AF598282

# Advances in Cryogenic Engineering

VOLUME 43, PART A

*A Cryogenic Engineering Conference Publication*

# Advances in Cryogenic Engineering

**VOLUME 43, PART A**

**Edited by**

**Peter Kittel**

NASA–Ames Research Center
Moffett Field, California

PLENUM PRESS • NEW YORK and LONDON

The Library of Congress cataloged the first volume of this title as follows:

Advances in cryogenic engineering, v. 1–
New York, Cryogenic Engineering Conference; distributed by Plenum Press, 1960–
v. illus., diagrs. 26 cm.
Vols. 1– are reprints of the Proceedings of the Cryogenic Engineering Conference, 1954–
Editor: 1960– K. D. Timmerhaus

1. Low temperature engineering—Congresses. I. Timmerhaus, K. D., ed. II. Cryogenic Engineering Conference

TP490.A3 660.29368 57-35598

Proceedings of the 1997 Cryogenic Engineering Conference,
held July 28–August 1, 1997, in Portland, Oregon

ISBN 0-306-45807-1


A Division of Plenum Publishing Corporation
233 Spring Street, New York, N.Y. 10013

http://www.plenum.com



10 9 8 7 6 5 4 3 2 1



Printed in the United States of America

## CONTENTS



## PART A

### Superconductivity Applications



### Superconducting RF Technologies

## Magnets: Design and Performance

## Magnets: Technology and Applications

## Large Hadron Collider – LHC



## Large Scale Systems

## Mechanisms and Machinery



## Instrumentation and Control

## Cryoelectronics



## Cryobiology and Medical Applications

## Space Applications



## Antarctica

## PART B

## Electric Power Applications

## Transportation Applications



## Thermal Properties and Mass Transport

## Helium Properties and Mass Transport

## Heat Exchangers, Cryopumps, and Heat Pipes

## Cryocoolers



## Cryocoolers: Regenerative

## Cryocoolers: Pulse Tubes

## INDICES

## FOREWORD

The Oregon Convention Center, Portland, Oregon, was the venue for the 1997 Cryogenic Engineering Conference. The meeting was held jointly with the International Cryogenic Materials Conference. John Barclay, of the University of Victoria, and David Smathers, of Cabot Performance Materials, were conference chairmen. Portland is the home of Northwest Natural Gas, a pioneer in the use of liquid natural gas, and Portland State University, where cryogenic research has long been conducted.

The program consisted of 350 CEC papers, considerable more than CEC-95. This was the largest number of papers ever submitted to the CEC. Of these, 263 papers are published here, in Volume 43 of **Advances in Cryogenic Engineering.** Once again the volume is published in two books.

## CEC PAPER REVIEW PROCESS

Since 1954 **Advances in Cryogenic Engineering** has been the archival publication of papers presented at the biennial CEC/ICMC conferences. The publication includes invited, unsolicited, and government sponsored research papers in the research areas of cryogenic engineering and applications. All of the papers published must (1) be presented at the conference, (2) pass the peer review process, and (3) report previously unpublished theoretical studies, reviews, or advances in cryogenic engineering.

Two or more specialists review each paper and assess its value to the community, and correctness in presentation. Reviewers return their comments to the technical editor who decides whether to accept, reject, or return the manuscript to the author for revision. The paper is accepted only after the positive recommendation of two reviewers. If one review is positive and one is negative, a third independent review is solicited. Authors must make changes deemed mandatory by the reviewers. To ensure objectivity, the names of the reviewers are withheld from the author and from each other.

**Advances in Cryogenic Engineering** resides in the libraries of most institutions throughout the world that conduct research and development in cryogenic engineering and applied superconductivity.

Peter Kittel
Ames Research Center
Moffett Field, CA
April 1998

## SAMUEL C. COLLINS AWARD

The Samuel C. Collins Award for outstanding contributions to cryogenic technology was not presented at the 1997 Cryogenic Engineering Conference.

The Collins Award was established in 1965 by the Cryogenic Engineering Conference to recognize outstanding contributions to cryogenic technology. The award is named in the honor of the late Samuel C. Collins, Professor of Mechanical Engineering at Massachusetts Institute of Technology, who was responsible for inventing the modern helium Liquefier. In addition to Dr. Collins, who received the award in 1965, this award has been presented to Professor Klaus D. Timmerhaus (1967), Dr. Edward F. Hammel (1973), Dr. Howard O. McMahon (1979), Professor Joseph L. Smith (1983), Professor William Gifford and Dr. Bascom Birmingham (1991), Professor Roger W. Boom (1993), and Dr. Ron Fast (1995). Since its establishment 32 years ago, the Collins Award has been presented only nine times.

## RUSSELL B. SCOTT MEMORIAL AWARDS

The Russell B. Scott Memorial Awards honor the first head of the Cryogenic Engineering Laboratory of the Boulder Laboratories of the National Bureau of Standards, now the National Institute of Standards and Technology. Mr. Scott was the founder of the Cryogenic Engineering Conference, the first of which was held in 1954 in Boulder, Colorado. He was author of the book: **Cryogenic Engineering**. He retired in 1965 after 37 years at NBS and died in 1967.

The objectives of the Scott Memorial Awards are to provide an incentive for high quality papers at the Cryogenic Engineering Conference and to provide recognition to authors who, in the judgement of the CEC board, presented the best papers at the proceeding conference. The award consists of a certificate and an honorarium.

The 1995 Scott Awards, for the best papers delivered at the 1995 CEC in the Hyatt Regency Hotel, Columbus, Ohio, and published in Vol. 41 of **Advances in Cryogenic Engineering**, were presented at the 1997 Awards luncheon in the Oregon Convention Center, Portland, Oregon

### Best Research Paper

J. Clausen, A. Hofmann, and M. Wanner

for their paper
"Novel Method for Measurement of the Para-Content of a Hydrogen Stream"

### Best Applications Paper

G.L. Mills, T.K. Kelly, and B. Brown

for their paper
"Integration of Thermoelectric Coolers into a Solid Nitrogen Dewar"

# STUDENT AWARDS

Student awards are presented by the CEC board to encourage the participation of students in the conference and to provide an incentive for students to write high quality papers. The awards are based of papers submitted prior to the conference.

The following students received
**Travel Assistance Awards:**

| | |
|---|---|
| Y.L. Ju | Chinese Academy of Science, Beijing, China |
| J.S. Panek | Florida State University, Tallahassee, Florida |
| A.K. Win | University of California, Los Angeles, California |

## Student Meritorious Paper Awards

Y.L. Ju
for two papers:
"Dynamical Experimental Study of the Multi-Bypass Pulse Tube Refrigerator with Two Bypass Tubes"
and
"Numerical Simulation and Experimental Verification of the Oscillating Flows in a Multi-Bypass Pulse Tube with Area Change"
authored by
Y.L Yu, C. Wang, and Y. Zhou

and

J.S. Panek
for the paper
"Liquid Level Control using a Porous Plug in a Two-Phase He II System"
authored by
J.S. Panek, Y. Zhao, and S. Van Sciver

## 1997 CRYOGENIC ENGINEERING CONFERENCE BOARD

| | |
|---|---|
| J.A. Barclay<br>Chairman | University of Victoria<br>Victoria, B.C., Canada |
| G.F. Green<br>Program Chairman | Naval Surface Warfare Center<br>Annapolis, MD |
| D.M. Coffey | Cryomagnetics, Inc.<br>Oak Ridge, TN |
| P.E. Gifford | Cryomech, Inc.<br>Syracuse, NY |
| K.T. Hartwig | Texas A&M University<br>College Station, TX |
| J.R. Hull | Argonne National Laboratory<br>Argonne, IL |
| J.P. Kelley | Babcock and Wilcox<br>Lynchburg, VA |
| P. Kittel | NASA-Ames Research Center<br>Moffett Field, CA |
| T. Kuriyama | National Institute of Standards<br>Boulder, CO |
| W.A. Little | Stanford University<br>Stanford, CA |
| J.C. Theilacker | Fermi National Laboratory<br>Batavia, IL |
| K.D. Timmerhaus | University of Colorado<br>Boulder, CO |
| A.R. Urbach | Ball Cryogenics<br>Boulder, CO |
| R.J. Witt | University of Wisconsin<br>Madison, WI |
| A. Zeller | National Superconducting Cyclotron Lab<br>East Lansing, MI |

**Awards Committee**

W.A. Little, Chairman  J.R. Hull  K.D. Timmerhaus

# ACKNOWLEDGMENTS

The Cryogenic Engineering Conference Board wishes to thank the Sponsors,

**Cabot Performance Materials,**
**Cryogas International, Inc.,**
**International Institute of Refrigeration,**
**Oxford Magnet Technology and Oxford Superconductor Technology,**
**Northwest Natural Gas,**
and
**Wah Chang,**

of the 1997 Cryogenic Engineering Conference.
These organizations greatly contributed to the success of the conference.

The Web Site for the 1997 conference was kindly provided by
W. Warnes and his staff
at the
Oregon State University, Corvallis, OR

Producing the **Advances in Cryogenic Engineering** requires the dedicated effort of many people. Of particular importance were those who took time from their busy schedules to review papers. Without their assistance the process would have taken even longer.

# EMERGING 21ST CENTURY MARKETS AND OUTLOOK FOR APPLIED SUPERCONDUCTING PRODUCTS

Carl H. Rosner

Intermagnetics General Corporation
Latham, New York 12110-0461 USA

## ABSTRACT

The spectacular discoveries in *superconductivity* that occurred during the 20th Century, as reported in the scientific literature and worldwide news media, have generally overshadowed the necessary and important underlying developments in *cryogenic* technology and equipment. The first and still only major commercial and economically viable application of superconductivity – MRI systems that have revolutionized medical diagnostics – owes much of its success not only to the availability of high-performance superconductor wire but also to advances in cryogenic technology.

Discoveries within the past decade of higher-temperature superconductive (HTS) materials have more recently generated much renewed interest in the need for reliable and cost-effective cryogenic systems. This historic interrelationship of the two technologies will become increasingly important in the early years of the 21st Century, when new multi-billion-dollar worldwide markets for superconductive products and equipment are expected to emerge.

HTS-based products, now under development, range from areas of applications not only in medical systems but also in electronics, communications, electric power, industrial processing and products, and transportation, among other likely markets. These potentially enormous opportunities are generally expected to utilize readily available liquid-nitrogen-based equipment, an assumption that seems overly simplistic and possibly naive. This presentation will review the prospects for new superconductive product areas and associated cryogenic equipment needs, as well as the corresponding technical and economic challenges faced by the superconductive and cryogenic industries.

## INTRODUCTION

The role that copper conductors and semiconductors have played in the 20th Century is likely to be expanded substantially by superconductor-based technology. Such technology

*Advances in Cryogenic Engineering*, Vol. 43
Edited by P. Kittel, Plenum Press, New York, 1998

has all of the hallmarks of becoming a pervasive engine for mankind's rapidly accelerating technological progress in the first few decades of the 21st Century, while also providing important energy-saving benefits. In making that prognostication, we must be aware that virtually all of the predictions over the past nearly 90 years that anyone has made – prophesying the near-term positive developments and beneficial commercial impact that superconductivity would have upon society – have, so far, been much too optimistic. Indeed, such predictions have generally been outright wrong as to the timeframe within which specifically defined benefits would be achieved.

Ever since the dawn of superconductivity – the absence of electrical resistance in metals discovered in 1911 by a Dutch physicist, H.K. Onnes[1] – high expectations of its usefulness were invariably replaced, time after time, by disappointments and letdowns. Yet the continued interest, decade after decade – by low-temperature physicists, theoreticians, and material scientists – has periodically led to breakthrough discoveries in understanding, as well as spectacular experimental demonstrations. Moreover, the award of several Nobel prizes in this area has stimulated further efforts towards the development of important applications.

Thus, each time a significant achievement in furthering the understanding of superconductivity was announced, more and more scientists and engineers became aware of the exciting and intriguing nature of the almost magical phenomenon of what some call "perfect conductivity." On the way, frustrations were encountered, largely because it took almost 50 years before an adequate theory was developed – the so called BCS theory[2] – to explain how a normal and what is typically a poorly conducting metal can suddenly become a "superconductor." Even today, when tens of thousands of scientists and engineers around the world are pursuing the promise of high temperature superconductors (HTS) in ceramic materials, the BCS theory does not provide guidance to those searching for new and better superconducting materials.

In very general terms, it is relatively easy to make strong arguments why superconductivity should become increasingly important, sooner rather than later. Early in the 20th Century, it was recognized that the increasingly widespread and worldwide introduction of electric power distribution – as well as related electrical equipment and devices – would accelerate much of the industrial revolution. Further, it was thought that the development and availability of ever better electrical conductors, including superconductors, would further expand the "electric age." Subsequently, it was assumed that semiconductor-based industries –information, communications, and computers – also could be expected to benefit greatly from the availability of superconductor-based elements.

Especially today, there exist strong convictions that all of these expectations are indeed still likely to be fulfilled. The questions are *when* and *how* and *what will it take* to have the prophecies replaced by real and fully convincing achievements.

There exists an overwhelming wealth of documentation and, in particular, a few recent excellent technical dissertations[3-8] that provide very thoughtful state-of-the-art reports. More important references are provided in the Biographical listing. Hence, this presentation, in a narrative form, will attempt to provide a brief overview of the enormous progress that has been made in the past few decades towards exciting goals for the large-scale commercialization of applied superconductivity. An attempt also will be made to identify the barriers that continue to stand in the way of worldwide application of superconductive devices and equipment for the benefit of the general public.

It should be emphasized that already a few fields of endeavor have benefited significantly from progress in superconductivity; high-energy physics research especially sees its future more and more tied to exploitation of the phenomenon. Also, fundamental developments in medical diagnostics – utilizing powerful and precisely functioning superconductive magnets to provide safe and enormously informative MRI "images" of human anatomy – have now become a multibillion-dollar-a-year business.

A number of entrepreneurial start-up companies are similarly positioning themselves to become beneficiaries of widespread applications of superconductivity for both the electric power and the communications industries.

Why has such a unique and attractive phenomenon not yet been fully harnessed? Why is it taking many decades to arrive at more definitive solutions? For answers, we need to take a quick historical look at the inherent and enormous complexities – and underlying challenges – of the science and technology of applied superconductivity. This challenge is continuing even though important milestones already have been surmounted. (Table 1.)

Equally important, especially to this audience, is the fact that development and progress in *cryogenics* has not only made possible the discovery of *superconductivity*; the two technologies also will continue to be interconnected for the foreseeable future, at least until room-temperature superconductors are discovered! Thus, the prospects for major cryogenic technology advances and equipment developments look equally exciting. They are deemed essential to achieve and exceed necessary economic and cost targets that have not yet been reached. Here lies the key argument for the major advantages attributed respectively to the development of higher-temperature (ceramic) superconductors (HTS) over the earlier lower temperature (metallic) superconductors (LTS). Refrigeration cost-savings expected to be achieved through the use of nitrogen-temperature refrigerators, rather than helium-temperature refrigerators, are principal assumptions in design decisions of superconductive systems which otherwise could utilize *either* LTS or HTS conductors. Aside from the relative cost of the superconducting materials themselves, the necessary cost of cryogenic refrigeration – as well as the cost of associated mechanical support structures and interface equipment – generally dominates system costs of virtually every superconductor-based application. Thus, the size and timing of the ultimate market will strongly depend upon the overall economic viability of such products in comparison to existing technology.

## KEY HISTORICAL EVENTS – THE FIRST 50 YEARS

The discovery of superconductivity by H.K. Onnes in 1911 was an unexpected seminal event. It followed several decades of success in the liquefaction of virtually all natural gases such as oxygen, nitrogen, hydrogen, air, and ultimately helium in 1908. The latter also was a first by Onnes and his co-workers. Their success was rooted in their ability to build ever more sophisticated apparatus towards the achievement of reaching the temperature of absolute zero, -273 Celsius. Within a few years thereafter they came remarkably close, to within one degree, of that ultimate goal. Also discovered, following soon thereafter, was the so-called "superfluid" phase of liquid helium, akin to the phenomenon of "superconductivity" in metals.

Thus, the successful liquefaction of helium by Onnes made possible investigations of the electric resistance behavior of pure metals at very low temperatures, leading to the discovery of superconductivity in mercury. This achievement was quickly followed by a search for superconductivity in more "practical" metals, such as tin and lead.

Since the period also coincided with the increasing development of electric machinery and power lines, Onnes envisioned high-field magnets that could operate without power losses. He actually demonstrated that possibility with coils made from lead wires; he also invented a "superconductive switch." Both were used to demonstrate "persistent" (non-decaying) current loops and "floating" magnets. His dream and desire to move towards superconductive transformers – and possibly motors, generators, and power transmission lines – was short-lived as further research showed that the initial superconductive behavior was lost abruptly when a critical and relatively low magnetic field was reached, where his pure metallic samples lost their superconductive behavior.

During the next several decades, establishment of low-temperature research laboratories in several countries led to important, yet mostly informal, scientific exchanges of communications and indirect collaboration in the nascent field of low-temperature physics.

Finally, in the mid-thirties, publication of explanations and experiments by Meissner and Ochsenfeld[9], as well as theoretical work by the London[10] brothers, seemed to explain the disappointing results encountered by Onnes, all of which led to a resurgence of interest in experimental superconductivity.

Progress in the field was interrupted to some extent by World War II. However, the more ready availability and potential usefulness of liquefied gases stimulated the development of commercial helium liquefiers. The most successful of these were first developed by Kapitza[11] and then by Collins[12], whose machine in 1947 ushered in an era of low-temperature physics research that could now more readily be conducted not only in universities but also in many industrial and government laboratories in several countries.

Renewed interest in the potential usefulness of superconductors was also enhanced by the discovery that wires made from alloys, in contrast to pure or "soft" metals, could sustain their superconductive behavior in higher magnetic fields. This activity was aided enormously by more complete theoretical descriptions provided by Abrikosov[13] and Landau[14], who in the late 1950's explained the superconductive behavior of metals and alloys and their subsequent classifications into so-called Type I and Type II superconductors. Progress in metallurgy and material sciences during World War II had encouraged Hulm[15] and Matthias[16] to pursue in the 1950's, independently and later collaboratively, approaches to finding and developing new superconductive materials with higher critical temperatures. They succeeded brilliantly and identified the key metallic superconductors that became the mainstay of today's commercially available superconductive wires, namely niobium-titanium (Nb-Ti) and niobium tin ($Nb_3Sn$). The combination of these efforts culminated in the dramatic announcement by Kunzler[17] at a 1961 MIT meeting, that $Nb_3Sn$ remains superconductive in a field of 88 kGauss, and that 67 kGauss had been achieved in a solenoid wound with wires of the intermetallic compound of niobium-tin.

The MIT event was, in fact, remarkably similar in its wide-ranging impact to the more recent discovery of higher-temperature superconductors. The "Kunzler announcement" set off a virtual stampede by individuals, laboratories, and companies around the world to seek the commercialization of products. It was what seemed to be an "era of renewed recognition" of the importance of superconductivity-based products. As noted, Table 1 lists the key events that stand out as having had major impact and influence on the course of the historical development of superconductivity.

**Table 1.** Key Enabling Events

| First 50 Years | |
|---|---|
| 1908 | Liquefaction of Helium |
| 1911 | Discovery of Superconductivity |
| 1913 | First Superconducting Magnet |
| 1933 | Meissner-Ochsenfeld Effect |
| 1935 | London Phenomenological Theory |
| 1947 | Collins - Helium Liquefier Machines |
| 1950 | Ginzburg and Landau Theories |
| 1954 | Matthias, Hulm - Materials with Tc's 10-18 K |
| 1957 | BCS Theory |
| 1957 | Abrikosov - Type II Superconductors |
| 1960 | Giaver/Josephson Tunneling |
| 1961 | Kunzler - 67 kG $Nb_3Sn$ Magnet |

## SUBSEQUENT EVENTS – THE NEXT DECADES

The confluence of discovery, understanding of physics, and unambiguous demonstration of the unique performance in an increasing number of small superconducting magnets – aided by progress in metallurgy and material sciences in general – led to the emergence of "applied superconductivity" as a commercial and industrial objective.

Rapid progress in the manufacture of long lengths of a variety of superconductive wires, such as molybdenum-rhenium, niobium-zirconium, niobium-titanium, and niobium-tin – all now called low temperature superconductors (LTS) – stimulated a race toward milestones that would confirm the industrial usefulness of superconductors. Teams of scientists and engineers seeking to exploit the new technology were formed at Westinghouse, AT&T, AVCO, RCA, GE, General Atomics and many other major corporations around the world. An early result was a large numbers of patent applications. Ready availability of government funding to national laboratories and universities, primarily via the Atomic Energy Commission and the Department of Defense, lent credence to the expectation that large-scale technical, economic, and social benefits could be achieved. To serve the rapidly growing need of R&D activities, also growing in numerous universities, commercial departments were established by industrial giants such as Westinghouse, RCA, GE and others to begin supplying wires, magnets, and measurement instruments. The demonstration of a 10 T (101 kGauss) magnet in 1963, and subsequent delivery to Bell Labs a few years later of a useful 10 T research system, including a "flux-pump," seemed to reinforce the conviction that superconductive technology truly was ready to be exploited by industry. [18]

In addition, spectacular achievements, such as one by AVCO[19] in building a large MHD magnet, also were hailed as important milestones, as were a number of other projects.

However, in the mid and late 1960's some doubt began to arise as to the magnitude of the market and commercial opportunities. It had been discovered that successful and loss-less direct-current (d.c.) performance could not be achieved under alternating-current (a.c.) conditions. Even devices for d.c. operating environments were plagued by sudden deviation from expected performance, called "flux jumps." Relatively high helium costs posed another problem. Disillusionment set in. Almost all of the large corporations in the U.S either abandoned their commercial efforts entirely and significantly retrenched their R&D activities.

A number of small companies were formed to continue these activities. Several of these have survived and prospered: Supercon and Intermagnetics General Corporation, along with a few others in the United States, and Oxford Instruments in the United Kingdom. By way of contrast, several large Japanese companies have exhibited continuous involvement and expanded activity, substantially abetted by strong and growing government funding. Their long-term perspectives now have allowed Japan to aspire realistically to a leadership position in the emerging HTS "technology transformation" that continues to be underway.

## MEGA DEMONSTRATION PROJECTS OF THE 1970's

The banner to pursue the potential of applied superconductivity now was assumed by the high-energy-physics research community, which recognized the importance of the availability of higher and higher magnetic fields for larger-scale research systems and for application to so-called "mega" projects. Likewise, agencies of the Department of Defense, such as the Air Force and the Navy, saw potential new applications that might deliver "cold-war" benefits. Specifically, naval ship-propulsion systems, based upon d.c. drives and very large-scale accelerator projects, among others, were pursued. These activities resulted in the successful launch of a Navy ship[20] outfitted with a superconductive motor as well as the completion of the TEVATRON[21] at the Fermilab National Laboratory.

**Table 2.** Key Enabling Events

| Subsequent Decades | |
|---|---|
| 1961 | Josephson Junction |
| 1962 | Commercialization of Superconductivity Begins |
| 1973 | Gevaler - $Nb_3Ge$ Tc of 23 K |
| 1980 | MRI Whole Body Medical Diagnostics |
| 1984 | Fermilab Accelerator Dedicated |

**Table 3.** Superconductive Based "Mega" Projects

| | |
|---|---|
| Fermilab I GeV Accelerator | United States |
| MagLev Train Project | Japan |
| Superconducting SuperCollider | (Canceled) |
| DESY Accelerator | Hamburg, Germany |
| KEK Accelerator | Japan |
| 70 MVA Generator | Japan |
| LHC Accelerator/CERN | Europe (France/Switzerland) |
| Large SMES Project | Alaska |
| 45 T Magnet System | Florida |

In Europe and Japan, similar "mega" project activities were pursued, ultimately leading to larger-scale construction of superconducting accelerator magnets at physics institutes such as KEK in Japan and DESY in Germany. Moreover, Japan embarked on a very ambitious magnetically levitated train program that is still ongoing.

Not nearly as visible, but also pursued with some vigor, were superconductive electronic devices, such as superconductive "tunneling" diodes, followed subsequently by "squids" and "Josephson" junctions, which became areas of intense research. IBM especially had embarked on a major program to develop superconducting computer elements utilizing lead.

However, toward the end of the 1970's, a further curtailment of industrial funding and government support severely dampened the outlook for major growth of superconductive-based industry, except for the mega projects that had been established. Table 2 lists some of the key enabling events in superconductor history over the years. Table 3 lists a number of large projects completed or currently underway.

## TURNAROUND IN THE 1980's

At the beginning of the 1980's, the remaining relatively small superconductive companies were increasingly pressed to find a modest and ongoing market in R&D equipment and superconducting-wire supply. Then, unforeseen by anyone, an opportunity emerged that seemed tailor-made for the application of superconductive magnets on a potentially large commercial scale.

While nuclear magnetic resonance (NMR) based research instrumentation had become a growing field for the application of small-bore research magnets that could be supplied with field strength to almost 18 Tesla, only a few very large-bore, lower-field magnets had been built. However, the successful demonstration, initially with resistive copper magnets, that

NMR could be useful for whole-body medical diagnostics[22] represented a major new opportunity for the fledgling superconductor industry. Although there was initial skepticism that NMR – renamed MRI – would be effective or truly useful for medical diagnostics, field stability features inherently offered by superconductive-magnet-based systems became an obvious solution. Large-bore, stable, and very uniform magnetic fields were judged to be economical and energy-efficient for this application.

Within a few years, the efficacy and economic viability of superconductive-magnet-based MRI was universally accepted, and the superconductor industry had finally found its first large-scale commercial application! Thus, it had taken about 70 years since the discovery of superconductivity before a product line could be judged to have achieved "commercial" status due to features and power-cost savings not possible with copper conductor systems. Nevertheless, aside from developments in fusion research and in MRI, research funding from industry and government departments in the USA had leveled off and was actually declining. There were serious debates as to whether the critical temperature of superconductors had a theoretical limit of around 25-30 K, since no new superconducting materials had been found in many years.

On the horizon, a bright spot appeared in the mid-1980's with the decision to proceed with an originally estimated $3-8 billion supercollider[23] project to be located in Texas. While a fledgling industry prepared itself to supply the large amounts of superconducting wires and magnets required, as well as refrigeration equipment, another major event played out in news headlines.

## DISCOVERY OF HIGHER-TEMPERATURE SUPERCONDUCTORS

The unexpected and initially low-key announcement in 1986 of a superconductivity discovery in ceramic-type materials by Bednorz and Müller[24] in the 30 K region re-ignited the worldwide race for higher-temperature superconductors (HTS).

At a meeting subsequently dubbed the "Woodstock of Physics," virtual pandemonium existed as several teams competed with reports that would confirm and possibly better the new high in Tc's achieved, now targeting the 40 K barrier, which was achieved soon thereafter.

A scant year later, teams headed by M-K Wu[25] at the University of Alabama and C-W Paul Chu[26] at the University of Houston, made additional and perhaps even more startling discoveries of materials having Tc's above the 77 K liquid-nitrogen barrier. This work led the way towards seemingly practical materials that could operate at that temperature, or higher. All restraint and skepticism, even by large industrial organizations, evaporated, and a new intensified race began. Thousands of multi-disciplinary teams of scientists and engineers around the world joined in the research and development of these and other materials that were expected to "revolutionize" our world within a few short years.

However, once again a similar disappointment reminiscent of the late 1960's set in. As *The New York Times* proclaimed in June 1989, "Hopes for Superconductivity Begin to Fade." The new materials could not carry much electric current.

The consequences were immediately apparent. Many large corporations in the U.S., having expected short-term commercialization opportunities and benefits, abandoned their efforts and curtailed financial support to others in the field.

Yet, many were not to be dissuaded and stubbornly persisted in their quest for understanding and discovery, again led by government-funded national laboratories and smaller businesses. The most visible vote of confidence came from Japan, where the formation of an International Superconductivity Technology Center (ISTEC) attracted nearly 100 of Japan's large companies. ISTEC has become the focal point for so-called "pre-

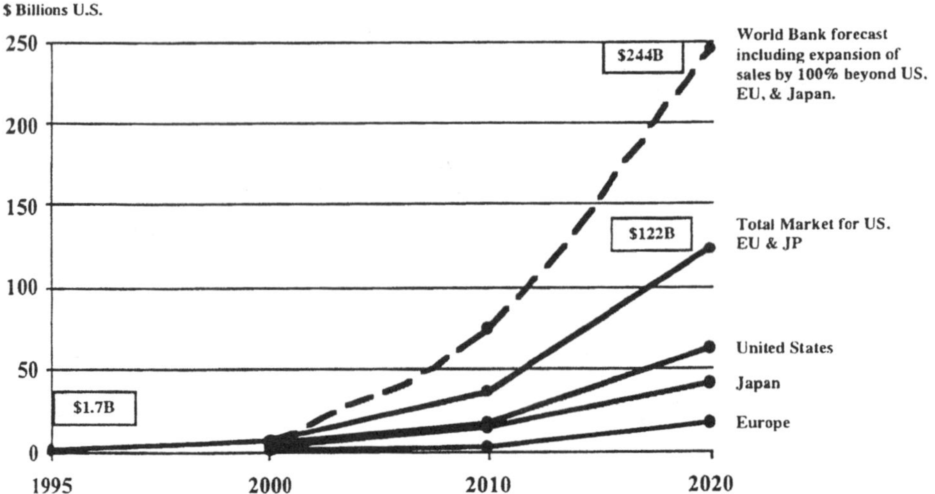


**Figure 1.** Forecast of Superconductivity Sales Opportunities. Fifth International Superconductivity Industrial Summit, May 1996.

competitive" research as well as long-term, systematically planned, and well-funded demonstration efforts in Japan. The door is wide open for any foreign organization to join ISTEC as a member. So far, only a handful of companies in the U.S. have done so.

## A DECADE OF PROGRESS

Within the past few years, careful analysis, experimentation, and innovative processing approaches have yielded improved current-carrying capacities in ever longer and longer lengths of various ceramic-conductor compositions. This has allowed construction of several feasibility demonstration devices in both electronic and electric power configurations. This more realistic, but equally intensive, effort appears to have turned the tide.

By the early 1990's, sufficient confidence was regained in the large-scale commercialization potential of superconductive technology that a group of leading experts from Japan, Europe, and the U.S. formed a committee called the International Superconductivity Industry Summit. ISIS has projected the sizes of various commercial markets and products. The 1996 projections are shown in Figure 1.

## CRYOGENIC REFRIGERATION

The role that cryogenic refrigeration technology has played in the advances achieved in superconductivity, over many decades, has been very much underrated and often taken for granted. Always in the shadow of the multi-billion-dollar liquefied-natural-gas industry, which also supplies liquid helium and nitrogen, smaller-size but sophisticated cryogenic refrigeration systems have played an important role. Specifically, improved MRI systems have relied on the availability of reliable two-stage 10-80 K shield coolers, thereby reducing the consumption of liquid helium by an order of magnitude in medical diagnostic MRI magnets, compared to the early MRI systems.

In the near future, two-stage 4 K shield coolers will reduce costs even further, but reliability may suffer in "cryogen-free" environments. The range of refrigerator equipment –

**Table 4.** Refrigeration Methods[27,31,32]

| | Refrigerator Classification | Temperature Range |
|---|---|---|
| 1. | Vapor-compression cycles | 240K - 300K |
| 2. | Vapor-compression Multistage & Cascades | 80K - 250K |
| 3. | Absorption | 250K - 300K |
| 4. | Hilsch or Ranque Tube | 250K - 300K |
| 5. | Thermoelectric | 200K - 300K |
| 6. | Expansion - Engine cycles | 10K - 200K |
| 7. | Multistage Joule-Thomson Refrigerators | 4K - 80K |
| 8. | Stirling Cycle* | 30K - 150K |
| 9. | Gifford-McMahon Cycle (GM cycle)* | 30K - 150K |
| 10. | Pulse Tube Refrigeration* | 30K - 150K |
| 11. | Adiabatic Demagnetization | 0.01 - 1.0K |
| 12. | Helium Three Dilution | 0.05 - 1.0K |

* By adding 2nd and 3rd stages to these cycles, their ranges can be extended to 4K.

and cycle of technologies – pursued, and to a large extent exploited successfully, are shown in Table 4. By adding 2nd and even 3rd stages to several of these cycles, their ranges can be extended substantially beyond those shown in the table.

An important parameter available to determine economic viability for cryogenic equipment is the ratio of power input at room temperature to that attainable at various low-temperatures. That relationship is shown in the accompanying figures for GM cycle equipment.[28]

Figure 2 shows the power input required to remove 100 W on the first stage of a good GM refrigerator. Figure 3 shows the same for removing 1 W on the second stage.

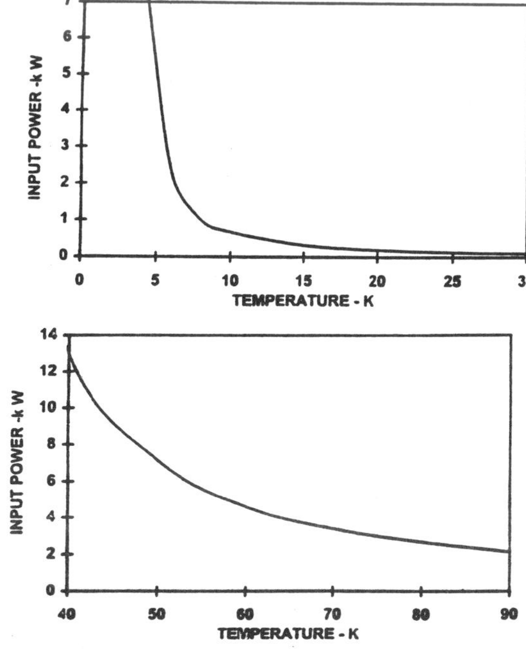


**Figure 2.** Electric power input for 100 W cooling on first stage of GM refrigerator with good efficiency.

**Figure 3.** Electric power input for 1 W cooling on second stage of GM refrigerator with good efficiency.

An MRI cryostat with a shield cooler having a load of 40 W at 45 K and 1 W at 11 K requires about 3.7 kW of power to cool the warm shield and 0.6 kW to cool the cold shield. Future challenges towards cost-reduction needs will require refrigerator developments that provide even lower equipment expenditures coupled with greater reliability.

In the past few years, pulse-tube refrigerator systems have been improved substantially and are likely to lead to enhanced reliability over a wide range of temperatures. Another approach that appears to have similarly great promise for much larger utilizations are refrigeration systems that use mixed gases, now beginning to be available commercially.

The ever-present performance/cost trade-off involved in operating superconductive devices at lower *vs.* higher temperatures will continue to be a dilemma, especially in the range between liquid helium, hydrogen, and nitrogen. The dilemma will cease to exist only if superconductors can achieve performance at nitrogen temperatures equivalent to that now achievable at helium temperatures. To truly achieve such a situation, superconductors with critical temperatures in the range of 150-200 K would be desirable. Can they be found? At this time, there are no theoretical or fundamental barriers to such discoveries. According to Little[29], even 1000-degree superconductivity may be possible in an "organically" based system.

## EMERGING MARKET OPPORTUNITIES

**Overview.** The development of superconductivity essentially is being driven by two forces. The first is related to the uniqueness of the phenomenon, which has attracted large numbers of scientists, principally from universities and research laboratories, whose goals are complete and fundamental understanding. Then, too, there is their hope of possibly winning a Nobel prize. The second driving force is generated by industry, searching for exclusive and proprietary new products and markets to address their needs for growth and profitability. For a long time, these forces seemed to be at loggerheads and were considered almost incompatible with their respective goals.

However, as research activities have reaffirmed how successful discoveries and results can meet – and be compatible with – the objectives of industry, a collaborative environment is proving to be exceedingly productive.

Since the 1960's, progress in superconductivity has benefited from such collaboration, which is generally more effective in Europe and Japan than in the U.S. However, government initiatives in the U.S., beginning with President Reagan's challenge to the superconductive industry ten years ago, are continuing to have a positive impact. What remains incompatible, however, is the generally unpredictable nature of when research projects will achieve success, which clashes with the desire by industry to obtain short-term results and quick payback of investments. Thus, the relatively slowly emerging market opportunities for superconductive products has caused much frustration, led to missed opportunities, and wasted talented work. In spite of the problems, though, the prospects of replacing electric power systems and of enhancing electronic devices with superconductors continue to be tantalizing indeed.

The very nature of superconductivity makes it seemingly easy to identify and pursue market opportunities (Table 5) already being served by emerging better performing superconductive products for which large markets exist. It is important to remember, however, that the energy-saving potential of superconductive products is not yet either well recognized or readily available.

It turns out that relatively few criteria can be established which in turn can guide development activities and yield potentially marketable products.

The opportunities for developing a range of superconductive products are shown in Table 6. It is particularly noteworthy that applications – whether based upon materials

**Table 5.** Major Areas of Market Opportunities

| | |
|---|---|
| HEALTH CARE: | Magnetic Resonance Imaging, Magnetic Resonance Spectroscopy, Functional MRI, Food Inspection. |
| TRANSPORTATION: | Magnetic Levitation, Ship Drives, Elevators. |
| ENERGY: | Energy Storage, Transmission Lines, Motors, Generators, Fusion, Current Limiters, Flywheels, Leads, Transformers. |
| RESEARCH: | Ultra High Field Magnets, Accelerators, Insertion Devices, Materials Characterization and Process Equipment. |
| ELECTRONICS: | Squids, Devices, Computer Cooling, Instrumentation, Communication. |
| DEFENSE: | Vibration Isolation, Mine Detectors, Detector Cooling, EM Launchers |

directly, on wires and cables, or on magnets – require a cryogenic temperature environment or suitable cooling equipment.

Two straightforward parameters must be achievable simultaneously if existing products are to be replaced. These requirements are: first, to provide much better performance; and second, to do so at significantly lower cost. Surprisingly, these have been and continue to be achieved relatively easily by the semiconductor and electronics industry, as well as in the electric appliance industry.

**Table** 6. Applied Superconductivity 1995 - 2020

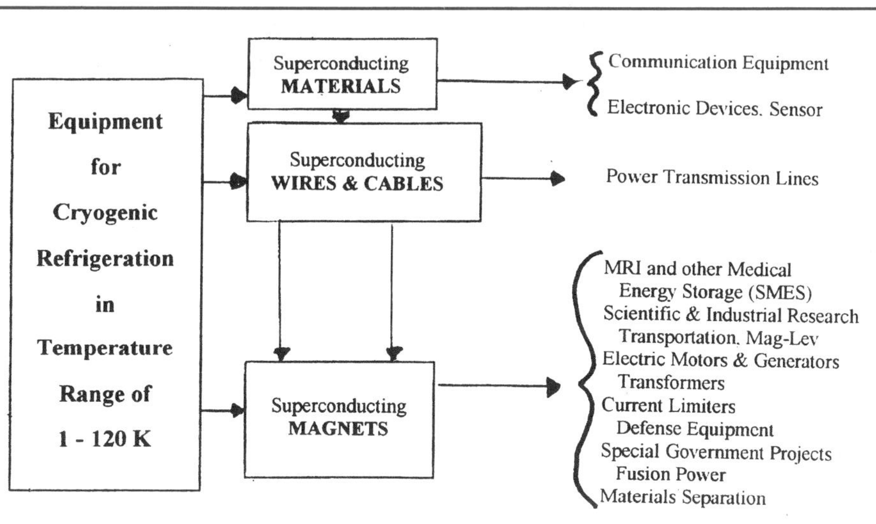

Can a superconductive-based product line achieve similar goals? The closest to attain such a characterization to date are research magnet systems, where a market of several hundred million dollar per year exists.[30]

**Magnetic Resonance Imaging (MRI).** An even more significant demonstration exists in products targeted for the medical diagnostic industry; i.e., superconducting magnets for magnetic resonance imaging (MRI). It is, however, sobering to realize that for nearly 15 years the utilization of superconducting magnets in MRI systems has been the only major and truly commercial application for superconductive technology. The level of annual revenues being generated is presently on the order of $2 billion per year.

What has allowed MRI magnets to attain such a position? The answer is simply that both economic and performance advantages over conventional products have been continuously demonstrated and extended. This had to be done in three key areas: superconducting wire, magnet design, and cryogenic equipment. Over the 15-year time span since the first delivery of a commercial MRI magnet, today's much better magnets cost less than one-half of those originally sold. Meanwhile, helium consumption has gone from about one liter per hour to less than one-tenth of that amount, due to advances in cryogenic engineering and the introduction of cryogenic "shield coolers".

Even though low-field MRI systems that utilize permanent magnets have achieved a significant market share, high-field-superconductor-based systems will continue to dominate MRI for the future. While the growth rates in that market have diminished from the early years, the efficacy of MRI as an important diagnostic modality will continue to grow substantially in the years ahead. Table 7 shows the growth pattern to date and forecasts for the next few years. These forecasts do not include a likely scenario where MRI, because of its safety compared to X-ray-based diagnostics, will become a universal screening tool for routine examinations, as well as for monitoring the effectiveness of pharmaceuticals, etc.

Further technical progress is also possible. For example, the increasing reliability of cryogenic refrigeration systems has allowed the introduction of "cryogen-free" MRI systems. Whether the use of HTS superconductors will lead to additional improvements and on-going demands for cost-reductions remains to be seen.

Superconductivity is expected to provide additional features to improve MRI images through new R.F. (radio frequency) surface coils, even though it may be difficult to dislodge

**Table 7.** Forecast: MRI Worldwide Installed Base

| Year End | Worldwide Installed [a,b,c] | Low Field [d] (Resistive/Permanent) | High/Mid Field [e] (Superconductive) |
|---|---|---|---|
| 1996 | 9,900 | - | - |
| 1997 | 11,088 | 1,775 | 9,313 |
| 1998 | 12,418 | 2,041 | 10,377 |
| 1999 | 13,908 | 2,347 | 11,561 |
| 2000 | 15,577 | 2,699 | 12,878 |

**Assumptions (approximated):**

a) 15% annual increase of all new MRI systems worldwide.
b) 3% annual decrease of all obsolete MRI systems worldwide.
c) 12% net growth of all MRI systems worldwide.
d) 15% annual increase of new low field MRI systems worldwide.
e) 11% annual increase of new high and mid field MRI systems worldwide.

**Rev.: June 1997**

**Table 8.** Examples of Ceramic HTS Superconductors with Commercial Potential

| Compound | | Critical Temperature (Tc) |
|---|---|---|
| $YBa_2Cu_3O_7$ | (YBCO) | 93 K |
| $Bi_2Sr_2CaCu_2Ox$ | (BSCCO) - 2212 | 90 K |
| $Bi_2Sr_2Ca_2Cu_3Ox$ | (BSCCO) - 2223 | 107 K |
| $Tl_1Ba_2Ca_2Cu_3Ox$ | (TBCCO) | 115 K |
| $HgBa_2Ca_2Cu_3Ox$ | | 134 K |

the very efficient conventional coils now in use. It is conceivable that superconducting electronics also may play a role in making MRI an even more useful and indispensable tool.

Beyond medical diagnostics, MRI's related technology is expected to be important in quality control of organic materials generally, and of food products in particular. The unrelenting requirements for improved productivity and higher quality, along with widespread adoption of international manufacturing standards and increasingly stringent government regulation, have driven industry to seek novel solutions to the automated inspection and control of a diverse array of manufacturing processes. MRI is therefore becoming more widely recognized as a possible unique solution to many inspection problems: in the analysis of foods and beverages, plastics and rubbers, petrochemicals, ceramics, explosives, narcotics, fuel propellants, and even timber. Already, successful demonstrations of such systems are used regularly for inspection of voids and sugar content in melons[31], the quality of fish suitable for sushi, as well as for millions of food containers that are being scanned on conveyor belts for quality deviations from exacting standards[32].

**Materials Developments.** While thousands of alloys and intermetallic compounds have been found to be superconductors, only a handful represent materials that are deemed useful enough to have generated technological and commercial interest. Thus, binary combinations of Nb-Ti, $Nb_3Sn$, Nb-Al, $V_3Ga$ and Nb-Ge have undergone substantial process developments, including additions of other elements that yield improved properties.

Similarly, in the area of HTS conductors, where already over 100 compounds have been discovered to have superconducting properties – principally higher Tc's than metallic materials – only a few are presently being developed with expectations of achieving commercially acceptable status. Table 8 shows a list of these, including their respective Tc's. Material developments are focused on the rather complex factors listed in Table 10.

**Table 9**. SC Wire Commercialization Requirements – Critical Characteristics

**Material Properties**

1. Control of Chemical Composition Achieved
2. Mechanical - Stress/Strain Characteristics Adequate
3. Controllable Thin-Film and Bulk Configurations
4. Compatibility with Other Materials
5. Acceptable Range of Thermal/Mechanical Behavior
6. Moderate Time/Temperature Processing Requirements
7. Isotropic Performance Characteristics
8. Acceptable Process/Materials Environment Behavior
9. Stable Temperature/Radiation Environment Behavior
10. Current Density > $5 \times 10^4$ Amps/cm$^2$ in High Fields
11. Low A-C Losses Achievable
12. High Critical Fields Attainable

**Table 10.** SC Product Commercialization - Critical Issues

**Device/Equipment Characteristics**

1. Favorable Marketplace Conditions
2. Superior Performance Achievable (compared to conventional approaches)
3. Favorable Economics Demonstrable in Each Case
4. Size/Weight Advantages Possible
5. Compatibility with Conventional Devices/Equipment
6. Large Scale Energy Savings Achievable
7. Engineering Possible with Existing Materials
8. Inventive and Innovative Products Demonstrated
9. Reliability and Safety Requirements Met
10. Continued Progress in Understanding/Developments

Since there is not yet a satisfactory theoretical explanation pointing to ultimately limiting parameters, it is assumed that the probability exists for discovery of newer and better HTS materials in the foreseeable future. Hence the search continues for both theoretical and practical reason. In a decade of such efforts, many materials have been discovered, but performance properties achievable at nitrogen temperatures have yet to match or exceed properties of metallic superconductors obtainable at helium temperature. Furthermore, the relatively brittle nature of HTS conductors requires very careful handling.

A range of material properties, required in combination, to meet the needs of most applications is shown in Table 9.

Once acceptable material properties are obtainable in wire or tape form, several important issues are likewise critically important to ensure commercial success for HTS device or equipment applications.

It long has been recognized that conditions for the economic viability of superconductive products require cost-effective conductors that can support large electric currents or have very large current-density characteristics. Hence, significant materials development is directed toward enhancing the current-carrying capacity of wires and tapes.

An outline of the specific targets for various applications is shown in Table 11. Since microstructure and grain-orientation have been identified as key elements in obtaining enhanced current-carrying characteristics, two promising tape-based and textured-deposition approaches have been pursued in Japan and the U.S. Recently, a "roadmap" (Table 12) has been generated that outlines and identifies important issues and milestones that must be pursued for the development of practical coated conductors.

**"MagLev" Train Systems.** Possibly the most intriguing and dramatically powerful demonstration of superconductivity evolves around the possibility to "float" devices, products, and even trains. (Indeed, the discoverer of superconductivity, H.K. Onnes, demonstrated and entertained his colleagues with "floating" superconductors.)

The idea of extending this capability to utilize superconductivity on a practical level was first developed in the United States by Powell and Danby, and later Kolm. Subsequently, their ideas were enthusiastically embraced both in Germany and Japan, where within a decade after the demonstration of LTS superconductivity, full scale prototype projects were underway. These projects are aimed at achieving remarkably high-speed performance and potential efficiency for human and cargo transportation over reasonable distances.

Following extensive multi-year testing and further exhaustive engineering analysis, both Germany and Japan decided to proceed with full-scale implementations of high speed

**Table 11.** High Temperature Superconducting Applications: Wire Performance Requirements

Commercial HTS Wire Requirements

| Application | $J_C$ ($A/mm^2$) 77K, self-field | Cost/tape ($/kA-m) | Field (T) | $Temp_{op}$ (K) | $I_C$ (A) 77 K self-field | AC loss (mW/A-m) | Bend Radius (m) | Strain (%) | Wire Length (m) |
|---|---|---|---|---|---|---|---|---|---|
| **Fault Current Limiter** | 10-100 | 30->10[a] | .3 - 3 | 40->65 | 100 | 0.4 | 0.15-0.05 | 0.2-0.4 | 200-1000 |
| **Motor** | 100 | 10 | 4 | >25 | 300 | NA[b] | 0.05 | 0.2-0.3 | 1000 |
| **Generator (100 MVA)** | 10 | 10 | 4-5 | 20->65 | 100-200 | NA[b] | 0.1 | <0.2 | 500-1000 |
| **Cable** | 10-100 | 10-100 | <0.2 | >65 | 30[c] | 0.15 | 0.01 | >0.4 | 100-1000 |
| **Transformer** | 10-100 | 20->5 | 0.15 | 20-65 | 200 | 0.25 | 0.1-0.2 | 0.1 | 250-3000 |
| **High Field Magnet** | 10-1000 | 5->1 | >20 | 4.2->65 | 300-500 | NA | 0.01 | 0.5 | 500-1000 |
| **Magnetic Separator** | 1 | 10 | 2-3 | 77 | 500 | NA | 0.5 | 0.2 | 1000 |

Table developed at 1997 DOE Workshop and Represents Consensus of Industrial Participants.

a Arrow represents cost decrease necessry from pre-commercial to final cost necessary to enable commercial device.
b For current field coil design.
c Current for individual wires; full cable assembly will carry 3-5 kA

Underlined items represent highest priorities for each application according to panel participants.

**Table 12.** Technical Roadmap for the Development of Practical Coated Conductors

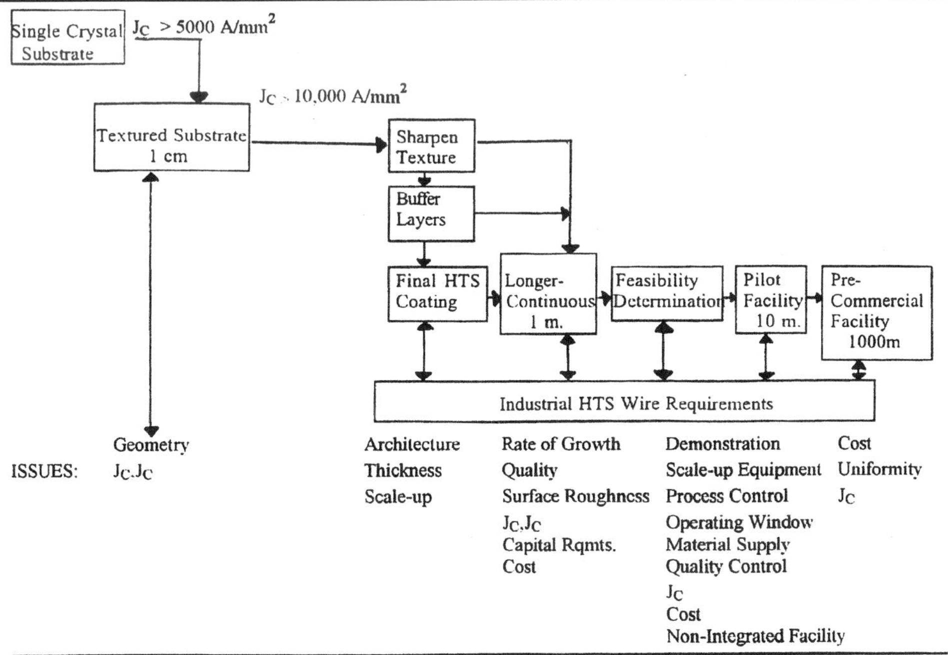

"MagLev" – magnetically levitated – train propulsion systems. There were, however, major differences in the approaches taken by the two nations. In Germany, where Siemens had taken the lead to pioneer this promising application, a debate ensued as to whether superconductive technology was mature enough to be exploited in a potentially risky environment. A decision was made fairly early to proceed with MagLev technology – now called Trans-Rapid – but to use conventional copper-based electromagnets for 1 cm levitation height and propulsion. Such a project is now slated to be operational with a link between Hamburg and Berlin sometime in the early part of the 21st Century. Unfortunately, and inexplicably, a first German superconductive prototype train demonstration project was dismantled and discarded in the 1970's.[35]

By way of contrast, the project in Japan was headed by Japan National Railroad (JNR), which enthusiastically and systematically extended their superconductive MagLev test program over two decades. Several increasingly sophisticated types of vehicles were developed. In 1980, the fastest of these achieved a record speed of 323 miles-per-hour on a 4-mile long test track. Levitation heights of up to 10 cm have been achieved. [36]

Following that successful demonstration, the Japanese transport ministry approved a program slated to begin with a $3 billion, 18-mile segment that is ultimately intended to connect Tokyo and Osaka through the interior of Japan. The initial segment is scheduled to be operational before the end of 1997. Sometime toward the end of the first decade of the 21st Century, travelers could be whisked between the two cities within one hour, whereas today's conventional high-speed "railroad" trains (Shinkansen) take almost three hours.

**Superconductive Magnetic Energy Storage (SMES).** There is considerable support for the idea that SMES may represent the next major opportunity for transition from feasibility demonstration to commercial status. The seeming simplicity of using d.c. currents to charge a superconductive magnet, within which the energy is then stored once it is placed into a "persistent" mode, makes it a quite logical "electromagnetic battery." A number of smaller demonstrations units, built in Europe and the U.S., have confirmed the feasibility of such a system.[37] One such LTS-based unit was installed in New Mexico,[38] performing well during a one-year test period. Plans for a major scale-up unit that could store up to 20 megawatt-hours were originally advanced towards a component test stage, but were then curtailed and abandoned.

A more successful approach to produce electric power applications was sponsored by the U.S. Air Force and resulted in the availability of commercial LTS-based "micro-SMES" systems from Superconductivity Inc., now a division of American Superconductors Inc. The unit's main advantages over lead-acid battery systems are rapid response capability and environmental attractiveness designed to improve "power quality" service.

Economic considerations will largely determine the extent to which scale-up to larger SMES sizes will take place in the foreseeable future. Power-grid stabilization needs, resulting from "deregulation" and power "wheeling" expansion, may accelerate the timetable for superconductive energy storage.

**Transformers.** Even though traditional transformers have historically been the most efficient of electric power devices, there appears to be a surprising – but soundly based – opportunity to introduce superconducting transformers into electric power grids. This actually may occur sooner than some may have thought possible or likely.

In addition to economic advantages, special considerations – including novel design approaches – may make superconducting transformers the earliest and most promising proving ground for HTS in the electric utility industry. Projects underway in Japan and Europe already have demonstrated the feasibility of HTS devices in electric utility environments. A consortium led by Asea Brown Boveri (ABB) has installed a transformer in the City of Geneva, where it is called upon to transform 18.7 kV to 420 volts for purposes of longer-term performance testing. Similarly, in Japan at Kyushu University, demonstration of a 550 kV single-phase project was also successful. In the U.S., a most ambitious project led by Waukesha Electric is implementing novel concepts through collaboration with Intermagnetics General, Rochester Gas & Electric, and the Oak Ridge National Laboratory. That project is nearing the test phase of a 1 MVA cryocooled single-phase transformer unit, rated at 13.8 kA/6.9 kV and operating at 20-35 K. This demonstration is to be followed by 5 MVA and 30 MVA units. Expectations are that reliable and economically attractive commercial units could then be available as early as 2003.

Key advantages and benefits expected are mainly in reliability, safety, and economic performance advantages. These are deemed to be particularly important needs in urban areas, where superconducting transformers require less space. Obviating the need to replace oil and nitrogen helps alleviate environmental concerns and reduces fire risks. Furthermore, such HTS transformers can handle excess peak loads, since they can withstand 200 percent overloads at no added cost, while maintaining acceptable a.c. losses.

**Electric Power Equipment.** It always has been assumed that the ultimate customers for superconductive electric power equipment would be the electric utility industry. A number of attempts – and a few successful demonstrations – were carried out over several decades, utilizing LTS superconductors. It occasionally was acknowledged that some benefits might be achievable. However, save for a few exceptions, the economic gains were judged to be insufficient to merit more serious consideration of introducing superconducting components into electric power systems. In part, this was because the overall advantages were considered to be marginal; electrical equipment already is remarkably efficient. Moreover, helium-based cooling requirements were considered to be potentially unreliable, especially for an industry where 20-30 year assurance of trouble-free life is an absolute necessity. Finally, the benign regulatory environment within which utilities operated made it relatively easy to pass on increased costs to their customers, hence there was little pressure to innovate or to pursue improved performance that might be achievable through the introduction of superconductors.

That environment now appears to be changing significantly. First, the possibility of operating in a liquid-nitrogen-temperature environment is much more amenable and likely to be more economical than using helium. Secondly, and possibly even more important, the impending deregulation of the electric utility industry, especially in the U.S. but also in other developed countries, has suddenly made it highly desirable to pursue cost-saving and efficiency improvements in both operational power equipment and transmission areas. There now seems to be an improved match between what superconducting technology can offer and what the electric power and utility industry needs.

Forward looking utilities and organizations – mainly in Japan, Europe and the U.S. – are funding feasibility projects by forming consortia involving superconductive manufacturing organizations. In the U.S., government laboratories, the Department of Energy (DOE), and the Electric Power Research Institute (EPRI) are strongly supporting and encouraging these activities.

**Rotating Machinery - Motors and Generators.** The technical feasibility of superconductive motors and generators has been successfully demonstrated in several countries around the world. In the 1970's and 1980's, both Westinghouse and General Electric successfully proved the feasibility of electric generators utilizing superconductive helium-cooled windings in prototypes, up to 20 MVA.

Subsequently, Japan initiated a collaborative 10-year national project with the objective of having a 70 MVA prototype generator in operation by the year 1997. With the prospects of HTS superconductive materials becoming available, the U.S. Department of Energy, in the mid 1990's, spearheaded a development project targeting the design and construction of a 100 MVA generator which was to follow demonstration of a 1.5 MVA unit, through collaboration of General Electric and Intermagnetics General Corporation. After successful demonstration of a race-track coil segment, utilizing HTS superconductive tape, the project is on hold for lack of funding. In the meantime, the development of superconductive motors, also utilizing HTS materials, and spearheaded by Reliance Electric, is proceeding with demonstration projects; commercialization is expected to occur early in the 21st Century.

**Electric Power Transmission Cables.** Attempts to demonstrate the advantages of superconductivity-based electric power components have been directed most extensively to the development of superconductive power cables. Successful and on-going demonstrations of first 1-meter, then 10-, 30- and even 50-meter cables, operating at reasonable current and voltage levels, are being pursued in Japan, Europe and the United States. Results so far have been sufficiently encouraging that plans for feasibility demonstrations for lengths of 100 meters – for both d.c. and 3-phase a.c. operations – are targeted for completion by 1999 and 2000 respectively. Should the ambitious objectives of acceptable low-loss and economical competitiveness be achieved, it seems likely that commercially-attractive 500- and 1,000-meter superconductor cable segments could be demonstrated by the year 2010. Thereafter, installation of such cables into congested urban areas, and possible replacement and upgrading of existing underground cables, appear as reasonable projections for subsequent decades.

**Superconductive Electronics and Space Applications.** The fundamental parameters underlying the development and subsequent growth of the electric power industry already were well established in the 19th Century and then extensively exploited in the 20th Century. The benefits of electric energy are now available in virtually every corner of the earth. Hence, that industry has grown substantially and continues to do so, albeit at much slower rates than during the first half of the 20th Century. As indicated earlier in this presentation, the availability of more practical superconductors will expand the reach, economic viability, and versatility of electric power equipment.

Much more dramatic has been the growth in the 20th Century of other areas of physics, most notably fields of atomic, molecular and electron physics, the latter commonly referred to as "electronics." The resultant emergence of the communications and computer industries has been so rapid and pervasive that it is difficult to envision how much further new developments and innovations can take us in the next century. Yet, it is precisely the requirements and demands for ever-increasing speed and higher performance of electronic circuits, signal processing, and miniaturization that offer such significant challenges to superconductive-based electronics as well as cryogenics. Already, cryogenically cooled "MASERS" are enhancing microwave communications. Similarly, much more sensitive sensors, probes, and measurement instrumentation utilizing superconductive elements are being explored and developed in a number of laboratories around the world.

The communications and semiconductor industries, having moved from vacuum tubes to transistors to integrated circuits, are now engaged in cutthroat, worldwide competition. Further progress will depend upon the availability of ultra-low-power devices that operate very efficiently under low-noise conditions.

Satellite technology and space communications will therefore embrace superconductive electronics because – as indicated above – it potentially offers all the keys necessary for progress in the next several decades. In particular, the "cool" environment of outer space appears to represent an ideal match for higher-temperature superconducting electronics. It is by no means a given that the determination and commitment exist today to make the development of superconducting electronics a high priority. The situation is made more difficult by the enormous capital-equipment expenditures required in today's semiconductor industry. Many manufacturers already are relying on cryogenics to enhance manufacturing quality, but they are not yet ready to engage in the larger-scale R&D required to expand into new and perhaps even more complex manufacturing processes required to exploit superconductive devices. Thus, it is my opinion that in the foreseeable future the enormously promising but very demanding area of superconductive electronics is more likely to be driven and developed by university laboratories than by industry. Nevertheless, niche applications – such as sensors, probes, microwave filters and very sensitive magnetometers – represent nearer-term commercial opportunities and should be available commercially within the next 2-5 years. The possibility for large-scale, consumer-type superconducting electronic products will have to await either the miniaturization of cryogenic coolers or the discovery of room-temperature superconductors.

**Fusion Power.** If the promises of utilizing "seawater" as fuel for fusion power generation is ever realized in a concept of "magnetic confinement" – also called magnetic energy fusion – superconducting magnets will play an indispensable role. For several decades, large-scale experiments in Russia, Japan and the United States have seemingly confirmed that TOKAMAK magnetic-confinement experiments can yield results that closely approach conditions necessary for fusion power generation. The next step is to design and build an engineering test reactor of sufficient size to demonstrate conclusively the technical feasibility of generating fusion energy.

The project's size and complexities prompted an international collaboration that provided a conceptual design, an effort that was completed in 1990. The main participants – the European Community, Japan, Russia, and the United States – have since collaborated on plans for an International Thermonuclear Experimental Reactor (ITER) with targeted completion before the year 2010. The magnet structures required for the design target of nominally 1,000 megawatts would be 30 meters high, having almost the same diameter.[39]

A project of this type would require very large superconductive magnets in order to generate magnetic fields approaching 11-15 Tesla. The time scale envisioned, and the enormous magnetic stresses that need to be supported, have made it necessary to utilize LTS superconductors, more specifically $Nb_3Sn$. In anticipation of – and as part of – project preparations, several tons of the required superconducting material have been fabricated in each of the member countries.

Should ITER proceed to the next step, an additional large quantity of superconductor materials would be needed for the fabrication of very large coils. Decisions as to whether, when and where to proceed is expected to be made before the year 2000.

**High Energy Physics.** The next largest existing market for superconductors is in the area of high-energy physics, where on-going research into the fundamental nature of the origin of the world and of life itself will continue to be pursued in very large-scale projects as well as in many smaller research activities.

The recent decision to proceed with the Large Hadron Collider (LHC)[40] project at CERN in Europe, will require huge quantities of LTS braided cables (Nb-Ti) and over 10,000 magnets, each 10-15 meters long. These will be augmented by the need for many hundreds of specialty coils. Such an enormous project clearly represents a major challenge to industry worldwide, probably extending over five years. However, the magnitude of it makes it "one-of-a-kind" and thus unlikely to represent an on-going commercial opportunity.

A similar and even more ambitious project, the "Supercollider" in the United States, was canceled in 1993 after expenditures of almost $2 billion. The aftershocks of that decision are still being felt by both the physics community and the superconductor industry in the U.S. The history of that project illustrates a major problem in the planning and implementation of long-term projects, especially those that require multi-year funding at significant levels. This situation carries over to investments required for the development of superconductive products, which need many years of funding before acceptance in the marketplace is assured.

## OUTLOOK AND CONCLUSIONS

The long-awaited transition from research to the successful commercialization of superconductivity-based technology, which began very slowly in the mid-1960's, has seen a gradual acceleration during the subsequent decades. Since the discovery of higher-temperature superconductors, and as a consequence of dramatic increases in worldwide HTS research and development activities, the likelihood for nearer-term commercialization has become a certainty.

Based upon developments to-date, and assumptions listed in Table 13, one can construct a reasonable time line for estimating when initial commercial availability of products will occur. Such a listing is provided in Table 14.

**Table 13.** Commercialization of Superconductivity

| Assumptions - Expectations |
|---|
| Expanded International Collaboration/Information Exchange. |
| Need for Energy Conservation Will Accelerate. |
| HTS Discoveries will be Extended. |
| Duplication of Efforts are Minimized. |
| Cryogenic Equipment Becomes Less Costly and More Reliable. |
| HTS Materials Will Become More Cost Effective. |
| Theory Will Provide Some Guidance to Materials Scientists and Engineers. |
| Significant Niche Markets Identified and Developed Beyond MRI. |

**Table 14.** Likely Superconductor Product Introduction Sequence

| | Type of Product | Year/Estimate of Initial Commercial Availability |
|---|---|---|
| 1 | Superconductive Wires and Cable | NOW - On-Going |
| 2. | High-Field Research Magnet Systems | NOW - On-Going |
| 3. | Cryotrons-Electronic Devices | NOW - On-Going |
| 4. | Squids, Sensors | NOW - On-Going |
| 5. | Computer Elements | NOW - On-Going |
| 6. | NMR Magnets | NOW - On-Going |
| 7. | MRI Magnet Systems | NOW - On-Going |
| 8. | Accelerator Magnets | NOW - On-Going |
| 9. | Material Separation Equipment | NOW - On-Going |
| 10. | General Purpose Magnets-Processing | NOW - On-Going |
| 11. | Magnetic Shields | NOW - On-Going |
| 12. | Current Leads | NOW - On-Going |
| 13. | Current Limiters | 1999 Onward |
| 14. | Flywheels | 2000 - 2005 |
| 15. | Superconducting Bearings | 2000 - 2002 |
| 16. | Transformers | 2002 - 2006 |
| 17. | High Speed Trains - MagLev | 2005 - 2010 |
| 18. | Ship Propulsion, Electro-Magnetic Launchers | 2010 + |
| 19. | Elevators | 2010 + |
| 20. | Power Transmission Cables | 2010 - 2020 |
| 21. | Motors | 2002 - 2010 |
| 22. | Generators | 2020 |
| 23. | Fusion Power Systems | 2040 to ? |
| 24. | Vehicles | 2020 |
| 25. | Unforeseen - New Products | 2000 - 2040 |

One must keep in mind, however, that products listed as "NOW-ON-GOING" involve primarily the use of metallic superconductors, and that HTS-based products are mostly projected for the future.

In making these projections, it should be recognized that ten years already have passed since July 28, 1987, the day President Reagan and his cabinet made the pursuit and rapid commercialization of superconductivity a "national objective". In our society, where rapid transformation of industry, as well as exploitation of short-term profitable opportunities, are mandated by shareholders ("Wall Street"), the longer-term route to market by superconductive products may be puzzling. Is the general – or investment – public intrigued with the magic and mystery that still surrounds superconductivity and magnetism? And is that community willing to take a much needed longer-term view? Certainly, the search to identify and serve markets that for many decades were intuitively judged to be "just around the corner," can be likened more to a century-long "marathon relay race." It is a contest in which many individuals and teams periodically made dramatic progress towards the goal, thereby gradually providing broader understanding of this forever fascinating phenomenon. To date, no team has arrived at the ultimate end, and the race continues. Previous characterizations of this situation have included terms such as "the prolonged adolescence of superconductivity." And some have said it is "a technology where solutions are looking for problems."

What is clear today is that "the race" will have to be extended well into the 21st Century. However, we are at a point in time where between now and the 100th Anniversary of H.K. Onnes' discovery (in 2011), substantial additional niche markets will most likely be identified periodically and confirmed in both the electric power segment and in the superconducting electronic sector.

Increasingly, that will require further advances in cryogenic engineering and equipment to the point where superconductive-based "turnkey systems" are the focal point, replacing the emphasis on individual component parts such as materials or magnets. Furthermore, since some superconducting materials provide much better performance at lower temperatures, the distinction between LTS and HTS conductors will be replaced by recognition for the need of "high performance superconductors" (HPS) configurations and reliably performing cryogenic systems that use *either* LTS or HTS conductors. Depending upon the overall limiting economics and optimization of application-oriented needs, cryogenic equipment costs – as well as "life-cycle" power costs – may determine the specific conductor preference. This is also dictated by the likely continued proliferation of superconductive wire developments, where various types and configuration display advantages and disadvantages, from a basic cost point of view. Hence, markets for superconductive-based products will emerge more gradually than we would like, especially when competition from other energy-related products are "moving targets." There is a need to more realistically assess the advantages and often unpredictable time horizons that surround developments in superconductivity. The designation "high temperature superconductivity" can reasonably be considered to be a misnomer and possibly misleading. Only through HPS superconductors that provide superior and economical performance at whatever optimal temperature is required for reliable and efficient performance will superconductivity have emerged from its science-based limitations.

Much of the initial excitement surrounding the discovery of HTS has abated. The remarkable fact is that as yet no satisfactory and universally acceptable theory for HTS has been established. This leads one to believe that eventually, when experiment and theory coincide, better materials can be developed and the possibility of room-temperature or true "high-temperature" superconductivity may be found. The area that will remain most challenging will be the usefulness of superconductors under alternating current (a.c.) conditions. Hence, here we may be up against a more formidable barrier. High-field/higher-current-density conductors will require more ingenuity from the equipment and device designers to overcome the propensity of high field/high current superconductors to generate losses under a.c. conditions.

I remain as convinced as ever that applied superconductivity, based upon the prototype developments now underway and examples cited here, has unequaled potential in its ability to deliver substantially improved performance, energy, and overall cost-saving benefits.

I am equally certain that applied superconductivity will be as important for 21st Century markets as conventional copper and semiconductor-based industries have been in the 20th Century.

## ACKNOWLEDGMENT

The views expressed in this presentation are based upon my own assessments. I am, of course, deeply indebted to the untold numbers of individuals who, for the better part of this century, have made contributions primarily to the understanding of metallic superconductors. Most of that knowledge is of inestimable value to the development of HTS – and HPS – technology. I am particularly indebted, for a lifetime of education, to my former colleagues at the General Electric Research Laboratory, especially the late Charles P. Bean, as well as my many colleagues at Intermagnetics General Corporation. Similarly, I have benefited greatly from much of the excellent work by the large international group of scientists and engineers, who individually and collectively have advanced this field of superconductivity. They have brought us to a point where we can realistically – and more confidently than ever before – predict and expect the advent of multi-billion-dollar annual market growth to occur throughout the 21st Century.

### REFERENCES

[1] H. Kamerlingh Onnes, Leiden Comm. 120b, 122b. 124c, (191 1).
[2] J. Bardeen, L.N. Cooper and J.R. Scbrieffer, Phys. Rev. 108, 1175 (1957).
[3] Paul Chu, High Temperature Superconducting Materials: A Decade of Impressive Advancement of Tc, IEEE TRANSACTIONS ON APPLIED SUPERCONDUCTIVITY, Vol. 7, No. 2, June 1997.
[4] D.C. Larbalestier. The Road of Conductors: 10 Years Make a Difference, IEEE TRANSACTIONS ON APPLIED SUPERCONDUCTIVITY, Vol 7, No. 2, June 1997.
[5] T. Van Duzer, Superconductor Electronics, 1986-1996, IEEE TRANSACTIONS ON APPLIED SUPERCONDUCTIVITY, Vol. 7, No. 2, June 1997.
[6] Paul M. Grant, Superconductivity and Electric Power: Promises, Promises ... Past, Present and Future, IEEE TRANSACTIONS ON APPLIED SUPERCONDUCTIVITY, Vol. 7, No. 2, June 1997.
[7] D. Bruce Montgomery, The Future Prospects for Large Scale Applications of Superconductivity, IEEE TRANSACTIONS ON APPLIED SUPERCONDUCTIVITY, Vol. 7, No. 2, June 1997
[8] A. Silver, Superconductivity in Electronics, IEEE TRANSACTIONS ON APPLIED SUPERCONDUCTIVITY, Vol. 7, No. 2, June 1997
[9] W. Meissner and R. Ochsenfeld, Naturwissenschaften 21, 787 - 788, 1933.
[10] H. London, Proc. Roy. Soc. A152, 650, 1935.
[11] P.L. Kapitza, Zhum. Tekhn, Fis 9, 99, 1939.
[12] S.C. Collins, SCIENCE 11 6, 289 (1952).
[13] A.A. Abrikosov, Soviet Phys. - JETP 5@ 1174 (1957).
[14] L.D. Landau, Phys. Rev. 75, 884, II,3, 1949.
[15] J.K. Hulm, R.D. Blaugher, Phys. Rev. 123, 1569 (1961).
[16] B.T. Matthias, Science & Technology of Superconductivity, Vol. 1, 263, 1973.
[171 J.E. Kunzler, E. Buehler, F.S.L. Hsu and J.H. Wernick, Phys. Rev. Lett 6, 89, 1961.
[18] General Electric - Private Communication.
[19] Z.J.J. Stekly, et al. A large superconducting magnet for MHD power generation, Boulder, June 1966.
[20] T.J. Doyle, H.O. Stevens. Inst. of Marine Eng., London, 1984.
[21] Dedication of the Energy Saver, U.S. Dept. of Energy, April 1984.
[22] P.C. Lauterbur, NMR in Medicine, Fourth Ann. Conf., Tokyo, Japan, (1981).
[23] Proceedings - International Industrial Symposium, New Orleans, 1989

[24] J.G. Bednorz and K.A. Müller, Z. Physic, Vol. B64, p. 189-193, 1986.
[25] M.K. Wu et al. Phys. Rev. Letter, Vol. 58. p. 908-910, (1987).
[26] C.H. Chu et al. Phys. Rev. Letter, Vol. 58, p. 405-410, (1987).
[27] W.E. Gifford, Proceedings of 1st International Cryogenic Eng. Conf., Tokyo and Kyoto, p.221, (1967)
[28] Adapted by R. Longsworth using Reference 27. See also M.J. Boiarski, V.M. Brodianski, R.C. Longsworth, Retrospective of Mixed-Refrigerant Technology and Modern Status of Cryocoolers Based on One-Stage, Oil-Lubricated Compressors, CEC '97/Portland, OR..
[29] W.A. Little, Journal of Polymer Science Pt. C, No. 29 (1970)
[30] Feasibility Study on Superconductivity Application Field, Exec. Summary Report, ISTEC, March 1997
[31] T. Miki. K. Saito, et al, Sugar Content Detection in Watermelon, Advances in Superconductivity VIII, 8th Inter-national Symposium on Superconductivity (ISS '95) Nov. 2, 1995.
[32] Intermagnetics General Corp., News Release, June 1997.
[33] R.F. Barron, Cryogenic Systems, 2nd Ed., 1985
[34] B.A. Hands, Ed.: Cryogenic Engineering, 1986
[35] Rail Systems Technology, Vol. 4, No. 21, Sept. 2, 1996
[36] H. Nakashima, "The Current State of Maglev Development in Japan," Inst. Phys. Conf. Ser. No.148, Applied Superconductivity, Edinburg, July 3-6, 1995.
[37] W. Hassenzahl, "Superconducting magnetic energy storage," IEEE Trans. Mag. Vol. 25, pp. 750-758, 1989
[38] J.D. Rogers, H.J. Boenig, R.I. Schermer, and J.F. Hauer, "Operation of the 30 MJ SMES system in the Bonneville Power Administration electrical grid," IEEE Trans. Mag. Vol. 21, pp. 752-755, 1985
[39] B.J. Green, M. Huquet, "The ITER Project, Status and Prospects," MT 14, 11-16 June, 1995, Tampere, Finland.
[40] L.R. Evans, The Present Status of LHC, IEEE Trans. Mag., vol. 32, no. 4, July 1996.
Ed., 1985

## BIBLIOGRAPHY - Further Reading

- D. Shoenberg, Superconductivity, C.U.P., 1952.
- F.E. Simon, et al. Low Temp. Physics, Four Lectures, Pergamon, London, 1952.
- S.C. Collins, R.L. Cannaday, Expansion Machines for Low Temperature Processes, 1958.
- R.B. Scott, CRYOGENIC ENG., D. Van Nostrand Co., Inc., NY. London, Canada, March 1959.
- K. Mendelssohn, Cryophysics, Interscience, London and New York, 1960.
- J. Bardeen, J.R. Schrieffer, Dept. of Physics, Univ. of Illinois, RECENT DEVELOPMENTS IN SUPERCONDUCTIVITY, Progress in Low Temperature Physics, Vol. III, April 1960.
- C.J. Gorter, ed., Progress in Low Temperature Physics, 4 Vols., North Holland, Amsterdam, 1964.
- C.P. Bean, "Magnetization of High-Field Superconductors," Rev. Mod. Phys. 36, 31-39 (1964).
- RCA REVIEW, Special Issue - Synthesis, Characterization, and Application of Superconducting Niobium Stannide (Nb3Sn), Vol. XXV, No. 3, September 1964.
- CRYOGENIC ENGINEERING - Present Status and Future Development, Proceedings of First International Cryogenic Engineering Conference, Japan, April 1967.
- D.B. Montgomery, Solenoid magnet design, New York: John Wiley & Sons Inc., 1969.
- M.N. Wilson, Superconducting Magnets, Oxford University Press (Oxford, New York, Toronto, 1983).
- W.J. Carr, AC Loss and Macroscopic Theory of Superconductors, (New York, London, Paris, 1983).
- E.W. Collings, App. SC, Vol. 1 Fundamentals, Vol. 2 Applications, Plenum Press (NY, London, 1986). Comp. review of properties of NbTi and excellent account of many topics pertinent to all Type II SC's.
- SUPERCONDUCTOR INDUSTRY, Vols. 1-10, 1988-1997.
- The Furukawa Electric Co., Ltd., SUPERCONDUCTIVITY 1988, Tokyo, Japan, December, 1988.
- B. Schechter, THE PATH OF NO RESISTANCE, 1989.
- M.W. Browne, Hopes For Superconductivity Begin To Fade, The NY Times, Section Cl, June 6, 1989.
- V.D. Hunt, Technology Research Corp., Superconductivity Sourcebook, John Wiley & Sons, Inc., 1989.
- S.T. Dale, S.M. Wolf, T.R. Schneider, ENERGY APPLICATIONS OF HIGH-TEMPERATURE SUPERCONDUCTIVITY, Vol. 1, Extended Summary Report, February 1990.
- Journal of Superconductivity, Special Issue: Bibliography of High-T. SC, Vol. 3, No. 1, March 1990.
- 2nd World Congress on Superconductivity - Prog. in High Temp. SC, Vol. 28, Texas, Sept. 1990.
- A National Program for the Superconducting Electric Power System of the Future, Prepared by the ad hoc Industry Working Group on Power applications of High-Temperature Superconductors, April 1991.
- J. Vranich, SUPER-TRAINS - Solutions To America's Transportation Gridlock, St. Martin's Press, NY, December 1991.
- K. Krishen, C.G. Burnham, WORLD CONGRESS ON SUPERCONDUCTIVITY, Proceedings of the 3rd International Conference and Exhibition, Part II, Munich, Germany, Sept. 15-18, 1992.
- SCIENTIFIC AMERICAN, Special Section - New challenges for 1994, December 1993.

- FEDERAL RESEARCH PROGRAMS IN SUPERCONDUCTIVITY, Materials Technology Subcommittee Communications Group on Superconductivity, June 1994.
- F.C. Moon, Cornell Univ., Superconducting Levitation - Applications To Bearings and Magnetic Transportation. JOHN WILEY & SONS, INC., 1994
- University of Cambridge, RESEARCH REVIEW 1994, Interdisciplinary Research Centre in Superconductivity, Cambridge Marketing Limited, 1994.
- T.P. Sheahen, Introduction to High-Temperature Superconductivity, Plenum, N.Y., 1994.
- K. Togano, "Bi-2212 Wires and Coils," ISTEC Journal, Vol. 8, No. 2, 1995, pages 48-50.
- S. Huang, D. Dew-Hughes, et al. "Highly reproducible high critical cur-rent density in partial-melt Bi2Sr2CaCU20y/Ag tapes fabricated by electrophoretic deposition," Super cond. Sci. Technology, Vol. 8 (1995), p 32-40.
- W.L. Carter, G.N. Riley, et al. "Adv. in the Development of Silver Sheathed (Bi,Pb)2223 Composite Conductors," IEEE Trans. on Applied Superconductivity, Vol. 5, No. 2, (June 1995), pages 1145-1149.
- R.H. Hammond, "Thick Film YBCO for Wires and Tapes: Scaie-Up Issues and Cost Estimates," Proceedings of ISS'95, Oct. 30th-Nov. 2nd, Hamamatsu, Japan, pages 1029-1033.
- 8th International Symp. on Superconductivity, Adv. in Superconductivity - VIII, Vols. I and 2, 1995.
- R.F. Giese, "The Status of Progress Toward High-Temperature Superconducting Bulk High-Amperage Conductors," report for this task - January 1996.
- 1996 INTERNATIONAL WORKSHOP ON SUPERCONDUCTIVITY, "High Temperature Superconducting Electronics: Fundamentals and Applications", Iwate, Japan, June 24-27, 1996.
- D.W. Hazelton, M.T. Gardner, J.A. Rice, M.S. Walker, C.M. Trautwein and P. Haldar, "HTS Coil for the Navy's Superconducting Homopolar Motor/Generator," Intermagnetics General Corporation, 1996.
- J.P. Voccio, et al. "125 HP Motor Field Winding Development," App. SC Conf, PA., August, 1996.
- C.W. Chu, W.K. Chu, D.U. Gubser and K.A. Miller, PROCEEDINGS of the 10th ANNIVERSARY HTS WORKSHOP on Physics. Materials and Applications, 1996.
- A.M. Wolsky, THE STATUS OF PROGRESS TOWARD HIGH-AMPERAGE CONDUCTORS INCORPORATING HIGH-TEMPERATURE SUPERCONDUCTORS, March 24,1997.
- IEEE Trans. Applied Superconductivity, Vol. 7, No. 2, June 1997.
- 1997 International Workshop on Superconductivity, June 15-18, 1997.
- J.W. Muehlhauser, R&D Roadmap to Achieve Electrical Wire Advs. from SS Coatings, July 1997.
- IEEE SPECTRUM, Superconductivity in Electric Power, July 1997.
- Hans J. Schneider-Muntau, High Magnetic Fields: Applications, Generation. Materials, 1997

# LARGE APPLICATIONS AND CHALLENGES OF STATE-OF-THE-ART SUPERCONDUCTING RF (SRF) TECHNOLOGIES

Quan-Sheng Shu [1, 2]

[1]Thomas Jefferson National Accelerator Facility
Newport News, VA 23606 and
[2]Northrop-Grumman Co., Bethpage, NY 11714

## ABSTRACT

Various applications of superconducting radio-frequency (SRF) accelerating structures in many fields around the world are introduced. These applications consist of high energy physics, nuclear physics, free electron lasers, energy amplifiers, nuclear materials and the treatment of radioactive wastes. A review of recent development of SRF technologies is presented. We also briefly discuss the future prospects of SRF technologies and applications.

## INTRODUCTION

Due to the great attraction of SRF cavities (quality value Q higher than Cu by $\sim 10^5$ and accelerating fields Eacc higher than Cu by an order of magnitude in continuous wave (CW) mode), tremendous efforts have been contributed to the development of SRF technology for four decades. In the 1960s, SRF cavities were basically the delicate devices of scientists pursuing potential applications at a few high prestige laboratories. In the 1980s, many laboratories and industrial partners joined the development efforts that resulted in the production of higher RRR (250) of Nb materials and overcoming of multipacting, field emission (FE) and thermal breakdown (TB). At the present time, a total of about 1,000 m of SRF structures with associated large cryogenic systems (multi-kW refrigerators both at 2K and 4K) have been operated successfully at many laboratories around the world.[1-5]

SRF cavities' applications are rapidly extended from the traditional high energy physics and nuclear physics to free electron lasers, [6,7] nuclear materials,[8] the treatment of radioactive wastes and energy amplifiers.[9] Several new applications with a total of more than 20 km of SRF cavities, such as TESLA,[10] APT, JAERI, etc. have been proposed.

The new proposed and ongoing projects have presented more challenges to the SRF cavities and associated systems that will definitely stimulate new momentum of SRF technology development.

In this paper, I will briefly introduce the advantages of SRF cavities and challenges facing the implementation of these advantages in applications. I will then present a variety of SRF applications with a summary of the existing, ongoing and proposed projects. Finally, I will discuss the status and future development of state-of-the-art SRF accelerating cavities in light of the applications.

## ADVANTAGES AND CHALLENGES

### Advantages of SRF Technologies

The technical advantages of superconducting RF (SRF) accelerating cavities over conventional RF approaches are: (1) the quality values Q of SRF cavities are about $10^9$–$10^{10}$ (surface RF resistance Rs $10^5$ times lower than that of Cu cavities) as shown in Fig. 1, and (2) accelerating gradients are around 10–25 MV/m in continuous wave (CW) mode, an order of magnitude higher than Cu cavities in CW, as shown in Fig. 2.[11] The benefits of SRF cavities in large scale applications can be summarized as follows:[12-14]

**Produce Very High Quality, High Energy Particle Beams**. The high Q value allows for us to use large aperture structures operating at lower frequencies (a few hundred MHz) with CW mode (or long macro-pulse length). The large aperture has a beneficial consequence of substantially reducing shunt-impedance, and thereby transverse and longitudinal wake-field effects, leading to relaxed accelerator alignment tolerances.

**Reduce Operational Cost.** Superconductors have a RF resistance (BCS contribution plus a residual RF resistance) below the critical temperature Tc. The high Q value of SRF cavities presents very low RF wall losses (lower peak power requirements) and low

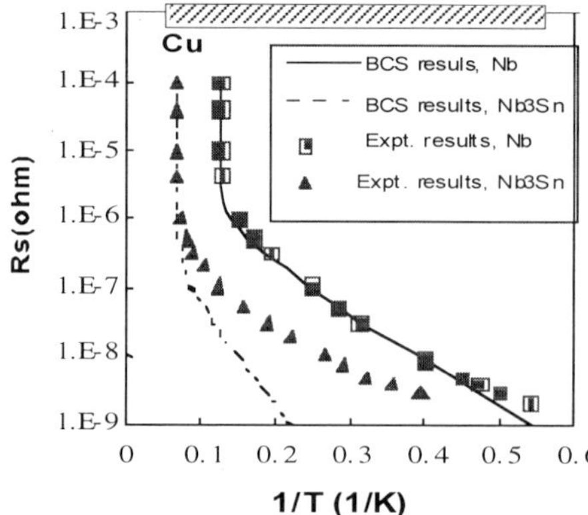


**Figure 1**. A comparison of surface RF resistances between Cu and Nb (also $Nb_3Sn$).

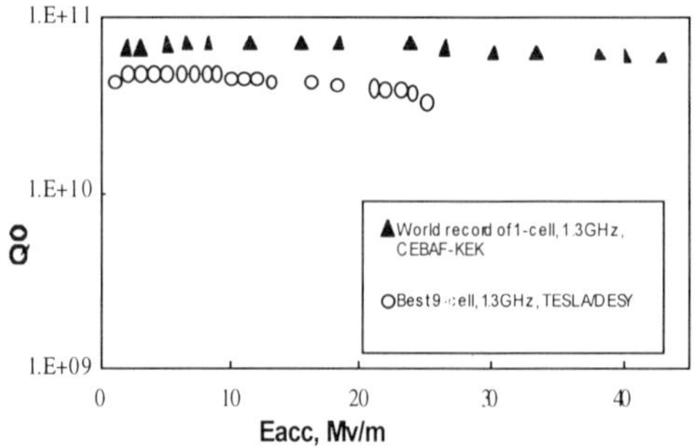


**Figure 2**. The overall RF performance of an one-cell and a 9-cell cavities.

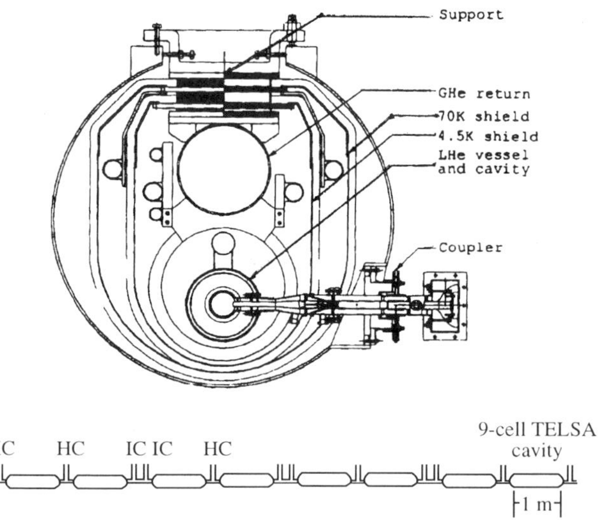


**Figure 3**. The cross-section of a TESLA cryomodule and arrangement of 8 9-cell cavities.

operational costs (net gain is still several hundred times after accounting for the refrigeration cost of an SRF cavity system).

**Reduce Accelerator Size and Construction Cost**. The high accelerating gradients make it possible to greatly reduce the construction cost and also the sizes of accelerators.

### Technical Challenges

**Continuously Push Performance to the Theoretical Limits.**

(A) Major technical efforts are focused on overcoming the most serious obstacles, field emission (FE) and thermal breakdown (TB), which prevent from reaching the theoretical performance limits (40-50 MV/m for Nb cavities[14,15]).

(B) Some applications require the SRF cavities to carry very high electron or proton beams (~1000 mA). The high input power (sub-MW/cavity) challenges designs of the power couplers and HOM couplers.[16,17]

(C) Procedures need to be identified for transferring the good performance of SRF cavities obtained in a test environment to an industrial mass production and regular operation.

**Demonstrate Cost Effectiveness.** A cost effective design of SRF systems must be demonstrated. The TESLA collaboration at DESY has a long-term goal of reducing the construction cost to from the current level of $40,000/MV to $2,000/MV by: (1) increasing Eacc, and (2) reducing cryomodule cost (Fig. 3).

## STATUS AND PROSPECTS OF SRF APPLICATIONS

The large-scale applications of SRF accelerating cavities have rapidly grown and have clearly demonstrated the technical advantages in long-term operations during the last ten years. Previously SRF cavities were mainly used in electron and heavy ion accelerators. The SRF cavities are now used to accelerate proton beams in many proposed large applications. The scales of SRF accelerators have developed from a few cavities to several hundred meters of SRF structure with hundreds of cavities. Based on these successes, some proposed applications, such as the TESLA 500-GeV $e^+ e^-$ collider, would utilize about

20,000 SRF cavities in two 10-km-active-length linacs with a superfluid liquid-He (LHe II) inventory of about 95,000 kg.[12] A brief summary of various SRF applications is presented herewith.

### SRF Structures in Successful Operation

**TRISTAN at KEK (Japan).** TRISTAN[2] was the first large scale application of SRF cavities in the world. It consists of 32 5-cell Nb cavities in 16 cryostats (Fig. 4A), 8 high power 508 MHz RF systems and one 6.5 kW LHe refrigerator. Energy gain per pass is 0.24 GeV. It has successfully served high energy research for 7 years.

**HERA at DESY (Germany).** Since 1992 the HERA electron storage ring has been equipped with 16 superconducting (SRF) and 84 normal-conducting cavities (both 500 MHz)[3] as shown in Fig. 4B. The cavities have a total of about 30,000 hours of beam operation time. Energy gain per pass is 0.08 GeV.

**LEP II at CERN (Switzerland).** Development of SRF cavities for more than 15 years yielded a gradual increase of the beam energy in LEP.[4] The majority of the 230 SRF cavities (352 MHz) are based on thin niobium films sputtered on a copper substrate. Energy gain per pass is 2.2 GeV. Fig. 4C shows a CERN SRF cryomodule.

**CEBAF at Jefferson Lab (USA).** The accelerator is designed for nuclear physics. It has a superconducting injector linac and two parallel superconducting main linacs containing 338 SRF cavities.[18] All cavities are a 5-cell design with waveguide couplers. The cavities operate at a frequency of 1.5 GHz and 2 K. The accelerator can simultaneously deliver beam to three experimental halls after 1-5 passes through the main linacs (re-circulation) for energies up to 4 GeV. Fig. 4D is the CEBAF cryomodule.

**S-DALINAC at Darmstadt (Germany).** It was designed for a free electron laser and has operated since 1990 with 8 20-cell SRF cavities (3 GHz) in 5 cryomodules.[6]

**ATLAS at Argonne (USA).** The heavy-ion linac of SRF quarter-wavelength SRF cavities began operation in 1978 and has operated and expanded continuously.[5]

**ALPI at L.N.L (Italy).** ALPI was completed in 1990 with 97 SRF quarter-wave resonators for heavy-ion acceleration.[19]

**Stanford University (USA), SUNY-StonyBrook[20] (USA) and Saclay (France)** have all contributed to the developments of SRF technologies for many years.

### SRF Structures Under Development or Construction

**TESLA-TTF at DESY (Germany).** The TESLA (TeV Electron Superconducting Linear Accelerator) Collaboration is an international R & D effort towards the development of an $e^+e^-$ linear collider with 500 GeV center of mass for multiple purposes: high energy physics, nuclear physics and laser technology. The TESLA Collaboration is building a prototype TESLA test facility (TTF) of a 500 MeV superconducting linear accelerator (~100 m) to establish the technical basis[12]. The collaboration will develop 20 km of active SRF accelerating structures at a frequency of 1.3 GHz and 25 MV/m if TESLA is finally approved (Fig. 5).

**B-Factory at KEK (Japan).** The B-Factory is designed mainly to search for the famous CP violation. It consists of 20 single-cell 500 MHz accelerating cavities as shown in Fig. 6. The very attractive feature of its SRF structures[21,22,23] is the capability of carrying very large beam current (300-1,000 mA), high input power (~300 kW per cavity) and very high HOM power extraction of 12 kW per cavity.

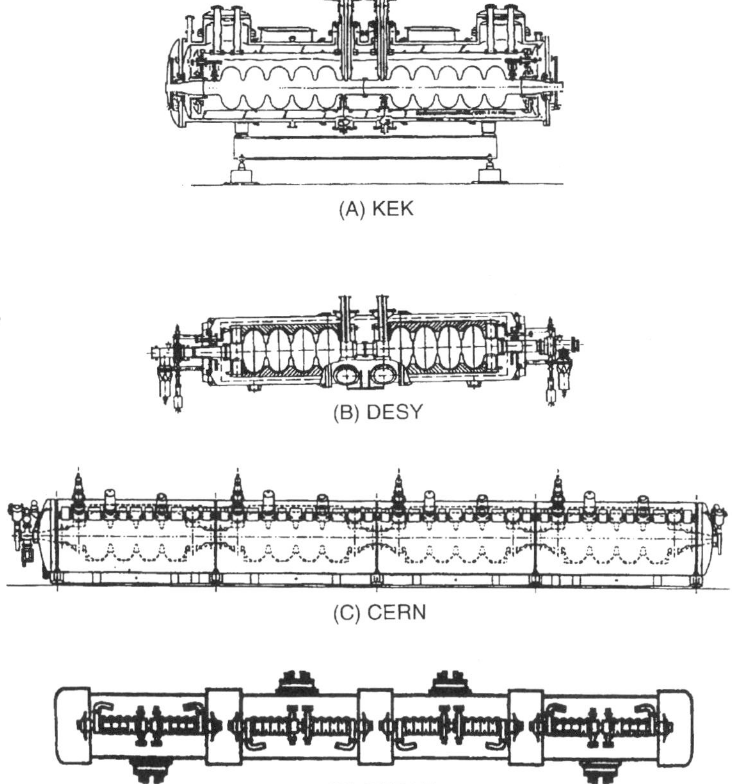


**Figure 4**. Cryomodules of the representative SRF structures in successful operation.

**CESR-II at Cornell University (USA).** Similar efforts for SRF development as KEK-B (but smaller scale) have been carried out at Cornell University for many years.[24]

**IR/UV-FEL at Jefferson Lab (USA).** A 1-kW demonstration continuous-wave infrared free electron laser[7] is being developed and will be operated in 1998. It uses the 1.5 GHz CEBAF cryomodules in both injector and linac shown in Fig. 7.

**Proton Linac at JAERI (Japan).** It will be the first large scale SRF proton linac in the world for both a nuclear waste transmutation facility and a neutron scattering facitity.[25] It will have 304 SRF cavities in 8 different β (0.45 - 0.89) regions of 699-m length as in Fig. 8.

## SRF Structures in Proposal and Consideration

**APT at LANL (USA).** The SRF accelerating system is designed for production of 3-kg/yr tritium. The linac will have 402 SRF cavities of about 1-km length and will be operated at 2 K. Prototype cavity was tested.[26]

**Energy Amplifier at CERN (Switzerland).** It was proposed by Bowman and Rubbia that it is possible to sustain a nuclear fission chain reaction under sub-critical conditions by providing the required balance of neutrons with a steady flow of neutrons from the spallation of an intense beam of protons on a solid target[9]. This method has a advantage of the safer operation since the chain reaction, if needed, can be easily controlled by acting on the proton accelerator. A superconducting linac has been proposed as the

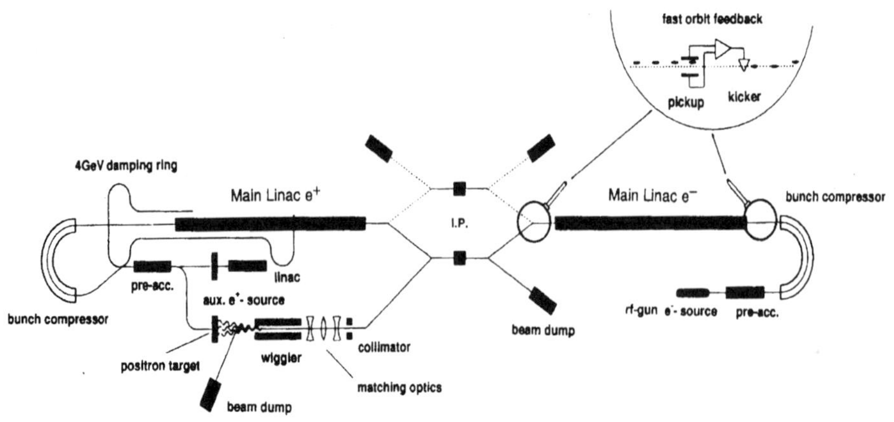


**Figure 5**. An overall layout of TESLA-500.

proton beam accelerator (Fig. 9) which drives a nuclear plant based on the concept of the energy amplifier. An example based on the net generation of 400 MW is described. This requires a proton beam energy of 1 GeV with a continuous beam current of 10 mA, corresponding to a beam power of 10 MW. Two frequencies, 360 and 805 MHz, have been considered for the linac design.

**Neutron Source (Europe)**[27] **and Muon Collider (USA)**[27] are other potential users of large scale SRF structures (not introduced here).

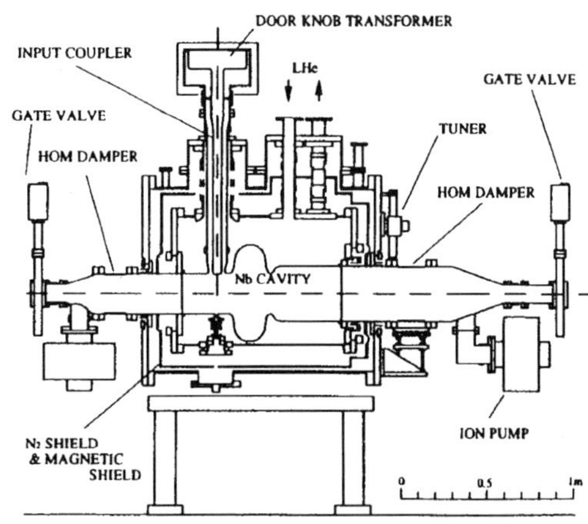


**Figure 6**. The layout of a cryomodule for the KEK-B SRF cavity.

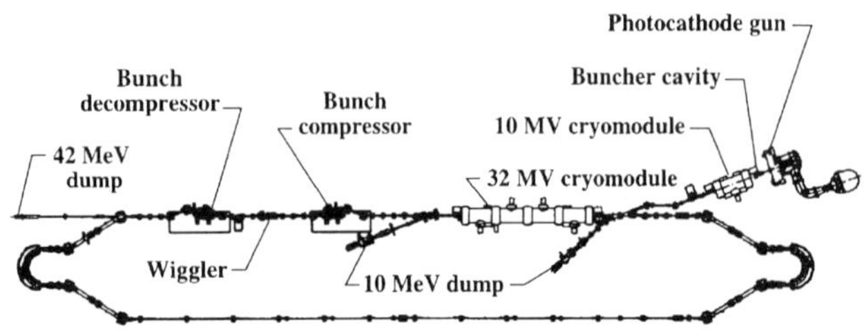


**Figure 7**. A schematic layout of the Jefferson Lab 1-kW IRFEL.

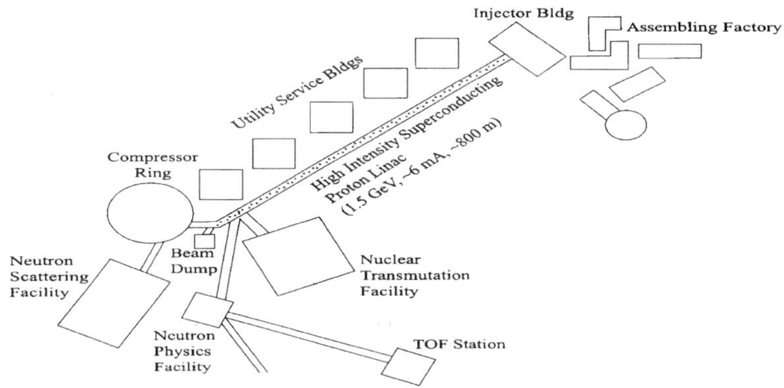


**Figure 8**. Plan of a facility layout for Proton Linac at JAERI (Japan).

## STATE-OF-THE-ART SRF TECHNOLOGIES

Two principal measures of cavity performance are the accelerating gradient Eacc and the unloaded quality value Qo. The performance highly depends on physical shapes, materials, cavity preparation and post processing, and the operational environment. Jefferson Lab's CEBAF and TESLA-TTF are used herewith as examples to introduce the representative achievements of the state of the art SRF technologies. Finally, table 1 summarized the status of SRF developments.

### Cavity Performance Before HPR, HPP & HT

The 338 5-cell cavities installed in CEBAF were fabricated and processed using the best SRF technologies developed in the 1980's, such as cavity design, mid-high RRR industrial NB sheets, cavity-fabrication by electron-beam welding and class 100 clean room assembly. Fig. 10 shows the overall statistics of the accelerating gradients reached in 324 5-cell SRF cavities[28] at Q0 above $3x10^9$. However, the high pressure rinsing (HPR), high RF power processing, (HPP) and high-temperature Ti-purification (HT) were not available.

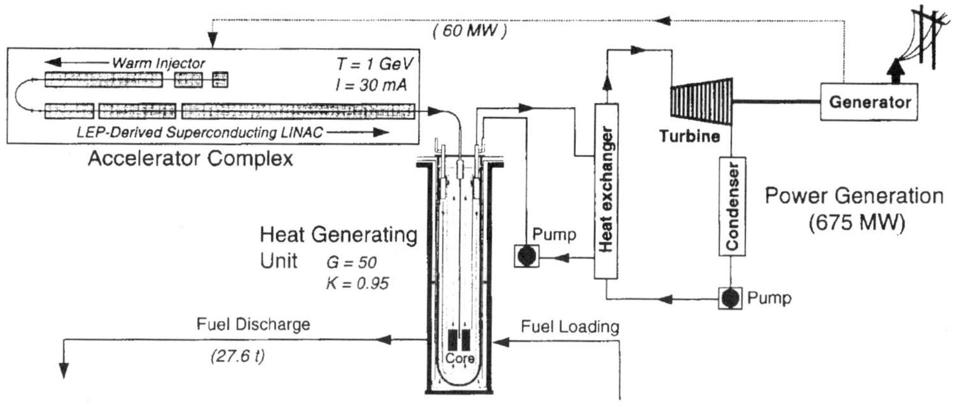


**Figure 9.** An energy amplifier with an SRF linac.

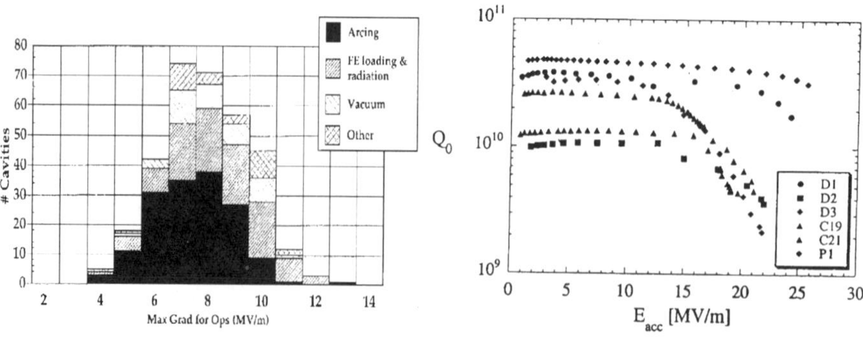


**Figure 10.** Performance of the CEBAF cavities in operation.

**Figure 11.** Performance of the TESLA cavities after HPR & HT.

## Cavity Performance After HPR, HPP & HT

The TESLA collaboration utilized almost all the technical experience developed to date. It adds the high pressure water rinse, a 5 MW klystron for high power processing on cavities and couplers and high temperature Ti-purification in an ultra-high vacuum oven into a new infrastructure.[28] The design goal for the gradient is Eacc > 15 MV/m at a quality factor of Q> $3x10^9$ for a pulse length of 800 µs. The majority of them exceeded the specification of 15 MV/m. Several cavities reached 25 MV/m at Q> $1x10^{10}$ in cw mode (as shown in Fig. 11) and up to 30 MV/m in pulsed mode.[11] The first cryomodule with 8 9-cell cavities in it reached Eacc > 12 MV/m). Cornell, KEK, Wuppertal and Jefferson Lab (through collaboration) have also conducted HT, HPP and HPR, and reached the Eacc region of 15-25 MV/m (multi-cell) and 20-40 MV/m (one cell).

**Table 1.** State of the Art SRF Cavities: Performance

| Laboratory | JLab | CERN | DESY | KEK | KEK | Cornell | Dmst. | TESLA |
|---|---|---|---|---|---|---|---|---|
| Project | CEBAF | LEP II | HERA | TRISTAN | B-Factory | B-Factory | Dalinac | (DESY, etc.) |
| f, MHz | 1497 | 352 | 500 | 508 | 508 | 500 | 3000 | 1300 |
| Structure | 5-cell | 4-cell | 4-cell | 5-cell | 1-cell | 1-cell | 20-cell | 9-cell |
| Materials | Nb | Nb/Nb-Cu | Nb | Nb | Nb | Nb | Nb | Nb |
| Cavities | 338 | 24/216 | 16 | 32 | 16/20 | 4 | 8 | 33 |
| T, K | 2 | 4.5 | 4.2 | 4.2 | 4.2 | 4.2 | 1.8 | 1.8 |
| Eacc (at D) | 5MV/m | 5-6 | 4 | 5 | 10/14 | 5-6 | 6-8 | 15/25 |
| Q at D-E | $2.4x10^9$ | $3-4x10^9$ | $2x10^9$ | $2x10^9$ | $1-2x10^9$ | $1x10^9$ | $1x10^9$ | $3x10^9$ |
| V(1-P), GeV | 0.845 | 2.2 | 0.08 | 0.24 | in AR | in CESR | 0.085 | 0.5 |
| Coupler, kW | 1.75/5 | 125/280 | 65/300 | 70/200 | 400/800 | 200/325 | 0.5 | 200, Pulse |
| I, mA | 0.2 | 6-7 | 30 | 15 | 570 | 220 | 1.7, Pulse | |

## Couplers for Very High RF Power and Beam Current

In several new applications (APT, B-Factory, sub-MW FEL, etc.), the SRF cavities need to handle with a very high beam current (300 - 1000 mA). Such a cavity requires its input coupler transferring several hundred-kW RF power into the cavity and each HOM

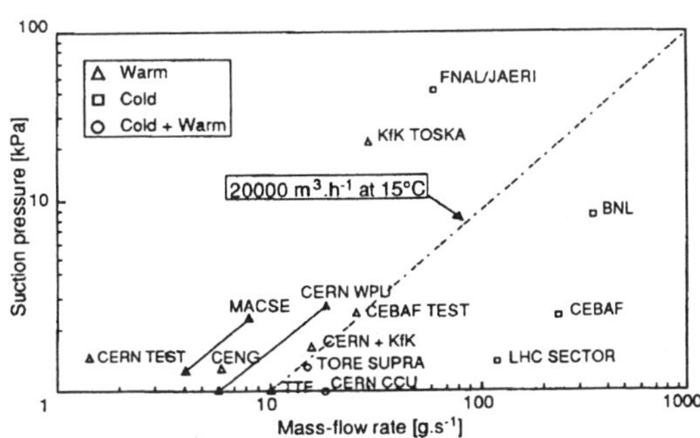


**Fig. 12.** Application range of the low-pressure He compressors.

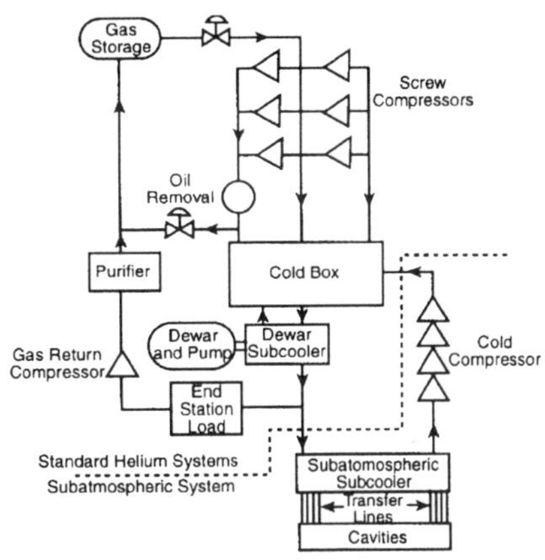


**Fig. 13.** A block diagram of the Refrigerator for CEBAF SRF Linac.

coupler extracting a 10-kW scale of HOM RF power out of the cavities. The coupler design, cooling arrangement, and the ultra-high vacuum are all challenging tasks. There have been two types of high power input coupler designs:[16,17] coaxial type design (CERN, DESY, KEK) and waveguide type (mainly Cornell). Both have successfully transferred several hundred-kW RF into cavities. A single cell cavity with a cylindrical large beam pipe was designed to propagate HOMs toward the beam axis and damp them by ferrite absorbers bonded on the inner surface of beam pipes (KEK, Cornell). The absorber was made by the HIP (Hot Isostatic Press) method and tested up to 12 kW RF power.

## Cryogenic Systems for SRF

The dissipated power in cavities has to be removed at 4 K. In order to further reduce the surface resistance Rs, SRF cavities with frequency larger than 1 GHz use He II as cryogen at 1.8K - 2K. It requires maintaining a pressure below 1.6 kPa on the heat sink of a 1.8 K cryostat system. Bringing the saturated vapor up to atmospheric pressure thus requires compression ratio exceeding 80, i.e. four times higher than that of refrigeration cycle for normal He at 4.2 K.

In a small-laboratory cryostat, this can be done by standard roots or rotary vacuum pump handling the very low pressure GHe. This technique may be pushed to larger scale. For instance, DESY has developed a large pumping station to reach 100-200 W at 2K for the TESLA-TTF SRF structure. However, 300 W at 2K is a commonly accepted limit for this technology. The alternative process is to perform compression of the vapor at low temperature, i. e., at the highest density. The pumps and recovery heat exchangers get smaller in size and less expensive. The pumping must be non-lubricated and non-contaminating, which seriously limit the choice of technology. Hydrodynamic compressors of the centrifugal or axial-centrifugal type have been used in large scale systems. Their limited pressure ratio imposes the arrangement of multistage configurations. Fig 12 shows the practical ranges of application of the different pumping techniques[29]. Fig. 13 is a block diagram of the 2 K and 4.8-kW (plus 12 kW of 50-K shield refrigeration) refrigerator at Jefferson Lab for the CEBAF SRF linac system.[30] It also delivers 10 g/s LHe to the end station. Four cold compressors with an inlet temperature of 3 K and magnetic bearings are utilized in the refrigeration cycle. It is the first multi-kW superfluid LHe cryogenic system in the world, and it has been successfully operated for three years.

## APPROACHES TOWARD THEORETICAL LIMITS

SRF cavities have successfully operated at many accelerators. However, they still do not reach the theoretical limits. Using a rule of thumb that 40 Gauss (4 mT) is equivalent to 1 MV/m accelerating gradient in cavities with the TESLA shape, the theoretical limit for Nb given by Bsh is then about 50 MV/m. Great efforts have been contributed to reaching theoretical limits in many laboratories around the world.

### Increased Cavity Thermal Conductivity

**Cavities with Thin Sputtering Nb Film on a Cu Substrate.** The main advantage of thin sputtering Nb film on a Cu substrate is the much higher thermal conductivity of Cu than Nb that significantly reduces the risk of cavity thermal breakdown.[4,14] The 230 cavities for LEP II with the techniques have been constructed and reached Eacc ~6 MV/m. However, the Q-value of these cavities decreases faster with RF power than with conventional niobium sheet cavities. To eliminate Q-degradation, the coating process parameters such as layer thickness (1-2 μm), noble gas mixture, coating temperature, and substrate treatment are studied.

**Cavities with Very High RRR Solid Nb Sheets.** Industry is now able to provide Nb sheet with RRR (RRR = R(300K)/R(4K)) of 250 (thermal conductivity ~ $RRR^{1/2}$) for cavity fabrication. Cavities using these sheets produce a range of TB at ~15 MV/m. Higher RRR of > 500 is desired for reaching a Eacc > 25 MV/m. Currently the way to increase RRR (>500) is to employ Ti-solid state gettering.[31, 32, 33, 34]

### Defeating FE and TB

**Understanding and Locating FE & TB.** The main approach to understand and locate the FE and TB of cavities is to study the hot spots (T) and X-ray (R) generated on the cavity surfaces during RF operation. Several fast and high resolution T-R mapping systems[35, 36] for cavities have been developed and successfully identify and characterize the TB and FE (Fig. 14). These cavity tests are backed by many sample tests: SEM, SIMS, TEM, scanning Auger, RRR, Tc, and AFM (Fig. 15). Examinations of Nb sheets become very important before making cavities.

**Preventing FE and TB Sources.** Cleaning techniques similar to those utilized in the semiconductor industry are used to remove potential contamination for FE and TB from the cavity's RF surfaces. Cleanliness during chemical etching, water rinsing (high purity water of 18 MΩ-m) and clean room have played an important role in achieving higher Eacc. The very high pressure (80 bar) water rinsing (HPR) device developed by CERN[37] is very helpful in removing foreign particles and is adopted by many labs.

**Destroying FE Sites in Situ.** Despite how well a job is performed to eliminate FE, there is always a possibility that particles escape removal and stay on cavity surfaces. Therefore a technique that can eliminate the emitters in situ is highly desirable. The techniques are high power processing (HPP) developed at Cornell University[38] and He-processing[34, 39]. During HPP a high power RF pulse is applied to the cavity in situ and eliminates the FE through an explosive process which destroys the remaining FE emitters. During the processing the cavity operates in the FE region in the pressure of a partial He gas just below discharge, causing a reduction of the electron loading. At DESY an HPP

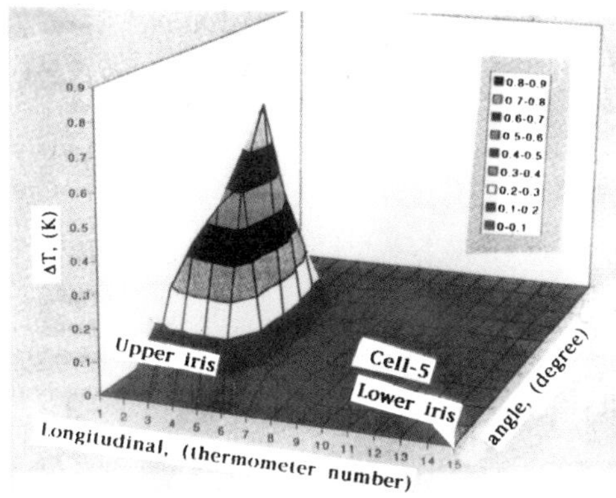


**Figure 14**. Identification of a TB location m at 12 MV/ (Eacc) on a TESLA cavity surface.

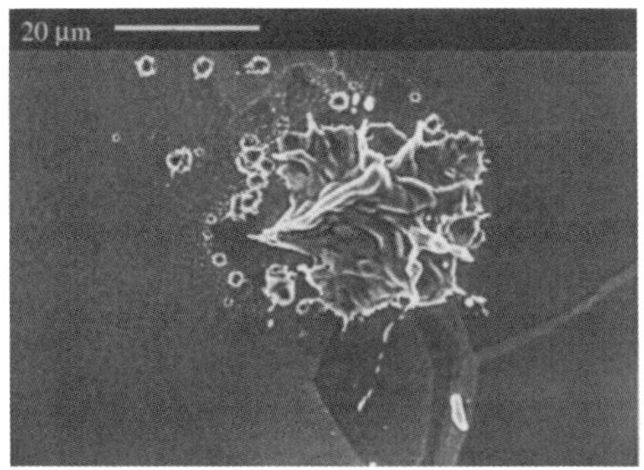


**Figure 15**. An emitter remained active at the field of Epk=17 MV/m.

facility (4 MW, 2 ms) has been used successfully in raising Eacc. At Jefferson Lab He processing is currently applied to SRF cryomodules in the CEBAF linac to gain Eacc.

## New Materials Other Than Nb for SRF Cavities

The limit magnetic field in RF is Bsh. The Bsh of $Nb_3Sn$ is about 400 mT, which represents 100 MV/m of Eacc. Several cavities with $Nb_3Sn$ coating on solid Nb sheet have been developed and tested by a collaboration of Wuppertal Univ., CEBAF, etc. Fig. 16 is the initial results[11]. It shows some encouraging results that an $Nb_3Sn$ cavity can be operated at 4 K with an acceptable Q and Eacc of CEBAF design while Nb cavity must be operated at superfluid He (2K).

Materials with a high-enough BCS energy gap 2Δ would allow a considerable reduction of the BCS surface resistance. According to the BCS theory, Δ is related to the critical temperature $\Delta = 3.5\ KT_c$. Therefore, superconductors with a higher critical temperature than niobium are possible candidates for cavity applications as shown in Fig. 17. Among them are the well-known "old" high-$T_c$ superconductors NbN (17 K), $Nb_3Sn$ (18.5 K), and "new" high-$T_c$ superconductors such as $YBa_2Cu_3O_{7-\delta}$ ($T_c$ = 93 K), which have also been studied or discussed for possible RF applications.[40]

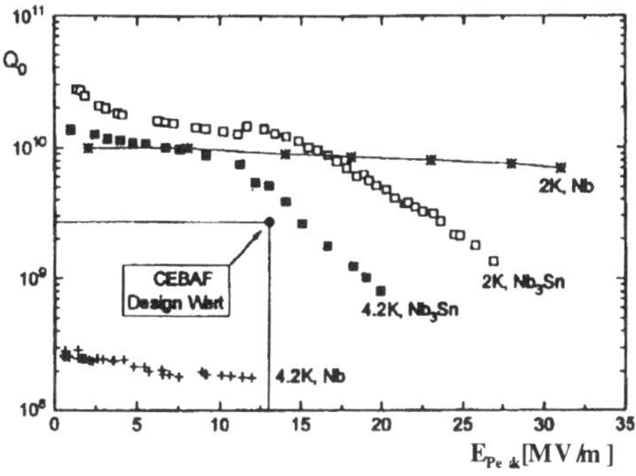


**Figure 16.** Initial results from a $Nb_3Sn$ cavity.

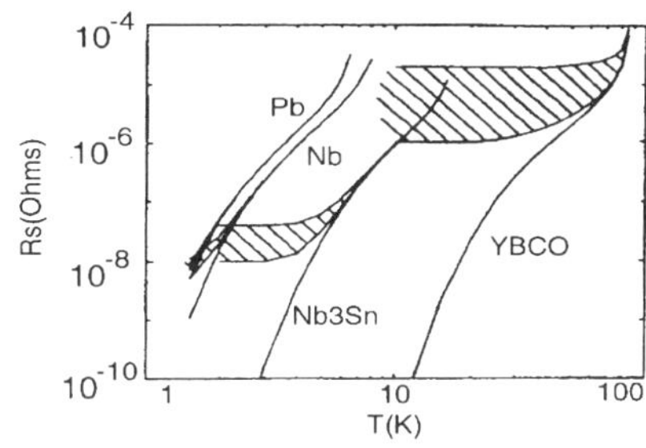


**Figure 17.** Surface resistances as functions of the temperatures for several LTc and HTc materials.

## ACKNOWLEDGMENT

Appreciation is given to colleagues in the field of SRF technologies for many fruitful discussions and information. I also sincerely thank S. Corneliussen, P. Kneisel, and C. Bohn for carefully reviewing this paper, and S. Spata and T. Wang for editing it.

## REFERENCES

1. C. Reece et al. Proceedings of PAC-95 conf. Dallas, TX. USA, (IEEE), p. 1512, 1996.
2. S. Noguchi, KEK, Proceedings of the 7th Workshop on RF Superconductivity, Saclay, France, p. 163. 1995.
3. B. Dwersteg, et al., Proceedings of the 7th Workshop on SRF, Saclay, France, p. 151. 1995.
4. G. Geschonke, CERN, In this Conference.
5. K.W. Shepard, Proceedings of the 6th workshop on SRF, Ed. by R. Sundelin, CEBAF, USA, 1993.
6. S. Döbert, et al. Proceedings of the 7th Workshop on RF Superconductivity, Saclay, France, p. 57. 1995.
7. C. Rode, In this Conference.
8. D. Chan, In this Conference. (withdrawn)
9. C. Rubbia, et al. CERN/LHC/1996-11, (EET).
10. H.T. Edwards, TESLA Parameters Update, DESY/TESLA 94-22, 1994.
11. W.D. Möller and P. Kneisel, Proceedings of the 7th Workshop on RF Superconductivity, Saclay, France, p. 163, 1995.
12. D. Edwards, TESLA-TTF Design Report, TESLA 95-01, DESY, 1995.
13. CEBAF Design Report, Newport News, VA. May 1986.
14. W. Weingarten, CERN Accelerator School (Report), May 1995. (CERN 96-03).
15. H. Padamsee, et. al., Proceedings of PAC-95 conf. Dallas, TX. USA, p. 1515, May 1995.
16. Kirchgessner, Proceedings of PAC-95 conf. Dallas, TX. USA, p. 1469, May 1995.
17. T. Tajima et al. Proceedings of PAC-95 conf. Dallas, TX. USA, p. 1620, May 1995.
18. J. Delayen, In this Conference.
19. V. Palmieri, Proceedings of the 7th Workshop on SRF, Saclay, France, p. 237, 1995.
20. I. Ben-Zvi, et al. Proceedings of the 7th Workshop on RF Superconductivity, Saclay, France, p. 55, 1995.
21. S. Mitsunobu, et al. Proceedings of PAC-97 Vancouver, Canada, 1997.
22. K. Hosoyama, In this Conference.
23. T. Tajima, et al. Proceedings of PAC-97, Vancouver, Canada, 1997.
24. S. Belomestnykh, et al. Proceedings of PAC-97, Vancouver, Canada, 1997.
25. N. Ouchi, Private Conversation, JAERI, Japan, 1997.
26. B. Rusnak, Private Conversation, LANL, USA, 1997.
27. (A) H. Lengeler, Proceedings of 6th workshop on SRF, CEBAF, USA, 1993.
    (B) Q. S. Shu, et al., PAC-97, Vancouver, Canada, 1997.
28. (A) C. Reece, Private Conversation, CEBAF, USA, 1997.
    (B) S. Wolf, PAC-95 conf. Dallas, TX. USA, (IEEE), p. 680, May 1995.
29. Ph. Lebrun, Proceedings of CERN Accelerator School, p. 321, 1995.
30. C. Rode, Proceedings of PAC-95 conf. Dallas, TX. USA, (IEEE), p. 1994, May 1995.
31. Q.S. Shu et al. Proceedings of ASC-96, Pittsburg, Pa. USA, 1996.
32. P. Kneisel, J. Less Common Metal, 139, 94, 1973.
33. H. Padamsee, et al. IEEE Tran. Mag.-21, 1007, 1985.
34. Q. S. Shu et al. IEEE transaction on magnetics, Vol. 25, No 2, 1989.
35. J. Knobloch, Proceedings of PAC-95, Dallas, TX. USA, p. 1623, 1995.
36. Q. S. Shu, Proceedings of the 7th workshop on SRF, Saclay, France, p. 163, 1995.
37. Ph. Bernard et al. PAC-92, Berlin, Germany, p. 1269, March 1992.
38. J. Graber, Ph. D. Dissertation, Cornell University, 1993.
39. H. A. Schwettman et al. Journal of Applied Physics, Vol. 45, No 2, p. 914, Feb. 1974.
40. B. Bonin, CERN Accelerator School (Report) , CERN 96-03, Hamburg, Germany, 1995.

# THE JEFFERSON LAB SUPERCONDUCTING ACCELERATOR*

J. R. Delayen

Jefferson Lab
12000 Jefferson Avenue
Newport News, VA 23606 USA

## ABSTRACT

The Continuous Electron Beam Accelerator Facility (CEBAF) at the Jefferson Lab is a 5-pass recirculating superconducting linac capable of delivering beam simultaneously to three experimental halls. The accelerator contains 338 niobium cavities operating at 1497 MHz in superfluid helium and was designed to accelerate 200 μA of electrons to 4 GeV. Each cavity is individually powered by a 5 kW klystron and controlled in phase and amplitude. First beam delivery to halls C, A, and B occurred in November 1995, April 1996, and December 1996 respectively. During 1996, 6 nuclear physics experiments were successfully completed in hall C. An upgrade program to raise the maximum energy of the accelerator is under way, and single-pass energy of 1 GeV has already been achieved. Jefferson Lab is also applying its superconducting rf and accelerator technology to other applications such as a Free Electron Laser for Navy and industrial applications.

## SYSTEM OVERVIEW

CEBAF was designed and constructed to accelerate an electron beam to 4 GeV by recirculating five times through two, 1497 MHz superconducting linacs, each providing 400 MeV per pass at the design accelerating gradient of 5 MV/m. As shown in Figure 1, the two antiparallel linacs are connected by 180 degree isochronous, achromatic recirculating arcs. The design maximum beam current is 200 μA CW (corresponding to a beam loading current of 1 mA). This beam can be divided among three interleaved 499 MHz bunch trains with each bunch train having individually adjustable current to match the requirements of the three experimental halls. The 499 MHz beams can be individually extracted after each pass thus simultaneously providing each experimental hall with a beam of different energy end current. The extraction is provided by room-temperature 499 MHz cavities which impart separate transverse kicks to the three beamlets.[1,2] The transverse kicks are amplified by quadrupoles and septum magnets to either deflect a beamlet to an experimental hall or a recirculating arc for further acceleration. This geometry allows only a single beam of a given energy to be sent to an experimental hall, except for the highest energy pass, where all three beams can be sent to the three halls simultaneously.

* Supported by U.S. DOE contract DE-AC05-84ER40150

A large fraction of the planned experiments will require polarized electrons. The polarized gun installed, at present, in the injector uses a bulk gallium arsenide cathode illuminated by a highly stable 1497 MHz rf gain-switched diode laser operating at 780 nm.[3] The quantum efficiency of the cathode declines over time but stabilizes around 1%. The increased lifetime of the cathodes at high currents is the focus of further development. Typical currents produced by the polarized gun are 75 μA with a measured polarization of 36%. A three-laser system will be installed next year so that each of the three 499 MHz beamlets will be generated independently of the others by its own laser.

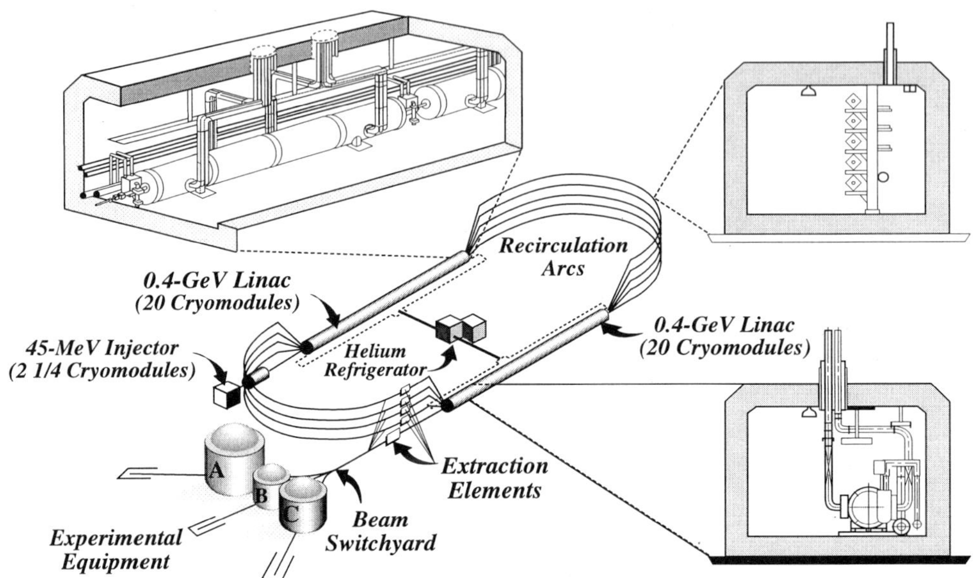


**Figure 1.** Recirculating, cw, superconducting linac.

The superconducting RF (SRF) system consists of 20 cryomodules in each of the two linacs and 2 1/4 cryomodules in the injector. Each cryomodule contains 8 accelerating cavities for a total of 338 cavities. The cavities, which are 50 cm long, were designed to operate at a gradient of 5 MV/m, which corresponds to an energy gain of 20 MV for each cryomodule. The cavity performance has greatly exceeded the design goal.[4] Each cavity is individually powered by a 5 kW klystron and controlled by a low-level rf control system to stabilize the rf field amplitude and phase. Having each cavity individually controlled allows for precise and careful optimization of the system and results in excellent phase and amplitude stability of the accelerating fields in the cavities. The entire rf system can be operated and monitored from a single control screen that provides the operators with the ability to quickly diagnose and locate system faults.[5]

The central helium refrigerator provides 2 K helium to the cavities with 95% availability. The system, capable of providing 4800 W of refrigeration, presently operates with a constant heat load of 3100 W, of which 1500 W is the actual rf heat load.

During commissioning, CEBAF replaced its in-house developed control system with EPICS (Experimental Physics and Industrial Control System), which is the product of an international, multi-laboratory collaboration. EPICS performs the task of monitoring and controlling the numerous systems which comprise the accelerator. With over 40,000 control points and 120,000 database records, CEBAF has become one of the largest installations of

EPICS worldwide. The open system architecture provided by EPICS allows access to the control system using a large variety of programs such as MATLAB, Tcl/Tk, UNIX, and C.[6]

The beam transport system is comprised of more that 2200 magnets, which were individually characterized and mapped prior to installation. Because of the required accuracy of the fields generated by the magnets, each magnet is tracked through a hysteresis loop to guarantee reproducibility, and the dc power supplies for the magnets have relative accuracy of $10^{-5}$.

Additional details on the design, commissioning, and operation of CEBAF can be found in previous conferences.[7–9]

## PERFORMANCE AND IMPROVEMENTS OF THE SRF SYSTEM

The design energy for CEBAF was 4 GeV, which corresponds to an average accelerating gradient in the cavities of 5 MV/m. The actual performance of the cavities far exceeded this value immediately after commissioning.[4] Nevertheless, it was lower than what had been achieved with "bare" cavities on a test stand, and much lower (by a factor of 5–10) than the ultimate potential of the superconducting material, niobium.[10] Furthermore, there is and will always be a desire and a push from the nuclear physicists to achieve higher and higher energies.[11]

Many phenomena limit the performance of superconducting cavities below what could be expected from the fundamental properties of the superconductor. Some are material related (field emission, x-ray radiation, quenches, excess power dissipation, *etc.*), some are system related (waveguide vacuum, beamline vacuum, arcing, *etc.*).[10]

Two different approaches have been used to improve the performance of the superconducting cavities *in situ.*[12] The first one, helium processing, is used to reduce field emission and x-ray production. This technique has been used extensively in the past, but mostly as an R&D tool. It involves the introduction of a small amount of helium gas in a cold cavity such that the gas pressure is lower than that needed to generate an rf discharge. The cavity is operated at high fields and the electrons emitted from a defect ionize the helium gas, and the positively charged helium ions stream back and bombard the emission site. Most of the benefit from this technique is obtained in less than an hour. Subsequently the cryomodule is warmed to about 20 K to remove the helium gas from the cavities.

The second technique, waveguide vacuum processing, is a form of pulsed rf power processing used to clean the thermal transition region between the warm and cold rf windows. Depending on the gradient in each cavity and the current being accelerated, the rf field profile in that region changes, leading to a change in the thermal profile. This shifting thermal profile can produce an outgassing of condensed gases, leading to arcing and waveguide vacuum trips.

To date, we have applied helium-rf processing at 2 K to 90 cavities in 12 of the 41 installed CEBAF cryomodules. We have also applied pulsed-rf processing to condition the input waveguide region in 10 cryomodules.[12]

Progress against field emission was observed by the reduction of x-radiation produced by the field-emitted electrons. Figure 2 illustrates this shift as measured by the Geiger-Muller tubes outside the cryomodule.

Many of the cavities subjected to helium processing were previously limited by arcing at the cold window. Following processing, the cavities could be operated stably at higher gradients. Figure 3 shows the distribution of maximum gradients for the processed cavities according to their performance-limiting constraint before and after helium processing.

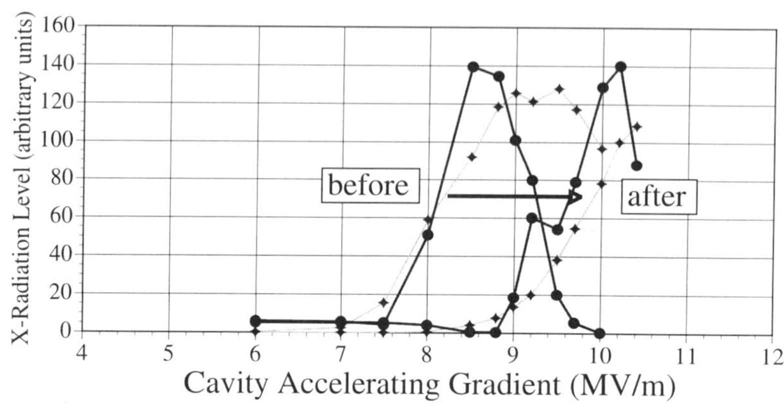


**Figure 2.** Radiation reduction with He processing.

**Improvement of Cavity Performance with Helium Processing**

**Distribution of Maximum Gradients by Type of Limitation**

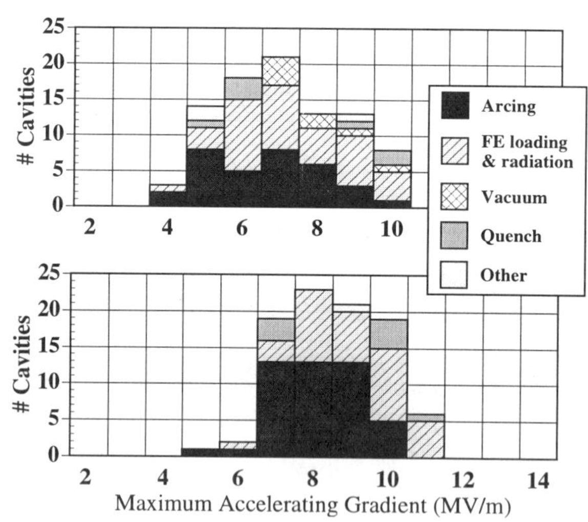


**Figure 3.** Distribution of maximum cavity gradients before (upper) and after (lower) helium processing.

**Table 1.** Improvement of cavity $Q$ due to helium processing

| Cavity # in NL03 | $E_{acc}$ (MV/m) | $Q_0 (\times 10^9)$ | |
|---|---|---|---|
| | | Before | After |
| 2 | 7.7 | 1.5 | 3.3 |
| 4 | 8.7 | 2.5 | 4.1 |
| 6 | 6.8 | 2.1 | >5.3 |
| 8 | 8.8 | 2.85 | 4.7 |

Careful measurements were made of the unloaded $Q$-factor for cavities in the first cryomodule processed. All cavities which were limited by field emission showed improvement. An example of this improvement is shown in Table 1.

The waveguide vacuum in 10 cryomodules was conditioned by applying low duty cycle rf pulses to each cavity, first detuned, and then tuned. The forward power was gradually increased to 4-5 kW, subject to maintaining the waveguide vacuum better than $1\times10^{-7}$ torr. The process requires ~4 hours. In many cases, the vacuum stability improved markedly. This stability reduces a complication that arises when a cavity is slightly mistuned. The relative merits of this compared with a thermal cycle to 20 K or higher remain an open question.

Using the two processing techniques on 11 of the cryomodules has resulted in an increase of the installed voltage of 60 MV, corresponding to an increase in beam energy of 300 MeV for five passes. This gain is shown in Figure 4.

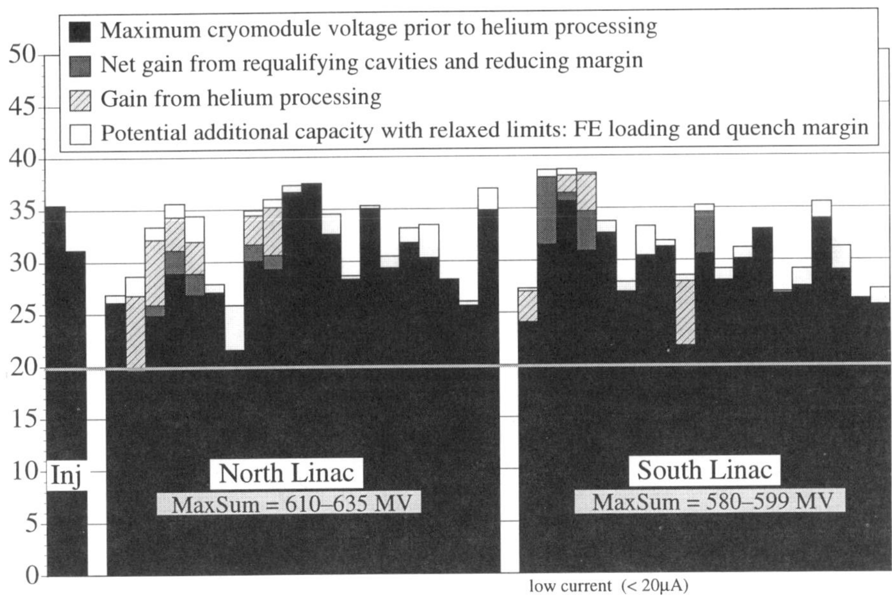


**Figure 4.** Maximum SRF cavity voltage per cryomodule in CEBAF.

Earlier this year, during machine development time, an attempt was machine to run the accelerator at the highest energy possible.[9] Because of limitations due to magnet power supplies in the recirculation arcs, only single pass operation was possible. A 90 μA beam was accelerated to 1.16 GeV in a single pass (equivalent to a 5.6 GeV, 18 μA 5-pass beam). The average accelerating field in the active cavities was 7.8 MV/m.

The maximum current required, and delivered, so far to the users has been 100 μA. During high-power tests currents of 180 μA have been accelerated to 4 GeV, just slighly below the nominal design value of 200 μA.[9]

## FUTURE PLANS

The first priority, at present, clearly is to achieve reliable, routine, 4 GeV, 3-hall operation, and complete the approved nuclear physics program. At the same time, there is an increasing interest in experiments requiring 5 to 6 GeV, and we have embarked on a program to incrementally increase CEBAF's energy from 4 to 6 GeV. The *in situ* processing of 11 cryomodules mentioned previously is the first step in that direction, and we plan to complete the processing on the remaining cryomodules during the next year. Experiments at 5 GeV are scheduled to take place in 1998. Loking further to the future, just beyond 2000, our user community has recommended to upgrade the energy to the 10 GeV range in an evolutionary manner, *i.e.* without extensisve impact on operation.

The recirculating arcs have been designed to accommodate a 20 GeV machine, and exploratory R&D is taking place in that direction. Such an upgrade would probably require removing and replacing the existing acceleration and transport systems.

Jefferson Lab is also applying the SRF accelerator technology to high-power, cw free electron lasers for the production of infrared, and later ultraviolet, light. This new program is presented in detail somewhere else in these proceedings.[13]

## REFERENCES


1. C. Leemann, C. G. Yao, "A Highly Effective Defelcting Structure", Proc. 1990 Linear Acc. Conf. 232 (1990).
2. C. Hovater, *et al.*, "The CEBAF RF Separator System", Proc. 1996 Linear Acc. Conf. (1996).
3. M. Poelker, "High-Power Gain-Switched Diode Lasr Master Oscillator and Amplifier", Appl. Phys. Lett., **67**, 762 (1995).
4. C. Reece, M. Drury, C. Hovater, J, Mammosser, T. Powers and J. Preble, "Performance Experience with the CEBAF SRF Cavities", Proc. 1995 Part. Acc. Conf., 1512 (1995).
5. J. Karn, B. Dunham and M. Tiefenback, "Operational Monitoring of the CEBAF RF System", Proc. 1995 Part. Acc. Conf. 2271 (1995).
6. C. Hovater, *et al.*, "Operational Experience with the CEBAF Control System", Proc. 1996 Linear Acc. Conf. (1996).
7. R. Legg, "Operational Experience at CEBAF", Proc. EPAC 1996.
8. B. Dunham, "Jefferson Lab, a Status Report", Proc. 1996 Linear Acc. Conf.
9. A. Hutton, "Operation of CEBAF with Heavy Beamloading", Proc. 1997 Part. Acc. Conf.
10. P. Kneisel, these Proceedings.
11. H. Grunder, "CEBAF Commissioning and Future Plans", Proc. 1995 Part. Acc. Conf.
12. C. Reece, M. Drury, M. Rao, V. Nguyen-Tuong, "Improvements of the Operational Performance of SRF Cavities via *In Situ* Helium Processing and Waveguide Vacuum Processing", Proc. 1997 Part. Acc. Conf.
13. C. Rode, these Proceedings.

# JEFFERSON LAB HIGH-POWER FREE-ELECTRON LASER*

C. H. Rode and the Jefferson Lab FEL Team

Thomas Jefferson National Accelerator Facility
12000 Jefferson Avenue
Newport News, VA 23606-1909 USA

## ABSTRACT

In partnership with the U.S. Navy, high-technology corporations, and research universities, Jefferson Lab is building a superconducting radio-frequency (SRF) accelerator–driven free-electron laser (FEL) and is outfitting an FEL user facility. The 1 kW, 3 μm infrared (IR) laser—which was being installed in the newly constructed user facility as of summer 1997—is the first step in a program to develop high-average-power SRF-based IR and ultraviolet (UV) FELs for multiple manufacturing applications as well as for defense-related applied research and basic scientific research. This initial FEL is driven by a 42 MeV, 5 mA recirculating SRF linac closely similar to the much larger SRF linac in Jefferson Lab's 4 GeV, 200 μA Continuous Electron Beam Accelerator Facility (CEBAF). The FEL will demonstrate 75% energy recovery. Its linac will be cooled by the existing CEBAF cryogenic system. At Jefferson Lab, an infrastructure of facilities and people already supports the advance of SRF and closely related technologies in the furtherance of the lab's primary mission of nuclear and particle physics. This paper, after describing that infrastructure, summarizes the opportunity represented by SRF-driven FELs and reports on the program now underway to develop them.

## INTRODUCTION

The Thomas Jefferson National Accelerator Facility, a $600 million U.S. Department of Energy national laboratory, serves basic science by carrying out a primary mission of nuclear and particle physics research. To enable this basic science, Jefferson Lab developed technology and constructed, commissioned, and is now operating the world's pioneering large superconducting radio-frequency (SRF) electron accelerator. A technologically related secondary mission now also exists for Jefferson Lab: applied research to develop SRF-based free-electron lasers. Powerful, multipurpose free-electron lasers (FELs) driven by SRF electron accelerators prospectively represent substantial, cost-effective new manufacturing capabilities for industry.

To conduct its primary mission, Jefferson Lab uses its Continuous Electron Beam Accelerator Facility (CEBAF),[1,2] in which the laboratory's 4 GeV SRF accelerator cost-effectively provides high-quality, CW electron beams to three experiment halls. Such beams are the key to the basic research conducted at Jefferson Lab. At more modest energies, such beams are also the key to producing coherent, single-wavelength light—that is, laser light—with much higher average power than is typically available from

* Supported by U.S. DOE contract DE-AC05-84ER40150

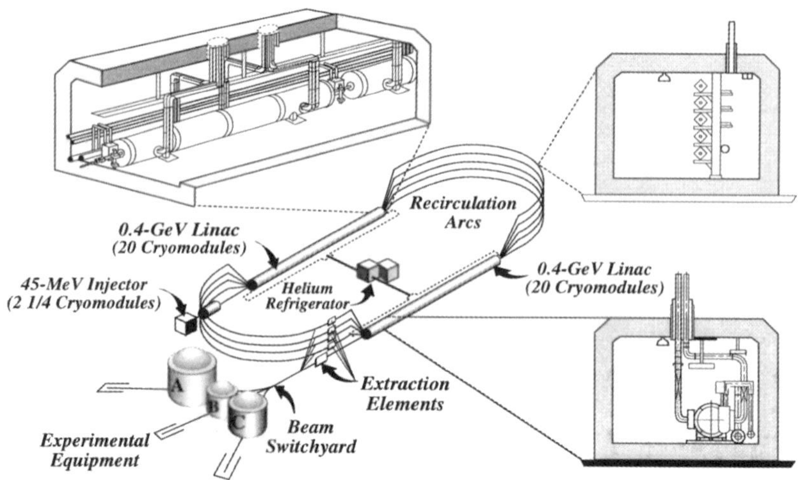


**Figure 1.** Schematic view of Jefferson Lab's SRF-based, 4 GeV Continuous Electron Beam Accelerator Facility (CEBAF). Electron beams originate in the injector, cycle as many as five times through a pair of antiparallel 400 MeV linear accelerators (linacs), and are split off for simultaneous experiments in each of three nuclear and particle physics experiment halls.

conventional lasers, and, also unlike conventional lasers, with tunability to any of a wide range of wavelengths. Industry has defined a clear need for a cost-effective source of such light, and has identified Jefferson Lab's SRF technology as the key technology for achieving it.

At Jefferson Lab, an infrastructure of facilities and people already supports the advance of SRF and closely related technologies in the furtherance of the lab's primary mission. This paper, after describing that infrastructure, summarizes the opportunity represented by SRF-driven FELs and reports on the cooperative industry-university-government-laboratory program now underway at Jefferson Lab to develop them.

## SRF INFRASTRUCTURE AT JEFFERSON LAB

CEBAF at Jefferson Lab (Figure 1) was constructed in the late 1980s and early 1990s and is based on SRF accelerating cavities (Figure 2) originally developed at Cornell University. The cavities operate at 2 K in cryomodules (Figure 3) cooled by liquid helium from a 4.8 kW Central Helium Liquifier (CHL, Figure 4).

**Figure 2.** A pair of five-cell, 1497 MHz CEBAF SRF accelerating cavities undergoing assembly in the Class 100 Clean Room in Jefferson Lab's SRF technology-development facility. Such cavities are the heart of the accelerator, and have substantially exceeded performance specifications in both accelerating gradient and $Q$, the cavity quality factor.

**Figure 3.** The SRF cryomodules of the CEBAF south linac. Each cryomodule contains four pairs of SRF cavities. In the twenty cryomodules of each accelerator and in the two and one-quarter cryomodules of the injector, the CEBAF accelerator uses a total of 338 cavities—over an order of magnitude more than will be needed in an FEL driver accelerator.

In addition to CEBAF itself, and in addition to a staff of physicists, engineers, and technicians who are experts in SRF technology—as well as a cadre of laser and FEL experts—Jefferson Lab's assets for SRF-based FEL development include an SRF technology-development facility, the "Test Lab." During the late 1980s, the laboratory renovated and equipped this pre-existing 57,000 square foot building to develop SRF and related technology specifically for application in CEBAF, and to process, assemble, and test the cavities, cryomodules, and other SRF technology hardware that went into CEBAF. The Test Lab now serves as a development facility for the incremental, evolutionary advances in SRF that are both foreseen and needed for future upgrades of the CEBAF accelerator (and for other applications elsewhere as well). The Test Lab has its own cryogenics plant as well as facilities such as clean rooms, cavity acid-treatment and processing equipment, cavity pair and cryomodule assembly shops, and cavity and cryomodule testing apparatuses, as well as beam-transport magnet measurement tools and an electron beam welder. The Test Lab also contains facilities for development of electron-source technology—that is, for development of the crucial accelerator subsystems that originate electron beams for further acceleration. Electron-source technology advances are vital both for future upgrades of CEBAF and for FEL development.

**Figure 4**. The 2 K and 4 K cold boxes of the 4.8 kW CEBAF Central Helium Liquifier.

## THE OPPORTUNITY

### Why FELs?

In principle, laser light offers an efficient, spatially and chemically precise, and environmentally benign way to process materials, as in surface-modifying polymers and metals and in micromachining. However, cost, capacity, wavelength, and pulse-length constraints have limited or even stymied progress with conventional lasers. Therefore a number of high-technology corporations (some of which are listed below) and research universities, believing in the potential of SRF-driven FELs to overcome these constraints, have formed the Laser Processing Consortium, and have joined with Jefferson Lab to develop the needed laser technology. Consortium members plan a range of industrial applications. In the area of polymer surface processing, they intend to develop:

- Amorphization to enhance adhesion (3M)
- Fabric surface texturing (DuPont)
- Enhanced food packaging (DuPont/TetraPak)
- Induced surface conductivity (Xerox)

In micromachining:

- Ultrahigh-density CD-ROM technology (3M, Aerospace)
- Surface texturing; micro-optical components (3M)
- Micro-Electrical Mechanical Systems (MEMS) (Northrop Grumman, Aerospace)

In metal surface processing:

- Laser glazing for corrosion resistance; adhesion pre-treatments (Ford, GM, Newport News Shipbuilding, Armco, Virginia Power)

In electronic materials processing:

- Large-area processing (flat-panel displays); laser-based "cluster tool" for combined deposition, etching, *in situ* diagnostics (IBM, Xerox, Lucent)

Concerning these prospective applications, an independently convened government review panel reported that the "potential commercial value of the technology is significant, impacting several multibillion-dollar markets."[3] Moreover, significant additional applications exist in basic and applied research. One key research area is the potential usefulness of future FELs for shipboard antimissile defense—the motivation for the bulk of the funding for the FEL development program's initial stage, which is being provided by the U.S. Navy.

### Why SRF-Driven?

In a free-electron laser, unbound—that is, free—relativistic electrons are transported from a linac to undulate transversely in the sinusoidal magnetostatic field of a "wiggler." (See Figure 5.) The resulting light output is initially spontaneous emission, but the light bounces back and forth inside an optical cavity until it is amplified to saturation. A conventional laser typically produces a single, set wavelength, but light from an FEL can be tuned to wavelengths throughout a wavelength range by varying the input electron energy or the magnetostatic field. With a CW input beam, the FEL can produce its wavelength-tunable light at high average power. The electron accelerator that provides such a CW electron beam is the technological heart of an FEL, and using an SRF accelerator will make the FEL notably cost-effective. SRF's high accelerating gradients translate into a compact FEL design with lower capital cost than for one with a corresponding conventional accelerator. Moreover, SRF's negligible RF losses allow a high rate of recovery of the unspent energy in the beam after it transits the wiggler. The energy-recovery process, with the beam returned to the SRF cavities in the main, post-injector part of the linac for conversion back to RF power, translates into a lower operating cost. An additional advantage of the SRF approach is beam apertures some three times larger than in a conventional accelerator, thereby minimizing beam instabilities and beam-loss-induced radiation.

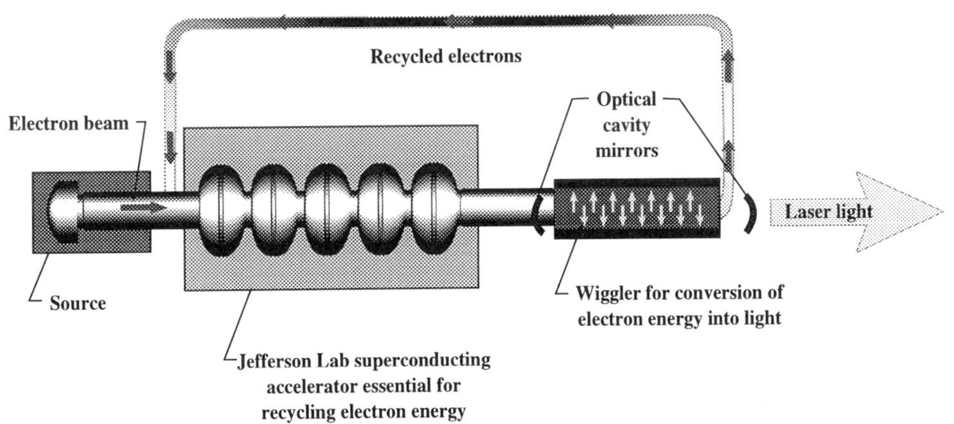


**Figure 5**. Basic concept of an SRF-driven FEL.

## FEL DEVELOPMENT

The potential of Jefferson Lab's SRF technology for industry was identified in the early 1990s by several high-technology corporations. By 1993 the Laser Processing Consortium had coalesced; by 1995 the U.S. Navy had become quite interested as well. A plan was formulated for beginning FEL development with a kilowatt-scale infrared (IR) device—about two orders of magnitude more powerful than previous, non-SRF FELs. Experience gained with this initial "IR Demo" FEL, funded mainly by the Navy, is planned to lead to FELs at still higher powers not only in the IR but in the ultraviolet (UV), where many of the most compelling potential applications lie. After the IR Demo and an upgrade of the IR Demo, the consortium intends to build a kilowatt-scale ultraviolet demonstrator FEL (the UV Demo), to be followed by scale-up and further development eventually leading to a 100 kW prototype device for cost-effective manufacturing use at industrial sites. Figure 6 shows the power output of the IR and UV Demo FELs at the various wavelengths through which they are tunable, and, for contrast, shows capabilities of typical conventional lasers as well.

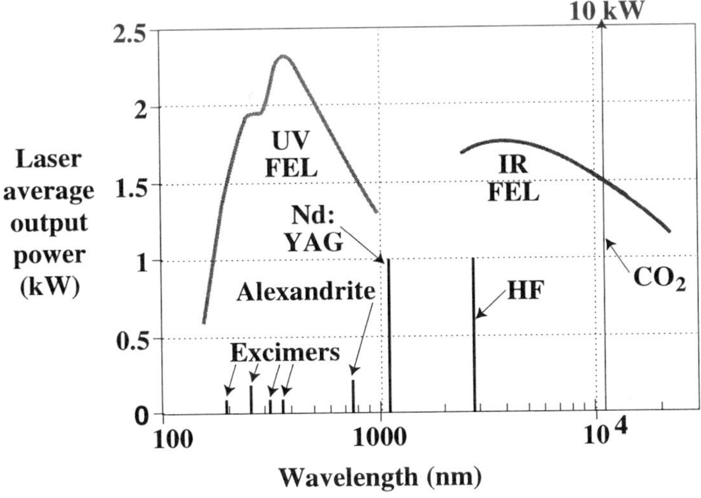


**Figure 6**. Power vs. wavelength for the IR and UV Demo FELs.

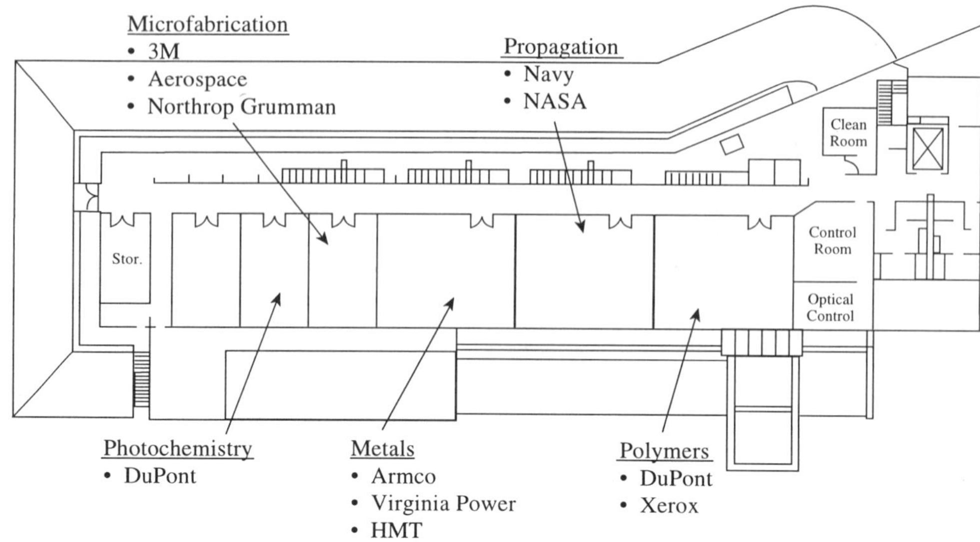


**Figure 7**. Layout of the Jefferson Lab Free-Electron Laser User Facility's user lab spaces.

In this effort the stakeholders' interests overlap: The Navy seeks a technology demonstration of a power-efficient, high-average-power, wavelength-tunable IR light source for potential extrapolation for shipboard self-defense. Consortium industrial members seek a power-efficient, high-average-power, wavelength-tunable, picosecond-pulse-length laser light source for materials processing. University members seek to participate in technology development, and also to exploit the basic science opportunities that these FELs will represent. Jefferson Lab seeks to profit from synergistic development in SRF-related technologies crucial for future CEBAF accelerator upgrades for advancing basic physics research. The industrial stakeholders in particular have committed substantial resources to the outfitting of the FEL User Facility. Figure 7, the user facility second-floor plan, shows the arrangement and initial assignments of the six laser-light applications-development labs.

However, the application of CEBAF SRF technology in the FEL driver accelerators, though clearly a nearly ideal example of a laboratory-to-industry technology transfer opportunity, is by no means a simple transference. The FEL-driver beam current, for example, is two orders of magnitude higher than in CEBAF. Nonetheless the FEL task is in some respects substantially smaller than was the task of creating CEBAF. Table 1 takes a "parts count" approach to contrast the relatively high complexity of CEBAF with that of each of the two demo FEL drivers.

**Table 1.** CEBAF vs. IR and UV Demo Driver Accelerator "Parts Count"

| | CEBAF | IR FEL | UV FEL |
|---|---|---|---|
| Number of cryomodules | 42.25 | 1.25 | 3.25 |
| Number of magnets | | | |
| Large magnets | 1303 | 75 | 150 |
| Correctors | 1015 | 51 | 75 |
| Power supplies | | | |
| Large supplies | 22 | 4 | 6 |
| Trim supplies | 1820 | 100 | 150 |
| Beam path (km) | 6.5 | 0.12 | 0.25 |
| Number of control variables | 140,000 | ~10,000 | ~15,000 |

**Table 2.** CEBAF vs. IR and UV Demo Driver Accelerator Cryogenic Loads

| | Units | Capacity | 4 GeV CEBAF | IR FEL | UV FEL |
|---|---|---|---|---|---|
| 2 K | kW | 4.8 | 2.5 | 0.2 | 0.5 |
| 50 K | kW | 12 | 10.0 | 1.0 | 2.4* |
| Liquid | g/sec | 10` | 0 | 0 | 0 |

* Will add 4 kW of 50 K refrigeration

Table 2 takes a similar approach to contrast the cryogenics technology involved. Because the CEBAF cavity $Q$ (heat load) is a factor of 2 better than specification, the CEBAF CHL has significant excess capacity, as the table shows. To tap this excess capacity, the FEL user facility building was located adjacent to the CHL[4]—which has a set of transfer lines each for the two CEBAF linacs—and a third set of transfer lines were added to connect to the FEL. About 10% of the CHL's excess capacity will be used for cooling the IR Demo's driver linac.

## First Step: "IR Demo" FEL

Figure 8 shows the IR Demo FEL schematically. The 1 kW device is designed to provide laser light at wavelengths from 3.0 to 6.6 μm, with optical beam quality two times the diffraction limit, and with electron beam losses of less than 5 μA at energies above 25 MeV. In this first FEL, about 99% of the energy from acceleration in the 32 MeV cryomodule will be recovered. (When the 10 MeV of pre-acceleration from the injector is accounted for, this actually represents only about 75% of the beam's overall energy. In planned future FELs, with higher ratios of accelerator energy to injector energy, this difference will lessen, with a corresponding rise in the benefit of energy recovery.) Figure 9, a July 1997 photo, shows an early stage of FEL installation.

FEL R&D is being conducted as an activity synergistic with the lab's main mission of running—and over time improving—the main SRF electron accelerator for physics experiments. Wiggler and beam-transport optics requirements for the IR Demo are substantially within the state of the art. The IR Demo effort's key technical challenges lie in the continuing development of high-gradient, high-reliability SRF accelerating cryomodules and high-brightness electron sources.

The IR Demo FEL's driver accelerator and its injector require, respectively, energy gain of 32 MeV (8 MeV/m gradient) in a single cryomodule and 10 MeV (10 MeV/m

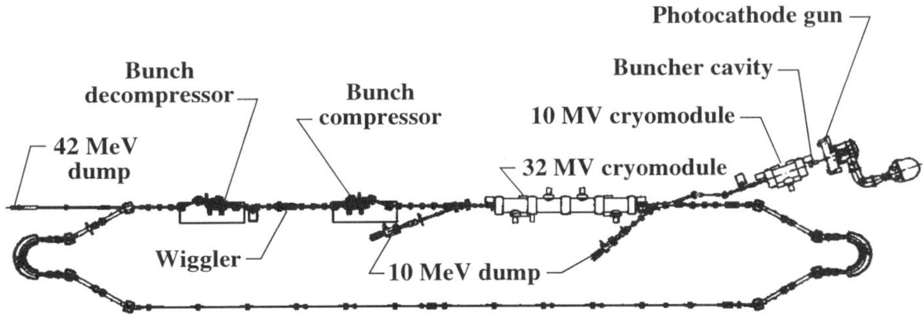


**Figure 8.** In the infrared demonstrator (IR Demo) free-electron laser, electrons from a photocathode source will attain 10 MeV in a two-SRF-cavity quarter-cryomodule before injection into a full cryomodule containing eight SRF cavities. At 42 MeV the electrons will transit a wiggler, where they will yield about 1% of their energy for the production of light in an optical cavity. Following recirculation to the injection point, most of the electrons' unspent energy will be recovered via deceleration and conversion back into RF power.

**Figure 9.** The beginnings of IR Demo FEL installation. In the foreground are quadrupole magnets of the driver linac's recirculation beam line; visible in the background are elements of the cryogenic and other subsystems of the linac.

gradient) in a quarter-cryomodule—specifications considered attainable based on experience operating well above the 5 MeV/m originally specified for CEBAF. Average quality factor $Q$ at 8 MeV/m in CEBAF has been twice that needed for the FEL driver. Some of the cryomodule components are being modified to handle the FEL's higher peak current and its 5 mA average current. The higher beam currents required two minor changes to our standard cryomodule. First, the interconnecting beam pipe and beam valves were increased in diameter to reduce impedances. Second, a redesign was carried out for the higher-order-mode (HOM) load system that dampens unwanted, beam-performance-degrading resonances. For the CEBAF accelerator the dissipated power was absorbed at 2 K since it was only 0.3 W per cavity, or 100 W total. For the UV Demo per cavity it will be 200 times higher, and therefore will be absorbed by the 50 K system. This is achieved by using a vacuum-insulated rod to conduct the heat and a labyrinth vacuum break between the beam and insulating vacuums (Figure 10).

The FEL injector's most technically demanding component is the 500 kV DC gallium arsenide photoemission gun. It was designed in collaboration with the University of Illinois, based on experience with an earlier gun at Stanford Linear Accelerator Center in California. DC was chosen because room-temperature RF structures require very high-

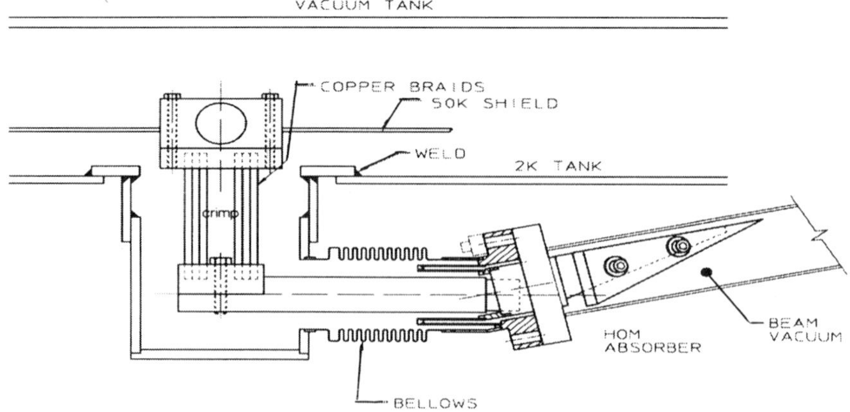


**Figure 10.** FEL higher-order-mode (HOM) load system.

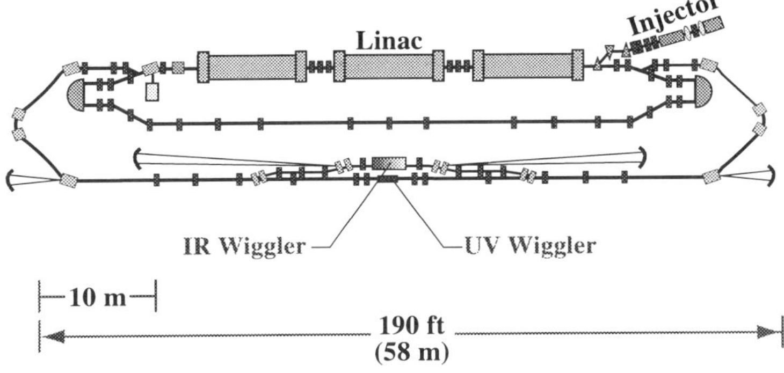


**Figure 11**. UV Demo layout concept.

average-power CW RF power sources and are thermally limited in gradient capability, and because the technology of CW RF guns based on superconducting cavities is not mature. The photoemission cathode was chosen to attain short pulse length and low emittance. Moreover, a thermionic emitter, with needed beam conditioning, would lead to emittance growth and would be very hard to scale later for higher-average-current beams. Use of gallium arsenide integrates with nuclear and particle physics program goals. To overcome the gun's technical challenges requires excellent vacuum conditions in the vicinity of the photocathode and limited field-emission currents from the electrode structures. The gun has been successfully tested at 350 keV.

**Status and Fiscal Year 1998 Plan**

As of July 1997, injector components have been tested, with the 350 keV operation of the photocathode gun in the Test Lab's Injector Test Stand. The injector 10 MeV quarter-cryomodule was also operated in the Test Lab this spring to full gradient. In June the injector was moved to the FEL building. With the completion of the infrastructure and cryogenic system, on July 10 the FEL cryogenic system and the 10 MeV cryomodule were cooled down to 2 K.

The 32 MeV cryomodule is being assembled; the four cavity pairs are assembled and qualification is complete, with one of the cavity-pair-containing quarter-cryomodules already assembled. The cryomodule is scheduled to be installed and cooled down in August.

The RF systems for the injector were fully tested at the Test Lab and are now in the process of being reinstalled in the FEL together with the installation of the 32 MeV cryomodule RF. Checkout of the RF system is scheduled to be complete in September.

The wiggler has been delivered and is in the process of being installed. The last two systems, beam transport and optical transport, are about 25% installed and are scheduled to be fully installed by the end of September.

The commissioning schedule is aggressive:

| | |
|---|---|
| October 1997 | Start FEL commissioning |
| February 1998 | First light |
| May 1998 | Energy recovery |
| June 1998 | High-power light |
| July 1998 | Start experiments |

A development path for proceeding beyond this initial IR Demo FEL has been outlined. Near-term steps will be incremental IR Demo upgrades. In the medium term, perhaps in Fiscal Year 2000, would come the UV Demo. Figure 11 shows the concept for the UV Demo, in which a 200 MeV SRF linac will alternately drive an IR and a UV wiggler, with IR output of 3.0–6.6 μm at 10 kW and 1.0–2.0 μm at 2 kW, and with deep ultraviolet (DUV) output down to 0.2 μm at 1 kW. In the more distant future, although precise ultimate end user goals can be expected to shift somewhat based on experience with the Demo and later FELs, it is expected that industry will ultimately want cost-effective, low-maintenance, easily operated 100 kW FELs operating in the IR at 5 μm and in the DUV down to 0.2 μm. And it is expected that the Navy will want a compact, robust, warship-compatible, and combat-reliable 1 μm device with a megawatt of power.

## ACKNOWLEDGMENTS

This paper represents the efforts of the Jefferson Lab Accelerator Division and the Jefferson Lab FEL Team. Special acknowledgment to S. Corneliussen for assistance in preparing it.

## REFERENCES

1. H. Grunder, "CEBAF Commissioning and Future Plans," *Proceedings of the 1995 Particle Accelerator Conference.*
2. J. R. Delayen, these proceedings.
3. Report of the Review of the LPC/CEBAF Free Electron Laser Proposal, May 10–12, 1995 (David W. Cheney, Associate Deputy Under Secretary of Energy for Technology Partnerships, Chair).
4. C. H. Rode, "CEBAF Cryogenic System," *Proceedings of the 1995 Particle Accelerator Conference.*

# SUPERCONDUCTING RADIO–FREQUENCY TECHNOLOGY: UNDERSTANDING AND IMPROVEMENTS OF LIMITATIONS THROUGH APPLICATION OF CRYOGENIC INSTRUMENTATION

P. Kneisel
Thomas Jefferson National Accelerator Facility
Newport News, Virginia 23606

## ABSTRACT

A large number of diagnostic methods have been developed during the last two decades to gain insight in limiting phenomena occurring in superconducting rf cavities during operation. Temperature mapping in subcooled and superfluid helium and x -ray mapping have successfully been applied to understand and subsequently prevent or eliminate loss mechanisms present in these devices. Newer techniques specific to problems such as rf window arcing are under development.

This paper reviews various diagnostic techniques and discusses the impact of their application on progress made in SRF technology.

## INTRODUCTION

Superconducting Radio –Frequency Technology has been applied successfully in the last decade in several large scale particle accelerator projects such as TRISTAN, HERA, LEP and CEBAF and is being considered seriously for future applications such as B-factories, proton accelerators for spallation neutron sources or for the production of tritium, TeV linear colliders and free electron lasers. To a large extent this multitude of applications and the renewed interest in this technology are a result of significant progress in understanding and overcoming the limitations encountered since the 1970's. The systematic application of sophisticated diagnostic methods such as temperature mapping and radiation mapping at cryogenic temperatures in conjunction with computer simulation calculations and surface analytical investigations led to the elimination of the most severe performance limitations of superconducting cavities caused by resonant electron loading (" multipacting") and was instrumental in identifying the causes for field emission loading and thermal magnetic breakdowns in cavities. Since the first use of a chain of carbon resistors by C. Lyneis at Stanford University in 1972[1] to detect the location of a quench area on a cavity, a variety of diagnostic methods have been "invented" to gain more insight into phenomena taking place in a superconducting cavity operated at high gradients.

Thorough reviews of the "state of the art" between 1981 and 1985 have been given in [2-4]; this paper will mainly concentrate on refinements of diagnostic methods in the last decade, new developments and the progress made in SRF technology by application of these methods.

## BEHAVIOR OF AN SC ACCELERATING CAVITY

A superconducting cavity as typically used in particle accelerator application is usually fabricated from niobium or niobium sputtered on copper. Its response to excitation by radio-frequency energy is described by two parameters, its Q-value and its achievable accelerating gradient.

The Q-value of a cavity is defined as the ratio of the stored energy (W) in the cavity to the power (P) lost in the cavity walls per rf cycle ($\omega = 2\pi f$)

$$Q_o = W / ( P / \omega ) \tag{1}$$

The $Q_o$-value is inversely proportional to the surface resistance (R) of the material of the cavity walls. The proportionality constant is called the geometry factor (G) and depends only on the geometry and the electromagnetic field configuration in the cavity. It is typically of the order of 270 $\Omega$ for structures designed to accelerate velocity of light particles:

$$Q_o = G/R \tag{2}$$

The accelerating gradient $E_{acc}$ is defined as the maximum energy a charged particle will gain in the time-varying rf fields by traversing an accelerating gap divided by the gap length. The accelerating gradient is proportional to the square root of the stored energy in the cavity.

$$E_{acc} \propto \sqrt{PQ_o} \tag{3}$$

The accelerating gradient is related to the peak surface electric and magnetic fields in the cavity through Maxwell's equations. Since superconductors go from the superconducting state to the normal conducting state, if a critical magnetic field $H_{crit,sh}$ is exceeded, there are fundamental limitations to the achievable accelerating gradients in a superconducting rf cavity. In the case of niobium as the superconducting material the fundamental magnetic field limit is app. 2400 Oe, which in a typical accelerating cavity corresponds to a gradient of $E_{acc} \approx$ 50 MV/m. Such gradients are by far beyond the present state of the art and are typically a factor of 3 to 6 higher than present achievements.

The experimentally observed behavior of a niobium cavity deviates from the theoretically expected behavior in four distinct features as shown schematically in the second curve in figure 1:

(a) The observed Q-value is significantly lower than predicted by the microscopic theory of superconductivity (BCS theory) due to the residual surface resistance caused by anomalous losses and defects in the material.

(b) At certain distinct fields the Q-value might drop to lower values caused by resonant electron loading (" multipacting")

(c) Above a certain field level in the cavity—typically 5 MV/m $\leq E_{acc} \leq$ 15 MV/m— the Q-value decreases exponentially due to non -resonant electron loading ("field emission")

(d) The experimentally observed field, at which the superconducting state disappears ("Quench") is significantly lower than the theoretically predicted field as already mentioned above.

All the above listed deviations from the ideal behavior are to a large extent caused by the surface conditions of the superconducting material. This knowledge has been gained through systematic application of diagnostic methods such as prominently temperature mapping or X-ray mapping in conjunction with scanning electron microscopy and elemental surface analysis as well as computer simulation calculations. Whereas the $Q_o$ vs $E_{acc}$ behavior of a cavity gives a "global" picture of the cavity as a whole—some conclusions of phenomena such as multipacting, field emission or quenching can be deduced from the rf-signal response of the cavity—the application of diagnostics led to significant progress in understanding of localized phenomena in these cavities.[5]

## DIAGNOSTIC METHODS

The diagnostic methods, which have been developed during the last two decades can be categorized into three main groups as follows:

a). **Thermal methods**: temperature mapping and use of second sound signals in the helium bath.

b). **Detection of ionizing radiation**: x-radiation mapping, measurements of x-ray energy spectra, electron currents , x-ray photography.

c). **Optical methods**: measurements of optical spectra of light emitted from areas of a cavity via fiber optic cables or the use of "warm" and "cold" optical devices such as a miniature CCD camera.

In the following several of these methods will be described and their usefulness in advancing the understanding of limiting phenomena encountered in these accelerating cavities will be discussed.

### Thermal Methods

As described by the BCS theory, an ideal superconductor in a high frequency electromagnetic field has a finite surface resistance, which vanishes exponentially as the temperature approaches absolute zero. In practice however, every superconducting cavity contains on its rf surface small areas, which are either normal conducting or are only weakly superconducting and become normal conducting in electromagnetic fields well below the critical field of the superconducting material. They can also be produced by the impact of high energy and high intensity electron currents during the operation of a cavity.

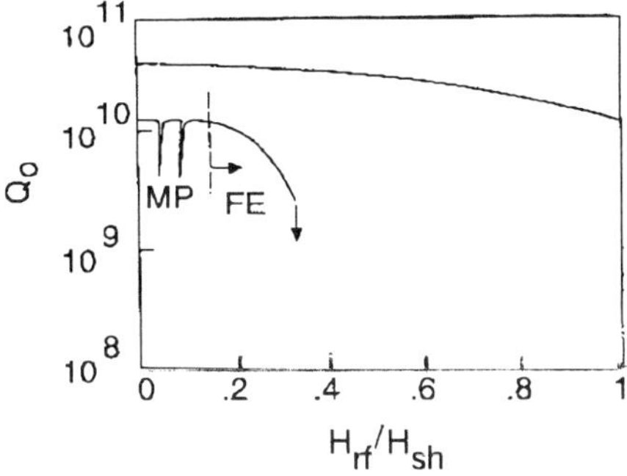


**Figure 1**. Schematic behavior of a sc cavity in electromagnetic fields .

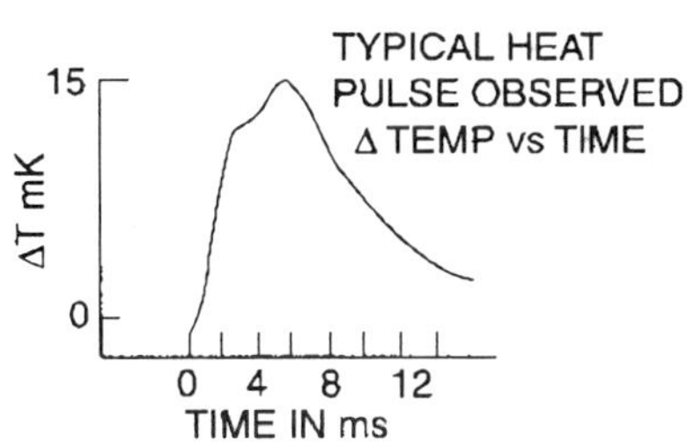


**Figure 2**. First observed heat pulse from a sc cavity[1].

These areas with increased losses will produce an increasing heat flux as the rf field level in a cavity is increasing and will lead to a temperature increase of their surrounding areas above the critical temperature or to the onset of film boiling on the outer wall of the cavity. Thus, each energy loss mechanism will finally lead to an increase of the temperature of the cavity wall, and therefore temperature measurements and the "mapping" of the outer cavity surface are of prime importance to identify the causes for performance limitations in the cavities.

The earliest application of a chain of rotating Allen-Bradley carbon resistors mounted a few millimeters away from the surface of a cavity (100 Ω, 1/8 Watt) for detection of the quench area was reported in ref. 1, and figure 2 reproduces the first published heat pulse detected near a quench area as captured by an oscilloscope trace. During a quench all the stored energy in the cavity is set free instantaneously. A very significant heat flux develops, leading to film boiling and a marked increase of the temperature of the helium film close to the area of thermal instability. This temperature rise can be detected easily in both normal and superfluid helium even if the thermometer is not in intimate contact with the cavity wall. In subsequent years this method has been used by many different groups working in SRF technology; the developments to improve the method concentrated mainly on two areas:

a) to increase the sensitivity of the thermometers above and below $T_\lambda$ so that loss mechanisms far below the actual "quench" could be investigated and

b) to increase the speed of data acquisition by different electronic schemes and data acquisition systems as well as by employing a large number of stationary resistors, which can be read out basically simultaneously, giving a "momentary" temperature map of a cavity and allowing the investigation of time-varying phenomena in a cavity such as multipacting, rf processing or helium processing.

**Increase in sensitivity.** In an experiment conducted in 1978 at CERN[6] it was shown that a temperature signal from a heated niobium plate immersed in helium at 2.3 K increased by one order of magnitude, if the bath conditions changed from saturated to subcooled (most favorable conditions $T_{Bath}$ slightly above $T_\lambda$, bath pressure ≈ 1000 mbar). This significant change in sensitivity can be explained by an increase of the heat transfer resistance from the metal surface to the He bath. In a subcooled bath nucleate boiling on the heated surface is absent and so is the very effective cooling mechanism of microconvection usually produced by bubbles. Therefore the surface temperature of the cavity wall increases compared to saturated helium and the thermometers are thermally decoupled from the surrounding bath, both effects leading to the observed higher sensitivity.

As a consequence of these findings, a rotating temperature mapping system with spring-loaded and encapsulated thermometers sliding on the outer cavity wall as shown in figure 3 was implemented at CERN and used to take temperature maps of 500 MHz niobium cavities. Examples of T- maps showing multipacting, Q-degradation due to hydrogen precipitation[20], field emission and a quench are shown in figure 4.

Calibration of the temperature increase of the outer niobium wall against the heat flux density and its dependence on bath temperature and orientation of the heated wall in the helium bath was subsequently carried out in different laboratories, permitting the conversion of "raw" temperature maps into maps of surface resistance distribution. However, temperature mapping in subcooled helium has its limitations. Because of the temperature dependence of the BCS surface resistance there exist upper limits to the surface magnetic fields above which global heating dominates the temperature response of a cavity wall and localized heating at troublesome areas is "buried" in this background. These limits are ≈ 250 Oe at S-band frequencies and ≈ 500 Oe at L-band frequencies [3], values which are quite often exceeded. In addition, obviously any local residual surface resistance, which is smaller than the BCS resistance above $T_\lambda$ will remain undetected and

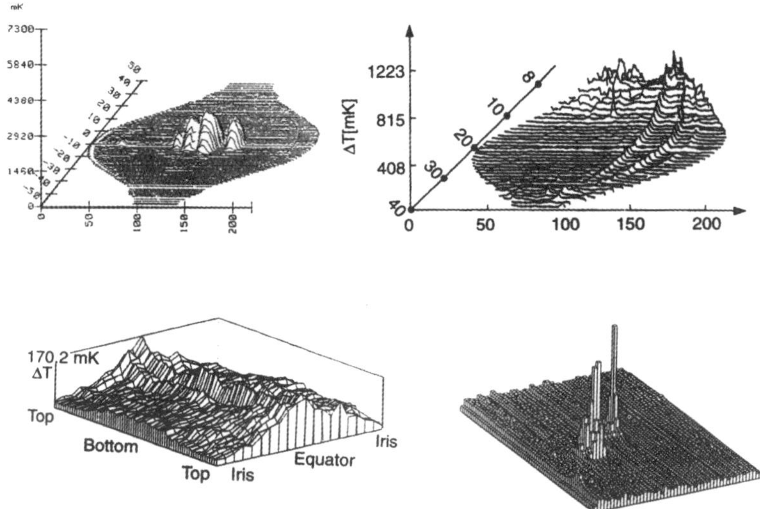


**Figure 4.** Examples of T - maps: Multipacting[15], Q-disease[19], field emission[2], quench[27].

investigations of low residual resistance distributions in a cavity become impossible under subcooled bath conditions. Both limitations require surface temperature measurements in superfluid helium. Two requirements for the sensor to be useful in measuring increased surface temperatures are a direct result of the extremely high thermal conductivity of superfluid helium [4,7,8]:

a) the sensors must be in good thermal contact with the metal surface; this is accomplished by spring loading and the use of a thermal bonding agent such as Apiezon N grease and

b) the sensors must be efficiently thermally insulated against the surrounding superfluid helium bath; this is accomplished by a thermal insulation housing from Stycast 2850 FT and the use of very thin manganin wire leads.

Such or very similar thermometers have successfully been used for investigation of the spatial surface resistance distribution in niobium cavities and for field emission studies.

Further improvements in sensitivity (≈ a factor of 3) have been gained in placing the sensitive thermometer part in a vacuum[9,10]. Non-stationary - (rotating) systems- have been developed, which feature spring-loaded sensors encapsulated in pointed copper housings and are routinely used at various labs, with decreased efficiency however due to the small contact area and the lack of a thermal bonding agent and a strong dependence of the

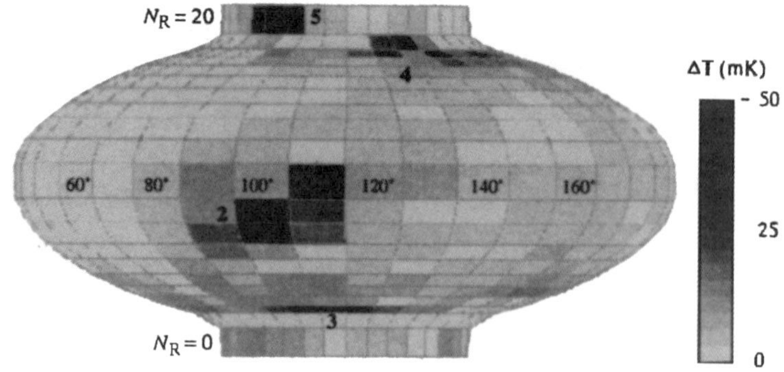


**Figure 5.** T-map of a single cell 1.5 Ghz cavity taken with a high speed system[12].

accuracy and reproducibility on the heat transfer conditions at the surface. They have been very useful in finding problem areas on multicell cavities.[11]

**Increase in speed.** Some of the limiting phenomena in superconducting cavities such as multipacting or field emission are time dependent and change during operation of a cavity. For the study of the dynamics of such process, very high speed data acquisition systems are needed and the application of mechanical , rotating mapping systems is not adequate. At Cornell University a high speed system has been developed in 1985 consisting of 686 high sensitivity, stationary thermometers, which could be read out in 15 sec.[8] Further improvements in the system reduced the read out time to ≈ 1 sec, permitting the investigation of processing of field emission sites, giving new insights in the FE process in superconducting cavities.[12] Figure 5 shows a T-map of a single cell 1.5 GHz cavity taken in superfluid helium with the high speed system of Cornell University.

In summary, temperature mapping in subcooled and superfluid helium has successfully been used to understand and eliminate limitations in sc cavities due to design flaws[13], multipacting[14,15], field emission [16,17], quenches caused by defects [5,18,19], Q-disease [20], and processing.[21,22]

**Second sound.** For reasons of completeness it should be mentioned that the detection of second sound waves generated by the release of large amounts of energy during a quench has been used at Stanford University more than a decade ago to localize specific troublesome cells in a string of accelerator structures assembled into a horizontal cryostat arrangement.

At Argonne germanium thermometers have been used in a split ring resonator to reconstruct from second sound signals the location of a quench area within 1 to 2 $cm^2$. At DESY sound propagation in helium has been used to reconstruct quench locations from temperature signals received by a few sensors distributed over the surface of a 1 GHz niobium cavity.[4]

## Detection of Ionizing Radiation

As mentioned above, in most cases superconducting cavities are limited in their high field performance by the onset of field emission loading. Electrons present in the cavities are accelerated by the high electric fields and produce bremsstrahlung when colliding with opposing cavity walls. X-ray intensities and bremsstrahlung spectra have been measured outside the cryostat with pin-hole cameras, ionization chambers or NaI detectors, giving information about electron trajectories and spatial distribution of radiation and leading to the conclusion that besides the primary process of electron impact—this would give a strong azimuthal peaking of the x-ray intensity—secondary multiplication processes play an important role.[23]

Surface barrier detectors in the helium bath were first used at CERN in 1981 for surface scanning of a cavity when mounted on a rotary frame.[2] These experiments clearly could separate signals generated by primary and secondary impact processes. This diagnostics method was further refined in the following years at the University of Wuppertal [17] and is being used in combination with temperature mapping to shed light onto electron loading processes encountered in the cavities during operation at high fields.[11] An example of a x-ray map is shown in figure 6.[17]

Recently, we have developed at Jefferson Lab a x-ray spectroscopic probe, which is located in the 2 K helium bath in close proximity to the cold rf window of the cavity. It consists of an NaI crystal and a Hamamatsu photomultiplier; it is thermally isolated against the helium bath by means of Stycast 2850 FT, uses a heated copper cylinder for temperature stabilization and is housed in a vacuum can. It is hoped that by direct measurement of the energy of the x-rays penetrating through the ≈ 1/8 in. thick alumina rf window and simultaneously observing arcing behavior at the window more insight into the charging mechanisms at CEBAF's rf windows can be gained.

The direct observation of electron currents using a biased electron probe usually assembled to an auxiliary port at a cavity is another diagnostic method which has been used to gain knowledge about phenomena occurring during the operation of a cavity at high field levels. The bias voltages used vary between +50 V and +100 V depending on the location of the probe in the cavity, the geometry of the cavity and the probe, and the energy of the electrons. Field emission currents collected this way follow quite well a Fowler-Nordheim law and from the slope of this curve a "field enhancement factor" can be extracted. This factor has been used as an indication of the severity of the field emission and the level of cleanliness of the superconducting surface.[24,25] An example of a electron current measurement is shown in figure 7.

**Optical Methods**

Glowing particles in cavities under operation have been seen as early as 1975.[2] These microparticles (dust), which have a rather poor thermal contact to the cavity walls, were heating up in the rf fields and started glow emission of electrons. Spectral analysis of glowing microparticles in rf fields has been carried out recently on intentionally contaminated surfaces of a room temperature cavity, which was used as a test vehicle to simulate electron loading behavior in superconducting cavities and to gain knowledge about conditioning effects. This study showed that dielectric particles strongly interact with rf fields depending on their size, leading to luminescence and explosive destruction of the luminous spots. The spectral intensity from various luminous spots peaked at different wavelengths between 700 and 800 nm.[26]

In the CEBAF superconducting cavities an alumina rf window coated with chromium oxide is placed very close to the beam line providing a hermetically sealing of the cavity in conjunction with gate valves at the beam pipes. These alumina windows are often subject to arcing phenomena, especially in the presence of ionizing radiation emanating from the cavities at higher gradients. To investigate the arcing behavior of the windows a fiber optical system was developed: the light produced during an arc at the window was monitored through a sapphire viewport mounted close to the rf window, then transmitted through a commercial grade acrylic fiber to a photo multiplier tube outside the cryostat and through a 300 nsec optical delay line to a spectrometer with a 990 element intensified photodiode array, which allowed the available spectrum to be recorded with 0.6 nm resolution.

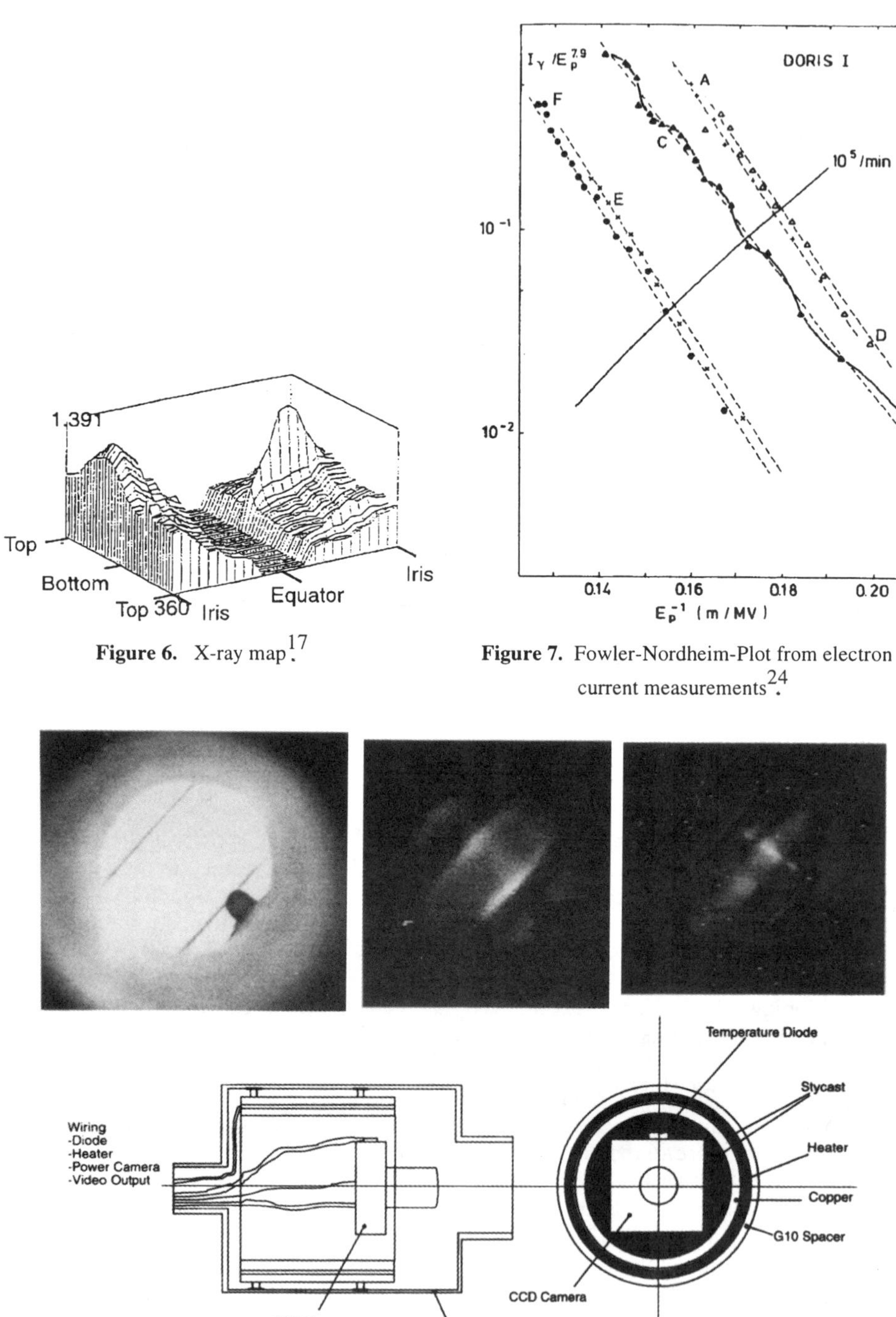


**Figure 6.** X-ray map[17]

**Figure 7.** Fowler-Nordheim-Plot from electron current measurements[24]

**Figure 8**. Arcs on an alumina window detected with "cold" CCD camera .

In several occasions distinct spectral lines for aluminum, hydrogen and chromium could be identified as indicators of a possible arc initiating mechanism; however more and complementary experiments are needed to understand the complexities of the arcing process.[27]

For one of such additional experiments we have developed a "cold" camera system, which is directly located at the window in the helium bath. It employs a miniature CCD camera, thermally isolated as shown in figure 8 and located in a vacuum can. In initial experiments with this camera we were able to record different types of arcs at the rf window originating from different locations. Also, to our surprise, the window started to luminance, when the cavity was operating in the field emission region at higher rf power levels. Luminescence of alumina rf windows as a result of multipacting electronic activity at higher rf power levels has been reported by Y. Saito [28] and one could speculate that CEBAF's cold rf windows are subject to similar phenomena. However, x -radiation could also lead to luminescence phenomena, and more data are needed to make the distinction.

## SUMMARY

Over the last two decades a variety of diagnostic methods have been developed to investigate limiting phenomena in superconducting cavities. Temperature mapping in subcooled and superfluid helium in conjunction with x-ray mapping have proven to be powerful tools and considerable understanding of loss mechanisms in these cavities has been gained, leading to significant performance improvements. A reflection of this progress in SRF technology is the increasing number of large scale accelerator projects, which either use the technology or are contemplating to equip their accelerators with superconducting cavities.

## ACKNOWLEDGMENT

Many thanks to P.Kushnik and J.Mammoser for their support of the "cold" camera tests, to S. Corneliussen, Q.S. Shu and J.Delayen for carefully reviewing this paper and to S. Spata for editing it.

This work was supported by the U.S. DOE under contract number DE-AC05-84ER40150.

## REFERENCES

1. C.M. Lyneis, M.McAshan, and Nguyen Tuong Viet, Recent Measurements of S-Band and L-Band cavities at Stanford, *Proc. 1972 Proton Linear Acc. Conf, 98 (Los Alamos 1972)*
2. H. Piel, Diagnostic Methods of Superconducting Cavities and Identification of Phenomena, *Proc. of the Workshop on RF Superconductivity, KfK Report 3019, 85 (Kernforschungszentrum Karlsruhe 19880)*
3. G. Mueller, Diagnostic Techniques and Defect Classification, *Proc. of 2.Workshop on RF Superconductivity, 377 (CERN 1984)*
4. C.Reece, Progress in Diagnostic Techniques for SC Cavities, *Proc. 3. Workshop on RF Superconductivity, Report ANL-Phy-88-1, 545 (Argonne Nat. Lab 1987)*
5. H.Piel, Fundamental Features of Superconducting Cavities for High Energy Accelerators, *Report WUB 86-14 (University of Wuppertal 1986)*
6. H.Piel and R.Romijn, Temperature Mapping on a Superconducting Rf Cavity in Subcooled Helium, *Report CERN/EF/RF 80-3 (CERN 1980)*
7. G.Müller and P.Kneisel, Development of Sensitive Thermometry for Calibrated Surface Temperature Measurements of Niobium Cavities Immersed in Superfluid Helium, *Report LNS SRF-851201 (Cornell Univ 1985)*
8. H.Padamsee,C.Reece et al., Field Emission Studies in Superconducting Cavities, *1987 IEEE Part.Acc.Conf, 1824 (Washington DC, 1987)*
9. M.Fouaidy, T.Junquera and A.Caruette, Surface Temperature Measurements on Superconducting Cavities in Superfluid Helium, *Proc. 5.Workshop on Rf Superconductivity, Report DESY M-92-01, 547 (DESY 1991)*

10. Ch. Liang,L.Phillips,and R.Sundelin, A new Method of Surface Resistance Measurement with a Niobium Triaxial Cavity working at 2K,*Rev.Sci. Instr. 64(7),1937 (1993)*
11. Q.S.Shu, G.Deppe et al., An Advanced Rotating T-R Mapping and its Diagnosis of TESLA 9-cell Superconducting Cavity, *Proc. 1995 Part.Acc.Conf., 1639 (Dallas 1995)*
12. J.Knobloch ,H.Muller and H.Padamsee, Design of a High Speed,High Resolution Thermometry System for 1.5 GHz Superconducting Radio Frequency Cavities,*Rev.Sci Instr., 65(11)3521 (1994)*
13. P.Kneisel, J.Åmato et al, Performance of Superconducting Storage Ring Cavities at 1500 Mhz, *IEEE Trans.Magn.21(2), 1000 (19850*
14. C.Lyneis, Electron Loading - Description and Cures, *Proc.of the Workshop on RF Superconductivity, Report KfK 3019,119 (Kernforschungszentrum Karlsruhe 1980)*
15. W.Weingarten, Electron Loading, *Proc.2.Workshop on RF Superconductivity, 551 (CERN 19894*
16. Q.S.Shu ,W.Hartung et al.,A Study of the Influence of Heat Treatment on Field Emission in Superconducting RF Cavities,*Nucl.Instr. and Meth. A278,329 (1989)*
17. R.W.Röth, H.Heinrichs et al., Suppression of Field Emission in Superconducting S-Band Acceleration Structures, *Proc. 2. Europ.Part.Acc.Conf., 1097 (Nice 1990)*
18. D.Moffat, P.Barnes et al.,Studies on the Nature of Field Emission Sites,*Part.Acc. 40, 59 (1992)*
19. M.Pekeler,Test results on the 9-Cell 1.3 GHz Superconducting RF Cavities for the TESLA Test Facility Linac, *Proc. 1997 Part.Acc.Conf.(Vancouver 1997), to be published*
20. B.Bonin and R.W.Röth, Q-Degradation of Niobium Cavities Due to Hydrogen Contamination, *Part.Acc40,59 (1992)*
21. J. Graber, High Gradient Superconducting RF Systems,*Proc. 1995 Part.Acc.Conf.,1478 (Dallas 1995)*
22. Q.S.Shu,K.Gendreau et al.,Influence of Condensed Gases on Field Emission and Performance of Superconducting Cavities,*IEEE Trans on Magnetics, 25(2),1868 (1989)*
23. I.Ben-Zvi,J.F.Crawford, and J.P.Turneaure,Electron Multiplication in Cavities,*IEEE Trans.NS-20(3),54 (1975)*
24. e.g.Sh. Noguchi,Y.Kojima, and J.Halbritter, Measurement of a Superconducting 500 MHz Nb Cavity in the $TM_{010}$-Mode, *Nucl.Instr. and Meth.179,205 (1981)*
25. R.J.Noer, Electron Field Emission from Broad Area Electrodes, *Appl.Phys. A28, 1 (1982)*
26. T.Junquera,S.Maissa et al.,Study of Luminous Spots Observed on Metallic Surfaces subjected to High RF Fields,*Proc. 1995 Part.Acc.Conf.,1632 (Dallas 1995)*
27. T.Powers,P.Kneisel, and R.Allen, Arcing Phenomena on CEBAF RF-Windows at Cryogenic Temperatures,*Proc. 1995 Part.Acc.Conf.,1645 (Dallas 1995)*
28. Y.Saito, Breakdown Phenomena in Vaccum,*1992 Linear Acc. Conf. Proceedings Report AECL-10728,575 (Ottawa 1992)*
29. M.Peiniger, Experimental Investigations on Superconducting $Nb_3Sn$-Cavities at Microwave Frequencies,*Report WUB-DIS 89-1 (University of Wuppertal 1989)*

# THERMAL STABILITY ANALYSIS OF SUPERCONDUCTING RF CAVITIES

T. Junquera, J. Lesrel, M.Fouaidy, and S. Bousson

Institut de Physique Nucleaire
(CNRS-IN2P3) 91406-ORSAY France

## ABSTRACT

Future linear accelerators using superconducting RF cavities (TESLA proposal), requires accelerating gradients Eacc=25 MV/m to be achieved reliably at large scale. This high gradient level is mainly limited by electron emission and thermal instabilities (quench). Impressive improvements have been recently accomplished by pushing further the onset threshold of electron emision using careful surface cleaning techniques. On the other side micron size resistive defects embedded in the niobium walls of the cavity continue to induce thermal breakdowns for Eacc in the range 15 to 25 MV/m. In this paper the thermal stability of SRF cavities is analysed using analytical and numerical simulation models. The effects of the most relevant parameters (i.e. defect size, RF frequency, thermal conductivity, cooling conditions, etc.) having an incidence on the cavity quench are studied. The steady-state and the transient solutions are presented in two cases : the defect free surface, and the micron size defect on the cavity surface. Experimental observations of the thermal events occuring during the quench have been obtained with the help of two diagnostic systems: surface thermometers working in superfluid helium and RF measurements. All the proposed experimental and modelling methods could contribute to get a more complete insight on the thermal effects taking place in the cavity wall.

## INTRODUCTION

In this paper we will consider only the thermal effects originated by lack of cooling in large surface areas and those induced by the presence of defects on the cavity wall. The electron emission can also give important thermal effects and some of the results can be applied to this particular case. The cavity wall (Fig. 1) behaves in a simple way from the point of view of heat transfert: a metallic slab cooled by helium on one side is sufficient to start the study of the thermal effects.

Most of the time, the thermal models give the steady state solution of the heat equation which allows to study in detail the influence of the main parameters of the cavity and its cooling conditions. It gives also a good starting point to simulate the operation of CW SRF cavities and to optimize the operating conditions in an accelerator. Several papers have covered this subject during the last past years [1,2,3]. One of the most interesting result was to evaluate the role played by the thermal conductivity of the niobium in the thermal quench process. The study of temperature transient is rather unusual, only a small number of papers have proposed an initial approach to this subject [4]. Some numerical codes calculate the time evolution of the cavity temperature field, offering the possibility of investigate the propagation of the normal zone during a quench. The TESLA project proposes to operate cavities in RF pulsed mode and modelling the transient regime could be of great interest in the work for searching ultimate performances.

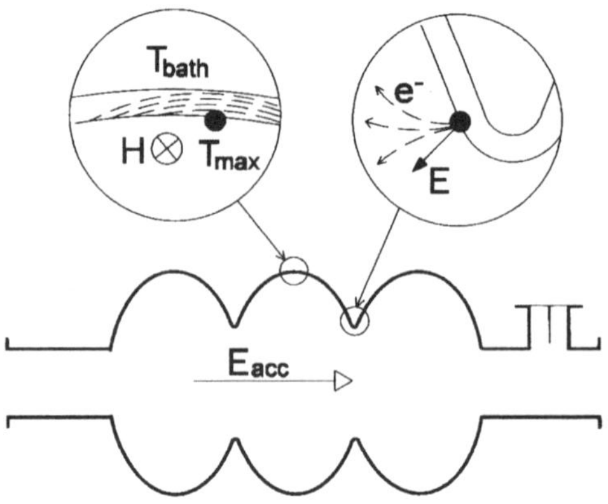


**Figure 1** Cavity wall thermal instabilities created by micron-size defects and electron emission.

In this paper we give some examples of the analytical and numerical solutions of the heat equation applied to the cavity problem with a systematic search for simple calculations. Some modelling procedures are proposed leading to a straightforward approach of the thermal quench calculations. Concerning the thermal response of cavities we present some experimental observations showing the quench detection with surface thermometers and the related temperature measurements.

## BASIC EQUATIONS AND MAIN PHYSICAL PARAMETERS

The thermal behavior of a cavity wall can be represented with a good accuracy by a slab of SC metal (in general niobium) where the temperature T in any point is governed by the heat equation:

$$\rho c_v \frac{\partial T}{\partial t} = \nabla \cdot (k \nabla T) \quad (\rho \text{ density, } c_v \text{ specific heat capacity, } k \text{ thermal conductivity})$$

Using cylindrical coordinates $(r,z)$, the boundary conditions can be easily introduced:

1) on the RF side of the cavity wall: $[-k(T)\ \partial T/\partial \tau]_{z=0} = q$ , where $q$ is the heat flux density : $q = (1/2)R_s H^2$ ($R_s$ is the surface resistance which depends on frequency and temperature, and $H$ the surface magnetic field). In our calculations we will introduce the accelerating field $E_{acc}$, which is proportional to the magnetic field $H$, the proportionality factor depending only on the cavity geometry. $E_{acc}$(MV/m)=$(4.\pi.10^{-7}/4.456.10^{-3}).H$(A.m) for the spherical shape 3 GHz cavities used in our experiments.

2) on the helium side of the cavity wall: $[-k(T)\ \partial T/\partial \tau]_{z=e} = h_k(T-T_{bath})$ where $e$ is the wall thickness and $h_k$ is the Kapitza conductance.

Analytical solutions of the heat equation are possible when all the parameters are constant, or in some cases when they are approximated by a linear temperature dependent fonction.The difficulty arises when considering the real temperature dependence of these parameters:

1) The thermal conductivity and heat capacity have to be treated like a set of measured points $k_i=f(T_i)$ or $c_{vi}=f(T_i)$ or by some modelling representation which in general depends on T, $T^2$, $T^3$. There is a large variety of values for this parameter. Commercial niobium sheets have values of RRR (Residual Resistivity Ratio $RRR=\rho_{295K}/\rho_{4.2K}$) in the range 40 to 400, which corresponds to values of $k$ in the range k(4.2K) = 10 to 100 $Wm^{-1}K^{-1}$, with k(4.2K) ~ RRR/4. Using different post-purification methods the final thermal conductivity of the cavity can be improved up to values close to RRR=1000. The specific heat capacity of the niobium is taken from a data book [5].

2) The Kapitza conductance is well represented by an experimental law: $h_k=h_{k0}f(T-T_{bath})$ where $h_{k0}=H_kT^n_{bath}$ is the low heat flux conductance, and $f$ a polynomial fonction of the temperature jump at the interface. In the following calculations, $H_k$ and the exponent $n$ are taken from different papers [6,7] (i.e. $H_k$=359 $Wm^{-2}K^{n+1}$, $n$=4.28) . The measured values are quite dispersed confirming the strong influence of the surface preparation

3) The surface resistance is represented by the following law:

$$R_s = (A/T)f^2 exp(-B/T)+R_{res}$$

where $f$ is the frequency, $R_{res}$ the residual resistance and, $A$ and $B$ are fit parameters which are easily deduced from RF measurements. We have adopted the fit parameters proposed by Weingarten [8] : $A$=10-4, $B$ = 18 (with $f$ in GHz) .

To solve the heat equation with the real temperature dependent parameters, a numerical method is needed. The FINITE DIFFERENCE method can handle simple geometry problems: this is the case of the study of a surface defect in cylindrical coordinates using simple uniform mesh or more elaborate progressive mesh with higher density in the vecinity of the defect. For more complicated geometries the FINITE ELEMENT method using an adapted triangular mesh gives quite accurate solutions.

## SOLUTIONS OF THE HEAT EQUATION

### Uniform RF heating of the cavity surface (Defect-free case)

The heat equation can be analytically solved when all the physical parameters of the cavity are taken as constants. Applying a classical heat transfert analytical treatment [9] we can obtain the temperature on the RF side of the cavity wall for a step heat input q:

$$T = T_{bath} + \frac{eq}{kL}\left[1 + L - \sum_{1}^{\infty} \frac{2L(\beta_n^2 + L^2)}{\beta_n^2(L^2 + L + \beta_n^2)} \exp(-\beta_n^2 \alpha \frac{t}{e^2})\right]$$

where $\beta_n$ are the positive roots of the equation: $\beta_n \tan \beta_n = L$, (L is the Biot number: $L=eh_k/k$, $\alpha$ is the thermal diffusivity : $\alpha = k/\rho c_v$). The steady state solution is simply given by: $T=T_{bath}+q(e/k+1/h_k)$ which shows the two contributions to the thermal barrier: the conduction through the wall $\Delta T_{cond} = q.e/k$ , and the convective term at the cavity wall-helium interface : $\Delta T_{Kapitza} = q/h_k$.. For wall thickness in the range $e$ = 0.5 to 3 mm and typical values: $k$(1.8K)=10 $Wm^{-1}K^{-1}$ , $h_k$(1.8K)=4000 $Wm^{-2}K^{-1}$ we obtain : $L$ = 0.2 ... 1.2. For low values of $L$ (close to 0.1) the cavity wall behaves like a "thin wall " where the cooling by convective heat transfert (Kapitza conductance) is the main governing mechanism. For high values, both mechanisms (convection and conduction) must be considered. The comparison between the analytical solution (constant parameters) and the numerical solution using Castem2000 (finite element code [10]) for typical cavities and physical parameters shows a good agreement (within 10 %) between the calculated values for both the temperatures and the time constants.

For a 3 GHz cavity ($T_{bath}$= 1.8 K, RRR=40, $e$=2 mm, $E_{acc}$=20 MV/m), the calculated time constant is $\tau$ = 6.3 $10^{-5}$, the RF and the helium side wall reach its steady state value in ~ 0.3 ms. The " thin wall " approximation ( values of $L$ close to 0.1) gives a simple solution depending on the Kapitza conductance and the diffusivity : $\tau_{thin\ wall} = \rho c_v e/h_k = 3.9\ 10^{-5}$.

In the steady state case, a more precise result can be obtained by a simple iterative method taking into account the thermal conductivity integral of the cavity wall between the temperatures of the two sides of the wall. The difference between the results obtained with this iterative solution and those obtained with Castem 2000 depends on the temperature gradient, and hence it is more important for higher fields. Nevertheless, the quench limit calculated by numerical method for the same example was 29.8 MV/m as compared to 32.6 MV/m using the iterative analytical method.

Using the iterative method we have compared the quench limit between cavities operating at different frequencies: 1.3 GHz and 3 GHz in the case of a defect-free surface (Fig 2). The calculations clearly show the thermal runaway at the quench limit, the thermal instability is reached at moderate wall temperatures ( $T$ < 2.3 K ). This result reflects the dominant role of the heat generation at the RF side of the cavity wall and the lack of available cooling at this temperature level (called by some authors : global thermal instability). The best result obtained at $f$ =1.3 GHz , was expected due to the quadratic dependence on frequency of the surface resistance. For this frequency the quench will be originated by the critical magnetic field limitation which is reached before the thermal instability threshold.

A few experimental proofs of this instability have been obtained [11] . They concern exceptional cavities prepared with careful conditioning techniques. Most of the time the cavities are limited by quenchs originated by defects, or by strong field emission. Nevertheless, these observations have shown some typical trends which can be easily modelled: heating of large cavity areas, detection of temperature increase which shift from the quadratic law dependence on field, and overall losses detected by RF measurements (slight decrease at high field of the cavity quality factor $Q_0$).

The short range of temperatures covered in these quench calculations offers a simple way to evaluate the field limit. By considering the thermal conductivity and the Kapitza conductance in the range 1.8 K - 2.3 K we are able to evaluate this limit in the defect-free case. If we look at the typical thermal conductivity curves of the niobium employed for cavity fabrication, $k(T)$ ranges from 5 to 15 $Wm^{-1}K^{-1}$ depending on the niobium purity (RRR), and the improvement of this value does not strongly contributes to the cavities performance which becomes finally limited by the Kapitza conductance.

## Surface defects on the cavity wall

An analytical solution can be obtained in the steady-state case [12]. The temperature at the center of a defect having the following characteristics (radius $r_d$, surface resistance $R_{sd}$) located on the RF side of the cavity wall, is given by

$$T = T_{bath} + r_d \cdot q_d \cdot \int_0^{\infty} \frac{J_1(x \cdot r_d)}{x} \left[ \frac{(k \cdot x - h_k) + (k \cdot x + h_k) \cdot \exp(2 \cdot k \cdot e)}{k \cdot \left[(k \cdot x + h_k) \cdot \exp(2 \cdot k \cdot e) - (k \cdot x - h_k)\right]} \right] dx$$

where $J_1$ is the Bessel fonction of first degree and $q_d$ is the heat power density in the defect area: $q_d = (1/2)R_{sd}H^2$. The RF power dissipated on the S.C. surface is neglected : we have taken $R_{sd} \sim 5.10^{-3}$ ohm which is largely more important than the superconducting surface resistance $R_s \sim 10 - 100 .10^{-9}$ ohm.

The calculated temperature at the center of the defect using this analytical solution has been compared to a numerical solution obtained with the Modulef finite element code [13] . The agreement is quite good, it also depends on the cavity field: 8 % to 13 % for cavity fields ranging from 6 MV/m to 18 MV/m ($f$=3 GHz, $T_{bath}$ = 1.8 K, $r_d$ = 20 µm, $e$ = 2.5 mm, RRR= 200).

One interesting point to be studied is the influence of the cavity wall thickness: the reduction of fabrication cost and the mechanical stability must be considered with respect to the thermal stability optimization.Using the analytical solution and for small power levels we have studied the influence of the thickness for different defect sizes: 20 µm to 1mm.

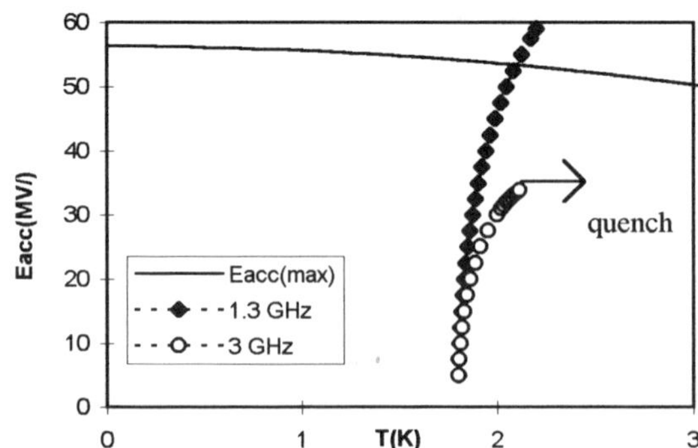


**Figure 2** . Quench limit for two different frequencies : 1.3 Ghz and 3 Ghz ( RRR=194, Rres=20 nohm, thickn.=2 mm). Eacc(max) : critical magnetic field of niobium.

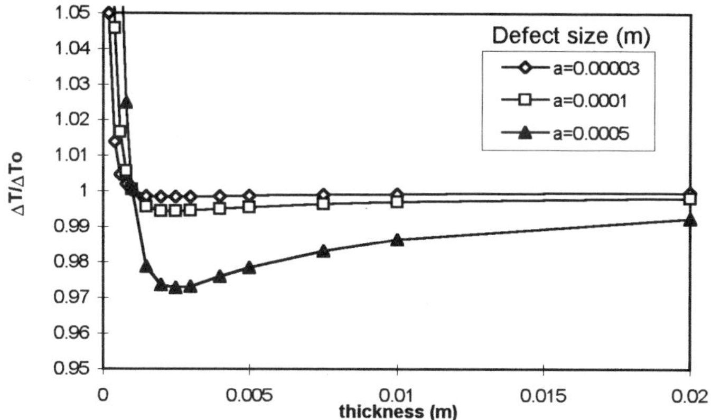


**Figure 3**. Influence of the wall thickness on the defect temperature.

The plot in Fig. 3 presents the relationship between *(ΔT/ΔTo)* and the wall thickness for different defect sizes. *ΔTo* is the temperature increase at the center of the defect size when the thickness becomes much larger than the defect size. This case corresponds to the semi-infinite solid case ($e \to \infty$) having a simple solution:

$$\Delta T = \frac{q_d \cdot r_d}{k}$$

It can be observed that all the curves tend to the semi-infinite solid solution for large thickness values. One important conclusion is the small influence of the cooling bath for small size defects. We can consider that for defect size < 0.1 mm the cavity wall behaves like a semi-infinite solid and the temperature increase at the center of the defect is totally governed by the thermal conductivity of the cavity wall.

This conclusion leads to a simple relationship between the power input into the defect and the conductive cooling available around the defect: $(1/2)R_{sd}H^2 = \Delta T.k/r_d$, so we can deduce a simple expression of the quench field $H_q$ for small size defects, as a function of the critical temperature $T_c$ (at $H=H_q$):

$$H_q = \sqrt{\frac{2 \cdot (T_c - T_b) \cdot k}{R_{sd} \cdot r_d}}$$

This result offers a simple way to evaluate the defect temperature evolution as a function of the accelerating field Eacc (Fig 4): taken into account the real thermal conductivity curve *k(T)* and iterating to solve the equation. We have found the quench values with an accuracy of 10% as compared to the solution using a finite element code.

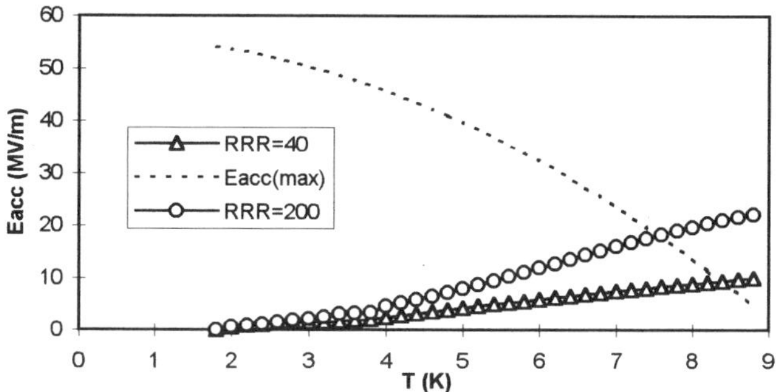


**Figure 4** Evolution of the defect temperature for two different types of niobium (thickn.=2.5 mm, f=3 Ghz, rdef=20 μm).

Experimental evidence of quenches originated by defects is a classical observation in superconducting RF cavities. Temperature maps using special rotating thermometers have been used to locate the quench with a good precision [14]. Detailed studies of thermal effects on cavities using fixed thermometers have also clearly shown the heating effects on cavity walls offering the possibility of interesting analysis using the model calculations. More recently a clear correlation between hot spots location in temperature maps and the presence of foreign metallic inclusions on the cavity RF surface has been proven with the help of sophisticated analysis methods ( eddy currents, neutron activation,...)[15].

The semi-infinite solid approach offers an analytical solution for small size defects:

$$T_{RF} = T_b + \frac{r_d \cdot q}{k} \int_0^{\infty} \frac{J_1(x \cdot r_d)}{x} \cdot erf(x \cdot \sqrt{\alpha t})dx$$

For larger defects the study of the time constants for wall thickness in the range 2-3 mm, requires the use of a finite element code. We have used a code developed at Cornell University by Tom Hays [16].

In the Fig. 5 are shown the temperature transients at the center of the defect corresponding to two different defect sizes: $t = 0.71$ µs to reach 0.9 $\Delta T_{max}$ for $r_d = 100$ µm, compared to $t = 99$ µs for $r_d = 5$ mm. For larger defects we obtain time constants which are close to the defect-free case.

## EXPERIMENTAL STUDY OF THE QUENCH

The following results have been obtained at the IPN Orsay (France) with 3 GHz monocell cavities. In addition to the RF measurement devices giving a global characterisation of the cavity performances, we have developed a surface thermometric system for thermal analysis of the quench. About 60 fixed thermometers [17] are placed around the cavity equator covering the high magnetic field zone. Eight photodiodes used for X-ray detection give a simple way to control the field emission activity. The thermometers and the RF detectors are connected to a fast acquisition system (amplifiers, multiplexor and digital oscilloscope) controlled by a Labview program. This system allows to locate the quench position, and is well adapted to follow the evolution of the temperature field before and after the quench, with time resolution of 5 µs.

The Fig. 6 shows the behaviour of one of the prototype cavities. The quench is detected at $E_{acc}$=15 MV/m and simultaneously a group of 4 thermometers (#25 to #28) measure important heatings (0.5K-4K) as displayed in Fig. 7. One of these thermometers (#26) is compared to the thermometer #2 located at the opposite side of the cavity. Thermometer #26 starts to detect some heating at 12.5 MV/m with a very sharp increase for more important fields, while the #2 has a more classical behaviour with a temperature increase dependence proportional to $E^2_{acc}$.

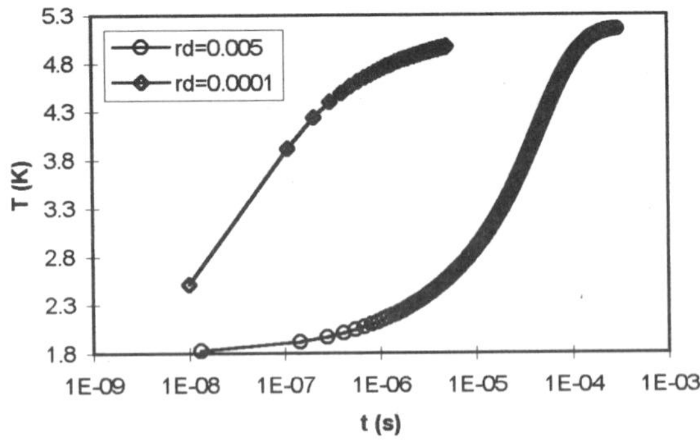


**Figure 5** . Temperature transients for two different defect sizes.

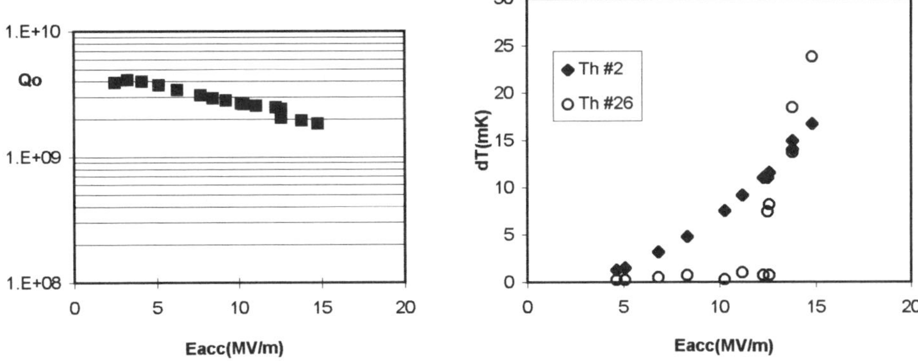


**Figure 6** . RF and temperature measurements in a 3 GHz cavity.

The Fig. 7 shows the time evolution of several thermometers close to the quench location. The temperature increase starts in synchronism with the sharp fall of the cavity field. The strong heating is observed during a long period (90 ms). Such a long period can be partially explained by the time response of the thermometers. In a separate experiment we made a measurement of this time constant The thermometers used in the 3 GHz experiments show a quite slow response: $\tau = 10$ ms. Applying short heat pulses of 1 ms, we have obtained temperature curves which are very similar to those obtained with the cavity. A different type of thermometer with improved thermal contact have a shorter response time: $\tau = 4.5$ ms. Our first conclusion is that the sensors are too slow to follow the thermal transient induced by a quench, however they are good and reliable for locating the quench and for quite accurate measurement of the steady state temperature field.

More precise RF measurements confirm a fast field decay (1 ms), and the thermal parameters (diffusivity), could explain a time constant of the same order of magnitude. We can also observe in the Fig. 7 how the field in the cavity can increase a few milliseconds after the quench, then our best guess is for long thermometer time constant to explain such long delays. The hypothesis of film boiling onset can be discarded for different reasons: the maximum heat flux density is lower than the critical value (~5 W/cm$^2$), no temperature jump is measured, and finally the recovery time are longer than those observed.

From the RF measurements we can follow the evolution of the field in the cavity during the quench. In the Fig. 8 we have plot the transmitted power signal Pt, which is proportional to $E_{acc}^2$, and the instantaneous $Q_0$, extracted from the $P_t$ curve with the help of a signal processing proposed by the group of Wuppertal University [18].

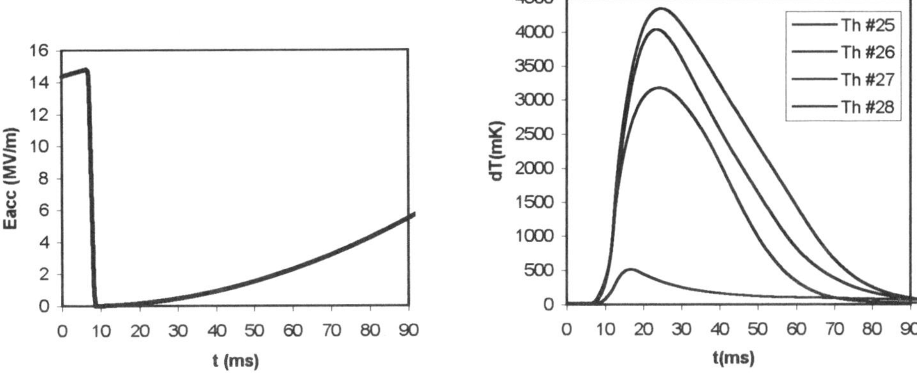


**Figure 7** . Temperature measurements during the quench .

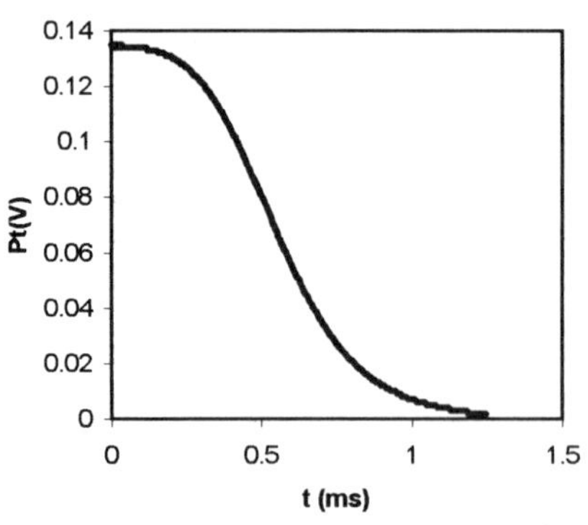

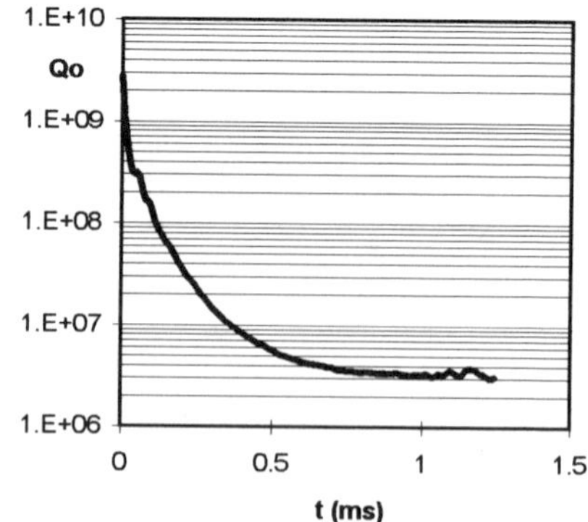


**Figure 8** . Cavity field (transmitted power) and Q0 transients during a quench

In a time t < 1 ms, $Q_0$ decreases from $3.10^9$ to $3.10^6$. This time constant is closer to the thermal model calculations presented in the above paragraphs either in the case of defect-free surface, or for large defects of several millimeters. From the initial and final $Q_0$ values, an estimation of the normal zone radius reached at the end of the quench can be obtained: $r$~7mm ( $R_s$ of normal niobium : 2. $10^{-3}$ohm).

## CONCLUSION

The models and methods presented in this paper show a simple way to calculate the thermal instabilities in cavities in both steady-state and transient regimes. The propagation rate of the normal zone on the cavity surface is not modelled yet either by analytical or a numerical model, but the RF measurements supply experimental data which can be analysed with the help of thermal model calculations. The surface thermometers give a good mean to study the thermal effects in the steady-state but their time constants are for the moment too large to supply an accurate measurement of the transient thermal effects during the quench.

## ACKNOWLEDGEMENT

This work was performed in the frame of a collaboration between the IPN - LAL (CNRS-IN2P3 Orsay) and the DAPNIA/SEA (CEA Saclay) for the study and development of SRF cavities. We would like specially thank J. Marini and J.L. Borne (LAL Orsay), J. P. Poupeau and H. Safa (CEA Saclay) and A.Caruette, A.Le Goff and N. Hammoudi (IPN Orsay).

## REFERENCES

1. G. Muller. Proceed. 3rd. Workshop on RF Superconductivity. Report ANL-PHY 88-1, Argonne USA
2. J.Tuckmantel. Report CERN/EF/RF 84-6, CERN, Geneva 1984
3. H.Safa. Proceed. 7th Workshop on RF Superconductivity. October 1995, Gif-sur-Yvette, France
4. X. Cao and D. Proch. IEEE Particle Conference. May 1991, San Francisco USA
5. Handbook on Materials for Superconducting Machinery. Nov. 1974, Batelle's Columbus Lab. USA
6. M. Fouaidy et al. Proceed. 5th. Workshop on RF Superconductivity. Report DESY M-92-01, Germany
7. Boucheffa et al. Proceed. 7th Workshop on RF Superconductivity. Oct. 1995, Gif-sur-Yvette, France
8. W. Weingarten CERN report 92-03 vol II. June 1992, CERN, Geneva
9. "Conduction of Heat in Solids" H.S. Carslaw and J.C. Jaeger, 2nd edition. Oxford Univ. Press 1959
10. Castem 2000. Report DEMT 88/176 . CEA Saclay, France
11. J. Graber et al. Report CLNS 91-1061 Cornell Univ. Ithaca NY, USA
12. M. Pekeler. Ph.D. Thesis. Report DESY M-96-16, Hamburg, Germany.
13. Modulef (1988). INRIA-Simulog, Rocquencourt Le Chesnay , France
14. Q.S.Shu et al. Applied Superconductivity Conference (1996). Pittsburgh, PA, USA
15. TESLA Collaboration Report 97-05. DESY, Hamburg, Germany.
16. Tom Hays: « http://w4.lns.cornell.edu/~trh/fondue/ »
17. M. Fouaidy et al. Proceed. 7th Worksop on RF Superconductivity. Oct. 1995, Gif-sur-Yvette, France
18. D. Reschke and R. Roth. Proceed. 6th. Workshop on RF Superconductivity. CEBAF 1993, VA, USA

# INFLUENCE OF THE RRR OF NIOBIUM ON THE RF PROPERTIES OF SUPERCONDUCTING CAVITIES

**H. Safa**[✧]

C.E. Saclay
91191 Gif-sur-Yvette, FRANCE

## ABSTRACT

The maximum accelerating field achievable in a superconducting niobium cavity is directly related to the thermal conductivity of the material, the Kapitza conductance between the niobium and the superfluid helium bath and the exponential behavior of the surface resistance with temperature. It is here highlighted that the RRR, an intrinsic physical measure which is related to the amount of impurities in the bulk, might affect these three parameters and hence the quench field. A new way of improving the niobium RRR by heat treatment is discussed as well as its impact on actual superconducting cavities.

## I INTRODUCTION

As superconducting radiofrequency (SCRF) cavities technology is improving in performance and reliability, the need for higher accelerating fields in these structures is hitting a severe limitation. Increasing the maximum magnetic field on the surface will induce a transition of a macroscopic part of the cavity surface from superconducting to normal conducting called "quench". As the drop in the corresponding Q value is a factor of $10^5$ between the two phases, this will immediately drop the accelerating field, making the cavity inoperable. In theory, this quench field for niobium should be higher than 150 mT in superfluid helium. The first cavities made of reactor grade material in the early seventies were quenching around 10 mT. The progressive use of higher purity niobium have subsequently improved this figure. But even then, when using nowadays RRR 400 (or higher) niobium, actual cavities still exhibit quenches around 80 mT. Showing the relationship between RRR and superconducting RF performances, the present study will focus on whether an adequate heat treatment might improve these performances to approach the theoretical values.

## II RRR

### Definition

The Residual Resistivity Ratio (RRR) is the ratio between the resistivity at room temperature (generally 295 K) to the resistivity at helium boiling temperature (4.2 K) in the normal state.

---

[✧] On leave for sabbatical at Los Alamos National Laboratory, Los Alamos, NM, 87545, USA

$$RRR = \frac{\rho_{295K}}{\rho_{4.2K}} \qquad (1)$$

The resistivity of all metals at low temperature can be expressed as

$$\rho = \rho_0 + \rho_{ph}(T) \qquad (2)$$

where $\rho_0 = \rho_{imp} + \rho_{gb} + \rho_{disl}$ is the residual resistivity at T = 0 K mainly due to impurities, grain boundaries and dislocations (In presence of an external magnetic field, there is an additional term due to magnetoresistivity). $\rho_{ph}$ is the phonons contribution which is temperature dependent. For niobium at low temperatures (T < 25 K), a good approximation is given by :

$$\rho_{ph}(T) = 6.5\,10^{-14}.\,T^3 \quad [\Omega.m] \qquad (3)$$

The impurities contribution is by far the most important one in the common case, using typically RRR200 niobium for SCRF cavities. Therefore, the RRR value gives an indication of the purity of the metal. An ideally pure single crystal niobium free of dislocations will have a RRR of 30 000 (only due to phonons).

There are two ways for measuring the RRR. One can either place a sample in boiling liquid helium (at T = 4.2 K) and apply a static magnetic field strong enough to perform the transition to the normal state. Or, staying at zero magnetic field, the resistivity is measured above $T_c = 9.25$ K, and extrapolated to 4.2 K using equation (3).

It has to be stressed here that the RRR value is an *average* measure of the bulk material over a macroscopic volume (typically hundreds of $mm^3$). Therefore, one microscopic inclusion (a few microns size) will not affect this value although, as it will be shown later, it may have a huge impact on the quench field.

## Purification

It has been shown that one way to purify a niobium piece is to heat treat it in a vacuum furnace with a solid state getter (Ti, Y or Hf) deposited on its surface [**1**,**2**]. This process of purification is based on the diffusion of light impurities (C, O & N) in the bulk and the more favorable free energy reaction of these with the getter, forming carbides, oxides or nitrides. Ideally, after purification, the niobium bulk would be free of all light elements (including hydrogen, which is, by the way, responsible of the Q-disease effect). The material should be also free of dislocations. The only remaining resistivity contribution should be the one due to metallic impurities (almost not diffusing) and to the grain boundaries. They will determine the maximum RRR value achievable using this purification process. Experimental RRR as high as 2000 can be obtained using a commercial niobium sheet having a starting RRR of 165 [**3**]. This confirms that the light elements are generally responsible for most of the residual resistivity. On reactor grade niobium samples (RRR = 25 to 40), maximum values between 250 and 800 are obtained depending on the initial metallic impurities contents, namely tantalum [**4**].

In practice, it has been shown that the furnace residual vacuum will tend to pollute the material. Heat treating at a too high temperature will finally results in a lower RRR [3]. The highest RRR were obtained at intermediate temperatures (1000°C-1200°C). Therefore, the final RRR after a given heat treatment will be a resultant function of temperature, time and vacuum. An optimized temperature-time cycle can be deduced for a given material type (impurities contents) and a given furnace to achieve the highest RRR [4].

One important drawback of the high temperature heat treatment (at temperatures over 1000°C) is the softening of the niobium due to the increase of its grain size. The yield strength (defined at 0.2% of elongation) might decrease from 70 MPa to 40 MPa. Some cavities shapes might then collapse while evacuated under their own weight if not stiffened (either by increasing the sheet thickness or by external stiffeners). When annealing below 1000°C, not only this softening is avoided, as the grains will have very little growth, but good mechanical properties will be restored as the cavity will be freed from the dislocations induced by the cold forming of the sheets.

# III RF PROPERTIES OF SUPERCONDUCTING CAVITIES

## Q-value

The standard way to show the RF properties of a superconducting cavity is to plot the quality factor Q of the resonator as a function of the accelerating field Eacc (integrated along the

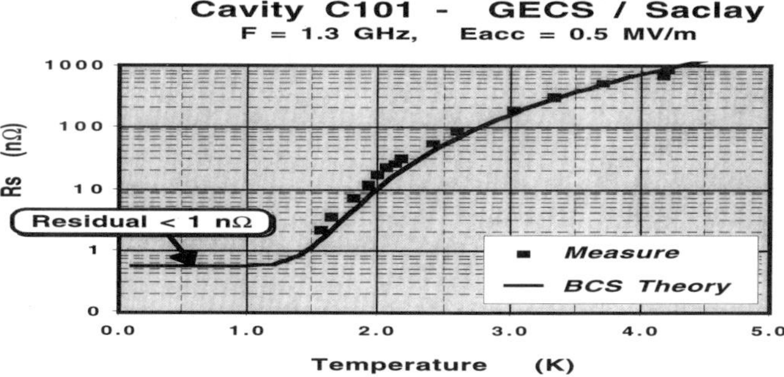


**Figure 1** *Surface resistance of a cavity with high purity niobium. The residual is lower than 1 nΩ, indicating a Q-value higher than $10^{11}$.*

particle path across the cavity). The Q value is equal to ($G/R_s$), where G is a geometrical constant (depending on the shape of the cavity), and $R_s$ the surface resistance of the niobium. At high temperatures (ratio to critical temperature $T/T_c > 0.3$), $R_s$ is dominated by the well known BCS surface resistance and can be pretty well predicted (function of temperature, frequency and resistivity). It *increases* with RRR. Therefore, for a cavity running at T = 4.2 K for example, a *lower* RRR will give a lower dissipation in the helium bath (provided the quench limit is not exceeded). At the other hand, a too low RRR will start degrading the superconducting properties (reducing the gap, the critical temperature and critical field). Consequently, for a given temperature, there is an optimum value for the RRR giving the lowest surface resistance.

At lower temperatures (typically below 2 K), the surface resistance will approach a non-zero value called residual resistance as the BCS part will vanish. Recent progress have been made in the understanding of this residual term. The more important effects have been shown to be due to the Q-disease [**5**,**6**] (or "100 K effect", hydride phase formed at the niobium surface between 100 and 160 K), to the granularity of the polycrystalline material [**7**], to the static magnetic flux pinned during cooldown [**8**] and to the residual impurities in the niobium [**9**]. A higher purity of the material will give a lower residual $R_s$. Reducing these effects, surface residual resistance lower than 1 nΩ (at a frequency of 1.3 Ghz) has been obtained on real cavities, giving Q values higher than $10^{11}$ at low field (figure 1).

## Quench Field

The quench field is mainly a thermal process. The surface resistance $R_s$ induces on the inner surface of the cavity a heating flux $Q = \frac{1}{2} R_s .H^2$ per unit area, where H = B/μ is the surface RF magnetic field. This heat has to be conducted through the bulk to the outer surface to be removed by the helium bath at the temperature $T_b$. The thermal equilibrium is reached when the inner temperature surface is equal to

$$(T - T_b) \quad = \quad Q \left[ \frac{t}{\Lambda} + \frac{1}{h_k} \right] \qquad (4)$$

where t is the sheet thickness, Λ the thermal conductivity of niobium and $h_k$ the thermal conductance at the niobium-helium interface (Kapitza in the case of a superfluid bath). As the surface resistance $R_s$ and the thermal conductivity Λ are both strongly varying with temperature, the above equation may get no stable solution above a given field in the superconducting state. Of course, as soon as either the magnetic field B exceeds the critical field $B_c(T) = B_{c0} \left[ 1 - \left( T/T_c \right)^2 \right]$, or if the temperature T exceeds $T_c$, the cavity will undergo a transition to the normal state. This thermal runaway will be defined as the quench field.

The thermodynamic critical magnetic field for niobium is $B_c$= 198 mT at zero temperature. But as niobium is a type II superconductor (even though very weak), its first critical field is

$B_{c1}$=163 mT. It drops to 155 mT at 2 K and to 129 mT at 4.2 K. Although it is not quite completely clear, one can assume that this will be the theoretical limit as, in the mixed state ($B > B_{c1}$), additional losses will sooner or later induce the thermal runaway. RRR value should have no influence on this critical field at least inside the grains. The intergrain critical value might be lower, reducing also the RRR. Generally speaking, a region having a low RRR (for example the cold work layer) exhibits also a lower critical magnetic field $B_{c1}$.

The thermal conductivity $\Lambda$ is directly proportional to the RRR, as normal electrons are mainly responsible for both current and heat transfer at these temperatures, even in the superconducting state. If one excepts the 1.8-2.2 K region, where large grains size can induce a "phonon heap" in $\Lambda(T)$, one can almost compute the thermal conductivity knowing the RRR [**10**].

The Kapitza conductance at superfluid helium temperatures ($T < T_\lambda = 2.17K$), have been shown to be much more sensitive to the external surface preparation (cleanliness, roughness) than to intrinsic properties of the niobium as the RRR [**11**]. More precise work is still needed to determine whether a higher purity niobium may exhibit better conductance.

The above analysis, valid for a uniform and homogeneous material, will give the uniform thermal instability of a cavity. This will be the maximum ideal field achievable, given a RRR niobium material, its thickness, surface resistance and the bath temperature. Figure 2 show an example of such value as a function of bath temperature.

Unfortunately, in most experimental cases, the quench field measured on actual cavities is somewhat lower than the calculated one given by the uniform solution. This inclines to believe that the real material is not uniform and that most probably, localized defects are responsible for the quench. More complete analysis including surface defects either analytical [**12**] or using computer codes [**13**,**14**,**15**] have been performed showing that this indeed could be the case. Microns size defects (assuming normal conducting local resistance) might actually explain the experimental results. In most cases, the origin or the nature of the localized defect inducing an increase of the local surface resistance (and experimentally identified by an increase of local heating at the quench location as pointed out by the temperature mappings of the external surface of the cavity) remains unexplained.

Although these defects are not yet quite well identified, what is well proven is that cavities made with higher RRR niobium will perform higher quench fields [4,**16**]. Part of this improvement can be attributed to the highest thermal conductivity giving hence a better cooling. But the main point is probably that the process of production of high purity niobium is, *in fine*, also producing a material with lower sizes defects or a lower number of them. As a matter of fact, a statistical analysis can be done assuming a random probability for having a given defect size on a given niobium surface. This analysis will predict a decrease in the quench field value if the number of cells (or the total cavity surface) is increased and is in pretty good agreement with the experience. This supports the above statement of defect induced quench field.

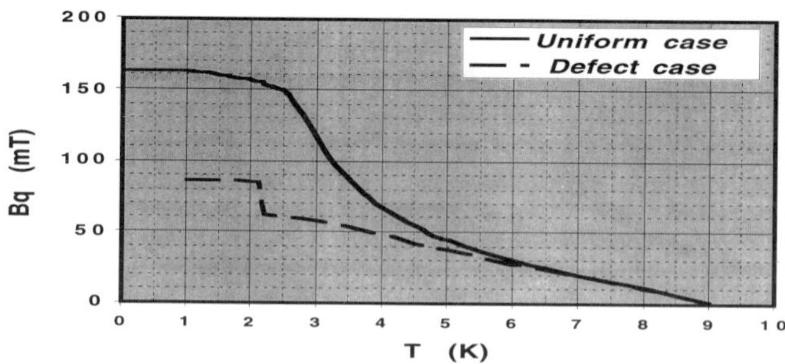


**Figure 2** Magnetic quench field as a function of bath temperature in the uniform case and in the case of a defect 30 μm size with a normal resistivity of $10^{-7}$ Ω.m.

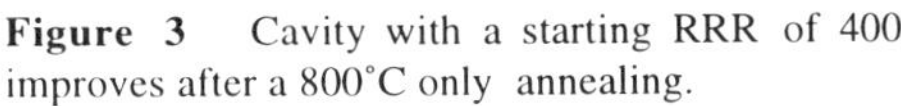

**Figure 3** Cavity with a starting RRR of 400 improves after a 800°C only annealing.

**Figure 4** Cavity with a starting RRR of 250 does not improve after a 800°C heat treatment but improves after a 1000°C one.

## Heat Treatment

Heat treating SCRF cavities will generally result in better performances, especially as regards to the quench field. As pointed out above, some of this improvement is due to the higher RRR obtained after a high temperature gettering. But even though the annealing is done at lower temperatures (800°C-1000°C) where no purification process can occur, clear benefits are obtained. Aside from the stress relieve pointed out above leading to better mechanical performances and the hydrogen free cavities, getting rid of the Q-disease problem, improvement in quench field might also be significantly observed after this annealing. Homogenization of the material might be the driving physical process here where small defects can locally dissolve in the bulk, making them less harmful. It seems that the higher the starting RRR material, the lower the annealing temperature can be. In the following graphs (fig. 3 &4), Q vs Eacc curves of two single-cells cavities tested at Saclay are shown. The first one is made out of niobium sheets having RRR400, while the second one is made with sheets having RRR250. Although they quenched at almost the same field level before annealing, the first one improved after a 800°C annealing, whereas the second needed a 1000°C one to perform better (no improvement was seen on the second one after only a 800°C heat treatment). Supporting this trend, cavities with reactor grade niobium (RRR 34) were tested following different heat treatment steps. These needed a high temperature heat treatment (1300°C) to start improving (figure 5). Even though, they still show a slope in the Q vs. Eacc curve, and probably need more tougher heat treatment to get through and perform better.

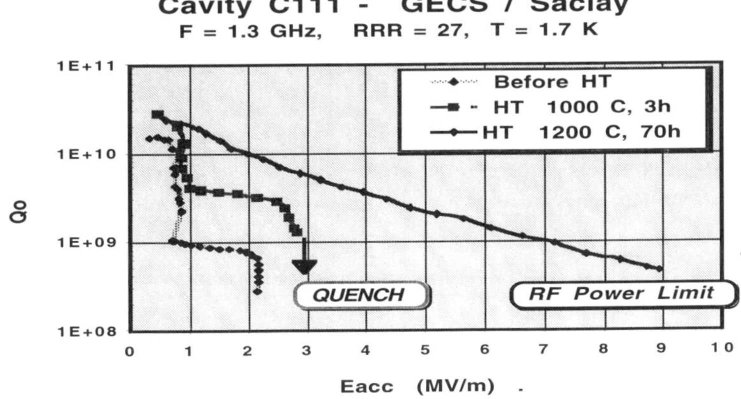


**Figure 5** Cavity made of reactor grade niobium. The high temperature heat treatment improves the performance but the final curve still exhibits a strong slope.

## IV CONCLUSION

In conclusion, high purity niobium is generally required for achieving high fields in SCRF cavities. But, as has been shown by KEK results **[17]**, cavities with RRR 200 might already achieve fields as high as 120 mT. The RRR, an intrinsic physical parameter easy to measure, can already give a good hint for the final RF performances. But as it has been demonstrated, local defects, while not affecting the RRR, may severely inhibit the quench field. Therefore, additional diagnostics able to measure local properties are needed to qualify the niobium. One way to override the quench limitation is to go for a high temperature heat treatment which gives several other benefits (Q-disease healing, recrystallization, removal of cold work). The higher the initial RRR, the lower the temperature annealing needed. One drawback of a too high temperature heat treatment is the softening of the material. Consequently, the use of a RRR 200 material, combined with a moderate temperature annealing (<1000°C) can lead to a good compromise between mechanical properties and RF performances.

## REFERENCES


**1** H. Padamsee, IEEE Trans. Mag., *21* (2) [1985] 1007

**2** P. Kneisel, J. Less Common Metals, *139* [1988] 179

**3** " Advances in the purification of niobium by solid-state gettering with titanium ", H. Safa, D. Moffat, B. Bonin, F. Koechlin, Journal of Alloys and Compounds, 232 (1996) p. 281-288

**4** "Nb purification by Ti gettering", H. Safa, D. Moffat, F. Koechlin, E. Jacques, Y. Boudigou, Proc. of the 7th Workshop on RF Superconductivity, Gif-sur-Yvette, France (Oct. 1995), p. 649

**5** " Degradation of niobium superconducting RF cavities during cooling time ", B. Aune & al, LINAC Conference, Albuquerque (USA, 1990)

**6** B. Bonin, R. Roth, Particle Accelerators, *40*, [1992] 59

**7** "Power Dissipation at high fields in granular RF superconductivity ", B. Bonin, H. Safa, Superconductor Science & Technology, *4* (1991), p. 257-261

**8** "Flux trapping in Superconducting cavities ", C. Vallet, M. Boloré, B. Bonin, H. Safa, Proc. of EPAC, Berlin (1992)

**9** "Surface Resistance of a superconductor ", H. Safa, Proc. of the 5th Workshop on RF Superconductivity, Hambourg (August 1991), p.711

**10** "Parametrization of the niobium thermal conductivity in the superconducting state", F. Koechlin, B. Bonin, Proc. of the 7th Workshop on RF Superconductivity, Gif-sur-Yvette, France (Oct. 1995), p. 665

**11** A. Boucheffa, Thesis at University of Pierre et Marie Curie, Paris VI, France [1994]

**12** "An analytical approach for calculating the quench field in superconducting cavities", H. Safa, Proc. of the 7th Workshop on RF Superconductivity, Gif-sur-Yvette, France (Oct. 1995), p.413

**13** G. Muller, "Superconducting niobium in high RF magnetic fields", Proc. 3rd Workshop on RF Superconductivity, Argonne, Ill., USA [Sept. 1987], p. 331

**14** H. Padamsee, "Calculations for breakdown induced by large defects in superconducting niobium cavities", CERN/EF/RF, 82-5 [1982]

**15** R. Roth, Proc. of EPAC, Berlin (1992)

**16** H. Padamsee, "Influence of thermal conductivity on the breakdown field of niobium cavities", IEEE Trans. Mag. *21* [1985] 149

**17** E. Kako et al., " Characteristics of the results of measurement on 1.3 GHz high gradient superconducting cavities ",Proc. of the 7th Workshop on RF Superconductivity, Gif-sur-Yvette, France (Oct. 1995), p.425

# STUDIES OF RF-SUPERCONDUCTIVITY PROPERTIES OF NIOBIUM FILM-COATED CAVITIES AT CERN

C. Benvenuti, S. Calatroni, I. E. Campisi,[1] P. Darriulat, C. Durand,[2] M. Peck, and R. Russo

European Organization for Nuclear Research, CERN, CH-1211 Geneva 23, Switzerland

[1] Visiting Research Associate from the Thomas Jefferson National Accelerator Facility, Newport News, Virginia, 23606, USA

[2] Presently with the Laboratori Nazionali di Legnaro, Istituto Nazionale di Fisica Nucleare, Legnaro (Padova), Italy

## ABSTRACT

Extensive tests have been performed on several cavities, operated at 1.5 GHz in the $TM_{010}$ mode, obtained by sputter-coating niobium on copper, in order to study the physical phenomena responsible for their ultimate performance. Cavities manufactured with the same technique, but resonating at 352 MHz, are in current operation in the CERN Large Electron-Positron Collider LEP.

The measured RF data have been analyzed to extract information about: 1) the BCS resistance properties of each film, 2) the residual resistance of sputtered niobium deposited onto copper cavities manufactured with various techniques, and 3) the influence of trapped magnetic fluxons on the film's superconducting RF properties.

Several tests have been made also on bulk niobium cavities at 1.5 GHz manufactured through different methods in order to compare and extend the results of the film studies.

Analyses confirm that the sputtered niobium films provide: 1) lower BCS resistance than the bulk, 2) lower sensitivity of the surface resistance to external applied magnetic field, 3) residual resistance comparable to the bulk, once the substrate's properties are well controlled.

## INTRODUCTION

Superconducting materials are nowadays established as a reliable and economical alternative to copper in the construction of microwave structures for charged particle accelerators[1]. Machines have been built which, by the use of superconducting RF structures, can provide beams with greatly enhanced properties (higher energy, lower emittance, high duty cycle) [2].

Most of these accelerators have used bulk niobium as the material for the construction of the RF resonators, and more are planned which will use that technology. The only exceptions to the use of bulk niobium are lead, electroplated onto copper low-beta structures for heavy ion machines, and niobium films, sputter-coated onto a copper substrate [3]. The latter technology has been successfully adopted and implemented for the Large Electron Positron (LEP) collider and it is now an established alternative to the use of the bulk metal [4].

The use of thin superconducting niobium films provides several advantages with respect to that of bulk niobium, and it has some additional complications as well. In this paper we will try to point out such advantages and complications and clarify the basic differences between niobium film and bulk, as they have been elucidated by the present research.

Bulk niobium cavities for practical accelerator use have been now manufactured for over 25 years and the factors which affect their performance (such as material purity, thermal conductivity, the manufacturing processes, the chemical surface treatments and the assembly cleanliness) are understood well enough that large scale, high gradient machines can now be designed using parameters (unloaded quality factor $Q_o$, accelerating gradient $E_a$) which are within reach of the present bulk niobium technology [5].

On the other hand, the experience accumulated in producing niobium films on copper is more recent, but the success of the technology as applied to LEP is as thorough as for bulk niobium. During the production of the more than 250 four-cell, 352 MHz cavities for the machine, care was taken to maintain a uniform production quality which would guarantee the original design performance to the design level of $4x10^9$ at 6 MV/m at 4.2 K.

As it is often the case in many large scale applications of new technologies, cost and scheduling considerations have driven the developers' efforts in the direction of the production of thin films according to a recipe which was well under control and understood, at the onset of the machine construction, rather than toward deepening the physical comprehension of the various phenomena which affect the performance of the niobium films in general.

In parallel with the industrial production of the LEP cavities, an effort has been undertaken at CERN to understand at a deeper level the factors governing the properties of niobium films. Even though the insight brought by this research program will not affect directly the production or performance of the LEP machine, which is now reaching its completion, it will clarify the limits of the installed cavities and it will provide recipes for improved and better controlled niobium films for future machines. This effort will also clarify, by comparison, some of the parameters and manufacturing of bulk niobium cavities.

In effect, one of the clear advantages of the niobium on copper technology over that of bulk Nb, is the possibility of tailoring the film properties to the specific needs of the accelerator application, by modifying impurity content and other parameters, without endangering the thermal grounding to the bath, which is provided by the high thermal conductivity copper.

An additional point to be made in favor of the thin film technology is that improvements to the cavity performance can be attained with a modest effort of recoating existing structures, should new methods or films prove to have better properties, without having to reinvest into major reconstruction efforts.

Along this line, the thin film technology allows an optimized use of a whole accelerator complex, including the cryogenics installations, since a tight control of the niobium film parameters can lead to an optimal performance of the cavities, which are usually coated as one of the last steps of the construction schedule, long after all the other machine parameters have been frozen and even after the mechanical fabrication of the copper structures has started.

The present work has concentrated on measurements performed on cavities operated in the $TM_{010}$ mode at 1.5 GHz. The choice is mostly due to practical reasons: working with an accelerating mode the results can be easily translated into some information directly applicable to real accelerator systems, and at this frequency the resonators are of a manageable size in terms of handling, preparation and testing, making a rapid turnover of results possible.

## SUBSTRATE MANUFACTURING

The resonators are constructed out of Oxygen-Free Electronic grade (OFE) copper, which provides low level of impurities and relatively high thermal conduction to the Helium bath, even at these low temperatures [6]. However, even though the starting material is the same, the cavity construction techniques have a very profound effect on the final performance of the film. In the LEP cavity production, a method of spinning half cells with subsequent welds was implemented. It is thought, however, that regions of welds must require a special attention and post manufacturing rework due to the different nature of a molten section of the substrate. To avoid these uncertainties, in the present work only seamless cavities were used, produced by three different methods:

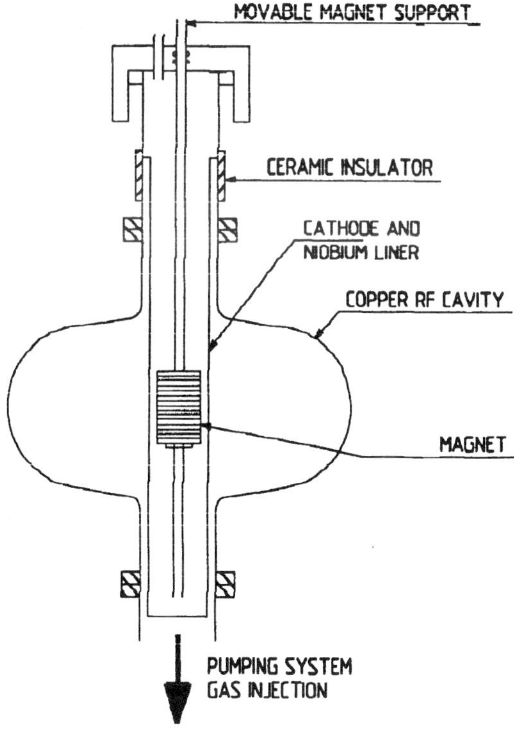


**Figure 1.** 1.5 GHz single-cell cavity sputtering configuration. For standard Ar discharge, a 3 A current at 360 V is maintained in the gas kept at a pressure of $1.5 \times 10^{-3}$ mbar.

1) Hydroforming [7]: the cavities were manufactured from seamless tubes and with three intermediate annealing steps at 600 °C. This fabrication method yields structures which exhibit fairly large grain size (hundreds of microns), due to the annealing. It is thought that the resulting roughness could render the adherence and the uniformity of the deposited niobium film less predictable and might be a cause of diminished performance.

2) Spinning [8]: this is a relatively novel procedure, developed for the seamless case at the Legnaro Laboratory (INFN, Italy), by which a sheet of copper (or other metals as well) of 3 mm thickness is spun around a specially keyed mandrel (which can be removed afterwards from the inside) so that a flat metal can be formed around an axially symmetric structure. Since there is no intermediate annealing in the process (a preliminary annealing of the sheet copper at 250 °C is performed), the original grain size is maintained.

3) Electroforming: This process involves the use of a disposable mandrel (usually aluminum, or, as in our case, glass) over which a thin copper layer is sputter-coated. Subsequently, a 3 mm layer of copper is deposited using pulsed electroplating techniques.

In principle, this technique should provide a very homogeneous and stress-free surface layer, onto which a smooth film could be deposited.

It is generally true, both from inner surface microscopic inspections and from the systematic differences in the results of cavities coated with films deposited with a similar recipe, that the substrate has a major importance in the final results and that as much care must be given to this aspect of the manufacturing as for the film deposition techniques themselves.

## SURFACE PREPARATION AND FILM DEPOSITION

The cavities are prepared for deposition by removing the inner layer (120 μm for spun cavities) via a sulfamic acid, hydrogen peroxide, n-butanol and ammonium citrate bath. Other methods for obtaining a smooth surface are now being investigated, such as electropolishing. High pressure water rinsing of the inner surface at 100 bar with ultra high purity water is performed before each deposition and each RF measurement.

Once the inner surface has been cleared of manufacturing defects and impurities, it is coated with a film of niobium, usually 1.5 μm thick [9]. This thickness is much larger than the penetration depth of the RF in the superconductor, and it is considered more than sufficient to ensure that all the surface irregularities of the cavities are properly coated. Coatings as thin as 0.5 μm and as thick as 7.5 μm have also been studied.

The coating method makes use of magnetron sputtering, as illustrated in Figure 1. In a noble gas atmosphere at a pressure of about 1.5 x $10^{-3}$ mbar, a discharge at 360 V and 3 A is maintained. The temperature of the substrate is held at a constant value of 150 °C during the discharge. It has been demonstrated that different deposition temperatures can have large effects on the properties of the film. With the above parameters the deposition rate is approximately 100 nm / minute. The standard coating thickness of 1.5 μm for the cell can be thus achieved in 15 minutes of treatment. The cutoff tubes are sputtered first (with slightly different discharge parameters: 320 V, 1 A, $10^{-2}$ mbar)), to make use of the gettering action of the freshly deposited niobium, to further improve the vacuum conditions when the cell is coated.

Two sputtering apparatuses exist, one with a single cathode and one with a second cathode. The main difference between the two systems consists in that, with the second copper cathode operated as an anode, it is possible to remove the oxide layer from the copper surface prior to deposition by reverse sputtering, providing a control of the nature of the interface between the copper substrate and the niobium layer.

## MEASUREMENT METHODS AND RESULTS

The RF measurements are performed with a relatively standard RF system. Fixed antennas are used to couple the power in and out of the cavity and the output antenna is also used as a pick up probe for electron current emitted from the cavity surface.
A phase lock loop maintains the generator's frequency locked to the cavity's frequency and measurements of incident, reflected and transmitted power are done at variable temperatures, from 4.2 K down to about 1.5 K. The measurements are done with a spectrum analyzer, which provides a linear output via a linear receiver and which filters out spurious harmonics. Power meter measurements are done for consistency check, but have usually a lower accuracy than the spectrum analyzer ones.

Calibrations of the antennas' external Q's are done for each set of measurements as a function of fields at 4.2 K and at the lower equilibrium temperature.

Measurements performed between the two temperatures yield information about the temperature dependence of the various components of the losses and about the residual resistance beyond the equilibrium temperature.

In Figures 2 plots of quality factors versus accelerating gradients are given for a niobium on copper cavity and for a bulk niobium cavity. Both cavities were manufactured by the spinning method.

With very good approximation, the losses at 4.2 K can be ascribed to the BCS losses (predominant at that temperature) whereas the losses at 1.8 K or lower are mainly dominated by the residual losses.

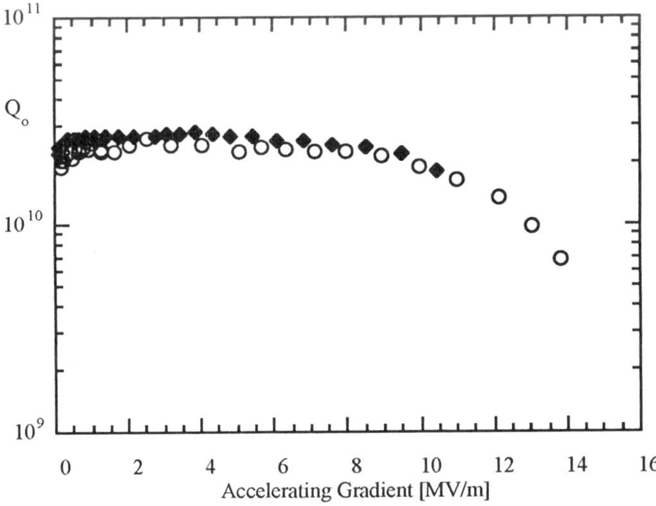


**Figure 2.** Unloaded Q versus accelerating gradient for a Nb/Cu cavity (◆) and for a bulk cavity (O) at 1.6 K. The copper cavity was coated with Nb and with standard Ar noble gas discharge. Several L-band Nb/Cu cavities have been produced with residual resistance below 10 nΩ up to the 10 MV/m range.

Each cavity is also tested with a variety of trapped magnetic fields, to determine the importance of the fluxon-induced losses of the films. It turns out that these losses are very different in films of different purity and their measurements can help establishing relative levels of defects in various films without resorting to direct destructive analysis of the cavity. Moreover, in cavities where surface defects are present, an initial independence of the losses from the magnetic field is observed, a signature which indicates that fluxons are affecting first areas which are already normal conducting (inclusions or film pinholes) and only subsequently the losses increase with increasing magnetic field, when the fluxons start to cover the surfaces of good superconducting quality.

The critical temperature is measured by observing the transition from the Meissner state. At the same time the trapping efficiency can be measured by measuring the relative amplitudes of Meissner, trapped and applied fields. Film cavities always exhibit complete trapping, whereas bulk cavities (also tested for comparison) show full trapping for as worked niobium and lesser degree of trapping (possibly by Meissner expulsion within purer grains) after having been fired at moderately high temperatures (~1000 °C).

By measuring the frequency shift as a function of temperature near transition, one can determine the penetration depth of the superconductor. This has been done in a number of cavities, giving numbers of approximately 37 nm in the clean (infinite mean free path) limit.

## DISCUSSION

In order to better understand the physical phenomena involved in the surface losses of the superconductors under study, it is convenient to separate them into three major contributing factors, each of which can be to some extent controlled and which yields information about different physical phenomena responsible for the various losses. These are the BCS losses, due to incomplete pairing at finite temperatures, the residual losses related to the material properties and surface conditions of the superconductor and which, to some extent, are out of the control of the experiment, and the fluxon losses which contribute by maintaining normal areas within the fluxons and by the motion of the unpinned or weakly pinned ones.

### BCS resistance

Extensive data exist on the properties of superconductors (in particular niobium) which produce substantial agreement between the experimental data obtained in various laboratories and conditions, and the predictions of the BCS theory [10,11]. It is certainly true that the temperature dependence, and to almost the same extent the frequency dependence, of the BCS losses have yielded excellent agreement between theory and experiment [12]. It is therefore important that for niobium films as well a thorough analysis be performed of the BCS loss term.

It is also true that there are no systematic data, nor comprehensive calculations, which describe the RF field dependence of the BCS resistance. From this perspective, it is important to carefully determine the differences between the behavior of bulk niobium and film, as future theoretical analyses will have to produce predictions in agreement with both sets of data as well as describe the differences in terms of basic physical parameters.

The data have been analyzed in terms of their compliance with the BCS theory. No major deviations from it have been observed, fact that indicates that the BCS fraction of the losses is consistent with it. In Figure 3, the temperature dependence of the BCS losses of all the data for film cavities is plotted. The energy gap at any temperature can be obtained from the ratio of the BCS resistance at that temperature and that at 4.2 K. These

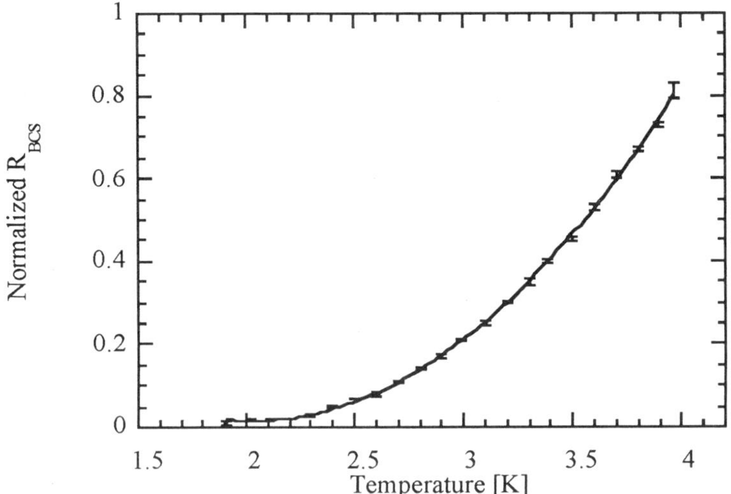


**Figure 3.** Temperature dependence of the BCS surface resistance (normalized at 4.2 K) for all the niobium film coated cavities.

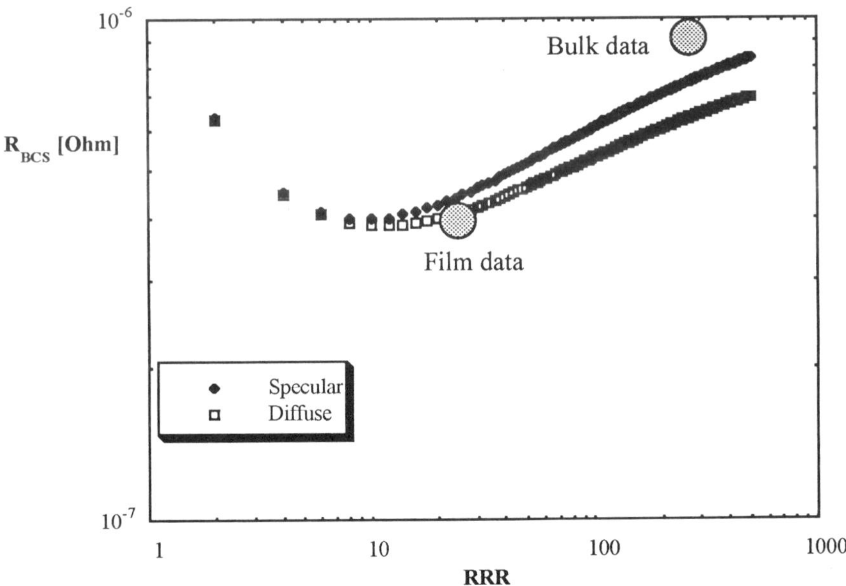


**Figure 4**. BCS resistance as a function of RRR for niobium film and bulk. The calculations were performed using a numerical code (Ref. 14) and for $T_C$=9.55 K, $f_0$=1.5 GHz, a coherence length of $\xi$=40 nm, a reduced gap $\Delta$/kT = 1.95. The difference in impurity concentration explains the different BCS resistance for film and bulk niobium.

experimental data give a value of 18.6 K, corresponding to $\Delta/kT=1.95$ in good agreement with earlier measurements. [13]

The actual values of the BCS resistance for films and for bulk niobium are considerably different (Figure 4). This difference can be explained in terms of purity (thus RRR).[14] For high RRR niobium, the BCS resistance can be more than a factor of two higher than for a film of RRR of about 20. These differences can be important in applications which require lower frequency resonators, which can be operated at higher temperatures and for which the BCS resistance is already sufficiently low.

### Residual resistance

The study of the BCS surface resistance is useful in confirming that the basic physics of the film's superconducting properties is well understood. On the other hand, attaining a low residual resistance is the ultimate goal of the study, with the understanding of the factors that lead to such values. Repeatability of the residual resistance is presently under control at the few nΩ level for the best films, and a significant progress has been made lately in understanding the phenomena which contribute to this term at the tens of nW level. Until recently, and still occasionally, problems with film adherence at the microscopic level have yielded cavities with residual resistance of several tens of nΩ which were not suitable for systematic studies of film properties. More importantly, the surface roughness of the copper substrate, brought about by the manufacturing or surface treatment procedures, were preventing a clear understanding of the film properties themselves. Standard coating procedures on spun copper cavities give now residual resistance of a few nΩ in the best cases and provide a good basis from which to start in order to study the effects of various parameter of substrates, interfaces and films.

### Magnetic flux resistance

Magnetic fields can be trapped in the cavity by using a solenoid external to the cryostat. Fields of up to a few Gauss can be generated and their effect on the RF losses studied.

The results of these measurements indicate that there is a very large difference in sensitivity to the external fields for bulk and films: in bulk niobium additional losses of the order of 200 nΩ/G are observed, whereas in standard niobium film contributions of 3-7 nΩ/G are most common. Within films, different sputtering parameters (noble gases, sputtering system cathode conditions, etc.) may give very large differences, so that significant differences in magnetic trapping behavior give a clear indication of film purity.

## SUMMARY AND CONCLUSIONS

Given the fact that reproducible standard coating procedures give excellent results, it is now possible to explore the phase space of various parameters which are suspected or known to alter the properties of the film. Although a complete study of all the possible permutations is not possible, a significant effort has been made and continues to be made to optimize the parameters and procedures in order to achieve in the most reliable and reproducible way the best result.

Changes of substrates have already provided suggestions that copper substrates with smoother surface finish give better results. The surface smoothness and the niobium film adhesion have an important role in the final performance of the cavity.

The role of the Cu/Nb interface, its oxidation condition and the migration of impurities at the boundary and within the niobium have proven to be important. This part of the study is only beginning now but it has indicated, for instance, that the lack of a barrier may produce in some cases films of poor performance, due possibly to the migration of impurities from the copper to the niobium. It seems therefore important to find a suitable barrier to prevent such migration, thus maintaining the film properties.

The controlled purity of niobium can both lead to an optimized BCS resistance and to an understanding of the role of the impurities themselves. Along these lines, the use of noble gases for sputtering other than argon has provided a deep insight into the nature of the impurity implantation and its consequences and possibly an optimal discharge gas could be found which provides the best film performance.

From the performed studies on 1.5 GHz $TM_{010}$ mode copper cavities coated with thin niobium films, it is possible to draw several conclusions about the role played by the films, the substrates and the interface in obtaining a successful performance of niobium sputter-coated cavities. A global clearer picture will become available in the near future.

## ACKNOWLEDGMENTS

The authors would like to thank V. Palmieri of LNL-INFN for providing spun copper and niobium cavities; P. Kneisel of TJNAF for discussions, measurements and for providing cavities for the study; S. Bauer for the chemical treatment of all cavities; G. Cavallari, E. Haebel, E. Mahner, W. Weingarten for useful discussions; S. Marsh, S. Sgobba for metallurgical analyses; N. Rensing for her participation in the early part of this study. The support of the Cryogenics Group at CERN, with J. M. Rieubland and his staff, is gratefully acknowledged.

## REFERENCES

1. C. Benvenuti, S. Calatroni, E. Chiaveri, G. Orlandi, and W. Weingarten, Superconducting cavities for particle accelerators, Cryogenics, 34:57 (1994)
2. H. Padamsee, K. W. Shepard and R. Sundelin, Physics and accelerator applications of RF superconductivity, Ann. Rev. Nucl. Part. Sci. 43 (1993) 635
3. W. Weingarten, Progress in thin film techniques, Part. Accel. 53:199 (1996)
4. C. Benvenuti, P. Bernard, D. Bloess, G. Cavallari, E. Chiaveri, E. Haebel, N. Hilleret, J. Tückmantel, and W. Weingarten, Superconducting niobium sputter-coated copper cavity modules for the LEP energy upgrade, 1991 IEEE particle accelerator conference San Francisco, CA, USA ; 6 -9 May 1991 (1023-1025 ).
5. R. Brinkmann, Status of the design for the TESLA linear collider, 16th Particle Accelerator Conference - PAC 95 Dallas, TX, USA ; 1 - 5 May 1995 ( 674-676).
6. C. Benvenuti, V. Palmieri, and R. Vaglio, Construction materials for superconducting RF accelerating cavities, 4th Topical Symposium on Superconductivity and Superconducting Materials Technologies, Florence, Italy; 28 June - 4 July 1994 . P Vincenzini, ed., Faenza (1994), p.637.
7. C. Hauviller, Fully hydroformed RF cavities, 1989 IEEE particle accelerator conference Chicago, IL, USA; 20 - 23 Mar 1989, p. 485.
8. V. Palmieri, R. Preciso, V. L. Ruzinov, S. Yu. Stark, I. I. Kulik, J. P. Bacher, J. P. Brachet, H. Fritz, F. Kuttl, and G. Lion, Recent experience with spinning of 1.5 GHz seamless copper monocells, Proceedings of the 7th Workshop on RF Superconductivity, Saclay, 1995, p. 571.
9. G. Orlandi, C. Benvenuti, S. Calatroni, M. Hauer, M and F. Scalambrin, Niobium coatings for 1.5 GHz RF cavities, Part. Accel. 46: 1 (1994).
10. D. C. Mattis and J. Bardeen, Phys. Rev. 111:412 (1958).
11. A. A. Abrikosov, L. P. Gor'kov and I. M. Khalatnikov, Zh. Eks. Teor. Fiz. 35 (1958) 265 and JETP 35:182 (1959).
12. J. P. Turneaure, J. Halbritter and H A. Schwettman, The surface impedance of superconductors and normal conductors: the Mattis-Bardeen theory, J. Supercond. 4:341 (1991).
13. J. P. Carbotte, Properties of boson-exchange superconductors, Rev. Mod. Phys, 62:1027 (1990) and references therein.
14. J. Halbritter, On surface resistance of superconductors, Z. Phys. 266:209 (1974).

# CONSTRUCTION, COMMISSIONING, AND CRYOGENIC PERFORMANCES OF THE FIRST TESLA TEST FACILITY (TTF) CRYOMODULE

C. Pagani[1], J. G. Weisend II[2], R. Bandelmann[2], D. Barni[1], A. Bosotti[1], G. Grygiel[2], H. Kaiser[2], U. Knopf[2], R. Lange[2], F. Loeffler[2], P. Pierini[1], O. Peters[2], B. Petersen[2], D. Sellmann[2], S. Wolff[2]

[1]INFN Milano-LASA, Via F.lli Cervi 201, I-20090 Segrate (MI), Italy
[2]DESY, Notkestrasse 85, D-22607 Hamburg, Germany

## ABSTRACT

A principal goal of the ongoing TESLA superconducting linac project is the production of affordable cryostats that still meet stringent requirements for alignment, vibration and heat leak. Each cryostat contains 8 superconducting RF cavities cooled to 1.8 K, a quadrupole magnet package cooled to 4.5 K, thermal shields cooled to 70 K and 4.5 K, active and passive magnetic shielding, cryogenic service pipes and all associated instrumentation. The axes of the 8 cavities must be aligned to the ideal beam axis to within ± 0.5 mm and those of the quadrupoles to within ± 0.1 mm. Additionally, the vertical mid plane of the quadrupole package must be aligned to the vertical direction to ± 0.1 mrad. These alignments must remain fixed after cool down and during operation. The cryostat must be designed so that there are no resonant vibration modes near the 10 Hz operating frequency of the accelerator. Although dynamic loads associated with the operation of the RF and the beam dominate the heat load, reasonable efforts to reduce the static heat leak into the cryostat are necessary.

The first of these cryostats has been tested in early June, 1997. This paper reports on the assembly of the cryostat and its performance during cryogenic and beam testing. Alignment, heat leak and vibration data are among the results discussed. The impact of the cryostat on the performance of the superconducting cavities is reviewed. Planned improvements in the cryostat design are also covered.

## INTRODUCTION

The cryomodule is the principal building block of the proposed TESLA superconducting linac[1]. The cryomodule contains 8 superconducting RF cavities plus, in many cases, a superconducting quadrupole package. Cavity support, alignment, cooling and thermal insulation are all provided by the cryomodule. In addition, the cryomodule must provide feedthroughs for the RF power and instrumentation, as well as connection to adjacent cryomodules or to cryogenic supply lines. One of the goals of the TESLA project is to reduce the cost of the cryomodules. Thus, the design must be made as cost effective as possible, while still meeting its operating requirements. These factors implied the need for building and testing a prototype cryomodule.

In the first half of 1997, the first cryomodule was assembled and tested. By mid July, 1997, the cryomodule had been cooled down twice. It will remain at cryogenic temperatures until September, while additional tests on the superconducting cavities, as well as additional cryogenic measurements, are made. Figure 1 shows a cross section of this cryomodule. The superconducting cavities and quadrupole package are attached to a 300 mm outer diameter gas return line that acts as the structural backbone for the cryomodule. This pipe is attached to three vertical posts, which in turn connect to the top of the vacuum vessel. The cryomodule also includes two actively cooled thermal radiation shields at approximately 70 K and 4.5 K. The cavities are bath cooled in He II at 1.8 K and the baths are fed by a two–phase He II flow in the pipe connecting the cavity helium vessels. The quadrupole is cooled by 4.5 K helium. Details of the cryomodule design have been previously published[2].

The cryomodule is tested in the TESLA Test Facility (TTF) linac. The front of the module is attached to a feedcap which provids the link to the cryogenic supply. The module is closed off at the downstream end by an endcap which seals off the vacuum space and turns around all the cryogenic flows.

## CONSTRUCTION EXPERIENCE

The construction of the first cryomodule was challenging, as it brought together components from five different laboratories in four different countries. A great deal of planning was done to ensure that there were no mechanical interferences between the various pieces. The assembly of the first cryomodule began in January, 1997 and lasted 8 weeks. The work was performed by technicians and engineers from INFN and DESY. Typically, five people worked on the assembly at one time with additional specialists called in as needed. In order to avoid problems, the assembly was carried out using a detailed written procedure. This procedure was altered as the circumstances dictated. The assembly was extensively documented using written records, photographs and videotapes.

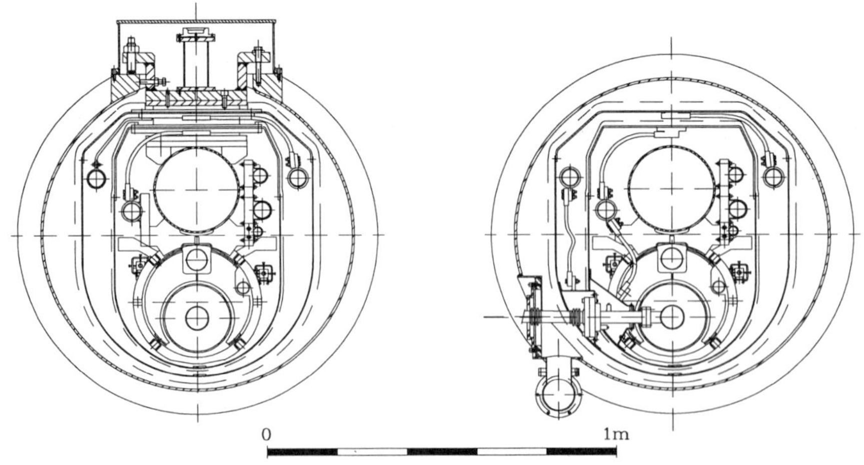


**Figure 1.** Two cross sections of the first TESLA Test Facility (TTF) cryomodule.

Work began after the string of eight cavities and one quadrupole package was brought out of the clean room. At that point, the beam line components had been connected together and filled with clean Argon gas, at a pressure slightly above 1 bar. The first step was to pre-align the cavities and quadrupole to within 2 mm of their desired position. This was done to facilitate welding of the cryogenic two–phase line connecting the cavity helium vessels. After this welding was complete, the line was tested and found to be leak tight. Next, thermometers, accelerometers, cables, and alignment targets were added to the cavities. Multilayer insulation (MLI) and magnetic shields were added to half of the cavities. Two different designs of magnetic shielding have been used in this first cryostat. This means that half of the magnetic shields were installed before the cold mass was attached to the cavity string, and half after the attachment.

Next, the cold mass was placed on the assembly tooling above the cavity string. The three support posts of the cold mass were aligned so that the He Gas Return Pipe (HeGRP - a 300 mm tube which provides the structural backbone for the cryomodule) was level, straight and aligned with the reference beam axis. The cold mass was then carefully lowered onto the cavity string. The cavities and quadrupole were attached to the cold mass using either the final supports or temporary supports in the case of those cavities whose magnetic shielding had not yet been installed. The assembled cold mass and cavity string was then lifted off the string support carts. A precision screw gear linkage system in the assembly tooling enabled this movement to be done without any damage to the cavities. At this point, the cavity tuner linkages and motors, remaining sensors and cables, and remaining cavity MLI and magnetic shielding were installed. The final supports were attached to the cavities and all temporary supports removed. The system was now ready for final alignment.

It was decided early on, that for this first cryomodule, we would have been satisfied if the cavities could have been aligned to within 0.5 mm of their ideal position and the quadrupole to within 0.2 mm. For all components, we required that the roll about the beam tube axis be zero. This decision was based on a desire not to damage the cavities by trying to meet a alignment tolerance that was not needed for this first prototype. The cavities and quadrupole were aligned one by one to the ideal beam axis. It was found that the tightening of the supports after the alignment alters the alignment. Thus, it was necessary to measure the alignment after the supports were tightened and realign the components as needed. The entire alignment process took about 3 days. Experience with the first cryomodule gives us confidence that the initial alignment of components in future cryomodules will meet the TESLA alignment tolerances (0.5 mm for the cavities and 0.1 mm for the quadrupole in the horizontal and vertical directions and a roll of 0.1 mrad for the vertical midplane of the quadrupole). Adjustment screws with finer threads will be used in future modules to simplify the alignment. The impact of transport, vacuum pumping and cooling on alignment is discussed below.

After the alignment was finished, the beam tube vacuum was leak checked to insure that the alignment procedure did not break the beam tube vacuum. No leaks were found.

The cold mass assembly was then transported to another assembly position where the INFN developed wire alignment system (WPM)[3] was installed and tested. This system was developed to measure alignment changes during operation of the cryomodule. Also at this location, the 4.5 K cooling lines were welded to the quadrupole package and this circuit was leak checked.

The cold mass assembly was then moved to its final assembly position. Here, the remaining 2 K MLI was installed along with the 4.5 K thermal shields, the 4.5 MLI, the 70 K thermal shields and the 70 K MLI. Next the 300 K vacuum vessel was slid over the cold mass assembly and the cold mass support posts were attached to the vacuum vessel. The support posts were then realigned in such a way as to bring the HeGRP, and thus the cavities and quadrupole, back into alignment with the ideal beam axis. The support post references were next transferred to references on the vacuum vessel so that aligning the vacuum vessel on the linac brought the cavities into the proper position.

Lastly, the warm portion of the main couplers were installed and leak tested. All instrumentation cables were brought out the appropriate ports and an instrumentation check was performed. At this point, the module was ready to be installed in the linac.

After aligning the module in the linac, the beam tube and cryogenic lines of the module were connected to the feedcap and endcap. All lines were vacuum leak checked. 70 K and 4.5 K thermal shields and MLI were added to the interconnect spaces between the module, feedcap and endcap. The outer vacuum jackets of these interconnects were closed and the isolation vacuum was pumped down and leak checked. The main couplers were attached to the 300 K RF wave guide and the current leads from the quadrupole package were attached to the power supplies. All instrumentation cables were connected to the module and a final instrumentation check was made. The module was then ready for cooldown. Figure 2 is a photograph of the complete cryomodule while inserting in the linac, to be connected to the feedcap and endcap. The endcap is also visible in the right side of the photograph.

The module assembly went quite smoothly and no major problems were seen. It was necessary to adjust the holes in the magnetic shields and in the thermal shields to fit the actual equipment used. These changes were recorded so that in the next module the proper holes will already be in place. The majority of temperature sensors that were placed between the cavity vessels and the magnetic shielding were not working after the completion of the assembly. This was thought to stem from the tight clearance between the vessel and the shielding. In future modules, the sensors will be moved to locations outside the magnetic shield which should fix this problem. Minor tooling changes, such as the way in which the couplers are supported during assembly, are planned to simplify the cryostat assembly .

**Figure 2.** Photograph of the complete cryomodule during placement into the linac. The cryomodule will next be connected to the feedcap and endcap. The endcap is visible in the right side of the photograph.

## COOLDOWN BEHAVIOR

Once installed into the linac, connected to the ancillary equipment and sealed, the cryomodule pump down started on May 14. The insulation vacuum had been pumped for a week before starting the cool down procedure on May 22. As planned, cavities were cooled through the cool down - warm up tube, in parallel with the 4 K shield, the fluid temperature being controlled to limit the maximum temperature difference inside the structure to 50 K, as suggested by the FEM calculation. A decreasing fluid temperature is obtained by mixing warm He gas to the cold box 4 K output. Cooling of the so called 70 K shield is performed via a second cold box output.

In the first cryostat design, the two radiation shields are cooled by copper braids, connected between the shield cooling tubes and the upper center line of the shields. This scheme prevents the development of high gradients in the shields regardless of the coolant temperature, because the heat flow is limited by the braids thermal conductivity. Nevertheless a cool down procedure, based on a coolant flow with decreasing temperature, is mandatory for a safe cooling of the cavity string.

Figure 3 shows the effect, as detected by the Wire Position Monitors (WPM), of a fast cool down performed on the cavity string, to check the possibility of preventing hydrogen driven Q disease in unbaked cavities, by a fast crossing of the 100 K region. The large bending was produced by the thermal conductivity of the cavity supports versus the lower part of the He gas return pipe (HeGRP), whose cool down was much slower. FEM simulations demonstrated that an average difference of 8 K between the lower and the upper temperature of the HeGRP is enough to produce this effect. This experience showed that a precooling to 100 K (instead of 130 K) of the HeGRP would have prevented this effect that, in any case, has been completely recovered once the final temperature is reached.

From the experience gained during the first cool down, a better procedure has been defined for the module cool down and warm up. This procedure was successfully followed at the beginning of July, after a good month of linac operation to set the RF system, accelerate the beam and measure the cryogenic performance of the cryomodule.

A general overview of the warm up and cool down procedure, as applied at the beginning of July 1997, is presented in Figure 4. In particular the two upper curves show, as a function of time, the input temperatures of the cooling Helium, after the mixing with the warm gas and before the entrance in the cryomodule, respectively for the 70 K (left) and the 4K (right) circuit. In the lower part of the same Figure 4, the behavior of the two worst temperatures in the shields are presented, to be respectively compared with their upper curves, to have a figure of merit of the instantaneous maximum gradient in the shields.

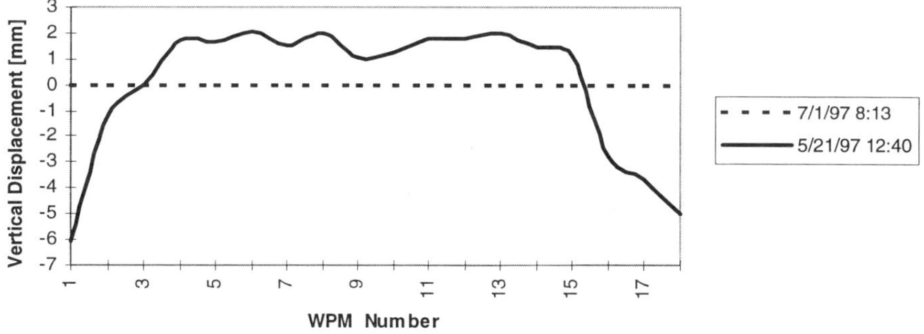


**Figure 3.** Taking as a vertical reference the stable cold position after the first cooldown (7/1/97), the solid line (5/21/97) shows the maximum vertical bending of the cavity string during the fast cool down.

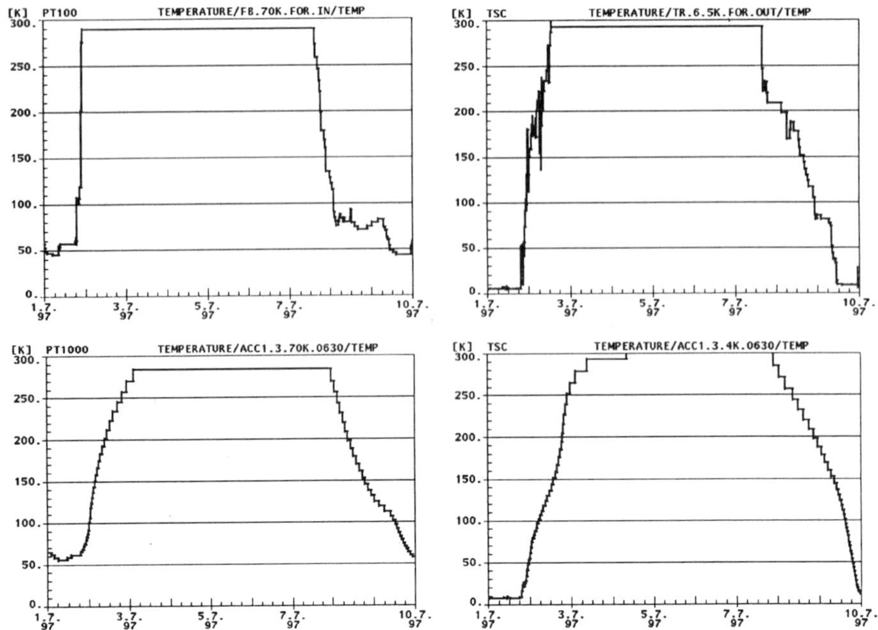


**Figure 4.** Temperature curves during the warmup and cooldown performed at the beginning of July. The two upper curves show, as a function of time, the input temperatures of the cooling Helium, after the mixing with the warm gas and before the entrance in the cryomodule, respectively for the 70 K (left) and the 4K (right) circuit. The two lower curves, to be compared with their upper curves, show the behavior of the worst temperature in the two shields respectively.

Given the coolant temperature, the instantaneous distribution of the temperature on the two radiation shields was verified to be in accordance with that anticipated by the FEM calculations[5], performed with the code ANSYS. As a consequence we conclude that, during warm up and cool down, the two shields behave as computed. The local stresses and overall deformations stay well inside the safety limits, both for tensile stress and geometrical interference.

The maximum vertical displacement of the cavity string during warmup and the second cooldown is shown in Figure 5, the curves being referred to the stable cold position. The final position, once cold, is also shown to have a figure of merit of the alignment reproducibility after a warmup - cool down cycle.

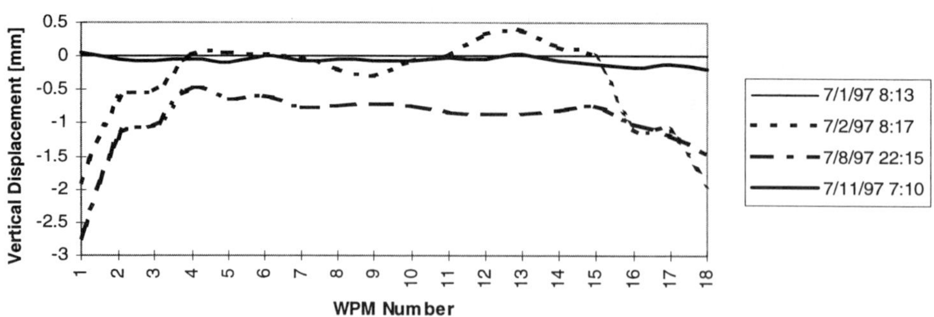


**Figure 5.** Maximum vertical displacement of the cavity string during warmup (7/2/97) and the second cooldown (7/8/97), referred to the stable cold position (7/1/97). The final cold position (7/11/97) is also shown to have a figure of the alignment reproducibility after a warmup - cool down cycle.

## HEAT LEAK RESULTS

Table 1 compares the measured and expected static heat leak values for the 70 K, 4.5 K and 1.8 K temperature levels. The 70 K and 4.5 K results were found by measuring inlet and outlet temperatures and pressures, calculating the change of enthalpy and multiplying by the measured mass flow rate. The 1.8 K result was found by measuring the mass flow rate at the pumps and then correcting this by the amount of vapor that was generated by the J - T expansion at the inlet to the two–phase line. This adjusted mass flow rate is then multiplied by the latent heat of helium at 1.8 K.

**Table 1.** Comparison of Measured and Predicted Static Heat Leaks

| Temperature Level | Predicted Static Heat Leak ( W ) | Measured Static Heat Leak ( W ) |
|---|---|---|
| 70 K | 76.8 | 90 |
| 4.5 K | 13.9 | 23 |
| 1.8 K | 2.8 | 6 |

It is seen that in all cases the measured static heat leak is significantly higher than predicted. However, most of the discrepancy can be traced to the additional instrumentation necessary for this first cryomodule. For example, the 150 coax cables needed for the stretched wire system contribute about 1.5 W total, most of which goes into the 1.8 K space. There is also the heat leak to 1.8 K from two quartz windows that connect the 300 K space with the 1.8 K space. These windows are required for the optical alignment measurements and will not exist in future modules. These windows also allow some additional heat leak onto the 4.5 K shields. Sensors mounted on the 4.5 K shields show a local hot spot where the several hundred instrumentation cables are heat sunk. All these cables also will not exist in the final design. This effect is shown in Figure 6. It can be seen that, on the right side, the temperature of the 4.5 K shield is substantially higher, with respect to the average temperature. The additional heat leak to the 70 K level is in part due to the manner in which some of the MLI at the main couplers was installed. This procedure will be changed in the next cryomodule. Moreover we note that the operating temperature for the 70 K shield is lower (see Figure 6).

The heat leak measurements will be repeated later in this run and additional measurements will be made with later cryomodules to ensure that the heat leak meets the design requirements.

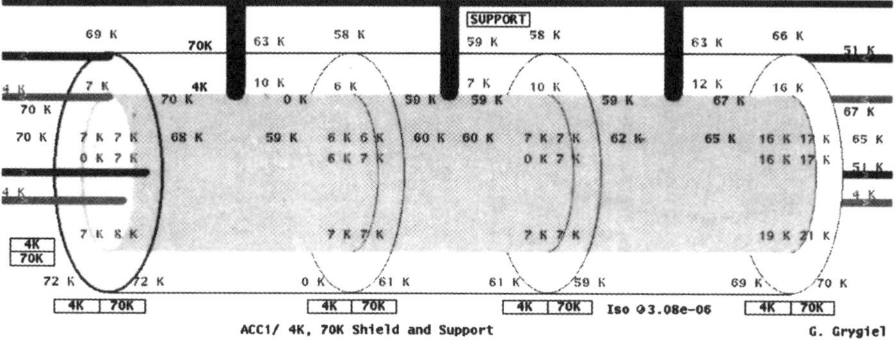


**Figure 6.** Control system screenshot, relative to the two shields when cold and stable. On the right side, the temperature of the 4.5 K shield is substantially higher, with respect to the average temperature.

## ALIGNMENT RESULTS

The general criterion adopted for TTF to have the active elements (quadrupoles and cavities) aligned once cold is to align them warm, when attached to the HeGRP, the last being pre-aligned and referred to the three Taylor-Hobson spheres on the suspension posts. In this way, in principle, the cavity string (including the quadrupole package) is referenced to the external reference, given by the Taylor-Hobson spheres, and the alignment can be finally recovered just aligning the external reference. Before pumping the insulation vacuum the posts are no longer accessible and the external reference is transferred to a similar one, attached to the vacuum vessel.

This philosophy is based on three main assumptions, supported by extensive FEM calculation, namely:

1. all the connections and fixtures among components are stable against movements and thermal cycles;
2. no asymmetric forces act on the supporting HeGRP when the system is under vacuum and cold.
3. all the strains in the components, which perform a thermal cycle, produce stresses well below the yield limit and no hysteresis is expected;

The experience gained so far, analyzing the data from the WPM system and comparing them with those given by the measurements with the optical targets, shows that in this first cryomodule the alignment reproducibility, from warm to cold, while acceptable for TTF, has to be improved to meet the TESLA requirements.

In particular, no hysteresis on the HeGRP has been detected, but some improvement on the next cryomodules has to be included in respect to the first two assumptions. Namely the fixture and support settling was not complete when the cavity string was connected to the HeGRP and an asymmetric force is produced, when cold, by the cryogenic connection to the HeGRP at the feedcap region.

As a consequence of the first of these two effects, we detected a small change (< 0.5mm) in the distance between the beam axis and the external Taylor-Hobson reference, associated with a minor waving, after the cold mass transportation. A similar permanent effect, but smaller, was also detected after the first thermal cycle, which includes pumping and cooldown.

The effect on the alignment due to the presence of asymmetric forces acting on the HeGRP, in the feedcap region, when cold, is shown in Figure 7 and 8, where the cavity string line is compared in different situations, respectively for the horizontal (X) and vertical (Y) plane. The presented curves are obtained by the WPM data and include the expected parallel vertical displacement, induced by the thermal shrinkage.

Conversely, as stated above, the reproducibility between two successive cooldowns has been checked to be very good, as can be seen, for vertical displacement, in Figure 5, where the bold solid line has to be compared with the reference one.

An all internal and simplified WPM system will be designed soon to be installed in the next two cryomodules string, to check the effect of the planned improvements to preserve the warm to cold alignment.

A preliminary analysis of the data from WPMs[3] shows a very small amplitude system vibration at 13.4 Hz. This vibration has a stable frequency, but its amplitude varies at each measurement, while being always less than 10 µm. Further analysis are needed to better understand the phenomenon, even if negligible.

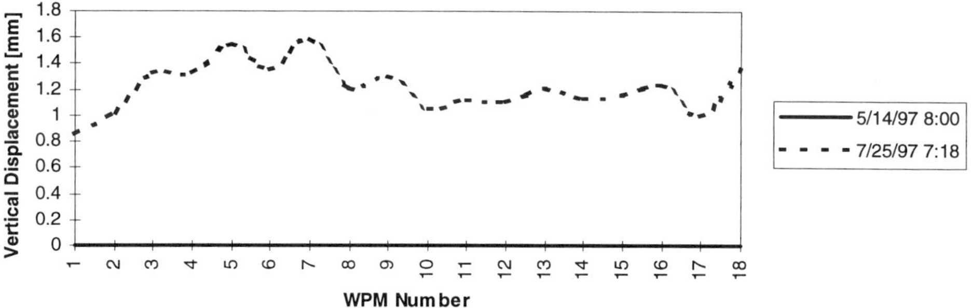


**Figure 7.** Displacement, in the vertical plane, of the cold cavity string (7/1/97), referred to the position measured before the first pumping and cooldown (5/14/97).

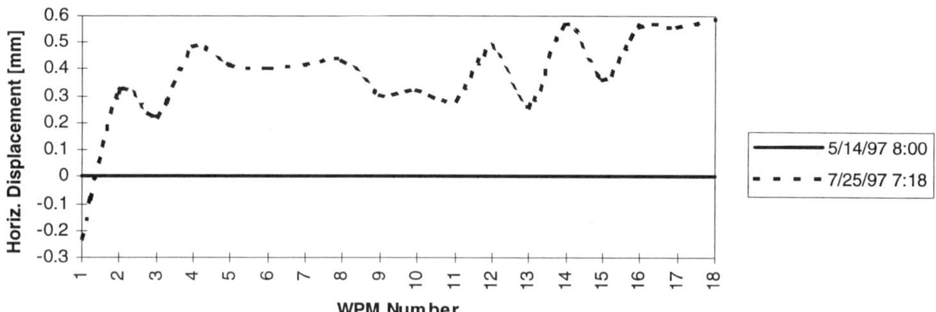


**Figure 8.** Displacement, in the horizontal plane, of the cold cavity string (7/1/97), referred to the position measured before the first pumping and cooldown (5/14/97).

## PLANNED CHANGES IN FUTURE CRYOMODULES

The experience in the construction and operation of the first cryomodule has led to changes in the design of the future cryomodules. Some of these changes are relatively minor, such as slightly changing dimensions in the magnetic shields and thermal radiation shields to avoid interferences, adjusting the dimensional tolerances and using finer adjustment screws for cavity alignment. Other changes, however, are quite significant.

The most important design change for the next two cryomodules occurs in the thermal shields. In the first cryomodule, the shields are connected together with screws and cooled by copper braids attached to the cooling pipes. This method is both expensive and labor intensive. In future cryomodules, the shield cooling will be done by aluminum cooling tubes welded to the shields and the shield pieces will, for the most part, be welded together. Further details on this design change are reported elsewhere in this conference[4].

In cryomodules number 4 and beyond an additional design change is envisioned. The 76 mm O.D. two–phase line will be eliminated and the 300 mm O.D. line will contain a stratified two–phase flow providing both liquid supply and gas return. This change will further reduce the cost of the cryostat. Conceptual design of these cryomodules is underway now. Theoretical and experimental investigations of He II two–phase flow in support of this design are also planned.

## CONCLUSIONS

The assembly and performance of the first TTF cryomodule has been quite gratifying. The assembly went smoothly and performance of the module has permitted the necessary superconducting RF cavity and beam experiments planned for the first phase of the TTF linac. The instrumentation in the cryomodule has indicated two areas of concern. The most important is the change of cavity alignment with transport, vacuum pumping and cooldown. Experiments are planned on the second and third cryomodules to try to understand and eliminate this problem. The higher than expected static heat leaks are also of concern. While it is expected that the additional heat leak is mainly due to the presence of various diagnostics, that are not planned for the final cryomodules, this hypothesis needs to be experimentally verified.

Additional modifications of the cryomodule design to simplify production and reduce the fabrication cost are also planned[4].

## REFERENCES


1. TESLA Test Facility linac - Design Report, DESY Report, March 1995, TESLA 95-01.
   Conceptual design of a 500 GeV e+e- linear collider with integrated X ray laser facility, R. Brinkmann et al., editors, DESY Report, 1997-048, 1997.
2. F. Alessandria, G. Cavallari, M. Minestrini, T. H. Nichol, C. Pagani, R. Palmieri, S. Tazzari, G. Varisco, Design manufacture and test of the TESLA-TTF cavity cryostat, Adv.Cryo. Engr. 41A:855 ( 1996 ).
3. D. Giove, A. Bosotti, C. Pagani, G. Varisco, A Wire Position Monitor (WPM) system to control the cold mass movements inside the TTF cryomodule, PAC97, Vancouver.
4. Pagani, C., Barni, D., Bonezzi, M., Weisend II, J. G., Design of the thermal shields for new improved TESLA Test Facility, Paper # FF 3, 1997 CEC/ICMC, Portland.
5. D. Barni, C. Pagani, P. Pierini, M. Todero, Cooldown simulations for the TESLA Test Facility (TTF) cryostat, Paper # FF 4, 1997 CEC/ICMC, Portland.

# THE PERFORMANCE OF 1.3 GHz 9-CELL SUPERCONDUCTING CAVITY SYSTEMS FOR THE TESLA TEST FACILITY

W.-D. Möller and M. Pekeler

for the TESLA collaboration

Deutsches Elektronen Synchrotron DESY, 22607 Hamburg, Germany

## ABSTRACT

For the Tesla Test Facility (TTF)[1,2] a new infrastructure has been established at DESY to prepare and test superconducting cavities. It consists of automised chemistry, large clean room facilities, high-vacuum furnace, high pressure water rinsing system and vertical as well as horizontal test stands. A 5 MW klystron is available for High Power Processing on cavities and for coupler conditioning. The design goal for the gradient is $E_{acc}$> 15 MV/m at a quality factor of $Q > 3 \cdot 10^9$ with the cavities operated in pulsed mode (800 μs constant gradient with 10 Hz repetition rate). More than 20 cavities fabricated by different companies were measured vertically. The majority of them exceeded the specification. Several cavities reached 25 MV/m at $Q > 1 \cdot 10^{10}$ in cw mode and up to 30 MV/m in pulsed mode. Eight cavities were tested in the horizontal cryostat fully equipped with helium vessel, high power coupler, HOM-couplers and tuning system. The superconducting capture cavity was installed and tested in the linac in Jan. '97. In addition a module consisting of 8 cavities has been installed recently and the first beam has been accelerated. Test results like gradient, quality factor and their limitations as well as the performance of the first cryomodule are presented.

## INTRODUCTION

TESLA (TeV Superconducting Linear Accelerator) is one of the proposed future TeV scale $e^-e^+$ linear colliders. It differs from other projects in its choice of superconducting accelerating structures and of low RF frequency (1.3 GHz).

In order to prove the technical basis of TESLA the Tesla Test Facility was established at DESY within the frame of an international collaboration. It has to show that accelerating gradients of 15 MV/m are reproducibly possible and can be kept also after assembly to the linac together with the necessary auxiliary systems.[3]

To develop the fabrication procedures and surface treatment methods required to produce high gradient multicell cavities, a rich infrastructure is being built at TTF. It consists of processing and handling systems as well as of high and low power test stands. Twenty-three 9-cell cavities, manufactured by four different companies have been processed and measured so far. The first cryomodule of the TTF linac containing eight superconducting 9-cell cavities has been assembled. Processing of couplers and cavities has been done. The first beam is accelerated by the module up to 125 MeV at a beam current of 8 mA, a macro pulse length of 40 μs and 2 Hz repetition rate.

*Advances in Cryogenic Engineering,* Vol. 43
Edited by P. Kittel, Plenum Press, New York, 1998

## CAVITY PRODUCTION AND TREATMENT

All cavities are made of RRR 300 (Residual Resistivity Ratio) niobium with a sheet thickness of 2.8 mm. After forming the half cells by deep drawing they are e-beam welded from the inside at the irises. In addition a stiffening ring is welded to provide mechanical stability. The equator welds are made from outside. One type of cavities is completed by a fully welded version of HOM-couplers whereas the second type has a flange for a separate HOM-coupler. Later on only the nine cells are housed in a titanium-helium-tank. The HOM-couplers are cooled by thermal conduction.

The standard cavity treatment is as follows:

1 Visual inspection and dimensions check;
2 Removal of the inner damage layer by 80 µm Buffered Chemical Polishing (BCP). This is done by a closed loop etching facility with active cooling in order to avoid heating of the acid to temperatures above 12° C.
3 Removal of 10 µm from the outer surface by BCP;
4 Heat treatment with titanium getter at 1400 °C, 4 h in order to improve the RRR of the niobium from 300 to 500;
5 90-150 µm BCP from inside to remove the titanium from the superconducting surface and 40 µm BCP from outside in order to improve the Kapitza conductance;
6 Field flatness measurement and tuning to correct frequency;
7 Final inner BCP of 20 µm;
8 High Pressure Water Rinsing (HPR)[4] at 100 bar, drying on clean room air class 10;
9 Assembly, HPR 100 bar, drying by pumping;
10 Vertical acceptance test and possibly High Power Processing (HPP);[5,6]
11 Welding of titanium-helium-vessel;
12 Inner BCP 20 µm HPR and drying on clean room air class 10;
13 Assembly of high power- and HOM-couplers;
14 HPR at 100 bar, drying by pumping;
15 Assembly of the tuning mechanism and magnetic shielding;
16 System measurements in horizontal test stand;
17 Ready for module assembly.

## MEASUREMENT RESULTS

### Vertical tests

Twenty-three 9-cell cavities produced by four European companies have been tested so far in a vertical teststand. Fig. 1 shows the Q(E)-behaviour of seven cavities. All of them reached gradients above 20 MV/m. The main limitation was thermal instability. Field emission could be strongly suppressed by means of careful high pressure water rinsing. Two cavities reached 25 MV/m at a quality factor $Q > 10^{10}$ without any field emission loading.

In the linac the cavities are operated in the π-mode with equal gradients in all cells. By excitation of the other coupled modes a non-uniform field distribution can be achieved which allows to determine the high field performance of individual cells. In five cavities of lower gradients a single cell was identified to be responsible for the limitation while the other cells could be excited to much higher fields.

### Defects in the bulk niobium

Three niobium cavities from one manufacturer, produced from the same ingot, showed quenches at gradients from 8 to 14 MV/m. They also exhibited a so called "Q-switch", which is a sudden drop in quality factor with increasing incident power (see figure 2). Localized areas with excessive heating were found by temperature mapping. In one cavity the defective cell was cut out and investigated with X ray fluorescence spectroscopy. A tantalum grain close to the RF surface was found. A sensitive eddy current scanning system, developed at the Bundesanstalt für Materialforschung, Berlin, was able to detect the inclusion. In the future this type of material defects will be excluded by eddy current scan of all niobium sheets. 5 % of recently produced niobium sheets had to be rejected.

## Defects in the equator welds

All cavities produced by another manufacturer exhibited a thermal breakdown in the 11 to 15 MV/m regime combined with a so called "Q-slope", a continuous reduction of the quality factor from approximately 3 x $10^{10}$ at low gradient to about 1 x $10^{10}$ near the quench field (figure 3). No field emission was observed. Three of the cavities were investigated with temperature mapping. Several locations with field-dependent surface resistance were found in the equator welds which caused the Q-slope and also the quench. An analysis showed that the welding procedures at this company were not adequate for gradients in the 20-25 MV/m

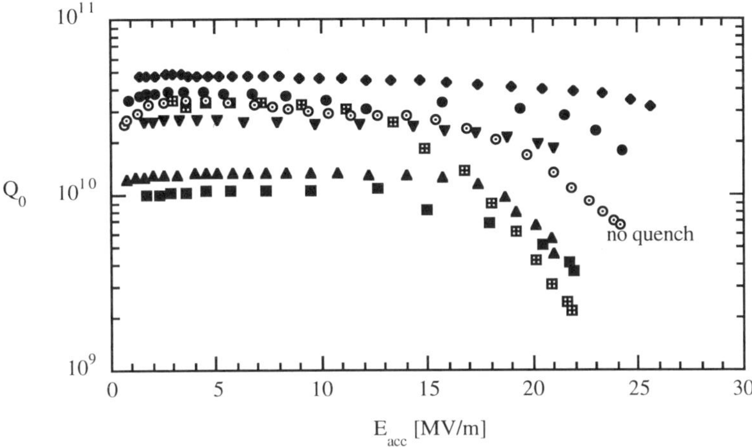


**Figure 1.** Q(E)-behaviour of TESLA 9-cell cavities which reached more than 20 MV/m. One cavity was limited by the available incident power, all others showed a thermal breakdown, some under field emission loading.

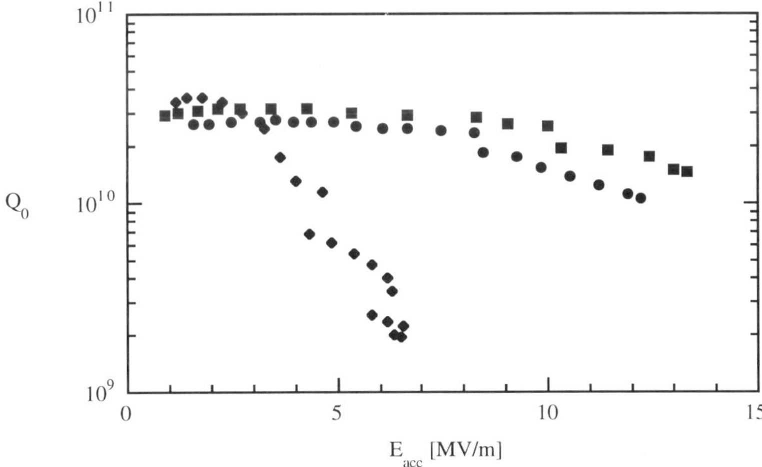


**Figure 2.** Q(E)-behaviour of 9-cell cavities which showed a "Q-switch". All cavities were produced by one manufacturer from the same ingot material. The Q-switches and quenches are probably caused by foreign material inclusions as tantalum.

regime. A cavity with improved welding technique has recently been fabricated by this company. The test result is shown in figure 4. A remarkable improvement is observed. The maximum gradient of 25 MV/m is not limited by quench but by the available incident power. There is no Q-slope visible up to gradients of 17 MV/m, however field emission sets in at 18 MV/m. This cavity will be high pressure water rinsed soon and retested again.

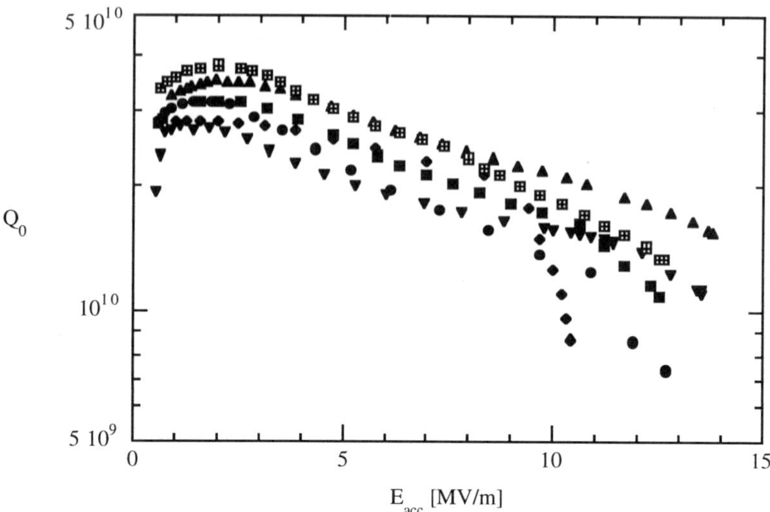


**Figure 3.** Cavities showing a "Q-slope" and quench in the 11-15 MV/m regime. There was no field emission present. All cavities were produced by the same manufacturer.

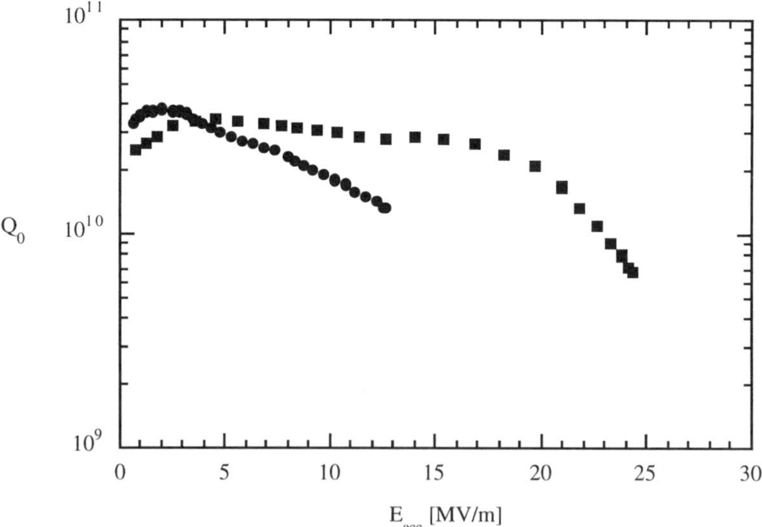


**Figure 4.** Q(E)-behaviour of a cavity produced with new welding technique. The cavity showed field emission above 18 MV/m. The maximum gradient is not limited by quench. For comparison one of the previous cavities (figure 3) is shown.

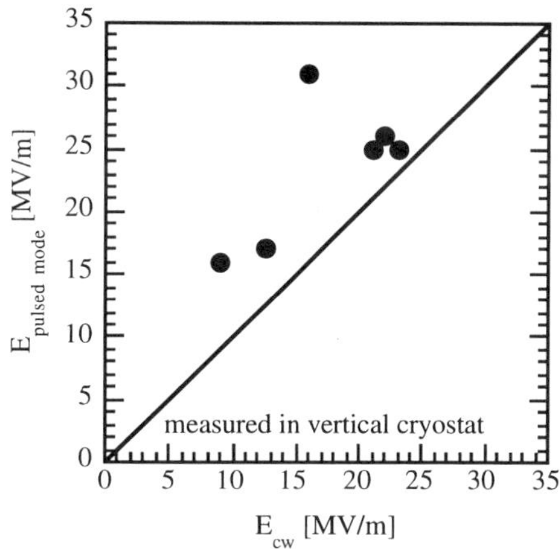


**Figure 5.** Comparison of cw and pulsed operation of cavities tested in the vertical cryostat.

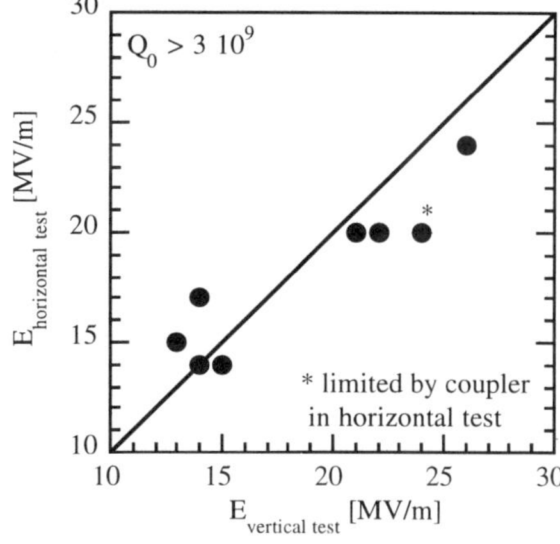


**Figure 6.** Comparison of the performance of cavities tested in the vertical cryostat in cw mode and in the horizontal test cryostat in the pulsed mode.

### Test in pulsed mode

A number of cavities were tested in the vertical cryostat also in the pulsed mode as foreseen in the linac operation. For this purpose the cavity was equipped with a high-power-coupler tunable from $Q_{ext} = 10^{10}$ (needed in cw operation) to 3 x $10^6$ (needed in pulsed operation). In pulsed mode the gradient is kept constant for 800 µs at a 10 Hz repetition rate. The gradients achieved in the pulsed mode are always higher than in cw operation (figure 5).

## HORIZONTAL TESTS

Eight cavities equipped with all auxiliary components (helium-vessel, high-power-coupler, HOM-couplers, tuning mechanism and magnetic shielding) were tested in a horizontal cryostat. Because of the low $Q_{ext}$ of the high power input coupler only measurements in the pulsed mode were possible.

The cavities reached an average gradient of 17.5 MV/m at Q > 3 x $10^9$ (see figure 6) compared to an average gradient of 18.6 MV/m at Q > 3 x $10^9$ in cw mode in the vertical test stand. In contrast to the vertical tests, field emission has been observed more frequently. The reason is probably particle contamination due to the more complicated assembly and the processing of the main power coupler. The higher field emission loading is responsible for the slight degradation in performance.

Cavities showing strong field emission can be further improved by applying HPP. For this purpose, short pulses of high instantaneous power (≈ 500 µs, up to 1 MW) are applied to destroy field emitters. The power coupler has to be carefully processed up to 1 MW, which may take several days. Due to a tight time-schedule HPP was not applied during the horizontal test.

## HIGH POWER INPUT COUPLER

### Design

The two types (Fermilab[7] and DESY design) of high power couplers consist each of two aluminum oxide ($AL_2O_3$) RF vacuum windows. One of it is a coaxial window located near the cavity and thermally connected to the cryostat 70 K heat shield. The other is a wave-guide window (Fermilab) mounted in front of the wave-guide to coax transition or a coax window (DESY) which is a part of the wave-guide to coax transition. Both types are outside the cryostat at room temperature. All ceramics are coated on the vacuum sides with a ≈100 Å layer of titanium nitride (TiN) to reduce the secondary emission coefficient and so suppress multipacting. The couplers feature bellows in the coaxial lines to allow for coupling adjustment and to accommodate cavity motion during cool down and warm up of the cryostat. The input coupler instrumentation includes charged particle pick ups upstream and downstream of the coax window, a photo-multiplier tube between both windows and temperature sensors on both windows. The coupler vacuum of all eight couplers on the module is connected and pumped by a common ion getter pump.

### Processing

The TESLA design parameters call for a transfer of 206 kW peak RF power to the beam with a pulse length of 1.3 ms and a repetition rate of 10 Hz. In addition it is also foreseen to apply HPP to the cavities in situ which requires 1 MW at a maximum pulse length of 500 μs and at a repetition rate ≤ 1 Hz.

The coupler processing procedure starts with the cavity and coupler at room temperature so that gases released by RF stimulated desorption can be pumped away. In the beginning of the conditioning the cavity is not on resonance. The first power rise is done with a pulse length of 20 μs, then the pulse length is doubled for the next power rise and so forth. At the full pulse length the power is sweeped for some hours. After this the cavity and coupler are cooled down and the same processing procedure is carried out again. The last part of processing is done with the cavities on resonance.

### Processing results on module I

Due to the limited time the goal of processing was restricted to 200 kW and 600 μs. The processing of the coupler on the module at room temperature was done at two couplers at a time in order to distinguish the vacuum reactions in the common coupler vacuum. It was always observed, that initially it is possible to rise the power very quickly without being limited by outgassing, electron signals or light. After a sudden burst of outgassing on the coupler side of the cold window at a typical power level of 50 - 100 kW many hours are needed to once again reach the earlier achieved power level. Once the maximum power of 200 kW is reached for the shortest pulse length the power rises for longer pulses are straight forward. After about 48 to 72 h only one multipacting level at 125 kW is visible at the charged particle detectors. But the signals are smaller by a factor of 10 compared to the first processing hours. A second multipacting level at 65 kW is processed away.

When the module was cooled down to the operating temperature of 1.8 K the same processing procedure took place. All multipacting levels were processed after 30 h. The temperatures of the Fermilab cold window increased due to dynamic losses by about 5 K. Due to a fabrication failure during brazing metallization of the cold DESY windows occurred. This causes a much stronger temperature rise of about 100 K which does not effect the coupler behaviour but creates higher losses at the 70 K shield circuit of the module.

When the cavities are on resonance the field distribution in the coupler differs from the off resonance full reflection condition. An additional processing is necessary. After only a few (1-5) more hours the couplers were fully operable without limiting the cavities. No interlock events were observed up to the TTF operating power of 125 kW in more than 30 days of operation time. After a few days power off time no new processing was needed to operate the module again.

## CAPTURE CAVITY

The first horizontal test of the capture cavity 'fully dressed' with the high power- and HOM-couplers, tuner and magnetic shielding gave a maximum gradient of 19 MV/m for the

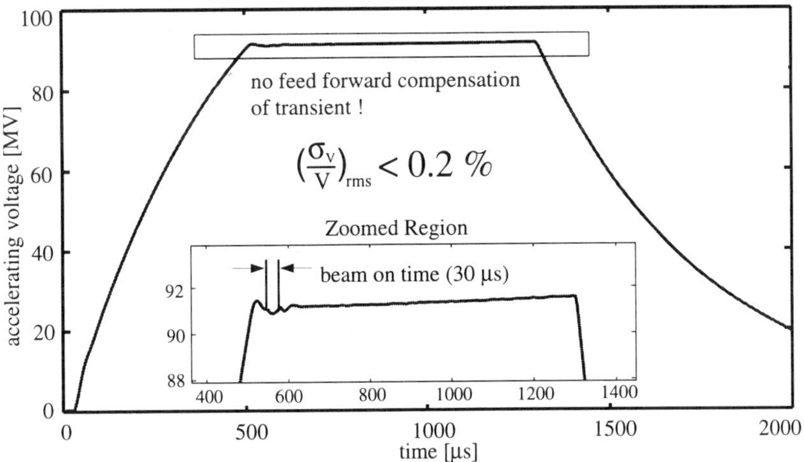


**Figure 7.** The accelerating voltage is controlled by a digital feedback loop. The stability of the voltage vector sum is better than $(\sigma V/V)_{rms} < 0.2$ %. Beam on-time is 30 μs whereas the RF pulse length in this case is 800 μs.

full TTF pulse length of 800 μs. After assembly of the cavity into the single cavity cryostat at Saclay (Paris), transporting it back to Desy and mounting it there into the injector, the cavity performance was degraded. The maximal operating gradient is now 12.5 MV/m, limited by field emission. The dark current at 14.8 MV/m measured on a Faraday cup (Ø = 25 mm) at 1.2 m away from the cavity is only $I_{peak} = 0.3$ μA. At this high field cryogenic losses are much too high to operate the cavity.

In order to suppress field emission it is planned to apply HPP to the cavity.

## MODULE I

The cryo module houses eight cavities and a superconducting quadrupole. All cavities are driven by one 5 MW klystron. A linear RF distribution system is branching off identical amount of power for each cavity by directional couplers. An individual three stub wave-guide tuner for phase matching of each cavity provides the possibility of ±30° RF-phase adjustment. Circulators for each cavity avoid a high field enhancement in the wave-guide which would be caused by the standing wave during filling time of the cavity.

## FIRST TESLA TEST FACILITY ELECTRON BEAM

On June 10th a first 125 MeV beam was delivered by the Tesla Test Facility. The electron gun delivered an 8 mA, 250 keV beam with a macro pulse length of 40 μs and a repetition rate of 2 Hz. In the capture cavity the beam energy was increased to 9.8 MeV by an RF pulse length of 1.3 ms and an accelerating gradient of 11 MV/m.

In the module seven cavities were under operation at an average gradient of 16.7 MV/m and an RF pulse length of 300 μs rise time and 100 μs flat top. The beam energy measured at the spectrometer was 125 MeV with a ΔE/E less than 1%. The vector sum of the seven cavities was stabilized by a digital feedback loop to better than $(\sigma V/V)_{rms} < 0.2\%$ and the phase better than $(\sigma_\varphi)_{rms} < 0.2°$ (see figure 7 and 8).[8] The maximum accelerating field is limited by the worst cavity due to the uniform power distribution.

The next step is the optimization of the RF distribution system in order to run every cavity at its individual limit. A feed forward compensation will be used to compensate the transient even more in field and phase during the beam on time. Also HPP is planned to suppress field emission.

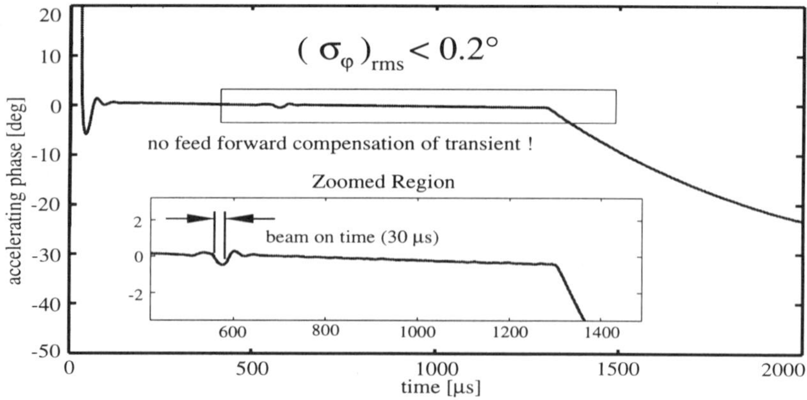


**Figure 8.** The stability of the digital feedback controlled phase is better than $(\sigma_\varphi)_{rms} < 0.2°$. The feed forward compensation of the transient during beam-on time is under development.

## CONCLUSION

Most of the first TESLA 9-cell cavities for the TTF Linac exceeded the specifications of 15 MV/m at $Q = 3 \times 10^9$. Seven cavities reached gradients above 20 MV/m, three reached 25 MV/m with $Q > 10^{10}$.

In cavities with poorer performance, two types of limitations were found: tantalum inclusion in the bulk niobium and imperfect equator welds.

Cures have been found for both types of imperfections. Using a dedicated eddy current scanning system defective niobium sheets will be rejected from the next cavity production. The second limitation has been overcome with a new welding technique. The first cavity built with this new welding procedure showed no quench at 25 MV/m. With better material control and better welding technique, gradients higher than 20 MV/m are expected to be routinely possible now.

In the horizontal tests, a slight reduction in gradient caused by higher field emission loading compared to the vertical test is observed. Work is in progress to improve the cleaning and mounting procedures prior to the horizontal test.

The first beam has been successfully accelerated. The average gradient in the first module is 16.7 MV/m which is more than the TTF goal of 15 MV/m. Further improvement of the gradient is possible by optimizing the RF distribution and applying HPP in situ to the cavities.

## REFERENCES


1. The TESLA (TeV Superconducting Linear Accelerator) R&D effort is carried out by a number of institutions which includes IHEP Beijing, TU Berlin, Max Born Institut Berlin, Cornell Univ., Krakow Univ., TH Darmstadt, DESY, TU Dresden, DSM/DAPNIA Saclay, JINR Dubna, Fermilab, Univ. Frankfurt, IN2P3/LAL Orsay, IN2P3/IPN Orsay, INFN Frascati, INFN Roma II, INFN Milano, FZ Karsruhe, INP Novosibirsk, Polish Acad. of Science, IHEP Protvino, SEFT Finland, UCLA Dept. of Physics, Warsaw Univ., BGH Wuppertal
2. D.A. Edwards: `TESLA TEST FACILITY LINAC - Design Report', TESLA 95-01, 1995.
3. B. Aune: `Results from the DESY TESLA Test Facility', PAC 97, Vancouver, to be published
4. P. Kneisel, B. Lewis, Part. Acc. **53**, 1996
5. J. Graber, PhD thesis, Cornell University, 1993
6. C. Crawford et al.: `High gradients in linear collider superconducting accelerator cavities by high pulsed power to suppress field emission', Part. Acc. **49**, pp. 1-13, 1995
7. M. Champion, Proc. LINAC 96, CERN 96-07
8. Th. Schilcher, S. Simrock, private communication

# SUPERCONDUCTING RF CAVITIES AND MAGNETS FOR A 4-TeV ENERGY MUON COLLIDER

Quan-Sheng Shu [1, 2], M. Green[3], D. Neuffer[4], D. Summers[5]
S. Simrock[6], E. Willen[7]

[1]Thomas Jefferson National Accelerator Facility, Newport News, VA 23606
[2]Northrop-Grumman Co. Bethpage, NY 11714
[3]Lawrence Berkeley National Lab, Berkeley, CA 94720
[4]Fermi National Accelerator Lab, Batavia, IL 60510
[5]University of Mississippi, Oxford, MS 38677
[6]DESY, Hamburg, D-22603 Germany
[7]Brookhaven National Lab, Upton, NY 11937

## ABSTRACT

The accelerators must take the muon beams from ~100 MeV to 2 TeV energies within the muon lifetime for a 4 TeV energy muon collider. These constraints plus the muon decay heating seriously challenge the designs of the superconducting RF (SRF) cavities and magnets in the accelerators and collider ring. The multiple superconducting recirculation linac and the very rapid-cycling superconducting synchrotron approach are both studied. We briefly introduce the technical considerations and preliminary designs of the SRF systems and magnets.

## INTRODUCTION

Muon colliders can provide strong potential advantages in high-energy physics, but also present serious technical challenges, as described in the Snowmass feasibility study[1]. A muon collider would use the superconducting RF (SRF) cavities for accelerating and superconducting (SC) magnets for bending the μ-beams throughout the machine. The central difficulty in the $\mu^+$-$\mu^-$ collider is the muon decay, with a lifetime of $2.2\times10^{-6}\gamma_\mu$ s (where $\gamma_\mu = E_\mu/m_\mu c^2$) that implies a requirement for very rapid increases in muon energies. At the design luminosity, the SC dipoles and quadrupoles must contain muon beams (up to $2\text{x}10^{12}$ muons per bunch) that will deposit over 1.8 kW of energy per meter into the storage ring.

First, we introduce a 20-Tesla (T) hybrid capture solenoid system[2] (combining a Bitter type water-cooled solenoid with a superconducting solenoid) that captures the pions and transfers the pions to a decay and phase rotation channel producing muons. Muons will then be accelerated in SRF accelerating structures. To do so, we have developed two accelerating scenarios. One of the options is a Rapid Cycling Synchrotron (RCS)[3], where two hybrid rings (2200m radius) of fixed SC magnets alternating with iron magnets ramping at 200Hz and 330 Hz between full negative to full positive field bending the muons. Muons are given 25 GeV of RF energy (700 MHz) per orbit by accelerating cavities. As an alternative we also discuss another option to RCS, where three recirculating superconducting linacs (respectively with energies reaching 70 GeV (with 350 MHz SRF), 250 GeV (800 MHz) and 2000 GeV (1300 MHz)) are used in a sequence of four recirculating linacs (RLAs)[4], each of which increases beam energy by ~ an order of magnitude, they accelerate beams up to 2 TeV for injection into a collider ring.

At 2 TeV the mean muon lifetime is 41.6 ms. To achieve high luminosity (higher number of turns and shorter orbit), the dipole should be over 8 T carrying muon over 1000 turns in a short-diameter orbit[5].

## SC MAGNETS FOR MUON CAPTURES

The pions produced by a target[1] must be captured by solenoid magnets during their decay into muons. A possible layout of a 20-T hybrid solenoid magnet with a clear bore of 150 mm is shown in Fig. 1, which is designed to capture particles with a transverse momentum of 225 MeV/c or less. The hybrid magnet option was selected for the following reasons: 1) The operating power for the water cooled solenoid is substantially lower if a superconducting outer-magnet is used. 2) The current density in the water-cooled Bitter solenoid can be low enough to insure that its lifetime will be long enough. (A reasonable lifetime goal might be 25000 hours.) 3) Additional space inside the Bitter solenoid can be made available for a heavy metal water-cooled shield. This reduces the incident energy from the target into the water-cooled solenoid and the surrounding superconducting solenoid.

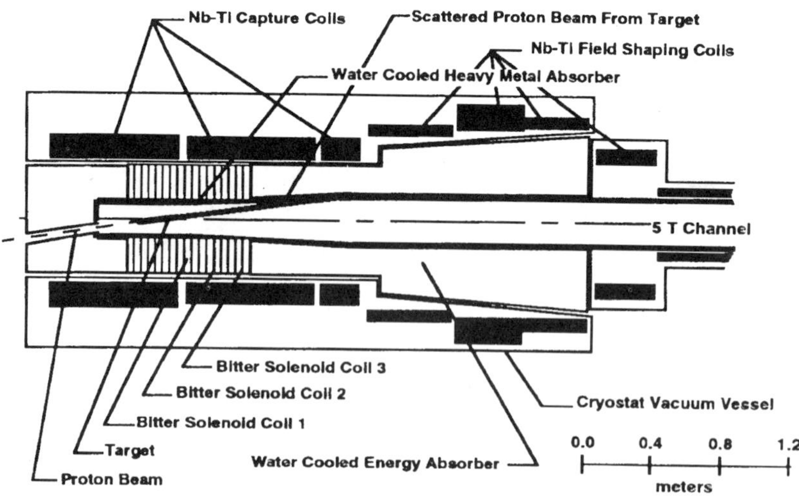


**Figure 1.** A 20T hybrid capture solenoid with a transfer solenoid system of 5T in phase rotation section (inside is a tilted target).

The water-cooled Bitter solenoid insert has an inner radius of 120 mm. The extra inner bore radius allows a heavy metal (tungsten or some other high Z, high density metal) shield that is 30mm thick to be inserted around the target region. The Bitter solenoid insert outer radius has been set at 345 mm. The cryostat of the SC outer-magnet (solenoid) starts at a radius of 370 mm.

The SC outer-magnet coil alone is designed to produce a central field of about 7.5 T. Operating under this condition, the outer-magnet solenoid can be made from Nb-Ti operating at 2.0 K. The outer-magnet has three coils with an inner radius of 400 mm. The outer radius of these coils is about 540 mm. The peak magnetic field in the outer-magnet coil is about 8.3 T when the coils are fully powered.

Other superconducting solenoids downstream from the outer-magnet coils form the transition region that shapes the magnetic field between the target and the phase rotation system. The decay of the pions to muons will occur at the 5 T region.

## RECIRCULATION SRF ACCELERATOR

In a muon accelerator the decay and acceleration rates require a small decay loss that means $dE/ds >> m_\mu c^2/L_\mu = 0.16$ MeV/m (including all lengths), which is relatively large, but can be reached in multipass systems with moderately high gradient.

### SRF Acceleration Systems

The basic accelerating unit in this scenario is the recirculation linear accelerator (RLA), which consists of two linacs with return arcs in a racetrack configuration. The beams are accelerated and returned for several passes in the same linacs, but with separate return arcs (10-20 turns). There are four RLAs, each of which increases beam energy by ~ an order of magnitude, and which accelerates beam up to 2 TeV for injection into a collider ring[4]. Figure 2 shows a conceptual overview of a 4-RLA system. The RLA permits economic multipass

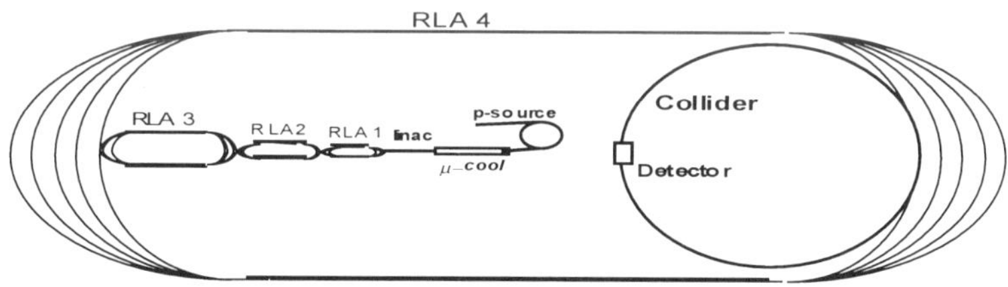


**Figure 2.** Overview of a $\mu^+$-$\mu^-$ collider system.

**Table 1.** Parameters of a 4-RLA μ-accelerating system

| | RLA 1 | RLA 2 | RLA 3 | RLA 4 |
|---|---|---|---|---|
| Beam energy (GeV) | 1→9.6 | 9.6→70 | 70→250 | 250→2000 |
| RF frequency (MHz) | 100 | 350 | 800 | 1300 |
| N turns | 9 | 11 | 12 | 16 |
| $V_{rf}$ /turn | 1.0 | 6 | 16 | 11.2 |
| $L_{turn}(2L_{linac}+2\pi R)$(km) | 0.26 | .095 | 2.32 | 12.6 |
| Beam Survival | 91% | 94.8% | 97.6% | 96.4% |
| $\sigma_{z,beam}$ (cm) | 8.3→4.8 | 1.3 | 0.6 | 0.3 |
| Cavity temp. K | | 4.2 | 2 | 2 |

acceleration, but the separate transport for each turn with cost and complexity limits the number of turns to ~10–20 per RLA. In this scenario, RF systems at 350, 800 and 1300 MHz are SRF cavities. Table 1 gives the main parameters of the SRF system.

If the total RF voltage and beam current are fixed, the use of pulsed RF can reduce the 'cryogenic cost' and allow a higher $E_{acc}$.

## SRF Cavities and SC Magnets

In RLA4, the muon energy increases from 250 GeV to 2000 GeV. As a baseline design 25 MV/m ($Q_0$=5 x $10^9$) at 2 K and 16 turns are chosen that requires about 112 GV per turn from cavities, or 4.5 km of active linac. Encouraged by the pulsed test results of the TESLA cavities, it is possible to use a higher $E_{acc}$. If 35 MV/m becomes realistic, the linac could be reduced to 3.2 km. For the 1300MHz SRF, 16 passes and 15Hz-RF pulses, the HOM power is ~300 W/m. One alternative will be to enlarge the aperture of the cavity from existing 70 mm to 102 mm. That will reduce the HOM mode damping requirements by a factor of 50%[6]. Fig. 3 shows a modified arrangement of the TESLA type cryomodule and Fig. 4 shows a cross-section of the multipass superconducting magnets.

Because of larger apertures and longer bunches, HOM power is expected to decrease as ~ $1/\lambda^3_{RF}$ so HOM excitation at 800 MHz and 350 MHz should be much less (respectively, ~60 W/m and 2 W/m in this scenario). Decay losses at this intensity are ~16 W/m. We require that the cryogenic system tolerate ~10% of decay losses at 2 K (1.6 W/m) with the remainder absorbed at higher temperature.

The 800 MHz RLA3 requires 16 GV of SRF or 1.07 km of linac (12 turns) at 15 MV/m ($Q_0$=5 x $10^9$) and 2 K. The 350 MHz RLA2 requires 6 GV of SRF cavities, or 600m (11 turns) at 10 MV/m and 4K. Such values have been achieved in smaller testing structures. Figure 5 shows a candidate of the 350 MHz SRF cryomodule from CERN-LEPII[7].

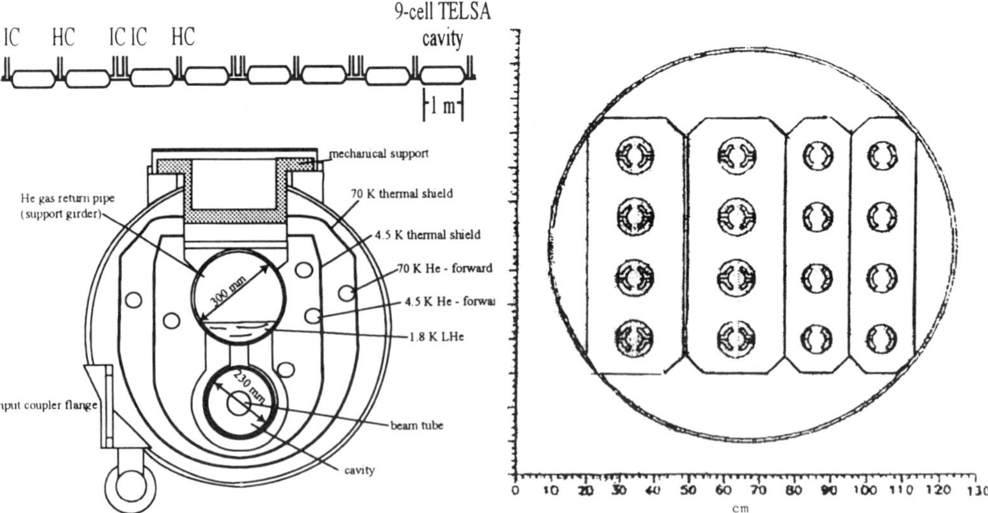


**Figure 3.** A cryomodule based on the TESLA cavities (couplers - IC and HOM couplers - HC).

**Figure 4.** A cross-section of the multipass superconducting magnets.

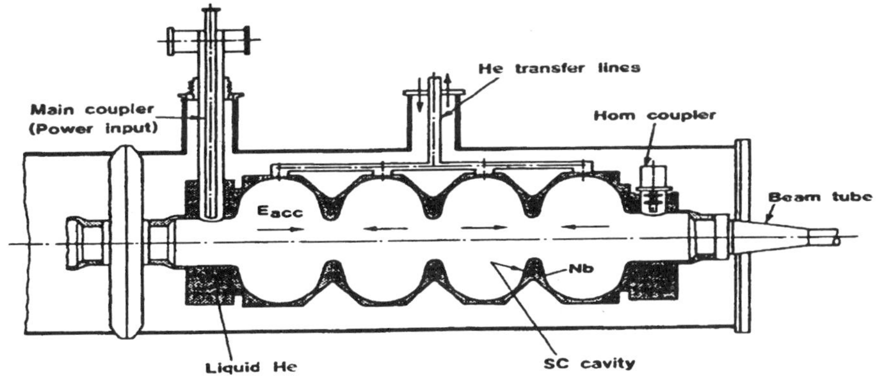


**Figure 5.** A candidate of the RLA2 cryomodule, based on a CERN 350 MHz SRF cavity.

**Table 2.** RF System and RF Power

| | RLA2 | RLA3 | RLA4 |
|---|---|---|---|
| RF pulse length, μs | 35 | 84.2 | 672 |
| Loaded Q to min. P | 1 x $10^6$ | 1 x $10^6$ | 1 x $10^6$ |
| ΔEacc/Eacc | 0.018 | 0.027 | 0.033 |
| Average I, mA | 100 | 45.6 | 7.6 |
| Available RF Power, kW/m | 200 | 200 | 400 |
| Voltage drop | 0.086 | 0.12 | ~0.0 |
| RF on - cryogenic loss, W/m | 119 (at 4K) | 71 (at 2K) | 78 (at 2K) |
| Ave. wall power for RF, kW/m | 5.2 | 2.6 | 5.25 |

## RF Power Requirements

The RF power requirements are dominated by the power needed for the acceleration of the beams. Additional power is required for RF controls. The control power needed depends on the magnitude of perturbations to be controlled to pass the muons 10 times through a pair of linacs. Table 2 lists the RF powers needed.

## RAPID CYCLING SYNCHROTRON

Rapid-cycling approaches require innovations in magnet designs and layouts, but could be more affordable than a RLA version. Fig. 6. is a schematic layout of RCS and table 3 summarizes the parameters of the sub-systems.

**100 MeV - 2 GeV Using RF=2GV:** A single pass 2 GV linac is used.

**2GeV - 25GeV Using RF = 2.5 GV/turn:** This is the first recirculating ring and has 2.5 GV of 100 MHz RF. The superconducting magnets with 10 bores (each with a different fixed field) are used.

**25 GeV - 250 GeV Using RF=6 GV/turn:** This stage uses a single ring of fast ramping *cos* $\theta$ dipoles. Thin stranded copper conductor is used at room temperature to achieve a 4 Tesla field. The low duty cycle is exploited to keep the $I^2R$ losses reasonable. A 6 GV of 350 MHz SRF system is distributed around the ring and accelerates the muons from 25 GeV to 250 GeV in 40 orbits.

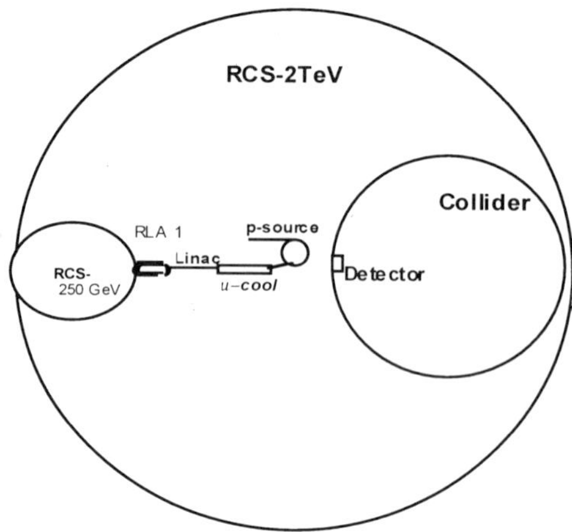


**Figure 6**. The Rapid -Cycling scenario of a muon accelerator and collider system.

**Table 3**. The main parameter of the RCS system

| | Linac | RLA | RCS1 | RCS2 |
|---|---|---|---|---|
| E (GeV) | 0.1→2 | 2→25 | 25→250 | 250→2400 |
| $f_{rf}$ (MHz) | 10→100 | 100 | 350 | 800 |
| $N_{turns}$ | 1 | 10 | 40 | 86 |
| $V_{rf}$(GV) | 2 | 2.5 | 6 | 32 |
| $C_{turn}$(km) | 0.4 | 0.36 | 2.3 | 14 |
| $\tau$(ms) | 0.0013 | 0.012 | 0.307 | 4.0 |
| $\sigma_{beam}$(cm) | 100→10 | 3 | 1 | 0.5 |
| Survival | 93% | 92% | 85% | 82% |

~60% survival over full cycle

**260 GeV - 2 TeV Using RF=25 GV/turn:** In the stage, we have two hybrid rings (2200m radius) of fixed superconducting magnets alternating with iron magnets ramping at 200 Hz between full negative and full positive field to bend the muons from the lower injection-energy to the higher extraction energy. Muons are given 25 GeV of RF energy (800 MHz) per orbit (tatol: 96 orbits). The SRF cavities are divided into multiple sections as at CERN-LEPII, so that magnetic fields and beam energies will match around the rings.

## SC MAGNETS FOR COLLIDER RING

### Special Considerations

Two factors govern the design of the collider ring magnets for a muon collider. (1) The decay of the muons within the ring can deposit up to 2 kW of beam energy per meter into the magnet bore in the form of high energy electrons, positrons and gamma rays. One must keep the muon decay products from depositing their energy into the superconductor of the collider ring magnets. (2) The luminosity at the collision point is inversely proportional to the ring circumference squared and it is inversely proportional to the beam size function squared. In order to get the circumference of the ring as small as possible, dipoles with a high central field are desirable. The strength of the quadrupoles should in some way be proportional to the collider ring dipole central field. A small beam size means that the beam beta function at the collision point must be very small. As a result, the focusing quadrupoles around the collision point must be strong with a large aperture. (The beam beta function in these quadrupoles is often very large in at least one direction.) Table 4 lists the muon decay parameters[5].

## Magnet Design with an Absorbing Bore Tube

As shown in Fig. 7, to attenuate the decay product power by three orders of magnitude, about 65 mm of tungsten shield is required. Only about 1.8 W per meter will then be deposited into the superconducting coil while the remainder of the beam decay power will be absorbed by a cooled metal bore (shield) tube[5]. A copper bore tube would be twice as thick. The primary problem with the large aperture magnet design is the large forces because of the high magnetic fields and the large inner coil radius. These forces scale as the square of the central field times the inner coil radius. Test dipoles with a coil radius of 25 mm have been built and operated at central field as high as 10.2 T at a temperature of 1.8 K. Fig. 7 is a schematic layout of the magnet.

**Table 4.** Muon Decay Parameters in Various Components of a RLA scenario

| Components | Peak Energy (GeV) | No. of Turns | Decay Rate ($\mu s^{-1}$)x$10^{13}$ | Decay Power (kW) | Decay Power per unit L (W $m^{-1}$) |
|---|---|---|---|---|---|
| First Ring | 9.6 | -NA- | 1.9 | 0.6 | -NA- |
| Second Ring | 79 | 11 | 1.2 | 3.6 | 1.64 |
| Third Ring | 250 | 12 | 0.8 | 19.7 | 1.75 |
| Fourth Ring | 2000 | 16 | 0.4 | 36.8 | 1.26 |
| Collider Ring | 2000 | 1000 | 13.1 | 14600 | 1840 |

## The Split Coil Design

In order for the muon decay products to leave the storage ring without colliding with the coil or a beam tube liner, there must be no material on the mid-plane of the magnet[5]. The muon decay products must pass the superconducting coil without interacting with any material. When the decay products are absorbed, they have to be absorbed in a way that prevents back scatter into the superconducting coils. A dipole using the split coil

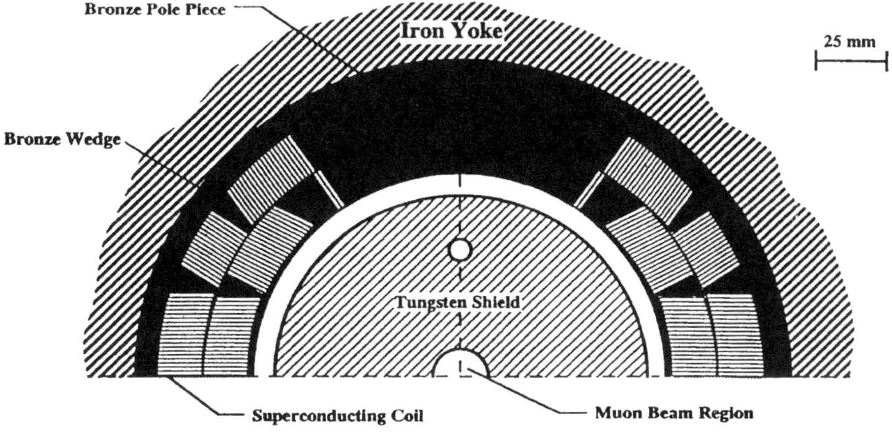


**Figure 7.** A cosine theta coil structure (muon decay energy is absorbed by a tungsten shield (65mm-thick).

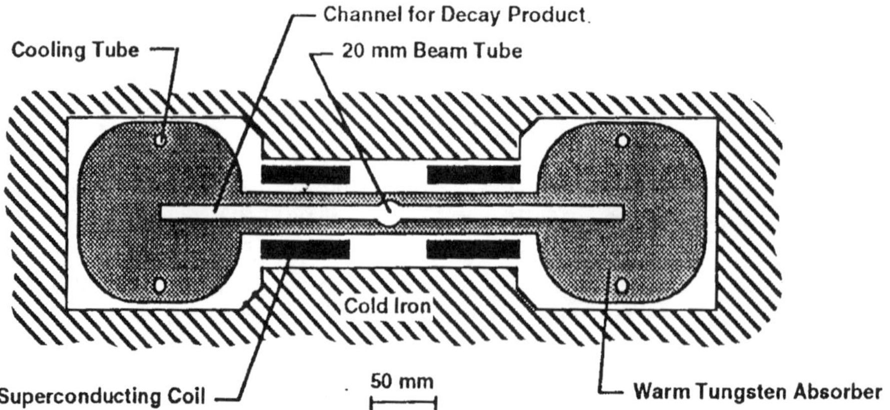


**Figure 8.** Separated coils on the mid-plane with a warm absorber for the muon decay products (an H magnet).

configuration can be fabricated in either a C or H configuration. Configured as an H magnet (see Fig. 8), the cold iron dimensions would be about 1060 mm wide by 700 mm high. The design shown in Fig. 8 has large forces that pull the coils together (across the mid-plane)[5]. The forces also push the coils apart parallel to the x-axis. In a cold iron configuration, the coil must be attached directly to the iron pole. Fig. 8 does not show this connection.

## ACKNOWLEDGMENT

Many thanks are given to colleagues of the muon collaboration. Q. S. Shu also sincerely appreciate S. Corneliussen and P. Kneisel for carefully reviewing this paper, and S. Spata and T. Wang for editing it.

## REFERENCE

[1] (A) $\mu^+\mu^-$ Collider - A Feasibility Study, BNL-52503, Fermi-Lab-Conf.-96-092, LBNL-38946, Muon Collider Collaboration, presented at the Snowmass workshop (1996).
(B) D. Neuffer, to be published in NIM A (1996).
[2] M. A. Green, et. al., LBNL-39107, Lawrence Berkeley National Lab, Berkeley CA 94720, 1997.
[3] D. Summers, et al., International Particle Accelerator Conference - 97, Vancouver Canada 1997.
[4] (A) Q.-S. Shu, et al., International Particle Accelerator Conference - 97, Vancouver Canada 1997.
(B) Q-S. Shu for the TESLA Collaboration, in Advances in Cryogenic Engineering, P. Kittel, ed. Vol. 41A, p. 827 (1996).
[5] M. A. Green, et al., LBNL-39106, Lawrence Berkeley National Lab, Berkeley CA 94720, 1997.
[6] J. Sekutowicz, Private communication, DESY, Hamburg Germany, May 1997.
[7] W. Weingarten, Private communication, CERN, Geneva Switzerland, September 1996.

# EXPERIENCE WITH THE LEP SUPERCONDUCTING RF ACCELERATING SYSTEM

G. Geschonke

CERN, SL Division
CH 1211 Geneva 23, Switzerland

## ABSTRACT

CERN is presently upgrading the large Electron Positron Collider (LEP) to higher energy by installing superconducting RF accelerating cavities. For a total installed circumferential voltage of about 2800 MV, 272 cavities operating at 352 MHz will be needed, representing an active length of 462 m and a cold surface of more than 1600 $m^2$. The series production cavities are made out of copper, sputter-coated with a thin layer of niobium and cooled with liquid He to 4.5 K. The cavities are produced by industry and the acceptance testing is done at CERN. In 1996 176 cavities had been installed and run successfully at their design gradient of 6 MV/m during physics at a beam energy of 86 GeV. As RF power sources 36 klystrons will finally be installed with a nominal RF output power of 1 MW each. In this paper the superconducting accelerating system in LEP will be described and experience gained during operation for physics as well as new developments will be presented.

## INTRODUCTION

The energy of CERN's biggest accelerator, the Large Electron Positron Collider (LEP) is presently being upgraded from 45 GeV per beam, where physics was done around the $Z_0$ mass, to above 95 GeV where experiments will be done above the threshold for $W^{\pm}$ production.[1] The circulating beam of energy E loses some of its energy in each revolution due to emission of synchrotron radiation, proportional to $E^4$. The loss per turn is 120 MeV at a beam energy of 45 GeV and 2.3 GeV at 95 GeV. This radiated energy must be replaced continuously to keep the beam circulating.

This was done initially by an acceleration system consisting of 128 normal conducting copper accelerating cavities, powered by klystrons.[2] This system was designed for a nominal circumferential voltage of 400 MV, with an installed RF power of 16 MW at a frequency of 352.2 MHz.

The energy upgrading requires an increase of the circumferential voltage to nominally 2830 MV. With the existing copper technology, this would require more than 900 cavities and the dissipated power in these cavities would be more than 100 MW. Therefore superconducting (SC) cavities are being used, which require less RF power and run at gradients of 6 MV/m as compared to 1.4 MV/m for the copper system. For compatibility reasons the frequency and the general layout of the RF power installation are the same for both the SC and the copper systems. In its final stage, planned to be completed at the end of the winter shutdown 1997/1998, 272 four-cell SC cavities will be in operation, in addition to 48 of the original Cu cavities.

## SYSTEM OVERVIEW

LEP is a circular accelerator/storage ring for counter-rotating electrons and positrons, traveling in a common vacuum chamber. It is housed in a circular tunnel, 60 to 100 m below the surface with a circumference of 26.7 km. The circulating electrons and positrons are each concentrated into four equidistant groups (bunchtrains) of up to four short bunches with an r.m.s. half-length of about 8-10 mm. The particles are allowed to collide in four places (IP) around the ring, equipped with physics detectors. On both sides of each of the four IPs the accelerating system is located in 200-m-long straight sections.

### Cavities and Cryostats

The accelerating cavities consist of four cells of elliptical shape, coupled to each other electrically via the beam hole. 16 of the cavities consists of solid sheet niobium; the other 260 are made out of copper, sputter coated with a thin layer of a few μm of Nb. This technology was developed at CERN, motivated by several advantages over solid Nb:

a) The higher thermal conductivity of the Cu substrate leads to a better stability against thermal quenches.
b) The thin Nb film is much less susceptible to degradations of the quality factor $Q_0$ due to external magnetic fields as compared to solid material.
c) No shielding for the earth magnetic field is necessary.
d) Lower cost.
e) Advantage can be drawn from ongoing R&D on different films of other materials and alloys. If better films become available, the cavities can be recoated.

This technology is now very well established; in total a surface of more than 1600 $m^2$ has been successfully sputter-coated. The cavities are made out of OFHC copper, by spinning and electron-beam welding. All vacuum connections are made with stainless steel knife-edge flanges. Any exposure of the inner surface to air, after cleaning and sputtering, has to be done under clean room conditions of better than class 100. The cavities are cooled by immersion into a bath of liquid He at a temperature of 4.5 K; the stainless steel He container is directly welded onto the cavity structure.

The cavity and its He vessel are flexible; therefore the dynamic frequency tuning during operation can be done by slightly changing the overall length. The cavities are enclosed in a cage formed by three nickel bars connected to the beampipes. The length of these bars can be varied by heating/cooling and via magnetostriction of the Ni by applying a magnetic field. The thermal tuner has a range of 50 kHz; the magnetostrictive one is used for fast variation, within a range of about 1.6 kHz.

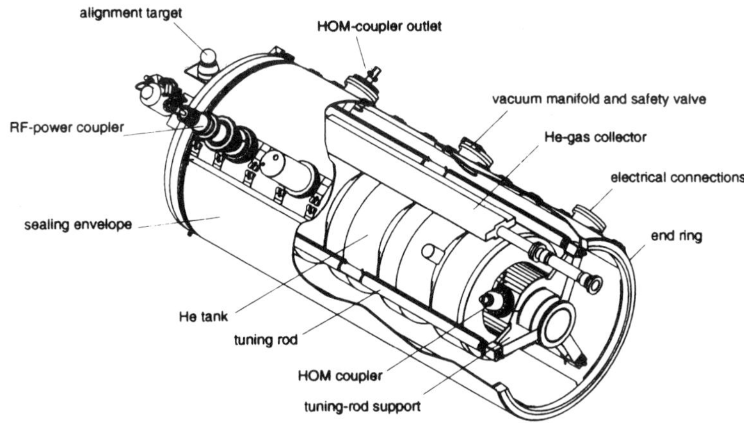


**Figure 1.** One cavity section of the cryostat. Four such sections are connected together to form a 12 m long "module". The He tank follows the shape of the cavities, in order to limit the volume of liquid He.

Four cavities are grouped together in one common cryostat of about 12 m length, called a "module". One cavity section of a module is shown in Figure 1.

Liquid He is fed into the first cavity and the gas is evacuated from the last one. Thermal insulation is done only via multiple layers of superinsulation foils. Apart from mechanical suspensions, several connections are necessary between components at 4.5 K and ambient temperature: coaxial RF and control cables, rigid coaxial links between Higher Order Mode (HOM) couplers and RF absorbers, RF power couplers and end cones bridge the full temperature difference. The heat influx through these large cross section

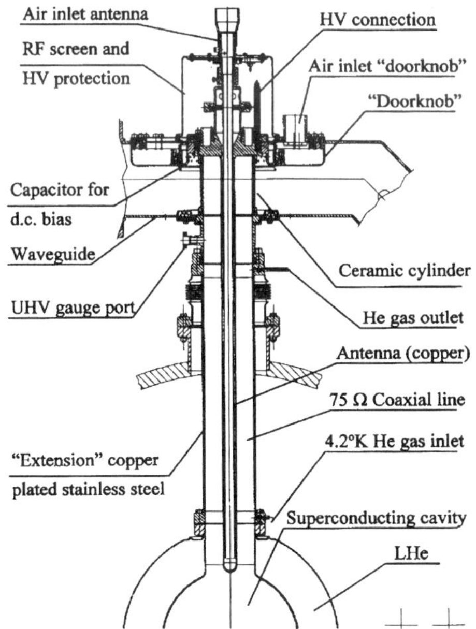


**Figure 2.** Schematic view of the RF power coupler.

links is minimized by cooling them with cold He gas. The total static heat losses amount to 20 W per cavity.[3] This has to be compared to the dynamic RF losses of 70 W at 6 MV/m.

The complete modules and all accessories were manufactured by industry. The know-how of technologies developed at CERN, such as for example the sputtering of the Nb film, was successfully transferred to three European companies. Contrary to previous practice, industry guaranteed the RF performance of the finished product; the $Q_0$-value of bare cavities was specified to be $3.4 \times 10^9$ and of the finished module $3.2 \times 10^9$, both at a field of 6 MV/m. After acceptance tests the power couplers and Higher Order Mode couplers were fitted at CERN.

Production in industry is finished by now, all cavities are accepted at CERN. A total length of 462 m of superconducting structures have been manufactured, giving an accelerating voltage of 2700 MV, the world's biggest superconducting RF system.

All bare cavities were systematically measured before assembly into modules and mounting of accessories. More than 98% of them reach fields above 7 MV/m after He processing, which is part of the acceptance test [4]; however, operation in LEP is presently limited to 6 MV/m, as will be shown later.

After contractual acceptance tests at CERN and fitting of power couplers and HOM couplers the finished modules are tested and conditioned in a test stand at the surface, which is an exact replica of the installation in LEP. In most cavities conditioning is mainly necessary for the power coupler. Conditioning was done in cw, as well as in pulsed mode. In some cases the cavities emitted strong X-radiation due to the presence of dust on the surfaces leading to field emission, and also showed increased thermal losses. In such cases He-processing was applied.

## Power Couplers and Higher Order Mode (HOM) Couplers

The power couplers have a double task: they transfer radio-frequency power up to 125 kW into the cavities by coupling to the electric field of the acceleration mode and at the same time they have to provide an RF transparent seal between the cavity vacuum and the atmospheric pressure in the waveguides. They consist of an all-metallic gas-cooled coaxial line, bridging the temperature difference between 4.5 K at the cavities and ambient temperature through the insulating vacuum. The connection to the waveguides is done via a doorknob transformer, matching the waveguide to the coupler. The RF window is a ceramic cylinder, brazed onto the outer conductor in the doorknob transformer. A drawing of the coupler design [5] is shown in Figure 2. Initial problems with multipacting in the coaxial line have been solved by applying a bias voltage of +2.5 kV to the inner conductor and by an *in situ* bakeout of the window before cooling. In tests on a dedicated cavity fitted with two couplers, 300 kW of RF power have been transmitted through the cavity, and in pulses of 1.5 msec duration 500 kW have been achieved.

The short bunches of the beam excite resonances in the cavities outside the acceleration mode (HOM). They have to be damped, otherwise they build up in amplitude if they are synchronous with the bunch repetition frequency and cause problems with beam stability or lead to electric breakdown or a quench.

The cavities are equipped with two HOM couplers mounted on the beam pipes, one at each side of the cavities. They couple to the electromagnetic fields via a loop with an integrated rejection filter for the fundamental accelerating mode. Previous problems were solved by geometry changes and increased cooling. The power they extract from the cavities is transported to RF absorbers outside the cryostats through rigid coaxial lines. The whole system was tested up to 850 W cw RF power per coupler at 630 MHz and beam currents up to 8 mA. Initially two commercially available 200 W RF absorbers were

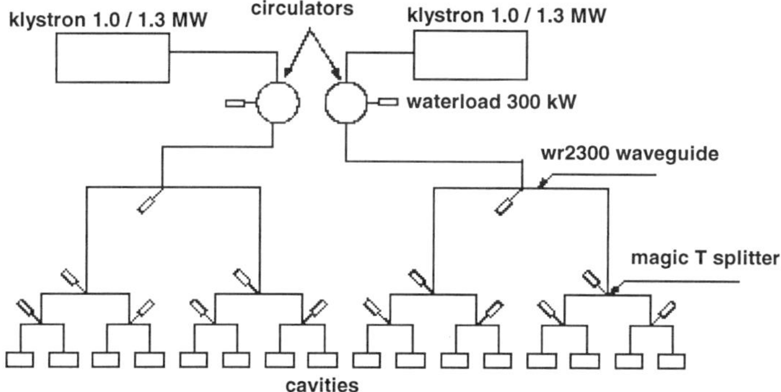


**Figure 3.** Schematic view of RF supply for two units of SC cavities connected to one common DC power supply.

connected in parallel to each HOM coupler. They were found not to withstand the high peak power which may occur during operation and are being replaced now by long, lossy enough coaxial cables.

## Cryogenic Plant

The cavities around each one of the four IPs are cooled by a separate cooling plant,[6] specified for a cooling capacity of 12 kW, out of which 10 kW are available at 4.5 K, and 2 kW are required to liquefy warm gas returning to the compressors. Because LEP is between 60 and 100 m under ground and for space restrictions, the cold boxes are split in two parts, at the surface and in the tunnel.

## RF Power Plant and Controls

Eight cavities are grouped together into one RF unit. The RF power[7] is provided by 1.3 MW klystrons and transported to the cavities via an aluminum waveguide system (WR2300). The splitting is done with magic Ts. Each klystron is protected against reflections by a circulator. In the case of a sudden switch-off of the RF power, most of the cavities' stored energy empties via the power coupler, and the reflected peak power can be as high as 1 MW, four times the zero-current forward power required for the operating field. Therefore the coaxial components on the circulator's RF absorbers were modified for higher power transmission. Two RF units share one DC power converter with a rating of 100 kV, 40 A. A schematic view is shown in Figure 3. The RF power plants run very reliably, so far, 11 of the LEP klystrons have accumulated between 27000 and 30000 operating hours.

The klystron output power is controlled via the modulating anode (slow, limited to about 10 Hz) and via a fast RF feedback, acting on amplitude and phase of the RF drive into the klystron. The RF feedback is based on the vector sum of the fields in all eight cavities driven by the same klystron. This fast control was introduced recently for the following reasons:

a) To assure beam stability at high beam loading.
b) To protect the beam from the effect of cavity field oscillations in amplitude and phase.

c) To increase operational reliability. It had been observed that, due to beam loading, a trip of one RF unit reduced the fields as seen by the beam in the other units, leading to a large loss of circumferential voltage and often to beam loss.

d) To protect the cavities against overvoltage in case of sudden beam losses.

A fast detuning of the cavities via the magnetostrictive tuner was introduced in order to avoid large beam-induced fields in cavities after an RF trip.

## STATUS OF THE UPGRADING PROJECT

During the winter shutdown 1996/1997 16 new modules (64 cavities) were installed in LEP, bringing the total up to 240 cavities for 1997 operation. The remaining 32 cavities will be installed during next winter.

In 1996 LEP operated successfully first at 80.5 GeV, near the threshold for $W^{\pm}$ pair production; in the second part of the year after installation of more cavities, 85 and 86 GeV were achieved.

In 1997 a maximum voltage of 2640 MV will be available. The beam energy will be chosen such as to leave some margin in voltage: experience has shown that in previous years LEP was running with about 95% of the nominal voltage, because of practical reasons in dealing with "sick" cavities and RF units, which cannot always be repaired instantly. Also the quantum lifetime has to be sufficient to survive a trip of at least one whole RF unit with two klystrons, which represents a loss of 160 MV. This gives an energy for 1997 between 91.8 and 92.4 GeV, depending on the optics parameters.

The currents accelerated so far were routinely 5.5 mA in physics, during machine development sessions 8 mA were achieved at low energy.

## PERFORMANCE OF THE SC SYSTEM IN LEP

### Statistics

The reliability of the LEP RF system is characterized by the number of fills which did not go into physics or were lost. This can be seen in Table 1 which gives the statistics over the last two years. A typical physics fill takes about one hour for filling, accelerating and adjustments, followed by storing stable circulating beams for typically 6-8 hours. 82 % of all RF faults caused a loss of 80 MV, 18 % tripped the high voltage, causing a loss of 160 MV. It should be noted, that the number of interlock circuits is very large, in 1996 we had 4500 acting on the RF and 720 acting on the HV.

### Cavity Performance

The cavities were routinely operated at their nominal gradient of 6 MV/m, the usable total circumferential voltage being 95% of the maximum, for reasons explained above.

Each module has a radiation monitor permanently mounted on each end. At nominal field the radiation levels went up to 50 kRad/hour for some modules; the majority were running at 10 kRad/hour. Once during LEP operation a degradation of a cavity was observed, which resulted in abnormal heating and high radiation. Apart from this one case, the cavity performance was stable. The $Q_0$ values as measured by cryogenic losses show no degradation; also the maximum fields are at design level. Some cavities are in LEP since 1994.

**Table 1.** Performance of the LEP RF system in the last two years

| | 1995 | 1996 |
|---|---|---|
| fills not leading to physics due to RF faults | 31 % | 16 % |
| down time due to RF | 37 % | 22 % |
| successful physics fills | | 111 |
| lost physics fills | | 39 |
| lost physics fills due to RF | | 24 |

***In situ* Processing.** After installation in LEP, conditioning of the couplers is again necessary for about 6-12 hours. After transport and installation some cavities had degraded in field emission performance above 4-5 MV/m. This resulted in abnormally high thermal losses, leading to a loss of liquid helium level, or in sharp pressure rises in the He bath. The degradation was accompanied by a considerably increased level of X-radiation emanating from the cavities. These cavities and also the one, which had degraded during operation, could be recovered *in situ* by RF processing with high power pulses or He processing or a combination of both. The pulses used were 5-6 ms long, with a repetition time of 200 ms, and the peak pulse height was 7.5 MV/m. The maximum peak power is presently limited by the reflection peak at the end of the pulse. He-processing was done at a He pressure of $6.5\times10^{-6}$ mbar with the klystron, on the modules as they are installed in LEP. The evolution of radiation for both processes is shown in Figure 4.

## Microphonics

Two types of mechanical cavity oscillation have been identified, leading to modulation of both cavity fields and phases. The first one, a thermo-acoustic oscillation, has been identified to be associated with the He gas return manifold and could be cured by filling an empty volume with Kapton™ foils.

The other type is a ponderomotive instability which appears if the cavities are driven at a frequency below their resonance. In this case, coupling between mechanical and stored electromagnetic energy occurs and the cavities start to oscillate at their mechanical resonant frequency around 100 Hz.[8] This effect is proportional to the square of the accelerating field. It occurs in operation when with higher beam current the tuning system detunes the cavities to compensate for the beam loading. Oscillations of field amplitudes up to ± 40% have been observed. The solution which could be implemented so far was to compensate

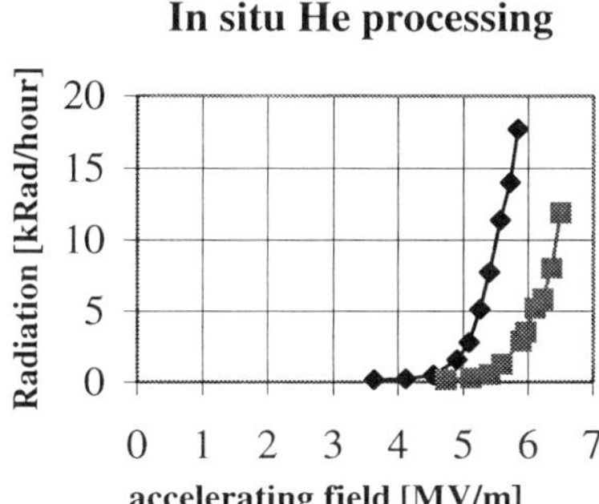


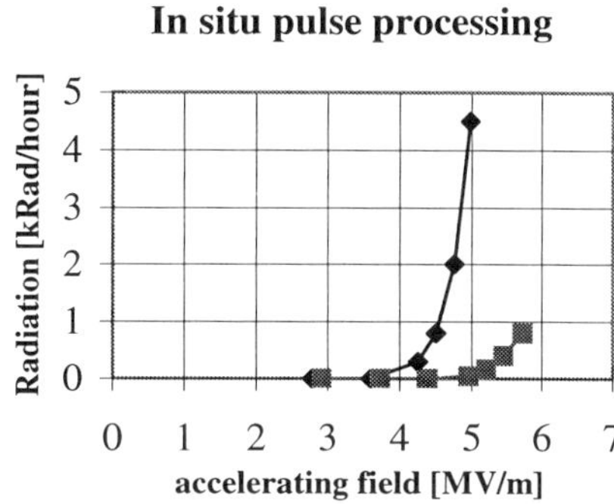


**Figure 4.** Effect of *in situ* He- and pulse processing of different modules in LEP. The diamonds represent the radiation from the modules before processing; the squares show the radiation after processing. In both cases a substantial factor could be gained.

the beam-induced detuning by introducing an offset into the tuning system by hand, so that at maximum current and field the cavities are driven at their resonance. This might eventually lead to a loss in beam stability which will have to be counteracted by other means. Automatic methods are under development, acting directly on the fast magnetostrictive tuners.

### Performance Limitations

Presently the SC cavities run very successfully at their nominal gradient and at beam currents of 5.5 mA. Running at higher beam currents or increasing the energy further by higher gradients, is presently limited by several effects:

**Cryogenic power.** During operation in 1996 an effect of the circulating beam on the cryogenic consumption was observed,[9] depending on the beam current, the bunch length and the beam energy. The bunch length and current dependence can be explained by higher order modes excited by the beam and heating normal conducting elements, such as inter cavity bellows and conical transitions at the end of the modules. So far the energy dependence cannot be explained. This effect might impose a limitation on beam current at higher energy, but the exact scaling laws will have to be established as soon as LEP starts again.

**Oscillations.** The above-mentioned oscillations make higher beam currents a delicate operation. Since there is a spread in external $Q_{ext}$ values of the cavities with an associated spread in cavity fields, control of these oscillations is difficult in practice, because the required tuning offset has to be adjusted individually for each cavity.

**Cavity fields.** Since eight cavities are driven in parallel from one klystron, the accelerating fields in the cavities vary because of the different $Q_{ext}$. The beam loading is different because of different cavity fields, tuning errors and asymmetries in the RF distribution. This means in practice that in order to achieve an average accelerating field of 6 MV/m, some cavities have to run as high as 7.5 MV/m, whereas others run only at 5 MV/m. Increasing the field risks to drive these cavities outside their operating range provoking an RF trip.

## CONCLUSION

The upgrading of LEP with superconducting cavities is nearing completion with 240 installed cavities. Initial difficulties with RF power couplers and HOM couplers have been solved and operation at nominal gradient has become routine up to beam currents of 5.5 mA during physics coasts in 1996. Presently limitations to further increase the performance beyond the specifications are in ponderomotive oscillations, the unequal field distributions and in high radiation levels.

## ACKMOWLEDGEMENT

This work is the result of continuous efforts from many groups at CERN. Without the dedication and motivation of all the people involved in engineering and development, manufacturing, installation, commissioning and operation the success of this large enterprise would not have been possible.

## REFERENCES

1. The LEP Design Report, vol. III, LEP2, C. Wyss, editor, CERN, AC/96-01, LEP2
2. G.Geschonke, S.Hansen, Status and Performance of the LEP RF System, *Part. Acc.*vol. 30, 779(1989)
3. M.Barranco Luque, D.Güsewell, Thermal loss analysis of cryostats and accessories for the superconducting cavities of the LEP energy upgrade, 4th EPAC, London 1994, p.2455
4. E. Chiaveri, Large scale industrial production of superconducting cavities, 5th EPAC, Sitges 1996, p. 200
5. E. Haebel, Couplers for cavities, in Proc. CAS, Hamburg 1995, CERN 96-03 (1996), p. 231
6. M.Barranco-Luque, S.Claudet, Ph.Gayet, N.Solheim, G.Winkler, Operation of the cryogenic system for superconducting cavities in LEP, proc. ICEC/ICMC, Kitakyushi, 1996
7. H. Frischholz, The LEP2 RF power generation system, Proc. IEEE PAC, Washington 1993, p.1247
8. D. Boussard, P. Brown, J. Tückmantel, Electroacoustic oscillations in the LEP SC cavities, 5th EPAC, Sitges (1996), p. 2097
9. Ph.Gayet, D.Kaiser, G.Winkler, Beam effects on the cryogenic system of LEP 2, this conference

# DESIGN AND PERFORMANCE OF KEKB SUPERCONDUCTING CAVITIES AND ITS CRYOGENIC SYSTEM

K. Hosoyama, S. Mitsunobu, and T. Furuya

KEK-High Energy Accelerator Research Organization
Tsukuba, Ibaraki, Japan

## ABSTRACT

The installation of the superconducting rf acceleration cavities and deflection crab cavities to the KEKB have been proposed and the R&D studies have been carried out extensively at KEK. The satellite refrigeration scheme is considered for the cryogenic system for the superconducting crab cavities. The R&D studies of the satellite refrigeration system and the high performance transfer line for this system have been started at KEK.

## INTRODUCTION

At KEK- High Energy Accelerator Research Organization, 32x5-cell 508MHz superconducting rf cavities in 16 cryostats were installed in the TRISTAN electron-positron collider and a large scale cryogenic system[1,2,3] with 8kW cooling capacity at 4.4K was constructed for upgrading of the electron and positron beam energy from 27GeV to 32GeV. Operation of the system was started in 1988 and continued until end of the TRISTAN physics run in June 1995. During this period superconducting rf cavities and the cryogenic system were operated very stably and reliably [4]. The total operation time of the cryogenic system was about 38,000 hours in 7 years.

After the TRISTAN project, construction of a high luminosity double-ring 8x3.5GeV asymmetric electron-positron high beam current collider KEK B-Factory[5] (KEKB) in the existing TRISTAN tunnel was started in 1994 as a five year project. Figure 1 shows the configuration of the KEKB. The two ring of KEKB are build in the existing 3km circumference TRISTAN tunnel. The design beam currents of electrons and positrons are very high to attain high luminosity of $10^{34}$ $cm^{-2}s^{-1}$ for detailed studies of CP-violation and other topics on decays of B mesons.

Installations of 10 single cell superconducting higher-order-modes (HOMs) damped acceleration cavities to KEKB have been proposed[6]. The R&D studies of these superconducting cavities have been carried out extensively at KEK. High beam current test of the prototype model was performed successfully at TRISTAN accumulation ring AR in 1996[7]. Construction of 4 Nb single cell superconducting HOMs damped cavities with design acceleration field gradient Eacc of 6MV/m (acceleration voltage Vc = 1.5MV) has been started. These cavities will be installed in the straight section of Nikko area of HER in summer of 1998 and will be operated from the commissioning phase of KEKB in autumn of 1998. A finite crossing scheme of 2 x 11 mrad has been adopted in KEKB to reduce the background rates and to simplify beam optics at interaction region. Installation of 4 Nb 508

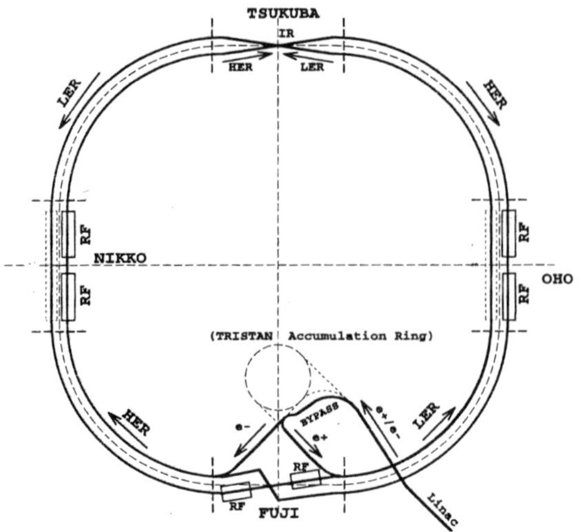


| | LER | HER | |
|---|---|---|---|
| Energy | 3.5 | 8.0 | GeV |
| Beam current | 2.6 | 1.1 | A |
| RF frequency | 508.887 | | MHz |
| RF volt | 5 -10 | 10 - 20 | MV |

Figure 1. Configuration of the KEKB.

MHz superconducting rf deflector crab cavities at Tsukuba straight section has also been proposed to eliminate the possibility of beam-beam instability due to the finite crossing angle scheme. The R&D studies of the KEKB superconducting crab cavities and cryogenic system of it have been started at KEK for the commissioning in autumn of 1999.

## KEKB SUPERCONDUCTING CAVITIES FOR ACCELERATION

For high beam current electron positron storage ring such as KEKB the coupled-bunch instabilities arising from the HOMs of cavities and the accelerating mode must be sufficiently suppressed to attain high luminosity operation such as $10^{34}$ $cm^{-2}s^{-1}$. In order to solve this problem, a new normal conducting accelerating cavity structure with an energy storage cavity, so called ARES[5], was proposed and extensively studied at KEK. Another solution is a superconducting cavity. An advantage of the superconducting cavity is, due to high

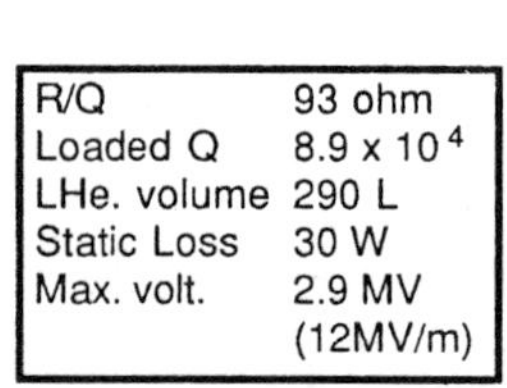

| | |
|---|---|
| R/Q | 93 ohm |
| Loaded Q | $8.9 \times 10^4$ |
| LHe. volume | 290 L |
| Static Loss | 30 W |
| Max. volt. | 2.9 MV (12MV/m) |

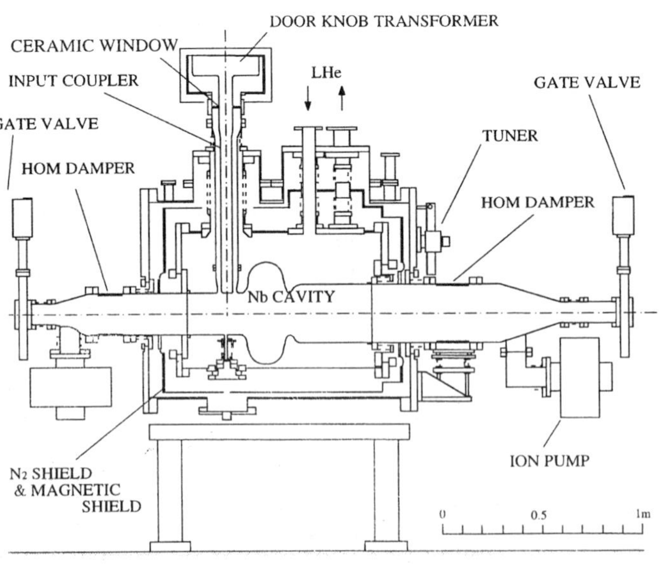


Figure 2. Superconducting HOMs damped acceleration cavity for KEKB.

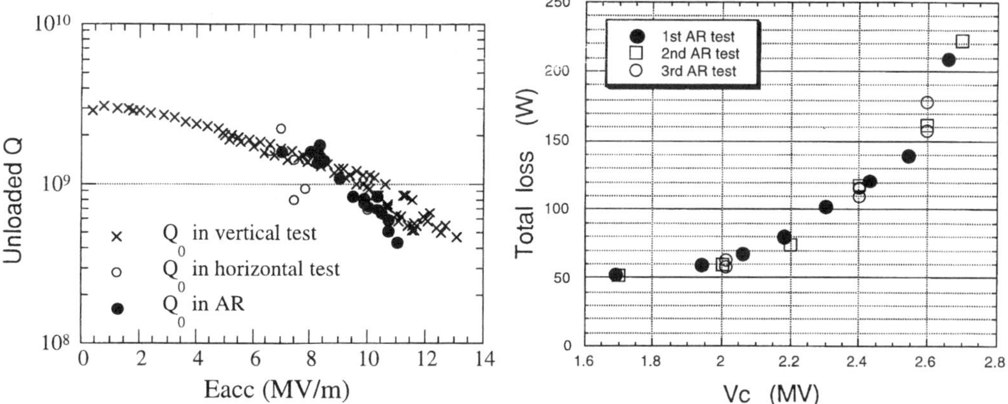


Figure 3. Unloaded Q-value vs. Eacc for KEKBsuperconducting HOMs damped cavity.

Figure 4. Total heat loss vs. Vc for KEKB superconducting HOMs damped cavity.

acceleration field of the superconducting cavity, a relatively small number of the cavities can provide the required rf voltage without greatly increasing the impedance of higher order mode and it stored energy is very large. Figure 2 shows the KEKB superconducting acceleration cavity installed in the cryostat. A spherical cell shape Nb cavity has large aperture beam pipes - 220 mm and 300 mm in diameters - and rf absorbers made of ferrite are attached at room temperature part of both end of the beam pipes. All HOMs of the cavity excited by the high current beam are extracted out through the large aperture beam pipes and absorbed by the rf absorbers. The high rf power of about 500kW must be supplied through a coaxial type input coupler to maintain a beam of 1.1A in HER.

For the feasibility study of the KEKB superconducting HOMs damped cavity, a prototype superconducting HOMs damped cavity and the horizontal cryostat were constructed and a cryogenic system with helium refrigerator (TCF200, 600W at 4.4K) and newly developed multi-transfer line was also constructed. After a bench test at Nikko test stand, this cavity was tested in high current beam of TRISTAN accumulation ring AR in 1996. The measured Qo-values of the cavity as a function of the acceleration field gradient Eacc. in vertical cryostat test, horizontal cryostat test and in AR are shown in Fig.3. The total heat load to liquid helium vessel including 30W of the cryostat static heat loss are shown as a function of the acceleration voltage Vc in Fig.4. The maximum current of 573mA was accelerated very stably in 2MV acceleration voltage and the HOMs in the cavity were damped sufficiently to the rf absorber. We could feed 280kW of the rf power to the input coupler and 160kW was transferred to the beam. The results of this high beam current test in AR verify the application of the superconducting HOMs damped cavity to KEKB.

## SUPERCONDUCTING CRAB CAVITIES

The crab crossing scheme[7,8], shown in Fig.5, will be adapted in KEKB to eliminate luminosity reduction due to geometrical effect and the possibility of beam-beam instability by synchrotron-betatron coupling resonances in the finite crossing angle collision. In this scheme bunches of electrons and positrons to the interaction point are tilted by time-dependent (508MHz) transverse kick in rf deflectors so called crab cavities and make the bunches head-on collision. After the collision these bunches are kicked back to the original orientation in another crab cavities. The R&D program for the KEKB superconducting crab cavity was started at KEK[9] in 1994. For the base line design of the KEKB crab cavity, we have adopted the squashed cell type cavity which was designed and studied extensively at Cornell[11] in 1991 and 1992 for CESR-B under KEK- Cornell collaboration. Figure 6 shows conceptual drawing of the KEKB crab cavity with damping scheme for all lower and higher frequency modes. Unwanted modes are extracted from the cavity through a large aperture beam pipe and a coaxial beam pipe and absorbed by rf absorbers attached on the beam pipes at room temperature. The required kick voltages are 1.4 and 1.44MV for KEKB LER and HER respectively which correspond to the maximum surface electric field of 21MV/m.

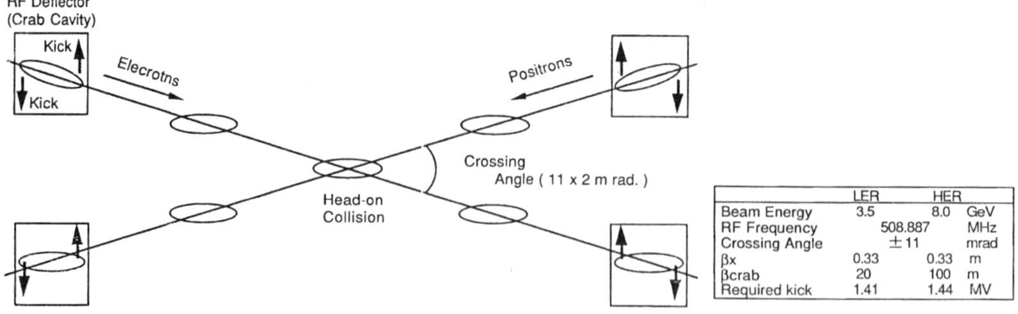


| | LER | HER | |
|---|---|---|---|
| Beam Energy | 3.5 | 8.0 | GeV |
| RF Frequency | 508.887 | | MHz |
| Crossing Angle | ±11 | | mrad |
| βx | 0.33 | 0.33 | m |
| βcrab | 20 | 100 | m |
| Required kick | 1.41 | 1.44 | MV |

Figure 5. Crab crossing scheme for KEKB.

## OVERVIEW OF TRISTAN CRYOGENIC SYSTEM

We want to discuss briefly about the TRISTAN cryogenic system[1,2,3] which has been designed and constructed for 32 x 5 cell superconducting cavities in 16 cryostats installed in Nikko. The helium refrigerator with 5 turbo-expanders has cooling capacity about 8 kW at 4.4 K without liquid nitrogen precooling. The helium compressor unit consists of 6 oil flooded screw compressors and 4 stage of oil filters. Liquid helium produced in a 12,000 L liquid helium dewar was distributed to 16 superconducting cavities cryostats installed in about 200m long straight section of underground TRISTAN tunnel through large size about 250m long multi-channel transfer line. A liquid nitrogen circulation system of 6.5kW cooling capacity at 80K with a turbo-expander and 2 screw compressors could supply liquid nitrogen to 80 K thermal shield of the cryostats in parallel through multi-channel transfer line. For the helium gas storage 9 x 100 $m^3$ medium pressure (1.9 MPa ) helium gas tanks are connected to the system. The whole cryogenic system was controlled by means of a process control computer system.

## CRYOGENIC SYSTEM FOR KEKB SUPSERCONDUCTING CAVITIES

The total of about 10 KEKB single cell superconducting HOMs damped cavities will be installed in the high energy ring HER of KEKB in the straight section of Nikko area. For cooling of these cavities we can use the existing cryogenic system with header transfer line and about 8kW at 4.4K helium refrigeration system and liquid nitrogen circulation system. Figure 7 shows the layout of the KEKB superconducting cavities in the tunnel. Because we must use the existing header transfer line designed for TRISTAN superconducting cavities the locations of the connection boxes were not optimized for KEKB superconducting cavities. We will require long sub-transfer lines for the connection of the KEKB superconducting cavities and the connection nearby boxes. To reduce the heat loss of the

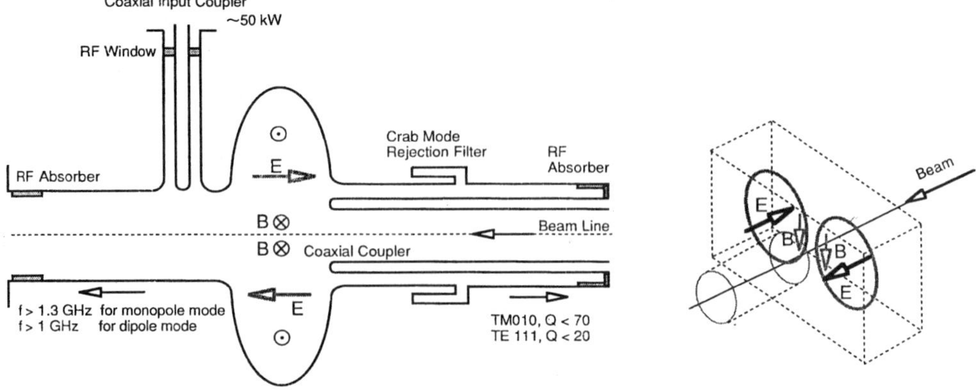


Figure 6. Conceptual design of KEKB crab cavity.

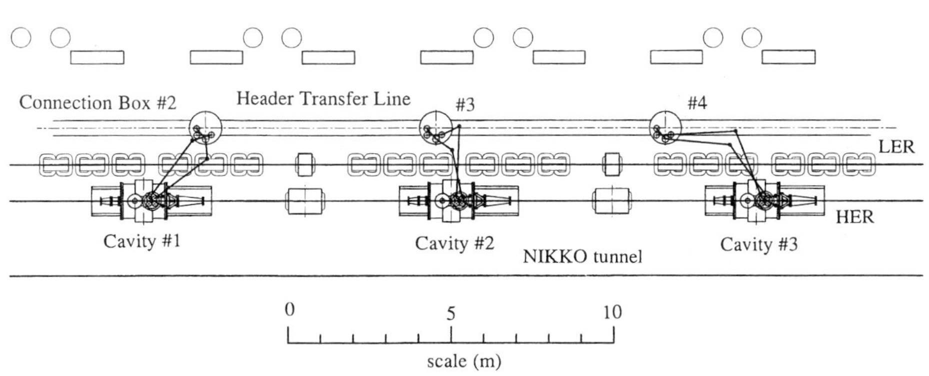


Figure 7. Layout of the KEKB superconducting cavities in the Nikko tunnel.

sub-transfer lines, we will use the newly developed high performance transfer lines with 80K liquid nitrogen thermal shield. The total heat load of the 10 single cell KEKB superconducting cavities will be about 3 kW including the transfer line loss of about 1kW.

Figure 8 shows the layout of the KEKB crab cavities in Tsukuba area. Four superconducting crab cavities will be installed in pair to LER and HER at both end of the collision point and the distance between these cavities is about 150m apart. The estimated heat load of the superconducting crab cavity at $Q_0=0.6x10^9$ is about 200W for each cryomodule including about 50% safety margin. The total heat load of the KEKB superconducting crab cavities is about 800W at 4.4 K. There are many possibility of cooling scheme for the KEKB crab cavity. We consider following cases.

1. Two independent cryogenic systems with about 400W cooling capacity at 4.4K placed near the 2 crab cavities of both end of the straight section of the Tsukuba area.
2. One cryogenic system with about 800W cooling capacity and distribution system for 4 crab cavities with long high performance transfer lines.
3. The 4 crab cavities will be connect directly to the Nikko cryogenic system through high performance long helium transfer line.
4. Two satellite refrigerators about 500W cooling capacities will be placed near the 2 crab cavities of the both end of the straight section of the Tsukuba area and liquid helium required for the satellite will be supplied from the Nikko helium refrigerator through about 1 km long transfer line.

From the stand point views of the operation and maintenance of the whole system and special reason that we have enough surplus cooling power in large cryogenic system in Nikko, case 3 and 4 seem very attractive. In case 3, in addition to the 1 km long transfer line for liquid helium supply, we must construct the return line for cold gas from the crab cavities to the helium refrigerator at Nikko and the inner diameter of this return line must be large enough to keep the pressure drop of the cold return gas very small. On the other hand in the

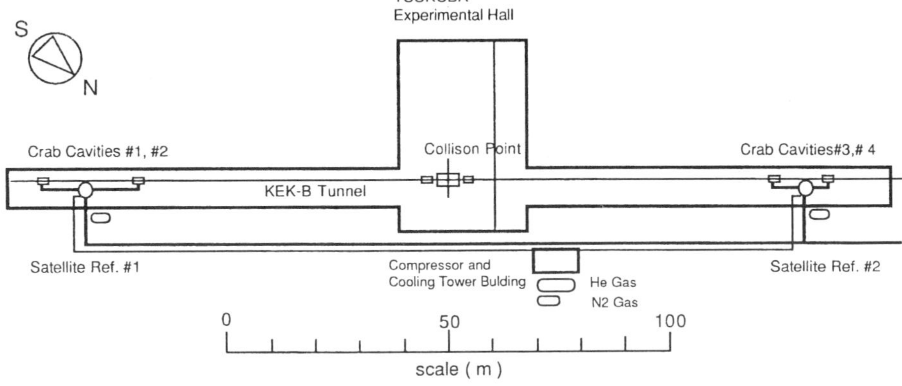


Figure 8. Layout of the KEKB crab cavities in the Tsukuba tunnel.

satellite refrigeration cooling scheme of case 4, only one way supply liquid helium transfer line will be required and return helium gas will be sent back to Nikko cryogenic system through room temperature line. The heat loss and the construction cost of the long transfer line will be reduced and flexibility of the system operation will be increase by this satellite cooling scheme. The satellite refrigerator scheme for helium refrigerator has been studied and used in Fermilab Tevatron cryogenic system[11,12] for cooling about 1000 superconducting magnets. This system have worked very stably for long time. The fundamental concept of the satellite refrigeration system for KEKB superconducting crab cavity is completely same to Tevatron system in Fermilab. Figure 9 shows the system flow of the satellite refrigerator for KEKB crab cavities. In our satellite scheme the heat exchanger and helium gas flow are used to utilize the cooling power of the cold gas but we do not use expansion engine because sufficient liquid helium about 150L/hr can be supplied from the large refrigerator in Nikko due to enough surplus cooling capacity of it. That is, all the cooling power required at satellite refrigeration systems in Tsukuba is produced by the large refrigerator which has high efficiency of the % Carnot of about 20% and very high reliability at Nikko. This satellite refrigerator scheme is very attractive compared to independent refrigeration system even if we take into account the heat loss of about 200W from 1km long transfer line, because we don not need expansion turbines and so that required helium gas flow rate of this system is less than half of the independent helium refrigerator and same time its configuration is very simple and the reliability of this system is very high. Two liquid nitrogen circulation systems will be used for 80K radiation shield of the crab cavity cryostats and liquid helium transfer line. The liquid nitrogen used in this system is supplied from Nikko cryogenic system through 80K radiation shield pipe of the 1 km long transfer line. We have started the system design of the KEKB satellite system in collaboration with industry.

## HIGH PERFORMANCE TRANSFER LINE

For construction of the KEKB cryogenic system for superconducting cavities, the high performance transfer line is a key component. The important factors for the KEKB transfer line is not only the performance of it but also the cost of it. Figure 10 shows the cross section of the typical transfer lines developed at KEK - (a) main transfer line of the cryogenic system for high beam current test in AR and (b) sub-transfer line for KEKB. In our transfer line design we adopted the 80K aluminum radiation shield cooled by liquid nitrogen to reduce the heat leak into helium line. The heat transfer between the radiation shield and the liquid nitrogen stainless steel piping is guaranteed by good fitting of it due to very precise dimension and fine surface of the groove made by extrusion molding. The whole structure of

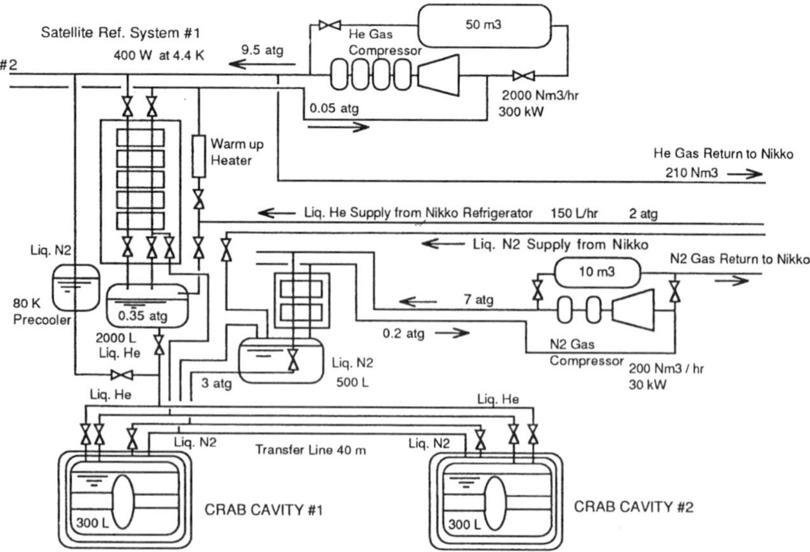


Figure 9. System flow of a satellite refrigeration system for KEKB crab cavities.

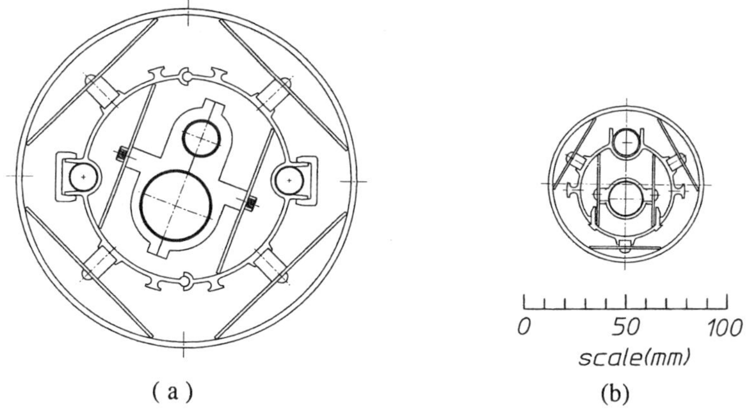


Figure 10. Cross section of (a) Main transfer line and (b) sub-transfer line.

the shield is designed to easy to assemble the transfer line. The helium lines are supported by the 80K shield with glass fiber reinforced polyester collars and G-10 plates. The 80K shield is supported by vacuum pipe with G-10 plates bolted to the shield. The helium pipes and the 80K shield are wrapped by 30 layers of aluminized polyester film with polyester interlayer net. The performance test of the main transfer line at 13 m long section of D10 cavity test stand showed the heat loss of the helium lines are 0.05W/m and 0.06W/m for supply and return line respectively. In the sub-transfer line shown in Fig.10 (b) a helium line and a liquid nitrogen line are integrated into one multi-transfer line. Figure 11 shows the joint part of this sub-transfer line. The helium line of part A joined downward through bayonet joint to that of part B and the liquid nitrogen line going upward and connected through bayonet joint on the same axis. By this configuration the length of the transfer lines, i.e. the distance between the connection box of header transfer line and cryostat, can adjust easily by changing the angles of the joint part.

## SUMMARY

The installation of the superconducting rf acceleration cavities and deflection crab cavities to KEKB have been proposed and the R&D studies have been carried out extensively at KEK. The construction of the superconducting HOMs damped acceleration cavities have been progressed for the installation into KEKB and commissioning in autumn 1998 in time for the commissioning of the KEKB. The installation and commissioning of the crab cavities

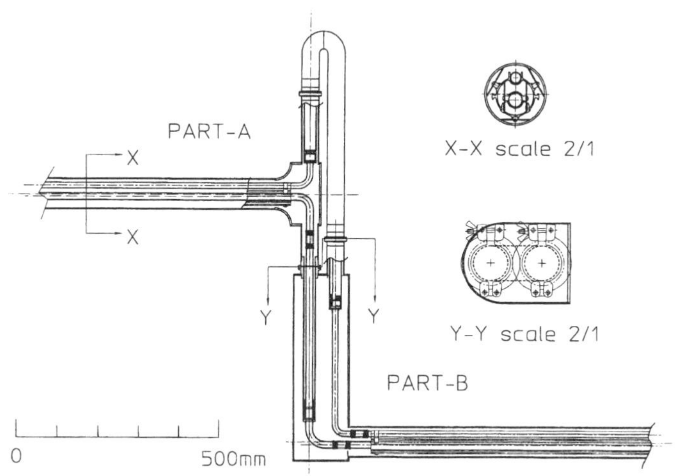


Figure 11. Joint part of sub-transfer line.

are scheduled in 1999. For the cooling of the superconducting HOMs damped acceleration cavities, existing cryogenic system constructed for TRISTAN superconducting cavities will be used. And for crab cavities cooling the satellite refrigeration scheme will be adopted. In this scheme liquid helium produced by large helium refrigerator at Nikko about 1km away from the satellite refrigerators will be supplied to the satellite refrigeration systems at Tsukuba through long transfer line. By using this supplied liquid helium the satellite refrigerators produce required cooling power for the crab cavities. We have started the system design of the satellite system and the R&D efforts of cost effective high performance transfer line to realize our superconducting cavities project.


## ACKNOWLEDGMENT

The authors wish to thanks Professors S.Kurokawa, E.Yamazaki and A.Ezura for their continuous support and encouragement and the staff of the superconducting cavities and cryogenic groups for their continuous support and many useful discussion. The authors also wish to express their thanks to Mr. K.Matsumoto of Hitach Ltd. for his process analysis of the satellite refrigeration system.


## REFERENCES


1. K.Hara et al., Cryogenic system for TRISTAN superconducting RF cavity, in: "Advances in Cryogenic Engineering", Vol.33, Plenum Press, New York(1988),p.615
2. K. Hosoyama et al., Cryogenic system for the TRISTAN superconducting RF cavities: performance test and present status, in: "Advances in Cryogenic Engineering", Vol.35, Plenum Press, New York, (1990), p.933
3. K. Hosoyama et al., Cryogenic system for TRISTAN superconducting RF cavities: upgrading and present status, in: "Advances in Cryogenic Engineering", Vol.37, Plenum Press, New York,(1992), p.683
4. K. Hosoyama et al., Cryogenic system for TRISTAN superconducting RF cavities: Description and Operating Experience, in: "ICEC 16 / ICMC Proceedings", Elsevier Science, New York,(1992), p.183
5. "KEKB B-Factory Design Report", KEK Report 95-7, August 1995, A
6. T. Furuya et al., Beam test of a superconducting damped cavity for KEKB, in: "Proceedings of Particle Accelerator Conferencce", May 1997 Vancouver
7. R. B. Palmer, SLAC-PUB 4707 (1988)
8. K. Oide and K. Yokoya, Phy. Rev. A40 315 (1989)
9. K.Hosoyama et al., " Crab cavity for KEKB", Proc. 7th Workshop on superconductivity (1995).
10. K. Akai, J. Kirchgessner, D. Moffat, H. Padamsee, J. Sears and M. Tigner, Proc. IEEE Part. Accel. Conf. (1993)
11. C. Rode et al., Energy Doubler refrigeration system, in: "IEEE Trans. Nucl.Sci., Vol.NS-27, No.3, (1977), p.1328
12. P.C.Vander Arend, Helium refrigeration system for Fremilab energy doubler, in "Advances in Cryogenic Engineering", Vol., Plenum Press, New York,(1977), p.420

# BEAM EFFECTS ON THE CRYOGENIC SYSTEM OF LEP 2

Philippe Gayet, Dieter Kaiser, and Gunter Winkler

LHC Division , CERN
CH-1211 Geneva, Switzerland

## ABSTRACT

The LEP collider was operated during 1996 for the first time with superconducting cavities at the four interaction points. During operation for physics it was observed that the dissipated heat in the cavities is not only a function of the acceleration gradient, but depends also on beam characteristics such as intensity, bunch length and beam current. These beam effects had not been foreseen in the original heat budget of the LEP cryogenic system.

The observations indicating the beam effect and its origin are presented. The available capacity of the refrigerators demonstrates that cryogenics might become a limiting factor for the performance of the LEP collider.

## INTRODUCTION

The CERN Large Electron Proton (LEP) collider is a 26.6 km circumference e+ e- storage ring. Originally equipped with a room temperature RF-system, it underwent an energy upgrade (LEP2) during the last years by installing superconducting (sc) cavities. At the end of the 1996 run, 176 installed sc cavities permitted operation for physics at 86 GeV per beam. During the shutdown 1996/1997 64 new cavities were installed.

The cryogenic system at each of the four acceleration points of LEP consists of up to 18 modules of 4 cavities each, and a cryoplant with an equivalent cooling capacity of 12 kW at 4.5 K. This power is required for cooling the sc cavities proper as well as various sub-systems. A summary of the refrigeration capacity available at the different points and the various sources of thermal load is given in Table 1, which shows that at 1996 operating conditions, about 6 kW were available for compensation of dynamic heat loads during operation of LEP.

*Advances in Cryogenic Engineering,* Vol. 43
Edited by P. Kittel, Plenum Press, New York, 1998

The main contribution to the dynamic heat load in a module is due to surface resistance of the sc material. The dissipated heat is proportional to the square of the acceleration gradient (figure 6). At the nominal gradient of 6 MV/m acceleration field the specification predicts losses of less than 280 W per module, which leaves a reserve of 55 W per module at the more heavily loaded cryoplants.

During last year, additional, unexpected dynamic losses were seen on the cryogenic system in the presence of beam. These losses, related to the beam intensity, particle bunch length and energy, show clearly both in the remaining cooling capacity available in the cold box and in the power dissipated in the modules.

This paper will review the different contribution and the foreseen solutions to achieve the maximum possible energy of LEP machine.

## HEAT LOAD MEASUREMENT PRINCIPLES

The cavities are cooled in a liquid helium bath at 4.45 K. Four cavities assembled in one cryostat form a module. Once filled, the modules have to be kept under stable conditions with minimum variation of helium level and pressure. For this task two feedback control loops and one feed forward calculation are performed.

The level is maintained with a PID control loop acting on the inlet valve. The pressure is kept constant at 1.25 bar with a PID control loop acting on the outlet valve. In order to achieve satisfactory stability, the changing dynamic RF loads must be compensated within the module. This is done using an electrical heater in each cavity, to which a pre-determined power (50-100W) is applied at zero RF. During operation, a feed-forward calculation reduces heater power as a function of the acceleration gradient in order to keep the total thermal load of the module constant.[1]

**Table 1.** Refrigeration capacity and heat load budget of 12kW cryoplant

| Specified duty | Power @4.5K |
|---|---|
| Cooling power @4.5K | 10000W |
| Liquefaction 13g/s (1g/s ≡ 125W @4.5K) | 1625W |
| Shield cooling 6700W @70K | 400W |

| | P2 | P6 | P4 | P8 |
|---|---|---|---|---|
| Measured cooling power @4.5K before 1997 tuning | 9000W | 9200W | 9200W | 9000W |
| | | | | |
| Load composition | | | | |
| Nb-sheet modules | 4 | | | |
| Nb-film modules | 12 | 16 | 18 | 18 |
| SC-quadrupoles | 2 | 2 | | |
| | | | | |
| Static heat load | | | | |
| Module connections (27W/module) | 432W | 432W | 486W | 486W |
| SC-quadrupoles (4g/s) | 500W | 500W | | |
| Transfer lines | 340W | 340W | 500W | 500W |
| Remaining power in pot for control | 400W | 400W | 400W | 400W |
| Static load of modules (80W/mod) | 1280W | 1280W | 1440W | 1440W |
| TOTAL | 2952W | 2952W | 2826W | 2826W |
| | | | | |
| Available power for dynamic RF-load | 6048W | 6048W | 6174W | 6074W |

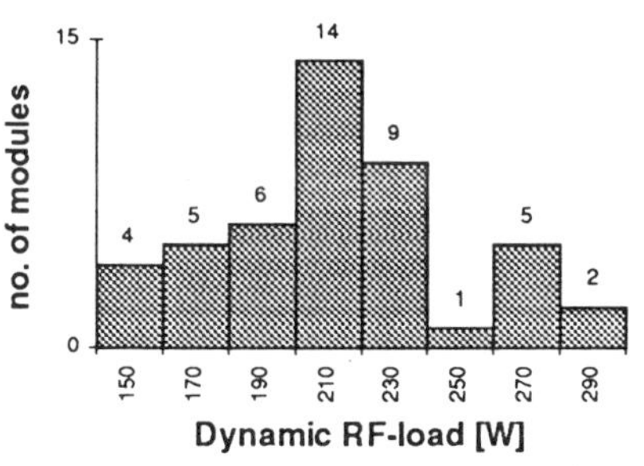


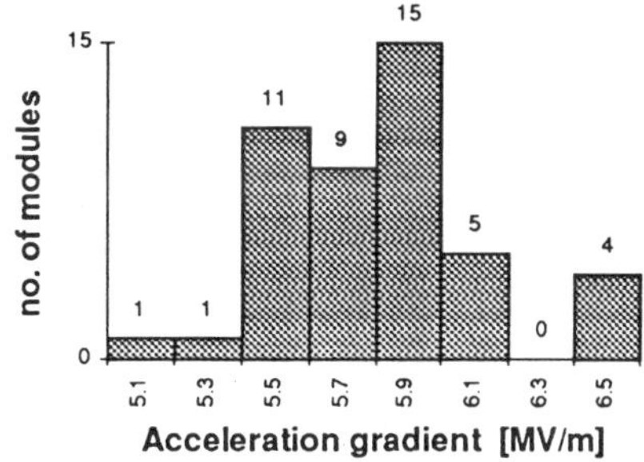


**Figure 1** Variations of dynamic RF-load and acceleration gradient measured before the 1997 run.

Each of the cryoplants runs in steady state. The high and low pressure sides of each plant are maintained constant by compensating load variations with an electrical heater. The helium is supplied to the modules at constant vapour quality, since the overall heat load of the plant is almost constant.

The inlet and outlet valve positions of the modules are therefore proportional to the heat loads and their variations provide a measurement of the power dissipation.

## DYNAMIC RF-LOAD MEASUREMENTS WITHOUT BEAM

The measurements done before the 1997 run show that the measured losses are generally lower than the predicted ones. However, the variations among modules are large. Figure 1 shows the distribution of RF-load and acceleration gradient over 46 modules. The average load is 210 W with an overall acceleration gradient of 5.7 MV/m instead of 240 W given by the specification for that gradient.

### Beam current

In Figure 2 it is shown that after a stabilisation period of the thermodynamic parameters following injection and increase of beam energy there is a strong correlation between heat load in the module and beam current. This was confirmed by similar data from a large number of modules.

The magnitude of the heat load variation induced by the beam current is about 15 W per module at a beam energy of 86 GeV and for a total beam current decreasing from 3.5 mA to 2.5 mA. It is proportional to the square of the beam intensity.

### Bunch Length

In November 1996, tests with variable bunch length and constant beam current were performed showing an increase from 30 W to 50 W of the power dissipated in the modules with a bunch length decreasing from 7.2 mm to 4.5 mm (Figure 3).

### Beam Energy

A preliminary cross-check done on different modules shows that at high beam energy, the variation in heat load with beam intensity is higher than at low beam energy. This result is consistent with temperature measurements on the bellows of the accelerator vacuum chamber outside the modules, which show a linear dependence on the energy.[2]

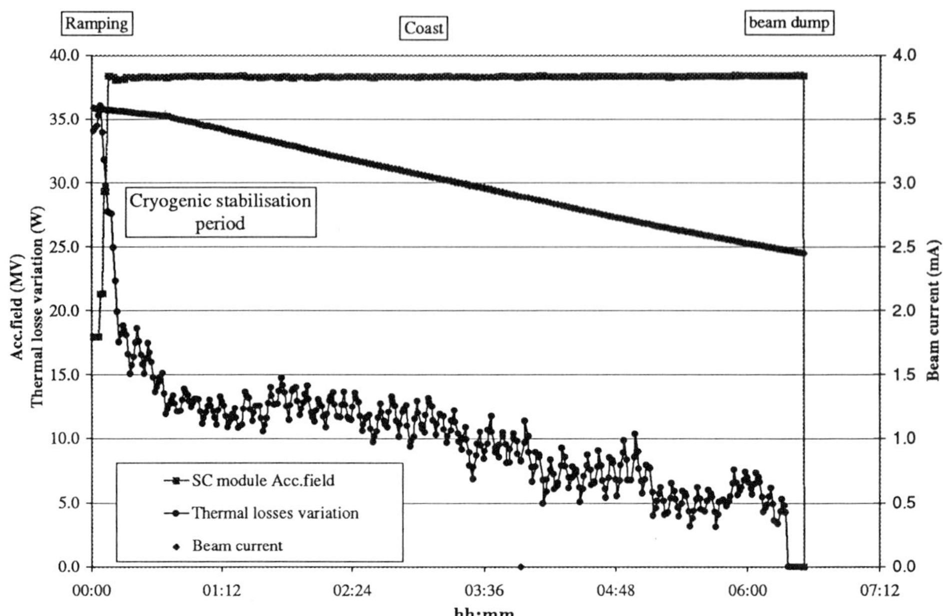


**Figure 2** Effect of beam current on heat load of module 6332.

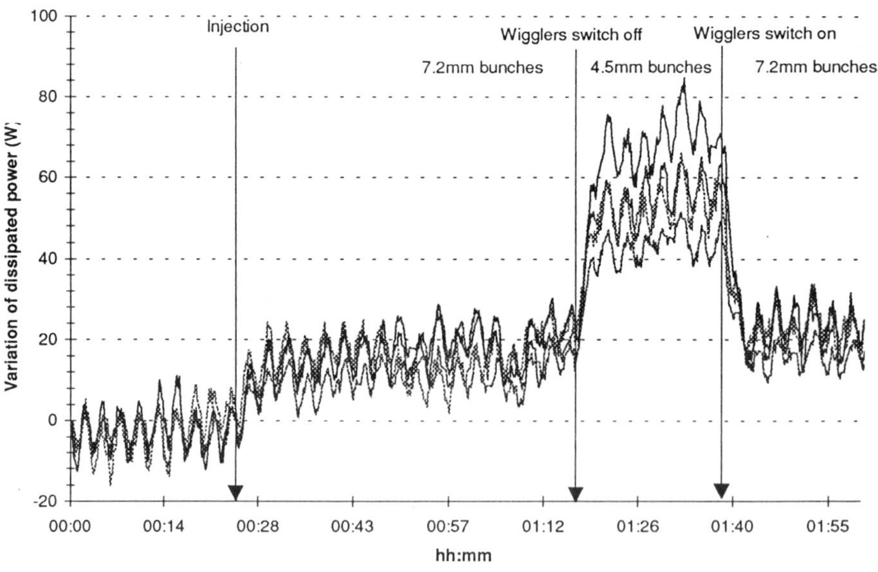


**Figure 3** Beam heat load on four modules with varying bunch length at 22 GeV without accelerating field.

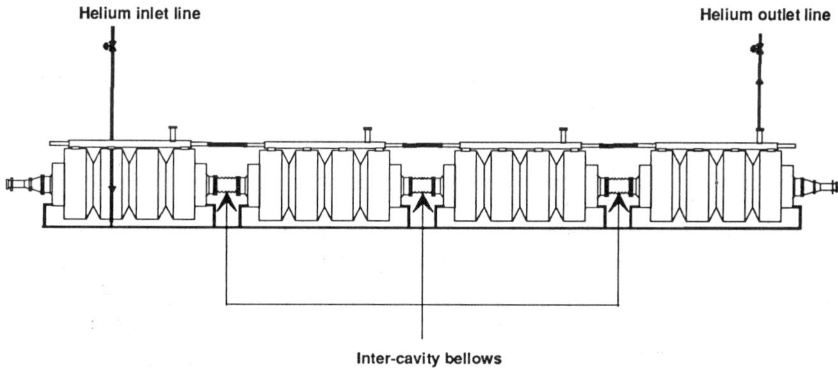


**Figure 4** Main components of the helium tank in a module.

## ORIGINS OF BEAM EFFECT

Thermometric measurements done inside the modules show that a beam load is detectable on the normal conducting bellows, which link together adjoining cavities in a module (Figure 4).

The power dissipation calculated from the temperature rise of the bellows shows the same general dependency on beam current as the module heat load seen by the cryogenic system (Figure 5).

The origins of this power dissipation are not yet completely understood. They may be caused by beam excited field, synchrotron radiation or radiation issued from other modules. A series of measurements will be made in the 1997 run to understand the causes and to scale the effect with the beam energy and total current.

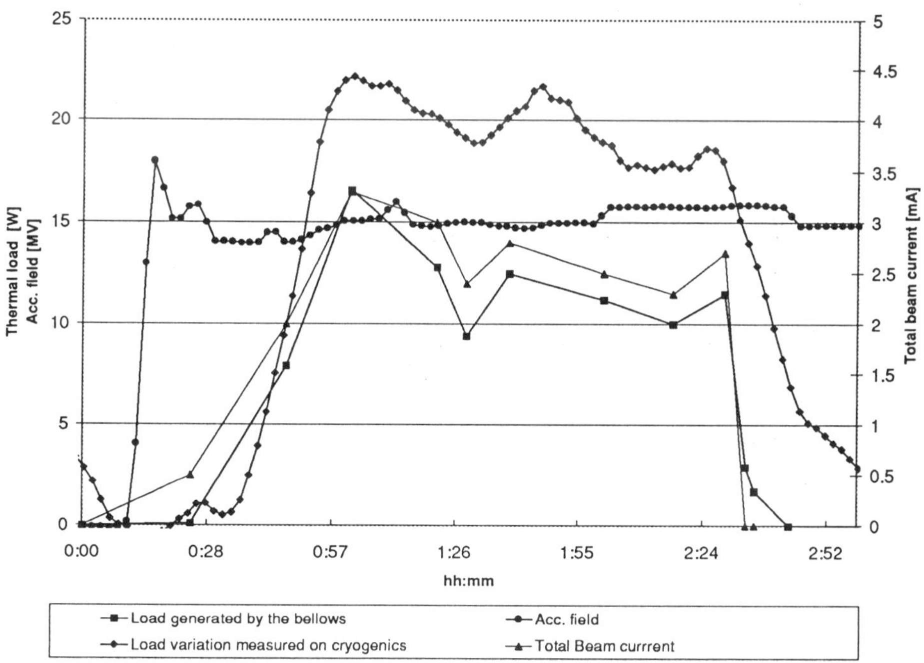


**Figure 5** Correlation between cryogenic load measurements and power dissipation on bellows.

## CONSEQUENCES

The unexpected heat loads on the superconducting cavities of the LEP2 collider produced by particle beams were observed at the end of the 1996 run, but could not be precisely measured. There are several possible mechanisms explaining the heat deposition in the modules, but a reliable prediction for losses at higher energies, higher beam currents or shorter bunches has not yet been made.[2]

In order to increase the limited reserve in cooling capacity, a tuning of the operating condition of the cryoplants was done before the 1997 run. By reduction of the outlet pressure of the last turbine from 2.5 to 2 bar and with a 2% increase of mass flow achieved by rising the low pressure of the cycle from 1.01 bar to 1.03 bar, which also allowed an increase of the high pressure of the cycle, a gain of 600W per cryoplant was achieved. Another 200 W were gained by improving the control parameters of the phase separator heater used to stabilise the remaining power, and by a reduction of cooling flow through ancillary RF-equipment.

However, the LEP2 program foresees a push to higher beam energy to the limit of the existing RF installation. This implies the use of accelerating gradients above 6 MV/m and the installation of four additional modules. Therefore a study of possible upgrades of the cryoplants was launched. Two alternatives are possible:

- a partial upgrade using the additional flow given by the recently commissioned redundant compressors, addition of an oil removal unit and replacement of some turbines,
- a complete upgrade using the additional flow given by the redundancy compressors and the addition of one extra compressor per cryoplant, and of an oil removal unit and the exchange of the turbines

The refrigeration power available to absorb dynamic heat loads is increased from 6.8k W to 10 kW in the first case and to 13kW in the second case. The upgrades would have to be implemented before the 1999 run.

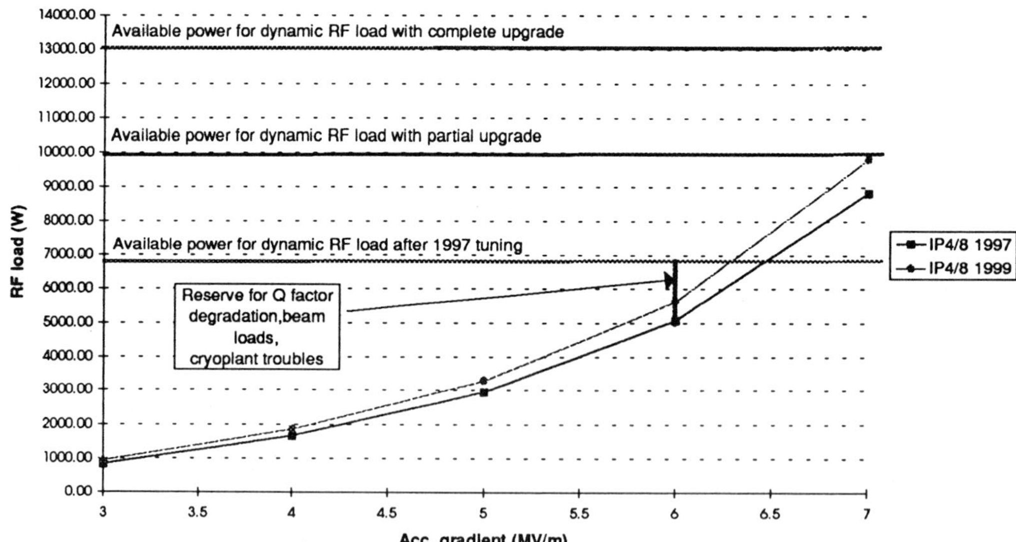


**Figure 6** Dynamic heat loads due to RF system and cooling capacity limits for P4 and P8 cryoplants .

Figure 6 shows the limitation of the existing plants by the higher loads at points 4 and 8 as well as possible upgrades. The curve show the variation of dynamic heat load as a function of acceleration gradient for the currently installed number of modules as well as for the 1999 configuration. The observed beam effects have not been added to the curves.

## REFERENCES

1. C. Wyss et al. LEP design report vol. III LEP2, CERN-AC/96-01(LEP2) (1996)
2. G. Cavallari, P. Gayet, G. Geschonke, M. Jimenez, S. Myers, Beam related thermal losses in the LEP SC RF system, CERN LEP2 note 97-40 (1997)

# A CRYOGEN-FREE 0.5 TESLA MRI MAGNET FOR HEAD IMAGING

J.M. van Oort, E.T. Laskaris, P.S. Thompson, B. Dorri, and K.G. Herd

GE Corporate Research and Development Center
1 Research Circle, Niskayuna, NY 12309

## ABSTRACT

A compact, cryogen-free 0.5 T superconducting Magnetic Resonance Imager (MRI) magnet for heads and limbs has been developed using conductively-cooled NbTi coils in conjunction with a 4.2 K Gifford-McMahon (GM) refrigerator.

The bell-shaped magnet configuration fits over the shoulders of the patient and provides a 20 cm spherical volume to image the head. The magnet is also suitable to image human limbs.

The lightweight magnet cartridge is designed for rapid cool-down, quench recovery, and field ramping, as required for clinical applications in small offices, intensive care units, and mobile vans. The NbTi coils are wound on a copper coil form, surrounded by a single aluminum thermal shield and a stainless steel vacuum envelope. The GM cold-head stages are in direct contact with the coil form and thermal shield through flexible connections to provide maximum vibration isolation.

Descriptions of the design, fabrication, and test results of the magnet are presented.

## INTRODUCTION

In MRI procedures, 50 % of the conventional examinations are of the head and 15 % of limbs. Nowadays, for MRI examinations of heads and limbs, regular whole-body scanners are used in most cases.

A few dedicated head, limb[1,2] or breast MR imagers exist, some of which are based on permanent magnet technology. A dedicated compact system with a field strength comparable to whole-body scanners[3] is the ideal solution for head and limb examinations in small offices, intensive care units, and mobile applications.

In order to achieve a compact and low-cost solution, a bell shaped configuration was selected for the system, as is illustrated in Figure 1. The larger inner diameter allows for

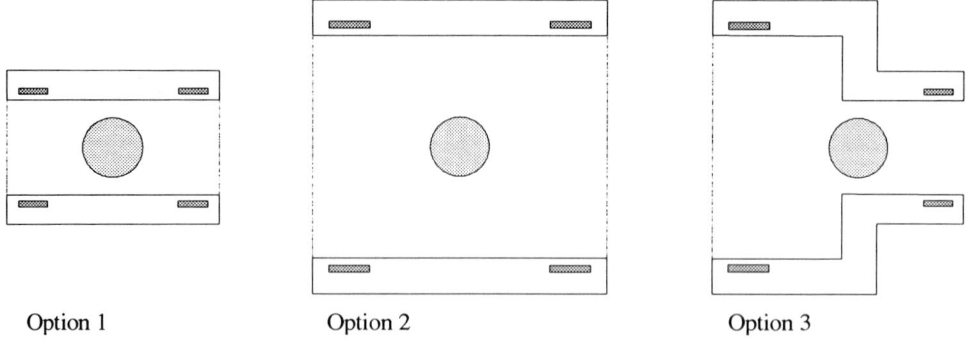


**Figure 1.** Overall system layout tradeoff. Option 1 has a small inner diameter to fit the head with a resulting short system length, but too long to clear the shoulder. Option 2 has a large inner diameter to fit both the head and the shoulder, but the system size becomes equivalent to a whole body system. Option 3 has a small diameter to clear the head, and a larger diameter at the shoulders. This results in a shorter and more economical system.

clearance around the shoulders of the patient, with the center of the 20 cm diameter spherical volume (DSV) at the transition to the smaller diameter.

The magnet and cryostat are supported by a cabinet that houses the cryo-cooler. The cryo-cooler is mounted on a plate which is bolted directly to the floor to minimize vibration. This configuration provides a small overall system size while placing the magnet bore at the right height for easy patient handling. An added benefit is the possibility to add sound insulation to minimize noise from the cold-head.

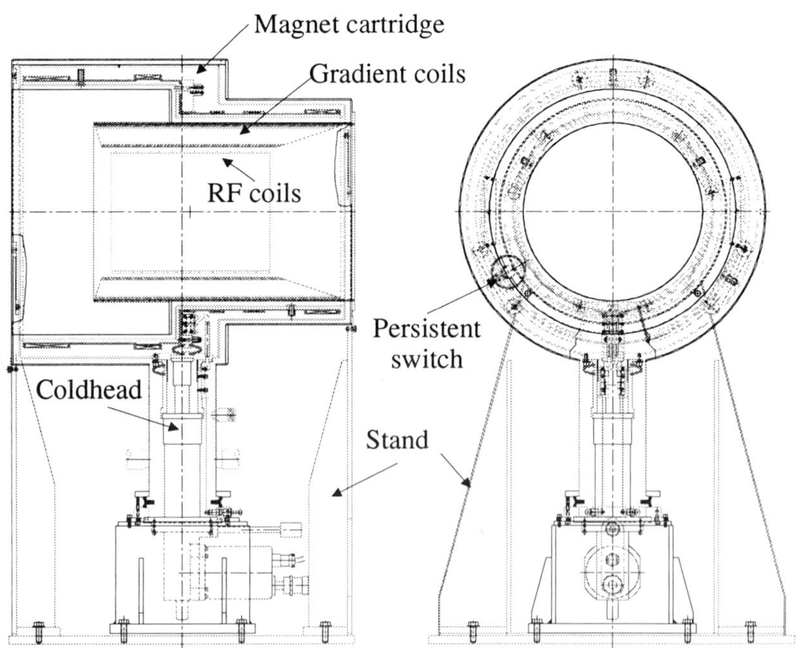


**Figure 2.** Detailed magnet system and support stand layout. The figure shows the location of the stand, cold-head, persistent switch, magnet cartridge, RF coils and gradient coils.

Table 1. Head magnet overall system design parameters

| PARAMETER | Value/Type |
|---|---|
| Field strength | 0.5 T |
| Superconductor | NbTi |
| Imaging volume | Sphere |
| Diameter | 20 cm |
| Inhomogeneity | < 10 ppm |
| Field stability | < 0.05 ppm/hr |
| Shimming system | Passive |
| Size | |
| Patient diameter | |
| Body | 56 cm |
| Head | 27 cm |
| Outside diameter | 70 cm |
| Length | 80 cm |
| Weight | 227 kg |
| Cooling | |
| Cryogen | none (conduction cooled) |
| Cool-down time | ≤ 2 days |
| Quench recovery | < 1 hour (from quench to next image) |
| Cryo-refrigerator | GM cryo-cooler |
| Stray field (5G line from FOV center) | |
| Axial distance | 3.75 m |
| Radial distance | 2.75 m |
| Room shield/RF shield | |
| Room width × height × length | 3 × 3 × 4 m |
| Room shield weight | < 2275 kg |

## SYSTEM DESIGN

The primary requirements for the magnet system are a 20 cm DSV imaging volume and comfortable positioning of the head. In order to achieve both requirements while keeping system size and cost at a minimum, a bell-shaped configuration is chosen. If one chooses a magnet bore size that just fits the head, the overall system length becomes too large. This results in the shoulders hitting the edge of the bore tube. A larger magnet bore solves this problem, but would essentially result in a system size equivalent to a whole body imager. Choosing an asymmetrical system that clears the shoulders of the patient but is smaller at the head allows for a shorter overall system length, lower conductor length, thus lower cost.

The magnet cartridge is built using two copper cylinders with different diameter. The two copper cylinders are joined with a flange at the middle. The two copper cylinders are thermally joined using a low temperature solder. A set of machine screws provides mechanical support to the solder joint. The magnet cartridge is cooled by conduction, with a single connection to the cold-head at the lower section of the copper flange. The overall system parameters are listed in Table 1, the system is illustrated in Figure 2.

## PATIENT HANDLING

The patient is positioned lying flat on an extension table which is placed in front of the large side of the magnet. Figure 3 illustrates the patient orientation for imaging heads. For applications as a limb imager for knees , the patient can be positioned with both legs inside

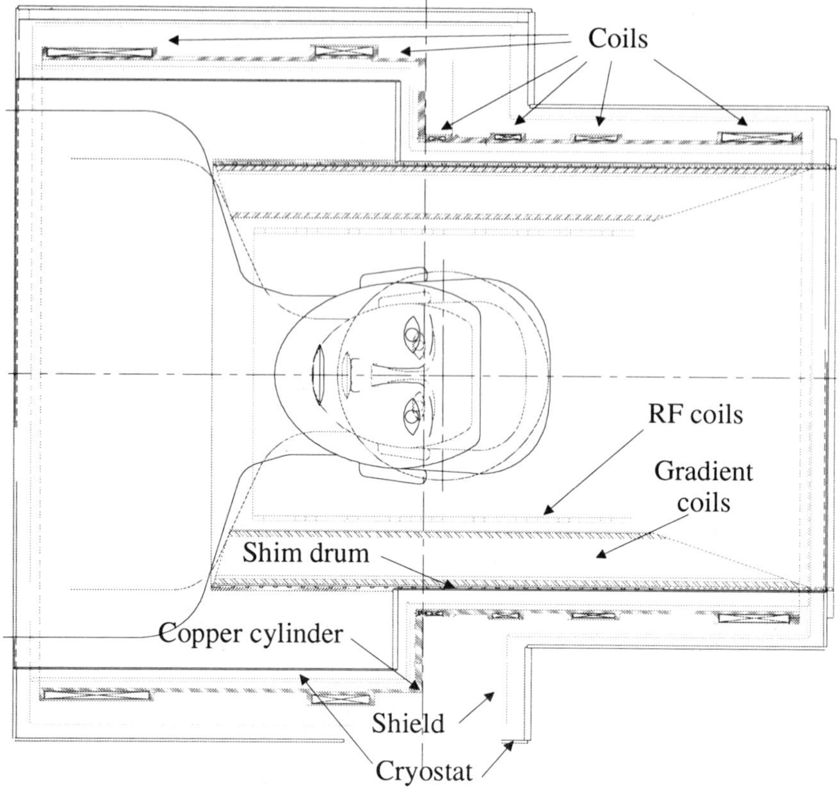


**Figure 3.** Patient orientation for use as a head imager. The shoulders clear the larger inner diameter on the left and rest against the gradient coil enclosure. Shown are a 99 percentile male and a 1 percentile female.

the magnet, or one leg in the bore of the magnet, and the other leg on the outside. For imaging elbows and wrists the arm can be placed within the imaging volume with the patient sitting in a chair next to the magnet.

## MAGNET DESIGN

The magnet cartridge is constructed using six coils, wound into fiberglass insulated coil grooves on the two copper cylinders. The coils are wound from a single length of 7890 filaments NbTi superconductor. The wire is an IGC 1.3:1 Cu/SC ratio SSC type conductor flattened to 0.0508 mm. A persistent switch is wound from the same superconductor. Two superconducting joints are made and positioned in a low field space. The coils are potted with a four part cryogenic stable epoxy.

Table 2. Magnet specifications

| Parameter | Value/Type |
|---|---|
| Number of coils | 6 |
| Total conductor length | 4814 m |
| Center field | 0.5 T |
| Conductor peak field | 1.41 T |
| Coil current | 158 A |
| Coil inductance | 2.65 H |

In order to minimize the heat leak to the magnet cartridge, BSSCO superconducting current leads are used. The current leads are stationed between the first and second stages of the cryo-cooler. The BSSCO leads are connected to copper pads on the first stage and flexible copper leads on the second stage.

The flexible copper leads are connected to the current leads on the coils via a copper pad. The copper pads are thermally anchored to first and second stages using nickel-plated beryllia (BeO) chips to provided electrical insulation. The copper cylinders are connected to the second stage of the cryo-cooler using indium sheet clamped tight with bolts to a flexible connection soldered to the cold-head.

The persistent switch is thermally anchored to the second stage of the cryo-cooler using copper rods.

The single radiation shield is constructed of 1100 aluminum. The shield is connected to the first stage of the cryo-cooler with flexible copper leads. All black holes are covered using adhesive backed aluminum foil. Forty layers of single sided Multi Layer Insulation (MLI) enclose the radiation shield.

The copper cylinders are supported by the aluminum shield using six reentrant glass-epoxy (G-10) tubes for radial support and two G-10 tubes for axial support.

The reentrant tubes provide adequate thermal stand-off within the confines of the magnet/vacuum envelope space. The shield is supported by the stainless steel vacuum envelope using six G-10 radial legs and two stainless steel axial supports. The magnet specifications are listed in Table 2.

## SHIMMING

The magnet is shimmed using passive steel shims placed on a Fiber Reinforced Epoxy (FRE) drum. The drum is made as a single piece and inserted between the inner cryostat wall and the gradient coils. There are 12 shim positions per ring, and 30 rings in the axial direction.

The length of the shim drum matches the length of the gradient coils. The maximum shim thickness per slot is 1 mm.

The optimum shim locations to minimize the field inhomogeneity are determined by a linear programming (LP) optimization algorithm[4,5]. An initial (virgin) field map is taken without steel shims in the shim drum. The field map used for the optimization algorithm uses 314 field points on a 20 cm diameter sphere. The points are spaced in circles of 24 points on 13 planes and two end points on the poles of the sphere. The pole points coincide with the longitudinal axis of the magnet. The mapping sequence starts at the first pole, then the second pole, and then alternating between the front and back planes inward to the central plane. A shim calibration file is generated analytically by computing the contribution of the magnetization of a single shim in each ring position to the field on the sphere.

Based on the measured virgin map and the computed calibration file, the optimum location of a set of steel shims is computed to minimize the inhomogeneity of the field on the sphere. The field inhomogeneity is minimized to a requested value while keeping the total volume of steel required at a minimum. The process generally has to be iterated between three to four times. This is due to the fact that interaction between shims is not taken into account in the calibration file, and exact placement of the shims is not guaranteed.

Figure 4 shows the initial inhomogeneity of 888 ppm in the field map before shimming and the final inhomogeneity of 8.7 ppm after four shimming iterations.

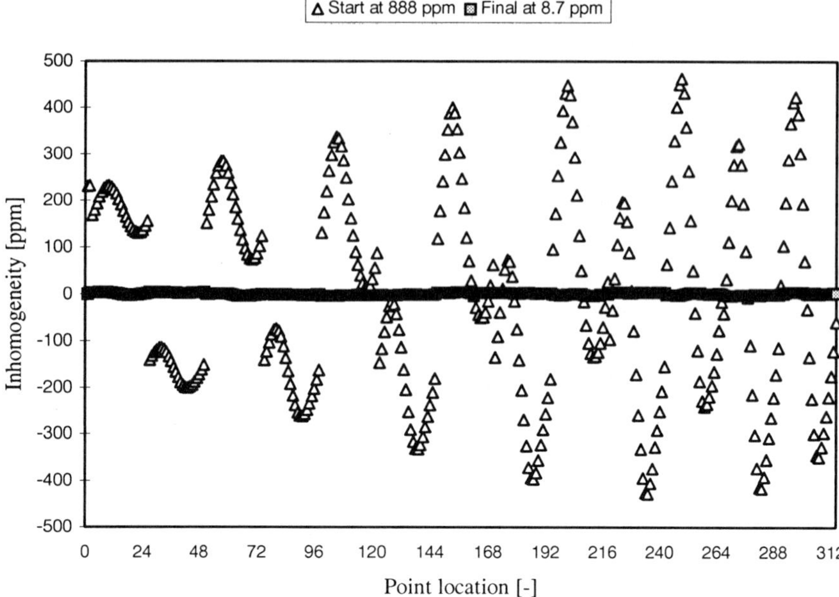


**Figure 4.** Initial and final shim iteration field inhomogeneity. The graphs show the measured inhomogeneity at the field points in the 314 points map. From left to right, the first point is the front pole location , followed by the back pole location. Thereafter, the front and back circles with 24 points each are plotted alternated, ending in the central plane of the sphere.

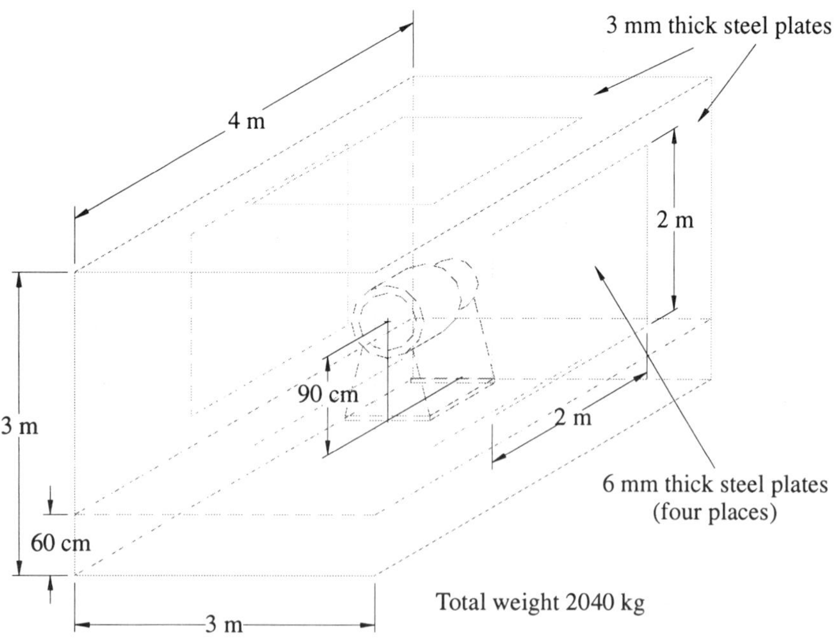


**Figure 5.** Room shield configuration with steel walls and additional plates.

Table 3. Magnet performance summary

| Parameter | Design goal | Measured value |
|---|---|---|
| Field strength | 0.5 T | 0.5 T |
| Magnet temperatures | | |
| NbTi coils | 6 K | 5.1 K |
| NbTi switch | 6 K | 5.2 K |
| Thermal margin to quench | > 1.5 K | 2.4 K |
| Thermal shield | 50 K | 44.8 K |
| Cryo-cooler first stage | 36 K | 35.4 K |
| Cryo-cooler second stage | 5 K | 4.8 K |
| Magnet drift rate | < 0.1 ppm/hr | < 0.01 ppm/hr |
| Cool-down time | 2 days | 1.5 days |
| Time to quench, cryo-cooler off | - | 3.5 min |
| Quench recovery time | 60 min | 45 min |
| Magnet power time | 60 min | < 20 min |
| Magnet weight | 227 kg | 273 kg |

## ROOM SHIELD

In order to protect the adjacent rooms from the stray field of the magnet, a steel room shield is required. The center of the imaging volume is positioned approximately 90 cm above the floor. The room shield is dimensioned to keep the 5 Gauss line at the wall in the longitudinal direction at $z = 1.5$ m, and at $r = 1.3$ m in the radial direction.

The required room shielding consists of room walls and ceiling shields of 3 mm steel, with an additional four steel plates positioned on the side walls, and on the floor and ceiling at the points closest to the magnet. The plates have a dimension of $1.80 \times 1.80$ m with a thickness of 6 mm. Figure 5 illustrates the location and size of the shield. The total weight is approximately 2040 kg.

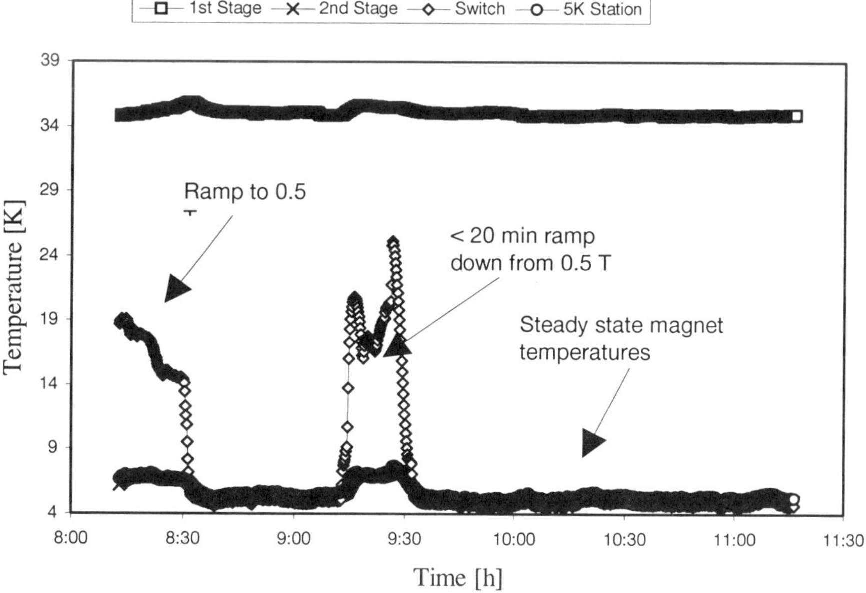


**Figure 6.** Temperature profile in a fast ramp-up to 0.5 T and ramp-down. Shown is the temperature increase in the system during the ramp-up to 0.5 T, up to 8:32, followed by the subsequent steady state temperatures. The ramp-down is started at 9:13, and ended at 9:32. The total time for ramp-down is less than 20 minutes, with the coils reaching a maximum temperature of 7 K.

## TEST RESULTS

Table 3 lists the performance of the magnet. The system cools down to the operating temperature in 1.5 days, exceeding the design value of 2 days. Also, the final coil temperature at 5.1 K is lower than the design goal of 6 K. The magnet ramps up to field within 20 minutes, with coil temperatures remaining below 7 K. The faster than predicted ramp speed is due to the lower coil temperature and associated increase in thermal margin. The switch reaches a maximum temperature of 25 K. The coils and switch reach the equilibrium temperature again after 6 minutes when the operating current of 158 A is reached.

Figure 6 shows the temperature profiles during and after ramping. Shown are the temperature rise in the first stage and second stage of the cryo-cooler, switch temperature and 5 K station temperature. During the ramp-up, the temperature in the coils remains steady at 7 K, while the switch temperature rises to 19 K. After reaching 0.5 T, the temperatures stabilize within 6 minutes.

Figure 7 shows the temperature rise and recovery during a thermally induced quench. The quench is induced by turning of the power to the cryo-cooler. After 3 ½ minutes, the magnet quenches, and the coil temperatures rise to 32 K. After turning the cryo-cooler back on, the temperatures are back to normal within 45 minutes. The total time from quench to the next image will thus be 1:05 hour (time for quench recovery, magnet ramp and temperature stabilization).

## SUMMARY

A compact, cryogen-free 0.5 T superconducting MRI magnet for heads and limbs using conductively-cooled NbTi coils in conjunction with a 4.2 K Gifford-McMahon (GM)

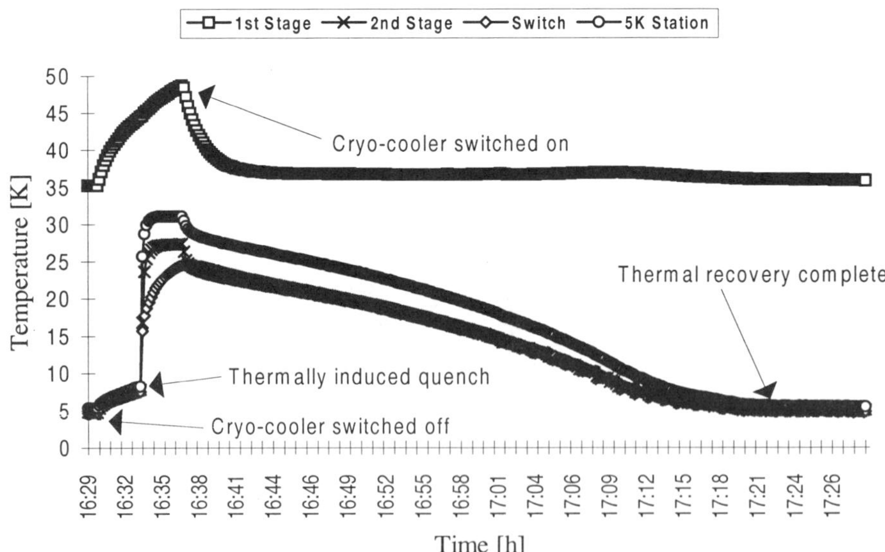


**Figure 7.** Temperature profile of a thermally induced quench. The graph shows the temperature at different locations in the magnet as a function of time. At 16:29:30 the cryo-cooler is switched of. The magnet quenches 3:30 minutes later, and after turning the cryo-cooler back on the temperature is stable after 44 minutes. The coils reach a maximum temperature of 32 K. The temperature of the shield rises to 49 K.

refrigerator has been successfully demonstrated[6]. The system can be cooled down from room temperature to 5 K in approximately 1 ½ days. Once at 5 K, the coils remain superconducting for 3 ½ minutes when the power to the cryo-cooler is interrupted. The system recovery time after a quench is 65 minutes. The magnet has an initial inhomogeneity of 888 ppm on a 20 cm imaging volume, and a final inhomogeneity of 8.7 ppm after four shimming iterations.


## ACKNOWLEDGMENTS

The authors wish to thank Michele Ogle for help on the coil optimization, Al Cooper for CAD support, Pete Morgan for technical support, and Harold Rothbell and Sam Walsh for shop support.


## REFERENCES


1 Laskaris, E.T., Herd, K.G. and Dorri, B., "A compact 0.8 T superconducting MRI magnet", *Cryogenics*, vol. 34, suppl. issue:65-8, 1994.

2 Laskaris, E.T., Dorri, B. and Herd, K.G., "Superconducting magnet for imaging human limbs", U.S. Patent 5,304,934, April 1994.

3 Laskaris, E.T., Ackermann, R. Dorri, B., Gross, D., Herd, K. and Minas, C., " A cryogen-free open superconducting magnet for interventional MRI applications", *IEEE Transactions on Applied Superconductivity*, vol. 5, no.2, pt.1:163-8, June 1995.

4 Dorri, B. and Vermilyea, M.E., "Method of optimizing passive shim placement in magnetic resonance magnets", U.S. Patent 5,045,794, September 1991.

5 Dorri, B. Vermilyea, M.E. and Toffolo, W.E., "Passive shimming of MR magnets: algorithm, hardware and results", IEEE Transactions on Applied Superconductivity, V3, N1, March 1992.

6 Dorri et al., "Over-shoulder MRI magnet for human brain imaging", U.S. Patent 5,416,415, May 1995.

# DESIGN, FABRICATION, AND TESTING OF THE ALISS SUPERCONDUCTING MINE-COUNTERMEASURES MAGNET

M. Heiberger [1], A.R. Langhorn [2], W.P. Creedon [1], S. Chapelle [1], R.A. Guzman [1], T.E. Harris [1], D. G. Morris [1], J. E. Medrano [1], J. Tannenbaum [3], M.E. Golda [3], T.H. Fikse [3], D. Waltman [3], G.F. Green [3], J.N. Chafe [3], J.D. Walters [3], C. Prenger [4]

[1] General Atomics, San Diego, CA 92186
[2] Startech Inc, Solana Beach, CA 92075
[3] Naval Surface Warfare Center, Carderock Division, Annapolis Detachment (NSWCCD), Annapolis, MD 21402
[4] Los Alamos National Lab, Los Alamos, NM 87545

## ABSTRACT

A conduction-cooled lightweight 7.5 MJ superconducting (SC) magnet capable of sustaining 100 g shock loads has been built and has undergone initial testing in support of the U.S. Navy's mine warfare program[1]. The solenoid magnet coil uses Cu:NbTi (4:1) superconductor wire and operates at a steady state temperature of less than 5 K while being cooled by two Gifford McMahon rare earth cryocoolers. The cryocoolers are detachable via a cold sleeve. Oxygen and ethane heat pipes thermally link the coil to the cryocooler's first stages to reduce cooldown time. The coil's cold-to-warm support structure comprises a series of re-entrant cylinders made of fiberglass reinforced epoxy composite. The vacuum vessel and thermal shields are constructed from formed aluminum isogrid panels which maximize the strength-to-weight ratio. The magnet uses grain-aligned YBCO high-Tc current leads operating between 30 K and 4 K, and the maximum design current is 147 Amps. The magnet has demonstrated stable operation up to 6.7 K at 60% of rated current. Operation of the magnet and its ancillary systems along with monitoring of all critical performance parameters is accomplished via a laptop computer. Additional tests are planned during which the magnet and its control system will be readied for participation in a joint Army-Navy Advanced Technology Demonstrator program involving mine sweeping exercises in 1998. During these exercises, the SC magnet system will be mounted on a small Naval craft designed to demonstrate new shallow water mine countermeasures technologies at sea.

## INTRODUCTION

Lightweight conduction-cooled SC magnets are being applied to the magnetic mine countermeasures effort under development by the U.S. Navy as part of the Advanced Lightweight Influence Sweep System (ALISS) program[1]. These devices are intended to trigger shallow water magnetic influence mines at safe stand-off distances from a small craft that carries the SC magnet and its auxiliary equipment. A key parameter that determines the effectiveness of a SC solenoid magnet in triggering a magnetic influence mine is the magnetic dipole moment, $M_d$:

$$M_d = N I \pi (r_i^2 + r_o^2) / 2, \quad (1)$$

where NI represents the coil's Amp-turns, and $r_i$ and $r_o$ are the coil's inside and outside radii respectively. It is the goal of the ALISS program to demonstrate a SC magnet design suitable for shipboard applications that is scalable in field strength to both larger and smaller devices. A design value for $M_d$ of 5 x $10^6$ Amp-m$^2$ was chosen as being the optimum value to demonstrate scalability. The planned at-sea demonstration will require 3 x $10^6$ Amp-m$^2$, and to date the system has been tested to a level of 3.3 x $10^6$ Amp-m$^2$.

The handling and use of bulk liquid cryogens is not considered practical for U.S. Naval applications, so the SC magnet is conduction-cooled using a pair of Gifford McMahon rare earth cryocoolers. In order to demonstrate robustness in a hostile environment, all structural elements and interfaces are designed to sustain loads and deflections from a 100 g static load equivalent from any direction. The system and its major components are illustrated in Figure 1.

## DESIGN TRADE-OFFS

A strong relationship between the magnetic dipole moment and coil diameter is obvious from Eq. (1). A larger coil diameter requires less Amp-turns which reduces current density, coil weight, and peak coil magnetic field, thus allowing the conduction-cooled coil to superconduct at a higher temperature. However, the larger diameter coil incurs an increasingly

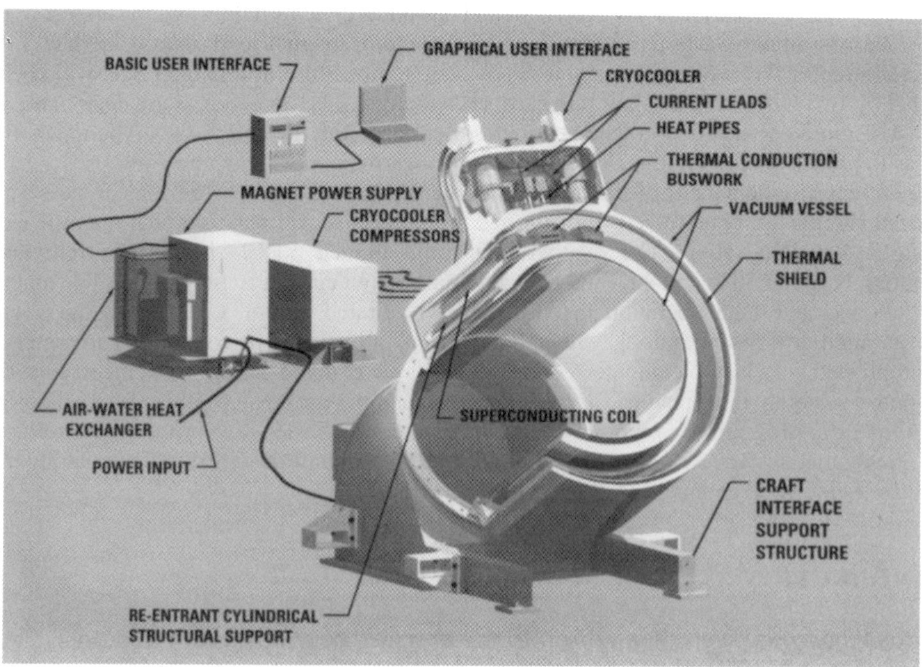


**Figure 1**. ALISS SC magnet system.

significant weight penalty from the vacuum vessel and other structural elements, which are also designed to withstand the high g loadings. The optimum coil diameter minimizes total cryostat weight.

The selection of coil design parameters to achieve a desired dipole moment is closely driven by the conductor critical transition temperature, $T_{crit}$, as a function of operating current and peak field. The theoretical relationship between these parameters can be determined based on conductor short sample data supplied by the manufacturer, IGC Advanced Superconductor, and extrapolations using equations from Ref. 2. Figure 2 is the load line for the coil design used and the conductor selected, where operating current density in the NbTi is designated $j_{op\ NbTi}$ At a magnetic dipole moment of 5 x $10^6$ Amp-$m^2$, the predicted critical transition temperature is 6.35 K.

For a given design, the coil's operating temperature can be estimated knowing the characteristics of the cryocooler used, the conductances of the heat paths in and around the coil, and the heat leaks into the coil. In order to ensure stable operation of the SC magnet, a design temperature margin between the value of $T_{crit}$ and the warmest point on the coil of 1 K was selected. Therefore, the temperature of the coil in the region of the highest field must not exceed 5.35 K. After iterating the design by exercising the relationships between these parameters, while striving to minimize the total weight and maintain acceptable Lorentz stresses and thermal-induced quench stresses within the coil, the coil design parameters were chosen. The final as-built coil design parameters are listed in Table 1.

**Table 1.** Key ALISS SC magnet design parameters

| | | | |
|---|---|---|---|
| Max. design dipole moment | 5 x $10^6$ A-$m^2$ | Peak field in coil at 147 Amps | 4.85 T |
| Nominal dipole moment | 3 x $10^6$ A-$m^2$ | Max. stored energy | 7.46 MJ |
| Max. Operating current | 147 Amps | Coil inductance | 690 H |
| Number of coil turns | 18,640 | Predicted coil operating temp. | 5.25 |
| Coil inside diameter | 148.4 cm | Predicted $T_{crit}$ at peak field | 6.35 K |
| Coil length | 44 cm | Shock load static equivalent | 100 g |
| Coil thickness | 4.1 cm | Vibration (broadband spectrum) | 2 g |
| Conductor size | 0.7 x 1.2 mm | Coil mass with epoxy | 634 kg |
| Conductor composition | NbTi:Cu (4:1) | Total cryostat mass incl. mount | 2270 kg |

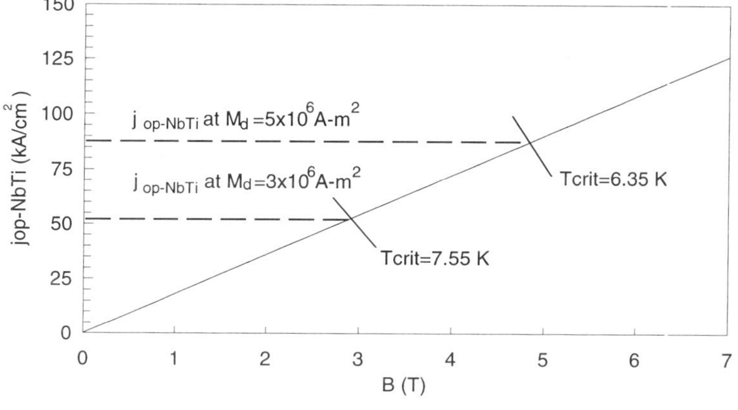


**Figure 2.** Coil load line indicating predicted values of $t_{crit}$.

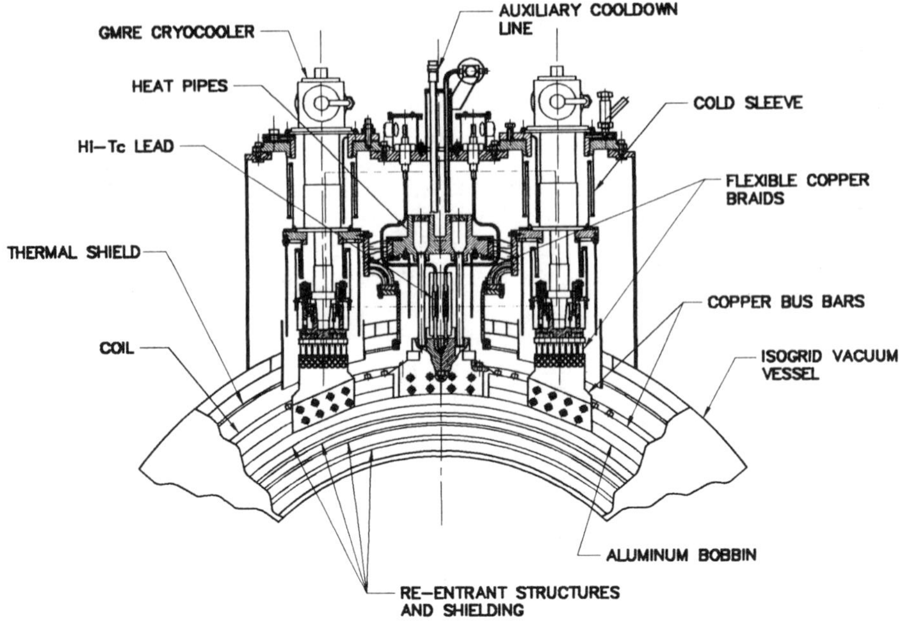


**Figure 3.** ALISS cryostat tower region showing cryocooler and thermal bus interfaces.

## DETAILED DESIGN

The ALISS magnet assembly consists of the SC magnet cryostat and its support pedestal, cryogenic refrigeration support equipment, and the power supply and controls components (Figure 1). Additional details describing the thermal interfaces in the tower region are shown in Figure 3 and details of the cold-to-warm-support (CTWS) structure are shown in Figure 4.

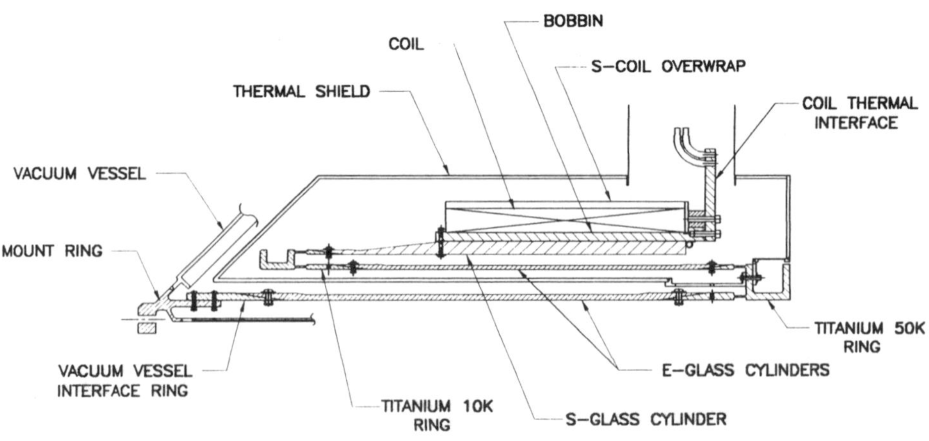


**Figure 4.** Cross section through ALISS cryostat showing the cold-to-warm support structure details.

## SC Coil

The coil was wet-wound from NbTi wire with a Cu-NbTi ratio of 4:1. The wire was manufactured by IGC Advanced Superconductor, and short sample tests on the wire performed by them indicated a typical critical NbTi current density of 285 kA/cm$^2$ at 4.2 K and 5 T. The epoxy used was CTD-521™, with 40% by weight silica filler. The use of filler is believed to strengthen the epoxy matrix and reduce the possibility of cracking. Table 1 lists some of the key coil parameters. The coil required about 87,000 meters of wire. Four splices were made inside the coil, plus two for the lead-in and lead-out. The wire was wound over a 6061-T6 aluminum bobbin which was adhesively bonded to the outer re-entrant S-glass cylinder (Figure 4) which makes up part of the CTWS. The aluminum bobbin contains an eddy current break, which consists of an Inconel interlocking "tongue-in-groove" style splice joint running the length of the bobbin. A 0.15 mm thick sheet of epoxy-impregnated fiberglass cloth was placed between the Inconel and the aluminum for electrical isolation. Without the eddy current break, several watts of resistive heating would result from eddy current flow in the bobbin while the magnet was being charged or discharged. The aluminum bobbin is needed to help distribute thermal energy across the coil during a quench and also to improve the thermal conductance in the path between the coil and the cryocoolers. After the coil was wound on to the aluminum bobbin-CTWS structure assembly, it was wet-overwrapped with about a 1.6 cm thick layer of highly tensioned S-glass fiber in order to drive the coil further into precompression. This step was necessary in order to prevent the coil from delaminating from the bobbin and becoming overstressed from Lorentz forces after it is energized. The completed coil-CTWS assembly is shown in Figure 5.

The coil SC current leads are connected to a shock and vibration hardened Hi-Tc lead assembly fabricated from grain-aligned YBCO. These leads interface with conventional copper leads at a thermal junction close to the cryocooler first stage (30 K typical operating temperature). The copper leads feed through hermetic plugs at the vacuum vessel boundary at the top of the tower.

## Structures

The structures for this system are required to withstand 100 g loads from any direction. Earlier studies conducted under the program have concluded that the re-entrant cylinder CTWS structural support method shown in Figure 4 offers the lowest structural weight impact compared with other methods meeting this requirement. The cylinders were wound from glass fiber in a lay-up pattern that was optimized to give the best combination of strength and thermal expansion coefficient, with special attention given to the joint interfaces. The interfaces between the cylinders and titanium rings shown in Figure 4 use a scarf joint that is adhesively bonded and riveted. The fiber lay-up in this zone was optimized to match the thermal expansion coefficient as closely as practical while maintaining adequate stiffness. These types of joints were structurally tested earlier in the program at cryogenic temperatures and were found to provide a suitable design margin under all anticipated load conditions, including fatigue[3].

The thermal shield and the vacuum vessel were fabricated from aluminum alloy 1100-H14 and 2219-T851 respectively. In order to minimize weight, these components were machined into an isogrid configuration. The components started out as flat panels of aluminum plate. The isogrid pattern was machined on a CNC mill. Next, the flat panels were brake-formed to the required shape and then welded to form subassemblies (Figure 6).

## Cryogenic Refrigeration

The magnet is cooled using two Gifford McMahon rare earth (GMRE) cryocoolers built by Sumitomo Heavy Industries Ltd. The coolers have been performance-mapped at NSWCCD Annapolis and each have produced 1 W of cooling power with the second stage running at 4 K,

**Figure 5.** Completed coil-CTWS structure.

while the first stage produced 23 W of cooling power at an operating temperature of 32 K. The cryocooler performance and the design heat loads on the magnet permit the coil temperature to run cool enough for stable superconducting operation with ample thermal margin. At full current, the predicted heat load on each first and second stages are 28 W and 0.7 W respectively, with the first stage temperature assumed to be at 40 K. Actual results are discussed below.

**Figure 6.** Finished isogrid vacuum vessel subassemblies.

Table 2. Key temperatures and heat loads within the ALISS cryostat

| | **TEMPERATURE (K), HEAT LOAD (W)** | |
|---|---|---|
| | **0 Amps** | **97 Amps** |
| Cryocooler first stage temp.* | 28.6 | 29.0 |
| Cryocooler second stage temp.* | < 3.5** | 3.9 |
| Average thermal shield temp. | 42.8 | 42.9 |
| Coil cold end temp. | 4.0 | 4.5 |
| Coil warm end temp. | 4.9 | 5.3 |
| Est. 1st stage heat load (each) | 19 | 21 |
| Est. 2nd stage heat load (each) | < 0.05 | 0.9 |

* There was no significant difference in operating temperatures between the two cryocoolers.
**Sensors are not calibrated below 3.5 K.

The cryocoolers interface with the magnet via cold sleeves that are designed to permit easy cryocooler replacement. Figure 3 shows details of the first and second stage cryocooler interfaces. The cold sleeve is designed so the second stage engages first, and a mechanism at the bottom of the cold sleeve deflects downward and squeezes the cooler's cold foot as downward pressure on the cooler is applied[4]. The cryocooler hard-stops when it seats at the first stage. The second stage interface surfaces consist of indium-plated copper and relies on precision machined surfaces and high pressure for good thermal contact. Conductance measurements made at NSWCCD Annapolis across this interface have indicated values of 4.4 W/K at 4.7 K and 6.6 W/K at 7.0 K. The first stage interfaces also use indium-plated copper on both mating surfaces, but in addition, a 0.5 mm thick indium gasket is used between them. Conductance measurements made on the first stage interface indicated values of 32 W/K at 40 K and 25 W/K at 50 K.

To improve the cooldown rate, four heat pipes connect between the coil thermal bus and the first stage of both cryocoolers (Figure 3). Two are filled with ethane gas pressurized to 4.14 MPa (600 psi) at room temperature, and two are filled with oxygen gas pressurized to 13.9 MPa (2000 psi) at room temperature. The heat pipes thermally short out the coil to the cryocooler first stages during cooldown from room temperature, thereby making the first stage power available to the coil mass. Two different gasses were chosen to improve the collective heat pipe performance over a wider range of temperatures[5]. When the coil temperature drops below about 50 K, the heat pipes freeze out, thermally isolating the coil from the first stage. The use of heat pipes reduces the cooldown time from room temperature by an estimated 50%.

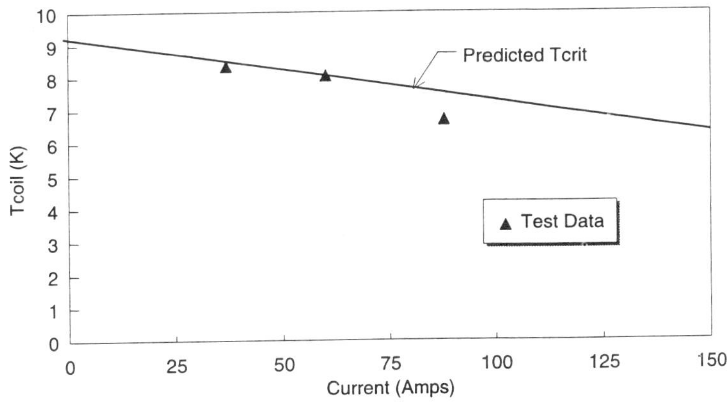


**Figure 7.** Coil current vs $T_{crit}$, theoretical and data.

## FACTORY TEST RESULTS

The magnet reached suitable operating temperature in about 200 hrs from the start of cooldown. Initially, the cryocooler cold head speed was increased by 60% over the normal operating speed to reduce the cooldown time. An array of Cernox™ resistors and GaAlAs diode thermal sensors monitored all key temperatures in the cryostat. The cryocooler speed was returned to normal when the coil cooled down to about 10 K. Prior to energizing the magnet for the first time, the warmest point on the coil measured 4.8 K and the average temperature of the thermal shield and structural thermal intercept was at about 43 K. Table 2 lists the key temperatures inside the cryostat, along with heat load estimates based on the cryocooler performance maps established earlier by NSWCCD. The higher first stage temperatures recorded at 97 Amps were the result of the added heat input from the Hi-Tc current leads, residual eddy current and hystersis losses, and a slight degradation in cryocooler performance due to the ambient magnetic field effects on the cryocooler regenerator.

The coil was energized to a magnetic dipole moment level of 3.3 x $10^6$ A-$m^2$ (97 Amps) with no training quenches. At the nominal operational level of 3 x $10^6$ A-$m^2$, the cryocoolers were turned off and the magnet was allowed to warm up and quenched. When the cryocoolers were shut off, the coil temperature at the warmest point rose from 5.2 K to 6.7 K in 11.4 min. A thermal sensor located near the warmest and highest field point in the coil recorded the temperature at the moment the quench initiated. This point is plotted in Figure 7 along with data from other test runs.

Subsequent magnet tests were performed. By allowing the temperatures to fully stabilize in the cryostat, the coil temperature can drop to as low as 3.9 K at its warmest point at zero current. The maximum charge rate of the magnet has been determined to be about 1.5 Amps/min. Attempts to charge faster sometimes resulted in quenches due to coil heat-up from internal eddy current and hysterises losses. Tests have also been conducted using quench heaters that are installed in the magnet that are able to force a magnet quench if an emergency discharge is ever needed. If an emergency shutdown is required, the heaters can be activated locally by engaging a switch mounted near the cryostat which diverts some of the magnet's current into the quench heaters.

The ALISS magnet has proven that conduction cooling of relatively large magnets using NbTi conductor is feasible using GMRE cryocoolers. The magnet is able to operate at the desired field levels with ample thermal margin. The system will be shipped to the Navy Mine Warfare test site in Panama City, Fl, where it will be integrated with a sea-going craft for tests planned in 1998.

## ACKNOWLEDGMENTS

The ALISS project acknowledges the efforts of P.S. Associates of San Diego for their work in isogrid structural analysis and design.

## REFERENCES

1. Program supported by the DOD, Naval Surface Warfare Center, Carderock Division, Annapolis Detachment, contract N61533-95-C-0080 with General Atomics.
2. R.X. Maix, D. Salathe, "Practical Scaling Formulas for the Determination of the Critical Currents in NbTi Superconductors, "Proceedings at the 9th International Conf. on Magnet Technology, Zurich, Switzerland", 1985
3. M. Heiberger, et al. "A Lightweight Rugged Conduction-Cooled NbTi Superconducting Magnet for U.S Navy Minesweeper Applications", Advances in Cryogenic Engineering, Volume 41, Plenum Press, New York, 1996.
4. A.R. Langhorn, J.D. Walters, M. Heiberger, "Development and Testing of a Demountable Cryocooler Thermal Interface", Proceedings from the 9th International Cryocooler Conf., Waterville Valley, New Hampshire, 1996.
5. C. Prenger, et al. "Heat Pipes for Enhanced Cooldown of Cryogenic Systems", Proceedings from the 9th International Cryocooler Conference, Waterville Valley, New Hampshire, 1996.

# MANUFACTURING OF LIQUID HELIUM FREE SUPERCONDUCTING MAGNETS FOR INDUSTRIAL USE

T.Kobayashi, Y.Sato, T.Sasaki, and S.Mine

Keihin Product Operations
Toshiba Corporation
Tsurumiku, Yokohama, Japan

## ABSTRACT

About ten kinds of liquid helium free superconducting magnets (HFSM) have been fabricated in the last three years by Toshiba. These magnets are used for research, magnetic separation and a single crystal growers. We have shipped about thirty sets of HFSMs. The cold mass which is cooled by GM refrigerator down to 4 K level has been increased up to nearly one ton recently.

## INTRODUCTION

In the last three years, about thirty superconducting magnets without liquid helium have been manufactured and shipped. Most of them are installed in universities and national laboratories, and are used for various studies with high magnetic fields. Some magnets are also for industrial use, where the are expected to reduce thermal convection of melted materials and so on. Helium free superconducting magnet can eliminate complicated liquid helium handling and helium gas exhaust piping. These eliminations just meet user's expectations.

This paper shows some typical HFSMs which can generate 0.31, 4, 6 and 10 Tesla magnetic fields. We also discuss some issues on future magnets.

## MANUFACTURED HFSM

Table 1 summarize our typical HFSMs. Quantity of the manufactured magnets are three for TM-6 and four for TM-4. Besides the Table 1, seven large bore (1300 mm in diameter) HFSMs for single crystal growers and five small HFSMs which have each three small right angled bores for physics experiments have been made. Moreover there are several special uses of HFSMs. One is the 2 meters long 4 Tesla dipole magnet for particle

**Table 1.** Main parameters of typical HFSMs

| | Mag. type | | TM-10 | | TM-6 | TM-4 | TM-0.31 |
|---|---|---|---|---|---|---|---|
| Dimension | Height | mm | 470 | | 600 | 505 | 600 |
| | Width | mm | 700 | | 875 | 1000 | 1000 |
| | Length | mm | 470 | | 550 | 1060 | 1060 |
| | Operation space | mm | 50 dia. | | 180 dia. | 400 dia. | 600x600x590 |
| Weight | 4K part weight | kg | 70 | | 260 | 190 | 110 |
| | Thermal shield | kg | 20 | | 30 | 30 | 60 |
| | Vacuum vessel | kg | 150 | | 200 | 430 | 500 |
| | Total weight | kg | 240 | | 490 | 650 | 670 |
| Coil | Super conductor | | NbTi | Nb3Sn | NbTi | NbTi | NbTi |
| | Central field | T | 6 | 4 | 6 | 4 | 0.31 |
| | Max. field | T | 6.6 | 10.4 | 6.4 | 5.81 | 5.9 |
| | Operation current | A | 132 | 132 | 135 | 118 | 95 |
| | Inner dia. | mm | 170 | 90 | 220 | 450 | 290 |
| | Outer dia. | mm | 240 | 157 | 400 | 532 | 400 |
| | Length | mm | 240 | 130 | 351 | 266 | 60 |
| | Stored energy | kJ | 183 | | 392 | 710 | 217 |
| | Current density | $A/mm^2$ | 184 | 126 | 71 | 166 | 213 |
| | Number of turns | | 11704 | 4148 | 16560 | 15340 | 7392x2 |
| | Inductance | H | 21 | | 43 | 102 | 48 |
| | Critical temperature | K | 5.5 | | 5.8 | 6 | 6 |
| | Ramp time | min. | 60 | | 60 | 60 | 60 |
| Heat leak (Design) | 4K Part | | | | | | |
| | Excitation | W | 0.41 | | 0.29 | 0.4 | 0.21 |
| | Non excitation | W | 0.13 | | 0.13 | 0.15 | 0.16 |
| | 40K Part | | | | | | |
| | Excitation | W | 17.3 | | 19.2 | 18.4 | 18.7 |
| | Non excitation | W | 10.7 | | 12.6 | 12.5 | 14.8 |

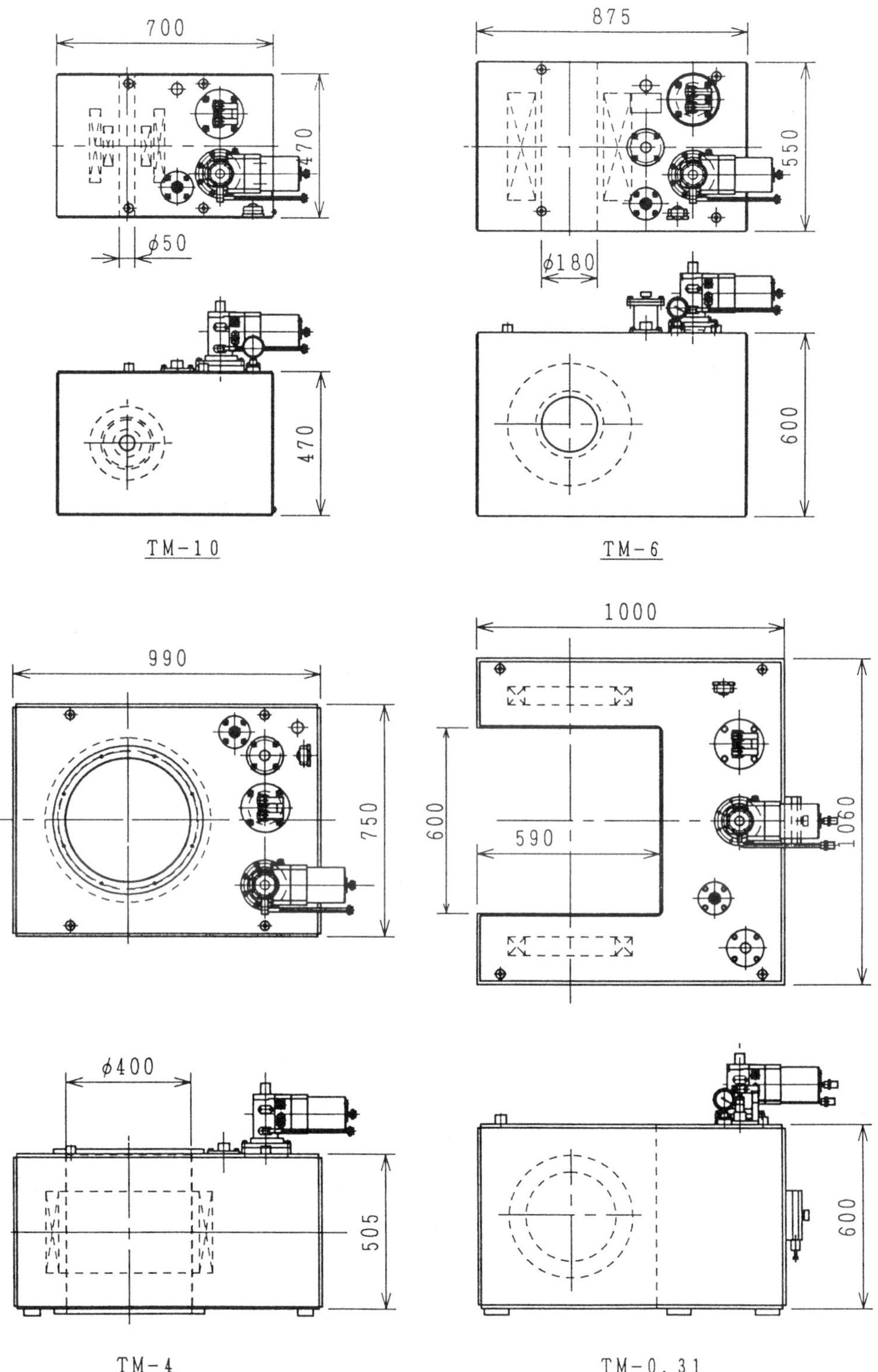


Figure 1. Outer shape of typical HFSMs.

**Figure 2.** Outer view of TM-4 and its application to the magnetic separation system.

**Figure 3.** The HFSM applied to the crystal growther (TM-0.31).

physics which is operated in persistent current mode. The other is a split magnet which will be set on the right and left sides of single crystal growers to suppress the thermal convection in liquid silicon. The single crystal pulling method is common for large sized crystal production named as MCZ (Magnetic field applied Czochralski method).

Figure 1 shows the outer shape of HFSMs which are listed in Table 1. Outer views of some HFSMs are shown in Figure 2 and Figure 3.

(1) TM-10

This system consists of a NbTi coil and a $Nb_3Sn$ coil and generates 10 T at the center of a 50 mm bore. This is the first 10 T HFSM. We confirmed the so called Moses effect using the horizontal bore. Recently our laboratory members achieved 11.5 T at 100 mm bore with the same kind of HFSM[1].

(2) TM-6

This coil is wound with two kinds of rectangular NbTi wires. Cold mass for the refrigeration load is about 260 kg, which was the heaviest HFSM at that time. Thus cooling time from room temperature without liquid $N_2$ is about 10 days. The pre-cooling tube for liquid $N_2$ was effective to shorten the cooling time.

(3)TM-4

The magnet has a wide bore (400 mm in diameter). It generates a central field of 4 T and is used for magnetic separation. The cooling time from room temperature was a few days with liquid $N_2$ pre-cooling.

(4) TM-0.31

This magnet has a large open space 600 mm x 600 mm x 590 mm for easy approach to the magnetic field. The field of 0.31 T at the center of this space is generated by a split coil. This magnet is used for the research of MCZ silicon single crystal fabrication.

## DESIGN CONCEPT

There are various types of HFSMs, but the composition and the design concept are same. Figure 4 is schematic cross-section of HFSM. In this case a NbTi coil is used, but if high magnetic field (>7 T) is required, a NbTi coil and a $Nb_3Sn$ coil are assembled.

Coils are designed to keep below about 6 K of the normal transition temperature under the operating current and field. To suppress the temperature rise during charging, finer filament superconductors are preferable. We have experience using 80 μm filament diameter wire.

The cooling of the coil is made as follows. The HFSM is cooled by a two stage G-M refrigerator which uses magnetic regenerative materials to achieve 4 K. Recently the refrigerator attained 20 W for 1st stage and 1 W for 2nd stage. This G-M refrigerator on HFSMs has been running continuously for about 10,000 hours. By the 1st stage of the refrigerator, radiation shield and copper current leads are cooled. And by the 2nd stage of the refrigerator, coils are cooled via the thermal conductive sheets. These thin high thermal conductive metallic sheets are glued on the outer surface of the winding. Coils are wound with epoxy. The sheets are laid directly to the 2nd stage cold head of the GM refrigerator with bending and twisting. The cross-section of the conductive sheet is not large, but there are only few joints on the thermal path which prevent heat flow.

As the current leads between the coil and the copper lead low temperature end, 10 mm dia. x 60 mm long Bi2223 pipes [2,3] are used. This lead is strong enough for shocks up to 30 G. To suppress the heat load to the 1st stage of the refrigerator, operating current is limited less than about 150 A. So operating current of the magnet is about 100 A.

Glass Fiber Reinforced Plastic and titanium support rods are used for sustaining coils and radiation shields. This support system is designed to endure transportational shocks.

For industrial use of the magnet, response during electricity failure is important. The magnet DC power supplies are backed up with the urgent power supply (UPS) of less than 10 KVA capacity. These are composed of an AC/DC/AC converter and Pb cells. The compressors of the refrigerator consume higher power than the DC power supplies. The magnet itself can be operated up to 10 minutes even in the refrigeration failure condition. In this case, temperature rise of the Bi lead and the coil are below the critical temperature. For protection of the coils, diodes are connected in parallel to coils. The power supply current can be cut off by the circuit breaker in case of Bi lead failure and the refrigerator stops.

Roughly speaking two GM refrigerators are mounted and operated in parallel for above 500 kg coils or larger than 1m coils.

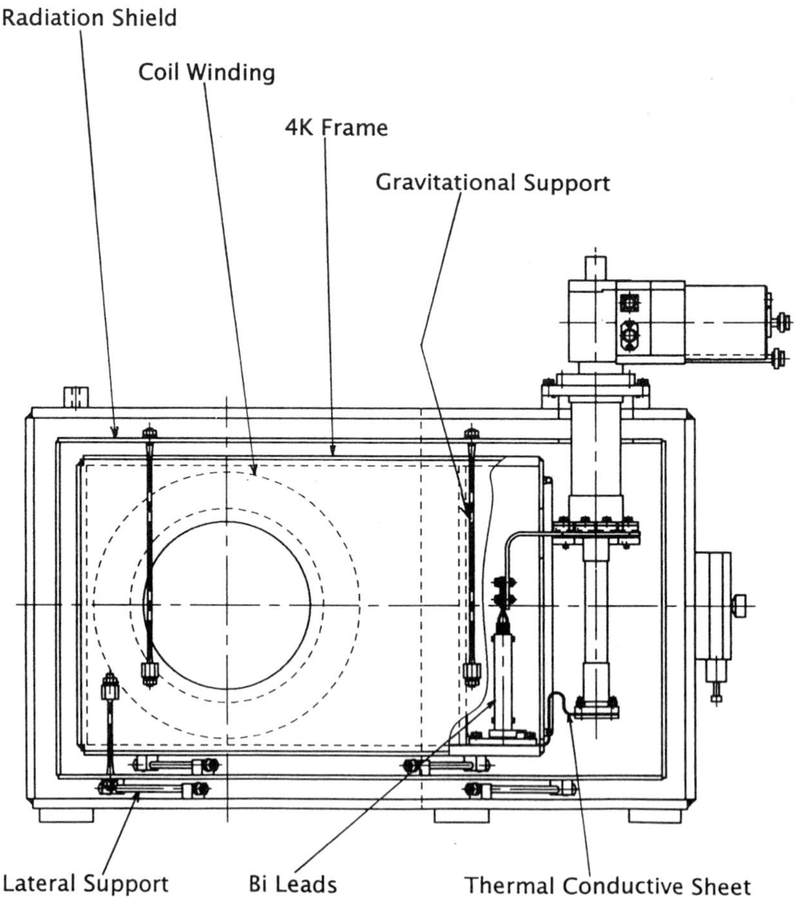


**Figure 4.** Structure of HFSM.

## DISCUSSIONS

There are some items to be solved for reliable magnet systems for industrial use. Superconducting coils of the liquid helium free magnet system are cooled by a 4 K GM refrigerator only. In other words, the GM refrigerator is one of the most important key parts, and the durability and reliability of the magnets sometimes depend on it. Unstable temperature of the GM refrigerator directly causes the coil temperature to rise. Stability of GM-refrigerator is expected.

Shortening of cooling time is the second item. GM refrigerator can cooldown the magnet by itself, however it takes several days. Various ideas are on study and trial. Pre-cooling channels on the coil and/or radiation shield is one of the ideas and can help to shorten the cooling time easily.

Refrigeration capacity of the GM refrigerator is also an important item, especially for a big magnet which has a big cooling mass. Large capacity of reliable 4 K GM refrigerator is expected for the HFMS.

In future, high temperature superconducting coils will be applied for HFMS. Their temperatures can be maintained low enough with traditional GM refrigerators which are already used for many cryo pumping systems or radiation shields. At that time, numbers of HFMSs will show up in industrial use.

## REFERENCES


1. Y.Ohtani et al., Development of a 11.5T Liquid Helium Free Superconducting Magnet System, Proceeding of the 17$^{th}$ International Cryogenic Engineering Conference, part2 (1997) p1113-1116
2. T.Honjo et al., Effects of Ag Addition in Bi-2223 Bulk Using Current Leads, IEEE Trans. Appl. Supercond.5 (1995) p1486
3. T.Honjo et al., Mechanical and Superconducting Properties in Bi-2223 Bulks for Current Lead, Proceedings of the 9$^{th}$ ISS(1996) to be published.

# RAMP RATE TESTING OF AN HTS HIGH GRADIENT MAGNETIC SEPARATION MAGNET

M. A. Daugherty, E. W. Roth, D. E. Daney, D. D. Hill, and F. C. Prenger

Los Alamos National Laboratory
Los Alamos, NM 87545 USA

## ABSTRACT

We report on the ramp rate testing of a prototype high temperature superconducting (HTS) high gradient magnetic separation (HGMS) magnet. HGMS magnets are ramped from full field to zero field to clean the separation matrix. The time spent ramping the magnet is unavailable for processing and must therefore be kept to a minimum. Existing commercial low temperature superconducting HGMS magnets are immersed in a liquid helium bath and are designed to ramp from zero to full current in one minute. The HTS magnet in our system is conductively cooled and operates in a vacuum at a temperature of approximately 30 K.

Heat generated during ramping is not as readily removed from the conductively cooled magnet as a bath cooled magnet. To verify that the conductive cooling can adequately remove heat generated during ramping we recorded magnet temperatures and voltages while ramping at rates of up to 4.8 A/second. The magnet can accommodate ramps from zero to 100 A (1.6 Tesla) in 21 seconds with no degradation in performance. The average magnet temperature rises a maximum of 1 K during ramping. Using temperatures recorded during ramps we have made rough estimates of the losses generated in the magnet.

## INTRODUCTION

High gradient magnetic separation uses large magnetic field gradients to separate ferromagnetic and paramagnetic particles from solids, liquids or gases. Fifty-nine elements have paramagnetic properties in some form and are thus candidates for HGMS. HGMS separators are built around a high-field solenoid magnet, the bore of which contains a fine-structured, ferromagnetic matrix, typically magnetic stainless steel wool. The design of our HTS HGMS system is shown schematically in figure 1 and has been discussed in two earlier papers.[1,2]

The matrix locally distorts the magnetic field and creates large field gradients, which trap contaminates on the matrix surface. As separation continues the contaminates build up on the matrix. Periodically the magnetic field must be turned off to clean the matrix and remove the accumulated contaminates. Then the magnet is turned on and separation resumes. In commercial magnetic separation systems the cycle times are typically on the order of twenty minutes of separation, one minute to ramp the field down, five minutes to flush the matrix, and one minute to ramp back up to the operating field.[3] Approximately seven minutes out of every twenty-seven are therefore not available for processing. Minimizing magnet ramping time is therefore an important issue in the design of HGMS systems.[4]

Existing low temperature superconducting HGMS systems operate with the magnet immersed in a bath of liquid helium.[4,5,6] The heat generated during ramping is relatively efficiently removed by convection of the liquid helium, ensuring approximately uniform magnet temperatures. In a system such as ours, which operates in a vacuum, the heat must be removed by conduction to the magnet bore tube and mounting plate. Significant temperature gradients may therefore arise between the outer sections of the magnet and the cooling surfaces at the inner bore and magnet bottom plate. If these temperature gradients are too large the superconductor will dissipate more energy than can be removed by conduction and the magnet will quench.

On the positive side the HTS material allows us to operate with a significant margin between the magnet operating temperature and the critical temperature of the superconductor. We also benefit from the much larger heat capacities of the magnet material at the magnet operating temperature, which greatly reduce the temperature rise for a given heat input. This work was carried out in order to examine the effect of these competing considerations and to estimate the magnitude of ramping losses in this conductively cooled HTS magnet system.

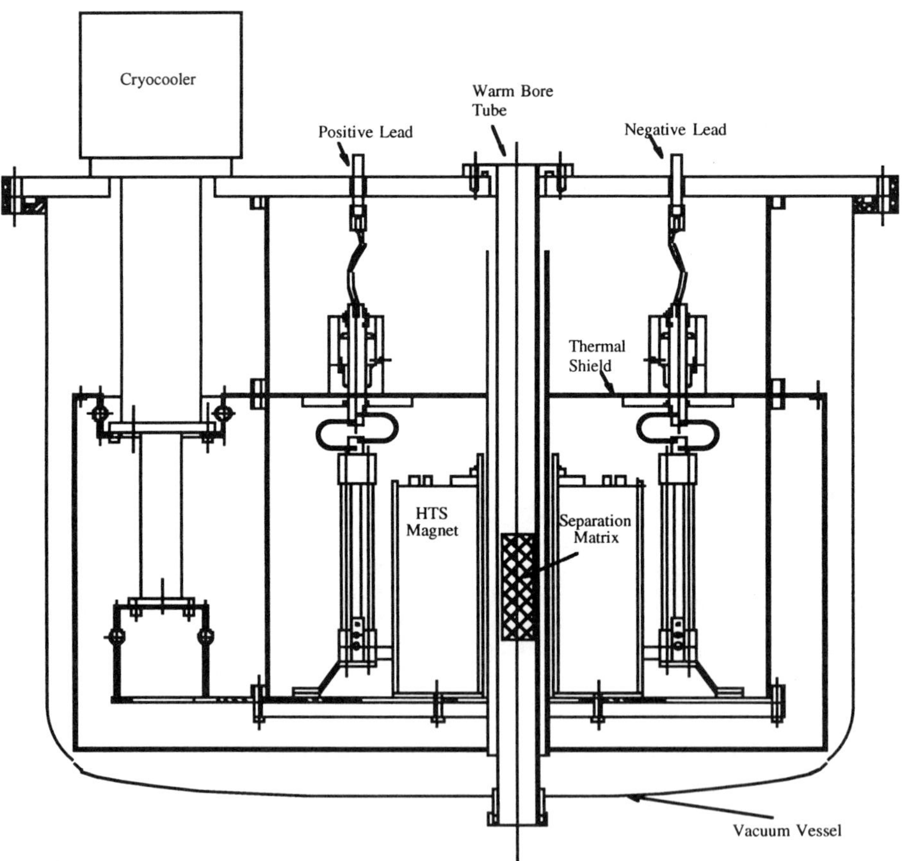


**Figure 1.** Schematic of the HTS high gradient magnetic separation system.

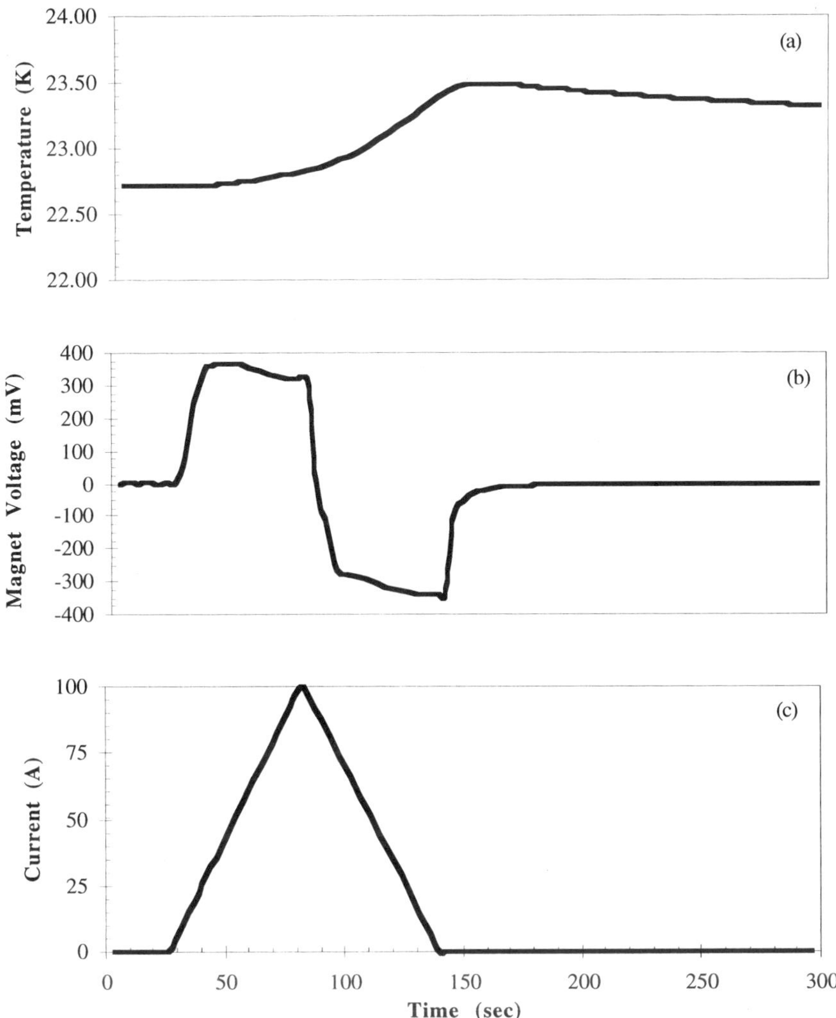


**Figure 2.** Time history of one magnet ramp cycle at 1.8 A/s showing average magnet temperature, magnet voltage and magnet current .

## EXPERIMENTAL PROCEDURE

The HTS magnet temperature was measured by three silicon diodes located at the top, middle and bottom of the magnet. The reported magnet temperature is the average of the readings from the three diodes. The response of the diodes is affected by the magnet field generated during a ramp, so the temperatures before and after the ramp, when the magnetic field was zero, were used to estimate the ac losses. The magnet voltage was measured by voltage taps located on the magnet terminals. The magnet current was measured using a 100 mΩ shunt resistor.

Prior to the start of a ramp the initial magnet temperature was set using a heater wound on the thermal link between the second stage of the cryocooler and the magnet cooling plate. Initial temperatures were varied from approximately 22 K to 41 K. After the magnet temperatures had stabilized a programmable power supply was used to ramp the magnet up to the maximum current desired at rates varying from 0.6 A/s to 4.8 A/s. Magnet current, voltage and temperature were measured before, during and after the ramp. Data from a representative ramp are shown in figure 2.

The magnet temperature, shown in figure 2a, peaks soon after the current ramp ends. The slight time delay between the end of the ramp and the peak magnet temperature is attributed to the time for the heat generated within the windings to conduct out to the diodes, which are located on the external surface of the magnet. The magnet voltage is shown in figure 2b and the current in figure 2c. Using the measured voltage of approximately 375 mV and the ramp rate of 1.8 A/s the magnet inductance is estimated to be 0.2 H. At the conclusion of a ramp the magnet temperatures were allowed to return to the initial temperature prior to the start of the next ramp.

## RESULTS

Magnet temperature, current and voltage were the primary quantities of interest measured in these experiments. To calculate the losses we used an energy balance given by:

$$\dot{Q} = C_p M \frac{dT}{dt} \tag{1}$$

Where $\dot{Q}$ is the heat being generated in the magnet due to losses, Cp is the temperature dependent specific heat of the magnet, M the total mass of the magnet and dT/dt the magnet's average rate of change of temperature. M was calculated from the dimensions and a materials list taken from the engineering drawing of the magnet assembly. While the determination of the magnet mass is believed to be accurate to within approximately 5 percent, the magnet is thermally coupled to the current leads, whose mass was not used in the calculation. This introduces a source of uncertainty because while the thermometers measured the magnet temperature, the magnet was thermally coupled to the rest of the system making accounting for the total relevant thermal mass difficult. dT/dt was approximated by taking the maximum average temperature change during a ramp cycle and dividing by the total elapsed time of the ramp cycle. These measurements and loss calculations were performed at various values of magnet temperature, current ramp rate, and peak current as indicated in the following figures.

Figure 3 shows the maximum average temperature rise of the magnet during the ramp cycle at various starting temperatures and ramp rates. The figure indicates the temperature rise increases with the faster ramp rate as expected but decreases at higher starting temperatures because of the higher heat capacity of the magnet. Note in all cases the temperature rise is less than one kelvin.

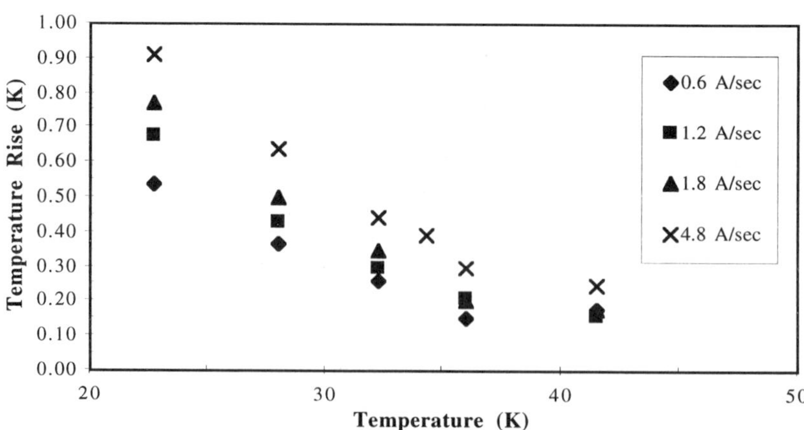


**Figure 3.** Average magnet temperature rise versus initial magnet temperature for various ramp rates.

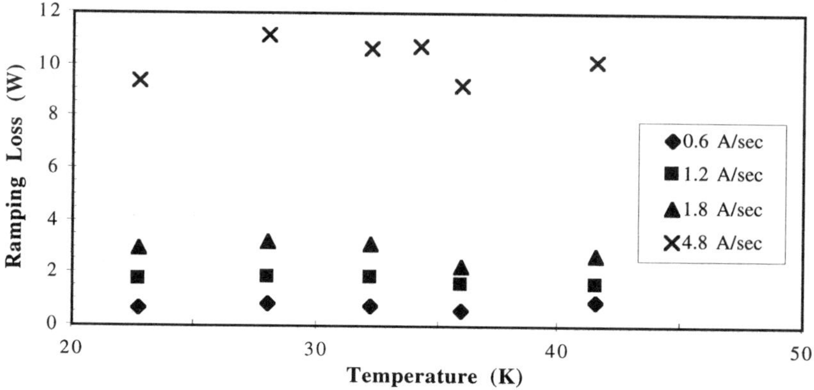


**Figure 4.** Ramping loss versus initial magnet temperature for various ramp rates.

Figure 4 shows the average power dissipated in the magnet during the ramp cycle at various starting temperatures and ramp rates. The figure shows increased dissipation at the higher ramp rates as expected. There does not appear to be much variation of dissipation with temperature. The average power dissipated is shown again in figure 5 as a function of ramp rate. The losses appear to scale with ramp rate to the 1.25 power.

Figure 6 shows the average power dissipated in the magnet as a function of peak current, which is equivalent to peak magnetic field. The ramp rate and starting temperature are fixed at 1.5 A/s and 30.5 K respectively. A peak current of 100 A corresponds to a peak magnetic field of 1.67 T. The data shows the average losses are independent of the peak current.

As we discussed in the introduction, commercial magnetic separators typically operate at full current for approximately twenty minutes, ramp down in one minute, dwell at zero current for five minutes to clean the matrix, and finally ramp back up to full current in one minute. In order to determine if this HTS magnet could tolerate such an operating cycle we allowed the magnet temperature to stabilize at full current, ramped down to zero current in one minute, held at zero current for one minute, and ramped back up to full current in one minute.

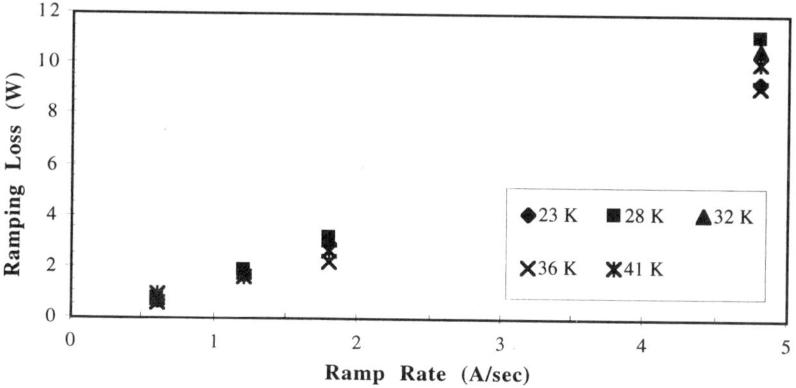


**Figure 5.** Ramping loss versus ramp rate for various initial temperatures.

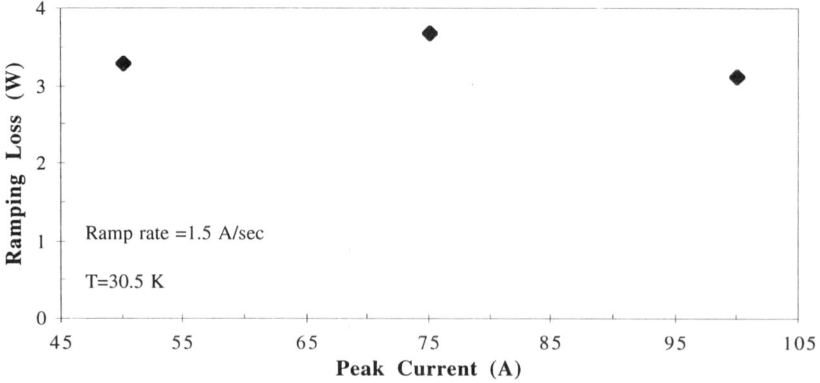


**Figure 6.** Ramping loss versus peak current (peak magnetic field) at a ramp rate of 1.5 A/s.

The results, shown in figure 7, clearly indicate the magnet survived the ramp and is beginning to cool back down. The cooling rate after the ramp is not sufficient for the magnet to reach its initial temperature in twenty minutes with this operating cycle. Note, however, that the dwell time at zero current here was only one minute as opposed to five minutes for a commercial separation cycle. The magnet recovery time can be improved by using a larger capacity cooler or improving the internal thermal conductivity of the magnet.

## CONCLUSIONS

We have examined the response of a prototype HTS high gradient magnetic separation system magnet to ramps between zero current and a current of 100 A. Ramp rates were varied from 0.6 A/s to 4.8 A/s and the initial magnet temperature was varied from approximately 22 K to 41 K. The largest measured temperature rise during a ramp from zero current to full current and back to zero current was 0.9 K. Measured magnet

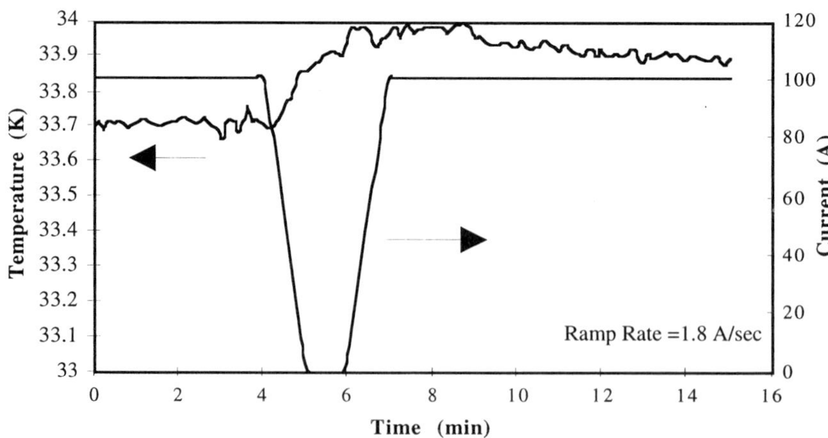


**Figure 7.** Time history of a magnet ramp cycle at 1.8 A/s to simulate a typical high gradient magnetic separation system matrix cleaning cycle.

temperature rises were used to estimate ac losses during a ramp cycle. At a ramp rate of 4.8 A/s losses were estimated to be approximately 11 W. Ramping losses were found to be largely independent of operating temperature in the range from 22 K to 41 K. Losses appear to scale with ramp rate to the 1.25 power in this same temperature range. Losses appear to be independent of peak magnetic field at a temperature of 30.5 K and a ramp rate of 1.5 A/s.

## ACKNOWLEDGMENTS

The authors gratefully acknowledge the technical assistance of W.L. Hults, J.Y. Coulter and D.E. McMurry. This research was supported by the U.S. Department of Energy, Office of Energy Efficiency and Renewable Energy.

## REFERENCES

1. M. Daugherty, F. Prenger, D. Hill, D. Daney, L. Worl, A. Schake, and D. Padilla, Application of high-temperature superconductors to high-gradient magnetic separation, 29th Intersociety Energy Conversion Engineering Conference, Monterey, CA, August 7-11, 1994, pp. 1502-1507.
2. M.A. Daugherty, J. Y. Coulter, W. L. Hults, D. E. Daney, D. D. Hill, D. E. McMurry, M. C. Martinez, L. G. Phillips, J. O. Willis, H. J. Boenig and F. C. Prenger,A. J. Rodenbush and S. Young, HTS High Gradient Magnetic Separation System, *IEEE Tran. on Applied Superconductivity*, Vol. 7, No. 2, Part 1, pp. 650-653, 1997.
3. J.A. Selvaggi, Magnetic Separation Consultant, Pittsburgh, PA, private communication, 1996.
4. J.H.P. Watson, Status of superconducting magnetic separation in the minerals industry, *Materials Engineering*, Vol. 7, Nos. 5/6, pp. 737-746, 1994.
5. J. A. Selvaggi, P. Vander Arend and J. Colwell, An industrial superconducting high gradient magnetic separator, *Advances in Cryogenic Engineering*, Vol. 33, pp. 53-60, 1988.
6. J. A. Selvaggi, T. Kranyecz and S. Bell, A 124" warm bore superconducting ironclad high gradient magnetic separator, *Applied Superconductivity*, Vol. 1, No. 1/2, pp. 13-18, 1993.

# Bi-2212 COILS FOR 1 T CLASS INSERT COILS

H.W. Weijers,[1] D. Hazelton,[2] L. Cowey,[3] Y.S. Hascicek,[1] I.H. Mutlu,[1] U.P. Trociewitz[1], and S.W. Van Sciver[1]

[1] National High Magnetic Field Laboratory
Tallahassee, Florida, 32310, USA
[2] Intermagnetics General Corp.
Latham, New York, 12110, USA
[3] Oxford Instruments
Abingdon, Oxon, OX13 5QX, England

## ABSTRACT

We present the construction and in-field testing of three insert coils using Bi-2212 conductor with pure silver and alloyed silver stabilizer. We use a stacked double pancake configuration, where double pancakes are wound from one piece of conductor. Insulation, impregnation, and winding techniques are discussed as well as characteristics of the coils in background fields up to 17 and 21 T respectively at 4.2 K. Stability of the coils against overcurrent is also investigated.

## INTRODUCTION

As part of the development of a 5 T HTS insert coil which will be part of a 25 T, 1.1 GHz, NMR system[1], the NHMFL set an intermediate goal to build an HTS magnet that generates 1 T as an insert magnet in a 50 mm bore and 20 T background field. This 1 T project is carried out in co-operation with industry, specifically Oxford Superconducting Technology and Intermagnetics General Corporation, and focuses on Bi-2212 Wind-and-React (W&R) coils. The final 5 T coil is currently envisioned as 5 concentric layer wound coils using possibly both W&R and R&W HTS technology. The choice for Bi-2212 is based on the expectation that, of the conductors available in sufficient lengths, it will have the highest engineering critical current density ($J_e$) at 4.2 K in magnetic fields between 20 and 25 T. The choice of the magnetic field level at which to change from applying $Nb_3Sn$ or similar conductor to applying HTS conductor in the 25 T system will depend on the average current densities that can be achieved in the windings ($J_{ave}$) and is subject to change. Where the benchmark design for the 5 T coil calls for a $J_{ave}$ of 100 A/ mm$^2$ at 25 T, the two stacks of double pancakes discussed in this paper require a $J_{ave}$ of 110 and 67 A/mm$^2$ respectively to generate 1 T. So, while size and stress during operation are limited for the stacks, the current density requirements are comparable to the benchmark design. One of our existing magnets, a 52 mm cold bore superconducting magnet, that can generate up to 17.5 T at 4.2 K, serves as the standard testing facility for the insert coil. In this paper we discuss the construction and in-field testing of two stacks of pancakes tested at the NHMFL and a single pancake tested in the $21^+$ T superconducting magnet at the Tsukuba Magnet Laboratories (T.M.L.) of the National Research Institute for Metals, Japan.

**Figure 1.** Picture of Coil A (right) and a separate unit.

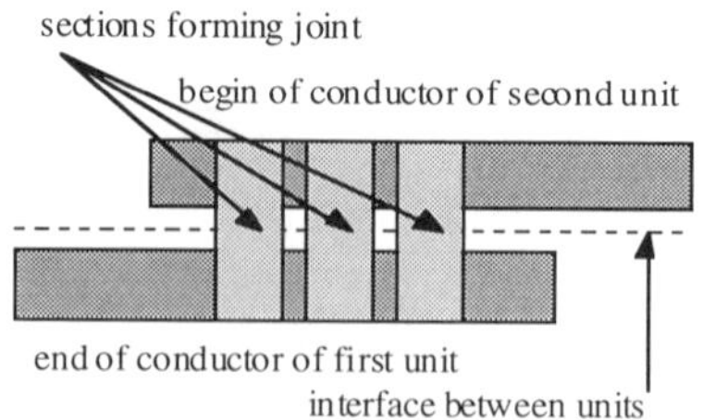


**Figure 2.** Coil A joint layout.

## CONSTRUCTION

We discuss three coils, A,B and C, in this paper. The double pancake units used to construct them have no internal joints as they are wound from one continuous piece of conductor. $ZrO_2$ is used as insulator, applied before winding as specified in Table 1. All units were heat treated in a small flow of $O_2$ at atmospheric pressure. The heat treatment consists of an outgassing phase of 24 hours at 820 °C, 0.4 hours at the peak temperature to create a partial melt and 14 hours of sintering at 825 °C, with controlled temperature ramps in between. The peak temperature that optimizes the critical current varies around 890 °C, the exact value differs per batch of conductor. Vacuum impregnation with Stycast 1266 followed after heat treatment. All impregnated units are 7.6 mm high. The units are then tested at 4.2 K in self field. A G-10 stack tube and flanges support the units and the leads. A second vacuum impregnation creates a robust and integral structure. Some other details differ per coil and are discussed below or listed in Table 1.

### Coil A

Although short samples from the conductor source used for Coil A typically carry over 200 A in the superconducting state, the specific batch used was limited to about 68 A. The insulation is a refractory paint from a commercial source and is sprayed on one wide surface plus the edges of the conductor. It suffices to create the conductor stand off needed for pancake winding but the adhesion on the edges is insufficient for layer winding. A sheet of 0.3 mm $ZrO_2$ cloth provides the separation between the pancakes. The units were joined by soldering three short sections of superconductor as indicated in Figure 2. Copper leads connect directly to the top and bottom units.

**Table 1.** HTS coil parameters

| Coil | A | B | C |
|---|---|---|---|
| Conductor size [mm*mm] | 0.25*2.54 | 0.2*3.1 | 0.21*2.95 |
| number of filaments | 84 | 19 | 19 |
| Matrix material | Ag+Ag alloy | Ag | Ag |
| Conductor length per unit [m] | 7 | 11 | 12 to 12.5 |
| number of turns per unit | 2*33 | 2*57 | 2*63 |
| number of units in stack | 4 | 1 | 6 |
| Winding height [mm] | 26.7 | 6.7 | 40.2 |
| Insulation application | spray | spray | sol-gel |
| Insulation thickness [μm] | 70 to 80 | 50 | 30 |
| Bore tube material | stainless steel | stainless steel | Macor |
| Bore tube OD [mm] | 22.2 | 15.7 | 16.3 |
| Free bore [mm] | 19.1 | 12.7 | 12.7 |
| Winding OD [mm] | 44 to 45 | 44.5 | 46 |
| Total OD [mm] | 49 | 49 | 50.2 |
| B/I ratio [mT/A] | 7.9 | 5.2 | 18.6 |
| Packing factor | 56% | 72% | 78% |

**Figure 3.** Picture of Coil B. **Figure 4.** Picture of Coil C.

## Coil B

Coil B is a double pancake unit prepared much like the units of Coil A, but with a smaller free bore and a slightly thinner layer of insulation.

## Coil C

$ZrO_2$ was applied on the conductor with a recently developed sol gel method[2,3], resulting in a thinner insulation (layer thickness is 15 μm on all sides) with better adhesion. This allowed a tighter packing and therefore a longer length of conductor in the same volume and correspondingly higher coil constants. A $ZrO_2$ sheet to separate the pancakes was omitted for four of the six units, relying solely on the insulation covering the edges of the conductor. The double pancake units were electrically joined using 7 to 9 parallel pieces of conductor in a layout similar to the one shown in Figure 2. The stacked units were pre-heated to 130 °C in a furnace before soldering the joints in place. The same technique was used to connect the top and bottom pancake to the copper current leads. Then the coil was insulated where necessary with Kapton tape and cast inside a brass tube with Stycast 2850 FT.

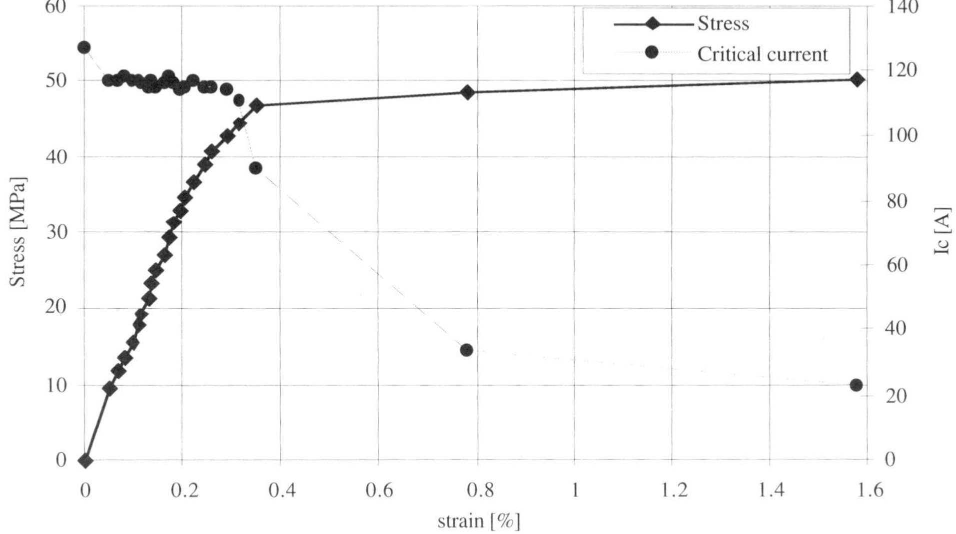


**Figure 5.** Stress-strain characteristic and strain dependence of the critical current (on left vertical axis) of unsupported conductor from the batch used for Coils B and C.

Lorenz-force loading of samples of the conductor used for Coils B and C resulted in the characteristics shown in Figure 5. These measurements are performed on unsupported circular samples in the bore of a magnet at 4.2 K. Details of the stress, strain, and subsequent critical current measurement can be found in references 4 and 5. A rapid decrease in critical current occurs around 44 MPa, corresponding to 0.31% strain.

## PERFORMANCE

All critical currents are determined with the conductor at 4.2 K and in self field with the 1 μV/cm criterion unless noted otherwise.

### Coil A

The units of Coil A, when first tested individually, had $I_c$ values of 57.4 to 58.5 A which is more then 80 % of the best short sample values. N-values varied around 13. Unit 4 is discussed in more detail in the section "Thermal Cycling and Aging". The increased self field after stacking reduced the critical current of the individual units, most noticeably for the end units that experience the highest radial self field. This spread in $I_c$ is reduced to (22.6+/-0.5) A in a 17 T background, where the axial field determines $I_c$. This corresponds to 45 % of the self field $I_c$, and about 1/3 of the short sample $I_c$. Figure 6 illustrates the dependence of the engineering current density $J_e$ on the total central magnetic field (central self field plus applied background field). A slow thermal runaway occured during the 17 T measurement at a total heat dissipation of 3.7 W in the coil.

The joints performed adequately at self field. Although the joints have a relatively small effect on $I_c$, the voltage across them was considerable in background fields of 8 and 17 T. Their resistance in field was 1 to 7 μΩ for low currents, but significant heating and an associated rise in resistance occurred around the critical current. We believe this is caused by the soldering technique we used for soldering the joints. Due to the thermal conductivity and total mass of conductor, a large heat load had to be applied locally to create a proper melt of the Pb/Sn 60/40 solder used.

### Coil B

After an initial 4.2 K characterization at the NHMFL, coil B was tested in the 21 T superconducting magnet at the Tsukaba Magnet Laboratories of NRIM, Japan. A background field of 21 T reduced the $I_c$ to 49 % of the self field value as measured at NRIM. A slight hysteresis is present in the critical current, with values for decreasing background fields being slightly higher, as seen in Figure 6. Table 3 gives the chronology of self field $I_c$ measurements. A rough estimation of the Lorenz-force induced hoop stress at the highest coil current yields $\sigma$=44 MPa for the outer turns at 21 T. We used $\sigma$ =J.B.r, where J is the current density, B is the magnetic field, and r is the radius.

### Coil C

The units of Coil C had critical currents between 136 and 148 A and N-values around 9 when tested individually. As with Coil A, the units of Coil C showed a reduced $I_c$ after stacking, most noticeably for the outer units because they are subjected to the largest radial self field. The $I_c$ of the entire coil including all joints and leads was 118 A. Figure 6 shows the dependence of $J_e$ on the total central field. For low applied fields the radial self field is the main factor determining the critical current, explaining the weak dependence on applied axial fields compared to typical short samples. See for example reference 6.

The slope of $dJ_e/dB$ between 15 and 17 T is slightly larger than the same slope for coil B, which used the same type of conductor. Also, the critical currents after a reduction of the background field are 2 to 5 A lower than after the initial increase. This indicates the coil sustained some damage as a result of the measurements at the highest fields. Using $\sigma$ =j.B.r gives an estimation of 45 MPa maximum hoop stress at 15 T and 49 MPa at 17 T. The power supply used tended to overshoot when programmed to supply a higher current level, causing the hoop stress to be momentarily even higher. These stress levels would reduce the critical current of an unsupported turn of conductor to less then 30% of its initial value (see

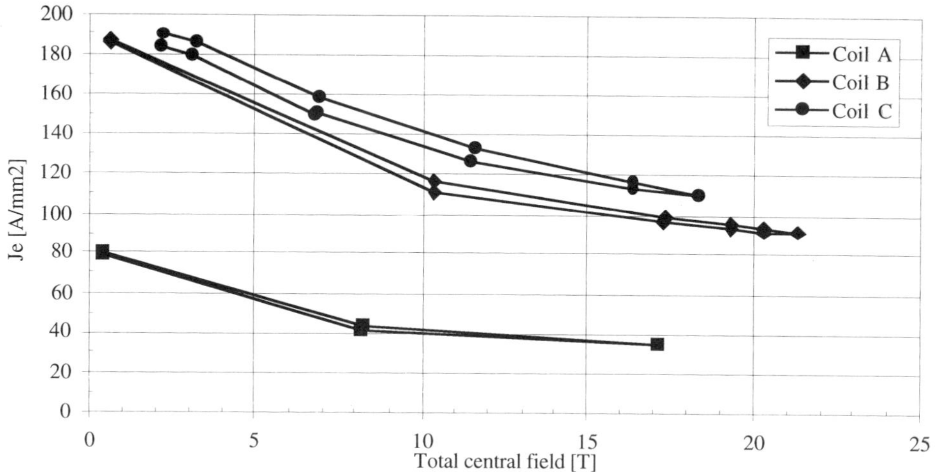

**Figure 6.** Engineering current density as a function of the total central magnetic field. Lines connecting the datapoints are a guide to the eye only.

figure 5). Therefore, it appears that the impregnated structure of the coil has prevented major damage, but not all damage.

At the first self field run Coil C was just stable at a total heat dissipation of 1.44 W in the coil at 122 A. For currents equal or larger then 123 A a thermal runaway occured. In the following measurements the maximum heat dissipation was limited to 1.2 W and no further thermal runaways occurred.

The joints between the units have resistances ranging from 20 to 160 nΩ at 17 T, which has no significant effect on $I_c$.

## Field Generation

All coils generated fields deviating less then 2% from the nominal values up to the critical current. Figure 7 gives an overview. Presented are the generated fields normalized to the nominal value at $I_c$ versus the current in the coil normalized to the critical current.

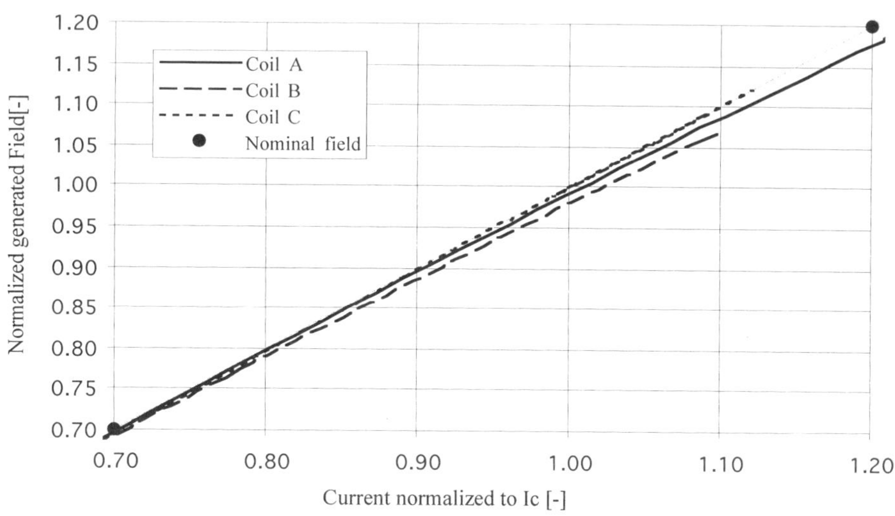

**Figure 7.** Generated field of the coils divided by the nominal field at $I_c$ vs. Coil current divided by the critical current.

**Table 2.** Generated central field at 4.2 K

| Coil | at self field [T] | in background field [T] | |
|---|---|---|---|
| A | 0.40 | 0.17 | in 17 T background |
| B | 0.61 | 0.29 | in 21 T background |
| C | 2.10 | 1.20 | in 17 T background |

Ideally all curves would fall on the line connecting (0,0) and (1,1). Figure 7 shows that there is a slight deviation from nominal behavior for the coils insulated with sprayed $ZrO_2$, where the units of coil C generate the nominal field. Investigations have shown that there is no metallic (Ag-Ag) contact between the turns of the units. We believe core material leaking through the matrix of the conductor during the partial melt phase of the heat treatment are causing the observed shorted behavior in Coils A and B. The same types of leaks are present in Coil C, but do not affect the field generation at 4.2 K. The conductor of coil C was insulated using a sol-gel technique.

### Thermal Cycling, Handling and Aging

The possible effects of age, handling, thermal cycling, and repeated charging and discharching on the properties of HTS coils is a relevant issue. Both double pancakes presented in Table 3 were thermally cycled three times and tested to 17 and 21 T respectively in the second run. Each time leads were soldered directly to the conductor, except for the third run of Coil B. Coil B was mounted on a stack tube, fitted with Cu leads and impregnated a second time before the second run. For the most part, warming up after the in-field measurements took place in ambient air, causing ice and water to deposit in the units.

Both coils show a small degradation after each run. The amount of data is limited but suggest handling (i.e. soldering) and/or thermal cycling are more important factors then aging since there is no clear relation between time and the measured critical current.

## DISCUSSION

With Coil C generating 1.2 T in a 17 T background field we have achieved our goal of building a 1 T Class insert coil. $J_e$ at 18.2 T central field was 110 A/mm$^2$ corresponding to a $J_{ave}$ of 85 A/mm$^2$ in the windings. This is considered a useful current density for NMR magnets[7] and close to the target value of 100 A/mm$^2$ for the 5 T HTS insert magnet envisioned to be part of a 25 T, 1.1 GHz, NMR system. The sol-gel technique provides excellent insulation in a thin, well adhering layer and is`resistant to core leaks. This insulation is expected to tolerate layer winding. The structure of Coil C has prevented major damage from Lorenz-force induced stresses. For coils subject to larger stresses reinforcement will have to be considered.

## CONCLUSIONS

- We have built an Bi-2212 HTS coil generating 1.2 T in a 17 T background field. The corresponding average current density in the windings is 85 A/mm$^2$.
- Double pancake units exhibiting only 1 to 4% variation in $I_c$, when using conductor from one batch, were built.
- Reversible in-field behavior with little hysteresis in $I_c$ was observed for Coils A and B.

**Table 3.** Chronology of $I_c$ measurement results at self field

| condition: | first run | second run | third run |
|---|---|---|---|
| Unit 4 of Coil A | 58 A | 57 A (after 4 days) | 55 A (after 151 days) |
| Coil B | 119 A | 113 A (after 32 days) | 112 A (after 71 days) |

- Field generation at 4.2 K is very close (within 2 % at $I_c$) to the nominal values for coils insulated with sprayed $ZrO_2$. Coil C, insulated using a sol-gel application technique, showed nominal behavior.
- Two double pancakes were repeatedly tested, thermally cycled and handled over periods of several months, showing a few percent degradation of the critical current.
- When joining double pancake units by soldering sections of superconductor between them, it appears necessary to preheat the stack in order to obtain acceptable joints.

## ACKNOWLEDGMENTS

Research supported in part by Grant 5-R44-CA60319-03 from the National Institutes of Health / National Cancer Institute. The authors also like to thank K. Marken and L. Motowidlo regarding conductors, H. Kitaguchi at NRIM for the measurements on coil B at the Tsukuba Magnet Laboratory, M.I. Godfrey for spray insulation of the conductors, S. Boutemy for the heat treatment of some of the units and L.Y. Xiao for the stress-strain-$I_c$ measurements.

## REFERENCES

1. W.D. Markiewicz, I.R. Dixon, J. Schwartz, C.A. Swenson, S.W. Van Sciver, and H.J. Schneider-Muntau, 25 T High resolution NMR Magnet Program and Technology, *IEEE Transactions on Magnetics 32:4:2586 (1996)*
2. I.H. Mutlu, R.E. Goddard and Y.S. Hascicek, ESEM Hot stage evaluation of Sol-Gel insulation coatings for high field HTS magnets, *Proceedings of Microscopy and Microanalysis '97 Conference*, (August 1997)
3. I. H. Mutlu and Y.S. Hascicek, Insulation for W & R HTS Coils, ibid.
4. L. Y. Xiao, D. K. Hilton, Y. S. Hascicek and S. W. Van Sciver, *Mechanical tests on the surface-coated Bi 2212 superconducting tapes*, Submitted to Cryogenics (1997).
5. D.K. Hilton, H.W. Weijers, Y.S. Hascicek, S.W. Van Sciver, and J. Schwartz, *Lorenz-Force Tensile Stress-Strain and Fatigue Apparatus*, Submitted to Reviews of Scientific Instruments
6. D.W. Hazelton, M.T. Gardner, J.M.Weloth, J.A. Rice, L.R. Motowidlo, Y.S. Hascicek, W.D. Markiewicz, S.W. Van Sciver, HTS Insert Coils for Ultra High Field NMR Spectroscopy, IEEE Transactions on Applied Superconductivity 7:2:885 (1997)
7. T. Kiyoshi, K. Inoue, M.Kosuge, H.Kitaguchi, H. Kumakura, H. Wada and H. Maeda, *NRIM R&D Program on HTS coils for 1 GHz NMR Spectrometer*, Proceedings of the CEC/ICMC conference of May 1996)

# COMPUTER SIMULATION OF QUENCH PROPAGATION IN QUELL

R.Zanino[1], L.Bottura[2], and C.Marinucci[3]

[1]Dipartimento di Energetica, Politecnico, I-10129 Torino Italy
E-mail address: zanino@alphaz.polito.it
[2]CERN, Div. LHC-MTA, CH-1211 Geneve 23, Switzerland
[3]EPFL, CRPP-Fusion technology Div., CH-5232 Villigen PSI Switzerland

## ABSTRACT

In the Quench Experiment on Long Length (QUELL), in the SULTAN facility, quench propagation was studied in a well instrumented sample wound using a Nb3Sn cable-in-conduit conductor (CICC) with central cooling channel, cooled by supercritical helium I. A few selected runs have been analyzed here by means of two different computer models - Mithrandir and Gandalf, using the same set of common input parameters, helium and material properties, heat transfer and fluid dynamic correlations. The predictions of the two codes are compared in detail with each other and against experimental data. Both codes are 1-D tools for the description of thermal-hydraulic transients in CICCs with cooling channel; Mithrandir differs from Gandalf mainly because it does not assume, as the latter does, the same thermodynamic state for the helium in the cable bundle region and the helium in the cooling channel. It turns out that although the 1-fluid model (Gandalf) gives a reasonable agreement with the experiment, 2-fluid (Mithrandir) modeling is more accurate both qualitatively and quantitatively.

## INTRODUCTION

In QUELL a series of over 60 experimental runs was conducted to study quench initiation and propagation in a ~ 100m long CICC with central cooling channel .[1] The channel mainly provides a low impedance path for the coolant, and a reservoir to limit pressure increase during a quench.

The cooling channel is lined by a thin (~ 0.7mm) strip, wound into a helix, with pitch and width such as to leave a free passage between the annular region where the cables are present (so-called cable bundle, $\mathcal{B}$ in the following) and the central channel of circular cross section (so-called hole, $\mathcal{H}$ in the following).

This type of conductor originally motivated the development of the 2-fluid code Mithrandir[2] , which describes supercritical helium flowing in the $\mathcal{B}$ and $\mathcal{H}$ regions at different speeds, and with different pressures p and temperatures T, coupled to heat conduction in cables and jacket (see next Section for more details). Mithrandir extends the 1-fluid model implemented in Gandalf [3], where p and T are assumed to be the same in the $\mathcal{B}$ and $\mathcal{H}$ regions.

A detailed validation of Gandalf against the experimental results of 17 selected runs has already been presented elsewhere [4], showing that the 1-fluid model is a reliable tool for engineering predictions, but also indicating that in some cases the details of quench initiation and propagation were not accurately described.

A first comparison[5] between the 1- and the 2-fluid models showed that instead the 2-fluid model implemented in the Mithrandir code was able to well reproduce the details of a single quench in QUELL by using suitable values of its free input parameters (see next Section); furthermore, the 2-fluid model was shown to be certainly needed for the description of slow transients ("heat slug" experiments) with weak coupling between the $\mathcal{B}$ and $\mathcal{H}$ regions.

The same two tools have also been used [6] for a first analysis of the effects of the central channel on stability, and the prediction of stability margins in ITER; finally, Gandalf and Mithrandir were recently extended to treat also helium II. [7]

Here we compare the two models against the experiment in six new QUELL runs, which we consider representative of a relevant subset of the quench parameter space.

## OVERVIEW OF THE MODELS USED FOR THE SIMULATION

Since details of the codes have already been presented elsewhere [2,3] we only review here the major features, emphasizing common aspects and differences.

Both codes describe 1-D (x coordinate along the conductor) thermal-hydraulic transients in supercritical HeI, flowing in CICCs; jacket (conduit) and conductor are at different temperatures $T_{jk}$ and $T_{co}$ respectively.

The helium is treated by the usual set of Euler equations with viscosity modeled by a friction factor. For both codes the He flow velocities in the two regions are different ($V_B \neq V_H$) but *in Mithrandir the He temperatures and pressures are treated separately* ($T_B \neq T_H$ *and* $p_B \neq p_H$), *whereas in Gandalf* $T_B = T_H = T$ *and* $p_B = p_H = p$. In other words, Gandalf assumes perfect (i.e., infinitely fast) thermal and hydraulic coupling between $\mathcal{B}$ and $\mathcal{H}$ regions, whereas in Mithrandir this coupling occurs at a finite rate, determined by a set of "free" input parameters to be discussed below in more detail.

Both codes use the same set of correlations for the heat transfer coefficients between helium and solids, and for the friction factor in the two regions.[1] Additionally, Mithrandir needs a relation for the heat transfer coefficient $h_{HB}$ at the no-wall interface between helium in $\mathcal{B}$ and helium in $\mathcal{H}$. This is a critical and rather underdeveloped subject, and more experimental/modeling work is possibly needed.

Both codes use the same set of initial and boundary conditions (in Mithrandir the same initial and boundary conditions are used for both the $\mathcal{B}$ and the $\mathcal{H}$ region). Conductor current as a function of time (with the related self-field), and external magnetic field in the SULTAN facility as a function of space [1], are also the same.

Both codes use an adaptive finite element method (linear elements) for space discretization, and the (implicit) Crank-Nicolson scheme with adaptive time step.

**Free input parameters in the 2-fluid Mithrandir model**

In Mithrandir there are essentially four free input parameters[2], each corresponding to a different uncertainty in the coupling between He in the $\mathcal{B}$ region and He in the $\mathcal{H}$ region:

- perforated fraction F of the helix wall between $\mathcal{B}$ and $\mathcal{H}$;
- multiplier $1/\sqrt{\kappa}$ in the mass flux $\Gamma$ through the perforation ($\Gamma$ is assumed to be proportional to $\sqrt{|p_B - p_H|}$);
- multiplier $\lambda_v$ in the radial flux of x-momentum convected through the perforation ($\lambda_v$ mainly depends on the geometry of the interface);
- multiplier $h_{nowall}$ of the energy flux conducted through the perforation (to compensate for the limitations of the model which we use for $h_{HB}$).

As a brief comment on these parameters (a detailed sensitivity study is underway and will be presented elsewhere) we can observe that the nominal perforation of the helix wall (about 14% in the case of QUELL) is not necessarily related to the actual unknown (!) perforation; the latter is determined among others by the spaces left open between cables after the manufacturing process. We chose F=1%, together with $\kappa = 1$, $\lambda_v = 1$ and $h_{nowall} = 10$; these values have been consistently used for all the runs presented here.

**Computer requirements and performance**

Mithrandir has been run on a DEC Alpha Open VMS Workstation with 100MHz processor and 96Mbyte RAM, Gandalf on a Sun ULTRA Workstation with 200Mhz processor and 256Mbyte RAM. On the Alpha, about 7 CPU seconds per Mithrandir time step are typically needed, with a standard 2000 element mesh.

In a single controlled run of both codes on the Alpha, Mithrandir needed about 3 times as much CPU time as Gandalf. This can be understood because of the increased number of unknowns (8 instead of 6 per node) and wider band of the system matrix, the larger number of He property calls, and the fact that the time stepping routine of Mithrandir is written in double precision FORTRAN (vs. single precision in Gandalf). [5]

## ANALYSIS OF QUELL QUENCHES

A detailed list of QUELL conductor data and experimental run characteristics is given in Ref.1. In all runs, and for both codes, the experimental pressures, as measured at x=0.12m and x=90.84m (x measured from the inlet), were used as boundary conditions, together with inflow temperatures equal to their initial (flat) values. The resistive heater is between x=40.93m and x=43.23m.

We compare between codes against experimental data of total resistive voltage $\Delta V$, total normal zone length NZ (processed), helium pressure p (at x=40.96m (P2), x=41.76m (P3), x=43.28m (P4), x=65.78m (P5)), conductor temperature $T_{co}$ (processed).

As it will become apparent below, uncertainties in material properties and other effects can become very important in the case of low temperature margin ($\Delta T < 1K$) or large current density (corresponding to $I > 8kA$). For the sake of clarity we define therefore two possible choices of current&field parameters, which will be used for the comparisons in the following:

| | | |
|---|---|---|
| α) $C0 = C0^* = 1.0138\ AT^{0.5}m^{-2}$ | and | nominal [1] magnetic self-field; |
| β) $C0 = 1.1 \times C0^*$ | and | magnetic self-field reduced by factor 0.66. |

where C0 is a scaling constant entering the expression for the critical current density. [1]

**Table 1.** Summary of Mithrandir results

| Run # | Features | Choice | $\sigma$ ($\Delta$V) | $\sigma$ (NZ) | $\sigma$ (p) | $\sigma$ ($T_{co}$) |
|---|---|---|---|---|---|---|
| 2 | **See Ref.4** | $\alpha$ | **6.6e-2** | **3.7e-2** | **1.5e-2** | **0.16** |
| | | $\beta$ | **9.5e-2** | **3.4e-2** | **2.2e-2** | **0.20** |
| 4 | **$\Delta$T =0.85** | $\alpha$ | **0.60** | **0.65** | **0.46** | **0.14** |
| | | $\beta$ | **7.2e-2** | **3.5e-2** | **2.8e-2** | **0.17** |
| 5 | **$\Delta$T=0.35** | $\alpha$ | **0.88** | **1.96** | **0.38** | - |
| | | $\beta$ | **0.45** | **0.42** | **0.15** | - |
| 6 | **I=6kA** | $\alpha$ | **8.7e-2** | **3.6e-2** | **6.5e-3** | **0.22** |
| | | $\beta$ | **0.11** | **1.9e-2** | **6.3e-3** | **0.30** |
| 8 | **I=12kA** | $\alpha$ | **0.39** | **0.42** | **0.33** | **8.8e-2** |
| | | $\beta$ | **9.7e-2** | **7.7e-2** | **2.7e-2** | **0.11** |
| 9 | **Zero flow** | $\alpha$ | **8.4e-2** | **5.6e-2** | **1.1e-2** | **0.20** |
| | | $\beta$ | **9.0e-2** | **8.3e-2** | **2.0e-2** | **0.25** |

The results of the simulations performed with Mithrandir are summarized in Table 1. As a measure of the global agreement with the experimental results we report for each of the quantities mentioned above the values $\sigma$ of their standard deviations integrated in time, normalized to their maximum during the transient. Run # 2 is characterized by I=8kA, $\Delta$T=1.85K and 4g/s nominal (i.e., initial) mass flow, and has been analyzed in detail in Ref.4.

**General discussion of the 2-fluid results**

The agreement in standard quenches (runs # 2,6,9) between Mithrandir and experiment is typically very good with choice $\alpha$, whereas choice $\beta$ is needed to find a good agreement in most nonstandard quenches (runs # 4,8), while run # 5 cannot be well reproduced with the presently used values of the free input parameters, even with choice $\beta$.

If choice $\beta$ is used for standard quenches the agreement does not deteriorate too much, the $\sigma$'s typically staying well below the 10% threshold for all quantities except $T_{co}$. The disagreement in the latter can be qualitatively justified by the several uncertainties in the processing of this quantity, which was not directly measured: among these we can mention the neglect of temperature gradients between voltage taps, the strong (fifth power) dependence of Cu resistivity on temperature, and the fact that not the whole current probably goes to the copper (as is assumed).

**"Standard" quench 1: zero flow at t=0 [run # 9]**

For this run both Mithrandir and Gandalf provide a good agreement with the measured resistive voltage drop. As to the normal zone length, however, Gandalf predicts a decelerating quench, whereas Mithrandir correctly predicts an accelerating quench, in agreement with the experiment (see Fig.1), and this turns out to be a rather ubiquitous feature of the simulations we performed, independently of the specific experimental run. Mithrandir also predicts local pressure evolutions in much better agreement with the experimental ones than Gandalf (see Fig.2).

**"Standard" quench 2 [run # 2]**

This case behaves similarly to run # 9: the predicted voltage is in good qualitative agreement with the experiment, but again the quench is predicted by Gandalf to decelerate,

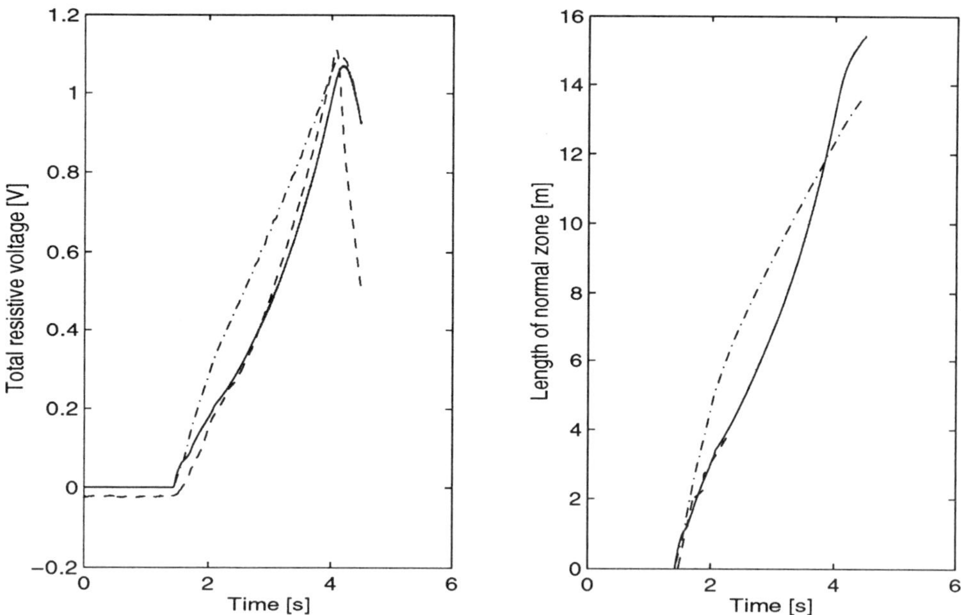


**Figure 1.** Run # 9: experiment (dashed), Mithrandir (solid), Gandalf (dot-dashed).

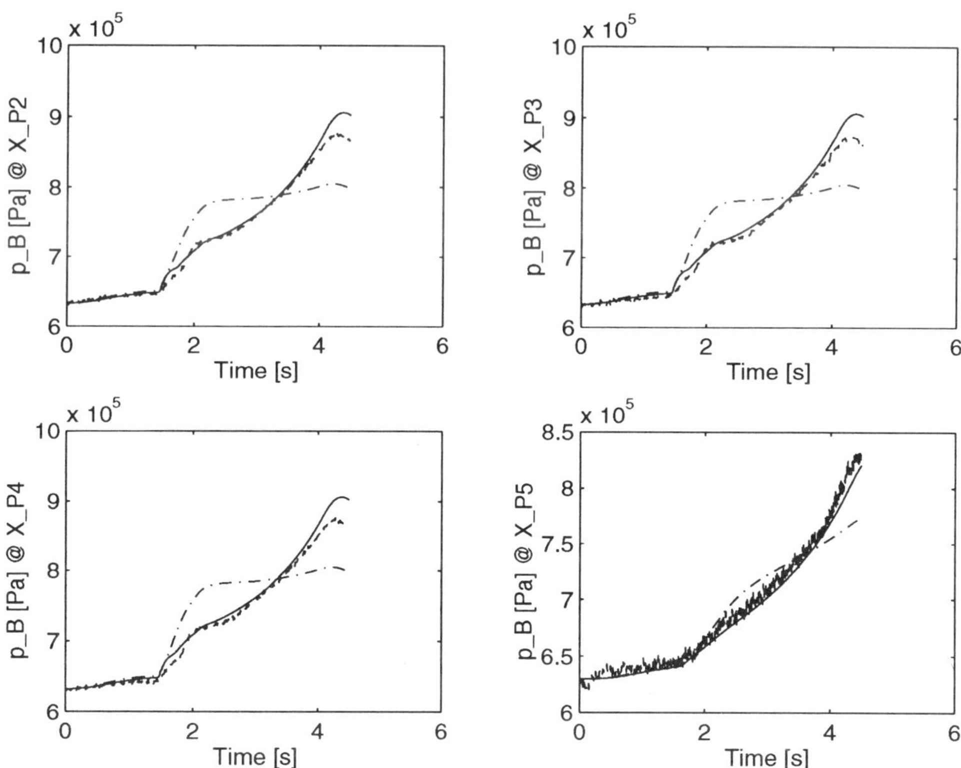


**Figure 2.** Run # 9, pressure at sensors: experiment (dashed), Mithrandir (solid), Gandalf (dot-dashed).

while Mithrandir predicts an acceleration as it happens in the experiment; the pressures are shown in Ref.4.

**"Standard" quench 3: low current = 6 kA [run # 6]**

With respect to the previous case we notice that here both codes show a good agreement with the experimental data (see Fig.3). In particular, the agreement of Gandalf is much better than in run # 9. This can be qualitatively explained by observing that: 1) the transient is now much slower, because the lower current gives a lower input power, and 2) the resulting overpressure at comparable times is also lower, almost impeding the attainment of flow reversal (not shown); the mass flow is therefore always near the nominal one, which is much larger than in run # 9. Both effects mentioned lead to a much better coupling between the $\mathcal{B}$ and the $\mathcal{H}$ region than in run # 9, so that the 1-fluid prediction now almost coincides with the 2-fluid.

**"Nonstandard" quench 1: low temperature margin = 0.85K [run # 4]**

This case is a first example of the sensitivity of code results to current&field parameters: with choice α above, Gandalf shows a significant disagreement with the experiment, whereas Mithrandir predicts a thermal-hydraulic quenchback (THQB) which is however not seen in the experiment (see Fig.4); with Mithrandir, choice β (not used with Gandalf) leads on the contrary to good agreement with the experiment, both with respect to voltage and normal zone, and with respect to the local pressures (Fig.5).

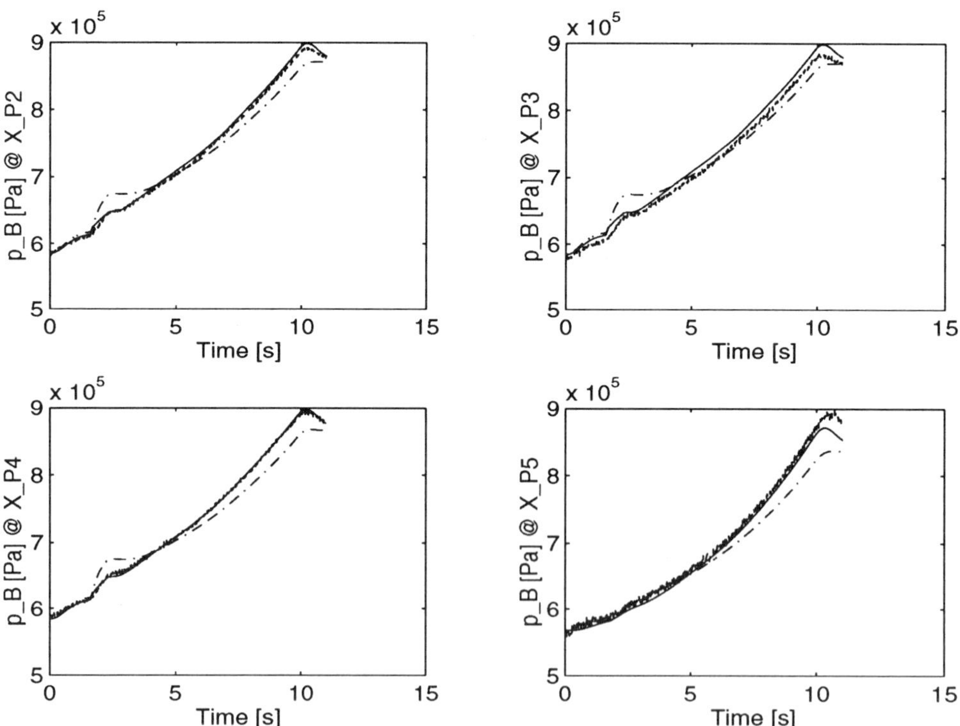


**Figure 3.** Run # 6, pressure at sensors: experiment (dashed), Mithrandir (solid), Gandalf (dot-dashed).

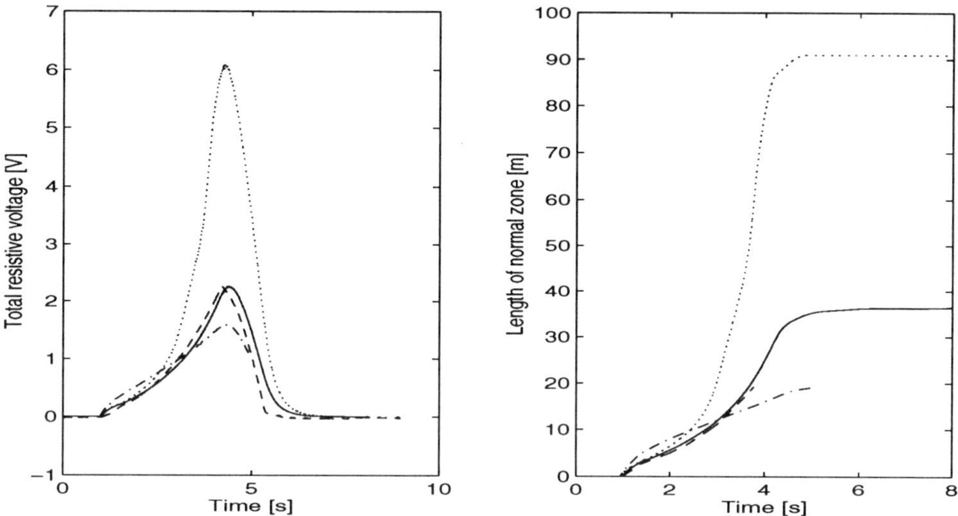


**Figure 4.** Run # 4: experimental (dashed), Mithrandir choice α (dotted), Mithrandir choice β (solid), Gandalf choice α (dot-dashed, simulation stops at start of current decay).

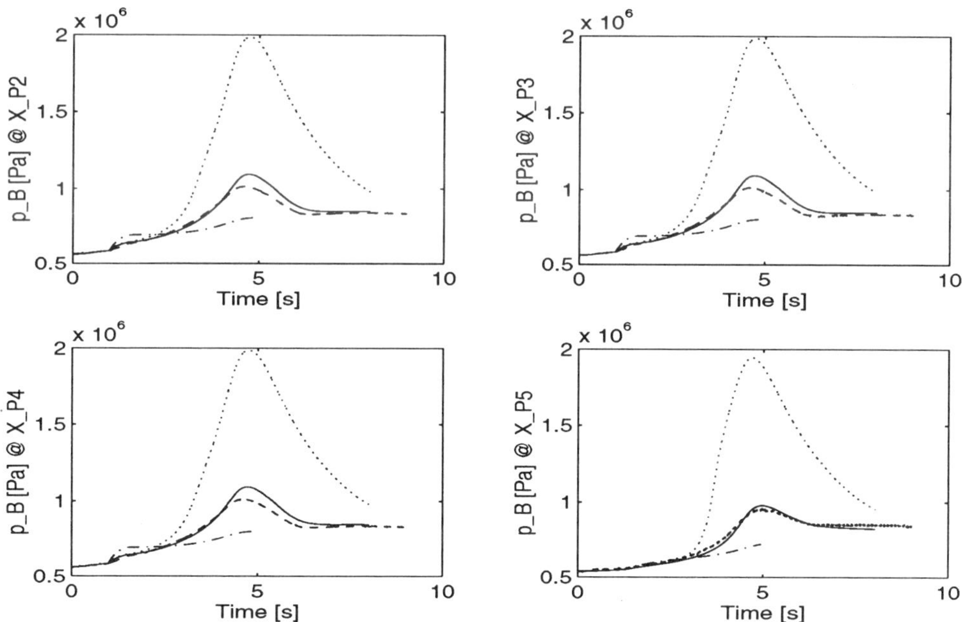


**Figure 5.** Run # 4, pressure at sensors: experimental (dashed), Mithrandir choice α (dotted), Mithrandir choice β (solid), Gandalf choice α (dot-dashed, simulation stops at start of current decay).

### "Nonstandard" quench 2: high current = 12kA [run # 8]

This case behaves somewhat similarly to the previous one (run # 4), although no THQB is predicted: with the nominal choice α (again the only one with which Gandalf was run in this case) a significant disagreement exists between the predictions of both models and the experiment (see Table 1 and Ref.4). With Mithrandir, choice β leads on the contrary to a reasonable agreement (see Table 1).

### "Nonstandard" quench 3: very low temperature margin = 0.35K [run # 5]

In this case the effect of external field modulation [1] is such that both codes, with nominal current&field parameters - choice α above, predict the onset of a THQB, which is not seen in the experiment. A reduction of the magnetic self-field by 0.66 is sufficient to obtain a fair agreement between Gandalf and the experiment. [4] On the contrary, even with choice β the disagreement with Mithrandir is reduced but still remains relatively large (see Table 1).

## CONCLUSIONS AND PERSPECTIVE

Interpretation of some standard QUELL quenches with a 1-fluid model (Gandalf) gives a reasonable agreement with the experiment, but 2-fluid (Mithrandir) modeling is more accurate both qualitatively and quantitatively, at the expense of a somewhat larger CPU time request.

For both approaches, runs with low temperature margin (< 1K) and/or high current density (corresponding to I > 8kA) can be roughly reproduced only with a suitable adjustment of some uncertain quantities, among which we chose the critical current constant C0 and the contribution of the self-field to the total magnetic field.

Predictive use of these codes will require a self-consistent treatment of the full hydraulic circuit and, for Mithrandir, a detailed analysis of sensitivity to its presently free input parameters.

## ACKNOWLEDGMENT

This work has been partially financially supported by the European Community under contract NET/96-424 to R.Z. and contract NET/96-425 to C.M..

## REFERENCES

1. A.Anghel, et al., The Quench Experiment on Long Length QUELL - Final Report, EPFL-CRPP, JAERI, MIT-PFC and SINTEZ-NIIEFA Report (1997).
2. R.Zanino, S.DePalo, and L.Bottura, A two-fluid code for the thermohydraulic transient analysis of CICC superconducting magnets, *J.Fus.Energy* 14:25 (1995).
3. L.Bottura, A numerical model for the simulation of quench in the ITER magnets, *J.Comput.Phys.* 125:26 (1996).
4. C.Marinucci, L.Bottura, G.Vecsey, and R.Zanino, The QUELL experiment as a validation tool for the numerical code Gandalf, submitted to *Cryogenics* (1997).
5. R.Zanino, L.Bottura, and C.Marinucci, A comparison between 1- and 2-fluid simulations of the QUELL conductor, *IEEE Trans. Appl. Supercond.* 7:493 (1997).
6. S.DePalo, C.Marinucci, and R.Zanino, Stability estimate for CICC with cooling channel using one- and two-fluid codes, to appear in *Adv. Cryo. Eng.*, Vol. 43.
7. R.Zanino, L.Bottura, L.Savoldi, and C.Rosso, Mithrandir+: a two-channel model for thermal-hydraulic analysis of cable-in-conduit super-conductors cooled with helium I or II, submitted to *Cryogenics* (1997).

# NUMERICAL EVALUATION OF THE QUENCH BEHAVIOUR OF THE ITER TOROIDAL FIELD MODEL COIL

R. Heller[1], J.L. Duchateau[2], S. Nicollet[2], and F. Prat[2]

[1]Forschungszentrum Karlsruhe, Institut für Technische Physik, 76021 Karlsruhe, Germany
[2]Association Euratom/CEA Cadarache
13108 St Paul Lez Durance Cedex, France

## ABSTRACT

In the frame of the ITER program, a Toroidal Field Model Coil is being built in Europe. Calculations about the thermohydraulic behaviour of this coil are presented in detail using the codes SARUMAN and GANDALF. The Helium temperature increase along the cooling circuits, the pressure drop and the mass flow partition inside the cable during normal operation will be discussed. One of the most crucial points is the behaviour of the coil during quench with a safety discharge. The general behaviour of the coil will be studied. Particular emphasis will be put on the pressure increase in case of a quench which in fact is strongly dependent on the dimensions of the joints and the external cryogenic circuit. The calculations and the resulting design choices for these parts are presented and discussed. An experimental program will be carried out to assess the validity of these choices as a necessary contribution to the ITER design.

## INTRODUCTION

The magnet system of the International Thermonuclear Experimental Reactor, ITER, consists of 20 toroidal field (TF) coils, a central solenoid (CS) and six poloidal field (PF) coils wound of superconductors. Despite different designs for the different coil systems, all conductors are designed as a cable-in-conduit (CIC) conductor having a central cooling channel surrounded by an annular region where the superconducting strands are located[1].

In case of the ITER TF coils, the conductor is encapsulated by a thin tube made of stainless steel or Incoloy 908 to form a Helium tight conductor. Then the conductor is put into grooves which are milled in a thick stainless steel plate (so called shear plate) which increases the stiffness of the pancake winding. The whole winding pack is then embedded in a stainless steel casing.

Within the ITER Engineering Design Activity (EDA) Phase, it was decided to construct a TF Model Coil (TFMC) in conjunction with European industry and test it in the TOSKA facility of the Forschungszentrum Karlsruhe beside the EURATOM LCT coil which delivers the background magnetic field to get the attraction forces for the TFMC. The TFMC is a Racetrack coil consisting of ten single pancakes having the joints at both the inner and outer sides of the coil. The conductor used for this purpose is a subsize model of the original TF conductor of type HP-I. In Table 1, the main parameters are summarized. For comparison, the parameters of the ITER TF conductor are given, too. The cross section of the TFMC conductor, of the winding pack as well as a perspective view of the TFMC is shown in Figure 1[2].

In parallel to the construction phase, calculations about the thermohydraulic behaviour of the TF Model Coil have been started. These calculations have to cover

- the real helium mass flow in the conductor, i.e., pressure drop and friction factor,
- the magnetic field pattern of the Model Coil,
- the pancake connections together with the helium supply lines, and
- the heating effect of the shear plates during discharge of the coil.

The knowledge of the quench behaviour of the coil is important for the design of the safety systems, i.e., the Helium supply system and the quench detection system. For the evaluation of the friction factor, it was decided to measure the pressure drop characteristic of a short piece of the conductor which was fabricated in industry using copper strands. The experiment is being performed at CEA Cadarache.

## EVALUATION OF THE FRICTION FACTOR OF THE TF MODEL COIL CONDUCTOR

In the frame of the ITER program, a quench experiment QUELL was done in the SULTAN facility at CRPP/Switzerland using a subsize ITER CS conductor[3]. The experiment had the intention to characterize the quench behaviour of a two-channel conductor and to compare the measurement data to calculation results obtained with numerical codes. This requires knowledge of the friction factor of the conductor used in the experiment. Two analytic approaches have been used for the central channel and for the

**Table 1.** Parameters of the ITER TF Model Coil Conductor. The data of the ITER TF conductor are shown for comparison

| Parameter | ITER-TF MC | ITER-TF |
|---|---|---|
| Type of conductor | HP-I | HP-I |
| Conductor hydraulic length [m] | 87.85 | 500 |
| Operation current [kA] | 80 | 60.4 |
| Maximum field [T] | 7.5 (+2.24 LCT) | 13 |
| Copper cross section [mm$^2$] | 408.1 | 500 |
| Non-Copper cross section [mm$^2$] | 148.4 | 315.7 |
| Jacket cross section [mm$^2$] | 196.5 | 299 |
| Strand perimeter [m] | 2.2903 | 2.68 |
| Helium annular cross section [mm$^2$] | 359.87 | 492.3 |
| Helium tube cross section [mm$^2$] | 78.5 | 132.7 |
| Tubular hydraulic diameter [mm] | 10 | 13 |
| Annular hydraulic diameter [mm] | 0.5651 | 0.6853 |

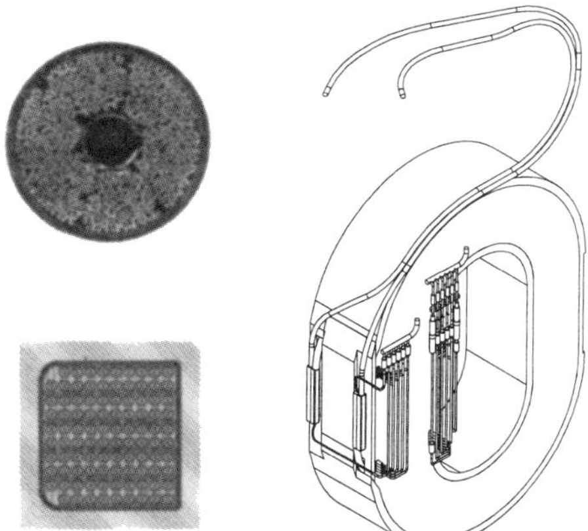

**Figure 1.** Cross section of the ITER TFMC conductor, of the winding pack and perspective view of the TFMC

annular region, i.e., the Blasius relation times a correction factor accounting for the roughness of the spiral used for the central channel, and an empirical relation published by H. Katheder which connects the void fraction of the CIC conductor, void, directly to the friction factor[4]. Using the He mass flow $\dot{m}$, density $\rho$, cross section A, hydraulic diameter $D_h$, and friction factor f, the pressure drop $\Delta P$ along a channel of length L is defined as

$$\Delta P = \frac{fL\dot{m}^2}{2\rho\, D_h A^2} \tag{1}$$

Then, one gets for the central spiral and for the annular region in case of the QUELL conductor sample[5]

$$f_{annular} = \left(\frac{1}{void}\right)^{0.72} (0.051 + \frac{19.5}{Re^{0.88}}),\ \text{and} \tag{2}$$

$$\frac{1}{4} f_{spiral} = 2.5 \cdot \frac{0.046}{Re^{0.2}}. \tag{3}$$

Here, the correction factor of 2.5 counts for the larger roughness of the spiral compared to a smooth tube. Re denotes the Reynolds number.

But for the ITER-TF Model Coil conductor, one has to find the appropriate correction factor. For this, an experiment is being done at CEA Cadarache to measure the pressure drop of a 5 m long test conductor piece as a function of mass flow rate, using nitrogen at room temperature.

## ESTIMATION OF THE MAXIMUM PRESSURE DURING QUENCH USING AN ANALYTIC APPROACH

### Dresner's approach

The most pessimistic assumption is that the whole length of the conductor gets normal and there is no energy release (adiabatic boundary conditions). With the copper density

$\rho_{Cu}$, cable space current density $J_{cs}$, and copper and Helium fractions $f_{cu}$ resp. $f_{He}$, one gets the following relation[6]

$$P_{max} = 0.14\left[\frac{\rho_{Cu}J_{cs}^2}{f_{cu}f_{He}}\right]^{0.72}\left(\frac{L}{2}\right)^{1.08} D_h^{-0.36} \tag{4}$$

Using the TFMC conductor parameters, one gets a maximum pressure of $P_{max}$=9.3 MPa.

**Shajii/Freidberg's approach**

A more realistic model takes into account a given length of the quench zone $L_q$. Considering the universal diagram presented in this approach, the TFMC is situated in the "long coil - high p" region, if the initial normal zone is 1 m long. One of the important input parameters is the initial temperature increase rate dT/dt. Using dT/dt=52 K/s (which is deduced from adiabatic hot spot evaluation taking into account the heat capacities of jacket and helium) and a characteristic time of $t_M$=5 s (which is related to the dump time constant of the coil), one gets a maximum pressure of about 2.2 MPa.

The latter number is much lower than the result obtained by Dresner's formula which is not surprising because this approach takes into account the quench propagation. So the results of numerical calculations should be more in the range of this formula than of Dresner's one. The reason for doing numerical calculations is the possibility to model the real magnetic field pattern in the Model Coil as well as the helium supply lines together with the pancake joints. Last but not least it is possible to take into account the influence of the eddy current heating of the radial shear plates.

## NUMERICAL CALCULATIONS

For the evaluation of the quench propagation of the ITER-TF Model Coil conductor, the two computer codes, SARUMAN[8] and GANDALF[9] developed for NET and ITER respectively are used. SARUMAN is a quasi 3-D code which allows modeling of the helium supply lines including pumps and reservoirs as well as an electric network to account for mutual coupling between different pancakes within the coil and/or between different coils including power supply or switches. The disadvantage of this code is the fixed mesh used which leads to the necessity of using a large number of nodes to get a representative and numerically stable result. GANDALF is a 1-D code which uses a very fine mesh at the quench fronts, so-called moving mesh. This allows the use of a marginal number of nodes and reduces the computer time considerably. The disadvantage is that in the actual version there is no possibility for introducing a hydraulic and/or electric network as it is possible in SARUMAN.

A second difference, which is of great importance for modeling the ITER type conductor, is that SARUMAN allows only one cooling channel whereas GANDALF offers the possibility of using two separate channel with same pressure and temperature but different velocities. The consequence is that one has to define so-called homogenized hydraulic quantities of the ITER conductor if SARUMAN is used. This means that one has to compute equivalent hydraulic diameters, friction factors and heat transfer coefficients of a conductor consisting of one cooling channel but with the thermohydraulic properties of a two-channel conductor. This method was proposed first by L. Bottura and used for analysing fault conditions for the ITER CS and P5 coils[10]. The computation of such homogenized quantities can be done in two different ways, either by adjusting the hydraulic

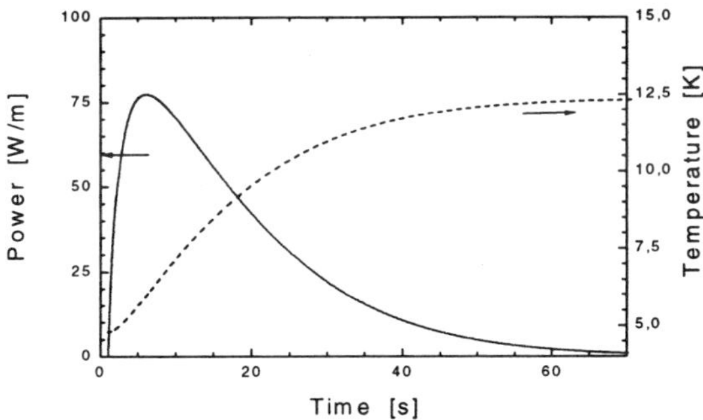


**Figure 2.** Heat flux from radial shear plates to the helium conductor of the ITER TFMC and resulting helium temperature increase as a function of time

diameter of the homogenized channel using the friction factor of the annular region or by adjusting the friction factor of the homogenized channel using the total hydraulic diameter.

## Calculation of quench propagation without hydraulic network

First the magnetic field in the winding pack of ITER-TF Model Coil was computed neglecting the background magnetic field of the LCT coil. The magnetic field calculations were done using the computer code EFFI[11]. The high magnetic field point of $B_{max}$=7.5 T was found to be near to the entrance of the inner pancake. Even at this location, the energy margin is expected to be very large because the maximum field for an operation current of 80 kA is much lower than the corresponding field of the ITER-TF coils at 60 kA. For the quench analysis, a perturbation of 4.5 J/cm$^3$, 0.5 m in length and 5 ms in time was put on the conductor. After 1 s of quench propagation, an exponential coil discharge with $\tau$=4 s was initiated. A total quench propagation time of 10 s was considered which is enough to obtain the essential results. For the moment, no hydraulic network was used and the heating of the shear plates was neglected, too. The boundary conditions were those of infinite helium reservoirs, i.e., constant helium pressure and temperature. Besides a pressure peak at the beginning of the quench of about 2 MPa, no dramatic pressure rise is observed. The maximum conductor temperature is about 135 K and the normal conducting length is very short, i.e., it is about 5 m after 10 s.

## Effect of radial plates heating on the quench propagation

Due to the current decrease initiated after the quench, eddy currents are induced in the stainless steel radial shear plates and in the coil casing[12]. Since the insulation wrapped around the TFMC conductor has a low thermal conductivity, the heat transfer between the plates and the helium conductor is strongly delayed. This delay will avoid any global quench of the coil. The heat flux through the outer surface of the jacket and the resulting temperature increase of the helium in the cable space as a function of time is presented in Figure 2. The calculation of the temperature increase was done neglecting Helium circulation.

The heat flux through the jacket is used in GANDALF to study the effect of the eddy current heating of the radial plates during the safety discharge following the quench of the inner pancake of the ITER TFMC. The result is an overall increase of the temperature inside the pancake and an increase of the He pressure, as expected. But the surprising result is a decrease of the normal length compared to the calculation without external heating of the radial plates. This can be explained by the heat flux from the jacket to the helium and the conductor leading to an increase of the mass flow resulting in an improved heat removal from the quenched region of the conductor. In addition, cold helium flows from the inlet in the conductor due to the boundary condition of an infinite volume both at inlet and outlet. Figure 3 shows the normal conducting length, the helium pressure and the conductor temperature as a function of time. For comparison, the results of the calculations without heating of radial shear plates are shown.

## Calculation of quench propagation with hydraulic network

To calculate the quench propagation of the TFMC, the helium reservoirs and supply lines have to be modeled. In addition, the effect of the pancake connections, i.e., the joints, have to be investigated. Due to fabrication and thermal reasons for the joints, the central spiral will be replaced by a tube with a larger wall thickness resulting in a smaller cross section for the helium flow. In addition, the void fraction of the annular region in the joint will be reduced to 25 %. The helium supply lines to the joints have an inner diameter of 8 mm, 13.6 mm and 32 mm respectively. So, the joint as well as the first pipe will act as a bottleneck for the helium in case of a quench, and the impact of it on the thermohydraulic performance of the TFMC has to be checked. For the calculations, the supply lines will be modeled with the hydraulic diameter being the average inner diameter of the joint and the three pipes. Figure 4 shows the hydraulic network used in the calculations.

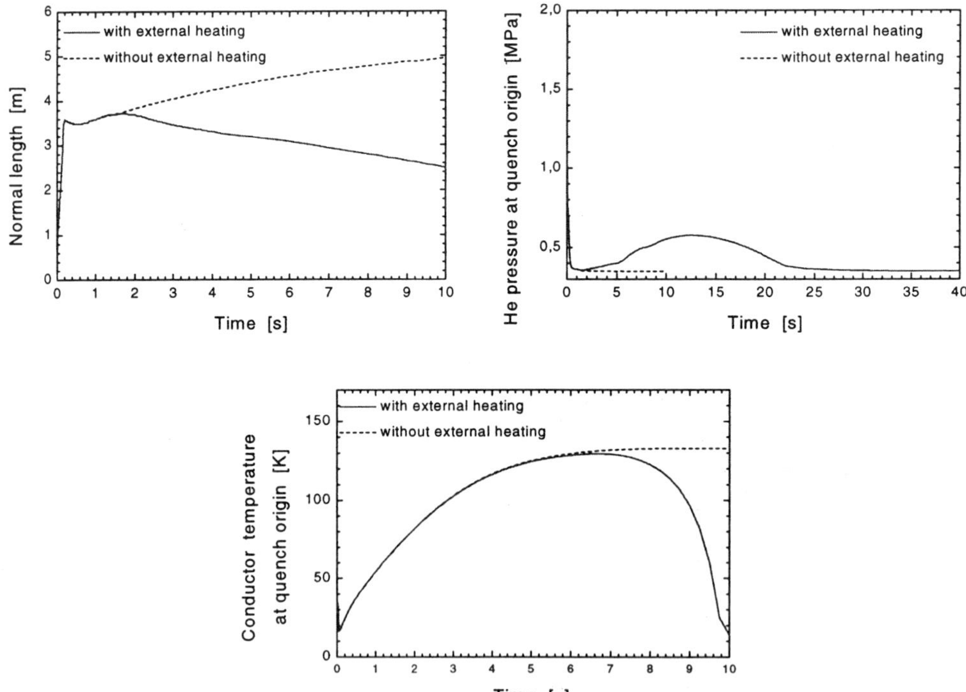


**Figure 3.** Normal conducting length, Helium pressure and conductor temperature at the quench origin of the ITER TFMC as a function of time with external heating. For comparison, the results of the calculation without external heating are shown, too.

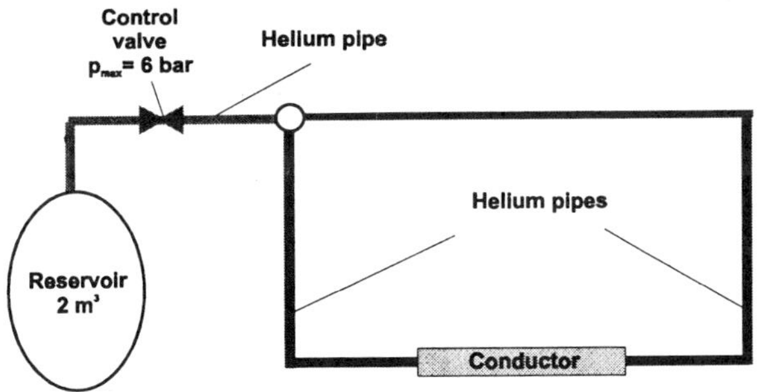


**Figure 4.** Model of the hydraulic network used in the calculations.

As already mentioned, the introduction of the hydraulic network implies the use of SARUMAN, i.e., the definition of the homogenized quantities. To study the effect of homogenization, two calculations were done with GANDALF using both the two-channel and the one-channel model. The differences are marginal, which validates the use of the one-channel model. Two calculations have been performed using the one-channel model and SARUMAN, i.e., without and with hydraulic network. The quench was initiated 2.5 m after the joint in the high field region of the inner pancake. No adjacent pancakes were modeled. Perturbation power and time were the same as for the GANDALF calculations done before. The effect of the eddy current heating of the radial shear plates and of the casing of the TFMC has been neglected. For comparison, one calculation was done using a new release of GANDALF and the same hydraulic network as for SARUMAN[13].

Figure 5 shows the pressure and temperature evolutions over the length of the pancake as well as the pressure at the joints as a function of time, calculated with GANDALF. In Table 2, the main results of the calculations are summarized.

## Discussions of results

From Table 2 it can be concluded that the effect of the hydraulic network doesn't change the overall quench behaviour of the ITER TFMC if averaged hydraulic channels are used, except the pressure head at the end of the coil discharge which rises up to the relief pressure of 0.6 MPa. All other numbers are in the same range. It is remarkable that the addition of the hydraulic pipes leads to an increase of the normal length by about 50 % which could be due to the fact that the boundary condition without pipes assumes an infinite helium reservoir, i.e., enhanced mass flow of cold helium from the ends occurs.

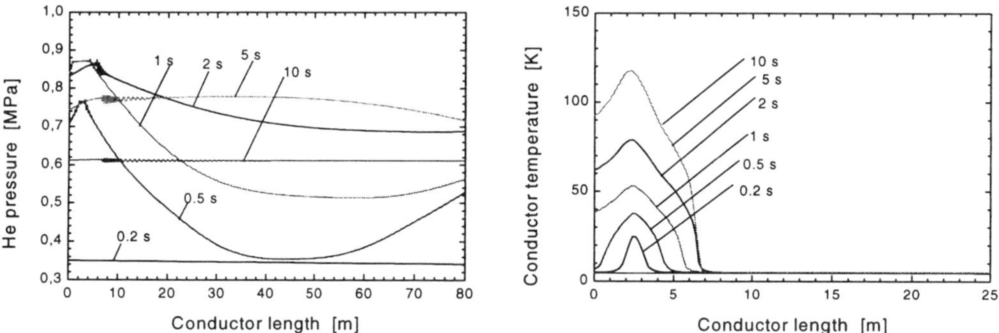


**Figure 5.** Evolution of the pressure and conductor temperature over the pancake length at different times as calculated with GANDALF.

**Table 2.** Main results of the calculations of the quench propagation of the inner pancake of the ITER TF Model Coil using GANDALF and SARUMAN

| Parameter | GANDALF two-channel | GANDALF one-channel | GANDALF ext. heating | SARUMAN no pipes | SARUMAN hydr. pipes | GANDALF hydr. pipes |
|---|---|---|---|---|---|---|
| Max. cond. Temp. | 132 K | 135 K | 130 K | 125 K | 134 K | 125 K |
| Max. He pressure | 1.92 MPa | 1.75 MPa | 1.92 MPa | 1.38 MPa | 1.87 MPa | 0.88 MPa |
| Max norm. length | 4.4 m | 4.3 m | 3.8 m | 3.7 m | 5.35 m | 6.7 m |
| Resistive Voltage | 1.61 V | 1.56 V | 1.51 V | 1.37 V | 1.42 V | 1.85 V |
| Max. Pressure at joint | n.a. | n.a. | n.a. | n.a. | 0.81 MPa | 0.84 MPa |
| Pressure after 10 s | 0.35 MPa | 0.35 MPa | 0.59 MPa | 0.35 MPa | 0.35 MPa | 0.6 MPa |

## CONCLUSIONS AND FUTURE WORK

The numerical evaluation of the behaviour of the ITER TFMC during quench with a safety discharge has been started. The effect of the eddy current heating of the radial shear plates during safety discharge as well as the introduction of an averaged hydraulic network results in no qualitative changes of the quench behaviour of the coil. Future work will cover a detailed analysis of the quench evolution using a more sophisticated hydraulic network which requires some modifications in the code GANDALF. The effect of eddy current heating with a longer discharge time constant of the coil has to be studied as well.

## ACKNOWLEDGEMENT

The authors acknowledge the kind help of L. Bottura who did one test run using the code GANDALF. This work has been performed within the frame of the European Fusion Technology Programme.

## REFERENCES

1. B.J. Green, M. Huguet, "The ITER Project: Status and Prospects", IEEE Trans. on Magn., vol. 32, No. 4, (1996), 2224-2229
2. P. Libeyre et al., "Conceptual Design of the ITER TF Model Coil", IEEE Trans. on Magn., vol. 32, No. 4, (1996), 2260-2263
3. The Quench Experiment on Long Length - Final Report, Villigen, Naka, Cambridge, St. Petersburg, (1997)
4. H. Katheder, "Optimum Thermohydraulic Operation Regime for Cable in Conduit Superconductors (CICS)", Cryogenics vol 34 ICEC Supplement 1994, 595-598 (1994)
5. A. Anghel, private communication
6. J.R. Miller, L. Dresner, J.W. Lue, S.S. Shen, H.T. Yeh, "Pressure rise during quench of a superconducting magnet using internally cooled superconductors, Proc. ICEC8, Genova, (1980), 3291-3297.
7. A. Shajii, J.P. Freidberg, "Quench in superconducting magnets II: Analytic solution", J. Appl. Phys. 76(5), (1994), 3159-3171
8. L. Bottura, O.C. Zienkiewicz, "Final Report on NET contract 90-245", (1993)
9. L. Bottura, "A Quench Simulation Model for the ITER Magnets", NET Internal Report N/R/08231/54/A, (1994)
10. R. Heller, "Quench Analysis for Magnet Safety Assessment", Final report Task D42, (1995)
11. S.J. Sackett, "EFFI, a code for calculating the electromagnetic field, force and inductance in coil systems of arbitrary geometry", LLNL, Livermore, California, UCRL-52402, (1978)
12. P. Hertout, J.L. Duchateau, P. Libeyre, B. Turck, Y. Stephan, "Prediction of Eddy Currents in the ITER TF Model Coil Experiment using the Code Corfou", presented at the ISTET Conference, June 1997
13. L. Bottura, private communication

# THERMAL AND HYDRAULIC MEASUREMENT IN THE ITER QUELL EXPERIMENTS

K. Hamada,[1] Y. Takahashi,[1] N. Koizumi,[1] H.Tsuji,[1]
A. Anghel,[2] B. Blau,[2] A. Fuchs,[2] B.Heer,[2] G. Vecsey,[2]
S. Smith,[3] S. Pourrahimi,[3] and M. Zhelamskij [4]

[1]Japan Atomic Energy Research Institute
801-1, Muko-yama, Naka-machi, Naka-gun, Ibaraki, Japan

[2]EPFL-CRPP, CH-5232 Villigen PSI, Switzerland

[3]Massachusetts Institute of Technology, Cambridge, MA 02139,USA

[4]Scientific Center, SINTEZ, St. Petersburg, Russia

## ABSTRACT

The Japan Atomic Energy Research Institute has a collaboration with the European Union, the United States of America and the Russian Federation for the International Thermonuclear Experimental Reactor (ITER). In the engineering design activity for ITER, a test coil named QUench Experiments on Long Length (QUELL), using 91 m and 1/5-size ITER superconducting conductor, was fabricated by JAERI. The performance tests were carried out at the SULTAN facility in Switzerland where quench propagation, thermal and hydraulic characteristics were determined and development and test of new quench detection system were conducted. The thermal and hydraulic behavior was not known well. This conductor has a central channel to reduce the pressure drop. In order to investigate the thermal and hydraulic characteristic of the conductor, the pressure drop has been measured at 5-13 K and 2-11 g/s, and the friction factor of the central channel was calculated. In heat slug propagation, an inductive and resistive heater on the conductor has been used and the velocity of the heat front and input energy are estimated from the temperature change of conductor.

## INTRODUCTION

Japan Atomic Energy Research Institute (JAERI) has participated in the International Thermonuclear Experimental Reactor (ITER) project[1]. The development of superconducting magnets for ITER is one of the large hardware R&D works in the engineering design activity (EDA) of ITER. ITER superconducting magnets use a cable in conduit conductor (CICC) with a central channel to reduce the helium pressure drop and to improve the cooling performance in the conductor. However, there is little experimental data on pressure drop and friction factor and also the thermal and hydraulic behavior were not understood clearly. The QUench Experiment on Long Length (QUELL)[-2-], which has an ITER sub- size and 91- m conductor, was planned in order to clarify the thermal, hydraulic, electrical, and quench characteristic[-3-] of CICC with central channel, to test new quench detection systems[-4-] and to validate numerical codes, with the purpose of establishing the design criteria of ITER superconducting magnets.

QUELL coil has been fabricated by JAERI and the experiment has been performed in collaboration with the European Union(EU), the United States of America and the Russian Federation, using a large scale superconducting test facility "SULTAN" at the Centre de Recherche en Physique des Plasma (CRPP) in Villigen, Switzerland.

The measurement of thermal and hydraulic characteristic using an ITER-like conductor with a central channel is one of main items in this experiment. The pressure drop has been measured and the friction factor of the central channel was estimated. QUELL coil has an inductive heater (IH) and a resistive heater (RH) in order to initiate a quench propagation and to carry out a stability experiment. Using these heaters, heat slug propagation in the conductor has been performed and hot helium flow velocity and input energy have been measured. These results are discussed in this paper.

## QUELL COIL

QUELL coil and cross section of the conductor are shown in Figure 1. Outer diameter and winding length of the QUELL coil are 507.2 mm and 735 mm, respectively. The coil has a two layer winding structure and the number of turns is 30 turns / layer. In order to reduce the mutual inductance between QUELL and background field coil (SULTAN), a non-inductive winding structure is used.

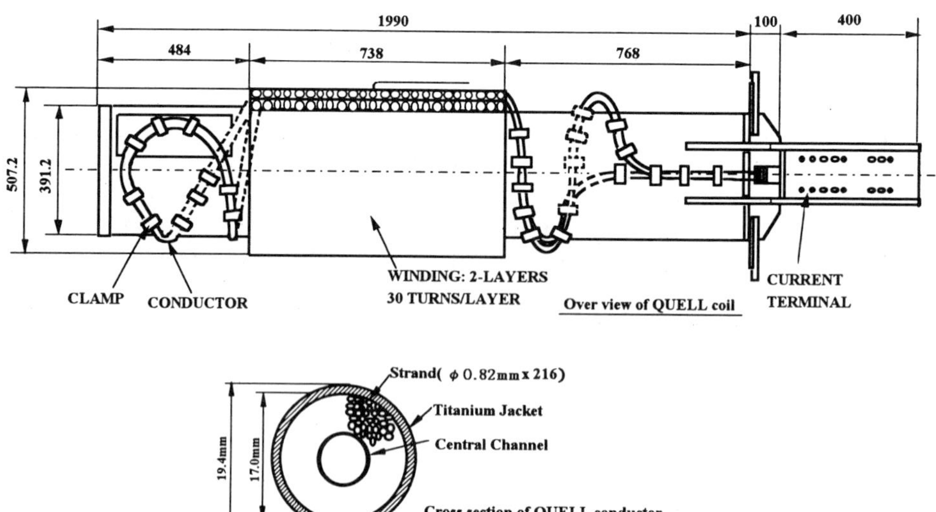


**Figure 1.** Over views of QUELL coil and cross-section of the conductor. Dimensions are in mm.

**Table 1.** Specification QUELL conductor

| | |
|---|---|
| Conductor inner / Outer diameter | 19.4 / 17.0 mm |
| Number of strands | 214 |
| Strand Diameter | 0.81 mm |
| Strand performance | 640 A/mm$^2$ at 12 T, 4.2 K(non copper) |
| Conductor rated current | 8 kA at 12 T |
| Central channel outer/inner diameter | 7 / 6 mm |
| Void fraction | 34 % |
| Conductor Length | 91 m |
| Hydraulic diameter / helium cross section at cable section | 0.47 mm / 66 mm$^2$ |

The major specifications of the conductor are listed in Table 1. The superconducting cable is composed of 216 - $Nb_3Sn$ strands, cabled around a central channel and inserted in a titanium jacket 91- m long. The QUELL conductor is a 1/5 scalded-down version of the ITER Center Solenoid (CS) Coil conductor. The strand has been fabricated in the ITER R&D to accomplish a mass predictability of ITER superconducting strand satisfying with the ITER specification. Figure 2 shows positions of pressure sensors, thermometers and heaters on the QUELL coil. In order to monitor the jacket temperature, nine carbon glass thermometers are inserted into a copper block attached to the jacket by ultrasonic solder. The heaters, IH and RH are placed in the middle of the conductor length. The length of RH and IH are 2.4 m and 0.12 m, respectively.

## PRESSURE DROP MEASUREMENT

Pressure drop characteristic of QUELL coil provides important information to estimate the hydraulic performance of a conductor with a central channel for determining cooling efficiency and pump work. Although the flow friction factor of CICC has been well investigated[5], the friction factor of the central channel made of a spiral tube has not been studied at cryogenic temperature. The friction factor of the conductor, which has two flow paths, a central channel and a cable space, is estimated from the experimental data. The measurement has been performed in the temperature range of 5 K to 13 K and the flow rate range of 2 g/s to 11 g/s. The measured pressure drop, dependent on mass flow rate, is shown in Figure 3. In order to predict the pressure drop characteristic of the ITER CICC, a parallel two-flow model consisting of central channel and cable space is assumed.

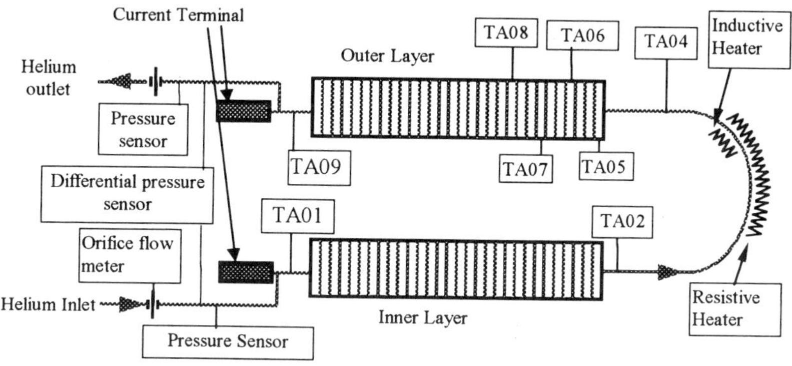


**Figure 2.** Thermometer and heaters position of QUELL coil. TA means a carbon glass thermometers.

The pressure drop can be calculated by the Fanning equation.

$$\Delta p = 4 \cdot f \cdot \frac{\rho \cdot v^2}{2} \cdot \frac{L}{D} \tag{1}$$

here, $\rho$ : density, $v$ : flow velocity, $L$ : conductor length, $D$ : hydraulic diameter and $f$ : friction factor

The friction factor of the cable void is estimated to be as a typical CICC and a general correlation formula[5] can be applied.

$$f_S = \frac{1}{4 \cdot v^{\alpha}} \left[ \frac{\beta}{\mathrm{Re}^{\gamma}} + \delta \right] \tag{2}$$

Here $\alpha$, $\beta$, $\gamma$ and $\delta$ :constant and Re: Reynolds number.

According to ITER design recommendation, the friction factor of the central channel should be calculated by the modified Blasius equation.

$$f_C = 0.046 \cdot \frac{N}{\mathrm{Re}^{0.2}} \tag{3}$$

Here, N-factor is equal to 2 for a central channel. However, the equation is not confirmed with experimental results using an ITER like conductor. In order to correct the N-factor to correspond to a practical conductor, the friction factor of the central channel is calculated from the experimental data and the deviation is studied. Solving two Eq. (1) for the central channel and the cable void, the friction factor of central channel $fc$ is obtained as follows:

$$f_C = f_S \frac{{v_S}^2}{{v_C}^2} \cdot \frac{D_C}{D_S} \tag{4}$$

Here, subscription $s$ and $c$ express the cable void space and the central channel, respectively.

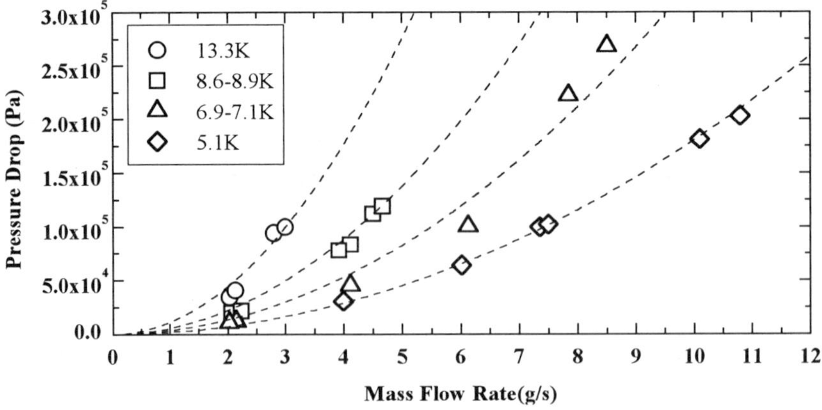


**Figure 3.** Pressure drop dependent on mass flow rate results in QUELL experiments. Broken lines are fitting curve by quadratic polynomial expression.

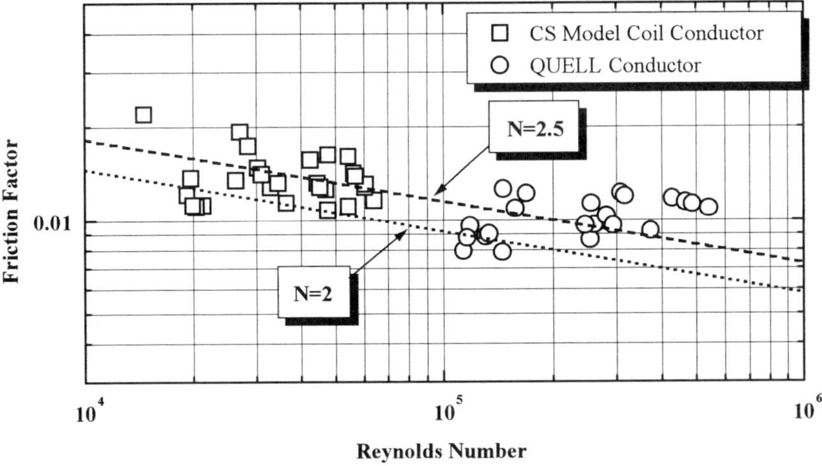


**Figure 4.** Friction factor dependent on Reynolds number of central channel.

Firstly, according to experimental condition of pressure and temperature, mass flow rate of the cable void space is calculated using Eq.(1) and (2). Then the mass flow rate in the central channel is calculated from the measured total mass flow rate and calculated flow rate in the cable void space. Finally, using Eq.(4), $fc$ can be obtained. A flow distribution between the central channel and the cable space is estimated as 70-80% and 20-30%, respectively. On the other hand, the friction factor of the central channel is also calculated, using the measurement data of ITER Central Solenoid (CS) model coil conductors[6], which are the real conductor for ITER and the length is around 160 m, at room temperature with nitrogen gas. These data are compared and the friction factor is estimated as shown in Figure 4.

Although the measurement conditions for the two kinds of conductor are completely different, the friction factor using N=2.5 is well-fitted to the experimental data.

## FLOW VELOCITY AND HEATER INPUT ENERGY IN CONDUCTOR

A heat slug propagation experiment has been performed to estimate the heat input energy and the propagation velocity. Heating energy is applied to the jacket and is conducted into the conductor. Thermometers located down stream from the heater detect the propagating temperature wave. A typical temperature wave propagation using IH is shown in Figure 5. Thermometer TA9 indicates a higher temperature than other thermometers because TA9 is located near the current terminal and is affected from a heat leak from current terminal. Helium flow velocity at each point is calculated from the time difference between the temperature peak of the thermometer and neighboring thermometer. In another way, flow velocity is calculated from the mass flow rate and overall helium flow cross section. The measured results at both 5 g/s and 10 g/s and predicted flow velocity using results of pressure drop measurements, are summarized in Table 2. The flow velocity based on the mass flow rate is in a good agreement with the results of propagation velocity. It means that the helium flow in the central channel contribute to the thermal diffusion from cable area to central channel.

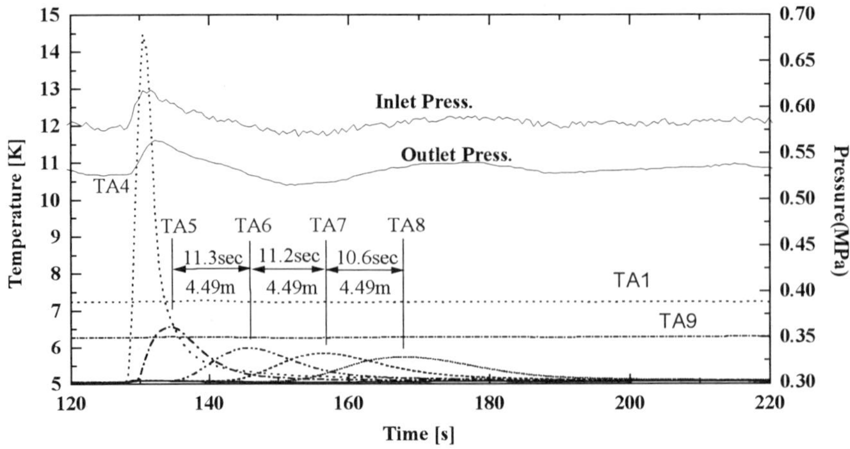


**Figure 5.** Hot wave propagation in the QUELL conductor at 5g/s, 5.1K and 0.6MPa.

In the section between TA4 and TA5, the velocity is dependent on the input energy and is observed to be 10-20-% slower than other velocities. The distance between TA4 and the heaters are 30 mm for RH and 180 mm for IH, respectively, and it takes a few seconds for the hot helium flow to pass TA4. Consequently, there is the possibility that the input energy does not diffuse enough in the radial direction of conductor in the first few seconds after heat pulse.

Heater input energy can be estimated from an integration of Enthalpy increment calculated from the temperature at each time. The RH has been operated with 300-ms pulse duration. The IH has been operated at 1 kHz and 590 Hz and the pulse duration was 40 ms. The result of RH is compared with joule heating as shown in Figure 6. The results for each thermometer (TA4-TA8) are in good agreement with the joule heating energy.

The calculation results for inductive heating are also compared with calibration using the calorimetric method[7] as shown in Figure 7. In Figure 7, x-axial represents the heating factor, which is proportional to an inputted energy to conductor by an IH. The input energy of IH, therefore, is estimated to be as a function of the heating factor. Although TA4 has shown relatively agreement with calorimetric method, the calculated value shows the effect of insufficient of thermal diffusion, as we already mentioned. TA5, TA6, TA7 and TA8 indicate around 30-% lower value than the calibration.

The diffusion of calculated propagation energy along the conductor length in the down stream direction is not observed clearly. This fact indicates that a thermal conduction between a turn and neighboring turn is negligible.

**Table 2.** Flow velocity estimation

| | Measured flow velocity | Flow velocity calculated from mass flow rate | Predicted flow velocity in central channel | Predicted flow velocity in cable void |
|---|---|---|---|---|
| 5 g/s | 0.41 m/sec | 0.43 m/sec | 1.0 m/sec | 0.1 m/sec |
| 10 g/s | 0.83 m/sec | 0.83 m/sec | 2.1 m/sec | 0.3 m/sec |

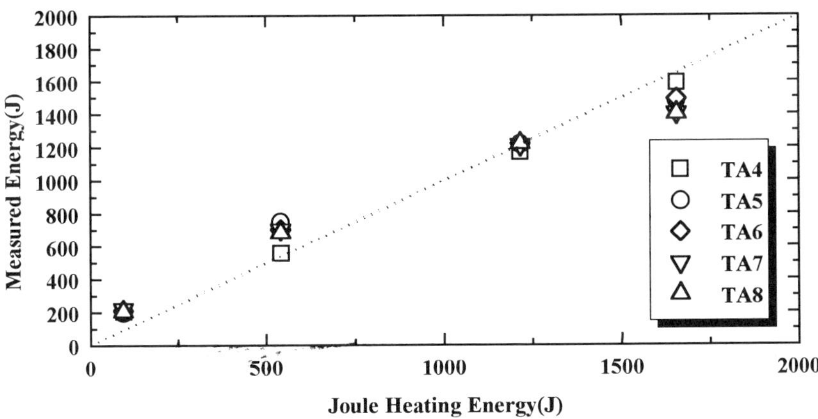


**Figure 6.** Input energy of resistive heater.

The main reason for the difference is considered to be the adiabatic compression of the helium due to the heat pulse. In case of 1 kHz operation, inlet and outlet pressure of QUELL coil has been increased about 0.05-0.1 MPa and the conductor temperature slightly increased to around 0.1 K when the heater was operated. The temperature increment corresponds to the change of the pressure, assuming adiabatic compression. From the deviation between the measured heat energy and calibration, the compressed helium volume is calculated to be from 5 liters to 6 liters. In the other hand, QUELL has 8.4- liters helium inventory and 4 or 5-liters helium, which is initially located down stream of each thermometer and compressed and is not included in above input energy calculation. Accordingly, it is considered that an adiabatic compression work due to heating occupies a large portion of heating energy. Especially, IH can input directly a high energy in the short length and short pulse duration, in comparison with RH. It is expected that IH can induce the adiabatic compression due to the high energy density, which is around 300-times larger than that of RH.

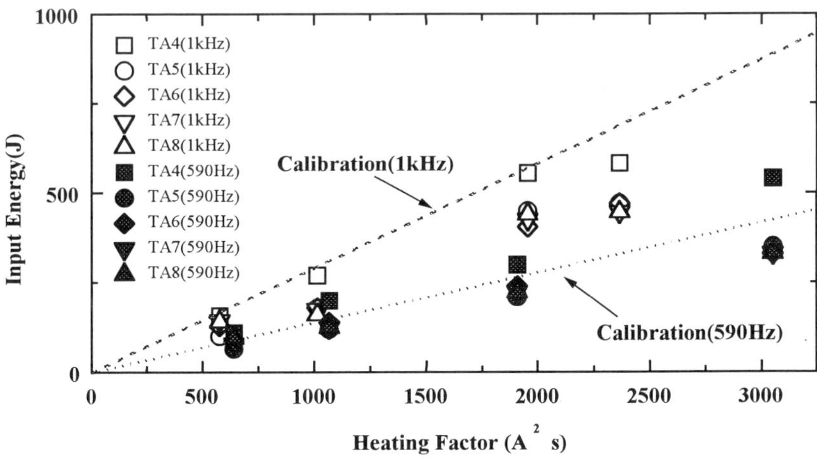


**Figure 7.** Input energy of inductive heater.

## CONCLUSION

In the QUELL experiments, thermal and hydraulic characteristics of a CICC with central channel similar to the ITER conductors have been measured. The experimental results are summarized as follows;

(1) Friction factor is estimated and the results have a good agreement with the ITER CS model coil conductor. N-factor of the central channel in the modified Blasius equation, is estimated to be 2.5.
(2) Heat slug propagation in the conductor has been observed and the flow velocity corresponds to that of mass flow measurement with a total helium cross section in conductor. The distribution ratio of helium flow is estimated to be as 70-80% for central channel and 20-30% for cable space. Central channel contributes to the thermal diffusion from strand to central channel.
(3) Input energy of RH estimated by enthalpy increment has good agreement with joule heating energy. Because IH induces an adiabatic compression of helium in the conductor due to high energy density, the work can not be neglected for estimation of input energy.

These data were used for establishing the design criteria for the ITER superconducting magnet.

## ACKNOWLEDGEMENT

The authors would like to acknowledge Drs. M. Ohta, T. Nagashima, and S. Matsuda for supporting this work. The contribution of this work from Hitachi, Ltd (QUELL coil fabrication) is grateful acknowledged.

## REFERENCES

1. B. J.Green and M. Huguet, The ITER Project : Status and Prospects, MT-14, IEEE Trans. On Magnetics . Vol. 32, No.4 2224 [1996]
2. A. Anghel, The Quench Experiment on Long Length Cable in Conduit Conductor (QUELL) in SULTAN Journal of Fusion Energy,14, 129 (1995)
3. A. Anghel et al. THE ITER -QUELL, a Quench Propagation Experiment on Long Length CICC with Central Channel, Proceeding of 19th SOFT, [1996]
4. S. Pourrahimi et al. Performance of the US Quench Detection System in the QUELL Experiments, IEEE Trans. on Applied Superconductivity. [1996]
5. H. Katheder, OPTTIMUM THERMOHYDRAULIC OPERATION REGOME FOR CABLE IN CONDUIT SUPERCONDUCTOR, Cryogenics, Volume 34 ICEC15 supplement [1994]
6. T. Ando et al. Design of the ITER-CS Model Coil- Outer Module, Proceeding of the ICEC16/ICMC, Vol. 2, 767 [1996]
7. T. Ito et al. Evaluation of Inductive Heating Energy of Quench Experiment on Long Length (QUELL) Conductor with the Calorimetric Method, Proceeding of the ICEC16/CEC Vol. 2, 1301 [1996]

# REARRANGEMENT OF SUPERCONDUCTING WIRES IN SOLENOIDAL MAGNETS

H.Moriyama,[1] F.Sawa,[1] H.Mitsui,[1] M.Arata,[1]
S.Nishijima,[2] and T.Okada[2]

[1]Toshiba Corporation
Yokohama, 230, Japan

[2]ISIR Osaka University
Ibaraki, Osaka, 567, Japan

## ABSTRACT

The strain behavior of the superconducting solenoidal coil form has a close relationship with the training behavior of the quench current. This relationship can not be explained by means of models of wire motion, friction and resin cracking, which are known as mechanical disturbances, as the cause of quench. Therefore, a wire rearrangement model was proposed, where the accumulation of the microscopic displacements of the multiple turns of superconducting wire manifests itself in the strain behavior of the form. Based on this model, the strain energy of superconducting wire released by rearranging the turns was investigated theoretically and experimentally. As a result it was confirmed that the released energy exceeds the minimum quench energy, MQE, resulting in the instability of a sample coil. From this result, it is expected that the wire rearrangement model is useful for estimating the stability.

## INTRODUCTION

In a superconducting magnet showing the training effect, it is expected that the winding of the magnet would reflect the training behavior.[1] It has been clarified that it is the solenoidal magnet form that shows the change of strain with training.[2 ~ 4] This study was conducted to enable the estimation of the stability of a magnet during the design stage by means of the strain behavior.

The strain behavior can not be explained by means of models of wire motion, friction and resin cracking, which are known as mechanical disturbance models, for the training.[4] Therefore, a rearrangement model, which is a model of the microscopic displacement of multiple turns of superconducting wire, was proposed and the friction and resin cracking models were devised.[4, 5]

When the superconducting wire is rearranged, the tension of the superconducting wire changes and part of the strain energy of the wire is released and converted into heat. When this released energy exceeds the minimum quench energy (MQE) of the magnet, the magnet quenches. Therefore, by calculating the tension of superconducting wire and inferring the relationship between the strain of the form and that of superconducting wire, the energy released by the rearrangement was studied theoretically. Further, by measuring the strain behavior of the former, the energy released by the rearrangement was obtained experimentally. By comparing this released energy with MQE, the released energy was determined to be valuable as a measure of the stability.

In this paper, the results of the superconducting wire tension calculation are described and the rearrangement model based on the load-strain line of superconducting wire is proposed. Further, the released energies obtained theoretically and experimentally are compared and the validity of the rearrangement model as a measure for stability is discussed.

## METHODS

### Sample Coil

To investigate the rearrangement model experimentally, a dry coil, with winding turns not bonded by resin, was fabricated.[4] Figure 1 shows the coil construction, where the axes r, $\theta$ and z are the cylindrical coordinates. The 5 mm thick cylindrical coil form was made from an aluminum alloy. The ground insulation was made of glass-fiber reinforced plastic (GFRP) attached to the outer surface of the form. The polytetrafluoroethelene (PTFE) sheet, which has a small friction coefficient, was glued on the outer surface of the ground

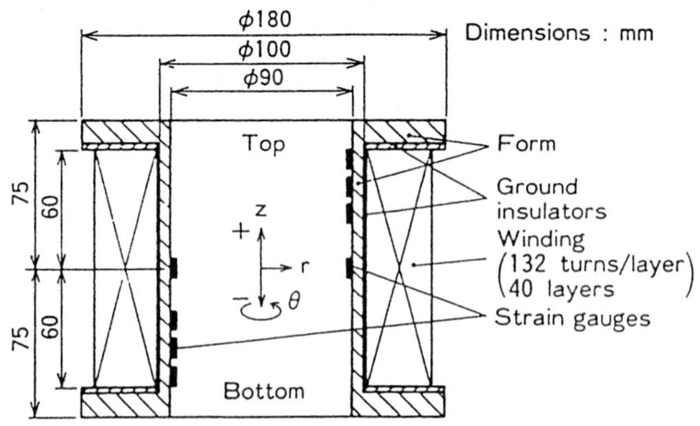


**Figure 1.** Construction of sample coil.

**Table 1.** Specifications of superconducting wire

| Items | Specifications |
|---|---|
| Superconductor | Nb-69%Ti |
| Copper to S/C ratio | 3 |
| Insulation | Polyvinylformal |
| Thickness of enamel | 50 $\mu$m |
| Diameter of bare wire | 0.8 mm |
| Diameter of enameled wire | 0.9 mm |

insulation. Table 1 shows the specifications of the superconducting wire. The diameter of enameled superconducting wire is 0.9 mm. The winding tension was set to 88 N.

## Theoretical Study of Rearrangement

**Calculation of Coil Form Strain and Wire Tension** The winding can be considered as a multi-layer cylinder of superconducting wire. When the cylinder is sufficiently long in the z direction, the tension of superconducting wire in each layer on the r- $\theta$ plane can be calculated using the multi-layer cylinder model. However, the strains of the form and the superconducting wire in the $\theta$ direction were not considered accurately in the previous papers. The wire tension $F_\theta$ in the $\theta$ direction was calculated by using the multi-layer cylinder model paying attention to the strain of the form and that of superconducting wire in the r- $\theta$ plane in this work.

When it is assumed that the form is a thin cylinder with an infinite length, the relationship between the inward compressive stress $\sigma_r$ and the strain $\varepsilon_\theta$ of the cylinder in the $\theta$ direction can be expressed by equation (1),

$$\sigma_r = 2\varepsilon_\theta E d / \phi \qquad (1)$$

where E is the Young's modulus of the cylinder, d is the thickness of the cylinder, and the $\phi$ is the average diameter of the cylinder.

The multi-layer cylinder model, in which layers of superconducting wire are wound on a cylinder and each layer has an infinite length, was assumed and a relation between the contact stress $\sigma_{r,n}$ at the ground insulation and the form strain $\varepsilon_{\theta,n}$, was obtained based on equation (1). The relational expression is shown in equation (2),

$$\sigma_{r,n} = 2\varepsilon_{\theta,n} E d / \phi$$
$$\varepsilon_{\theta,n} = ((\Sigma k_i(\omega_{i,n-1} + \varepsilon_{\theta,n-1} - \alpha(i-1)) + k_n \omega_{n,n}) / (1 + \Sigma k_i) \qquad (2)$$

where $\omega_{i,i} = \omega_{1,1}$

$$\omega_{i,n} - \omega_{i,n-1} = \varepsilon_{\theta,n-1} - \varepsilon_{\theta,n} - \alpha(i-1)$$
$$k_i = F_T / \omega_{1,1} \cdot 1/(E d g \phi) \cdot (\phi + d + h(2i-1)) / (1 + \beta(i-1))$$
$$i = 1 \sim n-1$$

and the winding tension, $F_T$, of the sample coil is 88 N, $\omega_{i,n}$ is the strain of the superconducting wire in the i-th layer, n is the number of the wound layers, g is the diameter of superconducting wire coated with polyvinylformal, 0.9 mm, and h is the average thickness of each layer, 0.755 mm.

The initial strain of superconducting wire, $\omega_{1,1}$, was $1.5\times10^{-3}$, obtained from the rule of mixtures for NbTi and Cu. The $\alpha$ is the coefficient for the strain transfer between the form and the winding, $\beta$ is the coefficient for the tension transfer of superconducting wire in the i-th layer to the form. When $\varepsilon_{\theta,n}$ decreases and approaches a constant value with n, $\alpha$ and $\beta$ were approximately $1\times10^{-6}$ and 0.1, respectively.

By substituting the superconducting wire strain $\omega_{i,n}$ which was obtained from equation (2), the superconducting wire tension $F_{i,n}$ of each turn was obtained as

$$F_{i,n}=F_T/\omega_{1,1}\cdot\omega_{i,n} \tag{3}$$

**Relationship between Rearrangement and Change of Strain Energy** Even if the winding tension of the superconducting wire is constant, the tension of the superconducting wire in the coil varies with layer and with position in the z direction.[4] Hence the stored strain energy differs from one turn to another. When the strain energy stored in one turn is converted into thermal energy and exceeds the MQE, a quench will occur. The strain energy change $\Delta E$ accompanied by the rearrangement was investigated based on the load-strain line of one turn of superconducting wire.

The superconducting wire moves in the r and z directions due to the rearrangement. Particularly, movement in the r direction is accompanied by a change $\Delta\omega_\theta$ of the wire strain. Because the inner layer is compressed by the outer layers, the movement in the outward direction due to the Lorentz force, is accompanied by the release of the strain energy $\Delta E_R$. The relationship between the wire strain change $\Delta\omega_\theta$ in the $\theta$ direction and the movement $\Delta r$ in the outward direction, can be expressed by equation (4). On the other hand, the relationship between $\Delta r$ and the released energy, $\Delta E_R$, can be expressed by equation (5),

$$\Delta r / r = \Delta\omega_\theta \tag{4}$$

$$\Delta E_R = 1/2\,(\delta\ F_\theta - \delta'\ F_\theta') = 1/2\,(L\ \Delta\omega_\theta\ F_\theta - L'\ \Delta\omega_\theta'\ F_\theta') \tag{5}$$

where the L is the length of one turn, 323 mm.

## Experimental Study of Rearrangement

**Measurement of Training and Strain Behavior** Training tests were performed by immersing the coil in liquid helium, and the training and the strain behavior measured as the form strain change $\Delta\varepsilon$ was observed. The energization rate of the coil was set to 10 A/min.

The active-dummy gauge method was used for measuring the form strain change $\Delta\varepsilon_\theta$ in the $\theta$ direction at z = 0 mm and z = 55 mm (see Figure 1).

**Energy Released by Rearrangement** When the superconducting wire in the innermost layer contacts the form through the thin ground insulators, the wire strain change $\Delta\omega_\theta$ in the $\theta$ direction and $\Delta\varepsilon_\theta$ can be thought to be equal to each other. Therefore, the $F_\theta$ - $\varepsilon_\theta$ line for one turn in the innermost layer can be obtained from the form strain $\varepsilon_\theta$. The energy $\Delta E_R$ released by rearrangement can be obtained from the $F_\theta$ - $\varepsilon_\theta$ line and the residual former strain $\Delta\varepsilon_\theta^R$. The $\varepsilon_\theta$, $\Delta\varepsilon_\theta$ and $\Delta\varepsilon_\theta^R$ were measured at z = 0 mm and z = 55 mm and $\Delta E_R$ was obtained from those measurements.

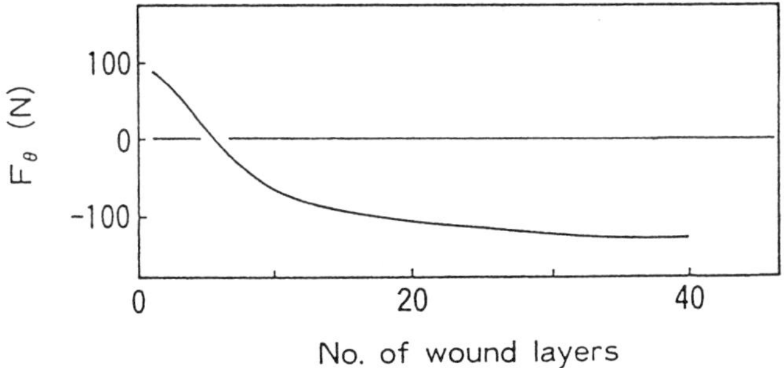


**Figure 2.** Relationship between innermost wire tension $F_\theta$ and number of wound layers.

## RESULTS AND DISCUSSION

### Theoretical Study of Rearrangement

**Tension of Superconducting Wire** Figure 2 shows the relationship between the wire tension $F_\theta$ of the innermost layer and the number of wound layers. The $F_\theta$ decreased from positive values to negative values as the number of wound layers increased. This means that $F_\theta$ changes from tensile to compressive as the number of layers becomes large.

**Relationship between Rearrangement and Change of Strain Energy** Figure 3 shows schematically the strain behavior of the form. The initial value of the form strain change $\Delta\varepsilon_\theta$ is set to zero. When the coil is energized, $\Delta\varepsilon_\theta$ increases from point A to point B. The rearrangement occurs at point B and it momentarily moves to point C, causing a quench. The $\Delta\varepsilon_\theta$ changes largely due to the abrupt change of the Lorentz force and the joule heating of the coil.[2] When the coil is cooled sufficiently, $\Delta\varepsilon_\theta$ may recover to point D. The changes from point D to point G in the second quench take place in the same manner as from A to D in the first quench. The residual strain $\Delta\varepsilon_\theta^R$ at point D and the accumulated residual strain $\Sigma\Delta\varepsilon_\theta^R$ at point G sometimes become negative.[2~4]

Figure 4 shows the relationship between the wire tension $F_\theta$ and the wire strain $\omega_\theta$ ( the $F_\theta$-$\omega_\theta$ line) of one turn in the innermost layer described by the rearrangement model. Points A to G in the figure correspond to the respective points in Figure 3. Because the value of $F_\theta$ is always negative (see Figure 2), the energy represented by the hatched area in the figure is the energy $\Delta E_R$ released by rearrangement. When the value of $\Sigma\Delta\varepsilon_\theta^R$ is positive, point D comes to the right of point A. On the other hand, when the value of $\Sigma\Delta\varepsilon_\theta^R$ is negative, point D comes to the left of point A.

From Figure 4, it was found that when the value of $\Sigma\Delta\varepsilon_\theta^R$ is positive, $\Delta E_R$ decreases with the quench number. On the other hand, when the value of $\Sigma\Delta\varepsilon_\theta^R$ is negative, $\Delta E_R$ increases with the quench number. According to this rearrangement model, stability is high when the value of $\Sigma\Delta\varepsilon_\theta^R$ is positive because the $\Delta E_R$ decreases with a quench, and the stability is low when the value of $\Sigma\Delta\varepsilon_\theta^R$ is negative.

### Experimental Study of Rearrangement

Figure 5 shows the change of strain behavior measured at z = 0 mm and z = 55 mm of the coil. The $\Sigma\Delta\varepsilon_\theta^R$ was positive at z = 0 mm and negative at z = 55 mm. Further, $\Sigma\Delta\varepsilon_\theta^R$ at z

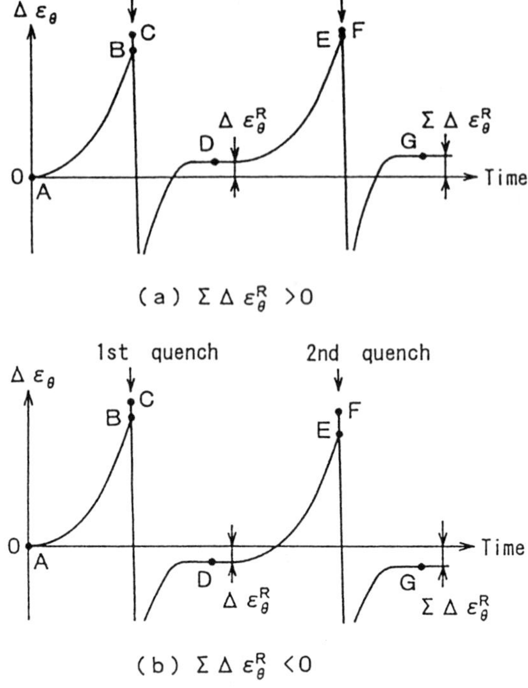


**Figure 3.** Schematic behavior of form strain.

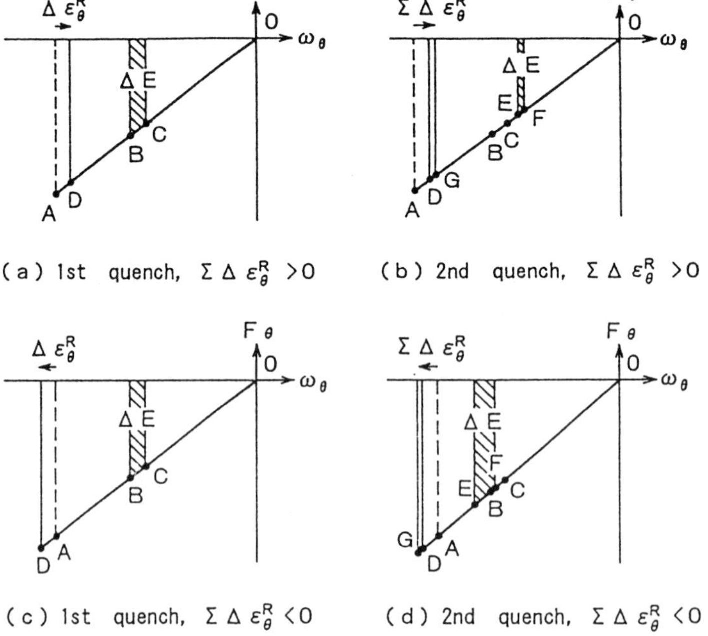


**Figure 4.** Relationship between tension $F_\theta$ and strain $\omega_\theta$ of innermost turn (Rearrangement model based on $F_\theta$ - $\omega_\theta$ line).

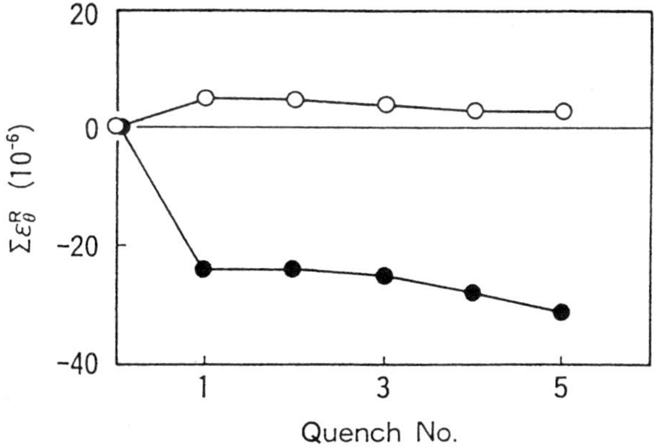


**Figure 5.** Accumulated residual strain $\Sigma\Delta\varepsilon_\theta^R$ induced by training.
○ : z = 0 mm、 ● : z = 55 mm

= 55 mm remained negative for all quenches. From the relationship between the $\Sigma\Delta\varepsilon_\theta^R$ and the stability (see Figure 4), the coil was evaluated to be unstable. This evaluation agreed with the training behavior.[4]

Figure 6 shows the load-displacement line ($F_\theta$ - $\delta$ line) and the released energy $\Delta E_R$ of one turn in the innermost layer. The $\Delta E_R$ at z = 0 mm and z = 55 mm in the first quench were estimated to be 190 $\mu$ J and 170 $\mu$ J, respectively. The released energy exceeded the MQE of 10 $\mu$ J/cm$^3$.[6] Although the $\Delta E_R$ at z = 0 mm at the second quench was 0 $\mu$ J, the $\Delta E_R$ at z = 55 mm was 170 $\mu$ J and was still larger than the MQE. These results suggest that rearrangement caused quenches in the coil.

## CONCLUSIONS

A rearrangement model consisting of the load-strain line of the superconducting wire in the innermost layer was proposed as a mechanical disturbance model for the superconducting

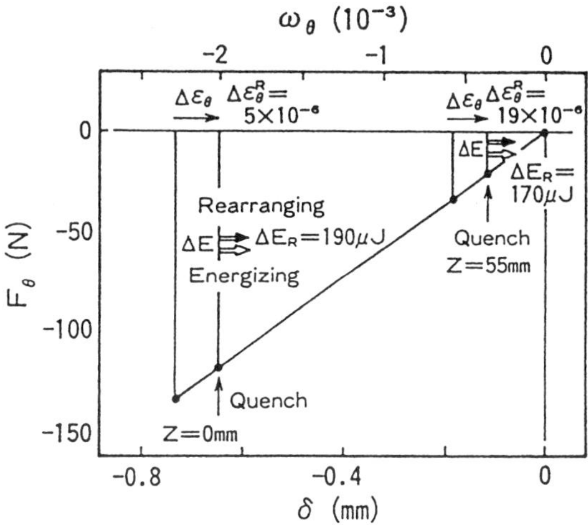


**Figure 6.** Load-displacement line and released energy $\Delta E_R$ of the wire.

solenoid magnet. Based on this rearrangement model, the strain energy released from the sample coil was investigated not only theoretically but also experimentally. The released energy of the sample coil was found to be larger than the minimum quench energy. This result accorded with the fact that the sample coil was actually unstable. It has been found that the rearrangement model is useful for estimating the stability of the superconducting solenoidal magnet at the designing stage.

## REFERENCES


1. J.Chikaba, F.Irie, K.Funaki, M.Takeo and K.Yamafuji, *IEEE Trans. Magn.* Vol.MAG-23 No.2: 1600 (1987)
2. H.Moriyama, H.Sekiya, H.Mitsui, A.Iwamoto, S.Nishijima and T.Okada, *Cryo. Engr (Jpn)* Vol.29: 653 (1994)
3. H.Moriyama, H.Mitsui, J.Ohmori, S.Mine, S.Nishijima and T.Okada, *IEEE Trans. Magn.* Vol.32 No.4: 3028 (1996)
4. H.Moriyama, F.Sawa, H.Mitsui, M.Arata, S.Nishijima and T.Okada, Proceedings of the 16th ICEC/ICMC Part 2 Kitakyushu, Japan, T.Haruyama, T.Mitsui and K.Yamafuji ed., Elsevier Science, (1996), p.1273
5. T.Kushida, S.Nishijima, Y.Honda and T.Okada, Proceedings of the 16th ICEC/ICMC Part 2 Kitakyushu, Japan, T.Haruyama, T.Mitsui and K.Yamafuji ed., Elsevier Science, (1996), p.1277
6. Y.Iwasa, *IEEE Trans. Magn.* Vol.28 No.1: 113 (1992)

# COMPLETION OF HELICAL COILS FOR LHD

S. Imagawa,[1] H. Tamura,[1] A. Nishimura,[1] T. Satow,[1] J. Yamamoto,[1] O. Motojima,[1] The LHD Group, K. Nakanishi,[2] O. Honma,[2] M. Sudou,[2] T. Kouda,[2] T. Yamagiwa,[2] and K. Asano[2]

[1]National Institute for Fusion Science
Nagoya 464-01, JAPAN
[2]Hitachi Ltd.
Tokyo, JAPAN

## ABSTRACT

Helical coils of LHD are large scale pool-cooled superconducting coils. In order to produce a fine magnetic surface, the accuracy of the coil position is very important. The required tolerance for manufacturing is $\pm$ 2 mm. Against large electromagnetic forces, the conductors were designed to be packed into thick cases which are supported by an outer shell structure. In order to suppress the maximum tensile stress of the conductors below the yield stress, the average compressive Youngs' modulus of the coils should be high. We developed insulator with a high compressive modulus, and we have managed the average gaps between layers within the aimed value while winding. After winding, the top cover of the coil case should be installed without leaving the residual gaps on the coils and without deforming the accurate shape. In order to reduce the assembling gap, we have machined adjusting spacers that were inserted between the coils and the top covers. After that, we tried to diminish the assembling gaps by utilizing the shrinkage by welding.

## INTRODUCTION

The Large Helical Device (LHD) is a fusion experimental apparatus under the final stage of construction, which equips with a pair of superconducting helical coils and three pairs of superconducting poloidal coils.[1] Since the amplitude of varying magnetic field is small in the helical coil except shut off of the coil current, the restriction for coupling losses is not severe. For attaining high accurate helical shaping and winding, we selected pool-cooled medium size conductors, which are directly wound on high accurate thick cases (HC cans). A pure aluminum stabilizer was adopted to perform high recovery current. Since it is impossible to withstand large electromagnetic forces by the conductors themselves, they are packed into the HC cans which are finally supported by an outer shell structure 100 mm thick. The can is used as a bath for liquid helium.

In order to produce a fine magnetic surface, the manufacturing error of the helical coil is demanded to be within $\pm$ 2 mm. In addition, the average elastic modulus of the coil in the minor radius direction should be high to keep the stress in the conductors within elastic region. At first, we have assembled the HC cans with absolute accuracy of 1.5 mm including setting errors. Next, on-site winding was carried out in day and night time from January of 1995 to May of 1996. The average accumulating gaps in the coil was successfully attained within 0.90 mm.[2] The position of conductor in the overturning direction was controlled by adjusting thickness of the spacers at both sides, and the errors were suppressed within ± 0.5 mm. The standard deviation of the errors of minor-radii of the coil was attained within 0.35 mm. After winding, top covers of the HC cans with arms have been set on the coils and welded. Since the electromagnetic forces on the conductors are supported by the cover, the gap between the coils and the covers should be small. We planned to insert adjusting spacers between them, and we tried to diminish residual gaps by utilizing the shrinkage of welding. Furthermore, this welding should be carried out very carefully not to damage the insulation in the coil. This paper is intended to summarize the design and method for assembling the top covers of the helical coils.

## STRUCTURAL DESIGN OF HELICAL COILS

### Requirement of mechanical rigidity

The cross-sectional view of the helical coil and surrounding structures are shown in Figure 1. From the aspects of plasma experiments, the asymmetrical deformation of the coil by electromagnetic forces should be small to produce accurate fields. The helical coils, therefore, were designed to be firmly supported from an outer cylindrical shell structure. In spite of the severe requirement for the tolerance, all the joints are welded to obtain enough strength. The thickness of main structures, which are the supporting shell, top covers of the HC cans and shell arms, were determined to be 100 mm from structural analyses to suppress the deformation of the current center of each coil within 3.4 mm in 4 T operation.

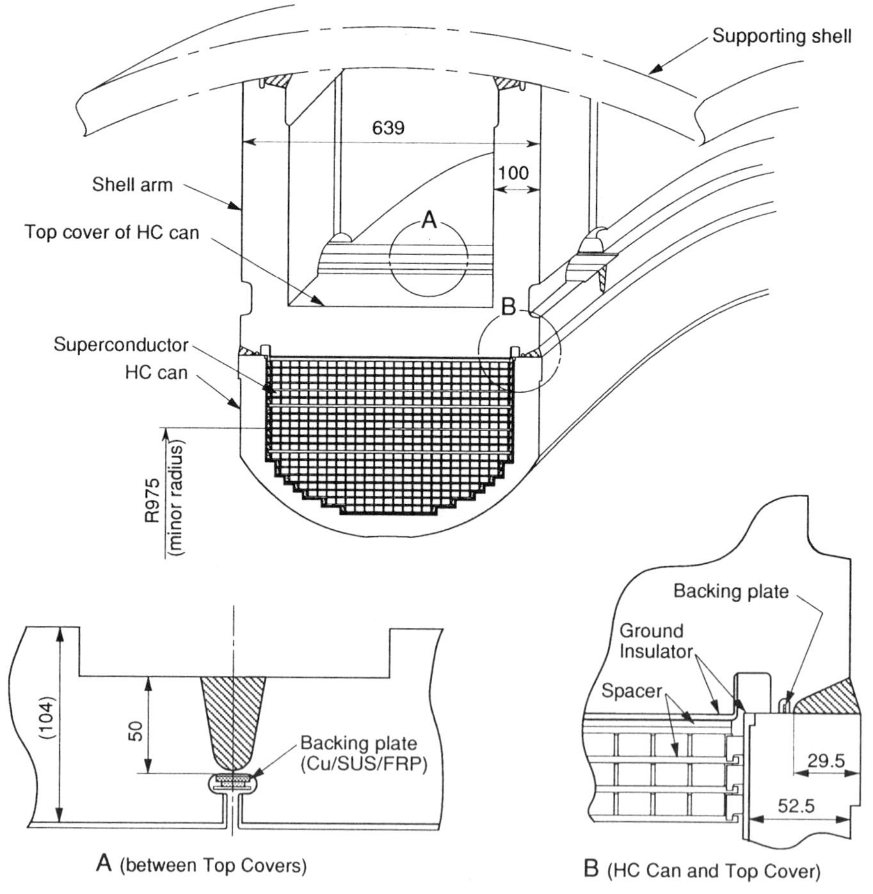


**Figure 1.** Cross-sectional view of helical coils and surrounding structures and design of the joint.

From the aspects of stability and reliability as a superconducting coil, the motion of the conductors due to electromagnetic forces must be small, that is, the rigidity of the coil must be high. Then, we have developed a glass-fiber-reinforced plastics with the Youngs' modulus higher than 22 GPa and the thermal contraction lower than copper. Nevertheless, the fabrication gaps between the conductor and the insulator decrease the equivalent rigidity of the coil. Since the coil will be enlarged in the minor-radius direction by the electromagnetic forces, the effect can be evaluated by assuming that the total gaps are equal to the increase of the displacement of the lowest layer. By approximating the helical coil to a circular coil with a equivalent curvature, the movement and stress in each conductor were calculated by using finite element method.[3] In the case of the rigidity of 22 GPa and the spacer ratio of 0.5, the movement of the lowest layer and the maximum stress in the conductor were calculated to be 1.1 mm and 173 MPa, respectively. In the case of 2.5 GPa, these were to be 2.4 mm and 288 MPa, respectively. From these results, the average fabrication gap was specified to be within 65 μm/layer, that is 1.3 mm/coil, to keep the stress in the copper sheath of conductors under the yield strength of 290 MPa. By attaining this condition, we will be able to avoid the coil quench caused by plastic deformation of the conductors.

### Design of top covers of HC cans

The electromagnetic forces on the helical coil are in the order of 10 MN/m at 4 T operations, and the minor-radius-hoop component is dominant in normal operation modes. When the helical coils are only excited, overturning forces become large, and the bending stress in the shell arms become large. The HC cans and shell arms are designed to have enough rigidity for keeping the position of the helical coils against the large electromagnetic forces. Since these structures are curved with minor radii of 1.7 to 2.2 m, the rigidity against the forces in the minor radius direction is estimated to be equivalent to the supporting shell. The space inside them are widely opened for workers to enter. It is used as a working space while welding the shell arms to the supporting shell and while maintaining the cooling pipe and sensor cables. Furthermore, we made maintenance holes in the shell arms per 30 degree in poloidal angle to check the clearance between the bottom of the HC cans and thermal shields of the plasma vacuum vessel.

The top covers were divided into thirty pieces and finished by the numerically controlled machinery with high precision of 0.2 mm. Since the accuracy of the coil surfaces were attained to be almost ± 2 mm, we planned to control the gap below 0.5 mm by inserting adjusting spacers between them. Besides, we planned to diminish the assembling gaps with utilizing the shrinkage by welding. The design of welding part is also shown in Figure 1. In order to decrease heat input and the deformation by welding, the depth of welding was determined by the stress limit. Then, the depth is almost half of the thickness of the structure. In order to protect the coil from the welding arc and spatter for the welding between the HC can and the top cover, a groove was machined behind the root face and a backing plate of stainless steel 1 mm thick were installed in the groove. For the welding between the top covers, a groove was also machined and a copper plate with a backing plate of stainless steel and reinforced plastics was inserted from the side hole. The hole was finally varied by build-up welding. The grooves play the roll of relaxing the stress concentration. A tungsten inert gas (TIG) welding is adopted for the first and the second layer to prevent the temperature rise of the nearest insulator. A metal arc gas (MAG) welding and a manual welding (SMAW) with flux-cored welding rods were adopted for the higher layers.

## METHOD OF ASSEMBLING THE TOP COVERS OF HC CANS

### Mock-up of the helical coil and trials of welding

Before actual installation of the top covers of HC cans, a real scale straight mock-up of length of 1.05 m was prepared. The materials and dimensions are same as the actual coil except that the conductors were made of copper only. Before trials of welding, an average Youngs' modulus was measured at a room temperature by using a 1000 ton press machine. A flat plate 120 mm thick and 700 mm long was set on the coil, and the plate were gradually pressed. The displacements were measured at the four corners and the center of the four sides. The stress of the insulator against the average displacement is shown in Figure 2. The stress was calculated by dividing the pressing force by the apparent contact area between the spacers and the conductors. The Youngs' modulus becomes higher with the higher stress. The reason is understood to be that the actual contact area increases at a higher compressive stress.[4] The measured data in a higher stress region are well fitted to the stress-displacement curve measured in a straight conductors package that was composed of real superconductors and spacers in six turns by six layers by length of 180 mm. We can predict the Youngs' modulus of the helical coils at the higher stress region.

After the measurement of Youngs' modulus, two trials of welding were carried out. One of the purpose is to predict the deformation by welding, and another is to demonstrate the temperature rise of the insulator in the helical coils. At the first trial, the gap between the welded surface of the HC can and the flat metal surface of the cover behind the root face was adjusted to be 1 mm before welding. In this case, the flat metal surfaces was contacted strongly after welding. At the second trial, the flat metal surface of the cover was shortened by 2 mm to estimate the effects to the deformation of the top cover. Before welding, the top cover was fixed on the coil by bolts. The temperature at the nearest insulator was measured with thermocouples while welding, and the shrinkage of welded part and deformation of the shell arms and the HC cans were measured with micrometers after each step of welding. A TIG welding was adopted up to the second layer, and a MAG welding was adopted after that. The length of the tack welding was 5 mm, and the bead length of the first layer was 50 - 60 mm. During the welding, the temperature of an inner-pass was kept below 150 degree C to prevent a hot cracking. The maximum temperature of the insulator was observed at the tack welding, which was successfully controlled below 180 degree C in the condition that the TIG welding current is from 80 to 90 A.

The deformations of the mock-up by welding are shown in Table 1. The difference among two trials is not small especially in the opening deformation of the shell arms. It is considered that the opening deformation was reduced by the counter-force from the metal contacted area behind the weldment. Nevertheless, the displacement of the center of the cover was almost same in two trials. Since the displacement of the center region of the cover is considered to be determined by the balance between the total welding force and the counter force from the coil, the displacement is mainly dependant on the rigidity of the coil in the case of same amount of welding by same method. Therefore, the gaps between flat metal surfaces are not important as compared with the welding method from the view point of diminishing the assembling gaps by welding.

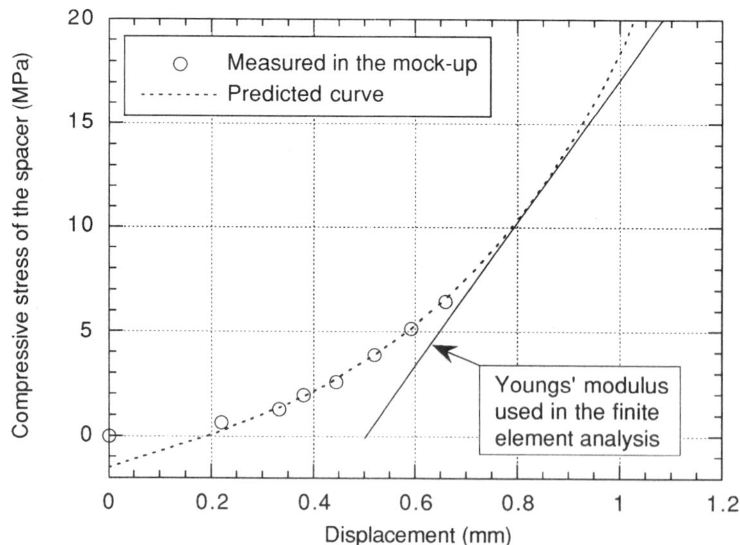


**Figure 2.** Stress versus displacement of a straight mock-up of the helical coil. The measured values are averages of four times measurements in loading process. The dashed line is the fitting curve in loading process of a 6 × 6 conductors package that was pressed up to 110 MPa. It is shifted to fit the Youngs' modulus to the measured data at the high stress region.

**Table 1.** Deformation of the straight mock-up of the HC cans by welding the top cover

| items | 1st trial | 2nd trial | calculated by FEM |
|---|---|---|---|
| displacement of the top cover at the center (mm) | 0.88 | 0.98 | – |
| bending of the top cover (mm) | 0.24 | 0.37 | 0.29 |
| Opening deformation of the shell arms (mm) | 3.83 | 4.73 | 4.38 |
| Shrinkage of welded part of the cover (mm) | 1.99 | 2.40 | – |

## Structural analysis of the top cover of the HC can

In order to estimate the stress of the spacer in the coil and to predict the deformation of the actual shell arms, structural analyses were carried out. The finite element model is shown in Figure 3. Each conductor and spacer between the conductors are represented by the four node solid. In order to estimate the force caused by welding between the HC cans and the top cover, the welded part was disconnected, and the force was applied on the nodes at the facing surface. The combined Youngs' modulus of the conductor and spacer was assumed to be 11 GPa to fit the predicted modulus in Figure 2 at the stress after welding. By considering the average spacer ratio of 0.5, the average Youngs' modulus of the coil was set to be 5.5 GPa. For the analyses of the straight mock-up, it was treated as plain strain problem. As the results of surveying the welding force to realize the measured deformation of the top cover and the shell arm, it was estimated to be 3 kN/mm. Since the depth of weldment is 30 mm, the force is corresponding to 100 MPa of residual stress.

For the analyses of the actual HC cans, the finite element model was treated as axisymmetrical problem. The radius was set to be equivalent curvature of three dimensional shape. By considering the curvature, the opening deformation of the shell arms becomes smaller, and the stress in the spacer also becomes lower as shown in Figure 4. The latter is undesirable for diminishing the assembling gaps. In order to increase the stress, we can utilize the shrinkage of the welding parts between shell arms. The effects of the shrinkage were shown in Figure 4 and 5. The compressive stress of the spacers become higher towards the side, and the average stress become higher with larger shrinkage between the top covers. From the aspect of keeping accuracy of helical coil position, it is desirable that the shrinkages between the top covers and between the top covers and HC cans become equivalent. In considering the allowable stress, the desirable value of shrinkage between the top covers is around 2 mm.

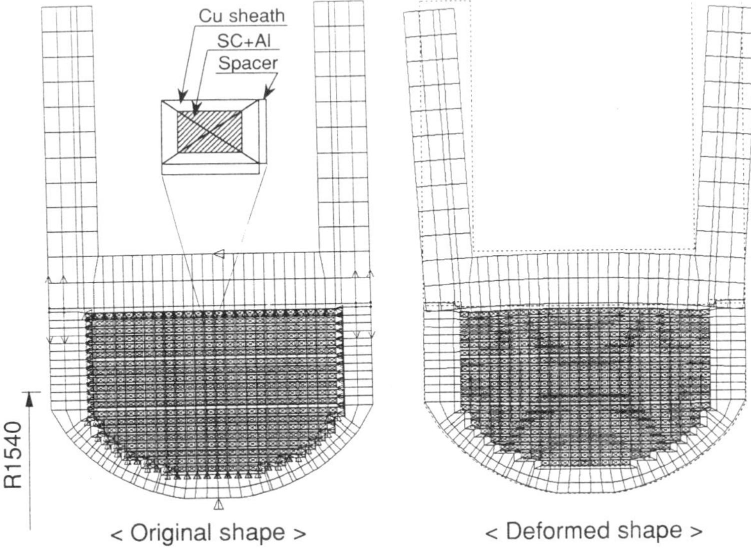


**Figure 3.** Finite element model for welding the top cover to the HC can. "SC + Al" represents the superconducting strands and a pure aluminum stabilizer.

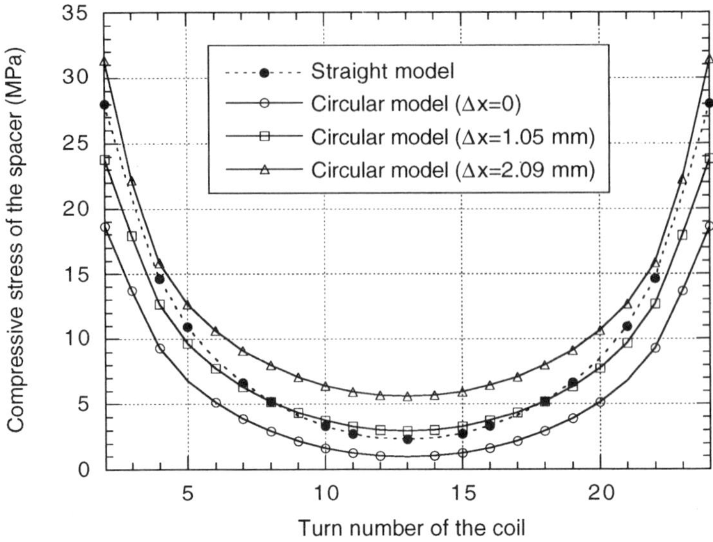


**Figure 4.** Calculated compressive stress of the spacer nearest the top cover. $\Delta x$ is the shrinkage of each welded part between the top covers, and $3\Delta x/2\pi$ is corresponding to the displacement of the top cover in the minor radius.

## MANUFACTURING

After winding the final layer of the helical coils, the relative depths of the conductors from the flat surfaces of the HC cans were measured at all the position of insulating spacers. All the spacers on the final layer were machined with the measured data for leveling the height. After pasting the spacers, the relative depths were measured again by a laser displacement meter (data A). Before setting the top covers of HC cans, the profile of the touching surface was measured by the laser displacement meter from the flat surface near the

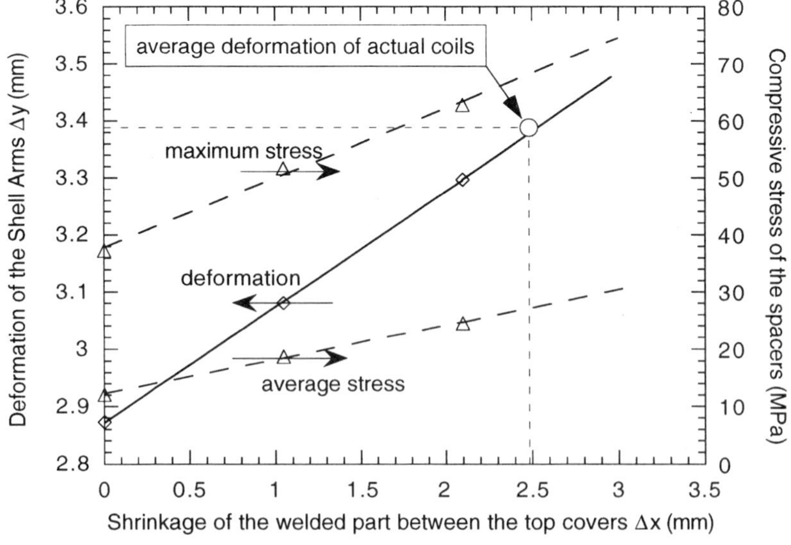


**Figure 5.** Deformation of the shell arm and compressive stress of the spacer in the coil. $\Delta y$ is opening deformation of both shell arms.

**Table 2.** Deformation of the actual shell arms and HC cans by welding the top cover

| items | average | deviation | maximum | minimum |
|---|---|---|---|---|
| displacement of the top cover at center (mm) | 0.86 | 0.19 | 1.4 | 0.2 |
| displacement of the top cover at end (mm) | 1.37 | 0.20 | 2.1 | 0.8 |
| Opening deformation of the shell arms (mm) | 3.39 | 0.43 | 5.1 | 2.0 |
| Sideway shift of the HC cans (mm) | 0.06 | 0.16 | 0.55 | -0.30 |
| Shrinkage of welded part of the cover (mm) | 2.48 | 0.43 | 3.5 | 1.4 |

root face (data B). Next, all the top covers were once set up on the HC cans, and we measured the gaps between both based surface at all the position of the spacers (data C). The adjusting spacers were machined with the data created by the data A, B and C. After pasting the adjusting spacers, the top covers were installed. In order to check the assembling gaps, five top covers were once installed and removed with inserting a silicon rubber 0.65 mm thick on each spacer and a pressure sheet on the top cover. The pressure sheet changed color at all position of the silicon rubbers, and we confirmed the assembling gaps were controlled within 0.5 mm.

The welding were carried out very carefully not to damage the coil. A TIG welding was adopted up to the second layer, and a MAG welding and SMAW was adopted after that. The length of the tack welding was controlled within 5 mm, and the bead length of the first layer was from 50 to 60 mm. The current of TIG welding was controlled from 80 to 90 A. According to the periodicity of the helical coil, ten working teams were organized, and each team was in charge of all welding in each pitch. It took 40 days by two shift working. The deformations by welding were measured at each 10, 45, 90, 135, 175, 185, 225, 270, 315, 350 degree in poloidal angle. The average deformations after each step of welding are shown in Figure 6. It shows that the deformation proceeded with amount of weldment. The deformations after welding were summarized in Table 2. The displacement of the top cover at the center and the shrinkage of welding between the top covers were successfully controlled from 0.2 to 1.4 mm and from 1.4 to 3.5 mm, respectively, without excess deformation of the helical coils. The average opening deformations of the top covers were well accorded with the results of structural analyses, as shown in Figure 5. It will prove that the assumption for the welding force is appropriate. The expected average and maximum compressive stress of the inserted spacers are estimated to be 28 and 68 MPa, respectively. The maximum stress is within allowable value of 100 MPa, and the average stress satisfies our target.

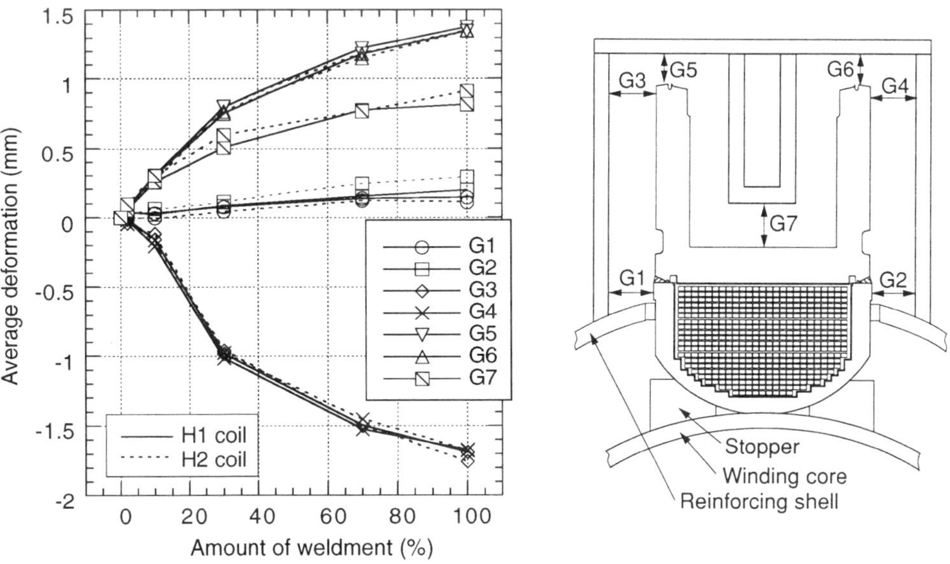


**Figure 6.** Average deformation of the actual helical coil cans and the shell arms.

## CONCLUSION

The helical coils are designed to be supported from the outer shell structure finally. After winding all the conductors into the HC cans, the top covers with shell arms were installed on the coil by welding. In order to suppress the maximum tensile stress in the conductors, the average compressive Youngs' modulus of the coils should be high. We developed insulating spacers with high compressive modulus, and we have managed the average gaps between conductors within 50 μm/layer while winding.We planned to install the top cover without remaining the residual gaps on the coils and without deforming the accurate shape. Before actual welding, we demonstrated a straight mock-up, and we carried out structural analyses to estimate the welding forces and compressive stress of the spacer in the coil. As the results of the analyses, the welding force per unit area was estimated to be 100 MPa, and desirable displacement of the top cover and the shrinkage of welded part between the top covers were estimated to be around 1.0 mm and 2.0 mm, respectively. By machining all adjusting spacers between the coils and the top covers with measured profile data, we have managed to keep the gap less than 0.5 mm. The welding were carried out very carefully not to damage the coil, and the displacement of the center of the top cover and the shrinkage of welded part between the top covers were successfully controlled from 0.2 to 1.4 mm and from 1.4 to 3.5 mm, respectively, without excess deformation of the helical coils. The average residual compressive stress of the spacers between layers is estimated to be 28 MPa.

## ACKNOWLEDGMENT

The authors are indebted to Dr. A. Iiyoshi, Director General of National Institute for Fusion Science, and Prof. M. Fujiwara for their continuous encouragement. We are also grateful to many staffs of Hitachi Ltd. who joined this work.

## REFERENCE

1. O. Motojima, N. Yanagi, S. Imagawa, K. Takahata, S. Yamada, A. Iwamoto, H. Chikaraishi, S. Kitagawa, R. Maekawa, S. Masuzaki, T. Mito, T. Morisaki, A. Nishimura, S. Sakakibara, S. Satoh, T. Satow, H. Tamura, S. Tanahashi, K. Watanabe, S. Yamaguchi, J. Yamamoto, M. Fujiwara, and A. Iiyoshi, Superconducting magnet design and construction of LHD, *Plasma Physics and Controlled Nuclear Fusion Research 1996*, F1-CN-64/G2-4..
2. S. Imagawa, S. Masuzaki, N. Yanagi, S. Yamaguchi, T. Satow, J. Yamamoto, O. Motojima, LHD Group, K. Nakanishi, K. Uchida, T. Yamagiwa, T. Yamamoto and K. Asano, Accuracy of superconducting helical coils for LHD, *Proceeding of the 19th Symposium on Fusion Technology*, Lisbon, Portugal, Sep. 16-20 (1996) (to be published).
3. S. Imagawa, N. Yanagi, S. Yamaguchi, T. Satow, J. Yamamoto, O. Motojima, LHD Group, T. Senba, S. Suzuki, K. Nakanishi, K. Asano, and T. Yamagiwa, Construction of Helical Coil Winding Machine for LHD and On-site Winding, *IEEE Transactions on Magnetics*, Vol. 32, No. 4 (1996) 2248-2251.
4. H. Tamura, S. Imagawa, A. Nishimura, J. Yamamoto and Y. Shindo, Mechanics of a composite structure of GFRP insulators and superconductors with a gap, *Advances in Cryogenic Engineering* 42 (1996) 197-204.

# DEVELOPMENT OF A SUPERCONDUCTING SOLENOID MAGNET SYSTEM FOR THE B-FACTORY DETECTOR ( BELLE )

Y. Makida, H. Yamaoka, Y. Doi, J. Haba, F. Takasaki, and A.Yamamoto

KEK, High Energy Accelerator Research Organization
Tsukuba, Ibaraki, 305, Japan

## ABSTRACT

A large superconducting solenoid magnet has been designed and fabricated for the BELLE particle detector, which will be installed in the KEKB colliding beam ring. It provides a central magnetic field of 1.5 T at 4000 A in a cylindrical volume of 3.4 m in diameter and 4.1m in length. The stored energy is calculated to be 42 MJ. The coil is wound with aluminum stabilized superconductor, after which aluminum strips are laminated on the coil for quench protection. The coil cold mass is approximately 7 tons with an estimated thermal load of 22 W at 4.2 K. A refrigerator of 240 W at 4.2 K is attached. The coil is indirectly cooled by forced flow of 2 phase helium through a cooling tube on the coil support cylinder. This method is preferred in order to avoid explosive boiling which may happen if a pool cooling method was used. A line in the solenoid is designed to allow the stored helium in the top buffer tank to circulate gravitationally in case of a suspension of cryogen, producing slow discharge without induced quench. This report presents the design and fabrication of the BELLE superconducting solenoid magnet.

## INTRODUCTION

KEKB is an asymmetric electron-positron collider ring under construction in Tskuba, Japan. The accelerator ring consists of two independent beam lines with one interaction point, where the $e^-$ ( 8GeV ) and $e^+$ ( 3.5 GeV ) beams collide with a finite crossing angle of 11mrad[1,2].

The BELLE detector is installed at the interaction point of the accelerator ring and will measure violations of symmetry between matter and antimatter ( CP violation ) in the B meson system, which is predicted by the Standard Model in particle physics[1,3]. Figure 1 shows the overall detector layout. The detector has an axisymmetric configuration and includes a solenoid magnet, an octagonal iron yoke structure, calorimeters and inner detecting devices, using techniques similar to these employed by several other collider experiments[4,5,6,7,8,9,10]. The central drift chamber ( CDC ) contained within the solenoid cavity provide pattern recognition and momentum measurement[11].

As shown in Fig. 1, four additional superconducting magnets are to be installed as accelerator elements near the interaction point[2]. The electromagnetic effects induced by the additional magnets on the magnetic force, inductance and other features of the solenoid are not considerably large.

## MAGNET DESIGN

### Design Outline and Boundary Conditions

Figure 2 shows the overall layout of the BELLE solenoid magnet. The magnet configuration is based on a superconducting solenoid surrounded by an outer iron yoke structure and a chimney port for cryogen and electric power service. The main parameters are summarized in Table 1.

The experiment requires the BELLE solenoid produce a magnetic field of 1.5T with a field uniformity within 2 % inside the CDC. The size of the solenoid magnet has to be kept within 4.0 m in outer diameter, 3.4 m in inner diameter and 4.44 m in overall length.

In most detectors used in the high energy experiment, an electromagnetic calorimeter is located outside the magnet. The magnet wall should therefore be as thin as possible in terms of the radiation thickness and be transparent enough to allow particles to pass though. However, in case of the BELLE detector, the solenoid can be robust and massive, as all the detectors except the muon detector in the yoke are located inside the solenoid cavity. This loose requirement on transparency allows for the use of a conservative coil and an inexpensive cryostat. The value of E/M ( stored energy / coil mass ) which gives a measure of transparency, is below 6.4 kJ/kg for the BELLE solenoid, while a transparent SDC prototype coil at KEK has an E/M of 9.6 kJ/kg[12]. Furthermore the vacuum vessel dose not need to minimize material, which allows all parts in the vacuum vessel to be made of plain stainless steel ( JIS-SUS304 ).

An electric power source and the bus line for the TOPAZ[7] solenoid in the TRISTAN collider are to be utilized for the BELLE solenoid, the capacity of which sets an upper limit on the excitation current of less than 4500 A.

### Magnetic Field and Iron Structure

The iron yoke structure consists of an octagonal barrel-yoke and two end-yoke sections. The overall height is about 9 m and the depth is approximately 8 m. The structure will be made from low-carbon steel (JIS-S10C), with a total weight of approximately 1300 tons. The end yokes are separated from the barrel yoke by a distance of 80 mm to keep a gap for signal cables from the inner detectors. The barrel yoke is composed of 14 layers of 47 mm thick steel plates with 40 mm thick air slots for calorimeters and an outermost iron layer 200 mm in thickness. Each end-yoke structure is a combination of a similar multi layered composition and a solid pole piece. The magnetic field produced by the coil is returned through these iron elements.

The field uniformity and the forces on the coil are strongly influenced by the proximity of the iron elements, the effects of which are calculated in a series of axially symmetric finite-element analyses using the computer code ANSYS. The configuration of the yoke and

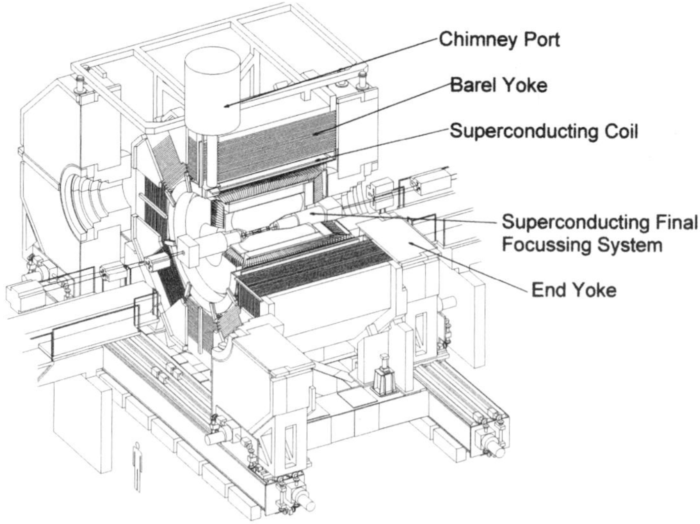


**Figure 1**. Layout of the BELLE detector .

**Table 1**. Main parameters of the solenoid

| Dimensions: | | |
|---|---|---|
| Cryostat | inner radius of vacuum shell | 1,700 mm |
| | thickness of inner shell | 6 mm |
| | outer radius of vacuum shell | 2,000 mm |
| | thickness of outer shell | 20 mm |
| | material | SUS304 |
| | inner radiation shield radius | 1,733 mm |
| | outer radiation shield radius | 1,940 mm |
| Coil | inner radius @ 4.5 K | 1,782 mm |
| | mean radius @ 4.5 K | 1,800 mm |
| | length @ 4.5 K | 3,910 mm |
| Conductor | cross section | $3\times33$ $mm^2$ |
| | winding pitch | 3.3 mm |
| | total length | 13.4 km |
| | SC core cable cross section | $1.15\times5.8$ $mm^2$ |
| | SC strand No. | 17 |
| | SC strand diameter | 0.67 mm |
| | SC cable twist pitch | 63.2 mm |
| | NbTi : Cu | 1 : 1.09 |
| | Critical current @ 4.2 K, 4 T | > 8000 A |
| | Purity of Al stabilizer | 99.998 % |
| Support cylinder | thickness | 25 mm |
| | material | A5083-H321 |
| Quench Protection | pure Al propagator thickness | 1 mm |
| | purity of Al | 99.999 % |
| Chimney line | length | 2150 mm |
| | maximum diameter | $\leq$ 350 mm |
| Control dewar | vacuum vessel diameter × height | 1.4 $m\phi\times2.12$ m |
| | Capacity of buffer tank | 200 ℓ |
| **Electrical parameters:** | | |
| Central field | | 1.5 Tesla |
| Peak field in the conductor | | 1 Tesla |
| Nominal current | | 4,000 A |
| Total ampere turn | | $4.7\times10^6$ A·Turn |
| Inductance | | 5.2 H |
| Stored energy | | 42 MJ |
| E/M | | 5.3 kJ/kg |
| Typical charging time | | 0.5 hour |
| **Mechanical parameters:** | | |
| Coil weight | | 7.9 ton |
| Total mass | | 27 ton |
| Radial magnetic pressure | | 0.9 MPa |
| Axial compressive force | | 3.4 MN |
| Axial decentering force | | $3.1\times10^5$ N/cm |
| Radial decentering force | | $2.6\times10^4$ N/cm |
| **Cryogenic parameters:** | | |
| Effective cold mass | | 8 ton |
| Thermal load @4.5 K | | 22.4 W |
| | current leads | 10.4 W |
| Radiation shield heat load @80 K | | 150 W |
| He cooling pipe | outer diameter × thickness | 31 $mm\phi$ ×3mmt |
| $LN_2$ shield cooling pipe | outer diameter × thickness | 18 $mm\phi$ ×2mmt |
| Cool down time ( 300 K → 4.5K ) | | 5 days |
| Recovery time from quench | | < 8 hours |

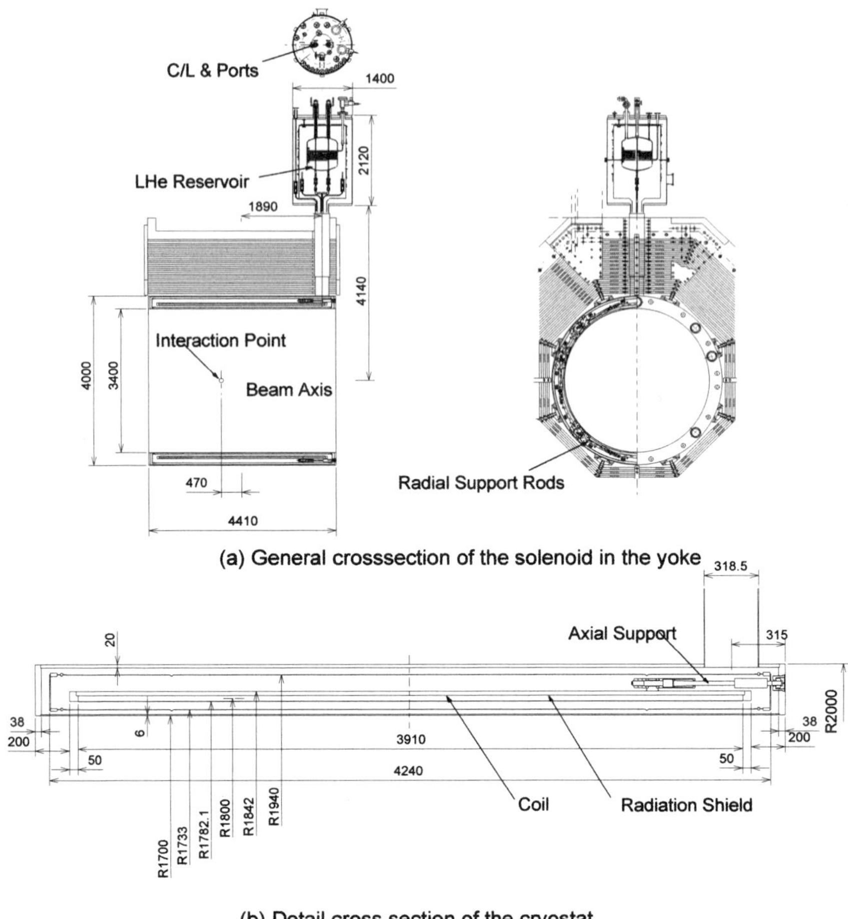


**Figure 2.** Cross section of the solenoid with the iron yoke. The end yokes are omitted in this figure.

the coil used in the calculation is shown in Fig. 4. The field uniformity in the CDC volume has been finally calculated to be 2.0%. The axial and radial decentering forces on the coil are taken into account in the solenoid design. The solenoid can withstand a force of $3.1\times10^5$ N/cm in the axial direction and $2.6\times10^4$N/cm in the radial direction.

## Coil Thickness and Stress Analysis

A crossectional drawing of the coil is shown in Fig. 2. The coil consists of a single layer edge winding of aluminum stabilized NbTi / Cu superconductor within an outer cylinder made of aluminum alloy.

The coil thickness is given by using the following way. When the magnet is quenched and a protective circuit fails to extract stored energy, the whole energy is dissipated into the coil itself. A conductor winding mainly suffers this energy as a heat load, but the outer support cylinder can't contributes a heat capacitance to the load, because of low thermal conductance between them. In case of allowable temperature rise of 90 K, a required thickness of the conductor winding becomes 33 mm according to its heat capacitance ( enthalpy ), as described in Eq. (1).

$$t_{cond} = \frac{E}{2\pi RL(h_{90K} - h_{4.5K})} \tag{1}$$

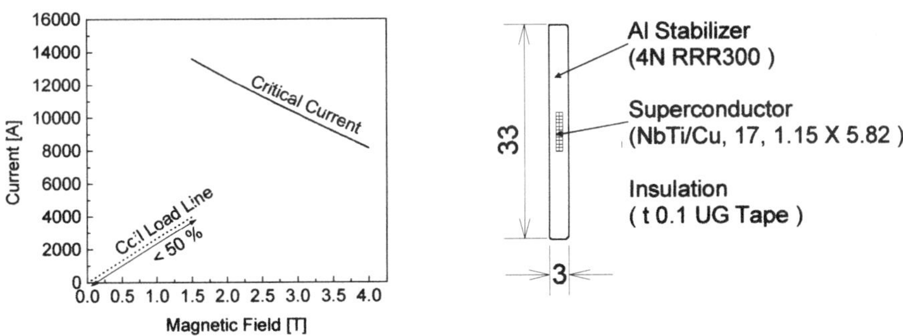


**Figure 3**. Superconducting characteristics and cross section of the conductor.

where $t_{cond}$ is the thickness of the winding, E is the magnetic energy, R is the main coil radius, L is coil length, $h_{90K}$ and $h_{4.5K}$ is the volumetric enthalpy at 90 K and 4.5 K respectively.

Next, the coil has to withstand hoop force estimated by Eq. 2 on the basis of magnetic pressure.

$$F_{hoop} = P_m R = \frac{B_z^2}{2\mu_0} \tag{2}$$

where $F_{hoop}$ is hoop force on unit coil length, $P_m$ is magnetic pressure and $B_Z$ is field. The hoop force is supported by the cylinder in the combination with the winding.

$$F_{hoop} = \sigma_{cy} t_{cy} + \sigma_{cond} t_{cond} \tag{3}$$

where $\sigma_{cy}$ (= 64MPa) and $\sigma_{cond}$(= 20MPa) is allowable stress at the cylinder and the winding respectively, and $t_{cy}$ is the cylinder thickness. According to Eq. 3, the cylinder thickness becomes 15 mm. Finally the practical cylinder thickness of 25 mm is decided in consideration of handling during manufacture.

## Fast Quench Propagation for Protection

Even if, in case of quench, the stored energy is not extracted from the coil, the uniformity of the energy dump into the whole coil eliminates the possibility of hot-spots. This is possible if the quench is propagated to the entire coil faster than the power decay time constant. A technique to increase the quench propagation velocity is to laminate pure aluminum strips on the winding surface or to put them between layers axially, which increases the net thermal conductivity across the conductors. This idea was developed and has been experimentally verified in the recent thin solenoid research in KEK[12,13]. In the BELLE coil pure aluminum strips with a thickness of 1 mm and a residual resistance ratio ( RRR ) ≥1000 corresponding to a thermal conductivity of > 5000 w/m·K at 4.2 K and 0 T are laminated onto the inner surface of the windings.

A quench simulation using program code "QUENCH" was performed to check the maximum temperature, which will be 90 K

## Superconductor

Figure 3 shows a cross section of the superconductor, which consists of 17 central compacted strands of $\phi$0.67 mm clad with aluminum stabilizer by using the co-extrusion method. The operation current of 4000 A at the maximum field of 1.5 T in the coil will be less than 50 % of the critical current at 4.2 K along the coil load line.

The coil is a single layer winding with aluminum stabilized NbTi superconductor. The conductor height is referred to consideration of the coil thickness. The conductor width of 3 mm is determined by considering an operation current of about 4000 A, which is within the operational capacity of the existing power supply and the bus line for the removed TOPAZ solenoid.

The purity of the aluminum stabilizer is not 99.999 % ( 5N ) but is sufficiently high at 99.99 % ( 4N ). As demonstrated by the SDC prototype, aluminum with RRR of 500 corresponding to 4N purity is sufficient if the stabilizer has a large enough cross section[12]. This type of conductor, which is less expensive than the 5N purity aluminum, was used for the CLEO-II and the ALEPH solenoids with no technical problems. The stability of the superconductor parametized by the minimum quench energy ( MQE ) is 1.3 J[14].

### Electric Insulation

Figure 4 shows the electrical ground insulation and the turn-to-turn insulation of the coil. The bonding strength is one of the fundamental points in the indirectly cooled coil both mechanically and thermally. The production and designs have been developed along;

- Polyimide film for the most reliable electrical insulation,
- Glass/epoxy-resin composite cloth for reliable mechanical bonding,
- Preprocessed hybrid lamination of the composite film and cloth.

The major characteristics of the insulation are summarized in Table 2.

The mechanical performance of the coil depends strongly on winding technique and quality controls. A prototype coil fabricated by using the same process and the same winding machine has achieved shear strength in bonding beyond 25 MPa at room temperature.

## COOLING SYSTEM

### Coil Cooling

The coil is cooled indirectly by using the forced flow of LHe, in the same way as "thin" solenoids. A serpentine tube with a bore diameter of 25 mm is welded on the outer surface of

**Table 2**. Mechanical characteristics of film / cloth composite electrical insulation

| Materials | | GUG-212 | UG-Epp |
|---|---|---|---|
| Insulation used for | | Ground | conductor |
| Composition | 1st layer | Glass - Epoxy | Epoxy coating |
| | 2nd layer | Upilex | Upilex |
| | 3rd layer | Glass - Epoxy | Glass - Epoxy |
| Thickness 1st/2nd/3rd | | 0.07/0.025/0.07 mm | 0.01/0.025/0.04 mm |
| Layers / insulation | | 2 | 2 |
| Breakdown voltage | | >> 2 kV | >> 2 kV |
| Shear strength | @300 K | 32 MPa | |
| | @77 K | 31 MPa | |

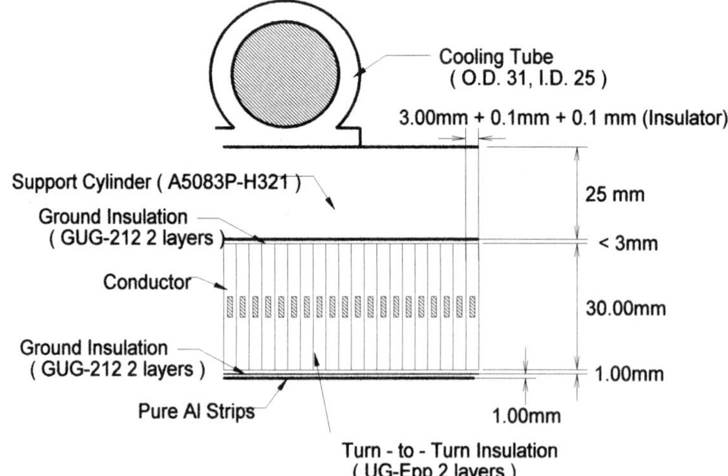


**Figure 4**. Detail cross section of the coil. Ground insulation and turn - to - turn insulation are pointed out.

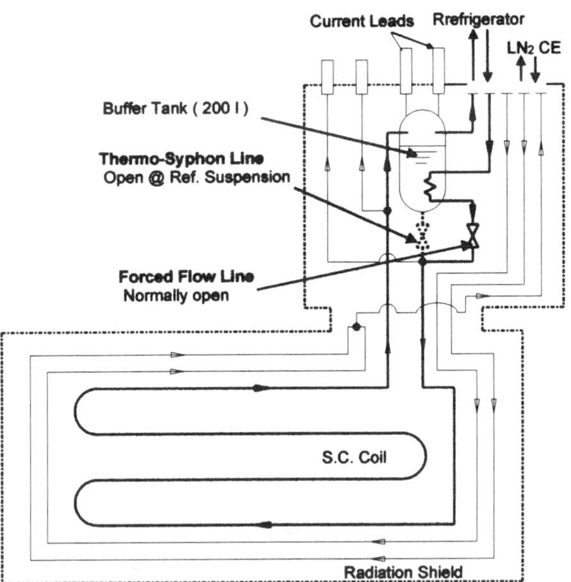


**Figure 5.** Flow diagram in the solenoid. Normally the coil is cooled by forced LHe flow. The line is switche to thermo-syphon circulation at the refrigerator suspension.

the support cylinder. One of main reasons that the pool boiling method is unfavorable is possibility of an explosive boiling off resulting in the rapid loss of cryogen in a quench. The force cooling method also has the advantage that a smaller quantity of cryogen is required.

A new scheme for the coil cooling is introduced in order to improve cryogenic reliability. When the LHe supply from the refrigerator is suddenly stopped, a pass from the bottom of the buffer tank to the coil part is opened to make the LHe in the tank circulate through the coil pipe gravitationally, as shown in Fig. 5. The route through the coil is not optimized for this thermo-syphon cooling mode, but it may be possible for the coil to be kept at a sufficiently low temperature until the discharge is completed.

### Chimney Port and Control Dewar

As shown in Fig.2, the control dewar, located outside the barrel yoke, provides the interface between cryogenic transfer lines and the solenoid. It has couplings for the helium and nitrogen supply and return lines. A liquid helium reservoir of 200 ℓ is fed from the helium return line and serves as the heat exchanger for the supply line to improve the quality of the two phase helium. The superconducting current bus lines from the coil enter the reservoir tank through insulated feedthroughs and are connected with helium-gas-cooled Cu current leads.

The chimney port containing all the cryogenic and electrical lines from the coil crosses the barrel yoke, where the superconducting current leads are indirectly cooled through thermal conduction from the helium return line.

### Refrigeration System

A helium refrigerator of 240 W at 4.2 K is installed to circulate liquid helium through the coil and a line from a cold evaporator with a capacity of 10000 ℓ is laid to transfer liquid nitrogen into the radiation shields.

## FABRICATION & SCHEDULE

Table 3 depicts the progress and schedule of the BELLE solenoid construction.

**Table 3.** Schedule of the BELLE solenoid construction

| | |
|---|---|
| 1994 | Conceptual Design, |
| 1995 | Engineering Design, Conductor fabrication |
| | Cryostat fabrication |
| 1996 | Coil winding, Magnet manufacture |
| 1997 | Installation into the detector |
| | Performance test, Field mapping |

## SUMMARY

The design principles, as well as an outline of the engineering design of the BELLE solenoid are described. The loose requirement, that the solenoid need not be transparent, allows for the use of a conservative coil and an inexpensive cryostat. Effort is focused on optimizing reliability and economy. The main features of this design and technique for achieving this goal are as follows:

- the employment of 99.99 % aluminum stabilizer without the necessity for the use of a 99.999 % purity.
- massive stainless steel vacuum vessel instead of studying stiff and transparent material.
- the application of pure aluminum quench propagator.
- the switch from forced flow to thermo-syphon circulation upon refrigerator outage.

## REFERENCES


1. "KEKB B-factory Design Report", KEK Report 95-7, (1995).
2. K. Tsuchiya, The superconducting magnet system for KEKB B-factory, in: "Proceedings of the 5th European Accelerator Conference ( EPAC 96 )", (1996 ), p.2287.
3. I. S. Hughes, "Elementary Particles", Cambridge University Press, New York ( 1972 ).
4. H. Desportes et al., Construction and test of the CELLO thin-wall solenoid, in: "Advances in Cryogenic Engineering Vol. 25", D-2(1980), p.194.
5. H. Minemura et al., Construction and testing of a 3 m diameter × 5 m superconducting solenoid for the Fermilab Collider Detector Facility ( CDF ), *Nuclear Instrument and Method in Physics Research* A23:18 ( 1985 )
6. M. Wake et al., A large superconducting thin solenoid magnet for Tristan Experiment ( VENUS ) at KEK, *IEEE Transaction on Magnetics*, MAG-21 No.2:494 ( 1985 )
7. A. Yamamoto et al., Test results of TOPAZ thin superconducting solenoid, in: "Proc. of 9th Int..Conf. on Mag. Technology ( MT-9 )", Zürich ( 1985 ), p. 167.
8. J.M. Baze et al., Design, construction and test of the large superconducting solenoid ALEPH, in: "Proc. of 10th Int. Conf. on Mag. Technology ( MT-10 )", Boston ( 1987 ), p.1260
9. P. Clee et al., Towards the realization of two 1.2 tesla superconducting solenoids for particle physics experiments, in: "Proc. of 11th Int. Conf. on Mag. Technology ( MT-11 )", Tskuba ( 1989 ), p. 201.
10. A. B. Oliva et al., Zeus magnet constructions status Report, in: "Proc. of 11th Int. Conf. on Mag. Technology ( MT-11 )", Tskuba ( 1989 ), p. 229.
11. J. Haba, KEKB and BELLE experiment, *Nuclear Instrument and Method in Physics Research*, A 368:74 ( 1995 ).
12. A. Yamamoto et al., Development of a prototype thin superconducting solenoid for the SDC detector, *IEEE Transaction on Applied Superconductivity*, Vol. 5 No. 2:849.
13. Y. Makida et al., Performance of a thin superconducting solenoid for particle astrophysics, in: "Advances in Cryogenic Engineering 37 A", (1991), p. 401.
14. M. N. Wilson, "Superconducting Magnet", Oxford University Press, New York ( 1983 ).

# PRELIMINARY TESTING OF THE DØ SUPERCONDUCTING SOLENOID

R. P. Smith[1], H. E. Fisk[1], K. Krempetz[1], D. Markley[1], R. Rucinski[1], B. Squires[1], R. Yamada[1], S. Mine[2], K. Kimura[2], T. Kobayashi[2], H. Kozu[2], and W. Odashima[2]

[1]Fermi National Accelerator Laboratory
Batavia, IL, 60544, USA
[2]Toshiba Corporation
2-4, Suehiro-Cho, Tsurumi-Ku
Yokohama, 230 Japan

## ABSTRACT

Toshiba Corporation has fabricated a thin superconducting solenoid [1–3] for the upgraded tracking system of the DØ Detector at the Fermilab Proton-Antiproton Collider. The magnet was successfully tested at the Toshiba Keihin Works in Yokohama before it was shipped to the US. The results of the Keihin tests are presented.

## INTRODUCTION

The DØ solenoid was specified[4] by Fermilab to cool down easily, charge rapidly, operate stably at either polarity, quench safely without a protection resistor, and provide a 2.0 Tesla solenoidal field with enhanced uniformity which is accurately aligned to the axis of the cryostat. The major parameters of the magnet are listed in Table 1.

Because the magnet must fit into the existing DØ detector the cryogens and electrical current are supplied to the magnet via a 14 m long narrow chimney which connects the magnet cryostat to the system control dewar. The control dewar contains the vapor cooled leads, a subcooler to recool the inlet two phase helium, control valves, bayonet connections, flow instrumentation, and relief valves.

The tests at Keihin demonstrated that the magnet fully met all specifications. When the magnet is later operated in the DØ Detector it will be the first thin superconducting solenoid to operate at 2.0 T in a particle physics detector.

**Table 1. Major Parameters of the DØ Solenoid**

| | |
|---|---|
| Central Field | 2.0 T |
| Operating Current | 4820 A |
| Cryostat Warm Bore | 1.067 m |
| Cryostat Length | 2.729 m |
| Integrated Field Homogeneity | $\pm 5.0 \times 10^{-3}$ |
| Stored Energy | 5.6 MJ |
| Inductance | 0.48 H |
| Cooling | Indirect, 2-phase forced flow helium |
| Cold Mass | 1500 kg approx |
| Conductor | 18 Strand Cu:NbTi, Cabled |
| Conductor Stabilizer | High purity Aluminum |
| Transparency | 0.9 $X_o$ |
| Cooldown Time | $\leq$ 120 hours |
| Magnet Charging Time | $\leq$ 30 minutes |
| Fast Discharge Time Constant | $\leq$ 15 seconds |
| Slow Discharge Time Constant | $\leq$ 330 seconds |
| Total Operating Heat Load | $\leq$ 18 W plus 1 g/s liquefaction |
| Operating Helium Mass Flow | $\leq$ 4 g/s |

## MAGNET COOLDOWN

The magnet system was precooled to 80 K in approximately 70 hours using an available refrigerator at Keihin. Cooldown from 80 K to 4.5 K was completed in slightly less than four hours by the transfer of approximately 600 liters of liquid helium into the system.

The rate of cooldown during refrigerator operation was limited by the requirement that the temperature gradient between the support cylinder and the coil not exceed 30 K at any time. This restriction was imposed to limit the axial thermal shear stress in the epoxy bond between the coil and the support cylinder during cooldown. It was found this restriction could be satisfied by limiting to 25 K the temperature difference between the inlet and exit helium flows to the coil. During refrigerator operation the circulating helium gas was maintained at a pressure of approximately 1.6 atma.

The amount of liquid required to complete the cooldown was quite close to that expected from considerations of the heat of vaporization of the liquid and the enthalpy of the cold gas.

The magnet resistance and residual resistivity ratio (RRR) of the conductor during the cooldown are shown in Figures 1a and 1b.

A one-Ampere current source and microvoltmeter were used to measure the resistance of the coil during cooldown. The measured data clearly showed that RRR = 500 was surpassed and observation of the behaviour of the microvoltmeter after RRR = 1000 was reached strongly suggested that RRR exceeded 1000 before the coil made the final transition to the superconducting state. The lack of an easily observed change of slope in the cooldown data beyond RRR = 1000 was likely due to the coil not cooling everywhere at precisely the same rate during the relatively rapid final cooldown.

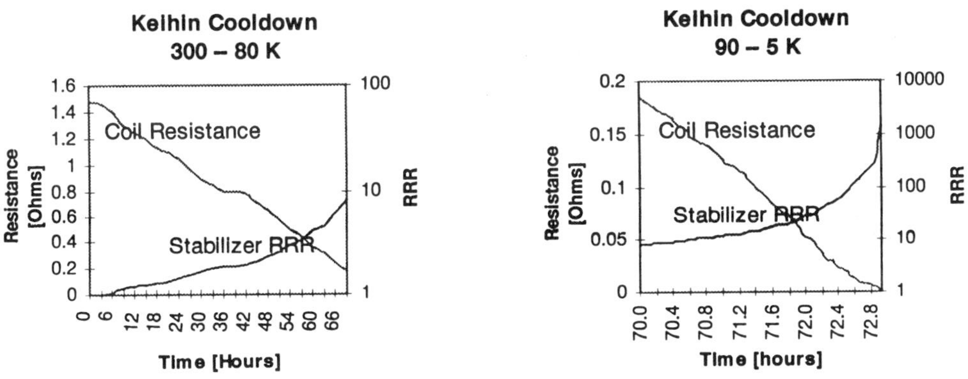


**Fig. 1.** Magnet Resistance and Stabilizer RRR During Cooldown.

## CRYOGENIC PERFORMANCE

During the Keihin tests the radiation shields were supplied with liquid nitrogen from a 100 liter supply dewar and the magnet system was supplied with liquid helium from a 1000 liter supply dewar. Both dewars were manually repressurized as required using helium gas cylinders and the mass flow from the helium dewar was calculated from changes in the liquid level in the dewar.

A programmable logic controller (computer system) operated the flow control valve which regulated flow of two phase helium into the magnet cryostat. The gas flow from the vapor cooled current leads was measured and regulated by automatic flow controllers. In the event of a quench a bypass valve was provided which permitted excess return flow from the magnet to be vented to the atmosphere instead of being returned to the subcooler heat exchanger. The steady-state discharge of cold gas from the subcooler vessel was regulated with a pressure regulating valve and measured with a flowmeter.

Because the temperature of the cold gas leaving the subcooler vessel was not measured it was not possible to accurately measure the instantaneous total heat leak to the system during operation.

Typically the flow of helium and nitrogen was stopped at the end of each day and the system was allowed to warm overnight. During a nominal eight hour standby period the radiation shield warmed to about 125 K and the cold mass to about 30 K. From the enthalpy change of the cold mass and the exact duration of the standby period the 4K heat leak to the cold mass can be obtained. An average value of about 6 W was obtained from these calculations.

## VAPOR COOLED LEADS

The vapor cooled leads were designed Toshiba for trouble-free service. Each consists principally of a round ETP copper rod with a helical helium channel machined into its periphery then encased in a tubular stainless steel jacket. The length and cross sectional area of the copper was selected so that in terms of standard descriptions[5] of ideal "self-sufficient" leads the DØ magnet leads were optimized for service at about 1.1 times nominal magnet operating current.

During bench tests of the vapor cooled leads made before they were assembled into the control dewar it was shown that after coolant flow was stopped the leads reached 114 C after approximately 13 minutes steady operation at 5000 Amperes. The tests demonstrated that the leads would not be damaged during even a slow discharge of the magnet after loss of coolant flow. The same margin of performance ensured their thermal recovery without current decrease after shorter transient periods of reduced coolant flow.

The behaviour of the leads during steady-state magnet operation at varying coolant flow rates was measured. Figure 2a shows that at higher flow rates the voltage drop on a lead stabilizes at a lower value almost immediately after the flow is increased, but at lower flow rates the voltage drop stabilizes more slowly at higher values.

Under ideal "self sufficient" conditions[5] the leads would be expected to operate stably with approximately $1.1 \times 1.04 \times 10^{-3}\,W/A \times 4820A/20.4\,J/g = 0.27\,g/s = 92$ standard liters per minute boiloff gas flow. Furthermore under these conditions the leads would develop a voltage drop of approximately $80/1.1 = 73mV$. The Toshiba leads are seen to operate stably at flow rates of 100 slpm per lead (0.3 g/s) or less at 4820 Amperes with a voltage drop of about 70 mV.

Figure 2b shows the helium supply dewar ullage increase during the tests. The slope of the curve is perhaps ≈20% smaller (43 l/hr, or 1.5 g/s) during high lead flow (1 g/s for two leads) than it is at the smallest lead flow (0.6 g/s for two leads) when it is approximately 55 l/hr (1.9 g/s).

The test data demonstrate that the system operates within the specified cryogenic budget as indicated in Table 1: a maximum of 1 g/s warm gas flow from the vapor cooled leads plus a maximum of 18 W refrigeration, or a total mass flow of 1.9 g/s.

## ELECTRICAL OPERATION

The magnet was specified to discharge at a rate corresponding to the slow discharge time constant without quenching, and to sustain no damage if it quenched without a protection resistor. To ensure the magnet operated stably at design field it was required to operate at 105% of design field for several hours as well.

Prior to energization the cold magnet coil was tested to ±2 kV, demonstrating DC isolation from ground greater than 100 Mohm. Energization testing then proceeded

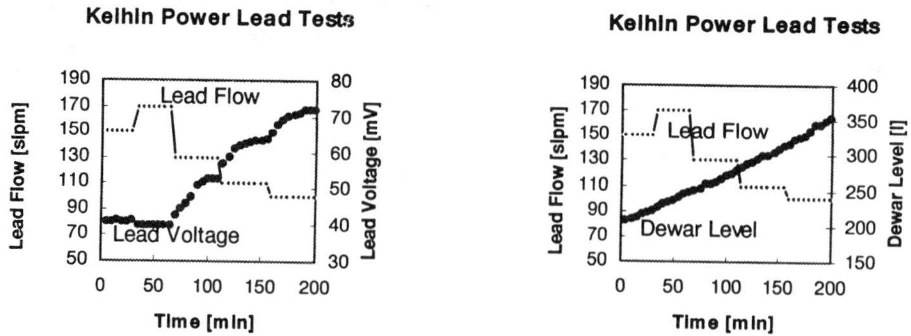


**Fig. 2.** (a)Lead Voltage and (b) Dewar Level (Total System Flow) vs. Lead Flow at 2.0 Tesla

in steps to increasingly higher final currents. At each step the magnet was discharged in the fast or slow discharge mode. On the eighth energization the magnet reached full field at 4820 A. The magnet was operated at 5000 A for several hours, operated at opposite polarity, and forced to quench without a protection resistor by artificially rapid charging. At no time did the magnet quench unexpectedly or show any training behaviour.

## QUENCH TESTING

Carbon glass resistors on the inside of the coil and on the support cylinder, and potential taps at the ends and at the 1/4 and 1/2 points of the coil were monitored during operation of the magnet. The signals from the potential taps were converted to engineering units and recorded by a fast hybrid recorder onto paper strip charts. The signals were also sent to the programmable logic controller which logged them at rates up to about 0.5 Hz. This computer also conditioned the signals from the thermometers and recorded them at 30 second intervals.

Calculations indicated that slow discharges would not quench the coil but fast discharges from currents greater than about 2000 Amperes would cause the coil to be driven normal by AC loss heating in the coil and support cylinder.

The tests showed that slow discharges from all currents equal to or less than 4820 Amperes did not heat the coil above the superconducting transition temperature. A slow discharge from 4820 Amperes caused the coil temperatures to reach only about 7.1 K between 30 and 60 seconds after the initiation of the discharge by which time the current was less than 3570 Amperes. Slow discharges from lower currents resulted in essentially negligible temperature elevations of the coil.

The first fast discharge, from 1500 Amperes, also evidently did not cause quenching in the coil. At some time during the first 30 seconds after the initiation of the discharge the coil temperature reached about 8.4 K (at the thermometer near the center of the inside layer) and it decreased thereafter. No resistive behaviour was seen on any of the potential taps.

The fast discharge from 3000 Amperes caused the potential tap pair which spans the layer-to-layer transition on the far end of the coil to show resistive behaviour about 3.2 seconds after the initiation of the discharge. The coil temperature reached 20.0 K at this end of the coil (and a few K less at the center and near end of the coil). Resistive behaviour was observed on the potential taps on the inside layer of the coil about 10 seconds after the initiation of the discharge.

The fast discharge from 4000 Amperes also caused resistive behaviour to be seen on the potential taps much as was seen in the 3000 A fast discharge (except the onset of this behaviour was at about 1.8 seconds into the discharge). The coil temperatures reached approximately 27.9 K at the far end of the coil (and a few Kelvins less in the center and near ends of the coil) very much like the 3000 Amperes fast discharge.

The fast discharge from 4820 Amperes warmed the coil to 29.9 K at the far end and the potential tap behaviour was much as in the previous fast discharges.

The quenching was cryogenically very benign. The 0.41 MPa (60 psi) relief valves provided on the helium inlet and outlet pipes to the coil did not open (the supply dewar valve closed and the bypass valve in the return line from the magnet opened to vent excess pressure to atmosphere, bypassing the subcooler vessel) and the coil smoothly and rapidly recooled after each test as soon as the valve settings were restored to normal.

## UNPROTECTED QUENCH TESTING

To simulate quenching of the coil with no protection resistor the slow discharge resistor was blocked from the power supply voltage with a large diode. The coil was charged rapidly to induce spontaneous quenching. It was found that charging the magnet at approximately 6.25 V (five times greater than the specified minimum charge voltage) caused it to quench at approximately 4650 A (96% of full current). The very small slow discharge resistor left in the circuit removed only a few percent of the stored energy from the magnet during the quench so the excercise achieved the desired simulation.

A similar unprotected quench induced by a charge voltage of 9.4 volts quenched the coil at 3900 Amperes.

The quenches induced by fast charging were very different than those induced by fast discharging. The fast-chargeup quenches heated the near end of the coil (where the current buses enter and exit) most rapidly and left the final coil temperatures much higher – 135 K in the case of the 6.25 volt chargeup. The potential taps likewise corroborated the fact that these quenches began at the near end of the coil instead of the far end as was the case with the fast discharge quenches.

The peak fields in the conductor occur at the ends of the coil (where high-current density enhancements are located) so it is perhaps not surprising that quenches begin at one end or the other of the coil.

## OVERCURRENT OPERATION

At the conclusion of the test where the magnet was operated at 5% overcurrent the coil was discharged into the slow discharge resistor. About 50 seconds after the initiation of the discharge, resistive behaviour was seen on the potential taps at the near end of the coil. Thirty five seconds later the quench wave reached the far end of the coil as signaled by the potential taps at the layer-to-layer transition. The coil temperature reached 43 K at the near end of the coil and 20 K at the far end.

## MAPPING THE FIELD

Toshiba carefully recorded conductor placement during coil winding, and the dimensions of the finished coil were measured. The location of the coil within the cryostat was surveyed before the vessel was made vacuum tight. In general the winding dimensions of the coil, and the expected changes upon cooldown, were rather well understood.

A field mapper was prepared by Fermilab to survey the field of the magnet to verify that the desired shape and position of the field had been achieved. The field mapper measured $B_r$ at two radial coordinates and $B_z$ at two radial coordinates using precision hall probes mounted on a movable arm. The movable arm could be manipulated under computer control to record the 4 field values stepping in $\phi$ at a fixed z or stepping in z at a fixed $\phi$. Sweeps in $\phi$ were made near each end of the coil, near the center of the coil, and at one intermediate location. Four sweeps in z at $\phi = 0$, 90, 180, and 270 degrees were made.

The measurements of $B_r$ near the coil ends enables the radial location of the coil with respect to the field mapper to be determined. The data showed a substantial $\phi$

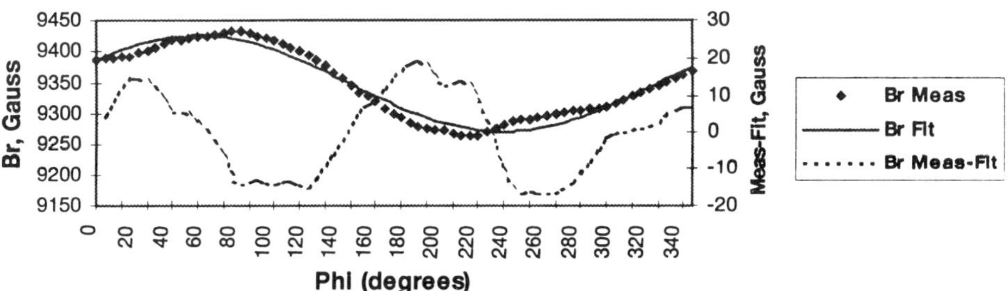


**Fig. 3.** Measuring the Radial Alignment of the Coil within the Cryostat.

variation which was fitted with a single sine function. The fit provided a measurement of the direction and extent of the radial offset. For the far end of the magnet, the results are shown in Figure 3. Shown are the measured data, the fitted curve, and fit residuals, as a function of $\phi$. The fitting results indicate the radial offset is about 0.95 mm.

A similar fit to the measured data at the near end of the cryostat yield a similar result – an offset of nearly 1.5 mm. The fits indicate that the axis of the magnet is misaligned with respect to the axis of the cryostat by only 0.21 mrad. This alignment accuracy implies that the magnet is completely suitable for use in the DØ detector.

From z-sweep data at a fixed $\phi$ the field derivative $dB_z/dz$ was calculated. The peaks at the ends of this quantity maximize at the electrical ends of the coil. By fitting the peaks with a polynomial to locate the maxima, the ends of the coil are found. The differences between these two locations then corresponds to the length of the coil. The results of the fits yield a value of 2548 mm for the (cold) length of the coil. This is to be compared with a value of 2543 mm derived from the as-fabricated coil length corrected for cooldown.

Figure 4 shows the measured values of $B_z$ at a radius approximately 1 cm from the vacuum vessel, and the values of $dB_z/dz$ calculated from them. The shape variations in

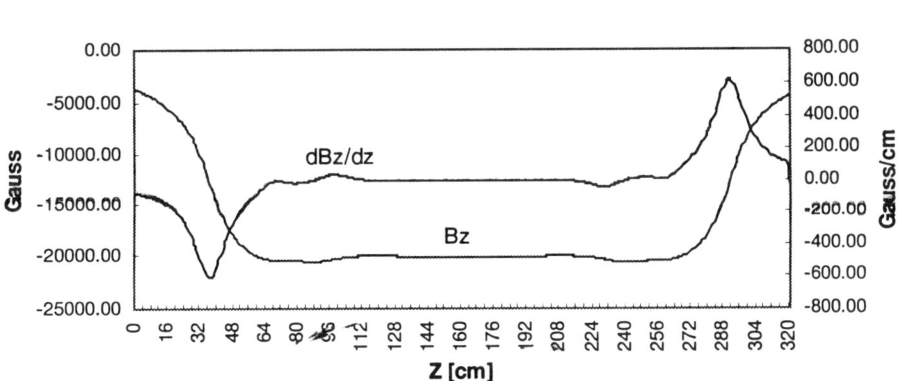


**Fig. 4.** Measuring the coil length with $B_z$ and $dB_z/dz$ vs. z at $\phi = 90$ degrees.

the quantities intermediate between the ends of the distributions reflect the enhanced current density windings at the ends of the coil used to improve overall field uniformity in the bore of the magnet.

## CONCLUSIONS

The DØ magnet fabricated by Toshiba meets all aspects of the detailed specification required for its use in the DØ collider detector at Fermilab. When later installed as part of the DØ Upgrade the magnet is expected to operate easily, efficiently, and safely during physics data taking at the Fermilab Collider at collision energies of 2 TeV/$c^2$ in the center of mass and at luminosities of up to $10^{33}cm^{-2}s^{-1}$.

## ACKNOWLEDGMENT

This work was supported by the U.S. Department of Energy under contract No. DE-AC02-76CH03000.

## REFERENCES

1. B. Squires, *et al.* Design of the 2 Tesla Superconducting Solenoid For the Fermilab DØ Detector Upgrade, *Advances in Cryogenic Engineering* 39:301, Plenum, New York (1993).
2. R. P. Smith, *et al.* Electrical Safety of a Thin Superconducting Solenoid in an Iron Yoke, *Advances in Cryogenic Engineering* 41a:1889, Plenum, New York (1996).
3. R. P. Smith, *et al.* The Aluminum Stabilized Conductor for the Fermilab DØ Solenoid, "Proceedings of the Sixteenth International Cryogenic Engineering Conference/International Cryogenic Materials Conference", p 863, T. Haruyama *et al.* editors, Elsevier Science, Oxford, UK (1997).
4. J. Brzezniak, *et al.* "Conceptual Design of a 2 Tesla Superconducting Solenoid for the Fermilab DØ Detector Upgrade", FERMILAB-TM-1886, Fermilab, Batavia (1994).
5. M. N. Wilson, "Superconducting Magnets", p. 256 ff, Clarendon Press, Oxford (1983).

# PERFORMANCE AND OPERATING EXPERIENCE OF THE RHIC FIRST SEXTANT TEST

J. Sondericker, M. Iarocci, G. McIntyre, H-C. Hseuh,
K.C. Wu, G. Ganetis, M. Anerella, J. Benson, Y Farah,
R. Grandinetti, E. Killian, C. Lac, A. Morgillo, J. Rank,
A. Nicoletti, E. Quimby, M. Rehak, and D. Zantopp

Brookhaven National Laboratory
Upton, N.Y. 11973-5000

## ABSTRACT

Installation and testing of the first sextant of RHIC magnets has been completed. The tests consisted of temperature cycling, quenching magnets, and measuring important engineering parameters of a string of magnets totaling a length of over 500 meters. This represents one sixth of the 3.8 kilometer circumference of the RHIC machine and serves as a basis for extrapolation of parameters for full machine operation. This paper gives an overview of the entire system, a detailed discussion of startup problems, operations and cryogenic system reliability over the period of the test and details of some of the important cryogenic operating parameters.

## INTRODUCTION

Brookhaven National Laboratory (BNL) has a long history of cryogenic experience dating from the mid-fifties, from operation of the first liquid hydrogen bubble chamber particle detectors through the transition to liquid helium operations and superconducting magnet development. By the late 70's BNL's ISAbelle Project was funded. A large 24.8 kW single refrigerator, with selected subsystem redundancy, was designed by the Cryogenics Group to cool the superconducting magnets of the machine. Midway through 1983, most of the architectural construction at the BNL site was near completion and the refrigeration plant was in the process of being installed by a contractor. Soon after, word was received that the High Energy Physics Advisory Panel (HEPAP) voted to recommend termination of CBA in favor of a larger undertaking to be called the Superconducting Super Collider.

Funding was continued to complete construction of the worlds largest helium refrigerator in order evaluate the design and gain experience in the operation of such a large plant. Compressor system acceptance tests commenced March '85 and were completed by mid April. Refrigerator tests began the following August and after four cool downs, the system met specifications and was accepted in March of 1986. Following the acceptance tests, the refrigerator was "mothballed".

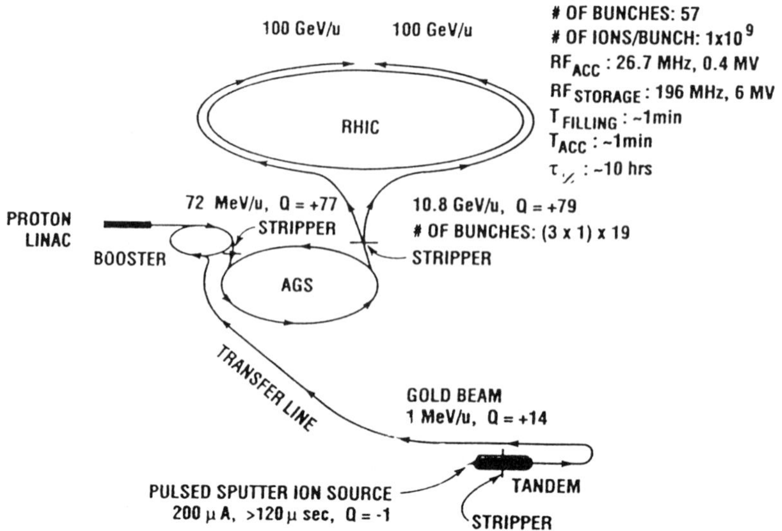


**Figure 1**. RHIC accelerator configuration and operating scenario.

## RHIC OVERVIEW

In 1991 funding was secured from the Department of Energy (DOE) to build a Relativistic Heavy Ion Collider (RHIC) at BNL which used to advantage the existing 3.8 km tunnel, experimental halls and the 25 kW helium refrigerator. In addition, an existing accelerator complex consisting of a Tandem van de Graff, the AGS Booster Synchrotron, and the Alternating Gradient Synchrotron (AGS) is to be used as the RHIC beam injector. Figure 1 shows the accelerator configuration of the RHIC facility and the operation scenario.

The collider is composed of two identical quasi-circular rings of superconducting magnets arranged to intersect with each other at six experimental areas. Figure 2 shows the RHIC ring

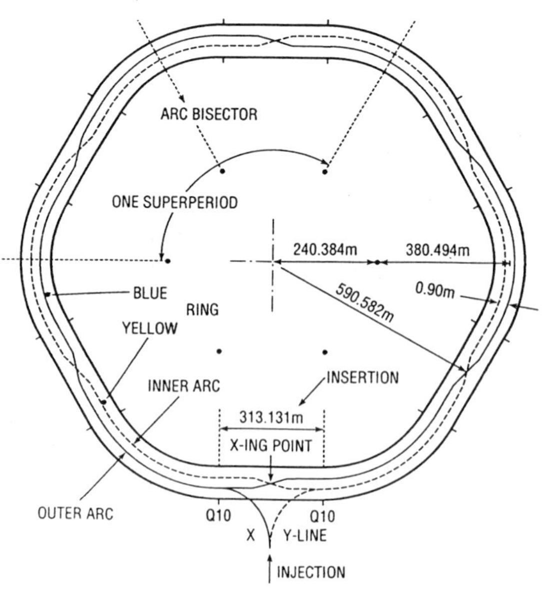


**Figure 2**. RHIC ring layout and injection scheme.

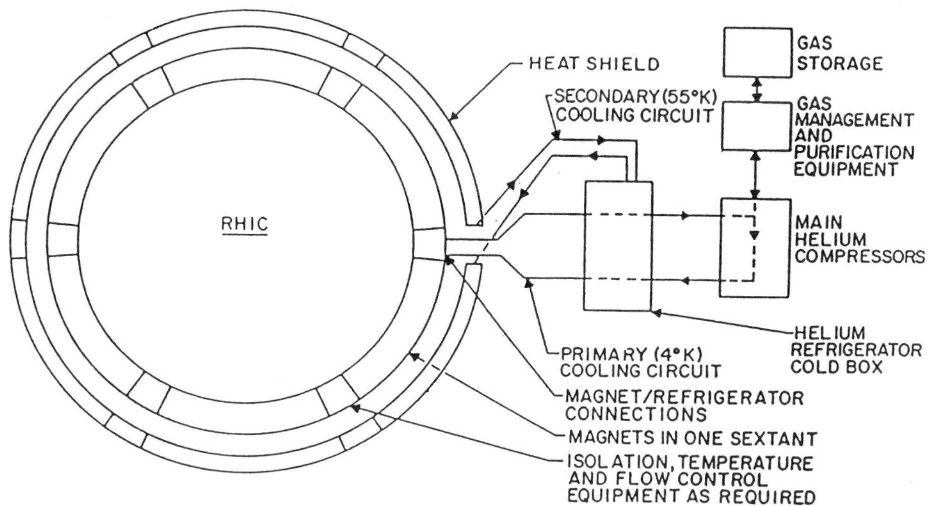


**Figure 3**. RHIC cryogenic subsystems.

layout and the injection scheme. Each ring consists of three inner and three outer arcs (each 355 m long), separated by six (283 m long) insertion regions. Each arc consists of 11 **FODO** cells, with each half cell consisting of a single dipole and a CQS assembly containing a quadrupole, sextupole, and concentric correction elements. Beam collisions occur at each crossing point in the insertion regions.

Dipoles and quadrupoles in both rings are separately powered in series from main supplies located in the 4 o'clock service building. There is only a short length of copper connecting the power supplies to the cryogenically cooled leads of the valve box.

Superconducting busses are connected to the 4K side of these leads and carried inside the magnet helium supply pipe to the neighboring sextant valve box. Here the conductor again goes warm to connect to the energy extraction system.. The cold bus scheme is used to power not only the arc dipoles and quads but all insertion magnets as well as the triplet, Dx, D0 and trims. Arc and Insertion sextupoles and correctors are powered locally via a12 X 50 ampere lead bundle at each cryostat location.

## THE RHIC CRYOGENIC SYSTEM

Performance specifications which have a major influence in the design of the cryogenic system include the following:

1. All magnets must be maintained at or below their nominal operating temperature of 4.6K during steady state operations.
2. Each sextant of each ring must be isolatable for independent warm-up and cooldown for repairs if necessary.
3. The worst case design heat load is set to be twice the calculated heat load for the magnets in each sextant. Under these conditions the temperature of the warmest magnet must remain below 4.6K
4. Reliability of the cryogenic system shall be such that its availability (at nominal design load conditions) is greater than 90%.

The RHIC cryogenic system (Figure 3) is made up of three large subsystems:

1. Helium compression and gas management system
2. The 24.8 kW refrigerator
3. The load; ring magnets, valve boxes and transfer lines

The compression system includes the main helium compressors, gas storage, gas management and purification equipment which supplies clean room temperature helium at operating pressure to the refrigerator. There are a total of 25 ea. Howden 321 mm rotary oil lubricated screw compressors installed, configured such that 20 ea. 600 h.p. driven units are paralleled for first stage use and the remaining 5 ea., 2250 h.p. are connected to use as second stage compressors. This system was designed to match ISAbelle requirements and is capable of supplying 4400 g/s of helium gas at 16.5 atm.. At full load, the compressors draw 14 Megawatts of electric power. RHIC magnets run at higher operating temperatures than ISAbelle and are overall, a more efficient design which permits reducing compressor flow to 2100 g/s.

The refrigerator system[1] fills a second building placed next to the compressor room. This design utilizes redundant subsystems to gain increased reliability and is capable of delivering a peak capacity of 24.8 kW at about 4K plus providing 55 kW of shield cooling flow at 55 K. BNL's many years of refrigerator operating experience were put to good use in placing redundancy in systems that are most sensitive to off design operation or apt to fail. The warm end of the refrigerator contains redundant top end heat exchangers which are the first cold surfaces that compressor helium sees in the refrigerator and has the greatest chance of becoming contaminated with condensables. Liquid nitrogen precooling was not used in this design because in the past, strikes and/or inclement weather has disrupted truck deliveries to Brookhaven.

Instead, a redundant train of expanders eliminates the need for nitrogen precooling and removes the problem from consideration. Redundancy continues in the second expander train, turbo oil bearing pumps and in line purification. The fifth vessel houses the low temperature end of the machine which includes the final expander, liquid pots, a cold vacuum compressor and connections to the RHIC ring magnets.

## PREPARATION FOR THE SEXTANT TEST

The Sextant Test was a RHIC Project DOE milestone scheduled for February 1997. The test was a full systems and beam trial whose goals were to:

- transport beam through sextant 5/4 to a dump
- verify cryogenic system performance
- verify the design of the magnet quench protection system
- test ramping and storage power supplies
- test vacuum system performance
- observe cold mass mechanical motion during thermal cycles

Magnets began to be installed in the tunnel in November 1995; first Grumman built arc dipoles and then arc quads, insertion elements and finally triplets[2]. By December 1996, installation of magnets in sextant 5/4 was completed and pump down commenced in preparation for cooldown. After a decent vacuum was established, individual helium circuits were pressurized to 1.5 MPa (225 psi). Leak detection showed the presence of one small helium leak in each of the two ring arcs but many large leaks were found in the final focus triplet assemblies at both 5 and 4 o'clock ends of the sextant. Upon opening the vacuum tanks it was found that pin hole leaks along the length of the flex hoses were the source of the problem. The holes were caused by the ultrasonic welder used to close multilayer insulation (MLI) blankets around the lines. In order to get a good weld, force has to be applied to the ultrasonic welder tip to locally compress the MLI. The effect was present only when the welder tip happened to land on the top of a bellow convolution and enough force was applied to spread the braid so that the 0.2 mm stainless wall was exposed to the energy of the welding process.

Leak location and repair continued into the month of January. As days went by it became clear that we wouldn't be able to find and repair all the leaks and still leave ample time for cool down and to get beam to the dump before the end of the gold beam AGS run which was scheduled to turn off at the end of the month. By the third week in January, it was felt that all the large

**Figure 4**. Operator display of magnet and recooler control variables.

leaks of the outer ring were repaired and since this was the path of beam injection we should close up and start cool down as soon as possible. Inner ring magnets would be pumped and but not cooled. The gamble was in hoping that the remaining leaks were warm and that available pumping could bring insulating vacuum pressure down low enough to cool down and reach operating conditions. Cool down commenced on Monday, January 20th.

## COOLDOWN, BEAM AND MAGNET TESTING

By Wednesday, January 22, the temperature of the sextant was down to 4.6K in spite of the leaks located in the 5 o'clock triplet cryostat. Figure 4 is a cryogenic control system screen display of magnet and recooler operating variables of the sextant under test.

The next two days were spent in checking power supplies, the quench protection system, completion of tests of the Personnel Access Safety System (PASS) and related safety reviews for accelerator readiness. By Saturday morning beam was being sent down the injection line and by 2:00 on Sunday, January 25th, signals indicated that beam reached 4 o'clock and the beam dump. For the remaining week of beam, tests of beam monitoring systems, beam permit links, controls and machine data bases were carried out. After beam was turned off, tests continued through the month of February in order to prove the soundness of collider subsystems.

Cold bore beam tube charcoal sorption pump He pumping capacity was measured and insulating vacuum system operating conditions were documented.

Power supplies and magnets were exercised over 100 times by repeatedly ramping magnet currents up to 5500 Amperes at a rate 10% faster than normal and switched to flat top storage mode for a duration of a few hours to over 12 hours without problem. Quench protection and detection systems were tested by quenching a portion of the cold crossing bus via spot heaters. Arc dipoles were found difficult to quench even at reduced flow and emptying the 50 Watt recoolers. Eventually, warm up heaters were used to force magnet quenches and at full current there was little impact on the cryogenic system.

## CRYOGENIC SYSTEM TESTS

Test and measurement of cryogenic system parameters included ring pressure drop at design He flow, heat loads of magnet strings, transfer lines, heat shield circuits and valve boxes. The refrigerator[3] was tested for maximum capacity against a calorimeter and cold vacuum compressor and circulating compressor data was acquired.

Some results of the tests are listed below and compared to design values:

### Pressure Drop

Pressure drop across one sextant of magnets, cooled with single phase helium at 5 atm., 100 g/s flow and 4.5K was measured to be .06 atm. This corresponds to 0.36 atm. for the entire six sextants of the machine. Design value is 0.42 atm.

### Heat Loads

Primary heat load of the sextant of magnets (Figure 4) was measured in four ways by two different people. In all cases the heat loads ranged from 650 Watts to a high of 900 Watts, for a 775 Watt rough average. Design value is 725 Watts.

### Shield Heat Load

For this test, heat shield cooling flow came through the 6 o'clock valve box, cooled the supply transfer line shield, all the sextant outer ring magnet shields and on to the 4 o'clock valve box. A temporary jumper was installed to the inner ring valve box to provide a return path for the shield flow back to the refrigerator. The inner ring of magnets were warm, so a sizable heat load was picked up by the shield flow on its return trip to the refrigerator. The outer ring shield load was measured at 1.0 kW and the total for the two rings at 3.7 kW, making the shield load 2.7 kW for the warm magnets. Checking the measurements, the warm magnets of the inner ring cooled down at the rate of about 3K per day and by specific heat calculation the heat load is 2.6 kW. The design value of shield heat load for a sextant of magnets is 1 to 1.25 kW.

Subsystem testing and data acquisition continued through February. Last on the task list was a warm up and cooldown cycle to further stress the magnet and cryogenic systems before turn off. Warm-up of the magnets was accomplished in 24 hours via built in electrical heaters. The following cooldown and warm-up cycle was uneventful and the Sextant Test officially ended on March 2nd.

## CONCLUSIONS

Overall the Sextant Test was a great success. From the systems point of view, all fundamental design concepts of RHIC were validated: components, mechanical aspects, electrical systems, cryogenics and beam systems. Some small problem areas were discovered but nothing even close to a magnitude large enough to impact the overall scope and plans for the project.

## ACKNOWLEDGMENTS

This work was performed under the auspices of the U. S. Department of Energy.

## REFERENCES

1. M. Iarocci, et al., "RHIC 25 kW Refrigerator and Distribution System, Construction, Testing, and Initial Operating Experience", CEC vol. 43, These Proceedings
2. R. Grandinetti, et al., "Design Considerations and Experience of the RHIC Dual Magnet Cryostat Installation", CEC vol. 43, These Proceedings
3. K C Wu, et al., "Status, Operation and Performance of the RHIC Helium Refrigerator", CEC vol. 43, These Proceedings

# MAGNETIC ELEMENTS FOR THE A1900 FRAGMENT SEPARATOR AT THE NSCL

A. F. Zeller, J. C. DeKamp, D. Johnson, F. Marti, D. J. Morrissey, B. M. Sherrill, J. Wagner, X. Wu, R. C. York,and R. Zink

National Superconducting Cyclotron Lab
Michigan State University
E. Lansing, MI 48824 USA

## ABSTRACT

A high rigidity, large acceptance fragment separator, the A1900, is being built at the NSCL. The device consists of twenty-four quadrupoles, four dipoles, sixteen sextupoles and sixteen octupoles. There are five sizes of the quadrupoles, three of which will contain sets of nested multipoles. The quadrupoles and the dipoles are superferric, with maximum pole tip fields of 2.5 T and 2 T, respectively. The quadrupole coils and the multipoles are random wound, potted using conductor from 0.85 to 1.35 mm in diameter. The dipole is layer wound with insulated conductor with a cross section of one by two mm. All coils are self protecting in case of a quench.

## INTRODUCTION

The National Superconducting Cyclotron Lab (NSCL) is in the process of upgrading[1] the capabilities of the lab for the production of secondary beams and higher energy heavy ions. This upgrade will couple the two superconducting cyclotrons and construct an upgraded fragment separator, the A1900. The A1900 will be able to produce secondary beams that are more rigid than the primary beam. Additionally, it will have a large acceptance in both solid angle and momentum range. This is achieved by using high gradient, large aperture quadrupoles and large gap dipoles. To obtain the desired resolution, sextupoles and octupoles are required. Because the site size is fixed, the separator must be made to fit within the present NSCL's building. The only rearrangements will be internal walls and moving of equipment that will be displaced by the required radiation shielding. The original liquid helium refrigerator will be moved to a location next to another refrigerator that was installed to accommodate the higher refrigeration loads that will be required for the A1900. The space constraint required that we design dipoles that operate at somewhat higher fields than our present dipoles[2]. The K1200 cyclotron and the A1900 magnetic element layout are shown in Figure 1. The NSCL will stop experimental operation in July 1999 for eighteen months for installation of the A1900, the coupling line between the two cyclotrons, and the necessary building changes.

**Table 1.** Parameters of the A1900 quadrupoles

| Type | Number | Iron length(m) | Effective length(m) | Pole tip radius(cm) | Warm radius(cm) | Multip-oles |
|---|---|---|---|---|---|---|
| QA | 4 | 0.65 | 0.723 | 13.3 | 10 | no |
| QB | 12 | 0.325 | 0.400 | 15.0 | 10 | yes(all) |
| QC | 4 | 0.715 | 0.790 | 15.0 | 10 | yes(all) |
| QD | 2 | 0.381 | 0.486 | 21.0 | 17 | no |
| QE | 2 | 0.625 | 0.700 | 15.0 | 11.6 | no |

## QUADRUPOLES AND MULTIPOLES

Based upon the success of the superferric quadrupoles used in the S800 superconducting spectrograph[3] we will use the same approach for the A1900. In general, however, the quads needed for the A1900 run at higher gradients, have larger bores and are longer than those used in the S800. This lead to several significant changes in the magnet parameters and in the cryostats. Because the quads are packaged at triplets, with most having imbedded multipoles, the bores were enlarged and the helium vessel and nitrogen shield designed with greater clearances to facilitate assembly. In order to maximize the transmission of secondary beams and at the same time fit the system into the available space, we are building five different size quads. The different magnets are listed in Table 1. The effective magnetic lengths are based on similar magnets which have been built in the lab. Effective lengths decrease with increasing field, but only the value for the maximum field is listed in the Table. Those sixteen magnets with multipoles have both sextupole and octupole windings. The lengths and warm bores have been chosen by balancing the two conflicting requirements of keeping the pole tip field below 2.5 T and making the magnets as short as possible to give the largest secondary beam transmission. Table 2 lists the appropriate wire, coil and magnetic parameters for the quads. The maximum required gradients are those which will produce the desired optical conditions. The design gradients are the magnetic parameters which correspond to the other parameters listed in the Table. The coils are wet wound with epoxy on a semiautomatic winding table. They are then cured and end forms added which position the coils on the ends. The coils are shimmed in place with respect to each other and the pole tips. A complete QE coil is shown in Figure 2. A QD coil is shown if Figure 3.

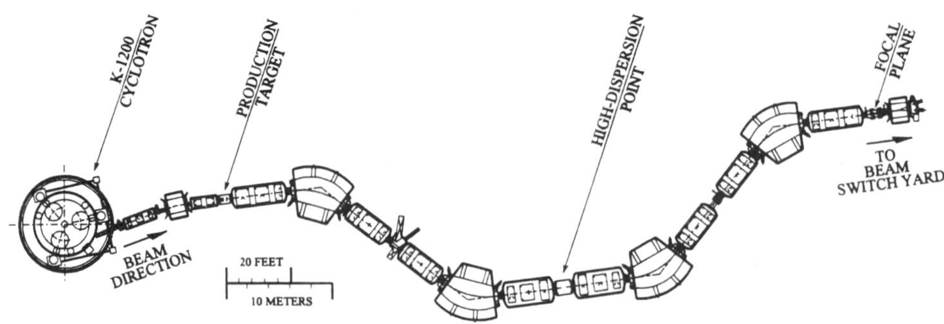


**Figure 1.** K1200 cyclotron and the A1900 Fragment Separator.

**Table 2.** The magnetic, coil and conductor parameters for the A1900 quadrupoles

| Type | Max required gradient (T/m) | Design Gradient (T/m) | Induct-ance (H) | Stored energy (kJ) | Max current (A) | Turns | Wire diameter (mm) |
|---|---|---|---|---|---|---|---|
| QA | 18.7 | 19.4 | 39.1 | 118 | 77.5 | 1950 | 0.85 |
| QB | 13.6 | 15.8 | 41.5 | 78 | 61.3 | 2820 | 0.80 |
| QC | 14.4 | 15.8 | 47.7 | 154 | 80.4 | 2150 | 0.95 |
| QD | 11.0 | 12.2 | 5.08 | 372 | 404.5 | 890 | 1.35 |
| QE | 14.4 | 15.8 | 43.8 | 135 | 78.6 | 2200 | 0.95 |

**Figure 2.** Complete coil for a QE size magnet. The meter stick shows the scale.

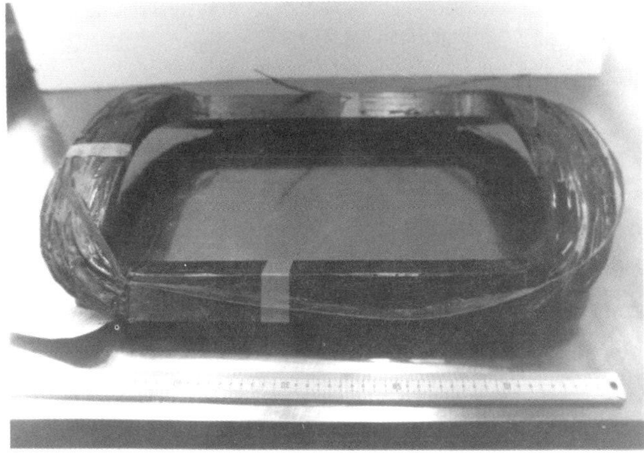

**Figure 3.** A QD coil.

**Figure 4.** The iron return yoke for a QD magnet.

The yokes and pole tips are machined out of low carbon iron (SAE 1006). After machining a radius on the back of the pole tip block the hyperbolic surface is cut in about ten hours per pole piece with a numerically controlled milling machine. The profile is cut. The yokes are cut from large billets and machined to the final size. A QD yoke is show in Figure 4.

Figure 5 shows a schematic of one of the triplets which contains a QD element (large bore) a QE and a QB (with multipoles). The magnets are positioned with respect to each other with stainless steel spacers. This is similar to what was done with the large quadrupoles for the S800 spectrograph, but with much larger sized magnets. The multipole inserts are shown in Figure 6. The inserts are similar to those used in the S800[3] and at CEBAF[4], which use the iron to enhance the field, but are essentially air core. The field uniformity of the multipoles becomes poor when the poles saturate, but resulting change in resolution in the secondary beam is small. At field levels

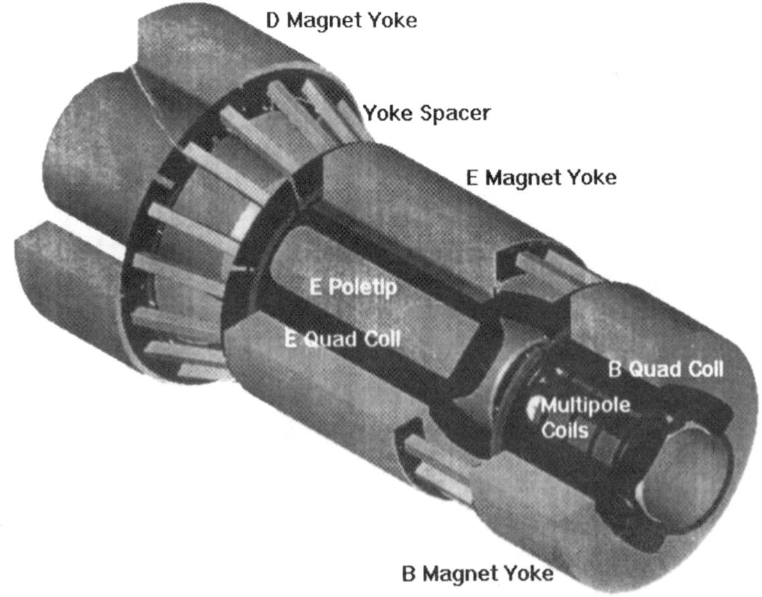


**Figure 5.** CAD view of a triplet which has a large quad (QD).

**Table 3.** Parameters of the multipole inserts

| Type | Max required field | Design field | Max current(A) | Turns | Wire diameter(mm) |
|---|---|---|---|---|---|
| Sextupole | 1.23 T/m$^2$ | 1.58 T/m$^2$ | 37.5 | 270 | 0.50 |
| Octupole | 0.54 T/m$^3$ | 0.62 T/m$^3$ | 19.2 | 260 | 0.50 |

**Figure 6.** The multipole assembly on a test bore tube. The octupole is on the inside.

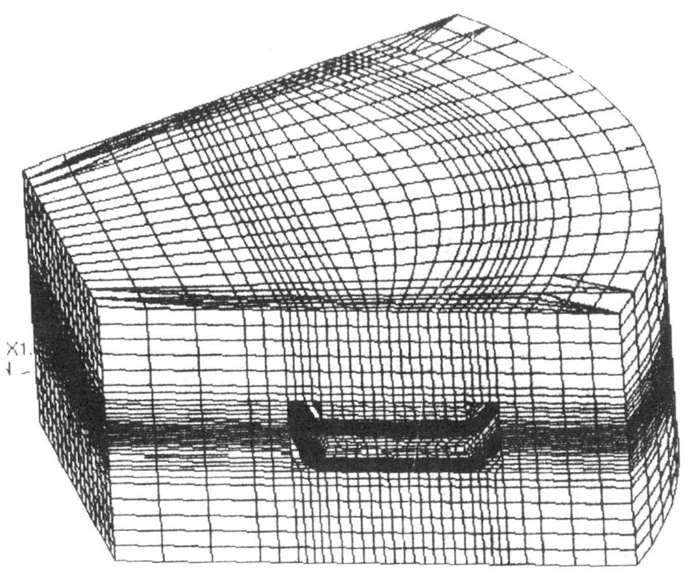

**Figure 7.** TOSCA view of the dipole showing the iron and coil shapes.

**Table 4.** Parameters of the A1900 dipoles

| **Conductor** | |
|---|---|
| Bare wire size | 0.898 X 1.898 $mm^2$ |
| Cu/Sc ratio | 9:1 |
| Maximum operating current | 161.3 A |
| Critical current at 2 T | 500 A |
| Coil | |
| Number of turns | 600 (25 turns by 24 layers) |
| Current density | 6450 $A/cm^2$ |
| Maximum quench internal voltage | 413 V (calculated) |
| Peak hot spot temperature | 216 K (calculated) |
| Magnet | |
| Bρ | 6 T-m |
| Gap | 9 cm |
| Central field | 2 T |
| Maximum stored energy | 0.53 MJ |
| Weight | 50000 kg |

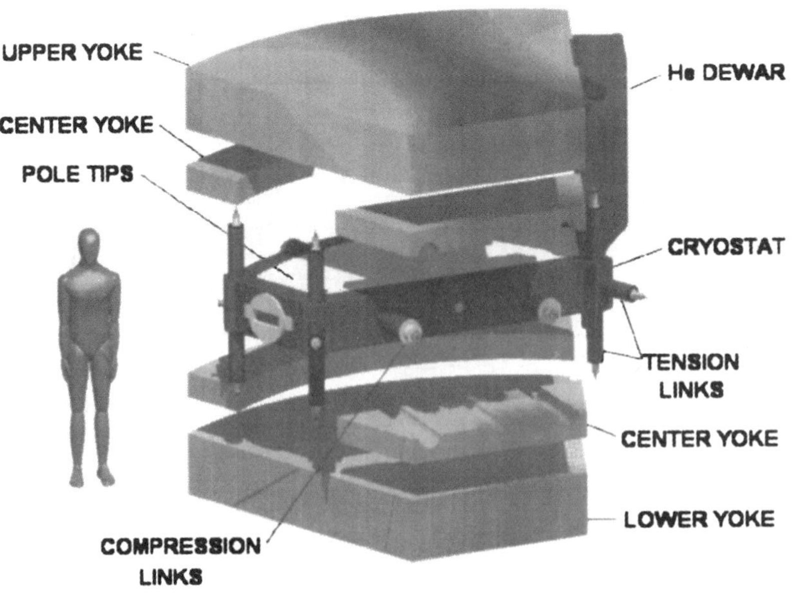


**Figure 8.** Exploded view of the A1900 dipole.

where the pole tips are unsaturated the transfer function is linear and the error terms remain constant. The parameters of the multipoles are given in Table 3. We use the normal procedure of placing the highest order multipole at the smallest radius, the octupole was placed at the smallest radius. This was because it operates at the highest pole tip field. Both multipoles operate at a maximum current density of 3 kA/cm$^2$ at the design fields. The multipoles are stacked onto the liquid helium bore tube with the necessary shims and spacers. The coils are shimmed to each other and then potted preserve the spacing. The final assembly is then held in place with stainless steel bands.

## DIPOLES

The standard NSCL dipoles are designed to operate at 1.75 T. Beyond this field, the uniformity begins to be worse than 0.1% in the volume occupied by the beam because of the compact design of the magnets. In order to fit the space constraints imposed on the A1900, a field of 2 T is necessary. For optical reasons, the dipole bend angle required is 45 degrees. Currently, a 45 degree bend is achieved by using a pair of 22.5 degree magnets. This option is precluded by space requirements. A minimum mass dipole would have the coils follow the beam sagitta, but this would result in one side of the coil having a negative curvature. While these coils can be wound, as was done for the S800 dipoles, they are very difficult. Therefore, we have eliminated this negative curvature by replacing it with a straight section. To further aid winding, a small positive curvature has been added, that is there is a 6 mm displacement at the center of the coil. Because the coil shape, and consequently the return yoke, can no longer be calculated with a two dimensional magnet code, such as POISSON, calculations with the three dimensional code TOSCA[5] were done to determine the pertinent parameters. The final parameters are given in Table 4. The area of the return yokes has to be adjusted to keep the saturation at a minimum and to keep the magnetic center aligned with the physical center. The configuration of the yokes and coils is shown in Figure 7. An exploded view of the entire dipole is shown in Figure 8. There are a total of four dipoles in the A1900.

A further requirement for the dipole was that they be modular. The first dipole is exposed to very intense radiation doses, both neutrons and charged particles, the coils must be readily replaceable in case of a failure. This means that the down time is small as well as keeping the radiation exposure to workers at a minimum. The top yoke assembly will be lifted off in a single piece to expose the cryostat and pole tip assembly. The entire assembly will then be removed to shielded storage and a spare assembly put in place. All the very active components are in the pole tip assembly so radiation exposure will be kept to a minimum. The calculated lifetime of the coils is approximately ten years.

## CONCLUSIONS

Magnetic elements for a new fragment separator have been designed are being constructed with a planned installation date of late 1999. Two of the five quadrupole coils types have been wound and sufficient multipole coils for the first triplets have been fabricated. A test wind of the dipole coil with aluminum wire has been done to proof the coil fabrication method. Machining of the quad iron assemblies and the dipole pole tips is underway. Tooling and fixtures for magnet assembly are almost complete. Quench calculations for all elements demonstrate that the coils will be self-protecting. The relatively low operating currents of all the magnets allows a safety margin in the availability of liquid helium from the refrigeration plants.

## ACKNOWLEDGMENT

This work is supported in part by a grant from the US National Science Foundation and from Michigan State University.

## REFERENCES

1. R. C. York, et al. "Proc. Of the Fourth EPAC", London, 1994, p. 554 and R. C. York, et al."Proc. Of the 1995 PAC", IEEE Press, Dallas, TX, p. 345.
2. A. F. Zeller, et al. *IEEE Trans. On Magnetics* 27:1951(1991).
3. B. Zhang, et al. *IEEE Trans. On Applied Superconductivity* 7:610(1997) and B. Zhang, et al. Adv. In Cryo. Eng. 41A, 1996, p. 375.
4. S.R. Lassiter, et al. *IEEE Trans. On Applied Superconductivity* 3:118(1993).
5. TOSCA ver. 6.3, Vector Fields, Inc. Aurora, IL

# DESIGN AND CONSTRUCTION OF COIL SUPPORTING STRUCTURE AND CRYOSTAT VESSEL FOR LHD

H. Tamura,[1] A. Nishimura,[1] S. Imagawa,[1] T. Satow,[1] H. Hayashi,[1]
O. Motojima,[1] K. Asano,[2] Y. Takahashi,[2] T. Kohda,[2]
M. Sudo,[2] O. Honma,[2] S. Suzuki,[2] and LHD group[1]

[1]National Institute for Fusion Science
Oroshi-cho, Toki-shi, Gifu 509-52, Japan
[2]Hitachi, Ltd.
Hitachi-shi, Ibaraki 317, Japan

## ABSTRACT

The Large Helical Device is a fusion experimental machine with all superconducting coils. Electromagnetic force generated by the coils is supported by the torus shaped "supporting structure" of 100 mm thick 316 stainless steel. This structure consists of 20 fan shaped parts that were welded together on site. The coils needed to be positioned within 2 mm accuracy in the absolute coordinates so that the supporting structure had to be assembled within 5 mm accuracy prior to the coil setting. After completion of the welding, the overall accuracy of < 5 mm was achieved.

The cryostat vessel for the cryogenic structures of LHD has also been assembled on site. The material used in this vessel is 304 stainless steel. The thickness is 100 mm for the base plate and 50 mm for the outer and inner cylinder parts and the upper cover. Calculations for the deformation of this vessel under atmospheric pressure were carried out by using the finite element method and the displacement of ports for plasma diagnostics was confirmed to be less than 4 mm.

## INTRODUCTION

The construction of the Large Helical Device (LHD) started as an 8-year project in 1990.[1] Figure 1 shows the schematic view of LHD including the components. LHD is a superconducting device that has a pair of helical coils and three sets of poloidal coils. The electromagnetic force generated by the interaction among these coils reaches up to 10 MN/m for the 4 T operation in the second experimental phase. The coils subjected to the electromagnetic forces are supported by the torus-shaped "supporting structure". The coils and the supporting structure will be cooled down to the liquid helium temperature in the first experimental phase (the helical coils will be further cooled by using superfluid helium in the second phase).

Minimizing the deformation during the excitation of the coils is extremely important for LHD to get a good confinement condition for plasma. The maximum allowable deformation was set to be 3.4 mm in the design concept according to the accuracy of the magnetic surfaces for the plasma confinement. Analytic estimation was done to decide the basic shape and the thickness of the supporting structure to satisfy the required deformation tolerance.[2] By these studies, the supporting structure was chosen to be a torus shape with the major and minor radius of 3.9 m and 1.85 m, respectively. It is made of 100 mm thick 316 stainless steel. The carbon and nitrogen contents of this stainless steel are controlled to increase strength at low temperature and to be strong against welding. Some mechanical tests of this stainless steel with welded sections have been carried out to investigate the fracture toughness.[3,4]

The weight of the supporting structure after assembly is about 400 tons. This structure will be cooled down to liquid helium temperature together with the coils. The total mass that will be cooled is about 850 tons. An adiabatic vacuum condition is realized by the cryostat vessel whose cross section is a bell shaped torus structure. There are radiation shielding plates near the inside surface of the cryostat vessel. The shielding plates will be cooled down to 80 to 40 K to reduce the radiation power input from the room temperature.

Both the supporting structure and the cryostat vessel have been assembled in LHD experimental hall since the transportation in its final form was impossible. They were transported in pieces and assembled on site. In this paper, the assembling procedures of these structures are described focusing on the welding processes. The analytical calculation results are also discussed for the deformation of the cryostat vessel.

## SUPPORTING STRUCTURE

The supporting structure primarily consists of two components, the upper and lower hemispheres. The only difference between them is that the lower part has the base for the cryogenic supporting posts. The weight of the supporting structure and the coils are sustained by the cryogenic posts which are set on the base plate of the cryostat vessel.[5] Thermal insulation between the room temperature and the liquid helium temperature is given by these posts. They also absorb a thermal shrink of 12 mm of the components in the major radial direction during the cooling down. The lower and the upper part of the supporting structure fasten the helical coil between them through the shell arms which are connected to the coil cans. The lower part was completed in FY 1994 and the upper part in FY 1996. The upper one had been assembled with the upside down posture to obtain the symmetry in the assembly with the lower one. Then the upper one was turned in the final position.

During the process of welding, the average deformations were monitored to be less than 5 mm for the ten sections of each hemisphere. The deformation of the upper half showed a good symmetrical correlation with the lower one. After the completion of the lower and

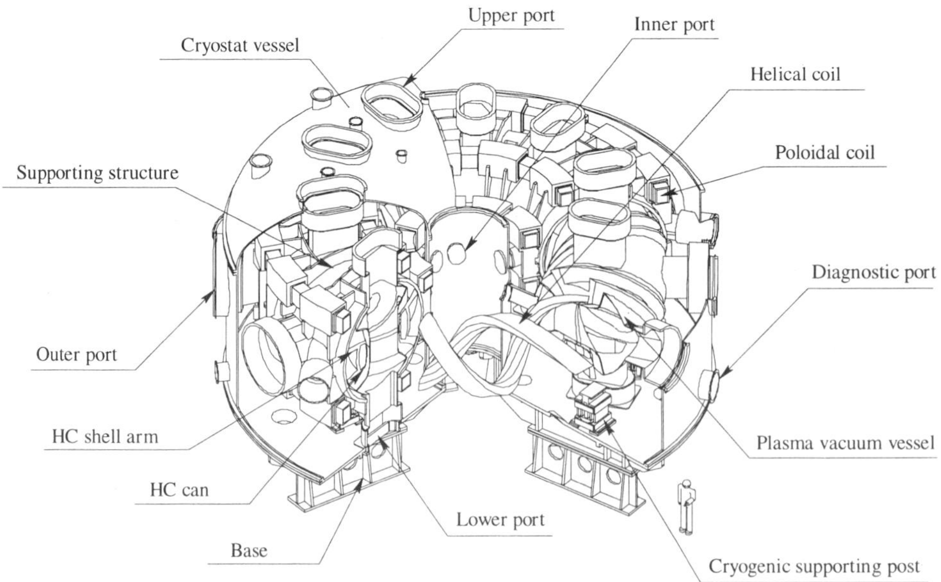


**Figure 1.** Schematic view of LHD.

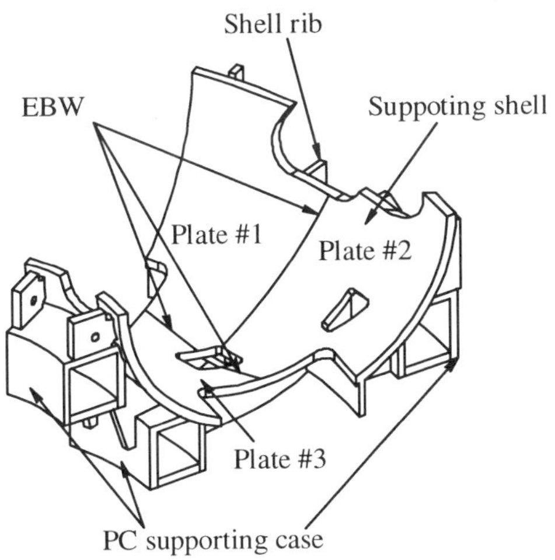


**Figure 2.** A fan-shaped section of the supporting structure. There are ten sections for each hemisphere.

upper parts, they were welded together at the equator, holding the helical coils in the torus shell. The shell arms were welded to the inner surface of the supporting shell simultaneously. The deformation after the completion of the supporting structure was less than 5 mm, which satisfied the required overall accuracy. Here follows the details of the construction of the supporting structure.

**Assembling procedure**

The 20 sections, manufactured in the factory, are 36-degree fan-shaped pieces as shown in Figure 2. They were assembled into the upper and lower hemispheres. Each section consists of three components. They are the "supporting shell", the "poloidal coil (PC) supporting case" and the ribs for strength. Three plates (plates #1 to #3 in Figure 2) were bent individually by hot press method and connected to each other by electron beam welding (EBW). Next, the PC supporting cases and the ribs were attached to the shell. Each of the 10 sections was mechanically joined on a machining center to finish the inner surface of the supporting shell and the inner wall of the PC supporting cases. The positional accuracy of these parts is important since the coils are installed inside. The target accuracy in this machining process was 0 to +2 mm considering the shrinkage during the welding process on site. The welding grooves for the joint section between the upper and lower hemisphere were also processed by the machining center. After the sectors were divided again, the grooves for sector-to-sector welding were processed and pipes for coolant were attached on the surface of the shell. Then every sector was transported to the experimental hall of LHD.

Figure 3 shows the details of the welding sections. Figure 3 (a) is for sector-to-sector welding and (b) for upper-to-lower welding. We adopted partial penetration welding to reduce the deformation during the welding process. In the case of (a), welding grooves were processed on both sides of the supporting shell and two holes were prepared in front of the grooves to reduce the stress intensity. The welding depth was decided by estimating the stress level in that area. On the other hand, there is one side groove in the case of (b), since there would be the shell arms of the helical coils inside the surface of the shell. The thickness around the welding zone was decreased to reduce the welding quantity and to be apparently full penetration welding. The stress level around the welding zone is multiplied by decreasing the thickness. However, the estimated stress level is below 540 MPa which is within the allowable stress level of the material. To minimize the deformation generated by the heat input, the welding lines were divided into ten sections and they were welded alternately.

After the supporting structure was assembled, the shell arms attached to the coil cans of the helical coils were connected to the inner surface of the supporting structure by welding. The upper and lower halves of the supporting structure were also connected by welding

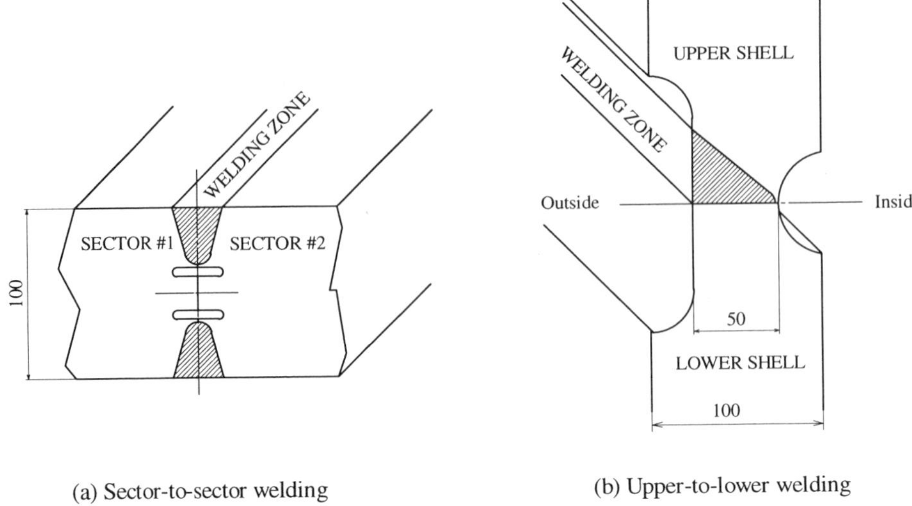


**Figure 3.** Schematic view of the welding zones of the supporting shell: (a) sector-to-sector and (b) upper-to-lower.

simultaneously. The helical coil had had a winding core before it was installed to the supporting structure to keep its shape. After this construction had finished, the winding core for the helical coils was removed.

**Result of on site assembling**

**Sector-to-sector welding.** The deformation during the welding process was measured at the following points: when the first layer of the welding was completed, after 40%, 70%, 100% of the welding, and when the finishing treatment had been completed. Important measuring points causing concern with the accuracy were inner surfaces of the supporting shell and the PC supporting cases. Figure 4 indicates the measurement locations during the process. RA, RB, RC, RD and RE are the minor radius of the inner surface of the supporting shell. IV/IS/OV-inner/outer R and IV/IS/OV-h are the major radius of the inside surface of every PC supporting case and the height of the ceiling of the cases, respectively. Figure 5 shows the measured deformation during the sector-to-sector welding. We measured every 36-degree of the poloidal cross section and the plots in the Figures 5 (a) to (d) are the average

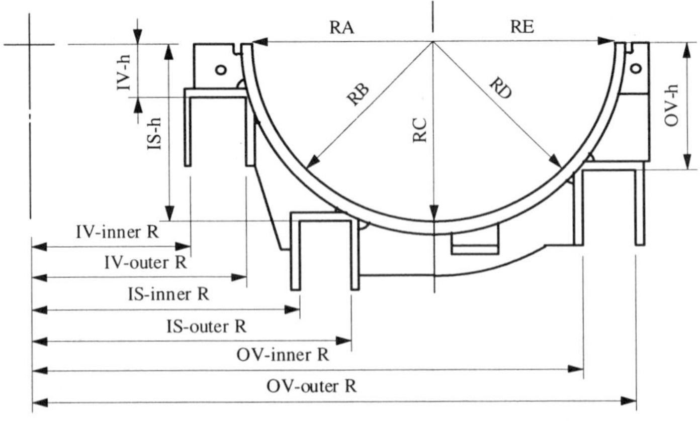


**Figure 4.** Measured points of the supporting structure during welding process.

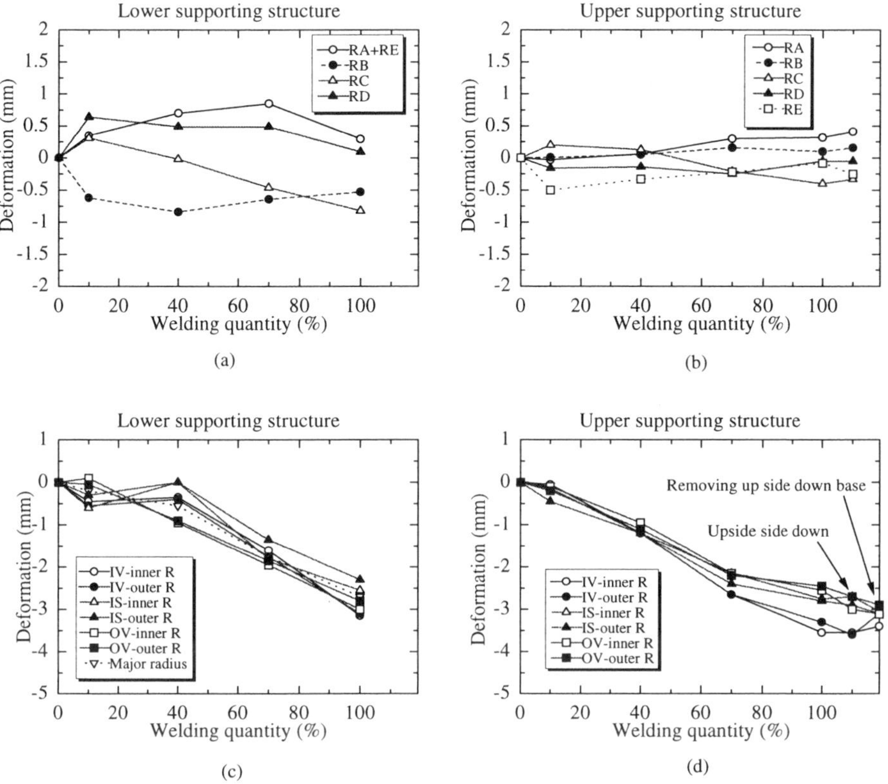


**Figure 5.** (a) Deformation of the minor radius (the lower hemisphere). (b) Deformation of the minor radius (the upper hemisphere). (c) Deformation of the major radius at PC supporting case (the lower hemisphere). (d) Deformation of the major radius at PC supporting case(the upper hemisphere).

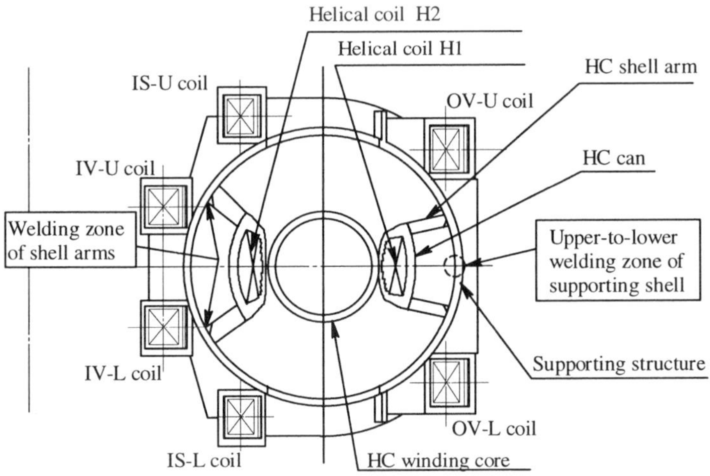


**Figure 6.** Cross section of the supporting structure and the helical coils with their welding zones.

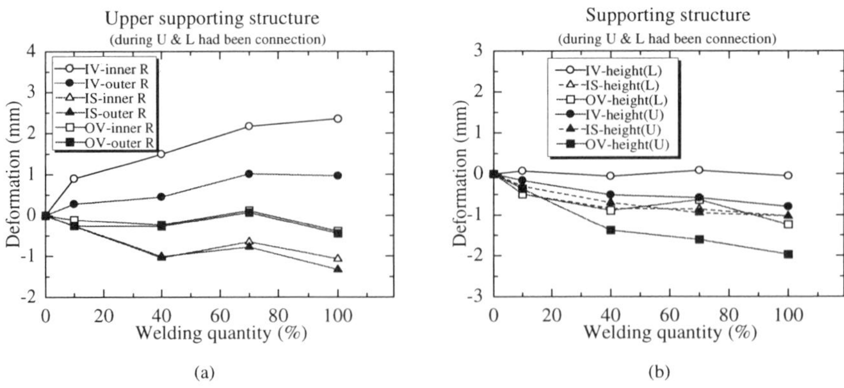


**Figure 7.** (a) Deformation of the major radius at PC supporting case during the upper and lower hemispheres were connected (the upper hemisphere). (b) Deformation of the height at PC supporting case (the upper and lower hemisphere).

values among them. The first layer was welded by using TIG welding and the rest of the zones were done by MAG welding. The deformation after the first layer TIG welding is indicated at 10 % of the welding quantity in the figures.

As shown in Figures 5 (a) and (b), the minor radius RA increases as the welding quantity increases while the radius RC decreases. As for the "inner R", the deformation increases almost linearly from the second layer to the end of the welding. The deformation of "IV-inner/outer R" and "IV-outer R" are a little larger than others are. We believe it is due to the influence of the expansion of the minor radius RA. The height measurement did not show as much deformation during the welding of each hemisphere.

**Upper-to-lower welding.** After completion of each hemisphere, the helical coils were set inside the supporting shell. The equator of the supporting shell was welded. The shell arms were also welded to the supporting shell as shown in Figure 6. These welds were done simultaneously. Figures 7 (a) and (b) show the measured deformations after the upper and lower hemisphere of the supporting structure were connected. The minor radius of the IV supporting case rather expanded with the welding. We suppose that a local deformation might have occurred in the region of the welding between the supporting shell and the helical coil shell arms. The PC supporting case might have inclined to the center direction of the minor radius of the supporting shell.

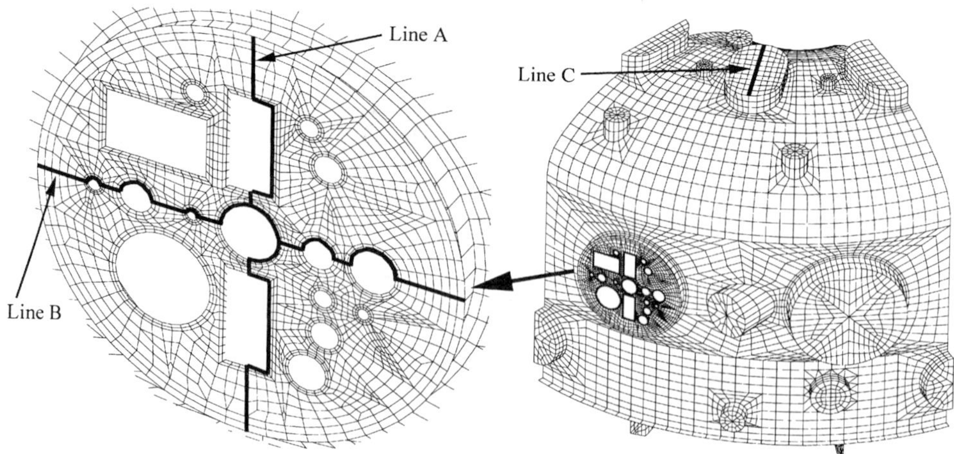


**Figure 8.** FEM model for the cryostat vessel.

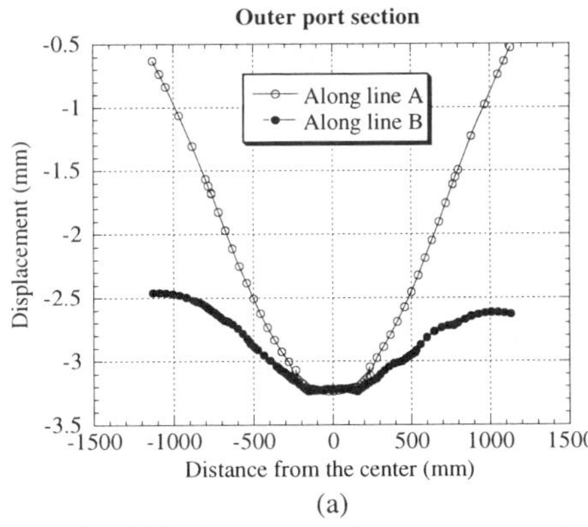


(a)

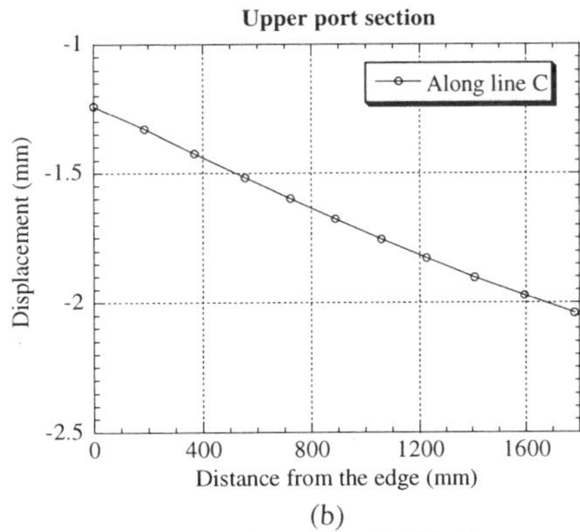


(b)

**Figure 9.** (a) Displacement in the outer port section along the line A and B in figure 8. (b) Displacement in the upper port section along the line C in Figure 8.

## CRYOSTAT VESSEL

In parallel with the assembly of the coils and the supporting structure, the assembly of the cryostat vessel for the cryogenic structures has progressed. The cryostat vessel of LHD is a 13-m-diam torus with a bell-shaped poloidal cross section. This vessel consists of four parts: the base plate, the outer wall cylinder, the inner wall cylinder, and the upper cover. Except for the inner wall cylinder, each part was divided into ten sections for transportation from the factory to the site. The vessel is made of 304 stainless steel with a thickness of 100 mm for the base plate and 50 mm for the cylinder wall and the upper cover.

One important design requirement for LHD cryostat vessel is to incorporate a sufficient number, 98, of access ports, for plasma diagnostics, vacuum exhaust, superconducting bus lines, safety valves, helium coolant plumbing, etc. High positional accuracy is required for these ports to ensure high diagnostic precision. When the device is operated, the outer cryostat vessel is subjected to the atmospheric pressure of 0.1 MPa and the vessel will be deformed by this pressure. Our target accuracy for these port sections was below 2 mm in the early design study. The global deformation of the outer cryostat vessel has been investigated to decide the basic shape and thickness.[6] An advanced analytical model was adopted to investigate the detailed mechanical behavior under the atmospheric pressure, especially for the port sections.

### Analytic model

Although the outer cryostat vessel does not have symmetrical shape, the analytic model was selected to form a typical 1/5 section of a toroidal angle. Figure 8 shows the model we prepared. This model has two outer ports, two upper ports, two lower ports, one tangential port, one inner port and nine other utility ports. One of the outer ports has some “sub ports” and the other is blind. We adopted 4-nodes shell elements for this model and calculated by using ANSYS 5.1. In the analysis, axis-symmetry was assumed, so that the cyclic boundary conditions were introduced which means the zero displacement is defined at each toroidal edge. The fixed boundary conditions were given to the ribs under the base plate since they contact the floor.

The Young’s modulus and Poisson’s ratio are 200 GPa and 0.3, respectively for stainless steel. The mechanical force subjected to the model is the external pressure of 0.1 MPa. The thickness of the port flange and its cover are valued as 80 mm and 60 mm, respectively.

### Calculation Results

Figure 9 shows the displacements of the outer port in the radial direction along the vertical line of the outer port (line A in Figure 8) and along the horizontal line (line B in

Figure 8). The vertical/horizontal location of zero means the center of the port. The maximum displacement in the major radial direction of the cryostat vessel appears at the center of the outer port and the value is –3.3 mm. Although it seems that the bending deformation compared with its original plane is large, the in-plane deformation is small except for the central area. As is seen in Figure 9 (b), the deformation in the upper cover is small. The pressure is transformed to the compressive load for the outer cylinder section. The maximum stress which appears at the edge of the outer port was 104 MPa. In the operational phase, every child port will have a pipe and this strengthens the rigidity so that the displacement of actual ports will be less than these analytical results.

## SUMMARY

The construction of the coil supporting structure of the Large Helical Device has been successfully completed. The deformation generated by the welding process of 100-mm-thick 316 stainless steel was measured and confirmed to be less than 5 mm, which meets our target accuracy. The cryostat vessel for cryogenic structures of LHD has also been assembled on site. Calculations for the deformation of this vessel under the atmospheric pressure show that the displacement of the ports for plasma diagnostics is within 4 mm.

The Construction of LHD is in its final stage. We have finished installation of all the superconducting coils into the supporting structure. After completing the plasma vacuum vessel, the radiation shield and some parts of the inner plumbing, the cryostat vessel will then cover the whole components. The wall cylinder and the upper cover of the cryostat vessel will be put on the base plate by the beginning of October 1997. The completed figure of LHD is going to appear at the end of the year.

## ACKNOWLEDGMENT

The authors would like to thank Prof. A. Iiyoshi, Director General of National Institute for Fusion Science for his helpful encouragement.

## REFERENCES

1. M. Fujiwara, et al., Large Helical Device (LHD) Program, *Journal of Fusion Energy*, Vol.15, 7-154 (1996)
2. S. Imagawa, et al., Structural analysis of the Large Helical Device, *Advances in Cryogenic Engineering*, Vol. 39, 309-316 (1994)
3. A. Nishimura, et al., Fracture Toughness of Partially Welded Joints of SUS316 Stainless Steel at 4 K by Large Bend-Bar Tests, *IEEE Transaction on Magnetics*, Vol. 32, No. 4, 2941-2944 (1996)
4. A. Nishimura, et al., Fracture Toughness of Partially Welded Joints of SUS316 in High Magnetic Field at 4 K, *Advances in Cryogenic Engineering*, Vol. 42, 315-322 (1996)
5. H. Tamura, et al., Structural and mechanical design of cryogenic support system for LHD, *Fusion Technology* 1996, 1019-1022 (1996)
6. H. Tamura, et al., Structural design of the cryostat for the LHD, *Fusion Engineering and Design* 20 97-101 (1993)

# PERFORMANCE OF MAGNET LEADS FOR RHIC[1]

A. Nicoletti, M.L.F. Rehak

Cryogenic Systems Group
Relativistic Heavy Ion Collider
Brookhaven National Laboratory
Upton, NY, 11973-5000, USA

## ABSTRACT

Three different conventional gas cooled power leads have been developed at Brookhaven National Laboratory for use in the Relativistic Heavy Ion Collider (RHIC). The RHIC cryogenic system will contain over 500 leads, which will be responsible for as much as 24% of the total 4.5 K refrigerator heat load. Leads optimized for 6300 A and for 1600 A are constructed with variable heat transfer segments and conductor cross sections so as to operate frost free over a wide range of currents without compromising safety or efficiency. A second type, which consists of a 12 lead bundle each carrying 150 A, is an improvement of an existing design, where parallel conductors are pressed into the perimeter of a plastic core. Each type of lead was optimized for minimum refrigeration load while maintaining a margin of safety in the event of coolant flow loss. The 6300 A and 1600 A leads will operate at 0.06 g/s-kA up to and including their design currents, and the 12x150 Amp lead bundle was stable at 0.09 g/s-kA. A mathematical model of each type of lead, which includes warm end components connecting it to the power supply and cold end superconductor, has been developed. The design and test result for each type of lead is presented here, and comparisons are made to their theoretically predicted performance.

## INTRODUCTION

The decision to use conventional leads was primarily based on the high cost and insufficient development of high temperature superconductors (HTS) at the time. Another contributing factor was the preexistence of the *Isabelle* 25 kW refrigerator, which will provide excess refrigeration capacity. Furthermore, the RHIC cryogenic system was not designed to readily provide the secondary cooling system required to gain maximum benefits

[1] Work performed under contract with the US Department of Energy

of HTS leads. With the constraint that heat be removed at 4.4 K, modest gains of only 20 % could be attained by using HTS. This is understandable since the warm end of the lead, which cannot be replaced by HTS materials, generates the most Joule heating. Four types of leads are required to satisfy the various RHIC superconducting magnet requirements. These are listed in Table 1, where the current ratings and their functions are shown. Because of space constraints, the 12x50 Amp corrector leads will not be discussed here.

Magnet leads are designed to minimize the total heat load on the cryogenic system. This is done by choosing a shape factor defined as LI/A (L: length, A: cross-section, I: current) and a flow rate, m, such that the total refrigerator load, defined by $Q_{tot}=Q_{in} + 80m$, is minimized. $Q_{in}$ is the conducted load and 80 Watts/g/s is a factor that, for a theoretical Carnot cycle plant at 4.4 K, characterizes the refrigerator in refrigerator/liquefier mixed-duty operation. Since the optimal flow does not have a sufficient margin over the critical flow (minimum flow required for stability, theoretically 0.052 g/sec-kA), an operating mass flow 20%, or 0.06 g/sec-kA above the critical flow, is expected for the 1600 A lead and for the 6300 A lead. The bundled lead types do not achieve as good heat exchange and consequently require higher mass flow. An additional requirement is that the leads must be able to sustain the operating current for 30 seconds after interruption of cooling flow. It is also postulated for the 6300 A and 1600 A leads that frost buildup be kept to a minimum, since water cooling of the warm end copper blocks is not available.

A complete mathematical model[1] was used to develop lead conductors. It includes the copper lead, the equipment attached at both the cold and warm end of the lead, free convection in air, free convection in helium and effects of superinsulation. Parametric studies on heat exchange surface, conductor resistivity ratio (RRR), and geometry were performed to reach these designs. Additionally, a simplified analytical solution was used to develop guidelines for curves of operating mass flow and voltage drop as functions of current.

## 6300 A and 1600 Amp Lead Design

The 6300 A leads and 1600 A leads are constructed by shrinking a stainless tube over a copper conductor which has a helical thread cut into it. The helium flows spirally up the lead until approximately 10 cm from the warm end, where it goes internal to the conductor. The temperature of the helium flowing up the helix is closely coupled to that of the conductor but the temperatures diverge once the gas is inside. An increased copper mass compensates for this decrease in heat exchange efficiency. This feature is intended to keep the warm end of the lead frost free at higher current while providing mass for temperature stabilization in the event of coolant flow loss. The model predicts that above half of the design current, the lead temperature is above freezing in the section exposed to air, without the use of auxiliary systems such as heaters, fans or water cooling. This is achieved at negligible cost to lead efficiency. The 6300 A lead (L=60.5 cm + 9 cm, D=1.4 cm), has about the same length and larger cross section than the 1600 A lead (L=59 cm, D=1.0 cm) since Joule heating effects are more significant than conduction. Both leads are manufactured of ETP copper with a RRR of approximately 75.

**Table 1**. Power leads for RHIC

| Type | Quantity | Function |
|---|---|---|
| 12 x 50 Amps | 492 | Corrector/Trim Magnets |
| 6300 Amps | 16 | Main Dipole/ Main Quadrupole |
| 1600 Amp | 42 | DX/D0 bypass |
| 12 x 150 Amp | 63 | Insertion Quadrupole, Dipole trim |

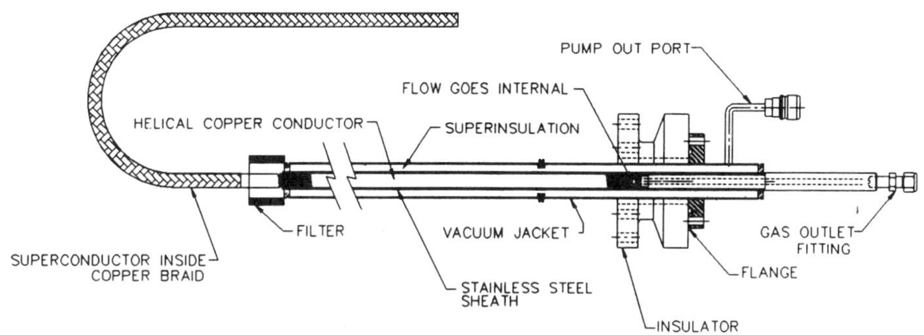


**Figure 1**. 1600/6300 Amp Lead design.

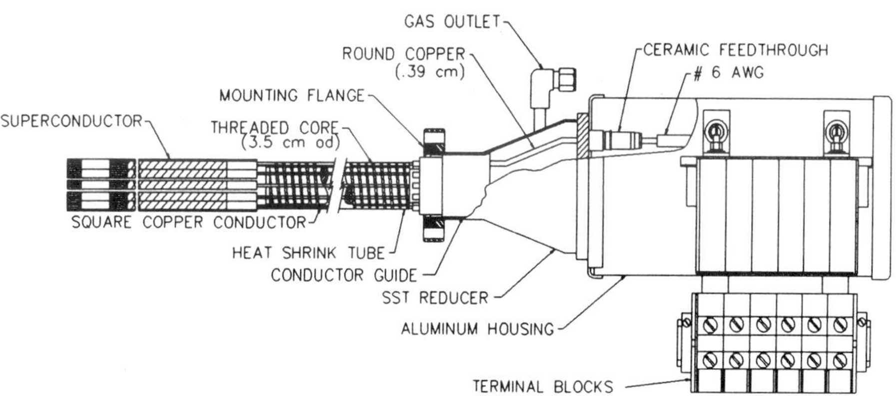


**Figure 2**. 12x150 Amp lead design

An Acme thread is cut into the conductor outer diameter. A stainless steel tube is shrunk around the conductor and serves as a helium containment jacket. A sintered metal filter prevents any particulate contaminants from plugging up the threaded helium passage. Pressure drop was not a major concern since inlet pressures are in the range of 4.5 - 5 atmospheres. The lead is shown in figure 1.

The cold end consists of two niobium titanium superconducting cables inside a copper braid. The braid provides an extended heat transfer surface and extra copper mass for cold end temperature stabilization. A vacuum jacket surrounds the entire length of the conductor, and terminates approximately 6 cm above the plastic (Ultem®) insulator. This keeps any frost from reaching sensitive 0-ring seals when the lead is operating at very low currents.

### 12 x 150 Amp Lead Design

These leads consist of 12 parallel conductors pressed into the perimeter of a plastic (Noryl®) mandrel core. The conductors are made of ETP copper and are square in cross section (.22 cm) to provide a tighter fit in the grooves cut axially into the mandrel, and thus minimize leakage flow. A plastic (Teflon®) jacket is shrunk over the diameter of the core, holding the wires in place and forcing the gas to flow spirally up the lead through the Acme thread. Helium gas traveling around the helix intercepts heat in the exposed section of each of the twelve wires in a cross-flow cooling mode.

The mandrel diameter is 3.5 cm and the square conductor is 70 cm long. The copper makes a transition to a round conductor (.4 cm) which passes through a stainless steel expander, before penetrating a hermetic seal through a ceramic insulator. #6 AWG welding cable carries the current from the ceramic insulators to a terminal block mounted on an aluminum housing. This housing protects the 12 ceramic insulators and also acts as a wiring point for cables coming from trays. At the cold end, each conductor transitions to a bundle of four NbTi superconducting cables, each of which is rated for 150 A at 4.5 K. The added superconductor helps insure cold end stability. These cables are subsequently spliced to the magnet bus. The lead is shown in figure 2.

## TEST RESULTS

### Problems Encountered

The leads were installed in a test fixture connected to a small refrigeration system. Initial results indicated excess heating at the warm end of the single conductor leads. This was eliminated when joint resistance in the warm copper bus assembly was reduced. Also, initially the superconductor on the single conductor leads would quench at less than half the design current. A redesign of the superconducting cable, which increased the heat exchange by a factor of over 100, ensued. It was subsequently impossible to quench the superconductor even during operation without coolant, as the leads would melt first. A third valuable fact learned was the need for the sintered filters at the cold end of the lead, as particulate contaminants partially plugged the helium passage cut into the threads on both the prototype 1600 and 6300 Amp leads.

### 6300 A Lead Test Results

Figure 3 indicates the minimum flow required for stability, and the corresponding voltage drop, as a function of current. Up to 2400 Amps, the mass flow rate is dictated by the conduction load and is a constant 0.12 g/s. From 2400 Amps to 5000 Amps, the lead

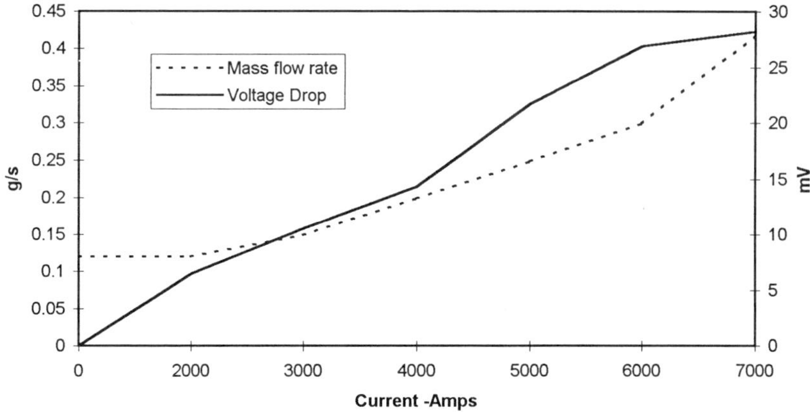


**Figure 3**. Minimum flow required for stability and corresponding voltage drop vs. current for the 6300 A lead.

was stable at 0.05 g/s-kA, and up to 7500 Amps at .06 g/s-kA. Above 5800 A, a fan was required to cool the copper bus at the warm end of the lead. Since only four of these leads will operate above 5500 Amps, the warm end bus will be scaled accordingly. The voltage drop across the lead was 25 mV at 6300 A and 28 mV at 7000 A (with the fan) at these minimum flows. This is close to the theoretical value of 40 mV for an ideal lead cooled at 0.06 g/sec-kA. This low voltage drop indicates that the lead is over cooled and thus has a good margin of stability. Temperatures reached 500 K during three minutes when the lead was powered at 6300 A without cooling. When installed, the lead will operate at 0.06 g/s-kA above 2400 Amps. At design current and mass flow, the pressure drop is 1 atm.

No frost was observed above 4000 A, and at lower currents, the vacuum jacket successfully kept frost off of the electrical insulator and its O-ring seal.

Figure 4 shows the actual temperature distribution along the lead as compared to the simulation at a current of 6300 Amps and a mass flow rate of 0.37 g/s. The model predicted the need for additional warm bus cooling, which is incorporated into the results shown. The

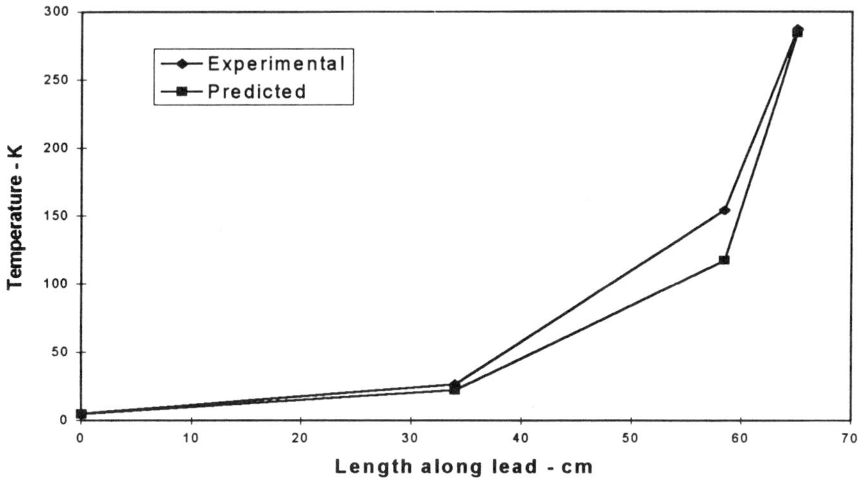


**Figure 4**. Actual and Predicted temperature distribution for the 6300 A lead at 6300 A and 0.37 g/s. lead.

measured values clearly show good agreement with the predicted values. This indicates the importance incorporating the warm and cold end connections into any lead simulation. Since the lead performs as predicted it is believed that the conducted heat load, which could not be measured accurately in the test setup, is as predicted by the model: under 2 W at design conditions.

## 1600 Amp Lead Test Results

The lead was stable at 0.06 g/sec-kA for currents up to 2000 A. Temperatures rose to 450 K after 4.5 minutes without coolant at design current. An operating mass flow of 0.05 g/s up to 800 A and 0.06 g/s-kA above 800 A will keep the cold end temperature below 6.5 K and therefore conducted heat minimal. A pressure drop of one atmosphere and a voltage drop of 33 mV at operating design and current were measured. Again the lead is slightly overcooled. Only a thin layer of frost was apparent above 1000 Amps. Figure 5 shows the operational mass flow rate and voltage drop as a function of current.

## 12X150 Amp Lead Test Results

The lead was instrumented with voltage taps and temperature sensors on two opposing conductors. Temperatures were read at the cold and warm ends of the square conductor, and at the upper end of the round conductor inside the reducer. The copper conductors do not have optimum cooling inside the reducer area so efficiency is lower for this type of lead and consequently the cold end temperature is higher. With all leads carrying 150 A, stability was achieved at 0.13 g/sec or 0.072 g/sec-kA, and the cold end of the lead was 9 K. When the lead was powered for two minutes without cooling, temperatures did not exceed 300 K and the superconductor did not quench.

Although the superconductor is heat sunk well enough to prevent quenching, there may be excessive heat conducted into the 4.5 K system. To keep the cold end temperature below 9 K an even higher flow rate would be needed An operating mass flow of 0.09 g/s-kA, which is based on the sum the currents in each of the conductors, will be used. For currents above 150 A, the required flow is no longer linear with current since at 200 A, 0.25 g/sec (0.104 g/sec-kA) is needed for stability. At low currents, heaters are required to prevent frost buildup on ceramic insulators. At operating conditions a very small pressure drop of the order of 0.2 atm was measured.

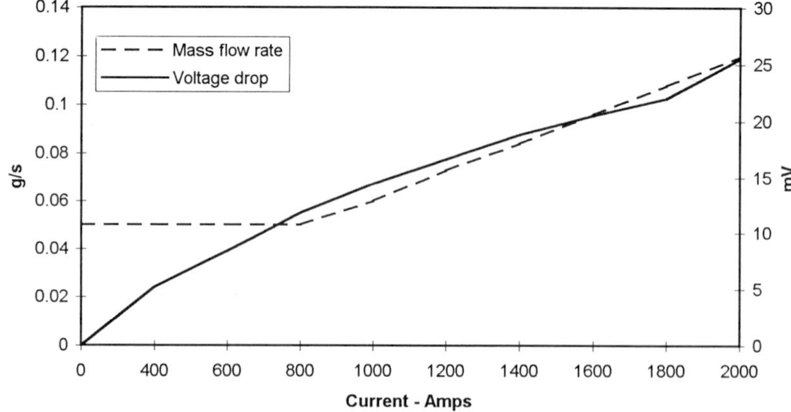


**Figure 5.** Operational voltage drop and mass flow rate vs. current for the 1600 A lead.

As there is a large amount of uncertainty in the heat exchange parameters, the computer model fails to predict the high temperatures at the cold end and thus the conducted heat. While absolute numbers cannot be precisely determined, relative values were valid and the model guided the choice of thread for improved heat exchange and conductor sizes. As these leads are adequate to operate in RHIC and any increase in their efficiency would have negligible impact on the cryogenic system, there is little justification for refining the model at this time.

## REFERENCES


1. M. Rehak, "A Simplified Theory of Current Leads", presented at CEC/ICMC, Portland, Oregon, 1997.

# MECHANICAL AND THERMAL CONSIDERATIONS FOR THE SUPERCONDUCTING MAGNET TO NORMAL CONDUCTOR MAGNET "WARM-TO-COLD TRANSITIONS" OF THE RHIC ACCELERATOR

James Rank, Ken Malm, Gary McIntyre, and John Sondericker

Relativistic Heavy Ion Collider (RHIC)
Brookhaven National Laboratory
Upton, N.Y. 11973-5000

## ABSTRACT

The RHIC accelerator, designed for maximum flexibility, allows for the insertion within the superconducting magnet lattice of specialty magnets of the normal conductor type. The beam character may then be customized at each of the experimental crossing points. This facility requires a beam tube transition from a cryogenic environment to ambient and back again. The "Warm-to-Cold Transition" is the system of mechanical and electrical components which permit this to occur. Discussed herein are: the systems mechanical degrees of freedom, shrinkage allowances, thermal and pressure induced stresses, cryogenic heat load, superconducting bus splices, and design considerations for each of the eight major assemblies.

## INTRODUCTION

Approximately forty Warm-to-Cold Transitions (W/C's) are designed into the RHIC lattice near the six intersection regions. Warm specialty magnets, some developed, some yet to be conceived, will occupy the region between opposite pairs of W/C's. The RHIC First Sextant Test of early 1997 included three varieties of W/C's. The particular type is determined by the superconducting (SC) magnet adjacent to the warm region onto which the W/C piggy backs. As will be seen the space allotted for the W/C is minimal.

The RHIC SC magnet shares a common heat shielded vacuum vessel with cryogenic process piping. These pipes or "RUSH" Lines: Return, Utility, Supply and Heat Shield serve to cool, recool, or even warm the magnet cold mass and superconducting bus. Normally these lines operate at 4K and from 1 to 5 atmospheres, except the Heat Shield Line which is at 55K and 15 atmospheres. During transitional or fault modes, though, some may run warm at 18.7 atmospheres. The passages through the cold mass which carry the helium converge at small pots or "end volumes" and outlet to two tubes known as the Magnet Lines with the same range of operating conditions. The end volume typically contains "omega" expansion loops in the SC bus work, named for their shape.

A Warm-to-Cold Transition serves multiple functions. Firstly, it shunts each of the process lines and Magnet Line from the ring, vertically, to a transfer line and around the normal conductor magnet. Secondly, it provides containment for splices between the hard bus of the SC magnet and the flexible cold crossing cable which is carried within each Magnet Line, while providing a means of access to the splices for servicing. Thirdly, it allows freedom for thermal contraction in the longitudinal and vertical directions simultaneously. Lastly and most important, it permits the passage of beam into a warm region.

All of these functions are served while two main design objectives are met. Firstly, heat load to the SC magnet is minimized. Secondly, the SC magnet maintains static equilibrium; i.e., pressure and thermal loads do not act on the magnet which is sensitive to position.

A W/C transition consists primarily of field assembled components which make up eight (sub)assemblies:

## THE ASSEMBLIES

**1. Superconducting Splice Can.** This is the containment for the splices which connect the hard to the flexible superconductor buses. Its design offers the maximum possible space and ease of access for convenient soldering of the splice and auxiliary copper conductor, as well as multiple signal and instrumentation splices, while providing for sufficient cooling of the less efficient joint.

The can itself allows servicing of splices by sliding up over the expansion joint (assembly eight). The upturned weld flanges at top and bottom can be ground back a number of times before the can must be replaced.

Also included in this assembly is the inverted dished head and shell which have penetrations for both Magnet Line tubes extending from the end volume. The upper (side) tube carries the bus and coolant for the CQS (corrector-quadrupole-sextupole) magnet and the lower (bottom) tube carries that for the dipole. Because the magnet tubes are of a very thin wall and have only a 0.06" root weld on the backside of a counterbore in the "end plate" of the end volume, they require reinforcing against bending moments that may be applied due to pressure or thermal loadings. This is accomplished in each case by an anchoring flange paired with an adaptor. The adaptor presses into the thin walled magnet tube bottoming against its shouldered extension which squares the built up assembly to the end volume and acts as a weld backing ring protecting the bus. The dipole adaptor is shouldered too, on the opposite end, where the split tee clamshells around it and the standoff (assembly two). The CQS adaptor is contoured to the ID of the shell into which it also welds.

The upper anchoring flange is split into two pieces so that it may be placed around the magnet tube for welding after the head and shell have been attached. The lower flange is one piece, having only one longitudinal cut. It slips over the adaptor before the split tee is assembled and presses into the end plate counter bore to insure squareness; this is necessary for proper alignment of the equilibrating standoff bellows (assembly two). Each flange welds to its adaptor and either welds or bolts to the "end plate," depending on magnet type. This

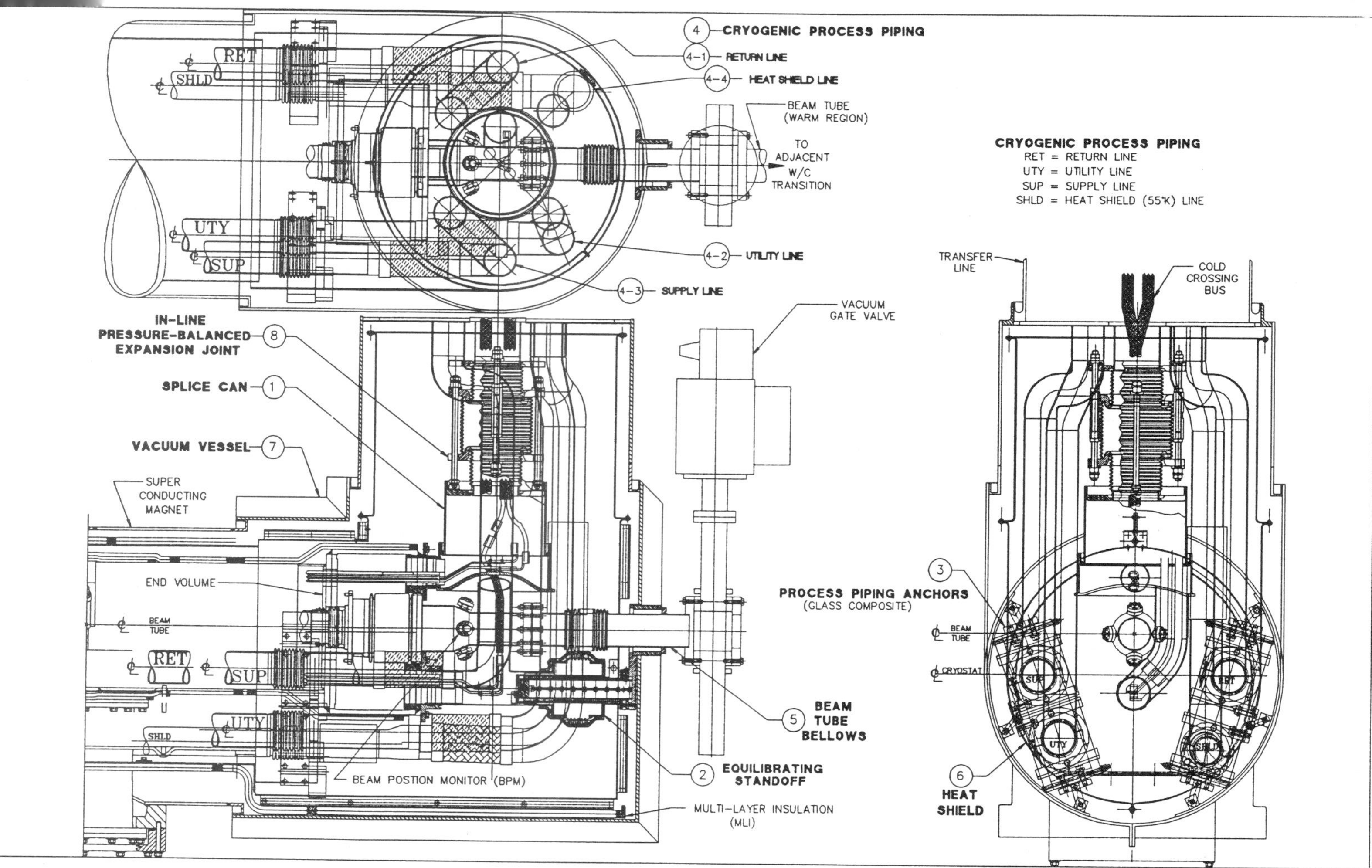


**Figure 1.** Warm to cold transition.

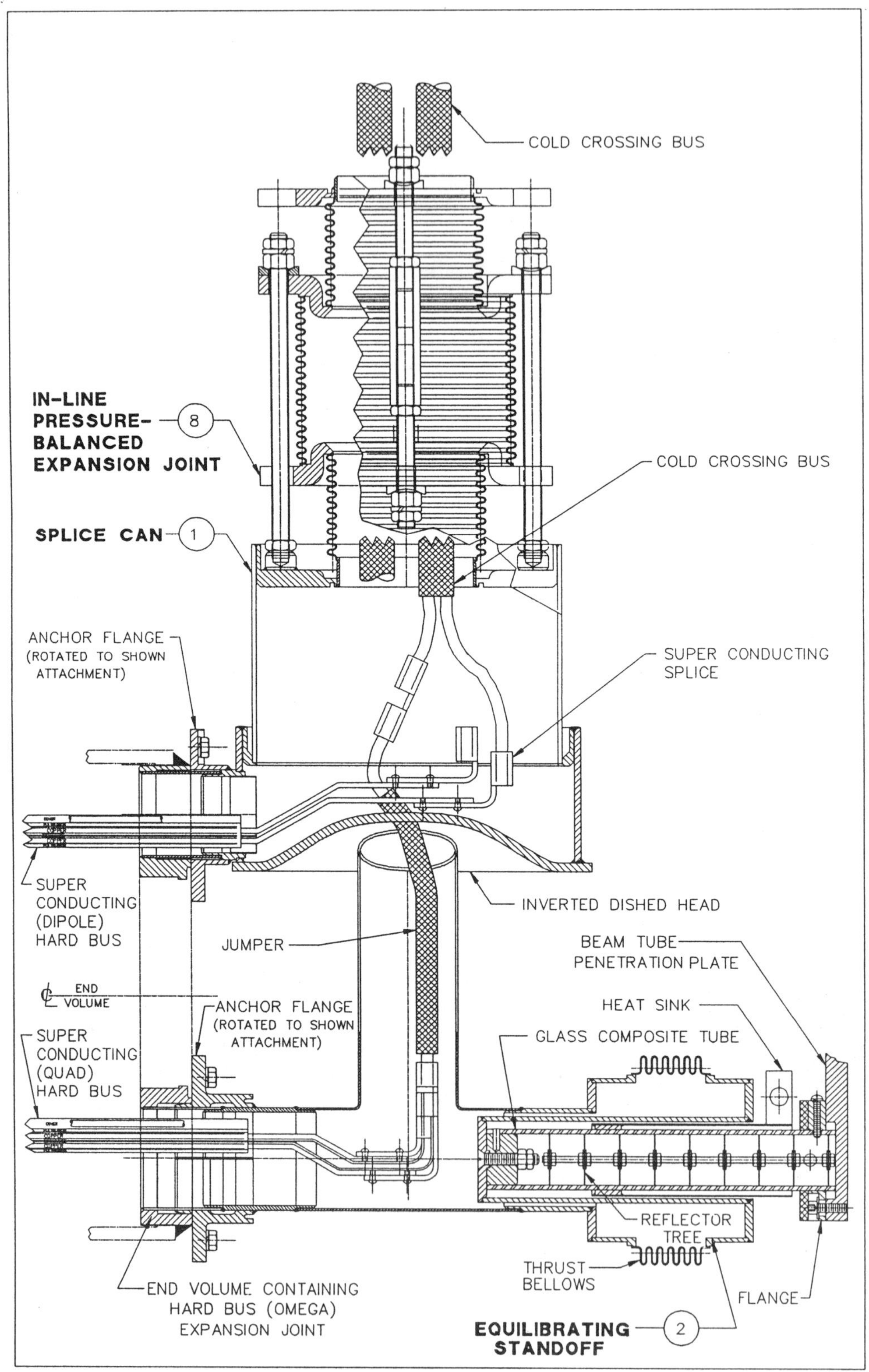


**Figure 2.** Magnet coolant line detail.

provides a sound structural connection of the splice can assembly to the magnets end plate eliminating stresses which would fail the magnet tube welds.

Each flange also employs a deep circumferential weld relief (where it welds to the adaptor) which is segmented. If a splice can is subjected to rapid cooling by a surge of colder gas, the less massive adaptor will cool faster in the flow than the heavy walled adaptor will by conduction, thereby shrinking at a faster rate. The deep segmented "fringes" act as soft cantilevered beams. They deflect radially preventing excessive stress in the adaptor (and welds) due to the difference in the rates of shrinkage. They will not, however, allow vertical or lateral deflection of the adaptor (or of the splice can assembly as a whole) since, regardless of the radial direction of any applied load, there will always be at least two "fringes" which resist deflection along their "tangential height," as would a stiff beam. This concept solves the problem of cold shock cracking of anchor ring welds which was witnessed with the prototypes of VJR (vacuum jacketed ring piping) transfer lines.

The splice can provides sufficient room for contraction of its two rigid splices and the split tee for its dipole jumper splice contained within its branch. The split tee is closed lastly by external weld strips that weld longitudinally along its run. The branch butt welds to a 45° elbow which feeds the dished head from underneath.

**2. Equilibrating Standoff.** This assembly connects to the run of the split tee (assembly one). It is thus exposed to dipole magnet tube flow directly. Its function is to equalize the pressure areas of the SC magnet. The bellows is sized such that the effective area is equal to the sum of that of the two regular interconnect bellows on the magnet tubes at the opposite end of the magnet. Thus, there is no net load in the longitudinal direction on the stands supporting the cold mass due to pressure. The pressure loading is transferred to the vacuum vessel walls of the W/C which are at room temperature. In order to minimize conductive heat load then, the standoff assembly makes use of a glass composite tube (internal to the bellows) which acts as a column in compression. This arrangement and material selection provides maximum structural stability with minimum heat load in the smallest possible package.

The cold end of the composite tube has a two-piece flanged plug pressed into the ID. This allows the tube to be bolted onto a stud after all stainless parts (the bellows, both of its end rings, the internal reentrant tube, the plug flange, and the stud) have been welded. The warm end has a split flange which clamps on the composite tube OD and bolts to a base plate. This plate rests flush against short leveling set screws which are ultimately plug welded from outside the vacuum vessel.

A reflector tree is inserted inside the composite tube to reduce the radiative load from the room temperature plate to the plug sitting in the flow. The tree is a multi shield package of aluminized Mylar discs strung along a light nylon threaded rod. The OD of the composite tube is stationed at 55K in the optimum location by means of a heat sink weldment, which also shields the ID of the stainless internal reentrant tube against radiation. The heat sink is a sleeve of rolled 20 mil pure aluminum sheet with ears, welded to a split ring which slips over the composite tube. At the opening a copper braid clamps to the ears on one end and the Heat Shield pipe on the other. The composite tube and the heat sink sleeve are each wrapped in MLI.

**3. Process Pipe Anchors.** The afore mentioned "RUSH" lines are anchored to the vacuum vessel of the W/C. The two anchor assemblies per W/C are angled to clear the magnet cold mass. One anchors the Return and Heat Shield, another the Supply and Utility Lines.

Each anchor is a structure of glass composite rectangular bar bolted with stainless hardware. Welds to the pipe OD are made with flanged rings having the same deep, segmented, radial weld relief as discussed in assembly one (fourth paragraph). Long bolts penetrate these flanges at the front and a keyed angle at the back, clamping the composite bar structure. The top and bottom of each anchor, as well as a midpoint for stabilization, are attached to a swivel link which welds to a rolled carbon steel shoe. Ultimately, after the vacuum vessel is mounted, each shoe is plug welded to the vacuum vessel from the outside.

**4. Cryogenic Process Piping.** The "RUSH" Lines which carry helium for service of the SC magnets must allow for contraction in both the vertical and longitudinal directions. In the horizontal, this is accomplished by means of four corrugated bellows and in the vertical by four braided flexible hose assemblies. These are separated by the anchors (assembly three).

Pairs of opposite facing elbows, by manipulation of their relative offsets and angles, correct any misalignments between the fixed components they interconnect; e.g., magnet and transfer line. The Heat Shield pipe has aluminum coolant sleeves, one horizontal and one vertical, that weld to the heat shield. The split sleeves slide on the pipe to allow for the different rates of contraction.

**5. Beam Tube Spool.** Typically the SC magnet adjacent to the W/C ends in a beam position monitor (BPM). A spool with a knife edged flange is carefully copper gasketed to the BPM to maintain ultra-high vacuum in the beam line. The base plate of assembly two and a vacuum vessel penetration sleeve, then finally a seal ring, are slipped over the open end of the beam tube spool before the larger flange which connects to a gate valve outside the vessel is welded on. The penetration sleeve fits and is welded inside a half hole in each clam shell of the vertical run of vacuum vessel. A low pressure bellows in the spool allows the magnet to move independently of the vacuum vessel. Critical adjustments are made and finally the seal ring is welded on the face of the penetration sleeve and to the beam tube OD.

A copper band brazed to the cold side of the bellows connects a copper braid to the Heat Shield pipe for heat sinking similar to that in assembly two.

**6. Heat Shield.** An eleven gauge pure aluminum radiative heat shield stationed at 55K envelopes the internals of the W/C. It is constructed in six major pieces. A pair of opposite flanged halves clamshell around the horizontal leg and another pair around the vertical leg of the W/C. The pairs are welded at the mitred seam where they meet. A removable clamshell pair is contained within the telescoping portion of assembly seven. It has a circular flange which mates with a matching flange on the two vertical halves. The removable shield flanges are crimped using an extruded aluminum channel. The internals of the telescoping vessel are then accessible by removing the crimp. Similarly a circular flange at the top crimps to a two piece cap which is notched to clear the "RUSH" lines. Two vertical seams are also crimped.

All parts of the shield assembly are wrapped in 64 layers of perforated MLI.

**7. Vacuum Vessel.** The entire W/C is contained in a vacuum shell of bumped 1/4" mild carbon steel. The vacuum space is not dynamically pumped, hence, it must be leak tight to $10^{-7}$ atm-cc/sec. The heavy wall is required to prevent buckling due to: external loads such as bending moments applied when the connected transfer line with its 20" vacuum jacket bellows is evacuated, and internal loads, such as RUSH anchor loads (assembly 3). The shell is further reinforced by extra long flanged returns on top and bottom. Besides stiffening the shell in the same manner as a structural tee section would, it reduces local deformation in the area where the pressure load is applied by the standoff (assembly 2).

No frosting or significant cold spots were witnessed, indicating that the heat load to 55K is not excessive.

The construction of the vacuum shell is similar to that of the shield with mitred horizontal and vertical clamshell pairs. The horizontal pair closes on the weld neck flange of the adjacent SC magnet. The vertical pair closes on the identical upturned lower flange of the telescoping vacuum vessel. This telescoping shell accepts the smaller upturned flange of the transfer line and, after grinding of both flanges, is free to slide up over the transfer line allowing access to the removable heat shield (assembly 6).

**8. In-line Pressure-Balanced Expansion Joint.** The expansion joint is the heart of the W/C. It permits the magnet cold mass to contract (relative to the anchor within the outer vessel of the transfer line) in every translational degree of freedom. And it does this without applying a pressure loading at the end of the unrestrained magnet which would otherwise tip. Furthermore, it allows an unimpeded, spacious, straight run for the superconducting cable which is free to contract. At assembly the expansion joint is fully contracted to maximize access to the spliced joint (assembly 1). It is a compact solution to a multitude of problems.

The design of the expansion joint includes three segments of bellows: the top and bottom, or outboard bellows, have large mounting flanges, or tiers, which accept the tie rods around their perimeter. The middle, or inboard bellows, has a tier at either end which also accepts tie rods. A given tie rod is fastened at one end to an outboard tier and at the other to the far inboard tier. For the first tier, at the bottom, the joint is a clevis and pin design; no penetration is made to the tier. The second through third tier has swivel joints from spherical nuts; the rods penetrate the tier.

The inboard bellows has, necessarily, twice the effective pressure area of the outboards. A conventional bellows in an in-line application allows a pressure load to be applied to the components it connects (here the magnet and transfer line) equal to the pressure times the effective area of the bellows. With the pressure-balanced expansion joint the same pressure load is also applied to the components in the opposite direction by tension in the tie rods. The source of this load is the pressure acting on the washer shaped ring that is the inner face of the second or third tier. The net load on each of the connecting components due to pressure is zero, regardless of pressure. The expansion joint spring rate determines the only load.

Longitudinal freedom of the magnet is accomplished by the swinging motion of the tie rods (up to 0.5") on their swivel joints. Vertical freedom of the magnet, required as the magnet support posts shrink downward while the Magnet Line piping shrinks upward, is accomplished by compression of the inboard bellows and extension of both outboard bellows.

## TESTING AND ACCEPTANCE

No prototype of the W/C or any of its (sub)assemblies was built. The first tests of any W/C were the eight which were included in RHIC's first sextant. Each was subjected to two thermal cycles for a total of sixteen successful thermal cycles. Every indication is that the performance of the W/C was exceptional, structurally, thermally and functionally.

# A GASEOUS-HELIUM COOLING SYSTEM FOR A HIGH-Tc SUPERCONDUCTING COIL

K. G. Herd, R. A. Ackermann, and P. S. Thompson

GE Corporate Research and Development
Schenectady, NY 12309, USA

## ABSTRACT

A gaseous-helium cooling system for a high-Tc coil operating at 20K is described. The system uses a two-stage Gifford-McMahon (GM) coldhead and counterflow heat exchangers to cool a pressurized flow of helium gas from room temperature down to approximately 19K. The cold helium gas circulates through a vacuum-insulated transfer line to a single-pass heat exchanger in contact with an epoxy-impregnated high-Tc coil The coil was wound using 2000 meters of Bi-2223 tape produced by Intermagnetics General Corporation (IGC). A water-cooled rotary compressor supplies pressurized helium gas to both the GM coldhead and the helium gas cooling circuit, in parallel. Descriptions of the coldbox configuration, the flow control and temperature monitoring systems, the high-Tc coil, and the test dewar are given. Test results, including the cooldown and energization of the Bi-2223 coil, are included.

## INTRODUCTION

This paper describes the development and testing of a closed-cycle gaseous helium cooling system for a prototype Bi-2223 high-Tc racetrack coil operating at approximately 20K. Cold helium gas is used to extract heat from the high-Tc coil via a single-pass heat exchanger bonded to the outer surface of the coil. The gas circulates back to a refrigerator coldbox, where the heat is extracted from the gas stream by counterflow heat exchangers and a two-stage GM cryocooler. The coldbox is located remotely from the coil, which is mounted in a separate test dewar. A vacuum-insulated transfer line carries the cold helium gas between the refrigerator coldbox and the test dewar.

The development and testing of the Bi-2223 racetrack coil was part of a DoE-supported Superconductivity Partnership Initiative (SPI) to evaluate the use of high-Tc windings in a 100 MVA utility scale generator. The Bi-2223 high-Tc racetrack coil design and fabrication details have been reported previously.[1-3] The coil is layer-wound with approximately 2000 meters of Ag-sheathed Bi-2223 tape produced by IGC, and vacuum-pressure impregnated with a low-viscosity epoxy. The coil assembly was suspended in a vacuum dewar and cooled to a temperature of 19K. The coil was energized with 30 amperes of current (generating ~35,000 ampere-turns of magnetomotive force) and the coil stabilized at a temperature of 20K. The coil current was subsequently increased to 34 amperes (~40,000 ampere-turns) and the coil stabilized at a temperature of 25K.

The following sections describe the refrigeration circuit (including the gas compressor, coldhead, coldbox, gas transfer line, coil heat exchanger, and test dewar) and the test results (including the cooldown, energization, and transient testing) of the Bi-2223 coil.

## DESCRIPTION

Figure 1 illustrates the basic features of the cooling system concept.[4] The gas supply flow leaving the compressor separates into two paths: one to the GM coldhead, and the other to the gas-cooling circuit. The distribution of the total gas flow between the coldhead and the parallel gas circuit was balanced to minimize the temperature of the gas leaving the coil heat exchanger (i.e. the coil operating temperature was minimized). Excess flow to the gas cooling circuit, at the expense of a reduced supply to the coldhead (hence reduced capacity), burdens the coldhead stages with an excessive heat load. Conversely, too little flow to the gas cooling circuit leads to an excessive temperature rise in the gas as it extracts heat from the coil heat exchanger, increasing the coil operating temperature. The optimum flow balance was determined by systematically varying the flow distribution and measuring the coil operating temperature. Approximately 2.7 grams/second flow to the coldhead and 0.3 grams/second flow to the coil circuit result in the optimum balance.

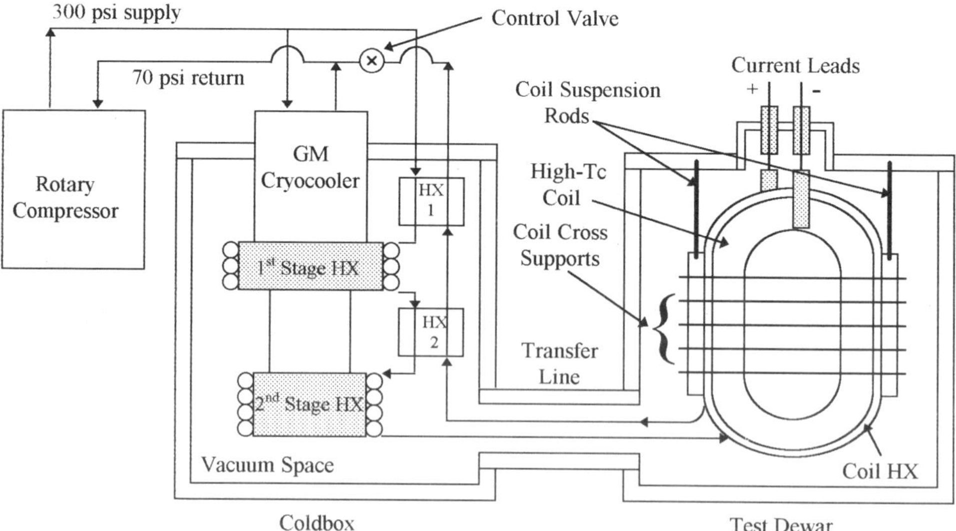


**Figure 1.** The basic features of the flow circuit are shown in the figure. A single compressor supplies the gas flow to both the coldhead and the parallel circuit to the coil heat exchanger. The coil is suspended in a separate dewar with a single-pass heat exchanger mounted to the outer surface.

## Compressor

The water-cooled rotary compressor, CTI model 8500, generates a supply pressure of 1930 kPa (300 psi) and a suction return pressure of 550 kPa (80 psi), at a mass flow rate of ~ 3 grams/second. The input power to the compressor is 5 kilowatts. An oil mist separator and a charcoal absorber are connected at the outlet of the compressor to minimize the oil entrainment in the helium gas stream entering the cooling circuit and the coldhead.

## Coldhead

The CTI two-stage GM coldhead is mounted in a vacuum-insulated coldbox, as shown in Figure 1. Heat sinks at both stages of the coldhead were fabricated by coiling and soldering the gas line around each stage. Ruthenium-oxide temperature sensors mounted to each stage were used for monitoring and control purposes. Five-watt heaters are also mounted to each stage to facilitate heat load mapping of the coldhead in-situ.

## Heat Exchangers

As shown in Figure 1, the incoming high pressure gas passes through a counterflow heat exchanger (HX1) prior to flowing to the first-stage heat sink. Low pressure gas returning from the coil heat exchanger flows through the cold side of the heat exchanger. The refrigeration efficiency of the cycle is increased by using the cold return gas stream to pre-cool the warm incoming gas. The contact length and flow area of the heat exchanger was sized to optimize its effectiveness. A second counterflow heat exchanger (HX2) pre-cools the gas stream after it leaves the first-stage heat sink, prior to flowing to the second-stage heat sink. Charcoal absorbers are connected downstream of each heat exchanger, prior to the stage heat sinks, to reduce the risk of contamination in the low temperature portions of the circuit.

## Insulated Transfer Line

Helium gas leaving the second-stage heat sink is at its lowest temperature in the circuit. In order to minimize the temperature rise of the gas flowing to the remotely-located coil, a vacuum-insulated transfer line is used to carry the gas to the coil heat exchanger and back to the coldbox. Fiberglass spacers separate the gas lines thermally and connect them together mechanically. Fifty layers of aluminized Mylar™ insulation are wrapped around the gas lines to minimize the heat leak from the warm outer tube. The theoretical heat leak to the cold gas lines is ~0.20 watts/meter. Small copper blocks are brazed to the stainless steel gas lines at both ends, each with a ruthenium-oxide temperature sensor bonded into a 1 mm hole, in order to monitor the temperature rise of the gas flowing through the transfer line.

## Test Coil

The heat exchanger is mounted on the outside surface of the coil. The heat exchanger assembly consists of a 0.16 cm thick by 10 cm wide sheet of copper with a single turn of stainless-steel tubing, 0.5 cm by 2.0 cm, brazed onto it. The heat exchanger conforms to the outer surface of the coil and is overbanded with 0.6 cm of fiberglass to provide mechanical integrity. Six calibrated ruthenium-oxide temperature sensors are epoxied to the coil assembly at the following locations: the inner surface of the coil end-turn at the mid-plane, the inner surface of the end-turn at the coil edge, on the start and finish lead busbars, and on the supply and return lines connecting to the heat exchanger.

### Test Dewar Setup

The test dewar consists of a single-walled vessel and a top flange mounted with an o-ring seal. As shown in Figure 1, the coil was suspended from the top flange of the test dewar using fiberglass rods. The stainless-steel heat exchanger tubing was joined to the gas supply and return lines using Cajon-VCR™ connectors with stainless-steel gaskets. The coil current leads were thermally anchored to the gas return line using dielectrically-insulated heat stations.[5] The leads were also heat stationed at an intermediate location to a liquid nitrogen reservoir (at a temperature of 77K). Warm-section leads (connecting 300K to 77K) and cold-section leads (connecting 77K to 20K) were made of drawn OFHC copper wires, optimized to minimize the heat leak to the heat stations at the maximum estimated operating current of 40 amperes. The coil assembly was wrapped with ~100 layers of aluminized Mylar™ in order to minimize the radiation heat load from the warm dewar wall to the coil. No additional thermal shielding was required. The test dewar vacuum space was rough-pumped down to approximately 100 millitorr, at which point a turbo-molecular pump was used to reduce and maintain the vacuum at a level of approximately $10^{-5}$ torr. Thirty-two instrumentation leads connect through the vacuum wall via a commercially available Ceramaseal™ connector.

The cooling circuit described above differs from the traditional GM pre-cooled Joule-Thompson (GM-JT) circuit in the following ways:

1) A single compressor supplies the pressurized helium gas to both the GM coldhead and the parallel gas circuit (compared to the two required for the typical GM-JT system).
2) No cold JT expansion valves are used. A throttling valve on the gas return line is used to regulate the flow through the parallel gas circuit.
3) Only two counterflow heat exchangers are required (four req'd in the GM-JT circuit).

## TEST RESULTS

Table 1 lists the coil temperatures, terminal voltages, and the coil heating for the steady-state tests. The lead heat stationing to the gas return line, described in the section above, was implemented for the testing of the coil at 20 amperes and above. The coil stabilized at 18.6K with no current flow. It can be seen from the table that the coil heating below 20 amperes is essentially negligible. The coil stabilized at 20K with a current of 30 amperes and a coil heating of 1.3 watts. Based on the in-situ load mapping of the refrigerator circuit prior to installing the coil (7.5 watts capacity at 20K), we estimate that the background heating through the MLI and down the current leads and support rods is approximately 6 watts.

**Table 1.** Summary of the coil test results

| Current (A) | Coil Temp. (K) | Coil Lead Temp. (K) | Coil Voltage (mV) | Coil Heating (W) |
|---|---|---|---|---|
| 0 | 18.6 | 25.1 | - | - |
| 5 | 18.6 | 25.1 | - | - |
| 10 | 18.6 | 25.3 | 1.5 | 0.015 |
| 15 | 18.6 | 23.5 | 3.0 | 0.045 |
| 20* | 18.7 | 25.5 | 8.0 | 0.16 |
| 25* | 19.1 | 20.1 | 10.0 | 0.25 |
| 30* | 20.2 | 21.3 | 50.0 | 1.30 |
| 34* | 25.1 | 26.5 | 100.0 | 3.40 |
| 40* | 20.3** | 21.5** | 210.0 | 8.40 |

*Lead heat stationing to the vent line implemented. **Coil and lead temperatures unstable.

The maximum stable current level was 34 amperes at a coil temperature of 25K. Currents in excess of 34 amperes led to a transient (unstable) temperature rise in the coil and leads. Transient testing up to 40 amperes indicated a coil heating level of 8.4 watts.

## SUMMARY

The development and testing of a closed-cycle helium gas cooling circuit capable of providing approximately 7.5 watts of refrigeration at 20K has been described. This gas cooling system has demonstrated, for the first time, a viable option to the standard approach of cooling high-Tc coils by solid conduction directly to the refrigerator, or immersion in liquid nitrogen at 77K. Several advantages are inherent to the approach. First, the approach facilitates the cooling of a rotating high-Tc coil with a gas stream at 20K, eliminating the need to rotate the refrigerator. Secondly, the approach can be used for applications which require the coldhead to be mechanically de-coupled from the coil for vibration or noise isolation purposes. Lastly, the gas cooling approach permits the cooling to be distributed over the coil, minimizing the temperature gradients that are typical of conduction-cooled magnets in which the coldhead extracts heat from the coil at a singular attachment point.

## ACKNOWLEDGMENTS

The authors wish to thank Trifon Laskaris, Richard Ranze, and Peter Morgan for technical advice and assistance in the experimental testing of the refrigeration circuit and the Bi-2223 high-Tc coil.

## REFERENCES

1. K.G. Herd, L. Salasoo, E.T. Laskaris, R. Ranze, C.G. King, P. Haldar, and J.G. Hoehn, Development and Fabrication of a Bi-2223 Racetrack Coil for Generator Applications, *IEEE Transactions on Applied Superconductivity* 7:531 (1997).
2. L. Salasoo, K.G. Herd, E.T. Laskaris, H.R. Hart, and M.V.K. Chari, Test Results for a Bi-2223 HTS Racetrack Coil for Generator Applications, *IEEE Transactions on Applied Superconductivity* 7:535 (1997).
3. C.G. King, D.A. Grey, A. Mantone, K.G. Herd, and E.T. Laskaris, Mechanical Stabilization of BSCCO-2223 Superconducting Tapes, *IEEE Transactions on Applied Superconductivity* 7:2046 (1997).
4. K.G. Herd, Remote Cooling System for a Superconducting Magnet, U.S. Patent 5,485,730 (1996).
5. K.G. Herd and E.T. Laskaris, High-Tc Superconducting Leads Assembly in a Cryostat Dual Penetration for Refrigerated Superconducting Magnets, U.S. Patent 5,260,266, (1993).

# A NOVEL THERMAL INSULATION METHOD FOR A NEXT-GENERATION CONDUCTION - COOLED SUPERCONDUCTING MAGNET

G. R. Chandratilleke, Y. Ohtani, H. Hatakeyama, M. Takahashi, M. Hiragishi, M. Okamura, and H. Nakagome

R &D Center, Toshiba Corporation
4-1 Ukishima-cho, Kawasaki-ku, Kawasaki, 210, Japan

## ABSTRACT

We have developed a new thermal insulation method for use with superconducting magnets or in applications where temperature needs to be maintained within a limit. Termed as Multi-shell-insulation (MSI) method, this method uses several thermal-energy storage shields to intercept heat flowing into the superconducting magnet. The shields are arranged in a vacuum vessel so that each shield is inside another slightly bigger shield and the superconducting magnet is inside the innermost shield. To use MSI, the shields are first cooled to about the same temperature as the superconducting magnet and then they are thermally separated from the cooling means. The temperature of the outermost shield would rapidly rise, but those of inner shields would rise only slowly; the closer the shield to the center, the slower the temperature rise. To demonstrate MSI effect, we used 10 copper shields, five of which were cooled to 30K and the other five to 4K by a two-stage Gifford-McMahon refrigerator. The 3kg-innermost shield could be maintained below 30K for a one-month period: a promising result for the realization of a cryocooler-detached, high-Tc superconducting magnet. It was also possible to maintain the inner shield below 10K for 8 days by coating the three inner copper shields with $HoCu_2$ and $Er_3Ni$.

## INTRODUCTION

Medium-sized superconducting magnets have progressed from early liquid-helium immersion types to cryocooler-cooled type magnets. Recent research has addressed some issues related to these types of magnets: for instance, use of a heat switch[1] can reduce magnet cool-down time from room temperature to operation temperature. Another issue remains to be addressed is elimination of interaction of the field of the magnet with the cryocooler's magnetic materials. Interaction of magnetic field with cryocooler will raise the operating temperature of cryocooler, lowering the temperature margin available for the magnet. Stopping the cryocooler will not only solve this problem but it will help save energy.

Even though stray heat flow to the magnet is only minimal, stopping the cryocooler will raise the magnet temperature in several minutes out of the operating temperature range of the magnet. This happens primarily because heat flows to the magnet through the cryocooler as it warms up. A new design concept is therefore needed to address the issue of separation of the magnet from the cryocooler and maintain the magnet within the operating temperature range.

We developed a new thermal insulation method, which makes it possible to maintain a magnet temperature within a limit for a long time. This paper reports on experimental demonstration of this insulation method, which we call multi-shell-insulation (MSI) method.

## MULTI-SHELL INSULATION METHOD

Typically how this insulation can be used is shown diagrammatically in Figure 1. The shields are arranged in a vacuum vessel so that each shield is inside another slightly bigger shield and the superconducting magnet or the object to be insulated is inside the innermost shield. The magnet is suspended from the shield just outside of it using a low thermal conducting material, and that shield is suspended from another just outside of it, and so on. The outermost shield is suspended from the flange of the vacuum vessel. To use MSI, the magnet and shields are first cooled to about the same initial temperature, and then they are thermally separated from the cooling means, which can be a cryocooler or anything else. The outermost shield would show a rapid temperature rise, mostly because of radiation heat from the vacuum vessel. The inner shields would rise in temperature very slowly because they are located between shields, which have nearly the same temperatures. Thus, heat conduction as well as radiation will be very small for the inner the shields, and closer the shield is to the magnet, the slower is the temperature rise.

Heat diffused from the outer shield will reach the inner shields as time passes. The propagation of heat is delayed because the shields act as thermal energy storage devices; the resistance to heat flow between the shields is also high. Heat propagation can be further delayed by using high heat capacity shields or increasing the number of shields.

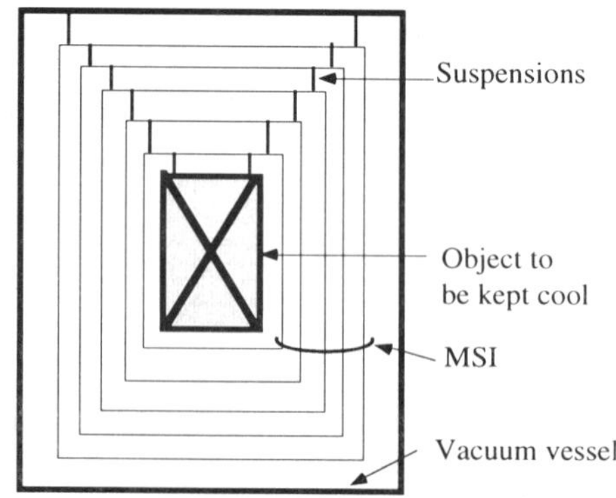


**Figure 1.** Application of multi-shell insulation to an object.

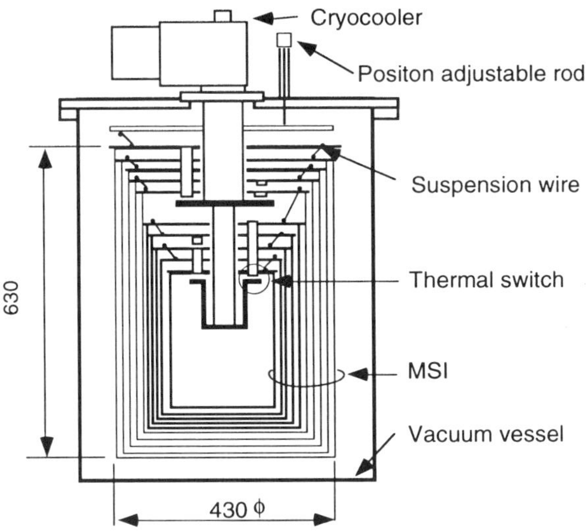


**Figure 2.** Schematic diagram of experimental apparatus for evaluating MSI effect.

## EXPERIMENTAL APPARATUS

To demonstrate the MSI effect, we used ten copper shields in the apparatus shown in Figure 2. No object was placed inside the innermost shield. For the suspension of the shields, the outer five shields used pin-jointed loops made of fiber-reinforced plastics (FRP) and the inner five shields used pin-jointed stainless-steel wire loops. Their lengths and cross-sectional areas were chosen so as to optimize heat conduction through them against their strength.(Given in Table 2) The shields were cooled by a two-stage Gifford-McMahon refrigerator, whose first-stage cooled the outer five shields roughly to 30K and whose second-stage cooled the inner five shields roughly to 4K.

A device was needed to thermally link the shields to the cryocooler for the initial cooling and then separate them from the cryocooler. We chose a mechanical contact on-off method as a thermal switch in the present apparatus because it was the easiest method at the moment. For connecting or disconnecting the shields with the cryocooler stages, which were fixed in position, the shields were required to be moved up and down. This was made possible by suspending the outermost shield from a FRP plate, which in turn is suspended by three rods that can be adjusted from outside the vacuum vessel. The shields were provided with short contact rods made of copper as shown in the Figure 2 for making the contacts with the cryocooler stages. Thermal resistance at these contact points is determined by the pressure at the points produced by the weight of the shields, and therefore it was necessary to select suitable contact areas for the contact rods; too small a contact area would result in longer cooldown time. The outer shields were directly brought into contact with the stages as shown in Figure 2, because it is where there are high heat flow rates and is therefore difficult to cool.

Temperatures of the shields were measured on the sides and bottom of shields at mid points. Sizes of the shields are tabulated in Table 1. The total weight of the shields is about 67 kg. Multi-layer insulation was wrapped on shield No 10, which is the outermost shield and on the shields up to the 5th one counted from the innermost shield; the number of layers on each shield is given in Table 1. Details of the suspensions are given in Table 2.

**Table 1.** Details of shields

| Shield | MLI | Dia(mm) | Height(mm) | Weight (kg) |
|---|---|---|---|---|
| 1 | - | 214 | 285 | 2.84 |
| 2 | - | 238 | 320 | 3.52 |
| 3 | - | 262 | 355 | 4.27 |
| 4 | - | 286 | 390 | 5.09 |
| 5 | 20 | 310 | 425 | 5.99 |
| 6 | 20 | 334 | 490 | 6.81 |
| 7 | 20 | 358 | 525 | 7.85 |
| 8 | 20 | 382 | 560 | 8.96 |
| 9 | 35 | 406 | 595 | 10.13 |
| 10 | 40 | 430 | 630 | 11.38 |

## EXPERIMENTAL RESULTS

Figure 3 shows the temperatures of shields as a function of time, whose origin is the time at which the shields were separated from the cryocooler by thermal switch operation. The temperature of the outermost shield shows quite a rapid increase while the other shields, in the order from outside to inside, show progressively lower rates of temperature increase. The innermost shield, weighing 2.8 kg, rose to only 30 K in one month's time. The cryocooler was not stopped after its thermal contact with the shields was broken because it was located in the innermost shield. For this reason, the innermost shield was somewhat cooled by the cryocooler and it was estimated that heat radiation to the cryocooler from the innermost shield was about 6% of the heat that was flowing into the shield from the shield next to it.

As the temperature alone is not enough to clearly show the insulation effect of the shields, we evaluated the thermal energy storage of the shields using specific heat values of copper for the temperature traces in Figure 3. The results are shown in Figure 4.

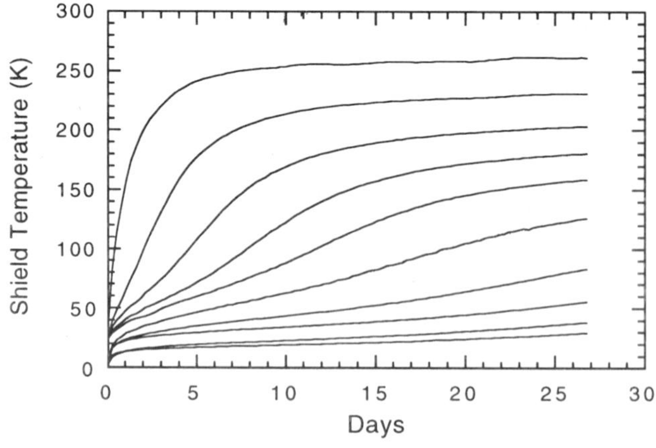


**Figure 3.** Temperature variation of shields after seperation from cryocooler.

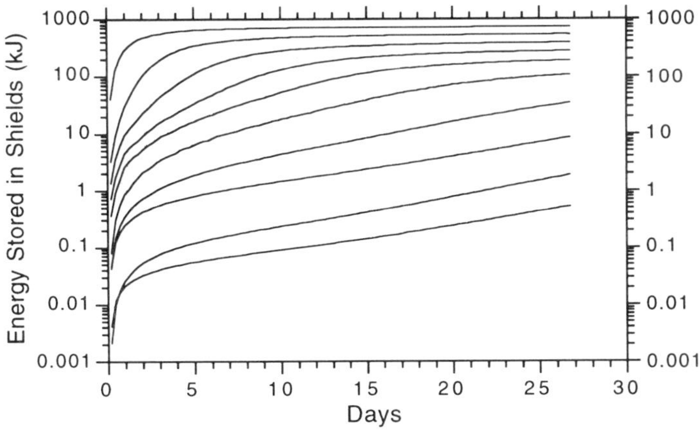


**Figure 4.** Variation of stored energy of shields after seperation from cryocooler.

The outermost shield is seen to reach a maximum energy storage capacity in about 5 days time, but the inner shields reach this state at progressively later stage. Once a shield has reached its maximum storage capacity, its function as an energy storage device is completed; then, whatever heat it receives is passed onto the next shield. The energy storage in shields 1 to 10 at the end of the experiment are respectively: 0.52,1.87, 8.59, 34.2, 107.1, 190.7, 278.7, 389.8, 541.1 and 736.8 kJ. The energy storage in the innermost shield or shield No. 1 is only 0.02% of the total energy storage of the shields. This clearly shows that the present insulation method is very effective in shielding heat.

Increasing heat capacity of the shields would lower the rate of temperature rise of the shield. To do that, we used magnetic materials, $HoCu_2$ and $Er_3Ni$, which have specific heat capacities respectively about 330 and 207 times the specific heat capacity of copper at 4K. We coated the three innermost copper shields with these materials: the innermost shield with 4.1 kg of $HoCu_2$ , the first shield from the innermost shield with 2.6 kg of $Er_3Ni$, and the second shield from the innermost shield with 3.4 kg of $Er_3Ni$. Figure 5 shows the temperature variation of the five innermost shields as a function of time.

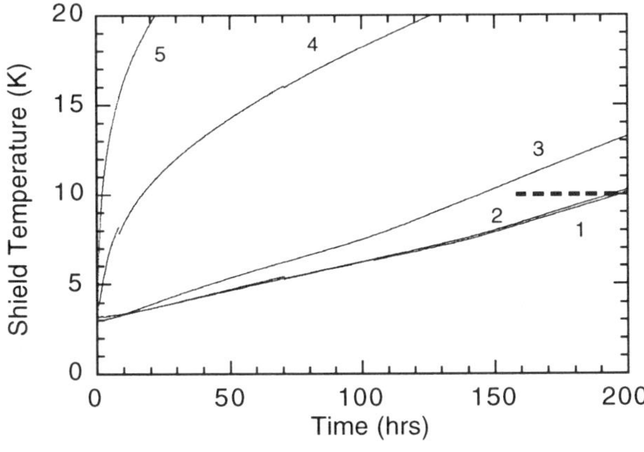


**Figure 5.** Temperature variation of shields when $Er_3Ni$ and $HoCu_2$ are used.

It took for the innermost shield (No.1) about 2.5 days to rise from 3 K to 5 K and 8 days to rise from 3 K to 10 K; as seen from the temperature traces 1 and 2 in Figure 5, the innermost shield had not been properly thermally isolated from the shield immediately outside of it. The experimental apparatus has not been particularly designed to keep the inner shield at 4 K for a long time; for instance, to optimize for 4 K use, it is required to determine the number of shields required at 4 K and where to place materials with large specific heats.

## ANALYSIS OF MSI METHOD

To analyze the transient heat propagation problem of the shields, the model of Figure 1 was considered. The following equations can be written for the i *th* shield:
Energy balance for the i *th* shield:

$$V_i C_i \frac{dT_i}{dt} = \dot{q}_{i+1} - \dot{q}_i \tag{1}$$

where, $\dot{q}_{i+1}$ and $\dot{q}_i$ are respectively the heat flow rates from the i+1 shield into the i*th* shield and from the i*th* shield to the i-1 shield. $V_i$ and $C_i$ are respectively the volume and volumetric specific heat of the ith shield respectively.

$$\dot{q}_{i+1} = \lambda_{i+1} A_{i+1} \frac{T_{i+1} - T_i}{l_{i+1}} + \varepsilon_{i+1} \sigma S_{i+1} F_{i+1} (T_{i+1}^4 - T_i^4) \tag{2}$$

where, the first term represents the heat conduction through the suspension wire between the shields i+1 and i, and the second term represents the radiation heat transfer between the i+1 and i shields. $A$ is the cross-section area of suspension wire, $F$ the geometric factor of radiation, $S$ the surface area of shield, $T$ the temperature of shield, $l$ the length of suspension wire, $\lambda$ the thermal conductivity of suspension wire, $\varepsilon$ the emissivity of shield and $\sigma$ is Stephan-Boltzman constant. $\dot{q}_i$ can be obtained by replacing i+1 in Eq. (2) by i-1.

**Initial and Boundary Conditions.** $T_i$=20 K; where i represents the shield number, which varies from 1 to 10 counted from the inside to outside. $T_{11}$=300 K; the vacuum vessel temperature.

**Assumptions.** Heat capacities of the suspension wires are negligible compared to those of the shields. Therefore heat conduction through them can be treated as steady heat conduction. The geometric factor, F was assumed to be 1 for each shield. The emissivity of all the surfaces was considered 0.03, and for $\varepsilon$ in Eq.(2), the effective value of emissivity for two infinite parallel plates was used.

**Analysis.** When i is varied from 1 to 10, ten simultaneously energy equations will result, and they are integrated numerically for the above initial and boundary conditions to obtain the temperature of shields as a function of time. Dimensions of shields are given in Table 1 and the sizes of suspensions are given in Table 2, where the i*th* suspension indicates the one between the shields i and i+1. Calculations were done for the configuration of the shields shown in Figure 1 for two cases: without an object inside the shields and with an object of 60 kg inside the shields. The object represents a high Tc superconducting magnet.

**Table 2.** Details of suspensions

| Suspension | Material | Total Cross-section($mm^2$) | Length(mm) |
|---|---|---|---|
| 1 | Stainless steel | 0.5 | 38.0 |
| 2 | Stainless steel | 0.5 | 38.0 |
| 3 | Stainless steel | 1.1 | 38.0 |
| 4 | Stainless steel | 1.1 | 38.0 |
| 5 | Stainless steel | 2.0 | 60.0 |
| 6 | FRP | 90.0 | 36.0 |
| 7 | FRP | 108.0 | 36.0 |
| 8 | FRP | 126.0 | 36.0 |
| 9 | FRP | 144.0 | 36.0 |
| 10 | FRP | 162.0 | 24.0 |

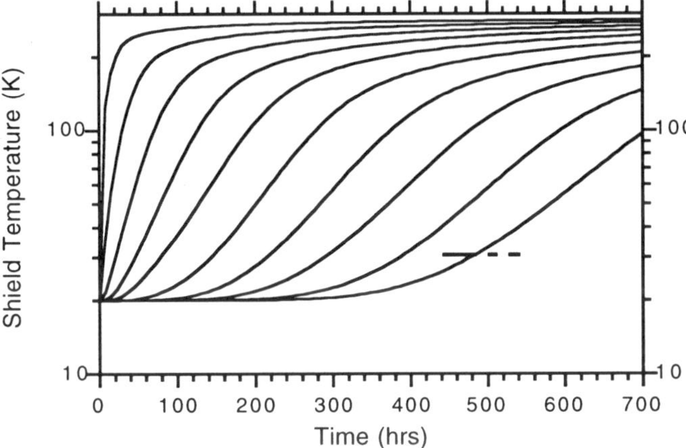


**Figure 6.** Numerical results showing temperatures of shields: no object inside shields.

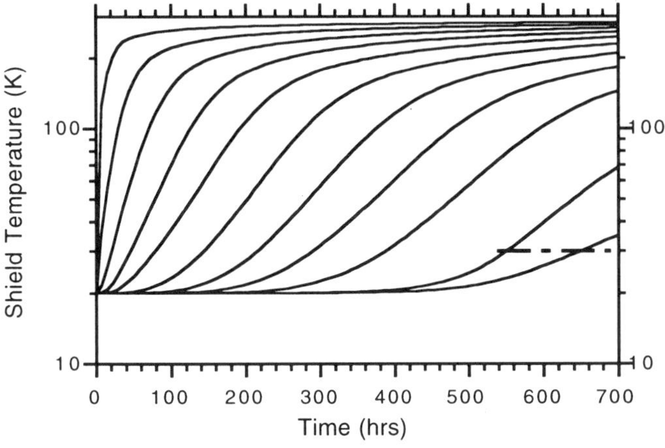


**Figure 7.** Numerical results showing temperatures of shields: 60 kg object inside shields.

**Analytical Results.** Figures 6 and 7 show numerical results for the two cases. For the innermost shield to rise to 30K from the initial temperature of 20K, it has taken 20 days as shown in Figure 6. The reason for shorter number of days compared to experimental results in Figure 3 is that in calculations no MLI was used on shields for reducing radiation heat. The case of MLI is difficult to model because the shield temperatures are affected by delayed diffusion of heat in MLI. If this effect is ignored, calculations show that it takes about 32 days for the innermost shield to rise from 4 to 20K and another 28 days from 20 to 30 K; experimentally, it took only 27 days for the temperature to rise from 4 to 30K.

In Figure 7, when there is a mass of 60kg inside the shields, the innermost shield has a slower rate of temperature rise than when the mass is not there, because some of energy from the inner shield has transferred to the inner mass. The rate of temperature rise of the 60 kg mass is seen to be lower than that of the innermost shield. This happens due to two factors. First, a smaller fraction of heat of the inner shield is transferred to the mass, and second, the mass has a larger heat capacity. It is also noted, comparing Figures 6 and 7, that the temperatures of shields other than the innermost shield have not changed due to the placement of the magnet inside the innermost shield. The effect of having an object inside the shields is localized. This is an indication of the strength of the MSI insulation method, because local changes are not propagated far from the locality where the change occurred. According to the figure, if MSI is used with a high Tc magnet of 60kg, the magnet can be maintained at the best about 26 days below 30K.

## CONCLUSIONS

Using the multi-shell insulation method, it was possible to maintain the 3kg-innermost shield below 30K for nearly a one-month period. Present work also has demonstrated that it is also possible to maintain a lower temperature for a reasonably long time by using high heat capacity materials with the shields. For instance, it was also possible to maintain the inner shield below 5 K for 2.5 days, or below 10K for 8 days by using $HoCu_2$ and $Er_3Ni$. These results are indicative of the possibility of realizing a liquid-free cryocooler-detached superconducting magnets at all temperature levels, although there are other technical barriers to overcome.

## REFERENCES

1. G. R. Chandratilleke et al., Gas - gap thermal switch for pre-cooling of cryocooler - cooled superconducting magnets, in: "Advances in Cryogenic Engineering, Vol. 41," Plenum Press, New York (1996), p. 139.

# CRYOCOOLER COOLED SUPERCONDUCTING MAGNETS AND THEIR APPLICATIONS

T. Hasebe,[1] J. Sakuraba,[1] K. Jikihara,[1] K.Watazawa,[1] H. Mitsubori,[1] Y.Sugizaki,[1] H. Okubo,[1] Y. Yamada,[1*] S. Awaji,[2] and K. Watanabe [2]

[1]Research and Development Center, Sumitomo Heavy Industries, ltd.
63-30 Yuhigaoka, Hiratsuka, Kanagawa, 254, JAPAN

[2]Institute for Materials Research, Tohoku University
2-1-1 Katahira, Aoba-ku, Sendai, Miyagi, 980-77, JAPAN

## ABSTRACT

Various types of cryocooler cooled superconducting magnets have been constructed and already used for some applications. An 11 T-52 mm room temperature bore magnet is used for a high-field heat treatment equipment, a 6 T-220 mm room temperature bore magnet is used for a new experiment on the electrochemical reaction in high fields, and a 5 T-50 mm bore with 10 mm gap split type magnet has been combined with an X-ray diffraction apparatus.

As a latest product, we have constructed a 10 T-100 mm room temperature bore cryocooler cooled superconducting magnet. The axis of the room temperature bore can be set vertically, horizontally or at any angle. The operating current was increased at a sweep rate of 0.25 A per second. It took about 11 minutes from 0 A to 162 A, which is the operating current for 10 T. The maximum central field of 10 T was held continuously for 9 hours at a constant coil temperature of 4.1 K.

## INTRODUCTION

A cryocooler cooled superconducting magnet directly cooled by a 10 K GM - cryocooler in vacuum was constructed using high-Tc supreconducting current leads for the first time in 1992.[1] There are many advantages of a cryocooler cooled superconducting magnet. These are easy operation, long-term continuous running and compact size, etc. These advantages stimulate high field applications in the field of chemical, biological and medical sciences. Vigorous developments of cryocooler cooled superconducting magnets have been carried out by many researchers and various types of cryocooler cooled superconducting magnets have been constructed.[2,3] We have constructed a high field magnet,[4] large bore magnet [5] and split type magnet with room temperature gap.[6] This paper describes some applications of these cryocooler cooled superconducting magnets.

*Present address: Faculty of Engineering, Tokai University
1117 Kitakaname, Hiratsuka, Kanagawa, 259-12, JAPAN

Furthermore, the latest result of a development is described in this paper. A 10 T-100 mm room temperature bore cryocooler cooled superconducting magnet has been constructed. The field sweep rate is high and it takes only 11 minutes from 0 T to 10 T. The direction of the room temperature bore can be set vertically, horizontally or at any angle. This magnet was designed to be more convenient and is expected to realize new applications of high fields.

## MAGNETS CONSTRUCTION

Parameters of typical cryocooler cooled superconducting magnets, which we have constructed and are used for some practical use, are listed in table 1.

### 4 T-38 mm room temperature bore magnet

A 4 T - 38 mm room temperature bore magnet was constructed in 1992 for a practical demonstration of conduction cooling of a superconducting magnet by a cryocooler and utilization of high-Tc $Bi_2Sr_2Ca_2Cu_3O_x$ superconducting current leads for the first time. The heat load to the second cooling stage of the GM-cryocooler is 0.55 W and the operating temperature at 4.6 T with a current of 465 A is 11.0 K. Because of the high critical current density at the operating temperature above 10 K, a multifilamentary Ti-added $Nb_3Sn$ superconducting wire which was fabricated by a Nb-tube process was used for the coil.

### 11 T-52 mm room temperature bore magnet

A high field cryocooler cooled superconducting magnet with a room temperature through bore was developed in the second step. A cross sectional view of the magnet is shown in Fig.1. Utilizing two 4 K GM-cryocoolers have enabled a cryocooler cooled high field $Nb_3Sn$ / NbTi hybrid superconducting magnet. A Nb-tube processed multifilamentary Ti-added $Nb_3Sn$ superconducting wire was used for an inner coil. The outer coil was wound using a multifilamentary NbTi superconducting wire. The heat load to each second cooling stage of the GM-cryocoolers is 0.53 W and the operating temperature at 10.0 T with a current of 150 A is 4.8 K.

**Table 1.** Parameters of cryocooler cooled superconducting magnets

| Superconducting coil | 4 T / 38 | 11 T / 52 | 6 T / 220 | 5 T / 50-10 | 10T / 100 |
|---|---|---|---|---|---|
| Conductor | $Nb_3Sn$ | $Nb_3Sn$ / NbTi | NbTi | NbTi | $Nb_3Sn$ / NbTi |
| Coil inner diameter (mm) | 59 | 86 / 187 | 260 | 80 | 127 / 248 |
| Coil outer diameter (mm) | 164 | 177 / 281 | 330 | 243 | 236 / 296 |
| Coil height (mm) | 188 | 222 / 225 | 334 | 41x2 with 60 mm gap | 230 |
| Room Temperature Bore diameter (mm) | 38 | 52 | 220 | 50 with 10 mm gap | 100 |
| Operating Current (A) | 465 | 150 | 144 | 68.5 | 162 |
| Central field (T) | 4.6 | 10.7 | 5.7 | 5.0 | 10.0 |
| Stored Energy (kJ) | 12 | 215 | 360 | 47 | 308 |
| Operating temperature (K) | 11.0 | 4.8 | 5.0 | 5.2 | 4.0 |
| Total cooling period (hr) | 23,800 | 15,100 | 14,100 | 7,500 | 2,900 |

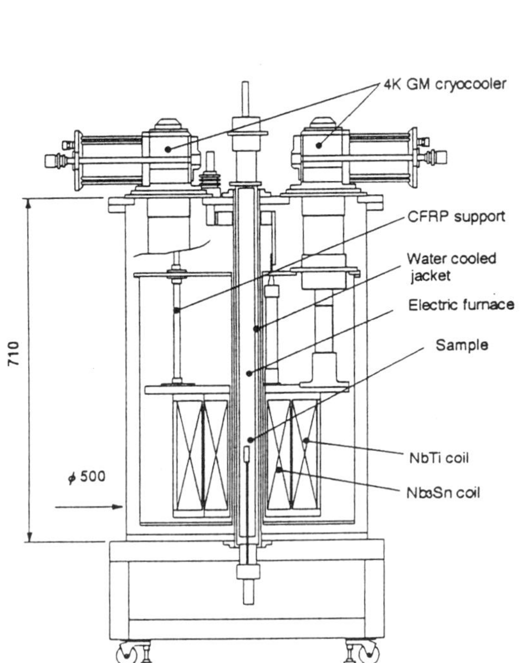


**Figure 1.** A cross sectional view of the 11 T - 52 mm room temperature bore cryocooler cooled superconducting magnet. An electric furnace is installed into a room temperature bore.[8]

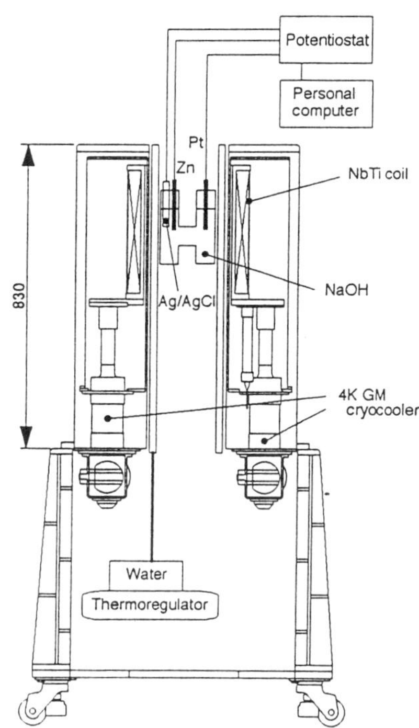


**Figure 2.** A cross sectional view of the 6 T - 220 mm room temperature bore cryocooler cooled superconducting magnet with an experimental set up for the electrochemical reactions.[5]

## 6 T-220 mm room temperature bore magnet

As a next step a large bore cryocooler cooled superconducting magnet was constructed utilizing two 4 K GM-cryocoolers and two grading NbTi coils. A cross sectional view of the magnet is shown in Fig.2. The heat load to each second cooling stage of the GM-cryocoolers is 0.31 W and the operating temperature at 5.7 T with a current of 144 A is 5.0 K.

## 5 T-50 mm bore with 10 mm gap split type magnet

A 5 T-50 mm bore with 10 mm gap split pair type cryocooler cooled superconducting magnet was constructed for an X-ray diffraction apparatus. Figure 3 shows a cross sectional drawing of the magnet system with the X-ray diffraction apparatus. It has a room temperature bore with a diameter of 50 mm for the sample cryostat and a 10 mm room temperature gap for the path of the X-ray beam. The magnet consists of a pair of split type coils wounded by NbTi wires with a gap of 60 mm. The heat load to the second cooling stage of the GM-cryocooler is 0.20 W and the operating temperature at 5 T with a current of 68.5 A is 5.2 K.

## 10T-100 mm room temperature bore magnet

As a latest product, we have constructed a 10 T-100 mm room temperature bore cryocooler cooled superconducting magnet. Figure 4 shows a schematic drawing of the developed magnet system. A hybrid $Nb_3Sn$ / NbTi coil is cooled by an improved 4 K GM-cryocooler. A bronze processed multifilamentary $Nb_3Sn$ superconducting wire was used for the inner coil. The diameter of filaments in the bronze processed $Nb_3Sn$ wire is 5 $\mu$m and the twist pitch is 25 mm. The outer coil was wound using multifilamentary NbTi superconducting wire. Coil support members are made of GFRP and placed at both ends of the coil. The direction of magnet system can be set vertically, horizontally or at any angle. Though the second cooling stage of 4 K GM-cryocooler should cool the large mass of superconducting coil, the cooling capacity of the second cooling stage is not enough above 20 K for initial cooling of the coil. A thermal switch which mechanically connects the second cooling stage of 4 K GM-cryocooler with the first cooling stage during initial cooling process is employed in this magnet system to shorten the cooling period. A calculated heat load to the second cooling stage of GM-cryocooler is 0.20 W at 10 T with a current of 162 A and the estimated operating coil temperature is 4.0 K.

## APPLICATIONS

### 4 T-38 mm room temperature bore magnet

In order to investigate the relation between the amount of the induced martensitic phase and the holding time in fields for stainless steel 304L,[7] an experiment was continuously performed for 1200 hours at 3.7 T which is generated by a current of 370 A.

### 11 T-52 mm room temperature bore magnet

The 11 T-52 mm room temperature bore magnet is used for high field, high temperature heat treatment equipment. An electric furnace with an outer diameter of 50 mm is installed into the room temperature bore of the cryocooler cooled superconducting magnet.[8] The temperature of a sample in the furnace was monitored and controlled using a thermocouple of Pt-Rh. This heat treatment system enables us to carry out a continuous experiment at 10 T and up to 1200 °C for a few weeks or months. As an example, a high field seeding process for $YBa_2Cu_3O_7$ oxide at 10 T and 1000 °C for 3 days was carried out.

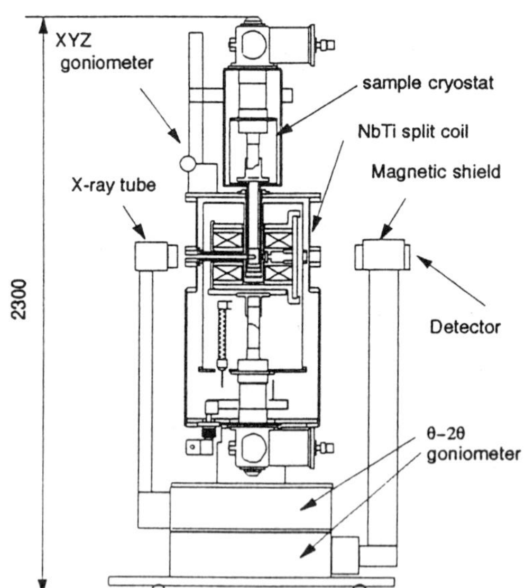


**Figure 3.** A cross sectional drawing of the 5 T 50 mm bore with 10 mm gap split type cryocooler cooled superconducting magnet system with an X-ray diffraction apparatus.

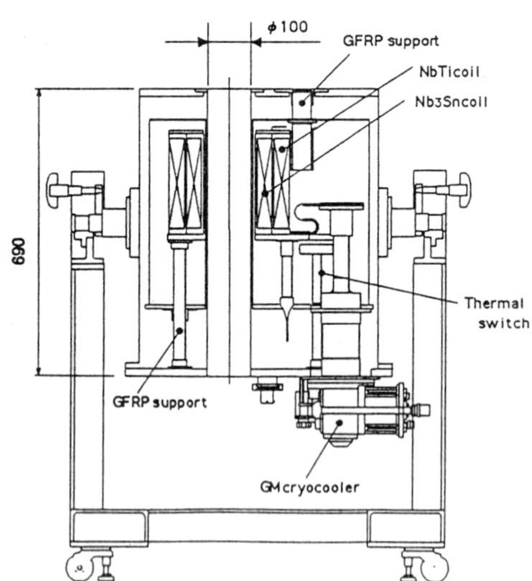


**Figure 4.** A schematic drawing of the 10 T 100 mm room temperature bore cryocooler cooled superconducting magnet.

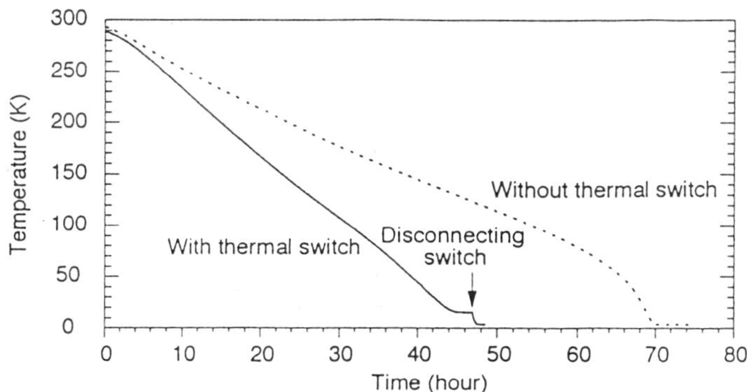


**Figure 5.** Initial cooling characteristics in the 10 T magnet with and without thermal switch.

## 6 T-220 mm room temperature bore magnet

The 6 T-220 mm room temperature bore cryocooler cooled superconducting magnet is used for a new experiment on the electrochemical reaction in high fields. The investigation on the field control of electrochemical reactions with a Zn electrode in a NaOH aqueous solution was carried out.[5] To keep the atmospheric temperature constant, a water circulator is installed in a room temperature bore.

The result of the experiment showed a significant difference with and without a field. In the absence of a field, the electrochemical oscillation was observed in this system. On the other hand, the current oscillation disappears in the presence of a field of 5 T.

## 5 T-50 mm bore with 10 mm gap split type magnet

The 5 T-50 mm bore with 10 mm gap split pair type cryocooler cooled superconducting magnet is placed on a $\theta-2\theta$ goniometer.[9] A scintillation counter is covered with a ferromagnetic magnetic shield. It is needed for photoelectrons with energy of few eV in the scintillation counter to eliminate an effect of a leakage field of 0.02 T when the superconducting magnet generates a field of 5 T. On the other hand, the energy of an electron beam in a X-ray tube is 40 kV, and the X-ray tube is not affected by the leakage field. A specimen is cooled by a GM cryocooler in vacuum down to 7.5 K. The temperature of the copper sample holder is monitored and controlled using a carbon glass resister (CGR) thermometer and a 20 W resistance heater in the temperature range from 7.5 K to 300 K with an accuracy of 0.1 K in the presence of a magnetic field. The cryostat for the sample has windows, made of CFRP on the vacuum vessel and of Be film on a thermal radiation shield, for the X-ray beam.

In an experiment, the field induced phase transformation for a perovskite $Nd_{0.5}Sr_{0.5}MnO_3$ was observed by the measurement of the lattice parameters of $Nd_{0.5}Sr_{0.5}MnO_3$ at the constant temperature of 140 K by applying a field.

# PERFORMANCE OF 10 T - 100 mm BORE MAGNET

The initial cool-down of the coil from room temperature was carried out in vacuum using a thermal switch within 48 hours as shown in Fig.5. The temperature of the coil saturated at 15 K, connecting the thermal switch between the first cooling stage and the second cooling stage, within 44 hours. It took an hour to an ultimate temperature after disconnecting the thermal switch at 15 K. The ultimate temperatures within each part of the cryostat are 3.6 K for the $Nb_3Sn$/NbTi coil and the second cooling stage and 43.1 K for the thermal radiation shield which is thermally anchored to the first cooling stage of the GM-cryocooler, respectively. Without using the thermal switch, it takes about 70 hours from room temperature to 3.6 K. Figure 6 represents the performance test result for a fast sweep of field

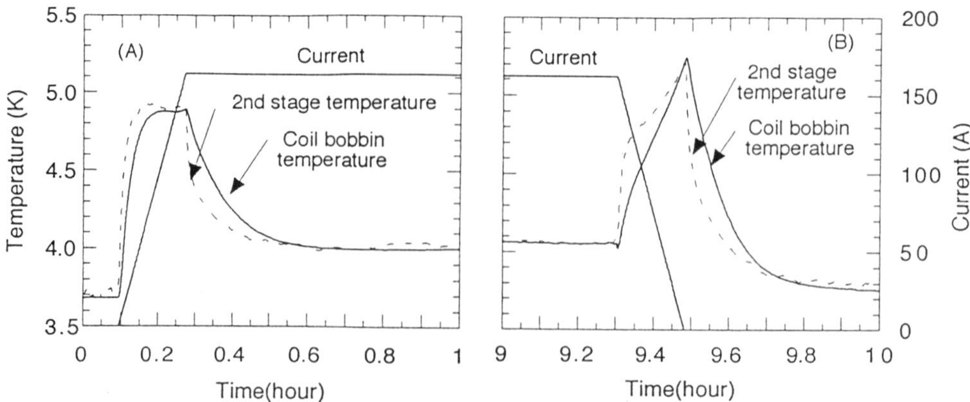


**Figure 6.** Result of a performance test of holding a field at 10 T with an operating current of 162 A for 9 hours. (A): at increasing current with a sweep rate of 0.25A/sec. (B): at decreasing current.

up to 10 T and a continuous hold of 10 T for 9 hours. The operating current was increased at a sweep rate of 0.25 A per second and it took about 11 minutes from 0 A to 162 A which is the operating current for 10 T. The coil temperature measured by a CGR sensor attached to the coil bobbin rises from 3.6 K to 4.9 K during the field sweep. The maximum temperature for the second cooling stage during field sweep is 5.0 K. The increased heat load to the second cooling stage of the GM cryocooler is due to AC losses in the superconductor. The diameter of the filaments in the employed $Nb_3Sn$ wire is 5 $\mu$m and the AC losses in this wire during field sweep is small with a high field sweep rate. After holding the field at 10 T for 20 minutes, the coil temperature fell to a constant temperature of 4.0 K. The maximum central field of 10 T was held continuously for 9 hours at constant temperatures of 4.0 K for the superconducting coil, 4.0 K for the second cooling stage and 46.1 K for the thermal radiation shield.

A maximum central field of 10 T is also generated in horizontal direction perpendicular to the direction as shown in Fig.4 without any change in the operating temperature.

## CONCLUSIONS

We have constructed various types of cryocooler cooled superconducting magnets and many application of high field have been realized in the field of chemical, material and other sciences.

A newly developed 10 T - 100 mm room temperature bore magnet can energize to maximum central field of 10 T in 11 minutes and change the direction of the through bore. Investigations of the frontier of magnetic science will start with this novel magnet system.

## REFERENCES

1. K.Watanabe, Y.Yamada, J.Sakuraba, F.Hata, C.K.Chong, T.Hasebe and M.Ishihara, $(Nb,Ti)_3Sn$ superconducting magnet operating at 11 K in vacuum using high-Tc $(Bi,Pb)_2Sr_2Ca_2Cu_3O_{10}$ current leads, *Jpn.J.Appl.Phys.* 32:L488(1993).
2. T.Kuriyama, M.Urata, T.Yazawa, K.Yamamoto, Y.Ohtani, K.Koyanagi, T.Masegi, Y.Yamada, S.Nomura, H.Maeda, H.Nakagome and O.Horigami, Cryocooler directly cooled 6 T NbTi superconducting magnet system with 180 mm room temperature bore, *Cryogenics*, 34:643(1994).
3. E.L.Laskaris, R.Ackermann, B.Dorri, D.Gross, K.Herd and C.Minas, A cryogen-free open superconducting magnet for interventional MRI applications, *IEEE Trans. Appl. Superconductivity*, 5:163(1995).

4. K.Watanabe, S.Awaji, J.Sakuraba, K.Watazawa, T.Hasebe, K.Jikihara, Y.Yamada and M.Ishihara, 11 T liquid helium-free superconducting magnet, *Cryogenics*, 36:1019(1996).
5. K.Watanabe, T.Masumoto, S.awaji, I.Mogi, M.Motokawa, K.Watazawa, J.Sakuraba, T.Hasebe, M.Ishihara and Y.Yamada, Cryocooler-cooled large bore NbTi Superconducting magnet using high temperature superconducting current leads, *Proceedings of the 16 the ICEC/ICMC*, 1121(1997).
6. K.Watanabe, S.Awaji, J.Sakuraba, K.Jikihara, K.Watazawa, T.Hasebe, F.Hata, C.K.Chong, Y.Yamada and M.Ishihara, Cryogen-free split-pair superconducting magnet with a $\phi$ 50 mm × 10 mm room temperature gap, in:"Advances in Cryogenics Engineering" , Plenum Press, New York(1996), 41:319.
7. K.Shibata, Y.Kurita, T.Shimonosono, Y.Murakami, S.Awaji and K.Watanabe, Effects of high magnetic fields on martensitic transformation at cryogenic temperatures for variously heat treated stainless steels, in:"Advances in Cryogenics Engineering Mterials" , Plenum Press, New York(1994), 40:1207.
8. K.Watanabe, S.Awaji and K.Kimura, Development of high field heat-treatment equipment and relevant applications, *Jpn.J.Appl.Phys.* 36:L673(1997).
9. K.Watanabe, Y.Watanabe, S.Awaji, M.Fujiwara, N.Kobayashi and T.Hasebe, X-ray diffraction investigation in high fields at low temperature for $Nd_{0.5}Sr_{0.5}MnO_3$, Submitted to :"Advances in Cryogenics Engineering Materials" , Plenum Press, New York(1997, in press).

# DEVELOPMENT OF TWO TYPES OF CRYOGEN FREE SUPERCONDUCTING MAGNETS ($5T$-$\phi$300mm AND $10T$-$\phi$100mm)

K. Shibutani,[1] S. Itoh,[1] O. Ozaki,[1] T. Takagi,[1] T. Miyazaki,[1]
R. Hirose,[1] S. Hayashi,[1] M. Shimada,[1] R. Ogawa,[1] Y. Kawate,[1]
K. Matsumoto,[2] N. Kimura,[2] and K. Takabatake[2]

[1]Advanced Products Development Center, Kobe Steel, Ltd.,
5-5, Takatsukadai 1-chome, Nishi-ku, Kobe, 651-22, Japan

[2]Japan Magnet Technology Inc.,
5-5, Takatsukadai 1-chome, Nishi-ku, Kobe, 651-22, Japan

## ABSTRACT

An Ic-B-T characteristic (critical current vs. magnetic field at various temperatures) for multifilamentary NbTi superconducting (SC) wire has been measured by using a conduction cooled critical current measurement apparatus up to 300A and up to 5T, in the temperature range from 5K to 7.5K.

Two types of cryogen-free (C-F) SC magnets, a large bore type and a high field type, have been designed and fabricated based on these Ic-B-T characteristics for the multifilamentary NbTi SC wire. The large bore type and the high field type magnets generated 5T and 10T stable magnetic field at the center of the room temperature bore of 300mm and 100mm diameter, respectively.

The average heat generation while energizing is obtained based on the temperature dependence of the cryocooler's cooling capacity. This average heat generation agrees well with the hysteresis loss of the whole magnet, calculated based on the hysteresis data for the multifilamentary NbTi short sample.

## INTRODUCTION

The conceptual design of an SC magnet using a cryocooler instead of liquid helium was reported in 1983.[1] The discovery of a high-Tc superconductor realized an oxide SC current lead[2] and a 4K-GM cryocooler has been developed by using magnetic regenerator material.[3] These technological breakthroughs have made it easy to design and realize the C-F SC magnet system.

*Advances in Cryogenic Engineering,* Vol. 43
Edited by P. Kittel, Plenum Press, New York, 1998

The difference of the design process between the conventional SC magnet and the C-F SC magnet is the treatment of the operating temperature. The operating temperature of C-F SC magnet is dependent on the cooling capacity of the cryocooler, alternating current (a. c.) loss from the SC wire, eddy current loss and the temperature gradient in the magnet.

For the magnet design, the reliable Ic-B-T characteristic data restricts the reliability of C-F SC magnet and controls the minimization of the material cost for an SC magnet. The reduced temperature (T/Tc) becomes larger, the importance of accuracy of Ic-B-T data becomes larger.[4, 5] However there have been few studies to optimize the C-F magnet design based on the measured Ic-B-T characteristic for the multifilamentary NbTi SC wire. In the temperature range from 4K to 10K, it is difficult to control the temperature while introducing the high current for the conventional measurement system. A conduction cooled critical current measurement apparatus for the temperature range from 20K to 90K has been reported by Hasebe et al.[6] In this study, we designed and fabricated the conduction cooled critical current measurement apparatus for the NbTi SC wire. We got the reliable measured Ic-B-T characteristic of several kinds of multifilamentary NbTi SC wire in the temperature range from 4.2K to 7.5K. Using these characteristic data, the large bore type magnet and the high field type magnet have been designed and fabricated. The large bore type magnet used NbTi wire only. The high field type magnet used both NbTi and Nb3Sn wires. The measured and the calculated Ic-B-T characteristics were employed for the NbTi magnet and Nb3Sn magnet, respectively. That is because the reduced temperature of NbTi wire is relatively high. These two types successfully generated 5T and 10T [7] stable magnetic field at the center of the room temperature bore of 300mm and 100mm diameter, respectively.

The main question is how to decide the operating temperature. In general, the SC magnet generates the a.c. loss depending field ramp rate. The important specifications, the maximum ramp rate is limited by the cooling capacity of cryocooler and acceptable heat generation. In this paper, we try to check the relation between the temperature rising of the magnet and the calculated hysteresis loss. After this relationship is clear, we will be able to design the optimum magnet; selection of the optimum SC wire and cryocooler for a required maximum ramp rate.

## Ic-B-T CHARACTERISTIC OF NbTi SC WIRE

### Apparatus and Measurement

The conduction cooled critical current measurement apparatus has been designed and fabricated in order to measure the Ic-B-T characteristic for NbTi SC wire directly. Figures 1 and 2 show the structure and a photograph of the conduction cooled critical current measurement apparatus, respectively. The sample is connected to the 2nd stage of a 4K-GM cryocooler with a deoxidized copper rod and is cooled directly. An electric heater is attached to 2nd stage of cryocooler in order to control the sample temperature. A carbon glass resistive (CGR) thermometer is attached to the removable electrode beneath the sample in order to measure the sample temperature. The sample current is introduced through the copper current lead from room temperature to the 1st stage of the cryocooler and through the oxide SC current lead from the 1st to the 2nd stage of the cryocooler. Table 1 shows the specifications of this apparatus. The temperature fluctuation of the sample are within 0.1K when the sample current increases from 0A to 300A in 1 minute. A C-F NbTi SC magnet with A thermally controlled NbTi persistent current switch (PCS) was employed as the external magnetic field. The measurements were done in order from low temperature to high temperature while the magnetic field was held constant.

### Results

Figures 3 and 4 show the examples of the result for the multifilamentary NbTi SC wire. Figure 3 shows the Ic-T characteristic of 0.65-1500-2.5 type NbTi SC wire (0.65mm bare diameter, 1500 filaments (corresponding to 10μm-class in filament's diameter) and 2.5 copper/non-copper ratio) in various external magnetic fields. In Figure 4, the Ic-B

**Table 1.** Specifications of conduction cooled Ic-B-T measurement apparatus

| | | | |
|---|---|---|---|
| Temperature stability | Zero current state | 4.5K< | <+/-0.05K |
| | Current sweep: from 0A to 300A for 1 min. | 5-6K | <+0.1K |
| | | 6K< | <+0.05K |
| Response time | Raising from 4.5K to 6K | | <5min |
| | Raising from 4.5K to 8K | | <15min |
| Sample exchange | Cooling from room temperature to 4.5K | | <6hours |
| | Raising from 4.5K toR. T. | | <3hours |

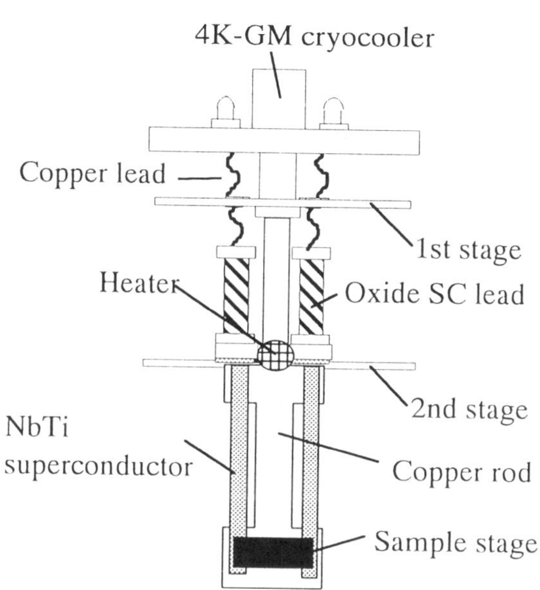


**Figure 1.** Structure of conduction cooled Ic-B-T measurement system.

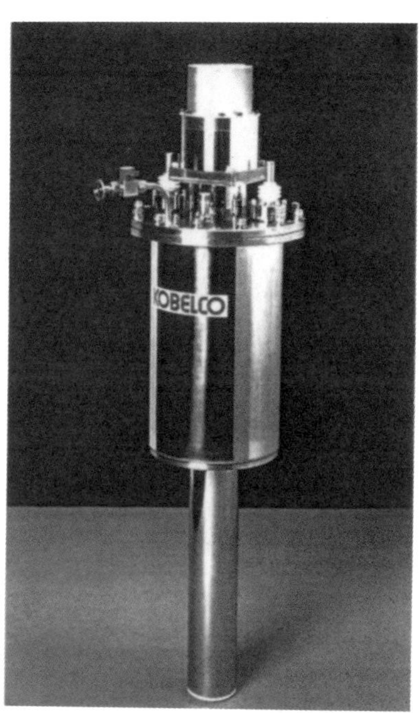


**Figure 2.** Conduction cooled Ic-B-T measurement system.

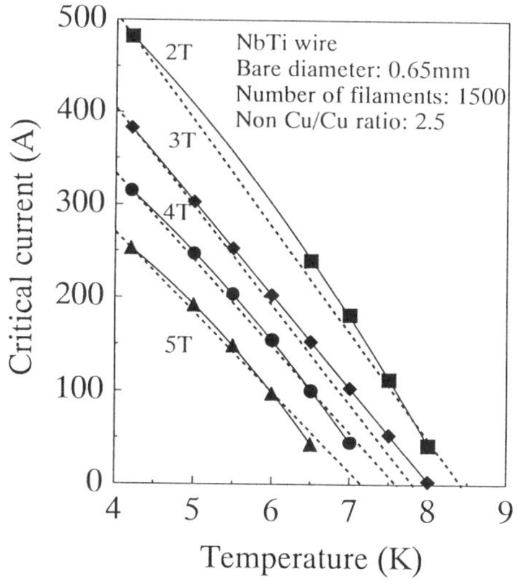


**Figure 3.** Ic-T (B) characteristic data of NbTi wire.

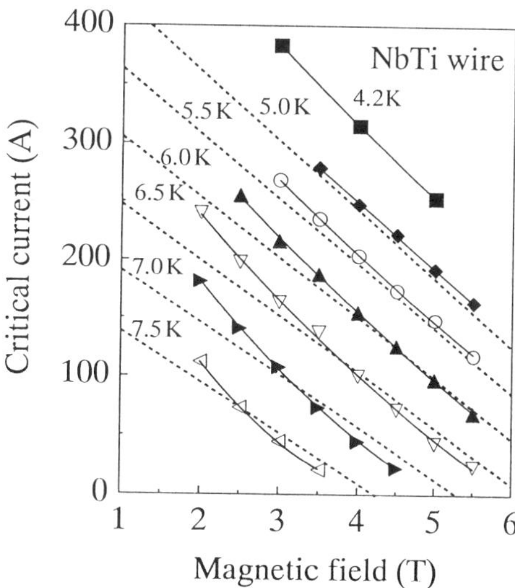


**Figure 4.** Ic-B (T) characteristic data of NbTi wire.

**Table 2.** Specifications of C-F 5T-ϕ300mm and 10T-ϕ100mm SC magnet systems

| | | 5T-ϕ300mm | 10T-ϕ100mm |
|---|---|---|---|
| Room temperature bore diameter | | ϕ300mm | ϕ100mm |
| Central field | | 5.0T | 10.0T |
| Field homogeneity | | <2% (ϕ100mmxH10mm) | <1.5% (ϕ10mm DSV) |
| Conductor grading | NbTi | ϕ0.65, ϕ0.75, ϕ0.85, ϕ0.95mm | ϕ0.60, ϕ0.65, ϕ0.70, ϕ0.85mm |
| | Nb3Sn | | ϕ0.70mm |
| Ramp rate | | 5T/60min. | 10T/60min. |
| Operating current | | 81.66A/5T | 79.37A/10T |
| Self inductance | | 157H | 114H |
| Initial cooling time | | <65 hours | <45 hours |
| Field direction | | Vertical | Vertical and horizontal |
| Input power of cryocooler | | 7.0kW (3ϕ200V-60Hz) | 7.0kW (3ϕ200V-60Hz) |
| Dimensions of OVC | | H944mmxW1090mmxD718mm | H832mmxf750mm |

characteristic of 0.65-1500-2.5 type NbTi SC wire at various temperatures is re-plotted from Figure 3. M. S. Lubell et al.[8] reported the experimental equation for Ic-B-T characteristic of NbTi SC wire. Lubell's experimental equation is given by

$$Ic(B,T) = Ic(B,4.2K) \times \frac{Tc(B)-T}{Tc(B)-4.2} \qquad \text{where } Tc(B) = Tc(0) \times (1-\frac{B}{Bc(0)})^{\frac{1}{1.7}} \tag{1}$$

where Ic is the critical current, Tc is the critical temperature, Bc is the critical magnetic field and Ic(B,4.2K) is the critical current as a function of magnetic field at 4.2K. The result of calculation from Ic(B,4.2K) data is also plotted as a dotted line in Figures 3 and 4 using typical values of Tc(0)=9.2K and Bc(0)=14.5T. It is found that the difference between measurement result and the calculation result from Lubell's equation is conspicuous around lower magnetic field and higher temperature region. However in the important region for the design of C-F SC magnet, higher magnetic field and lower temperature region (below 6K), measurement result agrees well with Lubell's equation. Every other kind of NbTi SC wire (range from 0.30mm to 0.75mm in bare diameter, 55 and 1500 in number of filament; corresponding to 10μm-class and 50μm-class in filament diameter respectively and 1.3 and 2.5 in copper/non-copper ratio) showed the same characteristic. These results suggest that the estimated Ic-B-T characteristic of NbTi SC wire based on the Ic-B data at 4.2K by using Lubell's equation gives sufficient reliability for cryogen free SC magnet design.

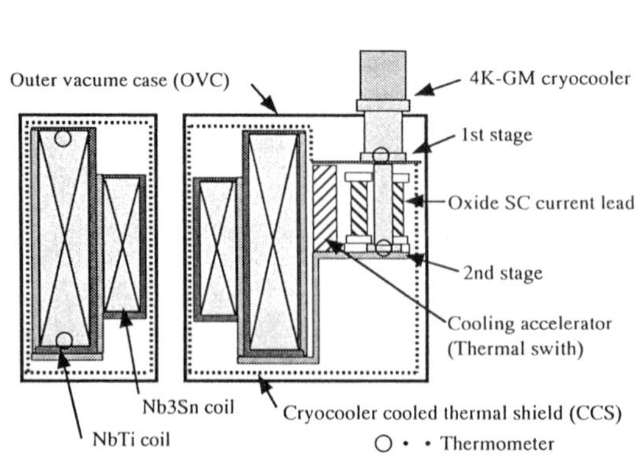


**Figure 5.** Structure of cryogen-free superconducting magnet.

**Figure 6.** 5T-ϕ300mm C-F SC magnet.

## CRYOGEN FREE 5T-$\phi$300mm and 10T-$\phi$100mm SC MAGNET

### System Design

C-F 5T-$\phi$300mm and 10T-$\phi$100mm SC magnet systems have the same structural composition. Figure 5 shows the structure of these systems. The main components of these systems are SC magnet, 4K GM cryocooler, Oxide SC current leads, neon gas conduction type thermal switch,[7] cryocooler cooled thermal shield (CCS) and outer vacuum case (OVC). Figure 6 shows the photograph of the 5T-$\phi$300mm magnet system. The specifications of these two systems are tabulated in Table 2.

An aluminum based alloy bobbin and a stainless steel bobbin are employed for the NbTi (for 5T-$\phi$300mm and 10T-$\phi$100mm) and Nb3Sn coils (for 10T- $\phi$100mm only), respectively. Each coil is covered with the cylindrical deoxidized copper plate and compressed by the outer stainless steel binding wire to keep the thermal contact even after the quench. The coil is cooled by thermal conduction through these plates. In order to protect the magnet in case of current lead shut down, eight and six pair diode shunt circuits are mounted on the bobbin of the 5T- $\phi$300mm and the 10T- $\phi$100mm, respectively. Three CGR thermometers are attached to the top and bottom flanges of the bobbin and the 2nd stage of cryocooler for each system in order to measure the temperature change. One Pt-Co thermometer is attached to the 1st stage of the cryocooler.

For both systems, the neon gas conduction type thermal switch that connects between the 1st and 2nd stage of the 4K GM cryocooler is employed as a cooling accelerator. It has become clear that this cooling accelerator reduced the cooling period by approximately two thirds when compared to cooling without an accelerator.[7]

### Initial cooling

Figure 7 shows the initial cooling characteristic (at the top flange of the bobbin) of two systems from room temperature to 4K. The cooling mass of 5T- $\phi$300mm and 10T-$\phi$100mm are 160kg and 130kg including the cooling parts, respectively. The initial cooling times of 5T- $\phi$300mm and 10T- $\phi$100mm are within 65 hours and within 45 hours, respectively.

### Average Heat Generation

Figure 8 shows the temperature change of the 5T- $\phi$300mm magnet while energizing. This is the temperature change of the CGR at the top flange of the bobbin. This point showed the highest temperature of the three CGR thermometers. The "top" was 1.2K and 0.4K higher than the "2nd" and the "bottom" all the time, respectively. The temperature of the 1st stage was about 30K all the time. The initial temperature of 2nd stage was 4.15K for all the case. The 5T-f300mm was energized three times on the various ramp rates, 5T/h ($2.3\text{x}10^{-2}$A/sec), 2.5T/h ($1.1\text{x}10^{-2}$A/sec) and 1.25T/h ($5.7\text{x}10^{-3}$A/sec). All the cases show the small hump just after starting ramp. The heights of these humps are 0.20K, 0.12K and 0.07K in the case of 5T/h, 2.5T/h and 1.25T/h, respectively. The height of these humps correspond to the a.c. loss from the superconductor, mainly.

Let us convert the height of the hump into the heat generation from the magnet. On this conversion, the measured load curve for 4k GM cryocooler (figure 9) is used. From the figure 9, the thermal load $P_{2nd}$(W) with $T_{1st}$=30K is given by

$$P_{2nd} = 0.124 \times T_{2nd}{}^{2} + 0.096 \times T_{2nd} - 1.582 \qquad (2)$$

where $T_{2nd}$ is the temperature of 2nd stage of the cryocooler. For example, the height of the hump is 0.20K in case of 5T/h. The same temperature, 0.20K-rising from 4.15K to

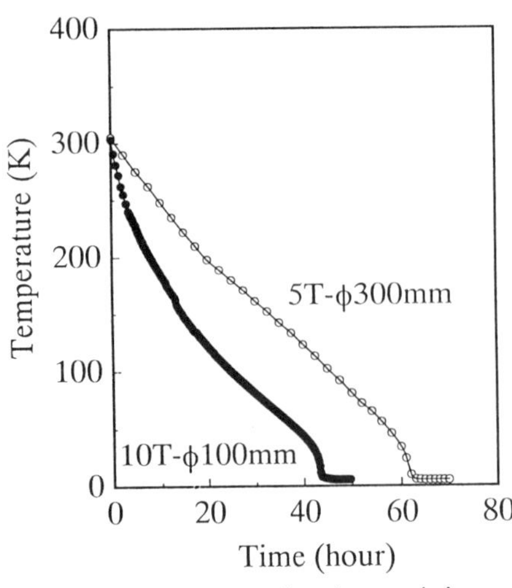


Figure 7. Initial cooling characteristic.

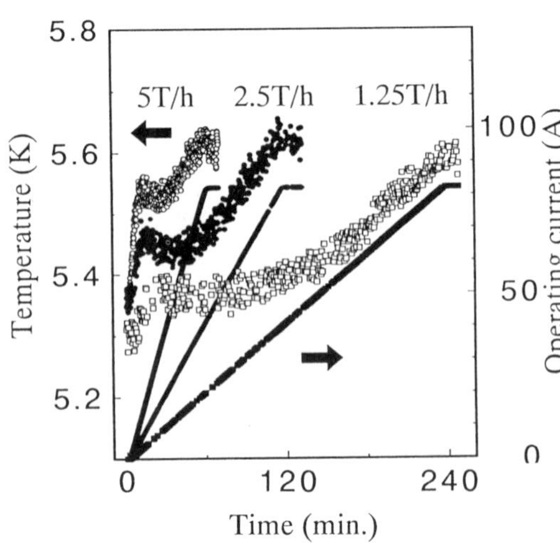


Figure 8. Temperature change of 5T- ϕ 300mm.

4.35K at the 2nd stage corresponds to 0.23W. From the same manner, the humps 0.12K and 0.07K correspond to 0.14W and 0.08W, respectively.

## Hysteresis Loss Estimation

Let us compute the hysteresis loss of whole the 5T- ϕ300mm magnet based on the result of a hysteresis loss of 0.65-1500-2.5 in multifilamentary NbTi short wire. Figure 10 shows the hysteresis loss of 0.65-1500-2.5 NbTi wire as a function of magnetic field. The loss measurement was performed using a.c.-magnetization methods.[9] Let us suppose that a hysteresis loss of NbTi wire with the same cross sectional arrangement is proportional to its filament diameter. The magnet was divided into 48 rings (or mesh in cross section) in order to consider the magnetic field distribution. The averaged hysteresis loss of the 5T-ϕ300mm; $P_{hys.}$(W) on the fixed ramp rate condition is given by

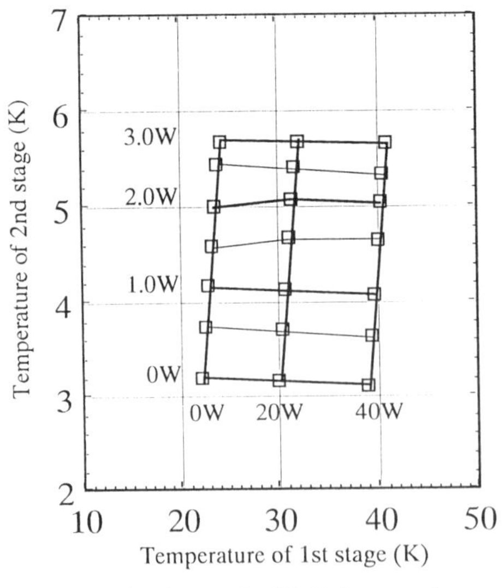


Figure 9. Load curve for 4K-GM cryocooler.

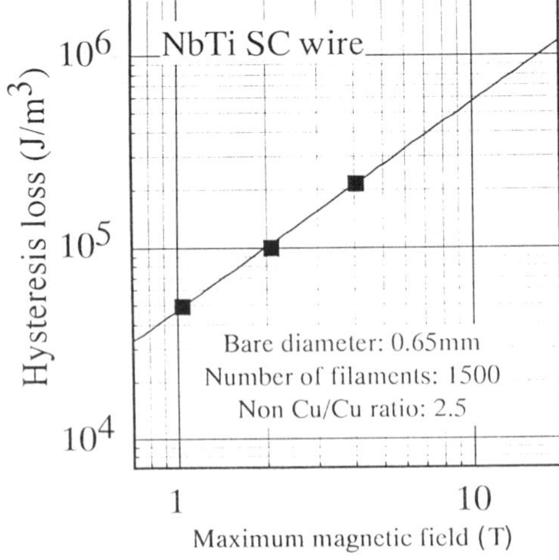


Figure 10. Hysteresisi loss of NbTi wire.

**Table 3.** Comparison between the converted average heat generation and the calculated hysteresis loss for whole magnet (5T- φ300mm magnet system)

| Ramp rate | 5T/hour | 2.5T/hour | 1.25T/hour |
|---|---|---|---|
| Rising temperature | 0.20K | 0.12K | 0.07K |
| $P_{2nd}$ (measured) | 0.23W | 0.14W | 0.08W |
| $P_{hys}$ (caliculated) | 0.21W | 0.11W | 0.05W |

$$P_{hys.} = \frac{1}{4} \times \frac{1}{t} \times \sum_{i=1}^{n} (Pi \times Vi) \qquad (3)$$

where Pi is the a.c. loss of the element i, Vi is the volume of the element i, n is the number of elements (=48) and t is the ramp period from 0T to 5T.

As the result, the hysteresis loss of the 5T- φ300mm magnet is estimated 0.21W, 0.11W and 0.05W in the case of 5T/h, 2.5T/h and 1.25T/h, respectively. The average heat generation ($P_{2nd}$) and estimated hysteresis loss ($P_{hys.}$) are tabulated in Table 3. Roughly speaking, the rising temperature of the C-F SC magnet for a specific ramp rate was estimated by the calculation based on the measured hysteresis loss data for NbTi short sample and the measured load curve for cryocooler. In other words, we will be able to take the optimum selection for the SC wire and the cryocooler under the required maximum ramp rate.

## CONCLUSION

Reliable Ic-B-T characteristic for the multifilamentary NbTi SC wire has been measured by using a conduction cooled critical current measurement apparatus. A 5T- φ300mm type and a 10T- φ100mm type C-F SC magnet have been designed and fabricated based on these Ic-B-T characteristic data. The two magnets successfully generated 5T and 10T stable magnetic field at the center of the room temperature bore. Converted average heat generation from magnets agreed well with the calculated hysteresis loss based on the measured hysteresis loss of the short sample.

## REFERENCES

1. M. O. Hoenig, Design concepts for a mechanically refrigerated 13K superconducting magnet system, IEEE Trans. Mag. 19:880 (1983).
2. J. L. Wu et al., Design and testing of a high temperature superconducting current lead, IEEE Trans. Mag. 27:1861 (1991).
3. T. Kuriyama et al., Development of 1 watt class 4K GM refrigerator with magnetic regenerator materials, in: Advances in Cryogenic Engineering, Vol. 39, Plenum Press, New York (1994), p. 1335.
4. Bu-Xin Xu et al., A cryogen-free superconducting magnet with 95 cm warm bore for whole body MRI, IEEE Trans. Mag. 32:2637 (1996).
5. K. Watanabe et al., Cryogen-free split-pair superconducting magnet with a f50 mm x 10 mm room temperature gap, in: Advances in Cryogenic Engineering, Vol. 41,Plenum Press,New York (1996), p. 319.
6. T. Hasebe et al., Critical current measurement unit utilizing Bi-based oxide superconducting current leads and cryocoolers, IEEE Trans. Appl. Supercond. 5:821 (1995).
7. K. Shibutani et al., Design and Fabrication of cryogen-free superconducting magnet, Proc. of the ICEC/ICMC16, Elsevier Science, (1996), p. 1129.
8. M. S. Lubell, Empirical scaling formulas for critical current and critical field for commercial NbTi, IEEE Trans. Mag. 19:754 (1983).
9. C. Schmidt et al. Second VAMAS a.c. loss measurement intercomparison: a.c. magnetization measurement of hysteresis and coupling losses in NbTi multifilamentary strands,Cryogenics 37:77 (1997).

# DESIGN OF THE THERMAL SHIELDS FOR THE NEW IMPROVED VERSION OF THE TESLA TEST FACILITY (TTF) CRYOSTAT

C. Pagani[1], D. Barni[1], M. Bonezzi[1], P. Pierini[1], J.G. Weisend II[2]

[1]INFN Milano - LASA, Via F.lli Cervi 201, I-20090 Segrate (MI), Italy
[2]DESY, Noktestrasse 85, D-22607 Hamburg, Germany

## ABSTRACT

One of the principal goals of the ongoing TTF project is to produce prototype low cost and reliable cryomodules meeting the stringent requirements for TESLA. The first cryomodule was assembled and tested at DESY. The experience gained during the design and the commissioning suggested new solutions to improving the technical design.

The most important improvement is the design of the new thermal shields at 4.5 K and 70 K. In this design the cooling helium pipes are directly welded to the aluminum shields. This solution simplified the technical design of the entire cryomodule and indicated some issues that needed study to verify the behavior of the new structure during cooldown and operation. The principal aspect studied has been the procedures necessary to avoid stresses and deformations produced during the cooldown which might damage the shields. The solution found also prevents the typical deformation induced by the welding and makes the production pre-alignment easier. Shields will be welded through small aluminum fingers that reduce the resistance section of the joint, while increasing the elasticity, unloading the welded joints and preventing possible damage during the cooldown.

## INTRODUCTION

One of the principal goals of the ongoing TTF project[1] is to study the production of low cost and reliable cryomodules meeting the stringent requirements for TESLA[1]. The first cryomodule[2] was assembled and tested at DESY[3]. The experience gained during the design and the commissioning suggested new solutions to improve the technical design.

In the new design of the TTF aluminum cryostat shields the cooling helium pipes are welded to the shields. To avoid stress breaks in the welded regions produced during the cooldown, we have studied a solution to unload the junction. This solution also avoids the typical deformation caused by the welding procedure. Results of analytical considerations have been tested with FEA calculations using the ANSYS code. Figure 1 shows two typical cross sections of the new cryostat design, to be compared with the old version[3], which was demonstrated to be considerably more expensive than expected due to the large number of special braids and threaded fasteners.

The original thermal shield design for the TESLA Test Facility (TTF) cryomodule[1-3] used aluminum shields cooled by copper braids connected to stainless steel cooling pipes. The shields themselves were split into parts connected together by hundreds of threaded fasteners. This shield design worked well in the first cryomodule[3]. However, the design has some drawbacks. The copper braids, with braised Cu and Al terminals, are expensive to produce and their thermal performance could not be consistently predicted. The large number (> 1000) of threaded fasteners are expensive and require a significant number of man-hours (and thus cost) for the drilling of holes, installation of threaded inserts and finally installation of the fasteners themselves. As part of the TESLA program to reduce the cost and complexity of the cryomodule, an improved shield design was developed together with a general reconsideration of the tolerances to reduce the fabrication cost of the major components (He Gas Return Pipe, Vacuum Vessel, etc.), while preserving and improving the quality of the final alignment. This new fabrication philosophy will be briefly sketched in this paper, concentrating on the "finger welding" solution.

## TOLERANCE SIMPLIFICATION

In the first cryostat design, in order to guarantee the needed clearance to connect the aligned cavity string to the supporting He Gas Return Pipe (HeGRP) through the suspension rings, most of the components, including the HeGRP, had to respect strict tolerances. These tolerances had to be preserved during the fabrication, and especially during weldings. Some expensive tooling, including those for the HeGRP optical measurement, were developed and used during the fabrication steps, thereby increasing the fabrication cost[2].

In the new cryostat design the above philosophy has been changed. In particular the general HeGRP tolerances were relaxed to meet those of a standard, high quality, welded pipe, i.e. ± 5 mm, including straightness. Once all the ancillary components and supports have been welded and the pipe straightness recovered, a 14 m milling machine is used to machine the upper three suspension posts flanges, to produce the reference axis. Figure 2 shows a scheme of the reference points that are used during fabrication and assembling.

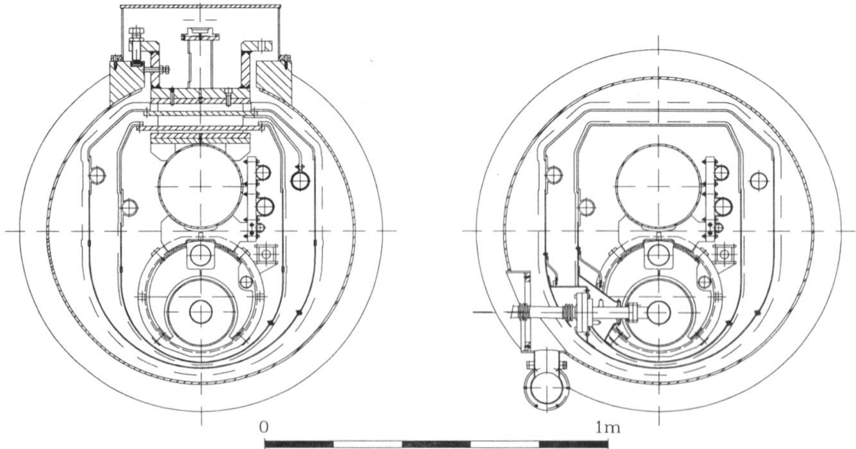


**Figure 1**. Two typical cross sections of the new cryostat design, to be compared with the old version[2].

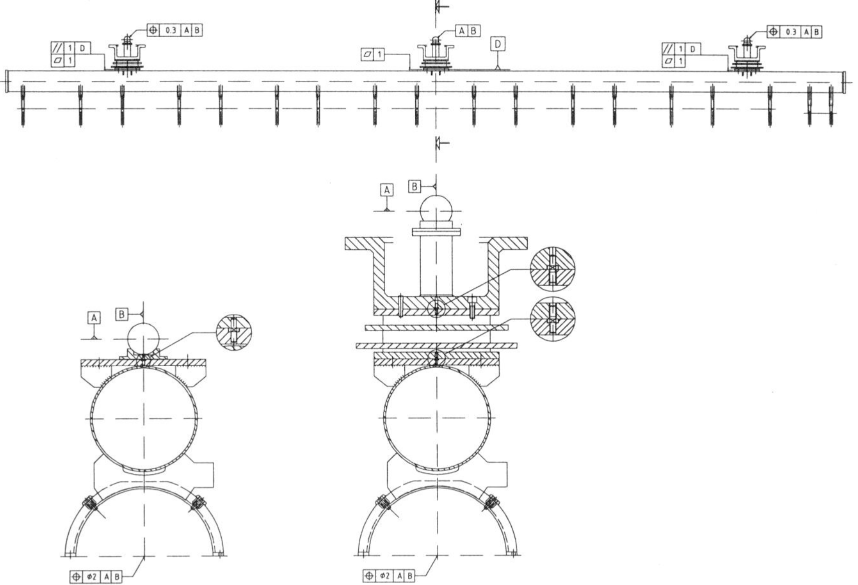


Figure 2. Schematic drawing of the references used during fabrication and assembly, in order to simplify measurement and relax the tolerances of the individual components.

As schematically shown in the lower left side of Figure 2 during all the fabrication steps the HeGRP can be easily referenced to its previously defined axis, by means of Taylor Hobson spheres (shown in the drawing) or pins, according to the precision required.

In order to further reduce fabrication costs, the tolerance specifications for the Vacuum Vessel has also been revised. In particular, all the components which need to meet strict tolerances, that is most of the previously welded flanges, are all machined on a 14 m long milling machine, without changing the vessel position.

## NEW THERMAL SHIELD DESIGN

A general view of the new shield design is shown in Figure 1, while some details of the 4.5 K shield are presented in Figure 3. Here the shields are cooled by aluminum pipes welded directly into the aluminum shields. Since stainless steel bellows are used in the interconnection between modules, the aluminum pipes require aluminum/stainless steel transition joints at each end. A total of 16 copper straps, fastened to the shields, are used in this design to connect the shields to the 4.5 K and 70 K heat intercepts of the main RF couplers.

As in the previous design, the shields are split into parts, but now all the thermal connections between shield parts are done by welds rather than threaded fasteners, which are just used for preassembling purposes. This new design, except for the "finger welding", is similar to that found on the magnet cryostats for the HERA, SSC and LHC accelerators.

As stated above, the major advantage of the new design is that of greatly reducing the number of copper braids and threaded fasteners needed for assembly. This results in a significant cost savings. The new design needs bi-metallic transition joints, which, however, are based on a technology that as been successfully used in the past.

The welding scheme, based on fingers, is discussed in the followings ections, together with its advantages and performances.

## THE WELDING SCHEME

As stated above, the new cryostat shields are directly welded to the cooling pipe. This connection represents the mechanical and the thermal link between the cooling pipe and the shields. This design leads to an asymmetric cooldown of the shields. We have checked through Finite Elements Analysis (FEA) the stresses and the general deformations produced by the asymmetric thermal fields, and have shown that a linear cool-down procedure from 300 K to 4 K, lasting two days, is completely acceptable for the shields[4]. We want to point out that, since in our cryogenic system the 4.5 K shield is cooled in parallel with the cavities, this cooldown procedure is, anyway, envisaged to prevent unwanted large movements of the cavity string[2].

The main aspects investigated in the following sections are those related to the welded joints between the cooling pipe and the shields, and, in the uncooled side, between the shield lower panels and its upper part. These two welding regions are the most critical from a thermo-mechanical point of view, so we have developed a solution to unload the joins from the thermally induced mechanical stresses. Figure 3 shows a simplified drawing of the welding between the pipe and the aluminum shields of the new TTF cryostat. As sketched in Figure 3, our solution consists of splitting each panel connection in two equal parts, symmetric with respect to the center of the shield panel. Each one of the two connections is performed by TIG welding, through properly shaped fingers which unload (as discussed below) the joint region. Each junction has 6 fingers, with nominal length of 50 mm and width of 23 mm.

## TYPICAL LOADS ON THE THERMAL SHIELD WELDINGS

The welded junction and its small cross-section represents a bottleneck for the heat-flow. This produces a thermal gradient between the shields and the cooling pipe. The difference in temperature between the two welded regions, through a differential expansion, causes stress on the junction. The same effect is also present in the welded junctions in the

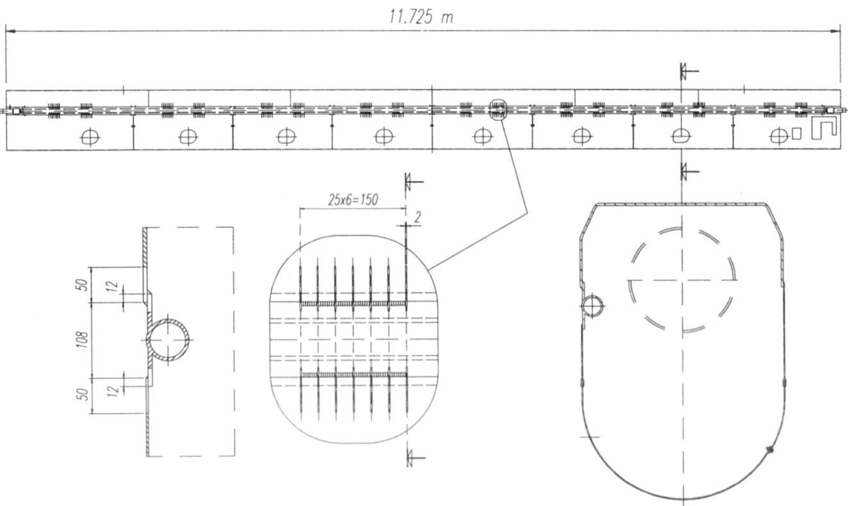


**Figure 3.** The welding scheme used to join both the cooling pipe to the aluminum shields and, in the opposite side, the shield panels to the upper part. All weldings are done on fingers to unload the welded join. Since the same scheme is used in both shields, just the 4.5 K shield is represented here.

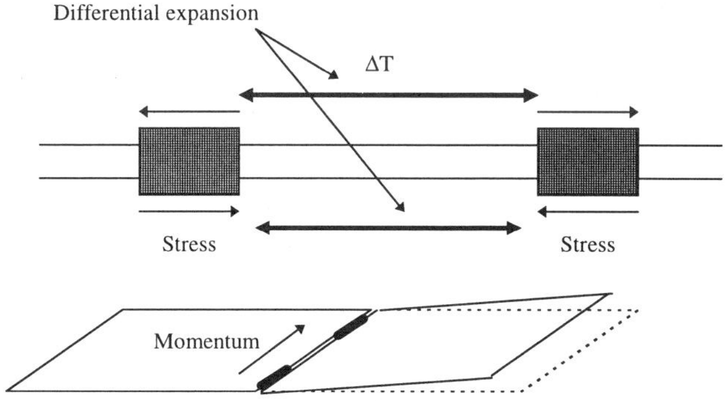


**Figure 4.** Schematic view of the different loads stressing the welded joints.

opposite side of the shield, i.e. with no cooling pipe. A schematic representation of this effect is sketched in the upper part of Figure 4. Moreover, since the helium pipe is connected only to one side of the shield, this causes an asymmetry in the temperature field. This last effect produces a global lateral bowing of the shield during the cooldown that is discussed elsewhere at this conference[4]. The lateral bowing induces a momentum in the direction of the welded joints, as shown in the lower part of Figure 4.

## Differential expansion

The introduction of the finger-cuts in the region of the welded joints modifies the loading of the joints. When two welded shells are kept at different temperatures the joint is essentially loaded by a shear stress status and it is well known that a welded joint is more critical in a shear that in a tension stress status. If the joints are welded over small finger this situation is radically changed. In Fig. 5 we show, as an example, the displacement field of the fingers due to the differential expansion induced by a temperature gradient of 40 K. The deformation is similar to that of an inflecting beam, constrained by two parallel planes at its ends. This situation may be studied analytically and the resulting stress state is shown in Fig. 6. It is interesting to note that the welding is very close to the region where the tensile stress is vanishing.

A more accurate calculation shows that the 6-finger solution is, from a tensile stress analysis, about six times better than a single finger solution. This means that for the same deformation in a 6-fingers case, six times less stress is produced.

The shear stress depends essentially by the section involved in the welding. The six cuts reduce the section by about 10% so the shear stress is increased by the same amount. Nevertheless, in this case the shear component is about one tenth of the tensile component, and can be neglected.

In Figure 7 we present the tensile stress values, as computed by ANSYS, for the case presented in Figure 4. It can be noticed that, due to the partially asymmetric deformation of the fingers, in the welding region the tensile stress, while strongly reduced, is not completely vanishing.

## Momentum on the weldings

Given the applied loads, the one large finger case is similar to the six small fingers case, because the moment of inertia depends linearly with the width of the finger. The significant parameter is the length of the fingers. For the same finger length, a more accurate analysis predicts a 25% momentum reduction for the six finger solution.

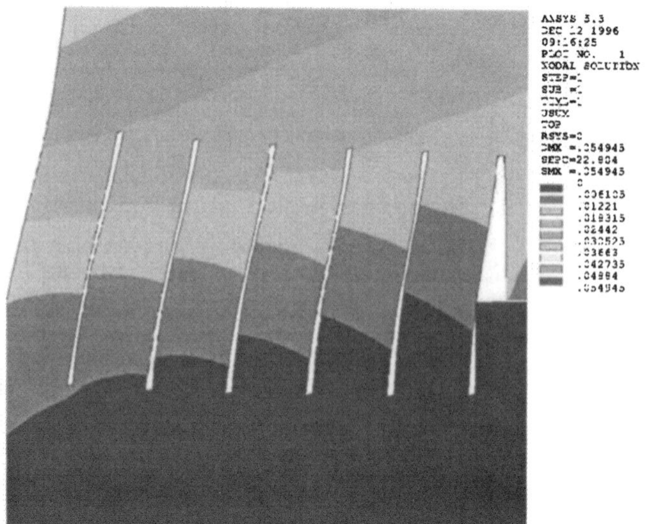


**Figure 5.** Finger deformation due to the differential expansion induced by a temperature gradient of 40 K.

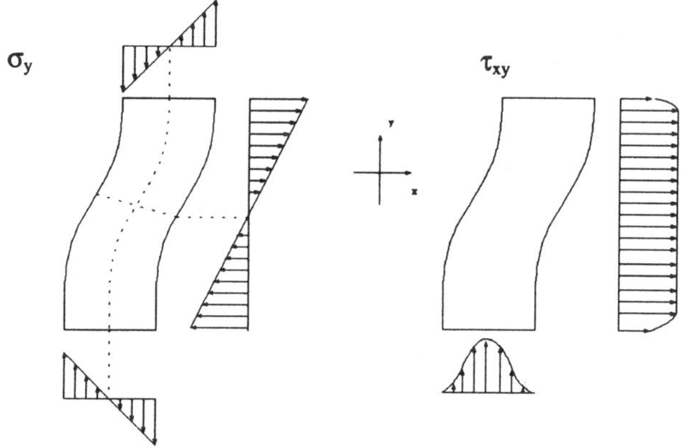


**Figure 6.** Stress state of a biflected beam. Note that the region of the welding is very close to the neutral region of the tensile stress.

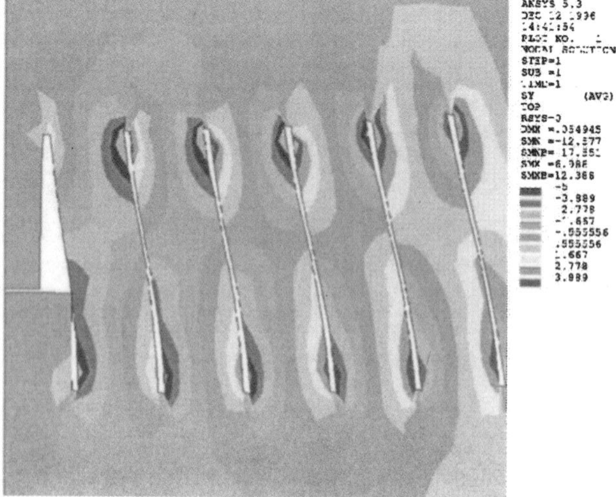


**Figure 7.** Tensile stress in the fingers for the loads of Figure 4. In the welding region the tensile stress is strongly reduced, but non zero.

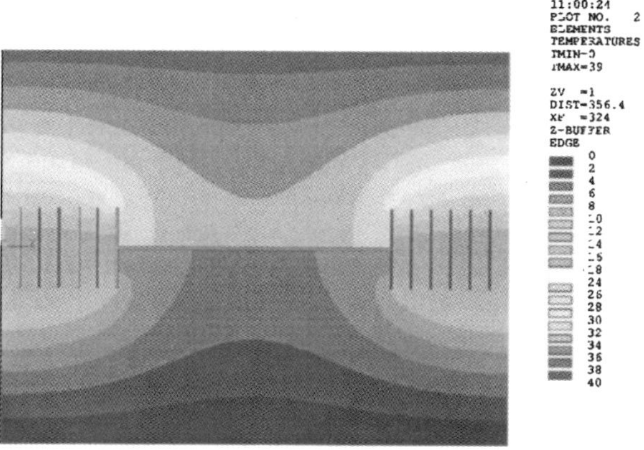


**Figure 8.** Real temperature field across the finger and the welding, in the worst case during the cooldown.

## FINGER SOLUTION IN ITS REAL THERMAL ENVIRONMENT

In another contribution to these Proceedings[4], cryostat shields FEA analysis have been used to determine the shield temperature gradients during the cooldown. In the models used for the cooldown simulations the details of the finger welding geometry were replaced by an equivalent heat transfer section. The temperature field obtained in this approximate model describes properly the finger welded shield at a distance from the fingers.

In order to check that the stresses arising in the finger region are compatible with the tensile characteristics of the Aluminum alloy used for the shield, we have performed a submodeling of this region, using the maximal temperature gradient developed during the cooldown, which has been determined to be less than 40 K for both shields.

In Figure 8 we show the temperature field, with a gradient of 40 K between the upper and lower part of the model, used for the local stress analysis of the joints. To evaluate the differential thermal expansion we have used a temperature independent (and overestimated, since referred to room temperature) expansion coefficient for the Aluminum 1050 alloy, $\alpha = 3 \cdot 10^{-5}$ $K^{-1}$. Fig. 9 shows the Von Mises stress in the finger region. The maximum stress is found to be lower than 2 $10^7$ $N/m^2$, well below the Aluminum alloy yield limit.

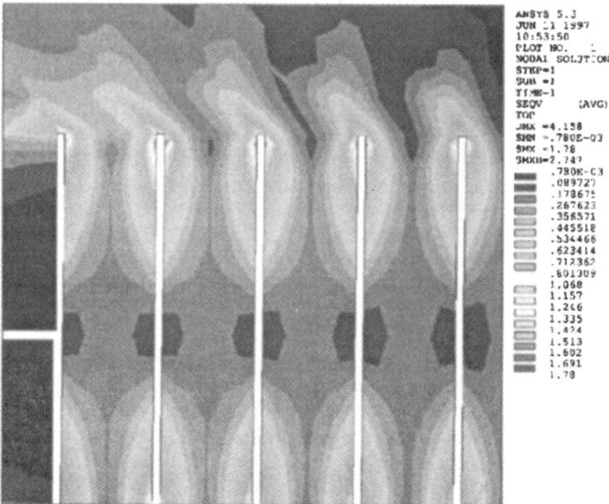


**Figure 9.** Tensile stress field in the finger region in the case of Figure 8.

## FINGER AND WELDING DEFORMATION

Another important aspect that led us to choose the finger solution is the well-known problem of deformations induced by the welding in large welded sheets. In our case this problem is crucial, since the deformation would occur during the final assembly of the shields, when the panels have to be welded. At this point the cavity string has been already attached to the HeGRP and aligned. As it can be easily understood and we have proved on welding samples, fingers have an active role in preventing a general deformation of the shields, induced by the local shrinkage of the welded seams. In fact, the moment of inertia of the fingers is very small compared to those of the connected components, so that the effect of the welded joint shrinkage stay confined into the fingers and is not able to induce global deformations to the rest of the structure.

## CONCLUSIONS

On the basis of the new design outlined above, two cryostats have been ordered, their fabrication is well under its way and the delivery to DESY of the first is due for next October (1997). A cost reduction close to 50%, with respect to the prototype operating in the TTF linac, has been reached. Experience gained at ZANON (in charge for the cryostat fabrication) has already shown a significant simplification in the handling and preassembling procedure.

As shown in this paper, the analysis on the characteristics of the welding procedures suggested a finger welding solution. This solution has been tested on significant samples and performed as expected.

The assembling and aligning procedure has been also revised to take advantages of the new design criteria. Further studies are under way to solve the problem of warm to cold alignment preservation.

## REFERENCES


1. TESLA Test Facility linac - Design Report, DESY Report, March 1995, TESLA 95-01. Conceptual design of a 500 GeV e+e- linear collider with integrated X ray laser facility, R. Brinkmann et al., editors, DESY Report, 1997-048, 1997.
2. F. Alessandria, G. Cavallari, M. Minestrini, T. H. Nichol, C. Pagani, R. Palmieri, S. Tazzari, G. Varisco, Design manufacture and test of the TESLA-TTF cavity cryostat, Adv.Cryo. Engr. 41A:855 ( 1996 ).
3. C. Pagani et al., Construction, commissioning and cryogenic performances of the first TESLA Test facility (TTF) cryomodule, Paper # DF 2, 1997 CEC/ICMC, Portland.
4. D. Barni, C. Pagani, P. Pierini, M. Todero, Cooldown simulations for the TESLA Test Facility (TTF) cryostat, Paper # FF 4, 1997 CEC/ICMC, Portland.

# COOLDOWN SYMULATIONS FOR TESLA TEST FACILITY (TTF) CRYOSTATS

D. Barni, C. Pagani, P. Pierini, M. Todero

INFN Milano - LASA
Via Fratelli Cervi, 201, I-20090 Segrate (MI), Italy

## ABSTRACT

The first TTF cryostat has been tested in June 1997 and monitored to study the deformations produced during the cool-down procedure. A numerical study of the thermal shields of the cryostat has been carried out with finite element analysis simulations. The results allowed to understand the effect of the cooldown on the TTF cryostat.

The new cryostat design, with the helium pipe directly welded to the thermal shields, needs a careful analysis of the cooldown procedure. To avoid damages to the cryostats during the cooldown a proper procedure has been devised and tested. According to the DESY cryogenic system capabilities, we have studied the possibility to use a mixture of cold and hot helium. By varying the relative percentage of the components, the shields will feel a smooth and progressive cooldown. Analysis results show that a single cryomodule can be cooled in about two days without damaging the cryostat.

This paper reports thermodynamic considerations and the results of the analysis performed for the first and second generation TTF cryostats.

## THE FIRST TTF CRYOSTAT INSTALLED AT DESY

The behavior of the aluminum shields of the first TTF[1] cryostat during cooldown has not been taken into detailed account up to now. Even if the entire cooldown procedure will be extremely slow, large thermal gradients in the structure could, in principle, lead to undesirable thermal displacements or intolerable stresses. In this section we show results of coupled thermal/mechanical ANSYS[2] simulations for the two shields (70 K and 4 K) of the TTF cryostats in the case of a fast cooldown, i.e. with the cooling pipe brought immediately at their final operating temperature. This is the worst situation from a mechanical point of view, so in the case of a slower cooldown procedure these results guarantee a safe cryostat operation. The two shields, including the copper braids responsible for heat conduction to the He pipes, has been modeled in a transient ANSYS analysis, using shell elements.

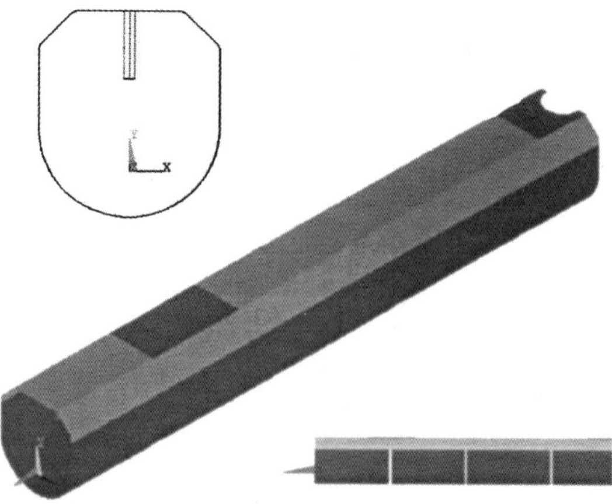

**Figure 1.** The FEM model of the 70 K shield, used in the ANSYS simulations. The figure in the upper left corner shows a front section, with the copper braids connected to the cooling He pipe. The figure in the lower right corner shows a longitudinal view.

## THE FEM MODELS FOR THE TWO SHIELDS

The FEM models for the two shields have been originated from the INFN drawings number 03.03.00 (70 K shield) and 03.01.00 (4 K shield). Longitudinally, only half of the screens has been modeled, for a total length of 5891 mm. Due to the relative length of the shields with respect to their thickness, shell elements have been used for the finite element analysis of the shield structure. The longitudinal positions of the heat conducting copper braids have been approximated with respect to the original CAD drawing, to allow for a mesh with a regular longitudinal step size of 6 cm. The difference of the actual braid positions with respect to the FEM model has been always kept smaller than 3 cm. Figure 1 shows the FEM model of the external 70 K screen.

The copper braids have been modeled in the transient thermal analysis as a series of 3D thermal links between the shield surface and the He pipes. On the He pipe surface a convective load has been applied, in contact with a fixed temperature (70 K or 4 K) bulk fluid. No temperature variation of the fluid in the pipes have been taken into account.

Radiation effects have been included in the FEM model. The outer 70 K shield experiences the thermal radiation of a room temperature environment at 300 K. Thermal losses to the inner 4 K shields have been neglected. The inner 4 K shield experiences the thermal radiation produced by the outer shield, during its cooldown. The thermal losses to the inner 2 K cavity vessel have been neglected. To take into account the effect of the super-insulation blankets a proper surface emissivity has been employed. This effective emissivity has been evaluated from the estimates of the steady state thermal load on the two shields after the cooldown: 0.9 W/m$^2$ for the 70 K shield and 0.1 W/m$^2$ for the 4 K shield.

## MATERIAL PROPERTIES AT CRYOGENIC CONDITIONS AND ASSUMPTIONS OF THE SIMULATIONS

The FEM model uses two different materials types for the thermal analysis: the aluminum alloy (6082) of the shields and the copper (ETP copper) used for the braids. Temperature-dependent material properties have been used for thermal conductivity,

**Table 1.** Temperature independent material properties

| Material Properties | Aluminum 6082 | ETP Copper | Units |
|---|---|---|---|
| Young Modulus | $6.85\times 10^{10}$ | $1.177\times 10^{11}$ | N/m$^2$ |
| Poisson's ratio | 0.3 | 0.3 | |
| Density | 2700 | 8960 | kg/m$^3$ |
| **Super-insulation** | **70 K shield** | **4 K shield** | |
| Surface emissivity | $1.97\times 10^{-3}$ | $5.02\times 10^{-2}$ | |

specific heat and thermal expansion coefficient, whereas density, elastic modulus, Poisson ratio and surface emissivity (for radiation effects) have been considered independent of the structure temperature.

In Table 1 we report the values used for the temperature independent material properties and in Figure 2 we plot the temperature-dependent material properties[3].

## TRANSIENT THERMAL ANALYSIS

A transient thermal analysis has been performed to study the cryostat cooldown when He at 4 K and at 70 K is flown in the LiHe pipe. The copper braids act as thermal filters in the heat diffusion from the shields to the pipe.

**Initial conditions and loads on the model.** The initial temperature for both models has been set to the room temperature of 300 K.

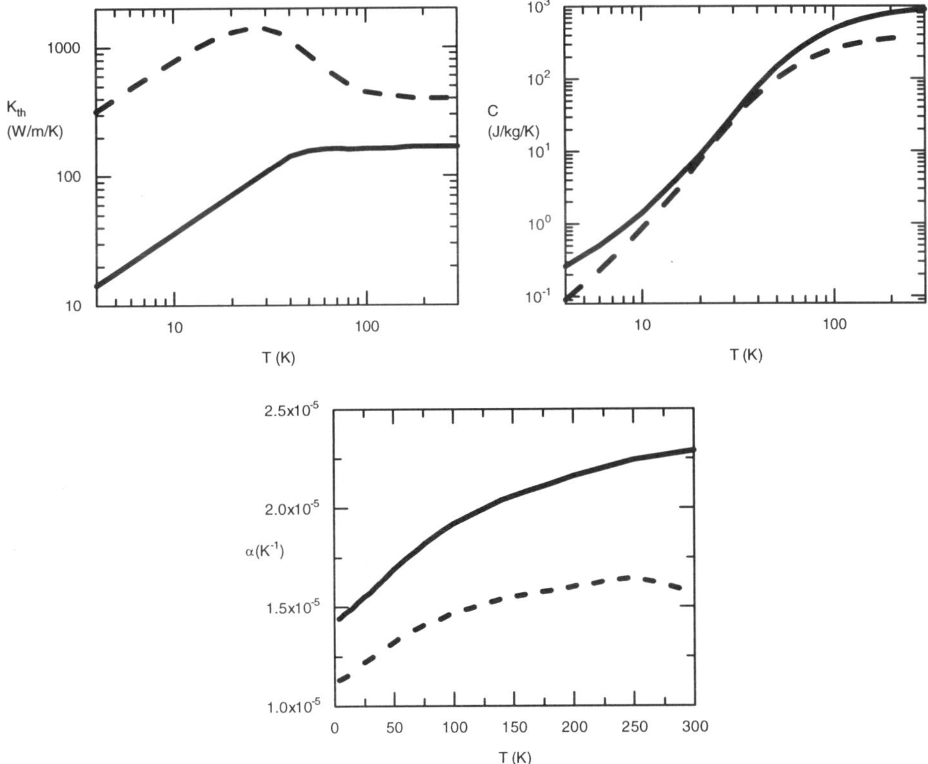


**Figure 2.** Thermal conductivity (upper left plot), specific heat (upper right plot) and thermal expansion coefficient for Aluminum 6082 (solid lines) and ETP copper (dashed lines), as a function of temperature.

**Convective heat exchange.** The heat exchange between the helium flowing in the pipe and the pipe walls is driven by convection effects. To compute the laminar coefficient ($H_f$), which drives the convective heat exchanges, we used the Dittus-Boelter relation:

$$\mathrm{Nu} = 0.026 \cdot \mathrm{Re}^{0.8} \cdot \mathrm{Pr}^{0.4} \tag{1}$$

where Nu is the Nusset number, Re is the Reynolds number and Pr the Prandtl number,

$$\mathrm{Nu} = \frac{H_f \cdot D}{K}, \quad \mathrm{Re} = \frac{\rho \cdot v \cdot D}{\mu}, \quad \mathrm{Pr} = \frac{C \cdot \mu}{K}, \tag{2}$$

$D$ is the equivalent pipe diameter, $K$ is the thermal conductivity, $\rho$ is the density, $\mu$ is the viscosity and $C$ the specific heat.

This relation is a good approximation if Re > 2000. Assuming a He mass flow of about 30 g/s and the viscosity varying according to the Keesom relation ($\mu = 5.023 \cdot T^{0.647}$) we obtain Reynolds numbers always over 2000 in the whole range of operating temperatures.

The convective heat exchange between He and the pipe is then followed by the conductive thermal drop in the braids. Assuming a minimum value of $H_f$=200 W/m$^2$/K (as shown in Figure 3) we found that the convective thermal drop is negligible with respect the conductive drop in the copper braids; thus it has been neglected in our model. Due to this assumption the laminar coefficient has been considered independent of the temperature.

**Loads on the 70 K shield.** The 70 K shield is cooled by 12 copper braids connected to the 70 K LiHe pipe. We assumed that all the room temperature pipe is instantaneously filled with the 70 K LiHe, neglecting any longitudinal temperature gradient on the steel pipe. Thus we have neglected a short (few minutes) time interval necessary for the pipe to reach a uniform temperature along its longitudinal axis. The assumption of a rapid cooldown procedure represents the worst possible operating case, and the results prove that the shields can sustain any smoother cooldown procedure. We used the laminar coefficient $H_f = 200$ W/m$^2$/K, as discussed above, for the heat flow from the braids to the LiHe.

The 70 K shield fits inside the room temperature vacuum chamber, thermally insulated from that by the super-insulation blankets. The radiation from the vacuum chamber, at a constant temperature of 300 K during the shield cooldown, has been taken into account. A steady state thermal load of 0.9 W/m$^2$ (radiation from the vacuum chamber heating the 70 K shield) has been used to derive the effective emissivity of the surfaces that correctly takes into account the super-insulation blankets. ANSYS computes the heat flow from one surface to another with the following formula:

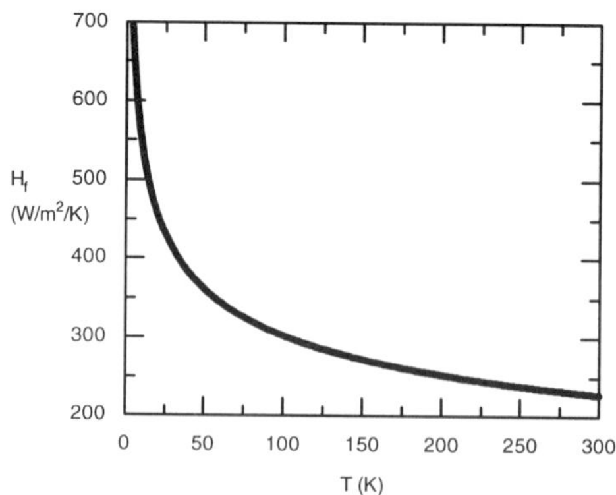


**Figure 3.** Convective heat exchange versus temperature. The coefficient is over 200 W/m$^2$/K which guarantee very small gradients over the surface between Helium and the pipe.

$$H_{12} = \sigma \cdot \varepsilon_1 \cdot F_{12}\left(T_1^4 - T_2^4\right) \quad (3)$$

where $\sigma$=5.6693 $10^{-8}$ W/m$^2$/K$^4$ is the Stefan-Boltzmann constant, $\varepsilon_l$ is the effective surface emissivity ($0<\varepsilon_l<1$) and $F_{12}$ is the view factor from surface 1 to surface 2 (defined as the fraction of total radiant energy leaving surface 1 arriving directly on surface 2). In our model the external vacuum chamber has been modeled as a single node with a constrained temperature of 300 K, corresponding to a surface with unitary view factor $F_{12}$=1. This node is the common extra node of the SURF22 elements used on the shield surface to consider radiation effects. The assumed steady state value of 0.9 W/m$^2$ when $T_1$=70 K and $T_2$=300 K leads to an emissivity of $\varepsilon$=1.97 $10^{-3}$. The emissivity (and hence the effect of the super-insulation blankets) has been considered independent of the structure temperature.

**Loads on the 4 K shield.** The 4 K shield is cooled by 11 copper braids connected to the 4 K LiHe pipe. We have used the same assumptions of the previous model regarding the initial condition of the LiHe pipe in the cooldown simulation. The same conservative value of the convection coefficient has been used also in this case. The computation of the radiation exchange with the 70 K shield needs to take into account the change of temperature of the external shield. For this purpose, the average temperature of the 70 K shield as a function of time has been used as a time dependent constraint for the single node used for the radiation effects in SURF22 elements. Assuming a steady state value of 0.1 W/m$^2$, we have evaluated an effective emissivity of $\varepsilon$=7.32 $10^{-2}$.

## RESULTS FOR THE 70 K SHIELD

Through the thermal analysis we followed the mechanical behavior of the shields during the cooldown. Three key points in the structure are shown in the historical temperature plots: the cold point at the copper braid end in contact with the cooling helium, the intermediate temperature at the top of the braid (by comparison with the cold point we determined the thermal flux through the braids) and the hot point, a reference for the hottest region in the lower part of the shield. It is important to remind that these points are representative of a complete thermal field in the structure. Fig. 4 shows the temporal behavior and the maximum temperature gradient in the shield.

The results show that the 70 K shields reaches the final temperature in about 24 hours. The cooldown is very smooth because the thermal properties of aluminum show a substantial improvement only below 80 K. The maximum gradient is reached after few hours (3.2 hours from the start of the cooldown) and is approximately 60 K. The temperature field in this situation has been used to compute the stresses on the structure, as discussed in the following sections.

## RESULTS FOR THE 4 K SHIELD

The results for the 4 K shield are similar to the results obtained for the 70 K shield and are summarized in Fig. 5. The maximum temperature gradient in the shield decreases faster than the 70 K shield, due to the reduction in the thermal capacity of the aluminum at low temperatures. The cold point temperature has an initial fast decrease, then it start feeling the influence of the thermal capacity of the braids and of the whole shield, and is forced to follow the average temperature of the shield. According to these calculation the cooldown of the 4 K shield will require approximately 10 hours under the assumption that the helium is injected in the cooling pipe immediately at the operating temperature of 4 K.

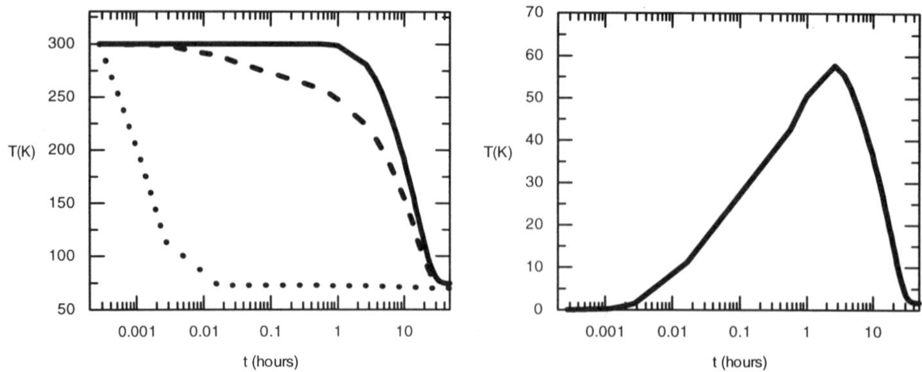


**Figure 4.** Left: Temperature vs. time at different positions in the 70 K shield. The fast decaying curve (dotted) corresponds to the He pipe surface temperature. The two other curves show the temperature at the shield end of the braid (solid) and at the "hot spot" at the bottom of the shield (dashed). Right: Maximum temperature gradient on the 70 K shield, between the shield end of the copper braids and the hot spot at the bottom.

## THERMAL-STRUCTURAL ANALYSIS AND RESULTS

Once the thermal transitory analysis of the shield was performed, we checked the mechanical effect of the temperature gradients on the structure. We performed a thermal-structural analysis of both shields at the time of the maximum thermal gradient. The temperature field from the transient thermal analysis has been applied as an external load to the structural analysis. The nonlinearity of the expansion coefficient of aluminum as been taken into account, feeding its temperature dependence to the simulation code (see Fig. 2 for the expansion coeffient of the materials used).

We were interested in two different aspects from the simulations. First of all, we wanted to know if the deformations produced by the temperature gradients could lead to thermal contacts between shields, causing power loss or, possibly, a mechanical damage to the shield and their superinsulation blankets. As a second aspect, we wanted to check the stress fields produced by the temperature gradient. In Fig. 6 we show the temperature map of the 4 K shield at the time when the maximum gradient develops on the structure.

The results are summarized in Table 2, where UY is the vertical component of the displacement, UZ is the longitudinal component and the last column indicates the

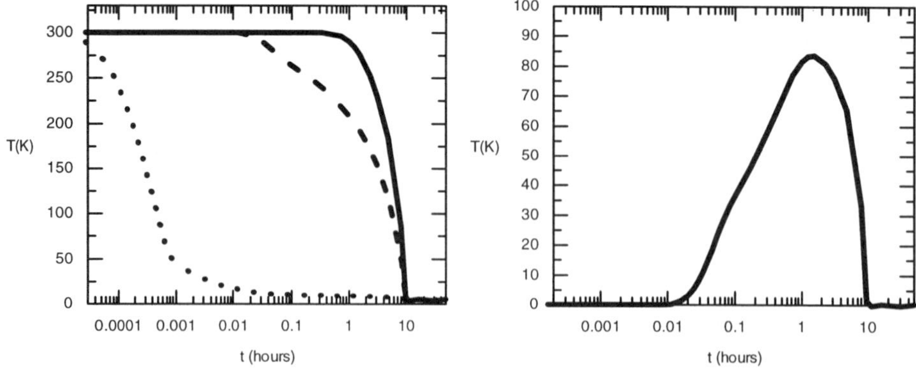


**Figure 5.** Left: Temperature vs. time at different positions in the 4 K shield. For a description of the curves refer to Figure 4. Right: Maximum temperature gradient on the 4 K shield.

**Table 2.** Summary of the results of the thermo-structural analysis of the first cryostat shields

| Shield | Time | UY | UZ | Von Mises Stress |
|---|---|---|---|---|
| 4 K | 3.2h | $6.9\ 10^{-3}$ m | $8.2\ 10^{-3}$ m | $\leq 30\ 10^{6}$ N/m$^{2}$ |
| 70 K | 4.8h | $6.5\ 10^{-3}$ m | $8.5\ 10^{-3}$ m | $\leq 30\ 10^{6}$ N/m$^{2}$ |

maximum Von Mises stress in the structure. The displacement vector has been split in its longitudinal and vertical components to isolate the thermal contraction from the bowing of the shield.

The results of the thermal-structural analysis show that the "stay clear" region around the shields are wide enough to avoid thermal shorts during the cooldown. The "bowing" effect due to the vertical asymmetric cooldown of the shield and induced by the thermal gradient along the structure does not lead to mechanical interference of the shields with the surrounding environment. The stress analysis has shown that the gradient induced by the thermal distribution along the cooldown does not lead to significant stresses in the structure

## THE SECOND GENERATION TTF CRYOSTAT

We then analyzed the second generation TTF Cryostat, as presented in another contribution to these Proceedings[4] Similarly to the first generation cryostats we performed thermal and mechanical analysis with ANSYS, to find the correct cooldown procedure and avoid undesirable deformation in the shields geometry.

In the cryostat for the second generation TTF cryomodule the cooling pipes are welded directly to the thermal shields. This solution reduced the complexity in the design of the cryostat, but has forced to study accurately the behavior of the shield and its welded joints during the cooldown. The first generation cryomodule has been developed and built to resist to any kind of fast cooldown procedure, by employing the copper conductive braids to act as thermal filters to the He pipe. For the second generation cryostat shields we checked a cooldown procedure that avoids undesirable deformations and leads to stresses which are compatible with the tensile characteristics of the materials.

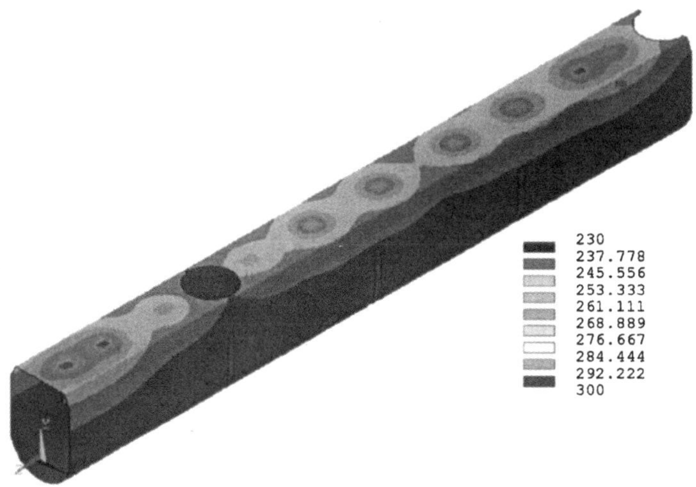


**Figure 6.** Temperature field in the 4 K shield at the time where the maximum gradient is developed in the structure.

## FINITE ELEMENT MODEL AND LOAD FOR THE SHIELDS

The finite element model for the thermal analysis has been derived from INFN drawings TTF2-03.01.00 (4 K shield) and TTF2-03.03.00 (70 K shield). Taking advantage from symmetry only half screen has been modeled, as in the previous case.

For the second cryomodule we have decided to employ the Aluminum 1050 alloy. Material properties at cryogenic conditions has been deduced from literature[3] and fed to the non linear code solver. Similarly to the first cryomodule investigations, we have neglected convective effects between the helium and the cooling pipe, and the heat diffusion in the steel pipes. The experience gained with first the cryomodule studies suggests also to neglect the radiation heat exchange between the shields and the surrounding environment. This assumption slightly offsets the steady state results, but does not influence the transitory history. According to our previous assumptions, we have investigated the thermal transitory by applying temperature constraints directly to the welded regions of the shield panel

In the actual shield design the panels are linked to the cooling pipe by small aluminum fingers, which allow to unload the welded junction from the stress generated during the cooldown. In our numerical model the fingers are not present, but the model considers the real heat transfer section. In this way we can derive some results for the finger region (these aspects are discussed elsewhere[4]). In the actual configuration it is necessary to avoid large gradients (more than 40 K[4]) in the welding regions.

We have decided to use a cooldown procedure where He is injected in the cooling pipe with a linearly decreasing temperature from the room temperature of 300 K to the final temperature of 4 K in 40 hours. Moreover, both shields are cooled simultaneously.

## RESULTS OF THE THERMAL ANALYSIS

The 70 K shield cooling stops when the final temperature of 70 K is reached. In this way the shield temperatures are always very close, and we can neglect radiation effects between the shields. In Fig. 7 and Fig. 8 we show the shield cooldown history, and the maximum temperature gradient in the structures. Our results show that once the cooldown has been started, both shields follow the decrease of the inlet He temperature.

The 4 K shield increases its cooling velocity when its average temperature decreases to approximately 80 K, due to the reduction of aluminum specific heat at temperatures lower than 80 K.

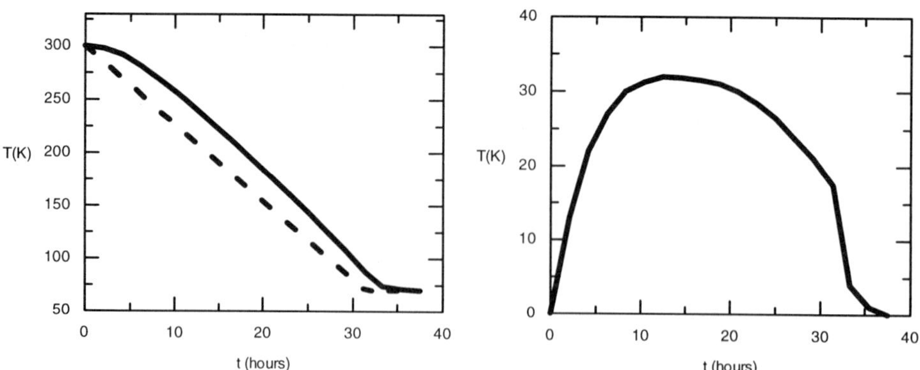


**Figure 7.** Summary of the cooldown for the second generation 70 K shield. Left: The lower (dashed) line represents the temperature of the inlet Helium. The upper (solid) line is the hottest spot in the shield temperature field. Right: the maximum gradient in the whole shield.

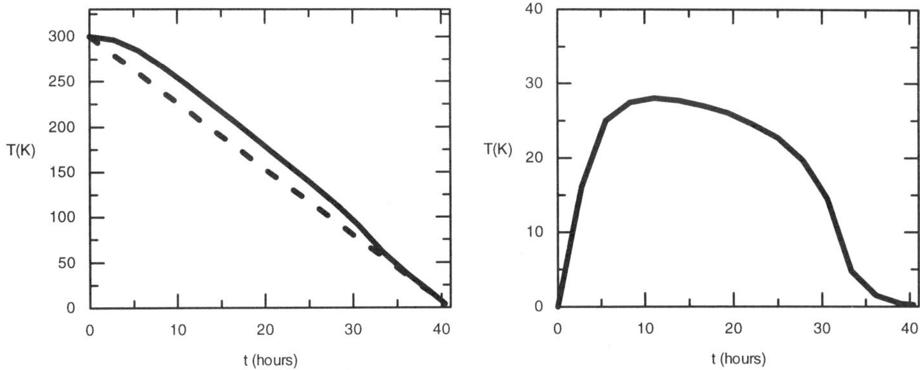


**Figure 8.** The 4 K shield has a similar transitory history as the 70 K shield until it reaches about 80 K. Then the decrease of the specific heat reduces the thermal inertia of the shield, which follows a faster cooldown rate.

The results in Figs. 7-8 show that the maximum temperature in the shields, due to the thermal inertia of the whole shields, decreases at a slower rate than the helium temperature until 12.5 hours from the beginning of the cooldown. Here the shield experiences the maximum temperature gradients. Then this gradient slowly decreases due to the fact that the system won its inertia and that the thermal properties of aluminum gets better.

**Table 3.** Summary of the results of the thermal-structural analysis of the second generation cryostats.

| **Shield** | **Time** | **UX** | **UY** | **Von Mises Stress** |
|---|---|---|---|---|
| 4 K | 12h | 4.8 $10^{-3}$ m | 5.5 $10^{-3}$ m | $\leq$ 30 $10^{6}$ N/m$^2$ |
| 70 K | 12.5h | 5 $10^{-3}$ m | 6.5 $10^{-3}$ m | $\leq$ 30 $10^{6}$ N/m$^2$ |

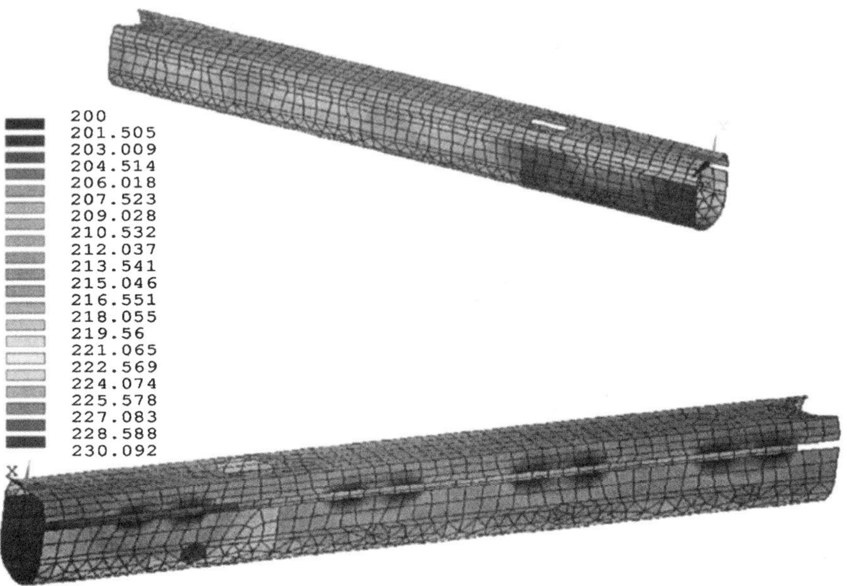


**Figure 9.** Temperature field on 4 K shield after 10 hours from the beginning of the cooldown.

### Temperature Field near the cooling pipe

The results shown in the previous paragraph summarize the behavior of the entire shields at two positions which are good representatives of the entire thermal field. However, the regions near the fingers linking the cooling pipe to the shield panels need to be analyzed separately, since these details are not included in the FEM model presented here. To understand the thermal effect of the fingers we examined the thermal fluxes flowing from the pipe to the shield. From the simulation results we computed the thermal fluxes through the finger regions. Taking into account the finger geometry, the computed heat flow of 5 W/mm$^2$ will cause a gradient of about 10 K across the finger. This gradient adds to the gradient evaluated in our simulation, which never exceeds 9 K, giving a good margin to the critical region of the finger (around 40 K).

## THERMAL-STRUCTURAL ANALYSIS OF THE SHIELDS

Once we guaranteed that the welding regions will be safe during a proper cooldown procedure (as the one described here), we checked that the gradient in the shields do not cause excessive deformations. We used a similar model to the one used for the thermal analysis, but we corrected the thickness of the shell elements in the region of the pipe to give the same moment of inertia of the real pipe. We then applied to both shields the thermal field at its maximum gradient (about 10 hours after the beginning of the cooldown) and calculated the deformations of the shields. In Fig. 9 we plotted the thermal field of the 4 K shield at its maximum gradient. The results obtained show an asymmetric thermal field which has a neutral region exactly at the joint of the two lower panel.

This asymmetric temperature distribution bows both shields sideways, without compromising the stay clear region for safe operation. In Table 3 we list the thermal-structural analysis results, indicating the maximum Von Mises stress, and the maximum deformations in the vertical (UY) and lateral (UX) directions.

## CONCLUSIONS

In this note we have described the results of thermal-structural calculations of the behavior of the first generation and second generation TTF cryostats.

We computed the thermal transient behavior of both the 4 K and 70 K shields and we found that the maximum gradient in the temperature distribution never exceeds 80 K (first cryostat) and 30 K (second cryostat). These gradients produce deformations which are compatible with the geometry of the cryostat and with the tensile characteristic of the aluminum used to build the shields. It is important to note that in the case of the first generation TTF cryostat we considered the situation of a very fast cooldown, which is the worst situation from a mechanical point of view. For the second generation TTF cryostats, where the cooling pipes will be welded to the shields, a smooth cooldown procedure has to be employed in order to prevent mechanical problems in the structures. Through the simulations we checked also the safety margins for the welded joints.

## REFERENCES


1. TESLA Test Facility linac - Design Report, DESY Print, March 1995, TESLA 95-01;
   C.Pagani et al., Construction, commissioning and cryogenic performances of the first TESLA Test Facility (TTF) cryomodule, Paper #DF 2, 1997 CEC/ICMC, Portland.
2. ANSYS is a registered trademark of SAS IP.
3. J.E. Jensen, R.B Stewart, W.A. Tuttle, Selected Cryogenic Data Notebook, BNL, 1967.
4. C. Pagani, D. Barni, M. Bonezzi, P. Pierini, J.G. Weisend II, Design of the thermal shields for the new improved version of the TTF cryostat, Paper #FF 3, 1997 CEC/ICMC, Portland.

# NOVEL CONCEPT OF CONCENTRATOR OF MAGNETIC ENERGY

V.R.Sobol, O.N.Mazurenko, A.A.Drozd, and B.B.Boiko

Institute of Solid State and Semiconductor Physics NASB
Minsk 220072, Belarus

## ABSTRACT

This paper presents a novel way of improving the magnetic system arrangement and efficiency via application of the Hall drift for the generation of additional magnetic field and an increase in the magnetic energy density. Such an increase may be realized with the use of a hyperconducting pure aluminum concentrator installed into a traditional solenoid. The Hall drift stimulated by the Lorentz force in crossed electric and magnetic fields has been investigated as a previously ignored source of magnetic field. The properties of the Hall drift of electrons and the generation of an associated self magnetic field are examined with aluminum conductors. An experiment on aluminum disk-like conductors with radial current between the inner and outer concentric contacts has been performed in an axial magnetic field. An analytical study has been made on the fundamental relations of charge flow in electric and magnetic fields under appropriate boundary conditions. The benefits and advantages of the application of the Hall drift magnetic field are discussed for the cryogenic temperature range.

## INTRODUCTION

The problem associated with the design of effective cryogenic energetic devices and systems further extends the requirement for a geometrical optimization for containing most of the magnetic field in the same volume for a given energy supply. The high conductivity of pure metals at low temperatures ensures a strong outside layer of alternating current that is already at low frequency. The skin depth in such a conductor fabricated of pure aluminum with an *RRR* of 10000 (resistivity at liquid helium is $2\times10^{-10}$ Ohm·cm) at a frequency of 1 Hz is only 0.1 cm. Thus, such metals as Cu, Al, Bi, and In may be utilized to increase the energy concentration of the nonstationary magnetic field by means of increasing the local magnetic field strength. Actually, a nonstationary magnetic field flux through a contour A with a cross section $S_A$ may be transformed into the field flux through contour $B$ with a smaller cross section $S_B$. The magnetic field magnitude $H_B$ in contour $B$ will be obtained from $H_B = S_A H_A / S_B$ where $H_A$ is a magnetic field in contour $A$. A cylindrical hollow conductor having a radial split through out its length is a concentrator of the magnetic field energy. The phenomenon of such a concentration may be applied in cryogenic asynchronous electrical motors for exciting the alternating magnetic field in stator coils. The weakest part of such a concentrator is a split which decreases the mechanical hardness of the active element.

For stationary magnetic fields there is still another mode of magnetic field increase, namely the phenomenon of the Hall drift in a transversely unlimited conductor. In this paper we describe a novel concept of hyperconducting concentrator fabricated of high purity normal metal. This concept is based on the Hall drift of carriers stimulated by the Lorentz force in crossed electric and magnetic fields. It has been shown that the transverse Hall drift may occur in a cylindrical conductor placed in a coaxial external magnetic field when the current is made to flow between the inner and outer concentric leads.[1-3] In this case the cylindrical active conductive element with no split may be represented as a magnetic field amplifier or concentrator because additional movement of the carriers generates excessive magnetic field for the same energy supply.

## EXPERIMENTAL AND ANALYTICAL PROCEDURE

The present investigation is both experimental and analytical. Aluminum was the material used in this study. The specimens used in the study were disks of different thickness (0.5 to 4 mm) and inner and outer diameters of 3 to 6 and 28 to 36 mm, respectively. The aluminum used had a polycrystal structure with an *RRR* of approximately 10000. The investigation were made in a regime of current supply where the maximum value of current density flowing through the sample between the inner and outer concentric contacts achieved a value of 1600 A/cm$^2$. An external magnetic field with a magnitude up to 85 kOe was directed normally to the end surface of the disk shape sample. Samples were placed in liquid helium with heat removal by vaporization of the liquid. A thermostat with automatic monitoring of the temperature was used for temperature measurements.[4-6]

Data acquisition involved determination of the integral magnetoconductivity, magnetoresistance, and self magnetic field in the disk center with an aid of Hall converters. Self magnetic field $H_s$ was obtained as a difference between the total magnetic field $H$ and the external field $H_0$ generated by the superconducting solenoid.

An analytical investigation has been made to provide basic relationships for current flow in a conductor having a cylindrical geometry. The main feature of this approach involves an equipotentiality of azimuthal contours due to the symmetry of the concentric inner and outer current contacts. This equipotentiality takes place both in the absence and in the presence of an external magnetic field for the polycrystalline material. To estimate a form of the current flow we used a charge discontinuity condition and a relation that connected the magnetic field with the current density as given by

$$\frac{1}{r}\frac{\partial}{\partial r}(rj_r) + \frac{\partial}{\partial \theta} j_\theta + \frac{\partial}{\partial z} j_z = 0; \quad \nabla \times \boldsymbol{H} = \frac{4\pi}{c}\boldsymbol{j}; \quad j_i = \sigma_{ik} E_k \tag{1}$$

Here $\boldsymbol{j}$ is a current density vector, $\boldsymbol{H}$ a magnetic field strength vector, $c$ a light velocity, $j_i$ and $E_k$ components of the current density and electric field in cylindrical coordinates, respectively. The conductivity and resistivity tensor components, $\sigma_{ik}$ and $\rho_{ik}$, were also used in the analysis.

## RESULTS AND DISCUSSION

### Theoretical Study

Results of the analytical investigation show that the Hall drift of carriers in such a cylindrical conductor is determined by the type of Fermi surface encountered, i.e., by the degree of electron and hole concentration, and by the quantity of carrier on open directions of the Fermi surface. Open directions indicate those electron trajectories where a movement is infinite and almost straight.

By assuming the absence of current flow in the $z$ direction, the current density of the Hall drift that is an azimuthal component $j_\theta$ is given by

$$j_\theta = -\frac{\rho_{\theta r} j_r}{\rho_{\theta\theta}} \tag{2}$$

For the compensated materials the conductivity tensor components are

$$\sigma_{rr} \sim \sigma_{\theta\theta} \sim \frac{1}{(\omega\tau)^2}; \quad \sigma_{r\theta} \sim \sigma_{\theta r} \sim \frac{1}{(\omega\tau)^2}; \text{ and } \sigma_{zz} \sim \sigma_0 \tag{3}$$

Here $\omega$ is a cyclotron frequency, $\tau$ the relaxation time, and $\sigma_0$ the conductivity in zero magnetic field. Under these conditions, the Hall drift current density is $j_\theta \approx j_r$.

For an open Fermi surface (for example, the open surface is along the $r$ direction) $j_\theta \sim \omega\tau j_r$ because $\sigma_{rr} \sim (\omega\tau)^{-2}$, $\sigma_{\theta\theta} \sim const$, and $\sigma_{r\theta} \sim (\omega\tau)^{-1}$. But, if the direction of the open surface is along the $\theta$ direction, then $j_\theta \sim j_r(\omega\tau)^{-1}$. In a polycrystalline structural material with an open Fermi surface due to averaging the different electron directions, the transverse resistivity component is $\rho_{rr} \sim \rho_{\theta\theta} \sim \omega\tau$. Thus, the transverse current Hall density drift is of the same order of magnitude as the radial current density; that is, $j_\theta \sim j_r$.

For a closed Fermi surface, $\sigma_{rr} \sim \sigma_{\theta\theta} \sim (\omega\tau)^{-2}$ and $\sigma_{r\theta} \sim \sigma_{\theta r} \sim (\omega\tau)^{-1}$. Accordingly, $j_\theta \sim \omega\tau j_r$, which makes this type of electron structure most suitable for developing a concentrator of magnetic energy. For this reason a high purity aluminum with a polycrystalline structure was chosen for the experimental modeling of magnetic field concentration.

### Approximation of Long Concentrator

The magnetic energy level depends on the total length of the cylindrical element.[7] Thus, an analytical study of energy concentration has been completed which involved two concentrator geometries, namely, the long concentrator and the short one. The long concentrator is a cylindrical element having a length that is much larger than the outer diameter. For such a conductor only the axial component of the magnetic field may be taken into account. Total magnetic field in the center of the concentrator is

$$H = H_0\left(\frac{r_2}{r_1}\right)^b, \quad b \approx \frac{2e\tau}{mc^2} I' \tag{4}$$

where $H_0$ is an external magnetic field, $r_1$ and $r_2$ the inner and outer radii, and $I'$ a linear current density per unit of length.

This relation is only valid for the uncompensated material exhibiting a closed Fermi surface. Strong dependence of the magnetic field on the radius is a result of the magnetic nonlinearity during strong Hall drift when the azimuthal current density is much higher than the radial one. For metal with a $\tau = 10^{-10}$ sec, parameter $b \approx 10^{-3}I'/3$, when $I'$ is given in A/cm.

### Short Concentrator

The other limiting case of concentrator geometry, namely, the short disk-like conductor is characterized with its small length when compared with its diameter. In this geometry, the azimuthal and radial magnetic field components and their derivatives cannot be ignored. The calculation of the axial magnetic field component occurring in the sample center has to be completed with a superposition of the magnetic fields for each one of the current elements. Thus, the self magnetic field in the disk center is

$$H_s = H_0 \frac{2I'R}{c\rho_{\theta\theta}} \left[ -\ln\left( \frac{r_1}{r_2} \frac{\frac{h}{2} + \sqrt{r_2^2 + \left(\frac{h}{2}\right)^2}}{\frac{h}{2} + \sqrt{r_1^2 + \left(\frac{h}{2}\right)^2}} \right) \right] \tag{5}$$

where $I' = I/h$, $h$ is the thickness of the short cylindrical conductor, and $R$ the Hall coefficient.

### Experimental Data

Experimental work has been completed for the conductor with the geometry closed to the short variant. Figure 1 represents the magnetic field increase in the conductor center as function of the ratio of the total magnetic field $H$ to the external magnetic field $H_0$. These data are shown as functions of the averaged radial current density $j_r$ given as

$$j_r = \frac{I}{\pi h(r_1 + r_2)} \tag{6}$$

Since radial current flows by means of an external supply, the current density is an independent parameter. The total to external field ratio indicates the degree of magnetic field increase. It is interesting that the triangles corresponding to higher external magnetic fields have less magnitude at other similar conditions. The reason for this phenomenon is due to a behavior of the transverse resistivity in the magnetic field whereas the Hall coefficient is a fixed parameter. Also, a nonmonotonic behavior at high external magnetic field is observed. The source of such a dependence has to be related to the additional increase of the transverse resistivity.

Self magnetic field energy density as a function of external field is shown in Fig.2. Self field energy is an additional energy similar to that of the external field. It is shown that the character of energy density growth at a radial current density of 600 to 700 A/cm$^2$ is the most promising at an external field of 80 kOe as noted by the higher slope of a curve. For an external field of 40 kOe, the prospective radial current density is 1200 to 1400 A/cm$^2$. The saturation and nonmonotonic character of the curves is observed in Figures 1 and 2. This indicates that some mechanism exists which restricts the magnitude of the self magnetic field.

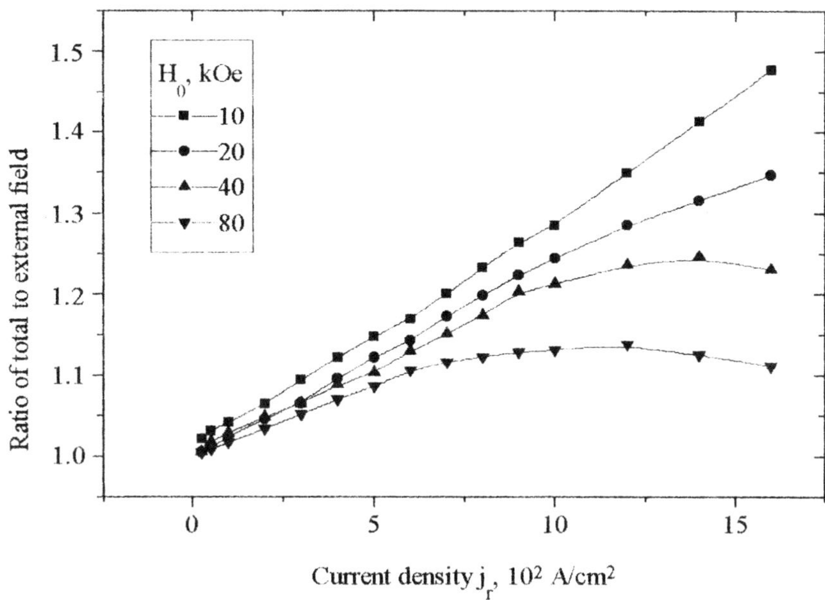


**Figure 1.** Ratio of the total magnetic field to an external one, $H/H_0$, versus the radial current density for the external field from 10 kOe up to 80 kOe.

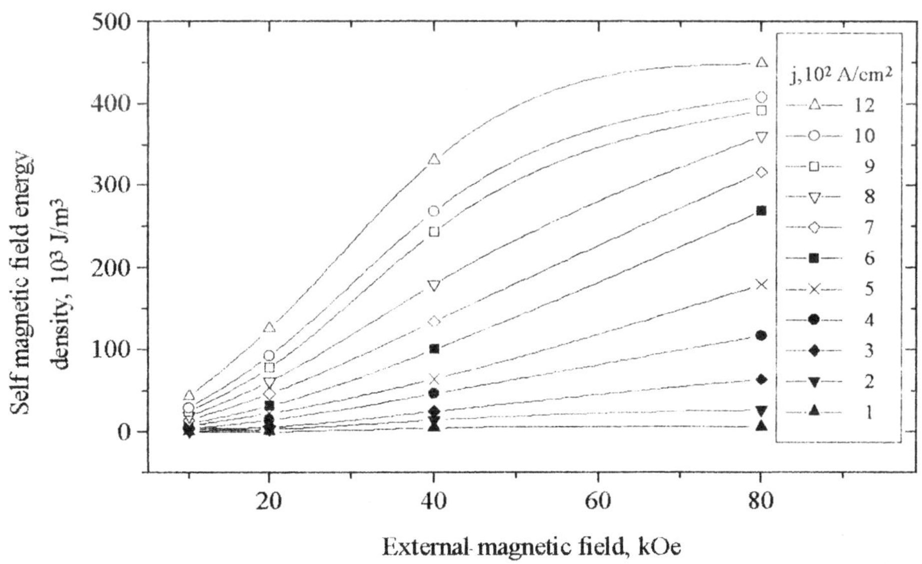


**Figure 2.** Self magnetic energy density in the disk center as a function of the external magnetic field for a radial current density from 100 to 1200 A/cm$^2$.

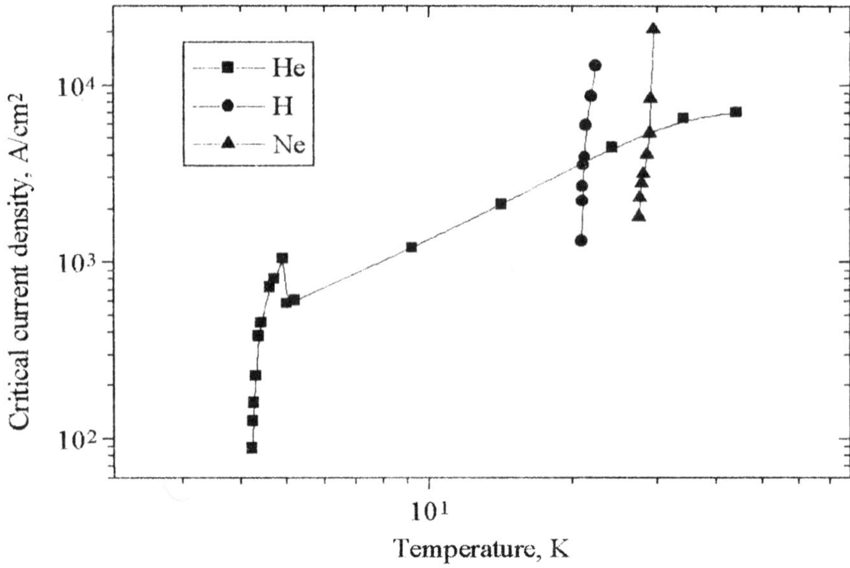


**Figure 3.** Critical current densities for different types of boiling for helium, hydrogen, neon.

Such limitation must be associated with the conductor thermal state, that is with dissipative processes in volume. In other words, an increase in conductor temperature occurs and as a result the azimuthal current density becomes smaller. The range of current density through the conductor is determined by the boiling curve of the cryogenic liquids. In this experiment we used liquid helium. The data show that at large current density the temperature of the conductor surface was higher than 4.2 K. Thus when the current density achieves a definite level, the helium can not provide a sufficient cooling and the temperature increases in accordance with the conditions of boiling. We have calculated a critical current density, based on our data, for three cryogenic liquids (helium, hydrogen and neon). Figure 3 shows the critical current density for the aluminum concentrator with the previously noted purity and structure versus the conditions of cooling. The data have been calculated based on the results obtained upon boiling He, H and Ne on the aluminum surface at different temperatures. The heat balance equation used is given by

$$\rho j^2 - \mathrm{div} q = 0 \tag{7}$$

where $q$ is the heat flux density vector through the conductor surface.

The critical current density denotes a maximum current density that may be supplied to the concentrator without a further temperature increase. The temperature dependence of the resistivity associated with a magnetic field magnitude of 80 kOe was taken into account in the calculation.[4]

## CONCLUSION

A steady magnetic field concentration may be realized at cryogenic temperatures based on the strong electron Hall drift occurring in crossed electric and magnetic fields in hollow cylindrical conductors with an axial external magnetic field in which radial current is forced to flow between the inner and outer cylindrical contacts.

The extent of the magnetic field concentration is determined by the geometry (the most favorable is the long cylinder approximation), dispersion law, crystal structure, temperature and purity of the concentrator material.

Dissipative processes in hyperconductors restrict the range of concentration; however, the experimentally observed level of magnetic field increases by 50% even for the more unprofitable short construction. An increase in magnetic energy density of about $5\times10^5$ J/m$^3$ has been obtained.

Large helical toroidal magnetic systems supplied with currents on the order of 10 kA/cm$^2$ are suitable for installation in the cavity of the long variant of the concentrator. As a result, the power law of magnetic field increase may be achieved under moderate heat conditions.

## ACKNOWLEDGMENT

This research is supported by the Belarusian Republican Fund of Fundamental Investigations (Grant No. F96-204).

## REFERENCES

1. D.A.Kleinman and A.L.Shawlow, Corbino disk, *J.of Appl.Phys.* 31:2176(1960)
2. E.K.Inall, A.E.Robson, and R.J.Turchi, Application of the Hall effect to the switching of inductive circuits, *Rev.Sci.Instrum.* 48:462(1977)
3. J.A.M.M.van Haaren, G.J.C.L.Bruls, A.P.van Gelder, H.van Kempen, and P.Wyder, Thickness variations and the Corbino effect, *Phys.Rev.* 34:6813(1986)
4. B.B.Boiko, V.I.Gostishchev, A.A.Drozd, V.S.Kuzmin, and O.N.Mazurenko, Corbino effect in aluminum, *Fiz.Met. i Metalloved.* (Russian). 63:1133(1987)
5. B.B.Boiko, V.R.Sobol, O.N.Mazurenko, and A.A.Drozd, Magnetosensitive conductivity of aluminum and the advantage of Corbino geometry, in: "Advances in Cryogenic Engineering (Materials)", Plenum Press, New York (1996), v.42, p.1063
6. B.B.Boiko, V.R.Sobol, O.N.Mazurenko, and A.A.Drozd, Magnetism due to Hall current in aluminum and copper conductors, in: "Proceedings of ICEC16/ICMC", Japan (1996) (in press)
7. B.B.Boiko, V.R.Sobol, and O.N.Mazurenko, Magnetic field of cyclic current in cylinder conductors, in: "Proceedings of ICEC16/ICMC", Japan (1996) (in press)

# STABILITY ESTIMATE FOR CICC WITH COOLING CHANNEL USING ONE- AND TWO-FLUID CODES

S. De Palo, [1] C. Marinucci, [2] and R. Zanino [1]

[1] Dipartimento di Energetica, Politecnico
I-10139 Torino, Italy
[2] EPFL, CRPP-Fusion Technology Div.
CH-5232 Villigen PSI, Switzerland

## ABSTRACT

Cable-In-Conduit Conductors (CICC) with central cooling channel have been recently tested in the QUench Experiment on Long Length (QUELL) at the Sultan facility, and are foreseen for the Toroidal Field (TF) and the Central Solenoid (CS) superconducting magnets of ITER (International Thermonuclear Experimental Reactor). A careful thermal-hydraulic analysis of this type of cables is needed in the design of magnets with a reasonable safety margin.

To the purpose of a more accurate description of the physics in the 2-channel topology, the 2-fluid code Mithrandir [1] was recently developed from the 1-fluid code Gandalf. [2] Mithrandir allows different thermodynamic state of the helium that flows in the cable bundle region and of the helium in the central channel. On the contrary, the homogenization intrinsic in a 1-fluid model can result, among others, in a significant overestimation of the stability margin.

Both Gandalf and Mithrandir have been very recently benchmarked against quench propagation data from QUELL. Here, we analyze the stability of QUELL, and a first comparison with experimental stability data for this type of conductor is presented, using a few measurements performed on QUELL.

Some stability calculations for an ITER CS conductor are then performed with the Mithrandir code, considering the influence of the 2-fluid coupling on the stability margin.

## INTRODUCTION

AC losses, cable movements, nuclear processes, release of mechanical energy through friction and cracking are some of the energy inputs that the cable must absorb without quench. The maximum quantity of energy that the cable can absorb defines the stability

*Advances in Cryogenic Engineering*, Vol. 43
Edited by P. Kittel, Plenum Press, New York, 1998

margin of the cable; a successful design of the magnet corresponds to the achievement of operating conditions that avoid quenches of the super-conducting cable.

In a CICC with an additional cooling channel, the latter mainly provides a low impedance path for the coolant, and a reservoir to limit pressure increase during a quench. No quench or stability data were available for this type of conductor until the QUELL experiment. In the meanwhile, numerical simulations for the ITER conductors were performed,[3] checking the effects of different physical (heat load distributions, heat transfer coefficients, etc.) and numerical (mesh size, numerical model, etc.) characteristics on the calculated stability margin. Very recently, finally, simulations with the POCHI code [4] were performed for the analysis of QUELL stability.

In the present stability study we use two different codes - Mithrandir [1] and Gandalf, [2] with the same set of common input parameters, helium and material properties, thermal-hydraulic correlations etc. Both codes describe 1-D transient thermal-hydraulics in a CICC with an additional cooling channel, solving Euler type equations for the helium coupled to heat conduction in the conductor and in the jacket. In both codes the helium flows at different speed in the region where the cables are present (so-called cable bundle, **B** in the following) and in the central channel (so-called hole, **H** in the following). The principal difference between the two codes is that the pressure p and the temperature T in the **B** and **H** regions are different in Mithrandir, whereas they are assumed to be the same in Gandalf. Therefore, only Mithrandir retains the finite equipartition times of temperatures and pressures in the **B** and **H** regions; Gandalf, on the contrary, assumes instantaneous thermal-hydraulic coupling between the two regions. Two major "free" parameters in Mithrandir control the **B/ H** coupling: the fraction F of the **B/ H** interface which is perforated, and the heat transfer coefficient $h_{HB}$ at the no-wall interface between the two regions. The latter parameter comes from a simplistic modeling of this complex phenomenon, [1] and we artificially enhance it by a multiplier $H_{nowall}$. The POCHI code mentioned above is a simplified 1-fluid code considering only the total mass flow **B+H**.

Both Gandalf and Mithrandir assume the current to be uniformly distributed among the strands. Since we mainly concentrate here on the effects of the coupling between **B** and **H**, the possible effects of a non-uniform distribution of the current on the stability margin [3] have not been considered.

Here we shall present a stability analysis of the QUELL conductor, which will be supplemented by a comparison with experimental data. It should be emphasized that this is not meant as a validation of the codes strictly speaking, because of the very limited amount of experimental data available. Still, we consider it valuable to analyze, from the point of view of stability, a conductor that was used already for a detailed validation both of Mithrandir and Gandalf in the context of quench initiation and propagation. [5-7]

In perspective, the interest in two-channel CICC is strongly related to the development of the ITER project. As a first step in the direction of a more detailed study, the effect of the perforated fraction F on the stability of an ITER CS conductor has also been analyzed with the Mithrandir code.

## STABILITY ESTIMATE FOR THE QUELL CONDUCTOR

In QUELL, a series of over 60 experimental runs was conducted to study quench initiation and propagation in a ~ 100m long CICC with central cooling channel. [4] A few data on the stability of the conductor were also collected, [4] some of which will be used here.

The cooling channel in QUELL is lined by a thin (~ 0.7mm) strip, wound into a helix such as to leave a free passage to and from the annular **B** region. A detailed validation exercise [5-7] against QUELL quench data has suggested to use F=1% and $H_{nowall}$=10 as reference values in Mithrandir, for which very good agreement with the experiment was found over a broad range of experimental conditions. These values, unless otherwise noted, will consistently be used also for the present study. All details on the QUELL conductor are given elsewhere.[4,6]

For all cases analyzed here the external magnetic field has a given modulation[4] and a maximum of 9T. The boundary conditions are constant in time, with inlet pressure of 0.6MPa, inlet temperature of 8.2K, inlet mass flow rate of $4gs^{-1}$. The inductive heater is 0.12m long, and is located at the middle of the cable length. Pulses of 20ms have been simulated, with different space and time shapes (see below).

The adaptive mesh used for the simulation starts with a uniform grid and a fixed refined zone (0.36m long) around the heater with a spacing $\leq$ 1.2 mm. The spatial convergence of the Mithrandir simulations has been checked doubling the number of elements in the refined zone, because in the simulations the normal zone has always been shorter than the fixed refined zone. Within an accuracy of some percent, the computed stability margin did not change.

## Results for the case of square wave external energy input

A first set of numerical simulations has been performed assuming a square wave energy input profile into the strands. Although this pulse shape was not used in the experiment (see next Section) it is still useful to compare also with this case because the square wave shape is typically used in the design of an experiment, e.g., ITER (see below). For this case an adaptive time step was used, starting from the value of 10μs.

Here and in the following, Q (W/m) is the linear external input power, $E_{ext}$ (J) is the total energy input into strands + jacket as computed by the code and SM (mJ/ccst) is the energy deposited into the strands divided by the effectively heated volume of strands. This volume obviously depends on the spatial shape of the external input power (see below), and is estimated as $L_h$ ($A_{sc}$ + $A_{Cu}$), where $L_h$ is the effective heater length, and the A's are the cross section areas of super-conductor and copper respectively.[4] For the case at hand, $L_h$ = 0.12m.

The results of the simulations are reported in Table 1 for Gandalf and in Table 2 for Mithrandir, where $Q_{st}$ = % of Q going into the strands, $Q_{jk}$ = % of Q going into the jacket, and $Q_{st}$ + $Q_{jk}$ = 100. In the fourth and fifth column the first figure is the upper limit for recovery, while the second figure is the lower limit for quench; the relative width of this window (i.e., the accuracy in the SM estimate) is of the order of some %. Different operating currents I were considered in the simulation.

We notice that for all current values I the 2-fluid results are well below the 1-fluid estimations, i.e., the 2-fluid estimation is conservative. The reason for this is that in the 2-fluid case the enthalpy of the helium in **H** is not immediately available for receiving the

**Table 1**. Gandalf stability margin (square wave input)

| Run # | I [kA] | $Q_{st}/Q_{jk}$ | $E_{ext}$ [J] | SM [mJ/ccst] |
|---|---|---|---|---|
| 1 | 8 | 100/0 | 13.3-13.48 | 972-985 |
| 2 | 10 | 100/0 | 9.46-9.62 | 691-703 |
| 3 | 12 | 100/0 | 6.10-6.22 | 446-454 |

**Table 2.** Mithrandir stability margin (square wave input)

| Run # | I [kA] | $Q_{st}/Q_{jk}$ | $E_{ext}$ [J] | SM [mJ/ccst] |
|---|---|---|---|---|
| 1M | 8 | 100/0 | 9.70-10.18 | 709-744 |
| 2M | 10 | 100/0 | 6.06-6.67 | 443-487 |
| 3M | 12 | 100/0 | 3.82-3.88 | 279-283 |

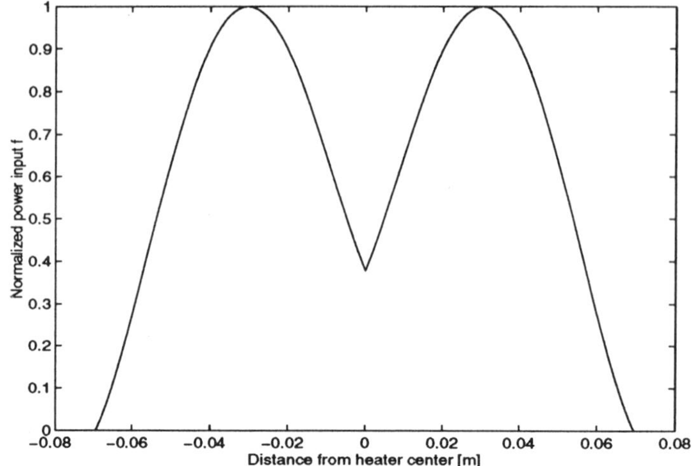


**Figure 1.** Spatial distribution of normalized external input power.

heat, because of the finite time constant in the B/H coupling. In the 1-fluid model, on the contrary, all the helium is assumed to be immediately available, and the stability margin artificially results to be higher. Both codes predict decreasing margin for increasing operating current, as expected.

## Results for the case of time and space dependent external energy input

A second set of numerical simulations has been performed using a space and time dependent external energy input $Q(x,t) \propto f(x)*g(t)$, where $g(t) = \sin^2(2\pi\omega t)$ with a frequency $\omega = 590$ Hz. For these simulations a constant time step of 10μs has been used during the heating period. The space dependence given by f(x) is shown in Fig.1, and was deduced from Ref.4. The resulting effective heated length is $L_h = 0.139$m. This pulse shape was used in the experiment, and both f and g have been measured in QUELL.[4] The results for this case have been summarized in Tables 3-5.

An additional parameter must be considered in the simulations of the case at hand, reflecting the fact that the external inductive heater will generally heat also the jacket. Calibration experiments in QUELL [4] indicated indeed that the power (W) in the jacket amounts to about 90% of the total external power into the cable. From the point of view of the comparison between 1- and 2-fluid codes, both conclusions, which applied to the square wave case above, are still valid. Additionally, one can observe by comparing runs 4M, 8M and 4 that for increasing perforation F the 2-fluid results tend to the 1-fluid, as expected. (Indeed, for F→100% and $H_{nowall} \to \infty$ Mithrandir typically reproduces the Gandalf results).

For the stability analysis of CICC with cooling hole the most severe operative conditions for each cable must be considered, and model uncertainties overcome introducing conservative hypotheses. One can, e.g., ignore the presence of the H helium in

**Table 3.** Gandalf stability margin

| Run # | I [kA] | $Q_{st}/Q_{ik}$ | $E_{ext}$ [J] | SM [mJ/ccst] |
|---|---|---|---|---|
| 4 | 8 | 10/90 | 54.18-55.10 | 342-348 |
| 5 | 10 | 10/90 | 43.40-45.46 | 274-287 |
| 6 | 12 | 10/90 | 34.66-34.89 | 219-220 |
| 7 | 8 | 100/0 | 13.06-13.14 | 824-829 |
| 8 (▼) | 8 | 10/90 | 30.12-30.41 | 190-192 |

(▼) Hole helium excluded (bundle mass flow as in run # 4)

**Table 4.** Mithrandir stability margin

| Run # | I [kA] | $Q_{st}/Q_{jk}$ | $E_{ext}$ [J] | SM [mJ/ccst] |
|---|---|---|---|---|
| 4M | 8 | 10/90 | 33.54-35.64 | 212-225 |
| 5M | 10 | 10/90 | 28.3-30.4 | 178-192 |
| 6M | 12 | 10/90 | 24.11-26.21 | 152-165 |
| 7M (▲) | 8 | 10/90 | 35.64-36.69 | 225-232 |

(▲) Modified C0 and magnetic self-field [7]

**Table 5.** Mithrandir stability margin (F=10%)

| Run # | I [kA] | $Q_{st}/Q_{jk}$ | $E_{ext}$ [J] | SM [mJ/ccst] |
|---|---|---|---|---|
| 8M | 8 | 10/90 | 44.03-46.12 | 278-291 |

the simulation, in order to reduce to a minimum the available enthalpy in the cable. This was done in run 8, consistently assuming that only the **B** mass flow computed in run 4 (about $1 gs^{-1}$) is now present in the cable. By comparing runs 4, 4M, and 8 we notice that the 2-fluid results correctly predict an intermediate stability margin, between the 1-fluid and the 1-fluid without **H** region. Neglecting the **H** helium amounts in fact to assuming an infinite time constant for the **B/H** coupling.

By comparing runs 4 and 7 one notices that the ratio $Q_{st}/Q_{jk}$ has a strong influence on $E_{ext}$. A much larger although less than proportionally increasing energy input can be absorbed by the cable without quench if the power fraction absorbed by the jacket is properly taken into account. Of course the situation is reversed when one reasons in terms of SM. The mentioned less-than-proportionality can be interpreted as the jacket playing a role in the stability of the cable through the helium temperature rise. [4]

By comparing runs 7 and 1, one notices that the square wave assumption is not strictly speaking conservative in this case from the point of view of $E_{ext}$. The larger difference in SM is essentially due to the difference in $L_h$.

Uncertainties in material properties and other effects have been shown to have important consequences on quench characteristics if the temperature margin is small and/or the current density is large. [7] These effects are considered here in case 7M, where C0 (a scaling constant entering the expression for the critical current density) is increased by a factor 1.1 and the magnetic self-field is reduced by a factor 0.66 with respect to the nominal computed value. All other input parameters are as in run 4M. This choice produces a slightly (about 5%) higher stability margin, as can be qualitatively expected.

We finally observe that the SM estimation of the POCHI code [4] lies below that of run 8 (most conservative here) in the case $Q_{st}/Q_{jk}$=10/90. It is possible that part of the difference between our results and those of POCHI be due to the fact that neither Gandalf nor Mithrandir presently implement the model [8] of POCHI for the transient heat transfer coefficient $h_t$. For the conductor of Ref.8, an analysis done with POCHI using the standard $h_t$ model [3] implemented here leads indeed to overestimation of SM with respect to the values computed with the $h_t$ model of POCHI. However, the estimation of POCHI is *above* that of run 1 (least conservative here) in the case $Q_{st}/Q_{jk}$=100/0; the reason for this behavior is not clear.

## COMPARISON WITH EXPERIMENTAL DATA FROM QUELL

The previous estimations can be compared with experimental values as given in Ref.4 (the transport current is 8 kA and the magnetic field has a maximum of 9 T):

| | |
|---|---|
| Recovery (experimental upper limit) | 97 mJ/ccst |
| Quench (experimental lower limit) | 108 mJ/ccst |

We notice that the 2-fluid SM estimations and the 1-fluid SM estimations obtained neglecting the presence of the **H** helium are within a factor of 2 from the experiment. In view of the several uncertainties in the experiment and in the modeling, and of the very limited amount of experimental data, we consider this agreement to be acceptable for the present. 1-fluid SM estimations that include both **B** and **H** helium are typically incorrect by larger factors.

## STABILITY ESTIMATE FOR ITER CS CONDUCTOR

The stability margin of ITER conductors is currently calculated with the conservative assumption that only the **B** helium is taken into account as enthalpy sink. [9] With the help of the two-fluid model, the validity of this assumption could be verified.

Here we present a preliminary analysis of the effect of the perforation F - a major coupling parameter between hole and bundle - on the stability margin for ITER CS conductors. For all the other "free" coupling parameters of Mithrandir we use the same values as for the QUELL study.

The main parameters of the CS conductor [10] are reported in Table 6. The relatively large ratio (~ 1/3) between **H** and **B** helium cross section areas indicates that the hole could, in principle, give a larger contribution to the stability margin with respect to the case of the TF conductors, where this ratio is much lower (~ 10%). [10]

Our present analysis is further restricted to the case of a mechanical disturbance as cause of external energy input to the cable, i.e., other input channels like AC losses are not considered here. This type of energy source is simulated with a square wave disturbance of 1ms duration, extending over a 1m length. [9] For this case we used an adaptive time step starting with the value of 1μs, and an adaptive mesh. The initial fixed mesh is refined over a length of 10m around the heated region, with uniform element size of 2cm in the refined zone.

The conductor is assumed to be immersed in a uniform magnetic field neglecting the self-field. Of all the copper, only that present in the strands is included in the model.

We can notice from Table 7 that the influence of F is smaller in the case of the ITER CS, compared to the case of QUELL. There, one notices by comparing, e.g., runs 4M and

**Table 6.** ITER CS conductor parameters used for the simulation

| Length [m] | 750 | Phtc [mm] | 2233.4 |
|---|---|---|---|
| Helium Hole [mm$^2$] | 113.1 | Phtj [mm] | 123.78 |
| Helium Bundle [mm$^2$] | 369.1 | Phtcj [mm] | 0 |
| SC [mm$^2$] | 262 | $\varepsilon$ | -0.0035 |
| Cu in strand [mm$^2$] | 262 | RRR | 100 |
| Cu extra strand [mm$^2$] | 65.7 | Inlet Temperature [K] | 5.3 |
| Jacket [mm$^2$] | 1381.8 | P inlet [MPa] | 0.6 |
| Insulation [mm$^2$] | 315 | P outlet [MPa] | 0.45 |
| Local void fraction | 0.36 | Op. Current [kA] | 42 |
| Dhh [mm] | 12 | Op. Field [T] | 13 |
| Dhb [mm] | 0.661 | | |

**Table 7.** Stability margin for the ITER CS conductor computed with Mithrandir

| Run # | Perforation F [%] | Input power Q [MW/m] |
|---|---|---|
| 1M-CS | 0 | 0.26-0.27 |
| 2M-CS | 1 | 0.26-0.27 |
| 3M-CS | 10 | 0.275-0.28 |
| 4M-CS | 99 | 0.28-0.29 |

8M an increase of ~ 25% of the stability margin when F is decreased by an order of magnitude. The stability margins of Table 7 are approximately in agreement with those computed previously, for slightly different input data.[9]

## CONCLUSIONS AND PERSPECTIVE

A stability study of cable-in-conduit super-conductors with cooling channel has been presented, based on the 1-fluid code Gandalf and the 2-fluid code Mithrandir. Both codes used the same set of common input parameters, helium and material properties, thermal-hydraulic correlations etc.

The study concentrated on the QUELL conductor because a detailed validation of the two codes was already performed, based on quench initiation and propagation data from QUELL, and because it is the first conductor of this type where some stability data were measured. Different time and space shapes of the external input power were considered.

The 2-fluid estimations for the stability margin turn out to be significantly more conservative, i.e., ~ 25% to 35% lower for the cases at hand, than the 1-fluid. If the helium in H is neglected, even more conservative estimations (i.e., ~ 10% to 20% lower than the 2-fluid) obtain, as expected. In the latter case, and with the 2-fluid code, the agreement with an experimental point is within a factor of 2.

Preliminary estimations with the 2-fluid Mithrandir code have also been presented for the stability margin of an ITER CS conductor following a mechanical disturbance. In perspective we aim at a detailed stability analysis of both TF and CS conductors in ITER, for different types of disturbance, to be compared with the predictions of the 1-fluid models.

## ACNOWLEDGMENTS

The European Community under contracts NET/96-424 and NET/96-425 has partially financially supported this work.

## REFERENCES

1 R.Zanino, S.DePalo, and L.Bottura, A two-fluid code for the thermohydraulic transient analysis of CICC superconducting magnets, *J.Fus.Energy* 14:25 (1995).
2 L.Bottura, A numerical model for the simulation of quench in the ITER magnets, *J.Comput.Phys.* 125:26 (1996).
3 M.Shimada and N.Mitchell, Stability estimations and quench simulations for the ITER conductors, *J.Fus.Energy* 14:59 (1995).
4 A.Anghel, et al., The Quench Experiment on Long Length QUELL - Final Report, EPFL-CRPP, JAERI, MIT-PFC and SINTEZ-NIIEFA Report (1997).
5 R.Zanino, L.Bottura, and C.Marinucci, A comparison between 1- and 2-fluid simulations of the QUELL conductor, *IEEE Trans. Appl. Supercond.* 7:493 (1997).
6 C.Marinucci, L.Bottura, G.Vecsey, and R.Zanino, The QUELL experiment as a validation tool for the numerical code Gandalf, submitted to *Cryogenics* (1997).
7 R.Zanino, L.Bottura and C.Marinucci, Computer simulation of quench propagation in QUELL, submitted to *Adv. Cryo. Eng.* (1997).
8 N.Koizumi, Y.Takahashi and H.Tsuji, Numerical model using an implicit finite difference algorithm for stability simulation of a cable-in-conduit superconductor, *Cryogenics* 36:649 (1996).
9 N.Mitchell, Private communication (1995).
10 ITER-DDD, Ch. 1 "Magnet Systems", Appendix C/ Annex II "Conductor database".

# A SIMPLE METHOD FOR MEASUREMENT OF AC LOSSES IN SUPERCONDUCTING WIRES UNDER CONFIGURATION OF A DOUBLE-LAYER NON-INDUCTIVE COIL

K. Funaki,[1] K. Kajikawa,[1] H. Tomiya,[1] M. Nakamura,[1] M. Iwakuma,[1] S. Miyake,[2] and T. Kumano [2]

[1] Research Institute of Superconductivity, Kyushu University 36
Fukuoka 812, Japan

[2] Showa Electric Wire & Cable Co., Ltd.
Kawasaki 210, Japan

## ABSTRACT

We measured AC losses in superconducting wires with NbTi fine filaments exposed to a transverse magnetic field of 60 Hz by a simple pickup coil. The samples have configuration of a double-layer non-inductive coil. Potential taps were attached to both terminals of the sample coil. In this configuration, the double-layer sample itself is a pickup coil to measure the AC loss. In this procedure, we can approximately estimate an averaged Poynting's vector on the side surface between the two layers by means of signals from a pair of the potential taps and a pickup coil for the external magnetic field. For comparison, we measured the AC losses both by a conventional electromagnetic method with coaxial pickup coils and by a calorimetric one. In the latter method, we obtain an equivalent AC loss by calibrating the amount of vaporized helium gas. Experimental results of the AC losses coincide with each other among the three methods in a wide range of the amplitude of external magnetic field. We also measured a transport-current loss in the sample coil and had a good agreement with results from the calorimetric method. Furthermore, we discuss the applicability of the proposed method to more universal types of AC loss measurement.

## INTRODUCTION

In superconducting pulsed and AC magnets, electromagnetic losses generated in the superconducting wires and windings are one of the most important factors for the basic design. AC losses have been studied by many researchers especially for applications of superconductivity to future fusion reactors, electric power devices and so on. They also proposed useful methods to evaluate the AC losses in various types of electromagnetic environment for the superconducting wires. They measured the electromagnetic losses in superconducting wires with AC transport current by means of a pair of potential taps attached along the sample wire, and in the wires exposed to transverse AC magnetic field with a pair of pickup coils. In cases where wires carrying a direct current are in the transverse AC magnetic field, a dynamic-resistance loss[1] was estimated by observing a terminal voltage from a double-layer non-inductive coil of the sample wire. Each method has been applied to measure the AC loss under restricted conditions for the electromagnetic environment.

Calorimetric methods[2] and so-called "coil-simulation" ones[3] were also used to measure AC losses of superconducting wires in a wider range of electromagnetic configurations. In the former methods, there are few restrictions in the electromagnetic configuration, if the system is immersed in liquid coolant. In the latter ones, on the other hand, we can measure the AC losses in a single-layer sample coil by observing Poynting's vectors individually on both inner and outer side surfaces with pickup coils and potential taps. This method gives the AC losses under the realistic environment equivalent to that of the winding in a superconducting coil.

We propose a simple method to measure the AC losses of superconducting wires in various types of electromagnetic configuration. In the proposed method, the sample has the configuration of a double-layer non-inductive coil. While the sample has the same structure as that in the method for the dynamic-resistance loss mentioned above, we can also estimate other types of AC losses, the transverse-field loss and the cyclic loss, for combined AC transverse magnetic field and transport current in the same manner as the coil-simulation method. In the present study, we considered fundamentals of measuring the transverse-field and transport-current losses in the sample coil. On the basis of these fundamentals, we measured the two types of AC losses in multifilamentary wires with NbTi fine filaments by the proposed method and compared the results to those from the calorimetric method and/or from the usual pickup-coil method. We also discuss the applicability of the proposed method to more universal conditions for AC loss measurements.

## FUNDAMENTALS OF AC LOSS MEASUREMENT

### AC Losses in Transverse Magnetic Field

We shall consider AC loss in a double-layer non-inductive coil of a superconducting multifilamentary wire, as shown in Figure 1 (a), where an AC magnetic field is applied parallel to the coil axis. Figure 1(b) schematically shows the distribution of magnetic induction in a pair of turns facing each other in the double-layer coil during a decreasing half cycle of the external transverse magnetic field. If the wire thickness is sufficiently small compared with the coil diameter, the distribution of magnetic induction is symmetric about the center of each layer.

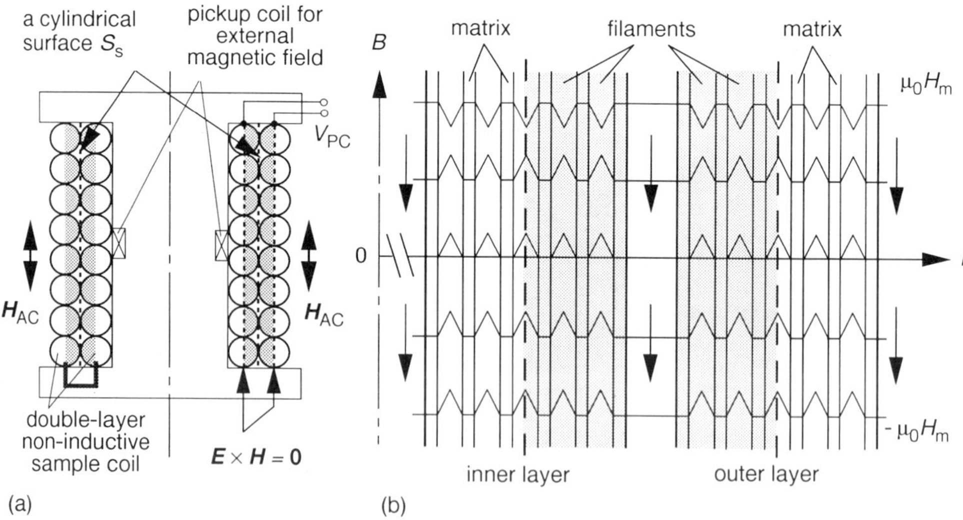


**Figure 1.** (a) A cross-sectional view of a double-layer non-inductive sample coil in an external magnetic field along the coil axis. The AC loss is estimated with a terminal voltage of the sample coil and a voltage signal from a pickup coil for the external field. (b) Distribution of magnetic induction in a pair of turns facing each other in the double-layer coil during a decreasing half cycle of the external transverse magnetic field. Poynting's vector at the center of each layer is equal to zero due to symmetry.

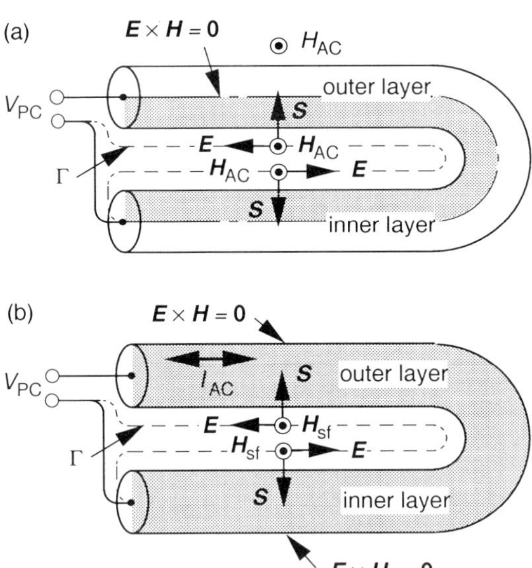


**Figure 2.** (a) A schematic explanation of fundamentals to measure the transverse-field loss in the sample coil. (b) A schematic explanation of the fundamentals to measure transport-current loss in the sample coil. In these figures, the terminal voltage $V_{PC}$ of the sample coil gives the electric field at a path $\Gamma$ along the wire on a cylindrical surface $S_S$ that is composed of an outer side of the inner layer and an inner side of the outer layer.

In general, we can estimate the AC loss in the superconducting wire by integrating Poynting's vector $\boldsymbol{S} = \boldsymbol{E} \times \boldsymbol{H}$ [W/m$^2$], the flux of electromagnetic power, on a closed surface surrounding the wire over a cycle of the external field. $\boldsymbol{E}$ and $\boldsymbol{H}$ are the electric field and the magnetic field on the surface. In the case of a double-layer non-inductive coil in an external magnetic field along its axis, the Poynting's vector is equal to zero at the center of each layer because of the symmetry in the distribution of magnetic induction. Therefore, we can estimate the AC loss in the shaded half region of the double layers in Figure 1 by considering the electromagnetic power flux through a cylindrical surface $S_S$ composed of the outer half of the inner layer and the inner half of the outer layer.

In the usual way, the AC loss density per cycle in the sample coil is given by the equation

$$W = -\frac{1}{V_S/2}\int_0^T \mathrm{d}t \int_{S_S} \mathrm{d}\boldsymbol{S}\cdot\boldsymbol{E}\times\boldsymbol{H} \tag{1}$$

where $V_S$ is the whole volume of the sample wire and $T$ is the period of the AC field. The direction of d$\boldsymbol{S}$ is outward from the shaded region. Let us assume that the external magnetic field $H_{AC}$ is much larger than that due to magnetic moments in the sample coil. Since the magnetic field on the surface $S_S$ is constant, Eq. (1) is changed to

$$W = -\frac{2d_S}{V_S}\int_0^T \mathrm{d}t\, H_{AC}\int_\Gamma \mathrm{d}\boldsymbol{s}\cdot\boldsymbol{E} \tag{2}$$

$$= -\frac{2}{nV_S}\int_0^T \mathrm{d}t\, H_{AC}V_{PC} \tag{3}$$

where $d_S$ is a diameter of the sample wire and is equal to the inverse of turn number per unit length, $n$, in each layer with tight winding, and $V_{PC}$ is the terminal voltage of the sample coil. If we take into account the distribution of electric field with a period of $d_S$ on $S_S$, we obtain Eq. (2) from Eq. (1) by integrating a mean electric field along a path $\Gamma$ parallel to the wire on $S_S$. The path is schematically shown in Figure 2(a). The terminal voltage $V_{PC}$ found along the center line of the wire, the solid line in Figure 2(a), is equivalent to that by the path Γ because the induced voltage due to the magnetic flux inside a closed loop composed of the above two paths is in-phase to the external magnetic field. Finally, we can approximate the AC loss by Eq. (3) with the external AC magnetic field and the terminal voltage of the sample coil.

## AC Losses due to Transport Current

We also consider AC loss in the double-layer non-inductive sample coil with AC transport current, as shown in Figure 3(a). Figure 3(b) shows a typical distribution of magnetic induction during a decreasing half cycle of the transport current, which flows in an outer region of the wire due to the self-field effect[4]. Since the distance between the two layers is usually negligible compared with the coil length, the distribution of magnetic induction is almost symmetric about the center of the double layers and the Poynting's vector on both side surfaces of the double-layer coil is nearly equal to zero, as shown in Figure 3.

In this case, we can estimate the AC loss in the whole superconducting wire by considering the Poynting's vector on the cylindrical surface $S_S$. In a similar procedure to the previous subsection, the AC loss density per cycle in the sample coil is expressed by the equation

$$W = -\frac{1}{V_S}\int_0^T \mathrm{d}t \int_{S_S} \mathrm{d}\boldsymbol{S}\cdot\boldsymbol{E}\times\boldsymbol{H} \tag{4}$$

In this equation, the volume to be considered is the whole of the sample coil as mentioned above. If we neglect terminal distortion in the distribution of magnetic field on the $S_S$, Eq. (4) is changed into

$$W = -\frac{d_S}{V_S}\int_0^T \mathrm{d}t\, H_{sf} \int_\Gamma \mathrm{d}\boldsymbol{s}\cdot\boldsymbol{E} \tag{5}$$

$$= -\frac{1}{nV_S}\int_0^T \mathrm{d}t\, H_{sf}V_{PC} \tag{6}$$

where $H_{sf}$ is a self magnetic field of the sample coil on the surface $S_S$. In the same manner as before, we obtain Eq. (6) as the expression of the transport-current loss with the self magnetic field $H_{sf}$ and the terminal voltage $V_{PC}$ of the sample coil.

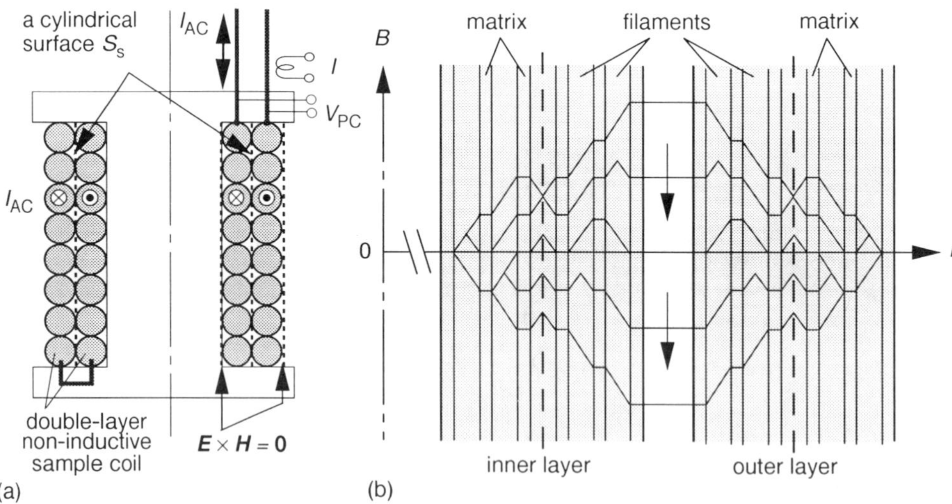


**Figure 3.** (a) A cross-sectional view of a double-layer non-inductive sample coil with AC transport current in measurement of the AC loss. The AC loss is estimated with a terminal voltage of the sample coil and a voltage signal from a pickup coil for the transport current. (b) Distribution of magnetic induction in a pair of turns facing each other in the double-layer coil during a decreasing half cycle of the transport current. Poynting's vector on both outer side surfaces of the double-layer coil is nearly equal to zero.

**Table 1.** Characteristics of sample wire, sample coils and pickup coils

<table>
<tr><th></th><th>pickup-coil method</th><th>proposed method</th><th>calorimetric method</th></tr>
<tr><td>Sample wire</td><td colspan="3">superconductor NbTi; filament number 21, 336<br>wire diameter 0.121 mm; CuNi / Cu / NbTi 2.17 / 0.16 / 1<br>filament diameter 0.45 μm</td></tr>
<tr><td>Sample coil<br>type<br>layer number<br>turn number<br>mean diameter<br>height<br>wire length</td><td><br>multi-layer coil<br>15 layers<br>2, 816 turns<br>30 mm<br>30 mm<br>278 m</td><td colspan="2"><br>non-inductive coil<br>double layers<br>394 turns<br>30 mm<br>30 mm<br>38 m</td></tr>
<tr><td>Pickup coils<br>turn number<br>mean diameter<br>height</td><td>(main / cancelling)<br>41 turns / 128 turns<br>40 mm / 23 mm<br>5 mm / 5 mm</td><td><br>394 turns<br>30 mm<br>30 mm</td><td></td></tr>
</table>

## EXPERIMENTS AND RESULTS

### Sample Coils and Pickup Coils

We prepared two types of sample coils using a NbTi multifilamentary wire. One is a double-layer non-inductive sample coil used for AC loss measurements both by the proposed method and by a calorimetric one, and the other is a multilayer coil used for the usual pickup-coil method. Characteristics of the sample wire and coils are listed in Table 1. The length of sample wire in the double-layer coil is about one seventh that in the multilayer one. This difference comes from an extremely low ratio of noise to signal in the proposed method compared with in the usual pickup-coil one. Parameters of pickup coils are also presented in the table.

### Transverse-Field Loss

Measurement procedure is the same as that in the usual pickup-coil method except for the structure of the main pickup coil. Figure 4(a) shows the electric circuit of AC loss measurement by the proposed method. The main pickup coil is, itself, the sample coil with a double-layer non-inductive structure. We measured the transverse-field losses of the sample

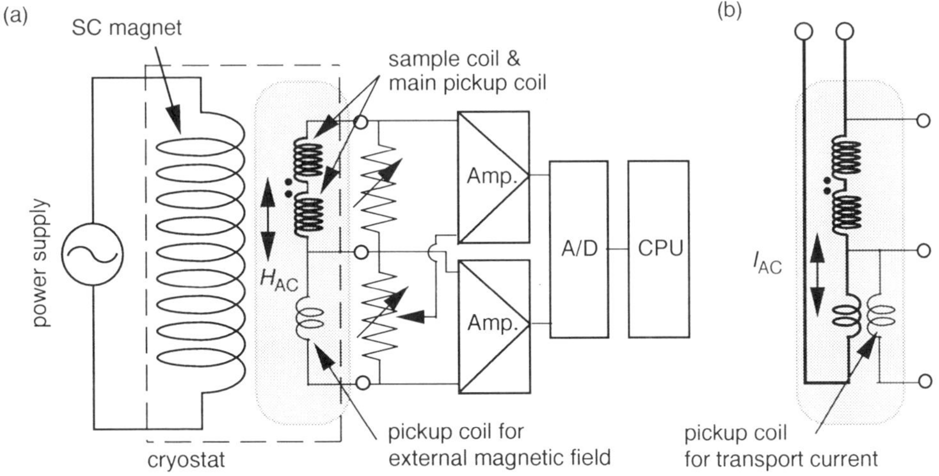


**Figure 4.** (a) Electric circuit to measure transverse-field loss in a double-layer non-inductive sample coil. (b) Partial electric circuit for the sample coil in measurement of transport-current loss. Whole circuit is made by replacing the part shaded in (a) with that in (b).

coil by computing Eq. (3) using the terminal voltages from the sample coil and the pickup coil for the external magnetic field. As seen in Figure 4(a), a component of $V_{PC}$ in-phase to the external field was canceled out in advance by using the signal from the latter pickup coil for a high accuracy in the computing process. Results measured in the external magnetic field of 60 Hz applied along the axis of the sample coil are plotted against the amplitude as open circles in Figure 5.

For comparison, we also measured the transverse-field losses by means of the usual pickup-coil method and the calorimetric one. In the pickup-coil method, we prepared another set of sample and pickup coils as shown in Table 1 and put the sample coil inside the main pickup coil. The procedure of processing signals from the pickup coils is in principle the same as that indicated in Figure 4(a) for the present method. In the calorimetric method, we directly monitored the volume of helium gas produced by AC losses of the sample coil through a slit of the cryostat by a video camera. In this procedure, we collected the helium gas by a glass vessel located upside down to cover the sample coil in liquid helium. We calibrated the AC loss by measuring the volume of helium gas produced by standard heater power. This configuration of the glass vessel gives an advantage in accurate measurement of AC loss compared with conventional types where helium gas is conducted to a gas flow meter outside the cryostat[2].

The results obtained by the above reference methods are indicated by closed circles and squares, respectively, in Figure 5. We have good agreement between the results of the present method and the reference ones over a wide range of amplitudes of external magnetic field. This agreement shows that the proposed method is applicable to the measurement of the transverse-field loss with an accuracy comparable to the conventional ones.

## Transport-Current Loss

Figure 4(b) shows the partial electric circuit for the sample coil used in measuring the transport-current loss. We obtain the signals for $V_{PC}$ and $H_{sf}$ in Eq. (6) from the sample coil and a pickup coil for transport current in Figure 4(b). In the measurement of transport-current loss, we set the circuit in Figure 4(b) into the shaded part in Figure 4(a). The procedure for the loss measurement is the same as in the previous subsection. While we supply no external magnetic field to the sample coil in this case, we can measure cyclic losses for combined AC external field and AC transport current in the present experimental setup as indicated below in the DISCUSSION section.

The losses were also measured by the calorimetric method for comparison. The results are given by open circles and closed squares for the present and calorimetric methods, respectively, in Figure 6. We have equivalent results for each other of the two methods.

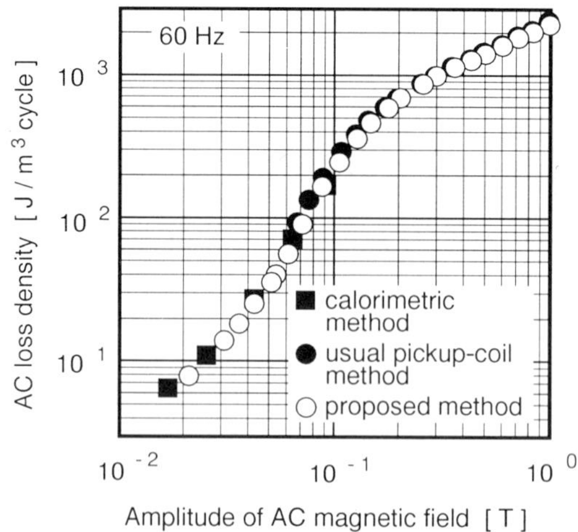


**Figure 5.** Dependence of transverse-field loss on the amplitude. The results by the proposed method are in good agreement with those both by the usual pickup-coil method and by the calorimetric method in a wide range of the amplitude.

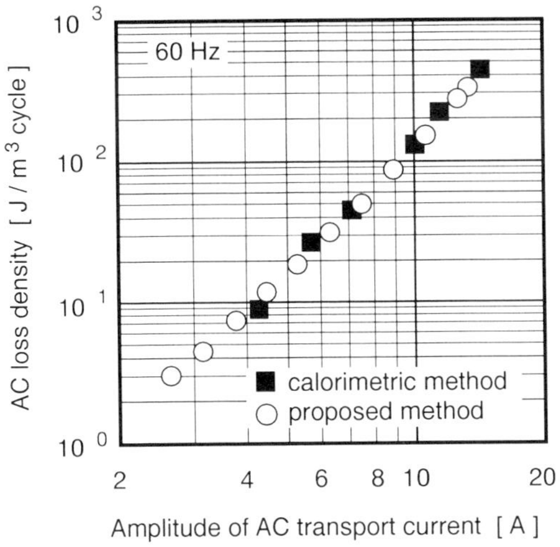


**Figure 6.** Dependence of transport-current loss on the amplitude. The results by the proposed method are in good agreement with those by the calorimetric method.

## DISCUSSION

It is in general difficult to estimate local cyclic losses of superconducting wires in AC multilayer coils. Because the wires with AC transport current are, at the same time, exposed to an AC magnetic field that has a distribution in the winding structure. Figure 7(a) shows the electromagnetic conditions for a layer in the superconducting coil. Subscripts 1 and 2 for a magnetic field $H$ and an electric field $E$ indicate that they are variables on an inner surface and an outer one of the layer, respectively. In Figure 7(a), the distribution of magnetic flux inside the layer of a monofilamentary wire is shown for a simplicity in explanation. The following consideration ought to be also available to the cases of multifilamentary wires.

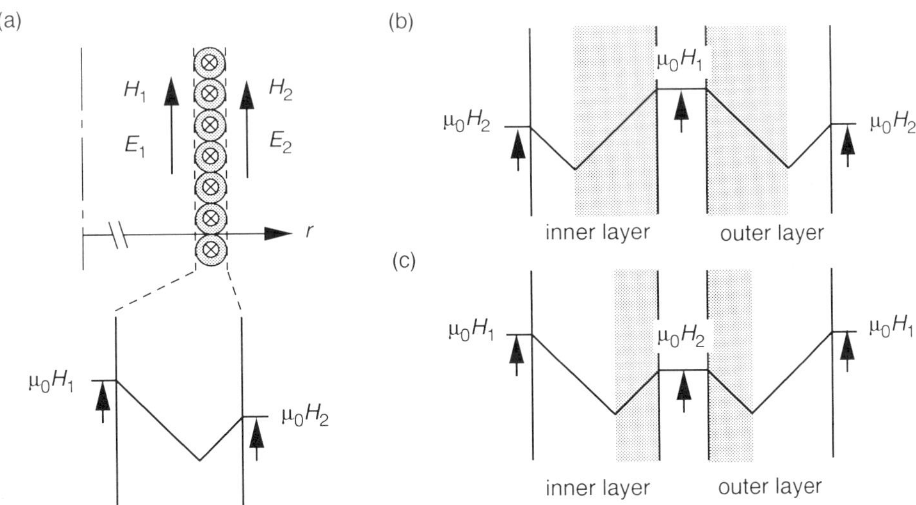


**Figure 7.** (a) General boundary conditions on both sides of a layer in a superconducting solenoid coil and the distribution of magnetic induction inside the layer wound by a monofilamentary wire. (b) and (c) Two types of distributions of magnetic induction in a double-layer non-inductive coil to simulate Poynting's vectors on the both sides of the layer in (a).

Supposing that we obtain the magnetic and electric fields on both side surfaces of the layer, we can estimate the cyclic loss in the layer by integrating the Poynting's vectors on the both sides over a cycle. In the so-called "coil-simulation" method[3], they measure the local magnetic fields individually inside and outside the layer by pickup coils, and the electric fields from two pairs of potential taps attached directly along both inner and outer sides of the wire over a whole length. In the case of the present method, we simplify the setup of potential taps in the coil-simulation method by using the sample coil itself as a potential tap.

In the coil-simulation method, we can observe the Poynting's vectors on the both sides of the layer at the same time by two sets of pickup coils and potential taps. In the present method, on the other hand, we basically observe the Poynting's vector only on one side of the layer as shown in Figures 2(a) and (b). So, in order to observe the same Poynting's vectors as those on the both sides of the layer, it is necessary to prepare two types of boundary conditions for the double-layer non-inductive coil, that are given in Figures 7(b) and (c).

For an example, we shall supply an external magnetic field $H_e = H_m \cos \omega t$ and a transport current $I_t = I_m \cos(\omega t+\phi)$ to the single-layer coil in Figure 7(a). In the configuration given in Figure 7(a), the boundary conditions of the magnetic field on the both sides are $H_1 = H_e + H_t$ and $H_2 = H_e$ with $H_t = n\, I_t$ for a turn number $n$ per unit length and per layer. The electric field on the boundaries are determined only with the present values of $H_1$ and $H_2$, their time derivatives and the structure of the superconducting wire. Therefore, we can realize the same conditions at the boundary between layers in Figures 7(b) and 7(c) by means of the following two types of conditions for the external magnetic field $H_e^D$ and the transport current $I_t^D$ to the double-layer coil, respectively;

$$H_e^D = H_e = H_m \cos \omega t\,, \quad I_t^D = I_t = I_m \cos(\omega t+\phi) \quad \text{for Figure 7(b)} \tag{7}$$
$$H_e^D = H_m^D \cos(\omega t+\alpha)\,, \quad I_t^D = -I_t = I_m \cos(\omega t+\phi+\pi) \quad \text{for Figure 7(c)} \tag{8}$$

In the conditions of Eq. (8), $H_m^D = H_m (1 + 2k \cos\phi + k^2)^{1/2}$ and $\tan\alpha = k \sin\phi / (1 + k\cos\phi)$ with $k = H_{tm} / H_m = n\, I_m / H_m$. In multilayer coils, inner layers with condition of $k \ll 1$ generally have a dominant part in the whole AC losses. Especially, for the condition of $k \ll 1$ and an in-phase condition $\phi = 0$ between $H_e$ and $I_t$, Eqs. (7) and (8) are simplified into

$$H_e^D = H_e = H_m \cos \omega t\,, \quad I_t^D = I_t = I_m \cos \omega t \quad \text{for Figure 7(b)} \tag{9}$$
$$H_e^D = (1+k)\, H_e = (1+k)\, H_m \cos \omega t\,, \quad I_t^D = -I_t = I_m \cos(\omega t+\pi) \quad \text{for Figure 7(c)} \tag{10}$$

In this way, we propose that we can experimentally estimate the total cyclic loss in the single layer of the superconducting wire shown in Figure 7(a) from a direct summation of the results obtained for the above two types of conditions in the present method.


## ACKNOWLEDGMENT

The authors are grateful to Dr. K. Kuroda for his cooperation in AC loss measurement by a calorimetric method. They also thank Dr. S. Nose for useful discussion on the method to measure the dynamic-resistance loss.


## REFERENCES


1. T. Ogasawara et al., Effective Resistance of Current-Carrying Superconducting Wire in Oscillating Magnetic Fields 1: Single Core Composite Conductor, *Cryogenics* 16 (1976), p.33.
2. K. Kuroda, Modified Boil-Off Method for Measuring AC Losses of Superconducting Composites, *Cryogenics* 26 (1986), p.566.
3. T. Ezaki et al., System for Coil Simulation Measurements of Losses and Instabilities in Superconducting Wires, *Cryogenics* 19 (1979), p.731.
4. J.T. Duchateau and B. Turck, Self-Field Degradation Effect in Adiabatic Condition, *Cryogenics* 14 (1974), p.481.

# HIGH-TEMPERATURE SUPERCONDUCTOR COIL SYSTEM FOR A PARTICLE DETECTOR ANALYZING MAGNET

R. C. Niemann,[1] L. R. Turner,[1] M. W. Morgan,[2] P. Haldar,[3] and J. G. Hoehn[3]

[1]Energy Technology Division, Argonne National Laboratory
Argonne, Illinois 60439 U.S.A.
[2]Ability Engineering Technology Inc.
South Holland, Illinois 60473 U.S.A.
[3]Intermagnetics General Corp.
Latham, NY 12110 U.S.A.

## ABSTRACT

The high energy physics experiment known as the Main Injector Neutrino Oscillation Search (MINOS), which is intended for the study of neutrinos, will incorporate a large particle detector system. The detector employs an analyzing magnetic field generated by magnetizing an array of steel plates, called absorbers, with a current-carrying coil. We have evaluated the feasibility of a high-temperature superconductor coil system that provides a 25,000 A-turn magnetizing force. An equivalent conventional coil system (i.e., water-cooled copper) was also designed for comparison.

## INTRODUCTION

The feasibility of using high-temperature superconductor (HTS) coils for high energy physics (HEP) detector magnets has been evaluated. The detector magnet type is that planned for the Main Injector Neutrino Oscillation Search (MINOS) experiment.[1]

The MINOS experiment is designed to determine if one kind of neutrino (electron, mu, or tau) can change into another, as it can if neutrinos have mass. This is not only a fundamental question in its own right, but bears directly on two major unresolved issues of contemporary physics and cosmology, the missing matter problem and the solar neutrino deficit. The MINOS collaboration has more than 100 participants from more than 20 institutions around the world. Magnet technology that would improve the MINOS experiment, or similar experiments, is clearly relevant to HEP.

Three benefits of using an HTS coil are foreseen. First, the HTS option allows operation at higher currents and thus provides insurance against two major uncertainties in the magnet design, i.e., the possible inability to achieve the design B-H curve of the magnet iron, and the possibility that the gaps within each lamination, or the field reduction due to the gaps, may be larger than anticipated. (Gaps in the laminations are unavoidable because of assembly constraints.)

Second, because the coil can operate at higher currents, the magnetic fields in the iron can be higher and more uniform. More of the detector iron will be at fields above 1 T. (Because the inner portion of the steel is already saturated, the field there will not increase much.)

Third, the use of a superconducting coil leads to savings in energy and operating costs. The location of an experiment in a mine deep underground (760 m [2500 ft] below the surface), as in the MINOS experiment, may increase the importance of these savings.

The application of HTS coils to HEP detectors will benefit from innovation in several technical areas. Among these are HTS conductors that perform effectively and reliably at the applied detector field; low-loss, reliable electrical junctions to and between HTS conductors; low-loss, reliable, field-assemblable cryostats; and effective, efficient, and reliable cryogenic refrigeration systems.

## THE MINOS DETECTOR

Figure 1 is an overview of the far detector. The total length of the detector is 36 m and total mass is 10,000 metric tons. The iron will be magnetized toroidally by running a coil through a hole in the center of the absorber plates to produce a reasonably uniform field. The magnetization allows measurement of high-energy muon momenta. The detector will have 600 octagonal cross-section steel absorbers.

The active detector elements consist of planes of streamer tubes. The total area of active detector planes to be instrumented is 32,000 $m^2$. The total number of instrumentation channels to be read out is 480,000.

The steel absorbers for the far detector consist of octagonal plates that are 4 cm thick and 8 m across from flat to flat on the sides of the octagon. The gap between plates is nominally 2 cm. Each steel plate has a total mass of 16.8 metric tons. The steel plates will be fabricated from AISI 1010 or equivalent steel; such steel is relatively inexpensive and provides good magnetic permeability.

The steel portion of the detector will have a central hole through which a current-carrying coil runs in order to magnetize the steel plates toroidally. Figure 2 shows the strength of the magnetic field as a function of position within a portion of one steel plate. For example, with $10^4$ ampere-turns running through the coil, the average field in the steel is 1.5 T and the field near the center is $\approx$2.0 T; average field 50 cm from the edge of the detector is $\approx$1.1 T. The field changes relatively smoothly and slowly across the face of the detector.

The magnet coil will be installed in three separate sections along the length of the detector, each section serving 200 plates of steel. Figure 3 shows conceptually how the coil will run around the ends of the detector segments. Sufficient gaps will be allowed between the steel plates at these locations to allow access for servicing the coil if necessary. Installation of the coil in sections will allow the first sections of the detector to begin normal data acquisition while the remainder of the detector is being completed.

The coil and iron will be assembled underground.

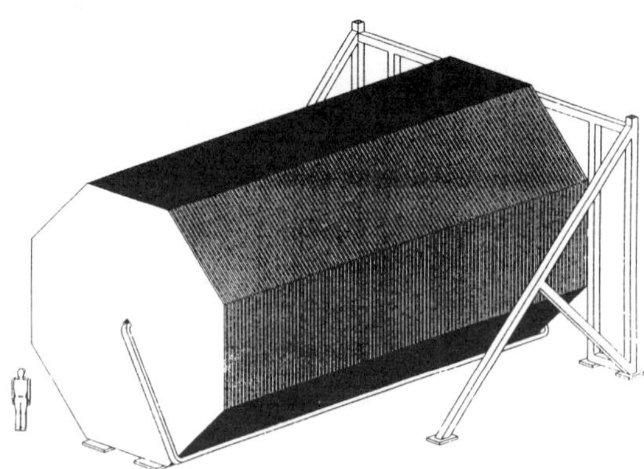

**Figure 1.** Overview of MINOS far detector (Figure from Lawrence Livermore National Laboratory, used by permission).

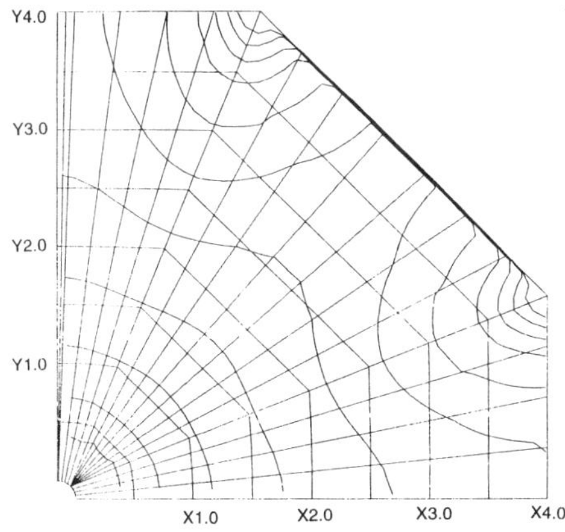


**Figure 2.** Magnetic field contour map of octagonal plates with 2.5 $10^4$ ampere-turns. Innermost field contour is 1.8 T, and outermost is 0.8 T (Figure from Lawrence Livermore National Laboratory, used by permission).

## COIL ASSEMBLY SPECIFICATION

### Geometry

The general layout of the coil is shown in Figure 4. Installation constraints include maximum cryostat OD = 0.30 m (12 in.) (outermost cryostat wall/flange), and maximum cryostat extension beyond detector end plates = 0.91 m (36 in.).

### Cryogenics

The working fluid is liquid nitrogen, and the coil will operate at 66 K +2/-0 K in single-phase fluid.

### Electrical

The magnetizing current is 25 kA-turns (DC). The power supply ripple ≤ 1%. A low ramp rate is employed during magnet charging.

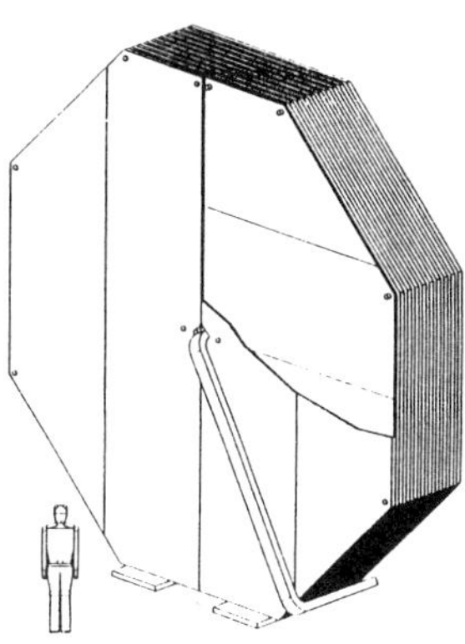

**Figure 3.** Magnet coil installation on the steel absorber plates will be in three separate segments along the detector.

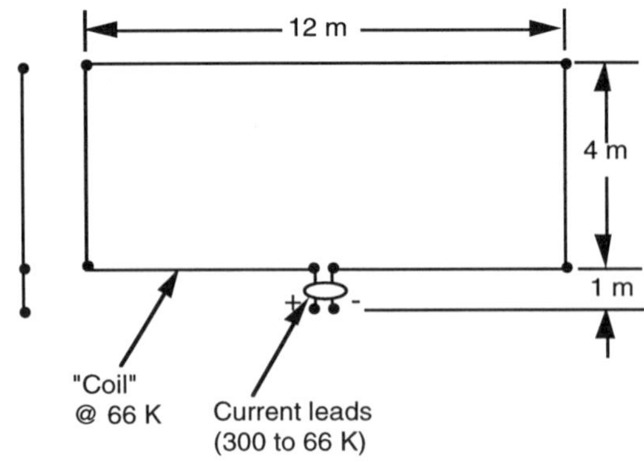


**Figure 4.** Coil assembly geometry.

## Magnetic

The coil will operate in an applied field generated by the magnetized steel plates. The applied field at the coil, as subdivided for magnetic field analysis, shown in Figure 5, assumed to be 10 × 10 cm square, is given by Table 1. Maximum Lorentz force on the coil is 670 N/m (≈4 lb/in.), occurring at the bends between the central and radial sections. The force is normal to the coil and is in the plane of the coil.

## Assembly/Installation

The three detector modules will be built sequentially and will operate independently of each other. Thus, there will be three separate coil assemblies, each with its own power supply, cryogenic system, and controls.

Because the coil will be assembled underground, components must fit within the mine shaft elevator, i.e., 1.2 m × 1.8 m (47 in. × 71 in.) cross-section; 8 m (315 in.) length; and 5442 kg (12,000 lb) maximum load.

# HTS COIL SYSTEM

## Design Assumptions

**Conductor.** The basic conductor element is HTS tape, grouped in either monolithic or transmission-line format. Operation is in the detector applied field.

**Coil.** The total number of coil turns is determined by the cryogenic load control, including current lead heat gain and joint ohmic heating, as well as assembly considerations. Joints are used at only one location, on the perimeter.

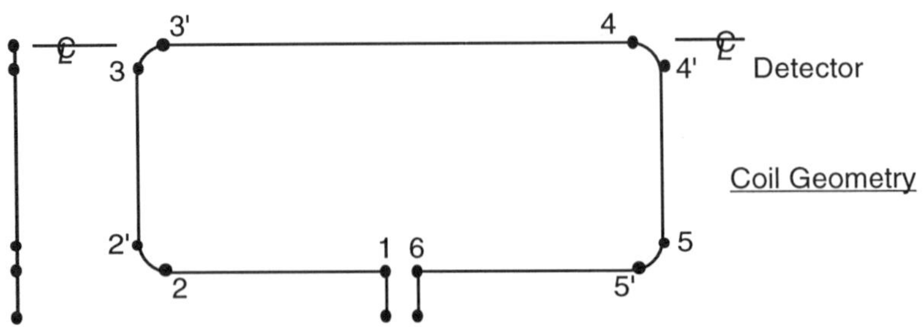


**Figure 5.** Coil subdivisions for magnetic field analysis.

**Table 1.** Applied field (in gauss) at conductor for NI = 25 kA-turns

| Coil Section | Along Coil | Normal to Coil in Coil Plane | Normal to Coil Normal to Coil Plane | Magnitude |
|---|---|---|---|---|
| Return (1-2) & (5-6) | 0.1 | 223 | 413 | 413 |
| Radial (2-3) & (4-5) | 3 | 670 | 448 | 747 |
| Central (3-4) | 108 | 770 | 710 | 850 |
| Outer Arc (5-5') | 20 | 175 | 618 | 618 |
| Inner Arc (4-4') | 18 | 662 | 825 | 852 |

**Cryostat.** Transmission line geometry is used, together with a single junction/current lead box.

**Cryogenics.** The working fluid is liquid nitrogen at an operating temperature of 66 K +2/-0 K. The coil will operate in single-phase fluid.

## HTS Coil

The coil consists of multistrand BSCCO-2223 multifilamentary cable. The operating current can be selected by grouping the individual cables together for a one-turn coil or grouping into several bundles for a multiple-turn coil. The coil will have only one joint, at the junction/current lead box.

To meet the 25 kA turn requirement for the specified conditions (assuming a 50% packing factor over the whole conductor cross section), the conductor must have an operating current density of at least 500 A/cm$^2$. The standard BSCCO-2223 for the given conditions of ≈66 K with a peak radial field of ≈0.85 T and peak axial field of 0.67 T can meet and exceed this requirement.

The forces on the cable are modest and occur only at the bends between the central and radial sections.

The conductor shown in Figure 6 comprises an annealed, perforated copper channel with the HTS conductor laid in lengthwise and soldered into one monolithic cable. Each coil consists of ≈250 cables packed into the total available coil cross section (≈10 × 10 cm). The annealed copper channels will facilitate handling and bending of the cable.

Because the conductor runs in DC mode, there is no need to helically wind the HTS conductor with pitch. The added benefits here are that the properties of the conductor can be maintained by avoiding a cabling operation and that less tape (length) is required.

If the piece lengths are limited to ≈40 m, each coil turn can be wound and delivered on a 1-m-diameter spool.

Each of the 250 cables will support the requisite 100 A plus a 10% margin between this operating current and the cable critical current.

Installation will not be a significant problem. As long as each turn can be assembled in the cryostat on site without making joints or having to bend the conductor around a radius of <0.5 m, the HTS conductor properties will be maintained.

Experiments[2] indicate that silver-sheathed HTS composite conductors are not prone to quench as low-temperature superconductors do, but rather exhibit current-sharing over a wide range of temperature and current. Consequently, for a conductor driven by a constant-current power supply, a local temperature perturbation or other triggering event would most likely lead to a detectable increase in power supply voltage. If the perturbation then disappeared, the voltage would decrease with no consequences. If the voltage continued to rise, the power supply could be programmed to shut down in an orderly manner, with no additional forces placed on the conductor. In a power failure at the site, the current in the conductor would decrease rapidly, but because there are no nearby current paths for inductive current transfer, there would be no additional forces on the conductor.

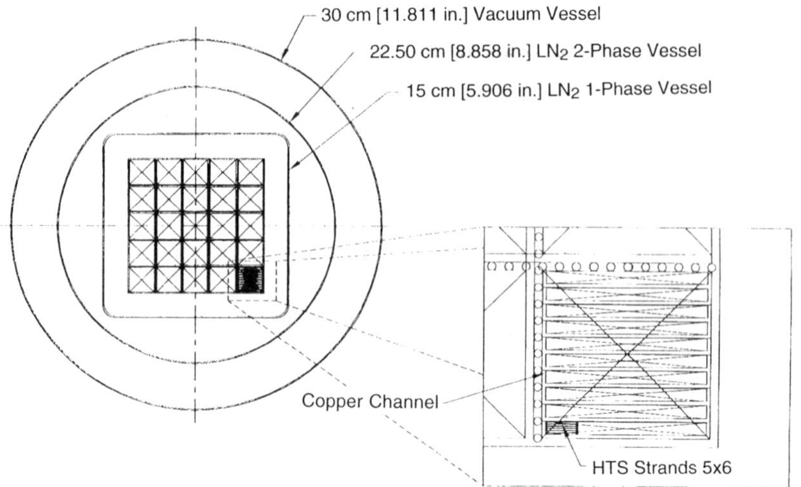


**Figure 6.** HTS conductor arrangement.

## Cryostat

The cryostat cross section is shown in Figures 7 and 8. The stainless steel inner line will contain the conductors surrounded by single-phase, subcooled liquid nitrogen. The annular space between the inner line and the surrounding tube will be filled with boiling two-phase liquid nitrogen. The wall between the two fluid streams provides a heat exchange surface to allow heat to be removed from the single-phase fluid along the entire length of the line.

The cryostat will incorporate a junction/current lead box that serves the dual purpose of containing all the necessary valving, instrumentation, and power leads, and of containing the single field joint.

## Cryogenics

The cryogenic system is shown schematically in Figure 9. Pumped pool boiling liquid nitrogen was chosen for simplicity, economy, and function. The heat load on the system is estimated to be 1700 W, due mostly to heat input from the power leads; 30 W has been budgeted for cryostat heat leakage. During operation, liquid nitrogen will be consumed at the rate of 8 g/sec. About 45 cfm of nitrogen gas will be vented into the tunnel or up a stack.

Liquid nitrogen will be purchased for this application and will be supplied to the cryostat from a storage vessel located outside the mine. The liquid nitrogen vessel will probably be located at ground level and the fluid can either be piped down or transported to the detector in dewars.

The vacuum pump will be located in a convenient area near the cryostat and will be sized to handle two cryostats in order to provide redundancy.

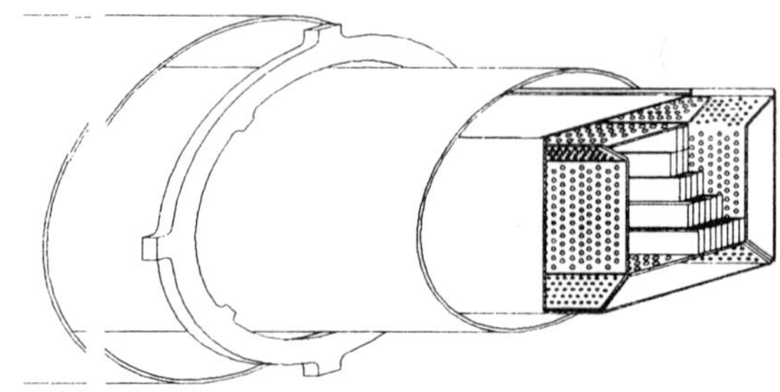

**Figure 7.** Cryostat cross section (isometric view).

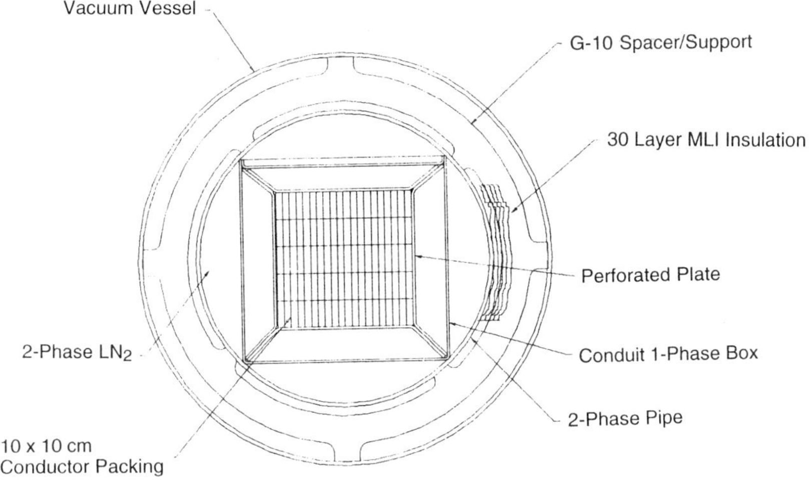


**Figure 8.** Cryostat cross section.

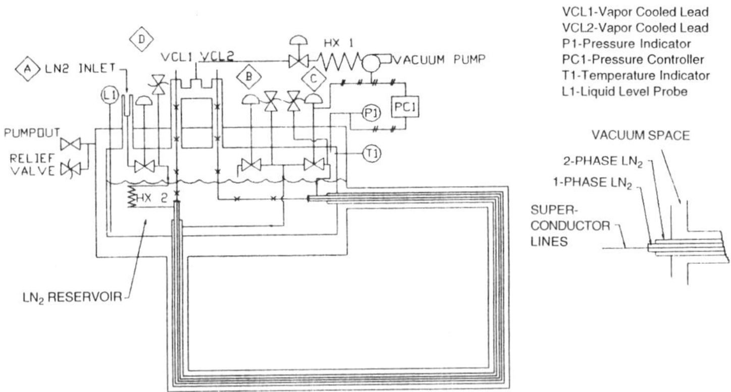


**Figure 9.** Schematic diagram of cryogenic system.

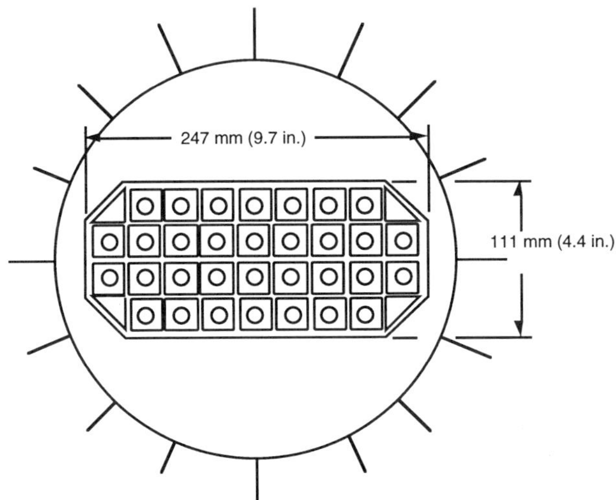


**Figure 10.** Conventional coil cross section.

**Table 2.** Coil assembly comparisons

| Characteristic | HTS | ANL Conventional |
|---|---|---|
| Coil assembly acquisition cost[a,b] ($$10^3$) | | |
| * Conductor | 1475 | 82 |
| * Tooling | 45 | 50 |
| * Components/hardware | 85 | 77 |
| * Assembly | 86 | 80 |
| | 1691 | 289 |
| Electrical power[c] | 4 kW | 24 kW |

[a]Cost for a single-coil system.
[b]1997 cost.
[c]For a single-coil system.

## CONVENTIONAL COIL ASSEMBLY

An equivalent conventional coil assembly was designed for comparison. The conductor is water-cooled copper, with a 2.54 cm (1 in.) square cross section and a 1.27 cm (0.5 in.) circular hole. The coil cross section is shown in Figure 10. The coil has 32 turns and operates at 781 A, with a 32 V drop. The cooling requirement is 24.2 kW/coil and requires 8 gpm cooling water flow.

## COST COMPARISON

The coil assembly acquisition costs and operating power requirements were estimated and are compared in Table 2.

## CONCLUSIONS

The specified magnetizing current of 25 kA-turns can be produced by both the HTS coil system and the conventional coil system with existing materials and technology.

The estimated 1997 acquisition cost of an HTS coil assembly (one of three required for the detector assembly) of $1691K is significantly higher than that for an equivalent conventional coil system of $289K.

The cost driver in the HTS coil assembly ($1691K) is the conductor cost ($1475K). Projected conductor cost reductions do not result in a cost-competitive coil option in the near future.

## ACKNOWLEDGMENTS

The authors gratefully acknowledge the engineering contributions of J. M Jagger and K. Thompson. The authors gratefully acknowledge the manuscript preparation skills of J. A. Stephens and editorial contributions of C. A. Malefyt. Work at ANL was supported by the U.S. Department of Energy, Energy Efficiency and Renewable Energy, as part of a program to develop electric power technology, under Contract W-31-109-Eng-38.

## REFERENCES

1. The MINOS Collaboration, "P-875: A Long-baseline Neutrino Oscillation Experiment at Fermilab," Proposal, Feb. 1995.
2. Y.S. Cha, S.Y. Seol, D.J. Evans, and J.R. Hull, Flux-flow resistivity of three high-temperature superconductors, *IEEE Trans. on Appl. Supercond.* 7(2):2122-2125 (1997).

# THREE-DIMENSIONAL NUMERICAL ANALYSES OF THE STABILITY OF Ag/BSCCO TAPE

E. E. Burkhardt[1] and J. Schwartz[2]

National High Magnetic Field Laboratory
Florida State University
Tallahassee, FL 32310-4005
[1]Department of Nuclear Engineering
University of Illinois at Urbana-Champaign
[2]Department of Mechanical Engineering
FAMU-FSU College of Engineering

## ABSTRACT

Practical design considerations for HTS magnets require detailed magnetothermal analysis. Here we report investigations of the stability of high-$T_c$ superconducting tapes. As a result of the broad range of temperatures during a transition and the strong temperature dependence and anisotropy of the material properties, the finite element method (FEM) is used to solve the three-dimensional heat conduction equation. The heat source is determined using the operating current, the resistivity of the Ag and the critical current density of $Bi_2Sr_2Ca_2Cu_3O_x$ (BSCCO). The minimum quench energy is determined for infinitely long uniform sources in the BSCCO.

## INTRODUCTION

The stability of superconducting magnets subjected to external thermal disturbances or flux jumps depends on the interplay between the electromagnetic and thermal characteristics of the system. As stability can limit the performance of a superconducting magnet, the stabilities of NbTi and $Nb_3Sn$ conductors have been well studied[1,2]. A recent study determining normal zone propagation in Ag/Bi-2223 superconducting tapes has shown the importance of heat dissipation along the length of the conductor[3], and a one-dimensional study of quench propagation has also been done[4]. Here, we use the finite element method to analyze the stability of a Ag/BSCCO superconducting tape[5,6]. The equations governing the entire system are Maxwell's equations and

$$\nabla \cdot \left(k(T)\nabla T\right) + E \cdot J + Q_{ext} = C_p(T)\frac{\partial T}{\partial t} \tag{1}$$

where $k(T)$ is the temperature dependent thermal conductivity, $E$ is the induced electric field, $J$ is the current density, $C(T)$ is the temperature dependent specific heat and $Q_{ext}$ is any external heat source (e.g. a crack in epoxy). The most general case of a high-$T_c$ tape conductor is shown in Figure 1.

Eq.1 must be solved in each region of the conductor. The boundary conditions are determined by the cooling mode: edge cooled, face cooled, or thermally insulated tapes. Although Eq.1 describes the thermal behavior of magnetothermal systems in general, the $E \cdot J$ term explicitly describes the superconductor and its dissipation. For example, for $Bi_2Sr_2Ca_2Cu_3O_x$ tapes operating at elevated temperature and in a significant background magnetic field, flux creep may be of most interest. In the superconductor, the electric field is exponentially increasing with respect to the ratio of the current in the superconductor, $I_{sc}$, and the critical current, $I_c$:

$$E_{sc} = E_c(I_{sc}/I_c)^n \tag{2}$$

where $E_c$ is the critical electric field that determines $I_c$ and $n$ is the index of transition (a higher $n$-value corresponds to a sharper transition from the superconducting state to the normal state), which is related at least in part to $U_o/T$, where $U_o$ is the pinning energy. In the Ag the electric field is determined by Ohm's Law:

$$E_{Ag} = \rho_{Ag} I_{Ag}/A_{Ag} \tag{3}$$

where $\rho_{Ag}$, $I_{Ag}$ and $A_{Ag}$ are the resistivity, current and cross-sectional area of the Ag, respectively. By knowing the total operating current, $I_{op} = I_{sc} + I_{Ag}$, and that the electric field must be constant across any $x$-$y$ plane, $E_f \equiv E_{sc} = E_{Ag}$, $I_{sc}$, $I_{Ag}$ and $E_f$ are determined. This will result in a current in an electric field being present at all times; however, when $I_{op}/I_c \ll 1.0$ the current in the Ag is insignificant.

A three-dimensional finite-element model has been developed. It uses the ITPACK 2C package developed by the Center for Numerical Analysis at the University of Texas at Austin. This package solves large sparse linear systems by adaptive accelerated iterative methods.

## THERMAL PROPERTIES OF HTS CONDUCTOR CONSTITUENTS

The material thermal properties are temperature dependent over a significant temperature range. As the temperature range of importance is expanded by HTS materials, it is necessary to include the following temperature and field-dependent physical properties in the stability analysis:

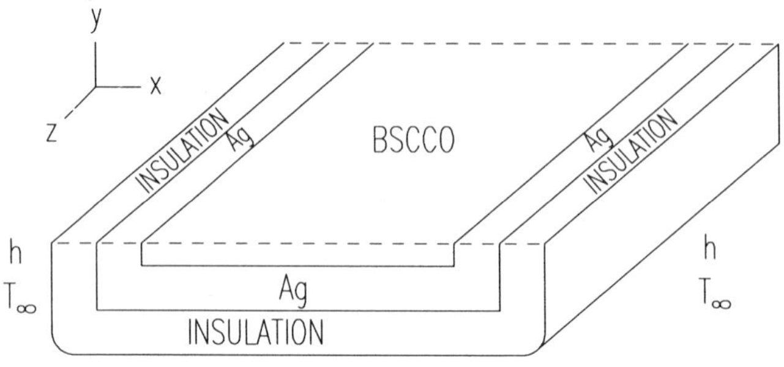


**Figure 1.** One-half cross-section of a Ag/Bi-2223 superconducting composite tape.

- specific heat
- thermal conductivity
- magnetoresistivity.

These properties and the significance of the temperature dependence have been discussed in detail previously and are summarized here[5,6].

At 4.2 K and below (where conventional superconductors operate) the specific heat of materials is quite low. As $T$ increases, the specific heat increases rapidly[7-9]. Thus the temperature dependence becomes more important for HTS systems.

For BSCCO, the thermal conductivities of pellets and tapes are expected to be quite different owing to the anisotropy of the crystal structure and varying degrees of grain alignment. This analysis uses the thermal conductivity of a pellet as the *a-b*-plane conductivity[10]. The thermal conductivity in the *c*-direction is estimated as one-fourth of these values, which is approximately the fraction of $J_c$ in the *c*-direction compared to the *a-b*-plane. Since this ratio is dependent on the grain alignment in the tape (if the *x-z*-plane in Figure 1 is equivalent to the *a-b*-plane), it will be varied in a future study. Although the thermal conductivity of Bi-2223 is much lower than that of Ag[11], it cannot be ignored because heat generated in the Bi-2223 requires a mode of transport to the Ag.

Small changes in purity due to the contamination by BSCCO can cause large changes in the magnetoresistivity of Ag. The magnetoresistivity of a silver tape that was processed with $Bi_{1.4}Pb_{0.6}Sr_2Ca_2Cu_3O_x$ has been measured at the NHMFL[12]. The silver tape was peeled off the BSCCO and cleaned with nitric acid to remove excess BSCCO. At lower temperatures, the magnetoresistivity of Ag increases as $B$ increases; however, the field dependence decreases as the temperature increases and becomes negligible beyond 100 K.

## FINITE ELEMENT METHOD

The finite element method is used to solve Eq. 1 in each region of the conductor. Due to the strong temperature dependence of the thermal conductivity, the first term on the right-hand-side of (1) does not simplify to $k(T)\nabla^2 T$. This introduces non-linearities and motivates the use of the finite element method instead of finite differencing. Note that the temperature dependence of $C_p$ and $J_c$ are maintained as well. The Crank-Nicolson method (central difference scheme) is then used to analyze the models. Four-node tetrahedral elements are used for the three-dimensional solution, and due to the temperature dependence of the material properties, average values are used for each element. The specific heats and thermal conductivities of BSCCO and Ag, and the magnetoresistivity of Ag have been fit to empirical equations which are incorporated into the program. The $E{\cdot}J$ source is determined using $J_c$, $I_{op}$ and $E_f$ determined from Eqs. 2 and 3. The $n$-value is maintained as a variable for parametric study.

To illustrate the FEM approach, the stability of the tape shown in Figure 1 is determined. The current propagates in the *z*-direction and anisotropy of the Bi-2223 occurs in the *y*-direction relative to the *x-z* plane. The tape considered has a 2.5:1 Ag/BSCCO ratio and is 3.56 mm wide and 0.23 mm thick. Here, an adiabatic boundary condition is used and the epoxy is only modeled at the top of the tape, consistent with the outer turn for a coil.

Starting with the tape at steady-state with a current $I_{op}$ operating in a magnetic field $B$, a heat pulse is introduced to drive the tape normal locally. The FEM can evaluate an infinite variety of sources in the stabilizer and/or the superconductor. The external disturbance must be large enough to create a substantial electric field. This is done by increasing the temperature which decreases $I_c$. Because $I_c$ is inversely dependent with $B$, operating in higher external field will decrease the initial $I_c$. To calculate $E_f$, $I_c$ is calculated by summing $I_c$ over the BSCCO elements at a given *x-y* cross-section. Due to the dependence of $E_{sc}$ upon $I_{op}$, heating occurs in the superconductor before $I_c$ decreases below $I_{op}$.

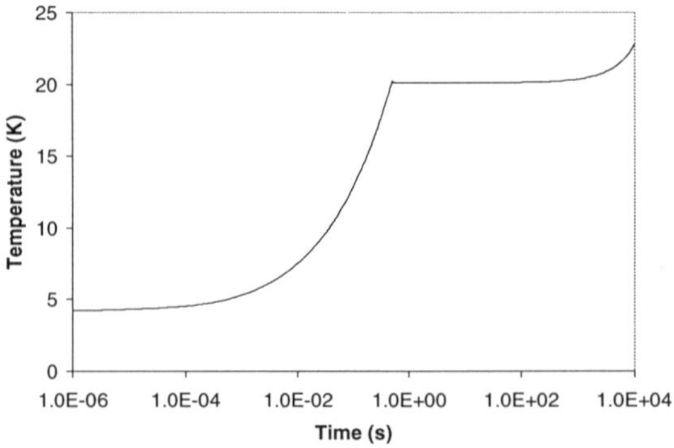


**Figure 2.** Temperature vs. time for an infinitely long tape (in *z*-direction), 0.5s pulse. $n$=20 and $I_{op}/I_c$=0.75.

The tape is considered unstable when the temperature rises such that $E_f$ does not vanish and the tape does not return to the initial state. If a pulse source is used, only the $E \cdot J$ source term remains after the pulse. If a constant heat source is applied, however, both it and the $E \cdot J$ source term exist (this case is only valid if convective boundary conditions are used, *i.e.* LHe or $LN_2$ are in contact with the tape, or the heat has some other method to leave the tape). Due to the nature of $E_{sc}$, heating will theoretically always be present, and an insulated (adiabatic) tape will never be stable. To eliminate this problem, when $E_f < 0.01$ μV/cm, it is assumed that all of the current flows in the superconductor and no heating occurs.

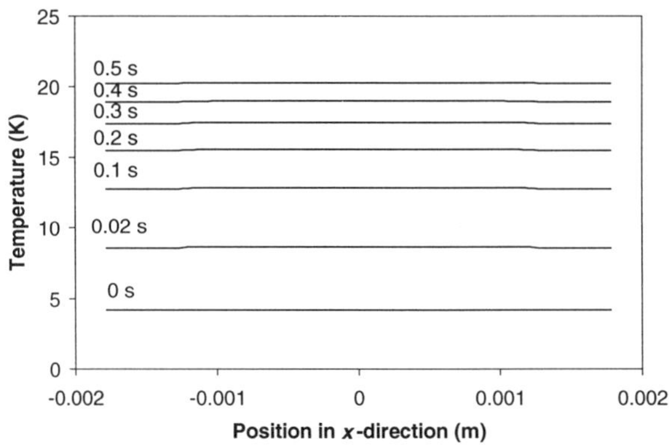


**Figure 3.** Temperature profile in conductor at various times for the same case as Figure 2.

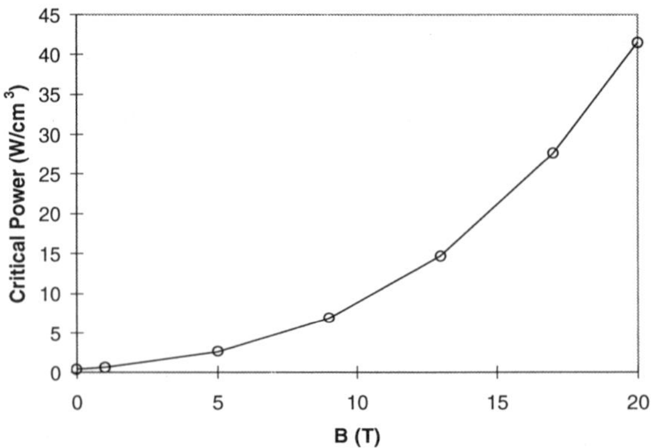


**Figure 4.** Critical power versus magnetic field for a uniform source in the BSCCO for a conductor operating at 4.2 K, $I_{op}/I_c = 0.9$ ($t$=0).

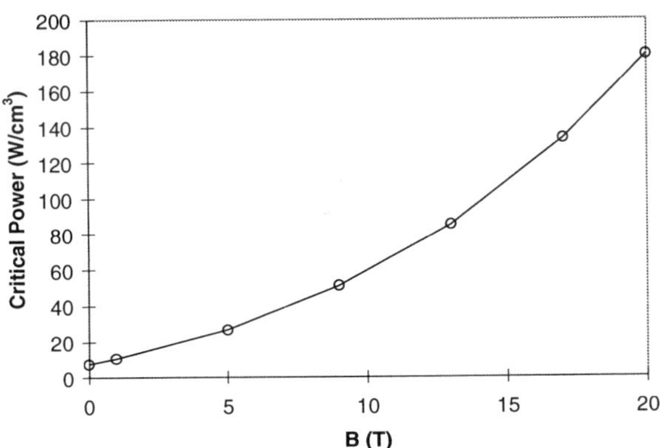


**Figure 5.** Critical power versus magnetic field for a uniform source in the BSCCO for a conductor operating at 20 K, $I_{op}/I_c$ ($t$=0)= 0.9.

The critical current density is found for each field at the operating temperature, with the operating current set to a (variable) fraction of $I_c$. A pulsed source is then applied which causes the temperature of the BSCCO to rise. If the pulse is sufficiently large, $I_c$ decreases to a value such that the amount of heat generated is sufficient for the tape to not return to a stable state, *i.e.* the temperature continues to rise. The magnitude of the pulse required to cause the tape to become critically unstable is determined and the values at various fields ranging from 0 to 20 T are calculated.

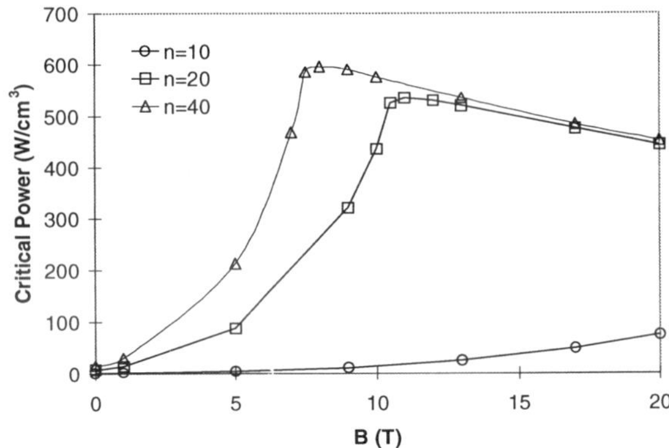


**Figure 6.** Critical power versus magnetic field for a uniform source in the BSCCO for a conductor operating at 4.2 K, $I_{op}/I_c$($t$=0)= 0.75.

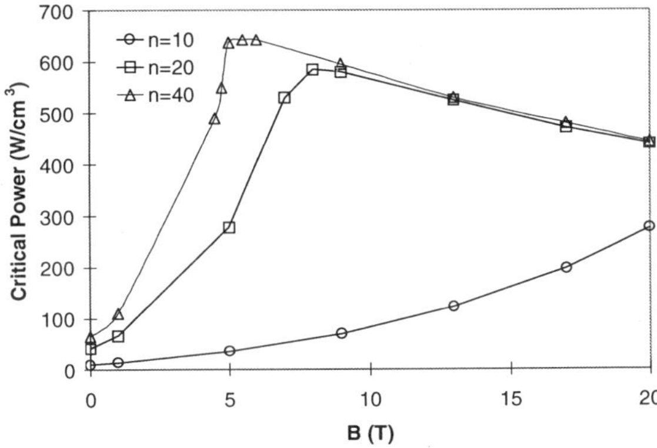


**Figure 7.** Critical power versus magnetic field for a uniform source in the BSCCO for a conductor operating at 20 K, $I_{op}/I_c$ ($t$=0) = 0.75.

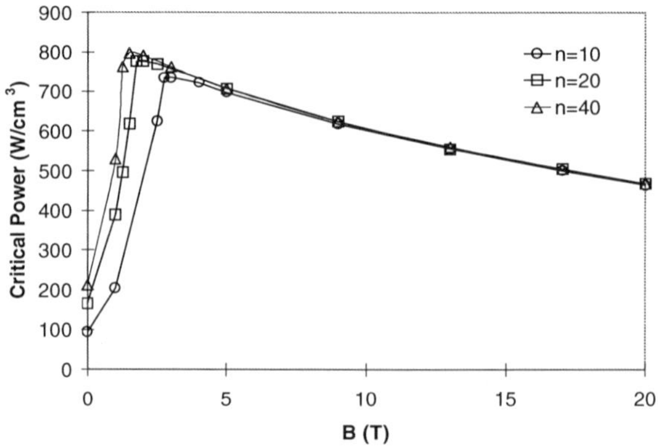


**Figure 8.** Critical power versus magnetic field for a uniform source in the BSCCO for a conductor operating at 4.2 K, $I_{op}/I_c$ ($t$=0) = 0.5.

## Uniform BSCCO Source

The results in this section are for an infinitely long (no variation in the $z$-direction) uniform source in the BSCCO that has a 0.5 s duration. A typical plot of the temperature vs. time for a point in the center of the tape for a slightly unstable case is shown in Figure 2. This is for the case at 4.2 K, 0 T, $n$ = 20, $I_{op}/I_c$ = 0.75, and a heat pulse of 10 W/cm$^3$ is applied for 0.5 sec.

Figure 3 shows the time progression of the temperature profile across the width of the tape during the heat pulse. Because $C_p$ and $k$ increase with temperature, $T$ will increase at a decreasing rate during the pulse. Also, as a result of the source being uniform in the BSCCO and the high thermal conductivity in the Ag, the temperature is almost constant across the tape at all times.

The results for the critical power needed to quench the tape for $I_{op}/I_c$ = 0.5, 0.75 and 0.9 at 4.2 and 20 K with $n$-values of 10, 20, and 40 have been determined and are shown in Figures 4-9. As the field increases the incremental reduction in $J_c$ with increasing temperature is proportionally smaller and less current is transferred to the Ag and the critical power required increases. Eventually, the critical current will fall to zero when the temperature is greater than $T_c$, and because $T_c$ decreases as $B$ increases, the critical power will decrease after this situation is reached. As expected, the tape initially becomes critical

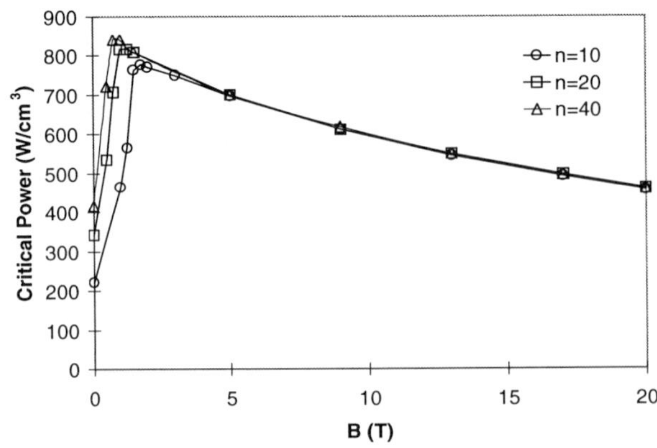


**Figure 9.** Critical power versus magnetic field for a uniform source in the BSCCO for a conductor operating at 20 K, $I_{op}/I_c$ ($t$=0) = 0.5.

as a result of $I_c$ decreasing to a value below $I_{op}$ and the critical power increases with temperature. Also, at the point where the tape becomes normal as a result of $I_c$ falling to zero, the critical power begins to decrease as $B$ increases. Because $I_c$ is not a function of $n$ in the region where $I_c$ decreases to zero, the critical powers are constant there. Note that for $I_{op}/I_c = 0.9$, only $n$=40 is used because for the lower $n$-values, the initial $E_f$ is greater than 0.01 μV/cm and heating will always be present.

## SUMMARY

A 3-D FEM technique for determining the stability of high-$T_c$ superconducting tapes has been established. This model allows temperature dependent thermal properties and the analysis of a variety of geometries. For the case of a uniform source in the BSCCO with varying $n$-values and values of $I_{op}/I_c$, the critical energy required to quench the tape behaved as expected, increasing when $J_c$ is the limiting factor and decreasing when $T_c$ is the limiting factor. Further development using three-dimensional models and including different cooling modes will provide a FEM capable of more comprehensive analysis.

## ACKNOWLEDGMENT

This work was supported by the Defense Advanced Research Projects Agency through the Naval Research Laboratory. Earle Burkhardt's research was performed under appointment to the MFET Fellowship Program administered by Oak Ridge Institute for Science and Education for the U.S. DOE.

## REFERENCES

1. S.Y. Seol and M-C. Chyu, Stability analysis for a composite superconductor subjected to a linear thermal disturbance, *Superconducting Science Technology*, 7:841-848 (1994).
2. A. Unal and M-C.Chyu, Three-dimensional stability analysis for quenching and recovery of superconductor tapes or films, *International Symposium on Safety of Superconductors and Related Heat Transfer at Low Temperatures*, August 28-31 (1994).
3. S.B. Kim, Y. Ueno and A. Ishiyama, Experiment and numerical analysis of normal zone propagation properties in Ag sheathed Bi-2223 superconducting tapes, submitted to *IEEE Transactions on Magnetics* and presented at the MT-14 conference.
4. R.H. Bellis and Y. Iwasa, Quench propagation in high-$T_c$ superconductors, *Cryogenics*, 34:129-144 (1994).
5. E.E. Burkhardt and J. Schwartz, Three-Dimensional Numerical Analysis of the Stability of Ag/Bi2Sr2Ca2Cu3Ox Tape Conductors, *IEEE Transactions on Applied Superconductivity* 7(2), 199-202 (1997).
6. E.E. Burkhardt and J. Schwartz, Two-Dimensional Finite Element Analysis of the Stability of Ag/BSCCO Tapes, *Advances in Cryogenic Engineering* 41, Cryogenic Engineering Conference, 545-552 (1996).
7. S.J. Collocott, R. Driver and C. Andrikidis, Specific heat of the ceramic superconductor $Bi_2Sr_2CuO_8$ from 0.4 to 20 K, *Physica C*, 173:117-124 (1991).
8. J.E. Gordon, S. Prigge, S.J. Collocott and R. Driver, Specific heat of $(BiPb)_2Sr_2Ca_2Cu_3O_{10-y}$ from 78 to 300K, *Physica C*, 185- 89:1351-1352 (1991).
9. L. Dresner, Stability and protection of Ag/BSCCO magnets operated in the 20-40 K range, *Cryogenics*, 33(9):900-909, (1993).
10. A. Jezowski, Thermal conductivity of Bi-based high-$T_c$ superconductors in the vicinity of superconducting transition," *Solid State Communications*, 75(10):779-783 (1990).
11. R.W. Powell, C.Y. Ho and P.E. Liley, *Journal of Physics and Chemistry*, Ref. Data 3, Suppl. 1, Vol. 606 (1974).
12. S. Nakamae, J. Schwartz, Magnetoresistivity of Ag-tape co-processed with $Bi_{1.4}Pb_{0.6}Sr_2Ca_2Cu_3O_x$ superconductor, *Cryogenics*, 36:395-397 (1996).

# DESIGN CONSIDERATIONS AND EXPERIENCE OF THE RHIC DUAL MAGNET CRYOSTAT INSTALLATION*

R. Grandinetti, J. Koehler, J. Sondericker, G. Ganetis, and G. McIntyre

Cryogenics Systems Group
Relativistic Heavy Ion Collider
Brookhaven National Laboratory
Upton, NY 11973

## ABSTRACT

The first RHIC Dual Magnet installation, of eight (8) cold masses in one vacuum vessel, has been completed. Design of the system included the requirement for assembly on location, the need for individual three-direction alignment capability at each cold mass cradle, a fixed position cradle for the entire assembly, longitudinal slide capability for all other cradles, two thermally isolated heat shields, accessible electrical splice boxes, removable interconnects, removability of individual cold mass sections, flexible supply lines for electrical power and cryogenic services and external loads imposed from the supply services. Initially two of the twelve total assemblies were fabricated for the Sextant Test of the RHIC ring. This paper covers an overview of that effort from initial design considerations, fabrication/installation_techniques, initial installation problems and the final design solutions.

## INTRODUCTION

The two counter rotating rings of the Relativistic Heavy Ion Collider (RHIC) cross each other in the horizontal plane at six interaction regions. On both sides of each interaction point (IP), the two rings (one ring designated as BLUE and the other YELLOW) converge together to allow particle collisions at the IP. Between the DX, (the last single bore magnet before the IP), and the two individual rings, the magnet cold masses were so close together that it was impractical to place them into individual cryostats. The D0 dipole ends were separated by approximately 0.30 inches apart. For the RHIC accelerator, this area consists of three quadrupole/corrector magnets (CQ1, CQ2 and CQ3) and one D0 dipole magnet for each ring for a total of eight cold masses designated for a common cryostat.

* Worked performed under the U.S. Department of Energy

Initial thoughts placed the dipole magnets in one cryostat and the quadrupole/corrector magnets in another. However, the separation distance between the D0 dipole and CQ1 quad was too close to consider separate cryostats. The final design has all eight magnets in one common cryostat. This component of the RHIC system is more commonly known as the Triplet Assembly.

The assembly of the cold masses for both rings in one cryostat is the first known installation of this type. The overall size and weight of the Triplet Assembly (4 feet diameter by 55 feet long and approximately 60000 lbs.) prohibit installation in the RHIC tunnel as one assembly. The Triplets were assembled on location. All subassemblies, vacuum vessels, heat shields, cold masses, etc. were installed on location.

Initially, only two Triplets were fabricated for the Sextant Test of the RHIC ring. This first article installation uncovered several leak problems. The leaks were predominately in the braided flex hoses. Some leaks were repaired and others required hose replacements. The balance were very small leaks that were allowed to remain for the Sextant Test.

## DESIGN CONSIDERATIONS

### 2.1 Basic Design Requirements

One cryostat
Field installation/assembly of Triplet Assembly
Individual magnet accessability and removal
Accessible splice boxes and cryo supply services
Individual magnet adjustability
Magnet Stand adjustability
Balanced cryogenic loads

### 2.2 Triplet Assembly - Final Design Overview

The Triplet Assembly consists of eight magnets, the D0 thru the CQ1, CQ2 and CQ3 installed in one horizontally split vacuum vessel (Fig 1). The upper sections are removable to allow individual magnet replacements. Longitudinally, there five removable interconnect sections that allow access to the interconnecting electrical splice boxes at the ends and between each pair of magnets. The electric and helium services enter and exit the Triplet Assembly above the D0/CQ1 interconnect and at the end of CQ3. The overall length of the vessel is almost 62 feet. The overall length of the magnet string is almost 55 feet. The estimated total weight is 60,000 lbs.

Each magnet is supported by two individually adjustable mounts. Therefore, there are 16 mounts for the 8 magnets in each cryostat assembly. Due to the alignment requirements of the individual magnets, the supporting mounts were each provided with individual vertical, lateral and roll adjustability. Longitudinal adjustability is provided by tie rods rigidly connecting the individual magnets together. The mounts are on top of plastic posts which thermally isolate the

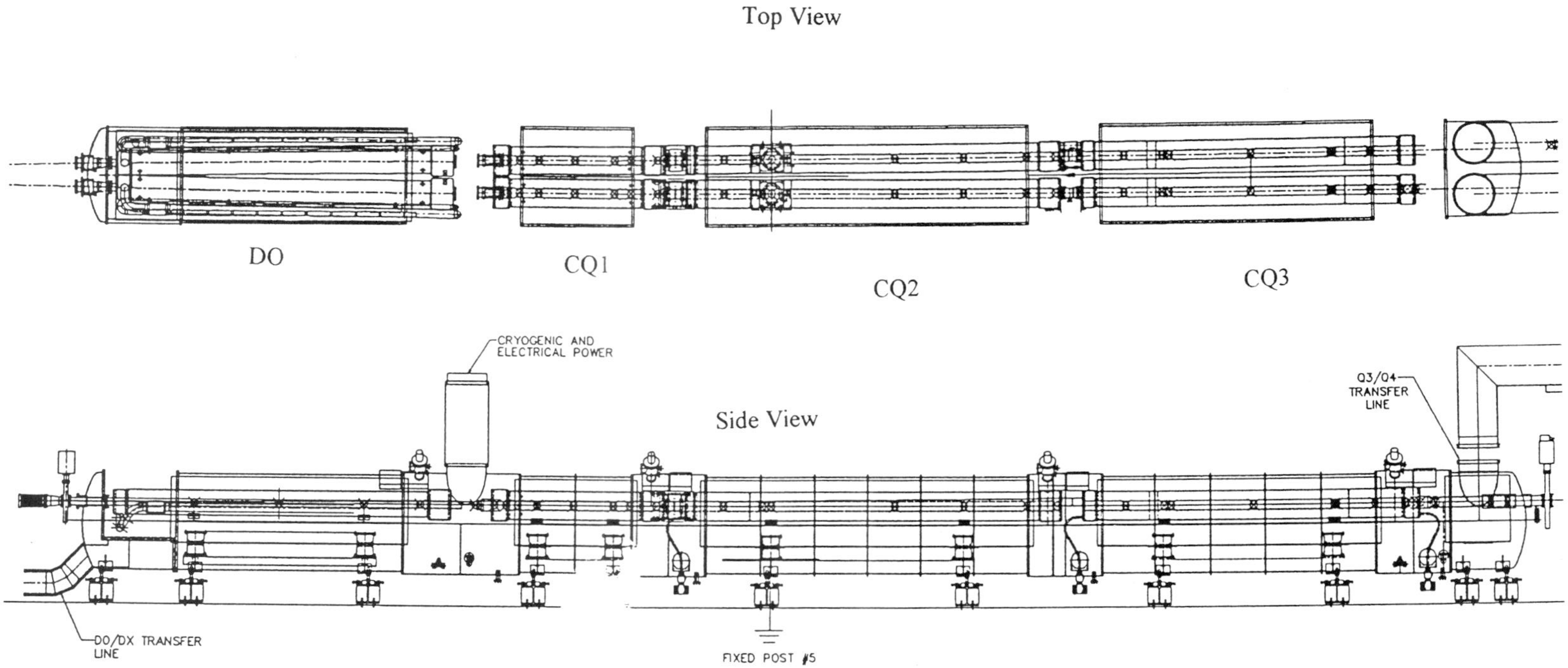

Top View

DO

CQ1

CQ2

CQ3

CRYOGENIC AND ELECTRICAL POWER

Q3/Q4 TRANSFER LINE

Side View

DO/DX TRANSFER LINE

FIXED POST #5

Figure 1 Triplet assembly .

cold mass from the heat shield and from the warm cryostat. Because the magnets were of various lengths, it was coincidental, but convenient, that one support of the CQ2 would be almost exactly the half point of the entire assembly. These mounts, there are two of them (one for each of the two CQ2 cold masses), are fixed to the warm cryostat while all the other 14 mounts slide toward the fixed mount during cool down. All contractions are toward the center. The ends contract almost and even amount, another convenient but coincidental item, with D0 at 1.042 inches and CQ3 at 1.029 inches. An important consideration during the design process was that the magnetic center of each magnet be in their design/operating position when at the cold operating temperature of 4° K. This requires that the magnets, which are installed and surveyed at a room ambient condition, consider the thermal contractions. This means the magnets will be offset when positioned warm and move to their correct position when cold.

The fixed CQ2 mount and plastic posts were strong enough to resist the unbalanced thrust loads imposed during cool down contractions, but to maintain the string on center during the contraction/expansion movements, additional straps were added to increase the longitudinal spring rate assisting the plastic posts to maintain the assembly in position.

All process lines and heat shields are also centered about the fixed CQ2 post. The process lines enter through a VJR line at the D0/Q1 interconnect and exit at the VJR line above the Q3 End. The lines are straight with out any bellows installed. Flex lines are used at both ends to absorb length changes due to contractions. The lines are in a balanced situation and there are no thrust loads imposed to the system with this design.

The heat shields for the assembly are made in four pieces to facilitate assembly and to thermally isolate the rings from each other. In the event that one ring needs to be warmed for maintenance, is would not be necessary to warm both rings. This requirement was based on the economic savings of refrigerator run time to re-cool only one ring as well as the re-cool time saved to achieve operating conditions.

## INSTALLATION EXPERIENCE

Initially, only two Triplets were fabricated to test one sextant of the RHIC ring. The installation started with the survey and anchoring of the cryostat stands. Next, the installation of the lower vacuum vessel sections and survey of the post platforms in the lower vessel sections were performed. The CQ3 end can and lower heat shields were the assembled in place. The MLI blankets, plastic posts, adjustable sliding mounts and then the eight magnet cold masses were then installed. The CQ2 cold masses are surveyed initially because the CQ2 cold mass cradle over post position # 5 is the fixed anchorage for the entire assembly. With the CQ2 location fixed, the remaining cold masses are then surveyed in their respective positions. Each of the sixteen mounts incorporated provides individual lateral, vertical and roll adjustments. Tie rods are installed between the cold masses to adjust the longitudinal position.

The next step was to electrically interconnect the cold masses using a welded clam shell splice box followed by the installation of the cryogenic pipes. The pipes were pressurized and leak checked after welding. This included the interconnect heat shield sections and all supporting instrumentation leads and conduits to the lower interconnect shells. The D0 end can was the last lower sectioned installed. At that point, the entire lower portion of the Triplet assembly had been installed with the exception of the interconnect sections. Installation of the vessel upper vacuum vessel and heat shield sections followed with the D0 end can MLI blankets and heat shields. The interconnect sections of the heat shields and vacuum vessel could then be installed. The two remaining open sections were the vertical interconnects to the cryogen and electric power supply pipes positioned directly above the D0/CQ1 interconnect and at the end can of CQ3. The connections at these locations are through flexible hoses because of the movements due to the cold temperature contractions (more than 1 1/4 inches at the CQ3 end). The flex hoses were installed after bench leak checks and their welded connections were pressure tested.

Upon completion of the assembly, insulating vacuum was obtained and the Helium lines were pressurized. Subsequent leak checking noted substantial leaks in the CQ3 end can and the CQ1/CQ2 interconnects. The interconnects in those areas were opened and leak checking was repeated. The leak checks revealed leaks in several hoses on the top or outer bellows convolution. Two other cases revealed leaks in the bottom or inner convolutions. Further investigation revealed that a stainless backing strip was not used or the welding gun slipped off the strip during the MLI blanket ultrasonic welding procedure, resulting in minute perforations to the bellows of the flex hose. The outer perforations were caused either by direct contact with the welder or by the welder pushing the braid against the bellows resulting in the same perforations.

This procedure was the cause of the perforated outer convolutions but did not explain the perforations on the inner convolutions. The inner perforations were possibly caused by improper cleaning of the hose after fabrication, creating a corrosion spot. Initial manufacturer vacuum leak checks did not reveal any leaks and the required pressure test was not performed. Receipt inspection for vacuum leaks conducted on all hoses at BNL were successful. It was not until the hoses were installed and pressurized did the hose bellows experience any stresses from internal pressure. It is believed that this initial stress excursion caused the corrosion spot to open and generate the inner convolution leaks. All hoses with outer convolution leaks were temporarily repaired by welding. The hoses with the inner convolution leaks were replaced. The Triplet was reassembled and the system pressurized in the vacuum environment. Additional leaks were evident but not serious enough to prevent successful completion of the Sextant Test.

After completion of the Sextant Test, the Triplets will be opened and the hose routing redesigned. All hoses will be replaced.

## CONCLUSIONS

Changes to the Triplet design, leak check procedures and hose procurement was necessary. The following is a summary of the changes to be made:

1. Redesign of the D0/CQ1 interconnect will use shorter hoses that are fully contained in the transfer line split clam shells to reduce disturbance to the cryostat during repairs. Overall, the use of shorter hoses, more pipe and pipe fittings were the guide lines.
2. The flex line path from the Triplet to the D0/CQ1 transfer line was a severely constricted space. The use of more pipe and pipe fittings that are of a smaller diameter than the flex hoses were the guide lines.
3. No ultra sonic welding of MLI blankets around the hoses will be allowed. The blankets will be taped and/or tied in position.
4. Leak check procedures will be revised to incorporate, as much as possible, isolated component leak checks. Systems previously exposed to Helium will be pressure tested and checked. Other systems could be initially vacuum-leak checked and then pressure checked.
5. Verification of the pressure testing of all flex lines prior to acceptance and installation.

## ACKNOWLEDGMENTS

The authors wish to acknowledge the assistance of Bob Duffin and Sue Norton in the Installation Section Design Group for their support of the Triplet Assembly project and for their participation in the preparation of detailed drawings, Dejan Trbojevic for his useful assistance in pre-alignment and positioning assistance discussions, Gary Mc Intyre and the Installation Group who assembled the Sextant Triplets and to Frank Karl and the RHIC Survey Team who surveyed and aligned the individual components that make up the Triplet Assembly.

# HIGH CURRENT DIFFUSION TYPE DIODES AT CRYOGENIC TEMPERATURES FOR THE LHC SUPERCONDUCTING MAGNET PROTECTION

L. Coull, R. Denz, D. Hagedorn

CERN, CH-1211 Geneva 23, Switzerland

## ABSTRACT

High-current by-pass diodes are required for the protection of the superconducting magnets for the Large Hadron Collider LHC at CERN. These diodes are at liquid helium temperature and will be exposed to irradiation. With the re-location of the by-pass diodes for the main dipoles underneath the iron yoke, and of those for the quadrupoles at the bottom of the cryostat the new estimations for the irradiation dose amounts to about 30 Gy with a neutron fluence of about 1.5 x $10^{11}$ n/cm$^2$ for the dipole diodes, and about 100 Gy and 5 x $10^{11}$ n/cm$^2$ for the quadrupole diodes over 10 years. These relatively low doses may allow the use of diffusion type diodes instead of epitaxial diodes. The electrical characteristics of several diodes were measured at temperatures in the range between 1.8 K and 300 K. Diffusion type diodes from three manufacturers were submitted to high current endurance tests in liquid helium. Electrical characteristics and temperatures were measured versus time and showed acceptable results. First irradiation tests show that modified diffusion diodes can be used at least for the dipole by-pass.

## INTRODUCTION

Originally it was anticipated that epitaxial diodes would be used for the high current by-pass at superfluid helium temperatures for the LHC superconducting magnets at CERN. This was due to the relatively high estimated radiation dose of about 35 kGy, and neutron fluence of about 1.5 x $10^{14}$ n/cm$^2$ over 10 years.[1]

With new detailed calculations for the irradiation dose, and with the re-location of the by-pass diodes for the main dipoles underneath the iron yoke, the most recent estimated 10-year doses and neutron fluences amount to about 30 Gy and 1.5 x $10^{11}$ n/cm$^2$ for the dipole by-pass diode. The 10-year irradiation dose for the quadrupole diode, now located in the bottom of the cryostat, is estimated at about 100 Gy with a fluence of 5 x $10^{11}$ n/cm$^2$.[2] These low doses may allow the use of diffusion type diodes instead of epitaxial diodes for the dipole by-pass. Epitaxial diodes may still be required for some quadrupoles.

## MAIN DIFFERENCES BETWEEN EPITAXIAL AND DIFFUSION TYPE DIODES

### Main Differences in Processing

**Diodes Based on the Epitaxial Technology.** To form a pn junction one starts to grow a thin (<100 μm) single crystal layer of silicon on a n-doped Si substrate in a special atmosphere, whereby a silicon compound is reduced to pure silicon on the surface of the substrate. The epitaxy provides a silicon layer with a very low defect concentration and allows an abrupt change of the doping concentration. Unfortunately, this technique is rather complicated and, especially for wafers of large diameter, not easy to control. The shaping of the pn junction is completed with the generation of a diffused p+ layer on the epitaxial layer. This is normally done by the diode manufacturer, whereas the epitaxy, because of the inherent difficulties, is performed only by a few specialized silicon suppliers. Therefore, small quantities of epitaxial diodes are very expensive and have a long delivery time.

**Diffusion type diodes.** For power diodes of the diffusion type, an n-doped base material produced in conventional manner, or a neutron transmutation doped silicon is used. In order to obtain the very high doping concentrations which are required for a power diode, a deep thermal diffusion of the dopant is performed on the silicon wafer. By this means the n+ zone of the diode is created. This high temperature process, which lasts several days, is carried out by the silicon supplier. For further processing the manufacturer grinds off one side of the silicon wafer. Subsequently, the p+ layer is produced by ion implantation followed by a thermal diffusion step. Since the diffusion process leads to a thicker base width than the epitaxial technology, these diodes are in general less radiation hard.

### General Characteristics of the Tested Diodes

The diodes from GEC-Plessey are standard high-voltage diodes from stock without any special passivation and locator for low temperature application, whereas the diffusion diode from WESTCODE is a specially developed thin wafer diode. The diodes from EUPEC are also specially developed. One diode is an epitaxial diode, and the two diffusion type diodes, one doped with gallium and the other doped with boron, are made with the deep diffusion technique applied to the basic material. With this kind of diffusion technique a thinner n-base region can be obtained, resulting in a lower forward voltage drop and higher radiation resistance but at the expense of reverse voltage blocking capabilities. The edge passivation is similar to the passivation applied on the epitaxial diodes. The wafers of all tested diodes have an outer diameter of 75 mm. Their general characteristics are shown in Table 1.

**Table 1.** General characteristics of the tested diodes

| Manu-facturer | Diode Type | Diode Name | Wafer thickness [μm] | Rated reverse Voltage [V] | Endurance Test at 4.2 K |
|---|---|---|---|---|---|
| EUPEC | Spec. develop. Epitaxial diode | E141 | 550 | 200 | yes |
| EUPEC | Spec. develop. Double Diff. (P, Ga) | EDD8 | 280 | 600 | yes |
| EUPEC | Spec. develop. Double Diff. (P, B) | E1024 | 280 | 600 | foreseen |
| GEC-Plessey | Standard Phase-Control, DS2101SY15 | M14 | 250 | 1500 | yes |
| Westcode | Stand. Rectif. Diode, SW14CXC32C | W046 | 234 | 1400 | yes |
| Westcode | Spec. develop. Rectif. Diode, SW05CXC32C | W003 | 184 | 500 | yes |

**Table 2.** Measured electrical characteristics on epitaxial and diffusion type diodes at different temperatures

| Electrical parameter | ( Initial ) Temperature $T_O$ [K] | Typical Epitaxial diode ( 22µm-base ) E141 | EUPEC diffused (Ga-doped) EDD8 | EUPEC diffused (B-doped) E1024 | GEC diffused M14 | WEST-CODE W003 |
|---|---|---|---|---|---|---|
| $V_{to}$ [V] | 4.2 | 1.2 - 1.5 | 12.8 - 16.5 | 6.4 - 8.8 | 7.0 - 8.0 | 5.7 - 7.5 |
| $10<dU_f/dt[V/s]>10^4$ | 1.8 | 1.3 - 1.7 | 15.5 - 17.2 | 6.9 - 9.4 | ~ 8.6 | 5.9 - 7.9 |
| $U_f$ | 300 | 1.0 | 1.4 | 1.08 | 1.2 | 1.2 |
| at | 77 | 1.25 | 2.5 | 1.59 | 2.6 | 2.2 |
| $I_f$ = 15 kA | 4.2 | 1.7 | 9.0 | 4.15 | ** | 4.3 |
| | 1.8 | 1.8 | ** | 4.15 | ** | 4.3 |
| | 300 | ~ 200 | > 720 | > 500 | > 1100 | ~ 900 |
| $U_{r\text{-}max}$ | 77 | ~ 150 | > 560 | > 370 | > 1100 | ~ 900 |
| [V] | 4.2 | ~ 100 | ~ 170 | ~ 370 | ~ 640 | ~ 600 |
| | 1.8 | ~ 100 | ** | ~ 370 | ** | ~ 300 |
| $\Delta U_f/\Delta I_f$ [µΩ] | 300 | ~ 14 | ~ 35 | ~ 16 | ~ 21 | 19 |
| $\Delta U_f/\Delta I_f$ [µΩ] | 77 | ~ 15 | ~ 95 | ~ 31 | ~ 85 | 70 |
| $\Delta U_f/\Delta I_f$ [µΩ] | 4.2 | * | * | * | * | * |
| $\Delta U_f/\Delta I_f$ [µΩ] | 1.8 | * | * | * | * | * |

* not measurable  ** not measured

## Main Electrical Differences

**Forward Characteristics.** The main electrical characteristics for 22 µm base-width epitaxial diodes and various diffusion type diodes are given in Table 2 for different temperatures. The forward voltage measurement at a forward current $I_f$ = 15 kA was carried out with a 200µs half sinusoidal current pulse.[3]

In Table 2 the average differential resistance $r_f = \Delta U_f/\Delta I_f$, derived from the forward current-voltage characteristics in Fig. 1 in the current range of 5 kA < $I_f$ < 15 kA, is also presented. When lowering the temperature from 300 K down to 77 K, epitaxial diodes show an almost pure voltage shift of about 0.3 V and almost no change in differential resistance, whereas diffusion type diodes show additionally to the 0.3 volt-shift a significant increase of differential resistance at 77 K. At 4.2 K and 1.8 K the differential resistance is so high that at currents pulses above 1 kA the temperature in the wafer increases significantly due to the measuring pulse itself, and the $I_f$ - $U_f$ characteristic is deformed. The forward voltages, $U_f$, at 4.2 K and 1.8 K indicated in the table for the epitaxial diode and the diffusion diode from EUPEC are the values for higher temperatures than the initial temperature due to warm up during current pulsing.

In Fig. 1 the typical $I_f$ - $U_f$ characteristics at 300 K and 77 K are shown for three diffusion diodes with a typical epitaxial diode for comparison. In Fig. 2 are the measured If-Uf characteristics at 1.8 K and 4.2 K for two diffusion diodes, and for the epitaxial diode given. The deformation of the curves (shift towards lower forward voltage) due to heating with increasing current is clearly visible.

**Reverse Voltage Characteristics.** The reverse voltage was measured before the endurance tests and after each run of endurance test. No significant changes were observed. Typical values are given in Table 2.

During the reverse voltage tests at 1.8 K and 4.2 K on the double diffusion diodes from EUPEC in current control, the reverse voltage increased up to about $U_r$ = 370V as shown in Table 2. After exceeding a reverse current, $I_r$, of about 50µA, the reverse voltage of all double diffusion diodes from EUPEC dropped down to about 130V up to reverse currents

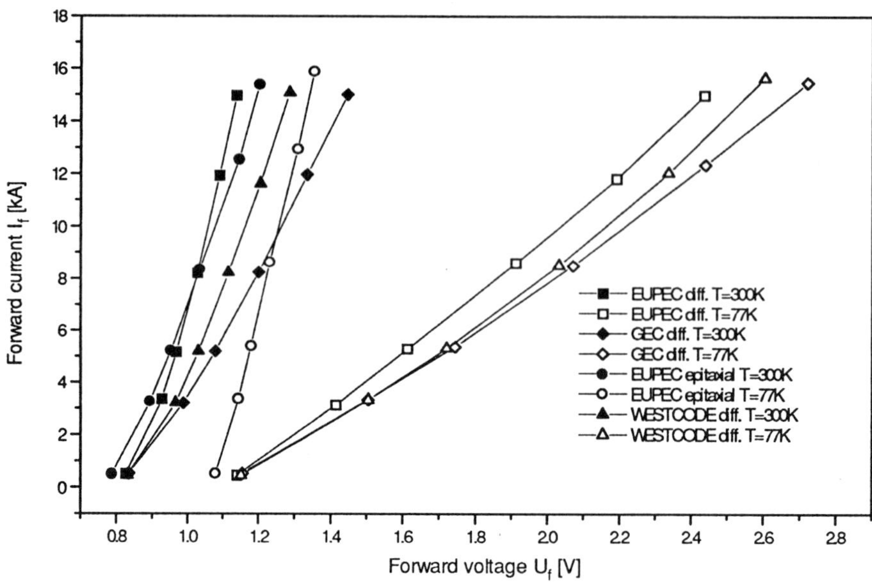


**Figure 1.** Measured forward current-voltage characteristics at 77 K and 300 K on three diffusion diodes and one epitaxial diode for comparison.

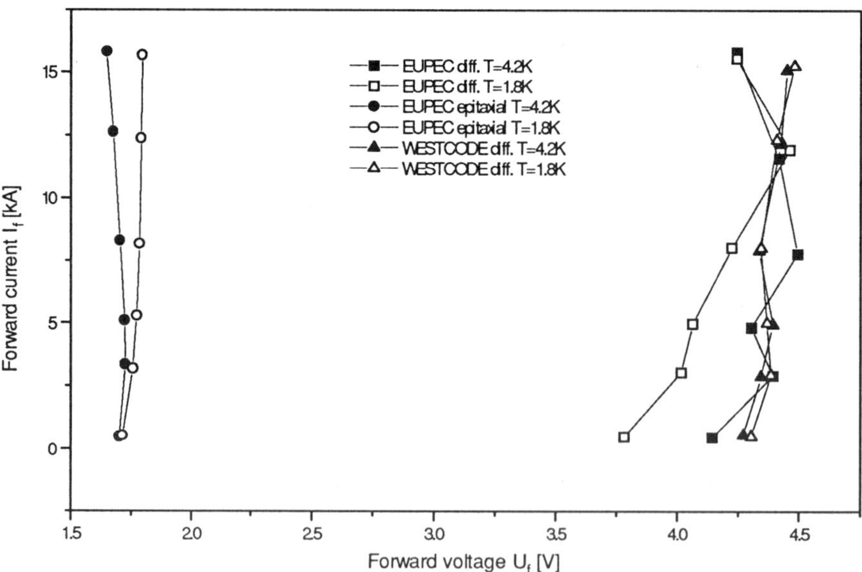


**Figure 2.** Forward current-voltage characteristic at 1.8 K and 4.2 K initial temperature for two diffusion type diodes and one epitaxial diode.

of about 1 mA. After warm up to 300 K, the reverse characteristics at 77 K and 300 K were found normal and unchanged. The same effect was observed on one of the diodes without passivation on the rim so that it cannot be attributed to the passivation. Some additional tests at several temperatures in the range between 4.2 K and 77 K lead to the conclusion, that the irreversible phenomenon starts below about 50 K and disappears at temperatures above 50 K. The observed behaviour is probably due to the impact ionisation of frozen out impurities leading to a generation of current filaments inside the depletion layer.[4]

**Turn-On Voltage Characteristics.** Turn-on voltages have been measured at 1.8 K and 4.2 K at different voltage rise rates in the range of 10 V/s < dU/dt < $10^4$ V/s.[5] The results in Table 2 allow the following general interpretation: the turn-on voltage $V_{to}$ increases with voltage rise rate, $dU_f/dt$, and $V_{to}$ at 1.8 K is higher than at 4.2 K. The lowest turn-on voltage for diffusion diodes is at least 3 times higher than for epitaxial diodes so that only one diffusion diode for the dipole by-pass needs to be used instead of two epitaxial diodes in series.

An abnormality has been observed on diode M14 each time after cool down to 4.2 K and 1.8 K: a turn on voltage of about $V_{to}$ = 60 V has been measured during the first $V_{to}$ measurement and this dropped down to reasonable values, as shown in Table 2, during subsequent measurements. This effect of excessive $V_{to}$ has already been observed on some of the HERA diodes tested at 4.2 K at CERN, when a $V_{to}$ of about 12V instead of 4-5V was measured. It seems to be a peculiarity of a few diodes of the diffusion type because it has never been observed on epitaxial diodes. This high $V_{to}$ could be dangerous for a magnet in the LHC by-passed since it will cause an additional delay before the diode is turned on.

## ENDURANCE TESTS

### Test Conditions

Diffusion type diodes from three manufacturers were submitted to endurance tests at liquid helium temperature similar to the procedure for epitaxial diodes.[6] Each diode was mounted under a force of about 44 kN between two copper disks of about 1600 $cm^3$ volume each acting as heat sinks. Prior to mounting, the quality of the contact surfaces between the diode electrodes and the heat sinks were checked with Fuji-paper prints.

Due to the rather high turn-on voltage, $V_{to}$, of about 4 - 17 V for the diffusion type diodes, compared to about 1.4 - 1.6V for epitaxial diodes, these diffusion diodes had to be tested individually in different runs ( power supply output voltage limitation ). A computer controlled power supply of 20 kA, 20 V provided the required current pulse consisting of a fast ramp to the peak current, $I_O$, within about 1 s and an exponential decay with a time constant of about 100 s.

During each current cycle, the current, the diode voltage, the heat sink temperatures, and the contact voltages were monitored versus time. Current and diode voltage measurements were used to calculate the junction diode temperature.[5] The heat sink temperature was monitored via Pt-100 sensors. Each diode was submitted to several successive current pulse cycles. After each cycle the reverse current-voltage characteristics of the diodes were re-measured after cooling down to 4.2 K.

### Endurance Test Results

**Forward Voltages.** The typical forward voltages of one epitaxial diode and three diffusion diodes during an endurance current cycle of 13 kA peak are plotted versus time in Figure 3. The insert is a zoom of the beginning of the cycle and shows the forward voltage just before turn on. As the resolution in time is 100 ms, the exact value of the turn-on voltage as shown in Table 1 is not always recorded. Immediately after this peak voltage, the forward voltage decreases rapidly within 10 s to about 1 V for the epitaxial diode and about 1.2 V for the diffusion diodes. Afterwards, up to 300 seconds, the average forward voltage is about 0.9 V for the epitaxial diode and about 1.1 V for the diffusion type diodes.

The high voltage of about 15 volt before turn on is typical for one of the firstly delivered diffusion diodes from EUPEC with gallium doping. The later diffusion type diodes from EUPEC are high level boron doped with a lower turn on voltage as shown in Table 1.

**Wafer Temperatures.** Figure 4 is a plot of the temperature of the diode junction itself versus time deduced from previous pulse current and voltage measurements. For all diffusion diodes, the maximum temperatures, reached after about 10-15 seconds, is between 300 K and 320 K. The temperature rise up to about 440 K for the epitaxial diode is due to a smaller heat sink volume (copper) foreseen for an installation in the magnet string at CERN. The slightly lower peak temperature of the WESTCODE diode, W046, can be attributed to the somewhat lower contact resistance between diode and heat sink which provides a better heat transfer to the heat sinks. The temperature decreases to about 200 K after 300 seconds. These wafer temperatures are only indicative, the two point calibration at 77 K and 300 K does not permit an absolute precision of better than about ± 20 K.

**Heat Sink Temperatures.** For all diffusion diodes the maximum heat sink temperature stays below 200 K, mainly due to the larger heat sink of about 1600 $cm^3$ of copper on each side of the diode, whereas the maximum temperature of the smaller heat sinks of the epitaxial diode of about 900 $cm^3$ reach about 220 K although the totally deposited energy is smaller than in all diffusion diodes. Using a larger heat sink volume also for the epitaxial diodes results in much lower temperatures for the wafer and the heat sinks.[6]

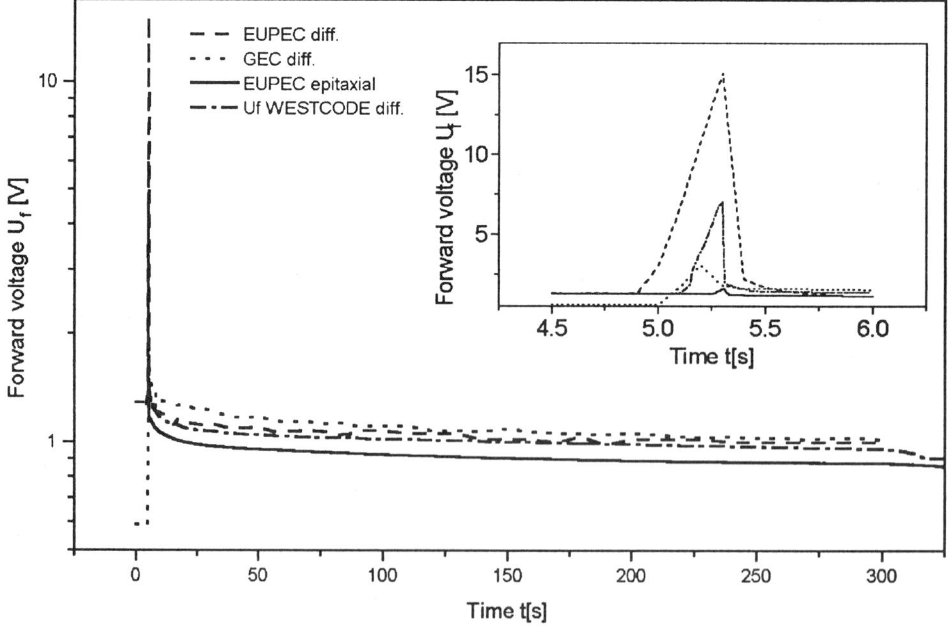


**Figure 3.** Forward voltages versus time of different types of diodes during an endurance test.

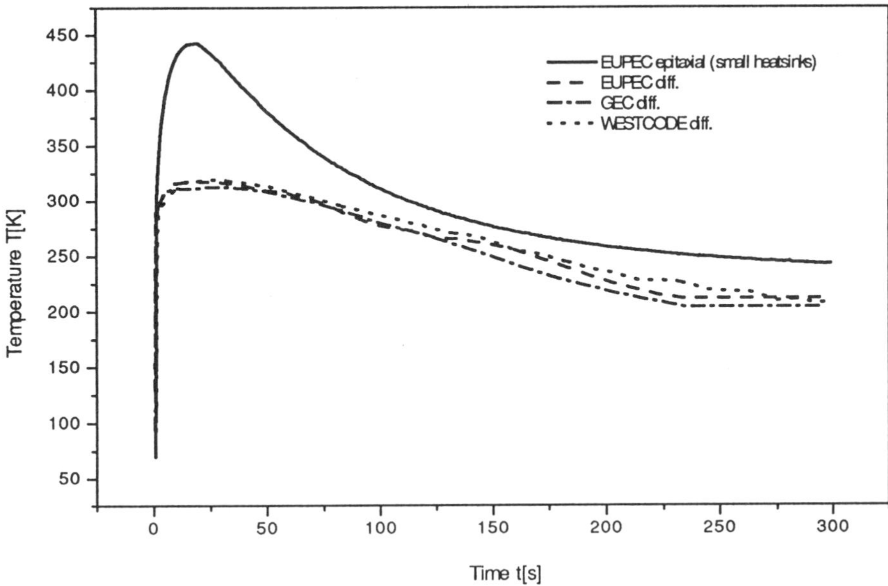


**Figure 4.** Wafer temperatures versus time of different types of diodes during an endurance test.

**Contact Resistances.** The electrical contact resistances between the heat sink and the electrodes of each diode were continuously monitored versus time. Average electrical contact resistances in the range between 0.3 μm and 4 μm were measured. The lower the electrical contact resistance the lower was the temperature difference between wafer and heat sink indicating a faster heat diffusion from the wafer to the heat sinks. In general, an average contact resistance value of a few μΩ is acceptable.

## PRELIMINARY RESULTS FROM IRRADIATION TESTS

The first results from irradiation tests at 77 K in an accelerator environment, just under way, show that the standard high-voltage diffusion diodes from GEC and WESTCODE are still too radiation sensitive, whereas the low voltage diodes from WESTCODE and especially the boron doped double diffusion diodes from EUPEC show sufficient radiation hardness up to a dose of about 300 Gy and fluences up to about $10^{12}$ neutrons/cm$^2$. A certain radiation hardness can be obtained only at the expense of lower reverse voltage. These irradiation test at 77 K in an accelerator environment will continue as well as irradiation tests at 4.2 K in a nuclear reactor.[7]

## CONCLUSIONS AND FUTURE EXPERIMENTS

The use of diffusion type diodes instead of epitaxial diodes for the high current by-pass, at least across the dipole magnets, will result in a significant cost reduction, since only one diode per double aperture dipole is necessary and diffusion diodes are less expensive than epitaxial diodes. The maximum wafer and heat sink temperatures depend on the copper heat sink volume and can be kept below critical values.

Occasionally, diffusion diodes show an abnormally high turn-on voltage at liquid helium temperatures. This will require systematic checking for elimination of those devices.

First irradiation results show that standard diffusion diodes are too radiation sensitive. Specially developed thin-wafer diffusion diodes are sufficiently radiation hard but at the cost of lower reverse voltage. The lower reverse voltage could be acceptable, at least for the quadrupole by-pass.

The following tests are required in order to validate completely the use of diffusion diodes for the high current by-pass in the LHC:

- irradiation and annealing of diode samples at 4.2 K
- establishing optimum annealing procedures after irradiation at 4.2 K
- over-irradiation of diode samples followed by endurance tests at 4.2 K
- testing of a small series of diodes from each manufacturer
  (up to now only a few prototypes have been tested).

## ACKNOWLEDGEMENTS

The authors would like to thank A. Gharib and O. Desebe for carefully carrying-out the measurements, J.M. Fraigne for the careful design of the sample holders, Mr. Rieubland and his team for their support during measurements at 1.8 K, and L. Walckiers and his team for their support during endurance testing.

## REFERENCES

1. L. Burnod, J. B. Jeanneret, and H. Schönbacher. Expected doses inside and around LHC dipoles, AC/DI/FA/ Note 93-06 (1993)
2. G. Stevenson, Private communication, 16.06.97
3. V. Berland, D. Hagedorn, A. Gharib, Power diodes submitted to radiation at 77 K in an accelerator environment - target T6 of TCC2 in the SPS -, CERN/LHC-ICP/VB 95-125, Internal note, July 1995.
4. R. E. Kunz, E. Schöll et al., *Solid-State Electronics*, Vol. 39, p.1155-1164, 1996
5. D. Hagedorn, W. Nägele, Quench protection diodes for the LHC at CERN, "Advances in Cryogenic Engineering (Materials) ", Vol.38, 1992
6. V. Berland, D.Hagedorn, A. Gharib, Comparison between two types of power diodes submitted to high current cycle endurance tests, CERN/LHC-ICP/VB 96-01, Internal Note April 1996
7. V. Berland, D. Hagedorn, H. Gerstenberg, Behaviour of the future LHC magnet protection diodes irradiated in a nuclear reactor at 4.6 K with intermediate annealing, RADiation Effects on Components and Systems conference (RADECS 95), Arcachon-France, September 1995.

# CRITICAL CURRENT DENSITY AND DISSIPATION IN SINTERED YBCO FILAMENTS

Y. S. Cha,[1] S. Y. Seol,[2] and J. R. Hull[1]

[1]Energy Technology Division, Argonne National Laboratory
Argonne, Illinois 60439 U.S.A.
[2]Chonnam National University
Kwangju 500-757, Korea

## ABSTRACT

Critical current density and dissipation of four sintered YBCO filaments were measured with a continuous DC power supply. Each sample exhibited somewhat different characteristics from the others even though three of the four samples were from the same batch. The data suggest that it is possible for the YBCO filament to be in a state of stable equilibrium when part of the filament is in the flux-flow state and part is in the normal state due to local hot spots. This behavior may be advantageous for resistive fault current limiters because the intermediate state (partly flux-flow and partly normal) has an overall resistance that is much higher than that of the flux-flow state but does not lead to thermal runaway (burnout) or excessive heating, unlike a YBCO filament driven into the normal state over its entire length. Two of the samples were broken as a result of local burnout.

## INTRODUCTION

In practical applications of high-temperature superconductors, it may be necessary to determine the dissipation characteristics of the superconductors beyond the critical current density usually defined by the 1 μV/cm criterion. For example, in predicting burnout of high-temperature superconductor current leads, information on flux-flow resistivity is needed.[1] In the design of a resistive fault current limiter, it is important to understand how the high-temperature superconductor responds to current above its critical current density.[2] In general, when calculating transient temperature fields in a conductor, the electrical resistivity of the superconductor in all thermodynamic states is required in order to determine the heat generation rate.[3] Relatively few reported data are available on dissipation (flux-flow) of various high-temperature superconductors well above its critical current density[4-6] because much of the reported data focus on critical current density. Data on burnout are even more scarce because burnout destroys the sample.

In this paper, we describe the results of measuring the critical current density and dissipation (up to local burnout) of four sintered YBCO filaments in liquid nitrogen. Implications of the results for application in resistive fault current limiters are discussed.

*Advances in Cryogenic Engineering,* Vol. 43
Edited by P. Kittel, Plenum Press, New York, 1998

## EXPERIMENTAL SETUP

The standard four-point technique was employed in the measurements. The current leads were made of copper braids that were silver-soldered to the ends of the YBCO filaments. Contact areas between the YBCO filament and the copper braids were relatively large because the filament was inserted into the braid. This helped reduce contact resistance between the YBCO filament and the current leads. Dissipation in the YBCO filaments was measured over a wide range of current far beyond the critical current density (1 μV/cm criterion). All tests were conducted at 77 K and with a continuous DC power supply. Four YBCO filaments were tested (samples 1 through 4); their dimensions and conditions are summarized in Table 1. The length of each YBCO filament was ≈77 mm. The diameter given in Table 1 is an approximate value because filament diameter was not uniform and could vary slightly along the filament length.

Sample 1 was a filament received more than 6 months before the testing date and stored in the laboratory with no special attention, such as storage in a desiccant chamber. Samples 2 to 4 were tested shortly after they were received and were also stored in the laboratory environment. All data reported here were obtained under steady-state conditions.

## EXPERIMENTAL RESULTS

Figure 1 shows the electric potential vs. current density (E/J) data for sample 1, which had a critical current density of 1,000 A/cm$^2$ (the critical current was 4.6 A), the highest among all the samples tested so far even though this sample had been stored in the laboratory for more than six months. As the current density was increased above 1,425 A/cm$^2$, which corresponds to a current of 6.5 A), the measured voltage suddenly jumped and then dropped, indicating some kind of instability. The test was then stopped and an attempt was made to remove the sample from the holder for visual examination. However, during removal, the filament was broken. Visual examination under a magnifying lens showed that the filament had broken because of mechanical stress (bending), not as a result of burnout, because both ends of the break showed that the filament structure (or texture) was not different from the bulk material (similar to that shown in Figure 4). This is supported by evidence that the current density was increased to a maximum of only 1.43 times the critical current density, and burnout is not likely to have occurred at this relatively small ratio of $J/J_c$.

**Table 1.** 77 K Test of YBCO filament

| | Sample 1 | Sample 2 | Sample 3 | Sample 4 |
|---|---|---|---|---|
| Diameter (mm) | 0.76 | 0.89 | 0.76 | 0.84 |
| Distance between voltage taps (mm) | 19 | 30 | 26 | 26 |
| Critical current (A) | 4.6 | 2.3 | 4.0 | 0.1 |
| Critical current density, 1 μV/cm criterion (A/cm$^2$) | 1000 | 370 | 875 | 18 |
| Date received | 12/16/94 | 7/10/95 | 7/10/95 | 7/10/95 |
| Pre-test condition | Stored in lab environment | Stored in lab environment | Stored in lab environment | Stored in lab environment |
| Post-test condition | Broken mechanically | Broken due to burnout | Intact | Broken due to burnout |
| Burnout current density (A/cm$^2$) | >1,425 | 820 | >1,575 | 2,100 |
| Burnout current density ratio ($J/J_c$) | >1.4 | 2.2 | >1.8 | 116.6 |

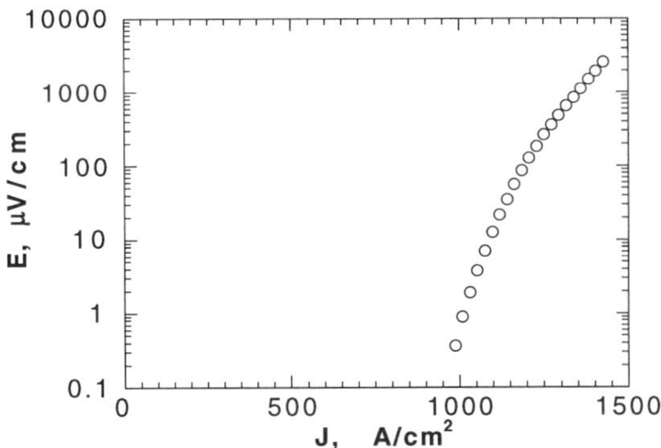


**Figure 1.** E versus J for sample 1.

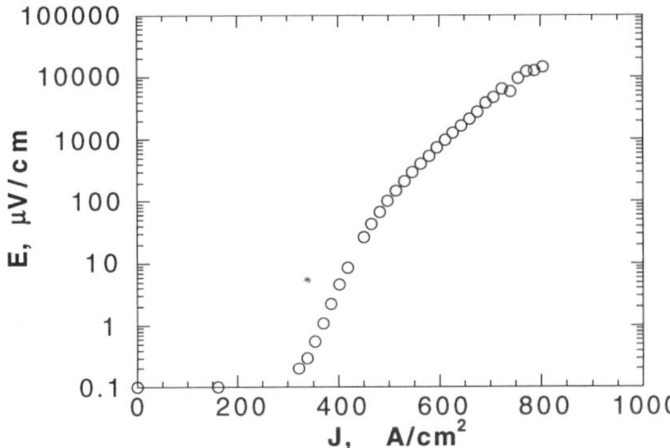


**Figure 2.** E versus J for sample 2.

Figure 2 shows the E/J data for sample 2. The critical current density of this sample was 370 A/cm$^2$ ($I_c$ = 2.3 A), considerably lower than that of sample 1. The data points shown in Figure 2 are average values because at currents >4 A, the voltage (electric potential) showed fairly large fluctuations (±10%). The test ended at I = 5.1 A because at this current, the electric circuit suddenly opened, indicating that the loop was broken. Post-testing examination showed that the YBCO filament was indeed broken near the solder joint of one of the braided current leads. Further examination under a magnifying lens revealed that the color of the YBCO filament on both sides of the break changed from black to gray, as shown in Figure 3. This color change at the break is quite different from that of a break due to bending (which exhibited no change in color, as shown in Figure 4). The grayish filmlike texture on the surfaces of the YBCO filament near the break is probably the result of local heating that eventually led to burnout. Local heating could be caused by inhomogeneity of material properties. A current of 5.1 A corresponds to a current density of 820 A/cm$^2$ and a ratio of $J/J_c$ = 2.2, which is higher than the ratio achieved for sample 1.

Figure 5 shows the E/J data for sample 3; critical current density for this sample was 875 A/cm$^2$. The numbers in Figure 5 indicate the order in which these data points were taken. For example, the experiment starts with point 1 and progresses to points 2, 3, etc., finally ending with point 21. By following the number, one can determine whether the data were taken in the direction of increasing or decreasing current; the arrows in the figure also help indicate the direction of increasing or decreasing current. Figure 5 exhibits interesting behavior for sample 3 that was not observed in samples 1 or 2. First, there is apparently a bistable region for current density between approximately 1,080 and 1,550 A/cm$^2$. Within this range, the electric potential can have two different values. When the electric potential of

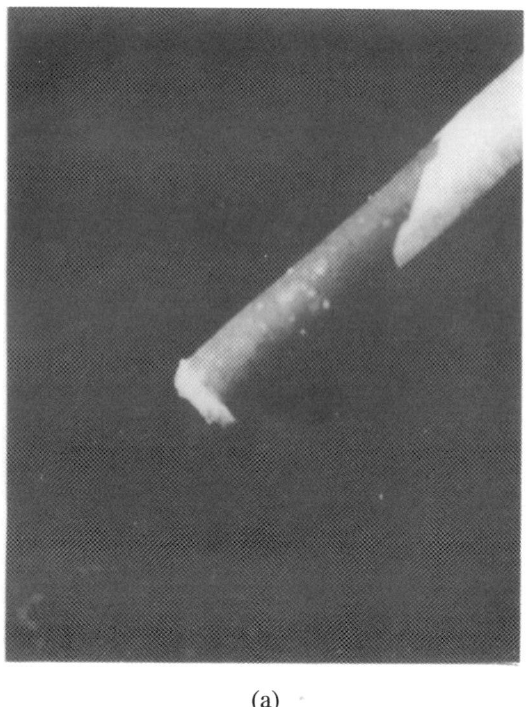

(a)

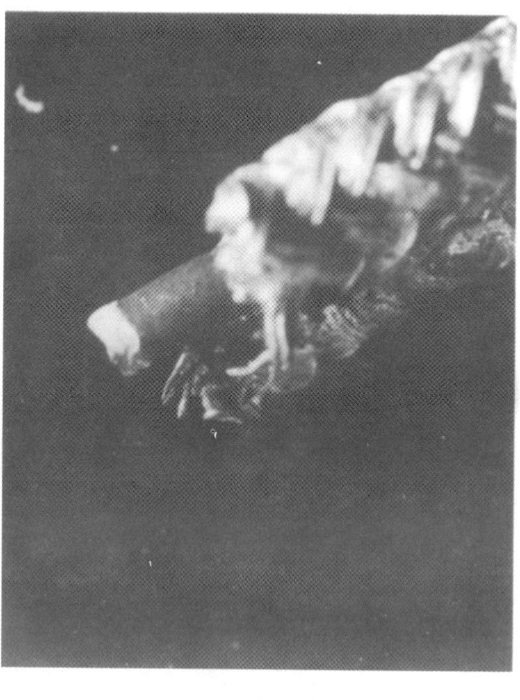

(b)

**Figure 3.** Photomicrograph of area near break (due to burnout) of sample 2: (a) on one side of break; (b) on other side of break.

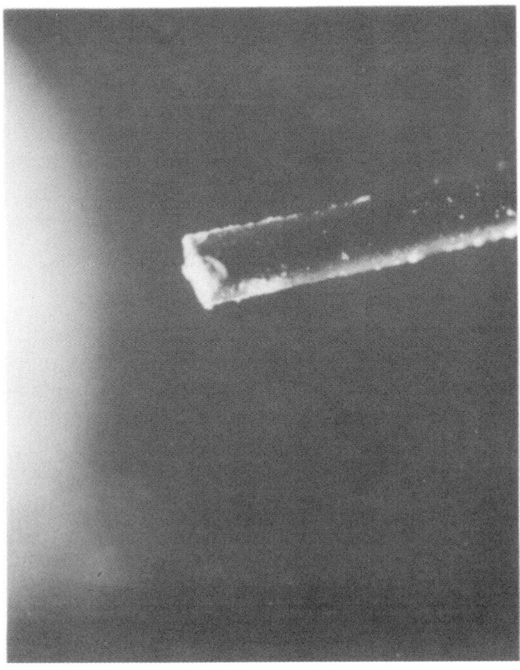

**Figure 4.** Photomicrograph of area near break (due to bending) of sample 2.

the YBCO filament is on the upper curve, we have observed a relatively large volume of vapor rising continuously toward the liquid/air interface from a location near one of voltage taps, indicating significant local heat generation. When the YBCO filament is on the lower curve, boiling is fairly uniform and there is no local concentration of boiling. Thus, the lower curve may correspond to the flux-flow state and the upper curve may correspond to a partially normal state when local hot spots become normal (near one of the voltage taps). It is also probable that the results shown in Figure 5 represent a transition between nucleate and film boiling.

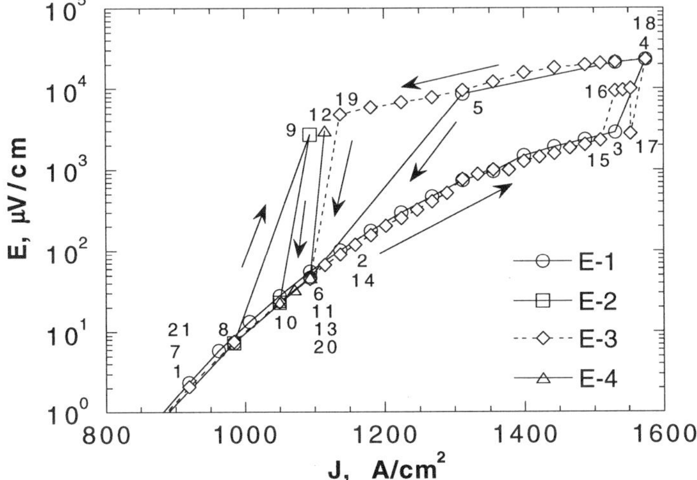


**Figure 5.** E versus J for sample 3.

Whether the electric potential will remain on the upper or the lower curve depends on the magnitude of ΔJ. For example, at point 8, a large increase in J (ΔJ ≅ 100 A/cm$^2$) caused E to jump from 7 to 3000 μV/cm (point 9). However, when a lower ΔJ is used, the data points move along the lower curve, as shown in Figure 5. The jump from lower to upper curve is likely due to a sudden large increase in resistive heating as a result of a large current increase. The YBCO heats almost adiabatically in such a short period that local temperature may exceed the critical value and the YBCO will then become normal, resulting in the voltage jump shown in Figure 5. The YBCO turns normal only locally at the hot spot (near one of the voltage taps) that may be present because the material is not homogeneous. Once part of the YBCO becomes normal, it remains in the normal state indefinitely, and locally it does not return to the flux-flow state. For lower ΔJ, local temperature anywhere never exceeded the critical temperature and the entire YBCO filament remains in the flux-flow state and on the lower curve. For J < 1,050 A/cm$^2$, the YBCO filament appears to be stable only in the flux-flow state and for J > 1,550 A/cm$^2$, the YBCO filament is apparently stable only in the partially normal state. As J increases near or slightly above 1,550 A/cm$^2$, the state of the YBCO filament becomes quite unstable and the voltage reading fluctuates up and down, as shown by data points 15-17 in Figure 5.

The tests of sample 3 were conducted up to a current ratio of $J/J_c \cong 1.8$. Even though part of the YBCO filament might have been in the normal state, equilibrium could still be achieved without experiencing thermal runaway up to $J/J_c = 1.8$. Sample 3 remained intact after the tests and could be used again if necessary.

Figure 6 shows E/J data for sample 4, which had a critical current density of only 18 A/cm$^2$ (1 μV/cm criterion). The flux-flow state existed over a wide range of current density up to J ≅ 1,400 A/cm$^2$ ($J/J_c \cong 78$). Between J = 1,400 and 1,900 A/cm$^2$, the YBCO is in an intermediate state that appears to be quite unstable and can jump up and down, as shown by the data points at J ≅ 1,630 A/cm$^2$. This intermediate state could correspond to a partially normal state as a result of a local hot spot similar to that of sample 3. When current density increases above 1,630 A/cm$^2$, there is another jump in voltage that could mean that another part of the YBCO also becomes normal. However, unlike that in sample 3, boiling appears to be fairly uniform and no local concentration of boiling was observed even though boiling intensity increases with current density. Further increase in current density to J = 2,100 A/cm$^2$ leads to thermal runaway and burnout because more and more parts of the YBCO become normal. The first evidence of burnout is, of course, detection of an open circuit. Post-testing examination showed that the YBCO filament had broken approximately 4-5 mm from one of the voltage taps. Figure 7 shows one side of the break (diameter of the YBCO filament is 0.84 mm); a large crack is clearly visible near the break. Similar to sample 2, the color of the YBCO changed from black to gray, indicating that the break is due to burnout rather than mechanical stress.

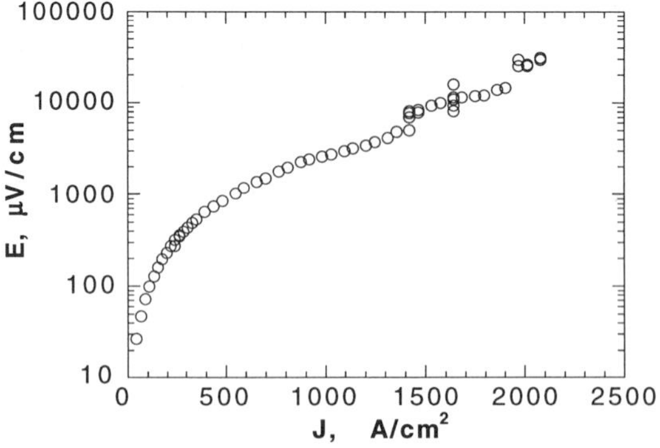


**Figure 6.** E versus J for sample 4.

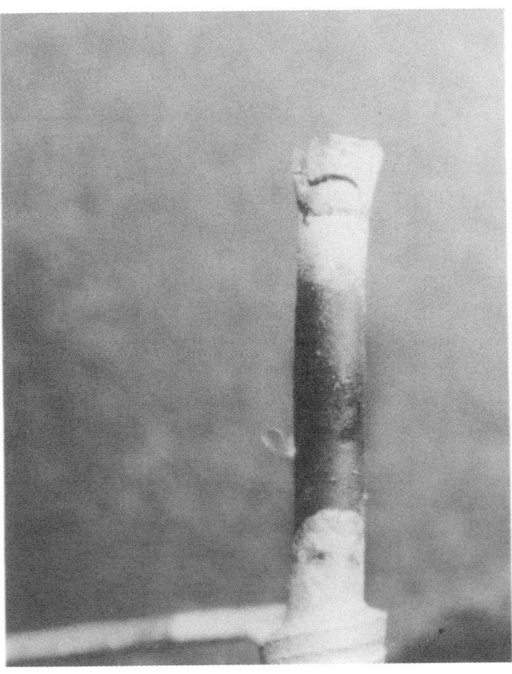

**Figure 7.** Photomicrograph of area near break (due to bending) of sample 4.

## SUMMARY

We tested four YBCO filaments (samples 1-4) at 77 K, and each sample exhibited somewhat different characteristics from the others even though samples 2-4 were from the same batch. It is not clearlyunder stood why the samples exhibited quite different critical currents. One possible explanation is that the diameters of the samples are relatively small (less than 1 mm), and local inhomogeneity may have a fairly large impact on the electrical properties of the samples. Samples 2 and 4 were broken as a result of burnout at local hot spots. The data from sample 3 seem to suggest that it is possible for the YBCO filament to be in stable equilibrium when part of the filament is in the flux-flow state and part of the filament is in the normal state because of local hot spots. As long as the number of local hot spots is limited, thermal runaway does not occur and the superconducting state can be recovered when the current is reduced. This behavior may be advantageous for resistive fault current limiters because the intermediate state (partly flux-flow and partly normal) has an overall resistance much higher than that of the flux-flow state but does not lead to thermal runaway (burnout) or excessive heating as it would in a YBCO filament driven into normal state over its entire length. The question of recovery time should be addressed and requires further experiments under transient conditions.

## ACKNOWLEDGMENTS

This work has been supported by the U.S. Department of Energy, Energy Efficiency and Renewable Energy, as part of a program to develop electric power technology, under Contract W-31-109-Eng-38. The authors thank Henry Muller of Illinois Superconductor Corp. for providing the YBCO samples.

## REFERENCES


1. S.Y. Seol, Y.S. Cha, R.C. Niemann, and J.R. Hull, Prediction of burnout of a conduction-cooled BSCCO current lead, *IEEE Trans. Appl. Supercond.* 7(2):696-699 (1997).
2. J. D. Hodge, H. Muller, D.S. Applegate, and Q. Huang, A resistive fault current limiter based on high temperature superconductors, *Appl. Supercond.* 3(7-10):469-482 (1995).
3. A. Abeln, E. Klemt, and H. Reiss, Stability considerations for design of a high temperature superconductor, *Cryogenics* 32:269-278 (1992).
4. Y.S. Cha, S.Y. Seol, D.J. Evans, and J.R. Hull, Flux-flow resistivity of three high-temperature superconductors, *IEEE Trans. Appl. Supercond.* 7(2):2122-2125 (1997).
5. Y.S. Hascieck and L.R. Testardi, Constant E-J relation in the current induced resistive state of $YBa_2Cu_3O_{7-x}$, *IEEE Trans. Magn.* 27:1186-1189 (1991).
6. T.R. Askew, J.G. Nestell, R.B. Flippen, D.M. Groski, and N.McN. Alford, Dynamic measurement of flux flow resistivity in $YBa_2Cu_3O_7$ wires, *IEEE Trans. Appl. Supercond.* 3:1398-1401 (1993).

# A POSSIBLE 1.8 K REFRIGERATION CYCLE FOR THE LARGE HADRON COLLIDER

F. Millet,[1] P. Roussel,[1] L. Tavian[2] and U. Wagner[2]

[1] DSM/DRFMC/SBT, CEA Grenoble
F-38054 Grenoble cedex 9, France

[2] LHC Division, CERN
CH-1211 Geneva 23, Switzerland

## ABSTRACT

The Large Hadron Collider (LHC) under construction at the European Laboratory for Particle Physics, CERN, will make use of superconducting magnets operating below 2.0 K. This requires, for each of the eight future cryogenic installations, an isothermal cooling capacity of up to 2.4 kW obtained by vaporisation of helium II at 1.6 kPa and 1.8 K. The process design for this cooling duty has to satisfy several demands. It has to be adapted to four already existing as well as to four new refrigerators. It must cover a dynamic range of one to three, and it must allow continuous pump-down from 4.5 K to 1.8 K. A possible solution, as presented in this paper, includes a combination of cold centrifugal and warm volumetric compressors. It is characterised by a low thermal load on the refrigerator, and a large range of adaptability to different operation modes. The expected power factor for 1.8 K cooling is given, and the proposed control strategy is explained.

## INTRODUCTION

The European Laboratory for Particle Physics, CERN, is working towards the construction of the LHC, a high-energy, high-luminosity particle accelerator and collider of 26.7 km circumference. The key technology of the LHC is the use of high-field superconducting magnets using Nb-Ti conductor operating in superfluid helium.[1] To maintain the some 25 km of bending and focusing magnets at their operating temperature below 2.0 K, the LHC cryogenic system will have to produce an unprecedented total refrigeration capacity of about 18 kW at 1.8 K, in eight cryogenic plants distributed around the machine's circumference. To reach this goal, several solutions of process design for the cooling loop of the 1.8 K refrigeration stage have been studied.

## CONSTRAINTS AND REQUIREMENTS

The cryogenic system as it will be installed for the LHC machine is facing several constraints. Of the eight refrigerators which are necessary to fulfil the cryogenic capacity demand, four will have to use the already existing installations for the Large Electron Positron collider (LEP). These LEP refrigerators must be upgraded in capacity and completed with a 1.8 K refrigeration stage.

To limit the number of design solutions, one single type of 1.8 K refrigeration stage shall be used both to complete the existing 4.5 K refrigerators of the LEP machine, and for the four new refrigerators. A suitable interface must therefore be found which can be adapted to the existing and to the new 4.5 K refrigerators.

In view of facility constraints, particularly in underground caverns, the four new 4.5 K refrigerators will be installed integrally at the surface,[2] thus differing from the existing LEP units which are of a split-cold-box design.[3] To limit the number of cryogenic transfer lines from the tunnel level, where the 1.8 K refrigeration stage must be located, to the 4.5 K refrigerators on ground level, it is desirable to connect the 1.8 K stage via the connections which serve as supply and return between the 4.5 K refrigerator and the LHC sector.

The demands in refrigeration capacity for the LHC, where all cooling which is to be provided by the 4.5 K refrigerators is non-isothermal, are totally different from those for LEP, where 80 % of the cooling is done isothermally at 4.5 K. The 1.8 K refrigeration stage should therefore complete the existing refrigerators in a way that their upgrade for LHC does not require excessive changes at the existing installations.

Due to the strong influence of the operating conditions of the LHC machine on the cryogenic cooling requirements,[4] the capacity at 1.8 K will vary by a factor of three between stand-by and full-load operation. The 1.8 K refrigeration stage must adapt to these variations in cooling capacity. Moreover, each 1.8 K refrigeration stage must cool down the sector of 3300 m length from 4.5 K to below 2.0 K without running into unstable operation of the cold compressors.

## DIFFERENT SOLUTIONS FOR THE 1.8K REFRIGERATION STAGE

A wide range of possible design solutions for the 1.8 K stage have been studied at CERN and in collaboration with the CEA Grenoble, in order to define a reference solution which satisfies all requirements mentioned in the chapter above. Common to all solutions is the suction pressure of the cold compressors, which is fixed at 1.5 kPa. Two principle solutions for the 1.8 K refrigeration stage were identified :

a) cycles without ambient-temperature compressors
b) cycles with ambient-temperature compressors

Figure 1 shows the principle block diagrams for these two solutions. Cycles including ambient-temperature compressors can further be "active" i.e. expand high-pressure helium in turbo expanders and thus create cooling capacity. In both cases, the cold compressor box (CCB) receives helium at 4.5 K from the main cold box, supplies the 1.8 K cooling loop, and feeds helium back at a certain temperature level to the main cold box.

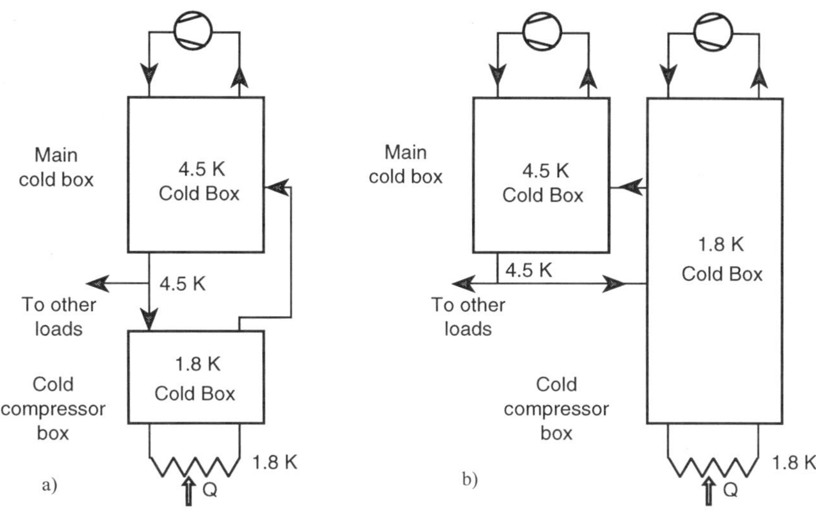


**Figure 1.** The principle choice for the 1.8 K refrigeration stage; a) cycles without ambient-temperature compressors, b) cycles with ambient-temperature compressors.

## Possible Interface Conditions

The possible interface conditions between the CCB and the main cold box, which apply equally to the existing and to the new 4.5 K refrigerators, are listed in Table 1.

**Table 1.** Possible interface conditions between CCB and main cold box

| | Temperature [K] | Pressure [MPa] |
|---|---|---|
| Supply | 4.5 | 0.30 |
| Return A | 20 | 0.13 |
| Return B | 50 | 1.85 |
| Return C | 75 | 1.60 |

## Cold Compressor Cycles without Ambient-Temperature Compressors

Cold compression to more than 0.13 MPa is excluded by the design of the cold hydrodynamic compressors as well as by the excessive exergetic losses it would create. For solutions without ambient-temperature compressor stages, the only possible return interface is at 20 K and 0.13 MPa as shown in Figure 2. A comparable cycle was realised for the CEBAF refrigerator system.[5] Unlike the CEBAF example, these solutions would need additional cooling capacity to be provided by the 4.5 K refrigerator in order to reach the 20 K limit, as cold compression from 1.5 kPa to 0.13 MPa leads to discharge temperatures between 40 K and 50 K for compressors with an isentropic efficiency between 75% and 60%, respectively. All CCB cycles without ambient compressors therefore increase the capacity demands on the main cold box, which is particularly undesirable for the existing installations.

## Cold Compressor Cycles with Ambient-Temperature Compressors

Cold compression with ambient-temperature compressors could have any return interface at temperature levels between 300 K and 4.5 K, and pressure levels up to the supply value from the ambient-temperature compressors.

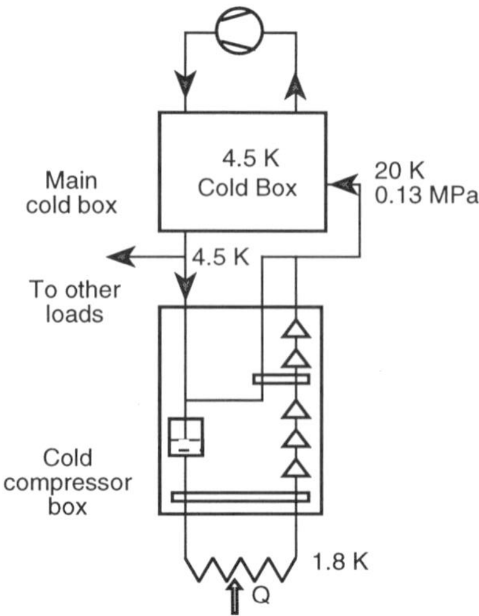


**Figure 2.** The 1.8 K refrigeration stage without ambient-compressors connected to a 4.5 K refrigerator via the possible interface connections (Return A)

The block diagrams of four different solutions including compression at ambient temperature are shown in Figure 3 and Figure 4. The two cycles in Figure 3 are characterised by the fact that no cooling capacity is created by expansion of helium gas in the cold compressor box. The helium returned to the main cold box at high pressure is expanded in turbines located therein. The difference between the two cycles in Figure 3 is given that in one case five stages of cold centrifugal compressors pump the helium from 1.5 kPa to above atmospheric pressure, whereas in the other case cold compression is only used to a pressure below atmospheric level, thus requiring warm machines with sub-atmospheric suction. A cycle with two stages of cold centrifugal compressors and sub-atmospheric suction was already realised for the refrigerator of the Tore Supra superconducting tokamak.[6]

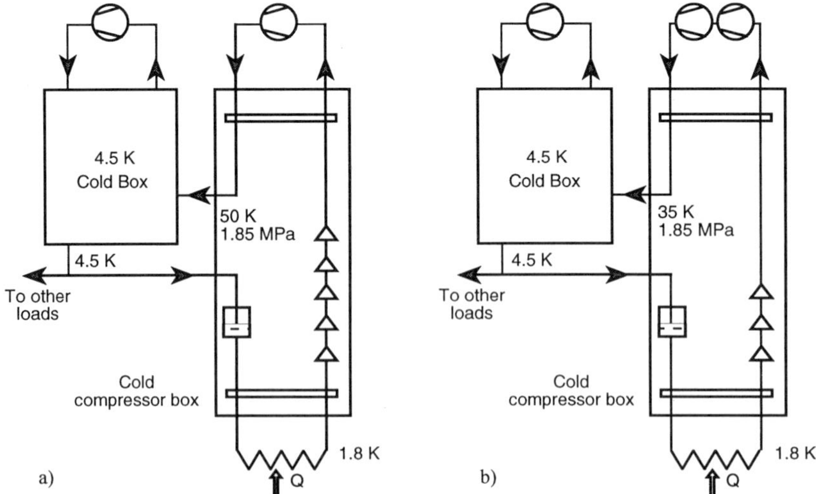


**Figure 3.** 1.8 K refrigeration stages including ambient-temperature compressors with no creation of cooling capacity via expanders; a) with cold compression to pressures above 0.1 MPa, b) with cold compression to pressures below 0.1 MPa.

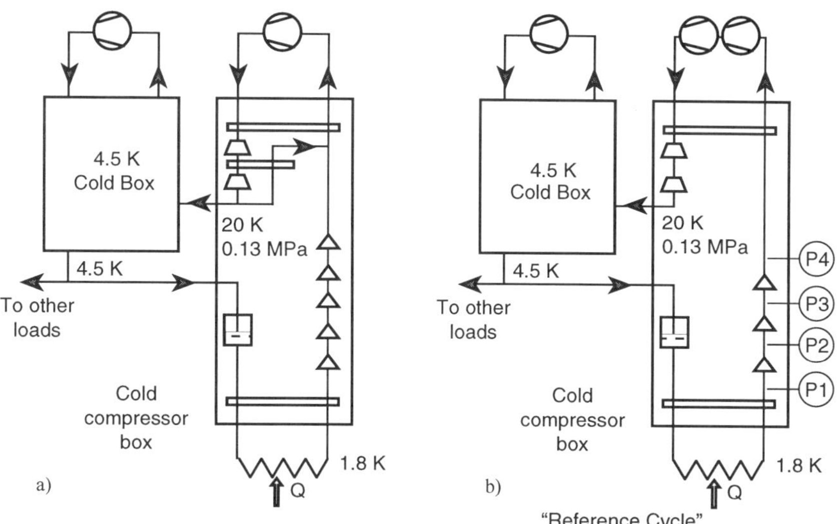


**Figure 4.** 1.8 K refrigeration stages including ambient-temperature compressors with creation of cooling capacity in turbo expanders; a) with cold compression to pressures above 0.1 MPa, b) with cold compression to pressures below 0.1 MPa.

The two cycles in Figure 4 are characterised by the fact that the high-pressure helium is expanded in turbines inside the cold compressor boxes. The helium gas is in both cases returned to the main cold box at 20 K and 0.13 MPa.

## THE 1.8 K CYCLE PROPOSED BY CERN AS REFERENCE SOLUTION.

The cycle which is represented on the right-hand side of Figure 4 was chosen by CERN as a suitable reference for comparison with other solutions, after evaluating its possible advantages and disadvantages. It uses three stages of cold centrifugal compressors, with a pressure ratio of three for each stage, up to a pressure level of about 40 kPa. At this pressure, efficient heat exchange in standard aluminium-plate-fin heat exchangers is possible. For ambient-temperature compressors, either a combination of liquid ring pumps up to 0.11 MPa and helical screw compressors for the high pressure level, or only helical screw compressors can be chosen. The overall efficiency of this cycle depends strongly on the efficiency of the warm machinery, which would favour the use of screw compressors. Even with isentropic efficiencies of 60% for the cold compressors, which is considered as conservative for the size of compressors to be installed for the LHC, a return temperature of 20 K or lower can be reached without additional cooling flow inside the CCB. During reduced capacity operation, the suction pressure of the warm volumetric compressors can be left floating, which eases considerably the control of the centrifugal machines and expands the operation range of the whole cycle. Moreover, this cycle contributes to the total refrigeration capacity, which decreases the demand on the main cold boxes. This is specially of interest for the reuse of the existing LEP refrigerators.

### Advantages of the Reference Solution

The advantages of a CCB cycle with a mixture of cold centrifugal and warm volumetric compression are:

– good cycle efficiency,
– lowering of the capacity demand for the main refrigerator,
– good adaptability to reduced-capacity operation,
– good controllability during cool-down of a machine sector below 4.5 K.

### Disadvantages of the Reference Solution

The only disadvantage of the 1.8 K cycle proposed as reference solution is that the first stage of the warm volumetric compressors has a sub-atmospheric suction pressure. This is generally avoided in helium refrigerators as it increases the risk of air contamination in the process gas. The experience of the French CEA with the Tore Supra refrigerator shows however, that a carefully designed warm sub-atmospheric system allows reliable operation. Moreover, a cryogenic system the size of one LHC sector has a multitude of inherent sources for contamination. An efficient purification system has to be installed, which can also be designed to cope with possible impurities resulting from the sub-atmospheric suction of the warm compressors.

## REDUCED-CAPACITY OPERATION

The 1.8 K cycle must be capable of operating continuously with flow rates down to a third of its nominal value. As operation at low flow can occur for extended time periods, it is not desirable to compensate by electrical heating in a 1.8 K bath.[7] Hydrodynamic cold compressors cannot easily match this flow variation at constant pressure ratio. In the reference solution for the 1.8 K cycle, it is envisaged to let the suction pressure of the warm volumetric machines decrease proportional with the decrease of flow, from e.g. 40 kPa at full load to 13 kPa at low capacity. A feasible and rather simple control strategy for three cold compressors in series is given in Table 2. The identifiers for the pressure points refer to those shown in Figure 4. The lowest-pressure stage is controlled to a fixed suction pressure of 1.5 kPa. The second stage is controlled to have a third of the total pressure ratio between suction of the first stage and discharge of the third stage. The third stage is controlled to a fixed pressure ratio of three in all load cases. The control is done for each cold compressor by varying the rotational speed. Following this control strategy, all centrifugal compressors stay well within their possible field of operation with a comfortable distance from the stall and choke limits.

## POWER FACTOR

To calculate the expected power factor for 1.8 K cooling, assumptions must be made concerning the efficiency of cold compressors, heat exchangers, turbo expanders and warm volumetric compressors, as well as of the future 4.5 K refrigerators.

**Table 2.** Control strategy for the cold compressors of the reference cycle during turn-down operation

| Cold compressor | Control strategy | Control variable |
|---|---|---|
| First stage | P1 = 1.5 kPa | Rotational speed first stage |
| Second stage | $P3/P2 = (P4/P1)^{1/3}$ | Rotational speed second stage |
| Third stage | P4/P3 = 3 | Rotational speed third stage |

For the cold centrifugal compressors, an isentropic efficiency of 60% seems reasonable, while that estimated for the turbo expanders is 70%. The heat exchanger efficiency is assumed to be 96%. For warm screw compressors with sub-atmospheric suction, an isothermal efficiency of 30% could be envisaged; for screw compressors operation above 0.1 MPa suction this value can be assumed to 50%. The Carnot efficiency of the cold box for the main refrigerator is further estimated to be 50%.

Using these rather conservative efficiency figures, the power factor for the 1.8 K cooling would be about 1060 W/W in the case of the proposed reference cycle. If for comparison one uses the same assumptions to calculate the power factor of a cycle as it is represented in Figure 4 a, the resulting value would be 1170 W/W.

## CONCLUSION

The cold compressor cycle which is proposed by CERN to be combined with the 4.5 K cold boxes of the LHC cryogenic system, is adaptable to the existing refrigerators of the LEP machine, which are of split design, as well as to the all surface installed new 4.5 K refrigerators, which will have to be installed in the coming years. No additional cryogenic transfer line connections are necessary besides those already foreseen to supply the LHC machine from the 4.5 K refrigerators. The cycle allows reduced capacity operation to a third of the nominal mass flow, and continuous pump-down of the LHC sector from 4.5 K. Moreover, it shows a promising potential for good overall efficiency. The sub-atmospheric suction of its warm compressors demands careful design and provisions to remove air contamination from the process helium.

## ACKNOWLEDGEMENTS

This work was performed in the framework of a collaboration agreement between CERN and CEA, Grenoble and Cadarache, France.

## REFERENCES

1. The LHC study group, "The Large Hadron Collider, Conceptual Design," CERN Report/AC/95-05(LHC)
2. U. Wagner, "The LHC refrigerators with surface-located cold boxes for the temperature range 300-4.5 K," LHC Project Note 70 (1996)
3. S. Claudet, W.K. Erdt, P.K. Frandsen, Ph. Gayet, N.O. Solheim, Ch. Titcomb and G. Winkler, Four 12 kW / 4.5 K cryoplants at CERN, *Cryogenics* (1994) 34 ICEC Supplement, p. 83-86
4. Ph. Lebrun, G. Riddone, L. Tavian and U. Wagner, Demands in refrigeration capacity for the Large Hadron Collider, in: "Proceedings of ICEC 16," T. Haruyama, T. Mitsui, K. Yamafuji, ed., Elsevier Science (1997), p. 95-98
5. C.H. Rode, D, Arenius, W.C. Chronis, D. Kashey and M. Keesee, 2.0 K CEBAF cryogenics, in: "Advances in Cryogenic Engineering," R.W. Fast, ed. Plenum Press, 35A (1990), p. 275-286
6. G. Claudet and R. Aymar, Tore supra and helium-II cooling of large high-field magnets, in: "Advances in Cryogenic Engineering," R.W. Fast, ed. Plenum Press, 35A (1990), p. 55-67
7. B.S. Bevins, W.C. Chronis, and M.S. Keesee, Automatic pumdown of the 2 K cold compressors for the CEBAF central helium liquefier, in: "Advances in Cryogenic Engineering," Plenum Press, N.Y. (1996), p. 663.

# A SIMPLIFIED CRYOGENIC DISTRIBUTION SCHEME FOR THE LARGE HADRON COLLIDER

M. Chorowski, W. Erdt, Ph. Lebrun, G. Riddone, L. Serio, L. Tavian, U. Wagner, and R. van Weelderen

LHC Division, CERN, CH-1211 Geneva 23, Switzerland

## ABSTRACT

The Large Hadron Collider (LHC), currently under construction at CERN, will make use of superconducting magnets operating in superfluid helium below 2 K. The reference cryogenic distribution scheme was based, in each 3.3 km sector served by a cryogenic plant, on a separate cryogenic distribution line which feeds elementary cooling loops corresponding to the length of a half-cell (53 m). In order to decrease the number of active components, cryogenic modules and jumper connections between distribution line and magnet strings a simplified cryogenic scheme is now implemented, based on cooling loops corresponding to the length of a full-cell (107 m) and compatible with the LHC requirements. Performance and redundancy limitations are discussed with respect to the previous scheme and balanced against potential cost savings.

## INTRODUCTION

In the former conceptual design of the LHC[1], the cryogenic distribution is based on a four-point feed scheme, with eight cryogenic plants distributed in pairs at each even point and each dedicated to the cooling of a complete 3.3-km sector. Within a sector, a separate cryogenic line feeds elementary cooling loops corresponding to the length of a half-cell (53 m). For a maximum of standardization, the layout of functional components (bus-bar plugs, cooldown-and-fill valves, safety relief valves, 1.8-K cooling loop components, etc.) in all elementary cooling loops is identical. The subcooling of helium for the superfluid helium cooling loop is centralized in the cold compressor boxes and its distribution requires a dedicated header for liquid helium at 2.2 K and 0.13 MPa.

The optimization of the distribution scheme by rearranging the cryogenic plant location and reducing the number of components and distribution headers as well as by increasing the cooling loop length to a full-cell, leads to major cost savings, but forces us to abandon the concept of complete standardization.

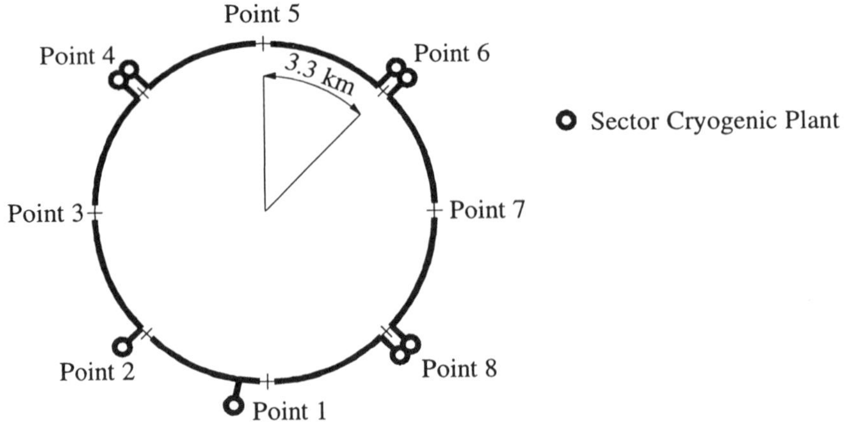


**Figure 1.** Location of LHC cryogenic plants.

## CRYOGENIC ARCHITECTURE

In view of installation and space constraints, particularly in underground caverns, and following a technical-economical study[2], the general architecture of the cryogenic system[3] has been modified. The four new cryogenic plants, which have to be added to the four existing split-coldbox helium refrigerators of LEP, are now integrally located at surface level. In addition, the decision of grouping two cryogenic plants at each even point was overruled for point 2 which will only keep the upgraded LEP refrigerator, since the other cryogenic plant is now foreseen in the area 18 of the point 1. Figures 1 and 2 show the location and the new architecture of the LHC cryogenics.

As a consequence of this new architecture, the cryogenic plant at point 1, located at the lowest point of the sector 1-2 gives a better cooling performance of this sector with respect to the previous configuration. In point 1, the cryogenic plant will first be used as a source of liquid helium for component tests. However, the interconnection between two neighbouring plants is no longer possible at point 2.

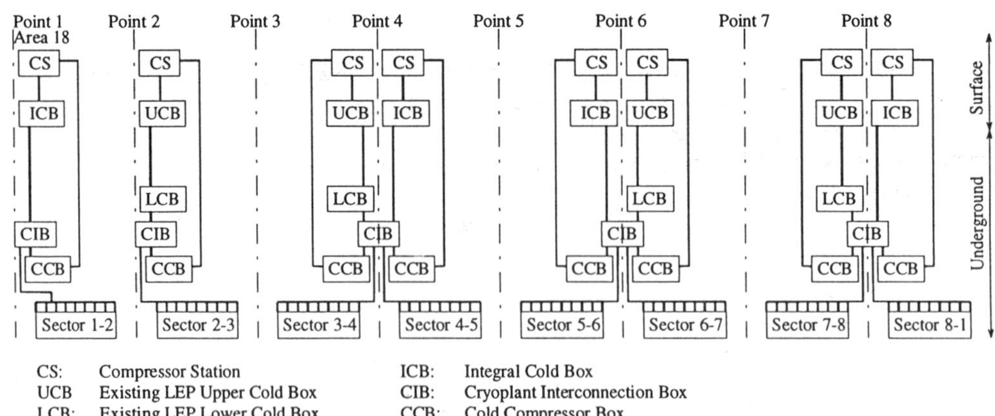


**Figure 2.** Cryogenic architecture of LHC.

## REDUCING THE NUMBER OF DISTRIBUTION HEADERS

In order to reduce the vapour mass fraction produced in the Joule-Thomson expansion of the superfluid cooling loop, a subcooled helium supply is required. The subcooling can be performed centrally in the cold compressor box of the corresponding sector, or distributed in each local cooling loop. The first case requires a single heat exchanger for the total flow of about 120 g/s and a distribution header in the cryogenic distribution line. In the second case, the need for this header disappears, but one smaller heat exchanger is needed for each cooling loop. The major difference between these two cases is the temperature of the returning low-pressure vapour in the pumping header, which has opposite consequences on the hydrostatic and frictional terms of the pressure drop. In the first case (central heat exchanger), the very-low-pressure vapour is pumped at 1.8 K, when it is dense, which minimizes frictional pressure drop. In the second case (distributed heat exchanger), the very-low-pressure helium is returned to the CCB at higher temperature, thus reducing hydrostatic head. Figure 3 shows the magnet temperature profiles which directly depend on the pressure profile of the pumping line, for nominal operating conditions.

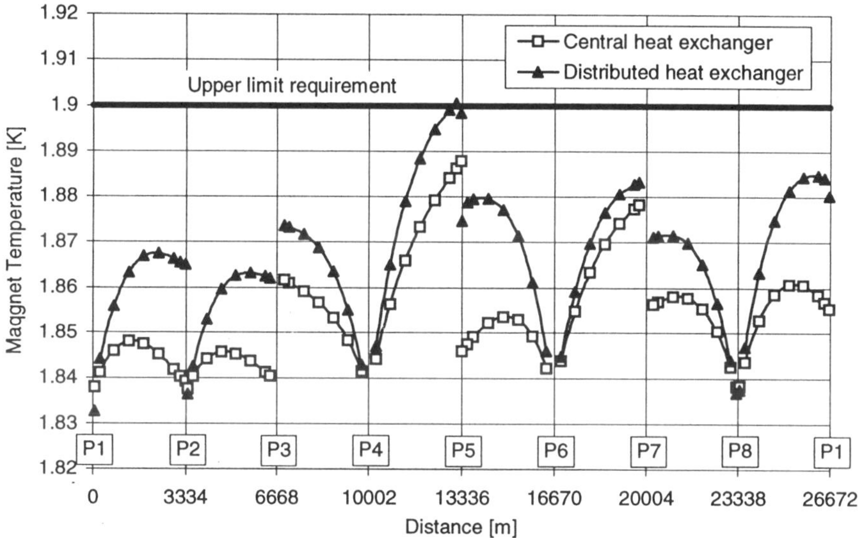


**Figure 3**. Magnet temperature profiles at nominal conditions (with uncertainty factor).

With distributed heat exchangers, the maximum magnet temperature is slightly higher (10 mK) but is still below the upper limit requirement of 1.9 K. A distribution scheme using distributed heat exchangers is acceptable. In addition, the validation of the design and construction of small distributed heat exchangers for a flow of about 5 g/s, is easier.

## REDUCING THE NUMBER OF COMPONENTS

The optimization of the cryogenic distribution scheme by reducing the number of components will decrease costs and heat inleaks, and increase the reliability of the cryogenic system.

## Bus-Bar Plugs

The bus-bar plugs segment the pressurized helium volume in the cold mass, which contains the magnets and the electrical bus bars connecting them in series. This hydraulic segmentation is required for the following reasons:

- limitation of the hydrostatic head due to the tunnel slope: the cold mass must remain above the atmospheric pressure to prevent air inleaks into circuits,
- possibility of short interventions on the cold mass (e.g. exchange of diodes, repair of instrument feedthrough, etc.); for such interventions, the warmup and cooldown time must be minimized and are only limited by the maximum allowed flow-rate per cell,
- possibility of magnet removal which require a sub-sectorization[4] of the sector.

The previous design was based on one segmentation at every half-cell, i.e. 53 bus-bar plugs per sector. Based on the above requirements, the number of bus bar plugs can be reduced to 14 elements per sector. Table 1 gives the bus-bar plug requirement for the previous and simplified schemes. Figure 4 shows the bus-bar plug spacing for the simplified scheme.

**Table 1.** Requirements of components number

| Component Type | Previous scheme | Simplified scheme |
|---|---|---|
| Bus-bar plug | 53 | 14 |
| Cooldown and fill valve | 54 | 15 |
| Safety relief valve | 108 | 42 |

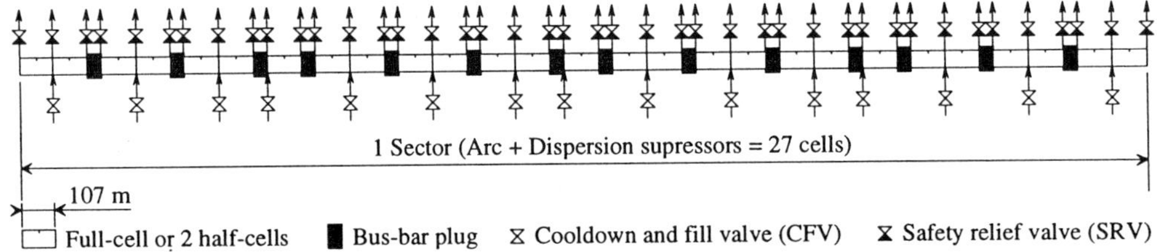


**Figure 4.** Requirement of component spacing of the simplified scheme.

## Cooldown-and-Fill Valves (CFV)

Due to pressure drop limitations, a maximum string length, corresponding to one full-cell, is acceptable for cooldown. With the bus-bar spacing defined for the simplified scheme and by using parallel cooling of two full-cells adjacent to a cooldown-and-fill valve, it is possible to cool down and warm up the sector without adding other bus-bar plugs between cells. Consequently, with respect to the previous scheme, the number of valves can be reduced from 54 down to 15 (see Table 1 and Figure 4). For parallel cooling, the flow balance and the cooldown time depend on differences in mass and hydraulic impedance between the two adjacent cells. Possible unbalanced cooldown could be corrected by pulse-duration modulation type control on the outlet valves.

### Safety Relief Valves (SRV)

The lengthening of the cooling loop to a full-cell increases the minimum distance between two safety relief valves from 53 up to 107 m. Figure 5 shows simulations of full-quench discharges of a cell for normal protection (simultaneous discharge through two SRV's) and for an accidental case in which one of these safety relief valve remains closed. These simulations were performed using a numerical model[5,6] based on quench discharges measurements of the Test String[7], in operation at CERN. The normal protection of the simplified scheme has been validated on the 50-m Test String by discharging it at one of its extremities only.

During a full-cell quench, the maximum pressure in the cold mass evolves around 1.3 MPa, which represents an ample safety margin with respect to the 2-MPa cold-mass design pressure. The number and spacing of SRV's, given in Figure 4, therefore conforms with the simplified scheme.

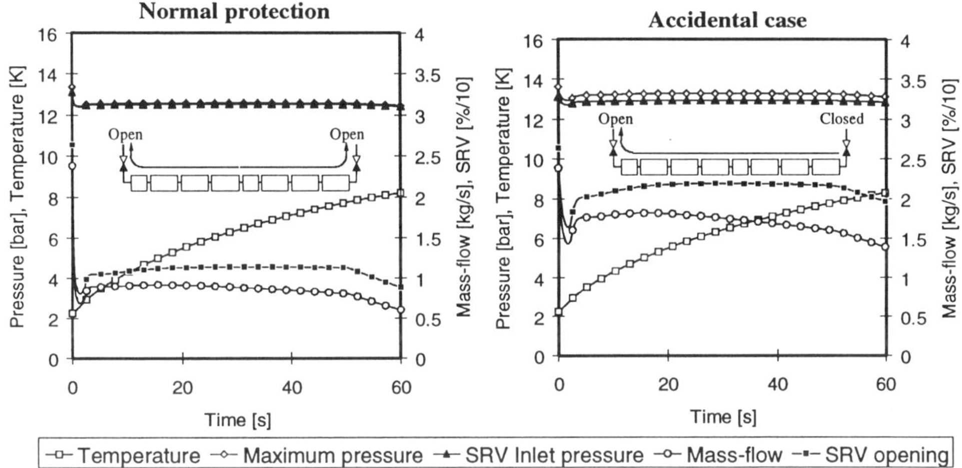


**Figure 5.** Simulations of quench discharge of full-cells.

## INCREASE OF THE LENGTH OF ELEMENTARY COOLING LOOPS

The component spacing defined above provides the possibility to increase the length of the elementary cooling loops to 107 m, and consequently reduce the number of cryogenic modules feeding the superconducting magnets from the cryogenic distribution line. To achieve this, two local loops, corresponding to the cold mass and the beam screen cooling, have to be modified.

### Supercritical Helium Beam-Screen Cooling Loop

The pressure drop in this circuit due to the very high hydraulic impedance of the beam screen channels does not allow the increase of the loop length to 107 m. Left-right supply is the only way compatible with a larger spacing between cryogenic modules. Figure 7 shows the new arrangement of the beam-screen cooling loops with two outlet temperature control valves (TCV2) per cryogenic module.

## Superfluid Helium Cold-Mass Cooling Loops

Saturated liquid helium in the cold-mass heat exchanger always flows downslope. Consequently, left-right supply circuits are not a viable option for the cold-mass cooling loops, which therefore need to be doubled in length.

The thermal performance of the cold mass heat exchanger[8] depends partly on its pressure drop. For given pipe geometry and linear heat load, doubling the length of this heat exchanger would lead to an increase in the corresponding pressure drop by a factor 8 which is unacceptable for the magnet cooling. The length increase must thus be compensated by modifying the pipe geometry using a combination of a change of friction factor and of hydraulic diameter. The outer diameter of the heat exchanger is limited by the 60 mm diameter aperture in the magnet yoke. Figure 6 shows two heat exchanger cross-sections as well as their main characteristics for the previous and simplified schemes. Taking into account the higher pressure level in the supply header, the external diameter of the inner pipe can be reduced from 14 mm to 12 mm. For the simplified scheme, the friction factor is improved by a factor 4 and the hydraulic diameter by a factor 1.2, the combination of which gives a pressure reduction by a factor 10. The hydraulic performance of the simplified scheme heat exchanger is better than that of the previous one.

The thermal and mechanical performance is determined by the wall thickness, which must be mechanically designed to withstand an external over-pressure of 2 MPa, while being as thin as possible for limiting transverse thermal impedance. The best thermo-mechanical optimization is to use OFHC copper with 20 % of cold work which increases the mechanical properties and requires a wall thickness of 2 mm only. The use of bellows may be required in each cold mass to compensate for thermal expansion.

The cooling loop length of 107 m allows the number of expansion valves TCV1 to be halved. The decision was made to add spare valves (TCV1') parallel to the expansion valves TCV1 and their subcooling heat exchangers. Such an arrangement permits a faster cooldown of the subcooling heat exchanger, a greater dynamic range for faster recooling after a magnet quench, and provides redundancy of the cooling loop in case of TCV1 valve failure or subcooling heat exchanger plugging.

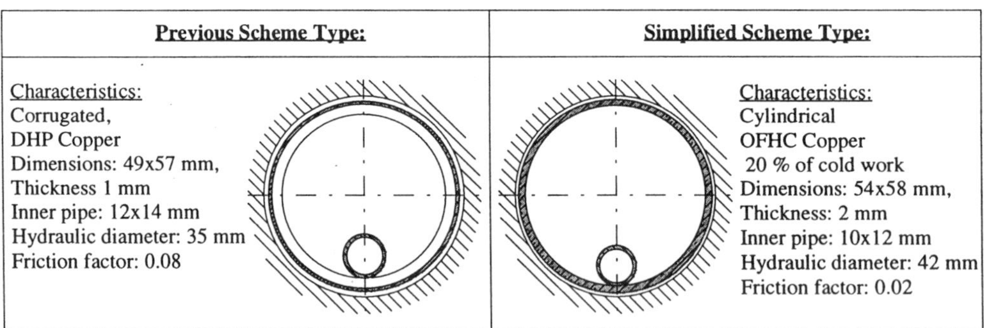


**Figure 6.** Cold mass heat exchanger cross sections and characteristics.

## THE SIMPLIFIED DISTRIBUTION SCHEME

Figure 7 shows the simplified scheme of the cryogenic distribution, which takes into account the component spacing and the cooling loop modifications.

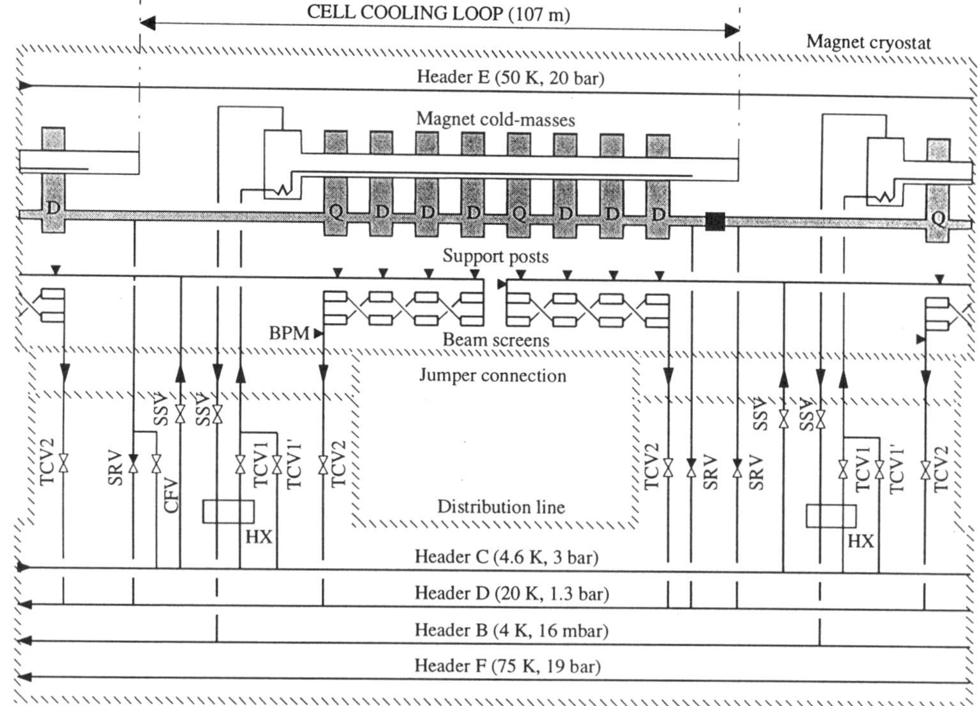


**Figure 7.** Simplified distribution scheme.

## COMPONENT INVENTORY AND COST SAVINGS

Table 2 gives the variation of component numbers with respect to the previous scheme for the complete machine, yielding a total reduction of 1260 valves, 312 bus-bar plugs and 216 cryogenic modules. The reduction of the number of components increases the reliability of the cryogenic system and gives corresponding cost savings of about 20 M$. The cost of extra subcooling heat exchangers is largely compensated by the cost saving corresponding to the suppression of the distribution header of subcooled liquid.

**Table 2.** Component requirements and variation with respect to the previous scheme

| Component type | Number/type | Variation |
|---|---|---|
| Distribution header | 5 | -1 |
| Subcooling heat exchanger | 231 | +223 |
| Cryogenic module | 236 | -216 |
| Bus bar plug | 124 | -312 |
| Phase separator sump | 227 | -217 |
| Control valve | 1028 | -300 |
| Safety relief valve | 360 | -528 |
| Sub-sector valve | 456 | -432 |

## CONCLUSION

The simplification of the cryogenic architecture is based on a four-plus-one feed scheme using four new cryogenic plants located integrally at ground level surface in addition to the four existing LEP2 plants. This new architecture results from a technical-economical optimization taking into account the cooling requirement of LHC and the high cost of cavern excavation.

The simplification of the cryogenic distribution scheme is based on the reduction in the number of active components and of distribution headers with the suppression of subcooled liquid distribution and the addition of distributed subcooling heat exchangers. These simplifications improve the reliability of the cryogenic system, preserve the operation and safety requirements, reduce heat inleaks and increase the redundancy level of the superfluid helium cold-mass cooling loop. The improvements far outweigh the inconvenience of some loss in standardization and in addition represent a cost reduction of about 20 M$.

## ACKNOWLEDGEMENT

We would like to acknowledge the contributions of our colleagues J. Casas-Cubillos, B. Hilbert, T. Kurtyka, P. Rohmig, F. Savary and B. Skoczen of CERN.

## REFERENCES

1. The LHC Study Group, "The Large Hadron Collider, Conceptual Design," CERN Report AC/95-05(LHC) (1995).
2. U. Wagner, "The LHC refrigerators with surface located cold boxes for the temperature range 300-4.5 K," LHC Project Note 70 (1996).
3. V. Benda, A. Bézaguet, J. Casas-Cubillos, S. Claudet, W. Erdt, Ph. Lebrun, G. Riddone, V. Sergo, L. Serio, L. Tavian, B. Vuillerme, R. van Weelderen and U. Wagner, Conceptual design of the cryogenic system for the Large Hadron Collider, in: "Proceeding of EPAC96", S. Meyers, A. Pacheco, R. Pascual, Ch. Petit-Jean-Genaz and J. Poole, editors, Institute of Physics Publishing, Bristol and Philadelphia (1996), p. 361-363.
4. M. Bona, P. Cruikshank, W. Erdt, J.L. Périnet-Marquet, A. Poncet, P. Rohmig, U. Wagner and T. Wikberg, "Cryogenic and Vacuum Sectorization of the LHC Arcs," LHC Project Report 60 (1996)
5. B. Hilbert, "Modeling and Experimental Analysis of the Thermohydraulics of Resistive Transitions on the LHC Prototype Magnet String," LHC Project Note 86 (1997).
6. M. Chorowski, B. Hilbert, L. Serio and R. van Weelderen, Thermohydraulics of Resistive Transitions of the LHC Prototype Magnet String: Theoretical Modelling and Experimental Results, paper presented at this conference.
7. A. Bézaguet et al, The LHC Test String: First Operational Experience, in: "Proceeding of EPAC96", S. Meyers, A. Pacheco, R. Pascual, Ch. Petit-Jean-Genaz and J. Poole, editors, Institute of Physics Publishing, Bristol and Philadelphia (1996), p. 358-360.
8. Ph. Lebrun, L. Serio, L. Tavian and R. van Weelderen, Cooling Strings of Superconducting device below 2 K: The Helium II Bayonet Heat Exchanger, paper presented at this conference.

# CRYOGENIC SYSTEM AND CRYOSTAT DESIGN FOR THE LHC IR QUADRUPOLE MAGNETS

Y. Huang, J. Kerby, T. Nicol, and T. Peterson

Fermi National Accelerator Laboratory
P O Box 500
Batavia, IL, 60510, USA

## ABSTRACT

The LHC interaction region (IR) inner triplet cryostats are under development at Fermilab. The inner triplet consists of four 70 mm aperture superconducting quadrupole magnets manufactured at Fermilab and KEK. These magnets are cooled with pressurized static superfluid helium at 1.9 K which transports the heat to saturated 1.8 K He II via a He II heat exchanger. This paper discusses the requirements for the inner triplet cryostat, development of the cryogenic system, and He II heat exchanger arrangement. Cooldown and quench pressure issues with the external He II heat exchanger are also discussed.

## INTRODUCTION

There will be four interaction points in the LHC machine at CERN: two high-luminosity insertions and two experimental insertions. A total of eight inner triplets will be installed, one on either side of each of four interaction points. Each inner triplet consists of four superconducting quadrupoles; named Q1, Q2a, Q2b, and Q3 starting at the interaction point. The magnetic lengths are 6.3 m for Q1 and Q3, 5.5 m for Q2a and Q2b.[1] Fermilab and KEK will provide the inner triplet quadrupoles, 16 quadrupole magnets each, not including spares. Fermilab is responsible for the cryogenic system integration of all the inner triplets, including the correction magnets and instrumentation. As part of the US-LHC Collaboration, Fermilab is also responsible for the development of a common cryostat for all IR quadrupoles, while LBNL is developing the cryogenic feed box for the inner triplet.

A detailed description of the LHC cryogenic system is beyond the scope of this report but can be found elsewhere in the literature.[2,3] With one exception, the cryogenic plants are installed at four points equally spaced around the LHC machine, and the refrigeration power is transported over a length of a 3.3 km octant by a separated cryogenic distribution line.

The thermal design requirements and heat loads at various temperatures are listed in Table 1. A thermal shield at 50 to 75 K is required to minimize the static thermal conduction and radiation heat loads.

**Table 1.** Inner triplet heat loads vs. temperature

| Temperature levels | 50 to 75 K | 4.5 to 20 K | 1.9 K |
|---|---|---|---|
| Static heat loads (W) | 170.7 | 3.2 | 10.0 |
| Dynamic heat loads nominal (W) | 0 | 0 | 146.5 |
| Total heat loads (W) | 170.7 | 3.2 | 156.5 |

## INNER TRIPLET CRYOGENICS

The inner triplet quadrupole magnets are cooled in static baths of pressurized He II at 1.9 K. The heat load is transported from pressurized He II to the flowing, saturated two-phase He II at 1.8K via a He II heat exchanger. This heat exchanger could be arranged either inside of the cold mass, the internal He II heat exchanger design, or outside of the cold mass, the external He II heat exchanger design.

Figure 1 shows a simplified flow schematic of the cryogenic system for the inner triplet with an external He II heat exchanger design. The refrigeration power is transported by the cryogenic distribution line over a length of 3.3 km and is available for the inner triplet magnets in a standard way, like many parallel cells in the octants. Supercritical helium (3 bar and 4.2 K) from the sector refrigerator, distributed through Line C, is cooled within the counterflow heat exchanger by the returning low pressure cold helium vapor to 2.2 K and expanded through a J-T valve. The saturated helium after J-T valve is then fed to the other end of the heat exchanger tube, absorbing heat and gradually vaporizing as it flows back. Line C also provides cooling for the current leads in the electrical feed boxes. The low pressure helium vapor is pumped via Line B, after giving off its cooling power to the incoming supercritical helium in the counterflow heat exchanger. Line D, maintained at 20 K and slightly above atmospheric pressure, collects the helium discharged from the magnets during a quench. Lines E and F circulate high pressure helium gas to cool the thermal shield and cold mass support post to intercept the heat leak from ambient by radiation and conduction.

Figure 1 presents the cryogenic system for the inner triplet on one side of interaction point in which the feed box is located in the lower position. The cryogenic system for the inner triplet located on the other side of the interaction point is identical except the feed box is now on the higher position, a pump line is connected to the lower end of the He II heat exchanger to pump the helium vapor. The pump line will replace the location of saturated helium supply tube in Figure 1.

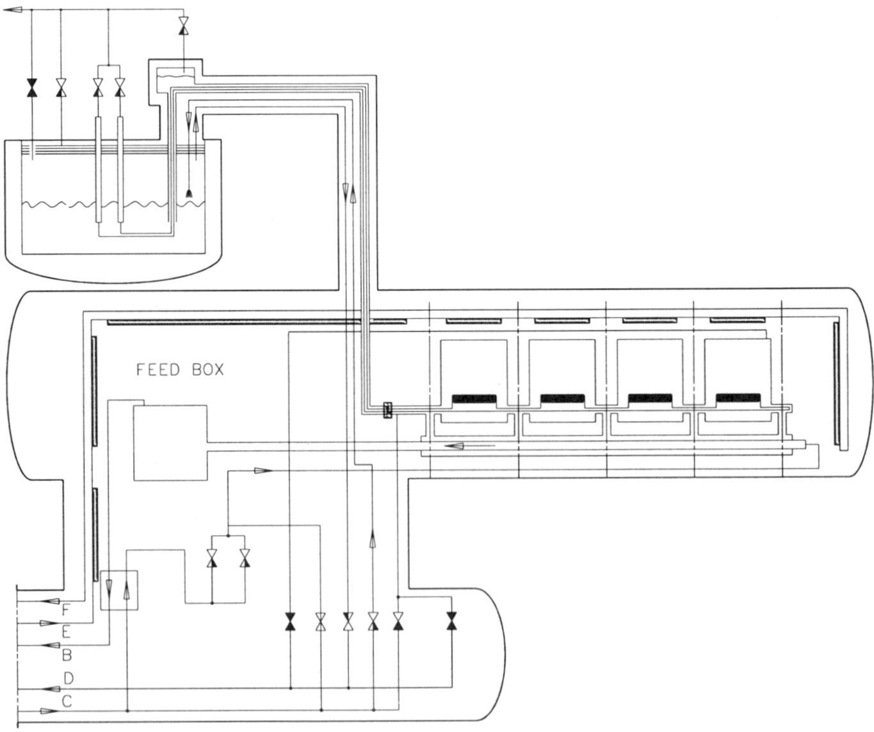


**Figure 1.** A simplified flow schematic of cryogenic system for the LHC IR inner triplet.

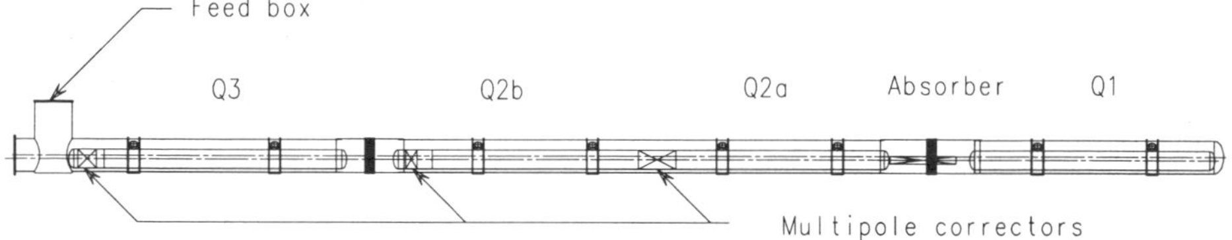


**Figure 2.** LHC interaction region triplet layout.

## CRYOSTAT

Figure 2 is a simplified illustration of an assembled LHC interaction region inner triplet. Each inner triplet consists of four cold mass assemblies, three multipole correctors, two beam position monitors, and supplementary heat absorbers. The overall length of each triplet is approximately 30 m long not including the feed box.

It is currently thought that quadrupoles Q1 and Q3 will each be assembled into separate vacuum vessels and that Q2a, Q2b, and two multipole correctors will be assembled into a single cold mass and a single vacuum vessel for a total of three distinct cryostats. These three individual cryostats will be joined using welded connections to form the complete inner triplet.

Figure 3 shows a typical cross section through any of the inner triplet quadrupoles at a cold mass support location. Each cold mass is supported at two places along its length by top-mounted supports oriented at ±45° from the cold mass centerline. Using two supports produces a much more rigid cold mass mounting system than one which uses a single support and may be an advantage in trying to meet the stringent alignment requirements for interaction region magnets. Also, supports mounted as shown in Figure 3 provide sufficient radial clearance for the external He II heat exchanger and 1.8 K saturated He II supply tube or a pump line. One of the two support locations per cold mass assembly will serve as the axial anchor. The other will have a sliding interface between the support and cold mass to allow for thermal contraction during cooldown.

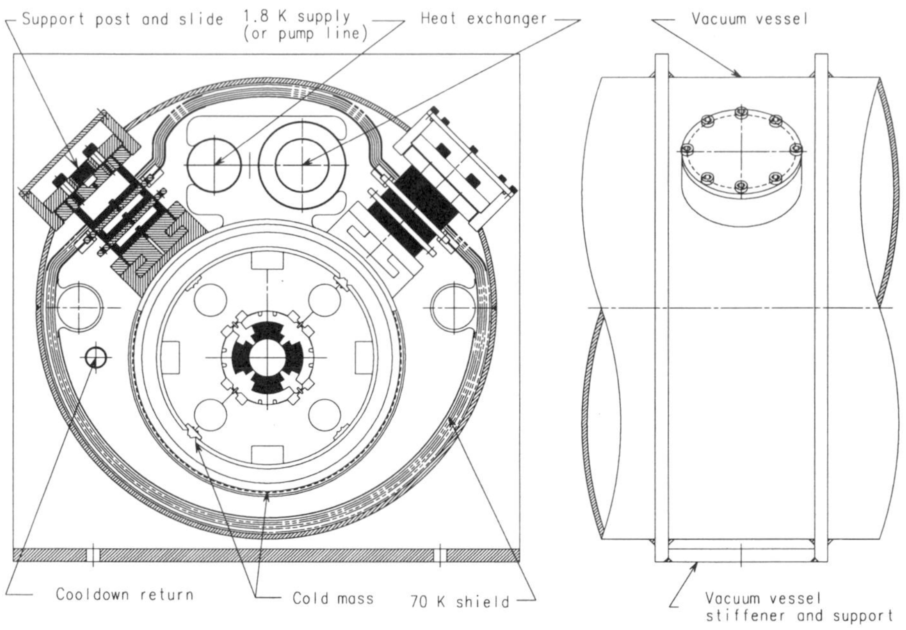


**Figure 3.** LHC interaction region quadrupole cryostat cross section.

The internal cold mass supports are coincident with the external vacuum vessel support rings. These support rings stiffen the vacuum vessel at the cold mass support points and serve as the interface between the vacuum vessel and the tunnel mounted support and alignment hardware. In the current conceptual design, the vacuum vessel is 864 mm outside diameter.

Ambient thermal radiation is intercepted by blankets of multilayer reflective insulation (MLI) immediately inboard of the inner vacuum vessel surface and by a 50-75 K helium cooled thermal shield. There is an additional MLI blanket on the cold mass.

There are two different outer diameters for the quadrupole cold mass helium vessels. Those being fabricated at KEK are 510 mm diameter while those being fabricated at Fermilab are 416 mm diameter. The current plan is to use an adapter on the Fermilab assemblies to increase their diameter at the cold mass attachment locations so that all cryostats will be identical regardless of the cold mass manufacturer. Any individual inner triplet uses cold masses with a single diameter, i.e. either all KEK or all Fermilab cold masses due to their having different operating currents. Figure 3 shows both cold mass diameters for reference.

## He II HEAT EXCHANGER

The He II heat exchanger is a key component in the successful cooling of the inner triplets. The main function of this heat exchanger is to extract both dynamic and static heat loads in the cold mass and maintain its temperature below 1.9 K during normal operation. Several experiments have been carried out at CERN to study the thermal performance of the heat exchanger in which the heat is transported from static pressurized He II to the flowing saturated He II through either a smooth tube or a corrugated pipe.[4-6] These studies show that the heat exchanger exhibits a stable two-phase boiling liquid helium flow and performs as an efficient heat sink when vapor and liquid flow in the same direction. The He II heat exchanger, however, could be arranged either internally (within the cold mass) or externally (outside of cold mass) as long as the thermal design requirements are fulfilled.

The design work for both internal and external He II heat exchanger arrangements is under way and the thermal analysis between the two is being compared. The final decision, depends on several issues, such as the inner triplet cooldown from room temperature, normal operation under various heat loads, the comparison analysis between the internal and external He II heat exchanger arrangements, peak quench pressure, and magnet cooldown after quench.

There are four 60 mm holes in the iron yoke for the Fermilab designed quadrupole magnets, the correction dipole magnets, and KEK fabricated quadrupole magnets as shown in Figure 4. The internal He II heat exchanger arrangement requires that the heat exchanger pipe be inserted all the way through from end to end of 30 meters in length. Thus, the positions of the 60 mm holes in the correction dipole magnets would be constrained to line up with those in the quadrupoles.

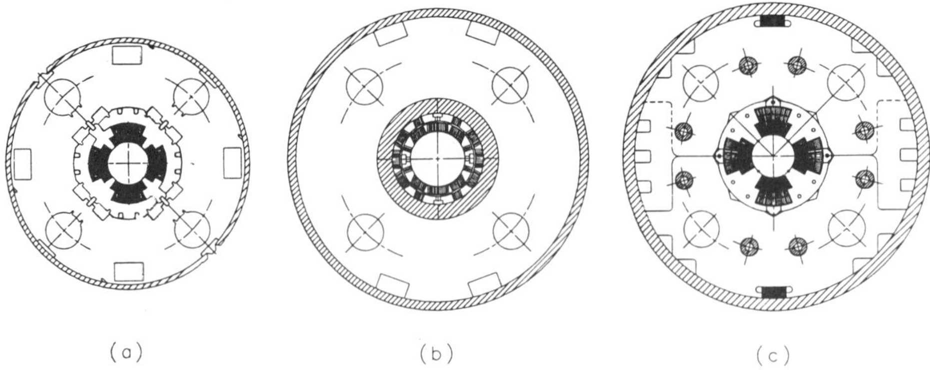


**Figure 4.** The cross section of inner triplet quadrupole magnet and correction magnet. (a) Fermilab designed, (b) the correction magnet, and (c) KEK designed.

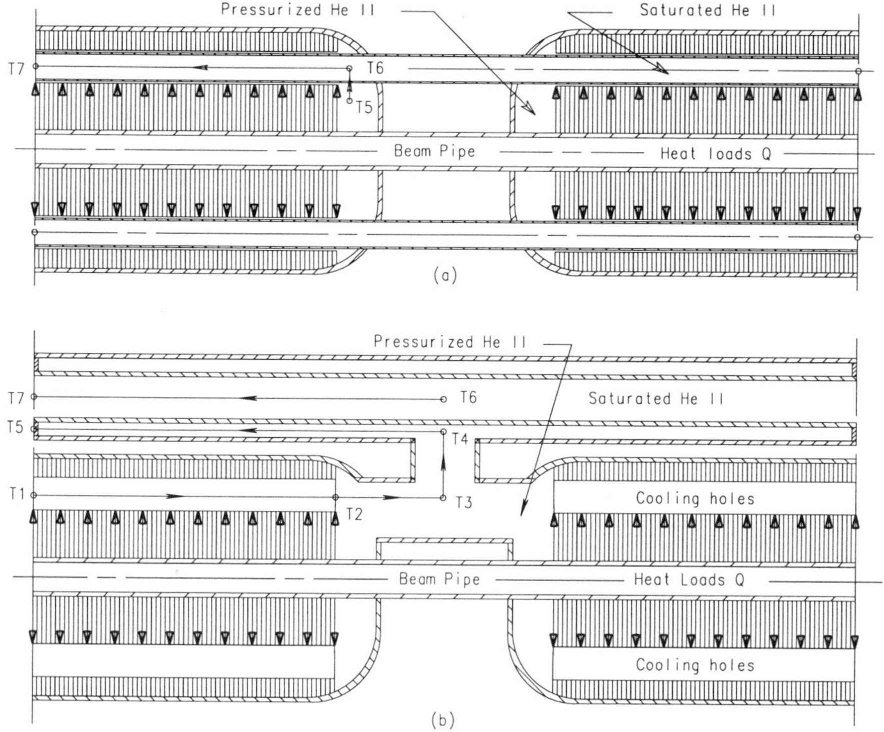


**Figure 5.** He II heat exchanger arrangements, (a) internally arranged; (b) externally arranged.

The He II heat exchanger arranged externally will eliminate this problem if the thermal design requirements are satisfied. The total temperature drop allowable is limited by the LHC cryogenic system, which is about 100 mK. For the externally arranged He II heat exchanger, the temperature drop must be maintained within 100 mK or comparable with that of internally arranged He II heat exchanger.

## Steady state performance

A schematic of the He II heat exchanger, arranged either internally or externally, is shown in Figure 5. The total temperature drop from cold mass to saturated He II at pump site consists of two parts for the internal heat exchanger design: across the heat exchanger wall, and caused by helium vapor flow pressure drop. The largest temperature drop is caused by the vapor pressure drop due to the fact that the heat exchanger pipe is limited by the size of cooling hole in the iron yoke. For the external heat exchanger arrangement, however, the total temperature drop consists of three parts: within the pressurized He II, across the heat exchanger wall, and caused by the vapor pressure drop. Even though an additional temperature drop within the pressurized He II is added, the temperature drop caused by the vapor pressure drop is minimized by the adequate sizing of the heat exchanger pipe.

The temperature drop within the pressurized He II is calculated by using the Gorter-Mellink equation for the constant heat load distribution and cooling channel dimensions. The temperature drop across the heat exchanger wall can be determined by assuming a constant heat conductance between the pressurized and saturated He II.[4-6] The pressure drop within the heat exchanger pipe is calculated based on the correlation given in the literature.[7] The parameters used in the calculation and results are summarized in Table 2. The temperature drop $\Delta T6$ is due to the vapor pressure drop in the 30 m long heat exchanger pipe(s). The heat load of 234.8 W is 1.5 times predicted heat load of inner triplets.

**Table 2.** Temperature drop of internal and external He II heat exchangers

| Parameter | Internal heat exchanger | Internal heat exchanger | External heat exchanger |
|---|---|---|---|
| Heat load Q (W) | 234.8 | 234.8 | 234.8 |
| No. of pipes | 4 | 2 | 1 |
| Pipe length (m) | 30 | 30 | 30 |
| Bellows tube I.D. (mm) | 40 | 49 | 110 |
| Bellows tube O.D. (mm) | 46 | 57 | 118 |
| Bellows pitch (mm) | 5 | 5 | 5 |
| Shell pipe O.D. (mm) | N/A | N/A | 160 |
| ΔT1 (T1-T2) (mK) | 0 | 0 | 1.6 |
| ΔT2 (T2-T3) (mK) | 0 | 0 | 4.1 |
| ΔT3 (T3-T4) (mK) | 0 | 0 | 4.0 |
| ΔT4 (T4-T5) (mK) | 0 | 0 | 6.4 |
| ΔT5 (T5-T6) (mK) | 23.8 | 38.9 | 34.7 |
| ΔT6 (mK) | 37.1 | 55.4 | 0.5 |
| Total ΔT (mK) | 60.9 | 94.4 | 51.3 |

## Cooldown

The inner triplets cooldown procedure consists of three steps: first, the inner triplets are cooled from 300 to 80 K with helium gas; once the system reaches 80 K, liquid helium is used to cool the system to 4.2 K; finally, He II heat exchanger starts to operate and cool the system to 1.9 K.

The first step cooldown can be simulated by a computer model as shown in Figure 6. The supercritical helium flows in the 60 mm cooling hole and the gap between the beam pipe and coils. The heat transfer coefficient is estimated for a constant mass flow rate, fluid properties, and cooling channel dimensions. For an initial temperature of cold mass at 300 K, inlet temperature of 80 K, a constant specific heat of 465 J/kg•K of iron yoke, and a heat transfer coefficient of 50 W/m$^2$•K, the cold mass reaches the temperature of 100 to 110 K after 15 hours of cooling.

The cooldown, from 4.2 to 1.9 K, can be estimated based on the cooling capacity of He II heat exchanger and the heat capacity of liquid helium in that temperature range. Our calculation shows that the cooldown from 4.2 to 1.9 K will be completed within 2 hours for a He II heat exchanger with a cooling power of 234.8 W.

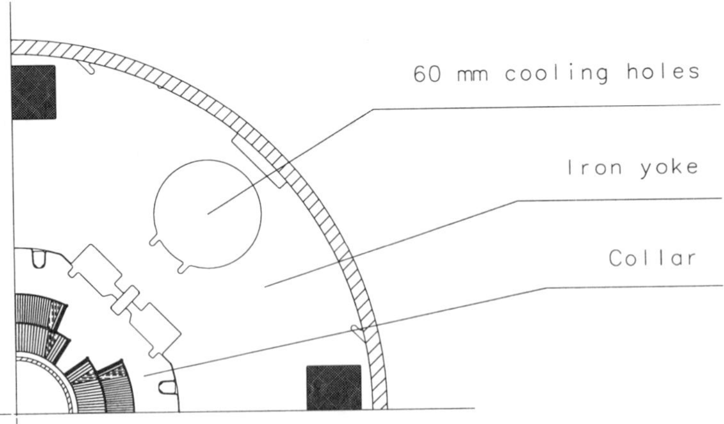


**Figure 6.** Thermal model to simulate the cooldown procedure of inner triplet magnets.

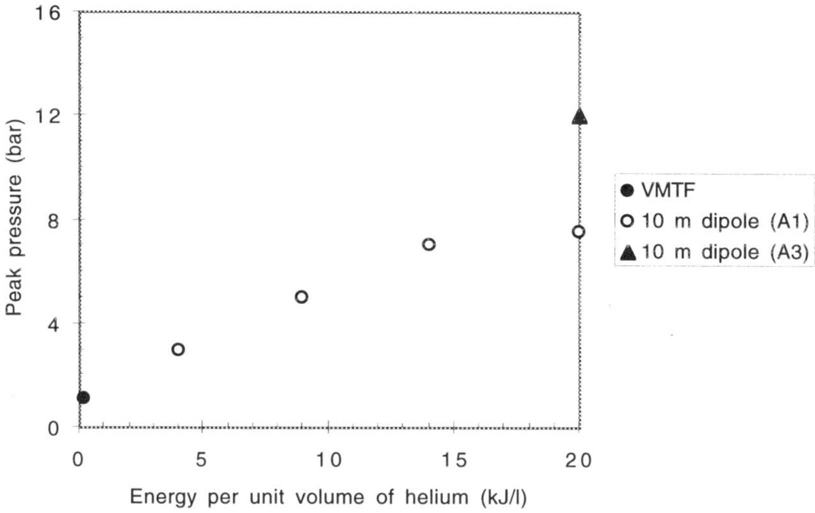


**Figure 7.** Peak quench pressure as a function of stored energy per unit volume of He II.

## Quench pressure

The stored energy of superconducting magnets will be dumped into the coils in less than a second. The peak quench pressure depends on the stored energy and amount of liquid helium within the cold mass. The measurement of quench pressure of dipole magnets as a function of time shows that pressure starts to rise gently between 0 and 60-80 ms and sharply afterwards. The quench relief valve starts to open 120-130 ms after the quench detection signal and reaches 90% opening within 210-250 ms. The longer the valve opening time, the higher the final pressure. A typical peak pressure is 12-14 bars if the valve opening time is 210-250 ms for a dissipated energy of 4732 kJ per 10 meter long dipole magnet.[8]

Figure 7 summarizes peak quench pressure as a function of stored energy per unit volume of liquid helium for the magnets cooled with He II. The data taken at CERN were from the 10 m long dipole magnets and data taken were on a 1.37 m long quadrupole magnet tested in the Vertical Magnet Test Facility (VMTF) at Fermilab. The experimental data show that quench pressure is proportional to the stored energy per unit volume of He II. It is worth noting that even for the same designed dipole magnets, the peak quench pressure may vary as much as several bars as shown in Figure 7 for dipole magnets of A1 and A3.

The stored energy of Fermilab designed inner triplet magnets is around 330 kJ/m at the operating current of 14 kA. The He II volume within the cold mass is about 20 liters. The stored energy per unit volume of He II is 16.5 kJ/l. From Figure 6, the peak quench pressure should be less than 12 bars. For the external He II heat exchanger arrangement, additional amount of He II will be added to the system and quench pressure will be reduced even further.

## SUMMARY

We discussed the requirements for the LHC IR inner triplet cryostat, development of the cryogenic system, and He II heat exchanger arrangement in this report. Cooldown and quench pressure issues with the external He II heat exchanger are also discussed. The He II heat exchanger arranged externally eliminates the problem caused by the unmatched cooling holes in the iron yoke for quadrupole magnets and multipole correctors. The total temperature drop can be maintained within the cryogenic system limit with a single He II heat exchanger pipe arranged outside of the cold mass.

## REFERENCES


1. R. Ostojic, T.M. Taylor and S. Weisz, Systems Layout of the Low-β Insertions for the LHC Experiments, Presented at PAC 1997, May 1997, Vancouver, Canada.
2. The LHC Study Group, The Large Hadron Collider, CERN/AC/95-05 (LHC), 20 October, (1995)
3. Ph. Lebrun, Superfluid Helium Cryogenics for the Large Hadron Collider Project at CERN, Cryogenics 34, ICEC Supplement (1994) p. 1.
4. A. Cyvoct, Ph. Lebrun, M. Marquet, L.Tavian and R. van Weelderen, Heated Two-Phase Flow of Saturated Helium II over a Length of 24 m, CERN-AT/91-28 (CR), LHC Note 169, (1991)
5. J Casas, A. Cyvoct, Ph. Lebrun, M. Marquet, L.Tavian and R. van Weelderen, Design concept and first experimental validation of the superfluid helium system for the Large Hadron Collider (LHC) Project at CERN, Cryogenics, vol. 32, ICEC Supplement, (1992), p. 118.
6. A Bezaguet, J. Casas-Cubillos, Ph. Lebrun, M. Marquet, L. Tavian, and R. van Weelderen, The Superfluid Helium Model Cryoloop for the Cern Large Hadron Collider (LHC), CERN AT/93-21 (CR), LHC Notes 233, (1993).
7. J.G. Weisend and S.W. Van Sciver, Pressure drop from flow of cryogens in corrugated bellows, Cryogenics, Vol. 30, (1990), p.935.
8. G. Gerin, B. Vullierme, and R. van Weelderen, Measurement of the thermo-hydraulic behavior of LHC dipole prototypes after quench, Adv. Cryo. Eng., 41A, (1996), p. 811.

# THE NEW SUPERFLUID HELIUM CRYOSTATS FOR THE SHORT STRAIGHT SECTIONS OF THE CERN LARGE HADRON COLLIDER (LHC)

W. Cameron,[1] Ph. Dambre,[3] T. Kurtyka,[1] V. Parma,[1] T. Renaglia,[1] J.M. Rifflet,[2] P. Rohmig,[1] B. Skoczen,[1] T. Tortschanoff,[1] Ph. Trilhe,[1] P. Vedrine,[2] and D. Vincent[3]

[1]CERN, European Organization for Nuclear Research
1211 Geneva 23, Switzerland
[2]CEA SACLAY DAPNIA, 91191 Gif sur Yvette, France
[3]CNRS, IPN, 91406 Orsay Cedex, France

## ABSTRACT

The lattice of the CERN Large Hadron Collider (LHC) contains 364 Short Straight Section (SSS) units, one in every 53 m long half-cell. An SSS consists of three major assemblies: the standard cryostat section, the cryogenic service module, and the jumper connection. The standard cryostat section of an SSS contains the twin aperture high-gradient superconducting quadrupole and two pairs of superconducting corrector magnets, operating in pressurized helium II at 1.9 K. Components for isolating cryostat insulation vacuum, and the cryogenic supply lines, have to be foreseen. Special emphasis is given to the design changes of the SSS following adoption of an external cryogenic supply line (QRL). A jumper connection connects the SSS to the QRL, linking all the cryogenic tubes necessary for the local full-cell cooling loop [at every second SSS]. The jumper is connected to one end of the standard cryostat section via the cryogenic service module, which also houses beam diagnostics, current feedthroughs, and instrumentation capillaries. The conceptual design fulfilling the tight requirements of magnet alignment precision and cryogenic performance are described. Construction details, aimed at minimizing costs of series manufacturing and assembly, while ensuring the high quality of this complex accelerator component, are given.

## INTRODUCTION

The design and construction of the first prototype Short Straight Section,[1] carried out in the years from 1992 to 1994, allowed its successful integration and operation in the LHC test string.[2] However, major changes of the machine layout, cryogenic distribution and magnet powering schemes since 1994 imposed a review of the conceptual design of the Short Straight Sections. Industrial techniques, in view of series production and cost optimization (reductions), have been implemented. Following the approval of the LHC project, a new collaboration for the design, prototyping, and industrial series production follow-up was concluded between CERN and two French Institutes, CEA and CNRS. Due to the special contribution of the host country to the LHC project, the existing collaboration with CEA/Saclay has been extended to the CNRS laboratory in Orsay and covers the design of the quadrupole magnets, the complete cold mass assembly (CEA) and the integration into the SSS cryostat (CNRS).

## NEW SSS LAYOUT

Figure 1 presents a drawing of the new LHC SSS with associated equipment. A major change from the previous design centers the quadrupole between neighboring dipole magnets. This places the corrector magnets on both sides of the quadrupole. Furthermore, the two quadrupole families (focusing and defocusing) are powered independently from the dipoles, thus reducing the integrated length of the tuning quadrupoles.

The integration of the cold mass (helium enclosure) within the quadrupole assembly now permits a vertical assembly with a common inertia tube, similar to CEA's design for the HERA quadrupole units. This considerably improves alignment precision for the magnets when compared to the first SSS prototype.

The implementation of a separate cryogenic feed line, QRL,[3] simplified the cryostat layout of the SSS. The cryogenic control valves, formerly located within the technical service module, are now housed in a dedicated pre-tested valve module. In addition, all cryogenic supply lines for the machine cryostat have been transferred to this feed line with exception of the thermal shield cooling lines. The jumper connection cryogenically links every second SSS to the QRL via the cryogenic service module, QQS. This permits the operation of a 107 m long cell. In addition to the cryogenic interconnect with its liquid helium phase separator, QQS of the SSS also contain beam position monitors, the protection diodes of the quadrupole, conduction cooled current leads and the cold mass instrumentation feedthroughs.

The LHC machine requires 12 different types of cryogenic jumper connections, depending on the slope of the machine and their location within the cryogenic loop.

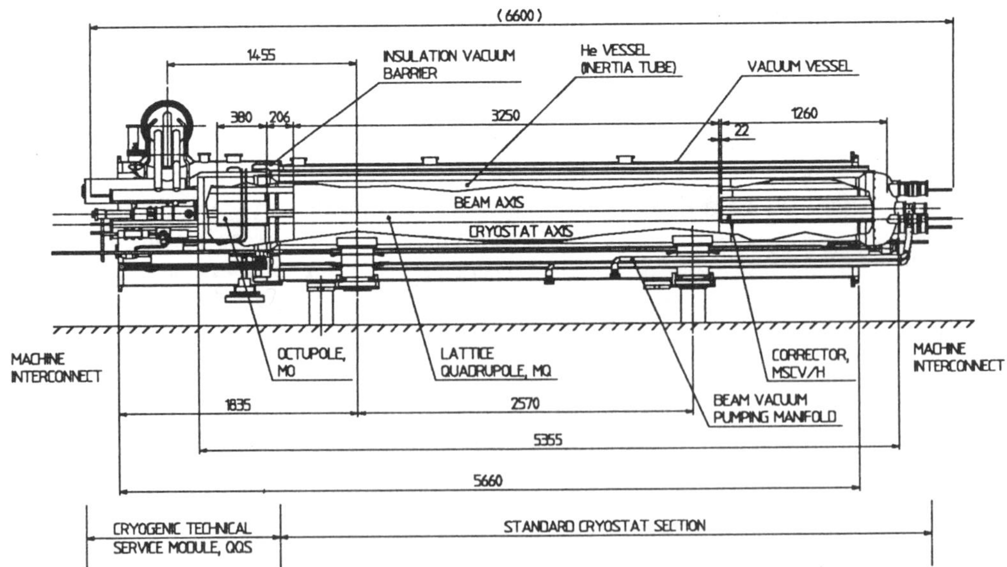


**Figure 1**. LHC Short Straight Section (arc) with jumper connection, cryogenic technical service module and vacuum barrier.

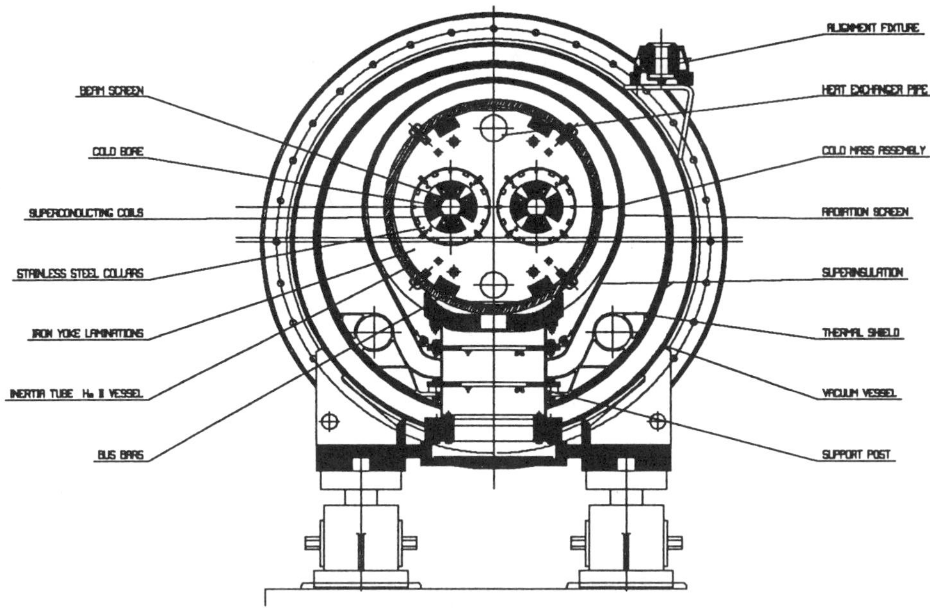


**Figure 2.** Cross-section of the LHC Short Straight Section (arc) with its quadrupole and cryostat .

To cope with possible failure of machine components and to minimize machine down time, a sector isolation of the cryostat insulation vacuum, the beam lines and the cryogenic circuitry has been proposed.[4] If finally adopted, this will be accomplished with cryostat vacuum barriers and bus bar pressure plugs in every second cell (214 m), and with the help of isolation valves in the beam vacuum and thermal shield line at 3 points in each of the eight 2.9 km long machine sectors.

## DESIGN IMPROVEMENTS FOR SERIES MANUFACTURE

Based on the experience gained with the first prototype, and to allow implementation of the conceptual changes mentioned, the LHC cryostat design and, in particular, that of the SSS has been reviewed.

### Cold mass and helium enclosure

The quadrupole magnets, the corrector magnets, and all standard cryostat section bus-bars are contained in the inertia tube. This 5.36 m long stainless steel tube provides not only the stiffening element for this assembly but also serves as the helium pressure vessel for the SSS. The tube and all it contains is called the cold mass. The feasibility of centering the magnets inside this inertia tube by means of a series of keys positioned by cylindrical pins and blocked by bolts, was confirmed with the first two prototypes. One of the two end covers will be of a dished type and identical to those of the dipole. The fixation of the BPM supports, aligned accurately to precisely drilled reference plugs on the inertia tube, imposes the requirement that other end cover be flat, with all passage tubes for bus bars, diode housing and current feedthroughs welded onto it.

### Standard Cross-section of the Cryostat

The Short Straight Section cryostat incorporates many design features identical to the dipole.[5] The thermal insulation system (Figure 2) consists of two actively cooled shields, the radiation screen at 4.5-20 K, and the thermal shield at 50-75 K. Extruded aluminum profiles of symmetric section have been adopted for both thermal shield bottom trays. Integrated cooling channel and clamp-in grooves for the upper pre-formed sheet segments, are included in the extrusion. Compared to earlier designs, this avoids mechanical instability due to asymmetric welding of the cooling tubes onto the bottom tray and to the pressure forces created in the interconnect bellows. This construction also provides a low manufacturing cost.

The cold support system[6] consists of two low heat leak, fiber glass reinforced, epoxy resin molded, support posts with integrated heat sink plates at the 4.5-20 K and 50-75 K levels. These plates also provide mechanical support for the thermal shields. With one fixed support post and a second allowing sliding for compensation of the thermal contraction between cold mass and vacuum vessel, a mechanical reproducibility, after thermal cycles, of the magnet position of better than 0.2 mm r.m.s. is expected.

All heat sinks, namely those of the fiber glass reinforced support posts and those of the shield segments, are obtained by welding. To separate the heat sink and anchoring functions, the support post heat sinks and supporting plates are linked by flexible aluminum multi-foil straps to the thermal shield bottom trays.

The choice of prefabricated Multilayer Superinsulation (MLI) blankets with Velcro™ fixtures has been confirmed by the first prototype as the most economical solution (reproducibility of quality and minimization of assembly time). Forthcoming tests with a Cryogenic Thermal Model (CTM), simulating different operating conditions, will validate the final type and set up of MLI for the 4.5-20 K radiation screen. For the 50-75 K thermal shield, blankets of MLI containing 30 layers of double aluminized Mylar™ (DAM) or single aluminized Mylar™ (SAM) type with polyester spacers will be used.

The vacuum vessel for the standard cryostat section of the SSS will be made out of Ø 914 (36") standard carbon steel tubes, as for the dipole cryomagnet. A comparison of costs quoted by possible future suppliers showed cost reduction of up to 30% for surface protected carbon steel tubes versus stainless steel tubes.

The same study showed that the choice between cast steel support modules versus welded ones, both integrating all functions for supporting and lifting, are strongly dependent on the labor costs of the manufacturer.

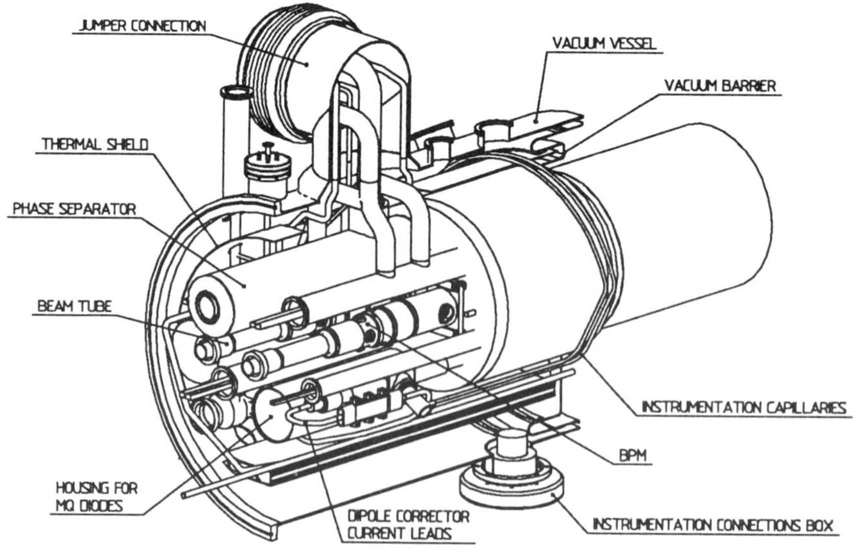


**Figure 3.** Cryogenic Technical Service Module (QQS) jumper connection type with phase separator.

### Vacuum Barrier

Vacuum barriers isolating the cryostat insulation vacuum every 2 cells (214 m) have been designed using two approaches that give similar mechanical and thermal performance. In the first prototype SSS, the vacuum barrier was made of corrugated cylindrical stainless steel segments (bellows), while a second unit, made from pre-impregnated fiber glass reinforced epoxy resin panels,[7] has been installed for two years in a dedicated module integrated into the LHC test string. The reliability of the composite/metal bonding for the welded interface to stainless steel cryogenic tubes on the composite barrier showed weakness but not degradation of leak tightness was observed. New prototypes and associated cost comparison analysis between the technologies will lead to a final choice before the end of this year.

### Cryogenic Technical Service Module (QQS)

With the introduction of the separate cryogenic feed line, all cryogenic control valves (Joule-Thomson, quench relief and other fill and control valves) formerly placed within this module are now located in a dedicated feed line valve box. Nevertheless, every second SSS is linked to the cryogenic distribution line by a jumper connection, equipped with bellows, housing up to eight supply lines (see Figure 3). Special attention has been paid to obtaining a low force compensation for any mechanical offset correction, in order not to interfere with the SSS magnet alignment.

There are 12 different types of QQS. The types are partially governed by the need to integrate a helium phase separator with its incorporated heat exchanger loop and control instrumentation. Depending on the SSS position along the tunnel slope, the separators are connected such that cooling of the upward cell is provided.

Conduction cooled 60 A current leads for the individual dipole correctors, as well as the cold mass instrumentation capillaries, are integrated in the QQS.

To cope with failures of the quadrupole protection diodes or corrector, their electrical connections are accessible via the cryostat interconnect. This permits the exchange of faulty diodes, or the use of spare bus-bars for by-passing the faulty correctors in a local intervention.

The two beam position monitors are fixed to, and thus integrated with, the SSS beam screens. These sub-assemblies must be absent during the cold magnetic measurements and will be inserted after all cold tests have been finished. They require pre-aligned supports attached to the cold mass end cover to permit easy assembly.

Three SSS in each arc, with beam vacuum sectorisation functions, will have manual operated valves integrated into the beam tubes upstream of the BPMs.

## CRYOGENIC PERFORMANCES

Based on the stringent cryogenic specification for the heat loads at different temperature levels, the thermal performance of the SSS was analyzed. The thermal insulation system, in particular the active 4.5-20 K radiation screen, needs further experimental validation in the CTM before a final choice for the LHC cryostat system can be made. Table 1 shows the estimated cryogenic performances of an LHC Arc Short Straight Section containing jumper connection and vacuum barrier under different operation modes. Beam-introduced heating, such as photoelectron deposition and beam gas scattering, is not considered in this table.

## DESIGN STATUS AND FUTURE PROGRAM

Design studies are proceeding for the cold mass mechanical design as well as for the Technical Service Module. Based on a similar design with the dipole cryostats, the standard section cryostat of the SSS is well advanced and prototype tendering is under way.

**Table 1.** Estimated cryogenic performances of a LHC Short Straight Section (arc) containing jumper connection and vacuum barrier under different operation modes in Watts

| Component | Static 50-75 K | 4.5-20K | 1.9 K | Nomina l 50-75 K | ( without 4.5-20K | static ) 1.9 K |
|---|---|---|---|---|---|---|
| **Standard cryostat section** | | | | | | |
| - Cold support post (2) | 9.5 | 1.4 | 0.04 | | | |
| - Thermal shield | 13.2 | | | | | |
| - Radiation screen | | 0.52 | | | | |
| - Beam pumping manifold | 0.8 | | 0.26 | | | |
| - Cold mass | | | 0.06 | | | |
| - Beam screens | | | 0.1 | | | |
| **QQS with jumper connect.** | | | | | | |
| - Thermal shield | 4.44 | | | | | |
| - Radiation screen | | 0.07 | | | | |
| - Beam Position Monitors | | 0.83 | 0.3 | | | |
| - Dipole correct. current leads | 5.2 | 1.1 | 0.11 | 0.34 | 0.08 | 0.15 |
| - Phase separator and piping | | | 0.02 | | | |
| - Cold mass instrumentation | | | 0.53 | | | |
| - Insulation vacuum barrier | 5.1 | 0.24 | 0.37 | | | |
| **Resistive heating** | | | | | | |
| - Quad. + correctors splices | | | | | | 0.62 |
| TOTAL | **38.24** | **4.16** | **1.79** | **0.34** | **0.08** | **0.77** |

Close contacts with industry allowed early integration of economical manufacturing technologies in the SSS design, with a view to low cost series production. Following a technical review within the three collaborating institutes in October, 1997, components for two new prototypes, one with jumper connection and vacuum barrier, the other without will be ordered from industry.

Cold mass assembly at CEA/Saclay and final cryostating at CERN with the assistance of CNRS/Orsay is planned to start in the second half of 1998. It is expected to have a first SSS cold-measured in 1999 and both SSS integrated in the new LHC Test String before the year 2000.

## CONCLUSIONS

Based on the experience gained with the first prototype built, which has been operating successfully since 1994 in the LHC Test String, a new generation of LHC Short Straight Section cryostats has been designed. Apart from conceptual changes, the integration of low cost components (pipeline tubes) and technologies (e.g. extruded thermal shield trays and cast supporting modules) was carried out. Considerable effort was invested to optimize all heat sinks (100% welded), a weak point of the first prototype. Attention was also given to push standardization of the 364 SSSs of the LHC arc as far as possible in the manufacturing process, and to achieve component failure repair within short machine down time.

## ACKNOWLEDGMENTS

The authors wish to thank the members of the Short Straight Section Working Group at CERN and the teams in the three institutes involved in the design of the SSS. Thanks also to our colleagues of the LHC-ACR group for the help on the heat load estimation and the fruitful discussions which lead to the present cryostat design.

## REFERENCES


1. W.Cameron, B. Jenny, G. Riddone, P. Rohmig and R. van Weelderen "Design and construction of a prototype superfluid helium cryostat for the Short Straight Sections of the CERN Large Hadron Collider (LHC)", Advances in Cryogenic Engineering, Vol. 39 (1994).
2. T. Tortschanoff et al., "The Short Straight Sections for the LHC", Particle Accelerator Conference, Vancouver (1997).
3. V. Benda et al. "Conceptual design of the cryogenic system for the Large Hadron Collider (LHC)", Eurpean Particle Accelerator Conference, Sitges (1996).
4. M. Bona et al. "Cryogenic and vacuum sectorisation of the LHC arcs", LHC Project Report 60 (1996).
5. J.C. Brunet, V. Parma, G. Peon, A. Poncet, P.Rohmig, B. Skoczen and L.R. Williams, "Design of the second series 15 m LHC prototype Dipole magnet cryostats", presented at this conference.
6. M. Mathieu, V. Parma, Th.Renaglia, P. Rohmig and L.R. Williams, "Supporting systems from 293 K to 1.9 K for the Large Hadron Collider (LHC) cryo-magnets", presented at this conference.
7. B. Jenny, P. Rohmig and J.M. Uriarte, "A composite vacuum barrier for the LHC Short Straight Section", Advances in Cryogenic Engineering, Vol. 41A (1996).

# COOLING STRINGS OF SUPERCONDUCTING DEVICES BELOW 2 K: THE HELIUM II BAYONET HEAT EXCHANGER

Ph. Lebrun, L. Serio, L. Tavian and R. van Weelderen

LHC Division, CERN
CH-1211 Geneva 23, Switzerland

## ABSTRACT

High-energy particle accelerators and colliders contain long strings of superconducting devices - acceleration RF cavities and magnets - operating at high field, which may require cooling in helium II below 2 K. In order to maintain adequate operating conditions, the applied or generated heat loads must be extracted and transported with minimum temperature difference. Conventional cooling schemes based on conductive or convective heat transport in pressurized helium II very soon reach their intrinsic limits of thermal impedance over extended lengths. We present the concept of helium II bayonet heat exchanger, which has been developed at CERN for the magnet cooling scheme of the Large Hadron Collider (LHC), and describe its specific advantages as a slim, quasi-isothermal heat sink. Experimental results obtained on several test set-ups and a prototype magnet string have permitted to validate its performance and sizing rules, for transporting linear heat loads in the W. $m^{-1}$ range over distances of several tens of meters.

## INTRODUCTION

The basic functions of beam guiding, focusing and acceleration in high-energy particle accelerators and colliders are now performed by superconducting devices - magnets and RF cavities - in order to achieve high performance at minimum investment and operating costs. As a consequence, large-scale helium cryogenics has become a key technology in accelerator construction and operation[1]. The drive for higher-field, more compact machines requires to push the performance of the superconducting materials by operating at lower temperature, in particular below 2 K in superfluid helium, where the specific transport properties of the cooling medium can also be exploited for more efficient heat removal and better stabilization of the superconductors[2]. Historically, this argument was first applied to the cooling of large superconducting magnets, e.g. high-field solenoids and tokamak toroids, which have been successfully built and operated over the past decades[3].

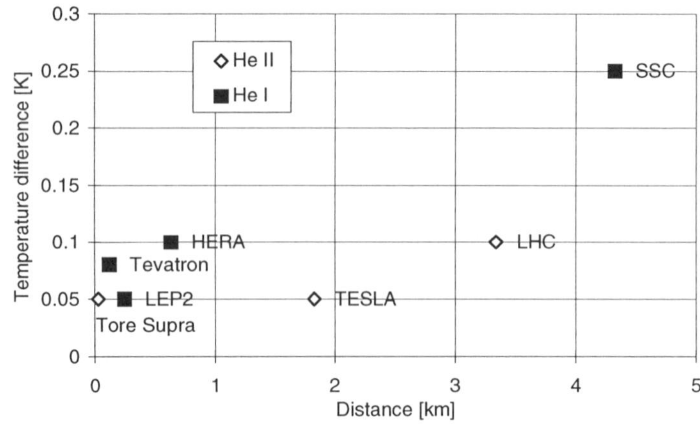


**Figure 1.** Transport of refrigeration in large distributed cryogenic systems.

Application of helium-II cooling to particle accelerators brings an additional constraint, that of large geographic extension in one dimension, the longitudinal position along the accelerator tunnel. In both circular and linear machines, the main problem is to ensure quasi-homogeneity of temperature in long strings of superconducting devices, i.e. to extract and transport heat efficiently from the furthest magnet or RF cavity, to the refrigeration plant serving the sector (Figure 1). In such cases, the helium II bayonet heat exchanger, developed at CERN for cooling the high-field magnets of the Large Hadron Collider (LHC)[4], represents an interesting alternative to schemes using conduction in pressurised helium II.

## BASIC PRINCIPLES AND DESIGN CHALLENGES

### Limitations of Conduction Cooling

Conduction cooling in static pressurised helium II, which works well for distances in the ten-meter range, very soon reaches its limits for extended geometries. Consider a uniformly heated tubular conduit of length L, cross-section A, filled with static pressurised helium II, representing an accelerator string with linear heat load q, and transporting heat Q by Gorter-Mellink turbulent mutual friction - equivalent to a non-linear bulk conduction - between temperatures $T_w$ and $T_c$ (Figure 2.a).

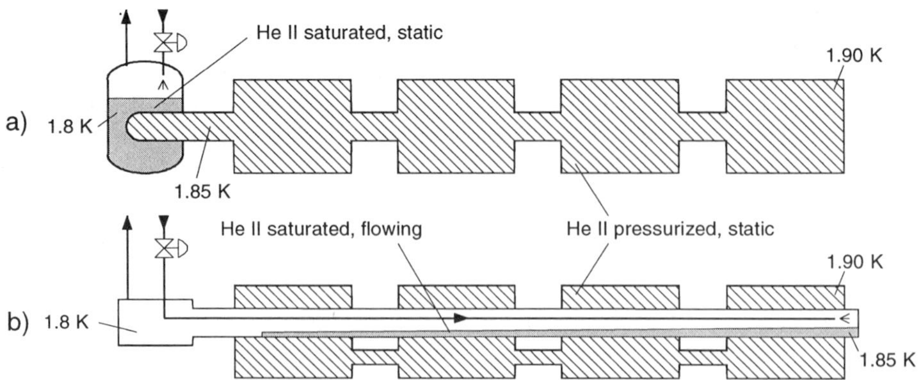


**Figure 2.** String cooling by (a) conduction in pressurised He II and (b) He II bayonet heat exchanger.

From energy conservation:

$$dQ = q \cdot dL \tag{1}$$

From the practical conduction law:

$$(Q/A)^n \cdot dL = X(T + dT) - X(T) \tag{2}$$

where the best experimental fit for n is 3.4, and X(T) is a tabulated function of temperature[5], equivalent to a conductivity integral.

After integration,

$$L^{n+1} \cdot (q/A)^n = (n+1) \cdot [X(T_c) - X(T_W)] \tag{3}$$

With a linear heat load q of 1 $W.m^{-1}$, and taking $T_w$ and $T_c$ respectively at 1.90 and 1.85 K, the diameter of the tubular conduit, which scales as $L^{0.65}$, is given in Figure 3 as a function of the conduction length. Another 0.05 K is allocated to the temperature drop across the lumped pressurised-to-saturated heat exchanger, so that the overall temperature difference between the warmest (furthest) device and the saturated heat sink is 0.10 K.

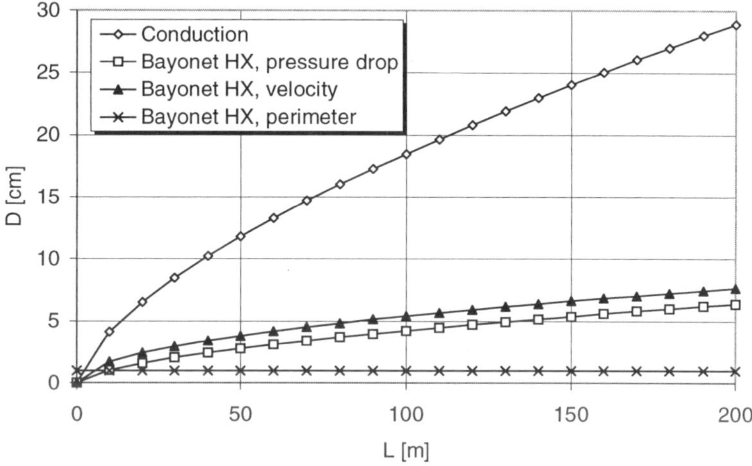


**Figure 3.** Sizing of conduit for string cooling by conduction in pressurised He II and bayonet heat exchanger, with linear heat load of 1 $W.m^{-1}$ between 1.9 and 1.85 K.

## Two-phase Flow of Saturated Helium II

To overcome the above limitations, the basic idea is to distribute the pressurised-to-saturated helium heat exchanger along the length of the accelerator string (Figure 2.b). In this fashion, a quasi-isothermal heat sink is provided at every longitudinal position along the string to be cooled, and the conduction distance in pressurised helium II is typically reduced to the transverse dimension of the cryostat. Subcooled helium is expanded in the Joule-Thomson valve and transported by the small-diameter feeder pipe to the end of the heat exchanger tube, from where it flows back and gradually vaporizes as it gathers heat along the way. The string heat load is extracted across the wall of the heat exchanger tube, which provides large developed surface area and hence low transverse thermal impedance, even when its wall is partially wetted. The device constitutes a bayonet heat exchanger, requiring access only from one end, and slim enough to be easily integrated in the cold mass of the superconducting devices in the string.

The basic question to be answered for ensuring adequate operation of this device, is that of stability and pattern of heated two-phase flow of helium II in horizontal or slightly inclined pipes, a subject on which no experimental data was available a few years ago. Application of predictive methods for flow-pattern maps, scaled from experimental correlations on other fluid systems[6] or from simple mechanical models[7], permitted to infer that the helium II heat exchanger tube would mostly operate in the stratified-flow regime, thus yielding a very partial wetting of the wall surface area.

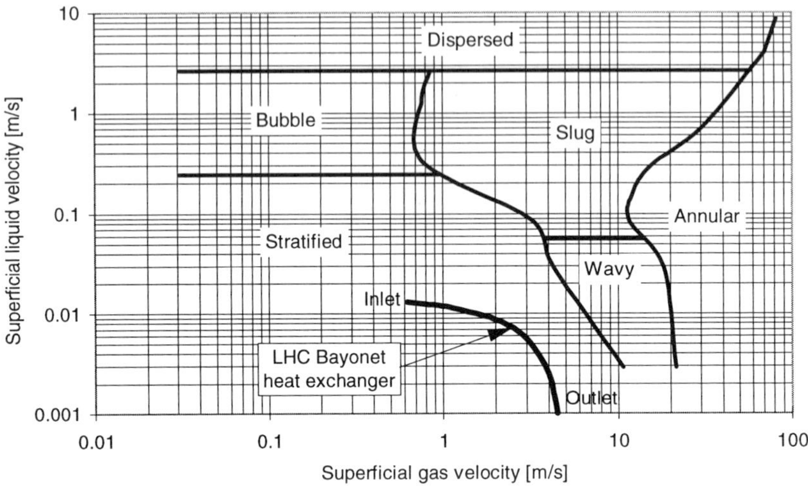


**Figure 4.** Predicted working line of helium II heat exhanger tube on a scaled Mandhane, Gregory and Aziz flow pattern map (saturated helium at 1.8 K, 1.64 kPa)

A first attempt to validate the concept on a 25-mm diameter, 24-m long cylindrical pipe[8], showed that horizontal and downward-sloping flows were very stable, the two-phases flowing clearly separated, and that excellent heat transfer at the tube wall could be obtained as long as liquid dryout did not occur. It also gave a first flavour of the control issues, in view of the slow (about 0.1 $m.s^{-1}$) velocity of the liquid phase.

A real helium II bayonet heat exchanger was then built and installed into two 10-m long modules containing pressurised helium II, connected in series and simulating LHC cryomagnets[9]. The heat exchanger consisted of a corrugated DHP copper tube, 39/44 mm in diameter and with 0.5-mm wall thickness. Tests on this set-up confirmed the stability and stratification of flow, as well as the excellent heat exchange, with conductances in the 100 $W.K^{-1}.m^{-1}$ range, up to linear heat loads of 2.4 $W.m^{-1}$. Measurement of mass flow-rates and estimation of flow velocities gave access to liquid holdup, void fraction, and hence wetted area, in reasonable agreement with the heat transfer data. The bayonet heat exchanger was originally operated wetted over its whole length, with excess liquid overflowing into a phase separator vessel, where it was vaporized to subcool the incoming helium. The liquid level in this vessel, which was used to control the position of the Joule-Thomson valve, responded with considerable delay due to the slow velocity of the liquid flow in the heat exchanger tube, thus resulting in a difficult control problem. It was then discovered that the conduction in the pressurised helium II, although insufficient to extract the applied heat load, could be used to convey control information with little delay, and subsequent tests were performed by controlling the Joule-Thomson valve position on the temperature of the warmest magnet-simulating module, and operating the bayonet heat exchanger with its length partially wetted. Specific control tools and methods were later developed along these lines, to cope with the strong non-linearities in the system[10].

In complement of this application-oriented tests, more thorough investigations of two-phase flow of saturated helium II were conducted at CEA Grenoble, France, on a test loop featuring 40-mm and 20-mm diameter heated cylindrical pipes, comprehensively equipped with diagnostics equipment[11, 12]. The diagnostics include wetted-perimeter indicators, as well as visual observation sections featuring CCD video cameras, which enabled to check the flow patterns inferred from hydraulic and thermal measurements. This work permitted to establish the validity of the stratified-flow model for predicting pressure drop, which eventually results in lack of isothermality along the flow. The liquid flow geometry - and hence the wetted perimeter - is adequately predicted by the model, which then coincides with the simple, open-channel flow description, up to vapour velocities of 4 to 5 $m.s^{-1}$. Beyond this range, an enhancement of heat transfer is observed, which cannot be explained within the stratified flow model, and is attributed to atomization of liquid droplets. This work has established a sound basis for sizing of helium II bayonet heat exchangers.

## Transverse Thermal Impedance of the Heat Exchanger Tube

All the extracted heat has eventually to flow across the tube wall, where it encounters three thermal impedances in series: the limited solid conduction across the metal constitutive of the wall, and the Kapitza resistances produced by the refraction of thermal phonons at the metal-to-helium interfaces. The overall transverse impedance was measured on fully wetted test samples at varying temperature, so that the different temperature dependence of the solid conduction and Kapitza terms enabled to resolve them[13]. For tubes with a wall thickness up to about 1 mm, the Kapitza resistance largely dominates below 2 K, and the use of high-purity, cryogenic-grade copper is not required.

## Corrugated versus Cylindrical Tubes

At equal free inner diameter, corrugated tubes exhibit, in fully developed turbulent flow, a friction factor typically 4 times larger than cylindrical, round tubes[14], which results in a strong increase in pressure drop, and hence temperature inhomogeneity along the flow. This drawback is partially balanced by the fact that the particular geometry of corrugated tubes yields a higher developed heat exchanger area, and an increased wetted perimeter at low liquid level. Moreover, they show better resistance against buckling by external overpressure, and thus may be built with a thinner wall. Another interesting feature of low-level, open-channel flow in such tubes is that the pressure drop is produced essentially by the series of *venae constrictae* at the corrugations, rather than by wall friction. As a result, the average flow velocity, and hence the wetted perimeter for a given flow-rate, are independent of the slope[15], which continuously varies around the LHC circumference.

## Practical Sizing Rules for the Bayonet Heat Exchanger

Consider the case sketched in Figure 2.b, and assume the vapour flow-rate increases linearly as the liquid vaporizes along the length of the flow.

From energy conservation:

$$Q = q \cdot L = L_v \cdot (m_{tot} - m_0) \tag{4}$$

where $L_v$ is the latent heat of vaporization, $m_{tot}$ and $m_0$ are the total vapour flow at outlet and inlet, respectively. Introducing $x_0$ as the inlet vapour quality,

$$m_0 = x_0 \cdot m_{tot} \tag{5}$$

A first sizing criterion is given by the allocation of a limited budget for the temperature difference along the saturated helium II flow, which translates into a maximum pressure drop. Since the void fraction in the tube is large, the pressure drop can be taken equal to that of the vapour flowing alone:

$$dP = \psi \cdot \frac{m^2}{2 \cdot \rho \cdot A^2 \cdot D} \cdot dL \tag{6}$$

where $\psi$ is the pipe friction factor, and $\rho$ is the mass density of the vapour.

After integration, for round tubes:

$$\Delta P = \frac{8 \cdot \psi}{3 \cdot \pi^2} \cdot \frac{q^2}{\rho \cdot L_v^2} \cdot \frac{L^3}{D^5} \cdot \frac{1 - x_0^3}{(1 - x_0)^3} \tag{7}$$

As an example, allowing the same 0.05 K temperature drop along the string length, and with the same linear heat load of 1 $W.m^{-1}$, the minimum diameter requirement set by this criterion for a cylindrical tube scales as $L^{0.6}$ and is given in Figure 3.

Another sizing criterion comes from the maximum vapour velocity ensuring to remain in stratified two-phase flow, say $v_{max} = 5\ m.s^{-1}$. The tube cross-section must then satisfy:

$$A \geq \frac{q \cdot L}{L_v \cdot \rho \cdot v_{max}} \cdot \frac{1}{1 - x_0} \tag{8}$$

This criterion, which leads to a diameter scaling as $L^{0.5}$, appears slightly more demanding than the pressure drop limit (see Figure 3), but still shows the clear advantage of the bayonet heat exchanger over helium II conduction.

A final check of the transverse heat transfer impedance should be made, to verify that the above criteria impose a tube with sufficient perimeter. Keeping the same value of 0.05 K for the transverse temperature difference, and considering an effective wetting of 20 %, yields a minimum diameter requirement of about 1 cm for a copper tube.

The helium II bayonet heat exchanger is therefore an efficient way to extract heat while using up a minimum amount of transverse space, which allows its easy implementation in the cold mass of the LHC cryomagnets (Figure 5).

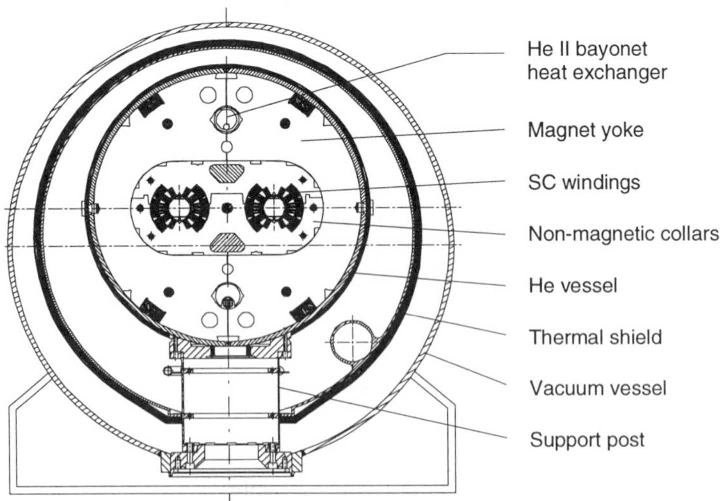


**Figure 5.** Transverse cross-section of LHC dipole cryomagnet, showing the helium II bayonet heat exchanger.

**Figure 6.** General view of the LHC Test String.

## EXAMPLES OF APPLICATION

### The LHC Test String

The first full-scale application of the helium II bayonet heat exchanger is the LHC Test String, a fully operational model of a half-cell of the future collider[16], which has been intensively tested since 1994. Over a length of 40 m, its cold mass is cooled by a helium II bayonet heat exchanger using a 41.5/50.5-mm diameter corrugated copper tube (Figure 6).

To cope with uncertainty in temperature measurements, as well as to allow for residual ripple in temperature control, the difference between maximum temperature in the pressurised bath of the cryomagnets, and saturated temperature at outlet was set at 35 mK The homogeneity of temperatures along the string under steady heat load can be seen on the left part of Figure 7, with a spread of a few mK. During a current ramp, which approximately results in doubling of the heat load, the bayonet heat exchanger contains the total temperature excursion to within 5 mK. The LHC test string has accumulated 3000 hours of operation at helium II temperature, as well as 2200 powering cycles, during which the bayonet heat exchanger has maintained the magnets below 1.9 K.

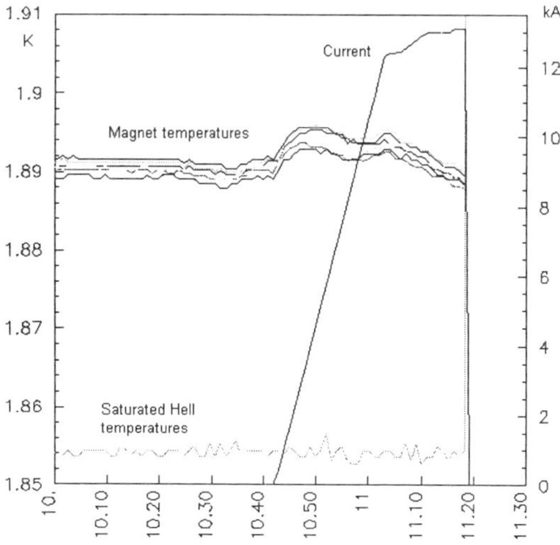


**Figure 7.** Response of the LHC Test String to a current ramp (horizontal scale: time in hours, minutes).

## The LHC Simplified Cryogenic Distribution Scheme

A recent revision of the LHC cryogenic distribution scheme has led to significant simplification, bringing the advantages of lower cost and higher reliability[17]. This was achieved by extending the length of the elementary cooling loops, including the helium II bayonet heat exchangers, from 53 to 107 m. Such a change meant doubling the length and helium mass flow-rate, which would have resulted in an unacceptable eight-fold increase in pressure drop, at equal tube diameter. Since the passage available in the cold mass of the cryomagnets could not be increased, the retained solution was to install a 54-mm inner diameter, 2-mm wall thickness cylindrical tube of OFHC copper, thus trading the advantages of corrugated tubes against the reduction in friction factor.

In total, the LHC will be cooled by some 230 helium II bayonet heat exchangers, distributed around its 26.7 km circumference.

## REFERENCES


1. Ph. Lebrun, Cryogenic systems for accelerators, in: "Frontiers in Accelerator Technology", S.I. Kurokawa, M. Month and S. Turner, editors, World Scientific, Singapore (1996), pp. 681-700.
2. Ph. Lebrun, Superfluid helium as a technical coolant, in: "Atti XV Congresso Nazionale sulla Trasmissione del Calore", Edizioni ETS, Politecnico di Torino (1997), pp. 61-77.
3. G. Claudet and R. Aymar, Tore Supra and helium II cooling of large high-field magnets, in: "Advances in Cryogenic Engineering" Vol. 35A (1990), pp. 55-67.
4. L.R. Evans, The Large Hadron Collider project, in: "Proceedings of ICEC16/ICMC Kitakyushu", T. Haruyama, T. Mitsui and K. Yamafuji, editors, Elsevier Science, Oxford (1997), pp. 45-52.
5. G. Bon Mardion, G. Claudet and P. Seyfert, Practical data on steady-state heat transport in superfluid helium at atmospheric pressure, *Cryogenics* 19 (1979), pp. 45-47.
6. J.M. Mandhane, G.A. Gregory and K. Aziz, A flow pattern map for gas-liquid flow in horizontal pipes, *Int. J. Multiphase Flow* 1 (1974), pp. 537-553.
7. Y. Taitel and A.E. Dukler, A model for predicting flow regime transitions in horizontal and near horizontal gas-liquid flow, *AIChE Journal* 22:1 (1976), pp. 47-55.
8. J. Casas-Cubillos, A. Cyvoct, Ph. Lebrun, M. Marquet, L. Tavian and R. van Weelderen, Design concept and first experimental validation of the superfluid helium system for the Large Hadron Collider (LHC) project at CERN, *Cryogenics* 32 ICEC Supplement (1992), pp. 118-121.
9. A. Bézaguet, J. Casas-Cubillos, Ph. Lebrun, M. Marquet, L. Tavian and R. van Weelderen, The superfluid helium model cryoloop for the CERN Large Hadron Collider (LHC), in: "Advances in Cryogenic Engineering" Vol. 39A (1994), pp. 649-656.
10. B. Flemsaeter, "Contribution to the Dynamic Analysis and Optimal Control of the Superfluid Helium Cooling Loop for the LHC Magnet String", Thesis, NTU Trondheim, Norway (1995)
11. A. Gauthier, L. Grimaud, B. Rousset, A. Bézaguet and R. van Weelderen, Thermohydraulic behaviour of He II in stratified co-current two-phase flow, in: "Proceedings of ICEC16/ICMC Kitakyushu", T. Haruyama, T. Mitsui and K. Yamafuji, editors, Elsevier Science, Oxford (1997), pp. 519-522.
12. B. Rousset, A. Gauthier, L. Grimaud and R. van Weelderen, Latest developments in He II co-current two-phase flow studies, paper presented at this conference.
13. M.M. Kado, Thermal conductance measurements on the LHC helium II heat exchanger pipes, CERN Report AT/95-34 (CR) (1995).
14. K. Kauder, "Strömungs- und Widerstandsverhalten in Gewellten Rohren", Thesis, Technische Universität Hannover, Germany (1971).
15. F. Déliot, private communication (1996).
16. A. Bézaguet, J. Casas-Cubillos, B. Flemsaeter, B. Gaillard-Grenadier, Th. Goiffon, H. Guinaudeau, Ph. Lebrun, M. Marquet, L. Serio, A. Suraci, L. Tavian and R. van Weelderen, The superfluid helium system for the LHC Test String. design, construction and first operation, in: "Advances in Cryogenic Engineering" Vol. 41A (1996), pp. 777-784.
17. M. Chorowski, W. Erdt, Ph. Lebrun, G. Riddone, L. Serio, L. Tavian, U. Wagner and R. van Weelderen, A simplified distribution scheme for the Large Hadron Collider, paper presented at this conference.

# SUPPORTING SYSTEMS FROM 293 K TO 1.9 K FOR THE LARGE HADRON COLLIDER (LHC) CRYO-MAGNETS

M. Mathieu, V. Parma, T. Renaglia, P. Rohmig,and L.R. Williams

CERN, European Organization for Nuclear Research
1211 Geneva 23, Switzerland

## ABSTRACT

The LHC machine will incorporate some 2000 main ring super-conducting magnets cooled at 1.9 K by super-fluid pressurized helium, mainly 15m-long dipoles with their cryostats and 6m-long quadrupoles housed in the Short Straight Section (SSS) units. This paper presents the design of the support system of the LHC arc cryo-magnets between 1.9 K at the cold mass and 293 K at the cryostat vacuum vessel. The stringent positioning precision for magnet alignment and the high thermal performance for cryogenic efficiency are the main conflicting requirements, which have lead to a trade-off design. The systems retained for LHC are based on column-type supports positioned in the vertical plane of the magnets inside the cryostats. An *ad-hoc* design has been achieved both for cryo-dipoles and SSS.

Each column is composed of a main tubular thin-walled structure in composite material (glass-fibre/epoxy resin, for its low thermal conductivity properties), interfaced to both magnet and cryostat via stainless steel flanges. The thermal performance of the support is improved by intercepting part of the conduction heat at two intermediate temperature levels (one at 50-75 K and the other at 4.5-20 K). These intercepts, on the composite column, are thermally connected to the helium gas cooled thermal shield and radiation screen of the cryo-magnet.

An overview of the design requirements is given, together with an appreciation of the system design. Particular attention is dedicated to the support system of the SSS where the positioning precision of the quadrupole magnet is the most critical.

## INTRODUCTION

After an intense R&D phase, which has lead to testing of several prototype supports, the main technological choices have been made[1,2,3]. The final layout of the support system is now defined within the cryostat design of the LHC Standard half cell. New prototypes for cryo-dipoles, manufactured by industry are presently undergoing testing at CERN.

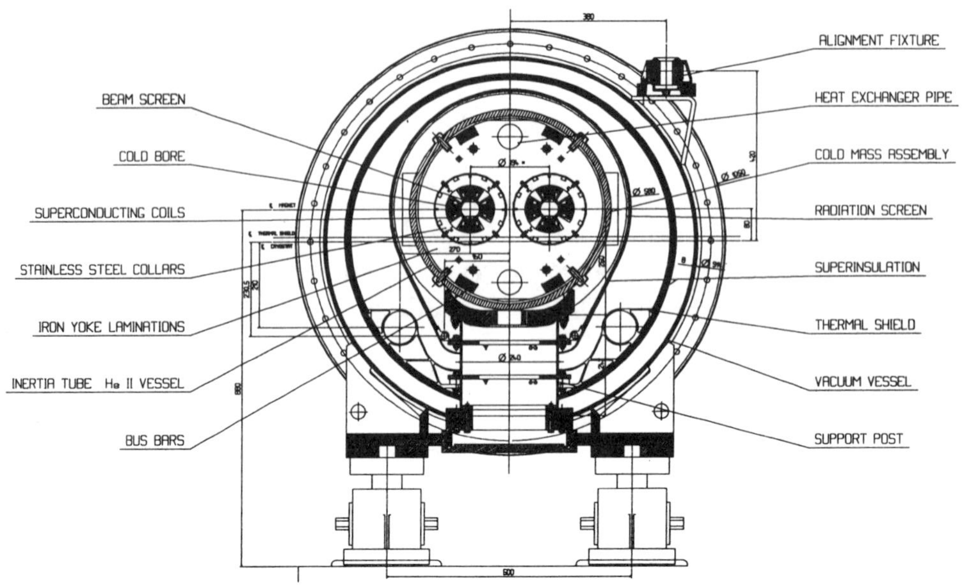


**Figure 1.** Transverse section SSS, fixed support.

Prototype supports for the SSS will be ordered in autumn 1997. Testing on dedicated facilities will validate the thermal and mechanical design, before they are mounted in the new generation of cryo-magnets. Tests will then follow on the magnet test bench facility and finally in a string of magnets representative of a complete cell of the LHC machine, which will become operational in the year 2000. Figure 1 shows the transverse section of the SSS and its fixed support.

## SUPPORT SYSTEM REQUIREMENTS

The accurate positioning of magnets to meet the alignment requirements for the LHC machine lattice imposes the most critical design constraint on the support system, as it is closely related to magnet performance and machine beam dynamics. Misalignments introduce perturbations similar to those from magnetic field errors resulting in a reduction of machine aperture and their correction would significantly increase the power requirements of closed-orbit correcting magnets[4]. In synchrotron machines, using traditional magnets, the alignment fiducials are mounted directly on the yoke. The LHC superconducting magnets housed in cryostats, inaccessible during operation, are aligned via external cryostat-mounted fiducials. This alignment, after machine installation, is therefore carried out "blindly", and implies that the positioning of the magnets is adequately guaranteed by the supporting system.

A thermo-mechanically optimized solution is therefore a compromise satisfying both the minimum stiffness required for precise alignment and the stringent heat load budget of the LHC.

The accurate positioning of the magnets inside their cryostats must be assured for machine operation in cryogenic environment during an estimated lifetime of 20 years.

**Table 1.** Main requirements for the support systems for cryo-dipole and SSS

| Requirement | Cryo-dipole | Short Straight Section |
|---|---|---|
| **Positioning accuracy:** | | |
| (after cryo-magnet assembly) | | |
| x (radial) | ± 1 mm | ± 0.5 mm |
| y (longitudinal) | ± 2 mm | ± 1 mm |
| z (vertical) | ± 1 mm | ± 0.5 mm |
| **Positioning reproducibility-stability:** | | |
| (in operation, during lifetime) | | |
| x (radial) | $< \pm 0.3$ mm ($3\sigma$) | $< \pm 0.3$ mm ($3\sigma$) |
| y (longitudinal) | $< \pm 1$ mm ($3\sigma$) | $< \pm 1$ mm ($3\sigma$) |
| z (vertical) | $< \pm 0.3$ mm ($3\sigma$) | $< \pm 0.3$ mm ($3\sigma$) |
| $\theta\ y$ (radial tilt) | $< \pm 0.3$ mrad ($3\sigma$) | $< \pm 0.3$ mrad ($3\sigma$) |
| **Thermal requirements** | | |
| (indicative heat loads, steady state) | | |
| 1.9K level | < 0.15 W | < 0.10 W |
| 4.5-20K level | < 1.5 W | < 1 W |
| 50-75K level | < 15 W | < 10 W |

Support system design must take into account long-term *position reproducibility* after repeated thermal cycling, and position *stability* with respect to *parasitic loads*. These loads may be caused by interconnect bellows and misalignment, jumper connection to the CDL line, *stick-slip* on the sliding interfaces, and to the 1.4% inclination of the LEP tunnel.

It is expected that all the SSS, demanding the most precise machine alignment, but only a fraction of the cryo-dipoles will undergo cold magnetic measurements and geometrical survey before installation in the machine. Therefore, all cryo-magnets and their support systems must be assembled to tight mechanical tolerances and demonstrate repeatable thermal contraction behavior. Manufacturing tolerances and positional *reproducibility* after thermal contractions of the series production units are main items to be addressed in the compilation of a strict Quality Assurance plan.

The main requirements for the cold support system are summarized in Table 1. *Positioning accuracy* and *reproducibility-stability* of the cold mass are given with respect to the cryostats. *Reproducibility-stability* are taken to be within the precision of re-alignment between cryostat vacuum vessels which is of 0.1 mm (*r.m.s.*) in the vertical-horizontal plane and 0.1 mrad (*r.m.s.*) radial tilt[5]. The maximum values are taken with 3 $\sigma$ confidence.

## SYSTEM DESIGN

### Layout

A three-point support configuration was chosen for the 15m-long, 25,000 kg dipole cold mass, where the inter-support spacing limits the maximum vertical sagitta, due to self weight, below 0.3 mm and the angular deflection of the extremities to acceptable values for magnet interconnect assembly[6]. Two support points are sufficient for the shorter (6 m) and lighter (6,000 kg) cold mass of the SSS to keep the maximum vertical self weight sagitta below 0.25 mm. Further details on the cryostat assemblies can be found in Reference[7,8].

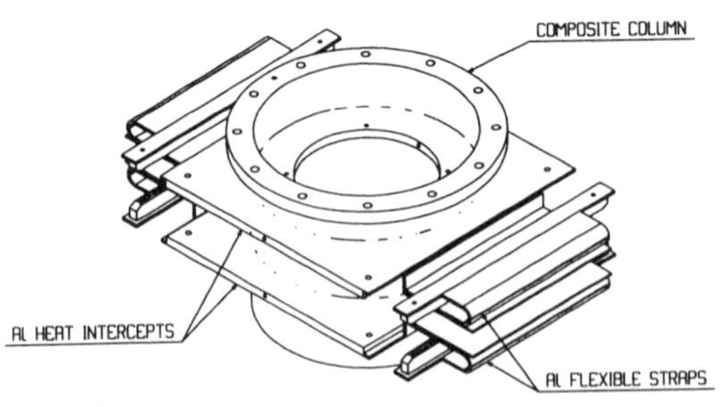


**Figure 2.** Support post assembly.

## Support post assembly and integration

Each support point is composed of a main non-metallic composite thin-walled column, bolted to stainless steel pads previously welded along the cold mass. The interface to the vacuum vessel allows the support to slide, leaving the thermal contraction of the cold mass unhindered. A simple support on the vacuum vessel and a key insert permits sliding in only one direction. The weight of the cold mass guarantees that lifting does not occur under any envisaged load case. A low-friction material plate (*Glacier DU*™) reduces stick-slip effects on the vessel. For the cryo-dipoles, the two lateral supports are free to slide longitudinally on the vacuum vessel, whereas the central support is longitudinally fixed but free in the radial direction, to allow possible changes in magnet curvature (due to excitation or repeated thermal cycling). For the SSS, one support is completely fixed, bolted to both the cold mass and vacuum vessel, as the longitudinal pressure force introduced by the insulation vacuum barrier (mounted in the SSS) would tend to lift the cold mass.

Moreover, an additional function requested to the columns of both cryo-dipoles and SSS is to support the thermal shield and radiation screen structures in the cryostat while allowing their differential thermal contractions with respect to the cold mass. The two aluminum heat intercept plates mounted on the composite column to thermalise the support at intermediate temperatures, additionally fulfill this function. An assembly view of a cold support is shown in Figure 2.

The estimated initial *positioning accuracy* of the SSS cold mass, with respect to the vacuum vessel interfaces of the supports, is less than ± 0.2 mm in the transverse plane and ± 1 mm in the longitudinal direction. The chain of tight mechanical tolerances achieves this precision. The *positioning accuracy* for the dipole cold mass, with respect to the vacuum vessel interfaces, is ± 1 mm in the transverse plane, achieved by shimming during cryostat integration. Angular errors between the vacuum vessel and cold mass interfaces, in the order of ± 0.3mrad, can be accommodated by the bending flexibility of the columns.

The assembly design for the support system is based on the use of standard tooling and on the simplicity of operations, with a view to maintaining the high assembly rate of the series production (2 cryo-dipoles per day) carried out by European industry. Road transport of the cryo-dipoles to CERN will be a critical phase for the integrity of the supports. Past experience from other similar accelerator machines has shown that blocking devices are required. Special reinforcing steel columns, mounted inside and concentric to the composite column are foreseen, unloading and stiffening the supports during transport, and finally removed when the cryo-dipole is positioned in the tunnel. An experimental program is underway to measure effects of road transport on a 15m cryo-dipole during delivery to CERN. The SSS, proposed to be assembled in the vicinity of CERN, will not mount reinforcing devices to avoid any mechanical perturbation of survey measurements. Also this strategy will need to be confirmed by tunnel transport testing.

### Composite technology

The composite columns are made of a 4-mm thick 240-mm external diameter tube with integrated top and bottom flanges. A glass-fiber/epoxy composite system is chosen for its high stiffness-to-thermal-conduction ratio, allowing minimization of the cross section and, as a consequence, of conduction heat loads. A long-fiber, woven fabric lay-up technology is chosen to achieve the highest stiffness. Either pre-impregnation or resin transfer mold (RTM) technologies are equally valid. Other materials and manufacturing methods such as short-fiber charged ULTEM™, ISARYL™ or ULTRAPEK™ thermoplastics have been exploited to improve thermal performance, but none of these provide a satisfactory combination of stiffness, and low thermal conductivity. Filament winding was also widely tested, however it requires glued flanges and due to the uncertain behavior of glued interfaces at cryogenic temperature over the estimated 20-year life of the accelerator, this technology has been abandoned for the SSS, which undergoes the highest shear loads. Filament winding is however interesting for its low cost of manufacture (up to 50% lower than other technologies), and is being reconsidered for the cryo-dipoles.

## MECHANICAL PERFORMANCE

During machine operation, the support posts are mainly subject to vertical load due to cold mass weight (10,000 and 3,000 kg per support on cryo-dipoles and SSS respectively). In addition, the supports may be subject to residual and unpredictable *parasitic* loads, which are estimated up to 2500 N per support in the radial direction and that lead to the minimum design stiffness to fulfill position *stability*. Under this load, a 0.27 mm radial displacement and 0.35 mrad tilt[9,10] are calculated for the axis of the SSS cold mass. Displacement *stability* is achieved but with little safety margin, whereas angular deflection is slightly too high. Mechanical testing on prototypes is needed to confirm these values.

The mechanical strength of the supports is not critical during machine operation. Testing on former prototypes has shown that the supports can withstand up to 200% of the nominal loads without failure. Calculation shows high safety margins with respect to static strength and buckling[11].

Although *creep* is a difficult phenomenon to master for composites, and extreme care must be taken for the LHC supports, testing of prototype cryo-magnets over 3 years (with 30% time at room temperature) does not show any evidence of viscoelastic effects (Figure 3). Also *relaxation tests* on compressed samples at room temperatures show a limited (6%) relaxation effect when loaded at 4 times the nominal compression load[11]. *Reproducibility* requirements are fulfilled with respect to thermal cycling; for the vertical contraction this is better than 50 μm, the dispersion observed is within the precision of the measurement system[12,13].

The thermal contraction of the supports (Figure 3) shows appreciable differences however (up to 0.2 mm). This is explained by the different technology applied by manufacturers (fiber/resin ratio, fiber orientation, etc.). The series production will need to be closely surveyed by a rigorous Quality Assurance plan to limit this uncertainty.

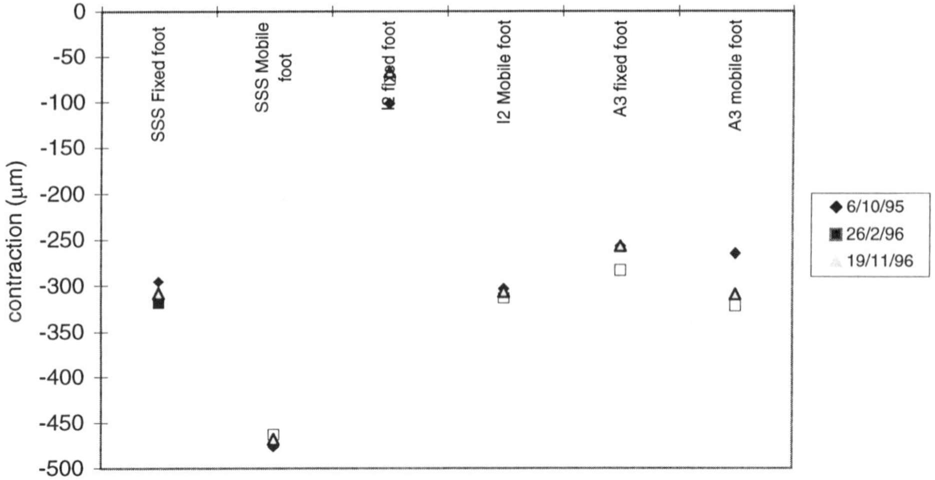


**Figure 3.** Supports vertical contraction, 1.9K.

## THERMAL PERFORMANCE

The thermal performance of the supports is improved by intercepting most of the residual conduction heat at two intermediate temperature levels, 50-75 K and 4.5-20 K, the precise temperature depending on the position of the cryo-magnet in the cryogenic loop of a machine sector. These intercepts consist of aluminum plates, integrated to the composite column and connected to the thermal shield and radiation screen via flexible aluminum straps. All-welded thermalisation assemblies are adopted, as past experience has demonstrated the unreliability of bolted connections for achieving low thermal impedance. Although cryostat integration and assembly limits the free choice of vertical position of the heat intercept plates along the composite column, these positions are chosen to minimize the exergetic cost of low-temperature refrigeration, according to the weighing function [14]:

$$Q_{tot} = Q_{1.8K} \cdot 1000 + Q_{5K} \cdot 125 + Q_{75K} \cdot 16$$

Where:

$Q_{1.8K}$, $Q_{5K}$ and $Q_{75K}$ are the heat loads at the temperature levels indicated, multiplied by their weight factors, and $Q_{tot}$ is the exergetic cost of refrigeration in Watts.

The optimization has been achieved for both the SSS and the cryo-dipole for arc-wise averaged temperatures at the heat intercepts (10 K for the 5-20 K range and 65 K for the 50-75 K range).

**Table 3.** Heat loads of Standard Half-Cell support systems (± 15%)

| Temperature | Cryo-Dipole | SSS | % of half cell static loads |
|---|---|---|---|
| **1.9 K** | 0.069 W | 0.038 W | 1.8 % |
| **8 K** | 2.58 W | 1.38 W | 62 % |
| **80 K** | 17.3 W | 9.5 W | 17% |

A review of this optimization criterion and of the position of the heat intercepts may be required, to reduce the cooling demands at the 4.5-20 K refrigeration, presently close to the budgeted limit for the re-conversion of cryogenic equipment from LEP.

Compared with a non-thermalised version of the same support of the SSS, the two heat intercepts allow a reduction of the heat loads to 1.9 K by a factor of 100. Sensitivity analysis indicates that the thermal efficiency is strongly influenced by the temperature at the heat intercepts[15]. The 4.5-20 K heat intercept is the most critical, where an off-design temperature of ± 2 K introduces a variation of ± 10 % on the exergetic cost of refrigeration. Off-design temperatures are encountered during cool-down (a transient study is being made to evaluate the impact on cool-down time and peak heat loads during this phase) but could also be caused by a faulty assembly of the heat intercepts to the support. Quality Assurance will play once more a crucial role in coping with this type of problems.

Inside the supports are mounted reflective aluminum disks for the interception and extraction of radiation heat via the heat intercepts. Additionally, aluminized *Mylar*™ reflective surfaces protect the part of the support column exposed to temperatures between 293 K and 50-75 K[16].

The heat loads for the cryo-dipole and SSS support systems, extrapolated from measurements on former prototypes of different geometry, with heat intercepts at 8 K and 80 K are shown in Table 3.

Measurements on a dedicated cryogenic test facility at CERN will be made to validate these results and will allow precise measurements of the heat loads across the working temperature ranges of each heat intercept. Validation of the new all-welded flexible thermalisation straps will be also possible.

## CONCLUSIONS

The system layout for the supporting of the LHC magnets inside the cryostats is retained and the main technological choices are made. An intense R&D phase has indicated the technology retained for the manufacture of the composite column (glass-fiber/epoxy pre-preg or RTM) although cost-reduction for the series production suggests continuing with further investigation on filament winding. The design of the supports is the result of a compromise between stiffness and thermal requirements. Calculations (to be validated by testing) show that stiffness is just within the positioning *stability* requirements; efforts must be therefore made towards the understanding and reduction of possible sources of mechanical misalignment. Thermal cycling over several years of operation on the first string of prototype cryo-magnets shows good position *reproducibility* with respect to thermal contractions of the cold mass inside the cryostats. However, an appreciable dependence of thermal contraction behavior on the manufacturing technology highlights the importance of limiting this dispersion by imposing a strict Quality Assurance program. Minor changes in the thermalisation design might still be introduced to reduce heat loads at the 4.5-20 K level.

New prototype supports have been manufactured by industry and are now being tested at CERN. Mechanical testing and evaluation of thermal heat loads on dedicated test facilities will validate the design. After being mounted in cryo-magnets, they will be tested first on the magnetic test-bench facility and finally in a string of magnets representative of a complete cell of the LHC machine, that will be operational in the year 2000.

## ACKNOWLEDGEMENTS

The authors would like to thank Ch.Disdier, T.Kurtyka and A.Pardon for their contribution to calculations, D.Missiaen and T.Dobers for the survey measurements. Finally, the team of the cryogenic lab at CERN for the heat loads measurements.

## REFERENCES

1. M.Mathieu, Th.Renaglia, Ch.Disdier, "Supportage de la masse froide du LHC", Technical Note EST-ESM/96-01, CERN - Geneva.
2. M.Mathieu, "Systeme de supportage du LHC: Point sur le development", Technical Note MT-ESM/94-3, CERN - Geneva.
3. M.Mathieu, "Développement de supports en matériau composite pour les aimants prototypes du LHC", Technical Note MT/93-06, CERN - Geneva.
4. J.F.Miles, "Concerning the Strength of Closed Orbit Correction Dipoles in the Arcs of LHC V4.2", LHC Project Note 43, CERN - Geneva.
5. P.Lefevre, for the Working Group on Mechanical Apertures, "Parameter & Layout Committee", minutes of the 9th meeting.
6. J.C.Brunet, "Extrémités des dipoles - Référentiel - Acceptabilité - Situation & Installation", LHC Project Note 50, CERN - Geneva.
7. W.Cameron, Ph.Dambre, T.Kurtyka, V.Parma, T.Renaglia, J.M.Rifflet, P.Rohmig, B.Skoczen, T.Tortschanoff, Ph.Trilhe, P.Vedrine, D.Vincent, "The new superfluid helium cryostats for the Short Straight Sections of the CERN Large Hadron Collider (LHC)", presented at this conference.
8. J.C.Brunet, V.Parma, G.Peon, A.Poncet, P.Rohmig, B.Skoczen, L.R.Williams, "Design of the second series 15 m LHC prototype dipole magnet cryostats", presented at this conference.
9. A.Pardon, "Verification of safety margins for the LHC cold support system", Technical Note under preparation, CERN - Geneva.
10. Ch.Disdier, "Validation by calculation of the thermo-elastic behavior of the LHC cold support system", Technical Note under preparation, CERN - Geneva.
11. Ch.Disdier, "Contribution a l'étude du décollement interfacial et des propriétés thermiques de composite unidirectionnels sous les aspects micro, meso et macro; applications", PhD Thesis, Université de Savoie - Chambery, France.
12. T.Dobers, G.Peiro, "Détection de mouvements de la masse froide par rapport au cryostat de dipole prototype I2 pour le LHC", Internal report AT/SU/TD, CERN - Geneva.
13. D.Missiaen, "Metrology of superconducting magnets: first experience and facts from the LHC test String", IWAA, International Workshop on accelerator alignment, Tokyo – Japan 1995.
14. L.Tavian, Private communication, CERN - Geneva.
15. V.Parma, "Thermalisations of cold supports: minimisation of conduction heat loads: results of calculation, implications on the design of thermal screens for the LHC half cell", minutes of the Cryostat Design Group meeting, 7 August 1996, CERN - Geneva.
16. B.Jenninger, "Influence of reflective surfaces on the LHC Support Posts", LHC Project Note 65.

# DESIGN OF THE SECOND SERIES OF LHC PROTOTYPE DIPOLE MAGNET CRYOSTATS

J.C. Brunet, V. Parma, G. Peón, A. Poncet, P. Rohmig, B. Skoczen, and L.R. Williams

CERN, European Organization for Nuclear Research
1211 Geneva 23, Switzerland

## ABSTRACT

An initial series of six LHC 10 m long prototype dipole magnets and cryostats have been manufactured by European Industry. The assembled cryo-magnets were tested singly and connected in series in a test string at CERN between March 1994 and December 1996. During the same period, an evolution in the requirements for LHC cryogenics distribution has lead project management to adopt a separate cryo-distribution line running parallel to the LHC machine.[1] The former standard LHC half-cell, made up of a short straight section unit and four 10 m dipoles, has been discarded and replaced with one composed of a short straight section unit and three 15 m dipoles. The new 15 m LHC dipole magnet cryostats are described. These units house the dipole magnet cold mass, which stands on three low heat in-leak support columns, and is enclosed within an actively cooled radiation screen operating at 4.5-20 K and an actively cooled thermal shield operating at 50-75 K.

## INTRODUCTION

Following project approval in December 1994, CERN commenced the design and construction of the Large Hadron Collider (LHC),[2] the next important step in its scientific program. This new facility, enabling the collision of protons at energies in the multi-TeV range, will mainly consist of a double ring of advanced technology super-conducting magnets which will be installed in the existing 27 km LEP tunnel. The dipole magnets are 15 m long with a twin-aperture design, producing bending fields in the 8 to 9 T range. For this purpose, they use Nb-Ti superconductors and operate at temperatures below 2 K. In order to test these magnets, the related cryogenics, and the other components of the LHC under realistic operating conditions, CERN has ordered a number of full scale super-conducting dipole cryo-magnets from European industry, and will install and operate them in a prototype string, 107 m long, representing an LHC full cell.

A cryo-magnet of this type operates in a bath of pressurised super-fluid helium at a pressure close to atmospheric and at temperatures below 2 K. The magnet string will allow the verification of magnetic performance, as well as the thermal and mechanical behaviour of a long cryogenic system operating with super-fluid helium.

The cryostat must provide a stable platform for the cold mass while maintaining heat in-leaks at a level matching the stringent heat load budget of the LHC. It must survive the thermo-mechanical transients associated with rapid cool-down or warm-up, a magnet resistive transition (quench), and a loss of insulation vacuum, while being simple and economic to manufacture in large numbers.

The dipole magnet[3] enclosed in its all welded helium vessel with heat exchanger and cold-bore tubes, forms the dipole cold-mass. The cryostat of the dipole magnet consists of three support posts to position the cold-mass, a radiation screen and a thermal shield both equipped with superinsulation, and a vacuum vessel. The dipole cold-mass assembled into the cryostat forms the cryo-dipole magnet. Figure 1 shows a cross-section of the LHC 15 m dipole cryo-magnet in the plane of an extremity support post. Similar cryostats have been designed for the LHC Short Straight Section (SSS) units.[4]

## DIPOLE COLD MASS

The cylindrical LHC dipole cold mass has a length of 15.2 m, a diameter of 0.57 m, and weighs about 25000 kg. During manufacture, it is curved about a vertical axis to a radius of 2700 m leading to a horizontal sagitta from end to end of 11.3 mm. Inside the cryostat, it is supported at three points along its curved axis, one at the mid length, the others being positioned as a compromise between minimizing vertical sagitta, rendering the cold mass approximately horizontal at its ends to ease interconnect assembly, and minimizing overhang outside the end supports to simplify magnet to magnet alignment. Maintaining the vacuum vessel wall thickness and mid length deformation under load within reasonable limits was also taken into account. These conditions require three internal supports located 5.4 m apart. The cold mass applies loads of 7800 kg at each end and 9400 kg at the mid-plane through the support posts to the vacuum vessel.

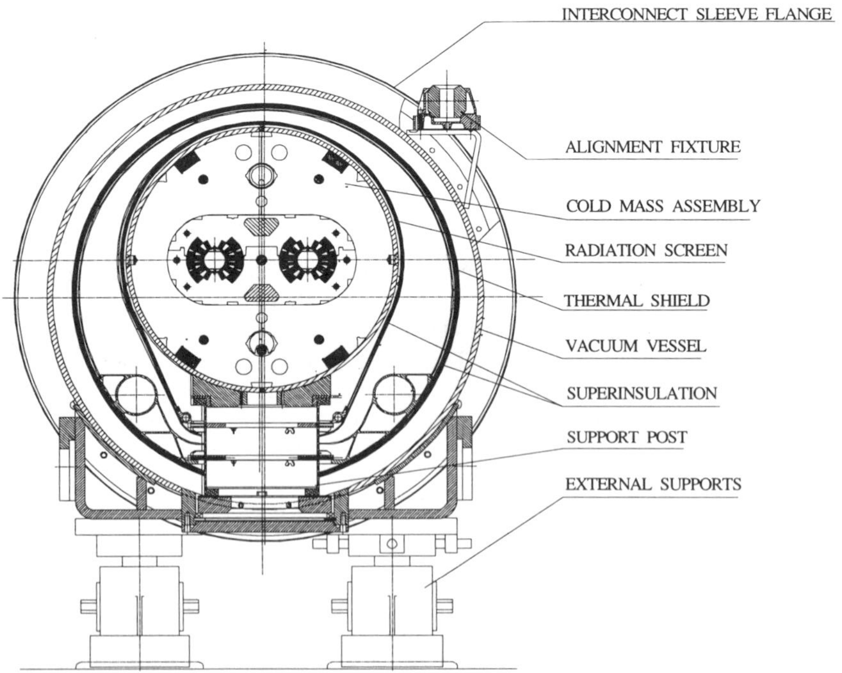


**Figure 1**. Cross-Section of the LHC 15 m Dipole Cryo-magnet Assembly.

To minimize heat transfer by radiation, the outer surface of the cold mass is closely wrapped with one layer of double aluminized Mylar™.

## INTERCONNECT

The interconnect regions between magnets are as short as possible to maximize bending magnet occupancy in the LHC. The electrical, cryogenic and vacuum circuits of successive dipole cryo-magnets are linked across the interconnect via stainless steel flexible bellows. These compensate for a thermal contraction of about 48 mm developed between the longitudinally fixed central supports of neighbouring units during cool-down. To permit cryo-magnet alignment in the LHC tunnel, the flexible components of each interconnect allow displacements of up to ± 4 mm in any direction perpendicular to the particle beam axes. Table 1 lists the interconnect temperatures and pressures associated with various operating modes.[5] All pipe-work installed in the interconnect region is in austenitic stainless steel with the exception of that linking the 15 mm internal diameter radiation screen cooling circuits, which at present terminate at one end with an Al 6060 single turn expansion loop instead of an aluminum to stainless steel transition and conventional bellows.

The final design will depend on the results of process validation, currently underway at CERN, for the helium leak-tight butt welding of these aluminum tubes in the tunnel. A specified fatigue life of 500 cycles combined with the operational pressure and temperature fluctuations has often imposed the use of multi-wall bellows. This leads to an overall interconnect stiffness of 890 N/mm longitudinally and 800 N/mm radially.

To avoid further complicating the interconnect, both the radiation screen and the thermal shield are attached to, and fixed laterally and vertically by, the two support posts at the cold mass extremities. On each side of the interconnect, these fixing points lie about 2.2 m inboard from the flexible bellows located in the thermal shield cooling circuit. Under the most unfavorable operating conditions, lateral and vertical stiffness with a safety factor of six to resist buckling instability of the flexible bellows has been incorporated into the thermal shield lower tray.[6]

Each interconnect requires 20 stainless steel and 1 aluminum tube welds and relies on compact orbital tube cutting and welding equipment for initial assembly and maintenance.

## VACUUM VESSEL

To contain the dipole stray field and reduce costs, the vacuum vessel shell is based on standard straight pipeline elements with an outer diameter of 914 mm (36"), and a wall thickness of 12 mm in mild steel, with end and auxiliary flanges of AISI 304L stainless steel. In case of a sudden loss of insulation vacuum, local cooling of the vessel shell to about 230 K may occur.[7] According to ISO standards, a minimum energy absorption of 28 J/cm$^2$ during a Charpy impact test at 240 K is required to qualify the vacuum vessel steel against low temperature embrittlement. These standard tube dimensions are chosen as a compromise between providing adequate resistance to bending under the cold mass loads, keeping the total cryo-dipole mass less than 35000 kg, and leaving sufficient space beneath the vacuum vessel to locate three adjustable floor mounted jacks to permit cryo-magnet alignment.

Three vertical through-penetrating flanges are positioned to locate the support posts, support the cold mass, and maintain its maximum static deflections to less than 0.3 mm. On each side of these three flanges, two welded annular reinforcements strengthen the vacuum vessel shell against ovalization. Lower 'U' shaped reinforcing elements, welded between the annular reinforcements, provide hard points for external supports and carry lateral lifting plates for handling. Two upper angle plates support fixtures into which Taylor Hobson targets can be located during alignment of the cryo-magnet with its neigbouring units.

As shown in figure 2, a captive flange behind the front end flange permits connection to a de-mountable sliding sleeve which encloses the interconnect region. This sleeve consists of two flanges separated by a tube incorporating two bellows, each of four corrugations.

**Table 1.** Operational pressures and temperatures

| | | Dipole cold mass | | | | Interconnect | | | |
|---|---|---|---|---|---|---|---|---|---|
| Component | DN | $T_0$ | $P_0$ | $P_Q$ | $P_C$ | $T_0$ | $P_0$ | $P_Q$ | $P_C$ |
| | [mm] | [K] | [bar] | [bar] | [bar] | [K] | [bar] | [bar] | [bar] |
| 50/75 K Circuit | 80 | | | | | 50/75 | 19 | 19 | 19 |
| He-He heat exchanger | 54 | 1.8 | 0.016 | 1.3 | 1.3 | 1.8 | 0.016 | 1.3 | 1.3 |
| Bus bar tubes | 80 | | | | | 1.9 | 1.3 | 19 | 10 |
| 5/20 K Circuit | 15 | | | | | 4.6 | 3 | 3 | 20 |
| He vessel | 570 | 1.9 | 1.3 | 19 | 10 | | | | |
| Beam tubes | 49 | 1.9 | 1e-14 | 1e-14 | 1e-14 | 1.9 | 1e-14 | 1e-14 | 1e-14 |
| Vacuum vessel | 914 | 293 | 1e-9 | 1e-9 | 1e-9 | 293 | 1e-9 | 1e-9 | 1e-9 |

$T_0$ is the normal operating temperature, $P_0$ the normal operating pressure, $P_Q$ the pressure during quench, $P_C$ the pressure during cool-down and DN the nominal tube diameter.

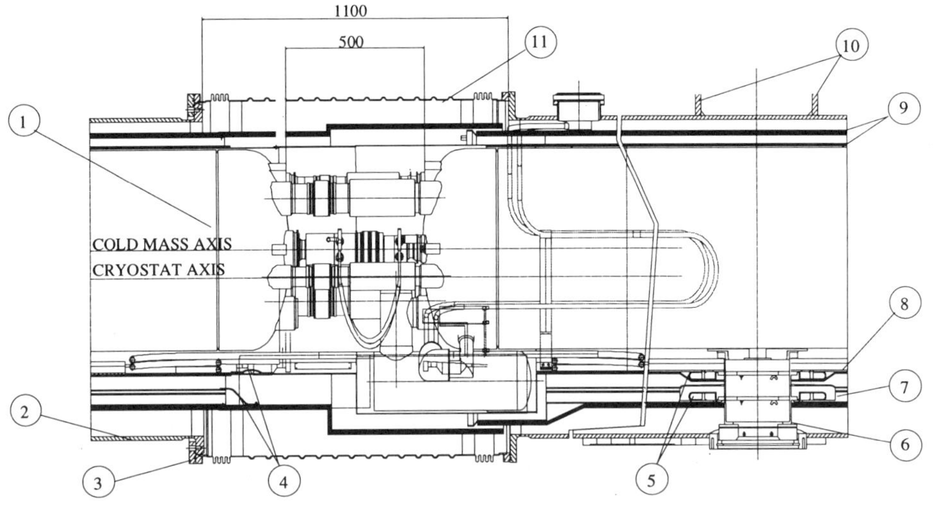


**Figure 2**. Beam-line Cross-Section of the LHC 15 m Dipole Cryo-magnet Interconnect Assembly.

1, Dipole cold mass. 2, Vacuum vessel. 3, Captive flange. 4, Interconnect thermalisation straps. 5, Support system thermalisation straps. 6, Extremity support post. 7, Thermal shield. 8, Radiation screen. 9, Superinsulation. 10, Annular reinforcements. 11, Interconnect sliding sleeve.

Four ISO DN100 and one ISO DN160 standard flanged ports are distributed around the vessel circumference at one end. During normal operation, the vessel contains a static insulating vacuum below $10^{-6}$ mbar and, in the event of a cryogenic leak, serves as a containment vessel up to a pressure of 1.4 bar at which point a pressure relief valve on the vessel wall opens.

## SUPPORT SYSTEM

The cold mass to vacuum vessel vertical eccentricity of 80 mm, together with the vacuum vessel inner diameter of 890 mm leads to an overall support column height of 210 mm. As a compromise between minimizing heat in-leaks and providing adequate stiffness, the support column consists of a glass-fiber/epoxy long fiber lay-up tube of 4 mm wall thickness with an outer diameter of 240 mm. Top and bottom flanges interfacing to the cold mass and the vacuum vessel are molded as an integral part of the composite column. These dimensions and composite technology have been selected to minimize shear and bending deflections and resulting cold mass lateral and tilt displacements due to parasitic forces induced by the interconnect bellows.

Each column is designed to withstand a worst case combined loading due to cold mass weight and unbalanced interconnect and vacuum forces of 125 kN in axial compression, and 40 kN in shear. The column, when equipped with two intermediate heat intercept plates, becomes a support post. The upper and lower plates support and are thermalized to the radiation screen and the thermal shield, respectively. The support system[8] consists of three identical support posts fixed to the cold-mass along its curved median plane. The middle post is blocked longitudinally and slides laterally on closely toleranced guiding devices located in the vacuum vessel. The other two posts, located on identical guiding devices, are blocked laterally and free to slide longitudinally. A plate in low-friction Glacier DU™ material reduces stick-slip effects at the sliding interfaces.

## THERMAL INSULATION SYSTEM

The thermal insulation system consists of two actively cooled shields, a thermal shield operating at 50-75 K and a radiation screen operating at 4.5-20 K.

### Thermal Shield

The thermal shield lower tray is extruded from Al 6060 in one 15 m length complete with two 80 mm diameter channels for the passage of helium coolant. Only one of these channels is used for cooling, requiring an aluminium to stainless-steel pipe transition at each end for connection to interconnect expansion bellows. The other provides vertical bending stiffness. Three machined features in the extrusion allow it to be blocked to the central support post lower flange and constrained to slide longitudinally on the end support lower flanges. The lower flange of each support post is thermalized to the lower tray via flexible aluminium straps welded at each end.

Divided into a minimum of five longitudinal sections, the roughly circular upper shell part allows ease of assembly and, during asymmetric cool-down and warm-up, it limits overall deflections and reduces stresses on the support posts. Industrially pure aluminium sheets 2.5 mm thick meeting ISO Al 99.0 standards, (equivalent to AA 1100), are used for the upper sections, providing adequate thermal conductivity. The upper sections are slotted into the bottom tray and then intermittently welded in place over 20 % of the total length to ensure adequate circumferential heat conduction under steady state conditions, and limit thermo-mechanical transient deflections during rapid cool-down and warm-up.[9]

The thermal shield, together with its thermalized extension pieces covering the interconnect region, is covered with superinsulation, made from 30 layers of double aluminized Mylar™ or an equivalent film with an adequate spacer material.

### Radiation Screen

Very similar to the thermal shield, the radiation screen is attached and thermalized to the upper flanges of the three support posts and is completed in the same manner, with at least 5 roughly circular upper sections 2 mm thick slotted into place and welded intermittently. These upper sections envelop the cold mass, leaving only a small (around 6mm) radial gap which, under normal circumstances, should remain at all times. As a precaution, each upper section is equipped with carbon composite studs to promote point contacts and minimise thermal shorts between the radiation screen and the cold mass.[10]

During a magnet quench, the close proximity of the radiation screen means that the rapid collapse of the dipole field will induce short term forces and displacements in its structure. The integrity of the radiation screen under this loading has been verified.[11]

The radiation screen together with its thermalized extension pieces covering the interconnect region are covered with superinsulation, made from 10 layers of double aluminized Mylar™ or an equivalent film with an adequate spacer material.

## THERMAL PERFORMANCE

Heat in-leak measurements[12] on a full scale cryostat thermal model simulating the previous prototype LHC dipole cryo-magnets[13] indicate that heat in-leaks to the cold mass at 1.9 K can be reduced by including a second "hard" radiation screen manufactured in high thermal conductivity aluminium alloy and actively cooled to temperatures between 4.5 K and 20 K.[14] Table 2 shows averaged heat loads estimated under a nominal insulation vacuum in the $10^{-6}$ mbar range for the cryostat configuration as shown in Figure 1 incorporating a thermal shield and a radiation screen, both actively cooled. The thermal performance of this cryostat will be fully measured under both nominal and accidental LHC operating conditions towards the end of 1997.

**Table 2.** Heat load estimates for one 15 m LHC dipole cryostat and interconnect

| Component | Averaged heat loads (Watts) | | |
|---|---|---|---|
| | 300 K to 50/75 K | 50/75 K to 4.5/20 K | 4.5/20 K to 1.9 K |
| **Cryostat** | | | |
| Thermal shield | 43.7 | | |
| Radiation screen | | 1.57 | |
| Cold-mass | | | 0.058 |
| Support System | 17.3 | 2.58 | 0.069 |
| **Total Cryostat** | **61.0** | **4.15** | **0.127** |
| **Interconnect** | | | |
| Thermal shield | 3.3 | | |
| Radiation screen | | 0.12 | |
| Cold-mass interconnect | | | 0.004 |
| Diagnostic feed-through | | | 0.39 |
| **Total Interconnect** | **3.3** | **0.12** | **0.394** |
| **Total** | **64.3** | **4.27** | **0.521** |

## STATUS

In December 1996, orders were placed with three firms in European Industry for the manufacture of five 15 m LHC dipole cryostats. These cryostats will be assembled with dipole cold masses and delivered to CERN between October, 1997, and March, 1998, for performance tests on the magnet test bench.

## CONCLUSIONS

Incorporating experience gained during manufacture and testing of the first series of 10 m LHC prototype dipole cryo-magnets, both on the single magnet test bench and when assembled into a half cell, a new generation of 15 m LHC dipole cryostats has been designed. This new design is fully adapted to the 15 m curved dipole cold-mass specified for the LHC machine and is compatible with the revised LHC cryogenic distribution scheme. To simplify the industrialization process that must precede the series production, a full quality control procedure has been established for the five units currently under manufacture, and emphasis has been placed on the selection of standard industrial components and materials.

LHC dipole and short straight section cryostat designs have evolved in parallel and have been constrained to use common components wherever possible.

## REFERENCES


1. V. Benda, A. Bezaguet, J. Casas-Cubillos, S. Claudet, W. Erdt, Ph. Lebrun, G. Riddone, V. Sergo, L. Tavian, B. Vullierme, R. van Weelderen and U. Wagner. "Conceptual design of the cryogenic system for the Large Hadron Collider (LHC)". EPAC 1996, Sitges (Barcelona).
2. The LHC Study Group, "The Large Hadron Collider Accelerator Project", CERN Report AC/95-05 LHC, 1995.
3. R. Perin, Superconducting magnets for the Large Hadron Collider. LHC Project Report 52. CERN Switzerland.
4. W. Cameron, T. Kurtyka, V. Parma, T. Renaglia, P. Rohmig, B. Skoczen, and Ph. Trilhe. "The new generation of superfluid helium cryostats for the short straight sections of the CERN Large Hadron Collider (LHC)." Presented at this Conference.
5. L. Tavian, U. Wagner, R. van Weelderen. "Dimensions, fluids, pressures and temperatures in cryostat and cryogenics distribution line piping for the LHC." LHC Project Note 48 06/06/1996. CERN Switzerland.
6. B. Skoczen, "Finite element analysis of the LHC thermal shield" private communication, October 1996.
7. P. Cruikshank, N. Kos, G. Riddone and L. Tavian. "Investigation of thermal and vacuum transients on the LHC prototype magnet string." ICEC/ICMC 16, 1996 Kitakyushu Japan.
8. M. Mathieu, V. Parma, T. Renaglia, P. Rohmig and L.R. Williams. "Supporting Systems from 293 K to 1.9 K for the Large Hadron Collider (LHC) Cryo-magnets." Presented at this Conference.
9. G. Peón, G. Riddone and L.R. Williams. "Analytical model to calculate the transient thermo-mechanical behaviour of long thin structures cooled from a pipe: Application to the LHC dipole thermal shield." ICEC/ICMC 16, 1996 Kitakyushu Japan.
10. G. Peón, B. Jenninger and B. Szeless. "Characterisation of net type thermal insulators at 1.8 K low boundary temperature." ICMC 1996, CERN Switzerland.
11. G. Peón and L.R. Williams. "Analysis of the LHC 15 m cryo-dipole radiation screen during a magnet quench." Presented at this Conference.
12. V. Benda, L. Dufay, G. Ferlin, Ph. Lebrun, J-M. Rieubland, G. Riddone, B. Szeless, L. Tavian, and L.R. Williams. "Measurement and analysis of thermal performance of LHC prototype cryostats." *Advances in Cryogenic Engineering* 1996, Vol. 41, pp. 785-792, CEC Supplement.
13. J-C. Brunet, J. Kerby, Ph. Lebrun, P. Rohmig, B. Szeless and L.R. Williams. "Design of LHC prototype dipole cryostats". *Cryogenics* 1992, Vol. 32, pp. 191-194, ICEC Supplement.
14. G. Ferlin, B. Jenninger, Ph. Lebrun, G. Peón, G. Riddone and B. Szeless. "Comparison of floating and thermalized multilayer insulation systems at low boundary temperature." ICEC/ICMC 16, 1996 Kitakyushu Japan.

# A FACILITY FOR EVALUATING SUPERCONDUCTORS ABOVE ATMOSPHERIC PRESSURE AT 1.8K

A. G. Prodell[1] and A. Arn[2]

[1] Brookhaven National Laboratory
Upton, New York 11973, USA
[2] CERN
Geneva 23, Switzerland

## ABSTRACT

A facility for evaluating superconductors above atmospheric pressure at 1.8K for the CERN LHC has been constructed and successfully operated at Brookhaven National Laboratory. The facility consists essentially of a modified vertical cryostat and a large pump station which have been incorporated into an existing helium liquifier and gas recovery system in a closed loop. The vertical cryostat has been equipped with a Lambda plate to separate a superfluid helium volume at 1.8K and 16.2 psia from a helium volume above it at 4.5K and 16.5 psia. The helium volume below the Lambda plate is cooled by a heat exchanger whose internal pressure is established and maintained by the large pumping station remote from the cryostat. Helium is injected into this heat exchanger from just above the Lambda plate through a counterflow heat exchanger above the plate and a J-T valve. A superconducting magnet suspended from the Lambda plate is energized to evaluate the current-magnetic field characteristics of LHC conductor samples separately mounted and powered on a removable holder which seals at the Lambda plate and extends into the bore of the magnet. The facility is the result of a joint effort between BNL and CERN. Details of the facility and its operation are described.

## INTRODUCTION

As part of a collaboration between CERN and Brookhaven National Laboratory (BNL), a test facility was established at BNL to evaluate superconducting conductors at 1.8K and atmospheric pressure for the Large Hadron Collider (LHC) to be built at CERN. This facility was constructed using some existing BNL components modified for operations at 1.8K and some components supplied by CERN based on the designs incorporated in the low temperature test facilities in operation at CERN.

The conductor test facility has several major components: 1) a helium cryostat modified to be capable of operation over a range of temperatures, 2) a large pump station, 3) a superconducting magnet in whose bore superconducting conductor samples are positioned, 4) conductor sample holders, 5) power supplies for the magnet and the conductor samples and 6) instrumentation for the cryostat and superconducting magnet.

## VERTICAL CRYOSTAT

An existing helium cryostat, shown in Figure 1, was modified by the addition of a ring to support a 6.35 cm thick G-10 Lambda plate which separates the 1.8K volume in the cryostat from the 4.5K volume above it. A copper thermal shield supported by a ring welded to the helium vessel above the Lambda plate encloses the low temperature volume below the plate and about half of the volume above it. The average temperature of the shield is calculated to be about 10K. Superinsulation is placed between the helium vessel and the shield and between the shield and the vacuum vessel. The heat load on the 1.8K helium volume is estimated to be less than 10 watts.

The helium vessel is filled initially by lifting a relief valve on the Lambda plate and flowing helium to the bottom of the vessel through a cold valve supplied by CERN until the operating level is reached, about 30 cm above the plate. This valve is then closed and the helium level is maintained by a so-called "trickle valve" which controls the set level above the Lambda plate to ±1.5 cm.

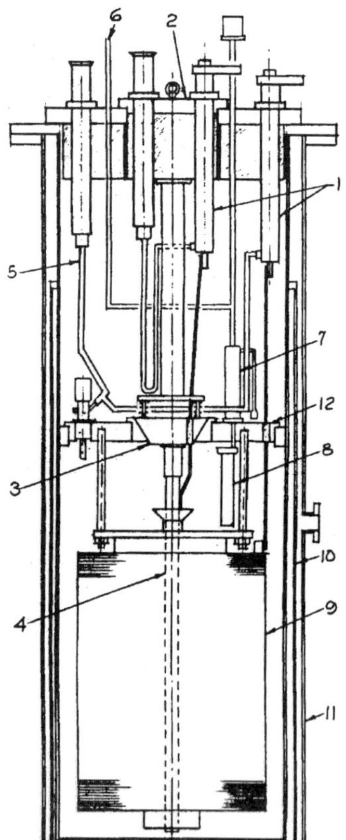


**Figure 1.** Vertical cryostat: (1) Liquid injection leads, (2) Sample holder, (3) Sample holder G-10 plug, (4) Conductor samples, (5) Fill line, (6) Pump line, (7) Upper counterflow heat exchanger, (8) J-T valve and lower heat exchanger, (9) Test magnet, (10) 10K shield, (11) Vacuumvessel, (12) Lambda plate.

Helium is supplied to the cryostat from two 3780 liter storage dewars connected to a 160 liter/hr (700W) liquifier/refrigerator. Because there was some interest in pumping the entire helium volume to below atmospheric pressure for some tests, the power leads for both the test magnet and the conductor samples, which are placed inside the bore of the magnet, are cooled by injecting helium from the supply transfer line directly into the bottom of the leads. The lead flows are manually controlled at this time but are being programmed to respond to current changes in the several leads.

The heat load on the total helium volume in the cryostat with tare flow on the power leads from the storage dewars is equivalent to about 32l/hr.

## HEAT EXCHANGER AND PUMP STATION

Once the helium level is established in the cryostat, the pumps in the pump station are turned on. These pumps pump on the helium in a 3 cylinder heat exchanger supplied by CERN mounted just below the Lambda plate as shown in Figure 1. Liquid helium drawn through a small tube from just above the Lambda plate passes through a vacuum insulated heat exchanger above the plate where it is cooled by exchanging with the helium being pumped. The helium then passes through a J-T valve into the lower heat exchanger, and back through the upper heat exchanger to the pumps and recovery. The heat exchanger assembly is shown in Figure 2.

The pump system is a two stage system assembled by Nash-Kinema, Inc. The first stage is a plugged rotor booster driven by a 74.6 kW (100 HP) three phase motor. The second stage is a NASH liquid ring pump, also driven by a 74.6 kW motor. Control of the suction pressure is achieved by the use of a bypass valve to control spoiling of the inlet gas flow. All active components are under computer control using a GE Programmable Logic Controller.

**Figure 2.** Upper and lower heat exchanger assembly.

The PLC actively controls the inlet pressure and monitors operating parameters critical to the functioning of the system. Should any operating parameters fall outside normal limits, the system is automatically shutdown to minimize damage to any component.

Operator control of the suction pressure is through a remote LCD panel. Functions provided at the remote panel are start, stop, crash and pressure set point in units of mm Hg. For operational pressures above 70mm Hg, the liquid ring pump is enabled. The booster lubrication loop is enabled also, and the booster is allowed to freewheel. For pressures below 70mm, the booster pump is enabled. The booster will remain enabled until the pressure set point is over 90mm. This hysteresis will be used to prevent chattering of the booster pump when the system is upgraded to allow pressure set point control using an analog signal from a remote computer control system.

## TEST MAGNET

The test magnet is an iron yoke magnet shown in Fig. 3 with superconducting coils weighing 1180 kg and is supported by three rods from the Lambda plate which, in turn, is hung by three rods from the top plate assembly. The second set of rods passes through sliding seals above the top plate to allow for differential contraction of the rods and the helium vessel when the system is cooled. The magnet has an 8 cm bore and an effective magnetic field length of 76 cm of which 61 cm is uniform to 0.1 percent longitudinally along the axis. It is capable of generating a magnetic field intensity of 7.5T with a current of 3400A at 4.5K and 10T with a current of 4850A at 1.8K. After the selected magnetic field intensity is achieved, a super-conducting switch across its terminals, is activated. The rate of field decay is then about 20 gauss/hr.

**Figure 3.** Test magnet.

## CONDUCTOR SAMPLE HOLDERS

A conductor sample holder is shown in Figure 4 suspended next to the test magnet assembly. The samples are clamped between the long bolted stainless steel bars which engage a stop on the bottom plate of the helium vessel and are also supported at their top to prevent rotation of the sample holder when the samples are carrying current in the field of the magnet. The G-10 plug seen above the clamped samples in Fig. 4 fits into the conical hole in the Lambda plate through which the samples are lowered to position them in the bore of the test magnet. When the top plate of the sample holder is bolted to the top plate of the test magnet assembly, the springs shown are compressed. As the helium vessel is cooled down, the springs expand to insure that the plug remains firmly seated in the Lambda plate to minimize leakage between the upper and lower helium volumes.

The threaded studs next to the springs are screwed down against the metal ring on the plug to give support when the sample holder is rotated to a horizontal position to change samples. Also seen in the figure are the stabilized superconducting power leads which penetrate the plug to the conductor samples mounted below.

## POWER SUPPLY SYSTEM

The power supply system shown in Figure 5 consists of two Rapid Electric Co. Air cooled supplies. The supplies are SCR controlled and each is capable of producing 15 kA DC at 24 volts. One power supply supplies current to the magnet (PS2) and the other (PS1), to the conductor samples. The magnet power supply has been derated and operates at only 6

**Figure 4.** Conductor sample holder with spring loaded G-10 plug.

kA, while the supply for the conductor samples is capable of supplying its full rated current of 15 kA.

Controls are provided by a BNL constructed control chassis through which a PC is capable of monitoring power supply status and operation. Manual programming of the current level through a simple control pot provides operator control of the current set point. An external input is also available to connect a ramp generator to control current level. Current feedback is provided by a DCCT which provides a current signal to the power supply regulator card and the data acquisition system. A link box allows sample lead configurations and polarity changes to be made easily.

## INSTRUMENTATION

The test facility is heavily instrumented with redundant sensors to monitor the temperatures and liquid levels above and below the Lambda plate. Diode and carbon-glass temperature sensors are installed near the bottom of the 1.8K volume and two carbon-glass sensors are located slightly recessed in the underside of the Lambda plate about 180 degrees apart. Two diode temperature sensors are located near the top of the test magnet and liquid level and carbon-glass temperature sensors are installed in the heat exchanger below the Lambda plate.

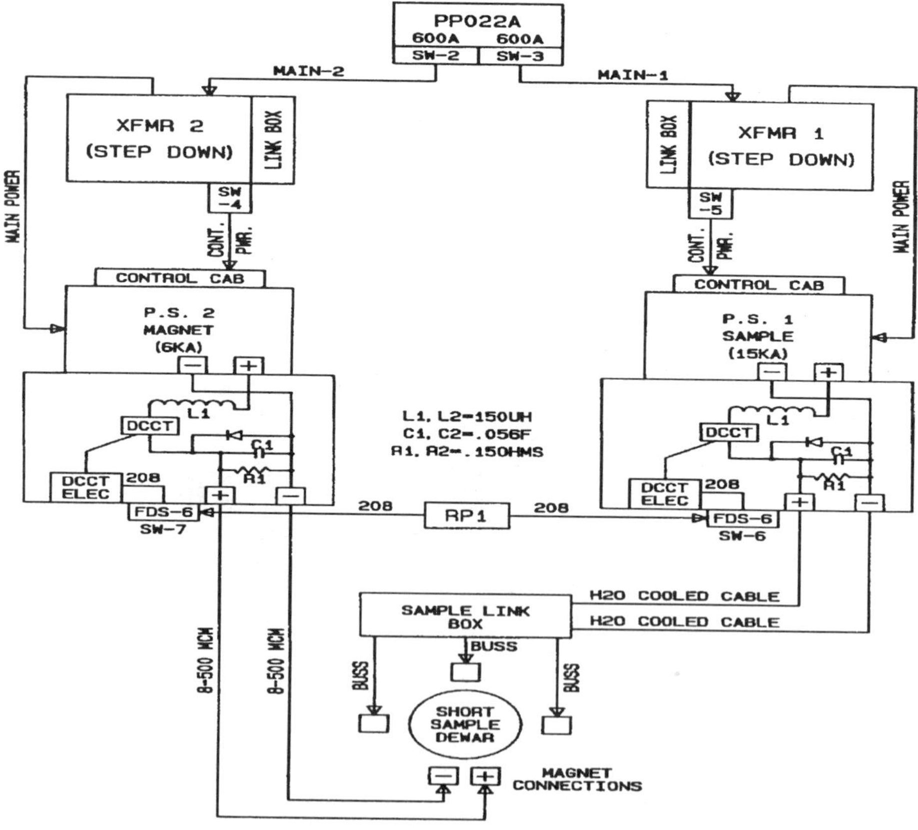


**Figure 5.** Short sample test facility. Power Supply Systems.

Carbon-glass and diode temperature sensors are positioned just above the Lambda plate and, in addition, diodes are located in the volume above the Lambda plate near the level where the helium liquid is controlled. Carbon-glass temperature sensors are placed near the bottom of the forced-flow liquid cooled power leads and voltage taps are installed bracketing each lead to monitor lead voltage.

The test magnet is instrumented with eight voltage taps, two carbon-glass temperature sensors and two Hall probes, and the magnet leads are connected to a superconducting switch with its associated heater to permit operation of the magnet in the persistent mode.

## TEST RESULTS

Once the helium vessel is filled as described above, the pump system is turned on and the pressure in the heat exchanger below the Lambda plate is reduced first to 70 mm and then to about 20 mm as measured in the pumpline just above the heat exchanger. Figure 6 shows the cool-down curves for the helium below the Lambda plate for three different settings of the J-T valve. The fastest cooldown time of 3 hours is achieved with the J-T valve fully open, i.e., at 30 percent. The kink in the curve with the J-T at 20 percent was caused by a temporary change in the J-T valve to a lower percent. Once the helium bath temperature is in equilibrium with that of the helium gas being pumped from the exchanger, helium liquid begins to accumulate in the exchanger and the level of that liquid is maintained by reducing the J-T valve setting to 10 percent.

Although the temperature of the helium gas being pumped from the exchanger is essentially the same for all three J-T valve settings as measured in the vacuum-jacketed section of the pump line above the cryostat, it is believed that the faster cooldown times for the larger J-T settings are due to a greater area of the heat exchanger surfaces being cooled by the larger flows of gas.

As expected, as the lower volume of helium is cooled to the Lambda point, all the temperature sensors rapidly converge to superfluid temperature.

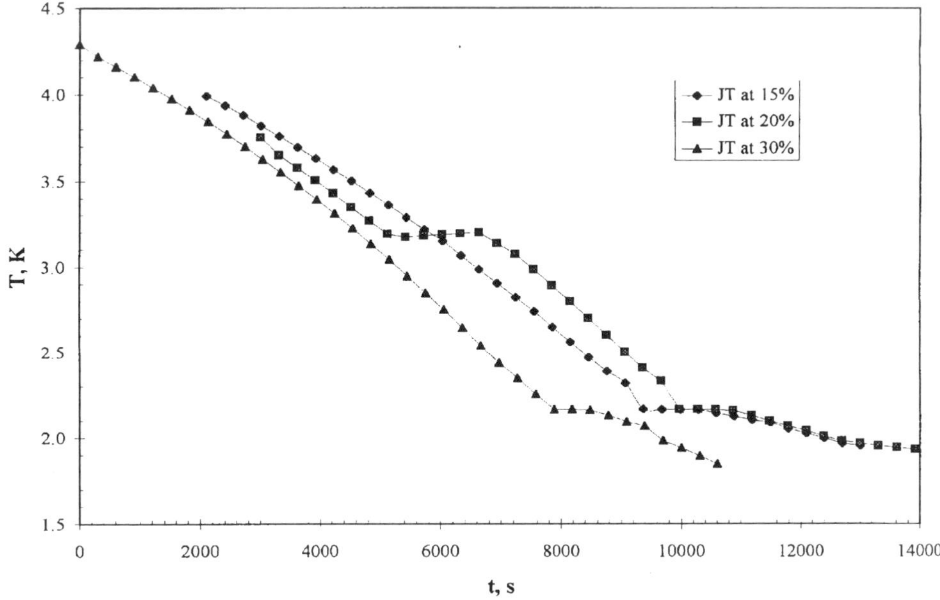


**Figure 6.** Cooldown curves for the helium below the Lambda plate for three different settings of the J-T valve.

## CONCLUSIONS

The facility for evaluating superconductors at 1.8K operates successfully and can be used for tests in the range from 1.8K to 4.5K. With a test magnet of 1180 kg and a helium volume of 260 liters below the Lambda plate, a bath temperature is reached in three hours. The system has also been operated at temperatures as low as 1.6K.

## ACKNOWLEDGMENTS

The authors would like to acknowledge the efforts of Charles Muller who worked on the design of the facility and of Frank Cullen, Joseph D'Ambra and Lon Werner, Jr. who were responsible for its assembly.

# CABLE MAGNETIZATION EFFECTS IN THE LHC MAIN DIPOLE MAGNETS

L. Bottura, M. Schneider, L. Walckiers, R. Wolf

CERN, CH-1211 Geneva 23, Switzerland

## ABSTRACT

Several short (1 m) and long (10 m) dipole models have been tested within the scope of the on-going R&D programme for LHC at CERN. We report here the results of measurements of field quality, focussing on the contribution of persistent current magnetization. We show that measurements of the dependence of allowed harmonics on magnetic field at 1.8 K can be computed with reasonable accuracy. The calculation is based on a model of the filament magnetization adjusted to fit available strand data. Low field measurements of the first allowed harmonics, sextupole and decapole, correlate satisfactorily with the filament diameter and the scaling obtained from the simulation program.

## INTRODUCTION

The Large Hadron Collider[1] (LHC) will accelerate proton beams from an injection energy around 450 GeV to the nominal coast value of 7 TeV. The superconducting magnets, operated in superfluid helium and forming the main part of the LHC ring, will need to be cycled in accordance to maintain the bending and focussing strength required. In particular, the main arc dipole magnets will receive the beam at the injection field of 0.54 T, and will follow its acceleration up to the nominal operating field of 8.36 T. From the experience on other superconducting accelerator magnets (Tevatron[2], HERA[3], RHIC[4] and SSC[5]), and from measurements of the field quality of LHC dipole prototypes, we expect that the field quality in the ring at the low field levels needed at injection will be dominated by the contribution of the magnetization of the superconducting filaments[6]. The LHC performance will be determined by field quality at injection, hence much attention has been paid till now to the prediction and measurement of magnetization effects[7,8].

The results reported here refer to several short (1 m) and one long (10 m) dipole models built and tested in the past 2 years within the scope of the on-going R&D programme for LHC. All magnets have the same coil geometry, with a 56 mm cold bore[1]. The coil is wound in 5 blocks and 2 layers with a 15 mm wide Rutherford cable. Most of the short models, the so called MBSSx series, have a single collared coil inside an iron yoke. A single short model, the MBST1, has two collared coils in a single vertically split

**Table 1.** Main features of the LHC short and long, 56 mm aperture dipole models tested

| Magnet | Length | type(*) | $I_c$ @ 4.2 K | | $D_{fil}$ | | Cu:NbTi | |
|---|---|---|---|---|---|---|---|---|
| | | | inner layer | outer layer | inner layer | outer layer | inner layer | outer layer |
| | (m) | | (kA@7T) | (kA@6T) | (μm) | (μm) | (–) | (–) |
| MBSS1 | 1 | S | 14.898 | 14.622 | 4.6 | 6 | >1.60 | >1.80 |
| MBSS2 | 1 | S | 14.917 | 14.700 | 4.6 | 6 | >1.60 | >1.80 |
| MBSS3 | 1 | S | 14.937 | 14.700 | 4.6 | 6 | >1.60 | >1.80 |
| MBSS4 | 1 | S | 14.465 | 15.555 | 4.6 | 6 | >1.60 | >1.80 |
| MBSS5 | 1 | S | 14.397 | 15.555 | 4.6 | 6 | > 1.60 | >1.80 |
| MBSS6 | 1 | S | 14.829 | 15.555 | 7 | 6 | > 1.60 | >1.80 |
| MBSS7 | 1 | S | 14.329 | 14.700 | 4.6 | 6 | 1.62 | >1.80 |
| MBSS8 | 1 | S | 14.829 | 15.528 | 7 | 6 | > 1.60 | 1.96 |
| MBSS9 | 1 | S | 14.660 | - | 7 | 6 | 1.63 | > 1.80 |
| MBSS10 | 1 | S | - | - | 7 | 6 | - | >1.80 |
| MBSS11 | 1 | S | - | - | 7 | 6 | - | > 1.80 |
| MBST1 | 1 | T | 14.829(†) | 15.351(†) | 7(†) | 6(†) | >1.60(†) | 1.96(‡) |
| | | | 14.829(‡) | 15.351(‡) | 7(‡) | 6(‡) | >1.60(†) | 1.96(‡) |
| MBL1N1 | 10 | T | 14.494(†) | 14.844(†) | 4.6(†) | 6(†) | 1.64(†) | 1.83(†) |
| | | | 14.638(‡) | 16.392(‡) | 4.6(‡) | 6(‡) | 1.63(‡) | 1.75(‡) |

(*) S = single aperture, T = twin aperture (†) first aperture (‡) second aperture

iron yoke that closes the magnetic circuit of both magnetic bores, or *apertures*. The MBL1N1 is a twin aperture 10 m model, the first long dipole with 56 mm aperture. The Rutherford cables used for these magnets have the same geometry, but the NbTi strand has evolved from magnet to magnet. In particular the magnetization (related to the critical current $I_c$ and filament diameter $D_{fil}$) is different for the various magnets. Typical filament diameter for the cables used range between 4.6 μm to 7 μm. The features of these magnets that are of interest here are summarised in Tab. 1, where we have reported in particular the average cable critical current (measured at 4.2 K at 6 or 7 T depending on the layer) and the filament diameter data for the inner and outer layer of each aperture.

## FIELD QUALITY DEFINITION AND THE EFFECT OF MAGNETIZATION

For accelerator magnets it is customary to consider the magnetic field as two dimensional, ignoring the length *z*, and to express the magnetic flux density **B** in the *x*-*y* plane normal to the beam using the following complex power series:

$$\mathbf{B}(x,y) = B_y + iB_x = 10^{-4} B_1 \sum_{n=1}^{\infty} (b_n + ia_n) \left( \frac{x+iy}{R_{ref}} \right)^{n-1} \quad (1)$$

where the $B_1$ is the dipole strength, $R_{ref}$ is the reference radius (10 mm for LHC), while $b_n$ and $a_n$ are the normal and skew 2n-pole coefficients. As given in Eq. (1), the multipole coefficients are expressed in so-called *units*, i.e. normalised and scaled by a factor 10,000.

One of the main source of field errors in a superconducting magnet is the magnetization *M* of the superconducting filaments. Screening currents, of persistent nature, are induced by any field change during operation of the magnet. These currents appear in first approximation as a trapped magnetization dipole $m=A_{fil}\,M$ with hysteretic behaviour, where $A_{fil}$ is the filamentary cross section in a strand. The multipoles generated by the plane magnetic moment $m=m_y + i\,m_x$ located at the position $R= x + i\,y$, including the effect of the cylindrical iron yoke with inner radius $R_{iron}$, can be calculated as follows:

$$\left(b_n + i a_n\right) = -10^4 \frac{n}{B_1} \frac{\mu_0}{2\pi R_{ref}^2} \left[ m^* \left(\frac{R_{ref}}{R}\right)^{n+1} + m'^* \left(\frac{R_{ref}}{R'}\right)^{n+1} \right] \tag{2}$$

where $m^*$ is the complex conjugate of $m$ and the mirror moment $m'$ and radius $R'$ are:

$$m' = -\frac{R'^2}{R_{iron}^2} \frac{\mu - 1}{\mu + 1} m^* \qquad\qquad R' = \frac{R_{iron}^2}{R^*}$$

Equation (2) shows that in principle all harmonic orders are generated. Symmetry conditions in a perfect dipole are such that only *allowed* harmonics are generated[6]. For a normal dipole magnet these are the multipoles $b_n$, with $n=2k+1$ and $k=[0...\infty]$. The model for the filament magnetization $M(B)$ is based on a fit[7] of measured magnetization data[9]. The fit consists of two formulae, one for the fully penetrated state and one for the penetration phase. The envelope of upper and lower magnetization branches in fully penetrated state is fitted as:

$$\left|M(B)\right| = \frac{\mu}{B^n} \tag{3}$$

where $B$ is the module of the magnetic field, while $\mu$ and $n$ are fit constants. The sign of $M(B)$ depends on the direction of the field change. The penetration phase, after a field reversal starting at $B_{rev}$ (corresponding to a magnetic moment $\left|M_{rev}\right| = \mu / B_{rev}^n$) is fitted as:

$$M(B) = M_{rev} + \left(h^2 - 2h\right) \Delta M\left(B_{rev}\right) \tag{4}$$

where $h$ is the normalised variable:

$$h = \frac{B - B_{rev}}{B_p\left(B_{rev}\right)} \tag{5}$$

and $\Delta M(B_{rev})$ is the width of the hysteresis loop from the field reversal point $B_{rev}$ to full penetration at $B=B_{rev} + B_p(B_{rev})$. The value of $B_p(B_{rev})$, the penetration field starting from the field reversal point $B_{rev}$, is computed scaling the experimental value of the penetration field $B_p(0)$ and the width of the magnetization loop $\Delta M(0)$ measured at zero field:

$$B_p\left(B_{rev}\right) = B_p(0) \frac{\Delta M\left(B_{rev}\right)}{\Delta M(0)} \tag{6}$$

Note that Eq. (6) is non linear because $\Delta M(B_{rev})$ can be calculated (using Eq. (3)) only when $B_p$ is known. To give orders of magnitude, the values of the parameters used to fit the magnetization of a LHC strand with 7 μm filament diameter are $\mu \approx 0.027$ ($T^{1+n}$), $n \approx 0.5$, $B_p(0) \approx 0.15$ (T), and $\Delta M(0) = 0.105$ (T). As we are interested in the magnetization at low field, where the critical current is much larger than the transport current, we do not make any transport current correction. Finally, all strands are assumed to be in a fully penetrated state at the end of a field sweep. This assumption limits the use of this model to major magnetization loops. We checked *a posteriori* that this causes a negligible error.

## MEASUREMENTS OF FIELD QUALITY AT INJECTION

To perform measurements of field quality we have used a radial rotating coil system. The short (1m) models were measured in a vertical cryostat, with a cold measurement coil covering 60 cm (the central part) of the dipole length. The 10 m-long dipole model was tested on a horizontal bench. In this case the rotating coil was in a warm bore, measuring a length of 75 cm at the center of the straight part of the magnet. To identify the contribution of the magnetization of the superconducting filaments, we have defined a dedicated measurement procedure. The procedure starts precycling the magnet to a maximum current of 11.75 kA (about 8.4 T), to simulate a collision run, followed by a ramp-down to a minimum current of 50 A (corresponding to a field of 35 mT). This cycle brings the magnet into a well defined magnetization state, on the upper magnetization branch, where most of the superconducting filaments in the cables are fully penetrated. We then ramp slowly, with a ramp rate of 2 A/s (1.5 mT/s), up to 2 kA (about 1.5 T), well above the nominal injection level (at approximately 800 A). During this slow ramp measurements are taken continuously (*on-the-fly*). The magnet is then ramped up to maximum current and down to 2 kA. From here a downward ramp at 2 A/s is started, where measurements are again taken continuously.

The purpose of the two slow ramps is to have a measurement of both upper and lower magnetization branches. The ramps are sufficiently slow that ramp-rate effects (due to coupling currents in the Rutherford cables) are negligible for most of the magnets tested. The reason for taking measurements on-the-fly, rather than stopping at several currents, is that we wish to avoid the decay of magnetization observed during current plateaus. We have shown elsewhere that a continuous magnetic field change can indeed prevent decay[8].

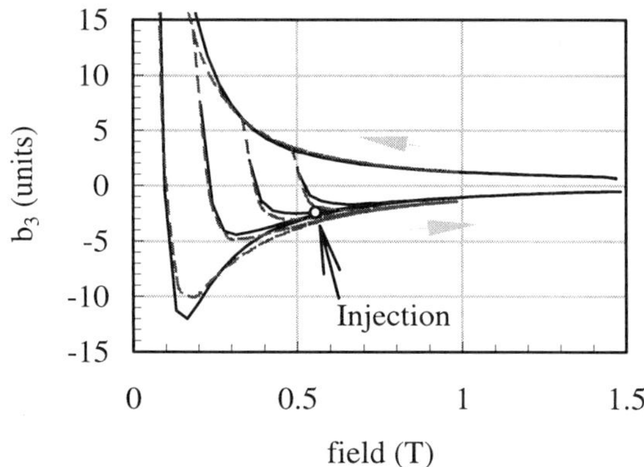


**Figure 1.** Normal sextupole ($b_3$) measured in the MBL1N1 long dipole model (thick continuous line). The arrows mark the ramping directions. The magnet has been cycled taking the minimum current as a parameter to investigate precycling influence at injection. The computed value (broken line) is reported for comparison.

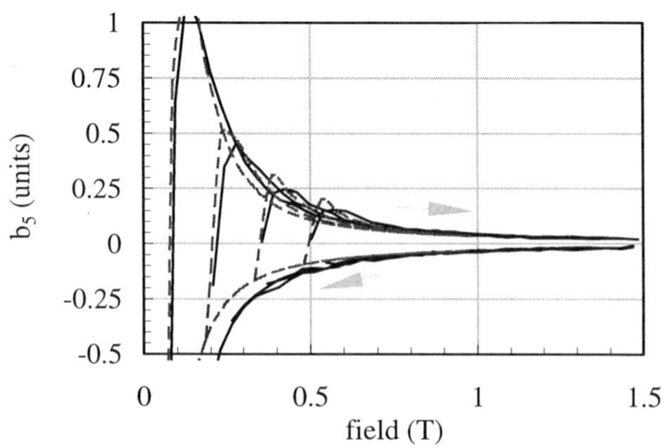


**Figure 2.** Normal decapole ($b_5$) measured in the MBL1N1 dipole (solid line) and computed (broken line).

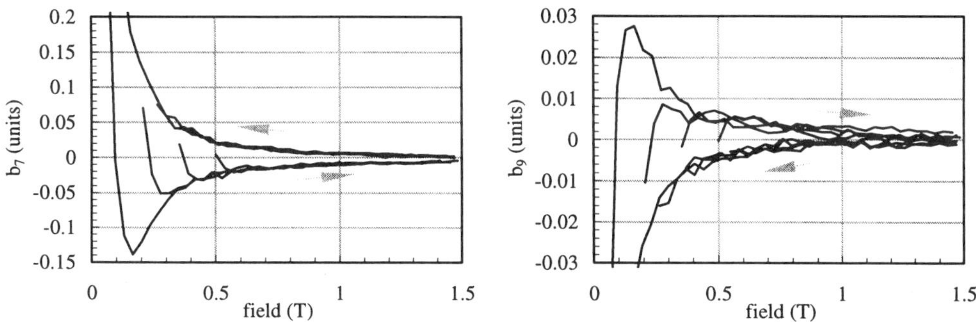


**Figure 3.** Higher order multipoles ($b_7$ and $b_9$) measured in the MBL1N1 dipole.

In the measurements reported here the minimum current has been taken as a parameter in the range of 50 to 650 A, with the aim to investigate the influence of the precycle procedure on the field quality at injection. All measurements were taken in superfluid helium, at stable bath temperatures ranging from magnet to magnet from 1.65 K to 1.85 K.

## Discussion of measurement results

Figures 1, 2 and 3 show an example of the detailed result of the measurements performed on the MBL1N1 dipole, the first 10-m long, 56 mm aperture model. We have plotted there the low field behaviour of the first allowed harmonics ($b_3$ to $b_9$), as measured with the procedure described above. The geometric component of the field error has been removed, shifting the measured curves so that they have zero average at middle field. The upward and downward ramps are marked by the arrow direction. Magnetization effects are clearly visible on these allowed multipoles. The steep portion of the curves, at low field in the upward branches, corresponds to the penetration phase for the majority of the filaments. After penetration the filament magnetization starts decreasing according to the $J_c(B)$ dependence. We see this effect as an extremum in the upward branch followed by a decrease as the field is increased. The upward branches with the more pronounced extrema corrrespond to a minimum current of 50 A. The subsequent cycles with minimum currents of 250, 450 and 650 A can be identified by the shift of the penetration phase towards higher field and the less pronounced extrema. As expected, for all allowed multipoles the penetration phase is significantly affected by the minimum current (i.e. minimum field) reached in the previous cycle. An interesting feature is that in all cases the upward branches with higher minimum current tend to overshoot slightly those with smaller minimum current. This small effect, common to all multipoles, can be most clearly seen on the decapole ($b_5$) curves of Figure 2. We attribute this effect to the difference in magnetization status of the different coil blocks, as discussed later. The downward branch is not affected by the minimum current, which is consistent with the fact that we cycle the magnet up to high field values, so that most of the filaments are in a fully penetrated state.

## Effect of minimum current

The effect of the minimum current during a precycle has been summarised in Figure 4, for the normal sextupole and decapole. We plot there the amplitude of the hysteresis cycles $\Delta b_3$ and $\Delta b_5$ at injection field (difference of the values on the upward and downward branches) as a function of the minimum cycle current. The effect is significant, and monotonous on $\Delta b_3$, while on $\Delta b_5$ only a small effect can be seen, with a peaking at minimum current of 450 A. The peak is due to the overshoot feature discussed earlier. The curves are flat below 250 A. If we consider this minimum cycle current and ideally change it by ±50 A the effect on the normal sextupole, the allowed multipole with the tightest relative tolerance for the machine optics, is less than 3 %. We consider this range of minimum current to be a safe compromise between the necessity to bring the magnets into a reproducible state and power supply design considerations.

### Statistics on measurements

As we stated in the introduction we have measured up to now several (12) 56 mm aperture short models in addition to the MBL1N1 long model discussed in detail in the previous sections. The general features of the measurements are similar to the ones presented for the MBL1N1. To summarise the statistic of these results, we have reported in Figure 5 the measured amplitude of the hysteresis on the sextupole and decapole for a minimum cycle current of 50 A. We see that the points are grouped, with two exceptions, in two clusters (marked to aid the eye). These two clusters correspond in Tab. 1 to the magnets wound with inner cables with small (4.6 μm, filled symbols) and large (7 μm, open symbols) filament diameter. The two outliers are magnets MBSS4 (filled symbol with exceptionally small $b_3$) and MBSS10 (open symbol with exceptionally large $b_3$). The first magnet had an extremely low interstrand resistance (around 3 μΩ), and therefore already at 2 A/s ramp-rate had significant cable eddy currents effects[5] invalidating our hypothesis of quasi-steady state measurement. There is no apparent explanation for the second outlier point. The residual scatter of the points in the two clusters can be attributed to the scatter in the magnetization properties that can be deduced from the variations in the cable critical current of Tab. 1, and in different bath temperatures from magnet to magnet during the measurements.

## CALCULATION OF MAGNETIZATION EFFECTS

The effect of magnetization can be calculated using Eqs. (2)-(6). A pre-computed map of the magnetic field is used to calculate the local field at each strand for the same current cycle used in the measurements. The contribution of the magnetic moment is then computed on a strand basis, using the local field value in the strand to determine $M(B)$, and summed over the complete coil cross section. The $b_3$ and $b_5$ computed with this procedure are plotted in Figures 1 and 2 (dashed lines), and they show an excellent agreement with the measurements. We attribute the accuracy of the calculation to the fact that we have calibrated $M(B)$ on a measured magnetization curve for a strand similar to those actually used in the coils of MBL1N1[9]. For $b_3$ both the penetration phase and the peaking as the filaments start reaching full penetration are matched very well. A small deviation is found on $\Delta b_3$ at injection, of the order of 20 %, that could be due to small residual differences

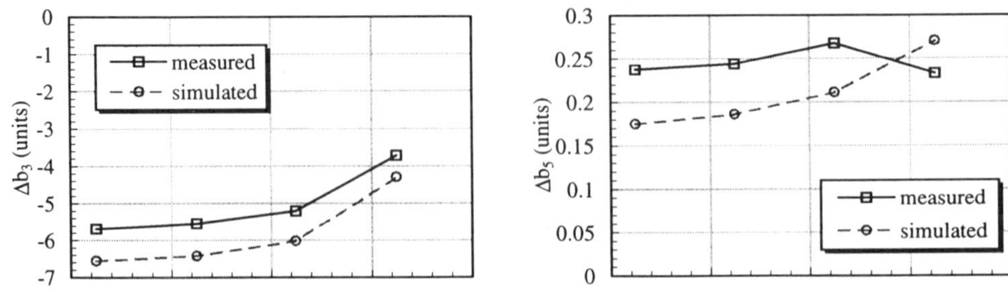


**Figure 4.** Amplitude of the sextupole and decapole hysteresis cycles at injection in the MBL1N1 dipole as a function of the minimum current during precycle as measured (squares) and simulated (circles).

**Table 2.** Layers contributions to $\Delta b_3$ and $\Delta b_5$, as computed for MBL1N1 at injection field

| | Δb3 (units) | Δb5 (units) |
|---|---|---|
| outer layer | -4.92 | -0.090 |
| inner layer | -1.62 | 0.261 |

between the fit $m(B)$ and the actual strand magnetization. In the case of $b_5$ the simulation correctly predicts the overshoot at increasing minimum cycle current, as observed experimentally. The width of the peaks of $b_5$ is slightly different, narrower in the simulation than in the measurement.

We have compared the simulations to the measurements on the effect of minimum cycle current in Figure 4. The $\Delta b_3$ curves, measured and simulated, are parallel. The shift is possibly due, as we said above, to residual inaccuracy in the fit of the strand magnetizations at injection. In the case of $\Delta b_5$ the initial behaviour, up to 450 A, is similar (apart again for a rigid shift). Because the simulated $b_5$ peak is slightly sharper than the one measured (compare to the details at injection in Fig. 2), we *miss* the maximum at 450 A minimum cycle current. We judge this inaccuracy to be a detail, rather inessential as compared to the overall good agreement found.

The simulation gives additional insight in the measurements, and in particular on the reason why the $b_5$ overshoot takes place. As already shown in Ref. 7, the different blocks in the coil have different weight on the overall contribution to the multipoles, depending on the complex interplay of their geometric location and magnetization strength. We show this fact in Table 2, where we report the partial contributions of the inner and outer coil layers to the $\Delta b_3$ and $\Delta b_5$ at injection, after a cycle at minimum current of 50 A. We see there that while for $b_3$ both contributions have the same sign, in the case of $b_5$ the inner layer has a strong positive contribution, as opposed to the weak negative contribution of the outer layer. These contributions change in magnitude, and for $b_5$ also in sign, depending on the operating field. In particular the outer layer gives a positive contribution to $b_5$ for small field changes $\Delta B \leq 0.25$ T after field reversal, and becomes negative for larger field changes. The peak in $b_5$ is found within the first range of field, where the two contributions add to each other. An increase of the minimum cycle current shifts the region where both

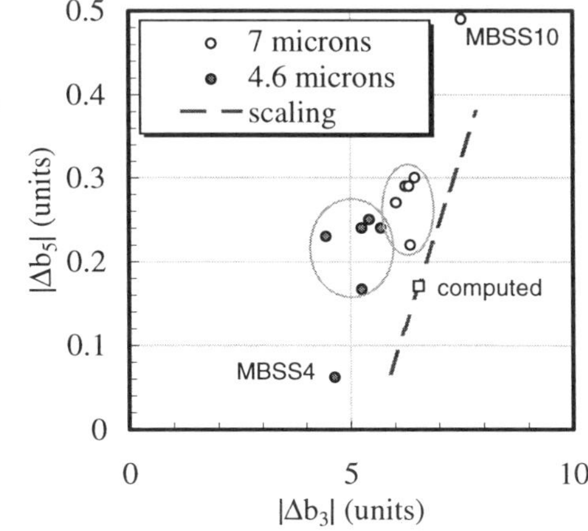


**Figure 5.** Summary of measured amplitudes of the hysteresis on $b_3$ and $b_5$, for 12 short models and one long model. The filament diameter $D_{fil}$ of the inner layer cables is reported in legend. All magnets have the same outer layer $D_{fil}$ = 6 μm (see Table 1). Also reported the calculated point for MBL1N1 (square) and the scaling with the magnetization of the inner layer (see text).

inner and outer layer give positive contribution to $b_5$ towards higher field. This explains the presence of the overshooting peaks at increasing minimum cycle currents.

We have finally used the computed weights of inner and outer layer as given in Table 2 to examine the scaling of the clusters of measurements of Figure 5. As seen from Table 1, all magnets tested had the same filament diameter for the outer layer cables, and different filament diameters for the inner layer cables. Therefore, if we neglect the contribution of the changes in critical current, we can expect that these two groups should differ just by the contribution of the magnetization of the inner layer. We have taken the computed point for MBL1N1, the square in Figure 5, and scaled the inner layer contribution according to the weights in Table 2. The result is the dashed line in Figure 5. Apart for a rigid shift, the magnets seem to broadly follow the scaling line, including the apparently anomalous MBSS4 and MBSS10. Hence we can interpret these two apparent outliers as having either an anomalously low (MBSS4) or high (MBSS10) inner layer magnetization.

## CONCLUSIONS

We have reported details and statistics on low field magnetic measurements on 12 short (1 m) and one long (10 m) LHC dipole models. We have shown how a relatively simple computer model is capable to reproduce the experimental results presented here with a very satisfactory agreement. The key to this good agreement is the use of directly measured magnetization data to produce an empirical fit used by the computer model. In addition the model gives information which is not available from the measurements, such as the weight of different coil layers to the total multipoles observed, used here to give a possible interpretation to the anomalous behaviour of two dipoles.

From a practical standpoint we could determine experimentally a safe value for the minimum precycle current to be used to bring the LHC dipoles into a known and stable magnetization state, in the range of 250 A. Relatively large variations in this range (±50A) produce small effects at injection field (below 3 % on the sextupole). A value in the range 250 to 300 A is thus a safe compromise between reproducibility and power supply design considerations, and can be taken as a standard for series measurements of the LHC dipoles.

## REFERENCES

1. N. Siegel, Status of the Large Hadron Collider and magnet program, IEEE Trans. Appl. Sup., **7**, 2, 252-257, 1997.
2. D.A. Herrup, et al., Time varying sextupole corrections during the Tevatron ramps, Proc. of 13th Particle Accelerator Converence, Chicago IL, 518-520, 1989.
3. H. Brueck, et al., Persistent current effects in the superconducting HERA magnets and correction coils, Proc. 2nd Eur. Part. Acc. Conf., 1990, 1160-1162.
4. A. Greene, et al., The magnet system of the relativistic heavy ion collider (RHIC), IEEE Trans. Mag., **32**, 4, 2041-2046, 1996.
5. Y. Zhao, et al., Current dependence of harmonic field coefficients of 5-cm-aperture, 15-m-long SSC dipole magnet prototypes, IEEE Trans. Appl. Sup., **3**, 1, 674-677, 1993.
6. K.H.Mess, P. Schmueser, R. Wolff, Superconducting accelerator magnets, World Scientific, 1996.
7. R. Wolf, Persistent currents in LHC magnets, IEEE Trans. Mag., **28**, 1, 374-377, 1992.
8. L. Bottura, L. Walckiers, R. Wolf, Field errors decay and "Snap-Back" in LHC model dipoles, IEEE Trans. Appl. Sup., **7**, 2, 602-605, 1997.
9. Magnetization measurements on LHC strands performed by A. Wisniewki, R. Puzniak, Institute of Physics, Polish Academy of Sciences, Warsaw and H Fikis, Technische Universität Wien, Vienna.

# THERMOHYDRAULICS OF RESISTIVE TRANSITIONS OF THE LHC PROTOTYPE MAGNET STRING: THEORETICAL MODELING AND EXPERIMENTAL RESULTS

M. Chorowski, B. Hilbert, L. Serio, and R. van Weelderen

LHC Division, CERN
CH-1211 Geneva 23, Switzerland

## ABSTRACT

In preparation for the Large Hadron Collider (LHC) project, a 40 m-long prototype superconducting magnet string, representing a half-cell of the machine lattice, has been built and operated. The superconducting magnets which comprise this string normally operate in a pressurized static bath of superfluid helium at a pressure of 1 bar and at a temperature of 1.9 K. At 13.1 kA they have about 15.3 MJ of stored magnetic energy. A series of tests was performed to assess the thermohydraulics of resistive transitions (quenches) of the string of magnets. These measurements provide the necessary foundation for describing of the observed pressure rise as the combination of two processes, each acting on a different time scale. The measurements are presented and an explanatory model description of the events is given.

## INTRODUCTION

With the magnet "cold mass" we designate the coils, the aluminum collars and iron yokes, the helium in which the whole is immersed, and the stainless-steel vessel around the assembly. When a magnet in the string quenches, the coils in the whole string are driven resistive by dedicated heaters and the current is ramped down with a decay constant of 250 ms. Consequently its full stored magnetic energy is dissipated by ohmic heating in the coils.[1] This heat will eventually end up in the cold mass, and unless the helium contained inside the cold mass is expelled, it will drive the pressure an order of magnitude beyond the 2 MPa (20 bar) design pressure of the vessel. Therefore a quench relief valve (QRV) must be opened, to allow the helium to discharge and to limit the pressure rise to below the 2 MPa limit.

A series of tests was performed on a 40 m-long prototype superconducting magnet string, representing a half-cell of the LHC machine lattice. These tests were aimed particularly at obtaining an understanding of the quench thermohydraulics, in order to enable the numerical evaluation of possible other magnet string configurations, and to permit the sizing of crucial cryogenic safety equipment and helium recovery systems.

We will discuss the main measurements and deduce from them an explanatory model description of the observed pressure and temperature transients.[2]

## LHC TEST STRING CONFIGURATION

### Installation Layout

Figure 1 shows the schematic of the cryogenic configuration of the prototype string. It comprises one quadrupole magnet (Q) and three dipole magnets (D), a cryogenic feed box (SFB) on the left end, and a return box (SRB) on the right end. The whole is mounted on a slope of 1.4 %, corresponding to the maximum slope that will be encountered in the LHC machine tunnel. The magnets normally operate in a pressurized static bath of superfluid helium at a pressure of 1 bar and at a temperature of 1.9 K. The liquid helium content in the helium cold mass is about 20 ℓ/m, which amounts to a total inventory of 850 ℓ. The four magnets of the string are hydraulically connected and this forms a single hydraulic unit. In case of overpressure the helium is discharged via a quench relief valve (QRV), mounted at the upper part of the rightmost dipole magnet. The helium then enters first the cold line D (20 K, 1 bar), into an isolated recovery vessel and finally a warm storage balloon. The other circuits are for cooldown, 1.9 K temperature control, and cryostat heat intercepts.[3]

### Instrumentation

The string is equipped with a full range of cryogenic instruments: temperature and pressure sensors, liquid level gauges and mass flowmeters. Of particular interest for quench analysis are the temperature sensors (Allen-Bradley 100 Ohm carbon resistors) and piezoresistive cold pressure sensors (Siemens KPY16, Siemens KPY47), placed in the end-volumes of each of the magnet cold masses.

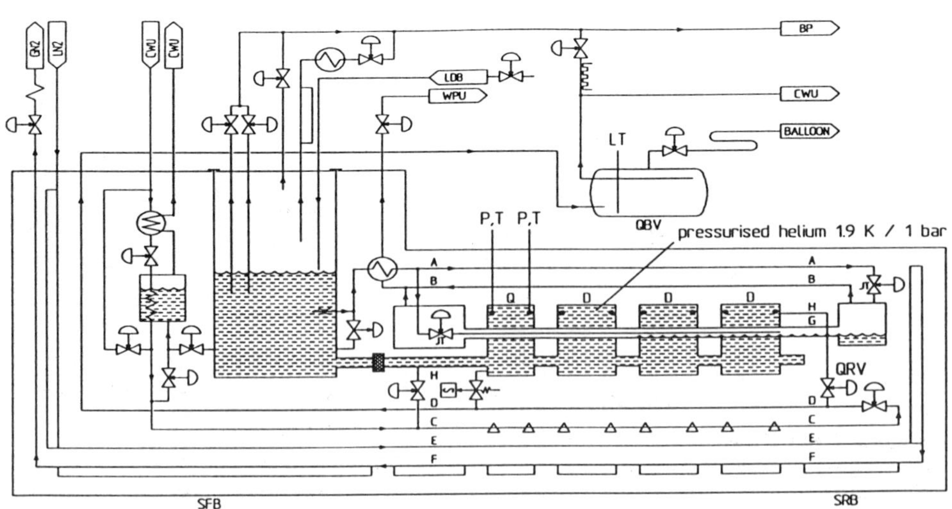


**Figure 1.** Schematic of the cryogenic configuration of the LHC prototype string.

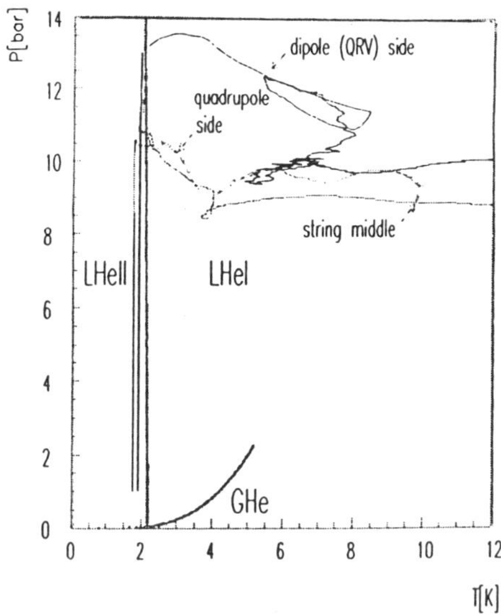


**Figure 2.** Phase diagram of the string cold mass helium after a quench.

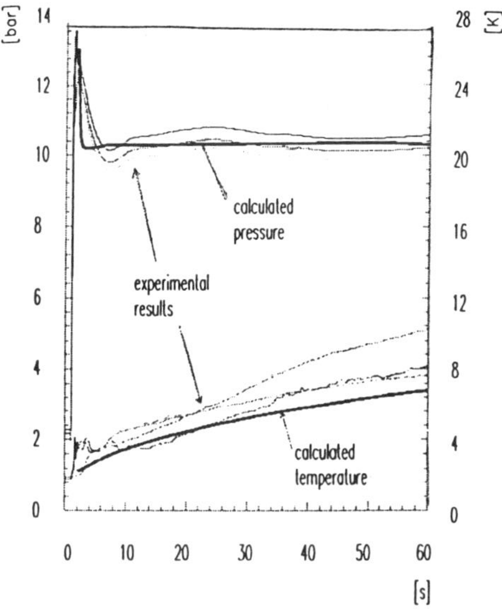


**Figure 3.** Measured development of the string cold mass helium after a 13.1 kA quench, as well as the model simulation.

Data was recorded at a sampling rate of 1 ms with transient recorders using pre-trigger and post-trigger recording.[4] The electrical quench detection signal was taken as trigger.

### Pressure Distribution after a Quench

Prior to their mounting in the string each magnet has been quench tested seperately in a stand-alone configuration.[5] Temperatures were measured in the end-volumes and midlength in the collar and yoke. Pressures were measured in the end-volumes, and at three locations along the length of the magnet in the collar noses. These measurements showed that the pressure development after a quench is homogeneous over the bulk of the magnet. In the string test only the pressure and temperature sensors in each of the end-volumes of each of the magnets are retained. Based on the previously demonstrated homogeneity of the pressures, they provide sufficient information to perform the analysis of the pressure rise processes and to establish the energy balance between magnets, helium inside and helium discharged from the cold masses.

## EVOLUTION OF THE HELIUM THERMODYNAMIC STATE AFTER A QUENCH

### Pressure and Temperature

Figures 2 and 3 show the development of the helium in the string cold mass after a quench. In the phase diagram of figure 2 the helium starts at the working point of 1.9 K, 1 bar, it then moves steeply (within about 150 ms) up in pressure up to about 14 bar while remaining at about the same temperature. The pressure drops then to about 10 bar and the temperature gradually increases. This last development where the helium evolves in the supercritical region is partly dictated by our choice of operating mode of the QRV. It serves as a pressure-regulating device, set at 10 bar.

In some experiments we have let the QRV function as a one-shot opening valve. In those cases, instead of evolving in the supercritical region, the working point moves to the saturation line. Two-phase flow might then limit the discharge capacity of QRV's by sonic constriction, complicates the analysis and makes any scaling of the results rather difficult. In addition, by remaining in the supercritical range, the overall heat exchange between the helium and the magnet structure is improved, since the uniformity of the supercritical phase allows the helium to make use of the maximum available area for heat exchange. More heat is then extracted from the cold mass, which shortens recooling times for the magnets. On these grounds we have therefore chosen, for further string tests as well as for the future LHC, to operate with the QRV's as pressure regulating devices only.

### Pressure Peak Dependence on Energy Dissipated in the Magnet Coils

The height of the pressure peaks (fig. 3) which occur about 150 ms after the quench are related to the energy dissipated in the magnet coils (fig. 4). The same kind of relation, showing some slight differences between the different magnets, was observed when the magnets were tested in stand-alone configuration.[5] In the string configuration, the heights of pressure peaks of neighbouring magnets tend to equalize. The extent to which magnets which are far apart influence each other, is determined by their separation in terms of sound propagation in helium.

### Pressure Peak Dependence on Valve Opening Time

Figure 5 shows the height of the pressure peak as function of the QRV opening time. This time was varied from 120 to 220 ms, with an additional measurement taken at 0 ms when we opened the QRV valve before triggering a quench. This clearly shows that the height of the pressure peak is inherent to the magnets and cannot be reduced to zero by any means of fast discharge. The influence of helium discharge is weak, and becomes appreciable only for opening delays greater then about 100 ms.

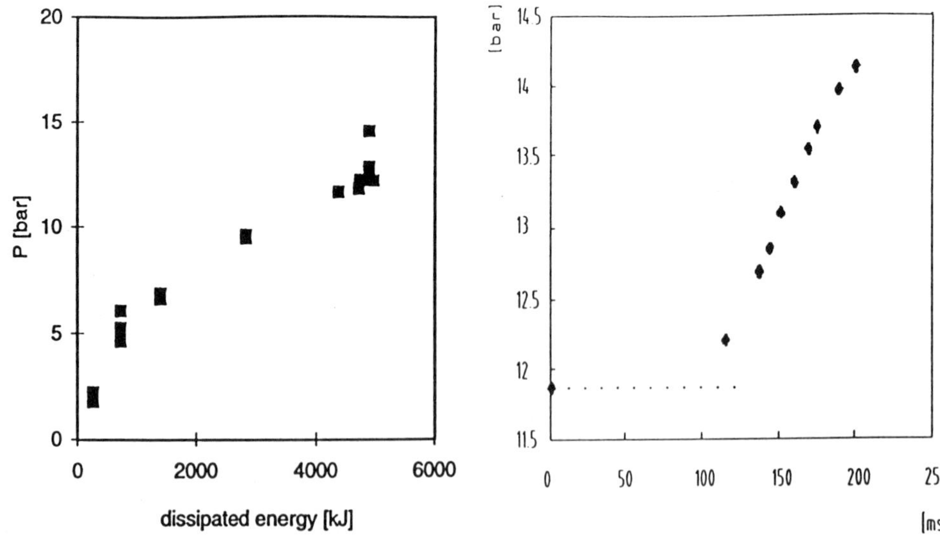


**Figure 4.** Pressure peak values from string measurements as function of dissipated energy in the coils.

**Figure 5.** Pressure peak dependence on QRV opening time.

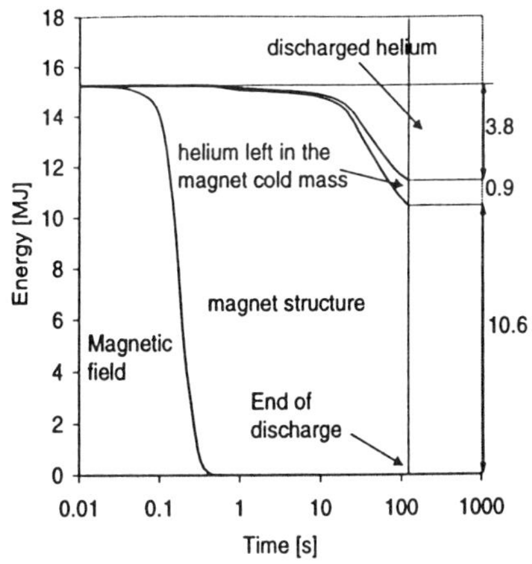


**Figure 6.** Time evolution of the distribution of energy over the magnetic field, magnet structure, cold mass helium, and discharged helium.

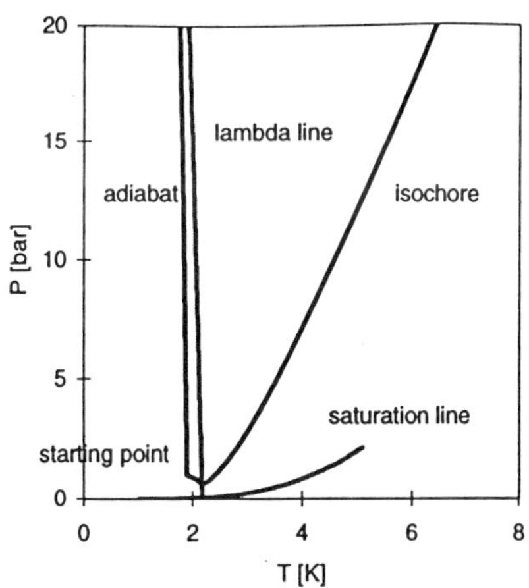


**Figure 7.** Ideal lines of evolution for isochoric heating, and adiabatic compression of helium.

## Energy Distribution

The magnetic field energy is calculated from the measured current decay, common to all magnets, and the respective magnet inductances. The internal energy of the helium inside each of the cold masses is calculated from the temperatures and pressures measured in the end-volumes, whereby it is assumed that the helium thermodynamic state is homogeneous over half the cold masses adjacent to the measurement locations. The helium mass discharge and corresponding enthalpy flow is calculated from the QRV $K_v$ value, and the temperature and pressure measurement before and after the QRV. The energy needed to complete the sum of these energies up to the stored magnetic energy at the quench current level is attributed to the magnet structure.

Figure 6 shows the time evolution of the energy distribution. From the initial magnetic field energy at quench current, about 6% ends up in the helium in the cold mass, about 25% leaves the cryostat with the discharged helium, and the remainder ends up in the magnet structure.

The heating power of the helium inside the cold mass is calculated as the differential of its energy development minus the enthalphy outflow (fig. 9).

## Process Identification

The foregoing observations have led us to identify the two main processes, each acting on a different time scale, which govern the helium thermohydraulics after a quench.

On a short time scale, up to roughly 150 ms after the quench, the bulk of the helium is adiabatically compressed. This follows directly from the measured evolution of the helium in the phase diagram as compared with ideal lines of evolution for isochoric heating, and adiabatic compression (fig. 7).

On a long time scale the bulk helium is heated at a relatively lower rate by the warm coils and the magnet structure, while simultaneously being allowed to discharge from the cold mass.

The existence of these two processes becomes clearly evident when in addition to the forementioned measurements, one observes the pressure evolution at much lower than usual quench energies. The two distinct rates of pressure rise associated with respectively the short time scale (high rate) and long time scale (low rate), as well as the changeover between these two processes can easily be recognized in figure 8.

## MODELS

### Short Time Scale Model

During about the first 150 ms, the bulk of the helium ($\approx$ 200$\ell$ per magnet) undergoes an adiabatic compression. At this stage the heat originating from the dissipated magnetic energy is still concentrated in the magnet coils. The coils exchange heat efficiently only with the relatively small amount of helium with which they are in close thermal contact: roughly 2% of the total helium contents as estimated from micrograph photographs of assembled coil cross-sections. This helium would attain very high pressures if not given the possibility, via a sufficient radial venting connection to the bulk, to expand into the region occupied by the bulk helium.[6] As a consequence the bulk helium gets adiabatically compressed. The, by now vapourized, helium surrounding the hot coils has become a limit to the heat transfer, which determines the end of the fast pressure rise.

One can thus distinguish: the (moderately) "confined" helium as being the fraction of helium in close thermal contact with the coils, the bulk helium, and the radial venting connection between the two. This connection must be sufficient, to prevent the development of high local pressures near the coils.

The adiabatic compression process is schematically depicted as a two-volume model[7] in figure 10. The peak pressure is very sensitive to the final temperature of the expanding "confined" helium. For the prototype magnet string, the maximum coil temperature after a 13.1 kA quench is about 140 K. Simply assuming no temperature difference between the confined helium fraction, 2%, and the coil would lead to peak pressure values of more than twice the measured values (fig. 11).

Assuming a 2% confined helium fraction we find that the heat and corresponding power necessary to reproduce the observed pressure peaks are quite small, a few hundreds of kJ

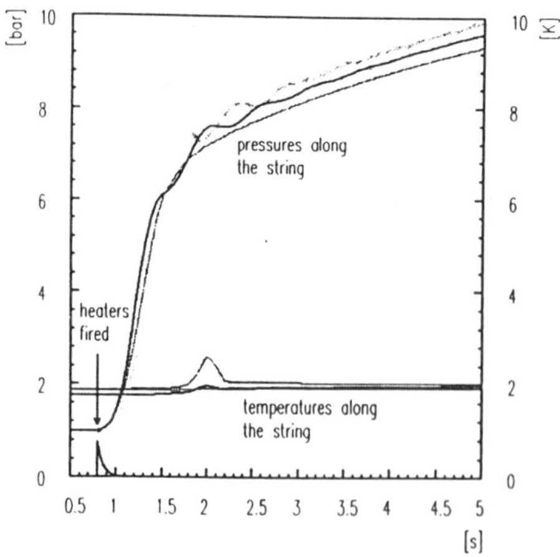


**Figure 8.** Development of the string cold mass helium after a low energy quench (at 5 kA).

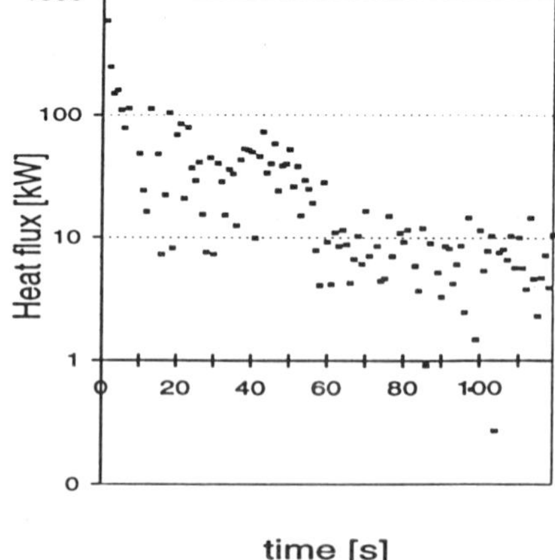


**Figure 9.** Power transfered from the magnet to the helium for a high energy quench (at 13.1 kA).

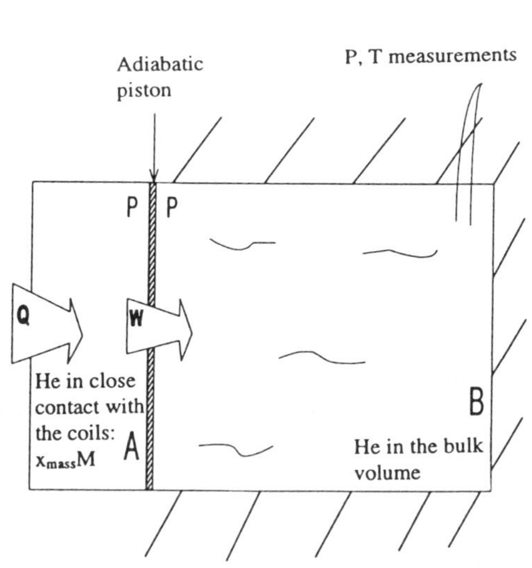


**Figure 10.** Two volume model of the adiabatic compression of the bulk helium (B) by the expanding confined helium (A).

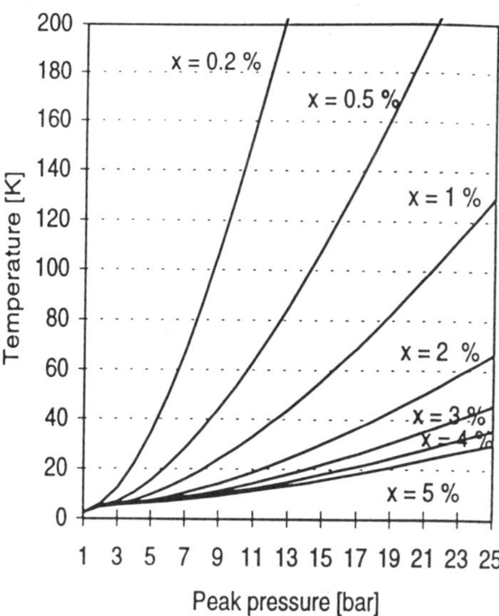


**Figure 11.** Calculated confined helium temperature necessary for compressing adiabatically the bulk helium to a given peak pressure; $x$-confined helium fraction.

(1-2% of the total magnetic energy) and about 1 MW, which is attained in our simulations with a heat exchange coefficient of the order of a few $\mathrm{W/cm^2}$ and a temperature difference between the coils and the expanding helium of about 110 K.

The height of the pressure peak depends on the transferred heat. This heat depends on the coil temperature, which in turn depends on the dissipation of magnetic energy. These links can explain the measured dissipated energy dependence of the pressure peak (fig. 4).

## Long Time Scale Model

After the adiabatic compression has ended, the expanded helium and the compressed bulk helium mix, the pressure thereby drops from its peak value down to a few bar and the temperature increases by about 0.5 K. At the same time heat in the coils is gradually being transferred by solid conduction to the collars and yoke. This way, the bulk helium gets exposed to an increasingly larger heat exchange area. At this stage we can consider the magnet structure as a heat source for the bulk helium. The helium warms up and pressurizes at a rate low enough such as to enable to keep its pressure below the cryostat design limits, by discharge through a QRV.

The heat and corresponding power necessary to reproduce the observed pressure and temperature development are quite small and can be easily accounted for by thermal convection.

## Model Comparison with Measured Data

Based upon these two models, linked through a mixing phase, we are able to reproduce the measured data (fig. 3) using as input a curve fit of the measured heating power as shown in figure 9.

## CONCLUSIONS

The thermohydraulics of resistive transitions of the LHC prototype magnet string is governed by two main processes: initially by a fast adiabatic compression of the bulk of the helium by the rapidly expanding confined helium, and finally by a global convective heating of the bulk helium. The transition between these processes is formed by a mixing between the expanded helium and the compressed bulk helium.

This overall behaviour remains valid as long as the radial venting connection between the confined and bulk helium remains sufficient to allow the confined helium to expand into the region occupied by the bulk helium.

The initial pressure peaks due to the adiabatic compression of the bulk helium are inherent to the magnet design. According to the adiabatic compression model they will be especially sensitive to the amount of confined helium. These pressures are mediated by sound waves. Mass transport is thus not involved, which is why the initial peak pressures cannot be influenced below their inherent value by any means of fast discharge. The influence of helium discharge is weak, and becomes appreciable only for opening delays greater then about 100 ms.

The measured quench pressure developments can be simulated with sufficient accuracy to justify extending the simulations to different string geometries.[2]

## REFERENCES


1. F. Rodriguez-Mateos, G. Gerin and A. Marquis, Quench Protection Test Results and Comparative Simulations on the First 10 Metre Prototype Dipoles for the Large Hadron Collider, *Fourteenth International Magnet Technology Conference (MT-14),* Tampere, (1995).
2. B. Hilbert, Modeling and Experimental Analysis of the Thermohydraulics of Resistive Transitions on the LHC Prototype Magnet String, LHC Project Note 86 (1997).
3. A. Bézaguet, J. Casas Cubillos, B. Flemsäter, B. Gaillard-Grenadier, Th. Goiffon, H Guinaudeau, Ph. Lebrun, M. Marquet, A. Suraci, L. Tavian and R. van Weelderen, The Superfluid Helium Cryogenic System for the LHC Test String: Design, Construction and First Operation, *Advances in Cryogenic Engineering* **Vol. 41** (1996), pp. 777-784.
4. D. Brahy, J. Casas Cubillos, M. Grippeling, D. Lavielle, G. Leo, L. Madaro, A. Rijllart, R. Saban, M. Skiadelli, The Control and Data Acquisition of the LHC Test String, *International Conference on Accelerator and Large Experimental Physics Control Systems* (ICALEPCS'95 Chicago, 1995).
5. G. Gerin, B. Vullierme and R. van Weelderen, Measurement of the Thermohydraulic Behaviour of LHC Dipole Prototypes after a Quench, *Advances in Cryogenic Engineering* **Vol. 41** (1996), pp. 811-818.
6. Ph. Lebrun, Th. Wahlström, R. van Weelderen and L. Williams, Investigation of Quench Pressure Transients in the LHC Superconducting Magnets, *Cryogenics* **34** ICEC Supplement (1994), pp. 705-708. Genova, (1994).
7. Ph. Lebrun, Private Communication, (CERN).

# HELIUM RECOVERY IN THE LHC CRYOGENIC SYSTEM FOLLOWING MAGNET RESISTIVE TRANSITIONS

M.Chorowski, B.Hilbert, L.Serio, L.Tavian, U.Wagner, and R. van Weelderen

LHC Division, CERN
CH-1211 Geneva 23, Switzerland

## ABSTRACT

A resistive transition (quench) of the Large Hadron Collider magnets provokes the expulsion of helium from the magnet cryostats to the helium recovery system. A high-volume, vacuum-insulated recovery line connected to several uninsulated medium-pressure gas storage tanks, forms the main constituents of the system. Besides a dedicated hardware configuration, helium recovery also implies specific procedures that should follow a quench, in order to conserve the discharged helium and possibly make use of its refrigeration capability. The amount of energy transferred after a quench from the magnets to the helium leaving the cold mass has been estimated on the basis of experimental data. Based on these data, the helium thermodynamic state in the recovery system is calculated using a lumped parameter approach. The LHC magnet quenches are classified in a parametric way from their cryogenic consequences and procedures that should follow the quench are proposed.

## THE LHC HELIUM RECOVERY SYSTEM

A LHC sector (1/8 of the machine circumference) consists of 46 half-cells in the arc plus a number of special cryomagnets. In the following for this analysis the whole sector will be taken as composed of 54 half-cells. The magnetic energy stored in the magnets that form a LHC half-cell will be released during a quench and may be as high as 23 MJ[1]. This stored magnetic energy is dissipated in the cold mass within a few tenths of a second following a quench. A fraction of this energy is subsequently transferred to the helium, causing a fast pressure increase and heating of the helium bath. Helium is finally discharged to the helium recovery system through a quench relief valve. The purpose of the recovery system is the preservation of the helium inventory and if possible, the recovery of its refrigeration capacity after a quench. As shown conceptually in Figure 1, the system for one LHC sector will consist of the following nodes:

1. 54 half -cells, each containing 850 litres of helium,
2. quench relief valves,
3. a vacuum-insulated line D having a length of about 3400 m and volume of 60 $m^3$ per sector, connected via quench relief valves (2) to the LHC half-cells (1),
4. an uninsulated vertical line having a length ranging from 42 to 145 m, depending on the sector, connecting line D (3) via valve (4a) with medium-pressure gas storage tanks (5),
5. medium-pressure (2 MPa) gas storage tanks with a volume of 250 $m^3$ each, forming a quench buffer volume of 2000 $m^3$,
6. piping and auxiliaries interconnecting the system with the sector refrigerator.

## LUMPED PARAMETER APPROACH TO THE RECOVERY SYSTEM

We assume that for the purpose of configuring and sizing of the quench recovery system a lumped-parameter approach is adequate. Therefore spatial variations of density, temperature and pressure can be neglected within a system node. For each node of the system, the following energy balance equation can be written:

$$dU = dQ + \sum_i h_i dM_i \tag{1}$$

where dU is the energy change of a node, dQ is the heat transferred to a node, $h_i$ is the enthalpy of the i-th stream of helium leaving or entering a node, $dM_i$ is the change of the mass of helium in a node due to the i-th flow.

### Estimation of Enthalpy and Mass Input Data

The enthalpy and mass of helium leaving the cold mass after a quench are basic inputs for any further calculations. Experimental data from the prototype magnet string[2], thermo-hydraulically equivalent to the LHC half-cell, were used to estimate:

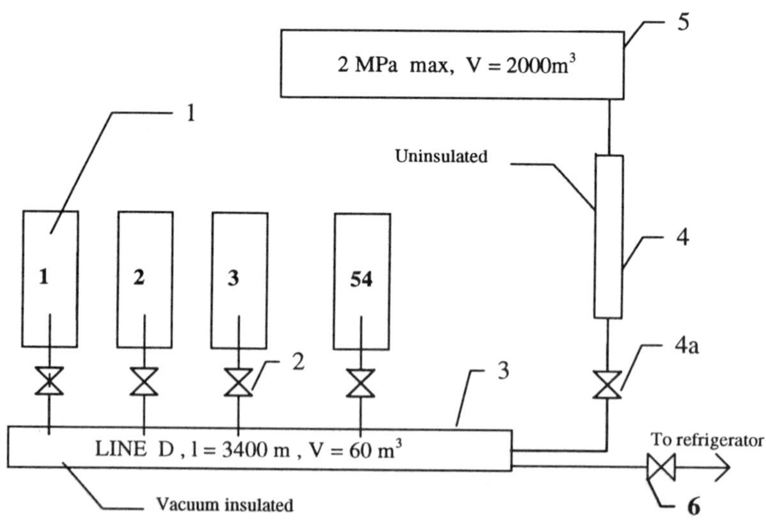


**Figure 1.** Helium recovery system, scheme for one sector: 1 - cryogenic half-cells ; 2 - quench relief valves; 3 - vacuum insulated line D; 4 - uninsulated vertical line; 5 - medium-pressure gas tanks; 6 - auxiliaries

- the final distribution of the magnetic energy released after a quench, between cold mass, helium left in the cold mass and helium expelled; this enables the calculation of the average enthalpy $\bar{h}_{He}$ of the expelled helium,
- the enthalpy, $h(t)$, and mass, $\dot{m}(t)$, flows of helium leaving the cold mass.

Table 1 gives the final magnetic energy distribution averaged over eleven quenches of the prototype string. The initial conditions of helium temperature, pressure and enthalpy were 1.9 K, 100 kPa and 1864 J/kg respectively. The stored magnetic energy was 15.3 MJ and an estimated amount of helium expelled from the cold mass was 120 kg. The cold mass temperature was measured 30 minutes after each quench. An average helium enthalpy of 33 J/g can be calculated from Eq. (2) on the assumption that the process of heat transfer to helium expelled from the cold mass is isobaric.

$$\bar{h}_{He} = (h_{He-ini}\Delta M_{He-ex} + \Delta E_{He-ex}) / \Delta M_{He-ex} \qquad (2)$$

where $h_{He\text{-}ini}$ is the initial enthalpy of helium, $\Delta M_{He\text{-}ex}$ is the mass of helium expelled from the cold mass and $\Delta E_{He\text{-}ex}$ is the energy transferred to helium expelled from the cold mass.

**Table 1.** Measured energy balance following a full current quench of the prototype string

| | Energy , MJ | Energy , % |
|---|---|---|
| Energy transferred to the helium expelled from the cold mass | 3.8 | 25 |
| Energy transferred to the cold mass | 10.0 | 65 |
| Energy transferred to the helium left in the cold mass | 1.5 | 10 |
| Coil magnetic energy | 15.3 | 100 |

Figures 2 and 3 show helium enthalpy and mass flow measured during discharge from the cold mass. Helium temperature, pressure and pressure drop at the quench relief valve were measured. An average helium enthalpy of 32 J/g can be calculated from Eq. (3), which is in good agreement to the value of 33 J/g obtained from Eq. (2). It confirms the validity of the helium mass flow $\dot{m}(t)$ estimation.

$$\bar{h}_{He} = \int_0^\infty h(t)\dot{m}(t)dt \Big/ \int_0^\infty \dot{m}(t)dt \qquad (3)$$

## CALCULATION OF HELIUM STATE IN LINE D AFTER A QUENCH

We use the measured helium enthalpy $h(t)$ and flow $\dot{m}(t)$, shown in Figures 2 and 3, as an input to calculate the helium thermodynamic state in line D after a quench. The heat capacity of line D is negligible in comparison to that of the helium. Therefore for a given number $n$ of half-cells quenched, Eq. (1) takes the form:

$$dU = n \cdot h(t) \cdot \dot{m}(t)dt \qquad (4)$$

The calculated time evolution of helium temperature, pressure and vapour quality in line D after a whole sector (54 half-cells) quench is shown in Figures 4 - 6.

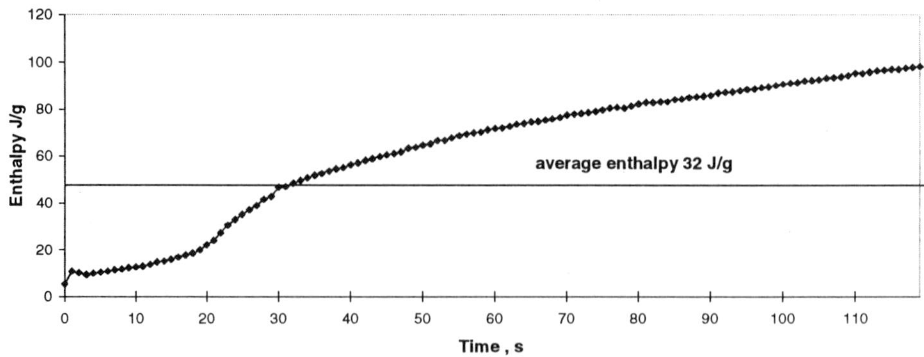


**Figure 2.** Enthalpy of helium leaving the cold mass after a full-current quench of the prototype string (measured).

## Steady-state Thermodynamic Parameters of Helium in Line D

The thermodynamic parameters of helium in line D will reach steady state several minutes after a quench. The temperature and pressure will then depend on the number of half-cells quenched, with the enthalpy $\overline{h}_{He}$ and mass of the discharged helium from one half-cell $\Delta M_{He-ex}$ as parameters - see Figure 7. The pressure in line D is limited to 2 MPa as it was designed for[1] and established by previous work[3]. Line D forms a buffer which can accumulate expelled helium if the number of half-cells involved does not exceed 44. For a higher number of half-cells quenched some helium must be discharged from line D.

## LHC Quench Classification

According to the number of half-cells involved, LHC quenches can be classified from their cryogenic consequences as:

- Limited quench (2-12 half-cells quenched). After a limited quench no liquid helium is present in line D during the cold mass discharge, maximal pressure in line D is equal to 0.42 MPa after a quench of 12 half-cells. A limited quench can be fully buffered by the capacity of line D.
- Medium quench (13-44 half-cells quenched). A medium quench is characterised by liquid helium present during discharge in line D for some time and maximal pressure 2 MPa (for 44 half-cells quenched). Due to the quench propagation considerations a medium quench is assumed to be a highly improbable event.

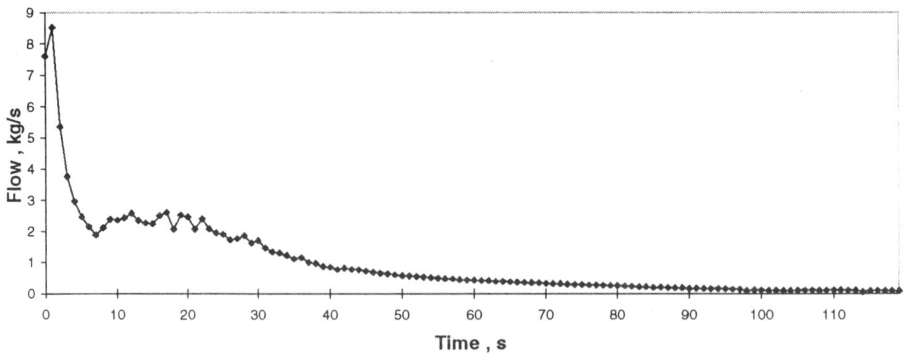


**Figure 3.** Helium mass flow out of the cold mass after a full-current quench of the prototype string (measured).

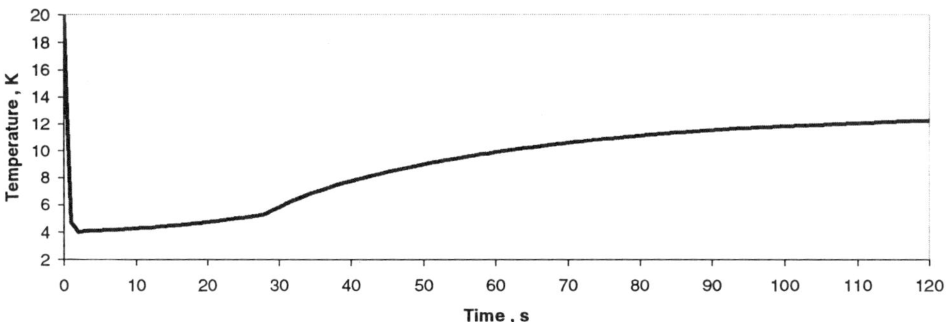


**Figure 4.**Temperature in line D after a whole sector (54 half-cells) quench (calculated).

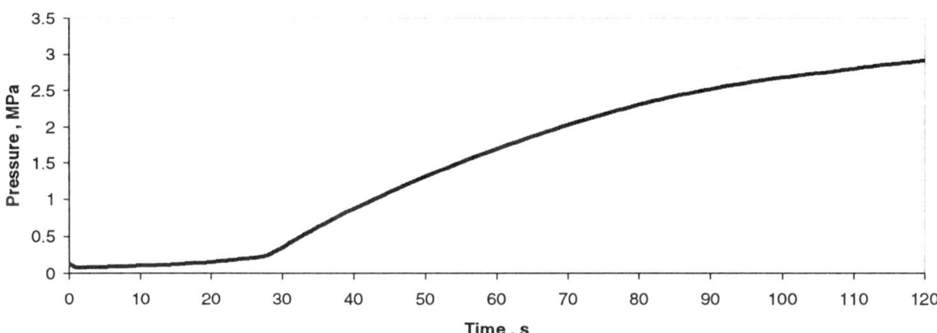


**Figure 5.** Pressure in line D after a whole sector (54 half-cells) quench (calculated).

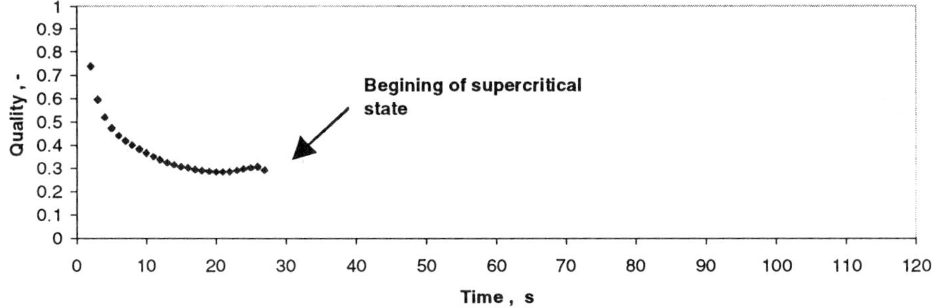


**Figure 6.** Vapour quality in line D after a quench of 54 half-cells (calculated).

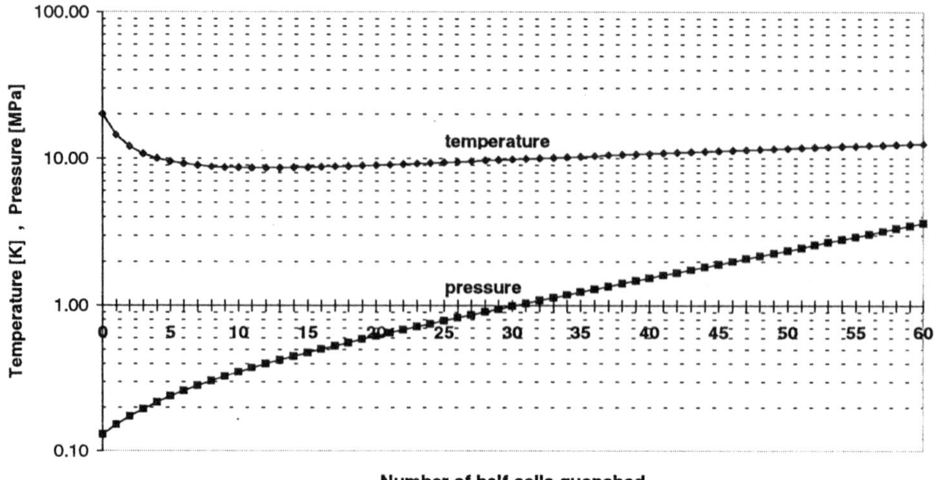


**Figure 7.** Calculated temperature and pressure in line D as a function of the number of half-cells quenched, average enthalpy of the discharged helium is 33 J/g, mass of helium discharged from one half-cell is 120 kg.

- Whole sector quench (45-54 half-cells quenched). A sector quench cannot be fully buffered by the capacity of line D and a fraction of helium must be discharged to medium-pressure gas tanks (Fig.1.5).

## HELIUM RECOVERY IN MEDIUM-PRESSURE GAS TANKS

As follows from the calculation of the helium parameters in line D following a quench, only in the case of a whole sector quench is the capacity of line D too small to contain (even for a short time) all the helium expelled from the cold mass and a fraction of helium must be buffered in the medium-pressure tanks. The medium-pressure tanks are made of carbon steel, which constrains the temperature of the wall to be higher than - 40 °C. This means that the in-flowing helium stream should be warmed up. We have considered two convenient passive sources of heat:

- Convective heating on the line connecting line D to the MP gas tanks (Fig.1.4),
- Heat capacity of the medium-pressure gas tanks (Fig.1.5).

Figure 8 shows the temperature of the helium inside the tank and the tank wall calculated for the following conditions:

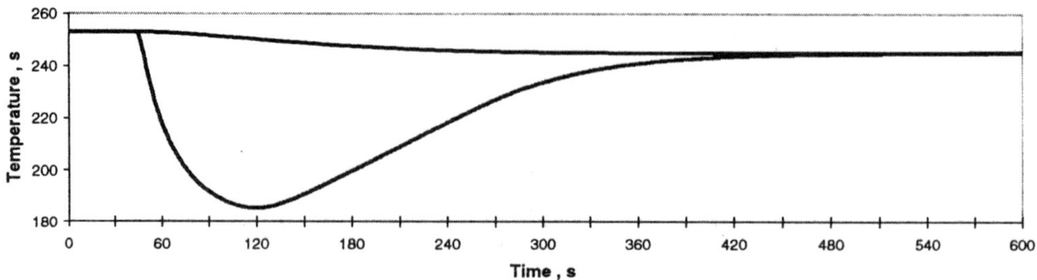


**Figure 8.** Calculated wall and helium temperature in medium-pressure tank during discharge of line D, initial temperature and pressure 253 K and 1 bar respectively.

1. Total quench buffer volume is $2000 m^3$ and its heat capacity is 203 MJ/K.
2. Initial helium temperature in is 253 K and initial pressure is 100 kPa.
3. Line D is discharged through a valve opening on pressure and set at 1 MPa, integrated helium flow to medium-pressure tank after a whole sector quench is 2340 kg, maximal instantaneous flow is 20 kg/s, time evolution of helium temperature and pressure in line D is given in reference[4].
4. Convective heating power on the vertical uninsulated line is estimated at 17 kW.
5. Heat transfer coefficient between tank wall and helium is estimated at 58 $Wm^{-2}K^{-1}$.
6. Heat transfer coefficient between the tank wall and ambient air is estimated at 10 $Wm^{-2}K^{-1}$.
7. Ideal mixing of helium occurs inside the tanks and no temperature gradient is developed across the tank wall.

The temperature of the tank wall decreases by 8 K only, reaching 245 K, and remains well above its lower limit (233 K). This means that the tank heat capacity is high enough to warm up the in-flowing helium and to prevent the tank wall against dangerous temperature drop with a 150 % safety margin. Nevertheless it must be stressed that the heat capacity of the tanks can be fully used only if a good mixing of helium inside the tanks is achieved e.g. by use of an appropriate tank inlet pipe-work, preventing jet formation and spot cooling of the tank wall.

## HELIUM INVENTORY AND EXERGY RECOVERY AFTER A QUENCH

From the LHC operation point of view, there are two probable sizes of quenches to be considered:

- Limited quench - it starts in one magnet and propagates to the adjacent magnets due to the inter-magnet quench propagation (up to 8 half-cells involved). A limited quench is considered as a part of the collider operation.
- Whole sector quench - all sector magnets quench simultaneously (46 - 54 half cells involved). A whole sector quench is considered as an accidental mishap.

In case of a limited quench, line D can fully buffer helium expelled from the LHC cold mass. Both helium inventory and its refrigeration capacity are then recovered.

In case of a sector quench a fraction of the expelled helium (up to 30%) must leave line D, be warmed up and stored in medium-pressure gas tanks. The temperature of helium leaving line D is below 12 K and the fluid is characterised by high exergy which is lost during warm up. As the sector refrigerator will be ready to start blowing down vacuum insulated line D about 10 minutes after the quench, the line will store the remaining helium for that period of time. The maximum flow capacity to blow down line D to the refrigerator may vary from 780 g/s to 260 g/s. If we assume that about 5000 kg of helium should be recovered from line D, it would be necessary to run the refrigerator compressors for 2 to 6 hours. After a sector quench it is then possible to recover all the helium inventory and about 70% of its exergy.

## CONCLUSIONS

1. For the purpose of configuring and sizing the LHC helium recovery system a lumped-parameter model based on energy balance is applicable.

2. The probable LHC quenches can be classified from their cryogenic consequences as:
   -limited quenches, up to 12 half-cells involved, no liquid helium present in line D during discharge, can be buffered by the capacity of line D, will be part of the machine operation,
   - sector quenches, 45-54 half-cells involved, considered as accidental mishap.
3. In case of a sector quench line D is not able to accommodate all the helium expelled, and a fraction of the helium (up to 30%) will have to be warmed up and stored in the medium-pressure gas tanks. The exergy of this helium is lost.
4. The heat capacity of the medium-pressure tanks is high enough to warm up the in-flowing helium and to prevent the tank wall temperature from droping below the design value.
5. To avoid local temperature drop of the tank wall below a design limit (-40 °C), a good mixing of helium must be achieved by use of a appropriate pipe-work inside the tanks.

## ACKNOWLEDGEMENTS

The authors would like to thank Philippe Lebrun for many helpful suggestions and his thorough proof-reading.

## REFERENCES

1. The LHC Study Group, The Large Hadron Collider Conceptual Design, CERN/AC/95-05(LHC).
2. A. Bézaguet et al., Cryogenic Operation and Testing of the Extended LHC Prototype Magnet String, Proc. ICEC 16/ICMC, Kitakyushu, Japan, May 1996.
3. L. Brue, Modélisation Thermohydraulique des Ecoulements Transitoires d'Hélium Cryogénique induits dans la Ligne de Récupération du Demi-Octant par les Transitions Résistives des Aimants du LHC, LHC Note 262, CERN AT/94-03 (CR).
4. M.Chorowski, Transient behaviour and helium recovery in the LHC cryogenic system following magnet resistive transitions, LHC Project note 77, CERN/LHC (1997).

# ANALYSIS OF THE 15 m LHC CRYO-DIPOLE RADIATION SCREEN DURING A MAGNET QUENCH

G. Peón and L.R. Williams

CERN, European Organization for Nuclear Physics
1211 Geneva 23, Switzerland

## ABSTRACT

The cost of extracting 1 Watt from a temperature of 1.9 K is about 8 times higher than that of extracting it from the 5 K to 20 K temperature range. In the LHC prototype half-cell cryostats,[1] having only one thermalised screen at 70 K, about 70% of the heat inleak to the 1.9 K temperature level comes from radiation. To enhance the cryostat performance, the second generation of prototype cryostats incorporates a second actively cooled screen operating in the 5 K to 20 K temperature range. This screen, which is manufactured in industrially pure aluminium alloy sheets 2 mm thick, is situated very close to the dipole magnet cold mass. The rapid collapse of the dipole field during a resistive transition (quench) will induce forces and displacements. An analytical, 2 D study to estimate this effect is presented.

## INTRODUCTION

Recent estimates based on measurements performed on the CTM [2,3] indicate the use of an actively cooled radiation screen at 5 K - 10 K to be beneficial, in particular, in limiting the heat inleaks to the cold mass at 1.9 K. Space restriction inside the cryostat requires the radiation screen to closely envelope the cold mass at a radius of about 5 mm from its outer skin. The cold mass and the radiation screen are prevented from touching by means of spacers.[4] The radiation screen is therefore situated in an environment of high stray magnetic field. A change in magnetic field induces eddy currents in the screen and these, in turn, interact with the magnetic field in the screen, to create forces and deformations on the screen structure. Large displacements of the screen may damage fragile internal cryostat parts and permanent deformations may lead to thermal shorts to the cold mass.

A study has been undertaken to estimate the effect of a resistive transition (quench) in the dipole magnet on the 5 K radiation screen. The study follows the following steps:

- Calculation of the magnetic field $(B_r, B_\phi)$ on a circumference concentric to the cold mass at a radius of R = 290 mm at the nominal dipole field of 8.36 T in the beam tube. The modification of the magnet field due to the eddy currents flowing in the radiation screen has not been considered.

- Calculation of the product $\vec{B}\,(dB_r/dt)$, 0.22 s after the quench initiation. (Point where $\vec{B}\,(dB_r/dt)$ is maximum).
- Calculation of the induced current density in the screen, J.
- Calculation of the forces on the screen ($F_r$,$F_\phi$).
- Calculation of the displacements (v,w) due to the calculated forces on the structure. Estimated displacements are calculated on the basis of a static system of forces, and therefore acceleration effects on the shield are neglected.

The type of insulating spacers to be used between the screen and the cold mass as well as the decision to stiffen or not stiffen the screen structure depends on the results of this work.

## DEFINITION OF COORDINATE AXES

The problem defined in 2 dimensions is solved in cylindrical coordinates as indicated in Figure 1. u, v and w denote the components in the x, $\phi$ and r directions of the displacement of a general point on the screen, approximated as circular, at an angle $\phi$. The supporting conditions of the screen along its length are not constant. The sections at the supports have their movement constrained at the bottom, whereas the sections furthest from the supports are likely to rest on the cold mass.

In this note, the problem is analysed in the plane of the supports. For this case, the maximum displacements are higher. For symmetry reasons, only the first 180° will be analysed.

## MAGNETIC FIELD ON THE SCREEN

This task was performed using the finite element program Roxie.[5] The simulation model included the magnet, the cold mass and the vacuum vessel. The initial values of the magnetic field seen by the radiation screen, $B_{rnom}$ and $B_{\phi nom}$, corresponding to a nominal dipole magnet field of 8.36 T in the beam apertures, are shown in Figure 2.

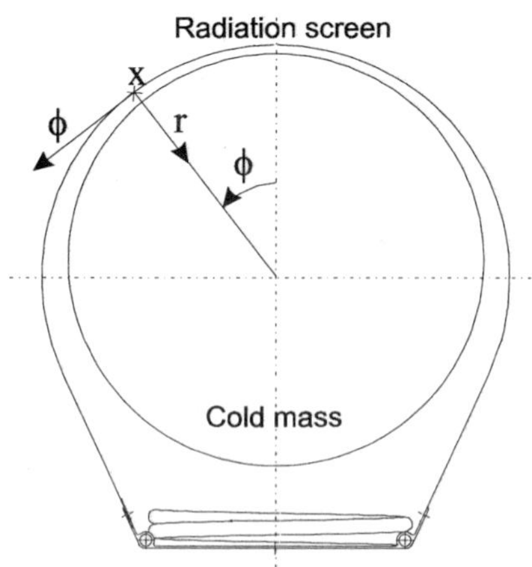


**Figure 1**. Definition of coordinate axes, x, $\phi$ and r.

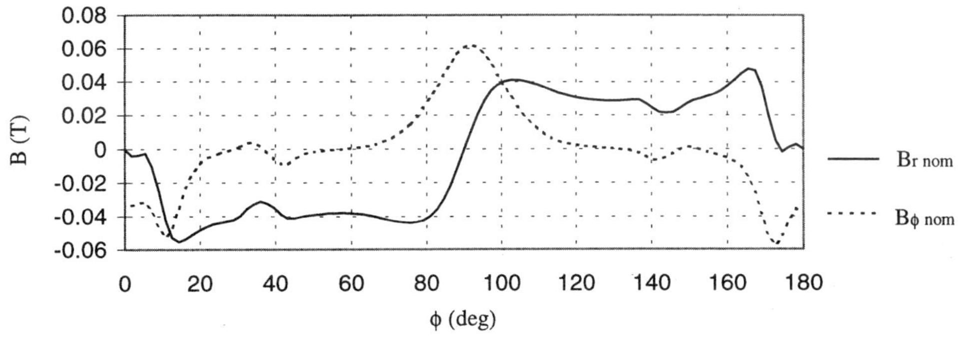


**Figure 2**. Magnetic field seen by the radiation screen for a magnetic field in the beam aperture of 8.36 T.

In a typical quench (simulated with QUABER [6]) of a 14.2 m long LHC dipole magnet where the heaters are effective 80 ms after the quench starts, the product of the current in the coil and its derivative with time, Ic(dIc/dt), presents a maximum at 0.22 s after the quench initiation.

The magnetic field, $\vec{B}$, if only the contribution of the coil is considered, is related to Ic as follows:

$$\vec{B}(t,\phi) = \vec{B}_{nom}(\phi) \cdot \frac{Ic(t)}{Ic_{nom}} \tag{1}$$

The maximum in $\vec{B}$ ($dB_r/dt$) is approximated to coincide with the maximum in Ic(dIc/dt), which is true only when iron is not present. The forces in the r and $\phi$ directions are proportional to the product $\vec{B}$ ($dB_r/dt$) and their maximum will also occur 0.22 s after the quench initiation.

At this time, Ic has decreased from 11.5 kA to 8.8 kA and the slope of the current decay, dIc/dt, is -41 kA/s. The field gradient, $dB_\phi/dt$, induces a change in current in the radial direction. These radial changes are considered negligible across the thin-walled structure of the radiation screen.

## INDUCED CURRENT DENSITY IN THE SCREEN

The change of radial magnetic field, $B_r$, with time induces a current density in the screen. As the total current must follow a closed loop, the sum of all currents flowing in the screen must be zero. For this dipole magnetic field, 4 current loops symmetric with respect to the vertical axis appear in the screen. The total current must, therefore, be zero in each half. This is expressed in Eq. (2) where t is the screen thickness, $\rho$ the material resistivity, R is the radius of the screen and $\phi_0$ an angle where the current density, $J(\phi)$ is 0.

$$I_{TOT}\Big|_{0^o}^{180^o} = \frac{tR^2}{\rho} \int_{0^o}^{180^o} \left( \int_{\phi_0}^{\phi} \dot{B}_r(\phi) d\phi \right) d\phi = 0 \tag{2}$$

From this equation, it is found that the eddy current density, $J(\phi)$ changes sign in the first quarter at $\phi_{01} = 48.6^o$ and, in the second quarter, at $\phi_{02} = 138^o$. The total current changes sign at the angle $\phi_1 = 99^o$.

Due to a lack of information on the residual resistivity ratio (RRR) of the aluminium alloy Al 6063-T5 (98.5% purity), this has been approximated to that of Al 5005 (99.3% purity) which represents the worse case analysis. Hartwig and Saffari-Kermani [7] quoted the following value:

$$\mathrm{RRR(Al\ 5005)} = 5.4 \tag{3}$$

The Al 6063-T5 resistivity at room temperature is:

$$\rho(300\mathrm{K}) = 3.37 \cdot 10^{-8}\,\Omega\mathrm{m} \tag{4}$$

and so

$$\rho(5\mathrm{K}) = 6.24 \cdot 10^{-9}\,\Omega\mathrm{m} \tag{5}$$

The current density in the screen can be determined from Eq. (6).

$$\mathrm{J}(\phi) = \frac{\mathrm{R}}{\rho}\int_{\phi_0}^{\phi} \dot{\mathrm{B}}_{\mathrm{r}}(\phi)\mathrm{d}\phi \tag{6}$$

## FORCES INDUCED ON THE SCREEN STRUCTURE

The interaction between the current induced in the screen and the dipole stray field seen by the screen structure generates forces in the $\phi$ and r directions. These forces can be calculated using the Lorentz formula.

$$\vec{\mathrm{F}} = (\vec{\mathrm{J}}\Lambda\vec{\mathrm{B}})\mathrm{t} \tag{7}$$

For this particular case, the maximum force distribution in $\phi$ and r, $F_\phi$ and $F_r$, as a function of $\phi$ are shown in Figures 3 and 4. In order to facilitate the calculation of displacements, the actual force diagram is modified according to the following simplifications:

- The force for each section is constant and equal to the maximum force in the actual diagram.
- The angle $\phi$, over which the simplified force is applied, is determined so that the area below the force line remains the same for both curves. In this way, the total force applied to the structure is the same both, for the simplified and the real case. The lengths over which the distributed forces are applied are smaller for the simplified case.

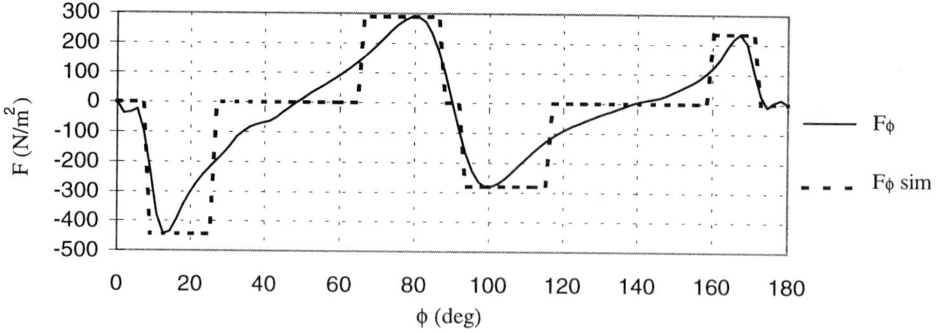


**Figure 3**. Force per unit of surface on the screen in the $\phi$ direction as a function of $\phi$. Calculated (continuous line) and simplified (dashed line).

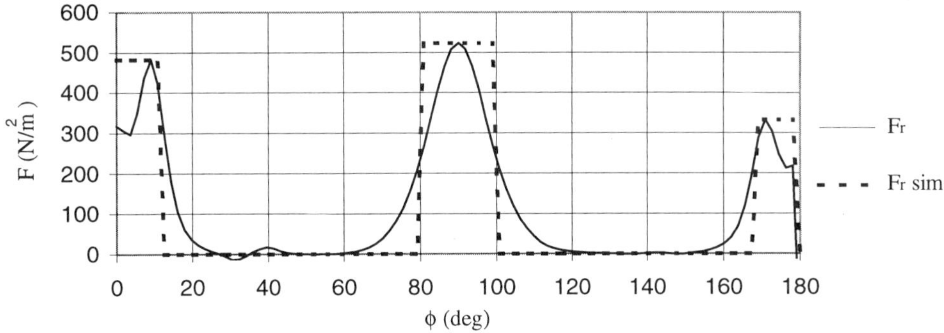


**Figure 4**. Force per unit of surface on the screen in the r direction as a function of $\phi$. Calculated (continuous line) and simplified (dashed line).

## DISPLACEMENTS CAUSED BY THE FORCES INDUCED

The forces induced during the decay of the dipole magnet field after a quench cause the screen structure to deform and also to accelerate. These forces change rapidly with time. In the present study, the structure is assumed to have zero mass and therefore, the deflections calculated represent a pessimistic estimate.

If the deformation of a cylindrical shell is to be inextensional, the three strain components in the middle surface must vanish:

- In the x direction,

$$\varepsilon_x = \frac{\partial u}{\partial x} = 0 \tag{8}$$

- In the circumferential direction,

$$\varepsilon_\phi = \frac{\partial v}{R\partial\phi} - \frac{w}{R} = 0 \tag{9}$$

- The shearing strain,

$$\gamma_{x,\phi} = \frac{\partial u}{R\partial\phi} + \frac{\partial v}{\partial x} = 0 \tag{10}$$

These requirements are satisfied for displacements in the following form:

$$u = 0 \tag{11}$$

$$v(\phi) = R\sum_{n=1}^{\infty}(a_n\cos(n\phi) - a'_n\sin(n\phi)) \tag{12}$$

$$w(\phi) = -R\sum_{n=1}^{\infty}n(a_n\sin(n\phi) + a'_n\cos(n\phi)) \tag{13}$$

for $a_n$ and $a'_n$ being two series of numbers.

In calculating the inextensional deformations of a cylindrical shell under the action of a given system of forces, the energy method is used. The strain energy, V, for a cylinder of a length L submitted to a load constant along the cylinder length can be expressed as a function of $a_n$ and $a'_n$ in the following form:[8]

$$V = \pi D \frac{L}{2} \sum_{n=2}^{\infty} \frac{(n^2-1)^2}{R^3} n^2 R^2 (a_n^2 + a'^2_n) \tag{14}$$

A theoretical treatment of the deformation of circular thin walled tubes subjected to constant-along-the-length radial and tangential forces symmetric with respect to the vertical axis is presented in a reference paper.[9] The displacements under radial and tangential forces are determined separately and the solution of the whole problem is found applying the principle of superposition.

## Displacements caused by the radial forces

For each pair of radial distributed forces symmetric with respect to the vertical axis, q, the displacement in the radial and tangential direction, $w_r$ and $v_r$, as a function of the angle, $\phi$, is given by Equations (12) and (13) with $a_{rn} = 0$ and:

$$a'_{rn} = \frac{-2qR^3}{\pi DL(n^2-1)^2 n^2} (\sin(n\phi_{fi}) - \sin(n \cdot \phi_{in})) \tag{15}$$

D is the flexural rigidity of the plate, L is its length, $\phi_{fi}$ is the angle where the distributed load finishes, and $\phi_{in}$ is the angle where it begins. The term n=1 refers to the displacement of the body in its plane as a rigid body and therefore is not taken into account.

## Displacements caused by the tangential forces

The displacements caused by distributed tangential forces symmetric with respect to the vertical axis, $w_t$ and $v_t$, are calculated using Eqs. (12) and (13), where $a_{tn} = 0$ and:

$$a'_{tn} = \frac{2qR^3}{\pi DL(n^2-1)^2 n^3} (\cos(n\phi_{fi}) - \cos(n \cdot \phi_{in})) \tag{16}$$

# RESULTS

The results of displacements in the tangential direction, $v_r$ and $v_t$ , are shown in Figure 5. The displacements in the radial direction, $w_r$ and $w_t$ , are found in Figure 6.

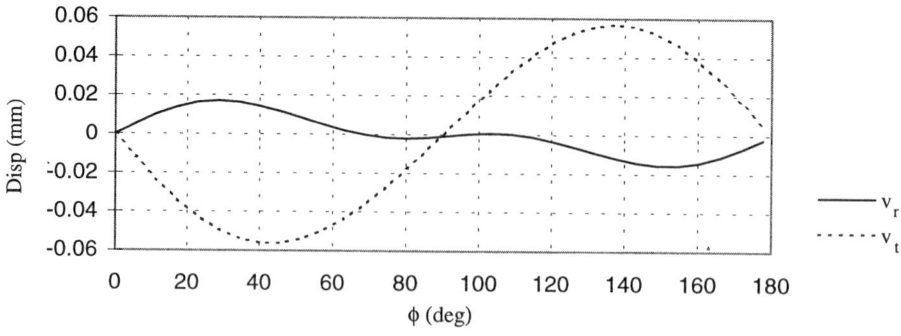


**Figure 5**. Quench induced radiation screen displacements in the $\phi$ direction by radial and tangential forces.

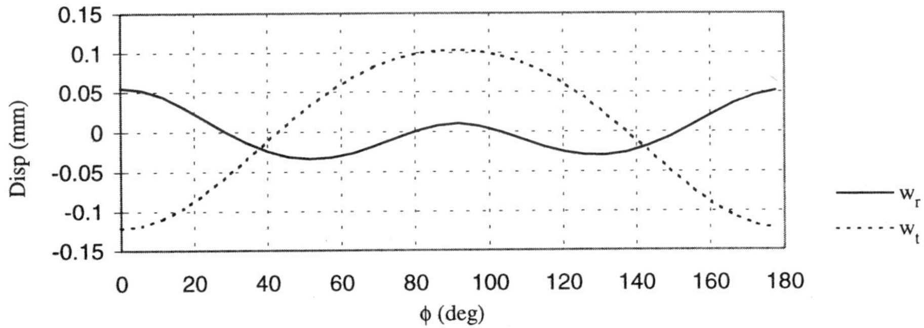


**Figure 6**. Quench induced radiation screen displacements in the r direction by radial and tangential forces

The total displacements, if the screen is fixed at the bottom part, $w_t$ at $\phi = 180^o$, are those shown in Figure 7.

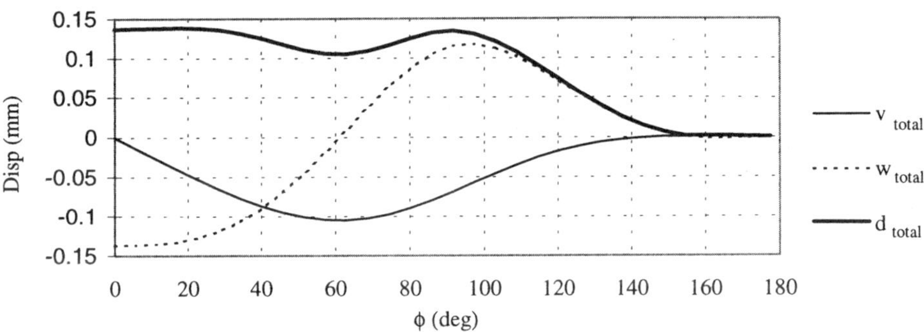


**Figure 7**. Displacements of the screen for a section anchored at $\phi = 180^o$.

## CONCLUSIONS

The displacements of the radiation screen induced by a quench on the LHC dipole magnets for a screen supported at the bottom part are 0.12 mm inwards at $\phi = \pm 90^o$ and 0.14 mm outwards at the azimuthal position, $\phi = 0^o$ .

The present report shows that displacements are small even for a worst case analysis and therefore more detailed study is not needed. The choice of separators between cold mass and radiation screen (net type spacers or nails) is not affected by the described phenomenon. Stiffening of the screen structure is not considered to be necessary.

## ACKNOWLEDGMENTS

The authors would like to thank C. Paul for the calculation of the magnetic field at the screen position and to S. Russenschuk for his advice and guidance in the first stages of the work. The authors also acknowledge F. Rodriguez-Mateos for the supply of the information related to quench in the half-cell dipole magnets and to J. Lucas for his careful proof-reading.

## REFERENCES


1. J. Brunet et al., Design of LHC prototype dipole cryostats, Cryogenics, ICEC Supplement, 32:191-194 (1992).
2. Dufay et al., A full-scale thermal model of a prototype dipole cryomagnet for the CERN LHC project, CERN AT-CR/IC/MA, 94-15 (1994).
3. G. Ferlin et al., Comparison of floating and thermalized multilayer insulation systems at low boundary temperature, International Cryogenic Engineering Conference (1996).
4. B. Jenninger, G. Peón and B. Szeless, Characterisation of net type thermal insulators at 1.8 K low boundary temperature, Presented at the Topical Conference on Nonmetallic Materials and Composites at Low Temperature VIII, September (1996).
5. C. Paul, K. Preis and S. Russenschuck, 2D Finite-element calculation of superconducting magnets applying a reduced vector potential formulation, CERN LHC Project Report 77 (1985).
6. D. Hagedorn and F. Rodriguez-Mateos, Modelling of the quenching process in complex superconducting magnet systems, IEEE Transactions on Magnetics, 28, no. 1:366-369, January (1992).
7. T. Hartwig and A. A. Saffari-Kermani, Characterization of several aluminum alloys for moderate-strength high conductivity applications in "Advances in Cryogenic Engineering", 26, (1980) p 621-624.
8. S. Timoshenko and S. Woinowsky-Krieger, "Theory of plates and shells", McGraw-Hill Book Company, Inc., (1959).
9. G. Peón and L. Williams, Analytical estimate of the dipole quench effect on the radiation screen, CERN LHC Project Note 82 (1997).

# STATUS, OPERATION AND PERFORMANCE OF THE RHIC HELIUM REFRIGERATOR

K.C. Wu, M. Iarocci, J. Sondericker, J. Benson, Y. Farah, C. Lac, A. Morgillo, A. Nicoletti, E. Quimby, J. Rank, M. Rehak, and D. Zantopp

RHIC Department
Brookhaven National Laboratory*
Upton, N.Y., 11973

## ABSTRACT

The Relativistic Heavy Ion Collider RHIC helium refrigerator was modified from the refrigerator originally constructed fifteen years ago for the ISABELLE accelerator. The modifications include a new cold vacuum compressor and relocating the cold circulating compressor to the cryogenic distribution valve boxes, which are located near the cryogenic feed to the collider rings. Prior to start up, an extensive effort was spent on improving conditions of the compressor system, turbine skids, valves, instrumentation and control apparatus. In February, 1996, the RHIC refrigerator was successfully cooled to liquid helium temperature with 10 kW of heat input at 4.5 K, 53 kW at 60 K and 44 g/s of liquefaction, using approximately two-thirds of the installed compressor capacity. In July, 1996, the refrigerator was used to cool a 350 meter vacuum jacketed transfer line and two cryogenic distribution valve boxes. In February, 1997, the refrigerator was used to cool the first RHIC sextant for over one month of operation. At the end of the first sextant test, an electric heat input of 25 kW at approximately 5.5 K was introduced to simulate a full refrigeration capacity mode. The capacity and reliability of the refrigerator have been demonstrated. The performance of the refrigerator including turbines, heat exchangers, cold vacuum compressor and circulating compressor is given.

## INTRODUCTION

The cryogenic system for the proposed ISABELLE accelerator was designed to provide 4000 g/s of 3.8 K helium to twenty-four parallel cooling passages to the magnets.[1] A four stage cold vacuum compressor and a cold circulating compressor were used. A description of the helium refrigerator is given in [2].

The RHIC cryogenic system utilizes recooler heat exchangers to maintain the temperature of the supercritical helium below 4.6 K.[3,4,5] A total of seventy-two recooler heat exchangers are used. The design flow rate for the cold helium through the RHIC magnets is 100 g/s at 5 atm between 4.28 K and 4.5 K. The cold vacuum compressor is designed for a suction pressure of 0.92 atm. The cold circulating compressor is designed to circulate up to 150 g/s of helium through each RHIC ring. The vacuum compressor and the two circulating compressors were purchased in 1993. Separate cold boxes were constructed for each cold compressor. The piping modifications for the refrigerator were completed in 1994.

*Work performed under the auspices of the U. S. Department of Energy.

## FLOW SCHEMATIC

The flow schematic for the RHIC refrigerator is given in Figure 1. It is a Claude type cycle with five expansion stages. There is no liquid nitrogen precooling. The main refrigerator is shown in the left half portion of Figure 1. The subcooler section, the supply and return header, and typical cooling loop is given in the right half portion of Figure 1.

The new cold vacuum compressor, identified as WCO, is installed in the subcooler section of the refrigerator. The circulating compressors, identified as WCIRC, are located in the cryogenic distribution valve boxes located near the cryogenic feed to the collider rings. The refrigerator is connected to the RHIC ring through a 350 meter vacuum jacketed transfer line.[5] One of the goals in the refrigerator modification was to minimize the piping alterations by using existing components of the refrigerator. The main compressor station, heat exchangers and turbines were not altered.

As shown in Figure 1, there are seven turbines WEX-1 through WEX-5. Both WEX-1 and WEX-2 consist of two turbines in series. Six redundant turbines, arranged in parallel to WEX-1 through WEX-4, are not shown. As shown in Figure 1, there are fifteen heat exchangers, HX-1 through HX-15. Two redundant heat exchangers for HX-3.2 and HX-7.2 are not shown. HX-1 and HX-2 are constructed together in three parallel cores. The refrigerator and the cryogenic distribution system are operated by a process control computer.[6] The oil skids for the turbines are operated through local control panels.

## COLD VACUUM COMPRESSOR

The baseline design for the cold vacuum compressor is given in Table 1. Lower suction pressures can be obtained by increasing compressor speed. The lowest design pressure and temperature are 0.65 atm and 3.8 K at 13,000 rpm. The unit is a centrifugal compressor utilizing ball bearings with low vapor pressure grease as a lubricant. The mean time between maintenance for these bearings is estimated at 8,000 hours by its manufacturer. A summary of the design features is given in Table 2. The cold vacuum compressor is mounted vertically on its own cold box. The unit can be operated from a local control box or by the process control computer.

**Table 1.** Nominal design condition for RHIC cold vacuum compressor

| | |
|---|---|
| Adiabatic Efficiency - % | 60 |
| Flow Rate - g/sec | 367 |
| Input Work - watts | 1747 |
| Inlet Conditions | |
| Pressure - atm | 0.92 |
| Temperature - K | 4.20 |
| Density - g/cc | 0.0149 |
| Volume Flow - L/s | 24.59 |
| Outlet Conditions | |
| Pressure - atm | 1.40 |
| Temperature - K | 5.22 |
| Density - g/cc | 0.0173 |
| Pressure Ratio - | 1.52 |
| Design Speed - rpm | 9,400 |

**Table 2.** Summary of RHIC cold vacuum compressor

| | |
|---|---|
| Manufacturer: | Barber-Nichols Inc. |
| Type: | Full emission centrifugal |
| Motor: | Hermetically sealed, 208V, 3- phase, 2 pole |
| Control: | Commercial variable frequency drive |
| Speed: | 9,400 - 13,000 rpm |
| Bearings: | Low vapor pressure grease ball bearing |
| Cooling: | Close loop water or glycol |
| Special Feature: | Rotating assembly can be removed from top of housing |

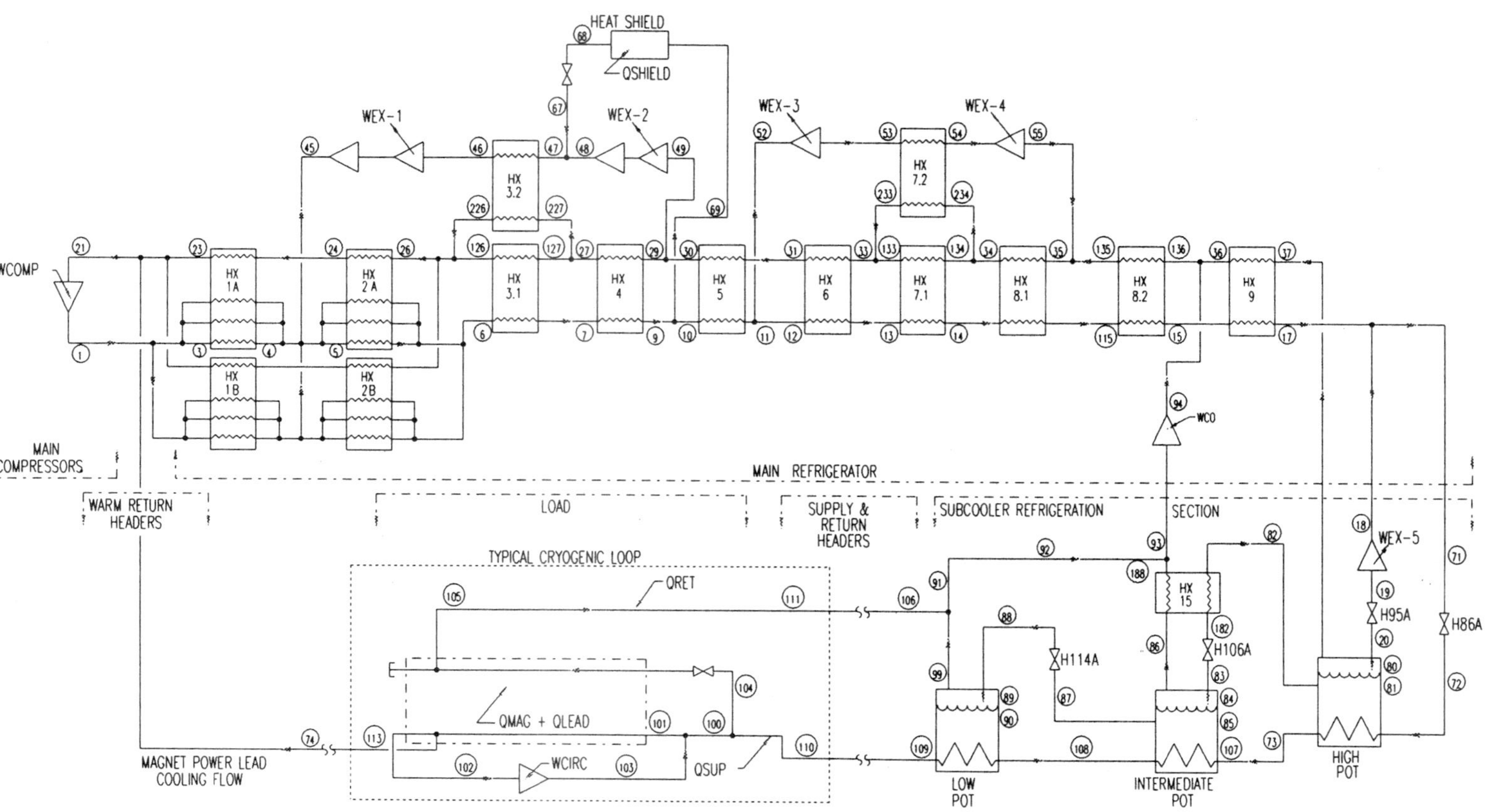


**Figure 1 Flow schematic for the RHIC refrigerator.**

## CIRCULATING COMPRESSOR

The baseline operating condition for the RHIC circulating compressor is 100 g/s of helium at 4.5 K and 5 atm for a pressure rise of 0.42 atm. The compressor is designed to handle a greater pressure rise of 0.84 atm at 100 g/s, and a high flow condition of 150 g/s and 1.76 atm pressure rise. The unit is a centrifugal compressor with a 7.5 centimeter wheel overhung on a 30 centimeter long shaft. There are two sets of ball bearing using low vapor pressure grease as the lubricant for the motor. One set of dry ball bearing is located in the cold end behind the impeller. The unit is mounted in its own cold box attached to the valve box and can be operated by a computer or through a local control panel. The circulating compressor is engineered and manufactured by Barber-Nichols, Denver, Colorado.

## RESTORATION AND IMPROVEMENTS OF THE REFRIGERATOR

The helium refrigerator was idle for over ten years since its acceptance test in 1984. Substantial efforts were spent in restoring and improving the condition of the refrigerator. The main compressors and motors have been realigned. All turbines have been disassembled for inspection and reassembled to manufacturer's specifications. The oil skids, the vacuum skid, and the gas regeneration system all have been inspected and restored. All leaky valves have been identified and the seats replaced. Every pressure and differential pressure transducer has been re-calibrated. The ranges for several differential pressure transducers in the cold end of the refrigerator have been adjusted for the new flow rates. A few bad temperature sensors have been replaced. New instrumentation has been added at key locations. All control and interlocks have been updated.

## OPERATING EXPERIENCE

In February, 1996, the RHIC Refrigerator was successfully cooled to liquid helium temperature using electric heaters as the load, and a cryogenic vaporizer to warm the gas for recovery. In July, 1996, the refrigerator was used to cool the 350 meter vacuum jacketed transfer line and two cryogenic distribution valve boxes, including the circulating compressors.[4] In February, 1997, the refrigerator was used to cool the first RHIC sextant, and the circulating compressor was used to circulate cold helium through the 600 meter sextant.[5] The system, including the turbines and the cold compressors, operated reliably. At the end of the first sextant test, the refrigerator was fully loaded to obtain the maximum refrigeration capacity without the liquefaction and the 55 K loads. With 25 kW of electric heat input, the temperature was about 5.5 K and the low pressure at the cold end of the refrigerator was 2.2 atm. The cold vacuum compressor was not used in the full refrigeration test.

Several operational procedures have also been exercised to prove the versatility of the refrigerator. These include the use of redundant turbines and heat exchangers, the warm-up of "contaminated" heat exchangers or turbines, the use of redundant absorber beds and helium gas regeneration skid, and the use of a back-up process control computer.

## PERFORMANCE OF THE TURBINES

There are thirteen turbines installed in the RHIC refrigerator. For normal operation, the RHIC refrigerator requires seven turbines. The six others are redundant. The capacity and the reliability of the refrigeration plant depend on these turbines. The turbines use oil bearing and each has a set of guide vanes for capacity control. For the dual purposes of process control and evaluation, these turbines are well instrumented with tachometers, and pressure and temperature sensors at their inlets and outlets. Venturi flow meters are used for flow measurements. The speeds of the turbines are controlled mainly by the guide vanes. The operation of the oil skid requires good training. The performance evaluation for these turbines is straight forward because these turbines reach "quasi" steady state quickly. The results for these turbines are consistent with the exception of turbine WEX 5, due primarily to its operation near 6 K where calculated results are very sensitive to measurement error.

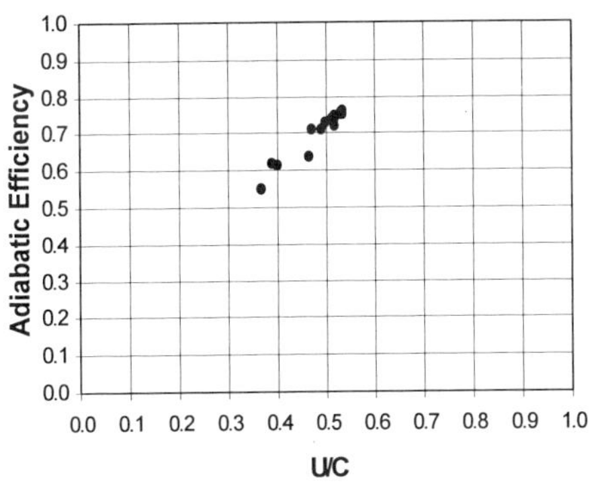


**Figure 2** Adiabatic efficiency of Turbine 1B2.

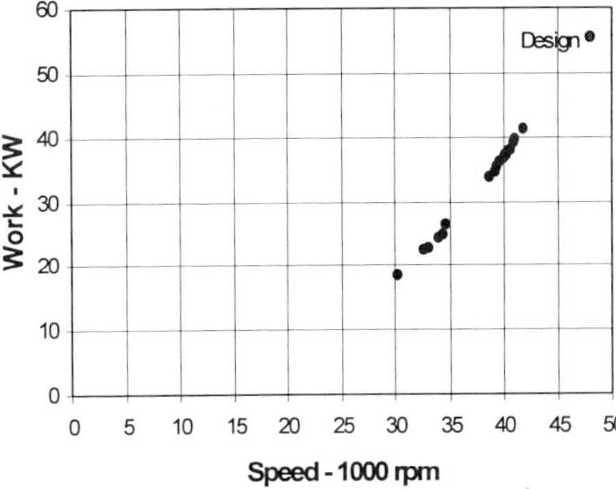


**Figure 3** Work extraction of Turbine 1B2.

The adiabatic efficiency for turbine WEX 1B2 is given in Figure 2 in terms of the parameter U/C, where U is the impeller tip speed and C is the speed evaluated from ideal expansion. The efficiency increases with U/C and approaches a peak value of about 80% at U/C near 0.6. The work extraction as a function of speed is given in Figure 3. The work extraction increases roughly to the third power of speed. All turbines have the same performance characteristic. The results for WEX 5 are not reliable and are not reported here. In fact, WEX 5 was originally sized for ISABELLE application and is slightly larger than needed for RHIC. Minor trouble in the labyrinth seal of WEX 5 occurred during the test and the turbine was not used in the runs of July, '96, and February, '97.

## PERFORMANCE OF THE HEAT EXCHANGERS

The performance evaluation of heat exchangers is difficult for the following reasons: 1) it is difficult to produce a true steady condition, 2) the terminal temperature differences are often small, 3) not all flow rates are measured directly, 4) the terminal temperature in a few exchanger is not available, 5) HX-8.1, HX-8.2 and HX-9 are packaged in one core, and 6) HX-1 and HX-2 consist of three parallel cores.

The effectiveness, AU and NTU, for HX-3 through HX-9 have been investigated. The effectiveness is defined as the ratio of the actual heat transfer rate to a thermodynamically limited maximum possible heat transfer rate. AU is the product of the heat transfer area and the overall heat transfer conductance. NTU, the number of transfer units, is defined as the ratio of AU to the minimum product of the specific heat and flow rate. The results vary greatly with the operating conditions in temperature, and with the flow rate ratio between the high pressure and low pressure streams. The uncertainties associated with the performance of the heat exchangers are large. Typical NTUs of these heat exchangers are close to the design requirement. Effectiveness are mostly greater than 0.93.

## PERFORMANCE OF COLD VACUUM COMPRESSOR

We have more than 1000 hours of operation on the cold vacuum compressor. The compressor unit is very robust and reliable with either steady or unsteady flow. The test condition covers the design flow range between 350 and 600 g/s. Suction pressures vary from 0.92 to 0.70 atm. The maximum operating speed in the test is 12,500 rpm. Safety interlocks were implemented to protect the compressor. When the compressor speed drops 20% below the set point due to an unfavorable operating condition, the computer closes the inlet valve and shuts down the compressor. The results are presented in the dimensionless form with the head and flow coefficients defined as

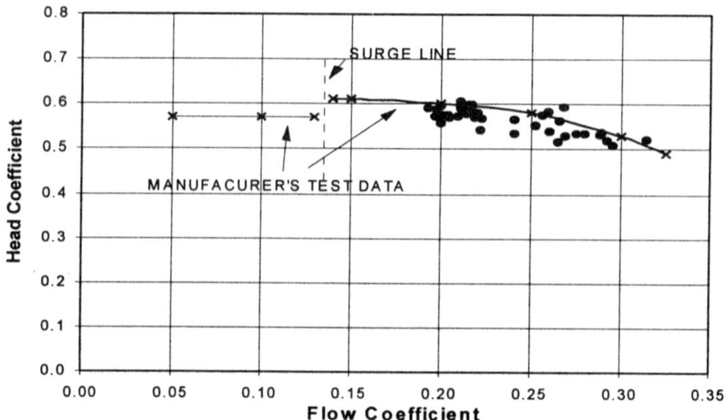


**Figure 4 Head and flow curve for cold vacuum compressor.**

$$\text{Flow Coefficient:} \quad \phi = \frac{Cm_2}{U_2}$$

$$\text{Head Coefficient:} \quad \psi = \frac{g \cdot H}{U_2^2}$$

$$\text{Adiabatic Efficiency:} \quad \eta = \frac{h_{o,i} - h_{in}}{h_{out} - h_{in}}$$

where
- $U_2$: Impeller tip speed
- $Cm_2$: Exit meridional velocity
- H: Head equals to pressure rise divided by density
- g: Acceleration due to gravity
- h: Enthalpy

subscripts
- in: at compressor inlet
- out: at compressor exit
- o,i: compressor exit for an isentropic compression

The head and flow coefficients, together with the results of an air test performed at the manufacturer, are given in Figure 4. The adiabatic efficiency is given in Figure 5. The efficiency is about 60 % with fluctuations between 40 and 80 % due to unstable conditions.

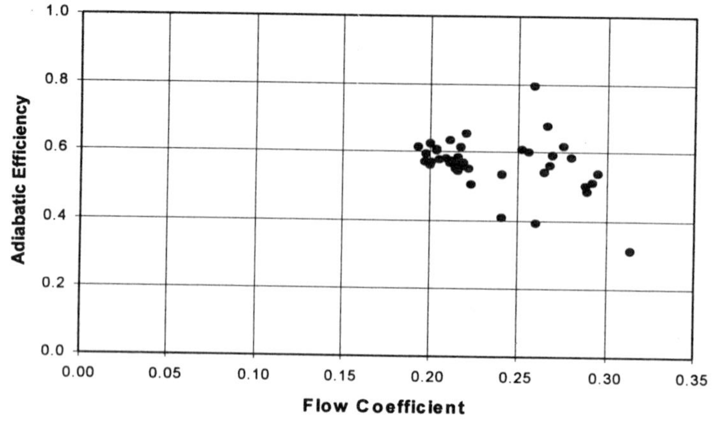


**Figure 5 Adiabatic efficiency of cold vacuum compressor.**

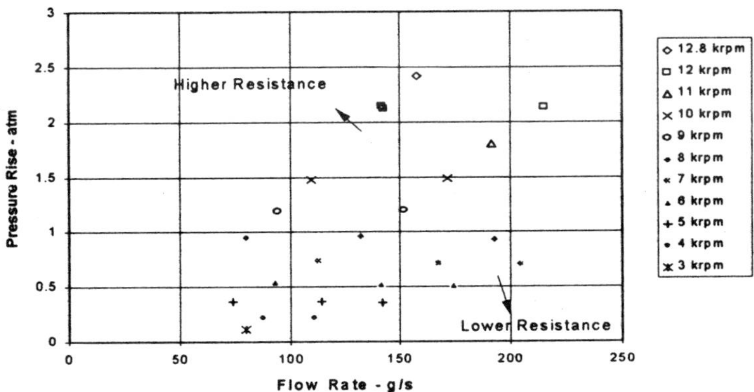


**Figure 6 Pressure Rise versus Flow at Various Speed and Resistance.**

## PERFORMANCE OF CIRCULATING COMPRESSOR

The circulating compressors were used in July, '96, and again in February, '97. In July, '96, a test loop inside the valve box was setup with the circulating compressor running on a bypass line. Several loop resistances were created using different amounts of shimming in the control valve. The compressor was operated over a speed range from 3,000 to 11,000 rpm. In February, '97, the compressor was used to circulate helium through the RHIC sextant. Total length of the cooling loop is more than 1200 meters.

Similar results were obtained between the July, '96 and the February, '97, tests. The flow rate increases linearly with speed at a given loop resistance. The pressure rise is proportional to the square of the pump speeds and is insensitive to the loop resistance. The pressure rise as a function of flow rate at different speed and loop resistance is given in Figure 6. The results are presented in a dimensionless form with head and flow coefficients given below. The pump curve, together with that obtained with methanol, is given in Figure 7. The agreement between the present results with the methanol test is very good. The head coefficient is rather flat for flow coefficients from 0.2 to 1.0. The efficiencies of the compressor were not available because the temperature sensors installed at the suction and discharge of the compressor were inaccurate.

Flow Coefficient: $\phi = \dfrac{V_{th}}{U}$

Head Coefficient: $\psi = \dfrac{g \cdot H}{U^2}$

where
- $U$: Impeller tip speed
- $V_{th}$: Pump diffuser throat velocity
- $g$: Acceleration due to gravity
- $H$: Head equals to pressure rise divided by density

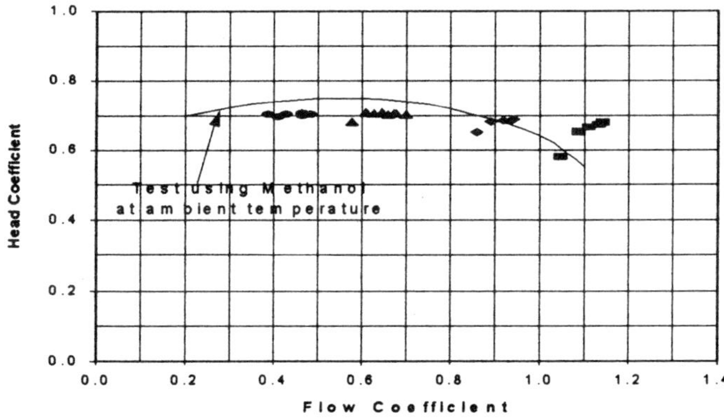


**Figure 7 Head Coefficient versus Flow Coefficient for the circulating compressor.**

## REFRIGERATOR PERFORMANCE

The refrigerator has two electric calorimeters, one for measuring the 4 K, and the other for the 55 K heat loads. The 4 K calorimeter was originally designed for heating supercritical helium in a closed circulation loop. Liquid helium can be extracted from the cold end of the refrigerator for a liquefaction load. In the refrigerator's modification process, we sacrificed the easy operation of the 4 K calorimeter for minimizing piping alterations. In the present configuration, helium flows through the calorimeter and becomes the vapor return to feed the suction of the cold vacuum compressor. While it has no effect on the operation of RHIC, flow through the 4 K calorimeter presented difficulties in flow control in the tests. The 4 K calorimeter was needed in the runs of July, '96, and February, '97, because the test heat loads were very small compared with the capacity of the refrigerator.

In February, 1996, electric heat input and liquid extraction were used as the load for the refrigerator. The purpose of the test was to demonstrate that the refrigerator was properly restored and modified, and has the capacity to cool the RHIC collider. It is not intended to push for the maximum output of the refrigerator or to optimize its efficiency. A heat load of 10 kW at 4 K, 53 kW at 55 K, and a liquefaction load of 44 g/s of liquid were obtained using approximately two-thirds of the 4,400 g/s installed compressors capacity. In this test, the lowest temperature achieved was 4.4 K. This is about 0.2 K higher than the design value because the pressure drop in the lines at the suction and discharge of the cold vacuum compressor was excessive. Both the suction and the discharge lines were enlarged for the subsequent test. In February, '97, the lowest temperature obtained was 3.8 K with a pressure of 0.71 atm.

## SUMMARY

The modification and restoration of the RHIC helium refrigerator was successful. The capacity of the refrigerator has been shown to meet the RHIC heat load requirement and the reliability has also been demonstrated. The cold vacuum compressor can easily meet the 4.2 K suction temperature with the possibility of reaching temperatures below 4.0 K. The circulating compressor is able to provide the necessary cooling flow for RHIC. The refrigerator is ready for RHIC operation in 1999.

## ACKNOWLEDGMENTS

The authors would like to thank the help provided from all personnel in the RHIC Cryogenic Section especially R. Picinich, W. Kollmer, F. Cullen, W. DeJong, W. DeVito, A. Gwilt, M. Sardzinski and A. Warkentein. The authors also like to thank S. Agnetti for the preparation of this paper.

## REFERENCES

1. D. P. Brown, M. Afrashteh, J. A. Bamberger, A. Fresco, A. P. Schlafke, W. J. Schneider, A. Werner and K. C. Wu, in "Advances in Cryogenic Engineering", Vol. 27, Plenum Press, N.Y. (1982), p. 509.
2. D. P. Brown, A. P. Schlafke, K. C. Wu and R. W. Moore, Cycle design for the ISABELLE helium refrigerator, in "Advances in Cryogenic Engineering", Vol. 27, Plenum Press, N.Y., (1982), p. 501.
3. A. Nicoletti and K. C. Wu, in "Advances in Cryogenic Engineering", Vol. 39, Plenum Press, N.Y., (1994), p. 1593.
4. M. Iarocci, etc., "RHIC 25 kW Refrigerator and Distribution System, Construction, Testing and Initial Operating Experience", presented in the 1997 CEC conference.
5. J. Sondericker, etc., "Performance and Cryogenic Operating Experience of the RHIC First Sextant Test", presented in the 1997 CEC conference.
6. J. Benson, etc., "Performance of the RHIC Cryogenic Control System During the First Sextant Test", presented in the 1997 CEC conference.

# TS DIAGRAM FOR ON-LINE MONITORING OF THE RHIC REFRIGERATOR

M.Rehak, G.Cornish, M.Iarocci, J.Sondericker, T.Wild

Brookhaven National Laboratory

Upton, New York, NY 11973

## ABSTRACT

The Brookhaven National Laboratory RHIC (Relativistic Heavy Ion Collider) 25 kW helium refrigerator[1] is now operating for the first time on the occasion of the first sextant test. A computer program, written in Visual Basic and running on a PC, has been developed as an on-line diagnosis tool. This piece of software overlays a theoretical and a live display of the refrigerator cycle on the Ts plane. The diagram is a graphical representation of the deviation from the theoretical state. This provides a visual feedback to refrigerator operators who can then make necessary adjustments. Any additional values of interest, such as work, efficiencies etc..,are also available on the diagram or as a table summarizing performance.

## INTRODUCTION

The RHIC refrigerator is presently operated from a control room where VAX workstations are running the CRISP operating system. Graphical static displays of P&ID's (process and instrumentation drawings), which show instrument readings such as temperatures, mass flows, pressures, are available on this system as well as process histories. However, it was found that performing calculations and having a live display of the refrigerator cycle is best executed on a PC using the Visual Basic language. Amongst numerous capabilities, the Visual Basic language allows file manipulation, issues commands to run Fortran executable code, transfers files, and has extensive graphics capabilities. The fourth author conceived the idea of generating a Ts diagram from instrument readings and overlaying it on a theoretical Ts diagram. Visual comparison would then allow operators to make proper corrections in order to reach the theoretical cycle. The availability of Visual Basic language made such a proposition a straight forward programming task. An existing code, developed by the second and last authors for monitoring magnet behavior during quenches, was adapted to the present task. At the inception of the project there were two unknowns: 1) would the theoretical and actual diagram be sufficiently distinguishable to provide feedback? 2) how useful would it be as a tool during operation?

## PROGRAMMING DETAILS

The Visual Basic code can run on any 486 PC equipped with FTP (File Transfer Protocol) capabilities and Fortran (Microsoft Fortran). The software has a graphical user interface with pull down menus offering choices such as frequency of readings, display of theoretical or actual diagrams and table of performance. The following is a description of the steps executed by the Visual Basic program.

### Raw data access

The raw data are stored as an ASCII file of about 2000 lines on the VAX in the control room. Each line contains one value corresponding to an instrument reading associated with a tag name. Tag names are names given to sensors on the refrigerator, magnets, valve boxes and P&ID's. A batch file containing VAX-VMS commands is sent from the VAX to the PC. This file instructs the VAX to transfer the raw data file to the PC. The first step of the Visual Basic code is to access the raw data in this manner.

### Performance calculations

The next step consists of executing a Fortran code, which generates the theoretical and actual values of process points. The position of valves indicates which paths of turbines and heat exchangers are in operation and consequently, which set of sensor readings are appropriate. Some of the raw data are used in a preliminary step to compute heat loads and other inputs necessary for calculations at the theoretical process points. These are: mass flows in leads and magnets, heat loads and pressure drops for various sections. All other values needed for those calculations such are efficiencies are assumed and fixed Temperature and entropy are calculated next for the actual cycle. These two arrays of values for the process points are then written to an output file. In addition, this Fortran code performs all calculations of interest since it has access to all input data from the refrigerator, valve boxes and tunnel piping. Examples of such calculations are heat loads for various sections of service lines, heat loads for magnets, across recoolers and heat shield. Heat exchanger effectiveness, turbine and compressor work and efficiencies, pressure drops, Carnot efficiency and entropy generation are all available in a pull down table from the main menu. Helium properties are obtained using the commercially available package HEPAK.

### Graphics display

The Visual Basic program reads the two data arrays and plots them. Zooming capabilities allow for close ups in areas where needed, which is mostly under the vapor dome.

### Data storing.

Raw data are saved for further processing, such as generating temperature time histories along magnets in the tunnel during cooldown and warm up. In order to avoid accidental termination of data acquisition, it was found to be more practical to separate the live Ts display from the data acquisition.

## USAGE DURING OPERATION

During cooldown, the table showing recooler levels, magnet temperatures and pressure drops was most useful, since most of the process points are all lumped in the high temperature region. After cooldown, the actual Ts diagram and theoretical Ts diagram are sufficiently close for visual comparison but distinct enough that zooming is needed mostly in the region under the vapor dome. There is no need to exaggerate deviations from the theoretical diagram with a multiplying factor. One must keep in mind that the logarithmic scale (without which process points would have been all lumped at low temperatures) exaggerates differences at the lower

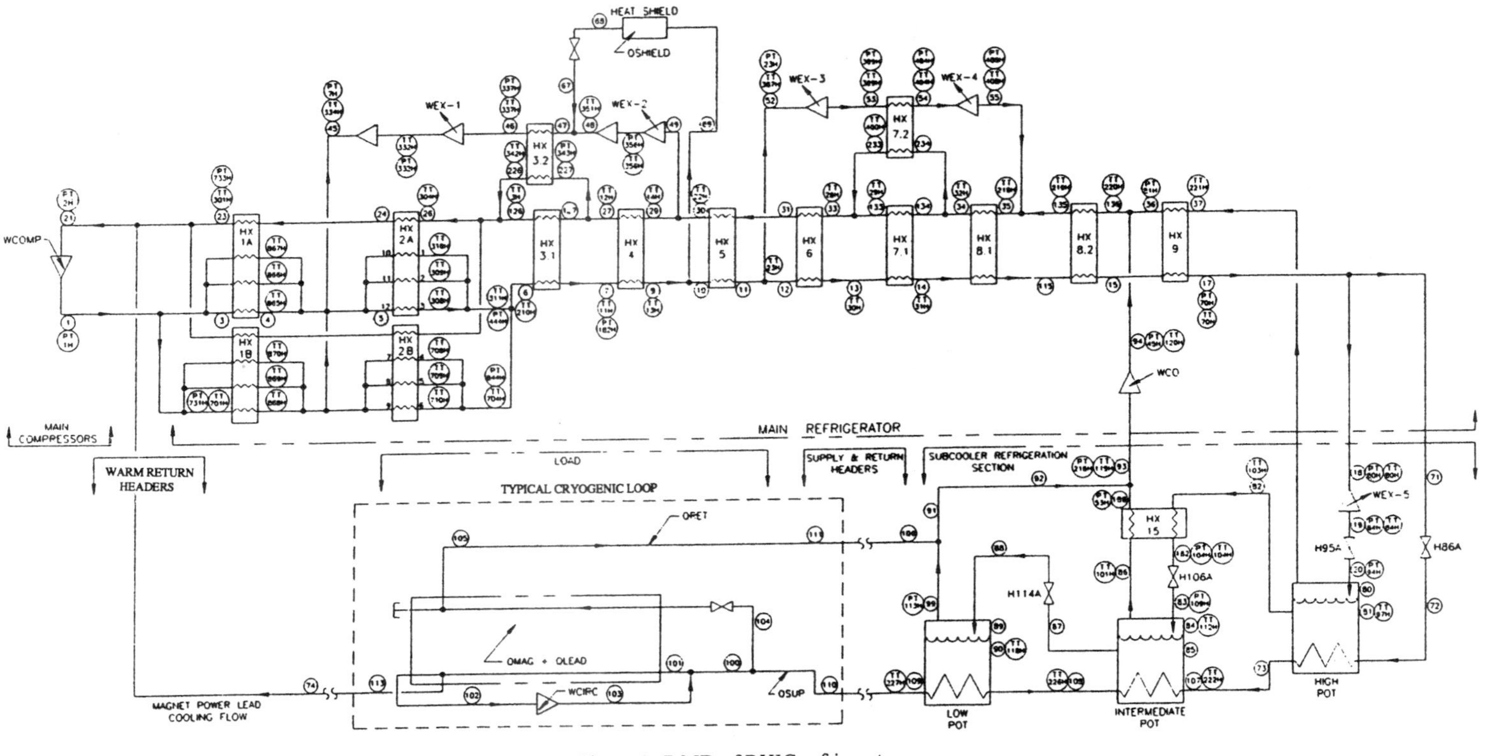


Figure 1. P&ID of RHIC refrigerator.

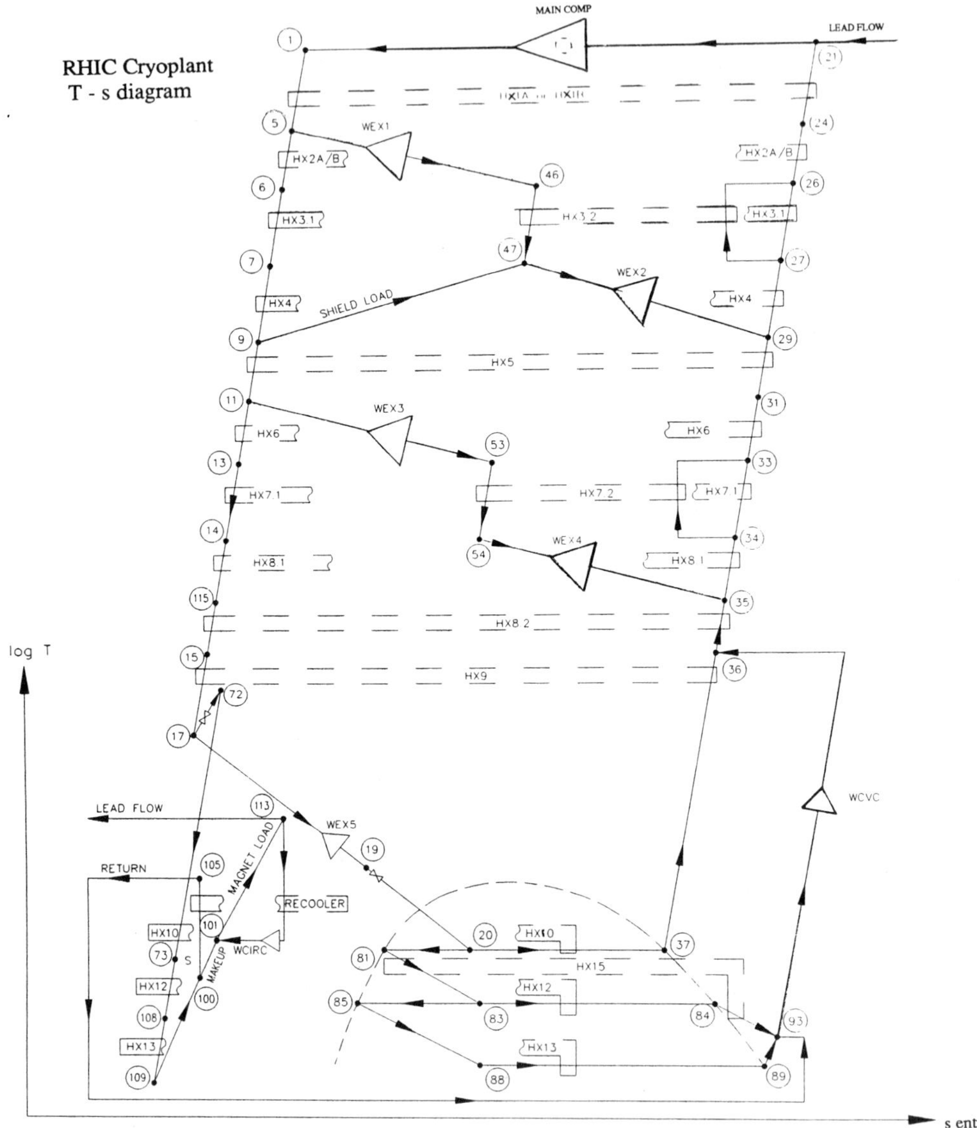


Figure 2. Schematic Ts diagram of the RHIC refrigerator cryoplant.

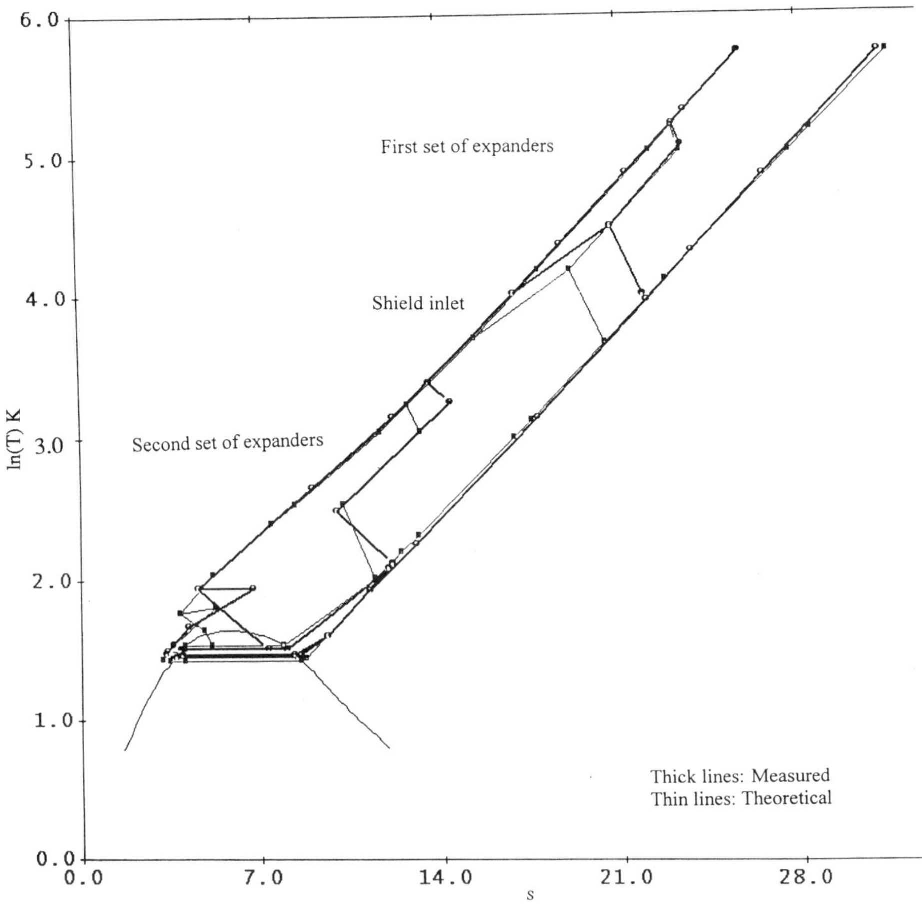


Figure 3. Live Ts diagram showing theoretical and actual process lines.

end and diminishes them at the higher end. For reference purposes, a P&ID and a schematic Ts diagram are shown in Figure 1 and Figure 2. The magnet cooling loop in the lower half corner of Figure 2 is a schematic and warrants a separate Ts diagram. It is not included in the live Ts diagram shown in Figure 3.

The thick lines in Figure 3 represent the measured refrigeration cycle, and the thin lines are the theoretical or objective to be reached. Colors for different sections of curves ease the reading. Numerical values (theoretical in black and actual in red) for efficiencies, work and effectiveness can be displayed at the physical location of the expanders or compressors as an option. Mass flow in all the branches, as well as pressure, temperature, enthalpy and entropy, are also available as options. Process points appear as dots (theoretical) or circles (actual) and correspond to the process points on the schematic in Figure 2.

As described above, the live Ts diagram is a tool that allows the trained thermodynamicist to analyze the refrigerator performance in real time and provide historical data, if desired. Also of great importance is the "visual feel" the graphical display gives the technicians assigned to operate the refrigerator. Even without an understanding of thermodynamic terminology, the technician is able to seek out information i.e., turbine efficiency, heat exchanger effectiveness, and most important curve shape comparisons to evaluate the performance of a steady state refrigerator operation. Some examples of the Ts diagram diagnostic capabilities follow.

## Expanders

Figure 3 indicates that the first and second set of expanders are running too slowly since temperatures higher than desired. Consequently, the input temperature to the heat shield is higher than anticipated. Efficiencies of turbines are graphically represented by the slope of the line between process points. For an isentropic process the line should be vertical. Hence, the steeper the line, the higher the efficiency. Therefore the set points should be changed to adjust turbine speeds.

It is apparent in the enlarged section shown in Figure 4 that expander WEX-5 (see Figure 2, between process points 17 and 19) has failed. Helium enters the vapor dome at a higher than desired pressure, and the quality of the first liquid pot is lower than desired, as can be seen from the fact that process point 20 is shifted to the right.

## Heat exchangers

The diagonal lines on Figure 3 running from the lower left corner to the upper right corner, which are lines of constant pressure, fall accurately on their theoretical location. The heat exchanger inlet temperature on the high pressure line should be as close as possible to the outlet temperature on the low pressure return. Thus these points should fall on an horizontal line for highly effective heat exchanger.

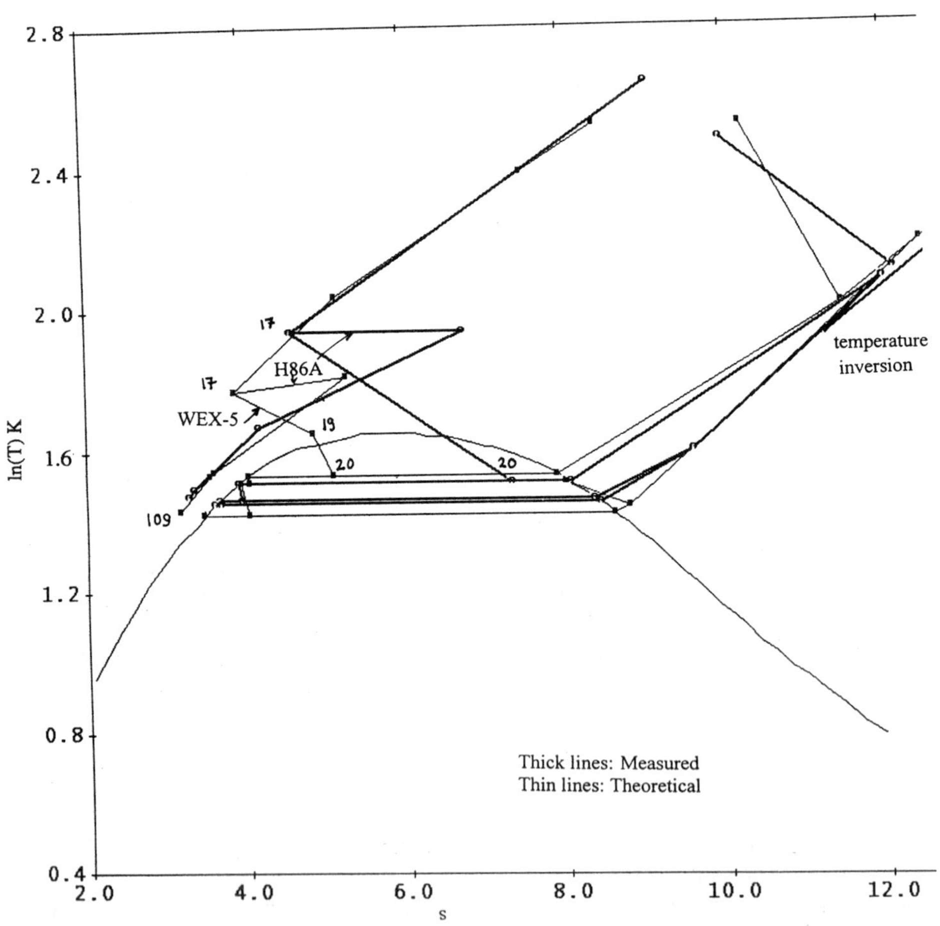


Figure 4. Close up of Figure 3 under the vapor dome, theoretical and actual process lines.

The jagged line on the left part of Figure 4 is the consequence that the actual piping is slightly different than indicated in Figures 1 and 2. The cold vacuum compressor outlet (with a lower temperature) joins the return line between heat exchangers 8.1 and 8.2, while the WEX-4 outlet ( with a higher temperature) falls between heat exchangers 8.2 and 9.

Modifications to the refrigerator piping will be made so as to reflect layouts in Figures 1 and 2, and have monotonous temperature distribution along the return path.

### Expansion valve

The expansion valve H86A, between process points 17 and 72 on Figure 2, is used to throttle the pressure from the incoming helium line at 15 atm to the 5 atm needed in the magnet supply loop. Theoretically, this isenthalpic expansion occurs to the left of the inversion curve on a Ts diagram, which leads to a temperature rise in the flow. Due to the lack of temperature sensor after the valve, the temperature was kept constant for display purposes. This and a few other missing sensors will be installed for the next test.

### Refrigerator cold end

Theoretically, there are three lines under the vapor dome, corresponding to the three liquid pots identified as low pot, intermediate pot and high pot on Figure 1. These lines are quite distinct in Figure 4 which is a close up of the region under the vapor dome of Figure 3. The temperature level of the third pot is higher than the theoretical value. Consequently, the Helium temperature in the high pressure make up loop is 4.4 K instead of 4.2 K as can be seen at the left most process point on Figure 4, corresponding to point 109 on Figure2.

## CONCLUSIONS

The concept of a live Ts diagram has proven to be a feasible idea. Its ease of use and the instant visual information it conveys have made it a practical diagnosis tool. Due to the portability of the software, complete information on the status of the refrigerator is conveniently available from any PC.

## ACKNOWLEDGMENT

Work performed under the Contract N0.De-AC02-76CH00016 with the U.S. Department of Energy.

## REFERENCES

1. D.Brown, A. Schlafke, K. Wu, "Cycle Design for the ISABELLE Helium Refrigerator", Advances in Cryogenic Engineering, Vol.27, Plenum Press, New York (1982).

# RHIC 25 kW REFRIGERATOR AND DISTRIBUTION SYSTEM, CONSTRUCTION, TESTING, AND INITIAL OPERATING EXPERIENCE *

M. Iarocci, J. Sondericker, K.C. Wu, J. Benson, Y. Farah, C. Lac, A. Morgillo, A. Nicoletti, E. Quimby, J. Rank, M. Rehak, and A. Werner

Cryogenic Systems Group
Relativistic Heavy Ion Collider
Brookhaven National Laboratory
Upton, New York 11973

## ABSTRACT

The installation and testing of most of the Relativistic Heavy Ion Collider (RHIC) cryogenic system is complete. The RHIC cryogenic system consisting of the modified ISABELLE refrigerator, and other new cryogenic components, has been completed from the 25 kW refrigerator through one sixth of the 3.8 km circumference RHIC accelerator up to the 4:00 valve box. This paper covers the progress to date with respect to construction and installation of the helium distribution system, including the vacuum jacketed piping and valve boxes, recoolers, and controls. In addition, the 25 kW refrigerator has been operated on three separate occasions over the past two years. These first experiences with the operation of the refrigerator and warm compressor system, along with some test results will be discussed here.

## INTRODUCTION

The cooling scheme used for RHIC, shown in fig. 1, uses a single Barber Nichols centrifugal pump per ring to circulate supercritical helium through the magnets at a nominal flow rate of 100 g/s. The temperature of the circulating gas rises as heat is absorbed from the magnets. This heat is then removed from the circulating helium as it passes through heat exchangers located periodically around the ring. These heat exchangers, called recoolers, are fed from the 25 kW refrigerator and represent the distributed cooling system for the RHIC magnets. More RHIC cryogenic system design parameters are shown in Table 1.

* Work performed under contract with the US Department of Energy

The cryogenic components required to adapt the existing 25 kW refrigerator for use with RHIC have been in the design and construction phase since 1991. The delivery of the first CVI Valve Box, in early 1995, marked the installation phase of the first large cryogenic device for this system. Followed in late 1995 by the receipt and installation of all Vacuum Jacketed Refrigerator to Ring (VJRR) piping and some Vacuum Jacketed Ring (VJR) piping, the system was ready for the helium distribution test with intent to verify all components up to the cryogenic entry point to the RHIC accelerator tunnel at 6:00. All major design efforts have been completed for approximately one year, and the current focus of effort is on the installation of the Ring Resident Control System [1], the balance of Valve Boxes, VJR supports and piping, minor refrigerator upgrades, and miscellaneous process piping and control hardware. During the first two years of construction and installation, tests of major cryogenic subsystems were conducted. These tests culminated with the most extensive test of the first sextant.[2]

## CRYOGENIC SYSTEM VERIFICATION

Over the years since acceptance tests [13] of the 25 kW refrigerator, [12] and associated equipment left over from the ISABELLE project, and the beginning of the RHIC project, the system was idle and mostly unattended. The warm Howden compressors, intercooler and aftercooler heat exchangers were left drained of cooling water and charged with helium in a heated, non-air conditioned environment with occasional inspections and maintenance such as, motor heater operation and compressor rolling. Compressor rolling was performed at six month intervals by rotating the compressors by hand with the oil pumps running. All other major compressor room components, such as oil pumps, backup process controls, cryogenic purifier, and instrument air compressors, were subjected to the same environment during the idle years. The exception is the I/O that had controlled temperature and humidity. The oil removal system, consisting of staged Monsanto coalescing filters and adsorbers and the compressor oil storage, makeup, and recovery system are found outside with full weather exposure. These items were not externally maintained but were kept pressurized with helium.

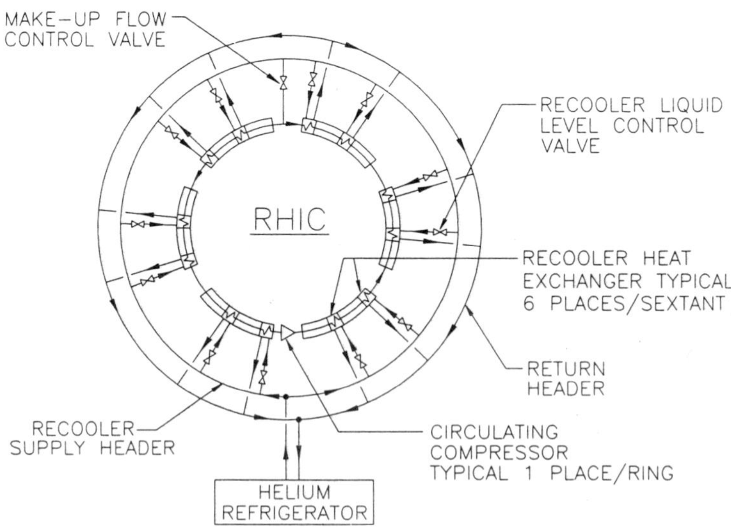


**Figure 1.** RHIC cooling scheme.

**Table 1.** Cryogenic system parameters

| Refrigerator Parameters | |
|---|---|
| Maximum Refrigeration @ 4 K, ~1200 g/s | 25 kW |
| Maximum Liquefaction (all expanders, no refrig.) | 270 g/s(8100 l/hr) |
| Heat shield max capacity, delivery @ 50 K, 400 g/s, 16-9 Atm. | 55kW |
| Maximum Compressor | 4200 g/s @ 17 Atm. |
| Refrigerator Operating Helium Inventory | ~3,000 kg (600 kSCF) |
| Refrigerator Return/ Low Pot Pressure | .8- .9 Atm. |
| Upper end turbines, 4 stage (with redundancy) | 180 to 40 K, 16 to 1 Atm., 750 g/s |
| Lower end turbines, 2 stage (with redundancy) | 25 to 7.5 K, 16 to 1 Atm., 1300 g/s |
| Final turbine | 5.03 to 4.98 K, 16 to 2.5 Atm. @ 800 g/s Max. |
| Refrigerator sub-cooled section is comprised of three pool boiling pots with heat exchangers. | |
| Load Parameters | |
| Cold end delivery requirement, Steady State | 4 K, 400 g/s, 5-6 Atm. |
| Liquefaction (lead flow) | ~ 50 g/s (equivalent to~4 kW) |
| Nominal heat load @ 4 K | ~ 8 kW |
| Combined refrigeration requirement, Steady State ( Q+ Liq.) | ~ 12 kW |
| Magnet Loop Pressure | 5 Atm. |
| Circulating compressor, magnet loop | 100-150 g/s |
| Cold return pressure | 1 Atm. |
| Process design/relief pressure | 18 Atm. |
| Heat Shield requirement, Steady State @50 K, ~300 g/s, 16 Atm. | ~38 kW |
| Total Helium Inventory @ full operating condition | ~ 35,500 kg (~7 MSCF) |
| Quantity of 50 Watt Recoolers per ring | 30 |
| Quantity of 225 Watt Recoolers per ring | 6 |
| Ring Circumference | ~3.8 km |
| Magnets quantities, arc and insertion ( two rings) | |
| Dipoles | 396 |
| Quads | 492 |
| Trim Quads | 72 |
| Sextapoles | 492 |
| Correctors | 492 |
| Magnet iron mass per sextant | 180,000 kg |
| Cooldown time per sextant, 300 to 4 K @100 g/s | 24 hr |
| Warm-up time per sextant, 4 to 300 K | 24 hr |

The refrigerator cold boxes extend partially into a building with the same environmental conditions as the compressor room. Most of each of the cold boxes and most of the process control valves resides outside and is subjected to all weather conditions. Besides the leading edge of each cold box, all turbo expanders, expander oil and gas control, cold adsorbent beds, regeneration skids, insulating vacuum systems, and instrumentation feedthroughs reside inside. The refrigerator control I/O racks reside in a temperature and humidity controlled environment. The refrigerator insulating vacuum was maintained during the idle period and the process piping was pressurized with helium. Failures due to system idleness and neglect will be discussed later.

During the last two and one-half years the cryogenic system, under construction for the RHIC project, has been tested to some degree on four separate occasions. The first test was conducted in July 1995 and was focused primarily upon the warm helium compressor system. Preparation for this test included compressor, motor, and electrical checkout including process control systems and 4160 VAC power. This test was the first full operation of the warm helium compressor system since its acceptance tests in 1984. Test results verified maximum advertised flow, pressure, and adiabatic efficiency. Only marginal system reliability was attainable during this test and was attributed to vibration failures of temperature and oil level probes, compressor motor bearing failures due to misalignment, dirty load/unload valves causing sticking, defective pressure transducers, inadequate high voltage terminal connections, slow interstage control valve response, contact corrosion on local compressor monitors and interlocks, and clogged or stuck compressor oil control valves.

The second test intent was the cooldown of the refrigerator. As with the warm helium compressor system, the last time the refrigerator ran was during the acceptance tests. We prepared for the latest phase of tests by modifying the ISABELLE refrigerator cold end for compatibility with RHIC operation, and reviewed and/or rebuilt all of the dynamic refrigerator components. The insulating vacuum controls were designed with analogue meters and were plagued with relay contact failures, this system was repaired but is scheduled for an interface with the refrigerator control computer. The cold turbines were mishandled over the idle years and most of the cold turbine lead seals were damaged. All turbine assemblies were serviced by Rotoflow and BNL technicians were trained in turbine maintenance. Turbine control skids were throughly tested for functionality and some controls were found working inversely. Valve actuators placed outside were damaged by water and ice causing seizure and occasionally breakage. Solutions included better sealing or draining of water traps. Some valve actuators placed inside were damaged from internal corrosion due to an ill maintained air dryer. A better practice would have included eliminating the need for instrument air for the long idle period. Approximately three man years went into preparation and/or repair of the refrigerator, and before the first cooldown attempt, many existing subsystems were tested and verified. Even with this formidable repair effort and careful attention paid to problem areas of the idle refrigerator two attempts were necessary to accomplish the refrigerator cooldown. Shortcomings that limited our goals, were again attributed to equipment trips and failures in the compressor room, of the same type listed above, and when compounded by a new set of problems with the refrigerator, the decision was made to stop the cooldown. Examples of critical refrigerator problems included a defective labyrinth seal in expander five causing flow that exceeded the capacity of the seal gas system, an upper end turbine failure due to over speed, a redundant upper end turbine with its seal gas control valve inversely configured, various cold leaks through valve stems and seats, and the interruption of insulating vacuum pump function due to clogged water filters. The cooldown was finally ended before we condensed liquid helium in the subcooled section of cold box 5, with the cold end at ~ 7 K. The problems identified were cleared with the exception of the clogging water filters. The cooling water for the insulating vacuum pumps is tied directly to the large tower system used for the compressors. These large systems tend to circulate much sediment and chronically clog the small filters. A better design will isolate the compressor water from the vacuum pump cooling process by incorporating a water to water heat exchanger. We were ready for another attempt at cooling this refrigerator after two months and the second cooldown, in February '96, was successful as we accomplished much of the proposed test plan.

After the compressor test and two cooldowns, and with most obvious repairs completed, we began to uncover subsystem flaws that were previously masked by the many failures mentioned above. To catagorize we found the seal gas compressor to be undersized, the regeneration skid inoperable due to wiring and instrumentation errors, and chronic leakthrough of Posi-Seal butterfly valves due to seal breakage. Clearly, the 12-year-old system was still in its infant state.

The helium distribution test was conducted in July '96 and included the first major articles of new cryogenic equipment (see fig. 2), including the VJRR piping, the 6:00 Valve Boxes, one section of VJR piping, and some ring resident controls. The final test in this series, the sextant test, was conducted from January 20, 1997 to March 5, 1997 and included the initial cooldown, steady state operation, one complete thermal cycle, and a final controlled warmup. These last two tests are discussed in the next section.

## INITIAL OPERATION EXPERIENCES, INTEGRATED SYSTEM

### Distribution System Test

After the series of tests and subsequent reworks of the existing refrigerator a test of the helium distribution system was performed. The installation of the cryogenic distribution system between the

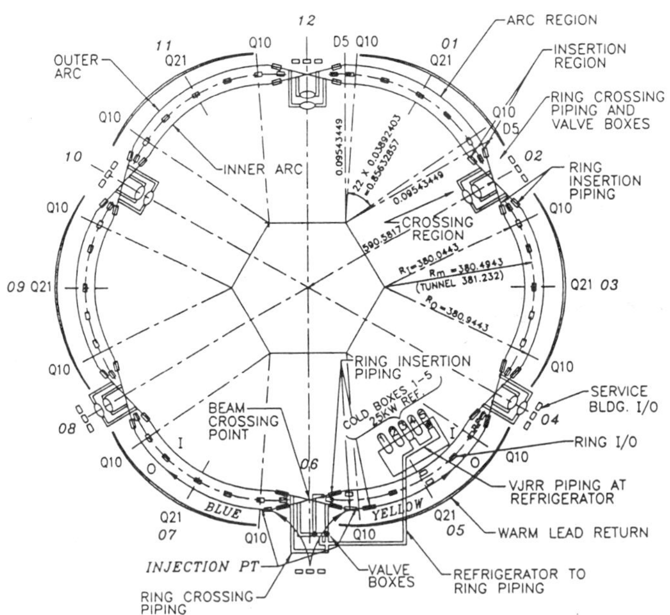


**Figure 2. Cryogenic component location**.

refrigerator and accelerator tunnel consisting of approximately 350 meters of multiple process, vacuum jacketed pipe and cryogenic valve boxes at the 6:00 accelerator location, was completed by May '96. The basic components of ring resident control system, except some final design hardware, were also in place for this test. See fig. 2 for component locations. We were ready in May '96 for this test, but delayed in the startup due to the inability for simultaneous operation of RHIC cryogenics with the Alternate Gradient Synchrotron (AGS). Simultaneous running was not possible due to electrical power limitations, so we commenced testing of the distribution system in July.

The test plan included verification of valve box process piping, measurement of VJRR and valve box static process heat loads, measurement of the 225 Watt recooler effectiveness, functionally tests of all new equipment, and categorization the cold circulating compressor. In addition, the section of CCB was powered to 7000 Amperes. During this test, a plan further to study the refrigerator cold vacuum compressor was not pursued due to a piping modification error. At the valve boxes, all instrumentation was functional, process valves and piping positions did in fact agree with design, but some valves lifted when subjected to near maximum differential pressure. The latter problem was caused by inadequate actuator spring rates and was avoided during this test with administrative procedures. A plan to replace the springs is in place.

The distribution test was considered successful and we were able to measure and test all mentioned above, nevertheless we were plagued the problem of excessive motor bearing temperature in the compressor room causing constant compressor trips. As in the past some typical problems like I/O connections, and temperature and oil level sensor breakage were also causing compressor trips, but were less frequent than before. One discovery here was the importance of exposing running equipment to all possible environmental conditions. The refrigerator and associated equipment ran very reliably during this test, causing only minor interruptions in the test program.

Throughout the cold period many tests and measurements were conducted on the newly integrated system, including the control system[3], cold vacuum compressors and cold circulating pump[4], Vacuum Jacketed Piping[5], Valve Boxes, and the existing refrigerator and refrigerator subsystems. Most of the test results are reported in a later section of this report.

## Sextant Test

After the distribution test, construction and installation of cryogenic and control components in the test sextant continued. By December '96 all magnets were in place, process connections were completed, and the final pressurization and leak testing were underway. Upon completion of leakchecking and leak repair, and after many hours of operating crew training we started cooldown of the refrigerator on January 20, 1997. A few hours after startup the refrigerator was cooling the distribution system and magnets. The system turned on this time without faults and the initial cooldown of the refrigerator, distribution system, and magnets were accomplished in approximately two and one-half days. The sextant cooldown is shown in fig. 3.

The first two weeks of testing were generally dedicated to accelerator physics because of the availability of the AGS heavy ion beam. By the end of January we achieved successful beam transport to the end of the sextant. The balance of the test focused upon cryogenics, magnet electrical, and vacuum. An extensive test plan including measurements and component functionality was established for the cryogenic system and is shown in table 2.

The simplicity of the initial cooldown and operation was outstanding when compared with all other test periods. As we continued through the cold test period, reliability was apparent, the cryogenic system maintained the magnets at RHIC parameters, with a parallel load of 6-8 kW applied at the refrigerator calorimeter, and without any failures. Any fault occurrences with the cryogenic system were incidental during the sextant test and finally allowed us to focus on system measurements.

## Summary of results from Refrigerator, Distribution, and Sextant Tests

Heat load and test results for critical cryogenic components are shown in table 3. During the later part of the test we induced a quench in a single Corrector Quadrupole Sextapole (CQS) magnet by introducing heat just upstream of the CQS, via one circuit of the magnet warmup heaters, to verify the quench detection and protection systems.

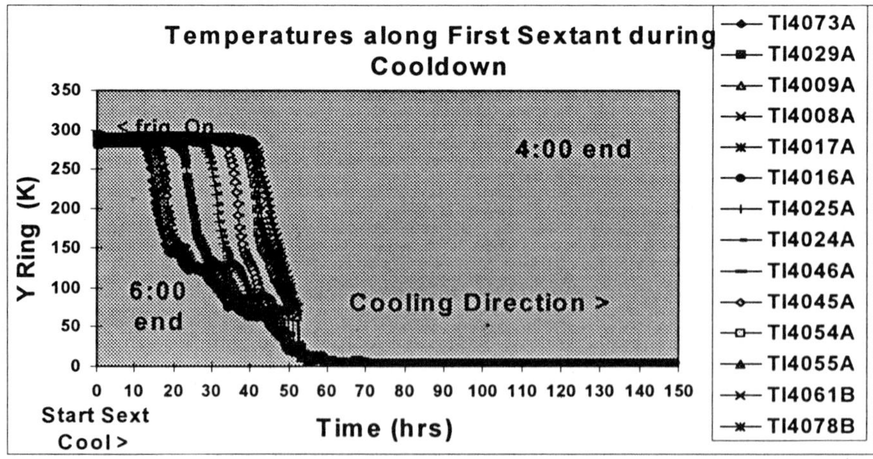


**Figure 3. Sextant cooldown.** Legend at the right shows temperature sensors located along the magnet string at the recooler inlet and outlet positions.

## Component Development

Development of associated equipment has been ongoing in house during the past few years. These major efforts include of diode calibration, [8] CQS lead flow instrumentation, [9] 50 Watt recooler testing, [7] and electrical power lead developement[11] and testing.[6] Although the testing of these sub-components has not required the operation of the large refrigerator, the research and development has required considerable effort and iteration to achieve optimized devices.

**Table 2.** Cryogenic System Sextant Test Plan

Measurements
Heat Load Measurements
Primary Load (4K)
- Recooler boil-off method
- Energy balances method with and without all recoolers operational.
- Recooler boil-off method with heat generation via distributed magnet warm-up heaters.
- Magnet loop boil-off at reduced pressure.

Secondary Load (55K)with energy balance method.
Warm Pressure Drop Measurements, Magnet Loop.
Cold Vacuum Compressor performance.
Cold Circulating Compressor performance.
Instrument air usage and redundancy configuration test.
Cooldown and warm-up rates using proposed methods.
Magnet (single) quench effect on the system.
Temperature sensor calibration and absolute accuracy.
Full refrigerator capacity.
Insulating vacuum readings for discrete vacuum spaces for VJR and VJRR.
Functionally Tests
Functional tests of existing systems:
Redundant Expander operation, Heat Exchanger and Adsorber switch over.
Regeneration skid operation.
Compressor system reliability.
Helium Management.
Cooling towers.
$N_2$ Instrument Compressor.
CRISP Control System Redundancy.
Seal Gas Recovery Compressor.
Refrigerator operational issues, full refrigeration.
Rotoflow turbine operation.
Air ventilation for the compressor room
Functional tests of New Systems:
Verified the integrity of major cryogenic devices such as VJ piping, Valve Boxes, Valves, etc.
Dynamic insulating vacuum systems used with the Valve Boxes.
Magnet warmup heater power distribution.
Cold vacuum compressor and other refrigerator cold box 5 modifications.
Cold Circulating Compressor at 6:00 Valve Box.
Ring Resident Control System and all associated communications hardware and software.
Temperature Sensor shield requirements.
Gas storage portion of the gas handling system.
Verify the helium inventory code, and live TS diagram display. [10]

**Table 3.** Cryogenic System Heat Load/ Performance Summary

| Process | Measured Range |
|---|---|
| Primary Heat Load, entire (boil-off) | 700- 900 W |
| Primary (4K)Load, entire with dist. cal. (boil-off) | 650- 780 W |
| Energy balance with all recoolers | 680 W |
| Energy balance with 50 W recoolers off line | 850- 950 W |
| Weighted Average of the above | 750- 850 W |

**Table 3 Cont'**

| Process | Measured Range | Design Value |
|---|---|---|
| Subtractions | NA | NA |
| Circulating Compressor | 50 W | NA |
| 4:00 VB and leads | 30W | NA |
| Extra VJR pipe | 30 W | NA |
| Insulating vacuum 5:00 triplet | 50 W | NA |
| Total subtractions | 160 W | NA |
| Estimated sextant heat load | 590- 690W | 550 W (Ave) |
| Projected primary load to full machine | 7.1 kW-8.3 kW | 6.6 kW |
| Secondary load one VB and yellow ring | 2.5 kW | 2.5 kW |
| Projected secondary load to full machine | 30 kW | 31 kW |
| Projected heat shield pressure drop | 4.8 atm | 6.0 atm |
| Magnet circuit pressure drop projection | < 0.5 atm | < 0.5 atm |
| Cooldown time per sextant | 24 hr | 24 hr |
| Warm-up time per sextant | 27 hr | 24 hr |
| Temperature sensor accuracy | +/- 20 mK | +/- 12 mK |

Cryogenic system single magnet quench response

| Quench Current | Tmax@recooler | Comments |
|---|---|---|
| 5357A | 6.5 K | Q21- Q6 recoolers empty |
| 5500A | 7.0 K | Q21 empty, Q14 down slightly |
| 5500A | 6.3 K | Q21 empty, Q14 down slightly |
| 5500A | 6.8 K | Q21 empty, Q14 down slightly |
| | Pressure rise < 1 atm. | |


## ACKNOWLEDGMENT

The authors thank members of the cryogenic section technical staff for their excellence in installing and operating the cryogenic system thus far. We also thank D. Brown for his planning, and guidance during the design stage of the project.


## REFERENCES


1. Y. Farah, et. al., "The RHIC Cryogenic Control System, Advances in Cryogenic Engineering," Vol. 39B, Plenum Press, New York, 1993.
2. J. Sondericker, et al., "Performance and Cryogenic Operating Experience of the First RHIC Sextant Test," to be published in these proceedings, July 1997 CEC.
3. J. Benson, et al., "Performance of the RHIC Cryogenic Control System During the First Sextant Test," to be published in these proceedings, July 1997 CEC.
4. K.C. Wu, et al., "Status, Operation, and Performance of the RHIC Refrigerator, to be published in these proceedings, July 1997 CEC.
5. E. Quimby, et al., "VJR/VJRR, Design, Construction, Installation and Performance," to be published in these proceedings, July 1997 CEC.
6. A. Nicoletti, M. Rehak. "Performance of RHIC Power Leads," to be published in these proceedings, July 1997 CEC.
7. A. Nicoletti, M. Iarocci, "Performance of RHIC 50 Watt Recoolers," to be published in these proceedings, July 1997 CEC.
8. J. Benson, et al., "Fully automated Silicon Diode Temperature Sensor Calibration System," to be published in these proceedings, July 1997 CEC.
9. A. Morgillo, et al., "Forced Convection Cooled Thermistors used as Magnet Lead Flow Sensors," to be published in these proceedings, July 1997 CEC.
10. M. Rehak, et al., "T.S. Diagram for On-Line Monitoring of the RHIC Refrigerator," to be published in these proceedings, July 1997 CEC.
11. M. Rehak, "A Simplified Theory of Current Leads," to be published in these proceedings, July 1997 CEC.
12. D. Brown, et al., Cycle Design for the Isabelle Helium Refrigerator, "Advances in Cryogenic Engineering," Vol. 27, Plenum Press, New York (1982), p 501.
13. D. Brown, et al., "Operational Tests of the BNL 24.8 kW, 3.8 K Helium Refrigerator, "Advances in Cryogenic Engineering," Vol. 31, Plenum Press, New York (1986), p 627.

# RING WARMUP AND HELIUM RECOVERY FOR THE RHIC ACCELERATOR: THERMAL ANALYSIS OF THE WARMUP OF A SUPERCONDUCTING MAGNET STRING

James Rank, Mike Iarocci, Margareta Rehak, and K.C. Wu

Relativistic Heavy Ion Collider (RHIC)
Brookhaven National Laboratory
Upton, N.Y. 11973-5000

## ABSTRACT

RHIC design criteria require that a faulty string be warmed rapidly to a serviceable temperature. Integral electric heaters warm the superconducting magnets. The magnets are very sensitive to temperature peaks and local gradients. Helium gas is circulated during warmup so that all points are warmed at nearly equal rates. An optimized string warmup procedure gives a continuous, uniform temperature profile, that is flat and rises steadily to room temperature at a rate which allows manageable recovery of the liquid helium content of the magnets. The ideal is not easily obtained. Various warmup schemes are studied. Limitations on the process, including those of the RHIC helium recovery system are discussed. A finite difference code (Fortran) is used to model a magnet string with flowing helium and electric heat generation. Temperature dependent properties of metal and gas including variable density are accounted for. Predictions are compared to physical test data.

## INTRODUCTION

The minimum string of vacuum insulated magnets which can be isolated for warming is one-sixth of one of the two 2.5 mile rings used in RHIC. The expense of downtime requires that this string be warmed from 4K to ambient within 24 hours for quick repair. Stress and recovery issues prohibit spoiling vacuum at any time during warmup. Warming is accomplished with discrete electric heating circuits looped in parallel through 30 m of magnets. With both circuits powered 395 W/m is generated.

Furthermore, there exists frequent magnet interconnecting components and transfer lines which have significant heat capacity, yet are not electrically heated. These are warmed by convection. Up to 100 g/sec of 293°K helium gas at 5 atm is available for this purpose.

## PHYSICAL ARRANGEMENT

In cross-section the cold mass has nearly quarter symmetry with "bypass holes" to carry helium and warmup heaters arranged around the central beam tube. A thin annular passage between the coil and beam tube also passes flow. The cold mass is constructed of 1/4" thick iron plates stacked within a 10.5" diameter cylindrical stainless steel shell.

The cold mass shares a radiative heat shield equipped vacuum jacket with four cryogenic process pipes. An aluminized mylar insulation (MLI) 16 layers thick wraps the cold mass, while 32 layers blanket the pipes and cold mass collectively, except the 55K pipe which conducts to the heat shield. The whole of these is enveloped by the shield wrapped in 64 layers.

Arc magnets with identical cross sections are interconnected throughout the arc portion of each of RHIC's two rings. Only the insertion magnets, which divide the rings into "sextants," deviate slightly. Helium flows from one end of the sextant in the bypass holes, returning through an insulated pipe. Sensors on tubes interconnecting certain cold masses indicate gas temperature. Iron temperature is determined by electrical resistance measurements of the superconducting bus.

## LIQUEFACTION AND GAS RECOVERY

Efficient operation requires that the 20,000 L liquid content of the warming sextant be reliquefied for use again upon cooldown. Heating raises the pressure of the sub-cooled liquid in the recoolers (pots which recool the magnet flow) to a state point from which a high quality mixture can be throttled into each of three liquid storage dewars totaling 122,000 L. High quality should be maintained for at least four hours, after which the JT efficiency will fall off; the returning gas is then diverted to the $1.7 \times 10^6$ SCF gas storage system of RHIC at 300 psi.

The design rate of liquefaction is 100 g/sec (up to 250 g/sec maximum during a power outage). This limits the rate of warming of a sextant; gas which cannot be reliquefied will be vented. Thus, heaters must be pulsed initially to control the recovery.

The thermal response of the system as a function of the heater duty cycle was studied using a 2-dimensional finite element analysis (FEA). A significant lag occurs between the power-on period and the transfer of heat to the fluid contained in the bypass holes. This is due entirely to the high thermal diffusivity of iron. Thus the effect of an over aggressive rate of warming will not be immediately realized and gas may evolve faster than can be recovered.

## PERMISSIBLE TEMPERATURE RANGE

The need for fast warming is established. The low temperature limit by definition of "warmup" is 293K. Peak temperature limits also exist; no magnet may exceed 305K as shifting of the superconducting coils due to creep in the plastics which house them will destroy the magnetic field quality.

An acceptable procedure gives a profile that is everywhere within this permissible range, never exceeds the limit, and induces gradients in any magnet of not more than 20K at all times. It is important then to know the effects of varying the controllable parameters:

$$q_L = \textit{rate of internal heat generation per unit length of magnet}$$
$$\dot{m}_g = \textit{mass flow rate of gas}$$
$$T_{g_0} = \textit{inlet gas temperature}$$

## THE FINITE DIFFERENCE METHOD

The transient heat transfer analysis of a sextant warmup with flowing helium and electric heat generation is modeled using a finite difference method. The one-spatial dimension, x, represents axial position along a magnet string measured from the end into which the helium enters. Solutions are sextant temperature profiles, or snapshots in time, t.

Some assumptions are made and justified. Firstly, assume that fluid and solid temperatures are uniform at a given axial coordinate, x; i.e., they do not vary radially. This reduces the problem to 1-D spatially. Evaluation of the Biot Number (Bi) gives a measure of the resistance to the transfer of heat by conduction within the solid to the resistance of transfer of heat to the solid by convection through the laminar fluid boundary layer from the turbulent flow field at uniform temperature. For the current arrangement Bi<0.1 proves the validity of these "lump mass" assumptions.

Secondly, neglect axial conduction. This supposes that transfer of heat in the axial direction is dominated by fluid transport. For the physical problem, the cold mass construction of laminar plates offers effectual resistance to axial conduction due to multiple thermal interfaces or gaps.

Thirdly, neglect thermal radiation and convection to the surroundings. As stated earlier, the cold mass is insulated against radiation by both the heat shield and three levels of MLI. The insulating vacuum is much less than 100 μ atmospheres, prohibiting convective warming.

Fourthly, assume the string has a uniform mass per unit length. In fact, it varies slightly for some components. Lastly, assume helium is an ideal gas at constant pressure, valid here above 10K.

**Nomenclature**

- $v$ -Flow velocity of gas.
- $\Delta x$ - Elemental length of solid.
- $\Delta t$ - Time interval.
- $h$ -Convective heat transfer coefficient.
- $A_L$ -Area per unit length of solid/fluid interface at boundary layer.
- $q_L$ -Rate of internal heat generation per unit length.
- $M_L$ - Mass per unit length of magnet (iron).
- $C_S$ - Specific heat of solid.
- $m_g$ - Mass flow rate of gas.
- $C_g$ - Specific heat of gas.
- $T_g$ -Temperature of gas.
- $T_S$ -Temperature of solid.
- i, n -Time and spatial coordinate indices, respectively.

**The governing partial differential equations.** The differential equations are developed by applying the laws of conservation of mass and energy:

$$hA_L(T_g - T_s) + q_L = M_L C_s \frac{\partial T_s}{\partial t} \tag{1}$$

$$-hA_L(T_g - T_s) = \dot{m}_g C_g \frac{\partial T_g}{\partial x} \tag{2}$$

In non-dimensional form (*) these become,

$$\frac{\partial T_s^*}{\partial t^*} = -(T_s^* - T_g^*) + Q^* \tag{3}$$

$$\frac{\partial T_g^*}{\partial x^*} = (T_s^* - T_g^*) \tag{4}$$

with initial condition of uniform temperature solid and boundary condition of constant temperature inlet gas.

$$T_s^*(x^*, 0) = 1 \tag{5}$$

$$T_g^*(0, t^*) = 0 \tag{6}$$

These equations are approximated by a forward difference scheme, with initial and boundary conditions, as

$$T_{s\,n}^{\,i+1} = (1 - \Delta t)\, T_{s\,n}^{\,i} + \Delta t\, T_{g\,n}^{\,i} + \Delta t\, Q \tag{7}$$

$$T_{g\,n+1}^{\,i} = \Delta x\, T_{s\,n}^{\,i} + (1 - \Delta x)\, T_{g\,n}^{\,i} \tag{8}$$

$$T_{s\,n}^{\,0} = 1 \tag{9}$$

$$T_{g\,0}^{\,i} = 0 \tag{10}$$

The numeric problem is a marching type. The approach is the method of successive intervals wherein a mass of fluid is transported incrementally from one solid element to the next, exchanging heat at each interval. The fluid temperature at the end of the first interval of time, serves as the initial temperature for the following interval. The mass flow rate is constant. The velocity of the fluid determines the relationship between the elemental length of solid, Δx, and this time interval, Δt, subject to the criteria of numeric stability. The number of solid elements which communicate with the fluid mass over the interval varies with the temperature dependent density. Wide ranges in fluid temperature demand flow velocity accountability.

## RESULTS OF ANALYSES

Controlled parameter settings for the study are: inlet pressure and temperature of gas, 5 atm and 300K; mass flow rate, 100 g/sec; heat generation 334 W/m (85% of full power).

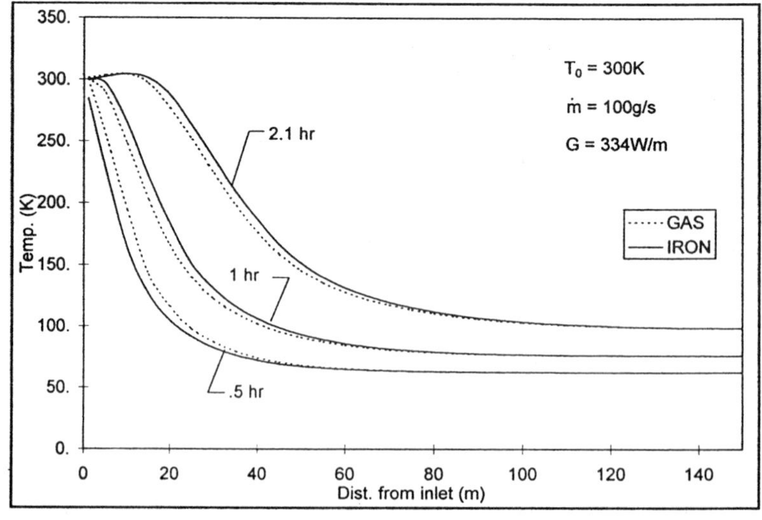


**Figure 1.** Full throttle warmup.

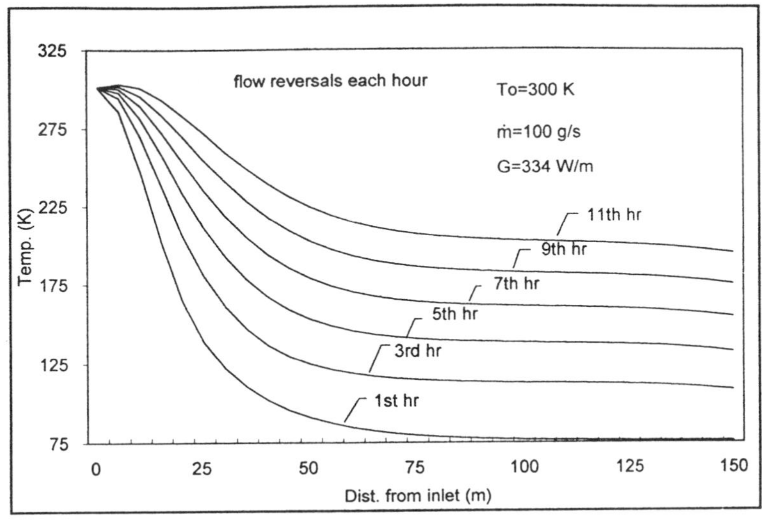


**Figure 2.** Periodic flow reversals.

**'Full Throttle'.** [Fig.1] Simply use continuous and maximum heat and flow.

The iron and gas temperature profiles are shown for the first 150 m of magnet when the allowable 305K is reached in the iron (t=2.11 hr, x=8.05 m). Allowed to continue, the peak propagates toward the end of the string as it grows. Far downstream of the peak, temperatures remain well below 293K. A more sophisticated warmup procedure is required than the nominal 'full throttle' warmup. The following options are studied:

**Flow Reversals.** [Fig.2] 1. Upon reaching 305K , reverse the direction of flow through the string. Continue flow reversals whenever a peak is reached. 2. Alternatively, flow reversals might occur periodically even in the absence of a peak.

1. The peak remains stationary, dancing left then right, and prematurely exceeds the limit. The flat middle portion of the profile continues to rise steadily. 2. Plots for periodic flow reversals are for each forward flow cycle. Limits are reached after the eleventh hour.

**Thermal Buffer.** [Fig.3] Install, before the string inlet, a "sacrificial" mass (also cooled to 4K); a length of unheated magnet might also act as a buffer.

The time to 305K is greatly increased (t=8.2 hr), though a significant length remains cold. It would be possible to determine a length of leading "buffer" magnet such that the peak 305K occurs when all downstream magnets are above 293K.

**Pre-cooling Inlet Helium.** Peaks move upstream. Flow rate regulates the time to 305K.

**De-Energize Heater Circuits.** [Fig.4] 1. As 305K is reached in any magnet, shut off its heaters. 2. More conservatively, shut the zone of heaters downstream of any sensor reading 300K.

The plots show snapshots at the occurrence of a limit. 1. This is promising. No over heating occurs. Local peaks develop at heater junctions but "wash out" after de-energizing. However, the method requires steady, repeatable controlled parameters, or a continuous distribution of temperature sensors. This is impractical. 2. Peaks flatten with time then redevelop at the next zone. After 15 hours the peak remains below 305 K. The criticality of sensor location is reduced. In both cases, the gradients downstream of the peaks are large.

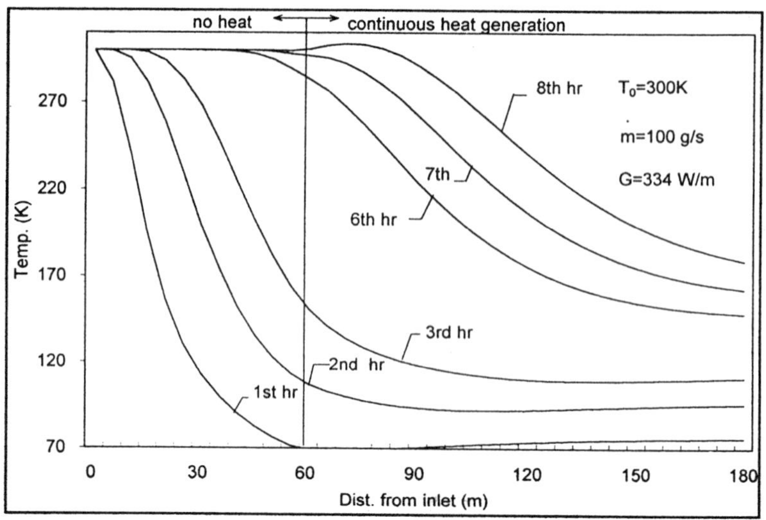


**Figure 3.** Thermal buffer.

**Inlet Counter-Flow Heat Exchanger.** [Fig.5] The inlet gas temperature tracks that of the string by heat exchange with the return gas.

Since the magnets can be warmed to ambient by their heaters alone in 22 hours, and they comprise most of the strings mass, heat added by inlet gas is unneeded. The upward slope is due to the time of travel for gas returning through an unheated process line within the magnet vacuum vessel to the inlet heat exchanger. The method is foolproof but costly.

**Delayed Flow.** Heaters are powered initially with no flow. A final flush with helium warms cold spots. A 3-D FEA shows large radial and longitudinal gradients which may produce high thermal stresses prior to the flush. Unlike continual flow methods, failure of a single heater will cause localized cold spots within magnets which will be subjected to thermal shock at the start of the flush.

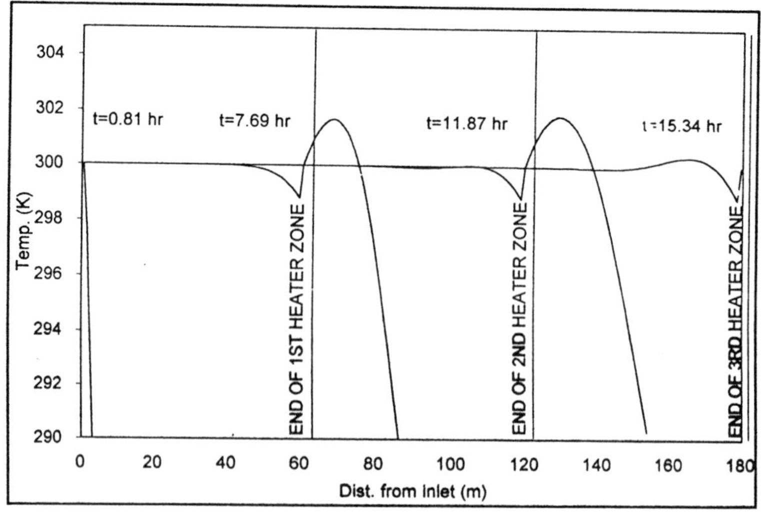


**Figure 4.** De-energize heaters at 300K.

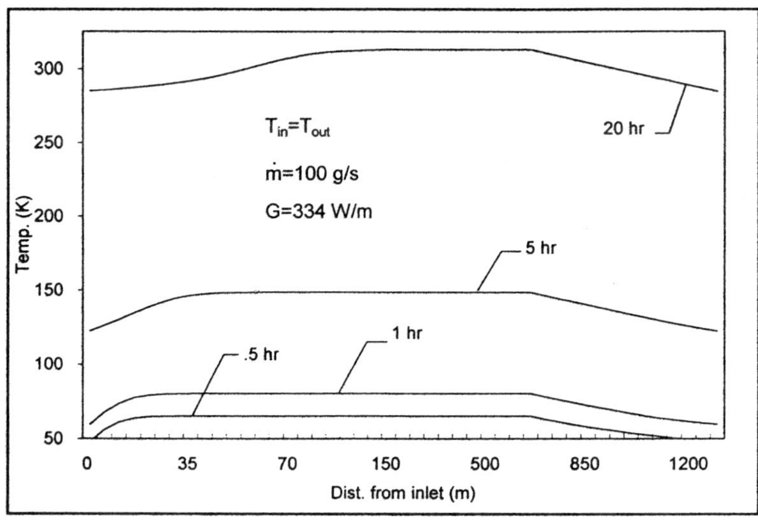


**Figure 5.** Inlet counter-flow heat exchanger.

## COMPARISON WITH TEST DATA

Many tests of RHIC's first sextant were conducted in early 1997, including two sextant (string) warmups. Data, however irregular, were recorded for both.

The procedure followed for the first warmup was one of delayed flow primarily because focus was on the recovery process which was done conservatively. It provided a valuable point of reference for the study of the second warmup which, though still conservative, included a modest flow rate of precooled gas. Resistance readings convert to temperature of magnets in the second warmup. Similar data was scarce for the first. Sensors external to pipes measure helium temperatures; however, for the no flow case they equate to magnet temperatures.

Much of the test is not representative of warmup of a single sextant of an otherwise cold ring. The 25 kW refrigerator and compressor were dedicated to a single sextant. Heaters were

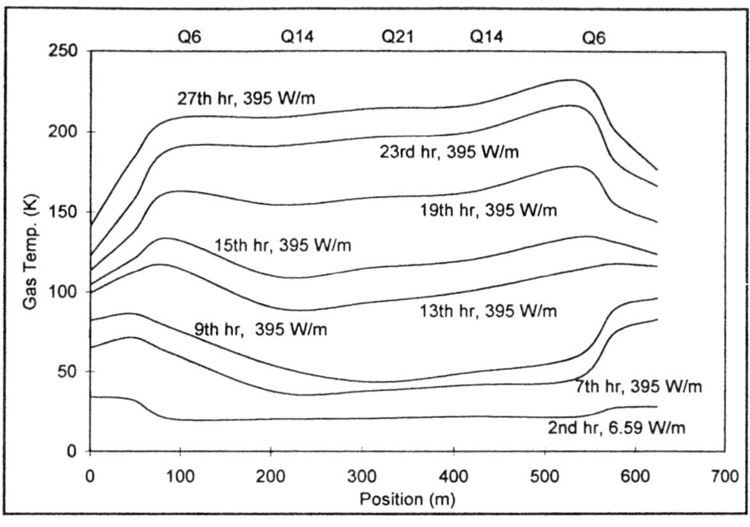


**Figure 6.** Warmup start 2/27/97 09:00. No gas flow.

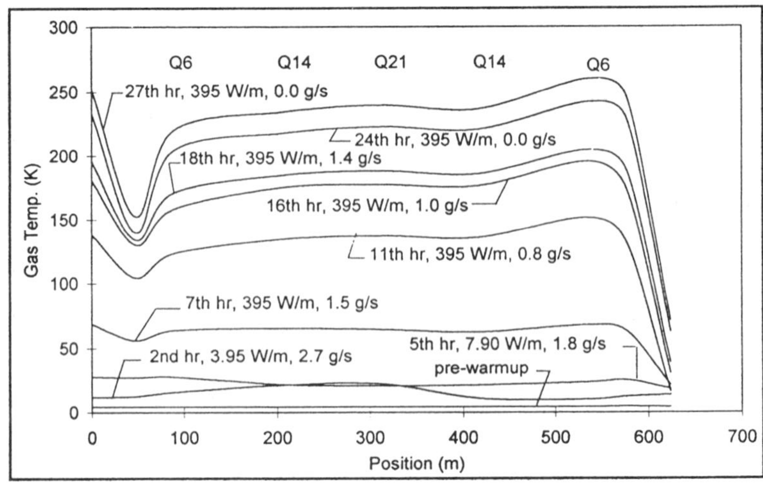


**Figure 7.** Warmup start 3/5/97 13:00. Decreasing gas flow.

pulsed at very low duty cycles during extended recovery periods. Complete warmup times were ~30 hours. The temperature profiles for the first and second warmups are shown [Figs.6, 7, 8]

The warmup beginning 2/27/97, with no flow [Fig. 6], plainly shows regions of magnets at either end of the sextant (intersection region) which lagged behind. The assumption of uniform heating is suspect here. Earlier periods show the same magnets leading the warmup. A contributing factor to this occurrence may be that, in the insertion region, magnets from both rings share a common cryostat; in fact, the cold masses are separated by ~6". During warmup of a single ring the warming magnet (in MLI wrap) is in line of sight (with a very high view factor) of the adjacent cold magnet. The injection region magnets are massive and, being thermally isolated and under insulating vacuum may, by radiation, act as a significant heat source (or sink). For example, a mass equal to that of a magnet with surface temperature of 50K absorbs 5 W/m from its surroundings at 100K (80 W/m from 200K) and has a time constant of 3 hr. (0.4

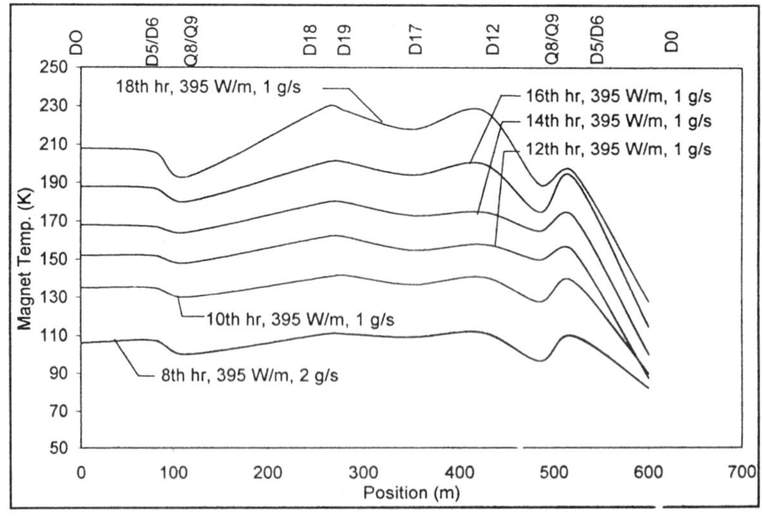


**Figure 8.** Warmup start 3/5/97 13:00. Decreasing gas flow.

hr). Since here only one ring is cooled, the adjacent magnets are between 4K and 55K (heat shielded) and act as a source, initially when the warming magnet is colder, and as a sink afterward. At any rate the symmetry one would expect in the no flow case is present.

The warmup of 3/5/97 with flow [Fig. 7, 8], on the other hand, does not possess this symmetry. The inlet temperature started lower than the iron temperature and, after time surpassed it. The beginning of the sextant clearly shows the short region of influence of the modest warming flow. Downstream of this region the profile is similar to the no flow case until more unheated/under heated components are encountered where the profile drops off drastically. This is due, in part, to the reasons already discussed, as well as to upstream flow being cooled by unheated transfer lines.

For the purposes of comparison, the finite difference code was customized to model the actual test hardware, procedures and controlled parameter settings. Modifications included modest, decreasing flow, initially pulsed heaters, variations in string mass per unit length, regions of higher heat load due to poor insulating vacuum and non-uniform heating.

## CONCLUSION

While it is futile to describe actual profiles entirely with limited sensors, the customized code output concurred with some notable trends over the first 150 m. Firstly, a negative slope at the entrance to the string is evident and is very dependent on the flow rate and inlet temperature as is predicted by the nominal study. Gas and iron temperatures converge shortly. Further upstream an inflection point and a localized upturn lead to a gradual climb to the peak temperature of the string. The gas changes much more abruptly. The center of the sextant is more stable. These features are coincident with code predictions. True verification requires substantial flow rates.

## ACKNOWLEDGMENTS

The author would like to thank Dr. Samuel Irvine, State University of New York at StonyBrook and Donald P. Brown , Brookhaven National Laboratory for their contribution.

## REFERENCES

1. RHIC Design Manual, Brookhaven National Laboratory, October, 1995.

2. R.P. Shutt, Cooldown Times for ISA Magnet Rings, Technical Note No.17, Isabelle Division, Brookhaven National Laboratory, April, 1976.

3. H. Lierl, P. Schmuser, Cooldown of a HERA Octant Equipped with Cold Iron Dipoles, Deutsches Elektronen-Synchrotron DESY, Hamburg, Report DESY HERA 84/16, July 1984.

4. P. Cruikshank, N. Kos, G. Riddone, L. Tavian, CERN, Investigation of Thermal and Vacuum Transients on the LHC Prototype Magnet String, paper presented at CEC '96, Portland, OR.

# PERFORMANCE OF RHIC 50 WATT RECOOLERS*

A. Nicoletti and M. Iarocci

Cryogenic Systems Group
Relativistic Heavy Ion Collider
Brookhaven National Laboratory
Upton, NY 11973-5000, USA

## ABSTRACT

Brookhaven National Laboratory has designed and tested cryogenic heat exchangers for use in the Relativistic Heavy Ion Collider (RHIC). These units, called recoolers, are located periodically in the superconducting magnet string and maintain the accelerator at design temperature by removing heat from the closed Helium circulating loop. Each is required to transfer 50 Watts from supercritical Helium, nominally flowing at 100 g/s to a Helium bath boiling at 4.25 K. With a 50 Watt heat load applied, there is a .028 K differential temperature between the Helium exiting the recooler and the recooler bath, and a heat exchange effectiveness of approximately 80%. The recoolers met design requirements during first article testing and then during the RHIC first sextant test. This paper describes the design and testing of the recoolers as well as their performance during the RHIC first sextant test.

## INTRODUCTION

### RHIC Recooler Function

The RHIC superconducting magnets are cooled by closed loop forced flow supercritical Helium. Other than lead flow makeup, the circuit is isolated from the main refrigerator process streams. The Helium is circulated by a single centrifugal compressor in a valve box near one of the six beam crossing regions. A 225 Watt heat exchanger[1] is located at each crossing region and 50 Watt recoolers are installed periodically in RHIC Corrector-Quadrupole-Sextupole (CQS) magnet cryostats, for a total of 30 per ring. They maintain the ring at design temperature by removing heat from the Helium circulating in the

* Work performed under contract with the US Department of Energy.

magnet coolant loop. The recooler shell is fed from the 25 kW refrigerator through a supply line to a J-T valve, which maintains a constant liquid Helium level. The boiloff flow is returned to the refrigerator via a cold return line. The supply and return lines are both located in the magnet cryostat. Figure 1 shows the 50 Watt recooler design parameters.

In this cooling scheme, heat from the magnets gradually warms the Helium in the circulating loop, until a recooler is encountered and the coolant temperature sharply drops. The result is a sawtooth temperature profile around the ring, with the upper and lower limits indicated by the recooler inlet and outlet parameters, respectively, in figure 1. Each recooler removes the heat accumulated from an average of 15 magnets and 30 meters of transfer line.

## Recooler Design

The magnet interconnect for the Helium coolant loop consists of a 2 ½ and a 3 inch tube in parallel. The tubes contain the main quad and dipole bus as well various power and instrumentation leads. The magnet coolant flow is diverted from the main loop into a recooler by a dam built into this magnet interconnect assembly at each recooler location. This dam consists of the magnet bus epoxy potted in a plastic tube approximately 25 cm long. This tube is in turn epoxied to a bellows, which is welded directly to the magnet end volume. This design provides for thermal movement of the bus relative to the cold mass, while keeping the leak rate significantly less than 1 % of the magnet coolant flow, at the relatively small differential pressure normal to RHIC operation.

## Recooler Testing

Testing was done in BNL's MAGCOOL magnet testing facility. Initially background heat load was established, then up to 60 Watts was applied via a calorimeter. At each heat

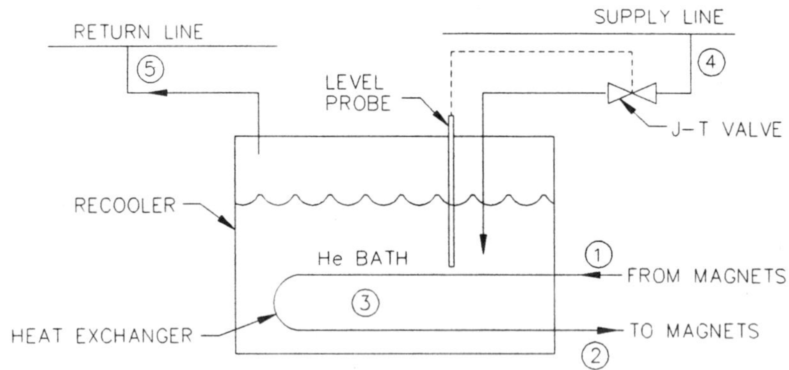


| Point | Location | Pressure (Atm.) | Temperature ( K) |
|---|---|---|---|
| 1 | Magnet coolant inlet | 4.5 | 4.40 |
| 2 | Magnet coolant outlet | 4.5 | 4.28 |
| 3 | Helium Bath | 1.024 | 4.25 |
| 4 | Supply Line | 4.8 | 4.30 |
| 5 | Return Line | 1.023 | 4.30 |

**Figure 1.** 50 Watt Recooler design parameters.

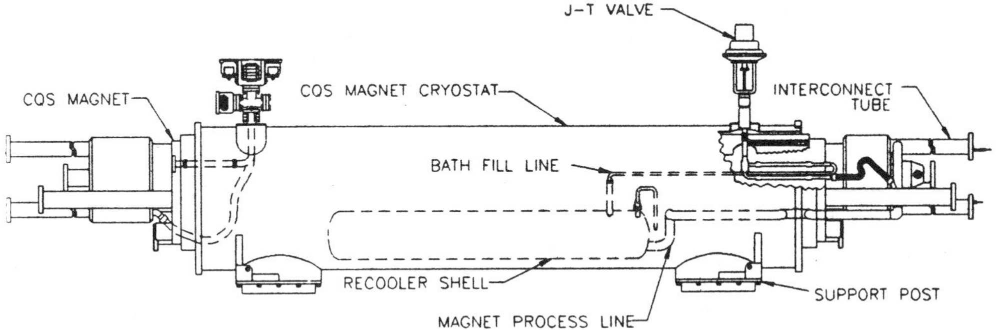


**Figure 2**. Recooler installed in RHIC magnet cryostat

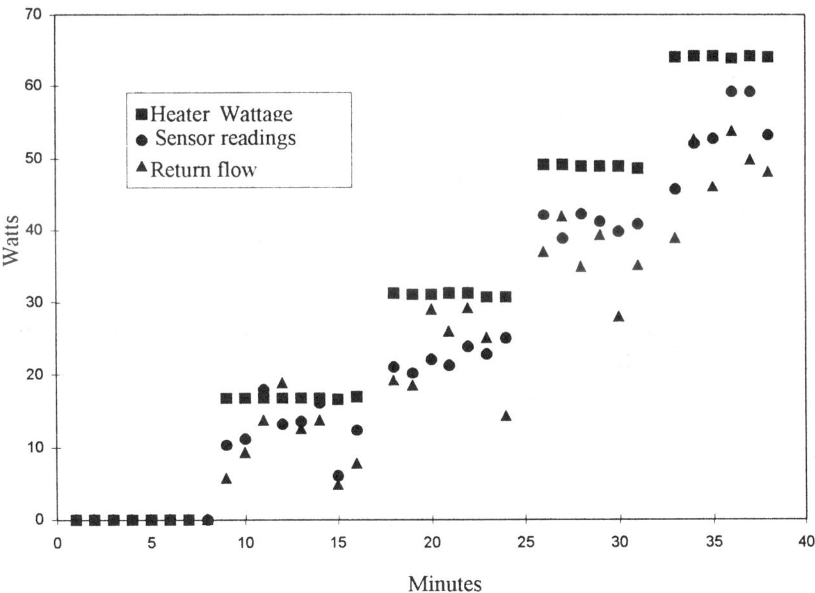


**Figure 3**. Heat transfer at heater settings of 0 to 63 Watts, as calculated from temperature instrumentation and boiloff flow.

load setting, the heat transferred in the recooler was measured by closing the J-T valve and monitoring the liquid Helium decay. In addition, boiloff flow was measured using a venturi flow meter in the return line. These measurements were verified with an energy balance using the recooler mass flow rate and inlet and outlet temperatures. The latter method was susceptible to uncertainty in temperature measurements and proved to be less reliable at lower heat load settings. At 10 Watts of heat transferred, the magnet process line temperature drops by only 0.025 K across the heat exchanger, so temperature measurement errors in the range +/- 0.005 K significantly impact energy balance calculations.

## TEST RESULTS

The steady state background heat load on the recooler, as determined from the liquid boiloff rate without any heat applied, was in the range of 10 to 15 Watts. The majority of the background load was due to the CQS magnet and other components in the MAGCOOL circulating loop, which were in series with the heat exchanger. Calculations indicate the heat load the recooler itself contributes is less than 5 watts, however because of limitations in temperature sensor accuracy, this exact number could not be determined. Figure 3 shows the net heat transfer, as calculated from the temperature instrumentation as well as the boil off flow rate, for different heater wattage settings. Any background heat load is subtracted from the data, so the results indicate the performance of the recooler with respect to the heat input from the calorimeter only. At each setting, after the system was allowed to stabilize, sensor readings were taken, then the JT valve was fully closed and the flow rate in the return line was recorded. As the heat is increased, of course, the percentage of heat removed decreases due to the reduced heat exchanger effectiveness. Although the return line flow rate varies considerably at each wattage setting, the two methods show good agreement if average values are used. At a heat load of 50 Watts approximately 80 % of the calorimeter heat is transferred to the boiling Helium bath.

The temperature of the Helium exiting the recooler is a critical parameter, as it defines the minimum magnet temperature. Figure 4 shows the effect of increasing heat load on the recooler inlet, outlet and bath temperatures. The recooler outlet temperature tracks the bath temperature, as is evident in the return pressure rise, which occurred at a heater load of 30 Watts. At a 30 Watt applied heat load, return temperature has risen a average of .020 K above the bath, at 50 Watts, .028 K and at 60 Watts the differential has spread to .032 K. The expected differential between the Helium exiting the recooler and the bath temperature at a 50 Watt heat load, from table 1 is .020 K. Temperature increases of this magnitude are inconsequential to the operation of the accelerator.

Other observations:

- With a 50 Watt heat load, the recooler refilled in less than 20 minutes.
- The JT valve can maintain a constant liquid level at heat loads in excess of 100 Watts.

## FIRST SEXTANT TEST PERFORMANCE

One sixth of one of the RHIC rings was cooled down to operating temperature in January 1997. This included a string of 81 magnet assemblies and five 50 Watt recoolers. The recoolers behaved as expected as they maintained the magnets at or below their design temperature, and contributed a pressure drop of approximately .008 atmospheres each to the magnet coolant loop. They also absorbed a considerable portion of the energy of an

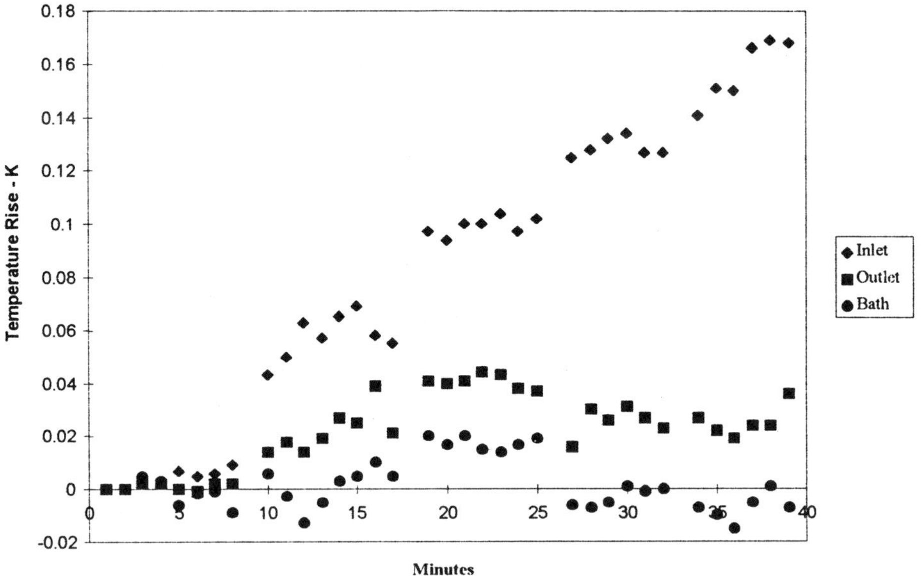


**Figure 4.** Effect of heat load on recooler inlet and outlet temperatures.

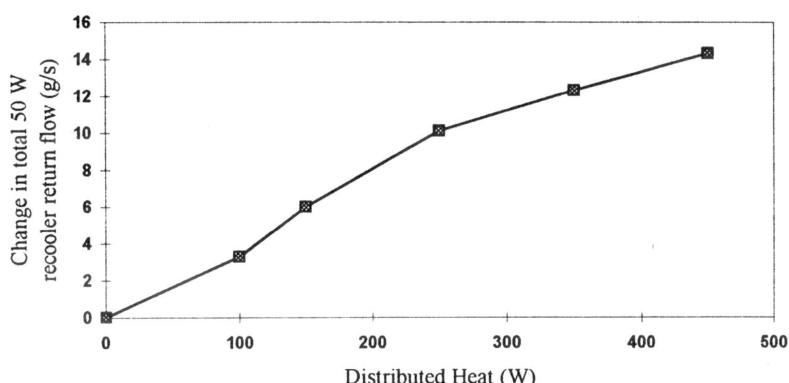


**Figure 5.** Change in total 50 Watt recooler return flow vs. distributed heat load, during the first sextant test.

artificially induced single dipole magnet quench and thus aided in decreasing the system recovery time.

Ring recooler testing was conducted similar to MAGCOOL testing, with magnet warm-up heaters substituting for the calorimeter as the artificial heat source. Since all magnets have these heaters built in, any heat they generate is evenly distributed throughout the sextant. However, roughly 15% of this load is applied to magnets not directly seen by any of the 50 Watt recoolers. 100, 150, 250, 350 and 450 Watts were applied to the magnets, and the resultant increase in boil off flow from the recoolers, with all J-T valves fully closed, was recorded. This heat is in addition to the 150 Watt background load. Since the five recoolers are spaced evenly throughout the sextant, this test applied up to 75 Watts above background to each of them. The results are shown in figure 5.

At slightly above one atmosphere, the heat of vaporization of Helium is approximately 20 J/g, thus figure 5 indicates roughly 80% of the 100 Watt distributed heat load applied to magnets seen by recoolers is removed by the recoolers. This increases to 90% at intermediate heat load settings, then reduces to approximately 75% at 450 Watts. The initial low reading could be attributed to higher susceptibility to measurement uncertainty at these smaller flow rates. At all settings, however, the applied heat would have had no impact on accelerator performance, thus indicating the considerable excess capacity of this portion of the cryogenic system.

## CONCLUSIONS

- The RHIC 50 Watt recoolers will meet heat exchange efficiency and process line pressure drop design parameters.
- The recooler system allows the accelerator to operate at a heat load/magnet of up to three times that seen during the first sextant test.

## REFERENCES

1. Nicoletti, A. and Wu, K.C., Testing of a Cryogenic Recooler Heat Exchanger at Brookhaven National Laboratory, *Advances in Cryogenic Engineering*, Vol. 39 (1993)

# PERFORMANCE OF THE RHIC CRYOGENIC CONTROL SYSTEM DURING THE FIRST SEXTANT TEST*

J. Benson, Y. Farah, and D. Zantopp

Cryogenic Systems Group
Relativistic Heavy Ion Collider (RHIC) Project
Brookhaven National Laboratory
Upton, New York 11973-5000

## ABSTRACT

The present RHIC cryogenic process control system is a hybrid of centralized and distributed processing. The setup marries together a centralized sub-system that was designed, installed, and in operation since 1984 with six identical distributed sub-systems, whose installation began in 1996 and will be completed in 1998. For the RHIC First Sextant Test, which took place from January 21, 1997 to March 5, 1997, it was necessary to have two new distributed cryogenic control sub-systems installed and tested.

During the First Sextant Test, gold ions were successfully injected into the superconducting collider at the 6 o'clock region, where they then traveled down Sextant 4/5, and hit a beam stop at the 4 o'clock region (Figure 1). The 6 o'clock region, 4 o'clock region, and Sextant 4/5 are the locations where the two new distributed sub-systems were installed and tested. The thrust of this paper is to discuss these new installations and the overall cryogenic control system performance during the First Sextant Test.

## REFRIGERATION AND COMPRESSOR CONTROL SUB-SYSTEM

The original centralized control sub-system, installed in 1984, handles all the I/O processing for the current RHIC refrigerator and compressor system. It was originally designed in the early 1980s for the now defunct Isabelle Project. It consisted of two PDP11® computers which ran CRISP® process control software and was hooked up in a hot-backup configuration. The two PDP11 computers have since been replaced by two VAX® 4000/400 computers, which now run an updated version of CRISP software (Figure

*Work performed under contract with the United States Department of Energy.

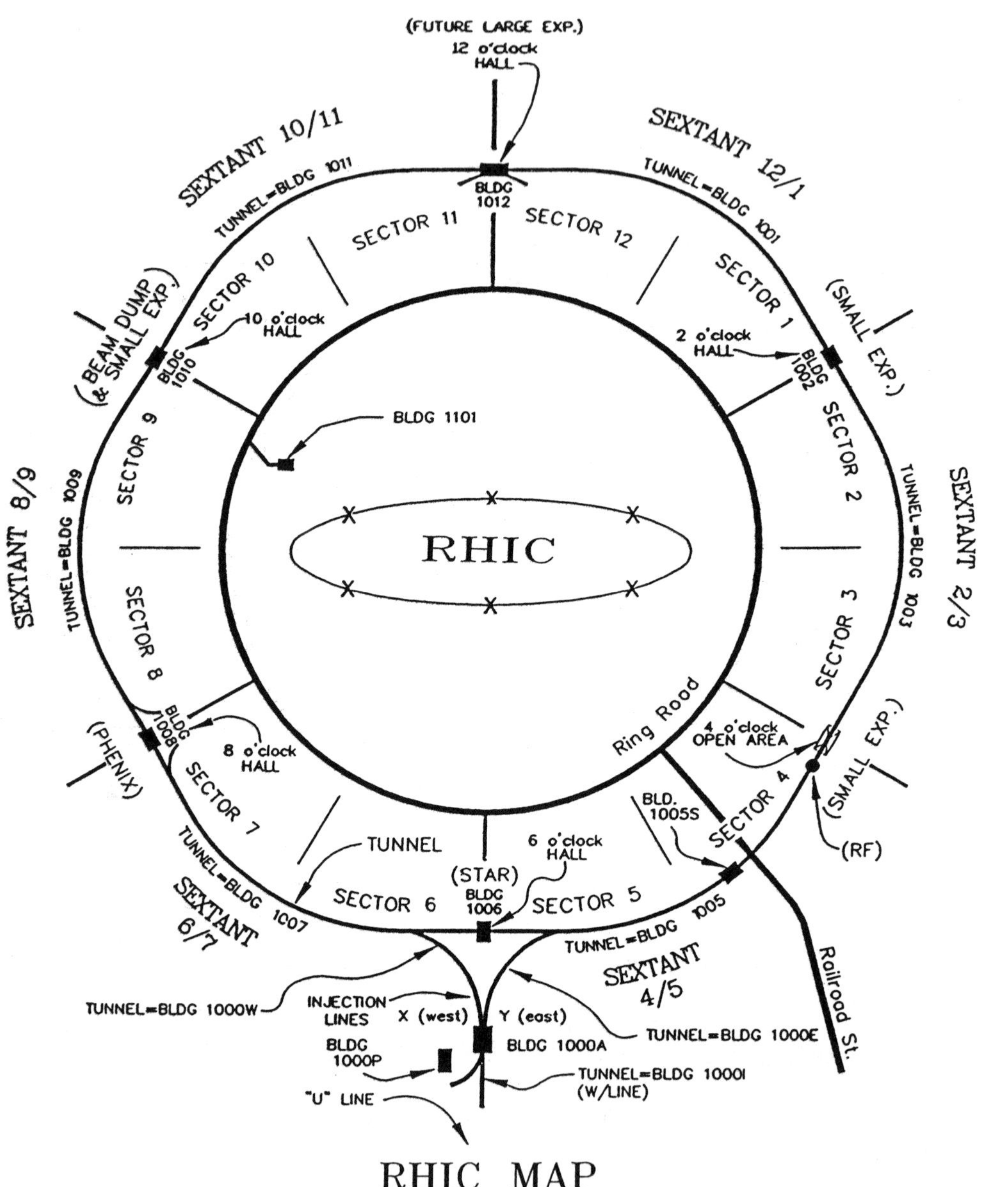


RHIC MAP

Figure 1. Region names and main buildings.

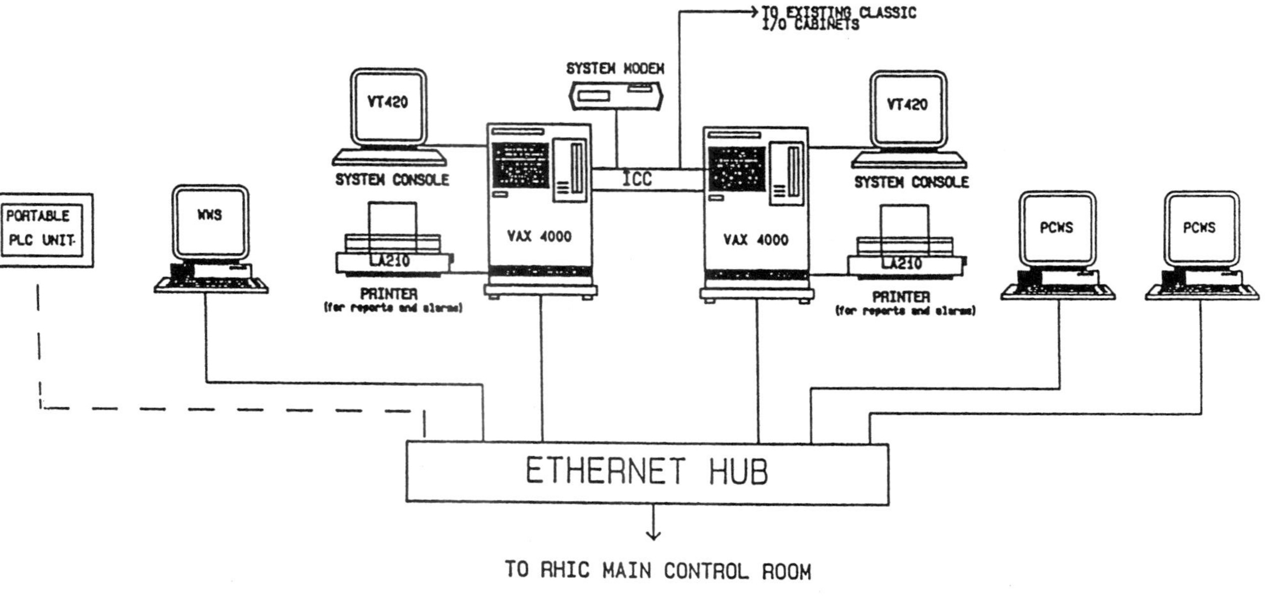


Figure 2. Cryogenic control room.

2). The I/O for the refrigerator and compressor system consists of memory mapped CRISP hardware. The CRISP I/O interfaces a total of about 1700 points consisting of analog and digital input/output points and temperatures. Through these points and with PID loops, the CRISP software controls compressors, turbine expanders, refrigerator and gas management, purifiers, and the cryogenic vacuum system.

## SERVICE BUILDING AND TUNNEL CONTROL SUB-SYSTEMS DISTRIBUTED AROUND THE RHIC RING

### Distributed Control

The six new distributed control sub-systems, when completed in 1998, will handle all the I/O processing for the cryogenic equipment that will be spread around the RHIC ring. These new sub-systems are "off the shelf" Programmable Logic Control (PLC) Systems manufactured by the Square D Company. PLCs are reliable, industrial grade control systems which require simple ladder logic programming to establish control.

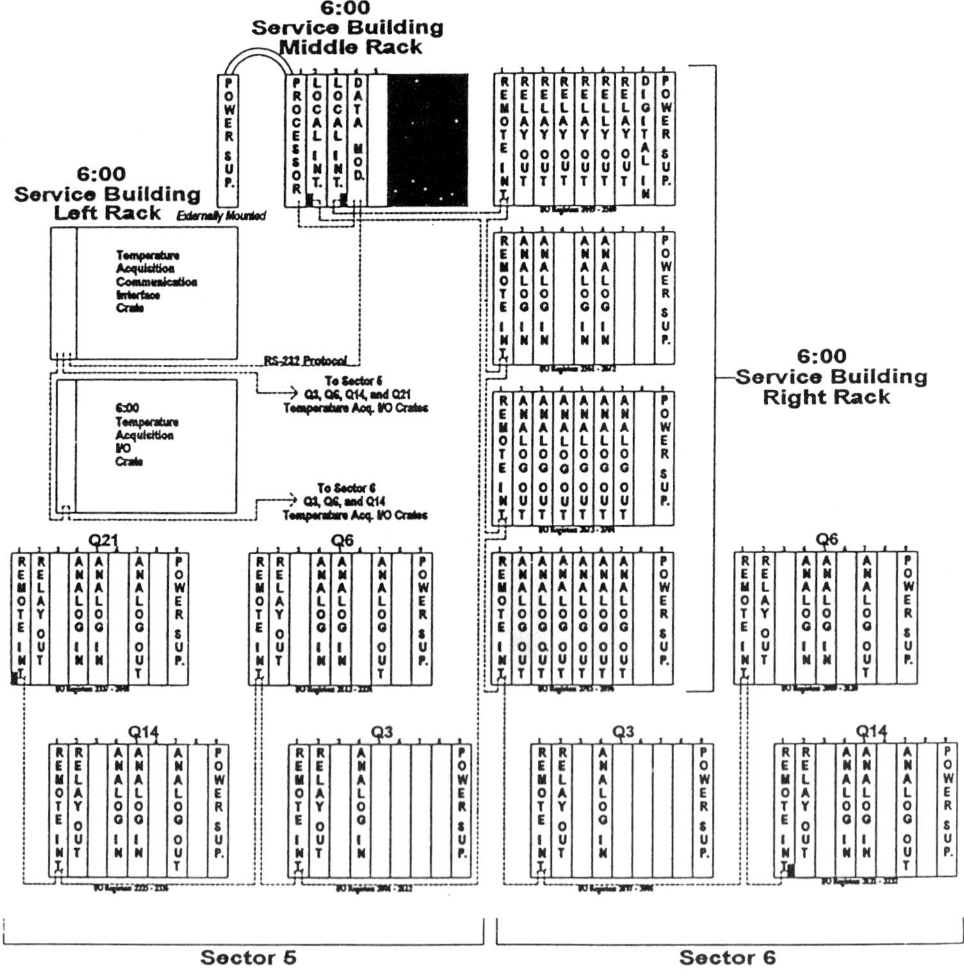


Figure 3. PLC equipment layout - 6:00 processor.

There are a couple of obvious advantages to distributed control. First, with distributed processing, a single-point failure will not take down the entire system. This is an especially critical factor for large control networks, such as the one used for RHIC's cryogenic refrigeration/distribution system. A second advantage is that distributed control lightens the processing load on each individual processor. This enables faster local processing response in carrying out such tasks as PID loop control.

**Distributed Sub-System Layout and Interconnection**

Each distributed ring sub-system contains an independent PLC processor, which is located at one of six RHIC ring service buildings (12 o'clock, 2 o'clock, 4 o'clock, 6 o'clock, 8 o'clock, 10 o'clock). Each PLC processor interfaces to the cryogenic system using remote PLC I/O crates, which are interconnected using a Square D proprietary networking protocol. Every PLC processor is responsible for controlling all of the cryogenic equipment located in the service building where it is physically located, and in the tunnel's two adjacent sectors, one on each side of the service building. The PLC equipment layout for the 6 o'clock processor can be seen in Figure 3. The six PLC processors, which are evenly distributed around the ring, the two VAX 4000/400 computers, and several CRISP PC workstations are all interconnected using a Star fiber optic network configuration, whose hub is located near the cryogenic control room.

**PLC Processor Tasks**

The PLC processor in each service building performs four primary tasks. First, it is responsible for transferring all the raw input data, via Ethernet, to the two VAX 4000/400 databases where it can then be downloaded, also via Ethernet, to the graphical user interfaces of the CRISP PC workstations. When RHIC is completed, there will be one CRISP PC workstation located in each of the six service buildings and five workstations located in the cryogenic control room. These PC workstations, which all run identical software, enable the system operators to both monitor and control the entire system from a single terminal. Second, the PLC processor is responsible for receiving all of the information sent to it by the VAX 4000/400 computer, via Ethernet, and performing the appropriate output function(s), such as shutting a valve. Third, it is responsible for handling all local processing tasks. Such tasks include PID loops that are necessary to control recooler Helium liquid levels and also the tasks required to control the RHIC magnet warm-up heaters. Fourth, it is responsible for handling all of the local interlocking that is necessary to protect certain cryogenic equipment from damage. For example, monitoring the differential pressure across a helium gas circulator and controlling the inlet valve position can be used to prevent over-spinning damage to the circulator.

**Service Building I/O**

In each of the six service buildings around the ring, there are two cryogenic valve boxes, one for each collider magnet string. A typical service building has four remote PLC I/O crates. Each of the twelve valve boxes requires approximately 110 points consisting of analog and digital input/output points and temperatures. For the valve boxes, the PLC system is responsible for reading inputs such as pressures, liquid Helium pot levels, Helium gas flow, and vacuum status. It is also responsible for sending output signals to the valve boxes to control digital/analog valves and to set gas flow controller set-points.

**Tunnel I/O**

There are five cryogenic process lines running throughout the twelve tunnel sectors of the RHIC ring. These five lines are used for distributing cold Helium gas around RHIC. The PLC control system in the RHIC tunnel is primarily responsible for monitoring the state of these process lines. A typical sector can have either three or four remote PLC I/O crates, which are located under certain designated quadrapole magnets (Q3, Q6, Q14, and Q21). There are approximately 150 I/O points monitored in each sector consisting of analog and digital input/output and temperatures. The inputs consist of such things as pressures, liquid Helium pot levels, Helium gas flow, and magnet warm-up heater status. The outputs consist of analog signals to control valves, relays to control power to liquid Helium level probes and Helium gas flow thermistors, and output signals to control the magnet warm-up heaters.

## TEMPERATURE ACQUISITION SYSTEM

The temperature acquisition system[2] is an in-house engineering design. The basic unit which comprises the system is a 19-inch rack mountable EUROCARD chassis which houses several functionally different electronic boards. Each crate, which is linked serially to other crates, has the capacity to read 84 cryogenic sensors. This multiplexed data acquisition system has an overall accuracy of 20 PPM, which translates into a temperature accuracy reading of less than 1 mK for temperatures below 10 K.

The system reads temperature of pre-calibrated transducers such as germanium, silicon diode, or platinum temperature sensors. The electronic system then linearizes the raw data by the use of corresponding polynomial coefficients (eighth-order). The temperature data is then shipped to a PLC Data Controller Module via RS-232 protocol. The Data Controller Module was programmed in BASIC language to interpret the serial information (address and corresponding temperature) and to appropriately direct the data to its pre-assigned PLC registers. It is then transferred, via Ethernet, to the VAX 4000/400 database.

## COMMUNICATION BETWEEN THE CRYOGENIC CONTROL SYSTEM AND THE RHIC MAIN CONTROL SYSTEM

The state of the cryogenic system is needed for the operation of some of the other systems of the collider, such as the power supply and vacuum systems. The opposite is also true. The state of other collider systems is needed for the proper operation of certain components of the cryogenic system. Therefore, to meet these requirements, the RHIC Controls Group has set up a database where key RHIC status data can be collected from the various different collider systems. The Controls Group is responsible for coordinating the essential data that must be collected from and delivered to the various different collider system's databases.

For the cryogenic system, the RHIC main control system has access to the VAX 4000/400 database through a router connected to the cryogenic Ethernet network. For example, on the cryogenic side, a directory is created under a VMS® operating system in

---

[2] Y.Farah, J.H. Sondericker, "A Precision Cryogenic Temperature Data Acquisition System", Advances in Cryogenic Engineering, Vol. 31, Plenum Press, New York, 1985.

which an ASCII input/output file is written. Critical information sent to the RHIC main control database that indicates overall cryogenic system status includes: liquid Helium pot levels, turbine expander status, flow to the load (magnets), and certain critical temperatures. The cryogenic system is also interested in monitoring certain critical status information from other RHIC systems including: the magnet lead current, the vacuum system status, and the Oxygen Deficiency Hazard (ODH) safety system status.

## CONTROL SYSTEM PERFORMANCE DURING THE FIRST SEXTANT TEST

From a cryogenic standpoint, the basic goal of the First Sextant Test was to cool down the sextant magnet cryogenic process line to approximately 4.5 Kelvin. To accomplish this, the flow of cold Helium gas was established from the refrigerator to the 6 o'clock valve boxes, via a vacuum jacketed piping system. From there, the gas was circulated into the tunnel, via more vacuum jacketed piping. It then traveled down Sextant 4/5 where it finally exited the ring, via vacuum jacketed piping, and entered the 4 o'clock valve boxes. A return path for the gas was set up at the 4 o'clock valve boxes.

The First Sextant Test called for two cool-down and two warm-up periods. For the majority of the time during the First Sextant Test, except during engineering studies, the cryogenic system was held in a steady-state condition, with the magnets being held at a temperature of approximately 4.5 Kelvin. During every phase of the First Sextant Test, all aspects of the RHIC cryogenic control system operated smoothly. Both cool-down periods, from room temperature to 4.5 Kelvin, took approximately 2.5 days. The gas recovery and warm-up procedure took the anticipated 30 hours. During steady-state operation, the control processors maintained the desired system parameters, such as the liquid Helium pot levels using PID loop control. The temperature acquisition system and its interface to the PLC system performed successfully. The hot-backup arrangement of the VAX 4000/400 was also intentionally tested as the switch over was successfully carried out while the entire cryogenic system was on-line. The communication between the two distributed PLC processors, the two VAX 4000/400 computers, and the CRISP PC workstations operated without a glitch. In addition, the communication between the PLC processors and their remote I/O crates operated without a problem. Despite being in its early stages, the transfer of data between the cryogenic control system and the RHIC main control system was successful.

The overall success of the Cryogenic Control System during the First Sextant Test gives confidence that when the remaining four distributed PLC systems are completed in 1998, the complete RHIC Cryogenic Control System will be capable of coming on-line for RHIC without a hitch.

------------

PDP11, VAX, and VMS are Digital Equipment Trademarks.

# VJR/VJRR DESIGN, CONSTRUCTION, INSTALLATION, AND PERFORMANCE*

E.C. Quimby[1], C.M. Lac [1], M. Iarocci [1], R. Sallash[2], A. Varghese [2]

[1]Cryogenics Systems Group
Relativistic Heavy Ion Collider
Brookhaven National Laboratory
Upton, N.Y. 11973-5000
[2]Gardner Cryogenics
2136 City Line Ave., Bethlehem, PA 18018

## ABSTRACT

The Brookhaven National Laboratory (BNL) Relativistic Heavy Ion Collider (RHIC) is a colliding beam accelerator consisting of two interlaced rings of superconducting magnets in an underground tunnel, 3.86 km (2.4 miles) in circumference. The superconducting magnets are cooled by compressed helium at 4.5 K. The cold helium is delivered from the refrigerator to two valve boxes at the magnet ring by way of the approximate 350 m long, 60.96 cm (24") diameter, Vacuum Jacketed Refrigerator to Ring (VJRR) piping run. The helium is then distributed around both rings through approximately 2,400 meters of 50.8 cm (20") diameter Vacuum Jacketed Ring (VJR) piping, working in conjunction with the magnet cryostats. The VJRR and VJR piping are both multi-line, 55 K helium heat-shielded, transfer lines, however, each differs in its fundamental design concept and component construction. This paper discusses VJR and VJRR design concepts, as well as certain aspects of component construction, field and tunnel installation, and thermal performance.

## PROJECT OVERVIEW AND RHIC RING TERMINOLOGY

Brookhaven National Laboratory (BNL) is presently constructing the Relativistic Heavy Ion Collider (RHIC) as shown on Fig.s 1 and 2. This colliding beam accelerator is capable of accelerating and colliding a wide variety of particles (ions) from protons to gold nuclei. The two rings of superconducting magnets will guide and focus particle beams circulating in opposite directions. The particles can be made to collide at six equally spaced beam crossing points, located around the circumference of RHIC, where experimental research in nuclear physics will be performed. At each of these six beam crossing locations (Figs. 1 and

* Work performed under Contract with the U.S. Department of Energy

**Table 1**. Summary of specifications for VJR/VJRR piping

| | | | | | |
|---|---|---|---|---|---|
| Design pressure (MPa): | helium process lines | | 1.89 | | |
| | vacuum jackets | | 0.152 | | |
| Design temperature helium (K): | primary coolant (process lines) | | 4.2 (4.5 avg.) | | |
| | secondary coolant (heat shields) | | 40 (55 avg.) | | |
| Heat loads totals max allowable (W): | | primary (4.5K) | secondary (55K) | | |
| | VJR | 1200 | 5000 | | |
| | VJRR | 200 | 800 | | |
| Operational Cycles: | mechanical | 5000 | | | |
| | thermal | 200 from ambient to operating temperature and back | | | |
| Design loads external supports (kN): | | | axial | lateral | dead wt. |
| | VJR | anchors | 35.6 | 13.3 | 13.3 |
| | | guides | 2.7 | 6.7 | 13.3 |
| | VJRR | anchors | 178 | 17.8 | 13.3 |
| | | guides | 2.7 | 13.3 | 13.3 |
| Design Life (yrs): | 20 | | | | |

2) an experimental area or hall exists. The straight tunnel sections shown immediately adjacent to each crossing point are referred to as insertion regions. The six curved sections of tunnel connecting consecutive insertion regions are known as the arc regions. The portion of the rings that lies between two crossing points is referred to as a sextant. The crossing points are individually identified by their clock position around the ring. For example, the Wide Angle Experimental Hall located near the beam injection tunnels from the AGS to RHIC, is at the 6 o'clock crossing.

The helium Cryogenic Distribution System (CDS) consists of approximately 2,750 meters of VJR/VJRR vacuum jacketed piping and twelve cryogenic valve boxes which contain all required valves, circulating pumps, lead pots, and 225 Watt recoolers. Additional recoolers (50 Watt) are built into the magnet cryostats themselves, uniformly distributed around the rings. The existing RHIC 24.8 KW Helium Refrigerator will provide the refrigeration effect for cooling the magnets. This system is designed to provide refrigeration at two nominal temperatures: a primary load at 4.5 K and a secondary load at 55 K. The function of the secondary circuit is to provide cooling for heat shields and heat stationing.

## VACUUM JACKETED RING (VJR) PIPING SYSTEM

The CDS transports the helium refrigerant from the refrigerator to the RHIC ring (6 o'clock valve boxes) by means of the VJRR piping system (Fig. 2). From the valve boxes, the VJR piping then distributes the cold helium to the magnets and around both magnet rings. The magnet rings, however, are not continuous, but contain gaps where non-superconducting ring components, such as RF cavities, are located. VJR piping is used, in the tunnel (Fig. 5), to carry the cold helium across these gaps, by-passing these additional ring components. The ring magnet gaps occur in the ring insertion regions. VJR piping is also used to by-pass the helium refrigerant around the crossing regions and through the valve boxes located there. Since the valve boxes are located in service buildings that exist outside the tunnel, the VJR piping must exit the tunnel in these regions. Because the

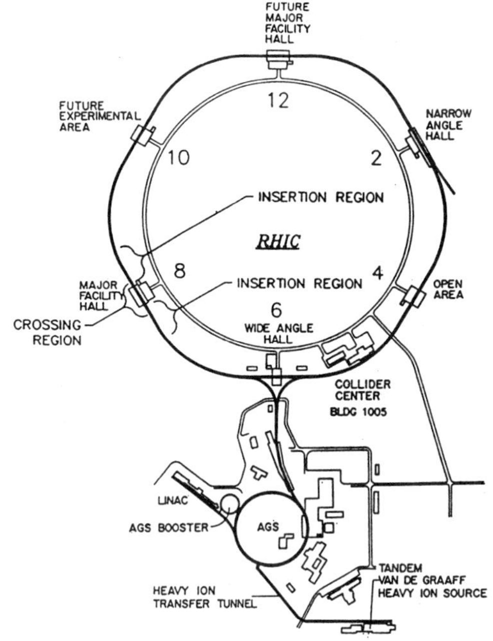


Figure 1. RHIC complex plot plan showing injectors.

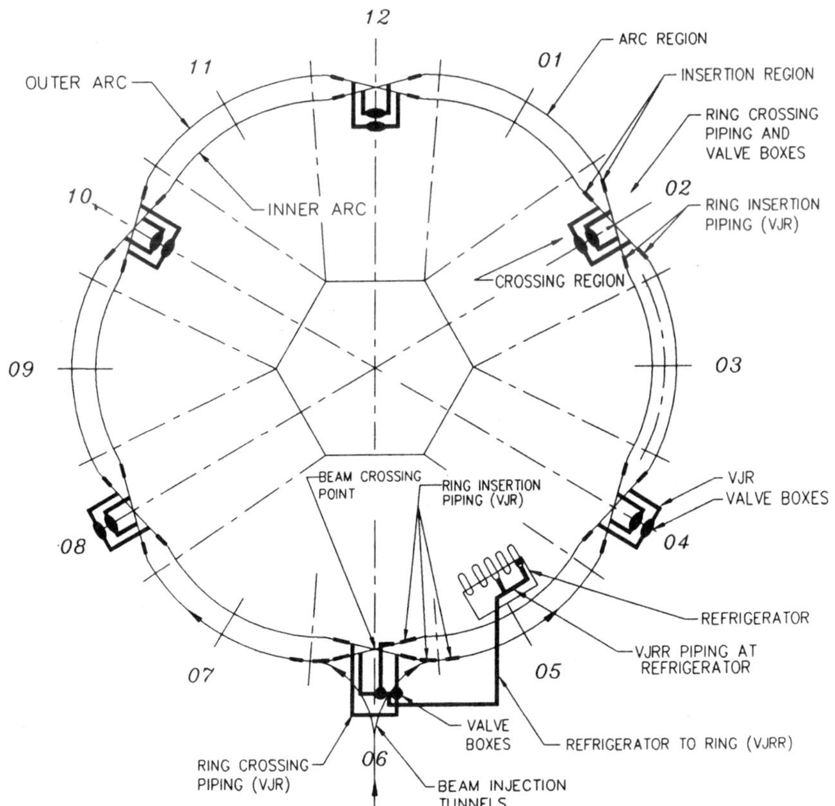


Figure 2. VJR/VJRR piping locations on RHIC accelerator.

outdoor terrain and RHIC facilities and construction differ at each crossing region, the piping runs and supports also differ considerably. Figure 2 shows all of the locations, requiring either VJR or VJRR piping drawn in heavy black lines.

The VJR piping system includes five (5) helium lines (Fig. 3) for each of the two (2) RHIC rings, including: Heat shield (40-70 K), Supply, Return, Utility, and Magnet lines (all 4.5 K). All five lines are contained within a common 50.8 cm diameter vacuum jacket and a 40-70 K helium Heat Shield. The 10.16 cm diameter magnet line will carry two 2.5 cm diameter flexible superconducting bus cables for the dipole and quadrupole magnets. The other four process lines, are 7.30 cm diameter and contain no cables.

## VJR Design Concepts

Since the VJR piping must exit the RHIC tunnel and connect to the valve boxes located in the crossing region's service buildings, the optimum piping run configurations would be significantly different in each region. Time, resources, and bidding constraints prevented us from providing an optimum custom design for each region early in the project. We therefore took the conceptual approach of creating each piping run from a standard limited set of basic piping components: 90° Elbow Assembly, Standard Straight Assembly, Vertical Elbow Assembly, Interconnect Assembly, etc. This approach minimized the piping design effort and maximized the amount of manufacturing repeatability of components and subassemblies. It also allowed us to design the piping runs for all six crossings in a relatively short time, and to early on determine the requirements for the whole project. Since the piping runs were now fixed, the pipe support design could proceed. This would be a separate effort from the pipe component contract.

The most basic and common VJR pipe component is the Standard Elbow (Fig. 3). The main variables for this elbow are: the Length ($L_1$), the Heat Shield orientation, and whether the outer jacket contains a Bellows (13), or not. All of the other components and subassemblies (items 1 through 13) remain the same. A brief description of some of these items follows. The heat shield orientation for a specific pipe is dictated by the process line orientations that exist at the valve box (Fig. 2), at one end of the VJR pipe run, and the RHIC magnet at the other end. A bellows is required in most pipe runs that are extended between two external anchor supports. The vacuum jacket bellows (13), vacuum break bellows (4)(5), and the flexible hoses (1)(2)(3), are all manufactured and tested by Senior Flexonics, Inc. Each of these components is subjected to pneumatic pressure testing, liquid nitrogen cold shock (except item 13), and He mass spectrometer leak testing.

The fixed (9) and sliding (10) process line supports are constructed of an aluminum inner pipe retainer with G-10 spider vanes that extend to the vacuum jacket wall. The fixed support vanes engage stainless steel brackets, which are welded to the jacket inner wall. These supports are spaced at a maximum of 4.6 m apart. As shown in the inset for the fixed support (9), G-10 tubing is used to provide low heat leak between the process line anchor ring and the main support structure. A 5 mm (0.20") anchor ring gap is provided to allow for process line contraction across the interconnect region. The process line anchor ring must have a good deal of radial flexibility so as not to induce excessive stress in the process line wall during welding and during operational thermal transients. Axial loads on these rings can approach 3.5 kN. Split rings or tabs are best here to keep the stresses in the pipe wall low. The weld size should always be less that the process line wall thickness. A finite element stress evaluation of this design was performed. The same thing applies to the vacuum break bellows (4)(5) collar welds to the process lines (Fig. 3). In this case a conical collar was designed to provide for the needed radial flexibility. However, in this case, the axial force on the weld does not exceed 70N due to the extreme softness of the vacuum break bellows design.

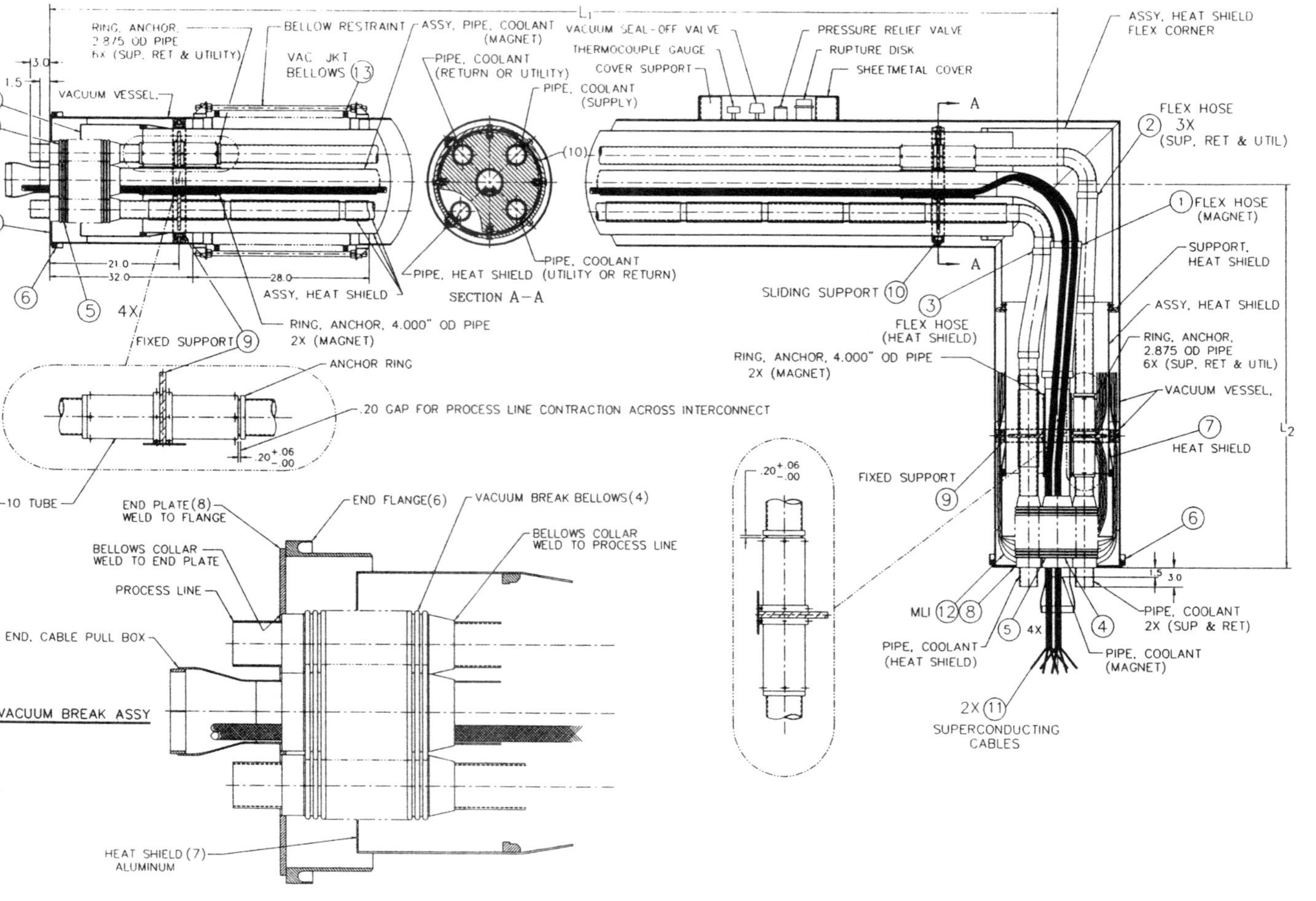


Figure 3. VJR standard elbow assembly.

## Expansion Elbow Section of VJR

Because of the physical size of the crossing region, the remote locations of the service buildings containing the valve boxes, the required height of the valve boxes to allow for vertical removal of the lead pots, and the length of the gaps between magnets in the insertion regions of the tunnel, the maximum straight piping runs for the VJR piping did not need to exceed 39 m (128'). The coldest helium process lines for this length would contract 11.6 cm. A maximum of 11.6 cm between pipe bends means that each bend must accommodate a maximum contraction of 5.8 cm. This result made it possible for us to handle all process line thermal displacements, including the independent displacements resulting when each line is at a different temperature, by means of the lateral displacement of flexible hose in the 90° elbow section of the VJR pipe. The use of flexible hose rather than in-line pressure bellows has the advantage of eliminating the very high hydraulic end loads on VJR piping internal components and external anchor supports, as well as eliminating sophisticated process line guide supports.

The VJR vacuum jacket is a 50.8 cm, 304/304L SST pipe. This diameter was determined by the number, size, and arrangement of process lines required (Fig. 3), as well as the spacing between lines required to be able to use automatic pipe welding and cutting equipment in the interconnect region, where the VJR components are connected together. It turned out that the 50.8 cm jacket was just large enough to accommodate the 5.8 cm flexible hose offset required for process line contraction, thereby making it possible to incorporate the expansion elbow section neatly into the basic elbow design.

## Independent Insulating Vacuum Space

The insulating vacuum for the VJR pipes is established, baked-out, and tested at the manufacturing plant prior to shipment. The pipes are then shipped under vacuum and checked upon receipt. Typically, 2 to 3 months may pass before a pipe is ready for installation. If the vacuum is still good, we know we are not likely to have a problem with that pipe. This method, which is preventative in nature, was chosen primarily to absolutely minimize the occurrence of VJR pipe leaks in the field after installation. Pipe leaks are very operational time consuming, and expensive to find and repair. This requirement helps the manufacturer to focus on leak tightness, since he cannot ship a leaky pipe. We benefit from this approach in that we experience very few leaks, we get a superior "baked-out" vacuum which we would not be able to produce at the RHIC site, and we do not have to pump the pipes down after we install them. Other advantages of the vacuum breaks become clear with additional knowledge of the piping system. The VJR system contains 166 pipes (1/3 of which are less than 9 meters long) and 160 interconnects 1.2 meters long. Therefore, half of the 320 vacuums spaces are 1.2 meters long. Leaks in the interconnects, which are much more likely because they must be field assembled and welded, are much less troublesome. They are isolated to a small vacuum space that is easy to find, and relatively easy and quick to fix, because the interconnect is designed to be opened and closed to allow access to the superconducting cables. Moreover, since the length of the interconnect vacuum space is small the thermal impact may be insignificant enough to allow the leak to continue until shutdown for yearly maintenance. Even more significant leaks may be allowed to continue since they are isolated by the vacuum breaks and do not affect the thermal performance of large portions of the VJR piping system.

The insulating vacuum is retained by a vacuum break assembly at each end of the pipe assembly. The vacuum break assembly is composed of a .95 cm thick end plate (8) (Fig. 3) and five vacuum break bellows (4)(5). The five process lines pass through holes in the end plate. Each bellows is welded at one end to the end plate and at the other end to the process line. The bellows is formed from a 0.203 mm wall stainless steel tube about 11.4 cm in diameter and about 2.3 m long. The final formed bellows is about 22.9 cm long and contains about 5 convolutions per centimeter of length, providing a high resistance to conduction heat leakage (about 1/8 Watt per bellows) from the 300 K end plate to the 4 K process line.

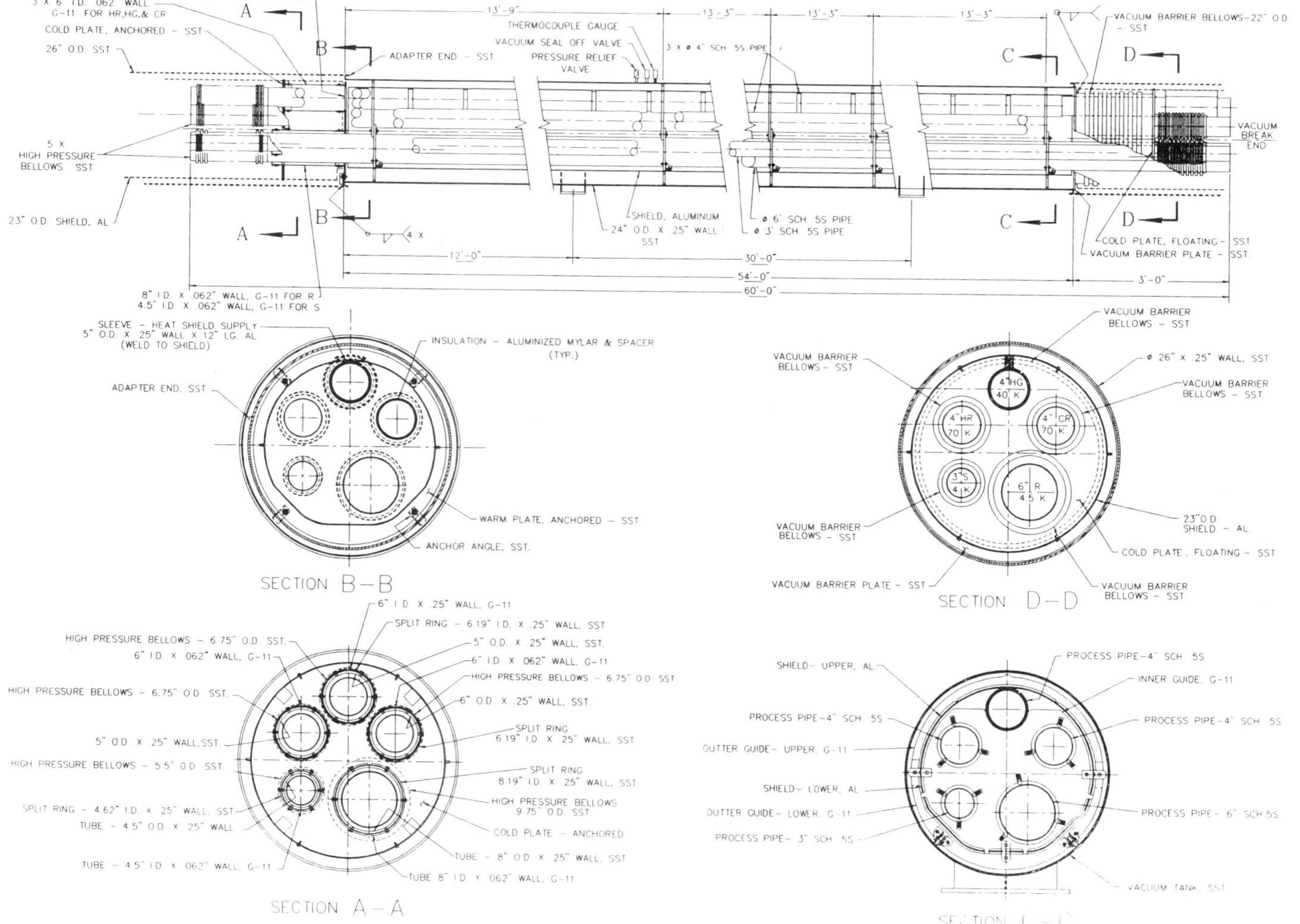


Figure 4. VJRR piping assembly.

The spring rate of this bellows is only a few pounds per centimeter. This flexible vacuum break allows process line movement during thermal contraction and expansion while maintaining its leak tightness.

### Vacuum Jacket Thermal Displacement Compensation

In the VJR piping runs, external main anchor support points occur near the pipe elbows. These anchor points help to insure that the VJR pipes do not experience bending due to lateral displacements at the ends. Lateral displacements could potentially cause binding of the internal process lines, which are designed to independently move axially as a result of temperature changes. The anchors also keep the vacuum jacket at a constant length, which is necessary to insure proper functioning of the expansion elbow section. However, a bellows is required in the vacuum jacket between any two-anchor points to avoid large thermally induced loads from breaking the anchors.

The VJR Vacuum Jacket Bellows (13) with bellows restraint rods is shown Fig. 3. This bellows has a .76 mm thick wall and an 1891 N/cm spring rate. The vacuum jacket bellows is restrained with an initial pre-compression of 5.1 cm, which provides a force of 9,608N that acts in opposition to the atmospheric end force resulting from the insulating vacuum. Therefore, the restraining rods see a net force of about 12,633N compression instead of the full 22,241N vacuum load. This allows us to reduce the rods from 25.4 mm to 19.05 mm diameter, resulting in a low profile bellows restraint assembly. The benefits of this design are threefold:

1) We obtain a low profile bellows restraint design, which is needed for the 4.9 m diameter corrugated steel tunnel, as well as for other low clearance areas.

2) In addition to lowering the load on the restraint rods, the 5.1 cm pre-compression in the 1891 N/cm bellows is designed to equalize (and therefore minimize) the magnitude of the worst case anchor pull and push loads, resulting from the various operational and failure modes which occur during operation when the bellows restrains would not be in place. This minimizes the strength and stiffness required by the structural steel supporting the VJR pipes, thereby reducing cost.

3) The higher stiffness bellows, being thicker (.76mm), are much less susceptible to accidental damage during shipment, storage and installation.

The bellows restraint rod assemblies are all stainless steel. The assemblies are bolted to the bellows collar to make them easily removable and replaceable. The restraints are to be removed immediately after the VJR pipes have been completely installed. They are then shipped back to Gardner for reuse. At the end of the project, about 30 sets of restraints will be returned to BNL. The restraints must be replaced, prior to removal of the pipe from the system or opening of the adjacent VJR Interconnects.

## VACUUM JACKETED REFRIGERATOR TO RING (VJRR) PIPING SYSTEM

### VJRR Design Concepts

The VJRR piping system consists of a single, mostly outdoor, above-ground, piping run of approximately 350 meters (1,148') between the helium refrigerator and the 6 o'clock valve boxes (Fig. 2). The VJRR (Fig.4) includes five (5) helium lines: Heat Shield Supply (11.43 cm, 40 K), Heat Shield Return (11.43 cm, 70 K), Recooler Supply (8.89 cm, 4 K) and Recooler Return (16.83 cm, 4 K), and Cool Down Return (11.43 cm, 70 K). All five lines are contained within a common 60.96 cm vacuum jacket and the 55 K Heat Shield. Since the VJRR piping run is characterized primarily by long straight runs, the flexible hose (expansion elbow) concept, used in the VJR piping, could not be used. Thermal contraction is compensated for by the use of high pressure guided bellows (Fig. 4) placed directly in the helium lines. The basic piping component for most of this system is a straight section approximately 18.3 m long. The design pressure and temperature for these lines is 18.9MPa

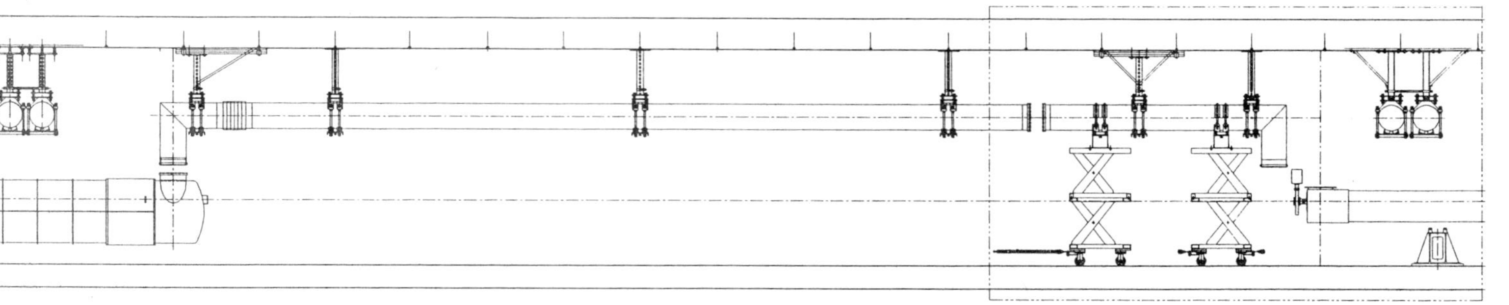

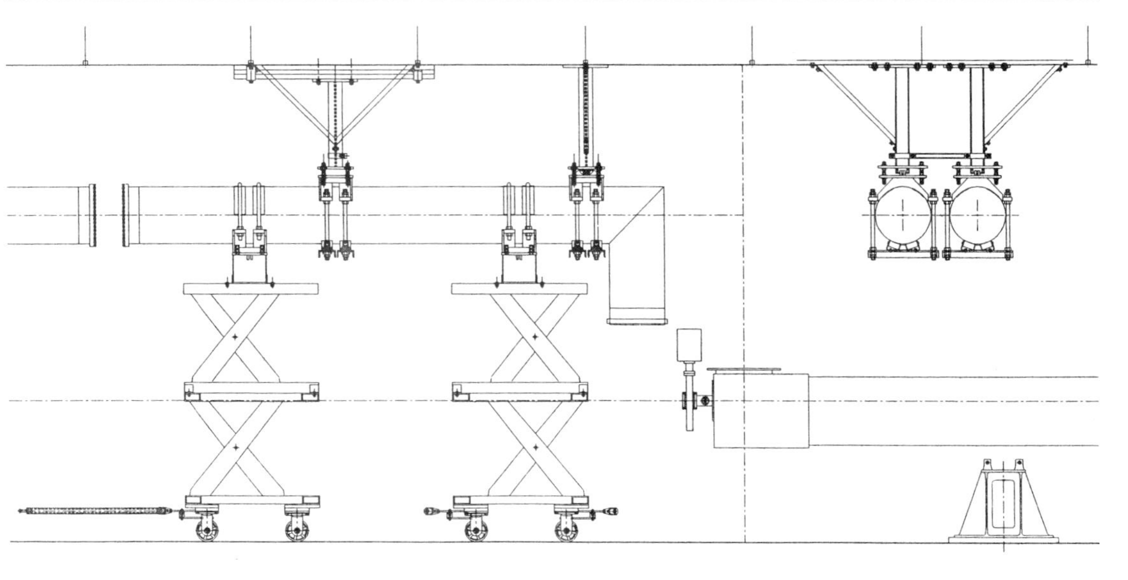

Figure 5. VJR Tunnel piping installation

**Table 2.** Thermal Performance

| Heat Load Category | Piping System | MLI (W/m) | Supports (W/sup) | End (W/end) | Calculated Heat Load (W) | Allowable Heat Load (W) |
|---|---|---|---|---|---|---|
| Primary | VJR | 0.1818 | 0.0954 | 0.238 | 621 | 1200 |
| (4.2K) | VJRR | 0.2556 | 0.0344 | 0.412 | 95 | 200 |
| Secondary | VJR | 0.738 | 1.082 | 0.238 | 2986 | 5000 |
| (55K) | | | 1.424* | | | |
| | VJRR | 0935 | 1.015 | 1.811 | 460 | 800 |

*fixed support

and 4.2 K, respectively. MLI is used throughout to minimize radiation heat leak. This piping system contains no electrical cabling.

The VJRR piping components contain one vacuum barrier per pipe. As a result, it is necessary to pump out each pipe vacuum space after field installation is completed. This vacuum barrier is based on the same principle as that of the VJR, however, due to the process line sizes and arrangement, it is a much more complicated assembly. The vacuum jacket bellows is included in the interconnect clamshell. The bellows is not pre-compressed, but is installed at its free length. The support system consists of numerous guide supports and a few large main anchors located at the ends and at the few bends existing in the piping run. The main anchors must be adequate to carry both the process line hydraulic end loads, as well as the atmospheric end load, because of the use of the inline pressurized bellows in the process lines.

## TUNNEL INSTALLATION VJR

During tunnel installation, two VJR elbow assemblies are each supported by roller bearing supports attached to the ceiling (Fig. 5). No ceiling anchor supports are used. The main anchors for the VJR pipes in the tunnel are provided by the magnet cryostats.

Two scissor-jack type transporters were designed and built by BNL. A pair of these units transport the VJR piping around the tunnel arcs to their installation points, and lift them into position, as shown. Once lifted into the support cradles and adjusted into position, the superconducting cables (11) (Fig. 3) must be installed before the interconnect can be closed.

Upon completion of the superconducting cable installation and the VJR pipe-to-magnet cryostat interconnect assemblies, the VJR to VJR interconnect is closed and pumped down. Finally, the bellows restraints are fully removed, thereby transferring the combined bellows pre-compression and atmospheric end load from the restraint rods through the VJR/magnet interconnects to the magnet cryostat's main anchors. All VJR piping loads, from this point on, are restrained by these anchors. This completes the installation.

## SUMMARY

The First Sextant Test, the major milestone test of the RHIC project, was preformed from January through March of 1997. The test included all 350 meters of VJRR piping, from the refrigerator to the 6 o'clock valve boxes, and all of the VJR piping from the 6 o'clock valve boxes through the tunnel to the 4 o'clock valve boxes (Fig. 2). The total VJR length, including 24 pipes and 18 interconnects, was approximately 365 meters.

The mechanical performance of both the VJR and VJRR was without fault, except for one small cold spot on a VJRR elbow and one leak in a VJR interconnect assembled by BNL. We were unable to obtain the actual thermal performance (Table 2) of the VJR/VJRR piping, however all indications were that the piping was performing very well.

# A 1400 LITER 1.8 K TEST FACILITY

T. J. Peterson, R. J. Rabehl, and C. D. Sylvester

Fermi National Accelerator Laboratory*
Batavia, Illinois 60510, USA

## ABSTRACT

A double bath superfluid helium dewar has been constructed and operated at Fermilab's Magnet Test Facility. The 1.8 K portion of the dewar is sized to contain a superconducting magnet up to 0.5 meters in diameter and 4 meters long in a vertical orientation in 0.12 MPa pressurized superfluid. The dewar can also provide a subcooled Helium I environment for tests; the entire temperature range from 4.4 K to 1.8 K at 0.12 MPa is available. This paper describes the system design, lambda plate, heat exchanger, and performance.

## INTRODUCTION

A double bath[1,2] superfluid helium system for testing superconducting accelerator components has been constructed and operated at Fermilab's Magnet Test Facility. Two unusual features of this system are its large volume—1450 liters below the lambda plate—and the external heat exchanger, which is designed to promote sufficient natural circulation to allow operation with subcooled Helium I, and give us the full range of temperatures from 4.4 K to 1.8 K at 1.2 bar pressure.

Figure 1 illustrates the dewar. The volumes are shown in Table 1, below.

**Table 1.** Volumes in the dewar

| | |
|---|---|
| Total volume above the lambda plate | 333 liters |
| 4.3 K liquid volume above the lambda plate | 100 liters |
| Total volume below the lambda plate | 1450 liters |
| Magnet volume during last run | 82 liters |
| Displacer volume during last run | 449 liters |
| Net helium volume below lambda plate during the last run | 919 liters |

* Work supported by the U. S. Department of Energy under contract No. DE-AC02-76CH03000

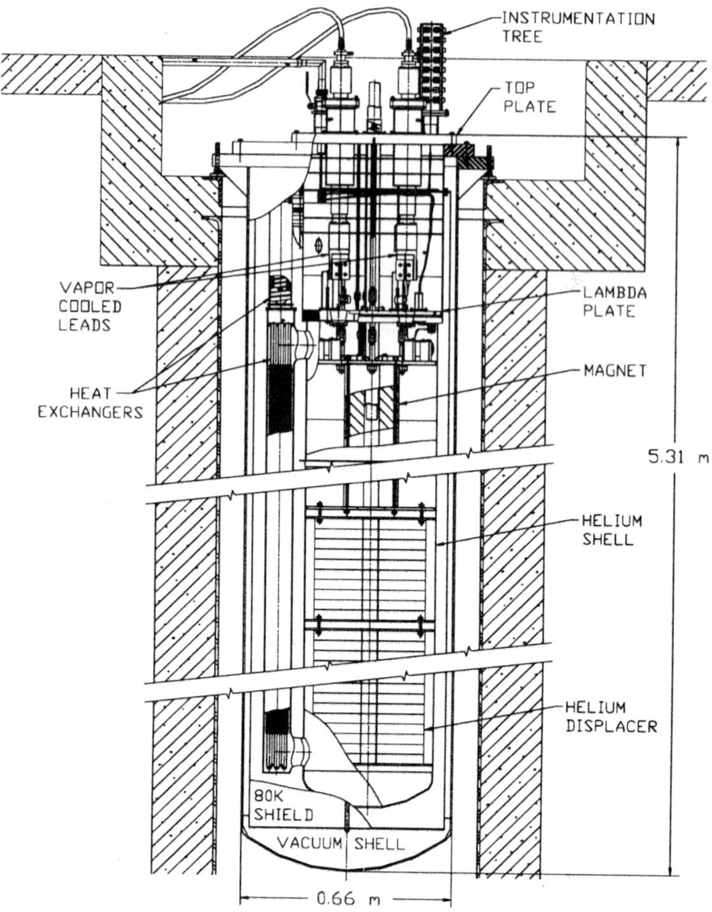


**Figure 1.** The dewar assembly.

The inner vessel, heat exchanger, and tubing are surrounded by a liquid nitrogen-cooled thermal shield and contained in a vacuum vessel. As is indicated in Figure 1, the vacuum vessel flange rests on the top of the concrete-lined pit and serves as the vessel support. The magnet assembly (which includes instrumentation tree, top plate, current leads, lambda plate, magnet, and Rohacel displacer) is inserted as an assembled unit into the inner vessel from the top.

## HEAT EXCHANGER DESIGN

The low temperatures below the lambda plate are produced by means of a heat exchanger, shown just to the left of the inner vessel in Figure 1, consisting of seven parallel vertical copper tubes, each 4 m long and 38 mm diameter with 3.0 mm thick walls. During normal operation, the tubes, which are connected together with a manifold at top and bottom so as to have the same liquid level in all seven tubes, are nearly filled with liquid helium at saturation pressure. Liquid level is measured by means of a probe, which extends down the length of one of the seven tubes. Pressurized 0.12 MPa liquid helium from the portion of the dewar below the lambda plate circulates by free convection via an upper and lower port over the outside of the seven copper tubes. The four-meter height allows a small density difference to provide large flow rates and keep temperature differences small in normal fluid. The large ports and heat exchanger surface area are more than adequate for heat transport during superfluid operation.

The boiling helium inside the seven tubes is supplied from the 4.3 K fluid above the lambda plate via a small Hampson-style, helical tube-in-shell heat exchanger and a control valve. The tube-in-shell heat exchanger precools the liquid to the lambda point (2.17 K) during operation at temperatures below the lambda point. This heat exchanger is similar to the analogous one in our horizontal superfluid test stand (Stand 5)[3].

## LAMBDA PLATE DESIGN

In a typical double bath liquid helium test dewar, an insulator is used to separate the 4.2 K liquid from the superfluid helium. For the Vertical Magnet Test Facility at MTF, this feature is derived from a 50 mm thick G-10 plate. Matching conical tapers machined in a type 304 stainless steel ring which is bonded to the G-10, provide a reliable seal between the two liquid volumes, while assuring ease of assembly and disassembly. For the outer perimeter seal the plan was to rely on matching machined conical tapers between the lambda plate and the dewar wall. Imperfections from manufacturing resulted in a mismatch between these two surfaces, requiring the addition of some filler material. The stainless steel ring and tapered plugs are attached to the G-10 by an adhesive bond between the G-10 disk and the stainless steel. Stycast 2850FT (Emerson Cumming Company) with catalyst 24LV was used for this application, and for the filler material. Cab-o-sil was added to the mixture used for the filler. Stycast 2850FT is a highly filled, epoxy system possessing excellent electrical and insulative properties. In addition, this system has excellent resistance to chemicals and solvents, high thermal conductivity and low thermal expansion. To test the effectiveness and strength of this design, and the performance of the Stycast adhesive over a few thermal cycles from room temperature to 4.2 K, a prototype lambda plate was developed and tested at 4.2 K prior to completing the final design of the VMTF unit.

## OPERATIONAL RESULTS

### Cool-down from 300 K to 4.3 K

The initial cool-down is accomplished by flowing liquid nitrogen through copper tubing which is wrapped around the magnet, with a positive pressure of helium gas actively maintained in the dewar. We do not attempt to make good thermal contact of the copper tube to the magnet; cool-down is primarily by free convection of the cooled helium gas around the warm magnet. Nevertheless, cool-down to under 100 K took less than 24 hours.

Figure 2 shows a flow schematic for the system. When the thermometry indicates less than 100 K throughout the dewar, the nitrogen is pumped out of the copper tube, liquid helium is supplied through control valve V3 to the bottom of the dewar, and helium vapor is vented through the current leads. A valve in the lambda plate is open during the cool-down process, allowing vapor and liquid to flow from the lower part of the dewar into the upper part. Liquid is supplied through valve V3 to the bottom of the dewar until the level rises above the lambda plate. Then valve V3 is closed, and valve V1 controls liquid level above the lambda plate for normal operations.

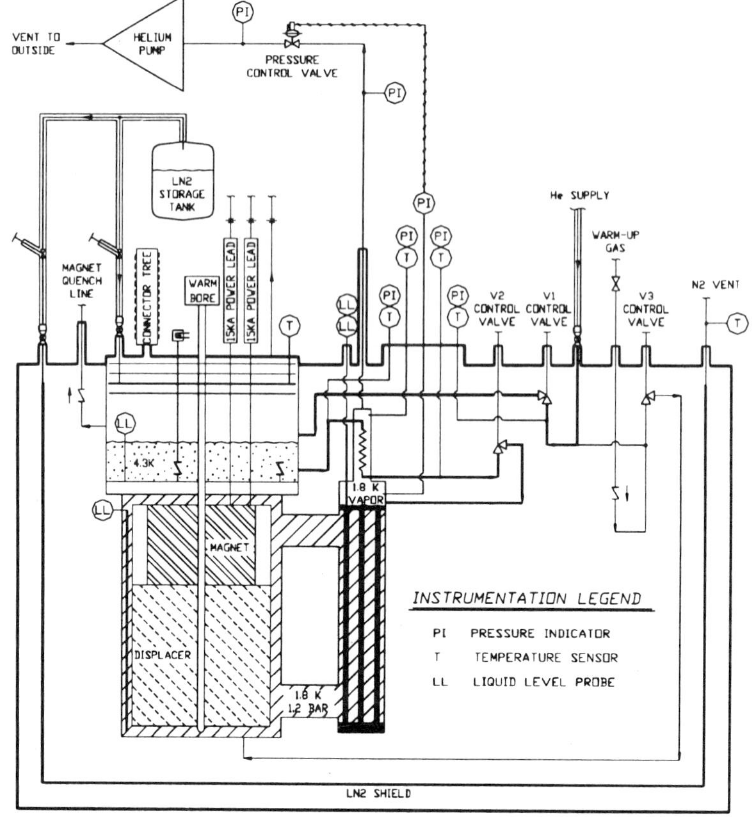


**Figure 2.** Simplified flow schematic of the vertical dewar system.

## Operation below 4.3 K

For operation below 4.3 K, valve V2 (see Figure 2) maintains the liquid level inside the seven tubes of the heat exchanger. Vapor is pumped away by a room-temperature helium pumping system consisting of an oil-injected blower, Kinney model KMBD-3201, followed by a two-stage liquid ring pump, Kinney model KLRC-951. This is the same pumping system, which has been used since 1988 to reach sub-lambda temperatures in our horizontal magnet test stand.[3]

Cool-down from 4.3 K to 1.8 K is shown in Figure 3. Pump-down from 4.3 K to the lambda point in this case took about 4 hours, with another 2 hours to 1.8 K. Temperatures were subsequently bumped back up to 2.0 K by two magnet quenches, and quenching continued with recoveries to 1.9 K.

One interesting feature of these cool-downs from 4.3 K to below the lambda point is the behavior of the thermometer located just below the lambda plate. Its temperature remains between 4.2 and 4.3 K until all the others have reached the lambda point, then it suddenly drops directly to 2.17 K. The vertical line in figure 3 is not just an artifact of the slow time plot; the temperature drop occurs within a few seconds. This dramatic effect is caused partly by the fact that the helium just below the lambda plate remains very effectively stratified and that the normal liquid is a poor thermal conductor. The maintenance of the 4.2 K temperature is also aided by the fact that liquid is drawn down from above the lambda plate during the cool-down between 4.2 K and 2.17 K.

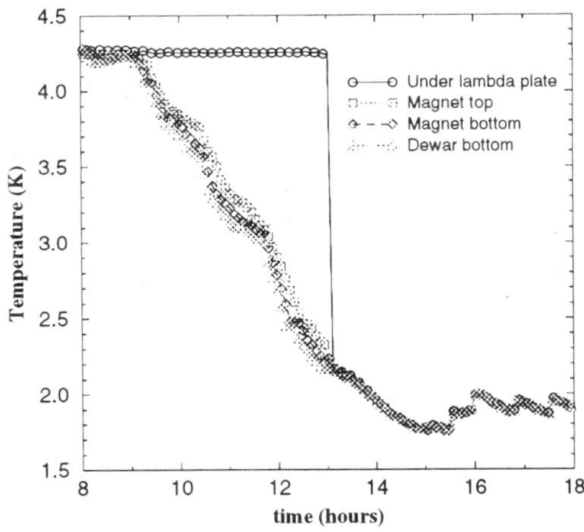


**Figure 3.** Cool-down from 4.3 K to 1.8 K followed by magnet quenches and recoveries.

Figure 4 shows a plot of helium mass below the lambda plate versus temperature. In cooling from 4.22 K to 2.167 K (lambda point) the density increases from 125 g/liter to 148 g/liter. So an additional mass of helium equivalent to 169 liters of 4.2 K liquid was pulled into the 919 liter volume through leaks in the lambda plate. This influx of 4.2 K helium helps to hold the temperature of the thermometer just below the lambda plate at 4.2 K.

An odd phenomenon which is caused by helium density changes is the response of pressure in the dewar to a quench in superfluid. Although there is an initial pressure increase, it is followed immediately by a sharp decrease to less than the starting pressure. Quench valves do not open. The initial upward spike is probably due to film boiling and vapor generation in the superfluid around the magnet coil. This vapor then collapses, and the superfluid bath returns to essentially isothermal conditions at a higher temperature. The helium density is a maximum at the lambda point, and warming from 1.9 to a temperature still below 2.17 K increases density, so pressure in the fixed volume decreases.

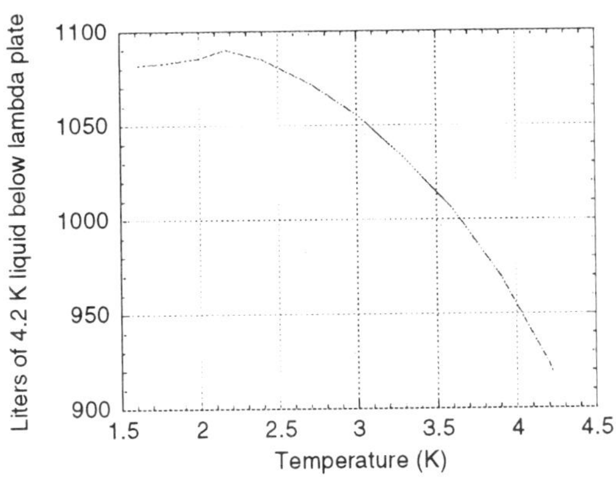


**Figure 4.** Helium mass below the lambda plate expressed as 4.2 K liquid liter equivalents for our last run, with 919 liters of helium volume under the lambda plate.

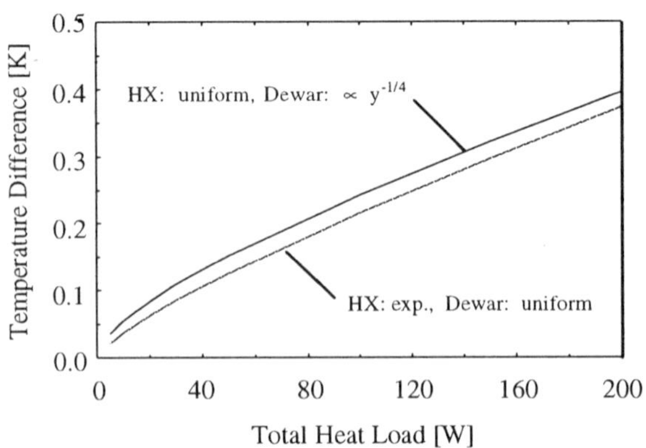


**Figure 5.** Calculated temperature difference between the top and bottom of the dewar vs. total heat load for operation of the dewar with 3 K normal helium for two possible cases of heat transfer distribution.

## Operation in subcooled normal fluid

An interesting aspect of the dewar operation that has been studied is that of the effect of natural convection on the temperature difference between the top and bottom of the dewar. A finite difference model was constructed to simulate the buoyancy-driven flow of 3 K normal helium through the dewar and the heat exchanger. Specified heat transfer distributions in the dewar and in the heat exchanger are required, as is the helium state at one node. The model is then solved with the constraint that the pressure drop through the closed helium circuit must be zero. The pressure drop through each node is calculated, and an energy balance is performed on each node. The solution yields the helium mass flow rate and the helium state at each node. Figure 5 is a plot of the calculated temperature difference between the top and bottom of the dewar as a function of the total dewar heat load for two possible cases of heat transfer distribution. Using the heat transfer distributions of the two curves with a 2.7 K bath and a 10 W total heat load, predicted temperature differences of 0.065 K and 0.043 K are in good agreement with the measured 0.05 K temperature difference.

## Thermal performance with superfluid

Heat load to the superfluid portion of the dewar was checked by two methods. The first was by estimating the boil-off rate from the heat exchanger. Room-temperature tests in which flow was metered into the dewar and pumped by the Kinney pumps indicate that they pump a mass flow (g/sec) = 2.1 x P/1600, where P is the pressure measured in Pascal at the pump inlet. This is somewhat less than the nominal rating of 2.5 g/s at 1600 Pa, probably due to some volumetric inefficiency.

With this estimate of pumped mass flow, we can estimate the heat removal rate from the dewar below the lambda plate. With 0.12 MPa, 2.17 K helium expanding into the low pressure region of the heat exchanger, at 1.7 K about 13% of the flow flashes directly to vapor with the remaining 87% liquid vaporizing at 23.2 J/g. So the total cooling at 1.7 K is 20 Joules per gram of pumped flow.

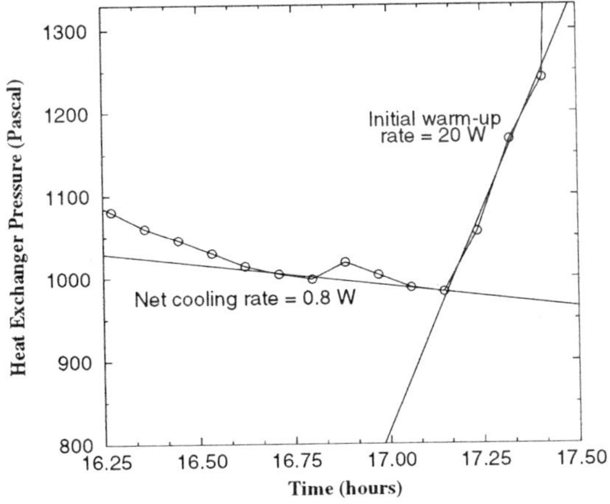


**Figure 6.** Pressure during pump-down to 1.7 K.

Figure 6 shows a plot of pressures as we pumped down to 1.7 K. Nearly leveling off at slightly below 1.7 K (980 Pascal), our heat load as indicated by pumped flow (705 Pascal at the pump, so 0.93 g/s mass flow) is 18.6 W - 0.8 W of cooling = 17.8 W.

Upon warm-up a heat load estimate is possible while temperatures remain below the lambda point since the dewar is still isothermal (Figure 7). The initial warm-up rate was 20 W, in good agreement with the pumping rate. The boil-off from the heat exchanger is just enough to maintain the vapor pressure of the helium on the low-pressure side of the heat exchanger as it warms up, which requires only a very small mass due to the low density of the vapor at low pressure. Only about 0.2 W is absorbed by the vaporizing helium on the low-pressure side during the initial warm-up from 1.7 K. So the heat exchanger has a negligible effect on this estimate of 20 W heat load.

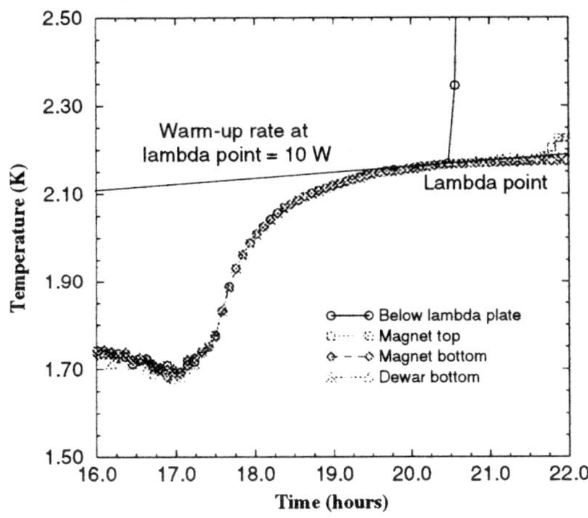


**Figure 7.** Temperatures below the lambda point during warm-up from 1.7 K to 2.2 K.

**Table 2.** Heat loads to the dewar below the lambda plate

| Source or mechanism for heat flow to 1.8 K | Heat (W) |
|---|---|
| Current leads | 2.50 |
| Conduction down magnet supports | 0.02 |
| Conduction down vessel walls | 0.00 |
| Cond through G-10 lambda-plate | 0.94 |
| Cond thru stainless lambda seal ring | 0.50 |
| Thermal radiation from sides | 0.72 |
| Warmup/fill line | 0.88 |
| Valves | 1.30 |
| Instrument wires (approx. 600 wires) | 0.18 |
| Heat flow thru s.f. in voids in inst. wire | 0.14 |
| Heat to He II via lambda plate "leaks" | 10.00 |
| **TOTAL HEAT LOAD (WATTS)** | **17.18** |

Just below the lambda point, the temperature difference through leaks in and around the lambda plate to normal fluid is very small. So this heat load includes only the heat loads other than those associated with the superfluid heat transfer. The result is a heat load at 2.17 K of 10 Watts (Figure 7). The difference in these two estimates represents the amount of additional heat that enters via superfluid at 1.7 K, about 10 Watts. Table 2 shows our estimates of heat loads to the dewar below the lambda plate. Values are calculated except for "Heat to He II via lambda plate leaks."

## CONCLUSIONS

A 4 meter tall, 1450 liter dewar capable of operating at temperatures from 4.4 K to below 1.8 K has been commissioned at Fermilab. With normal subcooled liquid helium between 2.2 K and 4.2 K, free convection provides temperatures uniform to within 0.050 K. Low heat loads and good sealing at the lambda plate have allowed operation down to 1.7 K.


## ACKNOWLEDGEMENTS

The authors wish to thank several reviewers who assisted with some early design critiques: Eric Haggard, Moyses Kuchnir, Al Lietzke, Peter Mazur, Steven van Sciver, and Jay Theilacker. Clark Reid was responsible for the design/drafting for this project. Meyer Tool & Manufacturing, Inc., contributed many design details and manufactured the dewar. The technical staff of Fermilab's Magnet Test Facility completed the system assembly, instrumentation, software, and operated the system. Their expertise and experience contributed greatly to the success of this project.


## REFERENCES


1. G. Claudet, et al., The design and operation of a refrigerator system using superfluid helium, in: "Proceedings of the 5th International Cryogenic Engineering Conference," Kyoto (1974), pp. 265-267.
2. G. Claudet, et al., Superfluid helium bath for superconducting magnets, in: "Proceedings of the 6th Cryogenic Engineering Conference," Grenoble (1976), pp. 159-162.
3. P. O. Mazur and T. J. Peterson, A cryogenic test stand for full length SSC magnets with superfluid capability, in: "Advances in Cryogenic Engineering," Vol. 35, Plenum Press, New York (1990), pp. 785-792.

# PERFORMANCE AND CONTROL OF A CRYOGENIC SYSTEM COOLING 1152 VLPC CHANNELS

T. H. Gasteyer and P. D. Wheelwright

Fermi National Accelerator Laboratory *
Batavia, IL, 60510, USA

## ABSTRACT

A liquid helium supplied cryogenic system has been designed, built, and operated continuously since September 1996. The system cools nine cassettes (128 channels each) of visible light photon counter (VLPC) chips to 6.5 K. Thermal stability of the chips has been excellent. Liquid helium is supplied from a modified CTI 1400 helium liquifier. The liquid flows into a vacuum insulated liquid nitrogen shielded cryostat where it cools the VLPC chips. Some of the helium boil off is used to reduce the cryostat heat leak. All boil off gas is returned to the compressor suction header for the Tevatron. The exceptional control of the cassette temperatures is achieved with two levels of PID control. The coarse control (All cassettes within 0.5 K) is reached by maintaining constant cryostat liquid level, cryostat pressure, and heat intercept flow. The fine tuning (All cassettes within +/- 0.010 K) is achieved with automatically controlled heaters in each cassette. The operating temperature range for the chips can be set anywhere between 5 to 15 K. The cryostat heat leak is about 2.0 W at 6.5 K.

## INTRODUCTION

Visible light photon counters are an emerging type of electro-optic sensor used in a high-energy physics particle detector called a scintillating fiber tracker. During particle collisions a large amount of sub atomic particles are generated. The scintillating fiber tracker is a tool for measuring the trajectory of these sub atomic particles. The VLPCs read the optical signals from the fiber tracker and convert them into electrical signals that are read into the data acquisition system. The trajectories are then computed from this data. Fermilab experiment E835 has implemented a small size detector of this type. The

* Work supported by the U.S. Department of Energy under contract No. DE-AC02-76CH03000

VLPCs are designed to operate at one specific temperature (6.5 K). Deviations from this nominal operating temperature point are undesirable since they affect the gain, efficiency, and noise rate in the VLPCs. It is therefore desired to minimize these temperature deviations. This paper discusses the liquid helium supplied cryostat designed to cool the devices to their operating temperature, the control system for achieving the desired level of temperature control, and the resulting control system performance.

## CRYOGENIC SYSTEM DESIGN

A closed cycle liquid helium refrigeration system has been developed for the purpose of cooling cassettes containing VLPCs to a temperature of 6.5 K. The system is comprised of a compressed pure helium supply at 2.0 MPa (290 Psia), a CTI 1400 helium liquefier, a 450 L storage dewar, vacuum jacketed transfer line, a phase separator, a cryostat for cooling the sensors, and a suction header for returning helium to the compressor system (See figure 1). Our system is tied into the main accelerator (Tevatron) helium system. This provides us with a source of high-pressure helium and a place to return the helium exhaust from our system. We simply liquefy the compressed helium with a CTI 1400, use it, and return it. Since the experiment this device is used on will run for more than one year we have added a 450 L storage dewar to increase the long term system reliability. This provides us with a backup liquid helium supply that will last for several days. The storage dewar has been used when the Tevatron high pressure helium system is off for maintenance, during power outages, and during CTI 1400 engine maintenance. It has reduced the number thermal cycles on the cryostat and the experimental down time. To lower the heat leak into the liquid helium, liquid nitrogen cooled shields are used throughout the system in the vacuum space. Control of the liquid level in the cryostat is of critical importance to control of the VLPC temperatures. Since there are about 35 meters of transfer line between the 450-L storage dewar and the cryostat we have added a phase separator directly upstream of the helium inlet valve. This provides essentially a 100 % liquid inlet condition to the control valve. By doing this, the sizing of the control valve can be done more accurately and the LHe level stability is better because the control system does not have to fight changing inlet conditions which are flow

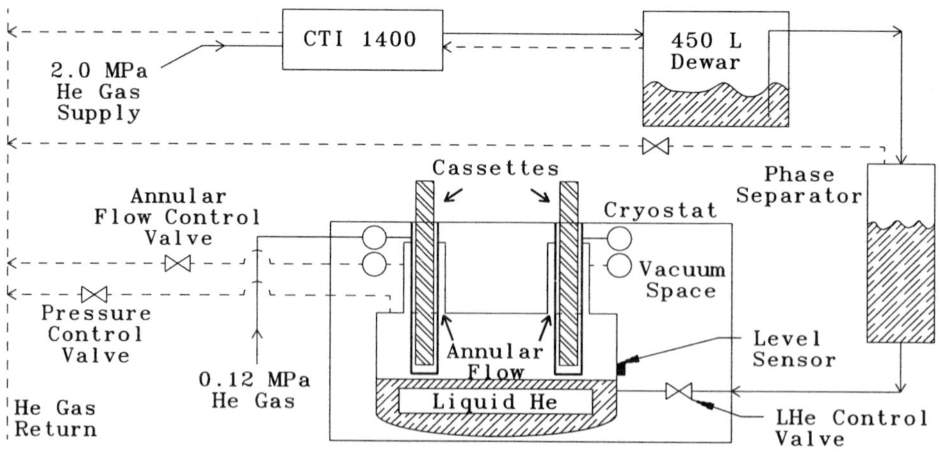


**Figure 1.** System schematic.

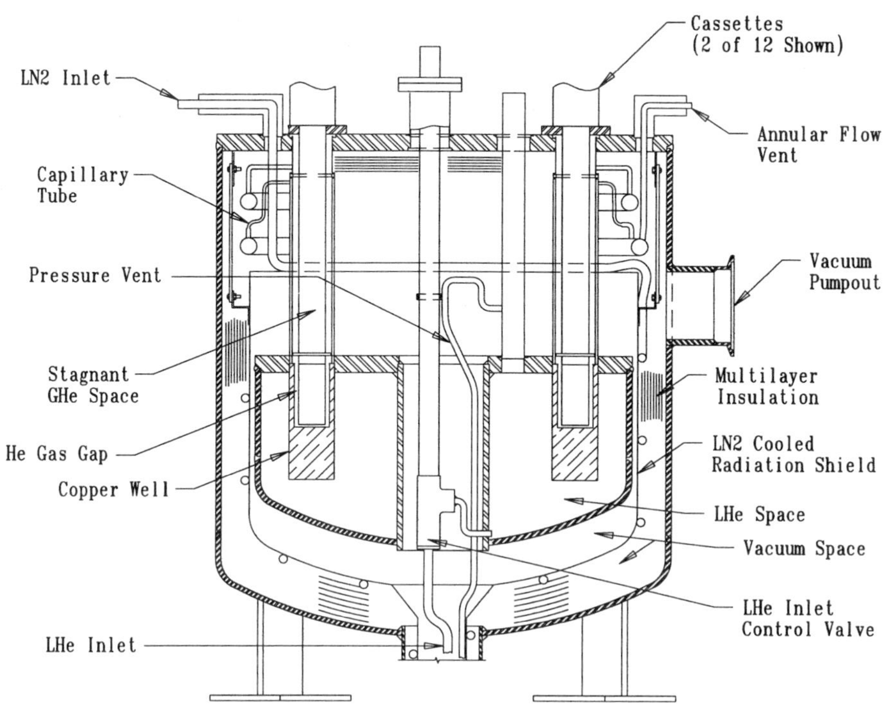


**Figure 2.** E835 VLPC cryostat.

rate and heat leak dependent. The LHe flows through the cryostat and is boiled off. This boil off is warmed up and flows through two control valves back into the Tevatron helium gas return.

The cryostat (See figure 2) contains a series of copper wells that hermetically seal the liquid helium space from the stagnant gas helium space inside the cassette volume. The liquid helium is controlled to a level just at the bottom of these wells. It cools them to roughly the saturation temperature of the liquid helium. On the inside of the well there is a hollow space which the cassette plugs into. The liquid helium inlet control valve establishes the level of liquid helium to a constant value. Boil off from the cryostat flows out two paths. There is a pressure vent whose flow rate is adjusted to maintain a constant pressure in the cryostat. And there is an annular flow vent whose purpose is to reduce the heat leak to the liquid helium by cooling the tubes that connect the vacuum vessel to the liquid helium vessel. These tubes are of double wall construction and the cold gas flows

**Table 1.** Cryostat and cassette thermal specification

| | Specification | Actual |
|---|---|---|
| Cryostat heat leak | 5 W | 2 W |
| Nominal VLPC operating temperature | 6.5 +/- 0.050 K | 6.5 +/- 0.010K |
| VLPC operating temperature range | 5 to 15 K | 5 to 15 K |

through the annular space between the tubes. The cryostat contains penetrations for twelve cassettes (Only 9 are currently used). The parallel gas outlets of the annular tubes are connected into a manifold with 3.2-mm (1/8 in.) capillary tubing. The purpose of this is to insure even flow distribution from tube to tube. Both the pressure vent and annular vent flows are then warmed to room temperature and go through mass flow control valves. (Figure 1) Since we have a closed cycle system, contamination is very important and must be maintained at a low level (Less then 1-PPM nitrogen). To monitor the contamination level, part of the helium exhaust flow goes through an arc cell that measures the PPM level of nitrogen in the return gas. The cassettes themselves are not hermetic and can introduce contamination into the stagnant gas space. This is the main reason the gas helium inside this space is separate from the liquid helium. The helium in the stagnant gas space is only there to provide a uniform heat transfer medium with the proper thermal resistance so that the VLPCs will operate close to there nominal operating temperature. Table 1 displays a list of the basic thermal specifications for the cryostat and VLPC temperature control.

One of the design obstacles to overcome in controlling nine different cassettes at the same temperature is to make sure that the thermal conductance from the cassette cold end to the copper well is consistent and repeatable. Conduction type contact joints at cryogenic temps can be problematic and were not used in the current cassette design. The solution was to use a thermal conduction gas gap[1]. The heat dissipation through the cold end is relatively small (Approximately 0.1 W). This heat is transferred across the annular helium gap (Between the cassette cold end and the copper well in the cryostat) by conduction and natural convection. The required gap size is on the order of 1 mm. This type of joint allows for cassette thermal contraction, is easy to fit, and provides a relatively consistent thermal joint from cassette to cassette. The cassettes themselves are not made to a high degree of accuracy and there are significant variations from cassette to cassette. If the cassette is bent when inserted into the cryostat it could touch the side of the copper well. This not a catastrophic event but will cause this particular cassette to run colder.

## CRYOSTAT CONTROL SYSTEM DESIGN

Accurate and repeatable control of VLPC temperatures is accomplished through good thermal and mechanical design of the cryostat and the implementation of a stable control system. The basic control system diagram is shown in figure 3. A description of each of the control loops in the system is provided in table 2. Since the VLPC devices generate only a minuscule amount of heat and this heat is roughly constant if we can establish a stable cryostat liquid level, pressure, and annular flow rate then the VLPC temperatures

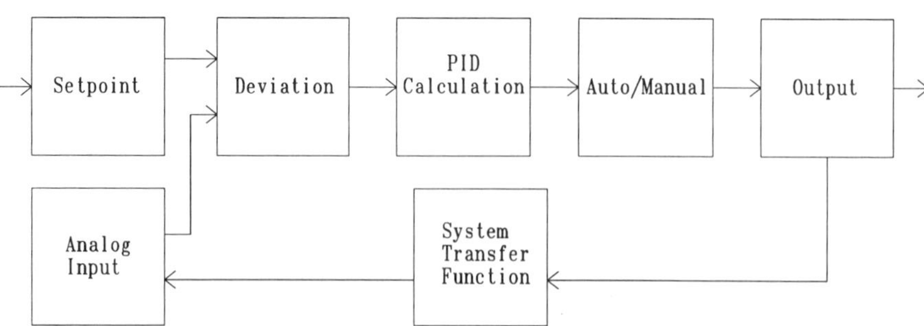


**Figure 3.** E835 VLPC cryostat control system schematic.

**Table 2.** Control system loop descriptions

| Loop | Setpoint | Input | Control Type | Auto/Manual | Output |
|---|---|---|---|---|---|
| LHe Level | Temperature | Cryostat Wall Temperature | PID | Auto | LHe Inlet Valve Position |
| Annular Flow | Mass Flow rate | Annular Flow | NA | Manual | Setpoint to Flow Control Valve |
| Cryostat Pressure | Pressure | Cryostat Pressure | PI | Auto | Setpoint to Flow Control Valve |
| Cassette Temperature | Temperature | LPC Temperature | PI | Auto | Current to Cassette Heater |

should be stable. In general there are two layers of control used to establish temperature control of the VLPCs.

The first layer consists of the three control loops that control the inlet liquid flow and outlet gas flow of helium through the cryostat. The first loop controls liquid helium level. This is controlled by a pneumatically actuated control valve (Cv = 0.32) which looks at a temperature sensor on the wall of the cryostat or a liquid level probe. We found that the temperature sensor on the outside of the cryostat was able to control the level better with a lower heat leak than a level probe. As the liquid level in the cryostat rises, this sensor reads a lower temperature. The sensor used is an Allen Bradley 100-ohm carbon resistor. At the control set point (Approximately 6.8 K) this sensor has a sensitivity of about 100 ohms/K. It took a full PID digital control loop to get this loop to work well. This is due to the thermal time delay of the signal as the heat travels through the stainless steel wall of the cryostat.

The second loop controls the annular flow to be constant. A commercially available, low pressure drop, thermal mass flow meter/controller is used for this loop. The controller range is 0 to 30 standard liters per minute (SLPM) helium with an accuracy of +/- 1% of full scale. This is a self-contained control system. A constant DC voltage setpoint signal is sent to this device. The controller then maintains a constant flow even when the upstream and downstream pressures are changing. The control circuitry in the flow controller proportionally opens or closes the valve so that the flow setpoint is achieved. The effect of this flow is to reduce the heat leak of the system from 42 W with no flow to about 2 W at the design annular flow. The constant nature of this flow is important to the overall control because it has a very strong effect on the cassette temperatures.

The third loop is for cryostat pressure. Its purpose is to maintain the cryostat pressure constant to avoid temperature changes induced by the effect of pressure on the saturation properties. A control loop which looks at the cryostat pressure and sends setpoint values to a mass flow controller is used to regulate the vent flow out of the cryostat. The type of mass flow controller used on this loop is the same as the controller used on the annular flow but the range is 0 to 80 SLPM helium. The boil off gas in both the vents is warmed up and passes through about 35 meters of tubing before reaching the flow controllers. Since the cryostat pressure is roughly 0.136 MPa (5 psig) and the vent pressure is usually about 0.115 MPa (2 psig) these controllers have to operate some times with as little as 0.007 MPa (1 psi) pressure drop due to line pressure drops. The purpose of all three of the loops is to achieve VLPC temperatures on all cassettes within 0.5 K.

The second layer of control consists of loops for trimming all the cassette isotherms to the same temperature. One hundred and twenty eight VLPCs are mounted on a highly conductive copper piece suspended in the bottom of each cassette. This piece is referred to as the copper isotherm. Each isotherm has two calibrated temperature sensors and a heater imbedded into it. A simple PI control loop for each isotherm then adjusts the heater output to maintain a constant isotherm temperature. With the heater on 100 percent the isotherm temp is raised about 0.5 K. The temperature set point of all the cassettes are set to the same value. The control circuit then adjusts the heater current from 0 to 20 mA to achieve the setpoint temperature. The sensors are driven by a 30 micro amp constant current source. The voltage across each isotherm sensor is then read into the control system with a temperature resolution of about 3.9 mK.

Since there are several loops that control parameters which affect the VLPC temperatures it is necessary to insure that there are no instabilities caused by loop to loop interactions. Specifically, the two loops which control the liquid helium level and the cryostat pressure must not interact. The philosophy used to prevent loop to loop interaction is to have the two loops have very different time response (One fast and one slow). For our present case the fast loop controls the liquid level and the slow loop controls cryostat pressure.

## CONTROL SYSTEM PERFORMANCE

Over the last 11 months the temperature stability of the VLPC chips has been excellent. Due to the large quantity of data recorded the figures shown in this section are from the first four days in May and are considered to be typical of control during the entire run. A temperature versus time plot of the VLPC temperatures is shown in figure 4. This plot shows the VLPC temperatures in a single cassette. One of the effects of being tied to the Tevatron helium system is that there are minor temperature upsets during magnet quenches. The suction pressure will rise from its normal value of 0.115 MPa (2 psig) to about 0.136 MPa (5 psig) or more. This causes the annular flow to stall and the cassettes

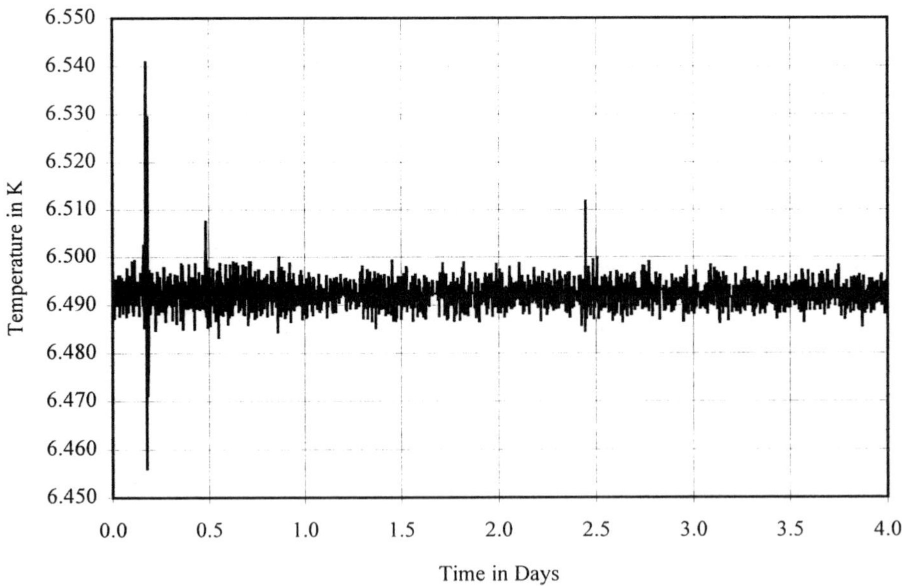


**Figure 4.** VLPC temperature stability during the first four days in May.

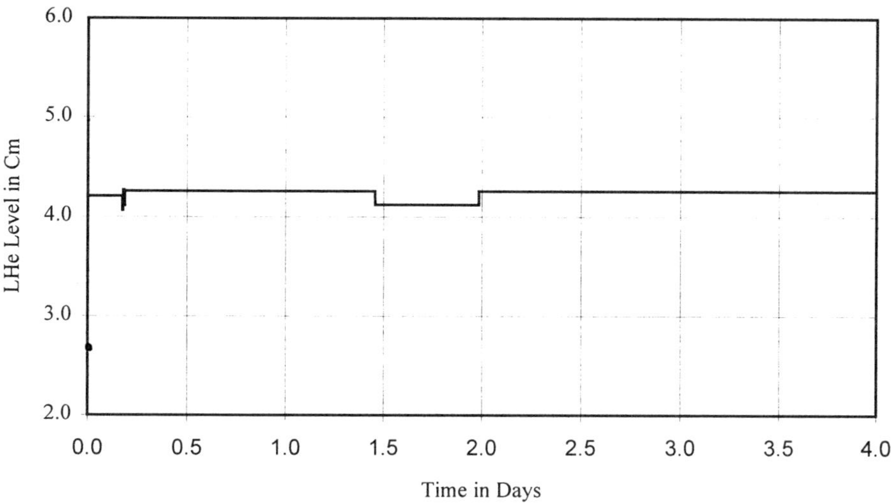


**Figure 5.** Typical cryostat level stability during the first four days in May.

begin to warm over their set point. Typically the maximum temperature rise is on the order of 30 to 40 mK (See temperature spikes in figure 4). This is within the allowable for the VLPCs. To get a better idea of how well the temperatures were controlled I have analyzed the data from four cassettes for the entire month of May using statistical analysis. During this period there were about 33,500 data points taken per cassette and 15 quenches. The results of this analysis are shown in table 3. The stability of the level (Figure 5) and pressure (Figure 6) in the cryostat was also excellent. In addition to having very stable temperatures the cryostat also has a low heat leak to liquid helium. The total cryostat heat leak is measured to be approximately 2.0 W. This data comes from the boil off flow rate measured by the mass flow controllers. For the most part, tuning of the control loops was accomplished with ease. All the loops were tuned using a trial and error method. The proportional gain was increased to a value below the point where system oscillation occurred. The integral gain was then increased until an acceptable level of offset was achieved.

**Table 3.** VLPC temperature control results for May

| | Cassette 1 | Cassette 2 | Cassette 3 | Cassette 4 |
|---|---|---|---|---|
| Control Setpoint | 6.492 K | 6.492 K | 6.492 K | 6.492 K |
| Average of Temperature | 6.4923 K | 6.4923 K | 6.4922 K | 6.4923 K |
| Standard Deviation of Temperature with Quenches | 0.0034 K | 0.0025 K | 0.0038 K | 0.0046 K |
| Standard Deviation of Temperature without Quenches | 0.0025 K | 0.0020 K | 0.0020 K | 0.0035 K |

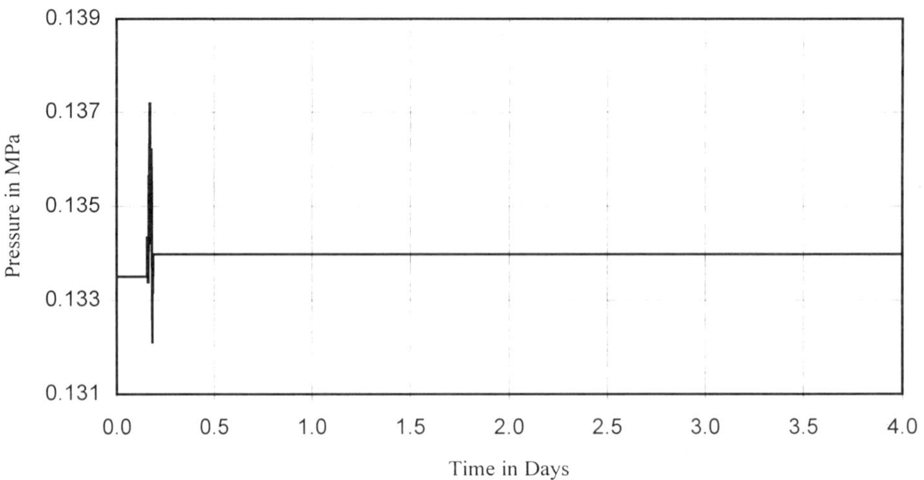


**Figure 6.** Typical cryostat pressure stability during the first four days in May.

## CONCLUSIONS

We have designed, fabricated, and tested a low heat leak cryostat for cooling a multitude of devices very accurately for long periods of time. The cryostat cools 1152 VLPCs to a temperature of 6.492 K +/- 0.010 K except during system upsets which happen infrequently. It has a total heat leak to liquid helium of approximately 2.0 W and has been operating for 11 months. The VLPCs have been controlled at up to 15.0 K by adjusting cryostat parameters (Liquid helium level, cassette insertion depth, cryostat pressure) and could be run at even higher temperatures.

## REFERENCES

1. T. H. Gasteyer, K. J. Krempetz, A. Lee, R. A. Rucinski, and A. M. Stefanik, "Heat transfer design and performance of a helium cryostat operating at 6.5 K", Advances in Cryogenic Engineering, Vol. 39A, P. 619, Plenum Press, New York, 1994

# THERMODYNAMIC ANALYSIS AND OPERATING EXPERIENCE OF THE NUCLOTRON HELIUM REFRIGERATORS IN THE "SATELLITE" MODE

N.N.Agapov, H.G.Khodzhibagiyan, and A.D.Kovalenko

Joint Institute for Nuclear Research
Dubna, Moscow region, 141980, Russia

## ABSTRACT

The cryogenic system for the superconducting synchrotron Nuclotron consists of 3 - 1,600 W refrigerators operating at 4.5 K. Since the Nuclotron only requires the output of 2 of these refrigerators, a distribution system based on the "satellite" refrigerator concept was developed to maximize reliability of the overall refrigeration system while simultaneously maintaining a high level of system efficiency. This paper presents a thermodynamic analysis of, and the experimental data for, the Nuclotron refrigerators operating in the "satellite" mode.

## NOMENCLATURE

| | |
|---|---|
| $e_0$ | specific exergy of liquid helium, J/g |
| F | surface of the heat exchanger, $m^2$ |
| G | compressed gas flow, g/sec |
| g | compressed gas flow per unit refrigeration capacity, (g/ sec)/W |
| i | enthalpy of high pressure stream at a certain position, J/g |
| $i_n$ | enthalpy of helium at the cycle position "n", J/g |
| k | heat transfer coefficient, $W/(m^2 K)$ |
| $p_3$ | compressor inlet pressure, MPa |
| $p_4$ | compressor outlet pressure, MPa |
| Q | refrigeration capacity, W |
| R | gas constant of helium, J/(g K) |
| r | heat of boiling, J/g |
| $T_a$ | ambient temperature, K |
| x | mass vapor content of the stream |

| | |
|---|---|
| $\alpha$ | liquid helium flow per unit compressed gas flow (coefficient of surplus return flow ) |
| $\Delta i_s$ | enthalpy reduction of the isentropic expansion, J/g |
| $\Delta T$ | temperature difference between high and low pressure streams at a certain position, K |
| $\varepsilon$ | total power expense per unit of refrigeration capacity |
| $\varepsilon_\alpha$ | liquefaction work of the independent liquefier |
| $\varepsilon_g$ | compression work |
| $\eta_c$ | isothermal efficiency of compressor |
| $\eta_{LHe}$ | efficiency of the independent helium liquefaction unit |
| $\eta_s$ | isentropic efficiency of the expander |

## INTRODUCTION

The superconducting synchrotron Nuclotron[1,2] has conducted 11 physics experiments since 1993. Its cryogenic system includes three parallel helium refrigerators. Each refrigerator has a nominal capacity of 1,600 W at 4.5 K. These refrigerators cool the accelerator ring, which has a perimeter of 250 m and a cold mass of about 80 tons. The total cooling power of 2 refrigerators is sufficient to handle the loads of the Nuclotron. The third unit is redundant. Therefore, thought was given to a distribution approach that permitted effective use of the additional capacity.

The "satellite" mode for helium refrigerators is characterized by liquid helium consumption without using expanders. The designers of the Tevatron[3] were the first to use this operating mode to improve the reliability of a cryogenic system. We have anticipated the need to deliver liquid helium from one to any other refrigerator in our system. This delivery requirement suggests applying the "satellite" mode to the Nuclotron. While this will permit cooling of the accelerator ring even when all of the expanders in a cold box are being replaced, the reason for applying the "satellite" mode is not just to improve reliability. This operating mode is also used to regulate the total system capacity. This is important because the heat load of the accelerator varies with the repetition rate of the accelerating cycles.

During normal operation of a helium refrigerator, piston or turboexpanders are used to cool a portion of the compressed gas, which is then used to precooling the remaining high pressure flow. Final liquefaction is accomplished with a J-T valve and/or a "wet" expander. There are three dry turboexpanders and a wet expander in the basic Nuclotron refrigerator cycle. With the injection of sufficient liquid helium from an independent liquefier into the cold end of the system (Fig. 1), the precooling performed by the expanders is no longer need. The excess gas boiloff generated from the injected liquid eliminates the need for expanding a portion of the high pressure stream, and the turboexpanders can be shut down. Full compressor flow can now be cooled and used to produce liquid, thus boosting the capacity of the individual refrigerator. The refrigerator cold box is now just a large unbalanced heat exchanger with greater flow in the low pressure pass than in the high pressure pass. Since this configuration can not be maintained without the injection of liquid, it was given the name "satellite".

Grassman and Kopp[4] have studied the optimal choice of temperature difference, $\Delta T$, between the high pressure outlet and the low pressure inlet streams of heat exchangers in cryogenic refrigerators. They showed that to most efficiently use the surface area of a heat

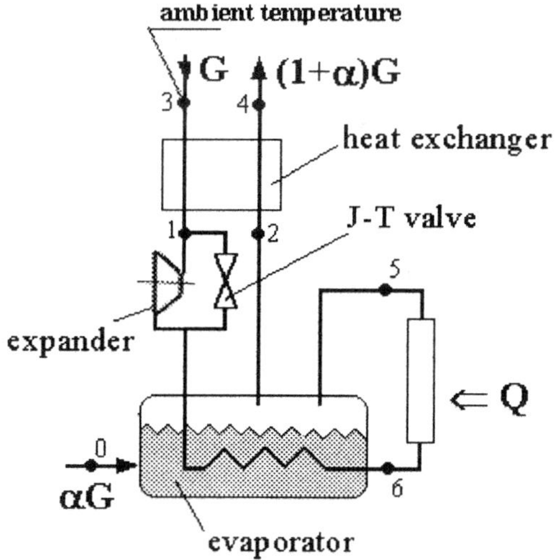


**Figure 1.** Flow diagram of the 'satellite" refrigerator.

exchangers, this temperature difference must be nearly proportional to the absolute temperature, T, of the low pressure stream inlet ($\Delta T/T \sim$ constant). Multiple expanders in the refrigerator cycle aid one in achieving this optimization through adjustment of the massflow and its impact on the absolute temperature at the low pressure inlet to the various heat exchangers. For the "satellite" configuration, where there is essentially one heat exchanger, the injected liquid performs the same task as the expanders, permitting optimization of the available heat exchanger area and thus, optimizing refrigeration efficiency.

The idea of the "satellite" refrigerator allows one to meet such contradictory requirements as high reliability and efficiency. This operation mode was a basic element of the system used to test the superconducting magnets for the Nuclotron[5], and is currently used to operate the Nuclotron. This paper presents a thermodynamic analysis of the "satellite" mode and describes special properties of its operation.

## OPTIMIZATION OF CYCLE

The refrigeration capacity of the "satellite" refrigerator (Fig.1) is given by:

$$Q = G[i_2 - i_1 + \Delta i_s \eta_s + \alpha(i_2 - i_0)] \tag{1}$$

which was derived using the assumptions:

- The liquefier injects ($x_0$=0.0) quality liquid into evaporator (100% liquid).
- Constant liquid level in the evaporator.
- The helium exits the load (5) at the same saturation temperature that exists in the evaporator. This flow can be 2 phase.
- There is no heat load from the transfer lines.

Compressed gas flow per unit refrigerating capacity is given by:

$$g = G/Q = 1/[i_2 - i_1 + \Delta i_s \eta_s + \alpha(i_2 - i_0)] \quad (2)$$

The value of efficiency is $\eta_s = 0$ for the case of using the J-T valve instead of the wet expander. Under this condition Eq. (2) is reduced to the following expression:

$$g = 1/[i_2 - i_1 + \alpha(i_2 - i_0)] \quad (2a)$$

To solve Eq. (2) or (2a), $i_1$ must be found. This is done by considering the equation for determining the surface area of a heat exchanger applied to the "satellite" heat exchanger conditions:

$$kF = \int_1^3 \frac{G}{\Delta T} di \quad (3)$$

Eq. (3) per unit of refrigeration capacity is:

$$\frac{kF}{Q} = g \int_1^3 \frac{di}{\Delta T} \quad (4)$$

One needs to use identical values of kF/Q to compare variants of refrigeration cycles. This is the only way to make a correct optimization. Fig. 2 shows the calculated results from Eq. (2) and Eq. (2a) for $kF/Q = 100\ K^{-1}$, with an ambient temperature of $T_3 = 293$ K, and a refrigeration temperature of $T_2 = 4.4$ K. $T_1$ was found by an iterative solution of Eq. (4) and Eq. (2) or Eq. (2a).

The total power expense for the "satellite" refrigeration system per unit refrigeration capacity (figure of merit) is given by:

$$\varepsilon = \varepsilon_g + \varepsilon_\alpha \quad (5)$$

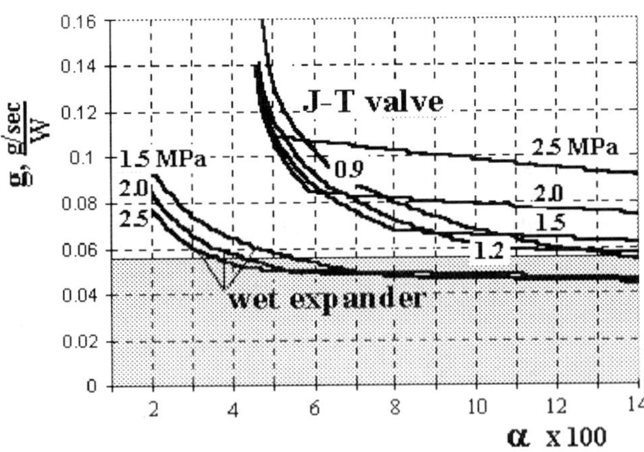


**Figure 2.** Compressed gas flow versus coefficient of surplus return flow, given as percent of the compressor flow, for various values of compressor outlet pressure.

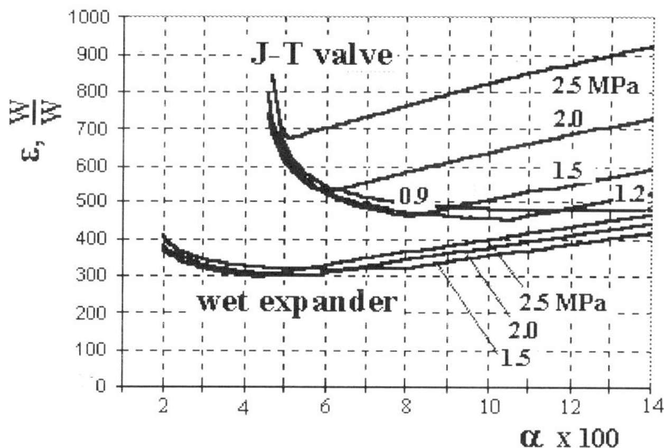


**Figure 3.** Figure of merit versus coefficient of surplus return flow for various values of compressor outlet pressure.

The compression work per unit of refrigeration capacity is:

$$\varepsilon_g = gRT_a \ln(p_3/p_4)/\ \eta_c \qquad (6)$$

The liquefaction work is given by:

$$\varepsilon_\alpha = \alpha g e_0/\ \eta_{LHe} \qquad (7)$$

Fig. 3 shows the calculated results from Eq. (5) for kF/Q=100 $K^{-1}$, $\eta_s$=0.7, $\eta_c$=0.5, $\eta_{LHe}$=0.15, $p_4$=0.105 MPa, an ambient temperature of 293 K, and a refrigeration temperature of 4.4 K. There are minima for the figure of merit for both the J-T valve and the wet expander. When using the J-T valve, the optimum parameters of operation are a high pressure of 1.2 MPa, and a coefficient of surplus return flow of 10.5 %. These parameters change significantly when using a wet expander. In this case they are $\alpha$= 4...6 % and $p_1$= 2.5 MPa. The minima of $\varepsilon$ are not very sharp. This allows one to have an efficiency close to the best value even if the system operates over a wide range of parameters. For example, with variations of $3\% < \alpha < 8\%$, and $1.5 \text{ MPa} < p_1 < 2.5 \text{ MPa}$ there is little impact on overall system efficiency. This is very important if one needs to change the refrigeration capacity of the "satellite" refrigerators.

Fig. 4 shows calculation results for $\eta_s$= 0, $\eta_t$= 0.5, $\eta_{LHe}$ = 0.15, $p_4$=0.105 MPa, an ambient temperature of 293 K, a refrigeration temperature of 4.4 K, for various values of kF/Q. Fig. 4 reveals that the optimum point is not a function of $\alpha$ if kF/Q $\geq$ 100 $K^{-1}$. As a rule this inequality is correct for helium refrigerators. So the results of this paper are useful not only for the Nuclotron cryogenic system, but for optimizing any other system including "satellite" refrigerators.

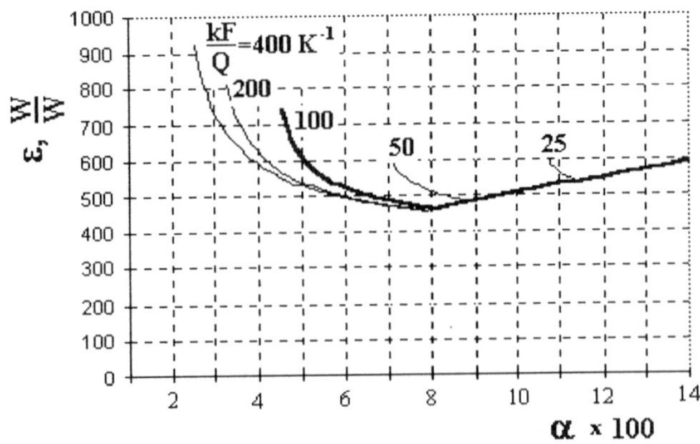


**Figure 4.** Figure of merit for various values of kF/Q.

## PERFORMANCE

The most valuable property of the Nuclotron magnets is their capacity for very fast cycling.[6] They can operate at up to 1 Hz. Dynamic heat releases depend on the repetition rate of the accelerating cycles. Table 1 shows that we have to regulate the refrigeration power from the value for static heat leaks, 1.7 kW, to more than 4.7 kW with the increasing repetition rate.

Fig. 5 illustrates a flow diagram of the Nuclotron cryogenic system. The magnets are piped in parallel resulting in about 100 channels in each half-ring returning two-phase helium flow. Therefore, it is necessary to be quite sure that there is only the liquid at the inlet of each of the magnets. Liquid helium passes, thus, through the subcoolers and phase separators (5,6) before being fed into the supply headers.

As shown in Fig. 5, two of the three refrigerators (1,2) have connection only to their half-rings and can operate independently. The third (3) is intended for increasing the total capacity for the cases of peak repetition rates and maximum design energies of the accelerator. It is also very useful for improving cooling reliability in situations where one of the refrigerators is stopped. The third unit has a connection to both of the other units through the liquid helium transfer line (4). This line allows the delivery of liquid helium from one to any other refrigerator.

The data of Fig.2 and Fig.3 allow one to optimize the system. However, there are some stipulations in our case that can be explained by the following way. The value $x_5$, mass vapor content of the stream at the outlet of the magnets, mustn't be more than 90 %. The compressed gas flow per unit of refrigeration capacity for the "satellite" refrigerator (Fig. 1) may be given by:

$$g = 1/(i_5 - i_6) \approx 1/rx_5 = 0.057 \tag{8}$$

**Table 1.** Heat load for the Nuclotron at 4.5 K versus the repetition rate of accelerating cycles

| Frequency, Hz | 0 | 0.1 | 0.2 | 0.3 | 0.4 | 0.5 |
|---|---|---|---|---|---|---|
| Head load, kW | 1.7 | 2.3 | 2.9 | 3.5 | 4.1 | 4.7 |

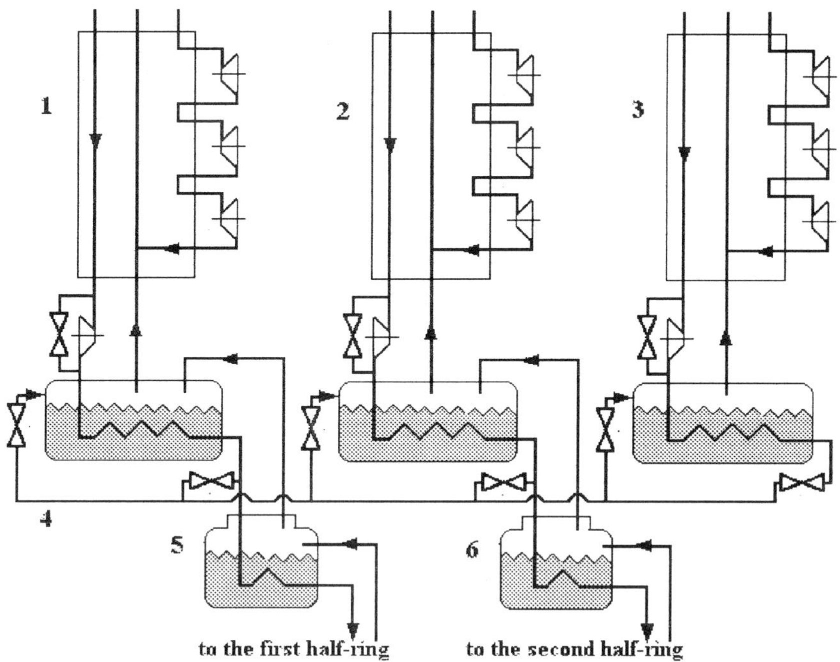


**Figure 5.** Flow diagram showing connections between refrigerators and Nuclotron half-rings. 1,2,3 - refrigerators 1600 W / 4.5 K; 4 - liquid helium transfer line; 5,6 - separators.

Applying Eq. (8) results in a requirement of $g < 0.057$ where the inequality $x_5 > 0.9$ holds. This is shown in Fig. 2 by the gray region at the lower end of the figure. Unfortunately, Figs. 2 and 3 indicate that the optimum points for the wet expander cycles are within this region. So, it is not possible to operate at optimized conditions. The permitted maxima for $\alpha$ are 3.6 % for 2.5 MPa, 4.4 % for 2.0 MPa and 5.5% for 1.5 MPa, but the efficiency is only slightly lower at these points than at the optima. It is clear from Fig.2 that there is no problem with the mass vapor content for the case of using a J-T valve.

During the 11 physics runs of the Nuclotron there have been no accelerator failures due to the cryogenic system. Usually there are sufficient reserves of liquid helium to switch each of helium refrigerators to the "satellite" mode. That is why cooling of the accelerator ring can be maintained even when all of the turboexpanders are broken.

Very often it is necessary to run with a refrigeration capacity of no more than 2.0 kW. Usually two refrigerators operate in this situation. The first one operates in its original mode: three gas expanders and a wet expander. It refrigerates the first half ring in addition to producing liquid helium. The refrigeration characteristics of the "mixed" mode: refrigeration + liquefaction were presented in [2]. The second refrigerator, which receives liquid helium from the first, operates in the "satellite" mode with either a wet expander or a J-T valve.

## SUMMARY

The cryogenic system of the Nuclotron has been developed by using the "satellite" mode of helium refrigerators. It provides high reliability with good thermodynamic efficiency, and the ability to regulate refrigeration capacity. The calculated results for thermodynamic optimization of the system have been presented. Special requirements for using the "satellite" refrigerator concept for the Nuclotron have been discussed.

## REFERENCES


1. A.M. Baldin, et al., Cryogenic system of the Nuclotron - a new superconducting synchrotron, in: "Advances in Cryogenic Engineering," Vol. 39, Plenum Press, New York (1994), p. 501.
2. N. Agapov, et al., Development & operating experience of the Nuclotron cryogenic system, in: Proc. of the 16th Int. Cryogenic Eng. Conf., Kitakyushu (1996), p. 139.
3. J. Theilacker, et al., Commissioning of the Tevatron satellite refrigeration system, in: "Advances in Cryogenic Engineering," Vol. 29, Plenum Press, New York (1984), p. 429.
4. P. Grassmann and J. Kopp, Zur gunstigsten Wahl der Temperaturdifferenz und der Warmeubergangszahl in Warmeaustauschern, Kaltetechnik. 10:306 (1957).
5. E.I. D'yachkov, H.G. Khodhzibagiyan, and V.N. Kuzichev, A satellite refrigerator for testing superconducting magnets with a forced cooling system, in: Proc. of the 10th Int. Cryogenic Eng. Conf., Helsinki (1984), p. 735.
6. A.M. Baldin, et al., Superconducting fast cycling magnets of the Nuclotron, in: IEEE Transactions on applied superconductivity, v. 5, No 2 (1995), p. 875.

# EXPERIENCE IN FOUR YEARS OPERATION OF THE 2KW-HE I-REFRIGERATOR AT FZK/KARLSRUHE

G. Perinić

Institute for Technical Physics
Forschungszentrum Karlsruhe
D-76021 Karlsruhe, Germany

## ABSTRACT

The 2kW-4.4K-refrigerator has been put into operation in 1993. It has altogether run a total of 11,000h in 4.4K refrigeration and liquefaction mode and ca. 3,000h in combined 3.3K and 4.4K operation mode including the test runs during commissioning.

The paper presents the experience gathered with the hardware components, the software and the operation of the plant as a whole since the acceptance tests of the refrigerator.

## INTRODUCTION

For development and testing of superconducting model coils, especially for the European and international fusion program, at the Forschungszentrum Karlsruhe (FZK), the magnet test facility TOSKA (toroidal coil test facility) [1] is available. In the framework of the TOSKA upgrade the cryogenic supply from an older 300W-He II-refrigerator has been upgraded by adding the more powerful up-to-date 2kW-He I-refrigerator. Besides supplying the TOSKA UP facility the new refrigerator does furthermore provide refrigeration power to the STAR experiment (superconducting test arrangement) and supply more than 120,000L of liquid helium per year to other users inside and outside the institute on site of FZK.

In the following sections first the refrigerator design, the operation history and the operation conditions are described and secondly the experience with individual components is summed up.

## DESCRIPTION OF THE 2KW-HE I-REFRIGERATOR SYSTEM

The process design, layout and the first tests of the 2kW-refrigerator are described elsewhere [2,3]. The refrigerator is specially adapted to the cooling requirements of the TOSKA UP facility which comprises controlled cool down of an object with up to 100t weight, cooling at 4.4K, cooling at 3.3K and simultaneous supply of liquefaction power. The cooling process is a modified Claude cycle utilizing two expansion turbines in the precooling circuit and one Joule-Thomson turbine. The liquefaction and refrigeration power can be increased with optional liquid nitrogen precooling or liquid helium precooling respectively.

The refrigerator plant is organized in seven subsystems: the compressor station, the coldbox, the valvebox, the LHe tank, the 4.6K calorimeter, the purge and vacuum systems and the cooling system. The compressor station comprises three screw compressors with their respective infrastructure. The coldbox contains three plate fin heat exchangers, one liquid nitrogen heat exchanger, nitrogen and neon adsorbers and the turbines. The supercritical helium gas is distributed inside the valvebox. Three rigid transfer lines connect the valvebox with the 10,000L tank, the TOSKA UP cryostat and the STAR experiment. The refrigerator is run fully automatically by an industrial process control system.

## OPERATION HISTORY AND OPERATION CONDITIONS

The on site assembly of the refrigerator started in August 1991. The first cool down was achieved in March 1992. In February 1993 the commissioning and the acceptance tests had been completed. Since then the refrigerator is operated by the cryoplant group of the Institute for Technical Physics.

In the first year of operation (1993) the fact that the TOSKA facility had not been connected yet was taken advantage of by running the plant in all possible operation modes. The aim was to find out all remaining deficiencies and to make the operation team familiar with the plant. The gradual connection of the TOSKA facility to the refrigerator was started in January 1994. It was completed in November 1995, after the test of the poloidal field coil (POLO) had been finished. The first coil test supplied by the 2kW-refrigerator was the test of the reinforced LCT-EU coil [4] which was performed in June/July 1996. The typical flow rate when supplying TOSKA is 40g/s in cool down and in operation (4.3K, 3.8bar).

In February 1996 the extension of the transfer lines to the STAR experiment was completed. The first cool down of the STAR cryostat took place in December 1996. The typical flow rate when supplying the STAR experiment is in the range of 20g/s during cool down and 6g/s during operation.

In the periods in between two experiments, the refrigerator is most commonly operated in liquefaction mode without liquid nitrogen precooling. In these periods the refrigerator is frequently started and stopped because it is only run as long as required. In the first nine months of operation after the acceptance test, the refrigerator has been started and stopped 87 times. In the first six months of 1997 it has been started an stopped 13 times. The longest operation period has been nine weeks during the LCT-EU coil test.

The refrigerator is supervised by the operators during cool-down, during operation mode changes and while the stop procedure is carried out. In the course of normal operation the monitors are checked periodically for warnings and alarms.

## EXPERIENCE WITH REFRIGERATOR COMPONENTS

A statistic of all faults and failures since the acceptance tests of the refrigerator is given in figure 1. The faults and failures comprise all problems noted by the operators from the error on the display up to the failure of the compressor motor. External failures such as power outage, cooling water loss etc. have been counted as faults of instrumentation or control origin. The external failures constitute approximately 1% of all faults and failures. The large number of problems observed in the first year of operation is due to the intensive test programme carried out during the warranty period of the refrigerator. The peaks in June/July 1996 and March/April 1997 coincide with the first and second coil tests in the TOSKA UP facility. The persistent basis is partly caused by a number of smaller problems which have not been resolved yet and which the operators have learned to handle.

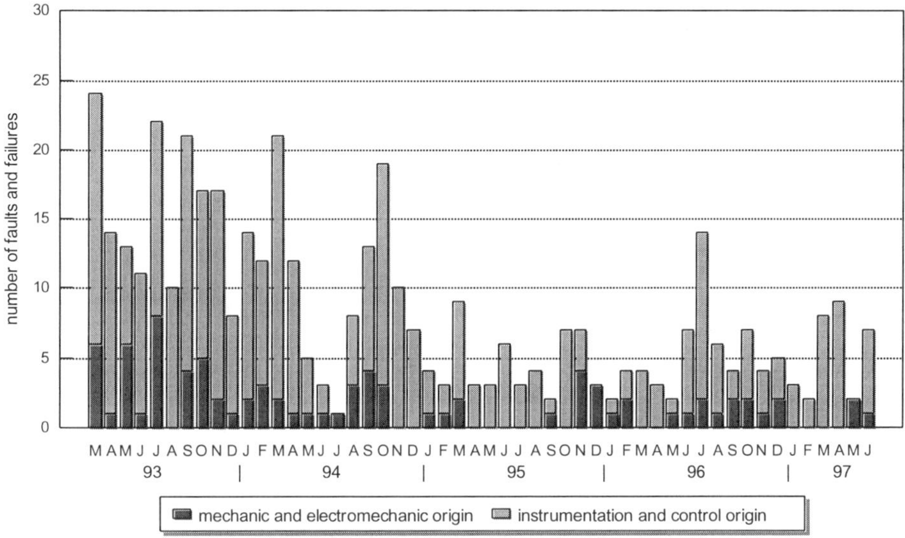


**Figure 1.** Statistic of faults and failures.

In the following paragraphs, the experience with individual components of the cryoplant is presented and major problems encountered with cryoplant components are described.

## Compressor station

**Compressors.** The compressors 2 and 3 have been running for 14,000h and the subatmospheric compressor 1 has been running for 3,000h altogether. Since the acceptance tests, two problems have been encountered with the compressors themselves. One of them was an oil leak on the last stage compressor caused by a broken lip seal of the slide vane actuator. The other problem was oil spray in the area of the motor clutch on all compressors caused by oil from the rotary face seals running along the driving shaft. The problem was solved by adding a splash ring on the shaft and modifying the shaft feedthrough.

**Motor.** In the compressor periphery, one of the compressor motors suffered a stator short-circuit after 5300 hours of operation. The insulation of the motors of the other compressors has consequently been altered as a preventive measure.

**Oil pump.** Due to repeated start-up problems one oil pump motor had to be replaced by a stronger motor. Furthermore one oil pump was destroyed due to a broken ball bearing after 13,000 hours of operation and 9,000 hours after the ball bearing had been replaced.

**Oil.** All three compressors are still running with the first filling of oil. The oil quality is analysed approximately every 2000 operating hours measuring the viscosity, the moisture content, and the concentration of various chemical elements. The only value that has changed notably compared with unused oil is the phosphor content which has increased up to 79 ppm. The phosphor most probably originates from parkerized surfaces.

Besides the oil quality the oil consumption of the rotary face seals is monitored in order to predict failures. The measured consumption of 4 seals is less than 0.6 litres per 1,000 hours, two seals have a consumption between 1 and 2 litres per 1,000 hours. All but one of the rotary face seals have a constant consumption.

## Coldbox

**$LN_2$ heat exchanger.** In the coldbox the plate-fin liquid nitrogen heat exchanger was replaced two years after the acceptance tests by a coiled tube heat exchanger. The heat exchanger was replaced as a preventive measure in order to avoid thermal stress in the first heat exchanger of the coldbox which could be caused by liquid nitrogen overflow from the liquid nitrogen heat exchanger. The new heat exchanger was also designed for an increase of the liquefaction capacity when utilizing liquid nitrogen precooling.

**Turbines.** Since the acceptance test, two failures of the first turbine have occured during operation after 1,900 and 4,800 turbine operating hours and one failure of the second turbine has occured during cool down after 12,800 operating hours. The first failure of the first turbine was attributed to temperature increases up to 98K at the turbine inlet during the cool down of the nitrogen adsorbers which caused the turbine speed to increase temporarily. The temperature increases were ironed out by a change to the cooldown process.

On the occasion of the two removals of the first turbine for replacement and one removal for inspection the rotor of the turbine was found to be coated with a considerable amount of black dust the first two times and still a little bit of carbon dust the third time. The carbon dust was either originating from the nitrogen adsorbers which are filled with active charcoal or from the abrasion of the vanes of the rotary vane regeneration blower. The turbine filter of the first turbine (mesh size 5μm) had to be cleaned twice and the filter of the second turbine once after the pressure difference had increased.

The second turbine had failed due to oxygen impurities in the helium freezing out on the turbine during cooldown.

**Rattling noise.** One yet unresolved problem concerning the coldbox is a loud rattling noise which originates from the region of the Joule-Thomson turbine. It appears if the inlet pressure of the turbine is reduced below 15bar and when a large flow of cold return gas is present. This is e.g. the case when the LHe-tank is full and the automatic power control decreases the compressor pressure and increases the heater power in the 4.4K calorimeter above 1kW. The rattling has a frequency of approximately 5Hz.

Even though one safety valve pipe inside the coldbox has been altered, some capillaries have been fixed, the turbine outlet non return valve has been blocked and various tests have been carried out, the source of the problem has not yet been tracked down.

**Regeneration blower.** The rotary vane blower used for the regeneration of the nitrogen and neon adsorbers has shown a strong abrasion of the graphite vanes in the range of 3mm per 50 operating hours from the very beginning. Notable is the fact that the abrasion could not be reproduced in a closed helium test loop with an identical blower outside the plant. One explanation might be the presence of moisture in the test loop and the total lack of moisture inside the adsorber regeneration loop of the refrigerator.

After a number of unsuccessful tests of various materials for the rotary vanes the blower was replaced by a scroll pump which is operating readily since.

### Instrumentation and control system

Significant result of the fault and failure analysis shown in figure 1 is the fact that more than four fifths of the reported faults and failures are related to instrumentation and control problems. Even though these problems have also led to considerable downtime, the majority of the instrumentation and control faults were minor bugs of the software and smaller problems with plant instrumentation that could be resolved by the specialists with minor capital costs.

During the superconducting coil tests in TOSKA no effects of the magnetic fields have been observed but the effect on the cathode-ray monitors. The magnetic field in the cryoplant control room and in the computer room has reached about 0.27mT. One monitor which was specially protected by a dual layer of high permeability metal has suffered only little distortion and could well be used for the plant control during pulses.

## FUTURE PLANS

For the supply of additional experiments requiring the refrigeration power of the 2kW-refrigerator the installation of a second valvebox is foreseen. Furthermore the control of the cold gas flow to and from the experiments needs to be improved as surges caused by the experiments have lead to problems in the past.

For the near future it is planned to optimize the refrigerator supervision. The first step is the installation of an automatic operator call system utilizing personal alphanumeric pagers. The second step will be the provision of remote access to the refrigerator by the means of a laptop equipped with a modem.

Furthermore the data collection of the refrigerator is to be improved by a PC-based data collection system and a remote control room is to be established for the operation during high field magnet tests.

## REFERENCES


1. A. Ulbricht et al., TOSKA: Testbett für Fusionsmagnete der Zukunft, *FZK Nachrichten* 29:5 (1997).
2. F. Spath et al., A 2kW He refrigerator for SC magnet tests down to 3.3 K, *ICEC 14 Proceed.* 32:56 (1992).
3. F. Spath et al., Performance test of a 2kW He refrigerator for SC magnet tests down to 3.3K, *Adv. Cryo. Eng.*, 39:563 (1994).
4. G. Zahn et al., Operation of the upgraded TOSKA facility and test results of the EU-LCT Coil cooled with forced flow helium II at supercritical pressure, *Proceed. 19th SOFT* (1996).
5. H. Lierl, H. Herzog, H. Berg, Die Heliukälteanlage für den supraleitenden Protonenring HERA - Bericht über Betriebserfahrungen und Zuverlässigkeit des Kältesystems, *DKV Tagungsberichte* 20 (1993).

# EXPERIENCE WITH A LARGE SCALE HE II REFRIGERATION SYSTEM AT TORE SUPRA

B. Gravil and B. Jager[*]

Association Euratom-CEA, Cadarache
13108 - Saint Paul lez Durance Cedex - France
[*]CEA, DRFMC/SBT/CEN
38054 - Grenoble Cedex - France

## ABSTRACT

A cryogenic system[1,2] which has a capacity of 300 W at 1.75 K, 1000 W at 4 K and 30 kW at 80 K has been operating for more than 10 years. It is used to cool the superconducting toroidal magnet of the TORE SUPRA Tokamak[3]. The 1.75 K refrigeration stage is constituted by 3 HeII saturated baths (cold sources) linked to the 18 HeII pressurized baths of the toroidal coils. The HeII saturated bath pumping system is achieved part at low temperature by two centrifugal compressors, part at room temperature by two oil ring pumps. This pumping system has a very large pumping range characteristic from 0.003 to 0.040 kg/s and a great operating reliability. The safety devices and protections against quench are peculiar.

## THE TORE SUPRA TOKAMAK

The TORE SUPRA Tokamak was built by the EURATOM-CEA association and established at the Cadarache Research Center which is located in Southeastern France. Its main vocation is the study of long-lasting pulses which required the use of a superconducting toroidal magnet. The constraints imposed by the use of superconducting coils come from the superconductor itself and from the environment created by a Tokamak. The performances of superconductors are limited by the maximum magnetic field in the conductor and by the maximum value of the current density. For TORE SUPRA a value of 9 T was chosen to make it representative of the technology of the future large devices. A second limitation comes from the Tokamak variable fields. The plasma current rises fastly to some pre-set level here from 0 to 1.7 MA at a speed of about $10^7$ A/s and later decays to zero. In addition, uncontrolled plasma behaviour in the present state of Tokamak plasma understanding may cause the plasma current to disrupt abruptly, thus producing extremely fast field variations. The eddy currents induced in the coil by the variable field could release enough heat to bring the conductor back to normal state, causing a quench in the superconducting coil system. Such problems imposed improved cooling by pressurized superfluid helium.

## HEII REFRIGERATION CIRCUITS

TORE SUPRA is the first machine of large size using a great volume of superfluid helium : about 5 $m^3$ of HeII essentially contained in the 18 toroidal field coils. The design and sizing of all the 1.8 K refrigeration circuits are the result of studies lead by the SBT Grenoble in the 70s and 80s. This was the case for the pumping of the saturated baths of which a part is done at low temperature by two levels of centrifugal compressors. This technique, used for the first time in the world, prevents the hazardous development of low pressure exchangers at very small pressure drop. The choice of warm oil ring pump is also the result of comparative studies both on the technical level and the level of investment and operating costs.

## SATURATED HEII CIRCUITS

All the 1.75 K refrigeration circuits are shown in Figure 1 starting with the saturated baths, they essentially consist of :

- Three saturated baths, one by satellite, make up the cold source. They are pumped in parallel and supplied by the liquid helium retrieved from the storage tank. Each saturated bath is made up of an annular vessel (volume of HeII per casing 175 liters). The upper part of these vessel is of reduced section to increase the sensitivity of the level regulation associated to the Joule Thomson valve. This reduction in section remains compatible with the limit flux in the HeII below the free surface.

- Three JT 4.5/1.75 K heat exchangers (ES2 ES4 ES6 Fig. 1) which are integrated in the pumping lines of the three satellites. They consist of a copper tube directly placed in the pumping tube. The temperature of the liquid upstream from the JT valves is slightly lower than $T\lambda$ (estimated $T \sim 2.015$ K). Each exchanger has a precooling by-pass which can also be used as a redundancy to the JT valve if the latter had an operating failure.

- Lines between the saturated baths and the refrigerator cold box. The central part of the cryogenic lines is reserved for the 1.75 K pumping. Depending on the position of the satellites, the length of the pumping tube varies between 25 and 45 m. The average total pressure drop between saturated baths and the suction of the first cold compressor is between 0.15 and 0.25 kPa. It is essentially located in the 4.5/1.75 K heat exchanger.

- Inlet cold compressors stabilizing exchanger. The purpose of this exchanger (E11 Fig. 1) is to set the inlet temperature of the cold compressor at 4.4 K whatever the flow rate or the thermal losses on the pumping line. The power retrieved to heat the low pressure helium allows to increase the liquefaction rate at the outlet of this expansion device of the 4.5 K refrigerator.

- Cold centrifugal compressors CC1 CC2. The two cold centrifugal compressors are located in a box under vacuum separated from the cold box. They are driven by electric motors cooled at 80 K. These motors are supplied by variable frequency converters (speed can be adjusted between 100 and 600 rot/s). The levitation is obtained by active magnetic bearings also at 80 K. The motor/bearings set was manufactured by the S2M company. The technology used for these machines has several advantages :

- limitation of the heat load towards the colder parts (motor + bearings at 80 K)
- no pollution risk (set under vacuum)
- possibility of adjusting the position of the wheel in relation to the bladed stator under operation (reaching the optimal efficiency or compression rate)
- great reliability, no parts in friction, no wear, reduced maintenance

These two compressors introduce an additional heat load in the refrigerator between 4.5 and 20 K. This heat load is compensated by an increase in the flow rate for the coldest turbine, therefore by a flow rate for the cycle compressors. However they reduce the volumetric flow rate of the warm pumping thus the size and the electrical consumption. It had been requested that on the level of electrical power the gain on the warm pumping at least compensate the addition on the level of the cycle compressors. The minimum characteristics imposed for the cold compressors are:
global compression ratio $\tau > 4.5$, isentropic efficiency $\eta s > 40$ %
At the tests, the measured characteristics were :

CC1 : $\tau_{max} = 3.3$, $\eta_{max} = 53$ % CC2 : $\tau_{max} = 2.3$, $\eta_{max} = 56$ %

- Low pressure exchangers: One of the main advantages of low temperature pumping is to avoid the development of low pressure exchangers at very small pressure drop. Consequently, the 20/80 K exchangers (E5 E6 E7 Fig. 1) are classic exchangers with brazed aluminium plates commonly used for refrigerators. The 80/300 K exchanger (E10 Fig 1) is an exchanger with a stainless steel tube on the HP side and finned on the low pressure side. The overall pressure drop between the outlet cold compressor and the inlet warm pump is of about 0.5 kPa for 0.015 kg/s of mass flow rate.

- P1 oil ring pump: During the study of warm pumping, we had the choice between :
- either a pumping group consisting of multi-stage Roots or a single liquid annular pump (no experience in the pumping of helium). After tests on helium pumping with a oil ring pump of small size, we finally selected this type of machine for TORE SUPRA. This choice was made with the following criteria in mind :

- simple and "rustic" machine commonly used in the chemical industry
- low maintenance costs (no parts in friction)
- low investment compared to the Roots
- machine well adapted to the pressure to be pumped
-possible adjustment of the pumping characteristic by modifying the transmission pulley. These pumps are equipped with double mechanical wear rings with oil injection on each shaft passage. This guarantees a tightness versus the ambient air. The choice of this type of machine was validated by 73000 cumulated operating hours for the two P1 and P2 pumps without any intervention on these machines.

The sizing of the P1 pump was done so that it could alone ensure the pumping at 2.1 K (cold compressors shutdown and by-passed) for the night hours. This night time operating mode guarantees that the HeII circuits of the coils are maintained at a temperature lower to

$T\lambda$, therefore full of superfluid helium without bubbles. In this operating mode, the flow rate to be pumped is reduced by half (set loss of about 130 W). The P1 pump discharges at a pressure of between 40 and 60 kPa, the P2 pump takes over the flow from the P1 pump and the flow from the 4 K cooling thick casings of the thermal buffer.

### 1.75 K Pumping Characteristics

In Figure 3, the characteristic of pumping Q=f(P) measured at the inlet of the first cold compressor is recorded. The latter is very large, the higher maximum flow rate at 0.040 kg/s (P inlet > 2.5 KPa) is not measurable by our system (Tcoils > $T\lambda$). It is the fact of having placed at the discharge of the cold centrifugal pumps, a hot volumetric pump which allows this range. Indeed, depending on the suction pressure of the CCs, the global characteristic of these CCs moves as indicated on the Q=f(P) characteristic (Figure 4). The discharge pressure adapts itself to the P1 volumetric pump characteristic and there is always a possible operating point. The experience gained from this for future machines (LHC, Tesla) is to be taken into account in order to obtain a large flexibility and an easy fine tuning of the pumping system. For TORE SUPRA, we note that the CCs/P1 pump adaptation was not optimized. The global compression rate of the CCs is not in conformity with the measurements performed during the CCs tests. The P1 pump is over-sized for pumping at 1.75 K. The consequence of this is to reduce the compression rate of the CCs. This was moreover confirmed during tests at 300W at 1.75 K by regulating the P1 inlet with the PCV02 valve. One can then obtain much higher compression ratio for the CCs and obtain a outlet discharge pressure of CCs of about 10 kPa to be compared with 4.5 kPa obtained with PCV02 valve (Fig 1) wide open. However, the operation is more unstable and we are getting closer to areas in which there is not necessarily an operating point (Fig 4).
The over sizing of this pump has two essential causes : too great a margin of the constructor and too drastic a condition imposed on this pump to ensure alone the 2.1 K operating mode. We are considering in a near future to equip this pump with a variable frequency converter allowing the rotation speed of this pump to vary and thus to adapt its characteristic according to the demand ; high speed to ensure alone the 2.1 K regime, low speed for operation with CCs at 1.75 K. We should also reduce the electric consumption of the system. A study of the costs is underway.

## PRESSURIZED HE II CIRCUIT

All the 1.75 K pressurized circuit are shown in Figure 2. The independence of T and P is obtained using a heat exchanger which ensures the thermal connection between the saturated helium bath which makes up the cold source and the circuit of pressurized helium to be cooled, maintained at P=Ctt.(0.125 MPa). The thermal insulation is performed by a cold safety valve which ensures a thermal and hydraulic tightness so that the interface 1.75 K $\Rightarrow T\lambda$ occurs at the level of the cold safety valve.
Above the valve can be found a homogeneous layer in temperature at $T=T\lambda-\varepsilon$ surmounted by a gradient layer of which the thickness depends on the heat flux which goes through it and thus of the thermal tightness of the cold safety valve. The level of He I in the cryostat is regulated by the LCV16 valve (Fig 2). Several detectors are placed in the gradient layer level. One of them gives us information on the thermal tightness of the cold safety valve. If

the valve is properly placed on its seat, the temperature is of about 2.25 K. On the contrary if it is not well placed the thermal flux is higher and the temperature decreases towards Tλ.
**Sizing of TORE SUPRA pressurized HeII circuits**: The heat exchanger between saturated HeII and pressurized HeII is made up of tubes of OFHC Cu which represents about 1 $m^2$ of exchange surface. This exchanger was sized to remove a power in night regime without shot of about 7 W per coil. In normal operation this power increases and can reach 16.6 W. This power corresponds to a flux still far from the critical flux.
**Analysis of thermal gradients:** The different temperature gradients calculated along the circuit, for different temperatures of saturated HeII bath, during idle and normal operation are given in the following table.

| T(Fig2) | T5 | T4 | T1 | T0 | Tsat | ΔT tube | ΔT kapitza heat exch |
|---|---|---|---|---|---|---|---|
| Idle 7W/coil | 1.622 K | 1.621 K | 1.615K | 1.614K | 1.6K | 7 mK | 14.7 mK |
| N Op 16.6W/coil | 1.821 K | 1.811 K | 1.781 K | 1.778 K | 1.75K | 40 mK | 28 mK |

In idle mode we can note that the highest ΔT is due to the thermal resistance of Kapitza, in normal operation the Kapitza ΔT is lower than the ΔT in the tube.

## Cooling of the Circuits

The startup of the cooling of the saturated HeII cold source causes the condensation of the gaseous helium contained in the pressurized circuit (Fig 2). In order to maintain the pressure in the coil circuit, we periodically admit a mass of He by opening the cold safety valve which is operated by an thrustor. The UV01 connection valve (Fig 2) with the 20000 l tank of LHe is open. Very quickly a gradient is established in the coil tube and the coil is cooled by removing a heat flux corresponding to the critical flux in the tube. The filling is finished when one is certain that the temperature in the coil is lower than Tλ which corresponds to the maximum of helium density. In order to do this we control either the T3 (fig 2) temperature under the current leads for most of the coils or the temperature on the vertical junctions T4 (fig 2) when the detectors have been added (case for 6 coils). When all the T3 and T4 temperatures (Fig 2) are lower than 2 K and all the coils have a pressure higher than 0.12 MPa, we consider that the filling is completed. The periodic opening of the cold valve is at this time stopped. The refrigeration power is then exclusively reserved for the temperature setting of the coils.

## Circuit Protection

The most important risk is linked to the quench of one or more coils. In this case one must be able, whatever the appearance dynamic of the normal areas, to remove the flow rate while limiting the maximum pressure of the enclosure. In our case the maximum acceptable pressure is about 2.5 MPa.
The protection is ensured by three additional devices (Figure 2) which are :

- a fast safety valve large diameter set at 0.35 MPa
- a magnetic valve set at 1.2 MPa

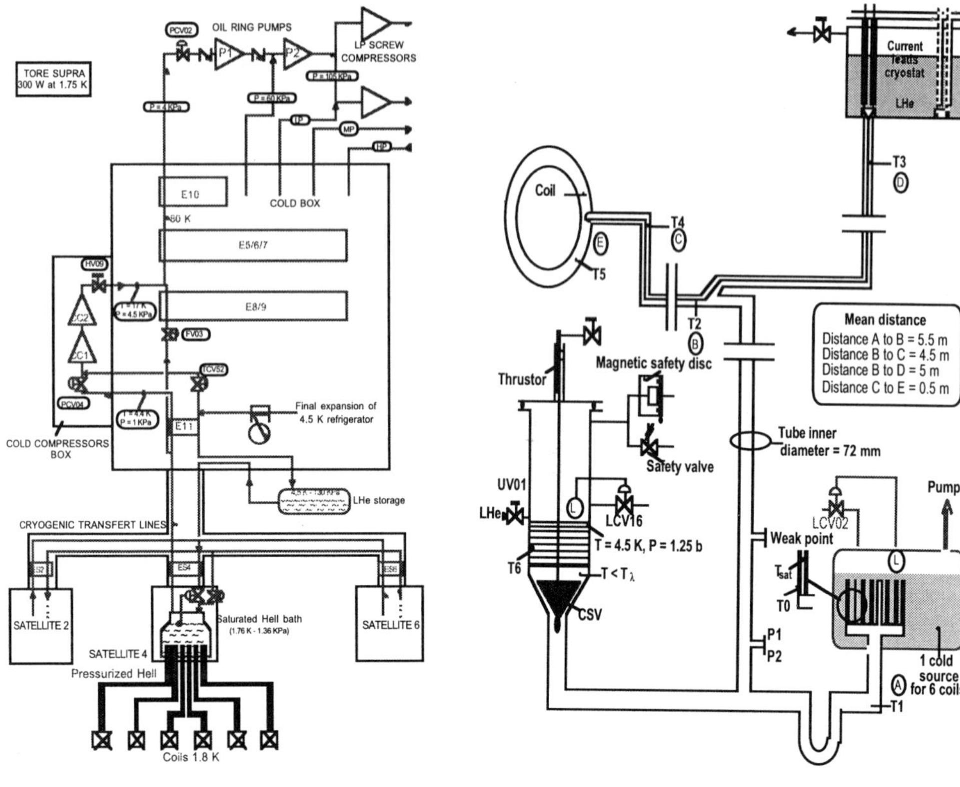


**Figure 1.** 1,8 K Refrigeration circuits.

**Figure 2.** 1,8 K Pressurized HeII circuits.

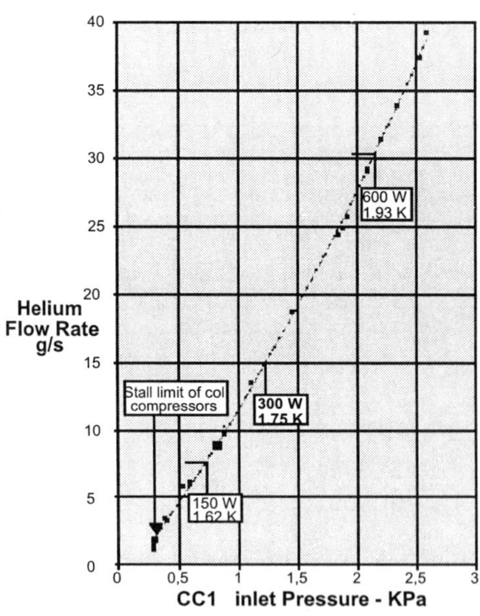


**Figure 3.** 1,8 K Pumping characteristic.

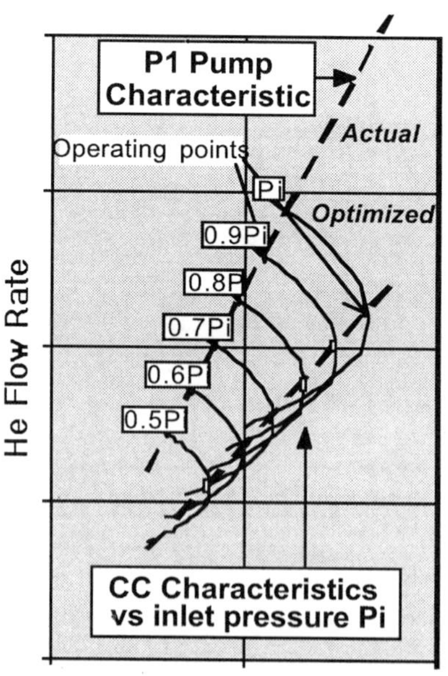


**Figure 4.** P1 and CCs Pumping adaptation.

- a weak point of this cold tube located in the satellite of which the bursting pressure is of about 2.2 MPa. This weak point limits the pressure in the thin casings in case of quench with blocking of the cold valve. The magnetic safety valve[4] has the particularity of opposing, once opened a very low resistance to the passage of fluid, the passage area is then equal to the tube area. It also ensures the tightness once the over-pressure is finished. For each satellite the safety devices, fast valve and magnetic valve of the 6 coil circuits discharge into a cylindrical collector (Figure 5) connected to the atmosphere by tube of very large diameter $\phi$=1 m. Each weak point (Figure 6) on the cold tube is connected to a lock chamber under vacuum protected in case of over-pressure by a magnetic safety valve. This lock chamber is itself protected by another, under controlled gaseous helium atmosphere. This last lock chamber is linked to the collector which is common to all the safety devices of the 18 coils.

## Principle of Active Protection of the Magnet[5]

During operation under field any operating dysfunction is detected and causes a discharge of the toroidal coil current. Depending on the severity of the incident this discharge can be slow or fast. The slow discharges are required in case of operating dysfunction of the cryogenic system. In this case the current of the toroidal coils is controlled by its electric supply. The lowering of current is of about 20 min. The fast discharges are required in case of severe problems. The time constant is set in this case at 130s. The energy stored under magnetic form in the coils is then dissipated in the external resistors.

## Safeties Used to Detect Quenches

Four different types of detectors are installed to detect quenches. In case of problems, they cause the rapid discharge of the toroidal current. These different detectors are :

- <u>$\Delta$V detectors</u>: The appearance of a normal area causes the appearance of a resistive tension to the coil connection which cause a potential difference between the faulty coil and a normal coil. To increase the safety each coil is compared to 2 others. The rapid discharge is performed and the quench information is given if at least 1 $\Delta$V is higher than 2 V.

- <u>Pressure detectors</u>: It is important to detect a pressure increase in the pressurized HeII circuit caused by the appearance of bubbles in the coil. In order to limit the response times and to be as close to the coil as possible the pressure detectors P1 and P2 (Figure 2) used are cold detectors of which the temperature is identical to that of the coils. The rapid discharge is caused and quench information is given if both detectors indicate a rapid increase in pressure > 0.2 MPa in less than 0.5 s.

- <u>Fluid movement detectors</u>: A carbon thermometer T6 (fig 2) placed in the gradient layer between $T\lambda$ and 4.2 K (in the cryostat of the cold valve) detects any liquid expulsion. Indeed if a hot zone with gas bubble appears in the coil, the HeII present in the connection tube not yet heated is going to be expulsed through the cold valve causing a temperature decrease of the thermometer placed in the gradient layer. The temperature under nominal operation indicates 2.25 K. As soon as its value is lower than 1.95 K the rapid discharge is ordered and the quench information is given.

-<u>Temperature detector</u>: The carbons installed under the current leads are also used in the safety logic. Any temperature higher than 1.98 K for more than 2 s is analyzed as a severe problem which must cause the rapid discharge of current.

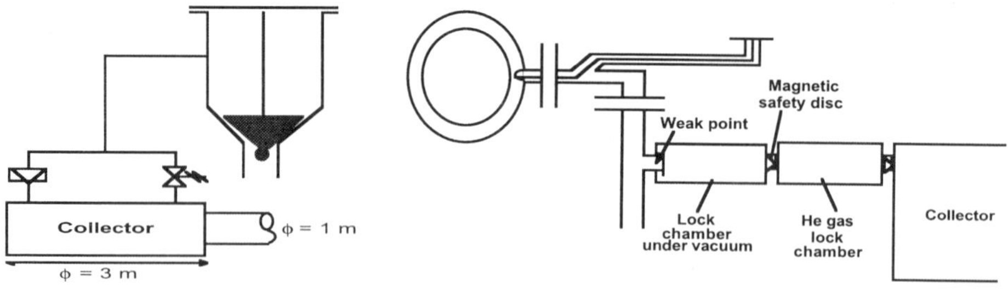


**Figure 5.** Normal safety devices circuits

**Figure 6.** Ultimate safety device circuits.

## Particular Problems Encountered

Reliability problems in the temperature measuring times have appeared.
The temperature measurements are all the temperature measurements under vacuum, the measured temperature is the temperature of the external wall of the tube which transports HeII. The probes used are all calibrated Carbon resistors placed in thermometric blocks specially developed for an easy industrial application. They ensure :

- a thermalization of the supply wires at a temperature close to that to be measured.
- a shielding of the thermometer block in order to eliminate thermal radiation.
- a good thermal coupling between the Carbon thermometer and the tube of which the temperature must be measured.
- a great thermal resistance between the thermalization block and the thermometer.

Each temperature detector is supplied by a 10µA current generator delivered by a conditioner placed as close as possible to the measurement.
All these measurements are not perfect and we note that it is very difficult to appreciate the temperatures in absolute terms. The relative values are however very easy to appreciate after yearly calibration.

## Noise of the Measurements During the Shots

The plasma heating by radio frequency of high power (several MW) induces noise on the temperature and pressure measurements. For the moment we have no solution to eliminate totally all these phenomena. The safety of the magnet depends for a large part on the dependability of the measurements. In fact the noise can either trigger a rapid current discharge in the case in which the noise corresponds to an increase of temperature or of pressure, or mask any real elevation in the opposite case.

## CONCLUSION

The TORE SUPRA cryogenic system totalizes more than 65000h of operation in which 24000h at $T<T\lambda$. All new components such as cold compressors, oil ring pumps, integrated in this system have been operated with success. The acquiescent behaviour of a large scale HeII circuit filled with stagnant pressurized helium has been demonstrated.

## REFERENCES


1. G. Claudet et al. Design of the cryogenic system for the TORE SUPRA tokamak, Cryogenics 26:443 (1986)
2. G. Bon Mardion et al.,Normal Operating of the 1.8 K TORE SUPRA Cryogenic system 1. Proc 13th ICEC Beijing CRYOGENICS (1990)30S 566-570
3. R. Aymar et al., IEEE Trans. Magn (1979) MAG 15 5426545
4. G. Bon Mardion et al. A Magnetic safety valve cryogenics 79:425 (1979)
5. J.L. Duchateau et al. Monitoring and controlling TORE SUPRA toroidal field system. IEEE Trans.Magn, 27:2053 (1991)

# OPERATION CHARACTERISTICS OF THE HELIUM REFRIGERATOR FOR THE LARGE HELICAL DEVICE WITH A DUMMY LOAD APPARATUS

R. Maekawa[1], T. Mito[1], S. Satoh[1], S. Yamada[1], A. Iwamoto[1], T. Baba[1], S. Moriuchi[1], K. Ooba[1], H. Sekiguchi[1], K. Nakamura[2], T. Fukano[2], H. Yamada[1], I. Ohtake[1], J. Yamamoto[1], and O. Motojima[1]

[1]National Institute for Fusion Science
322-6 Oroshi, Toki, Gifu 509-52, JAPAN
[2]Nippon Sanso Corpration
6-2 Kojima, Kawasaki, Kanagawa 210, JAPAN

## ABSTRACT

The helium refrigerator/liquefier operation for the Large Helical Device (LHD) with a dummy load apparatus (DLA) has been conducted under four different refrigeration conditions: no heat load mode; precooling mode; steady state mode and extra heat input mode. The stable operation of each mode, which simulates the refrigeration requirements for the LHD, was successfully demonstrated under semi-automatic control of the refrigerator. To establish the sequential control of the system, historical and real time trends of PID controlled electro-pneumatic actuators for valves and its corresponding thermal-hydraulic behavior of helium gas in the refrigeration system were collected as a data base. According to the operating experiences, the instabilities in the refrigeration system were observed. These instability mechanisms are considered to be quite complex: they are assumed to be coupled and/or combinations of heat exchangers in the refrigerator with approximately 90 to 100 m long supercritical helium hydraulic paths. To suppress these instabilities, adjustments of PID controlled valves and modifications of pressure control for the compressors were conducted. The paper describes operation characteristics of the helium refrigerator/liquefier with a DLA. The instabilities in the refrigeration system are also discussed.

## INTRODUCTION

The LHD, a large helioron torsatron type experimental fusion device, is now at the final stage of construction. A 10 kW class helium refrigeration system for the LHD, has been developed to provide reliable operation of superconducting coils: a pair of helical coils and three sets of poloidal coils; as well as superconducting bus-lines. To ensure the safe operation of the superconducting apparatus, it is important to scrutinize the characteristics of refrigeration system before starting the operation of LHD. We have been conducting several experiments with a Dummy Load Apparatus (DLA), which should be able to simulate all refrigeration requirements as a subscale model. This paper will present the experimental results which were conducted in October, December 1996. These results are compared with the latest experimental results which were taken after construction of a new helium buffer tank and modification of the pressure control scheme for the compressors.

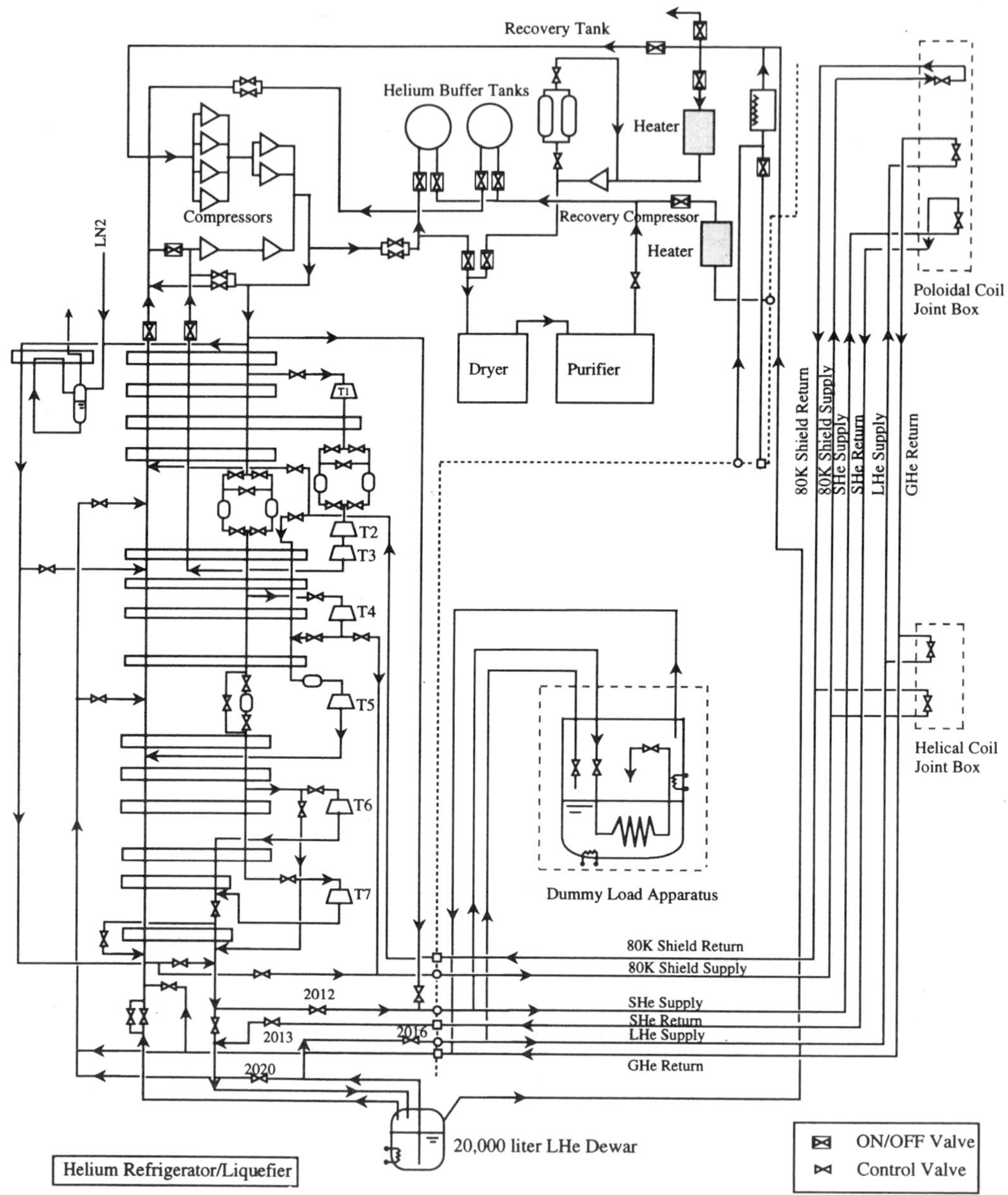


Figure 1. Flow diagram of a helium refrigerator/liquefier with a Dummy Load Apparatus.

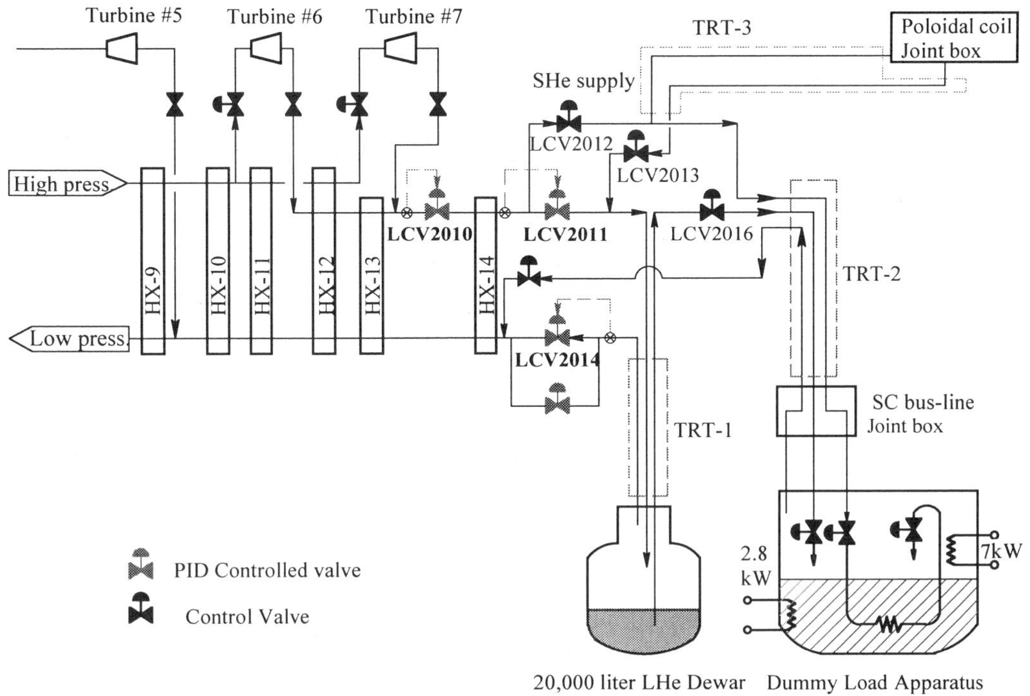


Figure 2. Flow diagram of the refrigeration system at low temperature environment.

## EXPERIMENTAL DESCRIPTION

Figure 1 shows flow diagram of the refrigeration system. The helium refrigerator has refrigeration capacity of 5.67 kW at 4.4 K and 679 l/h of liquefaction rate. A description of the helium refrigerator/liquefier has been reported.[1] The DLA was developed to test the refrigerator with various dummy heat inputs.[2] Two refrigeration schemes, forced-flow supercritical helium and pool boiling, can be simulated with two heaters at the supercritical helium (SHe) line and the liquid helium (LHe) bath, respectively. A cold box and the DLA is connected with approximately 43 m of TRT-2 containing multiple transfer lines. All cryogens, SHe, two phase helium and gaseous helium (GHe), are supplied and returned through TRT-2. The refrigeration systems are controlled by the cryogenic control system which consists of VME (Versa Module Europe) controllers, workstations and LAN (Local Area Networks).[3] Software packages have been developed and are now being modified for the LHD operation.

Figure 2 illustrates the flow diagram for low temperature regime in detail. There are primarily three PID controlled valves; LCV2010, LCV2011 and LCV2014. The LCV2010 maintains the line pressure above 0.93 MPa because of protection for turbine #7. The LCV2011, connected in series with LCV2010, maintains the SHe pressure above 1.08 MPa. LHe is supplied to the DLA through LCV2016 by controlling 20,000 liter dewar pressure with LCV2014. To supply the SHe to the DLA, LCV2012 has to be opened. The SHe line has two parallel paths, one is for a poloidal coil joint box (PCJB) via TRT-3 (~125 m long), the other is for a superconducting bus-line joint box. Since there is no control valve after the separation of each TRT, LCV2013 has to be opened slightly to avoid any substantial temperature gradients in SHe lines.

The experiments were conducted with four different refrigeration-conditions: Mode I, II, III & IV. Mode I simulates precooling process, SHe (5.7 K, 100 g/s) is supplied to the DLA and the heater mounted on SHe line provides approximately 7 kW to increase the SHe temperature to ~15 K. Mode II balances the mass flow rates of SHe supply to the DLA and GHe return to the refrigerator. Mode III is steady state operating condition: supplying

100 g/s SHe and 120 g/s LHe that are approximately 1/3 of overall mass flow rate for the LHD. Heaters at SHe and LHe lines provide 1.6 kW and 1.3 kW, respectively. In this mode, the total mass flow rates are also balanced as in Mode II. Finally, Mode IV simulates transient thermal disturbance (2.8 kW heat input to the LHe bath) to the refrigeration system, i.e., AC losses due to discharge of superconducting coils. Although the supply refrigerants are the same as Mode III, this condition induces 1.3 times larger mass flow return to the refrigerator. Each mode is considered to be simulating the refrigeration requirement for the LHD as a subscale basis.

## Modification of Discharge and Suction Pressure Control for Compressors

**Conventional Pressure Control.** Figure 3a shows simple diagram for the control scheme before modifying the pressure lines to the helium buffer tank. The discharge and suction pressures for the compressors were controlled using the discharge pressure. If the discharge pressure is higher than the setting value (SV) at PIC1001, a control valve 1001B opens and relieves gas to the buffer tank. For the case of lower pressure than SV, the 1001A valve opens and refills the gas to the compressor. PIC1002 provides additional gas to the suction line in the case of lower process value (PV) than SV. The buffer tank could not serve as a buffer with this control scheme since both suction and discharge pressures were controlled with the same line.

**New Control Scheme.** An added 700 $m^3$ buffer tank was constructed beginning January 1997 and completed at the end of March. The pressure control scheme of the compressors was also changed to suppress pressure and/or mass flow oscillation of the refrigerator. Figure 3b shows a new pressure control scheme. As one can see, the suction and discharge pressures are controlled independently through PIC1002 and PIC1001. PIC1005 has larger gain than that of PIC1001. Therefore, PIC1005 provides sensitive response to the transient pressure change in the suction line and to the pressure decreasing in the buffer tank. According to these modifications, the buffer tank would absorb pressure instabilities in the refrigeration system.

## RESULTS and DISCUSSION

### Four Refrigeration Modes

Four refrigeration modes were successfully demonstrated under semi-automatic control of refrigeration system in October 1996. Figure 4a & 4b illustrate the results obtained from DLA. Each mode was sustained about two hours to ensure stable operating condition. Mass flow-rates and temperatures oscillated at transitions of one mode to the next. In some cases, pressure oscillation amplitude became so large that the SHe pressure exceeded the relief valve setting. It was just impossible to suppress the oscillation with automatic control of LCV2011

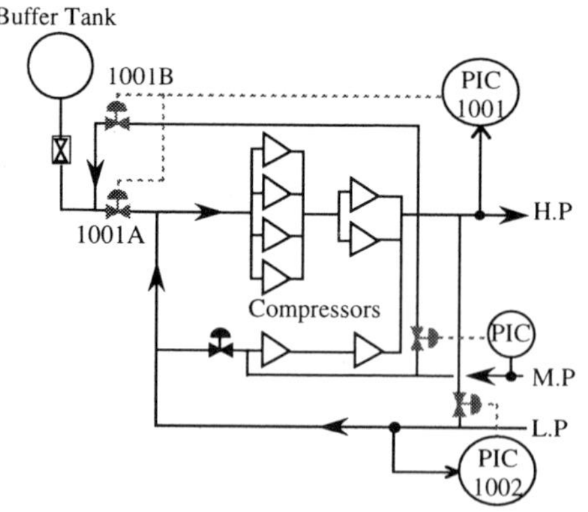

Figure 3a. Conventional pressure control scheme.

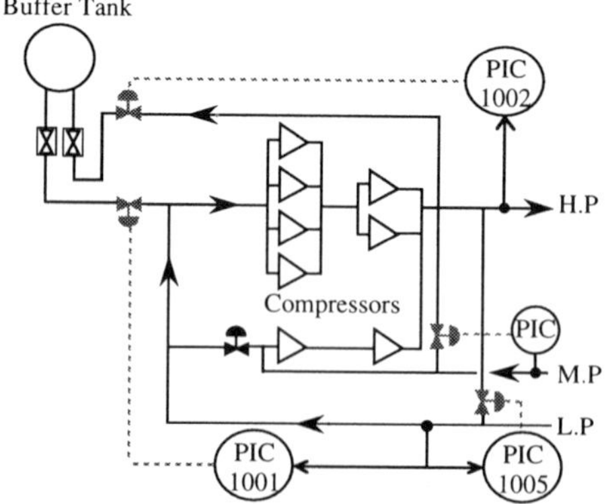

Figure 3b. New pressure control scheme.

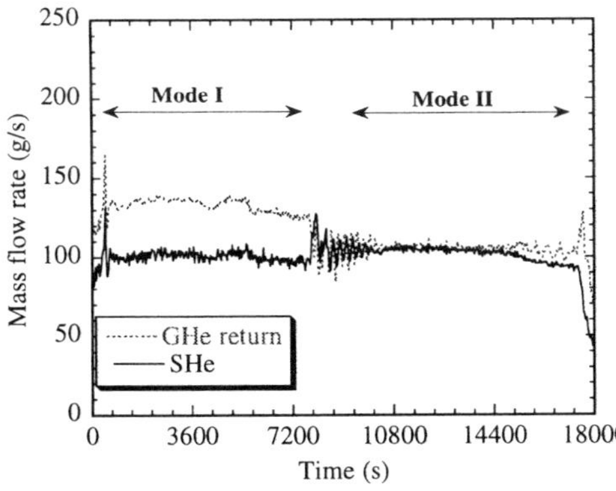


Figure 4a. Experimental result of Mode I and II.

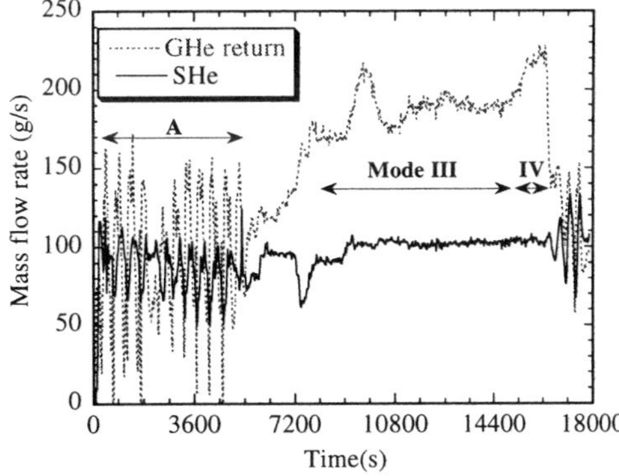


Figure 4b. Experimental result of Mode III and IV.

and LCV2014, SHe pressure control and LHe dewar pressure control, respectively. At that point, the interaction of control valves located in the low temperature region jeopardize the stable operation. To reduce the oscillation, the control of LCV2014 was changed from PID control to manual operation.

To understand the onset of instability, 'A' in Figure 4b, historical trends (one data point per minute) of refrigerator were reviewed. Figure 5 shows temperature profile as a function of time in the cold box. 'SHe_supply' temperature was measured at the down stream of LCV2012, while 'SHe_return TRT3' was measured at the up stream of LCV2013. Further, 'HX11_out' was SHe line temperature at the outlet of HX-11. The data were taken when LCV2012 was opened. Since the SHe line was closed during the night, transfer lines were warmed up because of heat in-leak.

According to Figure 5, 'SHe_supply' temperature decreased from 135 K to 6 K when supplying the SHe through LCV2012. At the same time, warm gas in the TRT's was returned to the low pressure side of a cold box through the DLA. As SHe reached to the warm transfer lines, the SHe absorbs heat and expands longitudinally. This induces the increasing heat transfer coefficients in the heat exchangers since some low temperature helium gas in the coid box was accelerated to the down stream side by the warm gas like a piston. This causes temperature decreasing 'HX-11_out' at t=2700 s. Finally, warm GHe reached to the heat exchangers and the temperature of HX-11 increased from t=2700 s to t=4000 s. This process might cause some off-design operation of heat exchangers so that the thermally induced flow and pressure oscillation were observed. It appeared that the oscillation was suppressed as decreasing the SHe temperature in TRT-3.

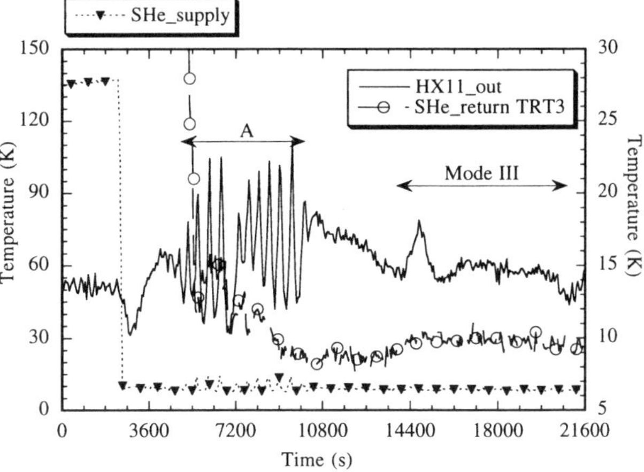


Figure 5. Temperature profile in the refrigerator during the experiment in Figure 4b.

In this case, the large pressure oscillation was induced by the warm GHe return to the cold box. Thus, the instability was created by transient thermal disturbance to the refrigerator. However, pressure oscillation was also observed right after Mode IV experiment as shown in Figure 4b. Unfortunately, the experiment had to be stopped at this point because of time restriction.

## Mode IV Experiment

To investigate the pressure oscillation in detail, real time trends (one data point per second) for the refrigerator were taken as a data base. Experiments were carried out to focus on the Mode IV operation because of drastic mass flow change in the refrigeration system. Figure 6a shows the results and hydraulic behavior of helium gas in the refrigerator in December, 1996. The pressure oscillations were observed in the entire refrigeration system. It seems like the amplitude of pressure oscillation was amplified through the compressors and traveled to the down stream. The buffer tank could not suppress the pressure oscillation at

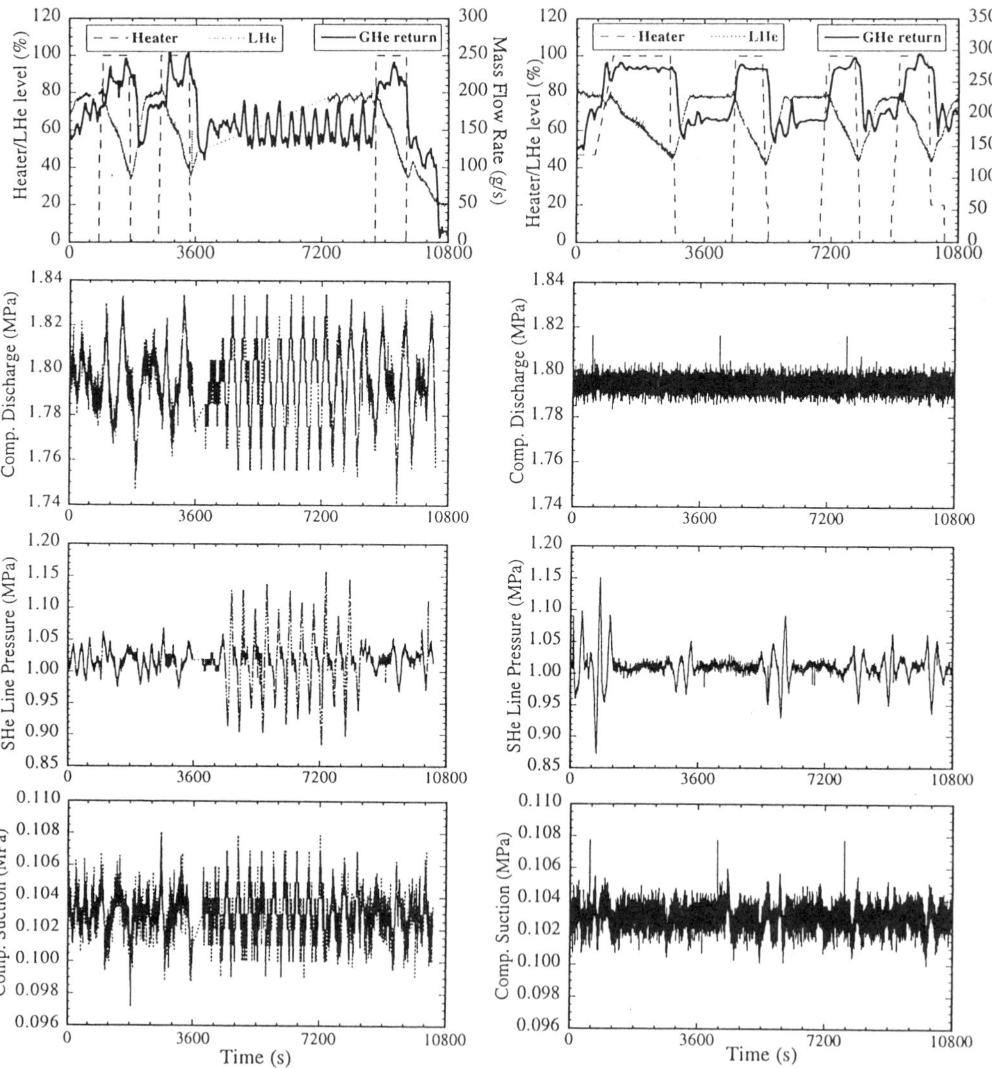


Figure 6a. Mode IV experimental results in December, 1996.

Figure 6b. Mode IV experimental results which were taken with a new pressure control scheme.

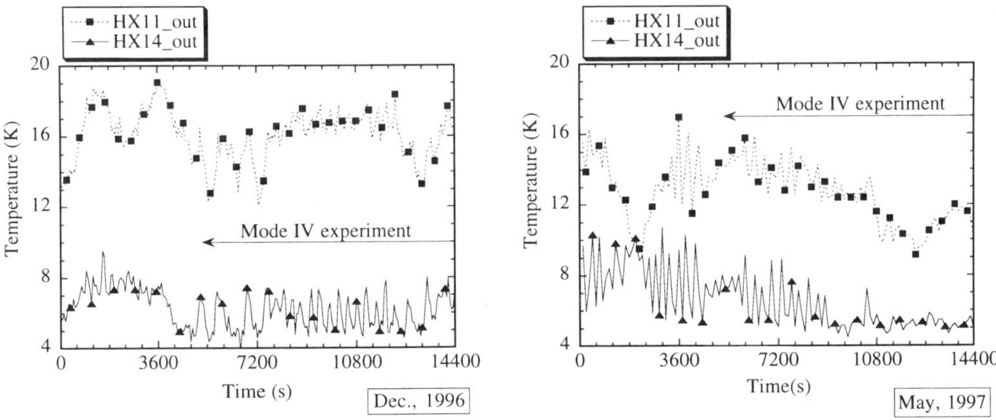


Figure 7. Comparison of temperature profile of heat exchanger 11&14 outlet in Dec. 1996 and May 1997. Heat exchanger 11 outlet shows lower temperature and the heat exchanger 14 outlet temperature is fairly stable in May 1997.

the discharge and suction of compressors, which was an intrinsic problem in the system. Pressure oscillations between the experiment could not be controlled with adjustment of PID parameters. In other words, the oscillation was self-sustained.

After the modification of pressure control line, the experiments were conducted from April to May 1997. Figure 6b shows the pressure of helium gas in the refrigerator as a function of time. Note that the improvement of pressure control at the compressor side. Although the mass flow and pressure were still oscillating at the low temperature region, discharge and suction pressures were fairly stable. The modified pressure control scheme isolates the disturbance from the low temperature environment in the refrigerator. Unfortunately, some pressure oscillations were still observed in the low temperature regime. In spite of PID parameter adjustments for LCV2011 and LCV2014, the system could not reach the stable condition. We found that the increasing gain for PID control induced the large oscillation amplitude in the system.

According to the experimental results, the refrigeration system sometimes showed oscillations in temperature, pressure and mass flow rate. The scenario of oscillation is as follows. Right after the Mode IV experiment, GHe return decreases from 300 g/s to 200 g/s. This mass-flow rate reduction affects the heat exchanger performance of HX 10~12. Therefore, the temperature of incoming helium gas will increase which also increases its pressure. Then, the LCV2011 would open to reduce its pressure. This affects the LCV2014 response through the DLA. The oscillation period is approximately five minutes. Once this process occurred, it is not easy to reach the equilibrium since time constant of the heat exchanger may be long. It will take some time to reach their temperature to the equilibrium state. As shown in Figure 7, these oscillations are closely associated with temperature at HX-11 outlet: the higher the temperature, the larger the oscillation amplitude.

## Troubles in the Refrigeration System

A new version of process control software was installed when the pressure control scheme was changed. The new software has serious problems such as, malfunction and/or slow-down of operating graphic consoles (man-machine interface), malfunction of data acquisition for the refrigeration system. These troubles were primarily caused by the substantial process load to a CPU for the VME controllers. Modification of a system controlled software was performed to continue the experiment. Another problem arose from refrigeration hardware. A brake control valve of turbine #5 showed improper behavior: a valve movement was completely opposite in direction as manipulated from the controller. A brake valve for turbine #7 was installed on the turbine #5 because of time restriction for the experiment. Therefore, turbine #7 was not operated during the experiment conducted in May, 1997.

## CONCLUSION

The helium refrigeration system successfully demonstrated four refrigeration modes with semi-automatic operation. Instabilities in the refrigeration system are primarily caused by the transient thermal disturbance to the system, which in turn induces oscillation in the heat exchangers. These oscillations are enhanced by the interaction of PID controlled valves: LCV2011 and LCV2014. The instabilities can be suppressed as improving heat transfer coefficient in the heat exchangers by substantial helium gas return to the refrigerator. This shows that the heat exchangers in the refrigerator were under unstable operating conditions. Since the DLA operation simulates the LHD as a subscale basis, the total mass flow rate was about 1/3 of that for LHD. Presumably this operating condition causes off-design operation of helium refrigerator, especially for the heat exchangers. Further study is still necessary to identify the coupled and/or resonant oscillations in the refrigeration system.

## ACKNOWLEDGMENT

We would like to dedicate our gratitude to Prof. Yamamoto who passed away in January, 1997. This work is supported by the ministry of education with a grant # 07555391.

## REFERENCES

1. S Satoh, et al., "Construction and Commissioning Tests of a 10 kW Class Helium Refrigerator for the Large Helical Device," Adv. Cryo. Engr., 41A: 727(1996).
2. R. Maekawa et al., "Study of Optimum Operating Condition for the Helium Refrigeration System of the LHD with a Dummy Load Apparatus," Proc. ICEC16, 1:87(1997).
3. T. Mito et al., "Cryogenic Control System for the Large Helical Device," Proc. ICEC16, 1:79(1997).

# DEVELOPMENT OF A CRYOGENIC SYSTEM FOR THE LARGE HELICAL DEVICE

T. Mito, S. Satoh, R. Maekawa, S. Yamada, K. Takahata, N. Yanagi, S. Imagawa, K. Watanabe, H. Tamura, A. Iwamoto, T. Baba, S. Moriuchi, K. Oba, H. Sekiguchi, I. Ohtake, T. Satow, J. Yamamoto, O. Motojima , and LHD Group

National Institute for Fusion Science
Oroshi, Toki, Gifu 509-52, JAPAN

## ABSTRACT

The construction and the cooling schemes of the cryogenic system for the Large Helical Device (LHD) are reported. The LHD is the first fully superconducting heliotron type experimental fusion device, which is under construction at the National Institute for Fusion Science (NIFS). The cryogenic system for the LHD consists of the helium refrigerator, the superconducting helical and poloidal coils and the peripheral equipment, such as superconducting bus-lines, control-valve-boxes and cryogenic transfer-lines. Three different cooling schemes are utilized for each cooling object; a pool boiling for the helical coils, a forced flow supercritical helium for the poloidal coils, a forced flow two phase helium for the coil supporting structure and the superconducting bus-lines. Many technical know-how were applied to develop the complicated cryogenic system for the LHD, which we could gather during the engineering research and developments such as the single coil test of the IV poloidal coil, the development and tests of the full-scale model of superconducting bus-lines and the test operation of the helium refrigerator with a dummy load apparatus.

## INTRODUCTION

The Large Helical Device (LHD) is a fully superconducting heliotron type experimental fusion device, which has the feature of current-less and steady state plasma confinement [1]. The construction of the LHD started in 1991, and is now at the final assembling stage toward the first plasma experiment in 1998. The LHD project is watched by all the world not only as the largest heliotron type experimental fusion device, but also as an actual application of a large scale superconducting coil system which is one of the key technologies for the realization of a fusion reactor.

## CRYOGENIC SYSTEM FOR THE LHD

### Cooling Objects of the LHD

Figure 1 shows a main part of the LHD cryogenic system in the LHD experimental room and the He refrigerator room; the LHD cryostat in which the superconducting helical coils, the poloidal coils and the coil supporting structures are installed, the helical and poloidal valve boxes by which the flows of the cryogen to the helical and poloidal coils are controlled, and the cold-box of the He refrigerator. The He refrigerator are separated from the LHD cryostat with a concrete wall of 2 m thickness for the neutron radiation shield in phase II operation, and the cryogenic multi-channel transfer lines connect them by way of the basements. The cold-box of the He refrigerator is divided into two stages of temperature due to a size limit for land transportation.

Figure 2 shows inside of the LHD cryostat. The superconducting coils for the LHD consists of two helical coils (H1 and H2) and three pairs of poloidal coils. The magnetic fields of the plasma center are 3 T in phase I operations and 4 T in phase II operations. The magnetic stored energies become 0.9 GJ in phase I and 1.6 GJ in phase II.

The major and minor radii of the superconducting helical coils are 3.9 m and 0.975 m, respectively. The helical coils were wound with an aluminum stabilized NbTi/Cu composite superconductor. The coils are designed to satisfy fully cryostable condition. The coils are cooled to 4.4 K with liquid helium in phase I operation and are planed to be cooled to 1.8 K with superfluid helium in phase II. The maximum magnetic field in the helical coils are 6.9 T in phase I and 9.2 T in phase II, respectively.

Three pairs of poloidal coils are named the Inner Vertical (IV) coils, the Inner Shaping (IS) coils and the Outer Vertical (OV) coils. The coil average diameters are 3.6 m for the IV coils, 5.6 m for the IS coils and 11.1 m for the OV coils. All poloidal coils are wound with NbTi/Cu cable in conduit conductors (CICC) and are cooled by forced flow supercritical helium. The coils have been designed to be operated with pulse modes in phase II operation to control the magnetic axis and shape of the plasma dynamically. The changing rate of the magnetic field in the poloidal coils is less than 0.5 T/s with time duration of 5 seconds and time interval of 5 minutes.

The helical and poloidal coils are mounted on the supporting structure to sustain large electromagnetic forces and to suppress deformations of each coil. The main part of the supporting structure is a toroidal-shape outer supporting shell. The helical coils covered with the stainless steel coil cans are connected to the outer supporting shell through the shell arms. The poloidal coils partly covered with the PC sleeve are mounted on the coil supports, which are built in the outer supporting shell. The supporting structure is cooled with two-phase helium flowing through cooling pipes attached on its surface. The supporting structure and all SC coils must be cooled to 4.4 – 4.5 K and are surrounded with the outer and inner 80 K radiation shields. The total cold mass of the LHD becomes 822 tons, and its breakdown is listed in Table 1.

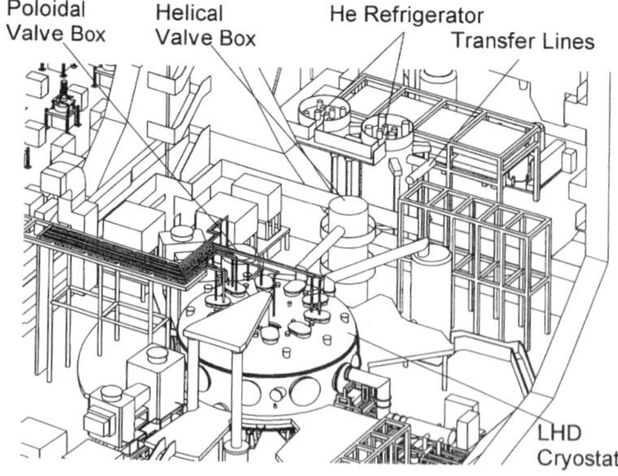


**Figure 1.** Bird's-eye view of main parts of the LHD cryogenic system.

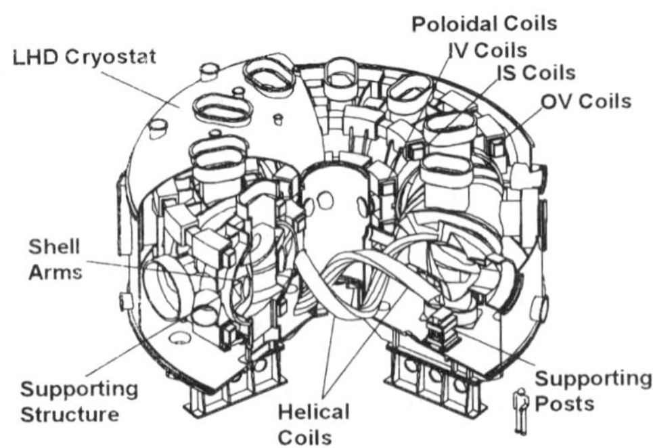


**Figure 2.** Inside of the LHD cryostat.

**Table 1.** Cold mass of the LHD

| Items | Cod mass |
|---|---|
| Helical coils | 240 ton |
| Poloidal coils | 182 ton |
| IV coils | 32 ton |
| IS coils | 50 ton |
| OV coils | 100 ton |
| Supporting structure | 390 ton |
| Piping etc. | 10 ton |
| Total | 822 ton |

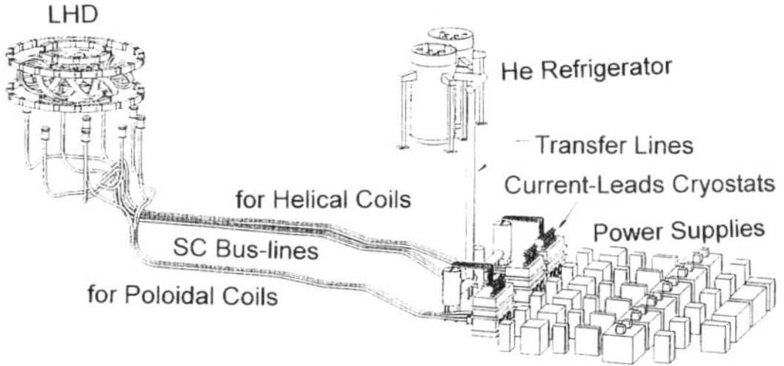


**Figure 3.** Bird's-eye view of the SC bus-lines.

Figure 3 shows a bird's-eye view of the superconducting bus-lines. For the LHD, the distances between the SC coils and the power supplies become 45 – 65 m. We have developed a flexible SC bus-line, which can connect SC coils and power supplies without energy consumption. One helical coil is electrically separated to three blocks named the inner, middle and outer blocks, and there are six blocks of the helical coils (H1-i, H1-m, H1-o, H2-i, H2-m, H2-o). On the other hand, each pair of upper and lower poloidal coils (IV, IS and OV) are connected in series inside the LHD cryostat. The total number of SC bus-lines is nine, of which six are for the helical coils and three are for the poloidal coils. The total length of the SC bus-line is 497 m. The operating current of the helical coils and its bus-lines is 13 kA in phase I operation and 17.3 kA in phase II operation. The operating current of the poloidal coils and their bus-lines IV, IS and OV are 20.3 kA, 21.3 kA and 31.3 kA, respectively.

## Requirements for the LHD Cryogenic System

Safety and stable operations of the SC coils should be required because the total magnetic stored energy of the LHD becomes 1.6 GJ in phase II operation. Long-term operations are necessary because a warm-up of the total cryogenic system including the SC coils will be done only once or twice per year. Full automatic control, corresponding to the various cooling modes such as, a cool-down, a steady-state operation, a warm-up, a power failure, a coil quench, is necessary for the cryogenic system to eliminate human errors of the operators due to the complicated cooling schemes of the LHD.

Figure 4 shows flow diagram of the LHD cryogenic system. The helium refrigerator/liquefier consists of the cold-box, the 50 g/s helium gas purifier, the 20,000 liter liquid helium reservoir, the 50,000 liter liquid nitrogen reservoir which is used only for the first stage of the cool-down and for the helium purifier. Eight oil-injected screw compressors (960 g/s, 1.9 MPa) and a recovery compressor (50 g/s, 1.8 MPa) are installed in the compressor room, which is 50 m apart from the He refrigerator room, and are connected with pipes running in the underground tunnel. Two 700 $m^3$ (2.0 MPa) spherical holder are used as helium buffer tanks to stabilize inlet and outlet pressures of the helium compressors and are also used for balancing the helium inventory in the LHD cryogenic system. A 100 $m^3$ (2.0 MPa) cylindrical holder is used as a supplying tank of impure helium gas, which must be purified by the purifier before supplying to the He buffer tanks.

The heat loads of the LHD, calculated from the final design of each component, are listed in Table 2. From the extensive efforts of designing and developing each component [2], the total heat in-leak to 4.4 K level becomes about half of the initial design value. For example, the heat load of the superconducting bus-line at 4.4 K was 3 W/m (total heat load of 1.5 kW) at the first design but it is reduced to 0.3 W/m (150 W) by inserting 80 K shield channel in their flexible coaxial transfer lines. The conductor joints in the helical and poloidal coils were also improved to reduce heat generation from 70 W to 10 W and to increase the coil stability, and their performances have been confirmed during the short sample tests of the jointed conductors and the coil.

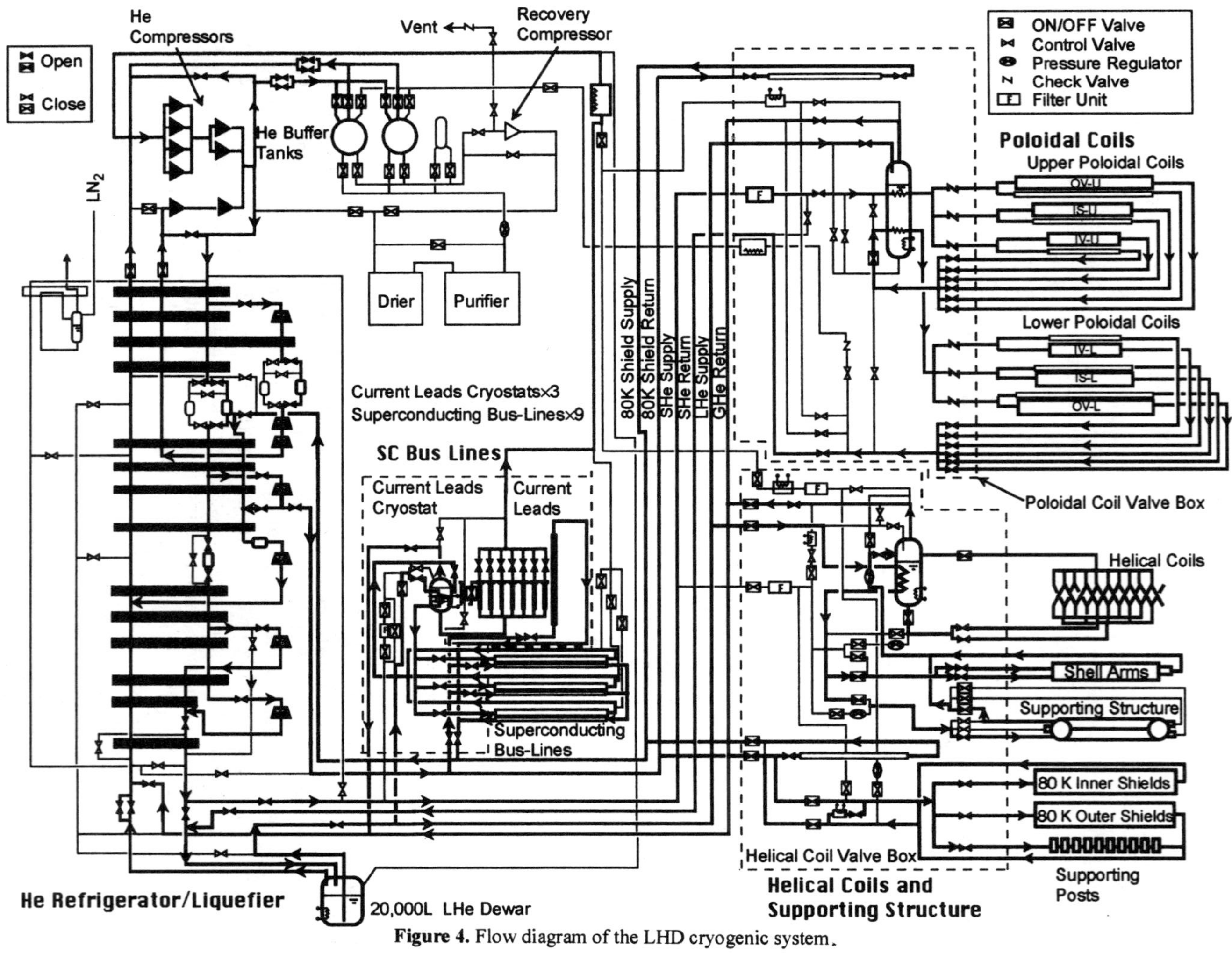


**Figure 4.** Flow diagram of the LHD cryogenic system.

**Table 2.** Heat loads of the LHD

| Items | | 4.4 K Refrigeration | 4.4 K Liquefaction | 80K Refrigeration |
|---|---|---|---|---|
| Helical Coils | | 70 W | - | - |
| Poloidal Coils | | 210W | - | - |
| H/C & P/C valve box | | 23 W | - | 765W |
| Supporting Structure | | 278W | - | - |
| 80 K Shield | | - | - | 8227 W |
| Superconducting bus-lines | | 150 W | - | 1500 W |
| Current leads (CL) | | - | 650 L/h | - |
| CL cryostats | | 310 W | - | 620 W |
| Transfer lines | TRT-1 | 335 W | - | - |
| | TRT-2 | 320 W | - | 700 W |
| | TRT-3 | 355 W | - | 800 W |
| Total | | 2051 W | 650 L/h | 13075 W |

The heat loads of the transfer lines were also improved. The transfer lines TRT-3 connects the cold-box and the helical and poloidal valve boxes with the length of 68 m. The TRT-2 connects the cold-box and the current leads cryostats with the length of 50 m. The TRT-1 connects the cold-box and the liquid helium reservoir. The TRT-2 and TRT-3 are sophisticatedly designed as a multi-channel transfer line with 80 K shield to reduce the heat leak to 4.4 K level. The heat loads to 80 K are also reduced from initial design values by improving supporting-structure of the radiation shields. The design cooling capacities of the helium refrigerator/liquefier are 5.65 kW at 4.4 K, 20.6 kW from 40 K to 80 K and 650 L/h liquefaction. We have an enough margin of the cooling capacities, which enable us to add a centrifugal supercritical helium (SHe) pump unit for the poloidal coils to increase SHe flow rate in the future. Figure 4 also shows cooling scheme of the LHD cryogenic system, in which active lines at a steady-state operation are marked with bold lines. The cooling schemes for each cooling object are explained below.

## Cooling Scheme of the Poloidal Coils

Each poloidal coil consists of 8 double pancake coils and has 16 parallel cooling paths for the forced flow SHe. The designed mass flow rate of the IV, IS and OV coils are 80 g/s, 65.6 g/s and 80 g/s, respectively. The total mass flow rate of six poloidal coils becomes 451.2 g/s. We separate the poloidal coils to two groups; the upper coils and the lower coils and connect them in series to reduce the total mass flow rate to 225.6 g/s. The total mass flow rate of the poloidal coils including the PC sleeves (supporting covers of the poloidal coils which are attached to the supporting structure) is 300 g/s. The inlet pressures of the upper and lower poloidal coils are 1.00 MPa and 0.84 MPa, respectively. The pressure drop of each coil is less than 0.1 MPa.

The SHe from the helium refrigerator is supplied to the poloidal valve box, in which it is cooled by a heat exchanger, goes through the upper poloidal coils and returns to the poloidal valve box and is cooled again by a heat exchanger, then is supplied to the lower poloidal coils. The SHe from the lower poloidal coils returns to the helium refrigerator and liquefies in 20,000 L Dewar through the JT valve.

## Cooling Scheme of the Helical Coils and the Supporting Structure

The liquid helium from the 20,000 L Dewar is sub-cooled by a heat exchanger in the helical valve box and goes through the cooling pipes of the coil supporting structure as the forced flow two-phase helium, then is separated to liquid and gas at a phase separator in the helical valve box. The cooling paths of the supporting structure are 80 parallel paths during a cool-down and are reduced to 20 paths in a steady operation. Total mass flow rate of the two-phase helium is 75.8 g/s. The helical coils are connected to the phase separator and are cooled by the pool boiling liquid helium without any active components for the reliable and safety operation.

### Cooling Scheme of the Superconducting Bus-lines

The liquid helium from 20,000 L Dewar is sub-cooled by a heat exchanger in the current-leads cryostat, then is supplied to the superconducting bus-lines. The returned two-phase helium from the bus-lines is separated at a phase separator in the current-leads cryostat. The liquid helium baths of the current leads are connected to the phase separator, and the current leads are cooled with helium vapor from the liquid helium baths. The flows of the bus-lines for the helical coils are the same as those of the poloidal coils except for the number of the SC bus-lines becomes 6 instead of 3. The total mass flow rate of 9 bus-lines is 109 g/s, and each of them is 12 g/s.

## DEVELOPMENTS OF THE CRYOGENIC SYSTEM

### Cooling and Excitation Tests of the Single IV coil (EXSIV)

We performed cooling and excitation tests of the single IV coil to confirm the designed characteristics of the IV coil as representative of the poloidal coils before installing them into the LHD cryostat [3, 4]. The basic configuration of the conductors for the IV, IS and OV coils are almost identical. The main parameters of the IV coil are listed in Table 3. The IV coil consists of 8 double-pancake coils with 16 parallel cooling paths and a length of each cooling path of 170 m. The IV coil was installed in a single test cryostat and was connected to the test facility with a supercritical helium centrifugal pump, a full-scale R&D model of the superconducting bus-line and a helium refrigerator.

The coil was cooled by the temperature controlled helium gas, mixing cold and warm gases in the helium refrigerator. According to the programmed refrigeration mode of about 1 K/h, the cool-down of the coil was successfully completed within 250 hours. The temperature difference between the coil inlet and the coil average was maintained less than 50 K until the coil temperature became lower than 100 K. There is no active flow control for 16 parallel paths of the coil, it was concerned that the flow balance becomes worse according to a decrease in the coil temperature. Figure 5 shows the flow balance of the IV pancake coils as a parameter of the coil temperature. The flow rates of each double pancake-coil (numbered 1 to 8) were normalized with their average flow rate. There was some deviation for the flow rate for No. 1 and No. 2 double-pancake coils even at a room temperature, however there was no serious unbalance for all temperature regions. Which was also confirmed from the temperature distribution in the coil; there was no temperature difference between each pancake coil.

The pressure drop of the coil was measured during cool-down and steady state operation. The pressure drop was transformed into the friction factor $\lambda$ by Darcy's equation:

$$\Delta P = \lambda \frac{\rho \cdot u^2}{2} \cdot \frac{L}{D_h}$$

$$u = \frac{\dot{m}}{A_{He} \cdot \rho} \qquad D_h = \frac{4A_{He}}{P} \tag{1}$$

where $\rho$ is a density of helium, $u$ is a velocity, $L$ is a length of the path, $A_{He}$ is a flow area of helium, $D_h$ is a hydraulic diameter and $P$ is a cooled perimeter.

Measured friction factor $\lambda$ agree well to the Katheder's formula [6]:

$$\lambda = \frac{1}{v^{0.72}} \left[ \frac{19.5}{R_e^{\ 0.88}} + 0.051 \right] \qquad R_e = \frac{D_h \cdot \dot{m}}{A_{He} \cdot \eta} \tag{2}$$

where $v$ is a void fraction of the conductor (=0.38 for the IV, IS and OV coils), $R_e$ is the Reynolds number and $\eta$ is the viscosity.

**Table 3.** Parameters of the IV coil

| *IV coil* | | *Conductor* | |
|---|---|---|---|
| Cooling method | Forced-flow of SHe | Type | Cable-in-conduit |
| Coil inner/outer radii | 1.6 m / 2.1 m | Superconductor | NbTi |
| Coil height × width | 0.47 m × 0.39 m | Critical current | 62.4 kA at 6.5 T |
| Number of turns | 15 × 16 = 240 turns | Conduit dimension | 23.0 mm× 27.6 mm |
| Operating current | 20.8 kA | Strand diameter | 0.767 mm |
| Maximum field | 5.2 T | configuration | $3^4 \times 6$ (=486) |
| Total mass flow rate | 80 g/s | Void fraction | 38 % |
| Cooling path length | 170 m | Strand insulation | none (bare copper) |

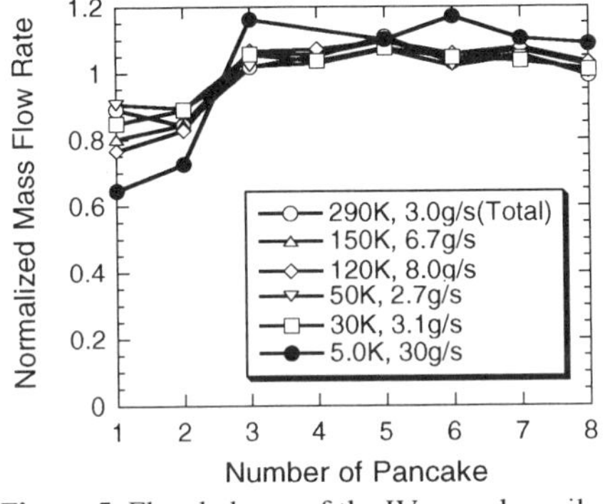


**Figure 5.** Flow balance of the IV pancake coils normalized with the average flow rate.

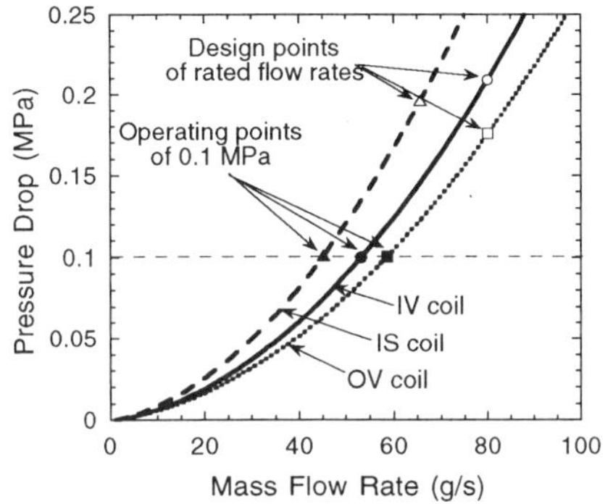


**Figure 6.** Pressure drop of the IV. IS. OV coil.

From above results, we can estimate the pressure drop not only for the IV coil but also for the IS and OV coils. Figure 6 shows calculated pressure drops of the IV, IS, OV coils as a function of the flow rate of the coil. The pressure drops at the nominal flow rate become 0.21 MPa for the IV coil (with a flow rate of 80 g/s), 0.20 MPa for the IS coil (65.6 g/s), 0.18 MPa for the OV coil (80 g/s). The pressure drops of all coils become higher than the initial design value because there was few database for a CICC coil at that time. The pressure drop of the coils is restricted to 0.1 MPa for the present LHD cryogenic system because the upper and lower poloidal coils are connected in series. Therefore the operating flow rates should be revised as a flow rate corresponding to the pressure drop of 0.1 MPa, which are 53.1 g/s for the IV coil, 45.1 g/s for the IS coil and 58.6 g/s for the OV coil.

The full scale R&D model of the SC bus line of 20 m length was cooled by forced flow two phase helium with a mass flow rate of 6 g/s. The measured heat load of the SC bus-lines was 65 W (3.25 W/m), which was almost the same as the design value. However we improved the configuration of the actual bus-line for the LHD from a four-channel coaxial transfer line to a five-channel coaxial transfer line. We can reduce the heat load to 4.4 K channels from 3 W/m to 0.3 W/m by inserting the 80 K shield gas channel.

## Test operations of the He refrigerator

The construction of the He refrigerator was completed at the end of FY 1994. A series of commissioning runs was conducted and the performances were measured in 1995. The refrigeration and liquefaction capacities were confirmed to satisfy the designed requirements as listed in Table 4. Then the test operations of the He refrigerator have been conducted with a dummy load apparatus, which can simulate heat load of the LHD cryogenic system before the completion of the SC coil system [5].

**Table 4.** Designed & measured cooling capacities of the He refrigerator

| Items | Designed Values | Measurements |
|---|---|---|
| Refrigeration capacity at 4.4 K | 5650 W | 5670 W |
| Liquefaction rate | 650 L/h | 704 L/h |
| Refrigeration capacity from 40 K to 80 K | 20.60 kW | 20.77 kW |

**Table 5.** I/O number list of the LHD cryogenic control system

| I/O signal | He refrigerator | Helical coils & valve box | Poloidal coils & valve box | SC bus-lines | Integrated VME controller | Total |
|---|---|---|---|---|---|---|
| Analog Input | 134 | 268 | 194 | 381 | 0 | 977 |
| Analog Output | 66 | 32 | 26 | 56 | 0 | 180 |
| Digital Input | 279 | 64 | 64 | 68 | 64 | 539 |
| Digital Output | 79 | 64 | 12 | 49 | 64 | 268 |

### Developments of a Cryogenic Control System

The development of a new control system is crucial for the cryogenic system because of fairly complicated cooling scheme of the LHD. The cryogenic control system is designed expandable using standard hardware and operating software. The system consists of workstations, VME (Versa Module Europe) controllers, Local Area Networks (LAN), operating graphic consoles and peripherals. The control system is composed as a duplex system, which significantly improves the reliable operation with fault diagnoses of each component including analogue input/output boards and digital input/output boards. The software tool packages are developed base upon the operating software of the UNIX and the Vxworks, which enables easy construction of the complicated automatic control programs for the LHD. The number of the input and output signals for the LHD cryogenic control system are listed in Table 5. The control system for the He refrigerator was constructed and has been confirmed its performances during the test operation of the He refrigerator. The control system for the cooling objects and the integrated controller for the total cryogenic system are now being constructed and will be finished at the end of 1997.

## SUMMARY

The development of the LHD cryogenic system was successfully conducted and it is now at the final assembling stage toward the first plasma experiment in 1998. The heat in-leaks calculated from the final design of the cryogenic system becomes about half of the initial design values because of the extensive efforts of designing and developing each component. Many technical data were gathered during the research and developments, and were utilized to modify the configurations and operational conditions of the cryogenic system. From precise measurements of the pressure drop during the single coil test of the IV coil, the operational flow rates of SHe for all poloidal coils were revised to coincide with the allowable pressure drop of 0.1 MPa. During the test operations of the He refrigerator, the cooling capacities, which have an enough margin against the calculated heat in-leaks, and the high performances of the cryogenic control system were confirmed.


## ACKNOWLEDGMENTS

The authors wish to acknowledge the staff of the LHD group, the Nippon Sanso Co., Ltd., the Toshiba Corporation, the Hitachi Ltd., the Fuji Electric Co., Ltd. for their previous and continuing works to construct the LHD cryogenic system.


## REFERENCES


1. O. Motojima, et al., "Large Helical Device Project for SC Steady-State Fusion Experiment," Proceedings of the 16th ICEC/ICMC, Kitakyushu, Japan (1997), p. 725.
2. S. Satoh, et al., "Design and Construction of Cryogenic Components for LHD," Proceedings of the 16th ICEC/ICMC, Kitakyushu, Japan (1997), p. 63.
3. K. Takahata, et al., "Hydraulic Characteristics and Stability for the Experiments on a Single Inner Vertical Coil (EXSIV) for the Large Helical Device," Proceedings of the 16th ICEC/ICMC, Kitakyushu, Japan (1997), p. 739.
4. T. Mito, et al., "Test Facilities of the Experiments on a Single Inner Vertical Coil (EXSIV) for the Large Helical Device," Proceedings of the 16th ICEC/ICMC, Kitakyushu, Japan (1997), p. 743.
5. R. Maekawa, et al., "Operation Characteristics of the Helium Refrigerator with a Dummy Load Apparatus for the Large Helical Device," presented this conference, CPE-6.

# CRYOGENIC SYSTEM FOR TJNAF'S HALL C

P. D. Brindza, S. R. Lassiter, and R. R. Wines

Thomas Jefferson National Accelerator Facility
Newport News, VA 23606, USA

## ABSTRACT

A cryogenic system has been operational in TJNAF's Hall C for four years. The design of this system is presented and the operational and control philosophy is discussed. The system consists of conventional rigid transfer lines and a unique flexible transfer line. The features and heat load measurements of this flexible line are presented. Operational data are stored continously on computer disc. There are typical plots of this data presented to illustrate system stability. The cryosystem for Hall C is used to operate the superconducting magnets of the HMS spectrometer and a cryogenic target for H2, D2 and He.

## INTRODUCTION

The cryogenic system for TJNAF/Hall C was installed in May 1993 and commissioned in August 1993. The system has been in continuous operation since commissioning supplying cryogens to the High Momentum Spectrometer (HMS) (ref. 1), a 7.5 GeV/c superconducting spectrometer, a cryogenic target system (ref. 2) and superconducting beamline instrumentation. A system upgrade is in progress to permit connection of the G0 Superconducting Toroidal Magnet Spectrometer (ref. 3) and a polarized target. The system is integrated into the overall TJNAF cryo system through a network of transfer lines (ref. 4) and is directly served by an 1800 watt refrigerator. The system is capable of delivering Helium at 2.5 atm and 4.5 K to the JT valves in the Superconducting magnets to keep the local reservoirs full and Helium at 15 K for cooling the cryogenic targets (see Fig. 1). LN2 is supplied from a large storage dewar at 4 atm. Variable temperature Helium for cooldown of large devices is produced in the cryo distribution system from a local heat exchanger. The heat exchanger can be used for either cooldown or warm-up. All devices are connected in parallel via "U-tubes". These demountable devices provide the flexibility for maintenance off line as needed. The system is fully instrumented and has off line data logging. There is currently 4 years of operational data on disc.

* Work supported by Department of Energy contract #DE-AC05-84ER40150.

## Transfer lines

The Hall C cryo system consists of two rigid transfer lines, one fixed and the other moving with the spectrometer, connected by a third flexible shielded transfer line. This system allows the HMS to rotate from 12 degrees to 90 degree relative to the electron beam for Nuclear Physics experiments. The design of the system is such that the HMS can be rotated using remote automatic control through its full allowed angle range with the SC magnets powered.

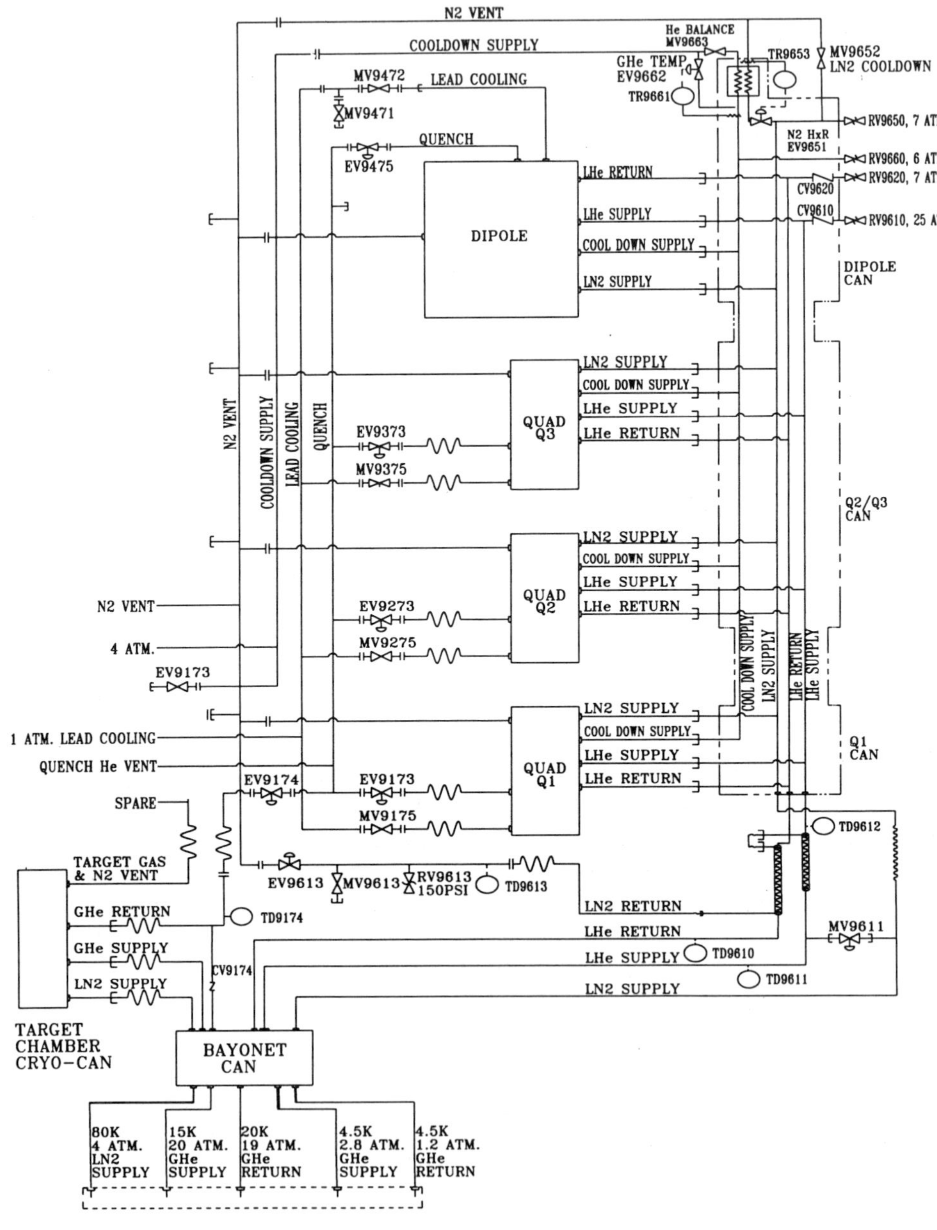


Figure 1. Hall C cryogenic flow sheet.

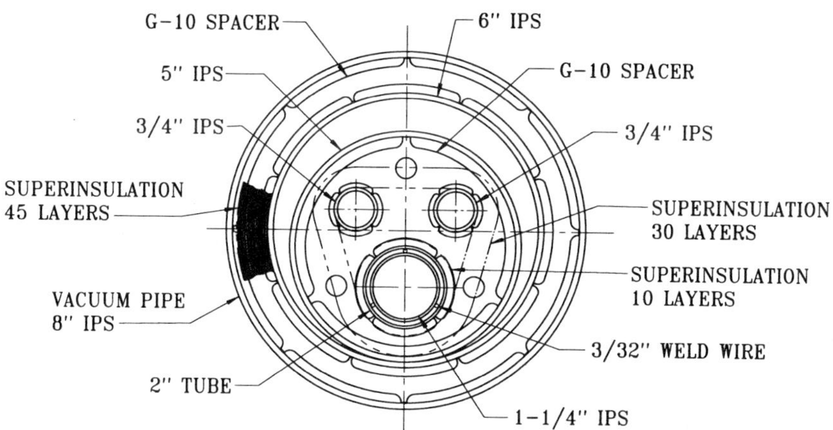


Figure 2. Hall C main transfer line section.

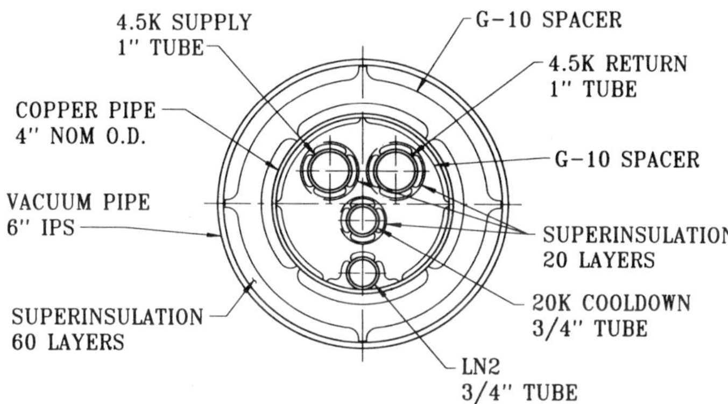


Figure 3. HMS transfer line section.

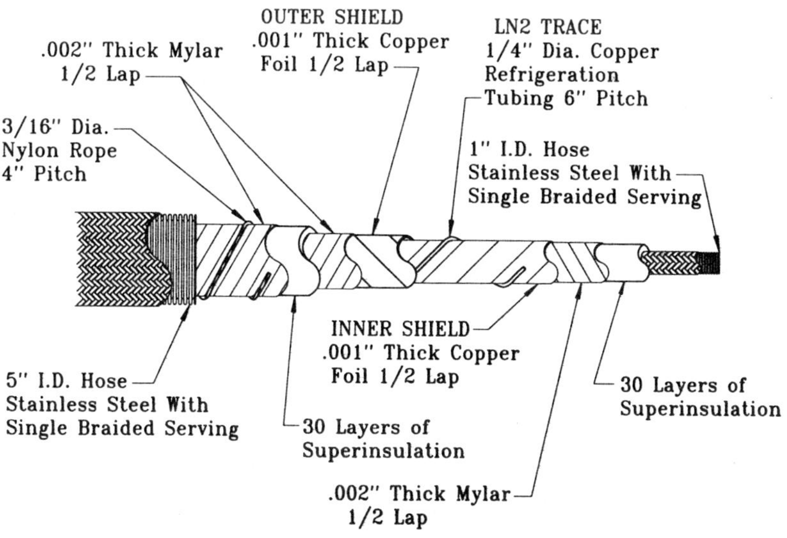


Figure 4. Hall C flexline cutaway view.

The main Hall C supply transfer line supplies both Gaseous He at 4.5 K, 3.0 atm and returns GHe at 4.2 K, 1.2 atm and Gaseous He at 15 K and 20 atm with the return at 20 K and 19 atm and LN2 at 4 atm and 80 K (see Fig. 5). The 4.5 K Helium supply is coaxial with the return and forms a 300 foot long heat exchanger to recover the superheat of the return helium. The 15 K supply and 20 K return for cryotarget use are separate pipes. These three lines are all mounted inside the LN2 annulus for good shielding. This line has been described previously in ref. 4 and a cross section of it is shown as figure 2.

The HMS transfer line is a rigid line with three isolated parallel lines mounted inside the LN2 cooled shield. The transfer line contains the cooldown/warm-up heat exchanger and cooldown Helium supply line. This line distributes the helium at 4.5 K to the magnet JT valves and it returns the boiloff from the magnet reservoirs at 4.2 K. Each of the SC magnets is cooled internally by the use of thermalsyphon circulation. The internal thermalsyphon loop starts and finishes inside the magnet reservoir and is driven by the magnets own heat leak. The inlet JT valve thus simply maintains liquid level in the reservoir. This maintains a degree of isolation between the four magnets as the function of the internal circulation does not depend on whether the reservoir is filling or not. The four independent JT valves are PID controlled from their respective magnet control systems. Operational stability is achieved by simply setting the valves integral time constant as long as possible, typically 10 to 30 minutes, and carefully choosing these time constants to be irrational ratios of each other. The valves have minimum and maximum openings that are chosen to not overdraw the supply and to not stop the flow to any magnet completely. This simple control philosophy avoids valve cross talk and coupling, avoids over stressing the supply during a group recovery from a system outage and avoids instability due to warm-up and cooldown of piping caused by interrupted flow. The system has operated for four years in this mode without difficulty not considering the very rare failure of components. A cross section of this line is shown as figure 3.

The system also contains the variable temperature helium supply line from the internal heat exchanger to the cooldown control valve in each magnet. The lines and magnets are completely parallel so that a magnet or magnets may be warmed up or cooled down while simultaneously the other magnets can be operated normally. It has been the case generally that maintenance has required that single magnets be warmed or cooled. A warm-up or cooldown of the entire spectrometer is possible by simply dividing the available 10 gm/sec cooldown flow.

The cooldown variable temperature Helium is produced in the HMS transfer line by precooling 4 atm Helium gas from room temperature to 80 K using a counterflow heat exchanger with boiler stage and LN2. This heat exchanger is sized to produce 10 gm/sec of 80 K helium and consumes 30 gm/sec of LN2. The variable temperature Helium is produced by blending a stream at 80 K with a 300 K stream. The variable temperature is monitored by a PT 100 thermometer mounted in the stream. The mixing valve and thermometer can be switched to any of the four SC magnets. The heat exchanger that is the heart of the cooldown system is a fintube counterflow design with a boiler stage and a gas to gas stage for maximum efficiency. The system can be used to cooldown multiple magnets by slaving the temperature control to the magnet which cools the slowest. The temperature control is formed by maintaining the variable temperature 50 K below the mean magnet temperature and keeping the maximum Delta T within the magnet also below 50 K. Each SC magnet has cooldown control logic that performs this task. This provides minimal thermal stress during cooldown. The typical cooldown duration ranges from three days for the HMS dipole up to 10 days for the Large HMS quads. The control of the LN2 consumption by the heat exchanger is achieved by maintaining the outlet N2 temperature at

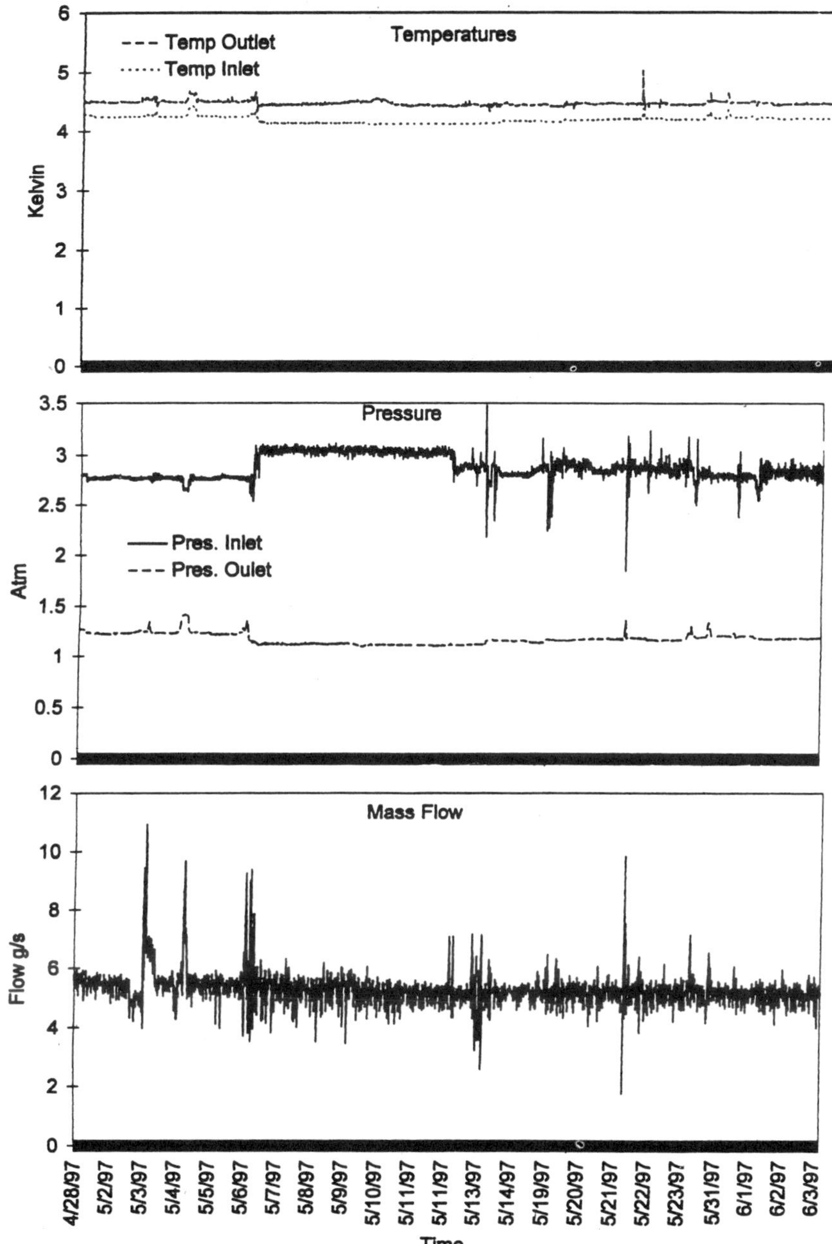


Figure 5. Time mass flow, pressure and temperature plot for transfer line.

typically 250 K. A local Honeywell controller and a Pt 100 thermometer is used for this function.

There is a parallel system of warm gas manifolds that handles the return of warm gas from current leads, the return of waste gas from cooldown and the supply of warm 4 atm gas for the cooldown heat exchanger. These helium lines are part of the sealed Helium refrigeration system. There is a line that vents the gaseous N2 from the magnets and the cooldown heat exchanger directly outdoors. The relief valves and rupture discs from the SC magnets vent directly into Hall C. The magnet relief valves were left vented to the Hall to increase safety during commissioning of the individual magnets. These valves have never opened during the four years of operation on Hall C. Connections exist to plumb these relief valves back into the appropriate gas return manifold should the need arise. The N2 gas vent is also used as an inverted relief line for the cryotargets which are typically H2 or D2.

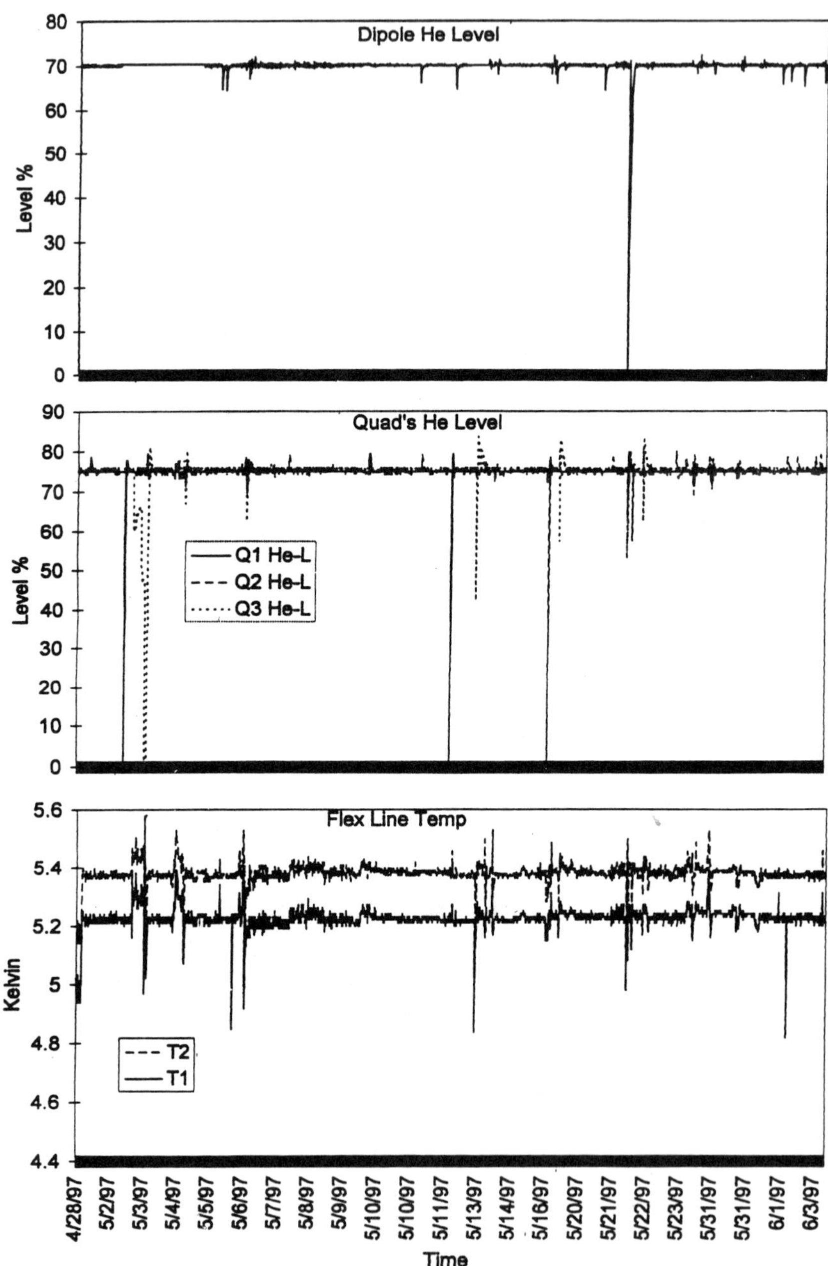


Figure 6. Time temperature, plot and liquid level in SC magnet.

**Shielded flexible transfer lines**

The HMS is required to rotate on demand automatically with the SC magnets at field during the normal course of running an experiment in Hall C. Therefore it was necessary to connect the HMS with a flexible transfer line system that would allow the 90 degree rotation. A unique shielded line was designed and built at TJNAF for this purpose. Figure 4 shows a cut back view of this line to illustrate its construction. The flex lines use a typical armored flex hose for the inner conduit and for the vacuum shell. The LN2 shield is built up from an inner and outer spiral wound copper foil with a spiral wound 0.25 inch copper tube to carry the LN2 in between. Spiral wound layers of mylar and clothesline are used to stabilize the superinsulation and allow the layers to slip past each other when the line bends during a rotation. The line was simple to fabricate and was less expensive than a comparable line made up from all flex hose.

The heat leak of the supply flex line is measured continuously as the inlet and outlet temperature, line pressure and mass flow are continuously recorded. Figure 6 shows the temperature difference for a 1 month period. The one year time averaged heat leak of this line is 10.5 ± .25 watts.

**Cryogenic target**

The cryogenic system supplies 15 K Helium or 4.5 K Helium and LN2 to the Hall C cryogenic target system. The cryotarget has three independent target pump loops which typically have LH2, LD2 and Helium in them. These loops are cooled through a miniature fintube heat exchanger with an inline axial flow pump. The pump circulates target gas at flow rates as high as 200 gm/sec for LD2 while the heat exchanger removes a measured 700 watts of beam heat and 100 watts of bias heater power. Optional operation at 4.5 K cooling temperature for high pressure HE3 targets and HE4 targets is included. Online switch valves allows the operation of HE3 at 4.5 K and then H2 and D2 calibration at 15 K. This helps prevent target gas freezing and lowers the average heat load on the refrigeration system.

The cryogenic target rotates and elevates to get in and out of the beam and to index the six cryogenic targets and two dummy targets into the beam respectively. This local motion is accomplished with flexible U-tubes. There is 1 meter of vertical motion and 90 degrees of rotation while at temperature. The same long flex lines allows both motions simultaneously. Automatic indexing is required during routine operation of the target for experiments and rotation is required to allow insertion of room temperature solid targets into the scattering chamber.

**SC Magnet connections.**

The four magnets of the HMS are connected to the cryo system by flexible U-tubes to permit magnet position adjustment. The HMS quadrupoles are mounted together on a Zoom Lens arrangement that allows the Quads to be moved forward by 16 inches to increase the spectrometer solid angle and back from the nominal position to decrease the minimum scattering angle that the spectrometer can view. These motions are routinely achieved while the magnets are cold but usually unpowered.

**Table 1.** Heat load summary

| | |
|---|---|
| Hall C total consumption | 5.3 gm/sec |
| Hall C Heat load | 109 watts |
| Hall C lead flow | 0.6 gm. sec |
| | |
| Flexline heat load | 10.5 watts |
| HMS Dipole heat load | 60 watts |
| HMS Quad 1 heat load | 10 watts |
| HMS Quad 2 heat load | 11 watts |
| HMS Quad 3 heat load | 12 watts |
| | |
| Residual system heat load | 5.5 watts |

## Conclusion

The Hall C cryogenic system has been in operation for four years. The system has simple control features that have resulted in stable operations. The system has proven to be reliable and rugged and has required no maintenance since commissioning. The flexible transfer line design is unique, has reasonable heat leak, is simple to build and is cost effective. The transfer system has continuous data logging and an off-line history exists for the first four years of operation. This makes possible a long term study of stability and performance.

The average heat load of the entire system, magnets included, is 109 ± 9 watts as shown in Table 1.

## REFERENCES


1 P. D. Brindza et al., Commissioning the Superconducting Magnets of the High Momentum Spectrometer (HMS) at TJNAF, IEEE trans. Mag Vol. 7, No. 2, June 1997, Pg. 755.

2 Cryotarget design for TJNAF, internal document.

3 P. D. Brindza et al. Superconducting Toroidal Magnet Design for the G0 Experiment at TJNAF, IEEE trans. Mag Vol. 7, No 2, June 1997, Pg. 618.

4 P. Brindza, W. Chronis, C. Rode, J. Kelley, M. Shea, CEBAF's Cryogenic System, Advances in Cryogenic Engineering, Vol. 39 Pg. 623.

# DESIGN OF THE FILL/TRANSFER STATION CRYOSTAT FOR THE OMEGA CRYOGENIC TARGET SYSTEM

C. R. Gibson,[1] C. M. Charmin,[1] J. V. Del Bene,[1] E. H. Hoffmann,[1]
G. E. Besenbruch,[1] and I. Anteby[2,3]

[1]General Atomics
San Diego, California 92121, USA

[2]University of Rochester, Laboratory for Laser Energetics
Rochester, New York 14623, USA

[3]Permanent address: Nuclear Research Center, Negev, Israel

## ABSTRACT

General Atomics is designing, testing and fabricating a system for supplying cryogenic targets for the University of Rochester's OMEGA laser system. A prototype system has demonstrated the filling of 1 mm diameter, 3 μm wall plastic spheres to 111 MPa (1100 atm) with deuterium and then cooling to 18 K to condense the fuel. The production design must be capable of routinely filling and cooling targets with a 50/50 mix of deuterium and tritium and transferring them to a device which places the targets into the focus of 60 laser beams. This paper discusses the design and analysis of the production Fill/Transfer Station cryostat. The cryostat has two major components, a fixed base and a removable dome. The joint between the base and the dome is similar to a bayonet fitting and is sealed by a room temperature elastomeric o-ring. Since the cryostat must be housed in a glovebox, its design is driven strongly by maintenance requirements. To reach the equipment inside the cryostat, the dome is simply unbolted and lifted. The inside of the cryostat is maintained at 16 K by a closed loop helium flow system. Gaseous helium at about 1.4 MPa (200 psi) flows through tubes which are brazed to the inner walls. Cooling is provided by several cryocoolers which are located external to the cryostat. Liquid nitrogen is used as a heat intercept and to precool the helium gas.

## INTRODUCTION

As part of the national Inertial Confinement Fusion Program, General Atomics is designing, constructing and testing the OMEGA cryogenic target system.[1] This system will be capable of filling, transporting, layering, characterizing, inserting, and positioning deuterium/tritium (DT) filled cryogenic targets. It will be part of the 30 kJ OMEGA laser at the University of Rochester's Laboratory for Laser Energetics in Rochester, New York.

The fill/transfer station's requirements are to fill the targets with DT, cool them to cryogenic temperatures and transfer them to the next device for transport to the target chamber. The process begins when four mounted targets, assembled into a rack, are placed inside a high pressure permeation cell. The targets are filled with DT at pressures up to 152 MPa (1500 atm) and then are cooled to cryogenic temperatures to condense the DT. Using precision vacuum manipulators, one target is removed from the target rack and inserted onto the stalk of a moving cryostat. The moving cryostat then transports the target to the center of the OMEGA target chamber. It is positioned very accurately and then, after rapid removal of the thermal shroud, is shot by the lasers.

The mounted target, shown in Figure 1, includes a 1 mm diameter plastic sphere. This sphere is suspended from a beryllium "C" by three spider webs. This assembly is attached to the top of a boron post. The entire mount is about 20 mm tall.

## CRYOSTAT DESIGN

As shown in Figure 2, the fill/transfer station consists of a large cryostat with several components attached. The inserter and target manipulator are precision vacuum manipulators for moving the targets. The permeation cell is a high pressure device capable of being uniformly cooled to 16 K. The moving cryostat port contains several pieces of equipment needed to hand off the target from the fill/transfer station to the moving cryostat system.

### Cryostat

The cryostat is of "belljar" design and consists of two distinct parts: a fixed base and a removable dome. The joint between the base and the dome is similar to a bayonet fitting and has a nominal 2.5 mm (0.10 in.) gap. This joint is sealed by a room temperature elastomeric o-ring at the joint between the base and dome. The cryostat is 1.25 m (49 in.) in diameter and 1.10 m (43 in.) high and is fabricated from 316 stainless steel.

Both the base and dome are vacuum insulated. These vacuum spaces are isolated to prevent tritium from contaminating the multilayer insulation (MLI). If this were to happen, it would be difficult to decontaminate the MLI due to its large surface area. The inner volume between the base and the dome operates with 6.6 Pa (50 mTorr) of helium. This gas removes heat from the target caused by decay of the DT.

The dome inner wall has an inside diameter of approximately 1.07 m (42 in.) and is a pressure boundary between the dome insulating vacuum and the inner volume. The inner

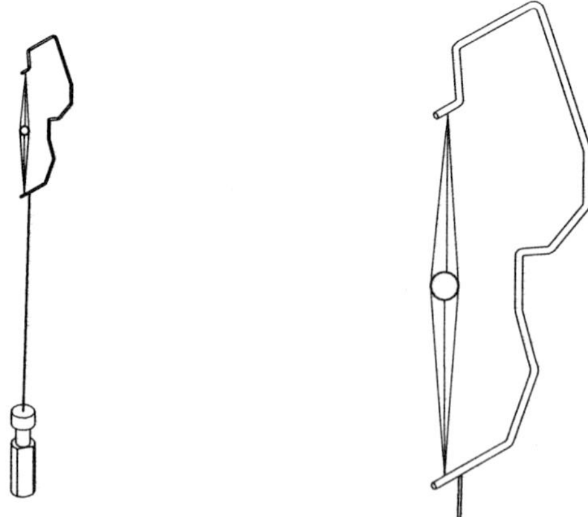

**Figure 1.** Sketch of mounted cryogenic target. Sketch on right is enlargement of beryllium "C", target sphere and spider webs.

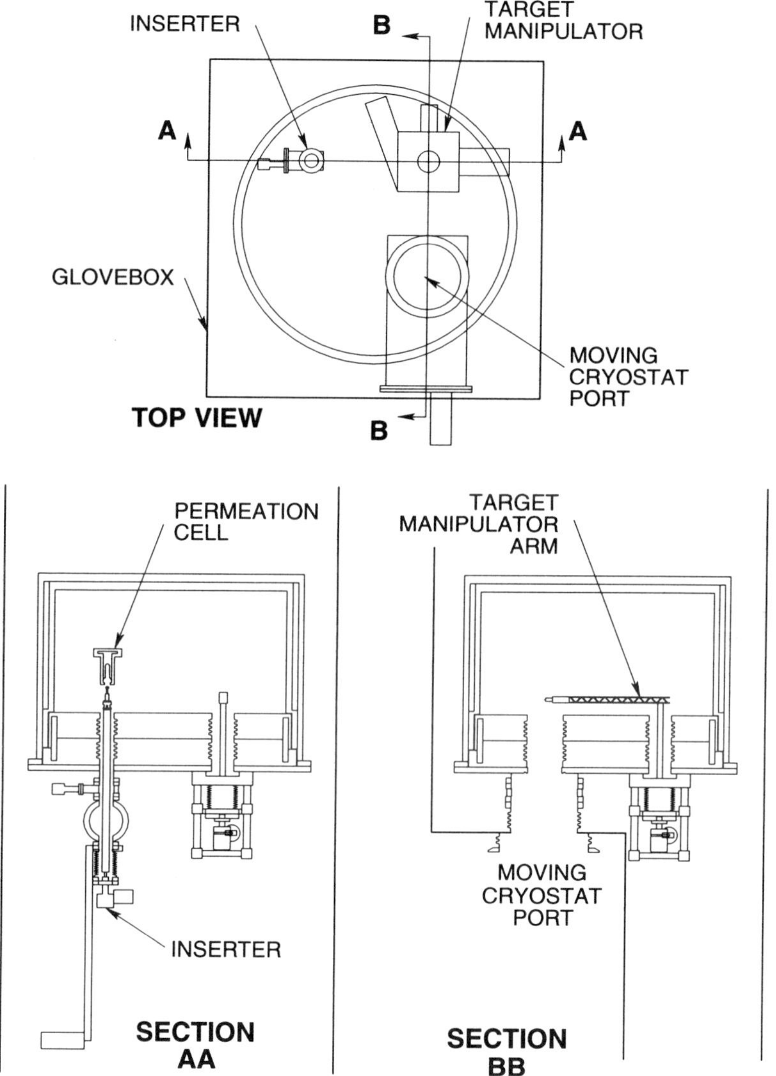


**Figure 2.** Sketch of Fill / Transfer Station showing cryostat, glovebox and major components.

wall has cooling tubes attached to it through which cold helium gas is circulated. This wall has a 16 K nominal temperature. Inside the insulating space, a liquid nitrogen cooled thermal shield is used to intercept both radiation heat transfer and conduction heat transfer coming up the mouth of the dome.

The cryostat base is built around the 32 mm (1.25 in.) thick bottom plate. This plate supports the cryostat during operation and also supports the vacuum manipulators hanging below it. Between the bottom plate and the top plate is a vacuum insulating space which includes a liquid nitrogen cooled thermal shield and MLI. The penetrations between the two plates are sealed with metal bellows to insure that the base insulating space is isolated. The top plate of the base has cooling tubes attached to its bottom and has a nominal temperature of 16 K. An elastomeric o-ring along the top edge of the base bottom plate is used to seal the base to the dome. A heating strip around the circumference of the bottom plate keeps the o-ring at room temperature.

For maintenance of equipment on the inside of the cryostat, the dome is simply unbolted and lifted vertically off the base. The lift mechanism is built into the glovebox. This design also allows warm alignment and checkout of all equipment, both outside and inside the cryostat.

## Cooling Module

The cryostat cooling is provided by the cooling module which is located directly below the cryostat. The cooling module is designed around eight closed-cycle cryocoolers, Sumitomo SRD-210W, grouped into five independent helium circulation loops.

One loop is used to cool the permeation cell and another is dedicated to cool the moving cryostat thermal shroud. These two loops each have one cryocooler and are designed to be frequently cycled between room temperature and cryogenic temperatures. Flowing compressed helium gas at approximately 0.3 g/s and 1.4 MPa (200 psi) connects the components to be cooled with the liquid nitrogen bath and cryocoolers. It first flows through a liquid nitrogen bath to cool to approximately 80 K. Next, it flows past the first stage of a cryocooler to bring it to about 50 K, then past the second stage to drop its temperature below 16 K. The helium then flows through tubes which are thermally connected to the items to be cooled.

The three remaining cooling loops are optimized to operate in a cryogenic steady-state mode, as shown in Figure 3. Each of the three loops have two cryocoolers. These loops operate at about 1.0 g/s and 1.4 MPa (200 psi). One loop is used to cool the cryostat base, one for the cryostat dome, and one for the equipment inside the cryostat. These loops each include three additional heat exchangers, HX1, HX2 and HX3, to maximize steady-state performance. The cooling module is also designed to raise the internal temperature of the cryostat to 473 K to help remove tritium that has been adsorbed by the inner wall.

## Thermal Analysis

A thermal analysis was performed to evaluate the transient and steady-state performance of the fill/transfer station cryostat and cooling module. The thermal analysis was performed using both hand calculations and the General Atomics' two-dimensional finite difference code TAC2D. This code has been used for many years and has been validated in General Atomics' helium gas cooled reactor programs.

The TAC2D[3] model of the cryostat is axisymmetric and includes both the cryostat dome, base, and internal components and the cooling module. FORTRAN subroutines added to the code include cryocooler performance maps, temperature dependent material properties, liquid nitrogen boiling heat transfer, pressure drop along the flow loops, and heat exchangers characteristics. Parametric studies were also performed to examine the sensitivity of the system to cryocooler performance, mass and heat load variations.

The TAC2D analysis, verified by hand calculations, gave the total steady-state heat input to the 16 K wall as 26 W. This heat load could be handled by two Sumitomo SRD-210W cryocoolers, however, for operational reasons five separate flow loops are required. In addition, the customer requested redundancy in the three steady state flow loops bringing the total number of cryocoolers to eight.

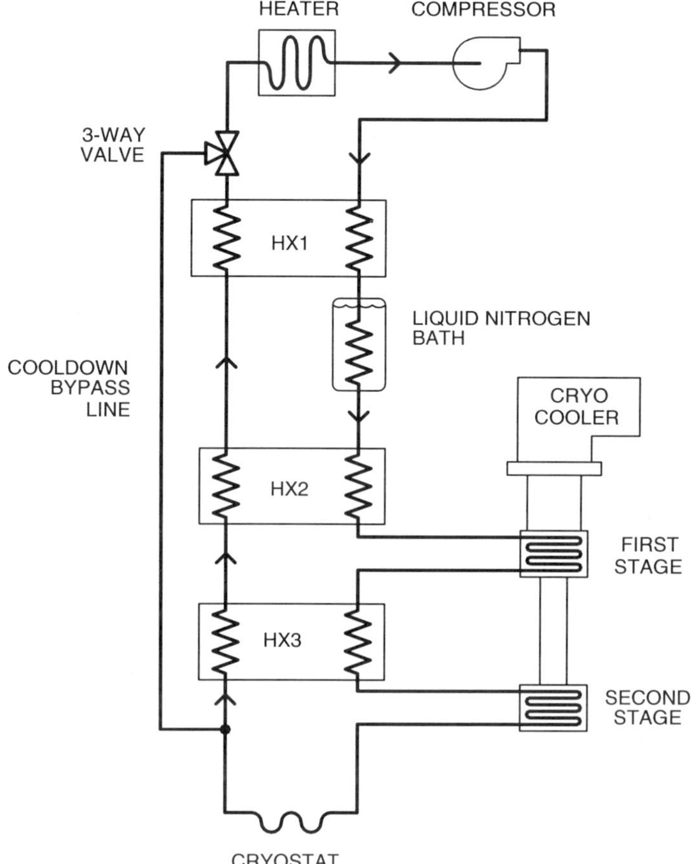


**Figure 3.** Schematic of steady-state flow loops (one of three).

The transient thermal analysis was also performed by TAC2D. The analysis gives the cooldown time to 16 K with eight cryocoolers as 10 hours. The stresses during this transient are all within allowables.

## Stress Analysis

The cryostat was designed and analyzed to the intent of the ASME Pressure Vessel Code. The stress analysis was performed using both hand calculations and the finite-element code ANSYS. The highest stressed region is the cryostat mouth, where the base and dome mate. This region is designed to be similar to a bayonet fitting. To minimize the heat leak, the two pieces must fit together with little clearance and must have minimum wall thickness. Using 1.50 mm (0.060 in.) thick 316 stainless steel for the walls satisfied both the structural and thermal constraints. However, to satisfy the buckling criteria, two stiffening rings were added to both the base and the dome.

The design is complicated by the fact that there are three independent vacuum spaces. In a fault condition, one of these spaces may be pressurized by a leak in a helium cooling loop or a liquid nitrogen tank. Evaluating all credible failure modes, gives large ranges of minimum to maximum stresses. This problem was solved by using very sensitive, 7 kPa (1 psig), pressure relief valves on each vacuum space.

## FILL AND TRANSFER EQUIPMENT DESIGN

In addition to the cryostat, the fill/transfer station includes components which actually fill the targets with DT and to transfer them to the moving cryostat. The design of this equipment is challenging since they must operate in environments which include cryogenic temperatures, high pressures, tritium, and with extreme mechanical precision.

### Permeation Cell

The permeation cell is located inside the cryostat inner volume. Although the cryostat interior walls remain cold, the permeation cell temperature is varied from 473 K to cryogenic temperatures. During target fill, the permeation cell is warm to allow the DT to permeate through the plastic target wall. Target fill is normally performed at room temperature, however increased permeation rates may be obtained at temperatures up to 473 K. Once the desired pressure is reached, up to 152 MPa (1500 atm), the permeation cell is cooled to trap the solid DT inside the target.

The permeation cell is manufactured from A286, a precipitation-hardened high strength stainless steel alloy. To minimize thermal gradients, a copper sleeve is brazed to the cell wall. Electric heaters attached to this sleeve maintain the cell temperature during target fill. To cool the cell, helium gas is flowed through tubes attached to the outside of the permeation cell. This helium gas is cooled by liquid nitrogen and a cryocooler in the cooling module.

After the targets are inserted, the permeation cell must be sealed to allow it to be pressurized. This is accomplished by means of a breech lock and a pneumatic actuator built into the cell. The actual pressure seal consists of mating sixty degree cones, similar in design to high pressure tube fittings.

### Inserter and Target Manipulator

These devices are used to manipulate the targets while they are at cryogenic temperatures. They are vacuum manipulators which consist of precision motion stages and bellows feedthroughs. Each axis of motion is driven by a stepper motor and controlled by a motion controller. The cold portion of these devices are actively cooled and are designed to minimize heat transfer.

The inserter is used to insert the targets into the permeation cell. It consists of a vertical thin-walled tube whose lower end is connected to an axial and rotary motion vacuum manipulator. Between the manipulator and the cryostat is a vacuum lock and a gate valve used to manually load the targets. In operation, the inserter raises the targets up into the cryostat and rotates sixty degrees to close the breech lock.

The target manipulator moves individual targets between the permeation cell and the moving cryostat. The movements are performed by a x-y-z and rotary vacuum manipulator. The portion of the target manipulator located inside the cryostat consists of a thin-walled stainless steel tube extending up from the manipulator. Attached to the tube is a horizontal arm with a end effector on the far end. This end effector is designed to pick up an individual target and place it on the moving cryostat.

### Moving Cryostat Port

The moving cryostat is the device which transports the target to the center of the OMEGA target chamber. It interfaces with the fill/transfer station cryostat at the moving cryostat port. This port includes equipment to remotely remove the thermal shield from the moving cryostat and cool it down to 16 K. The thermal shield is lifted through the use of a ball screw driven linear stage which operates at cryogenic temperatures. Also included is a mechanism which precisely aligns the moving cryostat with the target manipulator during the target transfer.

### Closed Circuit Television System

The closed circuit television system allows the operator to confirm the position of all of the components inside the cryostat. It consists of one fiberscope and two borescopes. The fiberscope is attached to the arm of the target manipulator and travels with it. There is one borescope near the inserter to confirm that the target manipulator has picked up the target. The second borescope is near the moving cryostat port to confirm that the target manipulator has successfully dropped off the target. The borescopes will be oriented at right angles to the fiberscope to give the operator orthogonal views.

### Instrumentation and Control

The instrumentation will consist of temperature, pressure, strain, tritium detection and alarm, and position sensors. All of the instrumentation will be interfaced into a local controller. This controller will be used to control the temperature and vacuum of the cryostat and to coordinate the motion of the vacuum manipulators.

## PROTOTYPE TESTING

General Atomics has demonstrated the filling, cooling and transporting of targets using prototypical equipment.[2] We have routinely filled 1 mm diameter mounted spheres with wall thickness as small as 3 μm to the nominal fill pressure of 111 MPa (1100 atm) with deuterium. These targets were then cooled to 18 K to condense the fuel and transported outside of the cryostat.

The prototype cryostat however, was not designed for use with tritium. For example, indium o-rings were used to seal all flanges which see cryogenic temperatures. It would be nearly impossible, however, to handle indium wire through heavy gloves.

Other key pieces of equipment have also been demonstrated. A full scale permeation cell has demonstrated operation at both 152 MPa (1500 atm) and at 16 K. A borescope has also been operated at cryogenic temperatures.

## STATUS

The fill/transfer station design is complete and fabrication drawings are currently underway. Purchase orders for key long lead items have been placed including the cryocoolers with Sumitomo, the cryostat with Ability Engineering Technology, and the target manipulator with Thermionics NW. The system will be assembled and operated with deuterium at General Atomics and then shipped to the University of Rochester for DT testing. The system is scheduled to be operational by summer 1999.

## ACKNOWLEDGMENT

The authors wish to acknowledge the dedicated and talented team which developed the design of the Fill / Transfer Station. Work supported by the U.S. Department of Energy under Contract No. DE-AC03-95SF20732.

## REFERENCES

1. N.B. Alexander et al., The Cryogenic target handling system for the OMEGA laser, Proc. of the 16th IAEA Fusion Energy Conf., Montreal, Canada 1996 (International Atomic Energy Agency, Vienna, to be published).
2. G. E. Besenbruch et al, Testing of the cryogenic target handling system for the OMEGA laser, Proc. of the 19th Symposium on Fusion Technology, PF52 (1996).
3. R.H. Boonstra et al, TAC2D a general purpose two-dimensional heat transfer computer code — user's manual, General Atomics Report GA-A14032 (1976).

# THE COLD NEUTRON SOURCE OF THE MUNICH RESEARCH REACTOR (FRM)

H. Gerstenberg, E. Krähling, D. Päthe, K. Schreckenbach, and W. Gläser

Fakultät für Physik E21, TU München
Reaktorstation, D-85747 Garching, Germany

## ABSTRACT

In order to satisfy the growing need for low energy ("cold") neutrons in various fields of science the Munich Research Reactor (FRM) has recently been equipped with a cold neutron source. The facility is based on the secondary moderation of thermal reactor neutrons by subcooled liquid hydrogen being contained in a moderator chamber of $V \simeq 0.9$ l. Due to the conflicting requirements of high transmission for low energy neutrons and sufficient mechanical strength the moderator chamber was fabricated in form of a thin-walled Al container being stabilized by a large number of internal Al support rods. It is installed within an evacuated beam tube in a distance of only few cm from the reactor core.

Special safety considerations had to be taken into account due to the use of hydrogen within a nuclear reactor. The formation of an explosive mixture of hydrogen and oxygen within the reactor building is prevented by means of a multi-barrier system. In addition in case of serious malfunctions the total hydrogen content of the cold neutron source is exhausted automatically by He inert gas into the open air.

The experimental areas are located outside the reactor building in a separate laboratory. First experimental setups, among these one for the neutron tomography of technical objects and a training experiment for students of the Technical University of Munich are in operation.

## INTRODUCTION

Neutron scattering is a powerful and well established experimental technique for the investigation of atomic and magnetic structures in solid state physics. Particularly, it is suited for the analysis of dynamical processes in solids and liquids.

The neutron spectrum provided by a research reactor is mathematically described by a Maxwell distribution corresponding to the moderator temperature. Wavelengths and kinetic energies of neutrons are related to this quantity via

$$\lambda(\mathring{A}) = 0.286/\sqrt{E(eV)} = 30.81/\sqrt{T(K)} \tag{1}$$

In standard neutron scattering experiments thermal neutrons with a kinetic energy of approximately 26 meV are used. For sensitive investigations of low energy atomic excitations or large structures (on an atomic scale) like e.g. biomolecules, however, subthermal neutrons to be generated in a cold neutron source (CNS) are required. Basically a CNS is a secondary neutron source which is physically based on the additional moderation of thermal neutrons within a cryogenic medium. Because of their low absorption probability and high stopping power for neutrons liquid hydrogen or liquid deuterium are the most suitable moderator materials for this purpose.

A CNS is available in most of the major German research reactors [1,2,3]. In order to satisfy the growing needs for cold neutron beams also the Munich Research Reactor (FRM) operated by the Technical University of Munich has recently been equipped with a CNS using subcooled liquid hydrogen as moderator. In thermal equilibrium with the cryogenic medium the Maxwell spectrum of the neutrons is shifted to lower temperatures, i.e. longer wavelengths. By means of special neutron guides the "cold" (low energy) neutrons are transported from the moderator chamber to the experimental areas, which – for reasons of low radiation background and enlargement of the useful space – are located outside of the reactor building.

## SETUP OF THE COLD NEUTRON SOURCE

### Installation of the CNS in the Munich Research Reactor (FRM)

The FRM is a light water moderated swimming pool reactor of the MTR type and operated at 4 MW thermal power. Its core is a rectangular assembly of fuel elements surrounded by a Be–reflector. The core dimensions including the Be–reflector are approximately $485 \times 685 \times 600\ mm^3$.

The setup of the CNS within the FRM is shown in Figure 1, which also indicates the main technical components of the facility:

- the hydrogen circuit including the moderator chamber, and the $H_2$/He heat exchanger,
- the He refrigeration system (coldbox),
- the off gas system,
- the evacuation system and vacuum spaces,
- the beam tubes providing the experimental areas with cold neutrons,
- the control and monitoring units including the power supply.

The moderator chamber containing the liquid hydrogen was integrated into the rectangularly shaped so-called QR beam tube which is directed perpendicularly towards the reactor core. For the insertion of the beam tube's thimble the thickness of the Be–reflector

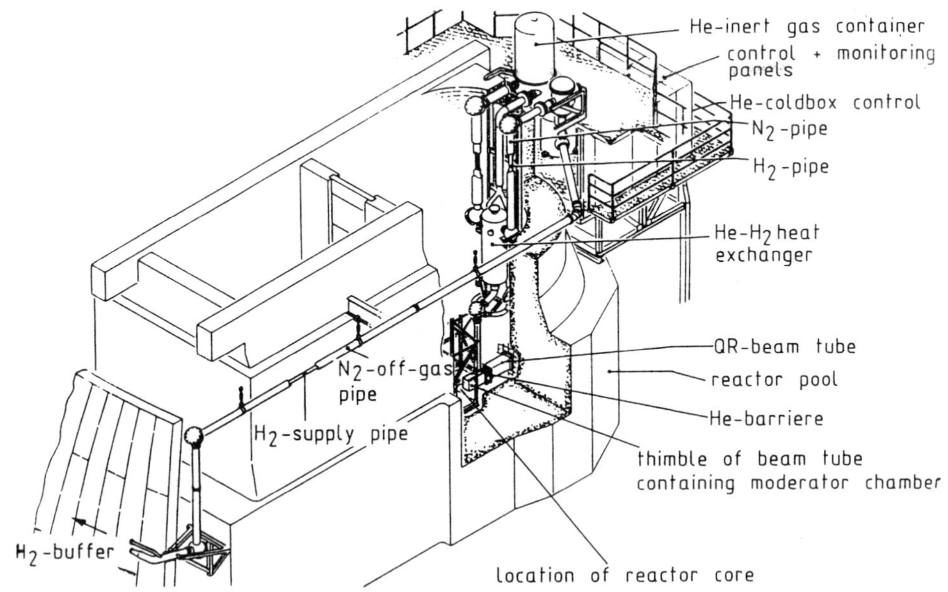


**Figure 1.** Schematic overview of the CNS installations within the reactor pool of the FRM.

was reduced by approximately 35 mm thus allowing the exposure of the moderator chamber to an intense primary thermal neutron flux density of $\Phi_{th} \simeq 2.5 \times 10^{13}\ cm^{-2}s^{-1}$.

## Moderator Chamber and Hydrogen System

The hydrogen circuit of the CNS includes the moderator chamber, the $H_2$/He heat exchanger, a $H_2$–buffer (V = 7.5 $m^3$) located outside the reactor building, and the connecting pipes with a total length of approximately 75 m.

The moderator chamber (see Figure 2) is the central component of the CNS. It was designed as a rectangular parallelepiped with outer dimensions of $200 \times 146 \times 28\ mm^3$. In order to minimize the intensity loss of the neutron beam within the moderator chamber the latter was manufactured from AlMg3, a material exhibiting a high transparency for cold neutrons, with a wall thickness of d = 1 mm only. During operation, on the other hand, the moderator chamber and the whole hydrogen circuit are loaded by an internal pressure of up to 4.5 bars. Therefore, due to the thin walls it was necessary to reinforce the moderator chamber by in total 178 Al support rods (square shaped with a = 2.5 mm) which cover only 4.2 % of its cross section. The stability of the design was proven in a loading test using an identical moderator chamber which showed up to be gas–tight up to a pressure of 61 bars. In addition the rectangular shape of the moderator chamber matches best the cross section of the QR beam tube and provides the optimal illumination of the neutron guides coupled to CNS.

During operation of the CNS, the hydrogen within the moderator chamber is a subcooled liquid at a temperature of $T \simeq 15$ K. The corresponding pressure is 3.5 bars. At full reactor power of 4 MW the near core installations of the CNS are subject to a nuclear heating power of $P \simeq 400$ W, resulting in a warm-up of the $H_2$ to $T \simeq 24$ K. The hydrogen which is kept as a nonboiling liquid within the entire cryogenic circuit is recooled in a $H_2$/He heat exchanger being mounted within the reactor pool in a distance of 2.5 m above the QR beam tube and the reactor core. Due to the thermosyphon mechanism the circulation of the liquid hydrogen is driven by natural convection only. The cooling capacity of the $H_2$/He heat exchanger is supplied by a He-refrigeration system (coldbox) which is installed on a separate platform at the periphery of the reactor pool.

## Safety Aspects

The installation of an experimental setup involving the use of hydrogen technology within a nuclear environment requires special safety measures.

In any case, the formation of an explosive mixture of hydrogen and oxygen by penetration of air into the $H_2$-circuit or by escape of $H_2$ inside of the reactor building has to be avoided. In the CNS this requirement is fulfilled even in the case of leakages by means of a triple barrier system.

In respect to the safety requirements the $H_2$-circuit has to be subdivided into the cryogenic components and those kept at ambient temperature. For reasons of heat insulation the cryogenic components of the $H_2$-circuit (forming the first barrier by themselves) are mantled by vacuum spaces, the pressure of which is monitored continously ($2^{nd}$ barrier). In addition these components are installed within the reactor pool. Thus the pool water forms the third barrier against penetration by air. The triple barrier concept is also obeyed with respect to a possible leakage of air into the evacuation system which is used for the heat insulation of the cryogenic components. In addition to the permanent monitoring of the vacuum, the corresponding out of pool installations are envelopped by a He-filled barrier. Within the $H_2$/He-heat exchanger both gas systems are separated by vacuum. The heat transfer takes place via heat conduction in the rectangular copper profile forming the walls of the vaccum space (see Figure 3). In addition, inside the reactor building the connecting pipes ($1^{st}$ barrier) between the cryogenic $H_2$-system and the gas buffer – being kept at room temperature

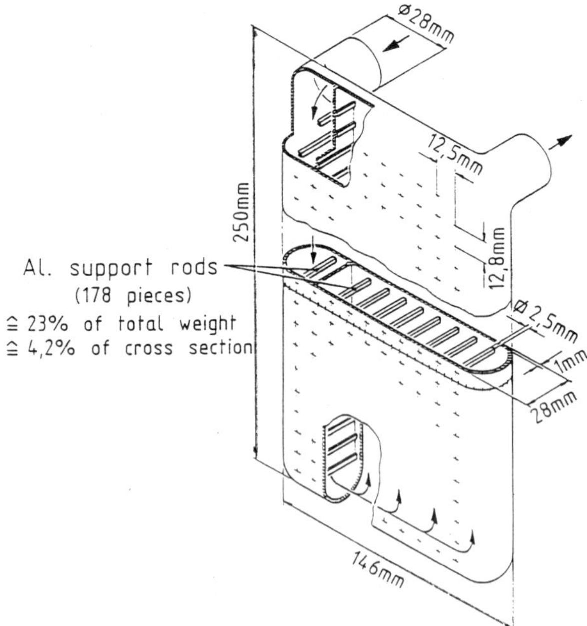


**Figure 2.** Moderator chamber of the CNS. Note the low wall thickness of d = 1mm and the lattice of internal support rods.

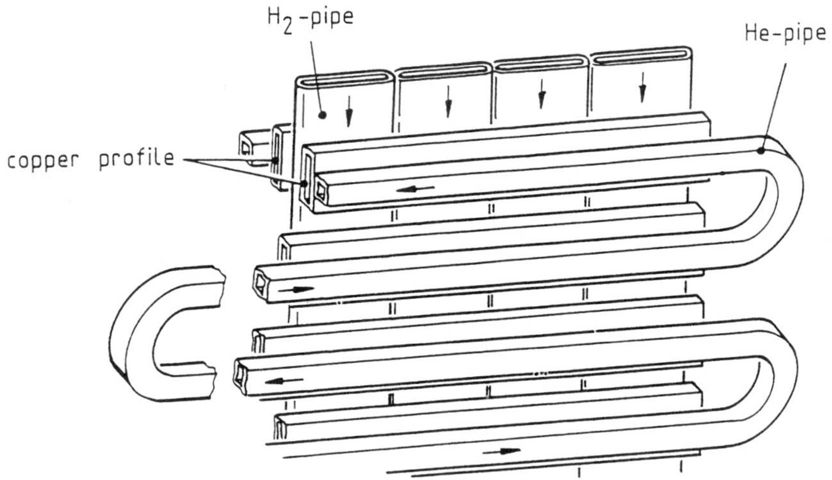


**Figure 3.** Inner structure of the He/$H_2$ heat exchanger exhibiting a multi barrier concept against mixture of the gases.

– are concentrically encapsulated by an off-gas pipe ($2^{nd}$ barrier) which is continously rinsed by a $N_2$–gas stream of $\dot{V} \simeq 3\ m^3/h$ ($3^{rd}$ barrier). A schematic overview of the decribed safety concept is given in Figure 4. Finally, outside closed buildings the German security standards for dangerous goods do apply. The hydrogen containing components are kept under an excess pressure which is monitored continously. In case of a leakage in this area the escaping $H_2$ is diluted immediately and a further protection against the formation of an explosive gas mixture is not to be required.

The front part of the QR beam tube containing the moderator chamber and pipes for the supply of liqid hydrogen is separated from the neutron guides and the experimental area by means of a special flange construction (see Figure 5): In a distance of 0.6 m from the thimble the beam tube is closed hermetically by the so-called He-barrier, a pillow shaped thin-walled (d = 1mm) AlMg3 component which is loaded by approximately 3 bars of He surpressure during operation. In case of an accidental penetration of pool water into the QR beam tube or a leakage of the liquid $H_2$ containing components into the insulation vacuum, the He–barrier withstands the corresponding excess pressures. On the other hand, the thickness of material within the neutron beam and consequently the neutron absorption is minimized by the design shown in Figure 5. Finally, this concept of a divided beam tube allows the repair or change of cryogenic components without dismanteling of the neutron guides.

## Protective Measures at Malfunction

Any malfunction of the CNS is handled according to a fail–safe principle. By activation of the protective measures "He–ventilation of vacuum spaces" and/or "Blow–out of $H_2$" the integrity of the CNS is guaranteed and a safe status is reached in any case. By means of an integrated monitoring– and control–system the reason for the activation of the above measures is registered automatically.

In case of loss of pressure within the $H_2$–system, the He–supply, or the He–barrier or in case of too low $N_2$–flow within the off–gas system, the hydrogen (no matter if in liquid or gaseous form) is blown out of the CNS into the open air automatically by

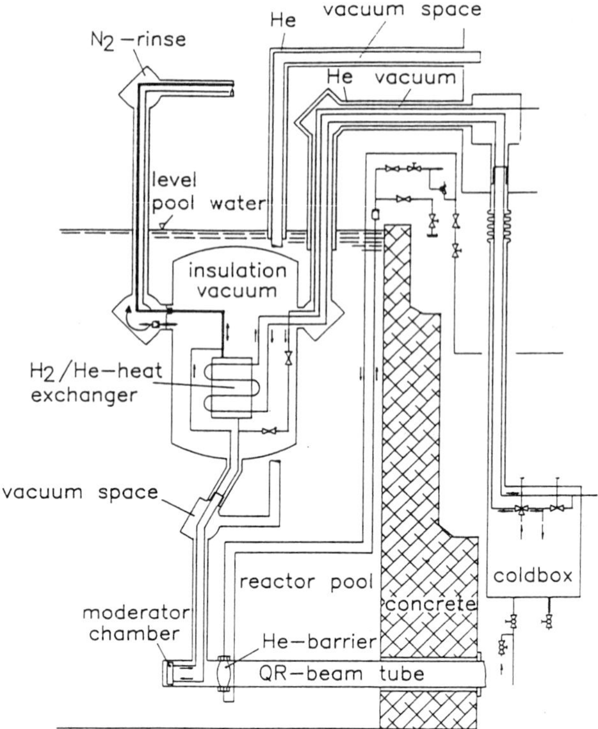


**Figure 4.** Sketch of the triple barrier system which prevents the formation of explosive mixtures of hydrogen and air within the reactor building or in the CNS components. See text for more details.

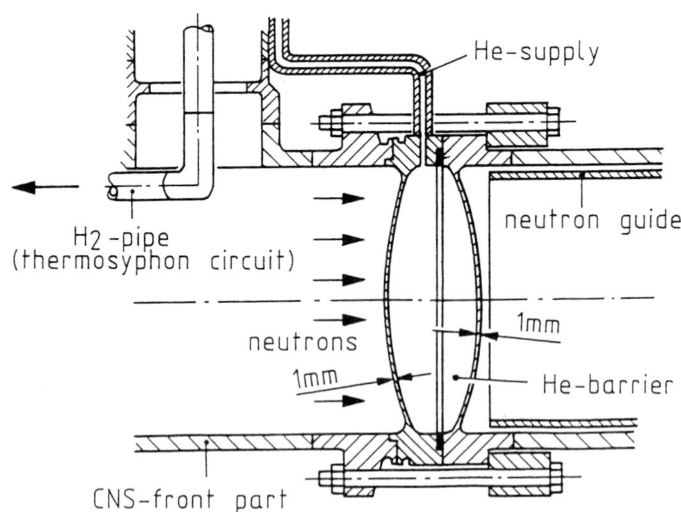


**Figure 5.** A He gas loaded barrier separates the front part of the QR beamtube containing the moderator chamber from the neutron guides connecting the CNS with the experimental areas.

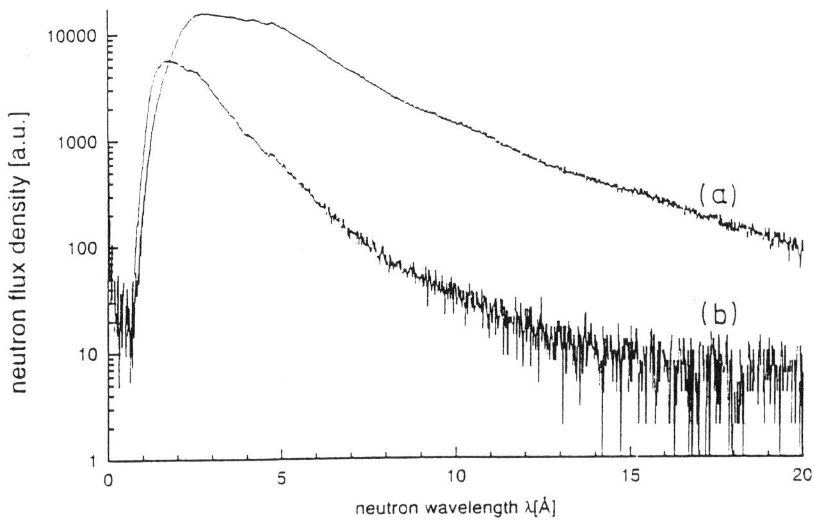


**Figure 6.** Spectral analysis of the neutron beam being available for experimental purposes. The different curves correspond to the moderator chamber containing liquid hydrogen at T $\simeq$ 24 K (a) or gaseous hydrogen at T $\simeq$ 320 K (b).

He–gas. For this purpose a He–buffer of approximately 1.5 $m^3$ at 10 bars is available. In a first step this gas is used in part for the He–ventilation of the vacuum spaces. Within about 40 s the corresponding heat input by thermal conduction leads to the evaporation of the liquid hydrogen contained within the cyrogenic components of the CNS. Simultaneously the off–gas valve of the $H_2$–cicuit is opened. In order to limit potential evaporation induced pressure peaks the $H_2$–system is still connected to the $H_2$–buffer via an one–way valve. Approximately one minute after activation of the blow out procedure, the $H_2$–buffer is completely separated from all further $H_2$–containing components and the pressure is decreased to a value allowing He inert gas supplied from the He–buffer to enter the $H_2$–circuit and to remove the hydrogen completely from all the CNS–components located inside the reactor building.

In contrast to various other research reactors the FRM may be operated at full power of 4 MW without cryogenic operation of the CNS. Under these conditions the moderator chamber and the in–pile components of the CNS are filled with gaseous $H_2$ whereas the vacuum spaces are filled with He–gas. Consequently during this mode of operation the temperature of the moderator chamber at 4 MW reactor power is limited to approximately 320 K.

It is noteworthy that none of the above scenarios does necessarily lead to a reactor shut down.

## EXPERIMENTAL AREA

The experimental area being supplied with cold neutrons is located in a separate laboratory outside the reactor building. It is connected to the CNS by a neutron guide the inner surfaces of which are Ni coated. The total length of the guides is 32 m and it exhibits a cross sectional area of 200 $\times$ 30 $mm^2$. In addition it is curved (R = 2850 m) in order to reduce the background of fast neutrons. In the laboratory the neutron guide is divided vertically into 3 separate guides.

By activation of thin Au-foils the integrated thermal neutron equivalent flux density being available for experimental purposes during operation of the CNS was determined to be approximately $2.5 \times 10^8\ cm^{-2}s^{-1}$. For comparison the measurement was repeated while only gaseous hydrogen was contained in the moderator chamber. The result was a flux reduction by a factor of $\simeq 10$. The corresponding neutron spectra were monitored by means of time of flight spectroscopy. The results are shown in Figure 6. It can be seen from the figure that at a neutron wavelength of $\lambda = 10\mathring{A}$ the CNS provides a gain factor of approximately 40.

The CNS of the FRM has been in routine operation since autumn 1996. So far, a variety of experimental setups was installed, among these a facility using cold neutrons for the tomography of technical objects. This technique offers the particular advantage to provide a high contrast for hydrogen rich materials like e.g. plastics which are almost invisible in classical x-ray scans. On the other hand, metallic structures become transparent to a high degree. In summary, neutron tomography offers complementary advantages in comparison to x–ray techniques. [4,5]

Besides this application, a test setup for so–called super mirrors was installed. A super mirror covers the inner surface of a neutron guide. It is coated by about 100 alternating layers of Ni and Ti which provide excellent reflectivity for neutrons beyond the critical angle of reflectivity. In addition training experiments for students of the Technical University of Munich are in use. The installation of further experiments using cold neutrons including a back scattering spectrometer is still in progress.

## REFERENCES


1. C. Doose and A. Tholen, Construction and operation of the cold neutron source at the reactor FRJ–2, Kerntechnik 13:539 (1971).
2. A. Axmann, and H.J. Didier, Ausbau und Betrieb des BER–II, atomwirtschaft 38:42 (1993).
3. W. Krull, A. Reymann, and H.U. Ruhnau, Aufbau und Betrieb der Kalten Neutronenquelle am FRG–1, atomwirtschaft 38:48 (1993).
4. B. Schillinger, 3D Computer Tomography with Thermal Neutrons at FRM Garching, J. Neutron Research 4:57 (1996).
5. B. Schillinger, R. Gebhard, B. Haas, W. Ludwig, C. Rausch, and U. Wagner, 3D Computer Tomography in Material Testing and Archeology, Proc. of the 5th. World Conference on Neutron Radiography, Berlin, June 17-20,1996 (in press).

# CRYOGENIC TRANSIENT HEAT TRANSFER IN THE g-2 CRYOSTATS AT THERMAL INSULATING VACUUM LOSS

L. X. Jia and L. H. Gong

Brookhaven National Laboratory, Upton, NY 11973

## ABSTRACT

A numerical simulation of transient heat transfer for thermal insulating vacuum loss in the g-2 cryostats was verified by the cold test results on one cryostat ring. The temperature profiles of the superconducting coil mandrel and outer shell were obtained for helium, nitrogen, and air leaks. Temperature drop and thermal contraction of the cryostat shell due to vacuum loss excited some design concerns. Heating methods for the cold cryostat shell, such as blowing natural air to increase thermal convection, were also simulated.

## INTRODUCTION

The muon (g-2) experiment at Brookhaven completed a 3-month run for initial data-taking in July, 1997. The g-2 cryogenic system[1], which supplies cooling to the superconducting magnet, has been satisfactorily commissioned since June, 1996. The three g-2 superconducting solenoids have an average diameter of fifteen meters - the world's largest superconducting solenoids. The superconducting coils are cooled indirectly by forced two-phase helium flow from a refrigerator of 625 watts at 4.5 K[2].

One of the special characteristics of cooling the large scale superconducting solenoids is its substantial thermal contraction of cold members of the cryostat. Each component of the cryostat, coil mandrel, heat shield, and outer shell, needs its own space for confined-free shrinkage. When cooled down from room temperature to liquid helium temperature (liquid nitrogen temperature for heat shield), the average thermal shrinkage of the g-2 coil mandrel is 31 mm in radius. Since it is difficult to cool every internal member of the cryostat evenly, certain clearance between each component is required for cool-down and warm-up. Figures 4 and 5 give the temperature limits and the movement boundaries of $LN_2$-cooled heat shield corresponding to the position and temperature of the mandrel, which must be maintained during cool-down and warm-up periods. At warm-up, see Fig. 4, the shield temperature can not be higher than 220 K, otherwise, mechanical interference between shield and mandrel will occur; for example, when mandrel is at 200 K there is only 3 mm clearance between shield and mandrel at certain positions.

Moreover, the outer shells of the cryostats could be very cold in some circumstances, which also require certain clearances for their thermal shrinkage between each cryostat and its surroundings, such as iron dipoles and particle beam ports, etc. In the case of thermal insulating vacuum failure in cryostats helium, nitrogen, or air may leak into the vacuum space and create a severe heat transfer environment between the cold superconductor mandrel (4.5 K) and the warm outer shell (300 K) in the pressure range of $10^{-3} \sim 10^{3}$ mbar. The large cold mass of the coil mandrel which contains 3,800 kilograms of aluminum at the temperature of 4.5 K, could become a large cold sink to the outer shell which is made of 3300 kilograms of aluminum. For example, the heat flux by conduction through leaking helium gas at internal pressure of one atmosphere could be as high as up to 900 $W/m^2$ between mandrel and outer shell. The possible lowest temperature of outer shell could reach 235 K in approximately 3.75 hours. Tests on outer cryostat has shown that its outer shell dropped 16 K from 293 K in one and a half hours when the coil mandrel was at 134 K and insulating vacuum was 5 mbar. It is obvious that the g-2 cryostats must be protected by providing either a confined-free structure design or a warming system to prevent the cryostats from severe interference with iron dipoles when thermal insulating vacuum fails in the cryostats.

This paper provides some theoretical analysis to simulate various cases of thermal insulating vacuum loss, which is compared with the test results using helium gas to spoil the vacuum in the outer cryostat. An air warming system for outer shell is also discussed.

## THE g-2 CRYOSTATS

A cross section of the superconducting magnet ring is shown in Fig. 1. Two inner cryostats, upper and lower, each of which contains a continuous 24 turns of SC cable loop along the ring, are mounted on the inner side of the iron yoke. Cross sections of the outer and inner lower cryostats are shown in Figures 2 and 3. One outer cryostat is buried in the iron yoke. The two SC cable loops (24 turns in each) are inlaid in the aluminum mandrel. Two helium cooling channels ($11 \times 30$ $mm^2$) are welded on the mandrel ring o.d. The LHe flow path starts at a point of the lower channel, travels counter clockwise around the mandrel, and moves vertically to the upper portion of the mandrel, then returns clockwise to the starting azimuth on the mandrel. On inner coils the fluid enters, loops around once and exits. The LN2 cooled aluminum shield consists of two cooling tubes to which aluminum sheet are attached forming a box structure around the mandrel. The MLI super-insulation is placed in the vacuum space between the mandrel, heat shield, and outer shell.

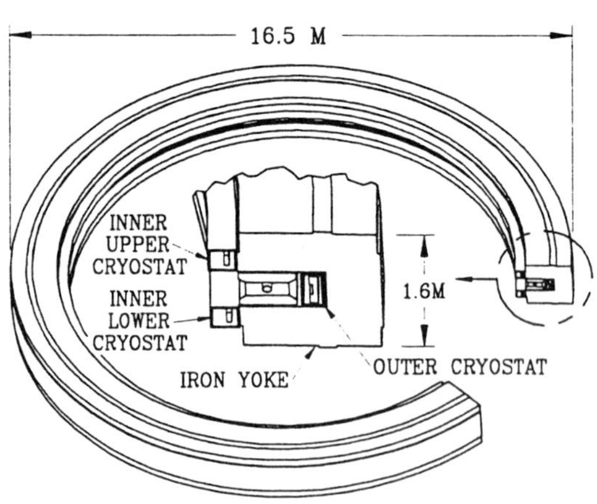


**Figure 1.** The g-2 superconducting magnet ring.

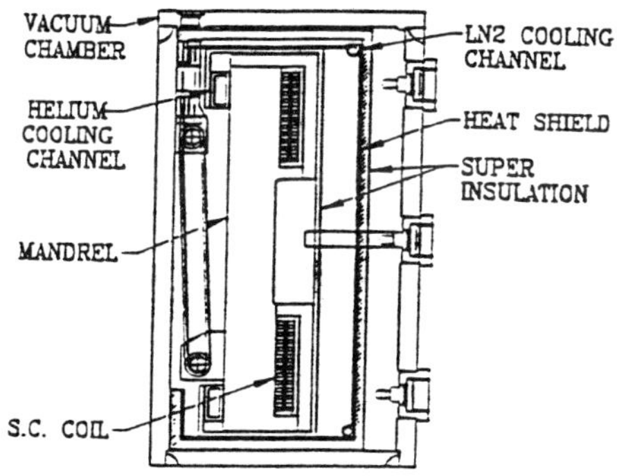


**Figure 2.** Schematic of the g-2 outer cryostat.

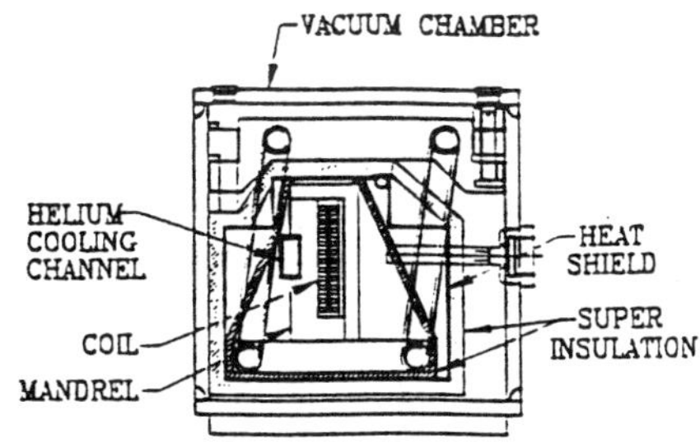


**Figure 3.** Schematic of the g-2 inner cryostat.

## EFFECTS OF VACUUM LOSS

Catastrophic failure of thermal insulating vacuum could cause serious damage because the large cold coil mandrel mass absorbs tremendous heat from the surrounding cryostat shell, resulting in substantial deformation of the shell due to thermal contraction.

There are three possible leaking sources, helium, nitrogen, or air at vacuum failure. LHe and $LN_2$ are used as cooling medium from refrigeration system through cryogenic valves, fittings, and transfer lines to cooling tubes which are welded onto the coil mandrels and heat shields. Although it has been highly addressed in the entire design and fabrication, cryogenic leak still cannot be fully avoided. The leaking rate varies and depends on the nature of the leak. It could be pipe weld cracks, valve or fitting seal failures, electrical isolator breaks, etc. Air leak is not uncommon in cryogenics either. In the case of the g-2 cryostats, to prevent air leaks is particularly difficult because the large weld area and large surface of the outer shell is a challenging leak-checking operation. The outer shells withstand all kinds of forces during magnet power-up or quenches. Some flexible joints and O-ring seal flanges are other possible sources of leaks.

Once thermal insulation fails, a severe heat transfer environment will be created. Since the heat transfer between the cold mandrel and the warm outer shell is dominated by heat conduction through each layer of the MLI superinsulation, helium leak will be the worst case in terms of keeping the temperature of outer shell from dropping too low. Indeed, if the cryostat can withstand the helium leak without problems nitrogen and air leaks, due to their physical properties, will be less serious.

When air or nitrogen leaks into the cryostats, the frost formation occurs on all cryosurfaces. Gas molecules of air components collide or interact with the cryosurfaces. A free molecule in the gas phase near cryosurface may experience a relaxation process in the potential well and attain thermal equilibrium with the solid. The residence time $t_r$ of gas molecule may be expressed as $t_c exp(E_d /kT_m)$,[3] which depends on the surface temperature $T_m$ of the coil mandrel. The $t_r$ could be very long because of large cold mass of the aluminum mandrel. In this case the gas molecules would be absorbed and reside at the bottom of the potential well with a small statistical chance of receiving enough energy from the cryosurfaces. Gas molecules would condense on the cryosurfaces in a solidified frost form. If air continuously leaks into the cryostat, it could be continuously captured as long as the frost surface is maintained at or below the capture temperature. Nitrogen gas could also be condensed or sublimated on cryosurfaces if surface temperature is at liquid helium temperature.

Since the thermal conductivity of the frost is more than 100 times smaller than aluminum, frost could behave like thermal insulation layer. However, the global heat flux may not be reduced much because heat transfer between the cold mandrel and the warm outer shell is dominated by heat conduction through each layer of MLI superinsulation. On the other hand, the latent heat of phase change may contribute to increase warming rate of the cold mandrel without increasing the cooling rate of the outer shell. As a matter of fact, gas condensation in the cryostat can only reduce the cooling rate at early stage of the process. As soon as the temperature of the mandrel surface rises to above a certain level, vaporization starts, which also takes heat from the outer shell and increases its cooling rate. In summary, the effect of gas condensation is flattening the temperature profile, and raising the lowest temperature point of the outer shell which occurs during the early cooling period of the outer shell.

## NUMERICAL MODELING

Numerical simulation of transient heat transfer for the g-2 cryostats at thermal insulation vacuum loss has provided a guidance for designs and operation to protect the system in case of catastrophic vacuum failure. The computations simulated various possible cases, such as that for helium leak, nitrogen leak, and air leak. It considered all modes of heat transfer: conduction, convection, condensation, and radiation, where applicable. Variable material properties as functions of temperature for gas and solid were also used.

All calculations start with initial conditions of one atm of internal pressure, 300 K of outer shell's temperature, 4.5 K of coil mandrel's temperature. Relative humidity is 60%.

### Transient Heat Transfer Equations

$$(M_m \cdot Cp_m) \cdot \frac{dT_m}{d\tau} = \frac{A_{cm} \cdot k_g}{\delta_{cm}} \cdot (T_c - T_m) + H \cdot R \tag{1}$$

$$(M_c \cdot Cp_c) \cdot \frac{dT_c}{d\tau} = A_c \cdot h_a \cdot (T_a - T_c) + \frac{A_{ac} \cdot k_a}{\delta_{ac}} \cdot (T_a - T_c) + A_c \cdot \sigma \cdot \varepsilon \cdot [(T_a)^4 - (T_c)^4] - \frac{A_{cm} \cdot k_g}{\delta_{cm}} \cdot (T_c - T_m) + q \tag{2}$$

The first equation describes the energy balance of the coil mandrel. The second term on its right-hand side is the latent heat flux which could be either for liquid or solid phases. The second equation represents the balance of each energy contribution to the cryostat shell. The first term on its right-hand side is the heat convection between outer shell and external fluid which performs either free convection or forced convection with outer shell. The second is heat conduction through external fluid which could be air or water contacting the surface of the outer shell. The third is the heat radiation from closed iron yoke surface. The fourth is heat conduction through internal gas between warm outer shell and cold mandrel. The last one is the external heat source which may be provided by an electrical heater.

### Material Properties

For the sake of convenience, physical properties of helium and nitrogen gas have been formulated as functions of temperature by using data base of the CRYODATA's GASPAK package, which are given in Table 1. For heat transfer calculations of the natural air, the formulated material properties are given in Table 2.

Table 1. Formulation of material properties of helium and nitrogen

| Gas | Thermal Conductivity [W/m-K] | Specific Heat [kJ/kg-K] |
|---|---|---|
| Helium | $k(T) = (3.4 \times 10^{-3}) \cdot T^{0.67}$<br>5 K ~ 300 K | $Cp(T) = 5.21 + 0.998 \cdot (T - 4.2)^{-0.86}$<br>5 K ~ 300 K |
| Nitrogen | $k(T) = (1.55 \times 10^{-4}) \cdot T^{0.9}$<br>77 K ~ 300 K | $Cp(T) = 1.034 + 0.355 \cdot (T - 77)^{-0.75}$<br>77 K ~ 300 K |

### Case I - Helium Leak

Case I assumes that the coil mandrel is at 4.5 K and the cryostat shell is at 300 K when helium gas suddenly leaks into the vacuum space. Temperature history of cryostat shell and coil mandrel calculated for Case I is given in Fig. 6. The lowest temperature of the cryostat shell happens to be 235 K at the 3.75 hour. The corresponding heat flux is 902 W/m$^2$. The coil mandrel warms up to room temperature in about 50 hours.

### Case II - Nitrogen or Air Leaks

Case II assumes that the coil mandrel and cryostat shell are at the same starting temperatures as that in Case I when nitrogen gas leaks into the vacuum space. Temperature history of the cryostat shell and the coil mandrel calculated for Case II is given in Fig. 7. The lowest temperature of the cryostat shell is 258 K at about 10 hours. The corresponding heat flux is 316 W/m$^2$. The coil mandrel warms up to room temperature in about 12 days.

## TEST RESULTS

A cold cryostat test was carried out on one of the g-2 cryostats (outer cryostat) at three different vacuum pressures and coil mandrel temperatures. The purpose of the test was to study the potential interference between the cold cryostat shell and the dipoles when the cryostat shell becomes cold and shrinks in radius due to vacuum failure. Thermal insulation vacuum was spoiled by introducing helium gas into the vacuum space. Six thermocouples were mounted on the surface of the cryostat along its 50 meter long circumference. Sixteen silicon diodes provided temperature readings around the coil mandrel ring. To avoid any possible mechanical damage to the cryostat during the cool down of the outer shell, the tests were conducted at relatively high coil temperature ranges. The three groups of the tests results are given in Table 3 and Fig. 8. The corresponding numerical data are also presented in Fig.8.

Table 2. Formulation of material properties of air

| Prandtl Number | Thermal Diffusivity [m$^2$/s] | Viscosity [m$^2$/s] | Thermal Conductivity [W/m-K] |
|---|---|---|---|
| $Pr(T) = 0.81 \cdot \exp(-4.6 \cdot 10^{-4} \cdot T)$ | $\alpha(T) = 2.5 \cdot 10^{-10} \cdot T^2$ | $\nu(T) = 5.4 \cdot 10^{-10} \cdot T^{1.8}$ | $k(T) = 0.105 \cdot 10^{-3} \cdot T^{0.97}$ |

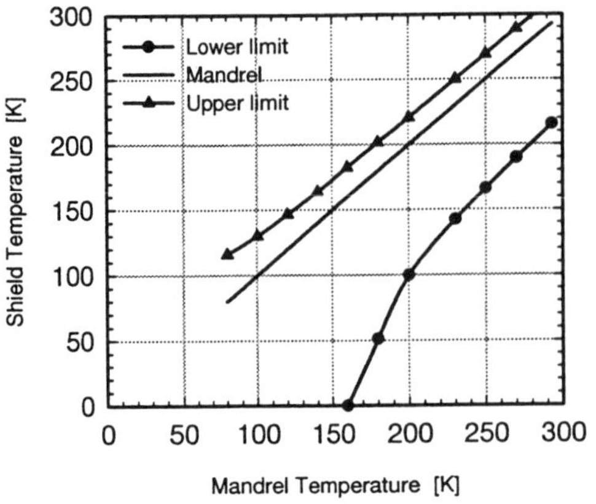


**Figure 4.** Corresponding temperature limits of heat shield with mandrel temperature.

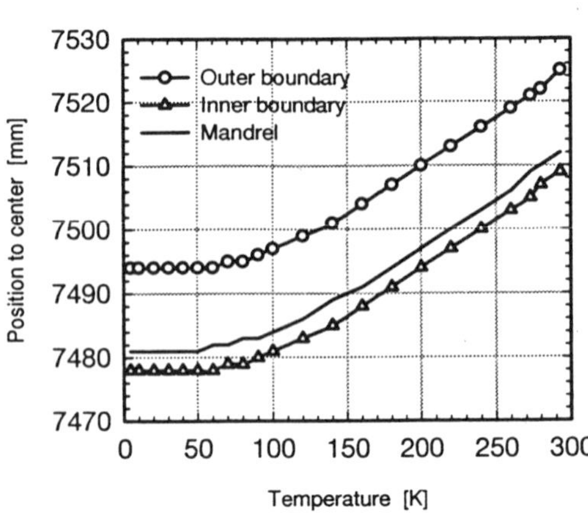


**Figure 5.** Inner and outer boundaries of coil mandrel.

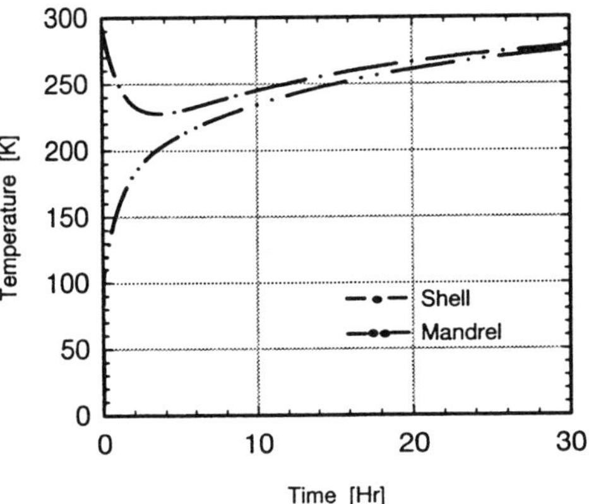


**Figure 6.** Temperature profiles of the coil mandrel and the cryostat shell with helium leak.

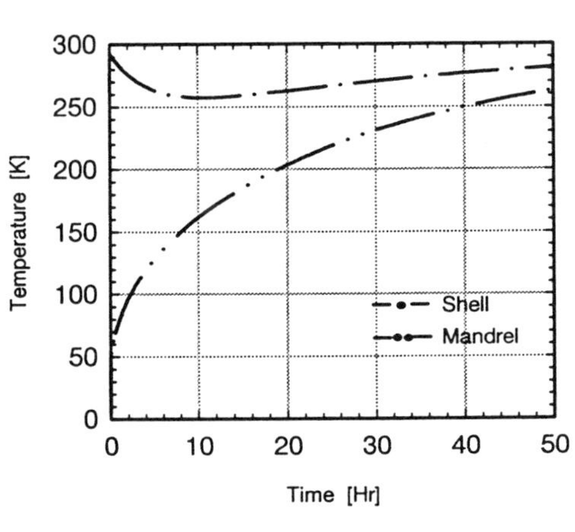


**Figure 7.** Temperature profiles of the coil mandrel and the cryostat shell with nitrogen leak.

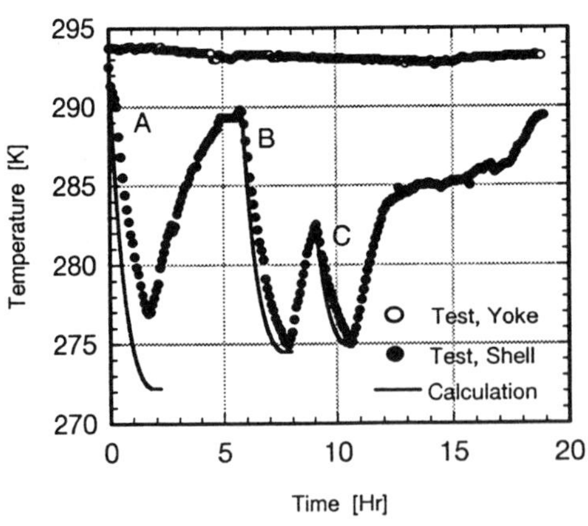


**Figure 8.** Measured temperature history of the cryostat shell when helium is introduced into vacuum space.

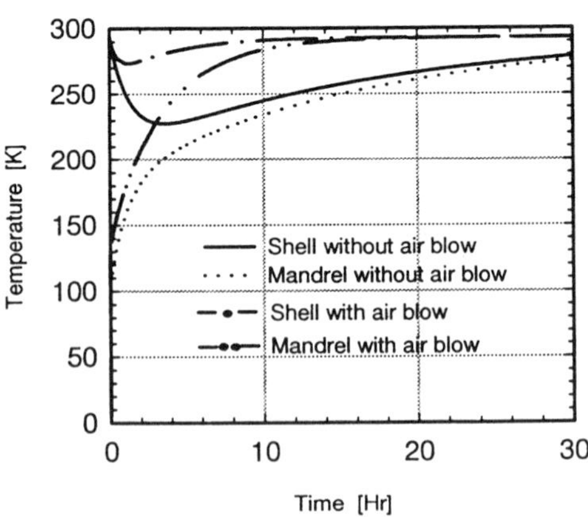


**Figure 9.** Comparison of temperature profiles under helium leak with and without air blow.

**Table 3.** Measured temperature drops of outer surface of the cryostat (outer)

| Test No. | Thermal Insulation Vacuum [mbar] | Average Temperature of Coil Mandrel [K] | Temperature Drop of outer shell [K] | Cool Down Time [minute] |
|---|---|---|---|---|
| A | 5 | 134 | 293-277 | 90 |
| B | 22 | 154 | 290-275 | 120 |
| C | 760 | 180 | 283-275 | 110 |

## AIR WARMING FLOW

In order to prevent a severe collision of the cryostat shell with the dipole ring which is placed close to the cryostat ring i.d., in case of losing vacuum and over-cooling of outer shell, some warming methods for the cold cryostat shell have been investigated by using this numerical modeling. A natural air-warming system was developed for this purpose. A model of forced convection in narrow slots (vertical and horizontal) between the shell and the iron yoke was used in computing equations 1 and 2. The results are presented in Fig. 9.

To design the air warming system, the air flow parameters were calculated based on a generalized one-dimensional, compressible, steady flow gas dynamics model[4]. This model can deal with an isothermal compressible flow with mass reduction and flow friction. The differential equation is given in Eq. 3.

$$\frac{dM}{M}=\frac{1+\frac{\gamma-1}{2}\cdot M^2}{1-M^2}\cdot\left[-\frac{dA}{A}+\frac{\gamma\cdot M^2}{2}\cdot\left[\left(\frac{4\cdot f\cdot dx}{D}\right)+\frac{2\cdot\delta D}{\gamma\cdot M^2\cdot p\cdot A}\right]+\frac{1+\gamma\cdot M^2}{2}\cdot\frac{dT}{T}+\left[(1+\gamma\cdot M^2)-y\cdot\gamma\cdot M^2\right]\cdot\frac{dm}{m}\right] \tag{3}$$

The heat transfer parameters used for the design is determined by the maximum allowable temperature drop of 20 K and heat flux of 520 [W/m$^2$] at the cryostat shell. The mass flow distribution around the cryostat ring which provided the gas dynamics parameters for design of the air warming system is shown in Figs. 10 and 11.

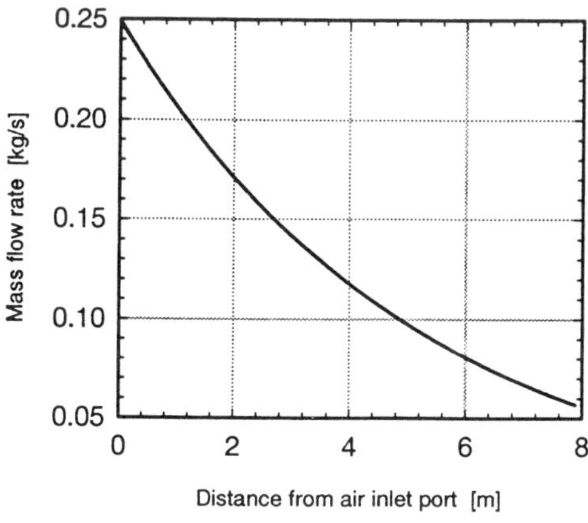


**Figure 10.** Air mass flow rate through nerrow gap between cryostat shell and iron yoke.

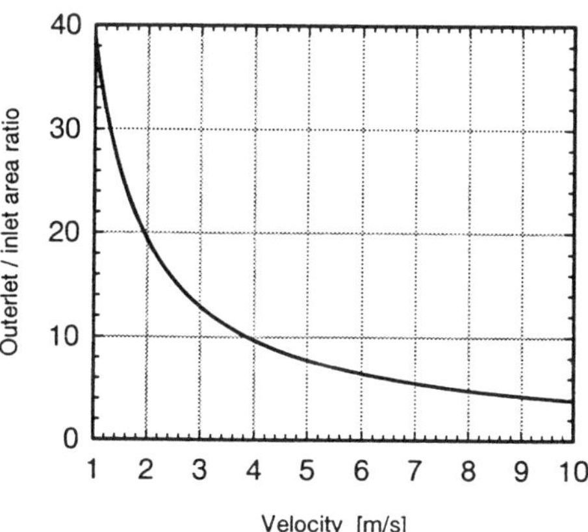


**Figure 11.** Ratio of outerlet area over inlet area at different venting veloccities.

## CONCLUSION

Design and operation of large scale superconducting solenoids require highly automatic controlled cryogenics, because of the thermal contraction of the cold cryostat members. Another concern is the failsafe design for vacuum loss in cryostats, which was one of the critical operation issues in commissioning the g-2 cryogenic system. The numerical simulation revealed the scale of thermal shrinkage and provided guidance to the system operation. A natural air warming system for the cold cryostat shell has shown its advantages over other methods, such as electrical heating. However, like other methods, the air warming system still requires the AC power to operate, which means that the vacuum loss in combination with AC power failure represents the worst scenario.

It takes about three weeks to warm-up entire cryostat rings at the ends of each experiments because of the limited clearances between each cold members. The air warming system can provide with a safe warming procedure for fast warming-up of the outer cryostat.

## NOMENCLATURE

| | | | |
|---|---|---|---|
| $A$ | = area, [$m^2$] | $A_c$ | = surface area of outer shell of cryostat, [$m^2$] |
| $A_{cm}$ | = surface area of mandrel, [$m^2$] | $Cp_c$ | = specific heat of outer shell material, [j/kg-K] |
| $C_{pm}$ | = specific heat of mandrel material, [J/kg-K] | $E_d$ | = cold well potential, [J] |
| $f$ | = friction coefficient | $H$ | = latent heat, [J/kg] |
| $h_a$ | = coefficient of heat convection, [$W/m^2$ -K] | $k$ | = Boltzmann constant, [J/K] |
| $k_a$ | = thermal conductivity of air, [W/m-K] | $k_g$ | = thermal conductivity of leaking gas, [W/m-K] |
| $m$ | = mass flow rate [kg/s] | $M$ | = mass, [kg] |
| $M_m$ | = mass of mandrel, [kg] | $M_c$ | = mass of outer shell, [kg] |
| $p$ | = pressure, [Pa] | $q$ | = heat source, [W] |
| $R$ | = condensation or solidification rate, [kg/s] | $T$ | = temperature, [K] |
| $T_a$ | = temperature of atmosphere, [K] | $T_c$ | = temperature of outer shell, [K] |
| $T_m$ | = temperature of mandrel, [K] | $t_c$ | = characteristic vibrational time of particle, [sec] |
| $t_r$ | = residence time, [sec] | $x$ | = distance from inlet, [m] |
| $y$ | = ratio of mass reduction stream to main stream | $\gamma$ | = specific heat ratio, |
| $\delta_{ac}$ | = thickness of air gap from shell to yoke, [m] | $\delta_{cm}$ | = thickness of insulation, [m] |
| $\delta D$ | = generalized body force, | $\epsilon$ | = emissivity |
| $\sigma$ | = Stefan-Boltzmann constant, [$W/m^2$-$K^4$] | $\tau$ | = time, [sec] |

## ACKNOWLEDGMENT

This project was supported by Office of Energy Research, Office of High Energy and Nuclear Physics, High Energy Physics Division, United States Department of Energy, under Contract DE-AC02-76CH00016.

## REFERENCES

1. G. Bunce, et al. "The large superconducting solenoids for the g-2 Ring," IEEE Transactions on Applied Superconductivity 5, No. 2, (1995).
2. L. X. Jia, et al. "Cryogenic System for the Muon g-2 Superconducting Magnet," Proc. of the 16th ICEC, Kitakyushu, Japan, (1996), p. 871.
3. W. Frost, "Heat Transfer at Low Temperatures," Plenum Press, New York, (1975), p. 213.
4. M.J. Zucrow and J.D. Hoffman, "Gas Dynamics," John Wiley & Sons, Inc. (1976), p. 493.
5. S. Kakac, R.K. Shah and W. Aung, "Handbook of Single-phase Convective Heat Transfer," John Wiley & Sons, New York, (1987), p. 12.27.

# SAFETY DESIGN, OPERATION, AND CONTROL OF A LIQUID HYDROGEN TARGET AT BNL

L. X. Jia, L. Addessi, V. Castillo, L. H. Gong,
J. Leskowicz, R. Meier, G. Miglionico, J. Scaduto

Brookhaven National Laboratory, Upton, NY 11973

## ABSTRACT

A liquid hydrogen target for high energy physics experiment, Crystal Ball Spectrometer, was recently built and operated at the Alternating Gradient Synchrotron (AGS) particle accelerator at the Brookhaven National Laboratory (BNL). The system safety design, operation, and control of the target was analyzed and evaluated during its two-month beam-taking experiment. This paper reports on this cryogenic project.

## INTRODUCTION

At the AGS each year for the past twenty five years an average of two liquid hydrogen or deuterium targets have been built for high energy physics experiments. Because of the usage and handling of large volumes of condensed flammable gas, the practice in this cryogenic field must adhere to the guidelines for safety design and operation, which involves extensive review of: material selection, target structure, condensation method, system test, operation procedure, control logic, and failsafe modes. A $LH_2$ target system for the Crystal Ball Spectrometer at the AGS was recently analyzed to evaluate its safety design and performance.

The $LH_2$ Mylar flask for the E913 is refrigerated by the helium G-M coolers in a closed cycle thermosyphon architecture. Hydrogen is stored, circulated, and recovered in a well defined space under various operating conditions. The PLC control is used for system operation and analysis. The safe and reliable operation of the target is complemented by upgrading the control technology and decreasing the complexity of the system. Safety design criteria and operation analysis for the E913 $LH_2$ target are discussed in this paper.

## LIQUID HYDROGEN TARGET SYSTEM

A schematic of the E913 $LH_2$ target system is shown in Figure 1. It consists of six basic components, (1) helium G-M coolers with compressors; (2) hydrogen gas supply and storage tank; (3) liquid hydrogen condensers; (4) liquid hydrogen flask; (5) vacuum chamber with valves; (6) PLC control system. A Mylar flask (0.25 mm in wall thickness) is positioned at the center of the Crystal Ball detector inside a vacuum beam tube (15 cm in diameter). There are two Mylar windows (0.25 mm in thickness) at each end of the vacuum beam tube. Two G-M

*Advances in Cryogenic Engineering,* Vol. 43
Edited by P. Kittel, Plenum Press, New York, 1998

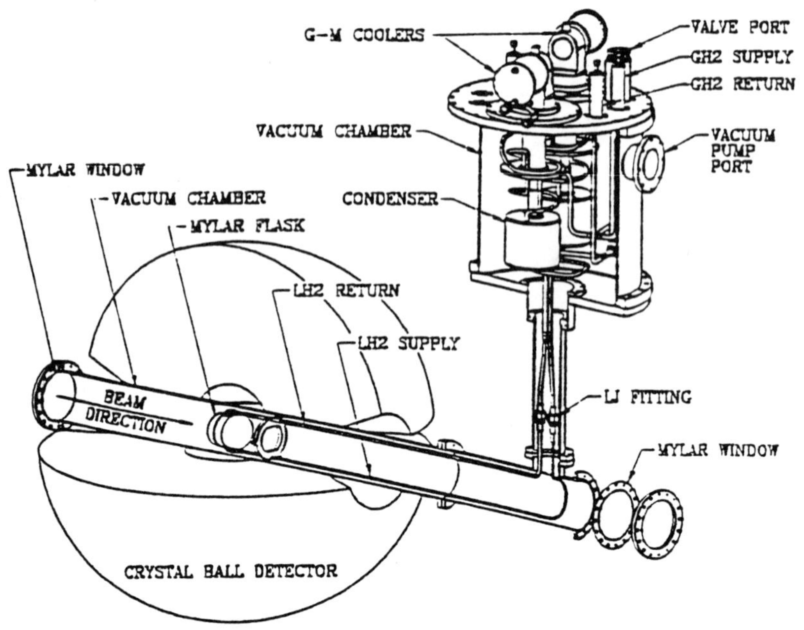


**Figure 1.** The schematic of the E913 liquid hydrogen target system.

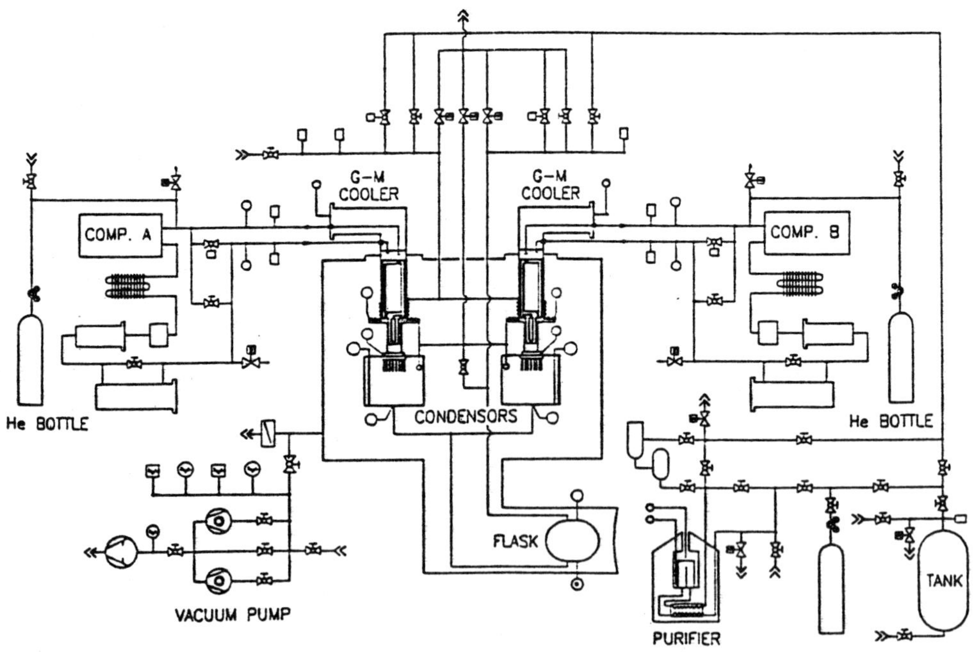


**Figure 2.** The cryogenic flow diagram of the E913 liquid hydrogen target system.

coolers with condensers attached at their second stages are installed in a vacuum chamber. Two hydrogen transfer lines, supply and return, from condensers to the Mylar flask are connected by two CESSC cryogenic tube fittings[1]. A cryogenic auto valve is installed at the hydrogen return sides of the flask for a special-purpose operation. There is only one hydrogen transfer line connecting the condensers to the ballast tank used for hydrogen supply as well as recovery. The cryogenic system flow diagram is shown in Figure 2. The hydrogen flow circuit adopts the closed cycle thermosyphon scheme. Hydrogen gas is initially transferred from a ballast tank, with a well-defined volume, and circulates between the condenser and liquid flask under various operating modes. The pressure in the entire hydrogen circuit is maintained even, except for a small driving potential during thermosyphoning. The two helium G-M coolers with their compressors are used to provide the cooling power to condense the hydrogen gas into liquid in the condensers. The liquefied hydrogen gravitates from the bottom of the condenser to the Mylar flask below and vaporized hydrogen, due to thermal load, returns in two-phase from the top of the flask to the condensers. Once the saturated gas flows back to the condensers, it is then re-liquefied by the cold fingers which are attached to the second stages of the G-M coolers. This ensures that heat is intercepted at any point along the 20 K circuit, while the hydrogen can only flow in one direction around the target, which does not upset the thermosyphon operation. Once an equilibrium state is reached, a small amount of hydrogen circulates between the condensers and the flask, being gravity-driven while constantly changing its phases.

### System Safety Design

Experience has shown that the safety devices added to the hydrogen target system may not improve the security unless the system itself is designed safe in principle. Safety design philosophy of the E913 hydrogen target system touches upon the following basic safety rules,

- avoid losing hydrogen gas to the uncontrolled area;
- keep hydrogen gas within a well-defined volume;
- provide appropriate confined space for condensed gas to expand;
- keep the storage pressure as lower as possibly needed;
- reduce the inventory of hydrogen gas to minimum;
- keep operating pressure as close as possible to the atmospheric pressure;
- reduce the maximum differential pressure in the entire system;
- do not run the system below the atmospheric pressure - avoid any possibility of sucking air into the hydrogen circuit;
- do not run the system at high pressure - avoid any possibility of blowing up the flask;
- simplify the flow path, especially, for hydrogen flows - minimize the possibility of hardware failures and contaminant plugs.

These safety design rules are chosen and emphasized from many other rules for a cryogenic system using flammable gas because the author intends to compare the current design with the conventional $LH_2$ target.

**Table 1.** Basic parameters of hydrogen flow in close cycle thermosyphon target system

| Thermal Heat Load [W] | $H_2$ Mass Flow Rate [g/sec] | Fluid Quality in Supply Line (6 mm i.d.) | Fluid Quality in Return (10 mm i.d.) | Differential Pressure [Pa] | Supply Flow Velocity [m/s] | Return Flow Velocity [m/s] |
|---|---|---|---|---|---|---|
| 8.11 | 0.10 | 0 | 0.16 | 500 | 0.043 | 0.023 |

**Table 2.** Hydrogen gas distribution, storage, and total inventory

| Status | Ballast Tank [g] | Hydrogen Condenser [g] | Transfer Lines [g] | Hydrogen Flask [g] |
|---|---|---|---|---|
| Storage | 390 | 0.4 | 0.24 | 0.58 |
| Operating | 305 | 22 | 14 | 50 |

### Thermosyphon Analysis

Operating parameters of a closed cycle thermosyphon system can be easily computed for hydrogen circulation if a homogeneous two-phase flow is assumed. Liquid flow between the flask and condenser is treated as inviscid and incompressible. For flows in the supply and return transfer lines to the flask, the momentum equations are used to calculate the wall friction. Hydrostatics is taken into account along vertical pipes. Mass continuity is applied to calculate the velocities of the fluid in each flow section. The thermal heat load determines the mass flow rate, which involves the energy equation in a form representing fluid density and viscosity of two-phase flow. The numerical results for the E913 target system under thermosyphon is given in Table 1.

### Hydrogen Distribution, Storage, and Inventory

One of the system design goals is to keep the operating pressure of hydrogen gas close to one atm everywhere: the warm ballast tank, cold condenser, and liquid Mylar flask. To achieve this a maximum storage pressure of 1.34 bar is needed when all liquid hydrogen in the system is vaporized and warmed up to room temperature. A hydrogen ballast tank of 3.7 $m^3$ is used for the hydrogen storage and recovery. The ballast tank is located outside of the experimental building and there is no need for hydrogen gas bottles in the experimental area. A stainless steel pipe of 2.5 cm i.d. and 20 meters in length is used as the transfer line between the hydrogen condenser and ballast tank. The total inventory of gaseous hydrogen in the E913 target system is limited to 391 grams. The design parameters, see Table 2, are given under both warm storage and cold operation conditions.

## CONTROL SYSTEMS

The control system consists of the operation controls and the safety controls. The operation control system is completely separated from the safety control system. The safety control furnishes the target system with a simple and reliable safety system using conventional relay switches. Further, only safety components such as fire detectors, hydrogen sensors, etc. are classified in the high quality assurance (QA) category. The operation control can be categorized in low QA category avoiding the expense of building a complex electronic system.

### System Operation Controls

The operation control provides the operator with an Allen-Bradley Programmable Logic Control (PLC) rack for remote operation of the E913 target system. The PLC process transmitters perform the monitoring, controls and digital or analog communications with various pressure transducers, temperature sensors, solenoid switches and valve actuators. The control software provides the operator with the computer interface, real-time supervisory control, data acquisition, and logging in both graphic and text formats. The hardware consists of an A-B PLC 5/20 enchanced processor, two A-B 12-bit analog input modules, two A-B 12-bit analog output modules, one A-B power supply, eight Lakeshore Model 231 temperature transmitter modules, eight Sensotec pressure transducers, redundant 24 VDC power supply, Win Pentium 585 pci 150 MHz personal computer, and Fortress Model L12.5K uninterruptible power system. The operating software, RSView, consists of 12 pages of Graphical User Interface (GUI). The target system is broken up into three pages, helium G-M cooler page, hydrogen flow circuit page, and hydrogen gas supply and storage page. They provide for the manipulation of the valves and the readout of the pressure and temperature sensors. There is a historical trend and a real time trend for the pressures and temperatures in both helium and hydrogen circuits. The $LH_2$ flask is depicted on a page with animation that shows the liquid level in the target. An information summary page of the system's running status is also created for the computer network communications.

### System Safety Controls

The $LH_2$ target system and Crystal Ball detector are located inside a 4×4×6 $m^3$ dry room enclosure at the experimental area. The dry room maintains the controlled temperature and humidity for beam counters (scintillators and phototubes). A herculite enclosure in the dry room is constructed around the target system to monitor hydrogen gas leaks while providing continuous venting with fresh air. The enclosure also provides a barrier between the Class I/Division II target area and the rest of the dry room. In the event of a Mylar window failure and subsequent liquid spill, the exhaust fan will switch to high speed facilitating faster venting of the enclosure. The herculite enclosure is extended to enclose the downstream vacuum Mylar window. This area is also continually vented with fresh air. The upstream vacuum Mylar window is protected from accidental puncture by adding a second Kevlar window clamped on the front face of the flange. An aluminum box is built around the upstream Mylar window and connected to the venting pipe.

Crystal Ball detectors around the vacuum window area inside the dry room are protected by a fire-wire protection system, which is interlocked to trip all experimental power to detectors and initiate an automatic building evacuate. Copper meshes serving as flame arresters are installed to cover the ignition sources in the dry room. Detectors for explosive gas are installed in the target enclosure exhaust trunk. Any indication of hydrogen gas would simultaneously trip all experimental power to the dry room and initiate alarms. Smoke detectors tied to the building fire alarm system and hence the BNL Fire Department are also installed in the dry room enclosure.

## OPERATION OF TARGET SYSTEM

A complete operation cycle of the hydrogen target includes the system cool-down, liquid condensation, flask empty and refilling, and system warm-up. The temperature and pressure profiles for a cycle during the experiment E913 are shown in Figures 3 and 4.

### Refrigeration Control

Automatic refrigeration control is achieved by synchronizing the helium gas mass flow rate and differential pressure between compressor discharge and suction with the pressure in the $LH_2$ flask. This is accomplished by the feedback control of two automatic by-pass valves in the helium flow circuits (one for each G-M cooler) with the pressure in the $LH_2$ flask which is held at the liquefaction range. In the E913 target, the maximum $LH_2$ pressure fluctuation in the flask is maintained within 1.04 ~ 1.08 bar. The corresponding temperature and density vary in the small ranges of 20.37 ~ 20.49 K and 70.70 ~ 70.55 kg/$m^3$, which contribute to the high quality of physics experiment.

### Empty and Refill

The physics experiment often requires the flask target be emptied and refilled with liquid hydrogen during normal operations. The target system must be capable of handling this process in a safe and timely manner. In the present E913 system this is achieved by using an automatic valve on the return side of the hydrogen circuit downstream of the liquid flask. Once this valve is closed, a small portion of the liquid in the return line and flask will be vaporized by natural thermal heat load and the slightly increased pressure would push the remaining liquid in the flask back to the condenser. When the valve is opened, liquid in the condenser will gravitate to refill the flask. Figures 5a and 5b show the process for the typical empty and refill operations. It took about ten minutes to empty the flask once the valve is closed and about two minutes to refill once the valve is opened.

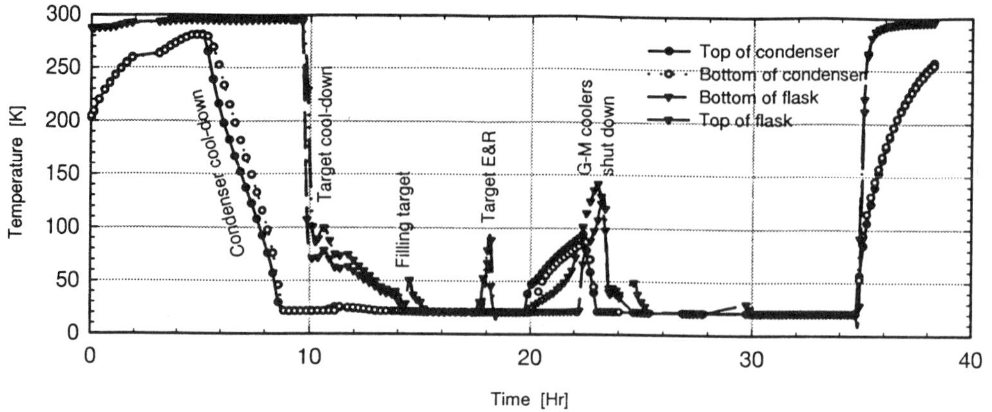


**Figure 3.** Temperature history of condenser and flask at various operating processes.

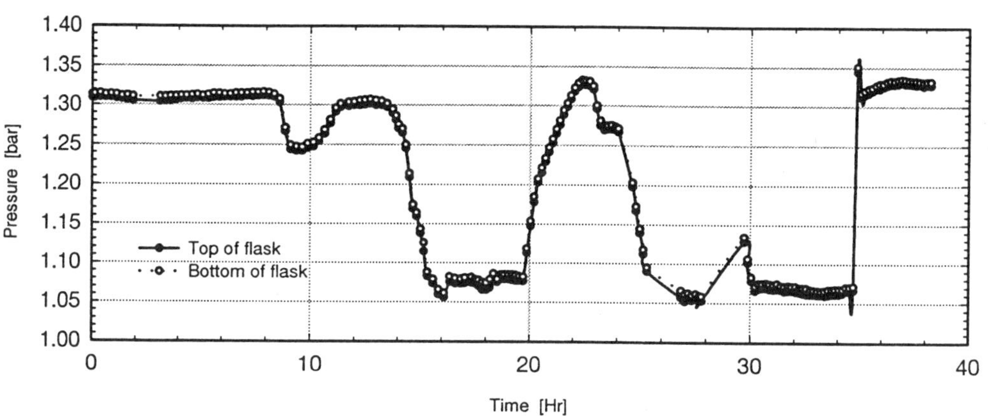


**Figure 4.** Pressure history in hydrogen circuit at various operating processes.

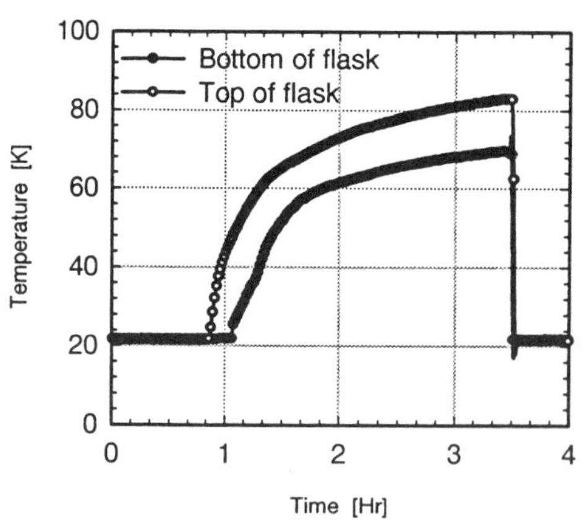


**Figure 5a.** Temperatures of liquid flask at emptying and re-filling.

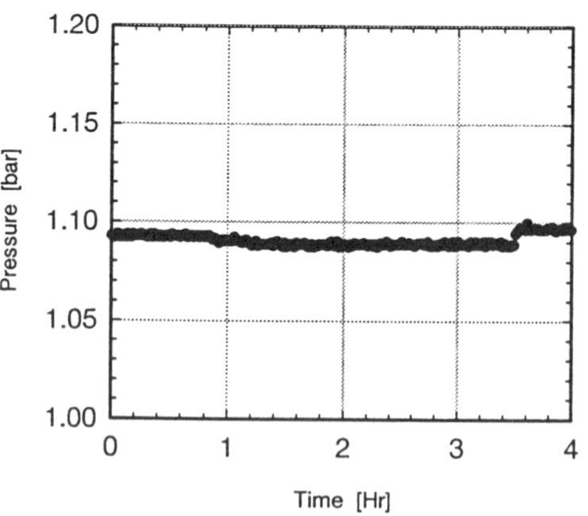


**Figure 5b.** Pressure at emptying and re-filling.

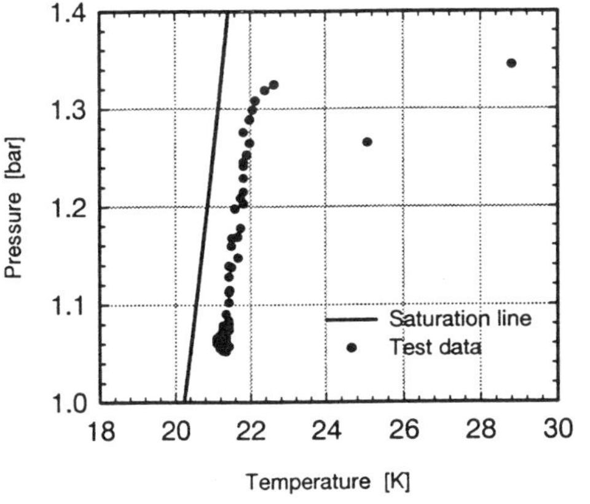


**Figure 6a.** $H_2$ P-T diagram in Mylar fask at bottom.

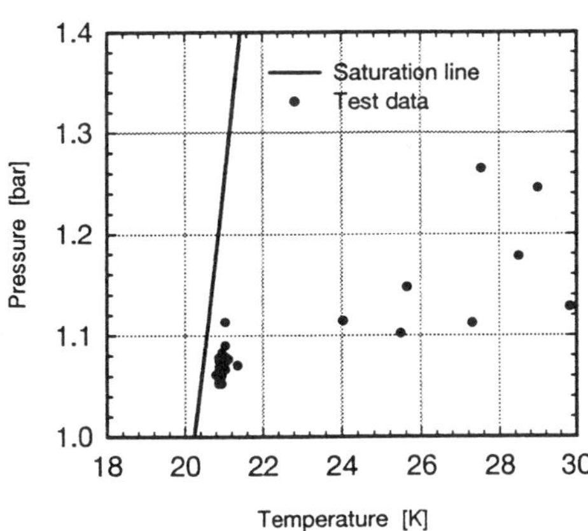


**Figure 6b.** $H_2$ P-T diagram in Mylar flask at top.

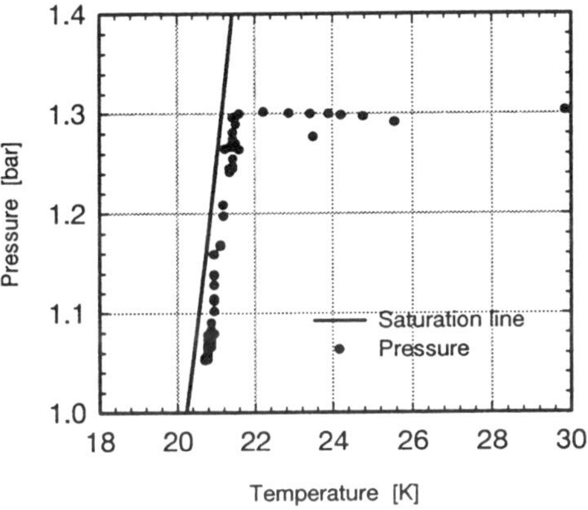


**Figure 6c.** $H_2$ P-T diagram in condenser at top.

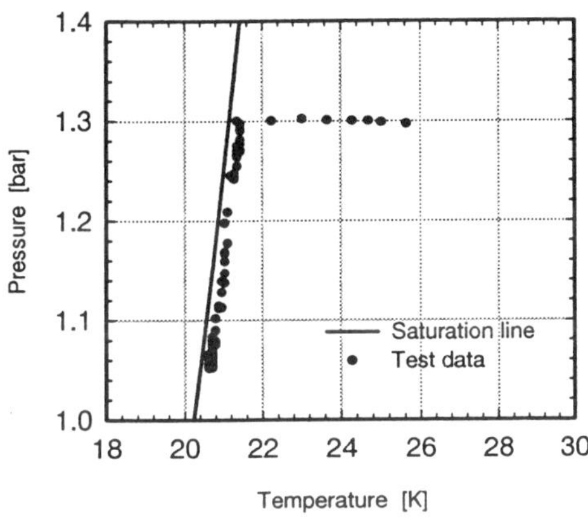


**Figure 6d.** $H_2$ P-T diagram in condenser at bottom.

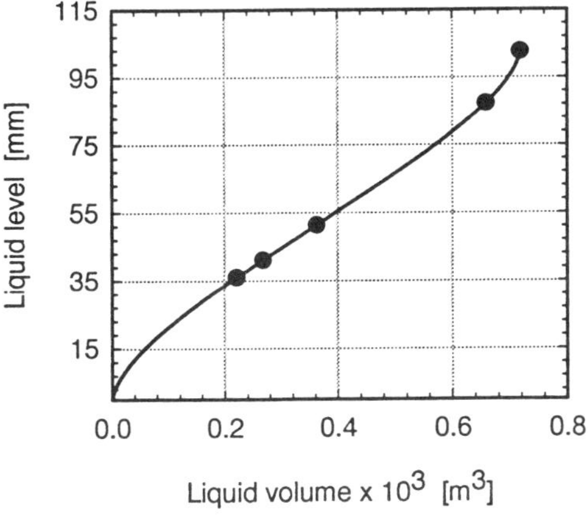


**Figure 7a.** $LH_2$ level vs. $LH_2$ volum in flask.

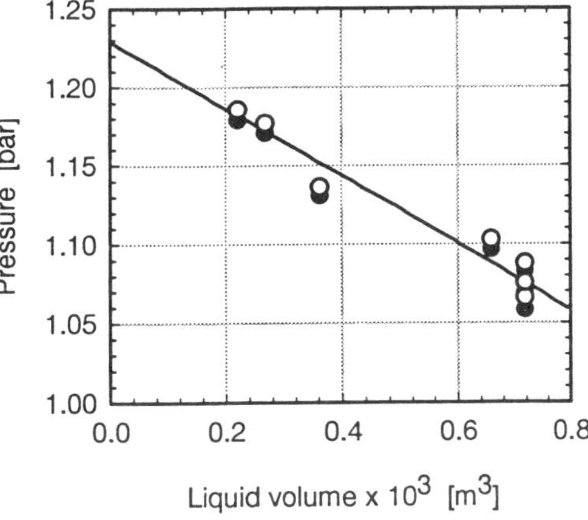


**Figure 7b.** $LH_2$ pressure vs. $LH_2$ volume in flask.

## Liquid Level Control

Liquid hydrogen level in the flask is a critical parameter in system operation. A full flask of $LH_2$ must be guaranteed for the valid physics experiments during the beam-taking. High precision temperature and pressure sensors are used for this purpose. Two silicon diodes are installed at the top and the bottom of Mylar flask. Figures 6a and 6b show the P-T diagrams of hydrogen at these two locations. These diagrams were used to verify a full flask. Theoretical saturation lines are also presented in these diagrams for references. The deviation of the test data was due to the mounting position of temperature sensors. The P-T diagrams of hydrogen in the condenser are given in Figures 6c and 6d. Since the information for middle levels is sometimes required, there is a scaled liquid level chart obtained by converting the flask pressure to liquid volume and applying flask geometry to scale the levels. Figures 7a and 7b show the level curves from this exercise. The test data was obtained by using a video camera during a system test run.

## CONCLUSION

The AGS E913 $LH_2$ target system successfully completed its first two-month experiment at BNL in June and July, 1997. The system design, manufacture, operation, and control of the E913 target have provided experience which furthers the establishment of reliable safety design guidelines for liquid hydrogen and deuterium target systems.

## ACKNOWLEDGMENT

This project was supported by Office of Energy and Nuclear Physics, High Energy Physics Division, United States Department of Energy.

## REFERENCES

1. X. Jia, "Contracting/expanding Self-sealing Cryogenic Tube Seals," *U.S. Patent No. 5,628,517* (1997).
2. D. H. Allspach, et al. "Design performance of liquid hydrogen target systems for Fermillab fixed target program, *Advances in Cryogenic Engineering*. 37:1495 (1992).
3. E. F. Hammel and J. D. Rogers, "Cryogenics and Nuclear Physics ---- Part 1," *Cryogenics*. 10:5 (1970).
4. L. M. Vail'ev, Yu. P. Dmitrevskii, et al, "A Universal Liquid Hydrogen Target for Experiments with High Energy Particles," *Cryogenocs*. 10:495 (1973).
5. K. Sh. Agababyan, R. T. Mina, et al, "A Liquid Hydrogen-deuterium Target," *Cryogenics*. 20:455 (1980).
6. Yu. T. Borzunov, V. F. Chumakov, et al. "A Cryogenic Target within an Electric Field," *Cryogenics*. 23:151 (1983).
7. L. M. Vasil'ev, I. A. Radkevich, et al. "A Mylar Liquid Hydrogen Target with Thin Mylar Side Walls," *Cryogenics*. 12:387 (1972).
8. M. Vasil'ev, Yu. P. Dmitrevskii, et al. "The Production of Lawsonite Film Liquid Hydrogen Targets," *Cryogenics*. 13:180 (1973).

# OPERATIONAL EXPERIENCE WITH A CRYOGENIC AXIAL-CENTRIFUGAL COMPRESSOR

L. Decker[1], A. Kündig[1], K. Löhlein[1], W. Purtschert[1], B. Ziegler[1],
Ph. Lebrun[2], L. Tavian[2], I. Brunovsky[3], and L. Tucek[3]

[1]Linde Kryotechnik AG
8422 Pfungen, Switzerland

[2]CERN, LHC Division
1211 Geneva 23, Switzerland

[3]PBS
59512 Velká Biteš, Czech Republic

## ABSTRACT

The Large Hadron Collider (LHC), presently under construction at CERN, requires large refrigeration capacity at 1.8 K. Compression of gaseous helium at cryogenic temperatures is therefore inevitable. Together with subcontractors, Linde Kryotechnik has developed a prototype machine. This unit is based on a cryogenic axial-centrifugal compressor, running on ceramic ball bearings and driven by a variable-frequency electrical motor operating at ambient temperature. Integrated in a test facility for superconducting magnets, the machine has been commissioned without major problems and successfully gone through the acceptance test in autumn 1995. Subsequent steps were initiated to improve efficiency of this prototype. This paper describes operating experience gained so far and reports on measured performance prior and after constructional modifications.

## COLD COMPRESSOR DRIVE

### Design Approach

Within the last decade ball bearing technology has improved significantly thanks to the introduction of new materials like ceramics beside metals to form mixed material systems, so-called hybrids. Hybrid ball bearings offer considerably higher operational speeds at longer service life than conventional types. In addition they are extremely well suited for permanent grease lubrication. Applications are found in grinding spindles and, close to cryogenics, vacuum machinery, specifically turbo-molecular pumps. In contrast to cryogenics these industries produce huge numbers of pieces with corresponding investment potential, highly developed products and excellent knowledge of components in terms of behavior and service life supported by statistics.

When Linde Kryotechnik resumed cold compressor development in 1994 it was decided to make use of this potential rather than stick to singular technology. A decision was taken to use ball bearing technology for the following reasons:

- they show comparably low costs to all other alternatives
- they are impressively simple and reliable
- they feature extremely small clearance thanks to pre-stressed bearing arrangement

Therefore a partnership with an industrial supplier for high speed spindles with excellent reputation was established. The basic concept was:

- technological transfer from a branch with high repetition to a one with low repetition
- design of a machine with standard components as far as possible with the advantage of detailed knowledge of the applied components' operating limits in advance
- low manufacturing costs for a special application
- establishing a number of standard frame sizes in order to cover multiple stage applications like CERN-LHC, DESY-TESLA, NIFS-LHD
- service friendliness by using standard components.

The key features of the system are:

- Asynchronous motor with frequency converter for variable speed working at ambient temperature
- Maximum speed: 700 rps
- The voltage at defined maximum load was selected to be 130 V in order to minimize the risk of motor damage due to reduced dielectric strength of He at low operating pressure. In addition special isolation material was chosen.
- Hybrid ball bearings working at ambient temperature, axial and radial with fixpoint of axial bearing next to compressor wheel
- Grease lubrication in order to minimize the risk of process contamination (total amount of grease is a few grams)
- Excellent suitability of ball bearings: high stiffness and extremely precise radial and axial positioning (unlike oil, gas or magnetic bearings during start/stop phase) >> with respect to clearance most suitable type for this application (shaft position during operation and off-duty identical)
- No emergency bearings are necessary
- No complex utilities as bearing gas or uninterrupted power supply are required.
- The control system requirements are as for standard machinery. Therefore all functionality can easily be integrated into a superior control system („open" system)
- The total speed range is clearly below the whipping speed of the rotor including prolongation with wheel
- Protection from air leakage is secured by a guard vacuum
- Easy access to drive for maintenance thanks to integration of guard vacuum into housing

- Minimum number of electrical penetrations (11 wires: 4 for electrical power, 4 for temperature sensors, 3 for speed pick-up; for magnetic systems typically this number is in the range of 50 -60)
- All welded water cooling passages
- Outer dimensions: ø 200 mm, 400 mm in height

### Warm tests

Prior to "cold" operation short- and long-term test series at the Linde workshop were performed with the following aim:

- Measurement of vibrations and imbalance of dummy wheel and vibrations of housing
- Verification of calculated first resonance frequency (nondestructive testing)
- Dielectric strength at low operating pressures
- Suitability of lubrication with grease at low operating pressures
- Verification of predicted ball bearings' service life of more than 8'000 hours

A test stand with recipient for low pressures and the necessary instrumentation was built. The results were:

- The complete assembly showed extremely low vibrations, which are in accordance with highest class of precision standards
- The maximum speed is far below first resonance frequency
- No disruptive breakdowns were observed
- After more than 8'000 operating hours the condition of bearings and grease was analyzed with excellent results.
- Potential for further improvements was identified.

## COLD STAGE

### Design Approach

The unit realized is based on CERN's specification. The main design data are:

| | | |
|---|---|---|
| Flow rate | 18 | g/s |
| Suction pressure | 10 | mbar |
| Discharge pressure | 30 | mbar |
| Pressure ratio | 3 | |
| Suction temperature | 3.5 to 4.4 | K |
| Isentropic efficiency | $\geq 60$ | % |

Thermodynamic, fluid dynamic and mechanical design, as well as manufacture of the cold stage, were performed by PBS, Velka Bites, Czech Republic. There has been a close collaboration between Linde and PBS for years. The characteristics of the cold stage are:

- Wheel diameter: 118 mm
- Wheel shape: backswept
- Diffusor: bladed
- Thermal intercept: None

The machine was designed to have a comfortable working field aiming for multistage application for LHC. Emphasis therefore was put on operational flexibility rather than efficiency alone.

## Tests at CERN

In autumn 1995 the system was installed and successfully commissioned and tested at CERN. As reported in [1] the efficiency at design conditions was 57 % or 3 % points lower than predicted. The operating field is wider than expected.

One main problem ocurred during commissioning, namely electrical breakdown in the motor housing at the leadthroughs, which at the end was easy to correct.

Based on the results it was obvious, how to improve efficiency:

- Reduction of heat inleak through housing (measured value: 51 W)
- Optimized bladed diffuser

Therefore in spring 1997 another series of tests were performed with a modified housing and diffuser. Reduction of heat inleak from 51 to 41 W resulted in an efficiency increase by 2 % points, the better adapted diffuser's contribution was 5 % points. In total the machine as it is now shows an efficiency of 64 % (Figure 1). The working field is somewhat reduced but still very comfortable.

Due to testing of other cold compressor prototypes and mainly usage of the facilities for magnet tests unfortunately the number of operating hours achieved are still small.

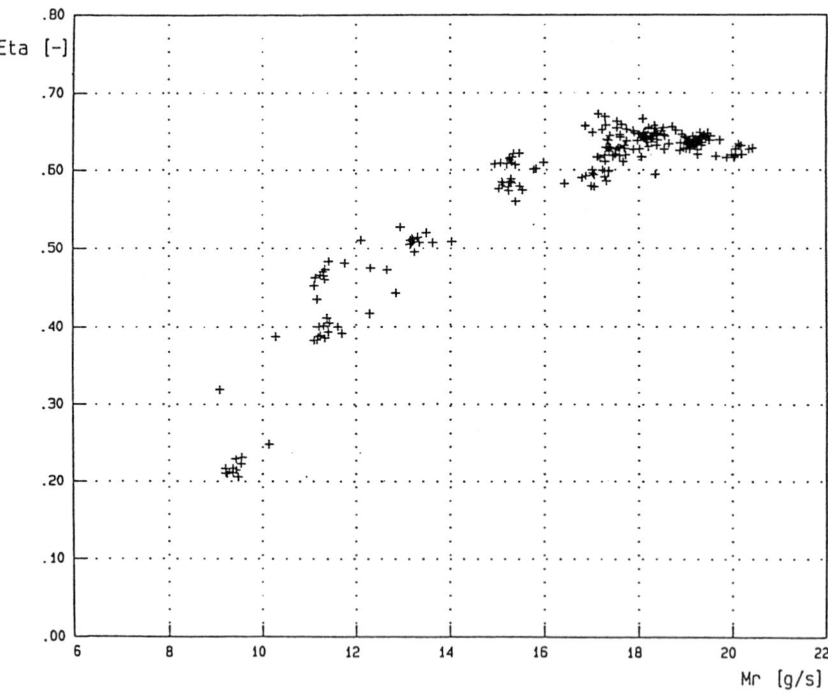


**Figure 1.** Efficiencies based on measurements during test series in spring 1997.

## PROSPECTS

In December 1996 Linde Kryotechnik was awarded a contract for the supply of a helium refrigeration system providing 200 W at 1.8 K for the Forschungszentrum Rossendorf, near Dresden, Germany. The plant will be realized with two-stage cold compression combined with a warm vacuum screw compressor system. The cold compressor system will be based on the experience with the CERN prototype. Plant operation is scheduled to be late 1998.

## REFERENCES


1. L. Decker, K. Löhlein, P. Schustr, M. Vins, I. Brunovsky, L. Tucek, Ph. Lebrun, and L. Tavian, A Cryogenic Axial-Centrifugal Compressor for Superfluid Helium Refrigeration, International Cryogenic Engineering Conference 16 (1996), 195-198

# A HIGH RELIABILITY GAS-DRIVEN HELIUM CRYOGENIC CENTRIFUGAL COMPRESSOR

M. Bonneton[1], L. Tavian[2], G.M. Gistau-Baguer[1], F. Turcat[1], P. Viennot[1]

[1] Air Liquide
Sassenage, 38360, France
[2] CERN
Geneva, Switzerland

## ABSTRACT

A helium cryogenic compressor was developed and tested in real conditions in 1996. The achieved objective was to compress 0.018 kg/s Helium at 4 K @ 1000 Pa (10 mbar) up to 3000 Pa (30 mbar). This project was an opportunity to develop and test an interesting new concept in view of future needs. The main features of this new specific technology are described. Particular attention is paid to the gas bearing supported rotor and to the pneumatic driver. Trade off between existing technologies and the present work are presented with special stress on the bearing system and the driver. The advantages are discussed, essentially focused on life time and high reliability without maintenance as well as non pollution characteristic. Practical operational modes are also described together with the experimental performances of the compressor. The article concludes with a brief outlook of future work.

## INTRODUCTION

In order to boost the capacity of the 1.8 K test stand at CERN, a single centrifugal cryogenic compressor called CCU (Cold Compressor Unit) was manufactured. It enabled to triple the available cooling power and flow rate from 0.006 to 0.018 kg/s. It was also a good opportunity to test new technologies for future use in LHC project. Therefore, the approach to this work was to develop and test all potentially interesting ideas for LHC. In other words, the design was chosen more to investigate future options rather than strictly answering at best profit and efficiency the needs of the test stand.

**Table 1.** Specified performance in nominal condition

| | Inlet | Outlet |
|---|---|---|
| Temperature (K) | 3.5 to 4.4 | to be minimized |
| Pressure (Pa) | 1000 | 3000 |
| Flow rate (kg/s) | 0.018 | |
| Isentropic efficiency | ≤ 60% | |

**Table 2.** Specified performance at reduced flow

| | **Inlet** | **Outlet** |
|---|---|---|
| Temperature (K) | 4 ± 1 K depending on flow and operation | to be minimized |
| Pressure (Pa) | 1000 | see curve attached on Figure 1 |
| Flow rate (kg/s) | 0.006 to 0.018 | |

## SPECIFICATIONS

The nominal design point of the compressor was specified to reach the values of Table 1. Moreover, reduced flow cases were considered and specified according to Table 2.

## MAIN DESIGN OPTIONS

The compressors such as those Air Liquide earlier delivered to CEBAF (now TJNAF), are still a reference of cryogenic centrifugal compression and have been giving full satisfaction since 1993. These machines, pumping up to 240 g/s from 2.8 kPa at 4.32 K up to 115.4 kPa, use electrical drive and bearings. The journal and thrust supports are active magnetic bearings and the drivers are electrical motors. Such a design normally results in relatively costly electronic cabinets and hardware while the time between maintenance is equivalent to good electronic quality standards.

The options for the CCU were very different. It is designed as an essentially pneumatic machine. The bearings are externally pressurized bearings from a standard Air Liquide type C6 expansion turbine. The driver is a centrifugal expansion turbine with 3 tangential outlets. Another major difference is that the machine has a cartridge design (i.e. : it can be extracted from the cold box relatively easily, without breaking the vacuum insulation.). These two features are normally found on all Air Liquide expansion turbines. However, they are totally innovative on cold compressors.

A priori, the following advantages can be foreseen. First, high rotational speeds are available with little practical limitations. Second, there is a significant reduction in investment cost: gas bearings are certainly cheaper than active magnetic bearings together with their electronics, and so is the expansion turbine compared to the motor together with its frequency converter. The reliability is already proven to a very large extent since the core of the machine is identical to that of existing expansion turbines (see Figure 2), which have accumulated years of continuous operation without any need for maintenance. The operation is also very simple. It is only necessary to make sure there is pressure inside the gas bearings (turn a valve on) and the compressor is ready to run at any speed without any special additional precaution. Of course, the gas bearings themselves do not introduce lubricant of any sort (as ball bearings could do for example) since they operate with the helium of the cycle.

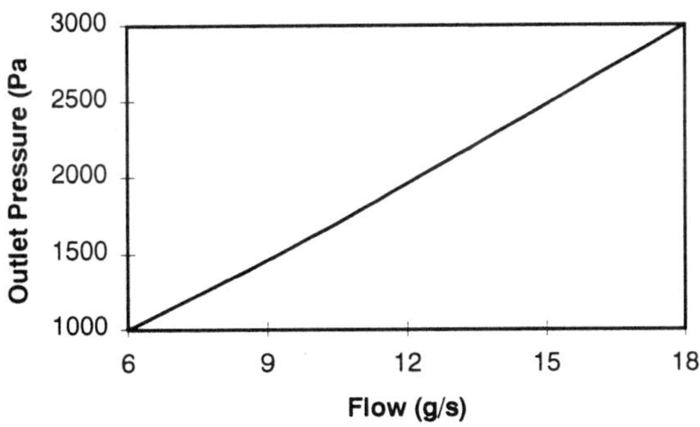


**Figure 1.** Outlet pressure versus flow rate.

However, these technical choices also bring some drawbacks. With pressurized bearings, the low pressure on one side of the shaft is about 100 kPa, a lot higher than the back-wheel pressure, which is close to 2 kPa. Leaks of warm 300 K helium from the bearings to the 10 K wheel would have a catastrophic effect on the outlet temperature of the cold compressor. A special non-contact seal has therefore been designed. The compressor drive makes uses of helium at high pressure as utility, which must be made available at reasonable cost.

The use of a pneumatic machine was not strictly necessary for the CCU. The maximum operating speed (below 400 Hz), together with the low requirements on reliability, were still in the range which can be covered by ball bearings and electrical motor. However, in view of the development goals of such machinery for the LHC project, pneumatic drive and bearings have been chosen.

## COMMISSIONING

### Measurements

Commissioning of the compressor took place at CERN. The quality of pressure and temperature measurements were first assessed. It was not so trivial to measure very low temperature with such a low density fluid 1000 Pa (10 mbar). We give in Table 3 our estimate of the precision of the measurements at compressor inlet.

### Compressor Performance

It became very soon clear that the thermal performance of the supporting structure at low temperature was insufficient. This resulted in larger heat leaks than the available envelope and to negative blade tip clearance. This problem was solved by insertion of a vacuum insulation space inside the compressor itself and reevaluation of the blade tip clearance. This internal vacuum insulation was an improvement to the standard Air Liquide expansion turbines, which

**Table 3.** Estimated precision of the measurements

| | Nominal value | Precision around nominal | Range |
|---|---|---|---|
| Pressure (Pa) | 1000 | 0.5% | 0-5000 |
| Temperature (K) | 4 | ± 0.05K | 3-10 |

normally do not request such techniques. This improvement lowered the heat leaks to the cold helium to 35 W in total, an adequate value for such a machine.

In search of compactness, the inlet pipe and general dimensions of the machine were reduced to minimum. The tests showed that the design had gone a little too far in this direction : in some places, sonic conditions were reached, thus preventing to reach good efficiency and pressure ratio. We implemented some minor modifications which enabled to reach the desired pressure ratio of 3 at 0.018 kg/s. The maximum efficiency reached 57% despite the fact that we could change neither the general dimensions of the machine nor the inlet pipe. To give an idea how stringent the constraint on the inlet pipe size has been, we foresee an improvement of the efficiency of the machine of approximately 10% with a larger inlet pipe.

The non-contact seal performed remarkably and no influence of the leak on the efficiency could even be detected. This corresponds to leak of the order of $10^{-6}$ kg/s warm helium.

## OPERATION

The operation of the compressor proved straightforward. To achieve cool down, the bypass valve was progressively closed while more and more helium was sent into the free-wheeling compressor. The shut down of the by-pass valve was controlled so that moderate quantities of gas were passed through the compressor : its free-wheeling rotational speed was kept below 120 Hz. Once the compressor was cold and the by-pass valve shut, the rotational speed was simply increased by opening the control valve of the pneumatic drive. The pressure ratio then increased continuously together with the flow. This was repeated until the nominal design point was reached. The user interface for this operation is a simple pressure set point : the system automatically adjusts the speed by means of the control valve.

Warm up was also simple : the speed set point is set to 0, the by-pass valve is opened, the inlets and outlets of the compressor are closed and warm helium is circulated through the machine.

The surge line was relatively close to the nominal operating point, due to the fact that the modifications described above had to be implemented to compensate for the over-compactness of the machine. However, there has not been any unexpected surge nor stability problem during the operation of the compressor. The compressor has shown good stability when submitted to fast change of its operating point. It was tested to automatically withstand changes from 0.016 to 0.01 kg/s and would probably show similar easy behavior over the whole flow range (0.006 to 0.018 g/s).

The bearings have shown excellent availability and robustness to all operating and test conditions. Whatever the temperature of the compressor wheel, the bearings could withstand any sort of positive or negative acceleration. There was a wide margin towards higher speeds if needed. There was no forbidden, so called "critical", speed to be specially avoided during operation. No maintenance of the bearings was needed after operation, and we do not foresee any in the long term. In fact, the bearings being identical to standard expansion turbines, there is no reason their life time is not identical. (The standard turbines have <u>proven</u> life time above 40000 hours and they have not met any hard limitation yet). The helium consumption in the bearings is of the order of 0.0018 kg/s. In order to save high pressure helium gas as utility, the driver (0.006 kg/s at maximum output torque) and the bearings can be put on the same line rather than in parallel.

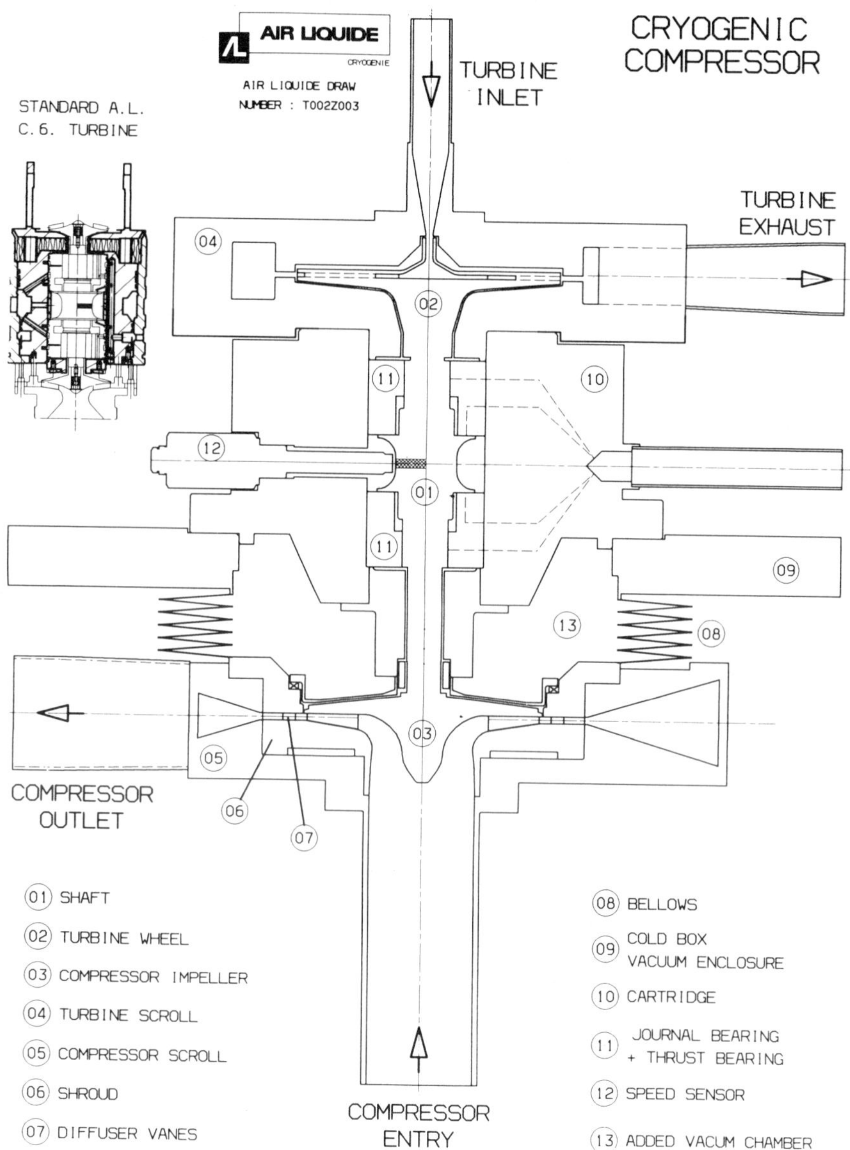


**Figure 2.** Schematic view of cold compressor.

## FURTHER DEVELOPMENTS

### Non Contact Seal

The seal between the bearings and the compression wheel is critical. The present design of the seal makes use of an intermediate-pressure chamber, evacuated by a vacuum pump. This is not in the general idea of no-maintenance compressors, although such maintenance (less than every 10 000 hours anyhow) would be easy and would never endanger the system. As an interesting alternative, a special, non-contact seal requiring no external pumping has been designed. It is now under construction and will be installed for tests. It should ensure good tightness through small auto-adaptive clearance. The residual leak will be of the order of a few $10^{-6}$ kg/s warm helium (corresponding to a few Watts at 4K)

### Tolerant Diffuser

The actual design has a high-density diffuser with numerous, thin blades. This promotes efficiency against flexibility. The compressor will be tested with bladeless diffuser.

### Drive

For more powerful machines, a different expansion turbine drive must be used. The centrifugal expansion turbine is well adapted for low power, such as for this CCU. For higher power and better efficiency, normal centripetal turbines are better suited. For this kind of machines, the more powerful the turbine, the easier to design and build.

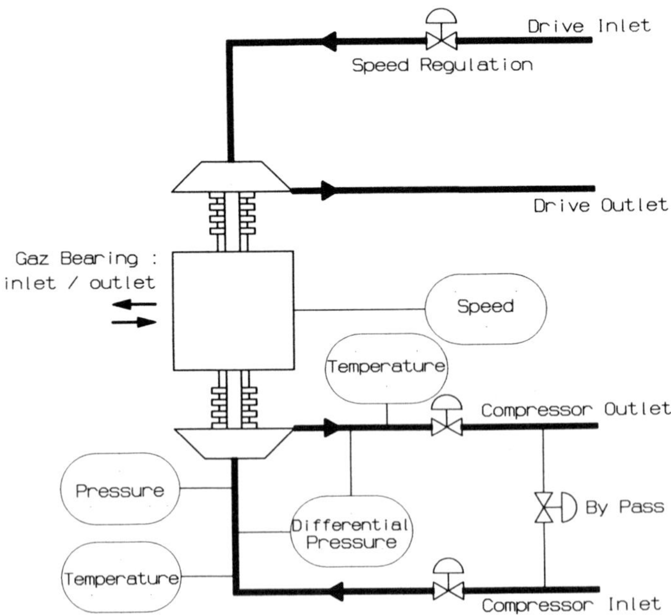


**Figure 3.** Overall system scheme.

### 3D wheel

The use of axial-centrifugal wheels can be well suited for this kind of machine. Although they often require higher rotational speeds (which the gas bearings can provide easily), they offer significant reduction of size and therefore, minimization of heat leaks. Such a wheel should be implemented and tested on the CCU. Design of future machines for LHC could also make use of such wheels.

## CONCLUSION

This type of pneumatic centrifugal cryogenic compressor is very well suited for high-reliability systems that require moderate investment cost. Many years continuous or intermittent use can be expected. It is suitable for use at higher rotational speeds than either advanced ball bearings or magnetic bearings can provide today. Wherever both pneumatic or alternative technology can be used (i.e. : lower speed and lower reliability), trade off shall be carried out to determine whether it is economically interesting to choose lower investment (gas bearings) with more helium utility consumption in the long term or more expensive electrical systems that do not require helium utility at all.

## REFERENCES


1. Gistau and G. Claudet, The Tore Supra 300 W - 1.75 K refrigerator, in : « Advances in Cryogenic Engineering » Vol 31, Plenum, New York (1985), p.607.
2. Gistau, High power refrigeration at temperatures around 2 K, in : « ICEC16/ICMC proceedings » Part 1, Elsevier Science, Oxford (1996), p. 189.
3. Parish, G.M. Gistau, C. Hood, K. Kreinbrink and W. Appleton, Design of a large 2.0 K refrigerator for CEBAK, in : « Supercollider III », Plenum, New York (1991).
4. The LHC Study Group, The Large Hadron Collider, Conceptual Design, CERN Report AC/95-05(LHC) (1995).
5. Book : Cryogénie, Ses applications à la Supraconductivité, by Institut International du Froid, Paris, France. ISBN 2 903 633 770
6. B.S. Bevins, W.C. Chronis and M.S. Keesee, Automatic pumpdown of the 2 K cold compressors for the CEBAK helium liquefier, in : « Advances in Cryogeninc Engineering » Vol 41A , Plenum, New York (1995), p. 663.

# OPERATING A TRAIN OF CRYOGENIC CENTRIFUGAL COMPRESSORS AT LOWER FLOW RATES AND CONSTANT COMPRESSION RATIO

Guy M. Gistau-Baguer

Air Liquide, Advanced Technology Division
BP 15, 38360 Sassenage, France

## ABSTRACT

Cryogenic centrifugal compressors are presently the only « industrial » machines used to reach temperatures lower than 4.4 K (Tore Supra, Fermilab, TJNAL)[1,2,3]. When the full cryogenic power is not needed, the processed flow is reduced. But the flexibility of a centrifugal compressor train is limited : reducing the processed flow rate leads to troubles (stall) and subsequently, surge. One of the easiest way to avoid such a situation is to inject electrical power into the liquid helium bath in order to provide the compressor train with a constant flow rate. This is not a very satisfactory procedure on a thermodynamical point of view... Other solutions have been proposed which generally lead to the operation of some parts of the room temperature helium circuits at sub atmospheric pressure, which is not consistant with a very long term reliable operation. Operating two parallel trains, each one processing half the nominal flow is another solution not very much flexible but which investment is roughly doubled.

## INTRODUCTION

When using a centrifugal compressor train, the question is : how to process a lower mass flow rate when keeping the same liquid bath temperature, i. e. the same compression ratio ? Our solution to cope with such a requirement is the addition of one more stage of compression.

## A REMINDER ABOUT CENTRIFUGAL COMPRESSORS

Figure 1 is a customary method of presentation of a typical performance map for a centrifugal compressor [4,5] : the pressure ratio across the whole machine is plotted as a function of « reduced » flow : $\dot{m} * \sqrt{T}$ / P ($\dot{m}$ : mass flow rate, T : inlet temperature, P : inlet pressure), for fixed values of « reduced » speed : N / $\sqrt{T}$ (N : rotational speed). At the

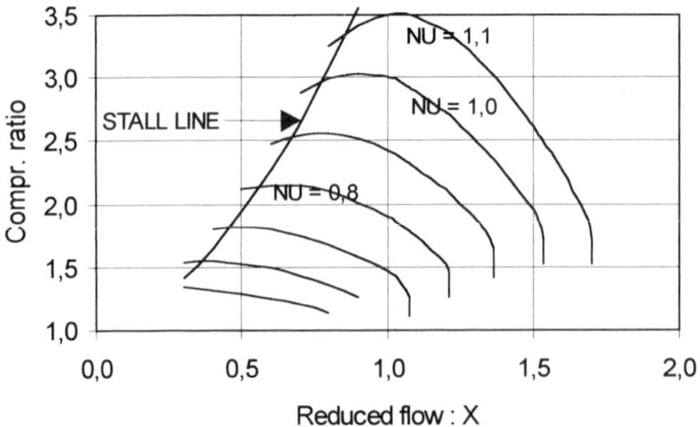


**Figure 1.** A centrifugal compressor typical performance map.

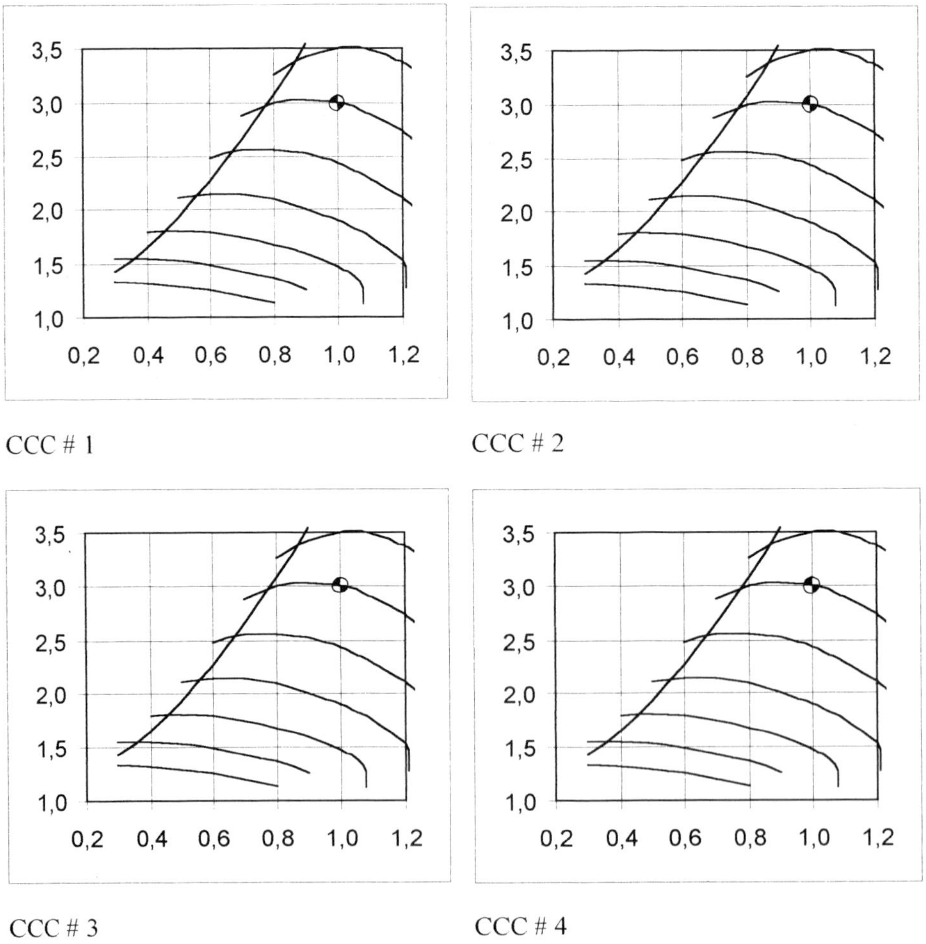

CCC # 1

CCC # 2

CCC # 3

CCC # 4

**Figure 2.** Operation of a CCC train at design conditions.

**Table 1.** Operation of a CCC train at lower mass flow rate

| DESIGN | Comp. ratio | Pressure | Temperature | Mass flow | X |
|---|---|---|---|---|---|
| | | (kPa) | (K) | (kg/s) | |
| Suction CCC # 1 | 2,51 | 3.0 | 3,3 | **0.240** | 1,00 |
| Suction CCC # 2 | 2,51 | 7.5 | 5,6 | 0.240 | 1,00 |
| Suction CCC # 3 | 2,51 | 19.0 | 9,4 | 0.240 | 1,00 |
| Suction CCC # 4 | 2,51 | 47.7 | 15,8 | 0.240 | 1,00 |
| Suction CCC # 5 | 1,00 | 120.0 | 26,6 | 0.240 | 1,00 |
| Discharge CCC # 5 | | 120.0 | 26,6 | 0.240 | |
| | Comp. ratio | Pressure | Temperature | Mass flow | X |
| Suction CCC # 1 | 1,96 | 3.0 | 3,3 | 0.200 | 0,83 |
| Suction CCC # 2 | 2,51 | 5.9 | 4,9 | 0.200 | 1,00 |
| Suction CCC # 3 | 2,51 | 14.8 | 8,2 | 0.200 | 1,00 |
| Suction CCC # 4 | 2,51 | 37.2 | 13,8 | 0.200 | 1,00 |
| Suction CCC # 5 | 1,28 | 93.6 | 23,3 | 0.200 | 1,00 |
| Discharge CCC # 5 | | 120.0 | 27,1 | 0.200 | |
| | Comp. ratio | Pressure | Temperature | Mass flow | X |
| Suction CCC # 1 | 1,57 | 3.0 | 3,3 | 0.170 | 0,71 |
| Suction CCC # 2 | 2,51 | 4.7 | 4,3 | 0.170 | 1,00 |
| Suction CCC # 3 | 2,51 | 11.8 | 7,3 | 0.170 | 1,00 |
| Suction CCC # 4 | 2,51 | 29.7 | 12,2 | 0.170 | 1,00 |
| Suction CCC # 5 | 1,61 | 74.8 | 20,6 | 0.170 | 1,00 |
| Discharge CCC # 5 | | 120.0 | 27,2 | 0.170 | |
| | Comp. ratio | Pressure | Temperature | Mass flow | X |
| Suction CCC # 1 | 1,32 | 3.0 | 3,3 | 0.150 | 0,63 |
| Suction CCC # 2 | 2,51 | 3.9 | 3,9 | 0.150 | 1,00 |
| Suction CCC # 3 | 2,51 | 9.9 | 6,6 | 0.150 | 1,00 |
| Suction CCC # 4 | 2,51 | 25.0 | 11,1 | 0.150 | 1,00 |
| Suction CCC # 5 | 1,91 | 62.8 | 18,6 | 0.150 | 1,00 |
| Discharge CCC # 5 | | 120.0 | 27,1 | 0.150 | |

left upper part of the map, the stall line shows the limit of stable operation, unstable operation being characterized by a severe oscillation of the mass flow rate through the machine. The right part shows the choked region where the constant speed lines tend to the vertical. It means that no further increase in flow is possible since sonic velocity is attained through some section of the machine.

In order to make such compressor performance maps comparable between different compressors, the « reduced-reduced » parameters are used : « reduced-reduced » pressure : $X = \frac{\dot{m}*\sqrt{T}/P}{\dot{m}_0*\sqrt{T_0}/P_0}$ and « reduced-reduced » speed : $NU = \frac{N/\sqrt{T}}{N_0/\sqrt{T_0}}$ ($\dot{m}_0$: nominal mass flow rate, $T_0$ : nominal inlet temperature, $P_0$ : nominal inlet pressure, $N_0$ : nominal rotation speed). For reasons of simplification, let us now talk of reduced flow as X and reduced speed as NU.

## WHAT HAPPENS ON A CCC TRAIN WHEN THE PROCESSED MASS FLOW RATE DECREASES ?

Let us suppose that a four stage CCC train is operating at its design point (see figure 2). It means that, for each stage, the parameters : reduced flow X and reduced speed NU are both equal to unit and the compression ratio is nominal. (In order to ease explanations, let us assume that the isentropic efficiency of each CCC and the CCC # 1 pressure and temperature do not change.) With such assumption, when the mass flow rate

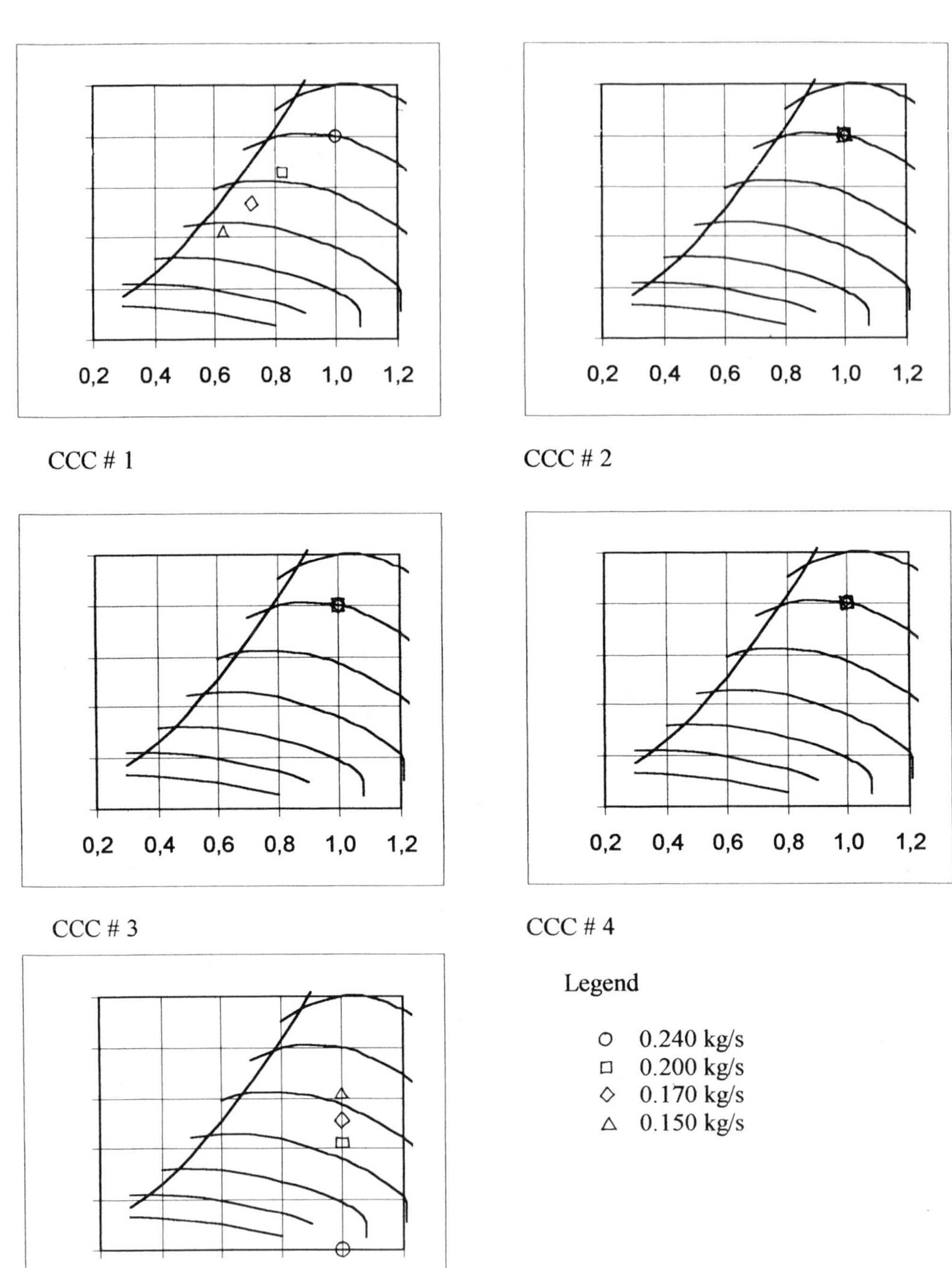


**Figure 3.** Operation of a CCC train at lower mass flow rate.

decreases, it is obvious that the reduced flow X of CCC # 1 is reduced by the same ratio. If the compression ratio of CCC # 1 does not change, a time will happen when its operating point hits the stall line : this situation is unbearable. It is necessary to decrease the compression ratio of CCC # 1 in order that its operating point stays on the right side of the surge line. In order to keep the nominal compression ratio, it means that the following CCCs have to compensate for the lack of compression ratio due to the new operating conditions of CCC # 1. Obviously, a limitation like speed, power or torque on CCC # 2 or # 3 or # 4 will be found.

## A SOLUTION

As a consequence of the reduction of the compression ratio of CCC # 1, the inlet conditions at CCC # 2 change : the pressure decreases a lot and the temperature decreases also (see table 1). It is possible to find a compression ratio value which makes the reduced flow X of CCC # 2 equal to unit. Therefore, if the compression ratio of CCC # 2 is kept at design, it means that this compressor will stay at its nominal operating point. Same reasoning can be made for CCC # 3 and # 4 which operating points stay at design. As a consequence of lower compression ratio on CCC # 1, the discharge pressure of CCC # 4 is lower than requested.

The solution simply consists in adding a fifth stage in order to perform the compression ratio which has not been taken care by CCC # 1 (see figure 3). The sizing of such a stage is done according to the mass flow rate which is to be processed.

## CONCLUSION

There is a thermodynamically efficient solution to process a lower mass flow rate at constant pressure ratio by adding an « n + 1 » centrifugal compressor stage downstream a compressor train.

## REFERENCES


1. G. M. Gistau, The Tore Supra 300 W - 1.75 K refrigerator, in : "Advances in Cryogenic Engineering," Vol 31, Plenum, New York (1985), p. 607.
2. J. Fuerst, Selection of cold compressors for the Fermilab Tevatron, in : "Advances in Cryogenic Engineering," Vol 39A, Plenum, New York (1993), p. 863.
3. H. Parish, G. M. Gistau, C. Hood, K. Kreinbrink and W. Appleton, Design of a large 2.0 K refrigerator for CEBAF, in : "Supercollider III", Plenum, New York (1991).
4. S. L. Dixon, "Fluid Mechanics, " Pergamon Press, Oxford.
5. G. M. Gistau, J. C. Villard, and F, Turcat, Application range of cryogenic centrifugal compressors, in : "Advances in Cryogenic Engineering," Vol 35B, Plenum, New York (1989), p.1031.

# PREDICTING PERFORMANCE OF EXPANSION TURBINES USING DIFFERENT WORKING FLUIDS BASED ON THE ARTIFICIAL NEURAL NETWORK

Liqiang Liu, Yanzhong Li,and Chunzheng Chen

Institute of Cryogenic Engineering
Xi'an Jiaotong University, Xi'an 710049, China

## ABSTRACT

In order to obtain the efficiency curve of a helium Expansion Turbine (ET) in the factory, a method which can be used to transform the performance of an ET with one kind of Working Fluid (WF) to another must be investigated because of the lack of the helium in the factory. A performance prediction program based on an one-dimensional analysis of ETs has been developed and has proved valid. On the basis of the program , the Artificial Neural Network(ANN) (Back-Propagation Algorithm) has been used to deal with the transformation problem. The method has proved effective by the efficiency experiments of an ET using the $CO_2$ and the air as the WF respectively.

## INTRODUCTION

When an ET using special WF (such as helium or $CO_2$ ) is developed, it is difficult to obtain the thermal performance of the ET under the design conditions in the factory. This performance is very important for users. A feasible method to solve the problem is to run the ET with an easily available WF (such as air) and then transform the performance of the ET from air to the special WF. Because of the difference of the ratio of specific heats, $\kappa$, the relations between using different WFs have intense nonlinearity and the transforming process is very complicated. We have developed a method of selecting similarity criteria to overcome the difficulty.[1] With the aim of dealing with the problem more effectively, the ANN technique which is very suitable for the nonlinear problem is used in this paper.

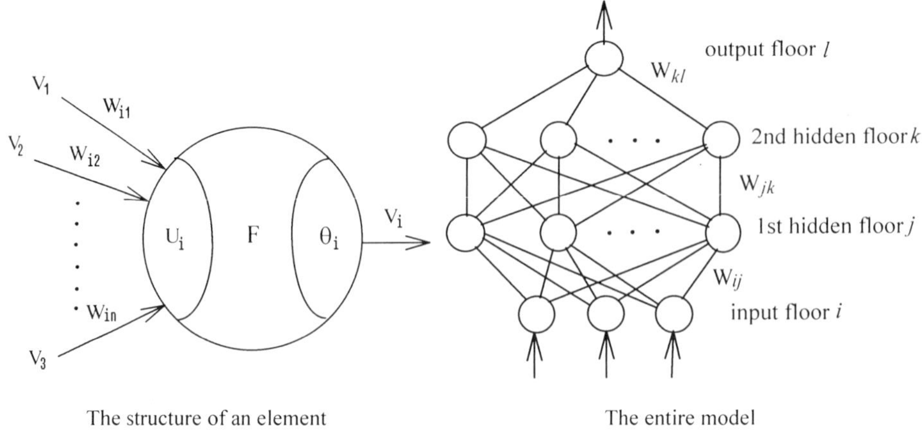

Figure 3. The model of the ANN.

with air or the special WF, keep the equal $p_0/p_3$), and the $U_1 / C_S$ of air. The only output parameter is the $U_1 / C_S$ of the special WF under which the ET has the same $\eta_S$ with the condition of the $U_1 / C_S$ of air which is one of the input parameters. When the $\eta_S$ and $U_1 / C_S$ of air are known by tests and the $U_1 / C_S$ of the special WF is acquired by the ANN, two curves of $\eta_S$ can be plotted, thus the ET performance relations between with air and the special WF can eventually be built up.

The Back-Propagation Algorithm is applied as the learning algorithm in the paper. In order to acquire the satisfactory precision, it is usually needed millions of learning circles. Fig. 4 gives the error curve of the learning process.

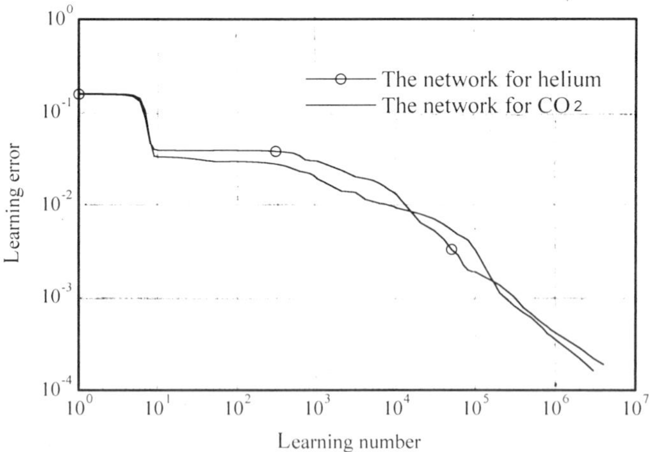

Figure 4. The learning error curves.

## PREDICTING PERFORMANCE PROGRAM

For the purpose of providing the base for the ANN method, an ET thermal performance prediction program based on an one-dimensional analysis has been developed. Under the assumptions, in which the turbine is perfectly isolated with no heat transfer from the surroundings and the flow in the turbine is one-dimensional steady viscous flow, the ET performance can be achieved by applying the law of continuity, conservation of energy, and the impulse-momentum principle to the nozzle, wheel, and diffuser of ET. In the program, the reaction of ET is assumed to be known in the beginning and is solved by iteration at last. When the reaction is known, the coefficients of nozzle loss, wheel loss, diffuser loss, incidence loss, internal leakage loss and disc windage loss are determined, consequently the adiabatic efficiency $\eta_S$ can be found.

Based on the structure of two air ETs made in China, we have determined the performance of the machines using the developed program. The calculation results and test data are both shown in Fig. 1 and Fig. 2 (the 6000ET design conditions are: the outer diameter of the wheel $D_1$ = 190 mm, the expansion ratio $p_0/p_3$ = 4.07, the velocity ratio $U_1/C_S$ = 0.66; the 150ET's are: $D_1$ = 36.5 mm, $p_0/p_3$ = 3.33, $U_1/C_S$ = 0.62). In the figures, E.R. means the expansion ratio. From the two figures, we can conclude that the program is valid

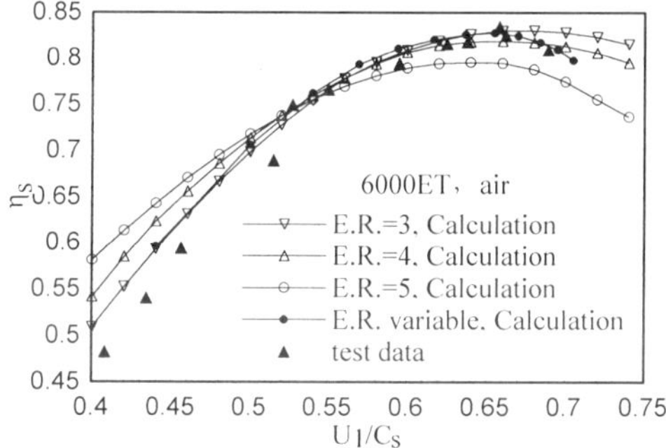


Figure 1. The 6000ET $\eta_S$ curves.

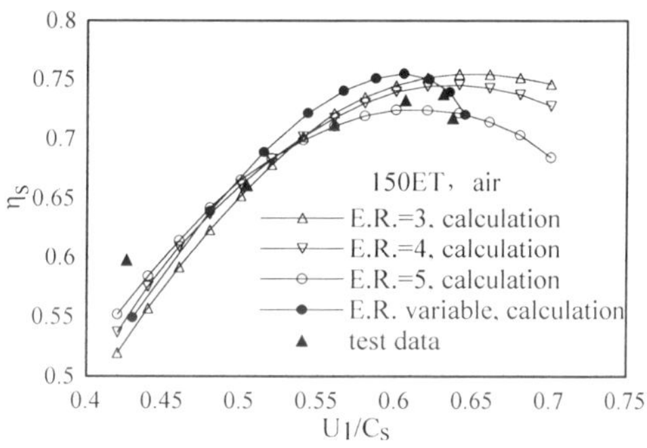

Figure 2. The 150ET $\eta_S$ curves.

## THE TRANSFORMING METHOD BASED ON THE ANN

Having developed the ET performance prediction program by which the ET performance with different WFs can be predicted, we can establish the relations between the ET performance using air and the performance using the special WF based on the ANN. The model of the ANN is shown in Fig. 3. As the figure shows, a three floors forward model is adopted here.

In the model, there are 9 elements in both the first and second hidden floors and 3 elements in the input floor. The three input parameters are $D_1$, $p_0/p_3$(when perform ET

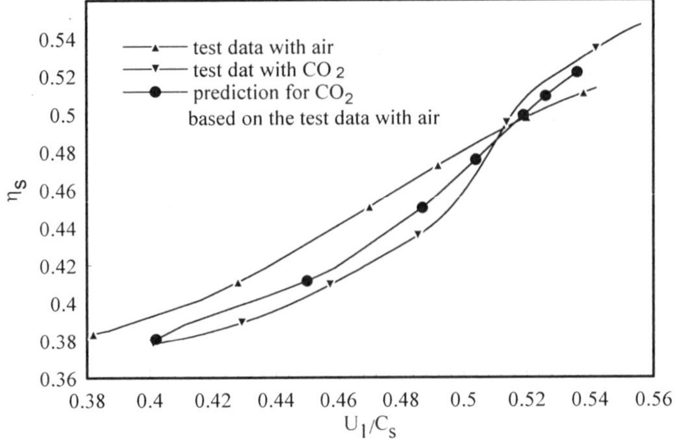


Figure 5. The comparing curves of $\eta_S$ of the small ET running with air and $CO_2$.

## COMPARING THEORETICAL RESULTS WITH EXPERIMENTAL DATA

In order to inspect the validity of the transformation method based on the ANN, we have set up a multifunction test rig for testing the performance of a small ET designed by ourselves, on which, the tests using different kinds of WF and operated under different range of inlet temperature can be done stably. On the test rig, the thermal performance of the ET was first tested using air. Then by controlling $p_0/p_3$, the thermal performance using $CO_2$ was tested. The test data both with air and $CO_2$ and the theoretical prediction for $CO_2$ based on the air test data are all shown in Fig. 5. From Fig. 5, it is obvious that the transforming method based on the ANN is effective, the error of prediction of $\eta_S$ is less than 1%.

## REFERENCE

1. Liqing Liu and Chunzheng Chen, Predicting performance of helium expansion turbines using similarity principles, ICEC16 Proceedings, Elsevier Science (1996), Part 1, 225-228
2. Whitfield, A. and Baines, N. C. In: Design of Radial Turbomachines, John Wiley Sons Inc., New York, USA(1990): 26-41
3. Misao, H., A one-dimensional analysis and performance prediction of subsonic radial turbines, Bulletin of the JSME (1980) 23 2064-2070
4. Ji, G. H., In: Expansion turbine, Jixie Gongye Press, Beijing, China (1989) 161-163

# A FEASIBILITY STUDY ON THE USE OF NEW GAS FOIL BEARINGS IN CRYOGENIC TURBOEXPANDER

Lian-You Xiong, Gang Wu, Chun-Zheng Chen and Yan-Zhong Li

Institute of Cryogenic Engineering
Xi'an Jiaotong University, Xi'an 710049, China

## ABSTRACT

Experiments have been conducted with two type of gas lubricated foil journal bearings and a small cryogenic turboexpander. One of these bearings is named as a spirally-wound foil bearing and the another is named as a multi-leaves foil bearing. Both of them have much simpler construction and lower cost as compared with the conventional foil bearings. The rotor of cryogenic turboexpander, 103 mm long and weighing 0.9 N, has the working speed up to 230,000 rpm. The main purpose of the investigation is a preliminary examination of stabilizing characteristics of those foil bearings under several different clearances and foil stiffness as well as the durability of the foils. Experimental results indicate the foil bearings presented here offer system stability and high durability, and it confirmed that the application of gas foil bearing to the cryogenic turboexpander is feasible and hopeful.

## INTRODUCTION

With the wide use of gas bearing turboexpander for small air separation plant and other cryogenic system, the requirements for the bearings become severer. The foil bearing, offers many advantages such as high speed, high stability and high impulse capacity, seems to be suitable for small high speed cryogenic turboexpander.

Up to present, many studies have been done for several types of foil bearings, such as Tension Type, Cantilever Type and Multi-wedged Type[1]. All of them have been successfully applied to turbomachinary. However, these types of foil bearings generally have complex construction and high cost, So it is necessary to develop a simple and high performance foil gas bearings applying to cryogenic turboexpander.

## NEW TYPE OF FOIL GAS BEARING

In recent years, a new type of foil journal bearing has been developed by Professor K. Hayashi[2]. The spirally-wound foil bearing consists of one foil element and a housing while the multi-leaves foil bearing consists of three foil elements and a housing. The foil element is a 0.06 mm thick phosphor bronze plate that has some number of 0.2 mm diameter copper wires on its backside and a 5 mm diameter steel pin at its end. This foil element is reversely wound double against the rotation of shaft and is set in the housing with its pin fixed into a pin hole in the housing. The nominal radial clearance c is measured with zero foil deflection. This bearing clearance can be varied by using the housing of different diameter while the stiffness of foil element is controlled by the number of copper wires fixed behind the foil.

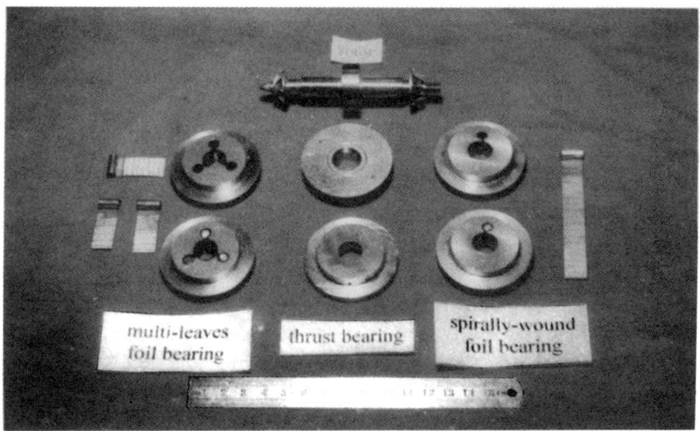


**Figure 1.** Rotor and test bearings .

## EXPERIMENTAL APPARATUS

The foil bearing tests have been done on a small turboexpander. Figure 1 shows the rotor of turboexpander and some test bearings. The rotor is driven by a reaction wheel with the diameter of 20 mm, and the output work is absorbed by a brake wheel with the diameter of 22 mm. The journal diameter is 12 mm, and its mass is 90 g. The working speed of this turboexpander is 230,000 rpm. A pair of test foil journal bearing is used to support the rotor and the externally-pressurized gas thrust bearings are used to equalize the axial force.

A pair of eddy-type displacement probes is mounted separately on expander housing in x-y direction to measure the radial vibration of the rotor. All probe outputs were recorded and displayed on computer. The rotational speed as well the vibrations of the rotor can be analyzed in real time by a computerized acquisition system.

## EXPERIMENTAL RESULTS AND DISCUSSION

Several test bearings with different clearance and different number of copper wires have been tested. The clearance and the number of copper wires of test bearings are as follows:

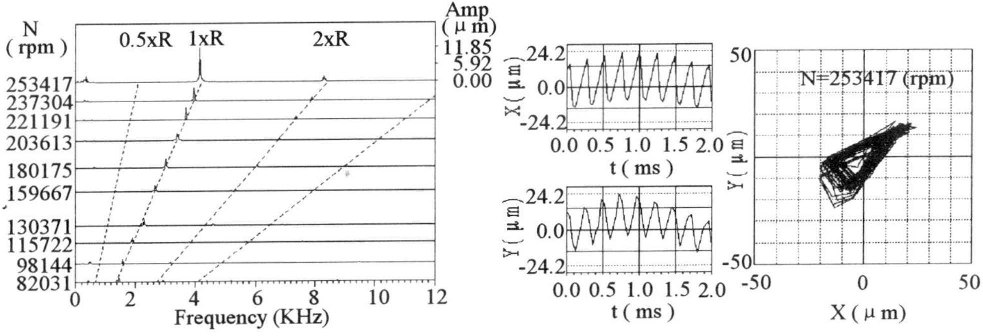


**Figure 2.** Start-up waterfall diagram and Time-base record of rotor vibration at maximum speed (spirally-wound foil bearing: n=21, c=10 μ m ) .

| | | |
|---|---|---|
| spirally-wound foil bearing: | n=15,17,19,21,23 | c=0, 10, 20 μ m |
| multi-leaves foil bearing: | n=6,7 | c=0, 10, 20 μ m |

The rotation characteristics of test bearings with the different number of copper wires under the clearance c=0 μ m have been fully discussed in the previous paper[3]. Accounting for the restriction of pages, only the result of spirally-wound foil bearing with n=19,21 and the multi-leaves foil bearing with n=7 under c=0, 10, 20 μ m are presented here typically.

The start-up waterfall diagram and the time-base records of the rotor vibrations at the maximum rotational speed are shown in figure 2, 3. The main portion of the rotor response is synchronous. The subsynchronous resonance is not significant and its frequency is very low (about 200Hz). The orbits of rotor center are clear and the amplitudes of vibration are small even when the subsynchronous resonance occurs, which indicate that both the spirally-wound foil bearing and the multi-leaves foil bearing show the good damping characteristics.

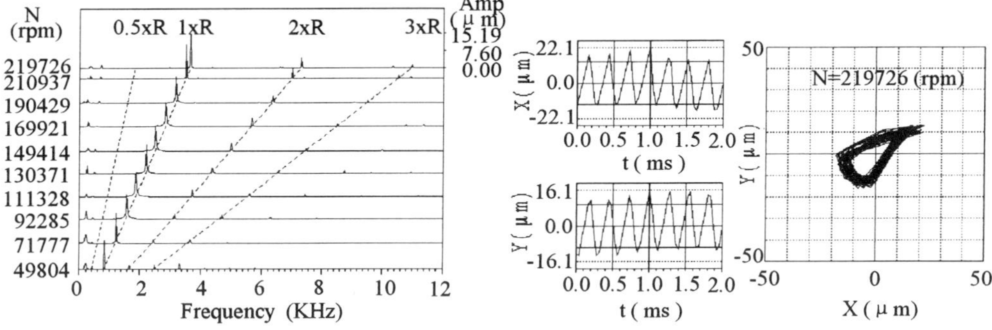


**Figure 3.** Start-up waterfall diagram and Time-base record of rotor vibration at maximum speed (multi-leaves foil bearing: n=7, c=10 μ m) .

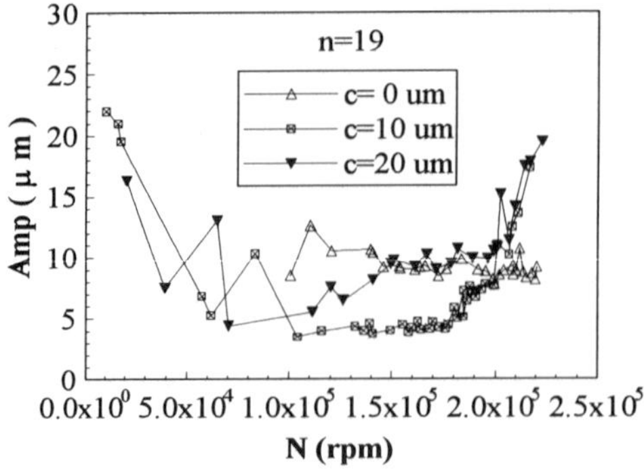


**Figure 4.** Variation of Rotor vibrations with Rotor speed (spirally-wound foil bearing: n=19) .

Figure 4 ~ 6 show a plot of peak-to-peak amplitudes as a function of shaft speed for various bearing clearances and various stiffness. The test results indicate that the amplitudes of rotor vibration are small in the middle speed region, and the choice of bearing clearance is very important to the amplitude of rotor vibration .

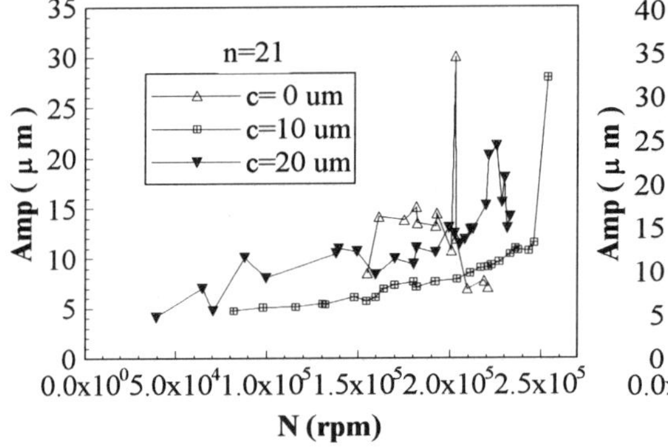


**Figure 5.** Variation of Rotor vibrations with Rotor speed (spirally-wound foil bearing: n=21) .

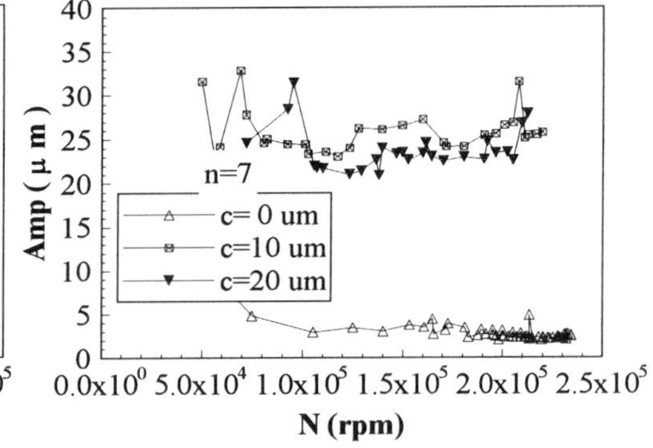


**Figure 6.** Variation of Rotor vibrations with Rotor speed (multi-wound foil bearing: n=7) .

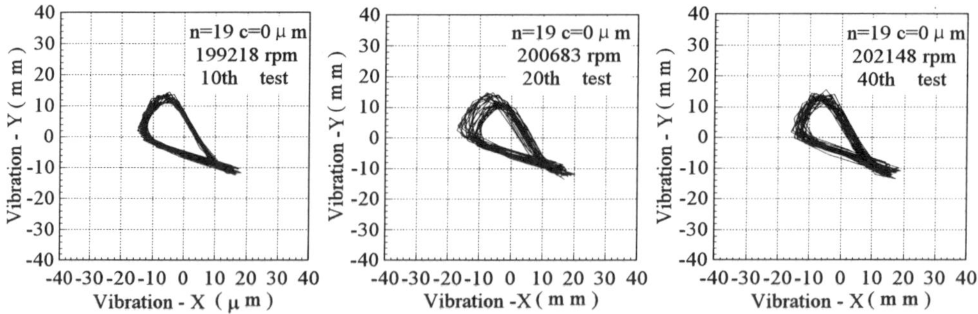


**Figure 7.** Start-Stop characteristics .

Figure 7 shows the durability test result of test bearing (n=19, c=0 μm). The start-stop cycle was repeated 40 times with the rotor speed up to 200,000 rpm, and the rotation characteristics change little during the tests.

## CONCLUSION

The foil stiffness and the bearing clearance play an important role in the dynamic performance of this new type of foil journal bearings. With the proper choice of the foil stiffness and the bearing clearance, the foil bearing presented here can achieve sufficiently high stability. Due to its simplicity and high performance, it is hopeful to apply to a small high speed cryogenic turboexpander.

## ACKNOWLEDGMENT

This project was supported by the Doctorate Foundation from the State Education Commission of China. The authors would like to thank Mr. Liu Guo jun et. al. of Shuzhou oxygen plant manufactory for their continued support.

## REFERENCE

1. S. Gray. Technology progress on compliant foil air bearing systems for commercial applications, *8th Int Gas Bearing Symposium* (1981)

2. K. Hayashi, Development of aerodynamics foil bearings for high-speed rotor, *21st Leeds-Lyon Symposium Tribology* (1994)

3. L. Y. Xiong , G. Wu et al . Development of Aerodynamic Foil Journal Bearings for High Speed Cryogenic Turboexpander, *Cryogenics*, vol. 37, n4, p221-230 (1997).

# 80K TURBO COMPRESSOR SYSTEM WITHOUT LN2 COOLING FOR HIGH RELIABLE AND EFFICIENT HELIUM REFRIGERATOR

H.Asakura[1], N.Saji[1], Y.Kaneko[1], S.Yoshinaga[1], M.Mori[1], H.Yamaguchi[1], T.Nogaku[1], and T.Umeda[2]

[1]General Machinery Div. Ishikawajima-Harima Heavy Industries Co.,Ltd., 3-2-16 Toyosu Koto-ku Tokyo 135 Japan
[2]Super-GM, 5-14-10 Nishitenma Kita-ku Osaka 530 Japan

## ABSTRACT

In Japan's national project 'Super-GM,' we developed a complete oil-free 80K turbo compressor system for the helium refrigerator which brought a remarkable reliability. This compressor has a good efficiency; nevertheless, the refrigeration system efficiency is almost same as that of a conventional refrigerator because of LN2 cooling. Then we are now developing a new conceptional cold compressor without LN2 cooling. The new compressor is driven by a coaxial expansion turbine instead of an electric motor, and the intercooling between compression stages is acted by cold helium gas which is produced by the expansion of a driving turbine. In order to make matching between compressor input and turbine output, an electric generator is provided coaxially. The efficiency of system with the new compressor is expected to improve very well. The study of system arrangement and the design of the cold compressor are reported in this paper.

## INTRODUCTION

The helium refrigeration system used for industrial machines such as superconducting generators requires much higher reliability, efficiency and space saving than that of experimental use.

Concerning high reliability, what is an important subject of development is to eliminate the contamination in the refrigeration system by adopting complete oil-free rotating machines and to reduce the number of rotating machines by managing the system arrangement.

In Japan's national project 'Super-GM', the advanced helium refrigeration system used for future 200MW superconducting generator has been developed, and we have already developed a 80K oil-free turbo compressor driven by high speed electric motor which is applied in the system as shown in Figure 1.[1] One of the features of this system is to realize a complete oil-free system by employing a turbo compressor which uses magnetic bearings for the main compressor. The other feature is to get a required pressure ratio with only 4 stages by increasing helium density by means of cooling the compressor inlet with liquid nitrogen (LN2). We have already carried out various tests of the compressor system and confirmed mechanical reliability, aerodynamic performance, multi-stage operation characteristics and their sufficient adaptability to the refrigeration system.[2,3]

The compressor itself has achieved a high efficiency. However, the total system efficiency is almost the same as that of the conventional refrigerator, because a lot of electric power is consumed to produce the LN2.

To solve this problem we have been developing a new conceptional refrigeration system by applying Brayton cycle to the compression unit for the temperature range from 300K to 80K and not using LN2 cooling. In this paper we report the arrangement of this refrigeration system and new cold compressor design.

## ADVANCED REFRIGERATION SYSTEM WITH CLOSED BRAYTON CYCLE

### Outline of Refrigeration Cycle

Figure 2 shows the new conceptional refrigeration system flow using Brayton cycle. The system constitution below 80K is the same as that in Figure 1 and Brayton cycle using helium gas is built in between 300K and 80K. The helium gas compressed by the compressor placed in 300K is introduced into 4 turbines in parallel through a recuperator. The gas after being expanded in the turbines whose temperature is cooled down to approximately 80K enters the heat exchanger to cool each stage compressor outlet gas and returns to 300K compressor passing through the recuperator. The 80K compressor is driven by the expansion turbine. The energy obtained through expansion is larger than the compressor driving power, so the remaining energy should be absorbed by the generator.

Therefore, the key component to establish this system lies in a rotating machine which has compressor, turbine and generator on the same shaft.

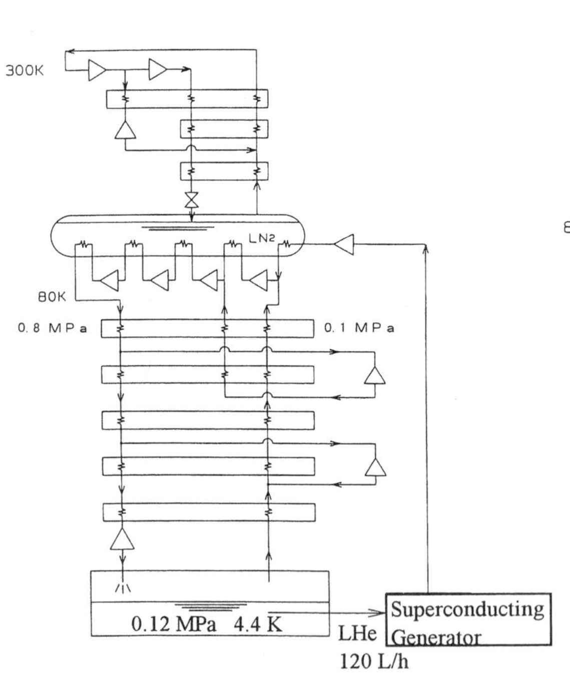


**Figure 1.** Advanced helium refrigeration system with $LN_2$ cooling. Eeach 80K oil-free compressor is driven by an electric motor.

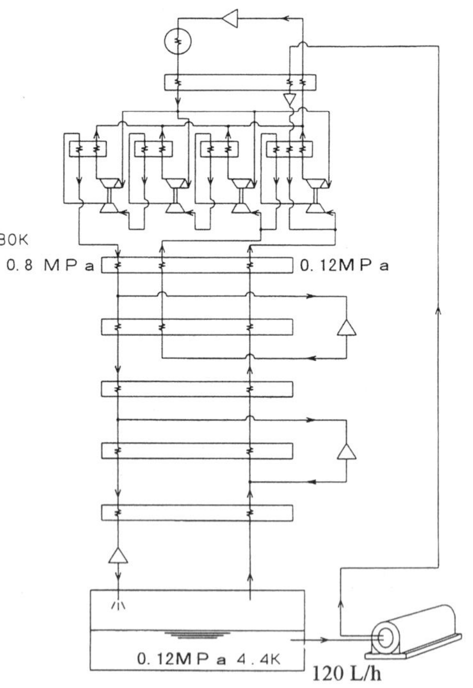


**Figure 2.** Advanced helium refrigeration system using Brayton cycle. The 80K cold compressors in Figure.1 are replaced with new compressors driven by turbines.

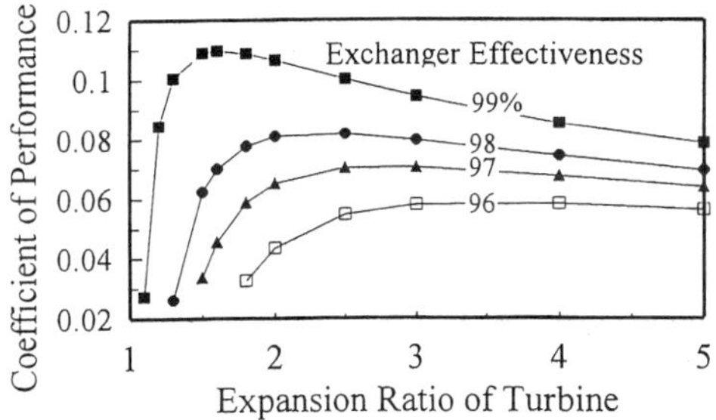


**Figure 3.** Efficiency of Brayton cycle as a function of expansion ratio of turbine. Coefficient of performance is defined as a refrigeration capacity at 80K per a power input.

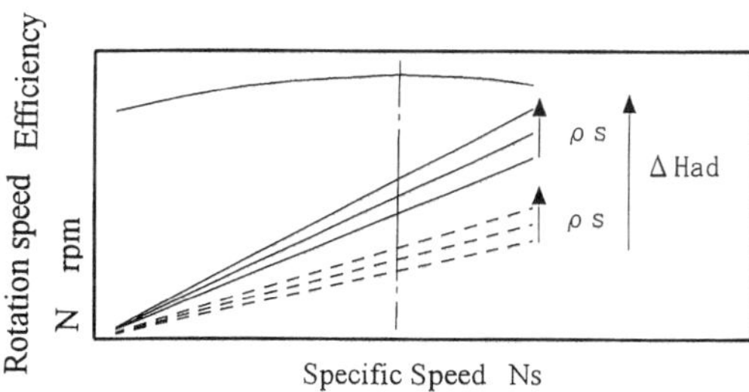


**Figure 4.** Specific speed vs. rotation speed and efficiency of centrifugal impeller. The larger ρ s and ΔHad is, the higher the rotation speed is at the optimum Ns.

## Measures for Attaining Higher System Efficiency

Brayton cycle has the optimum turbine expansion ratio with which the cycle efficiency becomes maximum depending on the exchanger effectiveness of the recuperator. As shown in Figure 3, the optimum turbine expansion ratio is approximately 2.0 at a 98% of exchanger effectiveness of a practical heat exchanger.

In order to increase the efficiency of the refrigeration system furthermore, the efficiency of the rotating machines should be increased. Especially efficiency of 80K compressors whose number is larger must be improved.

The efficiency of a centrifugal compressor impeller is roughly decided by the specific speed Ns which is a parameter of deciding the shape and it changes as shown in Figure 4. As shown in the figure, the optimum Ns to give the maximum efficiency is roughly decided. In designing, rotation speed and outer diameter of the impeller are decided so as to gain this optimum Ns as much as possible.

The specific speed Ns is defined as:

$$Ns=(N/60)\cdot Qs^{1/2}\cdot \triangle Had^{-3/4} \qquad (1)$$

where N: Rotation speed [rpm], Qs: Suction volume flow rate [m3/s], △Had: Adiabatic head [J/kg].

Rotation speed N is rewritten as:

$$N=60\cdot G^{-1/2}\cdot \rho s^{1/2}\cdot \triangle Had^{3/4}\cdot Ns \qquad (2)$$

where ρ s : Inlet gas density [kg/m3], G: Mass flow rate [kg/s].

When the flow rate is constant, degree of increase of N to Ns becomes larger with an increase of ρ s and △Had.(see Figure 4).

In the cold compressor, since the gas density becomes large and high head is required, considerably high rotation speed is required when the optimum Ns is desired. Meanwhile, when a motor or a generator is built on the same shaft with a compressor, maximum rotation speed is mechanically limited depending on output of the motor or generator.

The compressor developed for the refrigerator with LN2 cooling was designed to be driven by a motor, so the rotation speed limit was low and the Ns should be considerably lower than the optimum Ns. Breaking power of the generator installed in the turbine-drive compressor becomes low, so the rotation speed can be increased by 20%. In that case, the impeller Ns can reach near the optimum Ns and efficiency can be raised.

### Optimum Arrangement of Compressor Unit

The pressure ratio of each stage of 80K compressors in Figure 2 is approximately 1.7 and the one of a 300K compressor is 2.0 which is the same value as the turbine expansion ratio. Therefore, each one of the four 80K compressors can be used in common by a 300K compressor. In that case, flow rate of the 300K compressor increases by the amount of one-stage 80K compressor. However, because flow rate of one-stage turbine decreases inversely, the resulting flow rate does not change substantially. Therefore, with only change of the piping arrangement one rotating machine can be omitted and system efficiency increases due to omission of each conversion efficiency.

The optimum system flow is shown in Figure 5. Figure 6 shows comparison of each liquefaction efficiency and installation space of the system with LN2 cooling (Figure 1), 4 compressors (Figure 2) and 3 compressors (Figure 5) for the Brayton cycle system.

## COLD COMPRESSOR DRIVEN BY EXPANSION TURBINE

### General Features

The new cold compressor is driven by an expansion turbine on the same shaft instead of a motor. An electric generator is installed in the middle of the shaft to match compressor input with turbine output. That is, this machine is a high-speed rotating machine having three elements on one shaft.

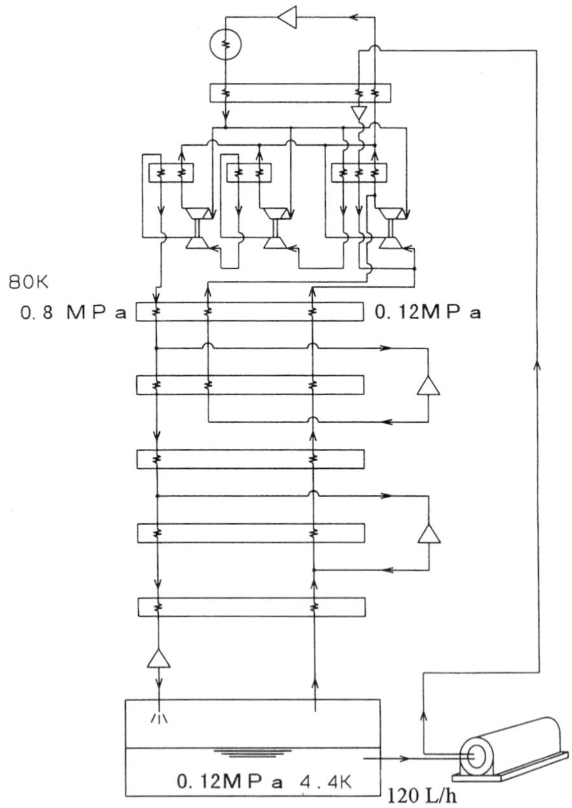


**Figure 5.** Optimum arrangement of compressor unit. The 2nd stage in Figure 2 can be omitted.

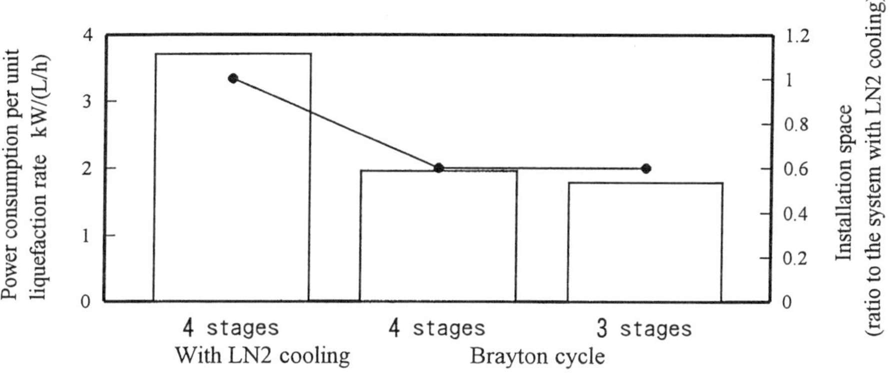


**Figure 6.** Comparison of liquefaction efficiency and installation space.

**Table 1. Specification of turbine and compressor**

| | | 1st stage | 2nd stage | 3rd stage |
|---|---|---|---|---|
| Turbine | Inlet pressure (MPa) | 0.38 | 0.38 | 0.38 |
| | Inlet temperature (K) | 100 | 100 | 100 |
| | Expansion ratio | 2.1 | 2.1 | 2.1 |
| | Flow rate (g/s) | 87 | 105 | 100 |
| | Rotation speed (rpm) | 120,000 | 120,000 | 120,000 |
| Compressor | Inlet pressure (MPa) | 0.12 | 0.38 | 0.56 |
| | Inlet temperature (K) | 80 | 80 | 80 |
| | Outlet pressure (MPa) | 0.18 | 0.56 | 0.8 |
| | Flow rate (g/s) | 50 | 66 | 66 |
| Generator | Absorption power (kW) | 4.3 | 4.8 | 4.3 |

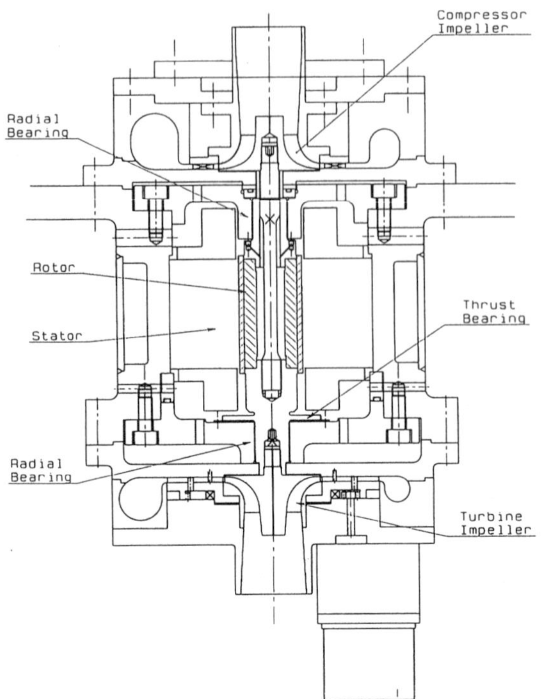


**Figure 7.** Cross-section of cold compressor driven by expansion turbine.

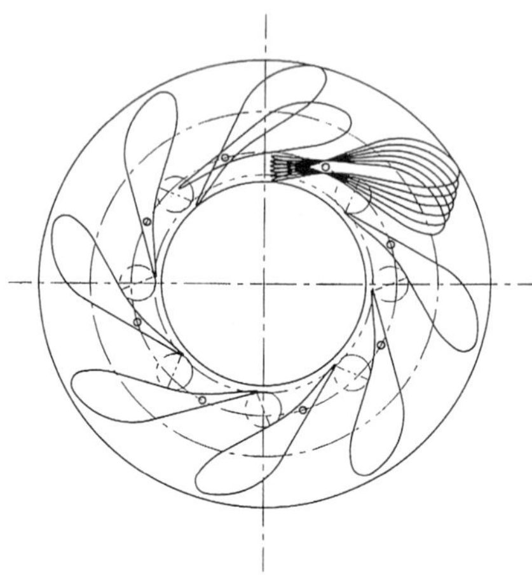

**Figure 8.** Mechanism of a variable geometry turbine nozzle.

Figure 7 shows the cross-sectional view of this machine and Table 1 shows the specification. The bearing must be an oil-free type. Because the rotor becomes compact thanks to an increase of rotation speed, dynamic gas bearings have been employed instead of magnetic bearings. As a control method of liquefaction rate, it is favorable to change the compressor rotation speed. In this system, rotation speed is controlled by changing turbine flow rate. And a variable geometry turbine has been adopted to keep high efficiency over a wide operation range.

## Variable Geometry Turbine

Expansion turbine flow rate has been conventionally controlled by inlet pressure adjustment with a throttle valve placed at the turbine inlet. However, with this method, gas pressure raised by 300K compressor is consumed uselessly and the expansion ratio of the turbine becomes lower. As a result this method has caused a large energy loss at part load.

Meanwhile, the variable geometry turbine which controls flow rate by changing the turbine nozzle throat area can control capacity from 0 to 120% without energy loss. The throat area is changed by changing the installation angle of the nozzle blades. Figure 8 shows a mechanism of the variable nozzles.

## Oil-free Gas Bearing

The oil-free bearing includes a gas bearing and a magnetic bearing. The gas bearing has demerits of low load capacity and weakness against external perturbation, but it has merits of a simple construction and no accessaries.

The dynamic gas bearing adopted for the present machine is a foil type. The radial bearing is very simple, that is a long stainless steel foil on which a bearing pattern is carved, and it is wound round the shaft three times. Since the foil bearing has a large

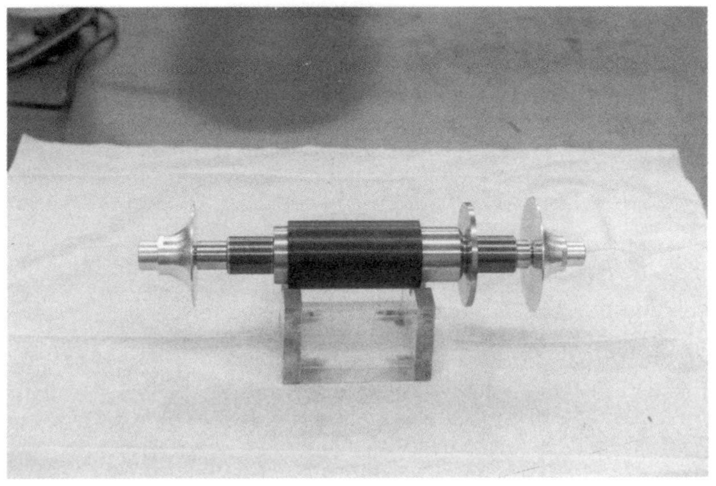

**Figure 9.** Rotor shaft. The central part is a permanent magnet reinforced by CFRP ring.

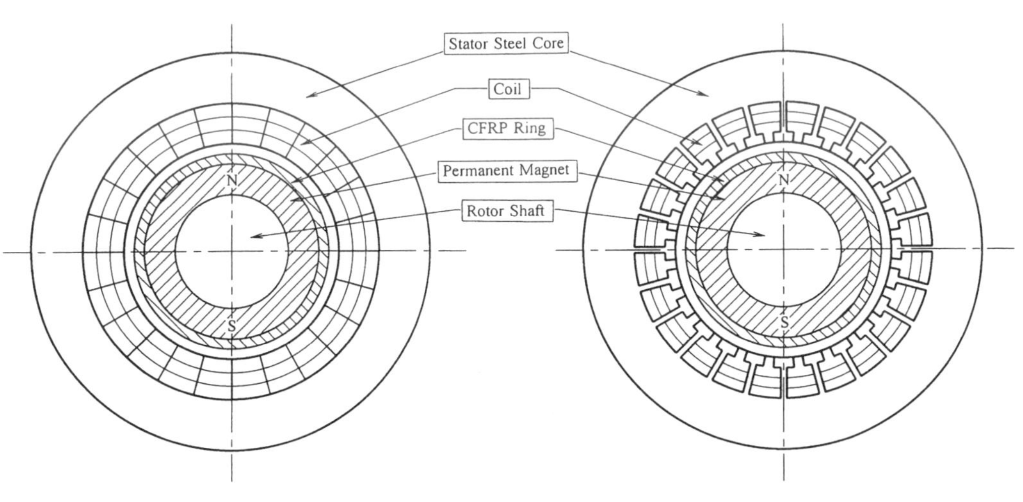


Toothless Stator

Conventional Stator

**Figure 10.** Toothless stator and conventional stator.

compliance with load, it has a certain self-center adjustment and characteristics to be able to resist a certain amount of impulse load. The surface of the shaft is coated with ceramics because the shaft touches the bearing foil at start and stop.

### High Speed Generator

The generator is required of a specification of 10 kW at 120,000 rpm. The highest powered Nd-Fe-B permanent magnet is employed on the rotor side, however its tensile strength is not strong. The CFRP ring is press-fitted on the magnet to make high hoop strength. Figure 9 shows a external view of the rotor shaft. If the center of the permanent magnet rotor is misaligned, a pull force into the stator works on the rotor. Such misalignment is disadvantageous to a gas bearing because of its low load capacity. The smaller the gap between the rotor and the stator steel core is, the larger the pull force becomes. To cope with this problem, the toothless stator shown in Figure 10 has been employed for the present machine.

The conventional generator stator has many slots on the steel core for the coil space. The tooth between the slots collects a magnetic flux from the rotor. Because the gap between the tooth and the rotor is small, the pull force becomes considerably large.

Meanwhile, in the toothless stator the steel core is placed outside the coil block in a cylindrical shape. Because the gap between the steel core and the rotor is large, the pull force becomes small. In this stator the magnetic resistance becomes large and the electromotive force per one turn of coil becomes small. However, because more number of coil turn can be installed due to reduction of the tooth space, generator output does not lower.

## CONCLUSION

The investigation of the refrigeration system for the superconducting generator has been carried, and it was found that the system using helium Brayton cycle instead of LN2 cooling is superior as to both efficiency and space saving. For this system a new conceptional cold compressor has been developed that is driven by an expansion turbine. The turbine is of a variable geometry nozzle type and is expected high efficiency over a wide operation range. The foil type gas bearing has been employed as an oil free bearing to ensure high reliability. For matching of the turbine output and the compressor input, a high-speed and high output generator has been installed on the same shaft. The tests of the compressor unit installing the present machine will be carried out in FY1998.

## ACKNOWLEDGMENT

This research has been carried out as a part of R&D on superconducting technology for electric power apparatuses under the New Sunshine Project of AIST, MITI, being consigned by NEDO.

## REFERENCES

1. H.Asakura, D.Kato, N.Saji, and H.Ohya, 80K Centrifugal Compressor for Helium Refrigeration System, in: "Advances in Cryogenic Engineering,"Vol.37B, Plenum Press, New York (1992), P.787
2. H.Asakura, D.Kato, N.Saji, H.Ohya, M.Kubota, Y.Kaneko, S.Nagai, and R.Toyama, Performance Test Results of 80K Centrifugal Compressor for Helium Refrigerator, in: "Advances in Cryogenic Engineering," Vol.39A, Plenum Press, New York (1994), P.893
3. H.Asakura, N.Saji, Y.Kaneko, S.Yoshinaga, M.Mori, J.Sato, A.Miyake, T.Iwasaki, I.Nishimura, T.Hosoya, and T.Umeda, Performance Analysis of Multistage 80K Centrifugal Compressors for Helium Refrigerator, Proceedings of ICEC16, P.203 (1997)
4. N.Saji, S.Nagai, H.Asakura, Y.Kaneko, Design of Oilfree All Turbo-type Helium Refrigerator, *Cryogenics* 34 ICEC Supplement, P.115 (1994)

# PROGRESS REPORT OF A CRYOMECHANICAL HYBRID VACUUM PUMP PROTOTYPE

J.P. Périn[1], J.J. Cordier[2] and R. Mathes[3]

[1]C.E.A. Grenoble / Département de Recherche Fondamentale sur la Matière Condensée/SBT, 17 rue des Martyrs, 38054 Grenoble Cédex 9, France

[2]C.E.A. Cadarache / Département de Recherche sur la Fusion Contrôlée/STEP
13108 Saint Paul lez Durance, France

[3]ALCATEL CIT, Division Technologie du Vide, 98 avenue de Brogny
BP 2069, 74000 Annecy, France

## ABSTRACT

A novel concept of a mechanical vacuum pump is presented and a prototype is described that is capable of operating at a low temperature of 25 K. Since gas density varies inversely with temperature, this pump delivers much higher mass flow rate at low temperature than at room temperature for a given size. Advantages of this concept are reduction in size and weight, when compared to a conventional pump scaled for the same mass flow rate at room temperature. A temperature gradient between the different pumping stages allows optimization of the exhaust pressure. This pump would be a solution for continuous tritium extraction and minimization of the mass inventory for fusion tokomaks fuel cycle control.

## INTRODUCTION

Next generation tokamaks will require the development of large tritium compatible high vacuum pumps. In large tokamaks, such as the International Thermonuclear Experiment Reactor (ITER), the plasma exhaust containing deuterium, tritium and a fraction of 5% helium must be removed in the burn and dwell regime. The mass flow rate is 2000 mbar-L/s (NPT) at a pressure of $5x10^{-3}$ mbar. To meet this requirement with present technology necessitates the use of cryopumps and turbomolecular pumps operating at room temperature.

This paper provides the technical description of a prototype developed from calculations and experimental results obtained previously[1,2] at low temperatures (80 K, 25 K) with a molecular drag pump (MDP) of 100 mm diameter operating with a few stages of a turbomolecular pump (TMP) running at the same temperatures. This pump would be an engineering solution for ITER which minimizes the scale-up and the tritium inventory. It will be also used for the future Toroidal Pumped Limiter of the next generation Tore Supra, where the particle control needs high pumping speeds.

## LOW TEMPERATURE PUMPING CHARACTERISTICS

The main results obtained from the calculations and experimental setup are the following:

(i) The rule of the flow rate being directly proportional to the gas density is verified within the performed experiments in the $1\text{x}10^{-3}$ mbar to $5\text{x}10^{-1}$ mbar range for TMP and MDP pumps and for different pure light gases like hydrogen, deuterium, helium and mixtures of hydrogen or deuterium with 5% of helium.

(ii) The maximum flow rate and the pressure ratio for MDP pump are higher if the gas is more viscous. Helium gas is more viscous than hydrogen or deuterium, so for mixtures the apparent viscosity is higher than for a pure gas as reported by the Mann law.

(iii) In MDP the compression ratio depends on the temperature and inlet pressure, so an optimum temperature level can be determined. This effect is predicted by the numerical model we have developed. At low inlet pressure ($1\text{x}10^{-3}$ mbar) the pump is working near the molecular flow range where the compression ratio ($\tau$) is in a first approximation inversely proportional to the average velocity of the molecules ($\tau \sim \exp(1/T^{1/2})$). In the viscous range at high pressure ($5\text{x}10^{-1}$ mbar) the pumping performances are determined by the variation of the dynamic viscosity of the gas with temperature ($\tau \sim T^a$, a is a dimensionless coefficient and its value ranges from 0.5 to 0.7). So the temperature can be used as an optimization parameter to obtain an optimum for the outlet pressure.

(iv) For TMP, working at low temperature is particularly useful because the mass flow rate and the pressure ratio are both increasing[3,4].

At low temperature the TMP is working in the transition regime, so the thermal losses due to friction are higher than at room temperature for the same working pressure. Indeed they are inversely proportional to the Knudsen number (ratio of the mean free path to the clearance between stages). They have to be taken into account in the design of the rotor blades. This point is very important for the cooling system capacity sizing.

Mechanical pumping of predensified gases at low temperature permits extraction of the light gas mixture at a very high flow rate while at a convenient pressure.

## PROTOTYPE DESCRIPTION

### Pumping System

To reach the desired requirement (mass flow rate 7 mbar-L/s NPT, inlet pressure $1\text{x}10^{-3}$ mbar, outlet pressure 1 mbar), a combination of a few turbomolecular stages and one molecular stage is needed. This vacuum hybrid pump (TMP and MDP association) increases the outlet pressure and provides a pressure ratio of $10^3$ (50 for TMP and 20 for MDP) (Fig. 1). With an inlet pressure of 1 mbar, dry fore vacuum pumps (roots or scroll pump) can be used to compress the gas to atmospheric pressure.

The prototype is designed from an industrial size pump supplied by Alcatel (290 mm diameter) and different pumping stages are optimized with computer codes that were developed and adjusted with previous experimental results. It includes 4 TMP stages : 3 compression stages (30° and 22° angle blade) and one impeller stage (45° angle blade). The axial clearance between the static and moving blades is 4 mm while the radial clearance for the same pieces is 1.5 mm. The axial gap has been determined to reduce thermal losses (evaluated at 10W) due to the friction torque.

The MDP stage is a radial disc pump type to avoid the problems associated with thermal dilatation, mechanical stresses and the increase in pumping speed. The channels are cut on the stator as a logarithmic spiral. The pumping channel in this design follows more closely the direction of the pumped gas. The geometric parameters of the MDP stage are summarized below :

Number of channels = 4 22
Inlet section of a channel = 108 mm$^2$
Developed length = 168 mm

Rotor diameter = 180 mm to 120 mm
Outlet section of a channel = 5 mm$^2$
Gap rotor /stator = 0.2 mm

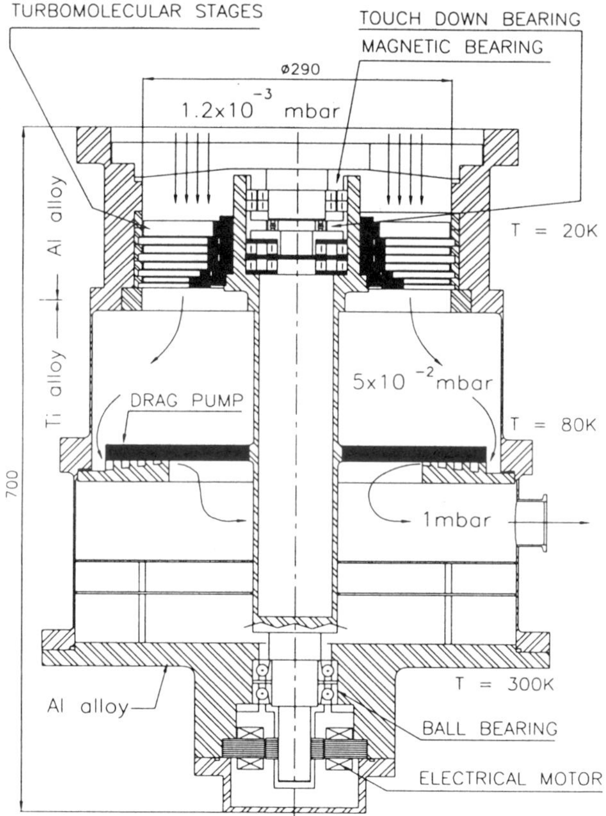


**Figure 1.** Schematic layout of the cryomechanical pump (dimensions in mm).

## Mechanical Structure

The rotating speed has been chosen as 21000 rpm to achieve the nominal pumping speed and to limit the mechanical stresses due to the centrifugal forces.

This pump is driven by an electrical motor provided with ball bearing at 300 K and a cold passive magnetic bearing at 25 K. This bearing selection provides a contactless bearing essential for running at low temperature and for a long life operation.

The two first natural modes of the rotating shaft are, a translation (25 Hz) and a rotation (precession mode at 85 Hz) of the spin axis. Operation in the super critical mode (350 Hz) has required the development of a 25 K dumping system since eddy currents induced by permanent magnets in a copper disc will provide a dumping ratio of about 0.02.

This system induces severe conditions for mass balancing. A 10 g-mm unbalance gives rise to a 0.15 mm shaft radial motion at critical frequencies. For emergency conditions, a touch down ball bearing is mounted near the magnetic bearing with a 0.25 mm radius clearance.

The material used for this prototype are essentially titanium and aluminum alloys. To prevent differential dilatation between the shaft and the pump body, both are manufactured of titanium alloy. This reduces the axial thermal flux from room temperature to the pumping stages. For the MDP stage or the TMP stage, aluminum alloys have been used for the moving and static parts; thus the radial thermal resistance is low and the pumping stages can be assumed isothermal.

## Cooling System

The cooling system is characterized by the two features. Liquid nitrogen is used for cooling the MDP stage and the thermal shields at 80 K. A closed thermosiphon hydrogen loop is used for the TMP cooling. The cold source is the second stage of a Gifford-Mc Mahon (GM) refrigerator fitted with a condenser. The condenser temperature is monitored at 15 K ( $H_2$ triple point is 13.8 K). The heat transfer is obtained by evaporation of liquid hydrogen in the evaporator which is attached to the TMP body (Fig. 2). This technique reduces the thermal gradient to 1.5 K between the cold source and the TMP. This gradient is due to the nucleate boiling.

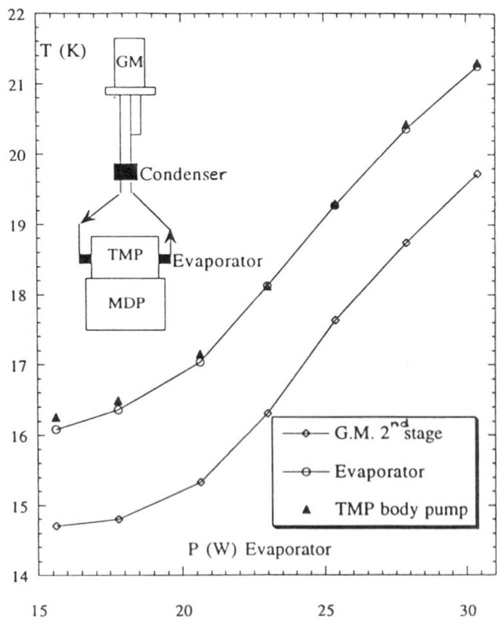


**Figure 2.** Temperatures of TMP versus heat power.

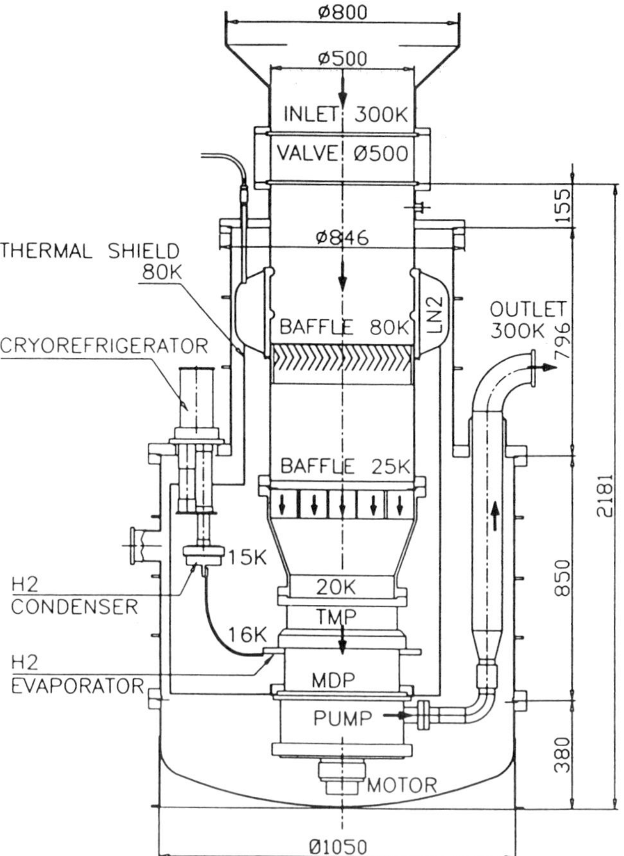


**Figure 3.** Cryomechanical pump located within the cryostat (dimensions in mm).

### Instrumentation and Process

This full scale prototype pump is instrumented with 6 displacement sensors (Eddy current probes), 5 absolute pressure gauges and 12 temperature carbon probes placed on the different stages of the pump. Different emergency procedures help us to prevent anomalous vibrations, leaks, or cooling system failure.

## PROTOTYPE TESTS

The prototype will be tested on a specific test bed which is a full scale model of the Tore Supra vacuum vessel. The pumped gas is first cooled down through the inlet 500 mm cylindrical pipe (300 K at its upper part), provided with two successive baffled heat exchangers (Fig. 3). The first one is cooled to 80 K while the second one cooled to 25 K is close to the entrance of the pump.

Various adverse operating conditions of the Tore Supra such as large air or water leaks will be simulated on the test bed. Protection for minimizing deviations in pump behavior are now under development including addition of a fast shut off valve.

The tests will be divided into three phases :

(i) The first one concerns only the pump performance (mass flow rate versus pressure and temperature), its mechanical characteristics (modal analysis, centrifugal stresses...).

(ii) In the second part on the test bed, the pump will be tested in the Toroidal Pumped Limiter configuration which is designed to pump neutralized gas as $D_2$, He , Ne, through the bottom triangular ducts of the Tore Supra vacuum vessel observation ports (Fig. 4). The expected gas flow rate pumped through the limiter throat is of 40 mbar-L/s (NPT). This gas flow is equally distributed through the ten ports, leading to a 4 mbar-L/s flow rate on each port. The expected pressure in the neutralizer part of the limiter is $2x10^{-3}$ mbar with a gas temperature of 500 K. The pressure to the right of the insulating slide valve is $1 \times 10^{-3}$mbar at 300 K.

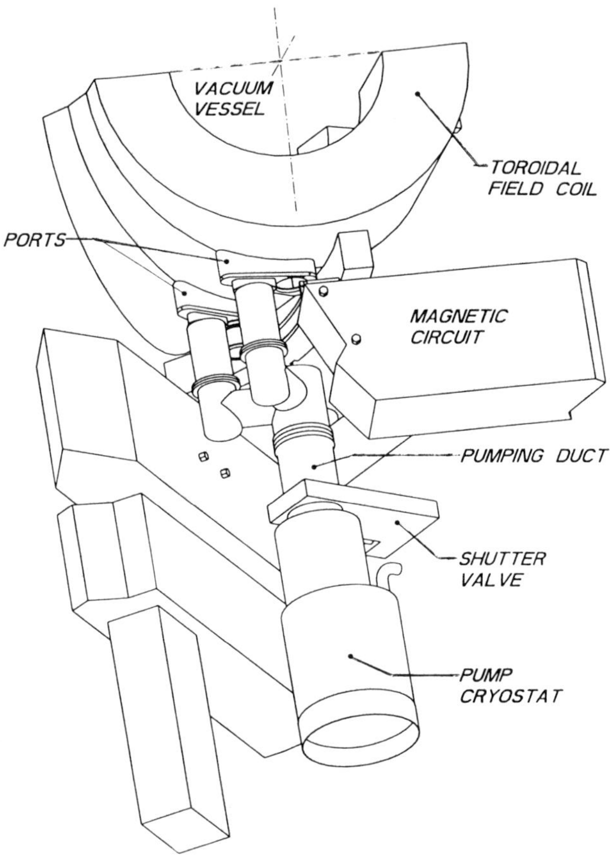


**Figure 4.** Pump location under Tore Supra.

The heat exchange between the static and moving parts of the pump is provided by the pumped gas. In order to keep the TMP rotor at a stationary 25 K operation temperature, a minimum quantity of gas has to be maintained in the pump. This effect resulted in a $5x10^{-4}$ mbar minimum inlet pressure. However, a servo-controlled valve located on the exhaust gas pipe at room temperature provides this minimum pressure when the insulating valve is shut and limits the ultimate pressure in the pump.

(iii) The pump (without mechanical pumping duct connection to the torus) will be placed in its expected final location under the machine. In this way, influence of vibrations, high frequency waves parasitic signals induced by ICRH antennas or magnetic field will be checked during plasma shots and disruptions. The pump placement in the Tore Supra environment is scheduled for the end of 1997.

## FIRST RESULTS

The pump already has been run at its nominal speed (350 Hz) with a measured deviation of the shaft is less than 30 microns peak to peak. At critical frequencies the shaft deviation is less than 140 microns peak to peak. At low temperature the copper disc resistance used in the dumping system is lower and this effect increases the eddy currents and improves the dumping factor from 0.06 at room temperature to 0.16 at 25 K.

The measured cooling power of the cryorefrigerator to cool the TMP stage at 20 K is 20 watts (corresponding to an electrical power input of 11 kW for the cycle compressor).

The equilibrium temperature of the MDP stage is 105 K in accordance with calculations of thermal losses due to friction torque (30 W).

The pumping speed for $1x10^{-3}$ mbar inlet pressure above the 80 K baffle is 9.3 $m^3/s$ (closed to the expected 10 $m^3/s$) for a gas temperature evaluated at 25 K at the TMP inlet. The corresponding mass flow rate is 3.7 mbar-L/s. In these conditions, the pressure ratio is 1000 (20 for TMP and 50 for MDP) corresponding to an exhaust pressure of 0.4 mbar.

The pressure drop due to the two baffles is about $6x10^{-4}$ mbar for an inlet pressure of $1x10^{-3}$ mbar. This pressure drop is mainly due to the 80 K baffle.

This baffle could be optimized or even suppressed. In this case, a mass flow rate of 10 mbar-L/s can be reached at $1x10^{-3}$ mbar inlet pressure without a large enhancement of the cryogenic load (40 W at 20 K) .

An ultimate pressure of $10^{-7}$ mbar as been achieved.

These experiments will be completed in the future to measure the equilibrium temperature of the shaft when no heat exchange occurs between the moving and static parts (ultimate vacuum regime). During these tests we shall measure the power dissipated in the turbomolecular stages due to the gas friction. These measurements will allow optimization of the cooling system before transferring the prototype to Tore Supra.

## DISCUSSION

The last key issue to be demonstrated is the capability to operate the running parts at high rotating speed regardless of the stray magnetic field influence around a tokamak machine [5]. Some calculations have been performed for the turbomolecular stages, where the cooling is the most difficult under different conditions for magnetic field :time transient (1T/s) or spatial gradient ($10^{-2}$ T/m). Both are perpendicular to the TMP blades. The results show that the time transient does not penalize the pump operation because the energy dissipated is in the range of 1 to 35 J for different scenarios of plasma disruptions. This power can be dissipated easily and the temperature rise is very low. The main problem is the spatial gradient, a dBz/dR of $10^{-2}$ T/m is equivalent to a permanent time variable magnetic field of 2 T/s. Under this condition a 10W continuous power has to be evacuated in the TMP stages. Some thought has been given to the possibility of locating or orienting the pump in the tokamak vicinity to obtain a lower spatial gradient or to reduce this problem by compensation coils.

## ACKNOWLEDGMENT

This work is been done for the French nuclear fusion program with the technical support of Alcatel for calculations and preliminary tests of the pump.

## CONCLUSION

The hybrid cryogenic vacuum pump has attained the nominal flow rate (7 mbar-L/s NPT at $1x10^{-3}$ mbar inlet pressure) with a pressure ratio of 1000. The reduced size of this technique seems very attractive for pumping a high flux of light gases like deuterium, tritium and helium.

After completing the prototype tests, the fabrication of 5 standard cryomechanical pumps for the Tore Supra Toroidal Pumped Limiter Project will be initiated. These pumps are scheduled to be operational by the end of 1999.

## REFERENCES

1. J.P. Perin, G. Claudet and F. Disdier, 18th SOFT, Karlsruhe (Germany) Aug. 22-26 1994.
2. J.P. Perin and R. Mathes, 13th International Vacuum Congress, Yokohama (Japan) Sept. 25-29 1995.
3. C.H. Kruger and A.H. Shapiro, 'The Axial-Flow Compressor in the Free-Molecule Range', Rarefied Gas Dynamics, Talbot, pp 141-160, New York, 1961.
4. L. Maurice, Jpn J. Appl. Phys.(suppl.), 2,21 1974,.
5. W. Becker et al. J. Vac. Sci. Technol., 15(2), 1978.

# CONDUCTION COOLED HIGH-TEMPERATURE SUPERCONDUCTING MAGNETIC BEARING MODULE

H. J. Bornemann, W. Fieger, A. Weindl, A.W. Kaiser, and R. Koch

Forschungszentrum Karlsruhe GmbH
INFP, P.O. Box 3640, 76021 Karlsruhe, Germany

## ABSTRACT

We have designed, built and tested a prototype of a conduction cooled HTSC bearing module. A superconducting ring, measuring ø 102 mm x 13 mm high was assembled from HTSC bulk material. Total mass is 547 g. Lifting force and stability at 77 K and 2 mm gap are approximately 150 N and 80 N/mm, respectively. The superconductor is integrated on a metallic cold plate. Modelling has been used for evaluation of thermal losses and optimization of the cryogenic design. For the present module, total losses at 0.01 Pa insulation vacuum are as small as 1 W, mainly due to radiation. Based on the thermal model the scaling behavior with respect to thermal losses as a function of maximum lifting capability was evaluated. The cooling power was found to be equivalent to about 400 mW electrical power per kg of levitated mass which compares quite favorably to the typical 0.1 ...1 W per kg for conventional active magnetic bearings. Details of both experimental work and modelling will be reported.

## INTRODUCTION

Superconducting magnetic bearings (SMBs) based on the high temperature superconductor (HTSC) $YBa_2Cu_3O_7$ (YBCO) can be used for frictionless levitation of rotating bodies. Applications for such bearings are seen for mechanical components in high-speed rotating machinery, in cryogenic systems, gyroscopes, transportation systems and vibration isolation. While SMBs are self-stabilizing and can be operated without any active control, the superconductor part has to be cooled below its critical temperature to be operational. Sufficient levitation properties are achieved at 77 K, so wet-cooling using liquid nitrogen has been a good solution for laboratory experiments in the past. In industrial systems, conduction cooling would be preferable because it would avoid hazardous risks such as cryo-leakage and convection thermal leaks.

In the past, a major concern has been the cooling power required to keep the bearing operational. Especially for applications such as advanced flywheels for energy storage, it was feared that excessive power consumption by the cryocooler would degrade the overall efficiency of the system. In this paper, we will present experimental data on the operation of a SMB and we will show, that with a carefully designed thermal system, the power consumption to operate such a bearing at the required low temperature is on acceptable levels.

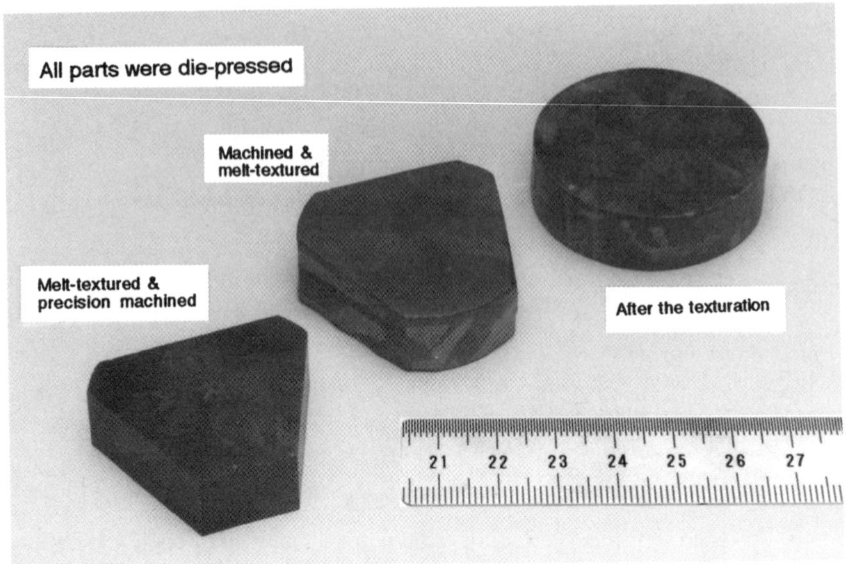


**Figure 1.** Fabrication of YBCO ring segments for assembly of a bulk superconducting ring.

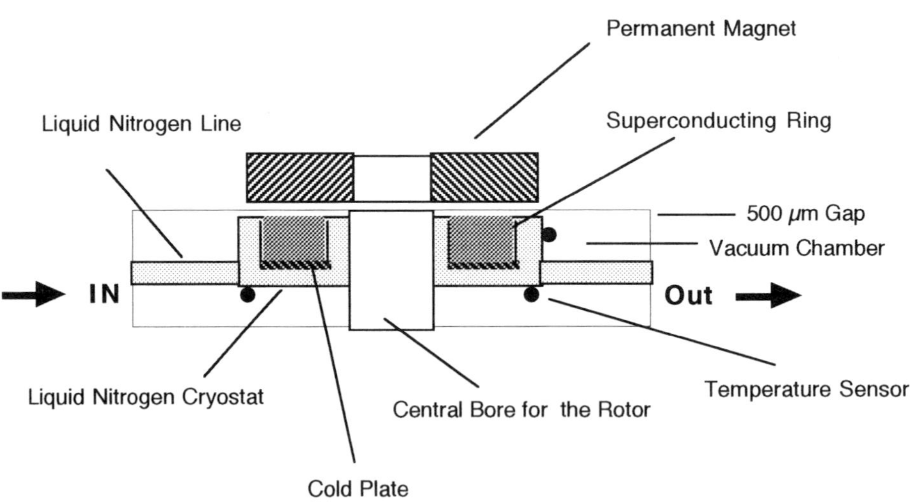


**Figure 2.** Schematic of the bearing module. For better thermal insulation, the vacuum chamber was typically operated at pressures in the $10^{-2}$ Pa range.

## EXPERIMENTAL

### Superconducting Bulk Materials

HTSC bulk parts are produced in a powder metallurgical process based on the melt-texturation process[1,2]. In this process YBCO powder is first mixed with stabilizing oxides and milled, pressed and compacted by cold isostatic pressing. A furnace treatment follows, including melting and resolidification in a specific mode. The desired form and surface finish is achieved with conventional methods such as cutting, milling and polishing. On a regular basis we are batch processing die-pressed bulk parts in the form of flat solid cylinders up to 70 mm in diameter, or as blocks up to 60 mm x 30 mm in size. Rods with a 10 mm x 10 mm cross-sectional area can be manufactured up to 100 mm in length. Components of more complex geometries can be manufactured as well. Depending on overall size and geometry, these parts are either die pressed, form pressed, cut from bulk blocks, or assembled from several pieces[2].

### Superconducting Bearing Module

The superconductor bulk ring used in the bearing measures OD 102 x ID 45 x 13 mm, total mass is 547 g. It was assembled from eight standard size ø 43 mm pellets, precision machined into segments, as illustrated in Fig. 1. Prior to assembly, the homogeneity of the material was checked by standard quality control measures, such as levitation force measurements and flux mapping. All eight pellets were single domain and of high quality. The variation in the maximum levitation forces of the eight pellets was measured to be constant within 2% of the average value.

A schematic of the superconducting bearing module is shown in Fig. 2. The superconducting ring is integrated onto the upper side of the cold plate of the cryostat. In contrast to earlier developments, where the superconductor is immersed in liquid nitrogen, cooling is provided by conduction only. The bottom side of the cold plate is in contact with the liquid which is forced through the cryostat from the storage vessel with slight overpressure.

To minimize losses and to prevent icing of the superconductor, the cryostat is placed in a vacuum chamber. A screw type mechanism allows precise positioning of the cryostat within the vacuum chamber, even during operation. For optimization of the lifting capabilities, this is extremely important since lifting capabilities of the bearing depend strongly upon both the distance between superconductor and magnet (see below) and the temperature of the superconductor. An increase in temperature of only 5 K above 77 K would result in a 50% reduced levitation force[3]. So the tricky task that had to be solved here was close placement of the cryostat below the cover of the vacuum chamber without thermal contact. Four sensors, placed near the inlet, outlet, bottom and sidewall are used to monitor the temperature. A cool down sequence is shown in Fig. 3. For clarity, only data of two sensors are presented. While the bottom part of the cryostat is cold after three minutes it takes about 10 min altogether until all four sensors indicate temperatures within the typical 79 K to 81 K range. The amount of $LN_2$ required to cool the module can be estimated according to Eq.(1):

$$\Delta E = m_{sc}\int_{80K}^{293K} c_{sc}(T)dT + m_{cr}\int_{80K}^{293K} c_{cr}(T)dT \tag{1}$$

Here, $\Delta E$ is the required energy to cool the module from 293 K to 80 K, $m_{sc}$, $c_{sc}$, $m_{cr}$ and $c_{cr}$ are mass and specific heat of the superconductor and cryostat, respectively. The specific heat of the superconductor is with values in the 300 ... 100 $^{J}/_{(mol\ K)}$ range considerably higher compared to the cryostat material. For the present design $m_{sc} \cong m_{cr}$, so most of the energy is used to cool down the superconductor. We find $\Delta E$ = 240 kJ, which is equivalent to an average cooling power of 400 W. Assuming complete vaporization of all $LN_2$ passing through the cryostat, the $\Delta E$ of 240 kJ equals 1.5 L of $LN_2$ in good agreement with our experimental observation.

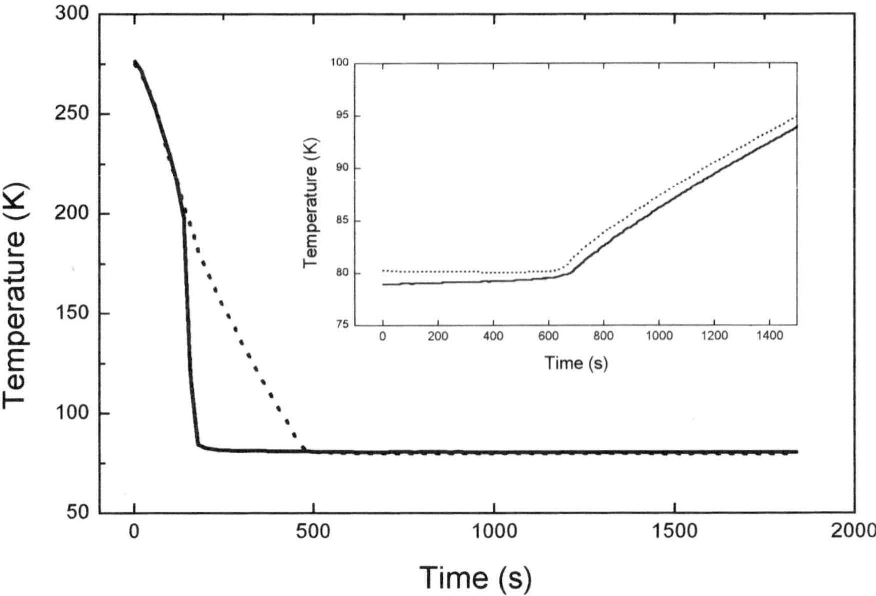


**Figure 3.** Typical cool down sequence. Data for two sensors are shown. Continuous line: Sensor at the bottom near the inlet. Dotted line: Sensor on the side. After 10 min, the superconductor is cold and the bearing can be operated. The insert shows temperature vs time after the liquid nitrogen supply tank was disconnected. Obviously, there is enough time for a controlled shut down of the component in case of failure of the cooling system.

**Figure 4.** Components of the bearing module prior to assembly: Cryostat with integrated cold plate shown with segmented superconductor ring made of YBCO bulk material and permanent magnet ring.

As shown in the insert of Fig. 3, after the supply tank was disconnected, the bearing stays at 80 K for about 10 min and then gradually warms up. This is an important feature with respect to the required safe operating conditions of such a device after failure of the cryocooler. In the present case, the bearing can be continued to operate for a considerable length of time before levitation is lost, during which measures can be taken for a controlled shut down of the system. Fig. 4 shows a photograph of the cryostat. In the next step of the development, cryogen free operation of the bearing module will be achieved by directly connecting the cold plate to the cold head of a closed cycle cryocooler.

## Levitation forces and stability

Levitation forces and stiffness under static load levels were analyzed using a $2^1/_2$ x-y-z motion stage with a three dimensional force sensor unit[4,5]. For these experiments, the vacuum chamber with cryostat was integrated onto the motion stage, while a flat, cylindrical can, holding high strength permanent magnet material, was connected to the force sensors. The minimum distance between top surface of the superconductor and bottom of the magnets was 2 mm, corresponding to zero visible gap. Depending on initial cooling conditions, i.e. superconductor is cooled in the field of the magnet (so called field cooled mode, FC) or outside the field (zero field cooled mode, ZFC), and the relative motion between magnet and cold superconductor, the interaction can be attractive or repulsive. Fig. 5 shows typical bearing characteristics. In Fig. 5 (a), the superconductor was centered below the magnets and cooled at a vertical distance of 5 mm. Then the distance was gradually reduced and a repulsive interaction force was measured, increasing from zero to 150 N, with a maximum stiffness of $8 \times 10^4$ N/m. In Fig. 5 (b), the superconductor was cooled at the minimum gap of 2 mm. It was then vertically moved away from the magnet, resulting in a strongly increasing attractive force with an initial slope of $8 \times 10^4$ N/m. In both cases, the maximum radial stiffness of $2.5 \times 10^4$ N/m is reached at minimum gap.

## Modelling of the thermal losses

For evaluation of the required power to keep the bearing cold, the thermal losses were modelled. In general terms, the overall heat transfer through the area A is a function of the temperature difference ($T_1$-$T_2$) and the heat transition coefficient k and can be written as[6-8]:

$$P_{loss} = kA(T_1 - T_2) \tag{2}$$

In our case, the total heat transfer from the liquid nitrogen bath to the outer surface of the vacuum chamber is a combination of individual processes:

$$P_{loss} = \left(\sum_{i=1}^{n} \frac{b_i}{\lambda_i A_i} + \sum_{j=1}^{m} \frac{1}{\alpha_j A_j}\right)^{-1} (T_{sur} - T_{LN}) + P_{rad} \tag{3}$$

Here, the first term denotes heat conduction (e.g. through the wall of the cryostat or through the superconductor), $b_i$ is the length of the path through the material with thermal conductivity $\lambda_i$ and area $A_i$. The second term describes heat convection (e.g. from the liquid nitrogen bath to the inner surface $A_j$ of the cryostat) with heat transition coefficient $\alpha_j$. The third term gives the heat transfer $P_{rad}$ due to radiation, which is a function of ($T_1^4$-$T_2^4$). Values for $\lambda_i$ and methods to calculate $P_{rad}$ and the $\alpha_j$ are given in the literature[6-9]. It was assumed that the system is in a stationary state, with the liquid nitrogen bath at a constant $T_{LN}$=77 K and the temperature $T_{sur}$ of the outer surface of the vacuum chamber fixed at 293 K. Only free convection was considered to be relevant, flow of liquid nitrogen was neglected. For the analysis, the total heat transfer from the liquid nitrogen bath to the outer surface of the vacuum chamber was divided into three components: (i) heat transfer in radial direction, (ii) via the top and (iii) via the bottom. Table 1 summarizes the results for three

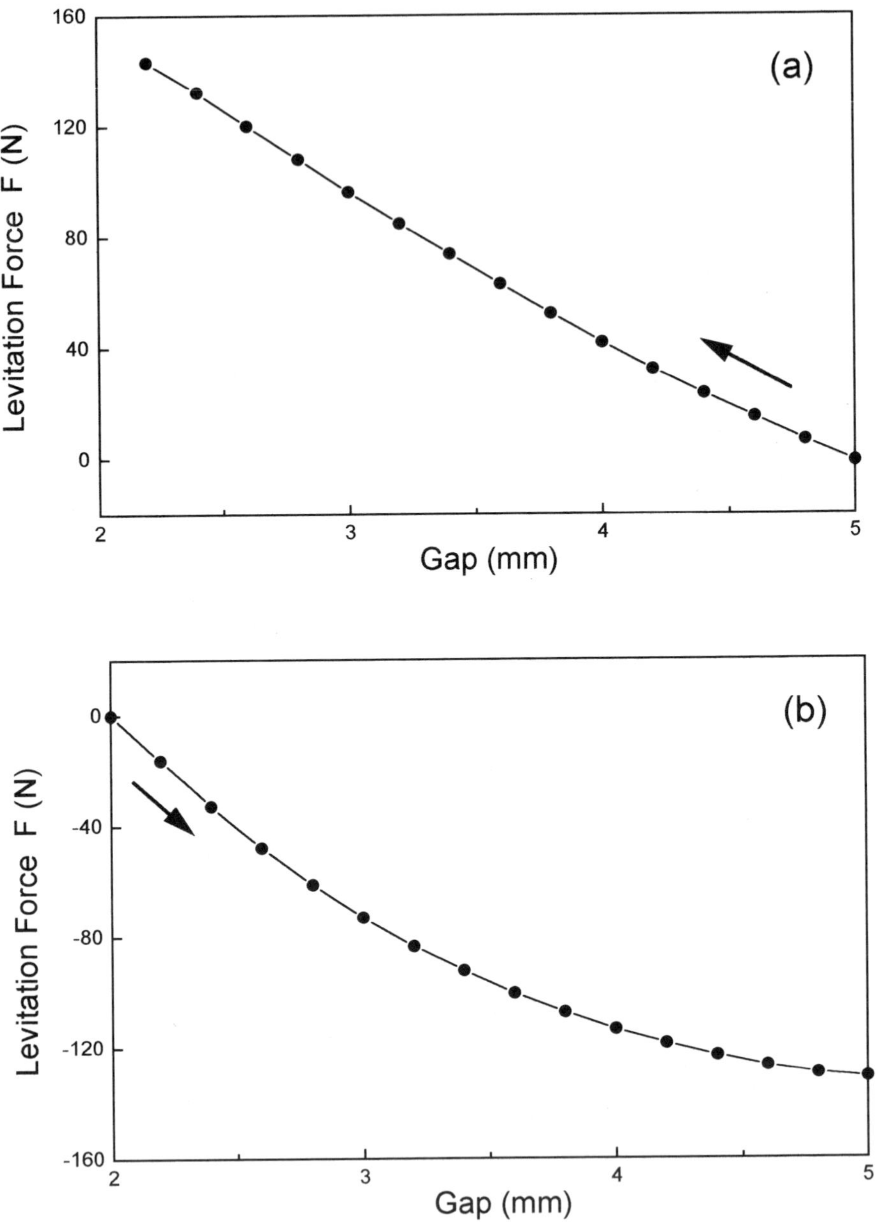


**Figure 5.** Levitation force vs gap between superconductor and permanent magnet under field cooled conditions: (a) initial gap was 5 mm, (b) initial gap was 2 mm.

**Table 1.** Calculated values for the heat transfer

| Insulation Vacuum | 0.01 Pa | 0.1 Pa | 1 Pa |
|---|---|---|---|
| Conduction + Convection | | | |
| Radial direction | 0.07 W | 0.69 W | 4.60 W |
| Via the top | 0.04 W | 0.47 W | 4.13 W |
| Via the bottom | 0.05 W | 0.43 W | 3.71 W |
| Total conduction + convection | 0.16 W | 1.59 W | 12.44 W |
| Total radiation | 0.85 W | 0.85W | 0.85 W |
| Total heat transfer | 1.01 W | 2.38 W | 13.29 W |

different pressures of the isolation vacuum. Obviously, there is not much difference between the three directions. At 1 Pa, conduction and convection are with about 12 W one order of magnitude higher compared to radiation. At 0.01 Pa, the total heat transfer is only 1 W, mainly due to radiation. A further reduction of about 50% can be expected when all cold surfaces are polished.

For practical use of the bearing, the scaling behavior with respect to thermal losses as a function of maximum weight of the levitated rotor is of particular interest. With a standard thickness for the superconductor of 1.5 cm, we typically measure lifting ratios of 1:20, mass of superconductor over levitated mass[1] (practical configuration with 3 mm minimum gap between superconductor and magnet, field cooled mode). So the lifting capability of the bearing can be scaled up by simply increasing the mass of the superconductor. Size of the cryostat and vacuum chamber were increased accordingly, while all distances between cryostat and superconductor were kept constant. The result is shown in Fig. 5 for 0.01 Pa, 0.1 Pa and 1 Pa insulation vacuum. For the lowest pressure of 0.01 Pa, the experimental

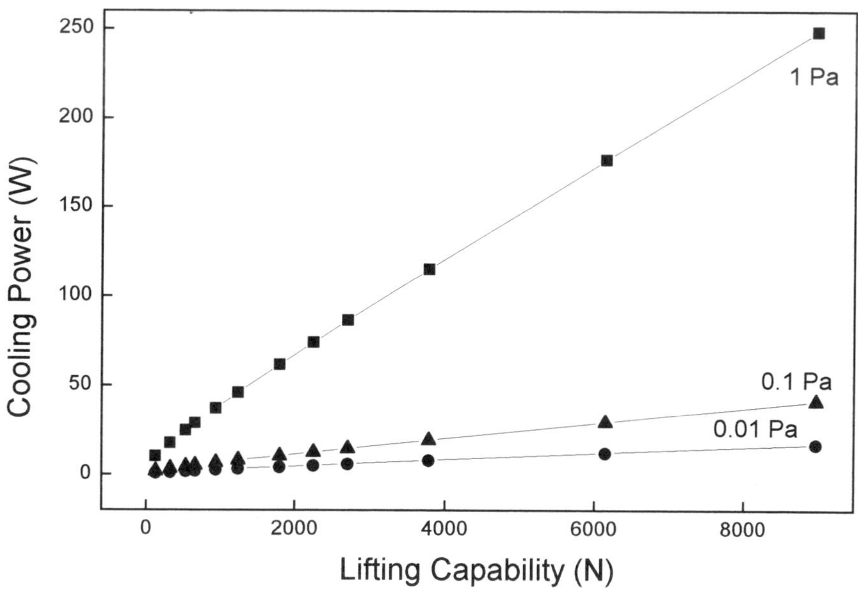


**Figure 6**. Scaling behavior of a superconducting magnetic bearing: Cooling power as a function of maximum weight of the levitated rotor and quality of the insulation vacuum.

data can be approximated by a straight line with a slope of only $2x10^{-3}$ W/N, i.e. per N of levitation force, only 2 mW of cooling power are required. Assuming an efficiency of 5% for the cryocooler, this translates into a consumption of electrical power of 0.04 W per N. For comparison, active magnetic bearings consume typically around 0.01 W ...0.1 W electrical power per N for lifting force and control, depending upon the specific design and number of controlled degrees of freedom.

As for the superconducting bearing being used in combination with a fast spinning rotor, what we have neglected so far is heat generated in the superconductor by magnetic friction. Magnetic friction results from flux motion in the superconductor due variations in the average magnetic field of the rotating permanent magnet[8] and is an additional heat load for the cooling system. In an earlier experiment with a superconducting bearing in a flywheel prototype[9], we measured about 50 mW per 10,000 rpm of heat generated in the superconductor due to magnetic friction. The maximum variation of the average field of the magnet used in that particular set-up was about 2% with respect to the maximum field.

## SUMMARY AND CONCLUSIONS

We have designed, built and tested a prototype of a conduction cooled HTSC bearing module. A superconducting ring, measuring ø 102 mm was assembled from HTSC bulk material. Total mass is 547 g. Lifting force and stability at 77 K and a working gap of 2 mm are approximately 150 N and 80 N/mm, respectively. The superconductor is integrated on a metallic cold plate. Modelling has been used for evaluation of thermal losses and optimization of the cryogenic design. The losses depend strongly upon the quality of the insulating vacuum. At 0.01 Pa, total losses were calculated to be around 1 W in good agreement with the experiment. Based on the thermal model, the scaling behavior with respect to thermal losses as a function of maximum lifting capability was investigated. The cooling power was found to be equivalent to about 0.4 W of electrical power per kg of levitated mass which falls right within the typical 0.1 ... 1 W per kg range of active magnetic bearings. Even operation at 0.1 Pa with about twice the power consumption might be acceptable in some cases. Here, the definite advantage would be a somewhat simpler vacuum system, because a conventional rotary pump would be sufficient to create and maintain the vacuum[12]. At any rate, our analysis has shown that even with the poor efficiency of today's available cryocoolers, the power consumption to operate a superconducting bearing at the required low temperature is acceptable.

## REFERENCES


1. H. J. Bornemann, Th. Burghardt, W. Hennig, and A. Kaiser, Processing technique for fabrication of advanced YBCO bulk materials for industrial applications, *IEEE Transact. Appl. Supercond.* 7:1805 (1997).
2. FZK Internet Online Product Information: www.infp.fzk.de/wheel/english/home.htm
3. S. Gauss and S. Elschner, Levitation and magnetic properties of melt processed high temperature superconductor cylinders, in "Proceed. 4th Int. Symp. Magn. Bearings," G. Schweitzer, R. Siegwart, and R. Herzog, eds, Zürich, Switzerland, (1994), p.435.
4. P. Boegler, C. Urban, H. Rietschel, and H. J. Bornemann, Standardized measurements of interaction forces in autostable superconducting magnetic bearings, *Appl. Supercond.* 2: 915 (1994).
5. FZK Test Bench Model TB-STML-ZM04
6. "VDI Wärmeatlas,"Verein Deutscher Ingenieure, Düsseldorf, Germany (1991)
7. "Handbuch Kryotechnik BW 41-07-03," VDI Bildungswerk, Düsseldorf, Germany (1975).
8. K. Gieck. "Technische Formelsammlung," Gieck Verlag, Heilbronn, Germany (1989).
9. C. Uher, Thermal Conductivity of High-$T_c$ Superconductors, *Journ. of Supercond.* 3:337 (1990)
10. V. V. Nemoshalenko, E. H. Brandt, A.A. Kordyuk, and B.G. Nikitin, Dynamics of a permanent magnet levitated above a high-Tc superconductor, *Physica C* 170:481 (1990).
11. H. J. Bornemann, T. Ritter, C. Urban, O. Zaitsev, K. Weber, and H. Rietschel, Low friction in a flywheel with superconducting magnetic bearings, *Appl. Supercond.* 2:439 (1995).
12. A. Chambers, R. K. Fitch, and B. S. Halliday. "Basic Vacuum Technology,"Adam Hilger, Bristol and New York (1989).

# COOL-DOWN DYNAMICS OF A CRYOSTAT WITH A CLOSED-CYCLE CRYOGENIC REFRIGERATOR

B.X. Zhang

Rockwell Automation
Cleveland, Ohio 44117, USA

## ABSTRACT

A cryostat designed for a large-scale superconducting motor application is to be cooled by a reversed Brayton cycle cryogenic refrigerator. A set of four high-temperature superconducting (HTS) field winding coils reside in the cryostat. A nominal operating temperature of 33 K for these HTS coils of the synchronous motor is maintained by the refrigerator through circulating cold helium gas through a pair of cooling channels in the cryostat. As part of the rotor of the HTS motor, the cryostat will be rotated at 1,800 rpm under normal operations. To reduce the thermal stress developed in the HTS coils and their supporting structure, the cool-down of the refrigerator and cryostat takes place simultaneously after being coupled by a pair of bayonets. Substantial variation of helium properties through the cool-down temperature range attributes to the dynamic behavior in hydrodynamics and heat transfer of helium gas in the cooling circuits. A numerical model is developed to simulate and characterize the cool-down process. Analytical results define optimal engine speeds at various temperature levels for maximum cooling capacity, which are compared to the design constraints of thermal stress level requirements. Potential hydrodynamic instability issues associated with the parallel cooling channels, typically during the cool-down, are also investigated and will be discussed in details.

## INTRODUCTION

The development of high-temperature superconducting (HTS) motors has been rapidly moving toward the design and testing of prototypes that possess the features required for commercially viable products. The current effort focuses on an air-core synchronous motor with HTS field winding rated at 1,000 horsepower. The nominal operating temperature of the HTS field winding coils has been set at 33 K. The cold space of the rotor cryostat will be cooled and maintained at the operating temperature by a closed-cycle cryogenic refrigerator. While an undersized refrigerator will certainly lead to impractically long cool-down periods, an oversized refrigerator will increase the overall cost of the HTS motor system and therefore may jeopardize the economical viability of the system. Results of this work contribute to the definition of design requirements for the ultimate refrigeration system for future commercial HTS motor products.

### System Features

The structure of the rotor cryostat resembles the one previously developed for a 125 horsepower HTS motor which was finally demonstrated at 200 horsepower with a few exceptions [1,2,3]. A vacuum and multi-layer insulation (MLI) system surrounds the rotor cold space and minimizes heat leak through thermal radiation. In order to reduce structural complexity, a secondary thermal shield in the vacuum space of the rotor cryostat has been excluded. A conductive cooling scheme is utilized to keep the HTS coils at their operating temperature. The coils are attached to a highly conductive support structure which is cooled by a pair of parallel cooling circuits. Instead of venting helium gas into the atmosphere with an open-cycle cryogenic refrigeration supply system, a commercial cryogenic refrigerator with a refrigeration capacity of 145 W on a 20 K return will be used to supply the required refrigeration to the rotor cryostat through a rotating transfer coupling.

### Cooling Requirements

Based on a preliminary design, the refrigeration load during motor normal operation is estimated at 60 W which includes all parasitic and dissipative heat leaks. For the given operating temperature and load, a custom-designed cryogenic refrigerator based on current technology will require approximately 100 W of compressor power to lift 1 W of heat leak at reasonable costs. It is imperative to minimize the heat leak into the rotor cold space, which has been a major cryostat design objective. In addition, it is also crucial to have a good understanding on the system cool-down dynamics that impose design constraints on the rotor cryostat structure and increase load requirement for the refrigeration system. The total heat capacity of the rotor cold space to be removed during the cool-down is estimated at $3 \times 10^7$ J. The refrigerator must be sufficiently sized to complete the cool-down within acceptable period of time.

As a typical issue associated with the cool-down of cryostats for rotating machinery, non-uniform thermal movement in the rotor cold space may cause local plastic deformation in rotor parts leading to dynamic imbalance of the rotor. The cool-down speed therefore has to be kept at certain level to ensure uniform thermal movement at each and every cross section of the coil support structure. The use of highly conductive construction material for the coil support structure lifts the cool-down speed limitation and reduces the chance for an unbalanced cool-down.

## DESIGN ANALYSIS

The rotor cooling system design is strongly coupled to the structural design of the rotor cryostat. Based on a preliminary structural design of the rotor cryostat, a refrigeration load was estimated. A computer simulation program was then developed for design iteration toward the optimization of the cryogenic process for the estimated refrigeration load. After selecting a refrigerator based on capacity requirements and budget constraints, the simulation program was expanded to reflect specific features of the refrigerator and rotor cryostat, as well as the dynamics during a simultaneous cool-down of the refrigerator and rotor cryostat. In an effort to determine the limit on cool-down speed, heat transfer characteristics during the system cool-down and at steady-state operating condition are

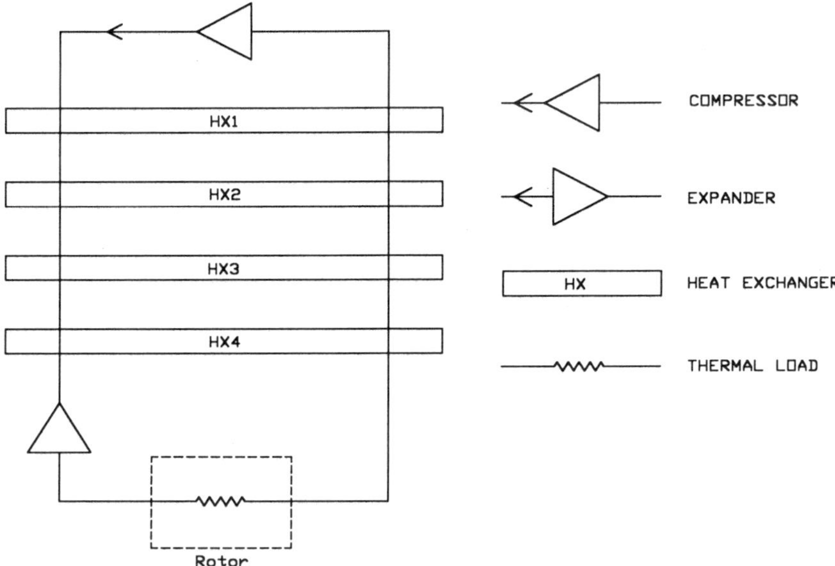


**Figure 1.** Single-loop reversed Brayton cycle flow diagram.

used as input for finite element analysis on thermal movements and thermally induced stresses in the rotor cryostat especially in the HTS coil mandrels and their support structure.

## Refrigeration Process Analysis

After comparing the merits of six design options based on current cryogenic refrigeration technology [4], a single-loop, reversed Brayton cycle was selected to provide flexibility in the first step of developing an ultimate refrigeration system for a large-scale industrial HTS motor. A simplified flow diagram for such a process is shown is Figure 1 where the entire rotor cryostat including the transfer coupling is depicted as the load to the refrigerator.

With reasonable assumptions on heat exchanger sizes and efficiencies for the compressor (36% isothermal) and expander (70%), the initial process analysis yielded a design that limited the compressor input power below 9 kW and a system efficiency close to 9% of the Carnot cycle [5].

The system cool-down simulation incorporated the actual performance data of the purchased refrigerator (Model 1420 from PSI) and the characteristics of the rotor cryostat structural design with expanded computational algorithm to maintain instantaneous system energy balance. As shown in Figure 2, in the simulation process the speed of the reciprocating expansion engine is initially fixed at 230 rpm allowing gradual increase of system mass flow rate till it reaches the full capacity of the compressor at 4.7 g/s (60 SCFM). The mass flow rate is maintained fairly constant thereafter resulting in a rapid slow-down of the engine speed as a consequence of increasing helium gas density at lower temperatures.

The overall heat leak into the rotor cold space is considered as a function of the cold space temperature and consists of two terms representing heat leaks through solid conduction and radiation. Starting with a zero value at ambient temperature, the heat leak gradually increases as the cold space temperature is reduced and eventually reaches 60 W when the system cool-down completes. There is a large swing in pressure drop across the rotor cryostat cooling circuits according to the underlying cool-down scheme in this study.

During the cool-down period of constant engine speed, the pressure drop in the cooling circuits increases with the increasing mass flow rate until it peaks. It subsequently decreases while the mass flow rate is maintained constant and reaches a final level at the end of the cool-down period when the engine speed is stabilized at 65 rpm. This procedure forces the engine to run at a constant maximum speed most of the time during the cool-down and substantially reduces the system cool-down time required otherwise.

## Cryostat Thermal Analysis

Upon entering the cold space of the rotor cryostat, the helium gas cooling stream branches into two parallel circuits with geometric symmetry. The parallel cooling circuit tubes are sized with an outer diameter of 0.00794 m (5/16") and each measures approximately 5 m long. They are integrated in the support structure of the HTS coils with intimate thermal contact. As plotted in Figure 3, the time history of temperature readings along each cooling circuit indicates that most of the heat transfer between the cold helium gas and the coil support structure effectively takes place within a short section of the cooling circuit tube near the entrance at the beginning of the cool-down. As the cool-down process progresses, heat transfer becomes less prominent in the entrance region and gradually spreads over the entire length of the cooling circuit. The convective heat transfer coefficient for these cooling circuits has a magnitude of over 1,000 W/m$^2$-K in the early stage of the cool-down and gradually reduces to approximately one-half of that value when the steady state is reached at the operating temperature level. The variation of heat transfer coefficient along each cooling circuit ranges from 11% at the beginning of the cool-down to less than 3% once the cool-down is complete.

The full potential of the M1420 system is also explored for the lowest operating temperature that could be possibly achieved. For the estimated load (60 W), the system can work at a return temperature of 14 K with the entire cooling circuit participating the heat transfer process. Temperature distribution along each cooling circuit at the full power of the refrigerator is shown in Figure 4. It represents a good indication on the length requirement for each cooling circuit.

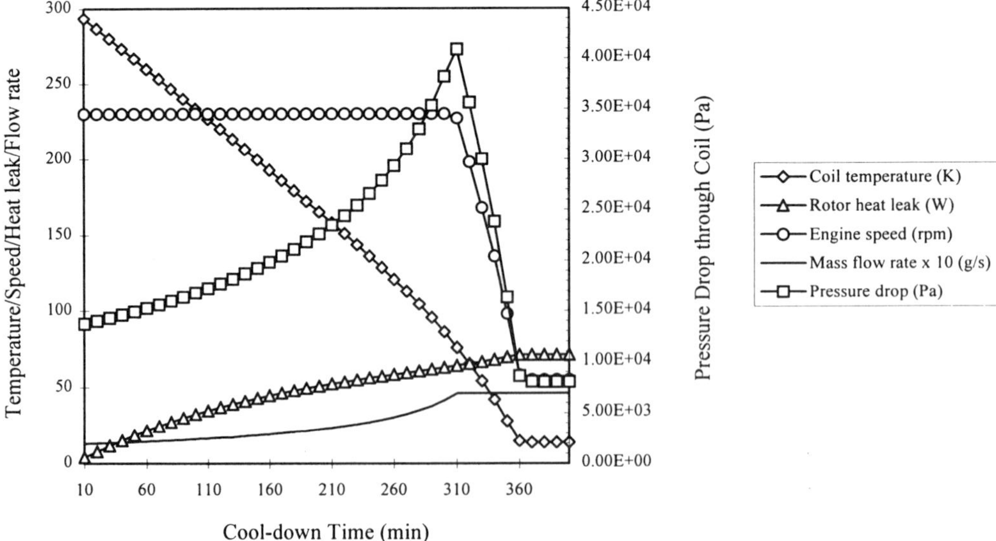


**Figure 2.** Cool-down dynamics of refrigeration system and rotor cryostat.

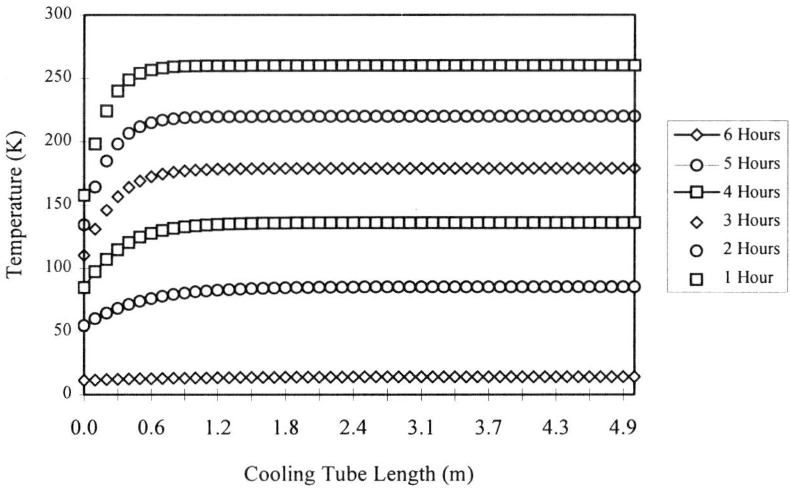


**Figure 3.** Temperature history of rotor cooling circuits during system cool-down.

At each time step of the cool-down simulation process, a constant temperature is calculated based on an energy balance of the total system. This approach offers a reasonable accuracy when the time step is reduced to approximately 10 seconds and is justified by the selection of a construction material structure with a high thermal conductivity for the support. Computed heat transfer coefficients for the cooling circuits at various cool-down times are then applied to a finite element analysis package ANSYS for further computation on temperature distribution in the HTS coil mandrels and their support structure, which in turn is used as input for computation of thermally induced stresses. The FEA results indicate a rather uniform temperature profile in the rotor cold space cross section during the cool-down with the maximum temperature differential in the axial direction limited to 10 K. The corresponding maximum thermal stress happens to be in the support structure and is computed at approximately $7 \times 10^7$ Pa. In the steady state of normal motor operation, the temperature drop in the helium cooling stream will be under 3 K and the temperature difference between the inlet helium gas and the coldest spot in the

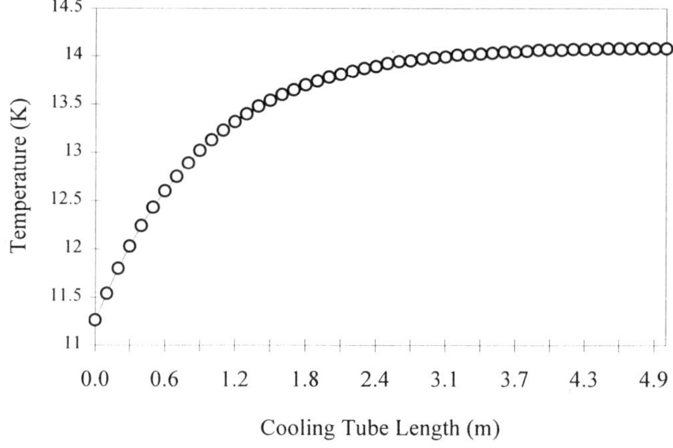


**Figure 4.** Temperature distribution along cooling circuit tube at maximum refrigeration capacity.

support structure will be close to 2 K. The worst-case scenarios of mass flow rate imbalance between the two parallel cooling circuits occurs when the temperature in the rotor cold space reaches approximately 70 K. Assuming a maximum temperature differential of 10 K across the cold space, the mass flow rate is estimated to split at a 48/52 percentage ratio.

## CONCLUSIONS

The simultaneous cool-down process of a rotor cryostat and a closed-cycle cryogenic refrigerator has been numerically simulated. The process analysis is based on an optimized refrigerator design for the particular application with modifications to accommodate features of a commercial system that will be actually used for rotor cryostat prototype testing. Temperature gradients in the HTS coils and their support structure have been minimized by selecting construction material with high thermal conductivity and moderate thermal diffusivity. The fairly uniform temperature distribution and the consequential low thermal stresses in the entire rotor cold space allow accelerated cool-down by keeping the expansion engine running at the maximum recommended speed till the maximum mass flow rate is reached.

## ACKNOWLEDGMENT

The program is funded in part by the U.S. Department of Energy under cooperative agreement DE-FC36-93CH10580 with Reliance Electric Company. The author also wishes to thank Richard Blaugher, Philip Eckels, and Fred Fickett for their support on verification of material properties.

## REFERENCES

1. Schiferl, R., et al., "Development Status of A 125 Horsepower Superconducting Motor," Adv. Cryo. Eng. Vol. 42, Plenum Press, New York, NY (1996).
2. Schiferl, R. et al., "High Temperature Superconducting Synchronous Motor Design and Test," Proc. of the American Power Conference, Chicago, IL (1996).
3. Driscoll, D. et al, "Construction and Testing of an Air-core Synchronous Motor with High Temperature Superconducting Field Winding," presented at the Applied Superconductivity Conference, Pittsburgh, PA (1996).
4. Zhang, B. et al. "Cryocooler Integration with High-Temperature Superconducting Motors," proceedings of the 9th International Cryocooler conference, Plenum Press, New York, NY (1996).
5. Zhang, B. and Driscoll, D., "Conceptual Design of Cryogenic Refrigeration System for Superconducting Motors," Corporate Research and Development Engineering Report No. 064, Reliance Electric Company, Cleveland, OH (1996).

# DEVELOPMENT OF 70MW CLASS SUPERCONDUCTING GENERATORS

Y.Imai,[1] K.Toyoda,[1] K.Kimura,[1] K.Inoue,[1] T.Ichikawa[1]
A.Sugawara,[2] M.Ogawa,[3] K.Yamaguchi,[4]
K.Suzuki,[5] and K.Miyaike[6]

[1]Super-GM, 5-14-10,Nishitenma Kita-ku,Osaka 530,Japan
[2]AIST,1-3-1,Kasumigaseki,Titoda-ku,Tokyo 100,Japan
[3]NEDO,3-1-1,Higashi-ikebukuro,Toshima-ku,Tokyo 170,Japan
[4]Hitachi Ltd., 7-1-1 Omika-cho,Hitachi,Ibaraki 319-12,Japan
[5]Mitsubishi Elec. Corp. 1-1-2,Wadasakicho Hyogo-ku,Kobe 652,Japan
[6]Toshiba Corp. 2-4 Suehirocho Tsurumi-ku,Yokohama 230, Japan

## ABSTRACT

This paper describes recent progress in an 11 year project for research and development (R&D) of 70MW class superconducting generators in Japan.

Three types of 70MW class superconducting generators (called model machines) have been developed. Two slow response excitation type rotors were completed and rotational excitation tests of them were conducted at each factory. A quick response excitation type rotor is in manufacturing and static excitation tests have been conducted at the factory. Also, the characteristics of a stator, which is to be commonly used for three rotors at the verification test facility, were measured by combining with one of the slow response excitation type rotors at the factory. Test run for verification tests on the first machine has been started in May 1997. Tests on the other two model machines will follow the test on the first one.

## INTRODUCTION

The application of superconductive technology to various power apparatuses in electric power system will be greatly important from the viewpoint of energy and resources saving. Superconducting generators using a superconductor in field windings, have many merits, e.g. increasing generation efficiency, improving power system stability and operation performance, and reducing size and weight in comparison with conventional generators.

The research and depelopment(R&D) conducted by Super-GM has been carried out as a part of R&D on superconducting technology for electric power apparatuses, under The New Sunshine Program of AIST, in MITI, being consigned from NEDO, based on an 11 year project started in 1988 [1].

The target of the R&D is to develop three types of model machines for verifying their performance and reliability and also to establish technologies for design and manufacture of 200MW class pilot machine.

**Table 1.** Major parameters of model machines

| Item | | Slow response type A (0.1pu/s) | Slow response type B (0.1pu/s) | Quick response type (1.0pu/s) |
|---|---|---|---|---|
| Specification | | | | |
| Capacity | (MVA) | 83 | 83 | 73 |
| Voltage | (kV) | 10 | 10 | 10 |
| Current | (A) | 4792 | 4792 | 4215 |
| Power Factor | | 0.9 | 0.9 | 0.9 |
| Rotation speed | (rpm) | 3600 | 3600 | 3600 |
| Synchronous reactance | (pu) | 0.35 | 0.35 | 0.45 |
| Field current | | | | |
| Rated | (A) | 3000 | 3000 | 3200 |
| Maximum | (A) | 3600 | 3600 | 4500 |
| Rotor structure | | | | |
| Warm damper | | Single layer type | Squirrel cage type | Three layper type |
| Cold damper/Radiation shield | | Three layer type | Single layer type | Single layer type |
| Thermal contraction mechanism | | Double bearing | Flexible support | Frexible support |
| Property of superconductor | | High stability | High current density | Low AC loss |
| Stator structure | | Air gap winding<br>Double transposed conductor<br>Water cooling | | |

The main objective of the R&D are as follows:

- development of elemental techniques for 70MW class medel machines,
- development of an assembly method for model machines by scalable rotor and stator model,
- demonstration and verification by means of tests of operational performance and reliability of model machines,
- verification of manufacturing techniques developed on scalable models,
- basic design of 200MW class pilot machine reflecting test results of medel machine[2].

Figure 1 shows the schedule of the R&D on the superconducting generators.

## Recent Progress in The R&D

### Main Parameters of Model Machines

Three types of 70MW class model machines, which have different in generator excitation response speed, superconductor type for field windings, damper structure and

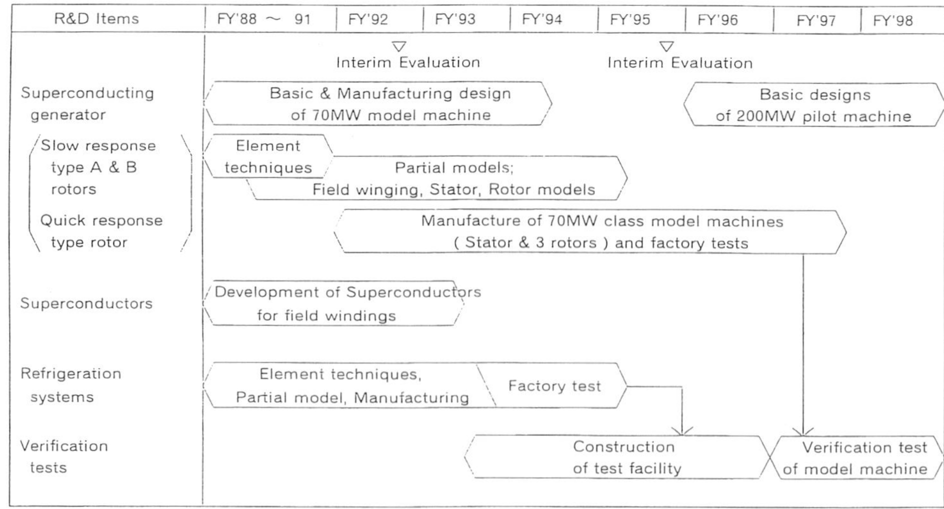


**Figure 1.** R&D schedule for superconducting generators.

Figure 2. Slow response excitation type A rotor finally assembled.

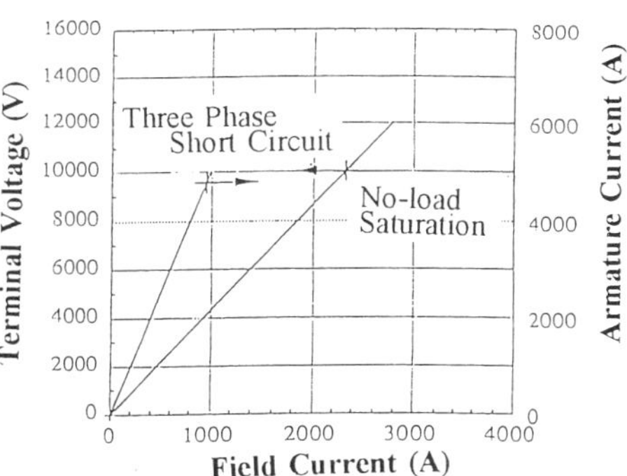


Figure 3. Electrical characteristics of type A machine.

mechanisms for absorbing thermal contraction between rotor components, have been developed as shown in Table 1 [3].

Development of Rotors

The recent developments of each rotor type are described as follows:

1) Slow response excitation type A rotor. One of the main features of the slow response excitation type A rotor is adoption of a cryostable superconductor. The stabilizer of the superconductor is aluminum having high electrical conductivity compared with copper at cryogenic temperature. Figure 2 shows this rotor finally assembled after static excitation tests. After final assembly, the rotor was combined with a stator, and rotational excitation tests at the factory were carried out. Figure 3 shows electrical characteristics of type A machine at the factory test. After the factory test, test run of the machine have been started in May 1997. Figure4 shows the machine installed in the verification test facility. The shaft vibration of the rotor up to the rated speed at liquid helium temperature is less than 40 μm p-p, whichis much the same amplitude as conventional rotor.

2) Slow response excitation type B rotor. The static excitation test of the field winding was carried out using a vertical type of helium dewar during the assembly process. Training, heater induced quench and pulsed excitation of the superconducting field winding were included in the test. The quench performance of the winding is shown in Figure 5. After the static excitation test, the model rotor was finally assembled and the cooldown test was carried out. From the results of cooldown test, it is estimated that the model rotor will be able to cooldown to liquid helium temperature within 72 hours using the refigeration system of the verification test facility. The external view of the rotor finally assembled with a magnetic shield is shown in Figure 6. After the rotational excitation test at the factory in 1996, it will be tested at the verification test facility in 1997.

3) Quick response excitation type rotor. As to the quick response type rotor, low AC loss type superconductor were wound into the winding support shaft, made of Ni-base alloy, as shown in Figure 7. And static excitation tests were carried out. Figure 8 shows the quench test results of the quick response type rotor. Manufacturing of other major components of the rotor, such as the warm damper (three layers; damper and its supports) and the radiation shield, was finished. This rotor will be assembled and tested at the factory in 1997, and it will be tested with the common stator at the verification test facility in 1998.

Development of Stator

After various tests using the stator winding model [4], a stator was manufactured in 1996. This stator is commonly used for three rotors in the verification test facility. Figure 9 shows the stator finally assembled. This stator and the slow response excitation type A rotor were assembled and tested as the factory test. Figure 10 shows the outer view of the factory test. After that, these were installed in the verification test facility in early 1997 as shown in Figure 4.

**Figure 4**. Type A machine installed in the verification test facility.

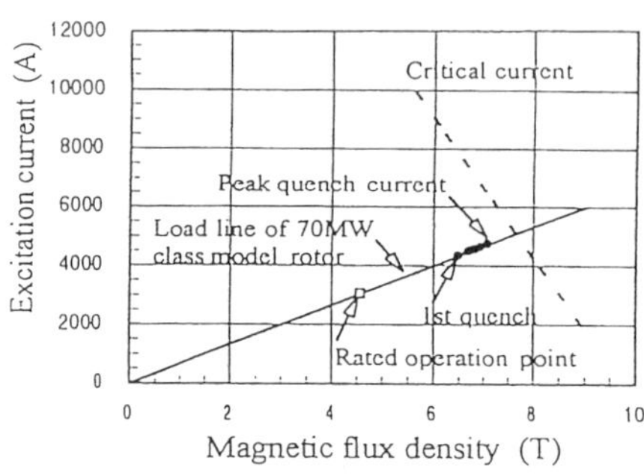


**Figure 5.** Static excitation test of field winding (type B rotor).

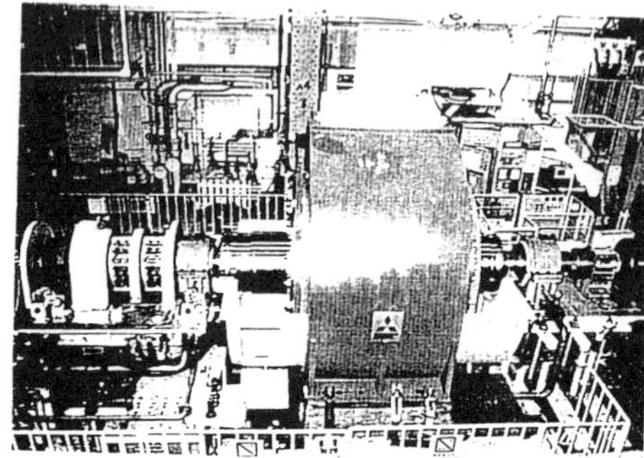

**Figure 6.** Slow response excitation type B rotor finally assembled with a magnetic shield.

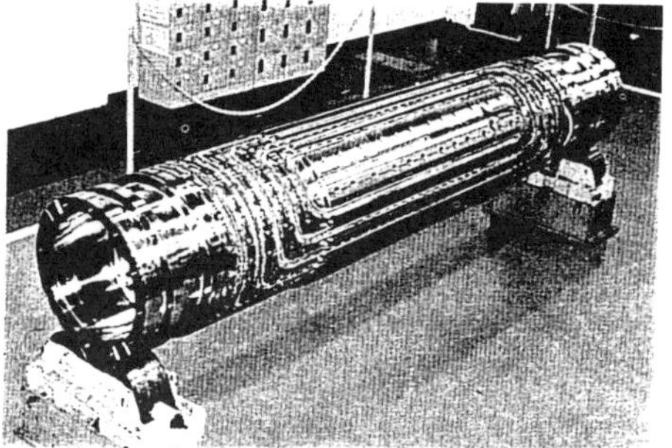

**Figure 7**. Quick response type rotor after winding the superconductor.

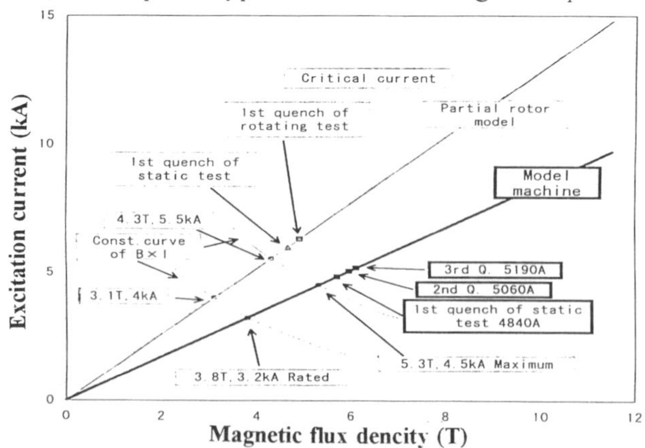


**Figure 8.** Quench test results of the quick response type rotor.

**Figure 9**. Common stator for three rotors.

**Figure 10.** Factory test of slow response type A rotor with a stator.

## VERIFICATION TEST

In order to establish advanced technologies for design and manufacturing of the 200 MW class pilot machine, operational performance and reliability of model machines should be tested in an associated test facility.

### Test and Verification Methods for Model Machines

Various techniques were developed through the project for design and manufacture of the 70MW class model machines. These techniques should be verified in order to establish technologies for design and manufacture of the 200MW class pilot generator. Table 2 shows the major subjects on model machines to be verified in the study. Various tests for obtaining static and dynamic data for the verification are planned to be conducted using the verification test facility. These tests are classified into four categories as shown in Table 3; basic tests for a synchronous machine, basic tests for superconducting generator, load tests, and severe tests. These contain more than 40 types of tests. The verification study is also planned to be performed not only through analysis of results from various tests at the factories and at the testing site, but also through analysis with computer codes utilizing the test results.

**Table 2.** Major verification subjects on model machines

| CLASSIFICATION | MAIN SUBJECTS |
|---|---|
| A<br>Establishment of SCG Design and Manufacturing Techniques | a. superconducting stability of the field winding<br>b. cool-down and warm-up characteristics, shaft vibration characteristics, of the multi-cylindrical rotor<br>c. electrical characteristics and mechanical strength of the air gap armature windings<br>d. effects of the damper system<br>e. increasing in unit capacity |
| B<br>Verification of a SCG Performance | a. enhancement of generator efficiency<br>b. improvement of transmission power limit and the system voltage stability |
| C<br>Verification of a SCG Operational Reliability | a. reliability of long term operation<br>b. soundness of the generator during system fault<br>c. same operability as conventional machines |

**Table 3.** Major verification tests on model machines

| |
|---|
| 1. Basic tests for synchronous machine |
| · No-load saturation curve, short-circuit characteristics<br>· Sudden short circuit test<br>· Efficiency test etc. |
| 2. Basic tests for superconducting generators<br>· Static excitation test<br>· Field current test<br>· Dumper test<br>· Vibration test<br>· Cool-down and warm-up test etc. |
| 3. Load tests |
| · P-Q characteristic test<br>· Longterm reliability test<br>· Excitation system voltage response test<br>· Daily Stop and Start test<br>· Load rejection test etc. |
| 4. Sever tests |
| · Overspeed test<br>· 3-phase sudden short circuit test<br>· Quenching test<br>· Excessive negative-phase-sequence current test |

Verification Test Facility

The construction of the verification test facility is almost completed. Using this facility, various tests will be performed to obtain static and dynamic data for the verification of performance and reliability of model machines and the refrigeration system. The verification test facility is located at the site of Osaka Power Station, Kansai Electric Power Company, as shown in Figure 11.

The simplified system configuration of the verification test facility is shown in Figure 12. Each of three 70MW class model machines is planned to be tested using a motor-generator (M-G) method. The model machine is not connected to the power network. The test facility consists of a M-G unit system, a refrigeration system and a

**Figure 11.** Whole view of the verification test facility.

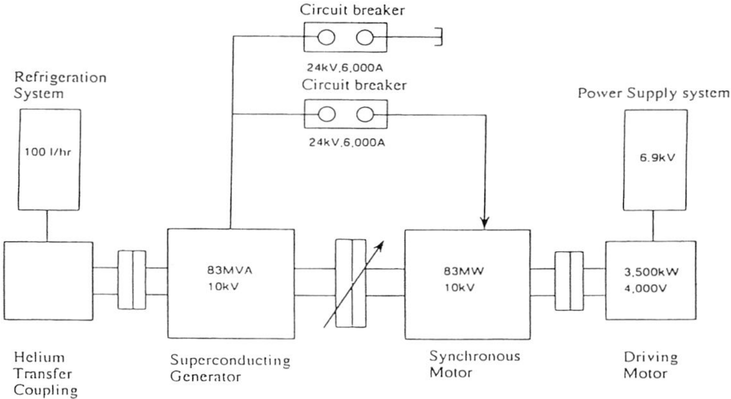


**Figure 12.** Test system for model machine.

control system for operating and monitoring these systems. The M-G unit consists of the model machine, a synchronous motor, a driving motor and turning equipment. These components are directly coupled. The model machine is driven by the synchronous motor and electric power generated by the model machine is fed back to the synchronous motor. The driving motor is used for start-up of the M-G unit and for compensating losses to maintain constant rotational speed of the model machine.

The test facility is designed so that various tests of the model machines and the refrigeration system under normal and severe transient conditions are available. For example, a M-G set foundation is designed to withstand against large mechanical load during severe tests such as the three-phase sudden short-circuit test.

A control building, a refrigeration building and a service building were already costructed. The synchronous motor and other auxiliary equipment were already installed. The refigeration system was also installed and a fine tune-up carried out. Thus, the construction of the test facility is almost completed.

## CONCLUSIONS

Super-GM has been conducting the 11 year R&D project mainly for developing superconducting generators. Three types of 70MW class model generators have been manufactured based on the fundamental technologies developed through the scalable rotor and stator models.

Verification tests on performance and reliability of model machines with the refrigeration system are one of the major items in the latter half of the project. The first model machine was already installed in the test facility and test run for the verification test has been started in May 1997, by combining with the refrigeration system.

The verification study will be performed not only through analysis of results from various tests at the factories and at the testing site, but also through analysis with computer codes utilizing the test resilts. Various technologies for design and manufacture of the 200MW class pilot generator will be established through the study.

In parallel with the research activities on the superconducting generator, Super-GM has been also conducting research works for developing conventional or low Tc superconductor and copper oxides or high Tc superconductors for DC and AC uses in power apparatuses, and for developing advanced oil-free refrigeration system to ultimately enchance reliability.

Super-GM will actively conduct the R&D for practical use of superconductive technology in the electric power field not only with cooperation and support from the utility companies, the manufacturers, university and national research organizations in Japan, but also with international cooperation and information exchange.

## REFERENCES


[1] "Feasibility study on superconducting machinery and materials technology to electrical power generations" prepared by Technova Inc. for Moonlight Project of AIST of MITI, March, 1988.

[2] T. Ageta "Recent Progress in Superconducting Generator Program in Japan ", International Symposium on Superconductivity 90,Nov.6-9,1990,Sendai,Japan

[3] N.Higuchi, et al. "Recent progress of development of 70MW class superconducting generators", IEEE Trans. on Applied Superconductivity Vol.3 pp.369-372, 1993.

[4] T.Ageta, et al. "Recent development progress of 70MW class superconducting generators (part2)", No.11-102,1996 CIGRE Session, Aug. 25 - 31, 1996, Paris, France.

# STUDY OF SUPERCRITICAL HELIUM CIRCULATION PUMP FOR A FORCED-FLOW COOLED MAGNET USING CICC

T. Kato, K. Hamada, K. Kawano, T. Shimba, K.Ootsu, H. Tsuji, and S. Shimamoto

Japan Atomic Energy Research Institute
Naka, Ibaraki, 311-01, Japan

## ABSTRACT

A fusion reactor, such as International Thermonuclear Experimental Reactor, should require a large scale forced-flow cooled cable-in-conduit conductor magnet. Such a magnet is operated by forced-flow cooling with a large amount of supercritical helium. A cryogenic pump is effective and reasonable to achieve such a forced-flow cooling for the magnet. A thermodynamic study shows a merit for the cryogenic pump compared to a conventional helium system in which the helium refrigerator supplies supercritical helium through the Joule Thomson line if the adiabatic pump efficiency of the cryogenic pump is more than 60%. A technical base for developing a large scale cryogenic pump is already established through continuous cryogenic pump developing at Japan Atomic Energy Research Institute as reviewed in the paper.

## INTRODUCTION

Japan Atomic Energy Research Institute (JAERI) has continuously developed a large scale of a cryogenic pump system that can supply a large amount of supercritical helium (SHe) to a large scale forced-flow cooled (FFC) cable-in-conduit conductor (CICC) magnet that is expected to be used for a fusion reactor, such as International Thermonuclear Experimental Reactor[1] (ITER). JAERI has so far developed a bellow type and a centrifugal type of the cryogenic pump with the pump head and the circulation SHe mass flow rate of 0.1 - 0.2 MPa and 100 - 500 g/s, respectively. Through the development and the operation experiment, it can say that the cryogenic pump should be effective and reasonable to achieve the SHe circulation to the FFC magnet. This paper try to clarify the features of the FFC magnet cooling condition in the view point of the helium refrigerator and the thermal efficiency comparison between the conventional circulation method and the cryogenic pump method by using a simple calculation. This paper also try to review the cryogenic pumps that have ever developed at JAERI.

## FUSION REACTOR AND FORCED-FLOW COOLED MAGNET

ITER is designed to have 20-superconcucting toroidal magnets, a superconducting central solenoid magnet, and 8-poloidal magnets. These magnets are required to be a large size, high current and high magnetic field. Thus, high strength and high rigidity are indispensable for the magnet structure. A FFC magnet, which has a cooling channel inside the conductor, can be availaþle to use a high strength structure material as a conductor jacket material, expecting to achieve a high strength and rigidity magnet. A voltage of more than 10 kV will generate in the magnet system during pulse operation by the poloidal magnetic field. Accordingly, high electric insulation is also required for the fusion magnet. There is only a FFC CICC magnet that will satisfy such requirements.

## FEATURES OF THE FFC CICC MAGNET COOING

Let's consider the cooling conditions on the FFC CICC magnet from the view point of a helium refrigerator.

(1) First, a required circulating mass flow rate for the magnet is not determined by the magnet heat load, meaning that the circulating mass flow is not designed by the following formula;

$$\dot{m}_m = \frac{Q_m}{h_o - h_i} \qquad (1)$$

Where, $\dot{m}_m$ is the magnet mass flow, $Q_m$ is the magnet heat load, and $h_o$ and $h_i$ are the outlet and the inlet enthalpy through the magnet, respectively.

The flow rate is practically designed to consider the following conditions:

(i) A transit time duration when the flow passes through the coil cooling channel

(ii) A required forced flow heat transfer coefficient.

The determined flow is generally much larger than the flow calculated by (1).

(2) Inlet and outlet helium conditions of the magnet are not determined from the view point of the magnet heat load but they are set up by considering the followings:

(i) Temperature performance of a used superconductor

(ii) Required temperature margin corresponded to the magnet operation scenario

For example, the inlet and outlet helium conditions for the ITER Central Solenoid coil are around 4.5 K, 0.6 MPa and 6.5 K, 0.45 MPa respectively, where the use of $Nb_3Sn$ superconductor and the temperature margin of 2 K are taken account of each other. Once a differential enthalpy through the cooling channel of the magnet, which is determined by the above conditions, is compared to the enthalpy difference related to a refrigeration power of the refrigerator which is defined as a enthalpy difference between the inlet enthalpy to the Joule Thomson (JT) valve and the enthalpy of a saturated liquid helium. The refrigeration power is generally much larger than the magnet differential enthalpy:

$$(h_{sv} - h_{BJT}) >> (h_o - h_i). \qquad (2)$$

Where $h_{sv}$ is an enthalpy of saturated vapor helium and $h_{BJT}$ is the inlet enthalpy to the JT valve, respectively.

(3) On the other hand, a refrigeration capacity defined by the magnet is ordinary designed to be equal or smaller than the refrigerator power to achieve refrigeration of the magnet:

$$\dot{m}_m(h_o - h_i) \leq \dot{m}_{JT}(h_{sv} - h_{BJT}) \qquad (3)$$

Where $m_{JT}$ indicates the mass flow rate through the refrigerator JT line, respectively. However, the mass flow rate of the magnet is generally larger than such JT flow rate:

$$\dot{m}_m >> \dot{m}_{JT}. \qquad (4)$$

Accordingly, if the JT flow is equal to the flow required by the magnet, which is usually used as a conventional SHe circulation system by a ordinal helium refrigerator, a produced refrigeration power is much higher than the heat load of the magnet:

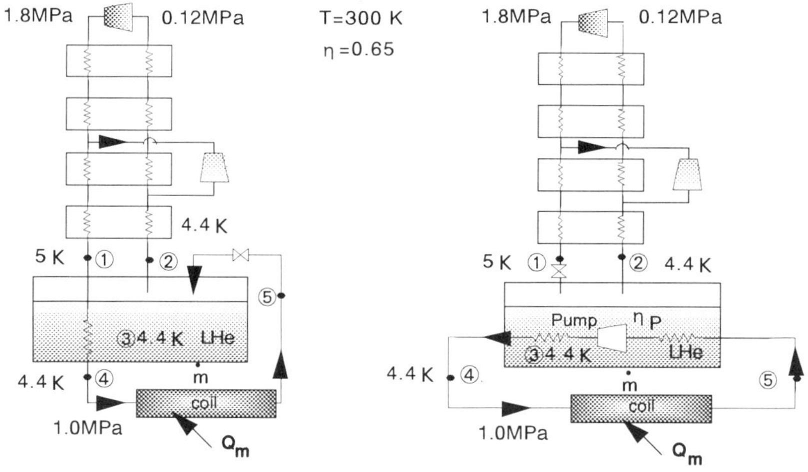


(1) Conventional circulation method (2) Cryogenic pump circulation method

Figure 1. A conventional SHe circulation method and a cryogenic pump SHe circulation method.

$$\dot{m}_m(h_{sv} - h_{sl}) >> \dot{m}_m(h_o - h_i) \text{ when } \dot{m}_{JT} = \dot{m}_m \quad (5)$$

It causes to produce a surplus refrigeration power in the cryogenic system. Then, to add an additional load, $Q_a$, to the magnet heat load, the heat balance will be established as:

$$\dot{m}_m(h_o - h_i) + Q_a = \dot{m}_{JT}(h_{sv} - h_{sl}). \quad (6)$$

Depending on the design conditions of both the magnet and the refrigerator, a reasonable heat balance will be achieved if the additional load will be used for increasing the magnet circulation flow. This is an idea to use a cryogenic pump which can increase the magnet circulation flow to be independent of the refrigerator JT flow.

## THERMAL EFFICIENCY COMPARISON ON THE CIRCULATION METHODS

A method that SHe circulation flow is directly supplied by the JT flow from the refrigerator is here defined as a conventional method. And a method that SHe flow is circulated by a cryogenic pump is then defined as a cryogenic pump method as shown in Fig. 1 (1) and (2), respectively.

Total heat load ($Q_T$) for the refrigerator in the conventional method and the cryogenic pump method is respectively:

$$Q_T = \dot{m}_{JT}(h_1 - h_4) + Q_m \text{ (Conventional)} \quad (7)$$

$$Q_T = W_{pump} + Q_m = \frac{1}{\eta_p}\frac{\dot{m}\Delta p}{\rho} + Q_m \text{ (Pump)} \quad (8)$$

Refrigerator power in the conventional and the pump is:

$$Q_R = \dot{m}_{JT}(h_2 - h_5) \text{ (Conventional)} \quad (9)$$

$$Q_R = \dot{m}_{JT}(h_2 - h_1) \text{ (Pump)} \quad (10)$$

If the heat balance is kept such as $Q_T = Q_R$, at the conventional method:

$$\dot{m}_{JT}(h_1 - h_4) + Q_m' = \dot{m}_{JT}(h_2 - h_5), \qquad \therefore \frac{\dot{m}_{JT}}{Q_m} = \frac{(\frac{\dot{m}_{JT}}{Q_m})(h_1 - h_4) + 1}{(h_2 - h_4) - (\frac{\dot{m}_{JT}}{Q_m})^{-1}} \tag{11}$$

At the pump method:

$$\frac{1}{\eta_p}\frac{\dot{m}\Delta p}{\rho} + Q_m = \dot{m}_{JT}(h_2 - h_1), \qquad \therefore \dot{m}_{JT} = \frac{1}{h_2 - h_1}(\frac{1}{\eta_p}\frac{\dot{m}\Delta p}{\rho} + Q_m) \tag{12}$$

Where $\dot{m}$ is a circulation mass flow, $\Delta p$ is a pump head, $\eta_p$ is an adiabatic pump efficiency, and $\rho$ is a inlet density of the circulating SHe, respectively.

A helium compressor power (Wc) is:

$$W_C = \frac{1}{\eta}\dot{m}_C RT \ln(\frac{p_o}{p_i}) = \frac{1}{\eta}\alpha\dot{m}_{JT} RT \ln(\frac{p_o}{p_i}) \tag{13}$$

Here, the compressor mass flow is assumed to be indicated as $m_C = \alpha m_{JT}$, and $\alpha$ is a constant. Where $\eta$ is a isothermal compression efficiency, R is a gas constant, T is temperature, and Po and Pi are an output pressure and suction pressure of the compressor, respectively.

When a Figure of Merit is defined as FOM=Wc/Qm, the FOM for the both methods:

$$FOM = \frac{1}{\eta}\alpha RT \ln(\frac{p_o}{p_i})(\frac{\dot{m}_{JT}}{Q_m}) = K(\frac{\dot{m}_{JT}}{Q_m}) \quad \text{Where,} \ K = \frac{1}{\eta}\alpha RT \ln(\frac{p_o}{p_i}) \tag{14}$$

Substitute (11) for (14) in the case of the conventional method:

$$FOM = K\frac{(\frac{\dot{m}_{JT}}{Q_m})(h_1 - h_4) + 1}{(h_2 - h_4) - (\frac{\dot{m}_{JT}}{Q_m})^{-1}} = \frac{Ax + K}{B - \frac{1}{x}} \tag{15}$$

Where $A = K(h_1 - h_4)$, $B = (h_2 - h_4)$, $x = (\frac{\dot{m}_{JT}}{Q_m})$, respectively.

Substitute (12) for (14) in the case of the pump method as well as above:

$$FOM = \frac{K}{h_2 - h_1}(\frac{1}{\eta_p}\frac{\Delta p}{\rho}(\frac{\dot{m}}{Q_m}) + 1) = Cx + D \tag{16}$$

Where $C = \frac{K}{h_2 - h_1}\frac{1}{\eta_p}\frac{\Delta p}{\rho}$, $D = \frac{K}{h_2 - h_1}$, $x = (\frac{\dot{m}}{Q_m})$, respectively.

As a result of the above simple calculations, FOM is indicated as a function of X in the both method. The shape of FOM along X for the conventional method can be assumed as a kind of a parabolic curve. On the other hand, the shape of the FOM for the pump is also assumed as a linear. Then, the thermal efficiency for each method can be evaluated by using the developed FOM functions when the values described in Fig. 2 are substituted for the both equation of (15) and (16). Evaluation results are shown in Fig. 2, in which the value of $\alpha$ is estimated to be 1.62 by the actual helium refrigerator data and the pump head of 0.2 MPa is used for calculation to correspond to the ITER magnet design.

Figure 2 shows interesting results as described in the following;

(1) The range of X is from 0.16 to 0.3 in the case of ITER magnets.

(2) If the pump adiabatic efficiency will be higher than 60 %, the cryogenic pump method is more effective than the conventional method.

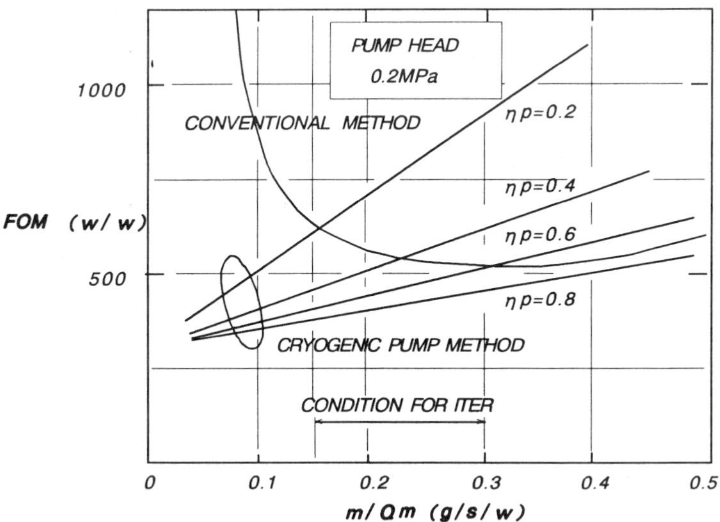


**Figure 2.** Evaluation results of the thermal efficiency comparison between the conventional SHe circulation and the cryogenic pump circulation.

(3) Especially, the higher pump efficiency such as more than 80 % will achieve that the pump is effective in spite of any X value.

Note that these results will be changed by the substituted values and the assumptions. Thus, it is difficult to say the results quantitatively. However, it can be qualitatively said that the pump method should be effective at some restricted X region or the higher pump efficiency.

## DEVELOPED CRYOGENIC PUMP AT JAERI

JAERI has continuously developed a cryogenic circulation pump to apply to the fusion magnet cooling since 1983. First technical goal was to develop the pump with a adiabatic efficiency of more than 60 % and the development of a bellow type pump was first started, considering to use the conventional techniques. Through the experiment of the bellow pump, it was understood that SHe should be treated as a incompressible fluid and the pressure head through the fusion magnet could be controlled by a low level such as 0.1 - 0.3 MPa in the magnet design. Thus, a centrifugal pump development was started after the success of the bellow pump development. Here, the developed bellow pump and the centrifugal pump are introduced.

### The bellow pump[2,3]

A bellow pump was developed to supply SHe to the Demo Poloidal Coil[4]. A goal of the pump was defined to simultaneously achieve a rated mass flow rate of 0.35 kg/s and a pump head of 0.1 MPa with an adiabatic pump efficiency of more than 60 % under a inlet helium conditions of 4.0 K and 1.0 MPa.

**Design and structure.** A view of the pump is shown in Fig. 3. A pump chamber is cooled at the cryogenic temperature in the vacuum atmosphere but a pump driving gear is set in room temperature. Both are connected by three columns including driving shafts. A pump space is separated into two rooms that are located top and bottom, providing a double acting mechanism, at which one is in suction process when the other is in exhaust process, to expect a large compressive ratio. Three suction valves and one exhaust valve are equipped in each pump room.

(1) Development of bellows: In order to assure a large mass flow and to reduce flow oscillation with the rated mass flow rate, a large diameter bellow with the diameter of around 430 mm is adopted and installed in top and bottom in the pump chamber. The

bellow is composed of five sheet of a SS304L with the thickness of 0.33 mm. Durability of the bellow was evaluated that the same bellow was tested to apply the stress ratio (generated stress / fatigue strength = around 0.42) and the test was repeated for $10^7$ cycles at room temperature.

(2) Driving gear: A oil pressure actuator with servo mechanism is selected as a pump driving gear. Adoption of the oil pressure type can be available for generation of large force required for driving of the large diameter bellow and provides a constant speed driving which act to reduce the flow oscillation. The servo mechanism can control the pump speed and the circulation flow at the control room. The actuator can produce the maximum pressure of 20.6 MPa, causing to be the pump force of around 3000 kgf.

(3) Suction and exhaust valve: An important point to get high volume efficiency for the volume type pump is to develop a high performance suction and exhaust valve. In the pump, a titanium alloy, SS316L, and SS304L are adopted as a material of the valve structure, the valve attachment, and the valve spring, respectively. It was experienced that a mechanical touch to seal a gas was more effective than the other type through the R&D.

**Achieved performance.** The pump could achieve the rated mass flow rate of 0.35 kg/s at several pump strokes and frequencies at the inlet helium condition of 4.5 K and 1.0 MPa. The maximum flow rate was measured to be 0.68 kg/s at the stroke of 12 mm with the frequency of 1.0 Hz. The adiabatic pump efficiency was also measured in the same test of the mass flow measurement. The adiabatic efficiency was determined by the enthalpy difference between the inlet and the outlet helium condition in the pump. The adiabatic efficiency of 60 % was verified at the rated operation condition. Furthermore, more than 60 % was achieved at the mass flow rate of 0.5 kg/s with the pump heat of 0.15 MPa. It was found that the adiabatic efficiency of more than 60 % was provided at the large pump work region of 100 - 600 W.

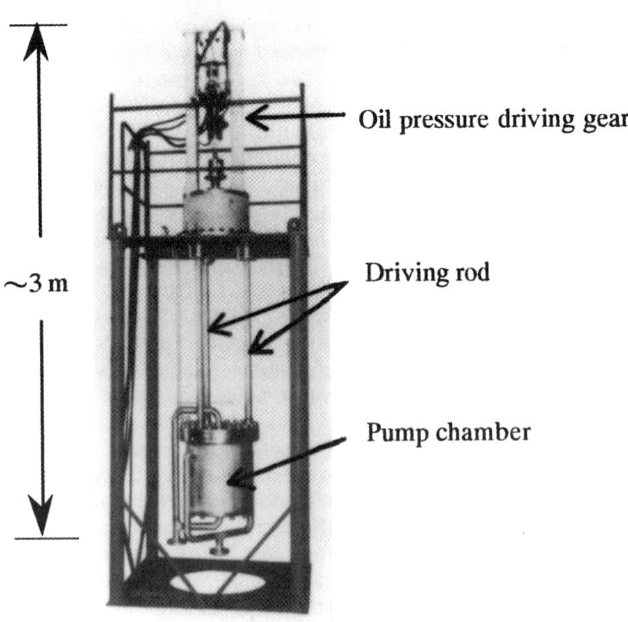


**Figure 3.** A view of the developed cryogenic bellow pump.

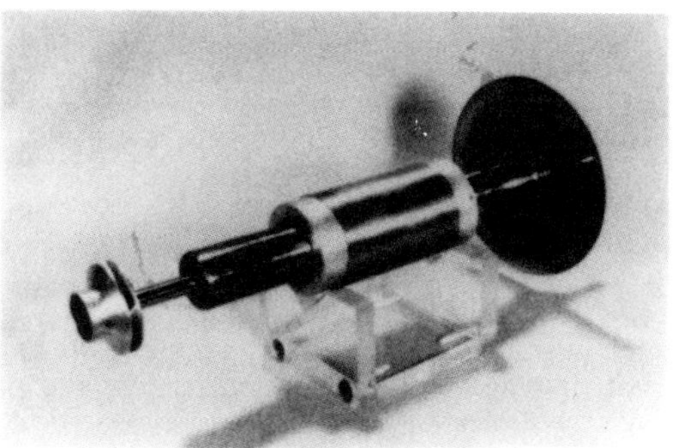

**Figure 4.** The view of the rotor part of the developed centrifugal cryogenic pump.

**Experienced merits and problems.** Merits of the cryogenic pump were experienced through the many operations. A list of merits are shown in the followings:

(1) Easy control for changing mass flow rate: It took only 10 minutes to change the flow from 0.1 to 0.3 kg/s, for example. Thermal matching between the refrigerator and the pump system was kept in a good condition.

(2) A large operation range: The pump could handle the mass flow rate from 0.1 to 0.6 kg/s. The minimum inlet temperature of 4.0 K and the maximum pump head of 0.2 MPa were attained each other.

What the problems for the pump is as follow.

It will be difficult to fabricate a larger pump because the pump driving force is almost in a limit. Though a adoption of larger oil actuator will resolve the issue, the strength of the pump installation flange and the driving shaft is also in a limit. A thick driving shaft will cause to provide a large heat leak from room temperature to the cryogenic temperature, bringing about a low pump efficiency.

## The centrifugal pump[5,6]

A centrifugal pump was developed with the same specification of the bellow pump except to the design pump head of 0.15 MPa. Development interests were set up how the high pump efficiency would be expected when applying the design base of the 90-% adiabatic efficiency water pump and how the pump size would be compacted.

**Design and structure.** A view of the developed centrifugal pump rotor is shown in Fig. 4. The pump is a single stage and adopts a radial-flow impeller. An inverter-controlled induction motor with 3.5 KVA is equipped, providing the maximum revolution of 33000 rpm. On the impeller design, a specific speed of 345 (rpm, m, $m^3$/min) is selected, expecting to produce an adiabatic efficiency of 80 % at the mass flow rate of 0.44 kg/s with the pump head of 0.15 MPa. A rated revolution and a impeller diameter are 26000 rpm and 37 mm, respectively.

(1) Impeller: Aluminum precision casting was adopted to fabricate a impeller. The impeller has four blades and a shroud cover is equipped around the impeller.

(2) Rotor: The center part of the rotor is a motor part which is composed of martensitic stainless steel. The other parts are made of austenitic stainless steel. The rotor diameter is 25 mm and the total length is 265 mm. The critical speed of the first bending mode is calculated to be 100000 rpm that has around four times margin for the rated speed.

(3) Bearing: A self-acting gas bearing is adopted as both a journal and a thrust bearing. The journal bearings are located at the both end of the motor part. A ring-type permanent magnet is equipped to lift up the rotor to support the thrust bearing force in a low speed.

**Achieved performance.** The designed performance, such as the mass flow rate of 0.35 kg/s and the pump head of 0.15 MPa, was achieved at the pump speed of 27000 rpm and inlet helium conditions of 0.47 MPa and 4.5 K. This is by 1000 rpm larger than the design speed. The measured adiabatic efficiency was around 86 %. This will be interpreted that the

SHe should behave as a incompressible fluid through the pumping process and the precision manufacturing should be achieved for the impeller and its casing. The developing goal was successfully accomplished. And it was demonstrated that the date base of the pump for using incompressible fluid such as water could be available to design the cryogenic centrifugal pump. A controllability of SHe circulation was also very smooth which was better than that of the bellow pump. The other feature was that the pump had a large operation range for the pump head and the mass flow rate. The pump can change the mass flow rate at a large region with the almost same pump head, which is a well-known feature of the centrifugal pump.

There were no problems for the developed pump. However, the minimum operation mass flow rate was limited in around 0.1 kg/s due to a limit of the thrust bearing force with a low speed. If the operation range will be more extended, the pump speed should be increased to provide a high thrust bearing force.

## CONCLUSION

(1) Cryogenic circulation pump is reasonable and effective device to supply supercritical helium circulation to the forced-flow cooled cable-in-conduit conductor magnet that will be applied to ITER.

(2) Technical base to develop a large scale cryogenic pump, which will be applied to ITER, is already established through the continuous developing work at JAERI.

(3) A 1-kg/s scale centrifugal cryogenic pump is now undertaken to develop at JAERI.


## ACKNOWLEDGMENT

The authors would like to express our gratitude to Drs. H. Kishimoto, M. Ohta, S. Matsuda, and T. Nagashima for their continuous encouragement on this work. The contribution to this work from Kawasaki Heavy Industries Co., Ltd. for the bellow pump and Ishikawajima-Harima Heavy Industries Co., Ltd. for the centrifugal pump is grateful acknowledged.


## REFERENCE


1. R. Aymar, Overview of the ITER project, Proc. of ICEC16/ICMC: 53 (1996)
2. E. Tada et al., Performance test results of cryogenic pump system for demonstration poloidal coil, IEEE Trans. on Magnetics, Vol 24: 1003 (1988)
3. H. Haraguchi et al., Development of a high capacity cryogenic pump for supercritical helium, Proc. of 12th International Cryogenic Engineering Conference: 103 (1988)
4. T. Kato et al., Operation performance of DPCF in the test of the Nb-Ti demo poloidal coils (DPC-U1,U2), Proc. of 11th International Conference on Magnet Technology: 1350 (1990)
5. T. Kato et al., Development of a large centrifugal cryogenic pump, *Advances in Cryogenic Engineering*, Vol. 37: 845 (1991)
6. T. Kato et al., Cryogenic helium pump system for the development of superconducting magnets in a fusion experimental reactor, Fusion Technology: 887 (1992)

# EXPERIMENTAL STUDY ON A THERMOACOUSTIC ENGINE WITH BRASS SCREEN STACK MATRIX

G.B. Chen, T. Jin, X. Bai, L. Zhao, and L. Fang

Cryogenics Laboratory, Zhejiang University
Hangzhou, 310027, P.R.C

## ABSTRACT

A thermoacoustic engine is a new type of pressure wave generator. With no moving parts, it is a simple, reliable, and long-term nonmaintenance machine. It has great potential in space application. An experimental thermoacoustic engine has been fabricated recently. A maximum pressure ratio of 1.12 has been obtained using a working fluid of nitrogen and 1.06 with helium, with corresponding frequencies of 25 Hz and 70 Hz, respectively. The influence of operating conditions and structural parameters on the performance of the engine was also tested.

## INTRODUCTION

With the wide application of cryocoolers, stability becomes a key problem for their further development. One of the effective means is to eliminate moving components. The development of the pulse tube refrigerator provides a mean to eliminate the moving displacer at low temperatures. But the moving piston at room temperature is still an obstacle to further development.

Recent research[1-3] shows that a thermoacoustic engine instead of a mechanical compressor to drive the pulse tube is a promising refrigeration concept. Two of the most important advantages are (1) there are no moving components except for the movement of the working fluid and (2) the thermoacoustic engine can be driven by heat energy which can utilize solar energy, fuel gas, etc. Since thermal energy is a low-grade energy, the thermal efficiency can be improved greatly. Furthermore, this energy is very useful in those areas where there is a lack of electric power.

*Advances in Cryogenic Engineering*, Vol. 43
Edited by P. Kittel, Plenum Press, New York, 1998

This paper presents an experimental apparatus of a thermoacoustic engine which is a promising driver for pulse tubes. Helium gas under a 2.0 MPa charging pressure oscillates when it is heated up to 400 ℃. The corresponding frequency and pressure ratio are 70 Hz and 1.06, respectively. Under the same conditions, the frequency and pressure ratio for nitrogen are 25 Hz and 1.12, respectively.

## THERMOACOUSTIC ENGINE

The thermoacoustic engine is a machine which converts heat energy into acoustic power under special conditions. Thermoacoustic oscillations can often occur when a gas-filled tube, closed at the hot end and open at the cold end, extends from room temperature to close to the temperature of liquid helium. It is often used as a liquid surface detector for liquid helium containers.

One of the essential conditions for spontaneous oscillations is to set up a sufficient temperature gradient, that is the temperature at the open end should be much higher than that at the closed end. The heat transfer from the hot end to the cold end is the driving force for oscillations, while the viscous resistance is the principal resistance. Only when the temperature gradient is larger than the critical one can the oscillations continue. With an increase in the pressure amplitude, the radial temperature gradient near the tube wall becomes so large that radial heat transfer increases rapidly and becomes the principal driving force.

Thermoacoustic engine consists of a thermoacoustic generator and resonator. The thermoacoustic generator includes a heater, (electric or gaseous), stack and cooler[4]. The heater and cooler provide the temperature gradient. The stack is similar in structure to a regenerator, but its mechanism and function are quite different from those of the regenerator.

In the regenerator, the thermal penetration depth of the working gas is much larger than the gap among the matrix, heat can be transferred continuously and smoothly. Considering there is only an axial temperature gradient, the gas is isothermal in the radial direction. However, in a thermoacoustic engine, the thermal penetration depth of the oscillatory gas is much smaller due to the high frequency and is almost equal to the gap among the acoustic matrix. There is a timing lag during the heat transfer through the boundary layer between this matrix and the gas. The thermal impedance results in a horizontal temperature variation. This results in a lag between the pressure and temperature waves. Since velocity is in phase with the temperature, i.e., a phase difference exists between the pressure and velocity waves, which results in thermoacoustic oscillations of the gas. Briefly, the stack has two main functions: (1) providing a working fluid with an axial heat source and (2) forming a suitable phase difference between the pressure and temperature waves. The temperature variation is partly caused by the adiabatic expansion and compression. The timing lag between the gas and matrix is also one of the decisive factors, and is essential to cycle operation.

In conclusion, in order to develop spontaneous thermoacoustic oscillations, it is essential that the actual temperature gradient of the acoustic stack be larger than the critical one . Of course, a selection of a suitable characteristic size of matrix is a sufficient condition.

## APPARATUS

The test apparatus of the thermoacoustic engine is shown in Fig.1. The heating unit of the apparatus includes two inner electric heaters, each with a 300W rating, and an outside heater of 400W, as well as a heat sink which is fabricated of copper. Matrixes of the stack and cooler are brass meshes of No.6 and No.10 stacked in a mixture of 1:2. The hot section of the apparatus is insulated with ceramic fiber material. The resonator is a stainless steel tube.

Main structure parameters of the apparatus are shown in Table 1. The characteristic dimension of the stack, i.e. the thermal penetration depth of the working fluids, given by $\delta_\kappa = (\kappa/\pi f \rho C p)^{1/2}$, is close to 0.5 mm. Length and diameter of the resonator are relatively larger, so as to obtain a rather low oscillating frequency and a small resistance loss that are suitable for pulse tube refrigeration.

**Table 1.** Dimensions of the test thermoacoustic engine.

| Item | Hot buffer | Heater | Stack | Cooler | Resonator |
|---|---|---|---|---|---|
| Diameter, mm | 140 | 52 | 56 | 56 | 36 |
| Length1, mm | (1.2L) | 50 | 288 | 50 | 4000 |
| Length2, mm | (1.2L) | 50 | 288 | 50 | 3000 |
| Length3, mm | (1.2L) | 50 | 288 | 50 | 2000 |

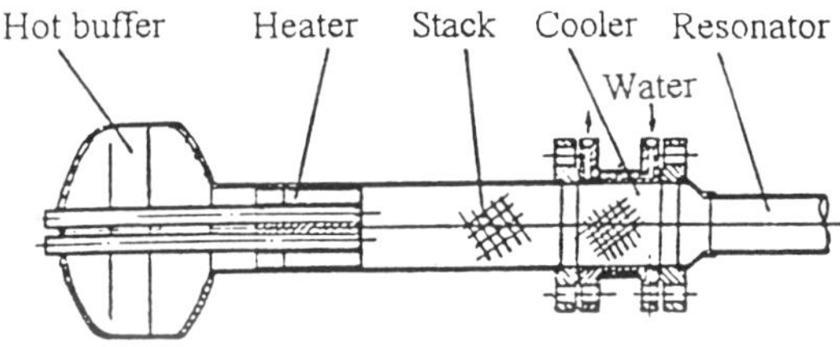


**Fig.1.** Schematic structure of a test thermoacoustic engine.

## RESULTS AND DISCUSSIONS

A number of tests have been performed on the thermoacoustic apparatus. The charging pressure is 2.0 MPa and the maximal heating temperature is 400 ℃. To initiate operation the system is heated slowly until the critical temperature is approached. When the gas in the resonator begins to oscillate spontaneously, a sine pressure wave is developed. A series of pressure waves under different heating temperatures and working pressures can be obtained by setting-up the heating temperature step by step.

According to the experimental results, the performance of the system is affected by both structure and operating parameters and is discussed in the following section.

### Oscillating Amplitude

Oscillating amplitude mainly depends on the heating temperature, charging pressure and working medium. Since temperature at the cold end of the stacks remains

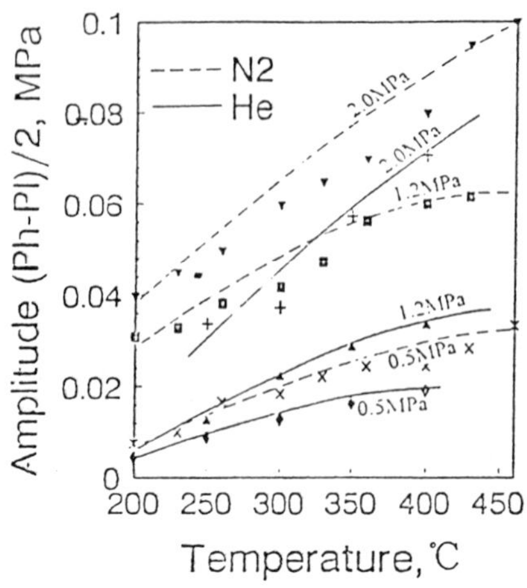


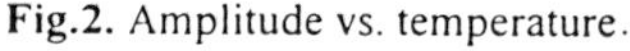
Fig.2. Amplitude vs. temperature.

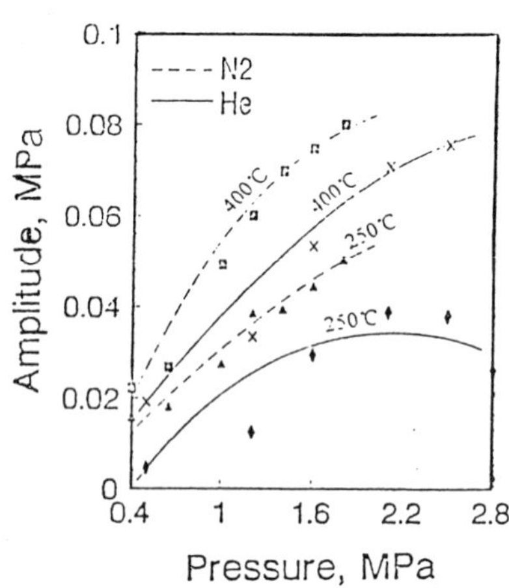


Fig.3. Amplitude vs. pressure.

approximately constant due to room-temperature cooling water, a rise in the heating temperature increases the temperature gradient on the stacks. As a result, the oscillating amplitudes are increased, shown in Fig.2. Figure 3 shows that the mean pressure rises with the charging pressure. This results in an increase of the amplitude. Furthermore, the oscillating amplitude also depends on the working medium. The amplitude of nitrogen is obviously larger than that of the helium under the same operating conditions. The oscillating amplitude obtained in this study is about 3 to 10% of the mean pressures.

## Pressure Ratio

Pressure ratio is closely related to the oscillatory amplitude and is primarily affected by the heating temperature, charging pressure and working medium. Figure 4 shows that the pressure ratio increases linearly with heating temperature. However, pressure ratio with a certain uncertainty increases with changing pressure, due to the influence of mean pressure

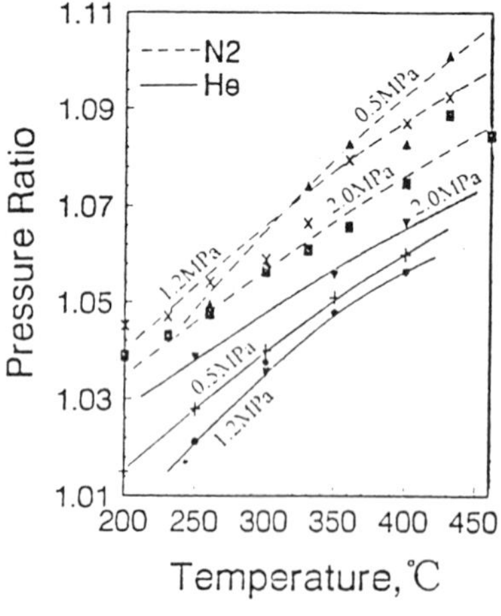


Fig.4. Pressure ratio vs. temperature.

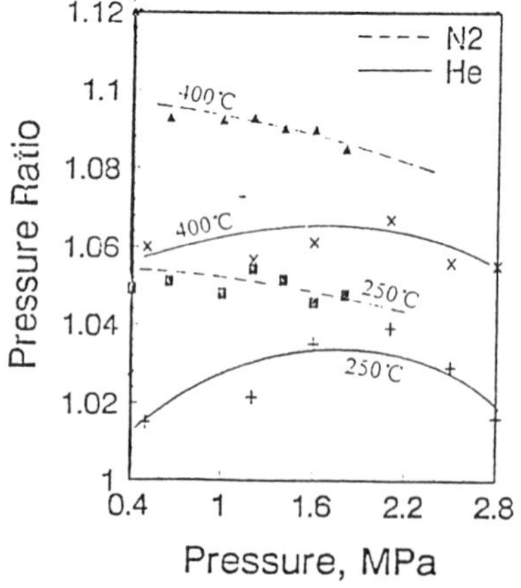


Fig.5. Pressure ratio vs. pressure.

(see Fig.5). Similar to the oscillatory amplitude, the pressure ratio of nitrogen is obviously larger than that of helium. In this experiment, the pressure ratios for nitrogen and helium observed are 1.12 and 1.06, respectively, under the conditions of a heating temperature of 400 ℃ and a mean pressure of 2.0MPa.

### Frequency

The oscillating frequency depends principally upon the size of the resonator and the working fluid. When the ratio of the diameter of the resonant tube to the stack decreases, and the length of the resonant tube increases, the oscillatory frequency drops. Furthermore the oscillating frequency increases with the sound speed of the working fluid. Experiments show that the oscillatory frequency of the nitrogen is about 25 Hz, while for helium it is about 70 Hz. The heating temperature and the mean pressure have little influence on the oscillating frequency.

Experiments also show that the length of the resonant tube has an influence on the characteristic form of the output wave. As the length increases, the pressure ratio and the oscillating frequency also increase (Figure 6 and 7), while the initial oscillating temperature decreases.

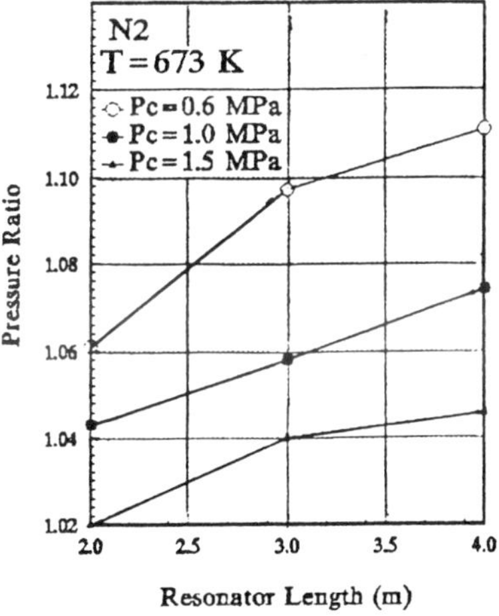


**Fig.6.** Pressure ratio vs. length of the resonant tube.

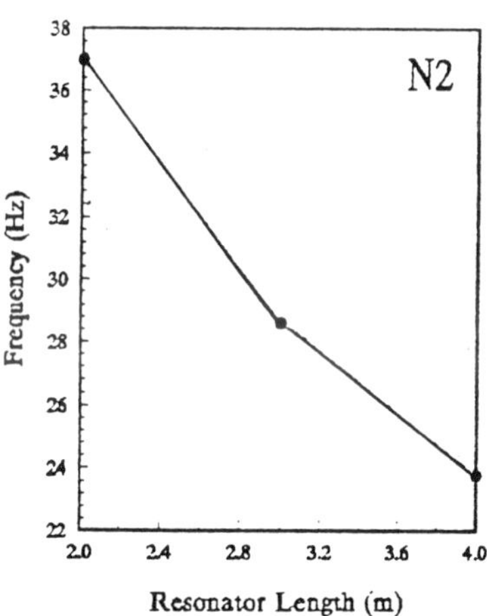


**Fig.7.** Oscillating frequency vs. length of the resonant tube.

## CONCLUSIONS

The thermoacoustic engine introduced in this paper successfully achieved thermoacoustic oscillations. The spontaneous oscillating temperatures of the system with a resonant tube with a length of 4m are 120 ℃ and 150 ℃, respectively, for nitrogen and helium with a charging pressure of 2.0MPa.

Maximum pressure ratios of 1.12 and 1.06 have been obtained for nitrogen and helium, respectively, under a heating temperature of 400 ℃ and a charging pressure of 2.0MPa.

Brass meshes can be used for matrixes of stacks and heat exchangers in the sonic generator. This new type of stack has advantages of structural simplicity, low cost and easy fabrication.

Characteristic form of the output pressure waves depend on the structure and operating parameters of the system. Pressure ratio is affected to a large extent by the length of the resonant tube, temperature gradient and mean pressure, while oscillating frequency mainly depends on the structure parameters and the working fluid. The length of the resonant tube also has an influence on the oscillating frequency and the initial oscillating temperature.

## ACKNOWLEDGMENTS

The project is supported by the Cao Guangbiao Hi-tech. Foundation of Zhejiang University and by the Natural Science Foundation of Zhejiang Province, P.R.C.

## REFERENCES

1. R.Radebaugh, Recent developments in cryocoolers, Proc. ICR'95, III(1995), p.973
2. G.W.Swift, Analysis and performance of a large thermoacoustic engine, J. Acoust. Soc. Am. 92(3), 1551 (1992)
3. S.L. Garrett and G.W. Swift, Thermoacoustic refrigeration and its relationship to the Stirling cycle, Proc. 7th ICSC, (1995), p.23
4. S.W.Zhu and Y. Matsubara, Theoretical and experimental study of thermal acoustic engine, Proc. ICSC'95, (1995), p.570

# INCREASING THE USEFUL LIFE OF QUENCH RELIEFS WITH INCONEL BELLOWS

W. M. Soyars

Fermi National Accelerator Laboratory*
Batavia, IL, 60510, USA

## ABSTRACT

Reliable quench relief valves are an important part of superconducting magnet systems. Fermilab developed bellows-actuated cryogenic quench reliefs which have been in use since the early 1980's. The original design uses a stainless steel bellows. A high frequency, low amplitude vibration during relieving events has resulted in fatigue failures in the original design. To take advantage of the improved resistance to fatigue of Inconel, a nickel-chromium alloy, reliefs using Inconel 625 bellows were made. Design, development, and testing of the new version reliefs will be discussed. Tests show that relief valve lifetimes using Inconel bellows are more than five times greater than when using the original stainless steel bellows. Inconel bellows show great promise in increasing the lifetime of quench relief valves, and thus the reliability of accelerator cryogenic systems.

## INTRODUCTION

The superconducting magnets of the Tevatron accelerator are protected from over pressurizing during a magnet quench (when the superconductor ceases to be superconducting) by a quench relief valve. These reliable, cost-effective cryogenic reliefs were developed by Fermilab. Figure 1 shows a cross-sectional drawing of these reliefs. The poppet seating force is provided by the remote pressurization of a bellows-driven actuator. When the pressure force from the magnet side, plus a small spring force, exceeds the force exerted by control pressure (that is, approximately the magnet pressure exceeding the control pressure), the valve opens for relieving with the poppet moving back counter to the flow.[1] This valve design has been in service since the beginning of the Tevatron era in the early 1980's.

While the quench relief valves have proven to be dependable for the Tevatron, their failures still contribute a great deal to accelerator cryogenics system downtime.[2] Valuable operating time could be gained if this portion of cryogenic downtime were reduced.

*Operated by Universities Research Association, Inc. under contract with U.S. Department of Energy

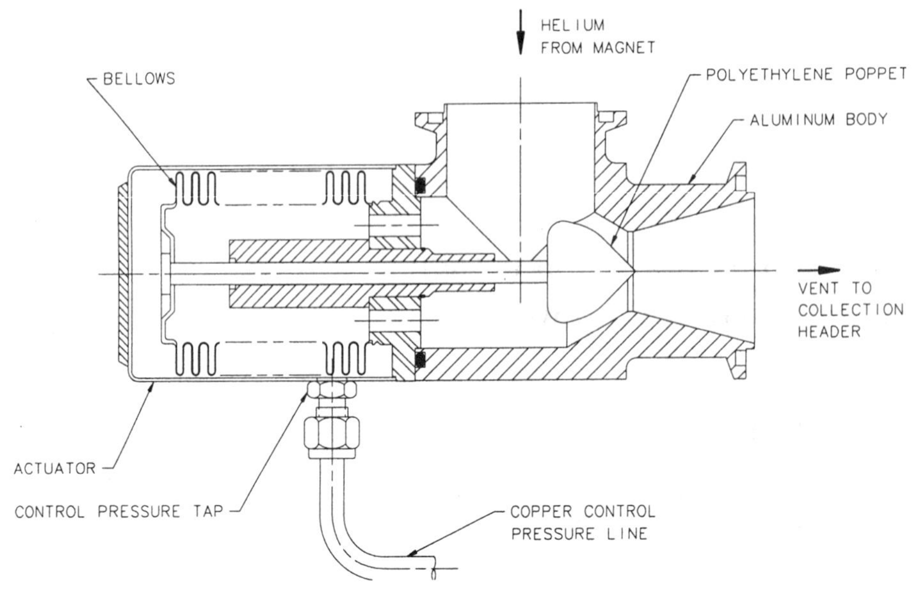


**Figure 1**. Fermilab's quench relief valve.

Let us consider the most common way a valve can fail, a cracked bellows. After many quenches, the thin bellows material can crack due to fatigue. This fatigue process is accelerated by a high frequency chatter of about 95 hz which a quenching bellows experiences.[3] Stress corrosion, such as from chlorides, is not an issue since the bellows is completely enclosed in a helium environment. Initially, a cracked bellows failure manifests itself as a loss of control gas pressure, which subsequently bleeds into the quench suction header. As the crack grows larger upon repeated relieving events, it eventually becomes very difficult or impossible to locally supply enough pressure to hold the cracked bellows closed. It is this group of failures that we intend to reduce with bellows assembly improvements.

## BELLOWS DETAILS

The relief valve bellows have previously been made of stainless steel 316. The effect of using bellows made of Inconel 625, a nickel-chromium alloy, will be studied. The intent is to extend the bellows lifetime due to the increased fatigue resistance of Inconel over stainless steel. For comparison, mechanical properties at room temperature of Inconel 625 (annealed sheet) and stainless steel 316 are shown in Table 1.[4,5,6] The impact strength shown is based on Charpy keyhole testing at room temperature. Inconel 625, like austenitic stainless steels, retains its ductility and toughness at low temperatures making it suitable for cryogenic service.

Other bellows actuator modifications were considered to improve the relief lifetimes, such as restricting actuator motion, throttling flow into the high pressure side of the bellows, or replacing the bellows with a sliding piston. Experiments with these schemes revealed that they either did an insufficient job of dampening chatter, were not conservatively safe, or

**Table 1**. Mechanical properties of competing bellows materials

| Material | Tensile Strength | Endurance Limit for Smooth Bar | Modulus of Elasticity | Impact Strength |
|---|---|---|---|---|
| Inconel 625 | 931 MPa | 620 MPa | 207 GPa | 66 J |
| Stainless 316 | 579 MPa | 269 MPa | 200 GPa | 95 J |

were too complicated. One helpful valve improvement was introducing opening/closing hysterisis by rapidly venting control side pressure upon a relieving event, which warrants further developement. For this study, we chose to focus on only improving the bellows material, which will avoid drastic change to the bellows assembly configuration and will retain the relieving characteristics of the original, proven, fast acting, operationally safe design.

Further lifetime improvements to a bellows potentially can be made by having more of the stroke in compression as opposed to tension. Compression is a more desirable way to load a bellows in that it does not promote growth of fatigue cracks the way tension does. The stainless steel bellows design had about 60% of its stroke in tension while the Inconel reliefs have about 35% of stroke in tension. Some lifetime variation between the two designs may be attributable to this. When observing stroke differences within a particular design, such variations appear to be small.

## THE WELDING EXPERIENCE

An important aspect of the bellows is how it is joined to the rest of the assembly. Poorly done welds can lead to large heat affected zones which deteriorate mechanical properties. Secure, leak tight joints are required. With an Inconel bellows, the joining of two different materials must be done. Furthermore, there is the potential difficulty of welding the thin bellows material to a thicker piece. These factors must be considered when developing a successful valve assembly.

If the weld joints are not properly designed, successful welding can be difficult or impossible. The key factor is to have a good, tight fit. Our final design calls for a slight interference fit between the straight extended neck of the bellows and its mating parts, the base flange or the cap. TIG fusion welds are used. The welding is done by the bellows manufacturer who prepares the bellows with a sizing die to expand and smooth out the bellows necks for a good fit. In our experience, it is helpful to have these critical bellows welds done by the bellows manufacturer for quality assemblies, utilizing their equipment and expertise.

Some further bellows assemblies were made using electron beam welded joints to study the benefit of improved heat control. The results, in terms of leak tightness, show no better performance than the fusion welds. If the fit is poor, neither welding process will be successful; therefore, the weld quality between the bellows ends and mating parts is ultimately determined by the fit. If the fit is good, the metal dissimilarities are not a factor. Insufficient data was gathered to determine whether or not an electron beam weld for the end assemblies would improve the bellows lifetime.

Another critical weld is on the bellows itself. The bellows we have used are mechanically formed with a longitudinal weld that joins the rolled sheet prior to formation of the convolutions. If an excessive weld bead is left behind, localized stress concentrations occur during bellows movement, causing cracks to initiate at the welded seam. This was learned by observing premature failures at the seam welds in the second batch of Inconel bellows which had unacceptable welds. Thus, our testing of Inconel bellows has been limited to the first batch of bellows with the superior welds, which clearly had smaller weld seams.

Future work involves setting the specifications to obtain more Inconel bellows, either with good welds or seamless bellows that do not have a seam weld.

## THE TEST FIXTURE

For testing purposes, room temperature compressed air is used to repeatedly cycle the relief valves. 0.7 MPa is available from a shop air compressor. It is accumulated in a vessel made of 15.2 cm pipe, 2.44 m long. This size allows the accumulated air volume to approximate the volume of helium relieved when a Tevatron dipole quenches. When a sufficient amount of pressure is allowed to build up, a solenoid valve will open exposing the vessels contents to the relief valve inlet. After the volume is expelled, the solenoid closes and pressure again builds for the next cycle. A counter keeps track of the number of cycles. There are pressure gauges on the accumulator vessel, the relief valve inlet, and the relief valve control pressure side. A photograph of the test setup is shown in Fig. 2.

The test fixture is looking for a bellows cracking failure, manifested by seeing high pressure on the control pressure side of the bellows which is connected to the pressure regulated bottle. The only way for high pressure to communicate with the control side is through a bellows crack. If such a rise in pressure is seen, the solenoid valve will cease cycling, freezing the cycle count.

The pressure transducer readings can be recorded with a fast data logging system that stores 16 seconds of readings at 1000 hz. These snapshots can subsequently be retrieved to study the features of one firing event. During a relieving event, most of the volume is carried away in the first two or three seconds.

Due to shop air compressor size and the restriction of the solenoid valve, the maximum pressure achieved in the test rig is not as high as peak pressures seen during magnet quenches, where peak pressures can be around 1.05 MPa.[3] Nonetheless, we are duplicating the motion that occurs and the stresses that result from this motion, giving a good simulation.

## RESULTS

Several relief valves were installed on the test fixture and cycled to failure. Pressures during one of these simulated relieving events is shown in Figure 3. How does this compare to an actual quench? For a magnet quench during current accelerator operating energy (900 GeV) peak pressure is 1.05 MPa, achieved 0.28 seconds after the quench, requiring 1.9 seconds to vent down to relief set pressure.[3] Another difference between this simulated quench and an actual quench is that here, as shown in Figure 3, the chattering starts immediately with the relief opening, while during an actual quench the chattering only begins after the relief valve inlet pressure has been lowered to near the set relieving pressure. This is explained by the fact that the test fixture room temperature flow is initially choked before the relief inlet by the 1.6 cm orifice of the solenoid valve. There is little pressure drop before the valve inlet when installed on magnets, which have a 3.0 cm tube for the cold flow leading up to the relief, so flow is not choked until the throat of the relief.

Since these data were taken at 1000 hz, a shorter segment of time can be closely looked at to better resolve the oscillations in relief inlet pressure that occur when the set pressure is approached. This is shown in Figure 4. A high frequency, low amplitude vibration of about 96 hz is seen. This duplicates valve flutter seen during actual quenches, which agrees with the second harmonic of the valve assembly's natural frequency as calculated assuming a lumped mass on a spring with mass.[3]

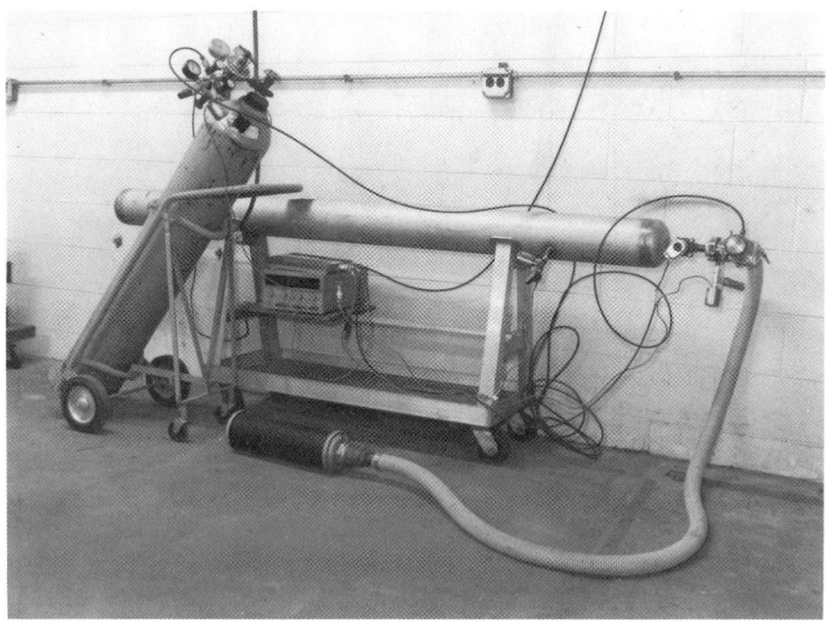

**Figure 2**. Test fixture for simulating quench relieving events.

In order to compare performance, we will keep track of the number of relief events before a bellows crack occurs. Notice that this count is not the same as the number of loading cycles on the bellows due to the high frequency oscillations that each relieving event brings. Ultimately the important design parameter for operations is how many quenches a relief can survive. While the test fixture does not exactly duplicate a quench, it allows for meaningful head-to-head comparison of competing designs.

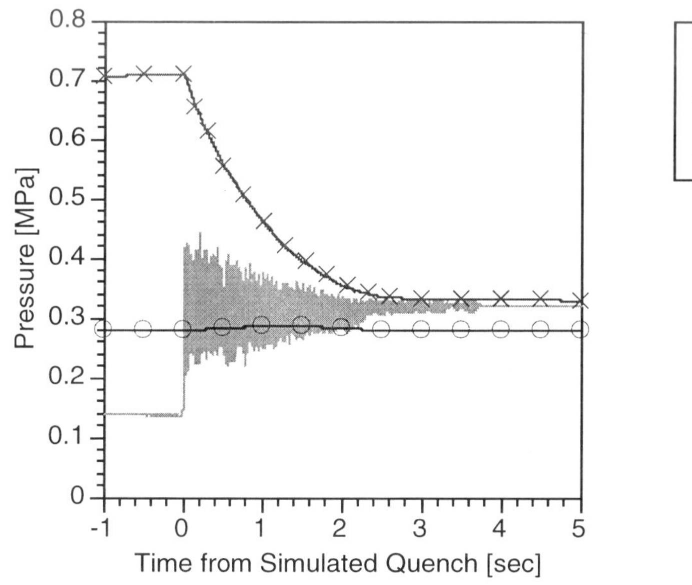


**Figure 3**. Pressures during one simulated quench .

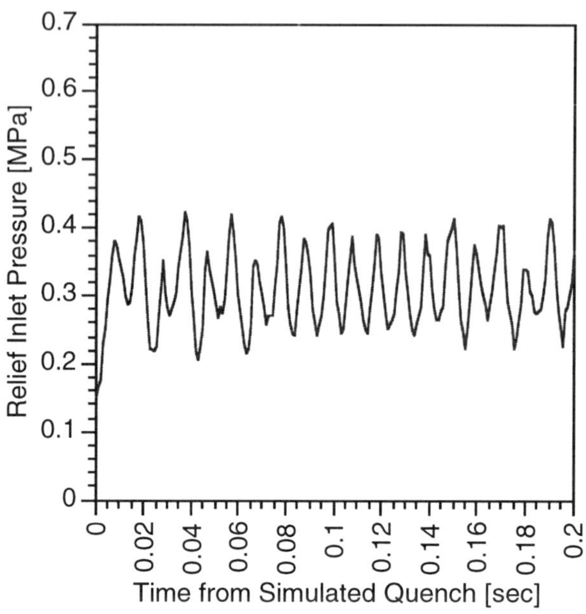


**Figure 4**. Oscillations in relief inlet pressure during one simulated quench event.

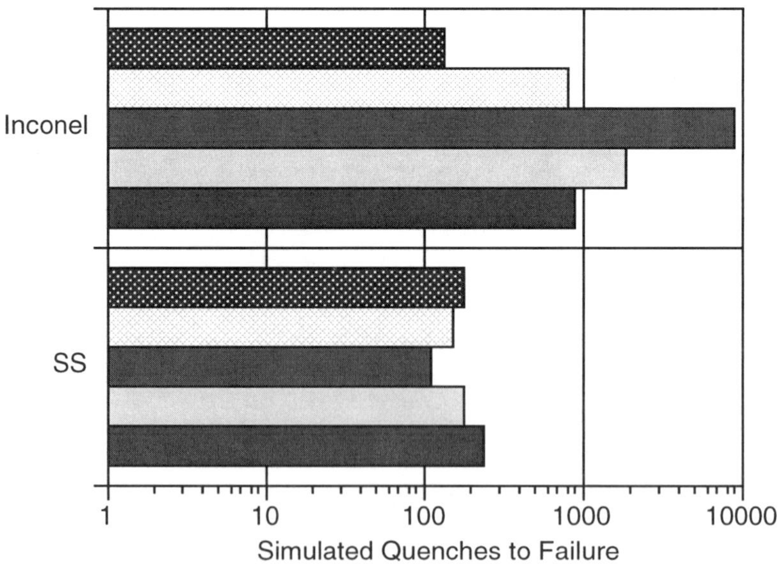


**Figure 5.** Lifetime of various relief valves before bellows failure.

The events-to-failure for tested relief valves are shown in Fig 5. Each article tested is represented with one bar. One can see how the Inconel bellows reliefs outperform those with stainless steel bellows. In one peculiar case, an Inconel bellows sample had a lifetime approaching 10,000 relieving events. This was attributed to the fact that in this case the pressure switch to detect bellows cracking and stop the count was not set with enough sensitivity, causing cycling to erroneously continue after failure until a larger crack had developed. The actual initial failure point was most likely more in line with the others of this type. In the limited testing of Inconel bellows done here, the scatter seen in the lifetime results will be addressed by future testing.

## CONCLUSIONS

Experience was gained in the design of improved relief valve bellows assemblies. Tests have shown that the new Inconel bellows relief valves have over five times the life (in terms of relieving events) than the original stainless steel style, lasting typically 900 relief events compared to about 170. Inconel bellows can increase the lifetime of quench relief valves, and thus, the reliability of the accelerator cryogenic system.


## ACKNOWLEDGMENTS

The author would like to thank the personnel of the Beams Division Cryogenic Systems group, especially Jeff Spencer, Darryl Richardson, Greg Johnson, Carl Pallaver, and Jim O'Neil for their help in assembling and testing new quench relief valves.


## REFERENCES


1. The Kautzky valve, Fermilab Reports, June 6-9 (1982).
2. B. L. Norris, Initial performance of upgraded Tevatron cryogenic systems, in: "Advances in Cryogenic Engineering," Vol. 41a, Plenum Press, New York (1996), p. 693.
3. J. C. Theilacker, B. L. Norris, and W. M. Soyars, Tevatron quench pressure measurements, in: "Advances in Cryogenic Engineering," Vol. 39a, Plenum Press, New York (1994), p. 469.
4. "Inconel Alloy 625 (Publication #T-42)", Inco Alloys International Inc., Huntington, West Virginia, (1988).
5. D. Peckner and I.M. Bernsein, "Handbook of Stainless Steels," McGraw Hill, New York (1977).
6. "Sourcebook on Stainless Steels," American Society for Metals, Metals Park, Ohio, (1976).

# MAGNETIC BEARING WITH MELT PROCESSED HTS

A. A. Kordyuk,[1] V. V. Nemoshkalenko,[1] and H. C. Freyhardt[2]

[1] Institute of Metal Physics
Kyiv, 252680, Ukraine
[2] Institut für Metallphysik, Universität Göttingen
Göttingen, D-37073, Germany

## ABSTRACT

We have studied the physical processes in the contactless magnetic bearings using melt processed YBCO as a stator. We have also designed and investigated the compound bearing with the additional components from the granular HTS. The high speed magnetic rotor on these bearings with self-centering and intrinsic damping was developed. The rotor's tolerance for its displacements, speed changes and phase disturbances was investigated. We have found the relations between the parameters of rotor's rotation and magnetic flux dynamics in the HTS bearings. We have studied the frequency dependencies of energy losses for rotors with non-ideal magnetic symmetry and found the optimization criterion for HTS bearing designing. We have also shown that such rotors can be used as a precision device for direct investigations of energy losses in magnetic materials in alternating magnetic fields.

## INTRODUCTION

The systems with the permanent magnets (PM) and high-temperature superconductors (HTS) have been under investigation since the HTS discovery[1-3] but the real opportunity to use them in the large scale applications appears only with the development of the melt-textured HTS technology.[4,5] The contactless magnetic bearings design has a significant place here, firstly, for energy-storage devices like flywheels. The different systems with such bearings are actively developed and investigated now.[6-9]

The increase of the critical current density in the bulk HTS materials is undoubtedly useful to increase the levitation force and stiffness and to decrease the hysteretic energy losses in HTS. The success in this direction makes the melt-textured HTS very close to the ideal hard superconductors that have no diamagnetic effect in the field cooled case (FC) and no hysteretic losses.[10] But for the contactless bearings it also results in a need to use the additional mechanical gadgets for centering of the rotor and for the damping of the low frequency parasitic vibrations.

In this work we studied the opportunity to design the compound bearing from the both melt-textured and granular HTS. The last component is used for the self-centering of the rotor and for intrinsic damping of its parasitic vibrations. We have performed the rotor optimization to decrease the energy losses in HTS at high frequencies. We have also obtained the relations between energy losses and rotor dynamical parameters. It gives an opportunity to use the rotor as a precision instrument for energy loss measuring.

## EXPERIMENTAL ROTOR SYSTEM

We use the configuration where the magnetic rotor is freely levitated and rotated in the superconducting stator. This configuration is presented in Figure 1 (*a*). The arrows show the directions of the PMs magnetization. The field from the rotating PMs in the bearings, in ideal case, is symmetric about the vertical rotational axis. In such a configuration, the energy losses of the rotor are mainly determined by air friction. The rotor is 50 mm in length and 7 mm in diameter. Its mass may be varied from 1.5 to 20 g by means of the non-magnetic inserts. Such a rotor and driving system were described in detail in our previous papers.[11,12] We investigated the rotors with different non-ideality in PM magnetization (see below).[12]

The results presented in this paper was obtained with the Y-123 material in the both granular and melt-textured HTS. All the PMs were made from $SmCo_5$.

## COMPOUND BEARING

### Self-Centering

Earlier[11,12] we had studied the rotor with the bearings from the granular HTS. The FC diamagnetic effect in these materials[13] was sufficient to rise the light 1.5 g rotor. So the main aim to use the granular components with melt-textured bearings was to obtain the self-centering effect.

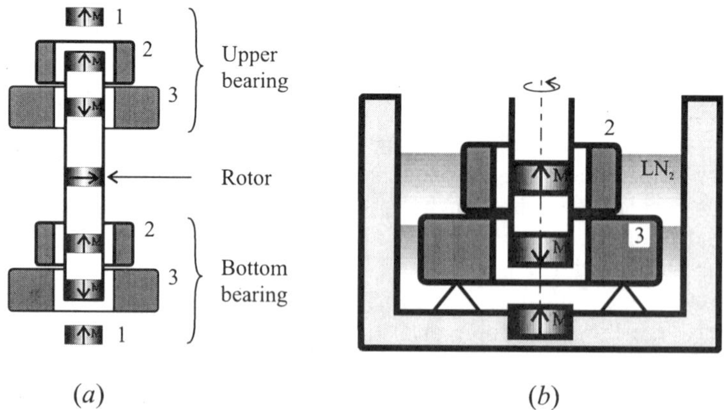


**Figure 1.** Experimental rotor configuration (*a*) and scheme of the compound bearing cooling (*b*): 1 – PMs in the stator, 2 – melt-textured HTS, 3 – granular HTS.

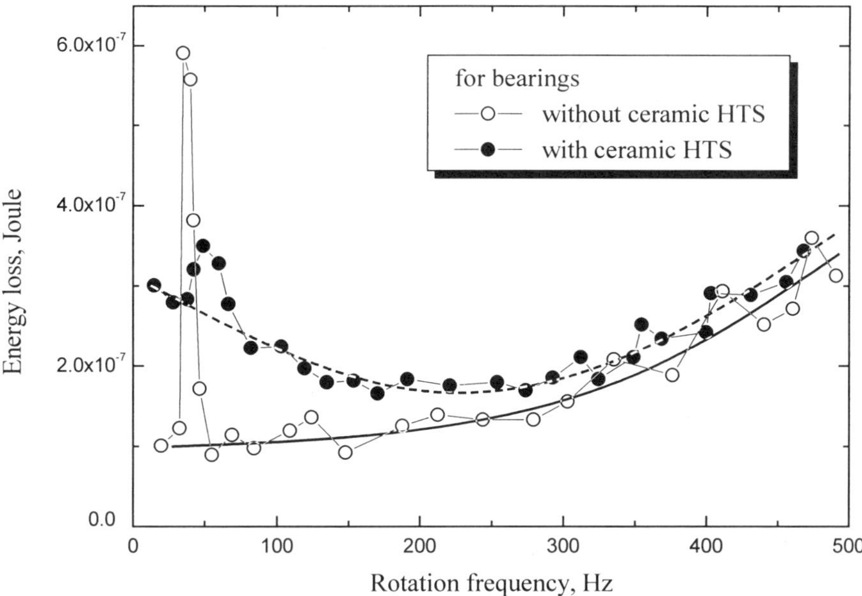


**Figure 2.** Energy loss per period *vs* rotation frequency of the rotor both with and without granular component.

Figure 1 (*b*) shows the bottom compound bearing from the both melt-textured and granular HTS. The PMs in the stator bear the main rotor weight. To obtain the self-centering effect, we use two stage cooling. In the first stage we cool the granular components and rotor is centered in the bearings. In the second stage we cool the melt-textured components that highly increase the stiffness of the rotor. The diamagnetic force is less than 5% from the lateral force due to rotor interaction with the melt-textured components but it was enough to achieve the self-centering effect for our light rotor.

## Intrinsic Damping

One of the problem in the HTS bearings is the damping of parasitic vibrations. They can be both the external vibrations and internal resonances. When the rotation frequency becomes near to one of the resonance frequency of the rotor-stator system the kinetic energy of rotation passes into the rotor vibrations. It results to the increase of energy losses and to the instability of the rotor.[6,8]

The use of the granular HTS component can resolve this problem. Figure 2 shows the dependencies of the energy loss per period on the rotation frequency of the rotor both with and without granular component. We obtained these dependencies in air by substracting the data for the best rotor (with no peak in energy losses) from data for the other non-ideal rotors. The granular HTS decreases the resonance peak and increases the rotor stability. The effective damping of the low frequency vibrations is due to high energy losses at intragrain magnetic flux motion in these materials,[14-16] but only for well optimized rotors there are no contribution of these granular components to energy losses at high frequency.[12]

## Rotor's Optimization

We have shown earlier[17] that there is a distinct crossover from viscous to hysteretic energy losses for granular Y-samples. This crossover takes place when the amplitude of external alternating magnetic field $h = h_c \sim 12$ Oe.[14] The losses arise because, in the real case, there are no ideal rotors with the ideally magnetized PM. The real PM always induces the a.c. magnetic field at its rotation. For very non-ideal rotors with a.c. field induced in the HTS bearings $h >> h_c$ we observed the considerable hysteretic energy losses that are frequency independent.[12] For slightly non-ideal rotors with $h < h_c$ the energy losses are much less and depend on the type of non-ideality, i.e. the position and direction of magnetic moments of the rotating PMs. When these moments are placed in parallel with the inertial axis of the rotor (see Figure 3), the longitudinal component of a.c. field is mainly generated. Such losses, as it follows from Figure 2, decrease with the speed of rotation and do not make a noticeable contribution to the rotor's energy consumption at higher frequencies. If these moments make some angle with the inertial axis, a transverse alternating magnetic field component is induced in the HTS bearings at the rotation. In this case the energy loss increases with frequency and reaches a peak at $\sim 10^3$ Hz.[12]

The nature of the different behavior of the rotors with the different type of non-ideality is in the different mechanisms of the longitudinal and transverse magnetic field component penetration into superconducting grains.[15,16] The longitudinal a.c. field have to penetrate through the grain boundary and the character time of this process is about 1 sec.[18] The transverse a.c. field component results in the intragrain vortex motion that can be described by the thermally assisted flux flow (TAFF) mechanism that has the character time about $10^{-3}$ sec.[12]

Thus, the use of the components from the granular HTS results in the rotor self-centering and effective damping of the parasitic vibrations and does not increase the energy losses at high frequencies for well optimized rotors.

## ROTOR AS AN INSTRUMENT

Low energy losses in melt-textured HTS bearings gives us an opportunity to use it as a precision instrument for energy loss measuring in magnetic materials in the presence of an alternating magnetic field. The spindown[6,8] measurement of the attenuation of the rotation speed is the express technique to get the dependence of the energy loss on field frequency $W(\omega)$. If the kinetic energy of the rotor $W_0 = J\omega^2/2$, the energy loss per period at the spindown experiments

$$W = -\frac{2\pi}{\omega}\frac{dW_0}{dt} = -2\pi J\frac{d\omega}{dt}, \tag{1}$$

where $J$ is the inertial moment of the rotor, $\omega$ is its frequency. We used this relation to obtain the data presented in Figure 2. The similar relation was used for the drag torque determination back in 1991.[6]

Apart from this technique, there are different dynamic parameters of the rotor that determined by the energy loss in the system and can be used for its measuring. At the forced

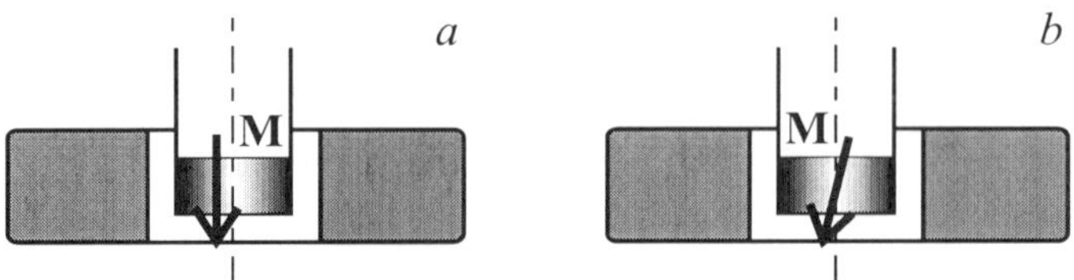


**Figure 3.** Two types of rotor non-ideality: *a* – the PMs magnetization **M** is parallel to the rotational axis *z* but does not coincide with it; *b* – there is some angle between **M** and *z*.

rotation of the rotor, if the driving field is $H = H_0 \cos \omega t$, the rotor's angle is $\alpha = \varphi + \omega t$. The phase of the rotor $\varphi$ is the function of time. It oscillates around its equilibrium position $\varphi_0$. For its small oscillations we can write $\varphi(t) = \varphi_0 + \delta(t)$.

The equation of rotor's rotation is

$$J\frac{d^2\alpha}{dt^2} + \eta\frac{d\alpha}{dt} = -\mu H_0 \cos \omega t \sin \alpha, \tag{2}$$

where $\mu$ is the magnetic moment of the central driving PM (Figure 1*a*). After its averaging over $\tau >> 2\pi/\omega$, we found the relations

$$W = 2\pi\eta\omega = W_m \sin \varphi_0, \tag{3}$$

$$\delta = C \exp\left(-\frac{\eta}{2J}t\right) \cos \Omega t, \tag{4}$$

$$\Omega^2 = \frac{16\pi W_0 \sqrt{W_m{}^2 - W^2} - W^2}{32\pi^2 J W_0}, \tag{5}$$

where $W_m = \pi\mu H_0$. In our experiment $\Omega$ was about 1–3 Hz and was decreasing to high frequency. Its value depends on $H_0$ and determine the phase rigidity $J\Omega^2$. The relation $\Omega > 0$ has been just the condition which determines the maximum speed of the rotor rather than the $\varphi_0 < \pi/2$. From (5) we can rewrite this condition as

$$\tan \varphi_0 < 8Q, \tag{6}$$

where $Q = 2\pi W_0/W$ is $Q$-factor (or durability) of rotor's rotation. The real maximum speed is less than obtained from (6) because the different phase disturbances appears at the speed increasing.

Thus, any relations of (1), (3)–(5) can be used for the energy loss determination and the choosing between them determined by the specific experimental conditions.

## REFERENCES


1. F.C. Moon, M.M. Yanoviak, and R. Ware, Hysteretic levitation forces in superconducting ceramics, *Appl. Phys. Lett.* 52:1534 (1988).
2. E.H. Brandt, Friction in levitated superconductors, *Appl. Phys. Lett.* 53:1554 (1988).
3. P.-Z. Chang, F.C. Moon, J.R. Hull, and T.M. Mulcahy, Levitation force and magnetic stiffness in bulk high-temperature superconductors, *J. Appl. Phys.* 67:4358 (1990).
4. K. Salama, V. Selvamanickam, L. Gao, and K. Sun, High current density in bulk $YBa_2Cu_3O_x$ superconductor, *Appl. Phys. Lett.* 54:2352 (1989).
5. M. Murakami, T. Oyama, H. Fujimoto, T. Taguchi, S. Gotoh, Y. Shiohara, N. Koshizuka, S. Tanaka, Large levitation force due to flux pinning in YBaCuO superconductors fabricated by melt-poweder-melt-growth process, *Jap. J. Appl. Phys.* 29:L1991 (1990).
6. B.R. Weinberger, L. Lynds, J.R. Hull, and U. Balachandran, Low friction in high temperature superconductor bearings, *Appl. Phys. Lett.* 59:1132 (1991).
7. J.R. Hull, T.M. Mulcahy, K. Salama, et al., Magnetic levitation and stiffness in melt-textured Y-Ba-Cu-O, *J. Appl. Phys.* 72:2089 (1992).
8. J.R. Hull, E.F. Hilton, T.M. Mulcahy, Z.J. Yang, A. Lockwood, and M. Strasik, Low friction in mixed-mu superconductig bearings, *J. Appl. Phys.* 78:6833 (1995).
9. T.A. Coombs, D.A. Cardwell, and A.M. Campbell, Dynamic properties of superconducting magnetic bearings, *IEEE Trans. Appl. Supercond.* 7 (1997) in print.

10. A.A. Kordyuk, Magnetic levitation for hard superconductors, *Appl. Phys. Lett.* to be published
11. V.V. Nemoshkalenko, A.A. Kordyuk, V.F. Los', A.D. Morozovskii, et al., 190 000 rpm low-power magnetic rotor with superconducting bearings, *Phys. Metal* 11:1113 (1993).
12. A.A. Kordyuk and V.V. Nemoshkalenko, High-speed magnetic rotor with HTS bearings for precision energy losses investigation, *IEEE Trans. Appl. Supercond.* 7 (1997) in print.
13. A.A. Kordyuk and V.V. Nemoshkalenko, Resonance oscillations of a permanent magnet levitated above granular superconductors, *Appl. Phys. Lett.* 68:126 (1996).
14. A.A. Kordyuk and V.V. Nemoshkalenko, Estimation of intragrain magnetic flux motion viscosity in high-temperature superconductors, *Journal of Supercond.* 9:77 (1996).
15. V.V. Nemoshkalenko and A.A. Kordyuk, Magnetic flux dynamics in granular HTS materials, *Low Temp. Phys.* 21:791 (1995).
16. V.V. Nemoshkalenko and A.A. Kordyuk, Interaction of HTSC with the longitudinal and transverse components of alternating magnetic field, *Phys. Metals* 14:1032 (1995).
17. V.V. Nemoshkalenko, E.H. Brandt, A.A. Kordyuk, and B.G. Nikitin, Dymamics of a permanent magnet levitating above a high-$T_c$ superconductor, *Physica C* 170:481 (1990).
18. A.A. Kordyuk and V.V. Nemoshkalenko, Intragrain magnetic relaxation in HTSC under variant magnetic field, *Czech. J. Phys.* 46:1251 Suppl. S3 (1996).

# RECENT ADVANCES IN THE REALIZATION AND DISSEMINATION OF THE ITS-90 BELOW 83.8058 K AT NIST

W. L. Tew, C. W. Meyer, G. F. Strouse, and G. T. Furukawa

National Institute of Standards and Technology
Gaithersburg, MD 20899

## ABSTRACT

Recent advances in our knowledge of the International Temperature Scale of 1990 (ITS-90), as realized and maintained at the National Institute of Standards and Technology (NIST), are briefly reviewed. As a result of these advances, the NIST disseminated version of ITS-90 has recently undergone small adjustments below 83.8058 K. These adjustments are all at the sub-millikelvin level and reflect the inclusion of recent data on ITS-90 realizations and intercomparisons of reference thermometers. Specifically, developments in the realization of ITS-90 fixed points, gas thermometry, and vapor pressure thermometry at NIST have significantly improved our knowledge of scale non-uniqueness and dissemination uncertainty in the lowest temperature ranges. We briefly describe the present status of ITS-90 realizations at NIST and which definitions are currently being disseminated. Our procedures for calibration of standard reference thermometers for cryogenic use are reviewed and our updated assessments of calibration uncertainties are presented. These include calibration of Standard Platinum Resistance Thermometers (SPRTs) of the capsule type between 13.8033 K and 273.16 K, and of Rhodium-Iron Resistance Thermometers (RIRTs) between 0.65 K and 83.8058 K. In addition, we preview a new Standard Reference Material® (SRM®) project which will make available "as defined" ITS-90 calibrated capsule SPRTs through the NIST SRM program. These continuing advances in scale realization and dissemination at NIST will improve the accuracy and availability of ITS-90 standards throughout the cryogenic engineering community.

## INTRODUCTION

The ITS-90 is the first international temperature scale ever adopted which fully integrates the cryogenic temperature ranges ($T_{90} \leq 83.8058$ K) with higher temperature ranges into its system of definitions[1]. A full realization of the scale in this range at the highest level of accuracy is complicated by its relatively complex form below 24.5561 K. A primary reason for this complexity is the gap between helium vapor pressure thermometry (with an upper limit at 5 K) and platinum resistance thermometry with a lower limit at 13.8033 K. In addition, the ITS-90 incorporates various overlapping definitions, particularly in the lowest temperature ranges. These overlaps, which were included in the scale to allow some flexibility on the part of various national laboratories to realize the scale in a manner best suited to their needs and abilities, also introduces some non-uniqueness into the scale.

At this writing, NIST is the only national laboratory worldwide that has fully implemented realizations of the ITS-90 over all defined ranges of contact thermometry. This has allowed us to disseminate an "as-defined" ITS-90 to the lowest limit of the scale (0.65 K). This version was not fully implemented prior to 1996. The version disseminated below 83.8058 K from 1990 to 1995 was an approximation, ITS-90(W), as maintained on reference SPRTs and RIRTs and commonly referred to as the "wire scale".[2] This full implementation has now allowed us to better evaluate calibration uncertainties, the small differences between the as-defined and wire versions of the scale, and the extent of non-uniqueness from overlapping definitions.

## ITS-90 REALIZATION

The NIST Low Temperature Realization Facility incorporates all defining instruments and fixed points necessary for the complete realization of ranges below 24.5561 K. The individual cells necessary for the ITS-90 realization in this range, as well as cells for the triple point (TP) realizations of oxygen and argon, are all contained within a single oxygen-free high-conductivity (OFHC) copper cylinder,[3] referred to here as the Integrated Realization Block (IRB). The IRB is cooled by a recirculating $^3$He refrigerator, capable of reaching temperatures below 0.65 K.

In addition, fixed-point cells for the realization of all SPRT sub-ranges above 83.8058 K are maintained in the NIST SPRT Calibration Laboratory. These include the triple points of Ar, Hg, and $H_2O$ in cells, referred to here as Long-Stem Immersion (LSI) type cells, designed for immersion from room temperature of long-stem SPRTs.[4] These fixed points are necessary for any SPRT calibration below 273.16 K on the ITS-90.

### 0.65 K to 5.0 K

Realizations of ITS-90 temperatures over the entire defined range of 0.65 K to 5.0 K have now been performed at NIST using $^3$He and $^4$He vapor pressure cells contained within the IRB. Vapor pressure measurements are accomplished via a system combining capacitance diaphragm gauges and absolute mode piston gauges.[5] Figure 1 shows results for both $^3$He and $^4$He with respect to the ITS-90(W), as previously maintained at NIST.

### 3.0 K to 24.5561 K

Realizations of the ITS-90 in the range 3 K $\leq T_{90} \leq$ 24.5561 K have now been achieved at NIST using an Interpolating Constant Volume Gas Thermometer (ICVGT).[6] The ICVGT is a 1 liter volume of $^4$He contained within the IRB. The gas pressure measurements are accomplished via a system combining capacitance diaphragm gauges and absolute mode piston gauges. Interpolation is accomplished via the ITS-90 defined equation of state for $^4$He. Results from 5 K to 24.5561 K are shown in Figure 2 with respect to the ITS-90(W).

### Fixed Points

Fixed-point realizations have been performed at NIST for all those fixed points defined on the ITS-90 including those in the cryogenic range. There are essentially three different techniques

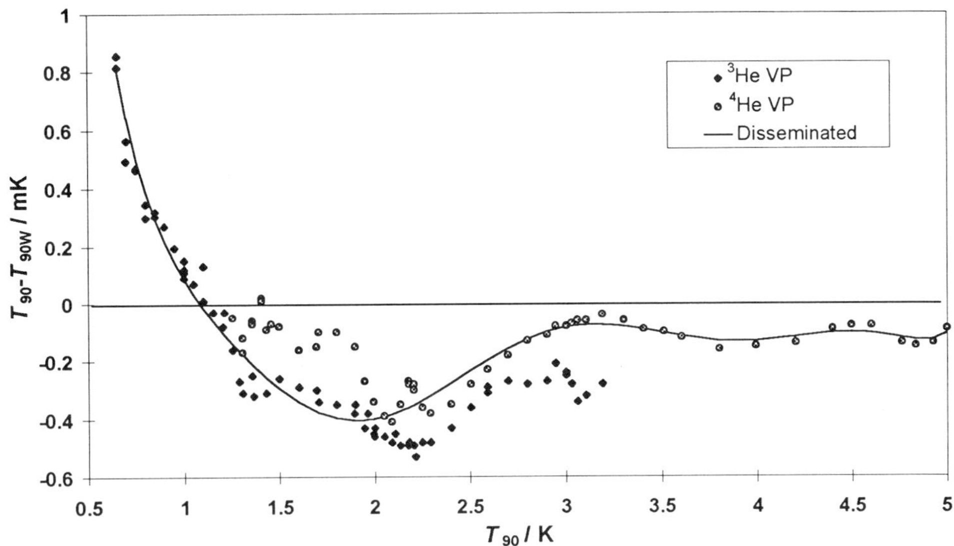


**Figure 1.** Differences in the indicated temperatures of the NIST implementation of the ITS-90 as defined by helium vapor pressures (VP) and the ITS-90(W) from 0.65 K to 5.0 K. The solid line shows a fit to the data for dissemination purposes.

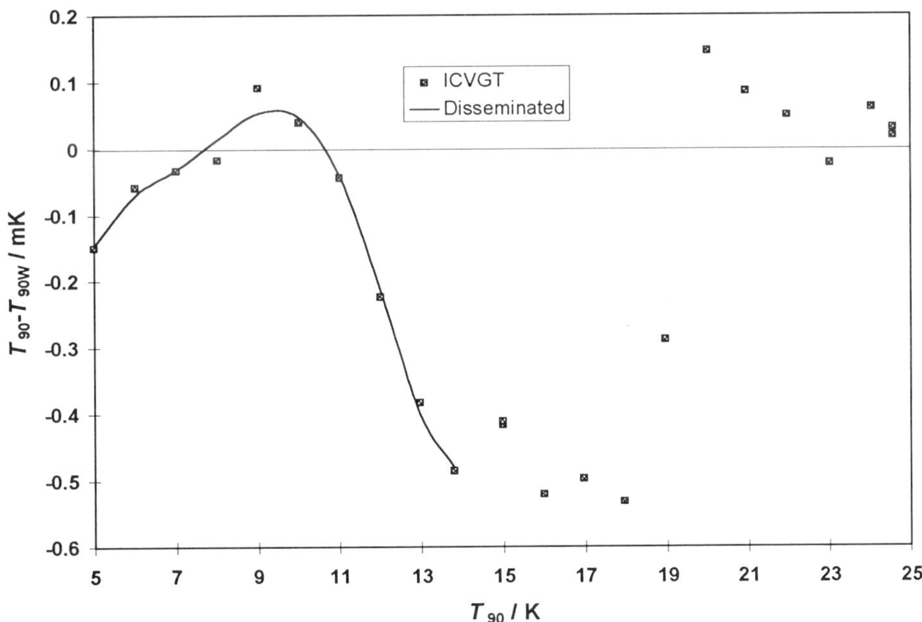


**Figure 2.** Differences in the indicated temperature of the NIST implementation of the ITS-90 as defined by the ICVGT and the ITS-90(W) from 5 K to 24.5561 K. The solid line shows a fit to the data for dissemination purposes from 5 K to 13.8033 K.

that may be used with capsule-type thermometers. Currently, the main technique for the cryogenic range uses isolated fixed-point cells, contained within the IRB, which are connected to room temperature manifolds and ballasts. Each cell manifold is dedicated to one particular species of gas with purities as specified in Table 1. The IRB contains six such cells for fixed-point realizations of: Ar TP, $O_2$ TP, Ne TP, e-$H_2$ vapor pressure points at 17.03 K and 20.3 K, e-$H_2$ TP, $^4$He vapor pressures, and $^3$He vapor pressures. The IRB can also be used to realize the SPRT calibration points at 17.03 K and 20.3 K via the ICVGT.

Another method, appropriate for triple point realizations only, is the sealed-cell (SC) technique, which enables the fixed-point cell to be permanently sealed and transported. To date, triple-point realizations using sealed cells with samples of Xe, Ar, $O_2$, and Ne have been performed at NIST.[7],[8]

In addition, an LSI-type Ar TP cell developed and maintained at NIST for simultaneous calibration of up to seven long-stem SPRTs can also accommodate capsule SPRTs when used with the appropriate type extension probes. The LSI-type fixed-point cells used at NIST for the Hg TP and $H_20$ TP accommodate capsule SPRTs as well, except that these cells are designed for only one thermometer at a time.

Table 1 lists all the ITS-90 defined fixed points used in the calibration of SPRTs below 273.16 K, as well as those necessary for calibration of the ICVGT, and reflects the current capabilities at NIST. The uncertainties listed are for the realizations alone, and do not include all the contributions to a calibration uncertainty. There is an observed disagreement of 0.6 mK at 17.03 K and -0.2 mK at 20.3 K in the realization of the NIST e-$H_2$ vapor pressure measurements with temperatures derived from the ICVGT. Some of the disagreement at 17.03 K is probably due to non-uniqueness in the definitions of the ITS-90 between 13.8033 K and 24.5561 K.

## ITS-90 DISSEMINATION

NIST maintains a variety of calibration services critical to the dissemination of the ITS-90 at the highest level of accuracy. These services are available to any customer with a suitable

**Table 1.** ITS-90 Fixed point cells for $T_{90} \leq 273.16$ K currently in use or pending at NIST

| Fixed Point | $T_{90}$ / K | Purity / % | Cell Type (ID, cm) | Uncertainty / mK ($k$=2) |
|---|---|---|---|---|
| $H_2O$ TP | 273.16 | | LSI (26.5) | 0.04 |
| Hg TP | 234.3156 | 99.999 99 | LSI (17.0) | 0.20 |
| Ar TP | 83.8058 | 99.999 99 | LSI (18.0) | 0.12 |
| Ar TP | 83.8058 | 99.999 99 | IRB | 0.06 |
| Ar TP | 83.8058 | 99.999 99 | SC | 0.10 |
| $O_2$ TP | 54.3584 | 99.999 99 | IRB | 0.06 |
| $O_2$ TP | 54.3584 | 99.999 99 | SC | 0.10 |
| Ne TP | 24.5561 | 99.999 | IRB | 0.18 |
| Ne TP | 24.5561 | 99.999 5 | SC | * |
| e-$H_2$ VP | 20.3 | 99.999 9 | IRB | 0.1 |
| e-$H_2$ VP | 17.0 | 99.999 9 | IRB | 0.1 |
| e-$H_2$ TP | 13.8058 | 99.999 9 | IRB | 0.09 |
| e-$H_2$ TP | 13.8058 | 99.999 9 | SC | * |
| $^4$He VP | 4.2 | 99.999 9 | IRB | 0.1 |
| $^3$He VP | 3.2 | 99.99 | IRB | 0.1 |

ID - Immersion depth of the fixed point material.

* - Realization results pending

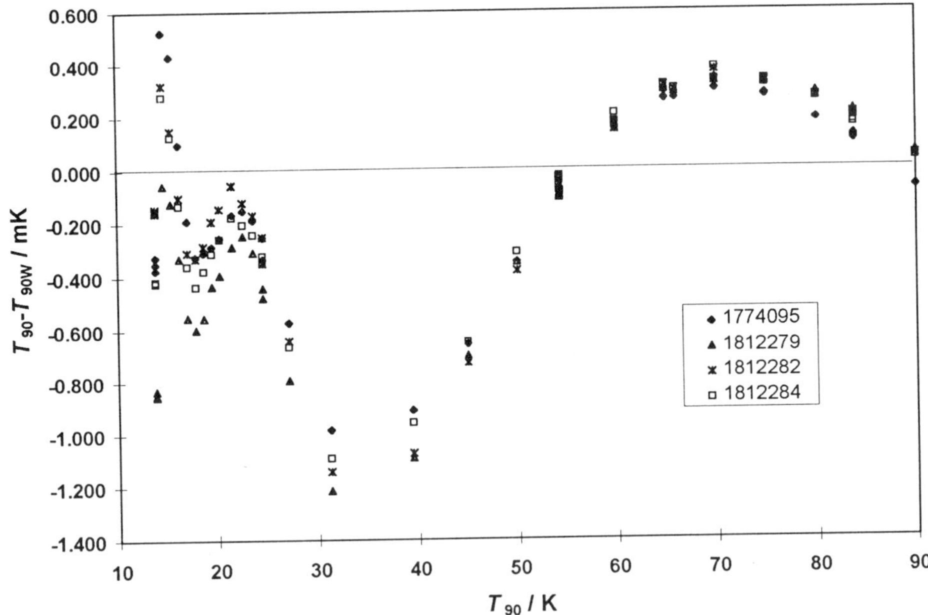


**Figure 3.** Differences in indicated temperature from 13.8 K to 90 K of the NIST implementation of the ITS-90 as defined by the SPRT sub-range from 13.8033 K to 273.16 K and the ITS-90(W). The four reference SPRTs which have been used to maintain the wire scale are identified by serial numbers as shown.

thermometer for various temperature ranges from 0.65 K to 1234.93 K for resistance-based thermometry. Of particular relevance to cryogenically oriented original equipment manufacturers, calibration laboratories, aerospace contractors and agencies, are those calibrations described under NIST Special Publication 250[9] with service identification numbers 33010C through 33141C for capsule type SPRTs and RIRTs ranging from 0.65 K to 505.078 K.

## Standard Platinum Resistance Thermometry: 13.8033 K to 273.16 K

Calibrations of customer SPRTs at temperatures above 83.8058 K are performed at NIST by direct measurements using LSI-type fixed-point cells. For calibrations points below 83.8058 K, the measurements are accomplished in a batch method via comparison to a reference SPRT which has been previously calibrated on the ITS-90 at the necessary fixed points. The batch calibration is performed with an OFHC copper comparison block, with a capacity for 24 four-wire devices, as many as 20 of which may be capsule-type thermometers.[10] These comparison calibrations are performed at NIST in the Low Temperature Calibration Laboratory, which also uses a recirculating $^3$He refrigerator for cooling.

Figure 3 shows the difference between the current NIST realization of the ITS-90 (as-defined) and the previous wire scale approximation, ITS-90(W), from 13.8033 K to 90 K. These data were obtained by comparison of four reference SPRTs, each calibrated according to the ITS-90(W), to a single reference SPRT recently calibrated on the ITS-90 as defined over the sub-range of 13.8033 K to 273.16 K. The wide scatter in the data below 20 K is almost entirely due to the imprecision in the original measurements used to construct the ITS-90(W) at NIST.

The SPRT calibration sub-range of 13.8033 K to 273.16 K is only 1 of 11 such sub-ranges defined on the ITS-90. While this sub-range is undoubtedly of the most interest for cryogenic thermometry, a capsule SPRT also may be calibrated in at least six other sub-ranges. Certain types are capable of being calibrated to temperatures as high as 505 K on the ITS-90, using other fixed points of pure metals. The implementation of the ITS-90 in these higher ranges has remained essentially unchanged since the inception of the scale in 1990.[11]

## Standard Reference Material 1750

While the SPRT is indispensable in achieving the highest level of accuracy on the ITS-90, the calibration of capsule SPRTs according to ITS-90 definitions is both costly and time consuming below 83.8058 K. Starting in 1998, NIST will introduce the Standard Reference Material 1750 which will significantly lower both cost and lead time for acquiring a calibrated SPRT. SRM 1750 will consist of a capsule SPRT, calibrated according to ITS-90 definitions over the range 13.8033 K to 429.75 K.

## Rhodium-Iron Resistance Thermometry: 0.65 K to 83.8058 K

While not a defining instrument of the ITS-90, the RIRT is the best suited device for disseminating the scale below 13.8033 K at the highest level of accuracy.[12] At NIST, a standard range calibration for an RIRT covers the range 0.65 K $\leq T_{90} \leq$ 27 K using 25 calibration points. An extended range calibration, up to 83.8058 K, also is available using an additional 13 points. Interpolation fits of resistance as a function of $T_{90}$ use break points at approximately 5 K and 26 K. Simple power series expansions of seventh order in $T_{90}$ are generally sufficient for interpolation over both ranges with inaccuracies of 0.1 mK to 0.2 mK.

RIRT calibrations also are performed via batch comparison to designated reference resistance thermometers maintained at NIST. For $T_{90} \leq$ 13.8033 K, a reference RIRT is used and for $T_{90} \geq$ 13.8033 K a reference SPRT is used.

NIST reference RIRTs have been calibrated on the ITS-90 using the following defined sub-ranges: the $^3$He vapor pressure scale from 0.65 K to 2.0 K; the $^4$He vapor pressure scale from 2.0 K to 5.0 K; and the ICVGT scale from 5.0 K to 24.5561 K. Reference SPRTs are calibrated on the ITS-90 using all fixed points within the sub-range of 13.8033 K to 273.16 K. Since 13.8033 K (e-$H_2$ TP) is a calibration point for both the SPRT sub-range as well as for the ICVGT, the two reference scales agree at this point to within the stated uncertainty for the calibration. While there is some disagreement in the overlapping range between 13.8033 K and 24.5561 K as mentioned above, the ICVGT scale over this range will not be disseminated from NIST unless by special request.

## Uncertainties

Along with these advances in realization and dissemination have come a commensurate reduction in the calibration uncertainties of resistance thermometers below 83.8058 K. Table 2 is a summary of the NIST disseminated calibration ranges, methods, devices and uncertainties for all calibrations below 273.16 K. The uncertainties given here reflect the most recent assessments at NIST for SPRT and RIRT calibrations. The basic statistical methodology underlying the uncertainty estimates has been adopted by NIST from documents defining the modern international practice for such assessments.[13],[14] This convention uses a coverage factor, $k$, to determine the statistical significance, and decomposes the uncertainties into Type A and Type B components. The Type A components are those derived from purely statistical methods and the Type B components are those known contributions which can not be directly measured. The expanded uncertainty, $U_k$, is then constructed according to the relation,

$$U_k = k(s^2 + \sum_j u_j^2)^{1/2}$$

**Table 2.** NIST expanded ($k$=2) calibration uncertainties for SPRTs and RIRTs .

| $T_{90}$ / K | Method (Cell Type) | Disseminated Definition | Reference | $U_2$ / mK SPRT | $U_2$ / mK RIRT |
|---|---|---|---|---|---|
| 273.16 | FP (LSI) | SPRT | $H_2O$ TP | 0.04 | |
| 234.3156 | FP (LSI) | SPRT | Hg TP | 0.20 | |
| 83.8058 | FP (LSI) | SPRT | Ar TP | 0.12 | |
| 83.8058 | C (IRB) | SPRT | SPRT | 0.22 | 0.22 |
| 54.3584 | C (IRB) | SPRT | SPRT | 0.18 | 0.18 |
| 24.5561 | C (IRB) | SPRT | SPRT | 0.24 | 0.24 |
| 20.3 | C (IRB) | SPRT | SPRT | 0.18 | 0.18 |
| 17.0 | C (IRB) | SPRT | SPRT | 0.17 | 0.17 |
| 13.8058 | C (IRB) | SPRT | SPRT | 0.18 | 0.17 |
| 5.0 to 13.8 | C (IRB) | ICVGT | RIRT | | 0.21 |
| 2.0 to 5.0 | C (IRB) | $^4$He VP | RIRT | | 0.08 |
| 0.65 to 2.0 | C (IRB) | $^3$He VP | RIRT | | 0.15 |

FP - fixed point calibration
C - comparison calibration

where $s$ is the Type A component and the $u_j$ are individual contributions to the Type B component. Unless stated otherwise, NIST uncertainty assessments are quoted with a coverage factor of $k$=2, yielding approximately 95% confidence for degrees of freedom of 50 or higher.

The SPRT calibration uncertainties in Table 2 are for the fixed point temperatures only. Each of these uncertainties propagate over the range 13.8033 to 273.16 K according to the plot shown in Figure 4. Further details concerning the uncertainties presented here may be found in the most recent NIST internal report on this subject.[15]

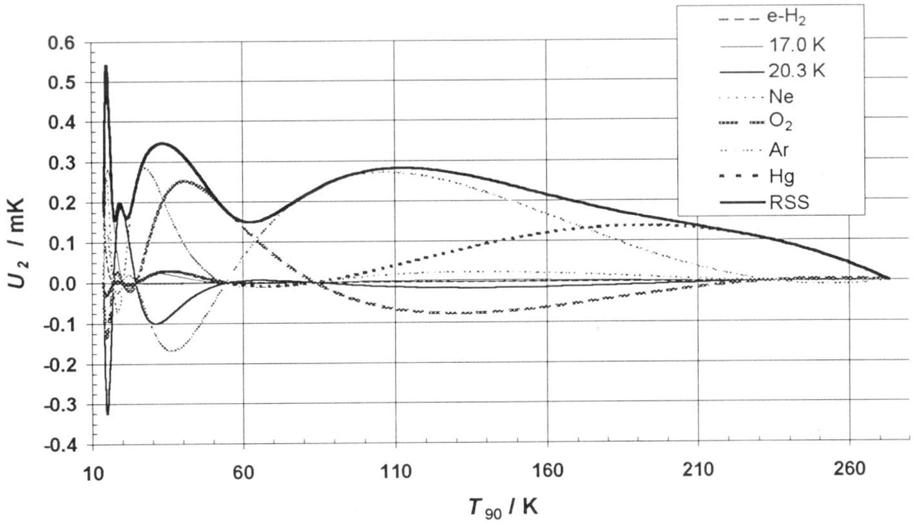


**Figure 4.** Expanded Uncertainties for SPRT calibrations over the sub-range 13.8033 K to 273.16 K as propagated from each of the ITS-90 fixed point temperatures in this range. The root-sum square (RSS) of the individual contributions as shown is the total calibration uncertainty over all temperatures in this range.

## CONCLUSIONS

The completion of the ITS-90 realization below 83.8058 K at NIST has verified that the previously maintained wire scale was indeed a reasonable approximation. The resulting necessary adjustments to the disseminated ITS-90 are small (≤ 1.2 mK) and should not have any practical impact on cryogenic engineering. Likewise, the small degree of non-uniqueness observed near 17 K will not affect the disseminated version of the scale. The most recent assessments of uncertainties in the disseminated ITS-90 represent a significant improvement in calibration accuracy and have enabled a level of assurance for low temperature thermometry that is the highest ever achieved at NIST. These advances should serve as the foundation for temperature measurement in cryogenic engineering for the foreseeable future.

## REFERENCES


1. H. Preston-Thomas, "The International Temperature Scale of 1990 (ITS-90)", *Metrologia*, **27**, pp. 3-10 (1990); ibid., p. 107 (erratum).
2. E.R. Pfeiffer, Realization of the ITS-90 below 83.8058 K at the NIST, *in:* "Temperature: Its Measurement and Control in Science and Industry", **6**, p.155, J.F. Schooley, ed., American Institute of Physics, New York (1992).
3. C. W. Meyer and M. L. Reilly, "A Progress Report on the Primary Realization of the ITS-90 from 0.65 K to 83.8058 K at the National Institute of Standards and Technology", Report of the Consultative Committee on Thermometry, BIPM, (1993).
4. B. W. Mangum and G. T. Furukawa, "Guidelines for Realizing the ITS-90", *NIST Tech Note 1265*, United States Department of Commerce, (August 1990).
5. C.W. Meyer and M.L. Reilly, " Realization of the ITS-90 in the range 0.65 K to 5.0 K using vapor pressure thermometry", *Metrologia*, **33**, p. 383 (1996).
6. C.W. Meyer and M.L. Reilly, "Realization of the ITS-90 in the range 3.0 K to 24.5561 K Using an Interpolating Constant Volume Gas Thermometer", *Proc. TEMPMEKO '96*, Torino, IMEKO TC 12, to be published, (1997).
7. G.T. Furukawa, *J. Res. Nat. Bur. Stand.* (U.S.), **91**, 255 (1986)., and G.T. Furukawa, *in*: "Temperature: Its Measurement and Control in Science and Industry", **5**, p.239, J.F. Schooley, ed., American Institute of Physics, New York (1982).
8. W. L. Tew, "Sealed Cells for the Realization of Triple Points at the NIST", *Proc. TEMPMEKO '96*, Torino, IMEKO TC 12, to be published, (1997).
9. J. D. Simmons, "NIST Calibration Services Users Guide", *NIST Special Publication 250*, United States Department of Commerce, (October, 1991).
10. W. L. Tew and B.W. Mangum, "New Capabilities and Proceedures for the Calibration of Cryogenic Resistance Thermometers at NIST." *Advances in Cryogenic Engineering,* **39 B**, p. 1019, Plenum Press, (1994).
11. G.F. Strouse, NIST implementaion and realization of ITS-90 over the range 83 K to 1235 K, *in*:"Temperature: Its Measurement and Control in Science and Industry", **6**, p.169, J.F. Schooley, ed., American Institute of Physics, New York (1992).
12. R.L. Rusby, The Rhodium-iron resistance thermometer, Ten years on, *in*:"Temperature: Its Measurement and Control in Science and Industry", **5**, p.829, J.F. Schooley, ed., American Institute of Physics, New York (1982).
13. ISO, "Guide to the Expression of Uncertainty in Measurement", International Organization for Standards, Geneva , Switzerland, (1993).
14. B. N. Taylor and C. E. Kuyatt, "Guidelines for Evaluating and Expressing the Uncertainty of NIST Measurement Results", *NIST Technical Note 1297*, 1994 Edition, 20 pp., U.S. Department of Commerce, (January, 1994).
15. W. L. Tew, G. F. Strouse and C. W. Meyer, "A Revised Assessment of Calibration Uncertainties for Capsule Type Standard Platinum and Rhodium Iron Resistance Thermometers", *NIST Internal Report*, (available in 1998).

# CRYOGENIC THERMOMETER CALIBRATION FACILITY AT CERN

C. Balle,[1] J. Casas[1], and J.P. Thermeau[2]

[1]CERN - division AT
CH-1211 Geneva 23, Switzerland
[2]IPN (CNRS-IN2P3)
91406 Orsay Cedex, France

## ABSTRACT

A cryogenic thermometer calibration facility has been designed and is being commissioned in preparation for the very stringent requirements on the temperature control of the LHC superconducting magnets. The temperature is traceable in the 1.5 to 30 K range to standards maintained in a national metrological laboratory by using a set of Rhodium-Iron temperature sensors of metrological quality. The calibration facility is designed for calibrating simultaneously 60 industrial cryogenic thermometers in the 1.5 K to 300 K temperature range, a thermometer being a device that includes both a temperature sensor and the wires heat-intercept. The thermometers can be calibrated in good and degraded vacuum or immersed in the surrounding fluid and at different Joule self-heating conditions that match those imposed by signal conditioners used in large cryogenic machinery. The calibration facility can be operated in an automatic mode and all the control and safety routines are handled by a Programmable Logic Controller (PLC). LabVIEW® is used both as the PLC operator interface and for configuring and reading the thermometric data sampled by the higher accuracy laboratory equipment. The isothermal support onto which the thermometers are mounted is thermally anchored through the wiring to a helium bath. The calibration procedure begins once the temperature of the support is stabilized. Measured data is presented and it is possible to infer that the absolute accuracy that can be obtained is better than ± 5 mK for the full temperature range.

## INTRODUCTION

The LHC project[1] will use about 1600 long superconducting magnets operating in pressurized liquid helium at 1.8 K and whose total cold length exceeds 25 km. The temperature of these magnets is a control parameter and as a consequence the cryogenic thermometers and their ancillary electronics are critical components that need to have adequate precision, redundancy and to be of industrial robustness; their total quantity is about 7000. This equipment will be installed inside the LEP tunnel, of 27 km circumference, and will be exposed to environmental conditions that are hostile from the point of view of ionizing radiation damage and electromagnetic interference.

The measurement of cryogenic temperatures is reputed to be difficult for various reasons like thermal anchoring of electrical leads, long-term stability of sensors, signal recovery and in the case of a particle accelerator, radiation hardness. The aimed accuracy is 0.25% of the measured temperature: this value is imposed on one hand by the maximum temperature at which the magnets can operate without losing the superconducting state and on the other hand by the cooling capacity of the cryogenic system[2]. In order to obtain such accuracy and a 20 year trouble-free operation, CERN has launched a program on cryogenic thermometry. This activity includes (a) the study of assembly techniques compatible with the factory floor[3], (b) design of calibration stations with high throughput, (c) procurement of temperature standards traceable to a national metrological laboratory, (d) sensor stability tests in existing machinery, (e) irradiation damage at low temperature, (f) creation of an Oracle® database for storing the sensor calibration data, making statistical analysis and studying long-term stability; and (g) selection of the most suitable sensor for the LHC project. This paper describes the calibration facility for cryogenic thermometers that satisfies the accuracy requirements of the LHC: it has the potential of processing about 240 thermometers per month, each batch being cycled and calibrated 3 times from ambient temperature down to 1.8 K.

At present the CERN temperature standard is maintained by comparing a set of secondary and working RhFe temperature standards; the apparatus for this purpose has been described elsewhere[4]. A primary temperature standard consisting of a neon triple point sealed cell is used to perform periodical checks on the secondary standards. The temperature is traceable in the 1.5 to 30 K range to the "Istituto di Metrologia G. Colonnetti", Turin, Italy. High quality RhFe standards of different origin (H. Tinsley and VNIIFTRI) are used in order to detect drifts that could be produced by the fabrication technique.

## APPARATUS

In order to get the best accuracy when installing a thermometer in the field, it is essential that its calibration be made in conditions as close as possible to those found in the cryogenic machinery. The main parameters are the thermometer's environment (vacuum or immersed in a fluid) and the Joule self-heating imposed by the type of signal conditioners used. The Joule effect seems to be critical with modern thin-film type thermometers probably because the heat evacuated through the wire bonds is not as efficient as that occurring in bulk-type sensors (for example an Allen-Bradley type) that use much thicker wires. The thermal anchoring of the wires and sensor with the object being measured can be reproduced in the calibration bench when using a thermometric block with built-in heat interception[3].

### Isothermal substrate and insert

An isothermal substrate (see Figure 1a) is used for comparing a set of industrial thermometers[3] with up to 5 working standards that are installed inside holes drilled at each end of the substrate, up to 60 thermometers can be calibrated simultaneously. The calibrations are done mostly under vacuum conditions by using a vacuum enclosure surrounding the substrate (Figure 1b). The vacuum tightness is obtained by using Helicoflex® metallic O-rings. The substrate is made of O.F.H.C. copper and its total mass is about 34 kg. The electrical connections are done by using manganin wire exclusively, there are 6 x 48-pin vacuum-tight connectors at room temperature and the wiring length is about 2 meters. The temperature of the substrate can be varied by either using an electrical heater or gaseous helium as heat exchange gas with the surrounding media.

The wires are thermally anchored at two different temperatures:

(a) at the low temperature end, a foil of kapton onto which all the wires are printed (319 copper strips) is used, this foil is pressed against the inner surface of a copper tube, heat is transferred to the cryogen through the tube wall and

(b) five disks fabricated by using printed circuit board technique are used for heating-up the wires when necessary. These disks are also used for distributing the electrical wiring to the thermometers.

Outside the vacuum enclosure, 60 thermometers are furthermore installed in order to increase the calibration throughput, these sensors are calibrated immersed in the surrounding liquid helium bath in the 1.6 to 4.2 K temperature range.

(a)

(b)

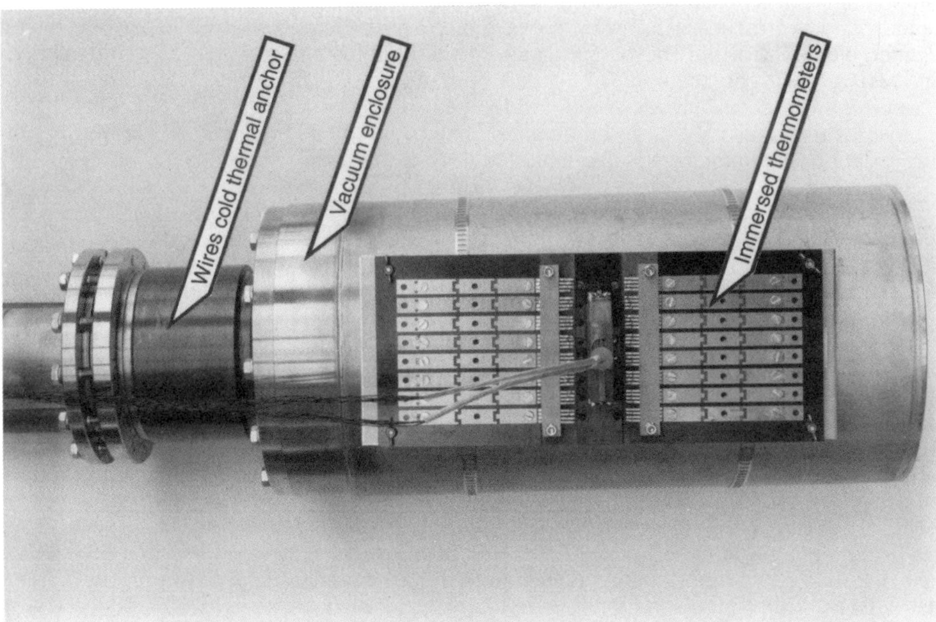


**Figure 1.** Insert (a) isothermal substrate with thermometers mounted and showing the wires heaters and (b) with vacuum enclosure and immersed thermometers.

## Process

The flow scheme of the installation is shown in Figure 2. The insert holding the isothermal substrate is immersed in either liquid helium or nitrogen. For transferring liquid helium, the gas pressure of a 500 liter storage dewar is increased by using a control valve FCV1 that regulates the gaseous helium flow FT1 at the outlet of the calibration cryostat, the transfer is interrupted after reaching a pre-defined liquid helium level or if any critical measurement exceeds a safety threshold.

The substrate temperature is lowered by using gaseous helium as heat exchange gas, its pressure is regulated by a pressure control valve PCV2; this valve is a programmable micro-valve capable of adjusting the pressure between $10^{-5}$ and 100 mbar. After reaching a pre-defined temperature level, the feeding valve is closed, the vacuum is re-established (typically better than 5 $10^{-5}$ mbar as measured by a room-temperature device - PIT3 in Figure 2) and the temperature of the substrate is stabilized (typically to within better than ± 1 mK) using an electrical heater.

The liquid helium bath temperature is varied using a butterfly valve PCV3; a simple PID control loop is not suitable for regulating the bath temperature because the optimum PID parameters depend strongly on the operating temperature.

After reaching the lowest calibration temperature, the system is warmed-up using various electrical heaters; during this procedure it is also possible to acquire calibration data.

## Control and acquisition equipment

A Programmable Logic Controller (PLC) is used for managing all the safety routines, the closed-loop controls, the sequence of temperature plateaus and the synchronization of the acquisition equipment. The man machine interface is provided by a Macintosh® computer running LabVIEW®. With a fully automated procedure a complete temperature cycle is estimated to have a duration of about one week.

The data acquisition equipment is of laboratory type; an IEEE-488 bus is used for communication between the instruments and the computer running LabVIEW®. The main components are a resistance bridge with dedicated scanner for measuring the working standards, a voltmeter, a programmable d.c. current-voltage source and a 400-channel scanner. For controlling the performance of the calibration equipment, various standard type resistors with an accuracy of about 5 ppm are used. At present the calibration data is stored in hard disk as flat files only and the post-processing of data is somewhat difficult because of the relatively large amounts of data to handle. This analysis should be easier once a data-base application is available for storing and retrieving this data.

# EXPERIMENTAL RESULTS

For performing a calibration it is necessary to have at least one temperature reference which is used for comparison with the thermometers attached to the isothermal substrate. In our substrate we typically install between 3 to 5 working standards that are calibrated in a different apparatus[4]. The working standards are compared periodically with a set of four secondary standards that are traceable in the 1.5 K to 30 K range to standards maintained in a national metrological laboratory. The working standards are not traceable above 30 K but the measurements suggest that the quality of the temperature sensor used for calibration in this range is sufficient for our application (e.g. better that 0.25 % on the measured temperature). Figure 3 shows the temperature spread between 3 working standards during one temperature cycle covering 1.6 K to 300 K, it can be seen that the maximum dispersion is below ± 2 mK. This demonstrates the good repeatability with the data obtained from our standard inter-comparison apparatus. The data presented assume that during a calibration scan there is no temperature drift of the substrate, better accuracy should be obtained by measuring this drift. This also shows that, within the measurement errors, it is not possible to detect a temperature gradient along the substrate (the working standards are installed at opposite ends).

The repeatability of the calibration for the 60 sensors installed in the isothermal substrate is assessed by performing two temperature cycles between ambient and 1.6 K;

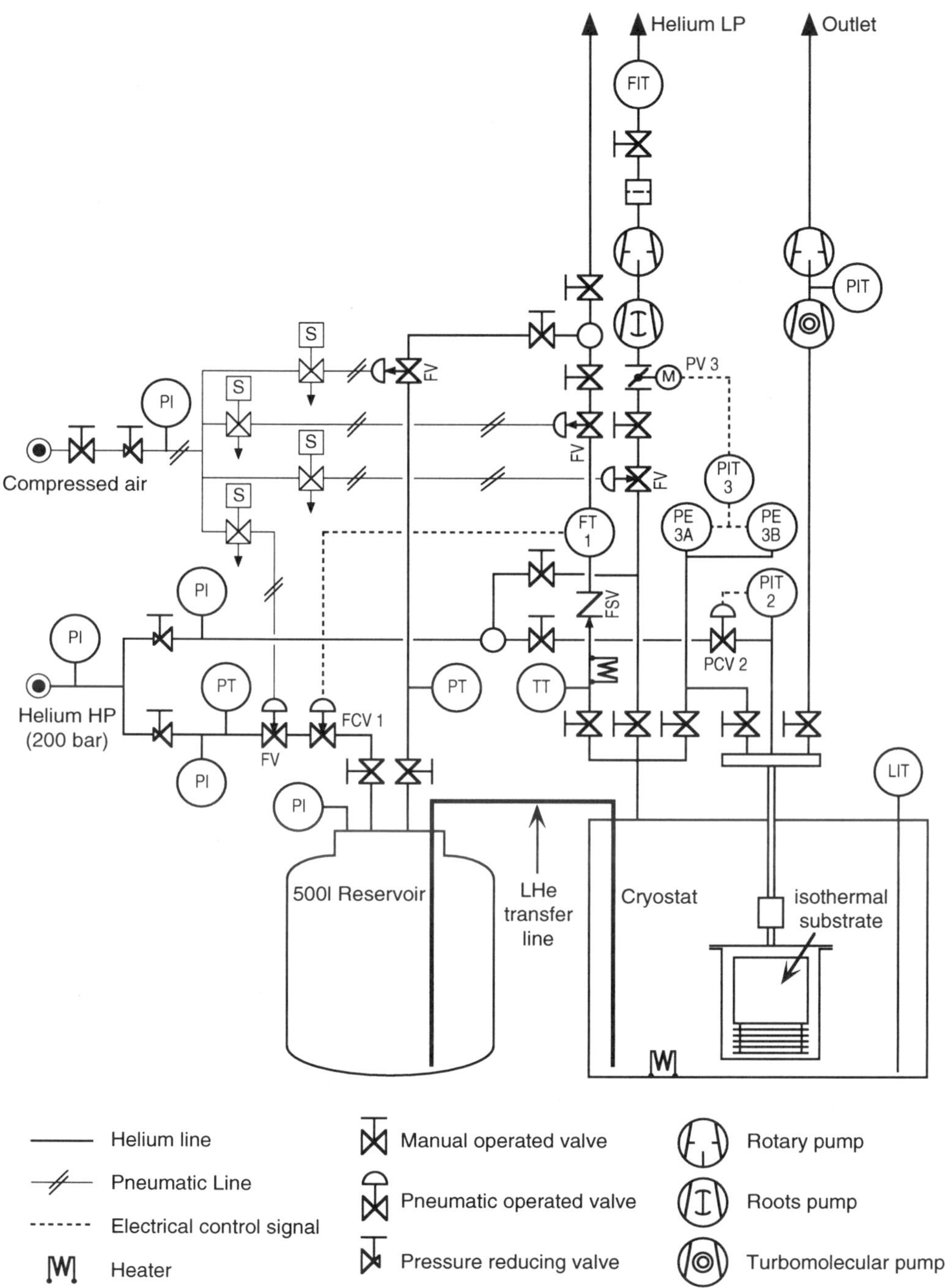


**Figure 2.** Flow scheme of the calibration facility. Letter code is C control, E element, F flow, I indicator, L level, M motor, P pressure, S solenoid, T transducer and V valve.

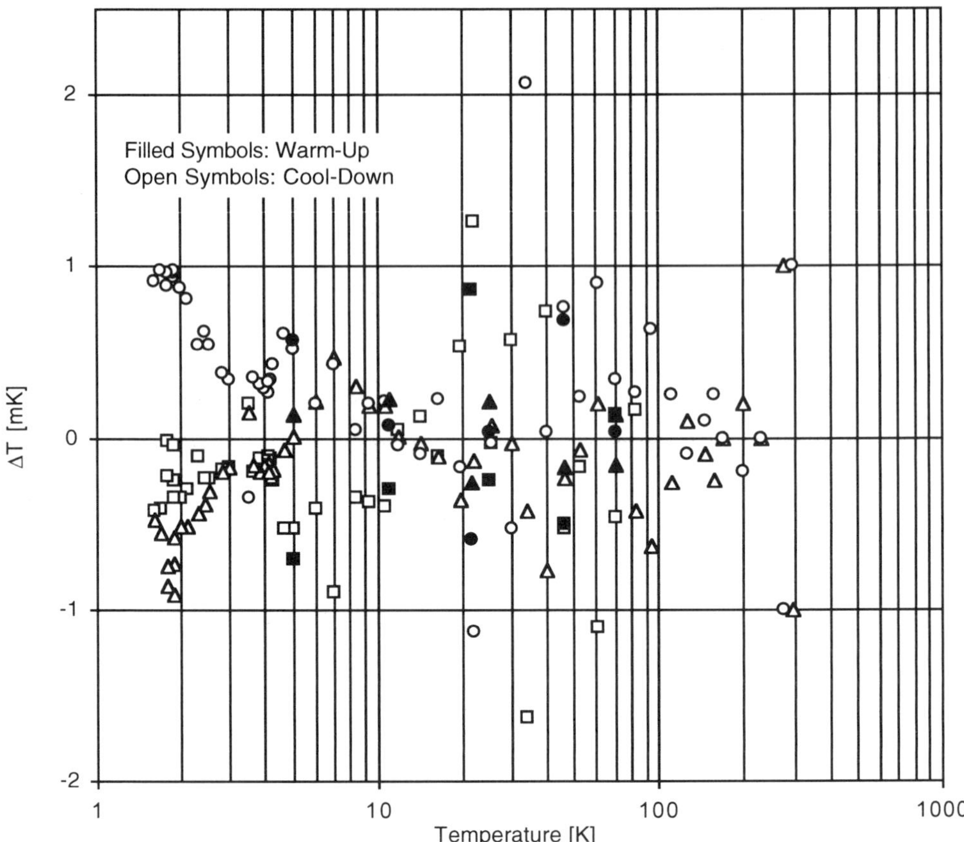


**Figure 3.** Temperature spread observed in the 1.6 to 300 K temperature range between three high quality RhFe temperature sensors used as working standards. One of the working standard (square symbols) is calibrated only below 90 K.

thermometers located on the lower (upper) sector of the cylindrical substrate are removed and installed in the upper (lower) sector before starting the second cycle. This measurement should reveal any temperature gradient or systematic error, within the stability over temperature cycles of the physical parameters determining the sensors characteristics. The temperature sensors examined during this test are: 58 CERNOX SD-1050, one Allen-Bradley type carbon sensor and one Russian TVO-type[5], all of them are mounted in CERN's industrial thermometer support[3]. Figure 4 shows the reproducibility of the measurements, it can be seen that above 150 K there seems to exist a temperature gradient along the surface of the isothermal support, this gradient is not observed with the working standards as shown in Figure 3. Below 100 K the repeatability of the measurements is better than ± 10 mK and Figure 4b suggest that it improves as the temperature is decreased. The spread in the repeatability at higher temperatures is probably the consequence of the type of sensors investigated: as the temperature is raised the variation of resistance with temperature decreases and a small variation on the sensor characteristics can account for a relatively large measurement error.

## CONCLUSION

A calibration facility capable of calibrating a large number of thermometers has been described. The self-consistency between the set of working standards is similar to what is observed in a high-accuracy metrological apparatus, showing that the facility described in

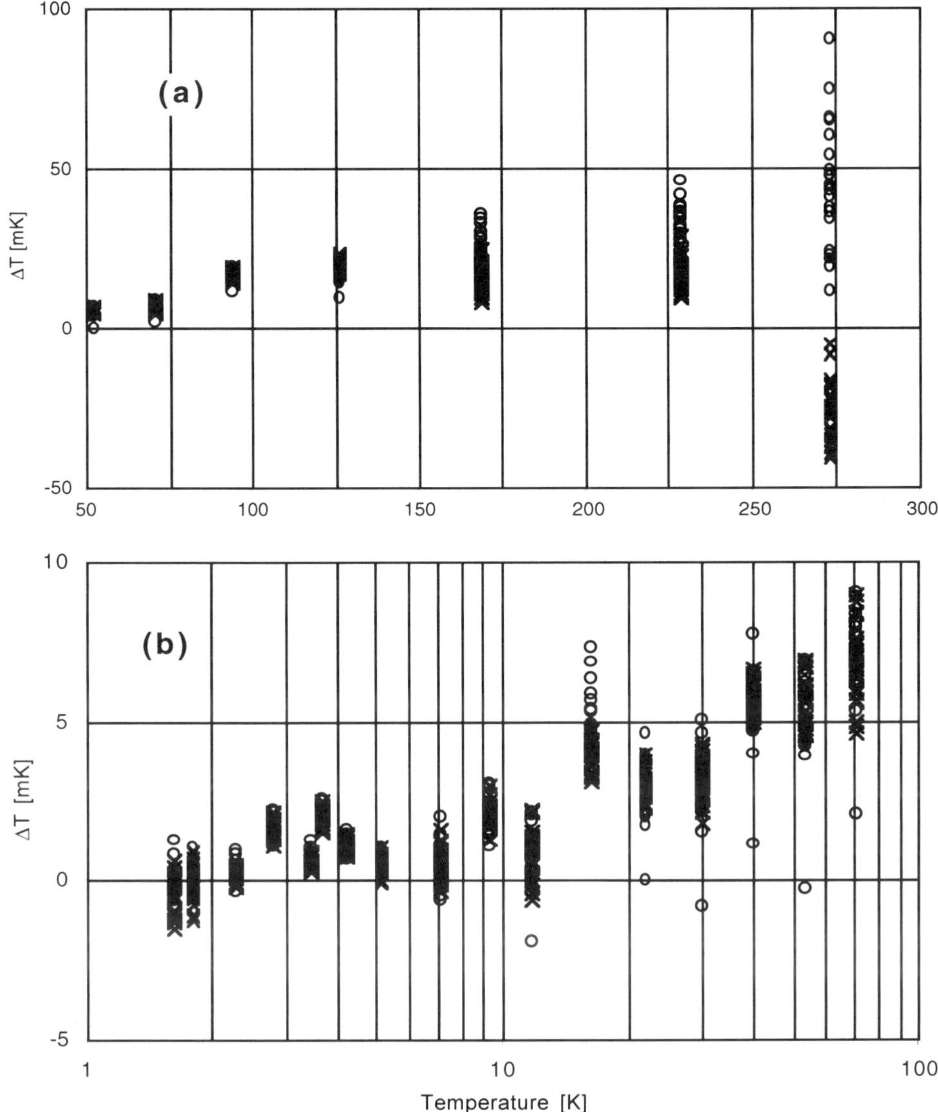


**Figure 4.** Reproducibility measured for 60 temperature sensors during two thermal cycles. "O" and "X' symbols correspond to sensors installed in the upper and lower sector respectively during the first temperature cycle. Their positions are inverted for the second cycle. (a) high temperature and (b) low temperature.

this paper has the potential of having an accuracy better than ± 2 mK in the 1.6 to 300 K temperature range.

It is seen however that the industrial-type thermometers have a much worse repeatability, although it improves with decreasing temperature. The data also suggest that a temperature gradient exist along the substrate for temperatures above 150 K. The accuracy of the calibration facility presented in this paper can be further improved by performing a new series of tests to assess the origin of this gradient. Nevertheless the results demonstrate that the installation absolute accuracy is much better than 0.25 % of the measured temperature that is the requirement dictated by the LHC cryogenic system.

In order to increase the overall throughput of this installation, an Oracle database for managing the calibration data is being designed and new automatic routines should be implemented to be able to work without the assistance of an operator.

## ACKNOWLEDGMENT

We would like to thank V. Datskov, visitor from JINR-Dubna (Russia) for many fruitful discussions.

This work has been performed in the framework of a collaboration agreement between CERN and CNRS-IPN Orsay, France.

## REFERENCES

1. L. R. Evans, The Large Hadron Collider Project, in: "proceedings of CEC 16-ICMC", Kitakyushu - Japan (1996), p. 45
2. V. Benda et al., Conceptual design of the cryogenic system for the Large Hadron Collider (LHC), EPAC-96, (1996), P. 361
3. C. Balle and J. Casas, Industrial-type cryogenic thermometer with built-in heat interception, in: "Advances in Cryogenic Engineering", Vol. 41B, Plenum, NY (1996), p. 1715
4. F. Pavese, V.M. Malyshev, P.P.M. Steur , D. Ferri and D. Giraudi, Routine calibration of cryogenic thermometers in the range 1.5-300K with an accuracy up to the milliKelvin level, and measurement of fixed points in sealed cells with a fully automated and self-contained modular apparatus, in: "Advances in Cryogenic Engineering", Vol. 41B, Plenum, NY (1996), p. 1683
5. V.I. Datskov and J.G. Weisend II, Characteristics of Russian carbon resistance (TVO) cryogenic thermometers, Cryogenics, Vol 34 (1994), p. 425

# MINIATURE RESISTANCE THERMOMETERS BASED ON Ge FILMS ON GaAs

V.F.Mitin

Institute of Semiconductor Physics of the
National Academy of Sciences of Ukraine
Prospect Nauki 45, 252650 Kiev-28, Ukraine

## ABSTRACT

Several types of miniature resistance thermometers based on Ge films have been developed and fabricated. They can operate in the 0.5 to 300, 77 to 400 and 200 to 500 K temperature ranges. The behavior of cryogenic thermometers has been investigated at low temperatures (1.7, 4.2 and 77.4 K) and in magnetic fields up to 14 T.

## INTRODUCTION

The variety of cases where temperature measurements are needed, as well as the diversity of operational requirements, have motivated the development and availability of a wide range of devices for measuring temperature. The present microelectronic thin-film technology opens fresh opportunities for developing miniature temperature sensors whose characteristics can meet the nowadays requirements.

The works [1-3] were devoted to development of the basic principles of designing sensors (for measuring temperature, strain, magnetic field, optical radiation) based on Ge films on GaAs substrates. Here we discuss the problem of developing miniature resistance thermometers. Such thermometers are suited to operate in the 0.5 to 500 K temperature range. We present thermometric characteristics of thermometers and discuss the conduction mechanisms in Ge films which are responsible for thermal sensitivity. Also we report on the behavior of cryogenic thermometers at low temperatures and in high magnetic fields.

## PRINCIPAL PREREQUISITES FOR DEVELOPING THE RESISTANCE THERMOMETERS BASED ON THE Ge-GaAs HETEROSTRUCTURE

The Ge-GaAs heterostructure seems to be promising for the temperature sensor development. The following reasons give grounds for such a statement. First, Ge and GaAs constitute an ideal heteropair. Their lattice misfit is extremely small (0.08 %). Their thermal expansion coefficients practically coincide over a wide temperature range. It is well known that absence of temperature stresses in a sensitive element is of great importance for devices operating in a wide temperature range. Second, during the process of epitaxial deposition of the Ge onto GaAs substrates, Ga and As atoms strongly diffuse from the substrate into the growing film and the solubility limits and diffusion coefficients of these impurities in the film turn out to be significantly (by $10^2$-$10^3$ times) higher than those in the Ge bulk. This interdiffusion disobeys the classical diffusion laws, and so it may be called anomalous. Interpenetration of both materials essentially depends on the technological conditions of the heterostructure formation. Ga (As) atoms serve as acceptors (donors) in Ge. Related to them are shallow impurity states in the Ge gap whose energies (~ 0.01 eV) are close. Ge serves as an amphoteric impurity in gallium arsenide.

Thin Ge films on the GaAs substrate are characterized by a complex spectrum of impurity states in the gap. This spectrum substantially depends on the film growth conditions. There exist three main sources of charge carriers in a Ge film (in addition to the intrinsic charge carriers), namely, Ga and As impurities and structural defects. Both the structural perfectness degree and the structural defect concentration are mainly determined by the substrate temperature during Ge deposition. The film structure may vary over a wide range - from poly- to single-crystalline. A quantity of structural defects (predominantly acceptors) exists in Ge films along with point defects - acceptors (Ga) and donors (As). A relationship between their concentrations substantially depends on the film growth conditions. Presence of the above defects results in the fact that Ge films on GaAs are both doped and compensated. The diffusion processes at the Ge/GaAs interface and in the transition region are determined by the technological conditions of the heterostructure formation. Among such conditions are the substrate temperature, deposition rate and the GaAs surface condition before the Ge deposition. Autodoping makes it possible to obtain Ge films of *p*-, as well as *n*-, type, and also of different doping levels (from $10^{17}$ to $10^{21}$ $cm^{-3}$) and compensation degrees. Under certain conditions the compensation degree may be close to unity for Ge films. Depending on the film formation conditions, the free charge carrier concentration (resistivity) in Ge films at room temperature may vary from $10^{14}$ to $5 \times 10^{20}$ $cm^{-3}$ (140 to $10^{-3}$ $\Omega$ cm).[2]

A distinguishing feature of diffusion processes in the Ge - GaAs heterostructure is that the anomalously quick diffusion that occurs during the film deposition significantly differs from the diffusion that occurs during the heat treatment of the formed hetetrostructure. Anomalous values of the impurity diffusion coefficients are immediately related to the very process of the epitaxial film formation. Heat treatment of the formed heterosystem does not lead to the anomalously quick diffusion of impurities. So the devices based on the Ge/GaAs heterostructure may be considered to be heat - resistant.

A variety of electrical properties exist for Ge films on the GaAs substrates. This enables using the Ge - GaAs heterostructure as a versatile sensitive material when developing different physical sensors (of temperature, strain, pressure, magnetic field, optical radiation).[2] To produce a wide-range resistance thermometers, it is necessary to provide a high thermosensitivity in a wide temperature range. This is commonly achieved through multicomponent doping and compensation of a semiconductor. Thus in a semiconducting material there exist several different mechanisms of conductivity, and each of them provides thermosensitivity in a definite temperature range. By varying the doping level and the compensation degree of Ge films, one can obtain the resistance thermometers characterized by different thermal sensitivity in different temperature ranges.

In this work the films have been prepared by thermal evaporation of Ge in a vacuum ($2\times10^{-4}$ Pa) onto semiinsulating GaAs (100) substrates. To produce sensitive elements, we have used a conventional microelectronic technology. It included the following procedures: (i) ohmic contacts deposition by thermal evaporation of AuGe in a vacuum; (ii) photolithography to form the topology of the sensitive region and terminal areas; (iii)

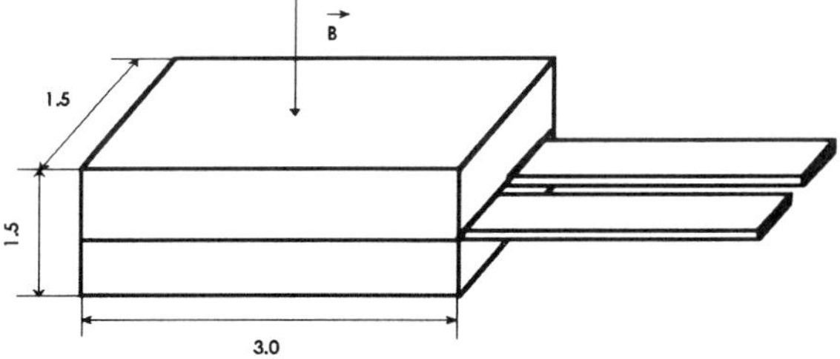


**Figure1.** The external view and dimensions (in mm) of the sensor.

wafer slicing; (iv) microassembly; (v) packaging. The sizes of sensitive elements were 0.5 × 0.5 × 0.2 and 1.0 × 1.0 × 0.2 mm for different types of sensors. The sensitive elements were placed into packs whose sizes were 3.0 × 1.5 × 1.5 and ∅1.6 × 4.0 mm.

## RESISTANCE THERMOMETERS FOR THE 0.5 TO 300 K TEMPERATURE RANGE

We have fabricated cryogenic wide-range resistance thermometers of two types, designated as TTR-1A and TTR-1B. They can operate in temperature range from 0.5 to 300 K. The external view and dimensions of the sensor are shown in Figure 1.

Figure 2 shows the characteristics of thermometers. The TTR-1A thermometers are more sensitive in 50 to 300 K range, while the TTR-1B thermometers are more sensitive at low (< 10 K) temperatures. The sensitive elements of the thermometers have been made under various technological conditions and thus differ in their thermometric characteristics. The thermal sensitivity of a thermometer depends on the technological conditions of film preparation, since they determine both the doping level and the compensation degree of Ge films. The thermometer resistance is determined by the sensitive element topology formed by lithography. By varying the film preparation conditions and the thermosensitive area topology, one can fabricate resistance thermometers differing in both resistance value and thermal sensitivity.

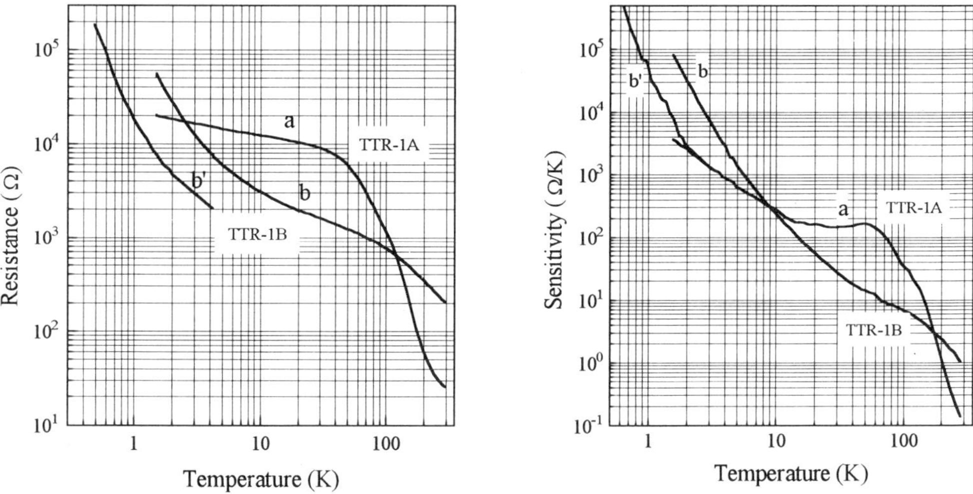


**Figure 2.** Thermometer resistance and sensitivity versus temperature curves: a - TTR-1A, b and b' - TTR-1B.

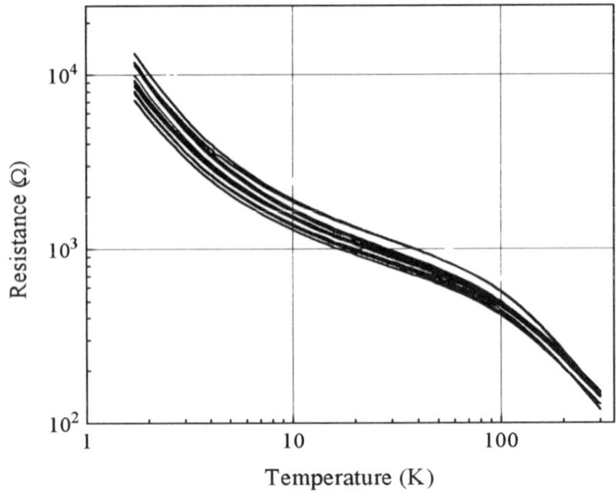


**Figure 3.** TTR-1B thermometer resistance versus temperature curves.

Figure 2b' shows the temperature dependence of the TTR-1B thermometer resistance in the 0.5 to 4.2 K temperature range. The resistance thermometers operating at low temperatures should not have high input resistance. Otherwise, even a small supply current will warm up the thermometer by Joule effect, which will result in a temperature measurement error. That is why we have chosen the TTR-1B thermometers with lower resistance for operating at the temperatures below 1.5 K. The TTR-1A thermometers also could be used at these temperatures, but their thermal sensitivity is lower than that of the TTR-1B ones.

Figure 3 illustrates difference of the characteristics of TTR-1B thermometers. Such thermometers have been produced from the same wafer of Ge/GaAs heterostructure. It is necessary to note, that about 10000 sensitive elements of thermometers may be produced from one wafer by means of microelectronic technology. In this sense a microelectronic technology is profitable and permits to manufacture cheap devices.

Figure 4 shows the dependencies of resistance on the supply current for two typical TTR-1B thermometers at 4.2 K. An essential change of resistance of thermometers has been observed at a current above 20 μA. Degree of constancy for resistance in the range of current up to 10 μA provides measurement of temperature with accuracy about 1 mK.

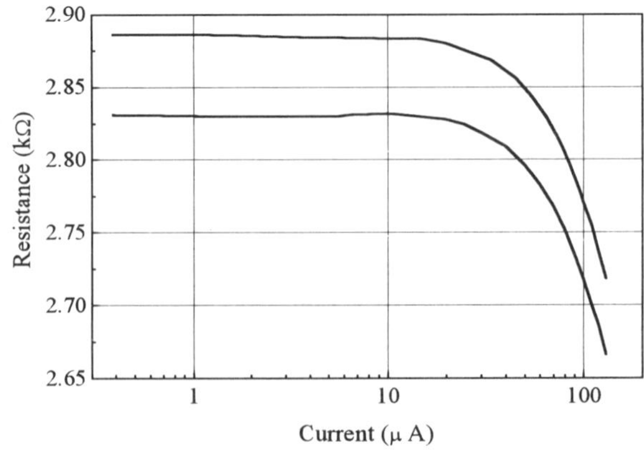


**Figure 4.** TTR-1B thermometer resistance versus current curves at 4.2 K.

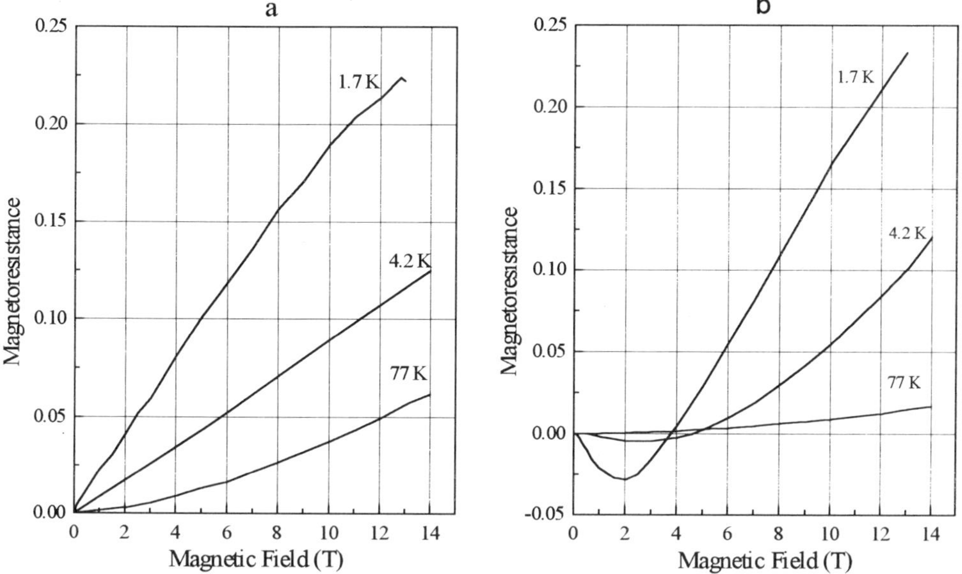


**Figure 5.** Dependence of the magnetoresistance (ΔR/R) on the magnetic field for two types of Ge film resistance thermometers: a - TTR-1A, b - TTR-1B.

## EFFECT OF MAGNETIC FIELD ON Ge FILM THERMOMETERS

In cryogenic engineering and experimental physics there is a demand for miniature temperature sensors that can operate in a wide temperature range and in high magnetic fields. Such sensors must have high accuracy of temperature measurements in magnetic fields.

The behavior of the Ge film resistance thermometers has been studied in high magnetic fields up to 14 T. The orientation of the magnetic field with respect to the sensor is shown in Figure 1. In that configuration the transverse magnetoresistance has been observed in the thermosensitive Ge films ( $\boldsymbol{j} \perp \boldsymbol{B} \,||\, \boldsymbol{n}$, where $\boldsymbol{n}$ is the normal to the film surface, $\boldsymbol{B}$ and $\boldsymbol{j}$ are the vectors of the magnetic induction and current, respectively). The magnetic field dependencies of magnetoresistance (ΔR/R) for thermometers are given in Figure 5. The TTR-1A thermometer shows a positive magnetoresistance in the whole range of magnetic fields and temperatures. On the contrary, the TTR-1B thermometer shows a negative magnetoresistance below 4.2 K in magnetic fields up to 4 T. But at 77 K its magnetoresistance is positive.

The magnetoresistance causes an error in the thermometer reading. The error can be given as a ratio $\Delta T/T$ (%), where $\Delta T = T(B) - T$, $T$ is the temperature measured at magnetic field $B = 0$, and $T(B)$ is the temperature measured at magnetic field $B$. The magnetic field dependent temperature errors, $\Delta T/T$ (%), for the Ge thin film thermometers at various magnetic fields and at 1.7, 4.2, and 77 K are displayed in Figure 6 and Table 1. At temperatures below 77 K the TTR-1B thermometer has the error in the temperature reading (caused by the magnetic field) much smaller than the TTR-1A thermometer has. At 77 K the temperature errors of both thermometers are close. The TTR-1B thermometer has small magnetic field-induced temperature errors, and can be used for temperature measurements at the presence of magnetic fields. The Ge film thermometers are better than typical bulk Ge thermometers if measurement in magnetic fields.[4]

**Table 1.** Magnetic Field-Dependent Temperature Errors $\Delta T/T(\%)$ at Magnetic Field

| Sensor Type | T(K) | Magnetic field (T) | | | |
|---|---|---|---|---|---|
| | | 2.5 | 4 | 8 | 14 |
| TTR-1A | 1.7 | -18.35 | -28.72 | -55.15 | - |
| | 4.2 | -9.39 | -14.82 | -30.94 | -54.77 |
| | 77.4 | -0.13 | -0.28 | -0.85 | -1.97 |
| TTR-1B | 1.7 | 1.04 | -0.21 | -4.62 | - |
| | 4.2 | 0.45 | 0.25 | -2.61 | -10.67 |
| | 77.4 | -0.11 | -0.23 | -0.81 | -2.25 |

## RESISTANCE THERMOMETERS FOR THE 77 TO 400 AND 200 TO 500 K TEMPERATURE RANGES

We have fabricated thermometers intended for operating in the 77 to 400 and 200 to 500 K temperature ranges, designated as TTR-2 and TTR-3, respectively. A high compensation degree in the Ge film is required to reach higher thermal sensitivity at the temperatures above 77 K. Shown in Fig.7 are the characteristics of the TTR-2 and TTR-3 thermometers. Conduction in these thermometers is of activation type. The conductivity activation energy for TTR-2 thermometers is constant in the 77 to 400 K temperature range and may vary from 0.07 to 0.15 eV (depending on technological conditions of film preparation).

To fabricate the resistance thermometers for the 200 to 500 K temperature range, we have used the heavily doped and highly compensated (HDHC) Ge films[5,6]. The resistivity versus temperature curve for a HDHC Ge film usually includes two exponential sections and may be described by the following expression:

$$\rho = \rho_1 \exp(E_1/kT) + \rho_2 \exp(E_2/kT).$$

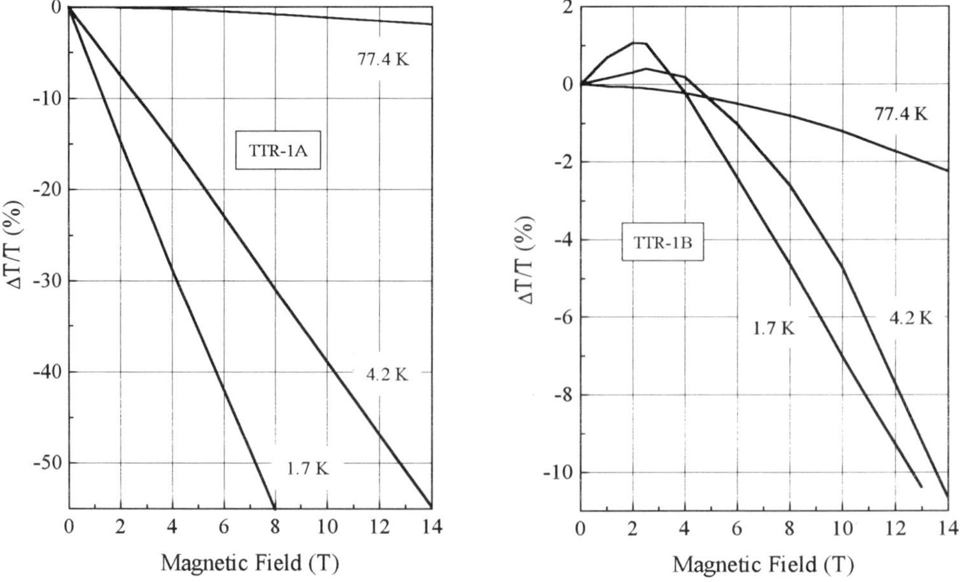


**Figure 6.** Temperature error in % versus magnetic field curves.

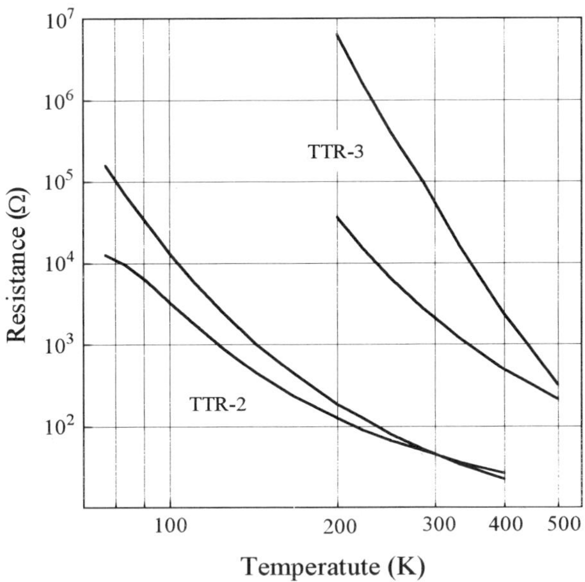


**Figure 7.** TTR-2 and TTR-3 thermometers resistance versus temperature curves.

A single-crystalline HDHC Ge film represents a complicated system of nonuniform composition. The properties of such a system are determined by the electrostatic potential fluctuations due to the nonuniform distribution of impurities.[5, 6]

At high temperatures the activation-type conduction in such films is determined by the thermionic emission of charge carriers from the Fermi level, $\varepsilon_F$, to the percolation level, $\varepsilon_p$. The activation energy is $E_1 = |\varepsilon_p - \varepsilon_F|$. Both $\varepsilon_p$ and $\varepsilon_F$ values depend on the compensation degree, dimensionality of the operating area and correlation in the impurity positions.[7] For the HDHC Ge films the activation energy of resistivity $E_1$ may reach half the band gap of Ge.[5, 6]

At low temperatures the conduction in HDHC Ge films is mostly realized through the localized states near the Fermi level and can be characterized by the activation energy $E_2 \neq E_1$. The temperature value at which a transition occurs from the conduction at the percolation level to that through the localized states depends on both the doping level and the compensation degree. It is determined by the two competing mechanisms, namely, thermionic emission and charge carriers tunneling in a potential relief. For the HDHC Ge films of the highest compensation degree the transition temperature is 250 - 290 K as a rule.[5, 6]

The resistance thermometers on the base of HDHC Ge films are characterized by a very high thermal sensitivity. Depending on the technological conditions of the Ge films preparation, this thermal sensitivity may be varied from 1.0 to 6.0 % $K^{-1}$ (at $T = 300$ K).

## CONCLUSIONS

Based on Ge films on GaAs substrates, several types of miniature resistance thermometers have been developed. Their operating temperature regions overlap the 0.5 to 500 K temperature range. A distinguishing feature in the fabrication technology of thermal sensitive elements is the possibility of obtaining the resistance thermometers of different both resistances and their temperature dependencies. Thermal sensitivities and resistances of such thermometers depend on the technological conditions of epitaxy. These conditions determine both the doping level and the compensation degree of Ge films. The thermometer resistance is also determined by the sensitive element topology. It is formed by photolithography and may be varied over wide limits. This enables one to fabricate thermometers which are intended for operating in different temperature ranges and whose operational characteristics can meet the user's demands.

The effect of magnetic field on some of there cryogenic thermometers has been studied. The Ge film thermometers designated as TTR-1B shows a small magnetic field induced temperature errors. Such thermometers can be used for temperature measurements under high magnetic fields.


## ACKNOWLEDGEMENT

The author is indebted to Prof. M.Oszwaldowski (Instytut Fizyki, Politechnika Poznanska, Poznan, Poland), and Prof. J.Klamut (International Laboratory of High Magnetic Fields and Low Temperatures, Wroclaw, Poland) for possibilities the investigations of Ge film thermometers in high magnetic fields. Also the author is very grateful to Dr. N.S.Boltovets (State Research Institute "Orion", Kiev, Ukraine) for the collaboration in the field of packaging. The author would like to thank Mr. V.Kholevchuk and Ms. V.Gavrilenko (Institute of Semiconductor Physics of the National Academy of Sciences of Ukraine, Kiev, Ukraine) for their technical assistance.


## REFERENCES


1. V.F.Mitin, Microsensors of temperature, magnetic field, and strain, in: *Proc. 15th Intern. Cryogenic Engineering Conf. (ICEC 15),* Genova, Italy, June 7-10, 1994, *Cryogenics* 34: 437 (1994) ICEC Supplement.
2. V.F.Mitin and Yu.A.Tkhorik, Production of sensors for different physical parameters on the base of all-purpose and unified technology, in: *Proc. 20th Intern. Conf. on Microelectronics (MIEL'95),* Nis, Serbia, September 12-14, 1995, Vol.2, pp.553-557.
3. V.F.Mitin, Yu.A.Tkhorik and E.F.Venger, New type of thin-film germanium resistance thermometer for use in a wide temperature range, in: *Proc. 16th Intern. Cryogenic Engineering Conf. / Intern. Cryogenic Materials Conf. (ICEC 16/ICMC),* Kitakyushu, Japan, May 20-24, 1996 pp.653-656.
4. L.G.Rubin, B.L.Brandt and H.H.Sample, Some practical solutions to measurement problems encountered at low temperatures and high magnetic fields, *Advances in Cryogenic Engineering*, Vol.31, pp.1221 (1986).
5. N.P.Garbar, L.A.Matveeva, V.F.Mitin, Yu.A.Tkhorik, R.Harman, Yu.M.Shvarts and Z.Stroubek, Heavily doped and strongly compensated heteroepitaxial germanium films, *Fiz. Tekhn. Poluprov.*, Vol.21, N 3, pp.393-399 (1987) (*Sov. Phys. - Semicond.*, Vol.21, N 3 (1987)).
6. V.F.Mitin, E.V.Mozdor and Yu.A.Tkhorik, Two-dimensional percolation in the heteroepitaxial films of germanium, *Fiz. Nizk. Temp.,* Vol.17, N 9, pp.1124-1126 (1991) (*Sov.* J. *Low Temp. Phys.*, Vol.17, N 9 (1991)).
7. B.I.Shklovskii and A.L.Efros, "Electronic Properties of Doped Semiconductors," Springer Verlag, Berlin (1984).

# HIGH RESOLUTION THERMOMETRY USING THE MAGNETIC PENETRATION DEPTH OF SUPERCONDUCTING FILMS

C.J. Yeager, P. J. Shirron, and M.J. DiPirro

Code 713, NASA/Goddard Space Flight Center
Greenbelt, MD 20771

## ABSTRACT

A Penetration Depth Thermometer (PDT) has been under development for potential use in microgravity critical point experiments. The active element is a superconducting film whose thickness is comparable to the magnetic penetration depth (5-20 nm). Field penetration is strongly temperature dependent just below the transition (~0.9 -1.0 $T_c$ ). Fields are generated and sensed by using excitation and pickup coils located near the film. Using a dc SQUID sensor for readout, temperature resolution on the order of 0.1 nK/√Hz can be obtained. Earlier work using aluminum films ($T_c$=1.69 K) demonstrated a maximum sensitivity of 1 nK/√Hz. Improvements in coil design have since yielded a factor of 3 improvement. PDT's have been made with aluminum films deposited on a planar sapphire substrate with transition temperatures greater than the lambda point in superfluid helium. For performance tests, a cryostat has been constructed where the PDT will be in an experimental cell filled with superfluid. This will provide thermal stability better than 0.1 nK/√Hz . Results of the current design are presented.

## INTRODUCTION

The study of critical phenomena and phase transitions is an active area of research. It also places great demands on high resolution thermometry as physical properties are probed near the critical temperature. This is particularly true for microgravity experiments where the limiting factor is now the thermometry. Since liquid helium provides an ideal system to test many aspects of critical phenomena, it is important to improve on the state of low temperature thermometry.

Current state of the art thermometry utilizes the temperature dependence of the susceptibility of paramagnetic salts to provide sensitivities less than 1 nK/√Hz. This was used in the Lambda Point Experiment to measure the specific heat near the superfluid transition. One of the limitations of this thermometer was its larger heat capacity and thus slow thermal response (fraction of a second). This was the primary limitation for the on orbit experiment, where cosmic ray events absorbed by the thermometer caused thermal spikes which limited the resolution to 1 nK/√Hz[1]

A PDT can be fabricated to have a small heat capacity and fast thermal response. Since it can have a simple geometry it could be easily incorporated into the experiment and would be useful in thermal conductivity measurement at the lambda point and second sound measurements.

Prior work has demonstrated the potential sensitivity of a PDT is on the order of 1 nK/√Hz[2]. This was seen in a 7 nm aluminum film with a transition temperature of 1.69 K. The absolute sensitivity could not be demonstrated due to much higher temperature fluctuations in the apparatus. It was also observed that the measured sensitivity was less than theoretically calculated due to the primary excitation coil generating magnetic fields near the surface of the film which are greater than the films critical field.

This paper presents results on three areas of development: The basic mechanism resulting in the loss of current carrying ability and efforts to optimize the PDT; the development of a PDT with operating temperature above the lambda point; and demonstration of the absolute sensitivity of the PDT.

## BASIC PRINCIPLES AND THEORY OF OPERATION

The magnetic penetration depth is the characteristic length a magnetic field penetrates into a superconductor. The superconductor generates screening currents on its surface to expel an external field. The Meissner effect can be expressed in general terms by the London equation

$$\nabla \times J = \frac{-1}{\mu_0 \lambda^2} B \tag{1}$$

The simple case of a bulk superconducting slab in a parallel field B field leads to an internal field as $B(x) = B_0 \exp(-x / \lambda)$. The temperature dependence of $\lambda$ is $\lambda = \lambda_0[1-(T/T_c)^4]^{-1/2}$; $T_c$ is the transition temperature and $\lambda_0$ is the penetration depth at T = 0 and is typically 5-20 nm.

A film with thickness of the order of the penetration depth will not be able to completely screen the applied field. The field that passes through the film is a measure of the penetration depth and is strongly temperature dependent. This attenuated field can be measured by a pick up coil where the induced currents are amplified by a dc SQUID.

A general method of operation is this: A persistent current is stored in a drive coil. A pick-up, or secondary coil, is initially nulled when the drive current is applied. This traps the initial flux of the drive coil-film system. As the temperature changes, the field seen by the secondary changes; to maintain constant flux, the current in the coil changes as its inductance changes. A SQUID with a 1 pA/√Hz. current resolution allows a potential temperature resolution better than 0.1 nK/√Hz.

The total inductance of the secondary-SQUID input loop system follows from Pippard[3] and is expressed

$$L_T = L_0 + L_s + L_n(\frac{2\alpha\lambda^2}{Dd}) \tag{2}$$

D is the length scale over which the applied field varies, d is the film thickness and $\alpha$ is a factor close to 1 for $d < \lambda$.

In general the flux seen by the secondary coil is the difference between the flux from the primary and the flux generated by the response currents of the film. From flux conservation, the current in the secondary can be calculated. The sensitivity is then expressed as

$$\frac{dI}{dT} = kI_p \frac{ct^3}{T_c(1-t^4)} \frac{L(s)}{(L_0 + L_s + L(T))} \tag{3}$$

where t is the reduced temperature.

The sensitivity is directly proportional to the applied current and is only limited by the critical current density of the film:

$$J_C = J_0(1-T/T_c)^{3/2} \tag{4}$$

where the current density at T=0 is $J_0$ and is $1 \times 10^{11}$ Amps/m$^2$ for aluminum films. Ideally, the critical current in the film is $I_c = J_c w d$, w is the width of the film (or radius for the case of the disk). The film will carry screening currents until it reaches the critical current density.

This assumes the current density is uniform along the film. For a long cylindrical film, this is a good approximation as the applied field is uniform and parallel to the surface of the film. Intuitively, the screening currents in the film will "mirror" the current distribution in the primary coil.

Due to the geometry of the disk system, the uniform current density approximation is not valid. J(ρ) can be calculated for a disk geometry. The self-consistent vector potential is

$$A_f(\rho) = A_p(\rho) + \frac{\mu_0}{4\pi}\int \frac{J_f(\rho')}{|\rho - \rho'|} d^3\rho' \tag{5}$$

where $\mathbf{A_f}$ and $\mathbf{A_p}$ are the vector potential of the film and primary coil respectively; ρ' is the coordinate in the film. Using London's equation the integral equation for **J**(ρ) is

$$J_f(\rho) = -\frac{1}{\mu_0\lambda^2} A_p(\rho) - \frac{1}{4\pi\lambda^2}\int \frac{J_f(\rho')}{|\rho - \rho'|} d^3\rho' \tag{6}$$

Instead of numerically evaluating the integral equation, a reasonable approximation can be made. To first order the current density of the film is assumed to have the same functional radial dependence as the vector potential of the primary coil. Calculations of $A_p(\rho)$ show a radial dependence that rises from the center and has a rounded maximum which peaks at half the coil radius. For a $J_f(\rho)$ with this radial dependence it is easily seen that one section of the film will reach the critical current before the rest. This limits the total current and thus the maximum sensitivity.

While a cylindrical film has a more uniform current response, the PDT sensitivity is limited by the need for the primary excitation coil to be close to the film. This can create large local fields at the surface of the film which reduce the total current that can be carried in the film.

## FILM FABRICATION

To develop a PDT that is useful for lambda point experiments it was necessary to pattern films with a $T_c$ slightly greater than 2.17 K. Films for the PDT were fabricated by vacuum deposition using aluminum and evaporated onto sapphire substrates. The films have been made on both cylindrical and disk geometries with film thickness ranging from 5 to 20 nm.

The small heat capacity of sapphire makes it an ideal substrate, providing a fast thermal relaxation to the thermal surroundings. The cylinder used was 3.81 cm long with a 1.27 cm diameter while the disk was a 2.54 cm diameter and 0.05 mm thick wafer. The estimated total heat capacity of the PDT is dominated by the sapphire and is approximately 6 μJ/K for the cylinder and 0.4 μJ/K for the disk. Most of the work in developing lambda point sensors was done using the wafers, to facilitate quick testing.

For some elemental superconductors, the transition temperature of a superconductor increases as its thickness decreases. With aluminum (bulk $T_c$ = 1. 14 K) a film 4.5 nm thick should have the desired transition temperature. However it was observed that films thinner than 7 nm showed marginal increases in $T_c$ and reached a maximum of 1.8 K.

One technique to achieve a higher $T_c$ was to form the film in the presence of oxygen[4]. During the evaporation process small doses of oxygen ($3\times10^{-4}$ to $16\times10^{-4}$ Pa) were admitted into the vacuum chamber. The oxygen bonds with the aluminum and has the effect of establishing small grains in the film. This reduces the size of superconducting domains relative to the coherence length. This is essentially the same as making thinner films allowing higher transition temperatures. Another result of the grains is to weaken the dependence of the transition temperature on the film thickness.

The rate of evaporation, which ranged from 0.3 to 0.6 nm/s, and the oxygen pressure determined the grain size and thus the transition temperature. Using a methodology where the thickness was held constant and the rate and pressure were varied, films with transition temperature greater than the lambda point were fabricated: 2.18 K and 2.25 K both with a film thickness of 10 nm.

The films were then tested in a liquid helium bucket dewar using an ac mutual inductance measurement. A sinusoidal current was passed through a primary coil located near the film. A secondary coil opposite the film was connected to a lock-in amplifier and the mutual inductance can easily be calculated.

Figure 1 shows the mutual inductance results for the $T_c = 2.25$ K film using excitation currents of 10 μA, 100 μA and 1 mA. The temperature where the inductance deviates from that measured with a lower current determines the critical current. It is evaluated from

$$J_c(0) = \frac{nI_c(T)}{d}(1-T/T_c)^{-3/2} \qquad (7)$$

where n is the turn density of the primary coil (184 turns/cm) and d is the film thickness. Using the 1mA data the critical current is reached at T =1.4 K. This gives a $J_c(0)$ of $3.9\times10^9$ A/m$^2$. This is 25 times smaller than the theoretical maximum of $1\times10^{11}$ A/m$^2$ and 2.5 times smaller than previous test performed on the cylindrical PDT.

As discussed previously, calculations show that the total current is expected to decrease based on the films non-uniform current response to the applied fields. Both this and the large local fields due to the individual turns, can reduce the effective critical current of the film. Using the estimation developed previously, the current is only carried in half the film which reduces the effective critical current by half.

A more important cause is the large local fields due to the primary coil. This was also observed in the cylindrical films when the coil was wrapped directly onto the film. Calculation of the field due to the primary show local fields can be 3 times larger than the

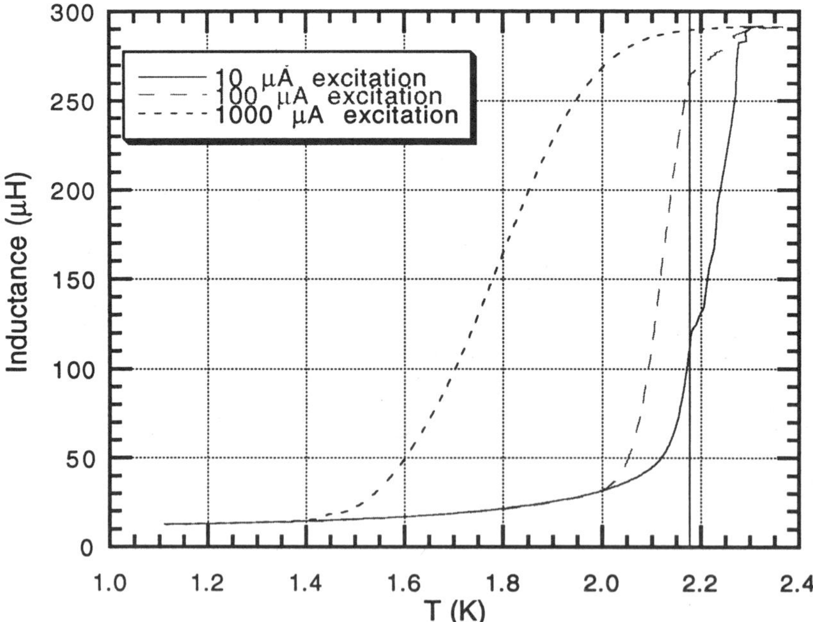


**Figure 1.** The mutual inductance vs. temperature for one film using various excitation currents. The transition temperature is greater than the lambda point which is marked by the vertical line.

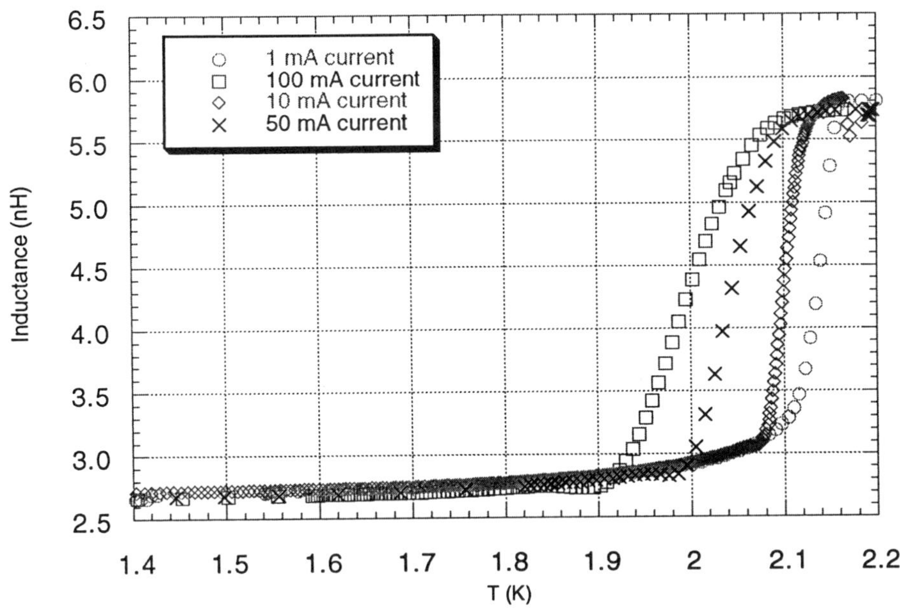


**Figure 2.** Mutual inductance verses temperature for one film using various excitation currents. These measurements were done using a different, single turn coil, and resulted in a factor of 3 improvement in the effective critical current.

average field, with large field gradients. To test this an alternate primary coil was designed that would eliminate the local peak fields. This was done by using a single turn coil, made from a copper disk with a radial slot. The copper sheet coil was made with the same outer and inner diameter as the pancake coil.

Mutual inductance measurements using this coil are shown in Figure 2. The higher background inductance at low temperature is the inductance of the leads which now become comparable with the inductance of the new coil. Noting the change in turn density, the effective critical current can be calculated. It is found the critical current is $1.3 \times 10^{10}$ A/m$^2$. This is a factor of 3 improvement compared to the wire wound coil. Using the approximation for sensitivity (I/L)dL/dT, the intrinsic temperature resolution using this film and coil arrangement is 0.3nK/√Hz. Currently, new coil designs and calculation of the current response film are being pursued for both disk and cylindrical geometries.

It should also be noted that it is the total current that determines the sensitivity of the PDT. The disk PDT, as currently fabricated, have one-half the sensitivity than the cylindrical PDT simple due to the loss of total cross-sectional area. An important result of the aluminum-oxygen films is the weakening dependence of film thickness vs. $T_c$. By making the film thicker some of the current carrying capability can be regained with out a loss of transition temperature.

## PDT TESTING

For resolution testing the previously characterized cylindrical PDT was used[2]. The primary and secondary coil were wrapped directly over the film and interspersed with each other. The wire used was 0.076 mm diameter Nb-Ti. The coils were wrapped as a solenoid with an estimated self-inductance of 60 µh for both secondary and primary coils.

Each coil was connected to a persistent switch: a small heater attached to the superconducting wire. To minimize the heat needed to activate the switch, it was constructed by wrapping resistive wire (Evanohm) directly over the superconducting wire.

This allowed the switch to be left on while cooling giving greater freedom to study the system.

To demonstrate the absolute sensitivity of a PDT it was necessary to design a cryostat that would allow an experimental platform to be controlled with a stability of better than 0.1 nK/√Hz with no thermal gradients and a time constant of several hours. A cell was designed so the PDT could be sealed in a bath of superfluid helium. This cell is attached to a series of isothermal platforms each weakly thermally linked to each other. The entire assembly is cooled by a pumped helium pot.

In order to eliminate heat leaks from the liquid helium fill line, it was necessary to make a cryogenic valve, which could be opened to fill the cell then closed to pump out the cell. The design that allowed the best thermal isolation was a hydraulically actuated needle valve using liquid helium as the actuating fluid[5]. The valve can be operated with a reasonable actuating pressure of 200 kPa. Its design allows it to act as a relief valve if the pressure in the cell exceeds 700 kPa. However, due to difficulties in developing a leak tight valve seat tests were done with no helium in the cell.

The thermal isolation stages in the cryostat filter out high frequency thermal noise in a manner analogous to a low pass filter where the ratio of thermal conductivity to heat capacity determines the "cutoff" frequency. Even with out the helium, regulating the isolation stages to a few μK allowed the cell to be stable to better than 1 nK/√Hz for frequencies greater than 20 Hz. Thermal noise was attenuated down to 0.1 Hz.

To demonstrate the absolute resolution of the film a current was placed in the primary and the cell was warmed to the 1.65 K region. The temperature of the cell was held stable while the SQUID output was sent through a low pass filter and then read by a Digital Multimeter (DMM). The SQUID output was also monitored by a spectrum analyzer.

To analyze the data, the long term thermal drifts on the order of several minutes were subtracted out. The data was then fourier transformed to determine the noise level. Figure 3 shows the results for two separate data runs. Over the range 0.2 to 1 Hz the noise is flat but in the 200-400 uV/√Hz. level which translates to a 50-100 nK/√Hz. There is also a noticeable 1/f noise below 0.2 Hz. This is most likely the result of long term temperature drifts .

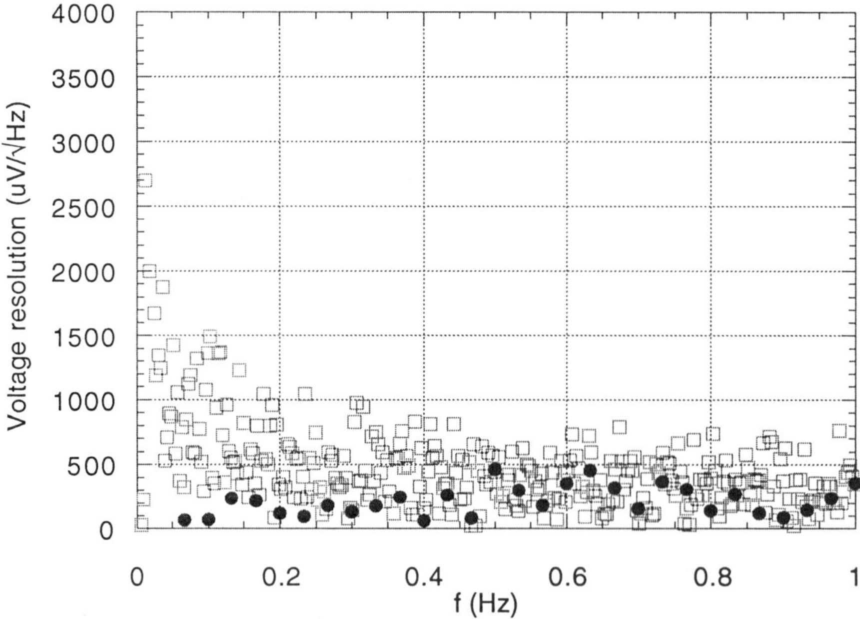


**Figure 3.** Voltage noise of PDT at constant temperature as a function of frequency. Two separate runs are shown. The calculated resolution of the PDT is about 50 to 100 nK/√Hz.

The noise at high frequency, as taken from the spectrum analyzer, show a noise level of 2 nK/√Hz over a frequency range 200 to 600 Hz. This is an important result which demonstrates the usefulness of the PDT as a bolometer with a fast thermal response. The noise level between 1 to 200 Hz could not be adequately determined since electromagnetic pick-up and the natural vibration mode of the cryostat dominated the noise spectrum.

A source of noise is true temperature fluctuations in the PDT and the cryostat. Estimating the statistical thermal fluctuations is done by evaluating

$$(dT)^2 = 4k_BT^2/C \tag{8}$$

for the low frequency thermal noise limit. C is the PDT's heat capacity and $k_B$ is Boltzmann's constant. For the case in which there is no helium in the cell and the PDT is assumed to be weakly linked to the thermal reservoir, the thermal fluctuation noise is no greater than 2 nK/√Hz. The other source of noise at low frequencies is thermal fluctuations in the cryostat that are not attenuated by the isolation stages.

## CONCLUSIONS AND FUTURE WORK

A PDT has been demonstrated to have an absolute resolution of 50-100 nK/√Hz for low frequency and 2 nK/√Hz for frequencies greater than 200 Hz. The high frequency results and the PDT's low heat capacity make it ideal for many bolometry applications.

A PDT has been made with an operating temperature near the lambda point. Because of the intrinsic geometry chosen, it did not have the current carrying ability as the cylinder PDT, but this problem can be addressed by the weaker $T_c$ dependence on film thickness. Further modeling, design and testing of various primary coil configurations is planned in an effort to reduce the peak field. Initial testing of a one turn sheet coil showed a factor of 3 improvement in the intrinsic sensitivity.

The resolution is above the projected level of 1 nK/√Hz. The limiting noise is primarily due to thermal fluctuation in the cell as well as statistical fluctuations in the PDT itself. Filling the cell with superfluid will eliminate both of these problems. By "anchoring" the PDT to the superfluid the statistical thermal fluctuation would decrease by a factor of 2000. The heat capacity of the helium would also change the external thermal time constant to the other stages making it much less susceptible to external low frequency noise.

The design of the test cryostat allows for testing the PDT's response to fast thermal pulses. Using the PDT in an ac mode allows the PDT to operate in the dissipative region increasing it's sensitivity an order of magnitude.

## ACKNOWLEDGMENTS

This work was supported by NASA's Microgravity Science and Applications Division through the Advanced Technology Development program.

## REFERENCES

1. J.A. Lipa, D.R. Swanson, J.A. Nissen, and T.C.P. Chui, Lambda Point Experiment in Microgravity, Cryogenics 34: 341 (1994).
2. M.J. DiPirro, J.G. Tuttle, and P.J. Shirron, Development of a High-Resolution Penetration Depth Thermometer, Adv. Cryo. Eng. 41: 1723 (1995).
3. A.B. Pippard, Magnetic field penetration through a superconducting film, Superconduction Science and Technology, p. 606 (1994).
4. R.W. Cohen and B. Abeles, Superconductivity in Granular Aluminum Films, Phys. Rev. 168: 444 (1968).
5. P.R. Roach, J.B. Ketterson, and M. Kuchnir, Hydraulically Actuated Valve for Very Low Temperatures, Rev. Sci. Inst. 43: 898 (1972).

# NEUTRON IRRADIATION TESTS OF CALIBRATED CRYOGENIC SENSORS AT LOW TEMPERATURES

T. Junquera[1], J. F. Amand[1], J.P. Thermeau[1], and J. Casas-Cubillos[2]

[1] Institut de Physique Nucléaire
CNRS-IN2P3, 91406-ORSAY, France
[2] CERN, CH-1211 GENEVA 23, Switzerland

## ABSTRACT

This paper presents the advancement of a program being carried in view of selecting the cryogenic temperature sensors to be used in the LHC accelerator. About 10,000 sensors will be installed around the 26.6 km LHC ring, and most of them will be exposed to high radiation doses during the accelerator lifetime. The following thermometric sensors : carbon resistors, thin films, and Pt, are exposed to high neutron fluences (> $10^{15}$ n/cm$^2$) at the ISN (Grenoble, France) Cryogenic Irradiation Test Facility in which a cryostat is placed in a shielded irradiation vault where a 20 MeV deuteron beam hits a Be target resulting in a well collimated and intense neutron beam. The cryostat, the on-line acquisition system, the temperature references and the main characteristics of the irradiation facility are described. The main interest of this set-up is its ability to monitor on-line the evolution of the sensors by comparing its readout with temperature references that are in principle insensitive to the neutron radiation (i.e. Argon gas bulbs when working at about 84 K, and below 4.5 K, either helium gas bulbs or the saturation pressure of the superfluid liquid helium bath). The resistance shifts of the different sensors at liquid helium temperatures are presented.

## INTRODUCTION

The next proton collider LHC [1] will include a large number of superconducting coils operating in superfluid helium at 1.8 K. Two proton beams (0.53 A , 7 TeV each) will circulate in opposite directions in the same magnet cold mass. The beam losses in the beam pipes (impacts on the beam screens and interactions with the residual gas) are estimated to $10^7$ protons m$^{-1}$ per second [2]. A detailed calculation gives the radiation levels at different distances of the cold mass axis (Fig. 1). In principle the cryogenic sensors will be mounted in the periphery of the cold masses (∅ = 0.57 m), where the integrated dose for ten years operation of the accelerator ranges from $10^2$ to $10^3$ Gy. The higher values corresponds to the intermagnet gaps where the shield effect of the metallic masses is reduced. These values apply for 95% of the machine, but in several regions of the accelerator, close to the interaction points where high luminosity beams are focused by an important number of superconducting magnets, the radiation doses can be 10 times stronger.

By taking a safety margin it has been decided to perform all the irradiation experiments at fluence values close to $10^{15}$ neutrons/cm$^2$ (corresponding to $2.10^4$ Gy).During the life time

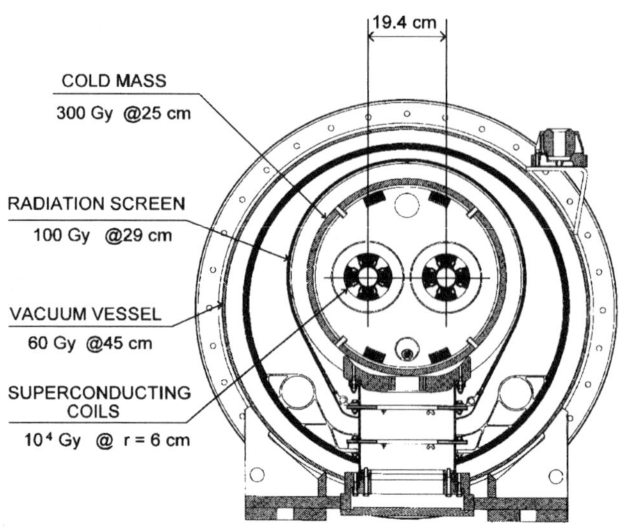


**Figure1.** Cross-section of a LHC magnet, the doses correspond to 10 years at nominal beam intensity.

of the accelerator several warm-up cycles for maintenance have to be considered, and some annealing of the damaged sensors could be expected.

The radiation resistance of cryogenic sensors has been performed at different dose levels [3, 4]. The aim of this work was to irradiate the sensors at high dose levels, and follow on-line the evolution of the sensors during the irradiation time.

We present in this article a description of the neutron irradiation facility which is installed in a beam line of a medium energy cyclotron in Grenoble (France). A special cryostat operating in liquid argon and helium has been constructed for the test of components of the detectors and accelerator systems of the LHC. A complete irradiation program of one year has started and we give here the first results obtained with different cryogenic temperature sensors.

## CRYOGENIC THERMOMETRY IN THE LHC

The s.c. magnet of the LHC accelerator will be operated at a temperature ranging from 1.8 K to 1.9 K, and their absolute value should be known better than ± 5 mK. The requirement for calibrated thermometers for the whole machine and the accuracies are presented in Table 1 [5].

A new thermometer block has been designed [6], it can be mounted on the vacuum side of the magnet cold masses in order to avoid leakages in the low pressure helium circuits. This thermometer incorporates a calibrated sensor and a microstrip printed circuit board for thermalization of the electric leads. The complete block should be in good thermal contact with the cold masses but it has to be located in hazardous regions from the point of view of the beam losses induced radiation.

**Table 1.** LHC cryogenic thermometry requirements

| Type of sensor | Range | Total | Accuracy |
|---|---|---|---|
| **Low temperature** | 1.6 K - 40 K | 2922 | 1% |
| **Low temperature** | 1.6 K - 300 K | 3696 | 0.25 % |
| **T > 50 K** | 50 K - 300 K | 2512 | 1 % |

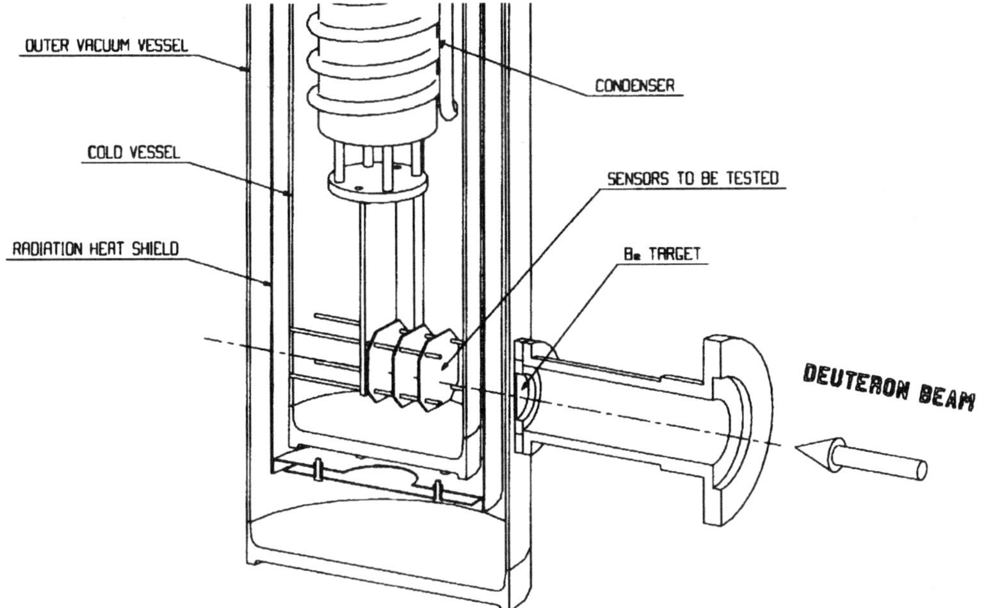


**Figure 2.** Cross section of the irradiation cryostat.

## NEUTRON IRRADIATION FACILITY

For irradiation tests of components for the detectors and the accelerator equipments of the future LHC, a neutron irradiation station has been recently installed at the SARA Cyclotron (CNRS-ISN Grenoble France) [7]. It is located at the end of one of its beam lines. A 5 μA beam of 20 MeV deuterons hits a 3 mm thick Beryllium target producing a high neutron flux. In the axis of the beam the neutron yield is $5.10^7$ $neutron.nA^{-1}sr^{-1}s^{-1}$, and its photon contamination is low (400 Gy for $2.10^{14}$ $n/cm^2$). The neutron energy spectrum is well adapted for studies of accelerator components ($<E_n> \sim 6$ MeV). As opposed to nuclear reactors this method produces very few thermal neutrons which drastically lowers the activity of exposed samples. At 11 cm from the Be target the measured dose was 4.3 $10^{14}$ $n/cm^2$ for an integrated charge of deuterons at the target of 1.04 C. To reach the final goal of $10^{15}$ $n/cm^2$ it was necessary to irradiate during a period of 5 days.

## THE IRRADIATION CRYOSTAT

A vertical cryostat (Fig. 2) is used to irradiate the sensors and is composed of a cold vessel, a radiation heat shield and an outer vacuum vessel :

a) The cold vessel is supported by four Epoxy fibre glass rods. One layer of active charcoal and 10 layers of superinsulation are wrapped around the cold vessel (heat losses 200 mW).

b) The copper radiation heat shield is wrapped in 25 layers of superinsulation and it is cooled with $LN_2$ circulation on the top. The cylindrical part is removable and is cooled by conduction. A pumping baffle on the bottom allows an efficient pumping (heat losses 7 W).

This cryostat can operate in two modes: 1) Liquid Helium, with automatic level regulation and temperature regulation by a valve, 2) Liquid Argon which is condensed by a $LN_2$ circuit and the bath pressure is regulated by the $LN_2$ circulation.

## EXPERIMENTAL METHODS

### The Temperature Reference

Three vapour pressure bulbs (Argon or Helium) immersed in the bath are used as temperature references. They are distributed along the vertical axis to measure any temperature gradient. These bulbs are composed of small copper cylinders ( 0.3 $cm^3$) and a stainless steel capillary tube (inside the cryostat: diam. 1 mm, length 1m, outside: int. diam. 6 mm, length 10 m) connecting the bulb to the pressure sensor (MKS pressure measurement system, calibrated at ±0.05 % of the reading).

### Irradiated Samples

During the irradiation experiment, we have tested four types of commecial cryogenic temperature sensors, and several test samples (Table 2). Six thermometers were placed in the vacuum, on the external face of the helium vessel, six centimeters in front of the neutron source. The 47 other sensors were soldered on three plates immersed in the helium bath of the cryostat, at a distance of 8 to 12 cm from the Beryllium target. The test samples are of two types : those used to confirm the behaviour of the sensitive materials of the thermometers (pure Platinium wires, stripped thermometers) and an Iron-Boron alloy sample which has been placed for control purposes : this material has an amorphic structure and a very small sensitivity to irradiation. Electrical resistance variation of such a sample may indicate problems with electronics or acquisition system.

### Measurement and Acquisition Systems

The measurement system is composed of a nanovoltmeter, a programmable current source (10 μA for thin film and carbon sensors, 1 mA for Pt and test samples) and the associated sensors for the cryogenic control system: helium level, pressures and temperatures in the cryostat components. This system is controlled by a Labview program running under Windows NT. The measuring cycle (reference temperatures, 50 sensors, and control devices) is 5 minutes. The measurement resolution is better than 0.2 mK at 4.2 K.

### Neutron Dose Measurement

One fondamental probleme in irradiation experiments is the measurement of the radiation dose. Three devices were used to calculate the neutron dose from indirect measurements :

1) The beam charge received by the Be target: it gives a precise on-line, not localized measure of the irradiation profile ( Fig 3)

2) Nickel foils attached to each temperature sensor: the neutron activation produces $^{58}Co$ isotope and its activity is measured after the experiment. It gives a non-precise (±15%), off-line, localized measurement (each foil being 4 mm ). A typical neutron dose map is presented in Fig. 4.

3) Nine silicon pin photodiodes were placed behind the cryostat at a distance of 42 cm of the Beryllium target. The radiation induced increase of the forward voltage give an on-line control of the neutron beam distribution.

By comparing the data shown in Fig. 3 and 4 we can estimate the received dose as a function of time.for each thermometer. This approximation is justified by the results of the silicon diodes, which indicate that the neutron beam had a good homogeneity during the irradiation.

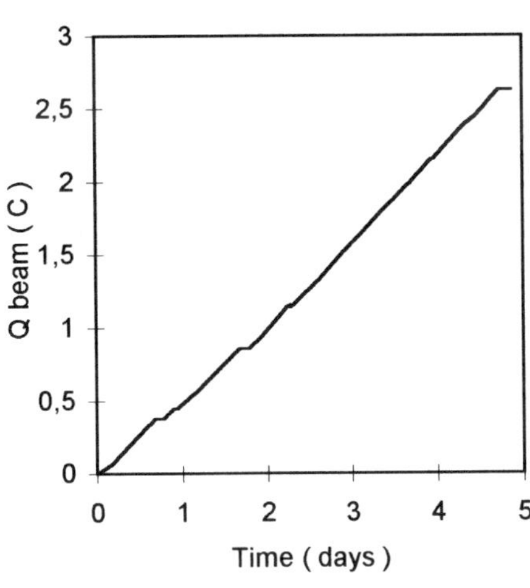


Figure 3. Deuteron beam charge on Be target.

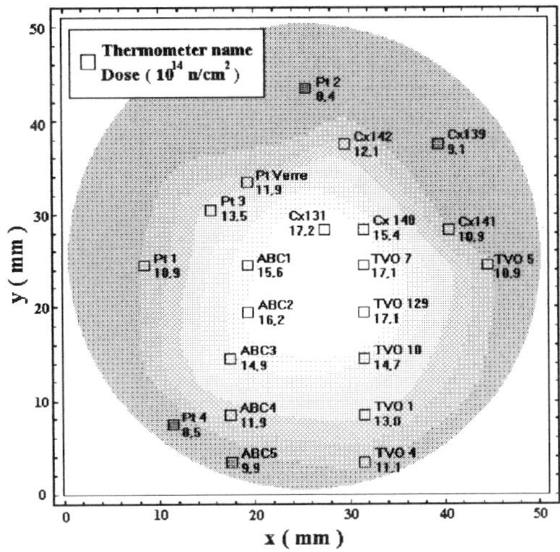


Figure 4. Neutron dose map (Nickel foil measurements).

## Bath Temperature Measurement during Irradiation

We maintained the temperature at 4.2K most of the time during the experiment, and some measurements were made at lower temperatures (1.6 K, 2 K, 3 K) to control the resistance of the sensors.

In fig.5 we have also plotted the reference temperature during a 3 hours period. This graph allows an estimation of temperature regulation quality. The noise on measured temperature is $<\sigma>= 0.085$ mK, and the typical temperature drift is about 1 mK per hour.

## Resistance Correction Procedure

As we want to estimate the irradiation induced resistance variations of our sensors, we must filter out the resistances changes due to bath temperature drift (Fig 6). This is done with the following correction procedure:

- We choose a correction temperature $T_{correction}$= 4.200 K.
- Rejected points: All the points outside the range 4.200 K $\pm$ 0.01K, and those considered not stable (i.e. have moved of more than 0.5 mK in one minute).

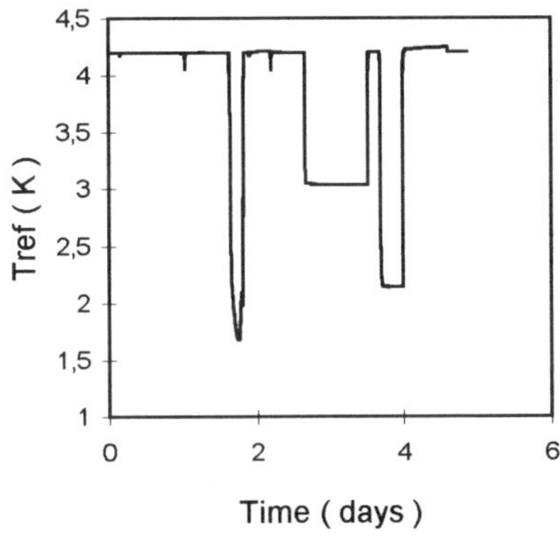


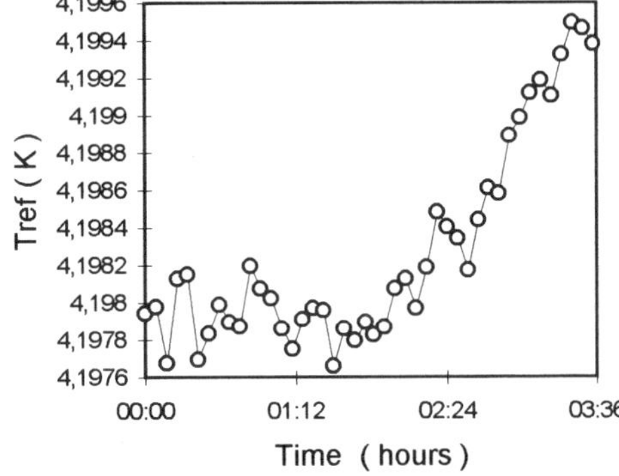


Figure 5. Measurement of the bath temperature (Reference Temperature) during irradiation.

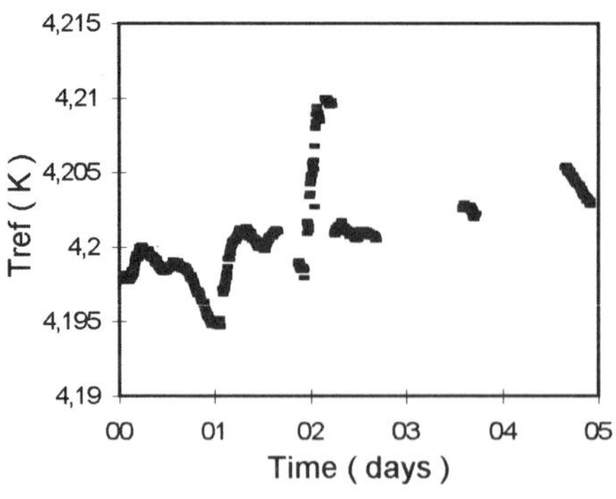

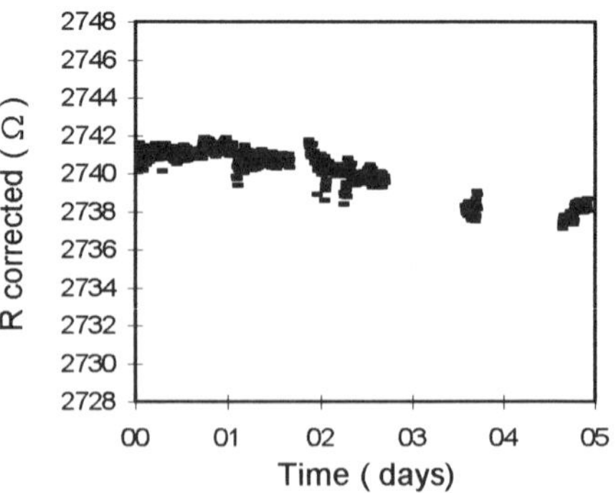


**Figure 6.** Resistance correction for a Cernox sensor ($T_{correction}$=4.2 K).

- For each measured point we calculate $\Delta T = T_{measured} - T_{correction}$
- We obtain : $R_{corrected} = R_{measured} \pm \Delta T.(dR/dT)$ , where dR/dT is the sensor sensitivity

## EXPERIMENTAL RESULTS

### Test Samples

We have used two kind of control samples: Pt and Fe-B alloy, in order to check the overall measurement and acquisition systems accuracy. The resistance variations of these samples are presented in Fig. 7 and 8:

- The irradiation produces a continuous resistance drift of the Pt sensors (including the wire) showing a classical response of irradiated pure metal samples at low temperature

- The Fe-B alloy sample exhibits a response in complete agreement with expectations: small influence of the temperature and near insensitivity to neutron irradiation.

### Cernox Sensors

The tested Cernox sensors came from two different batches (Fig. 9 and 10). Sensors from the first batch (Fig.9), which are less sensitive ($dR/dT \approx -1 k\Omega.K^{-1}$ @4.2K) have a continuous resistance decrease. For the whole experiment, they shifted of about 4Ω ($\Delta R/R$=-0.13%) and the equivalent measured temperature variation was of about 4mK ($\Delta T/T$=0.1% ). Sensors from the second batch (Fig.10) showed a resistance increase at the beginning of irradiation, and then a decrease. The resistance shifts for this batch was within a 3 ohm range, the equivalent measured temperature variation was of about 2.5mK. ($dR/dT$=1300 $\Omega.K^{-1}$, $\Delta T/T$=0.05% ).

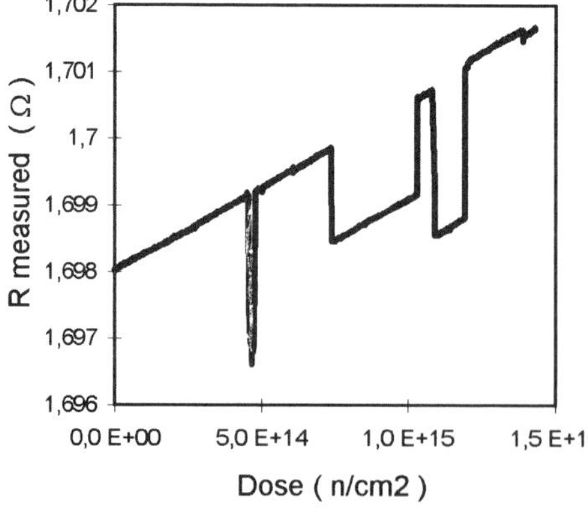


**Figure 7.** Irradiation of a Platinum sensor.

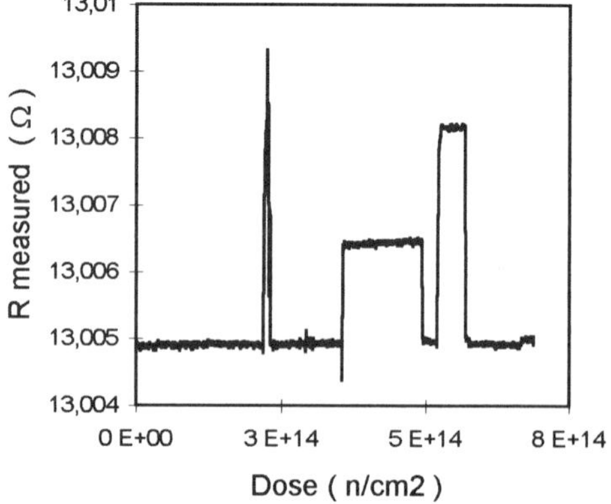


**Figure 8.** Irradiation of Fe-B sample.

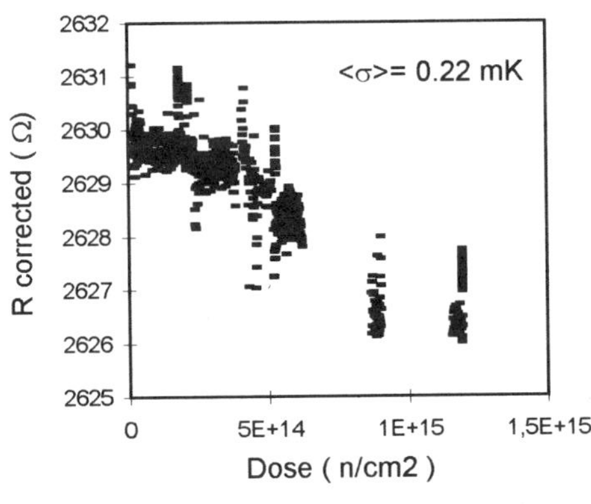

**Figure 9.** Cernox sensors ( batch 1).

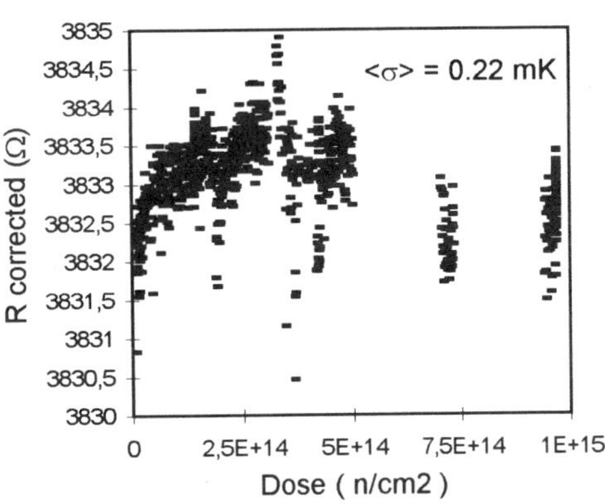


**Figure 10.** Cernox sensors (batch 2).

Note : the behaviour of sensors of each batch were almost the same, so the examples shown in Fig.9 and 10 are representative of the batches. The unencapulated Cernox sensors showed the same behaviour as sensors of first batch, with a total $\Delta R$ of 9 ohm ( $\Delta R/R$= 0.3 %), and a sensitivity dR/dT= 1300ohm.$K^{-1}$ , resulting in $\Delta T$= 7 mK ( $\Delta T/T$= 1.6 %).

A neutron irradiation at Argon temperature ( 85K ) of the Cernox sensors from batch #1 showed almost the same results : a continuous resistance decrease ( $\Delta R/R \approx 0.1\%$ for 1.5 $10^{15}$ neutrons/$cm^2$ )

## Carbon Sensors

Two carbon sensor types have been tested : TVO (JINR Dubna,Russia) and the classical Allen-Bradley 100 ohm resistors.

On Fig.11 is plotted a representative graph of Allen-Bradley corrected resistance during irradiation. For this family, we see very little resistance variations ( within a 0.6 ohm range ), which gives us an error on measured temperature of 1.5mK. On Fig.12 is reported the corrected resistance for a TVO sensor. We can see that this sensor has a continuous resistance drift, with a total amplitude of about 1ohm. This variation implies an error of 2mK ($\Delta T/T$=0.5%) on measured temperature. (Note : the TVO sensors came from two series, but all sensors had almost the same behaviour).

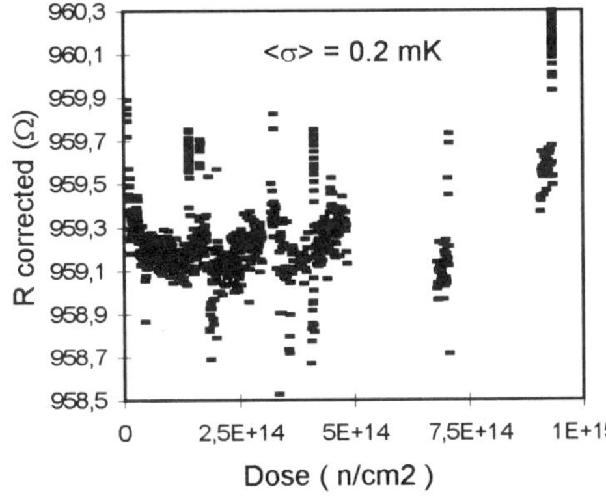


**Figure 11.** Allen Bradley Carbon resistor.

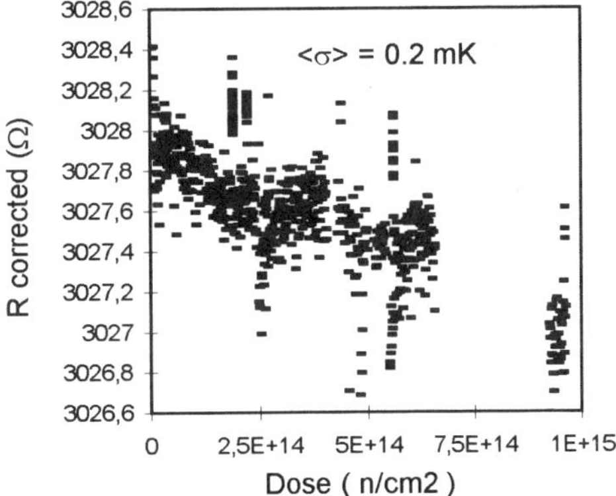


**Figure 12.** TVO sensor.

**Table 2.** Summary of Radiation Induced Resistance Shifts (all sensors receive neutron fluences between $8.10^{14}$ and $1.2 .10^{15}$)

| Family | Type | Quantity | R@4.2K ($\Omega$) | dR/dT@4.2K ($\Omega.K^{-1}$) | $\Delta R$ ($\Omega$) | $\Delta T$ (mK) |
|---|---|---|---|---|---|---|
| **CERNOX** | Serie 1 | 5+1 (vac.) | 3000 | -1000 | -4 | 4 |
| | Serie 2 | 4+1 (vac.) | 3800 | -1300 | -1.5 | 1 |
| | unencapsulated | 3 | 3200 | -1300 | -9 | 7 |
| **Carbon** | Allen-Bradley | 12+1 (vac.) | 960 | -370 | -0.5 | 1.5 |
| | TVO | 10+2 (vac.) | 3200 | -600 | -0.8 | 1.5 |
| **Platinum** | type 1 | 5+1 (vac.) | 1.7 | 0.0015 | $2.5\ 10^{-3}$ | - |
| | type 2 | 5 | 0.15 | 0.0012 | $2\ 10^{-3}$ | - |
| | Wire | 2 | 1 | 0.0018 | $2.5\ 10^{-3}$ | - |

**Neutron Beam Heating Effects**

During the irradiation experiments, a heating effect due to the neutron beam was observed on all sensors. When switching the beam off, an instantaneous resistance shift ranging from 0.2 to 1.5 ohm was observed on Cernox and Carbon sensors. This corresponds to temperature shifts ranging from 0.5 mK to 1.6 mK. During this transient no significant bath temperature shift was observed. In a separate experiment we have controlled the self-heating effect of the sensors by increasing the D.C. measurement current. It allows to evaluate the beam deposited power: it ranges from 0.2 μW to 0.6 μW depending on the sensor types and their distance to the neutron source.

## CONCLUSION

The whole irradiation set-up is now fully operational, most of the initial goals were reached in schedule and the first results (i.e. bath temperature measurement and stability, neutron beam reproducibility and stability, and the quality of the measurement and acquisition systems), confirm our expectations. The complete irradiation program (more than 200 samples) will continue in the next months, including systematic on-line control of the resistance shifts, comparison of calibrated sensors before and after irradiation, and finally the test of new sensor prototypes. During the next irradiation runs data will be sampled in superfluid helium temperatures in order to garantee a good bath temperature homogeneity, adding an extra control of the vapour pressure bulbs measurements against the bath saturation pressure.

## ACKNOWLEDGEMENTS

We would like to thank L.Thomé of the CSNSM (IN2P3-Orsay) for his continuous support and counsel on experimental and theoretical aspects of irradiation, and all the people at the 3 laboratories collaborating to this project: J. Collot, Ph. Martin, G. Mondin, M. Fruneau and all the cyclotron group at the ISN-Grenoble, F. Clapier, Ph. Bujard, Ch. Joly, D. Grolet, and specially S. Buhler, head of the Cryogenics group at the IPN Orsay, for his continuous support and fruithful discussions along the design phase, and finally, Ph. Lebrun, Ch. Balle, V. Datskov and all the LHC Cryogenics group at CERN.

## REFERENCES

1. L.R.Evans « The Large Hadron Collider », Particle Accelerator Conference (May 1995) Dallas, USA
2. K.M.Potter, H.Schonbacher, G.R. Stevenson. CERN-LHC report #18, November 1995.
3. D. Scott Holmes, S.Scott Courts, CEC-ICMC (July 1995), Columbus USA.
4. V.I.Datskov and J.G.Weisend II, Cryogenics, vol.34, p. 425.
5. LHC Conceptual Design Report, report CERN/AC/95-05, October 1995.
6. C.Balle, J.Casas-Cubillos, CEC-ICMC (July 1995), Columbus USA.
7. J. Collot et al. Nuclear Instruments and Methods A vol. 350 (1994) p. 525.

# EFFECTS OF HIGH INTENSITY CRYOGENIC IRRADIATION AND MAGNETIC FIELD ON TEMPERATURE SENSORS

Yu.P. Filippov, V.V. Golikov, E.N. Kulagin, and V.G. Shabratov

Joint Institute for Nuclear Research, Dubna 141980, Russia

## ABSTRACT

Four types of commercially available cryogenic thermometers were irradiated by gamma and/or neutron sources. These sensors are the well known rhodium–iron and platinum resistance thermometers, carbon resistance thermometer of TVO type and thin–film germanium resistance sensors; they can operate over the 1.5 K to 350 K range (14 K – for platinum sensor). The radiation source for experiments is the Fast Pulsed Reactor of Frank Laboratory for Neutron Physics, JINR. The total doses were about $10^{16}$ n/cm$^2$ and $10^5$ Gy at average energy of the fast neutrons of 1 MeV. The thermometers were calibrated at cryogenic temperatures and irradiated at 290 K, after that the comparison of the new calibration to the old one was made. Then the thermometers were irradiated and calibrated in–situ at 77.4 K. In addition, several tests of the thin–film germanium, platinum and TVO–carbon sensors were performed in the magnetic field. The obtained results are discussed.

## INTRODUCTION

Many devices of modern high energy physics installations will operate under conditions combining high radiation environment, magnetic field and cryogenic temperature. In this connection Lake Shore Cryotronics, Inc has performed some irradiation experiments with selected cryogenic temperature sensors. Thus, several types of sensors were calibrated, irradiated at room temperature by a gamma source up to the level of $10^4$ Gy or neutron and gamma source up to the fluence of $8.6 \cdot 10^{13}$ n/cm$^2$, and recalibrated as the first step to determine their suitability for use in radiation environment.[1] It is well known that the thermal annealing of sensors can occur upon warming to ambient temperature, and it is rather difficult to find the real reason of a possible shift in the calibration of the sensor after irradiation. To avoid these problems, thermometers were irradiated and calibrated in–situ to determine their suitability for cryogenic thermometry up to the total fluence of $2 \cdot 10^{12}$ n/cm$^2$ or gamma dose of $10^4$ Gy.[2] However, sometimes it is necessary to have an information concerning the effects of irradiation on sensors for significantly larger values of fluences. Thus, at the expected high luminosity of $L = 1 \cdot 10^{34}$ cm$^{-2}$s$^{-1}$ for ATLAS detector at LHC, the maximum year dose of 12 kGy/yr and 1 MeV equivalent neutron fluence of $1 \cdot 10^{15}$ n/cm$^2$/$yr$ are expected in the hadronic liquid argon end–cap calorimeter at the position closest to the beam pipe.[3] In this case an operation period can be of 10 years at high luminosity. Basing on the experimental data, one could expect that the TVO–sensor is suitable for use at the mentioned levels of irradiation.[4] However, the corresponding results were obtained at 300 K and a rather low average energy of neutrons of 0.1 MeV, and a possible damage caused by the radiation can depend on this value of the energy.

The aim of this work was to perform series of experiments with the commercially available cryogenic temperature sensors up to the total doses of about $10^{16}$ n/cm$^2$ and $10^5$ Gy for the fast neutrons and $\gamma$–rays, correspondingly. At first, the thermometers were calibrated against the rhodium–iron resistance thermometer at 1.9, 4.2, 77.4 and 290 K, then they were irradiated at 290 K, and the new calibration was obtained and compared to the old one again. Secondly, the cryogenic thermometers were irradiated and calibrated in–situ at 77.4 K. Besides, the tests of some types of the thermometers were performed under magnetic field up to 11 T at 1.5 K, 9 T at 4.2 K and 2.5 T at 78 K to verify their characteristics.

## EXPERIMENTAL SETUP

**Sensors and calibrator.** Four types of the commercially available sensors were used for tests: rhodium–iron (RIRT),[5] platinum (PRT),[5] carbon (TVO)[4] and germanium thin film (TTR)[6] resistance thermometers. The RIRT and PRT thermometers are designed and fabricated by Cryometrology Department (CRYOMET) of Russian National Scientific and Research Institute for Physical– Technical and Radiotechnical Measurements (VNIIFTRI). They are intended for calibration and precise laboratory measurements. The TVO–carbon resistors were originally designed for use in electronics under corrosive and wet conditions: they have a ceramic coat providing a hermetic seal.[4] The germanium thin film on GaAs substratum is used for TTR sensors. The contacts are made of Au, $\delta \approx 1$ $\mu$m, by vacuum deposition. The dimensions of the sensor crystal are $1 \times 1 \times 0.2$ mm.[6] The other characteristics of the thermometers are as listed in Table 1.[4,5,6]

To calibrate these thermometers, we use the calibrator AK–6.25 designed and fabricated by CRYOMET VNIIFTRI.[5,7] Its microcryostat is immersed into the liquid helium or liquid nitrogen dewar with the 50 mm nèck. The calibrated sensors are inserted into the comparison block which can be easily removed from the shield and vacuum insulation. The temperature is measured with reference to the rhodium iron resistance thermometer.

**Setup at the Reactor.** The irradiation has been carried out at the pulsed neutron reactor IBR–2[8] which has an operating frequency of 5 Hz, a half–width of the reactor pulse of about 220 $\mu$s and a peak power in the pulse of about 1500 MW. It produces a full spectrum of fast neutrons ranging from about $10^{-1}$ to 20 MeV; average energy, $E_n$ , is about 1 MeV. The reactor can deliver the neutron flux up to $10^{12}$ n/cm$^2$/$s$ over areas up to $20 \times 40$ cm$^2$. In addition to the neutron, gamma radiation is also produced

**Table 1.** Main characteristics of the tested thermometers

| Parameters | Types | | | |
|---|---|---|---|---|
| | RIRT | PRT | TVO | TTR |
| Temperature range, K | 0.35/373 | 13.8/303 | 1.5/423 | 1.5/350 |
| Resistance at 273 K, Ohm | 100 | 100 | 1000 | 25/125 |
| Biasing current, mA | 0.2/1 | 1/5 | 0.01 | 0.001/0.1 |
| Stability, mK per year | 2 | 3 | 2@4.2K | 10@4.2 K |
| Self heating, mK | $< 2$ | $< 2$ | $< 3$ | $< 2$ |
| Dimensions, mm | | | | |
| diameter | 3.7 | 3.7 | 1.3×2.4 | 1.5×1.5 |
| length | 40 | 40 | 8 | 3 |
| Mass, g | 2 | 2 | 0.1 | 0.04 |
| Accuracy, mK | ±5 | ±10 | ±10 @ 4.2 K | |
| | the whole range | | ±100 @ 77 K | |
| | | | ±1000 @ 300 K | |

in the nuclear reactions with maximal dose rates up to 10 Gy/s : the average energy is of about $E_\gamma \approx 1.5$ MeV. The ratio between the neutron fluence and the $\gamma$ dose can be varied by several orders of the magnitude by means of an appropriate beam filter and absorbers.[8] The flux of fast neutrons from $10^8$ to $10^{10}$ n/cm$^2$/$s$ was used for our measurements. The experimental setup includes a cryostat for the sensors, a beam filter, boron carbide layer, movable shielding platform, beam shutter, inner and outer reactor shields, cryogenic lines, measurement cables, nickel foil for both – monitoring of the total power and measuring of the homogeneity of the neutron fluence. In addition, the nickel foil covering the backside of the sensors was kept in the cryostat during the whole irradiation period. The induced activity in this foil was used for the monitoring of the total reactor power. To monitor the dose rates, the neutron and gamma dosimeters were mounted around the cryostat.[3] The accuracy to determine the fluence of the fast neutrons and $\gamma$–dose was ±10 %.

The $\gamma$ irradiation experiments were performed by using the Cesium–137 gamma source with the dose rate of 0.55 Gy/s at the average energy of about 0.67 MeV. In this case the tested thermometers can be irradiated at room and $LN_2$ temperatures.

**Magnetic Field System.** We use a solenoid superconducting magnet which consists of two coaxial sections. The dimensions of the magnet are as follows: height – 0.22 m, inner diameter – 0.038 m, outer diameter – 0.27 m. The maximum fields in the center of the solenoid are of about 9 T at 4.2 K and 11 T at 1.5 K. If permendur pole pieces are used, the maximum magnetic field can reach 13 T at 1.5 K within the gap width of 1.2 mm. The solenoid is supplied by a stabilized current source ($U = 6$ V, $I = 70$ A). Its integrator provides a smooth input/output of the current with the speed up to 0.5 A/s. The LMK Hall sensor is used to measure the magnetic field.

We had two possibilities to change the temperatures within the volume of the solenoid. If the cryostat is filled with liquid helium, the temperature can be regulated within the range from 1.5 to 5 K. In addition , we can use the cryostat inserted into the inner hole of the solenoid, and this cryostat might be filled with liquid nitrogen, for example. Magnetic field–independent temperature control was accomplished with the system to keep the pressure at the given value and measure the corresponding vapor pressure of the used cryogen.

## EXPERIMENTAL PROCEDURES AND RESULTS

**Radiation at Room Temperature.** First tests were performed with TVO and TTR 1B resistance thermometers radiated with the gamma rate of 0.11 Gy/s and neutron flux of $7 \cdot 10^9$ n/cm$^2$/$s$ at 290 K. After that the pre– and post–irradiation comparison was made at two or four temperatures: 1.87 K, 4.24 K, 77.4 K and 287.5 K. Nanovoltmeter Keithley 181 was used for all the measurements. For the tested sensors, the calibration data at 287.5 K and 77.4 K were precisely the same after doses of $1.5 \cdot 10^{13}$ n/cm$^2$ and 220 Gy; they were obtained in one week after irradiation.

The temperature shifts for the larger dose are presented in Table 2; they were obtained in two weeks after irradiation. Qualitatively, it looks like the behavior of the carbon-glass sensor of LSCI[1] with lower temperature shifts. Actually we were surprised at these comparatively good results for TTR 1B, at least. Respectively high radiation resistance for TVO is described in Reference 4, for example.

**Radiation in Liquid Nitrogen.** During these tests the thermometers were immersed in the dewar with liquid nitrogen whose level was kept constant with deviation of ±5 mm at the moment of measurement. To avoid the influence of oxygen concentration on saturated pressure of $LN_2$, the fresh liquid nitrogen was transferred into the cryostat from the laboratory 1000 L storage vessel to minimize contamination. For the presented data the problems with dissolved oxygen were neglected. To process the data, the following way was used: first, the basic line of the resistance of the sensor $R_0$ versus measured pressure $P_s$ without radiation, $R_0(P_s, 0)$, was built. Then the experimental points, $R(P_s, n+\gamma)$, under radiation were measured, and the differences $\Delta R = R(P_s, n+\gamma) - R_0(P_s, 0)$ were calculated. After that the values of the temperature shifts at each dose were calculated as $\Delta T(n+\gamma) = \Delta R/(dR/dT)$, where $dR/dT$ is the sensitivity of the sensor from the initial calibration data. It should be stressed that all the measurements with the thermometers were performed at the closed shutter of the beam which was opened again to increase the doses of radiation.

The first experiment in $LN_2$ was performed at the reactor with the n/$\gamma$ converter which allows one to maintain the parameters of the beam as follows: fluence of the fast neutrons is of $2.4 \cdot 10^{14}$ n/cm$^2$ and $\gamma$–radiation dose – $9.5 \cdot 10^4$ Gy. Two types of TTR sensors were used: they are of A and B types. The difference between them is that the resistance and sensitivity of the A–type sensor are higher with respect to the B–type by one order of magnitude , approximately, at 77.4 K.[6] The results showed that the deviations for TVO sensor are within the accuracy of its calibration, $\delta T = \pm 100$ mK, during radiation at 77.4 K. These values for TTR 1A sensors at the total doses are within the range of about ±100 mK too. After warming up to 290 K and subsequent cooling to 77.4 K in one day, no shift was observed. However, B–type sensor worked rather well up to $1.4 \cdot 10^{14}$ n/cm$^2$: the temperature shifts were within ±200 mK; but its shift at $1.6 \cdot 10^{14}$ n/cm$^2$ became several times larger – about 470 mK and reached the

**Table 2.** Temperature shift after $2 \cdot 10^{14}$ n/cm$^2$ and 3000 Gy dose

| Type | Shifts, mK, at selected temperatures | | | |
|---|---|---|---|---|
| | 287.5 K | 77.4 K | 4.24 K | 1.87 K |
| TVO | -130 | -85 | -12 | – |
| TTR 1B #147 | -530 | -20 | -10 | -10 |

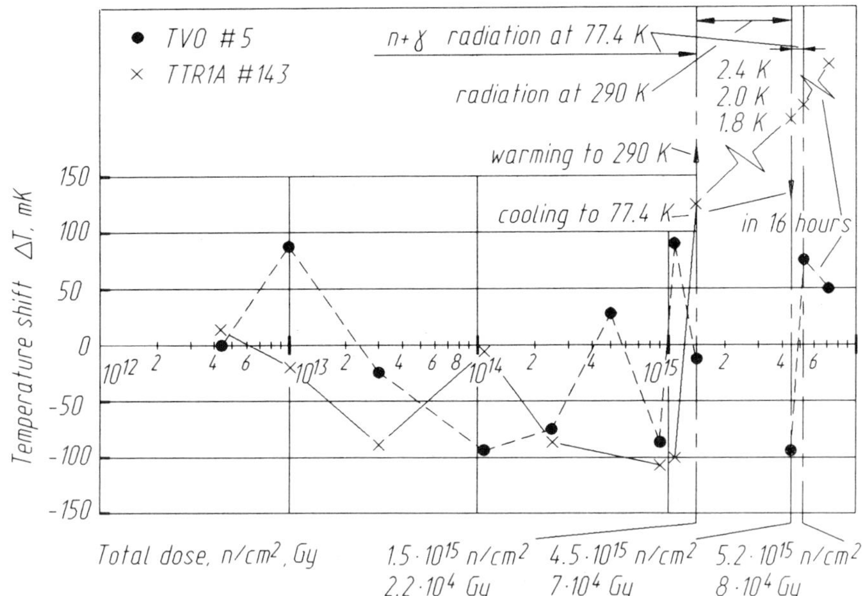


**Figure 1.** Temperature shift at 77.4 K due to neutron plus gamma irradiation ($7.5 \cdot 10^9$ n/cm$^2$/$s$ + 0.11 Gy/s) versus total dose for TVO and TTR thermometers.

value of 1100 mK @ $2.4 \cdot 10^{14}$ n/cm$^2$. After warming up to 290 K and cooling down to 77.4 K, the additional post radiation shift was revealed; its value was of 1.5 K. One can note here for comparison that the fluence of about $3 \cdot 10^{14}$ n/cm$^2$ is the margin of the stable performances for the GaAs preamplifiers developed in the framework of ATLAS project.[8]

The next experiment was performed at the neutron flux of $7.5 \cdot 10^9$ n/cm$^2$/$s$ and 0.11 Gy/s: only TTR 1A and TVO sensors were used in this case. The results are shown in Figure 1. It shows that the behavior of the TVO sensor is practically the same as in the previous case: the maximum shifts are within the range of $\pm 100$ mK during irradiation at 77.4 K, warming to 290 K, irradiation at this temperature, subsequent cooling to 77.4 K and additional cryogenic irradiation up to $5.2 \cdot 10^{15}$ n/cm$^2$ + $8 \cdot 10^4$ Gy. This thermometer showed the shift of 50 mK only after completing the irradiation in 16 hours. In its turn, TTR 1A thermometer worked successfully ($\Delta T \approx \pm 100$ mK) up to the doses of $1.1 \cdot 10^{15}$ n/cm$^2$. However, the shift of about $\pm 125$ mK appeared at the value of $1.5 \cdot 10^{15}$ n/cm$^2$. After warming to 290 K and irradiation at this temperature up to $4.5 \cdot 10^{15}$ n/cm$^2$, this sensor showed the shift of 1.8 K @ 77.4 K. Additional irradiation in $LN_2$ up to $5.2 \cdot 10^{15}$ n/cm$^2$ led to the new shift of 2.0 K. After completing the radiation, the sensor increased its shift to 2.4 K in 16 hours.

Another set of experiments was performed with the rhodium–iron, RIRT, and platinum, PRT, wire-wound standard resistance thermometers. First of all, they were tested using the Cesium–137 gamma source producing a dose rate of about 0.55 Gy/s. Sensors (RIRT #68 and PRT #226) were irradiated up to the total dose of $10^5$ Gy at 77.4 K. One can emphasize that the shifts and their oscillations for the both sensors showed to be less than 10 mK, perhaps, due to the change in liquid nitrogen because of its possible oxygen contamination and errors of saturated pressure measurements.

In the neutron plus gamma experiment, another pair of the thermometers (RIRT #106 and PRT #208) were irradiated at 77.4 K to the total doses of $7 \cdot 10^{15}$ n/cm$^2$

($2.4 \cdot 10^9$ to $8.5 \cdot 10^9$ n/cm$^2$/s ) and $9 \cdot 10^4$ Gy (0.11 Gy/s) by the reactor according to the time scheme shown in Figure 2. This figure shows that the first phase of radiation to $4.8 \cdot 10^{14}$ n/cm$^2$ ($2.4 \cdot 10^9$ n/cm$^2$/s) led to the maximum shifts of about ±15 mK for the both sensors. One can see that the majority of the temperature shifts had positive values. Qualitatively, it looks like the pictures showing the equivalent temperature shifts $\Delta T$ for platinum PT–103 and rhodium–iron RF–800–4 sensors irradiated at room temperature to the total dose of $10^4$ Gy using the Cesium–137 gamma source[1] : $\Delta T$ (in 2 weeks after irradiation) were about ±10 mK and ±25 mK for PT–103 and RF–800–4, correspondingly. The warming to 290 K and cooling to 77.4 K again do not practically influence upon the readout of the sensors: the shifts are less than 10 mK. The second phase of irradiation to $5.4 \cdot 10^{15}$ n/cm$^2$ ($8.5 \cdot 10^9$ n/cm$^2$s) gave a bit more shift - of about ±20 mK. The process of warming/cooling does not almost influence upon the resistance of the both thermometers again: the shifts are of about 5 mK for PRT and 15 mK for RIRT. The third phase of irradiation to $7 \cdot 10^{15}$ n/cm$^2$ showed approximately the same results. The warming to 290 K in two weeks and cooling to 77.4 K revealed the shifts of about 10 and 20 mK for PRT and RIRT, correspondingly, probably, due to some defects in the metal caused by irradiation. Several negative values of $\Delta T$ for the second and third phases of the experiment could take place because of possible electromagnetic interference in the experimental hall during the measurements.

**Tests in Magnetic Field.** Several experiments with the thermometers were performed in a magnetic field perpendicular to the sensor package.

First, the platinum thermometer irradiated to $7 \cdot 10^{15}$ n/cm$^2$ at 77.4 K was tested at 2.5 T, 78 K and 84 K. As expected,[9] the magnetoresistance $\Delta R/R$ was about 0.1 % at 78 K that gives the equivalent shift of ≈ 45 mK. The corresponding values at 2.5 T and 84 K were 0.08 % and 39 mK. The relative shift $\Delta T/T$ for TTR 1B sensor in the magnetic field of 2.5 T was about 0.1 % at 78 K.

One more experiment was performed with the TVO and TTR sensors at 1.52 K and 4.23 K. The results in the perpendicular magnetic field are presented in Figure 3.

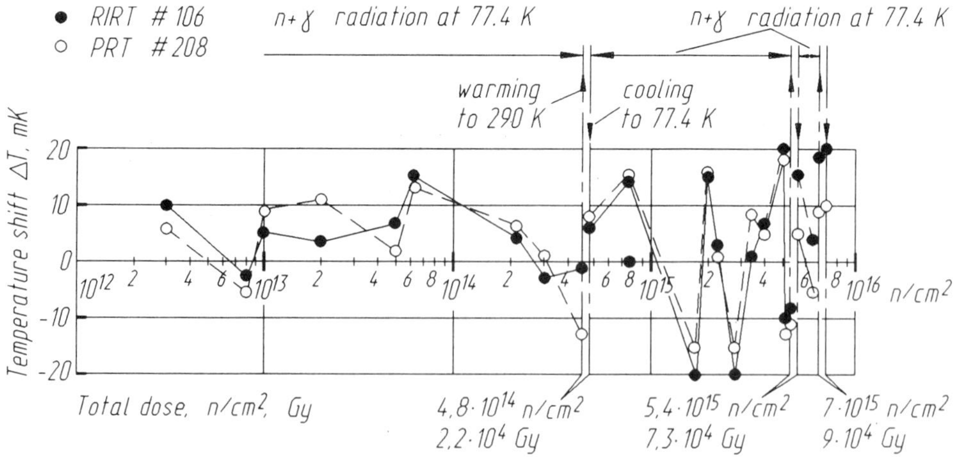


**Figure 2.** Temperature shift at 77.4 K due to neutron plus gamma irradiation (phase 1: $2.4 \cdot 10^9$ $n/cm^2/s$ + 0.11 Gy/s; phase 2 and 3: $8.5 \cdot 10^9$ $n/cm^2/s$ + 0.11 Gy/s) versus total dose for PRT and RIRT thermometers.

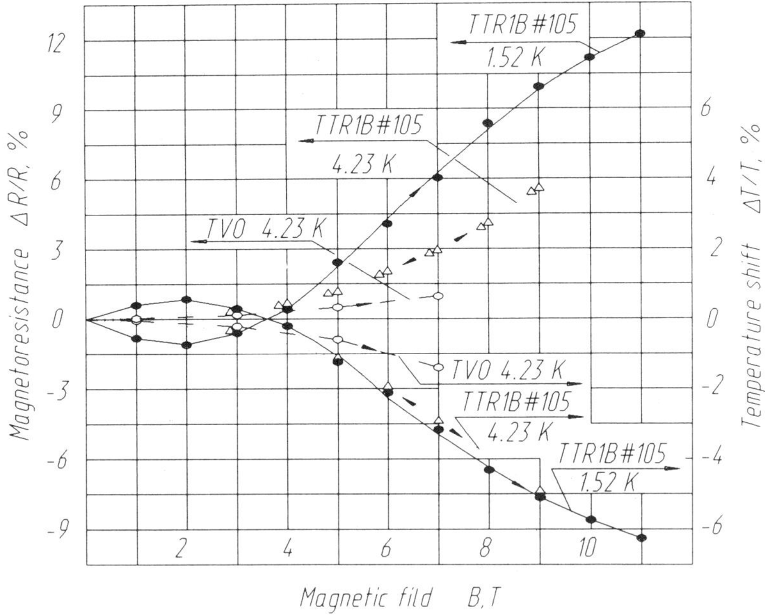


**Figure 3.** Temperature shift versus magnetic field for TVO and TTR thermometers.

The data processing has shown that the dependence $\Delta T(B, 4.23\ K)$ for TVO and TTR 1B can be expressed as $\Delta T = k_1 B^2$ and $\Delta T = k_2 B^{2.63}$ (for three sensors), correspondingly. These dependencies remained the same when the magnetic field was increasing or decreasing. One can see that the magnetoresistance and its equivalent temperature shifts are smaller for the TVO sensor with respect to the TTR 1B one. The dependence $\Delta T(B, 4.23\ K)$ for TTR 1B looks like the characteristics for the carbon glass resistance thermometers, approximately.[10] In the range 0 to $\approx$ 4 T, the temperature shifts for them due to magnetic field at 1.52 K indicate positive values.

## CONCLUSIONS

The rhodium–iron and platinum resistance thermometers (RIRT and PRT) of CRYOMET VNIIFTRI have exhibited the temperature shifts less than 10 mK during gamma radiation doses up to $10^5$ Gy, $E_\gamma \approx 0.67$ MeV, at 77.4 K: this value is within the accuracy of their calibration characteristics.

RIRT and PRT sensors have exhibited maximum shifts of about 20 mK during neutron+gamma radiation doses up to $7 \cdot 10^{15}$ n/cm$^2$, $E_n \approx 1$ MeV, and $9 \cdot 10^4$ Gy, $E_\gamma \approx 1.5$ MeV, for 77.4 K. The annealing effect was not revealed after warming from 77.4 K to the room temperature.

The carbon (TVO) resistance thermometers can shift up to 100 mK at 77.4 K during neutron+gamma radiation doses up to $5.2 \cdot 10^{15}$ n/cm$^2$, $E_n \approx 1$ MeV, and $8 \cdot 10^4$ Gy, $E_\gamma \approx 1.5$ MeV. The corresponding values for germanium film TTR 1A sensors are about 100 mK at 77.4 K during neutron+gamma radiation doses up to $1.5 \cdot 10^{15}$ n/cm$^2$, $E_n \approx 1$ MeV, and $3.3 \cdot 10^4$ Gy, $E_\gamma \approx 1.5$ MeV.

TTR thermometers have shown the stable reproducibility in perpendicular magnetic field up to 9 T at 4.2 K.Their magnetoresistance $\Delta R/R$ is about 1.2 % ($\Delta T/T \approx$

1.3 %), for example, for 4.2 K and 5 T. These characteristics for TVO sensors are better with respect to the TTR ones, and their magnetoresistance is less approximately by two times for the same values of the temperature and magnetic field ($\Delta T/T \approx 0.6$ % at 4.2 K and 5 T). The advantages of the TTR sensors are comparatively small dimensions, good calibration characteristics over the whole range from 1.5 K to 350 K and commercially available.

One of the best solutions for cryogenic thermometry in the 77.4 to 300 K range, magnetic fields up to 2.5 T and neutron+gamma radiation environments up to $7 \cdot 10^{15}$ n/cm$^2$, $E_n \approx 1$ MeV, and $9 \cdot 10^4$ Gy, $E_\gamma \approx 1.5$ MeV, is to use the platinum resistance thermometers: the additional known shift due to magnetic field can be of $\Delta T/T \approx 0.08$ % at the temperature of about 84 K and 2.5 T. The TVO–thermometers could be considered as competitor with PRT because of its rather attractive complex characteristics and significantly lower price.

To understand better the process of temperature measurements in radiation environments, the exploration with the platinum, rhodium–iron and TVO resistance thermometers will continue up to the total doses more than $10^{16}$ n/cm$^2$ and $10^5$ Gy at 77.4 K; the carbon glass resistance thermometers of VNIIFTRI will be added to the mentioned above in this case.

## REFERENCES


1. S.S.Court, D.S.Holmes and P.R.Swinehart, Neutron and gamma radiation effects on cryogenic temperature sensors, in: "Temperature, Its Measurements and Control in Science and Industry", J.F.Schooley ed., American Institute of Physics, New York, (1992), p.1237.
2. S.S.Court and D.S.Holmes, Effects of cryogenic irradiation on temperature sensors, in: "Advances in Cryogenic Engineering", Plenum Press, New York (1996), vol. 41B, p.1707.
3. ATLAS Collaboration, Technical Proposal, CERN/LHCC/94-43 (1993), M.Shupe, Private communication.
4. V.I.Datskov and J.G.Weisend II, Characteristics of russian carbon resistance (TVO) cryogenic thermometers, *Cryogenics* 34:425 (1994) ICEC Supplement.
5. D.I.Astrov, G.S Abilov, B.I Alshin et al. Investigations in the field of thermometry and thermophysical measurements, Proceedings of VNIIFTRI, Moscow (1967-1983), in russian.
6. V.Mitin, Y.Thorik and E.Venger, New type of thin–film germanium resistance thermometer for use in a wide temperature range, in: "Proc. ICEC 16, Part I", (1996), Kitakyuishu, Japan, p.653.
7. F.Pavese, V.M.Malyshev, P.P.M.Steur, D.Ferri and D.Giraudi, Routine calibration of the cryogenic thermometers in the range 1.5-350 K with an accuracy up to the millikelvin level, in:"Advances in Cryogenic Engineering", Plenum Press, New York (1996), vol. 41B, p.1683.
8. A.Cheplakov, V.Golikov, S.Golubyh et al., Radiation hardness of GaAs preamplifiers for liquid argon calorimetry at LHC, 80805 Munich, Forringer ring 6, MPI-PhE/96-15.
9. M.P.Orlova, O.F.Pogorelova and S.A.Ulybin. "Low Temperature Thermometry", Energoatomizdat, Moscow (1987), in russian.
10. T.Haruyama, N.Kimura, K.Tanaka and A.Yamamoto, Temperature measurement under high magnetic fields around 1.8 K by using CGR thermometers, in: "Proc. ICEC 16, Part I", (1996), Kitakyuishu, Japan, p.649.

# SILICON PRESSURE SENSOR FOR IN SITU PRESSURE MEASUREMENT IN A PRESSURIZED SUPERFLUID HELIUM ENVIRONMENT

T. Haruyama, N. Kimura , and T. Nakamoto

KEK, High Energy Accelerator Research Organization
Oho 1-1, Tsukuba, Ibaraki 305, Japan

## ABSTRACT

A silicon piezo-resistive pressure sensor is investigated for possible *in situ* pressure measurement in a pressurized superfluid helium environment. A small pressure sensor, PD116, from Toyoda Machine Works Ltd, has been reported as an applicable sensor for *in situ* pressure measurement up to the maximum pressure of 3 MPa at cryogenic temperatures down to 6 K. It has a low temperature coefficient of sensitivity around 200 ppm/K between 300 and 6 K. In order to evaluate the applicability in a pressurized superfluid helium environment, the PD116 sensor was set in a specific cryostat which can provide superfluid helium environment. The characteristics of the sensor were measured for the pressure range from 100 kPa to 170 kPa at 4.2 K and 2.0 K. A driving current of 1 mA was applied to the sensor in order to limit heat dissipation to less than 1 mW. The pressure dependence of the output voltage of the sensor also indicates high linearity at temperatures lower than 4.2 K. The experimental results of this study clearly indicate a possible use of this sensor for *in situ* pressure measurement in a pressurized superfluid helium environment.

## INTRODUCTION

*In situ* pressure measurement in a superfluid helium environment is important in understanding what is happening in the experimental system. For example, in the basic experiments on hydrodynamic characteristics of superfluid helium both under saturated and pressurized conditions, silicon sensors with a pressure range of 0-1 MPa have been used for many years. KPY series silicon piezo-resistive pressure sensors from Siemens AG have long been used for this purpose[1-3]. However, these sensors have a large temperature coefficient of sensitivity, and now the availability is very low. Recently, large projects require the use of many superconducting magnets at higher magnetic fields. To meet this requirement, magnets are operated at reduced temperature and with better heat transfer performance. For the case of a high field magnet quench, it has been reported that the pressure increased up to about 2 MPa[4] in the confined cryostat. Apart from this superfluid helium experimental research, *in situ* pressure measurements could be useful for obtaining information at low temperatures without time delay or signal distortion caused by using a long capillary pipe. For example, as these sensors are small enough to be set directly at the cold expansion space of cryocoolers, a pressure-volume (PV) diagram at the cold head can easily be obtained for understanding the cooling characteristics of crycoolers[5].

A small piezo-resistive silicon pressure sensor, PD116, from Toyoda Machine Works at Nagoya in Japan, was reported to be useful for *in situ* pressure measurement up to the maximum pressure of 3 MPa at temperatures down to 6 K[6]. This paper reports the results of performance tests of this sensor at 4.2 K and in a pressurized superfluid helium environment.

## SILICON PIEZO-RESISTIVE PRESSURE SENSOR

The main features of the PD116 sensor in this study are as follows:

- 3 MPa full scale range,
- small size (9 mm diameter and 17 mm length) and light mass (~1.5 g),
- heavily doped ($10^{-21}/cm^3$) piezo-resistive gauges on a silicon diaphragm,
- typical gauge resistance is 950 Ω ,
- 1 atm $N_2$ gas as the reference pressure,
- hermetically sealed by metal can.

These features indicate the possible use of this sensor for *in situ* pressure measurement in a superfluid helium environment. Figure 1 shows a photograph and a schematic diagram of this sensor. The low temperature characteristics of this sensor down to 6 K have already been reported by K. Nara et al.[6], concluding that it is applicable for use in a wide temperature range because of its small temperature dependence. The temperature coefficient of sensitivity is reported as small as about 200 ppm/K over the measured temperature range. To use this sensor in a superfluid helium environment, the important points to consider are the sealing performance against superfluid helium after thermal cycling and the effect of the nitrogen gas inside the metal can.

## EXPERIMENTS

### Set up

The schematic diagram of the experimental set up is shown in Figure 2. The liquid nitrogen shielded glass cryostat was used to maintain the superfluid helium temperature. The helium bath is connected to both the recovery gas-bag and the rotary pump with a

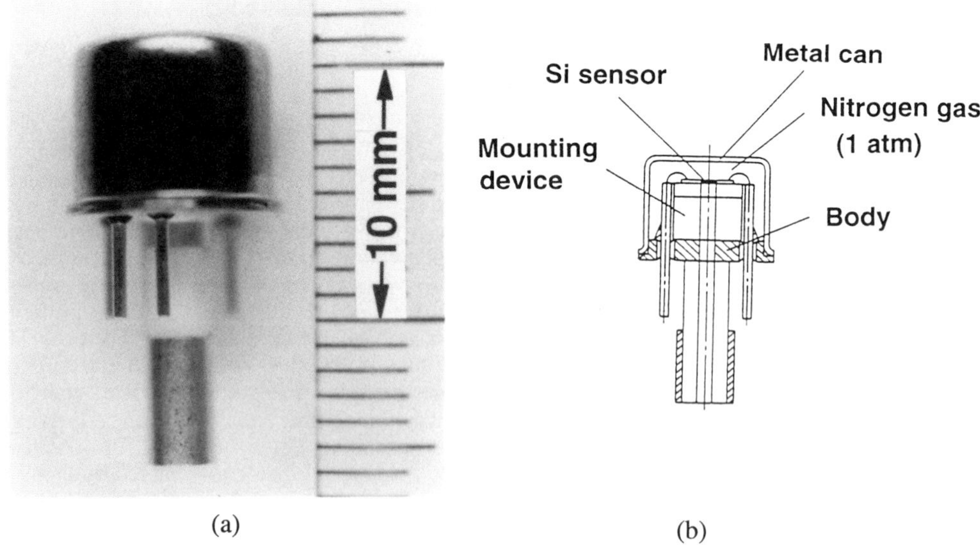


**Figure 1.** PD116 silicon pressure sensor (a) with a detailed schematic view (b).

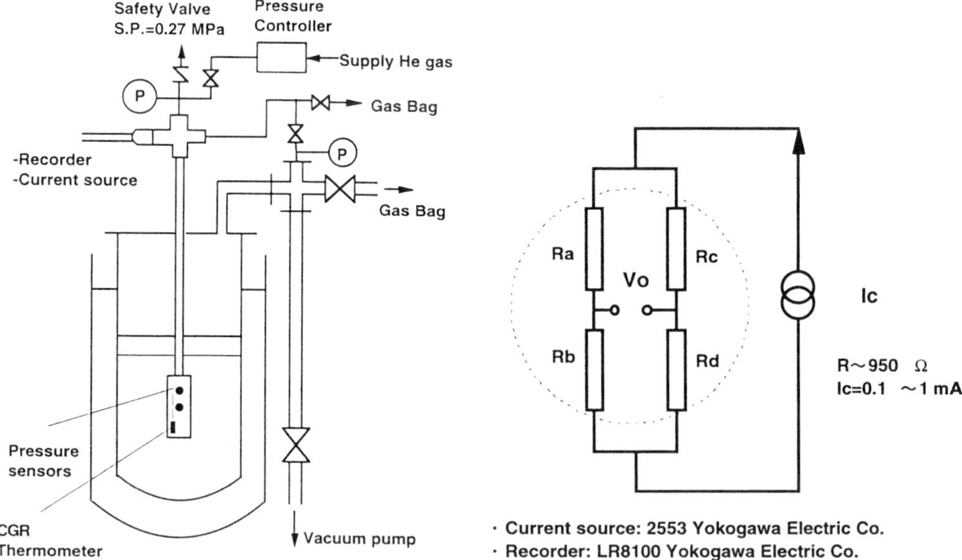


**Figure 2**. Experimental set up.

**Figure 3**. Measuring system .

pumping capacity of 1500 L/min. A sample holder 23.5 mm in diameter and 80 mm in length is suspended by the copper-nickel thin wall pipe from the top flange of the cryostat. Helium gas is supplied to the sample holder after being regulated by the hand-made pressure controller. The controller consists of a pressure monitor, solenoid valves and needle valves to regulate the pressure in the sample holder. Two precision pressure gauges, 2655 (Yokogawa Electric Co.), were used to monitor both the helium bath pressure and sample holder pressure at room temperature, as for the reference pressure. Five pressure sensors and one carbon glass resistance thermometer (CGRT, Lake Shore) are put in the sample holder and then sealed by soldering. The measuring circuit is shown in Figure 3. A D.C. constant current of 1 mA was supplied by the precision current source, 2553 (Yokogawa Electric Co.), and the output voltage of each sensor was recorded by the multi-channel pen recorder, LR8100 (Yokogawa Electric Co.).

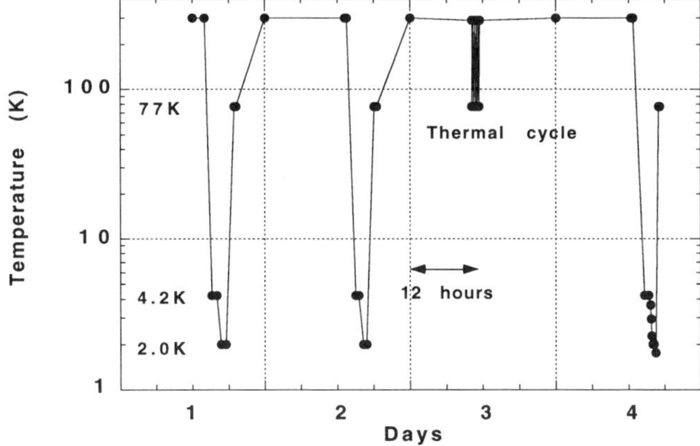


**Figure 4.** A profile of the experiments. Two sensors were tested between 297 K and 2.0 K three times and thermally cycled between 297 K and 77 K once.

## Procedure

The I-V characteristics and the pressure dependence of the output voltage were measured at 293 K. Then, the sensors were cooled down to 4.2 K with liquid helium. At 4.2 K, the same measurement was carried out. Then, the helium bath was pumped to reduce the temperature down to 2.0 K. Also, the I-V characteristics were measured and the sample holder was pressurized up to approximately 170 kPa. After completion of the measurement at 2.0 K, liquid nitrogen was poured into the cryostat for the 77 K measurement. In the present study, this set of measurements was carried out three times. The thermal cycle test between 293 K and 77 K was carried out once. Figure 4 shows the measurement profile in detail. During the measurement, the heat dissipation in the liquid helium is estimated to be less than 1 mW with a driving current of 1 mA.

# RESULTS

## I-V Characteristics

At null pressure, the output voltage dependence of the driving current was measured to check the ohmic resistivity of the differential resistance of the sensor at each temperature. Figure 5 shows the I-V characteristics at 293 K, 77 K, 4.2 K and 2.0 K. It indicates that the differential resistance of the sensor can be regarded as ohmic without any curious non-linearity over the wide temperature range.

## Pressure dependence

The pressure dependence of the relative change in output voltage (V-Vo; V = measured voltage, Vo = offset voltage) at 293, 77, 4.2 and 2.0 K is shown in Figure 6. The sensitivities at 77, 4.2 and 2.0 K are almost identical (0.5 mV/100 kPa at a driving current of 1 mA). This means that the sensitivity has almost no temperature dependence below 77 K. Also, the linearity between the output voltage and the applied pressure is as good as $10^{-2}$ to $10^{-3}$ % at 4.2 K and 2.0 K. The temperature dependence of the sensitivity normalized to 293 K is shown in Figure 7. The average temperature coefficient of the normalized sensitivity over the span is approximately 360 ppm/K. This result is almost consistent with the previously reported one. No apparent evidence of abnormal behavior by the solid nitrogen in the metal can was recognized during cooling from 77 K to 2.0K.

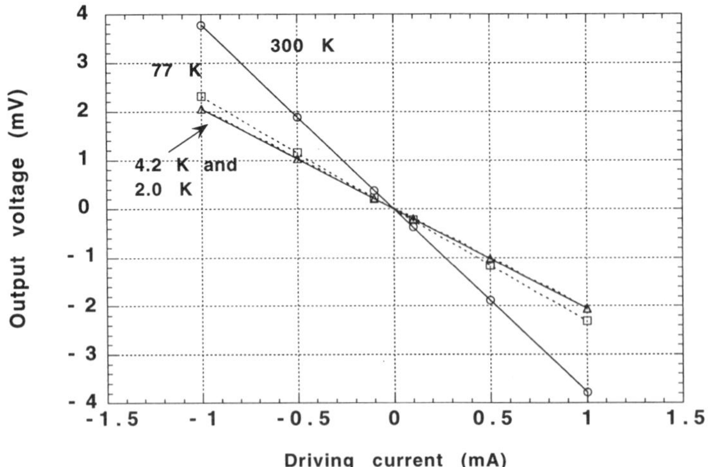


**Figure 5.** Driving current and output voltage (I-V) characteristics at each temperature. The I-V characteristics at 4.2 K and 2.0 K are almost identical.

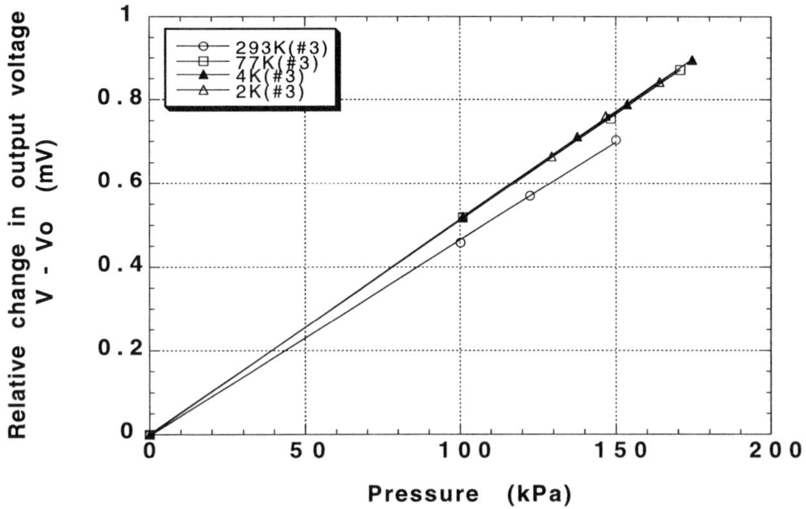


**Figure 6.** Pressure dependence of the relative change in output voltage at 293, 77, 4.2 and 2.0 K.

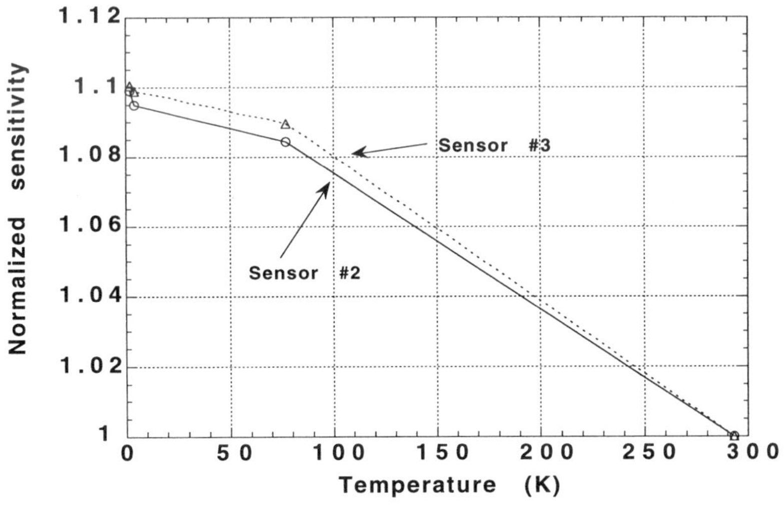


**Figure 7.** Temperature dependence of the normalized sensitivity.

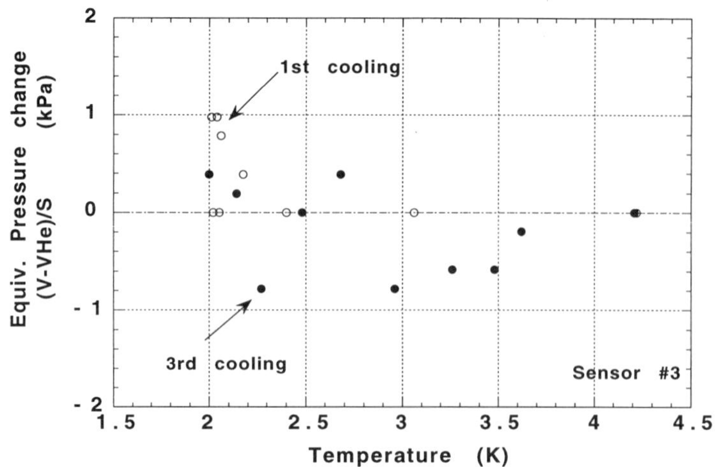


**Figure 8.** Equivalent pressure change during cooling to superfluid helium. Relative output voltage is divided by the sensitivity, S, and converted into the equivalent pressure change.

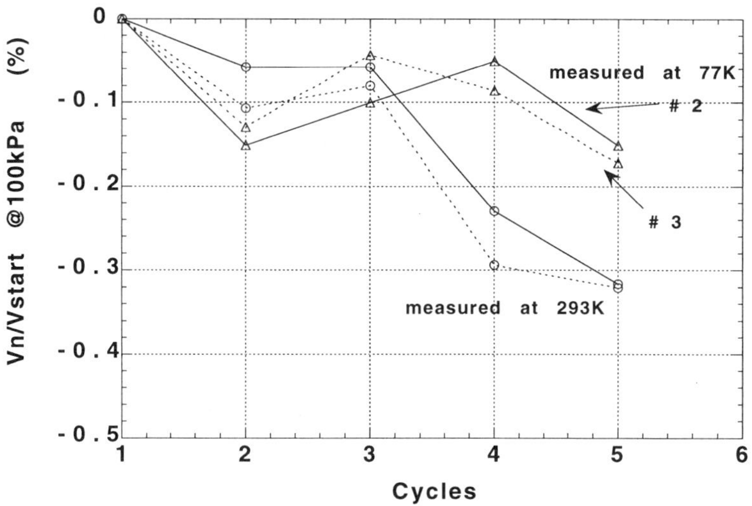


**Figure 9.** Relative output voltage shift after thermal cycling between 300 to 77 K for sensors #2 and #3.

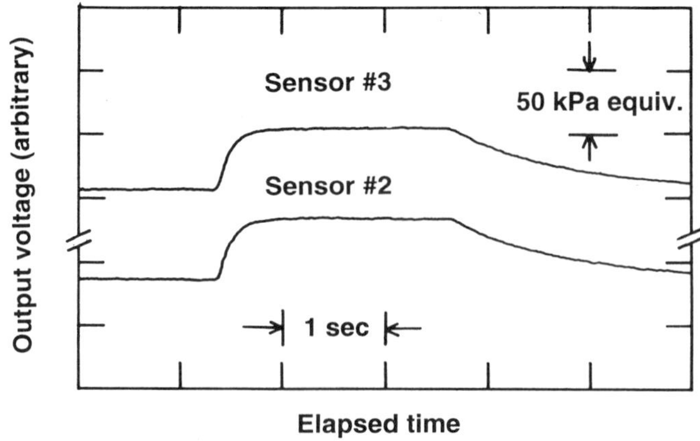


**Figure 10.** Typical response trace of the sensor .

**Table 1.** Summary of results

| Temperature | 4.2 K | 2.0 K | |
|---|---|---|---|
| Sensitivity | 0.5 mV/100 kPa | 0.5 mV/100 kPa | @1 mA |
| Linearity | 0.001 % | 0.027 % | |
| Reproducibility | 0.24 % | 0.20 % | @2K |
| Heat dissipation | < 1 mW | < 1 mW | @1 mA |

### Cooling to superfluid helium temperature

To evaluate the hermetic sealing performance in superfluid helium, the helium bath of the cryostat was pumped to reduce the temperature from 4.2 K to superfluid helium temperature keeping the pressure in the sample holder constant at nearly atmospheric pressure. During the cooling to 2.0 K, the output voltage and liquid temperature inside the sample holder were measured. The output voltage is converted into the equivalent pressure change. The results for the first and third cool downs are plotted in Figure 8. The five thermal cycle tests between 293 K and 77 K were carried out before the third cooling. The results indicate that the sealing performance of the metal can seems to be satisfactory in the superfluid helium environment.

### Thermal cycling

The thermal cycling test was carried out between 293 K and 77 K. Keeping the pressure inside of the sample holder at atmospheric pressure, the sample holder was immersed in a water bath and then in a liquid nitrogen bath. This was repeated five times. Each time, the output voltage was measured and compared with the initial voltage. Figure 9 shows the relative change of the output voltage at both 293 K and 77 K.

In addition to this liquid nitrogen thermal cycling test, these sensors experienced the thermal cycle between 293K and supefluid helium temperature for three times with no resulting damages, instabilities, etc. The output voltage of the sensor was measured at 2.0 K after the each cooling from 293 K. The average reproducibility is about 0.2 % at 2.0 K.

## TYPICAL OUTPUT TRACE AT PRESSURE CHANGE

The sensor response performance at 2.0 K was measured by changing the pressure in the sample holder quickly. The sample holder was pressurized by opening the supply valve. The pressure in the sample holder was changed from 100 kPa to 150 kPa. Figure 10 shows typical response traces of two PD116 sensors. As the sample holder is connected to the supply valve at room temperature with a long thin wall pipe, the exact response time of the sensor at 2.0 K could not be estimated quantitatively.

## SUMMARY

A small and convenient silicon piezo-resistive pressure sensor was evaluated for *in situ* use in a superfluid helium environment. The results summarized in Table 1 indicate possible use of the sensor in a superfluid helium environment. The nitrogen gas inside the metal can does not affect the performance, and the hermetic sealing of the sensor was unaffected after three thermal cycles between 293 K and superfluid helium temperature.


## ACKNOWLEDGMENTS

The authors wish to thank Profs. T. Shintomi and A. Yamamoto for their helpful comments. Also, they wish to thank the staff of the Cryogenics Science Center for their kind assistance during the experiments.

## REFERENCES

1. F. Breissuire et al, KPY12-A pressure transducer suitable for low temperature use, The 8 the International Conference on Magnet Technology (MT-8) Grenoble (1983) C1-671-674
2. R.M. Igra et al. Simultaneous pressure and temperature measurement on helium in the high speed rotating frame, *Adv Cryog Eng.* 31: 1005 (1986)
3. P.L. Walstrom et al. Use of Siemens pressure sensors at liquid helium temperatures, *Cryogenics* 27: 439 (1987)
4. F. Haug and A. McInturff, Measurement of pressure transmission in long capillaries, *Adv Cryog Eng.*.35: 1583 (1990)
5. T. Haruyama. and H. Inoue, Design and fabrication of a prototype Stirling Cryocooler for space borne applications, *Adv Cryog Eng.* 39: 1367 (1994)
6. K. Nara et al. Piezo-resistive pressure sensor applicable for in situ pressure measurement at cryogenic temperatures under magnetic fields, *Cryogenics* 33: 541 (1993)

# DEVELOPMENT OF AN ACCURATE DOUBLE-DIAPHRAGM SAPPHIRE CRYOGENIC CAPACITIVE PRESSURE TRANSDUCER

F.Pavese,[1] D.N.Astrov,[2] P.P.M.Steur,[1] D.Ferri,[1] and D.Giraudi[1]

[1]CNR, Istituto di Metrologia "G.Colonnetti" (IMGC)
Torino 10135, Italy
[2]VNIIFTRI
Moskow Region 141570, Russia

## ABSTRACT

Mechanical creep of the diaphragm is well recognized as a major problem for obtaining a reproducible cryogenic pressure transducer and it is mainly caused by the stress at the diaphragm edge, where this is rigidly constrained to the transducer body. The double-diaphragm design alleviates this problem. In this design, the rigid constraint of the diaphragm edge is replaced by a (non-ideal) hinge between two identical diaphragms, which allows for a considerable improvement of the reproducibility on thermal cycling. Sapphire has normally been used for this design because of its excellent elastic properties. However, the quality of the results critically depends on technological details of the fabrication.

The paper describes the design criteria used for the actual implementation of the double-diaphragm technique, making use of monocrystal sapphire disks of limited dimensions (∅ 25 mm). Results on the stability for $\Delta p = 0$ of a capacitive prototype ($\approx 0.1$ bar f.s., sensitivity $1.35 \cdot 10^{-3}$ pF $Pa^{-1}$), designed mainly for work in the range 3–30 K with a gas thermometer in the differential mode, showed a nearly zero temperature coefficient in that range, a zero shift decaying logarithmically with thermal cycling between 77 K and 300 K and limited to $\approx 0.025\%$, and a creep limited to 50 ppm in 8 days at $\approx$ 80–95 K.

## INTRODUCTION

One of the major problems in designing a reproducible cryogenic pressure transducer is the mechanical creep of the diaphragm, mainly caused by stress at the diaphragm edge, where it is rigidly constrained to the transducer body.[1] In this respect, the double-diaphragm design has proved to alleviate this problem and allow a considerable improvement as to creep and reproducibility on thermal cycling.[2] Metal diaphragms are generally used, but also materials with better elastic characteristics has recently been tested, such as pure silicon.[3] For the double-diaphragm design, only sapphire has been considered so far due to its excellent elastic properties.[2-4] However, the quality of results critically depends on technological details of the fabrication.

The paper describes the design criteria used for the actual implementation making use of monocrystal sapphire disks, and reports on preliminary results on a prototype with capacitive detection designed for work at cryogenic temperatures.

## TRANSDUCER DESIGN AND FABRICATION AND EXPERIMENTAL SETUP

Figure 1 compares the basic principle of a traditional diaphragm transducer and of a double diaphragm transducer. The traditional design (a), apart from technological variations or additions for improving the capacitive detection, basically involves a thin diaphragm *rigidly clamped* to a bulk external ring: two different pressures $p_1$ and $p_2$ (which can be zero) are applied to the sides of the diaphragm. In this geometry, the maximum *radial* stress is concentrated at the boundary ($r = R$) where the diaphragm is clamped to the ring and the displacement is null:

$$\frac{\sigma_{rmax}}{\Delta p} = \frac{3R^2}{4t^2} , \tag{1}$$

where $\Delta p$ is pressure difference, $t$ is thickness and $R$ is the plate radius. The normal stress, on the contrary, is maximum at the center ($r = 0$), and is: $\sigma_{zmax} = \sigma_{rmax}(1 + \nu)/2$ (for sapphire, $\nu \approx 0$).

Even for moderate displacements of the diaphragm, the elastic limit of the material may be exceeded, bringing to an anelastic behaviour (creep, hysteresis). By using pre-stressed diaphragms, the problem is limited, not cancelled. The double diaphragm design (b), involve the use of two identical diaphragms sealed together at the boundary, leaving a small internal space between them. If the internal space is kept evacuated during fabrication, the transducer can measure only an absolute pressure put on its outside. In this design, the *radial* stress is zero at the boundary if the hinge is ideal (edge freely supported). It has a maximum at $r = 0$:

$$\frac{\sigma_{rmax}}{\Delta p} = \frac{3R^2}{8t^2}(3 + \nu) , \tag{2}$$

(for sapphire, $\nu \approx 0$). The normal stress, too, has a maximum at the center ($r = 0$), and is: $\sigma_{zmax} = \sigma_{rmax}$. The maximum diaphragm deflection, that occurs at the center, in the freely supported case is about four times the value for the clamped edge case. If the hinge is not perfect, an intermediate situation occurs. In any case, stress concentration is minimized and so are, in principle, its consequences (creep, hysteresis).

If a hole is drilled across the center of one diaphragm, a small tube can be soldered to it, allowing access to the internal space and making the transducer a differential one.

The prototypes used in this paper were designed and fabricated in co-operation between IMGC and VNIIFTRI using sapphire disks.

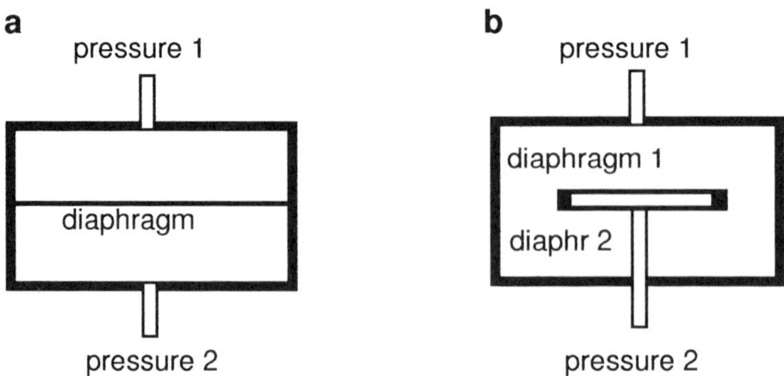


**Figure 1**. Design principle of pressure transducers: (a) traditional diaphragm, (b) double-diaphragm .

The prototype whose characteristics are reported in this paper (#2) has a small diameter (Ø 25 mm) and thickness ($s$ = 0.23 mm). Based on VNIIFTRI experience, single crystal disks were used with orientation of the $C_3$ crystallographic axis 60° to the surface and identically oriented. Soldering between the two disks was performed using a wire ring[a] of special glass whose thermal expansion best matches that of sapphire; the soldering area is restricted as much as possible to the boundary.

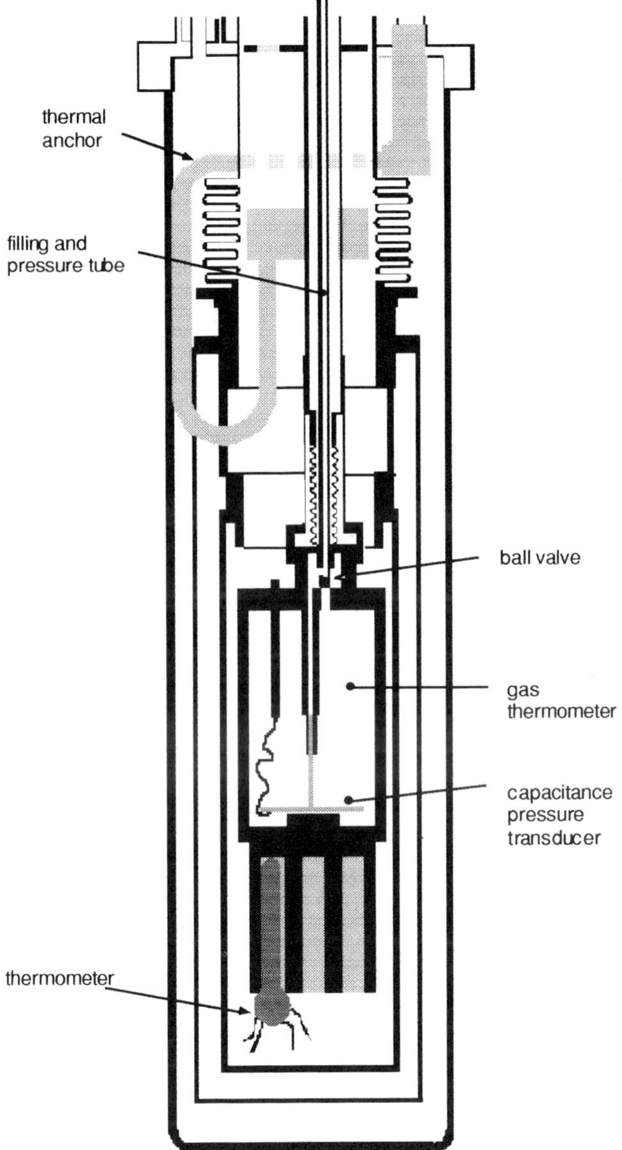


**Figure 2.** The zero-dead-volume ICVGT of IMGC (shown inserted in the lower part of the cryostat). Gas thermometer: ≈ 50 $cm^3$ gas container with appended thermometer block. The ball by-pass valve connects the ICVGT bulb (pressure 1 in Fig.1) to the pressure measuring tube (pressure 2 in Fig.1) to perform the transducer zero. The cryostat is immersed in a liquid cryogenic fluid.

---

[a] Instead of the glass powder used in Ref.2: IMGC had previously experienced on a prototype kindly supplied by the authors that such a sealing method may be quite brittle.

A mean gap of ≈ 0.03 mm was left between the disks. On each *external* surface of the transducer a platinum circular electrode was then evaporated for the capacitance measurement (without guard ring). At the center of one plate a hole was drilled and a small platinum tube was soldered to the plate with some glass. The prototype has a maximum *external* pressure (at which the plates contacts at zero internal pressure) of $\approx 10^4$ Pa and a sensitivity of $1.35 \cdot 10^{-3}$ pF $Pa^{-1}$ ($C \approx 39$ pF).

The transducer was measured using a precision capacitance bridge (General Radio mod. 1615A) with an external 100 pF standard capacitor (GR 1404-13) with a temperature relative coefficient of $10^{-5}$ $K^{-1}$ kept in a thermostat whose temperature was stable to better than ± 0.01 K. Measurements were performed supplying 7 V to the capacitor and using as a detector a lock-in amplifier EG&G mod. 9503 (10 nV noise limit) run by a computer. The transducer was connected to the bridge via 50 Ω coax cables (of the cryogenic type inside the cryostat), un-interrupted except at the cryostat outlet via a 2-pin LEMO connector.

A noise level of $\pm 2 \cdot 10^{-5}$ pF (1 Hz bandwidth) was consistently observed, corresponding to ± 0.5 ppm, or, in pressure, to $\pm 1.5 \cdot 10^{-2}$ Pa. The capacity of the open connections (transducer not connected to either cables) was measured to be $< (5 \pm 2) \cdot 10^{-5}$ pF.

After some preliminary tests in a simple cryostat aimed at checking the effect of thermal cycling from room temperature down to liquid nitrogen temperature, the transducer was mounted –suspended from its connecting tube– in an equipment aimed at realizing an interpolating constant-volume gas thermometer (ICVGT) for the realization of the International Temperature Scale of 1990 (ITS-90) in the range 3-26 K.[1] For this application, assuming a $^3$He gas filling at a density of $N/V = 300$ mol $m^3$, the sensitivity achieved is 2.5 kPa $K^{-1}$. Consequently, the aimed sensitivity of ± 0.1 mK requires a pressure sensitivity of 0.25 Pa, corresponding to ± 0.34 fF or ± 10 ppm of the capacitance value. Figure 2 shows the transducer mounted in the ICVGT: for further details see Ref.5.

## EXPERIMENTAL RESULTS AND DISCUSSION

Preliminary thermal cycling experiments were performed between 300 K and 77.3 K, since most of the thermal contraction takes place above the latter temperature. Figure 3 reports the results since the very first cooling. The shift appears to decay logarithmically on cycling, being sufficiently low for most needs. To attain the best accuracy, transducer zero must be checked *in situ* (by opening the provided by-pass valve).

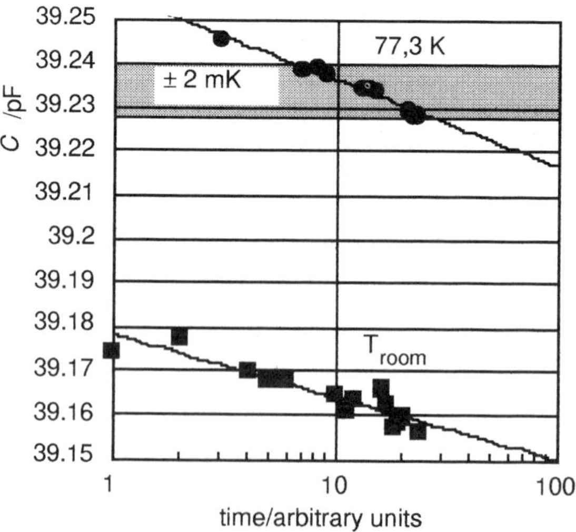


**Figure 3.** Thermal cycling of prototype #3 between 300K and 77.3 K. The shift is logarithmically decaying (the equivalent temperature error for the ICVGT –see text– is shown).

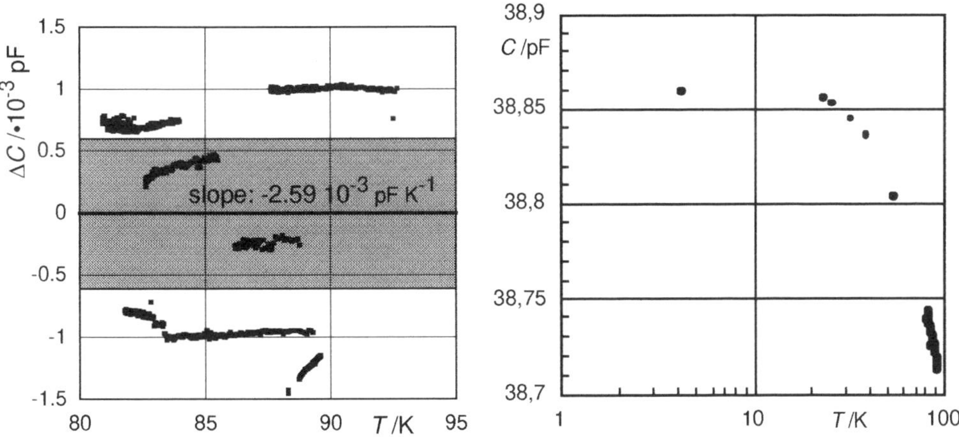


**Figure 4.** Residuals from a linear fitting with slope $-2.59 \cdot 10^{-3}$ pF K$^{-1}$.
Shaded area = ± 0.2 mK for the ICVGT.

**Figure 5.** Temperature coefficient in the range 4–100 K.

Then the extent of the possible creep has been checked at liquid nitrogen temperature in the cryostat of Fig.2. This information has to be extracted from the capacitance variation due to the transducer temperature coefficient. Several runs have therefore been performed with the by-pass valve open at different temperatures between 80 K and 93 K. Initially, the transducer was kept with both sides evacuated. Step shifts of the capacitance value occasionally were randomly observed. Checks showed that the capacitance readings were not affected by changes (within the noise level) when the transducer was subjected to small shocks or to the displacement of the cabling. Therefore, it was subjected to a few small pressure cycles, and eventually some gas (tens of pascal) was permanently left in both sides (by-pass open). The sudden shifts did not occur anymore. There are several possible explanations: one is that some impurities were "washed out", avoiding capacitance instability due to the adsorbed gas layer on the plates; another is that the plates were brought to a definite mechanical state, avoiding the occurrence of small inelastic displacements (the extent of hysteresis –which has the same origin– has not been measured so far).

The results of changing temperature are shown in Fig.4. A temperature coefficient of $-2.59 \cdot 10^{-3}$ pF K$^{-1}$ has been obtained from a linear fitting (a quadratic fitting is not necessary in this narrow temperature range). Over the 8-day period the measurements were taken, a total shift of $+ 2 \cdot 10^{-3}$ pF can be observed, equivalent for the ICVGT (see end of former Section), to a temperature uncertainty lower than ± 0.3 mK. It is attributed to creep. At shorter term, creep is not observed: changing speed and sign of temperature variation does not show any hysteresis within the noise level (the standard deviation of each series – up to 15 hours long– is few parts in $10^{-5}$ pF).

When the measurements were extended down to 4.2 K, the transducer showed the behavior of Fig.5. The capacitance shift was limited to $6.5 \cdot 10^{-3}$ pF in the range 4.2 K to 25.7 K, i.e. in the working range of the ICVGT ($T < 26$ K), equivalent to ≈ 2 mK, most of it above 20 K.

## CONCLUSIONS

A capacitance cryogenic transducer made of sapphire with the double-diaphragm technique was characterized by checking the creep effect due to thermal cycling on the zero reproducibility, for use as a differential diaphragm for an interpolating constant-volume gas thermometer (ICVGT) to be used in the range 3-26 K. Results on creep and thermal cycling showed that their occurrence is limited, at ≈ 77 K, to values lower than usually observed with metallic diaphragms, avoiding the need of a correction at all or, for an accuracy of ≈ ± 0.1 mK of the ICVGT, allowing an uncritical one.

The actual temperature coefficient in most of the working range of the ICVGT is null. Further measurements need to be performed for the off-zero calibration (linearity, reproducibility), for hysteresis, for the pressure coefficient of the zero and for checking these effects on a longer time scale.

## REFERENCES


1. F.Pavese, and G.F.Molinar, *Modern Gas-Based Temperature and Pressure Measurements*, Plenum, New York, 1992.
2. W.Griffoen, and G.Frossati, *Rev.Sci.Instrum.* 56: 1236 (1985)
3. P.M.Echternach, Inseob Hahn, and U.E.Israelsson, *Adv. Cryo. Engin.* 41B:1837 (1996)
4. D.N.Astrov, L.B.Beliansky, Y.A.Dedikov, S.P.Polunin, and A.A.Zakharov, *Metrologia* 26:151 (1989)
5. P.P.M.Steur, F.Pavese, D.Ferri, D.Giraudi, *Proc. TEMPBEIJING '97*, Chinese Society of Measurements, Beijing (1997), in press.

# MEASUREMENTS OF VOID FRACTION AND FLOW RATE OF LNG FLOW

A.M. Arkharov,[1] S.B. Glavatskikh,[2] A.G. Grechko,[1] and A.A. Zherdev[1]

[1]Cryogenics Division, Moscow State Technical University
Moscow, 107005, Russia
[2]Division of Machine Elements, Luleå University of Technology
Luleå, S-971 87, Sweden

## ABSTRACT

The paper describes the sensor and methodology to measure the void fraction and flow rate of liquefied natural gas two-phase flow. The design and calibration procedure of high frequency sensors as well as sensitivity and accuracy are discussed. High frequency method is used for void fraction measurements. Flow rate is measured by cross-correlation technique. The automatic measuring system to control two-phase cryogenic flows is described. Application of the system for testing fuel pumps of aircraft using cryogens as propellants is demonstrated. Results of calibration of the system on liquefied natural gas flow are presented.

## INTRODUCTION

Environmental problems such as the anthropogenic greenhouse effect, caused by the use of kerosene in air transportation, and limited resources of fossil fuels necessitate a selection of a suitable environmentally friendly energy carrier for future aircraft. Liquid hydrogen looks the most promising. However, for a country like Russia with large reserves of natural gas, the use of LNG may be a reasonable intermediate step towards liquid hydrogen.

The use of cryogenic fuels in aviation requires, besides engine modifications, the development of ground facilities for refueling a cryoplane. For the same time of pre-flight servicing as for kerosene aircraft, high refueling rates are required for the cryoplane. The ground facilities should be equipped with flowmeters. Conventional single phase flow meters are not applicable because of the vapor phase in the LNG flow and severe operating conditions. Thus, to avoid operational and safety problems such as tank damage in case of fuel level indicator failure or structural damage by temperature shocks, and to achieve appropriate fuel metering accuracy, a system for LNG flow monitoring during refueling must be designed for ground facilities.

Another problem arises at the development of suitable cryogenic fuel pumps for engine supply system. The cryogenic pumps, due to low fuel density, rotate at high numbers of revolutions to reduce the number of stages. Such pumps require a large cavitation margin. Their performance and anti cavitation feature should be experimentally investigated. A special sensor is needed to detect cavitation after the LNG fuel pump being tested. We have already developed a number of measuring instruments for two-phase cryogenic flows.[1-4] This article is a logical continuation of our previous work on vapor content sensors.[5-7] It is devoted to an automatic measuring system to monitor parameters of LNG flow. The system was made for the Tupolev Company and used during experimental investigation of fuel pump performance. In the process of system elaboration a new void fraction sensor was devised and patented. The objective was also to design a system not only for LNG flow monitoring, but for liquid hydrogen as well.

## METHODOLOGY

A fundamental flow parameter in vapor-liquid flow is the void fraction $\varphi$. This is the fraction of the pipe area at a given cross section occupied by the vapor phase. It can vary considerably, but usually the variations are at such high frequencies that a time-averaged value can be used. Due to the dielectric behaviour exhibited by LNG components this value can be expressed as follows:

$$\varphi = A_v/A = (\varepsilon_l - \varepsilon)/(\varepsilon_l - \varepsilon_v), \quad (1)$$

where $A$ is the total cross sectional area of the flow; $A_v$ is the cross sectional area occupied by the vapor phase; $\varepsilon_v$, $\varepsilon_l$ are relative dielectric constants of vapor and liquid phase respectively; $\varepsilon$ is the mean dielectric constant at a given cross section.

The extreme variations in void fraction in a two-phase flow create a considerable obstacle for single-phase flow meters. Cross-correlation techniques allow one to use these fluctuations for flowrate measurements. This technique is based on the measurement of the statistical similarity between two random signals, picked up from upstream and downstream sensors in a flow. The type of sensor depends to great extend upon the medium which is to be monitored. Liquid hydrogen and components of LNG are non-polar dielectrics (except propane and isobutane which are weakly polar fluids with small dipole moments[8-10]) with very low high frequency (HF) energy losses. So it seems reasonable to develop a sensor sensitive to changes in the mean dielectric constant of the flow. The dielectric constant can be measured by the capacitance method. But the HF method allows one to achieve higher accuracy of the dielectric constant measurement, as frequency can be measured more accurately and reliably than any other type of analog signal. There is also another advantage of this method: frequency measurement electronics do not drift as do electronics which measure capacitance.

The main idea of the HF method is to measure the resonance frequency of a resonator. At characteristic frequencies the electromagnetic waves in the resonator interfere with one another and produce a very narrow, high output power peak. By measuring the frequency of the resonant peak, one can measure the mean dielectric constant of LNG in the resonator with an extremely high degree of accuracy and resolution. The most important quantity of any resonant system is its quality factor which is in direct proportion to the resonance frequency and the total stored energy, and inversely related to the energy dissipated per second. The energy is dissipated in the metallic walls owing to their finite resistivity. At low temperatures the resistivity significantly decreases and it gives higher sharpness of the resonance peak and hence higher accuracy of the dielectric constant measurement.

Among a great variety of electromagnetic systems coaxial resonators look the most attractive as they allow one to get a concentrated electric field ( like in a capacitance) covering the whole cross section of the pipe and to electrically isolate this capacitance from the environment (important in operation with LNG or hydrogen since an explosion proof design is necessary). In other words, we can place a capacitance inside a volume resonator and get advantages of both methods: HF and capacitance.

### HF sensor

**Design.** The sensor or coaxial resonator is schematically shown on figure 1. It is a closed metal structure formed by a steel cylindrical housing 1, two coaxial flanges 2,3 and a ring 4 which is connected by a thin steel rod 5 to the housing. The sensitive element (a concentrated capacitance) is between the flange 2 and the ring. Thin plates 6 are installed inside the flange and the ring to create a uniform electric field and to increase sensor's resolution. To get higher quality of the resonator, the internal surfaces are covered by a thin layer of copper. The resonator cavity is filled with a dielectric material 7 to form the same cross-section of the sensitive element as in the pipe and to keep cryogens from entering the cavity which can influence the sensor's characteristics. Transmission and receiving of electromagnetic energy are effected by coupling loops 8,9 through two coaxial connectors 10,11.

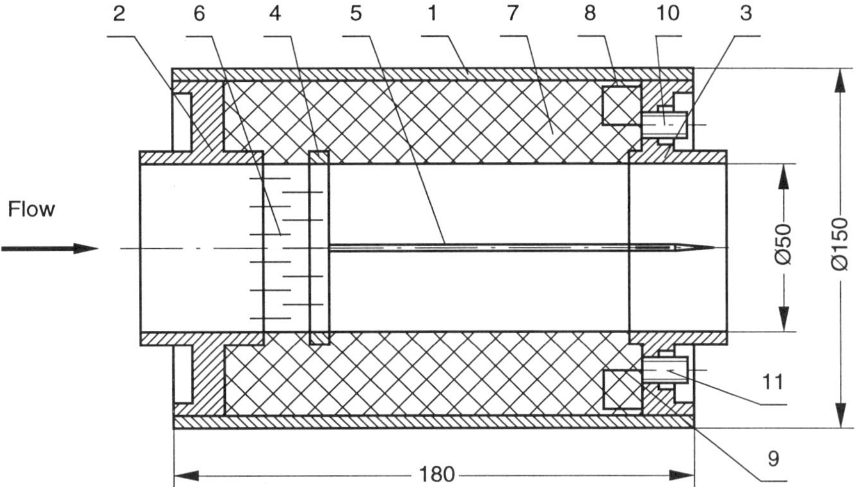


**Figure 1.** Scheme of the HF sensor.

A requirement for the LNG flow monitoring system is to place all electronic equipment at a distance of 110 meters from the sensor. Transmission of HF electromagnetic energy in coaxial cables to and from the sensor is associated with significant attenuation which is in direct proportion to the frequency. Power of the HF sweep generator is limited. It seems reasonable to decrease the resonance frequency of the sensor. It depends on inductance and capacitance. An increase in capacitance reduces the quality of the resonator, and an increase in inductance (which is in direct proportion to the length of coaxial tubes inside the resonator and the ratio of the housing inner diameter to the coaxial tubes outer diameter) makes the sensor grow in size. To solve this problem we installed the rod 5 instead of a tube. The rod, in contrast with the tube which was used in the original design of the sensor,[5] allows one to get four times lower resonance frequency of the sensor. Thus, with this sensor we can place electronic equipment significantly further away.

**Calibration.** The sensor calibration procedure can be separated into two parts: static and dynamic. The latter, which is related to flowrate measurement, will be described in the corresponding section. The static calibration consists of determining the dependence between resonance frequency $f$ and void fraction $\varphi$. This dependence can be derived from consideration of the electrical circuit of the resonator. It represents an oscillatory circuit with inductance $L$, measuring $C_m$ and parasitic $C_p$ capacitances, connected in parallel. Outer surfaces of the ring and rod serve as inductance. $C_m$ is concentrated between the plates and $C_p$ is formed by the dielectric filler. The resonance frequency of the circuit can be expressed as follows:

$$f = 1/\left[2\pi\sqrt{L(C_p + \varepsilon C_m)}\right], \tag{2}$$

where $\varepsilon$ is a relative dielectric constant of a medium filling the measuring capacitance. Using this formula one can get the following relationship

$$\varepsilon = (\varepsilon_r + D)(f_r / f)^2 - D, \tag{3}$$

where $D$ is calibration constant, $f_r$ is resonance frequency of the circuit when the measuring capacitance contains a reference medium of known dielectric constant $\varepsilon_r$. Using (1) and (3) one can obtain

$$\varphi = \left[(\varepsilon_l + D) - (\varepsilon_r + D)(f_r / f)^2\right] / (\varepsilon_l - \varepsilon_v). \tag{4}$$

Generally the choice of the reference medium is dictated by the practical situation. It is convenient to take $\varepsilon_r = \varepsilon_l$ and $f_r = f_l$ since we should know $\varepsilon_l$ to find $D$. The last relationship can be rewritten as

$$\varphi = (\varepsilon_l + D)\left[1 - (f_l / f)^2\right] / (\varepsilon_l - \varepsilon_v). \tag{5}$$

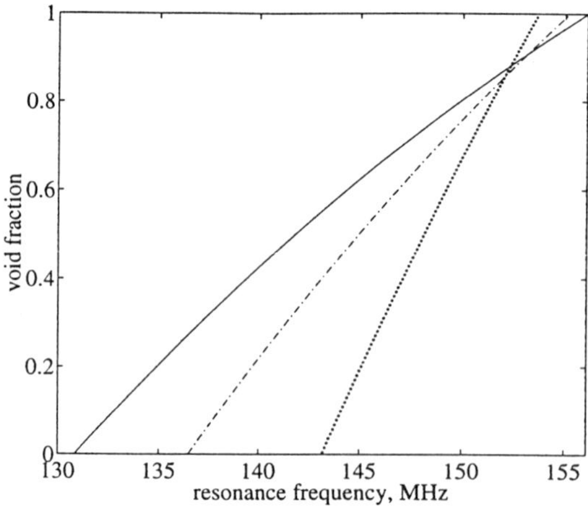


**Figure 2.** Calibration curves for saturated hydrogen (dotted line), nitrogen (dashdot line) and methane (solid line) at a pressure of 0.1 MPa.

Experimental verification of this dependence was carried out on vapor-liquid mixture of nitrogen in a special calibration test rig.[11] It was found that the experimental calibration curve was in good agreement with (5): the deviation (±1%) was within the limits of test rig error. Calibration curves for some cryogens at a pressure of 0.1 MPa are shown in figure 2.

The calibration constant $D$ is independent of the dielectric properties of the controlled medium, and determined by the sensor design. It can be easily calculated through measurements of two resonance frequencies, $f_1$ and $f_2$, when the sensor is filled with media of known $\varepsilon_1$ and $\varepsilon_2$. In the case of a pure substance like nitrogen or hydrogen these $f_1$ and $f_2$ are the resonance frequencies for the pure liquid and pure gas in the saturation line at given temperature. LNG is a multi-component mixture with unknown $\varepsilon_l$ and $\varepsilon_v$. Their values are measured by an auxiliary coaxial resonator with $D = 0$ and then they are used to calculate $D$ of the sensor to be calibrated.

**Sensitivity.** The void fraction sensitivity defined as $\partial f/\partial\varphi$ depends on $\varphi$ because of the non-linear dependence between $\varphi$ and $f$. It is more convenient to use a mean sensitivity which is expressed as follows:

$$\overline{S} = (f_v - f_l)/\Delta\varphi = \Delta f, MHz \tag{6}$$

where $\Delta\varphi=1$ (full range of void fraction change). There are several factors that influence the sensitivity, of which the type of cryogen and operational frequencies are the most important. For various cryogens the difference ($\varepsilon_l$ -$\varepsilon_v$) in the saturation line can significantly vary in magnitude. It is 0.224 for hydrogen (T=20.38 K) and 0.624 for methane (T=111.2 K). This means that the sensor has about 2.5 times higher sensitivity with methane than with hydrogen. In deciding on an appropriate dielectric material filling the cavity it is possible to increase the sensitivity with hydrogen. There is also another way to increase sensitivity, namely, magnification of the sensor resonance frequency. But as pointed out above, this results in higher attenuation in coaxial cables connecting the sensor and the electronic block. For the described sensor the sensitivity with methane is 25.23 MHz (at a pressure of 0.1 MPa). In other words, with electronic equipment resolution of 10 kHz the sensor can detect a bubble of ~1 mm in diameter in liquid methane flow. This bubble corresponds to a void fraction of 0.0004.

**Error.** An error in measurement of the void fraction arises due to inaccuracy in measurement of the resonance frequency, temperature and reference values used in (5), and in calculation of $D$ as well as due to non-uniformity of the electric field in the sensitive element. The maximum error of ±0.9% was obtained by theoretical estimation. This error correlates well with the result (±1%) of an empirical calibration of the sensor on the special test rig.[11,12] There are also some additional factors which can influence the total error. Variations in temperature and pressure cause thermal and mechanical deformations of the sensor. Furthermore, the dielectric constant of the material filling the cavity is a weak function of temperature. All result in a resonance frequency drift. The influence of these parameters was experimentally investigated on the same test rig[11,12] which was used for calibration. It was found that errors due to the pressure change from 0.1 MPa to 1 MPa inside the sensor were very small and could be neglected. However, the temperature error was considerably higher

$$\partial f / \partial T|_{\varepsilon} = 10 kHz / K.$$

To eliminate this error introduced by variations in temperature, it is important to measure temperature and void fraction simultaneously.

It should be noted that the error in measurement of the mean dielectric constant of a single-phase flow depends only on inaccuracy in resonance frequency measurement and calculation of D. That is the reason that it does not exceed ±0.05%.

## Density measurement

There is a simple and convenient method for cryogens density determination through a measurement of the dielectric constant. The dielectric constant is related to the density $\rho$ via the Clausius-Mossotti function (*CMF*):

$$\rho = (\varepsilon - 1)/(\varepsilon + 2)/CMF. \tag{7}$$

*CMF* has the property of being nearly a constant, generally varying about one percent over a wide range of density and temperature. For pure cryogenic substances like hydrogen the *CMF*s are known[10,13-16] and can be used at once in (7). Values of *CMF* for some cryogens are given in Table 1. LNG is a multi-component mixture and *CMF* is calculated according to the following formula:[17]

$$CMF = \sum_i x_i CMF_i, \tag{8}$$

where $CMF_i$ is the Clausius-Mossotti function of component *i* at the same temperature and pressure as the mixture, $x_i$ is the mole fraction of component *i*.

Thus, accurate and reliable measurements of single-phase flow density can be performed by the HF sensor. The error in measurement of liquid hydrogen density is better than ±0.2% and for LNG it is better than ±0.35%. Two-phase hydrogen flow density can be calculated from the equation:

$$\rho = \left[ \frac{\varepsilon_v - 1}{\varepsilon_v + 2} \frac{\varepsilon_l - \varepsilon}{\varepsilon_l - \varepsilon_v} + \frac{\varepsilon_l - 1}{\varepsilon_l + 2} \frac{\varepsilon - \varepsilon_v}{\varepsilon_l - \varepsilon_v} \right] / CMF. \tag{9}$$

In the case of a two-phase LNG flow this equation can also be used but compositions of both phases should be known to determine dielectric constants and *CMF*s of the phases.

## Flowrate measurement

As it was mentioned above cross-correlation technique is used to measure flowrate. If $f_1(t)$ and $f_2(t)$ are the two signals derived from upstream and downstream sensors respectively, the cross-correlation function of these signals in terms of the time delay $\tau$ is given by[18]

$$R_{f_1 f_2}(\tau) = \lim_{T \to \infty} \frac{1}{T} \int_0^T f_1(t) f_2(t - \tau) dt. \tag{10}$$

The time-of-flight is determined from the delay $\tau'$ corresponding to the abscissa of masspoint of the cross-correlation function in the peak region. The flow rate is calculated from this delay, cross section area *A* and the known sensor separation *L*:

$$Q = kAL / \tau'. \tag{11}$$

where *k* is flowmeter factor, defined as the ratio of the true mean velocity to the velocity measured by cross-correlation flow meter. It is determined by an empirical calibration.

**Table 1**. Values of *CMF* at 115 K (hydrogen at 20.268 K) for various cryogens

| Cryogen | $CH_4$ | $C_2H_6$ | $C_3H_8$ | $nC_4H_{10}$ | $iC_4H_{10}$ | $N_2$ | $H_2$ |
|---|---|---|---|---|---|---|---|
| *CMF* x$10^3$, $m^3/kg$ | 0.4093 | 0.3688 | 0.3673 | 0.3501 | 0.3660 | 0.1571 | 1.0051 |

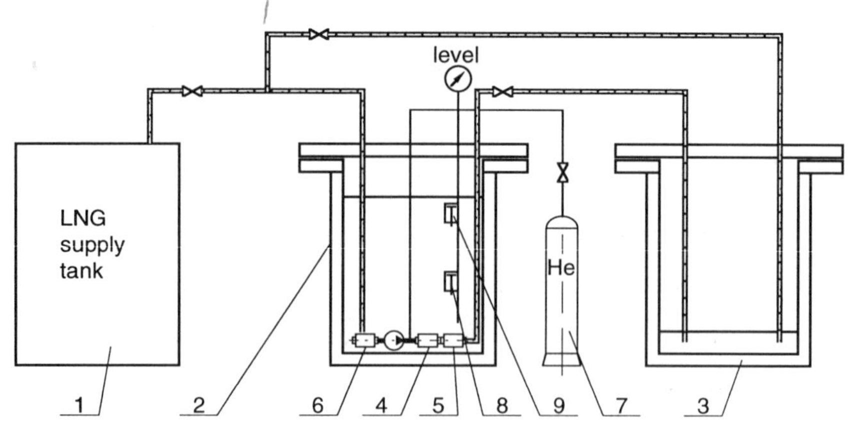


**Figure 3**. Test circuit used for LNG experiments.

## GROUND FACILITIES

Scheme of the circuit used for LNG experiments is shown in figure 3. The circuit includes LNG supply tank 1, two cryostats 2,3 and a loop facility. The pump to be tested is in cryostat 2. Cryostat 3 serves as a receiver for LNG. HF sensors 4,5 are installed in sequence after the pump to measure void fraction and flowrate. HF sensor 6 is required to measure dielectric constant of LNG for automatic compensation of a change in LNG composition and to determine LNG density if this composition is known. Compressed helium in cylinder 7 is used during calibration of cross correlation flowmeter and to create tags when there is no vapor phase in the LNG flow. Two auxiliary coaxial resonators 8,9 with $D$=0 are required in the process of HF sensors and cross-correlation flow meter calibration. They have adjustable end capacitance (sensitive element). A gap in the capacitance may be set to a very small value to decrease the error in level metering. There are several holes in the upper and lower flanges of each resonator. This allows LNG to flow through them.

## INTELLIGENT SYSTEM

A block diagram of the automatic measuring system is shown in figure 4. The system includes three HF sensors, temperature and pressure sensors, crate CAMAC with modules, a correlator and a computer. The HF electronic block sends and receives HF signals to and from the sensor, and maintains its resonance frequency. An output HF signal from the block

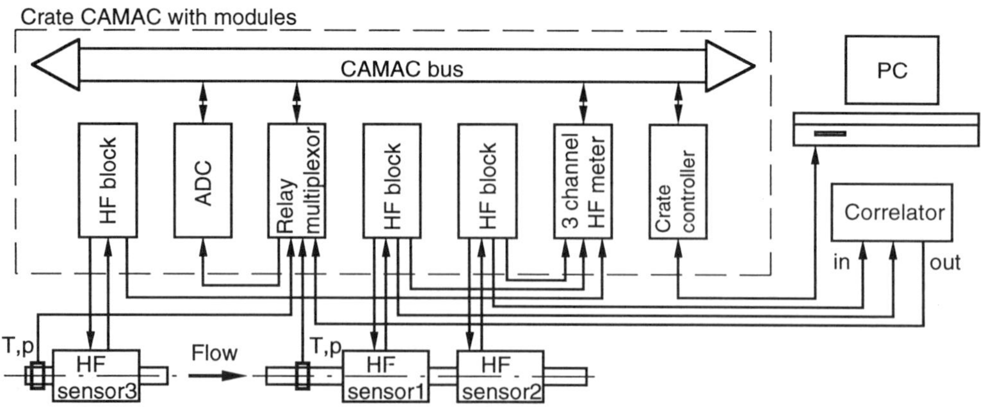


**Figure 4**. Scheme of an automatic measuring system.

goes to the three channel frequency meter and then through the CAMAC bus and cratecontroller to a computer. The second (low frequency) output of the block is connected to the correlator which produces a signal varied in direct proportion to the mean flow velocity.This signal is delivered to the input of the relay multiplexor. Other input signals come to the multiplexor from the temperature and pressure sensors. All sensors and the correlator are sequentially commutated by the relay multiplexor to analog-to-digital converter (ADC). Crate CAMAC feeds the temperature and pressure sensors and HF blocks. Temperature measurements are necessary to compensate influence of temperature on the sensor characteristics. Pressures are required on fuel pumps trials.

## EXPERIMENTS

The objectives of the experiments were to test fuel pumps for the presence of cavitation and to check system performance on a LNG flow. Before experiments the system should be calibrated at operating temperatures. At first the calibration of void fraction sensors was carried out. It involved determination of D, $\varepsilon_l$, and $\varepsilon_v$ appearing in (5). To simplify a process of calibration with LNG, an auxiliary resonator with D=0 (see figure 3) was used. The vapor phase of LNG was chosen as a reference medium for it. We set the dielectric constant of vapor phase ($\varepsilon_v$) equal to a dielectric constant of gaseous methane. This gave rise to a negligible error due to the fact that dielectric constants of all LNG components in vapor phase were very close to unity. Thus, we could calculate the dielectric constant of LNG liquid phase ($\varepsilon_l$) through measurements of the resonance frequencies of an auxiliary resonator filled with vapor and liquid phases of LNG. Then with known $\varepsilon_l$, $\varepsilon_v$ and measured $f_v$, $f_l$ of the sensor to be calibrated we obtained $D$.

To detect cavitation after the pump being tested it was necessary to know the dielectric constants of the phases in the flow. As in the calibration procedure, $\varepsilon_v$ was taken equal to the dielectric constant of gaseous methane. We also assumed that owing to the instantaneous character of LNG evaporation in the pump, the composition of liquid phase was not changed and corresponded to the composition of LNG in the cryostat. That is why we installed an additional HF sensor before the pump to perform continuous measurement of LNG dielectric constant (see figure 3). LNG which was used in experiments had the following composition: methane - 80-98%; ethane, propane, butane - 2-20%; nitrogen - 1-5%. The range of flowrates was 1-6 L/s. It corresponded to velocities from 0.5 to 3 m/s. At velocities higher then 2.8 m/s cavitation was detected. Void fraction was as much as 0.07.

Calibration of the cross-correlation flow meter was divided in two steps. It was made to estimate the error introduced by assumptions in calculations of vapor-liquid flow density. The first step was to calibrate the flow meter on single-phase LNG flow with velocities between 0.5 and 2.5 m/s. To ensure stable operation of the cross-correlation flow meter, a small amount of helium was injected in the flow. By doing so, fluctuations of the mean dielectric constant of the flow was achieved. The measured flowrate of LNG was compared with the flowrate determined by two auxiliary resonators (see figure 3). With known resonator separation, internal cryostat diameter and time it takes for the LNG level to be decreased from the first resonator to the second one, the flowrate was calculated. Data obtained in calibration showed that the difference between flowrates measured with auxiliary resonators and flowrates measured with cross-correlation flow meter did not exceed ±2%.

The next step was calibration of the flow meter on vapor-liquid LNG flow with velocities between 2.8 and 3.1 m/s. Unlike previous experiments, determination of the vapor-liquid flow density was required. Before experiments a composition of LNG was determined using a gas chromatograph. Knowing the LNG composition and based on the assumption of the invariance of phase composition after the pump, we could use equations (8) and (9) to calculate flow density. Comparison of mass flowrates measured by the flow meter with mass flowrates determined by the auxiliary resonators showed the difference of ±3%. The higher error in this case was related to the assumptions made in density calculations. Thus, the system calibration showed that the developed measuring methodology and HF sensors could give a comparatively high accuracy when monitoring LNG flows.

Finally, it is worth mentioning that when this system is used on hydrogen inaccuracies in void fraction, flowrate and density will be somewhat lower due to the known dielectric constants of vapor and liquid phases. The system will be also simpler since there is no need for the additional HF sensor before a hydrogen fuel pump.

## CONCLUSIONS

A HF sensor and methodology for void fraction and flow rate measurements of LNG flow have been developed and tested. Experiments have demonstrated an automatic measuring system operating reliably with LNG flow. The main restriction in measurements with LNG is the need to know liquid phase composition. In the case of liquid hydrogen, the system can be used without this restriction. The designed sensor contains no moving parts and electronics. It is simple and allows a comparatively high accuracy of measurements. The sensor can also serve as an accurate densimeter. Due to the specific sensor design, electronics can be now placed at the distance of 110 m from the sensor. Combination of two identical sensors allows one to measure the flowrate of cryogenic fuel flows. Experimental evaluation of the system will continue. Future work will include tests on hydrogen.

## ACKNOWLEDGEMENT

The authors are indebted to M.G.Emelyanov for his special assistance in installation of necessary electronic equipment.

## REFERENCES

1. A.M.Arkharov et al., Measurement of helium density and vapor content by means of microwave resonators, *Instruments and Experimental Techniques*, Vol 31 1:174 (1988).
2. A.G.Grechko and A.M.Arkharov, Use of radio frequency sensors for measurement parameters of cryogenic vapor-liquid flows, *Khimicheskoe i Neftekhimicheskoe Mashinostroenie*, 4:14 (1988) (in Russian).
3. A.M.Arkharov et al., Method for measurement of flowrate of vapor-liquid flows, Authors certificate SU No 1419271, *Invention Bulletin* No 31 (1988) (in Russian).
4. A.M.Arkharov et al., Flowmeter sensor for two-phase cryoproducts, Authors certificate SU No 1839234, *Invention Bulletin*, N37-38 (1993) (in Russian).
5. A.M.Arkharov et al., Apparatus for measurement of void fraction in cryogenic flows, Patent N 2001391 Russia, *Invention Bulletin*, N37-38 (1993) (in Russian).
6. A.M.Arkharov et al., "High frequency resonance sensors for single- and two-phase cryoproducts," Proc. 19th International Congress of Refrigeration, The Hague, The Netherlands, IIIb:1030 (1995).
7. A.M.Arkharov et al., Helium density monitoring system for QUELL experiment, Vestnik of Moscow State Technical University (1996) (in Russian).
8. W.M.Haynes, Measurements of densities and dielectric constants of liquid propane from 90 to 300 K at pressures to 35 MPa, *J.Chem.Thermodynamics* Vol.15 5:419 (1983).
9. W.M.Haynes, Measurements of densities and dielectric constants of liquid normal butane from 140 to 300K at pressures to 35 MPa, *J.Chem.Thermodynamics* Vol.15 9:801 (1983).
10. W.M.Haynes and B.A.Younglove, Dielectric constants of saturated liquid propane, isobutane, and normal butane, in "Advances in Cryogenic Engineering," Vol.27, Plenum Press, New York (1982).
11. A.M.Arkharov et al., Static method for calibration of microwave sensors, *Instruments and Experimental Techniques*, Vol 35 1:170 (1992).
12. A.M.Arkharov et al., Calibration of the vapor content HF sensors for cryogenic flows, *Khimicheskoe i Neftekhimicheskoe Mashinostroenie*, 6:27 (1992) (in Russian).
13. J.F.Ely and G.C.Straty, Dielectric constants and molar polarizability of saturated and compressed fluid nitrogen, *J.Chem.Phys.*, 61:1480 (1974).
14. L.A.Weber, Dielectric constant data and the derived Clausius-Mossotti function for compressed gaseous and liquid ethane, *J.Chem.Phys.*, 65:446 (1976).
15. G.C.Straty and R.D.Goodwin, Dielectric constant and polarizability of saturated and compressed fluid methane, *Cryogenics*, Vol 13 12:712 (1973).
16. R.D.McCarty, J.Hord and H.M.Roder, Selected properties of Hydrogen, National Bureau of Standards, Report MN 168 (1981).
17. W.M.Haynes and R.D.McCarty, Prediction of liquefied natural gas (LNG) densities from new experimental dielectric constant data, *Cryogenics*, Vol 23 8:421 (1983).
18. M.S.Beck and A.Plaskowski. "Cross Correlation Flowmeters - Their Design and Application," Adam Hilger, Bristol (1987).

# FORCED CONVECTION COOLED THERMISTORS USED AS MAGNET LEAD FLOW SENSORS*

A. Morgillo, M. Iarocci, A. Nicoletti, Y.Farah, J. Sondericker

Cryogenic Systems Group
Relativistic Heavy Ion Collider
Brookhaven National Laboratory
Upton, NY 11973-5000, USA

## ABSTRACT

Brookhaven National Laboratory (BNL) has developed a simple, inexpensive method for monitoring the flow, which cools the 12x50 Amp corrector power leads for the Relativistic Heavy Ion Collider (RHIC). Since there are more than 500 leads of this type distributed around the 3.8 km circumference accelerator, each requiring flow control and monitoring, the use of conventional mass flow meters is not economically feasible. The device consists of a negative temperature coefficient thermistor mounted inside a tube, through which helium gas flows from the power lead. The helium gas cools the sensor through convection causing a temperature change and therefore a resistance change in the sensor. The electrical circuit reacts by adjusting the current supplied to the sensor to maintain the resistance at a constant value. The change in current is used to calculate the helium mass flow rate. The flow meter monitors flow in the range of .03 to .05 g/s with an accuracy of plus or minus 10%. Presented here are the design calculations as well as the flow meter performance.

## INTRODUCTION

The RHIC being built at BNL, contains separately powered superconducting magnets for correction of beam field perturbations. These corrector magnets are individually powered through helium gas cooled warm to cold power leads. To monitor the helium mass flow rate, flow meters are required on each power lead. BNL has developed a thermistor flow meter to measure the mass flow rate in the range of .03 to .05 gms/sec. The temperature of the helium gas is about 290 K. The construction details, design calculations, electrical circuit design and test results are presented here.

*Work performed under contract with the US Department of Energy

## CONSTRUCTION DETAILS

The Thermistor Flow Meter (figure 1 below) consists of a 0.203 meter long by 0.0138 m outer diameter by 0.00924 m inner diameter stainless steel nipple. Positioned inside is a 0.2032 meter long Nema Grade G-10cr laminated plastic board assembly. The sensor is a thermistor soldered onto the board. The board assembly is held in position inside the stainless steel tube (nipple) by a Micarta plug positioned at one end of the tube. The thermistor is positioned on the board 0.165 meters from the entrance. The Tube assembly is located between the magnet lead flow return tube and a manifold consisting of either 4, 6 or 8 entrance ports and one exit port to return the room temperature helium at 0.1013 Mpa to the helium refrigerator.

The thermistor chosen is a standard negative temperature coefficient (NTC) bead type made up of a mixture of metal oxides with a suitable binder coated onto platinum alloy lead wires which are sintered into a ceramic body. The lead wires are welded to larger lead wires and the whole assembly is hermetically sealed in a glass tube for stability to form a glass probe type thermistor. The probe diameter is 0.00152 m with an overall length of approximately 0.0127 m.

The entrance and exit conditions were considered when determining the location of the thermistor sensor inside the tube assembly. The location of the thermistor was calculated to be positioned in a region far enough in the tube to be in a fully developed laminar flow region. For the flow range in this application of .03 to .05 gms/sec the hydrodynamic entrance length where laminar flow becomes fully developed is calculated to be about a minimum of 0.173 m (figure 2). Tests on the 0.203 m long assembly showed that the ideal entrance length is about 0.165 m. This is due to the exit conditions of the 0.203 m long tube affecting the development of the flow. Therefore the thermistor is positioned 0.165 m from the entrance point.

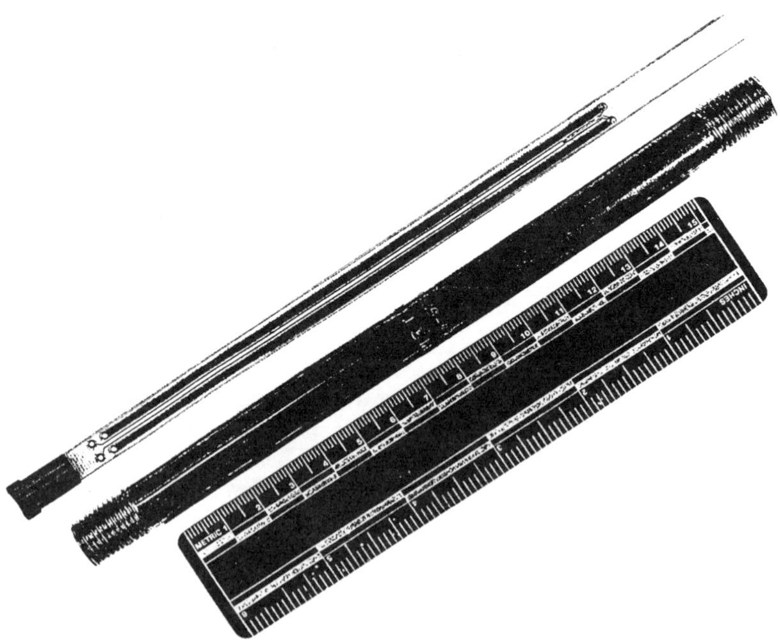


**Figure 1.** Photograph of thermistor board assembly and stainless steel tube.

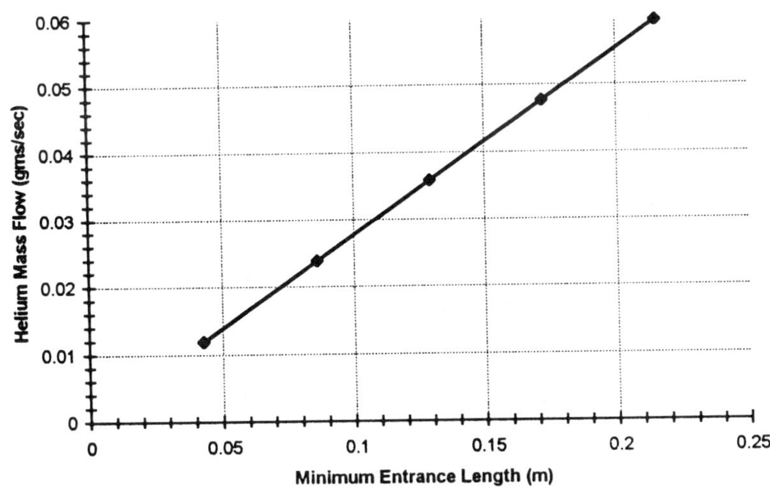


**Figure 2.** Helium mass flow vs. minimum hydrodynamic entrance length.

## DESIGN CALCULATIONS

The heat transferred from the thermistor at various mass flows was calculated prior to actual testing. Below are outlines the formulas used for this calculation: Some of the following formulas are empirical and practical relations developed for forced-convection heat transfer [1], [2].

First we calculate the velocity through the tube using the equation:

$$Vs = \frac{m_s}{A_t \rho_1}$$

where: $m_s$ the mass flow rate (gms/sec)

$A_t$ is the area of tube section calculated using tube hydraulic diameter (m)

$\rho_1$ is the density of helium (g/m3)

From the velocity we can calculate the Reynolds number across the sphere and through the tube to verify laminar flow conditions

$$\text{Sphere: Red}_s = \frac{\rho V_s D}{\mu_2} \qquad \text{Tube: Red}_t = \frac{\rho V_s D}{\mu_2}$$

where: $\rho_1$ is the density of helium (kg/m3)

$V_s$ is the velocity (m/s)

$D$ is the thermistor diameter (m)

$d$ is the tube hydraulic diameter (m)

$\mu_2$ is the dynamic viscosity at the sphere at thermistor film temperature T2 (kg/m-s)

where: $T2 = \frac{\text{Temp of Thermistor (Th)} + \text{helium bulk gas Temperature (T1)}}{2}$

The Nusselt Number ($Nud_S$) is calculated using the following empirical formula

$$Nud_s = 2 + \left[0.4 * (\mathrm{Re}d_s)^{.5} + 0.06 * (\mathrm{Re}d_s)^{.667}\right] * (\mathrm{Pr}_s)^{.4} * \left(\frac{\mu_1}{\mu_2}\right)^{.25}$$

where: $Red_S$ is the Reynolds number (sphere)
$Pr_S$ is the Prandtl number (sphere)
$\mu_1$ is the dynamic viscosity of helium at temperature $T_1$ (K)
$\mu_2$. the dynamic viscosity of helium at temperature of thermistor film $T_2$ (K)

From the Nusselt number and the thermal conductivity of helium gas we can calculate the film transfer coefficient ($h_S$) in W/m-K using the formula

$$h_S = Nud_S \cdot \frac{k_S}{D}$$

where $Nud_S$ is the nusselt number
$k_S$ is the thermal conductivity of helium gas (w/m K)
$D$ is the thermistor diameter (m)

Finally with the Film heat transfer coefficient (h) and temperatures $T_1$ and $T_2$ we calculate the heat transferred from the thermistor. The heat transferred is calculated as a function of mass flow using the formula:

$$q = h_s \pi D (T_2 - T_1)$$

where: $q$ is the heat transferred (W)
$h_S$ is the film transfer coefficient (w/m K)
$D$ is the thermistor diameter (m)
$T_2$ is the thermistor film temperature (K)
$T_1$ is the helium Bulk gas Temperature (K)

## ELECTRICAL CIRCUIT

Thermal Conductivity measurements NTC thermistors are usually designed using the wheatstone bridge circuitry where the thermistors are self heated to about 433.16K. In this application, where the thermistor is located as far away as 60 m from the electronic circuit, this is not possible because the resistance of the thermistor at 433.16K C is 24 ohms (for a 298.16K, 1000 ohm NTC thermistor), and the resistance of the wires connecting the thermistor leads to the bridge would introduce a significant error. Thus a 4 wire connection between the thermistor and the electronic circuit is used.

As shown in the block diagram (figure 3), a voltage ($V_{160}$) is applied at A3, which generates a current (I) to self heat the thermistor to about 433.16K

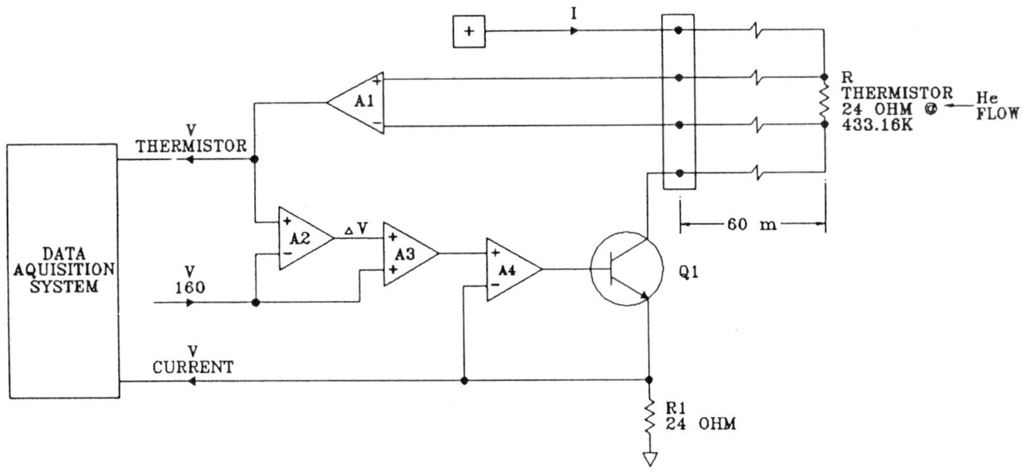


**Figure 3.** Electrical circuit block diagram.

At zero helium flow, $V_{160} = V_{therm}$ and $\Delta V = V_{therm} - V_{160} = 0$.

When the thermistor is subjected to helium flow, its temperature will decrease causing an increase in the resistance, which increases $V_{therm}$. This causes $\Delta V$ to increase which when added to the $V_{160}$ at A3 causes the current (I) to increase, resulting in the temperature returning to 433.16 K.

$V_{curr}$ and $V_{therm}$ are routed to the data acquisition system where the current (I) is

$$I = \frac{V_{curr}}{R_1}$$

Since the current in $R_1$ is equal to the current in the thermistor

$$R_{therm} = \frac{V_{therm}}{I}, \qquad R_{therm} = V_{therm} \frac{R_1}{V_{curr}}$$

Table 1 shows the recalculated electrical data for varying mass flow rates. Note that the Vcurrent equals Vthermistor and that Rtherm remains constant at 24 ohms representing a constant temperature of 433.16 K. The current in the thermistor is normalized to remove the ± 20% measurement variation possible between thermistors as stated by the thermistor manufacturer. Figure 4 is a graph of the normalized current (Inormal) in the thermistor vs the mass flow rate.

**Table 1**. Electrical Circuit Data

| Vcurrent (V) | I (ma) | Vtherm (V) | Rtherm (ohm) | Mass Flow (g/s) | Inormal (ma) | Power (mw) |
|---|---|---|---|---|---|---|
| 3.000 | 125.000 | 3.000 | 24.0 | 0 | 0 | 375.00 |
| 3.079 | 128.292 | 3.079 | 24.0 | 0.0119 | 3.291 | 395.01 |
| 3.176 | 132.333 | 3.176 | 24.0 | 0.0239 | 7.333 | 420.29 |
| 3.248 | 135.333 | 3.248 | 24.0 | 0.0358 | 10.333 | 439.56 |
| 3.303 | 137.625 | 3.303 | 24.0 | 0.0477 | 12.625 | 454.44 |
| 3.348 | 139.500 | 3.348 | 24.0 | 0.0596 | 14.500 | 467.05 |

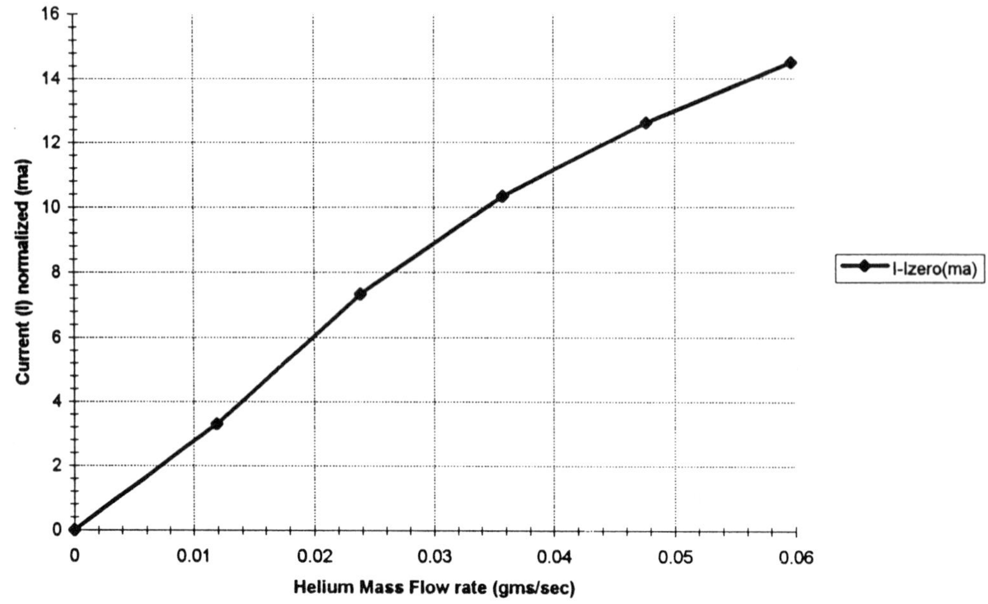


**Figure 4.** Normalized thermistor current vs helium mass flow.

## TEST RESULTS

The actual test results compare favorably with the calculated values. In the range of .03 to .05 gms/sec mass flow rate the difference between the calculated and actual heat transferred from the thermistor averaged less than 10%.

Table 2 shows the calculated data including the helium mass flow velocity, Reynolds numbers for the sphere and tube, and the calculated vs. actual heat transferred from the thermistor for the various mass flow rates tested. Figure 5 is a graph of the heat transferred from the thermistor (calculated and actual) vs mass flow rate.

## CONCLUSION

The forced convection cooled thermistor assembly used as magnet lead flow sensors offers a simple and inexpensive method for measuring helium mass flow through a tube. This design can be

**Table 2.** Calculated and Actual Heat Transfer Test Results

| Mass Flow gms/sec | Velocity m/s | Reds sphere | Calculated Reds tube | Calculated heat trans (W) | Actual heat trans (W) |
|---|---|---|---|---|---|
| .0119 | .361 | 27.27 | 95.445 | 0.283 | .342 |
| .0238 | .721 | 54.54 | 190.89 | 0.349 | .384 |
| .0358 | 1.085 | 82.04 | 287.14 | 0.402 | .421 |
| .0477 | 1.445 | 109.31 | 382.58 | 0.447 | .451 |
| .0596 | 1.806 | 136.60 | 478.03 | 0.487 | .467 |

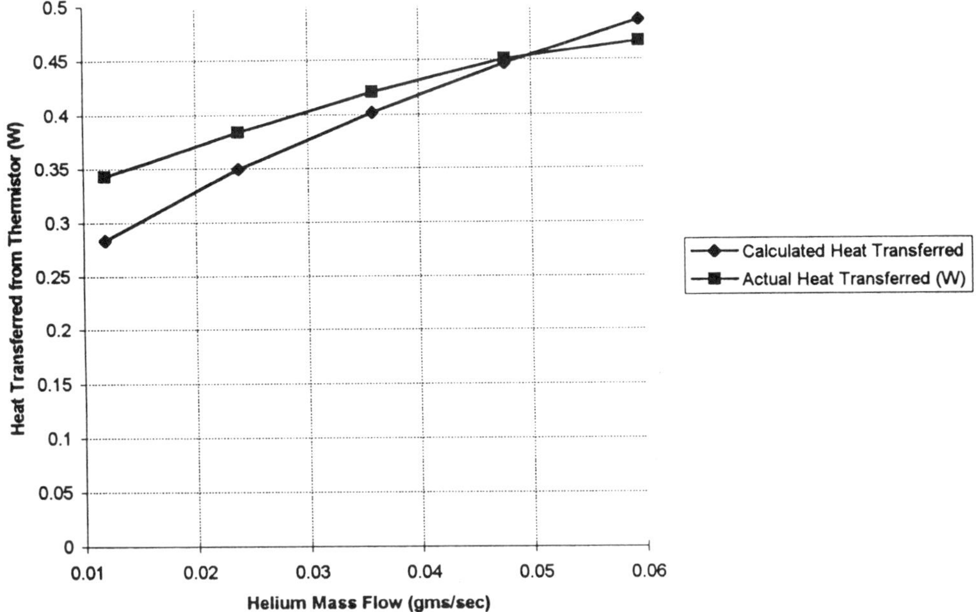


**Figure 5.** Heat transferred from thermistor (calculated and actual) vs helium mass flow rate.

adapted for various applications for measurements of the flow of gases through an orifice of varying shapes by choosing the proper thermistor and considering the entrance and exit lengths when positioning the thermistor within the orifice. Using 4 wire connection instead of a 2 wire connection eliminates the resistance impact of the length of the lead wires and thus improves the accuracy of the measurement. Reliability is expected to be excellent and will be verified once several hundred thermistor assemblies are installed and operating in the RHIC Machine. The RHIC machine is expected to be operational in late 1999.

## ACKNOWLEDGEMENTS


The authors would like to thank all personnel in the RHIC Cryogenic Section and is especially appreciative of the help from W. Hirzel.


## REFERENCES


[1] Holman, J.P., Heat Transfer, McGraw Hill, New York, 1981

[2] Dewitt, David P., and Incropera, Frank, P., Fundamentals of Heat and Mass Transfer, John Wiley and Sons, New York 1990.

# PRECISE WIDE RANGE HEATMETERS FOR 1.5 K UP TO 80 K

G. Ferlin, B. Jenninger, J-M. Rieubland

CERN-LHC division
CH-1211 Geneva, Switzerland

## ABSTRACT

Two heatmeters were designed at CERN for applications below 20 K with the option to work also at temperatures up to 80 K. The new calibration principle and design permits the construction of wide range heatmeters with precision in the range of milliwatts. The calibration function takes into account the temperature dependence of the thermal conductivity of the heatmeter material. The heat flow measurement is, therefore, independent of the base temperature, i.e. it is also independent on the temperature drop across thermal contact between heatmeter and the cold source. The simple calibration function makes the heatmeter a user-friendly portable diagnostic device. It is possible to quantify parasitic heat flow without a previous calibration, or to calibrate the heatmeter during a measurement with a specimen.

## INTRODUCTION

A heatmeter is a calibrated thermal resistance where the heat flow is measured as a function of its boundary temperatures. One of the main ideas is to use the heatmeter as a portable and interchangeable device. This implies that temperature sensors and heater for calibration are included in this device. This heatmeter principle was first developed and described by Kuchnir et al. at Fermilab 1985[1]. The heatmeters were designed for constant base temperatures (4.2 K and 77 K). Their calibration functions were, therefore, only valid for these base temperatures. A deviation from the base temperature would have required the application of correction factors.

Figure 1 shows two possible designs and the main features of heatmeters. In the original design at Fermilab, the thermal resistance of the heatmeters were determined by a stainless steel layer silver soldered between two copper blocks. This principle was adopted at CERN for a test bench to measure the heat load at different temperature levels of LHC prototype support posts [2]. The stainless steel was replaced by brass. Meanwhile, improvements have been made in both design and calibration principle, in order to adapt the idea of heatmeters in a better way to more general requirements, i. e. wide range applications for heat flow and temperatures[3,4]. The heatmeter of type 2 in Figure 1 shows the schematic design of the device that was developed at CERN. This time, the heatmeter is manufactured from a single piece of an aluminium alloy, i.e. a medium thermal conductor. The thermal resistance is determined by the ratio of section over length of the central part of the heatmeter. The tube-like design is to increase the aerial momentum of inertia for a given ratio. The bottom heater is an option for the control of the specimen temperature.

## CALIBRATION

### Theoretical calibration function

For wide range heatmeters the temperature dependence of the thermal conductivity needs to be taken into account. Equation 1 is a function that describes approximately the thermal conductivity of metal alloys in an approprate small temerature interval:

$$k = \alpha \cdot T^n \tag{1}$$

$\alpha$ and $n$ are considered constant within the temperature interval. This approximation function is practicable for regions where $n$ (i.e. the slope of the graph $k$ over $T$ in a double logarithmic scale) varies only little with the temperature. In regions where $n$ varies very fast, like for pure metals in the vicinity of 10 K where $k$ has a maximum, such an approximation function is not practicable. Pure metals are therefore no good choice for wide range heatmeters with working temperatures around this maximum in $k$.

For normal electrical conductors at about liquid helium temperatures the heat is practically only carried by electrons. In this region the thermal conductivity is proportional to the temperature ($n = 1$), unless the material becomes superconductive [5].

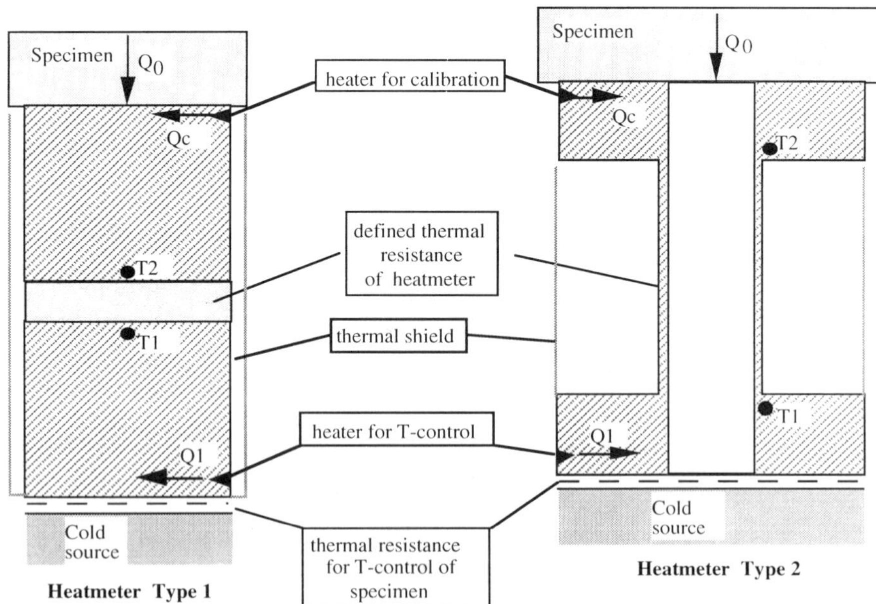


**Figure 1**. Two possible heatmeter designs.

The heat flow through the heatmeter is then:

$$\dot{Q} = \frac{S}{L} \cdot \int_{T_1}^{T_2} k \cdot dT \tag{2}$$

$$\dot{Q} = \frac{S}{L} \cdot \frac{\alpha}{n+1} \cdot \left(T_2^{n+1} - T_1^{n+1}\right) \tag{3}$$

$S$ and $L$ are the section and length of the thermal resistance between the boundary temperatures $T_1$ and $T_2$. Such a calibration function is only valid in the temperature interval where it is defined. The width of this temperature interval depends on the material chosen and the maximal systematical error accepted. For wide range heatmeters a material has to be chosen that varies only little in $n$, in order to reduce the number of intervals. For a well known material, $n$ can be estimated from the graph $k$ as a function of $T$.

$$n = \frac{T}{k} \cdot \frac{\Delta k}{\Delta T} \tag{4}$$

When $n$ is determined, $\alpha$ can be calculated from equation 1. In the real calibration function of a heatmeter the geometry and material parameters are unified in a single proportionality factor $C$.

$$C = \frac{S}{L} \cdot \frac{\alpha}{n+1} \tag{5}$$

Here, different calibration possibilities are proposed which may be applied to the same calibration data for cross-checking purposes.

- Calibration for $n$ with unknown $C$.
- Calibration for $C$ with known $n$ (parasitic heat load physically eliminated)
- In situ calibration for $C$ during a measurement with a specimen.

## Calibration for exponent $n$

In principle, $n$ can be chosen freely and an $\alpha$ be calculated from the calibration heating power $\dot{Q}_c$ and the corresponding boundary temperatures. The optimal fit, however, is the one that corresponds to the physical temperature dependence of the heatmeter material. From two calibration sets ($\dot{Q}_{c.1}$, $T_{2.1}$, $T_{1.1}$) and ($\dot{Q}_{c.2}$, $T_{2.2}$, $T_{1.2}$), the proportionality factor $C$ of the calibration function can be eliminated and $n$ can be calculated from the derivation of the calibration function, or, in other words, from the definition of the thermal conduction of a heatmeter.

$$\frac{\partial \dot{Q}}{\partial T} = \frac{S}{L} \cdot \alpha \cdot T^n \tag{6}$$

Interpreting $\partial \dot{Q}$ as the heat flow due to a temperature drop $\partial T$ along the heatmeter at the temperature mean $T$, a mean $n$ can be calculated from the two sets of calibration parameters:

$$n = \frac{\ln\left(\frac{\dot{Q}_{c.2}}{\dot{Q}_{c.1}} \cdot \frac{(T_{2.1} - T_{1.1})}{(T_{2.2} - T_{1.2})}\right)}{\ln\left(\frac{(T_{2.2} + T_{1.2})}{(T_{2.1} + T_{1.1})}\right)} \tag{7}$$

This formula gives a good estimation for $n$ in the temperature interval

$$T \in \left\{ \tfrac{1}{2}(T_{2.1} + T_{1.1}) \ldots \tfrac{1}{2}(T_{2.2} + T_{1.2}) \right\}$$

**Calibration for the proportionality factor *C***

If the exponent $n$ is known for a given temperature interval, either from literature or from a calibration like described above, it is quite easy to get the proportionality factor $C$ from one calibration set ($\dot{Q}_{c.i}, T_{2.i}, T_{1.i}$):

$$C_i = \frac{\dot{Q}_{c.i}}{T_{2.i}^{n+1} - T_{1.i}^{n+1}} \tag{8}$$

In the defined temperature interval, where $n$ was supposed to be constant, $C_i$ should be the same for each calibration set. The constancy of $C_i$ for different calibration sets is, therefore, a quality criteria of the calibration. During this calibration, parasitic heat load to the heatmeter needs to be physically eliminated. There is, however, a possibility to calibrate the heatmeter at the presence of parasitic heat flow or during a measurement with a specimen.

**Calibration at the presence of parasitic heat flow**

Supposing this parasitic heat flow $\dot{Q}_0$ to be constant during the calibration, equation (8) becomes:

$$C_i = \frac{\dot{Q}_0 + \dot{Q}_{c.i}}{T_{2.i}^{n+1} - T_{1.i}^{n+1}} \tag{9}$$

From two calibration sets (i = 1, 2), $\dot{Q}_0$ can be eliminated.
With $C_1 = C_2 = C$ and $\Delta_i = T_{2.i}^{n+1} - T_{1.i}^{n+1}$, $C$ can be determined with equation (10):

$$C = \frac{\dot{Q}_{c.2} - \dot{Q}_{c.1}}{\Delta_2 - \Delta_1} \tag{10}$$

**Precision of a heatmeter**

In the proposed design, the heat flow measurement is reduced to two absolute temperature measurements. Thus, the precision of the heat flow measurement is directly dependent on the precision of the temperature measurement. However, the thermal conduction of the heatmeter can be chosen in a way, that for the minimal expected heat flow at a given temperature, the temperature drop along the heatmeter is just big enough to be quantified with certainty. In the design of wide range heatmeters, one should also take into account the temperature dependence of the sensitivity of the temperature sensors. Equation (11) is a good estimation for the precision of a heatmeter at a given temperature level and for the case that the sensors are not calibrated together.

$$\Delta \dot{Q} = 2(n+1)CT^n \cdot \Delta T \tag{11}$$

Where $n$ and $C$ are the exponent and proportionality factor of the previous equations, $T$ is the temperature level and $\Delta T$ is here the uncertainty of the temperature measurement at this temperature level.

Using temperature sensors that were calibrated together, the precision of the heat flow measurement can be improved considerably, because systematical calibration errors of the sensors are partly eliminated.

## EXAMPLE

Two different heatmeters of type 2 were developed at CERN, both designed for operation below 20 K with the option to work up to 80 K. The measurements presented in this paper were made on a heatmeter that was initially designed to measure the heat load due to resistive radio frequency losses in coaxial cables. The required precision was 5 % of the measured value over a range of 20 mW to 600 mW at about 10 K [4].

Heatmeter material:
Aluminium alloy Al Mg Si 1 (Anticorodal 100)

Geometry:

Ratio Section over Length $\frac{S}{L} = 0.006\,[m]$

Temperature sensors:
Carbon-Ceramic sensors TVO from JINR, Dubna, Russia
Calibrated together from 1.5 K and 300 K.
Absolute precision: ≈ 0.1 % of the measured value

### Calibration below 20 K

For the calibration below 20 K, the heatmeter was mounted on a copper plate that was cooled by liquid helium at 4.2 K. The assembly was surrounded by an insulation vacuum of $10^{-6}$ mbar and shielded by a thermal screen at about 10 K.

In Figure 2 the proportionality factor $C$ is calculated for each set of calibration heat flow $Q_c$ and the corresponding boundary temperatures using equation (8). $C^*$ is the proportionality factor calculated with equation (10) from two adjacent calibration sets.

The precision of the heatmeter turned out to be about ± 1 mW up to a heat flow of 500 mW. At a heat flow of 1200 mW, the precision was still better than ± 6 mW. The heatmeter was removed several times from its mounting place and remounted again under slightly different conditions. The recalibrations showed no measurable changes in the calibration parameters. From the calibration data, $n$ could also be determined to be 1 ± 0.01. For $Q_c = 0$, $C$ becomes zero, because of a residual $\Delta T$ of 16 mK. This $\Delta T$ may be caused by parasitic heat flow of below 1 mW and an uncertainty in the temperature measurement. In $C^*$ the parasitic heat flow is eliminated using equation (10).

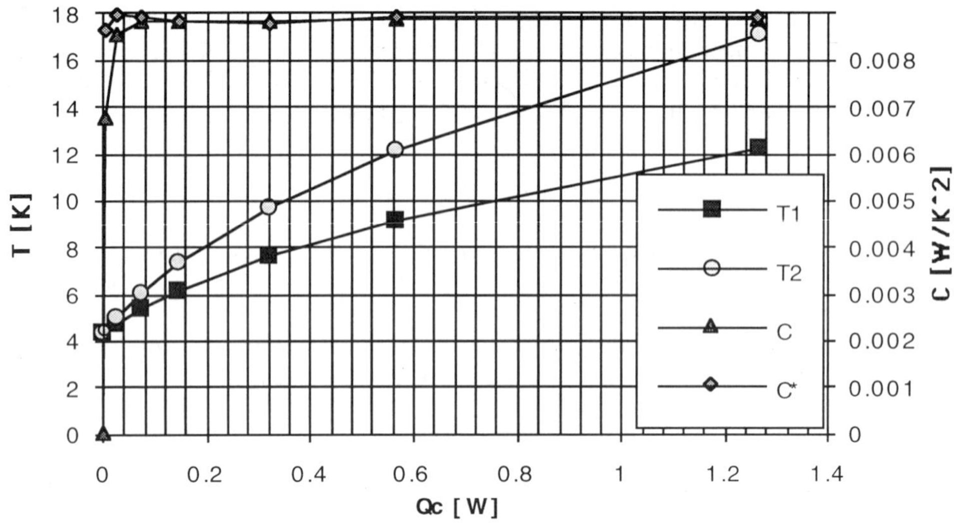


**Figure 2.** Calculated proportional factors $C$ and $C^*$ for sets of Qc, $T1$ and $T2$.

## Calibration from 10 K to 80 K

For the calibration in the temperature range from 10 K to 80 K, the copper plate was cooled with cold gaseous helium.

The calibration heater was adjusted to a fixed value ($\dot{Q}_C$ = 899 mW) and the temperature drop along the heatmeter was measured as a function of the temperature level. The temperature of the copper plate was adjusted by varying the GHe mass flow.
For a constant calibration heat flow equation (7) becomes

$$n = \frac{\ln\left(\frac{(T_{2.1} - T_{1.1})}{(T_{2.2} - T_{1.2})}\right)}{\ln\left(\frac{(T_{2.2} + T_{1.2})}{(T_{2.1} + T_{1.1})}\right)} \tag{12}$$

$$n = \frac{\ln\left(\frac{\Delta T_1}{\Delta T_2}\right)}{\ln\left(\frac{\overline{T_2}}{\overline{T_1}}\right)} \tag{13}$$

Table 1 shows the measured exponents $n$ at different temperature levels.

**Table 1.** Mean exponents $n$ of the heatmeter material

| $T$- interval | exponent $n$ |
|---|---|
| 1.5 K - 20 K | $1.00 \pm 0.01$ |
| 20 K - 40 K | $0.97 \pm 0.03$ |
| 40 K - 50 K | $0.55 \pm 0.05$ |
| 50 K - 75 K | $0.25 \pm 0.05$ |

**Table 2.** Parameters $n$ and $C$ of the calibration functions of the heatmeter

| | valid in $T$-interval | $n$ | $C$ [W/(m·K$^{n+1}$)] |
|---|---|---|---|
| Function 1 | 1.5 K - 40 K | 1 | 0.0085 |
| Function 2 | 40 K - 80 K | 0.4 | 0.11 |

From the calibration data, the calibration parameters were defined, covering two temperature intervals.

The calibration function is: $\dot{Q} = C \cdot \left(T_2^{n+1} - T_1^{n+1}\right)$, where the parameters $C$ and $n$ used for the corresponding temperature interval are given in Table 2.

In Figure 3 the calculated heat flow is compared to the constant reference heat flow. Function 1 covers well the interval from 1.5 K - 40 K and starts deriving above. Function 2 covers relatively well the interval between 40 K and 80 K. The measured accuracy in this temperature interval is better than 0.1 W. If a higher accuracy is required, the temperature range could be divided into three temperature intervals instead of two.

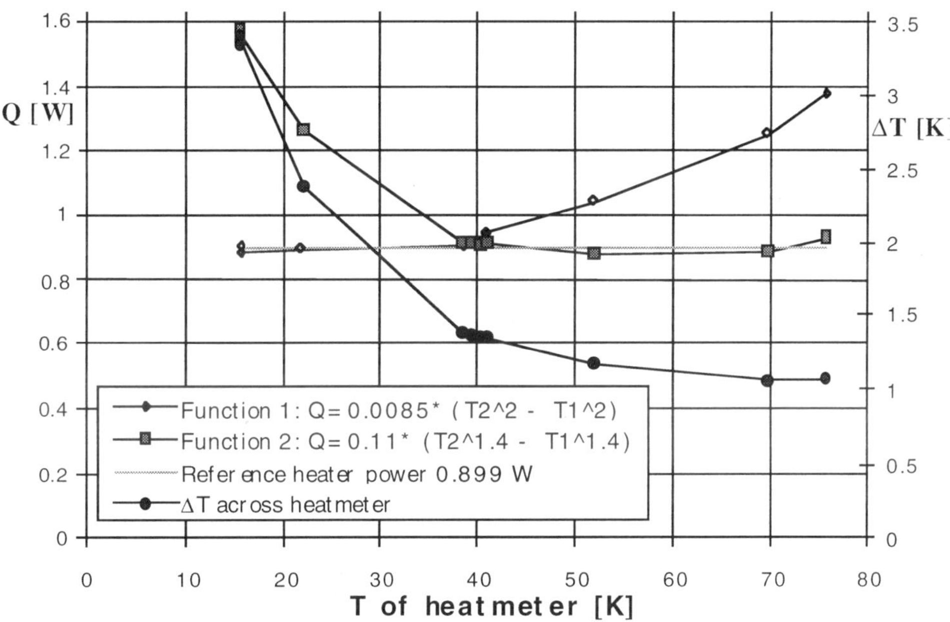


**Figure 3**. Two calibration functions for two temperature intervals.

**Table 3**. Estimated precision $\Delta\dot{Q}$ of the heatmeter

| $T$ [K] | $n$ | $C$ [W/(m·K$^{n+1}$)] | $\Delta T$ [K] | $\Delta\dot{Q}$ [W] |
|---|---|---|---|---|
| 4 | 1 | 0.0085 | 0.004 | 0.0008 |
| 20 | 1 | 0.0085 | 0.02 | 0.014 |
| 40 | 1 | 0.0085 | 0.04 | 0.056 |
| 40 | 0.4 | 0.11 | 0.04 | 0.056 |
| 80 | 0.4 | 0.11 | 0.1 | 0.14 |

Table 3 shows the calculated precision of this heatmeter due to the uncertainty of the temperature measurement using equation (11).

The real precision of the heatmeter is better than estimated above. This is due to the fact that in the calculation the uncertainties of the two sensors are considered uncorrelated. However, the sensors that are used were calibrated together which eliminates partly systematic calibration errors.

Note, that the calculated precision concerns the absolute precision of the heatmeter which is limited by the calibration of its temperature sensors. The heat sensitivity is mainly limited by the instrumentation (stability of current source, precision of voltmeter etc.) and can reach values much better than those calculated above.


## ACKNOWLEDGEMENTS

The authors would like to express their thanks to Vladimir Datskov for the many helpful discussions on precise thermometry and the supply of the calibrated carbon ceramic temperature sensors.


## REFERENCES


1. M. Kuchnir, J.D. Gonczy and J.L. Tague, Measuring heat leaks with a heatmeter, Proceedings R W Fast, Advancees in Cryogenic Engineering, Vol 31 (1985), 1285-1290
2. H. Danielsson, Ph. Lebrun, J.M. Rieubland, Precision heat inleak measurements on cryogenic components, Cryogenics 1992, Vol 32, ISEC Supplement, 215-218.
3. G. Peón, B. Jenninger, B. Szeless, Characterisation of net type thermal insulators at 1.8 K low boundary temperature, LHC Project Report 76, presented at International Cryogenic Material Conference ICMC'96 at CERN, published in Proceedings 1996
4. C. Bovet, B. Jenninger, J.P. Papis, J.M. Rieubland, Simulation and measurements of the thermal behavior of a prototype coaxial cable in a cryogenic environment, LHC Project Note 79, 1997, to be published.
5. Cryogenie et ses applications en supraconductivité, Institut International du Froid IIF, published by Techniques de L'Ingenieur T.I., 1995

# LOW TEMPERATURE PERFORMANCE OF A SUPERCONDUCTING ANGULAR ACCELEROMETER

E. R. Canavan[1], M. V. Moody[1], and P. J. Shirron[2]

[1]University of Maryland, Department of Physics
College Park, MD, 20742, USA

[2]NASA/Goddard Space Flight Center
Greenbelt, MD, 20771, USA

## ABSTRACT

An angular accelerometer has been developed that has an amplifier noise limited resolution of better than $10^{-10}$ rad $s^{-2}$ $Hz^{-1/2}$. The device consists of a superconducting niobium proof mass suspended by a torsional spring. A superconducting circuit and SQUID amplifier are used to measure angular displacement of the proof mass, which is proportional to the acceleration at frequencies below the natural resonance frequency. To prevent low-level linear accelerations from generating detectable angular accelerations, very precise mass balancing is necessary. Using room temperature measurements and wire EDM machining, the proof mass's center of gravity can be controlled to within 0.2 microns. In recent tests conducted at 4.2 K, performance better than $3\times10^{-9}$ rad $s^{-2}$ $Hz^{-1/2}$ has been observed. The noise floor can be traced to seismic and thermal inputs. Manufacturing details and performance tests are discussed.

## INTRODUCTION

An extremely sensitive Superconducting Angular Accelerometer (SAA) has been developed at the University of Maryland to complement the Superconducting Gravity Gradiometer (SGG). The SGG is a highly sensitive instrument capable of directly detecting small changes in the local gravity gradient with relatively high bandwidth. The SGG was developed with NASA support for a space environment, where it could provide a high-resolution global gravity map of the Earth[1,2] or other planetary bodies, and could be used for high-precision tests of fundamental physics.[3] Its capabilities have already been demonstrated in a precision test of the gravitational inverse square law.[4] The SGG consists of six linear accelerometers mounted to the faces of a precision cube. Pairs of accelerometers on opposing faces of the cube are coupled with superconducting circuits to form a three-axis in-line gravity gradiometer[2]. The instrument can be made insensitive to along-axis linear accelerations by tuning the excitation currents. However, misalignments of the accelerometers' sensitive axes allow accelerations in other degrees of freedom to couple into the gradient signal. In particular, angular accelerations are important if the accelerometers are not perfectly colinear. To remove the error, both the misalignments and angular accelerations must be measured. Hence, a precise angular accelerometer is necessary. The sensitivity requirement depends in part on the allotment of the error budget for the overall gradient measurement, but typically an input-referred noise with a power spectral density (PSD) of less than $3\times10^{-10}$ rad $s^{-2}$ $Hz^{-1/2}$ is desired.[5]

**Figure 1.** A completed angular accelerometer.

## DESCRIPTION OF THE INSTRUMENT AND TEST CONFIGURATION

The design of the angular accelerometer has been discussed in detail previously[5]; therefore, it will only be reviewed briefly here. A picture of a completed accelerometer is shown in Figure 1. Except for a cover, the mechanical elements of the device are all machined from a single disk of niobium 2.04 cm thick and 10.16 cm in diameter. The proof mass is formed by a single slot cut with a wire EDM (Electric Discharge Machine). The pivot, the small bridge between the proof mass and the housing, is defined by two holes at the ends of the slot. The minimum thickness of the bridge is 132 μm. The lighter circular region in which the pivot is located is a titanium alloy (Ti-6Al-4V) rod that is diffusion bonded into the niobium disk. Finite element analysis indicates that at maximum proof mass deflection, the peak stress in the pivot is less than 20% of the 0.2% yield strength of Ti-6Al-4V. This stress level is sufficiently low to avoid microyielding.

Elements of the displacement sensing circuit, shown in Figure 2, are housed in chambers machined into the disk. The two white rectangles along the horizontal diameter are the Macor glass-ceramic coil forms. Not visible in the picture are the niobium wire pancake coils on the upper surface of the coil form and the 75 μm gap between the coil and the lower surface of the proof mass. The sensing coils are single layer spirals formed from 52

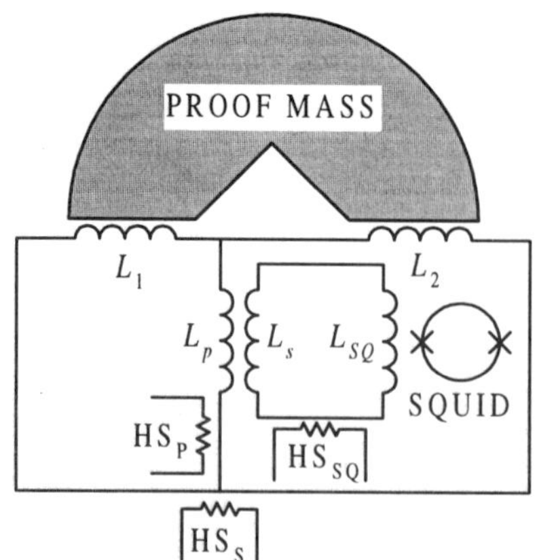


**Figure 2.** Schematic of the angular accelerometer sensing circuit.

turns of 145 μm (including insulation) wire. The inner and outer diameters of the coils are 4.8 and 19.8 mm respectively. The coil spacings are matched at room temperature by measuring the capacitance between the coils and proof mass, and lapping the back of one of the coil forms accordingly. The pockets in the lower half of the disk contain the heat switches (for storing current), the dc SQUID, and an impedance-matching transformer.

The sensing coils are connected in parallel to the SQUID through the impedance-matching transformer. Two heat switches allow flux to be trapped in either the same ($HS_S$) or opposite senses ($HS_P$). In the first case, persistent current initially flows only through the two sensing coils; in the second, it flows through the transformer primary then splits between the two coils. The first case is the primary operating mode, because an angular deflection of the superconducting proof mass increases the inductance of one coil and decreases that of the other (Meissner effect), forcing current to flow through the transformer, and generating a signal at the SQUID proportional to proof mass deflection. Trapping flux in the other configuration makes the SQUID output insensitive to proof mass deflection. As we shall see, this mode is useful for studying temperature sensitivity.

In the tests reported here, three angular accelerometers were integrated with the SGG as shown in Figure 3. Four tubes connect the SGG vacuum chamber to the cryostat top flange. Through each of the outer three tubes, which are equally spaced from the center tube and from each other, stainless steel cables run from the SGG suspension bracket to the bottom of rubber springs. These provide vibration isolation with a 2.3 Hz cutoff frequency (for vertical acceleration) and a relatively low Q (<10). A vertically-oriented permanent magnet on the adapter between the top of the suspension cable and the bottom of the rubber spring interacts with the field generated by a coil on the outside of the vacuum tube. These non-contact voice-coil actuators are used to shake the instrument for calibration and setup. A laser beam passes though the center tube and reflects off a mirror which is mounted perpendicular to the SGG cube diagonal. The deflection of the reflected beam is measured with an x-y photodetector to determine the tilt of the SGG in both degrees of freedom. The SGG is suspended so that one diagonal of the mounting cube is aligned with vertical. Thus, the three gradiometer axes make the same angle, $\arctan(2^{1/2})$, with respect to vertical. The three angular accelerometers are mounted on the three lower accelerometers of the SGG. The axes of the SAAs are, therefore, aligned with those of the SGG.

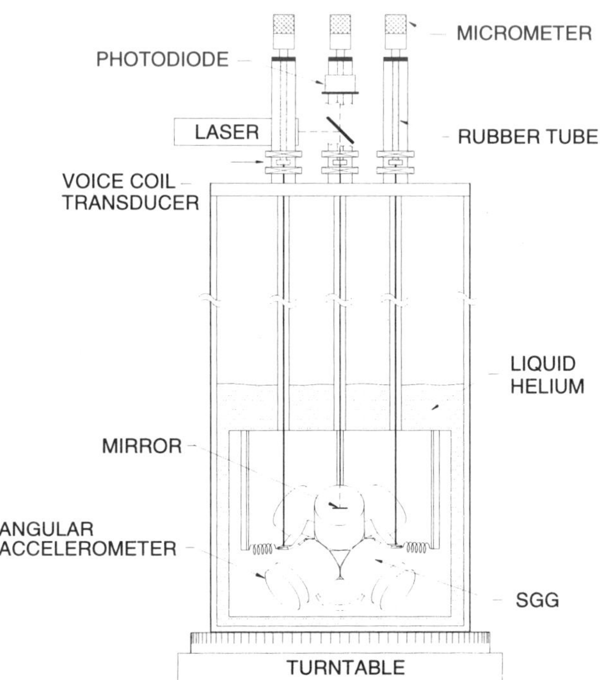


**Figure 3.** Schematic of the test cryostat.

## LOW TEMPERATURE MEASUREMENTS

### Sensing circuit inductances

By storing and dissipating small currents (0.1 A), measurements of certain combinations of inductances in the sensing circuit were made. The integral of the voltage across the heat switch while the current is decaying gives the product of the stored current and the inductance. Using $HS_P$, the series combination the sensing coils, $L_1$ and $L_2$ in Figure 2, was measured, while the parallel combination of $L_1$ and $L_2$ in series with $L_p$ was measured using $HS_S$ and $HS_{SQ}$. The series inductance values ranged from 3.4 to 4.2 μH, while the parallel values ranged from 4.8 to 5.1 μH. The transformer primary, $L_p$, and secondary, $L_s$, (measured before installation) were approximately 4.0 and 14 μH, respectively. These measurements indicated that the coil spacings had remained closely matched during the cool down.

### SQUID performance

Before storing current in the sensing circuits, the noise levels of the SQUID amplifiers were measured. White noise levels were approximately 1.3 pA $Hz^{-1/2}$ with all 9 SQUIDs (6 for the SGG and 3 for the angular accelerometers) operating simultaneously. The current noise was 1.7 times higher than the factory specifications. This appeared to be due, in part, to interactions between the different SQUID channels. The 1/f noise corner was at 0.1 Hz.

### Mechanical resonance frequency

In general, the resonant frequency of the mass-spring system will be a function of the current in each sensing coil due to the forces exerted by the screening currents in the proof mass. At low temperature, the mechanical resonance frequencies were measured with a sensing current sufficiently small to make the electromagnetic spring constant much less than the mechanical spring constant. They were found to be 13.78, 13.58, and 13.72 Hz, approximately 5.1% higher than at room temperature.

### Resonance frequency vs. sensing current

The resonance frequency of the accelerometers as a function of sensing current was measured for both series and parallel currents. The expected frequency shift can be derived by solving the equations for the circuit in Figure 2. For parallel current, the resonance frequency as a function of current is given by:

$$f_{P0} = \frac{1}{2\pi}\left[(2\pi f_0)^2 + \frac{I_P{}^2 c^2}{4\Theta}\left(2\frac{\lambda^2}{L_0} - \gamma\right)\right]^{1/2}, \tag{1}$$

where $I_p$ is the stored current, $f_0$ is the mechanical resonance frequency, c is the perpendicular distance from the plane of symmetry to the center of the coil (3.05 cm), and $\Theta$ is the moment of inertia about the pivot axis ($1.94\times10^3$ gm $cm^2$). To arrive at this equation the sensing coils are assumed to be matched. Also, the sensing coil inductance as a function of angle has been approximated as a Taylor series:

$$L(\theta) = L_0 + \lambda c\theta + \gamma(c\theta)^2/2. \tag{2}$$

where $L_0 = \mu_0 n^2 A d_0 D/(d_0 + D)$ ($n$ is the turns density, $A$ is the coil area, $d_0$ is the equilibrium spacing between the coil and proof mass, and $D$ is the spacing between the coil and the superconducting backplane), and $\lambda = L_0 D/[d_0(d_0 + D)]$ are determined from the coil geometry. The values of $L_0$ for the three accelerometers were 1.7, 1.8, and 2.1 mH, respectively, while the corresponding $\lambda$'s were 0.0168, 0.0168, and 0166 H $m^{-1}$. This leaves the $\gamma$'s which were determined from the experimental data to be 53, 64, and 75 H $m^{-2}$ respectively.

For series current, the resonance frequency as a function of current is given by:

$$f_{S0} = \frac{1}{2\pi}\left[(2\pi f_0)^2 + \frac{I_S{}^2 c^2}{\Theta}\left(2\frac{\lambda^2}{L_0 + L_{eff}} - \gamma\right)\right]^{1/2}. \tag{3}$$

where

$$L_{eff} = \left[L_{SQ} + (1-\kappa^2)L_s\right]L_p \,/\, (L_{SQ} + L_p) \;, \qquad (4)$$

is the effective inductance of the SQUID reflected through the transformer ($\kappa$ is the transformer coupling constant). A plot of $f_{0S}^2$ vs $I_S^2$ for one accelerometer is shown in Figure 4. As expected from Eq. (3), the data fits a straight line. The only free parameter in Eq. (3) is $L_{eff}$, hence, these measurements can be used to determine $\kappa$. They indicated $\kappa$ is approximately 0.99 for each of the three SAAs.

## Calibration of the SAAs and center of gravity measurements

The acceleration-to-current scale factor as calculated from the circuit equations is:

$$H_{\ddot{\theta}\to i_{SQ}}(\omega) = \left(\frac{\kappa\sqrt{L_P L_S}}{L_{SQ} + L_S}\right)\left(\frac{2c\lambda I_S}{L_0 + 2L_{eff}}\right)\left(\frac{1}{\omega_{S0}^2 - \omega^2 + \dfrac{j\omega_{S0}\omega}{Q}}\right), \qquad (5)$$

where $Q$ and $\omega_{S0}$ are the quality factor and angular frequency of the spring-mass resonance, respectively. The first two terms in Eq. (5) give the current through the SQUID input coil for a given proof mass deflection; the third term is the response of a simple damped spring-mass system. For frequencies of interest ($\omega << \omega_{S0}$), this term reduces to $1/\omega_{S0}^2$. Although sensitivity to deflection increases linearly with persistent current, $I_S$, $\omega_{S0}^2$ increases as $I_S^2$. From Eqs. (3) and (5), it is straightforward to show that optimum sensitivity occurs when $I_S$ is set such that the electromagnetic spring constant, $k_{em} = I_S^2c^2[2\lambda^2/(L_0 + L_{eff}) - \gamma]$, equals the mechanical spring constant, $k_0 = \Theta\omega_0^2$. The data plotted in Figure 4 indicate that a current of 3.3 A is necessary to achieve optimal coupling to the SQUID. However, the sensitivity at optimal coupling would result in frequent saturation of the amplifiers by the high frequency angular acceleration background. Since optimum sensitivity was not required for these tests, a sensing current of 0.500 A was used for normal operation. The resonance frequencies for this sensing current were measured to be 13.94, 13.74, and 13.88 Hz, respectively.

To measure the scale factors, the voice coil shakers were used to tilt the accelerometers in two orthogonal degrees of freedom. The outputs of the SQUID amplifiers versus tilt at various frequencies were measured and recorded. The slopes of the SAA outputs vs. frequency squared gave the acceleration-to-voltage scale factors, while the intercepts gave the offset of the proof mass c.g. from the pivot center, $\delta\mathbf{r}$. The magnitude of the tilt angle, $\theta$, was indicated using the laser and photodetector.

The magnitudes of $\delta\mathbf{r}$ as indicated by this data were 0.90, 0.88, and 1.11 μm, respectively. These show an increase of 0.8 to 0.9 μm over the room temperature values. The shifts are all in the same direction, to within several degrees. However, they are not along the axis of symmetry. This makes it unlikely that they were caused by the differential contraction between the Nb proof mass and the Ti-6Al-4V pivot.

A second method of calibration uses the SGG to measure the tilt angle. Though centrifugal acceleration introduces unwanted signals in the gradiometer output, in a constantly

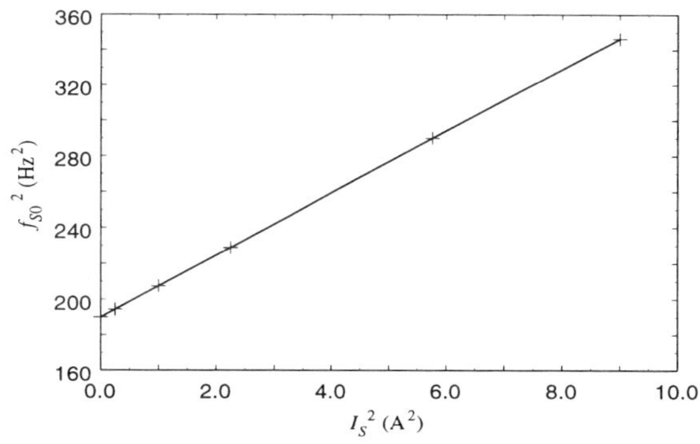


**Figure 4**. The square of the measured resonance frequency as a function of the sensing current squared.

rotating reference frame, such as the Earth, it can be used to simultaneously calibrate the SGG and the SAAs. The outputs of the three SGG axes are given by:

$$S_{\Gamma_i}(f) = H_{\Gamma_i} 2\left[\boldsymbol{\Omega}_{\mathrm{E}} \cdot \boldsymbol{\theta}\omega - (\boldsymbol{\Omega}_{\mathrm{E}} \cdot \hat{\mathbf{n}}_i)(\boldsymbol{\theta}\omega \cdot \hat{\mathbf{n}}_i)\right] , \qquad (6)$$

at $f$, where $H_{\Gamma_i}$ is the scale factor, $\boldsymbol{\Omega}_{\mathrm{E}}$ is the Earth's angular velocity and $\hat{\mathbf{n}}_i$ is the direction of the sensitive axis, and

$$S_{\Gamma_i}(2f) = H_{\Gamma_i}\left[(\boldsymbol{\theta}\omega)^2 - (\boldsymbol{\theta}\omega \cdot \hat{\mathbf{n}}_i)^2\right] , \qquad (7)$$

at $2f$. It is important to note that the acceleration signal from the SGG at $f$ is a velocity term, and, hence, is in quadrature with $\theta$ and $\theta\omega^2$. Thus, by doing phase sensitive detection, this signal can be separated from any residual misbalance term. Using Eqs. (6) and (7) and knowing the orientation of the SGG in the Earth's reference frame allows one to determine both the SGG scale factors and $\boldsymbol{\theta}$. Since the SAAs measure $\hat{\mathbf{n}}_\alpha \cdot \boldsymbol{\theta}\,\omega^2$ at $f$, where $\hat{\mathbf{n}}_\alpha$ is the direction of the SAA sensitive axis, knowing $\boldsymbol{\theta}$ permits determination of the SAA's scale factors. The scale factors measured in this manner were 9.5, 9.3, and 8.3 mA/(rad $s^{-2}$). Eq. (5) gives 11.9, 11.6, and 10.2 mA/(rad $s^{-2}$), respectively, for the SAA scale factors. The disparity from the measured values is due primarily to stray inductances which are not accounted for in the circuit equations. A plot of the SQUID output voltage versus the radial frequency squared, $\omega^2$, for a constant tilt angle is shown in Figure 5 for one accelerometer. The plot contains data for two orthogonal tilt directions.

The measured values indicate an amplifier noise limited sensitivity of $(1 + 0.1/f)^{-1/2}$ 1.4 $\times 10^{-10}$ rad $s^{-2}$ $Hz^{-1/2}$. At optimal coupling, the amplifier-limited noise would be a factor of 3.3 lower. A spectral density plot of the output is shown in Figure 6. The lowest observed spectral density in this test environment was approximately $3\times10^{-9}$ rad $s^{-2}$ $Hz^{-1/2}$ at 0.01 Hz. At frequencies above 10 mHz, the signal is the angular acceleration of the platform driven by seismic vibration. At lower frequencies the spectral density is dominated by coupling to temperature fluctuations.

### Brownian motion noise

In addition to amplifier noise, a second fundamental noise source in a spring mass accelerometer is Brownian motion noise. The magnitude of this noise is given by:

$$B^{1/2} = \left(\frac{4}{\Theta} k_B T \frac{2\pi f_{0S}}{Q}\right)^{1/2} . \qquad (8)$$

The quality factors of the three SAAs were determined by exciting the resonance with the shaker and measuring the decay time. The respective Q's were $2.3\times10^4$, $2.5\times10^4$, and $2.6\times10^4$ at a pressure of $8\times10^{-4}$ torr. The Q's were almost certainly limited by coupling to the platform suspension modes, which are highly damped. These values correspond to a Brownian motion noise level of $4\times10^{-11}$ rad $s^{-2}$ $Hz^{-1/2}$ at 4.2 K.

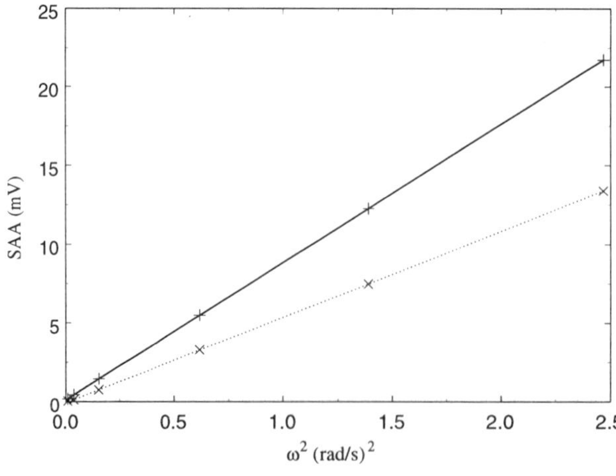


**Figure 5**. The output voltage of an SAA for nearly constant tilt angle in two orthogonal directions. The magnitude of the tilt angle is measured using the SGG.

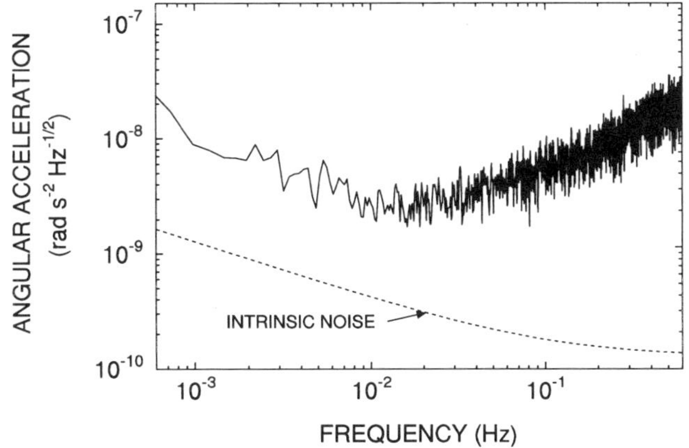


**Figure 6**. The spectral density of the SAA in the test cryostat. The intrinsic noise curve includes $1/f$ noise of the SQUID with a

## Temperature sensitivity

Sensitivity to temperature fluctuations is reduced in the normal operating mode since, to the degree that $L_1$ and $L_2$ are matched, any common change in their inductances will not induce a current in $L_{SQ}$. Even in the case of mismatched sensing coils, a small parallel current can be stored to balance out sensitivity to temperature fluctuations common to $L_1$ and $L_2$. In this case the SAAs are sensitive only to temperature gradients and thermal time constants become a critical issue.

To study sensitivity to temperature fluctuations, current in the sensing loops was stored in the parallel sense. This greatly reduced sensitivity to accelerations, while maximizing sensitivity to temperature. These measurements were performed using a manganin wire heater and a calibrated germanium thermometer attached to the SGG platform. In terms of a change in the coil-to-proof mass gap, $d_0$, the measurements indicated temperature sensitivities of $2.9\times10^{-8}$, $3.0\times10^{-8}$, and $2.5\times10^{-8}$ m $K^{-1}$, respectively, at 4.23 K. This sensitivity was independent of the magnitude of the sensing current within the measurement error. The temperature dependence of the sensing coil inductances was also measured at 3.21, 2.19, and 1.63 K. This data is plotted in Figure 7. As the figure indicates, there was a significant drop in temperature sensitivity from 4.23 K to 3.21 K; however, below 3.21 K, there was little change. The dependence of the temperature sensitivity on sensing current was also measured at 1.63 K, and again showed no significant variation. The thermal time constants between the accelerometer and the thermometer were also measured at 4.23, 3.21, 2.19, and 1.63 K The time constants were found to be 10.6, 4.8, 0.94, and 0.80 s respectively.

Previous experiments[6] led us to believe that the temperature dependence would be due primarily to the temperature variation of the superconducting penetration depth which is approximately:

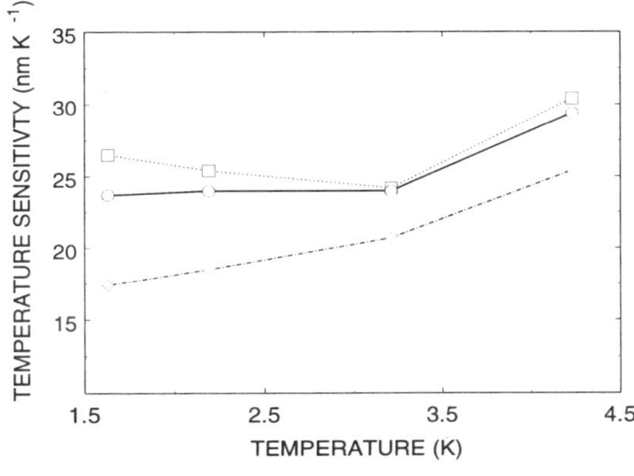


**Figure 7**. The temperature sensitivity for angular accelerometers with parallel current.

$$\lambda(T) = \lambda_0 / \left[1 - (T/T_c)^4\right]^{1/2} . \qquad (9)$$

where $\lambda_0$ is the penetration depth at $T = 0$ K, and $T_c$ is the critical temperature of the superconductor. However, this implies $dd_0/dT \cong 1.2\times10^{-9}$ m $K^{-1}$ at 4.24 K, well below the measured values. Futher, this effect should give a $T^3$ dependence for temperatures well below $T_c$, which is clearly not observed.

The thermal contraction and magnetic susceptibility of the Macor coil forms also contribute to the temperature sensitivity of the device. However, differential thermal contraction between the coil form and the Nb housing will give a displacement sensitivity of only $2\times10^{-11}$ m $K^{-1}$. The magnitude of the magnetic susceptibility effect is difficult to calculate, but it produces a temperature dependence proportional to current squared, contrary to what was observed. Furthermore, the magnetic susceptibility has a Curie law temperature dependence ($T^{-1}$), and the observed temperature dependence does not fit this.

## CONCLUSION

A monolithic angular accelerometer has been developed for use with the University of Maryland's Superconducting Gravity Gradiometer to compensate for angular and centrifugal acceleration errors. It is very robust, being machined from a single piece of niobium, and has been designed and proven to withstand the launch vibration loads of both the Space Shuttle and small expendable launch vehicles. The device is capable of resolving better than $10^{-10}$ rad/$s^2$ $Hz^{-1/2}$ in a single axis. Three units were integrated and tested with the SGG. Their characteristics are remarkably uniform and their performance correlated well with the model. Their resolution has been demonstrated to $3\times10^{-9}$ rad $s^{-2}$ $Hz^{-1/2}$, limited by the background angular acceleration of the platform.

Below 0.1 Hz, coupling to temperature fluctuations of the He bath was observed. The measured temperature sensitivities of approximately $3\times10^{-8}$ m $K^{-1}$ in the effective position of the proof mass are 20 times larger than would be expected from the temperature dependence of the magnetic penetration depth.

Although the proof masses were balanced to within 0.2 μm of the torsional spring's center of bending at room temperature, a subsequent shift of 0.8 to 0.9 μm upon cooling to 4.2 K was observed. Thermal expansion cannot account for either the magnitude nor the fact that it is not along the symmetry axis. Future work will focus on identifying the origin of the cg shift and the temperature sensitivity.


## ACKNOWLEDGMENTS

This work was sponsored by NASA Grant NAG-53104, and by NASA/Goddard Space Flight Center's Director's Discretionary Fund.


## REFERENCES


1. H.J. Paik, J.-S. Leung, S.H. Morgan, and J. Parker, Global Gravity Survey by an Orbiting Gravity Gradiometer, *EOS Trans.* 69:1601 (1988).
2. E.R. Canavan, M.V. Moody, H.J. Paik, P.J. Shirron, and M.J. DiPirro, Predicted Performance of the Superconducting Gravity Gradiometer on the Space Shuttle, *Cryogenics* 36:795 (1996).
3. B. Mashhoon, C.M. Will and H.J. Paik, Detection of the Gravitomagnetic Field Using an Orbiting Superconducting Gravity Gradiometer. I. Theoretical Principles, *Phys. Rev. D* 39:2825 (1989).
4. M.V. Moody and H.J. Paik, Gauss's Law Test of Gravity at Short Range, *Phys. Rev. Lett.* 70:1195 (1993).
5. P. J. Shirron, E. R. Canavan and M.V. Moody, Development of a Superconducting Angular Accelerometer, Proceedings of the SPIE: Cryogenic Optical Systems and Instruments VII, L. G. Burriesci and J. B. Heaney, eds., 2814:187 (1996).
6. Moody, H.A. Chan, H.J. Paik and C. Stephen,"A Superconducting Penetration Depth Thermometer," Proceedings of the Seventeenth International Conference on Low Temperature Physics, Edited by U. Eckern et al, (North Holland, New York, 1984), pp 407-408.

# NEW TRANSFORMER FOR POWERING SUPERCONDUCTING MAGNETS

O. A. Shevchenko, H. J. G. Krooshoop, H. G. Knoopers, and H. H. J. ten Kate

Faculty of Applied Physics, University of Twente
P.O. Box 217, 7500 AE Enschede, The Netherlands

## ABSTRACT

A new method to increase the efficiency of powering a superconducting magnet is presented. The powering system links a superconducting magnet and the mains. It contains both room and low temperature electrical components.

The use of a conventional power supply with a pair of high current leads or application of rectifier schemes based on superconductors or cold semiconductors shows certain limits. We propose to introduce an iron-core superconducting transformer between the conventional power supply and the magnet. This step leads to an optimized operation of the powering system and it is applicable with both low and high Tc superconductors.

In order to demonstrate the approach, a model of the proposed powering unit has been developed. The performance of the unit is demonstrated while ramping a current of a small SC magnet operating at 4.2 K. The characteristic in- and output currents are ± 20 A and ± 1 kA respectively.

## INTRODUCTION

Several large projects have designs in which superconducting magnets are employed with operating currents between 10 and 20 kA (like for example dipole and quadrupole and detector magnets). The unique features of superconducting magnets call for novel design configurations of their powering systems[1-3].

A number of different semi-conducting and superconducting devices can be used as components to build the system. A typical powering system of a superconducting magnet is sketched in Figure 1. It consists of a low temperature (cold) unit and a room temperature power control unit. The cold unit is electrically connected via high current leads to the magnet and via low current leads to the control unit. The purpose of the cold unit is therefore to link the magnet and the control unit. The magnet operates at relatively high direct current and low voltage.

When the room temperature power unit supplies just that, then a low output ripple and high current leads into the cryostat are required[1]. Note the contradiction that, in general, a superconducting magnet requires across the terminals an operating voltage of a few Volts or lower only, while a semi-conductor based supply is efficient when the voltage is well above one Volt.

When a superconducting converter is applied, the room temperature unit provides a low-amplitude current at a relatively high voltage. The drawback is the extra loss generated in the cold area due to the AC/DC energy conversion, especially important at higher operating frequencies. This will enlarge the cooling system.

Another option is to use a low-current and high-voltage power supply in combination with a step-down superconducting transformer. The method can be extended to even very high currents[4]. A superconducting transformer runs at a low frequency, which results in a high efficiency of the cold unit. A draw back of this method is that the self-inductance of the transformer (secondary coil) must be larger than that of the load magnet. This makes an air-core transformer to be comparable in size with the magnet to be powered. Therefore, this method is used almost exclusively for short low inductive samples.

The inductances of the transformer coils can be remarkably increased, when the transformer is equipped with an iron core. In order to prevent saturation of the iron, relatively large iron core will be obviously required. Nevertheless, when both power supply and cryogenics are included into consideration, the method shows reasonable parameters. It is proposed here to extend the application of the method to small size stand-alone magnets with high operating current.

The goal of the paper is to demonstrate the feasibility of powering small high current magnets using a single-cycle superconducting transformer equipped with an iron core. From the theoretical point of view the method is important, as it represents one of the extreme options available for powering a superconducting magnet (at the lowest frequency).

## PRINCIPLE OF OPERATION

The proposed transformer system is sketched in Figure 2. The cold unit of the powering system consists of the superconducting transformer. The primary coil of the transformer is connected to the programmable bipolar power supply $V$ placed at room temperature, while the secondary coil is attached to the magnet with the inductance $L_L$ via the joint resistance $R_j$. Both primary and secondary coils are superconducting. A small heater is attached to the section of the secondary circuit. When triggered, it forces the section to act as a switch. Due to the low operating frequency, a high efficiency of the switch can be achieved (if to make it for example from 0.5 m long and 0.03 m wide thin NbTi strip[5]). The transformer has an iron core.

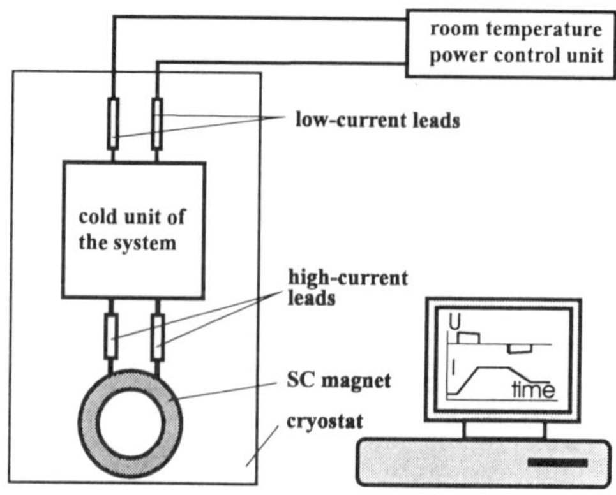


**Figure 1.** Schematic view of a magnet power system to control the current.

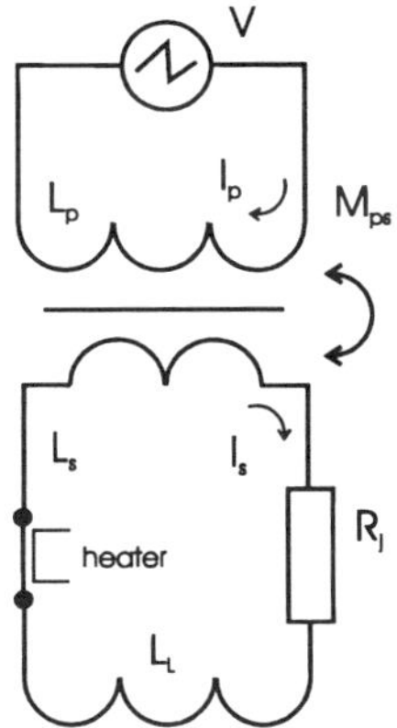


**Figure 2.** Basic electrical scheme of the transformer system: *V*- bipolar power supply; $L_p$, $L_s$ $M_{ps}$ are self-and mutual inductances of and between the primary and the secondary coils of the transformer; $R_j$ joint resistance.

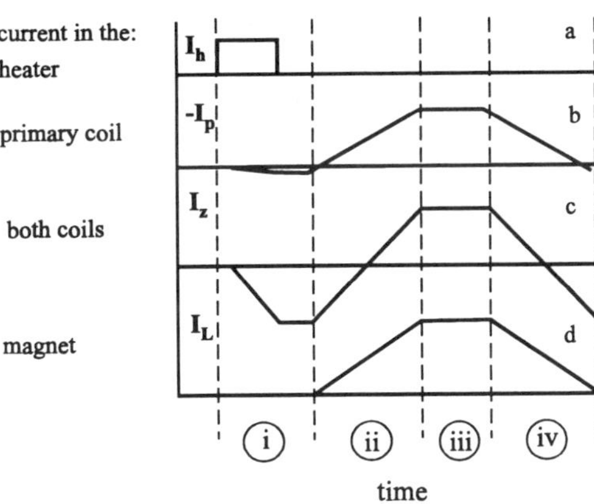


**Figure 3.** Time diagrams explaining the operation of the transformer while powering a sc. magnet: (i) preparation; (ii) ramping up the current; (iii) holding the current; (iv) ramping down the current.

Depending on the size of the magnet the core can be placed at room temperature (larger magnets) or at the same temperature as the coils (smaller magnets).

The diagrams in Figure 3 explain the cycle of operation. More details (like measuring the currents and quench protection) are given elsewhere[4]. The cycle consists of 4 stages: (i) preparation; (ii) ramping up the magnet current; (iii) control of the current at a certain level; (iv) ramping down the magnet current.

The preparation stage includes heating up the section of the secondary circuit. Furthermore, a voltage is applied to the transformer primary in order to pre-saturate the core as it is illustrated in Figure 3c. The primary current gets a small positive value, while the secondary (or magnet) current stays close to zero and equal to the leakage current of the switch, as it is explained in Figure 3b and 3d. After the heater is switched off, less than 0.1 second is required to recover the switch and to prepare the system to the next stage.

During stage (ii), a voltage with opposite polarity is applied across the transformer primary coil. Both primary and secondary currents rise to the required value. At the same time, the iron core of the transformer is magnetised in the opposite direction.

Stage (iii) starts when the desired value of the magnet current is reached. When the primary current is kept constant, the magnet current slightly decreases as the resistance of the joint (magnet) is small but not zero. When required, the current of the magnet can be kept constant. In any case the system is designed in such a way, that at the end of stage (iii) the iron core is still below the saturation value.

Stage (iv): to reduce the current of the magnet to zero, the voltage is applied to the primary coil. At the moment when the magnet current is zero, the primary current will be positive, as depicted in the Figure 3b. This state allows next ramping of the magnet current, when on request. Otherwise, the heater can be fired to return the system to the initial state.

## Theory of Operation

Consider an iron core with a magnetic cross-section $A_y$ and the path $l_y$. The core is made of the material with a high relative permeability μ, having applied magnetic field strength $H^*$ and an induction $B^*$ at the saturation point. When the closed core is placed around the coil(s) and the magnetic field outside the core is negligible, the inductance(s) can be calculated as follows:

$$M_{ij} = \mu\mu_0 \frac{A_y}{l_y} N_i N_j \tag{1}$$

with $\mu_0$ the permeability of the vacuum (indices represent the coil label). Therefore, when an iron core is used, the inductance of the superconducting coil rises drastically (up to ~$\mu$/10 times). Note, that at $\mu = 1$ the formula (1) gives wrong value of the air core inductance, as the magnetic field is present outside of the volume limited by $A_y$ and $l_y$. For example, for the coils of the transformer described below a correct value of an air core inductance can be obtained from (1) when to take $\mu \approx 9$.

In the case of an air core transformer, assuming a constant joint resistance $R_j$, a ramp rate of the primary current $\dot{I}_p$ and $L_s >> L_L$ a secondary current depends on time $t$ as[6]:

$$I_s(t) = \frac{M_{ps}}{L_s} \dot{I}_p \tau_d \left(1 - \frac{t}{2\tau_d}\right) \quad with \quad \tau_d = \frac{L_s}{R_j} \tag{2}$$

Because of the iron core, both mutual and self-inductances of the transformer depend on the applied voltage as sketched in Figure 4. This basically reflects the dependence of the core permeability on the induction. Note, that both scales are logarithmic; the shift between the curves is due to the air gaps present between the coils and the core. The superconducting transformer with the iron core in focus of this paper is shown in Figure 5.

A model of the transformer based on the constant inductances is not satisfactory. A simple non-linear model of the transformer assumes:

(a) primary coil self-inductance $L_p$ dependent on the applied voltage (Figure 4 and 6) and independent on the frequency (in the low frequency area of interest);

(b) constant transformation (winding) $k_{tr}$ and coupling $k_{ps}$ coefficients between the primary and the secondary coils;

(c) self-inductance of the secondary coil $L_s = L_p \times k_{tr}^{-2}$ ;

(d) mutual inductance between the coils $M_{ps} = L_p \times k_{ps} / k_{tr}$.

In this way equation (2) is still applicable for the iron core transformer, especially when $\dot{I}_p$ is constant. The amplification factor $M_{ps} / L_s = k_{ps} \cdot k_{tr}$ .

The status of the iron core can be evaluated based on the magnetisation current. Notice from Figures 2 and 3 that during the ramping stages the magnetic flux change in the magnet is equal to the flux change in the iron core. The field strength in the core is created by the total (or magnetisation) current $I_z$ of both primary and secondary coils having $N_p$ and $N_s$ turns respectively[7]:

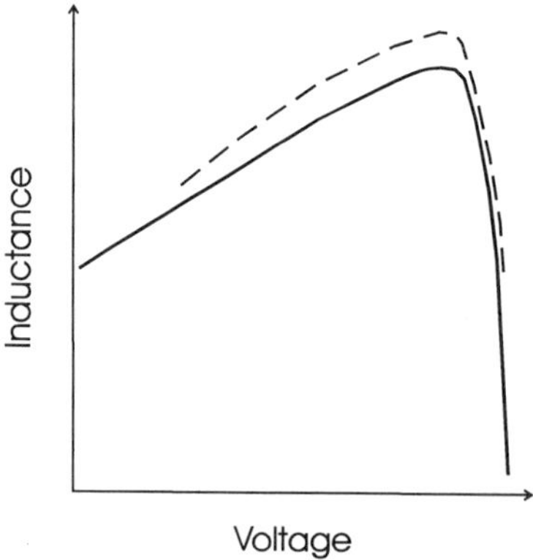


**Figure 4.** Inductance of iron core transformer coils versus applied voltage at a fixed frequency. Dashed line – calculated from the permeability data at room temperature, solid line – measured at 4.2 K[5].

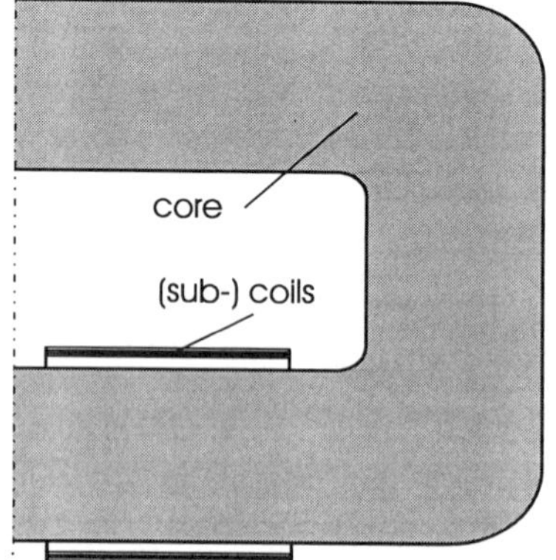


**Figure 5.** Cross-section of the transformer as used to simulate the magnetic fields: one half of the actual cross-section is shown due to the symmetry; each coil of the transformer has two equal sub-coils.

$$I_z = N_p I_p - N_s I_s, \quad or \quad I_z^s = k_{tr} I_p - I_s \quad with \quad k_{tr} = N_p / N_s \tag{3}$$

The last equation represents the magnetisation current $I_z^s$ of the transformer calculated from the secondary side. The last notion is convenient, as it allows direct comparison with the magnet current.
When the magnet current reaches its maximum value $I_{Lmx}$ (Figure 3), the core reaches the saturation point, so the following equations hold:

$$A_y = \frac{L_L I_{Lmx}}{B^* N_s}; \quad I_z^{s*} = \frac{H^* l_y}{N_s} \tag{4}$$

It may be clear from the Figure 3, when the preparation stage is executed prior to the ramping stage, the double value of the current $I_z^{s*}$ can be attained. In order to reduce the core cross-section, one has to increase the number of turns in the secondary coil. On the contrary, when a magnet operates at a high current (~ 10 kA), $N_s$ has to be minimised as a high current coil is rather expensive. Therefore, it is a subject for optimisation.

The magnet current $I_L$ is obviously equal to the current in a transformer secondary coil $I_s$, so one can write:

$$\frac{L_s}{L_L} = \frac{I_{Lmx}}{I_z^{s*}} = \frac{H^* l_y}{N_s} \tag{5}$$

The presence of the iron core limits the maximum duration of each ramping stages (ii) or (iv), see Figure 3. For a fixed ramping voltage $U_p$ or $U_s$ respectively across the primary or the secondary transformer coils, the limit is given by:

$$t^* = B^* A_y \frac{N_p}{U_p} = B^* A_y \frac{N_s}{U_s} \tag{6}$$

## DESIGN AND CONSTRUCTION

The main goal of the design of the model superconducting transformer with an iron core is to show the principle operation described above and to confirm the estimates made, rather than to create a device for a particular application. For this purpose a small model of the powering system is developed, able to operate inside a helium cryostat. The device uses a PC controlled power supply placed at room temperature, which is able to deliver ± 20 A and ± 20 V. It therefore acts as a current amplifier and provides a four-quadrant programmable operation of the magnet. The mass of the assembled cold unit is less than 1 kg, the volume is about 0.5 litres.

### The Transformer

The transformer consists of three concentric solenoids i. e. two primary coils in series and one secondary coil cooled with liquid helium at 4.2 K and wound around the iron core. The solenoids have two flat and two round sides in order to follow the shape of the core (parallel to the longer side) and to provide a reasonable bending radius for the secondary superconductor. In order to get a better coupling, the secondary coil is sandwiched by the primary coils. The specification of the used conductors and the dimensions of the coils and

Table 1. Specification of the conductors

| Applied in | Material | Bare size, mm | Number of filaments or laminates | Filament or laminate size |
|---|---|---|---|---|
| Primary coil | NbTi/Cu | ∅ 0.2 | 16819 | 0.8 μm |
| Secondary coil | $Nb_3Sn$/Cu | 4.3×0.2 | 1 | - |
| Magnet | NbTi/Cu | 2.3×1.1 | 161 | 65 μm |
| Iron core | TRAFOPERM | 14×45 | 35 | 0.3 mm |

Table 2. Dimensions of the coils

| Coil | Inner size, mm | Outer size, mm | Height, mm | Number of turns |
|---|---|---|---|---|
| Inner primary | 16 × 61 | 16.4 × 61.4 | 22+22=44 | 78+77=155 |
| Secondary | 16.5× 61.5 | 16.9× 61.9 | 22.5+22.5=45 | 6 |
| Outer primary | 16.9× 61.9 | 17.3× 62.3 | 22+22=44 | 77+78=155 |

the core are summarised in Tables 1 and 2 respectively. Each coil has a single layer, and it is divided in the middle in two equal and symmetric sub-coils of which one set is shown in Figure 5. The coils are wet-wound with STYCAST 2850 FT epoxy. The inner coil is wound over one layer of thin copper heat drains, separating the coil from the coil former.

The core has a magnetic path $l_y = 0.15$ m and operates at the same temperature as the coils. The magnet used in the experiments has an inductance of 4 μH. The presence of the joint $R_j$ is simulated by connecting a resistance of 0.1μΩ made of a phosphor-bronze strip, between the magnet and the secondary coil as depicted in Figure 2. The resistor is used as a shunt to measure the magnet current.

## RESULTS AND DISCUSSION

In idle mode the inductances of the transformer are measured at 50 Hz and 4.2 Kelvin both from the primary and the secondary side. The results are compared in Figure 6 using equations (c, d) of the non-linear model. One can see that the assumptions of the model are in fair agreement with the experiment in the practical range of the voltage (representing a fraction on the left branch of the curve in Figure 4). The maximum current is above 60 Amps when the primary coil is separated from the core.

In short circuit mode, the maximum current of the transformer (limited by the secondary coil) is found to amount 1.5 kA. The maximum current in the secondary coil is found to be above 1.2 kA even at operating frequency of 50 Hz (the sinusoidal shape). The test results are summarised in Table 3.

For powering the magnet, one set of the transformer sub-coils is used. The specifications of the transformer powering system are therefore as follows: $N_s$=3; $N_p$=155; $k_{tr}$ =51.5; $k_{ps}$ =0.997; $I_{Lmx}$ =1.4 kA; $\mu = 2000$; $B^* = 1.7$ T, $H^* = 500$ A/m; $A_y = 6.3 \cdot 10^{-4}$ m$^2$; $l_y = 0.15$ m; and finally $R_j = 0.1$ μΩ.

Table 3. Specification of the transformer

| Parameter | Dimension | Value |
|---|---|---|
| Primary self – inductance (at the voltage of 10 $V_{rms}$) | H | 0.245 |
| Maximum secondary current | kA | 1.4 |
| Maximum ramp rate | kA/s | > 10 |
| Transformation coefficient | | 51.5 |
| Coupling coefficient | | 0.997 |
| Max. inductance of the magnet to be powered | μH | 5 |

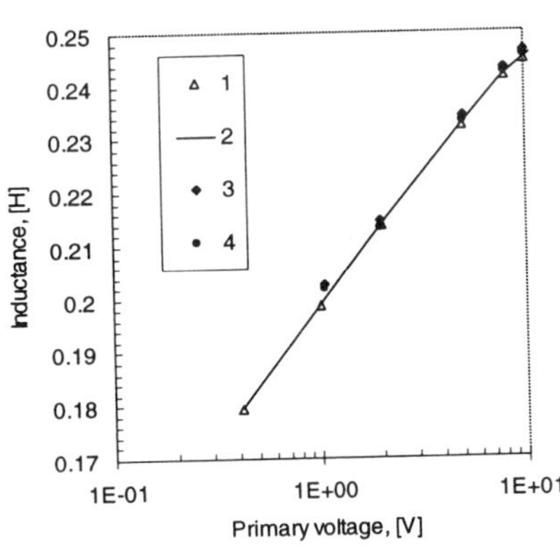


**Figure 6.** The transformer inductances versus the rms. voltage assuming $k_{tr}$=51.5; $k_{ps}$=0.997: 1- of the primary coil $L_p$; 2- the mutual $M_{ps}k_{tr}/k_{ps}$ ; the mutual $M_{sp}k_{tr} / k_{ps}$; of the secondary coil $L_s{\cdot}k_{tr}^2$.

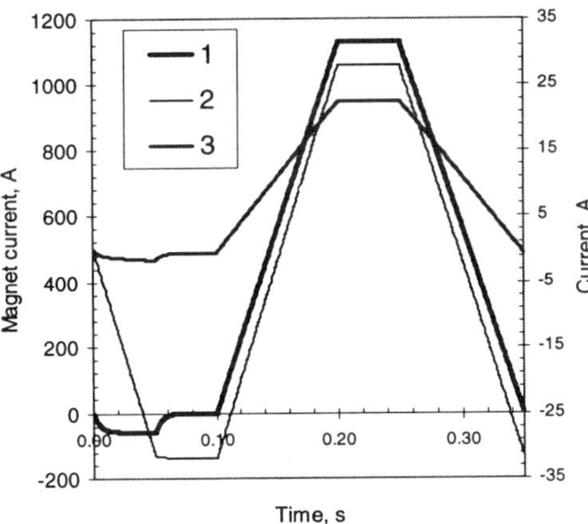


**Figure 7.** Current versus time dependence measured while powering the magnet: 1- in the magnet; 2 – in both transformer coils $I_z^s$, 3 – in the primary coil (both 2 and 3 use the right axis) .

Substitution in equations (2, 4, 5) gives the secondary magnetisation current $2\,I_z^{s*} = 50$ A; the inductance of the magnet to be powered $L_L = 4$ μH; the ratio between the inductances $L_s / L_L = 28$; and the time constant $\tau_d$ ~1000 s.
When to imagine an air core transformer made of the same amount of the superconductors, the secondary coil inductance $L_{sair} = 0.5$ μH, and the inductance of the magnet is found $L_L = 0.06$ μH, when typical for this case ratio $L_s / L_L = 8$ is used[4]. The time constant of the secondary circuit is also given by (2): $\tau_d = 5$ s.
Therefore, the iron core allows to gain a factor 67 in the inductance of the magnet to be powered and a factor ~ 200 in the time constant $\tau_d$.

## Operation

When the presence of the magnetisation current is accounted, the primary current is proportional to the magnet current: the amplification factor is found to equal 51.3. Both ramping up and down modes of the magnet current are available during the device operation. The preparation stage can be done only once for multiple ramps of the magnet current, as long as both magnet and primary currents are not required to be equal zero at the same time. Therefore, the actual loss in the switch is not of a primary importance.

During each operating cycle the primary current ramp rate was kept constant or zero at each stage. The characteristic dependence of the magnet current versus time obtained in a single shot following the algorithm is presented in Figure 7. For this particular run, the voltage across the primary coil is ± 3 or 0 Volts, the ramping time is 0.1 s.

A magnetic field calculation using the model arrangement depicted in Figure 5 shows, that field in the conductor area will reach the maximum of 0.1 Tesla in the secondary coil at the end of the stage (ii).

## Efficiency

Excellent efficiency of the system (the energy loss per cycle is estimated of about 0.3%, excluding the switch loss) is provided due to the lowest operation frequency of the cold unit of the device. A comparison of this transformer with a superconducting rectifier shows the following. The transformer scheme can be regarded as a single cycle superconducting rectifier consisting of the transformer and the switch. It operates with the lowest possible frequency (one cycle), set by the total ramping time of the magnet. When required, one solution to reduce the volume/mass of the iron core to the acceptable level is to increase the number of operating cycles. That is, to apply a full-wave superconducting rectifier with the iron-core transformer.

The limits of this approach have recently been investigated[8]. A magnet current of few kAmps is feasible up to 50-60 Hz. Note, that an increase of the operating frequency reduces the size of the core almost proportionally, but on the other hand it causes a rather fast increase of the cold unit loss. Therefore, the number of operating cycles is a subject for a careful optimisation in each case.

A converter using room temperature or cold semi-conducting switches[1,3] applied to power a magnet with a current of 10 kA will dissipate at least 10 kW due to the forward current voltage drop across the switch. Therefore, it only can be efficient when the ramping voltage across the magnet is well above 1 V. When additionally the low output ripple restriction, the current leads and the cooling system are included into consideration, the proposed transformer system may haver a good chance of application.

## CONCLUSIONS

The use of an iron-core transformer in a special scheme to power superconducting magnets is very simple and efficient. The conducted experiments confirm the estimates made. The method can be extended to laboratory scale superconducting magnets. The amount of iron required can be further reduced, when the number of the operating cycles of the powering system is optimised.

The method is equally applicable to low and high temperature superconductors, as long as there is no need for high frequency operation. For HTS one obvious application is to power prototype sections of DC superconducting cables.

## ACKNOWLEDGMENT

This work is supported by the Netherlands Organisation for Scientific Research (NWO) and by the EC, INTAS Program, grant N 95-0197.

## REFERENCES

1. J. McCarthy and C. C. Smith, "Power supply design for superconducting magnet applications", Superconductor Industry, pp. 22 (Spring 1992).
2. O. A. Shevchenko, H. J. G. Krooshoop, and H. H. J. ten Kate, "Switching properties of high quality superconductors", the paper to be published in the Proceedings of the EUCAS 97 ( held in June-July, 1997).
3. O. A. Shevchenko, M. A. Fedorovsky, and H. H. J. ten Kate, "Cold semi-conducting converter, a compatibility device for a superconducting magnet", Adv. in Cryogenic Engineering, Vol. 41, (1996).
4. H. J. Knoopers, S. Wessel, H. J. G. Krooshoop, O. A. Shevchenko, A. Godeke, and H. H. J. ten Kate, "Fast ramp 50 kA superconducting transformer for testing full-size ITER cable joints", Proc. of the 16-th Int. CEC/ICMC Kitakyushu, Japan 20-24 May 1996, Elsevier Science, part 2, pp.803-806 (1997).
5. O. A. Shevchenko, H. J. G. Krooshoop, M. A. Fedorovsky, and H. H. J. ten Kate, "Transformer and switch characteristics of a 50 Hz superconducting rectifier", Cryogenics Vol.34, ICEC Suppl., pp. 745-748 (1994).
6. G. B. J. Mulder, H. J. G. Krooshoop, A. Nijhuis, H. H. J. ten Kate, and L. J. M. van de Klundert, "A convenient method for testing high-current superconducting cables" ibid.[3], Vol. 35, pp. 763-770 (1990).
7. "Concise encyclopedia of magnetic & superconducting materials" / ed., J. E. Evetts Pergamon Press, Ltd. (1992)
8. O. A. Shevchenko, H. J. G. Krooshoop, M. A. Fedorovsky, and H. H. J. ten Kate, "A novel miniature superconducting converter for 1 kA magnets", IEEE Trans. On Magn., Vol. 32, N 4, pp. 2590-2593 (1996).

# CRYOGENIC CONTROL SYSTEM FOR THE g-2 MUON RING

T. Tallerico,[1] G. Bunce,[1] L. X. Jia,[1] W. Meng,[1] C. Pai,[1] and D. Konieczny[2]

[1]Brookhaven National Laboratory
Upton, New York, 11973, USA

[2]Princeton University
Princeton, New Jersey, 08544, USA

## ABSTRACT

The g-2 Muon storage ring consists of three superconducting solenoids 15 meters in diameter, and a beam Inflector solenoid 1.7 meters in length. All superconducting solenoids are indirectly cooled by forced two-phase helium. Cryogenic cooling is accomplished via a J-T circuit and a LHe control dewar. The control system for this cryogenic system is built around commercially available Programmable Logic Controllers (PLC's) and cryogenic hardware for monitoring temperature, pressure and flow control. The complexity of this system necessitated the use of a graphical user interface (GUI) which permitted operators to perform monitoring and control functions from a central control room. The graphical interface allowed for rapid operator training and, with the cooling circuit schematic shown graphically they can respond to critical situations promptly. Critical data points on the experiment are logged in the software and historical trends are provided for. Using a software based system allowed for rapid system revisions as they were required. Flow control can be performed manually by the operator or automatically by the software based on linear control algorithms. The system has been in use for two years with successful results.

## INTRODUCTION

The g-2 Muon experiment is an attempt to obtain a more precise value to the anomalous magnetic moment of the Muon subatomic particle. To achieve this end result requires three superconducting solenoid magnets connected in such a way that a dipole field is formed. The three solenoids and the Inflector will be cooled using forced two-phase helium in tubes. To keep such a large experiment superconducting for several months requires a cryogenic system that is simple to operate, can display over one hundred temperatures, 30 pressure transducers and allow for control and monitoring of a multitude of throttle and on/off valves. The system also had to monitor helium level in the dewar and operate the expansion engine speed control. Due to the system complexity and schedule constraints it became apparent an off the shelf computerized solution was required. This would also allow for a system design based on a central control room and would permit operators to monitor all sensors, transducers and valves located in remote parts of the experiment.

## PLC HARDWARE

For the last five years the Alternating Gradient Synchrotron (AGS) has been replacing old in house built control systems with Allen-Bradley (AB) Programmable Logic Controllers (PLC's). Historically PLC's have been used in an industrial environment where high-speed control and data collection is not a requirement. The hardware is controlled via a personal computer, which has the Allen-Bradley logic software on it. The logic software and hardware are a complete system in the sense that they are purchased as one unit. The graphical interface software (Controlview) is then purchased as an add-on package for operator interfacing.

Since the AGS has had a successful track record with these controllers we decided to consider these for our application. These controllers are built around a serial data bus known as DATA HIGHWAY PLUS (DH+). It allows PLC's to be distributed at remote locations around the experiment. Data is transmitted over the serial bus to the main PLC processor, which in turn displays the data on the personal computer in the cryo control room. Depending on the particular application, AB supplies interface modules that perform an array of functions. Since we were reading analog signals from the pressure transducers and temperature sensors 150 feet away we decided to use the eight channel differential input module. The throttle and on/off valves required analog output modules and TTL output modules respectively. The analog input and output modules resolve signals to 12 bits. The TTL output modules drive Opto-22 optoisolators to permit signal conversion to 120 volts ac. The analog input and output modules allow for an array of signal levels. You also have a choice of either voltage or current signals. Analog output modules permit an output current of 4-20 ma while the analog input modules permit a current level of 4-20 ma or 0-20 ma. A PLC-5/40E processor handles the central processing. A lithium battery backs up the program in the processor. An option is available to install an EEPROM for program security. This processor was purchased with an optional Ethernet communication interface to allow us to tie various systems to the experiment slow control system and allow for data packet downloading. Each PLC system communicates with the other via DH+ (i.e. cryo to vacuum or cryo to quench protection). All PLC systems not in the cryo control room (i.e. compressor room) have remote I/O adapters, which are scanned for data periodically by the central processor. Figure 1 shows the layout of the cryogenic control system.

## PLC SOFTWARE

We used four software packages to control the cryogenic system. Three were purchased from AB and the fourth system was purchased from FTP Software Inc. The AB software packages used were the logic software, the graphics builder software and the graphics runtime software. I will describe each package briefly. The logic software package is the main software where all the programming gets done. This DOS based package comes with a complete instruction set for writing complex programs to control the cryogenic process. The software has diagnostics and the ability to comment instructions and attach names to numerical addresses. This allows for ease of debugging. Typical processing cycle times (one complete cycle) are on the order of five to ten milliseconds and go up with program complexity. The graphics builder software is a windows based package where the graphic displays are created and the database of system tags is kept. Each tag is tied to a distinct address with a limit on the

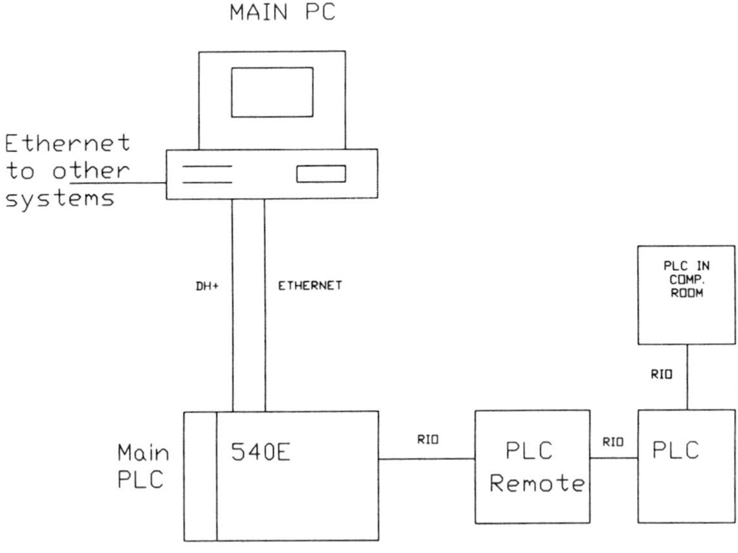


Figure 1

allowable number of tags. The limit is typically in the thousands of tags. The alarm database, data trending and data logging are setup from this software also.

The graphics runtime software is a DOS based package, which integrates the logic and graphics software and actually runs our cryogenic process. The software from FTP Software Inc. supplied the drivers, which allowed us to communicate over Ethernet. This permitted data transfers from the central processor to the experiment data acquisition system.

## TEMPERATURE SENSORS

Multitudes of cryogenic temperature sensors are installed in the g-2 storage ring and the Inflector. They were purchased from LakeShore Cryotronics Inc. and Chino. The storage ring sensors were selected and installed before the cryo control system was designed so we had to design around the selected devices. Having a software-based system afforded us the flexibility to do this. The storage ring contains silicon diode temperature sensors and Cernox thin film resistance temperature sensors (Ceramic Oxynitride on a Sapphire substrate). The Cernox sensors are not affected by the 1.5 Tesla magnetic field in the ring. The Inflector was built in Japan by KEK and they installed the temperature sensors during construction. Inflector sensors include Carbon Glass resistance temperature sensors form Lake Shore and Chino platinum sensors. Since the silicon sensors are affected by magnetic field dependent temperature errors we had to provide error correction in the software. The silicon sensors all followed a standard temperature curve, which permitted us to use a universal temperature transmitter. This transmitter provides a 4-20 ma output, which represents the temperature range from 1.4K - 325K. This allows us to read the transmitter output directly into the PLC. We had the capability to read 36 silicon temperature sensors directly. We then used temperature scanners to read 24 more silicon sensors which we downloaded to the PLC via a RS-232 port. The conversion to temperature is performed in the software using linear curve fitting. The Cernox and the Carbon Glass sensors have unique temperature curves so we had to download the data into each sensors associated transmitter module. These modules have 4-

20 ma outputs which we read into the PLC. Linear curve fitting is applied to these devices also. The platinum sensors in the Inflector did not have compatible transmitters so we had to do more work to read these out. We wrote a fifth order polynomial in the PLC software and curve fit the actual data coming back from the platinum sensors. Since a constant current was needed we used a one-milliamp constant current from a PLC output module to excite the sensor.

A typical equation for one platinum sensor is shown in equation one and the temperature versus resistor curve is shown in figure 2.

$$T = -306.807478 + 79.2429672R - 7.09216940\,R^2 + .291160906\,R^3 - 4.44431703\,R^4 \quad (1)$$

## PRESSURE TRANSDUCERS

We used a variety of pressure transducer ranges from Sensotec corp. The two types were absolute and differential. These devices have an internal amplifier built in. All transducers have 4-20 ma outputs which we read directly into the PLC. As with the temperature sensors each pressure transducer was scaled depending on its range in the PLC. Linear equations were used. The differential transducers were used to measure flow in four locations: 1) flow going to the J-T valve, 2) flow going to the expansion engine 3) flow to cool the magnet main leads and 4) flow to cool the Inflector leads. Absolute transducers were used in all other locations.

## HELIUM LEVEL

To measure LHe level in the dewar we employed a probe and controller from American Magnetics Inc. The maximum LHe level was 40 inches. The controller has an analog voltage output of 0 to 5 volts, which corresponds to 0-40 inches. This was read into the PLC, converted to liters and an animated display showing liquid level was employed.

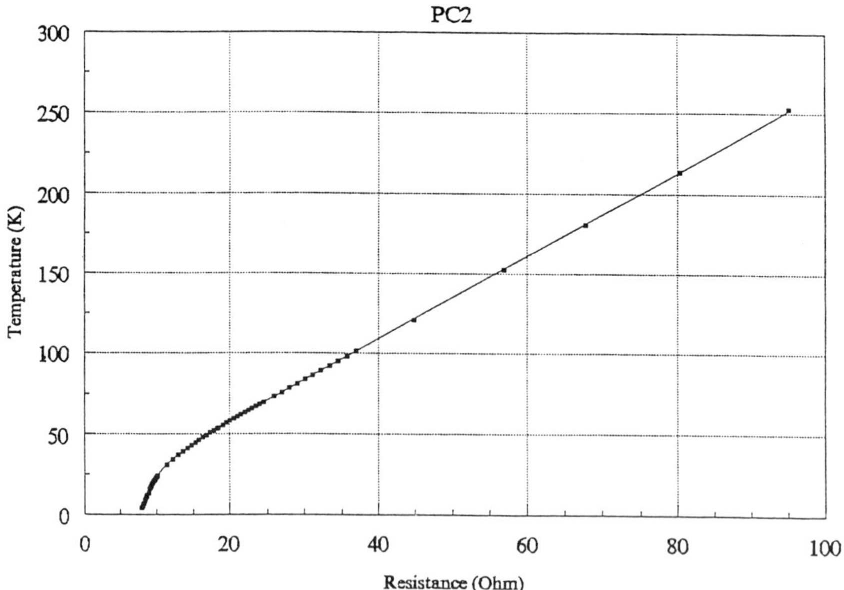


Figure 2 Platinum sensor temperature curve.

## VALVE CONTROL

Valves used on the experiment were throttle valves with 4-20 ma I/P controllers and on/off valves requiring 120Vac. We used valves from CVI and Baumann. The Joule-Thompson valve was purchased from Jordan. Valve sizes range from one to two inches.

Since the valves we use do not have any feedback elements we were not able to perform any PID control loops. We had to run an open loop system for each valve. The operator enters a percentage value from 0-100 on the computer console that he would like the valve to position to. 100 percent corresponds to a fully open valve. No feedback required us to watch pressures as valves were moved so we could get indication that valves were positioning properly. Utilizing software allows these valves to operate in a manual mode or automatic mode. In automatic mode each valve is put under control of a software algorithm. The valve is positioned as a function of pressure. Linear functions were used in all cases. In manual mode operators at the console had full control of valve position. A similar feature was performed with the on/off valves.

## DATA LOGGING

One of the requirements that the experiment has is a real time data logger for all processes that the system was running. The PLC has a data logger that allows us to set up multiple data log models and save data for defined periods of time. We logged the critical data every second (this was the maximum allowed) and all the noncritical data every 15 minutes. The total elapsed time for each model was a function of the file size specified. In the case of the one-second log we could store 16 hours of data. The 15-minute log allowed us to store data over several months. There are roughly 100 tags in each log model.

## DATA TRENDING

Real time data trending gives us on screen graphical displays of system data. You can also display trends of historical data stored in the data logger. This feature proved valuable when the magnet quenched and we needed to analyze temperature data expediently.

## EVENT DETECTOR

This checks for user defined events periodically. Events can be normal or abnormal. We used this feature to display a warning screen if any remote I/O communication failed. We also used this to print an hourly log of critical data points.

## ENGINE SPEED

The expansion engine speed was controlled manually due to problems with our automatic speed control. Closed loop control based on pressure proved erratic and it was determined that manual control was required. Closed loop control will be installed during the next major system upgrade.

## CONCLUSIONS

PLC's have proved to be an effective way to control cryogenic processes and software has given us flexibility in making system modifications as warranted.

## ACKNOWLEDGEMENTS

The authors wish to thank the members of the cryogenic target group for all their support. Their diligence and effort kept the experiment superconducting on a continuous basis.

# APPLICATION OF INDUSTRIAL CONTROL SYSTEMS TO A LARGE SCALE CRYOGENIC SYSTEM

R. C. Sanders, D. H. Allspach, R. L. Schmitt

Fermilab
Batavia, IL, 60510, USA

## ABSTRACT

Fermilab successfully installed and used off-the-shelf industrial controllers and software to operate five 600 Watt, one 90 Watt liquid helium refrigerators, seven strings of superconducting magnets and three liquid hydrogen targets and other miscellaneous subsystems. The equipment is distributed over a 3 km by 3 km area. The planning, wiring installation and programming was done entirely by mechanical engineers and technicians familiar with the processes being controlled, and in conjunction with their usual jobs. No computer programmers were involved. Sophisticated control schemes such as cascade PID, adaptive gains and automated quench recovery were employed on the Moore Products APACS and Siemens/TI controllers. Intellution FIX DMACS software running on PCs (with a Windows platform) provides alarming, data logging and interactive customized graphical displays of the processes for operators, engineers and managers. From any of 20 FIX DMACS nodes operators can change set points or other parameters on any of the subsystems. The FIX DMACS nodes communicate to each other over Ethernet using TCP/IP protocol and potentially can read/write to any other parameter in the controllers. Operational problems, operator man-hours and off hour call-ins have been dramatically reduced.

## OVERALL DESCRIPTION

Figure 1 shows an overview of the cryogenic control system for the Fermilab Fixed Target area spread out over a 3 km by 3 km area. Table 1 displays data on the control systems. The four largest cryogenic systems, MCC, PS4, PS1 and E835 include five 600 W and one 90 Watt liquid helium refrigerators, 39 superconducting magnets, one cryostat of helium cooled electronics that all use Moore Products APACS (Advanced Process

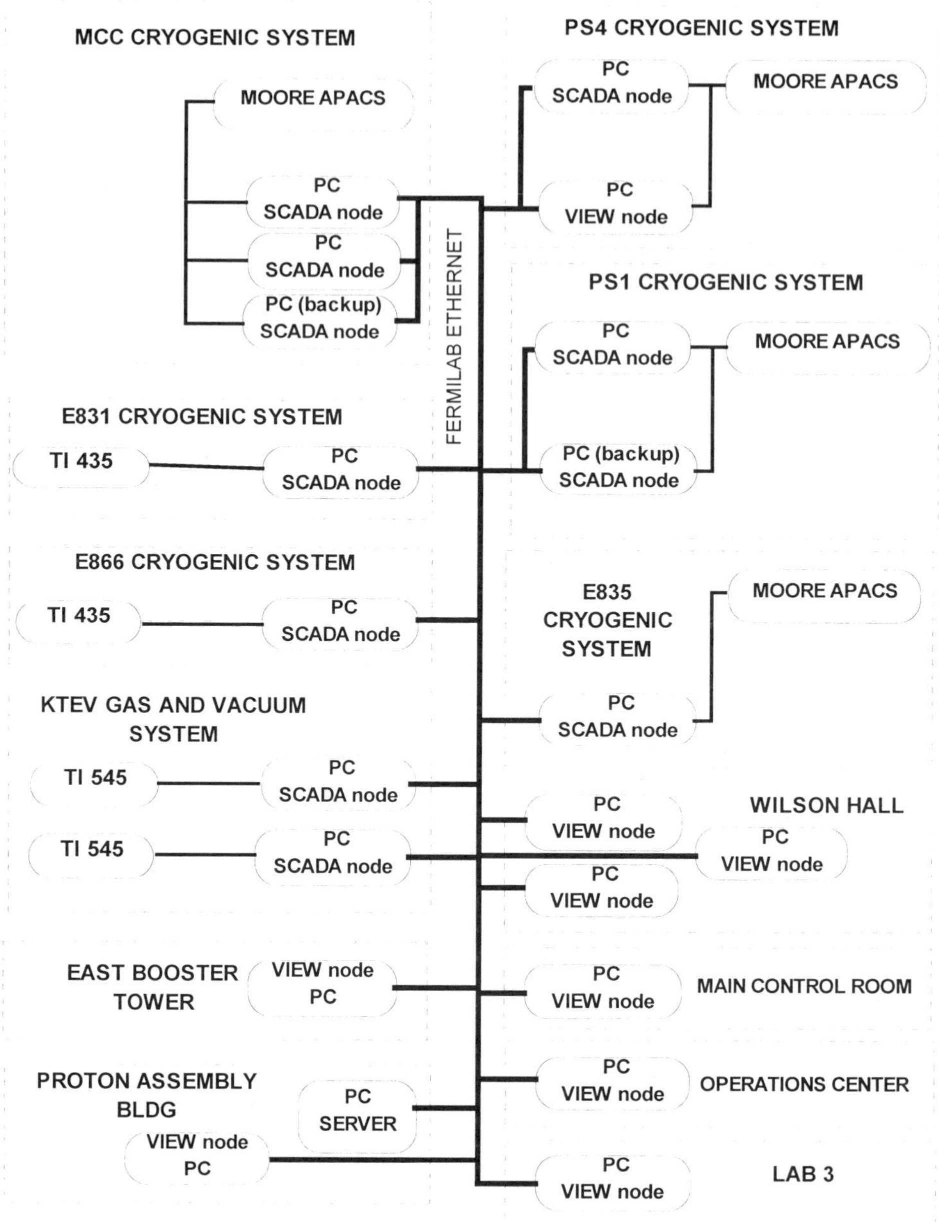


**Figure 1.** Schematic of Fermilab Fixed Target Area cryogenic control system. The PCs are connected over Fermilab Ethernet.

**Table 1.** Fixed Target Cryogenic Control System Data

| System | MCC | PS1 | PS4 | E835 | E866 | E831 |
|---|---|---|---|---|---|---|
| Controller | APACS | APACS | APACS | APACS | TI 435 | TI 435 |
| I/O modules | 37 | 11 | 17 | 6 | 8 | 14 |
| I/O Count | 644 | 171 | 304 | 131 | 84 | 149 |
| Active Controllers | 2 | 1 | 1 | 1 | 1 | 1 |
| Hot Backup Controllers | 2 | 1 | 0 | 0 | 0 | 0 |

Automation and Control Series) controllers. The two smallest cryogenic systems, E831 and E866, have three hydrogen targets cooled by five helium refrigerators and use Siemens/TI 435 controllers. The KTEV system has vacuum and warm gas processes and uses two Siemens/TI 545 controllers. All of these process systems are physically separate and operate independently of each other.

Using Intellution's FIX DMACS software, Fermilab personnel can monitor and operate any of the Fixed Target cryogenic systems from any of the PCs called SCADA (Supervisory Control And Data Acquisition) or VIEW nodes shown in Fig. 1. The PCs are either a desktop or rack mounted industrial type from Moore Products and communicate over the Fermilab Ethernet using TCP/IP protocol. This network uses a Windows NT server located in the Proton Assembly Building. Not shown in Fig. 1 is the D-zero control system, located outside the Fixed Target area that has a large collection of Siemens/TI controllers and SCADA and VIEW nodes using FIX DMACS software.

## APACS DESCRIPTION

Refineries, chemical plants, steel mills and other process industries use Moore Products APACS. It is suitable for complex cryogenic process systems where the primary need is to manipulate (real number) variables. Fermilab personnel configure the APACS controllers from the PCs connected directly to them. A Moore Products software package based upon the International standard IEC 1131 part 3 provides the configuration tools. The software integrates four programming languages: function blocks, ladder logic, sequential function chart and structured text. The configuration process is graphical and self documenting. APACS permits changing the configuration and testing the changes while operating. This is a big advantage. After an operational problem occurs, engineers can look at the historical data and make changes in the control scheme if needed. If necessary, very complicated algorithms can be implemented.

APACS has controller modules and I/O (Input/Output) modules that insert into a standard rack which can hold up to 10 modules. Each APACS I/O module has 16 to 32 channels that connect to field instruments and devices. The different I/O module types will handle a variety of functions including: 4-20 mA signals, analog voltage inputs, discrete voltage inputs, reading resistance thermometers or thermocouples. The modules communicate over two proprietary buses in the rack backplane. Larger control systems result from connecting the buses of individual racks. The SCADA nodes are PCs with a special card that connect to one of the backplane buses. The MCC and PS1 cryogenic systems have hot backup controllers that automatically take over the control in one scan cycle (100 to 200 mSec) if an error is detected in the active controller.

## SIEMENS/TI CONTROLLER DESCRIPTION

The Siemens/TI 435 PLCs (Programmable Logic Controllers) are appropriately used for the smaller control systems, E831 and E866, where primarily logic control is needed. A variety of discrete and analog I/O modules are available. The PLCs are configured with ladder logic. For the liquid hydrogen and deuterium target systems, the PLCs provide automatic pump down, semi-automatic cool down, safety interlocks and drive an operator control panel. Compressor interlocks remain with the compressor package, however a first out determination is being done by the PLCs. Each PLC interfaces with its SCADA node through a serial port connection. The PLC and SCADA node are supported by UPS's (Uninterruptable Power Supplies).

The Siemens/TI 545 PLCs are also programmed with ladder logic. They include more special functions and communication options than the Siemens/TI 435's. In addition to logic and instrument I/O the KTEV PLCs perform cascade analog control, receive serial data from high precision thermometers, and send serial data to physics data loggers. For the blockhouse temperature control the operation is completely automatic including recovery from power outages. The blockhouse PLC is equipped with a UPS. For the flammable gas and vacuum systems the PLC is used for interlocks and steady state operation and interfaces with the SCADA node by way of a special communication card.

## FIX DMACS DESCRIPTION

Intellution's FIX DMACS is a very powerful and flexible software package. FIX DMACS provides a graphical interactive display of pressures, temperatures, valve positions and other process variables. Using the graphical displays, personnel can change values of selected variables in the controller such as set points. The graphic objects can be made to rotate, change color, size and position based upon the real time value of process variables. Operators quickly and easily interpret graphical representations of dewar liquid levels, valve positions, and the on/off status of pumps and compressors. FIX DMACS can send alarm messages to desired locations when a process variable deviates unacceptably. Password based security privileges can be assigned to personnel to restrict access to particular systems or software features. Process data can be collected, stored and displayed. Archived data can be used for analyzing process performance. These features are available to operator from any VIEW or SCADA node

In each SCADA node, FIX DMACS continuously updates a database of real-time values of controller variables. The SCADA nodes connect directly to one or more controllers through serial ports or proprietary cards inserted into the PC. The communications path is two way; the values of measured variables such as temperatures and pressures pass from the controller to the SCADA node data base; while values of variables such as set points and desired valve positions pass the other way. The value of any variable in the controller can be made available to the SCADA node database. If a SCADA node shuts down or fails, the controller continues to run and will control the cryogenic process.

FIX DMACS does not maintain a database of controller variables in the VIEW nodes, which are usually located in remote control rooms and offices, and are not directly connected to controllers. When a VIEW node or remote SCADA node displays a variable, a request for the value of the variable is sent over Ethernet to the appropriate SCADA node. The value of the variable is then returned over Ethernet. This process is repeated continuously as long as the variable is displayed.

## PS4 CRYOGENIC SYSTEM SUCTION PRESSURE CONTROL EXAMPLE

A stable low compressor suction pressure is critical to the operation of a liquid helium refrigerator. The suction pressure affects temperatures and mass flow rates through the entire system. An unstable suction pressure causes control instabilities in many other process variables. The control system must respond properly to system upsets such as a magnet quench, or the sudden starting or stopping of a compressor or expansion engine. With improper control, suction pressure can easily go into a vacuum, pulling contamination into the suction header or go high and cause relief valves to open. Multiple devices, including the two compressors and the valves connecting the suction header, discharge header and storage tank provide suction pressure control.

The control algorithm for the valve that sends makeup helium gas into the compressor suction header from the helium gas storage tank uses adaptive gain; the bigger the deviation of suction pressure from its set point, the larger the proportional gain on the PID (Proportional Integral Derivative) loop. This algorithm results in a very stable suction pressure under normal conditions but is very responsive to significant deviations. The proportional gain is also automatically adjusted to compensate for changes in the gas storage tank pressure.

A sudden change in compressor loading can cause significant changes in suction pressure. The control algorithm for the compressors limits the rate of change of the loading. The compressors could load or unload only up to a certain maximum rate. The compressor mass throughput is a very nonlinear function of its loading; to compensate for this, the rate limit was made a nonlinear function of the loading so that the maximum rate of change of compressor mass throughput stays fairly constant. Normally the compressors operate at a fixed loading set by the operators; but they will unload if there is an excessively low suction pressure (to prevent pulling a vacuum) or an excessively high discharge pressure (to prevent lifting relief valves).

The control algorithm for the valve that sends helium gas from compressor discharge to the storage tank has two PID loops; one controlling discharge pressure and the other suction pressure. The valve position normally has a minimum of 10% and is the maximum of the outputs of the two PID loops. Either a high suction pressure or a high discharge pressure will cause the valve to open farther.

## MCC HELIUM FLOW BALANCING EXAMPLE

The MCC cryogenic system has three 600 Watt liquid helium refrigerators in parallel. They all connect to the same compressor suction and discharge headers and the same cryogenic transfer line. Historically it has been a major problem balancing the helium flow rates so that one refrigerator does not become too cold or too warm.

Helium returning from the six magnet strings is split between three heat exchangers. This low pressure helium cools the warm high pressure stream. Between the heat exchangers, relative flow rates are important, since low flow causes a heat exchanger to run warm, producing little or no liquid helium, and high flow causes a heat exchanger to run cold, which reduces the work the expansion engines can remove and wastes refrigeration. The problem is compounded since any of the six expansion engines can have poor efficiency or even shut down. Also it is possible to operate with just two of the three refrigeration systems when some magnet strings are not in use.

The control scheme determines whether a refrigeration system is running based on the operation of either engine. Then an average heat exchanger temperature is calculated. The average only includes operating units.

Each of the three low pressure helium return valves is controlled by a PID loop. The process variable is the difference between the individual heat exchanger temperature and the average. The set point is -0.1 K, which tends to drive one of the valves fully open, resulting in the least back pressure. Lower back pressure results in colder magnet temperatures.

This balancing control scheme is very stable during normal operations, and has compensated for unexpected engine trips automatically. The heat exchanger cold end temperatures are normally within 0.2 K of each other. Heat exchanger rebalancing during engine maintenance or adjustments is fully automatic.

## MAGNET QUENCH RECOVERY

In previous fixed target runs magnet quench recovery was done by operators manipulating valves. The operation was delicate and required expertise for timely completion and to avoid upsetting the refrigerator balance. This is due to the limited refrigeration capacity at these locations. Now quench recovery is fully automatic and requires no operator intervention. It is accomplished without refrigerator upset and enables a magnet power supply permit when completed.

Normally subcooled helium flows through the magnet, through a JT (Joule-Thompson) valve and back through an annulus around the magnet as two phase fluid. This helium returns to the heat exchanger(s) as described above and elsewhere.[1] Normally the JT valve is controlled by magnet temperature. Shortly after a quench the temperature rises and the JT valve would open. However if the temperature exceeds 6 K, returning this warm helium to the refrigerator would be deleterious. Also it is desirable to remove the heat quickly to resume operations. The magnet strings are equipped with a cool-down valve that returns this warmer helium directly to compressor suction.

Since a quench can deposit heat at either end of a long magnet string, the warmer of the two temperatures is selected to operate the quench recovery system. The normal temperature control of the JT valve is overridden quickly when temperatures exceed 6 K. The cool-down valve is opened using a cascade loop with temperature setting flow rate. The flow is important since high flow will unbalance the heat exchanger and since the conditions change dramatically across this valve. When the temperature falls below 5.5 K the JT valve is slowly opened and returns to normal temperature control.

The magnets must be refilled with subcooled helium and this is monitored with a liquid helium weir when so equipped.[2]

Quenches also caused the dry expansion engine to trip in the smaller systems since the large helium boiloff would overcool the heat exchanger. The trip often required local operator effort to restart. Logic now in use stops the engine following a trip which allows the compressor to recover suction pressure faster and prevents an engine trip. When temperatures and pressures return to normal the engine restarts. The logic for this combines engine inlet temperature, compressor suction pressure, and magnet two-phase pressure.

## COMPARISON TO IN-HOUSE DEVELOPED CONTROL SYSTEMS

It is important to compare industrial and in-house developed control systems. The Fermilab Fixed Target cryogenic systems were previously controlled with equipment and software developed by Fermilab over the last 15 years. During the last ten years, industrial control systems improved dramatically in cost, capability and ease of use.[3] Within the last 2 years, mechanical engineers and technicians, without the aid of professional programmers, installed, configured and put in service industrial control and monitoring systems in the Fermilab Fixed Target area. The advanced control algorithms and ability to make on-line changes of the industrial equipment and software allowed engineers, familiar with the thermodynamic and mechanical requirements, to easily automate the cryogenic processes without extensive programming time. The increased automation greatly reduced the labor required for start-up and operations, which more than offset the labor of installing the industrial equipment and software. The engineers involved were trained specifically in the implementation of industrial control systems and control dynamics rather than programming languages.

This is a sharp distinction with the practice of writing application programs for in-house developed control systems. Application programs are in principle completely flexible and can be written to perform any operation. However, in practice, considerable programming effort achieves only rudimentary control system features such as simple feedback loop control and the capability of remote set point changes. It is extremely unlikely if not impossible for a small number of in-house programmers to match the results of companies as Moore Products, Intellution or Siemens specializing in control system software. Usually few if any advanced control features are available from in-house developed control systems.[4,5,6,7,8] The reliability of in-house developed equipment and software is also less than industrial equipment and software.

## SUMMARY

Fermilab successfully installed industrial controls on to a large cryogenic system. The purchased hardware and software is very powerful, flexible and reliable. This demonstrates that industrial controls should be used for cryogenic systems.

## ACKNOWLEDGEMENT

Work supported by the U.S. Department of Energy under contract No. DE-AC02-76CH03000.

## REFERENCES

1. C. Rode et al., Energy Doubler Satellite Refrigerator Magnet Cooling System, in: "Advances in Cryogenic Engineering", Plenum Press , New York (1980), Vol. 25, pg. 326.

2. R. C. Sanders et al., Weir Flow Meter for Saturated Liquid Helium, “Advances in Cryogenic Engineering”, Plenum Press , New York (1996), Vol. 41, pg. 1797.
3. Elliott McCrory et al., Who gets to specify the Control System, “Proceedings of the 1995 International Conference on Accelerator and Large Experimental Physics Control Systems”, pg. 8.
4. Michael Oothoudt et al., Proton Storage Ring Control System Upgrade-Interim Status, "Proceedings of the 1995 International Conference on Accelerator and Large Experimental Physics Control Systems", pg. 218.
5. Marie Keesee et al., The CEBAF Control System for the CHL", “Advances in Cryogenic Engineering”, Plenum Press , New York (1996), Vol. 41, pg. 649.
6. private correspondence.
7. M. Barranco-Luque et al., Conclusions for Procuring, Installing and Commissioning Six Large Scale Helium Refrigerators at CERN, “Advances in Cryogenic Engineering”, Plenum Press , New York (1996), Vol. 41, pg. 761.
8. B. Lublinsky et al., New Tevatron Cryogenic Control System, "Proceedings of the 1993 Particle Accelerator Conference", pg. 1819.

# THE DESKSIDE SUPERCOMPUTER

John W. Peeples

KryoTech, Inc
714 Chris Drive
West Columbia, SC 29169

## ABSTRACT

State of the art cooling and electronic packaging technologies are distinguishing characteristics of supercomputers. The "deskside super computer" provides best-in-class computational performance by virtue of novel cooling and packaging technologies.

Chilling Complementary Metal Oxide Semiconductor (CMOS) circuits for performance enhancement in computing applications has been studied for decades. A synergistic evolution of critical technologies is occurring which promises to unleash this potential. Cost effective performance acceleration of workstation class computers via below ambient cooling is now commercially viable. Technical challenges and important program consideration of cooling enhanced computing are discussed in this paper.

## INTRODUCTION

CMOS is the semiconductor technology of choice for the computer industry. It was quickly embraced in the mid-1970's as a lower power alternative to the then dominant bipolar technologies. Subsequent investment in CMOS has increased, by many orders of magnitude, device densities and speeds. One ironic result is that today's high performance CMOS integrated circuits are as challenging to cool as were the 'hot' bipolar devices they displaced. Cooling can play a dual role with CMOS devices. As with bipolar devices, efficient cooling may be an operational imperative to keep devices from burning up. Unlike bipolar devices, cooling CMOS to lower temperatures can also enhance switching performance.

A few CMOS properties are disadvantaged at low temperature. These include a slight reduction in junction breakdown voltage, a higher risk of Hot Electron Injection (HEI) degradation (which can be more than offset by reduced operating voltage) and a reduction in the specific heat of the substrate material. Advantages of low temperature operation are numerous. Some are listed below as characterized by others at 77K.

- 2 to 6 fold increase in transconductance
- 5 fold increase in aluminum interconnect conductivity
- 6 fold increase in silicon thermal conductivity
- 6 orders of magnitude reduction in junction leakage
- Reduced soft error rate
- Bulk CMOS latch up suppression
- Well defined $V_{th}$ behavior and less $V_{th}$ variation
- Steeper sub-threshold slope
- No geometry effect degradation
- Better reliability
- Better digital performance

Performance enhancements of up to 300% can be achieved by operating CMOS computing devices at 77K. More modest performance gains are possible at higher temperatures. KryoTech currently cools CMOS microprocessors to 233K and achieves a 33% increase in switching performance.

Figure 1 plots modeled and measured performance increases that can be obtained from various technologies at lower junction temperatures.

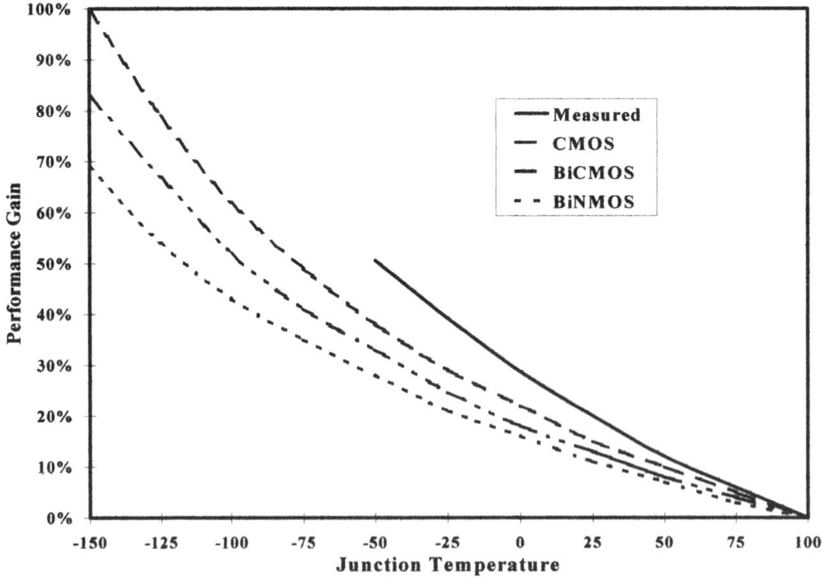


**Figure 1.** Cooling for performance, modeled and measured

**Deskside Supercomputer Opportunity**
*Worldwide Revenue ($ US)*

| | | 1997 | 1998 | 1999 | 2000 |
|---|---|---|---|---|---|
| Total Information Technology | | $540B | $578B | $618B | $661B |
| Workstation Hardware<br>% of Total Information Technology | | $ 25B<br>5% | $ 31B<br>5% | $ 37B<br>6% | $ 44B<br>7% |
| Deskside Supercomputer Market Segment | 20% | | | | $8800M |
| | 15% | | | $5550M | $6600M |
| | 10% | | | $3700M | $4400M |
| | 5% | | $1550M | $1850M | $2200M |
| Market Penetration | 1% | $ 250M | $ 310M | $ 370M | $ 440M |

**Figure 2.** Market opportunity for cooling enhanced workstations

## WHY 233K?

A lasting marriage of cryogenic technology with computing requires a strong value proposition. The benefits of cooling must obviously outweigh the costs. This simple economic truth is the Achilles Heel of cryo-computing. Investigations of cooled CMOS computing have often centered at temperatures well below 233K. Interesting prototype machines and several commercially unsuccessful ventures have resulted. The more moderate cooling target of 233K enables a much less costly cryo-computer capable of delivering a positive cost/benefit to a targeted market segment.

The cost of a cooler for computing application is driven by the cost of the cooling engine. Vapor compression benefits from an established manufacturing infrastructure serving white goods and thus holds a significant advantage in both cost and reliability. The lower temperature limit at which a single compressor can lift a 100 watt load is about minus 40°C or 233K. At this case temperature most merchant CMOS microprocessors can perform roughly one-third better than is possible when they are simply air cooled.

The performance of the latest generation microprocessors makes them low cost alternatives to the proprietary architecture machines which have dominated the engineering and scientific computing workspace. These open architecture alternatives for high performance computing require very precise implementations to realize all performance possible in order to run powerful CAD/CAE and graphical modeling programs. While a 33% performance increase is not compelling for office or transaction computing, it can deliver significant productivity gains in areas of product design and simulation and in graphical rendering. These compute-intensive segments of users will pay more for more performance. A single pass of a finite element simulation of the material creep of a turbine blade may run for seven days or more. Reducing run time to five days can have significant turnaround impact for the development program. Serving and rendering of graphical information is extremely consumptive of CPU bandwidth. A boost of 33% in this market segment at a reasonable price can make for good business. Lower temperatures will yield greater increase in performance and higher penetration into the workstation market segment. As can be seen from Figure 2, this can make for very good business.

## ENGINEERING CHALLENGES

Cryogenic containment in computing presents at least three serious challenges, all of which become more difficult as temperature decreases. These are (1) the development of a capable and cost effective cooling engine (2) the isolation of thermal transfer and of mass transfer of the chilled region (keeping heat and moisture out of the chilled volume) and (3) the cost effective and seamless integration of the cooling apparatus into a targeted computer. Commercial viability demands that all of these challenges are met.

The 233K cooling subsystem consists of two functional components, the chiller and the insulating cavity. The roles of these two components are distinctly different.

The chiller produces the cooling effect by lifting heat from the evaporator region and transferring that heat to ambient at the condenser. To maximize reliability, the chiller design should include minimal or no valves or switches. The compressor drives a closed refrigerant loop. Fan driven air flow cools the condenser and the compressor shell as depicted in the Figure 3 schematic.

The compressor and fans are thoroughly engineered and tested components. Compressor certification includes calorimetric (capacity) testing, noise, wear and lifetime tests. Some are certified for more than 500,000 on-off cycles. Testing includes transportation test and chemical and motor lab analysis. Test durations of a year or more are not uncommon in the compressor industry. The industry is vigilant in the maintenance of its high quality/high reliability image. Fan lifetimes are well understood in the electronics industry. Thus the chiller components represent a predictable and minimal risk to the computer. The main hazard associated with the vapor compression chiller is the loss of refrigerant. Secondary hazards include operational instability due to contamination. Thus the chiller component must be tested for refrigerant containment and for operating stability.

The insulating cavity blocks thermal transfer and mass transfer. The evaporative cold plate mates with the microprocessor. A coldplate for a Digital Equipment Corporation Alpha microprocessor is shown alongside the Alpha heatsink in figure 4. The heatsink is very close in size to the insulating cavity which encloses the microprocessor and coldplate as shown in Figure 5. The thermal performance of the insulating cavity is determined by its design. Cavity molding material, wall thickness and socket pin details are the major determinants of thermal performance of the cavity. Electrical pin counts of 500 and greater must be accommodated. The thermal performance is characterized by calorimetric

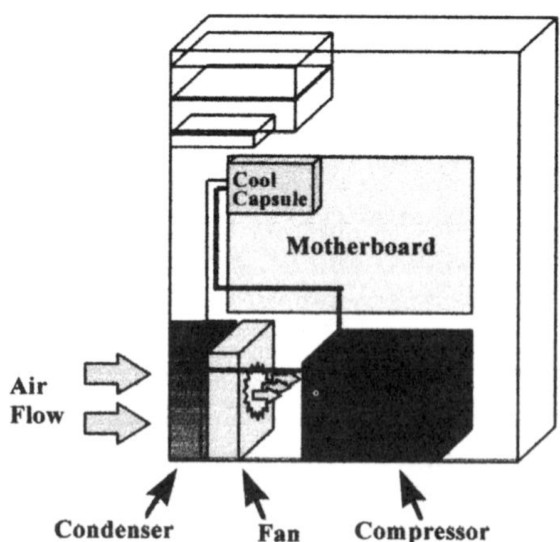


**Figure 3.** Schematic of cooled CMOS computer

testing and by thermal imaging. Mass transfer in the form of water vapor is a primary reliability concern. The reliability of the insulating cavity is dependent upon its wall permeability and seal integrity. The cavity is designed to minimize seal length and is constructed of low cost, low density and low permeability materials. An evacuated dewar type structure would be too expensive for this relatively warm application. The sealing character of the cavity is characterized through leak testing and high temperature/high humidity testing.

The containment problem in this application is complicated by the need to inexpensively insulate within a certain form factor and do so without interrupting the hundreds of high speed electrical signals involved. The 233K workstations employ reaction injected molded (RIM) foam cavities. These are generally "clamshell" in nature, normally having a bottom half that is attached in some way to the electronics board and a top half that assembles over the cold plate and processor to enclosed the chilled volume. Two methods are employed to encapsulate the microprocessor and coldplate.

One method encapsulates a portion of an electronic "daughtercard". The electrical signals are wired in the card which exits the insulated clamshell edgewise to interface with the un-chilled circuitry. This approach presents a relatively flat surface for sealing. It also can allow a reasonable wall thickness of insulation in the plane of egress. Novel methods are employed to seal the egress path and to force a proper thermal gradient to eliminate external frosting on the daughtercard.

A second method routes the actual pins of the microprocessor socket through the wall of the insulating clamshell. This approach is preferred for high speed signals and becomes mandatory as clock speeds increase. It eliminates the daughtercard and can be the more cost efficient approach. The confounding issues are that the socket pin approach results in hundreds of penetrations of the insulating shell which must be sealed against moisture ingress and that five hundred or one thousand phosphor-bronze IC socket pins comprise a low thermal resistance shunt into the chilled volume.

Condensation of moisture on any exterior surfaces of either the chiller or the insulating cavity will jeopardize the computer. Thorough insulation and, when required, spot warming are used to eliminate this hazard. Dew point testing of the cooling subsystem is conducted to insure adequate condensation control. A 233K insulating cavity is shown in Figure 5.

## PRODUCT CHALLENGES

The acceptance of cool CMOS workstations depends primarily upon price-performance, or value. Other cooling technology attributes, like size, weight and acoustical and electrical noise, reliability, condensation and moisture intrusion, can only diminish the product's appeal.

The size of the cooling subsystem includes its compressor, condenser, fans, distribution tubing, cold plate and insulating cavity. This is a substantial apparatus which must integrate into a well established deskside footprint without crowding out required computer functionality. The routing of the distribution tubes and the mounting of the cold plate and its insulating cavity to the microprocessor are extreme integration challenges. Figure 6 is a photo of a cooling pedestal which installs beneath a Digital Equipment Corporation Alpha workstation.

The weight of the cooling subsystem must be within reason and distributed in such a way as to survive normal transportation and installation abuses. 233K coolers capable of lifting 100 watts can be developed which weigh about 65 Kg. Mounting the bulk and most of the weight of the chilling apparatus in the base of the deskside unit serves to stabilize the tower case by keeping the weight low and allows for good front to back air flow over the condenser and compressor shell. Figure 7 shows a 233K workstation mounted on top of the cooling pedestal.

Acoustical noise requirements for workstation computers are becoming more stringent. Customers are more attuned to noise pollution and are demanding quieter machines. The introduction of a compressor changes the noise spectrum and may result in annoying spikes that must be designed out of the acoustic signature of the unit. Issues such as these, if ignored, become critical to customer acceptance.

Unacceptable electrical noise is more likely to originate in the computer than in the relatively low frequency compressor engine and fan motors. However, the cold plate and distribution tubes which attach to the high frequency processor can be very good antenna elements and can exacerbate the containment of electrical noise.

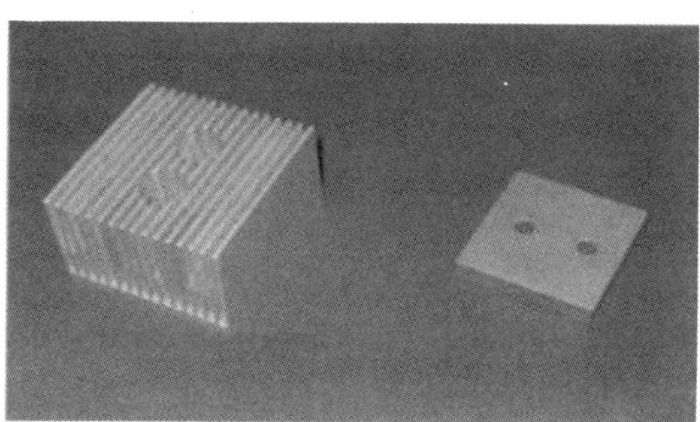

**Figure 4.** Heatsink (right) and evaporative cold plate (left) for Digital Equipment Corporation Alpha microprocessor.

**Figure 5.** 233K insulating cavity mounted in a Digital Equipment Corporation Alpha Microprocessor

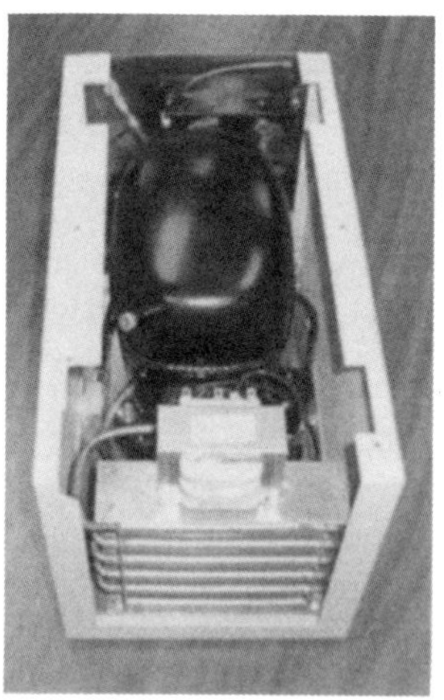

**Figure 6.** Cooling pedestal for a Digital Equipment Corporation Alpha Workstation

**Figure 7.** 233K DEC Alpha Workstation

## BENEFITS

The advantages of cool CMOS are numerous as indicated in the introduction. At the computer system level, the advantage is, first and foremost, increased performance. The actual performance gain to be realized is dependent upon the operating temperature supported by the cooling subsystem. That, in turn is a function of the sophistication and cost of the cooling engine and collateral components. The 233K subsystem has been installed in Digital Equipment Corporation Alpha microprocessor based systems and in Intel Pentium® Pro based systems. In both of these applications, standard CAD and graphics benchmarks were run to determine the actual, system level benefit observed in the target application. Performance gains of approximately 30% were experienced with Pro-Engineer by Parametrics, Softimage and with 3D StudioMax. This level of benefit can support the cost of the cooling subsystem described in this paper for the compute bound users. The obvious extension of this program is to drive to lower temperatures and higher performance enhancement without disproportional increase in the cost and price of the cooling subsystem.

## CONCLUSION

Commercially viable cool CMOS computer workstations can be developed employing vapor compression coolers. Compute bound workstation applications, including computer simulation and rendering, can cost effectively benefit from a 33% performance boost.

Critical program criteria include cost/performance as the number one purchase criteria, but reliability, size, and serviceability can severely restrict acceptance of such a product. Technical hurdles include cooler reliability and the thermal and mass isolation of the cooled region. Surface condensation can not be allowed in the system which must employ non-evacuated insulating cavities for cost reasons. Acoustic and electrical noise and subsystem weight are also important characteristics requiring sound engineering.

## ACKNOWLEDGMENTS

The author wishes to acknowledge the assistance of Rob DiGiacomo in the preparation of this paper and the work of Charles Wall and Terry Craps in the development of much of the technology reviewed herein.

# CRYOGENICALLY COOLED OPTICAL DYNAMIC RAM

A.E. Johnson, E.S. Maniloff, and T.W. Mossberg

Templex Technology
460 E. 2nd Ave
Eugene, OR 97401

## ABSTRACT

Optical Dynamic RAM (ODRAM) is the first entry in an entirely new class of memory devices, offering price and performance levels that are unachievable with any other technology. ODRAM will have the capacity of magnetic hard drives but will allow up to a three order of magnitude improvement in access time. ODRAM will provide this performance advantage at a fraction of the cost of silicon RAM. The extraordinary performance levels offered by ODRAM, which are the direct result of operation at cryogenic temperatures, will easily justify the expense of cryogenic operation. ODRAM offers a new and unique opportunity for cryogenic technology to address large markets and become an important technology in future computing and communications applications.

## INTRODUCTION

Data storage in modern computers is accomplished using two complementary technologies: semiconductor RAM and disk-based recording. RAM allows for very fast data access, however its high cost per unit capacity limits the amount of available memory in practical computing and data management systems. While hard drive storage units cost approximately 40 times less per gigabyte, their mechanical access mechanism results in data latencies approximately five orders of magnitude longer than RAM. The growth of networked computer systems over the last decade has created a situation in which multiple computers are frequently trying to access a single shared hard drive, thereby producing a bottleneck to the development of higher performance systems. In this paper, we discuss a fundamentally new memory technology that will address these data access bottlenecks. This new memory product, when used in conjunction with existing memory products, will enable significantly higher performance computing systems.

This new memory technology, which we term Optical Dynamic Random Access Memory (ODRAM) is based on the concept of a spectral memory: a memory which takes advantage of the optical properties of materials to store multiple ($10^3$-$10^7$) bits in a single spatial location.[1] Spectral memories depend on a specific and fairly general material property known as inhomogeneous absorption linebroadening. Materials exhibiting this property are comprised of atoms or molecules that individually respond to (absorb) very specific frequencies of light, but the collective response of all the material's absorbers covers a wide spectral region. Application of monochromatic light to such a material has the effect of affecting only a very small subset of active absorbers - those residing in the illuminated volume and being responsive to the light's specific frequency. If the frequency of the light is shifted, a different subset of active absorbers within the illuminated volume responds. The number of separately addressable subsets of active absorbers, which determines the potential capacity of the memory, increases dramatically at low temperatures.[1]

Figure 1 depicts a simple kind of spectral memory called "frequency-domain spectral memory". This type of spectral memory is not the type of memory that will be used in ODRAM, but is introduced here for ease of understanding. In frequency domain memory, specific frequency channels are allocated for each bit and the frequency simply becomes an additional addressing parameter. Frequency domain memory carries with it a fundamental speed limitation: accessing a very narrow frequency channel requires a very long interrogation time. This is obviously not the appropriate implementation of spectral memory when trying to build a fast device.

An alternative to the frequency-domain approach to spectral memories is the time domain approach, which is based on the realization that any arbitrary data stream can be represented by a unique Fourier decomposition into specific frequency and phase components.[2] Therefore, recording the Fourier spectrum of a complex, arbitrary data sequence is equivalent to storing the data sequence itself. In the implementation of time-domain spectral memory employed for ODRAM,[3-5] two beams, the data stream to be recorded and a reference beam, illuminate the storage material, directly recording the Fourier spectrum of the data beam in the recording material. Recall of the information is achieved by illuminating the storage material with a read beam that is identical to the reference beam, leading to the generation of an optical signal that is identical to the original data beam.

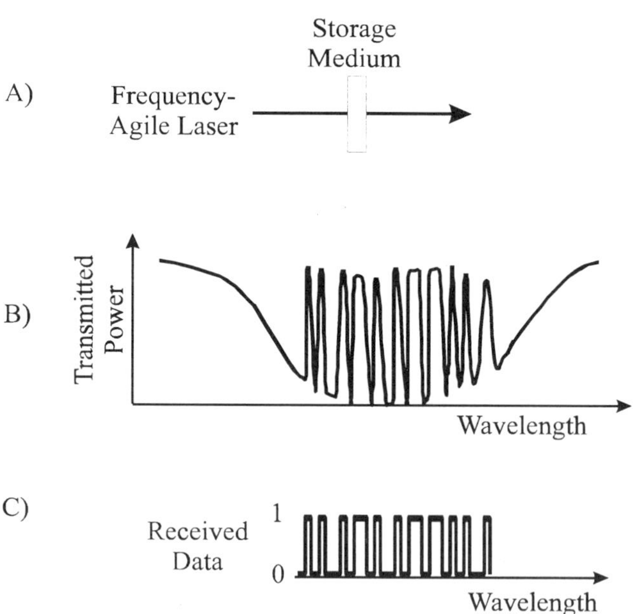


Figure 1. Schematic representation of frequency domain spectral memory. A) shows the frequency agile laser interrogating one spatial cell in the material. B) shows the transmitted power as a function of wavelength (frequency). C) shows the binary data corresponding to the transmission in B).

Clearly the implementation of spectral memories depends on the availability of viable storage materials. Cryogenics is an enabler of spectral memories as it greatly improves the memory performance of storage materials for spectral memories. The most notable performance characteristic that is enhanced by cryogenics is the potential storage capacity, which can be estimated by the ratio of the inhomogeneous linewidth to the homogeneous linewidth, $\Gamma_I/\Gamma_H$. While both quantities are temperature dependent, the homogeneous linewidth tends to have a much stronger temperature dependence. Materials at room temperature tend to exhibit one to ten spectral channels. However, at cryogenic temperatures, more than $10^7$ spectral channels can be available for data storage.[6] Therefore, to maximize capacity, ODRAM should operate at cryogenic temperatures.

ODRAM offers many new and exciting opportunities for researchers and manufacturers of cryogenic technology. Design and manufacture of a refrigeration device optimized for use as a subsystem for commercial ODRAM devices will offer enormous intellectual and technical challenges in the pursuit of innovation. In addition, the potential ODRAM market is large enough that cryogenic technology, as an enabler of ODRAM, could become a major commodity. Once that door is open, developers of many other mainstream technologies that could benefit from operation at cryogenic temperatures will see cryogenics as a cost-effective option, dramatically increasing the importance of cryogenics in a technologically advanced society.

## DEVICE DESCRIPTION

ODRAM is a memory device that offers simultaneous speed and capacity. The early implementations described in this paper are designed to take advantage of existing technologies and off-the-shelf components and subsystems. Focused development efforts will lead to improvement in many, if not all, subsystems and a concomitant improvement in overall system performance.

As discussed in the introduction, spectral memory, the ability to store multiple bits in a single location, is the enabling technology behind ODRAM. Storage of multiple bits in a single spatial location via spectral memory allows for non-mechanical access to large stores of data, overcoming the inherent slow response of mechanically accessed data devices. In particular, latency times can be as short as 10 microseconds, compared to 10 milliseconds for mechanically accessed data stores. In general, an optical RAM will store M bits per spatial location in an array of NxN nonmechanically accessed spatial locations. Initial devices will offer 10-100 Gbyte capacity by exploiting data storage on a single crystalline slab of storage material. Future implementations will exceed 1Tbyte of capacity by allowing mechanical access between multiple storage slabs.

In addition to latency, another important measure of speed is peak transfer rates. Initial efforts will pursue serial data transfer rates of 20-40 Mbit/s. Extension of serial technology to 32 bit parallel reading and writing will lead to bus-aggregate peak transfer rates on the order of 1 Gbit/s. This peak transfer rate is significantly faster than the peak transfer rate for hard drives and only slightly slower than the peak transfer rate for silicon DRAM.

The overall ODRAM system is depicted in Figure 2. The system consists of an optical unit, an archival backup unit, and an electronic interface and control unit. The archival backup unit will consist of a bank of hard drives, which will be used to mirror all data in case of power failure. The electronic interface and control unit will buffer data to/from the system data bus and perform electronic error correction. In addition, this unit will control a dynamic refresh cycle. This dynamic refresh cycle will rewrite data to the storage material during quiescent periods to alleviate decay of the data, which takes place

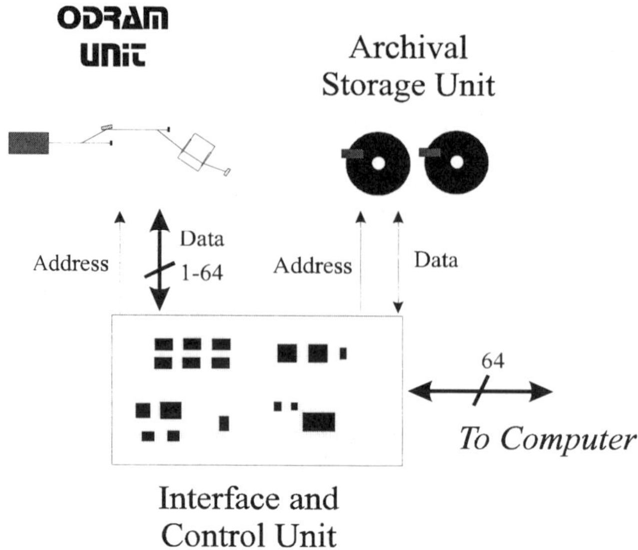

Figure 2. Overview of ODRAM architecture.

over several hours. Note that silicon DRAM also uses a dynamic refresh cycle, as the data storage time in silicon DRAM is of the order of milliseconds.

The optical subunit is depicted in Figure 3. The unit consists of a laser source, an optical data encoder, a spatial addressing unit, the storage material in its cryogenic chamber, and an optical detector. The laser source will be a frequency agile laser capable of tuning over 5 GHz at a rate of 1 GHz/μs. For single channel devices, the optical encoder will be an acoustooptic or electrooptic modulator. For multichannel devices, the optical encoder consists of an integrated array of electrooptic modulators. The spatial addressing unit will be a 2-axis acoustooptic beam deflector capable of resolving the requisite number of spatial locations. The detector will be a single photodiode for single channel devices and an integrated photodiode array in multichannel devices.

The initial storage material chosen for ODRAM applications is $Eu^{3+}$:$Y_2SiO_5$. At 10 K, $Eu^{3+}$:$Y_2SiO_5$ can store over $10^6$ bits per spatial location for several hours. The storage material size will be on the order of 75 x 75 x 1 $mm^3$. The heat load on the sample due to the data storage and reading processes will be less than 100 mW. Due to the requirement of optical access to such a large sample, the heat load due to radiation through ordinary optical windows would probably be significantly higher. Fortunately, the wavelength range over which optical access is required is very narrow (< 1 nm in the visible). Therefore, it will be possible to use windows designed as high reflectivity mirrors at all other regions of the spectrum, thereby greatly reducing the thermal load due to radiation. Because the storage density is a strong function of temperature, it is important to control the temperature of the entire material. In a cold finger type arrangement, it would be

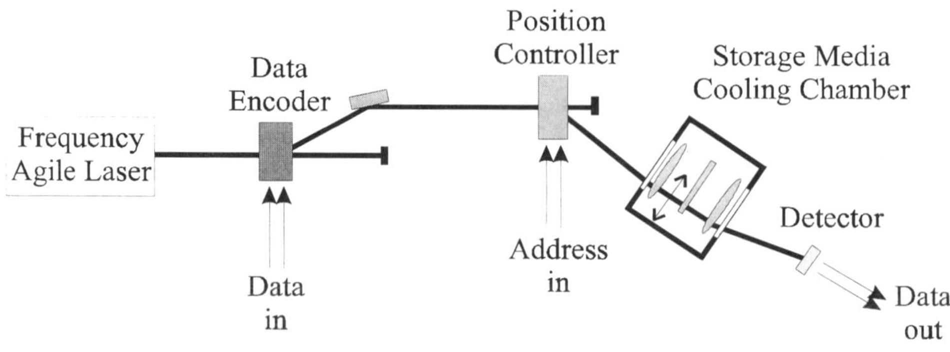

Figure 3. Optical subsystem for ODRAM

necessary to rely on the thermal conductivity of the crystal to transfer the excess energy to the cold finger, or to contact the crystal with another optically clear material that has a significantly higher thermal conductivity. Another possibility would be to house the sample in contact with cold helium gas used as a thermal conductor. In addition to issues directly related to cooling, the refrigeration unit must also have a long MTBF (mean time between failures) and minimal mechanical vibrations. (Note that minimization of mechanical vibrations is necessary to distinguish separate spatial locations whose size is of the order of 10-20 μm. Interferometric accuracy is not required in this application).

## COMPARISON TO EXISTING MEMORY TECHNOLOGIES

Modern computers contain a mix of several types of memories, each optimized to perform a specific function. Semiconductor memory is fast, volatile and expensive. Mechanically accessed memories, such as magnetic hard drives, are relatively slow, but have the advantage of low price and nearly permanent data storage. The large difference in cost and speed between semiconductor memory and hard drives provides the technological and business opportunity that ODRAM is designed to exploit.

In Figure 4, we compare average data transfer rates provided by ODRAM, magnetic hard drives, and silicon DRAM in a hypothetical application wherein a large number of random data records are accessed. It is assumed that the data within each record is contiguous but subsequent records are not. Thus the time required to transfer one record using a specific technology consists of the technology's latency time plus the time needed to transfer the record's contents. The average transfer rate, vertical axis, is the number of bits per record divided by the time needed to transfer each record. The horizontal axis shows the number of bytes contained within single records. Maximum record size in the various technologies is not accounted for in Figure 4. In the case of silicon DRAM, a 64 bit bus aggregate rate is used. Performance for ODRAM is shown for 1 bit serial and 32 bit parallel implementations at 30 Mbit/s single channel bandwidth. Also shown is projected late stage ODRAM-2 development employing 64 bit wide access and 1 Gbit/s single channel bandwidth. The magnetic hard drive is assumed to have a 100 Mbit/s single channel data bandwidth.

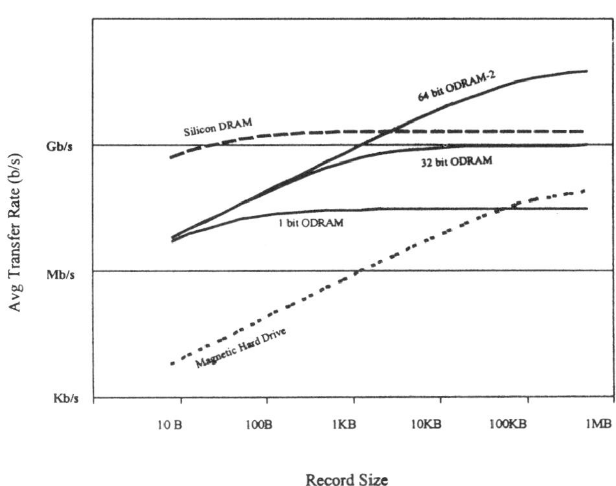


Figure 4. Average transfer rate as a function of record size.

Average transfer rate is a very important parameter. Any specific application scenario will be characterized by a required average transfer rate for a characteristic distribution of record sizes. For small records, 1-bit-wide ODRAM outperforms magnetic hard drives by approximately 3 orders of magnitude. As record size increases, the relative advantage of ODRAM decreases, but persists to record lengths of 50 kBytes. The performance advantage of 32-bit-wide

ODRAM persists throughout the entire range of record sizes considered. Late stage ODRAM (ODRAM-2) with 1 Gbit/s single channel bandwidth and 64-bit-wide access, outperforms even silicon DRAM for record sizes larger than a few kBytes. The data shown in Figure 4 shows that there is a considerable range of operational parameters where ODRAM offers dramatic improvement over magnetic hard drives.

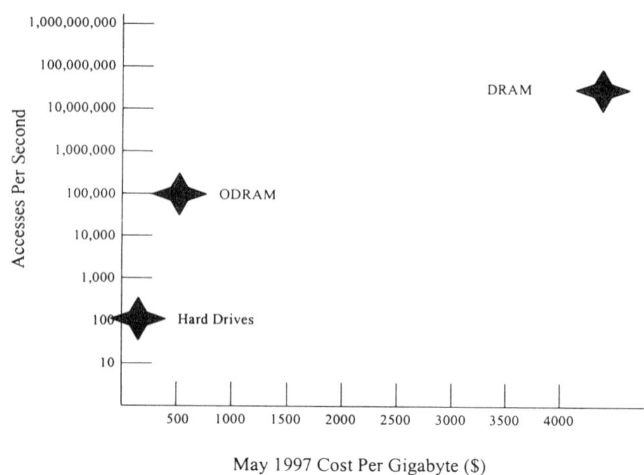


Figure 5. Number of small records that can be accessed per second for different technologies as a function of cost per unit capacity. Prices depicted are $4400 per Giagbyte for DRAM (Dataquest, May 21, 1997), $500 per Gigabyte for ODRAM (estimated 50 Gigabyte unit at an entry level price of $25,000) and $110 per Gigabyte for Hard Drives (Disk/Trend 1996).

The performance advantages of ODRAM relative to disk-based storage devices implies that the price of ODRAM does not have to match the price of magnetic hard drives. Figure 5 shows how the performance of ODRAM compares to magnetic hard drives and semiconductor memory as a function of cost per unit capacity. ODRAM offers a performance increase of 3 orders of magnitude for a factor of 5 in price. In comparison, semiconductor DRAM offers an performance increase of 5 orders of magnitude relative to magnetic hard drives for a factor of 40 in price.

## OPPORTUNITIES

The worldwide markets for data storage devices are enormous. For example, the worldwide sales of hard drives in 1995 was $26.7 billion[7] and the worldwide sales of silcon DRAM in 1996 was $25.9 billion.[8] As a new technology, ODRAM will capture a small portion of these markets. It is important to note, though, that ODRAM is not designed to replace any existing type of memory, but rather offer performance levels that will augment overall system performance when used in conjunction with existing memory devices.

ODRAM is expected to enhance those applications that are currently limited by the amount of fast memory available. In particular, ODRAM will allow for fast access to large databases, facilitating faster searching and manipulation of these same databases. End-users that will benefit from improved utilization of large databases include banks, airline reservation centers, credit card companies, telemarketing operations, sales departments, inventory tracking systems, internet and intranet servers, search engines, networked computing systems, *etc.* ODRAM will also be used as a high capacity, low latency swap space to accelerate and/or improve the convergence of scientific and engineering calculations, including computational chemistry and physics, rational drug design, atmospheric and meterological modelling, fluid and aerodynamic analysis and design, *etc.* Clearly, these applications are not desktop PC type applications. The initial users of ODRAM technology will be power hungry users with high-end workstations, mainframes, and servers. These users are willing to pay a premium for performance and they don't care "what's in the box" as long as the performance matches the specifications printed on the outside of the box.

Currently, large databases of the type that will benefit from storage on ODRAM devices are commonly stored in RAID systems, file cabinet size systems that are composed of large numbers of magnetic hard drives. Therefore, assessment of the RAID market, which currently targets database, client/server, and network computer applications will provide a more accurate indication of the potential market for ODRAM technology than the larger silicon DRAM and hard drive markets. In 1995, RAID systems sales reached $9.8 billion[9] and sales are expected to continue increasing rapidly, reaching $18.6 billion in 1999.[9] Capturing just over 1% of the value of the projected 1999 RAID market in our product entry year would correspond to first year sales of $187 million or 7500 units at $25,000 per unit. This assumption seems easily justifiable, given the dramatic performance enhancement in data access offered by ODRAM relative to RAID systems. Note that we cannot address the entire RAID market with our initial product entries because many RAID users purchase RAID systems with multiple redundant components to improve data security relative to hardware failure.

Cryogenics will be an important subsystem in commercial ODRAM devices. There are many challenges and opportunities for innovation in refrigeration technology, temperature sensing, temperature control, and many other aspects of cooling technology for application to ODRAM. As a short term cost-saving solution, we envision running multiple ODRAM devices with a single compressor. Projected into the future, we envision cryogenically equipped buildings, as depicted in Figure 6. Such a building would have one or two large compressors in the basement, and helium lines terminated with quick-connect nipples in other portions of the building. When users set up their computers, they plug a cable into the wall for electrical power, another cable into the wall for communications and a couple of hoses into compressed helium supply and return lines for device cooling. Obviously, once such a system is in place, the compressed helium would be available for cooling other computer components, including processors and silicon based memory.

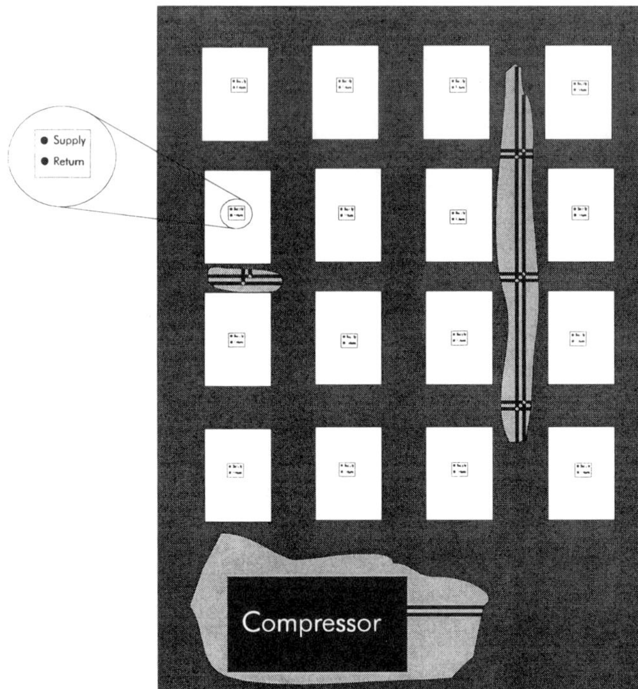

Figure 6. The cryogenically equipped office building.

## CONCLUSION

Optical dynamic RAM will enable higher performance computing through a dramatic improvement in data access rates. The potential markets are very large, and therefore the potential markets for the cryogenic subsystems are also very large. Initial systems will be individually cooled. Future developments may include cryogenically equipped office buildings, allowing the cold heads to be "powered" from wall outlets.

## ACKNOWLEDGEMENT

Partial support for the ODRAM development effort was provided by the BMDO SBIR program, under contract number F49620-97-C-0036, administered by Dr. Alan Craig of AFOSR.

## REFERENCES

1. W.E. Moerner, ed., "Persistent Spectral Hole-Burning: Science and Applications," Springer-Verlag, Berlin (1988).
2. T.W. Mossberg, Opt. Lett., 7:77 (1982).
3. T.W. Mossberg, Swept-carrier time-domain optical memory, Opt. Lett., 17:535 (1992).
4. H. Lin, T. Wang, G.A. Wilson, and T.W. Mossberg, Experimental demonstration of swept-carrier time-domain optical memory, Opt. Lett., 20:91 (1995).
5. H. Lin, T. Wang, and T.W. Mossberg, Demonstration of 8-Gbit/in.$^2$ areal storage density based on swept-carrier frequency-selective optical memory, Opt. Lett., 20:1658 (1995).
6. R.W. Equall, Y. Sun, R.L. Cone, and R.M. Macfarlane, Ultraslow optical dephasing in $Eu^{3+}$:$Y_2SiO_5$, Phys. Rev. Lett., 72:2179 (1994).
7. Disk/Trend 1996.
8. Dataquest, April 1997
9. Disk/Trend 1996.

# CRYOPACKAGING FOR REAL WORLD PRODUCTS

G. Lehmann, J. Ramsden, J. Sochor, and G. Beeck

Conductus
969 West Maude Avenue
Sunnyvale, CA, 94086

## ABSTRACT

Wireless products employing high temperature superconductors (HTS) and cryoelectronics are being sold in the commercial and military marketplace. Product requirements for reliability and cost are driving the configuration of the cryopackage. Engineering challenges for vacuum life, cryocooler reliability, cryoelectronics reliability, and HTS device reliability must be met within cost goals for the products to be viable in the long term. This paper discusses requirements and design approaches for wireless device cryopackaging.

## INTRODUCTION

High-temperature superconducting (HTS ) passive components (microwave filters) on the receive side are beginning to be employed in commercial and government wireless communication systems. The benefits of HTS filters combined with cooled low noise amplifiers (LNA) are associated with reduced noise figure and sharper filter skirts. In commercial base stations, this translates into larger range, increased selectivity, reduced size and weight for tower-mounting, and the potential for more channel capacity. In government systems, larger range and increased out-of-band rejection lead to clearer reception.

HTS circuits can deliver greater performance advantages in electronics applications if they can be packaged into a small volume to leverage the weight of the cooler over more components, provide immunity to vibration related de-tuning of the circuits, and maximize the performance gain allowed by their low power consumption. Both commercial and military applications in wireless communications favor operation relatively close to the material's transition temperature in order to reduce cooler costs. In that range, the electrical properties are a strong function of their temperature, hence uniform cooling and tight temperature control are essential for high performance

Significant progress has been made in the cryocooler industry in reducing size and cost and enhancing the reliability of cryogenic systems. As applications for cryocooled electronics, including HTS devices, emerge in the military and commercial marketplace, engineers are confronted with a series of tradeoff issues associated with the selection of the appropriate cryocooler and cryopackaging configuration.

Conductus' experience in commercializing HTS filters for the receive side indicate that the thermal design will have a large impact on the electrical design of the filter. Therefore packaginging considerations such as package material, package emissivity, parasitic heat loads (RF interconnects and dissipated heat) and radiation heat load must be taken into account.

## CRYOPACKAGING REQUIREMENTS

As the HTS microwave filters gain more and more market acceptance by commercial and military wireless customers, performance to specification of the cryopackage becomes increasingly important. There are three classes of application: commercial base station, commercial tower mount, and military avionics or surveillence. The most friendly environment for the cryopackage is the commercial base station installation such rural and urban cellular. Most severe for the cryopackage are the commercial tower mount and military applications. The base station applicatons generally have moderate ambient temperatures and units are in a protected, climate controlled building. Tower mount and military applications operate under severe environmental conditions. High ambient temperatures (>60°C), diurnal thermal cycles, wind loads, rain, snow, pollutants must be addressed in the design. Military applications have the added requirement of being portable and capable many startup cycles. Long term vacuum life and cryocooler reliability are key. Other requirements include stability and strength; Cool down time; Cooling Power at end of life; RF and electrical characteristics; and last but not least cost goals to compete in the market place.

### Vacuum life

At Conductus, our commercial wireless products are designed for 5 year vacuum life. End of life vacuum pressure is generally accepted as 0.0133 Pa ($10^{-4}$ torr). To achieve this requirement, careful selection of materials, adequate manufacturing processes, gettering, and good vacuum design practices are necessary.

### Cryocooler Reliability

Cryocooler reliability is similar to the vacuum life requirement. The cryocooler must operate continuously for 5 years without failure or maintenance. Reliability data from the cryopump industry suggests that mean time between failure of 80,000 hours are achievable for GM and Joule Thompson cryocoolers although some maintenance may be required to achieve this (oil adsorber replacement). Theoretical lifetimes for free piston Stirling cryocoolers using gas bearings and/or flexures or Pulse Tube cryocoolers are also of this magnitude, but do not have the operating history in numbers of units to back up lifetime claims.

### RF and Electrical Performance

Good RF and electrical performance and cryogenic design often conflict during cryopackaging design. RF interconnects from the warm vacuum vessel to the HTS device want to be long with small cross-sectional area for good thermal performance but short with large cross-sectional area and good surface conductance for low insertion loss. Allocation of the overall RF performance (insertion loss) and thermal performance (heat leaks, dissipated heat) are made to individual components of the cryopackage. DC interconnects must strike a balance between electrical resistivity and thermal conductivity to achieve acceptable heat loads. Cross talk between interconnects must be held to minimum.

### Thermal Performance

Thermal performance requirements include static heat loads at high ambient operating temperature, cryocooler end of life cooling capacity at high ambient operating temperature, cryopackage thermal mass, and cryopackage thermal profile. The cryocooler end of life cooling capacity at high ambient operating temperature is allocated to the individual components of the cryopackage such as RF interconnects, thermal radiation to the cryopackage, and dissipated heat in the cryocooled electronics (low noise amplifiers). For commercial markets, cool down time has not appeared as a requirement. However, for the military markets, cool down time is a requirement and can vary from minutes up to an hour. Cool down time requirements flow down to the cryocooler lift capacity and packaging materials and interfaces. An HTS device needs as near an isothermal environment during operation as possible. Temperature differences across the HTS device impact device performance.

### Structural Issues

Structural requirements for the cryopackage fall in two categories, thermomechanical loads (thermal shock, differential thermal contraction) and environmental loads such as transportation, handling, and shock loads. Thermomechanical loads are generated during cool down and operation and can impact device survivability depending on the cool down time and material property mismatches between components. Environmental loads frequently conflict with thermal performance requirements. Low stress or high stiffness do not in general go with low heat leak along structural members.

## CRYOPACKAGE DESIGN APPROACH

Many of the design challenges presented by cryopackaging of HTS devices for wireless application have already been addressed in one form or another by industry. The primary sources are from the military applications such as infra-red detector dewar, physics research applications such as accelerator technology, and the semiconductor industry. From these technology areas a design (and manufacturing) approach is developed.

### Vacuum Life

At Conductus, vacuum life is addressed in a three areas: good vacuum design practices (material selection, no virtual leaks, cleanable materials, etc.), gettering, and manufacturing processes. The cryopackage design has to be compatible with the vacuum requirement. Surface finish of metallic components such as the vacuum vessel and HTS device package is tightly controlled and held to minimum depending on cost so that outgassing of these components is minimized. Organic materials such as epoxy (some vacuum grades are acceptable), polyimides, and Teflon are avoided due to their propensity to outgass and absorb water. The number of fasteners used in the device packaging is held to a minimum or eliminated completely to minimize virtual leaks. If they are used, the threads are vented. Vacuum maintenance is provided by getters and/or activated charcoal. Getters have an advantage over activated charcoal in that it does not completely release any gas species back into the vacuum upon warming to room temperature and are regenerable. All vacuum components are cleaned to established cleaning procedures upon receipt and prior to storage. For example, ultrasonic bath followed by de-ionized water washes and inert gas oven bake for metallic components is typical. Vacuum components are generally stored in dry inert gas boxes. Elevated temperature bake out of individual components and the entire assembly prior to getter activation and final product shipment is also performed. Bake out duration and temperatures vary depending on the specific product. Bake out time and temperature are determined from accelerated life testing of representative test vehicles.

## Cryocooler Selection

For the commercial wireless market, Conductus uses the CTI Cryodyne Refrigerator. The Cryodyne is a Gifford-McMahon closed cycle cryocooler based on CTI M350 and 8F Cryopumps that are commonly used in laboratory, semiconductor, and vacuum industries with a wealth of operating experience to establish lifetimes. For the Cryodyne, the compressor and coldhead are separated allowing for distributed systems, remote operation, and siting flexibility. The end of life (5 years) cooling capacity is 6 watts at 60K. The Cryodyne is new design that is more compact and lighter weight compared to its precursors. Conductus products using the Cryodyne refrigerator have been operating at customer sites for about 1 year.

For the military markets, both the Cryodyne and Stirling cryocoolers are used. Stirling coolers are prefered when portability, low vibration, low acoustic signature, and lower cooling power are required. Conductus has units in the field that employ Stirling cryocoolers from CTI and Texas Instruments.

## RF Performance

Most standard RF components are not fully compatible with all of the cryopackage requirements such as vacuum life or thermal performance. For example, Teflon is very common dielectric present in most commercially available interconnects and coaxial cables. At Conductus, use of these materials is minimized or eliminated with custom proprietary components. Critical to the design of these components is maintaining impedance matching (50 ohm) and low insertion loss while insuring that thermal requirements are not violated.

## Thermal Performance

**Static Heat Load.** The static heat load from the cryopackage to the cryocooler can be broken down into four components: interconnect thermal conduction, radiation heat transfer from the warm vacuum vessel to the cold cryopackage, dissipated heat from the cryoelectronics, and residual gas conduction or convection depending on the vacuum pressure. For HTS microwave devices, there is a tradeoff between RF performance (insertion loss) and heat leak for the RF interconnects. The RF input interconnect wants to be short as possible for low insertion loss. Radiation heat load on the cryopackage is controlled two ways: gold plating on polished metal surfaces for low emissivity and multi-layer insulation where many HTS devices are packaged together. Multi-layer insulation or low emissivity heat shields are avoided due to negative impact on vacuum life (increased surface area). Dissipated heat from the cryoelectronics such as low noise amplifiers (LNA) and residual gas conduction/convection are generally minor players in the total heat load. The LNAs are low voltage and current devices with low dissipated heat.

**Thermal Profile.** The transient and steady state thermal profile of the HTS device and electronics are both very important to device reliability and performance. Although cool down times for the commercial products are not short (not considered to be a thermal shock), significant temperature gradients can exist during cool down. Steady state thermal profile of the HTS device can have a large impact on the device performance. At Conductus, we use a combination of finite element thermal and stress analysis using ANSYS to determine temperature gradients and resultant stresses of the cryocooled electronics during the design phase of product development. The HTS device and associated electronics are thermally cycled to insure device reliability during prototyping and production. Production cryoelectronics are tested and evaluated in a thermal environment that closely matches the actual product environment.

**Thermal Mass.** For the commercial market, thermal mass hence cool down time is not important right now. Cool down times range from 1 hour to 2.5 hours depending number of HTS devices being cooled. For our commercial products, we have more choices for material selection. Heat capacity and density are not as important allowing us to use low cost materials. For our military products, thermal mass is a major design constraint for the cryopackage and cryocooler. Cool down times of less than 30 minutes are typical of these applications.

## Structural Issues

The primary concerns on the structural side of cryopackage design are thermomechanical issues such as differential thermal expansion due to dissimilar materials or large thermal gradients during cool down and robustness of the cryopackage as a whole when it is not cold during transportation (shipping) and handling (and possibly operation for military products). Our approach is to match material thermal expansions as much as possible. Where epoxy or other adhesives are used in critical areas, the glue lines are kept to a minimum. Although immersion of our cryoelectronics in liquid nitrogen is not representative of the actual thermal environment the device sees during operation, it is a more severe test of the devices than repeated thermal cycling using the actual cooling device and arrangement of the cryopackage. Verification that the cryopackage is robust enough to handle shipping and handling loads is done using finite element analysis to perform stress analysis for shock loading and power spectral density analysis for random vibration analysis during the design phase and thorough testing to maximum environments during system qualification testing.

## SUMMARY

Cryopackage requirements have been presented and discussed. Our approach at Conductus to meeting those requirements has been discussed and key issues and approaches presented. Units designed and built at Conductus to the above requirements have been sold to commercial and military customers and currently operating in the field. Final verification of that all requirements will be met such as 5 year vacuum life and cryocooler life is ongoing in the field and at Conductus' facilities in Sunnyvale, CA.

# PACKAGING AND INTEGRATION ISSUES FOR CRYOELECTRONIC AND SUPERCONDUCTOR MATERIALS

T.H. Clynne

Infrared Components Corporation
811 Court Street, Utica, New York 13502 USA

## ABSTRACT

Device technology has matured to the point where necessary levels of performance in the cryoelectronic and superconductor components have been achieved, but now the limiting factor in producing products is the lack of high reliability, affordable packaging which allows the end user to treat the cryoelectronic subsystem as a procurable commodity. General packaging issues relating to the design, fabrication and integration of systems utilizing cryoelectronic and superconductor materials will be discussed by drawing upon historical work done in the Infrared field. System level performance topics will include limiters of thermal performance, enhancement of vacuum life and analysis methods, electrical & interconnect issues. Effects of environmental constraints and prediction tools will be reviewed. Problems to be anticipated when integrating devices into systems and methods of ensuring device survivability will be discussed.

## INTRODUCTION

In recent years, significant improvements have been made in basic and device technologies in the field of superconductivity. Performance requirements once considered to be only theoretically possible have been achieved, thereby enabling the application of superconductor materials and components into a wide variety of electronic and mechanical systems. This cycle of technology development is identical to most areas of advancement, and can be correlated directly with that of the innovations and accomplishments in the Infrared (IR) field. Initial attention is generally paid to the development of the device science and basic functionality, and measures relating to production and manufacturing are generally left to solve as a last requirement to gain commercial acceptance. The packaging and integration of the superconductor device into a product has been left as one of the last hurdles to overcome, but as the military Infrared (IR) technologists have found, packaging issues and the constraints associated with providing a long life product are not trivial, in fact, they can cause a reverse developmental effect, causing the redesign and possible iteration of the basic superconductor components themselves. The fundamental requirement of the packaging is that it be able to provide a reliable, efficient means of achieving the operational temperature of the device without degrading its performance characteristics. This is not limited solely to the function and performance of the cryocooler, but in fact can be attributed largely to the effect of the packaging and its ability to provide a controlled set of conditions for the cryocooler to operate under, namely careful control of the non-device related thermal loads and the means by which the device can operate unhampered by the constraints imposed by the packaging.

As previously stated, the IR industry has suffered through its share of packaging related product level issues and has been developing packaging techniques and materials for some 40 odd years, with the final solutions for some of the significant issues having been developed in only the past 5 years. Smaller scale military IR systems have been around for some time, but one of the major issues which prevented the commercialization and successful fielding of this technology were related to packaging. Figure 1 shows an example of ICC's Type II packaging, variations of which are currently used in a majority of the world's commercial staring IR systems. This type of packaging takes advantage of a standard set of design tools and common components which enable the rapid creation of new designs and configurations without having to go through the iterative testing and qualifications normally done to new products. As a case in point, ICC has built over 2500 of these systems over the past 7 years and has had no vacuum failure related returns.

The original IR systems were built in what was commonly referred to as a "slip-on" configuration, one whereby the device to be cooled was integrated into a sealed vacuum dewar, often manufactured from glass. This glass dewar integrated the device to be cooled onto the closed end of a tubular glass sleeve, or cold well, which allowed the dewar to be slipped onto the cryocooler cold finger, hence the name "slip-on". This was done primarily for two reasons. First, the glass dewars were manufactured along the same lines as vacuum tubes, with some of the basic components often built by some of the same companies which built millions of electronic vacuum tubes, and thereby allowed the producers of IR systems to rely heavily upon an established manufacturing capability and set of design tools and materials. Secondly, and perhaps most importantly, the cryocoolers of this generation of packaging had very short lives, often less than 250 total hours of operation. If the cryocooler failed, it could be easily replaced, thereby allowing

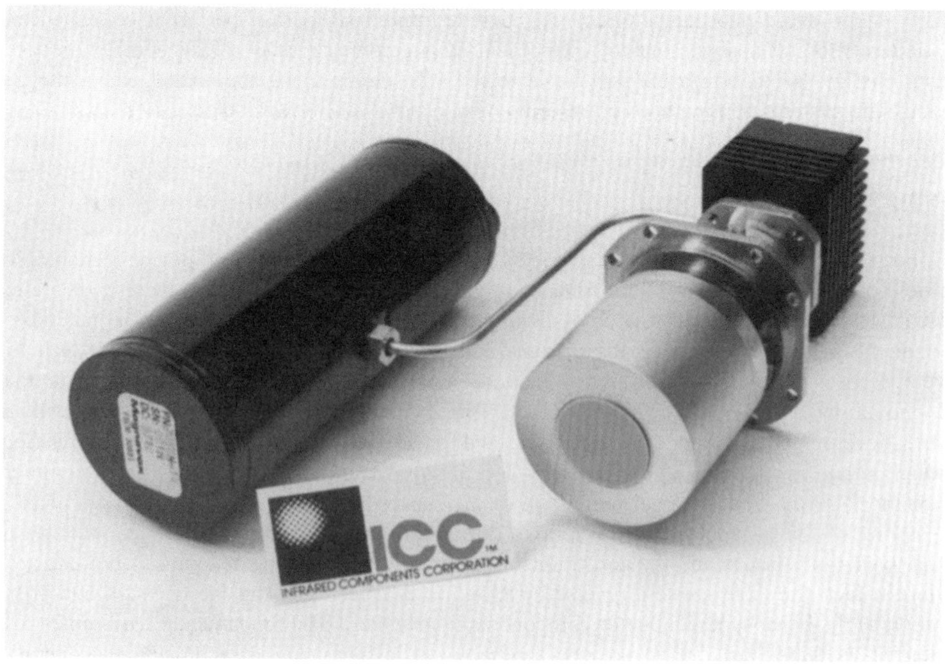


**Figure 1.** ICC Type II Integrated Detector Assembly.

the system to remain useful. While this sounds like an ideal situation, the additional thermal loads imposed by the glass cold well often consumed more than half of the available capacity of the cryocooler. Subsequent advances in cryocooler technology increased the expected lifetimes by up to 8000 hours, with extrapolated test data indicating potential lifetimes of 25000 hours or longer[1]. These advancements in cryocooler life allowed for the permanent integration of the cryocooler into the dewar as a structural and vacuum component, thereby eliminating the glass cold well and its associated thermal load. These latter day configurations are referred to as Integrated Detector Assemblies, or IDAs and constitute the great majority of staring class IR dewars built today.

## THERMAL PERFORMANCE ISSUES

The major role of the packaging in a cryocooled system, is to provide a high reliability, low thermal load environment. These thermal loads can be classified into three distinct categories, convection, radiation and conduction. Solutions to minimizing these losses are often interrelated, and one cannot ignore the subtle effect of a chosen material or configuration on the others.

### Convective Heat Loss

In a given system, convective heat loads can often be the largest. This type of heat load actually has two components, true convective heat transfer, where a substantial quantity of gas has accumulated in the dewar vessel to allow convection currents and gas conduction, where the density of the gas accumulated has not yet achieved a significant level to support mass flow, but instead, acts as a conductive thermal element. Gas convection and high losses from conduction typically occur at pressures above approximately $10^{-3}$ Torr. Pressures below $10^{-3}$ Torr are generally considered to be a thermally insulating vacuum and the remaining component of heat transfer is a linear relationship to the pressure of the gas. This conductive component is quite small, often less than a fraction of a percent of the potential convective load, and still statistically insignificant when measuring overall heat loads in the system. It is the achievement of pressures lower than $10^{-3}$ Torr that one strives for throughout the life of the product to be manufactured. The equation for convective heat load takes the form:

$$q = Ah(T_H - T_L) \tag{1}$$

Where A = Area of heat transfer surface, h = convective heat transfer coefficient and $T_H$ & $T_L$ are the respective temperatures of the surfaces exchanging heat. Determination of the convective heat transfer coefficient is rather complicated as it is a function of the fluid thermodynamic and transient properties of the gas and the geometry of the system. For a given configuration, one will strive to keep pressures thermally low enough, such that the resultant heat flow from this convection component can be negated. Methods of measurement of this component of load will be covered in a later section.

### Radiative Heat Load

The component of heat load associated with that of broadband radiative heat transfer is a function of the emissivities of each surface involved, their respective areas, and most importantly, their absolute temperatures raised to the fourth power. For a given system, the most profound way to minimize this component of load is to reduce the emissivity of one or both internal heat transfer surfaces involved, such that the equation of heat transfer no longer takes the form of a pure blackbody radiator, but instead takes one of the myriad forms of graybody radiation. Graybody radiation takes into account the differing emissivities of the hot and cold bodies in a system, as well as their spatial relationship to each other. These equations are well known and documented, but generally, the graybody radiative heat transfer component in a system can be simplified as that of heat transfer between concentric cylinders or some other simplified geometric relationship. Equations for these geometric relationships and forms of graybody heat transfer equations are many and can be found in any good text on heat transfer. Since the effect of the net temperature relationship between the two bodies under consideration is a function of the fourth power, the amount of heat transfer by increasing the ambient temperature from room (300K) to a higher ambient of say 71C (344K) can cause a corresponding 70+% increase in the amount of heat transferred from radiation. This consideration needs to be taken into account when specifying cryocooler capacity, as well as the net reduction in cryocooler capacity that generally occurs at higher rejection temperatures.

### Conductive Heat Load

Conductive heat load is that associated with direct conduction through a material from a hot source to a cold object, in the case of a cryocooled system, between the device being cooled and the outside of the dewar vessel, which is generally at or above the ambient temperature of the remainder of the system. Conductive heat transfer takes the form:

$$q = Ak(T_H - T_L)/L \tag{2}$$

Where A = cross sectional area of the conducting member, k = thermal conductivity of the material, L = length of the conducting member and $T_H$ & $T_L$ are the temperatures at each end of the conducting member. This is an area where a knowledge of materials and system level requirements is quite helpful. Often a material considered good for a room temperature system can cause a performance degradation in a cryogenic system. A typical example being that of the electrical leads in and out of a device. Room temperature system designers will often desire a high electrical conductivity, low impedance lead for interconnection purposes, but when traditional materials such as copper are considered in these applications, the associated thermal losses are generally too large. Knowledge of the joining processes available, their inherent manufacturability and reliability of the joints produced are also critical factors to consider when designing a cryocooled system. The thermal and electrical conductivities of a given material are related by the Wiedemann-Franz law:

$$k/\sigma T = \text{Constant} \tag{3}$$

Where k = thermal conductivity, $\sigma$ = electrical conductivity and T is the temperature under consideration. The thermal conductivity of most metals is proportional to the temperature up to about 60K, while for most dielectrics they are proportional to $T^3$. Above 60K, they will have a decreasingly lower value and dependence upon temperature. The thermal conductivities of alloys are generally at least an order of magnitude smaller than that of a pure metal and according to Eq. (3), there is a corresponding decrease in electrical conductivity as well.

## VACUUM LIFE

The issue of vacuum life and the methods to achieve substantial vacuum life are varied and complex. As stated earlier in the section under thermal convective heat loss, the maintenance of a good vacuum is absolutely critical in creating a reliable product. Long vacuum life is a complex function, relating materials science, processing and mechanical design guidelines into a resultant formulation for which there is a strict measure of it's success, that being the maintenance of vacuum pressure under $10^{-3}$ Torr. The ultimate lifetime of a dewar has less to do with the pinch-off pressure within the dewar when it is processed and more to do with the actual leak rate of gas into the dewar. For example, if a dewar is pinched off at $10^{-8}$ Torr, the amount of time to accumulate enough gas to cause the pressure to rise from $10^{-8}$ Torr to $10^{-7}$ Torr is only 10% of the time necessary to rise from $10^{-7}$ to $10^{-6}$ Torr, and likewise with each successive order of magnitude increase in pressure. When an end of life pressure of $10^{-3}$ Torr is desired, the time to rise from $10^{-4}$ to $10^{-3}$ Torr will consume almost 90% of the entire lifetime.

The units for leak rate of a specimen are expressed in Standard Cubic Centimeters per Second of Helium (SCCS He). Helium is usually chosen as a leak detecting gas due to it's small atomic diameter and chemical inertness. Leak rates are measured by pulling a vacuum on a test article and passing the gas pumped out over time across a Helium mass spectrometer. The highest sensitivity leak detector available today can detect leaks as low as $10^{-11}$ SCCS He, but when one calculates the leak rate required to ensure long product life, leak rates are needed which cannot be detected. This seemingly "no win" situation is eased by the use of vacuum maintenance components, such as getters and ion pumps, which will be covered in a later section. Vacuum loss comes from a combination of sources, real leaks and virtual leaks, details of which will be discussed below.

**Real Leaks**

Real leaks in a dewar are leaks into the dewar from the ambient surroundings. Typically these take the form of cracks in glass feedthrus, poor weld or solder joints, and other sources of physical damage which can create a gas pathway in any one of a hundred ways. Very small, undetectable leaks often appear in materials which are due to the permeation of gas through the bulk of the material itself due to microscopic pores or other defects within the components of the system. Plastics, rubber o-rings and epoxies are a common source of permeation style leaks, as all of these class of materials will permeate gas above a level which will eventually cause vacuum failure. Here again, the careful selection of the materials and knowledge of their joining processes is critical when designing a system. An understanding of the ambient conditions under which the system will operate must be taken into account, due to the buildup and potential relief of stress between dissimilar components. A system will often function quite well at room temperature, but when exposed to higher or lower temperatures, significant failures can occur, and many times, only after the system has been exposed to many cycles of heating and cooling. Typical joint materials such as solder, which maintain a degree of ductility after joining can cold flow out of a joint over time if the materials being joined are dissimilar enough to develop stress as a result of heating or cooling. As most joints are created using thermal processes such as welding, oven curing or solder reflow, the zero stress condition for a joint may be hundreds or even thousands of degrees above the ambient condition under which it is often tested. It is for this reason that low ambient conditions will likely cause more leak related failures than higher ambient temperatures, since the amount of stress in a joint is being increased with lower temperature.

**Virtual Leaks**

The term virtual leak covers a broad category of leak related causes, but is meant to imply any source of gas pressure accumulation not attributable to a real, direct leak. Sources for virtual leaks include trapped gas pockets within the vacuum space of the dewar, the use of volatile materials in the manufacturing or processing cycles, and the evolution of dissolved gas from within the materials of the system. The use of screw threads and large surfaces in contact with one another are typical sources of trapped gas pockets. While it may appear that the gas within a screw thread can escape, low pressure gas flow is largely statistical, and the apparent path that a given gas molecule must travel to be pumped away is often quite long on a molecular level. Long, tortuous paths create many opportunities for a gas molecule to bounce off of a surface and be deflected toward the direction from which it came. The longer a system can be pumped prior to pinch-off the better it will be, due to a combination of outgassing rates and this problem of statistical gas flow. Gas pockets may also be trapped within a epoxy joint or solder joint. Careful outgassing of epoxies before use can help to alleviate this problem, but no epoxy can completely eliminate permeation of gas from within itself. The designer should take this into account when using epoxies in a system. The smaller and thinner a joint, generally the better, as a gas pocket of just .020" in diameter releases enough gas into a 6 cubic inch volume to render it above the critical pressure level of around $10^{-3}$ Torr. A good test to perform on a sample epoxy is to prepare it for use and form it into a disc, approximately 1 inch in diameter and 0.1 inch thick. After curing, snapping the sample in half and examination under a microscope should reveal no entrapped gas pockets. If any pockets are detected, these should be considered as sources of virtual leaks, and will eventually cause failure of the system.

Volatile materials within vacuum systems can be a very sinister cause for vacuum life failure. A given system can be built, leak checked and pass all initial tests, but over time may fail due to the evaporation of volatile materials within the dewar space. Volatile materials in this context include oils left behind from cleaning processes, finger prints and skin oil from assemblers, and, more importantly, the use of improper materials in the design. Oils from cleaning and handling can be eliminated by better processes and housekeeping on the part of the assembly workers, but when a given material has a high outgassing rate associated with it, nothing can solve this problem except to design with a different material. As a general guide, room temperature outgassing rates for metals are approximately $10^{-9}$ to $10^{-7}$ mbar-liter/sec-cm$^2$, whereas non-metals will outgas at rates approaching $10^{-4}$. A prime example of this misuse of materials can be found in the case of a RF system. RF designers have come to rely upon the use of Teflon and other organic dielectric materials, which when subjected to the low pressure conditions in a dewar, outgas at unacceptably high rates. This outgassing, combined with the trapped gas between the metal sleeve and dielectric material, and the gas within an expanded dielectric virtually eliminates the use of traditional RF cables and most organic dielectric materials in a long life dewar system. Designers must search for materials and processes that achieve equivalent levels of performance found with high outgassing rate materials, but take into account the unique requirements for application into a cryoelectronic system. Along this line, ICC has developed a unique cabling approach for use in RF systems. ICC's Taylor Feed has been designed with the performance requirements of a long life vacuum system in mind, while maintaining the high performance specifications found with traditional RF cabling techniques at thermal load levels less than 5% of that of a traditional system.

The use of prepackaged components can also be a cause for failure in a dewar system. These prepackaged components, a hybrid package for example, are packaged with a gaseous atmosphere inside a "hermetic" package. The word hermetic is in quotes, because it has two meanings in the industry. Hermetic to an electronics packager indicates an acceptable leak rate many orders of magnitude higher than that needed for use in a dewar system and hence is to be considered a source of trapped gas within the system. Adequate means for venting internal gas cavities during the final bake out process must be allowed, along with an analysis of the materials used within the package for potential outgassing related failures from epoxies and the like.

While this assessment of virtual leaks and their sources sounds a bit gloomy, the designer should take heart. There are a host of materials and processes which are available to solve these problems. The research of the past 40 years of IR systems has resulted in a wide variety of epoxies and organic polymers which will likely find great application in the field of cryoelectronics once the design tools have been disseminated into the existing body of systems design knowledge.

Vacuum life can be enhanced via the use of scavenging devices such as getters and ion pumps. Traditional gettering materials take two forms, cooled and thermally activated. Cooled getters such as charcoal can provide a measure of pumping capacity only while cold. When allowed to warm up, they will release the gas which they have adsorbed, and potentially ruin the vacuum within the dewar. Thermally activated getters provide a measure of extra security as they will retain the gas pumped under all conditions, except that their ability to pump certain species of gas may be limited and careful selection is necessary to ensure maximum performance for a given dewar configuration.

## ANALYSIS TOOLS

The prediction of vacuum life from a strictly numerical standpoint is a very difficult and nearly impossible thing to do. While outgassing rates for many materials have been measured, their application into an actual system, and the handling and processing prior to pinch off have a greater effect on outgassing rates than anything. Overall cleanliness of materials in a system is of paramount importance as the elimination of oily residues is very difficult to do strictly from a baking out process. Analysis methods established by the IR industry can provide information on potential product life. Residual Gas Analysis (RGA) has proven to be a useful diagnostic and in-process tool for the design and manufacture of dewars. The RGA is basically a multiple channel mass spectrometer and can detect minute quantities of gases by their atomic mass numbers. Partial pressures of gas constituents can be measured directly, and these can be converted to outgassing rates by a simple experimental process. If the dewar being tested is valved off from the pumping system and an RGA scan is performed in real time, the partial pressures of the gases being evolved will necessarily increase, yielding a gas accumulation, or outgassing rate. When the outgassing rate of the system when taken without the dewar on it is subtracted, a net gas accumulation rate can be calculated. Since pressure is related to volume and the number of mols of gas by the ideal gas law, a maximum allowable amount of gas which can accumulate can be calculated.

Higher ambient lifetime predictions can be obtained by using a similarly simple test method. Design guidelines for vacuum maintenance techniques such as gettering are often given at room temperature. Similarly, RGA testing and other test results are generally given at room ambient temperature. When one considers that the rate of a reaction approximately doubles with each 10 degree Celsius rise in temperature, we can simulate lower ambient lifetimes by elevating the temperature by using the following equation:

$$t_H = 2^n \tag{4}$$

$$n = (T_H - 25)/10 \tag{5}$$

Where $t_H$ = number of hours at elevated temperature H, $T_H$ = elevated temperature in Celsius. It is assumed that the normalized values of outgassing and other design criteria are taken at 25°C. As an example, a dewar designed for a 10 year life at 25°C can be simulated in test by exposing it to a 1946 hour (81 days) test at 80°C.

Tests for accelerating lifetimes limited by real leaks can be done as well. A simple method to utilize involves increasing the ambient pressure on the outside of the dewar, which causes a corresponding factor of increase in the leak rate. Using Nitrogen as the test gas, increasing the pressure to 100 pounds per square inch accelerates the life test by a factor of 6.8. A 5 year life test can be reduced to a 268 day test in this method, and corresponding reductions in test time can occur by using successively higher pressures.

In all of these test methods, failure detection from direct measurement of the internal pressure will often be difficult to do. Other test methods for measuring internal pressure can infer, a pass/fail criteria, the primary one being the boil off test. The unit being tested will generally have Liquid Nitrogen ($LN_2$) poured into the cold finger onto which the devices are mounted internal to the dewar and the rate at which the $LN_2$ boils will be measured. This boil off rate is measured at intervals during the test cycle and when the pressure reaches the critical number around $10^{-3}$ Torr, a sharp increase can be detected. The rate of gas evolution during the boiling off process can be converted directly to a heat load by factoring in the Latent Heat of Vaporization for $LN_2$ and the number of mols of gas being evolved per unit of time. A basic guide is that if the boil off rate increases by any significant amount (>10%) the unit can be considered to have failed. In those systems for which the boiloff rate cannot be measured by the $LN_2$ method, a similar test can be done by monitoring the power consumption of the cryocooler. A given cryocooler will follow a predictable power consumption curve as a function of the amount of heat applied at the cold tip. An increase in power may indicate an increase in the ambient heat load attributable to an increase in gas pressure. A word of caution needs to be applied however. There are other causes for the power consumption of a cryocooler to increase, from mechanical breakdown, to contamination of the working gas within the cryocooler, so the idea of using the amount of power consumed as a sole indicator of vacuum pressure must be considered carefully. A power consumption test when combined with another test, the warm up rate test, can begin to achieve some degree of merit. The warm up test measures the temperature of the device being cooled over time after the means of cooling has been removed. If a warmup test measurement is taken of a given system when it is new, and the internal pressure levels are known to a high degree of certainty, a test at later date can be compared with its baseline initial test. Any deviation from the initial curve indicates a problem. Too rapid a warmup rate indicates a higher parasitic heat load, likely due to increased internal pressure, whereas a slower rate indicates a problem with the thermal path to the device being cooled. A slow rate can indicate a mechanical joint between the cryocooler and the device being cooled has somehow degraded, providing a greater thermal resistance in the thermal network.

## ELECTRICAL PERFORMANCE

In a cryoelectronics system, the designer strives to create an environment and package such that the performance of the device being cooled is not degraded as a result of being packaged. This requirement is made more difficult when the range and type of materials that can be used are restricted due to thermal and vacuum packaging issues. Typically, a system designer will specify a low electrical resistance connection be made to the device being cooled, but this can be difficult to do when materials such as Copper and Gold are ruled out due to their high thermal conductivity, which if used, would impose too high a conductive thermal load on the system. The relationship between electrical and thermal conductivity was given in Eq. (3) along with general guidelines for it's use. There exist several alloys which have been developed specifically for interconnections at cryogenic temperatures that attempt to balance the thermal conductivity and electrical resistance. Platinum-Iridium, Manganin and Constantan alloys are several examples.

Electro Magnetic Interference (EMI) and crosstalk between signals must be considered when providing an interconnect network within a dewar. Longer wires and interconnections are generally used for thermal reasons, but if individual wires are used within a dewar and they are

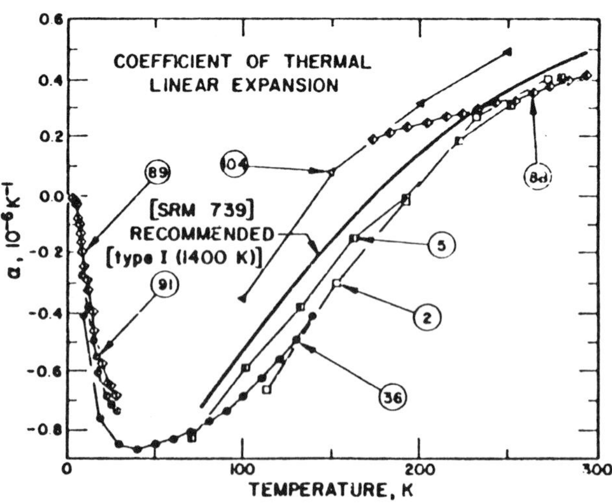


**Figure 2.** CTE chart for $SiO_2$ from 0K to 300K. See Reference 2.

allowed to run parallel to each other, crosstalk and interference can occur between signals. Additionally, the operation of the cryocooler can cause an effect on the performance of the remainder of the system. Linear Stirling cryocooler compressors are basically an AC magnet operating at a low (<60 Hertz) frequency and set up a large corresponding AC magnetic field during their operation. The electronic controllers used to drive them generally convert DC power to an amplitude controlled AC wave form, and as such, may inject large amounts of switching noise into the power buss of the system. Careful selection of these components and magnetic shielding can help to minimize the potential effect they may have.

## ENVIRONMENTAL ISSUES

As can be seen from review of previous sections, the environmental constraints under which a system must operate will dictate a great deal of the component level requirements. Environmental issues concern not only the temperature requirements of a system, but also include vibration and humidity. Military type environments have been used to develop a great deal of reliability and predictive tools, but many times commercial systems can impose a greater level of stress on a system, particularly high temperature and vibration environments. US military specification MIL-STD-810 can be considered an all inclusive type document, and includes a wide range of testing requirements. Of particular interest are the specifications for varying thermal ambients, called diurnal exposure requirements. Diurnal exposure scenarios are presented for a variety of applications and reflect the varying temperature a system will be exposed to during a typical 24 hour day. Compiling the diurnal requirements into an accelerated life test duration using Eqs. (4) and (5) above can be very useful when trying to qualify a system. Thermal shock and exposure limit test guidelines are also presented, as well as typical vibrational load profiles for a wide variety of applications. The actual requirements for a given system may vary, but this type of document can prove very useful as a guide.

## DEVICE SURVIVABILITY

The careful design of a dewar becomes a moot point if the operation of the system degrades or damages the device to be cooled. Most sources of failure can be traced to a co-efficient of thermal expansion (CTE) mismatch between adjacent or composite materials. Careful matching of the CTE values of materials must be taken into account, but of greater concern is the CTE matching analysis across the entire operational range of the system. CTE values tend to be non-linear across a wide range of temperatures, as can be seen in Figure 2.

Finite Element Analysis (FEA) techniques can prove very useful as an aid in predicting stress and design weaknesses in a particular configuration, but caution must be exercised in utilizing the ultimate values obtained to determine absolute survivability. The non-linearities, coupled with changing mechanical properties of most materials makes the absolute prediction of stress levels virtually impossible. Designs should seek to minimize mismatches and stress concentrations as a guide, but ultimate success can only be measured by building a representative system and subjecting it to the entire range of temperature values to be encountered. This can often be an involved process as it is critical to accurately simulate the operation of the cryocooler and subsequent components during test. Designers sometimes try to thermally shock a system by immersing it in $LN_2$, but this often times is not nearly as stressful as operating the system in its final configuration. The cooling of a device from a cryocooler is exceptionally non-uniform when considering that most systems remove heat from a single point or surface on the device, while leaving a great deal of thermal mass as yet uncooled and hence, at a higher temperature. Temperature differences can cause thermal stress even within carefully matched systems and often mechanical means such as sliding joints and unconstrained movement in one or more directions are adopted to relieve transient stresses built up during cooldown. Thermal stress will often relieve itself through a measure of bimetallic bowing, and can be detected via interferometric testing or careful dimensional measurement using other optical techniques. Many times the resultant stress in a system will be as a result of a component which may be connected indirectly to the device being measured. A good example of this is an often experienced source of electrical noise, sometimes referred to as microphonics. Microphonics are caused when an electrical device being cooled develops internal stress which is caused to change periodically as a result of the operation of the cryocooler. This piezoelectric stress can be difficult to remove and requires a very rigorous matching of the materials used to eliminate as much resultant stress in the device as possible.

## PROCESSING ISSUES

A fundamental requirement of vacuum packaging is that the dewar vessel be pumped down prior to pinch-off. This pumping down process is made more effective by elevating the temperature of the dewar to the maximum temperature allowable. This upper limit must be on the order of 20 to 30 degrees higher than the ultimate highest temperature the dewar will be subjected to in order to provide a significant margin of vacuum life. This is assuming that the system will be subjected to a typical diurnal type exposure, and is not planned to dwell for long periods at high temperatures. Ideal bake out temperatures are on the order of 100 to 105$^{\circ}$C for 7 to 10 days. The amount of gas evolved during this baking needs to be able to be pumped away by the vacuum system, and hence, it must be sized accordingly. When the bakeout temperature is limited below an ideal value, the duration of baking can be increased to provide a total equivalent number of ideal hours using Eqs. (4) and (5) as a guide. Bakeout is intended to remove surface contamination of a highly volatile nature, such as water vapor and those other adsorbed materials normally found in the atmosphere. Progress during the bakeout cycle can be ideally measured by taking periodic RGA readings of the gases being evolved, but at a minimum the total internal pressure of the dewar should be noted during the baking process. The internal pressure of a dewar will asymptotically approach the ultimate low pressure of the vacuum system doing the pumping, and when the heat is removed from the dewar and cooling takes place, a reverse pumping effect can actually occur caused by the cooler gas within the dewar vessel. The vapor pressures of some of the solid materials used for manufacturing the dewar must be considered when determining the highest acceptable temperature limit as their vapor pressures increase dramatically as a function of temperature.

## SUMMARY

A great deal of technology has been developed by the IR industry which is directly applicable to the packaging of cryoelectronics. This paper has attempted to provide a basic guide for the user to refer to when designing a cryoelectronics packge, but by no means is it all inclusive. There are many additional fields of technology and information available to the packager of cryoelectronics relating to cryocooler types, selection and reliability prediction models, again, largely derived from the IR field. Additional component sciences and techniques will necessarily have to be developed in order to truly create consumer level acceptance of cryoelectronic products.

## REFERENCES


1. G.R. Pruitt. Hughes Long Life Linear Stirlings: A Status Report, in: "Proceedings of the Eighth International Cryocooler Conference, held June 28-30, 1994 in Vail, Colorado," R.G. Ross, ed., Plenum Publishing, New York (1995), p. 113.
2. Author Unknown, taken from Linear Expansion of Silicon Dioxide (Fused), in: "Thermophysical Properties of Matter - The TPRC Data Series – Volume 13. Thermal Expansion - Nonmetallic Solids," Y.S. Touloukian, et al. ed., Plenum Publishing, New York (1977), p 359.

# A FULLY FUNCTIONAL CLOSED CYCLE CRYOSYSTEM THAT USES LESS THAN ONE WATT OF REFRIGERATION AT 4.5 K, FOR A 2.5 GHZ PER CHANNEL, 128 X 128 CHANNEL SUPERCONDUCTING SWITCH

E.L. Hershberg[1], T.J. Hendricks[2], D. Patelzick[1]

[1]Tektronix, Inc.
Beaverton, OR 97077
[2] Advanced Modular Power Systems, Inc.
Ann Arbor, MI 48108

## ABSTRACT

Tektronix has designed, built and characterized an extremely low loss cryosystem, incorporating a closed cycle cryocooler, that is capable of providing the thermal and magnetic environment required to operate a niobium Josephson junction based superconducting digital switch with 128 x 128 channels. The system is designed for serial data rates of 2.5 Gbps on a total of 1000 individual lines, necessitating a distance of only 15 cm from 300K to 4.5 K. Thermal radiation shields and independent magnetic shields, interleaved with optimized blankets of MLI, in combination with the conductive and dissipative load from the data I/O cables, deliver a 1.6 W heat load to the first stage of the cryocooler at 800 K. The 4.5 K cold finger incorporates a unique mounting stage that supports a ceramic multichip module and performs heat sinking for eight multi-channel I/O cables. The cable feedthroughs are demountable, and are assembled from inside the cryosystem.

The design of the cryosystem is optimized to the performance envelope of a single, closed cycle cryocooler, a Boreas B100. The design of the system is so precise, both thermally and mechanically, that primary data on thermal conductivity, yield strength and ultimate strength for certain critical materials were required to complete the design. State of the art CAD design and analysis tools were used to an extent that the system worked on the first assembly.

Characterization of the system using applied heat loads to mimic the presence of both I/O cables and a functioning MCM shows that the system meets the design goals without reaching the limits of the cryocooler.

## INTRODUCTION

As a participant in a federally sponsored program to develop a superconducting crossbar switch operating at 2.5 GHz, Tektronix undertook the task of designing, building and verifying the performance of a very high performance cryosystem to support the superconducting microelectronics. The electronics consist of a 128 channel x 128 channel superconducting crossbar switch, implemented in niobium Josephson junction technology, including both superconducting IC's on silicon, and a mixed, superconducting/normal metal multichip module (MCM) in multilayer ceramic. A key design requirement was that the system use a closed cycle cryocooler. The project therefore required fabricating and testing a uniquely-designed cryosystem, consisting of:

1. A cryogenic mount for the multichip module
2. A vacuum chamber with no more than 15 cm from the MCM to the chamber wall
3. Vacuum feedthroughs for eight I/O cables totaling over 1000 signal lines
4. Heat sinking system to cool the cables, shields and instrumentation
5. Thermal shielding
6. Magnetic shielding
7. Thermal insulation
8. Instrumentation
9. A closed-cycle cryocooler

## THE CLOSED CYCLE CRYOCOOLER

A Boreas B100 cryocooler was procured as the heart of the cryosystem. The remainder of the cryosystem was then designed to interface with the cryocooler. Characterizing the cryocooler along with the cryosystem was a critical part of the program plan for verifying the cryosystem's thermal and mechanical performance in the electronic cooling application. The Boreas B100 cryocooler was selected as the best cryocooler option for the cryosystem based on several selection criteria. These included the following factors:

1. First and third stage heat load capacity and temperature as a function of the load
2. Air-cooled compressor .
3. Low vibration characteristics.

Boreas, Inc., now a Division of Hypres, Inc., performed extensive testing on the Boreas B100 cryocooler during its development in order to establish its expected performance envelope. This performance was verified by Tektronix. Figure 1 and Figure 2 show the cold head heat load capacity vs. cold head temperature and cold head heat load capacity vs. first stage heat load, respectively, as tested by Tektronix. The cooling design requirement is close to 1.0 watt cooling capacity at 4.5 K, with a simultaneous first stage heat load of 2.0 - 3.0 watts. The data displayed in the two figures show that the Boreas B100 cryocooler has the thermal performance capability to satisfy the cryogenic system cooling requirements.

In these tests, the cryocooler took 16 hours to cool the cold head to the 4.2 K level from room temperature. The Boreas cryocooler was allowed to thermally equilibrate for a minimum of 2 hours at each individual test point to ensure steady state operation.

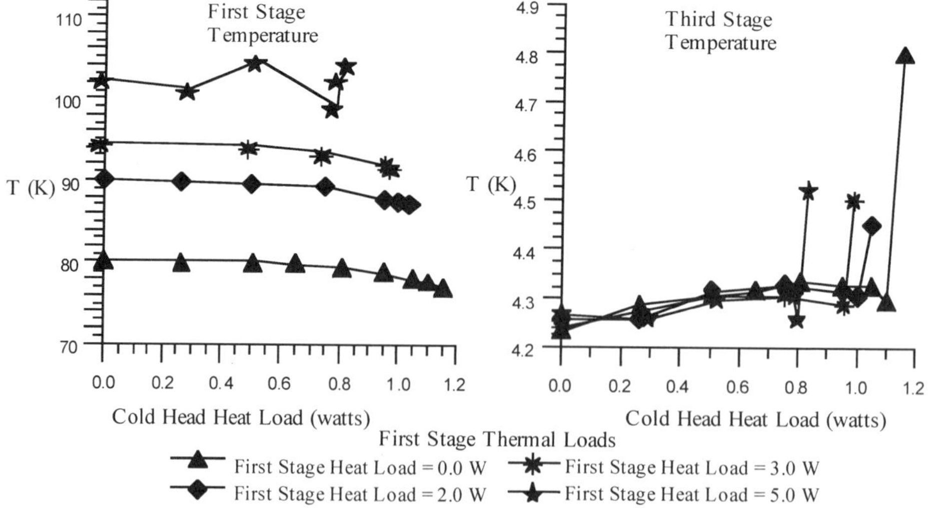


**Figure 1.** First stage temperature vs. cold head and first stage heat load.

**Figure 2.** Cold head temperature vs. cold head and first stage heat load.

## CRYOSYSTEM DESIGN

### Vacuum Chamber

The size and shape of the vacuum chamber is constrained by the length of the I/O cables, the size of the cryocooler and the need to bring eight flat flexible multi-channel cables onto the multichip module at 45 degree increments. The chamber's outer diameter is 457 mm (18 inches) and it's height is 353 mm (14 inches). The primary components are shown in Figure 3: the base, the mid chamber, an octagonal docking ring and a top plate. The base supports the entire system and is itself supported around its periphery. The Boreas cryocooler penetrates the bottom of the base through an ISO 100 flange and is held in place with clamps. Eight NW40 flanges are arranged around the bottom surface of the base. These provide feed through ports for instrumentation, heaters, vacuum gauges, vacuum pump out, and nitrogen purging. The octagonally shaped docking ring provides eight ports for the data cables. These ports are designed to allow the cables to be inserted or removed from either the interior or exterior of the vacuum chamber.

### Heat Load

The most significant heat load on the two cryocooler cold stages is from the eight I/O cables. These loads are the result of thermal conduction along the cables and joule heating in the cables. The conduction heat load has two major paths: from room temperature to the first heat sink point on the 80 K shield, and from the 80K shield to the 4.5K heat sink adjacent to the MCM. Cables with high thermal resistances are produced by using high thermal resistance materials with small cross sections. Joule heating loads occur in the electrical conductors of the cable as a result of current flow and ohmic resistivity. The combination of high thermal resistance with joule heating along the cables produce large thermal gradients that must be considered when the data cables are designed to fit this system.[1] The following cable related heat loads were used for this cryosystem design: 1500 mW at the first stage (80 K shield), 600 mW at the third stage (4.2 K coldhead). There is also a load of 25 mW on the MCM due to dissipation in the superconducting IC's.

### Thermal Radiation Shields

The thermal radiation shielding system is crucial to the performance of the cryosystem. Without proper shielding it is doubtful that the third stage could even reach 4.5 K. Two types of shielding are used: active and passive. Thermal network analysis of the shielding system showed that the active thermal shields needed to behave as isothermal entities. Thus the thermal and magnetic shielding could not be combined in one shield. The active shield, the 80 K radiation shield, is fabricated from 2.4 mm thick (0.093 inch) OFE[2] copper and is plated with gold to reduce the surface emissivity. The bottom surface is attached to the first stage (nominally 80 K) of the Boreas cooler with a copper interface ring. This shield also serves as a heat sink for the eight data cables. Using measured values of the copper conductivity, the shield has a calculated resistance of 0.22 K/W from the cable attachment ring to the first stage. The locations of the active shields are shown in Figure 4, which is a cross section of the complete cryosystem.

The passive shields are fabricated from multilayer insulation (MLI). There are two sets of passive shields. The first is located between the room temperature and 80 K magnetic shields and has approximately 50 layers. Thermal modeling predicts that the resulting heat load onto the active 80 K radiation shield is less than 0.3 W. The second passive shield is located between the active 80 K radiation shield and the MCM/cold head assembly, located on top of the third stage of the Boreas cryocooler. This shield extends down along the length of the cold head to insulate the section of the cryocooler between the cold head and the first stage. In comparison to the first MLI shield, there is much less thermal radiation between the 80 K shield and the third stage, so the second passive shield has only eight layers of MLI. The estimated radiation heat load on the third stage is 2.0 mW.

### Magnetic Shields

To attenuate both the earth's magnetic field and fields generated by the cryocooler and other machinery, a set of three magnetic shields were designed by another contractor. The location and orientation of the magnetic shields are shown in Figure 4 and Figure 5. They are concentric enclosures located within the vacuum enclosure. Each has a removable lid that incorporates pass-through slots for each of the eight data cables. The first magnetic shield, located immediately inside the vacuum chamber, is effectively at room temperature. The second shield surrounds the active 80 K radiation shield and is mounted on the same copper interface ring that connects the 80 K thermal shield to the 80 K cryocooler stage. The third shield is inside the 80 K radiation shield and surrounds the MCM. It is connected to the 4.2 K coldfinger at a single point of contact to reduce the possibility of induced currents in the shield due to thermoelectric effects.

## Multi Chip Module / Cold Head Assembly

The MCM/Cold head interface is designed to provide an isothermal mounting surface at approximately 4.5 K to mate with the bottom of the MCM substrate. The assembly, shown in Figure 5, consists of a circular MCM interface platter, a heat sink that reduces the heat load on the MCM, a magnetic shield, an adapter plate that provides both a conduction path and a support structure, and finally the associated hardware to attach the MCM to the cold head mounting surface. The MCM interface platter is an 8.9 mm thick (0.35 inch) disk of polished single crystal silicon 120 mm (4.7 inch) in diameter. It is attached to the mounting plate by a brass bolt passing through the center of the platter and threaded into a nut that sits a recess in the underside of the mounting plate. The MCM substrate is mullite, a ceramic that is specially formulated to have a thermal expansion similar to silicon. The MCM was bonded onto the silicon interface platter with Aremco 805[3], a thermally conductive epoxy.

We measured the thermal conductivity of the epoxy and the OFE copper used in the cryosystem. The data are shown in Figure 6 and Figure 7. For thin joints the epoxy has adequate thermal conductivity. The good low temperature bonding properties, coupled with the favorable coefficient of thermal expansion of the MCM and the silicon platter produced a functional bond between the two materials. This was verified with repeated cycles from room temperature to 4.5 K of an MCM bonded to a silicon platter. Temperature sensors bonded to the silicon platter and the MCM showed no evidence of increased thermal resistance at the platter to MCM bond.

Of great importance to the performance of this portion of the cryosystem is the thermal conductivity of the copper used for the heat sink, mounting plate and other elements in the thermal paths to the cold head. Very little margin for error was allowed for in the design, and

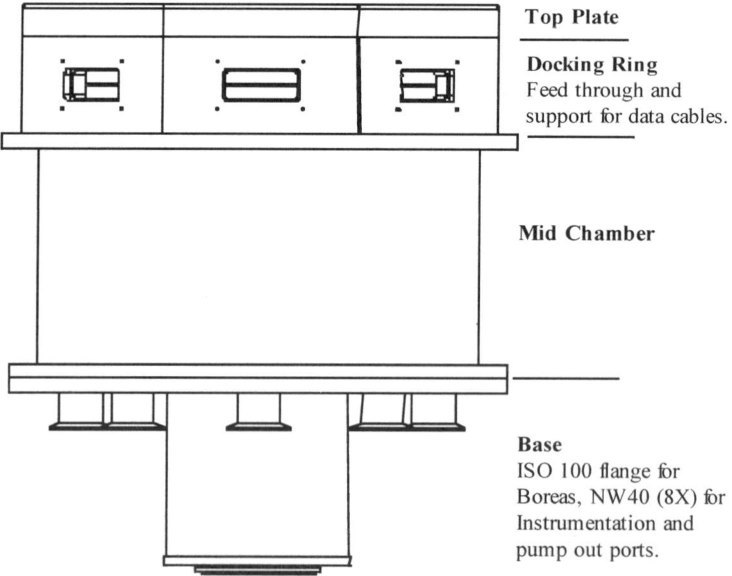


**Figure 3.** External view of the cryosystem vacuum chamber.

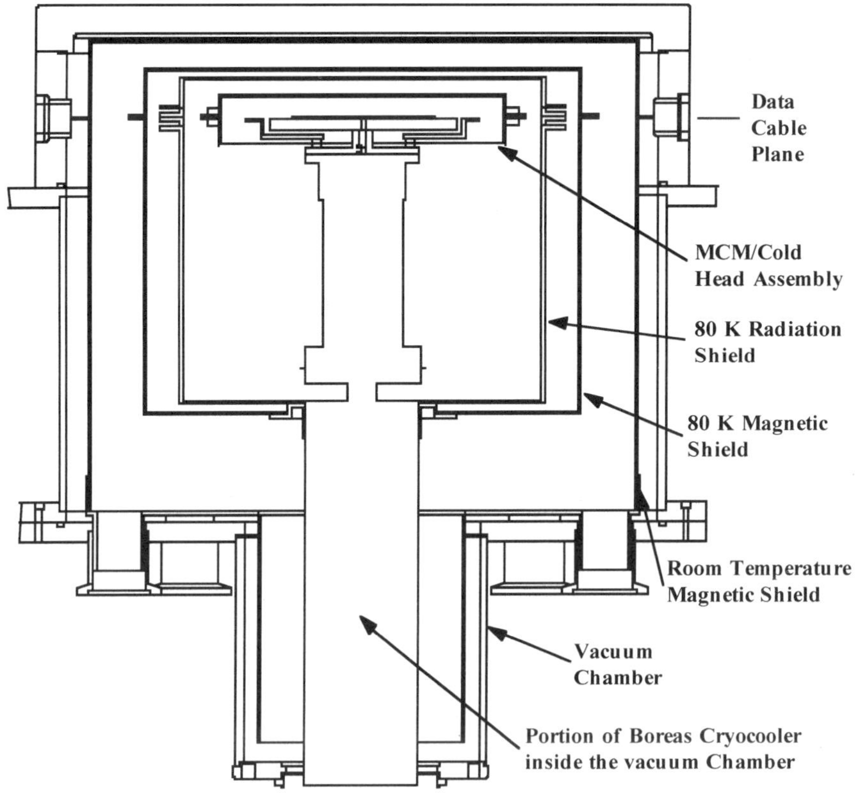


**Figure 4**. Cross section view of the cryosystem.

it was vital that the thermal models were precise and reliable. The data we obtained on the copper for this system were used in the thermal models for the thermal shields, the cable heat sinks, and cold head assembly, thus providing an unusually high degree of predictability.

## CRYOSYSTEM TEST

### Test Instrumentation

Ten Lakeshore Cryotronics model DT-470 series silicon diode temperature sensors[4] were used to monitor temperatures on the cryocooler cold head, first stage, and radiation shield. Three sensors are located on the MCM/platter assembly: one bonded to a corner of the MCM and the other two attached with screws onto opposing holes at the edges of the silicon platter. One sensor each was attached to: 4.5 K cable heat sink, 4.5 K magnetic shield lid, 4.5 K magnetic shield base, mounting plate, Boreas third stage, 80 K copper shield where the cables attach, and at the Boreas first stage.

Care was taken to minimize the thermal resistance between the temperature sensors and the mounting surface by using indium gaskets for the screw attached sensors and high strength thermal epoxy (Aremco 805) for the bonded sensors. The leads for each sensor located in a temperature zone of 10 K or below were carefully heat sunk away from the sensors to minimize heat conduction into the diode. All sensors were used in a four wire configuration. All sensors were calibrated using standard procedures, and two were factory calibrated sensors with a maximum error of 20 mK for temperatures below 10 K. One of the factory calibrated sensors was attached to the platter and one to the mounting plate.

The cable heat load was simulated with two sets of distributed electrical heaters located around the peripheries of the 80 K shield and the 4 K heat sink, with the individual heaters attached in roughly the same places as the cables are expected to attach. The heat load on the MCM was simulated using two Minco[5] film heaters bonded with epoxy (Aremco 805) to the

**Table 1.** Measured loads and temperatures

| Description/Location | | Units | Seq. 1 | Seq. 2 | Seq. 3 | Seq. 4 | Seq. 5 |
|---|---|---|---|---|---|---|---|
| 80 K Heat sink | - Heater | mW | 0.0 | 1520 | 4963 | 3535 | 3535 |
| 4.5 K Heat sink | - Heater | mW | 0.0 | 631.8 | 631.8 | 700.2 | 700.2 |
| MCM | - Heater | mW | 0.0 | 25.7 | 25.7 | 25.7 | 57.8 |
| Ch1 MCM | | K | 4.21 | 4.40 | 4.38 | 4.30 | 4.38 |
| Ch2 Platter edge | | K | 4.22 | 4.39 | 4.36 | 4.30 | 4.36 |
| ChA Platter edge | | K | 4.23 | 4.37 | 4.35 | 4.28 | 4.32 |
| Ch3 4.5 K heat sink | | K | 4.20 | 4.53 | 4.50 | 4.45 | 4.48 |
| Ch4 4.5 K mag shield lid | | K | 8.4 | 8.7 | 10.1 | 10.2 | 10.2 |
| Ch5 4.5 K mag shield | | K | 7.5 | 7.8 | 8.9 | 9.0 | 9.0 |
| ChB 4.5 K mount. plate | | K | 4.22 | 4.37 | 4.34 | 4.27 | 4.32 |
| Ch6 Boreas third stage | | K | 4.21 | 4.31 | 4.28 | 4.20 | 4.24 |
| Ch7 80 K copper shield | | K | 86.9 | 87.5 | 99.2 | 100.1 | 100.1 |
| Ch8 Boreas first stage | | K | 85.3 | 85.1 | 94.3 | 95.3 | 95.4 |

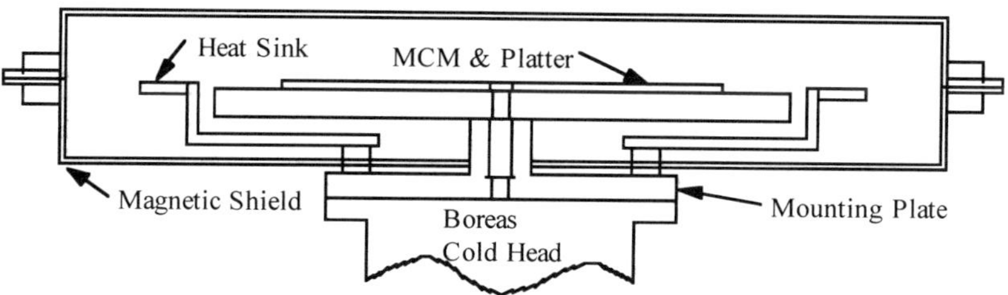


**Figure 5.** Cross section view of the MCM / Cold Head Assembly.

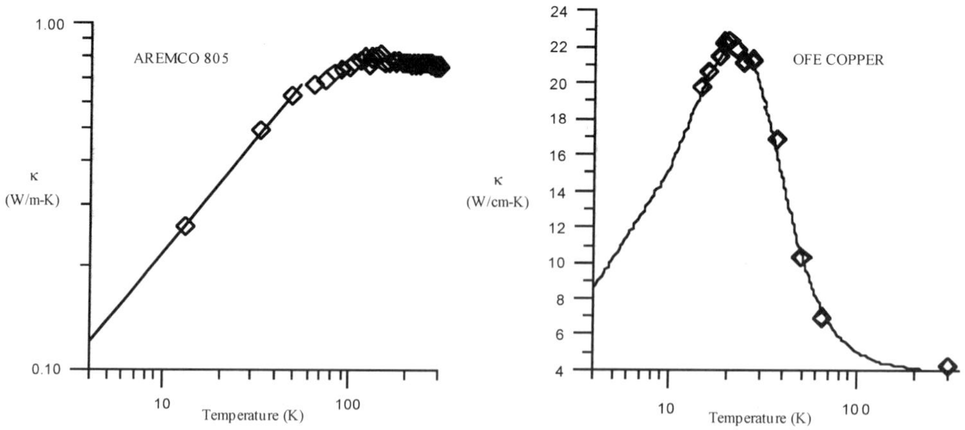


**Figure 6.** Measured thermal conductivity of the Aremco 805 epoxy.

**Figure 7.** Measured thermal conductivity of the OFE copper used in this program.

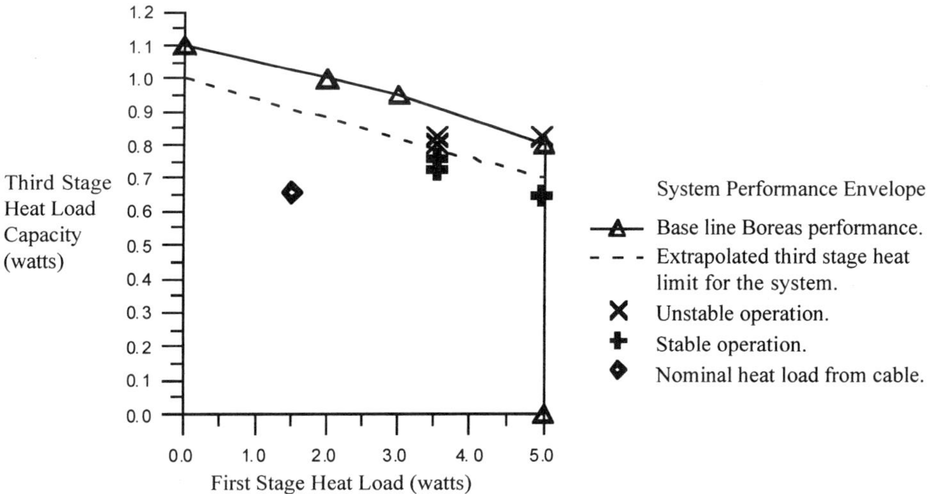


**Figure 8.** Measured cryosystem performance compared to the Boreas B100 baseline performance.

mullite MCM. The resistances of the heaters were selected so that joule heating in the phosphor bronze wires is less than six percent of the heat delivered to each location.

## Data Acquisition

Two computers were used to acquire and display the temperature data. A system using Labview software[6] monitored the two factory calibrated temperature sensors and was able to acquire accurate temperatures from each sensor in slightly less than a second. This resulted in three temperature readings per sensor for each expansion cycle in the Boreas cryocooler, providing a dynamic picture of the system behavior. The other data acquisition system, using Asyst software[7], took eight temperature measurements during a period of two minutes. This system averaged the temperature from each diode thus smoothing out the data.

## Results and Discussion

It takes 36 hours for the system to reach it's base no load operating temperatures. After reaching the operating temperatures, loads representing the cables and active electronics were applied to the first and third stages. The system was tested to the specified heat loads as shown in Table 1, Sequence 2. The system performed very well with the specified heat loads. With these loads the temperature on the MCM and 4.5 K heat sink are 4.40 K and 4.53 K respectively. With an actual cable in place the thermal connection between the MCM and the 4.5 K heat sink would be much greater so their actual temperatures should be almost identical, and will probably be between 4.4 K and 4.5 K.

Stable operation was established with the specified loads and we explored the limits of performance of the system to determine how much margin for error existed and whether the system was performing as the models predicted. We increased the heat load at a specific point and observed the system for temperature oscillations at the third stage. The results of this study are displayed in graphical format in Figure 8 and in tabular format in Sequences 3,4, and 5 in Table 1. These data were used to extrapolate a third stage "heat limit", thus providing a design boundary for the data cables.

The margin for adding additional heat load on the first stage is quite high. Comparison of Sequence 3 to Sequence 2 in Table 1 shows that an addition of 3500 mW on top of the design load of 1500 mW on the first stage does not increase the temperature of the cold finger. Sequence 4 shows that if the first stage is subjected to a heat load that is double the design load, and the third stage has a load 15% greater than design, the cold finger temperature will still be within specification.

The data in Table 1 illustrate the performance of the components mounted on the third stage of the Boreas cryocooler. The temperature of the MCM is essentially the same as the platter, indicating that the epoxy bond between the MCM and platter has a minor effect on the MCM temperature. The overall thermal resistance of the 4.5 K heat sink is 0.25 K/W, and most of that resistance can be attributed to the interface resistance between the heat sink and the mounting plate. By way of comparison, the thermal resistance between the mounting plate and the third stage of the Boreas cryocooler is 0.10 K/W.

Two sensors were mounted on the 4.5 K magnetic shield, one located on the lid and the other on the portion attached to the mounting plate. The magnetic shield temperature data in Table 1 are somewhat higher than predicted. This may be caused by the thermal load induced by the heater and sensor leads that have to pass through the cable slots in the shield. These are tied on one side to the 80 K copper shield and are sunk to the shield before passing on to the 4.5 K heat sink. Note that the shield temperature tracks the temperature of the 80 K copper shield. This additional heat load coupled with the low thermal conductivity of the magnetic shield material would have the effect of increasing the shield temperature.

## DESIGN APPROACH VERIFICATION

The very aggressive design goals fundamentally required optimized performance of all the system components, with very little margin for error or safety factor. We addressed the high risk aspect of this design challenge by using three basic design tools: thermal modeling, primary data on material properties, and mechanical modeling.

Thermal network models with up to 80 nodes were used, and many smaller models were used to refine the designs of certain components. Hierarchies of models were developed for establishing basic mechanical design parameters for the key components, and then more and more refined models were used to optimize the performance of the system.

Direct measurement of the properties of the actual materials to be used in the system was a critical factor in the success of the design effort. These measurements included the thermal conductivity of the epoxy used to bond the MCM to the silicon platter, the conductivity of the copper throughout the system, the expansion coefficients of the mullite used in the MCM, and the mullite yield and ultimate strengths at cryogenic temperatures. These data were invaluable for designing the system and for assessing in advance what approach to take in designing an effective thermal link between the MCM and the cold finger.

Three-dimensional models and assemblies were developed as the system was designed. This approach was very successful. The parts fit together on the first assembly.

The result of the approach we took to this project was that the first time the system was assembled and tested it came within 0.5 K of meeting the design goals. Improved flatness on the platter mounting plate was all that was required to achieve the results reported here. We are satisfied that the material property measurements were a deciding factor in our success.

## CONCLUSIONS

This cryosystem has the capability to cool superconducting microelectronics mounted on an MCM to 4.5 K when provided with data cables that perform within the design envelope shown in Figure 8. The thermodynamic performance of the Boreas B100 cryocooler was repeatable and did not degrade during the nearly 1000 hour course of system tests.

## ACKNOWLEDGMENTS

The authors are indebted to Dr. F. Bedard for his inspiration, guidance and support. Mr. N. Forrest did the mechanical design work and was a key contributor. This work is supported by the Maryland Procurement Office, Contract Number MDA904-92-C-5192.

## REFERENCES

1. T.J. Hendricks, M.W. Bruns, E.L. Hershberg, Thermal Transport and Electrical Dissipation in Ultra-Low Thermal Load, Multi-Gigahertz I/O Cables For Superconducting Microelectronics, in: "Advances in Cryogenic Engineering", Vol.41, P. Kittel, ed., Plenum Press, New York, 1996, p. 1761
2. N.J. Simon, E.S. Drexler, R.P. Reed, "Properties Of Copper and Copper Alloys At Cryogenic Temperatures", NIST Monograph 177, 1992, Chapter 1, p. 6
3. Aremco Products Inc., Ossining, New York
4. Lakeshore Cryotronics Inc., Westerville, Ohio
5. Minco Products Inc., Minneapolis, Minnesota
6. National Instruments Corp., Austin, Texas
7. Asyst Software Technologies, Inc., Rochester, New York

# A REVIEW OF CRYOBIOLOGY AND CRYOSURGERY

John Dobak, M.D

San Diego, CA 92121, CryoGen, Inc.

## ABSTRACT

Cryosurgery is the destruction of tissue by freezing. Modern cryosurgical techniques were first introduced in the mid 1960's, but, they only achieved modest application in selected medical specialties. This was in part due to limitations in cryosurgical devices which could not address the fundamental factors that influence cell death in response to freezing. It was also due to an inability to accurately assess or predict the growth of the ice ball within the tissue, and more importantly determine the zone of tissue death. Recent advances in liquid nitrogen cryosurgical devices and ultrasound guided cryosurgical techniques have addressed some of these issues, have significantly raised the interest in cryosurgery, and have expanded its application. Newer advances in cryosurgical devices and magnetic resonance imaging have great promise in making cryosurgery a primary therapeutic modality. This paper will review the biologic issues related to the freezing of tissue and their impact on the development of cryosurgical devices, as well as discuss the newest developments at the forefront of cryosurgical applications.

## INTRODUCTION

The goal of cryosurgery is to destroy cells within a limited diseased area and to allow the complete recovery of surrounding cells [1]. Cooling of tissue with a cryoprobe occurs via conductive heat transfer [2]. When a cryoprobe is in contact with tissue, an ice front advances in a radial direction away from the probe to form a cryolesion. The growth of the cryolesion is incremental and well controlled [3]. Frozen tissue is white in appearance and is hard to the touch, clearly delineating it from unfrozen tissue (Figure 1). It is common to create a cryolesion beyond the margins of the desired treatment area to ensure exposure of unwanted cells to the coldest possible temperatures [1].

A temperature gradient of gradually warming isothermal zones develops between the probe surface and the edge of the ice front which is at the freezing temperature of the tissue, or approximately -2 °C (Figure 2) [4]. The size of the lesion is proportional to the cryoprobe surface temperature, with lower temperatures creating larger cryolesions. Blood flow has a

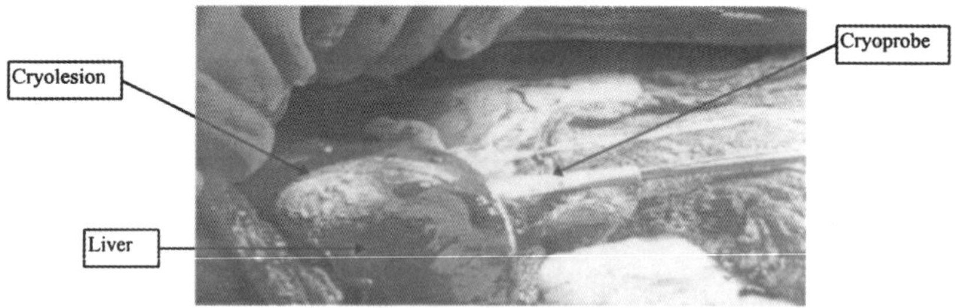


**Figure 1** Cryolesion formed in the medial lobe of goat liver. Note the sharp boundary between frozen and unfrozen tissue.

limited effect on the cryolesion, but may slightly reduce the size and gradient profile [5, 2]. Tissue thermal conductivity, which is related to water content, can affect lesion size; however, with the exception of fat and fibrous tissue, most tissues of the body have very similar thermal conductivities [6].

After freezing, the tissue is left in place to thaw. The tissue dies predominantly by aseptic necrosis and is absorbed or sloughed by the body. Multiple factors affect cell death during freezing.

## CELLULAR EFFECTS

The currently accepted physical events that occur with the freezing of cells are as follows. Reduction of the temperature to -2 to -5 °C results in supercooling of the extracelluar fluid, with gradual formation of extracellular ice crystals over the temperature range of -5 to -20 °C. Freezing of the intracellular water is initially prevented by the cell membrane. This creates an osmotic gradient in which water flows out of the cell and into the extracellular space. Eventually, the eutectic point of solutes within the cell is reached, and intracellular ice crystal formation commences. Studies have shown that there are three important factors related to the destruction and/or the recovery of cells frozen in culture: exposure temperature, cooling rate, and thaw rate [1, 7].

Exposure temperature has an affect on cell survival in culture. In the absence of cytoprotective agents, temperatures below -20 °C result in increased cell death. Cell death appears to reach a maximum at temperatures lower than -40 °C [8].

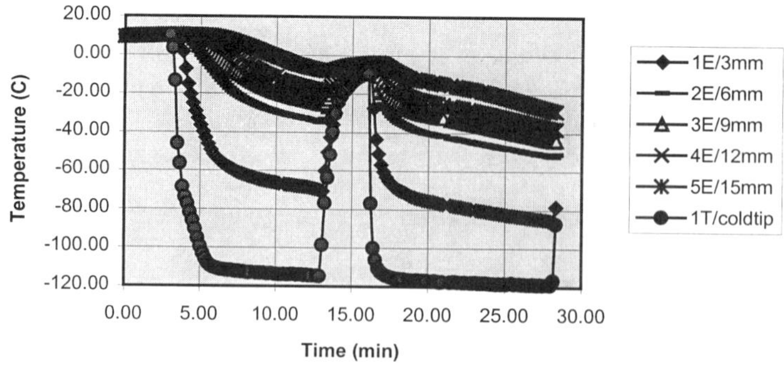


**Figure 2** Temperature gradients within a cryolesion. Type E thermocouples are spaced at 3 mm increments from the probe surface (1E-5E), and a type T thermocouple resides on the probe surface (1T) (Dobak, unpublished data). A thaw is allowed to occur between freezes.

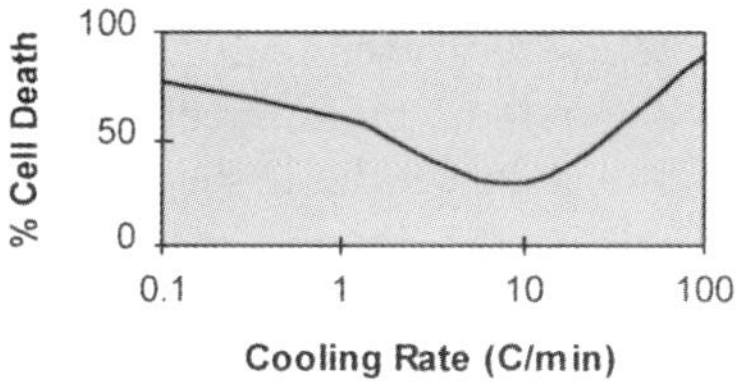


**Figure 3** Characteristic curve of cell death as a function of cooling rate.

Cooling rate has an even more profound effect on cell death. At slow rates of cooling (< 1-2 °C/min), the progressive loss of water from the cell can result in prolonged exposure of intracellular components to lethal high salt, or hypertonic, conditions [7]. At very high cooling rates (>15 °C/min), intracellular ice formation occurs with minimal cell dehydration, resulting in lysis of cell membranes and intracellular organelles. Cell lysis at high cooling rates has been shown to be the most lethal physical effect of freezing [9]. At intermediate rates of cooling, cell dehydration is balanced with intracellular freezing and cell survival is maximized. Figure 3 shows a typical cultured cell survival curve based on cooling rate.

Finally, thaw rate also has an effect on cultured cell death. Slower thaw rates (<10 °C/min) are more lethal to frozen cells than rapid rates [10]. This is believed to be the result of two factors. First, a slow thaw can result in prolonged exposure of cells to dehydration conditions encountered below -20 °C. Second, slow thawing is believed to result in intracellular recrystalization as water shifts across the cell membrane.

## TISSUE EFFECTS

The effect of freezing body tissues *in vivo* differs from what is anticipated by cell culture studies [4,11]. Just as a temperature gradient is established within a cryolesion, a cooling rate gradient is established, whereby decreasing rates extend radially from the probe. Cell culture studies would suggest necrosis would follow a well defined pattern, with a high percentage of cell death near the probe due to rapid cooling rates, a significant percentage of cell survival at intermediate distances from the probe where cooling rate is optimal, and another increase in percentage of cell death at the furthest distances from the probe where cooling rate is the slowest. In vitro studies would also suggest that maximal cell death would occur at the -40 °C isothermal zone. However, cell death in tissue is very clearly demarcated and nearly uniform even within the zone of optimal cooling rate. Further, uniform cell death has been shown to occur out to the -15 °C isothermal zone [4].

It is currently believed that a disruption of blood supply, or ischemia, promotes cell death in tissue. In the frozen region, the microcirculation may remain disrupted even after thawing. Additionally, the frozen area may swell after thawing which can also result in

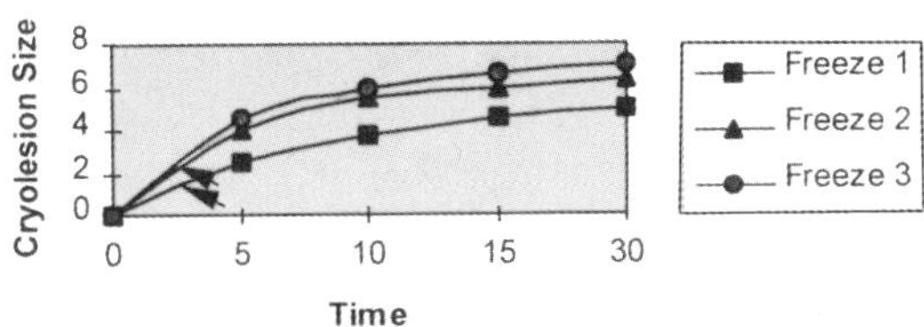


**Figure 4** Characteristic response of tissue to repetitive freezing. Note increased cryolesion size (arbitrary units) and rate of cooling (arrows) with each freeze. Adapted from Gill W, et al. (13).

disruption of microcirculation [2,12]. This ischemic effect is superimposed on the primary effects of cellular freezing, resulting in more uniform cell death within a cryolesion.

Repeated freeze-thaw cycles also affect the tissue response to cryosurgery. Repeated freezes have been found to improve tissue damage and eradication of tumor [11 13]. It was originally shown that repeated freeze thaws resulted in an increase in size of a cryolesion up to a maximum after 6 or 7 cycles [13]. Further, the cooling rate of the tissue was shown to increase with each repetitive freezing (Figure 4). This may account for the improvement in tumor or tissue destruction. The reasons for the increased cooling rate are not completely understood, however, they may be related to a disruption in microciruclation and changes in the tissue thermal conductivity due to the lysis of cell membranes [3].

Recent *in vitro* cellular effects of repeated freezing were investigated. It was shown that the temperature sensitivity of prostate cancer cells decreased on repetitive freezing [14]. Thus, cell death occurred at temperatures as warm as -19 °C after a second freeze. In addition, it was shown that this increase in cell sensitivity to warmer temperatures was further augmented by a cooling rate > 25 °C/min. These cellular effects may also account for improved clinical reports due to repetitive freezing and support the use of multiple freeze cycles.

## PROCEDURAL AND DEVICE IMPLICATIONS

Basic science studies have important implications for cryosurgical device and procedure development.

1. Cells in the treated region should be subjected to the highest cooling rate possible, or at least to a cooling rate higher than the optimum cooling rate for cell survival. For most invasive cryosurgical procedures, probes must be introduced into the body warm. To achieve the necessary cooling rates, cryosurgical devices should be developed with high cooling capacity to achieve a rapid cool down.

1. The minimum temperature achieved should be -20 °C or lower.

2. Thawing should be carried out at the slowest rate possible or the natural rate that can be delivered by the body.

3. Repeated freezes should be carried out with a complete thaw in between freezes. The combination of repeated freezing and rapid cooling rates may be very effective due to the increased sensitivity of cells to warmer temperatures. Further, this may prevent the need to create a cryolesion well beyond the margin of the desired treatment area in trying to reach low lethal temperatures at the edge. In this manner the cryolesion is much more focused, which can have important implications when the treatment area is near important body structures which must remain unfrozen, such as the rectum during prostate cryosurgery.

4. Because heat transfer occurs conductively, probe contact must be maximized to assure the highest possible cooling rate and deepest penetration of lethal temperatures. This can be accomplished through increased pressure and is further promoted by the presence of moisture or surgical gels between the tissue and probe [15]. Lubricants are noted to fill irregular crevices of tissue that might prove to be barriers to heat transfer.

5. Multiple probes may be placed within a single surgical sight to ensure exposure of the tissue to the greatest cooling rate possible. It should be noted, however, that it is not always possible to utilize multiple probes. In these cases, the temperature (-20 °C) achieved at the desired margin of the treated area becomes the most important guideline for therapy. This is due to the difficulty of achieving rapid cooling rates at the further distances from the probe. Because probe surface temperature is the most important factor related to the

**Table 1** Cryosurgical system comparison

| | | | | |
|---|---|---|---|---|
| **Technology** | Joule-Thomson | Phase Change | Joule-Thomson | Joule-Thomson |
| **Coolant** | Mixed Gas | Liquid Nitrogen | Argon | Nitrous Oxide |
| **Temperature (°C)** | -120 | -140 | -120 | -60 |
| **Pressure (psi)** | 350 | N/A | 3,000+ | 750+ |
| **Open or Closed** | Closed | Open | Open | Open |
| **Cost** | Moderate | High | Moderate | Low |

isothermal gradient development, cryosurgical devices must operate at an appropriate surface temperature to match the depth of penetration needed in a given procedure.

## CRYOSURGICAL INSTRUMENTATION

The two current methods by which low temperatures are achieved in cryosurgery are the circulation of liquid nitrogen to a probe or the expansion of high pressure gasses (J-T systems) such as nitrous oxide or argon [15]. Each system type has advantages, however, neither are ideally suited to meet the future needs for cryosurgical apparatus (Table 1).

Liquid nitrogen systems circulate the coolant to vacuum insulated probes. In early liquid nitrogen devices, Liedenfrost boiling often occurred in the tip which severely hampered heat transfer and probe surface cooling [1,16]. With Leidenfrost boiling, a warm gaseous film forms between the liquid nitrogen and the inner surface of the probe, effectively insulating the tip. Complex staged sub-cooling devices have been built that overcome this problem and allow colder probe surface temperatures (-130 to -140 °C), however, this significantly increases the cost and complexity of the device [16]. Further, because the coolant is liquid nitrogen, large storage tanks are required. This limits the ability to reduce the size of the device. Also, circulation of the cold liquid from the tank to the probe causes cooling of the entire conduit. This creates the need for bulky insulation and large inflexible hoses. Finally, the liquid nitrogen is consumed during use and must be continually replenished. This creates convenience problems related to storage and replacement of liquid nitrogen, and also creates the opportunity to run out of coolant during a procedure.

Other currently available cryosurgical systems operate on the Joule-Thomson (JT) principle [15]. These systems utilize a single compressed gas (e.g. nitrous oxide or argon) to

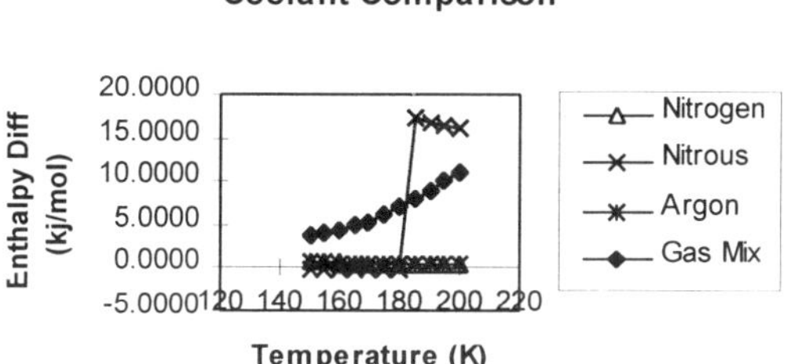


**Figure 5** Cryosurgical coolant comparison at an upper pressure of 300 psi and a lower pressure of 1 atm. Note that there is significantly more refrigeration potential in the gas mixture at 150 K (-120 °C) relative to the single gasses.

cool a probe through the JT effect. Because cooling occurs during the expansion process, the majority of the cooling can be focused at the tip, eliminating some insulation requirements. However, to generate probe surface temperatures below -100 °C under biological heat loads, high pressures (2000 + psi) are required which creates safety problems. Even temperatures as warm as -60 °C require pressures of 800 psi. Similar to liquid nitrogen the coolant is exhausted during use and must be continually replenished. Finally, the performance of these devices during use decays as the bottle pressure decreases.

### Mixed Gas Technology

Mixed gas Joule-Thomson devices may circumvent the problems associated with current technologies, and allow the development of a safe, low temperature, cost effective, and convenient device. By using a mixture of gasses, as opposed to a single gas, the non-ideal properties of the coolant are increased. This results in a greater JT effect, such that lower temperatures and greater power can be achieved at lower pressures (Figures 5)[5,17,18,19,20].

Thus, the advantages of JT based systems which include, low cost, compact size, focused cooling, and flexibility can be combined into a system that achieves the necessary operating temperatures for tissue destruction at safe pressures. With lower working pressures, a compressor can be incorporated into a mixed gas system, reducing replenishment needs and reliability concerns.

A novel mixed gas JT device has been built that incorporates a compressor. Operating pressures of only 300 to 350 psi are required to achieve temperatures below -120 °C with a biological heat load, such as freezing liver tissue at 37 °C [5]. Both a rigid probe, and a flexible catheter probe (100 cm length) that incorporates a unique mircominiature heat exchanger, have been developed [17]. The small size of the flexible catheter (3 mm outer diameter), combined with the added safety and flexibility of low working pressures, may open up new areas for cryosurgery such as intravascular treatment of arrhythmia or endoscopy.

## VISUALIZATION OF CRYOSURGERY

### Ultrasound

Freezing of tissue can be clearly visualized using ultrasound. The frozen tissue is a solid mass, and the ice front reflects a significant fraction of the acoustic waves. This creates a bright white line at the interface between frozen and unfrozen tissue. The remainder of the frozen tissue appears as a dark region behind the hyperechoic line and is referred to as a post acoustic shadow [21].

Resolution of the ice front is sufficient to allow cryosurgical treatment of liver and prostate tumors percutaneously with excellent safety and efficacy [22 23]. In studies comparing ice ball diameter as measured by ultrasound, with ice ball diameter as measured by calipers, less than 1 mm discrepancy is observed for ice balls less than 55 mm in diameter [24]. Consequently, the entire prostate can be frozen with little damage to the rectum despite the proximity of the two body structures. Although no information about the thermal history (ie. temperature and cooling rate) of the cells can be ascertained from ultrasound, treatment algorithms are being developed to ensure adequacy of therapy.

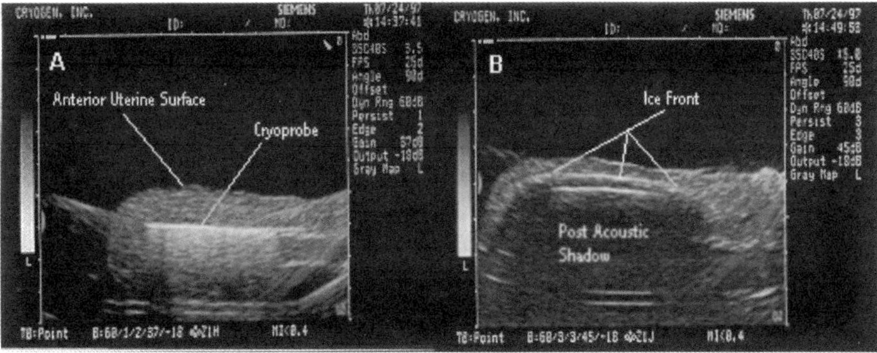


**Figure 6** Ultrasound characteristics of cryosurgery in a uterine specimen in a 37 °C water bath. Photo A is prior to freezing. Picture B is after 5 minutes of freezing.

## MRI

Like ultrasound, the growth of the ice ball can be clearly visualized using MRI [25]. There is no effective signal from the frozen region because the $T_2$ relaxation interval is sufficiently short. Subsequently, the cryolesion appears black.

Conventional MRI, like ultrasound, also fails to provide any information about the thermal history of the cells. However, it has been demonstrated that mathematical models for temperature distribution within a cryolesion can be utilized to obtain MRI assisted numerical solutions to the energy equations [26]. In this manner the thermal history within the cryolesion can be clearly mapped.

It is unclear where this form of cryosurgery is headed. The expense and complexity of MRI guidance may be inhibitory. In addition, special instrumentation and cryosurgical probes must be developed that are compatible with MRI. However, this technique may give the surgeon appropriate feedback and precision to treat a wide variety of tumors through minimally invasive approaches.

## REFERENCES


1. W. Bald, and J. Fraser, Cryogenic surgery, *Reports on Progress in Physics* **45:**1381 (1982).
2. R. Orpwood, *Phys Med Biol* **26**:555 (1981).
3. W. Gill, J. Fraser, and D. Carter, *Nature* **219:**410 (1968).
4. J. Smith and J. Fraser, *Cryobiology* **11:**139 (1974).
5. J. Dobak, X. Yu, and K. Ghaerzadeh, A novel closed loop cryosurgical device, in: "Advances in Cryogenic Engineering," P. Kittel, ed., Plenum Press, New York (1997)
6. T. Cooper and G. Trezec, *Aerospace Medicine* **42:** 24 (1971).
7. P. Mazur, *Science* **68**: 939 (1970).
8. I. McGann, *Cryobiology* **16:** 211 (1979).
9. J. McGrath, E. Cravalho, and C. Huggins, *Cryobiology* **12:** 540 (1975).
10. R. Miller and P. Mazur, *Cryobiology* **13:** 404 (1976).
11. A. Gage, K. Guest, M. Montes, J. Caruana, and D. Whalen, *Cryobiology* **22:** 175 (1985).
12. B. Rubinsky and D. Pegg, *Proceedings from the Royal Society of London (Biology)* **234:** 343 (1988).
13. H. Neel, G. Riggle, A. Ketcham, and W. Hammond, *Cryobiology* **8:** 501 (1971).
14. K. Tatsutani, B. Rubinsky, G. Onik, and R. Dahiya, *Urology* **48**: 441 (1996).
15. A. Gage and D. Torre, Cryosurgery, in: "Encyclopedia of Medical Devices and Instrumentation," J. Webster, ed. Wiley, New York (1988).
16. Z. Chang, J. Finkelstein, H. Ma, and J. Baust, *Biomedical Instrumentation and Technology* **28**: 383 (1994).

17. E. Marquardt, R. Radebaugh, and J. Dobak, A cryogenic catheter for treating heart arrhythmia, in: "Advances in Cryogenic Engineering" P. Kittel, ed., Plenum Press, New York (1997)
18. V. Alfeev, V. Brodyanski, V. Yagodin, V. Nikolsky, and A. Ivanstov, Patent (UK, 1973).
19. J. Jones, "Cryogenic mixed fluid application study and computer code development" *D79777* Jet Propulsion Laboratory, Pasadena, (1991).
20. R. Longsworth, M. Boiarski, and L. Klusmier, 80 K closed-cycle throttle refrigerator, in: "Cryocoolers 8" Plenum Press, New York (1995)
21. G. Onik, et al., *Cryobiology* **21:** 321 (1984).
22. G. Onik, et al., *Cancer* **72**: 1291 (1993).
23. G. M. Onik, D. Atkinson, R. Zemel, and M. L. Weaver, *Semin Surg Oncol* **9**:309 (1993).
24. C. Lam, S. Shimi, A. Cuschieri, *Arch Surg* **130**: 1068 (1995).
25. B. Rubinsky, *Cryobiology and Cryotechnology* **41**, 67-81 (1995).
26. J. Hong, S. Wong, G. Pease, B. Rubinsky, *Magnetic Resonance Imaging* **12**, 1021-1031 (1994).

# A NOVEL CLOSED LOOP CRYOSURGICAL DEVICE

John Dobak, Xiaoyu Yu, and Kambiz Ghaerzadeh

CryoGen, Inc., San Diego, CA 92121, USA

## ABSTRACT

A novel cryosurgical device has been developed that dramatically improves the cost, convenience, safety, and effectiveness of cryosurgical devices. The device incorporates a mixed gas J-T stage. The fluid mixture is non flammable, non toxic, non corrosive, environmentally safe, and therefore suitable for medical applications. The system is compressor driven utilizing a novel oil free motor. Working temperatures of -115 to -125 °C at pressures of 300 to 350 psi are achieved. Approximately 20 watts of refrigeration are delivered at -120 °C.

The compressor console is portable having a 4 cu-ft volumetric footprint. The cryoprobe has a handle 3.5 cm O.D. and 18 cm length with a probe shaft having a 4 mm O.D. and a 3.5 cm freeze zone. In bench top studies using tissue simulating media (2% gelatin, 98% saline) ice balls of 3.5 - 4.0 cm diameter and 5 cm length with a mass of 48 grams are created. Studies using calf liver tissue in a body temperature water bath show ice balls of similar size and mass (3.5 cm diameter, 48 g). In live goat animal studies, surface temperatures of -115 °C were achieved, and ice balls of 3.8 cm in diameter were created in 10 minutes in the liver. Isotherms below -20 °C were developed at 12 mm radial distance from the probe.

## INTRODUCTION

Cryosurgery is the destruction of tissue through the application of extreme cold [1]. The cryoprobe surface temperature is a critical determinant of tissue destruction [2]. Existing cryosurgical devices are based on the circulation of liquid nitrogen to a probe, or the expansion of high pressure gasses (J-T systems) such as nitrous oxide or argon [3]. While liquid nitrogen systems can achieve low surface temperatures (-120 to -140 °C), they are bulky, inflexible, and costly. J-T systems utilizing nitrous oxide tend to be smaller and less expensive, however, they have relatively warm surface temperatures (-50 to -60 °C) and high operating pressures (700 psi). Argon systems can achieve temperatures as low as -120 °C, however they utilize pressures in excess of 2000 psi. These high pressures raise safety concerns and limit use of these devices in some applications. Importantly, all currently marketed cryosurgical devices exhaust the coolant to the atmosphere and therefore require continual replenishment [3]. Thus, the convenience of these systems is reduced, and the risk of exhausting the supply during in the middle of a procedure is real.

A novel JT cryosurgical device has been developed that circumvents the problems noted above. The devices utilizes a gas mixture to achieve surface temperatures of -115 to -125 °C at pressures of 300 to 350 psi. Further, the system is closed, incorporating an oil free compressor to circulate the coolant.

## MATERIALS AND METHODS

### System

Design criteria for the system were 15-20 Watts of refrigeration at -120 °C with a maximum pressure of 350 psi. Two stages were required to achieve the desired cooling capacity. With no pre-cooling, given the limits of the pressure and flow provided by the compressor technology to be employed, the maximum refrigeration power yield was 9 Watts at -120 °C. However, with approximately 20 Watts of pre-cooling at -30 °C, this refrigeration power increases to 28 Watts at -120 °C. The first stage, or pre-cool stage, incorporated a commercially available refrigeration compressor and a commercially available freon. The second stage, or gas mix stage, incorporated a novel oil free compressor and a novel gas mixture. The compressor had a unique electromagnetic linear drive mechanism and gas bearings and a size of approximately 0.6 cubic feet. The overall size of the integrated pre-cooler compressor and gas mixture compressor is 2.3 cubic feet.

The power of the system was tested using a heater made of nichrome wire which was wound around the freeze zone to a known resistance. A DC power supply was used to provide current. The device was operated until the temperature reached a steady state and the power output of the heater was increased until the temperature began to rise.

### Gas Mixture

Gas mixture combinations were computer modeled using an algorithm that optimizes a unitless number termed the $(\Delta h^*)_{min}$. Conceptually this number provides a relative assessment of the amount of refrigeration power a gas mixture has available to extract heat from tissue for a given amount of heat exchange capacity. Since the heat exchanger capacity is limited by the size requirement for a cryosurgical probe, ideal gas mixtures are identified by this method. While this method produced potentially useful combinations and approximate percentage compositions, the mixtures had to be empirically optimized in the system.

### Cryoprobe

The cryoprobe was constructed by encasing heat exchangers in a metallic shell. The shell was subsequently evacuated to 10 E-6 to provide insulation. The metallic shell had a handle section and a shaft section. The heat exchangers of the two stages were integrated into the handle section which had an O.D. 3.5 cm and a length of 18 cm. The shaft was 4.0 mm outer diameter O.D. and 10 cm in length. and had a copper freeze zone, or tip, at the distal end 3.5 cm in length and 4.0 mm O.D. Braided stainless steel tubing was used to carry gas from the compressors to the cryoprobe.

Parallel tube counter flow heat exchangers were constructed by soldering of adjacent stainless steel tubes and mandrill winding. For the pre-cool stage, the high pressure gas mix tube was soldered to the low pressure freon tube. The high pressure gas mixture tube subsequently transitioned to a counterflow parrallel tube heat exchanger for the second stage. At the end of the second stage heat exchanger, the high and low pressure tubes transitioned into a coaxial construction that carried the gas down the shaft and to the freeze zone. A capillary tube restriction orifice was placed at the end of the high pressure coaxial line allowing the gas mix to expand within the tip.

### Gelatin Ice Ball Studies

Because room temperature gelatin simulates the thermophysical properties of tissue, gelatin studies are commonly used for testing and comparing cryosurgical instruments [4 5]. Gelatin was prepared by mixing 2.0% w/v of gelatin (Fisher Scientific) in normal saline. The gelatin was poured into glass beakers and allowed to set in a refrigerator at 10 °C for 6 hours. A thermocouple was attached to the surface of the cryoprobe freeze zone. A calibrated flow meter (MKS) was connected in series to the probe outlet line. A pressure transducer was connected to the probe inlet line. All data is recorded by a National Instruments Data Acquisition system.

The tip was inserted into the gelatin and the device was turned on. Ice ball growth was observed over a ten minute time period. The ice ball was measured using calipers after the beaker was removed from the beaker.

### Beef Liver Studies

Approximately 150 grams of beef liver were sealed in a plastic bag and placed into a 37 °C water bath. The liver was allowed to warm up to 37 °C as measured by a thermocouple placed into the tissue. The cryoprobe tip was subsequently placed into the tissue and the tissue was frozen for 10 minutes while it remained in the water bath. The ice ball was subsequently harvested from the tissue and the diameter measured with calipers. A multi-probe thermocouple array, with 3mm spacing between each probe, measured the isothermal gradients within the tissue.

### Animal Study

A single goat weighing 75 kg was utilized. The animal was pre-anesthetized with xylazine, intubated and kept under general anesthesia using isoflurane gas. The goat was placed in dorsal recumbency and the abdomen was prepped. Vital signs were monitored and recorded at regular intervals.

A midline incision was made and the liver was exposed. The probe was inserted into the medial and lateral lobes. The probe had a 15 Watt resistance heater wound around the tip. A thermocouple was attached to the outer surface of the probe. A multi-thermocouple probe with 3 mm spacing between the thermocouples was inserted into adjacent tissue parallel to the long axis of the probe. At each selected site, a double freeze was performed for ten minutes. Five minutes of thawing took place between each freeze. The heater on the probe was activated at full power until the probe surface thermocouple reached 0 °C. The ice ball size was measured using calipers.

## RESULTS

The computer modeling produced four potential combinations for the optimum gas mixture. Of these, two contained a component that was determined to have toxicity. Of the

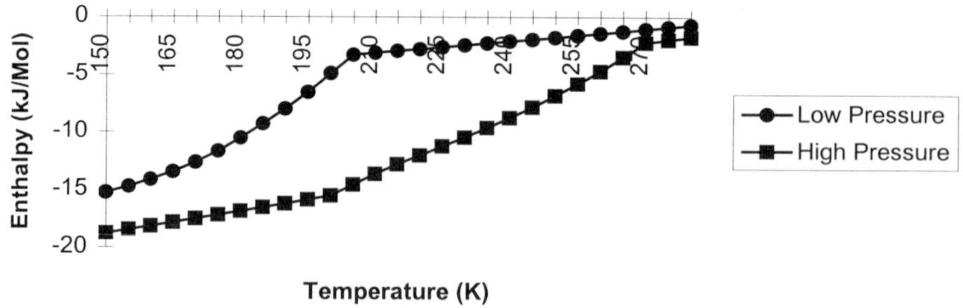


**Figure 1.** Enthalpy versus temperature curve for gas mixture used in the system.

two remaining combinations, some empirical computer modification was performed and the mixture with the highest $(\Delta h^*)_{min}$ was chosen for use in the system. The $(\Delta h^*)_{min}$ for this mixture was 0.3224. Figure 1 shows the enthalpy curve for the gas mixture used in the system. The overall hazard rating of this mixture was 1, and the DOT classification was non-corrosive, non-flammable, non-toxic [6]. Under ideal conditions (100% heat exchanger efficiency), the theoretical refrigeration power at flow rates of 10 liters per minute, a pressure ratio of 25:1 and 19 Watts of pre-cooling was 28 Watts at -120 °C. The measured cooling capacity of the device was approximately 20 Watts at -120 °C.

The oil free compressor was capable of producing 15 slm of flow with an outlet pressure of 2.5 MPa and an inlet pressure of 0.1 MPa. The maximum run time has not been determined, however, continuous run times of 180 minutes with no evidence of clogging or flow decay have been recorded.

The system was capable of making gelatin ice balls of 3.5-4.0 cm in diameter and 4 cm in length weighing 45 to 55 grams in a 10 minute freeze. Outer surface temperatures recorded ranged for -110 to -120 °C. Compressor outlet pressures ranged from 2.4-2.5 MPa with inlet pressures ranging from 0.12-0.16 MPa.

Ten minute freezes using beef liver at 37 °C produced similar sized iceballs as the gelatin test (average size was 3.45 cm diameter by 5.0 cm length, and average wgt. was 47 g). Average low surface temperatures were -122 °C (Figure 2).

In vivo goat liver freezes produced ice balls of 3.4 to 3.8 cm in diameter after ten minutes (Figure 3). Probe surface temperatures from -115 to -120 °C were recorded. Thermal gradients within the one ice ball as recorded with the multi-thermocouple probe show a temperature of -

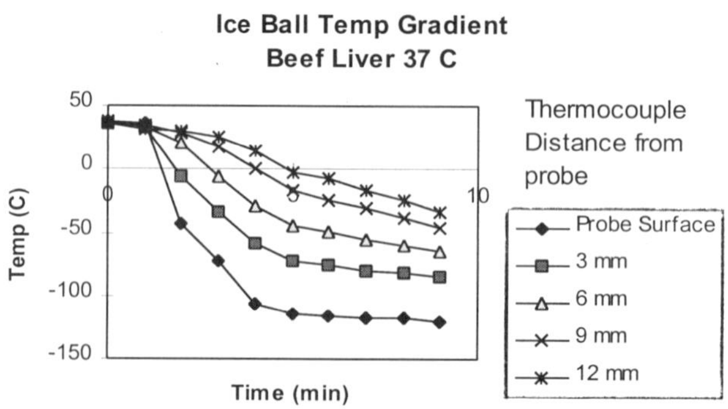


**Figure 2.** Isothermal gradients created during freeze of beef liver in a 37 °C water bath. Thermocouple distances are radial from the probe surface.

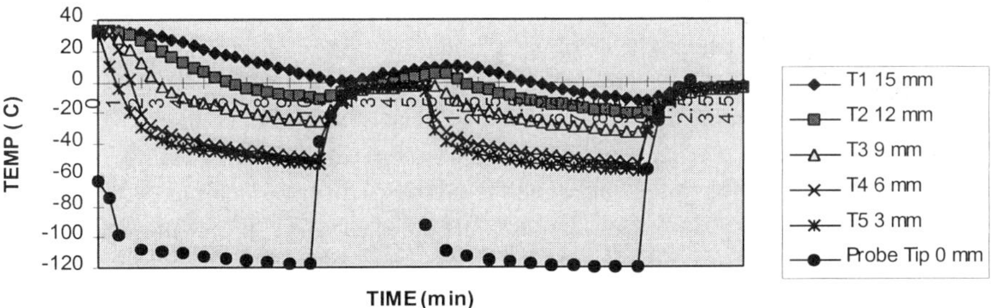


**Figure 3.** Isothermal gradients in freezing of liver in an anesthetized goat. Thermocouple distances are radial from the surface of the probe. A thaw was allowed to occur for 5 minutes between freezes.

20 °C at a distances of 9-12 mm from the probe (Figures 3, 4). It is important to note that multi-thermocouple probe was slightly misaligned with the ice ball, causing warmer than actual temperatures to be recorded at the more distal probes.

## DISCUSSION

In design of a cryosurgical apparatus, the two most important criteria are probe surface temperature and cool down rate [2]. The probe surface temperature determines the size of the ice ball created and the isothermal gradients within the ice ball. Most cells are destroyed at -20 °C and in some cancers a temperature of -40 °C is required [2]. When treating tissue, it is important to achieve a these necrotic temperatures at the desired depth of penetration or at the margin of a tumor, making surface temperature a critical factor.

The cool down rate is also relevant to cell death. A fast cool down rate results in the formation of intracellular ice crystals which increase the lethality of the technique. Different cells require different optimal cooling rates for destruction, but in general cooling rates above 25 °C/min are lethal [7,2]. Subjecting tissue to repeated freezing increases total cell death by increasing the ice ball size and the necrotic isothermal zone, and increasing cool down rate at any given point in the frozen area [8]. In addition, repeated freezing can lower the required necrotic temperature [7,8].

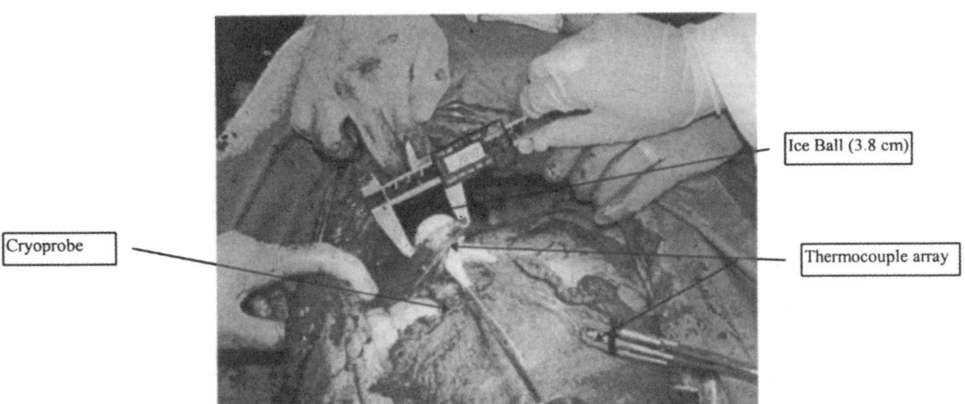


**Figure 4.** Photograph of ice ball created in goat liver. Ice ball diameter is 3.8 cm.

This newly developed device is capable of achieving lethal temperatures at radial distances of greater than 1 cm from the probe surface. Cool down rates of greater than 15 °C/min are seen within radial distances of 6-9 mm. This data indicates that the device, with a single probe, is capable of clinical treatment of small to moderately sized tumors. Further, it would be useful in other applications such as cervical disease or abnormal uterine bleeding. A multi-probe device could be developed for use in larger, more complex diseases such as prostate cancer.

The two stages in this system provide a number of advantages. First, the cool down rate is improved by increasing the overall cooling capacity. Second, a smaller second stage compressor can be utilized keeping the unit compact. Third, contaminants that could potentially freeze out at the JT orifice in the tip, may first condense in the pre-cooler and prevent clogging.

A small freon refrigeration loop was integrated into the system as the pre-cooler. Thermoelectric coolers meeting the pre-cooling requirement were found to be impracticably large. A split stream pre-cooler was considered, however, the overall capacity of the current oil free compressor was too limited.

An oil free compressor was chosen to improve reliability. Due to the small diameter tubing required to make the compact heat exchangers for the system, the low operating temperatures, and the nature of the gas mixture, oil could easily condense in the system and cause clogging. Using an oil based compressor, maximum cool time was less than 5 minutes, and continued to decline with successive freezes (unpublished data). With the oil free compressor, there is currently no limit on freeze time other than mechanical failure of the pump. Current run times of the system well exceed the needs to complete clinical procedures.

Life cycle testing of the compressor is currently underway, and it is anticipated that the mean time to failure will be greater than 5,000 hours. Because the duty cycle of the compressor is estimated at 15 % maximum, this translates to a minimum 4 year life time before major servicing.


## ACKNOWLEDGEMENT

The authors would like to thank Dr. Ray Radebaugh, Eric Marquardt, and Marcia Hubert, at the National Institutes of Standards and Technology for helping identify gas mixtures used in the system. The authors would also like to thank Terry Brown, Eric Ryba, and Dr. Ray Sarwinski for their assistance and insights during development.


## REFERENCES


1 I. S. Cooper, in: "*Cryogenics in surgery*," H. von Leden and W. G. Cahan eds., Medical Examination Publishing Company, Inc., Flushing, (1971), pp. 1-38 and p. 155.

2 R. Orpwood, Phys Med Biol **26:** 555 (1981).

3 A. Gage and D. Torre, "*Encyclopedia of Medical Devices and Instrumentation,*" Wiley, New York (1988).

4 H. Lenz and J. Eichler, Cryobiology **13:** 37 (1976).

5 T. E. Cooper and G. J. Petrovic, Journal of Heat Transfer **96:** 415 (1974).

6 "*Dangerous Properties of Industrial Materials*," Van Nostrand Reinhold, New York (1992).

7 K. Tatsutani, B. Rubinsky, G. Onik, and R. Dahiya, Urology **48:** 441 (1996).

8 W. Gill, J. Fraser, and D. Carter, Nature **219:** 410 (1968).

# A CRYOGENIC CATHETER FOR TREATING HEART ARRHYTHMIA*

E.D. Marquardt[1], R. Radebaugh[1], and J. Dobak[2]

[1] National Institute of Standards and Technology
Boulder, Colorado 80303

[2] CryoGen, Inc.
San Diego, California 92121

## ABSTRACT

Progress in the development of a cryogenic catheter to treat heart arrhythmia is discussed. This system uses a mixed-gas Joule-Thomson (J-T) refrigerator to cool the tip of a catheter that can be inserted into the body through the large veins leading into the heart. The cryogenic catheter is intended to treat heart arrhythmia characterized by an abnormally rapid heart rate, although the system has a wide variety of other medical applications. Approximately 2 million people in the U.S. suffer from rapid-rate heart arrhythmia. Catheter therapy has proven to be a more effective and less expensive method of treatment than alternatives such as drugs or surgery. A cryogenic catheter has significant advantages over existing catheters used in this form of therapy. The catheter has coaxial tubes for the high and low pressure streams with a miniature heat exchanger and J-T orifice at the catheter tip. The high pressure is maintained at 2.5 MPa. The largest diameter is 3 mm, the length is 90 cm, and all but the last 10-20 mm is flexible. The gas mixture has been optimized for the required operating conditions using non-flammable and low ozone depletion gases. Low cost techniques have been incorporated into the fabrication of the cold tip so that each catheter can be disposable. Several prototype catheters have been built. No-load temperatures down to 85 K were achieved with the cold tip exposed to ambient air. Using room temperature gelatin to simulate tissue heat loads, catheter tip temperatures of 160 to 175 K have been achieved, and ice balls about 26 mm in diameter weighing 11 g were created. We estimate that ice balls about 10 mm in diameter weighing 1.5 to 2.0 g are required to treat ventricular arrhythmia. Although the heat load in our experiments was less than the *in vivo* load, we think the current refrigeration power is sufficient to meet the clinical requirement.

---

## INTRODUCTION

Heart arrhythmia is a problem for over 2 million Americans. To date, such patients have been difficult to treat with conventional drugs or surgery. Catheter therapy has proven to be a more effective and less expensive method of treatment, but the electrosurgical catheters currently used are not very effective for heart arrhythmia treatment. They have limited destructive capability and are difficult to keep in contact with the heart. Lesions created with electrosurgical catheters are 3-4 mm diameter spheres, and lines of these spheres must be connected in order to perform some of the treatment. A cryogenic catheter has the potential to deliver greater destructive power to the tip, allowing larger lesions to be formed. Linear lesions are possible, and the catheter tip will adhere to the tissue during the cooling process.

To treat heart arrhythmia the cryogenic catheter must reach temperatures between 100 and 150 K. To be inserted through veins, it must be 3 mm in diameter or smaller and must be able to make a bend with a 10-15 mm radius. It must be about 1 meter in length and must have a surface temperature above 0 °C along its length. For safety reasons the maximum pressure should not exceed 3 MPa. It is estimated that 10 watts of cooling will be required. The refrigerant should be benign, nonflammable, and nontoxic, and have a low ozone depletion potential. The catheter also needs to be disposable for safety and sterilization reasons. These requirements have led us to choose a mixed-gas Joule-Thomson (J-T) refrigerator as the most practical solution.

The difficulty with existing J-T systems is that they are too large in size to be used for catheter therapy, the cold head is too expensive to be disposable, and the units require too high a pressure for catheter use. The cryogenic catheter we have designed is based upon the Joule-Thomson cycle shown in Figure 1. This is a closed-cycle system that does not require make-up gas at any time. The catheter has coaxial tubes for the high and low pressure streams with a miniature heat exchanger and J-T orifice at the catheter tip. The high pressure is 2.5 MPa. The largest diameter is 3 mm, the length is 90 cm, and all but the last 10-20 mm is flexible. The gas mixture has been optimized for the required operating conditions using non-flammable and low ozone depletion gases. Low cost techniques have been incorporated into the fabrication of the cold tip so that each catheter can be disposable.

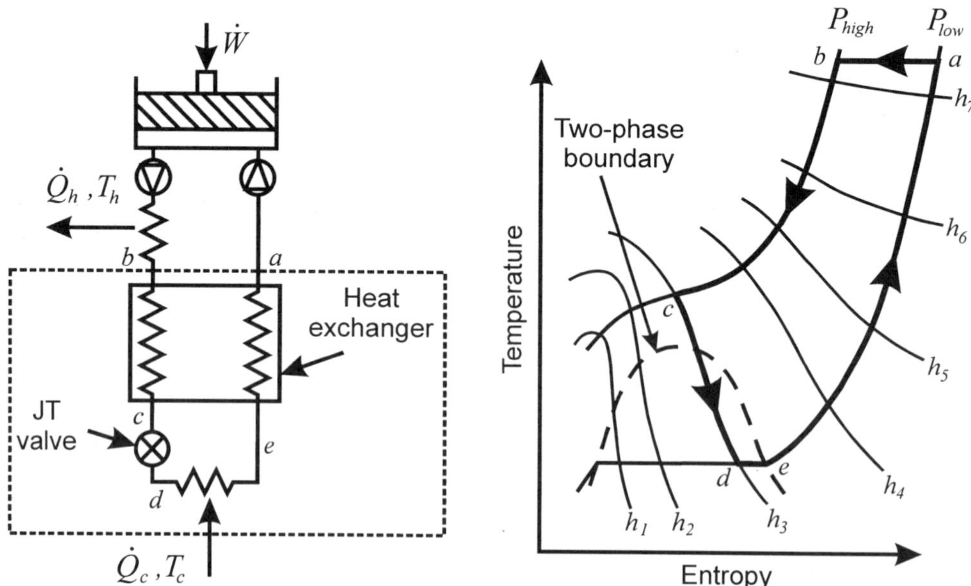


**Figure 1.** Schematic diagram of a Joule-Thomson system with temperature-entropy diagram.

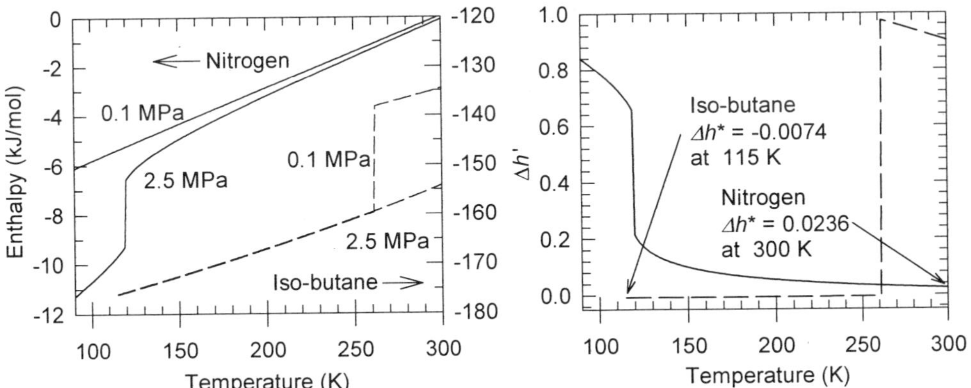


**Figure 2.** Enthalpy and normalized enthalpy for nitrogen and iso-butane.

## MIXED GAS REFRIGERANTS

Mixed-gas Joule-Thomson (J-T) refrigerators have been used since the mid-1930's when Podbielniak[1] received a U.S. patent on the first use of mixed gases in a single flow stream using a series of heat exchangers and phase separators. The phase separators allow only the liquid phase to be expanded at each stage, thus maintaining a high efficiency in the expansion process. Kleemenko,[2] Missimer,[3] and Little[4] have made subsequent improvements to this mixed refrigerant cascade cycle that uses phase separators. In 1969 Fuderer and Andrija[5] first used mixed gases in a Joule-Thomson cooler with no phase separators. In this simplified cooler the entire gas mixture flows through the heat exchanger and expands at the J-T valve or orifice. Fuderer and Andrija pointed out that because the mixture is in the two-phase region in most of the heat exchanger, much greater heat transfer coefficients are possible compared with pure gas streams. The mixture is usually 100% liquid at the cold end just prior to expansion. Alfeev et al.,[6] Little,[7] and Longsworth[8,9] have described subsequent improvements to this type of mixed-gas J-T system. We decided to use the mixed-gas J-T system without phase separators for the cryogenic catheter because the extreme space limitations at the cold tip would not allow for the use of phase separators.

Mixed-gas refrigerants allow for higher cycle efficiency by maintaining a more uniform enthalpy difference between the high and low pressure streams. Traditional J-T systems have used a pure fluid for the refrigerant. This results in a very inefficient system. J-T systems rely on the fact that there is a reduction in enthalpy with increasing pressure. An ideal gas has no change in enthalpy with pressure; therefore we must use a nonideal working fluid. Fluids are most nonideal in the liquid or near-liquid state. The vapor-compression cycle works well because the fluids are nonideal over the limited temperature range of operation. Cryogenic systems must operate over a very large temperature range, making the fluids very close to ideal over most of the temperature range, as can be seen in Figure 2 by the enthalpy of nitrogen.

The traditional analysis of a J-T system is very simple. We can draw a control volume, shown as the dashed line in Figure 1, around the cold end of the system. With the assumption of an ideal heat exchanger, the first law energy balance around the control volume can be written to find the refrigeration power, $\dot{Q}_c$, given by

$$\dot{Q}_c = \dot{n}(h_a - h_b) = \dot{n}\Delta h, \tag{1}$$

where $\dot{n}$ is the molar flow rate and the molar enthalpy, $h$, is evaluated at points $a$ and $b$ shown in Figure 1. From this simple expression, the refrigeration power can be maximized by simply maximizing the enthalpy difference at the warm end. Customarily these warm end enthalpies are evaluated at the same temperature when assuming a perfect heat exchanger. The problem with this analysis is that it also makes the assumption that the minimum enthalpy difference of the refrigerant occurs at the warm temperature; this assumption is rarely stated or even recognized. This traditional analysis works fine for a single fluid refrigerant, like nitrogen shown in Figure 2, but for a mixed gas we must recognize that the $\Delta h$ in Eq. (1) should be the minimum enthalpy difference over the entire operating temperature range.

Molar flow rate is used instead of mass flow rate because a constant volume compressor will give a constant molar flow rate but the mass flow rate will vary. Using molar flow rate makes it easier to compare gas mixtures with different compositions.

The enthalpy difference, $\Delta h$, needs to be normalized in order to compare different gas mixtures. The enthalpy difference can be normalized in a number of ways depending on what is critical for effective operation of the system. For the cryogenic catheter, the heat exchanger is the most difficult component to make effective due to the small size. The enthalpy difference should then be normalized by using the amount of heat transferred in the heat exchanger. Therefore, the normalized enthalpy difference, $\Delta h'$, is defined as

$$\Delta h'(T) = [h(T, P_{low}) - h(T, P_{high})] / [h(T_{high}, P) - h(T_{low}, P)]_{min}, \quad (2)$$

where $[h(T_{high},P)-h(T_{low},P)]_{min}$ is evaluated at either the high pressure or low pressure, whichever gives the smaller enthalpy difference. We now define $\Delta h^*$ as the minimum value of $\Delta h'$ over the operating temperature range.

By maximizing $\Delta h^*$ the gas mixture can be optimized for use in a cryogenic catheter.[10] Figure 2 shows the enthalpy, $h$, and the normalized enthalpy difference, $\Delta h'$, of nitrogen at pressures of 0.1 and 2.5 MPa between 90 K and 300 K. The figure shows that nitrogen has a very low change in enthalpy above 120 K and shows that $\Delta h^*$ is only 0.0236, which occurs at the highest temperature, 300 K. The large change in enthalpy at 120 K is due to the liquefaction of the nitrogen at the high pressure. All fluid property data were calculated using the mixture property database NIST14.[11] Figure 2 shows $h$ and $\Delta h'$ of isobutane; the liquefaction point can be seen at 261 K. The $\Delta h^*$ is slightly negative for pure isobutane at temperatures below 261 K. Figure 3 shows $h$ and $\Delta h'$ of a mixture of 80 mole% nitrogen and 20 mole% iso-butane, $\Delta h^*$ has been increased to 0.0317, a 34% gain. The $\Delta h^*$ occurs at a midpoint temperature of 168 K, whereas the temperature extremes have a high $\Delta h'$. $\Delta h^*$ can be increased by adding a gas to the mixture that has a liquefaction range about where a higher $\Delta h'$ is required. Figure 4 shows how $\Delta h^*$ is increased by adding methane (111-180 K liquefaction range) and ethane (184-275 K liquefaction range) to the mixture. Now $\Delta h^*$ is 0.0365. The component amounts are next

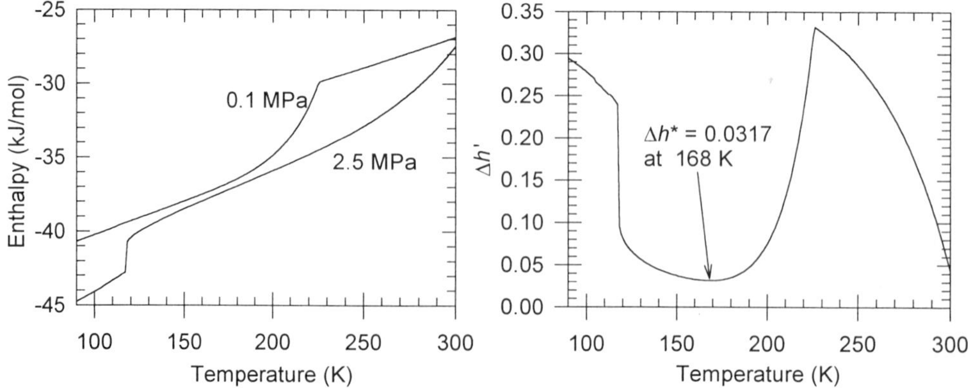


**Figure 3.** Enthalpy and normalized enthalpy for 80% nitrogen and 20% iso-butane.

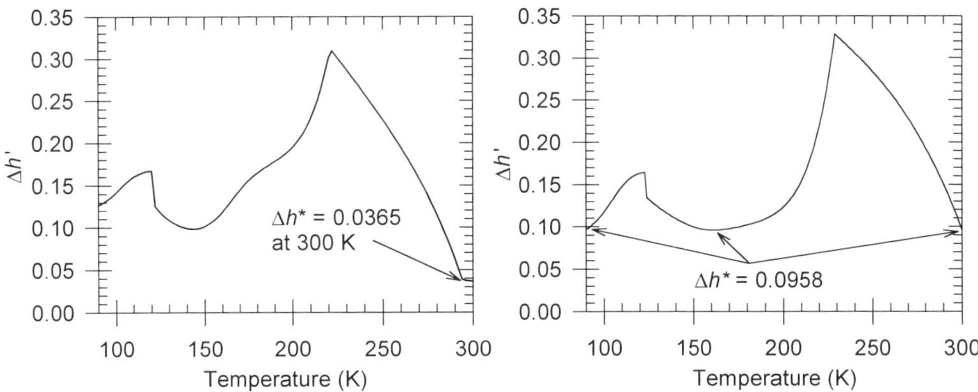


**Figure 4.** Normalized enthalpy for 45% nitrogen, 20% methane, 20% ethane, and 15% isobutane.

**Figure 5.** Normalized enthalpy for 38.5% nitrogen, 27.5% methane, 10.5% ethane, and 23.5% isobutane.

adjusted until the maximum $\Delta h^*$ of 0.0958 is achieved with a composition of 38.5% nitrogen, 27.5% methane, 10.5% ethane, and 23.5% isobutane by mole fraction shown in Figure 5. The new gas mixture has an improved $\Delta h^*$ more than 4 times that of pure nitrogen. Other gases could continue to be added to the mixture until $\Delta h'$ is nearly independent of temperature.

Precooling the gas mixture before it enters the cold end heat exchanger offers the possibility of greatly enhancing the gas mixture performance at the cold end. Figure 5 shows a typical optimized gas mixture has a steep negative slope in $\Delta h'$ at the highest temperature. If the high end operating temperature is reduced a small amount, the value of $\Delta h^*$ can be greatly increased by reducing the amount of the highest boiling point gas and increasing the amount of the other gases.

Gases other than hydrocarbons can be used and show great promise to create mixtures that perform better thermodynamically than hydrocarbons without being flammable. However, the advantage of hydrocarbon mixtures is that up to about 2% compressor oil can remain dissolved in it without freezing out at temperatures as low as 85 K.[12]

We can now write the refrigeration power $\dot{Q}_c$ of a Joule-Thomson refrigerator by

$$\dot{Q}_c = \dot{n}\Delta h_{min}, \tag{3}$$

where $\Delta h_{min}$ is the minimum enthalpy difference for the gas mixture. The ideal work of compression $\dot{W}_{ideal}$ is given by

$$\dot{W}_{ideal} = \dot{n}\{h(T_0, P_{high}) - h(T_0, P_{low}) - T_0[s(T_0, P_{high}) - s(T_0, P_{low})]\} = \dot{n} / \Delta g_0, \tag{4}$$

where $s$ is the entropy and $\Delta g_0$ is the change in the Gibbs free energy at the compression temperature $T_0$. The ideal coefficient of performance $COP$ of the refrigerator is given by

$$COP = \dot{Q}_c / \dot{W}_{ideal} = \Delta h_{min} / \Delta g_0 \tag{5}$$

If system efficiency is more important, $COP$ can be maximized by adjusting the mixture composition. The mixture discussed above which has a $\Delta h^*$=0.0958 gives a $COP$ of 0.2443, which for a temperature of 90 K is 57% of the Carnot $COP$.

Further studies are needed regarding the freezing point of gas mixtures, including gases with trace amounts of water and oil. There are very little experimental data on the freezing

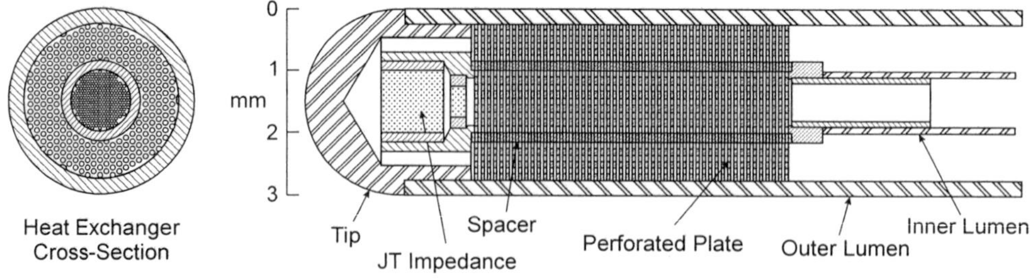


**Figure 6.** Schematic diagram of catheter tip.

point of gas mixtures, but it is clear that individual components that normally have a freezing point below the operating temperature may not freeze out of the mixture. Other areas that need further study include heat transfer effects of multi-component multi-phase mixtures.

## CRYOGENIC CATHETER SYSTEM

The schematic diagram of the catheter tip is shown in Figure 6 and details both the heat exchanger and the expansion orifice.

### Heat Exchanger

The miniature heat exchanger at the cold end is fabricated by diffusion bonding perforated plates of copper alternated with stainless steel spacers. Photos of the heat exchanger are shown in Figures 7 and 8. By diffusion bonding large metal sheets containing many individual heat exchanger layers, we are able to make large numbers of cold ends at one time, significantly reducing the cost in order to make them disposable. Prototype heat exchangers varied in length from 5 to 15 mm. The outer diameter of the heat exchangers was 2.5 mm. The enthalpy is more affected by changes in pressure at lower pressures, so the low pressure side of the heat exchanger must have a much lower pressure drop than the high pressure side.

### Expansion Orifice

The J-T impedance shown in Figure 6 was fabricated of sintered copper powder. This permitted many flow channels, limited plugging problems, and provided a large area for heat transfer. Later catheters were fabricated using a single knife-edge orifice and provided similar results with much less effort.

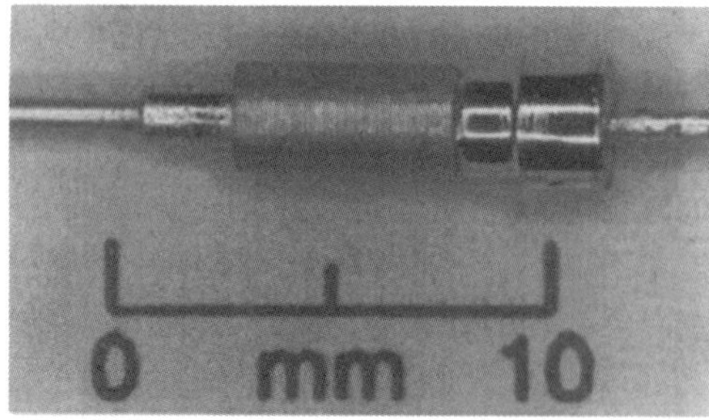


**Figure 7.** Catheter cold end. The tube on the right is only for making pressure measuments and is not part of the working catheter. The left tube is the high pressure inlet to the heat exchanger.

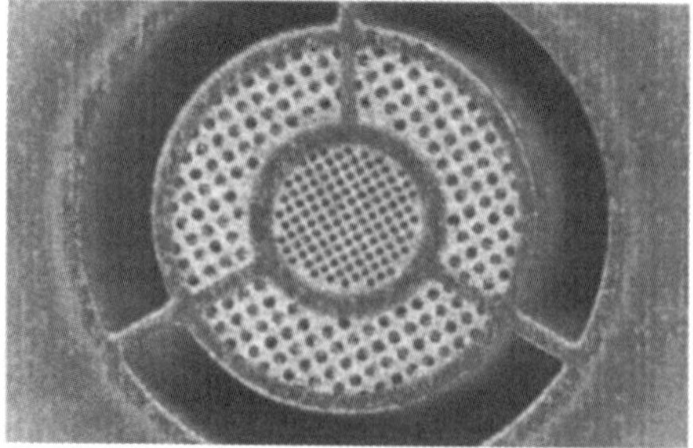

**Figure 8.** End view of heat exchanger. The top layer is a stainless steel spacer and the copper perforated plate can also be seen.

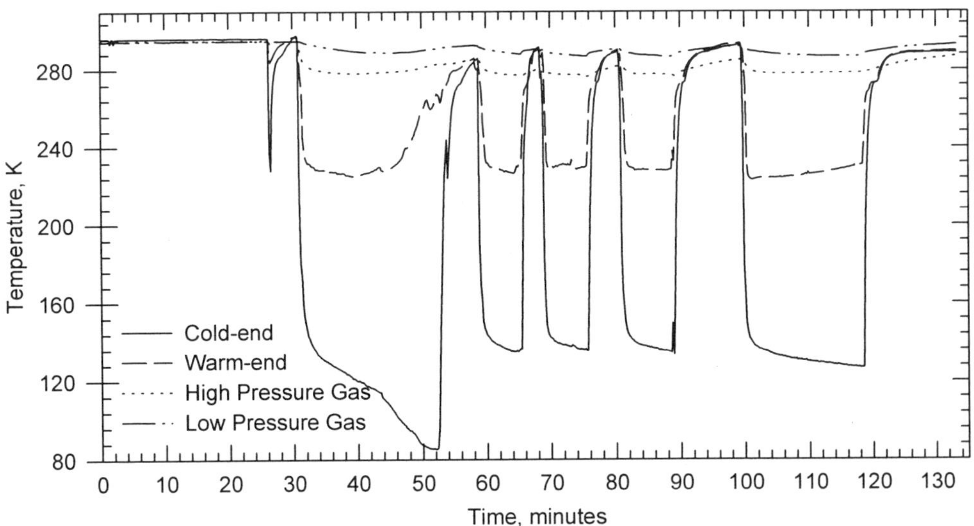


**Figure 9.** Catheter test results.The cold end is the catheter tip, the warm end is on the surface of the outer lumen at the warm end of the heat exchanger, high pressure gas is the gas temperature entering the inner lumen, and the low pressure gas is the gas temperature leaving the outer lumen.

## Lumens

A lumen is simply the tubing used for the flexible catheter pressure lines. The inner lumen was a 1 mm outer diameter polyimide tube with a stainless steel reinforcing braid. The reinforcing braid provided not only strength against the 2.5 MPa pressure but also helped to reduce the possibility of kinks in the tube. The outer lumen was a braided nylon-derivative tube with a 0.25 mm wall. For safety, the outer lumen was designed to withstand the highest pressure in the system, although a ballast volume was added to the low pressure side to reduce the average system pressure.

## Compressor

The compressor was a commercial single stage oil lubricated compressor that required input powers from 300 to 500 W. Some experiments operated two compressors in parallel to increase the mass flow rate. Later experiments were performed using a custom built oil-free compressor. This simplified problems with the gas mixture since higher boiling-point components are more soluble in the oil than the lower boiling-point components.

# RESULTS

Most experiments were performed with the catheter sitting on the bench top exposed to ambient air. A typical experimental result can be seen in Figure 9. The catheter is characterized by fast cool-downs and warm-ups. The lowest temperature achieved was 85 K with no load on the cold end. Typical operating conditions were 140 K with about 3 W of additional heat added to the cold end. Figure 10 shows the cryogenic catheter in operation.

A few experiments were performed with the cold end inserted into a room temperature gelatin to simulate biological heat loads.[13] Catheter tip temperatures of 160 to 175 K have been achieved and ice balls with diameters of about 26 mm and a mass of 11 g were created.

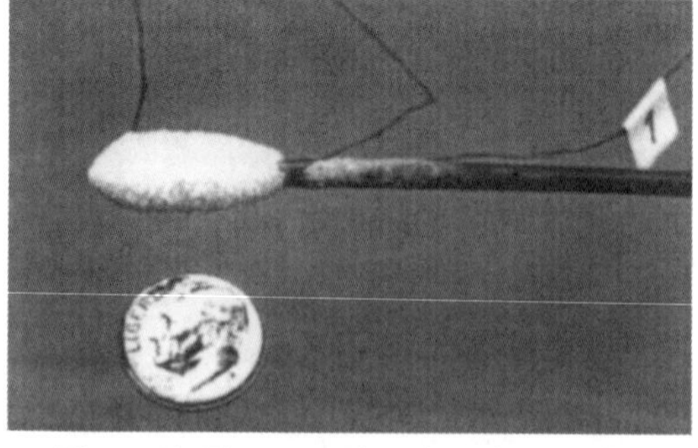

**Figure 10.** The cryogenic catheter in operation.

## CONCLUSIONS

We have developed a cryogenic catheter to treat heart arrhythmia. The required cold end temperature has been achieved, and ice balls with diameters of 26 mm weighing 11 g were created in a gelatin. We estimate that ice balls about 10 mm in diameter and a mass of 1.5 to 2.0 g are required to treat ventricular arrhythmia. Although the heat load in our experiments was less than the *in vivo* load, we believe the current refrigeration power is sufficient to meet the clinical requirement.

A procedure to optimize gas mixtures was developed. It was shown that optimization of a gas mixture can enhance the performance of a Joule-Thomson system by at least 4 times that of pure nitrogen. Low-cost techniques for fabricating miniature heat exchangers that yielded a heat exchange effectiveness of sufficient quality to reach a low temperature of 85 K were developed.

## REFERENCES


1. W.J. Podbielniak. "Art of Refrigeration," U.S. Patent 2,041,725 (1936).
2. A.P. Kleemenko, "One Flow Cascade Cycle (in schemes of natural gas liquefaction and separation)," International Institute of Refrigeration (1959), paper 1-a-6.
3. D.J. Missimer, "Auto-Refrigerating Cascade (ARC) Systems – an Overview," Tenth Intersociety Cryogenic Symposium, AIChE Spring National Meeting (March, 1994).
4. W. A. Little and I. Sapozhnikov, "Low Cost Cryocoolers for Cryoelectronics," Cryocoolers 9, Plenum Press, New York, 1997, p. 509.
5. M. Fuderer and A. Andrija, Verfahren zur Tiefkühling, German Patent 1426956 (1969).
6. V.N. Alfeev, V.M. Brodyanski, V.M. Yagodin, V.A. Nikolsky, and A.V. Ivantsov, "Refrigerant for a Cryogenic Throttling Unit," UK Patent 1,336,892 (1973).
7. W. A. Little, "Advances in Joule-Thomson Cooling", Advances in Cryogenic Engineering, Vol. 35, Plenum Press, New York, 1990, p. 1305.
8. R.C. Longsworth, "Cryogenic Refrigerator with Single Stage Compressor," U.S. Patent 5,337,572 (1994).
9. R.C. Longsworth, M.J. Boiarski, and L.A. Klusmier, "80 K Closed-Cycle Throttle Refrigerator," Cryocoolers 8, Plenum Press, NY (1995) p. 537.
10. R. Radebaugh, E. Marquardt, J. Dobak, and M. Huber, "Mixed Gas Refrigeration Method," U.S Patent number to be issued (1996).
11. NIST Mixture Property Program (DDMIX) (NIST14, Version 9.09), National Institute of Standards and Technology, Gaithersburg, MD, 1992.
12. Dr. Hans Quack, Techn. Universität Dresden, private communication (1997).
13. H.P. Lenz and J. Eichler, Cryobiology, Volume 13, p. 37-46, 1976.

# FAST JOULE-THOMSON CRYOCYCLING DEVICE FOR CRYOSURGICAL APPLICATIONS

B-Z. Maytal

Rafael Institute, P.O.Box 2250(39)
Haifa 31021, Israel

## ABSTRACT

A Linde-Hampson device has been developed, which incorporates the effects of positive and negative Joule-Thomson coefficients. Successive streams of pressurized gases, under and above their inversion temperatures (like nitrogen, argon, nitrous oxide on the one hand and neon or helium on the other), result in sequential cooling and heating. Fast cryocycling is obtained, because of the very low thermal inertia involved in a miniature Joule-Thomson "cooler". Consequently, a fast freeze and thaw complete cycle between 325 K and 105 K was accomplished within 20 s utilizing a 5.2 mm diameter surface heat loaded probe. A computer regulated fast electromagnetic valve and a thermocouple feedback have been used to maintain any steady state intermediate temperature and control temperature extremes and periods of cooling and warmup. The fast cycling capability provides a new dimension in the domain of cryo-interaction with living tissues. In this way the cryodestructive process of undesirable cells is intensified. Cold front depth of penetration may be controlled and calibrated in an open loop through frequency dependence. This may be extremely significant for delicate surgeries like those in the eye and brain. In addition, it enables frequency dependent selective cryodestruction.

## INTRODUCTION

Numerous cryosurgical devices utilize liquid nitrogen that is transferred into probes thereby interacting with living tissues for cryotherapy purposes. Liquid cryogen supply is charged and stored. There is also a group of devices that utilize the Joule-Thomson effect for cooldown of the pressurized gas.

The first cryosurgical apparatus utilizing the Joule-Thomson effect was proposed by Hood and Simons[1] in 1965, liquefying gas by the Linde-Hampson process. Nitrogen expanded isenthalpically through an orifice located in the probe. The associated temperature drop was enhanced with the use of a small recuperative heat exchanger until liquefaction is attained. In this development they took advantage of the miniature Joule-Thomson cryocoolers technology utilized earlier for infrared detection[2].

Bulat, Dray, and Sollami[3, 4] also used a nitrogen recuperative Joule-Thomson cryocooler. Their contribution was the addition of a bimetal temperature sensitive flow regulator to economize gas consumption. The ophthalmologist Amoils, focused on cryogenic cataract extraction[5]. He was the first to build a Joule-Thomson cryoprobe[6] employing a single isenthalpic expansion, but without a recuperative heat exchanger. Dobak[7], Radebaugh[8] and Marquardt[9] inserted a miniature Joule-Thomson cryocooler in a delicate catheter. The refrigerant was a hydrocarbon gas mixture pressurized by a closed cycle compressor. The idea to apply a mixed coolant extends the development of Joule-Thomson cryocooling from the work undertaken by Alfeev et al.[10].

The practice of cryosurgical therapy requires a means of thawing. Active warmup of the probe to release the tissue and minimize sticking is a highly desirable feature. Most of the devices mentioned above use electric heating directly applied to the probe. Another approach involves the flow of a coolant through the probe, maintained at room temperature and bypassing the expansion orifice to avoid cooling.

The present work[11, 12, 13] is another step in the development of recuperative Joule-Thomson cryosurgical devices. Its innovated feature includes both a recuperative warmup through a Joule-Thomson "cryocooler" and addition of the cryocycling. To the best of the author's knowledge, there is no other system with a fast alternating temperature drive over a wide range of temperatures. This technique may be useful in other applications as well.

## PRINCIPAL CONCEPT

A miniature Linde-Hampson liquefier, installed at the tip of the probe, creates a small bath of cryogen under the cold plate, shown in Figure 1. The refrigerant for this device is provided from a pressure vessel. The choice of argon over nitrogen or air, the normal choices for cryosurgical practice is justified solely for Joule-Thomson cryocooling. The trade off between the relevant thermophysical properties, on the one hand, and the boiling point of the gases on the other hand, indicate that the noble gases are the superior choice, as reflected in the study of Maytal[14]. The penalty of an elevated boiling point of 10 K relative to nitrogen, pays off.

The same Joule-Thomson "cryocooler" located inside the probe may be warmed up while being pressurized by helium gas instead of argon. Furthermore, it is a recuperative fast and intensive warmup. Murinets et al.[15], suggested warming up the probe with a gas exhibiting an inversion temperature below ambient. However, they used a nozzle without a recuperative heat exchanger, which resulted in only a single isenthalpic expansion. In this way the heat up will only develop a small temperature rise. For instance, room temperature helium, expanding from 40 MPa pressure will provide less than a temperature rise of 20 K. However, the self sustained recuperative warmup amplifies this small rise, what is by an order of magnitude higher temperature increase may be reached[11].

Cryocycling is achieved due to the main cooling and heating drives, on the one hand, and the low thermal mass (as 2 J) at the cold tip of a Joule-Thomson miniature cryocooler, on the other hand[12].

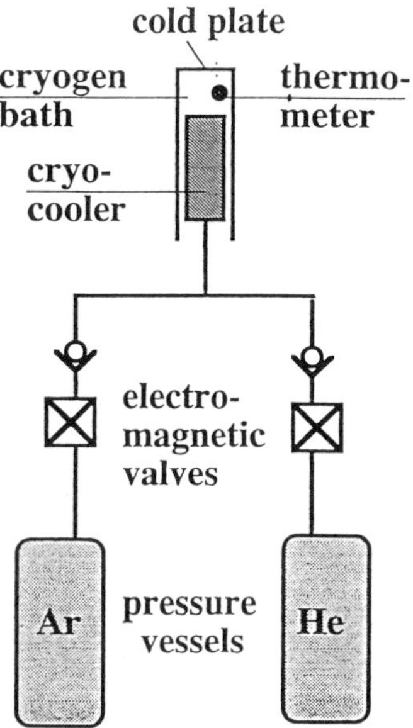


**Figure 1.** Principle of operation: a Joule-Thomson miniature cryocooler, installed at the tip of the probe, and successively pressurized by argon and helium, with a thermocouple feed back.

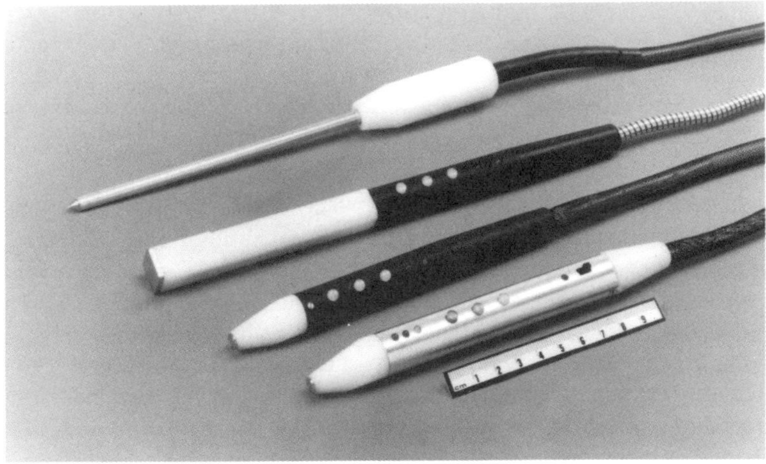

**Figure 2.** A variety of probes: 6 mm diameter, for liver (at the top) and the other for dermatology.

## CONSTRUCTION

The cryosurgical device incorporates three to eight simultaneously operated autonomous probes[13]. As displayed in Figure 1, Joule-Thomson cryocoolers, installed in each cryoprobe are exposed to a couple of pressure sources: an argon vessel and a helium vessel. Each supply channel is separately controlled by a fast electromagnetic on-off valve. Two check valves prevent bypassing flows between the two vessels. A thermocouple immersed in the cold/warm bath at the tip of the probe indicates the instantaneous temperature. This signal is fed into a computer and serves for further control decision making, either in termination of cooling or switching to heating to maintain desirable levels of upper and lower temperatures or automatically generate cryocycles.

Argon provides the fast cooldown refrigerating phase. Helium has a Joule-Thomson inversion curve much below room temperature; hence it is responsible for the warmup phase. This way, both cryocooling and warming are obtained solely by positive and negative Joule-Thomson effects.

Probes were constructed for the various applications in suitable and typical shapes. Bulk freezing as required inside a cancerous liver necessitates a long cylindrical probe of 6 mm diameter as shown at the top of Figure 2. Similarly shaped probes, but with a 3 mm diameter were prepared and tested for prostate cancer cryosurgery. Only about 2 cm of the tip is actually cryocooled, while the rest of it remains reasonably close to room temperature. Some of these probes are autonomously operated with the appropriate operating knobs. While inserting the “stick” knob, the temperature is reduced and stabilized at a level of about -30$^0$C, just to ensure that it resides in a position stuck to the tissues. “Freeze” and “Thaw” knobs initiate deep cryocooling and warm up, respectively. The “Stop” knob ceases any activity. Figure 2 also displays probes of different sizes and shapes.

## OPERATION AND PERFORMANCE

Evaluation of the advantages of these new devices should be made with the traditional ones, charged with liquid nitrogen. It should be noted that the new devices are cryogen-free. This basic feature manifests itself in many practical advantages. First, operation avoids the laborious need to charge the device before each surgical treatment. There is no need to store and circulate liquid cryogen. Since the cooling potential is preserved in the form of compressed gas, with no boil off, the device may be prepared and ready long before the planned surgery.

The hoses which connect the probes to the devices do not freeze; they permanently remain at room temperature and maintain complete flexibility during operation. Any hose conducting a stream of liquid nitrogen is much more bulky because of the required thermal insulation. Furthermore, it becomes stiff and must avoid contact with the patient.

Simultaneous multiprobe operation and further attachment of the probes is quite simply accomplished. Excellent controllability of each separate probe enables operation at any intermediate desirable temperature. The device is friendly to operate, silent, fast and easily accommodates shifting modes and set ups. An entire desirable protocol may be programmed.

The entire device is very compact, lacking the massive thermal isolation due to the absence of stored and circulating liquid cryogen. Handling of the liquid cryogen is much more bulky and demanding than maintaining a flow of ambient pressurized gas from a vessel. While the entire machine operates at room temperature, and the cryogen is

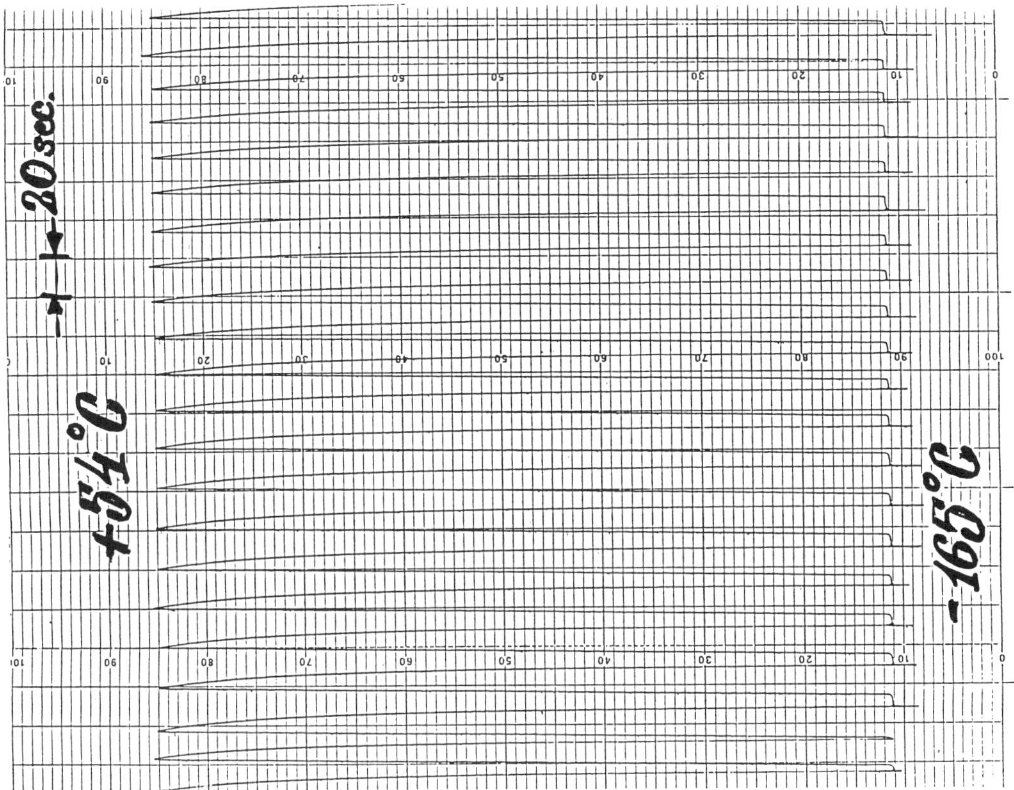


**Figure 3.** Fast cryocycling with a 5.2 mm flat surface probe, loaded by touching a tissue of a potato. Automatic cryocycling with 20 s period: 50% cooling and 50% heating. Temperature measured in the tissue adjacent to the probe.

produced solely at the tips of the probes, thermal losses are drastically reduced. Hence, a higher thermal efficiency is obtained.

The cylindrical 3.2 mm. diameter probe generates an ice ball in a water bath with dimensions of 41 mm diameter and 54 mm in length. This is a severe procedure for measuring the ice ball size, since in a liver tissue, for instance, it is significantly larger.

The active thaw is extremely intensive and fast. This probe is completely released within a second or two during a dermathological application. In the case of bulky freezing in a liver, sticking ceases after about 20 s. The low thermal inertia and the rapid switching from cooling to heating provide this useful feature.

## CRYOCYCLING POTENTIAL IN BIOLOGY

Beyond the previously mentioned operational advantages, the capability of such rapid changes over a relatively wide range of temperatures offers new cryobiological opportunities. The cryodestruction procedure for undesirable living tissues dictates that one follows a freezing phase by thawing. Often it is recommended to repeat this procedure once or twice. A cryocycling application should be regarded as a successive multiplicity of freeze and thaw cycles. This feature gains special significance in cancerous tissues, when its complete and undoubted removal is crucial. Figure 3 demonstrates a typical fast cycling from 327 K to 108 K. A 5.2 mm cryocooler within a nonevacuated encapsulation was pressurized by 30 MPa argon and helium and heat loaded by a tomato tissue. It was set for automatic cycling of 10 s cooling (while the cooldown lasts only 5 s) and 10 s heating, successively.

The process provides a method of stabilizing the effective depth of cold front penetration, as a steady periodic pattern. Furthermore, the process is controllable by adjusting the frequency of cycling and the extreme temperatures. Obviously, proper calibration is needed. This feature might be particularly beneficial in cases that do not require too deep or very delicate freezing requirements as in dermatology, eye and brain surgery or gynecology. Though periodically forcing very low temperatures, the depth of the cold front penetration remains limited and unaffected by the duration of exposure. It was verified by measuring the temperature variation at a depth of 2 mm in a potato tissue.

Cooling rates or cryocycling frequencies differently affect the various kinds of cells. While one kind may be damaged, another may be preserved. Hence, alternating the frequency may selectively preserve or destroy the cells that are targeted.

For a given ice ball size the cryodestructive diameter is larger than with the exposure to a constant low temperature. In practice, the ultrasound sensor is most often used, showing the ice ball size as an indication of the cryodestruction. Hence, in the case of cryocycling treatment the same information becomes even more relevant.

## SUMMARY

A cryosurgical multi probe device utilizing an argon Joule-Thomson cryocooler was built and successfully tested. The expected operational advantages such as ease of operation and enhanced controllability, were indeed realized, based on accumulated experience of liver surgeries. The fast defrost by the inverse Joule-Thomson warming effect was well verified. A few preliminary cryocycling experiments have been performed, to date, which basically demonstrate the cryocycling potential. More cryobiological basic experimentation will be conducted in the future, with focus on the cryocycling exposure affect on living tissues. It should verify the expected features in a practical environment.

Cryocycling exposure of living tissues has previously not been discussed as a cryobiological issue. The reason, it is believed, is the unattainability of such fast cryocycling by the prior state-of-the-art cryosurgical equipment.

## REFERENCES

1. Hood, C.B. and Simons, I.W., "Cryosurgical Apparatus", Patent No. US 3,477,377, November 11, 1969, CVI Corporation, Columbus, Ohio.
2. Geist, J.M. and Lashmet, J.M., "Miniature Joule-Thomson Refrigeration Systems", *Advances in Cryogenic Engineering,* Vol. 5, Plenum Press, New York, (1960), p. 324.
3. Bulat, T.J., Dray, W.L. and Sollami, B.J., "Cryogenic Surgical Instrument", Patent No. US 3,800,552, April 2, 1974, The Bendix Corporation, Delaware, U.S.A.
4. Solami, B.J., Bulat, T.J. and Dray, W.L., "Cryogenic Surgical Apparatus", Patent No. GB 1,422,445, Filed March 29, 1972, Patented January 28, 1976, The Bendix Corporation, Delaware, U.S.A.
5. Amoils, S.P., "A Cryothermic Cataract Extraction Technique", *South African Medical Journal,* Vol. 40, (1966), p. 797-801.
6. Amoils, S.P., "The Joule-Thomson Cryoprobe", *Arch.Ophthal,* Paris, Vol. 78, No. 8, ( 1967), p. 201-207.

7. Dobak, J., "Cryosurgical Instrument", Patent No.US 5,275,595, January 4, 1994.
8. Radebaugh, R. and Dobak, J., "Cryogenic Catheter", *Proceedings of the AIE Industrial Meeting of Military and Commercial Applications for Low Cost Cryocoolers,* San Diego, CA., (Jan. 1, 1996).
9. Marquardt, E. and Radebaugh,.R., "A Cryogenic Catheter for Treating Heart Arrhythmia", *Advances in Cryogenic Engineering,* Vol. 43, Edited by P. Kittel, Plenum Press, New York, (to be published).
10. Alfeev, V.N., Brodyanski, V.M., Yagodin, V.M., Nikolsky, V.A. and Ivanstov, V., "Refrigerant for a Cryogenic Throttling Unit", UK Patent 1,336,892, November 14, 1973, Nauchno-Issledovatelsky Institut Kriogennoi Elektroniki, Kiev.
11. Maytal, B-Z., "Fast Changing Heating-Cooling Device and Method Particularly for Cryogenic and/or Surgical Use", US Patent No. 5,522,870, June 4, 1996.
12. Maytal, B-Z., "Controlled Cryogenic Contact System", US Patent No. 5,540,062, July 30, 1996.
13. Maytal. B-Z., "Multi Probe Cryosurgical Apparatus", US Patent No. 5,603,221, February18, 1997.
14. Maytal, B-Z., "Noble Gases as Favorite Coolants for Joule-Thomson Cryostats", *Advances in Cryogenic Engineering,* Vol. 39B, edited by P. Kittel, Plenum Press, New York, (1994), p. 1935-1941.
15. Murinets, B.N., Nosov, M.E. and Bratslavsk, F., "Cryosurgical Instrument Head Reheating Method and Device by Using Gas of Lower Inversion Temperature than Ambient Temperature", Patent No. SU 1217377, March 15, 1986, Assigned for (AULO) AS UKR Low Temperature Physics.

# MAGNETOCARDIOGRAPHY AND MAGNETOENCEPHALOGRAPHY: NEW TECHNOLOGIES FOR FETAL SURVEILLANCE

R.T. Wakai,[1] A.C. Leuthold,[1] and C.B. Martin, Jr.[2]

Departments of Medical Physics[1] and Obstetrics and Gynecology[2]
University of Wisconsin-Madison
Madison, Wisconsin, USA

## ABSTRACT

The fetal magnetocardiogram (FMCG) and fetal magnetoencephalogram (FMEG) are among the most compelling applications of biomagnetism. For fetal surveillance the inherent signal transmission properties of the FMCG and FMEG are highly advantageous, compared to their electrical counterparts, and there is a paucity of alternative functional imaging techniques suitable for routine fetal evaluation. The need to develop new and improved techniques is considerable, as presently there is no single method with high predictive value for the assessment of fetal well-being, and there are no techniques whatsoever that can directly record human neurological activity *in utero*. While the first FMEG and FMCG studies were published more than ten and twenty years ago, respectively, there was surprisingly little activity in these areas in subsequent years. However, interest in fetal biomagnetism has reemerged recently, as a number of groups have reported findings at meetings and in the literature. This paper presents FMCG and FMEG studies performed by our group at the University of Wisconsin-Madison, and discusses the potential clinical utility of these techniques.

## INTRODUCTION

Since the advent of practical SQUID detectors in the 1970s, measurement of the extremely weak magnetic fields produced by the human body, i.e. biomagnetism, undoubtedly has been one of the most intriguing and challenging uses of cryogenic technology. Biomagnetism now encompasses the study of many organs and owes its appeal to the unique properties of magnetic signals, which in many situations can provide information about bodily function that cannot be obtained by other means.

Until recently, biomagnetism studies have focused almost exclusively on adults. However, cardiomagnetism and neuromagnetism studies of the human fetus have begun to attract greater attention. Unlike other areas of biomagnetism, where source localization is paramount, the rationale for fetal studies lies in the greatly improved signal quality afforded by magnetic detection. Given this and the paucity of alternative safe, noninvasive methods of obtaining functional information from the fetus *in utero*, it is somewhat surprising that fetal magnetocardiography (FMCG) and fetal magnetoencephalography (FMEG) were not pursued more vigorously following the initial reports[1,2] in 1976 and 1984, respectively. However, over the next few years we are almost certain to see greater activity, as more investigators are attracted to these areas. In this paper we begin by describing the favorable properties of fetal biomagnetic signals and discuss how these might be exploited to develop new clinical applications.

## THE FETAL MAGNETOCARDIOGRAM

### Signal characteristics of fetal electric potentials and magnetic fields

Electroencephalography (ECG) is the mainstay of cardiology; however, the fetal ECG (FECG) has never been used widely because the quality of the recordings obtained from electrodes attached to the maternal surface typically is poor. Signal-to-noise ratio is low and interference from the mother's ECG nearly always dominates any fetal signal that is present. The failure rate of FECG is especially high during the critical period 24-36 weeks' gestation, presumably due to the insulating effects of the vernix mucosa, a waxy substance which forms on the fetal skin around 24 weeks' gestation. In addition, FECG signal transmission is not well understood, and it has been shown that volume conductor theory does not apply after approximately 30 weeks' gestation[3]. Roche and Hon[4] proposed that the resistance of the fetal skin is much greater than that of the amniotic fluid and uterine muscle and that the electrical signals are conducted to the maternal surface through the umbilical vessel-placental pathway and the mucous-lined areas of the fetus. Consequently the configuration of the abdominal FECG is ambiguous, exhibits little spatial variation, and often shows little change even during labor or fetal movement.

The advantage of magnetic detection for fetal surveillance is manifest in the quality of the recordings. Figure 1 shows a comparison of the magnetic and electric signals recorded from one of our subjects at 36 weeks' gestation The FMCG data were recorded using a seven-channel, second-order SQUID. It shows much better signal-to-noise ratio than the FECG and is remarkably free of maternal interference. In many cases prior to 36

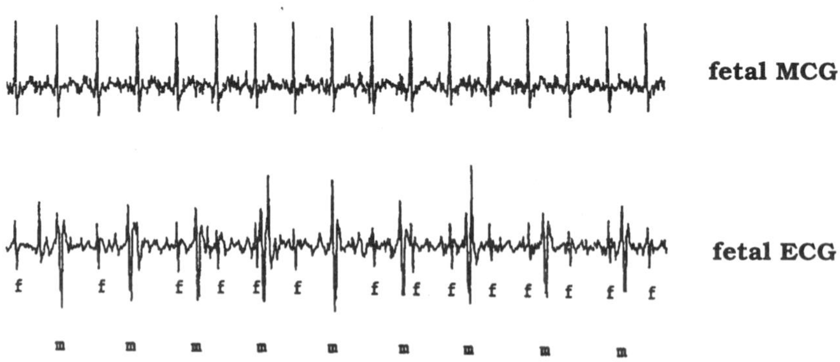


**Figure 1.** Simultaneous recording of the fetal magnetocardiogram (FMCG) and the abdominal fetal electrocardiogram (FECG) from a subject at 38 weeks. The FECG contains both fetal (f) and maternal (m) complexes, while the FMCG contains only the fetal complexes. From Wakai et al.[4].

weeks we are not able to detect any fetal electric signal, while the success rate of our magnetic recordings has been very high throughout the last trimester. We have been able to detect the FMCG as early as 16 weeks' gestation, which is considerably earlier than the FECG can be recorded. In addition, mapping studies[5] have demonstrated that, in contrast to FECG, FMCG shows signal transmission properties consistent with volume conduction, providing a basis for waveform interpretation.

## Potential clinical applications

One of the most important potential clinical applications of FMCG is fetal heart rate (FHR) monitoring for detection of hypoxic fetal distress. Each year in the US alone, 35,000 babies are stillborn, 40,000 die before age one, and 150,000 are born with deficits that severely affect their lives. Hypoxia is the primary factor in stillbirths and severe mental retardation and is the leading cause of preventable perinatal mortality and morbidity. A number of obstetrical conditions, e.g. growth retardation, preeclamsia, diabetes, and postdatism, place the fetus at increased risk of hypoxic insult due to impaired placental function. These high-risk pregnancies typically undergo twice weekly monitoring starting at 25-30 weeks' gestation. Presently, no one diagnostic test has high predictive value; thus the need to develop new techniques and/or improve the accuracy existing ones is real. The primary clinical method is FHR monitoring, which is one of the few techniques specific for hypoxia. Another widely used procedure that was developed to help improve the predictive value of FHR monitoring alone is the biophysical profile test--a composite test, consisting of FHR monitoring and assessment of amniotic fluid volume, fetal body movements, fetal breathing movements, and fetal tone, using real-time ultrasound.

FHR monitoring is performed by measuring the interval between each heart beat and plotting a continuous tracing of instantaneous FHR. Due to the high failure rate of FECG, Doppler ultrasound is the only technique used clinically during pregnancy. Interpretation of the tracings is subjective and based largely on baseline rate, the presence, absence, and timing of accelerations and decelerations, and variability in the baseline rate. For example, the presence of FHR accelerations and FHR variability (FHRV) is reassuring. The absence of these features and/or the presence of FHR decelerations that are late or variable with respect to uterine contractions is ominous.

One shortcoming of present technology is the inability during pregnancy to accurately assess short-term, or beat-to-beat, FHRV, which appears as small fluctuations in FHR, produced by asynchronous activity of the autonomic nervous system acting of the sino-atrial (SA) node. Clinically, a reduction in FHRV is associated with hypoxic or pharmacological depression. However beat-to-beat FHRV is technically difficult to assess during pregnancy. It is generally believed that Doppler ultrasound does not accurately measure beat-to-beat FHRV; only FECG and FMCG have adequate precision. One demonstration of the precision of the FMCG is the common appearance in our FHR tracings of small, rapid oscillations, known as respiratory sinus arrhythmia (RSA). RSA is associated with fetal breathing movements and is therefore a reassuring feature; however, it cannot be detected reliably by Doppler ultrasound.

The FMCG can also be used to infer fetal trunk movements. Due to its simple transmission properties, the amplitude and configuration of the FMCG is very sensitive to changes in the position and orientation of the fetal heart (Figure 2). Although these movement "artifacts" are undesirable for some types of measurements, they are useful indicators of fetal body movements in the assessment of fetal well-being. It is even possible to detect extremely small signal amplitude oscillations due to fetal breathing movements.

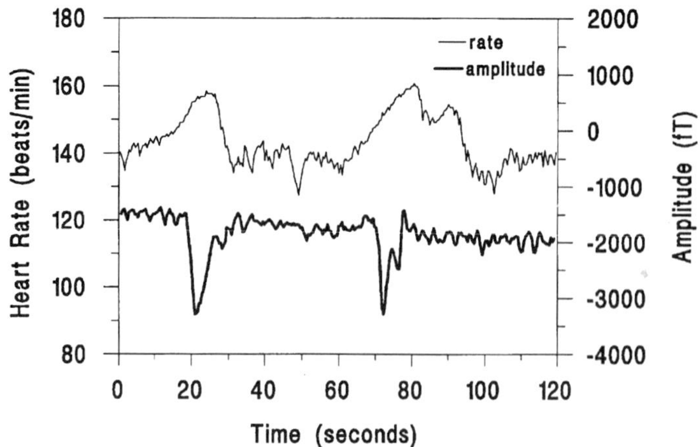


**Figure 2.** The upper tracing shows the fetal heart rate (FHR) ; the lower tracing shows the amplitude of the fetal magnetocardiogram (FMCG) signal, as measured by the amplitude of the QRS complex. .Several important aspects of the FHR are illustrated in this example. Two prominent FHR accelerations and considerable FHR variability (FHRV) are present. These features are reassuring. FHR accelerations are usually associated with fetal body movements, resulting in large changes in FMCG signal amplitude. Note that the FHR begins to accelerate before the fetus moves. The origin of FHR accelerations is not well understood. As seen here, FHRV typically is greatest when FHR is near baseline and is reduced significantly during accelerations. This is due to changes in autonomic nervous system activity that accompany fetal movement.

A common and very important use of the FMCG is measurement of waveform parameters. The heart signal consists of several components- the P-wave, QRS complex, and T-wave, which are due to atrial depolarization, ventricular depolarization, and ventricular repolarization, respectively (Figure 3). The timing and configuration of these components provide important diagnostic information. There is evidence from animal studies[6] and from human studies during labor[7] that measurement of waveform parameters can provide an early and sensitive indicator of hypoxia. Potentially this information can be used to improve the predictive value of fetal monitoring by resolving uncertainties that arise from the use of FHR monitoring alone.

Waveform parameters are also useful for diagnosis of fetal cardiac disorders, such as fetal arrhythmias. In one subject with a complicated heart malformation, MCG was used to detect an unusual fetal arrhythmia, where normal sinus rhythm predominated when fetal

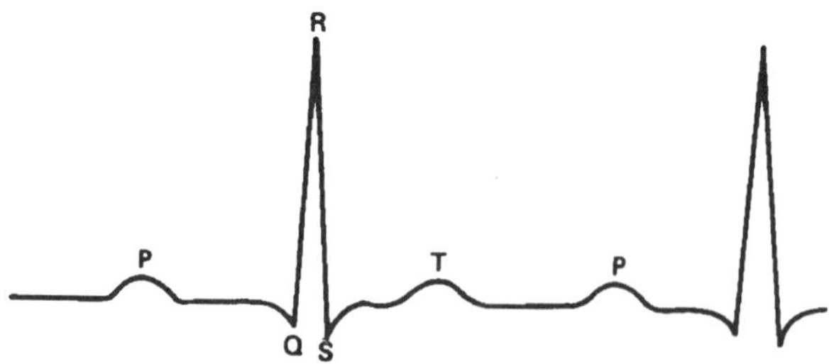


**Figure 3.** Schematic of cardiac waveform, consisting of P-wave, QRS complex, and T-wave. Cardiac rhythm analysis is based primarily on the temporal relationships among these waveform components.

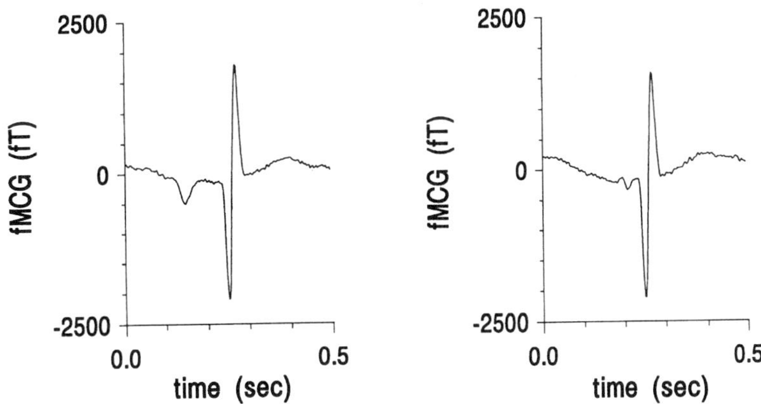


**Figure 4.** Fetal magnetocardiogram waveforms recorded from a subject at 34 weeks' gestation with an unusual rhythm disturbance that could not be detected by ultrasound. During periods of fetal breathing movements the waveform was consistent with normal sinus rhythm, as shown at left. In the absence of fetal breathing movements the P-wave configuration was altered markedly and PR interval was abnormally short, as shown at right. This waveform is consistent with an AV junctional rhythm. From Wakai et al.[8].

breathing movements were absent and a junctional rhythm predominated when fetal breathing movements were present[8]. RSA was used to infer the presence of fetal breathing movements. The junctional rhythm was manifested by a P-wave of reduced amplitude and altered configuration and a greatly shortened PR interval (Figure 4). It was hypothesized that changes in autonomic nervous system activity associated with fetal breathing movements were largely responsible for causing the rhythm alternations. This case was notable because it provided an example of a rhythm disturbance that was not detected by ultrasound, but could be analyzed in considerable detail using fetal MCG.

Another interesting case was a fetus with complete congenital heart block (CHB). In a normal heart the cardiac cycle is initiated by the sino-atrial (SA) node, which fires and causes the atria to contract. The impulse is then conducted to the ventricles through the atrio-ventricular (AV) node. In CHB, the impulse is blocked completely at the AV node. In this case, a secondary pacemaker, usually within the AV node, takes over as the pacemaker for the ventricles; however, the atria and ventricles beat completely independently of each other and the rate of ventricular contraction is much lower than normal. When no structural malformations are present, CHB usually is due to maternal anti-Ro (lupus) antibodies that attack the conduction system of the fetal heart, as was the case in this patient.

The patient was studied on several occasions from 24-38 weeks' gestation. A number of interesting behavior were documented that will be described in a forthcoming report. Among these was the observation that the atrial and ventricular FHR tracings showed prominent FHR accelerations, which is atypical in CHB patients, but was a reassuring sign. In addition, the degree of correlation observed between the atrial and ventricular FHR was remarkably high, given the dissociation between the two rhythms. The ability to analyze in detail both the atrial and ventricular heart rhythm and rate was ideal for study of this type of patient and would not have been possible using any other technique. In addition to diagnosis, FMCG may prove useful for assessing the efficacy and severity of side effects of various therapies used to treat fetal cardiac disorders. Given the increased use of early medical intervention, there is a definite need for such techniques.

## THE FETAL MAGNETOENCEPHALOGRAM

Despite the introduction of numerous methods of fetal evaluation- e.g. FHR monitoring, biophysical profile testing, assessment of fetal behavior- the incidence of cerebral palsy and serious mental retardation accompanying cerebral palsy remains virtually unchanged over the last few decades. Presently there is no technique with the specificity to be employed as a screening tool for detection of abnormal fetal brain function.

Direct recording of fetal neurological activity *in utero* has not been possible due to the inaccessibility of the fetal brain. While neuroscience has benefited greatly from the introduction of new functional brain imaging technologies- e.g. positron emission tomography (PET), single-photon emission computed tomography (SPECT), functional magnetic resonance imaging (fMRI), they pose potential hazard to the fetus and are not considered suitable for routine fetal evaluation. From this standpoint, magnetoencephalography (MEG) is an ideal technology in that it is completely safe and noninvasive, and records brain activity directly with high temporal resolution.

Recently, we recorded fetal brain responses to an auditory stimulus at 34-40 weeks' gestation[9]. The stimulus was a 20 msec tone burst of frequency 1.5 kHz and intensity 100 dB, delivered from speakers placed 1.5 m from the subject. Ultrasound imaging was employed so that the probe could be positioned as close as possible to the fetal scalp. One hundred trials were collected, using a random intertrial interval of mean duration 2-sec. The fetal auditory evoked responses (FAERs) were averaged and band-limited, using a 2-20 Hz passband.

Four of fourteen fetal subjects showed distinct FAERs with latencies near 200 msec (Figure 5). The largest of these signals was approximately 100 fT. Smaller deflections with latency 100-300 msec were also observed. Two of these subjects and four others were studied after birth and all showed neonatal AERs similar to the fetal responses.

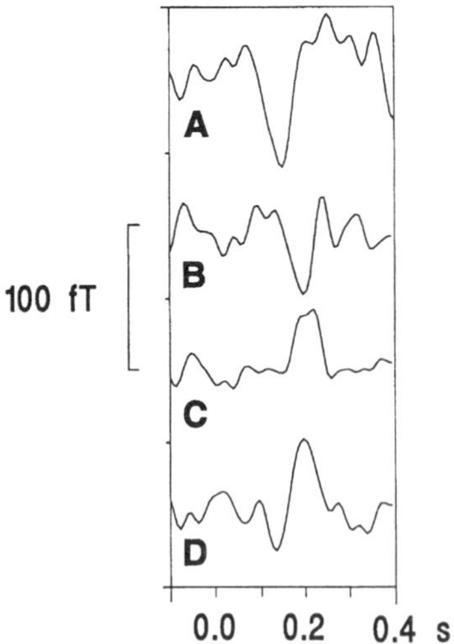


**Figure 5.** Fetal auditory evoked responses (FAERs) from four subjects, recorded at gestational ages A) 37 weeks, B) 34 weeks, C) 35 weeks, and D) 36 weeks. The stimulus occurs at time zero; the dominant component of the FAER has a latency of approximately 200 msec. As seen here, magnetic signals can have positive or negative polarity, depending on the placement of the probe. From Wakai et al.[9]

Previous to our study, Blum and co-workers recorded FAERs using a single-channel SQUID[2]. However, only a few subjects were studied and the quality of the recordings was not high. In our study, significantly better responses were obtained, but the overall success rate was only about 30%. Failure to detect a signal can be due to noise from various sources- instrumental, environmental, and biological- as well as factors beyond our control, including immaturity of the fetal brain, movement or unfavorable position or orientation of the fetal head, and attenuation of the sound intensity by the maternal tissues. Improvements in instrumentation and signal processing can help overcome these problems by increasing signal-to-noise ratio. Clearly, FMEG is a highly demanding application that will benefit from future advances in technology.

## DISCUSSION

It is still early to assess the potential of FMCG and FMEG for clinical use. Until recently, relatively little effort has been devoted to these areas, and technical advances in biomagnetism and competing technologies will continue to occur at a rapid pace. It must also be acknowledged that obstetrics is a difficult field in which to introduce sophisticated new technology, and the studies necessary to demonstrate clinical efficacy are difficult to perform due to the low percentage of pregnancies with poor outcomes.

For FMCG, significant progress has been made during the last few years. The quality of recordings has improved, allowing more extensive analyses to be performed. As a signal source for FHR monitoring, FMCG is precise enough to accurately measure short-term FHRV. It can also assess FMCG waveform parameters and signal amplitude changes associated with fetal movements. This is considerably more information than is provided by any other single method of FHR monitoring. We and others are working to demonstrate that these additional data can improve the predictive value of FHR monitoring.

For diagnosis of fetal rhythm disturbances it is already apparent that FMCG can be a valuable tool. We have shown in several subjects that the technique is capable of providing diagnostic information that cannot be obtained by any other means, including ultrasound. However, the number of patients at a typical medical center who might benefit from this type of examination is too low to justify the cost of an FMCG facility for this application alone.

Although FMEG has not been investigated as extensively as FMCG, it may have greater potential for scientific breakthroughs. Among existing functional brain imaging modalities, FMEG may be the only viable technology that can directly record human brain activity *in utero*. For this reason alone, it should continue to attract considerable interest.

In summary, FMCG and FMEG are among the few techniques well-suited for functional investigation of the fetus *in utero*. They offer compelling advantages over existing methods and can address important clinical and scientific problems that have been refractory for many years. Despite formidable challenges that must be overcome, we are optimistic that this field will come to be recognized as one of the most interesting and innovative applications of cryogenic technology.

## ACKNOWLEDGMENTS

This work was supported by grant R01-HL54579 from the National Institutes of Health. A.C. Leuthold was supported by NRSA grant CA09206 during the first few years of this project.

## REFERENCES

1. Kariniemi, V., Ahopelto, J., Karp, P.J. and Katila, T.E., The fetal magnetocardiogram, J. Perinat. Med. 2:214 (1974).
2. Blum, T., Saling, E. and Bauer, R., First magnetoencephalographic recordings of the brain activity of a human fetus, Br. J. Obstet. Gynaecol., 92:1224-1229 (1985).
3. Oostendorp, TF., Modeling of the fetal ECG, PhD thesis, University of Nijmegen, Nijmegen, The Netherlands, 1989.
4. Roche, J.B. and Hon, E.H., The fetal electrocardiogram. V. Comparison of lead systems, Am. J. Obstet. Gynecol. 153:402 (1965).
5. Wakai, R.T., Wang, M., and Martin, C.B., Spatiotemporal properties of the fetal magnetocardiogram, Am. J. Obstet. Gynecol. 170, 770-6 (1994).
6. Rosen, K.G. and Kjellmer, I., Changes in the fetal heart rate and EGG during hypoxia, Acta Physiol. Scand. 93: 59-66 (1975).
7. Murray, H.G., The fetal electrocardiogram: current clinical developments in Nottingham, J. Perinat. Med. 14: 399-404 (1986).
8. Wakai, R.T., Leuthold, A.C., and Martin, C.B., Association of fetal junctional rhythm and respiratory sinus arrhythmia detected by magnetocardiography, Ped. Cardiol. 18:201 (1997).
9. Wakai, R.T., Leuthold, A.C., and Martin, C.B., Fetal auditory evoked responses detected by magnetoencephalography, Am. J. Obstet. Gynecol. 74:1484 (1996).

# DEVELOPMENT OF THE LONG-LIFETIME SOLID NITROGEN DEWAR FOR NICMOS

C. D. Miller

Ball Aerospace Systems Division
Boulder, Colorado 80306-1062, U.S.A.

## ABSTRACT

The NICMOS dewar design was driven by requirements for long cryogen lifetime and minimal detector contamination. Several unique design features were implemented to meet the long-lifetime requirement, and are discussed. These include thermoelectric cooling, a single cryogen fill/vent line, and unique electrical cabling. Because the dewar annulus vacuum is not opened to space after launch, component outgassing was carefully evaluated during dewar integration. A series of tests was performed to evaluate the contributions of all non-metallic materials. Development of the outgassing rate specification and comparison to the measured rate at the end of dewar vacuum acquisition is presented. The methods used for long-term vacuum maintenance following pumpout, and long-term performance results are presented. The no-vent $LN_2$ fill technique and the fill efficiency achieved using this process are described.

## INTRODUCTION

The Near-Infrared Camera and Multi-Object Spectrometer (NICMOS) is a second-generation Hubble Space Telescope (HST) science instrument that was installed on orbit in February 1997. NICMOS is the first infrared instrument built for HST, and is designed to provide imaging and limited spectroscopic observations of astronomical targets between 1.0 and 2.5 microns. The instrument's three HgCdTe detectors are maintained at 58 K by a solid nitrogen dewar, shown in Figure 1. The dewar was designed for a cryogen lifetime greater than four years at an environment temperature of 270 K.

## DEWAR DESIGN DESCRIPTION

A cut-away view of the NICMOS dewar is shown in Figure 2. At the heart of the dewar are the three Boeing North American (formerly Rockwell) 256x256 pixel HgCdTe detectors, which are mounted to a cold bench. The cold bench is constructed of aluminum plates which are riveted together to form a stiff and thermally conductive bench. The cold bench is housed within a cold well formed into the cryogen tank. The 118-liter cryogen tank is filled with low-density aluminum foam, which acts both to conduct heat within the tank and to physically constrain the solid cryogen mass as it sublimes away from the tank walls.

**Figure 1.** NICMOS solid nitrogen dewar.

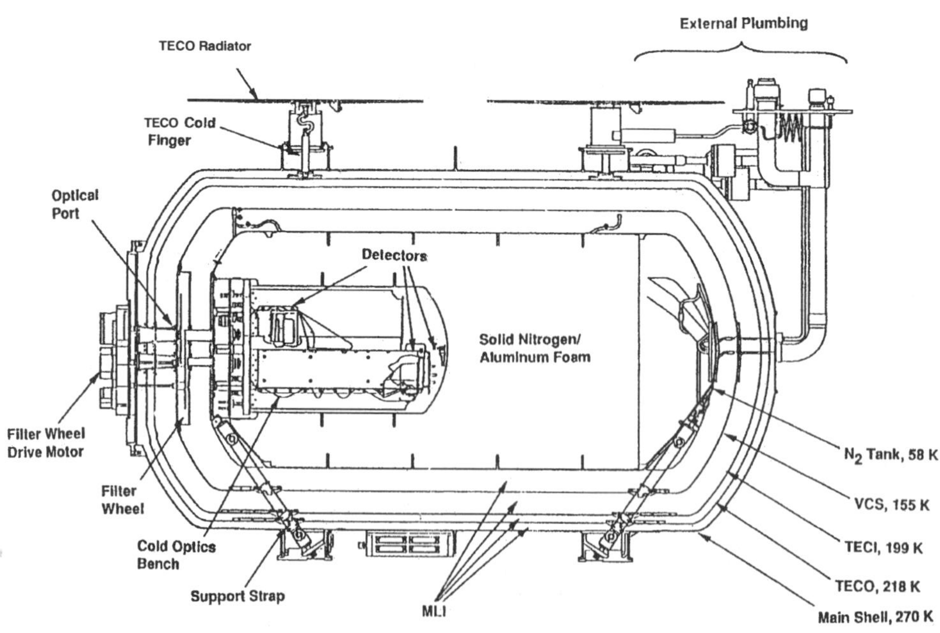


**Figure 2.** NICMOS dewar cutaway view.

The tank is surrounded by a vapor-cooled shield (VCS), which is cooled by sublimed nitrogen vapor. Surrounding the VCS are two additional thermal shields (TECI and TECO) that are each cooled by a set of two thermoelectric coolers (TECs). Each of the four TECs is mounted to the underside of a radiator that forms a part of the instrument enclosure. The cold tip of each TEC is attached to its thermal shield by means of a rigid cold finger and a flexible s-link.

The vacuum shell maintains annulus vacuum during ground operations and supports the cryogen tank and thermal shields. The support straps are anchored to two girth rings that form a part of the vacuum shell.

Over 100 layers of aluminized mylar/Dacron net insulation separate the various dewar stages. Six tensioned s-glass/epoxy straps support the cryogen tank and thermal shields. These supports were designed to retract during installation into the vacuum shell, resulting in greater length for better thermal isolation.

The plumbing system is shown schematically in Figure 3. A single fluid line serves as both the fill and vent path for the nitrogen. The on-orbit vent passes through a capillary line, creating a backpressure that yields the desired cryogen operating temperature. The emergency vent path contains two parallel burst disks to meet the two-fault tolerance requirement for STS payloads. The cooling loop at the aft end of the dewar facilitates nitrogen solidification and recooling using cold helium gas.

The optical path consists of separate optical ports for each of the three cameras. Each camera also has a separate 20-position filter wheel in a housing mounted to the VCS forward dome (Figure 4). Each wheel is driven by a stepper motor and gearhead mounted on the vacuum shell. The driveshafts are low-conductivity fiberglass tubes with bellows at each end to compensate for misalignment and thermal contraction.

Detector signal and clocking commands and instrumentation signals are carried from cold to warm regions by eight unique electrical cables routed along the support straps. These cables use small-diameter stainless steel leads, resulting in very low parasitic heat loads.

## DESIGN FEATURES

The chief design drivers for the NICMOS dewar were long cryogen lifetime and detector cleanliness. The long-lifetime requirement drove the need for certain unique fea-

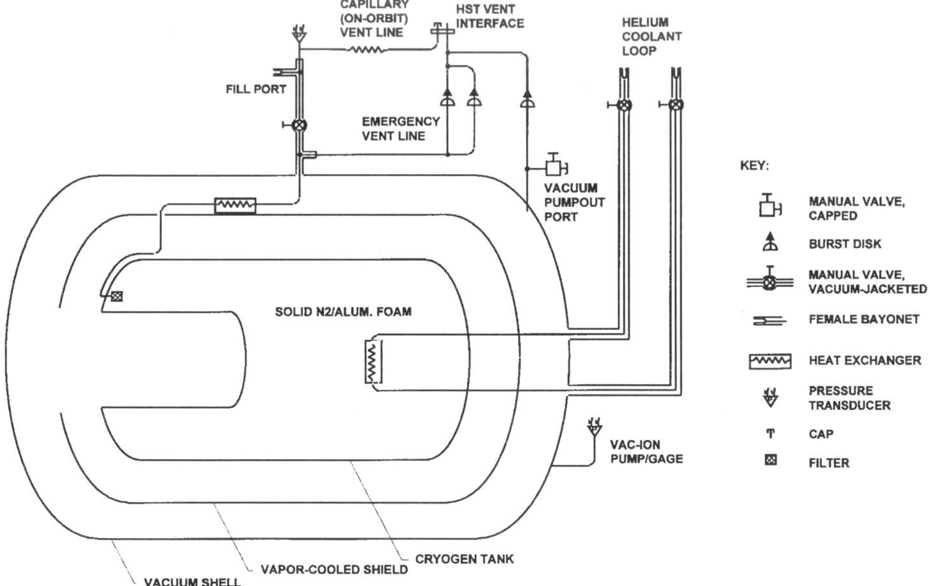


**Figure 3.** Dewar plumbing schematic.

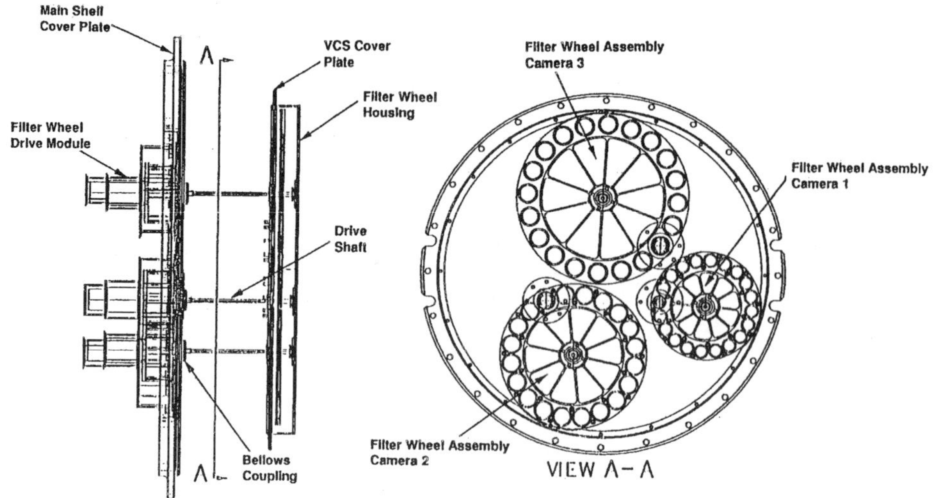


**Figure 4.** Filter wheel assembly.

tures, such as the thermoelectric coolers, the single cryogen fill/vent line, and low-conductivity electrical cabling. The detector cleanliness requirement forced careful evaluation of all materials used in the dewar, with an extensive test program implemented to evaluate outgassing characteristics of all non-metallic components including MLI. These design drivers are discussed below.

**Thermoelectric Cooling.** The use of thermoelectric coolers (TECs) to reduce parasitic heat loads has been estimated to increase the dewar lifetime by a factor of two[1]. TECs are compact and have no moving parts, which makes them very attractive for high pointing accuracy optical instruments that have environment temperatures greater than 200 K (they are now used in three of the four current HST science instruments). The TECs used in NICMOS are three-stage devices. Because the TEC couples are not very strong in tension or bending, the interfacing of the TEC to the coldfinger and radiator posed some unique challenges. Considerable thermal and structural testing produced an interface design which met mechanical loading, thermal performance and packaging requirements[2]. The TECs consume a total of 32.2 W of electrical power, and produced dewar thermal shield temperatures below 200 K and 230 K (TECI and TECO shields, respectively) during instrument performance testing with a 270 K vacuum shell temperature.

**Single Cryogen Fill/Vent Line.** Typically, solid nitrogen dewars utilize four plumbing lines, one each for fill and vent, and two which form a cooling loop for cryogen solidification. For NICMOS, one of these lines was eliminated by combining the fill and vent paths into a common line. The ability to perform a no-vent $LN_2$ fill process was demonstrated by testing a subscale model of the cryogen tank and plumbing. Results of the flight dewar fill are discussed below.

**Electrical Cabling.** A total of 360 electrical leads extend from the cold stages of the dewar. This large number of conductors dictated the development of low-conductivity electrical cabling. There are two cable types, both featuring small-diameter stainless steel conductors arranged to form a ribbon cable. The quiet cables, which carry detector and instrumentation signals, are laid up as a series of twisted pairs. The conductors are individually insulated with high-strength Teflon and embedded in a carrier film. The clock cables, which carry detector clocking signals, have conductors laid up in a parallel configuration. Both types of cables are overwrapped with a shield incorporating drain wires for noise reduction.

**Non-Metallic Material Evaluation and Testing**. To avoid unacceptable contamination of the detectors and degradation of annulus vacuum over the seven-year instrument life, it was necessary to budget the allowable outgassing rate of all materials placed inside the dewar. The outgassed product of primary concern was water, since it is absorbed to the greatest degree by non-metallic materials exposed to the atmosphere. The allowable quantity of water was determined by multiplying the ice thickness at which unacceptable IR light scattering would occur, approximately 2 microns, by the surface area of the cold well, the coldest portion of the cryogen tank, estimated at 10,000 $cm^2$. Assuming a safety factor of ten, a budget of 0.18 g of water was specified. The dewar would be evacuated during the two years prior to launch, while maintained at an ambient temperature of 20 °C, and during the assumed five-year mission life, while maintained at 0 °C. The resulting allowable 20 °C outgassing rate, determined from the Arrhenius equation using a water activation energy of 10 kcal/mole, was calculated at $1.65 \times 10^{-9}$ g/s.

To evaluate the outgassing contribution of the dewar non-metallic materials, a series of outgassing tests was performed in a 5-liter vacuum chamber that was capable of being vacuum-baked at 180 °C, and used metal seals to achieve a low background leakup rate. Materials tested for outgassing included the s-glass/epoxy support straps, urethane and epoxy adhesives, electrical cables, and flat black paint. The MLI outgassing rate was measured in a separate vacuum chamber constructed especially for that purpose. The test results showed an acceptable outgassing rate for all materials.

Of secondary concern were hydrogen desorbed from stainless steel under vacuum, and nitrogen and oxygen entering via leakage and permeation of elastomeric seals. To accommodate these loads, a small vac-ion pump was permanently installed on the vacuum shell. This pump was operated periodically during the two-year pre-launch period, and also served as an indicator of the vacuum level. Since the pump would not be operated on-orbit, its magnet was removed prior to launch. To help stabilize the vacuum level once on-orbit, an adsorbent was installed. The adsorbent container was a porous metal cylinder loaded with activated coconut shell charcoal. Capacity for nitrogen and hydrogen adsorption were calculated using data from Dushman[3] extrapolated to 60 K according to the method of Larin[4].

## VACUUM ACQUISITION

Vacuum acquisition was performed using an oil-less pumping station consisting of a rotary carbon-vane roughing pump, an $LN_2$-cooled sorption cryopump, and a refrigerated helium cryopump. The station was equipped with thermocouple and hot-filament ion vacuum gages, and a quadrupole mass spectrometer for gas specie identification.

The dewar was placed inside an air oven, with the pumping station outside. The oven temperature was carefully selected as the maximum at which the detectors and TECs, with their low-temperature solder interfaces, could be maintained for a period of several weeks.

The pressure measured at the pump manifold during vacuum acquisition is plotted in Figure 5. After six weeks of pumping, the pressure had dropped to $5 \times 10^{-7}$ torr, and the vac-ion pump reading, which better reflects the dewar annulus pressure, was $2 \times 10^{-5}$ torr. The leakup rate was measured at $1.1 \times 10^{-7}$ torr/sec, meeting the specification, but it was decided to extend the pumpout for an additional five days. The final leakup rate measurement was $9.2 \times 10^{-8}$ torr/sec, measured over a one-hour interval. The pumpout was terminated after dewar cooldown and a thorough leak check of all vacuum seals. Following vacuum acquisition, the vac-ion pump was operated for several weeks prior to loading the cryogen tank to achieve additional pumping, although at a considerably reduced rate.

A history of the dewar vacuum pressure prior to launch, as measured using the vac-ion pump, is shown in Figure 6. The vac-ion pump was operated during each recool, and at other times during instrument integration and test when convenient. The maximum pressure measured was $5 \times 10^{-7}$ torr, and the pressure was typically less than $2 \times 10^{-7}$ torr.

## CRYOGEN FILL

The cryogen tank was filled with its flight load of solid nitrogen in December 1995 to within 1% of the desired quantity. A no-vent fill process was used to load successive charges of liquid nitrogen into the tank. Each charge was frozen to solid prior to admitting the next charge of liquid. Each charge required precooling of the transfer lines and flight plumbing components with $LN_2$ that had been subcooled by pumping on the supply dewar

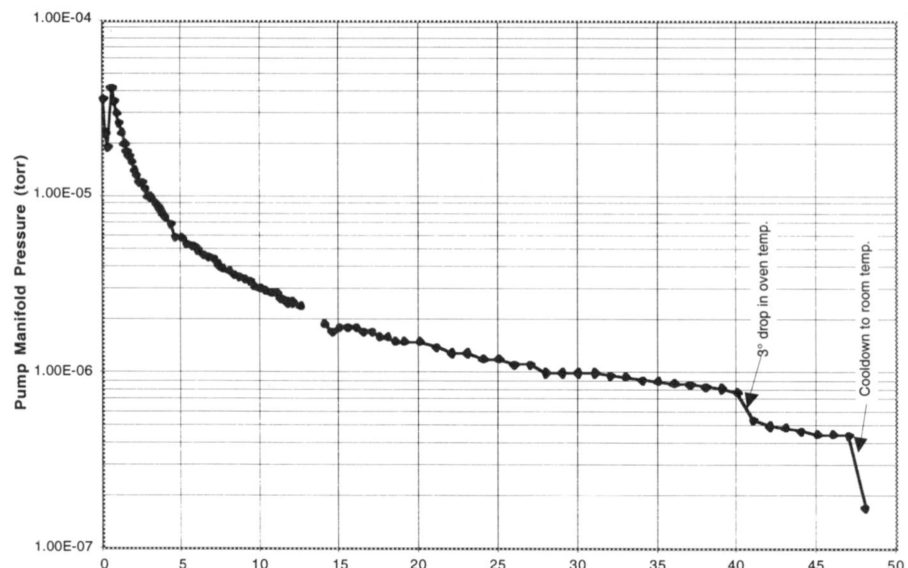


**Figure 5.** Vacuum acquisition manifold pressure history.

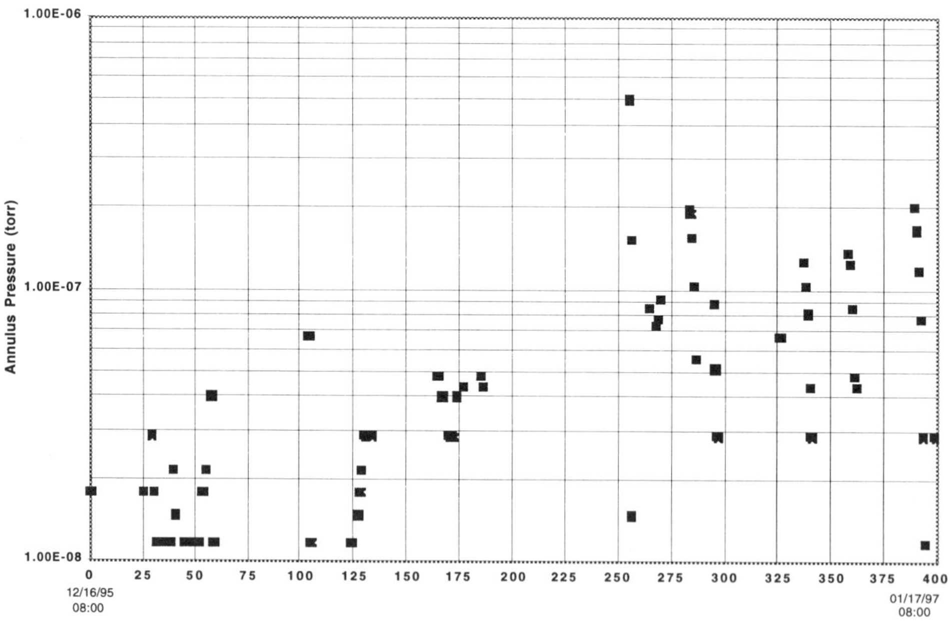


**Figure 6.** Dewar pressure history between cryogen fill and launch.

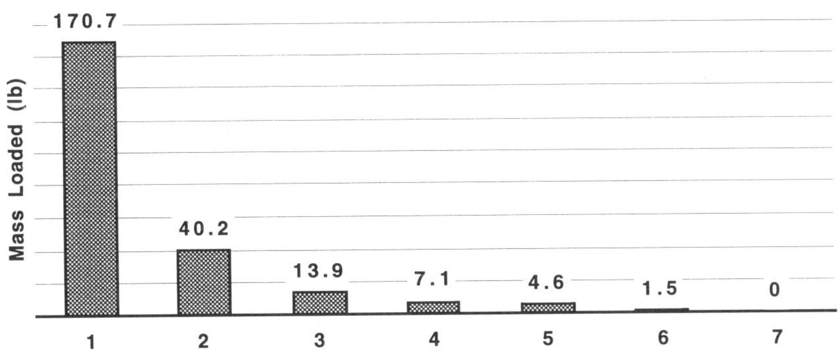


**Figure 7.** Cryogen mass loaded vs. fill attempt.

ullage. The subcooled $LN_2$ was then rapidly pressurized with $GN_2$ prior to opening the valve to the flight cryogen tank. After the liquid transfer rate dropped to near zero, the flight valve was closed, and the charge was frozen and subcooled to 58 K, the expected on-orbit operating temperature.

Prior demonstration testing had proven this no-vent fill approach for a similar cryogen tank, but with considerably different transfer lines and plumbing components, and using helium as the $LN_2$ supply dewar pressurant[5]. The demonstration test showed virtually 100% liquid fill on the initial charge, but the flight system achieved only about 70% fill for the initial charge. The difference in performance may be explained by larger parasitic heat load for the transfer lines and flight plumbing system, or may have been related to the use of $GN_2$ as pressurant, which would result in greater heat transfer rate to the supply cryogen due to condensation.

A total of seven $LN_2$ charges were loaded, with the final charge resulting in negligible mass added, as shown in Figure 7. The loaded cryogen mass was determined by suspending the flight dewar from a load cell as each charge was admitted. The total quantity of loaded cryogen indicated by this method was 108 kg, compared to the maximum possible mass calculated at 109 kg.

Following cryogen fill, the solid was subcooled to 40 K to begin the first ground hold interval. The cryogen was allowed to warm to 56 K to 58 K prior to recooling again to 40 K. At a warmup rate of 0.35 K/day, the recool interval could be as long as seven weeks, providing few interruptions for recooling during instrument integration.

## ACKNOWLEDGMENT

The NICMOS instrument was built by Ball Aerospace & Technologies Corp. for the University of Arizona under contract P172984. This work was supported by the NASA Goddard Space Flight Center through funding to the University of Arizona under contract NAS5-31289.

## REFERENCES

1. R. L. Oonk, Design of the solid nitrogen dewar for the near-infrared camera and multi-object spectrometer, "Advances in Cryogenic Engineering," Vol. 37B, p. 1385, Plenum Press, New York (1992).
2. G. L. Mills, T. K. Kelly, and B. Brown, Integration of thermoelectric coolers into a solid nitrogen dewar, "Advances in Cryogenic Engineering," Vol. 41, pp. 1121-1128, Plenum Press, New York (1996).
3. S. Dushman, Scientific Foundations of Vacuum Technique, second ed., chapter 7, 1962.
4. M. P. Larin, *Ultrahigh Cryovacuum Systems Cooled by Solid Nitrogen*, Soviet Physics-Technical Physics, 30(3), pp. 269-279 (March 1995).
5. C. D. Miller, Demonstration of no-vent fill and solidification of nitrogen for the NICMOS dewar, "Advances in Cryogenic Engineering," Vol. 39, pp. 145-152, Plenum Press, New York (1994).

# PRE- AND POST-LAUNCH PERFORMANCE OF THE NICMOS DEWAR

C. D. Miller

Ball Aerospace Systems Division
Boulder, Colorado 80306-1062, U.S.A.

## ABSTRACT

Approximately 14 months prior to launch, the NICMOS solid nitrogen dewar was filled with its flight load of cryogen and successfully completed subsystem verification testing. The dewar remained filled with solid nitrogen throughout instrument integration and testing, with periodic recooling of the cryogen to avoid venting. Thermal expansion and possible migration of the solid nitrogen during the recool-warmup cycles resulted in an unexpected shift in detector focus positions, revealed during an instrument-level optical system test. The results of an investigation of this anomaly are discussed. The instrument successfully completed system calibration and verification testing, and the cryogen sublimation rate measured during calibration was within 10% of the thermal model estimate. Subsequent recool-warmup cycles indicated that the expansion had stabilized at an acceptable level. However, shortly after launch and installation into HST, the solid nitrogen expanded further than before, moving one of the three detectors out of focus adjustment range. This expansion also caused physical contact of optical baffle segments, and a resulting large increase in the cryogen heat load and sublimation rate. Recent trends in the detector motion and heat load are discussed.

## INTRODUCTION

The Near-Infrared Camera and Multi-Object Spectrometer (NICMOS) was launched aboard space shuttle mission STS-82 and installed into the Hubble Space Telescope (HST) in February 1997. The NICMOS instrument is the first infrared instrument developed for the HST. It is designed for imaging and limited spectroscopy between 1.0 and 2.5 $\mu m$ wavelengths, using three Boeing North American (formerly Rockwell) 256 x 256 pixel HgCdTe detectors. The detectors are maintained at their operating temperature of 58 K by a subliming solid nitrogen dewar, shown in Figure 1.

## NICMOS DEWAR DESIGN

The NICMOS detectors are mounted to a conductive optical bench that is housed within a cold well of the dewar cryogen tank. Surrounding the tank are three thermal shields which reduce the parasitic heat loads reaching the cryogen. The innermost vapor

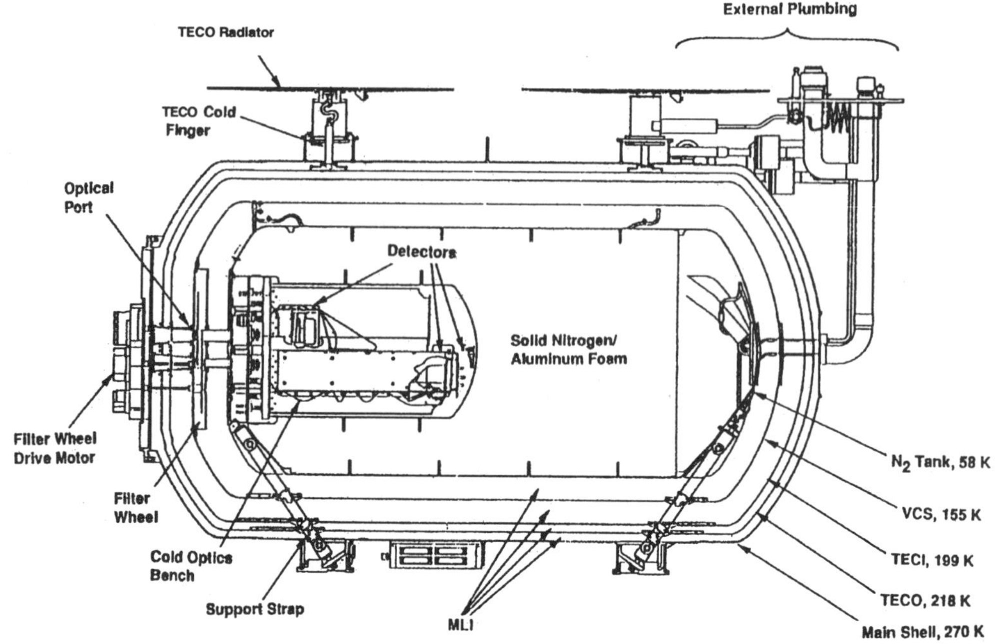


**Figure 1.** NICMOS dewar cutaway view.

cooled shield (VCS) is cooled by sublimed nitrogen vapor, and the other shields (TECI and TECO) are each cooled by a set of two thermoelectric coolers. The thermoelectric coolers reject their heat to radiators mounted on the instrument enclosure. The dewar vacuum shell supports the cryogen tank and shields from its two girth rings by means of six tensioned s-glass/epoxy support straps.

The cryogen operating temperature is determined by a flow restriction provided by the capillary vent line external to the dewar. The cryogen heat load results in a proportional sublimation rate of solid nitrogen. Flow of this sublimed vapor through the capillary line results in a back pressure inside the cryogen tank, which establishes the cryogen operating temperature according to the nitrogen vapor pressure vs. temperature relationship. The correlation between capillary inlet (cryogen) pressure and vapor flow rate was measured prior to launch. This pressure is now monitored on-orbit to provide a good estimate of the cryogen depletion rate, and the corresponding cryogen heat load.

The 118-liter solid nitrogen tank was designed to be filled with cryogen prior to integration with the rest of the instrument. Cooling coils located on the tank aft dome permit cryogen recooling with cold helium gas as needed. This approach allowed the dewar and instrument to be functionally tested at various points during integration without the need to fill and detank the cryogen, and also minimized deep thermal cycling of the detectors.

## DEWAR INTEGRATION AND VERIFICATION TESTING

The dewar integration, vacuum acquisition and cryogen fill was completed in 1995[1]. Following successful fill with 108 kg of solid nitrogen in December 1995, the dewar was subjected to random vibration and loads testing, and passed these tests successfully.

The cryogen warmup rate was monitored between recools and found to be 0.35 K/day, which agreed reasonably well with the dewar analytical model. This cryogen warmup rate was used as a metric for acceptable thermal performance. Due to program schedule constraints, dewar thermal performance testing at on-orbit conditions was performed after the dewar was integrated with the rest of the instrument.

## INITIAL DISCOVERY OF FOCUS ANOMALY

In June 1996, the integrated instrument underwent optical testing at Ball to verify alignment following acoustic testing. One of these tests uncovered an anomaly, in which one of the three detectors, Camera 3, was found to be out of focus by approximately 3.1 mm. The apparent motion was toward the forward end of the dewar, along the longitudinal axis. Because Camera 3 has a smaller depth of field than the other two cameras, an out-of-focus condition is more readily apparent and easier to quantify from its images. This discovery led NASA Goddard Space Flight Center management to form a NICMOS Anomaly Review Board (ARB) to investigate the problem. Their investigation concluded that the cause of the problem was the mismatch in CTE of the aluminum tank and solid nitrogen. This resulted in deformation of the forward and aft heads of the cryogen tank during the thermal cycling inherent to cryogen warmup and recooling. The problem may have been aggravated by the very high solid nitrogen fill quantity and the depth and speed of the initial cryogen recools. During recools, cryopumping of nitrogen into voids resulting from the CTE mismatch may have been a factor as well. The tank deformation was manifested as a forward motion of the coldwell, which carried the cold bench and detectors forward by as much as 4 mm. As part of the investigation, a non-linear finite-element structural model of the dewar was developed, and reasonable agreement between predicted and actual forward motion was obtained[2]. Also, several test efforts were carried out to obtain fatigue life data for the aluminum cryogen tank material[3], and to measure compressive strength and modulus of solid nitrogen[4].

The recommendations of the NICMOS ARB included extending the range of travel of the focusing mechanism known as the Pupil Adjust Mechanism, or PAM. The PAM was redesigned to provide twice the original range of travel, and the mechanism was relocated to offset the center of the focus range from the original position. Other ARB recommendations included performing recools more frequently to reduce the temperature range between recools, and reducing the temperature gradient across the tank during a recool.

Focus position was systematically monitored through final instrument integration. Subsequent recool and warmup cycles indicated that the focus position had stabilized, since the focus at a given cryogen temperature was repeatable during each cycle.

## INSTRUMENT VERIFICATION AND PERFORMANCE TESTING

NICMOS underwent system calibration testing in a thermal vacuum chamber at Ball in September 1996. The instrument was brought to on-orbit operating conditions and maintained there for 22 days while calibration was performed. During this time, the dewar thermal performance was evaluated, and cryogen sublimation rate was measured using a mass flowmeter. Extrapolation of the sublimation rate to steady-state conditions indicated a cryogen heat load of just under 200 mW, which would yield a lifetime of just over 4 years.

Plots of the cryogen and VCS temperatures measured during calibration testing are shown in Figure 2. At the end of the 22-day test, the dewar temperatures had not attained equilibrium. However, an extrapolation to steady-state conditions showed the heat load to be within 7% of the analytical model prediction for the extrapolated final VCS temperature.

In addition to its earlier acoustic testing, the instrument also underwent launch phase load simulation in the NASA-GSFC centrifuge facility to demonstrate launch survivability. These tests showed no effect of launch loads on focus or instrument operation. Based on this result, and the apparent stability of the focus after numerous recool/warmup cycles, the decision was made to launch NICMOS as scheduled.

## POST-LAUNCH PERFORMANCE

The NICMOS and STIS instruments were launched aboard STS-82, and installed into the HST on February 14, 1997. The dewar temperatures appeared nominal for the first 18 days following installation, with the shield temperature profiles similar to those observed during T/V calibration testing. The cryogen temperature was lower at launch than at the start of T/V calibration testing.

As shown in Figure 3, on day 63 (3/4/97), the VCS temperatures began to fall at a much faster rate, approximately -6.5 K/day, compared to the previous rate of -2.5 K/day.

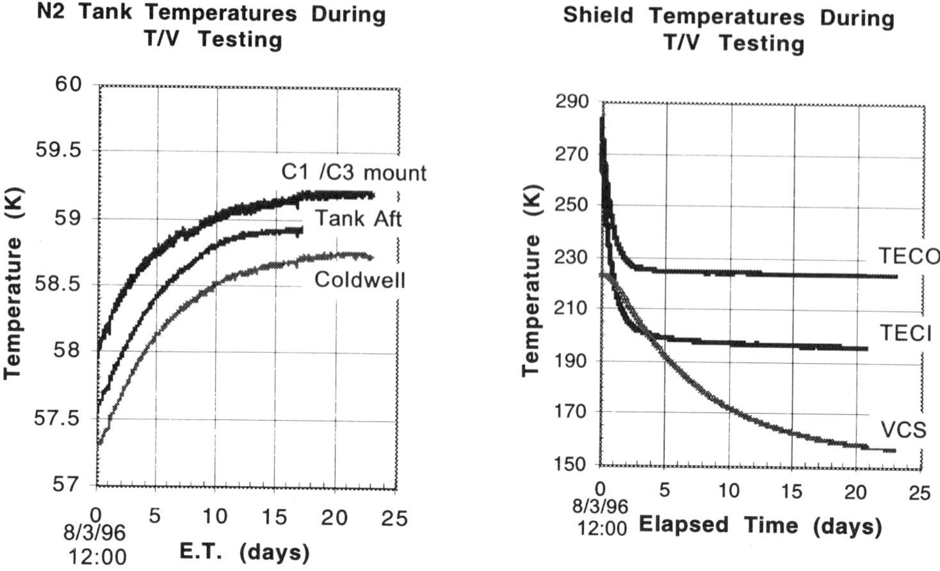


**Figure 2.** Cryogen tank and shield temperatures during T/V calibration testing.

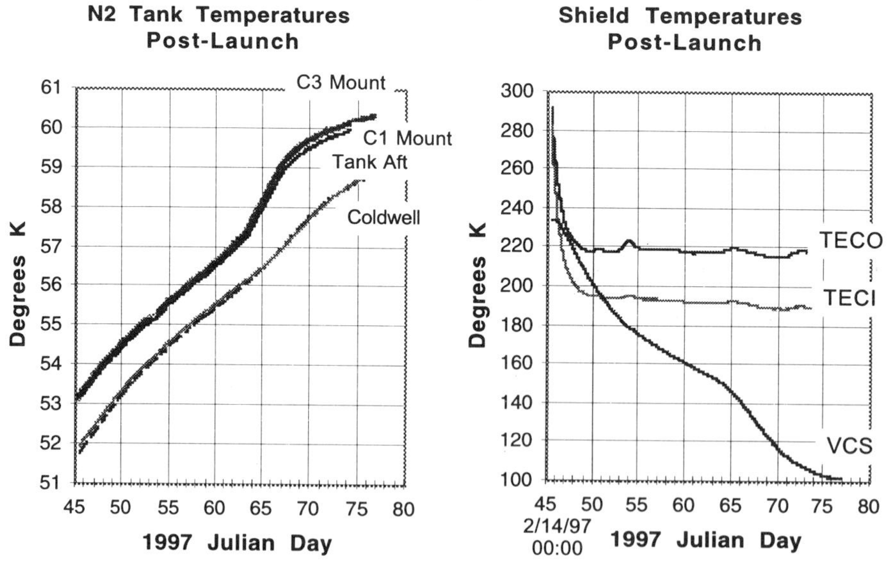


**Figure 3.** Cryogen tank and shield temperatures following HST installation.

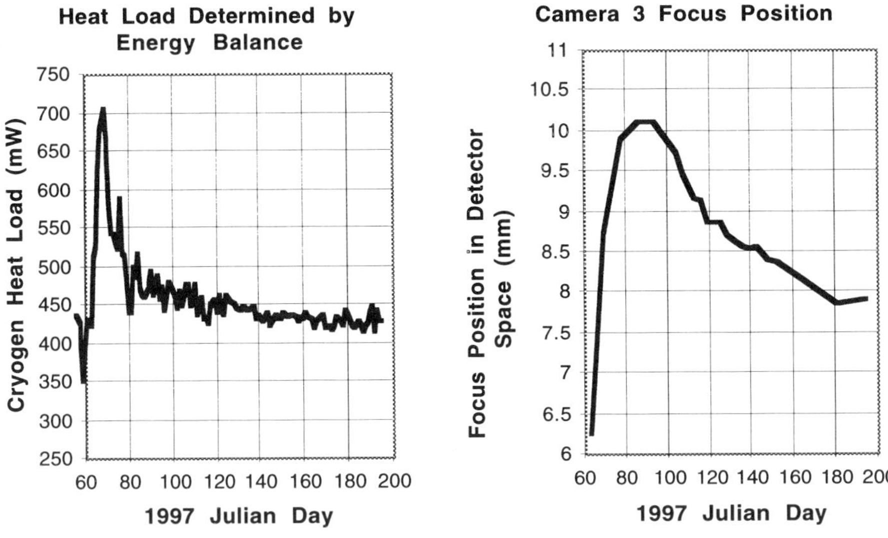


**Figure 4.** Cryogen heat load and focus history since onset of anomaly.

This was accompanied by a change in the rate of detector temperature increase to +0.50 K/day, compared to the previous rate of +0.22 K/day. The TECI and TECO shield temperatures did not appear to be significantly affected. The data indicated that a significantly higher rate of heat transfer between the VCS and cryogen tank had developed. An aft-to-forward temperature gradient of 1 K began to develop in the VCS, at the same time a 1 K forward-to-aft gradient developed in the cryogen tank. This indicated that the additional heat transfer was occurring near the forward end of the dewar, i.e., near the optical port area. A focus measurement made just prior to onset of this anomaly indicated a Camera 3 focus position approximately 6.3 mm forward of the original position. Structural analysis indicated that this position would, indeed, result in contact between the optical baffles of the VCS and those of the cold bench.

In the days following the anomalous behavior, the Camera 3 focus position changed dramatically from 6.3 mm to 10.1 mm over a period of 24 days as shown in Figure 4. Camera 3 was now out of focus, since the extended-range PAM could focus only to a position of 6.2 mm. Because of their longer focal length, however, Cameras 1 and 2 were still well within the PAM focus range. A second NASA-GSFC Anomaly Review Board was convened to evaluate the new situation. The ARB concluded that there was no action short of refocusing the HST secondary mirror that could be taken to improve the situation, but that the cryogen depletion over time would reduce the forces causing the tank deformation.

## CURRENT STATUS AND EXPECTED FUTURE PERFORMANCE

As expected, the focus position of Camera 3 has been improving over time, and the heat load has been decreasing in a similar fashion, as shown in Figure 4. This is most likely due to reduced pressure on the coldwell as the cryogen sublimes, resulting in reduced force exerted on the optical baffles in contact. Finite-element structural modeling of the system has provided a good understanding of the mechanisms that caused the anomaly. However, since there is no way to accurately specify the geometry of the deformed tank components, such modeling cannot conclusively determine if the thermal short will break before the cryogen is depleted. The most recent Camera 3 focus measurement showed a position of approximately 8.1 mm on June 30, 1997. A value of approximately 6.2 mm is needed to bring Camera 3 into diffraction-limited focusing range.

The current heat load is estimated at 430 mW and is slowly declining. Assuming the heat load continues at this value, the cryogen lifetime will be 1.9 years. The solid nitrogen temperature is stable, but has been slowly declining at a rate of .02 K every 30 days, probably due to the declining heat load. The detector temperatures have risen .03 K during the same period, likely due to sublimation of cryogen away from the forward end of the cryogen tank, where the cold bench is attached. The VCS temperature has increased at a rate of about 1 K every 30 days, with a reduction in the aft-to-forward temperature gradient. This is further evidence of a declining heat load. The thermoelectric coolers continue to operate nominally, with inner and outer shield temperatures of 190 K and 218 K, respectively.

## CONCLUSIONS

The NICMOS dewar performance measured in thermal-vacuum calibration testing would have yielded a cryogen lifetime of slightly more than four years. Had this performance continued after launch, NICMOS would have achieved the longest cryogen lifetime of any known dewar flown in space. However, unexpected expansion of the solid nitrogen after launch caused optical baffle segments of the VCS and cold bench to contact, increasing the heat load by more than a factor of two. The reason for the post-launch behavior being different than that observed during ground testing has not been determined. It is possible that additional cryogen tank deformation occurred in the 5 months between ground testing and launch, even with the modified recool procedures in place, and that this deformation was somehow masked in the focus measurements.

NICMOS is currently expected to accomplish more than 90% of its original science program using two of the original three detectors and an accelerated observing schedule. Cameras 1 and 2 are producing excellent image quality, due in part to the reduced VCS temperature, which is resulting in extremely low background noise from the filters and cold pupils. Camera 3 continues to move closer to diffraction-limited imaging capability at slow rate.

## ACKNOWLEDGMENT

The NICMOS instrument was built by Ball Aerospace & Technologies Corp. for the University of Arizona under contract P172984. This work was supported by the NASA Goddard Space Flight Center through funding to the University of Arizona under contract NAS5-31289.

## REFERENCES

1. C. D. Miller, Development of the long-lifetime solid nitrogen dewar for NICMOS, to be presented at the 1997 Cryogenic Engineering Conference, July 28 to August 1, 1997, Portland, OR.
2. R. E. Dame and E. Cofie, A final report on the thermal expansion analysis of the NICMOS cryogenic subsystem's dewar, prepared under Sub-Contract SC00891 for Fairchild Space Contract NAS5-32539, December 18, 1996.
3. R. E. Dame, Results of low cycle sample fatigue tests conducted on 5083-0 billet material used for the NICMOS dewar nitrogen tank, Mega Engineering Report, December 4, 1996.
4. R. L. Pederson, C. D. Miller, M. P. Schulze, and K. Blount, Mechanical properties of solid nitrogen between 40 K and 58 K, to be presented at the 1997 International Cryogenic Materials Conference, July 28 to August 1, 1997, Portland, OR.

# THERMAL PERFORMANCE OF THE XRS HELIUM CRYOSTAT

Susan R. Breon[1], Michael J. DiPirro[1], Brent A. Warner[1],
Peter J. Shirron[1], and James G. Tuttle[2]

[1]Cryogenics, Propulsion, and Fluid Systems Branch
Code 713
NASA/Goddard Space Flight Center
Greenbelt, MD 20771
[2]Hughes STX
4400 Forbes Blvd
Lanham, MD 20706

## ABSTRACT

The Astro-E/X-Ray Spectrometer (XRS) is required to operate for 2 years on-orbit with approximately 20 liters of superfluid helium. The total heat load on the helium must be less than 1 mW, which is significantly smaller than any previous spaceborne helium system. Consequently, the XRS cryostat incorporates several new techniques that have not been used before in spaceborne cryogenic systems, including high temperature superconducting leads, a film flow suppressor in the vent line, and kevlar suspension systems. The thermal design of the cryostat is described and results of the ground tests of both an engineering model cryostat and the flight cryostat are presented. Tests of the flight cryostat are incomplete at the writing of this paper, so only preliminary results are given.

## INTRODUCTION

Goddard Space Flight Center has produced two helium cryostats for the Astro-E/XRS project. The XRS instrument and the collaboration between NASA and Japan's Institute for Space and Astronautical Sciences have been described in earlier papers.[1-3] One cryostat, known as the Mechanical/Thermal Model (MTM), was shipped to Japan in July 1996 and has been in use in tests of the Engineering Model (EM) solid neon dewar since September, 1996. Assembly of the flight model (FM) XRS helium cryostat was completed June 31, 1997. In mid-June, the flight model cryostat had undergone a short-term test at cryogenic temperatures to ensure that everything was operational. A test is now underway to measure the thermal performance of the cryostat without the instrument heat loads; i.e., without the Adiabatic Demagnetization Refrigerator (ADR)[4,5] and the XRS detector assembly. In September, the cryostat will be integrated with the ADR and detector assembly to become the helium insert, which will undergo thermal performance tests and a vibration test. The final ground testing will occur after the helium insert is integrated with the neon dewar in 1998. Launch is scheduled for January 2000.

## CRYOSTAT DESIGN

### Flight Model (FM) Cryostat

The primary design driver for the cryostat is the lifetime: the helium must last at least 2 years with a goal of 2.5 years. In order to meet this requirement within the constraints of mass, envelope, and thermal interfaces that were fairly firmly established before the start of the cryostat design, we needed to develop a cryostat with a total heat load <970 μW, of which 340 μW comes from the ADR and the detector assembly. The cryostat was first conceived when XRS was cancelled from NASA's Advanced X-ray Astrophysics Facility (AXAF) and reconstituted as part of the Japanese Institute for Space and Astronautical Sciences' (ISAS) Astro-E mission. Work on the XRS detector assembly, ADR, and ADR magnet had been underway at Goddard Space Flight Center for over 5 years. Thus the magnet and ADR size, approximate detector location, and instrument heat load on the cryostat were already established. On the dewar side of the interface, ISAS used the design of the IRTS helium dewar[6] as the starting point for the XRS dewar, which defined the envelope available for the cryostat, ADR, and detector assembly as the annulus within the neon tank.

The XRS cryostat is shown in Figure 1. The cryostat, together with the ADR and the detector assembly, comprise the XRS helium insert which will be delivered to ISAS as a unit for integration into the neon dewar. The cryostat can be broken down into four subassemblies: support structure, helium tank, plumbing, and wiring and sensors.

The support structure consists primarily of a riveted titanium cylinder with titanium top and bottom rings. Longerons run axially and stiffeners circumferentially along the cylinder to provide the needed rigidity. Six inconel flexures, bolted to the top ring of the support structure, will be pinned and bolted to mate the helium insert to the mounting flange of the neon tank. The cantilevered design was chosen to simplify the interfaces between the neon tank and the helium insert and to shorten the time needed for integration of the two systems. The titanium has a high strength-to-weight ratio which makes it ideal for reacting the loads from the suspension system supporting the tank. However, on cooldown it shrinks only about half as much as aluminum and so the flexures are needed to accommodate the thermal contraction mismatch between the titanium support structure and the aluminum neon tank. The titanium cylinder is gold-plated to lower its emissivity, particularly to reduce the radiative exchange with the helium tank. The support structure is expected to be at nearly the same temperature as the neon tank, nominally 17 K.

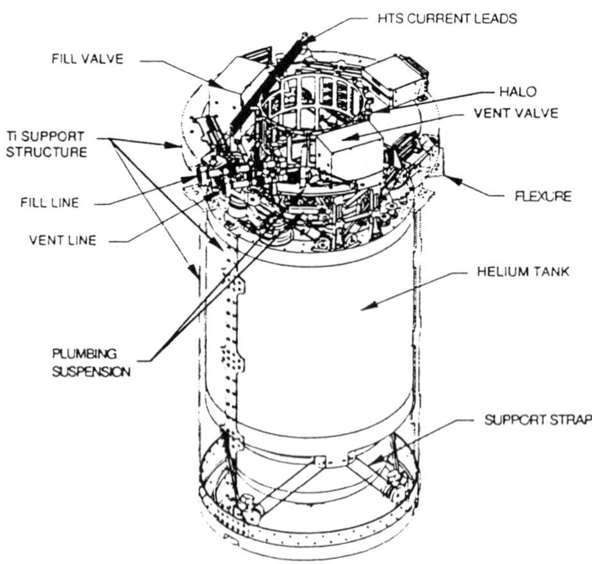


**Figure 1.** XRS helium cryostat.

The helium tank is welded aluminum. The cylindrical sides are polished to lower the emissivity. The tank is suspended from the support structure by 12 graphite-epoxy straps—6 fore and 6 aft. Copper wires connected between the fore straps and two heat exchangers on the vent line intercept most of the heat conducted along the straps. The fore straps have a cross-sectional area of 5.3 $mm^2$ and a length of 132 mm. The aft straps, which have a cross-sectional area of 4.2 $mm^2$ and a length of 173 mm, are not vapor-cooled. Inside the tank is the ADR magnet, a shielded superconducting magnet capable of producing a 2 Tesla central field with a current of 1.91 amps. With the magnet installed, the tank has a volume of 32.5 liters. The tank is protected against overpressurization by a 45 psid cold burst disk located on the aft end of the tank. A carbon-carbon composite getter is mounted on the tank to adsorb any helium gas that might be present in the vacuum space. The getter has been sized so that it could adsorb 1.5 standard liters of helium gas. The getter is covered with a 50 μ mesh stainless steel screen to ensure that no particles escape which could damage the XRS thin-foil filters.

The plumbing consists of a fill line and a vent line, with a cryogenic motor-operated valve for each line mounted on top of the tank. Both lines are assembled from welded combinations of 0.25 mm wall stainless steel tubing, 0.15 mm wall welded bellows, and customized fittings. Connections to the valves and to the neon dewar plumbing are made with conflat fittings. There are radiation baffles fitted into the fill and vent lines at the neon dewar interface to intercept radiation down the plumbing. The vent line has two heat exchangers which provide vapor cooling to the straps, the fill line, and the magnet and valve leads. The bellows provide adjustment capability in the plumbing interfaces and thermal isolation between the heat exchangers, the helium tank, and the 17 K interface. As a result, however, the plumbing is not self-supporting. Instead, a kevlar suspension system has been devised to support the vent and fill lines at two places without thermally shorting the plumbing

One of the most critical developments for the cryostat thermal performance was a method to reduce film flow and film recondensation downstream of the porous plug. A more detailed description of the film suppression technique is presented in a separate paper[7]. The concept relies on using a relatively small porous plug, 6.4 mm thick by 9.5 mm in diameter, and subsequently allowing the helium in the vent line downstream of the plug to exchange heat with the helium in the tank. Film flowing into the heat exchanger is evaporated, with the latent heat conducted from the helium in the tank. To evaporate the film completely, the porous plug would have to operate with a 50 mK temperature drop; the result is that the bath is 50 mK warmer than it would be otherwise. The temperature drop across the plug is determined by the plug size, the bath temperature, and the flow rate through the plug. As the temperature difference across the plug is decreased, some film will flow through the heat exchanger and evaporate somewhere along the vent line, thus losing much of the cooling power the helium could provide. For XRS, film losses must be traded off against the decrease in efficiency of the ADR, which will dissipate an extra 0.6 μW per mK rise in bath temperature. Figure 2 shows the same figure of merit for the reduction of film flow. Initially, the benefit is large because of the highly non-linear relationship between film thickness and temperature. That is, the film becomes highly unsaturated when its temperature is raised by only a few mK. Beyond that, film losses are reduced by at least 1.5 μW/mK until the film is completely evaporated (where the value is then zero).

The cryostat wiring appears to follow paths through every nook and cranny in the cryostat, but it has been laid out in ways that minimize the heat load on the cryogen tank. Wires that bridge the gap from the 17 K interface to the helium tank are optimized for a combination of thermal conductance, radiation, and joule heating. Wires carrying the lowest currents, $\leq$1 mA, are made of 42 AWG (0.06 mm diameter) stainless steel wire. Medium current wires, between 1 mA and 0.1 A, are made of stainless steel wire and the size varies as appropriate for the amount of current being carried. For the highest current wires, 0.5 to 2 A, that carry power for the valves and the magnet, low thermal conductivity leads were developed using filaments of the high temperature (HTS) superconductor YBaCuO[8]. The HTS leads are thermally anchored at the cold heat exchanger on the vent line. Copper clad NbTi wires, partially stripped of copper to decrease thermal conductivity, carry the current from the cold end of the HTS leads to the niobium blocks mounted blocks on the helium tank. The NbTi wires and the magnet leads, which are also

NbTi, are connected to the niobium block to form a superconducting joint. The cryostat is instrumented with 19 ruthenium oxide thermometers of which 7 will be read out on orbit, two liquid level detectors (for ground use only), and a mass gauge heater for use on orbit. There are also five calibration resistors mounted on the helium tank that are used to remove the effects of temperature drift in the thermometer read-out electronics box. The cryostat wiring is constrained by kevlar fibers strung on an aluminum structure known as the "halo." The purpose of the halo is to allow wires to be routed from the connectors at the 17 K interface to their positions on the cryostat without attaching them directly to the tank, which would be a thermal short. The halo also keeps the wires out of the area to be occupied by the ADR and the detector assembly.

**Mechanical/Thermal Model (MTM) Cryostat**

The MTM was built primarily to verify the structural and mechanical design of the cryostat, but has lower fidelity in the thermal design. The differences between the FM cryostat and the MTM are surface finishes, the getter, plumbing, and wiring. Neither the helium tank nor the support structure of the MTM have had any special surface preparation, and so the emissivity is higher than for the flight cryostat. The plumbing is made mostly of thick-walled (0.71 mm) stainless steel tubing. Bellows are only used to provide adjustment capability at the neon dewar interface and the plumbing is self-supporting: there is no elaborate kevlar suspension system on the MTM. The straps and the fill line are not vapor-cooled and there is no attempt to suppress the film flow in the vent line. The wiring on the MTM is much simpler than on the FM cryostat. The MTM does not have a magnet, ADR, or detector assembly and so no wiring is needed for these components. Furthermore, the restrictions on routing of the wires were much more relaxed, so the halo is not included in the design. Finally, the HTS lead assemblies are not present in the MTM: power to the valves is carried on 0.6 mm and 0.8 mm diameter stainless steel wires approximately 25.4 cm in length.

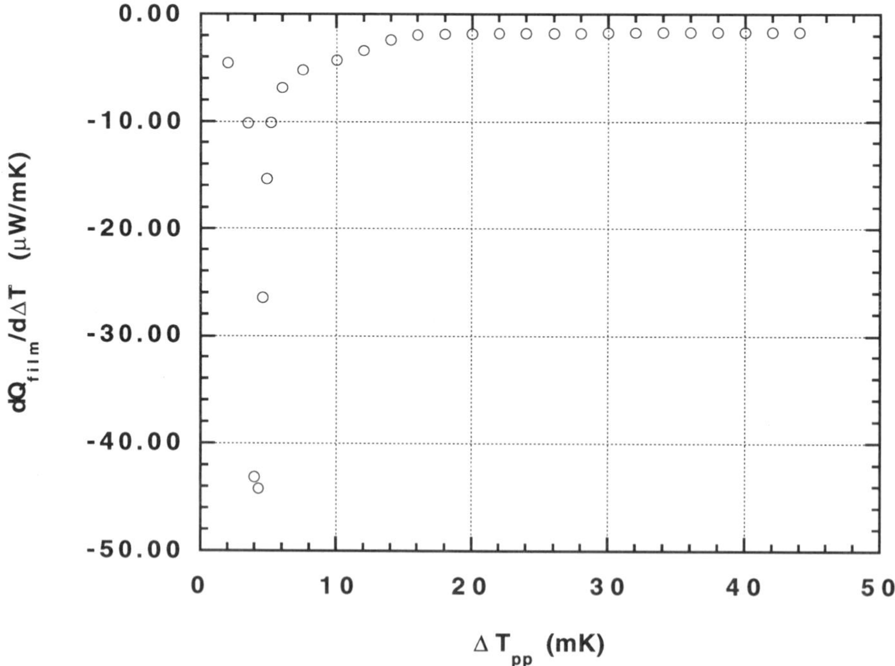


**Figure 2.** Effectiveness of changing temperature drop across the porous plug to reduce heat load on helium bath.

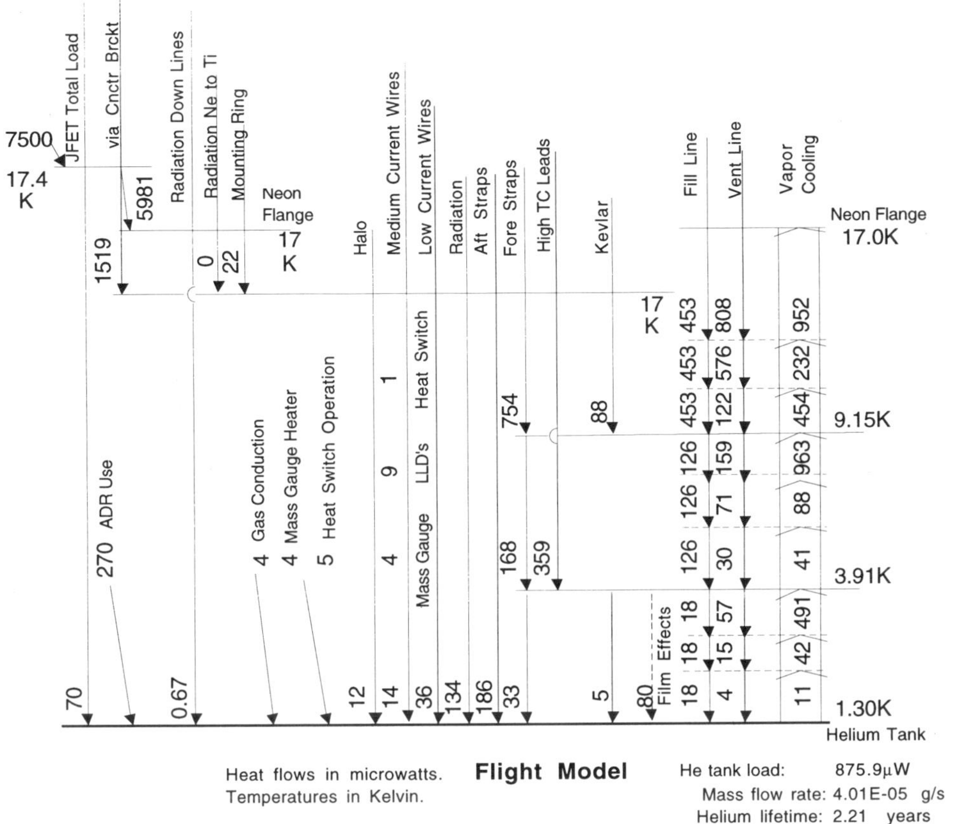


**Figure 3.** Helium insert heat map.

## THERMAL MODEL

### FM Cryostat

The helium insert heat map, which includes the ADR and detector assembly heat loads, is shown in Figure 3. The current predicted lifetime for the helium is 2.2 years with a heat load of 876 μW. There are several assumptions that are included in the calculation:

- Neon tank interface temperature of 17 K, effective heat sinking of wires and plumbing at the interface, effective radiation shielding from warmer dewar components
- Bath temperature of 1.3 K
- Detector assembly heat load on helium tank of 70 μW
- Time-averaged ADR heat load on helium tank of 270 μW
- Efficient low-temperature top-off resulting in 27.6 liters of helium available for on-orbit operations
- 30% margin (lifetime estimates based on 19.3 liters of helium)

With the exception of the need for a 30% margin, the rest of the assumptions should be verified over the course of ground testing at various levels of assembly. Once the cryostat testing is complete, the ADR and the detector assembly will be integrated with the cryostat and their respective heat loads will be measured. Preliminary measurements of the detector assembly heat load on the helium tank show that it should meet its 70 μW thermal budget. Measurements that have been made of the flight ADR performance separately would lead to an estimated ADR heat load[5] of 235 μW, pending measurements of the detector assembly heat load on the ADR.

### MTM Cryostat

The thermal model for the MTM was modified for the design differences outlined earlier in this paper. The predicted heat load for a 17 K neon dewar interface and a 4.2 K bath temperature is 1.6 mW. For a 24 K interface temperature, the predicted heat load is 3.3 mW. At 1.3 K, the predicted heat load would be on the order of 1 mW higher because of the film that will exist downstream of the porous plug.

## THERMAL TESTING

### FM Cryostat

The cryostat is tested in a dewar built specifically for this purpose. The dewar has an annular liquid nitrogen tank and a separate helium tank which is filled with normal helium. The mechanical and thermal interface to the neon dewar is simulated by a thermal shroud that is suspended from the nitrogen tank by kevlar straps. The shroud is cooled by the vapor boiling off from the test dewar helium tank. Heaters are mounted on the shroud as well. Using a combination of vapor cooling and heater control, the shroud can be maintained at any temperature over the expected range for neon operations, approximately 10 K to 24 K.

To date, the FM cryostat has successfully completed a short-term functional test which checked out the sensors, demonstrated operation of the valves and magnet, and included leak checks of the cryostat with superfluid helium. The seal across the fill valve seat was also leak-checked. A significant heat leak, as much as 14 mW, was observed on the cryostat when the fill valve was open. The excess heating may have been due to a small thermoacoustic oscillation, although the pressure pulses were too small to register on a compound pressure gauge on the line. The heat load dropped below 4 mW with the valve closed and the line evacuated.

A second test to characterize the thermal performance of the cryostat has been inconclusive. The heat load on the helium is determined by measuring the helium flow rate. At 4.2 K the measurements are affected by changes in atmospheric pressure. In future tests, the pressure will be better regulated to eliminate this effect. A wet-test flowmeter is being incorporated into the system to allow the flow to be integrated over a period of time. At present measurements are done as instantaneous mass flow readings. The most critical measurements, those at superfluid helium temperatures, have been hampered by disturbances in the system that have prevented the cryostat from reaching equilibrium: a 1 mK/hr change in bath temperature of 15 liters of helium at 1.5 K is equivalent to a heat load 688 μW, or nearly the entire cryostat heat budget. The disturbances are mostly due to an apparent leak into the guard vacuum. That problem is now being investigated.

Once the problem has been repaired, we will measure the thermal performance of the cryostat. The heat map for the predicted thermal performance of the cryostat alone is given in Figure 4. The parasitic heat load on the cryostat will be measured as a function of the interface temperature. In addition, the performance of the film suppressor will be characterized. The vent line impedance will be measured in order to estimate the expected on-orbit bath temperature, although the vent line impedance is dominated by the warm plumbing in the neon dewar and the exhaust plumbing external to the dewar. The heat load on the helium due to hysteresis and eddy current heating in the magnet will be measured. Additional tests will be performed to characterize the dewar and the cryostat in preparation for tests at higher levels of assembly. These include measurement of the conductance between the ADR cold plate and the helium bath; operation of the cryostat and dewar in a horizontal position, which is required in order to integrate the dewar with the equipment needed to calibrate the XRS detectors; and a low temperature top-off to develop procedures for final top-off on the launch pad.

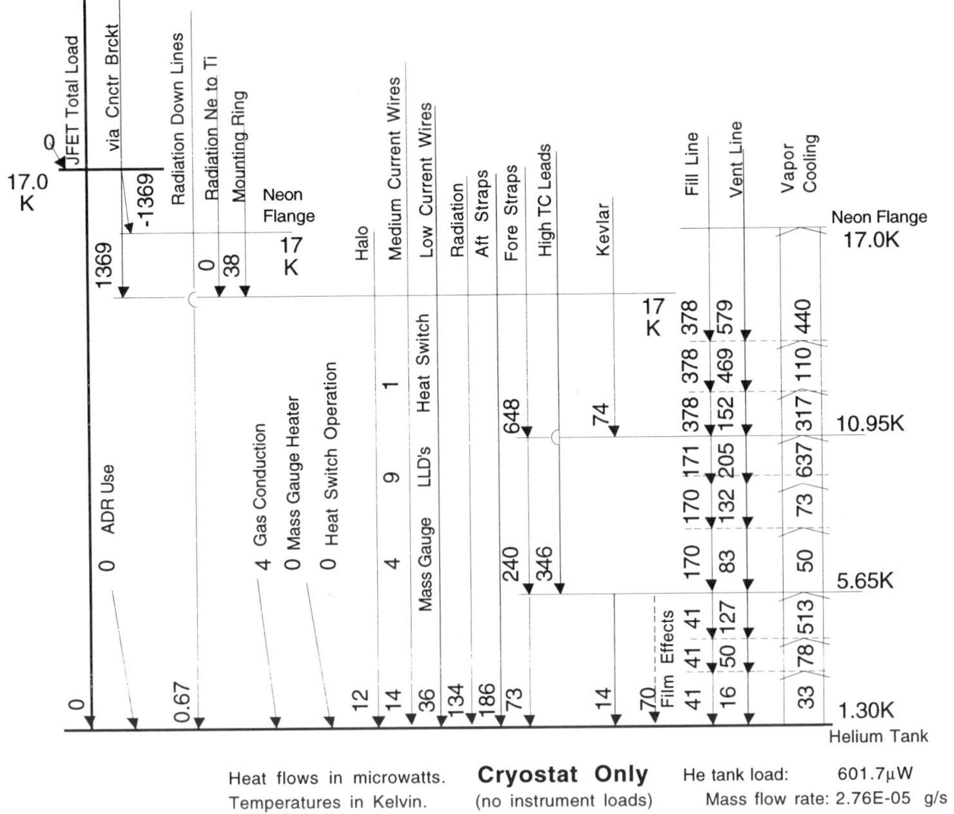


**Figure 4.** Cryostat heat map.

## MTM Cryostat

The MTM was vibration tested and thermally tested at Goddard Space Flight Center prior to shipment to Sumitomo Heavy Industries (SHI) in Japan. The vibration test was performed with the MTM cooled to approximately 80 K by a liquid nitrogen tank in the vibratile dewar. The helium vent line, which was a duplicate of the flight design but with a different support system, failed in vibration. It was replaced with a line very similar to the fill line, which was self-supporting and had survived vibration. The thermal test was done to verify operation of the sensors, heaters, and valves at helium temperature and to leak check the MTM helium tank when filled with superfluid helium. The duration of the thermal test at Goddard was not sufficient to determine the heat load on the MTM.

Once the MTM was integrated with the EM neon dewar, the dewar was evacuated for about a week and leak-checked. During this test the MTM burst disk was found to have developed a leak, which was eventually traced to the burst disk diaphragm. After much investigation, it was concluded that a hole in the diaphragm had been temporarily plugged when the diaphragm was cleaned with acetone. The leak did not appear on several leak checks performed both at Goddard and at SHI prior to integration with the EM dewar. The hole then reopened during the long evacuation period prior to cooldown of the EM dewar. We were able to reproduce this failure scenario with the offending burst disk and diaphragm during the investigation and no other explanations of the failure were uncovered. The MTM burst disk was replaced and the system was closed again and leak-checked. Once the system was confirmed to be leak-tight, the dewar was re-opened for final installation of the vapor-cooled shields and thermal blankets. Then the dewar was closed out and both the MTM helium tank and the EM dewar neon tank were filled with liquid helium.

The EM dewar has undergone a series of performance and vibration tests. During the first cooldown and freezing of neon, the heat load on the MTM was measured to be 2.76 mW with the helium at 4.28 K and the neon tank interface at approximately 16 K. This heat load is higher than that predicted for the MTM. Unfortunately, because of constraints on travel to Japan and other testing required for the neon dewar, the reason for the higher heat load has not been determined. The MTM tank has been pumped superfluid, to temperatures as low as 1.30 K. In a test of the effect of film flow in the plumbing, it was found that the heat load on the helium tank dropped by 710 μW when the fill valve was closed. The fill line was left open to vapor through a cross-over valve that connects to the vent line, which allowed us to determine that the effect was not due to thermo-acoustic oscillations or convection. The magnitude is also consistent with the expected value for film flow in the vent line.

The tests to date with the EM dewar show that the neon temperature should be at or below the nominal interface temperature of 17 K. The helium bath temperature is also expected to be at or below 1.3 K. Low-temperature top-off tests planned for late August 1997 will help establish the expected efficiency of the launch pad servicing operations. All of these factors will be included in establishing the lifetime estimate of the XRS cryostat during on-orbit operations.

## ACKNOWLEDGMENTS

The authors wish to acknowledge the recommendation from R. Trant of the University of Wuppertal to investigate the need to incorporate a helium getter in the design. Without a getter the resultant heat load could have represented a significant fraction of the total thermal budget for the XRS cryostat.

This work is supported by the National Aeronautics and Space Administration Office of Space Science.

## REFERENCES

1. S.R. Breon, H.D. Branch, G.J. Blount, M.L. Jackson, R.F. Boyle, J.G. Tuttle, Design of the XRS Helium Insert, *Adv in Cryo Engr* 41:1129 (1996).
2. S.R. Breon, J.A. Gibbon, R.F. Boyle, M.J. DiPirro, B.A. Warner, and J.G. Tuttle, Thermal design of the XRS helium cryostat, *Cryogenics* 36:773 (1996).
3. S.M. Volz, K. Mitsuda, H. Inoue, Y. Ogawara, M. Kyoya, M. Hirabayashi, The X-Ray Spectrometer (XRS)—A multi-stage cryogenic instrument for the Astro-E x-ray astrophysics mission, *Cryogenics* 36:763 (1996).
4. A.T. Serlemitsos, M. SanSebastian, and E. Kunes, Final design of the Astro-E/XRS ADR, To be presented at the 1997 Cryo Engr Conf.
5. A.T. Serlemitsos, M. SanSebastian, and E. Kunes, Design and performance of a spaceworthy ADR, To be presented at the 1997 Space Cryo Workshop
6. G. Fujii, S. Tomoya, M. Kyoya, M. Hirabayashi, M. Murakami, H. Okuda, H. Murakami, and T. Matsumoto, Thermal and mechanical performance of superfluid helium dewar for IRTS, *Adv in Cryo Engr*, 38:1359 (1992).
7. P.J. Shirron and M.J. DiPirro, A superfluid film flow suppressor and its use in spaceborne helium dewars, To be presented at the 1997 Cryo Engr Conf.
8. J.G. Tuttle, T.P. Hait, R.F. Boyle, and S.R. Breon, A high Tc superconducting current lead assembly for the XDS helium cryostat, To be presented at the 1997 Cryo Engr Conf.

# SUPPRESSION OF SUPERFLUID FILM FLOW IN THE XRS HELIUM DEWAR

P. J. Shirron and M. J. DiPirro

NASA/Goddard Space Flight Center
Greenbelt, MD, 20771, USA

## ABSTRACT

Superfluid helium dewars for space missions use a porous plug phase separator to retain the bulk liquid while releasing boiloff vapor. Although the thermomechanical force is quite effective in holding the liquid against the internal vapor pressure, van der Waals forces will lead to a leakage of helium through the porous plug to feed the flow of a superfluid film in the vent line. Since the latent heat is generally not extracted from the remaining liquid, the film poses a serious problem for space missions such as the X-Ray Spectrometer (XRS) which have extremely low heat load budgets. A design for the XRS porous plug and vent system is presented which keeps film losses below 2 μg/s through a combination of a heat exchanger, to evaporate the film, and a knife-edge device. Tests of the vent have been conducted in a minus 1g configuration. At the expected on-orbit tank temperature of 1.22 K, film losses are negligible, but rise to 4 μg/s at the maximum tank temperature of 1.30 K. Design parameters and performance tests are discussed.

## INTRODUCTION

The X-Ray Spectrometer (XRS) mission is one of a number of current and planned space missions requiring very high performance superfluid helium dewars. It contains a small, 30 liter helium tank shielded by a solid neon tank at 17 K and is designed for a 2.5 year lifetime. It is projected that as little as 18 liters will remain after reaching equilibrium on orbit, thus the average heat load on the helium must be less than 800 μW. Since this represents an improvement of two orders of magnitude over previous missions, many parasitic effects that could previously be ignored have been re-evaluated. One such effect, which is quite commonly encountered in helium-4 and dilution refrigerators, is a direct mass loss through the flow of a superfluid film up the vent line. Since the flow velocity can be as high as 250 cm/s, mass losses can exceed 50 μg/s in a typical 1 cm diameter vent line[1]. The heat of vaporization is not extracted from the remaining liquid, so this represents a cooling loss of 1.1 mW and would cut the XRS lifetime by more than half. The goal for XRS was to reduce film flow to an equivalent of 40 μW or less.

To see how this can be done, it is instructive to note that the mass flow rate of the film is

$$\dot{m}_{film} = \rho_s C v_c (d-d_0), \tag{1}$$

where $\rho_s$ is the superfluid density (which at the temperatures of interest is nearly equal to bulk density, $\rho$), $d$ is film thickness, $d_0$ is the inert layer thickness, and $C$ is the smallest perimeter of the vent line or flow channel. The critical superfluid velocity, $v_c$ is a function only of $d$ and temperature, $T$. $d_0$ is almost always 1 layer (0.36 nm) and is an important correction only for very thin films. It has been suggested[3] that surface roughness can make C effectively much larger, however, we have not seen this in any of our measurements on a variety of substrates.

At first glance, Eq. (1) suggests that any structure that significantly limits the film thickness or flow perimeter, such as an orifice, would be quite effective in suppressing film flow. In fact, orifices are routinely used in helium-4 and dilution refrigerators for this purpose, although the size that would be necessary for XRS to meet its 40 μW goal would be a rather small 0.03 cm diameter. Unfortunately, the situation is not that simple because of the rather complex behavior of helium films, and it turns out no single orifice or knife-edge device would be entirely suitable. In order to see why this is so, we must first discusses the dynamics of superfluid films in the context of a porous plug vent system. We then present the XRS vent system design and performance test results. Incidentally, all tests were conducted in a minus 1g configuration so that gravity effects could not give artificially beneficial results.

## SUPERFLUID HELIUM FILM DYNAMICS

Figure 1 is a schematic of a typical porous plug phase separator and vent line of uniform radius $R$. Under normal operating conditions, the porous plug is completely filled with liquid and evaporation occurs at the downstream surface. The temperature gradient across the plug generates a thermomechanical pressure[4] which exceeds the vapor pressure difference by a factor of 12-16. This imbalance causes the liquid to retreat slightly into the pores, increasing the surface tension pressure, until a pressure balance is achieved. The chemical potential of the liquid in the pores is

$$\mu_{liq}(T_p) = -\frac{m\sigma(T_p)}{\rho r_p} \tag{2}$$

where $m$ is the mass of a helium atom, $\sigma$ is surface tension at the porous plug temperature $T_p$, and $r_p$ is the radius of curvature of the liquid surface. The chemical potential of the vapor just downstream of the phase separator is given by the Frenkel-Halsey-Hill relation[5]:

$$\mu_{vap}(P,T_p) = -k_B T_p \ln\left[\frac{P_{svp}(T_p)}{P}\right] \tag{3}$$

where $P$ is the pressure and $P_{svp}(T_p)$ is the saturated vapor pressure. $k_B$ is Boltzmann's constant. Since the vapor and liquid are in equilibrium, $\mu_{vap}(P,T_p)=\mu_{liq}(T_p)$. For typical operating conditions[4], $r_p$ is tens of microns (and positive), giving a vapor pressure which differs from $P_{svp}$ by only 0.005%.

The vent line will be coated with a film of thickness $d$ whose chemical potential, in a general formulation, includes terms for the van der Waals potential, kinetic energy and surface tension:

$$\mu_{film}(T) = -\frac{\Gamma}{d^3} + \frac{1}{2}\frac{\rho_s}{\rho} m v_s^2 - \frac{m\sigma(T)}{\rho r_s} \tag{4}$$

where $\Gamma$ is the van der Waals constant, $v_s$ is the superfluid velocity, and $r_s$ is the radius of curvature of the surface. Viscous drag prevents flow of the normal component, so only the

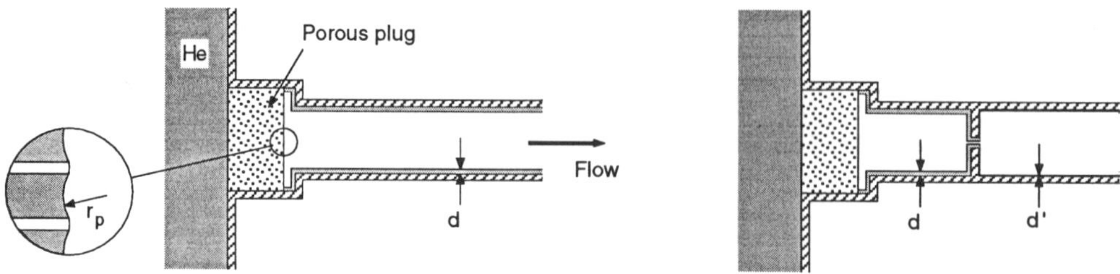


**Figure 1.** (a) Schematic of a porous plug and vent line coated with a film, and (b) with an orifice added.

superfluid kinetic energy appears. However, for all practical purposes, it is negligible. For most materials, the van der Waals binding energy of first layer atoms is on the order of 20 K, giving $\Gamma \sim 1.2\times10^{-50}$ J•m$^3$. Equating Eqs. (3) and (4) gives a general expression for the film thickness as a function of the local pressure, temperature, and, if significant, radius of curvature:

$$d = \left[ \frac{\Gamma}{k_B T \ln\left(\frac{P_{svp}(T)}{P}\right) - \frac{m\,\sigma}{\rho r_S}} \right]^{1/3} \tag{5}$$

In the vent line downstream of the porous plug, the surface tension term is negligible, so from the pressure computed above, we find $d$ is approximately 25 nm.

The flow that a film of this thickness can support can be computed from Eq. (1) provided that the critical velocity is known. Although early measurements of $v_c$ were not entirely consistent, more recent measurements, along with our own, demonstrate that $v_c$ is inversely proportional to $d$ except for extremely thin films ($d$<2.0 nm for $T$~1.3K)[1]. Below this limit, $v_c$ begins to drop rapidly. In the proportional region, we find

$$v_c d = 1.25\times10^{-4}\ \text{cm}^2/\text{s},\ \ 2\ \text{nm} \leq \text{d} \leq 35\ \text{nm} \tag{6}$$

This compares well with theoretical predictions[6] which give $v_c d = \hbar/2m = 7.9\times10^{-5}$ cm$^2$/s. Thus superfluid films possess the unique ability to support the same mass flow rate almost regardless of thickness. Therefore, changes in flow rate and temperature of the porous plug, which alter the radius of curvature in the pores, do not change the film flow losses.

In Figure 1a, as one travels downstream (to the right), the vent line temperature rises due to parasitic heat flow while the vapor pressure decreases due to the flow impedance. This will cause $\mu_{vap}$, and hence $\mu_{film}$, to become more negative. This potential gradient drives the film downstream, readily accelerating it to its critical velocity so that the flow rate is consistent with Eq. (1) with $C=2\pi R$. The gradient also causes the film to become thinner with a progressively increasing critical velocity. Downstream, the film eventually becomes so thin (<2 nm) that its critical velocity starts dropping and it begins to evaporate. Regardless of the distances involved, evaporation starts when the temperature has risen by only a few mK and is complete when it reaches 60-80 mK.

## Suppression of Film Flow

We now examine the effect of placing a film flow suppressor in the vent path as shown in Figure 1b. Assume it to be a small orifice of radius $R_o$. At the orifice, the situation is the same as above: the film flows at the critical velocity and its thickness, $d$, is simply that given by Eq. 5. Upstream of the orifice, the film's velocity must be less than $v_c$, so it flows without dissipation. This means there are no appreciable chemical potential gradients and that the film thickness, also equal to $d$, is uniform. Downstream, however, the situation is very different. Chemical potential gradients will accelerate the film to the critical velocity, but the film thickness can no longer remain in the regime where $v_c$ is inversely proportional to thickness since that would represent a higher flow rate than exists at the orifice. Instead, it will fall into the regime where the critical velocity drops with decreasing thickness. Using Eqs. (1) and (6), mass conservation requires that the thickness, $d'$, satisfy

$$v_c(d')d' = (1.23\times10^{-4}\ \text{cm}^2/\text{s})R_o/R. \tag{7}$$

$v_c$ exhibits a rather complex dependence on thickness and temperature, and it is easiest to estimate $d'$ from critical velocity data. As an example (see reference 1), if $R_o/R$ were 0.2, the thickness at 1.30 K is slightly less than 4 layers, or 1.3 nm, and $v_c$=110 cm/s. As the ratio $R_o/R$ is varied, d' will actually remain relatively constant, and $v_c$ will change proportionally. Therefore, it is always the case that the film will be thick on the upstream side of a flow suppressor, and thin downstream.

The effect of this is quite interesting: because the vapor must be in equilibrium with the film at all points, there must also be a temperature difference across the orifice. Looking at it in reverse, if the temperature were uniform, the film thickness gradient would represent a large chemical potential gradient that would cause the immediate condensation of vapor and a rise in temperature on the downstream side. Using Eqs. (3) and (4), if there is very little direct film flow (i.e. $R_0/R$ is small), the rise, $\Delta T$, must be on the order of 50 mK to restore equilibrium. Since heat will be conducted back to the upstream side, in steady state there will be a continuous evaporation of film on the upstream side and condensation of vapor on the downstream side, both at a rate equal to $K\Delta T/L$, where $L$ is the latent heat and $K$ is the thermal conductance of the orifice.

The conductance is therefore a very important parameter for the design of film flow suppressors. For XRS, a budget of 2 µg/s (equivalent to 40 µW) for film flow sets an upper limit of $K$~1 mW/K.

## XRS POROUS PLUG AND VENT SYSTEM DESIGN

Although the above discussion makes reference only to the film flow suppressor, the thermal conductance of any component in parallel with it must be included in the total. This is a particular problem for XRS since its vent line contains a cryogenic stepper-motor valve (identical to those used on the Superfluid Helium On-orbit Transfer flight demonstration and Gravity Probe-B[7]) that is used to bypass the porous plug. This is a standard arrangement and is necessary for ground servicing. The valve's conductance is estimated to be 80 mW/K, meaning the valve would completely short out any film flow suppressor. It was therefore necessary to employ other techniques for reducing film losses.

The approach taken was to try to evaporate the film instead. The XRS vent, shown in Figure 2, consists first of a relatively small diameter porous plug to obtain a temperature drop of approximately 50 mK for nominal operating conditions (temperature and flow rate). The plug is the followed by a small diameter vent line (0.15 cm I.D.), a copper heat exchanger anchored to the cryogen tank, then a knife-edge device. The small vent line provides a gross limit to film flow into the heat exchanger of about 9 µg/s. As the film rises by 50 mK back to the bath temperature, it becomes highly unsaturated and most, if not all, of it is evaporated. The 9 µg/s value is not critical, however it is important to keep in mind that as more film is evaporated in the heat exchanger, the same amount less is evaporated at the porous plug. Thus, the correct plug size depends on the tube's diameter and on how effective the heat exchange will be at eliminating film losses. It was judged best to make the tube as small as possible without introducing a significant impedance.

In sizing the plug, it was especially important to realistically estimate the final on-orbit helium tank temperature. Because the 50 mK drop is very close to, if not above, the "knee"[4] in the XRS operating temperature range (up to 1.3K), it will be very difficult to pump to much lower temperature if it is estimated too high. Because the temperature drop varies as the thirteenth power of temperature, estimating too low will result in less than 50 mK across the plug and higher film losses. Based on estimates of the total vent line impedance, and adding in the 50 mK, the expected operating temperature is 1.14 K. At the XRS vent rate of 40 µg/s less the 9 µg/s film flow, the proper plug size is 0.76 cm diameter.

The knife-edge device was originally planned as the sole means of suppressing film flow, but is now used as a secondary limit. The device works by introducing an extremely

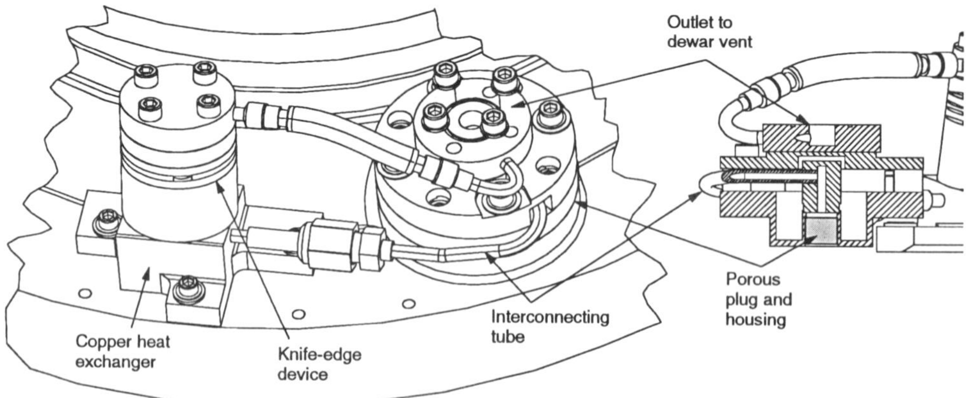


**Figure 2.** The porous plug, exchanger and knife-edge device are mounted to the top rim of the XRS helium tank. A cut-away of the porous plug housing shows the attachment of interconnecting tubes.

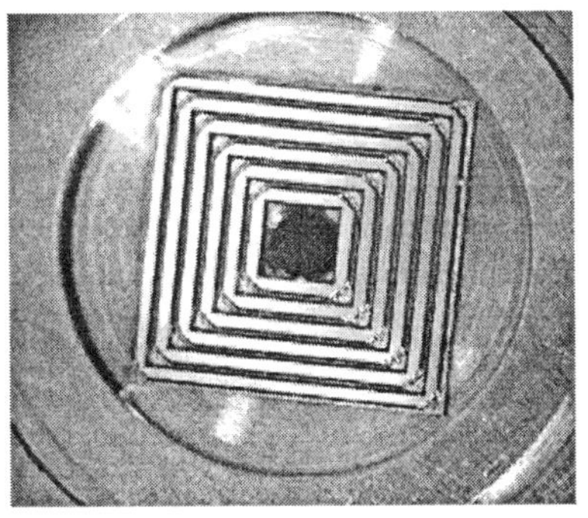
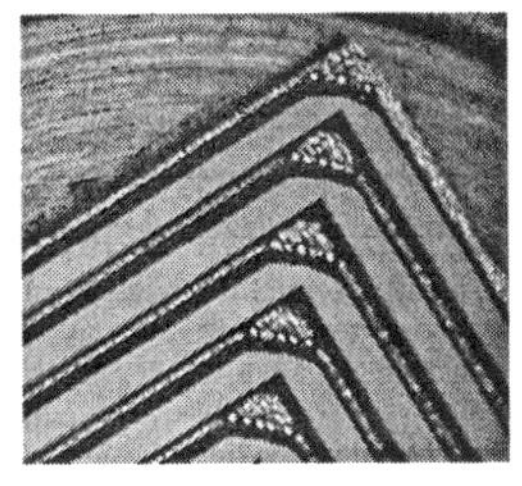

**Figure 3.** Two silicon wafers, each having 5 etched troughs and a center hole, are mounted onto a flange which has been machined from the edge to reduce its thermal conductance. The center picture shows the unintended etching pattern at the corners.

sharp edge into the film flow path whose small, negative radius of curvature gives rise to a large surface tension energy. The film at the edge will therefore be very thin. If thin enough, i.e. less than ~2 nm, the mass flow rate is reduced. We can use Eq. (5) to estimate just how small $r_{KE}$ must be to do this. The appropriate temperature to use is the cryogen tank temperature, which by design is $T_p$+50 mK, and the appropriate pressure is $P_{svp}(T_p)$. In the heat exchanger, where the surface tension term is negligible, the film thickness is about 1.4 nm. Since $v_c$ is a very strong function of thickness in this range, reducing $d$ by an angstrom or so yields a marked reduction in mass flow. This requires $r_{KE}$ to be on the order of 10 nm.

The knife-edge device is formed from a <100> oriented single crystal 0.3 mm thick, silicon wafer which is patterned photolithographically and etched as shown in Figure 3. The central hole is 0.15 cm square and is surrounded by 5 equally spaced, concentric square troughs. An anisotropic etchant, such as 20% potassium hydroxide or 25% tetra-methyl ammonium hydroxide, is used to etch the silicon preferentially along the <111> crystal plane, at an angle of 57° relative to the surface. In the case of the troughs, the etching is self-limiting where the two angled sides intersect. The width at the surface was precisely set to have this occur at a depth of 0.20 mm. Once this occurs, the etching is stopped to prevent the breakthru at the corners. Since the center hole is incompletely formed at that point, additional etching is required from the back side.

The final device is made by bonding one wafer to each side of an Invar flange, as shown in Figure 3, using Armstrong A-12 epoxy. Invar is used to minimize thermal contraction differences to prevent fracturing of the silicon, and the epoxy, which is flexible at room temperature, minimizes thermally induced strains.

In their final form, each wafer contains 10 knife edges in series. This configuration was chosen not for redundancy, but to obtain as low a thermal conductance as possible. Assuming for the moment that each knife-edge is capable of completely stopping film flow, the way this works is as follows: as film flow is blocked at the outer beveled edge of the upstream wafer, helium vapor will condense on the flat surface adjacent to the bevel. Since the edge is atomically sharp, the thermal conductance between the two regions is relatively large, even accounting for Kapitza resistance. The condensed film attempts to travel toward the central hole, but is blocked at the next edge where this process is repeated, and so on. Kapitza conductance is proportional to surface area, so each succeeding knife edge has a progressively smaller conductance.

The conductance of each wafer is estimated to be $K$=0.5 mW/K, using a value of 0.14 mW/cm$^2$•K for the boundary conductance at 1.3 K, and including the parallel conduction through the silicon. The latter was minimized by etching the troughs as deep as possible. By using two wafers, the conductance of the device is cut to 0.25 mW/K.

If the assumption that each knife edge is completely capable of eliminating film flow is not true, or does not hold for each edge, for example due to imperfections or damage, then the total film flow rate downstream of the device will be in excess of that given by $K\Delta T$. Unfortunately, it is not possible to independently measure $K$ since the film is an integral part of the heat transport. Performance tests can only give information about total flow rates, not the relative contributions of direct flow and condensation.

## PERFORMANCE TESTS

Performance tests of the complete XRS vent system and of its constituent components were conducted in a specially designed helium-4 refrigerator, shown in Figure 4. The porous plug is located at the lower end of the J-shaped vent line, and is followed by a knife-edge device, a copper test cell (bottom horizontal section), and a vent line out of the cryostat. The inlet to the porous plug is open to the liquid in the pot, while the remainder of the J-tube is surrounded by vacuum to simulate the flight vent. The test cell contains a heater mounted on the outside and germanium resistance thermometer (GRT) mounted on the inside. These were used to measure film flow rates by determining the power needed to evaporate it. During each test, the liquid level was raised above the inlet of the porous plug. In the upside down configuration, the porous plug must support the hydrostatic head of liquid so that film flow measurements will, at worst, *over*estimate what will be experienced on orbit.

For tests of the flight porous plug and vent system, the heat exchanger was bolted to an aluminum block to simulate its attachment to the flight tank. A heater and GRT on the block facilitated measurements of the joint conductance and allowed the temperature of the heat exchanger relative to the porous plug to be varied. This made it possible to determine film flow rates as a function of $\Delta T$ across the porous plug. Measurements also included $\Delta T$ across the plug as a function of flow rate at various temperatures between 1.15 and 1.3 K. These data, along with estimates of the total parasitic heat load, were used to predict the on-orbit tank temperature and film flow losses.

### Determination of Film Flow Rates

Film flow rates were determined by measuring the difference in temperature, $\Delta T$, between the test cell and the porous plug as a function of power into the test cell heater. Initially, the heat will go into evaporating the film as it flows into the cell and the temperature will show little effect. Once the power exceeds $\dot{m}_{film}$L, the film is completely evaporated and any additional power simply heats the flowing vapor. This results in a linear dependence of $\Delta T$ on $\dot{Q}$, with the slope given by $\dot{m}C_p$, where $\dot{m}$ is the total mass flow rate and $C_p$ is the specific heat. The $\Delta T$=0 intercept is the heat going into the cell which produces no temperature rise; i.e., the power needed to evaporate the film.

Results are shown in Figure 5 for a porous plug with (a) no film suppressor, (b) a 0.15 cm diameter orifice, and (c) a knife-edge device. Case (b) uses the same flange as (c), so

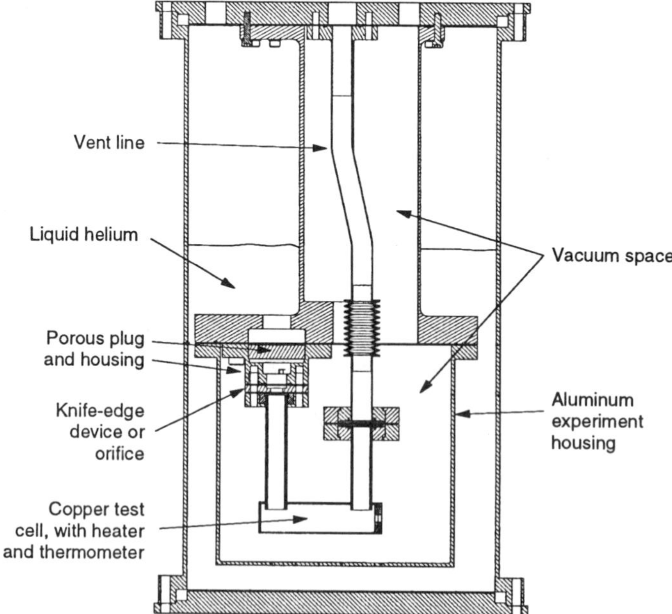


**Figure 4.** Cut-away of the helium-4 pot used for porous plug tests.

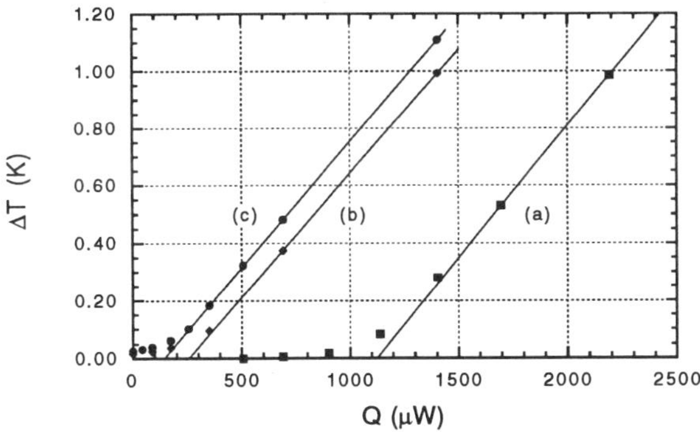


**Figure 5.** The temperature rise of the test cell as a function of heat input with (a) no film suppressor, (b) a 0.15 cm diameter orifice, and (c) an etched silicon knife-edge device. The solid lines are fits to the linear portions of each data set.

any differences can be attributed directly to the silicon knife-edges. In all cases, the vent line downstream of the porous plug is 0.90 cm I.D.and can support a maximum film flow of 50 μg/s. The mass flow rate out of the cryostat was measured with an external flowmeter to be approximately 0.1 standard liter/minute. The linear portions all can be fit with straight lines with less than 1% deviations and the slopes agree with the flowmeter to 5%. The intercepts give (a): $\dot{Q}_{film}$=1.12 mW, $\dot{m}_{film}$=50 μg/s, (b) $\dot{Q}_{film}$=240 μW, $\dot{m}_{film}$=11.0 μg/s, (c) $\dot{Q}_{film}$=155 μW, $\dot{m}_{film}$=7.1 μg/s. Repeated measurements do not vary by more than 5%.

We consistently find lower flow rates with the knife edge device present than without, and conclude that the radii of curvature are indeed on the order of 10 nm. Furthermore, despite the potential for contamination (solvent residue or frozen gases) to compromise performance of the knife edges, measurements on several devices show less than 20% variations in film flow. Because of the very small heat loads this represents, it was critical to allow the test apparatus enough time to thermally stabilize (~1 day) before taking any data.

For the flight vent system, film flow rates were determined in the same manner as a function of the temperature difference between the heat exchanger and porous plug. The data, in Figure 6a, show that raising the heat exchanger 35 mK above the porous plug evaporates nearly all (>85%) of the film. By extrapolation, at 50 mK, either no film remains or it is too thin to support superflow.

Figure 6b shows the heat flow, $\dot{Q}_{HX}$, that was required to raise the heat exchanger temperature by the indicated amount. As expected, there is a large initial growth in $\dot{Q}_{HX}$ without an equal decrease in the film flow rate. The reason is that the small tube leading to the heat exchanger can supply film at a faster rate than it flow through the knife-edge device. This excess flow must be evaporated before there is any noticeable change in $\dot{m}_{film}$. Beyond that point, there is an exact correspondence between the two quantities.

Finally, measurements of the porous plug temperature as a function of total vent rate are shown in Figure 7 at various bath temperatures between 1.15 and 1.35 K. These can be

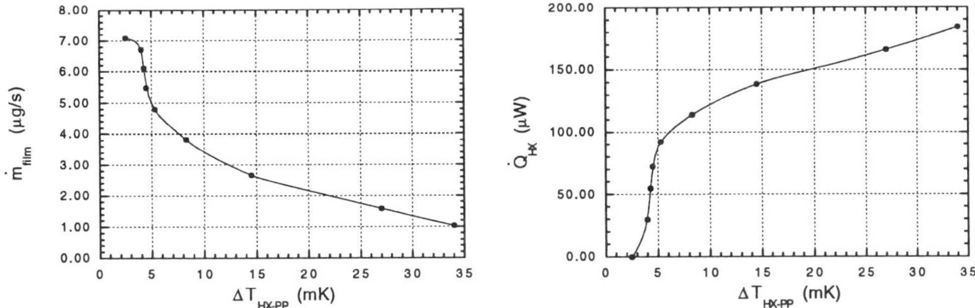


**Figure 6.** Performance data of the flight XRS vent system showing (a) film flow rates as a function of temperature across the porous plug and (b) the corresponding heat flow into the heat exchanger. The solid lines are interpolated fits.

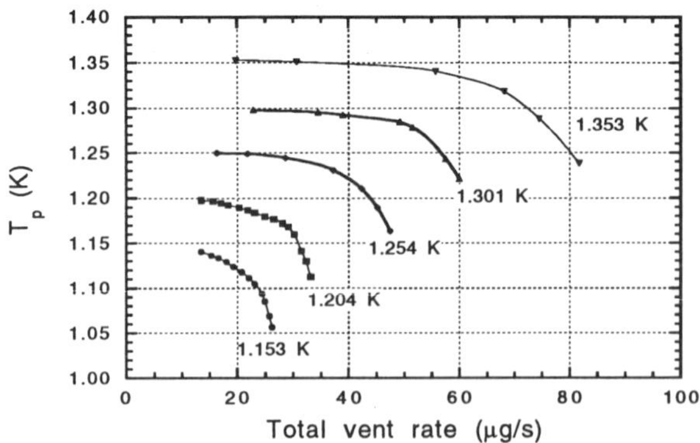


**Figure 7.** Downstream temperature of the porous plug versus flow rate for fixed bath temperatures.

used to estimate the final helium tank temperature. To vent at a rate of 40 μg/s, the driving pressure needed at the porous plug is estimated to correspond to saturated vapor at 1.09 K. The temperature of the curve that would include this point is estimated to be 1.22 K. This means the $\Delta T$ across the plug will be 0.13 K, and, from Figure 6a, that the film will be completely evaporated in the heat exchanger. Thus, it appears that the porous plug is sized smaller than necessary and will be operating above its knee.

## SUMMARY


The XRS porous plug and vent system has been designed to reduce or completely eliminate the direct loss of liquid helium from the tank which can result from superfluid film flow. Two techniques were implemented. First, a relatively small porous is used to achieve a relatively large temperature drop and the vent line is subsequently heat exchanged with the cryogen tank. Second, and as a backup, an etched silicon knife-edge device was developed and is located downstream of the heat exchanger. This vent line design has been built and tested, and has been installed on the flight XRS cryostat. Test results indicate that the film will be completely evaporated, so film losses are reduced to zero.


## ACKNOWLEDGMENTS


The authors wish to express their thanks to John Hendricks for identifying superfluid film flow as a potential problem for flight helium dewars, and acknowledge the contributions of Sridhar Manthripragada, Brent Mott, and Sanghamitra Dutta in fabricating the silicon knife-edge devices. This work was sponsored by NASA's Office of Space Sciences.


## REFERENCES


1. I. Rudnick, On the thickness, Doppler shift, superfluid density and critical velocity of a flowing unsaturated helium film, in: "Liquid and Solid Helium," C. G. Kuper, S. G. Lipson, and M. Revzen, eds. Wiley and Sons, New York (1975), p. 311
2. P. J. Shirron, K. A. Gillis, and J. M. Mochel, Third sound in thin helium films on Ar, Ne, and $CO_2$, *J. Low Temp. Phys.*, 78:157 (1990).
3. R. Torii, S. R. Bandler, T. More, F. S. Porter, R. E. Lanou, H. J. Maris, and G. M. Seidel, Removal of superfluid helium films from surfaces below 0.1 K, *Rev. Sci. Instrum.* 63:230 (1992).
4. M. J. DiPirro and J. Zahniser, The liquid/vapor phase boundary in a porous plug, *Adv. Cryo. Eng.* 35A:173 (1990).
5. See for example, D. M. Young and A. D. Crowell, "Physical Adsorption of Gases", Butterworths Publishing Company, Washington (1962).
6. A. Bijl, J. DeBoer, and A. Michels, *Physica* 9:655 (1941).
7. R. H. Haycock, Remote-controlled Stepper-motor-activated cryogenic valve: design, development, and testing, Proceedings of the SPIE: Cryogenic Optical Systems and Instruments VII, L. G. Burriesci and J. B. Heaney, eds., 2814:166 (1996).

# FINAL DESIGN OF THE XRS/ASTRO-E ADR

A. T. Serlemitsos*; M. SanSebastian, E. S. Kunes**

*Code 713.1, NASA/ Goddard Space Flight Center, Greenbelt, MD 20771

**HSTX Corporation, 4400 Forbes Blvd, Lanham, MD 20706

## ABSTRACT

GSFC has developed an X-Ray Spectrometer (XRS) to be flown aboard Astro-E, in cooperation with the Japanese Institute of Space and Astronomical Science (ISAS). XRS uses an array of 32 microcalorimeters capable of detecting X-rays in the energy range of 0.3 - 10 keV with a resolution of 12 eV. In order to accomplish this, the detectors must be operated at a temperature of 0.065 K. In space, an Adiabatic Demagnetization Refrigerator (ADR) must be used to cool the detectors to that temperature. A spaceworthy ADR has been developed at GSFC to be used in the XRS. Originally, the ADR was developed to be flown aboard the AXAF[1,2]. Budgetary constraints necessitated the move of the XRS to the Astro-E program and this resulted in a considerable reduction in weight and a much tighter thermal specifications for the ADR. The allowable average thermal load of the ADR to the LHe dewar was changed from 2.6 mW to 250 μW. Time constraints did not allow a complete redesign of the ADR. The original shape and size were left unchanged and the new specifications were met by streamlining the heat switch and lengthening the salt pill magnetization cycle time. Size and weight contraints forced us to abandon the original redundancy plan of having a heat switch on either side of the salt pill. We experimented to provide redundancy by replacing a single heat switch with two much slimmer heat switches occupying the same volume, each with almost half the capacity but also half the parasitic heat leak. They were independent from each other, thus providing redundancy, but could be operated simultaneously so as not to reduce the heat switch capacity significantly. During vibration testing we discovered that the considerable reduction in stiffness made these heat switches vulnerable to shorting. It was, therefore, decided that it was preferable to proceed with a single heat switch without redundancy. For a LHe bath temperature of 1.3 K the gas gap heat switch presently used has an on/off ratio of ~28000 and a parasitic heat leak of 2.4 μW/K.

## INTRODUCTION

One of the experiments on the Japanese Astro-E satellite is the X-Ray Spectrometer (XRS)[3]. The detector system in this instrument will be an array of 32 calorimeters (bolometers) capable of detecting X-rays in the energy range of 0.3-10 keV[3]. To obtain a resolution of 12 eV, the calorimeters must operate at a very low temperature (65 mK). In space, an Adiabatic Demagnetization Refrigerator (ADR) must be used to cool the XRS

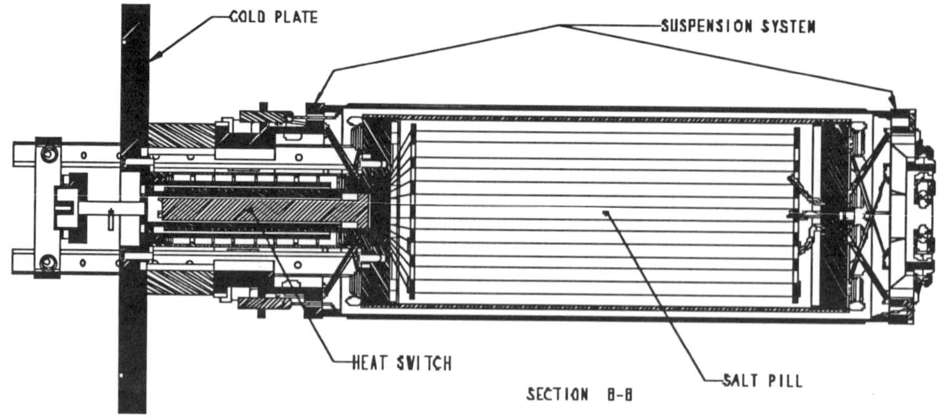


**Figure 1.** Schematic Diagram of the GSFC ADR.

calorimeters. The development of the Goddard Space Flight Center (GSFC) ADR started in 1980[4,5]. In the past several years the development of the ADR went through various stages in an effort to arrive at as efficient a unit as possible while at the same maintaining sufficient margins in its structural integrity. In this paper we shall concentrate on the final design and construction. A brief account of various methods used will be given, along with certain things one must avoid.

## CONSTRUCTION

A schematic diagram of the ADR is given in Figure 1. The main parts of the ADR are the paramagnetic salt pill, the gas gap heat switch and the suspension system. The detector assembly is thermally, but not structurally attached to the end of four thermal bus bars.

### Salt Pill

Ferric ammonium sulfate, $[Fe_2(SO_4)_3 \cdot (NH_4)_2SO_4 \cdot 24H_2O]$, is used for the salt pill of the GSFC ADR. The thermal bus, Figure 2, is made up of 4500 pure gold wires 200 μm in diameter. Sheets of wire are made and secured in a thermal bus frame spaced evenly apart. The ends of the wires are tightly bundled in a cylindrical shape, passed through a hole in the

**Figure 2.** Gold-wire Thermal Bus Harness on Gold-plated Brass Frame.

**Figure 3.** Harness and copper end piece assembly ready for welding.

copper cross piece, Figure 3, and welded to the copper piece. Four copper thermal bus bars for the detectors, 9 mm in diameter, are also welded to the cross piece. As seen from Figure 3, the central portion of the copper cross piece has as small diameter as possible in order to minimize the generation of eddy current heating during the adiabatic demagnetization of the salt. After all the welding is finished the two ends of the salt pill harness are plated with 12 μm of gold for protection from the highly corrosive properties of the salt solution. The G-10 fiberglass end cap, Figure 4, is now bonded to the copper cross piece with Stycast 2850FT epoxy[6]. Eight 2-56 stainless steel screws are used to join the G-10 end cap to the copper piece for structural purposes.

The leads of two Ruthinium Oxide thermometers are passed through a small hole in the G-10 end cap sealed with Stycast. The thermometers are well sealed to protect them from the salt solution during the crystal growth. One of the two will be used in flight during the magnetization cycle to control the maximum temperature the salt is allowed to reach. The second is there for redundancy. The G-10 end cap is threaded with #40 thread. A G-10 tube

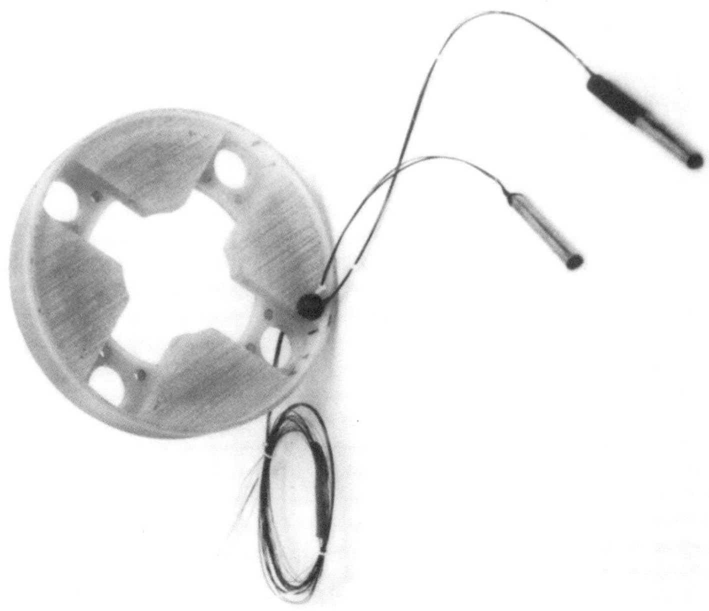

**Figure 4.** G-10 Fiberglass End Cap.

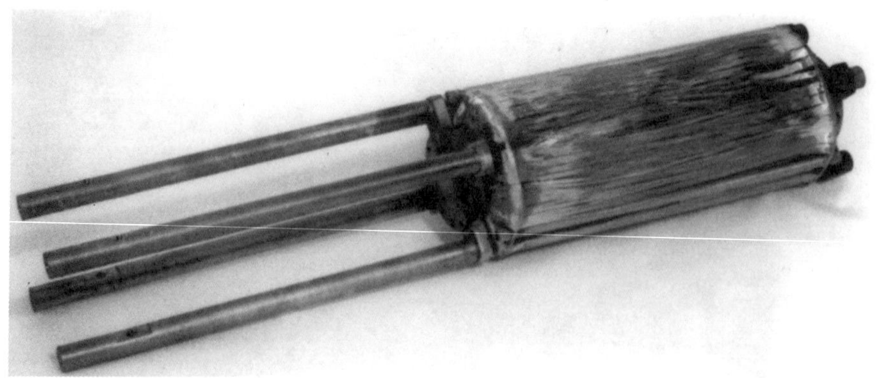

**Figure 5.** Salt Pill Harness welded at both ends.

threaded at the end is attached to the end cap using Stycast epoxy. In the original design the far end of the salt pill was identical to the near end except for the bars protruding only 1.5 mm from the copper cross piece, Figure 5. The purpose of that was to provide thermal contact to a redundant heat switch same as the primary heat switch.

The reduction in the weight limits necessitated by the cooperation with Astro-E, forced the elimination of the redundant heat switch so that the size of the magnet could be reduced. The flight ADR was build with a redesign of the far end that eliminated the cross piece. Figure 6 shows a schematic diagram of the new design. All except 60 of the 4500 gold wires were cut 1.5 mm above the wire harness frame. This eliminated the closed wire loops that the previous design had, thereby reducing the generated eddy current heating. The 60 gold wires that are not cut short, are used to provide a thermal link between the salt and the far end cap.

Once the salt pill container is completed, it is checked for leaks with a helium leak detector, and then the salt is grown. After an initial seeding using ~5 cc of solution saturated at 35 °C, the crystal is grown by filling the container to the top and maintaining a temperature gradient of 15 °C by cooling the bus bars to ~10 °C, keeping the other end to ~25 °C. Once a day the used solution is dumped and fresh solution saturated at 35 °C is added until the crystal is grown all the way to the far end of the thermal harness. Finally the far end is sealed again with Stycast and the salt pill is complete. The salt is coated with Epon epoxy which is allowed to set overnight before the final sealing with Stycast is made. This additional step was added when it was found that Stycast does not set properly if it comes in contact with the salt.

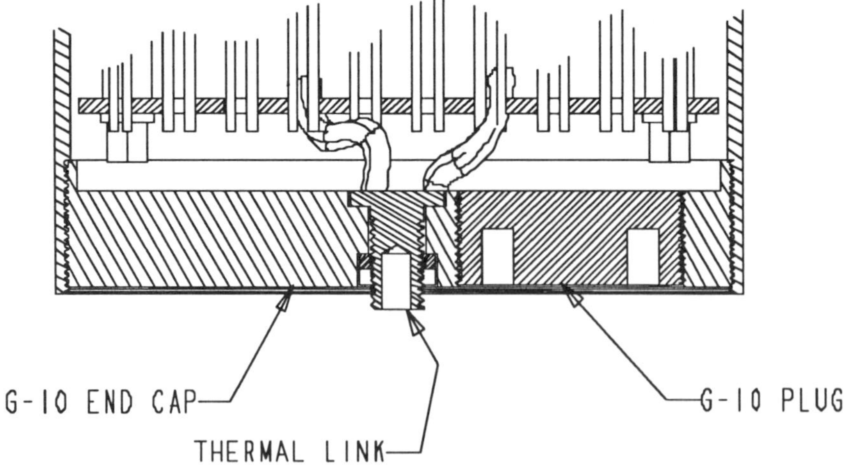


**Figure 6.** Schematic Diagram of the Far End Design of the Flight ADR.

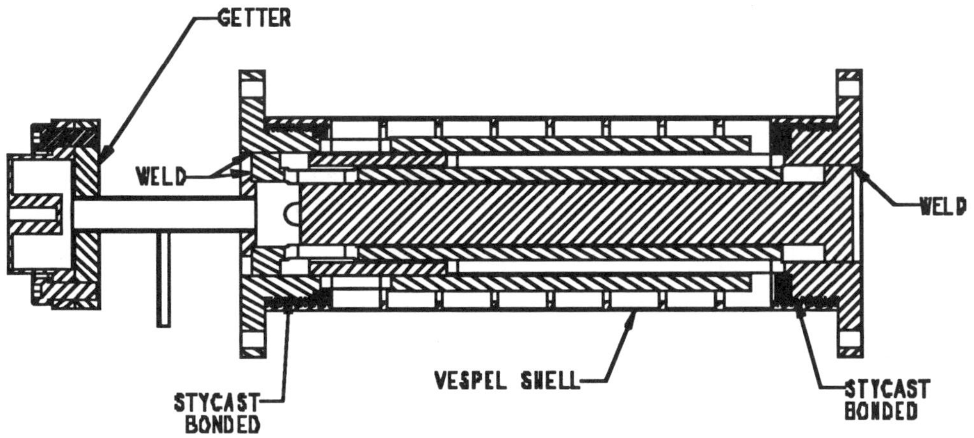


**Figure 7.** Schematic Diagram of the Gas Gap Heat Switch .

## Heat Switch

A gas gap heat switch is used for the GSFC ADR. Figure 7 shows a schematic diagram of the present gas gap heat switch. Each heat switch end is made up of two concentric copper cylinders. (The copper used both in the salt pill and the heat switch is OFHC copper.) The two cylinders for each end are welded using a steel mandrel to keep them concentric. After welding all the copper parts are plated to a thickness of 12 μm of gold. When interlaced they leave a nominal gap of 200 μm between the cylindrical surfaces. A cylindrical shell with a wall thickness of 500 μm is used to separate the two ends of the heat switch and to confine the helium gas which provides the thermal conductance across the heat switch. The material used for this shell is Vespel, a polyimide made by DuPont. A gold plated stainless steel tube, used to connect the heat switch to the getter, is soft soldered to the heat switch. A filling tube adapter has already been silver-brazed on the connecting tube.

A 10 mm length on either side of the Vespel shell is threaded to a #40 thread, so are the mating surfaces of the two ends of the heat switch. This provides a much stronger heat switch assembly, as well as a means of maintaining parallelism of the heat switch cylinders. Stycast is used for bonding the heat switch pieces to the Vespel shell and for sealing the heat switch. Helium gas diffuses through the Vespel shell at the rate of ~5E-5 standard cc/s at room temperature. At this rate the half life of the gas in the switch is only of the order of two weeks, which is unacceptable, since the ADR is expected to be warm and perhaps in vacuum for a few months. To remedy this problem, the Vespel shell is wrapped with a special titanium alloy with low thermal conductivity. Armstrong A-12 epoxy is used to do the wrapping, resulting in a diffusion rate of ~5E-8 scc/s. The necessary wrapping of the heat switch results in doubling the parasitic heat leak through the heat switch shell.

Zeolite is used as a helium getter. The getter pellets are silver-soldered to the bottom of the aluminum getter cup. The base of the getter is gold plated copper which has been soft soldered to the gold plated connecting tube. The getter is baked in vacuum at 380 °C. The getter cup is attached to the heat switch in a helium filled glove box and the assembly is sealed with an indium O-ring. The fill tube is used to pump the heat switch briefly and then fill it with helium at slightly above atmospheric pressure and seal it by pinching the fill tube then folding it twice and using soft solder for the final seal.

## Suspension System

The suspension system for the ADR must support the salt pill well enough to assure a safe launch. It must also be rigid enough to minimize the microphonic noise input to the calorimeters. The outline of the suspension system can be seen in Figure 1. A titanium near end support ring is rigidly attached to the cold plate using eight 8-32 screws. Four 2.4 cm long stainless steel spacers are used between the cold plate and the support ring. A similar titanium support ring is used at the far end. A titanium tube, 7.5 cm in diameter, and a wall thickness of 360 μm is utilized to keep the two support rings the right distance apart. Four 900 N-test Kevlar 49 ropes are tensioned between the two rings to a total of 1800 N in order to keep the support structure rigid during the suspension process. At low temperatures the support is provided by the magnet mandrel. At the near end the support ring is attached to the magnet flange via the cold plate. At the far end the support ring is very tightly held by the magnet mandrel due to differential contraction between the Aluminum mandrel and the titanium support ring. At each end of the salt pill eight 6.2 cm long Kevlar 49 loops made of 710N-test rope are used for supporting the salt pill. Eight #8-32 screws attached to each end of the salt pill are used to anchor the Kevlar loops. At the opposite end of each loop threaded hooks and nuts are used to tension the loops. A pretension of 180 to 220 N at room temperature is applied to each loop.

## RESULTS

Three complete ADR units have been completed recently and thoroughly tested. The flight unit (FM2) incorporated the far end redesign described earlier, while the other two (FM0 and FM1), one of which (FM1) is the flight spare unit, were built following the original design. What prompted the change in design was the realization that some fluid gets trapped in the near end of the salt pill. In the last engineering model, EM2, as well as in FM0 and FM1 tiny leaks developed at the near end gold wire welds, when the salt pill temperature was allowed to go up to 28 °C. It is believed that because of the convergence of 4500 gold wires at that point, salt is formed all around that volume thus trapping the used solution there. The solution to this problem was to drill a small hole at the center of the weld and with the help of vacuum to drain that space completely. The hole is then tapped for a #2-56 thread and is sealed with Stycast and a stainless steel 2-56 flat head screw. All the salt pills built up to that point, even the ones that did not show any leaks, had some fluid drained at the near end. It was simply a matter of the depth of the weld. For FM2 an effort was made to make the weld much deeper. Also the procedure for growing the salt was changed. Before the used solution was emptied, the temperature of the entire unit was raised to 25 °C and after the used solution was emptied, the salt pill was placed upside-down in vacuum sucking out any remaining used solution. Finally, the far end was redesigned as described earlier to create a gap of at least 5 mm between the top of the salt and the bottom of the far end cap. The salt pill had no signs of any leaks after 6 thermal cycles to LN2 temperature, and an elevation of its temperature to 28.2 °C at which it was left for at least two hours. Nevertheless, it was decided to drill a draining hole like the other salt pills. The following observations resulted. The weld was indeed much deeper than in the previous salt pills -- ~4.5 mm of copper and 1.5 mm of gold. Yet, ~4 cc of used solution was removed. However, whereas in the previous salt pills the trapped solution was under pressure, in FM2 there was no pressure presumably because of the buffer volume at the far end of the salt pill. After the draining and the subsequent sealing none of the salt pill showed any signs of a leak when subjected to thermal cycling, and to an elevated temperature of 28.2 °C.

The thermal performance of the ADR surpassed the original requirements. The weight of the salt compared to the expected weight from the measured volume was between 96% and 97% and the cooling power of five salt pills are within 3% of each other. With a bath temperature of 1.3 K and the salt pill at 64 mK the total parasitic heat leak was measured to be 3.6 μW, giving an estimated heat switch leak of 3 μW or 2.43 μW/K. With the salt pill at 1.8 K the conduction across the heat switch was measured at 68 mW/K giving an on-off ratio of 28000 for the conditions of this ADR.

Finally, the suspension used was shown to be providing considerable margin. The present form of the ADR was vibrated twice to the levels expected during launch with no obvious damage, or deterioration of its performance.

## REFERENCES


1 S.S. Holt, "X-Ray Spectroscopy of AGN with the AXAF 'Microcalorimeter'", Astrophysical Letters and Communications 26 61, (1987).

2 A.T. Serlemitsos, et al., Adiabatic Demagnetization Refrigerator for Space Use", Advances in Cryogenic Engineering 35 1431, (1989).

3 R.L. Kelly et al., "High Resolution X-Ray Spectroscopy Using Microcalorimeters", Proc. SPIE 219 (1988).
4. S.H. Castles, "Refrigeration for Cryogenic Sensors", NASA Conference Publication 2287 389 (1983).
5. S.H. Castles, S.R.Breon, B.A.Warner, A.T. Serlemitsos, S.M. Volz, and M.G. Ryschkewitsch, "The Cryogenic Subsystem for the X-ray Spectrometer on the Advanced X-ray Astrophysics Facility", Proc. SPIE 973, 110 (1988).
6. Stycast 2850-FT, Emerson and Cummings Inc., Canton MA 02021.

# A HIGH $T_C$ SUPERCONDUCTING CURRENT LEAD ASSEMBLY FOR THE XDS HELIUM CRYOSTAT

J. G. Tuttle,[1] T. P. Hait,[1] R. F. Boyle,[2] H. J. Muller,[3]
J. D. Hodge,[3] and S. R. Breon[2]

[1]Hughes STX Corp.
c/o Code 713, NASA/GSFC
Greenbelt, Maryland 20771

[2]Code 713, NASA/Goddard Space Flight Center
Greenbelt, Maryland 20771

[3]Illinois Superconductor Corp.
451 Kingston Court
Mount Prospect, Illinois 60056

## ABSTRACT

The X-ray Spectrometer Detector System (XDS) helium cryostat consists of a tank of pumped liquid helium at about 1.3 kelvin suspended inside a seventeen kelvin cylindrical support structure. The tank is a heat sink for an adiabatic demagnetization refrigerator (ADR) and its superconducting magnet. The cryostat's small initial helium volume and mission lifetime goal of 2.5 years require that the average total heat load to the helium be less than about 800 microwatts. During the mission the superconducting magnet requires a current of 2 amps with a three percent duty cycle. In addition, wires capable of carrying up to 1 amp are needed for cryogenic valve operations during the cryostat's ground servicing. The best optimized conventional current leads between the 17 kelvin stage and the magnet and valves would contribute an average heat load to the helium of about 3 milliwatts. An assembly of superconducting YBaCuO fibers bonded to a fiberglass tube and suspended by a Kevlar* braid was developed to conduct the current from the 17 kelvin support structure to a vapor-cooled 4 kelvin stage. NbTi wires provide a superconducting path from the 4 kelvin stage to the magnet and valves on the 1.3 kelvin helium tank. This paper describes the assembly's fabrication and suspension and presents the results of its performance and vibration tests.

* Kevlar is a Du Pont registered trademark

## INTRODUCTION

The X-Ray Spectrometer Detector System (XDS) helium cryostat[1] consists of a tank of pumped liquid helium at about 1.3 kelvin suspended inside a seventeen kelvin cylindrical support structure. The tank is a heat sink for an adiabatic demagnetization refrigerator (ADR)[2] and its superconducting magnet. The cryostat's small initial helium volume and mission lifetime goal of 2.5 years require that the average total heat load to the helium be less than about 800 microwatts. One challenge involved in reaching this goal was the design of wiring to provide electrical power to the ADR magnet and to two cryogenic valves atop the tank. The magnet will be operated on orbit with a maximum current of 2.0 amps with an estimated 3 percent duty cycle. The valves allow the fill and vent plumbing lines to be closed after the final ground servicing so that the on-orbit venting will be via a porous plug phase separator. The valves are only operated on the ground, but their stepper-motors require currents of up to 1 amp over a period of 20 seconds during each operation.

## WIRING DESIGN OBSTACLES

An attempt was made to design magnet and valve current leads from the 17 kelvin support structure to the 1.3 kelvin helium tank using conventional wire. An iterative thermal model was used to optimize the ratio of the wires' cross sectional areas to lengths. This model arrived at a steady state wire temperature profile by balancing the heat flow into and out of each of several nodes along the wire. It assumed that the wires were blackbodies exchanging radiation with a 17 kelvin shell. The calculations revealed that the best average heat load for two copper magnet leads with a 3% duty cycle would be about 2.7 milliwatts, exceeding the helium tank's entire thermal budget by more than a factor of three. The Wiedemann-Franz law implied that using another metal would not improve the situation much, as indeed thermal modeling of stainless steel (optimum is 2.7 milliwatts for two leads) and manganin (2.8 milliwatts) wires showed.

Each magnet current cycle includes a one-hour period of ramping up, holding at 2.0 amps, and ramping back down. This is followed by about 30 hours during which the current is low enough to be conducted efficiently by small conventional wires. Using a mechanical switch (solenoid or magnetostriction-driven) to open the high-current magnet leads thermally while the current is low would improve the average heat load due to two optimized copper leads to about 320 microwatts. This might have led to a solution in which the "current -on" heat was dumped to a vapor-cooled heat exchanger below 9 kelvin on the helium vent line, allowing NbTi superconducting wires to carry the current to the 1.3 kelvin tank. However, the overall XDS thermal model showed that the large heat load into the vapor while the current was 2 amps would have driven the heat exchanger's temperature well above the 9 kelvin superconducting transition temperature of NbTi. Thus a conventional wire would be needed to conduct the current from the heat sink to the tank. In such a scenario with a heat sink temperature of 8 kelvin, the time-averaged heat load due to two optimized leads would be about 1.2 milliwatts.

Each of the two valves requires four phase leads carrying 0.5 amps and one return lead carrying 1.0 amp. Since the valves are only operated on the ground, there will be no ohmic heating in the leads on orbit. If the valve power leads were made of a conventional wire, they would be designed to have a steady-state "current on" temperature profile with a maximum temperature below some reasonable value (such as 300 kelvin) to avoid damage to the wires or insulation. This temperature profile was modeled with various wire sizes, and the total resulting valve lead heat load (with I=0) was 230 microwatts. As with the magnet leads, this zero-current heat load could be absorbed by a vapor-cooled heat exchanger, but the enormous heat load when the current is on could not. Again, a set of conventional leads would be required to carry the current from the heat sink to the tank, resulting in a large heat load to the helium.

It might have been possible to count on the high heat capacity of stainless steel wires and the relatively short duration of the valve operation current to limit the maximum wire temperature. By not designing for the steady-state case it would have been possible to use smaller wires and have a lower on-orbit heat load. However, experience with launch-pad ground servicing and with these particular valves and their support electronics indicated the need for a design that accommodated multiple valve operations in a short time period.

In view of all these problems, the decision was made to use a high $T_C$ superconducting (HTS) material for the magnet and valve leads. One option considered was the use of thin film superconductors.[3] Unfortunately the technology was not quite ready in time to produce a lead assembly that would meet the thermal needs. Instead, Illinois Superconductor Corporation was consulted. Two of the authors produced a design and built a series of three prototype assemblies using $Y_1Ba_2Cu_3O_7$ (YBCO) filaments. These prototypes were performance-tested at Goddard Space Flight Center (GSFC). The actual flight assemblies were built and tested in-house at GSFC using YBCO filaments provided by Illinois Superconductor.

## HTS ASSEMBLY DESIGN REQUIREMENTS

The HTS lead assembly needs to satisfy several basic requirements. It must provide two leads to conduct the 2.0 amp magnet current and ten leads capable of carrying 0.5 to 1.0 amps to operate the two cryo-valves. The filaments need to have adequate critical currents at 17 kelvin, the maximum operating temperature of the leads. Obviously the assembly must fit in the available space on the helium cryostat. This limits its overall length to 25 cm and its cross section to about 0.75 cm by 2.5 cm. The assembly must conduct as little heat as possible to the helium tank and have virtually no electrical resistance. A vapor-cooled station at about 4 kelvin on the vent line is available to intercept some of the conducted heat. According to the cryostat's thermal model,[4] each additional 100 microwatts added at this stage will raise its temperature 0.39 kelvin and add 14.5 microwatts to the helium tank's heat load, trimming 11 days off the mission lifetime.

Individual HTS filaments of the appropriate size are quite fragile, and the entire assembly must survive the Japanese M-V rocket's launch vibration and shock loads, with low-frequency peak accelerations around 25g and an overall vibration environment of 8.5 grms.

## DESIGN DETAILS

The flight helium cryostat uses two HTS lead assemblies wired in parallel for redundancy. Each of the assemblies contains twelve 25-cm long, 0.25 mm diameter HTS filaments which are epoxied to a thin-walled 0.64 cm diameter fiberglass tube of the same length. A copper wire conducts current between to the warm end of each filament and the 17 kelvin stage. The bundle of 12 copper wires heat-sinks the HTS assembly's warm end to that stage. The wire diameters are 0.64 mm for the two magnet leads and 0.51 mm for the valve current leads.

Each assembly's cold end includes a copper braid epoxied into the fiberglass tube. This braid is soldered to a 0.64 mm diameter copper wire which is thermally anchored to the 4 kelvin vapor-cooled heat exchanger. For the electrical connections, a copper-clad niobium-titanium wire conducts the current from each filament's cold end to the valves or magnet on the 1.3 kelvin tank. The copper has been etched away from 2.5 cm of the wire length to reduce the thermal conduction. The iterative model indicates that each of these 0.051 mm diameter bare NbTi segments conducts less than 100 nanowatts. The NbTi leads have a critical current of about 5 amps at 4 kelvin and about 2.5 amps at 7 kelvin.

The NbTi wires used as valve leads are attached to the valves via a heat sink on the helium tank. This is necessary because the valve stepper motors heat up significantly during operation. In fact, performing two operations of a single valve in a short period of time heats its motor housing up to about 15 kelvin, much higher than the NbTi's transition temperature.

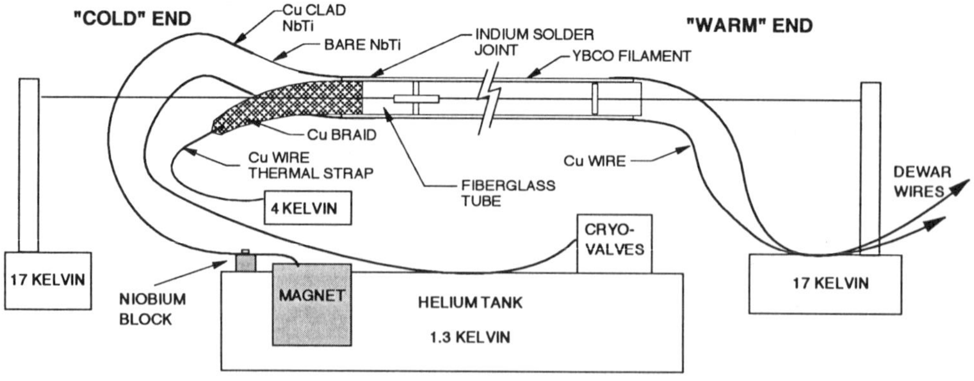


**Figure 1.** A schematic representation of one HTS assembly in its flight configuration.

To eliminate joule heating in the magnet circuit, the NbTi magnet wires are connected to the magnet leads via a superconducting niobium terminal block on the helium tank. The two wires in each connection are abraded first and then slightly crushed onto the block under a washer. In addition, a copper-clad portion of each magnet lead is soldered to its corresponding copper-clad NbTi wire from the HTS assembly in order to provide a backup joint. Figure 1 shows a schematic of one of the HTS leads assemblies as it is configured in the flight cryostat.

Each current lead assembly is supported on a Kevlar yarn running through the axis of the fiberglass tube. The yarn is tied to each end of an aluminum tension link which is bonded to a fiberglass disc, and two more fiberglass discs are glued along one length of the yarn. The tension link's disc is bonded into the cold end of the current lead assembly, and the other two discs assume their positions axially inside the tube when the yarn is tensioned. This supports the tube at one location axially, and with sufficient tension in the yarn, three locations laterally. Each end of the Kevlar is supported from a post mounted on the cryostat's 17K support structure. This support scheme is shown in Figure 2.

## FABRICATION OF CURRENT LEAD ASSEMBLY

### YBCO Filament Fabrication

The HTS filaments are composed of sintered polycrystalline ceramic, nominal composition $Y_1Ba_2Cu_3O_7$ (YBCO). They are extruded with a 0.25 mm diameter and cut to 25 cm lengths. Later they are coated with silver along 2.5 cm of length at each end and baked in an oven in order to fuse the YBCO. Each filament is electrically tested at 77 kelvin, and those having a critical current below 0.5 amps are discarded. Test data on this material show that the critical current increases by a factor of at least seven from 77 kelvin to 17 kelvin, so the filaments passing this test will easily carry the maximum current of two amps. The filaments are shipped to GSFC where they are immediately put into a dry box for storage. The desiccant in the box keeps the humidity well below 1% and ensures that the filaments are not damaged by moisture.

### Fiberglass Tube Fabrication

The first step in making the fiberglass tube is to slide a copper braid over the end of a Teflon mandrel. The mandrel's diameter is 0.64 cm except near one end where the braid is installed. At this end the mandrel is narrower so that the braid is flush with the rest of the

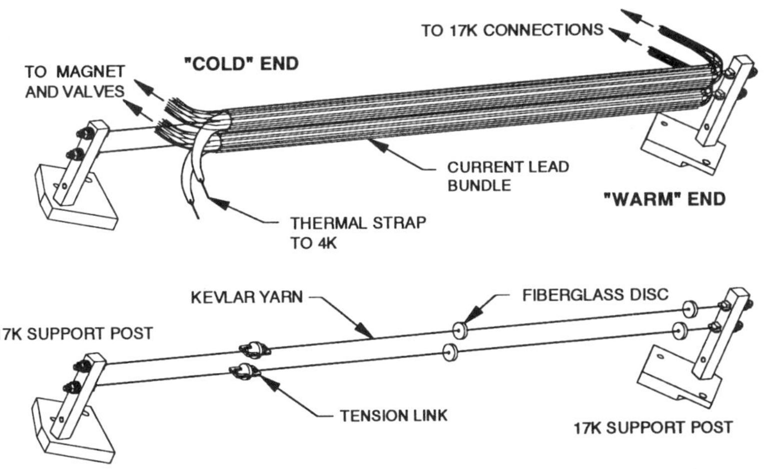


**Figure 2.** A pair of HTS assemblies shown suspended in their flight configuration .

mandrel. A 12.7 μm thick glass cloth is wrapped twice around the mandrel and braid and is impregnated with CTD-521 epoxy manufactured by Composite Technology Development, Inc. The manufacturer adds 20 percent alumina filler by weight to the resin so that the fiberglass tube's thermal contraction will match that of the YBCO filaments. This epoxy is allowed to cure at room temperature for at least 24 hours. The resulting fiberglass tube, with its copper braid, is removed from the mandrel and installed on a smaller mandrel to prevent it from flexing during the filament mounting process.

## Assembly of Components

The YBCO filaments are extremely fragile and must be handled with great care until they are safely glued to the fiberglass tube. First, a 0.64 or 0.51 mm diameter copper wire is indium soldered to the "warm" end of each YBCO filament. The filament's "cold" end is indium soldered to a formvar-insulated wire with a 51 μm diameter NbTi core and a 12.7 μm thick copper coating. The NbTi wire is first wound around the filament's silver-coated end so that about 3.2 cm of the wire length are in each indium solder joint. As was mentioned above, the copper has been etched off 2.5 cm of the wire's length to reduce its thermal conduction. After each joint has been checked under a microscope, a short length of very thin Teflon tubing is slid over the NbTi wire until it butts against the solder joint. This tubing insulates the part of the wire from which the Formvar has been stripped in order to etch the copper, and it helps keep the tiny wires from tangling or kinking. The YBCO filaments are tacked to the outside of the fiberglass tube with dots of Devcon 5 minute epoxy, taking care to keep them flat against the tube. Then a thin layer of the CTD-521 epoxy is painted over the filaments in order to protect them from moisture, which would degrade the HTS material's performance over time. In addition, the epoxy is applied under each filament to fill in any gaps between it and the tube. This work is done very carefully with a white sable brush. A thick layer of the epoxy is applied over the solder joints at the ends of the tube. About 48 hours later, when the epoxy has partially hardened, a second "touch-up" epoxy paint job is done. All of this epoxy is allowed to cure at room temperature for at least 24 hours. The assembly is then baked in a vacuum oven at 80°C for about 15 hours.

The suspension components are assembled separately from the current lead bundle. The tension link and fiberglass disc are bonded together with CTD-521 epoxy. Kevlar 49

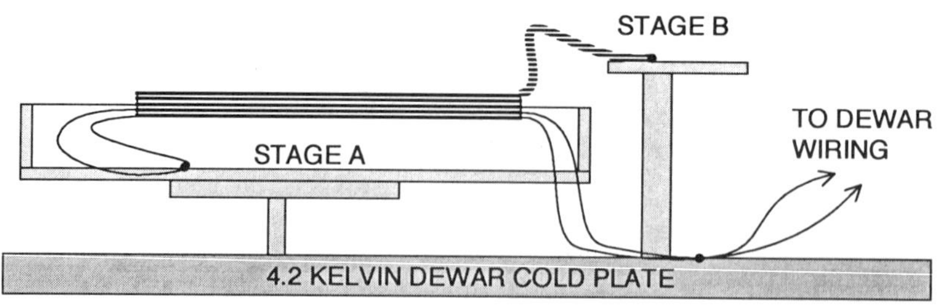


**Figure 3.** A schematic representation of the thermal performance test setup.

yarn with a 27.7 N rated strength is tied onto each end of the link, and the knots secured with cyanoacrylate adhesive. The other two discs are then bonded to the Kevlar, and the epoxy is allowed to cure for three hours at 80°C. These components are pull-tested to 5.5 N, and then the link disk is bonded into the fiberglass tube with CTD-521 epoxy, near the cold end of the current lead bundle. The other two disks are allowed to float axially, preventing the tube and superconducting filaments from carrying any of the axial load in the Kevlar yarn. After a room-temperature cure of the epoxy, eye screws are tied onto the ends of the Kevlar. The position of the eye screws is set to allow about 0.32 cm in for subsequent stretching of the Kevlar. The entire assembly is then baked again at 80°C for 12 hours, driving off all measurable amounts of volatile materials.

Prior to mounting the lead assemblies in the flight cryostat, they are suspended from two posts. A strongback fixture is then used to stretch the Kevlar 2.3 mm and position the posts at their nominal separation. This stretch corresponds to a load of about 4.8 N in each yarn, sufficient to give a first mode lateral frequency of about 100 Hz. Upon cooldown, the titanium of the cryostat support structure has a large contraction relative to the Kevlar yarn. However, the Kevlar's modulus has been found to increase around 60% with this temperature decrease, so there is no net loss of load in the yarn for this setup. Creep in the yarn, assumed to lengthen it by about 0.51 mm over five years of life,[5, 6] is monitored with a set of Belleville washers, and re-set to the initial tension after about one month. The preload in the yarn is low enough relative to its breaking strength that it will not rupture within the life of the experiment.

## TESTING

### Critical Current Test

For its critical current test each assembly is mounted in a helium dewar. Each filament's copper wire is connected to one dewar lead, and its NbTi wire is connected to another. The dewar is filled with liquid helium so that the assembly and its wires are submerged. Currents of one, two and three amps are then sent through each filament while the voltage across the entire circuit is measured. For undamaged filaments the zero-current circuit resistances are all about 0.5 ohms, and they change by no more than 2 milliohms when the current is raised from one to three amps. This small change is easily explained by heating of the copper wires in the circuit.

### Thermal Performance Test

Figure 3 is a schematic representation of the thermal performance test setup, which is mounted on another helium dewar's 4.2 kelvin cold plate. The HTS assembly is supported by its Kevlar string in the flight configuration. Its two support posts are mounted on an

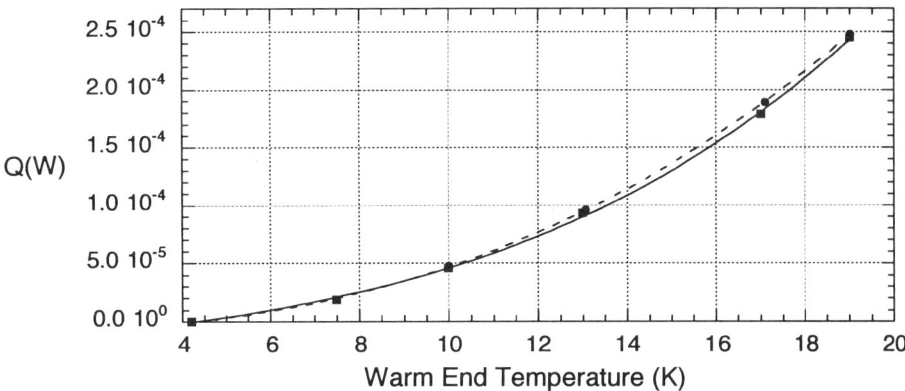


**Figure 4.** Thermal conductance data for each of the two flight HTS assemblies with their cold ends held at about 4.2 kelvin. With their warm ends controlled at 17 kelvin they each conducted about 185 microwatts.

aluminum bar attached to temperature-controlled stage A. The copper wires on the two magnet lead filaments are soldered together in a loop and heat sunk to this stage. The copper braid at the cold end of the assembly is attached to temperature-controlled stage B. The conductance between each of these stages and the 4.2 kelvin cold plate has been characterized. The NbTi wires on the magnet lead filaments are connected to the dewar leads via a heat sink on the 4.2 kelvin cold plate.

To measure the HTS assembly's thermal conductance, stage A is controlled at one of several temperatures between 7.5 and 19 kelvin. After a steady state is reached, the temperatures of stage B and the cold plate are measured. The temperature drop from stage B to the cold plate gives the heat load on stage B, allowing the assembly's zero-current thermal conductance to be calculated. Figure 4 shows this data for the two flight assemblies, each of which conducted about 185 microwatts with its warm end at 17 kelvin.

At this point two amps are applied through the two magnet filaments, with stage A controlled at 17 kelvin. The resulting change in stage B's temperature gives the ohmic heating in the YBCO-NbTi solder joints. The current is allowed to flow steadily for up to four hours during the testing. The average magnet filament solder joint resistance was 4.2 microohms per joint in one flight assembly and 8.6 microohms in the other. Average joint resistances as low as 1.4 microohms were seen in the prototype testing.

## Vibration Test

A prototype current lead assembly was vibrated at room temperature for four minutes at qualification levels, in a configuration similar to flight. There was no indication of any problem with the superconducting filaments, any change in the critical current, or any wear products between the fiberglass tube and support disks. A strobe light was used to confirm that the assembly's lateral mode was actually at 100 Hz.

The flight HTS assemblies were vibrated at room temperature, two-at-a-time, suspended from the flight support posts in the flight configuration. In their flight configuration the assemblies are oriented perpendicular to the spacecraft's thrust (z) axis. The vibration loads were applied along the axis parallel to the Kevlar suspension yarns, along the spacecraft's z axis, and along a third, orthogonal axis. The vibration program began with a 30 second duration random vibration in each direction. Then, in order to simulate a shock load, a 6-cycle, approximately 25 Hz sine burst was applied in each direction. After this test, the leads were again cooled to 4.2 kelvin for another critical current test. All of the filaments on the flight assemblies remained superconducting at currents up to three amps, so no damage was indicated. One of the assemblies was also retested for thermal performance, and no significant degradation was seen.

## CONCLUSION

Two HTS lead assemblies have been completely tested at the component level and qualified for space flight. They are installed on the XDS flight helium cryostat which is undergoing system-level testing. As this document goes to print they have met all of their performance requirements. They will experience a system-level vibration and more performance testing after that. If they continue to function as expected they will help the XDS cryostat meet its mission lifetime goal.

## ACKNOWLEDGEMENTS

This work is supported by the NASA's Office of Space Science.

## REFERENCES

1. S. R. Breon *et al.*, Design of the XRS helium cryostat, *Adv. Cryo. Eng.*, 41B:1128 (1996).
2. A. T. Serlemitsos, Final design of the ASTRO-E/XRS ADR, *to be presented at this conference.*
3. S. R. Foltyn *et al.*, Pulsed laser deposition of thick $YBa_2Cu_3O_{7-\delta}$ films with $J_C$ > $1MA/cm^2$, *Appl. Phys. Lett.*, 63:1848 (1993).
4. S. R. Breon *et al.*, Thermal Performance of the XRS helium cryostat, *to be presented at this conference.*
5. "Properties and Uses of KEVLAR in Fiber Optic and Electromechanical Cables," Bulletin K-506C, Du Pont Advanced Fiber Systems, Wilmington, Delaware (1995).
6. "KEVLAR for Fiber Optic and Other Cables - Data Manual", Du Pont Advanced Fiber Systems, Wilmington, Delaware (1991).

# CRYOGENIC/THERMAL SYSTEM FOR THE SIRTF OBSERVATORY

R. A. Hopkins[1], S. J. Nieczkoski[1], K. L. Russell[1], M. J. Lysek[2], A. Klavins[3]

[1]Ball Aerospace Systems Division, Boulder, Colorado 80306-1062, U.S.A.
[2]Jet Propulsion Laboratory
[3]Lockheed Martin Missiles and Space Company

## ABSTRACT

The Space Infrared Telescope Facility (SIRTF) Observatory will be launched in the year 2001 and will provide unprecedented sensitivity to explore the birth and evolution of the universe. The Science Instruments will be maintained at a temperature of 1.4 K inside a superfluid helium cryostat. The telescope is located outside the cryostat and will be warm at launch. Helium effluent will vapor cool the telescope and cryostat vacuum shell to their required 5.5 K operating temperature. This is made possible by a multistage thermal system that controls Spacecraft and Sun heat inputs to the telescope and cryostat to very low levels. The system achieves this by using heat rejection to space, low-thermal conductance alumina/epoxy truss supports, miniature electrical cables, and radiation shields to block radiation from the Spacecraft elements. The preliminary design and predicted performance of this unique cryogenic/thermal system are described.

## INTRODUCTION

SIRTF is the final component of NASA's Great Observatory Program, joining the Hubble Space Telescope, the Advanced X-ray Astrophysics Facility, and the Gamma Ray Observatory. The project is managed by the Jet Propulsion Laboratory. After more than two decades of Observatory concept definition and considerable detector development, NASA approved Phase B of the project in December 1996. Launch is planned in December 2001. The Observatory, shown in Figure 1, will carry three Science Instruments: the Infrared Array Camera (IRAC), the Infrared Spectrograph (IRS), and the Multiband Imaging Photometer for SIRTF (MIPS). These are housed inside a superfluid helium cryostat. These instruments will make measurements over the wavelength range 3.5 - 160 μm to improve understanding of protoplanetary and planetary debris disks, brown dwarfs and super planets, ultraluminous galaxies and active galactic nuclei, and the early universe.

The planned Delta 7920H launch vehicle limits the Observatory launch mass to 910 kg. Three major decisions have been made to reduce the Observatory mass while achieving the desired mission lifetime. First, the telescope will be warm at launch; i.e., the telescope is outside the cryostat, rather than inside as was the case in previously flown infrared telescope satellites. Mass is saved by not having a vacuum shell surrounding the telescope, but new issues related to telescope temperature and contamination control must be carefully considered. Second, the Observatory will be placed into an orbit which is

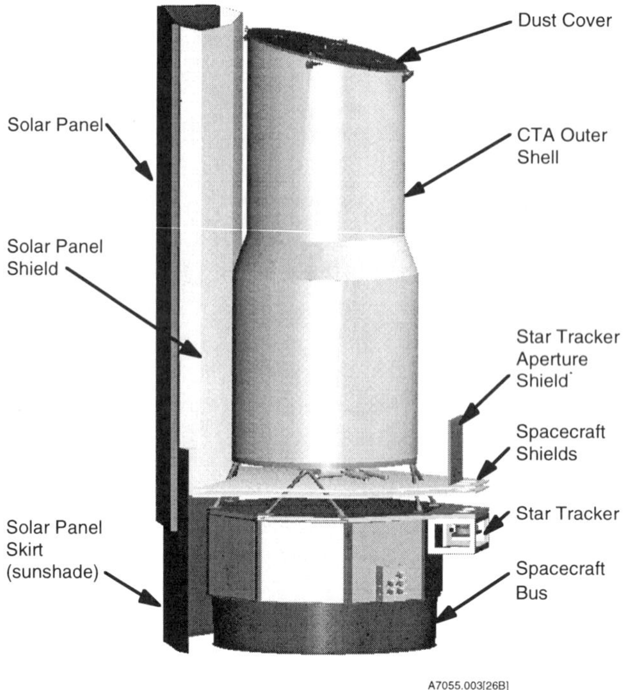


**Figure 1.** The SIRTF Observatory preliminary design concept.

extremely benign thermally, so that the cryostat vacuum shell can be passively controlled to a temperature so low that parasitic heating into the cryostat is negligible. This orbit is an Earth-trailing, Sun orbit with an average distance from the Earth of 0.3 AU, which makes heating of the Observatory by the Earth negligible. Third, Observatory pointing is limited to a small range: 80 to 120 degrees from the Sun line. This allows very efficient heat rejection from the Observatory thermal system to space.

The Observatory is comprised of three major systems: the Spacecraft, the Cryogenic Telescope Assembly (CTA), and the three Science Instruments. The cryogenic/thermal system discussed in this paper is a part of the CTA.

## DESIGN DESCRIPTION

The CTA (Figure 2) consists of five subsystems: (1) the multiple instrument chamber (MIC), which houses the Science Instruments and is bolted to the cryostat helium tank, (2) the 360 liter superfluid helium cryostat, (3) the 85 cm beryllium Ritchey-Chrétien telescope, including secondary focus mechanism, (4) the outer shell group, and (5) the warm electronics, which provides sensor readouts and controls mechanisms.

The outer shell group serves several major functions. It controls the heat load on the telescope and vacuum shell by providing thermal isolation from the Spacecraft bus and solar panel, by rejecting heat to space from strategic intermediate surfaces, some as cold as 40 K, and by using refrigeration available in the helium vented from the cryostat. It mounts the CTA to the Spacecraft bus with adequate alignment control and overall stiffness (14 Hz lateral and 40 Hz axial). It passes electrical cables from the cryostat to the Science Instruments and CTA warm electronics, that are located in the Spacecraft bus. It provides GSE interfaces for CTA servicing, monitoring, and handling. Finally, it encloses the telescope and cryostat in a purged cavity, i.e., the outer shell, for contamination control while on the ground.

As shown in Figure 1, radiation from the Spacecraft bus to the outer shell is blocked by two shields. They are made of aluminum honeycomb panel. Both surfaces of the shields

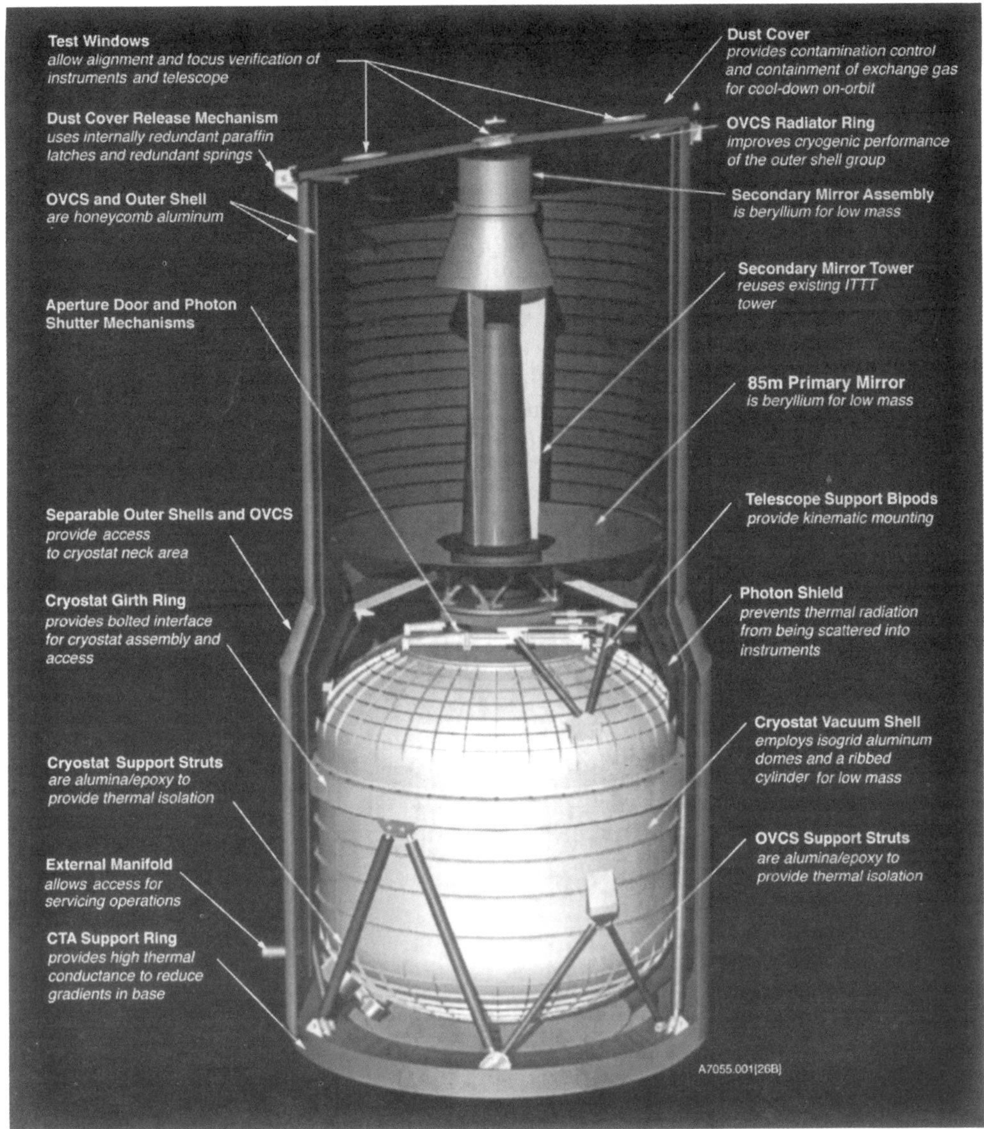


**Figure 2.** The Cryogenic Telescope Assembly (CTA) preliminary design concept.

have low emittance. The shields are supported from the truss that connects the CTA to the bus. The use of multilayer insulation blankets attached to the shields is being evaluated. Note there is a hole in each shield to accommodate the star tracker line of sight and a small shade attached to the colder shield to prevent radiation from the warm star tracker aperture to the outer shell.

The solar panel shield blocks radiation from the solar panel to the outer shell. This shield is also made of aluminum honeycomb panel. It is covered on the warm side with multilayer insulation. Its surfaces have low emittance, except for the wings along the long edges, which are painted black to reject heat to space. The solar panel shield is supported from the solar panel with thermally isolating fiberglass/epoxy brackets.

The outer shell consists of the machined aluminum baseplate, the aluminum honeycomb panel cylindrical section, and the aluminum honeycomb panel dust cover that closes out the top. The outer shell is a sealed unit to allow purging for telescope contamination control. The dust cover, that will be ejected several days after launch, includes a relief valve to vent the outer shell internal volume during launch. The half of the outer shell cylinder facing away from the solar panel is painted black to reject heat to space. The surfaces of the dust cover are also painted black to assist cooldown.

Between the outer shell and the telescope/cryostat is the outer vapor-cooled shield, as shown in Figure 2. It is the warmest element to be vapor cooled. It also is made of aluminum honeycomb panel.

The surface of the aluminum vacuum shell will be designed for overall low emittance. The telescope aluminum barrel baffle is an integral machined piece to assure the light tightness needed to provide the very low photon background for operating the long wavelength detectors. It is mounted to the vacuum shell dome with a bolted flange, and its outer surface has low emittance. The vacuum shell is vapor cooled, and the telescope is cooled by thermal straps connecting it to the vacuum shell.

The various aluminum surfaces that must have very low emittance will be either polished aluminum or gold coated (where the performance advantage warrants the cost).

The CTA uses four support trusses made of alumina/epoxy tubes. This material was chosen because of its superior combination of high stiffness, high strength, and low thermal conductivity over the temperature ranges of interest. The primary truss attaches the CTA to the bus; it will be at a temperature of ~300 K during launch and span the range 300 - 40 K during flight operation. The second truss attaches the cryostat to the outer shell baseplate; it will be at ~300 K during launch and span the range 40 - 5 K during flight. The third supports the outer vapor-cooled shield; it will be at ~300 K during launch and span the range 40 - 20 K during flight. The fourth supports the helium tank from the vacuum shell; it will span the range 300 - 2 K during launch and 5 - 1.4 K during flight.

Electrical cables pass about 1400 electrical conductors from the Science Instruments and CTA to the warm electronics mounted in the bus. These cables are a miniature ribbon configuration, some with overall shields, to meet difficult electrical requirements and thermal isolation requirements.

## PERFORMANCE REQUIREMENTS

The driving requirements from Table 1, lifetime and telescope/vacuum shell temperature, are related in a way that is not immediately obvious. If the heat load to the telescope/vacuum shell becomes too great, the helium effluent produced by the instrument power dissipation may not provide cooling sufficient to maintain the telescope at less than 5.5 K. Excessive telescope/vacuum shell heat load would result from the intermediate thermal system not performing as well as expected. Adding heat to the bath solely to produce more helium effluent for cooling the telescope would be highly undesirable because of the mission lifetime loss. In actuality the situation is not quite this simple because the

**Table 1.** Major cryogenic/thermal system performance requirements

| Parameter | Requirement | Comments |
|---|---|---|
| On-orbit cooldown time | <30 days | To cool down to operational temperatures after launch |
| Observational lifetime (following 30 days cooldown plus 30 days initial orbital checkout) | >2.5 years; 5 years goal (with instruments' average heat load of 6 mW nominal, 8 mW maximum) | To allow follow-up observations and provide overlap with NASA's other great observatories |
| Superfluid helium temperature | <1.4 K | To limit dark current in longest wavelength detectors |
| Bath temperature stability | <0.05 K over 30 minutes with 5% helium fill | For scan operation |
| Telescope and cryostat vacuum shell temperatures | <5.5 K | To limit photon background on longest wavelength detectors |
| Ground lifetime | >3 months | To accommodate testing |
| Pre-launch lockup capability (time from GSE disconnect until launch) | >1.8 days (for bath temperature to rise from 1.5 K following final topoff to 1.9 K at launch) | To accommodate 1 day launch delay; bath pumpdown from 1.9 to 1.3 K results in budgeted 5% mass loss |

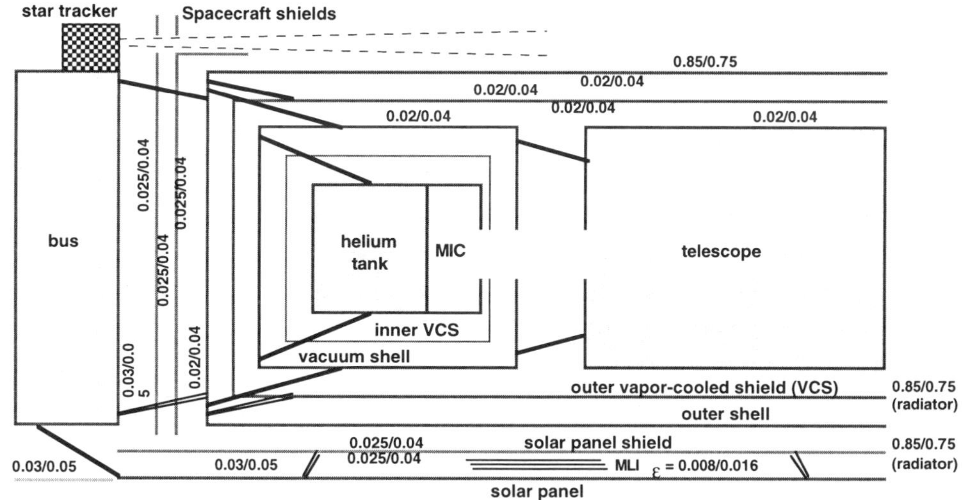


**Figure 3.** Thermal system block diagram showing surface finishes used for modeling.

allowable telescope temperature is considerably higher for the shorter wavelength detectors. However, to simplify our lifetime analysis, we have conservatively assumed the telescope must be less than 5.5 K at all times.

Because the vacuum shell is so cold, the heat load into the helium bath is almost solely due to instrument power dissipation. The Science Instruments operate one at a time for a duration of about 2 weeks. For simplicity we treat the instrument power dissipation as a constant for calculating helium loss rate and lifetime. In actuality it varies significantly, depending on which instrument is on and what operating mode it is in. Bath temperature stability is not a problem because of the large heat capacity of the liquid helium. However, since the helium bath pressure and flow rate respond very slowly to changing heat load, careful operational planning will be needed to maximize overall observing time while meeting the telescope temperatures actually needed by each instrument.

Since the helium bath parasitic heat load during flight operation is negligible compared to the instruments' heat load, ground performance requirements drive the cryostat thermal design. The cryostat uses two vapor-cooled shields, high-performance multilayer insulation blankets, alumina/epoxy struts for helium tank support, and miniature electrical ribbon cables. It is a high-performance cryostat by any standard.

## MODELING AND PERFORMANCE PREDICTIONS

During the concept definition phase of the program, modeling of the intermediate thermal system was performed cooperatively by Ball Aerospace, the Jet Propulsion Laboratory, and Lockheed Martin. This approach took advantage of the thermal network models developed independently during the pre-proposal and proposal phases to quickly evaluate the early thermal architecture trades. This process also validated the models. Design and continued modeling of the thermal system is the responsibility of Ball Aerospace. The Jet Propulsion Laboratory will continue their modeling efforts through the preliminary design phase to support the optimization process and validate the Ball analyses.

Figure 3 shows the surface finishes used in the model. Bec ause emissivity decreases with temperature, lower values are assumed for the colder surfaces. Notice that two values of emissivity are shown: nominal and worst case. Because of the complexity and interactiveness of the end-to-end thermal system, we chose to evaluate performance margin in part by predicting performance with both nominal and worst-case models. The worst-case model has all the worst possible radiative and conductive inputs stacked in the model at once; these inputs are based on uncertainty estimates. This is a very conservative approach to calculating worst-case performance. Our goal is to predict a 5-year observational lifetime with the nominal model. Our requirement is to predict at least 2.5 years of observational

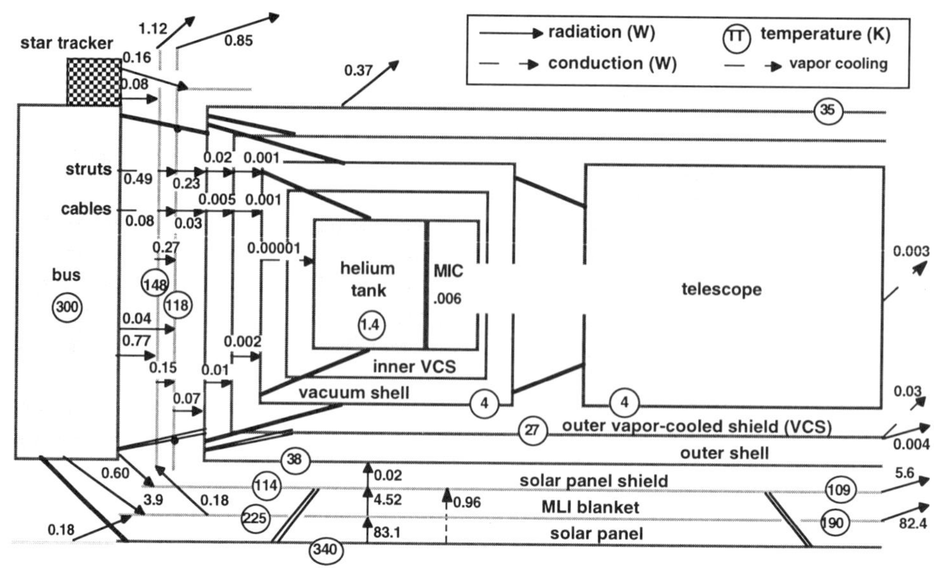


**Figure 4.** Predicted temperatures and heat flows for the nominal case.

lifetime with the worst-case model. As the design matures, some of the estimated uncertainties used for the worst-case model inputs will be reduced.

Figure 4 shows the predicted temperatures and heat flows for the nominal case. The large reduction in heat flow to each colder stage is evident, showing the effectiveness of the system in controlling heat input to the telescope and cryostat vacuum shell. The temperature of the outer shell is essentially independent of the thermal subsystem inside it, which permits us to plan a segmented, cost-effective approach to test verification of the end-to-end thermal system. In general, radiative and conductive heat flows are equally important. Conductive heating is dominated by the support trusses.

Observational helium lifetime is based on a helium mass equivalent to 80 percent of the tank volume. The remaining 20 percent is budgeted as 5 percent ullage following the final pre-launch topoff, 5 percent loss during pumpdown of the helium bath to operating temperature after launch, and 10 percent loss during the 2 months of system cooldown and initial orbital checkout before the start of observation. These budgets will be updated as the modeling effort becomes more mature and as test results become available. The calculated lifetime is based on an internal heat load of 6 mW (8 mW for the worst-case model) and the requirement to maintain the telescope/vacuum shell temperature less than 5.5 K.

**Table 2.** Predicted performance compared to requirements

| Parameter | Requirement | Nominal Prediction | Worst-case Prediction |
|---|---|---|---|
| Observational lifetime | >2.5 years | 4.9 | 2.7 |
| Helium bath temperature | <1.4 K | 1.2 - 1.3 | <1.4 K |
| Helium temperature stability | <0.05 over 30 minutes with 5 percent fill | <0.025 | not applicable |
| Telescope/vacuum shell temperature | <5.5 K | 3.7 | 5.5 |
| Ground lifetime | > 3 months | 4.3 | to be determined |
| Pre-launch lockup capability | 1.8 days (from helium bath temperature of 1.5 K following final topoff to 1.9 K) | 4.2 | to be determined |

Table 2 shows that all performance requirements are met by the model predictions, with comfortable margins in some areas. Our main performance concern is ground lifetime. For the worst-case model, the total helium bath heat load needed to cool the telescope to 5.5 K is 10.9 mW. In this case, the heat required in addition to that dissipated in the Science Instruments would reduce helium lifetime about 2 years. If no heat is added to the bath, the predicted telescope temperature is 13 K.

## CONCLUSIONS

An innovative design concept has been developed for the SIRTF thermal control system. It takes advantage of a solar orbit far from the Earth and restricted orientation to the Sun to provide the required Science Instruments and telescope operating temperatures with minimum mass and high reliability. Much work remains to finalize the design details and produce more mature performance predictions.

## ACKNOWLEDGMENT

The work reported here was conducted under contract to NASA as part of the SIRTF Program.

# A SMALL LONG-LIFE DEWAR FOR EXPLORER-TYPE SPACE EXPERIMENTS

A. R. Urbach, K. A. Gause, and S. J. Nieczkoski

Ball Aerospace Systems Division
Boulder, Colorado 80306-1062, U.S.A.

## ABSTRACT

Present and future space experiments are being designed to be small, lightweight, low cost, and if superfluid helium (SfHe) cooled, often will fly in a solar orbit. This paper presents the design of a 90-liter SfHe dewar that fulfills these requirements. The dewar is constructed from aluminum with stainless steel plumbing, has thermally efficient gamma alumina supports, multilayer insulation, and low thermal conductance electrical ribbon cables. With a vacuum shell radiantly cooled to <60 K , the anticipated lifetime is 2 years. The dewar is fully instrumented with mass quantity gages, and temperature and pressure sensors. It also has strain and accelerometer gages to permit detailed performance measurement which will be used to identify future design improvements. An instrument cavity is provided for housing instruments at <1.4 K.

## INTRODUCTION

This paper presents the design developed at Ball for a SfHe dewar of the size required for the Small Explorer (SMEX), the Mid-Sized Explorer (MIDEX), and the Space Infrared Telescope Facility (SIRTF) missions, which require cold temperatures and long life. This dewar will also be used to demonstrate and initiate concepts for cost and weight reduction which may be implemented in future space qualified hardware. The dewar design and test programs provide a knowledge base for the design of future space hardware used in heliocentric orbits.

## OPERATIONAL REQUIREMENTS

The following requirements were selected as representative of SMEX, MIDEX, and the SIRTF class missions. Of equal importance are the specific performance parameters, operational characteristics, and the ability to demonstrate performance in a straightforward manner. The Long-Life Dewar (LLD) was designed to the performance parameters and operational requirements shown in Table 1.

*Advances in Cryogenic Engineering,* Vol. 43
Edited by P. Kittel, Plenum Press, New York, 1998

**Table 1.** LLD requirements and predicted performance

| Item | Requirement | Performance |
|---|---|---|
| Weight | 100 kg | 102 kg |
| Volume | 90 liters | 95 liters |
| Lifetime: | | |
| 300 K environment | 3 months | 2.6 months |
| 60 K environment | 1.8 years minimum w/2 mW internal heat load | 1.9 years |
| Operating pressure: | | |
| Tank | 10 psig internal, 13.2 psia external | As specified |
| Vacuum shell | 10 psig internal, 15.2 psia external | As specified |
| Operating temperature | 1.4 K maximum @ 80 K shell temperature | 1.4 K |
| Launch environment | Pegasus XL, Taurus, Delta Lite, and Delta 2910 | As specified |
| Resonant frequencies | Axial–35 Hz minimum, Lateral–10 Hz minimum | 91 axial, 62 lateral |
| Instrumentation | Dewar to be fully instrumented to enable complete diagnostics of thermal and structural performance | As specified |

## DESCRIPTION OF DEWAR

The 86.2-cm-high by 95.2-cm-diameter dewar consists of six functional elements: cryogen tank subassembly, tank supports, insulation system, fluid management subassembly, outer shell, and instrumentation. A description of these elements and their specific requirements are listed in the following paragraphs and shown in Figure 1.

### Cryogen Tank Subassembly

The cryogen tank consists of two aluminum alloy ellipsoidal heads welded together at the girth and at the midpoint inside of the instrument cavity. The tank outside dimension is 44.4 cm high and 66.5 cm in diameter and contains 90 liters of SfHe with 5% of ullage space. The instrument cavity is 21.1 cm internal diameter. By providing mounting holes on

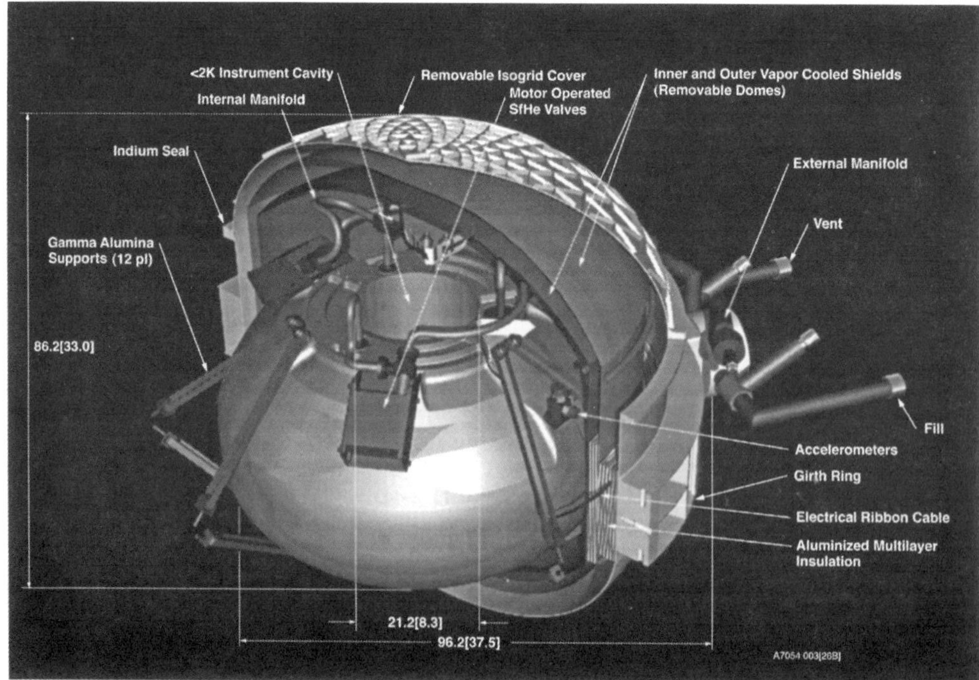


**Figure 1.** Long-life dewar.

each head an instrument inside the cavity can be mounted from either the upper or lower head.

### Tank Supports

The cryogen tank is supported from the girth ring by six bipod tension strap supports; three bipods are attached to the upper head and three are attached to the lower head. Each support is made of uniaxial filament wound gamma alumina bonded in an epoxy matrix to form a strap. First mode frequency for the dewar is 62 Hz lateral and 91 Hz axial. Gamma alumina fibers were selected because they provide the best combination of structural and thermal properties for a heliocentric orbit flight environment.

### Insulation System

The insulation system consists of multilayer insulation (MLI) spaced by two vapor cooled shields, which are supported through flexures by the support system. The flexure attachments are thermally efficient and provide for cooling the supports. Surrounding the tank under the inner vapor-cooled shield is a layer of aluminum foil which reduces the effective emissivity of the tank and internal plumbing manifold. The MLI is 1/4-mil mylar layers aluminized on each surface and spaced by dacron net. All penetrations are buffered to minimize radiation tunneling through the penetration. The insulation system is installed one layer at a time, except for the upper head insulation, which is made in 5-layer quilts. This is done to ease access into the instrument area. There are 3 quilts between the inner vapor cooled shield (IVCS) and outer vapor cooled shield (OVCS) and 5 quilts between the OVCS and outer shell. The VCS heads are installed with screws for easy access to the instrument. The completed insulation system is shown in Figure 2.

The VCSs are spun aluminum heads lined with aluminum foil to reduce the inner surface emissivity. The stainless steel vent line has a good thermal attachment to each vapor cooled shield to intercept the incoming parasitic heat. The fill line is also lightly attached to the vapor cooled shields to reduce parasitic heat from conduction along the fill tube wall.

### Fluid Management Subassembly

The plumbing components of the fluid management subassembly direct and control the helium in a safe manner and utilize the refrigeration in the effluent GHe to minimize

**Figure 2.** Insulated dewar.

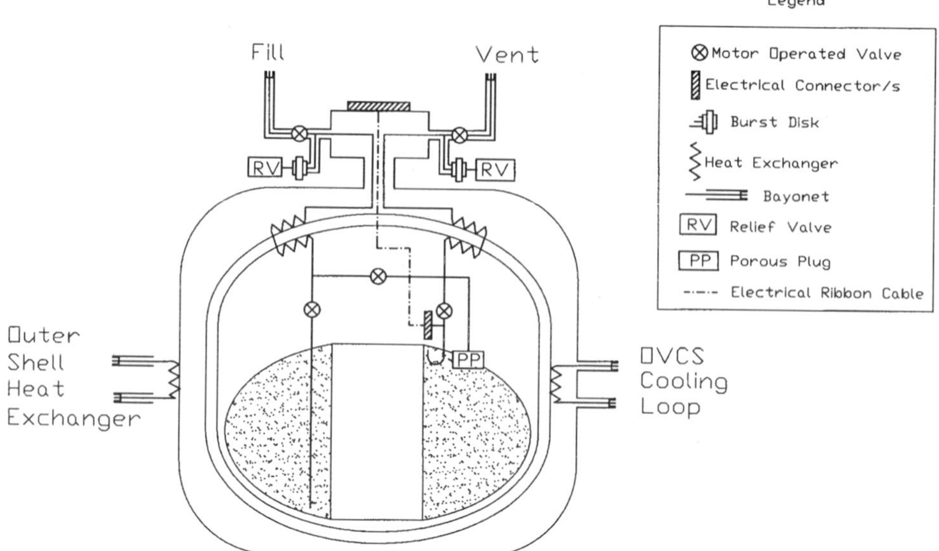


**Figure 3.** LLD fluid management scheme.

incoming parasitic heat. Figure 3 presents the scheme used for the LLD. Three motor operated valves located in the internal manifold (shown in Figure 4) are used during initial cooldown, fill, and topoff operations. For a flight system the external fill and vent valves would also be motor operated, but to save cost for the LLD, manual valves were used. SfHe phase separation is achieved by closing the internal valves and directing the flow through the porous plug and out the vent line to a vacuum system which maintains the internal tank pressure at less than 37.8 torr.

Burst disks are included on the fill and vent lines to prevent over pressurization of the system. A relief valve is placed downstream of each burst disk to prevent back flow of air into the cold lines which operate at less than one atmosphere.

**Figure 4.** Internal and external manifolds.

**Figure 5.** Removable, lightweight dewar cover.

## Outer Shell

The outer shell is constructed of aluminum alloy and consists of the upper isogrid head, a girth ring, and the lower head. The upper head (see Figure 5) is removable and sealed by an indium o-ring to the girth ring. Only part of the head was made into isogrid, purely for economic reasons. Because this was a research dewar, cost reduction was a constant companion and the critical structural element of the isogrid was clearly demonstrated by machining only the apex of the dome.

The girth ring shown in Figure 4 provides the rigidity to react the support system loads and consists of a closed box structure that is vacuum tight. The lower aluminum head is welded to the lower side of the girth ring. Connected to and supported from the girth ring is the external manifold, which is described and shown schematically in the fluid management system section.

## Instrumentation

Evaluation of a research dewar's performance is mandatory. Therefore, a large amount of instrumentation was installed on this dewar. A list of the instrumentation is included in Table 2. Low thermal conductance ribbon cables were used for all the instrumentation.

**Table 2.** Instrumentation list

| |
|---|
| **Germanium temperature sensors** |
| • Nine are included to monitor porous plug performance, internal mass gaging, internal cold flowmeter, insulation system performance, and tank temperature profile |
| **Silicon diode temperature sensors** |
| • Nine are included to monitor tank cooldown, insulation and support system performance, and outer shell temperature |
| **Pressure transducer** |
| • Monitors tank pressure |
| **Vacuum gage to monitor dewar vacuum level** |
| **Structural sensors** |
| • Accelerometers to monitor system resonant frequency |
| • Strain gages to monitor support preload |
| **Valve position indicators** |
| **Liquid-level sensor for fill operations** |

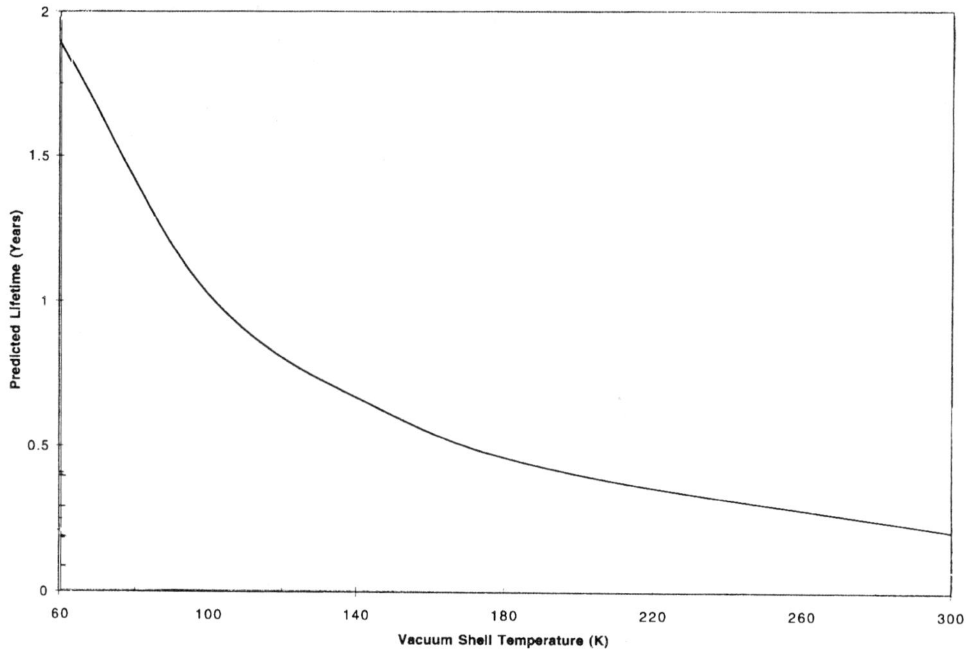


**Figure 6.** LLD provides long life for cold outer shell temperatures.

## THERMAL PERFORMANCE PREDICTIONS

Dewar thermal performance is predicted by a computer analysis tool having heritage and verification from both the Infrared Astronomical Satellite (IRAS) and Cosmic Background Explorer (COBE) programs. Computer models used to predict thermal performance account for all heat flows into the dewar. These models use nodal representations for each isothermal stage within the dewar, temperature dependent material properties, and an MLI radiation exchange correlation developed from testing our previously built cryogenic systems.

System lifetime is predicted using a computer model that calculates the steady state heat flow to the SfHe. Inputs to the model can be easily changed to evaluate the performance impact of configuration alternatives or the sensitivity to thermal boundary conditions. Figure 6 shows the predicted LLD lifetime over a range of vacuum shell temperatures. Predictions are based on 85 liters SfHe useable on orbit and 2 mW internal power dissipation.

Heat flow into the SfHe is greatly reduced by the two stages of vapor cooling. With a 60 K vacuum shell the heat flows in the dewar are dominated by solid conduction. This is not the case with a room temperature vacuum shell where radiation heat exchange dominates dewar performance. In the outer regions of the dewar, conduction through the straps are roughly 50% of the heat flow to the OVCS and IVCS. Inside the IVCS only 13% of the heat load to the helium is due to the support straps. The total predicted heat load to the helium is 4.6 mW under these conditions, of which only 57% is due to parasitic heating.

A separate computer model was constructed to predict the transient behavior of the dewar during a ground hold scenario. The ground hold capability of a SfHe dewar is defined by the time required to raise the helium bath temperature from its initial temperature, following final fill operations and closure of the vent valve to the lambda point transition temperature of 2.18 K. Allowing the helium to transition through the lambda point and become normal would compromise the phase separation function of the porous plug. The LLD has the capability of being in a hold mode for up to two days before the lambda temperature is reached. This assumes an initial bath temperature of 1.5 K at the time the vent valve is closed. This is an excellent capability for a system of this size.

## PLANNED PERFORMANCE TESTING

In mid-August 1997 a test program of the LLD will start. Planned tests include:

1. Lifetime at room ambient outer shell temperature. This test will compare data to the analytical model and demonstrate lifetime to be anticipated in the laboratory.
2. Ultimate bath temperature that can be achieved at both 300 K and 85 K outer shell temperatures. Besides instrument operating temperature capability this will also demonstrate dynamic range of the porous plug.
3. Cold outer shell temperature performance. A test with the outer shell cooled by $LN_2$ should result in an 85 K or less environment, which is to simulate a heliocentric orbit average temperature. Our analysis indicates its possible to achieve outer shell temperatures as low as 40 K with a properly designed external thermal system.
4. Vibration test with normal LHe contained in the tank. This test will demonstrate structural integrity of the dewar and provide actual resonant frequencies of the dewar for comparison to the structural analysis.

## CONCLUSIONS

It is possible to develop a relatively inexpensive dewar which will be of low enough weight to support SMEX and MIDEX instrument opportunities available today.

## ACKNOWLEDGMENTS

The authors wish to thank the large number of people at Ball Aerospace & Technologies Corp. for their contributions to this dewar development. This work was an internal research and development program supported by Ball Aerospace & Technologies Corp.

# LIFETIME ENHANCEMENT OF THE CHeX CRYOGENIC SYSTEM

T.S. Luchik, M.E. Larson, U.E. Israelsson, D. Pearson, D. Petrac, and S. Elliott

Jet Propulsion Laboratory, California Institute of Technology
Pasadena, CA 91109, USA

## ABSTRACT

The Confined Helium Experiment (CHeX) is manifested for flight on the Space Shuttle (STS-87) in November, 1997. The experiment was constrained to re-use a previously flown flight cryogenic system. However, the inherited system did not meet CHeX science requirements, nor current Shuttle launch requirements; enhancements of the system and procedures were required. We present model and data of the full flight system, before and after the enhancements were made. A discussion of procedural changes and their impact on the system are also presented. With the changes, all Shuttle servicing requirements were met. A net improvement in ground servicing of 30% and a net parasitic reduction of 15% were achieved. Predicted flight performance is also presented.

## INTRODUCTION

The superfluid helium cryostat which provides the thermal environment to the JPL Low Temperature Platform (LTP) has flown on two previous missions, SpaceLab 2 in 1985 and the Lambda Point Experiment (LPE) in 1992 (part of the first United States Microgravity Payload, USMP-1). For these missions, the 10-day cryogen lifetime was sufficient to meet the requirements of the experiments it was supporting, given the launch-hold constraints of the time. This cryostat will be used to provide the cryogenic environment for the Confined Helium Experiment (CHeX) which is part of USMP-4 currently scheduled for launch in November, 1997. This experiment will measure the specific heat of helium in the vicinity of the lambda point in a geometry confined to two dimensions (50 μm spaces).

This cryostat is described in detail elsewhere and will not be described here[1,2]. The essential features of this cryostat is that it has two vapor-cooled shields, has allowance for a 20 cm diameter x 90cm long instrument and holds up to 90 liters of liquid (set by the porous plug location) with an instrument of the size described above installed.

For CHeX, the requirements on the cryostat are significantly more rigid than they were for either of its previous two missions. For CHeX, the cryostat must have a safe life between helium servicings of at least 207 hours and provide 7 days on orbit[3]. A "safe" level for servicing is defined as 20 liters remaining in the cryostat at the start of servicing. It is at this volume of helium that there is sufficient coupling between the instrument and the bath that heating of the instrument is dissipated to the bath. If the helium level in the cryostat is less than 20 liters at the start of cryostat servicing then superconducting shields on the instrument's high resolution thermometers could go normal destroying the functionality of the instrument. The second requirement assumes a maximum launch hold of 96 hours and that the last servicing occurred 65 hours prior to nominal launch so that the total lifetime from full-to-empty for the mission is 329 hours.

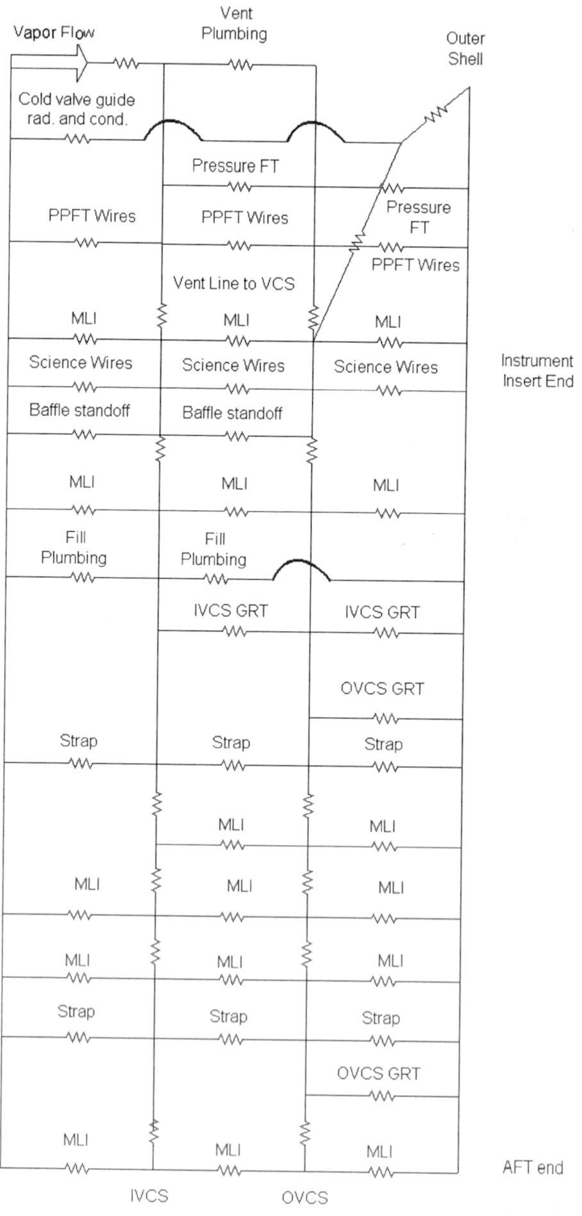


**Figure 1**. Thermal node Diagram for modified CHeX cryostat.

**Table 1.** Predicted and measured results for cryostat performance

| | No Instrument | | Instrument, Ground | | Instrument, Flight |
|---|---|---|---|---|---|
| | Measured | Predicted | Measured | Predicted | Predicted |
| Boil off (liters/day) | 6.2 | 6.3 | 6.8 | 6.8 | 5.7 |
| $T_{IVCS}$ (K), neck area | 32 | 37 | 33 | 36 | 32 |
| $T_{OVCS}$ (K), neck area | 80 | 86 | 88 | 88 | 78 |
| $T_{OVCS}$ (K), aft area | 105 | 100 | 100 | 101 | 89 |

Since the system was not able to meet the requirements for this experiment, it was modified so as to improve its thermal performance. Modifications of the cryostat included improving the thermal contact between the vent line and the vapor-cooled shields, improving the thermal anchoring of the cryostat-engineering sensor wiring and reducing the thermal leak from guide tubes for the cryostat cold valve actuators. Details of the modifications were presented elsewhere[3]. The system has been continuously filled with helium since June, 1996 with the flight instrument installed. Direct measurements of thermal performance on the ground will be presented. A thermal model of the system, which models the on-ground performance of the cryostat at 4K and 2K, has been developed and used to predict the on-orbit performance of the cryogenic system.

## THERMAL MODEL

The helium cryostat is a helium dewar with two vapor-cooled shields. The inner tank of the cryostat is suspended by six G-10 fiberglass straps (3 fore, 3 aft). Support wiring for cryostat maintenance and housekeeping have their own feed-throughs and are thermally anchored directly to the vent line. The cryostat has an opening approximately 20 cm in diameter to accommodate the CHeX science instrument. The instrument is closed off with baffles that are thermally anchored to the vapor cooled shields. All instrument wiring is anchored to the closure baffles. These baffles are conductively attached to the vent line of the cryostat by small copper straps. The fill line for the cryostat is thermally anchored to the inner vapor cooled shield only.

A simple thermal model for the cryostat has been developed to describe the cryostat. The model was initially developed and validated against the Lambda Point Experiment data. The model was used to identify areas where significant improvement could be achieved. Modifications to the cryostat were made and compared to the predictions. These results were described elsewhere[3]. This model has been updated to include axial variation in the temperature of the vapor-cooled shields and more detail on the instrument. A simplified node diagram for the thermal model is shown in Figure 1. The model assumes that the outer shell and the inner tank are at uniform temperature and calculates the temperature profile of the vapor-cooled shields and the parasitic heat into the cryostat.

Table 1 shows a comparison of predicted and experimental data for a variety of experimental conditions. The data with no instrument were used to validate the "cryostat only" portion of the model. A thermal model of the instrument was then included in the overall model and compared to experimental data. The agreement is very good and gives good confidence that the model is representative of the physics of the cryostat. The model was then run for the expected on-orbit conditions. The primary difference between ground operation and flight operation is that the outer shell of the cryostat is expected to cool to 273K as it did on the Lambda Point Experiment. On LPE, the system was found to be significantly effected by the shell temperature. Because many of the poor conduction leaks have been improved for CHeX, the system is predicted to now be more effected by the change in outer shell temperature.

## LAUNCH-SCRUB-TURN AROUND

As was mentioned in the introduction, the CHeX cryostat must provide a safe environment for the instrument even in the event of a launch scrub. Safe has been defined to be at least 20 liters of helium in the cryostat at the start of servicing. The minimum time between servicings following a launch scrub is defined by the Shuttle operations at Kennedy Space Center and has a value of 207 hours. This includes final servicing at L-65 hours prior to initial launch attempt, a maximum of 96 hours for all launch attempts before a launch is scrubbed, and 46 hours for the Shuttle to be safed and access to be provided through the Payload Change-out Room for a servicing to begin. We have repeatedly measured more than 27 liters of helium remaining in the cryostat after 207 hours. Typical data are shown in Figure 2. At this liquid content and our ground parasitic rate, we have more than one day of contingency of the servicing operation.

## FLIGHT TIMELINE

Performance of the cryostat on the ground is significantly enhanced on-orbit because the outer shell of the cryostat is colder (300 K on the ground, 273 K on orbit). A predicted performance history of the cryogenic system is shown in Figure 3. The performance history includes warming of the cryostat for 4 hours after launch in order to

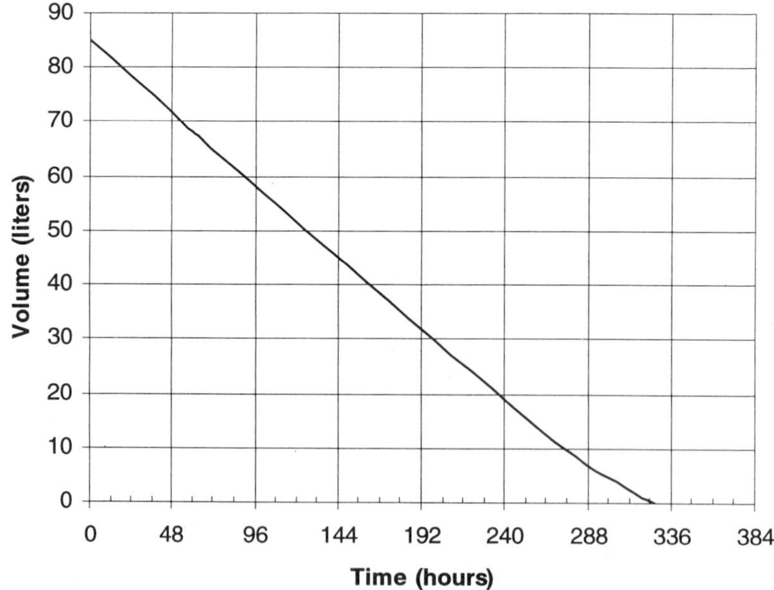


**Figure 2.** Ground hold time for cryogen bath at 1.95 K.

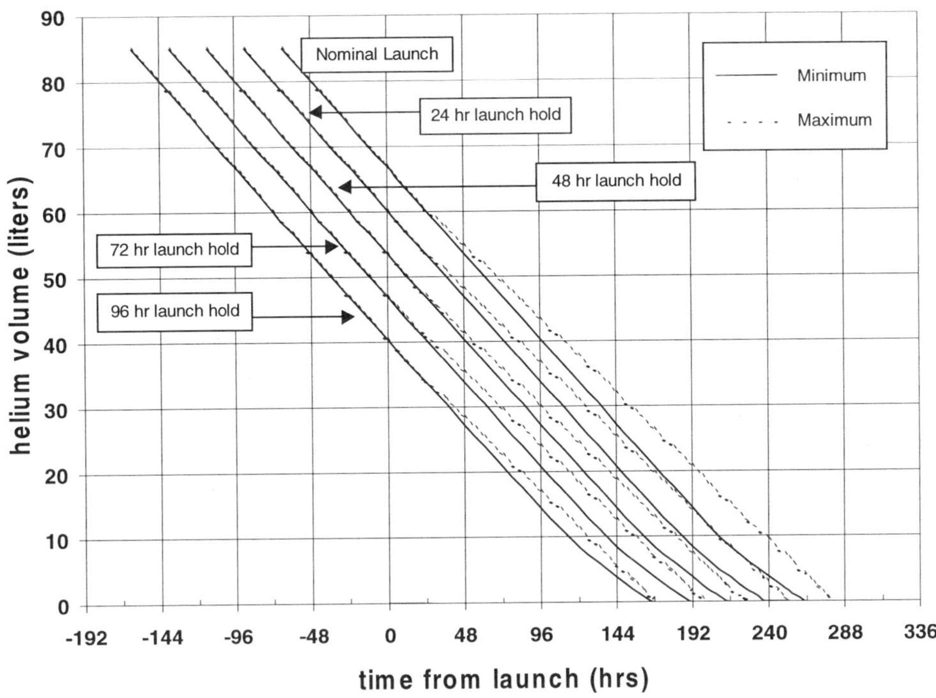


**Figure 3**. Flight timeline for various launch scenarios.

assist in the evacuation of the instrument guard vacuum. This space is filled with $^3$He prior to launch to provide a conduction link between the cryostat bath and the instruments high-resolution thermometers. This provides a cooling path for heat generated in the instrument due to its mechanical motion in response to the launch vibration. The curves in Figure 3 show the cryostat performance as a function of launch attempt. Maximum values for lifetime are obtained by using the predicted performance with an outer shell temperature of 273 K (as was measured for USMP-1) starting 24 hours after orbit is achieved and the ground parasitic rate for all other times. The minimum performance is obtained by using the as-measured ground parasitic rate. For all cases, the 7-day lifetime requirement is achieved if the maximum prediction is used. In the case of the minimum prediction, the 7-day lifetime requirement is met with the exception of the 96-hour launch-hold case, which falls short of 7 days, by 2 hours. In the case of nominal launch, the system will provide for more than 11 days of science gathering.

## CONCLUSION

A thermal model of the CHeX flight cryogenic system has been developed and experimentally verified against measured ground performance data. The model was used to obtain a prediction of the on-orbit performance of the cryostat where the shell temperature is expected to reach 273 K. A combination of measure ground performance data and flight predictions were used to develop a flight time history of the cryogenic system performance. These results showed that the system would provide at least seven days of experimentation on orbit. Experimental data were used to determine that the CHeX system could be safely serviced even in the event of a launch scrub.

## ACKNOWLEDGEMENT

This work was carried out by the Science and Technology Section of the Jet Propulsion Laboratory, California Institute of Technology, under contract to the National Aeronautics and Space Administration. NASA Microgravity Science and Applications Division funded the work.

## REFERENCES

1. D. Petrac, U.E. Israelsson, D. Otth, L.L. Simmons, J. Staats, and A. Thompson, "Technical Challenges Involved in Supporting the Lambda Point Experiment," *Cryogenics*, 30, September supplement (1990).
2. D. Petrac, U.E. Israelsson, and T.S. Luchik, "The Lambda Point Experiment: Helium Cryostat, Cryo-Servicing, Functions and Performance," *Advances in Cryogenic Engineering*, 38, (1993).
3. T. Luchik, U.E. Israelsson, and D. Petrac and S. Elliott, Performance enhancement of the CHeX flight cryostat," *Advances in Cryogenic Engineering*, 39, (1995).

# DESIGN AND TESTING OF AN IMPROVED CRYOPUMP FOR THE CONFINED HELIUM EXPERIMENT

M. J. Lysek,[1] U. E. Israelsson,[1] T. C. P. Chui,[1] M. E. Larson,[1] D. Petrac,[1] S. E. Elliott,[1] D. R. Swanson,[2] X. Qin,[2] and J. A. Lipa[2]

[1]Jet Propulsion Laboratory, California Institute of Technology
Pasadena, CA 91109, USA

[2]Stanford University
Stanford, CA 94305, USA

## ABSTRACT

A cryopump flown on the space shuttle for the Lambda Point Experiment[1] (LPE) was improved and redesigned for use in the Confined Helium Experiment (CHeX). The successful Lambda Point Experiment (LPE) flew on the space shuttle in October 1992. The Confined Helium Experiment (CHeX) will launch in October 1997. CHeX uses high resolution thermometers[2] to measure the heat capacity of liquid helium when confined within finely spaced parallel disks. To eliminate heat flow between the calorimeter and the cryogenic bath, a cryopump is used to reduce the pressure in the experimental probe. The LPE cryopump trapped activated charcoal behind a nuclepore filter. The new cryopump uses activated charcoal glued to copper plates to eliminate the nuclepore filter. Vibration tests verified that the cryopump does not generate charcoal dust contamination. Early performance tests indicated that the probe vacuum is significantly lower with the new pump, and that the pressure falls more rapidly than with the previous design. Recent tests with the CHeX experimental probe show a significant reduction in the heat leak between the calorimeter and the cryogenic bath.

## INTRODUCTION

LPE and CHeX use high resolution thermometers[2] (HRT) to measure the heat capacity of liquid helium samples with great accuracy. The HRT used for these missions measures temperature with noise less than 1 nK at 2 K. In order to achieve high accuracy, the calorimeter cell must be thermally isolated from its environment. The cell is isolated from conductive heat leaks by several temperature controlled stages. The final stage has HRT temperature control and encloses the cell within a radiation shield. A high vacuum is established to isolate the stages from the liquid helium bath and to isolate the cell from the radiation shield.

During environmental testing for the LPE, it was discovered that the launch vibrations would warm the support structure to above 9 K. Each HRT depends on a stable magnetic flux stored in a superconducting niobium tube. The flux is lost when the niobium warms above its superconducting transition temperature, 9.3 K. The HRT sensitivity would be greatly degraded if the magnetic flux were to be lost. To prevent the temperature rise during launch, 5 torr of $^3$He gas was introduced into the probe. The gas couples the stages to the cryogenic bath. To operate the experiment after launch, the majority of the gas was expelled from the probe to space through a mechanical valve. The remaining gas was trapped by a cryopump mounted on the upper flange of the instrument, between the instrument vacuum space and the cryostat guard vacuum.

Once the LPE data was analyzed, it was determined that, although the experiment was highly successful, the heat flow through the residual He$^3$ gas was one of several factors that limited the calorimetric accuracy. Therefore, the CHeX probe was modified to reduce virtual leaks, and an improved cryopump was designed, built and tested.

## LPE Cryopump Design and Operation

The LPE cryopump consisted of about 15 g of loose activated charcoal captured behind a 12 μm pore size nuclepore filter. The nuclepore filter prevented any charcoal dust from escaping out of the cryopump and adhering to the calorimetric cell. If a small particle of charcoal were to become attached to the cell, then the heat of adsorption generated by the changing pressure in the probe could disturb the heat capacity measurements.

The charcoal filled copper container was supported on a weak thermal link to the 1.7 K superfluid helium bath. An electric heater was used to warm the charcoal above 40 K to expel any adsorbed helium gas. A 2 K radiation baffle positioned across the pump throat limited the amount of thermal radiation from the charcoal container to the calorimeter stages during the 40 K bake-out.

Several days before launch, the 7 L instrument volume was filled with 1.3 Kpa (10 torr) of $^3$He gas. $^3$He heat exchange gas proved to be better than $^4$He because it yielded a faster evacuation and a lower ultimate pressure. This gas provided enough thermal conductivity between the HRT flux tubes and the liquid helium bath at 1.7 K to keep them below the superconducting transition during launch. Shortly after launch, mechanized valves opened to allow the probe to be evacuated through a 1 m long, 6.4 mm I.D. corrugated vacuum line. At the same time, a heater warmed the charcoal in the cryopump to above 40 K in order to expel the adsorbed $^3$He gas. After about five hours, the residual gas pressure was below 13 Pa (0.1 torr) and the evacuation valve was closed and the heater turned off. Then the charcoal cooled to 2.0 K by conduction through the weak thermal link to the liquid helium bath. The residual $^3$He gas in the probe passed through the nuclepore filter where it is trapped by the cold charcoal. For the first 30 hours after launch, a 3 W heater was cycled on and off to warm the liquid helium bath up to 2.0 K. Increasing the instrument temperature reduced the time required to reach operating pressure, about $10^{-7}$ Pa ($10^{-9}$ torr).

## CHeX Cryopump Design

The cryopump for CHeX was required to achieve a lower ultimate pressure and also produce a high vacuum in the probe more rapidly than the LPE pump. It retained the same mechanical and electrical interface. The cryopump must not allow any charcoal to escape into the instrument vacuum space, especially during launch. The pump must be vacuum

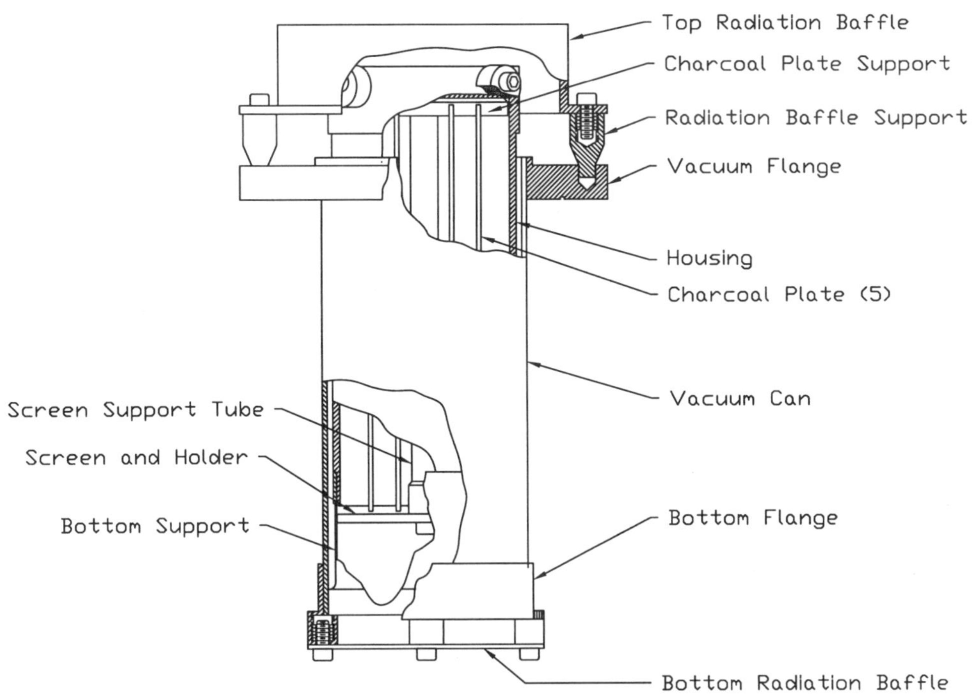


**Figure 1.** Cut-away view of the CHeX cryopump.

leak tight since its upper interface seals the instrument vacuum space from the cryostat vacuum space.

The main improvements of the LPE cryopump design were to increase the thermal contact between the charcoal and the 1.7 K bath and to reduce the flow impedance from the instrument vacuum space to the charcoal supply. The LPE cryopump used loose charcoal grains trapped behind a nuclepore filter, while most cryopumps[3] have charcoal glued to metal plates. A pump with charcoal bonded to copper plates offers several advantages. Each grain of charcoal is in good thermal contact with the substrate. The charcoal covered plates can be arranged to offer good gas conductance to the instrument. And, since the charcoal is fixed in place, it would not be a source of contamination, eliminating the flow impedance caused by the nuclepore filter used in the LPE cryopump. Considering these advantages, the LPE cryopump was redesigned with charcoal glued to metal plates using a high thermal conductivity epoxy.

The basic overall geometry of the CHeX cryopump is similar to the LPE pump design. Figure 1 and 2 show cut-away and cross section views of the new design. The housing holds five copper plates coated with activated charcoal.[4] The charcoal is embedded in a 0.8 mm thick layer of aluminum filled epoxy[5] on both sides of each plate. The total plate surface area of 180 $cm^2$ holds about 9 g of charcoal. This epoxy was also used to attach the plates in the charcoal plate support. The plate assembly is stiffened by four 1.6 mm diameter copper rods glued into holes in the plates. The screen support tubes hold the plate assembly in place inside the charcoal housing. A 100 mesh (per in) screen is soft soldered into a copper frame that fits tightly over the mouth of the housing. The screen retains any larger pieces of charcoal that might escape from the plates.

As with the LPE cryopump, the housing is brazed into the thin walled type 304 stainless steel bottom support. The length and cross section of this tube was selected to provide adequate cooling for the charcoal housing while limiting the heater power required

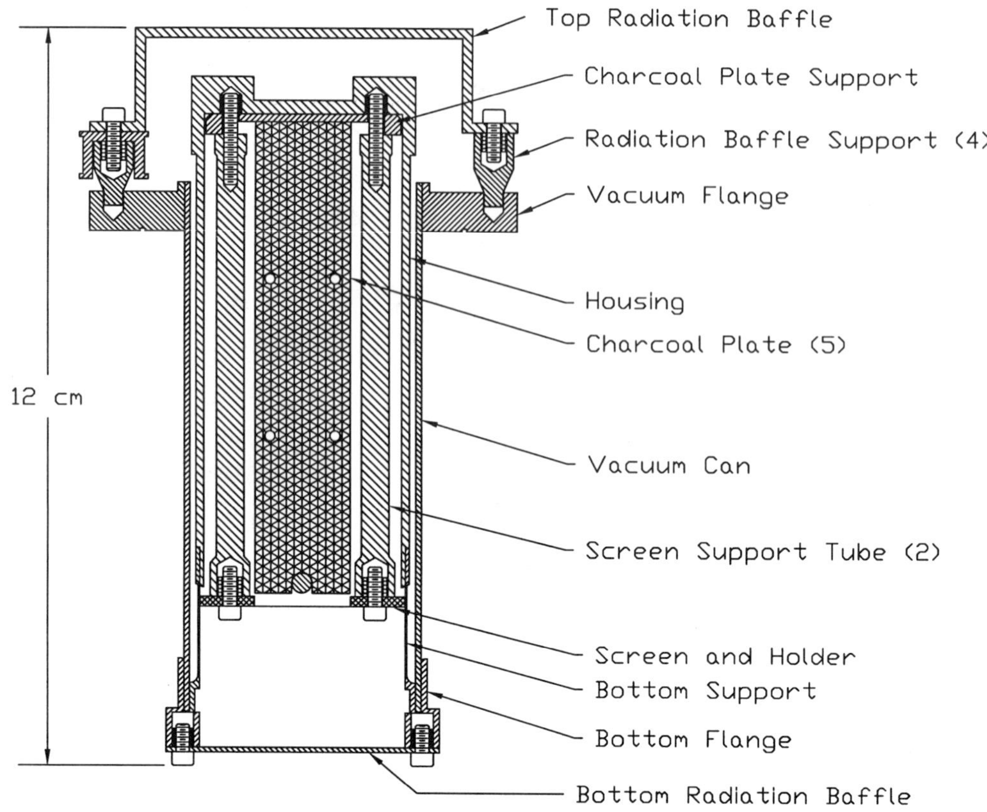


**Figure 2.** Cross section view of the CHeX cryopump.

to raise the housing temperature to 40 K during the post-launch bake-out. The bottom support is welded into the vacuum can. The bottom flange is a copper ring that is thermally strapped to the cryogenic bath to provide cooling for the pump housing. The bottom flange includes supports for the gold plated bottom radiation baffle. This baffle reduces the heat radiated into the instrument during the bake-out. The vacuum can is welded into the vacuum flange. The outer surface of the housing is always in a high vacuum so the charcoal is thermally isolated from the vacuum flange. The heat input during the bake-out is 0.25 W even while the instrument vacuum space is filled with $^3$He heat exchange gas. The radiation from the housing to the 1.7 K liquid helium bath is reduced by a gold plated radiation baffle attached to the top of the cryopump.

## ACCEPTANCE TESTING

### Mechanical design

The only component on the LPE and CHeX pump designs that is significantly stressed by the expected launch vibration is the thin walled stainless steel bottom support. Analysis indicated that a 0.24 mm wall thickness would offer a large safety factor. Sample pull tests were performed of the epoxy joint between the plates and the plate support and for the furnace brazed vacuum joint between the bottom support and the housing. The completed cryopump was exposed to random level vibrations in all 3 axes while at 77 K. No damage was observed to any of the components. The completed CHeX probe, including the cryopump, survived a 2 axis random vibration test at launch levels.

### Contamination

In order to verify that the pump did not expel any fine charcoal particles during launch, it was exposed to launch level vibrations while in a sealed chamber at 77 K. The chamber was cleaned prior to the test, and then washed with solvent after the test. The solvent was filtered to collect any charcoal particles. After the first qualification test, a total of 15 μg of dust was recovered from the test chamber. This level of contamination was considered unacceptable. When examined, it was determined that the pump plates were vibrating against the housing walls and the screen support rods. Four 1.6 mm diameter copper rods were glued into holes in the plates to stiffen the assembly. The plates were modified to provide adequate clearance between the charcoal and the pump body by masking off the areas adjacent to the plate edges during the charcoal coating process. The plates were scrubbed with a stiff brush and blown off with clean $N_2$ gas prior to assembly in order to eliminate any poorly bonded charcoal. After these changes, only seven charcoal particles totaling 160 ng in weight were expelled from the pump during the second vibration test.

### Vacuum Leak Testing

Both before and after the launch vibration tests, the pump assembly (without the charcoal covered plates installed) was tested for leaks with a mass spectrometer leak detector. In both cases no leak was detected within the sensitivity of the detector. In addition, the instrument vacuum integrity was verified after the system level vibration test.

### Pumping Performance

Two measurements were available to determine the pressure in the instrument. The $^3$He leak rate measured by a mass spectrometer leak detector (or a residual gas analyzer) connected to the instrument vacuum pump-out line can be correlated with the instrument pressure. The thermal coupling between the $^4$He bath and the calorimeter radiation shield depends on the pressure. Then the pressure can be estimated from measurements of the steady state heater power needed to maintain the shield at 2.0 K. The first method was available during the development of the CHeX cryopump, while both methods were used once the pump was integrated with the completed CHeX instrument.

Figure 3 shows a comparison between the $^3$He leak rate data for the LPE and CHeX instruments. The CHeX pump reaches a lower pressure more rapidly than the old design. While it is not possible to accurately correlate the pressure with the leak rate, these measurements are a good indication of the relative performance of the two pump designs. Although some of the improvement can be attributed to the elimination of virtual leaks in the CHeX instrument, early performance tests with the LPE prototype instrument demonstrated a significant improvement over the LPE pump.

Measurements of the heater power needed to maintain the calorimeter radiation shield at 2.0 K show that after about 35 hours the pressure in the instrument drops almost as low as when no heat exchange gas is used, about $10^{-7}$ Pa ($10^{-9}$ torr).

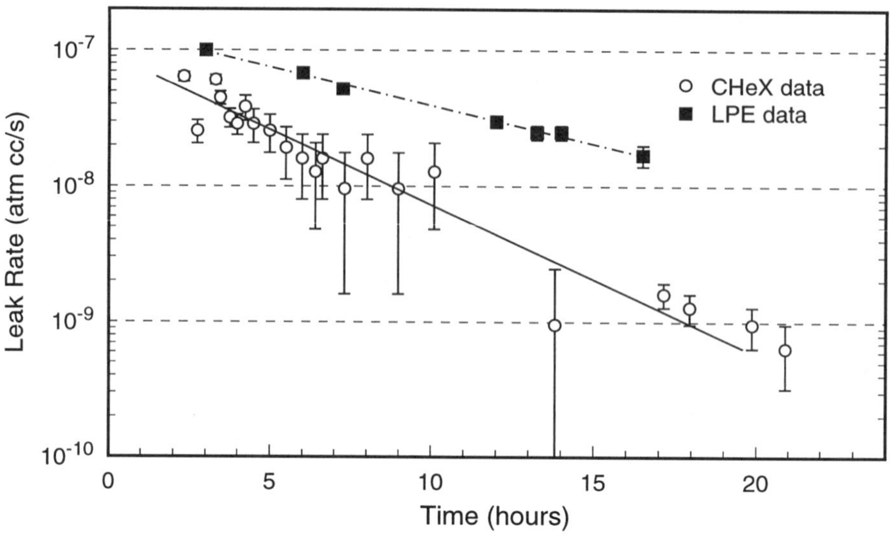


**Figure 3.** Plot of CHeX and LPE pre-flight leak rate data.


## ACKNOWLEDGMENT

D.R. Swanson and T.C.P. Chui designed and built the original LPE cryopump. S. Elliott and J. Hardy carefully assembled and tested the CHEX cryopump. Robert Galletly, Johnathan Lamb, and Gordon Wang provided mechanical engineering support. The design work benefited from conversations with D. W. Sedgley and G. Rao. This work was carried out by the Science and Technology Development Section of the Jet Propulsion Laboratory, California Institute of Technology, under contract to the National Aeronautics and Space Administration. The work was funded by the NASA Microgravity Research Division.


## REFERENCES


1. J.A. Lipa, D.R. Swanson, J.A. Nissen, and T.C.P. Chui, Lambda-point experiment in microgravity, Cryogenics 34(5):341 (1994).
2. T.C.P. Chui, D.R. Swanson, M.J. Adriaans, J.A. Nissen, Temperature fluctuations in the cannonical ensemble, Phys Rev L (21):3005 (1992).
3. D.W. Sedgley, T.H. Batzer, and W.R. Call, Helium cryopumping for fusion applications, J Vac Sci A 6(3):1209 (1988).
4. Type 12x30 PCB charcoal from the Calgon Carbon Corporation. Charcoal screened between tyler standard screens #16 (1.0 mm) and #18 (1.4 mm) to make the grain size more uniform.
5. Type 568 VLV aluminum loaded epoxy by Aremco Products Co., Ossining, NY, USA. All of the metal surfaces were glass bead blasted prior to glue application.

# DESIGN AND FABRICATION OF A ZERO GRAVITY LIQUID CRYOGEN COOLER USING SURFACE TENSION CONFINEMENT TECHNOLOGY

G.C. Driscoll, J.F. Maguire, and P.M. Winn

Applied Engineering Technologies, Ltd. (AET, Ltd.)
Woburn, MA 01801

## ABSTRACT

AET, Ltd. was contracted by NASA GSFC to design and fabricate a zero gravity liquid cryogen cooler for use in NASA's Hitchhiker Program. The cooler was designed to fit inside a standard Get Away Special (GAS) canister located in the cargo bay of the space shuttle. The main function of the cooler is to provide an inexpensive way for commercial customers and the research community to perform cold temperature testing in a zero gravity environment. Surface Tension Confinement Technology used in the design of the cooler, allows a liquid cryogen to be used as the cooling media without the traditional problems associated with the use of fluids in zero gravity. The paper discusses the design of the cooler, the testing results, and the epoxy joint design criteria that was developed during the contract, for a specific epoxy joint used in the design of the cooler.

## INTRODUCTION

Surface Tension Confinement Technology was initially developed by Stephen Castles and Michael Schien of NASA GSFC[1] in 1988. Surface Tension Confinement refers to the natural wicking property exhibited by open pore foams on liquids. This phenomenon can be particularly useful in zero gravity applications where containment of a liquid is difficult. While there are several potential applications of this technology for the space program[2], this effort focused on the integration of the Surface Tension Confinement Technology into the design of a zero gravity cryostat.

In 1992 AET, Ltd. was contracted by NASA GSFC to investigate the feasibility of providing a liquid cryogen cooler (LCC) for use aboard the Space Shuttle as a part of NASA's Hitchhiker Program. The goal of the Phase 1 (SBIR) contract was to demonstrate that an LCC was a viable method of providing cost effective cooling through the use of Surface Tension Confinement Technology. Specifically, the wicking and thermal performance of the foam needed to be tested to insure it would hold cryogens in zero gravity, and be capable of providing the necessary power (5 Watts) for typical sensor cooling applications. A 1/4 scale Proof of Concept (POC) cooler was designed and fabricated for that purpose. The foam chosen for testing was an HTP-6 foam developed by Lockheed Martin. The HTP-6 has excellent wicking properties (due to its high porosity) and is extremely light weight. HTP-6 has the added benefit of being readily available to NASA since it is the same foam used in the thermal tiles located on the exterior surface of the shuttle.

The POC testing demonstrated not only that the foam would wick in zero G but that it had enough wicking power to hold the cryogen in -1G. The foam's high porosity also led to large percentage fills (~84% when saturated in liquid nitrogen) , which would be necessary due to mission requirements as well as the volume restrictions of the shuttle's GAS canister for which the LCC would ultimately be designed. The cryogen volume of the LCC not only had to satisfy the 5 watt experimental heat load for a period of 5 days, but also a potential 30 day ground hold time. Therefore, a vent gas cooled radiation shield was designed into the POC cooler to keep parasitic heat losses to a minimum. The POC testing indicated that all thermal objectives could all be met in a full size LCC.

As the project progressed, AET, Ltd. proposed to establish some design criteria for the G-10/Aluminum epoxy joints that would be used in the fabrication of the LCC as an additional part of the Phase 2 of the contract. G-10 is a glass composite that has excellent structural and thermal isolation properties. It is commonly used by industry as a support mechanism in space flight (and other) cryostats, and is the main support tube in this LCC. Although G-10 support tubes are most often epoxied in place, there was no design criteria available for this epoxy joint.

## DESIGN REQUIREMENTS OF THE LCC FOR USE IN A NASA GAS CANISTER

In May of 1993 AET received a Phase 2 (SBIR) contract to design and fabricate a space flight qualified LCC. The unit needed to meet the thermal requirements listed above and still fit inside the standard Get Away Special (GAS) canister located in the cargo bay of the shuttle. Spatial limitations of the GAS canister required that the overall dimensions of the LCC be less than 50.16 cm in diameter and less than 71.76 cm long. NASA Hitchhiker specifications also required the unit be suspended and solely supported from the top plate of the GAS canister. In addition to the spatial and thermal requirements placed on the LCC, the flight qualified unit had to be designed to survive typical launch conditions seen inside the shuttle's cargo bay.

## LCC DESIGN

An Assembly drawing of the LCC is given in Figure 1 below. The main component of the LCC is the inner cryogen vessel that houses the foam containment structure (FCS). The FCS is fitted to the diameter of the inner vessel to insure good thermal contact. The FCS also has several heat fins passing through the foam to insure no unusable pockets of liquid nitrogen form in the middle of the foam. The FCS was designed to hold only enough nitrogen for the 5 day, 5 Watt heat load. Orientation of the foam containment structure was crucial, as it was necessary to insure that the foam was saturated in the excess liquid nitrogen during launch.

The void space above the FCS was filled with F-11 grade felt. The felt acts like a wick, similar to the foam. The vent line was then placed at the very top of the void space, above the felt. This configuration ensures that any excess liquid not confined by the foam after launch, is wicked (and held) by the felt, preventing the liquid from escaping out of

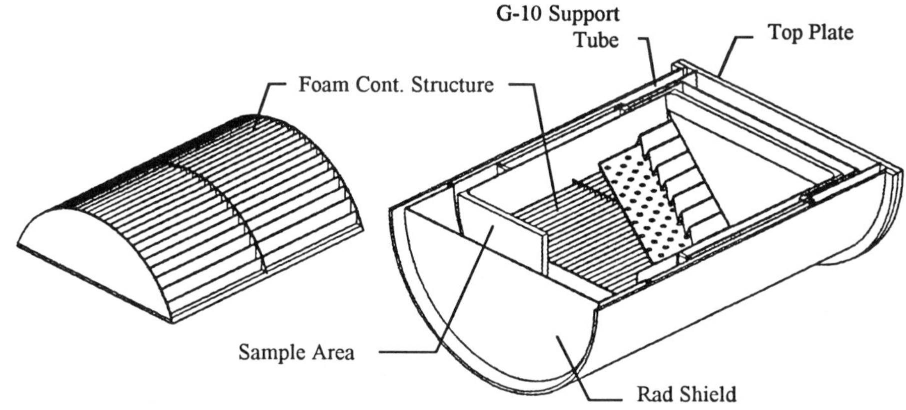


**Figure 1** LCC Design.

the vent. The felt was then secured in place with some perforated aluminum sheets, that were in turn welded in place to the inside of the pressure vessel.

The inner vessel is supported from the top plate via the G-10 Support Tube. This support tube is interrupted by the radiation shield that surrounds the inner vessel as shown in Figure 1. The radiation shield is vent gas cooled (as it was in the POC cooler) to keep its static temperature as low as possible to minimize the radiant heat load on the inner vessel. The radiation shield is also gold plated to keep its emmissivity as low as possible.

## DESIGN CONCERNS

The POC testing in the Phase 1 portion of this effort eliminated all of the thermal design concerns. The wicking capability of the foam was determined, so confining the liquid would not be a problem. The performance of the vent gas cooled shield also indicated that the mission and hold time requirements could be met within the volume requirements of the GAS Canister.

The only real concern once the flight unit was being designed was the structural integrity of the epoxy joints used on the aluminum and G-10 connections. The design of the LCC requires these joints to support the entire inner vessel (while full) and radiation shield assembly (80 Kg) during launch conditions. The epoxy most commonly used for this application is an Epibond Resin 1210-A with hardener 9615-10, manufactured by Ciba-Geigy Formulated Systems Group, Furane Aerospace Products. A finite element analysis of the LCC indicated that a joint stress of 6.16 MPa was expected during launch conditions.

### Epoxy Joint Design Questions

As reliable data was not available for bonding G-10 to Aluminum with this epoxy, AET, Ltd. developed a test plan to provide some design criteria for this (rather common) joint. The major questions which AET set out to address are as follows:

1. What is an appropriate allowable joint stress?
2. How much does the epoxy clearance in the joint effect the strength of the joint?
3. How effective is increasing the depth of the epoxy joint in increasing the strength of the epoxy joint?

## Experimental Plan

In an attempt to develop a test plan to specifically answer the design questions, it was decided that a series of G-10 rods would be epoxied into aluminum cylinders of varying joint depths and thickness. Figure 2 shows schematically the test articles. Three different diameter G-10 rods were inserted into 5 different joint depths. The joints had 3 different epoxy clearances (groove widths) for each of the varying joint depths. Joints at both 300K and 80K were tested. Table 1 lists the different rod diameters, joint lengths and epoxy clearances that were tested.

Twelve test articles were made for each possible configuration for a total of 540 test articles. Care was taken to insure that the test article epoxy joint procedure was noted and carefully followed during the construction of the actual hardware with one exception. To eliminate any tensile effects of the epoxy in the testing, the bottom surface of the-G-10 rod was not epoxied in the test articles. For the 12 articles in each configuration, 6 were tested at ~300K and 6 were tested at ~80K.

A joint depth of less than 6.4 mm was not chosen since it was felt that any depth less than that would be too shallow for practical applications, particularly in applications where the connection must carry significant forces. Once the 540 test articles were constructed, they were pulled axially to failure.

Three different sized rod diameter's were used to determine the effect of differential shrinkage between the two dissimilar materials during the cold joint testing. AET was more concerned with this parameter affecting the experimental test set up, than affecting the system design.

## Epoxy Joint Testing Results

All of the load vs. time plots for the 540 tests were analyzed after testing. The data obtained from those plots was reduced to study the effect of the various parameters as a function of the overall joint stress. Figures 3-6 show the reduced data for the 6.4 mm rod diameter test articles. The data from the 12.7 mm rod and 25.4 mm rod are similar, but the plots have been excluded for brevity. AET, Ltd. chose not to statistically reduce the raw data but to simply plot it. This was done to show not only the trends and results, but also the scatter. In this particular application the scatter in the data was as an important design concern as the data itself.

Quick inspection of Figures 3 & 4 immediately reveals a very clear trend of decreasing (ultimate) joint stress as the groove depth is increased for both warm and cold

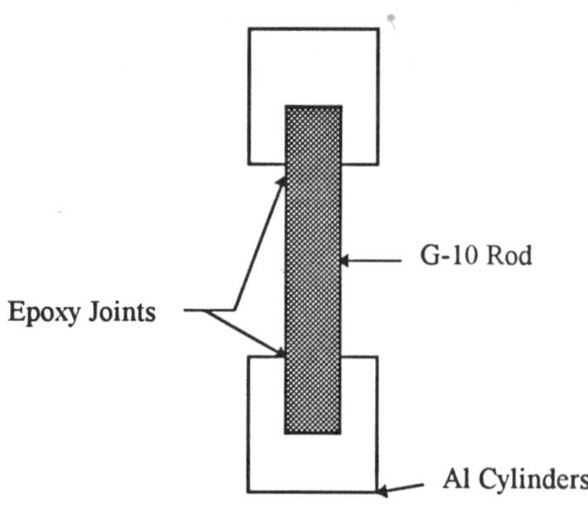

**Figure 2** Test Articles.

**Table 1** Test Article Parameters

| Rod Diam.) | Joint Length) | Epoxy Clearance |
|---|---|---|
| 6.4 mm | 6.4 mm | 0.076 mm |
| 12.7 mm | 9.5 mm | 0.178 mm |
| 25.4 mm | 12.7 mm | 0.279 mm |
| | 19.0 mm | |
| | 25.4 mm | |

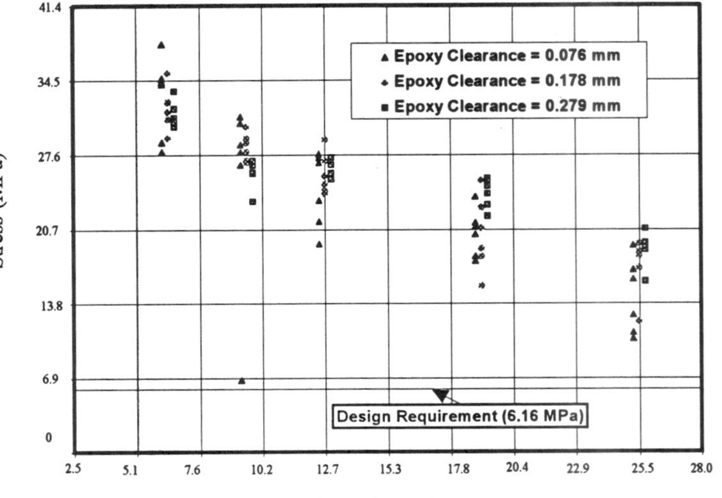


**Figure 3** Joint Stress Vs Depth For 6.4 mm Rod (T=300K).

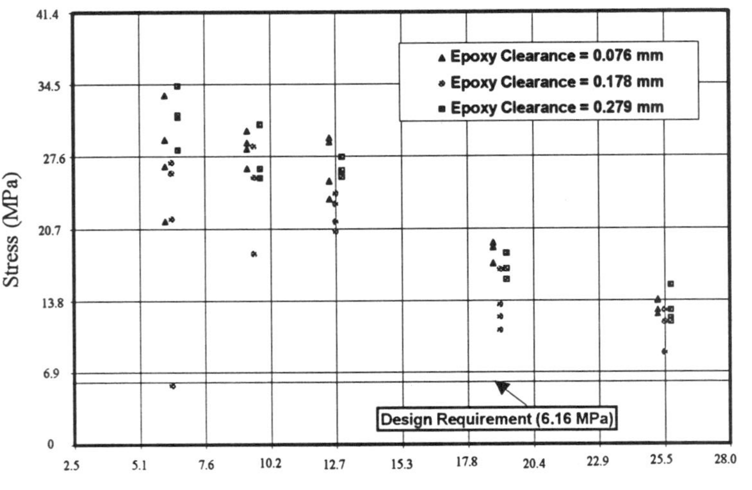


**Figure 4** Joint Stress Vs Depth For 6.4 mm Rod (T=80K).

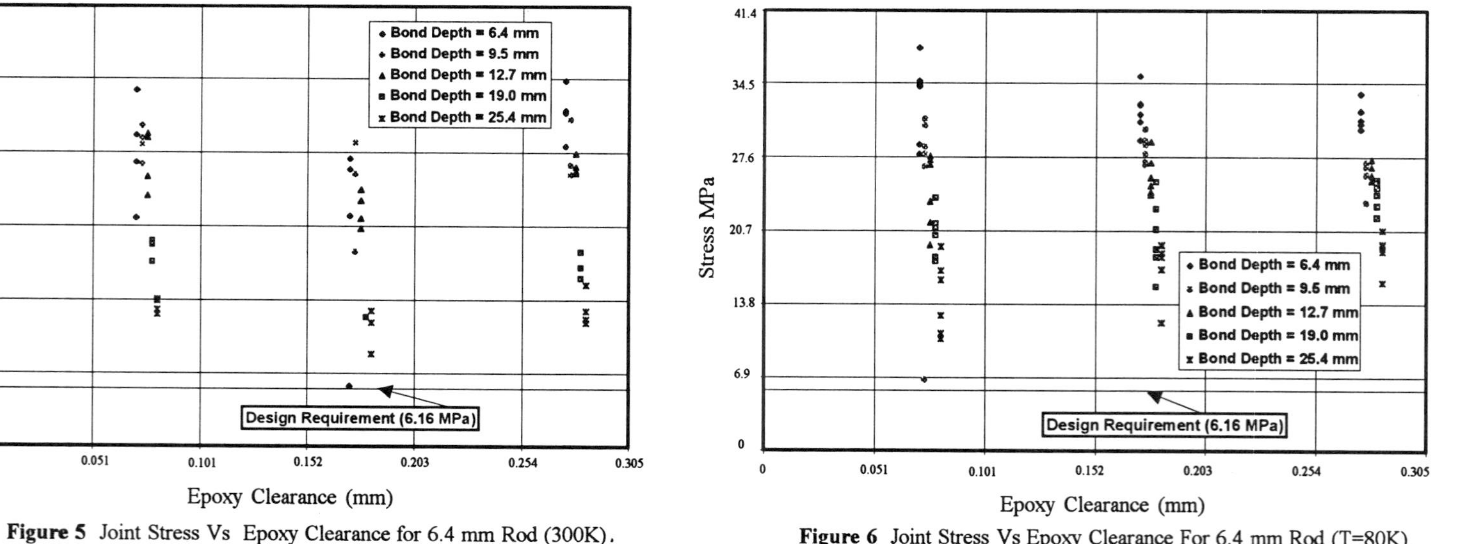


**Figure 5** Joint Stress Vs Epoxy Clearance for 6.4 mm Rod (300K).



**Figure 6** Joint Stress Vs Epoxy Clearance For 6.4 mm Rod (T=80K).

joints. Although the joint stress does decrease as the groove depth increased, it is important to note that the overall ***strength*** of the connection increased with increasing groove depth.

Careful examination of Figures 5 & 6, reveals no clear trends for either the warm or cold tests. The ultimate joint stress appears to be relatively independent of the width of the epoxy groove, at least over the range of groove widths tested (0.076 mm - 0.279 mm). If the groove width does effect the joint strength, it is not discernible from the scatter in the data. The 6.4 mm rod data does indicate that there is decreased scatter with increased groove width. However, this is only seen in the 6.4 mm rod data. When viewed alone, the 12.7 mm data indicates the opposite trend, and no trend is seen in the 25.4 mm data.

The pre-stress imposed on the connection due to differential shrinkage (during the cooling process) also did not appear to effect the ultimate joint stress. This is visible upon inspection of the 12.7 mm and 25.4 mm rod data. Again, the main concern here was whether the pre-stress would influence the cold temperature testing data.

Although, differential shrinkage due to cooling did not appear to effect joint stress, there may be a trend towards decreasing joint strength at liquid nitrogen temperatures. It is difficult to determine if it is a definite trend, or if it is simply that the cold temperature data has more scatter in it than the warm data. In either case, whether the joint is not as strong or not as reliable, caution should be used in picking an appropriate allowable joint stress for cold temperature joints.

## Application Of Epoxy Joint Testing To LCC Design

**Joint Depth.** It was decided that a 6.4 mm groove depth would be used for the critical epoxy bond in the LCC designed for the GAS canister. The test data did indicate (as expected) that the allowable joint stress decreases as the depth of the epoxy groove is increased. Although smaller groove depth may have yielded higher allowable joint stresses, it was decided that a 6.4 mm deep groove was the smallest practical joint size for a large diameter G-10 tube. 6.4 mm was therefore the smallest groove depth tested, and the groove depth that yielded the highest allowable joint stress both warm and cold.

**Joint Width.** Since no definite trends were seen in the varying of the epoxy clearance, it was decided that a 0.127 mm clearance would be used in the critical epoxy joints. Some variation to this clearance around the diameter of the epoxy joint was expected in the design due to the manufacturing tolerance of the G-10 tube. G-10 is fabricated by winding a series of thin fiberglass sheets on a mandrel with epoxy between the layers. Therefore, the outer diameter of G-10 tubes is often slightly oversized due to increased wall thickness. The most accurate dimension of a large G-10 tube is its inner diameter. AET choose a 0.127 mm groove width based on this manufacturing tolerance. AET anticipated that with the common oversized outer diameter of large G-10 tubes, the epoxy clearance would vary between 0.076 mm and 0.178 mm on average. Since, the testing indicated that the joint stress is independent of the groove width, design concerns stemming from the G-10 tube tolerancing were alleviated.

**Temperature.** Due to the possible decrease in joint strength/reliability mentioned earlier, caution was used in choosing an appropriate allowable joint stress for the cold temperature critical epoxy joint used in the design of the GAS LCC.

## Comparison Of Epoxy Joint Test Results To Joint Strength Required By Design.

Data furnished from the epoxy manufacturer indicates that the typical lap shear strength at room temperature is about 17.2 MPa for aluminum to aluminum bonding. The test data indicates that for a G-10 rod to aluminum groove, room temperature bond with a 6.4 mm epoxy groove depth, 17.2 MPa appears to be a conservative number for the allowable joint stress, regardless of the epoxy groove width chosen (inside the test range). In fact, 17.2 MPa appears to be a conservative number (although less conservative) for an identical cold temperature joint.

**Margin Of Safety**. The finite element analysis of the critical epoxy joint in the GAS LCC requires the joint stress of only 6.16 MPa for the warm and/or cold joint. NASA's safety documentation for the program defines the Margin of Safety as follows:

$$MS = (\text{Allowable Stress}/\text{Actual Stress}) - 1 \qquad (1)$$

The Hitchhiker Program requires that the Margin of Safety (for structural analysis) for any payload be greater than 0. Using an allowable stress of 17.2 MPa and an actual stress of 6.2 MPa and substituting in equation (1), the margin of safety for the critical epoxy joint of the GAS LCC is ~1.8

## GAS LCC TESTING

### Structural Testing

**Hitchhiker Requirements**. The launch conditions required the LCC to be capable of withstanding an 11 G shock load in both the axial and lateral directions, as well as a 0.044 $G^2/Hz$ random vibration (between 20 and 2000 Hz) in all axial directions. The LCC was also required to have its first natural mode higher than 35 Hz.

**Engineering Model and Flight Unit**. Prior to the flight unit being fabricated and tested, an Engineering Model Of the LCC was constructed and tested to levels higher than launch conditions (11G shock, 0.1$G^2/Hz$). After passing testing, the EM was disassembled and inspected for damage. As no damage was found, the flight unit was fabricated and tested. The GAS LCC passed all structural tests.

### Thermal Testing

Thermal testing on the LCC was not quite as successful. The parasitic heat loads on the inner vessel were higher than predicted. In order to meet the 5 day 5 Watt heat load requirement, the LCC was only capable of a 25 day ground hold in its 1G launch pad position. The high parasitic losses are attributed to physical and support constraints placed on the insulation material, to insure it would not degrade during shock and vibration. Further investigation into more effective attachment methods of the insulation is planned.

## ACKNOWLEDGMENTS

AET, Ltd. would like to thank Susan Breon and Rob Boyle of the NASA-Goddard Space Flight Center for their support of this SBIR effort (Contract Number NAS5-32502).

## REFERENCES

1. S. H. Castles and M. E. Schein. "Advances in Cryogenic Engineering," Vol 33, Plenum Press, New York, p. 819.
2. J. F. Maguire, P. M. Winn, and G. C. Driscoll, Long Term Storage of Liquid Cryogens in Space Using Fluid Surface Tension Confinement, ASME Annual Winter Meeting, (1993).

# LIQUID HYDROGEN DENSIFICATION HEAT EXCHANGER

G.E. McIntosh,[1] R.J. Stochl,[2] and T.M. Tomsik[2]

[1]Cryogenic Technical Services, Inc.
Longmont, CO 80501, USA
[2]NASA Lewis Research Center
Cleveland, OH 44135, USA

## ABSTRACT

Densification (subcooling) of liquid hydrogen and oxygen propellants offers potential reductions in size and mass of space launch vehicles. Oxygen densification is easily accomplished by heat exchange with a bath of boiling liquid nitrogen. Parahydrogen presents a more difficult problem because the cooling bath temperature is limited by its 13.803 K triple point value. To achieve a high degree of densification, the parahydrogen cooling bath must be held close to its triple point and the heat exchanger exit $\Delta T$ must be small.

We report on a parahydrogen heat exchanger designed to remove approximately 60 kW from a 0.8165 kg/s (1.8 lb/s) flow with an exit $\Delta T$ of 0.6 K with respect to a bath held at 14.4 K or lower. Details of the test system, physical design of the heat exchanger, and heat transfer analysis are given. Test results are presented and compared with design predictions.

## INTRODUCTION

Subcooled cryogenic liquid propellants for reusable launch vehicles (RLV's) have the potential to reduce vehicle gross lift-off weight by 15 to 32 percent[1] compared to fueling with normal boiling point (NBP) cryogens. These reductions start with smaller and lighter weigh propellant tanks and compound throughout the propulsion system. Similar advantages were recognized a few years ago when slush hydrogen technology was developed for the National Aerospace Plane. However, many of the technical problems[2] identified with slush hydrogen are eliminated or reduced by use of densified liquid.

Cryogenic densification is conceptually simple. As shown in Figure 1, the only required additions to a normal fueling system are transfer lines for the densification loop, a recirculating pump, the densification heat exchanger, and a vacuum pump. With the possible exception of the heat exchanger, all of the components are within the state of the

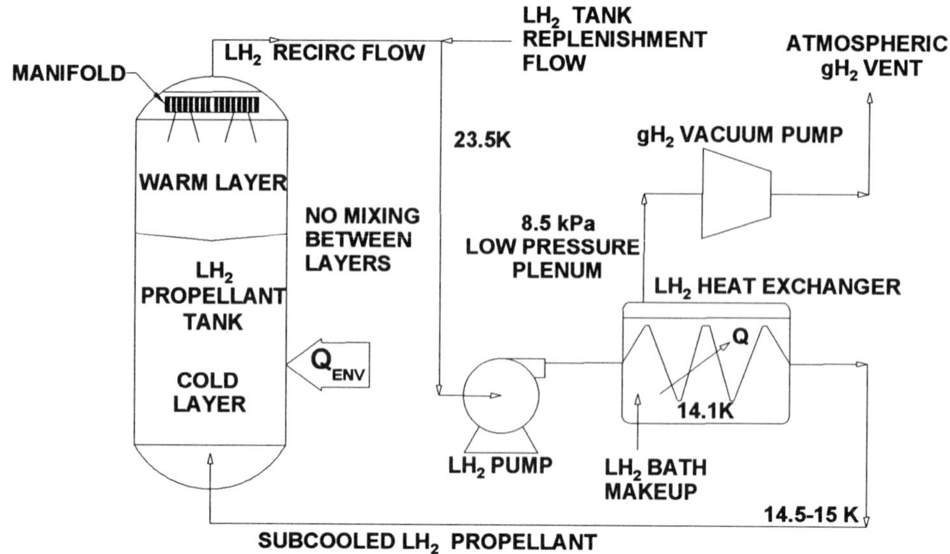


**Figure 1.** Integrated RLV propellant tank and $LH_2$ propellant densification unit.

art. Notably, either commercial vacuum pumps (with ambient warmup heat exchangers) or steam ejectors can be used instead of cold compressors.

Design of the heat exchanger is driven by the need for close temperature approach at the exit end. Also, there is some lack of design flexibility because pool boiling heat transfer is primarily a property of the fluid with limited geometric influence. Liquid hydrogen subcoolers are further constrained because boiling Q/A values decrease with falling bath pressures. Alternatives open to the designer include use of extended surface (fins) on the boiling side and selection of any feasible combination of passage size and number which achieves the desired heat transfer within limits of allowable pressure drop.

## HEAT TRANSFER ANALYSIS

Internal heat transfer calculations were based on the conventional Dittus-Boelter[3] equation for forced flow conditions in the turbulent range. Liquid hydrogen properties were taken from NIST 12[4]. The internal design problem was focused on obtaining sufficient surface area relative to the cross sectional flow area. As shown in Figure 2, heat transfer surface varies inversely with diameter for a given total cross sectional area. (i.e. for constant G in $kg/m^2$.) Thus, it is immediately evident that a large number of small flow passages is advantageous provided that the combination of flow and diameter does not cause Reynolds number to fall into the laminar range. A previously developed (for the SSC) machined 6063 aluminum extrusion was used to meet the internal flow criteria. As shown in Figure 3, these extrusions have eight 1/8 inch (0.003175 m) diameter internal passages which provide the necessary heat transfer area while reducing the number of leak-tight joints required. The heat exchanger consists of 150 of these 2.4384 m (8 ft) long extrusions running in parallel for a total of 1,200 individual passages.

External pool boiling heat transfer was calculated using the Kutateladze[5] correlation which is expressed as follows:

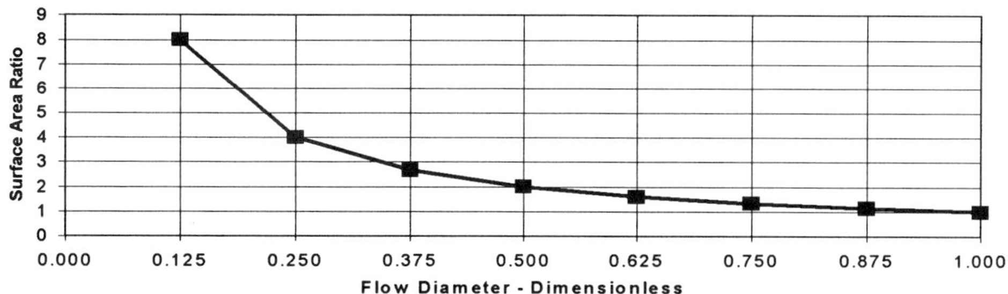


**Figure 2.** Surface area vs. flow diameter.

$$Q/A = 4.87E\text{-}11((C_p)_l/\lambda\rho_v)^{1.50}(k_l\rho_l^{1.282}P^{1.750}/\sigma^{0.906}\mu_l^{0.626})(\Delta T)^{2.50}$$

Where

l and v are liquid and vapor subscripts
$(C_p)_l$ = liquid specific heat at constant pressure, j/g-K
$\lambda$ = latent heat of vaporization at saturation, j/g
$\rho_v$ = vapor density, g/cm$^3$
$k_l$ = liquid thermal conductivity, W/cm-K
$\rho_l$ = liquid density, g/cm$^3$
P = boiling pressure, dynes/cm$^2$
$\sigma$ = liquid-vapor surface tension, dynes/cm
$\mu_l$ = liquid viscosity, g/cm-s

The Kutateladze equation only requires evaluation at the operating temperature of the bath. This leaves $\Delta T$ as the single variable. Having $\Delta T$ as a single variable facilitates a procedure for calculating the heat exchanger in one-degree K increments. This procedure involves evaluating the heat load for the one-degree step and equating the internal and external heat transfer totals based on the constraint of the total $\Delta T$ available. The actual length of heat exchanger required for each 1 K increment is calculated in this manner and the results are plotted as Figure 4. The predictable result is that very small lengths are required for larger temperature differences and nearly 39 % of the heat exchanger is used in the final increment.

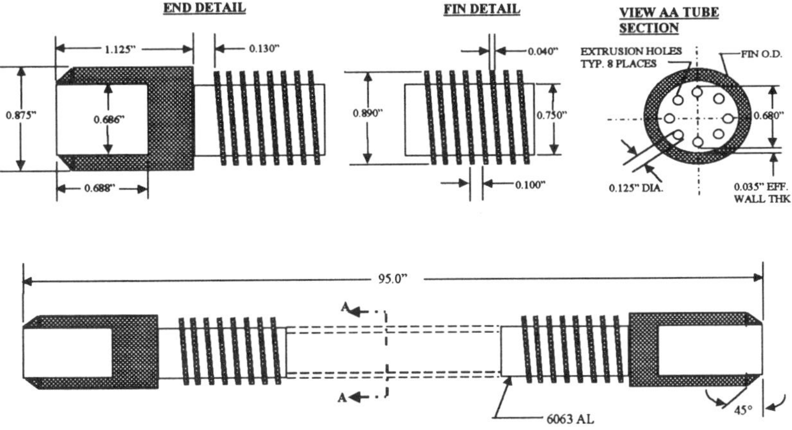


**Figure 3.** Extruded tube detail.

## HEAT EXCHANGER DESCRIPTION

NASA requirements for the liquid hydrogen densification heat exchanger are set forth in Table 1.

### TABLE 1

### HEAT EXCHANGER DESIGN REQUIREMENTS

| | | | |
|---|---|---|---|
| Mass Flow Rate | 0.8165 | kg/s | (1.8 lb/s) |
| Inlet Temperature | 23.667 | K | (42.6 R) |
| Inlet Pressure | 275.8 | kPa | (40 psia) |
| Outlet Temperature | 15 | K | (27 R) |
| Pressure Drop | 6895 | Pa | (1.0 psi) |
| Bath Pressure | 9.65±1.38 | kPa | (1.4±0.2 psia) |
| Bath Temperature | 14.39±0.28 | K | (25.9±0.5 R) |
| Coolant Flow Rate | 0.00145-0.0027 | $m^3/s$ | (23 – 43 gpm) |
| Coolant Supply Pres. | 241.3 | kPa | (35 psia) |
| Coolant Inlet Temp. | 23.61 | K | (42.5 R) |
| Comp. Bypass Flow | 0.0635 | kg/s | (0.14 lb/s) |
| Thermal Loss Rate | 2 %/day | | |

A concept section view and schematic of the heat exchanger designed to meet the above requirements is shown in Figure 5. The all-aluminum heat exchanger assembly is fastened to doubler bars attached to the bottom of the stainless steel hydrogen reservoir. Level in the dewar is indicated and controlled from a diode rake which drives the position of the coolant inlet valve. The $LH_2$ stream to be densified flows through the heat exchanger without any internal controls or restrictions. The six-inch butterfly valve on the vent line has pressure control capability but this function was later integrated into the cold compressor system. The reservoir is treated as a typical dewar and is insulated with approximately 50 layers of glass paper and aluminum foil MLI. The reservoir is supported by filament-wound fiberglass/epoxy trunnions at each end. The piping end trunnion is fixed and a warm sliding socket is incorporated at the other end to accommodate thermal shrinkage of the inner assembly. Physically, the heat exchanger assembly is 1.07 m (42 inches) in diameter and 5 m (16 ft) long. Active heat transfer areas are 29.19 $m^2$ (314.16 $ft^2$) for forced convection and 55.694 $m^2$ (599.5 $ft^2$) on the boiling side.

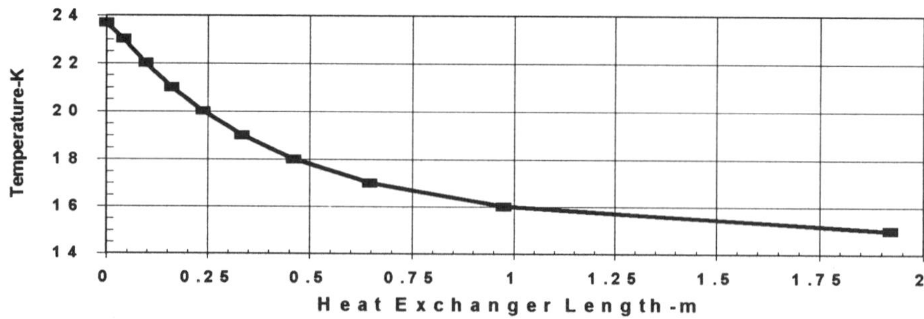


**Figure 4.** Heat exchanger temperature distribution.

## DENSIFICATION TEST SYSTEM

The densification test system set up at NASA Lewis Research Center, Plum Brook K-Site facility, is shown schematically in Figure 6. Key elements include a 49.21 $m^3$ supply dewar, dual circulating pumps, the densification heat exchanger, a four-stage cold compressor (vacuum pump), a second 49.21 $m^3$ liquid hydrogen receiver dewar, and associated controls. Except for the two liquid hydrogen dewars, all of the components shown were mounted on a skid fabricated at Plum Brook. The test system generally follows the basic concept shown in Figure 1 with the addition of the second hydrogen dewar and a gaseous hydrogen bypass line returning partial compressor flow to bubble back into the reservoir liquid volume. The entire test system was remotely controlled from a fully instrumented control room located some 250 m from the liquid hydrogen area. Multiple temperature, pressure, and flow measurement locations are shown on Figure 7 which is the Instrumentation and Control Schematic.

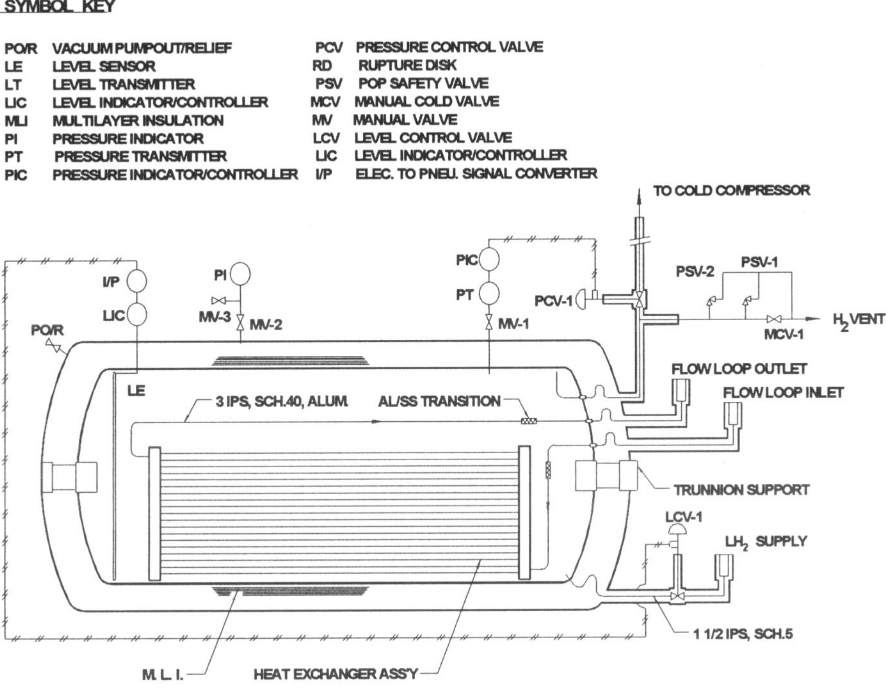


**Figure 5.** $LH_2$ subcooler schematic.

## TEST RESULTS

Some of the test objectives were obscured by difficulties with system components. In particular, neither of the $LH_2$ circulating pumps worked on either nitrogen or hydrogen and cold compressor performance was only partially successful. Several of the silicon diode temperature sensors failed and all of them gave initially faulty readings until the cold compressor variable frequency drive was shielded. Non functioning of the circulation pumps proved not to be a problem because supply dewar pressure proved sufficient to maintain the desired flows and pressures. The cold compressor system limited running time and did not achieve the desired 8.27 kPa (1.2 psia) design pressure although this value was closely approached several times. The test program was terminated by bearing failure of the second stage of the cold compressor.

Despite the problems noted above, resourceful NASA technical people were able to overcome most of the difficulties and obtain credible data. By working with the cold compressor, they developed a technique for bringing it up to speed in steps and to prevent surging by timely introduction of bypass flow. In its most successful run, the densification system cooled a 0.907 kg/s (2 lb/s) $LH_2$ flow from 21.67 (39) to 16.11 K (29 R) with a bath pressure of 20.68 kPa (3 psia). In this instance, $\Delta T$ between the bath and densification stream was 0.22 K (0.4 R) for a flow 11 % greater than design.

Aside from the exit $\Delta T$, measurements relevant to heat exchanger design yielded mixed results. In particular, 6 silicon diodes were distributed along the length of a centrally located heat exchanger tube with the aim of comparing the actual temperature distribution to that calculated and shown in Figure 4. As installed, these diodes measured the temperature of the aluminum extrusion (essentially that of the bath) and did not duplicate the calculated curve. Three additional diodes were installed in a vertical line near the bottom, center and top of the heat exchanger bundle. The results of these measurements were either inconclusive or showed very little vertical temperature gradient.

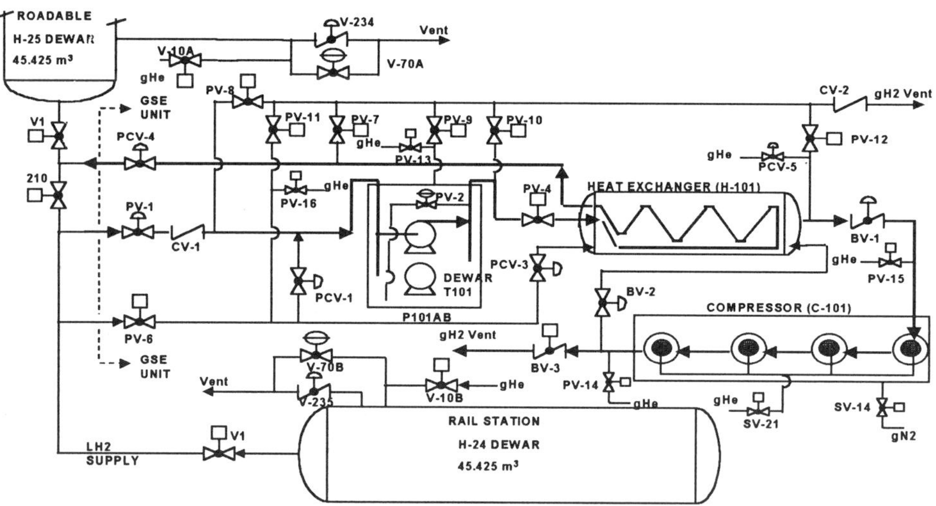


**Figure 6.** $LH_2$ GSE simplified flow schematic at K-site.

## CONCLUSIONS

Technical feasibility of liquid hydrogen densification was confirmed. The heat exchanger functioned properly and produced a smaller exit $\Delta T$, 0.22 K, than was predicted by design calculations. Turbulent boiling heat transfer from a large, closely spaced tube bundle was found to equal or exceed pool boiling Q/A values from a single tube.

Temperature measurement techniques on the aluminum extrusions were inadequate. The sensors must be placed in a well allowing direct contact with the densification flow.

There was no definitive evidence of vertical stratification.

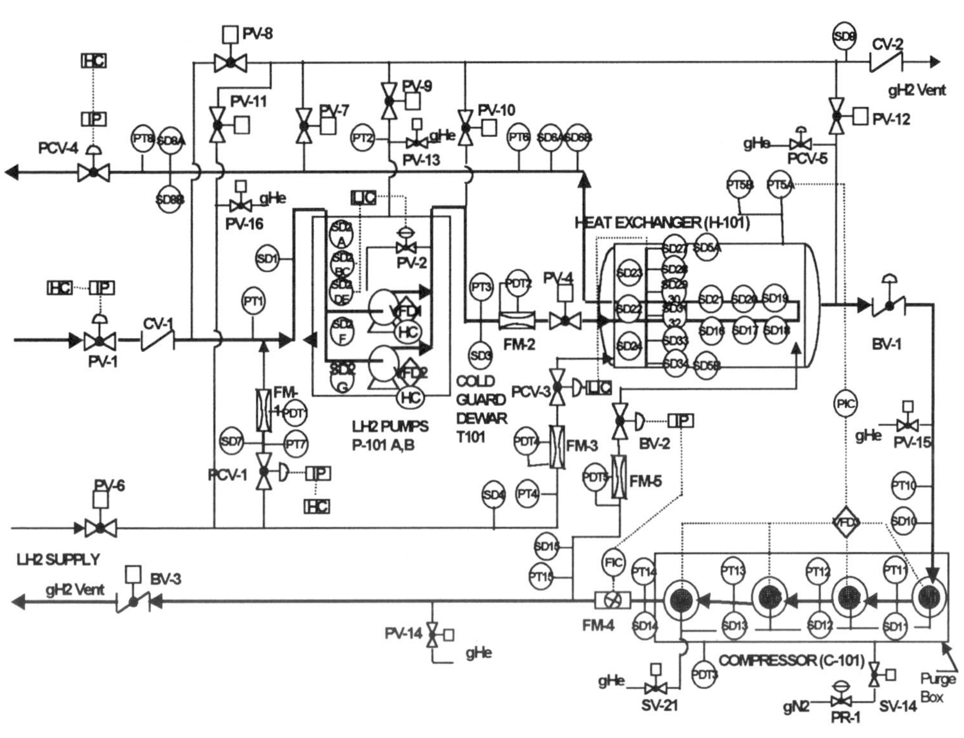


<u>Instrumentation Legend</u>

| | | | | | |
|---|---|---|---|---|---|
| PT | Pressure Xder | PDT | Diff Pressure Xder | IP | Current/Pressure Xder |
| SD | Silicon Diode | HC | Hand Control | VFD | Variable Frequency Drive |

**Figure 7.** $LH_2$ GSE instrumentation and control schematic.

Better vapor/liquid separation is needed to reduce liquid entrainment in the low pressure line to the cold compressor.

Densification system components should be kept as simple as possible to ensure reliable performance.

## ACKNOWLEDGEMENTS

The densification heat exchanger was designed and fabricated with support from NASA Lewis Research Center, Contract NAS3-27788. Mr. Herb Mundhenke made 300 aluminum extrusion to header welds without a single leak.

## REFERENCES

1. T.M. Tomsik, Performance tests of a liquid hydrogen propellant densification ground support system, AIAA-97-2976, 33rd AIAA/ASME/SAE/ASEE Joint Propulsion Conference and Exhibit, July 6-9, 1997 /Seattle, WA.
2. N.P. Hannum, Technology issues associated with fueling the National Aerospace Plane with slush hydrogen, NASA TM-101386, 1989.
3. F.W. Dittus and L.M.K. Boelter, Heat transfer in automobile radiators of the tubular type, in: "Univ. Calif. Publ. Eng.," Vol. 2, (1930) pp. 443-461.
4. NIST Standard Reference Database 12, U.S. Dept. of Commerce, National Institute of Standards and Technology, Gaithersburg, MD, September 1992.
5. E.G. Brentari and R.V. Smith, Nucleate and film pool boiling design correlations for $O_2$, $N_2$, $H_2$, and He, in: "Advances in Cryo. Eng.," Vol. 10B, Plenum Press, New York, (1965).

# POST-FLIGHT ANALYSIS OF A 10 K SORPTION CRYOCOOLER

R. C. Bowman, Jr., P. B. Karlmann, and S. Bard

Jet Propulsion Laboratory
California Institute of Technology
Pasadena, CA 91109 USA

## ABSTRACT

In May 1996, the Brilliant Eyes Ten-Kelvin Sorption Cryocooler Experiment (BETSCE) was flown aboard the Space Shuttle flight STS-77 in the first space-flight demonstration of chemisorption cryocooler technology. While on orbit, BETSCE successfully met its science objectives of cooling from 70 K to 10 K in less than two minutes, sustaining a 100 mW heat load below 11 K for over ten minutes, and demonstrating recycle times of approximately eleven hours. The hydride compressors were able to circulate and compress the hydrogen refrigerant fluid in a repeatable manner with no adverse microgravity effects being observed.

Since completion of the Shuttle flight, the BETSCE flight hardware was carefully inspected and a repair was made to a system valve which had developed an internal leak during flight. A series of successful cooldown cycles to below 9.5 K were completed during post-flight ground tests following repair of this valve. All observed compressor and cryostat parameters obtained in the measurements performed after flight were entirely consistent with the behavior seen in the pre-flight tests. These results verify that key functional performance of the BETSCE instrument is highly reproducible and did not degrade throughout the extensive ground tests and flight operations.

## INTRODUCTION

Hydrogen sorption cryocoolers utilize metal hydride sorbent beds to compress and circulate hydrogen fluid within a closed-cycle Joule-Thomson (J-T) system to provide active cooling in the 10 to 30 K temperature range. Sorption cryocoolers offer long-life, low-vibration solutions to meet the cooling needs of future instruments in astrophysics space telescopes[1,2], space superconducting electronics, and military surveillance satellites[3].

Johnson and Jones[4] recognized that a periodic 10 K sorption cryocooler permits repeated quick cooldowns from initial temperatures ~65 K and low average power consumption for those applications where intermittent operation is sufficient. The 10 K periodic cryocooler concept has been thoroughly described elsewhere[3,5,6]. Briefly, metal hydride sorbent beds are used to compress hydrogen gas to a nominal pressure of 10 MPa for storage in a tank between short duration cooling periods. On command, hydrogen gas flows through an upper stage precooler at ~65 K to a J-T expansion valve to form liquid hydrogen in a cryostat-reservoir with the non-condensed hydrogen gas being absorbed by a $LaNi_{4.8}Sn_{0.2}$ filled sorbent bed. Solid hydrogen at below 11 K is formed as the pressure in the J-T reservoir is reduced to ~100 Pa using a second hydride sorbent bed containing ZrNi powder. The entire cooling process from 65 K to below 11 K is completed within 120 seconds and the low temperature is maintained until all the solid hydrogen has been sublimated. The recharging of the hydrogen tank from the sorbent beds takes 6 - 12 hours before the rapid cooldown can be again performed.

The Brilliant Eyes Ten-Kelvin Sorption Cryocooler Experiment (BETSCE) project was undertaken to develop a complete closed-cycle periodic 10 K sorption cryocooler suitable for demonstration in earth orbit on the Space Shuttle. The primary objectives of BETSCE were: (1) provide a thorough end-to-end characterization and performance validation of a hydride sorption cryocooler in the 10 to 30 K temperature range, (2) acquire the microgravity database needed to provide confident engineering design, scaling and optimization, (3) identify and resolve interface and integration issues, and (4) provide hardware qualification and safety verification heritage. Detailed understanding of microgravity issues associated with the heat and mass transfer processes within the sorbent beds and the separation and retention of the liquid and solid phases of hydrogen within the J-T coldhead is necessary to eliminate excessive design margins in future periodic and continuous 10 K sorption cryocoolers. The present paper provides an overview of BETSCE behavior during flight and ground testing with a final assessment of vital lessons-learned for future design and fabrication of long-life periodic and continuous 10 to 30 K sorption cryocoolers.

## SYSTEM DESIGN AND OPERATION

The BETSCE instrument contains five integrated subsystems: Sorbent Bed Assembly (SBA), Cryostat Assembly (CA), Tank & Valve Assembly (TVA), Control Electronics Assembly (CEA), and the Structure Assembly (SA). The basic BETSCE operating cycle is briefly summarized in Figure 1 with the use of the fluid schematic. Since step-by-step descriptions of the cooldown-recharge cycles are available elsewhere[5,7], no additional details will be given here.

BETSCE functional ground tests[5] clearly demonstrated that all of the basic performance requirements had been exceeded by the hardware. Both the heat and mass transfer performance of the BETSCE Sorbent Bed Assembly exceeded the early predictions[3]. The Fast Absorber Sorbent Bed (FASB) was able to absorb the necessary hydrogen and dissipate the heat of reaction during the 2 minute cooldown phase while maintaining the back-pressure typically below 0.35 MPa. The Low Pressure Sorbent Bed (LPSB) demonstrated that it could solidify the hydrogen within the coldhead in less than 10 seconds and continue to maintain the coldhead at temperatures as low as 9.44 K. Additionally, the High Pressure Sorbent Bed (HPSB) repeatedly compressed hydrogen from pressures around 0.07 MPa to greater than 10 MPa.

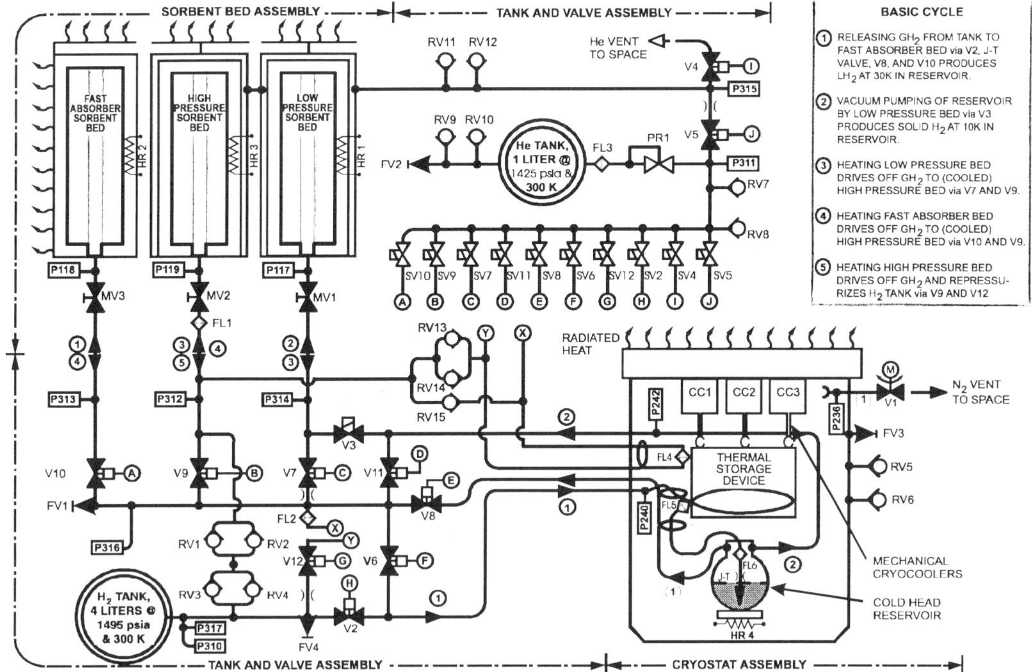


**Figure 1.** BETSCE fluid schematic with outline of 10 K cooldown cycle.

During initial tests, freezing of contaminants (e.g., $H_20$, $CH_4$, etc.) sometimes caused the J-T capillary tube to clog within the cryostat when cooling to ~26 K. To solve this problem two custom "cold trap" adsorption filter assemblies were fabricated and installed within the hydrogen system. Both assemblies are located on the Thermal Storage Device (TSD) which is cooled to 65 - 80 K within the cryostat. These cryogenic adsorbers are constructed from 316L stainless steel with 2 μm porous filter discs shrink fitted and welded at each end of the devices. One adsorber assembly (identified as FL5 in Figure 1) is filled with sodium aluminosilicate molecular sieve material and is located just upstream of the cold heat exchanger and J-T valve inlet. The second cold trap (denoted FL4) contains activated Saran carbon and is located in-line between the HPSB and the hydrogen tank as shown in Figure 1. Consequently, hydrogen from the HPSB is filtered and cleaned as it flows through this cold adsorber before entering the $H_2$ tank during the tank recharge process. Following installation of the two adsorber cold traps, no further difficulties with contaminant gas freezing was noted during ground and flight tests.

Solid particulates also caused problems during pre-flight testing. After the second occurrence of a particulate becoming lodged within the J-T capillary tube, a custom J-T valve filter assembly was fabricated and installed to reduce the recurrence of further particulate contamination of the J-T capillary. This J-T filter assembly (denoted as FL6 in Figure 1) fully encapsulates the inlet side of the J-T valve with a shrink-fitted 2 μm porous stainless steel filter disc installed ahead of the capillary opening. No further entrapment of particulates at the J-T valve were observed during subsequent ground and flight testing.

An important objective of the BETSCE project was to make comparisons of performance parameters predicted by a transient thermal simulation model[7] with experimental data from ground tests. In support of this objective, a series of 10 K cooldown experiments was conducted to assess the impact of several parameters on low

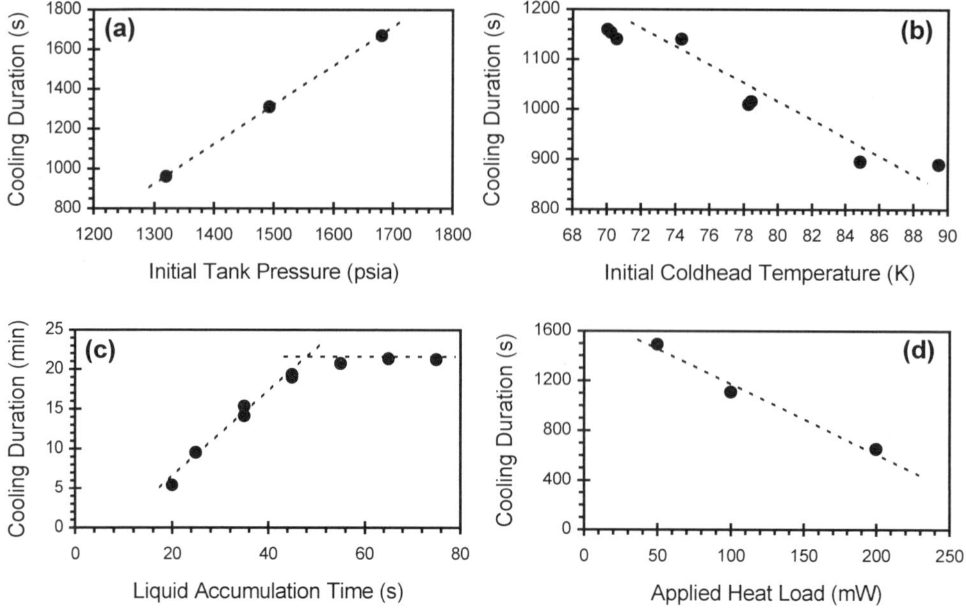


**Figure 2.** Impact of (a) initial tank pressure, (b) initial coldhead temperature, (c) liquid $H_2$ accumulation time, and (d) applied $I^2R$ heat load on the coldhead cooling duration below 11 K.

temperature cooling duration. These results are summarized in Figure 2. In accordance with thermal model simulations, the cooling duration below 11 K was observed to increase with tank pressure and liquid hydrogen accumulation time, while these times decreased as the initial coldhead temperature and applied heat load increased. The minimum coldhead temperature ($T_{min}$) was found to be insensitive to these parameters (i.e., 9.4 K < $T_{min}$ < 9.5 K) except for heater input power where temperature rose to 9.8 K at 200 mW. These tests also showed the cooldown time to increase with rising initial coldhead temperature and to decrease with rising tank pressure, which corresponded to trends predicted by the model simulations.

In summary, the pre-flight ground tests confirmed that BETSCE exceeded all performance requirements, was capable of cyclic operation with reproducible cooldown temperatures and duration, and satisfied all Shuttle safety and interface requirements.

## IN-FLIGHT PERFORMANCE

The BETSCE instrument was launched into orbit on Shuttle Mission STS-77 (Orbiter Endeavour) on May 19, 1996 with BETSCE mounted on the Shuttle orbiter payload bay side-wall. Waste heat from BETSCE radiated to space by flat plate passive radiators oriented out (+Z) of the Shuttle bay. In order to begin cycling operations, BETSCE initially completed preconditioning sequences that cooled the TSD to ~65 K and then desorbed the hydride sorbent beds to pressurize the hydrogen tank. As recently reported[8], the sorbent beds successfully brought the $H_2$ storage tank to an operating pressure of 9.8 MPa within 24 hours after the TSD had cooled below 70 K. These results demonstrate that the sorbent beds were able to effectively compress hydrogen to identical pressures as had been achieved in ground tests. Temperature profiles of the bed walls, outer shells, and radiators for the individual sorbent beds during their desorption heating-cooling steps are presented in Figure 3.

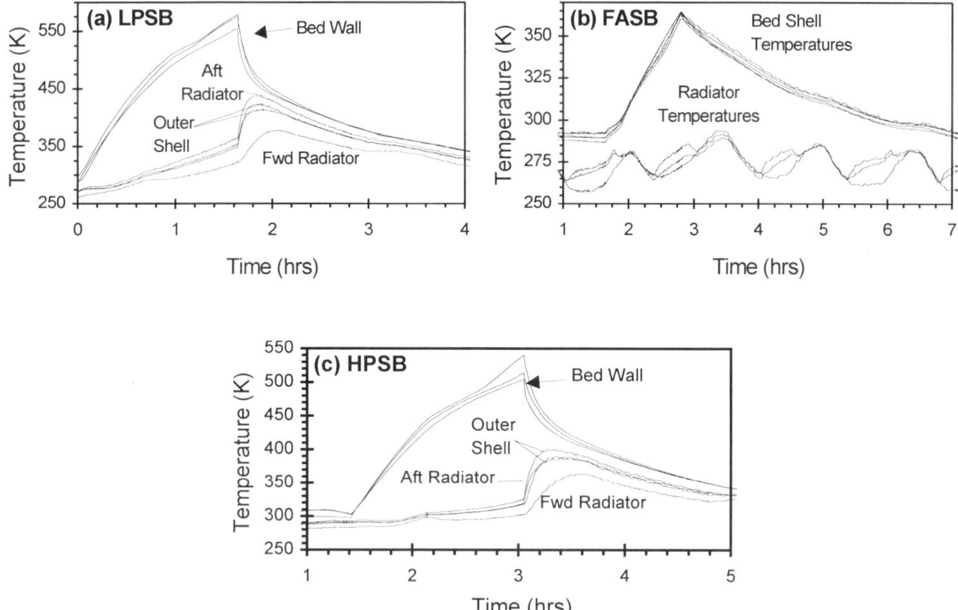


**Figure 3.** Temperature profiles of sorbent beds during on-orbit desorption and cooldown: (a) Low Pressure Sorbent Bed, (b) Fast Absorber Sorbent Bed, (c) High Pressure Sorbent Bed.

Although bed temperatures during heating in orbit were very similar to behavior found in ground testing[5], the rates of cooling were slower and the outer shell and radiators were hotter (e.g., peak temperatures greater by ~25 K and ~50 K for LPSB and HPSB, respectively) during flight. Both of these effects are attributed to the fact that beds were in vacuum during earth orbit which restricted heat transfer from the beds to the flight radiators. In contrast, the bodies and shell structures of the sorbent beds also lost heat via natural convection with the ambient air during ground tests, thus providing further cooling. Temperature oscillations due to Shuttle orbiter attitude changes and ~90 minute day-night cycles during earth orbit are quite distinct for the FASB radiator temperatures shown in Figure 3b. These cyclic fluctuations are less apparent in Figure 3a and 3c, but are also present for the LPSB and HPSB radiators.

Figure 4 shows the initial on-orbit 10 K cooldown for BETSCE. The cooldown from 70 K to 11 K was completed in under 2 minutes, and a 100 mW $I^2R$ heat load was sustained at below 11 K for 10 minutes, thus meeting the primary system performance objectives[3] for the BETSCE flight. After the cooldown was completed valve V2, which isolates the tank from the cryostat, did not reseal properly. This allowed hydrogen to leak from the tank to the cryostat[8].

Although the problem with valve V2 was to prevent additional 10 K cooldown cycles, valve control sequences were modified to allow continued cycling of the metal hydride compressors and BETSCE was able to produce liquid hydrogen. The coldest temperature achieved in these cycles was 18.4 K and the longest cooling duration was over 32 minutes. A total of eight liquid hydrogen cycles were achieved by the end of the mission[8]. The temperatures recorded during the initial hydrogen absorption by the FASB in these cooldowns confirmed that melting of the phase change material n-hexadecane while in microgravity remained capable of absorbing much of the heat released by the hydriding process[3,5,7].

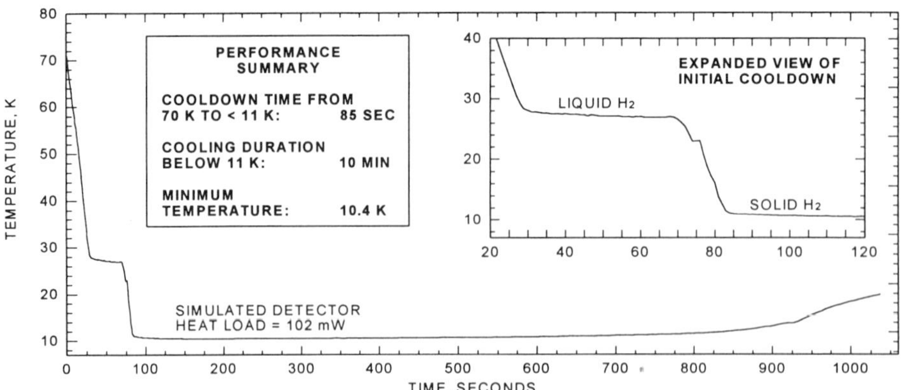


**Figure 4.** Summary of on-orbit 10 K cooldown.

## POST-FLIGHT EVALUATIONS

After the return of the BETSCE hardware to JPL, the first activity was to disassemble the leaking valve V2 for inspection. A microscopic examination of the valve seat revealed numerous metallic particles embedded in the sealing groove. While the specific sources of these particulates cannot be established with any certainty, their most probable origin is fabrication debris trapped in the 4-liter hydrogen storage tank or associated TVA plumbing. Although efforts were made to clean and rinse this vessel prior to welding it into the hydrogen manifold, the small (i.e., 0.46 cm) opening to the connecting tubing impedes not only removal of residual particulates but also permits entrapment of any particles generated during later assembly procedures. For ground tests in any orientation of the BETSCE system, the particles would remain trapped in the bottom of the tank. Since these particulates would be floating in the tank while in orbit, they were apparently swept out of the tank and lines towards the valve during the nominal 70 sL/min flow rate for the cooldown and liquid hydrogen accumulation steps of the first 10 K flight cooldown. After determination of the presence of particles on the removed valve stem, valve V2 was thoroughly cleaned and a new valve stem/seat was installed. Subsequent vacuum and pressure leak checks confirmed that the integrity of valve V2 had been restored and it was sealing properly when closed.

In order to establish whether the BETSCE hardware had retained its performance capabilities after flight operation and repair of the valve V2, a series of the ground tests were conducted. No contamination or clogging effects were seen in fifteen 10 K cooldown cycles. The minimum coldhead temperatures were consistently between 9.45 and 9.50 K with cold duration periods > 1100 seconds and all other sorbent bed properties during cooldown and recharging were nearly identical to the pre-flight ground test results. Figure 5 compares six post-flight 10 K cooldowns to a representative pre-flight run under slightly different test conditions. These results verified that no substantive alteration in cooldown performance remained after the Shuttle flight and subsequent repair of valve V2. No permanent damage from the particles, which had limited on-orbit testing, was detected with the other BETSCE valves or J-T components during post flight testing.

Over 130 periodic 10 K cooldown-recharge cycles were made with the BETSCE hardware during the ground tests in addition to the flight experiments performed aboard the Shuttle. No degradation in the storage capacities and reaction kinetics for the hydride sorbent beds was observed up through the final post-flight ground tests. All components in the cryostat, including the Stirling cryocoolers, TSD, heat exchangers, and J-T reservoir

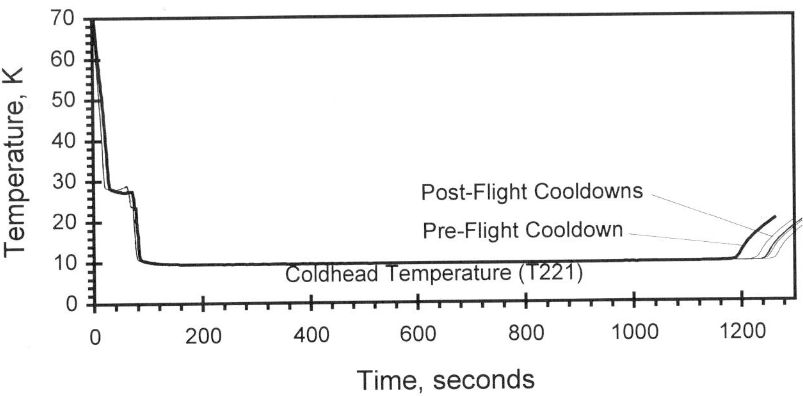


**Figure 5.** Comparison of several post-flight 10 K cooldown cycles to a pre-flight test.

functioned appropriately during the tests. The temperature and pressure sensors performed well and the TVA valves operated dependably (except for the leaking of V2 during flight from particle contamination).

## LESSONS LEARNED

### Future Design Improvements

Based on the BETSCE hardware development experience and analysis of the BETSCE ground and flight test results, a number of valuable lessons-learned can be applied to improve the design of future operational 10 K sorption cryocooler systems[1,2]. First, it is important to recognize that the size and weight of a 10 K cryocooler can be greatly reduced from that of BETSCE. The BETSCE instrument was essentially an adaptable test bed that allows extensive characterization testing of an experimental 10 K sorption cryocooler system within the highly constrained Shuttle safety and interface requirements. Hence, only nominal efforts were made to minimize the size, weight, or power requirement, as the BETSCE resources were focused on developing the important new technology elements of the system, particularly those that needed to be validated in microgravity. As described in References 5 and 9, an operational periodic 10 K sorption cryocooler should be achievable with a mass of ~30 kg including an upper stage 60 K mechanical cooler.

### Contamination Control Issues

For a long-life flight cooler it is necessary to completely eliminate the contamination problems encountered by BETSCE. It is crucial that the system be designed to minimize virtual leaks that trap contaminant gases internal to the gas system. Electropolished tubing and fittings should be used wherever possible, and use of organic materials (e.g. in valve seats) should be minimized. Welds are preferred to braze joints to minimize flux contaminants and trapped volumes. Although it was constructed of electropolished 316L stainless steel tubing, and carefully cleaned and verified to aerospace industry standards, the BETSCE hydrogen storage tank had a single 0.46 cm i.d. opening serving as inlet and outlet to the 4L volume. For future systems, it is recommended that all tanks contain both an inlet and an outlet to aid in more effective cleaning. It is important to pay special attention to filter manufacturing processes and cleaning procedures, especially for the filters contained within sorbent beds. These filters have very large surface areas that can trap adsorbed contaminants, and should be subjected to numerous fill/ flush and vacuum bake-outs prior to loading of sorbent material. In-line filters should be included to protect all valves and the J-T refrigeration loop from solid particulates. Detailed examination of the disassembled BETSCE valve V2 has confirmed that the leakage was caused by metallic particulates which were able to flow from the tank, fittings or plumbing only while in the microgravity conditions of flight. It should be noted that no problems with clogging of the J-T capillary line were encountered during flight or post flight ground tests. While it is likely that in microgravity conditions some particles reached the J-T inlet, this region was presumably protected by the J-T filter assembly FL6. Finally, to further reduce the risk of clogging, in-line cold trap getters should be installed to continually remove gas phase contaminants that may be generated in the closed-cycle cooler.

## SUMMARY AND CONCLUSIONS

BETSCE achieved its original flight objectives[3]. It demonstrated the ability to cooldown and produce solid hydrogen at 10 K in under 2 minutes on its first cooldown attempt on orbit while sustaining a simulated $I^2R$ detector heat load of 100 mW for 10 minutes, the ability to repeatedly produce liquid hydrogen below 19 K, and the ability to repeatedly recycle the sorption compressors. Microgravity operation was found to have no adverse effect on the ability to retain liquid and solid hydrogen, and the sorption compressors demonstrated similar heat and mass transfer characteristics as determined in ground testing[5]. The successful BETSCE space and ground test results have validated many critical technologies for both periodic and continuous configurations of hydride sorption coolers. The end-to-end characterization and the quantitative microgravity database should enable early insertion of 10 to 30 K sorption cryocooler technology into future long-life, low-vibration, space remote sensing missions.

## ACKNOWLEDGMENTS

This work was sponsored by the Ballistic Missile Defense Organization, the Air Force Space and Missiles Systems Center, and Air Force Phillips Laboratory through an agreement with the National Aeronautics and Space Administration. This work was made possible by the support provided by Erwin Myrick, Dr. Dwight Dustin, and Col. Ralph Gajewski.

## REFERENCES

1. L.A. Wade, in: "Cryocoolers 8", R.G. Ross, Jr., ed., Plenum, New York (1995), p. 845.
2. L. Wade, et al., in: "Cryocoolers 9", R.G. Ross, Jr., ed., Plenum, New York (1997), p. 577.
3. S. Bard, et al., in: "7th Int'l Cryocooler Conf.", PL-CP-93-1001, Kirkland AFB, NM (1993), p. 1107.
4. A. Johnson and J. Jones, in: "7th Int'l Cryocooler Conf.", PL-CP-93-1001, Kirkland AFB, (1993), p. 831.
5. S. Bard, et al., in: "Cryocoolers 8", R.G. Ross, Jr., ed., Plenum, New York (1995), p. 609.
6. J.J. Wu, et al., in: "Adv. Cryogenic Engineering, Vol. 39", Plenum, New York (1994), p. 1507.
7. P. Bhandari, et al., in: "Cryocoolers 8", R.G. Ross, Jr., ed., Plenum, New York (1995), p. 581.
8. S. Bard, et al., in: "Cryocoolers 9", R.G. Ross, Jr., ed., Plenum, New York (1997), p. 567.
9. P. Karlmann, et al., in: "Adv. Cryogenic Engineering, Vol. 41", Plenum, New York (1996), p. 1305.

# THE REDSTONE INTERFACE COOLING SYSTEM

J.M. Lester

Redstone Engineering Company
Under Contract to Ball Aerospace Systems Division
Boulder, Colorado 80306-1062, U.S.A.

## ABSTRACT

An adaptive interface for connecting conventional cryocoolers to space instruments has been designed, built and tested. Many advanced space cryo-cooling applications require tight temperature control in the presence of a widely varying cooling load. Also, often cooling loads must be removed from large distributed areas rather than from concentrated points. These requirements must be met, while at the same time, power is minimized. The Redstone Interface is a temperature control and thermal storage system designed to satisfy the above requirements. The system combines three well known principles into a useful hybrid: a conventional cryocooler such as a Stirling or Turbo-Brayton, a low pressure Joule-Thomson cooling loop and a thermal storage unit. A test system operating at 65 K was shown to maintain the temperature over an 0.5 m long by 50 mm wide simulated sensor to within 0.05 K while the thermal load varied from 0.65 to 3.60 W over an extended period of hours. Other useful system integration features such as ease of heat rejection and lack of sensor disturbances were demonstrated.

## INTRODUCTION

Cooled space instruments are becoming larger and more complex as new applications are pursued. New requirements are placed on the cooling systems to support these instruments. For instance, larger distributed area sensors need to be cooled. Temperature must be more precisely controlled than before. These temperatures must also be held constant even though the instrument (with its heat load) may be turned on and off. In addition, more than one cooling temperature may be needed. And finally, all of these things may be needed simultaneously. As always, the input power, weight and size must be kept to a minimum and the instrument must not be affected by vibration or other disturbances from the cooling system. The heat from the cooling system must be rejected efficiently and reliability must be high. Clearly, conventional cryocoolers cannot supply many of these features. In response to these new requirements the Redstone Interface has been developed in an IR&D program at Ball Aerospace. This paper describes the system and test results.

## THE REDSTONE INTERFACE

The Redstone Interface Cooling System (RICS) is a hybrid arrangement consisting of a conventional cryocooler (pre-cooler) and one or more Joule-Thomson (J-T) cooling loops. These J-T loops contain thermal storage units (TSUs). The objective of this arrangement is

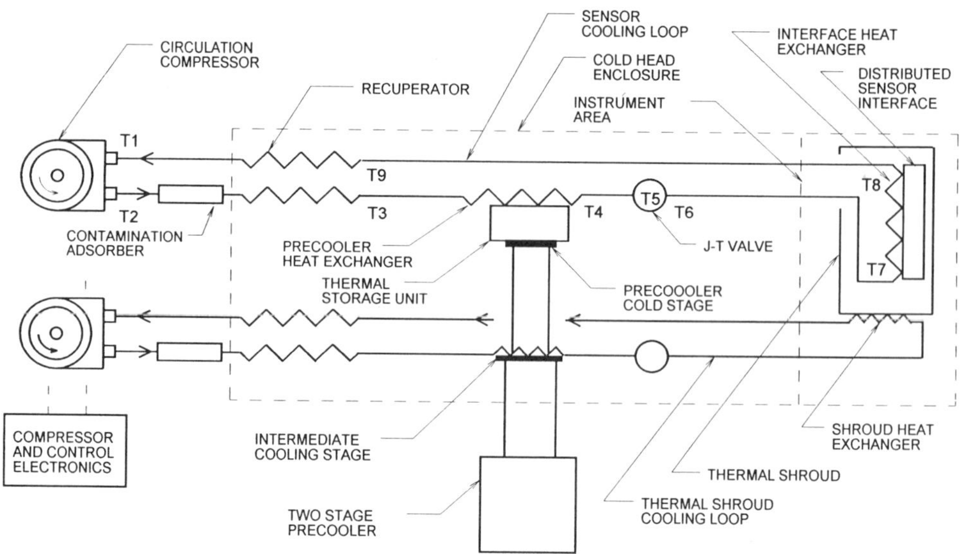


**Figure 1.** Redstone Interface System schematic shows two J-T cooling loops pre-cooled by a two-stage Stirling or Pulse Tube cooler.

not to produce lower cooling temperatures than the pre-cooler can provide, but to adapt the pre-cooler more effectively to the requirements of the advanced instruments.

The RICS system is shown schematically in Figure 1. Small, low pressure compressors circulate gas in cooling loops. The purpose of the loops is to transfer cooling from the pre-cooler and TSU to the instrument and to control instrument temperatures. The loops contain several heat exchangers which are contained in a vacuum-insulated space with the TSU. The low pressure compressors operate at ambient temperature and are outside the vacuum space. The RICS version shown in Figure 1 shows two cooling loops, one for a sensor at the lower temperature and one for a thermal shroud around the sensor at a higher temperature. The thermal load from the sensor is variable and a TSU is required. The thermal load on the thermal shroud is constant; therefore, no TSU is required. Looking at the sensor loop, gas flows through the incoming leg of the recuperator and is cooled to temperature T3 by the returning flow. Further cooling, condensation and liquid sub-cooling to temperature T4 is done by the pre-cooler and the TSU. At the J-T valve the sub-cooled liquid is expanded to the lower return pressure, creating a two phase mixture in this expansion. Since the purpose of the J-T valve is to create a two phase mixture for precise temperature control and not to produce refrigeration, the pressure and temperature are only slightly decreased. Next, the two phase mixture flows through the sensor interface heat exchanger and absorbs the sensor's heat output by boiling at a constant temperature and pressure. The boil-off vapor and some residual amount of liquid flow back to the compressor and pre-cool the incoming flow in the recuperator. The temperature in the sensor interface heat exchanger is nearly the same at all points as a result of the boiling process in the cooling tubes. The interface temperature is held constant by a feedback loop. Temperature information from T7 or T8 is delivered to a control circuit that varies compressor speed in order to adjust the pressure and temperature of the interface heat exchanger, thereby holding the sensor temperature constant.

The thermal storage function is performed by the TSU that is attached between the pre-cooler and the pre-cooler heat exchanger. The TSU is comprised of a thermal storage material in a small container. The TSU operates in the following manner: when the sensor heat load is low, the cooling transferred by the J-T loop is not fully used. As a result, the excess liquid flows out of the sensor interface heat exchanger at T8. Hence, the excess cooling decreases the temperature at T9; and in sequence T3 is also decreased. Furthermore the heat load on the TSU and the pre-cooler is reduced. Since the pre-cooler runs at a

**Table 1.** Features and benefits of the RICS system

| Features | Benefits |
|---|---|
| Controlled temperature interface as a result of constant temperature and pressure boiling process. | Precise temperature control of a distributed area sensor in the presence of a variable sensor heat load. |
| Thermal storage units level out the heat load on the pre-cooler. | Allows use of smaller, average-capacity pre-cooler which operates at its most efficient point. This reduces required input power. |
| Thermal storage temperature can vary as it stores and releases cooling. | De-couples temperature control function from TSU. |
| Allows use of solid thermal storage materials for most applications. | Eliminates effects of zero-gravity on performance. Eliminates heat transfer problems associated with phase change materials. |
| Sensor cooling is done by small tubes. | De-couples sensor from pre-cooler disturbances such as vibration. Allows maximum packaging flexibility for sensor. |
| Allows use of different types of pre-coolers. | Can select best or most available pre-cooler. |
| Heat producing components are located at the heat rejection interface. | Maximizes heat rejection efficiency which, in turn, minimizes required input power and weight. |
| One two-stage pre-cooler can cool two loads at two temperatures. | Simplifies cooling of thermal shields around sensor. Minimizes weight and power. |

constant capacity, i.e., a constant input power, its temperature decreases when both the heat load decreases and the temperature of the TSU decreases. The colder TSU now becomes a reserve of cooling for later periods of high heat load. When the sensor heat load is high, there is very little or no excess cooling at T8. Because there is no excess cooling, T9 and T3 increase and a higher heat load is applied to the TSU. The TSU is designed to have a high heat capacity so it will absorb the high heat load as it warms by only a few degrees. In this way, the constant capacity pre-cooler is adapted to the variable load sensor and the required pre-cooler input power is minimized. Any type of cryocooler that produces the average cooling capacity at the required pre-cooling temperature can be selected as the pre-cooler. Stirling, Pulse Tube, J-T or Turbo-Brayton coolers are pre-cooler candidates. Table 1 above summarizes the beneficial features described and includes several others.

**Table 2.** Example performance requirements

| Requirement | Value |
|---|---|
| Sensor cooling temperature | 63 K +/-0.25 K |
| Sensor heat load (during on periods) | 5 W |
| Duration of on period | 1 hour |
| Number of on periods per 24 hour day | 6 |
| Sensor heat load (during off periods) | 1 W |
| Duration of off period | 1 hour |
| Average sensor heat load | 2 W |
| Size of sensor | 0.5 m by 0.5 m |
| Thermal shield cooling temperature | 118 K +/- 1 K |
| Thermal shield heat load (continuous) | 5 W |

## AN EXAMPLE APPLICATION FOR THE RICS

The benefits of the RICS are further revealed by examining key performance values for the typical application defined in Table 2 below.

## LOOP CIRCULATION COMPRESSORS

The loop circulation compressors must be reliable and vibration free. Their light duty performance requirements shown in Table 3 make this possible. Note that only 15 W are needed for the sensor loop and 8 W for the shroud loop to drive these compressors. These values were calculated using a conservative efficiency of 30% for the combination of compressor, drive motor and power supply. The efficiency of these compressors can be low, since very little power is needed to produce the small pressure rise and flow. Other performance requirements, such as high reliability, are easier to satisfy when efficiency can be low. An oil-free compressor is preferable because zero gravity oil control features are avoided. A multi-unit reliability program can be done for reasonable cost because these compressors are inexpensive. Several types of compressors commonly used in industrial and aerospace applications can be adapted for the RICS. An IR&D project is currently underway at Ball to develop and verify the life of the loop circulation compressors.

## CONCEPT DEMONSTRATION UNIT RESULTS

A concept demonstration model of the RICS was built and tested in 1995 and 1996. Figure 2 shows the hardware. The object shaped like a top hat is the vacuum enclosure containing the TSU, heat exchangers and J-T valve. The finned object in the background is a commercial compressor used for loop circulation. The long tube contains a 0.5 m long

**Table 3.** Key performance values for a RICS design that meets the requirements listed in Table 2

| Factors | Values |
|---|---|
| Pre-cooler capacity at 65K | 2.2 W |
| Pre-cooler capacity at 120K | 5.5 W |
| Pre-cooler type | Two stage Stirling |
| Pre-cooler input power | 198 W |
| Pre-cooler weight | 14 kg |
| TSU weight | 2.5 kg |
| TSU thermal capacity | 28800 J |
| TSU temperature rise per sensor on period | 7 K |
| Sensor cooling loop fluid | Gas mixture |
| Sensor cooling loop compressor pressures | 13.8 kPa (2 PSIA) to 89 kPa (10 PSIA) |
| Sensor cooling loop compressor input power | 15 W |
| Sensor cooling loop compressor weight | 0.75 kg |
| Thermal shield cooling loop pressures | 103 kPa (15 PSIA) to 138 kPa (20 PSIA) |
| Thermal shield cooling loop compressor input power | 8 W |
| Thermal shield cooling loop compressor weight | 0.75 kg |
| Total system input power | 221 W |
| Total system weight | 23.0 kg |

**Figure 2.** This demonstration system validated the RICS concept in 1996.

simulated sensor. Other small objects in the picture are flow meters, pressure transducers and interconnecting tubing.

Figure 3 shows the simulated sensor which can be several feet away from the TSU. One small flexible stainless steel tube coming from the J-T valve feeds several smaller parallel tubes. Flow is collected and returned through the other tube. Wires are for temperature sensors and load simulation heaters. Note the small amount of hardware that is needed to cool the sensor. No heat pipes, flexible thermal straps, TSU containers, liquid holding load tanks nor additional structural supports are needed. The volume of attached hardware is small and weighs only a few ounces, allowing for maximum flexibility in packaging the sensor.

The demonstration model was extensively tested and the concept was validated. Figure 4 shows a typical test run wherein the sensor load was suddenly changed from an off

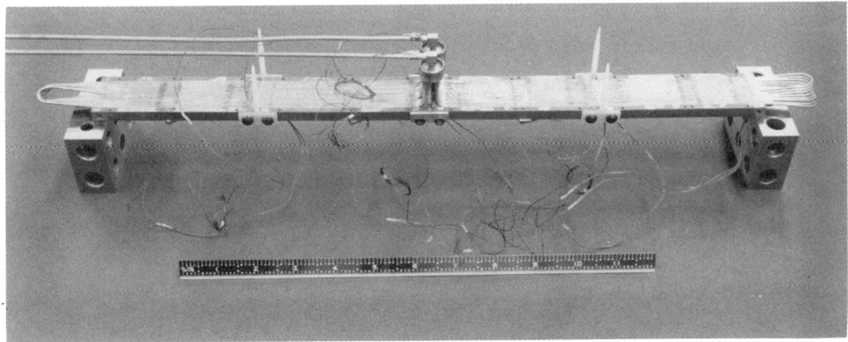

**Figure 3.** This 0.5 m long distributed area interface was housed in the long tube shown in Figure 2. Temperature stability tests were run for a variety of load change cases between 60 K and 70 K.

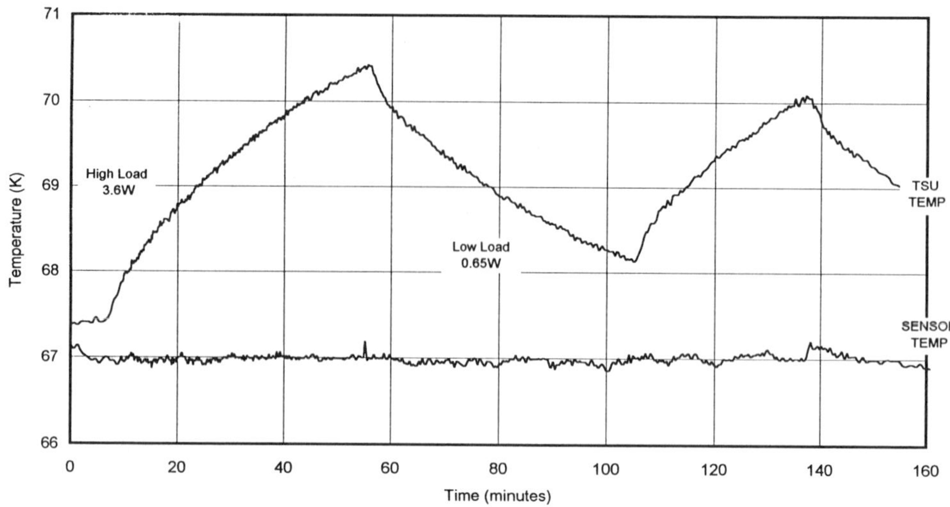


**Figure 4**. Redstone Interface operation. The sensor temperature was controlled by manually adjusting compressor speed. The TSU absorbed the high heat load as it warmed by 3 K.

condition of 0.65 W to on at 3.60 W. The lower curve shows that the temperatures measured along the simulated sensor do not change more than 0.25 K through the load variation. In addition, the temperature gradient over the entire sensor was less than 0.05 K. The upper curve shows the temperature rising as the TSU absorbs the high heat load and shows the temperature decreasing as the capacity of the pre-cooler exceeds the heat load when the sensor is turned off.

## CONCLUSION

The Redstone Interface has been shown to be an efficient way to adapt cryocoolers to space sensors that have variable heat output and precise temperature control requirements. The system has several other highly desirable features which aid in sensor and spacecraft integration. Work on this system is continuing and will include compressor development and experiments at other temperatures.

## PATENT CITATION

1. The Redstone Interface is a patented concept with additional patents pending.

# COALS TO NEWCASTLE - CRYOGENICS IN ANTARCTICA

A. W. Francis and J. Alcorta

Arthur W Francis Consulting
New City, New York 10956 USA
Antarctic Support Associates
Englewood, Colorado 80112 USA

## ABSTRACT

The National Science Foundation's Antarctic Program requires liquid oxygen, nitrogen and helium to support scientific research in Antarctica. Biologists employ liquid nitrogen refrigeration to freeze, store and transport cell samples. Astrophysicists cool detectors with liquid helium and liquid nitrogen to eliminate noise and enhance sensitivity to radiation in the microwave and far infrared. Gamma ray detectors are maintained at liquid nitrogen temperature. Meteorologists use cryogen cooled instruments to analyze pollutants in the atmosphere, including those believed to be responsible for loss of ozone in the stratosphere. Aviators breathing oxygen is stored aboard aircraft in the liquid state.

Cryogens must be supplied, thousands of miles from industrial sources, to an area which can receive ocean freight less than one month each year. Air freight is limited to three to six months and is often dependent on ski equipped aircraft of limited cargo capacity. Ambient temperatures vary from 200 to 280 K (-100 to +40 F). Wind and snow are frequent hazards. Supplying the program needs involves specially selected equipment and procedures to overcome these problems safely.

This paper describes some of these applications and the techniques developed to supply Cryogens along with descriptions of life and work in this most remote continent.

## INTRODUCTION

The title of this talk suggests a paradox, that the coldest continent is not cold enough so that we must bring cold to Antarctica. This has only happened recently. The earliest expeditions had extreme difficulty handling the low temperature provided by nature. They could do little more than establish that an ice covered continent existed and that the explorers had visited specific locations, such as the South Pole. By the 1930's Admiral Richard Bryd's repeated trips began to map the physical features and conduct weather and biological surveys. The United States conducted frequent and extensive exploration and scientific studies after World War II. A continuous United States presence on the continent has been maintained since the International Geophysical Year in 1957. This presence includes three currently operating

*Advances in Cryogenic Engineering*, Vol. 43
Edited by P. Kittel, Plenum Press, New York, 1998

stations, McMurdo on Ross Island, the southernmost port site available to large ships for a few weeks each year, Amundson-Scott station at the geographic South Pole, and Palmer, in the Antarctic Peninsula south of Terra Del Fuego in South America. Other locations have been repeatedly occupied during Austral summer seasons, roughly November through January.

The original buildings of 1957 at South Pole became snow covered and had to be abandoned in the 1970's when the central core of the current station was constructed. During this early period most of the scientific observations concerned weather, snow conditions and similar phenomena specific to local conditions. However, a few scientists began to take advantage of its unique properties to make observations difficult or impossible at other locations. For example, the six months of continuous sun light, attracted the attention of Solar Seismologists. Data from these experiments, conducted since the early 1960's, have provided much insight into conditions in the interior of the sun. South Pole Station is undergoing a major facelift because of its age, rising maintenance costs, facilities that are marginally able to provide for increasing research, by larger numbers of science personnel.

The new station at South Pole and new facilities built at other stations in the past twenty years, are intended to support the type of investigation that depends on characteristics of the locale which are optimum for observations of general scientific interest.

## LONG WAVE LENGTH ASTRONOMY

Astronomers want to locate their telescopes to minimize the effects of the earth's atmosphere. Higher elevations reduce the amount of air between the receiving telescope and the incoming radiation. Some chemicals, especially water vapor, can absorb certain wave lengths. In the Troposphere temperatures normally fall with increasing height which tends toward vertical convection currents that decrease the air's transparency.

Atmospheric density at the South Pole location is equivalent to a height of 11,500 feet above temperate zone sea level. Its very low temperature ,200 to 245 K assures very low levels of water vapor. Ground level air temperatures are so low that Stratospheric conditions of increasing temperature with increasing altitude occur at ground or slightly above.

These properties make the polar plateau attractive for observations of the far infra-red and micro-wave radiation. Viewing conditions in these wave lengths during the Austral winter can approach the quality of space based telescopes, and far exceed those of any other earth based observatories. The cost of constructing and operating equipment at South Pole is high, but less than one percent that of comparable equipment in orbit. Manual attention can be essentially continuous and major changes can be made every year.

Observations in the far infra-red and micro wave portions of the spectrum are of increasing interest to Astronomers to detect both relatively low temperature nearby sources, such as dust clouds and "Brown Dwarf" stars, and higher temperature radiators, such as "Main Sequence" stars and Galaxies, at very great distances with consequently large red shift. Radiation which is omnipresent throughout the sky in the microwave is believed to originate in the period shortly after the "Big Bang" during the separation of matter and energy as the earlier chaotic glop cooled enough for matter to condense into atoms.

Common to all these observations is the requirement that detectors of radiation in these wave lengths must be cooled well below the equivalent black body temperature of the radiation to be observed. This requires cooling at least to the range of liquid helium, i.e. less than five Kelvin. In the studies of anisotropy in the micro wave background radiation, the detectors have been cooled to about twenty millikelvin. This also requires liquid helium, vacuum pumped to about one degree, as a heat sink for salt demagnetization refrigerators.

The first attempt at such observation was made in the Austral summer season of 1986-87. The liquid helium supply logistics plan was based on flying a 500 liter dewar of liquid about every week. It was impossible to maintain this flight frequency. Also, the shipments often arrived with very little liquid . As a result there was not sufficient liquid for meaningful measurements. In the 1988-89 season a single large shipment of about 13,000 liters was made at the beginning of November. The 9,100 Kilogram , 2.4x2.5x6.0 Meter transport was flown from Point Mugu NAS via Honolulu , Pago Pago, and Christchurch to the ice runway at McMurdo in an Air Force C-141. At McMurdo it was transferred to a ski equipped C-130 for the flight to South Pole. The container was offloaded to a sled and moved about a mile away to the site of the microwave antennae.

From mid November liquid was transferred into 100 liter dewars as needed, every two or three days.This sufficed to maintain two experiments for 45 days of apparatus testing and observations. Supply technique, especially transfer efficiency,continued to improve and lower heat leak equipment became available, so that smaller shipments were able to supply growing activities throughout the entire summer season, i.e. to the end of January.

Finally the program began to operate through the winter season . Supplies for winter operations must arrive at South Pole before mid February when the temperature drops low enough to prevent further flights into or out of the South Pole runway. The temperature remains below safe operating level until the last week in October, about 250 days later. Most dewars have liquid helium evaporation rates from 0.5 to 2.0 % of capacity per day. They would be completely empty from evaporation alone in less than half that time. Nevertheless, in 1995, one telescope was kept operational throughout the entire year.

Currently, the program aims to maintain experiments on two different telescopes continuously. Each telescope consumes about five liters a day, more than 2,500 liters for the season. The initial inventory this year was 15,000 liters in two bulk containers. It is expected that with evaporation and transfer losses this should be adequate. Additional telescopes with still more cooled detectors are planned. Whether by imported liquid, by dedicated refrigerators, or by reliquefiers, it is likely that these growing demands will be supplied.

## LONG DISTANCE BALLOONING

High altitude balloons can also be effective in carrying instruments above most of the atmosphere that interferes with the desired measurements. Large balloons can carry thousands of pounds to heights of about twenty five miles. At this level less than one percent of the sea level atmosphere remains above the instruments. These balloons are capable of staying aloft for long periods, gradually descending as the buoyant helium diffuses through the envelope. At very great altitudes the air pressure is very low, so the force driving this diffusion is also very low. At sea level a toy balloon, made of Mylar plastic, can remain buoyant for several days.

High altitude balloons could theoretically maintain their altitude for several weeks, except for the effect of diurnal heating and cooling. As the sun warms the balloon, its helium expands and the balloon rises. After sun set the balloon cools, contracts and falls. This daily excursion, which may seriously interfere with the experiment, will after only a few cycles bring the balloon down to ground level. However, in the Polar zones, during their respective summer seasons, the sun shines through the entire twenty four hour day. With nearly constant solar heating a balloon will not experience the daily thermal cycle, expanding and contracting. As a consequence it becomes possible to maintain experiments at a nearly constant altitude up to three weeks.

During such a long duration flight a balloon can be expected to drift with the wind for thousands of miles. At the end of the flight the experimental apparatus may descend onto an inaccessible area of land or onto the open ocean where it may sink before recovery. In the Antarctic, a weather pattern develops around the time of summer solstice that includes steady circumpolar winds at 20-25 mile altitude. As a result, it is possible to float instruments at this height for a period in excess of one week on a course that roughly circles the continent and returns near to the initial launch area.

For more than a decade a cooperative effort between the National Science Foundation and the National Scientific Balloon Facility of NASA has made one or two of these flights each year from a launch site near McMurdo station. One of these flights carried a liquid helium cooled superconducting magnet for studies of cosmic rays. Several flights have carried liquid nitrogen cooled gamma ray detectors.

## BIOLOGY

There is little biological activity of any kind other than human in the interior of Antarctica. However some very interesting discoveries have been made in the ocean off the Antarctic coast and in certain valleys that remain free of snow and ice through very low local precipitation. In the dry valleys microbes have been found that live inside sandstone rocks. Summer sunlight penetrates the transparent quartz grains to provide barely enough energy to maintain cellular activity. In certain fish, indigenous to the Antarctic coastal seas, internal fluids are kept free of ice by a protein that acts as an antifreeze. These fish live in salt water at a temperature that averages 2-3 degrees below zero centigrade.

Liquid nitrogen is used for preservation of specimens in Antarctica and during transport back to the United States. For the latter, the use of "Dry Shippers" has been very helpful. These are vacuum insulated dewars whose inner vessel is lined with adsorbent material. Liquid nitrogen is filled into the container, and after rapid boiling stops, the remainder is poured out again. Sufficient nitrogen is retained by the adsorbent to maintain its temperature for several days, enough to make the flight north to Christchurch, New Zealand and on to the experimenters home base. The prior alternative was to pack the samples in dry ice (solid carbon dioxide) in insulated cardboard boxes. The nitrogen cooled "Dry Shippers" are safer, more effective, and have a lower overall cost.

Much of this activity is carried on from the ice breaking research ships operated by NSF in the waters near Antarctica. In addition, liquid nitrogen is used with instrumentation for detecting and analyzing atmospheric contaminants including those responsible for the destruction of ozone in the Stratosphere.

## LOGISTIC EFFICIENCY

Helium, nitrogen and oxygen are required in gas phase. Ocean transport to McMurdo is possible only in late January at the height of the annual thaw. Compressed gas supply to activities the following summer, November to January, must remain in storage for eleven months before use. Empty pressure vessels, returned to the United States for refill in February wait empty until the following December to return. That means in a two year period the compressed gas container makes one round trip. Container capital investment or rental expense becomes the major cost of gas supply. Long term storage through the Antarctic winter often causes major loss of product.

During the summer season at the time of gas use, the only transport available is cargo aircraft. From late October through early December a hard surface ice runway is available. Toward the end of this period the ice is beginning to soften and crack. Before it becomes dangerously weak the runway is closed. After that only ski equipped aircraft can land and take off from the airstrip in the permanent snow of the Ross Ice Shelf. At South Pole station the airstrip is snow the year round with ski planes used for all flights.

These conditions impose a very high cost per pound on any cargo imported to McMurdo by air cargo, with an even higher cost at South Pole. Transporting helium, nitrogen and oxygen in liquid phase reduces the container weight. For example, seven times more mass of liquid helium can be carried than compressed gas in equipment of equal total weight and volume. In recent years at South Pole, enough liquid helium has been brought in for refrigeration of long wave length detectors, that recovering and compressing the normal evaporation loss from liquid transports supplies all the gas required for daily launch of meteorology balloons.

Liquid oxygen is being used for the first time this year at McMurdo station. Fifteen hundred liter transports of LOX will be shipped to McMurdo twice during the summer. This LOX will be used on the Air National Guard C-130's for conversion to gaseous oxygen for breathing by the air crew. Previously these planes would have to return to Christchurch, New Zealand once each week for a LOX refill. Having LOX available in McMurdo will allow the aircraft to stay on the ice longer and fly more missions each season.

## LOCAL SUPPLY

All cryogenic shipments are considered hazardous cargo for aircraft and in the NSF program prohibit the flight from carrying passengers. Shipments of liquid nitrogen in 1500 liter transports were required frequently throughout the entire summer season. This caused extra difficulty in maintaining needed personnel movements. Several years ago the program acquired a small (10 liquid liters per hour) nitrogen liquefier, installed at McMurdo. This unit separates nitrogen from air in a Pressure Swing Adsorbent (PSA) system and liquefies the nitrogen with a GM cooler. During the summer season more liquid nitrogen is now produced at McMurdo than ten transport loads, the maximum previously imported. This has helped the overall logistics very significantly.

Despite having a much greater heat of vaporization than liquid helium (60 times greater per volume of liquid), there are no commercially available liquid nitrogen containers with normal evaporation rate low enough to provide liquid throughout the winter season. To meet this requirement a very small unit with capacity to liquefy one liter per hour was installed at South Pole station. This unit separates nitrogen from air by semi-permeable membrane and liquefies with a GM cooler. Both these units have had initial problems during their first operating seasons. However with more experience and some necessary alterations to meet the local environmental conditions, they both are performing very well this year.

## SUMMARY

The environment in Antarctica is quite exotic. The need for cryogenics is entirely normal. Deep cooling for superconductivity and for especially sensitive instruments, preservation of biologicals, and efficient transport of gases are very common uses anywhere else in the world. Providing for these needs safely and effectively in far from normal conditions challenges the skill of the engineers and technicians that have performed those services during the past decade.

## ACKNOWLEDGEMENT

The National Science Foundation division of Polar Programs, through its prime civilian contracting organization, Antarctic Support Associates, has provided financial, managerial and engineering support to the organization of cryogenic supplies in Antarctica.

# INVESTIGATION OF THE SELF DESTRUCTION OF A LIQUID NITROGEN CONTAINER AT SOUTH POLE STATION, ANTARCTICA

A. W. Francis

Arthur W. Francis Consulting
New City, New York 10956

## ABSTRACT

A 2200 liter liquid nitrogen container suffered a spontaneous self inflicted accident which resulted in separation of the carbon steel outer shell into two parts at the joint between the forward dished head and the cylindrical body. The cylindrical shell, rear head, and skid mounting were displaced about six meters from their original location without damage to other equipment or personnel. The inner vessel remained connected to the forward head by four unbroken lines and did not move appreciably from its former location. However, the stainless steel liquid fill and withdrawal line was separated completely at its connection to the pass through the forward head. The rupture disc which protected the outer shell from overpressure in the vacuum space was not ruptured. There was no indication that the inner vessel pressure relief valve had operated. It was not damaged, remained connected to the forward head, and was later shown to operate successfully at its rated pressure.

Analysis of the accident concluded that it was initiated by a brittle failure of the liquid fill and discharge line at a pressure estimated to be less than four bars, well below its rated safe operating pressure of seven bars.

## INTRODUCTION

The United States National Science Foundation Antarctic Program has used liquid nitrogen for a number of experiments at its stations and field camps for more than a decade. Liquid was obtained from an air separation plant in Christchurch New Zealand. It was transported to McMurdo Antarctica and from there to South Pole station by cargo aircraft in vacuum insulated 1500 liter dewars of a type used routinely by the United States Airforce. The National Science Foundation had obtained three units of this type and used them effectively during the several years prior to 1991. In planning for the experimental season of November 1991 through January 1992, nitrogen use at McMurdo, South Pole and Williams field, about ten road miles from McMurdo, would be large enough to justify locating a nitrogen tank at each location.

The time to send an empty to Christchurch, refill and return ranged from five to seven days, but could be extended by as much as an additional week if weather conditions delayed flights. It appeared that the three 1500 liter containers could not be expected to provide both on-site storage and transport for all three sites. It was decided to lease some tankage of similar size to provide supplemental storage. Two 2200 liter vacuum insulated liquid nitrogen dewars were found to be available for rental for the season.

## CONTAINER BACKGROUND

These containers were horizontal cylindrical, double wall, vacuum multi-layer insulated. The outer shell was carbon steel, 8 mm thick in the cylindrical section with 12 mm dished heads. The inner vessel was 304 stainless steel with design safe working pressure of 8 atmospheres absolute, (atm.a.). It was equipped with an automatic relief valve to prevent over-pressure.

One of these was delivered empty to South Pole station in November 1991. After a few days one of the 1500 liter liquid nitrogen transports arrived to make the initial fill to the storage tank. The storage tank was supposed to have been delivered to the Antarctic program in useable condition. However, prior to filling it was purged and filled with nitrogen gas at a slight pressure. Several lines at valves and gauges were found to leak. One of these was a one inch ips. threaded connection at the main liquid fill and withdrawal valve. It would not be possible to move the 2200 liter tank into a heated area of the station for several hours because of the press of other work. The necessary repairs were quite minor and in other circumstances might have been dealt with in a few minutes.

The tank location was a snow field among rows of timber blocks on which various cargo not harmed by freezing temperatures and snow accumulation was stored prior to use. The weather was clear, bright sunshine, wind less than 15 km/hr, at about 230 Kelvin temperature. It was possible to work for periods of only ten to twenty minutes before hands became sufficiently cold to prevent further activity. It was then necessary to seek warmth in a nearby heated building for five minutes before continuing the repairs. It required about five hours for two men to complete this task.

Nitrogen gas from the liquid transport was then transferred at about 3 atm.a. to check the effectiveness of the joint repairs. Pressure held steady for an hour which indicated that any residual leakage was at a tolerably low level. The storage tank was cooled by filling liquid nitrogen at a very low rate. The temperature of the venting vapor was observed. The tank appeared to be sufficiently cooled after transferring about 150 liters over a period of about one hour. The liquid addition rate was increased and 1200 liters transferred in about 30 minutes. The pressure in the source container declined during transfer to about 2.5 atm.a. Pressure in the 2200 liter storage tank rose to this level at the end of the transfer. With all manual valves closed the storage tank pressure was observed an hour later at about 2.6 atm.a. There was no detectable leakage from any valves or connections, or any other indication of abnormal status.

## THE INCIDENT

This observation was made at 4 PM station time. At around midnight there was a loud bang, a somewhat muffled explosive noise. The station manager was nearby and reached the cargo area less than one minute later. The area around the 2200 liter liquid nitrogen storage tank was enveloped in fog. Pieces of aluminum foil and glass fiber paper were scattered about.

Liquid nitrogen poured from a bare tank through a ruptured line. This tank was attached by several other tubes to a dished head resting against a cargo pallet about 30 cm. from the tank. About seven meters away, in the opposite direction, the rest of the outer shell (cylindrical section and dished head resting on small "I" beams) had come to rest without touching anything. No personnel had been in the area, and no one came in contact with the remaining liquid nitrogen, which completely evaporated in a few minutes. Damage to the cargo pallet which had been struck by the front dished head was inconsequential. The area was marked off and the pieces of storage tank were left untouched to be examined later.

A 25 mm. tubing size and two small diameter gage lines connecting to the vapor space at the top of the inner vessel remained connected to both the liquid vessel and the front head. A small diameter line connecting the bottom of the liquid vessel to the head and a drain valve was also intact. A fifth line of 25 mm. diameter tubing was connected to the main liquid fill and withdrawal valve on the head. It was completely separated by a brittle fracture in the darkened area of the weld affected zone next to the weld attaching to the pass through on the outer head. The break was nearly circular, with no evidence of tearing.

The frangible disc on the outer head was not ruptured. The break separating the front outer head from the rest of the outer shell was entirely within the weld metal. There was also a crack in the cylindrical section of the outer shell which extended from the lowest point of the failed weld back and up through the shell for about three feet. This crack crossed a stiffening ring which was completely separated.

One potential cause could have been overpressure in the liquid nitrogen tank. However, there was no indication that the relief valve had opened. It was intact and its connection to the inner vessel was unbroken. On subsequent bench test it performed normally. No one had heard any noise of the nature of gas discharging at high velocity, as would have been the case if the relief valve had opened.

## FAILURE SEQUENCE

It appears that the accident initiated with rupture of the liquid line. It is not known what pressure was reached in the liquid vessel, except that it did not exceed the setting of the relief valve. It is likely that the pressure was between 3 and 4 atm.a. from the apparent rate of pressure rise in the hour after filling. It is likely that the failure occurred through metal fatigue from thermal cycling during 23 years of use.

Subsequently, liquid nitrogen entered the insulation space, collecting at the lowest point of the outer shell. The outer shell was rapidly cooled in the area in contact with the liquid pool. Pressure in the annular insulation space changed from a vacuum to a rapidly rising positive pressure.

The combination of stress from uneven shrinkage from thermal contraction and the rapid change from external to internal pressure caused the cracks in the shell and the weld. The weld "unzipped". The positive pressure in the annular space propelled the two sections of the outer shell apart from each other.

The area of the circular cross section of the outer shell was approximately 13000 sq cm. The force separating the two sections would therefore be approximately fifteen tons for each atmosphere of positive pressure in the annular space. Nevertheless the front head had done very little damage to the cargo pallet against which it had come to rest. While the rear section had been propelled about twenty feet, it had slid along the surface of the snow on the "I" beam "runners" which were the base of its skid. This provided a rather good approximation of a sled with modest friction against the snow surface. It appeared therefore that the separation of the two sections occurred before the annular space pressure exceeded 0.3 atm.g.

## SAFETY CONCERNS

While this accident did not cause any loss to the program, other than 2200 liters of liquid nitrogen and the use of the two storage tanks, the management was reluctant to use other vacuum insulated cryogenic tankage until it was determined whether this incident was unique to this particular unit or whether a generic problem of environment, age, or other factor might suggest that similar incidents might occur. Of particular concern was the extreme low temperature of the South Pole. If this caused the accident, the use of other cryogenic tankage might have to be restricted at that location. In this case the break in the liquid line could not be attributed to a low temperature environment since the break occurred inside the insulation space, in a line intended to operate in a temperature range of 78 to 325 K. The atmosphere at South Pole station was well within this range.

It is possible that low temperature may have contributed to the crack in the outer shell and failure of the circumferential weld. Mild steel becomes brittle at temperatures below 260 K. When the pool of liquid nitrogen created internal stresses, due to uneven cooling of the head and cylindrical sections, the steel did not have sufficient ductility to deform to relieve the stress. On the other hand, portions of the shell were certainly subject to much lower temperatures from contact with liquid nitrogen, and may very well have cracked and failed irrespective of the environmental temperature.

A close inspection of all the welds on the destroyed container showed that the welds were made by hand, in sharp contrast with the machine welds of the supposed sister container. Weld penetration in the failed weld was generally poor. In spots there was less than one millimeter of effective weld. In addition the fit between the cylindrical shell and the dished head at the other end was found to be very poor. The cylinder was more than 8 mm. out of round. It became evident that the failed container had not been made to the exact same design, or by the same fabricating technique, as the other.

## CONCLUSION

Our final conclusion was that this incident was primarily the result of design and fabrication details unique to this unit and not at all typical of industry practice of the 1960's and certainly not in keeping with present day standards. It is understood throughout the industry that rupture of a liquid line in the insulation space of any vacuum insulated cryogenic liquid container will cause a serious incident. It is also well understood that use of liquid lines to fill and remove Cryogens, subjects these lines to large temperature excursions that cause shrinkage and expansion of the metal and movement of the lines. This could result in fatigue after many years of use. That the industry has generally escaped such incidents, is a tribute to its skill in design and fabrication of liquid lines and especially their connections through the outer shell.

The Antarctic program continues to utilize cryogenic storage and transport equipment without further incident. However, it is now standard practice to maintain a low pressure on all containers regardless of the design maximum working pressure. Before each trip via cargo aircraft, the pressure is reduced to near atmospheric. The intent of these measures is to reduce the likelihood and severity of any similar incident.

## ACKNOWLEDGEMENT

The author investigated this incident with the assistance of the engineering staff of the Antarctic Support Associates, prime contractor to the National Science Foundation Division of Polar Programs.

# CRYOGENIC SAFETY IN THE UNITED STATES ANTARCTIC PROGRAM

J. J. Alcorta
Antarctic Support Associates
Englewood, CO 80112 USA

## ABSTRACT

The United States Antarctic Program (USAP) has experienced an increase in the use of cryogenic fluids over the past decade. This increased use requires that the USAP purchase or lease suitable equipment, retain competent personnel, provide training to increase the knowledge base and find adequate storage for the inventory of equipment.

Safety is extremely important to the USAP because of the remote location of the work. Antarctica remains as a lonely reminder that accidents do happen and that help may not be immediately available. Communication between remote stations often relies on radio or electronic mail. The geographic location adds long flight times to travel between these stations and the rest of the general population. Most of the safety aspects of the USAP are common to any other typical research program in a typical laboratory setting. The USAP also faces unexpected safety problems not generally encountered elsewhere.

## AN ANTARCTIC OVERVIEW

The United States Antarctic Program (USAP) began a pattern of increased consumption of cryogenic fluids starting in 1987. These fluids consist of both liquid nitrogen and liquid helium. Liquid nitrogen is used at Palmer Station on the Antarctic peninsula, McMurdo Station on Ross Island and South Pole Station, Antarctica. Liquid nitrogen is also used on the research vessels' cruises of the southern oceans. Liquid helium is used almost exclusively at South Pole Station to support astrophysical research of the microwave and far-infrared wavelengths. The austral summer of 1997/98 will be the first year that liquid oxygen will be used by the program at McMurdo Station.

The USAP consumes approximately 37,000 liter of liquid helium and 15,000 liters of liquid nitrogen each year between McMurdo and South Pole Stations. Liquid nitrogen use on research vessels accounts for approximately 6,500 liters of liquid nitrogen and Palmer Station needs account for less than 5,000 liters each year.

In October, 1993, the USAP installed a liquid nitrogen plant at McMurdo Station. This plant has a capacity of 5-15 liters per hour and is operated on a continuous basis from the last week of August through the last week of April. This was in response to an accident that occurred at South Pole Station in the 1990/91 research season. A leased liquid nitrogen dewar experienced an unprecedented catastrophic failure resulting in self-destruction of the container.

In February, 1994, the USAP installed a small one liter per hour nitrogen liquefier at South Pole Station to support researcher needs of liquid nitrogen during the austral winter. This plant is used during the winter only because of it's limited capacity. Studies are being conducted to determine the feasibility of a larger plant that would meet year round needs.

## TRANSPORTATION

Antarctica has no natural helium deposits and mining is not permitted under various Antarctic treaties. All liquid helium brought to Antarctica must be flown in by military aircraft or brought in by sea. Sea transport has long been an established means of transporting vast quantities of liquid helium around the world. Only in the last two years has the USAP made greater use of this method of shipment.

Flying liquid helium or any other cryogenic fluid is much more common. This is due to the fact that most stations are unreachable by sea routes for ten months out of the year. Flying cryogenic liquids to South Pole Station is the only method available for delivery. The military aircraft used include the C-5, C-141 and C-130. Each flight requires a technical escort to accompany the cryogenic transport. The duty of the technical escort is to ensure the safety of the crew and aircraft before and during the flight.

Prior to the cryogenic transport being loaded onboard the aircraft the technical escort must check the transport, ensure that the gauges are operational, the pressure is low, that there is sufficient liquid for the users and that the transport is secured properly. An additional check for improper equipment was added last year. In a previous year, a transport was altered and valves were placed in the vent lines compromising the automatic safety system. This altered the normally safe condition of the transport to an extremely unsafe condition since the transport could not vent automatically.

The technical escort is sometimes called upon to assist in getting the transport in the aircraft. A full container of liquid nitrogen weigh 1700 kilograms and full liquid helium containers weigh 10,000 kilograms. Each aircraft is equipped with a winch but these are difficult to use when space is limited. Several people must get in the aircraft and push the container in until it is locked in place. The extreme weight of some containers can shift causing the container to roll backwards out of the aircraft. The USAP has not yet experienced this and is fortunate in this regard.

The technical escort must also connect the vent system from the transport to the off-board vent once the container is onboard the aircraft. Several lengths of flexible hose and a one-way check valve help accomplish this. There are two different styles of vents used and the escort must be familiar with each style so that appropriate adapters are available. The purpose of the vent system is to vent gas build up from the transport outside of the aircraft during flight. Not every aircraft used by the USAP has this feature.

The inner space of a C-130 is 283 cubic meters. The liquid nitrogen transports used by the USAP hold 1500 liters of liquid nitrogen, enough to inert the entire atmosphere of the aircraft in an accident scenario. The problem becomes more severe with the 12,000 liter liquid helium transports. In the event of a major failure of the transports, with concurrence of the aircrew, the technical escort may be called upon to eject the transports from the aircraft during a flight.

Occasionally, the USAP is called upon to support liquid nitrogen requirements of research vessels that have run out. This calls for a resupply mission by helicopter. A typical resupply of liquid nitrogen means flying 480 liters, in 160 liter dewars, suspended under the helicopter. The flight may take up to 1 hour to reach the ice edge where the research vessel is waiting. Prior to the flight the technician must make sure the pressure in the container is extremely low to prevent liquid being forced from the relief valves.

## WEATHER RELATED PROBLEMS

Antarctica is the highest, driest, windiest and coldest continent and thus presents a formidable challenge to cryogenic experts. The extremes in weather at McMurdo Station range from -40°C to 2°C, while at South Pole Station temperatures can reach a balmy -35°C during the austral summer. Winter temperatures can and do dip below -60°C at McMurdo Station while South Pole Station can drop below -90°C and frequently reach -100°C for days at a time.

These temperature extremes cause valve failure even though such valves were designed for cryogenic use. Seals composed of Buna-N™ or Viton™ shrink and vacuum can be lost during filling operations of a dewar. Equipment can become damaged or contaminated when left outside exposed to weather. Transportation can breakdown leaving the cryogenic technician to pull the dewar on a sled to the scientist; a distance of up to 2 kilometers. This alone is exhausting work at South Pole Station since the altitude is 3,000 meters above sea level. Cranes to haul dewars to the roof become inoperable at -80°C because most greases become solid masses at the extremely low temperatures.

Normal protective equipment, gloves, face shields or goggles and aprons become health hazards at times. Gloves are standard equipment for USAP participants and mittens are much preferred at times. Goggles can become fogged over and face shields are impossible to use outside because of the need to wear hats and ear protection for the cold weather. Most USAP cold weather gear is designed to help the user defend against the cold but does restrict mobility making movement clumsy and finesse impossible at times.

Many times, to protect major pieces of equipment, heat trace is used to defend equipment from extreme cold weather. Transports, cranes and critical areas of equipment are wrapped with heat trace and the plugged in to the electrical grid to warm the area of interest. Industrial heat guns are used to thaw areas such as joints and stuck valves to facilitate operations outside. In more remote locations a small, portable generator may be required to operate so equipment.

Sometimes weather affects USAP operations in other ways not normally thought of elsewhere. When a transport is loaded aboard the aircraft weather conditions can be excellent for flight. In the three hours to South Pole Station from McMurdo the weather can change making it difficult to land if not impossible. Flights can be delayed several days stranding a person at a station when they have tasking to complete at another station.

Winter is the most difficult time to supply liquid helium to the researcher at South Pole Station. The station closes on February 12th of each year with approximately 28 people. After that there are no flights in until the 1st of November. Liquid helium must be on station in quantities great enough to last through the end of October.

The last weather related item of concern is known as the "contrail". At -48°C and lower the exhaust from the aircraft condenses creating a visible trail and obscures the back of the aircraft. This mixed with blowing snow makes unloading large containers of cryogenic fluids very hazardous to the worker and the aircraft.

## TRAINING OF PERSONNEL

Training is primary to the USAP as is retaining highly qualified individuals. Every year a cryogenic technician is told that the dewar in the loading dock is making some kind of hissing noise. This is of great comfort to hear for several reasons but mostly because it means that the safety relief valves are working properly. However, it points out the lack of knowledge of even the simplest storage dewar.

Antarctic Support Associates (ASA) retained a knowledgeable cryogenic consultant early on. ASA took the opportunity to appoint individuals within their infrastructure to interface with this consultant and have learned a great deal regarding cryogenics. These people have set up small training classes to involve more of the station personnel. More people are capable of dealing with liquid nitrogen now than five years ago. The next step is to turn capable people into people with high comfort levels as well.

Liquid helium is still pretty much the territory of a small hand full of people between McMurdo Station and South Pole Station. Those that work with liquid helium on a more frequent basis are very comfortable with the work. Training of other personnel with liquid helium progresses much slower though.

There are precious few people that have experience with oxygen systems. The Diving Coordinator for McMurdo Station is probably the most experienced individual in compressed oxygen systems due to maintaining the high number of compressed oxygen cylinders for diving operations. People are now being put through an oxygen course provided by another professional organization. Liquid oxygen use is not expected to progress much beyond 6,000 liters and will be limited to breathing systems on aircraft.

Transportation of biological samples represents the greatest area of training for station personnel as well as USAP personnel overall. Some scientists need to transport and store samples at 77 K because of the biological activity that needs to be slowed. Equipping people with the knowledge of what can and cannot be done reasonably is primary. Ensuring that containers are stored, labeled, marked and shipped in a proper fashion is the basis of the training.

## OTHER CONCERNS

To resupply liquid helium in the most efficient manner, liquid helium is now brought in by sea in February. These transports are checked when placed on the resupply ships in the United States, New Zealand and as soon as possible during offloading of the ship at McMurdo Station. The transport can also be checked by the ship's crew while in transit. Transfilling operations are performed in a deserted area away from the general populace of McMurdo Station. This helps keep moving equipment to a minimum and helps prevent damage to containers. Proper ventilation is extremely important even when working outside with large containers. Lines are extended well away from the immediate working area. This transfilling operation can be performed by a single individual however two people are rotated in case of problems.

## CONCLUSION

The United States Antarctic Program has many of the same safety concerns as any lab setting. The unique environment presents some other safety issues that may not be encountered elsewhere. Weather, transportation and training make a big difference in how the USAP supports the needs of the researcher in Antarctica. Increased volumes makes for handling problems not seen in the lab or generally encountered in the production setting.

## ACKNOWLEDGMENT

This paper is presented by Antarctic Support Associates under the auspices of the National Science Foundation, Office of Polar Programs. Antarctic Support Associates is a private contractor to NSF-OPP.

# Advances in Cryogenic Engineering

VOLUME 43, PART B

*A Cryogenic Engineering Conference Publication*

# Advances in Cryogenic Engineering

**VOLUME 43, PART B**

Edited by

**Peter Kittel**

NASA–Ames Research Center
Moffett Field, California

PLENUM PRESS • NEW YORK and LONDON

The Library of Congress cataloged the first volume of this title as follows:

Advances in cryogenic engineering, v. 1–
New York, Cryogenic Engineering Conference; distributed
by Plenum Press, 1960–
v. illus., diagrs. 26 cm.
Vols. 1– are reprints of the Proceedings of the Cryogenic Engineering
Conference, 1954–
Editor: 1960– K. D. Timmerhaus

1. Low temperature engineering—Congresses. I. Timmerhaus, K. D.,
ed. II. Cryogenic Engineering Conference

TP490.A3 660.29368 57-35598

Proceedings of the 1997 Cryogenic Engineering Conference,
held July 28–August 1, 1997, in Portland, Oregon

ISBN 0-306-45807-1


A Division of Plenum Publishing Corporation
233 Spring Street, New York, N.Y. 10013

http://www.plenum.com



10 9 8 7 6 5 4 3 2 1



Printed in the United States of America

# ALASKA SMES: FORM AND FUNCTION FOR THE WORLD'S LARGEST MAGNET

S. F. Kral, V. Karasik, Y. M. Lvovsky, K. E. Grut, J. P. Kelley, K. D. Dixon, P. F. Ribeiro, and R. Kunz

BWX Technologies, Inc.
Naval Nuclear Fuel Division
Product Development Department
P. O. Box 785
Lynchburg, VA 24505-0785, U. S. A.

## ABSTRACT

A superconducting magnetic energy storage (SMES) system for utility applications is planned for installation at Anchorage Municipal Light & Power in 2000. The system's stored energy of 1800 MJ (0.5 MWh) will be delivered at power levels up to 31.5 MVA for spinning reserve, voltage support and electrical network stabilization.

A low aspect ratio solenoidal superconducting magnet will store and deliver energy. The magnet's maximum operating current is 10.8 kA and the peak field in the winding is 6.1 T. The maximum discharge voltage is 3.375 kV. The magnet will be cooled by pool boiling helium at 4.45 K. A review of our conductor configuration trade analysis and a description of the cryostable, copper stabilized, NbTi conductor is presented. Conductor characteristics including stability are discussed. The pancake winding pack is described. Magnet quench characteristics and protection are presented. Current lead selection criteria are presented and the leads are described. Cryostat design is described and overall magnet system thermal and structural characteristics are presented. Seismic conditions in the Anchorage area are identified and associated structural issues discussed.

Thermal loads are discussed and a description of the refrigeration system is provided. Implications due to the unavailability of $LN_2$ are discussed as well as the required recovery of the cryogenic system from discharge events.

## INTRODUCTION

In 1993, BWX Technologies (BWXT) joined with Municipal Light & Power (ML&P) of Anchorage, Alaska to build the first commercial utility-scale superconducting magnetic

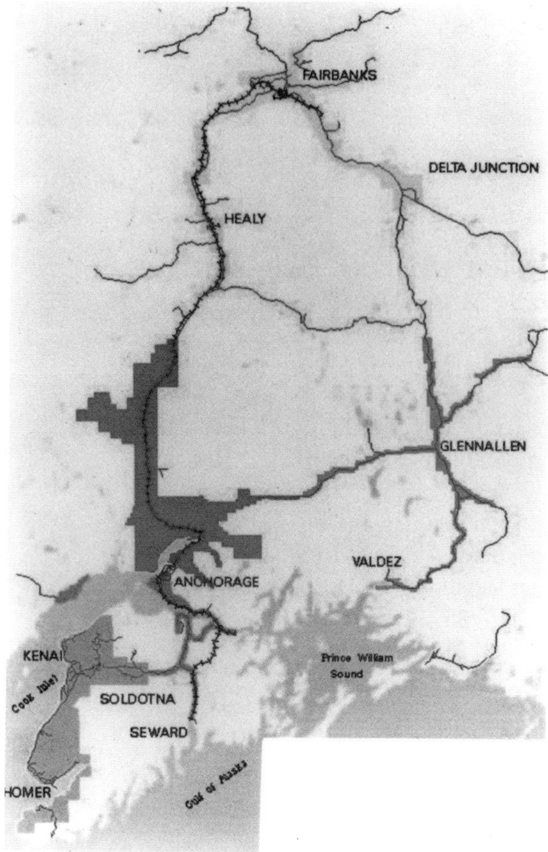


**Figure 1.** The Alaska railbelt intertie.

energy storage (SMES) system. BWXT, supported by ML&P, together with the Argonne National Laboratory, the University of Alaska, the Applied Superconductivity Center of the University of Wisconsin, Virginia State University, Norfolk State University and St. Paul's College, was awarded a Technology Reinvestment Project (TRP) grant for its project "Commercialization Demonstration of Mid-Sized Superconducting Magnetic Energy Storage Technology for Electric Utility Applications." Responsibility for the TRP program resides in the Defense Advanced Research Projects Agency of the U. S. Department of Defense. The TRP program requires industry cost sharing. Approximately two-thirds industry funds and one-third government funds support this project.

ML&P is one of seven utilities that share the Alaska railbelt intertie, a transmission system that joins load and generation facilities from Homer on the Kenai peninsula to Fairbanks at its northern terminus. Figure 1 shows this network. The intertie shares no interconnection with other utility systems and consequently is susceptible to serious disruption during unscheduled load or generation changes.

The primary ML&P SMES application enables the use of a slowly responding hydroelectric facility, Bradley Lake, as spinning reserve for system reliability improvement.[1] Spinning reserve must begin to supply power within seconds of generation contingencies because these contingencies can cause a rapid drop in power system frequency that can force undesirable underfrequency load shedding. Maximum Bradley Lake output may take as long as two minutes to achieve. The SMES system will operate so that its power output, combined with the response of other spinning reserve, will support system frequency above underfrequency load shed settings during the Bradley Lake ramp-up.

The stored energy and power output of the SMES system were established through economic and technical analyses of the spinning reserve application. Railbelt utility operating rules require ML&P to maintain, on average, 41 MW of spinning reserve capacity. About 10 MW is now carried by low-cost hydroelectric generation. The remaining spinning reserve requirement is met by spinning costly combustion turbines. Spinning reserve generation cost modeling shows that significant financial benefit from SMES delivering power up to 30 MW is realized through the reduction of combustion turbine fuel costs. Above 30 MW the additional financial benefit realized by ML&P is small. The system supplied to Alaska will provide up to 31.5 MW to accommodate characteristics of the SMES power electronics system.

Technical analyses show that 30 MW SMES power output supports system performance adequately even in the case of the most demanding intertie contingency. For example, in a "worst case" scenario the intertie system was studied under winter load conditions with system load exceeding 700 MW. Typically, this load is supplied by a combination of steam cycle, combined cycle, combustion turbine and hydroelectric generation.[2] The largest non-hydroelectric resources supplying power are two combined cycle plants. Additional power comes from many small combustion turbines; in total, more than 25 small combustion turbines operate to serve the winter load. A generation contingency, where a combined cycle combustion turbine trips off line and the heat recovery steam turbine spins down over about 45 seconds to about one-half its maximum output was studied. Loss of the combustion turbine represents a sudden 75 MW drop in available power. The combined cycle steam turbine output reduces from 50 MW to 25 MW. Without SMES, the intertie system experiences underfrequency load shed a few seconds after this contingency.

In the study assuming an installed SMES, 31.5 MW of combustion turbine spinning reserve is removed from the system. The SMES operating system is tuned to begin delivering power when the frequency falls below 59.8 Hz. The on-line combustion turbines will ramp-up quickly to maximum output as their governors sense dropping system frequency. Remaining spinning reserve assets will also respond within a few seconds. Not to be ignored is the use of SILOS (shed-in-lieu-of-spin) an alternative strategy used by some railbelt utilities. These utilities systematically shed load if frequency drop is sensed rather than maintain and pay for combustion turbine spinning reserve, sacrificing reliability for reduced cost. Between 10 MW and 20 MW of sacrificial load shedding may occur. SMES power can also be used to avoid SILOS. The SMES control system is tuned such that sufficient power is supplied so that the system frequency does not fall below a predetermined floor, about 59.1 Hz. However, the power output is controlled so that SMES energy is not expended to raise the frequency above the floor. By keeping the frequency at or near the floor, the SMES makes most efficient use of its stored energy by maintaining a condition where the governors of combustion turbines on the system will maintain their power output at maximum levels. The SMES does not have sufficient stored energy so that it alone could support the network. During the approximately 90 seconds that the SMES is able to deliver power under such conditions (albeit at constantly decreasing power levels) Bradley Lake output increases such that system frequency is restored to 60 Hz at about 2 minutes after the contingency. With appropriate control, even in this extreme case only about 1350 MJ of SMES stored energy is expended. To accommodate load growth and provide operating margin, 1800 MJ will be supplied in the ML&P magnet.

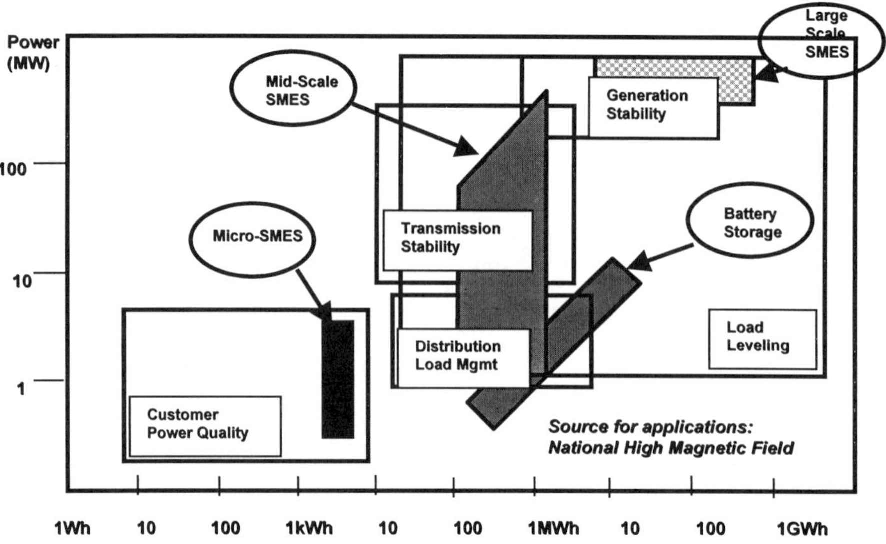


**Figure 2.** Applications of SMES. Commercialization effort focuses on Mid-scale SMES.

While ML&P and BWXT investigated the SMES system requirements, BWXT also assessed SMES as a potential commercial utility product. In the past, SMES application has focused on very large systems (perhaps 1000 MW, 3000 MWh) for diurnal load leveling and load following or very small, single circuit services. Excess generation capacity and increased economic impetus for power wheeling, including the U. S. Energy Policy Act of 1992, eliminate the possibility of economic benefit from large SMES use. However, wheeling power from excess generation introduces unplanned stresses into the power transmission and distribution (T&D) network. SMES systems that increase reliability through enhanced system stability, provide voltage compensation through reactive power service and increase power quality are effective in supporting the T&D network.[3] Figure 2 illustrates the full range of SMES applications and identifies the range of SMES energy and power appropriate to the T&D applications that are the focus of BWXT's effort.

BWXT's requirement that the ML&P SMES project fit commercialization plans established guidelines within which technical decisions were formed. These included the need for reliability consistent with utility expectations, including system availability not less than 96% and a 25-year system lifetime, and system costs that would be attractive compared to competitive technologies and consistent with the utility's expected financial benefits. Because the system would be supplied with commercial warranties, only technical and manufacturing approaches that, through operational experience or due to their inherent properties, had been demonstrated as low risk with proven reliability were permitted. Finally, anticipating an emerging market, design and manufacturing approaches that favored modular assembly and factory fabrication were required.

With increased definition of system costs and as fuller appreciation of the lack of a near-term market for SMES emerged, some design and manufacturing guidelines were relaxed. Most significant with respect to the cryogenic system was the abandonment of the requirement for SMES magnet factory fabrication. The high aspect ratio solenoid originally proposed to fit within the geometric envelope required for transportability is no longer required. In addition, the early design effort focused on an aluminum stabilized, cryostable conductor. Only with such a stabilizer could a SMES magnet with sufficient energy and low weight be factory manufactured and shipped at acceptable costs and with reliability and performance characteristics expected by utilities. With on-site fabrication, a copper stabilized conductor was chosen that can be fabricated at lower cost and less developmental effort. Because of the extensive experience with such conductors in many magnets, increased confidence of project success results.

To meet the needs of a commercial cryogenic system for utilities, guidelines were applied to the design and specification of other subsystems and components. Only refrigeration systems with extensive operating experience and components with demonstrated reliability were considered. An HTS lead development project was terminated when the hoped-for benefit of a refrigeration load reduction sufficient to permit purchase of a smaller refrigeration system with proven performance was not realized. The Alaska utilities can not support R&D projects nor will unproven technologies be accepted into regular service. No R&D is included in the project plan.

## SMES SYSTEM PERFORMANCE MODES

Three SMES system performance modes are identified that are beneficial to the utility. (Several other system operational modes serving system internal needs such as maintaining magnet charge also are defined but are not discussed here.) The first is the spinning reserve application, also called the frequency support mode. In this mode when network frequency is determined to have fallen below a threshold (probably 59.8 Hz) and the rate of frequency decrease is within utility defined bounds (probably less than 3 Hz/s) power will be supplied up to the maximum of 31.5 MVA until one of the following occurs: (1) The system frequency is restored to an acceptable level. (2) Available SMES energy is depleted to 5% of maximum storage. In this mode the system will deliver active and reactive power as determined by the system control algorithm. Because reactive power can be supplied without depleting magnet stored energy and because reactive power can also be effective in supporting system frequency, control system algorithms for this mode appropriately will husband stored energy. The rate of frequency change limitation in initiating this mode is necessary to avoid discharging the magnet in circumstance where the event affecting system frequency is so catastrophic that the SMES would not be effective.

The second performance mode, like the frequency support mode, will usually involve changes in the stored energy of the SMES magnet. This is the system stabilization or frequency stabilization mode. In power systems rapid changes of load, generation or other system properties can induce power oscillations involving the exchange of electromagnetic and mechanical energy between power network elements and generators. SMES is highly effective in dampening these oscillations. Typically, the power absorbed or delivered by the SMES system is similar to the amplitude of the oscillation. Operating in this mode in the Alaska system, the SMES will charge or discharge the magnet at power levels up to 10 MW. Control such that this power is applied as needed to stabilize the line frequency within a utility defined tolerance is provided. The control also must keep the magnet's energy within a maximum (1800 MJ) and a minimum (5% of maximum). Finally, signals from the helium refrigeration system can also terminate this mode if excess thermal load is generated.

The third mode does not involve the cryogenic system. This mode is called the voltage support or variable VAR (volt-ampere reactive) compensation mode. In this mode energy is drawn off the network by the SMES system and, through the action of the SMES power electronics subsystem, is used to synthesize a sinusoidal voltage signal of

**Table 1**. Operational Requirements of the Alaska SMES System

| Mode | Power Level | Duration | Expected Frequency | 1996 Experience |
|---|---|---|---|---|
| Spinning Reserve: Deep Discharge | 31.5 MVA | ~60 s | Monthly | 4 events |
| Spinning Reserve: SILOS Avoidance | <31.5 MVA | <60 s | Unknown | 23 events |
| VAR Compensation | ±31.5 MVAR | Continuous | Continuous | 15 ± 10 MVAR Daily Average |
| System stabilization | 10 MVA | < 20 s | Weekly | Unknown |

appropriate phase for injection into the network. Maintaining system voltage levels is a continuous task as inductive and capacitive load characteristics change constantly, giving rise to voltage changes that the SMES system can compensate.

Table 1 identifies the applications of the SMES system and also provides the expected duration of each incidence of service and the expected frequency of network need for the service. For comparison, an estimate of SMES usage based on the actual 1996 operational experience is provided. We note that the most severe demand on the cryogenic system because of the potential for magnet deep discharge, the spinning reserve application, occurs very infrequently. Indeed, the infrequency of this demand allows us to treat the magnet system as a DC system, considerably simplifying the requirements of the cryogenic system. The system is designed to exploit the infrequency of deep discharge by absorbing the thermal loads associated with discharge over a much longer time than the discharge itself. System stabilization too is infrequent. Also of interest is the expected continuous use of the system as a reactive power source in comparison to the rare use of the system as an active power source.

## SMES SYSTEM DESCRIPTION

The SMES system comprises three subsystems:

- The power conversion subsystem (PCS) which (a) converts stored energy into ac power for injection into the network, (b) absorbs ac power from the network and converts it to dc energy for storage and (c) absorbs ac power from the network, synthesizes an ac voltage of desired phase relative to the network and injects it into the network.
- The superconducting magnet subsystem (SMS) which is the SMES energy storage medium;
- The helium refrigeration subsystem (HRS) which establishes and maintains the operating environment for the SMS.

All system power electronics is included in the PCS, including the chopper. The Alaska SMES PCS uses a voltage source converter and a chopper is required. The chopper is a DC to DC converter that interfaces the magnet, a current source, with the voltage source converter.

Specification development for the SMES system and its subsystems began with definition of system power and energy requirements, as described in the Introduction. Several additional system requirements were imposed because of utility technical requirements and on the basis of commercial practice for utility systems. Subsystem specification focused on the SMS-PCS interface, specifically the interface current and voltage because of the significance of these parameters to our cost and reliability targets. Magnet current and voltage was specified for maximum utilization of the capability of the solid state devices that switch power in the PCS. Finally, the HRS specification was developed to meet the SMS requirements. Table 2 summarizes the results of this process

**Table 2.** SMES Ratings

| | |
|---|---|
| Rated energy | Not less than 1800 MJ at full charge<br>Minimum charge not less than 5% full charge |
| Rated power<br>Full four quadrant forced commutated service | Active power ±31.5 MW<br>Reactive power ±31.5 MVAR<br>Combined active and reactive power 31.5 MVA |
| Line frequency operating range | 60 ± 2 Hz<br>Operate for all rates of change < 3 Hz/s |
| Line voltage | 115 kV, 3 phase nominal at point of common coupling<br>Operating range: +10%, -20%<br>Phase unbalance < 3% |
| Magnet current | 10 kA dc maximum |
| Magnet terminal voltage | 3.375 kV dc maximum |
| Deep discharge limit | Not less than three deep discharges per month<br>Time between deep discharge greater than 24 hours |
| Operating life | 25 years |
| Operational availability | Not less than 96% |
| Operational limits | 3,000 deep charge/discharge cycles<br>5 cool down/warm up cycles<br>15 emergency discharge events |

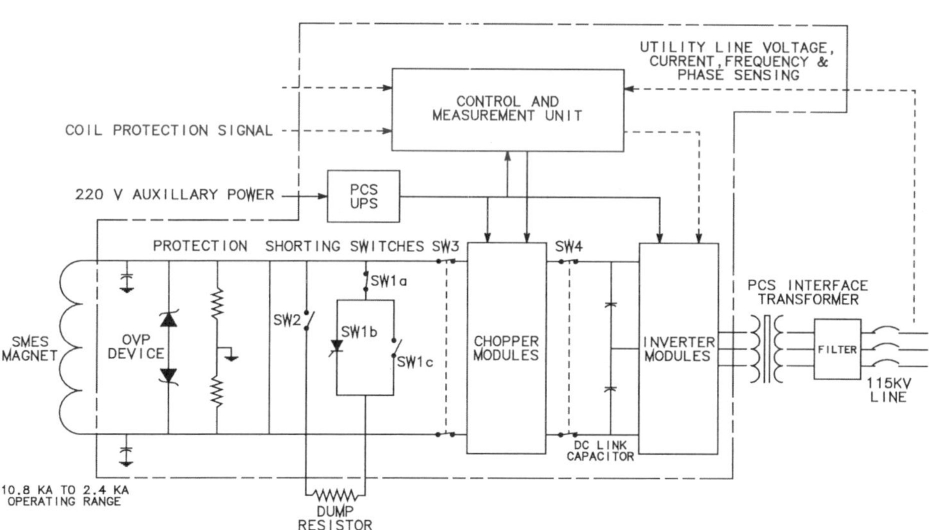


**Figure 3.** SMES schematic diagram. The switching configuration for magnet shunt, emergency discharge and safe PCS maintenance is shown. Utility network interface in Alaska is through a step-up transformer to 115 kV.

**Figure 4.** SMES site plan. The equipment building is located outside a 3 mT limit. Overall site dimensions limit magnetic field exposure outside the system boundary to less than 1mT.

Figure 3 is a schematic diagram of the SMS and PCS that illustrates the electrical circuit of the system. Of note is the bypass switching assembly, SW1a, SW1b and SW1c. These switches are used during normal operation of the system, when the magnet is charged and waiting for the call to service. SW1a and SW1c are mechanical switches that provide a lower loss coil-shunting path because of their inherently low resistance. SW1b is an electronic switch that has greater losses than the mechanical switches but also faster response. In typical operation, shunting will take place with SW1a and SW1b closed and SW1c open. This faster response, greater loss path permits the SMES system to commence power delivery to the network from the shunted mode well within the 100 ms time after a contingency that is the utility requirement. Also, the lifetime of the electronic switch will be much longer than that of the mechanical switch, reducing maintenance costs and downtime. Losses in these switches are yet to be determined. However, a typical mechanical switch resistance is a few micro-ohm while the voltage drop across solid state switches commonly used by utilities is about 2 V.

Figure 4 is a site plan that shows buildings for the SMS, and HRS and PCS. The SMS building will be constructed of non-magnetic materials. Climate control will be provided for all buildings. The site is a zone 5 seismic region and system specifications include protection against the affect of earthquakes. The site is in the outskirts of Anchorage in a reserve area with no nearby residential housing. Magnet fringe fields reduce to 1 mT at approximately 30 m from the magnet and at this site such a fringe field is acceptable. The magnet will be wound in the SMS building.

## SUPERCONDUCTING MAGNET SUBSYSTEM

The superconducting magnet subsystem (SMS) is a free standing cylindrical coil mounted on a thermally isolating, seismic resistant support structure. The coil is a series connected assembly comprised of 38 single pancake windings using an unconditionally stabilized conductor. The coil is cooled by pool boiling helium and the cold mass is enclosed in an annular cryostat. The coil general specification are given in Table 3.

**Table 3.** Superconducting Coil General Specifications

| | |
|---|---|
| Stored magnetic energy | 1800 MJ |
| Peak magnetic field | 6.1 T |
| Operating current | 10.8 kA |
| Normal operating temperature | 4.45 K |
| Number of single pancakes | 38 |
| Number of turns per pancake | 50 |
| Coil inside diameter | 6.7 m |
| Coil outside diameter | 8.44 m |
| Coil height | 2.44 m |
| Coil weight | 258,000 kg |
| Conductor length | 45,163 m |

## SMS Conductor Description

The conductor configuration and dimensions are shown in Figures 5a and 5b. The conductor includes a superconducting Rutherford cable, which is fabricated of NbTi filaments within a copper matrix. The cable critical current at 6.1 T and 4.45 K is 20 kA. ($I_{op}$ / $I_c$ = 0.54). This cable is soldered into a longitudinal groove on one side of the conductor's massive copper stabilizer. A second, symmetrically located stabilizer groove provides an open surface to improve stabilizer heat transfer to the liquid helium coolant. The copper stabilizer is also a coil structural member supporting the hoop and axial stresses of the magnet's Lorentz load. Fishbone spacers (see Figure 5b), providing open space for liquid helium flow separate adjacent turns. The spacers are manufactured from G10 glass-epoxy composite.

## Conductor Structural Analysis

Table 4 gives the design criteria for the copper stabilizer and other winding structural members. Because conductor turns are supported only intermittently in both the radial and

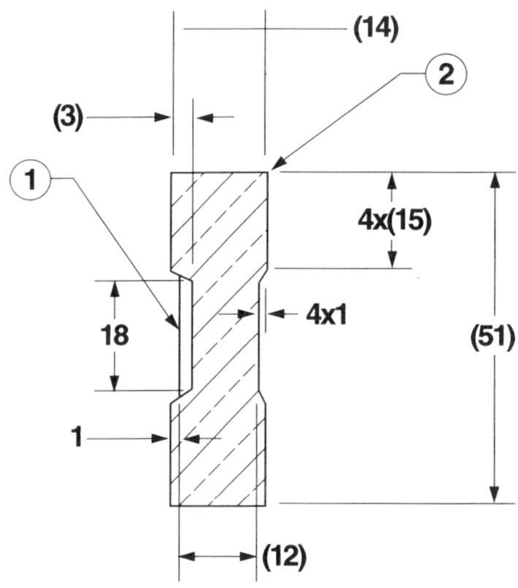


**Figure 5a.** SMES conductor cross section. 1-Rutherford cable. 2-Copper stabilizer. All dimensions in mm.

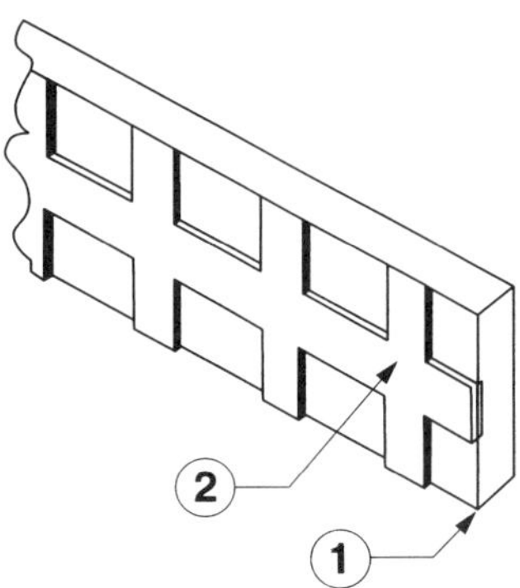


**Figure 5b.** SMES conductor turn assembly. 1. Copper stabilized conductor. 2. Radial spacer ( fishbone ).

**Table 4.** Structural design criteria

| Stress Type | Allowable Stress |
|---|---|
| Primary Membrane Stress, Sm | < 2/3 minimum yield<br>< 1/2 ultimate strength |
| Primary Bending Stress, $S_b$ | < minimum yield<br>< ½ minimum ultimate strength |
| Membrane, $S_m + S_b$ | $S_m + 2/3\ S_b$ < minimum yield<br>$S_m + 2/3\ S_b$ < ½ minimum ultimate strength |
| Peak Stress | < minimum ultimate strength |
| Cyclic Fatigue Allowable | < 4x operating cycles |

the axial directions, a three dimensional stress state consisting of bending stresses and cross sectional distortions must be examined in detail in order to establish stress levels and confirm design margins. Because of the large coil cross section and number of conductor turns, a correlation from a global structural analysis to the local conductor level was required. Furthermore, the turn-to-turn interaction is not uniform throughout the magnet pack and therefore an accurate method of predicting the boundary conditions acting on the individual conductors was also required. Also, because of the SMES coil's low aspect ratio, the Lorentz force direction is significantly different at discrete coil cross section locations and the assumptions for a long slender solenoid do not apply.

A finite element sub-modeling technique was developed for establishing details of the conductor stress based upon the global magnetic field distribution and the conductor turn-to turn interaction. The global magnetic field distribution and general displacements are estimated using a coupled magnetic/structural 2D axisymmetric finite element model. The field distribution is extracted and applied to a structural model of a single pancake. This model uses contact elements that simulate the interface between adjacent conductor turns. These contact element results, in turn, are applied to a detailed finite element model of the individual conductor turns.

The maximum magnetic pressure occurs at the coil inside diameter mid-plane. The magnetic field reverses at approximately the 35$^{th}$ turn and the Lorentz force at the coil outside diameter acts inward, tending to pull the Rutherford cable away from the stabilizer groove. Also, the turns at the coil extremes have the maximum axial Lorentz force acting on them. Finally, at approximately the center of the coil cross section, the cumulative axial and radial forces are a maximum and result in the maximum compressive stress acting on a conductor turn. Table 5 provides a summary of the preliminary structural analysis of stresses for the coil section. The table also identifies design margins for two typical conditions of copper work hardening. The measure of work hardening used is the reduction in cross section during extrusion. The preliminary analysis concentrated mainly

**Table 5.** SMES conductor design margins for different conditions of work hardening

| Stress category | Design value | Design allowable @ (20% Cold Work) | $S_{allowable}/S_{design}$ | Design Allowable @ (30% Cold Work) | $S_{allowable}/S_{design}$ |
|---|---|---|---|---|---|
| Membrane Stress ($\sigma_m$) | 197 MPa | 204 Mpa | 1.038 | 226 MPa | 1.149 |
| Mem.+Bend Stress ($\sigma_{m+b}$) | 205 MPa | 306 Mpa | 1.49 | 339 MPa | 1.65 |
| Local Stress | 301 MPa | 306 Mpa | 1.018 | 339 MPa | 1.127 |

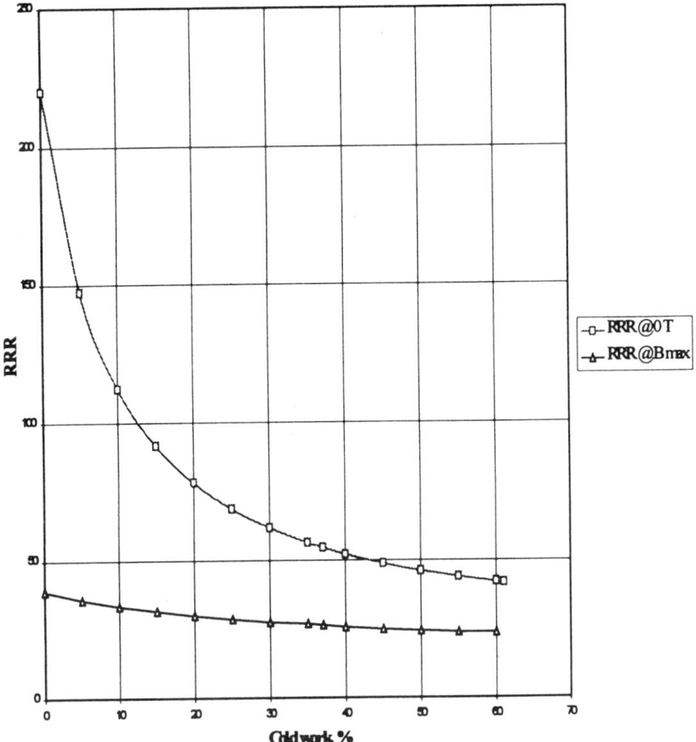


**Figure 6**. Copper RRR ($R_{300K}/R_{4.2K}$) as a function of cold work and magnetic field.

on the stresses at the coil middle plane because these turns experience the maximum loading conditions.

## Conductor Stability Analysis

The conductor stability analysis was performed in two steps. First, a parametrical analysis was done to determine the minimum stabilizer cross-section needed to guarantee unconditional cryostability with a 20% margin. Then a detailed stability calculation was completed for the chosen configuration with fixed parameters. The following approach and assumptions apply to the stability analysis:

- An unconditional steady-state cryostability requirement $Q_{min} > W_{norm}$ was applied. Q is heat transfer from the stabilizer surface, cooled by liquid helium, and W is heat generation inside the stabilizer.
- A design requirement with an explicit 20% margin on the $Q_{min}/W_{norm}$ ratio was set.
- Heat transfer calculations were based on test data obtained previously from the National High Magnet Field Laboratory for aluminum vertical and horizontal smooth surfaces in the channels.
- No credit was taken for the heat transferred from the conductor side facing downward.
- Conservative correlations for the electrical resistance and heat transfer were used.

Calculation of the copper resistivity at the maximum magnetic field included degradation due to cold work hardening and magnetoresistance.[4] Cyclic strain degradation of copper electrical resistance is negligible in the elastic strain range under consideration. The initial RRR of copper was assumed to be 210. Figure 6 shows the final $RRR_f$ of cold worked copper in the magnetic field. The minimum stabilizer area needed to obtain a 20% stability margin was calculated as a function of the cold work.

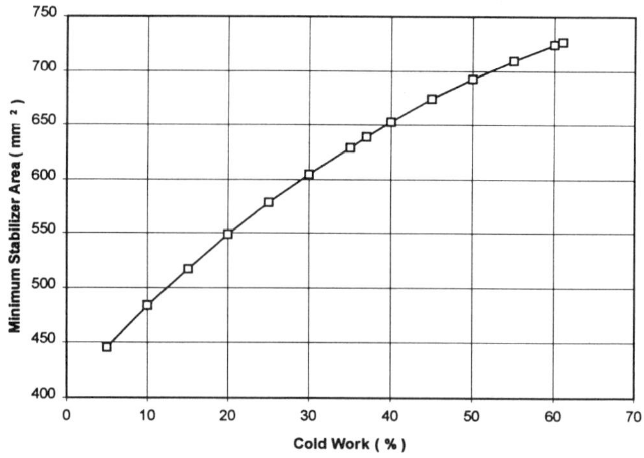


**Figure 7.** Minimum copper stabilizer area as a function of the cold work.

The radial spacers were assumed to cover 21.6% of the stabilizer area and the axial spacers were assumed to cover 29% of the area. The groove channels were assumed totally uncovered. Based on these results, and on the results of structural analysis, a manufacturing requirement that the cold work of the SMES conductor be bounded between 20% and 37% was established, the latter corresponding to the least stability but the best structural performance.

A stabilizer cross section of 657 $mm^2$ was used for further analysis. The structural criteria limit occurs for this cross section at about 20% cold work. This area can not be reduced further either from the stability or the structural standpoint.

The refined stability calculations were performed for copper parameters of initial RRR of 210; RRR after cold work of 58; final RRR at magnetic field 6 T of 28; stabilizer height of 51.25 mm; and stabilizer width of 14.25 mm. Figure 8 presents the conductor cooling versus heating curves. A linear transition in the boiling curves is assumed. The recovery current was calculated using both linear and conservative transition regions in the boiling

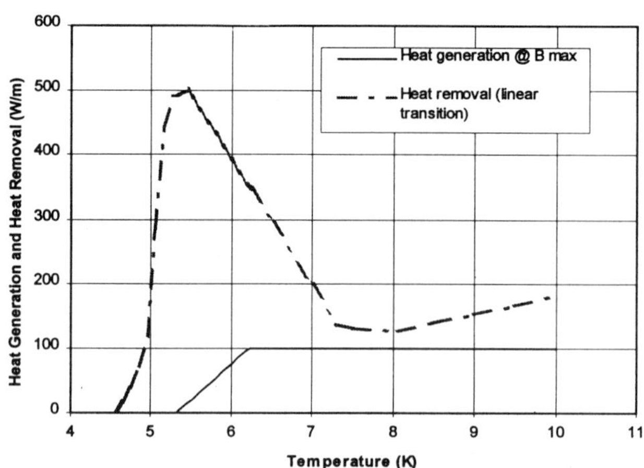


**Figure 8.** Heat generation and heat removal : copper stabilizer in magnetic field 6 T .

curves, using the Equal Area Theorem criterion. This calculation shows sufficient margin with respect to the cold-end recovery. For the linear approximation $I_{EAT\ linear}$ = 16700 A. For the conservative approximation $I_{EAT\ conservative}$ = 13900 A. In the first case the recovery current is 1.55 times greater than the SMES operating current; in the second case it is 1.29 times greater. The conductor unconditional cryostability is assured.

## Electrical Insulation

Helium liquid and possibly vapor provides electrical insulation within the coil. The dielectric strength of liquid helium is larger than the dielectric strength of helium vapor. Therefore, our conservative insulation analysis used the dielectric properties of 20 K helium vapor.

The SMS normally operates with an external shunting switch closed across its terminals. Except during charge and discharge, no voltage is applied to the coil. At the maximum discharge rate during deep discharge and emergency discharge, the normal maximum SMS terminal voltage is 3.375 kV.

During charge and discharge, with the shorting switch open, voltage transients can be applied to the coil terminals, including ripple produced by the PCS chopper and surges from the network passed through the PCS. These higher frequency disturbances could excite coil resonances whose frequencies are determined by the distributed capacitance and inductance of the coil. The SMS transient over-voltage is limited against potential disturbances from the PCS at 4.215 kV, 25% over the normal operating voltage limit. The following devices provide voltage limiting:

- Harmonic filters at the front end of PCS filter out the majority of network switching and lightning surges;
- Semiconductor over-voltage protection (OVP) devices at coil terminals clamp any let-through or other fast surges;
- Low pass filters at coil input filter out other transients.

Criteria chosen for design enhancement factors are typical for large superconducting magnets and include:

- Transient let-through enhanced with 25% margin ⇒ 4.2 kV
- Withstand voltage with margin of 1.5 times ⇒ 6.3 kV
- Peaking factor 3.2 greater than withstand( lower than turn to turn ) ⇒ 20 kV
- Safety factor of 2 applied to all margins ⇒ designed at 40 kV.

Electric field concentrations at conductor edges and corners lead to higher fields than average. An enhancement factor of 4 was established as a preliminary design criteria for turn to turn and pancake to pancake voltages. With an enhancement factor for normal operation of 4, the transient enhancement factor becomes 3.2.

The possibility of flashover and surface tracking was taken into consideration. In flashover an avalanche of electrons or ions bombarding an insulation surface causes breakdown. These charges are always present because of ionizing radiation and the large electric field gradient at the insulation's surface is conducive to bombardment. Experiments indicate that flashover can happen at half of the breakdown voltage. For preliminary design, to minimize the possibility of flashover, all surface to surface gaps were enlarged by a factor of 2. Surface tracking is caused by carbonization of contaminated or degraded insulator surfaces through leakage currents that eventually lead to flash-over or breakdown. Surface tracking is a slow and gradual process.

The insulation capability of the insulators was compared to the established criteria after the structural requirements of the insulators were defined and after helium flow and heat transfer concerns were addressed. In every case, the insulation system sized using these methods exceeded criteria.

### Turn to turn insulation design

The normal expected turn to turn operating voltage is 1.78 V. The turn to turn insulation capability is 85 V in consideration of possible breakdown and transients. Electric insulation between turns is provided by 1.19 mm thick G10 fishbone shaped sheets. In addition, the solid G10 material is resistant to punch through.

### Pancake to pancake insulation design

The normal expected pancake to pancake operating voltage is 89 V. The pancake to pancake insulation design criterion is established at 8.5 kV in consideration of possible breakdown and transients. G10 spacer bars of 5.56 mm thickness and 63.5 mm width are placed between pancakes in the radial direction. (Structurally, the bars are also sufficiently wide to take large axial load in the coil mid-plane ( 140 MPa ).) Finally, we note that the tracking field is less than 80 V/mm.

### Coil to ground insulation design

The center of the coil is grounded and the maximum normal operating voltage experienced is 1687.5 kV. In the case of a ground fault this voltage could be as large as 3.375 kV.

Coil to ground insulation is provided by appropriate electric insulation of coil supporting structure, cross-beams, axial tie bars, bus, current leads and instrumentation

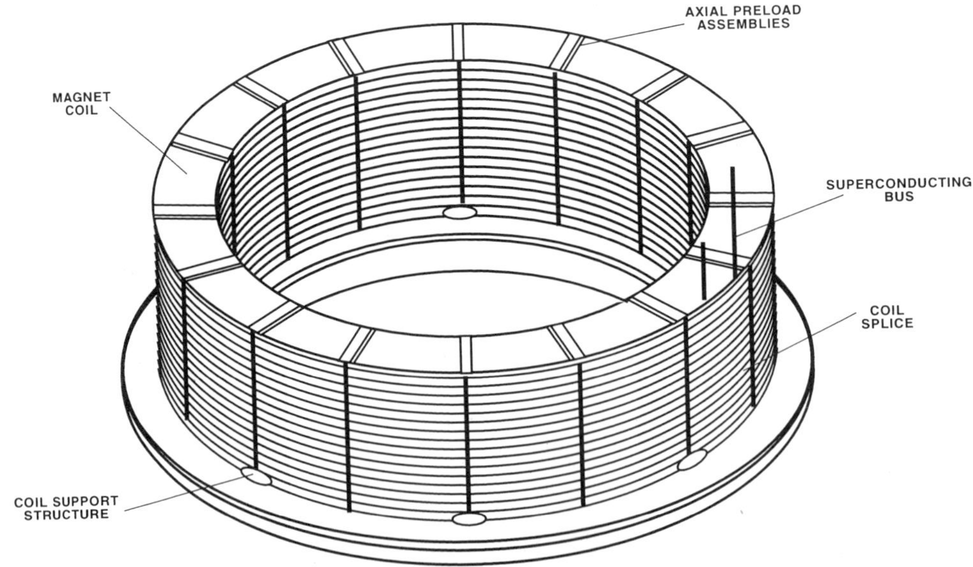


**Figure 9.** SMES magnet coil assembly.

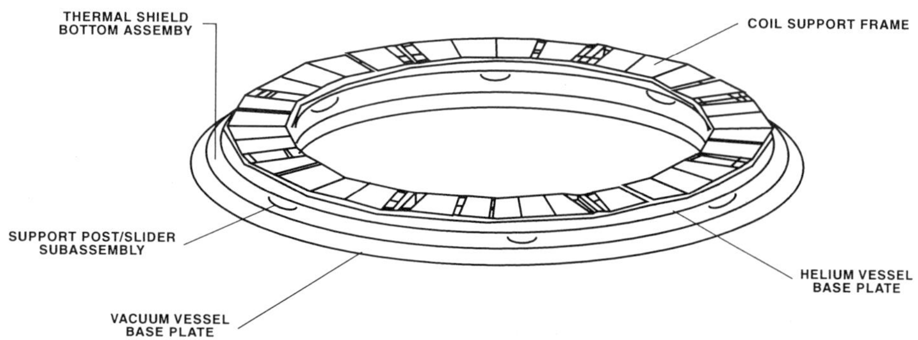


**Figure 10.** Coil support base .

leads. Also a clean manufacturing environment during coil assembly and in process testing will ensure sustained margins.

## Coil and Cold Mass Design

The SMES coil assembly is shown in Figure 9. The magnet comprises 38 single pancakes assembled on a support base. Axial spacers separate the pancakes mechanically and electrically. Pancakes are restrained axially by tie rods. They are connected electrically in series with splices and to the power leads through a bus.

Figure 10 shows the support structure for the coil assembly, the helium vessel and the thermal shield for combined gravitational, thermal differential, and seismic loading conditions. The composite support posts in the load path minimize heat transfer from the foundation to the cold mass. A radial sliding interface accommodates relative thermal motion between the cold mass at 4.5 K and the foundation at 300 K. A segmented box structure distributes the coil support load and provides bottom attachment points for axial tie rods. A total of 16 polymer support posts of high compressive strength and low thermal conductivity support the cold mass at eight locations around the coil. The posts are stacked three high, with a thermal shield intercept. Pretensioned tie rods inside and outside the coil at sixteen locations provide sufficient compressive normal force between the pancakes to prevent relative motion due the to lateral seismic accelerations or magnetic restoring forces acting on an out-of-round or decentered pancake. G10 spacers of thickness 31.75 mm will insulate the coil from the base support structure and provide passages for helium flow to the base of the coil.

A superconducting bus connects the conductor ends at the bottom and top of the coil to the power lead lugs. The superconducting bus uses the same conductor as the coil and is encased in G10 insulation. The bottom end of the bus is supported at the coil splices, and the top end at the bus-to-power lead splice. The Lorentz loads on the longer bus from the bottom of the coil are reacted with G10 stand-offs that are attached to an adjacent tie rod.

## Helium vessel design

To minimize eddy current heating in the structure, the helium vessel is as thin as can be safely achievable. Figure 11 shows the welded stainless steel annular vessel. The vessel has flat top and bottom heads and it is supported from below by the cold mass support

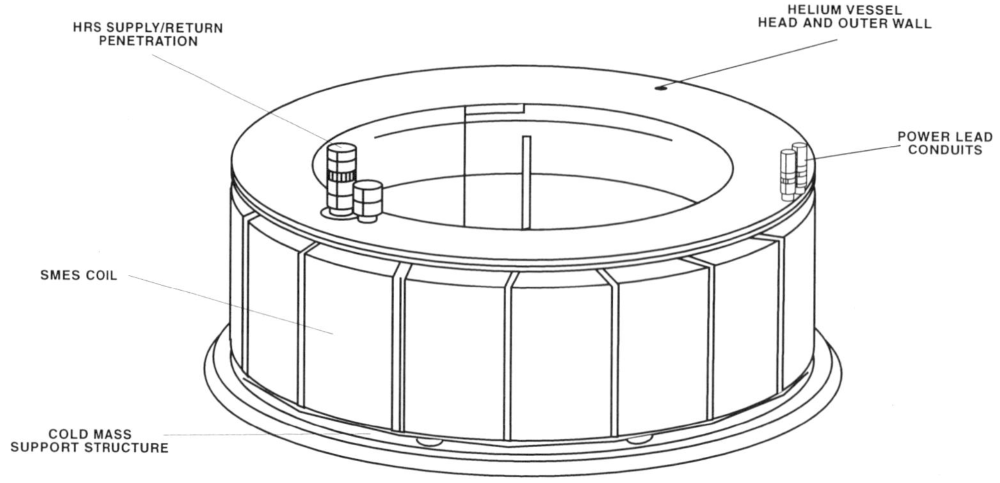


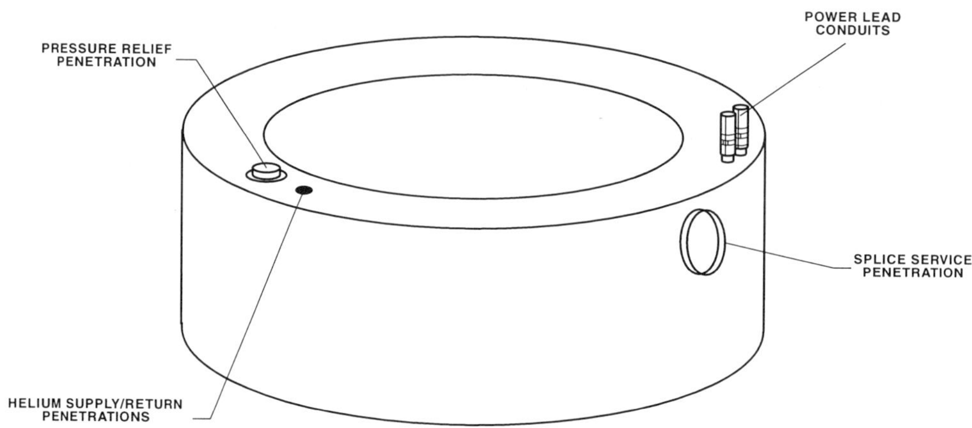


**Figure 11.** Helium vessel : above outer cylinder removed; below complete vessel.

**Table 6.** Design requirements for the helium vessel

| | |
|---|---|
| Design pressure (maximum allowable pressure) | 2 bar |
| Nominal operating pressure | 1.4 bar |
| Operating temperature | 4.45 K |
| Seismic loads | vertical 1 +/ - 0.5 g, lateral 0.7 g |
| Liquid helium volume (total) | 35,000 liters |
| Liquid helium above coil | 14,000 liters |
| Allowable stress | 2/3 yield stress |

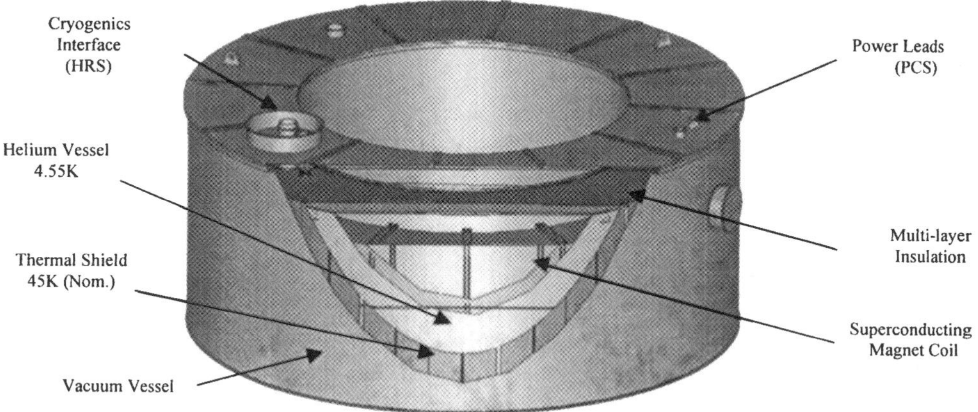


**Figure 12.** SMES magnet cryostat.

structure. The helium vessel has penetrations for the current leads, helium supply and return, and for instrumentation leads. Preliminary analysis indicates an outer helium vessel wall thickness of 10 mm. The inner wall of the vessel requires a thicker section (19 mm) since it is treated as an externally loaded cylinder and is potentially subject to buckling. The thickness of bottom head is 43 mm, and that of the upper head is 25 mm. Table 6 describes the operating conditions and requirements for design of the helium vessel.

## Cryostat design

Figure 12 shows a cut-away view of the cryostat: the annular stainless steel helium vessel, containing the cold mass; the thermal shield; and the multi-layer insulation.

The thermal shield is a thin-wall segmented aluminum cylinder supported by a network of tubing to minimize eddy current heating. Each segment is made of 3 mm thick aluminum sheet onto which tubing for the helium coolant is brazed. The tubes are welded to headers at the top and bottom ends at final assembly. The thermal shield is supported by the helium vessel with stainless steel struts that are attached to the frame at the middle of the thermal shield skirt. Eddy currents induced in the shield during magnet discharge interact with the magnetic field to produce a Lorentz load normal to the horizontal mid-plane of shield segment. This load is reacted in part by the shield curvature and in part by the tubing assembly.

Multi-layer insulation will be fabricated as blankets that are supported by the thermal shield within the 15 cm space between the shield and the vacuum vessel.

The vacuum vessel has 19 mm thick cylindrical walls with top and bottom flat heads. The bottom head plate thickness is 50.8 mm, which supports the weight of cold mass and cryostat internal. The top head is 19 mm thick and is reinforced by ribs and flanges at the inside and outside diameters.

The top end of the cryostat contains interconnections to the bus, cryogen transfer line, and instrumentation leads, as well as pressure relief and vacuum pump-out ports. A vacuum level of $10^{-9}$ bar will be maintained within the cryostat by getters sized to absorb leaks up to $10^{-9}$ bar cc/sec during 25 years of magnet life. Additionally, the cold mass will operate as a cryopump.

## Magnet protection and instrumentation

The magnet protection system must detect conditions that could cause a safety hazard to personnel or damage to the magnet. Magnetic energy must dissipate safely if:

- The conductor quenches;
- The liquid helium level falls below the top of the magnet coil;
- Cryostat vacuum is lost;
- Current lead liquid helium flow is lost;
- Magnet protection system power is lost.

The quench detection system must be capable of distinguishing between three types of normal zones:

- A stagnant normal zone, which is characterized by increasing length and a temperature rise in the conductor that could damage the magnet;
- A steady normal zone, which develops in a limited area with a peak temperature below the damage threshold due to heat transfer along the conductor. One source for such a zone would be the absence of a thermal and mechanical connection between a short length of cable and the stabilizer;
- A resistive zone that recovers after the causal disturbance disappears.

The quench detection system (QDS) should prevent any conductor temperature rise above 200 K. Temperature increase calculations were performed to determine for how long the QDS could wait before initiating dumping of magnet stored energy in the dump resistor. The maximum allowable hot spot temperature was taken as 200 K. The calculation give a conservative estimate of the maximum delay time, using an adiabatic approximation with no heat transfer to the helium, adjacent turn or along the conductor. The maximum delay time was found to be 296 seconds.

The primary method of detecting a stagnant normal zone employs voltage taps. Due to the large inductance of the magnet, the inductive voltage across the coil can be large, up to 3.375 kV, compared to the resistive voltage that must be detected. Digital inductive voltage compensation will be used. Inductive voltages measured during magnet testing will be stored and compared during normal operation to identify anomalies accompanying abnormal behavior.

Flow meters measuring magnet boil-off will provide a secondary quench detection system. A 30 mV resistive voltage generated by a stagnant zone can give rise 324 W of additional thermal load and increased boil-off from the helium vessel. An increased flow of up to about 6 g/s through the return line of helium refrigeration system is possible. The flow meters will be magnetic field insensitive, with an accuracy of at least 1% of the calibrated range.

Instrumentation will be incorporated into the magnet to monitor thermal, structural and electrical characteristics of the coil. The magnet's operating history will be stored. Temperature sensors will be installed at locations on the thermal shield, warm-to-cold supports, helium inlet and outlet lines, and on the winding pack to monitor the thermal behavior of the magnet system during cool down, warm up and normal operation.

Warm-to-cold structural supports will be equipped with strain gauges as will the other locations of maximum stress in the coil. Two type of liquid helium level sensors will be provided: continuous-reading superconducting probes, and discrete carbon resistors.

**Table 7.** Emergency discharge resistor specifications

| | |
|---|---|
| Energy capacity | 1,800 MJ |
| Maximum current | 10,800 A |
| Resistance | 0.31 Ohms |
| Maximum voltage during discharge | 3.375 kV |
| Discharge time (3 time constant) | 307 sec |
| Maximum rate of heat transfer to water | 0.5 $MW/m^2$ |
| Water temperature increase | 50 C |
| Water volume | 9 $m^3$ |
| Temperature gradient during discharge | 15 K |
| Stainless steel resistivity | 0.75 $\mu\Omega$-m |
| Mass of assembled plates | 97.6 kg |
| Overall plate length | 72.1m |
| Number of plates | 40 |
| Plate height | 0.914 m |

### Emergency discharge resistor

The emergency discharge (dump) resistor is composed of stainless steel plates welded into a non-inductive sinuous circuit. To withstand the magnetic forces, insulating spacer rods supports the plates. The resistor is installed within a plastic vessel filled with purified water. High voltage feed-throughs mounted on the can cover are connected to the resistor by copper current leads. The dump resistor is provided with a water purification system and sensors for monitoring the water level and water conductivity.

Specifications and design characteristics of the dump resistor are presented in Table 7.

## HELIUM REFRIGERATION SUBSYSTEM

Siting the helium refrigeration subsystem (HRS) in Anchorage, Alaska presents demanding requirements. In addition to meeting the static heat load and liquefaction needs of the magnet, it must handle large dynamic loads. Due to the large magnetic fringe field, the HRS must be installed a significant distance from the magnet, leading to additional costs and heat loads from unusually extended transfer lines. Availability balanced against system costs are primary considerations, and finally, system reliability is critical since utility technicians will handle most maintenance activities. The following describes an HRS that meets these needs.

### Requirements

The HRS is a mixed mode system designed to provide both refrigeration and liquefaction at 4.45 K. Conductor stability permits an operating range up to 4.55 K. A process flow diagram for the HRS is given as Figure 13. Table 8 presents the static 4.45 K and shield heat loads for both the magnet and the ancillary equipment (transfer lines, dewar, etc.). The major components of the HRS are located outside the 30 Gauss line of the magnet, 31.7 m (104 ft.) from magnet centerline, as is indicated in Figure 4, dictating long transfer lines with their associated heat leak. This is reflected in Table 8, which shows that the 4.45 K ancillary equipment's static heat loads are more than twice that of the magnet's. The shield loads for the ancillary equipment are also significant, being ~ 40% of

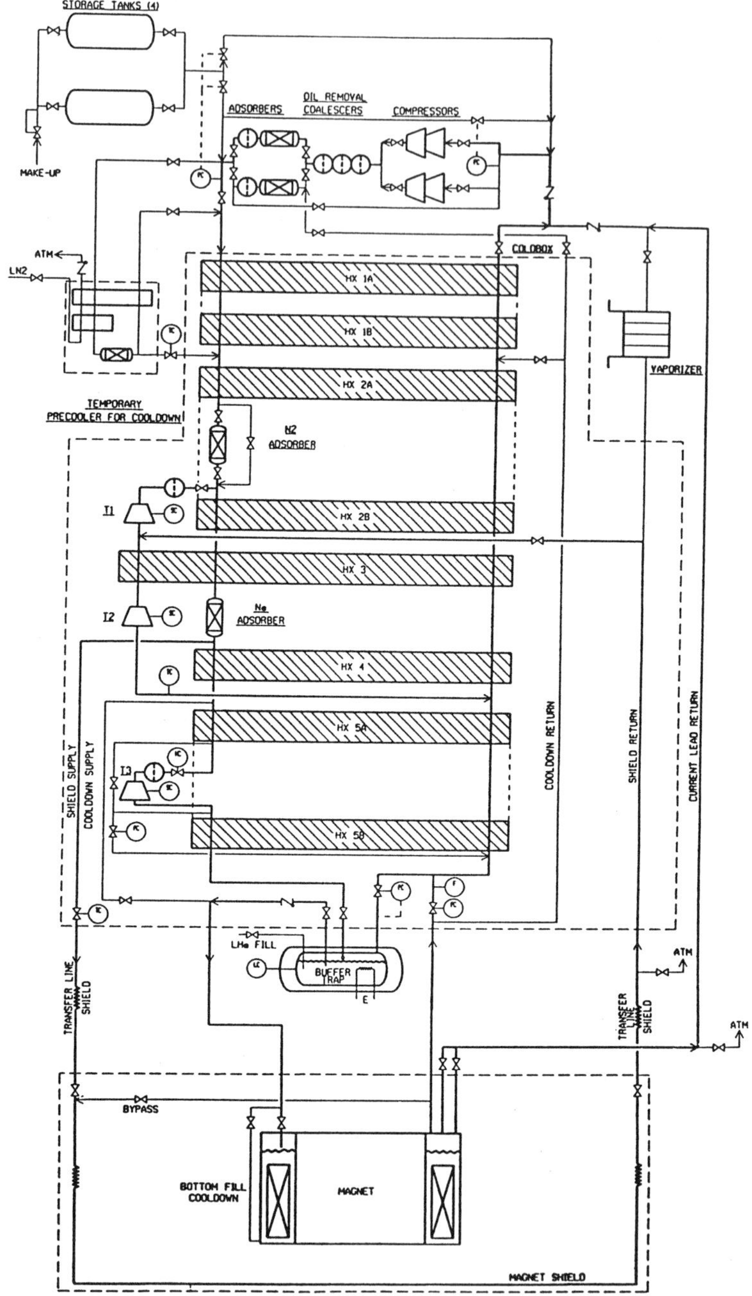


**Figure 13.** Helium refrigeration system process flow diagram.

**Table 8.** SMES static heat and liquefaction loads (with margins)

| Component | Load (W) |
|---|---|
| **4.45 K Magnet Static Loads** | |
| Posts | 3.40 |
| Penetrations | 4.16 |
| Splices | 7.35 |
| Radiation | 11.55 |
| Residual Gas Conduction | 16.50 |
| **Subtotal** | **43.0** |
| | |
| **Ancillary Equipment 4.45 K Loads** | **95.5** |
| | |
| ***Total HRS 4.45 K Static Heat Load*** | ***138.5 W*** |
| | |
| **Magnet Liquefaction @ 4.45 K** | **g/s** |
| Current Leads | 1.44 |
| Downtime Make-up | 0.46 |
| **Subtotal** | **1.90** |
| | |
| ***Total Liquefaction @ 4.45 K*** | ***1.9 g/s*** |
| | |
| **Magnet Shield (45 K) Loads** | **Watts** |
| Posts | 66 |
| Penetrations | 124 |
| Radiation/Gas Conduction | 460 |
| **Subtotal Total** | **650** |
| | |
| **Ancillary Equipment Shield Loads** | **275** |
| | |
| ***Total Shield Heat Loads*** | ***925 W*** |

those for the magnet. Dynamic heat loads (Table 9), when averaged, are similar in magnitude to the static magnet loads. Because of the large ancillary heat loads, which are either well understood or correctable, the potential impact of "underestimating" magnet loads is reduced. Total heat and liquefaction loads are given in Table 10. These are the "design" loads for the HRS.

Liquefaction loads (Table 8) are also significant, a consequence of current lead demands at 10,800 A and HRS downtime make-up requirements. Liquefaction make-up requirements are based on the availability demonstrated by the HERA refrigeration system after 3 years of operations.[5] Eliminating load based complaints, that system was off-line an average of 2.61 hours/week or 136 hours/year. It is recognized that the HERA system consists of multiple identical components and is maintained by dedicated personnel, while utility technicians with other responsibilities will maintain the HRS. Therefore, to meet availability requirements, skids with a mean time to repair of greater than 9 hours are made redundant and sufficient liquid helium inventory will be stored in the magnet to permit all SMES operations, including one deep discharge, without HRS provided make-up for an 18 hour period.

Refrigerator availability is driven more by cryogen logistics than SMES demands. Helium is unavailable in Anchorage and must be imported from the "lower 48" states. Rail or plane can deliver small quantities, but large quantities are only deliverable by barge. All transport modes can be adversely impacted by weather, and large unplanned needs have a

**Table 9.** SMES magnet dynamic (ac loss) heat loads to liquid helium (with margins)

| | Frequency | Energy Dissipated | Real Time Load | Ave. Heat Load |
|---|---|---|---|---|
| Deep Discharge | 1/Month | 3,520 kJ in 60 sec. | 58,670 W | 1.36 W |
| Recharge after Deep Discharge | 1/Month | 552 kJ/533 sec. (Est.) | 979 W | 0.21 W |
| Idling/VAR | Continuous | | 14.7 W (Est.) | 14.7 W |
| Combined Modes | 2/Day | 872 kJ/1,800 sec. (Est.) | 484 W | 20.2 W |
| **Total** | | | | **36.5 W** |

minimum delivery time of 2 weeks. Additionally, on-site warm storage capacity is expensive. When full, the magnet's helium inventory is ~31,500 L of LHe and 3,500 L of cold GHe (21,000 L is required to cover the coils). Cost constraints dictate that only a portion of this inventory can be stored as a warm pressurized gas. Liquid nitrogen ($LN_2$), though available locally in limited quantities, is expensive and reliable delivery is questionable. For these reasons, it will only be used during clean-up and cooldown operations. There are numerous helium refrigerator systems world-wide that operate without $LN_2$ precooling, and so this is not a technical challenge.[6,7,8]

## Description

A process flow diagram of the HRS is given in Figure 13. A baseline description of the major subsystems of the HRS is as follows. The compressors are redundant, compound, oil lubricated, rotary screw compressors (Mycom 2520 MSC with 500 H.P. motors) sharing redundant 90 gpm, 15 H.P., oil pumps. Under design conditions, each compressor will provide 83.2 g/s between 1.04 and 13.2 bar. Oil removal consists of a bulk oil separator with 2 coalescers to reduce the oil content to < 1 ppm (wt.), with a third "guard" coalescer. Redundant charcoal beds, each designed for a 5 year lifetime, have a goal of reducing helium stream oil content to 1 ppb (wt.) During HRS start-up and magnet commissioning, one charcoal bed will be charged with molecular sieve material and plumbed to compressor suction to perform the dryer function.[9] After commissioning, this bed will be isolated, and the molecular sieve removed and replaced with charcoal. Sufficient molecular sieve material will be inventoried on site for use in the event of inadvertent water introduction into the HRS. With this provision, since the system is closed and current plans are to provide liquid make-up (yearly helium losses to power outages, seal leaks, and bed regeneration are estimated to be < 1,000 L LHe), there is no need for a permanent dryer on-site. Oil and helium are air cooled. The outdoors installed, air cooled, heat exchangers are located to permit gravity draining of oil to the sump, located in-doors, if the compressor goes off-line to minimize the possibility of oil freezing at low environmental temperatures. During operations, switch-over between compressors, pumps and charcoal beds will be performed monthly. Compressor switch-over can be performed in 15-30 minutes.[10]

The cold box is a standard Linde Kryotechnik TCF-50, based on a modified Claude cycle, with minor alterations for shield cooling and the addition of a "wet" turbine. The high pressure and "wet" turbines are TGL-16s, the low pressure turbine is a TGL-22. A JT valve is mounted in parallel with the "wet" turbine for use during cooldown or reduced load conditions. Heat exchangers are aluminum plate finned, mounted vertically. Two charcoal adsorbers, one for nitrogen and one neon, are included. The nitrogen adsorber includes a bypass to permit regeneration during operations. Regenerating the neon adsorber requires a slight warming of the cold box. Connections are provided on the high

**Table 10.** Total heat and liquefaction loads (with margin)

| | |
|---|---|
| 4.45 K Static Heat Load and Dynamic Loads | 175 W |
| 4.45 K Liquefaction Loads | 1.9 g/s |
| Shield Heat Load | 925 W |

pressure pass below the neon bed for shield flow supply, which is returned downstream of the high pressure turbine. Work extracted by the turbines will be dissipated and the subsequent heat will be removed by one of two redundant chiller units.

Due to large magnet cold mass (~230,000 Kg), a temporary helium precooler using $LN_2$ will be installed for HRS start-up and magnet commissioning. The unit, which includes a nitrogen boiler, a 3 pass gas heat exchanger and a charcoal bed, is designed to cool and purify 30 g/s of room temperature helium to 80 K. Bed regeneration is accomplished by warming the vessel and purging with helium. The precooler can be removed and reused to commission future systems, or left in place to boost HRS capacity if required.

A 3,785 L (1,000 gal.) buffer dewar is provided. This dewar, when used for liquid filling of the system, serves as a contamination trap. Additionally, it is intended to commission the cold box and compressor systems prior to the completion of magnet assemble. This permits operator training, rectification of start-up failures, and elimination of these components as variables for consideration when troubleshooting during magnet/transfer line commissioning. The dewar is instrumented to verify cold box performance. Finally, the dewar plays a critical role during system recovery from a deep discharge. Table 8 notes that the real time heat load of a deep discharge is ~ 59,000 W over 60 seconds. The boiloff from this event can not be handled cost effectively in real time, nor do utility specifications require such treatment. However, the mass of liquid in the magnet is such that, although the pressure will exceed 1.342 atm (the saturation pressure corresponding to the conductor stability temperature limits of 4.55 K), the pressure does not approach the helium vessel's maximum allowable pressure of 2 atm. Since the primary purpose of the SMS is to handle deep discharge, and since this event is anticipated to happen on a monthly bases, a 24 hour recovery period is permitted for the HRS to return the magnet to operational conditions. The excess gas generated is returned to the cold box in a regulated manner, at flow rates that greatly exceed normal operating conditions. Turbine flows, which are controlled by monitoring discharge temperatures and pressures, are shutdown thus configuring the system as an Unbalanced JT Cycle. The buffer dewar is isolated from the magnet, and collects the liquefied helium. The system is maintained in this configuration until magnet pressure is returned to 1.23 atm.[11] The system is then reconfigured into the basic "design" mode (with turbines) and the collected liquid is discharged to the magnet by pressurizing the dewar with an internal heater. Once the magnet is filled, it can be returned to operations. While 24 hours is allowed for deep discharge recovery, under some conditions recovery can be accomplished in less than 2 hours.[12]

There will be 452 $m^3$ of warm gas storage available (4 - 30,000 gallon tanks). This is only sufficient to store the equivalent of ~ 8,400 L of 4.45 K LHe. During normal operations, the HRS will maintain the magnet's liquid volume nominally between 29,700 and 31,500 L ($\Delta V$ = 1,800 L). Due to the high density of the cold helium boiloff remaining in the magnet, due to the fact that a significant portion of the ancillary heat load at 4.45 K is removed sensibly, and due to the fact that liquid volume increases during a deep discharge, there is sufficient warm gas storage to handle the boiloff for a minimum of 18 hours of normal operations without HRS providing make-up to the magnet.

## Operating Modes

Initially pumping and purging with Grade A helium, and then circulating helium with a warm compressor accomplish magnet clean-up. A portion of the high pressure helium is purified in the precooling unit and then mixed with the remainder of the flow prior to entering the cold box. The turbines and nitrogen adsorber are bypassed and the flow provided either directly to the bottom of the magnet or, through the dewar to the magnet. Flow is returned to the compressor through the charcoal bed filled with molecular sieve material. A small portion of this flow can be directed to the shield and current leads in parallel.

Cooldown from 300 to 80 K proceeds similarly. Flow is directed to the precooler, cooled and purified and split. A portion is warmed to 300 K and directed to the cold box where it is mixed with remaining cold gas to maintain a temperature difference across the magnet of 50 +0/-5 K. Flow is then directed to the bottom of the magnet directly or through the dewar. Flow is returned to the compressor system directly or, as the magnet cools, through the high temperature heat exchangers. The shield is not actively cooled initially, but is allowed to cool by conduction and radiation exchange with the helium vessel. Active cooling of the shield is initiated when its temperature reaches 120 K. For 80 to 5 K cooldown, turbines T1 and T2 are activated and flow through the precooler reduced. Turbine T3 is bypassed and flow is directed to the bottom of the magnet, or through the dewar to the magnet, when it is then returned to compressor suction through the appropriate connection in the low pressure return circuit of the cold box.

For normal operations, T3 is activated and helium is liquefied into the dewar, which is maintained at ~ 1.34 atm. Liquid is then directed from the dewar to the top of the magnet. Boiloff from the dewar and a portion of that of the magnet is returned to the cold box low pressure pass, while the remainder of the magnet's boiloff is routed to the current leads where it is warmed to 300 K and returned to compressor suction. Helium gas is provided to the shield nominally at 5 g/s and 26 K, and returned to the turbine interstage at ~ 60 K.

While it is desired that the magnet never be warmed, if necessary, warm-up can be accomplished with the same configurations as cooldown. The precooler is not required but would be helpful for removing frozen out contaminates that are liberated during warm-up. A leased high pressure compression system could be used to recover the helium in tube trailers.

In the event of a cold box related failure, normal magnet operations can be continued for up to at least 18 hours. Eighteen hours is sufficient time to deryne a heat exchanger, change a valve actuator, or accomplish any number of minor repairs. The buffer dewar is isolated from the magnet and boiloff is routed both through the current leads and through the shield to compressor suction, where it is compressed and stored. If deemed necessary, a fraction of this flow can be routed to compressor suction through the return transfer line to minimize line warm-up. For power outage or compressor failure, boiloff is again routed through the leads and shield where it is then vented to atmosphere.

## CURRENT LEADS CONCEPTUAL DESIGN

The 10.8 kA rated current leads will be mounted toward the outer edge of the vacuum vessel top plate assembly and will connect the superconducting bus within the magnet to the ambient-temperature dc bus leading to the PCS. The heat conducted through the leads due to the temperature difference between the internal and external environments and due to Joule heating will be largely reduced by counter-flowing boil-off helium gas. These

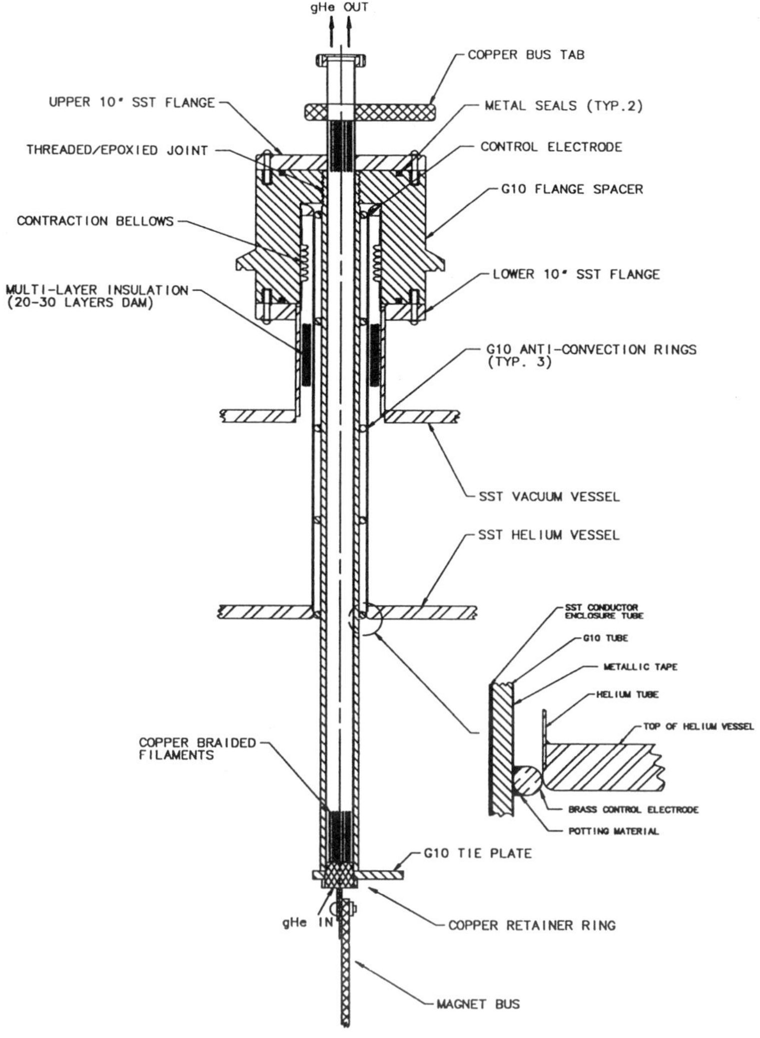


**Figure 14.** SMES power lead design .

leads must provide safe, reliable and efficient performance with minimal required maintenance.

## Vapor Cooled Lead

Vapor cooled leads of a proven design will be purchased from a commercial source. The leads will be modified to meet the high voltage, structural, and thermal requirements particular to the SMS.

The commercial vapor-cooled lead includes perforated solid copper end pieces bridged by copper braided filaments enclosed within a thin walled stainless steel tube. The filaments provide a large heat transfer area and a proprietary internal restraint prevents them from bunching together due to the Lorentz forces. The enclosure tube is thin-walled to reduce heat leak into the helium vessel. Although the guaranteed heat leak to current ratio is 1.13 W/kA, these leads reportedly perform closer to 1.04 W/kA[13,14], the ideal value for materials obeying the Wiedemann-Franz Law.

## High Voltage Insulation

During a ground fault, the leads could experience a 4.2 kV surge potential. They are conservatively designed for a breakdown voltage of 30 kV, which is derived from safety, testing, and local enhancement multipliers as described earlier. The leads, including high voltage components, will be tested for breakdown and flashover to nearly 10 kV (2 X operating voltage + 1 kV) under gaseous helium conditions to confirm the design.

A NEMA grade G10CR tube and G10 flange assembly electrically insulates the high voltage stainless steel conductor tube and the upper flange from the grounded stainless steel helium vessel tube assembly and lower vacuum vessel flange (see Figure 14). These dielectric materials are chosen for their availability and documented thermal and electrical characteristics at cryogenic temperatures. An electrical strength of 2.5 kV/mm, an industry standard for glass epoxy laminates, accounts for cracks and pin-holes that are inherent in these materials. Dielectric material thicknesses have been chosen to both eliminate flashover and to reduce electrical fields to acceptable levels.

The G10CR tube will be threaded and epoxied to the G10 flange spacer to promote good structural integrity and to prevent voltage breakdown across the interface between the G10 components. This insulating tube surrounds the stainless steel conductor enclosure tube and extends from the flange spacer to the G10 tie plate at the lead bottom. Strips of metallic foil bonded to the inner and outer surfaces of the G10 tube will ensure an equipotential; thus eliminating charge build ups along the length and diameter. Control electrodes that will be potted to the exterior of the insulator tube, will be placed at both the upper and lower helium tube rounded edges to minimize local field enhancements[15]. Similarly shaped G10 rings will be placed between the electrodes to break up significant convection currents.

The G10 flange spacer insulates the upper to the lower steel flanges. A protruding lip on the exterior of this spacer precludes any condensed water vapor near the top flange from bridging to the grounded, lower flange. A design incorporating screws was chosen over a flanged spacer to reduce the shorting potential between the metallic fasteners given a maximum manufacturing thickness of this spacer.

## Structural and Mechanical Considerations

The magnetically induced forces will be taken up in the lead at three points: the vacuum vessel outer tube weldment, the helium vessel shell, and the tie plate that connects the bottom of the two current leads. Lead components have been evaluated for these forces and also those due to flange bolting, seismic activity, thermal effects, and other reactions due to the warm and cold buses that are not taken up by there respective support systems.

According to a TOSCA model of the SMES coil, there will be a 1.29 T- 2.75 T field gradient across the leads in the vertical direction. The field is nearly perpendicular to the direction of the current and will give rise to a total force and moment of approximately 22 kN and 16 kN-m acting on each lead. The additional repulsive force due to the field generated by the current leads themselves is negligible when compared to the above forces. The loading on both the warm and cold temperature buses due to the field and other effects will be taken up by their own support systems and not by the leads.

The concentric tube assembly consists of two tube sub-assemblies joined to a top ring within the upper portion of the G10 flange spacer (see Figure 14). This design is chosen to reduce the heat leak to the upper potion of the lead where a large thermal gradient exists. Both the thin-walled inner (helium) and upper outer tubes provide a large thermal resistance to the cold helium environment and will guide the lead assembly into the vessel.

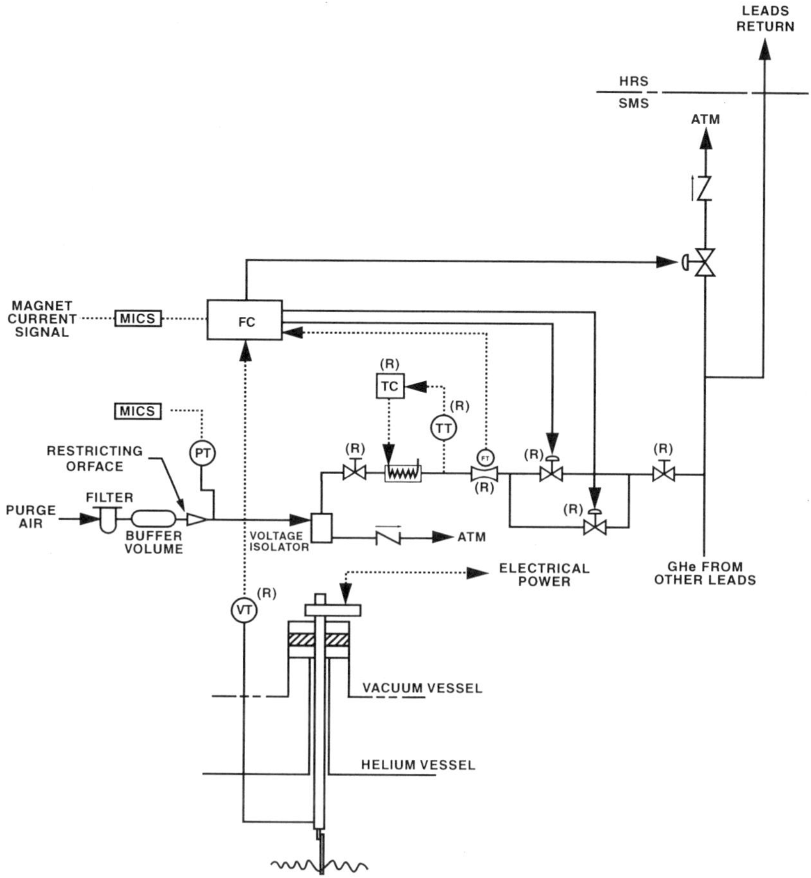


**Figure 15.** Power lead control schematic.

A welded bellows in the upper portion of this outer tube compensates for the differential contraction between the tubes. The thicker lower outer tube, which attaches the vacuum vessel to the lower stainless steel flange, bears most of the weight of the entire lead assembly and resists loading due to Lorentz forces.

Helium and vacuum environments will be contained within the current lead assembly by weldments, braze joints and mechanical seals. Brazing will be used to join dissimilar metals or those having large differences in thickness. Spring-loaded C-seals can endure large off-normal temperature drops and will limit the helium leakage to levels at or below those required for the helium vessel for the lifetime of the system without requiring replacement.

## Thermal Performance

The cross sectional area to length ratio of the lead will be thermally evaluated using a finite difference method given by Wesche and Fuchs[16]. Preliminary results are in good agreement using design charts given by Alff [17]and Locke[18]. This analytical method can be used to make slight alterations to account for the estimated effects of magnetic field and boundary temperatures different than those assumed by the supplier.

A liquefaction load of 0.72 g/s has been budgeted for each lead. This should account for the contribution of heat due to the surrounding support and insulating tubes (<1 Watt), the uncertainty of the magnetic field contribution, and an added margin of safety against burn-out. Since the liquefaction load is sensitive to the temperature of the upper copper bus tab, it will assume the warm bus temperature that will be maintained close to ambient. The thermal performance of the leads will be checked with liquid helium prior to installation.

**Flow Control**

The leads are self sustaining; i.e., the amount of heat that flows down the leads is nearly the exact amount required to boil enough helium gas off to optimally cool them. This works well in an open bath of helium; however, these leads are tied to a refrigerator having a fluctuating suction pressure. Therefore, the lead flow will be artificially controlled with the use of flow and resistance controlled valves as shown in Figure 15. This approach will work well with these vapor cooled leads since the flow impedance through them is a small fraction of the total gas return impedance to the compressor suction. The resistance across each lead is an excellent indicator of the axial temperature profile and it will govern the amount of flow that will be regulated using integral flow controllers or a combination of venturi flowmeters and control valves. The resistance signals will be determined by the voltage drop of each lead coupled with the current signal available from the PCS. Resistance heaters coupled with temperature sensors will ensure that the helium flowing from the leads into the flow controlling devices will be maintained within the proper operating range. A voltage isolator, surrounded by dry air, will eliminate the high voltage from the control piping and HRS system.

**Protection**

Voltage taps across the leads and at the bottom splice connections provide input to determine the resistance signal and are part of the primary protection for the leads and coil. The tap wires will be of the same type used for the magnet coil taps for the quench protection system. A pair of voltage tap wires (one redundant), enclosed in capillary tubing, will be soldered at the bottom copper lug and will run along the braided conductor strands within the conductor tube enclosure. They will join a pair of upper tap wires at the upper copper lug (surrounded by the copper bus tab) and will emerge out of a teed high voltage feedthrough connection at the top of the assembly. Inductive voltage noise is minimized by this routing and the use of twisted, shielded pairs at the near the feedthrough.

During normal operations, the resistance signal will control the flow through each lead that is expected to be between 0.36 to 0.72 g/s. Flow control bypass valves are set to open if the resistance across any of the leads exceeds 10 $\mu\Omega$ or if the flow meters indicate less than the minimum flow. If the resistance continues to rise or if the refrigeration system is down, the HRS control system will vent the helium return gas to atmosphere.

If the resistance continues to increase after these measures are taken or if the flow to the current leads is totally stopped, then the magnet protection system will dump the magnet current. A preliminary MIIts calculation indicates that the ratio of heat capacity to resistivity is large enough to withstand such a scenario without any damage to the leads; even if the time lag for initiation of the dump is several seconds.

## SUMMARY

A SMES system that meets utility requirements for commercial products has been presented. Applications that guide SMES system performance and design characteristics have been described. These include expectations for the magnet of rare deep discharge, infrequent shallow discharge and occasional shallow, oscillatory charge and discharge. Continuous reactive power exchange with the network is expected, but the cryogenic system does not participate in this service. To meet these requirements, and to meet utility demands of reliable operation for a 25 year system lifetime, a conservatively designed, low cost, superconducting magnet will be supplied. The central feature of the magnet is a cryostable, copper stabilized, NbTi conductor. The refrigeration subsystem contains only components with documented reliability. Also, redundant skids are provided when the mean time to repair is greater than 9 hours. Although no significant new technology is included in the SMES subsystems, the integrated system represents a challenging step forward in the application of superconductivity and power electronics.

The SMES system presented in this paper is appropriate to the infrequent needs of the Alaska electric power system. SMES systems to meet the high power, rapid discharge requirements called for in some power system stabilization and defense applications will generate significantly larger thermal loads during use. In addition, much more frequent use is expected. These high power, rapid discharge SMES systems will require conductors that minimize ac losses and refrigeration systems capable of absorbing considerably larger real time thermal loads. We have established guidelines for such systems and are beginning to define design criteria for subsystems and components. For us, the Alaska system provides invaluable experience in advancing utility and defense SMES technology to the next step.

## ACKNOWLEDGMENTS

This paper was prepared with the support of the U. S. Department of Energy, Grant DE-FC36-94G010011. However, any opinions, findings, conclusions, or recommendations expressed herein are those of the authors and do not necessarily reflect the views of the DOE.

The authors acknowledge the significant contributions of J. Waynert, X. Huang, M. Xu, G. Lehman and Moe Aslam to the project.

The authors also acknowledge the significant contributions to this work of J. Jurgen Clausen and Monte Schwartz of Lotepro, and of Armin Senn and Andres Kuendig of Linde Kryotechnik AG.

## REFERENCES

1. Spinning reserve refers to the utility practice of maintaining generators in an operational state so that they are prepared to deliver power to make up for the unanticipated loss of other generators. Spinning reserve resources are costly because of fuel, maintenance and other costs.
2. Combined cycle plants consist of a series combination of combustion and steam turbines. The thermal energy in the exhaust gases of the combustion turbine drives the heat recovery steam turbine.
3. Recent advances in high power electronic devices make possible power electronics systems that can provide similar services. The combination of power electronics and the T&D network is sometimes called the Flexible AC Transmission Systems (FACTS). The IEEE includes SMES as one potential system in the FACTS ensemble. SMES is the most sophisticated of these systems in that it is the only system with significant stored energy and active power capability.

4. N.J. Simon, E.S. Drexler and R.P. Reed. "Properties of Copper and Copper Alloys at Cryogenic Temperatures," NIST Monograph 177, U.S. Department of Commerce (1992).
5. H. Berg, M. Clausen, H. Herzog, G. Horlitz, H. Lierl, Report on operational experience and reliability of the HERA cryogenic system, in: "Advances in Cryogenic Engineering, Vol. 39A," Plenum Press, New York (1994), p. 75.
6. G.M. Gistau and M. Bonneton, A 6kW at 4.5 K helium refrigerator for CERN's cryogenic test station, in: "Advances in Cryogenic Engineering, Vol. 39A," Plenum Press, New York (1994), p. 547.
7. W.K. Erdt, P. Frandsen, D. Güsewell, K. Kurtcuoglu, K. Löhlein, A.E. Senn, N.O. Solheim, U. Wagner, G. Winkler, B. Ziegler, Performance of an efficient 6 kW helium refrigerator for the LEP 200 project at CERN, in: "Advances in Cryogenic Engineering, Vol. 39A," Plenum Press, New York (1994), p. 553.
8. J. Clausen, A. Patzelt, A. Stephan, M. Wanner, The Linde turborefrigerator for MR-tomographs, in: "Advances in Cryogenic Engineering, Vol. 39A," Plenum Press, NY (1990), p. 949.
9. J.J. Clausen, Lotepro, private conversations, 10/31/95.
10. Mohammad Gauhar, EDTI, private conversations.
11. A.E. Senn and A. Kuendig, Linde Kryotechnik, private conversations, 12/4/96.
12. J.P. Kelley, J.J. Clausen, A.E. Senn, A. Kuendig, Deep discharge recovery with the AML&P SMES helium refrigeration system, to be presented at the 1997 Cryogenic Engineering Conference.
13. M. Wilson, "Superconducting Magnets," Oxford, New York (1983).
14. K. Efferson, Helium vapor cooled current leads, *Rev.Sci.Inst.*, 38:1776 (1967).
15. S.W. Schwenterly, private communication.
16. R. Wesche and A.M. Fuchs, Design of superconducting current leads, *Cryogenics*, 33:714 (1994).
17. J.J. Aliff, Practical charts for designing gas-cooled current leads, in: "Proceedings of MT-9," SIN, Switzerland (1985), p. 800.
18. J.M. Lock, Optimization of current leads into a cryostat, *Cryogenics*, 9:438 (1969).

# RESEARCH OF A NEW COIL DESIGN FOR SUPER-CONDUCTING MAGNETIC ENERGY STORAGE

A. Mailfert, M. Takorabet, A.Colteu

GREEN INPL
Groupe de Recherches en Electrotechnique et Electronique de Nancy
Institut National Polytechnique de Lorraine
2 av. Forêt de Haye 54516 Vandoeuvre les Nancy FRANCE
Email : amailfer@ensem.u-nancy.fr

## ABSTRACT

Some superconducting devices like Superconducting Magnetic Energy Storage (SMES) need a small external stray field and sometimes a constant internal field (MRI). Both problems can be approached using special coil configurations. Basic forms like torus, solenoid, and alternate solenoids with parallel axis, are well known. The active shielding method is based on the cancelation of the lower order magnetic moments of a current loop. We propose two designs that handle both problems. One configuration is composed by two spherical concentric distributions of current where a special diameters ratio is determinated in order to realise good conditions for the superconducting windings. The other one is a polygonal structure. The technological approach of these two theoretical designs is considered. The study will focus on energy, flux density and geometrical data.

## INTRODUCTION

One question associated with storing magnetic energy with superconducting windings resides in the value of the external magnetic stray field that must be as low as possible, (the value proposed by NIH in USA is 0.5mT), [1,2,3,4]. Some aspects must be taken into considerations [5,6,7,8,9,10,11]: The confinement of the stray field reduces the eddy current losses appearing in a conducting cryostat.

Usual coil configurations include solenoid, torus and system of solenoids with parallel axes :

- For a simple solenoid, the higher the ratio of its radius to the length the greater is stray field[5]. High passive shield field solenoids need a great quantity of iron ; solenoids with active shielding, (concentric solenoids) present the most attractive shape.
- A torus is a configuration that theoretically produces no external magnetic field, but it is not easy to wind it. For these reasons it is constructed by putting together a discrete set of coils with the drawback of allowing stray field losses. Moreover it is expensive considering the quantity of wire required.
- The system of solenoids with parallel axes [2,3]., have a small magnetic stray field : the larger the number of solenoids, the lower the magnetic stray field. Such systems require much more

**Table 1.** Comparison between configurations [2]

| | | |
|---|---|---|
| Energy 11MJ | solenoid alone | 10086 Gauss at 1.5m |
| Energy 15MJ | bundle of solenoids | 2166 Gauss at 1.5m |

superconducting wire. Beyond a certain energy threshold of about 5MJ, the electromagnetic confinement tends to be less efficient. A comparison of the magnetic stray field is given for two configurations (simple solenoid and a system of four parallel solenoids)[2], Table1.

This paper proposes new designs of two low stray field superconducting coils : a spherical configuration and a polygonal configuration.

## THE SPHERICAL CONFIGURATION

An ideal spherical coil with a sinusoidal distribution of current creates a constant internal field and an external dipolar field vanishing at infinity. A system of two concentric spherical coils with sinusoidal distribution of well-chosen currents cancels the external field, keeping a constant field inside the internal coil and a non uniform field between the two coils. There is no particular problem with making a spherical cryostat. The difficulty resides in producing the spherical distribution of the current. So we propose a technological approach to the theoretical distributions by using several coils or solenoids, Figure1b.

### Analytical Study

A spherical frame of reference ( $\vec{u}_r \vec{u}_\theta \vec{u}_\varphi$ ) is considered. Figure 1a represents the three domains with different magnetic permeabilities delimited by the two spheres of a and b radii (a < b) : the internal domain (I), the domain between the two coils (II), and the external domain (III). The superficial density of the current on each sphere « i » is given by $\vec{K}_i = K_i \sin\theta\, \vec{u}_\phi, i = 1,2$.

The magnetic vector potential $\vec{A} = A(r,\theta)\vec{u}_\varphi$ that satisfies the Laplace equation and the boundary conditions will be :

$$
\begin{aligned}
A_1 &= C_1 r \sin\theta & r < a \\
A_2 &= \left[ C_2 r + \frac{C_3}{r^2} \right] \sin\theta & a < r < b \qquad (1) \\
A_3 &= \frac{C_4}{r^2} \sin\theta & r > b
\end{aligned}
$$

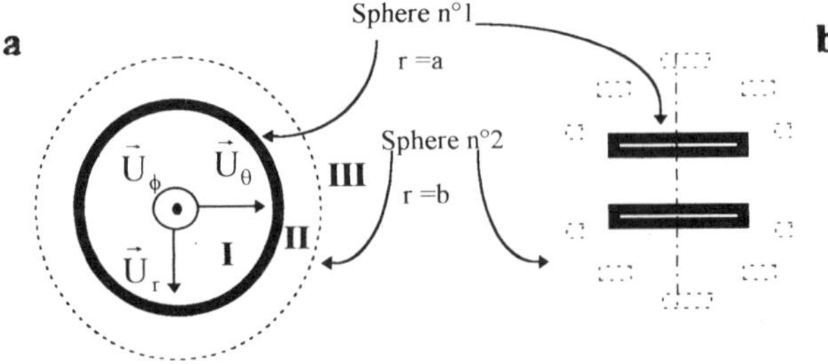


**Figure 1.** Suggested coil configuration for the solenoid approach.

**Table 2.** Example of a spherical system with $\mu_0$ internal core ($B_{max}$ on wire=5T)

| a [m] | b[m] | $K_1$[A/m] | $K_2$[A/m] | $B_{in} \in D_1$[T] | $V_{tot}[m^3]$ | Energy[J] | Capacity [J/m³] |
|---|---|---|---|---|---|---|---|
| 0.298 | 0.476 | 0.795 $10^7$ | -0.195 $10^7$ | 5 | 0.452 | 2.21 $10^6$ | 4.9 $10^6$ |

The boundary conditions

$$\begin{aligned} &A_1 = A_2 \\ &-\frac{1}{\mu_1}\frac{\partial}{\partial r}(rA_1) + \frac{1}{\mu_2}\frac{\partial}{\partial r}(rA_2) = r\,K_1 \sin\vartheta \qquad \text{for } r = a \\ &A_2 = A_3 \\ &-\frac{1}{\mu_2}\frac{\partial}{\partial r}(rA_2) + \frac{1}{\mu_3}\frac{\partial}{\partial r}(rA_3) = r\,K_2 \sin\vartheta \qquad \text{for } r = b \end{aligned} \tag{2}$$

allows to determine the coefficients $C_1$ to $C_4$.
In order to cancel the external magnetic field, $K_1$ and $K_2$ must be chosen in such a way that $C_4$=0. Thus, for $\mu_1 = \mu$ and $\mu_2 = \mu_3 = \mu_0$ we obtain the relation

$$\frac{K_1}{K_2} = \frac{-(2\,a^3(\mu - \mu_0) + b^3(2\mu_0 + \mu\,)}{3\,a^3\mu} \tag{3}$$

In the case where all permeabilities are $\mu_0$, the condition is :

$$a^3K_1 + b^3K_2 = 0 \tag{4}$$

The magnetic energy of this system can easily be calculated by :

$$E = \frac{1}{2}\left[\int_{S_1} \vec{A}_1\vec{K}_1 ds_1 + \int_{S_2} \vec{A}_2\vec{K} ds_2\right] \quad \text{where } S_i \text{ is the spherical surface « i ».} \tag{5}$$

**Results**

The system of two spheres behaves like a dipole without stray field outside the external sphere, Figure 2. The values of flux density are equal to zero too after this border, and are constant in the internal sphere, Figure 3. The maximum value of the flux density is located in and on the internal sphere and on the windings.

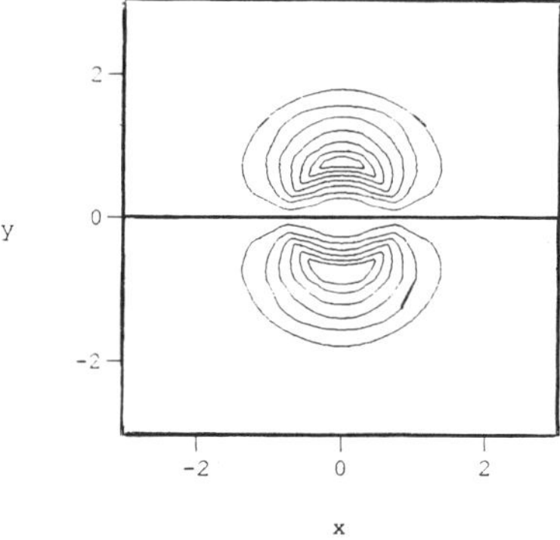


**Figure 2.** Dipole behavior of the system.

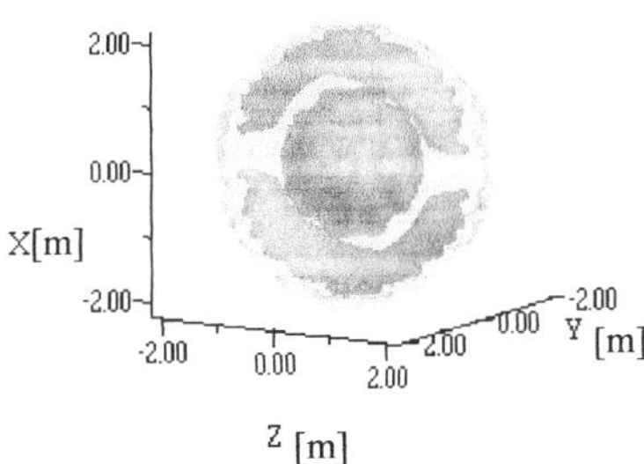


**Figure 3.** 3D values of the induction : a low and intense isosurfaces are simulated for visualising the two spheres.

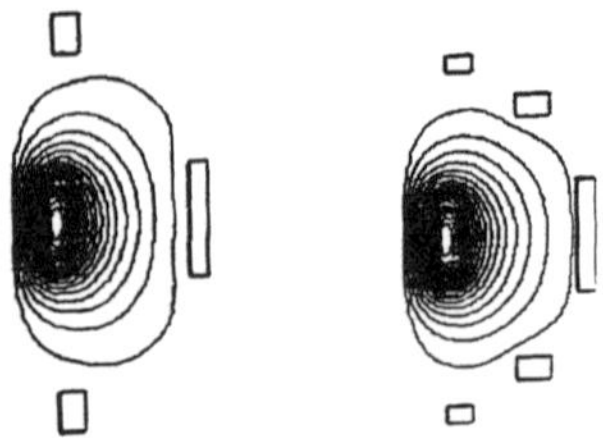

**Figure 4.** Simulating $S_1S_3$ , $S_1S_5$ by finite element method.

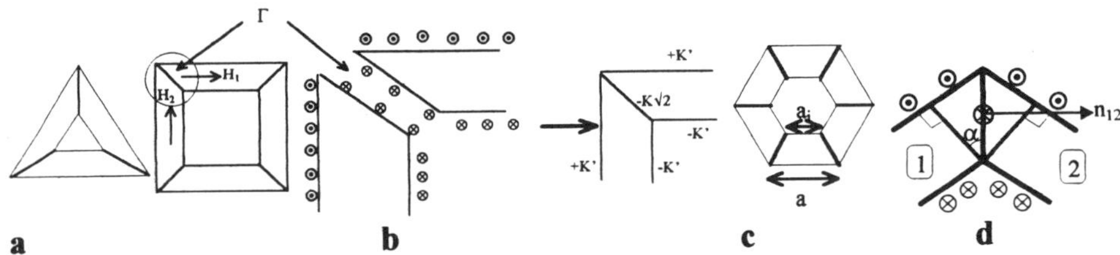


**Figure 5.** Family of Polygonal Structure ($a_i$ : internal side ; a external side).

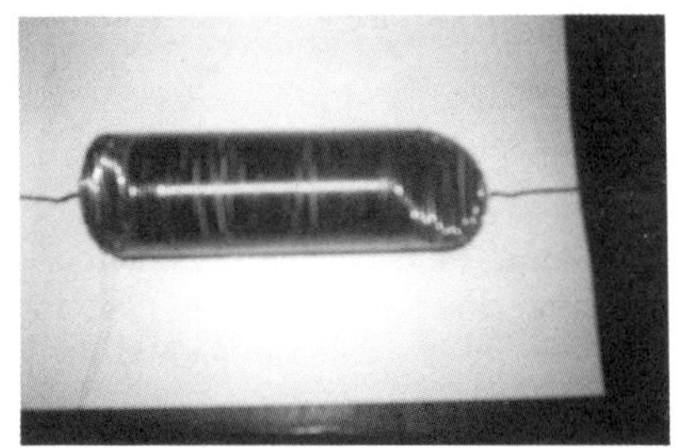

**Figure 6.** Demonstration model of technological feasibility for a polygon ( branch detail ).

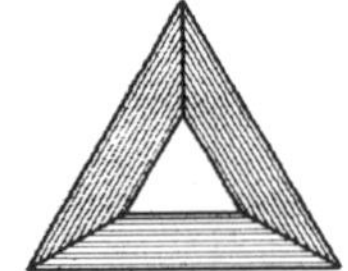

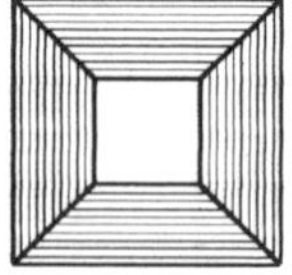

**Figure 7.** Magnetic field for triangle and square configurations .

In order to impose a constant value of the flux density modulus at every point of the superconducting windings of the internal sphere, the following relation between the current $K_2 = -K_1/4$ must hold. Thus, the optimal dimensions of a two sphere system in a vacuum is characterized by the special relation $b/a = 4^{\frac{1}{3}}$. The mean magnetic energy per unit volume (for external sphere) then comes to $\frac{E}{V} = \frac{1}{4\mu_0}(B_{in})^2$. Table 2 illustrates the optimization of such a system of maximum 5T of flux density.

**Technological Approach**

The theoretical system is simulated by coaxial solenoids. In Figure 4, $S_1S_n$ stands for $S_1$ internal sphere simulated by a single solenoid and $S_n$ the external sphere simulated by n coaxials solenoids. The first condition to be satisfied in order to approach a zero external flux density (r > b) is to make the global dipolar momentum equal to zero. This leads to a weak field at a relative small distance from the system.

$$\sum_{i=1}^{n} \vec{M}_i = -\vec{M}_0$$ where $M_0$ is the moment of the central solenoid

$M_i$ is the moment of the $i^{th}$ external solenoid

In practice, analytical method is used to compute the field distribution in the $S_1S_n$ configuration The currents inside the fine external solenoids have to be calculated in order to cancel the dipolar moment of the whole system and furthermore the first terms of higher degree in the development of the external vector potential. .

## POLYGONAL CONFIGURATION

The second structure proposed in this paper is based on a family of regular polygons. A « n - polygon » with n branches is defined. This family begins with the triangle (n = 3) and then with the square (n = 4) up to the n-polygon, Figure 5. A branch, Figure 6, is the basic unit for the family of polygons and will therefore be studied in detail. Each branch is made of two corners and one main part. The branches are wound such that the external side of the polygon receives the up- current while the internal sides receive the down- currents. The number of conductors per unit length is constant both along the main part and the corners. The continuity of the winding at the border insures the following relation between the superficial current densities on the main part and on the flat side of the corner, respectively K' and K :

$$K'\sin\alpha = K$$

where $\alpha = \pi/n$ is defined on Figure 5d. For instance, for the square (b) : $\alpha = \pi/4$.
Thus, the total superficial current density at the border is $2\,K'\sin\alpha$.

It is easy to verify that such a current distribution insures a constant field $\mathbf{H}_1$ inside the branch 1 and a constant field $\mathbf{H}_2$ inside the branch 2, the fields having the same modulus $H_1 = H_2 = K'$. These values allow the following boundary conditions to be fullfilled :

$$\vec{n}_{12} \wedge (\vec{H}_2 - \vec{H}_1) = \vec{K}_1 + \vec{K}_2 \qquad \vec{n}_{12}.(\vec{B}_2 - \vec{B}_1) = 0 \qquad (6)$$

Two exemples are given involving the triangle and the square, Figure 7. The field is calculated using a 2D finite elements method. The flux density is verified to be constant, and there is no stray field. So for the calculation of the energy, only the external volume of the polygon is needed. The energy density does not take into account the cross section of the branches. Let $N_1$ and $N_2$ be the number of turns (in the main part and in the corner). For one branch of the polygon, the length of superconducting wire is:

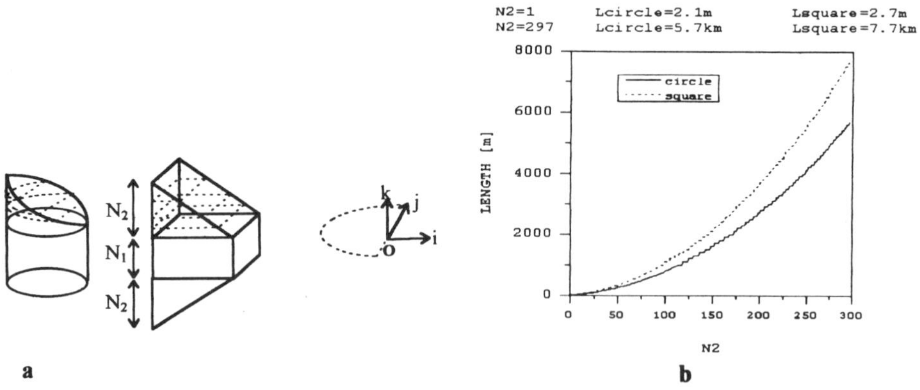


**Figure 8.** Comparison of circular and square cross-sections : $N_{1,2}$ is the number of turns.

for a square cross section :

$$L = 4.2R.N_1 + 2\sum_{i=1}^{N_2}(2.\ 2R + 2\left(1 - \frac{i}{N_2}\right)2R)$$

for a circular cross section :

$$L = N_1 2\pi R + 2\sum_{i=1}^{N_2}(\ 2R\ \sin(\frac{i\pi}{N_2}) + 2\pi R\ (1 - \frac{i}{N_2})\ ) \qquad (7)$$

For a fixed diameter 2R of the chosen solenoid and an 'n' polygon (with fixed angle $\alpha$), $N_2$ is known since $N_2 = {}^{2R}/_{d.\tan\alpha}$. Figure 8b shows the length variation of the windings for the two cases, $N_2$ varying from 1 to $N_{2max}$. For big energy, it is clear that $N_1$ must be low in comparison to $N_2$. Furthermore, if the comparison is made just with the basic cross section of each corner, the perimeter for a square is higher than that of a disk with a ratio ${}^{4}/_{\pi}$, i.e. a saving of 21% for the same winding. From a commercial point of view, the polygonal structure with a circular cross section is thus the most attractive. The study of mechanical forces has not been developed here but the advantage of a circle is a priori to have a better stress distribution than the square.

## Practical Approach

For a practical approach, the corner of the branches can be realised with a set of discrete coils. The flux density is than calculated using the Biot and Savart law in the solenoidal main part and at the two corners. For each corner, the flux density at a point M for a D-shape [12] coil in a (O,i,j,k) coordinate system is given by :

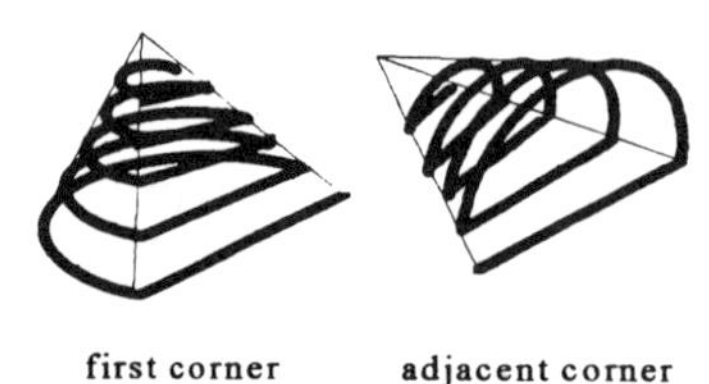


**Figure 9.** Technological liaison between two corners.

**Table 3.** Example of an Actual Case

| K[A/m] | B[T] | $a_e$[m] , $a_i$[m] | $V_{branch}[m^3]$ | $V_{tot}[m^3]$ | Energy[J] | Capacity[J/m$^3$] |
|---|---|---|---|---|---|---|
| $100\pi\ 10^3$ | 5 | 1.3 ; 0.1 | 0.113 | 0.4524 | $4.49\ 10^6$ | $9.947\ 10^6$ |

$$Bx = \frac{\mu_0 I}{4\pi}\int_{\alpha 1}^{\alpha 2}\frac{Z_0 R\cos\alpha}{\Delta} + \frac{Z_0 R\cos\alpha}{\Theta}\,d\alpha$$

$$By = \frac{\mu_0 I}{4\pi}\int_{\alpha 1}^{\alpha 2}\frac{Z_0 R\sin\alpha}{\Delta}\,d\alpha$$

$$Bz = \frac{\mu_0 I}{4\pi}\int_{\alpha 1}^{\alpha 2}\frac{-X_0 R\cos\alpha - Y_0 R\sin\alpha + R^2}{\Delta} - \frac{R\cos\alpha(X_0 - R\cos\alpha_1)}{\Theta}\,d\alpha$$

$$\text{where}\quad \Delta = \left[(X_0 - R\cos\alpha)^2 + (Y_0 - R\sin\alpha)^2 + Z_0^{\,2}\right]^{3/2} \qquad (8)$$

$$\Theta = \left[(X_0 - R\cos\alpha_1)^2 + (Y_0 - R\sin\alpha)^2 + Z_0^{\,2}\right]^{3/2}$$

I is the current and $(X_0, Y_0, Z_0)$, the coordinates of M in the local coordinate system (O,i,j,k). For each D-shape, (Bx, By, Bz) are expressed in the fixed coordinate system of the polygon and added in order to obtain the total components of the flux density. We consider the case where the chosen polygon is a square, current density (200 A/mm$^2$), wire diameter (d) of 2 mm, and a thickness (2r=0.6m), (results in Table3). The energy is calculated using the expression given below where : K is the superficial current density, $B_t$ is the flux density, V the volume, and a the external side, $a_i$ the internal side of a polygon :

$$E = \frac{B^2}{2\mu_0}V\ ; \quad a = a_i + 2R\tan(\alpha)\ ;$$

## CONCLUSION

Two new designs for SMES are described : two optimized spheres and a family of regular polygons. The proposed SMES don't give any external stray field and have a constant internal field. For the same field space, the same flux density, (see Table 2. and 3.), the system of spheres presents a lower storage capacity (E/V) than the family of polygons. Optimality criterion was established for the system involving 2 concentric spheres (radii a and b) in vacuum : $b/a = 4^{\frac{1}{3}}$. The two theoretical designs have been simulated technologically with different solenoids and coils. The polygonal design involves D-shape coils in the corners.

## REFERENCES


1. C.Levillain, « Minimal performances of high TC wires for cost effective SMES compared to low Tc's », IEEE Trans. MAG, Vol.32, N°4,pp.2308-2311, 1996.
2. Schoenung, S.M., Meier W.R., Hassenzahl W.V., « A comparison of large scale toroidal and solenoidal SMES systems », Journal IEEE Trans. MAG. Vol 27 Iss 2 pt3 p2324 - 2329, 1991.
3. W. V. Hassenzahl, « A comparison of the conductor requirements for energy storage devices made with ideal conductor geometries », Journal IEEE Trans. MAG Vol 25 p1799-1802, 1989.
4. W. V. Hassenzahl, « Considerations against a force compensated coil », Journal IEEE Trans. MAG Vol 25 p1854-1858, 1989.
5. Mostafa K. Abdelsalam, « MicroSMES Magnet Configuration for reduced Stray field applications », IEEE Trans. ASC, Vol.5, N°2, 1995.
6. K.Ishibari, K. Kamezawa, A. Katase, A. Terashima, Y. Sakakibara, Y. Funashi, H. Hirabayashi; « Design study on future accelerator magnets using largely keystoned cables », Journal IEEE Trans. MAG Vol 25 Iss 2 pt3, p1628, 1989.
7. H. Blosser, J. Bailey, R. Burleigh, D. Johnson, E. Kashy, T. Kuo, F. Marti, J. Vincent, A. Zeller, « Superconducting cylotron for medical applications », Journal IEEE Trans. MAG Vol 25 p1746-1755, 1989.

8. J.H. Schultz, D.B. Montgomery, H.J.Thome, E.S. Bobrov, and R.J. Fleming ,« Poloidal field coil and cryogenic systems for the FCX Tokamak », Journal IEEE Trans. MAG Vol 21 p1064-1066, 1985.
9. W.Weck, P.Ehrhart, A.Müller and G.Reiner, « Superconducting Inductive Pulsed Power Supply for electromagnetic launchers : design aspects and experimental investigation of laboratory set-up », IEEE Trans. MAG, Vol.33, N°1, pp.524-527, 1997.
10. F. Vivaldi, L.Ottonelo, « Study on possible constructive solution of SMES for different applications in the italian background. », Proceeding of the IEA Symposium on the use of superconductivity in energy storage, Editor World scientific-Juengst, Komarek, Maurer, 1994.
11. G. Fresacher, G. Schönwetter, J.Gerhold, «Active shielded transportable SMES systems », IEEE Trans. MAG, Vol.32, N°4 pp.2349-2352, 1996.
12. Y. Wachi, A. Miura, T. Hamyaima, T; Uchida ,«Developpement of a forced cooling D-Shaped superconducting coil by supercritical helium»,; Journal IEEE Trans. MAG Vol 19 Iss 3 p320-324 May 1983
13. E. Durand, Electrostatique II, Problèmes généraux conducteurs, Edition Masson

# TEST RESULTS OF ELEMENTALY COILS FOR TOROIDAL SMES

H.Hayashi[1], T.Imayoshi[1], H.Kanetaka[1], Y.Yamamoto[1], F.Irie[1], T.Ezaki[2], M.Takeo[3], S.Ioka[4], K.Asano[5], and S.Nose[6]

[1] Kyushu Electric Power Co., Inc., 2-1-47, Shiobaru, Minami-ku, Fukuoka, 815, Japan
[2] Oita University, 700, Dan-noharu, Oita, 870-11, Japan
[3] Kyushu University, 6-10-1, Hakozaki Higashi-ku, Fukuoka, 815, Japan
[4] Toshiba Corporation, 1-1-6, Uchisaiwai-cho,chiyoda-ku,Tokyo,102,Japan
[5] Hitachi Ltd., 1-1-3, Saiwai-cho,Hitach-shi,Ibaraki,317,Japan
[6] Fuji Electric Co., Ltd..,7 Yawata-kaigandori,Ichihara,chiba,290,Japan

## ABSTRACT

Our experimental SMES of 1kWh/1MW named "ESK" (Experimental SMES of Kyushu Electric Power Co., Inc.) is going to be completed in 1997. It has a toroidal coil system with 6 elementary coils (~ each 1/6 kWh) made by 3 manufactures independently, each of which has its own type of coil and conductor: Compacted Rutherford type for conductors, and pancake and layer-winding type for coils. The newly designed coil shape "modified D" has two straight parts in-between D parts.

Measurements were made for each of the 3 coil types in the 6 elementary coils. The results are: coils were stable up to the current corresponding to the rated 1000A of the toroidal arrangement. Losses were measured for the condition of repeated charging and discharging with high ramp rate were within the designed value. Strains have also been measured on the support structures of each coil, mechanical characteristics of each coil are compared.

## INTRODUCTION

Small or medium scale SMES will be in high demand for electric power system stabilization and load fluctuation compensation. We are advancing development to scale up SMES capacity(1). For scaling up the SMES, we must solve technical problems for each step. In order to expand the capacity up to a practical scale, high voltage and large current of the direct-current circuit of SMES are required. However, it is hard to realize in the present condition technology. A modular-type SMES, which is composed of normalized small units, is very effective at this point(2).

One of the severest problems for scaling up SMES is the mechanical force on conductors. Reducing leakage field is a goal as an environmental control issue. Therefore ESK, a toroidal coil system with 6 coils, is adopted to reduce a leakage field. In this system, for reducing stress on conductors as well as bending moment in coils, a modified D type is proposed for the coil shape[(3)].

Now, manufacturing and testing of the 3 elementary coil types have been finished. In this report, experimental results of the test are shown and discussed focusing on the mechanical force.

## MODIFIED D TYPE TOROIDAL COILS OF ESK

Each modified D type coil is composed of 4 sections, 2 D shape sections and 2 straight line sections, as shown in Figure1. Sections of straight line are connected to each other by support structures with backing plate. Approximately no bending moment exists in either the D sections or the straight sections.

These modified D type coils have the following characteristics.

1) The mechanical disturbance is smaller than the cases of circular and the existing D type coils. This is due to the smaller tension in conductors caused by support structures, and also due to the shape with no bending moment.
2) The quantity of conductor per unit stored energy is reduced from that of the D shape or circular coil.
3) The floor area and the leakage field are small because of the reduction of torus radius accompanied by the small coil width.

Fundamental design parameters of ESK and configuration of the modified D shaped coil are shown in Table 1.

**Table 1.** Fundamental design parameters of ESK

| | | |
|---|---|---|
| Maximum stored energy | 1 | kWh |
| | [3.6] | [MJ] |
| Maximum Power | 1 | MW |
| Rated current | 1,000 | A |
| Rated voltage | 500 | V |
| Number of modules | 2 | Units |
| Superconducting coils | | |
| - Type of coil | Toroidal type | |
| - Number of coils | 6 (3/module) | |
| - Coil core | FRP | |
| - Cooling system | Each coil is immersed in LHe each vessel | |

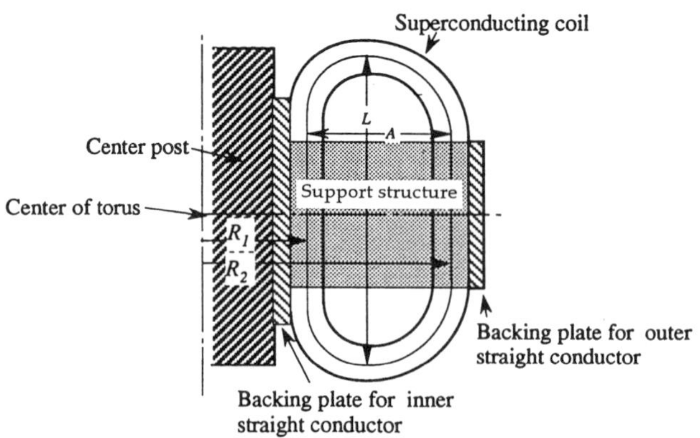


**Figure 1.** Modified D shaped coil.

**Table 2.** Design parameters of conductors

| Conductor | | A | B | C |
|---|---|---|---|---|
| Superconductor material | | NbTi | NbTi | NbTi |
| Current density | A/mm$^2$ | 40 | 40 | 40 |
| Cu/CuNi/NbTi | | 4.7/2.5/1 | 4.7/2.5/1 | 2.5/1.9/1 |
| Dimension of conductors | mm$^2$ | 2.13 × 8.4 | 1.90 × 7.60 | 1.98×7.15 |
| Twist pitch of conductors | mm | 91.7 | 85.0 | 55.0 |
| Number of strands | | 16 | 16 | 13 |
| Ic (5T,4.2K) | kA | 3.68 | 3.41 | 3.74 |

**Table 3.** Design parameters of the 3 elementary coil types

| Coil | | A | B | C |
|---|---|---|---|---|
| R1 | mm | 430 | 430 | 430 |
| R2 | mm | 794 | 794 | 794 |
| Radius(toroidal) | | 612 | 612 | 612 |
| Straight section length | mm | 364 | 364 | 370 |
| Aspect ratio | | 2.0 | 2.0 | 2.0 |
| Type of winding | | 10DP* | 12DP* | LW** |
| Turns | | 1,520 | 1,512 | 1,551 |
| Inductance | H | 1.00 | 1.03 | 1.04 |
| Length of conductor | m | 2,850 | 2,380 | 2,903 |

* DP:Double-pancake, **LW:Layer-winding

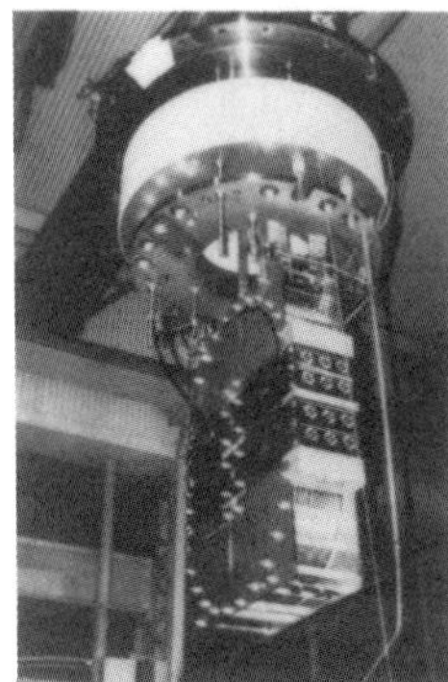
Coil A

Coil B

Coil C

**Figure 2.** View of 3 elementary coil types .

The design parameters of conductors are shown in Table 2. Since SMES system for power system control have to be operated in a pulse mode, pulse losses of conductors should be as small as possible. From this point of view, 3 sample NbTi conductors of Rutherford cable type are chosen. We measured the crossover resistances between strands of conductors by use of short sample conductors, and evaluated the resistances [4].

The design parameters and photographs elementary coil types made by 3 manufactures are shown in Table 3 and Figure 2.

## TEST RESULTS ON COMPLETED ELEMENTARY COILS

### (i) Inductance test

We obtained values of inductance from the induced voltage in charging and from the time constant in discharging. The results of test are shown in Figure 3 and Table 4.

The measured coil inductances are 1.00~1.07 H in both cases of charging and discharging, which are almost consistent with the design values.

### (ii) Electromagnetic force

We have succeeded in energizing a single coil up to 1220A. This is equivalent to the case of full rating (1000A) test for toroidal arrangement of 6 coils. The maximum forces on a coil in these two cases are equal to each other as shown in Figure 4.

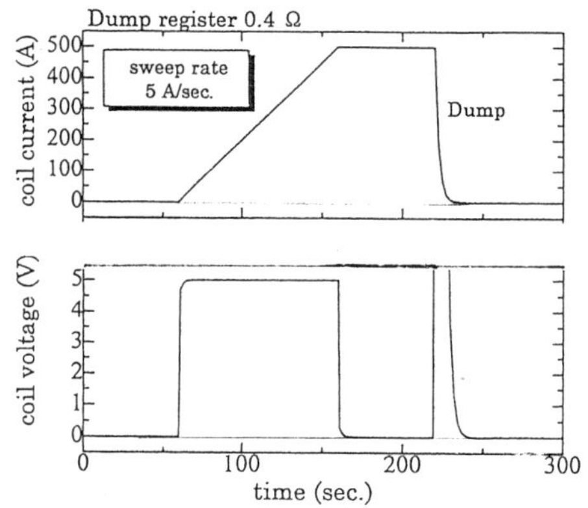


**Figure 3.** Dump test (coil A)
(The sign of the voltage is reversed for the case of dumping).

**Table 4.** Test results of inductances

| Coil | Inductance |
|---|---|
| A | 1.00 H |
| B | 1.04 H |
| C | 1.07 H |

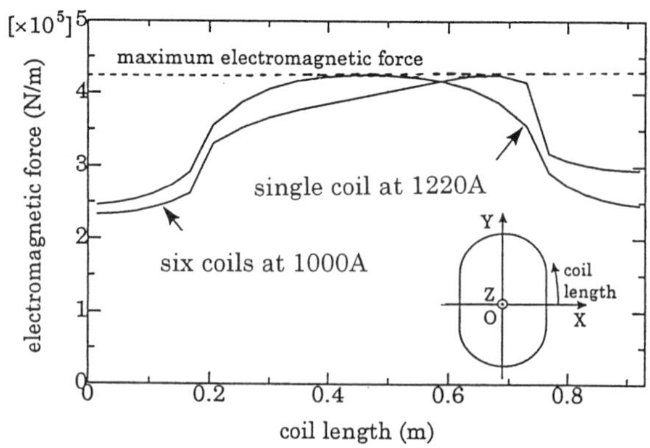


**Figure 4.** Electromagnetic force analysis (coilA).

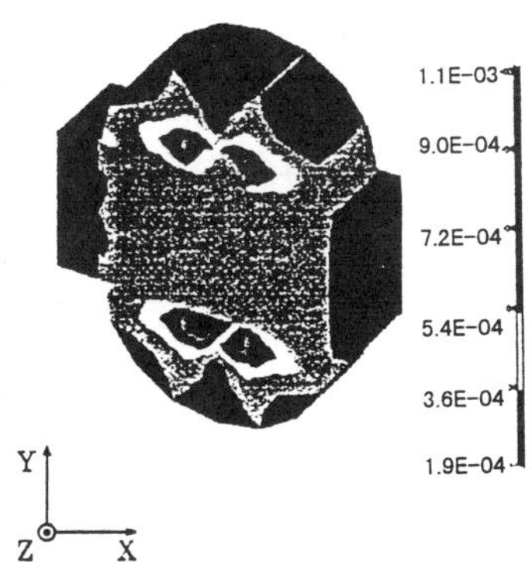


**Figure 5.** Analytical result of FEM.

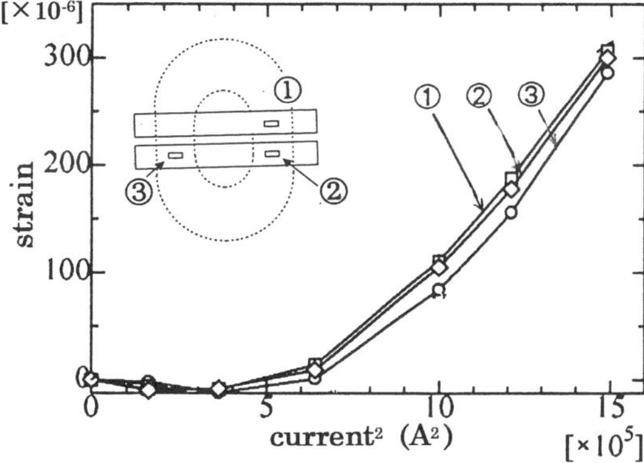


**Figure 6.** Strain on the support structures(coil A).

**(iii) Measurement of strain on the support structure**

We have analyzed strain distribution on support structure by finite element method (FEM). The result is shown in Figure 5. In this model, the coil is simulated as a race tack coil with its straight conductor section supported by flat support structure(SUS304) with 6 mm thickness.

In this experiment, we measured the strain on the support structures connecting outer and inner straight sections. Results are shown in Figure 6~Figure 8. These results show that the value of strain generated on the support structures were less than $300\times10^{-6}$ which is much smaller than $2{,}000\times10^{-6}$ (the value of yield strain of 304 stainless steel at 4.2 K). The estimated strain in the windings is also under the elastic limit, as shown in Figure 6.

We observed hysteretic characteristics in strain-(current)$^2$ curves, (Figure 7,8) where (current)$^2$ should be proportional to Lorenz force. They must come from stress-strain hysteretic characteristics of conductors, because the stresses of the support structures are well in the elastic region. These hysteretic characteristics are important because they are related to coil instability. In these Figures, some delayed rise of strain can be seen. It may have come from almost free motion of wires in a weak force region. The difference of strain between front and back sides in Figure 7 comes from backing of the support structures.

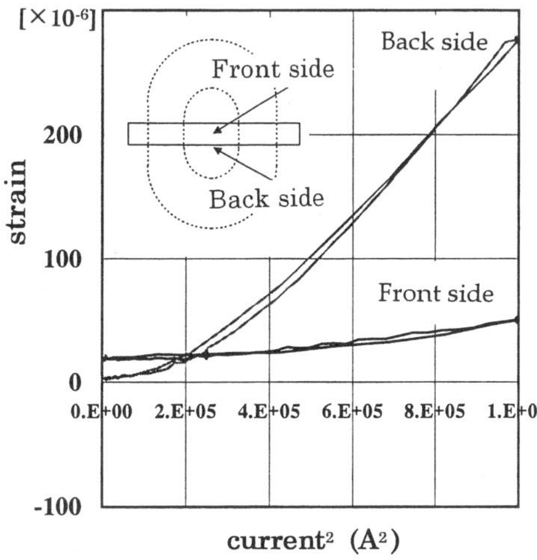


**Figure 7.** The strain on support structures(coilB).

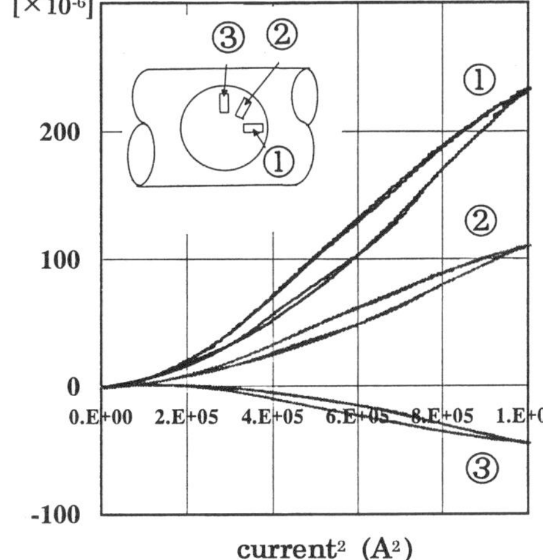


**Figure 8.** The strain on support structures(coilC).

**Table 5.** Estimated loss of each elementary coil for repeated pulse operation with the rate of increase of the current was 10A/s and that of decrease was 50A/s

| | Coil | | A | B | C |
|---|---|---|---|---|---|
| Case 1* | Theoretical values | W | 1.7 | 1.1 | 1.1 |
| Case 2* | Estimated values from short sample tests | W | 1.6 | 1.1 | 1.5 |
| Case 3** | Measured values for elementary coils | W | – | – | 1.8 |

*500-1000A pulse operation, **0-1000A pulse operation

We consider that the hysteretic characteristics, the mechanical losses caused by wire motion and the coil instability are connected to each other. Mechanical problems such as mechanical loss and resulting instability in the straight section of a coil were our primary concern. However, instability was not observed in the course of energization. The mechanical hysteretic loss was quantitatively estimated roughly from the measured strain of the support material sections. Those are going to be studied in detail.

### (iv) Pulse losses

Pulse losses of coils were estimated by the decrease of LHe (liquid helium) level in vessels during the cyclic energization between 500A and 1000A (Case1 and Case2), and between 0A and 1000A (Case3) where the rate of increase of the current was 10A/s and that of decrease was 50A/s. Results are shown in Table 5. Though measurements were somewhat simplistic, the measured results proved to be consistent with the preliminary calculation.

## COIL COOLING SYSTEM

The cooling system of 6 coils are shown in Figure 9. This system, which has a helium gas space, is our new proposal. Each coil is immersed in LHe in a separate FRP vessel which is not a cryostat. The outer space of the vessel is filled with GHe.

This cooling system is expected to have several merits. It does not have eddy current loss while the separate metal cryostat system has. Supporting structures for forces between coils are rather simple for the present system of FRP vessels in a cryostat, while the existing FRP cryostat system is not. This is because in the latter case, support structures should have thermal insulation in the parts that go through vacuum spaces. However, the original merit of the separate cryostat, i.e., no quench propagation between coils, holds even in this case. Besides, large GHe space in the

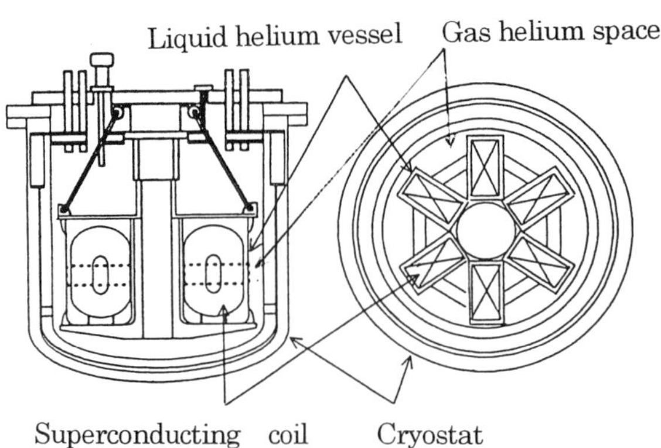

**Figure 9.** Systematic diagram of cooling system for 6 coils.

cryostat may be useful for recovering some quantity of helium without raising pressure too much in case of quenching.

The cooling system has been manufactured and preliminary testing is now going on. The results seem to be satisfactory and will be reported later.

## CONCLUSION

Three type elementary coils for ESK have been tested to confirm stable operation. The results of test are as follows ;

1) Rated operation of each coil was successful.
2) Strains on support structures of coils were measured.
3) AC losses of each coil were estimated by the decrease of LHe levels. The measured results are consistent with the preliminary calculation.
4) Cooling test of the whole system is now going on.

The whole ESK system will be tested in the power line system in Fukuoka beginning 1998. We are expecting the data to elucidate merits of SMES .

## REFERRENCE


1. T. Imayoshi et al., "Development of a 1kWh-Class Module-type SMES-Design Study," Vol.7, p844, ASC96, Pittsburgh, USA (1996)
2. H. Okada et al., " Study on Modular SMES ," IEE(J)Trans. B, 115, p52 (1995)
3. T.Ezaki et al., "Design of a Modified D Shaped Troidal Coil for 1kWh/1MW Experimental SMES," p1279, ISS'95, Hamamatsu, Japan (1995)
4. F.Sumiyoshi et al., "Coupling Losses in Rutherford Cable Designed for 1kWh/1MW SMES," p116, Cryogenic Engineering Japan (1996)

# STRUCTURAL DESIGN OF SUPERCONDUCTING ENERGY STORAGE SOLENOIDAL MAGNETS[1]

Yu Wang[2]

BWX Technologies, Babcock & Wilcox
Product Development, P.O.Box 785
Lynchburg, VA 24505, U.S.A.

## ABSTRACT

Mid- and large scale commercial superconducting magnetic energy storage (SMES) magnets have been actively studied recently. It has been found that cost is a major consideration in determining the magnet overall dimensions if the energy requirement is defined. Generally, conductor length has been considered as a critical parameter. Studies have shown that many other design requirements could be as expensive as the conductor length. This paper discusses the stress characteristics and some structural limitations for low aspect ratio solenoids. Literature and analytical relations are reviewed. Stress distributions relative to major dimensional parameters are presented. Other critical structural considerations, such as turn-to-turn contact, composite conductor configuration, are investigated.

## INTRODUCTION

Superconducting magnetic energy storage (SMES) systems have been studied for utility applications since the early 70s. The SMES system originally studied was an earth supported, large-scale, single layer solenoid used as a load-leveling device. General relations that correlate the stored energy and magnetic field to cost-related design parameters were developed for those thin wall solenoids to assist in design of an optimized magnet[1,2,3].

The recent SMES studies are more focused on transmission line stability, spinning reserve, voltage control, and power quality enhancement in general. These applications generally demand mid- or small scale, high power, self-supported thick wall solenoids.

Babcock & Wilcox (B&W) has been studying utility SMES applications since the late 80s and has developed several alternative designs for a 30MW, 1800 MJ SMES to be installed in Anchorage Municipal Light & Power (ML&P), Alaska.

Cost has been a major limitation for a magnet designed for commercial applications. The overall dimension plays a critical role in driving the raw material cost. Some of the major design parameters are

[1] Work supported by U.S. Department of Energy Grant No. DE-FG36-94GO10011.
[2] Now at GE Corporate R&D Center, P.O. Box 8, K1-EP 123, Schenectady, NY 12301

- Solenoid inside radius, $a$
- Shape factor, $\alpha = b / a$ ($b$ = solenoid outside radius)
- Solenoid aspect ratio, $\beta = l / a$ ($l$ = half length)
- Current Density, $J$
- Stored energy, E
- Magnetic field, B

If dimensional parameters β, α, and $a$ are fixed, two parameters among J, E, and B are then independent. The third parameter can be derived from the known. A cost optimization must also balance among the cost of raw material, fabrication, installation and operation. Configuration details are vital in the design. These details are likely affected by the following structural considerations:

- Stress distribution within the solenoid
- Stress distribution at mid-plane
- Turn-to-turn interactions
- Pancake-to-pancake interactions
- Local stresses within a composite conductor
- Conductor stabilizer properties
- Choice of structural materials
- Amount of structural materials
- Structural reinforcement method
- Pre-stress demand
- Stress/strain stability with respect to the winding process
- Long term reliability (material strength)

In addition, the magnet must meet performance requirements. Examples of the requirements are

- Conductor stability
- Quench detection and protection
- Operating degradation (RRR, material properties)
- Conductor disturbance

The objective of this paper is to demonstrate the structural characteristics of an energy storage solenoid. The stress distributions relative to the dimensional parameters are provided. Turn-to-turn contact behaviors of a low aspect ratio solenoid, which requires the shortest length of conductor for a given energy and field, are given. These characteristics may be the critical considerations for design engineers.

## THIN WALL SOLENOIDS

Boom, Moses, and Eyssa studied the thin wall SMES solenoids extensively at University of Wisconsin[1,2,3]. The following is a summery of the principal relations developed:

### The Ampere-Meter, Surface Area, and Radius(Moses 1976)

$$IS = Q_{IS}(\beta)E^{2/3}/B_M^{1/3} \tag{1}$$

$$A = Q_A(\beta)E^{2/3}/B_M^{4/3} \tag{2}$$

$$R = G(\beta)E^{1/3}/B_M^{2/3} \tag{3}$$

where IS = the product of current, I, and conductor length, S; A = magnet surface area; $\beta = l / a$, the aspect ratio; $a$ is the inside radius and $l$ is the half length; E = stored energy; $B_M$ = the maximum field at the mid-plane. The quality factors were given as

$$Q_{IS} = 4(\pi\beta K'/\mu_o K^2)^{1/3} \tag{4}$$

$$Q_A = 8(\pi\mu_o^2\beta K'^4/K^2)^{1/3} \tag{5}$$

$$G = (\mu o K'^2/\pi\beta K^2)^{1/3} \tag{6}$$

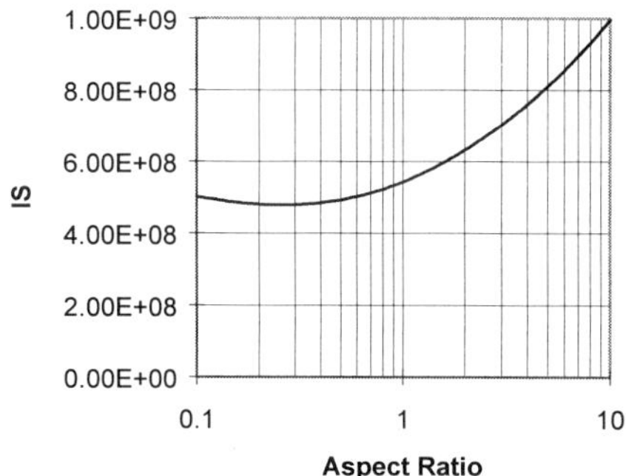


**Figure 1.** Ampere-Meter Relative to Aspect Ratio for ML&P Conceptual Solenoids.

K and K' are factors that approach unity for a long solenoid. For β < 1, K may be calculated as (Grover 1946)

$$K = \frac{2\beta}{\pi}[(\ln\frac{4}{\beta} - \frac{1}{2}) + \frac{\beta^2}{8}(\ln\frac{4}{\beta} + \frac{1}{8}) - \frac{\beta^4}{64}(\ln\frac{4}{\beta} - \frac{2}{3}) + \ldots] \tag{7}$$

and K' as (Eyssa and Hilal 1980):

$$K' = \frac{1}{2} + \frac{1}{2\pi}\int_0^{\pi} \frac{\beta d\theta}{[\beta^2 + 4\sin^2(\theta/2)]^{1/2}} \tag{8}$$

For β < 1, K' may be approximated as

$$K' = \frac{1}{2} + \frac{\beta}{2\pi}\ln(\frac{2\pi}{\beta}) \tag{9}$$

It is seen from Equations (1) through (3) that for a given energy and expected mid-plane field, conductor IS is directly related to the solenoid aspect ratio β.

To conceptually configure a 2000 MJ, 7 T peak field SMES solenoid for ML&P, IS vs. β relationship is plotted in Figure 1. It is shown that the IS is minimized when 0.2< β < 0.4. Given the current, conductor length would be the shortest in that aspect ratio range. For a given β, solenoid radius is defined, and therefore the solenoid half length $l$ will be fixed, see Figure 2 for relationships of β vs. $a$ and $l$ vs. $a$, respectively to meet ML&P SMES energy requirement.

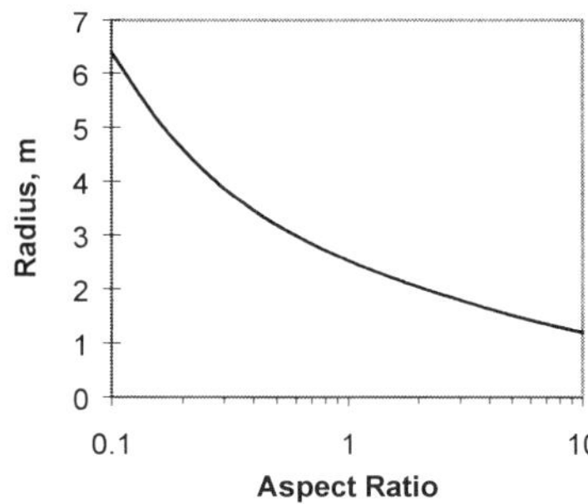


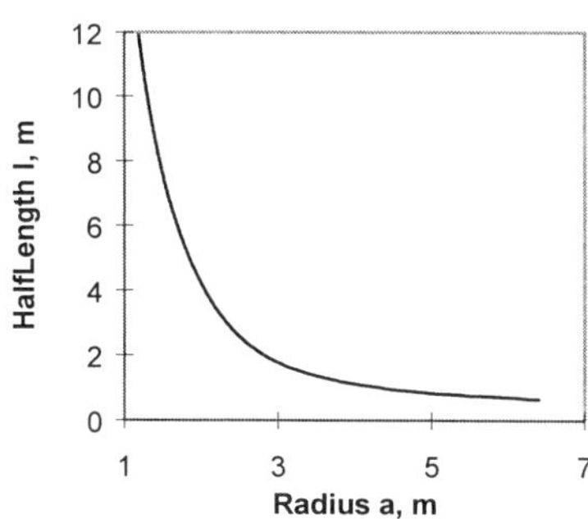


**Figure 2.** Radius to Aspect Ratio and Half Length to Radius Relation for ML&P Conceptual Solenoid.

## Virial Theorem

In a thin wall solenoid, sufficient structural materials must be provided to support hoop tension resulting from the radial pressure and the axial compression resulting from the axial force. The total mass required according to Tresca criterion, M, is

$$M = M_t + M_c \tag{10}$$

where $M_t$ is the mass under tension and $M_c$ is the mass under compression. It was shown (Boom et al 1974, Eyssa and Hilal 1980) that

$$M_t = Q_t\left(\frac{\rho E}{\sigma}\right) \tag{11}$$

and

$$M_c = Q_c\left(\frac{\rho E}{\sigma}\right) \tag{12}$$

where ρ is the density of the support mass and σ is the allowable stress of the material. The quality factors were given by the following equations

$$Q_t = 2\left(1 - \frac{\beta}{2K}\frac{\partial K}{\partial \beta}\right) \tag{13}$$

$$Q_c = \left(1 - \frac{\beta}{K}\frac{\partial K}{\partial \beta}\right) \tag{14}$$

Both $Q_t$ and $Q_c$ are increasing function of β. If the supporting mass is uniformly distributed along the conductor length, the conductor cross section areas required for both tension and compression can be obtained from Eqn (1), (10), (11), and (12) as

$$A_c = Q_{cA}\left(\frac{I}{\sigma}\right)(EB_M)^{1/3} \tag{15}$$

$$A_t = Q_{tA}\left(\frac{I}{\sigma}\right)(EB_M)^{1/3} \tag{16}$$

$$A = A_c + A_t \tag{17}$$

where $Q_{cA} = Q_c/Q_{IS}$, and $Q_{tA}=Q_t/Q_{IS}$. It is shown in Figure 3 that the required conductor cross section is peaked when $\beta \approx 1$. The ratio of $Q_{tA}/Q_{cA}$ is a decreasing function of β. For different size solenoids with identical conductor cross section, energy and current density, the stresses will follow the same trend of $Q_A$.

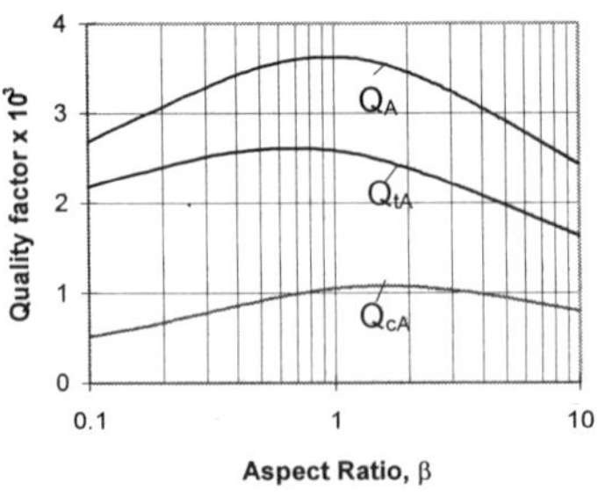


**Figure 3.** Quality Factors For Conductor Cross Section.

Based on the curves given in Figures 1 through 3, very preliminary configuration optimization for a ML&P SMES solenoid can be obtained. These formula were derived for a thin wall solenoid that has inside radius, $a$, and length, $2l$.

For the ML&P SMES application, thick wall, multi-turn solenoids were proposed. The relations for thin wall solenoids can only be used to provide preliminary estimations. Many additional characteristics of a thick wall solenoid were also considered during the ML&P SMES design.

## STRESSES IN THICK WALL SOLENOIDS

"Free turn" stresses could be obtained assuming each conductor turn act independently. Assuming the axial field varies linearly with radius, the field at radius $r$ can be obtained as

$$B = \frac{(\alpha - \kappa)B_a + (\kappa - 1)B_b}{(\alpha - 1)} \tag{18}$$

where $B_a$ = the field at the inner radius $a$; $B_b$ = the field at the outer radius $b$; $\alpha = b/a$; $\kappa = r/a$. The conductor free turn hoop tensile stress $\sigma'_\theta$ is

$$\sigma'_\theta = BJr \tag{19}$$

As we know, conductor turn does not act independently in a multi-turn solenoid. Turns are acting interactively. Adjacent turns will press on each other and develop a radial force which will modify the hoop tension. If the solenoid wall properties are assumed isotropic, the problem of distributed body forces in a cylinder can be solved analytically. The plane stress solution for hoop stress and radial stress were shown[4] as

$$\sigma_\theta = \frac{C(2+\nu)}{3(\alpha+1)}\left\{\alpha^2 + \alpha + 1 + \frac{\alpha^2}{\kappa^2} - \kappa\frac{(1+2\nu)(\alpha+1)}{(2+\nu)}\right\} - \frac{D(3+\nu)}{8}\left\{\alpha^2 + 1 + \frac{\alpha^2}{\kappa^2} - \frac{(1+3\nu)}{(3+\nu)}\kappa^2\right\} \tag{20}$$

$$\sigma_r = \frac{C(2+\nu)}{3(\alpha+1)}\left\{\alpha^2 + \alpha + 1 - \frac{\alpha^2}{\kappa^2} - (\alpha+1)\kappa\right\} - \frac{D(3+\nu)}{8}\left\{\alpha^2 + 1 - \frac{\alpha^2}{\kappa^2} - \kappa^2\right\} \tag{21}$$

where

$$C = \frac{(\alpha B_a - B_b)Ja}{(\alpha - 1)} \tag{22}$$

$$D = \frac{(B_a - B_b)Ja}{(\alpha - 1)} \tag{23}$$

It is shown that solenoid stresses are functions of shape factor $\alpha$, fields $B_a$ and $B_b$, and current density $J$.

The results from Eqn. (20) and (21) are slightly deviated from those of an actual finite length solenoid due to the facts that fields are not uniformly distributed along the length and the plane stress assumption ignored the axial force effects. However, the solution provides a simple and reliable alternative approach to benchmark the results from the advanced finite element modeling.

### Aspect Ratio Effects

A computer program SOLDIM was used at B&W to configure sizes for energy storage thick wall solenoids. To demonstrate the stress variations, five solenoids with different aspect ratio $\beta$ were proposed as examples. Double pancake configuration and identical conductor cross section (19mm x 55 mm) were assumed. Current density J = $10.53 \times 10^6$ A/m$^2$. In determining the dimensions, the shape factor $\alpha$ was fixed to 1.3 for stress comparison. The key parameters are given in Table 1. In Table 1, #DP=number of double pancakes; n=number of conductor turns within a pancake.

**Table 1.** ML&P SMES Solenoid Proposals

| Design | $a$, m | $b$, m | $2l$, m | $\beta=l/a$ | # DP | n | E, MJ | $B_{min}$,T | $B_{max}$,T |
|---|---|---|---|---|---|---|---|---|---|
| S1 | 3.8 | 4.94 | 1.65 | 0.22 | 15 | 60 | 2170 | -3.49 | 6.56 |
| S2 | 3.0 | 3.9 | 2.53 | 0.42 | 23 | 47 | 2015 | -2.80 | 6.91 |
| S3 | 2.5 | 3.25 | 3.85 | 0.77 | 35 | 39 | 2018 | -1.971 | 7.03 |
| S4 | 2.0 | 2.6 | 6.82 | 1.71 | 62 | 32 | 2025 | 0.083 | 7.00 |
| S5 | 1.5 | 1.95 | 17.38 | 5.79 | 158 | 24 | 1998 | 0.075 | 5.93 |

A 2-D axisymmetric homogeneous, anisotropic finite element model was developed using ANSYS to characterize the stress distributions within a solenoid. Electromagnetic analyses were performed as the first step to evaluate the magnetic filed distribution and to calculate the electromagnetic forces, which were generated and distributed in a non-uniform manner within the solenoid. The force matrix was then applied to the same model to perform stress analyses. Distributions of the homogenous equivalent stresses and the associated component stresses were obtained. When current density is uniformly distributed across the wall, the peak stress intensity always occur on the solenoid mid-plane, where the accumulative axial force was maximized, and adjacent to the inside radius, where hoop stress was peaked. Field and stress distributions across the thickness at the mid-plane are plotted in Figures 4 through 5.

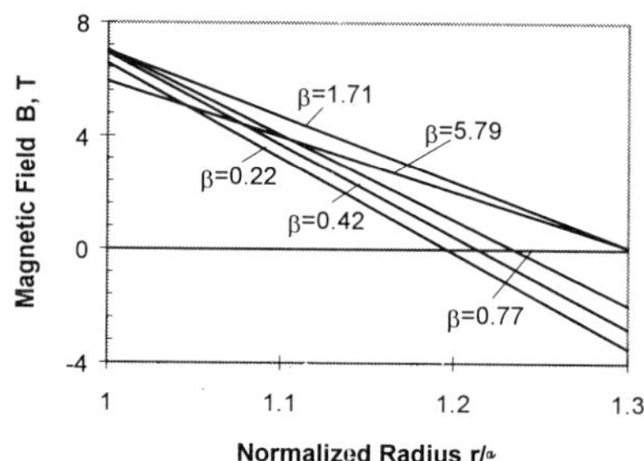


**Figure 4.** Magnetic Field Distributions Across The Wall Thickness.

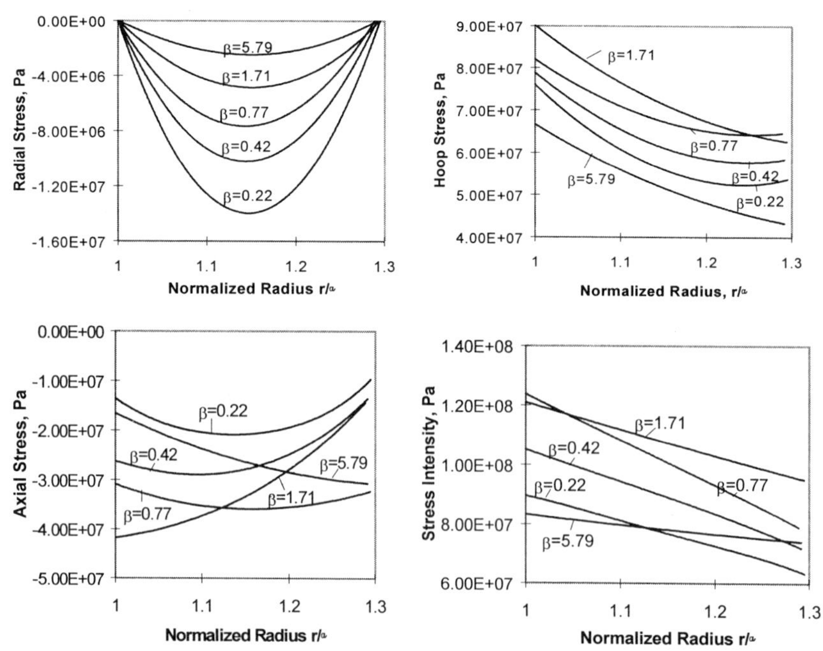


**Figure 5.** Stress Distributions Across The Wall Thickness at Solenoid Mid-Plane.

As shown in Figure 4 that negative B appears at the outside radius for β = 0.22, 0.42 and 0.77. As the result, the generated radial electromagnetic forces are acting inwards even though the net effect was still to expand the solenoid outwards.

It is shown in Figure 5 that radial stress is a decreasing function. Hoop stress, axial stress, and the stress intensity are increasing functions except for very large aspect ratio. This trend is consistent with the $Q_A$ function, as shown in Figure 3.

The hoop stress $\sigma_\theta$ was compared with $\sigma_\theta^{'}$ calculated for independent turns for different shape factor α in reference 4. It was reported that the difference between $\sigma_\theta$ and $\sigma_\theta^{'}$ was decreasing when α was increasing from 1.3 to 1.85. The radial stresses were compressive, thus beneficial in reducing hoop stress. When α = 1.85, $\sigma_\theta = \sigma_\theta^{'}$ at $r/a = 1.0$. When α was greater than 1.85, the radial stress was tensile and would make matters worse.

The difference between $\sigma_\theta$ and $\sigma_\theta^{'}$ is an indication of dependency between turns. A large difference represent a strong support between turns, which will help to share the loads and to make the stress distribution more uniform. However, it demands better contacts between turns. On the other hand, a small difference between $\sigma_\theta$ and $\sigma_\theta^{'}$ will represent a small radial stress; that is, it shows an insignificant coupling between turns.

From Eqn. (19) it is seen that $\sigma_{\theta\max}^{'} = B_{\max} Ja$. Noted that for fixed energy and current density, consistent peak field and shape factor, the inside radius $a$, and therefore, the difference between the peak $\sigma_\theta$ and $\sigma_\theta^{'}$ are reducing as the aspect ratio β is increasing, as shown in Figure 6.

## Non-Homogeneous Stress State

Actual conductor turns are often made of composite materials to include electrical insulator, superconducting cable, stabilizer, and/or additional structural members. Different materials may experience different thermal contraction during the cool-down, therefore may develop gaps between conductor turns. Different materials may have different mechanical properties, such as Young's modulus, yield point. Thus, stresses would be

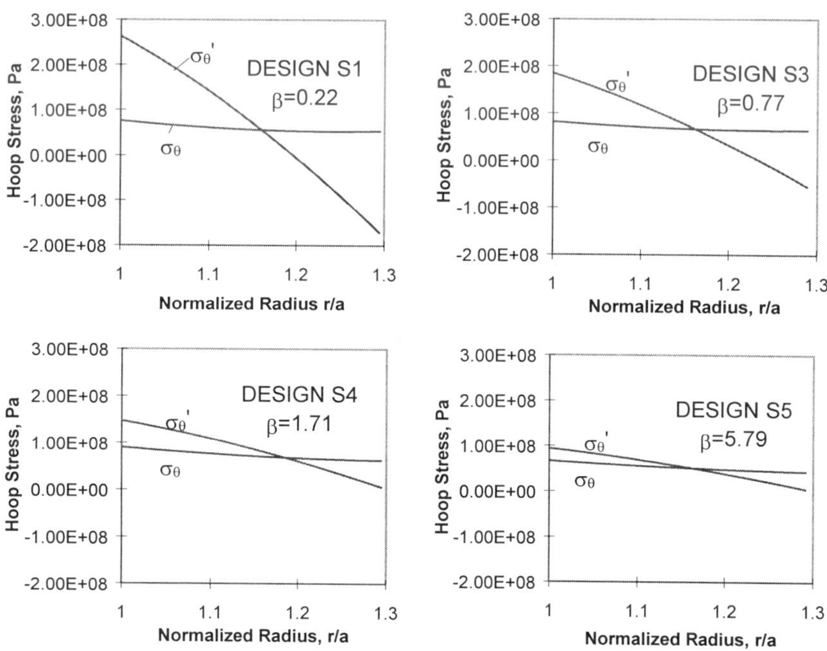


**Figure 6.** Hoop Stress Distributions Compared With The Independent Turn Solutions.

distributed unevenly among the composite members. A poor contact between conductor turns, use of radial spacers to create cooling channel, or use of significant thickness of electrical insulating materials, may develop "sponginess" in the winding. It would be much more involved to determine the stresses inside the winding for actual composite conductors.

Middleton and Trowbridge[5] considered the stresses in a helical wound pancake. Conductor turns were separated by picket fence type spacers served as insulating layer and coolant passage, as applied in the High Field Bubble Chamber magnet. It was reported that when the spacing between the pickets increase, the coil become more "sponge" and result in higher peak hoop stress at the inside radius.

## Composite Turn

If a conductor turn is constructed by two materials which have different moduli of elasticity, the assumption of consistent deformation is a quick way to estimate the stress ratio between the two materials. Assuming a composite conductor turn is thin, the hoop strain of the two materials are the same.

$$\varepsilon_{\theta weak} = \varepsilon_{\theta strong} \tag{24}$$

$$\frac{\sigma_{\theta week}}{E_{weak}} = \frac{\sigma_{\theta strong}}{E_{strong}} \tag{25}$$

and

$$\frac{\sigma_{\theta strong}}{\sigma_{\theta weak}} = \frac{E_{strong}}{E_{weak}} = m \tag{26}$$

If a more stiffer structural support is used to replace the structural material which has the same modulus of elasticity of the conductor, the conductor stress would be 1/m of that of the stiffer structural support. Note that the effective supporting area of the conductor is reduced by a factor of m when replacing the structural material. The reduction in total supporting cross section would result in higher base stresses unless the conductor is not counted as a structural member.

## Turn-To-Turn Contact

Among ML&P SMES solenoid candidates, we have witnessed very strong coupling between the turns in the low aspect ratio design S1. Therefore, the winding pack stress would be sensitive to the conductor turn contact quality. The contact quality is affected by the thickness and properties of the electrical insulating materials, winding pre-stresses, thermal contraction compatibility, conductor tolerances, winding precision, and the details of the conductor design.

A finite element pancake model was developed to illustrate the contact effect of a winding pack. The corresponding radial electromagnetic forces and maximum accumulated axial forces were applied on each turn. The shear effects between pancakes and between turns were neglected when determining the boundary forces for the mid-plane pancake. This assumption would slightly over predict the peak stresses since the shear between pancakes tends to uniform the radial deflection along the solenoid length. However, a pancake model will generally provide conservative peak equivalent stress. In the design of S1, the calculated peak stress intensity from pancake model was approximately 5% higher than that determined from the homogeneous solenoid model

An important parameter of a contact layer is its layer thickness δ. If the contact layer is extremely "soft", that is, a gap exists between turns, the inner turn can not be supported by its adjacent turn until they are in contact. Since all of the turns are deforming under the generated electromagnetic force from the current, it is the deferential movement between the turns creating a potential of interference which develops the turn-to-turn interactions. The potential turn interference may be expressed as

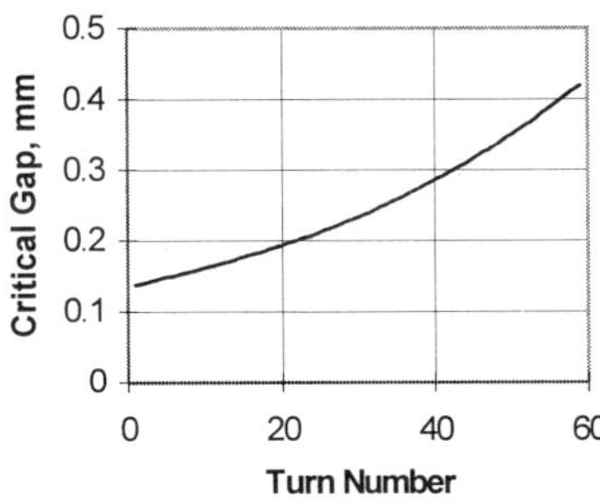


**Figure 7.** Critical Gaps for Design S1.

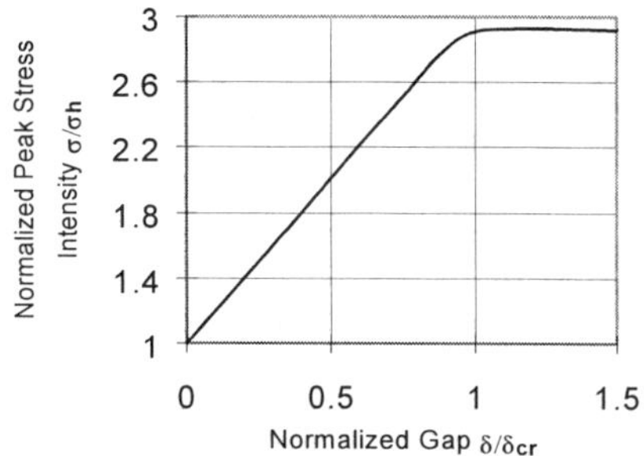


**Figure 8.** Effects of uniform gaps on the peak stress intensity of the mid-plane pancake.

$$\delta_{cr} = u_i - u_{i+1} \tag{27}$$

Therefore, if an initial gap $\delta$ is greater than $\delta_{cr}$, the critical gap, the inner turn i will behavior like a free turn since turn (i+1) would not able to support turn i during the whole loading process. $\delta_{cr}$ can be determined from thin wall cylinder theory,

$$\delta_{cri} = Q_x \frac{(B_i r_i^2 - B_{i+1} r_{i+1}^2)J}{E} \tag{28}$$

where factor $Q_x=[1+\mu(\sigma_z+\sigma_r)/\sigma_\theta]$, resulting from Poisson's effect. Typically, $Q_x=1\sim1.3$. All turns are assuming independent in Eqn (28) and $\delta_{cr}$ will vary across the wall. If gaps are uniformly distributed to each turn-to-turn interface, the effective critical uniform gap for the peak stress can be estimated considering those turns in the peak stress region, which would be highly interacted if there were no gaps between turns.

In the ML&P SMES case, if design S1 is evaluated, $(\sigma_z+\sigma_r)/\sigma_\theta \approx 0.1$ adjacent to the inner radius in the mid-plane from the independent turn solution, thus $Q_x \approx 1.03$. It is shown in Figure 6 that $\sigma_\theta'$ and $\sigma_\theta$ are intersected at approximately 1/2 of the coil thickness. That is, the inner 1/2 of the coil is the significant turn interaction region for the peak hoop stress which occurs at the inside radius. The calculated $\delta_{cr}$ for different turns are plotted in Figure 7. Therefore, average $\delta_{cr}$ for the inner half coil is approximately 0.2 mm.

Assuming turn-to-turn gaps were uniformly distributed, stress analyses were performed using the ANSYS pancake model. The equivalent stress was redistributed when different gap was assumed. When initial gap $\delta$ was less than the estimated average $\delta_{cr}$, the peak equivalent stress which also occurred at the inside radius was approximately linearly

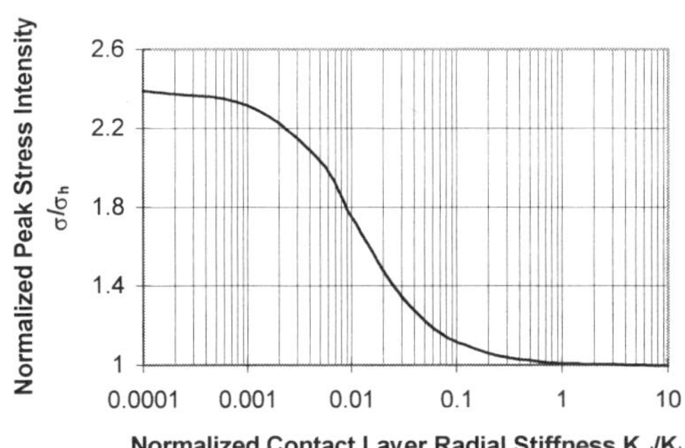


**Figure 9.** Effects of the contact layer radial stiffness on the peak stress intensity of the mid-plane pancake.

increasing with $\delta$, then reached the free turn solution. The peak stress intensity obtained from the solution without any gaps was $\sigma_h$. The normalized peak stress intensities $\sigma/\sigma_h$ relative to $\delta/\delta_{cr}$ is plotted in Figure 8.

If the contact layer developed between two conductor turns has certain stiffness, the turns will have interactions all the time. A "softer" contact layer will reduce the radial stiffness of the wall, thus reduce the load sharing capability of the winding pack. Following the springs in series principle, the effective combined radial stiffness is

$$\frac{1}{K} = \frac{1}{K_1} + \frac{1}{K_{c1}} + \frac{1}{K_2} + \frac{1}{K_{c2}} + ... + \frac{1}{K_n} + \frac{1}{K_{cn}} \quad (29)$$

where $K_i$ (i=1,...n) is the radial stiffness of conductor turn i, and $K_{ci}$ is the radial stiffness of the contact layer between conductor turns i and (i+1). It is typical that $K_{ci}$ is smaller than $K_i$. The conductor turn and contact layer stiffness can be presented as

$$K_i = \frac{EA}{\Delta} \quad (30)$$

$$K_{ci} = \frac{E_c A_c}{\delta} \quad (31)$$

where E is the conductor base material elastic modulus, A is the radial contacting surface area, and $\Delta$ is the conductor turn thickness. While $E_c$ and $A_c$ are the contact layer effective elastic modulus and radial surface area. From Eqn. (29) it is shown that the smallest $K_{ci}$ will dominate K.

In design S1, major dimensions were $a$=3.8m, conductor height h = 55 mm, $\Delta$=19mm, n=60, b= 4.94m, E=130 x $10^9$ Pa, and the average conductor turn stiffness $K_i$ = 10.4 x $10^{12}$ N/m. The contact layer thickness was assumed to be equal to 0.138mm, the 69% of $\delta_{cr}$. The normalized peak stress intensities $\sigma/\sigma_h$ relative to different $K_c/K_i$ is plotted in Figure 9.

It is shown from Figure 9 that peak equivalent stress will increase if the contact layer stiffness is less than the conductor, that is, the characteristic turn stiffness. The range of the stress variation is limited by the solution assuming clearance gaps for the contact layer. Due to relative small thickness $\delta$ of typical contact layers, the $E_cA_c$ could be reduced by a factor of $\Delta/\delta$, according to Eqn. (31), without elevating the primary stresses.

## CONCLUSIONS

An energy storage solenoid can be initially sized using simple thin wall solenoid relations. Final configuration must consider the structural characteristics among many other limitations. For fixed energy level, current density, and shape factor, stress distributions are plotted for solenoids with different aspect ratios, thus different radius, wall thickness, and length. Additional relations relative to composite conductor and contact qualities are also provided. These relations should be helpful to energy storage magnet designers to perform intelligent trade-offs and to pay attention to the appropriate parameters during the magnet design process.

## REFERENCE


1. Eyssa Y M, Design of Single Layer Superconductive Energy Storage Magnets, J. Phys. D: Appl. Phys., 13 (1980) 1719-26.
2. Moses, R.W., Jr., The Configurational Design Of Superconductive Energy Storage Magnets, Advances in Cryogenic Engineering, Vol. 21, 1975
3. Boom, R.W., et. al., "Wisconsin Superconductive Energy Storage Report", Vol. 1, University of Wisconsin at Madison, Engineering Experiment Station, 1974.
4. Wilson, M.N., "Superconducting Magnets", Oxford Science Publications, 1983
5. Middleton A. J., and Trowbridge C. W., "Mechanical Stress in Large High Field Magnet Coils", Proc 2nd Int. Conf. Magnet Technology, 1967

# ALUMINUM STABILIZED CONDUCTOR IN A LARGE CRYOSTABLE SMES MAGNET[1]

Yu Wang[2] and T. Brandsberg

BWX Technologies, Babcock & Wilcox, Product Development
P. O. Box 785, Lynchburg, VA 24505, U.S.A.

## ABSTRACT

An aluminum stabilized cryostable conductor concept for the 30Mw, 1800MJ superconducting magnetic energy storage (SMES) system to be installed at Anchorage Municipal Light and Power (ML&P) is presented. Mid- and large scale commercial SMES magnets for utility applications have been actively studied at Babcock & Wilcox. Co-extruded aluminum stabilized conductor has been considered as one of the major alternatives for the ML&P SMES application because of its maturity in extrusion process, high RRR, relative low raw material cost and light in weight. However, many significant technical issues have been identified associated with using aluminum conductor in high field and high energy magnets. Critical design considerations are discussed in this paper.

## INTRODUCTION

Babcock & Wilcox (B&W) has completed several phases of design efforts for the project of supplying the first commercial SMES system[1,2] to the power utility, Anchorage Municipal Light and Power (ML&P), for spinning reserve and to enhance the grid power quality. The base configuration of the magnet is a low aspect ratio solenoid assembled from multiple double pancakes. The conductor has been designed to be unconditionally cryostable and will be cooled by pool boiling helium. The main magnet parameters are summarized in Table 1.

**Table 1.** ML&P SMES Magnet Parameters

| | | |
|---|---|---|
| Energy | MJ | 1800 |
| Voltage | kV | 3.375 |
| Operating Current | kA | 10.8 |
| Peak Field | T | 6.5 |
| Inside Radius | m | 3.35 |
| Outside Radius | m | 4.3725 |
| Coil Height | m | 2.08 |
| Coil Type | | Double pancakes |
| Number of Double Pancakes | | 19 |
| Number of Turns per Pancake | | 50 |

[1] This work was supported by the U.S. Department of Energy under grant number DE-FG36-94GO10011.
[2] Now at GE Corporate R&D Center, P.O. Box 8, K1-EP123, Schenectady, NY 12301

Several design alternatives have been evaluated and traded-off for the conductor stabilizer. Aluminum stabilized conductor has been very favorably considered in the process. The NbTi/copper Rutherford cable can be co-extruded with a high purity aluminum (HPA) stabilizer in great length. The extrusion process for industrial application is proven and reliable. HPA has high initial RRR, therefore, the size could be minimized.

The major drawback of HPA stabilizer, however, is its very low mechanical strength. ML&P SMES design demands high field and high energy, which will significantly elevate the stress state within the winding pack. Significant plastic strain will be expected in HPA. This situation has imposed two major challenges. First, low strength aluminum conductor must be reinforced and confined by adequate structural materials to support the generated electromagnetic forces and to prevent excessive plastic deformation. Since the pool boiling cooling method requires the stabilizer to be exposed to helium fluid, picket fence type spacers were proposed to create channels for helium flow. The cooling exposure requirement has made the structural confinement task difficult. The second major challenge is that the plastic strain must be minimized to assure the conductor stability performance. During the alternating charge and discharge operation of SMES, the developed HPA plastic strain varies cyclically. Excessive cyclic plastic strain can cause significant RRR degradation [2]. Therefore, the cyclic plastic strain must be controlled to limit the RRR degradation.

In a companion paper of the principal author[4], the stress characteristics of a low aspect ratio energy storage solenoid has been demonstrated. It is shown that for low aspect ratio SMES solenoids, conductor turn-to-turn interactions are very strong. The high dependency between conductor turns defines a significant sensitivity of the stress/strain results to the quality of the conductor turn-to-turn contacts. Adequate pre-stress must be applied.

In this paper, an adequately reinforced and confined aluminum stabilized conductor configuration for ML&P SMES is described. The associated major structural considerations are presented.

## CONDUCTOR CONFIGURATION

There are four components in a conductor turn: co-extruded aluminum stabilized conductor, two pieces of radial support side bar, two layers of radial spacers, and a layer of backing strip. Shown in Figure 1, the aluminum stabilized conductor is integrated (brazed) at two edges with the two structural side bars, made of stainless steel. The conductor with side bar forms an integrated conductor, which is sandwiched by two layers of picket fence spacers, made of electrical insulating material, radially separating the conductor from the adjacent members.

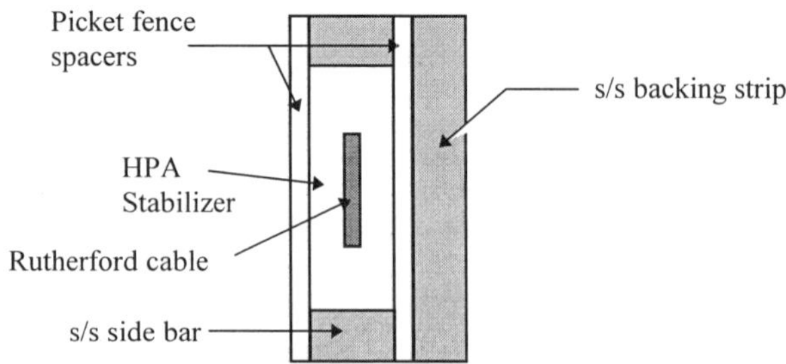


**Figure 1**. Reinforced and confined aluminum stabilized conductor for ML&P SMES.

**Table 2.** Conductor Parameters

| | | |
|---|---|---|
| Conductor Material | | NbTi |
| SC Rutherford cable | mm | 2x18 |
| Stabilizer Material | | High Purity Aluminum |
| Initial HPA RRR | | 1500 |
| Critical Current | kA | 18 |
| Side Bar Material | | SA 240-GRD 316L |
| Radial Spacer Material | | G 10 |
| Backing Strip Material | | SA 240-GRD 316L |
| Conductor Radial Surface Exposed | | 50% |

The primary structural member within a conductor turn is a backing strip, made of stainless steel. Key conductor parameters are summarized in Table 2.

## STRUCTURAL CONSIDERATIONS

### Design Criteria

Industrial codes and standards always require primary stresses in any loaded member to be less than the material yield strength. That is, gross sections of any structural member must not yield under normal and abnormal design conditions. There is significant quantity of high pure aluminum stabilizer in a conductor turn. HPA has very low strength, 17 MPa yield strength at 4.2K under annealed condition. It is impossible to economically design a magnet for this application to limit the HPA stress below its yield point. On the other hand, HPA can not be excluded from the load path since the electromagnetic forces are generated from the current in the embedded S/C cable. HPA must distribute the forces from the cable to the supporting members.

It is important to determine if HPA stabilizer is appropriate to be applied in this high field and high energy SMES magnet. To achieve the decision, HPA must be demonstrated to be fully confined. That is, its plastic deformation is restrained and limited by a sufficient containment surrounding the stabilizer. Unlimited plastic deformation should not occur, at any time under all design conditions. Total accumulated permanent deformation of HPA under alternating operating loads over the design life should be limited to an acceptable quantity. In addition, sufficient structural reinforcement must be provided to support the entire load.

If a composite conductor can meet the described confinement and reinforcement criteria for the low strength aluminum member, it would be acceptable.

### Cool-Down Thermal Stresses

The composite conductor is assembled from components made of different materials, which have different material properties, such as thermal contraction rate and Young's Modulus. The integrated linear thermal contraction from 300K to 4.2K of the component members are

**Table 3.** Thermal Contraction Rates

| | Average Contraction Rate |
|---|---|
| HPA | 0.42% |
| S/S | 0.33% |
| G10 (Normal) | 0.70% |

During the cool-down, interactive constraints will be developed between components with differential thermal contractions, thus thermal stresses are developed. Due to significant strength difference between HPA and s/s reinforcement members, the HPA stabilizer will easily reach the yield point and limit its restraints on the s/s members. Using thin wall cylinder under pressure formula, assuming yield stress of $\sigma_s$ for HPA, the residual hoop stress of the s/s reinforcement can be estimated as

$$\sigma_\theta = -\frac{\sigma_s A_{Al}}{A_{s/s}}$$

When $A_{Al}/A_{s/s} = 0.75$, $\sigma_s = 17$ MPa, $\sigma_\theta = -12.75$ MPa. It was showed in the ANSYS analysis that considering complex stress effect and stress distribution, the peak s/s hoop residual stress was -12.2 MPa, an excellent match to the estimate. If the strength of S/C cable was considered, the s/s residual stress would be slightly reduced to -10.3MPa. The residual hoop stress in the S/C cable would be -33.8 MPa.

The picket fence radial spacers are discontinuous in hoop direction, therefore, they do not create any constraints to adjacent members. Since radial spacers are in the radial load path, a uniform gap of 0.02 mm will be developed between each turn after the cool-down, as a result of differential contraction between stainless steel and G10 spacers.

### Structural reinforcement

Several finite element models were developed to confirm that adequate structure is provided to support the entire electromagnetic forces. Major structural models include (1) 2-D axisymmetric homogeneous, anisotropic solenoid model; (2) axisymmetric non-homogeneous single pancake model; and (3) three dimensional conductor local model.

The major objective of the 2-D solenoid ANSYS model was to calculate the magnetic field and electromagnetic force distributions. The results from this model was then applied to the non-homogeneous single pancake model and conductor local model.

In the axisymmetric pancake model, only structural members, which meet all stress limits, are modeled. Since the HPA stabilizer with S/C cable was not governed by the structural criteria, it was not included in the model. The pancake model was used to determine the primary stresses in the structural components. The calculated results are summarized in Table 4.

3-D local model was intended to demonstrate the stress concentration resulted from non-uniform details of the conductor configuration. Table 5 shows the peak Von Mises equivalent local stresses.

### HPA Confinement

Under the operating condition, the Nb/Ti Rutherford cable embedded within the HPA stabilizer could generate up to 31,000N/m radial force at the mid-plane inner first turn, and 42,900 N/m axial force, at the end-plane turn 31, which must be transferred and distributed

**Table 4.** Primary Stresses

| | Maximum Von Mises Equivalent Stresses (Assuming 0.02mm cool down gaps) | | |
|---|---|---|---|
| | Side Bars | Backing Strips | Allowable |
| Membrane Stress (M), MPa | 229 | 241 | 272 |
| Bending Stress (B), MPa | 1.68 | 14.8 | 408 |
| M+2/3B, MPa | 230 | 251 | 408 |
| Total, MPa | 230 | 254 | 1020 |
| Max. Radial Displacement | 3.66 mm | | |
| Peak Hoop Strain | 0.109% | | 0.122% |

**Table 5.** Local Stresses

| | Backing Strip Maximum | Side Bar Maximum | Allowable |
|---|---|---|---|
| Membrane Stress (M), MPa | 261 | 238 | 272 |
| Bending Stress (B), MPa | 7.98 | 110 | 408 |
| M+2/3B, MPa | 266 | 311 | 408 |
| Peak, MPa | 364 | 349 | 1020 |

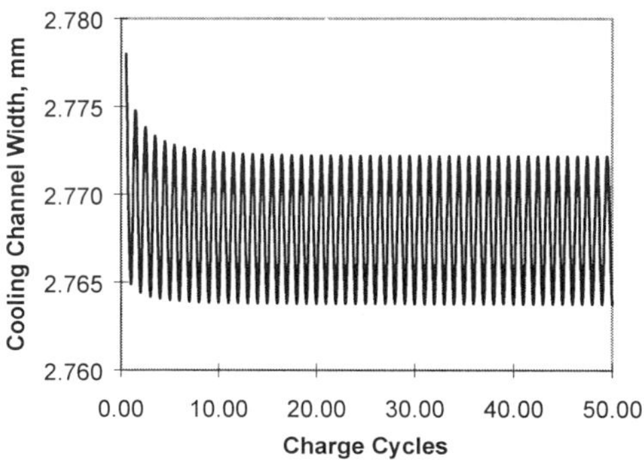


**Figure 2.** Dynamic cooling channel width with respect to charge cycles

to the structural reinforcement. When SMES is fully charged, total strain of HPA at all locations exceeds its yield point. The front and back faces of the aluminum stabilizer are supported by picket fence type spacers as the confinement members, which provide only a discontinuous support. Finite element analysis was performed to see if HPA under cyclic plastic deformation would have the potential to gradually close up the cooling channel. Figure 2 shows the dynamic cooling channel size under the full power cycles. It is shown that the strain hardening effect of HPA reduces the permanent plastic deformation of each cycle. Therefore, the cumulative total plastic deformation into the cooling channel cease to increase after approximately 15 charge cycles. In addition, the fact that the embedded Rutherford cable, which has significant higher strength and stiffness than HPA stabilizer acts like a "bone", increasing the conductor total stiffness. For ML&P SMES design, the cooling channel width would be reduced from 2.778 mm to a stable 2.764 mm after 50 charge cycles, a 0.014 mm reduction. It is evident that HPA is confined in radial direction.

Conductor is wound into a spiral coil with helix angle. Tangential shear force will be generated due to differential hoop forces between turns. The shear must be resisted by either friction or mechanical bonding to prevent unwinding the coil. Due to small helix angle in the design, the shear force is small, less than 100N/m. It was shown that the friction bond between HPA and its adjacent components is sufficient to prevent conductor from unwinding. The physical bonding between HPA and side bars assures that any relative motion between conductor members in tangential direction does not occur.

Another potential weak point of the confinement is also in tangential direction, but at the conductor splice location of the outer-most turn. There is a sharp tensile force reduction when structural reinforcement is separated from HPA at the splice. The longitudinal stress of HPA varies from its yielding point to free stress state, which results in a shear force of 11,300 N. In addition to a reliable conductor bonding, a mechanical lock was designed at the splice location to support the shear.

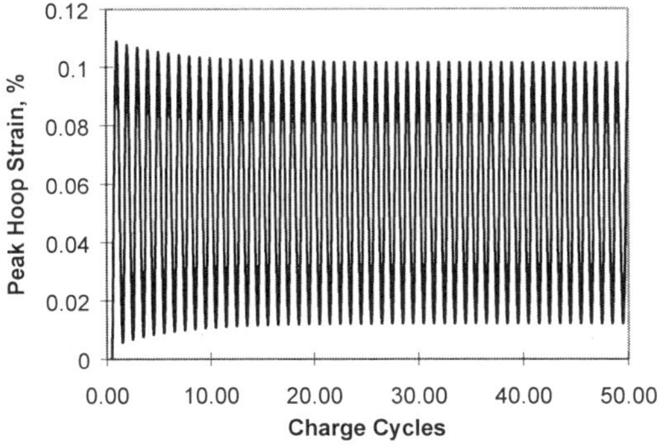


**Figure 3.** HPA peak cyclic hoop strain.

## HPA Operating Cyclic Plastic Strain

HPA can experience RRR degradation under alternating strain cycles[3]. The level of the strain range defines the magnitude of this degradation. Reducing the strain requires significantly increasing the structural material. The concept of using composite material was introduced. The structural stresses are defined by the amount of structural materials when solenoid dimension and power/energy specification are determined. It is independent of the type of structural material as long as the stresses are within the elastic range. From this observation, stainless steel which has a significant higher elastic Modulus than structural aluminum alloys was selected for the primary structural members. For the same stresses, the associated strain is reduced by a factor of $E_{s/s}/E_{Al}$.

HPA will initially yield at both fully charged and fully discharged operating points. However, due to strain hardening effect, HPA strength increases during the plastic cycles. Thus, the peak HPA strain would shakedown and the residual strain at 0 power would increase during operation cycles. The residual strain in HPA reflects the equilibrium point at the discharge mode. Figure 3 shows the cycle history of HPA peak hoop strain under 0 to full power cycles. It is shown that significant shakedown only occurs during the initial 20 cycles.

## Winding Pack Tightness Control (Pre-Stress)

The companion paper "Structural Design of Superconducting Energy Storage Solenoidal Magnets" presented in this conference has shown the significant turn-to-turn dependency of a low aspect ratio solenoid. Pre-stress is essential to provide a high quality contact if no bonding is used between contacting layers. There are three common ways of developing compressive hoop stress at the location where the operational hoop stress is significant: 1) take advantage of the differential thermal contraction rates; 2) apply tension during the winding; 3) apply mechanical preload clamping.

To wind a coil with tension, a coil form bobbin should be utilized to react the tension loads. A permanent coil form will assist to retain the prestress, however, the manufacturing cost and the operating AC losses during the SMES changing transients must be addressed.

# CONCLUSIONS

Aluminum stabilized conductor can be used in high field, high energy SMES magnets, provided the low strength aluminum stabilizer is sufficiently confined and reinforced. Due to significant strain hardening behavior, the cumulative plastic deformation of the aluminum stabilizer can converge to a stable amount after a few alternating charge cycles. The phenomena has significantly reduce the confinement requirement. A candidate aluminum stabilized conductor for ML&P SMES magnet is presented and analyzed. The applicability of such conductor has been demonstrated.

# REFERENCE


1. X. Huang, S.F.Kral, G.A. Lehmann, Y.Lvovsky, and M. Xu, "30MW Babcock and Wilcox SMES Program for Utility Applications', *IEEE Trans. Appl. Supercond.*, Vol.5, pp428-432, June 1995.
2. S. Kral, M. Aslam, P.F. Ribeiro, X. Huang, M. Xu, "Superconducting Power Delivery Systems For Transmission & Distributions", Presented at the 57th American Power Conference, Chicago, IL, April 18-20, 1995
3. K.T. Hartwig, G.S.Yuan "Strength and Resistivity Changes Caused by Cyclic Strain at 4.2K in Pure Aluminum", Cryogenic Materials'88, Vol.2, Structural Materials, 1988
4. Y. Wang, "Structural Design of Superconducting Energy Storage Solenoidal Magnets", Presented at 1997 CEC/ICMC Conference.

# DESIGN OF A MONOLITHIC 15 kA SUPERCONDUCTOR FOR SMES

O. R. Christianson,[1] D. T. Hackworth,[1] W. V. Hassenzahl,[2] T. Hordubay,[1] M. P. Krefta,[1] T. Kupiszewski,[1] G. T. Mallick, Jr.,[1] M. J. Supercynski, Jr. [3]

[1]Westinghouse Science & Technology Center
Pittsburgh, PA 15237, USA
[2]Piedmont, CA, 94611, USA
[3]Naval Surface Warfare Center
Annapolis, MD, 21402, USA

## ABSTRACT

The configuration and projected performance of a 15 kA monolithic superconductor designed for 0.36 to 72 GJ (0.1 to 20 MW-hr) SMES use for the Naval Surface Warfare Center is described. The conductor operates at 4.2 K and 1 atmosphere in a peak magnetic field of 5 T at 70% along the load line, which corresponds to a temperature margin of 1.7 K. The conductor consists of a Nb-Ti copper composite Rutherford cable soldered into a high purity aluminum stabilizer that has a nominal RRR of 1500. The conductor is cryogenically stable at the operating point. Pulsed devices and energy storage systems using superconductors require conductors with low AC losses. Internal eddy current barriers limit AC losses during a discharge. Extrapolation of the conductor configuration to further reduce AC losses is considered. A possible test configuration for the conductor is described as it is related to design specifications.

## INTRODUCTION

Westinghouse has designed a low AC loss 15 kA monolithic conductor for SMES installation from one to tens of megawatt hour rating for large-scale applications. This report summarizes the description of the conductor design. Design requirements are given in Table 1.

---

Work supported by Naval Sea Systems Command in support of the SMES program office at the Naval Surface Warfare Center - Annapolis, MD.

**Table 1.** Design Requirements

| Operating Point: | |
|---|---|
| Current | 10 - 15 kA |
| Magnetic Field | 5 T |
| Temperature, LHe bath | 4.2 K |
| Joint Heating, maximum | 5 W |
| SMES Stored Energy | 0.1 to 20 MWhr |
| Coil Discharge Duration, minimum | 10 seconds |

## CONDUCTOR SPECIFICATIONS

The proposed low AC loss conductor, shown in Figure 1 and described in Table 2, consists of 48 Nb-Ti copper composite strands that are 0.808 mm diameter. These strands are formed into eight subcables. These subcables form a Rutherford cable. The cable is ultrasonically soldered into a groove in the rectangular high purity aluminum stabilizer that has a RRR of 1500.

The monolithic superconductor is designed to have a margin of 70% along the load line and a temperature margin of 1.7 K at the operating point of 15 kA, 5T, 4.2 K, see Figure 2. Calculations indicate that the minimum propagating zone for the overall conductor is 28 cm and that the minimum quench energy is 0.41 J. Initial calculations for these SMES applications were for a 10 kA conductor. The conductor described here was initially designed for 10 kA, but found to have ample margin and to be cryostable at 15 kA.

## JOINT DESIGN

Several types of joints can be made in this conductor. The joint design chosen, called a quad overlap, consists of two superconducting shunts each made of a number of conductor subcables overlapping the ends of the conductor. The shunts occupy rectangular cross sectional volumes which have been cut out of the interior side walls of the stabilizer slot on the sides of the conductor cable. Therefore the joint outer dimensions are the same as the conductor. Details of the expected joint electrical performance calculated at 10 kA are included in Table 3. (Initial calculations assumed a 10 kA transport current. Margin and cryostability criteria allow 15 kA operation.)

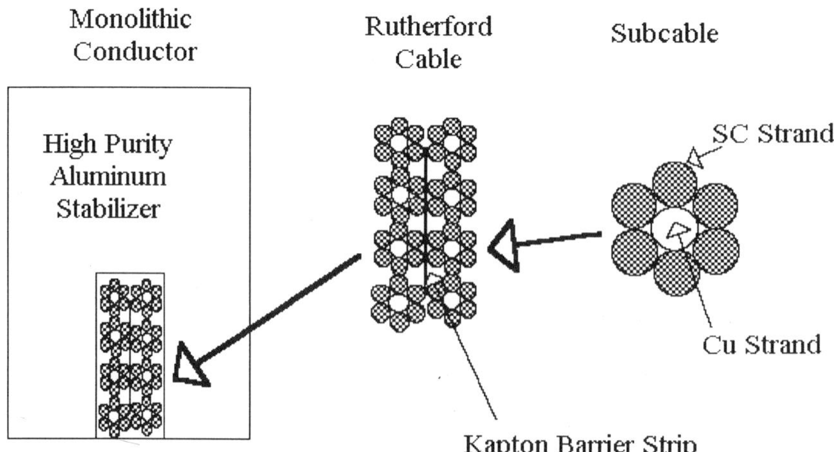


**Figure 1.** The superconducting cable showing a subcable and the internal Kapton barrier strip.

**Table 2.** Superconductor Specifications

| **SC Strand:** | |
|---|---|
| Strand Ic(A) @ 5T and 4.2 K | 700 (short sample) |
| Filament Diameter (μm) | 6.6 |
| SC Strand Diameter (mm) | 0.808 |
| SC strand Cu:nonCu ratio | 1.24:1 |
| Strand Cu RRR | 194 |
| Strand filament twist pitch (mm) | 18 |
| Strand twist pitch (mm) | 25.4 |
| "η" Factor 5T | 60 |
| **Cable:** | |
| No. of SC copper composite Strands | 48 |
| No. of pure Cu Strands | 8 |
| Cu Strand Diameter (mm) | 0.808 |
| No. of Subcables | 8 |
| Subcable diameter (mm) | 2.438 (w/solder) |
| Cable lay pitch (cm) | 7.6 +/- 0.5 |
| Insulating barrier width (mm) | 6.35 |
| Insulating barrier thickness (mm) | 0.127 |
| Cable width (mm) | 11.0 |
| Cable thickness (mm) | 4.7 ± 0.1 |
| Pb:Sn:Sb solder composition (wt%) | 39.47:60.08:0.39 |
| **Stabilizer:** | |
| Stabilizer dimensions (mm) | 23.75 x 16.25 |
| Stabilizer slot dimension (mm) | 11.5 x 5.0 |
| Wetted perimeter (mm) | 23.75 |
| Stabilizer area ($cm^2$) | 3.28 |
| HPAL RRR | 1500 +/- 600 |

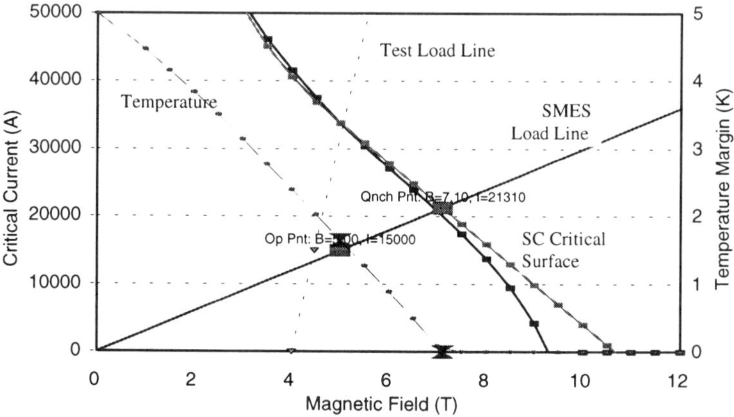


**Figure 2.** The superconductor critical surface and a SMES device load line. The test load line produced by the background magnet and the test specimen is also shown.

| Table 3. Quad Overlap Electrical Performance at 10 kA | |
|---|---|
| Length (mm) | 415 |
| Loss (Watt) | 0.78 |
| Resistance | 7.8 nOhm |
| Heat Flux (W/cm$^2$) | 0.008 |
| Max. Temperature (K) | 4.34 |

## CURRENT DIFFUSION EFFECTS DURING QUENCH

Current will diffuse into the high purity aluminum stabilizer during a quench, but will not flow uniformly in the stabilizer for a second or so. The redistribution of current results in an initially high loss in the superconductor and the aluminum stabilizer close to the superconductor, see Figures 3, 4, and 5. Figure 3 shows the non-uniform current distribution in the stabilizer at 0.2 s where sections of the stabilizer close to the superconducting cable initially carry more current than sections further away from the cable. Figure 4 shows the joule heating in the superconductor and stabilizer as the current redistributes. Figure 5 simulates voltage traces for non-inductive voltage taps on the surface of the stabilizer. The excess heating enhances the quench propagation velocity, see Figure 6.[1,2,3]

## HALL EFFECTS

Hall currents increase the effective magnetoresistivity of cryogenically cooled high purity aluminum in the presence of another dissimilar conductor.[4] This aluminum stabilized conductor is analyzed for the importance of the Hall effect using a finite element technique.[5] At a field of 5T, the effective magnetoresistivity increases by 33% due to Hall currents, which decreases the cryostable current by roughly 20%. Changes in the aluminum RRR with processing may obscure this slight increase in resistivity.

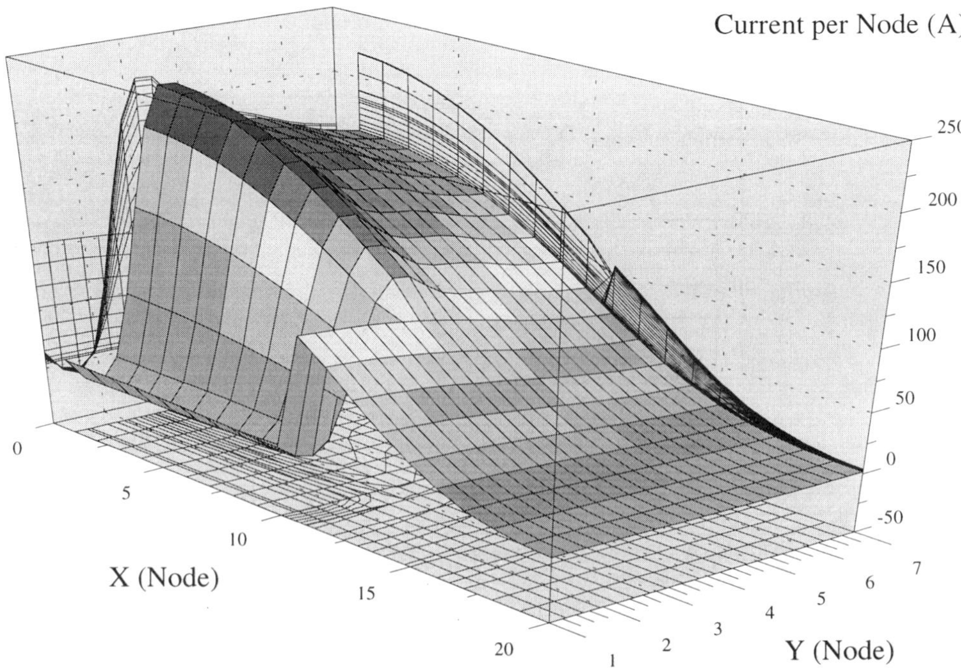


**Figure 3.** Current distribution due to current diffusion in the conductor at 0.2 seconds.

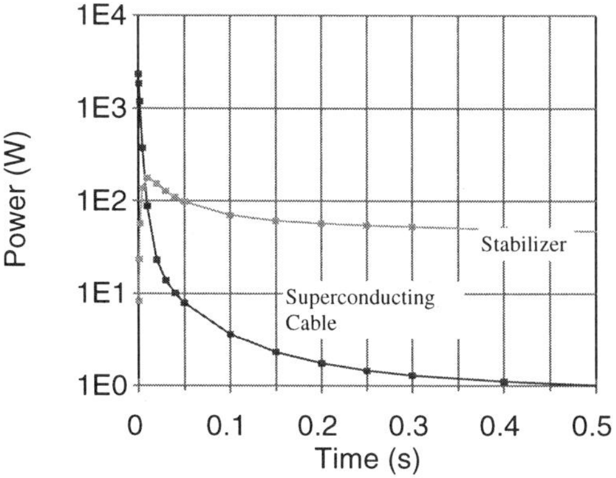


**Figure 4.** Losses in the superconducting cable and the aluminum stabilizer due to current diffusion during a quench.

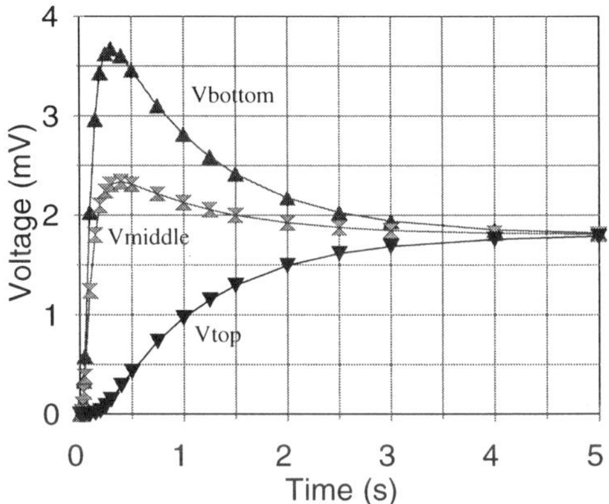


**Figure 5.** Simulated voltage traces for non-inductive voltage taps placed on the surface of the stabilizer showing current diffusion effects.

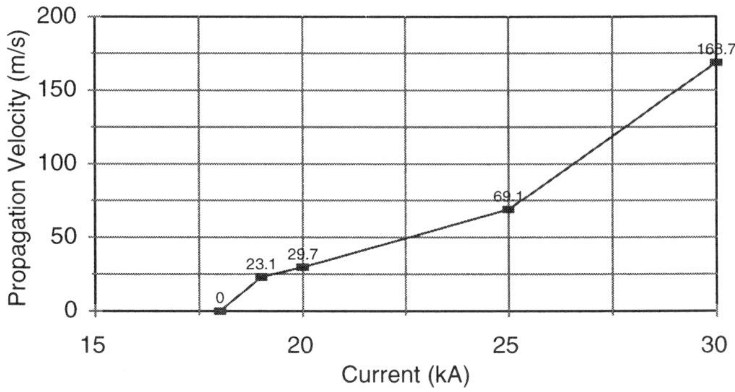


**Figure 6.** Quench propagation velocity enhanced by current diffusion.

**Table 4.** Losses for a 10 kA transport current at two different charge/discharge cycles

| Stored Energy | | Discharge Power | Charging Time | Discharge Time | Power Loss |
|---|---|---|---|---|---|
| MWhr | J | MW | s | s | W/m |
| 21 | 7.56E+10 | 400 | 1800 | 94.5 | 6.4E-03 |
| 0.5 | 1.8E+09 | 30 | 600 | 54 | 1.8E-01 |

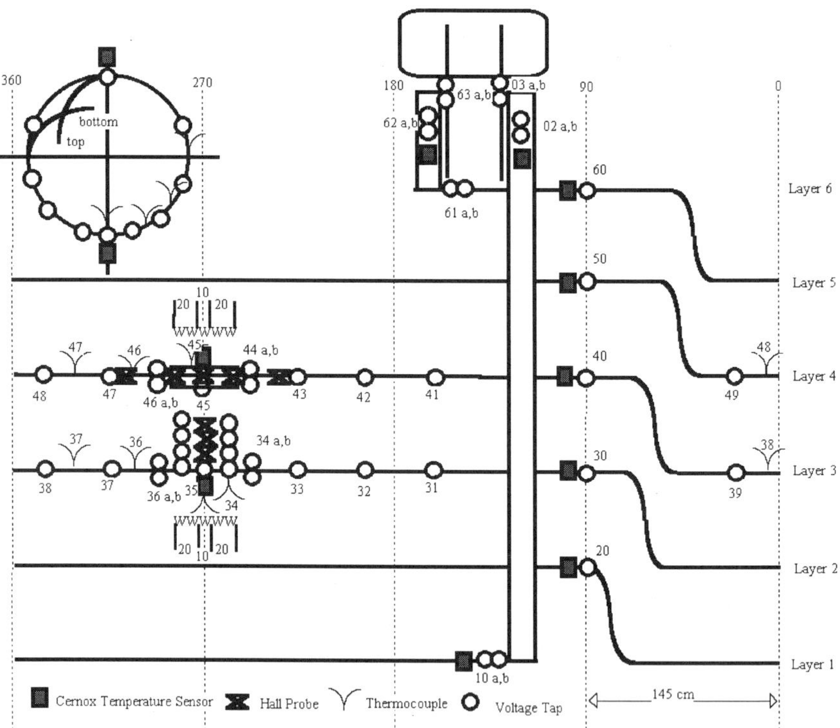


**Figure 7.** Possible test configuration to measure performance of the 15 kA SMES conductor.

## DESIGN VERIFICATION

To prove that this conductor will meet SMES performance requirements will require thorough testing. An experiment schematically shown in Figure 7 could test conductor design capabilities, including the superconductor critical surface, minimum quench energy, recovery currents, and quench propagation velocities. Resistive heaters would be used to initiate a quench. Measurements of quench propagation and current diffusion effects would require tens of meters of conductor length, equal to several turns of the conductor, and special arrangement of voltage taps and magnetic field probes. Joint performance could be measured by including a joint in another section of the conductor, which requires a couple of additional turns. A possible experiment would require five turns at a 1.5 to 2 meter diameter.

## AC LOSSES

AC losses have been calculated for the initial 10 kA operating conditions. AC losses are minimized by including a Kapton insulating barrier within the superconducting cable, soldering the cable in the aluminum stabilizer instead of coextruding, and by orienting the conductor such that the magnetic field is perpendicular to the narrow face of the aluminum stabilizer. Losses calculated at a 10 kA transport current for two charge and discharge cycles are listed in Table 4. Losses in the stabilizer dominate, and are given by

$$W/l = hw^3 \frac{\sigma \dot{B}^2}{12} \tag{1}$$

for a rectangular conductor where w is the dimension of the stabilizer perpendicular to the magnetic field, h is the height of the stabilizer, $\sigma$ is the conductivity, and $\dot{B}$ is the time rate of change of the magnetic field. Further reductions of the losses in the stabilizer can be affected by notching the surface ($W/l \Rightarrow 0.25\ W_o/l$) or by including internal barriers ($W/l \Rightarrow 0.10\ W_o/l$).

## PRESENT STATUS OF CONDUCTOR DEVELOPMENT

A short length of conductor, less than 100 m, is being fabricated. The superconducting strands and subcables have been fabricated by Oxford, and the New England Electric Wire produced the Rutherford cable. High purity aluminum was purchased from Sumitomo Chemical Company Limited and extruded by National Extrusion & Mfg. Co. All components of the conductor are available and are ready for final conductor fabrication.

## REFERENCES


1. O.Christianson, Normal Zone Evolution and Propagation in a Cryogenically Stable Superconductor, in: "Advances in Cryogenic Engineering," Vol. 31:383-390 (1986).
2. X.Huang, Y.M.Eyssa, and M.A.Hilal, Dynamic and Transient Stability of He II Cooled Conductors, in: "Advances in Cryogenic Engineering," Vol. 35:155-163 (1990).
3. J.M.Pfotenhauer, M.K.Abdelsalam, F.Bodker, D.Huttleston, Z.Jiang, O.D.Lokken, D.Scherbarth, B.Tao, and D.Yu, Test Results from the SMES Proof of Principle Experiment, *IEEE Transactions on Magnetics* 27(2):1704-1707 (1991).
4. P.W.Eckels, N.C.Iyer, A.Patterson, A.T.Male, J.H.Parker, Jr., and J.W. Coltman, Magnetoresistance: the Hall Effect in Composite Aluminum Cryoconductors, *Cryogenics* 29:748-752 (1989).
5. M.P.Krefta, O.R.Christianson, and J.H.Parker, Jr., Steady-State Hall Losses in Composite Cryogenic Conductors, *Cryogenics* 36:291-301 (1996).

# A METHOD TO BUFFER AC LOSSES IN A FORCED FLOW SUPERCONDUCTING MAGNET

B. K. Parsons and C. A. Luongo

Bechtel National, Inc.,
P. O. Box 193965,
San Francisco, CA 94119-3965, USA

## ABSTRACT

A SMES device is proposed for an application requiring repeated charging and discharging over a short period of time, followed by a long period of relatively low usage (standby). The magnet consists of a toroid wound with a cable-in-conduit conductor (CICC) cooled by forced flow supercritical helium, each coil module containing a limited helium inventory. The hypothetical operating scenario results in high energy deposition due to AC losses into the SMES magnet over the short (less than one hour) operating period on an approximately daily cycle. Removing this energy deposition in real time would require a refrigeration system of extraordinary size. To reduce the size of the necessary cooling system, while maintaining operability in the postulated scenario, the supercritical circuit includes a large separate helium inventory to buffer the AC loss energy deposited during the discharge cycle. A methodology is described to establish the required supercritical inventory for magnet pulsed operation, and to compare the present approach with steady-state removal of AC loss loads.

## INTRODUCTION

In Superconducting Magnetic Energy Storage (SMES) energy is stored in the magnetic field generated by a high-current superconductor wound to form a magnet. As energy is removed from the SMES the magnetic field decreases and this change in field generates eddy current flows within the metal parts of the conductor. These eddy and coupling currents generate heat within the magnet and are referred to as AC losses. The amount of AC loss generated in a magnet depends on the nature of the energy removal (rate and duration) and the characteristics of the conductor.

A common magnet design incorporates the cable-in-conduit conductor (CICC) cooled in forced-flow by supercritical helium at 4.2 K. The CICC concept offers two distinctive advantages: low AC losses in the cable relative to monolithic conductors, and a low helium inventory relative to bath-cooled magnets. However, this low helium inventory can actually become a drawback in heavily pulsating applications. Even though the AC losses per pulse are low in a CICC, there may not be enough helium inventory within the CICC to absorb the cumulative AC losses of frequent pulses. The design challenge is how to accommodate the AC losses of repeated pulsing without exceeding the current sharing temperature of the conductor (the point at which an irreversible quench would be initiated).

The conventional way of accommodating AC losses from a single pulse consists simply of (a) designing the CICC with sufficient operating enthalpy margin such that a single discharge will not quench the magnet, and (b) ensuring sufficient refrigeration capacity such that the energy deposited during a discharge event can be removed by the refrigerator, restoring the coil to design temperature, prior to the initiation of the next discharge event. However, in an operating scenario where a large number of discharges is postulated over a short period of time, followed by a long period without discharge activity, steady state removal of the losses is not feasible. Under those circumstances the enthalpy margin of the CICC helium volume is not enough to keep the temperature below critical, but some type of 'averaging' of the losses over the longer period may be feasible. This paper presents a refrigeration scheme in which the AC losses of pulsed operation in a CICC magnet are buffered to allow removal over a longer period by steady state operation of a smaller refrigeration system than would otherwise be needed.

## MAGNET DESCRIPTION

This description is provided to serve as an example. It should be emphasized that this refrigeration approach can be applied to a range of applications with similar operating considerations. The superconducting magnet of interest is a toroid wound inside a vacuum vessel cryostat with a single thermal shield. The conductor is a $Nb_3Sn$ CICC rated at 70 kA at a peak field of 8 T and operating at 4.2 K and 5 atm helium (supercritical). The CICC (shown in Figure 1) features a central cooling channel to provide a low pressure drop path for helium circulation during magnet pulsing (when flow rates have to be high). The central channel wall is perforated so that there is helium exchange between the channel and the cable space. The magnet characteristics are summarized in Table 1. The toroid (shown in Figure 2) is composed of 24 racetrack coils storing 535 MJ. All 24 coils are connected in series so that the magnet is a single electrical circuit that can provide the required power pulses (described in the next section) without exceeding 5.5 kV at the terminals. The coils are connected in parallel hydraulically (or even individual pancakes within the coils can be paralleled) to provide the short flow paths needed for cooling during pulsed operation.

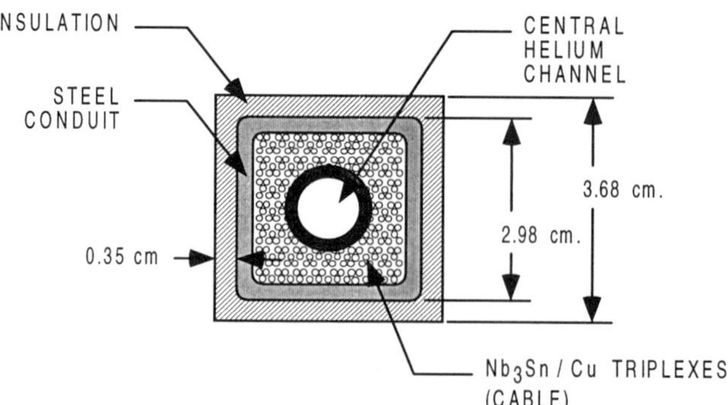


**Figure 1.** Cable-in-conduit conductor cross section.

**Table 1.** Magnet Characteristics

| Parameter | Value |
|---|---|
| Magnet Type: | Toroid |
| Number of Coils in Toroid: | 24 |
| Stored Energy: | 535 MJ |
| Peak Field: | 8.1 T |
| Conductor Current: | 70 kA |
| Peak Voltage: | 5.5 kV (terminal) |

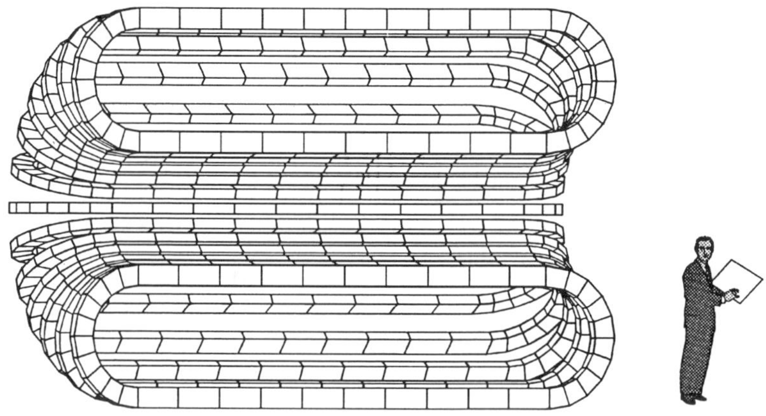

**Figure 2.** Schematic of the toroidal magnet (half of the racetrack coils are shown).

## MAGNET OPERATING SCENARIO AND HEAT LOADS

The magnet of interest is intended to store energy and service a pulsed load with a high degree of repetition. Power pulses start 45 seconds apart and comprise a 2-second triangular pulse with a peak of 150 MW, so that 150 MJ are withdrawn from the magnet with each pulse. The scenario comprises 50 consecutive pulses over roughly one hour once a day (so the system operates for 1 hour and is on standby for 23 hours). The magnet, storing 535 MJ, is therefore capable of providing three consecutive power pulses without recharging. The pulses can be delivered from a fully charged, 2/3 charged, or 1/3 charged magnet. In actuality the magnet pulsed operation involves a combination of charge/discharge scenarios such as the one depicted in Table 2; 32 of the 50 pulses are delivered from a fully charged magnet, 12 pulses are delivered at 2/3 charge, while 6 of the pulses are from 1/3 charge. This distinction is important because the AC losses per pulse are a strong function of the magnet state of charge at the beginning of the pulse.

The AC losses per pulse are calculated using the standard expression:

$$\dot{q}'''_{AC} = \frac{2\Theta}{\mu_o}\left(\frac{dB}{dt}\right)^2 \tag{1}$$

where $\dot{q}'''_{AC}$ represents the AC losses per unit volume (W/m$^3$) of conductor metal, dB/dt is the rate of change of the local magnetic field, and $\Theta$ is the AC loss time constant which combines all the effects of eddy currents and coupling losses at the different stages of the cable. The AC loss time constant was conservatively estimated as 150 ms. A magnetic field simulation code (*M'c*, part of the *SuperMagnet* suite by CryoSoft of Thoiry, France) is used to calculate the field distribution in the toroid. From the field distribution and the current profile as a function of magnet charge for the different pulse scenarios, it is possible to infer the instantaneous AC loss at the peak field location (maximum loss) and average loss over the conductor length for all cases of interest. Because (dB/dt) changes over time differently depending on the magnet state of charge, the AC losses are different for the three different types of pulses. Figure 3 shows the average instantaneous power dissipation over the 2-second pulse for each of the three cases under consideration: pulse begins from a fully charged, from a 2/3 charged, and from a 1/3 charged magnet (denoted as 1st, 2nd, and 3rd discharge respectively). The power dissipation shown in Figure 3 is expressed per unit length of conductor (rather than per unit volume). Similar curves can be produced for all locations along the conductor as the magnetic field profile is fully mapped. It is then possible to calculate the total AC loss energy deposition for the entire magnet for each type of pulse, as summarized in Table 2. It should be noted that as the charging time is more than an order of magnitude greater the discharge time (40 s compared to 2 s), AC losses during the charge cycle are negligible compared to the those from discharge and have not been included. The quantities of Table 2 are summed for the complete scenario of 50 pulses to yield a total energy deposition to the coil of about 11 MJ after about an hour of pulsed operation. This amount of energy deposited at 4.2 K needs to be removed over the ensuing 23 hours for the magnet to return to its initial state prior to pulsed operation the following day.

The next two sections present two refrigeration approaches to remove the AC loss heat load from the magnet, one based on steady state cooling, the second based on buffering the load and 'averaging' the load for removal over the entire 24 hour period.

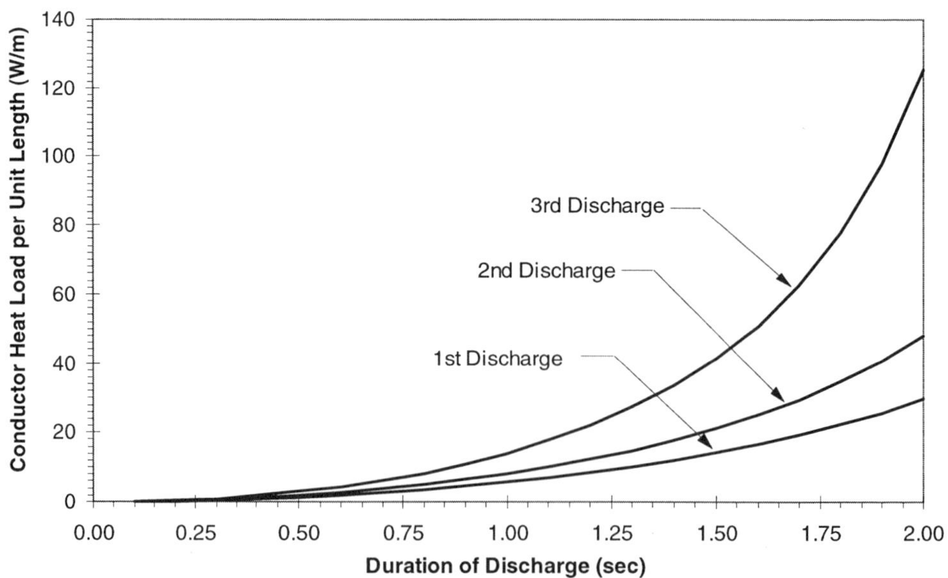


**Figure 3.** AC Loss vs. pulse duration for three different discharge pulses. 150 MJ of the 535 MJ total magnet stored energy are delivered with each discharge.

**Table 2.** Total AC Loss Energy Deposition

| | Number of Pulses | AC Loss / Pulse (kJ) |
|---|---|---|
| Pulses at full power (*i.e.*, 1st discharge): | 32 | 158 |
| 2nd Discharge Pulses: | 12 | 238 |
| 3rd Discharge Pulses: | 6 | 509 |
| Total: | 50 | 10,980 |

**Table 3.** System Heat Loads Exclusive of AC Losses

| Heat Loads (W) | | Current Lead Load | Refrigerator Compressor |
|---|---|---|---|
| At 4.5 K | At 45 K | (l/h) | Power (kW) |
| 240 | 348 | 75 | 235 |

## CONVENTIONAL REFRIGERATION APPROACH

Exclusive of the AC losses, the heat loads to the refrigerator (equivalent loads at the temperatures shown) for the subject system are estimated to be as shown in Table 3. The required refrigerator compressor power has been calculated assuming a liquefaction coefficient of 100 W/(g/s) and an efficiency of 15% of Carnot.

A conventional approach to refrigeration in this operating scenario would be to remove this AC loss energy in real time, *i.e.*, between pulses, so that the coil is returned to its initial operating temperature prior to the occurrence of the subsequent pulse. To quantify such a refrigeration capacity, one can consider an 'average' deposition energy of about 220 kJ per pulse at an 'average' frequency of once per minute during the pulsed operation period. This represents an equivalent heat load of about 3670 W at 4.5 K. If this load is added to those shown in Table 3, the refrigeration system is then sized at about 4.24 kW equivalent at 4.5 K and its compressor power requirement jumps to 1.85 MW. Using the estimating methods of Green *et al*[1], this equipment would cost on the order of $3.8 million.

## BUFFERED REFRIGERATION APPROACH

The buffered refrigeration approach uses a supplemental helium volume to increase the thermal mass of the supercritical system as shown schematically in Figure 4. The temperature of the entire supercritical volume is allowed to climb during the pulsed discharge period but the overall rise does not bring the temperature to its critical value. The system operates as follows. In idle operation, the base heat loads to the magnet (radiation from the enclosure, conduction through the support structure, pumping energy, and heat generated in conductor splices) are passed to the supercritical helium circulating through the conductor. The circulating helium carries this enthalpy increase to the buffer where heat is exchanged with the subcritical helium flow from the refrigerator. In this state, the load on the helium refrigerator is at some value below the refrigerator capacity, and the supercritical loop is maintained at its rated operating temperature. Also, the pumping energy can be kept to a minimum as the required helium flow rate through the magnet in this condition is small (approximately 11 g/s).

When the system operates in a rapid pulse discharge mode, the cooling of the refrigerator is increased to its maximum, but its capacity is not enough to keep the

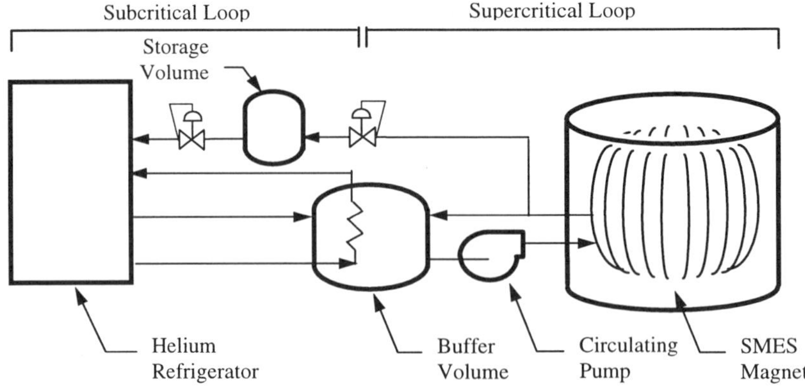


**Figure 4.** Cryogenic system schematic.

temperature in the supercritical loop at its design point. To keep the magnet at as low a temperature as possible, the helium flow as delivered by the circulating pump is increased during the discharge period to a rate which flushes the conductor between pulses, *i.e.*, in 45 seconds. The comparatively warm helium in the magnet after a pulse is returned to the buffer volume where it mixes with the existing inventory. As AC loss energy accumulates, the temperature of the supercritical loop is allowed to slowly rise. Pressure in the supercritical loop is maintained at a constant value by venting helium. By virtue of the helium mass in the buffer volume, the maximum temperature reached at the end of the discharge cycle remains below the current sharing temperature. Over the course of the discharge cycle the magnet looses temperature margin gradually, but retains enough to continue safe operation. After the cycle is complete, the refrigerator can begin to return the supercritical loop to its base temperature. The required capacity for the refrigerator is then determined by the total AC loss energy deposited and the time available between discharge cycles to remove it (23 hours).

The required buffer volume is determined as follows. For the conductor specified, the operating condition is 4.2 K at 5 atm, the coil helium volume is 2.23 $m^3$, the current sharing temperature at the operating point is 7.8 K. An allowable supercritical system temperature rise of 1.5 K is selected for a maximum conductor temperature at the start of a pulse of 5.7 K. That is, the magnet can execute the last pulse from any charge state with a starting temperature of 5.7 K and still have sufficient temperature margin during the final discharge. Due to the decrease in density with increasing temperature, if a constant pressure is maintained in the supercritical loop, helium will be expelled as the pulsing cycle progresses. Applying an energy balance to a control volume consisting of the supercritical system, the system volume can be approximated by:

$$Q_{in} = V_s(\rho_f u_f - \rho_i u_i) + V_s \left[ \int_i^f h \, d\rho \right]_{p = const} \tag{2}$$

or

$$Q_{in} \approx V_s(\rho_f u_f - \rho_i u_i) + V_s(\rho_i - \rho_f)\left(\frac{h_i + h_f}{2}\right) \tag{3}$$

where $Q_{in}$ is the AC loss plus pump heat input, $V_S$ is the volume of the supercritical system, $h$ is enthalpy, $u$ is internal energy, $\rho$ is density, and $i$ and $f$ indicate initial (4.2 K, 5.0 atm) and final (5.7 K, 5.0 atm) conditions respectively.

In pulsed operation, the required helium flow rate is determined by the need to completely flush the helium in the coil in the period between pulses. This ensures that the AC loss thermal energy is removed from the coil prior to the subsequent pulse. Based on the conductor flow path and flow area, and the time between pulses (45 seconds), the required coil helium flow rate in the pulse mode is about 1480 g/s at an estimated ΔP of 48.8 kPa. Applying a typical pump efficiency of 65%, total pumping power is about 80 W, which is treated as a heat load to the supercritical system during the period of pulsed operation. Including the AC loss energy of 10.98 MJ, the total input energy is 11.2 MJ, so the required supercritical loop helium volume is 11.3 $m^3$, and the buffer volume is 9.1 $m^3$ (subtracting the magnet helium inventory). For this example, the required buffer volume could be contained in a spherical vessel 2.6 m in diameter or in a cylinder 2 m in diameter by 3 m long.

The helium expelled from the supercritical loop during the discharge period, about 340 kg, must be captured either as warm gas compressed for storage, or at low temperature prior to recovery of cooling in the refrigerator. As the expelled mass is substantial, the low temperature option is more attractive both to limit the variation of flow rates in the refrigerator, and because of the large volume required for ambient storage even at elevated pressures. If this helium is released from the supercritical loop to a volume at intermediate pressure, it can be captured economically.

To accomplish this, a storage tank is provided which initially contains helium vapor at 4.2 K and 1 atm. To determine the required storage volume, an energy balance can be performed which determines the internal energy in the storage volume after an injection of cold helium from the supercritical system:

$$M_f^T u_f^T = M_i^T u_i^T + \Delta m^S \bar{h}^S \tag{4}$$

where $M^T$ is the mass in the storage volume, $\Delta m^S$ is the mass leaving the supercritical system, $h^S$ is the enthalpy of the expelled supercritical mass, $u^T$ is internal energy of the mass in the tank, and $i$ and $f$ indicate initial and final conditions respectively. By solving this equation stepwise with incremental mass flows from the supercritical system, for a given storage volume, the final conditions (internal energy and specific volume) of the storage tank can be approximately determined. The final temperature and pressure of the storage tank are then established from property tables. Using this approach, it can be shown that a storage volume of 5.0 $m^3$ will receive the expelled mass and end the pulse cycle at conditions of about 5.5 K and 3.0 atm. Over the remaining 23 hours of the daily cycle, this mass can be expanded at a comparatively constant rate to return to the refrigerator at about 1 atm and saturated conditions.

In comparison to the conventional approach, 11.2 MJ refrigerated over a 24 hour (86,400 sec) period represents an average load of about 129 W of refrigeration at 4.5 K. Adding this to the base load as shown in Table 3, a system of about 695 W equivalent at 4.5 K is required, with a capital cost of under $1.4 million, including an allowance for the buffer and storage vessels. This comparison is summarized in Table 4.

A related approach to refrigeration in pulsed operation is described by Slack[2] wherein a two-phase dewar with internal heat exchangers provides a buffering action through pressure variation. Several significant differences between these two approaches should be noted. The system described here has a distinct supercritical operating loop, which contains the buffer and storage volumes. The "averaging" of AC loss loads takes place entirely within this supercritical loop, so the thermodynamics of the process are quite different. Finally, this system is designed to function over a discharge period of considerable length, comprised of a large number of pulses, whereas the arrangement proposed by Slack averages pulse loads over the period between pulses.

**Table 4.** Comparison of Conventional and Buffered Systems

| Parameter | Conventional | Buffered |
|---|---|---|
| Base Load at 4.5 K | 566 W | 566 W |
| Equivalent AC Loss Load at 4.5 K | 3670 W | 129 W |
| Maximum Equivalent Load at 4.5 K | 4236 W | 695 W |
| Room Temperature Power | 1854 kW | 304 kW |
| Estimated System Cost | $3.8 million | $1.4 million |

## CONCLUSIONS

A refrigeration method is proposed for forced-flow superconducting magnets operating under heavily pulsating loads. The principle is to buffer the AC losses in a helium volume within the supercritical loop and outside the magnet cryostat. By "averaging" the loads over a time longer than the pulsed operation period, significant reductions in refrigerator cost and power requirements can be achieved. The cost of extra tankage required by this method is reasonable and justifies the approach.

## ACKNOWLEDGMENT

Work performed under contract P.O. 86-FNX-18795 with Westinghouse Electric Corporation, Machinery Technology Division.

## REFERENCES

1. M. A. Green, R. A. Byrns, S. J. St. Lorant, Estimating the cost of superconducting magnets and the refrigerators needed to keep them cold, "Advances in Cryogenic Engineering", Plenum Press, New York (1991), Vol. 37A, p. 637.
2. D. Slack, A novel scheme to handle highly pulsed loads with a standard helium refrigerator, "Advances in Cryogenic Engineering", Plenum Press, New York (1994), Vol. 39, p. 1223.

# INTEGRATED DESIGN OF CRYOGENIC REFRIGERATION SYSTEMS FOR SMES SYSTEMS

Brian J. Bowers and Joseph L. Smith, Jr.

Massachusetts Institute of Technology
Cambridge, MA 02139

## ABSTRACT

This study investigated techniques for specifically matching the designs of a SMES magnet and its cryogenic refrigeration system. Computer models estimated the heat that can be removed by a steady flow of helium through a CICC flow passage. This data was generalized for use with flow passages of different sizes. The transient response of a recirculator-loop cooling scheme was modeled in a manner consistent with the characteristics of the magnet, pump, heat exchangers, and buffer tank. Several buffers were compared and a novel buffering technique was proposed. The results showed that the pumping power required for cooling of CICC magnets can significantly increase the refrigeration load. The pumping power can be reduced by re-cooling the helium before it enters the pump and by using a higher pressure in the system. Parallel flow cooling schemes and high efficiency pumps can also decrease the pumping power and allow a smaller refrigerator to be used. These results show it is important to include pump characteristics when calculating SMES refrigeration loads.

## INTRODUCTION

The cryogenic refrigeration system is a significant part of any superconducting magnetic energy storage (SMES) system. A method for matching the designs of the magnet and refrigeration system could minimize the required cooling power and lead to a smaller, less expensive refrigeration system. This study modeled magnet/refrigerator interactions in the lowest temperature area of the SMES system since this area affects the performance of the entire refrigerator. Magnets made with a cable-in-conduit conductor (CICC) were considered.

## STEADY STATE MODELING OF HELIUM FLOW THROUGH A PASSAGE

A computer model[1] simulated steady flow of supercritical helium through a generalized CICC passage (length L, hydraulic diameter $D_h$, cross-sectional area A, friction factor f). The equations giving the changes in temperature and pressure per unit length (dT/dx and dP/dx) were found using energy conservation and standard pipe flow relations for one-dimensional flow without axial conduction. A finite difference program numerically integrated these governing equations to find the conditions throughout a particular conductor passage as a function of the distance from the inlet (x). The inlet conditions of the conductor were chosen along with the maximum allowable temperature ($T_{max}$) and the minimum allowable pressure ($P_{min}$). The core of the program calculated the conditions throughout the conductor assuming a constant heat load per unit length ($q'$) and a constant mass flow rate ($\dot{m}$). An iteration routine then found the maximum allowable heat load ($q'_{max}$) for a range of temperature and pressure constraints.

To generalize the results, the governing equations were non-dimensionalized by defining a dimensionless temperature ($\theta=(T-T_{ref})/\Delta T_{ref}$) pressure ($\Pi=(P-P_{ref})/\Delta P_{ref}$) and distance ($X=x/L$) [Eqns. (1), (2)]. All fluid properties, including density ($\rho$) and specific heat (Cp), are functions of the dimensionless distance from the conductor inlet (X). These equations were used to find a parameter that can define thermodynamically similar flow in passages of different dimensions. Flows were considered similar if each has the same fluid conditions at the same fraction of the distance through the passage (X). To equate the temperature and pressure, Eqns. (1) and (2) would have to be integrated for each passage. However, the integration is avoided by noticing that the slopes must also be the same since the distance has been normalized. Thus, for two conductors (1 and 2) to have the same dimensionless temperature-pressure profile, they must have the same inlet conditions and Eqn. (3) must be valid at any location X.

$$\frac{d\Theta}{dX}(X) = \frac{q' \cdot L}{\dot{m} \cdot C_p(X) \cdot \Delta T_{ref}} + \frac{\Delta P_{ref}}{C_p(X) \cdot \Delta T_{ref}} \left[ \frac{\Theta(X) + \frac{T_{ref}}{\Delta T_{ref}}}{\rho(X)^2} \cdot \frac{\partial \Pi}{\partial \Theta}(X)\bigg|_{\rho} \cdot \frac{\partial \rho}{\partial \Pi}(X)\bigg|_{\Theta} - \frac{1}{\rho(X)} \right] \frac{d\Pi}{dX}(X) \quad (1)$$

$$\frac{d\Pi}{dX}(X) = \frac{-f \cdot L \cdot \dot{m}^2}{2 \cdot \rho(X) \cdot D_h \cdot A^2 \cdot \Delta P_{ref}} \quad (2)$$

$$\frac{d\Theta_1}{dX}(X) = \frac{d\Theta_2}{dX}(X) \quad \text{and} \quad \frac{d\Pi_1}{dX}(X) = \frac{d\Pi_2}{dX}(X) \quad (3)$$

Eqns. (1), (2), and (3), are combined to find a group of parameters which must be a constant for conductors which are thermodynamically similar [Eqn. (4)]. This group contains only the passage dimensions and the heat load per unit length. It does not contain any fluid properties, the mass flow rate, or the arbitrary reference conditions. The group is a constant (C) that depends only on the inlet conditions and the constraints being used.

$$f \frac{L}{D_h} \left( \frac{q'_{max} \cdot L}{A} \right)^2 = C(P_{in}, T_{in}, P_{min}, T_{max}) \quad (4)$$

Once this flow constant (C) is found for a range of inlet conditions and constraints, the maximum allowable heat load per unit length can be estimated for various passages [Eqn (5)]. The total heat load throughout the conductor is $q'_{max} L$.

$$\boxed{q'_{max} = \frac{A}{L} \sqrt{\frac{1}{f} \frac{D_h}{L} C(P_{in}, T_{in}, P_{min}, T_{max})}} \quad (5)$$

Figure 1 shows the flow constant C for an inlet temperature of 4.5 K, a minimum pressure of 3 atm, and a range of inlet pressures and maximum allowable temperatures. Due to the changing Joule-Thompson coefficient, increasing the inlet pressure allows more heat removal up to about 7 or 8 atm but decreases the possible heat removal above 8 atm. The results also show that allowing higher temperatures always permits more heat removal.

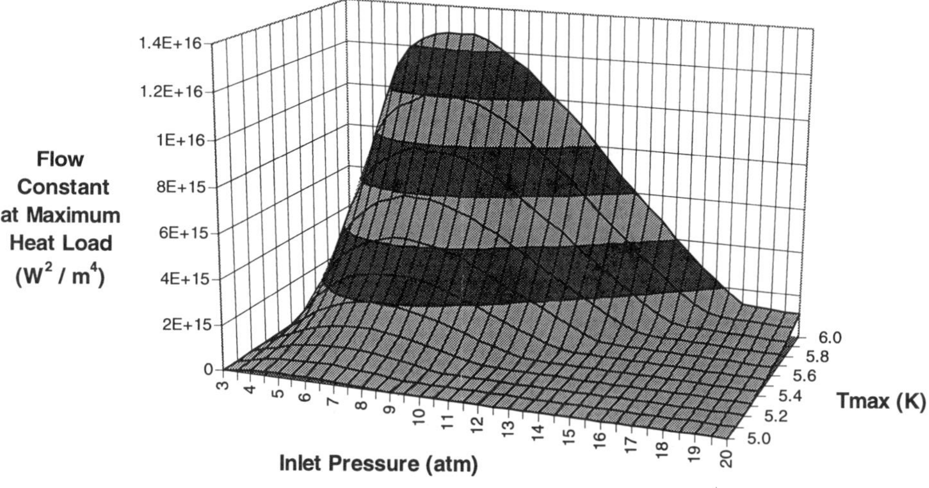


**Figure 1.** Flow constant at maximum allowable heat load for a range of conductor inlet pressures and maximum allowable temperatures.

## INFLUENCE OF LOW-TEMPERATURE PUMP ON REFRIGERATION LOAD

Figure 2a shows a possible SMES cooling scheme for use with CICC magnets. Supercritical helium circulates in a closed loop (ABCD). Helium exits the pump at location A and is then cooled in a heat exchanger that is inside a pool of liquid helium (the buffer tank). The helium enters the magnet at B and then absorbs heat from the magnet. After leaving the magnet at C, the helium returns to the pump to be recirculated. This configuration is referred to as a recirculator loop.

The refrigerator removes heat from the liquid pool so that there is no net heat load on the buffer tank during steady state operation. An energy balance on the system shows the refrigerator must remove both the heat load of the magnet ($\dot{Q}_{magnet}$) and the pump power ($\dot{W}_{pump}$). Therefore, the pump power has the potential to significantly increase the load on the refrigerator. Reducing the pump power will reduce the refrigeration load and allow a smaller refrigerator to be used.

Figure 2b shows a possible way to reduce the required pump power. A heat exchanger is added to the recirculator loop to cool the helium before entering the pump. The cooler, denser

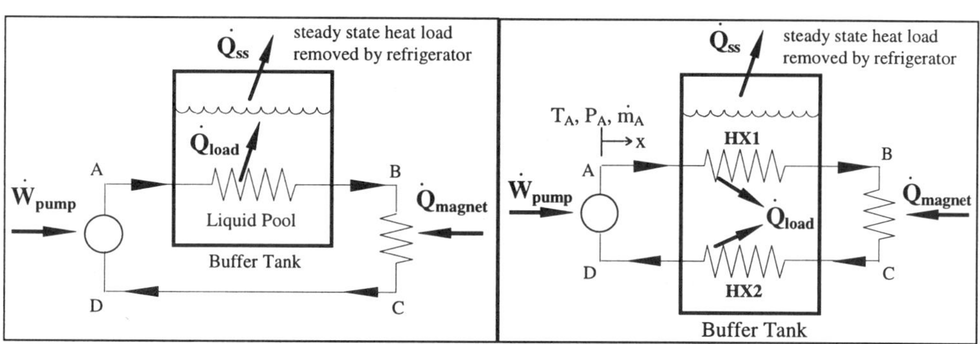


**Figure 2a.** Warm pump cooling scheme.

**Figure 2b.** Cold pump cooling scheme.

helium will require less pump power for the same pressure rise. This reduction can be explained by looking at the equation which determines pump work for an ideal, steady state pump[2]:

$$w = \int v \cdot dP \tag{6}$$

where w is the pump work per unit mass, P is the pressure, and v is the specific volume. Equation 6 shows that the ideal pump work is proportional to the specific volume of the fluid flowing through the pump. The work required to produce a given pressure rise can be reduced by pumping a cold, high density (low v) fluid.

The two systems of Figure 2 were compared using a computer model. The magnet was modeled using the technique described in the previous section. The temperature at B was chosen as 4.5 K. The temperature at D was found by either assuming it is the same as the magnet exit (warm pump) or by assuming HX2 is sized to re-cool the helium to 4.5 K (cold pump). To simplify this analysis, pressure drops in the heat exchangers were neglected and transfer line losses were also neglected.

Figure 3 shows that pumping the helium at a colder temperature will reduce the pumping power ($\dot{W}_{pump}$). The difference between warm and cold pumping becomes greater as the maximum helium temperature in the magnet is increased. This difference occurs because the warm pumping is done at approximately $T_{max}$, while the cold pumping is always done at 4.5 K. Therefore, the difference between warm and cold pumping becomes greater as $T_{max}$ is raised. But in each case, the refrigerator load decreases when cold pumping is used, allowing a smaller refrigerator to be used. Figure 3 also shows how the pump power has the potential to be as large as the magnet heat load ($\dot{W}_{pump} / \dot{Q}_{magnet} = 1$). Therefore, pressure drops in the magnet should be minimized to reduce the pump power and the required refrigeration load. Finally, allowing higher temperatures in the magnet can significantly reduce the pumping power due to the lower mass flow requirement.

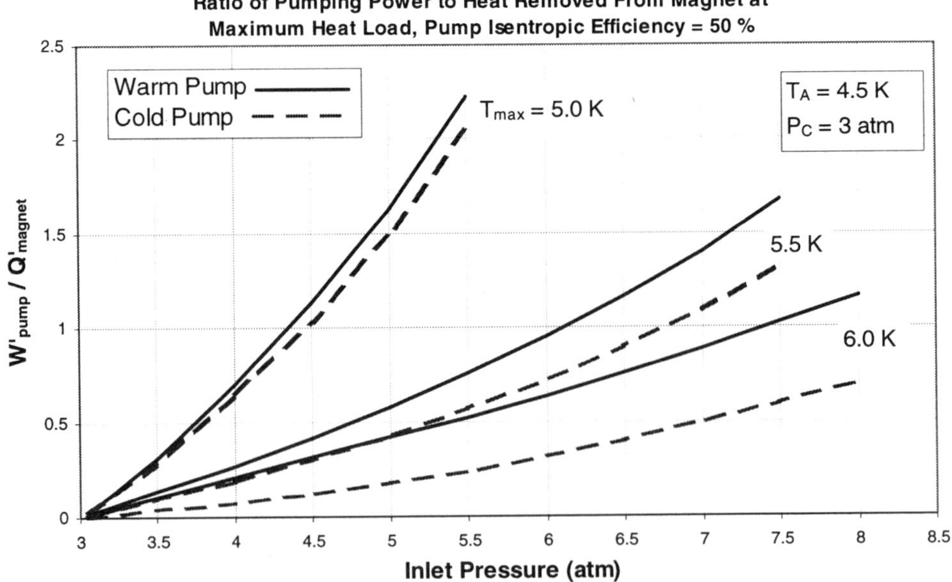


**Figure 3.** Ratio of pumping power to heat removed from the magnet at maximum heat load.

**Table 1.** Comparison of several buffer tank designs

| METHOD | SPECS | ENERGY STORAGE PER VOLUME ($J/m^3$) | PROS | CONS |
|---|---|---|---|---|
| 1. Boiling at 1 atm with an expandable external tank to hold vapor | Start with saturated liquid, end with saturated vapor. (No superheat allowed to keep at 4.2 K) | 344,000 (includes external tank volume) | Constant Temp of 4.2 K | Poor Q/V ratio due to large vapor volume. |
| 2. Heat compressed liquid in closed container | Start with compressed liquid at 4.2 K. Optimum $P_{start}$ varies with $T_{max}$ | 285,000 ($T_{max}$ = 5 K) (22 to 27 atm)<br>659,000 ($T_{max}$ = 6 K) (28 to 39 atm) | Simple | Pressure vessel required. Requires temp rise. |
| 3. Heat 2-phase helium in closed container | Start at 4.2 K, two phase liquid. Program searched for optimal initial quality. | 424,000 ($T_{max}$ = 5 K) ($x1_{init}$ = 0.37)<br>720,000 ($T_{max}$ = 6 K) ($x1_{init}$ = 0.07) | Simple. | Requires temp rise. |
| 4. Heat two-phase helium with a rigid external tank to hold the vapor | Start at 4.2 K. Two phase in main tank, saturated vapor in external tank. | 727,000 ($T_{max}$ = 6K) (V2/V1=0.36)<br>(V1 = main tank vol)<br>(V2 = external tank vol) | --- | More complex than 3, but no gain. |
| 5. Heat liquid helium in an evacuated closed container | Start with saturated liquid at 0.1 atm (T=2.5 K). Close tank and add heat at constant density. | 806,000 ($T_{max}$ = 5 K)<br>1,158,000 ($T_{max}$ = 6 K) | Very Good Q/V ratio | Difficult to maintain 2.5 K. Requires vacuum pump and extra components. |
| 6. Divided tank with energy transfer to the environment | Insulated float divides tank into a section at 4.2 K and a section at 300 K. $P_1$= 1 atm | 795,400 ($T_{max}$ = 6 K)<br>(conduction losses neglected) | Good Q/V ratio | Complicated design with moving parts. |

## BUFFER SYSTEMS

When a SMES magnet is charged or discharged, AC losses deposit heat energy in the conductor at a much higher rate than the steady state refrigerator load. Only a high power refrigerator could remove this energy "pulse" at the same rate as it is deposited. This refrigerator would be expensive and oversized for most times of operation.

A buffer system temporarily stores the pulse energy in some material or fluid (usually helium). After the pulse, the refrigerator can slowly remove the stored energy. In effect, a buffer system "averages" the transient loads over a long period of time so that a smaller refrigerator can be used.

Since some SMES applications have strict space limitations, several helium buffers were compared based on their energy storage per volume. The results are summarized in Table 1. The evacuated tank concept (method 5) can store the most energy per volume, but could be difficult to carry out in a real system. The divided tank (method 6) allows the second most energy storage, but an actual system may not perform as well as the ideal system that was modeled. The closed container with two-phase helium (method 3) has a good energy storage and is probably the most practical system.

A novel buffer concept[1] (method 6 and Fig. 4) was proposed and an idealized model was created. The system uses a rigid container that is divided by an insulated float into two compartments. Compartment A holds cold helium and is where the magnet energy is rejected to. Compartment B holds warm helium that is in thermal contact with the ambient environment. As heat is added to A, the helium in A expands and causes the float to rise upward, compressing the warm helium in B. Since B is held at a constant temperature via contact with the ambient, the compressive work done on B will be rejected to the atmosphere ($Q_{out}$). As compartment A is re-cooled, the float will move back down. Energy will be

**Table 2.** Iteration criteria for transient simulation of recirculator loop cooling scheme

| Parameter | Criteria for Iteration of Transient Solution |
|---|---|
| $\dot{m}_A$ | vary until $\Delta P_{system} = \Delta P_{pump}$ |
| $T_A$ | consistent with isentropic efficiency and conditions at location D |
| $P_A$ | vary until $\dot{m}_A = \dot{m}_D$ |

transferred back from the ambient to compartment B and then to A via a work transfer. In effect, the system uses the ambient environment to temporarily store some of the energy that the refrigerator cannot immediately absorb. Equation (7) gives the energy received per unit volume for this idealized model.

$$\frac{Q_{in}}{V_{tot}} = \frac{\rho_{A_1}}{1+\Gamma}(u_{A_2} - u_{A_1}) + \frac{P_1}{1+\frac{1}{\Gamma}} \ln\left(\frac{1}{1-\frac{\vartheta}{\Gamma}}\right) \tag{7}$$

where $Q_{in}$ is the amount of heat taken in at the cold end, $V_{tot}$ is the total volume of A and B, $\rho_{A1}$ is the initial density in A, u is the internal energy, $P_1$ is the initial pressure, $\Gamma$ is the initial ratio of the volumes ($V_{B1}/V_{A1}$), and $\vartheta$ is the fractional increase in the volume of A ($\Delta V_A/V_{A1}$).

A range of volume ratios were evaluated for an initial condition of 4.2 K in volume A. The optimum values found for $T_{A\ max}$ = 6.0 K are: $Q_{in}$ / $V_{tot}$ = 785,400 J/m$^3$, $\Gamma$ = 0.66 and $\vartheta$ = 0.5 with $\rho_{A1}$ = 135 kg/m$^3$, $P_1$ = 1 atm, and $P_2$ =4 atm. This is a 10% improvement in energy storage per total volume over a simple closed container (method 3) and a 32% increase in energy storage per helium mass. However, this analysis ignored loss mechanisms such as conduction and heat capacities of the walls and float. Therefore, further analysis is needed to see if a real system can obtain any of the gain shown in this idealized model.

## TRANSIENT MODELING OF A SMES COOLING SCHEME

During charging and discharging, transient heat loads in the SMES magnet will cause helium temperatures and pressures to rise. Modeling the behavior of the refrigeration system can help to understand the issues that affect this transient period. This section presents a method for modeling transient behavior of the coldest section of a SMES refrigeration system Sample results are discussed to show some of the information that the modeling can provide.

A computer model[1] analyzed the transient response of a recirculator loop cooling scheme (Fig. 2b). The governing equations for one-dimensional flow through a differential control volume (Eqns. 8) were derived from energy conservation, mass conservation, and standard pipe friction relations. Variables not previously defined are time (t) and enthalpy (h). These

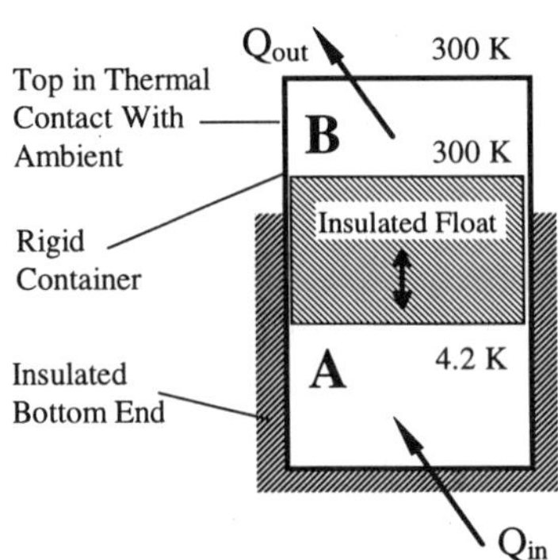


**Figure 4.** Divided buffer tank with energy transfer to environment.

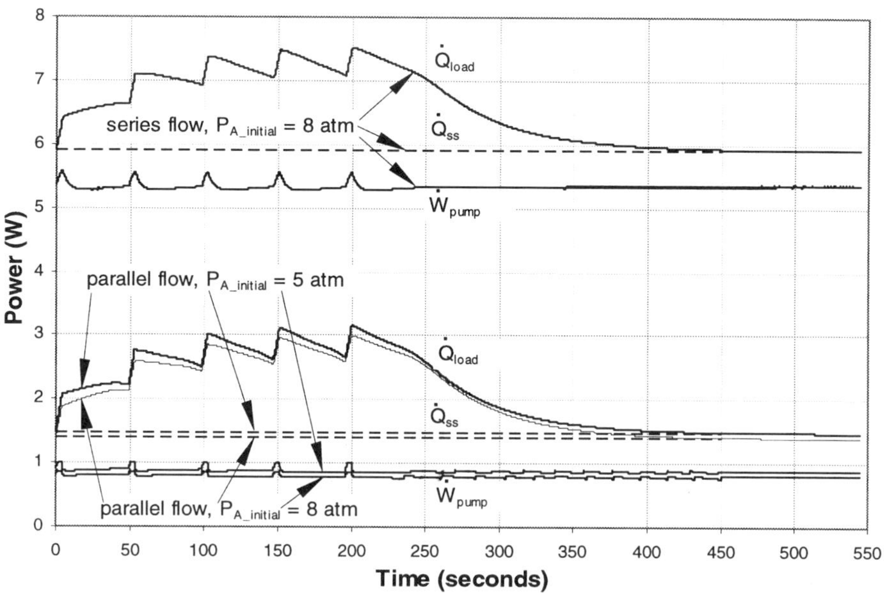


**Figure 5.** Buffer tank heat loads and pumping power versus time.

equations were used to model the flow through the magnet and two heat exchangers (HX1 and HX2). For the heat exchangers, the heat load per unit length ($q'$) was replaced with the relation $q' = ua(T\text{-}T_{buffer})$ where ua is the overall heat transfer coefficient per unit length and $T_{buffer}$ is the temperature of the buffer tank. These equations were written in matrix form and a routine was written to solve the equations for a given set of conditions at the pump exit ($T_A$, $P_A$, $\dot{m}_A$). An outer iteration loop then varied these conditions until the solution was consistent with the pump curve, the pump isentropic efficiency, and the mass conservation in the pump. The pump was assumed to have such a small internal volume that it could store no mass, making the mass flow rate the same at A and D. Table 2 summarizes these iteration criteria.

Computer simulations were run to investigate the influence of parallel or series flow through the magnet and the influence of the initial pressure in the recirculator loop. To study the effect of parallel or series flow, two flow configurations were simulated. The first simulation was taken to be the "series flow" case by considering one pass through a magnet in which all conductors are connected end-to-end. A "parallel flow" case was simulated by considering the same conductors to be split into two parallel paths, each half of the total length. To study the effect of the initial pressure in the recirculator loop, the parallel flow configuration was re-run using a lower initial loop pressure. The two pressures used were 8 atm and 5 atm.

The pump isentropic efficiency was assumed to be 60% and the pump curve was a rough fit of data for a pump at 100 rpm[3]. To simulate a magnet discharge, the magnet heat load ($\dot{Q}_{magnet}$) was increased to 100 W for 4 seconds and then returned to the steady state value of 0.6 W for 45 seconds. Five discharges were simulated with a 300 second final recovery period. These values, as well as the magnet and heat exchanger dimensions, were used only for illustrative purposes and are not based on an actual design.

Figure 5 shows that the transient heat load on the buffer tank ($\dot{Q}_{load}$) is related to the steady state heat load ($\dot{Q}_{ss}$). The steady load $\dot{Q}_{ss}$ is drawn in dotted lines to give a reference for the increased heat load during the transient period. The shape of the heat load curve is similar for the parallel flow and series flow systems. The curves are just offset by the difference in their steady state heat load, which is caused by a difference in steady state pumping power.

Figure 5 shows that the flow configuration strongly influences the steady state refrigerator load. The series flow configuration requires a refrigerator that is 4 times larger than that off the parallel flow configurations. The pumping power accounts for this difference. A larger pumping power is required to force the helium through the longer passage of the series system. Since the refrigerator is assumed to be sized for the steady state load, the series configuration will require a larger, more expensive refrigerator.

Figure 5 also shows that the initial pressure in the recirculator loop can slightly influence the buffer tank heat load ($\dot{Q}_{load}$). For the same parallel flow configuration, an 8 atm initial pressure results in a slightly lower heat load than a 5 atm initial pressure. Once again, this difference is due to a difference in pumping power. The 8 atm helium requires slightly less pumping power since it has a higher density than the 5 atm helium.

In addition to the heat load data, this modeling technique can provide temperature, pressure, and mass flow data for the recirculating loop as well as the buffer temperature and pressure. Due to space limitations, plots of these other parameters are not included. However, the heat load plot gives a feel for how the designs of the magnet and refrigeration system can be improved by using a model which includes several refrigeration components in addition to a magnet model.

$$A\cdot\rho\left.\frac{\partial u}{\partial T}\right|_P\frac{\partial T}{\partial t}+A\cdot\rho\left.\frac{\partial u}{\partial P}\right|_T\frac{\partial P}{\partial t}+\frac{P}{\rho}\cdot\frac{\partial\dot{m}}{\partial x}+\dot{m}\cdot Cp\frac{\partial T}{\partial x}+\dot{m}\left.\frac{\partial h}{\partial P}\right|_T\frac{\partial P}{\partial x}-q'=0$$

$$A\left.\frac{\partial\rho}{\partial T}\right|_P\frac{\partial T}{\partial t}+A\left.\frac{\partial\rho}{\partial P}\right|_T\frac{\partial P}{\partial t}+\frac{\partial\dot{m}}{\partial x}=0 \qquad (8)$$

$$\frac{\partial P}{\partial x}+\frac{f\cdot\dot{m}^2}{2\cdot\rho\cdot D_h\cdot A^2}=0$$

## CONCLUSIONS

Models of CICC SMES systems should include the pumping power when estimating the refrigeration load. The models should include characteristics of the pump, heat exchangers, and buffer tank to give a solution that is consistent with all components. This comprehensive type of model can then be used to study alternative magnet cooling configurations and better understand how the magnet design can be altered to reduce the refrigeration load. Other conclusions of the project are that: 1) steady state cooling of CICC passages with flowing helium can be characterized by a parameter named the flow constant, 2) re-cooling helium before pumping can reduce pumping power, and 3) a novel buffer system was proposed but further analysis is required to determine if the system is feasible.


## ACKNOWLEDGMENTS

The authors thank Westinghouse Electric Corporation and the U. S. Navy for their support and input throughout the project. Specifically we thank Hank Lokay and Don Hackworth of Westinghouse and Mike Superczynski, representative of the U. S. Navy, for their work in organizing the SMES community.

## REFERENCES


1. B. J. Bowers. "Integrated Cryogenic Refrigeration System Design For Superconducting Magnetic Energy Storage Systems". Masters thesis. Massachusetts Institute of Technology (1997).
2. J. B Jones and G. A Hawkins. "Engineering Thermodynamics," 2nd Ed. John Wiley & Sons, Inc., New York (1986).
3. J.W. Lue, et al. Test of a cryogenic helium pump. "Advances in Cryogenic Engineering, Volume 27," Plenum Press, New York, (1982), pg 785.0

# HE II COOLING SYSTEM PERFORMANCE IN THE EBASCO PROOF OF PRINCIPLE EXPERIMENT DURING LAMBDA TRANSITION, WARMUP, AND QUENCH

O. R. Christianson,[1] E. F. Daly,[2] and J. M. Pfotenhauer[3]

[1]Westinghouse Science & Technology Center
Pittsburgh, PA, 15235, USA
[2]Standford Linear Accelerator
Menlo Park, CA, 94025, USA
[3]University of Wisconsin - Madison
Madison, WI, 53706, USA

## ABSTRACT

Temperature measurements in a 700 L bath of helium during cool down, the transition to He II, and operation in subcooled He II taken during the Ebasco SMES proof of principle experiment, POPE, are reported. Cooldown proceeds at a uniform rate throughout the bath, but the lambda transition appears to propagate from the heat exchanger to the current leads, which provide the largest heat load to the helium II region, over ten minutes. During warm-up and quench significant temperature stratification is observed with the bath consisting of boiling helium at the top and subcooled helium with decreasing temperature below. Both regions consist of a constant thickness while the helium level decreases as helium is vaporized at the top of the bath. The region of helium at the bottom of the dewar is at the lambda point, decreases in thickness with time, and exists for up to six hours after the He II heat exchanger is turned off. Calculations based on a model incorporating temperature dependent density and thermal conductivity of helium matches the measurements. This behavior is also observed during quench at an accelerated time scale.

## INTRODUCTION

The Ebasco SMES proof of principle experiment has been conducted to demonstrate the stability and operation of the superconductor in the Ebasco team Engineering Test Model design. Details of the facility, experiment, conductor stability, and operation have been reported by Pfotenhauer.[1,2]

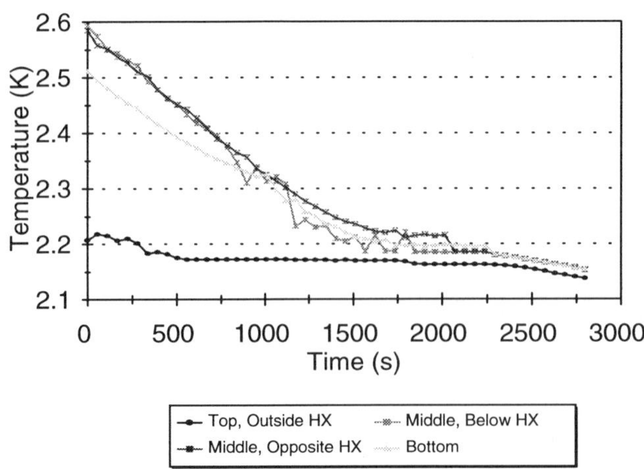


**Figure 1.** The lambda transition in the POPE. The bulk helium transitioned to He II 30 minutes after He II first appeared outside the heat exchanger. The transition occurred over a time period of less than 10 minutes for the bulk of the bath.

The experimental facility consists of a 100 kA DC power supply, a 4 T one meter bore background field split solenoid, a three turn one meter diameter test coil, a dewar for operation of the solenoid and test coil at 1.8 K and 1 atmosphere, and various support systems for vacuum, helium supply and recovery, and data acquisition.

The dewar consists of two parts, an annular inner dewar surrounded by a larger, cylindrical vacuum vessel. The test coil and background coil rest on the bottom of the inner dewar and are operated at 1.8 K and 1 atmosphere cooled by a heat exchanger mounted on a 12.7 cm thick G-10 isolator flange. Operation of the heat exchanger is similar to that described by Canavan et al.[3] A reservoir of saturated 4.2 K helium is maintained at a constant level above the G-10 isolator flange through the use of a liquid helium keep-fill system. Operation of the He II refrigeration system has been reported by Pfotenhauer.[2]

Temperature measurements in a 700 L bath of helium during cool down, the transition to He II, operation in subcooled He II, and subsequent warm-up and quench taken during the Ebasco SMES proof of principle experiment are reported in this paper.

## COOL DOWN AND THE LAMBDA TRANSITION

Cool down and He II operation occurred on two occasions, August and November of 1993. Ultimate temperatures of 1.94 and 1.85 K respectively were achieved with a heat load of 35 W.

Cool down of the 700 l helium bath in the POPE to He II occurred within several hours. Once He II appeared outside of the heat exchanger, the transition of the bulk fluid was complete within about 30 minutes. The time for a transition to occur in the bulk fluid was relatively rapid, less than 10 minuets, see Figure 1.

Several features of the cool down are noteworthy. Cool down proceeds relatively uniformly suggesting good heat transport. Once helium reaches the lambda temperature around the heat exchanger, temperatures do not decrease there until the entire bath has transitioned to superfluid helium. As the temperatures in the bath decrease toward the lambda point, temperature gradients or differences exist. The bottom of the bath is cooler than the middle of the bath below the current leads. Finally, the last region to reach the

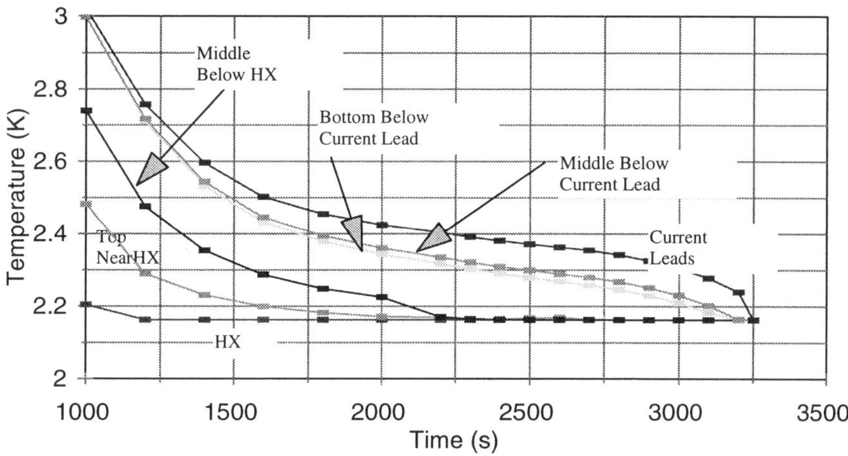


**Figure 2.** Calculational simulation of the cool down to and the transition through the lambda point.

lambda temperature is around the current feedthroughs, which are a continuous source of heat and are located opposite the heat exchanger.

These features can be explained based on the physical processes involved. The initial appearance of He II close to the heat exchanger occurs at the coolest point of the apparatus, whereas the last region to reach the lambda temperature occurs around the current leads which have the highest heat leak of 15 W and are situated opposite the heat exchanger. The presence of the background magnet eliminates or significantly reduces circulating convective currents, but also provides a thermal conductive path through the conductor copper. This has the effect of allowing a relatively uniform cool down throughout the bath. Once a section of the bath transitions to superfluid with its high thermal conductivity and larger density compared to normal liquid helium; the normal and superfluid helium exist simultaneously in the bath with an interface that propagates from the heat exchanger to the current leads. Similar observations of the existence and propagation of a normal-superfluid helium interface have been reported by Caspi.[4]

A simple two dimensional finite difference model simulates cool down and transition through the lambda point. The differential equation solved is

$$CV\frac{\partial T}{\partial t} = kA_{x,y}\nabla^2 T \tag{1}$$

where C is the specific heat, V is the cell volume, $A_{x,y}$ is the appropriate heat transfer surface in the vertical or circumferential direction, and T is the node temperature. The bath is approximated by 50 nodes, 5 vertical and 10 circumferential over half of the bath so that the heat exchanger and the current leads are included as boundary conditions on the top outside nodes respectively. The thermal conductivity of copper and the specific heat of helium is used in the model where the volume is divided between the magnet with the copper and the helium. The specific heat includes the lambda transition allowing model simulation of the transition through the lambda point. It is assumed that once a node transitions through the lambda point, the thermal conductivity increases such that the there is no longer a temperature gradient between transitioned nodes. This duplicates the normal-superfluid interface and the propagation of the interface from the heat exchanger to the current leads. Results are shown in Figure 2. Qualitative and rough quantitative agreement is found between the calculations and the experimental measurements. A time sequence of the

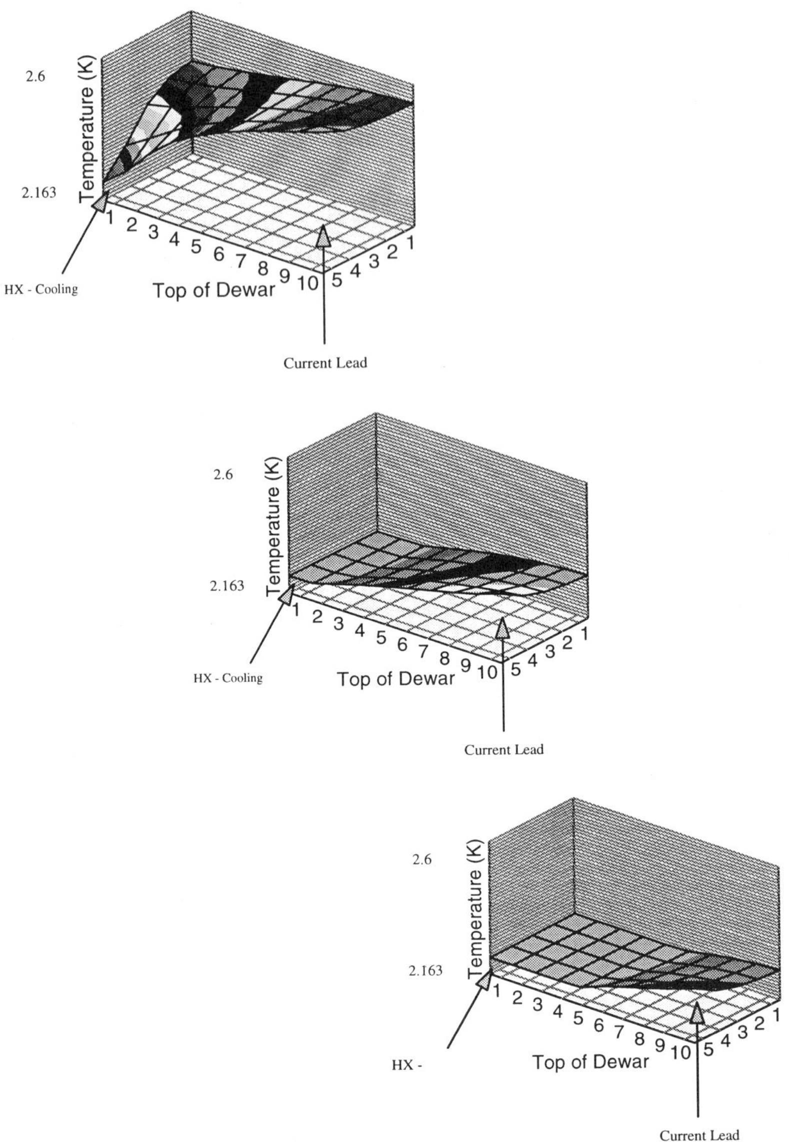


**Figure 3.** A time sequence of the temperature distribution in the helium bath during cool down and transition through the lambda point.

temperature distribution is shown in Figure 3, where the coolest spot is at the heat exchanger, the warmest spot is at the current leads, and the temperature at the bottom of the bath is colder than for some distances closer to the heat exchanger, but also closer to the current leads.

## WARM-UP AND HELIUM STRATIFICATION

After the experiment the heat exchanger was turned off, and the helium bath warmed up slowly. The transition of the superfluid to normal helium was observed. The temperature of the bulk fluid rose to the lambda point uniformly. The temperature of the fluid at the bottom

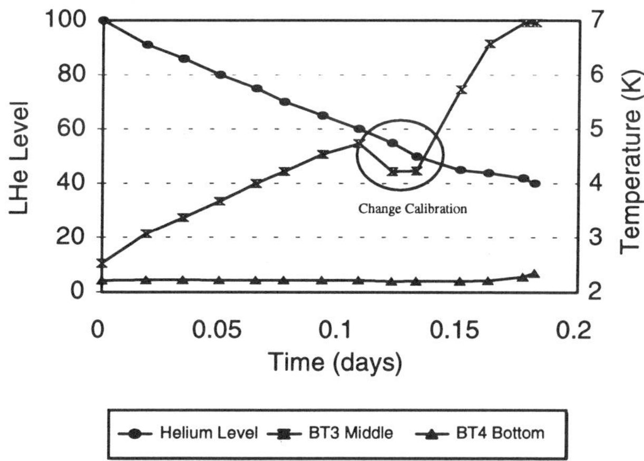


**Figure 4.** The transition from He II to normal helium. Cold helium remained at the bottom of the dewar for 4 hours while helium boiled at the top of the fluid and the level dropped. The temperature as a function of time indicates a temperature gradient existing in the helium.

of the dewar remained at or just above the lambda point for over 4 hours, see Figure 4. Boiling occurred at the top of the fluid, and the level of the fluid decreased. The temperature of the fluid at the middle of the dewar increased as the fluid level decreased. This indicates there was a region at 4.2 K which was boiling, a subcooled region where the temperature decreased from 4.2 to 2.2 K, and a region at or just above the lambda point.

A graphical picture of the spatial temperatures at different times can be obtained by transforming the temporal temperature measurements to a spatial temperature distribution, see Figure 5. The subcooled region exists for approximately 30% of the vertical height of the dewar.

A simple model gives approximate quantitative agreement with these measurements. It is assumed that the background magnet eliminates convective currents in the liquid helium bath, and that the heat load is deposited in the subcooled region by heat transport along the copper in the conductor. Since the density of liquid helium increases with decreasing temperature, the hotter fluid is located at a larger vertical height and stratification of the

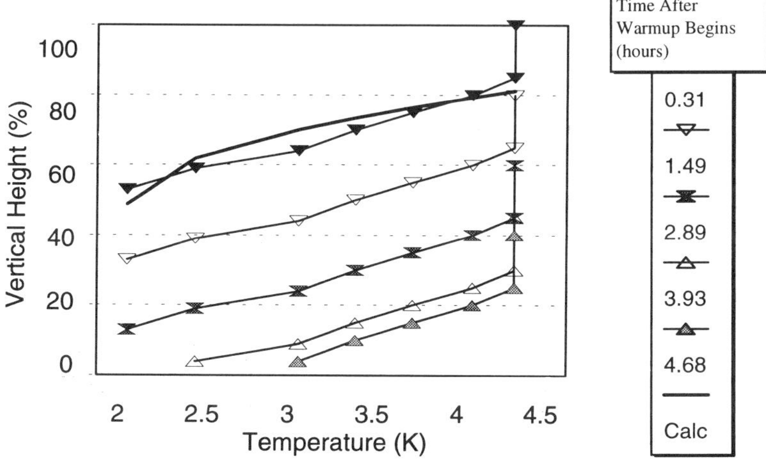


**Figure 5.** Spatial temperature distributions at different times obtained by a coordinate transformation from time to height.

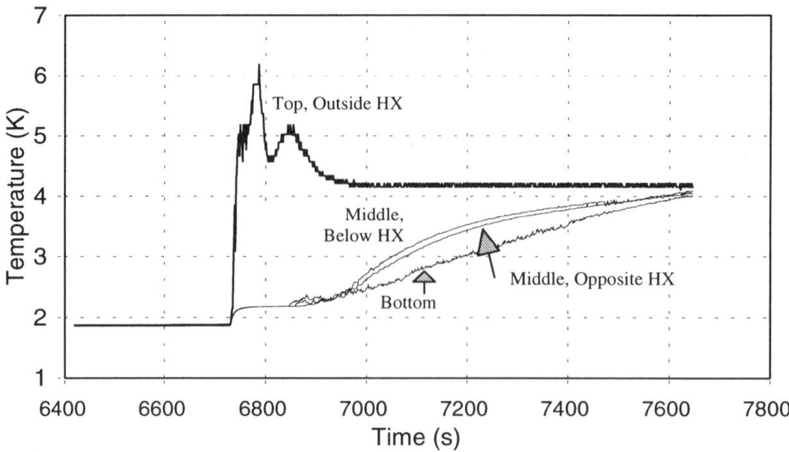


**Figure 6.** Bulk helium temperatures after a quench during the November 8, 1993 POPE experiment.

helium can result. The temperature distribution is given by one dimensional heat transport in the liquid helium with its poor thermal conductivity, k, and the heat load is modeled as a volumetric source, q''', such that

$$\frac{d^2T}{dz^2} = \frac{q'''}{k} \tag{2}$$

which has the solution

$$T(z) = T_\lambda + \frac{q'''}{2k}(z - z_o)^2 \tag{3}$$

where $z_o$ is the top of the helium that is at the lambda temperature. Results of the calculation are shown in Figure 5. Further attempts at comparison between calculation and measurement are not warranted because of the assumptions in both the analysis and the data portrayal.

## QUENCH

A quench occurred in the background coil during the November 1993 experiment. Significant vaporization of helium occurred, with the pressure relief valves opening. The temperature of the fluid is shown in Figure 6. The temperature at the top of the magnet increased to over 4.2 K. This is reasonable since the top of the magnet was no longer immersed in liquid helium. The temperature of the fluid at the bottom rose in temperature over a period of 15 minutes with some stratification occurring. This behavior is similar to the stratification during warm-up, but at an accelerated time frame.

## SUMMARY AND CONCLUSIONS

Several interesting features were observed in operating the He II refrigeration system in the POPE. These included a uniform cool down rate to the lambda point, creation of a normal-superfluid interface, and propagation of the interface throughout the bath of helium. This behavior can be understood by noting that circulating convective currents are eliminated or drastically reduced because the background magnet occupies most of the dewar volume, the copper in the magnet conductor facilitates heat transport in normal helium, and the superfliud helium has very good thermal conductivity and a density greater than normal helium allowing coexistence of normal and superfluid helium accompanied by the interface propagation. A finite difference calculation qualitatively, and roughly quantitatively, duplicates these features. Subsequent warmup of the bath results in an additional feature of stratification of helium such that boiling occurs in a layer at the top, a subcooled region of the bath occupying about 30% of the volume available, and a section of the bath at the bottom of the dewar residing at the lambda point.

## ACKNOWLEDGMENT

This work was supported by the Defense Nuclear Agency under contract #DNA001-92-C-0064.

## REFERENCES


1. J.M.Pfotenhauer, M.K.Abdelsalam, F.Bodker, D.Huttleston, Z.Jiang, O.D.Lokken, D.Scherbarth, B.Tao, and D.Yu, Test Results from the SMES Proof of Principle Experiment, *IEEE Transactions on Magnetics* 27(2):1704-1707 (1991).
2. J.M.Pfotenhauer, O.D.Lokken, O.R.Christianson, and E.F.Daly, He II Refrigeration System Performance in the SMES Proof of Principle Experiment, submitted to *Cryogenics*.
3. E.R.Canavan, D.C.Larbalestier, and S.W.Van Sciver, A 13T-1.8K NbTi Laboratory Test Coil, *IEEE Transactions on Magnetics* 24(2):1082-1085 (1988).
4. S.Caspi, Gravitational convection of subcooled He I and the transition into superfluid He II at atmospheric pressure, in: "Proceedings of the Eighth International Cryogenic Engineering Conference," C. Rizzuto, ed., IPC Science and Technology Press, Genova (1980), pp. 238-242.

# DEEP DISCHARGE RECOVERY WITH THE AML&P SMES HELIUM REFRIGERATION SYSTEM[1]

J.P. Kelley,[1] J.J. Clausen,[2] A. Kuendig,[3] A.E. Senn,[3] and K.D. Dixon[1]

[1]BWX Technologies, Inc.
Lynchburg, VA 24505, USA
[2]Lotepro Corp.
Valhalla, NY 10595, USA
[3]Linde Kryotechnik AG
CH-8422, Pfungen, Switzerland

## ABSTRACT

A full deep discharge event for the Anchorage Municipal Light & Power (AML&P) SMES magnet will dissipate 4,150 kJ into the magnet's helium bath in 60 s. Resulting bath pressures and liquid levels will exceed those that permit normal magnet operations. The helium refrigeration system (HRS) must stabilize bath conditions to permit normal operations within 24 h of event initiation. A 4-stage deep discharge recovery approach is outlined, and models developed for each stage are discussed. Model predictions of the HRS response are presented.

## INTRODUCTION

A full deep discharge of the AML&P SMES magnet in response to a frequency shedding event presents the HRS with its most challenging task. Fig. 1 is an HRS process flow diagram. HRS and magnet details are presented in [1]. AC losses deposit 4,150 kJ of energy in 60 s into the pool boiling magnet's helium bath, increasing pressure, and potentially the liquid level, beyond the magnet's normal operating regions ($4.45 \leq T \leq 4.55$ K, and nominally $29{,}700 \leq V_L \leq 31{,}500$ L). The total available magnet volume is 35,000 L. Levels between 21,000 L, which minimally covers the coils, and 29,700 L are reserved for helium losses to power outages, bed regeneration, and for operations when the HRS is off-line. The HRS, which is designed for 175 W and 1.9 g/s @ 4.45 K, and 925 W @ 45 K,[1] must return the bath to 4.45 K with sufficient liquid volume prior to magnet recharge. Recharging is also a significant event, dissipating 651 kJ in the bath in 533 s. The HRS is required to return the magnet to operations within 24 h. Additionally, it must handle 3 consecutive deep discharge events with a 24 h recovery period between each event.

Recovery from a deep discharge is a 4-stage process. The 1st stage is pumpdown. Helium molecules are removed from the magnet and either liquefied into a buffer dewar, or compressed and stored as a room temperature gas. Pumpdown ends when the magnet is at 0.125 MPa (4.45 K). The 2nd stage is a transfer of liquid from the buffer dewar to the magnet vessel by maintaining a pressure gradient between the two units. During the 3rd stage the magnet is recharged. This is handled in the same manner as pumpdown, with

---

[1] This work was supported by the U.S. Department of Energy under grant number DE-FG36-94GO10011.

helium molecules stored as liquid in the buffer dewar, or as gas in warm storage tanks. The 4th stage is a final liquid transfer simultaneous with magnet operations. This process is identical to Stage 2 except that magnet heat and liquefaction loads are greater.

Models of these stages were developed, and the input and output of the models tied together. They were then used to determine the final system recovery conditions, including recovery time, final magnet liquid volume and temperature, and warm and cold stored volumes, for a 4,150 kJ deep discharge for given sets of initial conditions. Ranges of initial conditions were identified where recovery is possible. It is not prudent in all cases to immediately initiate Stage 3 after completing Stage 2 since the lower loads existing prior to recharge permit quicker magnet filling than those that exist after recharge. Therefore, estimates of the liquefaction capability of the HRS were made for the lower load conditions, and a delay introduced before recharge initiation. Using the delay, new ranges of initial conditions were established for which recovery is possible for a 4,150 kJ deep discharge. Once these ranges were established, the possibility of sustaining 3 consecutive deep discharges, one every 24 h, was investigated.

## THE DEEP DISCHARGE EVENT

Dissipating 4,150 kJ in 60 s in a helium bath is equivalent to a real time heat load > 69.2 kW. Since the HRS can only handle a fraction of the boiloff in real time, the magnet's

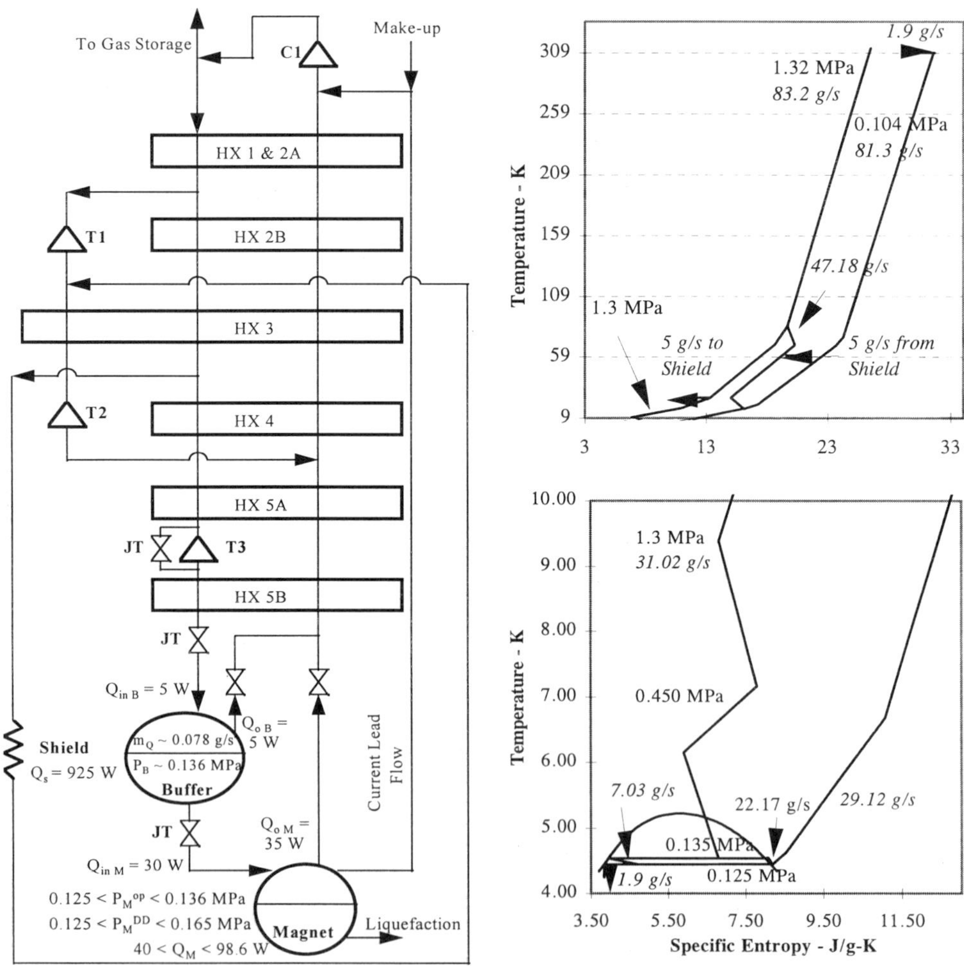


**Figure 1.** HRS Process Flow Diagram and Design T-s Diagram.

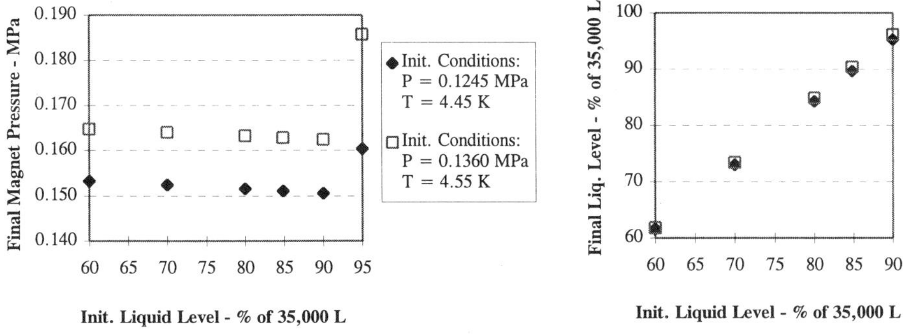


**Figure 2.** Final magnet pressures and liquid levels resulting from a 4,150 kJ deep discharge.

helium vessel can be treated as a closed system when determining final magnet pressure, temperature and liquid level. Conservation of Energy reduces to $\Delta Q = M^T \Delta u^T$. The solution details to this problem are given elsewhere and are not repeated here.[2,3] Fig. 2 presents the results for a 4,150 kJ deep discharge for various initial liquid levels, at the operating temperature range limits. Note that for an initial liquid level of 95%, the magnet is completely filled with liquid after a 4,150 kJ deep discharge (not shown). Since portions of the vessel's surface area will be at room temperature, this is an unstable situation that would cause continued vessel pressurization. For this reason, the maximum magnet fill level is 31,500 L, 90% of available volume. Then, the maximum calculated pressure is 0.165 MPa (1.63 atm), occurring with initial conditions of 4.55 K and 60% level.

To avoid magnet overfilling, helium molecules in the HRS are limited based on the following criteria: 31,500 L LHe @ 4.45 K in the magnet, 400 L LHe @ 4.55 K in the buffer, and room temperature helium @ 0.101 MPa (1 atm) in the warm gas storage tanks.

## DEEP DISCHARGE RECOVERY - PROCESS

### First Stage of Recovery - Pumpdown

When a deep discharge occurs, the magnet pressure climbs rapidly, exceeding the buffer's operating pressure, 0.136 MPa (1.34 atm) (Fig. 1).[1] Magnet loads drop with current from 62 to 40 W @ 4.45 K, and from 1.44 to 1.07 g/s for the leads. The large magnet return flow is regulated, maintaining buffer pressure and minimizing compressor loading. The magnet supply from the buffer is isolated. Turbine flows, which are controlled on discharge temperatures, are shutdown by cooling resulting from the excess return flow. Turbine T3's JT bypass is opened and the system is configured as a JT cycle (Fig. 3). This cycle is unbalanced, with return flow exceeding supply flow. Excess return flow is compressed and stored as a warm gas. The pressurized magnet acts as a cold gas source for 'supercharging' the cold box, boosting liquid production in the buffer. A T-s diagram for this operation (Fig. 3) with the magnet pressurized to 0.155 MPa indicates that 67.5 g/s of helium is removed from the magnet, 64.33 g/s @ 4.54 K is liquefied in the buffer, and 16.9 g/s is stored as a warm pressurized gas. Table 1 presents pumping massflows, liquefaction and storage rates as functions of magnet pressure with the buffer maintained at 4.54 K. Reduced massflows at 0.125 MPa are due to the desire not to drive compressor suction subatmospheric.

Pumpdown time is determined by writing mass and energy balances for the magnet, leading to the expression

$$u_f^T = \{[Q_M - m_o h_o]\, t + M_i^T u_i^T\} \,/\, [M_i^T - m_o\, t]. \tag{1}$$

The 2 unknown, $u_f$ and t, are determined by trial and error, with the constraint

$$[v_f^T - v_f^G] \,/\, [v_f^L - v_f^G] = [u_f^T - u_f^G] \,/\, [u_f^L - u_f^G]. \tag{2}$$

Here $h_o$ is taken as the average of the exiting flow's initial and final enthalpies. Pumpdown is evaluated in two steps. For magnet pressures > 0.135 MPa, $m_o$ is held at 67.5 g/s

**Table 1.** Pumping, liquefaction and warm storage massflows generated by the unbalanced JT cycle as functions of magnet pressure during magnet pumpdown. $P_B = 0.135$ MPa

| Magnet Pressure | Pumping Massflow | Liquefaction @ 4.54 K | Stored @ 300 K |
|---|---|---|---|
| 0.165 MPa (1.626 atm) | 67.51 g/s | 64.36 g/s (1,968 L/h) | 16.89 g/s |
| 0.135 MPa (1.332 atm) | 67.52 g/s | 64.33 g/s | 16.92 g/s |
| 0.125 MPa (1.23 atm) | 63.88 g/s | 60.63 g/s (1,854 L/h) | 16.26 g/s |

(pumping massflow) + 1.07 g/s (leads). For pressures < 0.135 MPa, the average value of the massflows for 0.135 MPa and 0.125 MPa (Table 1) is used. Once t is found, liquefaction and storage rates are handled similarly and the total accumulations determined. A pumpdown from 0.162 to 0.125 MPa (initial deep discharge conditions of 4.55 K and 90%) is predicted to take ~ 4,000 s ($t_1$). During this period, 2,164 L of 4.54 K LHe collect in the buffer. The equivalent of 559 L LHe @ 4.45 K are stored as a warm gas, resulting in 1,110 L available in warm storage (551 L result from the 4.55 K initial conditions with a full HRS). With no magnet make-up, its liquid volume decreases to 28,080 L (80.2%).

## Second Stage of Recovery - Liquid Transfer from the Dewar to Magnet

The second stage of recovery involves transferring the significant 4.54 K inventory of liquid to the magnet vessel. During this operation, the turbines are restarted. Transfer is effected by maintaining a pressure difference between the buffer and the magnet. By isolating buffer return, pressure is maintained by the buffer heat load (1.48 W), the gas

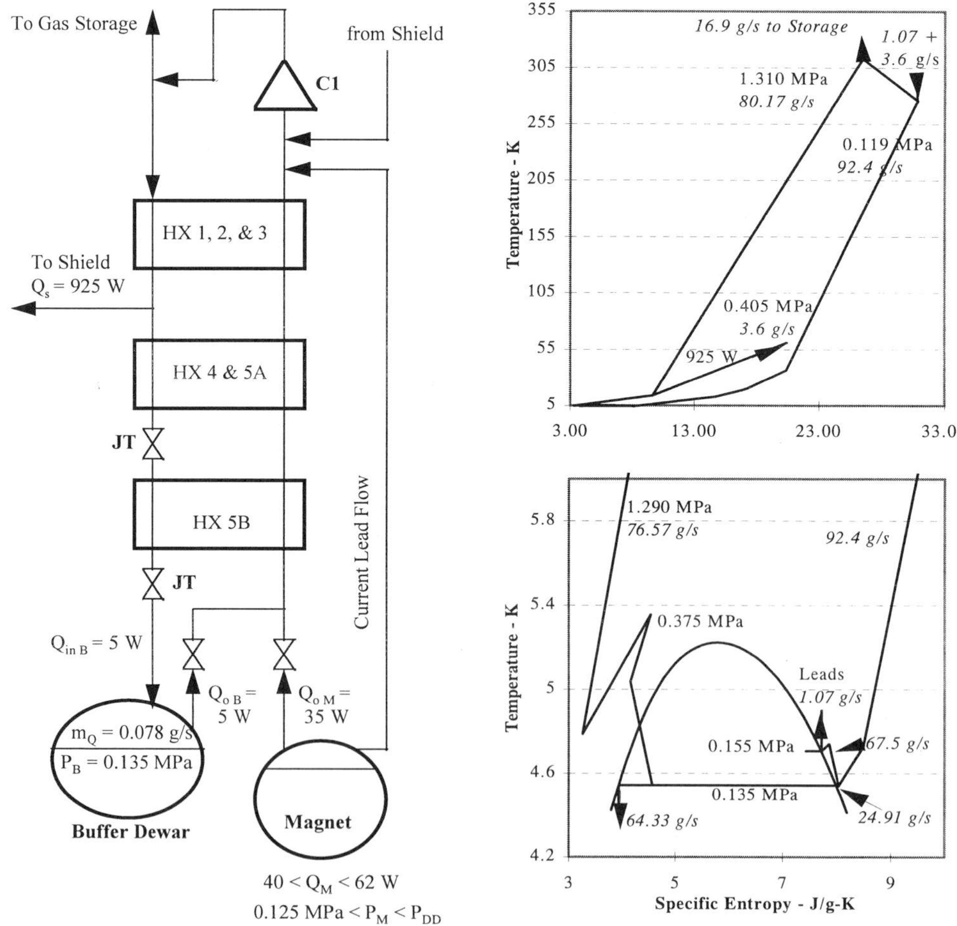


**Figure 3.** Process flow diagram and a T-s diagram for the HRS operating with an unbalanced JT Cycle.

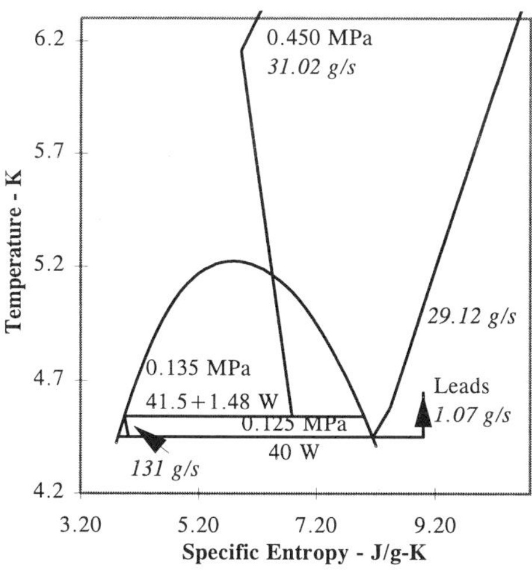

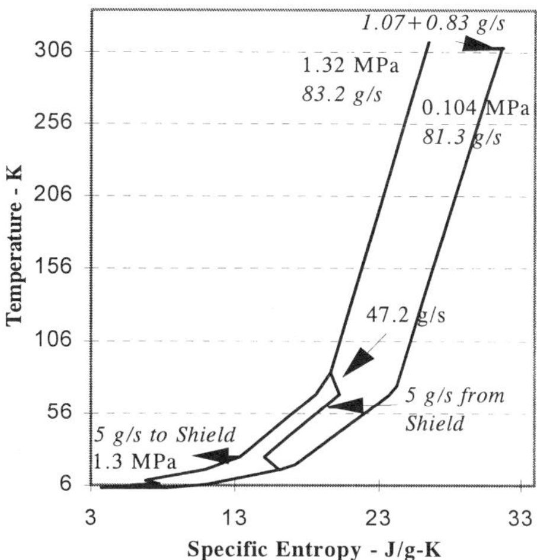


**Figure 4.** T-s Diagram foı Stage 2, Liquid Transfer.

generated by the JT process, and from an internal heater. The initial gas boiloff caused by cooling the warmed magnet supply transfer line was not considered. Fig. 4 presents a T-s diagram with the relevant parameters for this process. Using a 41.5 W heater input, liquid is directed to the magnet at a rate of 131 g/s.

Transfer time is determined by writing a mass balance for the buffer based on the known initial and desired final state (taken as 400 L LHe, ~ 10% of total volume, to preserve temperature gradients and prevent the release of any frozen contaminants).

$$t = [\, M_f^T - M_i^T \,]_B \,/\, [m_{in\,B} - m_{in\,M} - m_{o\,B}] \tag{3}$$

A mass balanced applied to the magnet vessel leads to the net mass collected in the magnet.

$$[\, M_f^T - M_i^T \,]_M = [m_{in\,M} - m_{o\,M} - m_{CL} - m_{LIQ}\,]\, t \tag{4}$$

The mass of collected liquid can be found with Eq. (5):[4]

$$\{[M_L]_f\}_M = [\, M_f^T - M_i^T \,]_M \,/\, [1 - \rho_G/\rho_L\,]_M \tag{5}$$

Applying these results to the case of pumpdown from 0.162 MPa, the magnet collects 2,090 L of 4.45 K LHe in 2,040 s ($t_2$), increasing its liquid volume to 30,170 L (86.2%).

## Third Stage of Recovery - Recharge

If the level in the magnet vessel is above the nominal operating minimum, recharging can be started. During recharging, 651 kJ is dissipated into the bath in 533 s ($t_3$), equivalent to 1,220 W in real time. Turbines are assumed to be shutdown, and the buffer is isolated from the magnet, initiating the unbalanced JT mode. Magnet loads of 51 W and 1.44 g/s were used for this stage's evaluation. In addition to lead flow, the heat loads produce a boiloff of $[(1{,}220 + 51)\text{W}/(19.23\ \text{J/g})][1-\rho_G/\rho_L] = 54.53$ g/s @ 4.45 K that must be handled by the cold box. Table 1 indicates that this boiloff could be handled in real time if mode switch-over could be initiated immediately. If the magnet vessel level is below the allowable minimum, recharging is delayed while the HRS provides make-up to the system.

The JT mode model predicts that the buffer liquid increases by 230 L, the warm gas volume increases by the equivalent of 66 L LHe @ 4.45 K, while magnet liquid volume falls by 300 L to 29,870 L (85.3%).

## Fourth Stage of Recovery - Liquid Transfer after Recharge

Stage 4, the transfer of accumulated liquid from buffer to magnet, is the same as the 2nd stage except that the magnet loads are taken to be 62 W and 1.44 g/s. The model predicts a lower buffer heater output, 20.2 W, with 222 L transferred in 228 s ($t_4$).

For recovery from a full deep discharge ending in a magnet pressure of 0.162 MPa, the magnet could be operational in the sum of the times for discharge and the 4 stages, 6,861 s (1.9 h), since its final conditions, 4.45 K and 30,092 L, are within operating parameters.

## IMPACT OF INITIAL CONDITIONS ON DEEP DISCHARGE RECOVERY

During normal operations, deep discharge initial conditions could be anywhere between 4.45 K and 4.55 K, and 84.84% and 90%. Pumpdown results predict large liquid losses in the magnet. Additionally, the system is required to handle 3 consecutive deep discharges, one per 24 h period. The models are systematically used to determine initial conditions where recovery from a single deep discharge is possible, and the regions over which the requirement for 3 consecutive deep discharges can be met.

### Immediate Recharge

The criterion used for permitting recharge immediately after initial liquid transfer is whether the magnet liquid level is ≥ 29,700 L (84.86%) after recharge. A condition of recharge is that the magnet be at 4.45 K prior to initiation. During recharge, the magnet liquid level decreases 300 L. Within the assumed constraints, this value is not a function of initial liquid level. Hence, the minimum liquid level must be at least 29,700 + 300 = 30,000 L prior to commencing recharge. Fig. 5 displays the models' predictions for the initial conditions that result in the existence of the minimum operating level at the end of recharge for a 4,150 kJ deep discharge. If the point defined by the initial conditions falls onto the left hand region, immediate recharge is possible. If the initial conditions are described by a point falling onto the right hand region, immediate recharge is not recommended and should be forbidden for all but the most unusual of situations. For these later conditions, the magnet must be maintained at low current until the HRS returns the liquid level to 30,000 L. Note that for any initial liquid level below 87%, immediate recharge is not recommended for any initial temperature.

### Deep Discharge Recovery in 24 Hours

The models indicate that the time consumed in executing the 4 recovery stages is < 2 h. Hence, there are ~ 22 h that can be used between Stages 2 and 3 to liquefy stored gas to increase the magnet's liquid level. During this period, with the current held at low levels, the performance of the HRS is maximized. At reduced currents, there is 58.5 W and 0.83 g/s of capacity available @ 4.45 K, treating the lead flow as pure liquefaction. Using an approach advanced in [5], the equivalent liquefaction capacity of a refrigerator operating between 4.45 and 309.2 K is calculated to be 102 W/g/s, which agrees conservatively with existing equipment performance.[6] Hence, there is ~ 0.83 + 58.5/102 = 1.4 g/s (41.98 L/h) of liquefaction capacity when the magnet is at low current levels. The following gives the final liquid level in the magnet assuming recharging is delayed until the last instant

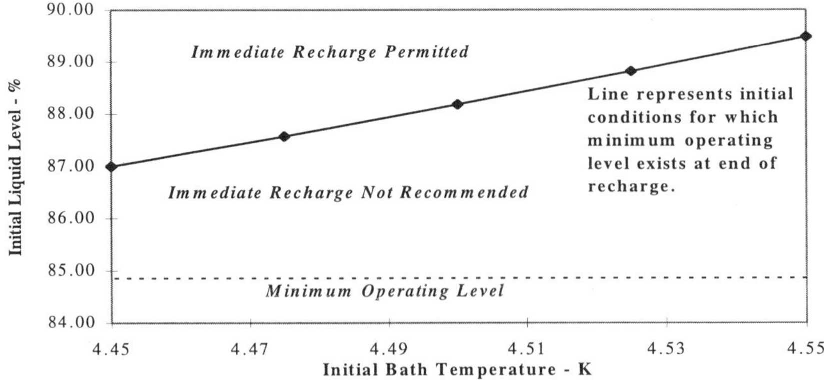


**Figure 5.** Regions where immediate recharge is permitted after a 4,150 kJ deep discharge.

**Table 2.** Magnet liquid volume at the end of initial liquid transfer, and pumpdown and initial liquid transfer time vs. initial liquid level and temperature for a full deep discharge

| Init. Temp. | 4.45 K | 4.45 K | 4.5 K | 4.5 K | 4.55 K | 4.55 K |
|---|---|---|---|---|---|---|
| Liquid Level, % | Liq. Vol. after Transfer, L | Pumpdown + Liq. Transfer Time, $t_{1+2}$, s | Liq. Vol. after Transfer, L | Pumpdown + Liq. Transfer Time, $t_{1+2}$, s | Liq. Vol. after Transfer, L | Pumpdown + Liq. Transfer Time, $t_{1+2}$, s |
| 85 | 29,299 | 4,136 | 28,915 | 5,086 | 28,508 | 6,015 |
| 86 | 29,649 | 4,136 | 29,257 | 5,089 | 28,841 | 6,021 |
| 87 | 29,999 | 4,136 | 29,598 | 5,092 | 29,174 | 6,026 |
| 88 | 30,349 | 4,135 | 29,940 | 5,094 | 29,507 | 6,032 |
| 89 | 30,699 | 4,135 | 30,281 | 5,097 | 29,840 | 6,038 |
| 90 | 31,049 | 4,134 | 30,623 | 5,100 | 30,173 | 6,044 |

$$V_f^L = \text{Liq. Vol. after Init. Transfer} + \{1.4\ \text{g/s}[(24\ \text{h})(3{,}600\ \text{s/h}) - (t_{1+2} + t_3 + t_4)]/\rho_L\} - 300\ \text{L} + 222\ \text{L}$$

Table 2 provides the values for $t_{1+2}$, and for the liquid volume after initial transfer, for a range of magnet initial conditions. The results of applying this equation are given in Figure 6, which is a plot of final liquid level vs. initial liquid level, with initial temperature as a parameter. For initial levels above ~ 89% of 35,000 L and initial temperatures below ~ 4.5 K, it is possible to not only recover to 0.125 MPa (4.45 K), but to also have the magnet filled to its maximum level in less than 24 h. For initial liquid levels above ~ 86% at all initial temperatures, recovery in 24 h to a point within the region of allowable liquid levels is possible. For initial temperatures less than ~ 4.5 K, recovery is always possible within 24 h. For initial liquid levels below 86%, and initial temperatures above ~ 4.51 K, recovery in 24 h is likely not to be possible if a 4,150 kJ deep discharge occurs. At lower deep discharge energies, recovery becomes more likely.

## Three Consecutive Deep Discharges in 72 Hours

It has been shown that recovery is possible within a specific range of initial conditions. These examples represent the first of 3 consecutive deep discharge events. Successful recovery from the first event by definition returns the magnet bath to a temperature of 4.45 K and to a liquid level ≥ 29,700 L. Hence, 4.45 K is the initial temperature for the second deep discharge, which is assumed to occur at the end of the first 24 h period, with an initial level of ≥ 29,700 L. For these initial conditions, Fig. 5 indicates that the system will always recover from a 4,150 kJ deep discharge, and the magnet will either be at its maximum liquid level at the end of the 2nd 24 h period, or end this period with more liquid in the magnet than existed at the start. By extension, the same can be said of the third

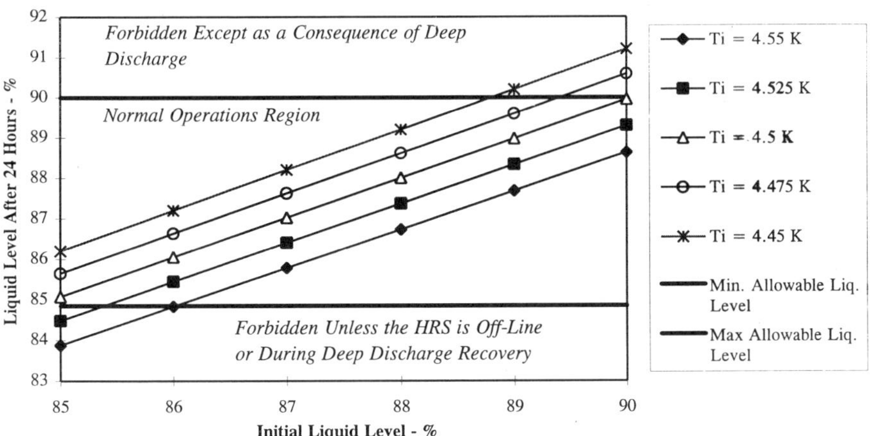


**Figure 6.** Recovery from a 4,150 kJ deep discharge: liquid level 24 h after event. Recharge delayed while liquefying at 1.4 g/s.

period. If the HRS can bring the magnet back on-line 24 h after the first deep discharge, it can always handle 2 (or more) additional events spaced at 24 h intervals.

Since recovery includes pumpdown to 4.45 K, it is possible for the system to handle 3 consecutive deep discharges in 72 h even if it does not recover from the 1st in 24 h, fulfilling part of the criteria. Consider a 4,150 kJ deep discharge with initial conditions of 4.55 K and 85%. Table 2 indicates that the liquid level after Stage 2 is 28,508 L. At a liquefaction rate of 41.98 L/h, it takes [(30,000 - 28,508) L]/(41.98 L/h) = 35.54 h. to reach the 30,000 L required for recharging. After Stage 4, there would be (30,000 - 300 + 222) L = 29,922 L of liquid in the magnet (85.5%). Using 4.45 K and 85.5% as the new initial conditions, Table 2 indicates (by interpolation) that after Stage 2 for the 2nd deep discharge, the level is 29,474 L. Filling to the minimum recharge level will take [(30,000 - 29,474) L]/(41.98 L/h) = 12.53 h. The total time required to handle both events is

$$\{[(6{,}015\text{ s})(1\text{ h}/3{,}600\text{ s})]+35.54\text{ h}+[(t_3+t_4)(1\text{ h}/3{,}600\text{ s})]\}_{\text{event 1}}+\{[(4{,}136\text{ s})(1\text{ h}/3{,}600\text{ s})]+12.53\text{ h}+[(t_3+t_4)(1\text{ h}/3{,}600\text{ s})]\}_{\text{event 2}}=51.31\text{ h}$$

where 6,015 and 4,136 s are $t_{1+2}$ for the 2 respective cases. Using the same initial conditions for the 3rd event, 3 consecutive full deep discharges can be handled in

$$51.31\text{ h}+[(4{,}136\text{ s})(1\text{ h}/3{,}600\text{ s})]+12.53\text{ h}+[(t_3+t_4)(1\text{ h}/3{,}600\text{ s})]\}_{\text{event 3}}=65.2\text{ h}$$

The HRS should be able to handle 3 consecutive deep discharges in under 72 h when the initial conditions of the first do not permit 24 h recovery, provided the period between each deep discharge is the time it takes to achieve the minimum recharging level of 30,000 L.

## SUMMARY AND DISCUSSION

The results of the models indicate that for most initial conditions, the HRS can handle a 4,160 kJ deep discharge and recover in 24 h, and handle 3 of these discharges at 24 hour intervals. The magnet is not overpressurized and no helium is lost during recovery. Transition times between modes were not addressed. However, with the large massflow increases generated by deep discharge and recovery, and the low material specific heat, reaction times will be on the order of minutes, not hours, minimally impacting final results.

Recently, the magnet's conductor stabilizer was changed from aluminum to copper. This resulted in reduced AC losses. Energy deposited as heat in the magnet for a full deep discharge is reduced from 4,160 kJ to 3,520 kJ with similar reductions expected for recharge. This will reduce ultimate magnet pressures, shorten recovery times, and is likely to permit 24 hour recovery for all initial conditions.

## NOMENCLATURE

| h - enthalpy | V - volume | **Superscripts** | **Subscripts** | i - initial | o - output |
|---|---|---|---|---|---|
| M - mass of helium | v - spec. volume | DD-deep discharge | B - buffer | in - input | Q - heat load |
| m - massflow | Q - heat load | G - gas | CL - current leads | L - liquid | s - shield |
| P - pressure | ρ - density | L - liquid | f - final | LIQ - liquefaction | 1,2, 3, 4 - stage number |
| t - time | ΔQ - energy dissipated | op - operating | G - gas | M - magnet | |
| u - internal energy | | T - total | | | |

## REFERENCES


1. S.F. Kral, V. Karasik, Y.M. Lvovsky, K.E. Grut, J.P. Kelley, K.D. Dixon, P. Ribeiro, R. Kunz, Alaska SMES: Form and Function for the World's Largest Magnet, Presented at the 1997 Cryogenic Engineering Conference, Portland, OR, July 28 - August 1, 1997.
2. L.E. Scott, R.C. Van Meerbeke, B.S. Kirk, and G.C. Nubel, Technology of Liquid Helium, R.H. Kropschot, B.W. Birmingham, and D.B. Mann, ed., NBS Monograph 111 (1968), p. 167.
3. M. Wiseman, R. Bundy, J.P. Kelley, and W. Schneider, CEBAF cryounit loss of vacuum experiment, Applications of Cryogenic Technology, Vol. 10, J.P. Kelley, ed., Plenum Press, NY (1991), p. 287.
4. L.E. Scott et al., op cit., p. 164.
5. T.R. Strobridge, Cryogenic Refrigerators - An Updated Survey, NBS TN 655, June 1974.
6. A.E. Senn, Linde Kryotechnik AG, Private Conversations.

# SIMULATION AND DESIGN OF A PROTECTION SYSTEM FOR SMES

F. Rosenbauer and M. Harke

Lehrstuhl für Elektrische Maschinen und Geräte
Technische Universität München
80290 München, Germany

## ABSTRACT

The protection system of a superconducting magnetic energy storage (SMES) prototype consists of a quench detection system, bypass switches, quench heaters and a discharge circuit. The bypass switches are semiconductors that are placed within the cryogenic area. They serve for bridging the coils in the case of a quench in order to obtain a faster discharge.

A complex numerical simulation system was developed to investigate the behaviour of this protection system because both the superconductor and the semiconductor are highly non-linear. Based on the results of the simulation, the components of the protection system can be rated and the process-oriented sequential control can be optimized.

## INTRODUCTION

Currently there is a 1 MJ/1 MW SMES prototype under test in our laboratory. The storage consists of six coils. Each coil has 385 turns of the LCT 10+1 NbTi superconductor which are arranged in seven layers. This results in an inductance of 190 mH. Between the 3rd and 4th layer the winding has a tap. The six coils are assembled as a torus (total inductance 1.486 H) to minimize the stray field. The coils are cooled by supercritical helium that flows in two cooling layers, which are inside and outside the winding. All cooling pipes are serial connected with several intermediate coolers between the coils.[1,2]

Special effort was made for the quench protection system as the superconductor is not cryo-stable. Thermal and electrical quench detection is employed. The first uses GaAlAs-diode thermometers, which are placed at the critical point of each coil. For the latter, the voltages across the coils are compared to the voltages across auxiliary coils.

In the case of a quench the whole system is de-energized into an external resistor (Fig. 1). Additionally, each coil has a switchable bypass. This allows a faster decrease of the current in the affected coil. To avoid losses caused by additional current leads the switch is placed within the cryogenic area. This switch is realized using IGBT modules and power diodes. The diode is essential, because the IGBT modules contain an inverse diode (Fig. 2).

To get the complete coil quenched and to minimize the temperature gradient within the quenched coil there is a quench heating or booster device. This fires directly into the center tap of each coil and quenches the whole coil first by exceeding the critical

*Advances in Cryogenic Engineering,* Vol. 43
Edited by P. Kittel, Plenum Press, New York, 1998

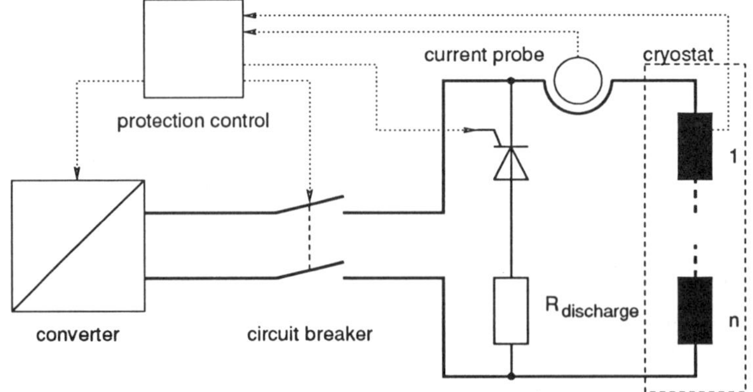


**Fig. 1.** Overview of a modular SMES with protection elements. Our prototype consists of $n = 6$ coils.

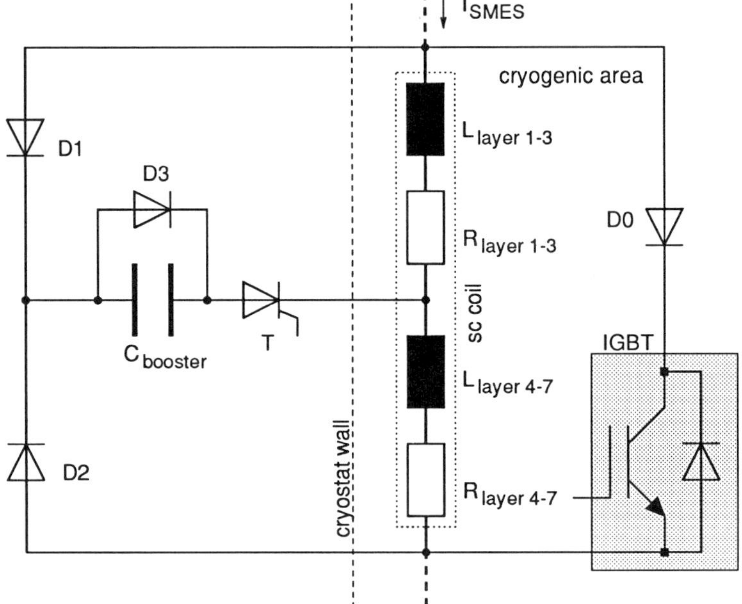


**Fig. 2.** Protection elements of a single coil. The mutual inductances between the layers and to the other coils are not shown for simplicity.

current and the critical induction and second by the hysteresis losses caused by the high $dB/dt$.

To investigate the behaviour of the coils and of the protection system during a quench, a simulation system was developed. The simulation model includes the thermal, electrical and magnetic behaviour of the coils and the thermal and electrical behaviour of the power semiconductors working at cryogenic temperature.

## SIMULATION MODEL

### Model of the semiconductor devices

It can be seen easily that the behaviour of the IGBT modules and power diodes, which are placed within the cryostat, influences protection system behaviour. The lower the on-voltage of the devices, the faster is the commutation from the coil into the bypass and the better works the protection. This on-voltage is highly dependent on the temperature of the semiconductor.[3] The on-voltage again causes losses and therefore a warming up of the device, especially as the thermal capacities and thermal

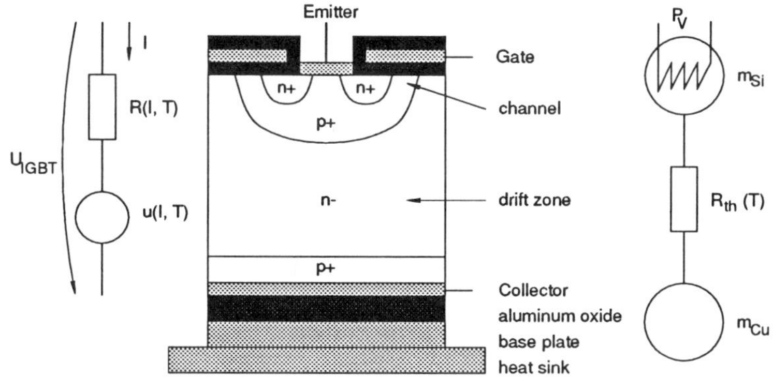


**Fig. 3.** Cross section of an IGBT with electrical (left) and thermal (right) simulation model.

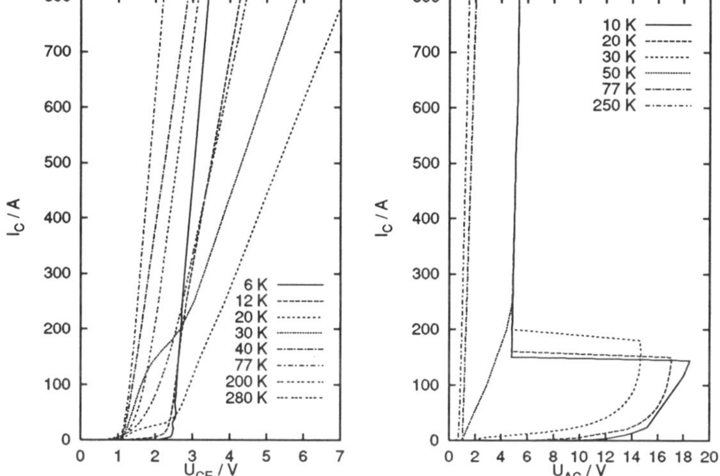


**Fig. 4.** Characteristics of the IGBT FZ 360 R 17 KF (left) and the power diode D 798 N 12 (right) at variable temperature.

conductivities are very small at cryogenic temperature. Thus, a simulation model must include both, the electrical and the thermal behaviour of the semiconductor devices.

**Electrical behaviour.** The high inductances of the coil and therefore the relatively small $di/dt$ in combination with the fast switching of the IGBT modules with their typical turn-on time of 1 $\mu$s allow us to neglect the dynamic behaviour of the IGBT modules. However, the static voltage drop has to be considered. It includes[4]

$$U_{\mathrm{IGBT}} = U_{pn} + U_{\mathrm{drift}} + U_{\mathrm{channel}} \tag{1}$$

$$U_{\mathrm{drift}} = \frac{W_{\mathrm{drift}}}{(\mu_n + \mu_p)\,\tau} \tag{2}$$

$$I_{\mathrm{C}} = I_{\mathrm{saturation}}\left[\exp\left(U_{pn}/U_{\mathrm{therm}}\right) - 1\right] \tag{3}$$

$$U_{\mathrm{channel}} = I_{\mathrm{C}} \cdot \frac{W_{\mathrm{channel}}}{A_{\mathrm{channel}} \cdot q_0 \cdot (\mu_n n + \mu_p p)} \tag{4}$$

The voltage drop depends on the charge carrier mobilities $\mu$, the charge carrier densities ($n$,$p$) and the charge carrier lifetimes $\tau$. All have different temperature dependencies. For example, the voltage $U_{pn}$ across the $pn$-junction near the Collector (Fig. 3) increases with falling temperature, but the resistance first decreases (down to a temperature of about 100 K) and then increases again.[5] Up to now there is no model, which is based on the geometry and the doping profile of the device, that describes the behaviour of an IGBT or other power semiconductors over the whole temperature range from 5 to 400 K. But even if there was such a model, it would require a complex and time consuming calculation.

The data of $U_{IGBT}$ as a function of current and temperature was already known (Fig. 4).[3] This data was used for the simulation. The on-voltage for intermediate values of the current or the temperature is obtained by linear interpolation of the measured characteristics. This procedure requires measurements of the characteristics of the devices to be simulated. The devices also had to be tested to determine if they work in cryogenic environment.

**Thermal behaviour.** When the bypass is switched on, the voltage across the coil is nearly zero. Therefore any switching loss can be neglected. After switching, the rising current in the bypass causes the dissipation

$$P_V = U_{IGBT} \cdot I \tag{5}$$

which can exceed 2 kW. The heat transfer into the high vacuum and the heat radiation from the device can be neglected. As the links are very long and have a small diameter, heat conduction via the terminals can be neglected also. Relevant heat conduction is only possible via the base plate into the heat sink.

The one-dimensional thermal model uses two lumped thermal masses: the silicon of the device, and the base plate (Cu) with heat-sink (Cu). The two masses are coupled with a thermal resistance that represents a part of the drift layer and the aluminum oxide insulation layer (Fig. 3). The heat sink may be connected to a cooling medium.

The dimensions were determined with an opened device and confirmed by the manufacturer. The data of the specific thermal capacities and thermal conductivities can be found in the literature.[6–8] It is interesting to note that, from the thermal point of view, operation at about 40 K would be ideal. Here the thermal conductivities of silicon and aluminum oxide have their absolute maximum. It is 30 times larger for Si, compared to the values at 400 K, and 180 times larger for $Al_2O_3$.

The electrical and thermal models of the power diode are based on the same assumptions as mentioned above for the IGBT (Fig. 4).[3]

## Model of the superconducting coil

For the thermal simulation the winding of a coil is discretized into 9625 lumped thermal masses that are connected by the appropriate thermal resistance. The temperature of the cooling layer serves as the boundary condition. Here two modes can be selected:

1. Fast Simulation: The temperature of the cooling layer is constant. Due to the symmetry only one coil has to be simulated thermally.
2. Complete Simulation: The temperature of the cooling layer is variable. This can be achieved if the complete thermal behaviour of the six (thermally in serial connection) coils including cooling plant is taken into account.[2]

The heat dissipated in a single thermal mass can be calculated with[9]

$$P_{sc} = \begin{cases} N_C \cdot (N_F \cdot P_{hyst} + P_{coup,\,composite}) + P_{coup,\,cable}, & \text{if sc} \\ R_{sc}(T, B) \cdot I \cdot (I - I_{crit}), & \text{if nc} \end{cases} \tag{6}$$

where $N_C$ denotes the number of composites (here 10) and $N_F$ the number of filaments (318) of the cable. These losses can only be calculated on-line, if the future peak values of the induction are known. Therefore several consecutive simulation runs have to be performed.

As the hysteresis losses $P_{hyst}$ are proportional to $dB/dt$ and the coupling losses $P_{coup,\,composite}$ and $P_{coup,\,cable}$ are proportional to $(dB/dt)^2$, a 3D semi-analytical magnetic field calculation that is based on Biot-Savart's formula was included. Special effort was made to have a fast calculation that considers the field distortion caused by a bypassed and boostered coil. Besides the modulus of the induction it is possible to calculate its perpendicular and parallel component with respect to the superconductor cable.

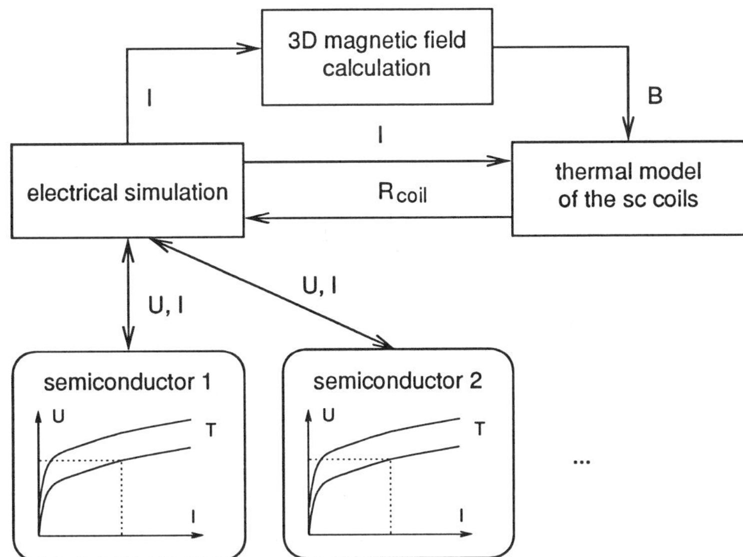


**Fig. 5.** Overview of the simulation system. The electrical model interacts with the thermal model of the coils and the models of the semiconductors. The latter are coded as objects that "know" their electrical and thermal state.

**Table 1.** Simulation results for selected protection strategies (D = disabled, E = enabled; $w$ denotes where the energy stored in the coil before the quench is dissipated)

| no. | | 1 | 2 | 3 | 4 | 5 |
|---|---|---|---|---|---|---|
| discharge | | D | E | E | E | E |
| booster | | D | D | D | E | E |
| bypass | | D | D | E | D | E |
| $w_{coil}$ | % | 99.6 | 21.2 | 11.6 | 33.4 | 15.0 |
| $w_{dump\ resistor}$ | % | 0.0 | 78.6 | 88.2 | 66.4 | 84.8 |
| $T_{coil,max}$ | K | 241 | 150 | 126 | 147 | 115 |
| $\max(T_{coil,min})$ | K | 63 | 27 | 16 | 78 | 57 |
| $\max(\Delta T)$ | K | 178 | 123 | 110 | 69 | 58 |

Every lumped thermal mass corresponds to a superconductor element. Once the temperature, the current and the induction are known for every element, its critical current can be calculated.[10] This allows to determine if this element is in superconducting or normal conducting state.

## Coupling of the models

The models of the semiconductors and of the coils join together in the electrical model (Fig. 5). It is based on the equivalent network shown in Fig. 2 which is combined with external elements like discharge circuit, power converter and protection logic (Fig. 1). Using the state analysis, a system of differential equations can be set up and solved using the well known methods. The electrical simulation and the simulation of the coils operate with different time steps that take into account the respective smallest time constants (electrical 4 $\mu$s, thermal 1 ms).

The interface between the electrical model and the thermal model of the coils are the currents that run in the two sections of the coil and the resulting resistances $R_{layer\ 1-3}$ and $R_{layer\ 4-7}$. In Fast Simulation Mode only one coil is simulated with the complete equivalent network and the other coils are combined into a single inductance. This is different for the Complete Simulation Mode: All coils are simulated with their thermal behaviour and the resulting resistance including the quench detection system at every coil, but only the coil that quenches first is simulated with the bypass and booster devices.

Table 1 shows the results of several simulation runs with different protection strategies using the Fast Simulation Mode. The progression of different currents and temperatures of run no. 5 can be seen at Figs. 6 and 7.

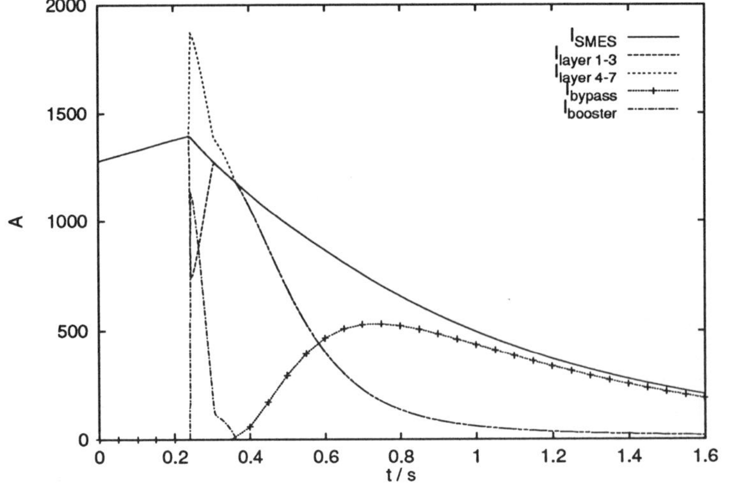


**Fig. 6.** Currents after a quench. As soon as the booster is extinguished, the current commutates into the bypass.

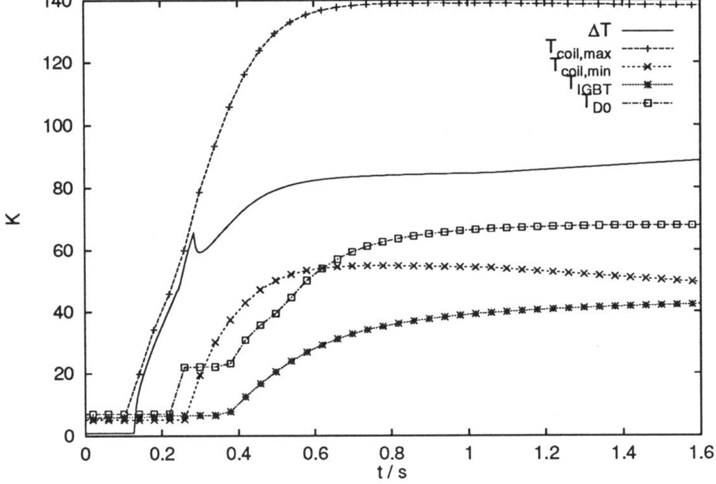


**Fig. 7.** Temperature rise during a quench. All protection elements are enabled. $\Delta T = T_{\mathrm{coil,max}} - T_{\mathrm{coil,min}}$

## DESIGN OF THE PROTECTION SYSTEM

### Switchable bypass

As already presented in a former publication[3] we chose the IGBT module FZ 360 R17 KF for our bypass switch. The rated current of this module is 360 A, the maximum possible current at 5 K is 800 A. Although the current through the bypass should never reach the maximum current of our SMES of 1360 A, we use two modules working in parallel to stay on the safe side.

To reduce the cooling power at the 5 K level, the whole SMES is placed within a liquid nitrogen screen. The IGBT modules are mounted within this screen in the gap between two coils above the actual torus (Fig.8). The maximum magnetic field at this place yields only 250 mT which is very small compared to the absolute maximum of 3.7 T within the cryostat.

The heat that is generated by the semiconductor has to be removed from the chip to prevent destruction of the device by overheating or too high thermal gradients within

**Fig. 8.** Paralleled IGBT modules (black) and compensation diode with clamping device (white) assembled on their heat sinks. The IGBTs are connected to the superconductor via a meandered metal sheet.

a module. A direct connection of the IGBTs to the liquid helium cooling circuit is not possible, as just in the case of a quench this one is highly loaded. On the contrary, the thermal conduction from the modules to the superconductor has to be reduced. That is why the distance between the Emitters of the IGBT modules and the superconductor has to be enlarged by the factor six using a meandered metal sheet. The link between the diode and the superconductor was already large enough. Another possibility was a connection to the $LN_2$ screen for cooling purposes. This is also impossible, because it would represent a thermal bridge between the 5 K and the 77 K temperature levels.

These circumstances lead to the decision to mount the IGBTs on a simple copper heat sink. This must be large enough to store the energy dissipated by the semiconductor during the bridging. On the other hand, the copper block must not be too large, as this would cause significant eddy current losses by the alternating magnetic field nearby the storage coils. The size of the block was determined with the simulation program. The calculations resulted in a thermal mass of 1.6 kg, whose eddy current losses are below 0.25 W. Although it is able to limit the maximum temperature of the modules to a moderate value and take advantage of the high thermal conductivities around 50 K.

The drivers of the IGBTs are placed outside of the cryostat. First this simplifies the electrical installation within the cryostat and second we do not need a driver that works at 5 K. A disadvantage of this layout is the long line (up to 4 m) between the driver and the modules. This should be no problem because the IGBT does not need to work at high frequencies. In fact it is switched on only once during a long period.

## Quench heater

The key cost factor of the quench heating device are the capacitors. The simulation results allowed to reduce the voltage of the heater capacitors from the originally chosen 1600 V to 1000 V. Additionally, the bipolar capacitors, that were proposed first, were replaced by electrolytic capacitors and diodes (labeled D3 in Fig. 2). Both measures reduced the overall cost (including diodes) to 35%.

The 13 copper lead wires (3 at each coil, 5 of which are common) have a graduated diameter for low thermal transfer from ambient temperature to 5 K and minimum resistance. Outside the cryostat the area is 4 $mm^2$, between the current leads and the $LN_2$ level it is 1.5 $mm^2$ and from there to the coils it is 0.75 $mm^2$.

## Measurement and Control

The control of the protection system is performed by a 16 bit microcontroller of the type 80C167. The voltages across the coils and the auxiliary coils and the signals from the temperature sensors and the SMES current probe are preamplified and handed over to the microcontroller. This performs an analog to digital conversion and evaluates the sensor characteristics. All analog input lines are galvanically insulated from the controller with insulation amplifiers.

The measured data is transfered via a RS-232 interface to a PC, that serves as user interface. Here the measured data can be displayed and commands to the microcontroller and parameters for the threshold values of the quench detection can be entered. Last but not least the data can be stored to hard disk.

If a quench is detected, i. e. one of the signals exceeds its respective threshold value, the controller can react by triggering the appropriate bypass and heater. In the same time the storage is insulated from the power converter and discharged. To protect the cooling system, a relief valve can be opened. All components are insulated from the microcontroller by optical fibres. Besides the galvanic insulation this avoids noise generated by the pulsing magnet.

## CONCLUSIONS

A practical model for the simulation of a SMES and its accessory protection was presented. This model takes into account the special electrical and thermal behaviour of both, the superconducting coil and the semiconducting protection elements.

Based on this simulation, a semiconductor switch for 5 K was developed, that takes the special thermal boundary conditions of our plant into account. During the first cooling tests of the SMES, the temperature sensors within the coils were already in use. All elements work satisfactorily, but the whole system has to prove is functionality.

## ACKNOWLEDGMENT

The authors wish to thank Prof. Dr.-Ing. H.-W. Lorenzen, Lehrstuhl für Elektrische Maschinen und Geräte, Technische Universität München for initiating this work and Deutsche Forschungsgemeinschaft for the provision funds to develop and build the SMES.

## REFERENCES

1. J.F. Kärner, H.W. Lorenzen, F. Rosenbauer, J. Schaller and R.M. Schöttler, A protection system for small high power SMES with power semiconductors working at cryogenic temperature, *IEEE Trans. Applied Superconductivity* 5:266 (1995)
2. R.M. Schöttler and H.W. Lorenzen: Temperature and pressure rise in supercritical helium during the quench of indirectly cooled sc coils, in: "Advances in Cryogenic Engineering", Vol. 41, Plenum Press, New York (1996), p. 325
3. F. Rosenbauer and H.W. Lorenzen: Behaviour of IGBT modules in the temperature range from 5 to 300 K, in: "Advances in Cryogenic Engineering", Vol. 41, Plenum Press, New York (1996), p. 1865
4. N. Mohan, T.M. Undeland and W.P. Robbins. "Power Electronics: Converters, Applications and Design", 2nd, John Wiley & Sons, New York (1995)
5. S.M. Sze. "Physics of Semiconductor Devices", 2nd, John Wiley & Sons, New York (1981)
6. R.B. Scott. "Cryogenic Engineering", D. van Nostrand Co., Princeton (1967)
7. P.D. Maycock, Thermal conductivity of silicon, germanium, III-V compounds and III-V alloys, *Solid-State Electronics* 10:161 (1967)
8. W.D. Kingery, H.K. Bowen and D.R. Uhlmann. "Introduction to Ceramics", John Wiley & Sons, New York (1976)
9. R.M. Schöttler. "Optimized Thermal Utilization of Small High-Power Superconducting Magnetic Energy Storage" (thesis, in german), Technische Universität München (1997)
10. M.S. Lubell, Empirical scaling formulas for critical current and critical field for commercial NbTi, *IEEE Trans. Magnetics* 3:754 (1983)

# POWER SYSTEM FEEDBACK FROM HIGH-Tc SUPERCONDUCTOR SHIELDED CORE FAULT CURRENT LIMITER

J. Gerhold

Technical University
Graz, A-8010, Austria

## ABSTRACT

The shielded core fault current limiter is attractive with high critical temperature superconductors. The device acts during normal operation as an ideally shorted current transformer; the secondary superconducting loop, preferably a superconducting ring or cylinder, prevents from any flux entering. In case of a fault, the large current in the secondary current quenches the superconductor, and the secondary current is limited due to a considerable voltage drop; flux can now enter the core. The flux induces a primary voltage which counteracts the driving system voltage; this limits the fault current efficiently.

Very fast action is an essential requirement. However, this can give rise to dangerous transient oververvoltages both in the system to be protected which incorporates various inductances, and across the fault current limiter. Transient voltage occurence is discussed by means of equivalent circuits. Some means to control these voltages are highlighted in the view of different characteristics of the superconductor: ideal critical state performance versus slow or fast quench, respectively. The findings are illustrated by two numerical examples. It is shown that the shielded core fault current limiter gives a good chance to limit transient overvoltages down to a tolerable amount by introducing an additional tertiary winding. At the same time, heat generation in the superconductor can be kept very low.

## INTRODUCTION

Fault currents caused by short circuits in a power system are normally switched off by circuit breakers close to current zero passage, and a delay of at least 50 ms must be accepted. The instantaneous short-circuit current peak cannot be reduced. The related electrodynamic forces can cause severe stress on the equipment.

The introduction of a Fault Current Limiter (FCL) as an automatically activated series impedance $Z_{FCL}\{t\}$, which is almost zero during normal operation but increases very rapidly with time $t$ in case of a fault, is an interesting option to reduce equipment stress [1-4]. Superconductors offer a unique feature by utilizing the fast transition from superconductivity into ohmic voltage drop (flux flow or resistive state) when reaching the critical current density $j_{sc}$. Figure 1 (top) may illustrate the basic FCL arrangement in a three phase power system to be protected. The sytem can be treated as a single phase circuit in the case of grounded systems fed with an rms-voltage $U$ of frequency $f$ (bottom).

A lot of work has been carried out with metallic superconductors, but the cost of liquid helium cooling has been found somewhat prohibitive [1, 2]. The new high critical temperature superconductors (HTS) can be cooled much cheaper with liquid nitrogen. Unfortunately, wires and cables matching resistive FCL specifications are hardly available at present; bulk materials such as rings or cylinders can be produced much easier, and may be readily purchased from industry.

## TRANSIENT PERFORMANCE

It is obvious that the fast activation of an FCL, i.e. switching $Z_{FCL}$ from zero to a defined final level which guarantees the allowed rms-fault current $I_{FCL}$

$$I_{FCL} = U / Z_{max} \tag{1}$$

gives rise to transient overvoltages. The fault may start at an arbitrary time $t_{st}$ where the driving voltage is just $u(t_{st})$. The fault current increases for the first time according to $u(t) = (L_{so} + L^*) \cdot \partial i / \partial t$. The momentary fault current $i_{tr}$ as flowing at the instant of FCL triggering, $t_r$, cannot change instantaneously; a transient voltage $u_L = (L_{so} + L^*) \cdot \partial i / \partial t \,\vdots_{t=tr}$ appears along the power system inductances $L_{so}$+$L^*$; the former term incorporates any stray inductance of the feeding equipment. The voltage $u_{so}$+$u_{L^*}$ is compensated for by an instantaneous voltage $u_z$ across the FCL impedance $Z_{FCL}$ $(t)$. Hence

$$u - (L_{so} + L^*) \cdot \partial i / \partial t - u_z = 0 \tag{2}$$

for any subsequent time instant.

Transients must not become excessive so serious constraints may be required for the time dependent $Z_{FCL}(t)$. A simple ohmic impedance $Z_{FCL}(t) = R_{FCL}(t)$ where the resistance $R$

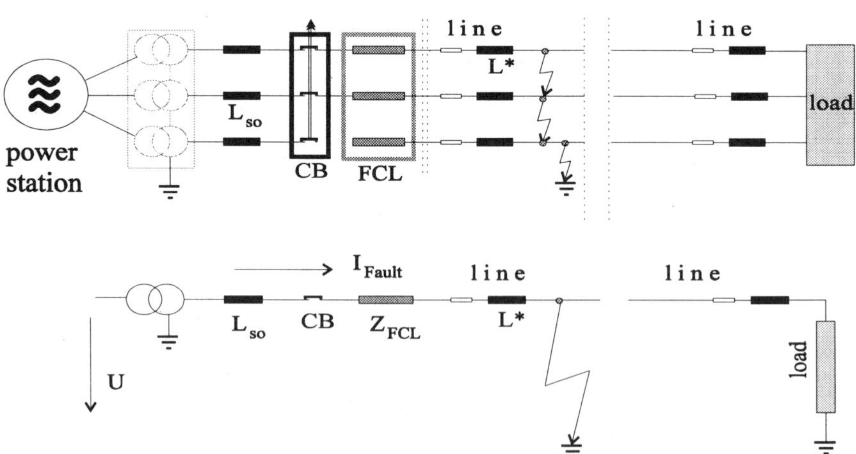


**Figure 1.** Three phase power system with integrated fault current limiter (top), and equivalent single phase circuit (bottom): $L_{so}$... inductance of power source and transformer; *CB*... circuit breaker; *FCL*... superconducting fault current limiter with impedance $Z_{FCL}$; $L^*$... inductance of transmission system; $U$...rms feeding voltage; $I_{fault}$...limited rms fault current

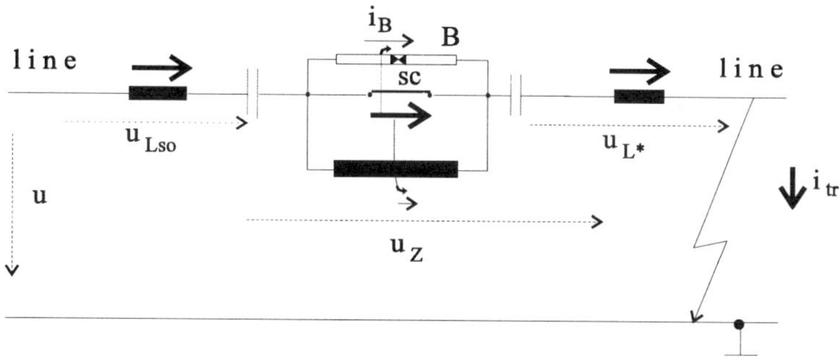


**Figure 2.** Transient voltages with inductive fault current limiter in case of activation: $u$... instantaneous driving system voltage at triggering time $t_r$; $u_{Lso}$... instantaneous voltage along source inductances; $u_z$... instantaneous voltage along fault current limiter impedance $Z_{FZL}(t)$; $u_{L*}$... instantaneous voltage along transmission system; $i_{tr}$... fault current at trigger level; $i_B$... balancing current across bypass $B$

is limited to $\leq R_{max}$ may cause a voltage spike up to

$$\hat{u}_{tr} \approx i_{tr} \cdot R_{max} \tag{3}$$

in the worst case. Overvoltages can in principle be controlled by providing a defined resistive path [4].

However, due to the large amount of heat being produced in any resistive device during a power system fault, inductive FCL's have been claimed to give a better performance [2]. But, transient voltages can be much more dangerous by the latter. Figure 2 may illustrate this in case of an inductance $L_{FCL} = Z_{FCL}$. This inductance is bypassed by the superconductor *sc* during normal operation. When a fault occurs, the superconductor may quench very fast at a current $i_{tr}$, up to a high resistance so almost no heat is dissipated. However, energy $1/2 \cdot i_{tr}^2(L_{so}+L^*)$ is already stored in the power system inductances immediately before $t_r$. This energy cannot jump. That means it has to be redistributed immediatly after $t_r$ between $(L_{so}+L^*)$ and the FCL-inductance $L_{FCL}$ according to $i_{tr}^2 \cdot (L_{so} + L^*) = (i_{tr} - \Delta i)^2 \cdot (L_{so} + L^* + L_{FCL})$. The related current jump $\Delta i$ would induce extreme voltage spikes $u_L$, $u_z$, which cannot be tolerated. A parallel path $B$ must be provided in order to allow a bypass current $i_B$. Special voltage surge protector devices have been proposed now and then, but this incorporates an additional risk of malfunction.

## SHIELDED CORE FCL

HTS cylinders can be used in a transformer type FCL only, i.e. as a secondary winding. This single turn secondary is coupled to the primary winding (turn number $n$), which carries the power system current at a high voltage, via magnetic flux in a ferromagnetic core. The superconducting secondary may tightly adhere to the core and allows no flux variation inside so the core is under near-zero flux condition during normal operation, i.e. it is shielded from flux entering. The transformer operates essentially as an ideally shorted current transformer where leakage flux only occurs in the primary winding. Core losses then are negligible. The primary leakage flux however causes an inductive voltage drop. Up to $0.05 \cdot U$ may be accepted in medium voltage distribution systems. Limitation of ohmic losses as well as the required high voltage insulation of the primary winding is a design challenge where wide experience from transformer design can be used.

The FCL is activated in case of a fault, i.e. an impedance $Z_s\{t\}$ is inserted into the secondary by triggering the superconductor into a voltage drop operation [5].

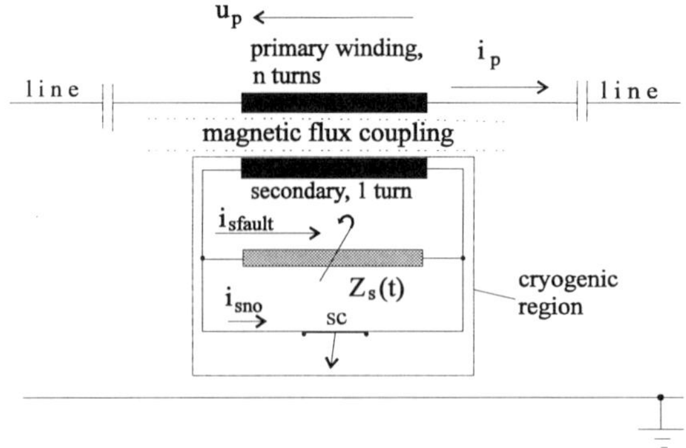


**Figure 3.** General scheme of transformer - type FCL; $u_p$...instantaneous driving voltage across the primary winding; $i_p$...instantaneous current through primary winding; $i_{sno}$... current across superconducting secondary *sc* during normal operation; $i_{sfault}$...instantaneous fault current across secondary impedance $Z_s(t)$ after FCL triggering in case of a fault

## Scheme of operation

The general scheme is illustrated in Figure 3. Figure 4 and 5 illustrate two different configurations. Figure 6 shows the T - equivalent circuit [6]. Inductances are involved.

The power system should see almost no series impedance during normal operation: the transformer primary stray inductance $L_{ps}$ and resistance $R_p$, must be very low; this is a challenge for the high voltage winding design. The superconducting secondary can be built almost without stray inductance [6]; a cold core may be advantageous since it can be housed together with the superconductor within a common thermal insulation envelope. No core losses occur during nomal operation where no flux enters the core.

The normal operation performance, the sustained short circuit performance and the transients can be described by a set of simple equations when neglecting $R_p$:

i) normal operation with $\Delta U$ << (rms): $\Delta U \approx 2\pi f\, L_{ps} \cdot I_{no}$ (4a)

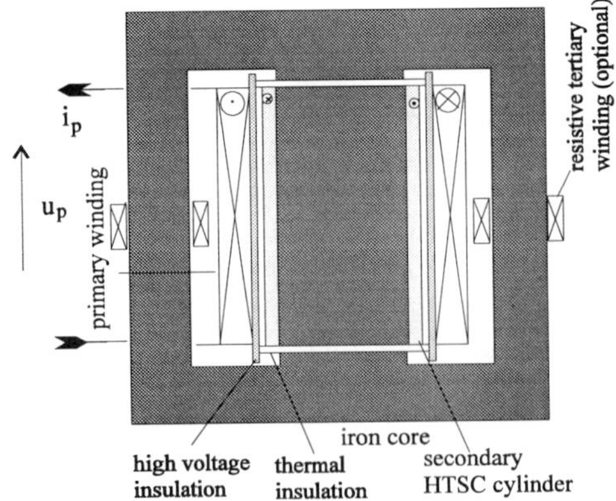


**Figure 4.** Shielded core FCL with closed flux.

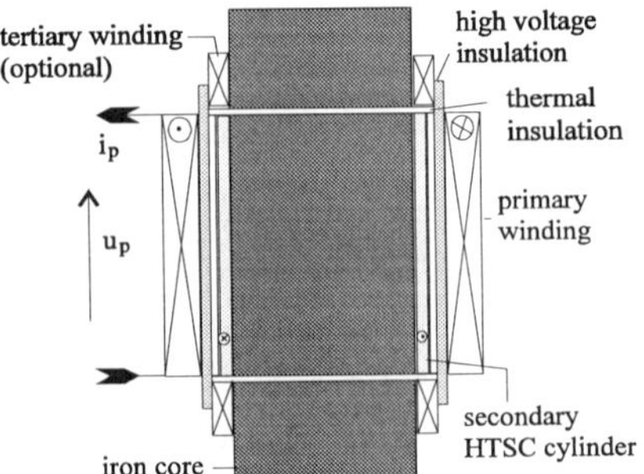


**Figure 5.** Shielded core FCL with open flux.

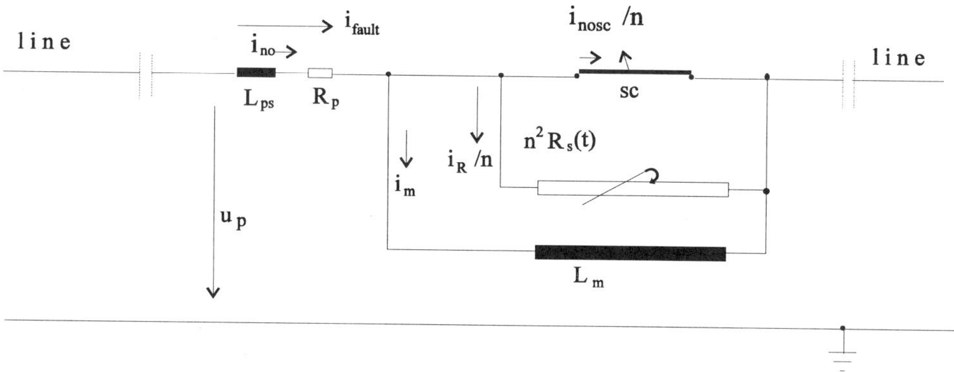


**Figure 6.** Equivalent T-circuit of shielded core FCL; $u_p$...driving voltage accross the primary winding; $i_{no}$... normal operation current; $i_{fault}$... limited fault current; $L_{ps}$...stray inductance of primary winding with turn number $n$; $R_p$...resistance of primary winding; $i_m$...current across main inductance $L_m$; $i_{nosc}/n$... equivalent current across superconducting secondary $sc$ during normal operation; $i_R/n$... current across equivalent resistance $n^2R_s(t)$ after triggering in case of a fault

ii) sustained short circuit operation (rms):

$$U_Z=((2\pi f\,L_{ps}\cdot I_{fault})^2+(n\,R_{max}\cdot I_R)^2)^{1/2};\; n\,R_{max}\cdot I_R-2\pi f\,L_m\cdot I_m=0;\; I_{fault}=((I_R/n)^2+I_m^2)^{1/2} \tag{4b}$$

iii) transient at $\geq t_r$: $u-(L_{so}+L^*)\cdot \partial i_{fault}/\partial t=n\cdot L_{ps}\cdot \partial i_{fault}/\partial t+n\cdot R_s\{t\}\cdot i_R$;

$$u-(L_{so}+L^*)\cdot \partial i_{fault}/\partial t=L_{ps}\cdot \partial i_{fault}/\partial t+L_m\cdot \partial i_m/\partial t;\; i_{fault}=i_m+i_R/n \tag{4c}$$

FCL activation occurs after fault begin when the secondary cross section $A_{sc}$ has been filled up with the critical current density $j_{sc}$, i.e. $i_{tr}=A_{sc}\,j_{sc}/n$. This triggers the superconductor into a resistive state at time $t_r$, i.e. flux flow or the occurence of a quench. The final equivalent resistance comes up to

$$n^2R_{max}=n^2\rho_{sc\,max}\,\pi D_{sc}/A_{sc}=n\cdot\rho_{sc\,max}\cdot j_{sc}\cdot \pi D_{sc}/i_{tr} \tag{5}$$

($\rho_{sc}$...resistivity of the quenched superconductor; $D_{sc}$...diameter of the HTS cylinder). A small cross section is required to ensure sufficiently high resistance. Hence a high critical current density is important. The resistivity may strongly depend on temperature which rises with time $t$ due to the dissipated heat.

It may be easy to adjust the FCL performance in a rather ohmic range by setting the main inductance $L_m > n^2 R_{max}$, or by selecting a rather inductive performance by providing a moderate $L_m$ value. A closed core configuration as illustrated in Figure 4 can be easily adjusted to a ohmic mode since $L_m$ is large [6]. An open core configuration [7] -see Figure 5- may enforce an inductive performance due to the rather large magnetic reluctance of the flux return path in open air, but requires a much less amount of iron and investment cost may be lower. The inductive voltage drop under normal operation will be higher on the other hand.

## Transient performance of shielded core FCL

The limitation of transient overvoltages is a key problem in the case of shielded core FCL's. Although an ideal Type II superconductor without quenching may give a very appropriate performance, a real superconductor may behave quite differently, e.g. due to temperature rise in course of heat dissipation. Hence additional means may be appropriate.

**Ideal Type II superconductor cylinder.** The secondary superconductive current is limited in between the boundaries $\pm(A_{sc}\, j_{sc})$. No flux can enter the core, and only stray - inductance voltage drop $\Delta U$ occurs in the primary winding. Any fault current increases at first according to equation (4c) whilst $R_s = 0$ up to the trigger level $i_{tr} = A_{sc} j_{sc} / n$. Then, flux flow which can be symbolized by a resistance $R_s$ enforces a constant current flow, according to the Bean - model. Magnetic flux $\Psi$ now enters the core, and any primary driving voltage $u_p$ must be compensated by flux variation in the core according to $u_p\{t\} \propto \partial\Psi / \partial t$ .

The flux variation induces a secondary voltage $u_s$ also. In the T - equivalent circuit of Figure 6 we find $u_s = \partial\Psi / \partial t \approx L_m \cdot \partial i_m / \partial t$. There is no reason for any transient voltage spikes since magnetic energy buildup begins just at $t_r$. Figure 7 illustrates, at a first sight, a very well FCL performance. Even in the cases where $i_m$ is comparable with $i_R/n$, i.e. in an open core configuration, no transient voltage spikes have to be encountered.

Unfortunately, there is an important heat dissipation issue in the secondary during the full fault duration. This heat must be stored almost adiabatically by enthalpy rise $\Delta h$ of the HTS, i.e. $q = j_{sc} \cdot \frac{1}{\pi\, D_{sc}} \cdot \int_t u_s dt = \Delta h$ . The enthalpy rise lifts the HTS temperature $T$, and the critical current density comes down to zero when approaching the critical temperature $T_c$.

**Type II superconductor with fast quench.** The temperature rise in case of heavy faults can be very rapid, and $T_c$ may be reached in the millisecond or even submillisecond range. Then the normal resistivity controls the secondary resistance $R_s$. There is still heat dissipation leading to further temperature increase. The resistivity course $\rho(T)$ where $T = T(t)$ now controls the appearance of transient voltages - see equation (4c). This scenario imposes serious restraints on the quench process of the HTS in order to guarantee limited transient voltages: the resistance has to remain limited. A satisfying performance may be hard to achieve since fault conditions can be quite different. An order or magnitude estimate yields an upper limit for the allowed critical current density of some $10^4$ A cm$^{-2}$. Heat dissipation will become important during the fault duration. This is not compatible with a fast reclosing after fault clearance.

**Resistive balance coil.** A high secondary resistance is preferable in order to keep the dissipated heat in the quenched HTS as low as possible. A high resistance can be provided when the HTS critical current density is high, i.e. > $10^5$ A cm$^{-2}$. However, a balance current is needed to limit transient voltages. A resistive shunt in parallel with the primary winding is not adequate due to permanent losses produced by the inductive voltage drop $\Delta U$ during normal operation, unless the shunt is combined with an arrester. The shielded core transformer however suggests an alternative solution: a low - voltage tertiary winding can be easily added. This has already been indicated in figures 4 and 5 as an option. The tertiary winding is definitely inactive during normal operation where no flux exists in the core, but provides a resistive current path parallel to the quenching HTS in the equivalent circuit under fault conditions: $R_s$ is now given by the tertiary and almost constant, the high HTS resistance being negligible. Heat extraction from the low voltage tertiary winding after fault clearance may be rather easy. The transient voltage spikes can be limited to a level as indicated in equation (3). Figure 8 illustrates the transient performance for an FCL assembly where the peak transient voltage $\hat{u}_z$ is limited to almost the crest driving voltage $\hat{U}$.

It must be pointed out that a tertiary winding also limits transient voltages in case of a rather inductive FCL, e.g. when using an open core. The maximum voltage peak is still in-

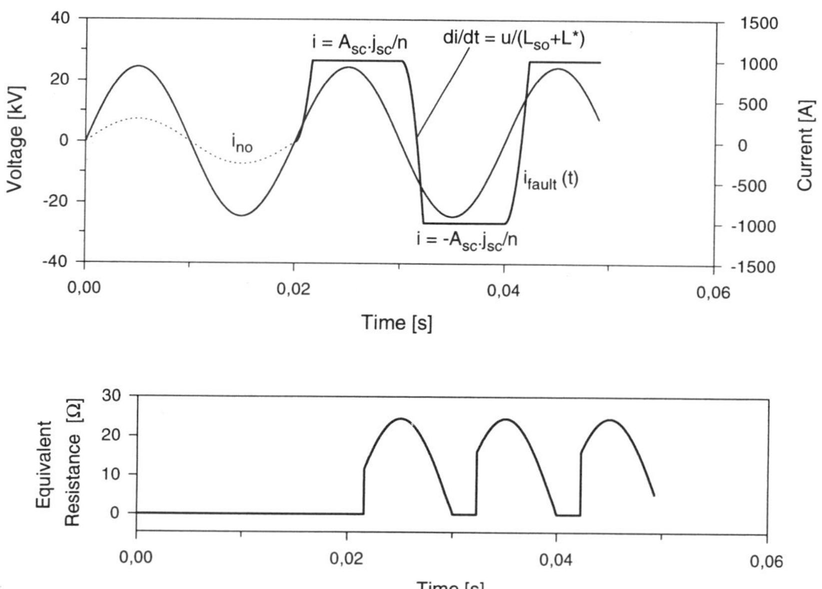


**Figure 7.** Limited fault current through an ideal type II superconducting FCL; no transient overvoltages; U = 30 kV, $(L_{so}+L^*) = 9.55$ mH, $L_{ps}$ <<
top...driving system voltage u and limited current $i_{fault}$;
bottom...trace of equivalent secondary resistance

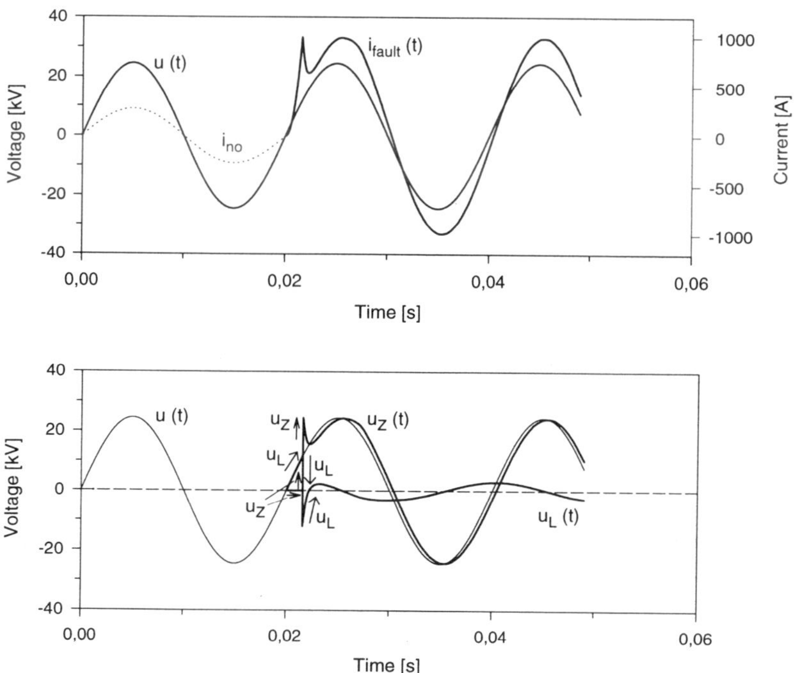


**Figure 8.** Limited fault current through an FCL with fast quenching secondary superconductor and resistive tertiary winding; constant equivalent resistance $R_s = 24.5\ \Omega$ for $t \geq t_r$
top... driving system voltage $u$, normal operation current $i_{no}$, and limited fault current $i_{fault}$;
bottom... driving voltage $u$, transient voltage $u_Z$ across *FCL*, and transient voltage $u_L$ response from transmission system

dicated by equation (3). However, there is an additional inductive current across $L_m$ which may be asymmetric. The total instantaneous fault current peak can be significantly higher.

The most important advantage of the tertiary winding is the decoupling of the transient FCL performance from the superconductor. The voltage rise or quench after triggering of an HTS cylinder is not crucial for the transient voltages. Hence the HTS can be designed mainly in view of a sufficiently fast activation in order to limit the instantaneous fault current for any fault conditions. This gives some margin to meet some other problems such as nonuniform cylinder activation and aging effects due to thermal cycling which have to be overcome.

## CONCLUSIONS

Shielded core fault current limiters can be operated with limited transient voltage spikes. To this end, a resistive tertiary winding is of great use. This winding decouples the transient FCL performance from the particular time course of voltage rise and quenching in the secondary HTS cylinder. Thus, much more space is available for the specifications of the latter. The heat being developed in the tertiary winding during a fault may be easily removed since neither high voltage, nor cryogenic temperature, is involved in this winding.

## ACKNOWLEDGMENTS

Important parts of this work have been carried out in course of a sabbatical stay at the Research Center Karlsruhe/Germany. Thanks are due to Professor Peter Komarek and his coworkers for support and helpful discussions. The assistance of M. Humann for calculations and layout of the paper is appreciated.

## REFERENCES

1. L. Salasoo, Comparison of superconducting fault limiter concepts in electric utility applications, *IEEE Transactions on Applied Superconductivity Vol.* 5:1079 (1995)
2. R.F. Giese and M. Runde. „Fault Current Limiters," Argonne National Laboratory, Argonne/Illinois (1991)
3. R.F. Giese. „Fault Current Limiters - a second look," Argonne National Laboratory, Argonne/Illinois (1995)
4. J. Gerhold, E. Telser, K. Piswanger and M. Hubmann. „Einsatzpotential von supraleitenden Kurzschluß-strombegrenzern," KfK-Primärbericht 13*03*04P 12B, Nuclear Research Center Karlsruhe/Germany (1991)
5. D.W.A. Willen and J.R. Cave, Short circuit performance of inductive high Tc superconducting fault current limiters; *IEEE Transactions on Applied Superconductivity Vol.* 5:1047 (1995)
6. W. Paul, T. Baumann and J. Rhyner, Tests of 100 kW high-Tc superconducting fault current limiter, *IEEE Transactions on Applied Superconductivity Vol.* 5:1059 (1995)
7. V. Meerovich, V. Sokolovsky, G. Jung and S. Goren, High-Tc superconducting inductive current limiter for 1 kV/25A performance, *IEEE Transactions on Applied Superconductivity Vol.* 5:1044 (1995)

# STUDIES ON THE CURRENT DISTRIBUTION IN HTSC POWER CABLES

S.F. Chen, L.Z. Lin, and P. Ye

Institute of Electrical Engineering
Chinese Academy of Sciences
Beijing, 100080, China

## ABSTRACT

A 2000A HTSC power cable 10m in length is under project. The HTSC cable is designed by helically winding a number of Bi-2223 tapes in 10 layers on a flexible stainless steel former of 32mm in outer diameter. And the Bi-2223 tapes 4.0mm in width and 0.2mm in thickness are designed to carry current of 15A at 77K in self-field. In order to utilize the current capacity of the multi-layer conductor effectively, it is necessary to consider the current distribution in the cable during the design stage. In this paper, the mechanism of the current distribution in multi-layer conductors is studied and a simplified current sharing model is presented. Based on this model, a program was developed to simulate the current distribution in multi-layer conductors. The simulation results of a 10m long power cable are presented and several measures to achieve uniform current distribution are discussed.

## INTRODUCTION

With the development of high-Tc superconductors, a project to develop a 2000A HTSC D.C. power cable is arranged in China. The Bi-2223 tapes with cross section of 4.0mm $\times$ 0.2mm are provided by our partners. The properties of these tapes are measured. We find that the I-V characteristics of the tapes which come from different production batches differ from each other dramatically, and that such characteristics of the tapes which come from the same batch have little difference. For example, the critical current of one batch ranges from 10A to 12A, while those of another batch range from 18A to 20A. When all these tapes are used to manufacture a multi-layer conductor for a power cable, the current distribution will be non-uniform. The current transmitted by some tapes may be small, while the current transmitted by others may be large. In order to utilize the current capacity of all these tapes effectively, it is necessary to investigate the factors that determine the current distribution in the cable.

We consider that the I-V characteristics of the tapes, the joint resistance and the self-fields determine the current distribution in these tapes. Based on this consideration, a current sharing model is proposed and a program is developed to simulate the real situation of the current distribution in every tape.

## CURRENT SHARING MODEL AND SIMULATION

Among the factors that affect the current distribution, the self-fields degrade the I-V characteristics of the tapes and their effect can be converted into the I-V characteristics of the tapes. Since the cable conducts D.C. current and there is no skin effect, we can assume that the current is distributed evenly over the cross section when considering the self-field effect. The self-fields at different layers are different. And the self-field effect is converted into the I-V characteristics of each tape according to which layer this tape belongs to. Thereafter, the current distribution is determined by the new I-V characteristics of the tapes and the joint resistance.

According to the I-V characteristics, the tapes will develop their own resistance, i.e., the flux flow resistance, while conducting current. And the influence of the I-V characteristics on the current distribution is embodied by the flux flow resistance that develops in the tapes. So the current distribution is determined by the joint resistance and the flux flow resistance. However, the flux flow resistance of each tape conversely depends on the current the tape conducts. While its transport current increases, the flux flow resistance will also increase. And then the transport current will decrease until a kind of dynamic equilibrium between the transport current and the flux flow resistance in this tape is reached.

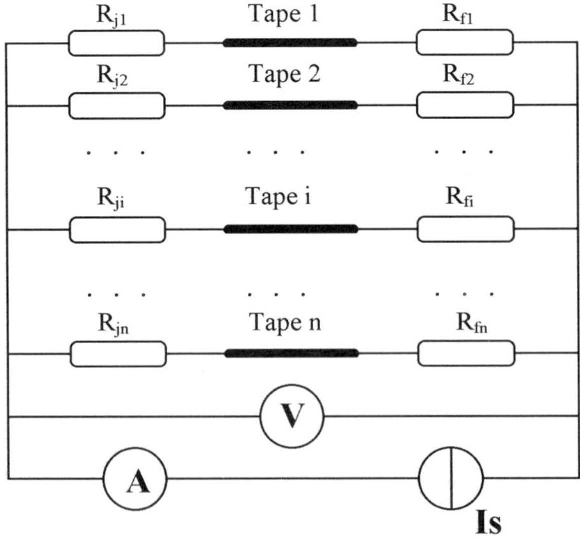


**Figure 1**. The simplified circuit of the current sharing model, where $R_{ji}$, $R_{fi}$ are the joint resistance and the flux flow resistance of tape i, respectively.

This kind of equilibrium of each tape should be reached in determining the current distribution in all the tapes when they are paralleled together.

We use a simplified circuit to simulate the real situation of the operation of the cable(see Figure 1). The circuit consists of a number of branches connected in parallel. Each branch is made up of a Bi-2223 tape, a joint resistor and a flux flow resistor. The joint resistor represents the resistance of the two joints at both ends of the tape, and the flux flow resistor represents the flux flow resistance developed in the tapes. A source is paralleled to this circuit, and every branch in this circuit has the same voltage.

It is assumed that the heat generated by the flux flow resistance will transfer to the coolant immediately when the transport current of the tapes is lower than their critical current. And thus the I-V characteristics of the tapes will not change.

In calculating the current distribution, we do not consider the turn-on transient process. During the transient process, many current loops will be induced and will increase or decrease the transport current of the tapes in the cable. And this is similar to the case in a multi-strand cable[1]. The time constants of these loops are estimated to be $\sim 10^2$s when all the tapes are 10m long. Therefore, the interval for the decay of the induced current loops is very short compared with the long term steady operation. This paper is primarily concerned with the steady current distribution in cables which are made up of tapes of non-uniform I-V characteristics and with non-uniform joint resistance, and we assume that the transport current of the cable is increased evenly and slowly enough for the transient process not to change the properties of the tapes irreversibly. The transient process preceding the steady state and its effects may be addressed elsewhere.

Assuming there are n pieces of branches tapes, we can find:

$$\begin{cases} I_1 \times (R_{j1}+R_{f1})=V \\ I_2 \times (R_{j2}+R_{f2})=V \\ \cdots \quad \cdots \\ I_i \times (R_{ji}+R_{fi})=V \\ \cdots \quad \cdots \\ I_n \times (R_{jn}+R_{fn})=V \end{cases}$$

where V is the voltage of the circuit model, $I_i$, $R_{ji}$, and $R_{fi}$ (i=1,2,...,n) denote the transport current, the joint resistance and the flux flow resistance of tape i, respectively.

Based on this model, we developed a program to simulate the current distribution in all the tapes. The program can randomly generate the joint resistance for every tape according to normal distribution $R \sim N(R_0,\sigma^2)$, where $R_0$ and $\sigma$ are the mean joint resistance and the mean square deviation, respectively. They can be changed through keyboard. Considering that the tapes which come from the same batch have almost identical I-V characteristics, the program can generate the I-V characteristics for each tape based on the I-V characteristics of a tape in the same batch. And the $I_c$ of all the tapes in every batch conforms to the normal distribution $I_c \sim N(I_0, \sigma^2)$, where $I_0$ and $\sigma$ are the mean critical current and the mean square deviation, respectively. They also can be changed through keyboard.

After the operation current of the cable is specified, the program can compute out the transport current and the flux flow resistance in every tape. And then, it will give the I-V characteristics of the whole cable with regard to joint resistance and self-field effect. All the simulation results can be viewed in graphical mode or in text mode.

## SIMULATION RESULTS

The Bi-2223/Ag tapes(multi-filamentary or single core) used to manufacture the cable are provided by our partners. We use the properties of these tapes to run the simulation program mentioned above. According to the measurement of the I-V characteristics, the criterion of 1μV/cm is used in determining the critical current. The average critical current of one batch of Bi-2223/Ag tapes is 22.23A, and that of the other batch is 13.28A. Therefore, the maximum allowable flux flow resistance generated in the first batch should be less than 4.5μΩ/m, and that generated in the second batch should be less than 7.5μΩ/m. On the other hand, we choose 85% for the ratio of the total transport current of the cable to the total critical current of all the tapes.

First, the current distribution in the tapes of the first batch which have different I-V characteristics is considered. Assuming that the joint resistance of all the tapes is uniform, we find that the transport current of these tapes differs little, although the critical current of these tapes varies considerably. The critical current of the 40 tapes we used ranges from 20.27A to 24.19A ($I_c \sim N(22.23,\ 1.0^2)$), with the average of 22.23A, and the joint resistance is 1.0μΩ. Assuming the total transport current of 40 tapes is 85% of the total critical current, i.e., 756.00A, the operation voltage is calculated to be 19.21μV. The simulation results show that the greatest difference between the transport current of every two tapes is less than 2.0A, and the flux flow resistance exhibited in the tapes is negligible when compared with the joint resistance.

Second, we consider the current distribution in one batch of tapes with the same I-V characteristics but different joint resistance. The joint resistance is assumed to conform to normal distribution $R_j \sim N(1.0,\ 0.8^2)$, ranging from 0.19μΩ to 1.90μΩ. When the cable is only 1m long, the simulation results show that when the total transport current of the tapes is 85% of the total critical current, the transport current in the tapes ranges from 13.30A to 20.52A. And there is no quenched tape(see Table 1). We also can see that the joint resistance dominates the current distribution in these tapes. However, if the length of the tapes is increased to 10m, the maximum allowable flux flow resistance in these tapes will also be 10 times greater than that of 1m long tapes. The simulation results show that the flux flow resistance dominates the current distribution(see Table 2), and that the current distribution in these tapes is more uniform.

On the other hand, as to a 1m long cable, whose maximum allowable flux flow resistance is about 4.5μΩ, when the joint resistance ranges from 1.12μΩ to 10.41μΩ ($R_j \sim N(6.0,\ 5.0^2)$), the transport current would range from 12.39A to 22.26A, which exceeds the critical current of the tapes. A few tapes would quench. But as to a 10m long cable, the transport current would range from 14.24A to 20.32A, and there is no quench. Therefore, we conclude that the current distribution in the tapes can be affected by the joint resistance obviously only when the joint resistance is comparable with the maximum allowable flux flow resistance. In order to achieve uniform current distribution, we should either equalize the joint resistance when it is comparable with the maximum allowable flux flow resistance of the tapes, or control the wide dispersed joint resistance to be much less than the maximum allowable flux flow resistance.

Third, we consider two batches of tapes which are paralleled together for transmission. Assuming the joint resistance and the I-V characteristics in every batch are identical, the simulation results show that the two batches mentioned above can be paralleled to transmit

**Table 1.** The simulation results of 10 tapes 1m in length with non-uniform joint resistance

| n | I (A) | $R_j$ (μΩ) | $R_f$ (μΩ) | n | I (A) | $R_j$ (μΩ) | $R_f$ (μΩ) |
|---|---|---|---|---|---|---|---|
| 1 | 20.071 | 1.000 | 0.337 | 6 | 20.316 | 0.800 | 0.521 |
| 2 | 20.073 | 0.999 | 0.338 | 7 | 14.809 | 1.812 | 0.000 |
| 3 | 18.999 | 1.413 | 0.000 | 8 | 20.606 | 0.469 | 0.833 |
| 4 | 20.358 | 0.900 | 0.418 | 9 | 20.358 | 0.900 | 0.418 |
| 5 | 20.335 | 0.696 | 0.624 | 10 | 16.156 | 1.661 | 0.000 |

85% of the total critical current, i.e., 604A, without quench. The transport current of every tape in the first batch is 19.82A, and that in the second batch is 10.38A. The flux flow resistance induced in the tapes of the first batch is 0.28μΩ/m, and in those of the second batch is 1.44μΩ/m. However, when we change the critical current of the second batch to 3.28 A, the simulation results show that the tapes in the second batch are near to quench as the cable transmits 85% of its total critical current. The transport current of each tape in the first batch is 20.32A, and that in the second batch is 1.40A. The flux flow resistance induced in the first batch is 0.526μΩ/m, while in the second batch is 21.63μΩ/m. Since the average critical current of the two batch is 12.78A, we conclude that the maximum range of the critical current should be 12.78 ± 9.5A, which is approximately 12.78A ± 75%. However, if the joint resistance and the I-V characteristics of every batch are not identical, the allowable range of the critical current should be narrower. We assume that the joint resistance conforms to $N(1.0,0.8^2)$, the critical current of the first batch conforms to $N(22.23, 1.0^2)$, and that of the second batch conforms to $N(I_{c2}, 1.0^2)$, where $I_{c2}$ is the average critical current of the second batch. The simulation results show that if $I_{c2}$ is lower than 5.28A, some tapes of the second batch may quench when the cable transmits 85% of the total critical current. Therefore, the average critical current is 13.76A, and the maximum range of the critical current is 13.76 ± 8.48A, which is approximately 13.76 A ± 60%. These results indicate that the electrical performance of the tapes can range considerably when these tapes are paralleled together in cables. However, the range of the critical current is preferably controlled to be ± 30% in order to achieve uniform electrical performance of the cable.

And finally, we achieved the simulation results of a 10m long power cable, which is made up of 100 tapes, 50 of the first batch, and another 50 of the second. The joint resistance conforms to $N(1.0, 0.8^2)$, and the critical current of the two batches conforms to $N(22.23, 1.0^2)$ and $N(13.28, 1.0^2)$, respectively. The tapes are assumed to be separated in 10 layers, and the self-field effect is also considered in this simulation. The simulation results

**Table 2.** The simulation results of 10 tapes 10m in length with non-uniform joint resistance

| n | I (A) | $R_j$ (μΩ) | $R_f$ (μΩ) | n | I (A) | $R_j$ (μΩ) | $R_f$ (μΩ) |
|---|---|---|---|---|---|---|---|
| 1 | 19.416 | 1.304 | 1.225 | 6 | 19.381 | 1.413 | 1.120 |
| 2 | 19.640 | 0.339 | 2.162 | 7 | 19.558 | 0.587 | 1.924 |
| 3 | 19.486 | 1.000 | 1.520 | 8 | 19.692 | 0.188 | 2.306 |
| 4 | 19.467 | 1.100 | 1.423 | 9 | 19.486 | 0.999 | 1.521 |
| 5 | 19.517 | 0.800 | 1.716 | 10 | 19.341 | 1.531 | 1.008 |

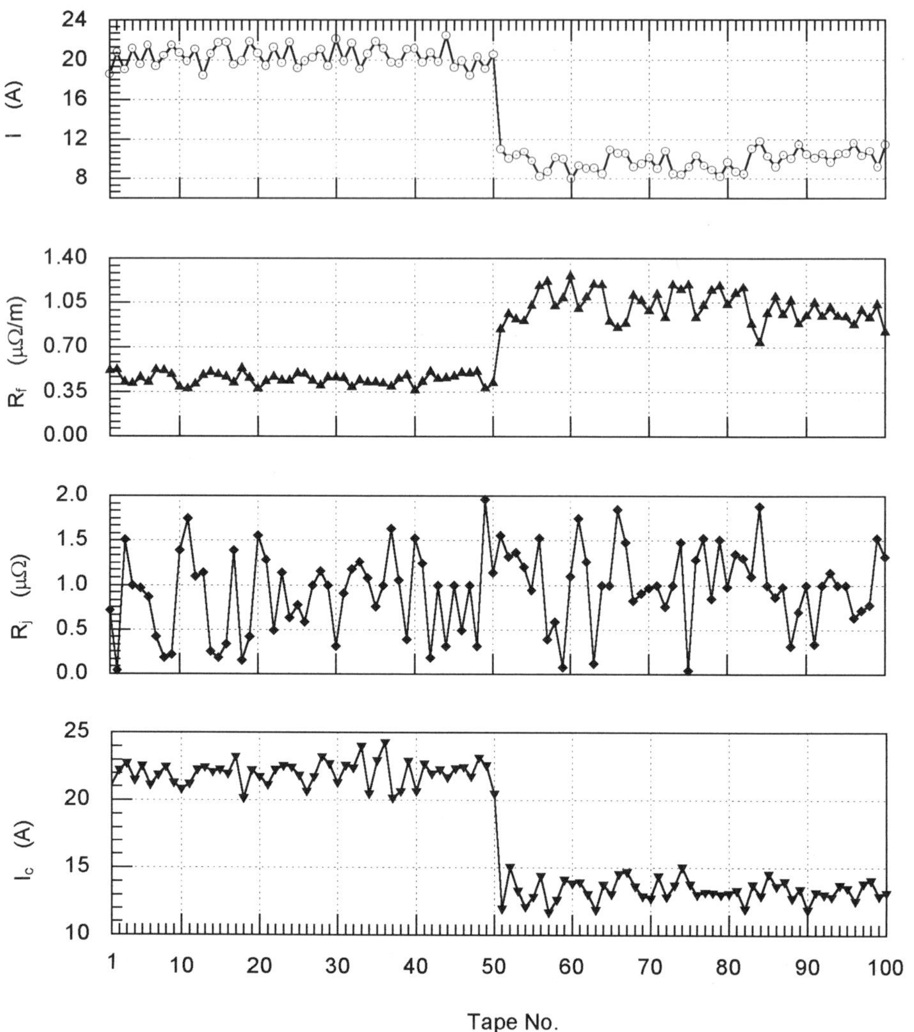


**Figure 2** The simulation results of 100 tapes, $I_c$, $R_j$, $R_f$ and I are the critical current, the joint resistance, flux flow resistance and transport current of the tapes, respectively. Tapes from No.1 to No.50 belong to the first batch, and those from No.51 to No.100 belong to the second batch.

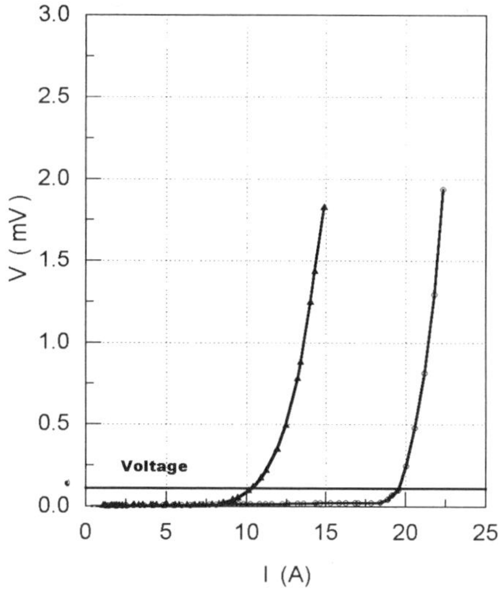


**Figure 3.** The I-V characteristics of 2 tapes in the 10m long cable with regard to the joint resistance, and the operational voltage is also presented.

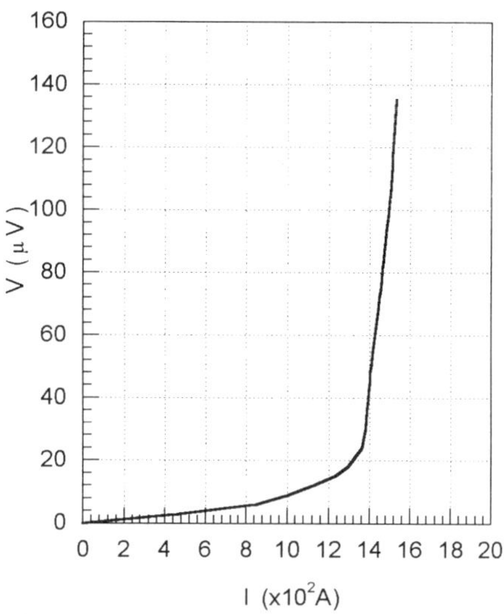


**Figure 4.** The I-V characteristics of the whole cable with regard to the joint resistance.

show that the operation voltage is 110.70μV when the transport current of the cable is 1509A, which is 85% of the total critical current of all the tapes, and that there is no quenched tape(see Figure 2). In addition, the simulation program enable its users to see the I-V characteristics graphically of all the tapes while using keyboard to change the operation voltage and thus the transport current of the cable(see Figure 3, only 2 tapes are presented). We also can view each tape's transport current, joint resistance and flux flow resistance which correspond to the operation voltage selected previously. In the end, the I-V characteristics of the whole cable is also available(see Figure 4). We can see that the transport current of the cable increases from 0A to about 1400A when the operation voltage varies from 0μV to 45μV. And as the operation voltage triples to 135μV, the transport current increases only about 120A, i.e., 8.6%. Therefore, it is more reasonable to run the cable at about 1400A, which is approximately 80% of the total critical current of all the tapes(Of course, if the criterion in determining the critical current of the tapes is changed to 0.1μV/cm, this percentage can be enhanced).

## DISCUSSION AND CONCLUSION

In this paper, the dynamic equilibrium between the transport current and the flux flow resistance in superconducting tapes was proposed and a current sharing model for D.C. high-Tc superconducting cables was presented. Based on this model, a simulation program was developed correspondingly.

The simulation results show that the non-uniformity in either the I-V characteristics of the tapes or the joint resistance will bring about non-uniform current distribution in the tapes of a power cable. Although their I-V characteristics differ considerably, Bi-2223/Ag tapes can be paralleled together in a power cable without quench while transport 85% of the total critical current. The current will distribute in all the tapes according to their capacities. But great difference will result in non-uniform electrical property of the cable. The influence on the current distribution caused by non-uniform joint resistance depends on the maximum allowable flux flow resistance of the tapes. If the joint resistance is comparable with the maximum allowable flux flow resistance, it is important to equalize the joint resistance because it dominates the current distribution in the tapes. Otherwise, the wide dispersion of joint resistance does not affect the current distribution obviously, and the flux flow resistance will dominate the current distribution.

In order to achieve more uniform current distribution in the tapes, thus utilize the transport capacity of the cable better, we recommend to

1. ensure the joint resistance to be far less than the maximum allowable flux flow resistance of the tapes or equalize the joint resistance as far as possible when these two kinds of resistance are close;
2. select the Bi-2223/Ag tapes according to the preferable range of the critical current, which is assumed to be no more than $\pm$ 30%; and
3. place those batches of tapes with better I-V characteristics around the outer layers in order to minimize the non-uniformity caused by the self-field effect because the self-fields in outer layers are greater than those in inner layers.

## ACKNOWLEDGMENT

This work is supported by the National Center for R&D on Superconductivity of China.

## REFERENCES

1. Vitaly Vysotsky, Makoto Takayasu, Sangkwon Jeong, Philip Michael, Joel Schultz and Joseph Minervini, Current Redistribution In Multistrand $Nb_3Sn$ CICC Due to Current Ramp, ICEC16/ICMC (Japan, May,1996)

# TEST RESULTS FOR DIFFERENT HIGH TEMPERATURE SUPERCONDUCTING TRANSMISSION CABLE PROTOTYPES

J. A. Demko[1], J.W. Lue[2], M. S. Lubell[2], U. Sinha[3], J. Tolbert[3], and L. Dresner[4]

[1]Oak Ridge Associated Universities/ORNL
P. O. Box 2009
Oak Ridge, TN 37831
[2]Oak Ridge National Laboratory
P. O. Box 2009
Oak Ridge, TN 37831
[3]Southwire Company
Carrollton, GA 30119
[4]Private Consultant
111 Stanton Lane
Oak Ridge, TN 37830

## ABSTRACT

Development has begun on high temperature superconductor (HTS) cable systems for power transmission. Many design factors affect the performance of a HTS transmission cable. Typical constructions will have multiple layers of superconductor. It is known that the current distribution among the layers has an effect on the performance of the cable. Measurements on two different prototype cable constructions have been performed. The prototypes are approximately 1 m long and were fabricated by helically winding Ag sheathed Bi-2223 high temperature superconductor tape on a stainless steel tube former. Both prototypes had four HTS layers consisting of 2 pairs of oppositely wound tapes. Comparisons of the measured performance of the two prototypes and the measured current distributions are compared and discussed.

## INTRODUCTION

The requirements for delivery of more electrical power through existing right-of-ways has brought about the need for increased current carrying capacity of electrical transmission cables. High temperature superconductors provide a means of achieving this goal. Southwire Company, through a CRADA agreement with Oak Ridge National Laboratory, has been engaged in development activities for HTS transmission cable systems.

Prototype cables using Ag sheathed Bi-2223 high temperature superconductor tape exhibit a power law cryoconductor behavior when transitioning to the normal state and when normal. A detailed description of this behavior is discussed by Dresner [1]. In addition, self field effects can be present in multiple layer constructions as in a typical transmission cable. Testing of several different 1 meter HTS transmission cable constructions is being conducted. The test results of two similarly constructed prototype cables, prototypes #2 and #5, are discussed. Details of the d.c. current voltage characteristics, a.c. loss measurements, and current distribution between layers are discussed.

## PROTOTYPE CABLE DESCRIPTIONS

The prototype cable #2 consisted of 4 layers of HTS tape with Kapton insulation separating each layer. The lay angle for the tapes was 15 degrees in all the layers. Each layer was independently connected to the power bus by a braided cable consisting of 164 strands of #32 AWG tin coated copper conductor. One end of each cable was soldered around the HTS tapes on the end plug and the other end was bolted to the power supply cables using a bus bar. The four braided cables on one end of the prototype cable had voltage taps placed 10 cm apart to measure the currents in each layer during the tests.

The prototype cable #5 consisted of 4 layers with a lay angle of 30 degrees. The inner pair of layers had no insulation between the layers, and the outer pair of layers were insulated by Kapton with the individual tapes separated by a thread. This prototype had voltage taps that enabled the voltage across the inner and outer pairs of layers to be measured. The pairs of layers were connected independently using copper bus bars (2 at each end of the cable) that clamped to the end plugs of the sample and bolted to the power supply cables. The HTS tapes in a pair of layers were soldered directly to the end plugs. During mounting of the second prototype in the test dewar, some of the tapes on the outer layer were damaged. This occurred while connecting the power cables to the copper end plugs. A slight twisting of the outer pair of HTS layers creased some of the tapes. A simple G-10 support was fabricated to prevent this from happening again during testing. The inner pair of HTS tape layers were not damaged.

## EXPERIMENTAL APPARATUS AND PROCEDURE

A 3000 ampere d.c. power supply and a 2000 ampere a.c. power supply were used to energize the cables. A schematic of the setup for the first sample is shown in Fig. 1. The power passes through a single main shunt where the total current supplied to the cable was measured. This was also used to measure the braid resistance during individual layer experiments of prototype #2. When multiple layer experiments were conducted, the total current measured at this shunt provided a consistency check of the individual layer currents. From the main shunt, the power was split through two similar shunts using equal size cables to try to match the resistance in each branch. The voltages from the shunts, braids, and voltage taps placed on the prototypes were read using a Keithly 2001 digital multi-meter using Labview 2.2 data acquisition software running on a Macintosh IIfx computer.

The second test schematic for prototype #5 differs from the previous one only in that there are not copper braids connecting each individual layer to the power bus. Instead, pairs of layers of HTS tapes were soldered to copper end plugs at each end of the sample that were electrically insulated from each other. Clamps from the power supply cables could be attached to either or both pairs of HTS layers in the prototype. In either case the main shunt would be used to measure the total current supplied to the sample. For prototype cable #5, the inner pair and outer pair shunts, as shown in Fig. 1, were used to measure the current distribution in the cable sample.

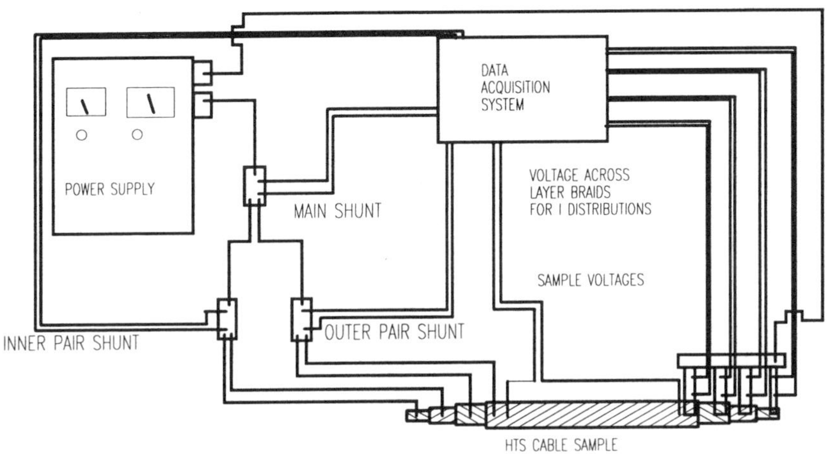


Figure 1. High temperature superconductor cable prototype test schematic.

For prototype cable #2, the d.c. voltage current characteristics were measured for each individual layer. Both d.c. and a.c. measurements were made with the inner pair of layers (#1 and #2), the outer pair of layers (#3 and #4), and all four layers energized together. Measurements of the current distribution in the layers of prototype cable #2 for d.c. and a.c. conditions are provided. The a.c. loss measurements have been reported previously [2].

For prototype cable #5, both d.c. and a.c. measurements were conducted for the conditions where the inner pair, the outer pair , and both pairs of layers were energized. Measurements of the d.c. characteristics, a.c. loss, and current distribution are shown. The a.c. experiments on the inner pair were conducted with (1) the outer pair in a open circuit, and (2) the outer pair of layers in a closed loop passing through a secondary shunt to measure the current induced in the outer layer, but disconnected from the main shunt.

A calorimetric technique was used to measure the a.c. loss in prototype cable #5. In this technique the temperature rise of the sample for different d.c. currents over Ic are measured for calibration. The power can be determined from voltage measurements at these conditions. The power dissipated under a.c. is determined from the measured temperature rise of the sample using the d.c. power calibration results.

## RESULTS FOR PROTOTYPE #2

The d.c. characteristics were measured separately for each individual layer of prototype #2. The d.c. critical current for each of the four layers are provided in Table 1 using the 1 $\mu$V/cm criterion. They show that the inner three layers have similar Ic values, and that the outer layer Ic is much higher. Also provided in Table 1 are the approximate n-values for the power law cryoconductor model of each layer. These values indicate that there exists a wide variation in the conductor properties in this prototype.

The d.c. voltage-current characteristics are shown for the first two layers of prototype #2 in Figures 2 and 3. The voltage at a given current and the power law exponent (n-value) for individual layers were observed to change when multiple layers are energized as compared to energizing the individual layer alone. A voltage increase is observed when all four layers are energized and in the outer layer of a pair when only a pair is supplied current.

Table 1. D.C. critical currents for first prototype

| Layer | Ic (Amperes) | Exponent n |
|---|---|---|
| 1 | 121 | 2.9 |
| 2 | 100 | 4.5 |
| 3 | 93 | 2.1 |
| 4 | 210 | 3.4 |

The current distribution of prototype cable #2 for d.c. is provided in Fig. 4. The d.c. current distribution is controlled by the series resistance of the cables from the power supply, joints, HTS cable, and shunts. The contact resistance between the HTS and the electrical connection was not measured.

Under a.c. conditions, as shown in Figure 5, the current is carried mostly by the outer layer, with decreasing amounts of current in the inner layers. This indicates that the resistive component of the cable impedance is small compared to the inductance

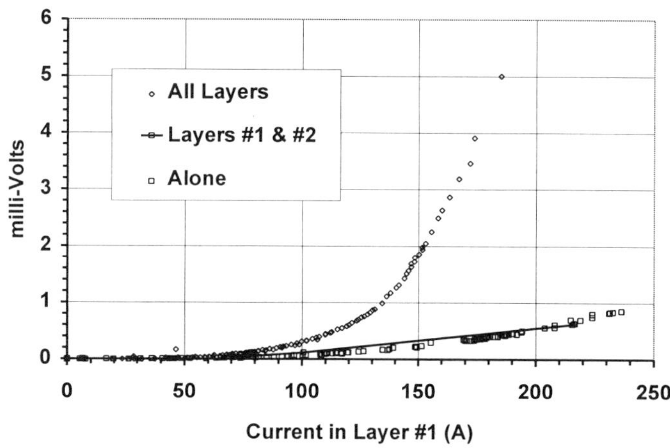


Figure 2: Voltage - Current characteristics for layer #1 (innermost) of prototype #2.

## RESULTS FOR PROTOTYPE #5

The d.c. voltage current characteristics for prototype #5 are presented in Figures 6 and 7 for the individual pairs of layers. For the inner layers, there is a significant change in the characteristics when both pairs of layers carry current. The Ic drops from around 1000 amperes when the inner layers alone carry current to around 700 amperes when both the

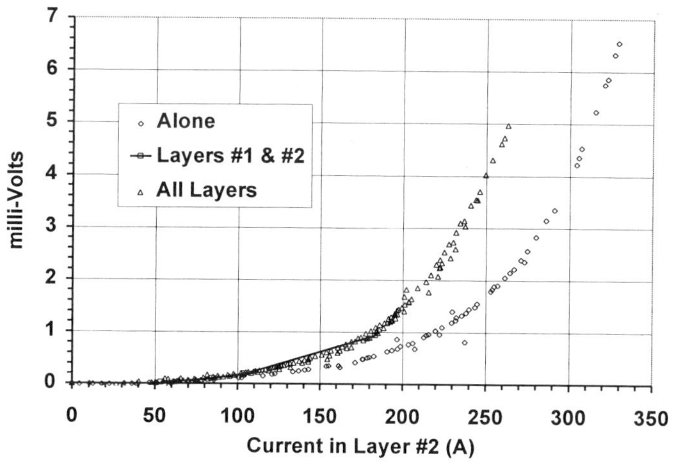


Figure 3. D.C. voltage current characteristics for second layer of prototype #2.

inner and outer layers carry current. This is similar to what was observed on prototype #2 and is attributed to magnetic field effects of the nearby layers.

The voltage measurements in Fig. 7 were for three different HTS tapes in the outer layer. The connected data are for current applied only to the outer pair of layers. The unconnected data are for both pairs of layers carrying current. No significant voltage difference was observed between these cases. The different voltages between tapes are from unequal currents in the tapes caused by variations in the tape characteristics and the joint contact resistance between the HTS tapes and the copper end plugs.

The measured current distribution for prototype #5 is shown in Figure 8 for both d.c and a.c. conditions. The combined series resistance was not measured but was almost the same for each pair of layers since under d.c. the current is almost the same in both pairs of

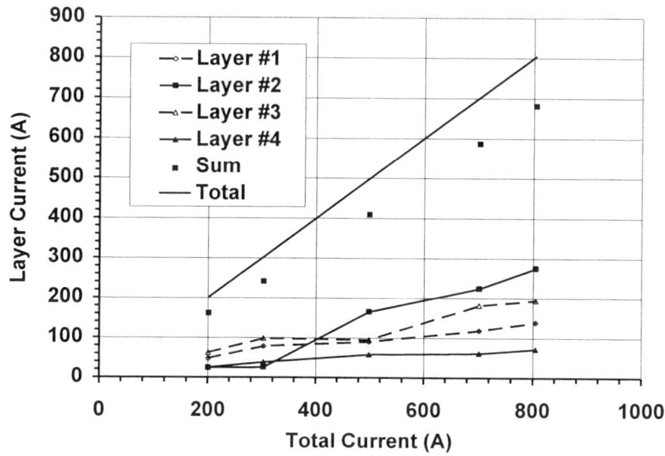


Figure 4. D.C. Current distribution in HTS transmission cable prototype #2.

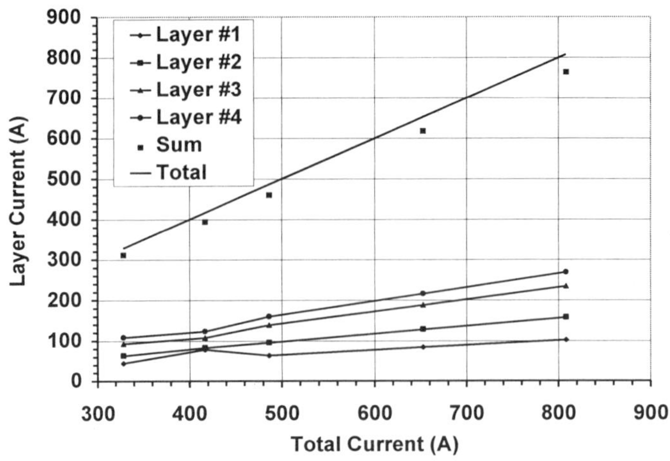


Figure 5. A.C. Current distribution in HTS transmission cable prototype #2.

HTS layers. For a.c. conditions, the outer layers carry a higher portion of the current. The ratio of the current in the outer pair to the total current was found to be around 0.51 for d.c. and 0.58 for a.c. These ratios were constant over the range of currents tested.

The calibration of the cable temperature rise with the power dissipated was accomplished by passing d.c. current through the cable at values higher than the critical current. This produces a resistive heating in the sample that can be determined from voltage measurements. Since the characteristics of each pair of layers are different, a calibration was obtained for the three conditions in which the a.c. loss was measured.

The a.c. loss measurements for prototype #5 are summarized for all conditions in Figure 9. For testing of the inner pair of layers, there were two configurations of the outer pair of layers as described previously. The measurements indicate that the a.c. loss for the inner pair of layers in a closed loop may be slightly higher than when the outer loop is

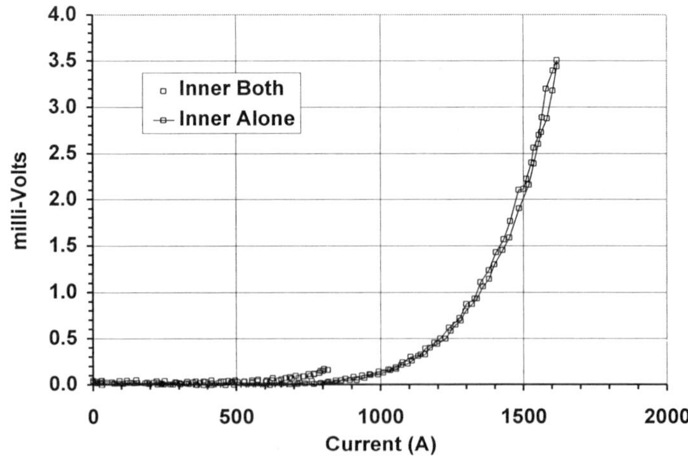


Figure 6. Voltage-current characteristics for inner pair of layers of prototype #5.

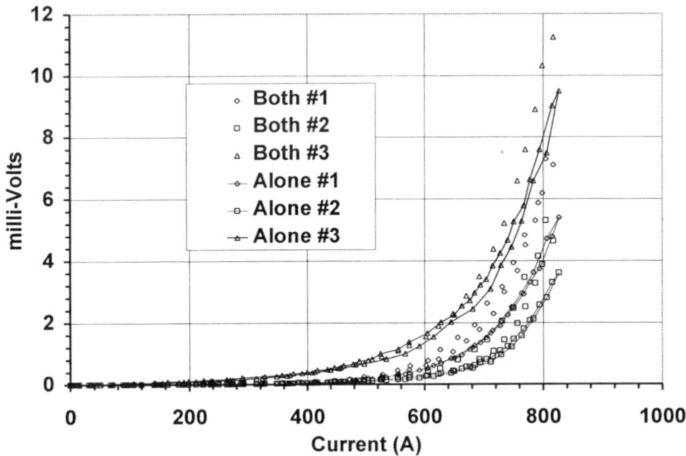


Figure 7. Voltage current characteristics for three voltage taps on outer pair of HTS layers of prototype #5.

open. When the inner loop alone is supplied current, the ratio of the current measured in the outer loop to the inner loop was nearly constant at 0.133.

The a.c. loss for the outer layers appears at significantly lower currents than for the inner layer most likely because of the damage present in some of the tapes. The outer pair is the predominant source for the a.c. loss when both pairs of layers carry current. When taking into account the current distribution among the pairs of layers from the previous figure, it becomes apparent that the inner pair is operating at currents where the a.c. loss is minor, while the outer pair, that is carrying the major portion of the current, is operating where significant a.c. losses are present. For example at a total current of 800 amperes, the inner pair carries around 320 amperes, which exhibits no measurable a.c. loss, while the outer layer carries the balance of current. At 480 amperes, the a.c. loss of the outer pair is comparable to that of both pairs at 800 amperes. A power law fit of the data is also shown on the figure for all three cases.

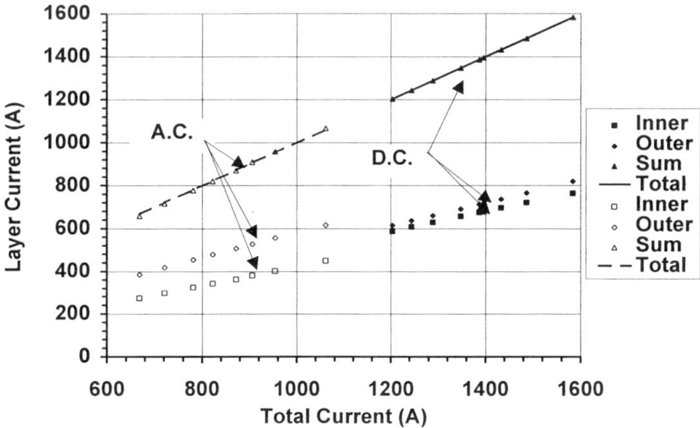


Figure 8. Current distribution in HTS transmission cable prototype #5 under a.c. and d.c. conditions.

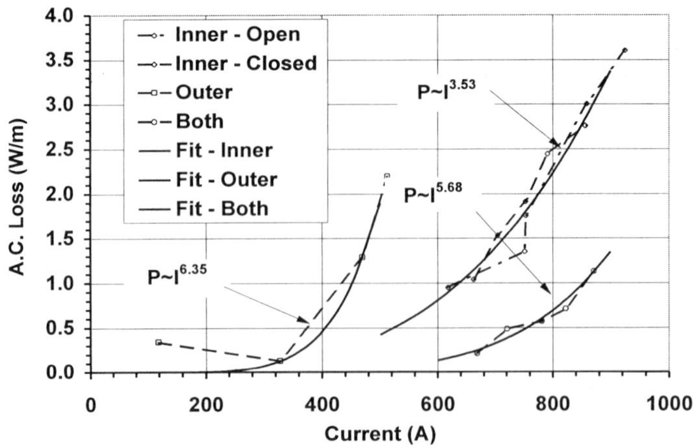


Figure 9. A.C. loss measurements for HTS transmission cable prototype #5.

## SUMMARY AND CONCLUSIONS

Testing of two different prototype HTS transmission cables has been conducted and provided some interesting results. The d.c. voltage-current (V-I) characteristics of a particular layer in a multi-layer transmission cable will depend on whether the current is flowing in neighboring layers. The magnetic field from the inner layer in a pair of layers caused the d.c. critical current of the outer layer of the pair to decrease.

The current distributed equally among the pairs of layers under d.c. conditions for prototype #5. For prototype #2 it was influenced by the layer V-I characteristics. Under a.c. conditions, the current distribution among the layers is dominated by the inductance. The current distributes so that the highest fraction flows in the outermost layer and decreases in the inner layers. This was demonstrated in both prototype cables. The a.c. losses in a four layer cable, protoype #5, were measured in four different operating configurations.

## ACKNOWLEDGMENTS

Research performed under a CRADA between Southwire Co. and ORNL, the latter is sponsored by the Office of Energy Efficiency and Renewable Energy, U.S. Department of Energy under contract DE-AC05-960R22464 with Lockheed Martin Energy Research Inc. This research was supported in part by an appointment to the Oak Ridge National Laboratory Postdoctoral Research Associates Program administered jointly by the Oak Ridge National Laboratory and the Oak Ridge Institute for Science and Education.

## REFERENCES

1. L. Dresner, Applied Superconductivity, Vol. 4, No. 4, (1996); Applied Superconductivity, Vol.4, Nos. 7/8, 337-341 (1996).
2. J. W. Lue, M. S. Lubell, E. C. Jones, J. A. Demko, D. M. Kroeger, P. M. Martin, U. Sinha, and R. L. Hughey, IEEE Transactions on Applied Superconductivity, Vol. 7, No. 2, p 302-305, (1997).
3. S. Muyokama, et al., IEEE Transactions on Applied Superconductivity, Vol. 7, No. 2, p 1069-1072, (1997).

# R&D OF HIGH RELIABLE REFRIGERATION SYSTEM FOR SUPERCONDUCTING GENERATORS

Y. Ikeda,[1] T. Umeda,[1] T. Ichikawa,[1] H. Yanagi,[2] H. Asakura,[3] K.Sano,[4] and H.Sugawara[5]

[1]Super-GM, 5-14-10 Nishitenma Kita-ku, Osaka 530, Japan

[2]Advanced Technology Lab., Mayekawa Mfg. Co., Ltd., 2000 Tatsuzawa, Moriya-Machi, Kitasooma-Gun, Ibaraki-Pref. 302-01 Japan

[3]General Machinery Div., Ishikawajima-Harima Heavy Industries Co., Ltd., 3-2-16 Toyosu Koto-ku, Tokyo 135, Japan

[4]Energy Conversion & Storage Dept., NEDO, 3-1-1 Higashi-Ikebukuro Toshima-ku, Tokyo 170, Japan

[5]AIST, 1-3-1 Kasumigaseki, Chiyoda, Tokyo 100, Japan

## ABSTRACT

Super-GM has been promoting R&D on 70MW class superconducting generators (called Model Machines) in addition to their refrigeration system. The helium refrigeration system for the SCGs must meet a stringent requirement of reliability higher than that of conventional systems. So far, its nominal performance has been successfully confirmed through a factory testing program. Subsequently, its system has been shifted to the verification site and a series of commissioning runs were made to verify the specifications. These runs include a performance of low temperature adsorber housed in a cold box , which is an influential factor for its continuous operation. For developing an advanced refrigeration system, an oil free turbine-drive turbo compressor have been tested and clarified the performance.

The present paper will report the test results of adsorber capacity measurements, cooling-down／warming-up the Model Machine and the recent progress in development of turbine-drive turbo compressors applied for the advanced refrigeration system using the helium Brayton cycle.

## INTRODUCTION

Super-GM (Engineering Association for Superconducting Generation Equipment and Materials ) was established in September, 1987 to develop technology for superconducting generation equipment and materials as a national project. Since then we have been developing two types of high reliable refrigeration system for the SCG ; one is a conventional refrigeration system using oil injected screw compressors and the other is an advanced refrigeration system using oil free compressors.

The main subject in developing its refrigeration system for verification tests on site of the SCGs are to develop technology for removing impurities in helium gas which largely affects its reliability and to further improve the reliability and efficiency of its components. Through factory testing in FY 1995 a developed refrigeration system has been confirmed to meet its requirements as a liquefaction capacity, a precooling capacity, an automatic control to different modes of operation ( cool-down, steady-state, etc. ) and then the system was installed at the verification site in FY 1996.[1,2] Since then a series of commissioning runs have been made to verify the specifications. These runs include a performance of low temperature adsorber housed in a cold box.

This paper represents the test results of adsorption capacity by measuring a breakthrough curve of low temperature adsorber, and those of the cool-down ／warm-up of the Model Machine which was conducted as one of verification tests on site.

For R&D of the advanced refrigeration system, so far the 80K oil free turbo compressor driven by high speed electric motor has been developed.[3] However it needs to increase the helium density by cooling $LN_2$ in order to get the necessary compression ratio, which results in increase of power consumption in producing the $LN_2$. Since then we are much concentrated on a development of turbine-driven turbo compressor to obtain a high efficiency, with which the new refrigeration system was proposed and analyzed, applying for the helium Brayton cycle instead of $LN_2$ cooling.

## REFRIGERATION SYSTEM REQUIREMENTS

The refrigeration system requirements are shown in Table 1. A cooling stream of 4 g／s of liquid helium is required for cooling of the three types of rotor. The liquid helium is fed to the rotor at a pressure of 0.13 MPa and a temperature of 4.5 K. The evaporated gas stream coming from a torque tube and power lead returns to the suction side of compressor at a pressure of 0.1 MPa and a temperature of about 300 K.

The cool-down time of the rotors is to be less than 3 days using $LN_2$ precooling. A temperature difference of 100 to 150 K between the reference temperature of the rotor and that of the cooling gas stream is permitted during the cool-down.

The reliability of the refrigeration system has to guarantee an uninterrupted liquid helium supply to the SCG for more than 10,000 hours, as required in power plant operation. Figure 1 shows a block diagram of the system and its main components.

**Table 1.** Refrigeration system requirements

| | | |
|---|---|---|
| Mass of rotor | ton | 4 ~ 5 |
| Heat load | l/h | 83 ~ 87 |
| Liquefaction | l/h | 100 |
| Mass flow | g/s | 68 |
| Precooling | | |
| Temperature | K | 90 20 |
| Flow rate | g/s | 56 10 |
| Power | kW | 240 |
| Figure of merit | kW/(l/h) | 2.4 |

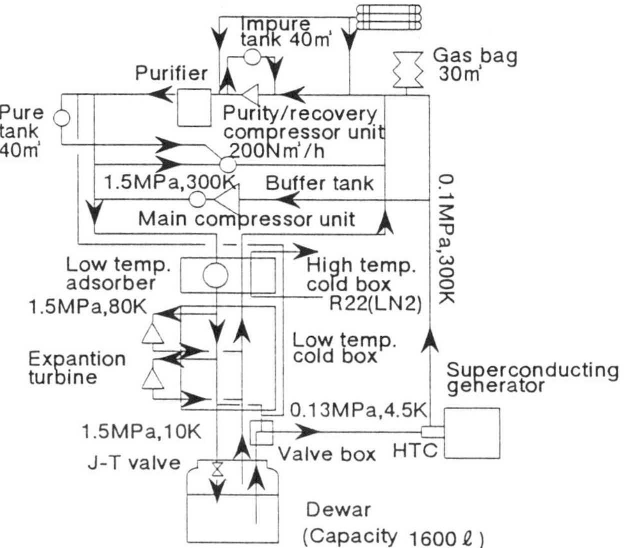


**Figure 1.** Conventional Refrigeration System Flow Diagram.

## VERIFICATION TEST RESULTS ON SITE

### Adsorber Capacity Measurements

The breakthrough curve measurement on the low temperature adsorber (24 kg activated charcoal ) was carried out under the nominal conditions as the throughput of circulated helium of 1,400 $m^3/h$, its inlet pressure of 1.5 MPa and the temperature of 80 K. An inlet concentration of nitrogen in helium gas to the adsorber was maintained at 160 ppm as an accelerated test by charging and controlling a mixed gas flow of nitrogen and helium from an outside cylinder vessel into the inlet of the adsorber.

The test result is tabulated in Table 2. The equilibrium adsorption capacity and the mass transfer zone was 122 cc/g N.T.P. and 10.5 cm, respectively. This results are plotted in the Polanyi characteristic curve as shown in Figure 2, which has been obtained through the factory testing using a 1/4 scale model.[4] Notice that the test result is quite consistent with the curve. Accordingly, it is confirmed that the adsorption capacity at the design point of 1ppm is possessed of 60.9 cc/g N.T.P., which is corresponding to 1044 hours continuos operation until reaching its breakthrough.

**Table 2.** Characteristics of the low temperature adsorber

| 10%breakthrough time(mini.) | 90%breakthrough time(mini.) | Mass transfer zone (cm) | Equilibrium adsorption (cc/g N.T.P.) |
|---|---|---|---|
| 1050 | 1165 | 10.5 | 122 |

## Cool-down and Warm-up

**Cool-down.** The cooling pattern of Model Machine is shown in Table 3. The first stage cooling from 303 K to 296 K was conducted by circulating 90 K helium gas over running 6 rpm for 5 hours just to cool the helium transfer line and transfer coupling ( HTC ) of the SCG. The second cooling to the last filling up stages were made by means of liquid helium directly from the liquid helium dewar via a valve box.

The helium vaporizing in the rotor passes through torque tube extensions and current leads, and leaves the HTC in three partial streams via flexible lines. According to measuring values obtained from the SCG, these partial streams were controlled individually. After streaming through electric heaters, the three streams are combined into one and then led to the suction side of compressor.

Figure 3 shows the test result of cooling of the SCG from 300 K to 4.5 K. The flow rate of HFC ( L/min ) in the figure represents that of the total three streams. During cooldown, the liquid helium has been sent to the SCG over liquefying. Notice that its cooldown was successfully done within the specification of 72 hours.

**Warm-up.** The warming-up pattern of Model Machine is illustrated in Table 4. Its warming-up was done by four stages as shown in the table. The 1st stage was done by controlling of liquid helium flow to the HTC. When its temperature is raised higher than 80 K, both of the third and fourth stages were done by feeding and circulating the heated pure helium at 300 K from a purifier. So that the compressor was switched to the purify/recovery compressor from

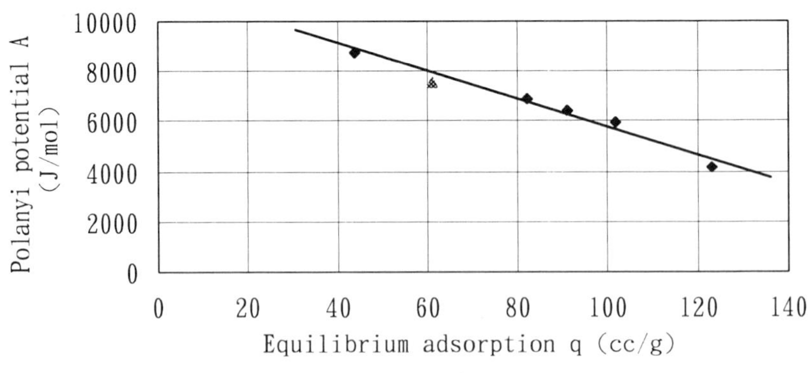


**Figure 2.** Polanyi Characteristics Curve.

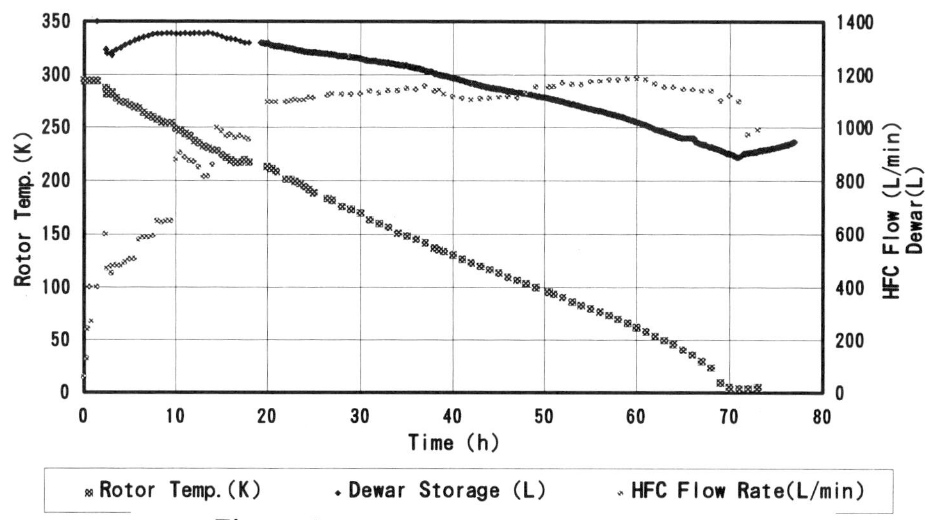


**Figure 3.** Cool-down of The Test Rotor.

**Table 3.** Cooling pattern of Model Machine

| Stage | 1st cool-down | 2nd cool-down | 3rd cool-down |
|---|---|---|---|
| Temp. of rotor | 303 → 296 K | 296 → 4.5 K | 4.5 K |
| Cooling rate | 1.4 K／h | 4 K／h | — |
| Refrigerant | 90 K gas | LHe | LHe |
| Rotor speed | 6 rpm | 500 rpm | 500 rpm |
| Flow rate | 13 $m^3$／h | 65 ～ 70 $m^3$／h | 60 $m^3$／h |
| Operating time | 5 h | 72 h | — |

that of liquifier. The impurity gas released from the rotor was simultaneously adsorbed by the low temperature purifier. The warming-up rate was controlled less than 10 K／h through all the stages.

Figure 4 demonstrates the test result of warming up of the Model Machine from 4.5 K to 300 K. It can be seen from the figure that at the third stage the warm up rate is increased due to exciting a 20 A to the field windings to accelerate the temperature rise. Notice that it takes longer in the warming- up process than in the cool-down process. This reason is due to a small diameter of HTC line so that the maximum flow rate is limited up to approx. 35 $m^3$／h.

**Table 4.** Warm-up pattern of Model Machine

| Stage | 1st warm-up | 2nd warm-up | 3rd warm-up | 4th warm-up |
|---|---|---|---|---|
| Temp. of rotor | 4.2 → 20 K | 20 → 80 K | 80 → 273 K | 273 → 300 K |
| Warm-up rate | 10 K／h | 10 K／h | 10 K／h | 10 K／h |
| Refrigerant | LHe | 90 K gas | 300 K gas | 300 K gas |
| Rotor speed | 6 rpm | 6 rpm | 6 rpm | 6 rpm |

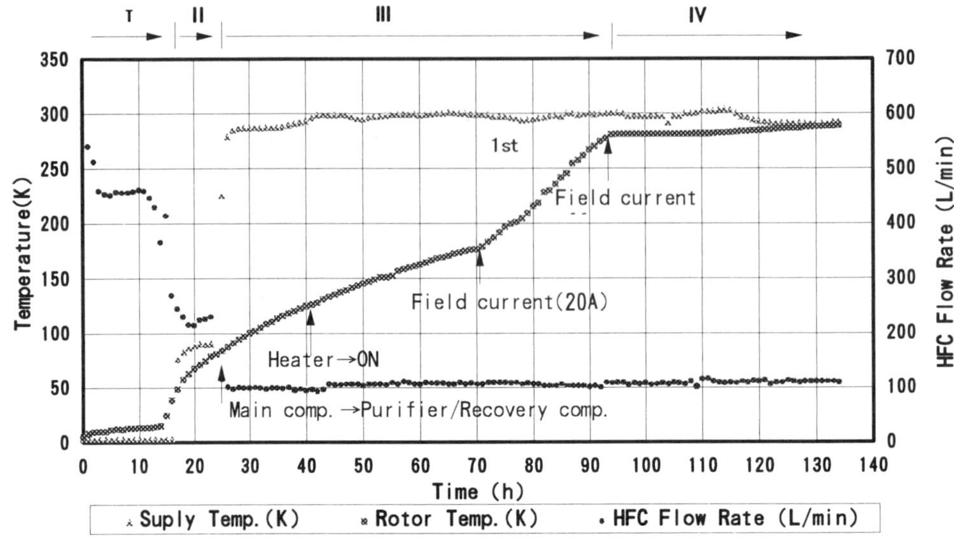


**Figure 4.** Warm-up of Test Rotor.

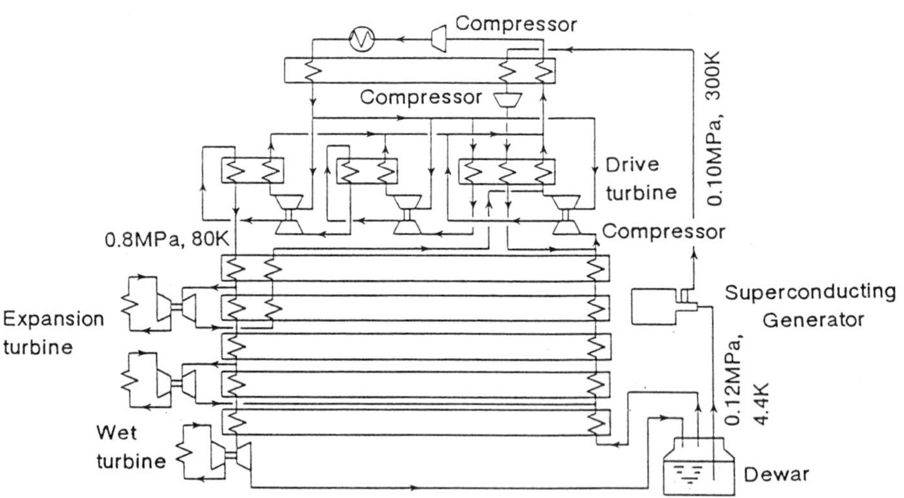


**Figure 5.** Advanced Refrigeration System Flow Diagram

## ADVANCED REFRIGERATION SYSTEM

### Proposed Refrigeration Cycle

Figure 5 shows the new refrigeration system flow diagram using a closed Brayton cycle. This proposed system can be expected to have a value of FOM as 1.8 kW／ ( L／h ).

**Table 5.** Specification of turbine and compressor

| | | 1st stage | 2nd stage | 3rd stage |
|---|---|---|---|---|
| Turbine | Inlet pressure(MPa) | 0.38 | 0.38 | 0.38 |
| | Inlet temperature(K) | 100 | 100 | 100 |
| | Expansion ratio | 2.1 | 2.1 | 2.1 |
| | Flow rate (g/s) | 87 | 105 | 100 |
| | Rotation speed (rpm) | 120,000 | 120,000 | 120,000 |
| Compressor | Inlet pressure(MPa) | 0.12 | 0.38 | 0.56 |
| | Inlet temperature(K) | 80 | 80 | 80 |
| | Outlet pressure(MPa) | 0.18 | 0.56 | 0.8 |
| | Flow rate (g/s) | 50 | 66 | 66 |
| Generator | Absorption power(kW) | 4.3 | 4.8 | 4.3 |

The helium gas compressed by a 300 K stage compressor is fed to 3 turbines in parallel through a heat exchanger for recuperation, and the each expanded exhaust gas of 80 K in the turbines passes through the heat exchangers over cooling the incoming outlet flow from the each stage compressors and then returns to the suction side of 300 K stage compressor via the recuperator. The 80 K compressors are driven by their own expansion turbines. The energy absorbed through the expansion is larger than that of compressing so that the remaining energy should be absorbed by means of a generator.

Hence the key component to realize this system lies in developing a rotating machine which is composed of compressor, turbine and generator on the same shaft.

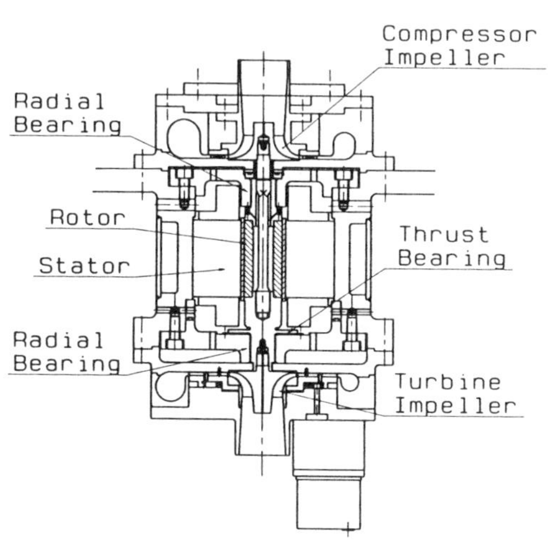


Cross-section of compressor unit

Drive-turbine impeller and nozzle

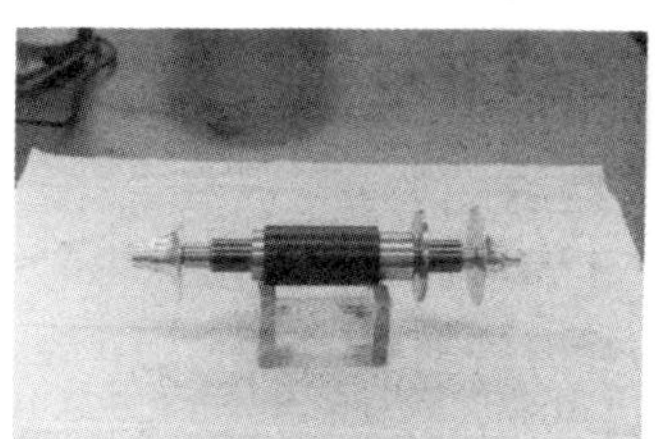

Rotor shaft

**Figure 6.** Developed Cold Compressor Driven by Expansion Turbine.

### Cold Compressor Driven by Expansion Turbine

Figure 6 demonstrates a developed cold compressor driven by expansion turbine on the same shaft applying for a foil type gas bearing, which allows to operate at 120,000 rpm. The employed turbine here features in variable turbine nozzles so as to be able to control a flow rate, which enhances an efficiency in partial load.

Table 5 represents the specification of turbine and compressor for each stages.

## CONCLUSION

### Conventional Refrigeration System

The verification test on site has been carried out successfully to confirm the specified capacity of the helium refrigeration system combined with the Model Machine. Its total operation time is now over 1000 hours continuously with no trouble.

### Advanced Refrigeration System

The new refrigeration system applying for the helium Brayton cycle has been studied. It was found that this new system is promising with a high efficiency. In addition the new cold compressor necessary for the system was developed, which is driven by expansion turbine on the same shaft using for a foil type gas bearing.

## ACKNOWLEDGMENT

This work has been performed as part of " R&D on Superconducting Technology for Electric Power Apparatus " under the New Sunshine Program of Agency of Industrial Science and Technology, MITI, being consigned by NEDO.

## REFERENCES

1. T. Hosoya, H. Yanagi. And et al., R&D of High Reliable Refrigeration System for Superconducting Generators, Proc. of CEC/ICMC '95 conference, Columbus, Ohio, USA (1995)
2. H. Yanagi, M. Ikeuchi, A. Machida and Y. Ikeda, Development of Helium Refrigeration Systems for 70MW Class Superconducting Generators, Ploc. of ICEC16, P.159 (1997)
3. H. Asakura, N. Saji and et al., Performance Analysis of Multistage 80K Centrifugal Compressors for Helium Refrigerator, Proc. of ICEC16, P203 (1997)
4. H. Yanagi and et al., The Planning of A Low Temperature Purifier for Super-GM helium Refrigeration, " Advances in Cryogenic Engineering," VOL.37 (1992), p771

# 1 ATM SUBCOOLED LIQUID NITROGEN CRYOGENIC SYSTEM FOR OXIDE SUPERCONDUCTING POWER TRANSFORMER

S.Yoshida[1], Y.Yagi[1], T.Umeno[1], T.Ishida[1], Y.Kamioka[1], H.Ogino[1], M.Iwakuma[2], K.Funaki[2], Y.Yasukawa[3] and M.Konno[3]

[1]Taiyo Toyo Sanso Co.,Ltd., Kawasaki 210 Japan
[2]Kyushu University, Fukuoka 812 Japan
[3]Fuji Electric Co.,Ltd., Kawasaki 210 Japan

## ABSTRACT

A subcooled liquid nitrogen cryogenic system was developed for an oxide superconducting power transformer. The performance of the transformer was improved in capacity and in reliability. The transformer is cooled by 1 atm subcooled liquid nitrogen supplied by a pump. The cryogenic system has the refrigeration power of 100 W at 70 K and the capacity of the transformer was improved from 500 kVA at 77 K to 800 kVA at 66 K. Because of the steep temperature gradient at the vicinity of the liquid surface, there is a saturated liquid nitrogen at the liquid surface and an ordinary open-dewar can be used as a cryostat for the transformer. Below the steep temperature gradient region, uniformly subcooled bulk liquid is found and there are no bubbles in the cryostat during operations. This feature makes a reliability of the superconducting power transformer increased.

## INTRODUCTION

Electric power devices using oxide superconductors have advantages against ones using metallic superconductors. Those advantages are not only a low refrigeration cost compared with liquid helium but also an improvement of an electric insulation and an increase of superconductor specific heat, because liquid nitrogen is used as a cryogen.

A 500 kVA class oxide superconducting power transformer was developed and some characteristic tests were carried out in 1 atm saturated liquid nitrogen[1]. In order to obtain higher performance of the transformer, a 1 atm subcooled liquid nitrogen cryogenic system was developed. The transformer is installed in 1 atm subcooled liquid nitrogen of 70 K, and a lower temperature improves the capacity of the transformer. Furthermore, an enthalpy margin of the subcooled liquid nitrogen prevents the transformer from being surrounded with bubbles. It makes a electric insulating strength improved.

The design specifications of the cryogenic system are the refrigeration power of 100 W at 70 k and no cryogenic pressure vessel.

## SYSTEM CONFIGURATION

Fig.1 and Fig.2 show a flow diagram and a photograph of the cryogenic system. The system consists of three main units and a control panel. Three main units are a main cryostat

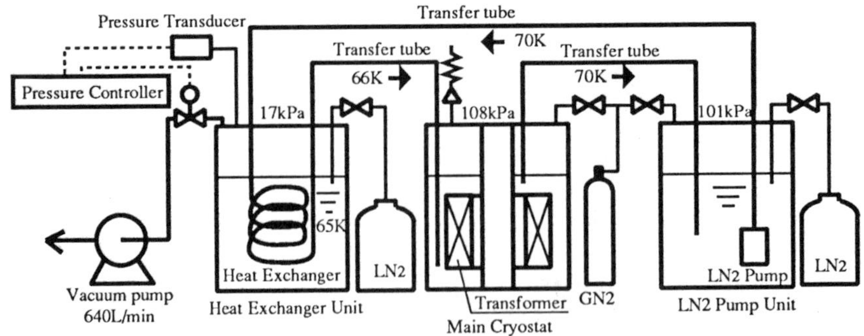


**Figure 1.** Flow Diagram.

which contains the transformer, a heat exchanger unit and a liquid nitrogen pump unit. Each unit are connected by means of flexible transfer-tubes.

## Main cryostat

A main cryostat is made of Glass Fiber Reinforced Plastics (G-FRP) and an oxide superconducting power transformer is installed in the main cryostat . The main cryostat has a height of 1210 mm, an outer diameter of 785 mm, and a room temperature bore of 337 mm through which an iron core is installed. The volume of the liquid nitrogen in the cryostat is 120 liters. Table 1 shows design parameters of the oxide superconducting power transformer.

There are current leads, a relief valve and transfer-tube ports on the top flange. An inlet of subcooled liquid nitrogen is near the bottom of the main cryostat, and an inlet of a transfer-tube for return to the pump unit is at the position of 120 mm under the top flange. The 1 atm subcooled liquid nitrogen at 66 K is transferred to the bottom of the main cryostat and takes heat away from the oxide superconducting power transformer. The subcooled liquid nitrogen is warmed up to 70 K and returned to the pump unit. A pressure of the main cryostat is kept 7 kPa higher than the dewar pressure of the liquid nitrogen pump unit by means of the relief valve. This pressure difference makes the subcooled liquid nitrogen in the main cryostat transferred to the liquid nitrogen pump unit. The system always maintains the main cryostat pressure a little higher than the atmospheric pressure and requires no complicated design such as cryogenic sealing devices for the main cryostat, so the system allows to use an ordinary open-dewar as a main cryostat.

## Heat exchanger unit

The heat exchanger unit consists of a stainless steel dewar and a coiled finned copper tube as a heat exchanger which should be set in the stainless steel dewar. The stainless dewar has a height of 1375 mm, an inner diameter of 270 mm and an outer diameter of 313 mm. Fig.3 is a photograph of the heat exchanger. The finned tube of the heat exchanger has an inner diameter of 7 mm and an outer diameter of a finn is 12.7 mm. The heat exchanger is totally 12 m long.

The stainless steel dewar is evacuated and the pressure inside the dewar is controlled

**Table 1.** Design parameters of the transformer

| | |
|---|---|
| Superconductor | Bi-2223 Ag-sheathed multifilamentary tape |
| Capacity at 77 K | 500 kVA |
| Capacity at 66 K | 800 kVA |
| Frequency | 60 Hz |
| Voltage (Primary/Secondary) | 6600 V / 3300 V |
| Current (Primary/Secondary) at 77 K | 76 A / 152 A |
| Current (Primary/Secondary) at 66 K | 121 A / 242 A |
| Core | Silicon steel plate |
| Winding diameter (Primary/Secondary) | 465,553 mm / 509,597 mm |
| Winding height | 748 mm |

**Figure 2.** Subcooled Liquid Nitrogen Cryogenic System.

**Figure 3.** Heat Exchanger.

to produce 65 K saturated liquid nitrogen by means of a pressure control device and a vacuum pump. The capacity of the vacuum pump is 640 liters/min. The 1 atm liquid nitrogen in the copper tube exchanges heat with 65 K saturated liquid nitrogen and it is cooled down to be 66 K. The 1 atm subcooled liquid nitrogen after the heat exchanger is transferred into the main cryostat.

A Pt-Co resistance thermometer is set at the outlet of the heat exchanger and checks a temperature of the subcooled liquid nitrogen. The exchanger unit is needed to replenish liquid nitrogen, when the 65 K saturated liquid nitrogen runs out and the subcooled liquid nitrogen temperature comes above 66 K . In order to refill liquid nitrogen, the system operation is stopped and the pressure of the stainless steel dewar should resume to be atmospheric pressure by supplying nitrogen gas. A circulation of a subcooled liquid nitrogen through the system is halted until the temperature of the saturated liquid nitrogen in the exchanger unit becomes 65 K again.

## Liquid nitrogen pump unit

A liquid nitrogen pump is installed in a stainless steel dewar, whose dimensions are the same as those of the heat exchanger dewar. The liquid nitrogen pump is a single acting bellows type and its plunger is driven by a pneumatic mechanism.[2] The capacity of the pump is 1 L/min and the 1 atm subcooled liquid nitrogen is circulated in the system. Table 2 shows characteristics of the liquid nitrogen pump and Fig.4 is a photograph of the liquid nitrogen pump.

**Table 2.** Characteristics of Liquid Nitrogen Pump

| Discharge Pressure (kPa) | Operation Frequency (Hz) | Flow Rate (L/min) |
|---|---|---|
| 101.3 | 1.2 | 1.04 |
| 111.4 | 1.2 | 1.03 |

## Control panel

A control panel has the pressure controller which controls a pressure in the dewar of the heat exchanger unit, a temperature indicator for the temperature of subcooled liquid nitrogen in the heat exchanger unit and liquid level indicators for the heat exchanger unit and the liquid nitrogen pump unit.

# EXPERIMENTS

## Refrigeration power test

A stainless steel dewar, which has an inner diameter of 432 mm, an outer diameter of 476 mm and a height of 1324 mm, is used in this experiment instead of the main cryostat. A heat load is supplied by an electric heater set in the stainless steel dewar and the heat load experiments are carried out. In this experiment the temperature distributions of the subcooled liquid nitrogen in the stainless steel dewar are measured at loading of 0, 50, 75 100 and 125 W. Temperature measuring points are indicated in Fig.5. Platinum resistance thermometers are used to measure temperatures and set every 5 mm in a vicinity of the liquid surface. Between the depth of 50 mm and 350 mm, platinum thermometers are set at every 50 mm. The total number of thermometers is 17.

Temperature distributions of the subcooled liquid nitrogen after halting a system operation are also measured.

## Cooling test

Initially the oxide superconducting transformer is in 77 K saturated liquid nitrogen and it is cooled down to 66 K. A cooling procedure is the following. The main cryostat, the heat exchanger unit and the liquid nitrogen pump unit are filled with 77 K saturated liquid nitrogen. The liquid nitrogen in the heat exchanger unit is evacuated and the liquid nitrogen pump is turned on. Then the subcooled liquid nitrogen is circulated in the system and the transformer begins to be cooled down. During the system operation, a pressure of the main cryostat is maintained 7 kPa higher than the atmospheric pressure by means of a relief valve. The dewar of the liquid nitrogen pump unit is open to the atmosphere through a flapper which

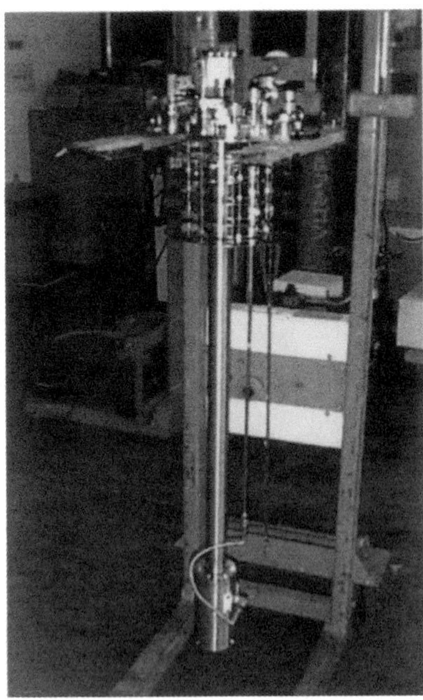

**Figure 4.** Liquid Nitrogen Pump.

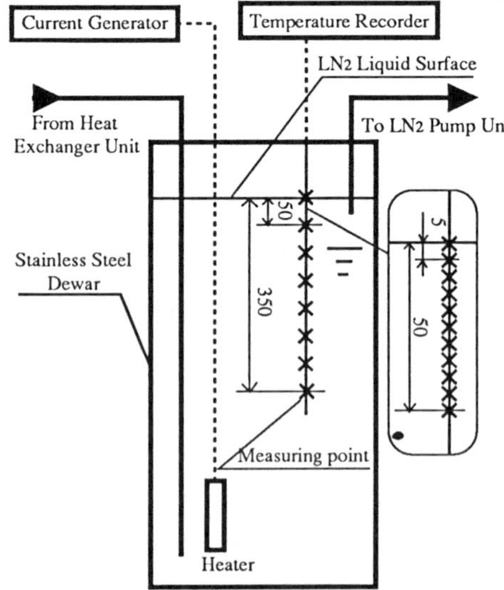


**Figure 5.** Setup for Refrigeration Power Test.

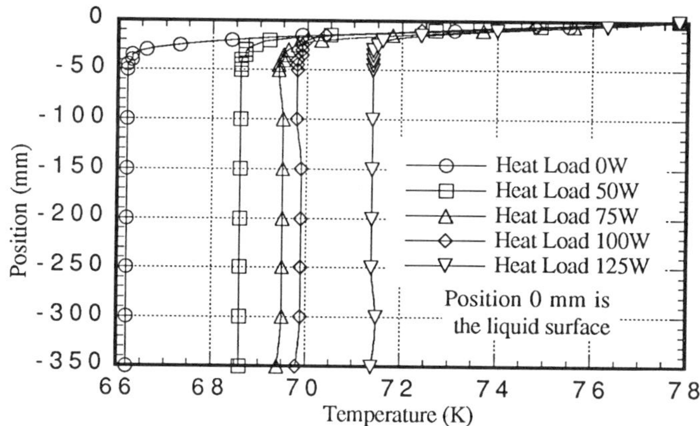


**Figure 6a.** Temperature Distribution Patterns of Liquid Nitrogen.

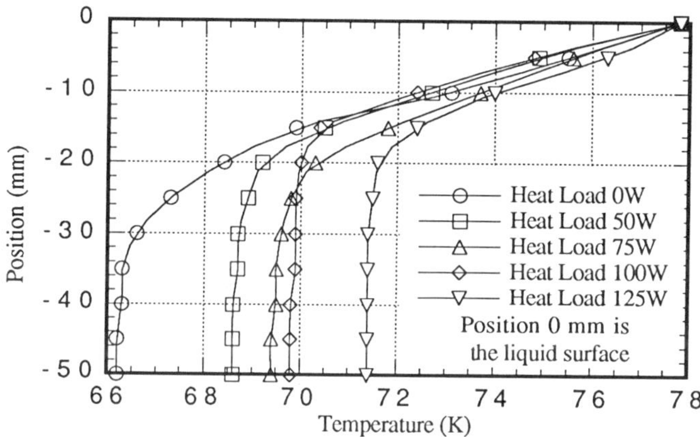


**Figure 6b.** Temperature Distribution Patterns of Liquid Nitrogen.

keeps air out of the dewar. The dewar is supplied by nitrogen gas if its pressure becomes below the atmospheric pressure. In the case that the pressure of the main cryostat is decreased, outflowing subcooled liquid nitrogen to the liquid nitrogen pump unit is reduced and a subcooled liquid nitrogen level in the main cryostat comes up. Finally a heat conducted into the liquid nitrogen in the main cryostat is increased and the pressure of the main cryostat is resumed. In a steady state a flow rate of the liquid nitrogen pump and a transferring rate of subcooled liquid nitrogen in the main cryostat to the liquid nitrogen pump unit are the same. Consequently a circulation rate is constant and there is no evaporation of the circulated subcooled liquid nitrogen in the system.

During cooling down no current was supplied to the transformer. After the temperature of the transformer reached at 66 K, the transformer was operated under a short circuit condition.

## RESULTS

### Refrigeration power test

Fig.6 shows temperature distributions of the subcooled liquid nitrogen . The subcooled liquid nitrogen under the depth of 30 mm had unified certain temperatures depending on heat loads. This phenomena were caused by the circulation of the subcooled liquid nitrogen and the heat loads made convective flows in the dewar and encouraged unifying the temperatures of the subcooled liquid nitrogen. The temperatures of the bulk liquid nitrogen at each heat

**Table 3.** Temperatures of Subcooled Bulk Liqiud Nitrogen at Heat loads

| Heat Load (W) | Temperasture (K) |
|---|---|
| 0 | 66.2 |
| 50 | 68.6 |
| 75 | 69.5 |
| 100 | 69.9 |
| 125 | 71.4 |

loads are listed in Table 3. The table shows the heat load at 70 K is more than 100 W. This refrigeration power is enough to keep the transformer at 70 K, because the total heat load for the system is estimated about 100 W (including AC losses and heat leaks into the main cryostat) for a 750 kVA operation which is the design specification.

Fig.7 shows temperature distributions of the subcooled liquid nitrogen after halting the system operation. A thermal stratification is formed just below the liquid surface because of no liquid circulation[3]. Fig.8 indicates that a heat load was put again and the stratification was vanished because of convective flows by a heat load.

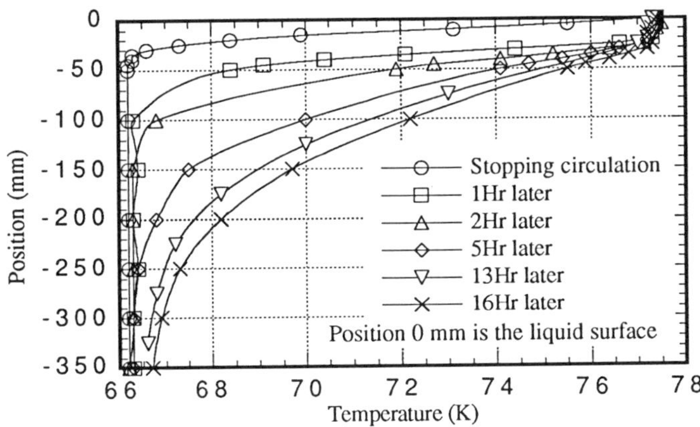


**Figure 7a.** Temperature Distribution Patterns of Liquid Nitrogen after Stopping Circulation.

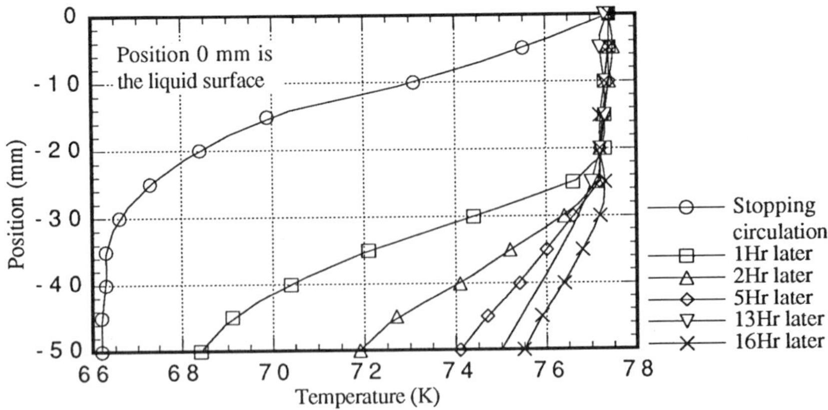


**Figure 7b.** Temperature Distribution Patterns of Liquid Nitrogen after Stopping Circulation.

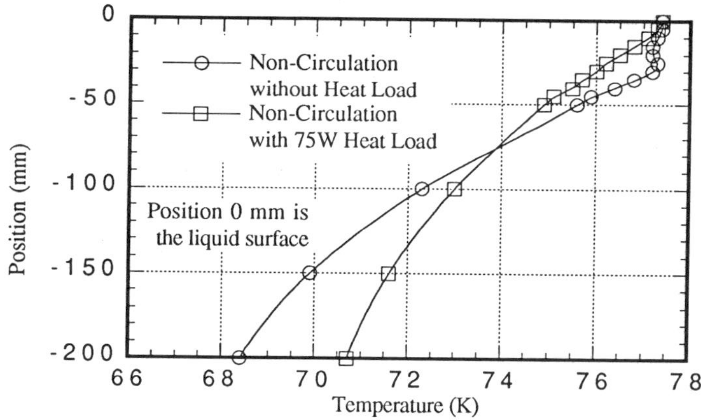


**Figure 8.** Temperature Distribution Patterns of Liquid Nitrogen at Non-Circulation State.

## Cooling test

Fig.9 shows a cooling test result. There was the first replenishment of liquid nitrogen into the heat exchanger unit at 400 minutes because the liquid nitrogen in the heat exchanger unit run out. It took 800 minutes for the transformer to be reached at 66 K. No current was supplied to the transformer during this test.

The transformer was kept at 66 K, and then a current which is equivalent to 800 kVA was supplied. The transformer was excited without any quench[4].

## CRYOGENIC SYSTEM WITH REFRIGERATOR

The present cryogenic system is a batch operation type. Replenishment of liquid nitrogen is required when liquid nitrogen in the heat exchanger unit runs out. It is inconvenient that a system operation must be interrupted to replenish liquid nitrogen and wait for cooling down the heat exchanger again . In order to solve this problem, a cryogenic system with a refrigerator is proposed.

A refrigerator is installed in the dewar of the heat exchanger unit and saturated liquid nitrogen of 65 K is produced by means of a refrigerator. This system can be operated continuously for a long period of time.

Another system is shown in Fig.10. A refrigerator is installed in the liquid nitrogen pump unit. This system produces directly subcooled liquid nitrogen in the liquid nitrogen pump unit and doesn't have the heat exchanger unit .

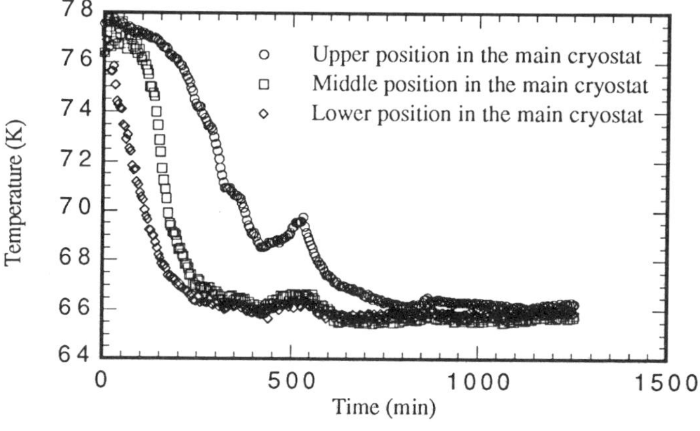


**Figure 9.** Cooling Test.

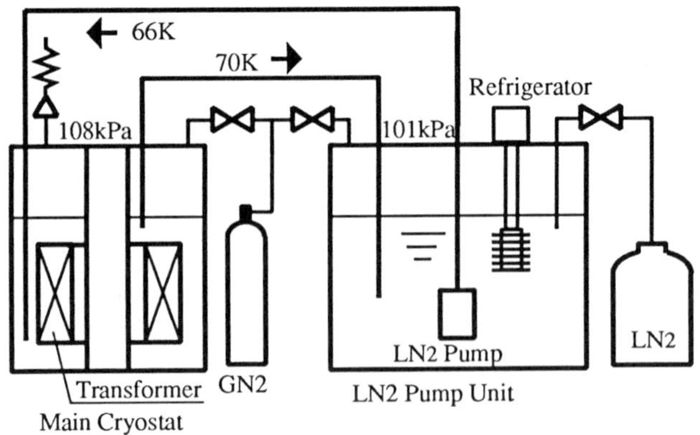


**Figure 10.** Cryogenic System with Refrigerator.

## CONCLUSIONS

A 1 atm subcooled liquid nitrogen cryogenic system for the oxide super conducting transformer was developed, and refrigeration power tests and a cooling test were carried out. The cryogenic system had a refrigeration power of more than 100 W at 70 K.

During operations there was a steep temperature gradient from the surface to the depth of 30 mm, and the temperatures of subcooled liquid nitrogen below this gradient are unified.

It took 800 minutes to cool the transformer down from 77 K to 66 K, and a current was supplied without a quench. It is found that the transformer has a capacity of 800 kVA.

The cryogenic systems with a refrigerator are presented for a continuous long operation cryogenic system.

## REFERENCES

1. K.Funaki et al., Design and Construction of a 500 kVA-Class Oxide Superconducting Power Transformer Cooled by Liquid Nitrogen, in: "Proc. of the 16th ICEC / ICMC," Elsevier Science, Oxford (1997), p.1009
2. Y.Kamioka et al., A Double Acting Bellows Type liquid helium pump, "Proc. of the 10th ICEC," Butterworth & Co, Guldford (1984), p.578
3. J.A.Clark, A Review of Pressurization, Stratification, and Interfacial Phenomena, in: "Advances in Cryogenic Engineering," Vol.10 Sec. M-U, Plenum Press, New York (1965), p.259
4. K.Funaki et al., Preliminary Tests of A 500 kVA-Class Oxide Superconducting Transformer cooled by Subcooled Nitrogen, in: "to be published in Proc. of ASC '96 Pittsburgh (1996)"

# OPTIMIZED LNG STORAGE TANKS FOR FLEET-SIZE REFUELING STATIONS WITH LOCAL LNG LIQUEFIERS

J. A. Barclay, A. J. Corless, and E. H. Nelson

University of Victoria, Victoria, B.C. V8W 3P6

## ABSTRACT

The capacity of a liquid natural gas (LNG) storage tank in a LNG fleet-size refueling station is determined in part by the desire to have fuel capacitance to meet uneven fuel-use patterns during the day/week, to allow for maintenance and unexpected equipment failure, and to take advantage of off-peak feedstock prices. These considerations drive the selection of the LNG storage tank size upwards. On the other hand, the cost of conventional LNG storage tanks, restrictions due to safety codes, the size of the refueling skid, the installation costs, fuel quality variations due to weathering effects, and site preparation costs all drive the selection of the LNG storage tank size downward. Conventional LNG storage tanks for local refueling stations with a capacity of ~1000 DEG/day are made of either 304L stainless steel or 9% Ni-steel inner shells with carbon steel outer shells (DEG = Diesel-Equivalent-Gallons, or approximately 1.7 L LNG). The tanks are double-walled with thermal isolation provided by vacuum and superinsulation or, in some cases, perlite. The tanks are extremely thermally efficient and therefore a relatively expensive component of a refueling station. We have performed a thermoeconomic analysis to determine the optimum level of tank technology to apply to a LNG storage vessel for a local refueling station. One key result is that a local liquefier allows the use of a much less efficient and less expensive LNG storage tank, thus reducing the cost of delivered LNG fuel.

## INTRODUCTION

North America's thirst for petroleum based fuels is becoming problematic. Environmental, geopolitical, and economic drivers are forcing the public to look for cleaner, cheaper, and locally abundant fuels. Several alternatives to petroleum-derived fuels have been identified, but natural gas (NG) has been shown to offer the most promise as a fuel capable of displacing gasoline and diesel, and to gain a significant share of the multi-billion dollar vehicular fuels market[1]. While NG currently supplies approximately 26% of the North American total energy requirements, it accounts for less than 0.05% of the transportation fuel market[2]. NG fuels only about 90,000 of the more than 200 million vehicles registered in North America, with 30% of those being in Canada. At present there are approximately twelve hundred compressed natural gas (CNG) and less than 25 liquefied natural gas (LNG) refueling stations versus nearly 200,000 gasoline and diesel station[3]. It is the cost and performance of existing CNG and LNG refueling equipment and the surcharges on NG vehicles (NGV's) that are impeding the rapid development of this infrastructure. These in turn are continuing to impede the rapid adoption of NG as a vehicular fuel.

*Advances in Cryogenic Engineering,* Vol. 43
Edited by P. Kittel, Plenum Press, New York, 1998

Current world liquefier technology falls into four categories of size:

1. base load plants, producing up to thousands of metric tonnes per day ,
2. peak-shaving plants, producing up to hundreds of metric tonnes per day,
3. fleet-size units, producing up to tens of metric tonnes per day, and
4. residential/laboratory units, producing up to 0.05 metric tonnes per day

To overcome the refueling barrier, high-performance and inexpensive refueling equipment is required. Two scenarios for fleet-sized refueling systems are in development. The first would utilize large-scale LNG production facilities, such as existing base-load plants, peak-shaving plants, or other LNG liquefier plants coupled with commercial transport-truck distribution to supply a network of local storage tanks. The economies of scale inherent in large-scale plants would be offset partly by the need for expensive, high-capacity storage systems, losses during transport due to sloshing and unrecoverable boil-off, as well as flashing losses during the transfer from transport vehicles to the storage tank. Studies[4] have shown that these types of losses in the handling of liquid hydrogen (NBP 20.3 K) can be significant.

An alternative model would have delivery of pre-purified NG feedstock through high-pressure pipelines to a network of refueling systems at the local level, which would then produce and dispense LNG. These smaller plants could exploit pipeline pressure as a source of mechanical energy for drive motors. Local fleet-sized liquefaction plants could generate LNG on an "as-needed" basis, which would require comparatively smaller storage capabilities. This has the potential to reduce the capital cost of the overall refueling infrastructure. Furthermore, there is the added benefit that boil-off gas could be recovered and re-liquefied. A system of this type is illustrated schematically in Figure 1. Study of the cost-effectiveness of a distributed LNG system shows that a major component is the storage tank itself. LNG with a NBP of 111.7 K must be stored in insulated containers. Because it is impossible to completely eliminate heat-leaks from the environment into the cryogenic liquid, some evaporation (or, boil-off) is inevitable, and this gas can either be vented, re-injected into the pipeline to be used for some other purpose, or re-liquefied locally. The conventional approach has been to develop superior, and more expensive, insulation technology as the solution. Advanced insulation methods, including cooled radiation shields, high vacuum and superinsulation (multi-layer insulation) can reduce boil-off significantly, but as will be shown, this adds considerably to the capital costs. It is worthwhile to carefully examine the cost-effectiveness of storage systems in the light of benefits gained overall.

The goal of this paper is to present a critical examination of the effects of: (1) LNG storage capacity, (2) quality of insulation, and (3) LNG storage pressure on the capital costs associated with a local LNG storage facility. Data on the capital cost, quality of insulation, and typical boil-off rates were obtained by vendor quotations provided by several commercial suppliers of insulated LNG storage tanks. Additionally, reasonable assumptions with respect to the operating costs of conventional gas cycle liquefiers were included in the analysis, because on-site re-liquefaction of boil-off gases is a potential option.

## LNG STORAGE CAPACITY

The optimum capacity of a local LNG storage vessel in a commercial refueling system is determined by several operational factors. Adequate fuel capacitance should permit uninterrupted supply of LNG to fleet customers during unavoidable fluctuations of feedstock supply, scheduled or unscheduled maintenance, etc. These tend to drive the anticipated capacitance higher. Each of these factors and their influence on storage capacity are listed below.

1. Uneven fuel use patterns during nighttime or weekends implies the need for some minimum capacitance. The size of relatively expensive liquefaction equipment can be

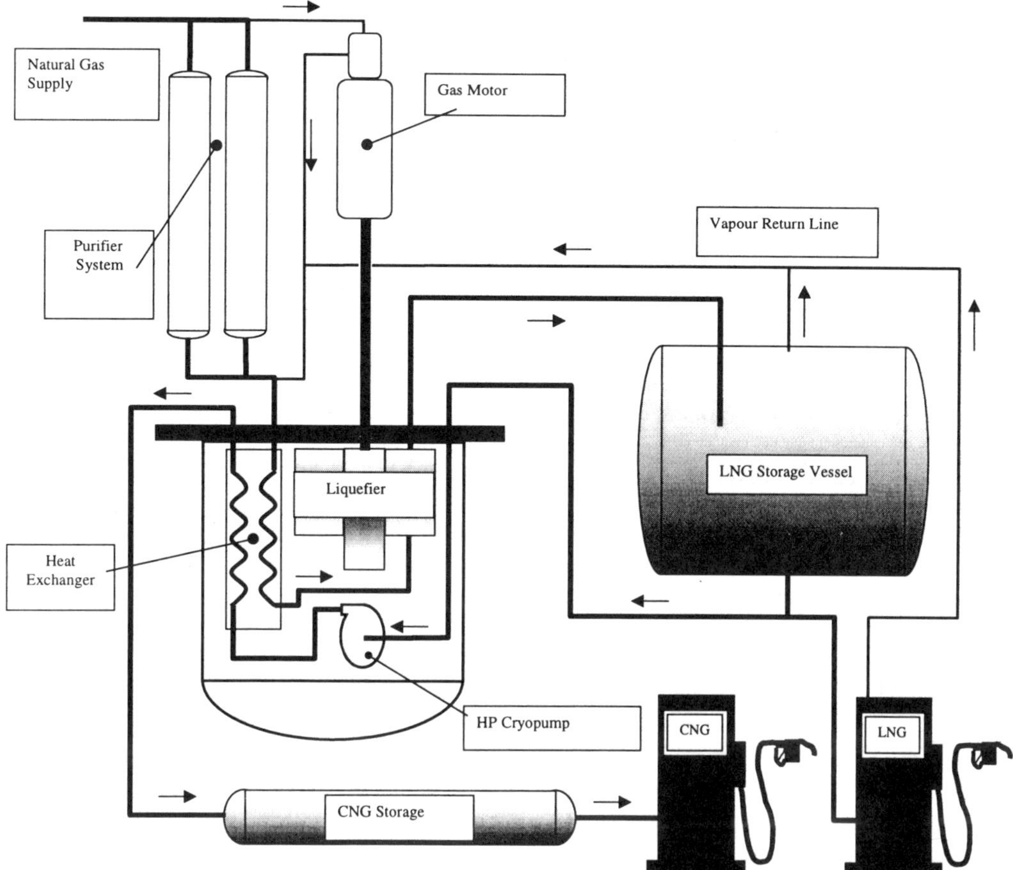


**Figure 1 LNG/CNG refueling system**.

reduced if there is sufficient time during low-demand periods to re-liquefy boil-off, and the capacity to store it.

2. Adequate capacitance is necessary to permit scheduled maintenance and accommodate unexpected break-downs of liquefaction equipment. It is unlikely that fleets converted early to LNG or CNG will be able to find alternate supplies of fuel on short notice. If we consider a fleet-sized refueling system with capacity of 1000 DEG's per day (7.45 x$10^{-5}$ $m^3/s$, or 1700 US gallons/day) we can set a minimum size requirement by considering the maximum number of days we could expect to be off-line in the event of a system failure. If we estimate somewhere in the range of 2 to 5 days, this implies a required storage capacity in the range of 12.87 to 32.17 $m^3$ (3400 to 8500 US gallons) of LNG.

3. A local LNG storage and supply facility should be able to accept off-peak or interruptible NG supply to negotiate better feedstock prices. In certain instances, NG utility companies will offer reduced feedstock prices to customers who purchase gas during low demand periods such as at night. Given sufficient savings in feedstock price, it may prove economical to install a liquefier with excess capacity to produce LNG during these periods.

However, there are also reasons to minimize the quantity of fuel stored on-site. These are listed below.

1. Safety codes (such as NFPA 57)[5] distinguish between large and small LNG storage tanks, mandating costly safety measures for the larger tanks. Currently, the division

between the two classes falls at 7.6 $m^3$ (2000 US gallons). Installations of LNG storage vessels under 7.6 $m^3$ (2000 US gallons) are not required to provide special containment systems. The Canadian standards currently under review are expected to parallel the NFPA standards.

2. Faster LNG turn-around minimizes the effects of LNG aging. Aging refers to increasing concentration of heavier components (ethane, propane, butane, etc., present in feedstock) during longer periods of dormant LNG storage, during which boil-off primarily consists of methane gas.

3. Smaller storage vessels may be rated for higher pressures without significant additional costs. This is very important when considering the production of high pressure LNG.

The overall capacity of the LNG storage system will most likely be dictated by a minimum storage capacity to allow for liquefier down-time. Security of fuel supply is the primary concern of fleet operators who are considering switching to an alternate fuel[6]. For a 1000 DEG per day system, this leads to a LNG storage requirement of at least 19.30 $m^3$ (5100 US gallon) for a three-day holding capacity which is well above the NFPA 57 designation for "small" LNG storage tanks. A 22.71 $m^3$ (6000 US gallon) storage system could consist of three 7.57 $m^3$ (2000 US gallon) vessels or a single 22.71 $m^3$ (6000 US gallon) tank. The capital costs of conventional LNG storage tanks as a function of size are easily determined. Figure 2 shows capital cost as a function of storage capacity for LNG storage tanks. The data come from a number of established suppliers, including Taylor Wharton, MVE, and Hydra Rig. One of the interesting features of this graph is that the relationship between capacity and capital cost is essentially linear.

The capital cost for storage of 22.71 $m^3$ (6000 US gallon) would be approximately $84,000 for three 7.57 $m^3$ (2000 US gallon) tanks, or approximately $55,000 for a single 22.71 $m^3$ (6000 US gallon) tank. This does not include the cost of integrating the tanks into the refueling system, which would be considerably higher for the three-tank option. The installation of LNG tanks over 7.57 $m^3$ (2000 US gallon) must include a spill containment wall to mitigate the dangers of a LNG release. A simple containment dike around the storage area capable of holding the contents

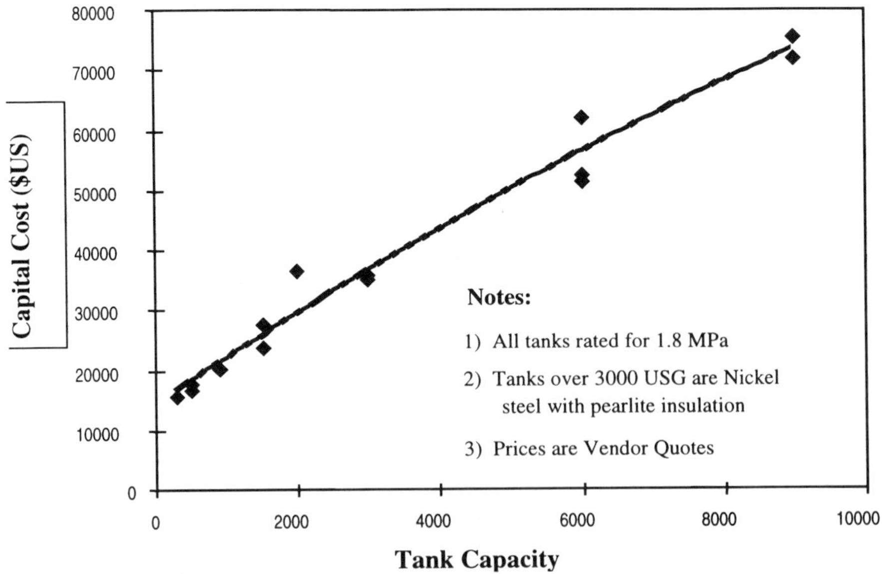


**Figure 2 Capital cost of LNG storage tanks**.

of the tank is all that is required. Based on the price differential of approximately $30,000 the economics would seem to favor the installation of a larger tank, even with the more stringent regulations.

## OPTIMIZATION OF LNG STORAGE VESSEL INSULATION EFFICIENCY

The insulation technology for a conventional LNG storage tank is largely dependent on the capacity of the vessel. For tanks under 11.36 $m^3$ (3000 US gallon), double-walled vacuum insulated tanks constructed of 304L SS and using super insulation are most common. These tanks are efficient and are produced in large quantities, particularly for common cryogenic industrial gases such as $LO_2$ and $LN_2$. For tanks ranging in size from 11.48 to 75.72 $m^3$ (3000 to 20,000 US gallon) double-walled tanks insulated with perlite under a rough vacuum are most common. 9% Ni-steel is the normal construction material. These tanks have high insulation efficiencies because the ratio of surface area to volume is smaller. Much larger tanks, such as those used at utility peak-shaving plants, are usually constructed in place and consist of a double-wall vessel containing perlite insulation in large quantities. These tanks are maintained at atmospheric pressure and usually have a floating lid.

The current trend in LNG storage tank development is to utilize more expensive, high-efficiency insulation in order to maximize holding times and reduce daily losses of usable fuel. This is particularly important when the LNG is produced at a central facility, then transported and stored on-site. To be able to accept a full LNG tanker load, local storage capacity must typically be 56.77 $m^3$ (15,000 US gallons). A fleet using 1700 DEG of LNG daily will have a fuel inventory of approximately 9 days. These longer holding times amplify the adverse economic effects of LNG boil-off. The cost of boil-off must be assessed as the cost of LNG from the supplier plus the cost of the capital equipment required to handle the boil-off, less the current market value of the gas itself which will be used for some other purpose. To determine the optimum level of technology for an LNG storage vessel in a distributed LNG refueling system, a thermoeconomic analysis is necessary. The details of this analysis are provided below.

When LNG is produced on-site a very different set of operating conditions and constraints exist. The primary differences which directly affect LNG storage are as follows:

1. LNG is produced continuously. The size of the storage tank is no longer linked to the size of a LNG transport truck. It can therefore be much smaller, as long as it meets the minimum fuel capacitance requirements as outlined in the previous section.
2. Active refrigeration is available on-site to handle storage tank boil-off. Tank boil-off can be fed directly back into the liquefier for re-liquefaction or LNG may be fed in to the storage vessel sub-cooled to collapse boil-off vapors and reduce the pressure in the storage vessel.

For a given size tank there are several technological choices available which greatly affect the delivered price of the vessel. An 11.36 $m^3$ (3000 US gallon) LNG storage tank was considered in this analysis because of the availability of cost data. Four insulation technologies were considered:

1. liquid-helium quality dewar with internal heat-shields and vacuum-superinsulation,
2. double-walled SS with vacuum-superinsulation,
3. double-walled 9% Ni with perlite insulation,
4. uninsulated single-walled SS vessel.

The vessels in this analysis were rated for 1.8 MPa operating pressure. The capital cost of an 11.36 $m^3$ (3000 US gallon) storage tank was determined as a function of insulation quality (boil-off rate) using vendor data. These four data points (% boil-off rate versus capital cost) were fitted to the power-law expression of Equation 1, which is shown in Figure 3. Here, C is the capital cost in US$ and B is the boil-off rate (% per day).

$$C = 23{,}038\ B^{-0.4251} \qquad (1)$$

The second part of the analysis involved consideration of the net present value of the LNG boil-off for a distributed LNG refueling system. The boil-off rate for cryogenic dewars is commonly expressed as a % of total capacity per day. The cost of boil-off is calculated to be the energy cost of recondensing the vapor, plus the capital cost of the additional refrigeration capacity. The costs of an increase in the marginal boil-off rate, which is expressed as $ per 1% change in the boil-off rate, are estimated in the following way, starting with these assumptions:

1. 3 day LNG storage capacity,
2. work input per DEG = 3.5 kW hr,
3. energy costs = $ 0.05 / kW hr,
4. boil-off NG is warmed to room temperature then re-liquefied, the additional liquefaction capacity can be accommodated by a slight increase in operating frequency, so there are negligible additional capital costs.

A marginal rate increase of 1% corresponds to $1.31 \times 10^{-6}$ $m^3/s$ (30 US gallons per day, ~18 DEG) of boil-off for an 11.36 $m^3$ (3000 US gallon) tank. Assuming a liquefier efficiency of 30%, a NG liquefaction pressure of 0.1 MPa, and an energy cost of $0.05/kW hr, the costs associated with the re-liquefaction of this boil-off are:

3.5 kW hr/DEG x ($0.05/kW hr) x (18 DEG/day/1% per day boil-off)
= $3.15/day/1% per day boil-off
= $1150/year/1% per day boil-off

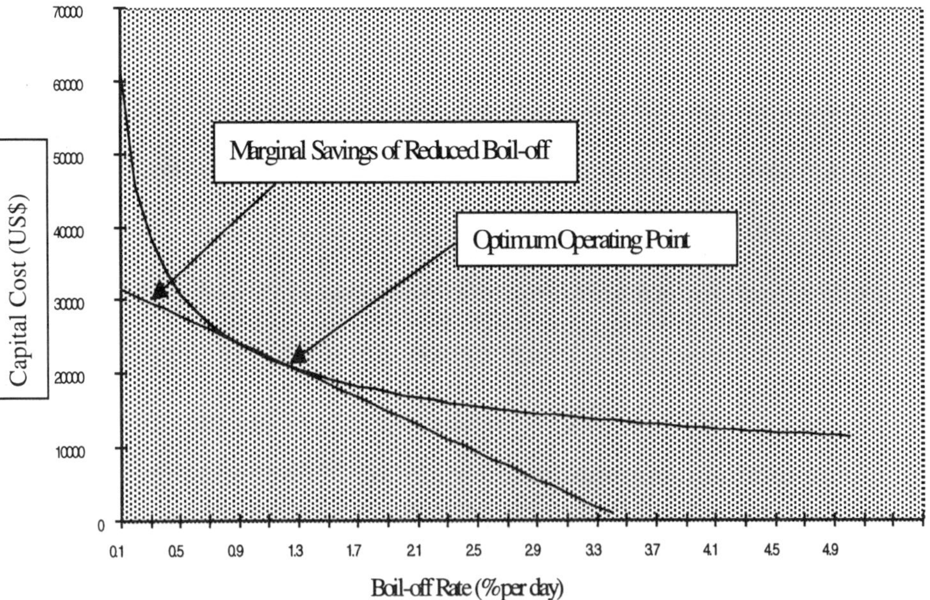


**Figure 3 Effect of LNG storage technology on capital cost .**

At a 10% discount rate over 10 years this equates to a net present value of \$7771 / 1% per day boil-off. This is the additional capital expenditure needed to reduce the boil-off rate by 1% per day (i.e. the marginal value, MV, of the added insulation). This is represented in Figure 3 as a line with slope of -(\$ 7771/1% per day boil-off).

The marginal cost (MC) of improving the insulation quality of the dewar is the slope of the capital cost versus boil-off rate curve, given by Equation 2. The optimum design point is obtained by equating MC to MV, which gives a value of B = 1.176%/day. Substituting this value into Equation 1 yields C = \$21,503. By reference to Figure 2, this indicates that the savings in the price of the 11.36 $m^3$ (3000 US gallon) LNG tank is ~\$12,000.

$$MC = d\{23{,}038\ B^{-0.4251}\}/\{dB\} = -9793.45\ B^{-1.4251} \qquad (2)$$

Unfortunately, this solution is based on the assumption that there exists a continuum of technologies which allow trade-offs between capital costs and efficiency. The analysis does, however, point out that inexpensive insulation should result in more favorable life-cycle economics.

To assess whether such a dewar is practical, we consider the simple solution of spraying a polyurethane expanding foam over a bare tank. We can calculate how much foam would be required to achieve a boil-off rate of 1.176%/day. A vertical 11.36 $m^3$ (3000 US gallon) LNG storage vessel would have approximate dimensions of 2 m (6 feet) diameter by 4.25 m (14 feet) tall.

For a 11.3 $m^3$ (3000 US gallon) tank:

1.1% boil-off = $1.44 \times 10^{-6}\ m^3\ s^{-1}$ (33 US Gallon/day) = 95.4 kg/day methane
= 0.0011038 kg/s (methane)

$h_{fg}$ (methane) @ 110 K = 510 kJ/kg

Heat leak, Q = boil-off x $h_{fg}$ = 562 W.

Surface Area, $A = \pi DL + 2\pi D^2/4 = 30\ m^2$.

Therefore, surface heat flux = $Q/A = 18.7\ W\ m^{-2}$.

A conservative estimate for the required thickness of foam is obtained by assuming a thermal resistance model which includes conduction across only the thickness of the foam coating. In this case:

Thermal conductivity of foam, $k = 0.026\ W\ m^{-1}\ K^{-1}$,

Temperature difference ($\Delta T$) = 300 K – 110 K = 190 K.

Heat flux, Q = k A dT/dx, therefore:

dx = k dT/(Q/A) = (0.026 x 190)/(18.7) m = 0.264 m.

We estimate the cost of such a modification to the tank would be \$1000-\$2000.

The conclusion drawn by this analysis is that, contrary to popular conception, it would be economically advantageous to reduce the quality of the LNG storage dewar for an on-site fleet-sized refueling system to minimize delivered fuel costs. It also illustrates yet another advantage of distributed LNG/CNG refueling systems over centralized systems.

## SUMMARY

LNG storage capacity is required to provide a buffer between LNG production and use and a margin of safety to the fleet owner in the event that the NG liquefaction equipment should fail. Consideration was given to the appropriate size of the system and it was concluded that the size should be determined by the maximum amount of time which could be required to arrange for an alternate supply of LNG. Conventional LNG storage vessels are very efficient and expensive. An analysis was performed to determine if this level of technology was appropriate for a distributed LNG refueling system. The analysis showed that a storage vessel with higher heat leaks and lower capital costs would result in a lower life cycle cost. The effect of storing LNG at higher pressures was considered. It was determined that the performance of the LNG storage vessel would improve at higher pressures, although the effect was small.

## ACKNOWLEDGEMENT

The authors gratefully acknowledge the support of the Natural Science and Engineering Council of Canada (NSERC), WestCoast Energy, Inc., through Centra Gas, B.C., and the University of Victoria.

## REFERENCES

1. Energy Information Administration, 1994, "Annual Energy Outlook 1994 with Projections for 2010", *U.S. Department of Energy, DOE/EIA-0383(84)*

2. Rogner, H.H., (1989), "Natural Gas as the Fuel of the Future", Annual Review of Energy, 1989, Vol. 14, pp 47-73, Department of Energy, April, 1995, *Alternative Fuels Data Center Update*, Vol. 3, Issue 4, pg. 5

3. Clean Energy Resource Institute, 1989, "Solar Hydrogen Energy System", T. Nejat Veziroglu (principal investigator), University of Miami

4. National Fire Protection Agency, 1992, "Proposed Draft of NFPA 57: Standard for Liquefied Natural Gas (LNG), Vehicular Fuel Systems"

5. Easton Consultants, Inc., 1991, Natural Gas Vehicle Fleet Market Study, pp, 15-18, published by the AGA and the NGV Coalition.

6. Adrian Corless, M.A.Sc. Thesis, University of Victoria, 1996

# A COMPARATIVE STUDY OF CNG PRODUCTION METHODS FOR A 1750 DEG / DAY VEHICLE FLEET

M.A. Barclay and T.C. Brook

CryoFuel Systems, Inc.
33301 12th Avenue SW
Federal Way, WA 98023

## ABSTRACT

Three approaches to the production of compressed natural gas (CNG) for a 3500 $Nm^3$/day [1750 diesel equivalent gallon (DEG/day)] transit bus fleet were compared using a life cycle cost analysis approach. Three currently available methods of CNG production were compared. The analysis considered direct compression of pipeline natural gas and the compression and vaporization of liquefied natural gas (LNG) supplied by either tanker truck or locally produced with a small-scale liquefier. Considered contributions to the cost of CNG production included equipment capital costs, maintenance costs, operating costs, and feedstock price. CNG production with the use of a local liquefier resulted in the lowest total cost of fuel production.

## INTRODUCTION

The development of a natural gas (NG) vehicle infrastructure requires creative solutions to a number of industry barriers. Compressed natural gas (CNG) is the most successful alternative fuel to date and the public has become increasingly comfortable with this fuel over the last five years. CNG is typically supplied from locally available pipeline natural gas with the use of a multistage inter-cooled compressor. This approach is the simplest and is a natural extension of well-established NG processing and handling technology. Compressing LNG with a cryogenic pump followed by vaporization may also produce CNG. CNG produced through this approach is typically called Liquid to Compressed Natural Gas (LCNG). LNG can either be truck-delivered from nearby large LNG facilities or may be produced on-site with the use of a local liquefier.

Several variables must be considered to evaluate the cost of CNG production:

1. Pipeline NG specific variables such as the feedstock price, line pressure, and purity of the locally available NG.
2. Fleet variables such as the size, growth, and maintenance strategy for the vehicle fleet.

*Advances in Cryogenic Engineering*, Vol. 43
Edited by P. Kittel, Plenum Press, New York, 1998

2. LNG trucking specifics such as the cost, and availability and delivery schedule of LNG.
3. Other related issues include the capital costs of the refueling systems, maintenance costs, operating costs, space available for refueling facilities, and local zoning.

A comparative analysis was completed for a bus fleet planning to incrementally convert vehicles to operate on CNG between 1997 and 2010. Feedstock price, capital costs, and operating costs are the three most significant factors for the transit fleet. The anticipated CNG fuel requirements are shown in Figure 1.

Figure 1 shows the linearized fuel requirments for the transit fleet increasing from 3500 $Nm^3$/day (1750 DEG/day) in 1997 to over 12000 $Nm^3$/day by the year 2010. To accomidate incremental increases in fuel consumption two stages of refueling equipement installation were planned. This comparisin assumes the following:

- The fueling station must have non-interruptable fuel supply in the event of major equipment failure.
- The pipeline NG is supplied at 3.10 Mpa (450 psig) and 300 K with a composition of 95% methane, 3% ethane, 1.5% nitrogen, 0.5% $CO_2$ and saturated water.
- The pipeline NG is supplied at a price of $0.084 / $Nm^3$ (2.37 / Mscf).
- The truck delivered LNG is supplied at a price of $121.50 per $Nm^3$ ($0.46 per LNG gallon)
- The outlet pressure of the gas is 31 MPa (4500 psi) and 300 K.
- NG engines supply the work input to the equipment.

The initial analysis included operating costs, maintenance costs, capital costs and feedstock costs for the three competative technologies.

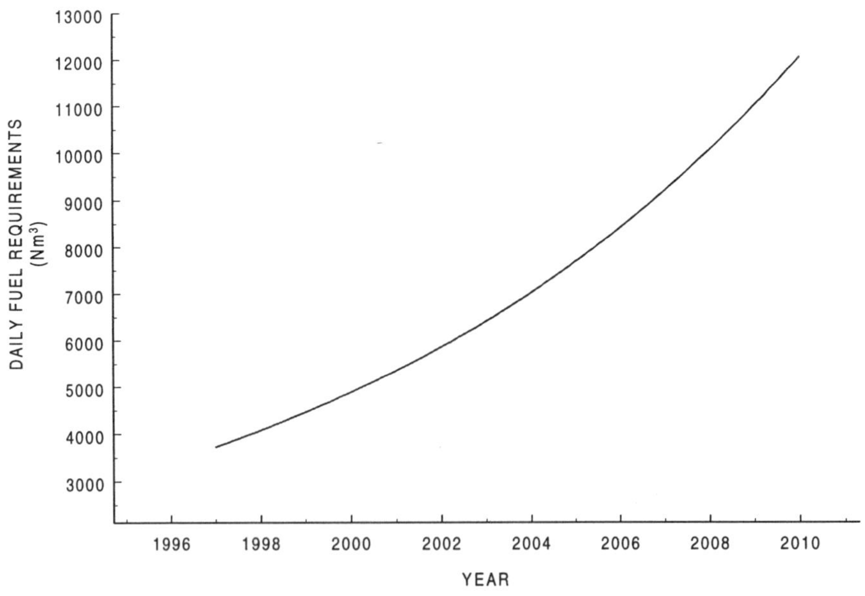


**Figure 1.** Natural Gas Fleet Requirements.

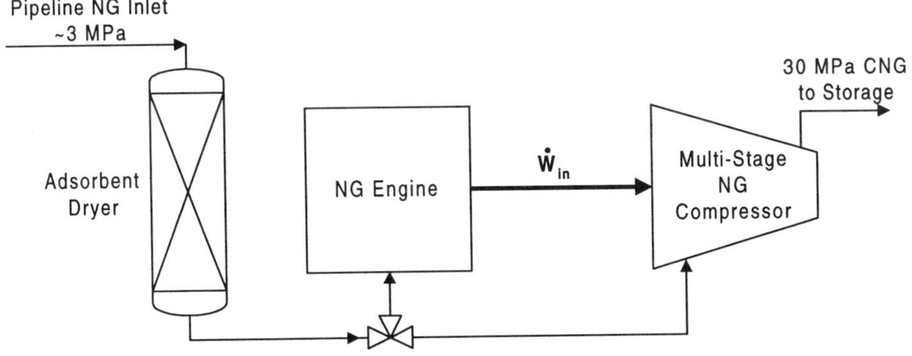


**Figure 2.** Multi-stage Compressor Approach.

## DESCRIPTION OF CNG PRODUCTION METHODS

CNG can be produced through direct compression of pipeline NG or two approaches using compressed and vaporized LNG. A description of each of these approaches follows.

### Compressor Approach

Vehicular CNG is usually produced with multi-stage compressors. Compressor stations have become widespread in the past decade and have a well-established service record as well as mature operating and manufacturing standards. A schematic of compressor production of CNG is seen in Figure 2.

Pipeline NG is first sent through an adsorbent dryer to avoid downstream hydrate formation or condensation. The NG then enters the inter-cooled multi-stage compressor where the gas is compressed to ~3 MPa. After exiting the compressor the CNg amy travel through oil separation and a coalescing filter prior to exiting the considered system to the storage tanks.

These compressors often operate with a pressure ratio of 100 or more with outlet pressures of 3 MPa. Station installation and costs of compressor units often favor air-cooled compressor designs. Problems with oil carryover into CNG buffer storage and vehicles has led to oil-free designs being favored. The harsh operating conditions coupled with design preferences have led to a history of seal problems, high maintenance costs, and early compressor failure. Uninterruptible supply with a compressor station can be provided only with the use of two compressor stations installed in parallel. This is reflected as increased capital cost in the final cost analysis. As expected, the maintenance costs for the compressor station are the highest of the three methods.

### Trucked LNG Approach

Alternatively LNG may be trucked into a fleet refueling system and used to produce CNG as seen in Figure 3.

A cryogenic tanker delivers LNG periodically to match the fuel consumption. The delivered LNG is stored in a 57 $m^3$ double-walled vacuum insulated cryogenic storage tank. The LNG is withdrawn from the storage tank and enters a cryogenic pump where it is compressed to ~30 MPa. The high pressure LNG is then run through a vaporizer to produce LCNG without any work recovery.

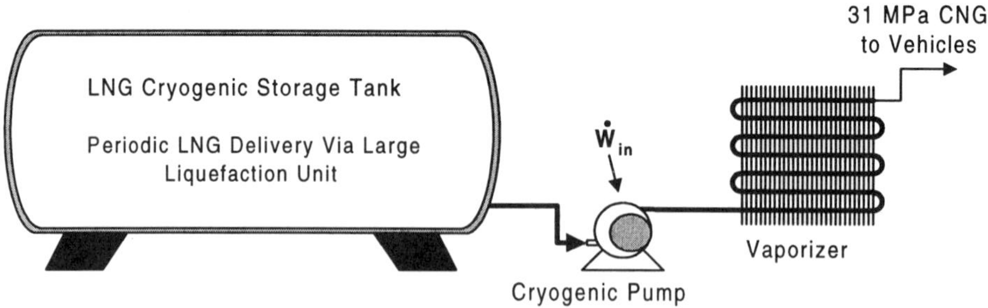


**Figure 3.** Trucked LNG Option.

This approach enjoys several benefits that have made it attractive to several fleets. It minimizes complexity and capital expenditure. The system consists of a cryogenic storage tank, the vaporizer, cryogenic pump, transfer lines, and dispensing equipment. Having only one moving part minimizes the system maintenance costs. The work input into the cryogenic pump is 0.7 % of the work required to compress the same mass of gas. This system has the ability to provide LNG if required and can match incremental fleet conversions by accepting more deliveries rather than increaing gas peocessing capacity. There is also the ability to temperature compensate the outlet gas to assure a full fill of the vehicle tank.

The disadvantages to this approach come in the form of feedstock supply. First, the price of the feedstock is higher than pipeline NG. This price differential increases as pipeline natural gas purity or inlet pressure increase. Additionally, localtion of large-scale LNG facilities can limit the viability of this appraoch. Although LNG has been delivered to end-users over 1,600 km, economcially feasible transport distances are usually limited to less than 300 km.

**Locally Produced LCNG**

CNG may also be produced by compressing and vaporizing locally produced LNG. A schematic of this approach is pictured below in Figure 4.

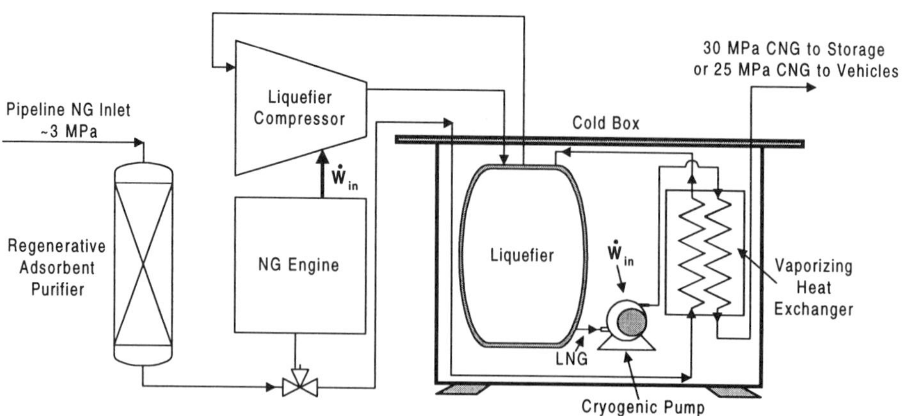


**Figure 4.** Locally Produced LCNG Approach.

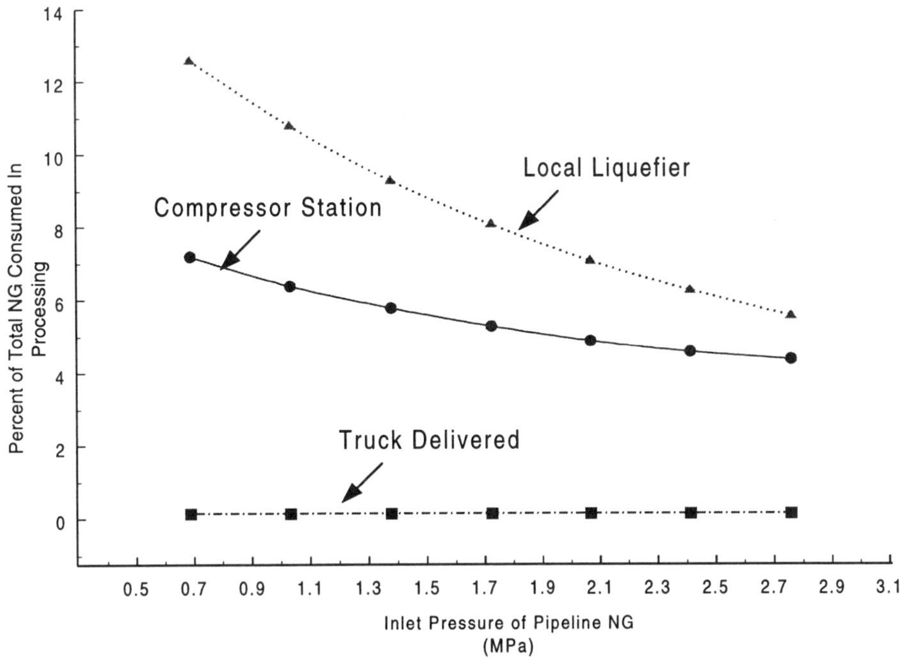


**Figure 5.** Operating Costs As Function of Inlet Pressure.

Low-pressure pipeline NG is supplied at ~3 MPa as in the case of the compressor system. The gas enters the regenerative adsorbent purification system that removes mercaptans, water, carbon dioxide, and other impurities to avoid freezing and corrosion problems in the liquefier. Next the gas enters the cold box and flows through the pre-cooling heat exchanger prior to entering the main liquefier. Following liquefaction, the LNG is treated similarly to the LNG in the truck-delivered LNG case. The LNG is compressed in a cryogenic pump to ~30 MPa and transferred to the vaporing heat exchanger. The vaporizing LNG flows in a counter flow configuration against the incoming room temperature NG. The LCNG is then transferred to the CNG storage tanks as the case with the compressor station.

Several advantages of this approach are immediately obvious. First inexpensive pipeline NG is used, reducing fuel costs and offering a long term secure supply. As with the delivered LNG option, a local liquefier offers greater operational flexibility and can act as an intermediate step to LNG or $LH_2$ in the future. Maintenance costs are significantly less than that of the compressor stations. Operating costs are only slightly higher than compressor systems. This approach also offers the highest quality fuel because impurities have been removed to very low levels and boil-off enrichment of less volatile components is not a problem.

Disadvantages associated with this approach include the highest capital and operating costs per unit capcity of the three approaches considered. Additionally, the local liquefier remains an unproven technology for fleet sized vehicle fleets.

## METHODOLOGY OF ANALYSIS

The total costs of the CNG from the transit operators point of view can be determined primarily by adding the operating costs of converting the NG supply in to CNG, the feedstock costs, and the capital costs of equipment required to convert the feedstock into

CNG. Maintenance cost includes capital costs of recommended spare parts for operation, service contracts, and other costs associated with operating a fleet. The total capital costs of an installed system is several hundred thousand dollars more than the figures used in the analysis due to the inclusion of dispensers, fuel management systems, civil engineering and site prep. These capital costs are similar in all cases and were not included in the comparison of CNG production methods.

Operating costs for the three approaches to CNG production were compared by isolating the various work inputs for control volumes drawn around the installed refueling systems as seen in the above schematics. For the compressor system work input into the dryer and the multi-stage compressor was considered. The truck delivered option has work input only into the cryogenic pump to compress the LNG. The local liquefier has work input into the purification system, the liquefier, and the cryogenic pump. Small power requirements for valves and control systems were ignored as insignificant.

As can be seen in Figure 5, the truck-delivered LNG operating costs are by far the lowest of the three appraoches considered. The operating cost of the local liquefier is higher than the standard compressor method. Liquefier operating costs approach the compressor station's as inlet pressure increases.

Capital costs of the various systems were determined through a combination of vendor quotations and engineering estimates where required. Equipment specifications were based on the expectations of the transit fleet engineering consultant. Reliability concerns with the compressor systems resulted in a decision to provide 100% redundancy of compressor equipment to ensure availability of fuel. LNG storage provided sufficient fuel capacitancy in the other two cases to ensure fuel supply security.

Fuel costs were determined by quoted prices from the Local Distribution Company (LDC) and LNG preliminary pricing from two LNG delivery companies. The price quoted by the LDC was \$0.084 per $Nm^3$ (\$2.37/Mscf), this is below the average price for pipeline NG in the United States. The delivered LNG was \$121.50 per $Nm^3$ (\$0.46/gal).

The capital costs were amortized over a 10-year period and assessed using quotes acquired in May 1997. Total Costs are seen below in Figure 6.

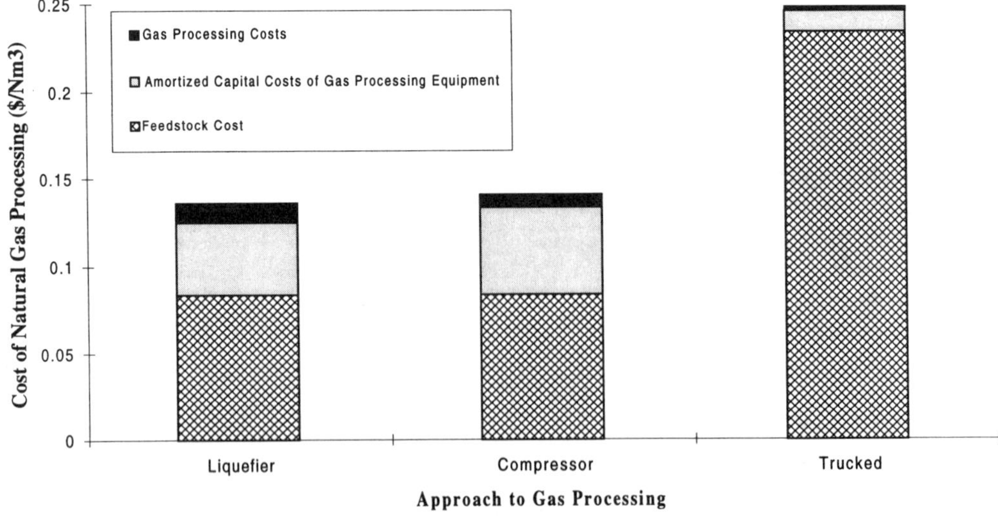


**Figure 6.** Total CNG Production Costs for the Analyzed Transit Fleet.

## SUMMARY AND CONCLUSIONS

A comparative cost analysis of three competing CNG production options was prepared for a transit bus fleet incrementally converting vehicles over the next thirteen years. The local liquifier provides fuel at a cost just below that of the conventional compressor stations. In this particular case truck-delivered LNG was a poor choice because of the relatively long delivery distance and high feedstock costs. Although both the compressor and local liquifier provide fuel at relatively the same cost, the local liquifier can offer significant advantages. Temperature compensated fill ensures desired vehicle range is achieved. The local liquifier offers lower maintenance costs and improved reliability. Unlike compressor stations, the local liquifier also can be installed underground to apprear more like a traditional refueling facilty. This analysis showed the production of LCNG with a local liquefier is a viable alternaitive to direct compression of NG or truck delivered LNG. Ultimately, the NGV infrastructure should develop with help from all three alternative approaches to CNG production.

## REFERENCES


1. J.A. Barclay, Cryofuels, Now and in the Future, Adv. Cryog. Eng., 41B, 1015-1031 (1996).
2. J. A. Barclay, D.W. Goudie, and C.E. Reid, Comparison of Natural Gas Liquefiers for Fleet-size LNG/CNG Refueling Systems, Proc. Of the 1995 Windsor Workshop on Alternative Fuels (1995).
3. C.F. Blazek and R.T. Biederman, LNG Plants in the U.S. and Abroad, Presented at the SAE LNG TOPTEC Conference, (1992).
4. M.A. Barclay and J.A. Barclay, Purification Techniques for Natural Gas Refueling Stations, Presented at the 14th National NGV Conference and Exhibition, (1996).

# TESTING OF A BURIED LNG TANK

J.M.Jurns,[1] G.M.Pease,[2] R.R.Tison,[3] R.J.Sprafka[3], R.F.Nigro[3]

[1]NYMA Inc.
Brook Park, OH 44142
[2]NASA LeRC
Cleveland, OH 44135
[3]East Ohio Gas
Cleveland, OH 44101

## ABSTRACT

Development of alternative fuels is a significant national priority. Liquefied natural gas (LNG) is an alternative fuel that has many attractive features. However, before LNG can be extensively exploited, the infrastructure for transferring, storing, and dispensing LNG needs to be developed. One important aspect of this development is addressing technology issues associated with installing LNG dewars at customer sites. The National Fire Protection Association (NFPA) has recently revised standards 57A and 59 covering installation of these dewars. This paper describes a test program that was conducted through a collaborative effort between the East Ohio Gas Company and NASA Lewis Research Center. Test results are shown for data that was collected on a LNG dewar that was buried at a site in Norfolk, VA. Data are used to validate predictions of vapor generated during normal operations and failure (loss of dewar vacuum) conditions. Data are also used to validate predictions of soil temperature surrounding the dewar. This testing provides information that will be useful in improving the NFPA code, and will lead to the safe and effective implementation of LNG as an alternative fuel.

## INTRODUCTION

East Ohio Gas (EOG) is currently working on developing an alternative fuel infrastructure. One important aspect of this activity is addressing technology issues associated with burying liquefied natural gas (LNG) tanks at customer sites. The National Fire Protection Association (NFPA) has recently revised standards 57A and 59 covering installation of these tanks. The objective of this experimental program will be to collect data on a buried LNG tank and provide information that could be used to argue for improvements to the NFPA Code. In addition, it is anticipated that the results of this test program will lead to the safe and effective implementation of LNG as an alternative fuel.

EOG currently owns a prototype 11356 L (3000 gal) LNG tank, and has installed it underground at a test site in Norfolk, VA. This paper describes the test program and results obtained from tests conducted on this buried tank.

## TEST PROGRAM

### Overview

The test program is designed to address technical and safety issues associated with the burial of an LNG storage tank[1]. The LNG tank is a double walled, vacuum insulated vessel, with a high performance insulation in the annular space between the outer jacket and inner liquid container. Tank failure as defined in this program is the loss of vacuum in the annular insulation space. Critical technical issues include determining the impact of tank failure on the LNG storage tank and the surrounding soil. Specifically, the following issues are addressed:

1. Does soil surrounding the LNG tank freeze with tank failure, and does it result in excessive pressures on the outer jacket of the tank?
2. Does tank failure result in extremely low temperatures in the outer jacket, and what problems could result from this?
3. Will the volume of boil off gas be a problem?
4. Will the operator know the failure has occurred?

To conduct this test, a 11356 L (3000 gal) tank designed for direct burial was buried at a test site in Norfolk, VA. Instrumentation was installed to measure vent gas flow rates, tank and annulus pressure, outer jacket temperatures, surrounding soil temperatures, soil pressure on the outer jacket, and liquid level in the tank.

The tank was filled with liquid nitrogen ($LN_2$) for this test. $LN_2$ was used instead of LNG for several reasons. The site did not have any means of accommodating vented natural gas; use of $LN_2$ eliminated the need for explosion proof test equipment; and $LN_2$ is colder than LNG, which would result in more severe test conditions. After the tank was filled with $LN_2$, the vacuum in the annular space was "broken" using nitrogen gas. The pressure in the annulus was increased to approximately 101 kPa (14.7 psia). Pressure was monitored and maintained at this pressure during the entire test using boil off nitrogen gas from the inner liquid container.

### Test Objectives

The test program has the following objectives:

1. Determine the amount of vapor generated due to loss of vacuum conditions (loss of vacuum boil off rate).
2. Determine the soil temperature profile surrounding tank under loss of vacuum conditions.
3. Determine outer jacket temperature under loss of vacuum conditions.
4. Determine soil pressure on the outer jacket of the tank under loss of vacuum conditions.
5. Determine the indicators that the failure has occurred.

**Loss of Vacuum Boil Off.** The purpose of this test is to measure the amount of vapor generated under loss of vacuum conditions. A potential source of failure for this tank would be loss of vacuum in the annulus due to either corrosion through the outer jacket, or failure of the outer jacket relief devices or evacuation valve. This failure was simulated by introducing nitrogen gas into the annulus to increase the annular space pressure to approximately atmospheric. Following completion of the loss of vacuum boil off test, the annulus was re-evacuated to the original vacuum level, and sealed off.

**Soil Temperature Profile.** Under normal conditions, the heat flux through the insulated vacuum jacket of the buried tank is very small, and the temperature of the outer jacket would remain in equilibrium with the surrounding soil. However, under failure conditions, the temperature of the outer jacket would be expected to decrease due to the increased heat flux through the annular insulation space. The temperature profile in the soil surrounding the test tank was measured and recorded all the while the loss of vacuum boil off test was being conducted. Based on representative soil thermal properites[2,3], it was anticipated that the soil immediately surrounding the tank would freeze, with the temperature of the soil increasing to ambient conditions at a distance of approximately 3 m (10 feet.).

**Outer Jacket Temperature.** Temperatures were measured on the outer jacket to determine the minimum temperature it would reach under failure conditions. As the outer jacket was fabricated from SA36 carbon steel, temperatures below it's ductile to brittle transition temperature (approximately -40 C) could be a source of concern.

**Soil Pressure on Outer Jacket.** A potential result of soil freezing around the tank is increased pressure on its external surface. Load cells installed on the top and sides of the tank monitored soil pressure on the outer jacket. It was anticipated that as soil froze, pressure may increase to a point that exceeded the ability of the jacket to withstand buckling.

**Failure Indicators**. It is important to determine what (if any) obvious signs would indicate that a failure has occurred. If any positive indication of failure of the storage system were observed, it would be cause to remove the system from service for repair or replacement. The tank used for this test was heavily instrumented, so any change in the condition of the system was easily noted. However, in normal installations, instrumentation is likely to be much less extensive, and some other means of determining failure is important.

## Test Hardware

**Test Article.** The LNG tank for this program is a commercial horizontal vacuum jacketed dewar manufactured by Minnesota Valley Engineering Inc. (MVE). Specifications for the tank are as follow:

| | |
|---|---|
| Capacity (gross liquid) | 11954 L (3158 gal) |
| Capacity (net liquid) | 11356 L (3000) |
| M.A.W.P. | 1207 kPa (175 psig) |
| Inner vessel mat'l | 9% Ni steel SA553 |
| Outer jacket mat'l | Carbon Steel SA36, 1.27 cm (0.5 in) th |
| Manway | 43.8 cm (17.25 in) minimum i.d. |
| Weight (empty) | 7258 kg (16000 lb) |
| Weight (full LNG) | 12066 kg (26600 lb) |
| Outside diameter jacket | 243.8 cm (96.0 in) |
| Overall length | 462.9 cm (182.25 in) |

152.4 cm (60 in) collar around manway for access
Multi-layer insulation (MLI) with composite filler insulation in annulus

The tank is provided with piping connections for liquid and vapor phase, as well as piping connections for pressure and level measurement. Piping components (valves and instrumentation) were field installed.

**Instrumentation.** The tank and surrounding soil were instrumented with 54 type "T" thermocouples to measure jacket and soil temperatures. The jacket thermocouples were mounted directly on the outside of the jacket. The soil thermocouples were mounted in arrays, and extended radially outward from the tank in several directions. One vertical thermocouple array was buried approximately 6.1 m (20 ft) away from the tank as a reference to measure undisturbed soil temperature. A diagram of the thermocouple locations is shown in Figure1. Four resistance temperature detectors (RTDs) were used to measure the liquid and vapor temperatures inside the tank. Four load cells were installed at various locations on the tank to record the various soil pressure on the tank during installation and testing. The load cells were 0-227 kg (0-500 lb) range, and had a live surface area of 126.6 $cm^2$ (19.63 $in^2$). Pressure transducers with an accuracy of ±0.11% FS were used to measure tank pressure, annulus pressure, flow meter pressure, and tank liquid level. A diaphragm flow meter with a pulse type output was used to measure gas flow rate. The flow meter had a capacity of 377 actual L/min (800 SCFH) with a 0.12 kPa (0.5 inch W.C.) pressure drop across the meter, and a capacity of 802 actual L/min (1700 SCFH) with a 0.5 kPa (2 inch W.C.) pressure drop across the meter. Accuracy of the flow meter was ± 1.0%.

**Data Acquisition.** The data acquisition system used a 133 MHz PC workstation. A 16 channel analog input module installed in the workstation was used to acquire data. Analog signals from the chassis were multiplexed into the workstation using one of the sixteen available analog inputs. There were two, 32 channel thermocouple input modules and one input module for analog signals. During the test, data were scanned every 15 seconds and held in a queue for averaging. Every ten minutes the data was averaged and stored to disk.

## Test Site

The 11356 L (3000 gal), 2.44 m (96.0 in) diameter, LNG tank was installed at a future Alternate Fuel Station site in the CD Area at the Navy Public Works Center in Norfolk, VA. The tank was buried approximately 0.6 m (2 ft) below grade. The excavation was backfilled with crushed stone to the midpoint of the tank and then with sand to within 0.3 m (1 ft) of grade. A plastic rock shield was wrapped around the entire tank to prevent damage to the outer jacket corrosion preventive coating. A 0.3 m (1 ft) thick concrete pad 4.6 m (15 ft) wide by 7.3 m (24 ft) long was poured over the buried tank. The top of the pad is 3.7 m (12 ft) above sea level. The fill extends approximately 0.9 m (3 ft) from the tank. Beyond the fill, the native soil is characterized as alluvial deposits consisting of sand, silt and clay. During the test, the ground water level was approximately 1.2 m (4 ft) above sea level. A small shed was erected on site to house the data acquisition computer and signal conditioning equipment.

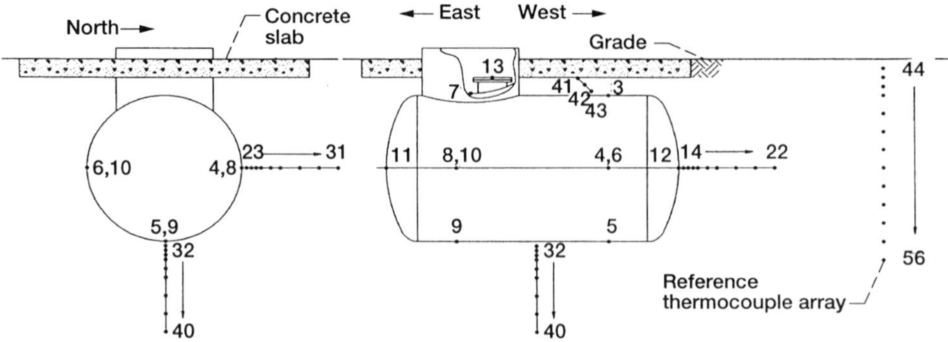


**Figure 1.** Thermocouple locations.

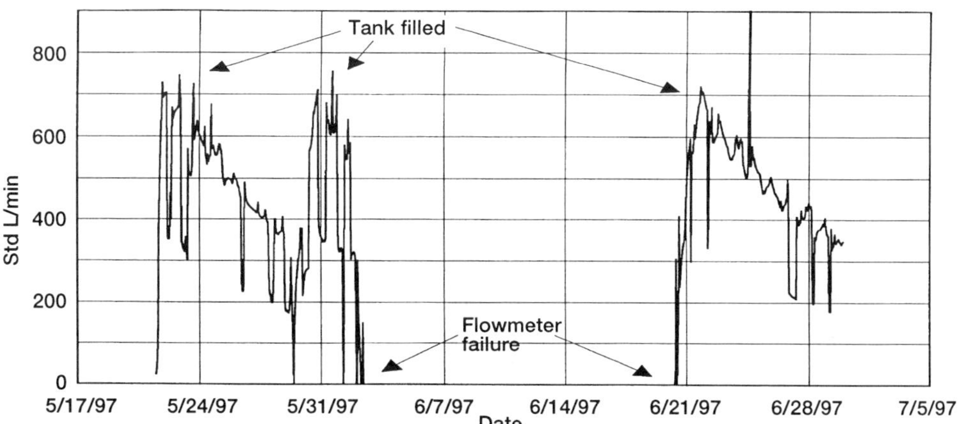


**Figure 2.** Boil off flow rate versus time.

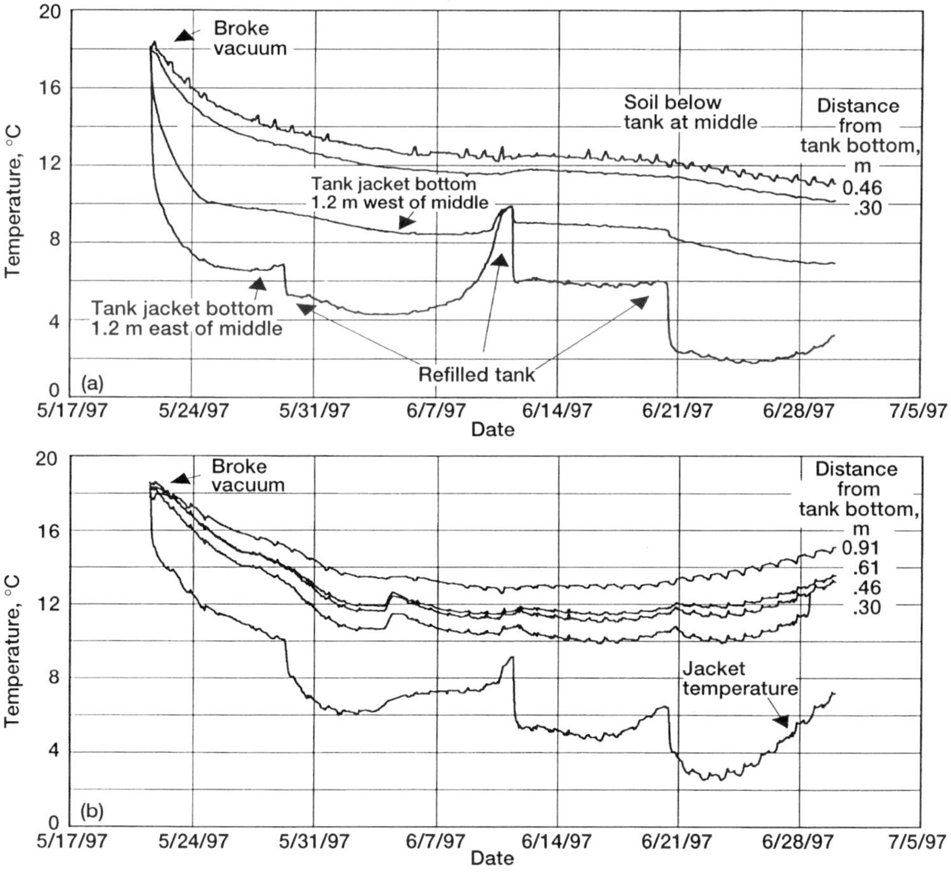

Broke vacuum
Soil below tank at middle
Distance from tank bottom, m
0.46
.30
Tank jacket bottom 1.2 m west of middle
Tank jacket bottom 1.2 m east of middle
Refilled tank
Temperature, °C
(a)
20
16
12
8
4
0
5/17/97
5/24/97
5/31/97
6/7/97
6/14/97
6/21/97
6/28/97
7/5/97
Date

Broke vacuum
Distance from tank bottom, m
0.91
.61
.46
.30
Jacket temperature
Temperature, °C
(b)
20
16
12
8
4
0
5/17/97
5/24/97
5/31/97
6/7/97
6/14/97
6/21/97
6/28/97
7/5/97
Date

**Figure 3.** Temperatures versus time for tank jacket and soil. (a) Underside. (b) End.

## TEST RESULTS

### Boil Off Rates

Theoretical boil off rate under normal vacuum conditions was estimated to be approximately 28.3 standard L/min (60 SCFH), based on a heat leak into the tank of 88 W (300 BTU/hr). Actual boil off rate was approximately 47 standard L/min (100 SCFH). This was larger than expected, but still within acceptable limits. This boil off rate is the equivalent of 0.8 % per day loss, which is typical for cryogenic storage vessels this size.

Theoretical boil off rates under loss of vacuum (failure) conditions was estimated to be approximately 472 standard L/min (1000 SCFH), based on a heat leak into the tank of 1464 W (5000 BTU/hr). The average actual boil off rate agreed with predictions, although it varied from 750 standard L/min when the tank was full, dropping to about 190 standard L/min when the tank was near empty (1600 to 400 SCFH). See Figure 2. The break in the data is the result of a flow meter pulse output failure.

### Temperatures

**Outer Jacket Temperatures.** With loss of vacuum, there was an increased heat flux through the insulation between the inner liquid container and the outer jacket. Consequently, jacket temperatures were expected to fall. The jacket temperatures dropped over time, but after 40 days and 3 LN2 refills, the coldest temperature recorded was only -16 degrees C (2.5 F), and average jacket temperatures were considerably warmer than that. Figure 3 shows jacket temperatures versus time.

**Soil Temperatures.** The soil temperatures surrounding the tank decreased, which was expected. Temperatures were coldest next to the tank, and increased with increasing distance from the tank. Soil temperatures were the same as for undisturbed soil at a radial distance of approximately 2.4 m (8 ft) from the outer jacket. The coldest soil temperature recorded was 3 C (37 F), and was recorded at a distance of 0.3 m (1 ft) from the bottom of the outer jacket (TC32, see Figure 4). Soil temperature versus time was plotted for each of the three buried temperature arrays as shown in Figure 3.

### Soil Pressure

Two load cells were mounted on the top of the tank, one was mounted at the 4 o'clock position, and one was mounted at the 8 o'clock position. The 8 o'clock load cell apparently failed during installation, as there was no signal output. The top two load cells indicated soil pressures on the jacket of anywhere between 28 and 90 kPa (4 and 13 psi). Daily swings in pressure were as high as ±21 kPa (± 3 psi). The load cell in the 4 o'clock position initially indicated a pressure of 90 kPa (13 psi), and steadily decreased to 42 kPa (6 psi) after 25 days. See Figure 4 for a plot of soil pressure versus time. The increasing pressure on the top cells along with the decreasing pressure on the 4 o'clock cell suggest that some soil freezing may have occurred along the bottom of the tank.

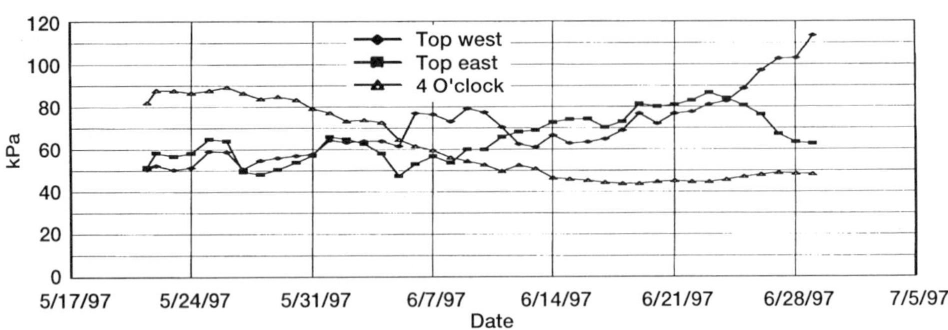


**Figure 4.** Soil pressure versus time.

### Failure Indicators

As predicted, boil off rates increased by a factor of 10. This increase caused a small increase in vent gas noise and an increase in the length of vent piping which was frosted. Since these changes were small and could be ignored by station operators, the loss of vacuum test was extended to 40 days. It was reasoned that within 40 days the product loss would become the failure indicator. The observed boil off would represent about a 25% increase in the stations delivery requirements assuming an active station that required a fuel delivery every three days.

## CONCLUSIONS

Data collected to date from this test program lead to the following observations and conclusions:

For these tests, loss of vacuum did not precipitate any type of catastrophic event. Loss of vacuum boil off rate is predictable and low enough to be accommodated with standard safety relief devices and vent systems.

For the period of time this test has extended, outer jacket and surrounding soil temperatures did not decrease to a point that could cause a potential failure in the outer jacket. Soil temperatures for the most part remained above freezing, and outer jacket temperature remained above the ductile to brittle transition temperature.

External soil pressures were not excessive. If there was any soil frozen, it only made minor contribution to soil pressure on the outer jacket.

If an operator does not notice the increased boil off rate, we believe the accountant will find it, thus limiting the period of a loss of vacuum failure.

Based on the above observations, it is reasonable to assume that no special precautions are necessary for underground installation of this type tank other than providing normal safety relief devices and vent piping typical for a cryogenic storage vessel.

## ACKNOWLEGEMENTS

Thanks are extended to the Navy Public Works Center in Norfolk, VA for providing a suitable site to conduct this test. Thanks are also extended to the Viccellio Construction Inc. of Virginia Beach, VA (general contractor), the Riggins Company of Hampton, VA (field piping), and Technical Piping and Service Corporation of Strongsville, OH (piping fabrication). Also, special thanks go to Mr. Chris Wrocklage and Mr. Dale Burcher, both of Virginia Natural Gas Company. This test program could not have been carried out without their able assistance. Finally, Pacific Gas and Electric is acknowledged as a funding partner for this test program.

## REFERENCES

P.A.Outtrim et al. Undergournd LNG Storage Developments, presented at LNG's Emerging Integrated Marketplace, Houston, TX, March 4, 1997.

F.D.Haynes et al. "Thermal Diffusivity of Frozen soil", CRREL Special Report 80-38, December, 1980.

O.T.Farouki, "Thermal Properties of Soils", CRREL monograph 81-1, December, 1981.

# REGIME OF UNIFORM HEATING OF LIQUID PHASE DURING STATIC STORAGE OF LIQUEFIED NATURAL GAS

Zh.A. Suprunova and V.E. Seriogin

Special Research & Development Bureau in Cryogenic Technologies,
Institute for Low Temperature Physics & Engineering
National Academy of Sciences of Ukraine
47, Lenin Ave., Kharkov, 310164, Ukraine

## ABSTRACT

The case of heat and mass transfer is investigated for static storage of liquefied gases, which, with exception of the near-wall region, occurs without temperature stratification of the liquid phase. The maintenance of vessels in a regime of uniform heating reduces losses by a factor of 2.5 - 3.5. The range of parameters for the process corresponding to this regime under self-pressurization are investigated.

## INTRODUCTION

This work is devoted to the development of techniques for static storage of liquefied gases with the most favorable uniform heating of the liquid. The static storage of cryogens is frequently used in practice. It is used in transport vessels, propellant tanks , and in stationary tanks. For storage of liquefied natural gas (LNG) both small (1-3 $m^3$) and large (50-180 $m^3$) vessels and tanks are used. The small vessels operate at pressures up to 2.5 MPa, while large tanks are used for storage at pressures close to atmospheric. Vessels with static storage are used in the production of liquid nitrogen, in trucks filling stations, and in river and marine ships.

The static storage of cryogens is accompanied by a pressure rise in the vessel, as well as temperature stratification of the liquid. In the latter case, elevated pressure in a vessel corresponds to an interface temperature which considerably exceeds the mass-body temperature. The evaporation rate in a vessel increases 2-3 times, and the rate of pressure increases by 2.4-3.5 times. To reduce these negative effects, various engineering solutions are used, such as intermixing devices, heat exchangers, cold fluid sprayers at the phase interface, insulation blankets at the phase interface, boil-off gas liquefiers etc.

At the same time, static storage has a bifurcation branch corresponding to uniform heating of liquid. This feature was revealed experimentally during the investigation of storage of oxygen in a spherical tank with diameter 0.3 m [1], and in a torus with innner and outer radii of 0.096 and 0.175 m, filled with liquid nitrogen [2]. In the present work, the results of research into singularities of the regime of uniform liquid heating under conditions of self-pressurization are presented, and the dimensionless parameters appropriate to this regime are identified.

## DISCUSSION

The results of research into static storage of liquefied gases have been considered previously [3,4]. Common regularities of such processes connected to the origin of this phenomena, are as follows. Under conditions of self-pressurization in the range of Raleigh numbers $Ra > 10^5$, interior flows at a heated vessel wall will develop thermal and hydrodynamic boundary layers around fixed or slowly circulating basic mass of fluid. The ascending flow in the upper area of liquid turns along the free surface and washes it by a jet flow, which then turns into of the mass of liquid, being intermixed with it. Under conditions of stratification this flow is unstable and actively interacts with the main mass of liquid. As the result, heat input from gas to the interface partially goes deeply into the body of liquid, forming a stratified field of temperatures. However, under certain conditions, when the heat flow through the fluid interface is insignificant, the jet flow gains a steady character. The heat flow through the interface, mainly is carried deep into the liquid, and being intermixed with it, aligns the temperature. As the result there is uniform heating with minor temperature stratification. The spatial and temporal distribution of temperatures in this regime is shown on figs.1 and 2. Data were obtained in a toroidal vessel (fig.1) and spherical vessel (fig.2), filled with liquid nitrogen. As can be seen from the figures, transition from uniform heating of liquid to temperature stratification is accompanied by the rise of liquid level, and the onset of temperature oscillations of at the surface of fluid. The more prolonged the stage of uniform heating, the more rough the subsequent transition to stratification, when the body of superheated liquid with loss of a stability of surface stratum penetrates to the interface and evaporates. Besides the influence of heat flow through the interface, the stability of the surface flow is affected by the type of liquid flow (turbulent or laminar), the character of phase transitions at the interface, and transport of heat in the walls of the vessel.

While analyzing and describing the experimental results, the following dimensionless parameters were used:

- $q_i / q$; - ratio of the heat flux through interface to the total external heat flux;
- $(\rho_{eff} / \rho_{cr})$ ,$(T_0/ T_{cr})$ - reduced density and temperature of the vessel content at the initial moment;
- $l = V/S$ - effective linear scale of the process , (having dimension of meters);
- $(T-T_0)^* = (T-T_0)\, c_p / L = \Delta u / L = (q\, \tau\, S) / (V\, \rho_{eff}\, L) = (q\, \tau) / (l\, \rho_{eff}\, L)$ - relative increase of the mass-body temperature or relative internal energy increase ;
- $Ra^*_{int} = (q\, g\, h4\, \beta) / (\nu a\, \lambda)$ - the modified Raleing criterion;
- $Pr = \nu/a$ - Prandtl criterion;
- $Fo = a\, \tau/ l^2$ - Fourier criterion (dimensionless time);
- $S_i / S_s$ - ratio of the interface area to the vessel dome area (interfaced to the gas ullage );
- $S_\mu / S_{i=}$ - ratio of the vessel envelope cross- section at the interface level to the interface area;
- $m = V_{liq} / V$ - vessel filling factor;
- $(C_w\, \rho_w\, \delta_w) / (C_p\, \rho\, l)$ - envelope thermal inertia parameter;

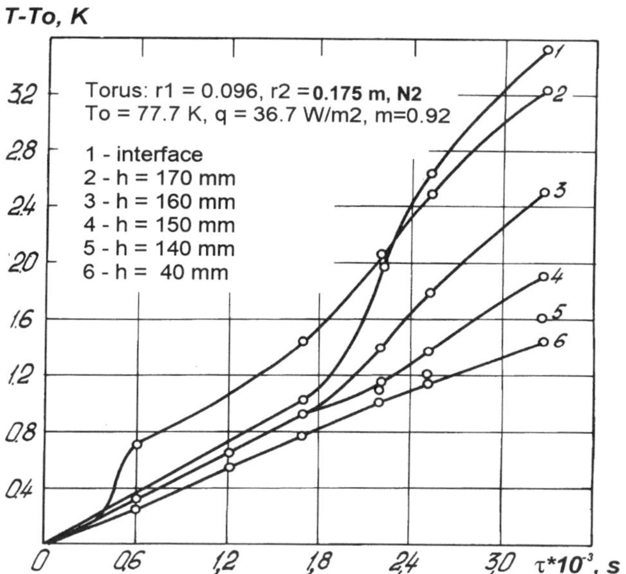


**Figure 1.** A time history of temperature in a volume of liquid at a stage of uniform heating, in a transient regime and at a stage of developed temperature stratification.

- $\lambda_w / \lambda$; $\delta_w / l$ - parameters, describing the vessel envelope thermal conductivity.

The subscripts designate: w - wall; i - interface of phases; s- a dome of a vessel; cr - critical filling of a capacity; eff - effective value; int - internal, liq - liquid, 0 - initial conditions.

In figures.3 and 4 the results of experimental work on a toroidal and spheric tanks are shown, in term of dimensional temperature increase **$(T\text{-}T_0)^*$** $= (q\,\tau) / (l\,\rho_{eff}\,L) \cdot 10^{-2}$ and variables $q_i/q$ and effective density $\rho_{eff.} / \rho_{cr.}$ Time $\tau$ in a parameter $(q\tau) / (l\,\rho_{eff}\,L)$ corresponds to the onset of temperature gradients in liquid. According to fig.3 the

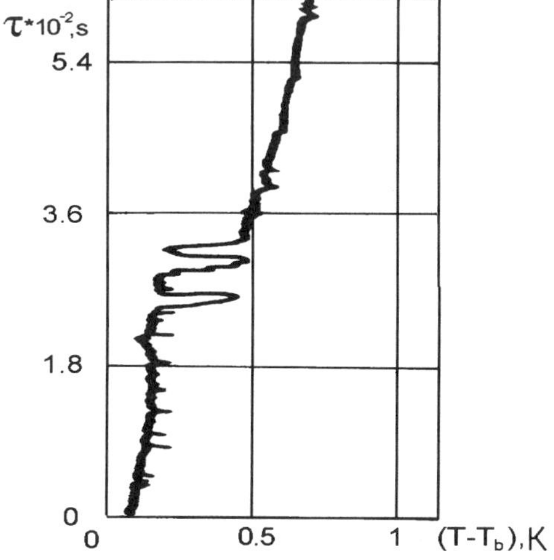


**Figure 2.** Temperature drop on a height of volume of liquid at a stage of transition from uniform heating to temperature stratification. The temperature sensor is located 1 mm below the phase boundary ( spherical tank with $LN_2$, r=0.15 m, filling factor m=0.78, q = 48 W/m$^2$ ).

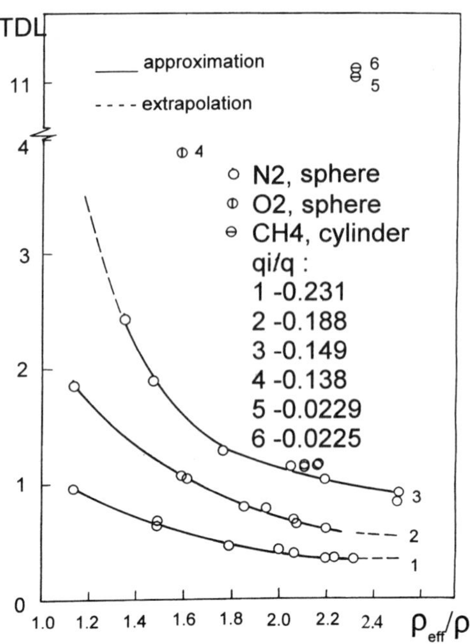


**Figure 3. Dependence of the energetic threshold of the temporal (existence) boundary of the isothermal regime on the thermodynamic state of a two-phase medium, and distribution of heat flows in a vessel. ( TDL = Temperature DimensionLess = $(T\text{-}T_0)^* = (q\ \tau) / (l\ \rho_{eff}\ L) \cdot 10^{-2}$)**

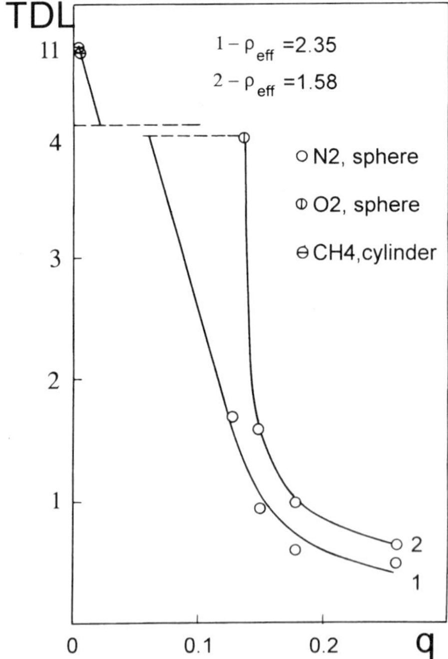


**Figure 4. Dependence of a energetic threshold of the temporal boundary of existence of an isothermal regime on distribution of heat flows in a vessel for two values of a parameter ($\rho_{eff}$ / $\rho_{cr}$ ) = 1.58 and 2.31 . (TDL - see fig.3, $q = q_i / q$ )**

**Table.1.** Performance characteristics of automobile methane tanks

| Parameters | Automobile model | |
|---|---|---|
| | ZIL-130 | GAZ-53 |
| Capacity of a tank, l | 190 | 160 |
| Overall dimensions | | |
| Length, mm | 1350 | 1200 |
| Diameter, mm | 620 | 620 |
| Mass, kg Empty | 83 | 78 |
| Full | 164 | 159 |
| Operating pressure, kPa | 140-300 | 140-300 |
| Evaporability, % per day | 3.3 | 3.3 |
| | Dimensionless variables | |
| $(q\ \tau) / (l\ \rho_{eff}\ L)$ | 0.11 | 0.11 |
| $q_i / q$ | 2.29.10-2 | 2.25.10-2 |
| $\rho_{eff} / \rho_{cr}$ | 2.31 | 2.31 |

power threshold for existence of an isothermal condition is increased with a decrease of the parameter $\rho_{eff} / \rho_{cr}$ (increase of initial pressure or lowering of filling factor of the vessel by liquid). However, the increase of temperature of the medium also increases the degree of flow turbulization, which is accompanied by lowering of stability of the flow, and shortening the uniform heating. The data for $q_i / q = 0.231$ (curve 1 on the graph) corresponds to maximum values of heat flows through a surface of liquid, in which the condition of uniform heating of liquid is obtained. The lower curves (1,2) correspond to process conditions with large external heat flows and more brief periods of uniform heating. The upper curves 3, 4 correspond to small heat flows and longer periods of uniform heating. Modifications to the distribution of heat flows in a system which depends on external heat flows is basically connected to the limited heat-conducting ability of a vessel walls. In fig.4 the experimental data are shown in the form $(q\ \tau) / (l\ \rho_{eff}\ L) = f(q_i / q)$ for two values of a parameter $\rho_{eff} / \rho_{CK}$ (1.58 and 2.31). As seen in fig. 4, reducing the heat flow through the interface the duration of unstratified heating quickly increases.

The isothermal heating of liquid under conditions of self-pressurization was realized in tests of automobile fuel tanks with liquid methane. The automobile tanks were developed and manufactured in SR&DB in CT of the ILTPh&E of the NAS of Ukraine for automobiles ZIL-130 and GAZ-53. The tanks were made of stainless steel with use of superinsulation and had internal bulkheads of aluminium alloy. The results are submitted in a Table. 1.

## CONCLUSIONS

Figs.3,4 show, based on calculations of distribution of heat flows in a system, how to select liquefied gas vessel parameters, in which uniform liquid heating will be assured. For example, the vessel cross section, according to this method of calculation, should be in the limits $(S_s - S\ \mu) / (S_i - S\ \mu) = 2.28 - 1.1$ for fillings m = 0.4 - 0.96 accordingly. ($S\ \mu$ - cross-section of a vessel wall). This imposes a restriction on the ratio of surface area of the dome of a vessel washed by gas, to the surface of free liquid.

These results were obtained for a spherical vessel with a radius r = 0.15 m, filled with $LO_2$, $LN_2$, a torus with radii $r_1$ = 0.096 m and $r_2$ = 0.175 m, filled with $LN_2$, and a horizontal cylinder with a radius r = 0.31 m, filled with $LCH_4$.

The data cover the following range of dimensionless parameters.

**Table.2.** Range of dimensionless parameters

| Parameter | Range of values |
|---|---|
| Modified Raleigh criterion Ra* | $10^{10}$ - $10^{13}$ |
| Fourier number Fo | $10^{-2}$ - 5 |
| initial filling factor of a vessel m | 0.4 - 0.96 |
| The enevelope thermal inertia parameter $(C_w \rho_w \delta_w) / (C_p \rho l)$ | 0.043 - 0.057 |
| The ratio of a thermal conductivity of a wall to that one of a liquid $\lambda_w / \lambda$ | 54 -350 |
| Relative thickness of a wall $\delta_w / l$ | 0.01 - 0.04 |
| Initial reduced temperature $T_o / T_{cr}$ | 0.56- 0.61 |
| Prandtl criterion Pr | 2.17 - 2.4 |
| A ration of the envelope cross-section to the surface area | 0.014 - 0.80 |

Experiments on industrial vessels are necessary to perfect this technique.

## REFERENCES


1. Yu.. A.Kirichenko, V.N.Schelkunov, V.N. Timon'kin, L.Yu. Radchenko. Peculiarities of free convective heat transfer in cryogens during their storage in static vessels. In Heat- and Mass Transfer - V. Proc. of Fifths All- Union Conference on Heat and Mass Transfer, Minsk, May 1976, vol.1, part II, p.314.
2. Yu. A.Kirichenko, Zh. A. Suprunova. Experimental investigation of convective heat and mass transfer in the closed vessel. In: Hydrodynamics and Heat Transfer in Cryogenic Systems, Kiev, Naukova Dumka, (1977), p.8-7.
3. J.A.Clark. Cryogenic Heat Transfer. In: Advances in Heat Transfer, Academic Press, New York, London, vol.5, (1968), p.361-554.
4. J.A.Clark. A Review of Pressurization, Stratification, and Interfacial Phenomena, In: International Advances in Cryogenic Engineering, (1965), Plenum Press, New York, p.259-283.

# THE ASSESSMENT AND COMPARISON OF ALTERNATIVE AND TRADITIONAL FUELS FROM A STUDY OF THE VEHICLE'S ON-ROAD PERFORMANCE AND EMISSIONS

S. De Maria

University of South Australia, Energy and Engines Research Group
Adelaide, South Australia, 5000, Australia

## ABSTRACT

A Vehicle and Engine Performance Monitoring and Analysis System (VEPMAS) has been under development since 1989 as part of the assessment phase of compressed natural gas buses in Australia.

Since then the analytical capabilities and features of VEPMAS have been enhanced to assist with optimisation of engine conversions and evaluation of fuels such as natural gas and diesel. VEPMAS provides engineers with information on the vehicle's performance whilst the vehicle is in actual use.

The results are based upon comprehensive engine data from dynamometer tests of the engine fitted to the vehicle or from a similar engine. By logging two independent engine variables when the vehicle is in service, cumulative data can be generated every second to provide total fuel consumption and emissions for a particular bus run, vehicle trip or selected cycle. The results obtained have been validated against actual measurements.

This paper is a non technical description of the concept and operation of the system. Typical results obtained from road trials are presented to demonstrate the output format of the data and to highlight the most importance application aspect of VEPMAS namely its ability to provide a quantitative basis for the comparison of fuels.

## INTRODUCTION

The Vehicle and Engine Performance Monitoring and Analysis System, VEPMAS was developed by the Energy and Engines Research Group (EERG) at the University of South Australia to predict the fuel consumption and exhaust emissions of vehicles powered by internal combustion engines. Comprehensive dynamometer engine data and the combustion equation form the basis of operation of VEPMAS.

*Advances in Cryogenic Engineering,* Vol. 43
Edited by P. Kittel, Plenum Press, New York, 1998

The accuracy of VEPMAS in terms of fuel consumption has been shown to be better than ±5% and in some cases as precise as ±2%. Validation of the emissions component is the current subject of research supported by national and state funding bodies.

VEPMAS to date has been used in various Australian cities to compare the on-road performance of compressed natural gas (CNG) public transport buses. In Malaysia it has been utilised by Petronas Research Scientific Services Sdn Bhd to determine the environmental benefits of using CNG buses in that country.

VEPMAS portends to have many useful applications from the point of view of OEM's, research organisations, fleet operators, and environmental protection agencies.

## BASIS OF OPERATION

VEPMAS utilises results obtained from comprehensive dynamometer engine tests[1] to predict the on-road performance of the vehicle in terms of fuel consumption and exhaust emissions.

### Engine dynamometer tests

An extensive dynamometer engine test is conducted to obtain the base engine performance data required by VEPMAS. Variables such as engine speed, torque, gas consumption and gas supply pressure are measured and logged at numerous points of the engine operating load-speed matrix using an automated data acquisition system[2].

The engine operating matrix may be described as a function of both engine speed and load. A selected number of equally distributed steady-state test points forms a grid over the engine operating matrix catering for all combinations of engine speed and power output. Engine performance data collected at these points is later adapted to VEPMAS by mathematically fitting a continuous three dimensional surface.

Once compiled, the possible engine conditions are divided up into 5 possible regions. These are :

- Idle
- Main Engine Map
- Wide Open Throttle (Full Load)
- High Engine Speed (Governor)
- Deceleration

These are indicated graphically in figure 1, below.

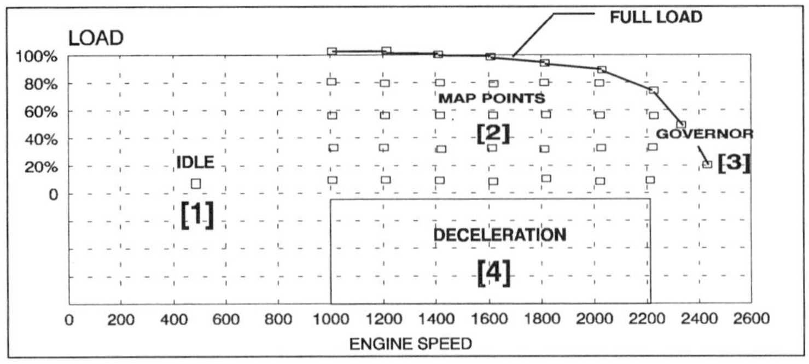


**Figure 1.** The five possible engine operating regions.

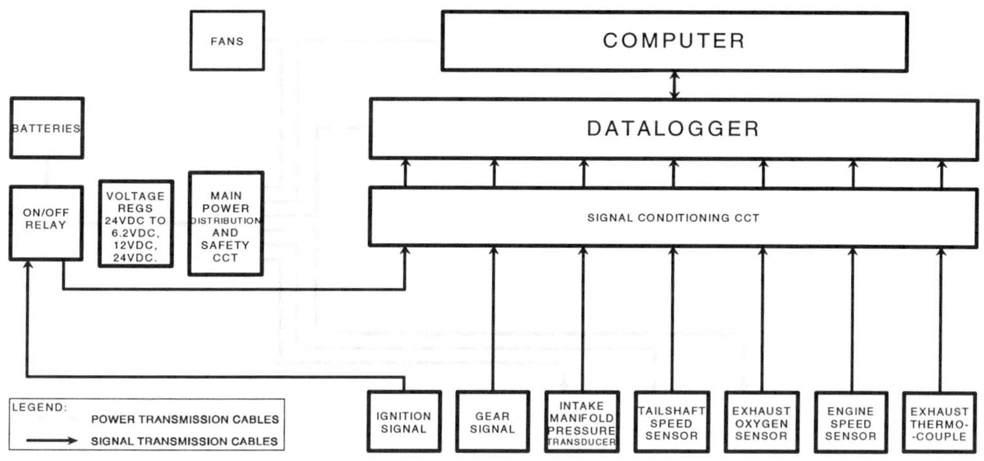


**Figure 2.** Vepmas on-road data acquisition system.

## On-road measurements

On-road measurements are performed in a manner similar to dynamometer measurements[2]. The system is capable of measuring the desired parameters at a frequency of approximately 1 Hz. Figure 2 indicates the components used in the latest version of VEPMAS[3,4].

A total of six sensors are fitted to the vehicle as shown in Figure 3.

The operating condition of the engine at a particular instant can be defined using two independent engine varables, in this case engine speed and intake manifold pressure. A proximity switch is used to determine the kinematics of the vehicle, whilst an oxygen sensor is added to check and compare the air/fuel ratio predicted by VEPMAS. The gear change and exhaust thermocouple are indicative of driver pattern and engine durability respectively.

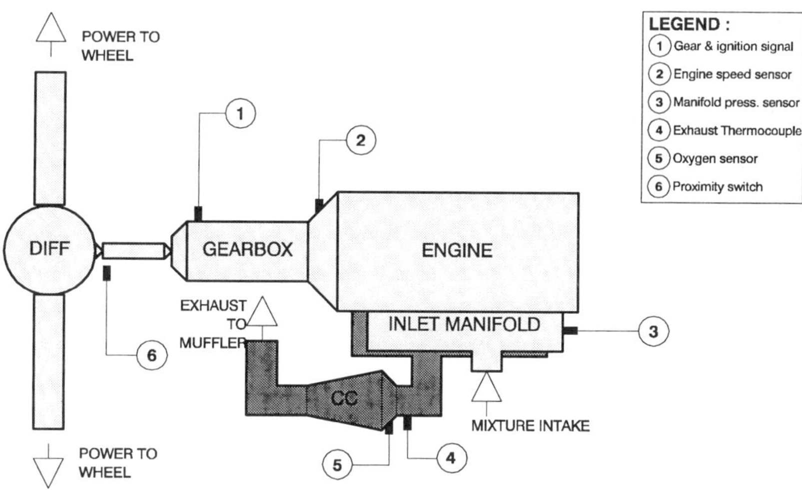


**Figure 3.** Vepmas sensors location on the engine.

When a test is conducted, the bus is driven over typical routes (flat, hilly, city, highway etc.) preferably by usual drivers and involving normal or simulated passenger operation. The data is logged and later either down loaded to a laptop computer or transferred via a telecommunication link to base where it is analysed and a hard copy of the results is produced as depicted in Figure 4.

## Combustion equation and emission output

The exhaust products analysis is derived from combustion theory. In this, the molar amounts of fuel and air are obtained from the engine load-speed matrix (Figure 1) via values of manifold pressure and engine speed logged from the on-road test.

The mass flow rate balance between the left hand side and right hand side of Eq. 1 is achieved by an iteration process which examines each of the different atomic elements, C, H, N and O to determine the molar fractions of the respective constituents of the exhaust products.

The concentrations of some of the exhaust products obtained during the dynamometer engine test is used to solve the set of simultaneous equations and hence the molar fractions of the exhaust constituents. This allows the calculation of the exhaust mixture mass flow rate and hence completes the combustion equation. This sequence is iterated until the sum of the emission fractions approximately equals 1.0 and the mass of reactants equals the mass of products.

These values are numerically balanced to satisfy the chemical equation of the combustion model. This then provides the corresponding amounts of the various emissions being produced by the engine at that instant in time. Combustion Eq.(1) forms the basis for the prediction of exhaust products and emissions by VEPMAS. A knowledge of the composition of the fuel and of the exhaust products concentrations is required.

$$\begin{aligned}&\dot{x}\left[0.9751\left(C_{1.0631}H_{4.1824}\right)+0.0135N_2+0.0114CO_2\right]+\dot{y}\left[0.20890_2+0.7911N_2\right]\\&\rightarrow\dot{\psi}\left[lO_2+pN_2+qH_2O+rNO_2+sCO+tCO_2+u\left(C_{1.0631}H_{4.1824}\right)\right]\end{aligned} \tag{1}$$

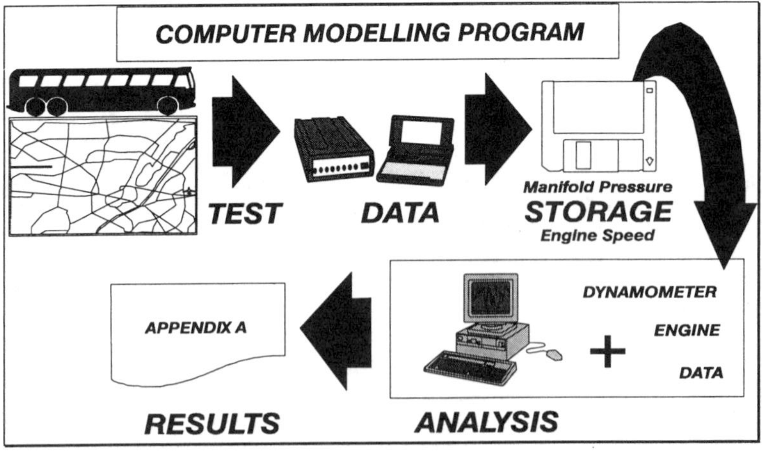


**Figure 4.** Diagram of data flow for vepmas.

**Table 1.** Raw summary output file (Malaysia data)

```
VEPMAS v3.0 Analysis Summary     10:57:27  Thursday 14-10-1993
----------------------------
Input file : 24_8_93.dat

number of hits : 2306   number of points : 2306

Emission Analysis
   total gas consumed  (g) :    6530
   total air consumed  (g) :  104862
   total reactants consumed (kg) =     111.39
   total O2 produced  (kg) :      10.07
   total CO produced   (g) :      34
   total CO2 produced (kg) :      17.15
   total H2O produced (kg) :       9.53
   total N2 produced  (kg) :      80.90
   total NO2 produced  (g) :       9
   total HC produced   (g) :       3
   total exhaust produced (kg)   =     117.69

Mass Balance Check :   5.79

Average A/F (weighted) from maps, etc.     : 15.80
Average Lambda value                       :   1.0

Total distance travelled (km) :      17.11
Average Road Speed (km/hr)    :       27.0

Gas consumption rate           (g/100km) :   38159
Air consumption rate           (g/100km) :  612767

O2 production rate            (kg/100km):      58.83
CO production rate            (g/100km) :     196
CO2 production rate           (kg/100km):     100.20
H2O production rate           (kg/100km):      55.69
N2 production rate            (kg/100km):     472.76
NOx production rate           (g/100km) :      49
HC production rate            (g/100km) :      14

Average Exhaust temperature     (Deg. C) :     514
Average Oxygen (mV)          (from Sensor) :     369.90

Gear Change Analysis
   Total gear changes       :    133
   Average gear changes/sec :   0.05
   Average gear changes/km  :   7.77

Average observation length (seconds) =  1.08
Total time (seconds) =    2493.68
```

## VEPMAS OUTPUT

Although VEPMAS can be customised to meet client wishes, normally two output files are produced.

The first file is a summary containing summations and averages of the various statistics of interest (Table 1). From this file the user can examine such items as total fuel consumption, totals of emissions produced, average road speed. Consumption and emission statistics are also shown on a distance basis, to allow comparisons of different routes.

This file is typically put into a spreadsheet format to summarise a series of tests at one time. An example appears in Table 2, below.

**Table 2.** Typical representation of vepmas route statistics (Brisbane data)

| Filename | Route No. | Atmospheric Temperature | Duration of Run | | Run Distance | Total Gas Consumption | Air Consumption | Exhaust Mass | Air/Fuel Ratio (mass basis) | Gas Consumption |
|---|---|---|---|---|---|---|---|---|---|---|
| | | °C | min | sec | km | gm | gm | gm | - | kg/100km |
| TRIP3 Garden City to City (Unladen) | 511 | 27 | 41 | 38 | 21.74 | 6471 | 142891 | 149390 | 22.1 | 29.8 |
| TRIP4 City to Garden City (Laden) | 511 | 28 | 39 | 51 | 21.9 | 6793 | 154276 | 161100 | 22.7 | 31.0 |
| TRIP4 City to Garden City (Laden) | 31 | 23 | 57 | 57 | 21.08 | 7223 | 160508 | 167760 | 22.2 | 34.3 |
| TRIP15 North Quay to Cannon Hill Unladen) | 31 | 22 | 43 | 26 | 21.43 | 7588 | 170733 | 178350 | 22.5 | 35.4 |

| Filename | Route | Total Emissions on a Mass Basis | | | | | | | Weighted Air/Fuel Ratio | |
|---|---|---|---|---|---|---|---|---|---|---|
| | | $O_2$ | $N_2$ | $H_2O$ | $NO_2$ | CO | $CO_2$ | HC | Program | Sensor |
| | | gm | gm | gm | gm | gm | gm | gm | - | - |
| TRIP3 Garden City to City (Unladen) | 511 | 8510 | 109790 | 13590 | 440 | 60 | 16970 | 30 | 19.82 | 17.49 |
| TRIP4 City to Garden City (Laden) | 511 | 9860 | 118510 | 14260 | 550 | 60 | 17820 | 30 | 20.38 | 17.31 |
| TRIP4 City to Garden City (Laden) | 31 | 9680 | 123310 | 15160 | 560 | 60 | 18950 | 30 | 19.6 | 17.7 |
| TRIP15 North Quay to Cannon Hill Unladen) | 31 | 10620 | 131150 | 15930 | 640 | 70 | 19910 | 40 | 19.8 | 17.6 |

| Filename | Route | Total Emissions on a Distance Travelled Basis | | | | | | | Exhaust Temp. |
|---|---|---|---|---|---|---|---|---|---|
| | | $O_2$ | $N_2$ | $H_2O$ | $NO_x$ | CO | $CO_2$ | HC | 'Y' Junction |
| | | gm/km | gm/km | gm/km | gm/km | gm/km | gm/km | gm/km | °C |
| TRIP3 Garden City to City (Unladen) | 511 | 391 | 5050 | 625 | 20 | 3 | 781 | 1 | 514 |
| TRIP4 City to Garden City (Laden) | 511 | 450 | 5411 | 651 | 25 | 3 | 814 | 1 | 540 |
| TRIP4 City to Garden City (Laden) | 511 | 459 | 5850 | 719 | 27 | 3 | 899 | 1 | 534 |
| TRIP15 North Quay to Cannon Hill (Unladen) | 511 | 496 | 6120 | 743 | 30 | 3 | 929 | 2 | 544 |

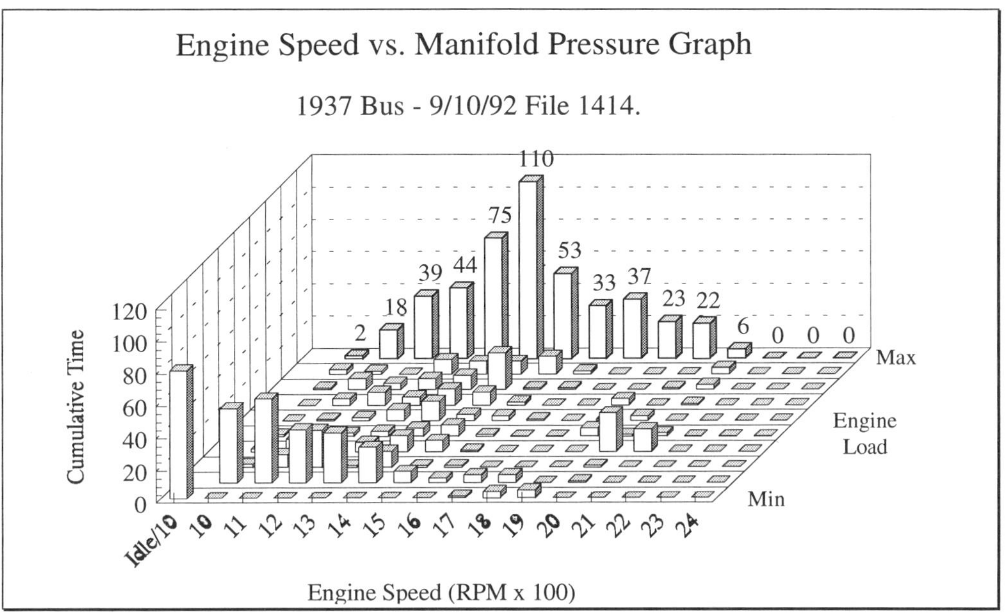


**Figure 5**. Harvard graphics graph of frequency output data (Adelaide data).

The second file is a table representing cumulative time spent at each of the points of a engine speed/load matrix, together with idle. This was designed to be imported into a graphic package and displayed pictorially, thus providing a picture showing the nature of the test.

Figure 5 is a graph produced by Harvard Graphics of the 'frequency file' output. Note that the idle value has been scaled down by a factor of 10 to maintain detail of the remainder of the figure.

This graph shows that the bus is idling for a very high proportion of the time followed by a high percentage of its time accelerating (full load) and decelerating. This would be typical of heavy traffic or city operation.

A comparison of such cumulative frequency files provide a valuable insight into

- traffic conditions
- driver behaviour
- state of tune of the engine
- suitability of engine selection

Such knowledge can assist with

- gear change strategy in the case of programmable automatic gearbox
- engine service and maintenance schedule
- driver education
- vehicle and engine selection

## SUMMARY

A vehicle and engine performance monitoring and analysis system has been developed because of the need to quantitatively assess the relative performance of vehicles operating on traditional and alternative fuels. The system can be utilised to optimise the performance of an engine and is capable of providing a realistic measure of a vehicle's fuel consumption and emission.

The system is ideally suited to provide the necessary data to fleet operators, engineers, research organisation and government bodies with relevant data for decision making.

## ACKNOWLEDGMENTS

Research supported by the Australian Research Council and the South Australian State Energy Research Advisory Committee.

## REFERENCES

1. S. De Maria. “Ngv program of the state transport authority of South Australia: the dynamometer evaluation of dual fuel and dedicated cng engines,” The Second Biennial IANGV International Conference and Exhibition, Buenos Aires, Argentina (1990), Vol.1, Paper 26.
2. S. De Maria. “Design and implementation of a real-time monitoring and logging system for compressed natural gas buses,” The Third Biennial IANGV International Conference and Exhibition on Natural Gas Vehicles, Goteborg, Sweden (1992), Part II, p.231
3. S. De Maria.. “Vehicle and engine performance monitoring and analysis system (vepmas),” European Conference On Natural Gas Vehicles, Amsterdam (1993), Paper 9
4. A. Klos and S. De Maria. “Vepmas for performance monitoring and analysis of internal combustion engines,” Third Australasian Instrumentation and Measurement Conference, Adelaide (1994), p.253.
5. C. J. Trowbridge and S. De Maria. “The emissions validation of a ‘vehicle and engine performance monitoring and analysis system '(vepmas),” The Fourth Biennial IANGV International Conference and Exhibition on Natural Gas Vehicles, Toronto (1994), Vol.2, p.593
6. S. De Maria. “Comparative evaluation of performance and emissions of a Scania OSC11 G01 compressed natural gas bus and a Scania DSC11-24 diesel fuelled bus based on dynamometer and on road tests,” The Fifth Biennial IANGV International Conference and Exhibition, Kuala Lumpur, Malaysia (1996), Vol.4, PS61.

# THE ELIMINATION OF OXIDES OF NITROGEN FROM THE EXHAUST OF A DIESEL ENGINE USING CRYOGENIC AIR SEPARATION

A. Manikowski[a], G. Noland[a] and M. A. Green[b]

a Procyon Power Systems Inc.
Alameda, CA 94501

b E. O. Lawrence Berkeley National Laboratory
University of California
Berkeley, CA 94720

## ABSTRACT

The high temperatures associated with combustion in diesel Engines produce oxides of nitrogen as a pollutant. Under the Clean Air Act, it is desirable to eliminate both the oxides of nitrogen and particulate matter from diesel engine exhaust gasses. This paper presents a method for eliminating oxides of nitrogen from the exhaust stream by eliminating the nitrogen from the incoming air entering the engine. The proposed cycle for a cryogenic separation system that separates the oxygen and heavy inert gasses from nitrogen in the air is discussed. Much of the power needed to operate the cryogenic separation plant would come from the heated exhaust gas stream from the engine. Since there is almost no nitrogen in the incoming oxidizer stream, the engine can be run much leaner to reduce carbon monoxide, hydrocarbons, and particulate emissions from the engine exhaust stream.

## INTRODUCTION

Recent changes in the Clean Air Act call for drastic reductions in ozone and particulate matter from the exhaust streams of internal combustion engines. The ozone in smog is formed by a reaction between unburned hydrocarbons and the oxides of nitrogen in the presence of sunlight. Gasoline and diesel engines produce both unburned hydrocarbons and oxides of nitrogen as by-products of combustion. The diesel engines used widely for trucks and other forms of transportation also produce particulate matter, primarily in the form of unburned carbon. Recent rulings from the Environmental Protection Agency (EPA) call for large reductions in the production of oxides of nitrogen ($NO_x$ nitric oxides) and particulate matter (soot), particularly from automotive diesel engines.

The usual way of reducing the production of the oxides of nitrogen is to reduce the temperature during combustion under a condition where the engine runs rich (fuel rich oxygen poor). Operating a diesel engine this way produces increased emission of carbon monoxide, unburned hydrocarbons and particulate matter in the form of soot. In addition, the engine runs less efficiently and requires catalytic converters on the exhaust stream. Running a diesel engine at the correct fuel to air ratio or slightly lean greatly increases nitrogen oxide production and thus ozone. This is a catch-22 because standard diesel engine air pollution control methods can no longer meet the new air quality standards and more fuel is used by the engine due to inefficient engine operation.

One way to get rid of the oxides of nitrogen is to get rid of the nitrogen in the incoming air stream. Air is 78 percent nitrogen, 21 percent oxygen, 1 percent argon and a number of other gasses (in order of abundance) that include: water, carbon dioxide, neon, helium, methane and other hydrocarbons, krypton, hydrogen, nitrous oxide ($N_2O$) and xenon. The nitrogen in the incoming air to a diesel engine could be replaced by recycled carbon dioxide and water from the diesel engine exhaust stream. The primary change in the incoming oxidizer stream to the engine would be a slight increase in the average molecular weight (from 29 to 31) and a reduction in the ratio of specific heats (from 1.4 to about 1.35). By removing the nitrogen from the air, the diesel engine can be operated at the stoichiometric point or slightly lean. Operation slightly lean greatly reduces the formation of carbon monoxide, unburned hydrocarbons and carbon soot. Little or no oxides of nitrogen are formed because there is very little nitrogen in the engine oxidizer stream that can be oxidized. Because combustion in the engine is hotter and the fuel is more completely burned, the thermodynamic efficiency of the engine should go up.

This paper presents a way of removing the nitrogen from the incoming air stream to a diesel engine by separating the nitrogen and light gasses from the oxygen and argon in the air. This separation is through a cryogenic air separator that is attached to and is part of the diesel engine cycle.

## THE CRYOGENIC AIR SEPARATION CYCLE

A number of cycles for the separation of nitrogen and oxygen from the air were investigated[1]. The goal was to minimize the amount of energy drawn from the shaft of the diesel engine. In order to do this, waste heat from the exhaust stream was employed and the energy extracted in the refrigerator expander was used to compress the separated nitrogen stream so that it could be used to generate power at the room temperature end of the cycle. Figure 1 below illustrates a promising cycle for extracting the nitrogen from the incoming air stream to the diesel engine. The cycle shown in Figure 1 gets the compressor power for the incoming air from a combination of the engine shaft and an expander for the nitrogen gas separated from the air. Exhaust gas from the engine is used to heat the nitrogen stream before it is expanded and ejected from the system. Table 1 presents the state point conditions for the cycle shown in Figure 1. The cycle shown in Figure 1 and Table 1 delivers a gross power of 75 kW (100 HP) and a net power of 69.8 kW (93 HP).

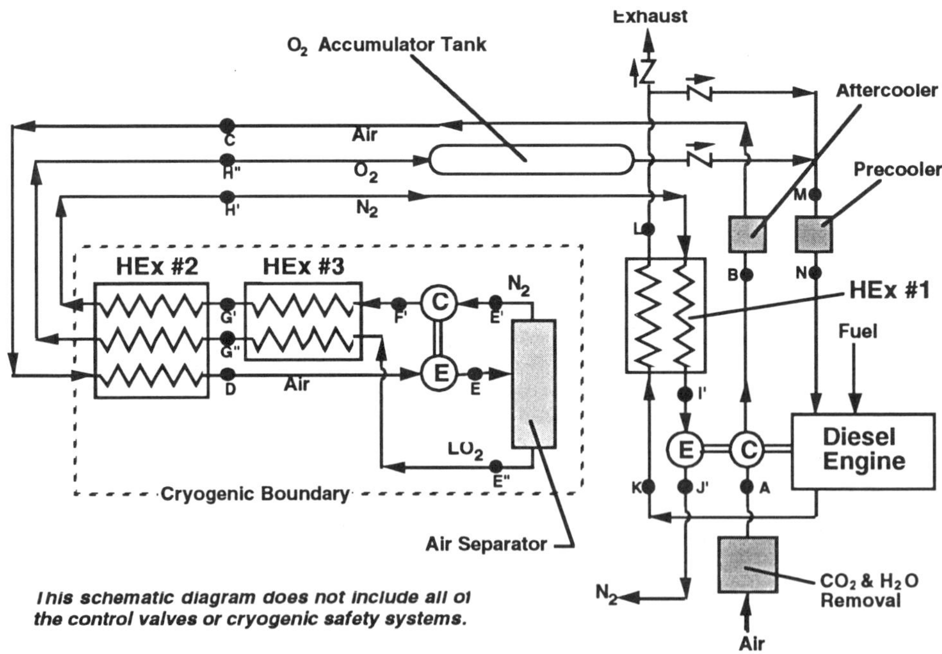


Figure 1 A Schematic Representation of a Cryogenic Air Separation System for Nitrogen Removal from Incoming Air to a Diesel Engine.

Table 1 The State Points for the Diesel Engine Air Separation Cycle shown in Figure 1 (The table includes Constituents, Mass Flow, P, T, H and S.)[2]

| State Point | mass flow ($kg\ s^{-1}$) | P (MPa) | T (K) | H ($J\ g^{-1}$) | S ($J\ g^{-1}\ K^{-1}$) |
|---|---|---|---|---|---|
| Air Stream | | | | | |
| A | 0.0630 | 0.101 | 300.0 | 462.4 | 4.536 |
| B | 0.0630 | 2.000 | 454.1 | 614.4 | 4.086 |
| C | 0.0630 | 2.000 | 330.0 | 486.5 | 3.754 |
| D | 0.0630 | 2.000 | ~120.0 | 246.1 | 2.554 |
| E | 0.0630 | 0.210 | >85 | 207.4 | 2.666 |
| $N_2$ Stream | | | | | |
| E' | 0.0484 | 0.200 | 83.4 | 231.1 | 2.840 |
| F' | 0.0484 | 0.800 | ~137 | 281.0 | ~2.890 |
| G' | 0.0484 | 0.800 | ~104 | ~222 | ~2.58 |
| H' | 0.0484 | 0.800 | ~310 | 470.1 | 3.821 |
| I' | 0.0484 | 0.800 | 880.0 | 1063.8 | 4.952 |
| J' | 0.0484 | 0.101 | 543.2 | 716.2 | 5.023 |
| $O_2$ Stream | | | | | |
| E" | 0.0146 | 0.200 | 97.0 | 71.6 | 0.945 |
| G" | 0.0146 | 0.200 | 97.0 | 277.5 | 3.090 |
| H" | 0.0146 | 0.200 | ~310 | 475.0 | 4.164 |
| Engine Exhaust Stream | | | | | |
| K | 0.0773 | 0.130 | 900.0 | 942.3 | NC |
| L | 0.0773 | 0.130 | 543.7 | 569.3 | NC |
| M | 0.0773 | 0.130 | ~504 | ~528 | NC |
| N | 0.0773 | 0.130 | ~330 | ~337 | NC |

| | |
|---|---|
| Shaft Power Generated by the Diesel Engine (kW) | 75.00 |
| Warm Air Compressor, Point A to B $\eta = 0.9$, Input Power (kW) | 22.02 |
| Cold Air Expander, Point, D to E, $\eta = 0.82$ Output Power (kW) | 2.44 |
| Cold Nitrogen Compressor, Point E' to F' $\eta = 0.85$, Input Power (kW) | 2.42 |
| Warm Nitrogen Expander, Point I' to J' $\eta = 0.9$, Output Power (kW) | 16.82 |
| Engine Shaft Power needed to Run the Air Plant (kW) | 5.20 |
| Net Diesel Engine Shaft Power to the Load (kW) | 69.80 |
| Fuel Burn Rate to Generate the Shaft Power ($kg\ s^{-1}$) | 0.0043 |

With the state point values given in Table 1, about seven percent of the shaft power is used to separate the nitrogen from the incoming air. If there were no recovery of energy within the cycle the power needed to separate the nitrogen from the incoming air would be about 30 percent of the shaft power of the engine. In order to reduce, the incoming shaft power needed to separate the nitrogen from the air one must recover the expansion engine power and use it to recompress the nitrogen. One must also use the heat from the exhaust gasses to help drive the input air compressor. The amount of shaft power saved depends on the efficiency of the compressors, the expanders and the heat exchangers. The state point values given in Table 2 are values that might be achievable if the right types of components are selected.

## Efficient Compressors and Expanders

The warm compressor is assumed to have an efficiency of 90 percent as compared to isentropic compression. The cold compressor is assumed to have an efficiency that is 85 percent of isentropic compression. Isothermal compression requires less power than isentropic compression, so the secret to achieving efficiency that is high compared to isentropic is to stage the compression with intercoolers to approach isothermal compression[3].

The single warm compressor shown in Figure 1 can be replaced with a staged compressor with intercoolers between stages and an after cooler after the final stage of compression. A three stage warm compressor is illustrated on the left hand side of Figure 2. Water cooled intercoolers are after stages one and two and a water cooled after cooler is after the third stage. This reduces the compressor input power needed to compress the gas and it reduces the temperature of the gas entering the cold box heat exchangers.

Inter cooling can also be applied to the cold compressor. A two stage cold compressor with intercooling and after cooling is illustrated on the right side of Figure 2. The liquid oxygen stream can be used to provide the intercooling and aftercooling for the compressed nitrogen stream.

Both the warm compressor and the cold compressor should be of the positive displacement type. Scroll compressors or a lubricated piston compressor should be considered. Some development is needed to achieve the level of compressor efficiency quoted in Table 2.

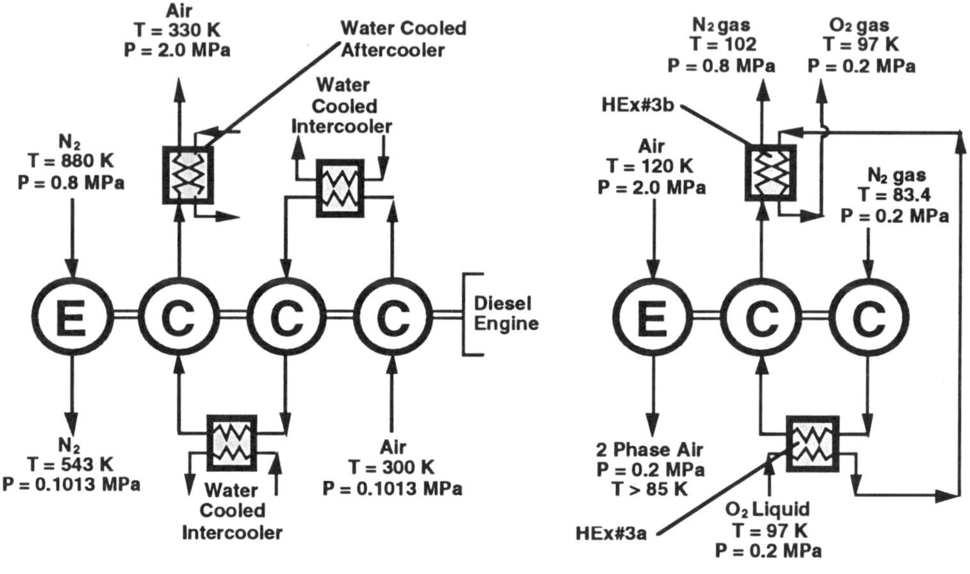


Figure 2 A Schematic Representation of Warm and Cold Multistage Compressors from the Cycle Shown in Figure 1 with Intercoolers and Aftercoolers.

The expander efficiency quoted may be achievable with positive displacement expanders now being used. The expander has to be compatible with the cold compressor that is being proposed. Both machines will share a common shaft and would operate at the same speed. A vane or scroll type of expander capable of operating within the two-phase region is needed. Considerable development work may be needed to achieve an efficient reliable expander that con be coupled to the cold compressor.

## Efficient Separation of Nitrogen from Argon and Oxygen

Efficient separation of the oxygen (plus argon and other heavy gasses) and nitrogen (plus the light gasses) reduces the amount of incoming air that needs to be processed. Efficient separation reduces the amount of nitrogen introduced into the engine and the amount of $NO_x$ that is produced by the engine. A simple cyclone separator is probably not adequate for efficient air separation. A fractionating column with a reflux condenser and a reboiler will do a nice job of separating the air gases so that the oxygen stream contains less than a few percent nitrogen. Conversely little of the oxygen is lost in the nitrogen stream.

A packed column separator with plenty of contact surface can be used for gas separation as long a there is an oxygen reboiler at the lower end of the column and there is a nitrogen reflux condenser at the upper end of the column[4]. A schematic for a packed column separator with reboil and reflux is shown in Figure 3.

Heat for the reboiler is supplied by using warm nitrogen from the cold compressor. The nitrogen can be reintroduced back into the nitrogen stream, between compression stages with some loss of compression efficiency. Only a small percentage of the nitrogen stream is used for reboiling the oxygen. Cold for the reflux condenser can be supplied by expanding liquid nitrogen at the G' end of heat exchanger HEx #3. This liquid nitrogen is expanded to about 0.12 MPa to provide the cooling to the reflux compressor. The nitrogen used to cool the reflux condenser can either be recompressed to the pressure at E' (with an oxygen cooled after cooler) or the cold from the reflux condenser cooling stream can be recovered provided an extra pass is added to heat exchanger HEx #3 and HEx #2. The flow in the reflux condenser cooling stream is expected to low.

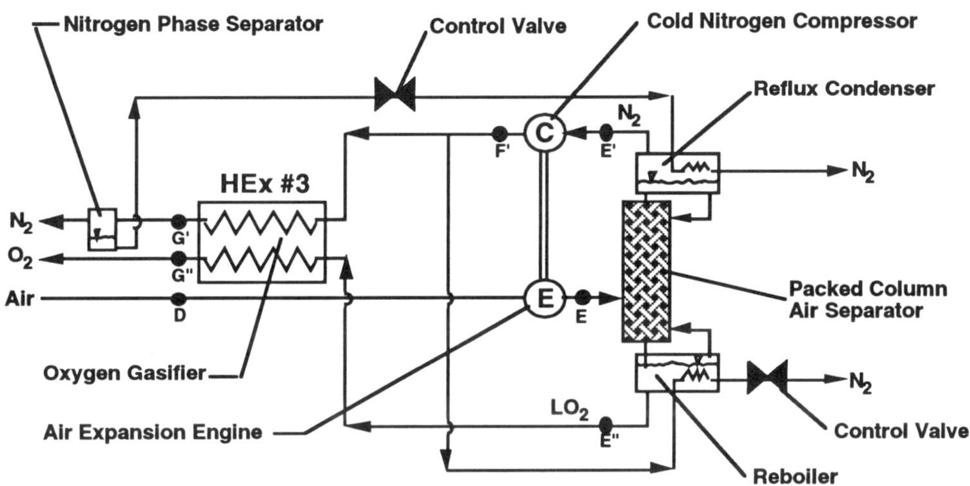


Figure 3 Reflux and Reboil to Improve Air Separation Efficiency.

## Heat Exchangers for the Air Separator

Ordinary tube and shell heat exchangers would be too large and too expensive for a compact air separation plant for a diesel engine. Compact heat laminar flow heat exchangers, that have boundary layer thicknesses from 0.5 to 1.5 mm appear to be attractive for this application. Table 2 shows the parameters for the laminar flow heat exchangers

Table 2 The Log Mean $\Delta T^5$, U Factor, and Area for the Heat Exchangers in Figure 1

| HEx # | Q (kW) | $\Delta T_{LM}$ (K) | $h_{c1}$ ($W\ m^{-2}\ K^{-1}$) | $h_{c2}$ ($W\ m^{-2}\ K^{-1}$) | U ($W\ m^{-2}\ K^{-1}$) | Area ($m^2$) |
|---|---|---|---|---|---|---|
| 1 | 2.88 | 87.4 | 97.1* | 20.4** | 16.8 | 19.6 |
| 2 | 15.25 | 20.1 | 33.4* | 33.4* | 16.7 | 45.4 |
| 3 | 3.01 | 12.9 | ~21.0* | ~360 | 19.8 | 11.8 |

* The boundary layer thickness in the laminar heat exchanger is about 0.5 mm.

** Due to fouling, the flow passage must be larger, the boundary layer thickness is 1.5 mm.

The laminar flow heat exchangers can be made from etched copper plates that have cooling passages that are 1 mm deep. The heat exchanger between the engine exhaust stream and the nitrogen gas returning from the cold box may have to be made from another material. The passages on the exhaust side of this heat exchanger would be about 3 mm deep. On the nitrogen side (not subject to fouling), the passage depth would be only about 1 mm. Depending on how the flow passages are shaped, extended heat transfer surfaces can be created. The laminar flow heat exchangers would have many passages in parallel so the pressure drop along the heat exchanger would be quite low. The estimated physical size for the three heat exchangers shown in Figure 1 are: HEx #1 would be 1.0 m long, 0.4 m wide, and 0.4 m thick. HEx #2 would be 1.0 m long, 0.5 mm, and 0.5 m thick. HEx #3 would be 0.2 m long, 0.5 m wide, and 0.5 m thick. Heat exchanger HEx #3 can be subdivided to provide the intercooling needed for the cold compressor. The whole cold box assembly including the expander, the cold compressor, and the air separation column could fit in a vacuum cylinder with dished heads that is 0.85 meters in diameter by 1.7 meters long.

### Heat Exchanger Fouling, Air Purification and Other Issues

Heat exchanger fouling is an issue on the exhaust side of HEx#1 where unburned carbon and hydrocarbons can be deposited in the heat exchanger passages. Fouling in this heat exchanger will be minimized if the particulate matter and unburned hydrocarbons are burned as fuel in an oxygen rich atmosphere.

Fouling may also be a problem on the air side HEx#2. The problem here is water vapor, carbon dioxide, hydrocarbons (ethane and above) krypton and xenon. Air at 305 K (32 C) can contain up to 29 grams of water for each kilogram of air[3]. The cycle shown in Table 1 processes about 227 kg of air per hour. This means that 6600 g of water, 113.5 g of carbon dioxide, 0.75 g of krypton, about 0.2 g of higher hydrocarbons and 0.09 g of xenon are also involved the process[6]. Krypton, higher hydrocarbons and xenon can be taken care of by molecular sieve purifiers that are in the middle of HEx #2. Methane will also be removed by these purifiers, but it does not have to be. Whether or not these purifiers have to be recharged depends on the number of hours the engine runs without being shut off.

The real problem is the water and carbon dioxide that enter the cycle with the air. Molecular sieve or silica gel desiccants can remove both water and carbon dioxide. How much carbon dioxide is removed depends on the temperature of the desiccant. Purifiers needed to remove the water can be at the entrance to the air compressor (see Figure 1), but to remove carbon dioxide a purifier may have to be somewhere in the upper end of heat exchanger HEx #2. Purifiers for removing water and carbon dioxide must be recharged periodically. The nitrogen coming from the expander at a temperature of 543 K (270 C) can be used to recharge both the water and carbon dioxide purifiers.

The cycle shown in Figure 1 and Table 1 shows a temperature after precooler on the mixed gas stream entering the engine (point N) of 350 K. It is desirable to have the gas entering the engine at as low a temperature as possible, but lowering this temperature below 345 K will result in the condensation of water in this stream. The mass flow of the oxidizer stream into the engine is reduced 14 percent compared to an air stream at 300 K. One can compensate for the reduced mass flow by increasing the percentage of oxygen in the oxidizer stream to about 25 percent by volume. The oxidizer stream (25 percent oxygen) that enters the engine will have a ratio of specific heats of 1.354 and a molecular weight of 30.4. Normal air (21 percent oxygen) has a ratio of specific heats of 1.4 and a molecular weight of 29. How one controls the amount of oxygen in the oxidizer stream may be an issue.

## CONCLUDING COMMENTS

It is desirable to remove oxides of nitrogen, particulate matter (carbon), carbon monoxide and unburned hydrogen from the exhaust stream of internal combustion engines. One can greatly reduce particulate matter, carbon monoxide and unburned hydrocarbons by burning fuel in the engine with a slight amount of excess oxygen. This greatly increases the amount of $NO_x$ produced by the engine.

Removal of nitrogen oxides from the exhaust of a diesel engine can be accomplished by separating the nitrogen and oxygen from the incoming air stream to the engine. The nitrogen is replaced by a mixture of carbon dioxide and water vapor. The engine can now run lean to get rid of particulate matter, CO and unburned hydrocarbons. The use of a cryogenic separation plant to separate nitrogen from the air appears to be feasible technically.

In order to reduce the shaft power needed to separate the incoming air one needs to recover waste heat from the exhaust and one needs to recompress the nitrogen separated at the low temperature lend of the cycle. The power for this recompression would be supplied by a cold expander used to produce liquid air at the low temperature end of the cycle. The amount of shaft power from needed to separate nitrogen depends strongly on the efficiency of the compressors, expanders and the heat exchangers. Multiple stage compression with intercooling and aftercooling is clearly desirable. The colder the gas is between stages, the more efficient the compression will be become compared to an isentropic compressor.

A cycle that has 100 percent efficient compressors and expanders will require no shaft power from the engine to separate the incoming air. (All of the power needed comes from waste heat in the exhaust stream.) For the cycle presented here with less than one hundred percent efficient expanders and compressors, the shaft power used to separate the incoming air represents about seven percent of the shaft power developed by the engine. The increase in efficiency due to higher combustion temperature may, in part, make up for the losses due to air separation inefficiencies.


## ACKNOWLEDGMENT

Over half of this work was done with funds from the private sector. The remainder of the work was performed with the support of the Director, Office of Energy Research, Office of High Energy and Nuclear Physics, High Energy Physics Division, U. S. Department of Energy under Contract No. DE-AC03-76SF00098.


## REFERENCES


1. R. B. Scott, *Cryogenic Engineering*, D. Van Nostrand Company Incorporated, New York, (1969)
2. V. J. Johnson editor, " Properties of Materials at Low Temperature (Phase 1), a Compendium," Part 1 Properties of Fluids, Pergamon Press, New York (1961)
3. J. B. Jones, and G. A. Hawkins, *Engineering Thermodynamics*, John Wiley and Sons Incorporated, New York (1960)
4. R. E. Traybal, *Mass-Transfer Operations*, McGraw-Hill Book Company Incorporated, New York, (1955)
5. F. Kreith, *Principles of Heat Transfer*, International Textbook Company, Scranton PA, (1961)
6. *Handbook of Chemistry and Physics*, 58th Edition, p F210, CRC Press, (1977)

# CRYOGENIC HEAT ENGINE EXPERIMENT

M. C. Plummer, C. P. Koehler, D. R. Flanders, R. F. Reidy, and C. A. Ordonez

University of North Texas
Denton, Texas 76203

## ABSTRACT

Experimental results on the operation of a "cryogenic heat engine" are presented. A cryogenic heat engine employs a cryogenic medium as a heat sink and the atmosphere as a heat source. Cold thermal storage by refrigeration to produce the cryogenic medium is thus equivalent to energy storage. Using liquid nitrogen as the cryogenic medium, a small cryogenic heat engine which utilizes a simple gas expansion process has been evaluated to experimentally provide 19 kJ of mechanical energy per kilogram of nitrogen exhausted. Theoretical modeling indicates that larger specific energy values are readily possible using more advanced cryogenic heat engine processes. Cryogenic heat engines are potentially suitable as power systems for zero emission vehicles. In order to prove that this is possible, an automobile has been converted for operation using liquid nitrogen as its fuel. In addition, a study has been conducted to assess the feasibility of using a cryogenic heat engine as a zero emission vehicle power system.

## INTRODUCTION

A "cryogenic heat engine" is a concept for using a cryogenic substance to produce useful energy.[1,2] One realization of the concept is the operation of an engine using liquid nitrogen as a combustionless fuel. Operation of an engine using liquid nitrogen is similar to operation of an open-cycle steam-driven engine. For an open-cycle steam-driven engine, fuel is burned to vaporize water under pressure to produce compressed gas (steam). When operating a liquid-nitrogen-driven engine, atmospheric heat is used to vaporize liquid nitrogen under pressure to produce compressed (nitrogen) gas. In both cases, the compressed gas is used to run a pneumatic motor or turbine with the exhausted gas released into the atmosphere. A thermodynamic analysis of using liquid nitrogen to run an engine in this way has recently been published.[1] The system considered was very simple, consisting of only three primary components: a pressurized

**Table 1.** Specific energy values for liquid nitrogen and battery systems

| System | Specific Energy (kJ/kg) |
|---|---|
| 1 MPa (145 psi) $LN_2$ | 200 |
| 3 MPa (435 psi) $LN_2$ | 300 |
| Ideal (Carnot) $LN_2$ | 760 |
| Lead/Acid Batteries | 130 |
| Mid-Term Battery Goal | 290 |
| Long-Term Battery Goal | 720 |

tank (which stores liquid nitrogen), a heat exchanger (which vaporizes and then heats the nitrogen under pressure to near atmospheric temperature using atmospheric heat), and a pneumatic motor. For such a system, possible specific energy values for two different system pressures are shown Table 1.[1] Also shown is the specific energy for an ideal Carnot process. (The Carnot value is obtained by using $T_{cf} = T_h = 298$ K in Eq. (12) of Ref. [2].) These values for liquid nitrogen systems can be compared to the specific energy values for electrochemical batteries. Along with the specific energy for commercially-available lead/acid batteries, the mid-term and long-term goals established by the U.S. Advanced Battery Consortium are shown in Table 1.[3] It should be mentioned that in these comparisons, the specific energies for liquid nitrogen and battery systems have slightly different definitions. The specific energy for a liquid nitrogen energy system is defined as the energy produced per unit mass of liquid nitrogen used. For batteries, the specific energy is defined as the total energy which can be produced by a battery on a single charge divided by its mass.

A possible use of a cryogenic heat engine is as a power system for zero emission vehicles. This is a rather appealing prospect since widespread use of liquid nitrogen energy systems provides a long-term potential to *reverse* environmental impacts associated with burning fossil fuels provided non-polluting (e.g. solar) energy sources are used to produce the liquid nitrogen. Undesirable airborne particulates are necessarily removed during the production of liquid nitrogen. In addition, carbon dioxide is separated in the liquefaction process and could be used in greenhouse agriculture. The more that liquid nitrogen powered automobiles are driven, the cleaner the air would become.

According to the U.S. Environmental Protection Agency (EPA), more than half the U.S. population lives in areas where pollution levels exceed established air quality standards. Beginning in the year 2003, existing legislation calls for 2% of new automobiles to produce zero emissions in California. Thus, strong incentives exist to develop zero emission vehicles for commercialization. With batteries presently the primary candidate for powering zero emission vehicles, the work presented here indicates that cryogenic heat engines should also be considered for powering zero emission vehicles. As with battery power systems, a disadvantage of using liquid nitrogen as a fuel to power a vehicle is that the vehicle will have a shorter driving range than existing vehicles because the specific energy for liquid nitrogen is small compared to that for fossil fuels. However, it may be possible to reduce this limitation by using liquid nitrogen in conjunction with liquid natural gas. After liquid nitrogen is expanded using atmospheric heat, it would be further expanded using the heat from low-emission natural gas combustion before being passed through a pneumatic motor. This combined approach would make the vehicle more practical by increasing the maximum range available for long-distance trips. For short-distance trips, the vehicle could be powered solely using liquid nitrogen.

**Table 2.** Experimental data

| Speed (rpm) | Torque (N·m) | Gauge Pressure (kPa) | Flow Rate (g/s) | Specific Energy (kJ/kg) |
|---|---|---|---|---|
| 1544 | 0.82 | 296 | 8.19 | 16 |
| 1678 | 0.79 | 283 | 8.19 | 17 |
| 1600 | 1.14 | 345 | 9.28 | 21 |
| 1671 | 1.24 | 407 | 10.4 | 21 |
| 1774 | 1.27 | 448 | 11.5 | 21 |
| 1516 | 1.33 | 482 | 11.5 | 18 |
| 1630 | 1.43 | 427 | 10.9 | 22 |
| 1442 | 1.08 | 379 | 9.28 | 18 |
| 1721 | 0.73 | 310 | 8.74 | 15 |

## EXPERIMENT

A bench-mounted, liquid-nitrogen-fueled cryogenic heat engine has been assembled and tested. The experiment incorporated a pressure-regulated liquid nitrogen storage tank which provided nitrogen gas to two external heat exchangers connected in series. The compressed gas from the heat exchangers was sent through a pneumatic motor which was used to drive a dynamometer (a pony brake arrangement assembled in-house with rotation speed measured using a phototachometer and torque measured using a spring scale arrangement). Flow rate, temperature, and pressure of the inlet gas to the air motor were measured using conventional laboratory equipment. The specific energy produced is related to raw experimental data by the relation,

$$w = \frac{2\,\pi\, f\, \tau}{\dot{m}} \tag{1}$$

where: $f$ is the rotation speed in rps (revolutions per second); $\tau$ the torque in N·m; and $\dot{m}$ is the mass flow rate in kg/s, obtained from application of the ideal gas law. The inlet temperature to the air motor was close to ambient temperature during the measurements (307 K). (It should be mentioned that, at the rates at which the nitrogen gas was withdrawn from the liquid nitrogen storage tank, the gas leaving the tank was also at near ambient temperature. Hence, the external heat exchangers were not actually necessary.) The power produced by the air motor during the measurements was between 131 W and 243 W. Table 2 shows the experimental data along with the corresponding specific energy values. The average of the specific energy values is 19 kJ/kg while the standard deviation is 2.5 kJ/kg. The average value for the specific energy is 2.5% of the Carnot value.

## LIQUID NITROGEN POWERED VEHICLE

In order to show that it is possible to power an automobile using a cryogenic heat engine, a 1973 Volkswagen has been converted for operation using liquid nitrogen as its fuel. Figure 1 shows a photograph of the completed conversion. A Taylor-Wharton XL-45 180 L tank which can hold 124 kg of liquid nitrogen was mounted in the vehicle along with two Thermax D8.3 vaporizers connected in parallel. The vaporizers provide

**Fig. 1.** The liquid nitrogen fueled automobile after a continuous 8 km (5 mi) drive at 9 m/s (20 mi/hr).

compressed nitrogen gas to a 9 hp Gast 16AM-FRV-13 vane-type pneumatic motor which was mounted in place of the gasoline motor. The outer shell of the car was removed and a pipe structure was welded into place in order to support the liquid nitrogen tank and the vaporizers. Pressure in the tank and vaporizers is maintained at 1.2 MPa by the tank's internal vaporizer circuit. Before the pneumatic motor, a pressure regulator reduces the pressure to 0.7 MPa. Between the pressure regulator and the pneumatic motor, a throttle consisting of a butterfly valve was mounted and connected to the driver's accelerator peddle. The mass of the vehicle with a full tank is 700 kg. It should be emphasized that the vehicle in Figure 1 is intended to prove that the concept is feasible and not that it is technically feasible to build a vehicle with practical operating characteristics. The pneumatic motor is small in comparison to typical gasoline motors. Nevertheless, it provides sufficient torque to accelerate the vehicle at 1.1 $m/s^2$ at low speed (at less than 4.5 m/s). This is due to the pneumatic motor's characteristic high torque at low rotation speed. The maximum speed of the converted automobile was found to be 11 m/s. The maximum speed is limited by constricted gas flow through two stainless steel transfer lines which connect the tank to the vaporizers. The driving range without refueling the automobile was found to be approximately 24 km while traveling at about 9 m/s. For one test, the vehicle was driven for 8 km at an average speed of 9 m/s without stopping. During the drive, ice formed on the tank as shown in Figure 1. However, only a small amount of ice formed on the external heat exchangers with most of the condensation blown off by the passing air. The temperature of the nitrogen gas provided by the heat exchangers never dropped below 273 K during the drive. In general, the driving characteristics were found to be very similar to those prior to conversion. Other than the limited speed, the most noticeable difference is that the motor does not idle while the vehicle is stopped.

## FEASIBILITY

Assume that the coefficient of performance ($COP$) associated with extracting heat, $Q_c$, from a quantity of nitrogen gas to convert it to liquid nitrogen at atmospheric pressure is a fraction, $\eta_{ref}$, of the Carnot coefficient of performance ($COP_{carnot}$). Thus,

$$COP = \frac{Q_c}{W_{in}} = \eta_{ref} \frac{Q_c}{W_{carnot}} = \eta_{ref}\, COP_{carnot} \tag{2}$$

Here, $W_{in}$ is the energy required by the cryogenic refrigerator for extracting heat, $Q_c$, while $W_{carnot}$ would be the minimum possible (Carnot) value for $W_{in}$. Rearranging provides,

$$W_{carnot} = \eta_{ref}\, W_{in} \tag{3}$$

For a cryogenic heat engine which operates using the atmosphere as a heat source and a cold substance as a heat sink, the amount of heat which can be absorbed by the cold substance is limited, but the amount of heat available from the atmosphere is not. Consequently, the inverse coefficient of performance ($ICOP$) is a more appropriate performance parameter for characterizing a cryogenic heat engine than the usual heat engine efficiency.[2] Suppose that the inverse coefficient of performance associated with producing an amount of energy, $W_{out}$, using a quantity of liquid nitrogen is a fraction, $\eta_{che}$, of the Carnot inverse coefficient of performance. Then,

$$ICOP = \frac{W_{out}}{Q_c} = \eta_{che} \frac{W_{carnot}}{Q_c} = \eta_{che}\, ICOP_{carnot} \tag{4}$$

Thus,

$$W_{out} = \eta_{che}\, W_{carnot} = \eta_{che}\, \eta_{ref}\, W_{in} \tag{5}$$

It is convenient to define a total efficiency as the energy produced by the cryogenic heat engine per unit energy required by the cryogenic refrigerator: $W_{out}/W_{in} = \eta_{ref}\, \eta_{che}$.

Existing refrigerators (e.g., the Philips model PPG-2500 cryogenic refrigerator operated at a pressure of 4 MPa) can be used to produce liquid nitrogen with a coefficient of performance of 40% of the Carnot value.[4] Modern liquid nitrogen production plants can exceed this even when they include the energy cost associated with the air separation process.[5] An alternative to using liquid nitrogen is to use liquefied air. Modern plants are capable of producing liquefied air at 50% of the Carnot value.[5] However, due to preferential evaporation of nitrogen during venting of typical storage tanks, liquefied air becomes enriched with oxygen. In order to avoid this problem, the tank's pressure relief valve can be located within the liquid as illustrated in Figure 2 such that liquid is released through the valve rather than as vapor. Upon passing through the valve, the liquid is subjected to a lower (atmospheric) pressure causing it to vaporize and absorb heat while passing through the remaining liquid. The gas, which is then vented, has the same composition as the liquid. An equilibrium is reached when the rate at which gas is vented is such that the heat absorbed by the gas from the remaining liquid equals the heat which enters the inner vessel and is absorbed by the remaining liquid. Without the problem of oxygen enrichment during the venting of storage tanks containing liquefied air, use of liquefied air for powering zero emission vehicles is much more attractive than the use of liquid nitrogen since the possibility of asphyxiation would be avoided.

For the present analysis, production of the fuel (liquefied air or liquid nitrogen) at 50% of the Carnot value has been assumed. With a cryogenic heat engine which operates at 25% of the Carnot inverse coefficient of performance (i.e. at a specific energy of 190 kJ/kg), the total efficiency is 12.5%. Using renewable energy at a cost

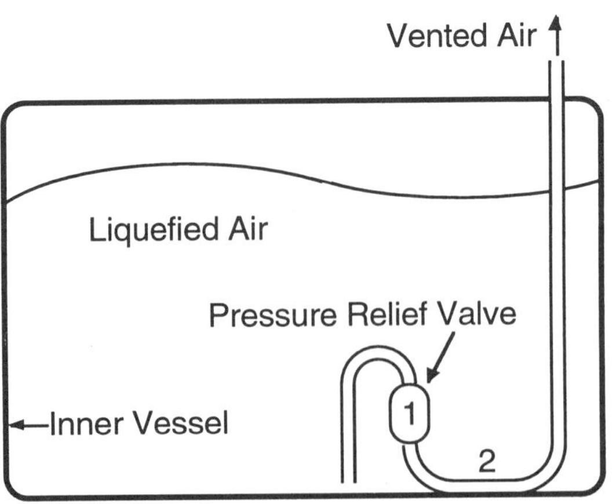


**Fig. 2.** An illustration of the pressure relief valve location (1) within the inner vessel of a tank used to store liquefied air. At location 2, a heat exchange with the inner vessel occurs such that the vented air leaves with a temperature essentially the same as that of the liquefied air contained in the tank. (The relative sizes shown are not to scale.)

of 1.7 cents/MJ (solar-thermal energy systems are expected to achieve this[6] ) to run the cryogenic refrigerator, the cost of energy delivered by the cryogenic heat engine is 14 cents/MJ. Alternatively, using excess off-peak energy from nuclear power plants at a cost of 0.42 cents/MJ to run the cryogenic refrigerator, the cost of energy delivered by the cryogenic heat engine is 3.4 cents/MJ. It is difficult to compare these values with the cost of energy from battery systems. The U.S. Advanced Battery Consortium long-term goal of achieving \$28/MJ is not the overall energy cost, but is essentially the cost of a battery divided by the maximum amount of energy produced on a single charge. In order to assess the conceptual feasibility of using a cryogenic heat engine to power a zero emission vehicle, the cost per kilometer and vehicle range are determined for a "typical" automobile (a 1993 Saturn SL2). First, the average speed and power for the automobile are needed. These are determined using the 1995 EPA Urban Driving Schedule. The driving schedule is a two-column list of speed versus time values (in equal increments) used for emissions testing. The average speed (simply the arithmetic mean of all the speeds) is found to be 8.7 m/s. The power is then calculated from the vehicle's mass, speed, acceleration, and frictional losses.

The total power needed to drive an automobile is the sum of the power required to overcome friction and the power required to increase speed: $P(a,v) = m\,v\,[a_f(v) + a]\,\Theta(P)$. Here $a_f$ is the speed-dependent force per unit mass needed to overcome friction and $a$ is the acceleration of the vehicle. The Heaviside step function, $\Theta(P)$, which is zero for negative values of $P$ and one for positive values, is added to require the power to be positive or zero thus avoiding the possibility of negative power values. Consequently, regenerative braking is not considered. The normalized force required to overcome deceleration due to friction, $a_f$, was determined empirically using the Saturn SL2. The mass of the vehicle including two passengers was approximately 1375 kg. Measurements of the time required to slow 2.25 m/s while in neutral gear were taken at speeds ranging from 2.25 to 24.5 m/s. The measurements were conducted on a relatively flat stretch of road in both directions. An average of the deceleration values obtained while driving in each direction was taken. A second order polynomial fit to the data, which is adequate due to the nature of the two main drag forces (tire to road resistance,

**Table 3.** Energy cost values for cryogenic heat engine and internal combustion engine vehicle power systems

| System | Cost of Energy |
|---|---|
| 12.5% Efficient $LN_2$/Solar | 6.3 cents/km (10 cents/mi) |
| 12.5% Efficient $LN_2$/Nuclear | 1.5 cents/km (2.4 cents/mi) |
| 1993 Saturn SL2* | 3.1 cents/km (5.0 cents/mi) |

*actual fuel cost.

which dominates at low speed, and wind resistance, which dominates at high speed) is in m/s$^2$: $a_f(v) = 0.122 + 0.00393v + 0.000191v^2$. This relation agrees to within 5% with the measured data. The average power is: $(1/N)\sum_{i=1}^{N} P(a_i, v_i)$ where the speeds, $v_i$, are taken directly from the driving schedule, and the accelerations, $a_i$, are determined by subtracting adjoining speeds and dividing by time elapsed. The average power is calculated to be 4.0 kW.

In the conceptual feasibility analysis presented here, it is assumed that a "typical" cryogenic-heat-engine powered vehicle will have frictional drag characteristics similar to that of the 1993 Saturn SL2 such that its average power consumption for city driving will be 4.0 kW. The effect of having to carry along a larger mass of fuel is neglected although this can be compensated some with an optimized design including the use of advanced materials. Multiplying the initial mass of fuel in the vehicle by both the fuel's specific energy and the ratio of the average vehicle speed to power provides an estimate of the vehicle driving range. With the fuel mass taken to be 500 kg (this mass is typical for a battery pack in comparable electric vehicles), and using 190 kJ/kg as the specific energy for liquid nitrogen or liquefied air (this assumes the vehicle's cryogenic heat engine operates at 25% of the Carnot inverse coefficient of performance), the vehicle range is 210 km. A comparison can be made with the Saturn EV1, the lead/acid-powered electric vehicle commercialized by General Motors Corporation, which has a typical driving range of 121 km. Assuming a total efficiency of 12.5% for the cryogenic-heat-engine powered vehicle, the cost of energy to produce the fuel used to propel the vehicle is 6.3 cents/km if energy provided by a solar energy plant at 1.7 cents/MJ is used. For excess off-peak energy provided by a nuclear energy plant at 0.42 cents/MJ, the cost of energy to produce the fuel used to propel the vehicle is 1.5 cents/km. In comparison, the 1993 Saturn SL2 gets 24 miles per gallon of gasoline consumed under EPA Urban Driving Schedule conditions. Considering gasoline to cost $1.20 per gallon, the cost of fuel to propel the gasoline-powered vehicle is 3.1 cents/km. For easy referral, these values are summarized in Table 3

## CONCLUDING REMARKS

In previous work,[1,2] the scientific basis of the cryogenic heat engine concept has been developed. In the work reported here, the conceptual feasibility of using a cryogenic heat engine to power a zero emission vehicle has been assessed. A bench-top experiment has been carried out and an automobile has been converted for operation using a proof-of-principle cryogenic heat engine power system. A preliminary assessment of the driving range and fuel cost for a "typical" automobile powered by a cryogenic heat engine has also been conducted. In future work, the technical feasibility of using cryogenic heat engines to power zero emission vehicles will be evaluated. The primary question to be addressed is whether a vehicle power system operating at 25% of Carnot performance can be achieved.

## ACKNOWLEDGMENT

The authors would like to thank P. R. Foster, M. E. Parker and D. M. Peterson for helpful discussions and technical assistance. This material is based in part upon work supported by the Texas Advanced Technology Program under Grant No. 3594-003.

## REFERENCES

1. C.A. Ordonez, Cryogenic heat engine, *Am. J. Phys.* 64:479 (1996).
2. C. A. Ordonez and M. C. Plummer, Cold thermal storage and cryogenic heat engines for energy storage applications, *Energy Sources* 19:389 (1997).
3. M. J. Riezenman, The great battery barrier, *IEEE Spectrum* 29(11):97 (1992).
4. G. Walker, "Cryocoolers," Plenum Press, New York (1983).
5. B. Mostello, private communication.
6. Solar Two, California, USA, *Power* 140(4):103 (1996).

# CRYOGENIC MARINE LOADING VAPOR RECOVERY SYSTEM

Fred H. Babet

Babet Engineering, Inc.
Pasadena, Texas, 77501, U.S.A.

## ABSTRACT

Liquid nitrogen is used to condense vapors evolved from liquid hydrocarbons or chemicals during marine loading operations. The paper describes the major components of the system and points out the controls and other criteria which must be installed to comply with 33CFR154, Subpart E. Advantages and disadvantages of the system are discussed.

## INTRODUCTION

The EPA and State air quality control boards have mandated the collection and destruction or recovery of vapors generated during the loading of some hydrocarbons and chemicals into marine vessels.

The potential for a disaster -- in excess of 500 Kilograms of flammable vapor -- is ever present during the loading of hydrocarbon liquids. Most transportation vessels have a very low pressure rating -- normally 100 mm Hg pressure and 50 mm Hg vacuum. Because human operations are always required for hook-up and because loading rates are relatively high, the element of error is also present. Most vapor control systems utilize an enclosed or open flare, adding yet another source of disaster -- ignition!! This system and any other "recovery" system does not utilize a combustion devise and is therefore inherently safer than a destruction system.

Recognizing the dangers involved in collecting hydrocarbon vapors, the United States Coast Guard formulated regulations covering the manner in which vapor control facilities must be installed and published them in the June 21, 1990 Federal Register under 33 CFR Part 154, subpart "E" - Vapor Control Systems.

The Coast Guard also realized that they did not have sufficient technically competent staff to oversee the design and installation of the control systems. They, therefore, invited qualified engineering companies to apply for the position of "Certifying Entity". There are now 23 engineering companies that are approved by the U.S.C.G. to design or certify Marine Loading Vapor Control Systems. (They cannot do both on the same system.)

*Advances in Cryogenic Engineering,* Vol. 43
Edited by P. Kittel, Plenum Press, New York, 1998

In general, a vapor control system design and installation must eliminate potential overfill hazards, over pressure and vacuum hazards, and sources of ignition to the maximum practical extent.

The vapor control system selected will probably be one of the following:

1. Incineration - enclosed flare (thermal oxidizer, incinerator, or combuster)
   - open flare stack

2. Absorption - lean oil for hydrocarbons
   - water for acetone or phenol

3. Adsorption - activated carbon beds
4. Cryogenics - liquid nitrogen

5. Mechanical Refrigeration

6. Vapor Balancing

This paper will discuss the design of a cryogenic vapor recovery system.

Referring to Figure 1, we can trace the flow of collected vapors as they pass through the piping system and the many safety features which are required by 33CFR154, subpart "E".

1. The vessel must be equipped with high level alarm instrumentation and high-high level shutdown instrumentation which are tied into the dock-side control system through a "Hubbell" 5-pin connector. The vessel high level instrument will activate the dock-side audio-visual alarm and the high-high level instrument will close the vapor valve, and normally, the liquid cargo loading valve. Any blowers which may be operating will be shut down.

2. The vapor collection hose or arm must be equipped with an insulating flange. It may be installed on the dock-side end of the hose, providing that it is impossible for the hose to ground out on the dock. Otherwise, an insulating flange must be installed on the vessel-side of the hose to ensure that the final "make or break" connection cannot produce a spark.

3. A pressure indicator must be installed as close to the vapor connection as possible, to allow the dockman to monitor the pressure in the marine vessel.

4. A high pressure alarm and shutdown must be installed in a redundant manner.

5. A strainer is installed to prevent plugging of the detonation arrester which contains very small openings.

6. A manual block valve is required.

7. A U.S. Coast Guard approved detonation arrester must be installed within 6 meters of the facility vapor connection. In most instances, temperature sensors must be installed on each side of the arrester to shut down the system if a flame is detected. A pressure differential sensor is recommended, but not required, to detect fouling of the tiny passages in the detonation arrester. The differential pressure device may also be used to detect reverse flow.

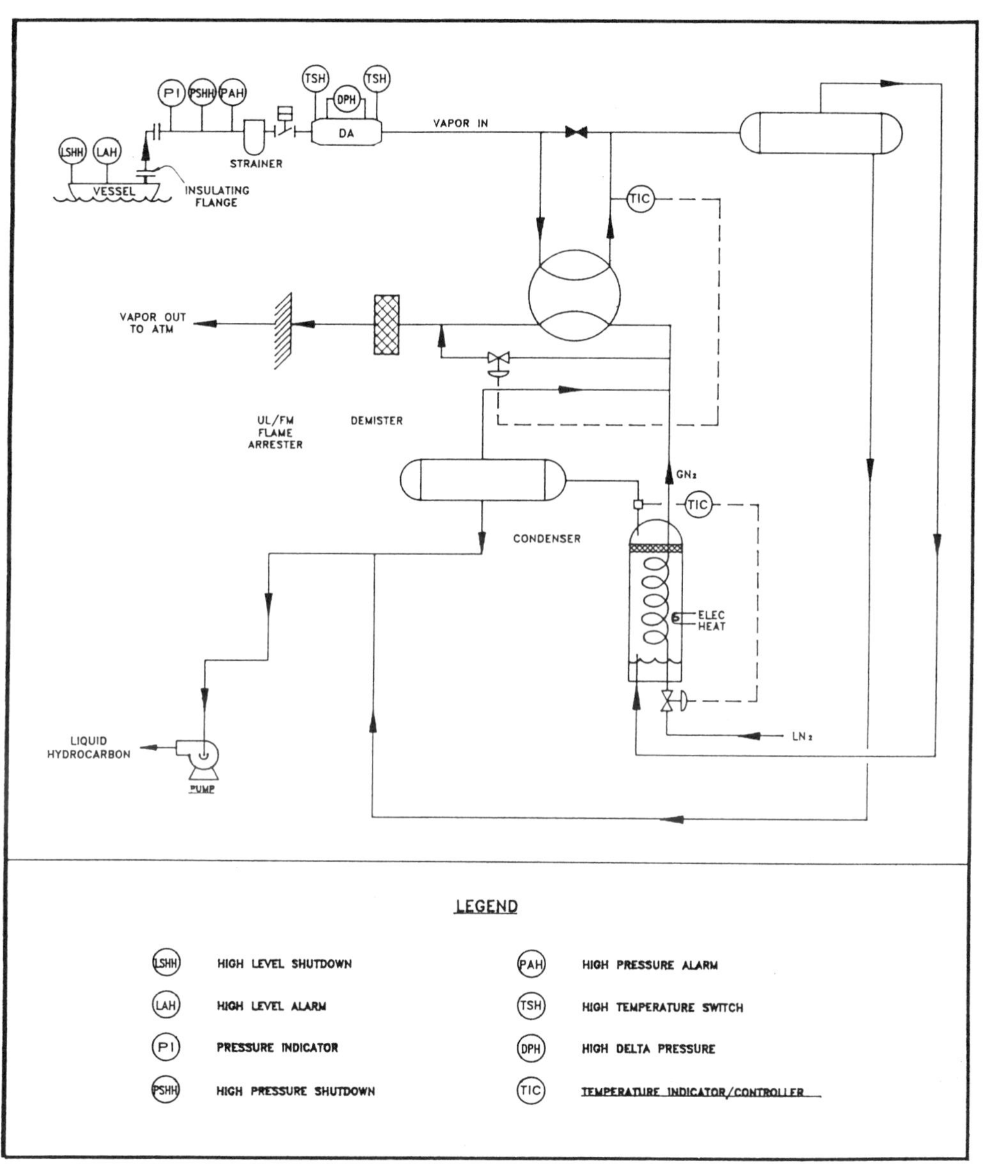


FIGURE 1 CRYOGENIC MARINE VAPOR RECOVERY SYSTEM.

8. The warm gas being displaced from the vessel is exchanged with the cold gas being vented to atmosphere.

9. Condensed water is knocked out of the gas stream in a separator at 274K.

10. The chilled gas enters the condenser where it is further cooled to 230K by liquid nitrogen. This condenses in excess of 98 percent of the hydrocarbon vapors which are discharged to

11. A collection tank which separates the liquid hydrocarbon from the water.

12. The cold, dry gas passes through the gas-gas exchanger, through the demister, and the flame arrester, and out to atmosphere. If the cold gas is not warmed up before passing through the demister, ice will form on it and plug it.

This system can be installed for about $350,000, compared to over $600,000 for a mechanical refrigeration system. Although the operating costs for the cryogenic unit are relatively high, this can be justified when the system is only operated about once per month and the barges loaded are 10,000 barrel or less, or if a use for the vaporized Nitrogen can be found. If the barges to be loaded are larger than 10,000 barrel, a two column condensing system would have to be utilized, because the condenser would freeze up due to the moisture in the vapors coming from the barge. (Most products are pumped off the barge with the vapor space open to atmosphere, allowing water saturated air to enter the barge.) The additional cost to install this modification would be about $100,000.

Significant developments have been made recently in the utilization of adsorption, absorption, and cryogenics in the design of marine loading vapor recovery systems. The EPA is currently studying ways to decrease the generation of Carbon Dioxide. One obvious method would be to require the recovery, rather than the destruction, of marine loading vapors. It is therefore advisable, if one is contemplating the installation of a marine loading vapor control system for single product loading, to consider the use of recovery, rather than destruction.

# ADSORPTION STORAGE OF GASEOUS HYDROGEN AT CRYOGENIC TEMPERATURES

R. Chahine and P. Bénard

Institut de recherche sur l'hydrogène
Université du Québec à Trois-Rivières
Trois-Rivières, Québec, Canada
G9A 5H7

## ABSTRACT

The storage of hydrogen on board vehicles is one of the most critical issues for the transition towards an hydrogen-based transportation system. The physical adsorption of hydrogen on activated carbon can reduce the pressure required to store compressed gases. Though an efficient adsorption-based storage system for vehicular use of natural gas can be achieved at room temperature, the application of this technology to hydrogen using activated carbon as the adsorbent requires its operation at cryogenic temperature. In this paper we present a comparative study of the efficiency of adsorption and compressed gas storage of hydrogen as a function of temperature and pressure. The isothermal hydrogen storage and net storage densities for passive systems at 77 K, 150 K and 293 K are compared and discussed.

## INTRODUCTION

Various alternative transportation fuels are being explored to increase the efficiency of existing internal combustion engines and reduce the impact of harmful emissions on the environment. Electrical vehicles using hydrogen fuel cells constitute, in the mid to long term, the only means to completely eliminate the harmful emissions of the transportation sector while satisfying market concerns on vehicle range and speed of refueling[1]. In the shorter term, the transition from a hydrocarbon-based to an hydrogen-based transportation system will most likely be done by internal combustion engines fueled by natural gas. The predominance of gasoline as the automotive fuel of choice is largely based on its storage

*Advances in Cryogenic Engineering*, Vol. 43
Edited by P. Kittel, Plenum Press, New York, 1998

efficiency. Its liquid state allows convenient storage in thin and light tanks which yield superior range, as well as fast and simple refueling of the vehicles[2]. Both hydrogen and natural gas require improvements in energy storage in order to achieve acceptable performances and to compete with established hydrocarbon-based fuels. The storage problem is more severe for hydrogen than natural gas[3], which can achieve higher energy storage performances. An electric vehicle powered by a hydrogen fuel cell will require 3.1 kg of hydrogen to achieve a range of 500 km. This amount, when stored in a typical gasoline tank, would correspond to a hydrogen density[4] of 65 $kg/m^3$ (including the storage system) and 6.5 % wt. Presently, only liquid hydrogen ($LH_2$) systems (with a density[5] of 51 $kg/m^3$ and 14% wt) is close to this target. However, the cost of using $LH_2$ as a transportation fuel is nearly twice that of a gaseous hydrogen ($GH_2$), due to the liquefaction process, increased fuel transportation costs and more complex manipulation of the fuel. The use of $LH_2$ is therefore restricted to certain niche applications. The other two standard storage systems, namely compressed hydrogen gas (3.1 % and 14 $kg/m^3$ at 250 bars and 293 K)[5] and metal hydrides (1.5% and 28 $kg/m^3$)[5] are far from the above stated goal for hydrogen, and any substantial improvement can only be done at the cost of more severe operating conditions and/or higher penalties on range and fuel consumption[6].

## STORAGE CAPACITY OF HYDROGEN ON ACTIVATED CARBON

The recent advent of activated single-walled carbon nanotubes (SWNT) has put adsorption in the forefront of research in hydrogen storage technology[7]. Physisorption on activated carbon offers a promising avenue for lowering the storage pressure of compressed gas fuels such as natural gas and hydrogen. For example, the storage density of adsorbed natural gas can be 3/4 of compressed natural gas at 1/6 the gas pressure, in addition to the advantage of allowing the use of on-board conformable tanks, and this at room temperature. However for hydrogen, the temperature must be lowered in order to attain this level of gain if currently available activated carbons are used. Experiments show that at 77 K and 17 bars, a density of 40% that of $LH_2$ at 20 K can be achieved, which corresponds to five times the density of gaseous hydrogen under the same conditions. The key to achieving this level of performance is the use of a densification process which can produce activated carbons with a high surface area, coupled with a high bulk density[8]. If these conditions are not met, the performance of adsorption can be lower[9,10] by as much as 28 to 40%.

The performance of hydrogen adsorption ($AH_2$) storage systems must be carefully analyzed in order to define specific operating guidelines. The gross storage capacity of such systems undergoes a dramatic loss of efficiency with increasing temperature. Furthermore in the case of passive $AH_2$ systems, the net storage densities can be lower than that of $CH_2$ due to the higher amount of residual gas stored in the tank below the discharge pressure, which is the operating pressure of the energy conversion system. Performance estimates of the storage efficiency of adsorption requires the characterization of the adsorption isotherms over a wide range of pressures and temperatures above the critical point of the adsorbate. One of the difficulties in achieving reliable performance predictions has been to develop a model for the adsorption isotherms under such conditions for microporous adsorbents. Though the Langmuir isotherms and the virial expansion approaches can be used at low pressures and on a limited temperature scale, the maximum of the isotherms must be described by an expression designed to handle excess adsorption isotherms at high pressures. The Ono-Kondo lattice gas approach, recently extended to supercritical gas adsorption can describe methane adsorption on activated carbon over a wide range of temperature and pressure[11,12,13]. In this approach, the adsorption process is directly mapped onto the

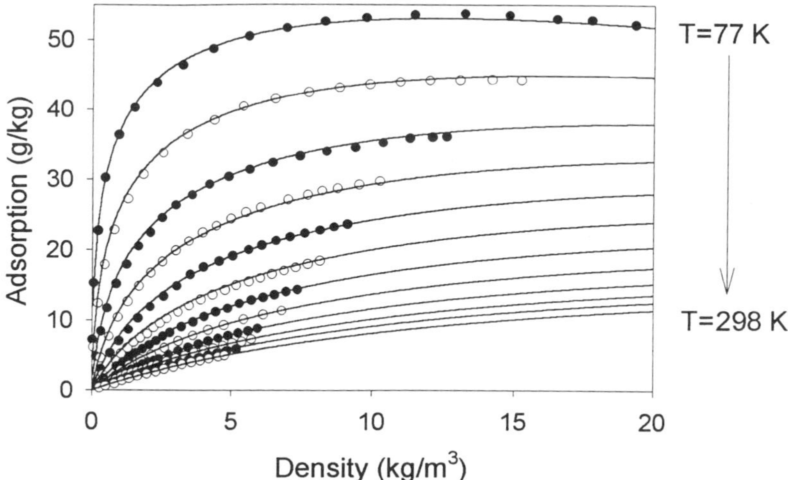


**Figure 1.** Ono-Kondo fit of the adsorption isotherms (lines) to the experimental data (points). The extreme values of the temperature are 77 K and 298 K respectively. The other isotherms are separated by 20 K intervals, from 93 K to 273 K. The arrow points towards increasing temperature.

hexagonal graphite planes, taking into account the Lennard-Jones interaction between adsorbed molecules on the graphite layers[13]. The molecular fraction per adsorption site on the i$^{th}$ layer $x_i=\rho_i/\rho_s$ (where $\rho_s$ is the molar density of a completely filled adsorption layer) can be calculated from

$$\ln\left(\frac{x_i(1-x_b)}{x_b(1-x_i)}\right)+\frac{z_0 E}{kT}(x_i-x_b)+\frac{z_2 E}{kT}(x_{i+1}-2x_i+x_{i-1})=0 \quad (1)$$

subject to the boundary conditions:

$$x_1=x_N \text{ and } \ln\left(\frac{x_1(1-x_b)}{x_b(1-x_1)}\right)+\frac{E}{kT}(z_1 x_1+z_2 x_2-z_0 x_b)+\frac{E_A}{kT}=0, \quad (2)$$

The excess adsorption is given by

$$N=C\sum_{i=1}^{N}(x_i-x_b), \quad (3)$$

where $x_b$ is the bulk density of the adsorbate $\rho_b$ divided by $\rho_s$, and M is the maximum number of layers in a typical microporous slit of the activated carbon. The prefactor $C$ takes into account the density of the active pores of the adsorbent and other structural properties of the adsorbent.

The Ono-Kondo equations are parameterized by the in-plane lattice coordination number z, the adsorbate-adsorbate interaction E and the adsorbate-adsorbent interactions $E_A$. The parameters used in the fit are z=6, E=43,89 K and $E_A$=-531.3 K. The adsorbent in the hydrogen adsorption experiments data used to fit the theory was the activated carbon AX-21™ with a specific area of 2778 m$^2$/g. The prefactor C of equation (3) was found to be temperature-dependent. The temperature dependence was fitted by a polynomial to cover the range 77 K-298 K. Figure 1 shows an excellent fit of the data over the whole pressure and temperature range. The hydrogen storage density for adsorption storage systems can calculated using[14]

$$d_s = Nd_b + \left(1 - \frac{d_b}{d_c}\right)d_g, \tag{4}$$

where $d_g$ is the bulk gas density at the storage conditions of P and T, and where $d_b$ and $d_c$ are respectively the bulk and the skeleton density of the adsorbent ($d_c$=2.3 g/cm$^3$ for graphite). Even if its surface area is the highest, AX-21™ has a low bulk density (0.3 g/cm$^3$). This leads to a relatively low storage density because its surface area per unit volume is only 840 m$^2$/cc. This value can be increased to 1400 m$^2$/cm$^3$ if the adsorbent is optimized by densification. Even if densification increases the bulk density to 0.7 g/cc, it lowers the surface area per unit mass from 2778 m$^2$/g to 2000 m$^2$/g. The excess adsorption of hydrogen depends linearly on the BET surface area[14]. At 77 K, comparing the experimental $H_2$ adsorption isotherm on densified AX-21™ to the $H_2$ isotherm on AX-21™ powder shows that the later is larger by a constant factor of 1.47 over the whole pressure range. This factor is within 6% of the ratio of the BET surface areas. The reduction in adsorption can therefore be accounted for by the change in the BET surface area. Because the temperature characterization of the adsorption isotherms of densified AX-21™ is not available for the range 77 K-298 K, we use equation (4) with the $d_b$=0.7 g/ cm$^3$ and the fit to the Ono-Kondo equations to simulate its storage properties. The fit was corrected to account for the lower surface area of densified AX-21™. The resulting storage density is, at 77 K, in good agreement with results of reference 8.

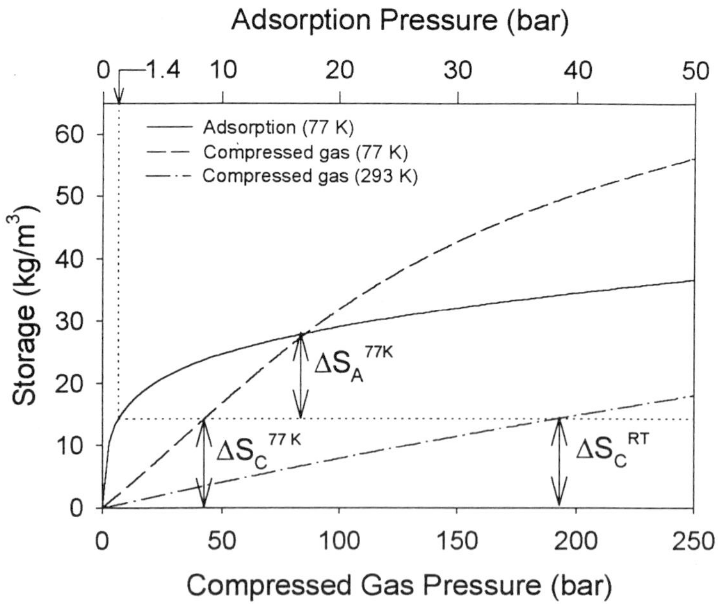


**Figure 2.** Intrinsic storage density of hydrogen adsorption on activated carbon at 77 K compared to $GH_2$ at 77 K and 293 K. The pressure scales for adsorption and compressed gas are at the top and the bottom of the figure respectively. The added volumes of the reservoir shells are not included.

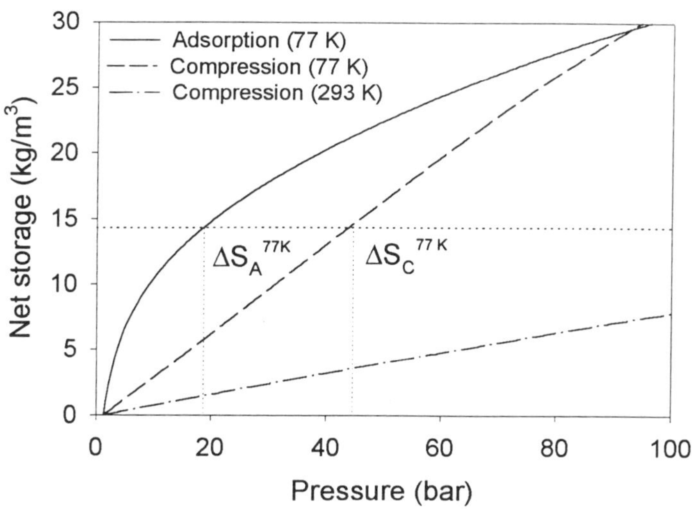


**Figure 3.** Net storage capacity of adsorption at 77 K compared to compressed gas storage at 77 K and room temperature. An operating pressure of 20 PSI (1.36 bar) is assumed. The added volumes of the reservoir shells are not included.

## RESULTS

The isothermal storage densities were calculated at three temperatures of technological interest (77 K, 150 K and 293 K). Since we compare adsorption and compresed gas storage densities for different temperatures, the calculations are limited to intrinsic densities, which do not take into account the weight and volume of the storage system. The gross storage density of hydrogen on activated coal can be calculated using expression (3) . The largest gain of adsorption over compression is achieved at low temperatures and pressures. At high pressures, compressed gas storage eventually becomes more efficient than adsorption. This occurs after the adsorbed density reaches its maximum value. For higher pressures, the volume occupied by the adsorbent in the reservoir reduces the available volume for bulk gas storage. Figure 2 shows the storage density of adsorption on activated carbons at 77 K compared to compressed gas at the same temperature and at room temperature (293 K).

For a passive storage system, the pressure of the reservoir must remain above the operational pressure of the energy conversion system, which is called the discharge pressure. The discharge pressure in this simulation was set to 1.36 bars (20 PSI), which would be a reasonable value for a hydrogen fuel cell. The net storage density ΔS delivered is equal to the difference between the density at a given pressure and the storage density at the discharge pressure, which is called the residual density. To discharge the tank beyond the residual density, hydrogen must either be pumped out or heat must be provided to the storage bed. As can be seen from figure 2, the residual density for adsorption at 1.36 bars is 14.3 kg/m$^3$, or half the $AH_2$ storage density at 18.8 bars. This is much larger than compression at 77 K (0.44 kg/m$^3$) and 293 K (0.11 kg/m$^3$). The pressures corresponding to a net storage density of $\Delta S = 14.3$ kg/m$^3$ are shown in figure 2 for $AH_2$ at 77 K, $CH_2$ at 77 K and $CH_2$ at 293 K. The values obtained were 18.8 bars for adsorption ($\Delta S_A^{77K}$), 43.8 bars for compression at 77 K ($\Delta S_C^{77K}$) and 190 bars for compression at 293 K ($\Delta S_C^{RT}$). A pressure gain of 4.3 (190 /43.8) can be achieved by reducing the temperature of compressed gas from 293 K to 77 K. This gain is multiplied by a factor of 2.3 ( 43.8/18.8)

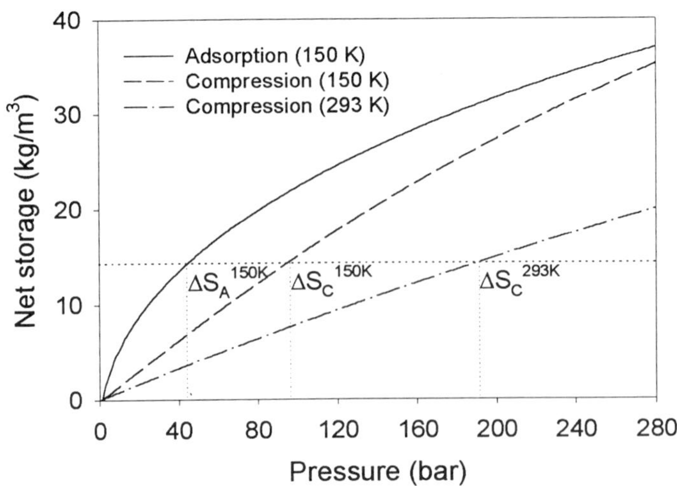


**Figure 4.** Net storage capacity of adsorption at 150 K compared to compression at 150 K and 293 K. The drop lines show the pressures for net storage densities of 20 $kg/m^3$. The added volumes of the reservoir shells are not included.

for a total gain of 9.9 for $AH_2$ at 77 K over compression at room temperature. To get the equivalent net storage capacity of $CH_2$ at 250 bars at room temperature (18 k/m3), the pressure in an $AH_2$ storage reservoir at 77 K must be 31 bars.

The pressure dependence of the net storage density is shown in figure 3. At 77 K, the large residual density of adsorption limits the gain over compression for passive systems to pressures below 100 bars. This range can be extended if a pumping device is used or heating is added. Adsorption at 77 K provides a substantial gain over compressed gas storage at room temperature over all pressures relevant to practical applications.

The net storage density ΔS for $AH_2$ at 150 K and $CH_2$ at 150 K and 293 K is shown in figure 4. A net storage density of 14.3 $kg/m^3$ is achieved at 44.5 bars for adsorption and

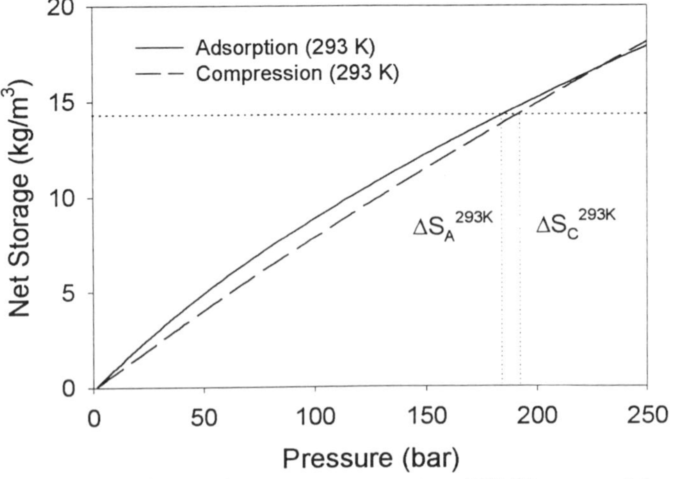


**Figure 5.** Net storage capacity of adsorption at room temperature (293 K) compared to compression. The added volumes of the reservoir shells are not included.

95 bars for compressed gas at 150 K. The pressure gain of adsorption over same temperature compression at 150 K is similar to the one observed at 77 K because of the much smaller residual density of $AH_2$ (1.37 kg/m$^3$). The residual density for $CH_2$ is 0 21 kg/m$^3$ and 0.12 kg/m$^3$ at 150 K and 293 K respectively. The consequences of the dramatic reduction of the residual density at higher temperatures for $AH_2$ can be seen by comparing its storage densities at 18.8 bars for T=77 K and T=150 K. The increase in temperature lowers the net storage density by 43% (from 14.3 kg/m$^3$ at 77 K to 8.2 kg/m$^3$ at 150 K), while the gross storage density is decreased by 66%, from 28.6 kg/m$^3$ to 9.6 kg/m$^3$.

Figure 5 shows the net storage of adsorption and compression at room temperature (293 K) for a passive system. The residual density at 1.36 bar was 0.1124 kg/m$^3$ for compression and 0.16 kg/m$^3$ for adsorption.

The pressure required to achieve a net storage of 14.3 kg/m$^3$ at 293 K is 185 bars for adsorption and 190 bars for compression, for a small pressure gain of 1.03 of adsorption over compression. Compression becomes more efficient than adsorption at room temperature for pressures beyond 225 bars.

## CONCLUSION

We have presented a model to calculate the net storage capacity of a passive adsorption storage system for hydrogen over a wide temperature and pressure range. For a passive energy conversion system, the large residual density in adsorption storage units at low temperatures must be taken into account in an efficiency comparison with compressed gas. Despite a large residual density, passive adsorption systems using densified activated carbon show a clear pressure gain over compression at 77 K and 150 K for a storage density of 14.3 kg/m$^3$. However, this gain vanishes at room temperature.

The net storage densities of $AH_2$ mentioned above could be doubled at 77 K for an active system by pumping or heating the reservoir. This would increase the pressure gain of adsorption over compressed gas storage even further. The substraction of the residual storage density in the calculation of the net storage density for a passive system modifies the pressure gain of adsorption over compression in a complex way. The residual density decreases exponentially as the temperature is increased for adsorption. This leads to a slow decrease in net storage efficiency for a $AH_2$ passive system as the temperature is increased from 77 K to 150 K. At 18.8 bars, the dramatic reduction of the residual density of $AH_2$ at 150 K leads to a decrease of 43% of the storage density of $AH_2$ at 77 K, despite the fact that the gross storage density decreases by a factor of almost 3.

An $AH_2$ system using activated carbon can be made more efficient than compressed gas at cryogenic temperatures. The key to the successful application of adsorption storage is the densification of activated carbon, without which the performance of the system may be comparable to compressed gas. The use of new materials, such as SWCNT, may lead to more substantial gains at higher temperatures.

## ACKNOWLEDGEMENTS

We acknowledge the support of the Ministère des ressources naturelles du Québec.

## REFERENCES


1. Conference report, *Basic Research needs for vehicles of the Future*, New Orleans, January 5-7 1997.
2. J. S. Cannon, *Harnessing Hydrogen, The Key to Sustainable Transportation*, Inform, 1995.
3. J. Corless and J. A. Barclay, in *Advances in Cryogenic Engineering vol 41b*, P. Kittel ed., p1033 (1995).
4. De Luchi, *Hydrogen Fuel Cell Vehicles,* Institute of Transportation Studies, University of California at Davis, 1992.
5. The storage densities cited in the text are for BMW (LH2), Daimler-Benz (NECAR II, compressed hydrogen gas) and Mazda-Myata (metal hydrides).
6. M. A. Dougherty in *Advances in Cryogenic Engineering vol 41b*, P. Kittel Ed, p1049 (1995).
7. A.C. Dillon et al, *Nature* 386:377 (1997).
8. R. Chahine and T. K. Bose, *International Journal of Hydrogen Energy* 19:161 (1994).
9. J. A. Schwartz, *Final Report for the Tasks XC-1-1108-1 and XAE-3-13346-01*, National Renewable Energy Laboratory, Golden, Colorado (1994).
10. S. Hynek et al, *International Journal of Hydrogen Energy* ***22***, 6:601 (1997).
11. S. Ono and S. Kondo, *Molecular Theory of Surface Tension in Liquids*, Springer Verlag, Berlin (1960).
12. G. L. Aranovitch and M. D. Donohue, *Carbon* 33:1369 (1995).
13. P. Bénard and R. Chahine, *Langmuir* 13:808 (1997).
14. R. Chahine and T. K. Bose, *in Hydrogen Energy Progress XI*, T.N. Veziroglu *et al* editors, p1259 (1996).

# A FUEL CELL POWER SUPPLY FOR LONG DURATION BALLOON FLIGHTS USING STORED CRYOGENS

M. A. Green[a], A. Manikowski[b], G. Noland[b], and R. L. Golden[c]

a E. O. Lawrence Berkeley National Laboratory
University of California
Berkeley, CA 94720

b Procyon Power Systems Inc.
Alameda, CA 94501

c New Mexico State University
R. L. Golden Particle Astrophysics Lab
Las Cruces, NM 88003

## ABSTRACT

Large balloon launched cosmic ray experiments can require up to 1.8 kW of power for the duration of the mission. Present day battery packs, which have a mass of 550 kg, limit the mission time for such experiments to less than 60 hours. Long duration polar balloon missions require a power supply that can deliver power at the rate of 1.5 to 2 kW for a period of time from 10 to 21 days. A hydrogen-oxygen fuel cell is an attractive option for a power supply because 20.6 kg (291 liters) of hydrogen and 165 kg (144 liters) of oxygen can provide the 1.42 kW of power for a 18 day mission at a fuel cell efficiency of 80 percent. If the water produced by the oxidation of the hydrogen in the fuel cell can be dumped during the mission, the required ballast needed for the mission can be reduced by almost 190 kg. The waste heat from the fuel cell can be used to preheat the fuel and oxidizer before they enter the fuel cell. The remainder of the waste heat must be transferred away from the balloon by radiation. This report describes a fuel cell power supply configuration.

## INTRODUCTION

Much of today's cosmic ray physics is being done on balloon platforms operating at a distance of 30 to 40 km (100000 to 130000 feet) above the earth's surface. At this altitude all but 0.3 percent of the earth's atmosphere is below the experiment. Modern balloon experimental payloads can have a mass of more than 3000 kg (6600 lb.). Included in a typical particle astrophysics payload mass would be the gondola, possibly a superconducting magnet, particle detectors such as drift chambers and calorimeters, the electronics needed to detect and record the particle tracks, the payload control electronics, the telemetry to connect the experiment to the ground, and some ballast (perhaps 200 to 300 kg) for altitude control.

As with ground based particle physics detectors, physics detectors carried in a balloon gondola require electric power to run the detectors and the electronics associated with these detectors[1]. In addition, electric power is needed for the telemetry used to communicate with the ground and temperature control within the gondola. In a typical balloon experiment, the power is supplied by lithium chloride batteries, which can store about 140 Whr (0.5 MJ) of electrical energy per kilogram of battery mass. (By comparison, lead acid batteries, which are rechargeable, can store only about 0.2 MJ per kilogram of battery mass.) Even short missions that require 1.8 kW of for only a few days require battery packs that have a mass of 450 to 500 kg. The battery pack can not be counted as part of the balloon ballast, because batteries are considered to be toxic waste. Batteries have to be taken up and brought down with the balloon gondola, so that they can be disposed of in a proper manner.

Long duration balloon missions are power limited. Several types of alternative power sources have been proposed over the last several years. These include: 1) photo voltaic solar cells that convert solar energy to electrical energy (8 to 12 percent efficient); 2) solar heat collectors with a heat engine that converts collected solar heat energy to electrical energy (20 to 30 percent efficient); 3) fuel cells that convert chemical energy to electrical energy at efficiencies from 65 to 90 percent; and 4) other battery options (i.e., nickel cadmium cells).

Power from solar cells has been used on some long duration balloon missions, but these missions have had power requirements that are less than a few hundred watts. The maximum electric power generated by solar cells is about 100 W $m^{-2}$ for cells pointed directly at the sun. For solar cells attached to the side of the balloon gondola (these cells do not track the sun) the average power generated will always be less than 20 W $m^{-2}$. The amount of solar collector needed becomes even more of a problem for low latitude missions that experience a day and night cycle. (At low latitudes, day and night are more nearly the same length.) In short, solar cells appear to be attractive for some polar missions where only a few hundred watts of power are needed for the balloon payload.

## A FUEL CELL POWER SUPPLY

A hydrogen-oxygen fuel cell can convert chemical energy to electrical energy at an efficiency of greater than 65 percent. Fuel cells are advantageous for the following reasons: 1) The available energy per unit mass is potentially much higher for a fuel cell than for the best batteries. For example, if the fuel cell power supply is 80 percent efficient, the available electric energy density is 3500 Whr (12.6 MJ) per kilogram of fuel and oxidizer. Even if the tankage and fuel cell weigh as much as the fuel and oxidizer, the power source generates twelve times more electrical energy per unit mass than lithium chloride batteries. 2) The water produced by the fuel cell can become part of the ballast for the balloon, which means less ballast has to be launched. The savings in the ballast can be applied to extra fuel, oxidizer and tankage. 3) Fuel cells generate their electrical energy at voltages that are compatible with existing balloon payload avionics. Fuel cell voltages in the vary with the output power. For example, a 2 kW fuel cell can produce 32 V at no load and 24 V when the cell is fully loaded. This means that a fuel cell power supply can replace the 24 V power supply modules now being used by New Mexico State and other users. 4) A fuel cell power supply can be located at any convenient location on the balloon gondola. This means that the power supply module can be separated from the experiment. The keys to the successful use of a fuel cell for powering a balloon payload are reducing the mass of the fuel cell, the cryogenic tankage, and the heat rejection system.

Table 1 presents some of the basic cryogenic parameters for hydrogen[2] and oxygen[3]. The parameters include the triple point temperature and pressure, the normal boiling temperature, the critical temperature and pressure, heat of vaporization at 1 atm, the average specific heat at constant pressure above 200 K and the enthalpy change from the saturated liquid condition at 1 atm to a gas at 1 atm at 323 K (50 C the operating temperature of the fuel cell). Liquid hydrogen is predominately in the para state; hydrogen gas at room temperature is about 75 percent ortho hydrogen. As a result, the heat of conversion from the para to ortho state is included in Table 1. (There is no equivalent change of state for oxygen.) The change in enthalpy from the saturated liquid hydrogen at 20.3 K to gas at 323 K reflects the para to ortho conversion energy. A catalyst is needed for the para to ortho conversion to occur in the flow stream between the liquid hydrogen tank and the fuel cell. Most of the para to ortho conversion will occur between 100 and 225 K. Since the total enthalpy change for hydrogen between 20.3 K and 323 K is ten times the heat of vaporization, hydrogen cooled shields and cold mass support intercepts would be very useful for reducing the heat leak into the hydrogen storage dewar. (The entire hydrogen flow from the tank to the fuel cell can be used to cool the tank shield and intercept.) The use of shield gas in the oxygen tank is hardly worth doing because the heat of vaporization is a large portion of the total enthalpy change.

**Table 1** Basic Parameters of Cryogenic Hydrogen and Oxygen

| Parameter | Hydrogen | Oxygen |
|---|---|---|
| Liquid Density at 1 atm (kg m$^{-3}$) | 70.6 | 1140 |
| Triple Point Temperature (K) | 13.95 | 54.4 |
| Triple Point Pressure (kPa) | 7.04 | 0.15 |
| Boiling Temperature at 1 atm (K) | 20.3 | 90.2 |
| Critical Temperature (K) | 33.0 | 154.7 |
| Critical Pressure (MPa) | 1.29 | 5.08 |
| Heat of Vaporization at 1 atm (J g$^{-1}$) | 441.5 | 222.5 |
| Heat of Conversion Para to Ortho at Boiling Point (J g$^{-1}$) | 527.1 | --NA-- |
| Average Specific Heat from 200 to 323 K (J g$^{-1}$ K$^{-1}$) | 14.5 | 0.92 |
| Enthalpy Change from Liquid to 323 K Gas (J g$^{-1}$)* | 4580.8* | 436.7 |

* includes the para to ortho conversion enthalpy

## The Fuel Cell Power Supply Cycle

A schematic representation of the fuel cell power supply is shown in Figure 1. Table 2 shows the constituents at each state point (Note: H is hydrogen, O is oxygen and W is water.), the mass flow of that constituent, the constituent temperature, constituent pressure and constituent enthalpy for the numbered state points in Figure 1, the fuel cell power supply gas circuit. The state point diagram, Figure 1 and Table 2 shows the fuel cell power supply as it operates while developing a gross electric power output of 1.5 kW. (The net power to the experimental load is over 1.4 kW.) The fuel cell power supply shown in Figure 1 would nominally deliver 2 kW at its full capacity. The Heat rejection panel would be sized to reject 0.7 kW of heat energy by radiation heat transfer (assuming a worst case fuel cell efficiency of 74 percent at the full 2 kW capacity). The volume of the stored cryogen needed depends on the experimental power needed and the length of the mission.

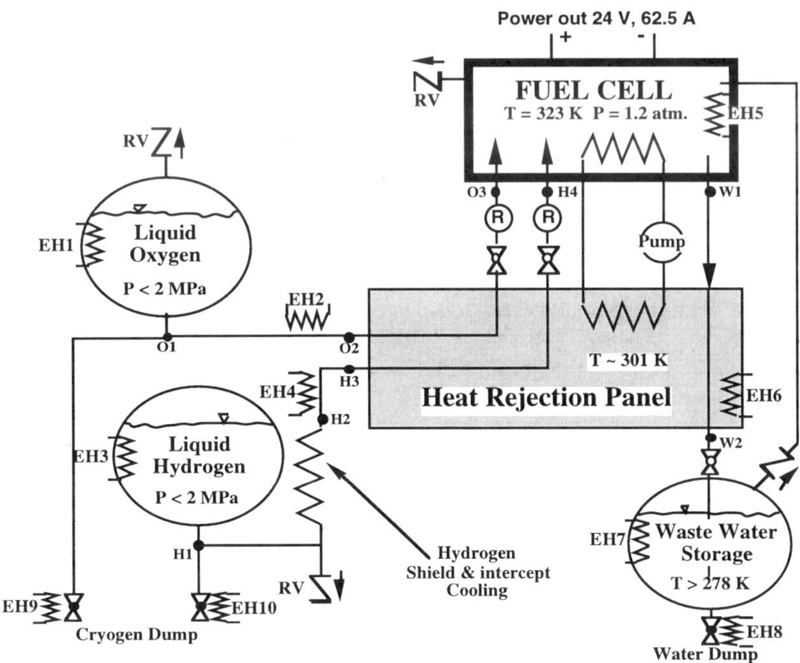


**Figure 1** A Schematic Representation of a Fuel Cell Power Supply Delivering 1.5 kW (Note: Electric heaters are EH, relief valves are RV, and pressure regulators are R.)

**Table 2** State Point Mass Flow, Temperature, Pressure, and Enthalpy for the Hydrogen Oxygen Fuel Cell Power Supply Cycle Shown in Figure 1

| State Point | Fluid Type | Mass Flow ($g\ s^{-1}$) | T (K) | P (MPa) | Enthalpy* ($J\ g^{-1}$) |
|---|---|---|---|---|---|
| H1 | Hydrogen | 0.0132 | 27.3 | 0.5 | ~84 |
| H2 | Hydrogen | 0.0132 | ~53 | 0.5 | ~618 |
| H3 | Hydrogen | 0.0132 | 278 | 0.5 | 3928.3 |
| H4 | Hydrogen | 0.0132 | 301 | 0.12 | 4261.8 |
| O1 | Oxygen | 0.1061 | 118 | 0.9 | ~69 |
| O2 | Oxygen | 0.1061 | 278 | 0.9 | 395.3 |
| O3 | Oxygen | 0.1061 | 301 | 0.12 | 416.5 |
| W1 | Water | 0.1193 | 323 | 0.12 | 209.3 |
| W2 | Water | 0.1193 | 301 | 0.12 | 117.2 |
| | Gross Power Output from the Fuel Cell (W) | | | | 1500.0 |
| | Heater EM1 Power (W) | | | 0.2 | |
| | Heater EM2 Power (W) | | | 34.6 | |
| | Heater EM3 Power (W) | | | 1.4 | |
| | Heater EM4 Power (W) | | | 43.7 | |
| | Circulating Pump Power $\eta = 0.05$ (W) | | | ~2.0 | |
| | Total Parasitic Power (W) | | | | 81.9 |
| | Net Power to the Balloon Experiment (W) | | | | 1418.1 |

* Note: For hydrogen and Oxygen the zero enthalpy point is saturated liquid at 1 atm. For water the zero enthalpy point is liquid at the triple point (273.16 K)

The cycle in Figure 1 and Table 2 shows a number of heaters labeled EH. The heaters represent the major source of parasitic power to the power supply. The heater power can be reduced by not putting any of the waste water in contact with the heat rejection panel. This means that the pump loop shown must contain another working fluid besides water. If methanol is used as a circulating cooling fluid for the fuel cell, the minimum temperature for hydrogen and oxygen entering the heat rejection panel can be 170 K. This results in a reduction of the parasitic power by thirty-eight percent. Heater EH6 can be eliminated.

Heater EH5 is a heater used to heat the fuel cell to its operating temperature. This heater is only used while the balloon is on the ground. Heater EH7 keeps the water in the storage tank from freezing. Most of the time this heater is not powered. If the waste water storage tank is insulated, this heater can be limited to 10 W of operating power (when it is operating). Heater EH10 keeps the dump valve from the water storage tank from freezing. If the waste water is considered part of the ballast, it must be dumped periodically. If the waste water is to be kept aboard the gondola, then the waste water tank must be large enough to carry all of the water generated during a mission. In either case it may be a good idea to dump the waste water during the balloon decent. The waste water dump valve should be located at the bottom of the waste water storage tank.

Heaters EH8, and EH9 keep the cryogen dump valves from freezing. The cryogen dump valves are used to dump the liquid hydrogen and liquid oxygen when the balloon is descending. The oxygen would be dumped first; then the hydrogen. The dump valves shown in Figure 1 can be replaced by rupture discs that are opened rapidly. The valves or rupture discs can be actuated by ambient pressure switches. A helium gas system (not shown in Figure 1) could be used to purge the hydrogen and oxygen storage tanks as the balloon gondola nears the ground.

At the full operating altitude of the balloon (30 to 40 km above the earth), the external air pressure is below the triple point pressure for hydrogen (7.04 kPa). The dump valve or rupture disc is located at the bottom of the tank so liquid hydrogen is dumped (when the tank pressure is below hydrogen's critical pressure of 1.29 MPa). Heater EH10 keeps the hydrogen from freezing in the valve or disc port. Heater EH9 may not be needed because the ambient pressure of the atmosphere at an altitude of 40 km (about 0.27 kPa or 2 torr) is above the triple point pressure for oxygen (0.15 kPa).

## The Fuel Cell

Hydrogen-oxygen fuel cells have been used by NASA since the Apollo program in the late 1960's. Fuel cells are currently in use as a power source for NASA Space Shuttle flight. While fuel cells have been used by NASA, they have not been used very much outside of the space arena. This could change, because there is serious work occurring in the automotive sector by a number of companies.

A fuel cell for balloon application can be quite small. A 2 kW fuel cell power supply could have as few 30 cells that are 6.4 mm (0.25 inch) thick. The total stack thickness is only 192 mm, but the total cell length would be about 305 mm when the rest of the system is included. The size of these cells can be as small as 152 mm by 152 mm (6 by 6 inches), but reducing the current density in the cell would increase its overall efficiency and the reduce amount of cooling that is needed. The fuel cell overall package size could be as small as 300 mm by 300 mm by about 500 mm. Much of the fuel cell mass is water in the electrolyte contained in the cell while it is operating. The optimum operating temperature of the fuel cell is probably in the 60 to 80 C range. Recent developments in fuel cell technology suggest that light weight regenerative fuel cells can be built[4]. For some missions this could be attractive.

The nominal voltage for the fuel cell power supply under no load (no current is produced) would be 32 volts. As the cell is loaded, its terminal voltage goes down (like a battery). At its full design load, the cell terminal voltage could be as low as 24 volts. The voltages required for the balloon payload at New Mexico State are $\pm$ 6 volts, $\pm$12 volts and $\pm$24 volts. Other types of balloon payloads have similar voltage requirements.

## Cryogen Tankage

The balloon fuel cell tankage is based on an existing 75 liter liquid oxygen tank produced by Essex Cryogenics. The Essex tank can carry 85.5 kg of liquid oxygen at an operating pressure up to 2.07 MPa (300 psig). The vacuum insulated tank and liquid withdrawal system is all stainless steel and has a mass of 31.8 kg (70 lb.). The Essex tank is 610 mm high and about 600 mm in diameter. The heat leak into the Essex oxygen tank is 6 to 7 watts and it will hold oxygen without venting up to 12 hours. The existing Essex tank is entirely suitable for oxygen storage for the balloon experiment.

The heat leak into the standard Essex tank is too large for the hydrogen storage tanks, but one should be able to modify the it so that hydrogen flowing from the tank intercepts heat coming down the cold mass supports and through the insulation. The heat leak in a modified Essex tank can be brought down to less than 1 watt when 0.004 g $s^{-1}$ of hydrogen is withdrawn from the tank. The modified Essex tank mass should be about 34 kg (75 lb.) and it should hold 5.3 kg of hydrogen. For every 75 liter tank of oxygen used, two 75 liter tanks of hydrogen must be used. The fuel cell, a water storage tank and six Essex cryogen tanks (the entire power supply for an 18 day mission using 1.42 kW of power) would fit into a space that is 1.9 meters in diameter and 0.7 meters high.

The Essex tanks are by no means the best tanks that can be used for cryogenic tankage. Replacing the stainless steel with aluminum would reduce tank mass by a factor of two. Replacing the stainless steel vacuum shell with an aluminum shell would reduce the tank mass by 8 to 10 kg. By comparison, the 590 liter Shuttle hydrogen storage tanks weigh 103 kg and they hold 41.8 kg of hydrogen. The shuttle tanks will store hydrogen for 30 days without venting.

The maximum mission life is determined by the quality of the cryogenic tankage used. If one uses the modified Essex tankage, the maximum mission time is 30 days even if no power is used by the payload. Tankage that is similar to the Shuttle tankage should permit mission times in excess of 90 days, assuming the balloon will stay up that long.

## The Heat Rejection Panel

Fuel cells operate best at a temperature above 323 K (50 C) The waste heat from the production of electrical energy is carried into the cell by the water formed when hydrogen and oxygen combine. As a result, the fuel cell must be cooled. A cell producing 1.5 kW of electricity at an efficiency of 65 percent will produce 808 W of waste heat. At 80 percent efficiency, the waste heat production is reduced to 375 W.

At an altitude of 40 km, the surrounding air is quite cold, but unfortunately, the heat transfer to the air is rather poor. Heat rejected from the power supply (and the balloon payload for that matter) must be by radiation. Heat transfer from the heat rejection panel is by long wave radiation heat transfer to deep space above (at 3 K) and the earth below (at 240 to 260 K). Heat is transferred to the payload and its heat rejection system through short wave radiation from the sun. This radiation may be direct or by reflection from clouds (about 30 percent is reflected) or snow (about 70 percent is reflected) on the ground. The power supply heat rejection panel is sized for the condition where the balloon is flying above the Antarctic ice pack in the sunshine. Under this condition the temperature of the panel should be no higher than 313 K (10 K below the minimum operating temperature of the fuel cell). When the payload is in the dark the heat rejection panel temperature will be lower.

In order to control the temperature of both the payload and the fuel cell power supply, selective coatings must be used on the payload and the heat rejection panel[5]. The selective coating must absorb a lower percentage of the short wave length (solar) radiation than the percentage of the long wave length radiation (characteristic of a body at 290 K) emitted. NASA white paint is such a coating. It has an $\alpha$ over $\varepsilon$ of 0.3 (where $\alpha$ is the absorption coefficient of solar radiation and $\varepsilon$ is the emissivity for thermal radiation at 290 K) NASA white paint is satisfactory except that the panel temperature variation between the highest solar load case (in the sunlight over an ice field) and the no solar load case (at night) is a little large for heat rejection panel in contact with liquid water. The temperature variation is less extreme if a Shelldahl silver Teflon coating ($\alpha$ over e of 0.12) is used on the heat rejection panel. The panel area for a maximum heat rejection of 0.7 kW is about 4.9 square meters. The heat rejection panel would fit around the six modified Essex cryogen tanks, the water tank and the fuel cell.

## A COMPARISON OF FUEL CELLS WITH BATTERIES

Table 3 compares a particle astrophysics experiment power supply based on lithium chloride batteries with a fuel cell power supply for the same experiment. In both cases the experimental payload is 1500 kg (3300 lb.). The mass of the payload recovery system, the balloon avionics, and the backup batteries is the same. The total ballast mass is the same for both cases, but the spent water from the fuel cell is assumed to be ballast. The power used by the experiment is 1.42 kW in both cases. Table 3 presents the mass of major constituents of the power supplies and the rest of the balloon payload. The maximum mission time and the available electrical energy per unit mass are compared for both cases.

**Table 3** A comparison of Mass and Operating Time for a Balloon Payload Using Lithium Chloride Batteries or a Fuel Cell Power Supply

| Component | Battery Powered | Fuel Cell Powered |
|---|---|---|
| Balloon Experimental Payload Mass (kg) | 1500 | 1500 |
| Balloon Avionics and Recovery System Mass (kg) | 100 | 100 |
| Iron shot Balloon Ballast Mass (kg) | 250 | 64 |
| Main Battery Power Supply Mass (kg) | 545 | ----- |
| Backup Batteries | 55 | 55 |
| Fuel Cell System Mass (kg) | ---- | 41 |
| Cryogen and Water Tankage Mass (kg) | ---- | 270* |
| Fuel Cell Heat Rejection Panel Mass (kg) | ---- | 20 |
| Fuel and Oxidizer Mass (kg) | ---- | 186 |
| Total Mass Launched (kg) | 2450 | 2236 |
| Main Power Supply Available Net Energy (kWh) | 72 | 613 |
| Maximum Mission Time @ 1.42 kW Net (hr) | 50.7 | 436** |
| Power Supply Mass including cryogens (kg) | 545 | 517 |
| Available Electrical Energy per Unit Mass (MJ per kg) | 0.48 | 4.27 |

* This includes 4 hydrogen tanks, 2 oxygen tanks, 1 water tank, piping and a skid to hold the tanks.

** The mission life is based on 80 percent efficient fuel cells.

## CONCLUDING COMMENTS

A fuel cell power supply with cryogenic storage appears to be a viable power supply option for long duration balloon flights that carry particle astrophysics experiments. It is clear from Table 3 that the fuel cell power supply can potentially supply power over eight times longer than a power supply based on lithium chloride batteries. In addition, the launch weight of the balloon can be reduced by over 200 kg because 186 kg of iron shot ballast can be eliminated. The fuel cell power supply described in Table 3 is capable of delivering 0.85 kW for 30 days. At that life time, the power supply is limited by the heat leak into the modified Essex tanks. As a comparison, the power supply based on lithium chloride batteries would deliver 0.85 kW for about 3.5 days.

One can argue that the batteries can be improved. (A factor of two improvement may be possible.) An equally persuasive argument can be made for the possible improvement of the fuel cell power supply over the power supply given in Table 3. Improved tankage alone will result in a lower mass power supply that can supply electrical energy over a longer period of time. Improved tankage would allow some kinds of payloads to be supplied with power for more than 90 days.


## ACKNOWLEDGMENT

The authors thank Steve Stochaj of the R. L. Golden Particle Astrophysics Laboratory at New Mexico State University for his help in obtaining information on the power needed of balloon born particle astrophysics experiments. Long duration balloon flight power supplies based on the use of fuel cells was a pet project of Bob Golden's before his death, hence he is an author of this paper. The authors also thank B. Kimball of New Mexico State University.

This work was performed at the Lawrence Berkeley National Laboratory with the support of the Director, Office of Energy Research, Office of High Energy and Nuclear Physics, High Energy Physics Division, U. S. Department of Energy under Contract No. DE-AC03-76SF00098.


## REFERENCES


1. R. L . Golden et al. "Performance of a Balloon-Born Magnet Spectrometer for Cosmic Ray Studies," Nuclear Instruments and Methods in Physics Research **A306**, p 366, (1991)
2. V. J. Johnson editor, " Properties of Materials at Low Temperature (Phase 1), a Compendium," Part 1 Properties of Fluids, Pergamon Press, New York (1961)
3. "Selected Properties of Hydrogen" NBS Monograph 168
4. Lawrence Livermore National Laboratory Science and Technology Review, May 1997 and private communication with F. Mitlitsky, Lawrence Livermore National Laboratory
5. E. M. Sparrow and R. D. Cess, *Radiation Heat Transfer*, Brooks/Cole Publishing Company, Belmont CA, (1966)

# CURRENT STATUS OF THERMODYNAMIC PROPERTIES OF CRYOGENIC FLUIDS

R. T Jacobsen[1], S. G. Penoncello[1], and E. W. Lemmon[1,2]

[1] Center for Applied Thermodynamic Studies
University of Idaho, College of Engineering
Moscow, Idaho 83844-1011, USA

[2] Current Address:
Physical and Chemical Properties Division
National Institute of Standards and Technology
325 Broadway, Boulder, Colorado, 80303, USA

## ABSTRACT

A review of the current status of mathematical models and correlations for the calculation of thermodynamic properties of cryogenic fluids is presented. This review provides the practicing engineer and interested scientist with information on the thermodynamic properties of air, argon, carbon monoxide, deuterium, ethane, fluorine, helium, hydrogen, krypton, methane, neon, nitrogen, oxygen, parahydrogen and xenon. We have included the most accurate formulation for each fluid available at this time. We have referenced the original works that contain details of the mathematical models and correlation methods, and provide the user with sources of information about using the computer formulations for applications. We have included information about Internet access to computer programs for thermodynamic properties of the fluids. Estimates of the accuracies of properties calculated using the available computer models are also included.

## INTRODUCTION

This paper reports on the current status of computer models for the calculation of thermodynamic properties of cryogenic fluids. Further details on these models are given in Jacobsen *et al.*[1] Studies of the importance of accurate thermophysical properties are given by Zudkevitch[2], Zudkevitch and Gray[3], Sengers and Klein[4], Chappelear *et al.*[5] and Haynes *et al.*[6] Table 1, modified from Haynes *et al.*[6], estimates the relative importance of accurate thermophysical properties in specific process areas. Haynes *et al.*[6] present a discussion

**Table 1.** The relative importance of thermophysical property data in cryogenic processes (Modified from Haynes *et al.*[6])

| Process | Pressure-Density Temperature | Phase Equilibria | Calorimetric Properties | Vapor Pressure Critical Properties | Transport Properties |
|---|---|---|---|---|---|
| Distillation | 2 | 1 | 2 | 1 | 2 |
| Compression and Expansion | 1 | 1 | 1 | 1 | 2 |
| Heat Exchange | 3 | 1 | 1 | 2 | 2 |
| Metering and Sales | 1 | 2 | 2 | 2 | 3 |
| Storage | 2 | 1 | 1 | 2 | 3 |
| Safety | 2 | 1 | 1 | 2 | 3 |

Type of Property: columns 2–6. 1=most important 2=very important 3=less important

including examples of specific processes and equipment which depend upon individual property information, including additional references to literature on this subject.

The demands of expanding technology during the past four decades have resulted in organized efforts to satisfy the needs for accurate thermodynamic property formulations, tables and graphs on a world wide basis. Historical numerical data projects on thermodynamic properties of cryogenic fluids are described in References 7 through 9.

## GENERAL REQUIREMENTS FOR THERMODYNAMIC PROPERTY FORMULATIONS

The term "thermodynamic property formulation" is used to define a group of equations required for the complete definition of the thermodynamic properties of a fluid. The equation of state is the major part of the formulation and generally requires extensive data for its development.

Current practice in the development of computer programs, property tables and charts for system design and analysis involves the correlation of selected experimental data for a particular fluid or mixture. The objective of the correlator is to ascertain the accuracy of the available experimental data for the particular fluid or system under investigation, and to develop a mathematical model capable of representing the data within the reported or estimated experimental uncertainty. The formulations described here for cryogenic fluids are considered by the authors to be the most appropriate for use in engineering applications at this time.

For comparisons of calculated values to measured properties, the single-phase thermodynamic regions shown in Figure 1 are arbitrarily defined as:

1. Liquid below the critical temperature,
2. Vapor below the critical temperature,
3. Fluid above the critical temperature, and
4. Critical region (within five percent of the critical temperature and 25 percent of the critical density).

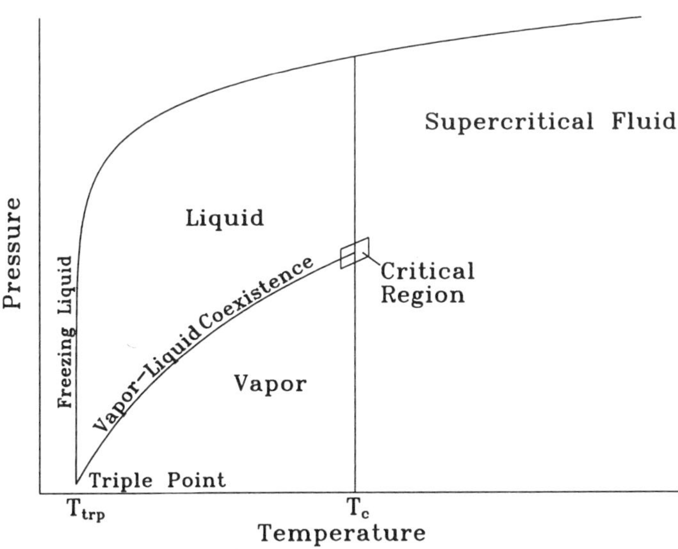


Figure 1. Thermodynamic surface regions.

The accuracies of calculated values of various properties are determined by comparing them to measured values. Graphs which illustrate such comparisons show differences in experimental data from alternate sources and facilitate the determination of random errors and relative consistency of data. Statistical analysis based on the percent deviation is also used to determine the overall estimated accuracy of various data sets, and to define the accuracies for various properties calculated with the formulation.

The percent deviation for any property, X, is given by

$$\%\Delta X = 100\left(X_{data} - X_{calc}\right)/X_{data}. \tag{1}$$

General estimates of the attainable accuracies of various properties are given in Table 2.

**Table 2.** Estimated accuracies of various measured thermodynamic properties

| **Calculated Property** | | **State of the Art Experimental Accuracy** | **Accuracy to be Expected from an Equation of State** |
|---|---|---|---|
| Density | | ±0.020 percent | ±0.1 percent |
| Pressure | P < 0.5 MPa | ±0.020 percent | |
| | P > 0.5 MPa | ±0.050 percent | |
| Temperature | | ±1.0 mK | |
| Isochoric Heat Capacity | $\rho > \rho_c$ | ±0.500 percent | ±0.5 percent |
| | $\rho < \rho_c$ | ±1.000 percent | ±1.0 percent |
| Isobaric Heat Capacity | $\rho > \rho_c$ | ±0.500 percent | ±1.0 percent |
| | $\rho < \rho_c$ | ±2.000 percent | ±1.0 percent |
| Speed of Sound | $\rho > \rho_c$ | ±0.100 percent | ±0.5 percent |
| | $\rho < \rho_c$ | ±0.001 percent | ±0.1 percent |
| Vapor Pressure | P < 0.1 MPa | ±0.050 percent | ±0.5 percent |
| | P > 0.1 MPa | ±0.010 percent | ±0.2 percent |

## THERMODYNAMIC RELATIONS AND FUNCTIONAL FORMS FOR EQUATIONS OF STATE

A number of different functional forms have been used for equations of state for cryogens. A common equation of state used by engineers and scientists is the pressure-explicit form, generally expressed in the form $P=f(v,T)$ or $P=f(\rho,T)$.

The term "fundamental equation" generally refers to an equation which contains calorimetric and datum state information so that absolute values of specified properties may be calculated directly by mathematical differentiation. In current engineering practice, some form of fundamental equation, explicit in Helmholtz energy, A, is commonly used for accurate practical thermodynamic property formulations in design and analysis of energy systems.

In the development of a thermodynamic property formulation for a pure fluid or fluid mixture, it is desirable to determine separate ancillary functions for the liquid-vapor coexistence properties, for the ideal gas heat capacity and for the freezing liquid line (often called the melting curve). Equations for the vapor pressure, saturated liquid density, and saturated vapor density are used to define the saturation (coexistence) states during the development of the equation of state and as initial estimates in the iterative calculation of saturation properties using computer programs for the completed property formulation. The equation for the freezing liquid line is used to determine the low temperature boundary of the liquid region, and the ideal gas heat capacity equation supplies the calorimetric information.

At the coexistence boundary of a pure fluid, the Maxwell criterion (equal Gibbs energies in the liquid and vapor phases at the saturation temperature and pressure) may be applied to the equation of state, *i.e.*, coexistence property values predicted by the equation of state should be consistent with those from the ancillary equations for the vapor pressure and the liquid and vapor saturated densities. Most of the equations used here are suitable for these calculations.

Correct behavior of the equation of state in the critical region is of concern to some users of property formulations. Not all of the formulations described in this paper are accurate near the critical point. Because of the nonanalytic behavior of certain fluid properties in the critical region, special correlations, (*e.g.*, the scaled equations of state presented by Tang *et al.*[11] and Sengers[12]), using models which account for the fluid behavior at the critical point have been developed for critical region properties. We suggest that if applications require calculation near the critical state, that the user investigate the availability of such a correlation.

## THE GENERAL WIDE-RANGE FUNDAMENTAL EQUATION FOR CRYOGENIC FLUIDS

The fundamental equation described in this section is a general form suitable for the calculation of properties of a large number of fluids. Although the original references reporting the equations of state for cryogenic fluids use a wide variety of equation formats, all equations have been transformed to the fundamental form of Jacobsen *et al.*[10] Details of the implementation of this equation as a generalized form for cryogenic fluids are given by Jacobsen *et al.*[1]

## The Fundamental Equation Explicit in Helmholtz Energy

The Helmholtz energy is given by

$$A(\rho,T) = A^0(\rho,T) + A^r(\rho,T), \tag{2}$$

where $A^0(\rho,T)$ is the ideal gas contribution to the Helmholtz energy and the term $A^r(\rho,T)$ represents the contribution due to the compressibility of the fluid. Other properties can be calculated similarly using accepted thermodynamic relations.

Practical equations of state generally use reduced variables to simplify computer calculations. Properties are reduced by accepted critical state values or other parameters to make the equations nondimensional. The variables used in this work are,

$$\alpha = \frac{A}{RT}, \; \delta = \frac{\rho}{\rho_c}, \text{ and } \tau = \frac{T_c}{T}. \tag{3}$$

Using these variables, Equation (2) is written as

$$\alpha(\delta,\tau) = \alpha^0(\delta,\tau) + \alpha^r(\delta,\tau). \tag{4}$$

The ideal gas contribution to the dimensionless Helmholtz energy is given by

$$\alpha^0(\delta,\tau) = \frac{H_0^0\tau}{RT_c} - \frac{S_0^0}{R} - 1 + \ln\frac{\delta\tau_0}{\delta_0\tau} - \frac{\tau}{R}\int_{\tau_0}^{\tau}\frac{C_p^0}{\tau^2}d\tau + \frac{1}{R}\int_{\tau_0}^{\tau}\frac{C_p^0}{\tau}d\tau. \tag{5}$$

The form of $\alpha^r(\delta,\tau)$ is based on theoretical and empirical considerations required to represent experimental data for a fluid. The general form used by Jacobsen *et al.*[10] is

$$\alpha^r(\delta,\tau) = \sum_{k=1}^{m} N_k \delta^{i_k} \tau^{j_k} \exp(-\gamma\delta^{l_k}), \tag{6}$$

where the $N_k$, $i_k$, $j_k$, and $l_k$ are determined by a least-squares fit to experimental data, and $\gamma$ is a coefficient which has a value of 0 for terms with $l$=0 and a value of 1 for terms with $l$ greater than zero.

## Thermodynamic Properties from the Fundamental Equation

Table 3 gives selected thermodynamic property relations for the fundamental equation summarized in the previous section. The functions in this table and those for other properties needed for applications are readily used in computer programs for the calculation of thermodynamic properties as given by Lemmon *et al.*[13,14] Details of the calculation of properties for specific fluids are given in Jacobsen *et al.*[1]

# THERMODYNAMIC PROPERTIES OF CRYOGENIC FLUIDS

This section provides information about the thermodynamic properties of fifteen cryogenic fluids. Many of the correlations used in this work are based upon comprehensive data bases which include P-ρ-T and other property data (isobaric heat capacity, isochoric heat capacity, speed of sound, and saturation properties). Table 4 lists the fluids detailed in this section along with the literature reference for each correlation. Fluid specific constants, coefficients, and exponents are presented for the fundamental equation for each fluid in Jacobsen *et al.*[1] The details about each fluid property correlation may be obtained by consulting the references in Table 4.

**Table 3.** Functions for calculation of selected thermodynamic properties

| Property | Relation |
|---|---|
| Compressibility Factor | $Z = \frac{P}{\rho RT} = 1 + \delta\left(\frac{\partial\alpha^r}{\partial\delta}\right)$ |
| Pressure | $\frac{P}{P_c} = \frac{\delta}{\tau Z_c}\left[1 + \delta\left(\frac{\partial\alpha^r}{\partial\delta}\right)\right]$ |
| Internal Energy | $\frac{U}{RT} = \tau\left[\left(\frac{\partial\alpha^0}{\partial\tau}\right) + \left(\frac{\partial\alpha^r}{\partial\tau}\right)\right]$ |
| Enthalpy | $\frac{H}{RT} = \tau\left[\left(\frac{\partial\alpha^0}{\partial\tau}\right) + \left(\frac{\partial\alpha^r}{\partial\tau}\right)\right] + \delta\left(\frac{\partial\alpha^r}{\partial\delta}\right) + 1$ |
| Entropy | $\frac{S}{R} = \tau\left[\left(\frac{\partial\alpha^0}{\partial\tau}\right) + \left(\frac{\partial\alpha^r}{\partial\tau}\right)\right] - \alpha^0 - \alpha^r$ |
| Speed of Sound | $\frac{W^2}{RT} = \frac{C_p}{C_v}\left[1 + 2\delta\left(\frac{\partial\alpha^r}{\partial\delta}\right) + \delta^2\left(\frac{\partial^2\alpha^r}{\partial\delta^2}\right)\right]$ |
| Isochoric Heat Capacity | $\frac{C_v}{R} = -\tau^2\left[\left(\frac{\partial^2\alpha^0}{\partial\tau^2}\right) + \left(\frac{\partial^2\alpha^r}{\partial\tau^2}\right)\right]$ |
| Isobaric Heat Capacity | $\frac{C_p}{R} = \frac{C_v}{R} + \frac{\left[1 + \delta\left(\frac{\partial\alpha^r}{\partial\delta}\right) - \delta\tau\left(\frac{\partial^2\alpha^r}{\partial\delta\partial\tau}\right)\right]^2}{\left[1 + 2\delta\left(\frac{\partial\alpha^r}{\partial\delta}\right) + \delta^2\left(\frac{\partial^2\alpha^r}{\partial\delta^2}\right)\right]}$ |

**Table 4.** Thermodynamic property formulations for cryogenic fluids

| Fluid | Reference | Temperature Range (K) | Maximum Pressure (Mpa) | Temperature Scale of Reference |
|---|---|---|---|---|
| Air | Jacobsen *et al.* [15] | 59.75 – 870 | 70 | 1968 |
| Argon | Stewart and Jacobsen [16] | 83.804 – 1200 | 1000 | 1968 |
| Carbon Monoxide | Friend *et al.* [17] | 68.127 – 300 | 30 | 1968 |
| Deuterium (normal) | Friend *et al.* [17] | 18.71 – 423 | 320 | 1968 |
| Ethane | Friend *et al.* [18] | 90.352 – 623 | 69 | 1968 |
| Fluorine | de Reuck [19] | 53.48 – 300 | 20 | 1968 |
| Helium | McCarty and Arp [20] | 2.1768 – 1500 | 100 | 1976 |
| Hydrogen (normal) | McCarty [21] | 13.95 – 500 | 40 | 1968 |
| Krypton | Juza and Sifner [22] | 115.77 – 800 | 300 | 1968 |
| Methane | Setzmann and Wagner [23] | 90.694 – 625 | 1000 | 1990 |
| Neon | Katti [24] | 24.562 – 700 | 700 | 1968 |
| Nitrogen | Jacobsen *et al.* [25] | 63.148 – 2000 | 1000 | 1968 |
| Oxygen | Schmidt and Wagner [26] | 54.361 – 300 | 82 | 1968 |
| Parahydrogen | Younglove [27] | 13.8 – 400 | 121 | 1968 |
| Zenon | Sifner and Klomfar [28] | 161.36 – 1300 | 100 | 1968 |

Property tables are included in Jacobsen *et al.*[1] to provide information for those who wish to estimate properties without using computer programs. They may also be used to verify calculations by those who wish to develop their own computer programs for specialized applications.

Pressure-enthalpy (P-h) and temperature-entropy (T-s) diagrams calculated using each formulation are also given in Jacobsen *et al.*[1] These property diagrams illustrate the surface of state for each cryogenic fluid as represented by the correlation referenced in Table 4.

## COMPUTER PROGRAMS FOR THERMODYNAMIC PROPERTY CALCULATION

Long-term research in the Center for Applied Thermodynamic Studies (CATS) has resulted in the development of computer programs for the calculation of thermodynamic properties of fluids and fluid mixtures. ICMPROPS is a set of computer routines developed specifically for cryogenic fluids. All necessary coefficients, constants, and iterative routines for a variety of input property pairs are provided in ICMPROPS. Details of the use of these programs are given in Jacobsen *et al.*[1] and the programs and licensing information for commercial users are available on the World Wide Web at <http://calvin.engr.uidaho.edu/~cats/cats.html>.

## REFERENCES


1. R. T Jacobsen, S. G. Penoncello, and E. W. Lemmon, "Thermodynamic Properties of Cryogenic Fluids," Plenum Press, New York (1997).
2. D. Zudkevitch, Imprecise data impacts plant design and operation, *Hydrocarbon Processing* 54:97 (1975).
3. D. Zudkevitch and R. D. Gray, Jr., Impact of fluid properties on the design of equipment for handling LNG, in: "Adv. Cryo. Engr.", Plenum Press, New York (1975), 20:103.
4. J. V. Sengers and M. Klein, ed., "The Technological Importance of Accurate Thermophysical Property Information," Special Publication 590 Nat. Bur. Stand. (U.S.) (1980).
5. P. S. Chappelear, R. J. J. Chen and D. G. Elliot, Pick K correlations carefully, *Hydrocarbon Processing* 56:215 (1977).
6. W. M. Haynes, A. J. Kidnay, N. A. Olien, and M. J. Hiza, Status of thermophysical properties data for pure fluids and mixtures of cryogenic interest, in: "Adv. Cryo. Engr.," Plenum Press, New York (1984), 27:919.
7. F. Din, "Thermodynamic Functions of Gases: Volume 2. Air, Acetylene, Ethylene, Propane and Argon, Volume 3. Ethane, Methane, and Nitrogen," Butterworths, London (1962).
8. R. B. Stewart and K. D. Timmerhaus, The correlation of thermodynamic properties of cryogenic fluids, in: "Adv. Cryo. Engr.," Plenum Press, New York (1964), 9:20.
9. D. E. Diller, Thermophysical properties data research on compressed and liquefied gases at the NBS Cryogenics Division, in: "Adv. Cryo. Engr.," Plenum Press, New York (1976), 21:522.
10. R. T Jacobsen, R. B. Stewart, M. Jahangiri, and S. G. Penoncello, A new fundamental equation for thermodynamic property correlations, in "Adv. Cryo. Engr.," Plenum Press, New York (1986), 31:1161.

11. S. Tang, G. X. Jin, and J. V. Sengers, Thermodynamic properties of 1, 1, 1, 2-tetrafluoroethane (R134a) in the critical region, *Int. J. Thermophys.* 12(3):515 (1991).
12. J. V. Sengers, Effects of critical fluctuations on the thermodynamic and transport properties of supercritical fluids, in: "Supercritical Fluids," E. Kiran and J. M. H. Levelt Sengers, eds., Kluwer, Dordrecht (1994), p. 231.
13. E. W. Lemmon, R. T Jacobsen, S. G. Penoncello, and S. W. Beyerlein, Computer programs for the calculation of thermodynamic properties of cryogens and other fluids, in: "Adv. Cryo. Engr.," Plenum Press, New York (1994), 39:1891.
14. E. W. Lemmon, R. T Jacobsen, S. G. Penoncello, and S. W. Beyerlein, "Computer Programs for Calculating Thermodynamic Properties of Fluids of Engineering Interest, Version 4.2 and 4.2W," CATS Report 97-1, University of Idaho, Moscow, Idaho (1997).
15. R. T Jacobsen, S. G. Penoncello, S. W. Beyerlein, W. P. Clarke, and E. W. Lemmon, A thermodynamic property formulation for air, *Fluid Phase Equilibria* 79:113 (1992).
16. R. B. Stewart, R. B. and R. T Jacobsen, Thermodynamic properties of argon from the triple point to 1200 K at pressures to 1000 MPa, *J. Phys. Chem. Ref. Data* 18(2):639 (1989).
17. D. G. Friend, M. L. Huber, and J. S. Gallagher, Thermophysical property computer packages from NIST, in: "Computerized Thermophysical Property Packages, American Society of Mechanical Engineers," S. G. Penoncello, ed., Symposium Proceedings, (1992), HTD Vol. 225:13.
18. D. G. Friend, H. Ingham, and J. F. Ely, Thermophysical properties of ethane, *J. Phys. Chem. Ref. Data* 20(2):275 (1991).
19. K. M. de Reuck, "International Thermodynamic Tables of the Fluid State-11 Fluorine," International Union of Pure and Applied Chemistry, Chemical Data Series, Pergamon Press, Oxford (1990).
20. R. D. McCarty and V. D. Arp, A new wide range equation of state for helium, in: "Adv. Cryo. Engr.," Plenum Press, New York (1990), 35:1465.
21. R. D. McCarty, "Hydrogen, Technological Survey - Thermophysical Properties," NASA SP-3089, National Aeronautics and Space Administration (1975).
22. J. Juza and O. Sifner, Modified equation of state and formulation of thermodynamic properties of krypton in a canonical form in the range from 120 to 423 K and 0 to 300 MPa, *Acta Technica CSAV* 1:1-32 (1976).
23. U. Setzmann and W. Wagner, A new equation of state and tables of thermodynamic properties for methane covering the range from the melting line to 625 K at pressures up to 1000 MPa, *J. Phys. Chem. Ref. Data* 20(6):1061 (1991).
24. R. S. Katti, "Thermodynamic Properties of Neon, From the Triple Point to 700 K with Pressures up to 700 MPa," M.S. Thesis, University of Idaho, Moscow, Idaho (1985).
25. R. T Jacobsen, R. B. Stewart, and M. Jahangiri, Thermodynamic properties of nitrogen from the freezing line to 2000 K at pressures to 1000 MPa, *J. Phys. Chem. Ref. Data* 15(2):735 (1986).
26. R. Schmidt and W. Wagner, A new form of the equation of state of pure substances and its applications to oxygen, *Fluid Phase Equilibria* 19:175, 1985.
27. B. A. Younglove, Thermophysical properties of fluids, *J. Phys. Chem. Ref. Data* 11(1):4 (1982).
28. O. Sifner and J. Klomfar, Thermodynamic properties of xenon from the triple point to 800 K with pressures up to 350 MPa, *J. Phys. Chem. Ref. Data* (1):63 (1994).

# THERMODYNAMIC PROPERTIES OF MIXTURES OF NITROGEN, ARGON, AND OXYGEN, INCLUDING AIR

E. W. Lemmon[1,2] and R. T Jacobsen[1]

[1]Center for Applied Thermodynamic Studies
University of Idaho, Moscow, ID 83844-1011
[2]Current Address: Physical and Chemical Properties Division
National Institute of Standards and Technology
325 Broadway, Boulder, CO 80303

## ABSTRACT

A mixture model explicit in Helmholtz energy has been developed which is capable of predicting the thermodynamic properties of mixtures containing nitrogen, argon, and oxygen, including air. The calculation for air assumes a ternary mixture of these three components and neglects the presence of water vapor, carbon dioxide and trace elements known to be present in atmospheric air. The Helmholtz energy of the mixture is the sum of the ideal gas contribution, the residual contribution, and the contribution from mixing. The contribution from mixing is given by a single generalized equation which is applied to all mixtures used in this work. The independent variables are the reduced density and reduced temperature. The model may be used to calculate thermodynamic properties of mixtures at various compositions including dew and bubble point properties and critical points. It incorporates the most accurate published equation of state for each pure fluid.

The model may be used to calculate the properties of mixtures generally within the experimental accuracies of the available measured properties. The estimated accuracy of calculated properties is ±0.1% in density, ±0.5% in the speed of sound, and ±1% in heat capacities. Calculated bubble point pressures are generally accurate to within ±1%.

## INTRODUCTION

Many engineering applications require the use of equations of state capable of accurate prediction of thermodynamic properties of nitrogen-argon-oxygen mixtures. These include air liquefaction and separation and the operation of aircraft gas turbines at extreme pressures and temperatures.

*Advances in Cryogenic Engineering,* Vol. 43
Edited by P. Kittel, Plenum Press, New York, 1998

The development of equations for the thermophysical properties of air and mixtures of nitrogen, argon, and oxygen has been a continuing project at the University of Idaho Center for Applied Thermodynamic Studies for the last ten years. An equation of state for air as a pseudo-pure fluid was published by Jacobsen *et al.*[1] In addition, an extended corresponding states model for calculating the thermodynamic properties of nitrogen-argon-oxygen mixtures was published by Clarke *et al.*[2] and a Lagrangian interpolation model for calculating vapor-liquid equilibrium properties for this system was published by Lemmon *et al.*[3]

The mixture model used to calculate properties of nitrogen-argon-oxygen mixtures in this work was developed by Lemmon[4] and is also reported by Lemmon and Jacobsen.[5] The formulation is based on corresponding states theory. Reducing parameters which are dependent on the mole fraction are used to modify absolute values of density and temperature. The Helmholtz energy for an ideal solution is determined at the reduced density and temperature of the mixture using accurate pure fluid equations of state for the mixture components.

The formulation for nitrogen-argon-oxygen mixtures given here is a modification of the generalized model developed by Lemmon[4] for a wide range of fluids including hydrocarbons and cryogens. It preserves the general nature of the previous work, but is more accurate for the calculation of nitrogen-argon-oxygen mixture properties than the prior formulation at low temperatures. By restricting the application to a particular class of fluids, the model was designed to represent the unique characteristics of this system. The model represents available measured data on all parts of the thermodynamic surface within their estimated experimental accuracy.

An advantage of the approach used here is that the portion of the model describing the Helmholtz energy contribution from mixing is the same for all binary mixtures, and relatively simple scaling factors are used to determine its magnitude for a particular mixture. Experimental data from several different mixtures were used to determine the generalized function. The model is capable of predicting mixture properties for fluids with limited experimental databases. In addition, all vapor and liquid thermodynamic properties, including density, energy, entropy, heat capacity, sound speed, vapor-liquid equilibrium, and the mixture critical temperature, pressure, and density, can be calculated accurately using this approach.

The mixtures studied in this work are:

- Air as a ternary mixture of 78.12% nitrogen, 20.96% oxygen, and 0.92% argon.
- Nitrogen-Oxygen
- Nitrogen-Argon
- Argon-Oxygen
- Nitrogen-Argon-Oxygen

Data for these mixtures were used to evaluate the behavior of the model, to calculate the coefficients of the generalized mixture equation and to calculate the individual parameters for each binary mixture. Table 1 lists the pure fluid equations of state used in this work.

**Table 1**. Pure fluid equations of state used in the mixture model

| Fluid | Author | Temperature Range (K) | Maximum Pressure (MPa) |
|---|---|---|---|
| Nitrogen | Jacobsen *et al.*[6] | 63.148 - 2000 | 1000 |
| Argon | Stewart and Jacobsen[7] | 83.804 - 1200 | 1000 |
| Oxygen | Schmidt and Wagner[8] | 54.361 - 300 | 82 |

## THE MIXTURE EQUATION

The Helmholtz energy for mixtures of nitrogen, argon, and oxygen can be calculated using

$$\alpha_m(\delta,\tau,\mathbf{x}) = \alpha_m^i(\delta,\tau,\mathbf{x}) + \alpha^E(\delta,\tau,\mathbf{x}), \tag{1}$$

where $\mathbf{x}$ is the concentration variable. The reduced values of density and temperature for the mixture are

$$\delta = \rho / \rho_j \text{ and} \tag{2}$$

$$\tau = T_j / T, \tag{3}$$

where $T_j$ and $\rho_j$ are the reducing values given by

$$T_j = \sum_{p=1}^{r} x_p T_{c_p} + \sum_{p=1}^{r-1} \sum_{q=p+1}^{r} x_p x_q \zeta_{pq} \text{ and} \tag{4}$$

$$\rho_j = \left[ \sum_{p=1}^{r} \frac{x_p}{\rho_{c_p}} \right]^{-1}. \tag{5}$$

The coefficient $\zeta_{pq}$ is used to define the shape of the reducing temperature line. These reducing parameters are not the same as the critical parameters of the mixture. They define the relationship between the critical lines of different binary mixtures and are determined simultaneously in the nonlinear fit with the generalized factor, $F_{pq}$, described below.

The Helmholtz energy for the ideal solution is

$$\alpha_m^i(\delta,\tau,\mathbf{x}) = \sum_{p=1}^{r} x_p \left[ \alpha_p^0(\delta,\tau) + \alpha_p^r(\delta,\tau) + \ln x_p \right], \tag{6}$$

where $r$ is the number of fluids in the mixture, $\alpha_p^0$ is the ideal gas Helmholtz energy of component $p$, and $\alpha_p^r$ is the residual Helmholtz energy of component $p$. Equations for the ideal gas Helmholtz energy and residual Helmholtz energy for the pure fluids are given in the references shown in Table 1.

The contribution to the Helmholtz energy from mixing used in this work is

$$\alpha^E(\delta,\tau,\mathbf{x}) = \sum_{p=1}^{r-1} \sum_{q=p+1}^{r} x_p x_q F_{pq} \sum_{k=1}^{7} N_k \delta^{i_k} \tau^{j_k}, \tag{7}$$

where the $N_k$, $i_k$, and $j_k$ are coefficients and exponents obtained from linear regression of experimental mixture data. The coefficients, $N_k$, of the equation are determined using a stepwise least-squares technique with a search and selection procedure which selects an optimum group of terms from a large comprehensive function based on statistical evaluation of the significance of individual terms. All thermodynamic properties can be calculated from the Helmholtz energy using differentiation with respect to density or temperature as described by Lemmon *et al.*[9]

The independent parameters of the equation for the contribution to the Helmholtz energy from mixing given in this work are reduced density and temperature and mixture composition. In addition, one other parameter, $F_{pq}$, is required which relates the properties of one binary mixture to those of another. This value is determined with a nonlinear fitting procedure that minimizes the sum of squares of the deviations between the equation and the data for a small selected set of measurements.

Details of the fitting procedures to determine the generalized factors and mixture parameters are given by Lemmon[4] and are not repeated here. The values of the coefficients and exponents in Eq. (7) are given in Table 2. The generalized factors and mixture parameters are given in Table 3.

**Table 2.** Coefficients and exponents of the mixture equation

| $k$ | $N_k$ | $i_k$ | $j_k$ |
|---|---|---|---|
| 1 | $0.110\,495\,938\,854 \times 10^{-1}$ | 1 | 1 |
| 2 | $-0.132\,116\,298\,744 \times 10^{-2}$ | 1 | -2 |
| 3 | $0.749\,312\,295\,074 \times 10^{-2}$ | 2 | 1 |
| 4 | $-0.674\,613\,233\,792 \times 10^{-3}$ | 4 | -1 |
| 5 | $-0.490\,864\,461\,893 \times 10^{-5}$ | 7 | 4 |
| 6 | $0.525\,114\,559\,789 \times 10^{-6}$ | 10 | 3 |
| 7 | $0.123\,175\,103\,037 \times 10^{-6}$ | 10 | -2 |

**Table 3.** Parameters of the mixture equation

| Binary Mixture | $F_{pq}$ | $\zeta_{pq}$ |
|---|---|---|
| Nitrogen-Argon | 0.459 571 | -2.483 638 |
| Nitrogen-Oxygen | 1 | 0 |
| Argon-Oxygen | 1.446 511 | 0 |

## COMPARISONS OF CALCULATED MIXTURE PROPERTIES TO EXPERIMENTAL DATA

Summary comparisons of values calculated using the mixture equation to $P$-$\rho$-$T$ data, isochoric heat capacity data, sound speed data and VLE data for mixtures of nitrogen, argon, and oxygen are given in Table 4, as well as the temperature range and composition range for the first component listed for each data set. Details of the property calculation methods including those for VLE states are given by Lemmon.[4]

The percent deviation in any property $X$ is defined as

$$\%\text{deviation in } X = 100\left(\frac{X_{\text{data}} - X_{\text{calc}}}{X_{\text{data}}}\right) \tag{8}$$

where $X_{\text{data}}$ is the experimental value and $X_{\text{calc}}$ is calculated from the equation of state. The comparisons in Table 4 report the following statistics:

$$\text{AAD} = \frac{1}{n}\sum_{i=1}^{n}\left|\%\Delta X_i\right| \tag{9}$$

$$\text{Bias} = \frac{1}{n}\sum_{i=1}^{n}\left(\%\Delta X_i\right) \tag{10}$$

where $n$ is the number of data points. Also given in Table 4 is the maximum deviation for each data set.

The comparisons of air data agree well with the equation of state for air (Jacobsen *et al.*[1]), *e.g.*, the AAD for the Howley *et al.*[11] data is 0.05% for this work and 0.04% for the equation of state of Jacobsen *et al.*[1] In the generalized model presented by Lemmon[4], experimental data for air at temperatures less than 70 K deviated from the model by as much as 0.5%. This was due to the incorrect extrapolation behavior of the nitrogen equation below its triple point. Comparisons of the nitrogen-argon-oxygen mixture model of Clarke *et al.*[2] showed nearly identical trends for temperatures below 70 K. Corrections for this behavior have been made in the Helmholtz energy contribution to mixing reported here, and comparisons of low temperature data agree well with the published air equation of state. Similar problems resulting from the extrapolation behavior of the nitrogen equation of state were observed in the calculation of the isochoric heat capacity in the original equation of Lemmon.[4] The equation reported here compares well to isochoric heat capacity data at all temperatures.

**Table 4.** Comparisons of mixture properties calculated from the model to air and mixtures of nitrogen, argon, and oxygen

| Author | No. Points | Temp. Range (K) | Comp. Range | AAD (%) | Bias (%) | MaxDev (%) |
|---|---|---|---|---|---|---|
| **Air -- $P$-$\rho$-$T$** | | | | | | |
| Blanke[10] | 110 | 61-170 | | 0.457 | -0.320 | -6.103 |
| Howley *et al.*[11] | 286 | 67-400 | | 0.049 | 0.004 | 0.889 |
| Kozlov[12] | 348 | 288-873 | | 0.063 | -0.020 | -7.093 |
| Michels *et al.*[13] | 157 | 273-348 | | 0.032 | 0.022 | 0.098 |
| Michels *et al.*[14] | 199 | 118-248 | | 0.202 | -0.175 | -5.247 |
| Rogovaya and Kaganer[15] | 23 | 173-273 | | 0.100 | -0.090 | -0.616 |
| Romberg[16] | 124 | 84-122 | | 0.046 | 0.042 | 0.158 |
| Vasserman *et al.*[17] | 109 | 77-199 | | 0.090 | -0.070 | -0.363 |
| **Air -- $c_v$** | | | | | | |
| Magee[18] | 227 | 66-299 | | 0.501 | 0.138 | 1.810 |
| **Air -- $w$** | | | | | | |
| Ewing and Goodwin[19] | 13 | 255 | | 0.013 | 0.013 | 0.030 |
| Van Itterbeek and De Rop[20] | 44 | 229-313 | | 0.213 | 0.142 | 1.766 |
| Younglove and Frederick[21] | 169 | 90-300 | | 0.286 | -0.160 | -1.655 |
| **Nitrogen-Oxygen -- $P$-$\rho$-$T$** | | | | | | |
| Blagoi and Rudenko[22] | 19 | 67-79 | 0.11-0.80 | 0.523 | 0.523 | 1.091 |
| Knaap *et al.*[23] | 7 | 77 | 0.29-0.91 | 0.150 | 0.150 | 0.212 |
| Kuenen *et al.*[24] | 153 | 135-293 | 0.25-0.50 | 1.039 | -0.893 | -8.692 |
| Pool *et al.*[25] | 7 | 84 | 0.25-0.74 | 0.153 | 0.153 | 0.223 |
| **Nitrogen-Oxygen -- VLE** | | | | | | |
| Armstrong *et al.*[26] | 70 | 65-78 | 0.03-0.94 | 2.411 | 2.313 | 7.532 |
| Cockett[27] | 62 | 81-91 | 0.07-0.81 | 2.378 | 2.328 | 7.969 |
| Din[28] | 119 | 79-116 | 0.10-0.89 | 1.420 | -0.634 | -26.182 |
| Duncan and Staveley[29] | 11 | 63 | 0.10-0.80 | 3.917 | 3.893 | 7.240 |
| Hiza *et al.*[30] | 63 | 63-100 | 0.07-0.88 | 0.996 | 0.486 | 3.817 |
| Pool *et al.*[25] | 11 | 84 | 0.10-0.90 | 1.137 | 1.021 | 2.552 |
| Wilson *et al.*[31] | 138 | 78-136 | 0.05-0.99 | 0.643 | 0.182 | -4.214 |
| **Nitrogen-Argon -- $P$-$\rho$-$T$** | | | | | | |
| Crain and Sonntag[32] | 264 | 143-273 | 0.20-0.80 | 0.322 | 0.264 | 7.500 |
| Kosov and Brovanov[33] | 201 | 293-353 | 0.16-0.81 | 0.204 | -0.194 | -0.810 |
| Maslennikova *et al.*[34] | 88 | 298-423 | 0.26-0.75 | 0.220 | 0.040 | 2.852 |
| Palavra[35] | 203 | 94-106 | 0.32-0.70 | 0.025 | 0.001 | 0.088 |
| Pool *et al.*[25] | 24 | 84 | 0.18-0.89 | 0.494 | -0.494 | -1.194 |
| Ricardo *et al.*[36] | 21 | 119 | 0.52 | 0.134 | 0.104 | 0.281 |
| Townsend[37] | 144 | 298-323 | 0.16-0.84 | 0.065 | 0.033 | 0.685 |
| Zandbergen and Beenakker[38] | 55 | 171-293 | 0.19-0.74 | 0.169 | 0.165 | 1.127 |
| **Nitrogen-Argon -- VLE** | | | | | | |
| Fastovskii and Petrovskii[39] | 56 | 79-103 | 0.05-0.90 | 1.600 | -0.164 | 24.495 |
| Hiza *et al.*[30] | 36 | 85-100 | 0.10-0.89 | 0.291 | 0.187 | 1.898 |
| Miller *et al.*[40] | 14 | 112 | 0.11-0.85 | 0.372 | -0.372 | -1.087 |
| Narinskii[41] | 98 | 80-120 | 0.02-0.96 | 0.430 | -0.031 | -2.913 |
| Pool *et al.*[25] | 12 | 84 | 0.06-0.88 | 0.651 | 0.531 | 1.201 |
| Thorpe[42] | 68 | 81-115 | 0.10-0.91 | 0.820 | 0.750 | 2.607 |
| Wilson *et al.*[31] | 183 | 72-134 | 0.05-1.00 | 1.273 | -0.984 | -23.968 |
| **Argon-Oxygen -- $P$-$\rho$-$T$** | | | | | | |
| Blagoi and Rudenko[22] | 36 | 70-89 | 0.10-0.87 | 0.389 | 0.263 | 1.304 |
| Knaap *et al.*[23] | 6 | 90 | 0.28-0.79 | 0.101 | 0.101 | 0.139 |
| Pool *et al.*[25] | 15 | 84-90 | 0.17-0.86 | 0.108 | 0.108 | 0.159 |
| Saji and Okuda[43] | 20 | 85-87 | 0.25-0.92 | 0.264 | 0.111 | 0.667 |

**Table 4**. Comparisons of mixture properties calculated from the model to air and mixtures of nitrogen, argon, and oxygen (continued)

| Author | No. Points | Temp. Range (K) | Comp. Range | AAD (%) | Bias (%) | MaxDev (%) |
|---|---|---|---|---|---|---|
| **Argon-Oxygen -- VLE** | | | | | | |
| Burn and Din[44] | 140 | 85-118 | 0.10-0.91 | 0.252 | 0.198 | -1.540 |
| Clark *et al.*[45] | 55 | 90-110 | 0.10-0.90 | 0.277 | -0.205 | -1.050 |
| Hiza *et al.*[30] | 16 | 95-100 | 0.11-0.79 | 0.765 | -0.746 | -1.866 |
| Narinskii[46] | 55 | 90-120 | 0.03-0.96 | 0.227 | -0.185 | -0.697 |
| Pool *et al.*[25] | 24 | 84-90 | 0.10-0.90 | 7.983 | 0.501 | -29.560 |
| Wilson *et al.*[31] | 200 | 87-139 | 0.00-0.98 | 0.448 | -0.231 | 12.406 |
| Yorizane *et al.*[47] | 58 | 89-92 | 0.00-0.76 | 0.852 | 0.168 | 6.052 |
| **Nitrogen-Argon-Oxygen -- VLE** | | | | | | |
| Funada *et al.*[48] | 60 | 90 | 0.0-0.004 | 0.282 | 0.083 | 1.457 |
| Narinskii[49] | 115 | 82-120 | 0.04-0.86 | 0.745 | 0.415 | -3.542 |
| Weishaupt[50] | 41 | 81-92 | 0.01-0.89 | 1.107 | -0.805 | -4.390 |
| Wilson *et al.*[31] | 1427 | 78-136 | 0.00-0.99 | 0.596 | -0.041 | -6.746 |

The Kozlov[12] data for air extend to temperatures of 870 K. Because of the generalized and predictive nature of the model, calculated properties of mixtures of nitrogen and argon and of argon and oxygen extrapolated to high temperatures up to 1000 K should be accurate to within ±0.5%. Although the equation of state for oxygen (Schmidt and Wagner[8]) is valid only to temperatures of 300 K, the original work of Lemmon[4] demonstrated that the oxygen equation extrapolates to high temperatures up to 1000 K within ±0.5% in density.

The Maslennikova *et al.*[34] data for the nitrogen-argon mixture extend to 800 MPa, and the model should extrapolate within ±0.5% in density up to 800 MPa for the nitrogen-oxygen and argon-oxygen systems. The magnitudes of the deviations for most of the nitrogen-argon data are similar to those for the extended corresponding states model of Clarke *et al.*[2]

Very few data exist for the nitrogen-oxygen and argon-oxygen mixtures. For the nitrogen-oxygen system, air data were used to determine the equation parameters. Although the temperature range of experimental P-ρ-T data for the argon-oxygen system is 70-90 K with pressures up to 0.2 MPa, the model should be valid for all temperatures and pressures for argon-oxygen mixtures within the ranges of experimental data for the nitrogen-oxygen and nitrogen-argon systems due to the predictive nature of the model. The accuracy of P-ρ-T calculations in the extended range for argon-oxygen mixtures is expected to be within ±0.5%. This claim is due in part to comparisons to the large amount of VLE data available for this mixture.

The deviations for the VLE data of Wilson *et al.*[31] for the nitrogen-oxygen, nitrogen-argon, argon-oxygen, and nitrogen-argon-oxygen mixtures are within ±1.0% and are nearly the same as those for the VLE model developed by Lemmon *et al.*[3] However, for the lower temperature data of Hiza *et al.*[30] and Armstrong *et al.*[26] the deviations for the nitrogen-oxygen system are about 1% and 2.5%, respectively, about 50% larger than those for the VLE model of Lemmon *et al.*[3]

## ACCURACY ASSESSMENT

An assessment has been made to determine the accuracy of the mixture model reported here. Generally, the equation is accurate to ±0.1% in density, ±1% in heat capacity, and ±0.5% for the speed of sound for both binary and ternary mixtures. Calculated bubble

**Table 5**. Regions of stated accuracy of the mixture model

| Mixture | Temperature Range (K) | Maximum Pressure (MPa) |
|---|---|---|
| Nitrogen-Argon | 70-420 | 800 |
| Nitrogen-Oxygen and Air | 60-870 | 300 |
| Argon-Oxygen | 70-90 | 0.2 |
| | 70-400* | 300* |

*No data are available to verify this range, however, the accuracy of the equation should be at least 0.5% in density for the range stated due to the predictive nature of the model.

point pressures are generally accurate to within ±1%. The mixtures and ranges for which calculated properties have been verified by experimental data to be within these accuracies are listed in Table 5. In regions where there are no binary mixture data, the accuracy is estimated to be of the same magnitude. However, this cannot be verified by the authors until experimental data are available to support these conclusions.

New measurements are continuously being made, and these measurements will confirm whether the equation is valid in regions not covered by the experimental data used in the development of this model. These data will enable continued evaluation and development of the model.

## REFERENCES


1. R.T Jacobsen, S.G. Penoncello, S.W. Beyerlein, W.P. Clarke, and E.W. Lemmon, *Fluid Phase Equilib.* 79:113 (1992).
2. W.P. Clarke, R.T Jacobsen, E.W. Lemmon, S.G. Penoncello, and S.W. Beyerlein, *Int. J. Thermophys.* 15(6):1289 (1994).
3. E.W. Lemmon, R.T Jacobsen, and S.W. Beyerlein, "Adv. Cryo. Engr." 37:1107 (1992).
4. E.W. Lemmon, "Generalized Model for the Prediction of the Thermodynamic Properties of Mixtures Including Vapor-Liquid Equilibrium," Ph.D. Dissertation, University of Idaho (1996).
5. E.W. Lemmon and R.T Jacobsen, Submitted to *Int. J. Thermophys.* (1997).
6. R.T Jacobsen, R.B. Stewart, and M. Jahangiri, *J. Phys. Chem. Ref. Data* 15(2):735 (1986).
7. R.B. Stewart and R.T Jacobsen, *J. Phys. Chem. Ref. Data* 18(2):639 (1989).
8. R. Schmidt and W. Wagner, *Fluid Phase Equilib.* 19:175 (1985).
9. E.W. Lemmon, R.T Jacobsen, S.G. Penoncello, and S.W. Beyerlein, "Adv. Cryo. Engr." 39:1891 (1994).
10. W. Blanke, *Proc. 7th Symp. Thermophys. Prop.*, A. Cezairlizan, editor, The American Society of Mechanical Engineers, New York (1977), p. 461.
11. J.B. Howley, J.W. Magee, and W.M. Haynes, *Int. J. Thermophys.* 15(5):801 (1994).
12. A. Kozlov, "Experimental Investigation of the Specific Volumes of Air in the 20-600°C Temperature Range and 20-700 Bar Pressure Range," Dissertation, Moscow Power Engineering Institute (1968).
13. A. Michels, T. Wassenaar, and W. van Seventer, *Appl. Sci. Res.* 4:52 (1954).
14. A. Michels, T. Wassenaar, J.M.H. Levelt Sengers, and W. de Graaf, *Appl. Sci. Res.* 4:381 (1954).
15. I.A. Rogovaya and M.G. Kaganer, *Russ. J. Phys. Chem.* 34(9):917 (1960).
16. H. Romberg, "Neue Messungen der Thermischen Zustandsgrossen der Luft bei tiefen Temperaturen und die Berechnung der Kalorischen Zustandsgrossen mit Hilfe des Kihara-Potentials," VDI-Forsch.-Heft 543, Dusseldorf: VDI-Verlag (1971).
17. A.A. Vasserman, E.A. Golovskii, E.P. Mitsevich, and V.A. Tsymarnyi, "Measurement of the Density of Air at Temperatures of 78 to 190 K up to Pressures of 600 bars," VINITI Deposition No. 2953, Odessa Institute of Marine Engineering, Ukraine (1976).

18. J.W. Magee, *Int. J. Thermophys.* 15(5):849 (1994).
19. M.B. Ewing and A.R.H. Goodwin, *J. Chem. Thermodyn.* 25:423 (1993).
20. A. Van Itterbeek and W. De Rop, *Appl. Sci. Res.* 6(A):21 (1955).
21. B.A. Younglove and N.V. Frederick, *Int. J. Thermophys.* 13(6):1033 (1992).
22. Yu.P. Blagoi and N.C. Rudenko, *Izv. Vyssh. Uchebn. Zaved. Fiz.* 6:145 (1958).
23. H.F.P. Knaap, M. Knoester, and J.J.M. Beenakker, *Physica* 27:309 (1961).
24. J.P. Kuenen, T. Verschoyle, and A.T. Van Urk, *Proceedings of the Royal Academy of Amsterdam* 26:49 (1922).
25. R.A.H. Pool, G. Saville, T.M. Herrington, B.D.C. Shields, and L.A.K. Staveley, *Trans. Faraday Soc.* 58:1692 (1962).
26. G.T. Armstrong, J.M. Goldstein, and D.E. Roberts, *J. Res. Natl. Bur. Stand.* 55(5):265 (1955).
27. A.H. Cockett, *Proc. R. Soc. London* A239:76 (1957).
28. F. Din, *Trans. Faraday Soc.* 56:668 (1960).
29. A.G. Duncan and L.A.K. Staveley, *Trans. Faraday Soc.* 62:548 (1966).
30. M.J. Hiza, A.J. Kidnay, and W.M. Haynes, "Liquid-Vapor Equilibria in Binary Systems Containing Nitrogen, Oxygen, and Argon between 63 and 100 K", NIST, Boulder, CO, unpublished data (1990).
31. G.M. Wilson, P.M. Silverberg, and M.G. Zellner, Paper Q-6 in: "Adv. Cryo. Engr." (1965), p. 192.
32. R.W. Jr. Crain and R. Sonntag, "Adv. Cryo. Engr." 11:379 (1966).
33. N.D. Kosov and I.S. Brovanov, *Inz.-Fiz. Zh.* 36(4):627 (1979).
34. V.Ya. Maslennikova, A.N. Egorov, and D.S. Tsiklis, *Russ. J. Phys. Chem.* 53(6):919 (1979).
35. A.M. Palavra, "Effect of Pressure and Temperature on Excess Properties in the System Argon-Nitrogen," M.S. Thesis, Instituto Superior Tecnico Universidade Tecnica De Lisboa (1979).
36. A.A. Ricardo, S.F. Barreiros, M. Nunes da Ponte, G.M.N. Albuquerque, and J.C.G. Calado, *J. Chem. Thermodyn.* 24:1281 (1992).
37. P.W. Townsend, "Pressure-Volume-Temperature Relationships of Binary Gaseous Mixtures," Ph.D. Dissertation, Faculty of Pure Science, Columbia University, New York (1956).
38. P. Zandbergen and J.J.M. Beenakker, *Physica* 33(2):343 (1967).
39. V.G. Fastovskii and Yu.V. Petrovskii, *Zh. Fiz. Khim., Russ. J. Phys. Chem.* 30:76 (1956).
40. R.C. Miller, A.J. Kidnay, and M.J. Hiza, *AIChE J.* 19(1):145 (1973).
41. G.B. Narinskii, *Russ. J. Phys. Chem.* 40(9):1093 (1966).
42. P.L. Thorpe, *Trans. Faraday Soc.* 64(549):2273 (1968).
43. Y. Saji and T. Okuda, "Adv. Cryo. Engr." 10:209 (1965).
44. I. Burn and F. Din, *Trans. Faraday Soc.* 58(475):1341 (1962).
45. A.M. Clark, F. Din, and J. Robb, *Proc. R. Soc. London* A221:517 (1954).
46. G.B. Narinskii, *Kislorodnyi,* 10(3):9 (1957).
47. M. Yorizane, S. Yoshimura, H. Masuoka, A. Toyama, Y. Nakako, and I. Funada, *Chem. Eng. Sci.* 33:641 (1978).
48. I. Funada, S. Yoshimura, H. Masuoka, and M. Yorizane, "Adv. Cryo. Engr." 27:893 (1982).
49. G.B. Narinskii, *Zh. Fiz. Khim.* 43(2):408 (1969).
50. J. Weishaupt, *Angew. Chem., Ausg. B* 20(12):321 (1948).

# MATHEMATICAL MODEL FOR BOILING HEAT TRANSFER FROM A THIN WIRE IN LIQUID NITROGEN

C. Fusco, E. Bodegom, J.S. Semura, and L.C. Brodie

Department of Physics and
Environmental Sciences and Resources Program
Portland State University
Portland, OR 97207

## ABSTRACT

In observations of steady state nucleate boiling, fluctuations in the temperature and heat flux might initially appear to be completely random. However, for a platinum wire immersed in liquid nitrogen, it has been shown that the fluctuations about the steady state exhibit an average counterclockwise circulation when the heat flux is plotted versus the degree of superheat. A model is constructed to simulate, by discrete time steps, the fluctuations in the boiling heat transfer process from a thin wire immersed in liquid nitrogen. The fluctuations in the wire temperature and heat flux obtained from the mathematical model are compared with experimental data. From the insight gained, several parameters are identified that seem to capture the essential features of the experimental observations.

## INTRODUCTION

The analysis of boiling heat transfer is very complex. The many factors that play a role in this physical transport phenomenon are difficult to identify and the interactions between them cannot be easily isolated. In general, the mathematical description requires the use of a sizable set of interacting nonlinear partial differential equations. Because of this complexity the investigation of boiling phenomena is usually carried out experimentally and by using phenomenological approximations. It is our purpose here to capture some of the essential features of boiling by implementing a few clear physical assumptions. The mathematical model here described, constructed using a formalism similar to that of the Ising model, simulates an experimental setting in which a thin platinum wire is immersed in liquid nitrogen.

In this type of experiment,[1,2,3] a current is passed through the wire while the resistance is measured. The orientation of the wire can be changed from horizontal to vertical. The

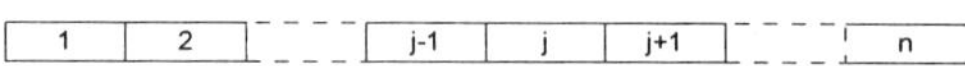


**Figure 1.** Model of the platinum wire.

fluctuation of wire temperatures and heat flux are then calculated using the fluctuation of resistance for both orientations. The aim of those experiments was to gain insight about stability and the transition to film boiling. The fluctuation in temperature and heat flux obtained from the mathematical model can be compared with the experimental data for validation. The mathematical model is useful to identify and classify all the parameters that could have an effect on stability and the transition to film boiling.

In a first approximation we recreate all the essential features of the experimental setting. The only parameter we manipulate is the power supplied. The heat extraction due to conduction and convection alternates between two states: bubbling and no bubbling. The heat extracted by the bubbling process is constant and is assumed to be proportional to the nucleation temperature. The qualitative physical picture obtained with the model is consistent with the heat flux and temperature fluctuations from experiment.[1,2,3]

## THE MODEL

The time dependence of the boiling heat transfer in liquid nitrogen is described by the trajectory in the heat-flux superheat temperature plane of the state of the system. To obtain this trajectory we need to look at the cycle of bubble formation and detachment. Once a bubble detaches from a nucleation site and rises through the liquid, a new thermal boundary layer forms. Then the nucleation site heats up. At a later time, and at a high enough degree of superheat, a new bubble begins to grow at the site. This process keeps repeating.[4,5,6]

We consider the thin wire to be a linear chain of n segments or elements as shown in Figure 1. To each element is associated a positive superheat temperature $T_{i,j}$, where $j$ identifies the position of the element at the time step $i$. There is an interaction only between neighboring elements of the wire, with the effects of the interaction showing at the next time step. If $T_{i,j}$ is the temperature of the element $j$ at the time $i$, then at the time $i$+1 the temperature of the j$^{th}$ element is given by:

$$T_{i+1,j} = T_{i,j} + \lambda\left(\frac{T_{i,j+1} + T_{i,j-1}}{2} - T_{i,j}\right) \tag{1}$$

In Equation (1), $\lambda$ is a constant representing the thermal conduction along the wire, and depends also on the time scale. At each time step, energy is supplied to the system from the external environment (power supply) while energy is lost by the cooling process (to the bath). The supply of energy results in a temperature increase proportional to the supplied power. We denote the heat gain by $\varepsilon_p$ where $\varepsilon_p \propto Pt$, where $P$ is the supplied power and $t$ is the time interval over which power is applied. The energy loss is represented by two different mechanisms. One part represents the heat loss due to conduction and convection, while the other part is the loss due to nucleate boiling or bubbling.

$\varepsilon_c$ represent the energy loss to the bath due to conduction and convection. This flow of heat between the element and the bath is proportional to the difference in temperature between them. In a first approximation $\varepsilon_c$ varies linearly with the cell's superheat temperature:

$$\varepsilon_c \cong \alpha \cdot T_{i,j} \tag{2}$$

The coefficient $\alpha$ is a function of the thickness of the liquid thermal boundary in contact with the j$^{th}$ element and varies with the orientation of the wire. The temperature change of the j$^{th}$ element at the time i+1 without bubble formation is proportional to $\varepsilon_p - \varepsilon_c$:

$$\left(T_{i+1,j} - T_{i,j}\right) \propto \left(\varepsilon_p - \varepsilon_c\right) \tag{3}$$

Let $T_n$ be the heterogeneous nucleation temperature and $T_q$ the quiescent temperature for the fluid. These are defined as follows: we consider $T_n(j)$ to be the temperature at which the liquid surrounding the j$^{th}$ cell starts bubbling and $T_q(j)$ the temperature at which the liquid surrounding the j$^{th}$ cell ceases to bubble. When the liquid surrounding a cell starts bubbling the temperature of that cell drops by an amount proportional to $\varepsilon_b$ until the temperature reaches the quiescent temperature $T_q$, where $\varepsilon_b$ is the heat lost to bubble formation. During bubbling, the combined cooling effects and the supply of energy result in a temperature change proportional to $\varepsilon_p - \varepsilon_c - \varepsilon_b$:

$$\left(T_{i+1,j} - T_{i,j}\right) \propto \left(\varepsilon_p - \varepsilon_c - \varepsilon_b\right) \tag{4}$$

When the cell temperature reaches the quiescent temperature, the liquid surrounding the cell ceases to bubble and its temperature starts to rise again according to Equation (3) until it reaches $T_n$ and the cycle is repeated. The energy $\varepsilon_b$ is related to the rate of formation of bubbles which we assume depends linearly on the nucleation temperature. Since the superheated liquid in contact with the wire will extract more heat per unit time if its temperature is higher, we make the assumption that $\varepsilon_b$ is proportional to $T_n(j)$. During bubbling the total energy lost by cooling depends on Equation (4). The cycle of the heat extraction is shown in Figure 2.

When the cell temperature is decreasing $\left[t(T_n) \text{ to } t(T_q)\right]$ we impose Equation (4); otherwise $\left[t(T_q) \text{ to } t(T_n)\right]$ Equation (3) applies. The change in temperature of the j$^{th}$ cell at a time $i$+1 is given by:

$$T_{i+1,j} = T_{i,j} + \lambda\left(\frac{T_{i,j+1} + T_{i,j-1}}{2} - T_{i,j}\right) + f(T,\dot{T}) \tag{5}$$

where $f(T,\dot{T})$ is given either by Equation (3) or Equation (4), and $\dot{T}$ is the rate of change in temperature.

Equation (5) is consistent with the heat equation, which can be seen as follows. By the Taylor formula:

$$T_{i+1,j} - T_{i,j} = k\frac{\partial T_{i,j}}{\partial t} + \theta(k) \tag{6}$$

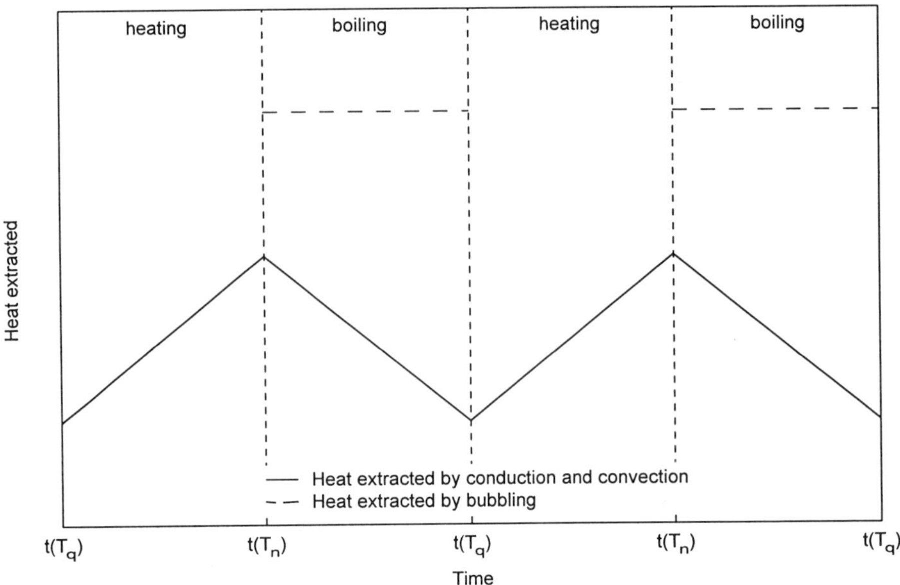


**Figure 2.** During heating without bubbling there is a loss of energy due to convection and conduction. During boiling there is an additional energy loss due to the bubbling. On the *x*-axis we have the time progression of the temperature change. This is not a real time graph and is for illustration purposes only.

where $\theta(k)$ represents the higher order terms, and

$$T_{i,j+1} + T_{i,j-1} = 2T_{i,j} + h^2 \frac{\partial^2 T_{i,j}}{\partial x^2} + \theta(h^2) \tag{7}$$

where h and k are small positive numbers and $\theta(h^2)$ represent the higher order terms. If we set $u = T_{i,j}$ and substitute Equation (6) and (7) into Equation (5) we obtain:

$$\frac{\partial u}{\partial t} = \xi \frac{\partial^2 u}{\partial x^2} + \eta f(u, \dot{u}) \tag{8}$$

where $\xi = \dfrac{\lambda h^2}{2k}$, and $\eta = \dfrac{1}{k}$.

Equation (8) is a nonlinear diffusion equation. $f(u, \dot{u})$ is a nonlinear term which takes into consideration the switch from cooling due to conduction and convection to cooling due to conduction and convection and boiling. While in the Taylor formula *h* and *k* are small positive steps for space and time respectively, in the experimental setting k would be associated with the sampling frequency and *h* with the size of a bubble, i.e., between 0.1 and 1 mm in diameter. If we neglect the second term on the right hand side of Equation (8) we notice that $\xi$ represents the diffusivity $\beta$. Therefore:

$$\xi = \beta = \frac{K}{\rho C} \tag{9}$$

where $K$ is the thermal conductivity, $\rho$ is the density and $C$ is the specific heat of the platinum wire.

## COMPUTER SIMULATION

We consider a linear chain of 100 elements where each element's length is comparable to an average bubble diameter. This is roughly equivalent to a wire of 1 cm in length. In our simulation we are interested in capturing the qualitative behavior of a thin wire immersed in liquid nitrogen under steady-state conditions. The program for the simulation was written in Visual Basic 4.0.

The function for $\varepsilon_c$ was derived from Sinha et al.[7] From their heat flux vs. superheat diagram it can be seen that for small heat fluxes $\varepsilon_c$ varies linearly with the superheat temperature and the constant $\alpha$ can be calculated. For larger heat fluxes, in the nucleate boiling regime, the transfer of heat is more efficient. The transfer can be attributed to the cooling due to conduction and convection ($\varepsilon_c$) and boiling ($\varepsilon_b$). A value for $\varepsilon_b$ can be derived in a similar manner. To each cell is attached a $T_q(j)$ and a $T_n(j)$. The distribution of these fixed temperatures along the wire is random within fixed lower and upper limits. Values for $T_q$ and $T_n$ can be obtained again by referring back to Sinha's heat flux vs. superheat diagram. The lower limit for both $T_q$ and $T_n$ is that value where the nucleate boiling curve meets the convective heat transfer curve. The upper limit is a value beyond the transition to film boiling.

Results for the fluctuations in a wire's average temperature about the liquid nitrogen boiling temperature for several heating power values are shown in Figure 3. The average temperatures are in agreement with the steady state average temperatures obtained experimentally with a platinum wire immersed vertically in liquid nitrogen.[7] The fluctuations in the temperature of the wire have similar frequencies and amplitudes as the experimental data.[1,2,3]

The characterization of the heat transfer properties of a boiling liquid is usually done by plotting the *steady-state* values of heat flux vs. superheat temperature. The heat flux into the liquid per unit wire surface area ($\dot{q}/A$) can be written from the conservation of energy relationship as:

$$\dot{Q} \equiv \frac{\dot{q}}{A} = \frac{P}{A} - \frac{C}{A}\frac{dT}{dt} \tag{10}$$

where $\dot{q}$ is the heat flow rate, $A$ is the surface area of the heater, and $dT/dt$ is the time derivative of the wire temperature.

Figure 4 shows a typical trajectory in the heat flux vs. degree of superheat plane for moderate power inputs under nucleate boiling conditions. The average superheat temperature and the average heat flux have been subtracted from the data to emphasize the fluctuations. The fluctuations in the degree of superheat are about 1 K. The average values and the range of the fluctuations depend on the applied power. The fluctuation diagrams (Figure 3 and 4) are qualitatively similar to the fluctuation diagrams obtained in the experiment.[1,3]

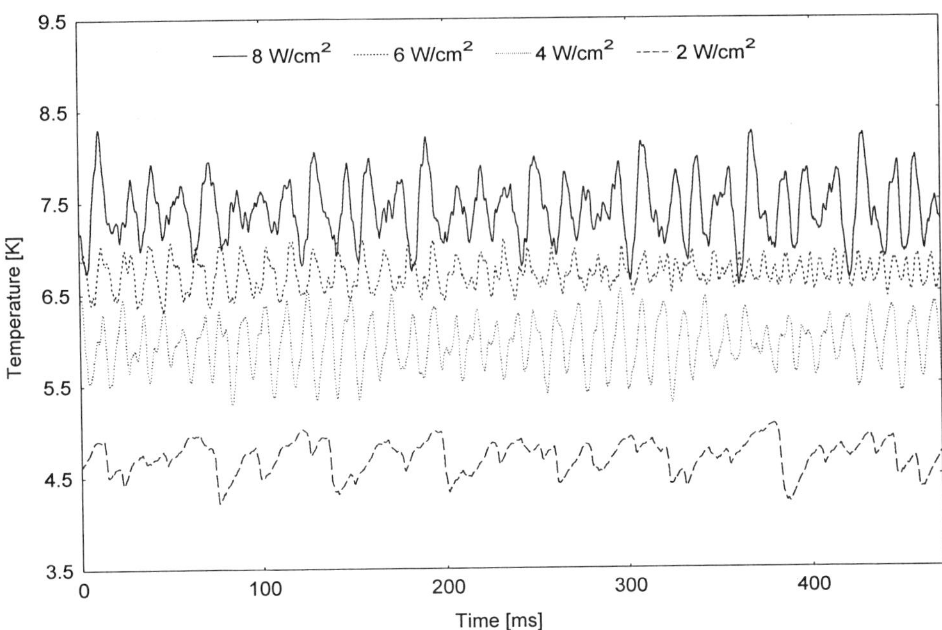


**Figure 3.** Temperature fluctuations in the mathematical model as a function of time for several power inputs.

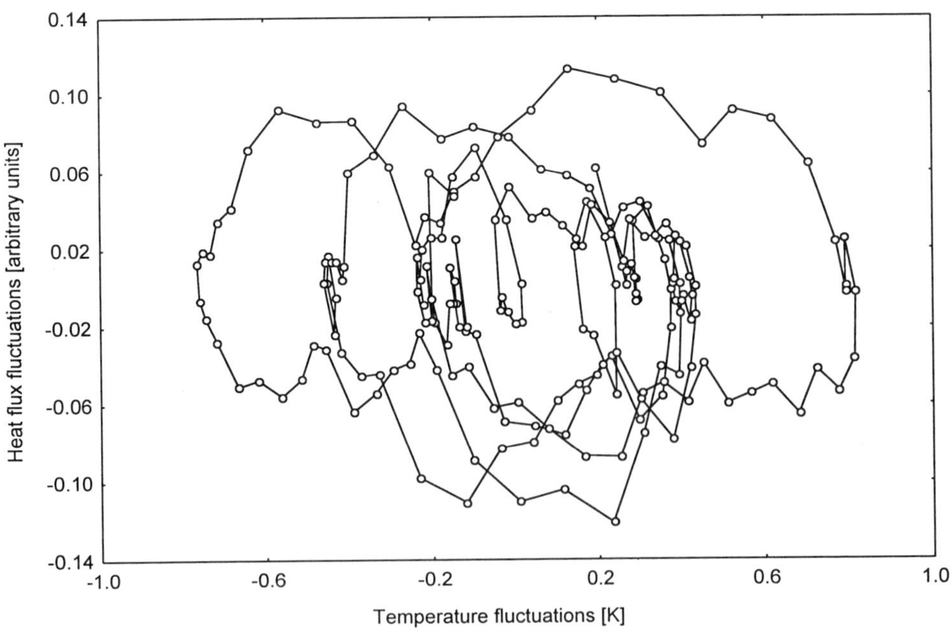


**Figure 4.** Fluctuations in the mathematical model about the average in the heat flux vs. the average superheat temperature.

## CONCLUSIONS

We have developed a simple mathematical model that reproduces the major features of the fluctuations in nucleate boiling from a wire immersed in liquid nitrogen. The results obtained with the computer simulation show the same behavior as the data obtained experimentally. The three main observations generated by the model, which are also supported by experimental observations, are as follows. First, using the calculated values for $\varepsilon_c$ and $\varepsilon_b$, we obtain the same heat flux as a function of degree of superheat. Second, the fluctuations in the superheat are reproduced. This can be seen most clearly in Figure 4. Very similar looking data were obtained previously by Osborne et al.[3] Third, just as in the experiments, the measure used to characterize these fluctuations, the circulation, has a linear dependence on the applied power. These observations suggest that in a first approximation, the assumptions made for the mathematical model are sufficient to capture the general characteristics of the boiling phenomena under consideration. The mathematical conditions are very simple and with only one variable, the power input, most experimental data can be adequately modeled. The functions describing the heat extraction due to conduction, convection and bubbling need more fine-tuning. Future development of the model will include more complex heat extraction functions together with a mechanism to differentiate the horizontal from the vertical wire orientation. We anticipate that the balancing mechanics between the vertical and horizontal functions should provide further insight in the understanding of boiling phenomena.

## REFERENCES


1. E.W Roth, Nucleation and heat transfer in liquid nitrogen, Ph.D. Dissertation, Portland State University, Portland, OR (1993).
2. E.W. Roth, J.S. Semura, L.C. Brodie, and E. Bodegom, Novel approach to the analysis of fluctuations in steady-state nucleate boiling, in: *Advances in Cryogenic Engineering*, **Vol. 39**, (1994) p.1741.
3. W.F. Osborne, E. Bodegom, L.C. Brodie, and J.S. Semura, Influence of heater orientation on fluctuations in steady-state nucleate boiling, in: *Advances in Cryogenic Engineering*, **Vol. 41**, (1996) p. 187.
4. M. Shiotsu, K. Hata, and A. Sakurai, Heterogeneous spontaneous nucleation temperature on solid surface in liquid nitrogen, in: *Advances in Cryogenic Engineering*, **Vol. 35A**, (1990) p.437.
5. S.T. Hsu and F.W. Schmidt, Measured variations in local surface temperatures in pool boiling of water, *J. Heat Transfer, Trans. ASME*, (August 1961) p.254.
6. T.F. Rogers and R.B. Mesler, An experimental study of surface cooling by bubbles during nucleate boiling of water, *A.I.Ch.E. Journal,* (September 1964) p.656.
7. D.N. Sinha, L.C. Brodie, J.S. Semura, and F.M. Young, Premature transition to stable film boiling initiated by power transients in liquid nitrogen, *Cryogenics* **19:225** (1979).

# NUMERICAL ANALYSIS OF HEAT TRANSPORT MECHANISM IN NITROGEN NEAR THE CRITICAL POINT

A. Nakano[1], M. Shiraishi[1], M. Nishio[1], F. Takemura[1] and M. Murakami[2]

[1] Mechanical Engineering Laboratory
Agency of Industrial Science and Technology
Ministry of International Trade and Industry
Namiki 1-2, Tsukuba, Ibaraki, 305 Japan

[2] Institute of Engineering Mechanics
University of Tsukuba
Tennodai 1-1-1, Tsukuba, Ibaraki, 305 Japan

## ABSTRACT

A heat transport mechanism in nitrogen near the critical point is investigated by means of a numerical calculation. The thermofluid equations are solved by using the finite difference method. Calculations confirm a piston effect in nitrogen near the critical point. These results show that thin thermal boundary layers form near the walls while the remaining bulk fluid exhibits a uniform temperature distribution. This suggests a typical feature of the piston effect which is a relatively new mechanism of thermal energy transfer ; i.e., the thermal energy propagates as acoustic waves rather than as heat conduction. The thermal boundary layers become thinner as the system approaches the critical point. The effect of gravity on the heat transport mechanism is also investigated in this report.

## INTRODUCTION

Fluids enter the supercritical state when their temperature and pressure are above the critical temperature and pressure. The liquid-vapor interface no longer exists and the density takes an intermediate state between the liquid and vapor. Near the critical point, the thermophysical properties exhibit a very strange behavior. The specific heat and the isothermal compressibility show a strong divergence at the critical point. The piston effect is one of several typical thermal phenomena that occur near the critical point, particularly since the thermal energy propagates as acoustic wave. This effect originates from the high compressibility and the low thermal diffusivity of supercritical fluids. Heat transport in dense fluids is normally achieved by the basic mechanisms of convection, diffusion and radiation. The piston effect is a fourth mechanism of heat transport that needs to be considered. This effect has been investigated by many researchers. Several experiments have been carried out under both terrestrial and microgravity conditions(1)(2)(3). As for theoretical studies, thermodynamic equations(4)(5)(6) and molecular dynamic equations(6) have been proposed and solved by means of numerical calculation. But most of these studies have only investigated substances with critical points above the normal temperature range. There is no study concerned with cryogens.

We have investigated the heat transport mechanism in cryogenic nitrogen near the critical point. A thermofluid dynamics analysis has been carried out and the governing equations have been solved directly in this study. An effect of gravity on the thermal energy transfer is also investigated by using two-dimensional numerical analysis.

## THEORETICAL CONSIDERATION

The following thermofluid equations(6) have been used to investigate the thermal energy transfer mechanism in nitrogen near the critical point :

$$\frac{\partial \rho}{\partial t}+\frac{\partial (\rho V_i)}{\partial x_i}=0 \tag{1}$$

$$\rho\frac{\partial (V_i)}{\partial t}+\rho V_j\frac{\partial V_i}{\partial x_j}=-\frac{\partial P}{\partial x_i}+\frac{\partial}{\partial x_i}\left(\zeta\frac{\partial V_k}{\partial x_k}\right)+\frac{\partial}{\partial x_j}\left[\eta\left(\frac{\partial V_i}{\partial x_j}+\frac{\partial V_j}{\partial x_i}-\frac{2}{3}\delta_{ik}\frac{\partial V_k}{\partial x_k}\right)\right] \tag{2}$$

$$\frac{dQ}{dt}=\frac{\partial}{\partial x_i}\left(\lambda\frac{\partial T}{\partial x_i}\right)+\sigma'_{ij}\frac{\partial V_i}{\partial x_i} \tag{3}$$

$$\frac{dP}{dt}=\frac{1}{\rho\alpha_T}\frac{d\rho}{dt}+\frac{\beta}{\alpha_T}\frac{dT}{dt} \tag{4}$$

Here, $\rho$, V, P, T and t indicate the density, velocity, pressure, temperature and time variables, respectively. The term Q is the thermal energy per unit volume and $\sigma'_{ij}$ is the viscous stress tensor. The terms $\eta$, $\zeta$ and $\lambda$ represent the dynamic viscosity, the volume viscosity and the thermal conductivity, respectively. The terms $\alpha_T$ and $\beta$ are the isothermal compressibility and the volume expansion coefficient, respectively. The term $\delta_{ik}$ is the unit tensor. Equations (1)-(4) represent the equation of continuity, the momentum equation of viscous fluid, the energy balance equation and the equation of state, respectively. The left term in the Equation (3) can also be expressed as

$$\frac{dQ}{dt}=\rho T\frac{dS}{dt}=\rho T\left\{\left(\frac{\partial S}{\partial T}\right)_P\frac{dT}{dt}+\left(\frac{\partial S}{\partial P}\right)_T\frac{dP}{dt}\right\} \tag{5}$$

$$=\rho C_P\frac{dT}{dt}-\rho T\left(\frac{\partial v}{\partial T}\right)_P\frac{dP}{dt} \tag{6}$$

where, S, v and $C_P$ indicate the specific entropy, specific volume and specific heat at constant pressure, respectively. Hence, the dQ/dt term can be written as

$$\frac{dQ}{dt} = \rho C_P \frac{dT}{dt} - \rho(C_P - C_v)\frac{\alpha_T}{\beta}\frac{dP}{dt} \tag{7}$$

Here, $C_v$ is the specific heat at constant volume. On the other hand, the second term in right hand side of Equation (3) can be written as,

$$\sigma'_{ij}\frac{\partial V_i}{\partial x_i} = \frac{1}{2}\eta\left(\frac{\partial V_i}{\partial x_j} + \frac{\partial V_j}{\partial x_i} - \frac{2}{3}\delta_{ik}\frac{\partial V_k}{\partial x_k}\right)^2 + \zeta\left(\frac{\partial V_k}{\partial x_k}\right)^2 \tag{8}$$

Substituting Equations (4), (7) and (8) into the Equation (3) results in

$$\frac{dT}{dt} + \frac{(\gamma - 1)}{\beta}\frac{\partial V_i}{\partial x_i} = \frac{1}{\rho C_V}\left[\frac{\partial}{\partial x_i}\left(\lambda\frac{\partial T}{\partial x_i}\right) + \frac{1}{2}\eta\left(\frac{\partial V_i}{\partial x_j} + \frac{\partial V_j}{\partial x_i} - \frac{2}{3}\delta_{ik}\frac{\partial V_k}{\partial x_k}\right)^2 + \zeta\left(\frac{\partial V_k}{\partial x_k}\right)^2\right] \tag{9}$$

where $\gamma$ is the ratio of the specific heats, $C_P/C_v$. The governing equations are therefore Equations (1), (2), (4) and (9). These equations are solved by the explicit Mac Cormack method with the FCT algorithm(7).

The thermophysical properties used in this calculation are shown in Table 1. The thermophysical properties $C_v$, $\eta$, $\lambda$ and $\alpha_T$ are obtained from references (8) through (11), respectively. Values of $C_P$ and $\beta$ are evaluated from following thermodynamic relations : $C_P = C_V + \frac{T}{\rho}\left(\frac{\partial P}{\partial T}\right)_\rho^2 \alpha_T$ and $\beta = \left(\frac{\partial P}{\partial T}\right)_\rho \alpha_T$. Here, $\left(\frac{\partial P}{\partial T}\right)_{\rho crit} = 1.715\times10^6$ dyn/cm²/K (11) is used in this study.

**Table 1.** Thermophysical properties of nitrogen

Critical Point of Nitrogen ($T_c$=126.2143K, $P_c$=3.398 MPa, $\rho_c$=314 Kg/m³)

| | Supercritical state | | | Liquid |
|---|---|---|---|---|
| To (K) | 126.2243 | 126.3143 | 127.2143 | 75 |
| To-Tc (K) | 0.010 | 0.100 | 1.000 | — |
| P (MPa) | 3.399715 | 3.415150 | 3.569500 | 0.1 |
| $\rho$ (kg/m³) | 314 | 314 | 314 | 817.05 |
| $C_v$ (J/kg K) | 3400.80 | 2485.77 | 1599.91 | 1076.43 |
| $C_P$ (J/kg K) | $2.0526\times10^7$ | $1.2084\times10^6$ | $7.2989\times10^4$ | 2027.86 |
| $\eta$ (Pa・s) | $210.4\times10^{-7}$ | $198.0\times10^{-7}$ | $185.0\times10^{-7}$ | $1.677\times10^{-4}$ |
| $\lambda$ (W/m K) | 0.0393 | 0.0385 | 0.0335 | 0.1369 |
| $\alpha_T$ (1/Pa) | $1.7358\times10^{-3}$ | $1.0192\times10^{-4}$ | $5.9910\times10^{-6}$ | $5.313\times10^{-8}$ |
| $\beta$ (1/K) | 297.690 | 17.479 | 1.028 | 0.0235 |

## THE CALCULATION MODEL

The one-dimensional (1-D) and two dimensional (2-D) models are shown in Figures 1-a and 1-b. In the case of the 1-D model, we consider a very narrow tube of supercritical nitrogen with a length of 1cm. Heat is applied in step functions from the left side wall. The opposite side wall is maintained at constant temperature. The heated wall is 5 mK higher than the other side wall. Three cases of initial temperature $T_0$, which are 10, 100 and 1000 mK (1K) above the critical temperature are examined in this study. For comparison with the supercritical states, the liquid state is also calculated here.

In the case of the 2-D model, we consider a square cavity with a single, horizontal, heated wall located at the bottom. The top wall is maintained at constant temperature while the two side walls are perfectly insulated. As for the velocity, no slip condition is applied at the walls. In this calculation, the $T_0$ is set 1000 mK above the critical temperature. In this temperature range, the thermophysical properties do not show a strong temperature dependence. Therefore, we set a 50 mK temperature difference between the top wall and the bottom one. The effect of gravity is investigated by using this 2-D model.

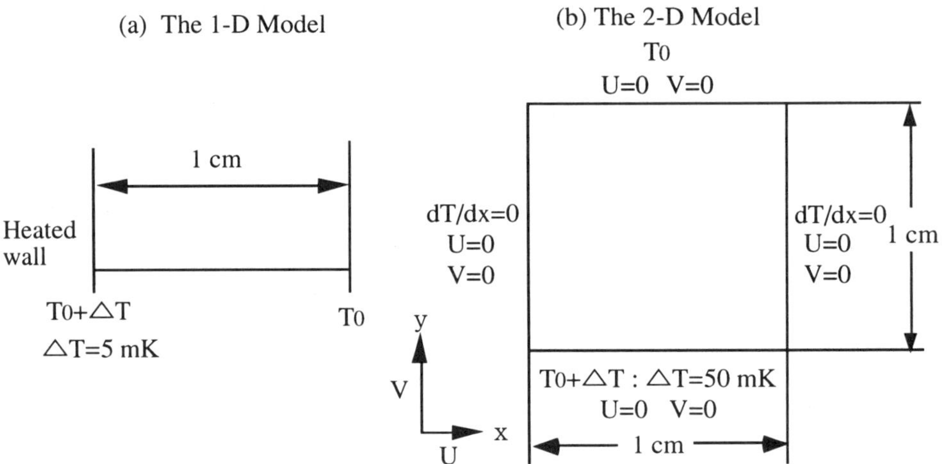


**Figure 1.** Calculation models for supercritical nitrogen.
(a) The 1-D model : The initial temperature of the system $T_0$ is set as 10mK, 100mK or 1000mK above the critical temperature $T_c$. The left sidewall temperature is 5mK higher.
(b) The 2-D model : A square 2-D cavity with a single, horizontal, heated wall located at the bottom. The wall temperature is 50 mK above $T_0$. The initial temperature of the system $T_0$ is fixed at 1000mK above the critical temperature $T_c$.

For the analysis of supercritical fluids, the thickness of the thermal boundary layer is an important factor because all of the physics of the problem are contained in the thermal boundary layer. The order of the thickness $\delta_{BL}$ is given by $\delta_{BL} \propto o(\varepsilon^{0.5} \cdot \mu^{0.25})$ [5]. Here, the $\varepsilon$ is a Knudsen number-like parameter defined by $\varepsilon = P_r \cdot t_a/t_d$, where the $P_r$ is the Prandtl number and $t_a$ and $t_d$ are the characteristic acoustic time and the characteristic time of heat diffusion. The mesh grid size is determined by taking into account the thickness of the thermal boundary layer.

## RESULT OF THE 1-D MODEL

Figures 2-a, 2-b and 2-c with $T_0$ values of 10 mK, 100 mK and 1000 mK above the $T_C$ of nitrogen show the time variations of the temperature distribution. The left-side heated wall is set 5 mK higher than the right-side wall which remains at a constant temperature of $T_0$. It is observed in these figures that a homogeneous temperature field is formed in the bulk of the fluid and steep thermal boundary layers are formed near both walls. This suggests that heat is transported by a piston effect in the nitrogen near the critical point. The increasing rate of the homogeneous temperature field becomes smaller and the thermal boundary layer becomes thinner as the system approaches the critical temperature.

It is concluded from this calculation that the piston effect is initiated by the high compressibility of the critical fluid. A diffusive thin thermal boundary layer formed at the heated wall-fluid interface expands due to the high compressibility. This acts as a piston, generating an acoustic wave which propagates into the bulk of the fluid and is reflected by the opposite side wall. Thermal conversion of this pressure wave can heat the fluid in an adiabatic manner and a spatially homogeneous temperature field is formed by this effect. In the case of Figure 2-c, the compressibility is not as large when compared with the cases of Figures 2-a and 2-b. This indicates that the effect of thermal conductivity becomes rather large. This is evidenced by the broadening of the thermal boundary layers near both walls in Figure 2-c. For comparison with the supercritical state, Figure 2-d shows the result for the case in the liquid state. This figure suggests that the major effect in the liquid state is the heat conduction. Near the critical point, the dominant heat transfer mechanism is the piston effect which is completely different from heat conduction.

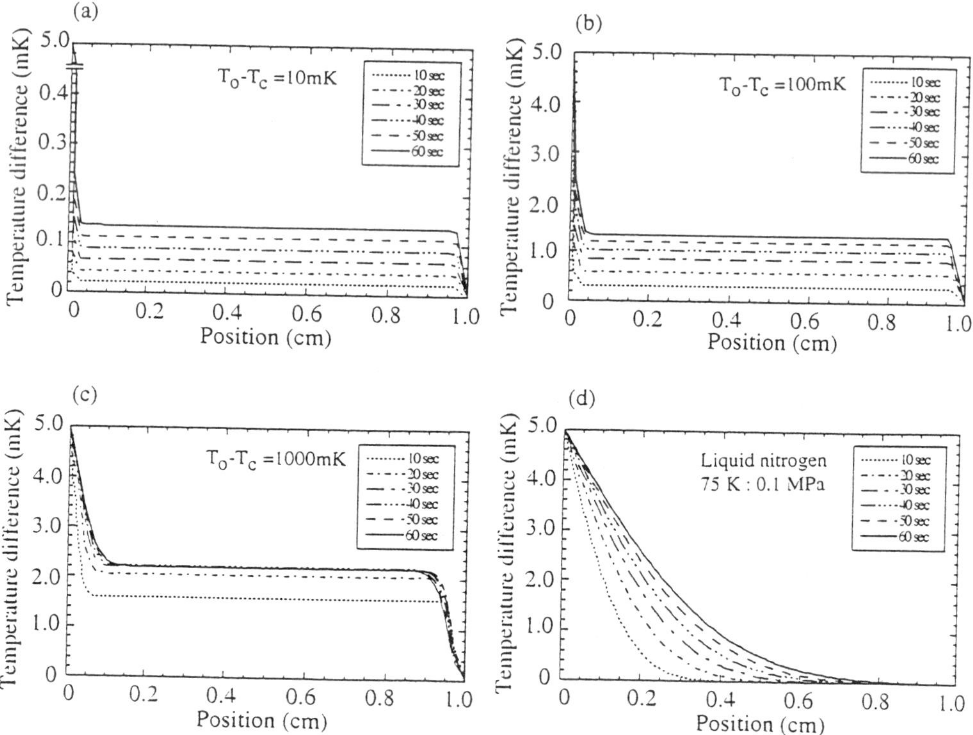


**Figure 2.** Calculation results of the 1-D model.

## RESULT OF THE 2-D MODEL

The 2-D model is used to investigate the gravity effect. Zappoli et al. [(5)] reported that the effect of gravity induces a high velocity field near the critical point. This might prevent the observation of the piston effect. Thus a suitable experimental condition needs to be developed in which the inference of the velocity field is negligibly small. For such a situation a system 1000 mK above the critical point has been selected. In this analysis, the temperature difference between the top and the bottom wall is set at 50 mK, which is ten times larger than the cases investigated in the 1-D analysis.

Under the terrestrial condition, the vertical (y) component of the momentum equation is written as

$$\frac{\partial(\rho V)}{\partial t}+\frac{\partial(\rho UV)}{\partial x}+\frac{\partial(\rho V^2)}{\partial y}=-\frac{\partial P}{\partial y}+\eta\left(\frac{\partial^2 V}{\partial x^2}+\frac{\partial^2 V}{\partial y^2}\right)+\frac{1}{3}\eta\left(\frac{\partial^2 U}{\partial x\partial y}+\frac{\partial^2 V}{\partial y^2}\right)-\rho g \qquad (10)$$

where U and V are the velocity components in the x and y direction, respectively, and g is the acceleration of gravity. As a simple way of considering the gravity effect, we introduce a Boussinesq approximation because the volume expansion coefficient, $\beta$, does not vary appreciably in this temperature range. The pressure P can be written as $P=P_0+P'$, where the $P_0$ is the static pressure and P' is the perturbation of pressure when the velocity is induced. The density is also written as $\rho=\rho_0+\rho'$. In the static condition, g can be written as

$$g=-\frac{1}{\rho_0}\frac{\partial P_0}{\partial y} \qquad (11)$$

Density is given by

$$\frac{1}{\rho}=\frac{1+\beta(T-T_0)}{\rho_0} \qquad (12)$$

From Equations (11) and (12), Equation (10) can be rewritten as

$$\frac{\partial(\rho V)}{\partial t}+\frac{\partial(\rho UV)}{\partial x}+\frac{\partial(\rho V^2)}{\partial y}=-\frac{\partial P'}{\partial y}+\eta\left(\frac{\partial^2 V}{\partial x^2}+\frac{\partial^2 V}{\partial y^2}\right)+\frac{1}{3}\eta\left(\frac{\partial^2 U}{\partial x\partial y}+\frac{\partial^2 V}{\partial y^2}\right)+\rho g\beta\,(T-T_0) \qquad (13)$$

The calculation results are shown in Figure 3. Figure 3-a shows the velocity field while Figure 3-b shows the temperature distribution 1.2 sec after the heater is powered on. The interval of the temperature contour lines is set at 0.5 mK. Figure 3-c shows the temperature profile at the center location on the x-axis in Figure 3-b. In Figure 3-a, the maximum velocity is $2\times10^{-3}$ mm/s, which is an extremely small value. Figures 3-b and 3-c suggest that the homogeneous temperature distribution is formed in the bulk of the fluid and steep thermal boundary layers appear near the top and bottom walls. This suggests that heat is essentially transported by a piston effect. Certainly, the temperature is weakly affected by the velocity field near the both side walls in Figure 3-b, but the variations are very small. In this calculation, the density of the fluid near the heater decreases due to the heating and the density near the other side wall increases because of the compression; however, the fluctuations in the density are within 0.2 % when compared with $\rho_c$. Therefore, the buoyancy induced by the density differences is very small. In fact, the induced velocity is also very small. This calculation shows that the effect of convection generated by gravity is very small at a temperature which is 1000 mK above the critical point and the piston effect can be observed under terrestrial condition, at least in the first few seconds.

## CONCLUSION

We have investigated a heat transfer mechanism in supercritical nitrogen near the critical point. The results of the 1-D numerical calculation suggest that a homogeneous temperature field is formed in the bulk of the fluid and steep thermal boundary layers are formed near both walls. This suggests that the thermal energy propagates as an acoustic wave rather than via heat conduction at temperatures near the critical point. Thus the heat is transported essentially by a piston effect in supercritical nitrogen near the critical point.

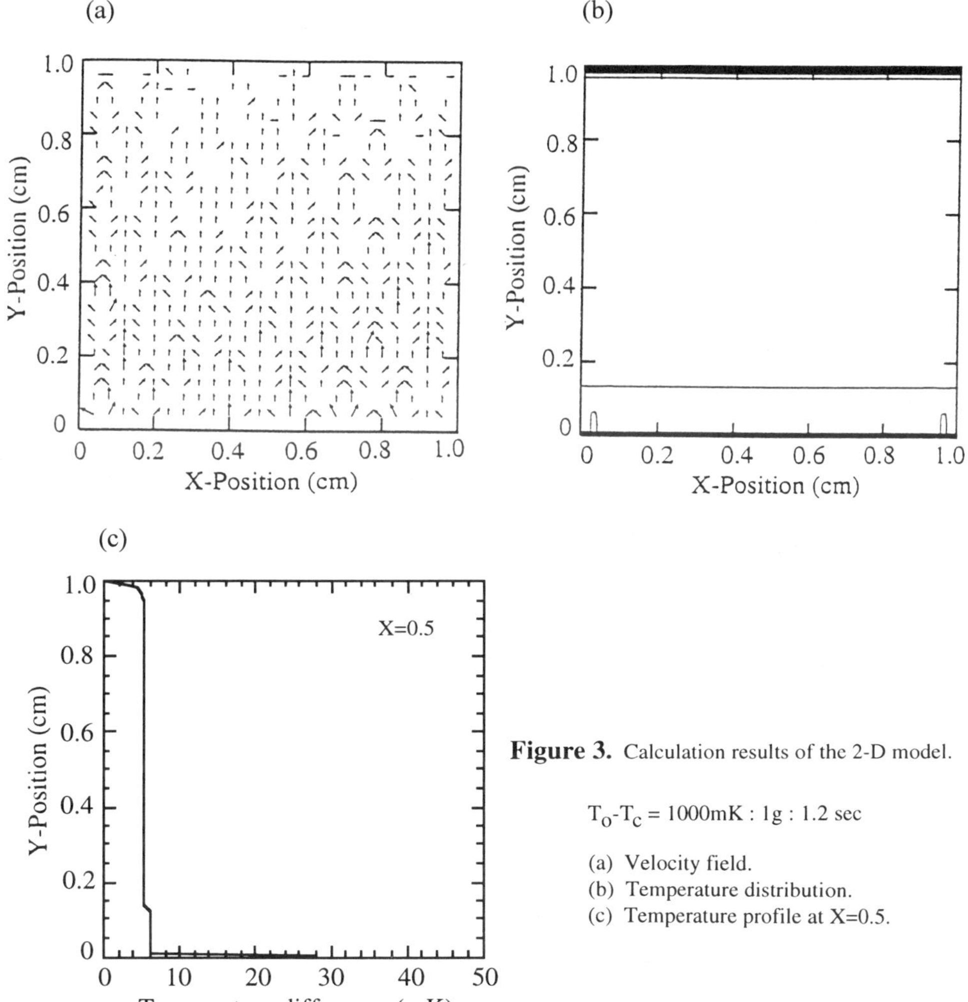


**Figure 3.** Calculation results of the 2-D model.

$T_0$-$T_c$ = 1000mK : 1g : 1.2 sec

(a) Velocity field.
(b) Temperature distribution.
(c) Temperature profile at X=0.5.

The results of the 2-D numerical calculation suggest that the effect of gravity is very small when the temperature is 1000 mK above the critical point. This provides a guide line for an experimental studies of the piston effect in nitrogen under terrestrial conditions.

## ACKNOWLEDGMENT

The authors acknowledge support from RIPS Center in the Agency of Industrial Science and Technology. This work was proceeded by using the Cray C90 in the RIPS Center. The authors especially wish to thank Mr. N. Itoga of the National Defense Academy for useful suggestions relative to the numerical calculation techniques.

## REFERENCES

1. Guenoun, P., Khalil, B., Beysens, D., Garrabos, Y., Kammoun, F., Neindre, B.L., and Zappoli, B., Thermal cycle around the critical point of carbon dioxide under reduced gravity, Phys. Rev. E, 47 (3) (1993) pp. 1531-1540
2. Beysens, D., New critical phenomena observed under weightlessness, materials and fluids under low gravity, 464 (1995) pp. 3-25
3. Straub, J., Eicher, L. and Haupt, A., Dynamic temperature propagation in a fluid near its critical point observed under microgravity during the german spacelab mission D-2, Phys. Rev. E, 51 (6) (1995) pp.5556-5563
4. Onuki, A., Ferrell, R. A., Adiabatic heating effect near the gas-liquid critical point, Physica A, 164 (1990) pp.245-264
5. Zappoli, B., Amiroudine, S., Carles, P., Ouazzani, J., Numerical solutions of thermoacoustic and buoyancy-driven transport in near critical fluid, Materials and Fluids under Low Gravity, 464 (1995) pp.27-40
6. Maekawa, T., Ishii, K., Temperature propagation in a single component critical fluid, Thermal Science and Eng. (The Heat Transfer Soc. of Japan), 5 (1) (1997) pp.15-23
7. Fletcher, C.A.J., Computational technique for fluid dynamics, Springer Verlag (1988)
8. Voronel, A. V., Gorbunova, V. G., Chashkin, Yu. R, and Shchekochikhina V. V., Specific heat of nitrogen near the critical point, Sov. Phys. JETP, 23 (1966) pp.597-601
9. Basu, R. S., Sengers, J. V., Viscosity of nitrogen near critical point, J. of Heat Transfer, 101 (1979) pp.3-8
10. Ziebland, H., and Burton, J. T. A., The thermal conductivity of nitrogen and argon in the liquid and gaseous state, British J. of Appied Phys., 9 (1958) pp 52-59
11. Pestak, M. W. and Chan, M. H. W., Equation of state of N2 and Ne near their critical points. scaling, corrections to scaling, and amplitude ratios, Physical Rev. B, 30 (1) (1984) pp.274-288

# EFFECT OF SURFACE CONDITIONS ON POOL BOILING CRITICAL HEAT FLUXES IN SATURATED AND SUBCOOLED LIQUID NITROGEN AT VARIOUS PRESSURES

A. Sakurai,[1] M. Shiotsu,[2] K. Hata[3] and K. Fukuda[4]

[1]Future Energy Research Assoc.
c/o Research Institute for Applied Science, Kyoto 606, Japan

[2]Dept. of Energy Science and Technology, Kyoto University
Uji, Kyoto 611, Japan

[3]Institute of Advanced Energy, Kyoto University
Uji, Kyoto 611, Japan

[4]Dept. of Nuclear Engineering, Kobe University of Mercantile Marine
Kobe 658, Japan

## ABSTRACT

Critical heat fluxes (CHFs) in the fully developed nucleate boiling (FDNB) regime were measured for a 1.2 mm diameter horizontal cylinder with commercial or rough surface (CS or RS) in a pool of liquid nitrogen for subcooling covering a range from 0 K to about 40 K at pressures ranging from 0.3 to 2 MPa. The CHFs for the cylinders with the two different surfaces were compared under the same experimental conditions. Measured all data of the CHFs for both surface conditions were well explained by the two different CHF mechanisms ; one is due to the hydrodynamic instability(HI), and another is due to the heterogeneous spontaneous nucleation in originally flooded cavities. The CHFs due to the HI obtained for both surface condition agree well with the values derived from the saturated and subcooled CHF correlations : the former was given by Kutateladze and the latter was given by the authors, respectively. It was clarified that the CHFs due to the HI are independent of the surface conditions. Little effect of the surface condition tested here was observed on the CHFs due to the HSN except the CHFs at a pressure of 2047 kPa for low subcooling near saturated conditions. An empirical correlation for the CHFs due to the HSN is presented.

## INTRODUCTION

Correct understanding and predictions for the critical heat flux in the fully developed nucleate boiling(FDNB) regime at which heat transfer crisis occurs on a solid surface in a pool of saturated and subcooled liquid nitrogen become important as the database for stability design of high-$T_C$ temperature superconductors increasing year by year.

Vapor behavior at the direct transition from natural convection regime to film boiling on a horizontal cylinder at atmospheric pressure due to a steadily increasing heat input was observed by the authors using high-speed video camera.[1,2] A large vapor tube around the cylinder was first formed by the explosive-like initial boiling which was clearly different from that due to active cavities entraining vapor, and then by the growth of a Taylor unstable wave on the top of vapor-liquid interface of the vapor tube, large vapor bubbles break away from the vapor tube due to existing buoyancy force. After the collapse of the vapor tube due to the detachment of the vapor bubbles, a new thin vapor film occurred. The thin vapor film was similar to that corresponding to steady-state film boiling.

This direct transition was well explained by assuming[1,2] that it occurs due to the heterogeneous spontaneous nucleation(HSN) in originally flooded cavities. There exists no active cavities entraining vapor for initial boiling on a horizontal cylinder in liquid nitrogen and helium I due to a steadily increasing heat input. The surface tension of liquids are so low that vapor is not entrained in the surface cavities, and there is no dissolved gases in liquid helium I or in liquid nitrogen except for possible trace amounts of helium, hydrogen and neon.

On the other hand, after the detachment of vapor bubbles from the vapor tube at pressures higher than around 300 kPa, the nucleate boiling from the cavities entraining vapor appears with a decrease in surface superheat down to that in the FDNB regime; thus, the heat transfer crisis occurs in the FDNB regime at a critical heat flux due to a steadily increasing heat input. The authors have suggested that the CHFs in the FDNB regime in saturated and subcooled liquid nitrogen, helium I, ethanol and water[1–6] at various pressures occur not only due to the hydrodynamic instability(HI)[7,8] but also due to the heterogeneous spontaneous nucleation (HSN) at the lower limit of the HSN surface superheat. The CHF due to the HSN is lower than the value supposed due to the HI. The process and its mechanism of transition from initial boiling to the FDNB regime is provided in detail elsewhere with photographs of the boiling behavior near the cylinder surface.

The purpose of this paper is first to present experimental data of the CHF in the FDNB regime on horizontal cylinders with commercial and rough surfaces in saturated and subcooled liquid nitrogen at pressures ranging from about 300 kPa to 2 MPa with subcooling from 0 K to 40 K, second to clarify two different CHF mechanisms based on the data obtained and the effect of surface conditions on the CHFs due to the mechanisms, and third to present the corresponding CHF correlations for the mechanisms.

## EXPERIMENTAL APPARATUS

A vacuum insulated cylindrical pressure vessel with an inner diameter of 20 cm and the height of 70 cm was used in the experiment as shown in Fig. 1. The maximum working pressure is 3.5 MPa. The vessel is connected by a valve to a liquid feed tank. The pressure of the vessel is automatically controlled to the desired pressure of $\pm 1$ kPa by a pressure controller consisting of a transducer and a control valve. A sheathed 1.0mm-diameter copper-constantan thermocouple is inserted in the vessel to measure the temperature of bulk liquid. The vessel has two sight ports to observe the boiling behavior on the test heater using a high-speed video camera with a speed of 200 frames/s and a rotary shutter exposure of 0.5 ms. Experiments were carried out using the two 1.2mm-diameter platinum horizontal cylinders with a length of about 50 mm provided with commercial or rough cylinder surfaces. The latter was finished by using Emery-3 paper to roughen the surface. The electrical potential taps of fine $30\mu$m-diameter platinum wires were spot-welded about 50 mm apart on the test cylinder surface. The heat transfer on the cylinder surface between the taps was measured. These test heaters were annealed and an electrical resistance versus temperature behavior was obtained for over range in the temperatures between 77 K and 453 K in both of liquid nitrogen, water and glycerin baths.

The test heaters were heated electrically by the power source with maximum capacity of 12V$\times$600A. The average temperature of the test heater between the potential taps was measured the resistance thermometry using the heater itself as a branch of a double bridge circuit. The signal voltage of the bridge circuit, together with the voltage drops across the potential taps and across a standard resistance due to the heating current,

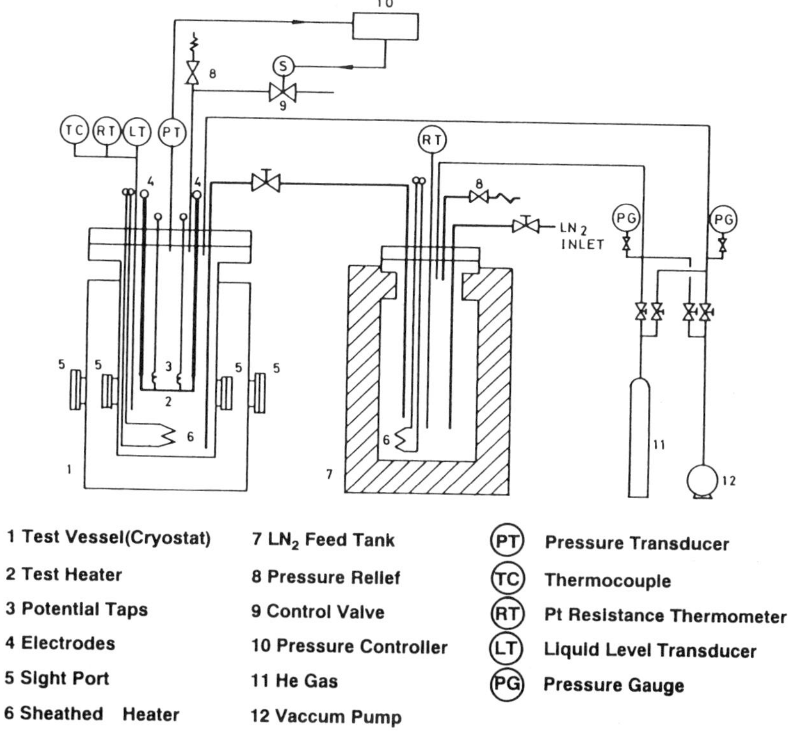


**Figure 1.** Schematic of experimental apparatus.

were amplified and sent to the digital computer through analog-to-digital converters. The average temperature is calculated from the signal voltages by using the previously determined resistance-temperature relation. Measurement errors for the surface temperatures and the heat fluxes were estimated to be about ±1 K and ±2 %, respectively.

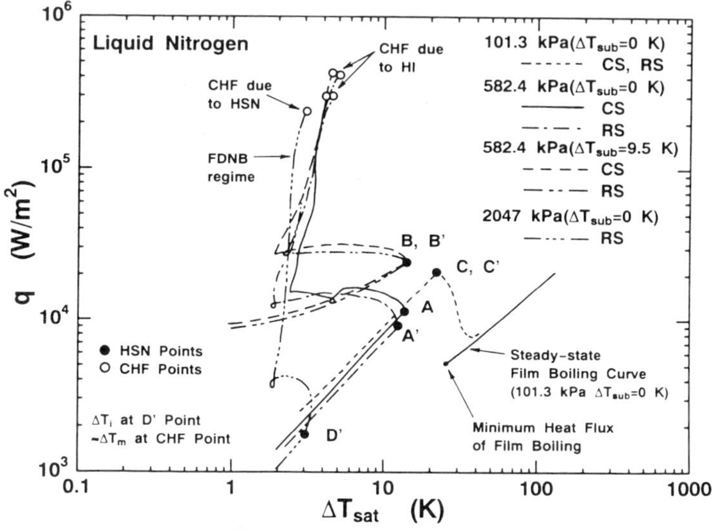


**Figure 2.** Heat transfer processes for pressures of 101.3, 582.4 and 2047 kPa.

## EXPERIMENTAL RESULTS AND DISCUSSION

### Typical Heat Transfer Processes up to CHFs in Fully Developed Nucleate Boiling Regime

Typical heat transfer processes resulting from an exponentially increasing heat input, $Q_0 e^{t/\tau}$, with a period, $\tau$, of 20 s on 1.2mm-diameter horizontal cylinder in saturated and subcooled liquid nitrogen are shown on the graph of heat flux, $q$, versus surface superheat, $\Delta T_{sat}$, for both surface conditions of commercial and rough surfaces in Fig. 2. The heat input corresponds to a steadily increasing one: the CHFs obtained almost agree with those obtained for the heat inputs with the periods longer than 20 s for the same conditions tested.

The processes for $\Delta T_{sub} = 0$ at atmospheric pressure for both surface conditions are represented by a single curve showing the direct transition from the steadily increasing natural convection regime (which agrees with theoretical natural convection curve) to film boiling. After the transition, the heat flux decreases, and then increases respectively with increasing surface superheat through around the steady-state film boiling curve as shown in the figure. The curve was given by authors' film boiling correlation.[9] The surface superheats at C and C' points almost agree with each other ; namely, there is no effect of surface conditions on the surface superheat at the direct transition. At pressures lower than atmospheric, the direct transitions were clearly observed for the heat inputs with the increasing rate ranging from steady to rapid ones. At pressures higher than around 300 kPa, the direct transition was not observed for a steadily increasing heat input, though the direct transitions from the transient conduction regime was observed for an increasing heat input. As already described in previous papers by Sakurai et al.[1-6], it was assumed that the direct transition occurs due to the explosive-like heterogeneous spontaneous nucleation in originally flooded cavities on the cylinder surface.

Typical processes up to the CHFs for the subcooling of 0K and 9.5 K for both surface conditions at a pressure of 582.4 kPa are also shown in Fig. 2. The initial boiling occurs at points A and B for subcoolings of 0 K and 9.5 K for the CS, and A' and B' for the same subcoolings for the RS. At the locations, the large vapor tubes around the cylinders rapidly appear similar to that at point C at atmospheric pressure due to the HSN. The vapor tubes then collapse rapidly and large vapor bubbles break away from the cylinder surface. After the detachment of the vapor bubbles, a nucleate boiling appears from active cavities entraining vapor with a decrease in surface superheat down to that of the FDNB regime. The initial surface superheat on the same surface for the $\Delta T_{sub}$ values of 0 K and 9.5 K almost agree with each other and are independent of the subcooling. Those for the CS and RS for the same subcooling, namely those for A and A' and those for B and B', also almost agree with each other, though the heat fluxes for the $\Delta T_{sub}$ of 9.5 K are larger than those for the $\Delta T_{sub}$ of 0 K. The CHFs in the FDNB regime for both surface conditions at a pressure of 582.4 kPa for each subcooling of 0 K and 9.5 K agree with each other respectively ; the CHFs are independent of different surface conditions since the CHFs are caused by the hydrodynamic instability.

It should be noted that the surface superheats at the direct transition points C and C' under saturated conditions almost agree with the minimum film boiling surface superheat as shown in Fig. 2. It was already assumed[10] that a minimum film boiling surface superheat at which the vapor film collapses, is determined by the lower limit of the HSN surface superheat at the same condition.

### Saturated CHFs at Various Pressures for Both Surface Conditions

The CHFs for both surface conditions under saturated conditions at pressures up to about 2 MPa were measured as shown in Fig. 3 and compared with the curve obtained from the saturated CHF correlation with a $K_1$ value of 0.17 in Eq.(1) as given by Kutateladze[7]. The curve agrees with the CHFs for both surface conditions at pressures lower than about 1 MPa. However, the CHFs at pressures higher than 1.5 MPa become lower than those predicted by Eq. (1); the data suggest that there is another

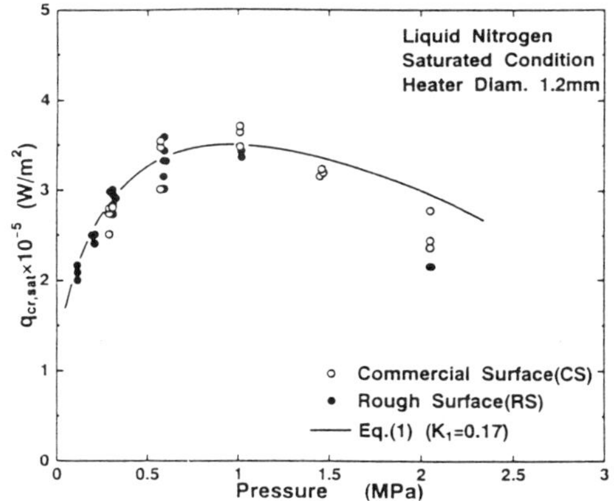


Figure 3. Critical heat fluxes under saturated conditions at various pressures.

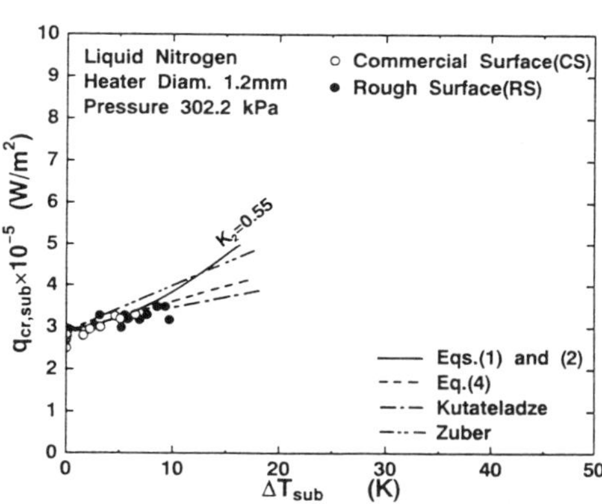


Figure 4. Critical heat fluxes for liquid subcoolings at 302.2 kPa.

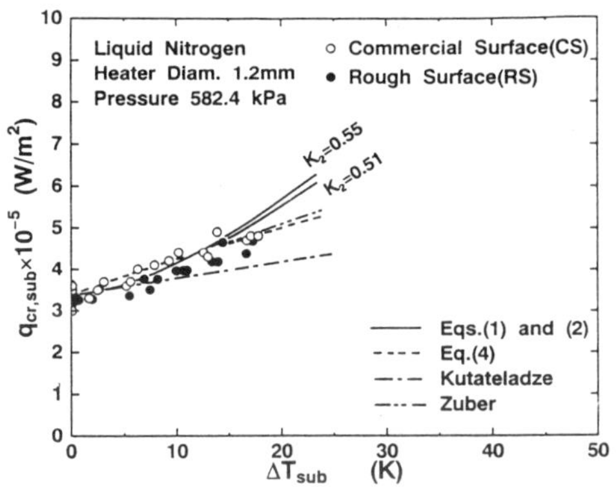


Figure 5. Critical heat fluxes for liquid subcoolings at 582.4 kPa.

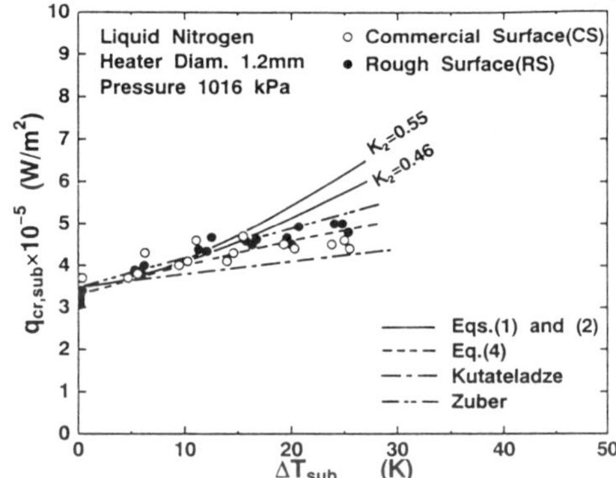


Figure 6. Critical heat fluxes for liquid subcoolings at 1016 kPa.

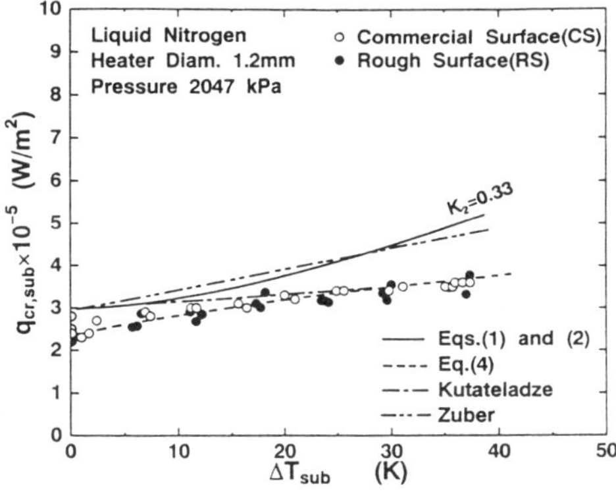


Figure 7. Critical heat fluxes for liquid subcoolings at 2047 kPa.

Table 1. Constants in Eqs.(1), (2) and (4) for Liquid Nitrogen

| Heater | Pressure | $K_1$ | $K_2$ (*) | A | B | C | $q^*_{cr,sat}$ |
|---|---|---|---|---|---|---|---|
| Horizontal Cylinder (1.2mm dia.) | 302 kPa | 0.17 | 0.55 | 0.378 | -0.104 | 0.0111 | $2.8\times10^5$ |
| | 582.2 kPa | ↑ | 0.51 | ↑ | ↑ | ↑ | $3.4\times10^5$ |
| | 1016 kPa | ↑ | 0.46 | ↑ | ↑ | ↑ | $3.3\times10^5$ |
| | 2047 kPa | ↑ | 0.33 | ↑ | ↑ | ↑ | $2.4\times10^5$ |

(*) $K_2$ values obtained by Eq.(3)

mechanism for the CHFs besides the hydrodynamic instability model given by Eq.(1). The effects of surface conditions on the CHFs were clearly observed at a pressure of 2047 kPa as shown in the figure.

### The CHFs in the FDNB Regime versus Subcooling for Both Surface Conditions at Various Pressures

The CHFs in the FDNB regime for various subcoolings, $\Delta T_{sub}$, at pressures of 302.2, 582.4, 1016 and 2047 kPa for both surface conditions are shown in Figs. 4, 5, 6 and 7. As shown in Fig. 6, the typical trend of the CHFs versus subcooling at a pressure of 1016 kPa is as follows: the increasing rate of CHF for high subcooling is lower than that of the CHF for lower subcooling. This trend means that two CHF mechanisms exist for lower and higher subcooling : it was already suggested by the authors[5] that the former occurs due to hydrodynamic instability and the latter occurs due to the HSN at the lower limit of the HSN surface superheat in previously flooded cavities in the FDNB regime. The lower limit of the HSN surface superheat for a certain subcooling at a specified pressure is measured as the incipient boiling superheat caused by an exponentially increasing heat input with a period longer than 20 s.

**The CHF Correlation based on the Hydrodynamic Instability Model.** The CHFs for subcooling on the horizontal cylinders with diameters of 1.2, 2 and 5 mm placed in water have been observed[5] by the authors for wide ranges of subcoolings up to about 170 K and pressures up to about 2 MPa; it was deduced[5] from the results that two CHF mechanisms exist due to the hydrodynamic instability and to heterogeneous spontaneous nucleation. The CHF correlation based on the hydrodynamic instability for lower subcooling expressed by Eqs. (1) and (2) was presented by Sakurai et al..[5]

$$q_{cr,sat} = K_1 h_{lg} \rho_v [\sigma g(\rho_l - \rho_v)/\rho_v^2]^{1/4} \tag{1}$$

$$q_{cr,sub} = q_{cr,sat}[1 + K_2(\rho_l/\rho_v)^{0.69}(c_{pl}\Delta T_{sub}/h_{lg})^{1.5}] \tag{2}$$

Equation (2) which includes Kutateladze's saturated CHF correlation was obtained by modifying Kutateladze's subcooled CHF correlation[7] taking into account the nonlinear effect of subcooling based on the experimental data obtained for wide ranges of subcooling and pressure on horizontal cylinders for water; his correlation did not provide a good correlation with the experimental data. The values of $K_1$ and $K_2$ were determined by the data obtained for the cylinder diameters of 1.2 mm, 2 mm, 5 mm by the authors and 5.8 mm by Ponter & Haigh[11], and for wide ranges of subcooling and pressure in water.

Figure 8 plots the constant $K_1$ and $K_2$ of Eqs. (1) and (2) for a dimensionless length $L'(= D/2\sqrt{g(\rho_l - \rho_v)/\sigma})$: the $K_1$ values for horizontal cylinders are close to 0.17 and are independent of the cylinder diameters tested and the $K_2$ values are expressed by the curve represented by Eq.(3). The curve for the $K_1$ derived from the existing correlation[12] which depends to large extent on short values of $L'$, is also shown in the figure. As shown in Figs.3 and 8, the $K_1$ value for liquid nitrogen at pressures of 302.2, 582.4 and 1016 kPa except the pressure of 2047 kPa is 0.17. The $K_2$ values of lower subcooling at pressures up to 1016 kPa are to be around 0.55 judging from the data obtained as shown in the figure. On the other hand, it is possible to derive the $K_2$ values from the correlation, Eq. (3), based on water data.

$$K_2 = 0.38(1 - log_e L') \tag{3}$$

The $K_2$ values 0.55, 0.51 and 0.46 are obtained for corresponding $L'$ values of 0.638, 0.707 and 0.808; those depend on pressures in liquid nitrogen of 302.2, 582.4 and 1016 kPa. On the contrary, the $K_2$ values for water are independent of the pressures, and only depend on the cylinder diameters as shown in Fig. 8. The curves derived from Eq. (2) with a $K_2$ value of 0.55, and the curves for $K_2$ derived from Eq. (3) are shown in Figs. 4, 5 and 6. It seems that the curves based on Eqs. (2) and (3) approximately agree

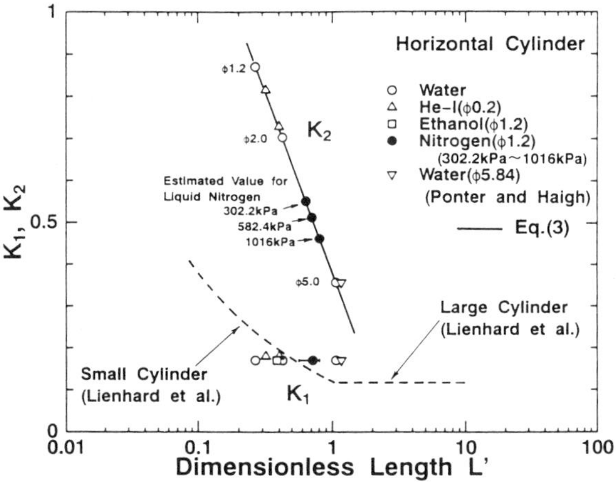


**Figure 8.** $K_1$ and $K_2$ values obtained by various liquids versus dimensionless length.

with the data for lower subcooling obtained for both surface conditions as shown in the figures. The curves derived from Kutateladze and Zubers' subcooled CHF correlation[7,8] are also shown in Figs. 4, 5 and 6. More research of the subcooled CHF correlation based on the HI model is necessary to confirm the values of $K_2$ for many other liquids.

**CHF Correlation based on the Heterogeneous Spontaneous Nucleation Model.** As shown in Figs. 4, 5 and 6, the curves derived from Eq. (2) almost agree with the data for lower subcooling. However, the data for higher subcooling are lower than the corresponding values of the curve. The CHFs for higher subcooling at which the transitions from the FDNB regime to film boiling occur are due to the explosive-like HSN at the lower limit of the HSN surface superheat and the CHFs are lower than those derived from Eq. (2). It was confirmed in the research that the surface superheat of the CHF, $\Delta T_m$, for large subcooling at high pressures agrees with the lower limit of the HSN surface superheat, $\Delta T_i$. The heat transfer process from incipient superheat, $\Delta T_i$, in the natural convection regime to the CHF surface superheat, $\Delta T_m$, in the FDNB regime at a saturated pressure of 2047 kPa for the cylinder with RS caused by a steadily increasing heat input is shown in Fig. 2 as typical ; the $\Delta T_m$ almost agrees with the $\Delta T_i$ which corresponds to the lower limit of the HSN surface superheat as shown in the figure, and the CHF at the $\Delta T_m$ is less than that derived from Eq. (1) with a $K_1$ value of 0.17 as shown in Fig. 3. The CHF due to the HSN occurs even under saturated condition at the high pressure. On the other hand, the $\Delta T_m$ values are clearly lower than the corresponding $\Delta T_i$ values under saturated condition at 101.3 kPa and 302.2 kPa, and under the subcooled condition at a pressure of 302.2 kPa for both surface conditions. This indicates that the CHFs occur not due to the HSN but due to the HI. The detailed results on the relation between $\Delta T_m$ and $\Delta T_i$ at various pressures for high subcooling including effect of surface conditions will appear elsewhere. It should be noted that as shown in Figs. 3 and 7 the CHFs at the higher pressures such as 2047 kPa occur due to the HSN in the FDNB regime at all subcooling including 0 K, namely at saturated conditions. The following empirical equation to express the CHFs was obtained:

$$q_{cr,sub} = q^*_{cr,sat}[1 + (\rho_l/\rho_v)^{0.5}\{A(c_{pl}\Delta T_{sub}/h_{lg}) + B(c_{pl}\Delta T_{sub}/h_{lg})^2 + C(c_{pl}\Delta T_{sub}/h_{lg})^3\}] \tag{4}$$

where $q^*_{cr,sat}$ is the CHF at saturated conditions due to the heterogeneous spontaneous nucleation. The constants $A$, $B$, $C$ and $q^*_{cr,sat}$ in the relation are obtained from the experimental data and shown in Table 1. The curves derived from the equation are shown in Figs. 4 to 7.

## CONCLUSIONS

Critical heat fluxes (CHFs) in the fully developed nucleate boiling (FDNB) regime on 1.2mm- diameter horizontal cylinders with commercial and rough surfaces in a pool of liquid nitrogen due to steadily increasing heat inputs were measured for subcoolings ranging from 0 K to about 40 K at pressures ranging from about 0.3 to 2 MPa.

Measured all data of the CHFs for both surface conditions were well explained by the two different CHF mechanisms ; one is due to the hydrodynamic instability(HI), and another is due to the heterogeneous spontaneous nucleation in originally flooded cavities.

The CHFs due to the HI obtained for both surface condition agree well with the values derived from the saturated and subcooled CHF correlations : the former was given by Kutateladze and the latter was given by the authors, respectively. It was clarified that the CHFs due to the HI are independent of the surface conditions.

Considerable effect of the surface conditions on the CHFs and the incipient boiling surface superheat due to the HSN was not observed except the CHFs for low subcooling at the pressure of 2047 kPa. The empirical correlation for the subcooled CHFs due to the HSN was presented.

## REFERENCES


1. A. Sakurai, M. Shiotsu, and K. Hata, Boiling heat transfer from a horizontal cylinder in liquid nitrogen, "Heat Transfer and Superconducting Magnetic Energy Storage," ASME HTD-Vol.211, (1992), p.7.
2. A. Sakurai, M. Shiotsu, and K. Hata, New transition phenomena to film boiling due to increasing heat inputs on a solid surface in pressurized liquids, in: "Instability in Two-phase Flow Systems," ASME(1993), p.27.
3. A. Sakurai, M. Shiotsu, and K. Hata, Mechanism of nucleate boiling on a solid surface in liquid helium I, "in Advances in Cryogenic Engineering," Vol.39, Plenum Press, New York(1990), p.1757.
4. A. Sakurai, M. Shiotsu, and K. Hata, Incipient boiling superheats and critical heat fluxes due to increasing heat inputs in subcooled He I at various pressures, "in Advances in Cryogenic Engineering," Vol.41, Plenum Press, New York (1996), p.203.
5. A. Sakurai, M. Shiotsu, and K. Fukuda, Pool boiling critical heat flux on a horizontal cylinder in subcooled water for wide ranges of subcooling and pressure and its mechanism, "in Proceedings of the 31st National Heat Transfer Conference," Vol. 4, ASME, New York(1996), p.93.
6. K. Hata, M. Shiotsu, and A. Sakurai, Critical heat fluxes in pool boiling of subcooled liquid nitrogen at elevated pressures, "in Proceedings of the 16th International Cryogenic Engineering Conference/International Cryogenic Materials Conference," Part 1, Elsevier, Tokyo(1996), p.585.
7. S.S. Kutateladze, "Heat transfer in condensation and boiling," AEC-tr-3770, USAEC(1959).
8. N. Zuber, "Hydrodynamic Aspects of Boiling Heat Transfer," AECU-4439, USAEC(1959).
9. A. Sakurai, M. Shiotsu and K. Hata, A general correlation for pool boiling heat transfer from a horizontal cylinder to subcooled liquid, Part 2: Experimental data for various liquids and its correlation, ASME Journal of Heat Transfer, 112:441(1990).
10. A. Sakurai, M. Shiotsu and K. Hata, Effect of system pressure on minimum film boiling temperature for various liquids, Experimental Thermal and Fluid Science, 3:450(1990).
11. A. B. Ponter and C.P. Haigh, The boiling crisis in saturated and subcooled pool boiling at reduced pressures, Int. Heat and Mass Transfer, 12:429(1969).
12. J. H. Lienhard and V. K. Dhir, Hydrodynamic prediction of peak pool-boiling heat fluxes from finite bodies, ASME Journal of Heat Transfer, 95:152(1973).

# HEAT TRANSFER THROUGH A MULTILAYER INSULATION SYSTEM AS A FUNCTION OF PRESSURE IN THE CRYOSTAT VACUUM SPACE

M. A. Green

E. O. Lawrence Berkeley National Laboratory
University of California
Berkeley, CA 94720

## ABSTRACT

When one builds a high quality cryostat, one strives to have a good vacuum (better than 1 mPa) in the space where there is multilayer insulation between the cryogen vessel and the outer vacuum vessel. When designing any cryostat, one should always ask the following question; What happens if the insulting vacuum is not good? This paper presents a technique for analyzing the performance of a cryogenic insulation system when the insulating vacuum is not so good. From the analytical calculations presented in this report, one can improve the insulation system so that acceptable insulation performance can occur even when the pressure in the vacuum space is as high as 100 mPa. Of particular interest is the calculation of insulation performance when the vacuum fails. The difference between an air or nitrogen leak, and a helium leak or a hydrogen leak is illustrated.

## INTRODUCTION

This report describes a method for calculating mixed heat transfer through the multilayer insulation used to insulate a cryostat. The method described here permits one to estimate the insulation potential for a multilayer insulation system from first principles. The heat transfer regimes in a typical cyrostat are: radiation, gas conduction, solid conduction through the insulation and free convection heat transfer (at high pressures) at the cryogen vessel wall and cryostat vacuum vessel wall. In most cases, heat transfer is dominated by a combined gas conduction and radiant heat transfer from the cryostat vacuum wall to the cold cryogen vessel wall. The equations for gas conduction are given for helium, hydrogen and nitrogen gas between the cryostat vacuum chamber wall and the cold wall.

The cryogenic multilayer insulation system is modeled as a stack of flat plates where the following assumptions apply: 1) The spacing between the plates is much smaller than the smallest surface dimension of the plate. As a result the form factor for radiation and free molecular heat transfer is one. 2) The spacing between the plates is larger than 0.6 times the predominant wave length of the emitted heat, so there is no radiation tunneling between the plates[1]. At the lowest temperature (say 4 K), the plate spacing must be greater than 0.42 mm. On the room temperature side (300 K), the plate spacing has to be greater than 0.006 mm. 3) The plates shall be low resistivity metal or they shall be coated with a low resistivity metal so that they are opaque to the wavelengths of infrared on the plates[1]. At 300 K, mylar with 300 angstroms of aluminum on it is opaque to thermal radiation. At low temperatures, the aluminum thickness has to increase (0.03 mm at 4 K). 4) The gas between the plates does not absorb nor emit thermal radiation[1]. 5) The spacing between the plates is small enough to prevent convection cells from forming during continuum gas conduction[2].

*Advances in Cryogenic Engineering,* Vol. 43
Edited by P. Kittel, Plenum Press, New York, 1998

Gas conduction between the plates falls in two general regimes: 1) The free molecule regime is where gas molecules carrying heat travel from plate to plate without colliding with each other. The spacing between the plates is much less than a mean free path for gas molecule collisions 2) The continuum gas conduction regime is where the gas molecules carrying the heat from plate to plate collide with each other in the process of traveling from one plate to the other. The spacing between the plates is much greater than a mean free path for gas molecule collisions There is a transition regime between the free molecular gas conduction regime and the continuum gas conduction regime.

The mean free path for collisions between gas molecules can be determined from the viscosity of the gas. The mean free path is defined as follows[3]:

$$\lambda(T_g) = 1.23 \frac{\mu(T^*)}{P} \left[\frac{R\,T}{M}\right]^{0.5} \quad (1)$$

where $\lambda(T_g)$ is the mean free path for gas molecule collisions for gas at a temperature T*; $\mu(T^*)$ is the gas viscosity at a temperature T*; T* is the average temperature of the gas between plates; R is the universal gas constant ($R = 8314\ J\ K^{-1}\ mole^{-1}$); M is the molecular weight of the gas; P is the pressure measured by a vacuum gauge looking at a space with a temperature T (use 300 K). Note: the equations here are all given in SI units. (A vacuum of 1 torr is 133.29 Pa.) Analytic expressions for gas viscosity $\mu(T^*)$ and thermal conductivity k(T*) (valid from the 1 atm boiling point to 400 K) are given as follows[4,5]:

$$\mu(T^*) = A\,T^{*B} \quad (2a)$$

$$k(T^*) = C\,T^{*D} \quad (2b)$$

where $T_g$ is the temperature of the gas. Equations 2a and 2b are useful between the gas normal boiling temperature and 400 K. Equation 2a and 2b are independent of pressure as long as the gas is near ideal. The numerical values of the coefficients given in Equations 2a and 2b can be found in Table 1 for helium, hydrogen and nitrogen.

In order to determine the gas heat conduction regime one can define a dimensionless number called the Knudsen number Kn[3];

$$Kn = \frac{\lambda(T^*)}{S} \quad (3)$$

where S is the spacing between the insulation layers. The free molecular heat transfer equations can be applied precisely when $Kn > 10$. In many cases, the free molecule equation can be used as Kn approaches 1, but the accuracy is reduced. The ordinary conductive heat transfer equation can be applied when $Kn < 0.003$. The region from $0.003 < Kn < 10$ is a transition region.

## COMBINED HEAT TRANSFER IN THE FREE MOLECULE, CONTINUUM, AND TRANSITION REGIMES

The heat flow through a stack of multilayer insulation can be approximated using a radiation heat transfer term plus a free molecular gas heat transfer term provided $Kn > 10$. This calculation can be done on a plate by plate basis or globally for the whole stack[6]. Global heat transfer through a stack of multilayer insulation can be approximated as follows:

$$Q_{RF} = \frac{H(T_m) - H(T_o)}{m} \quad (4)$$

where $Q_{RF}$ is the combined heat transfer through multilayer insulation that has an upper stack temperature of $T_m$ and a lower stack temperature of $T_o$. There are m layers of multilayer insulation in the stack. The values of $H(T_m)$ and $H(T_o)$ are defined as follows:

$$H(T_m) = E(T_m)\, T_m^4 + G(T_m,P)\, T_m \tag{5a}$$

and

$$H(T_o) = E(T_o)\, T_o^4 + G(T_o,P)\, T_o \tag{5b}$$

where E is the radiation heat transfer function and G is the free molecular heat transfer function.

If one assumes that E and G are temperature independent (to first order they are), the following expressions for E and G result:

$$E = E(T_m) = E(T_o) = \frac{\varepsilon\,\sigma}{2} \tag{6}$$

where e is the average emissivity of the multilayer insulation (use $\varepsilon = 0.02$); and $\sigma$ is the Stefan-Bolzman constant ($\sigma = 5.67\times10^{-8}$ W m$^{-2}$ K$^{-4}$).

$$G(P) = G(T_m,P) = G(T_o,P) = \frac{\alpha}{2}\frac{k+1}{k-1}\left[\frac{R}{2\pi\, MT}\right]^{0.5} P\,\Delta \tag{7}$$

where $\alpha$ is an overall accommodation coefficient ; k is the ratio of specific heats for the gas; R is the universal gas constant; M is the molecular weight of the gas; P is the pressure measured by a vacuum gauge looking at a plate at a temperature T. The $\Delta$ is the transition region function defined by Equation 8. The ratio of specific heats k and the average accommodation coefficient are given in Table 1 for helium, hydrogen, and nitrogen.

Heat conduction in the transition region may be characterized by a function $\Delta$ that relates the total heat flow $Q_T$ to the free molecular heat flow $Q_{FM}$ at a given pressure. Heat conduction through gas in the transition region may be obtained by extrapolation from either the continuum region or free molecular region. Lees' linearized four moment Maxwell molecule model[7] approaches the problem from the free molecular end (Kn > 10). Experimental measurements taken by Teagan and Springer using air at 300 K were found to agree favorably with the Lees' four moment model[8]. The ratio of transition region heat flow to free molecular heat flow $\Delta$ using Lees' four moment model can be stated as follows:

$$\Delta = \frac{\Xi\ Kn\,(2/\alpha - 1)}{1 + \Xi\ Kn\,(2/\alpha - 1)} \tag{8}$$

where Î is the transition region function; $\Xi$ is a fitting parameter (In most cases $\Xi$ is about 1.7) that fits the transition function into the continuum region. The value of $\Xi$ is shown in Table 1 for specific cases using helium, hydrogen and nitrogen. Kn is the Knudsen number; and $\alpha$ is the average accommodation coefficient between the plates (see Table 1 for $\alpha$). If the fitting parameter $\Xi$ is properly chosen, $Q_{RF}$ will equal the continuum heat transfer $Q_{RC}$ given by Equation 10, when Knudsen number is small.

It is of interest to calculate the pressure in the vacuum space where free molecular conduction heat flow equals heat flow by radiation. An expression for this pressure P* is given as follows:

$$P^* = \frac{\frac{\varepsilon\,\sigma}{2}\left[T_m^4 - T_o^4\right]}{\frac{\alpha}{2}\frac{k+1}{k-1}\left[\frac{R}{2\pi\, MT}\right]^{0.5}\left[T_m - T_o\right]} \tag{9}$$

where all of the constants in Equation 9 have been previously defined. It is of interest to note that P* is independent of the number of layers m. P* is lower for lower values of $T_m$. Values for P* are given for various cases in Table 1. From Equation 9, it appears that the vacuum inside of an 80 K shield can be an issue to consider in insulation design.

When the pressure is high, the heat flow through a stack of multilayer insulation can be approximated using a radiation heat transfer term plus a continuum gas conduction heat transfer term provided Kn < 0.01. Continuum heat transfer through a stack of multilayer insulation with gas in the vacuum space can be approximated as follows:

$$Q_{RC} = \frac{E(T_m)\,T_m^4 - E(T_o)\,T_o^4}{m} + \frac{C}{D+1}\,\frac{[T_m^{D+1} - T_o^{D+1}]}{S\ m} \tag{10}$$

where $Q_{RC}$ is the combined heat transfer through a stack of multilayer insulation that has an upper stack temperature of $T_m$ and a lower stack temperature of $T_o$. There are m layers of multilayer insulation in the stack. S is the distance between the layers. The values of $E(T_m)$ and $E(T_o)$ are defined by Equation 6.

## CALCULATION OF HEAT FLOW THROUGH MULTILAYER INSULATION

Table 1 shows the values for various constants used in Equations 1 through 10 for helium, hydrogen and nitrogen. The helium cases apply to heat flow between 80 K and 4 K and from 300 K to 4 K. The hydrogen case applies to heat transfer from 300 K to 20 K; the nitrogen case applies to heat transfer from 300 K to 77 K. Figure 1 shows the calculated heat transfer from 300 K to 4 K through a stack 25 aluminum plates with an average emissivity e = 0.02 and a plate spacing of 1 millimeter with helium as the gas between the plates. The calculations were made over a range gas pressures from $10^{-4}$ Pa ($7.5 \times 10^{-7}$ torr) to $10^5$ Pa (0.987 atm).

**Table 1 The Value of Various Constants in the Heat Flow Equations for Various Gases and Temperature Ranges**

| Parameter | Helium 80 - 4 K | Helium 300 - 4 K | Hydrogen 300 - 20 K | Nitrogen 300 - 77 K |
|---|---|---|---|---|
| $T_{boil}$ (K) | 4.22 | 4.22 | 20.3 | 77.4 |
| T* (K) | 50 | 160 | 170 | 190 |
| M | 4 | 4 | 2 | 28 |
| k | 1.67 | 1.67 | 1.4 | 1.4 |
| A | $5.03 \times 10^{-7}$ | $5.03 \times 10^{-7}$ | $8.43 \times 10^{-8}$ | $1.25 \times 10^{-7}$ |
| B | 0.65 | 0.65 | 0.78 | 0.87 |
| C | $3.83 \times 10^{-3}$ | $3.83 \times 10^{-3}$ | $1.03 \times 10^{-3}$ | $1.35 \times 10^{-4}$ |
| D | 0.65 | 0.65 | 0.91 | 0.92 |
| $\alpha$ | 0.10 | 0.14 | 0.14 | 0.19 |
| $\Xi$ | 1.69 | 1.72 | 1.82 | 1.70 |
| P* (Pa) | 0.0015 | 0.053 | 0.027 | 0.092 |

## DISCUSSION

The calculated heat fluxes per unit area through multilayer insulation shown in Figure 1 are achievable provided the insulation is properly applied. Multilayer insulation between an 80 K shield and the helium tank is a good idea unless one is absolutely convinced that the vacuum will always be good. Multilayer insulation between an 80 K shield and a helium tank (4 K) should have a thicker layer of aluminum (up to 0.01 mm thick) than the insulation used for temperatures above 80 K. The first few layers on the helium tank should be an insulation system using aluminum foil instead of aluminized mylar. This will reduce the radiation heat load between the shield and the helium vessel. The standard 300 angstrom layer of aluminum on aluminized mylar insulation becomes transparent to infrared radiation at about 40 K. Even standard aluminized mylar insulation has some value at temperatures less than 40 K because free molecular gas conduction is reduced by the this form of insulation. Layers of insulation between the shield and the cryogen tank also prevent the formation of convection cells when the gas pressure in the insulation is above 3000 Pa. Finally, layer of insulation on the helium tank will greatly reduce the heat flux during a catastrophic air leak in the vacuum tank.

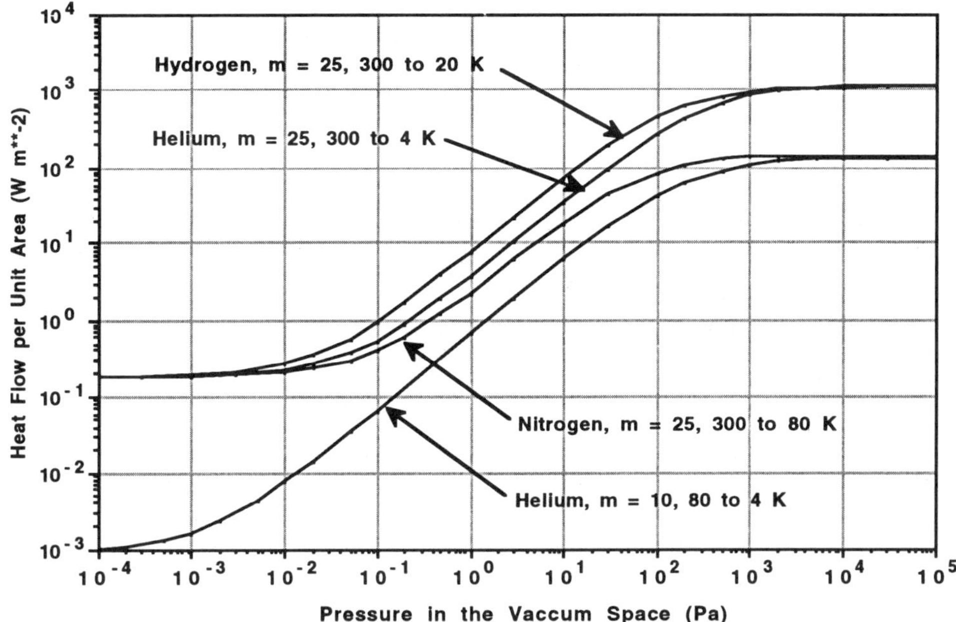


Fig. 1 The Heat Transfer per Unit Area through a Stack of Aluminum Plates 1 mm Apart as a Function of the Gas Pressure Between the Plates for Various Gases.

For cases where the highest temperature was 300 K, one finds that from 0.3 Pa to 100 Pa, heat flow through the stack of plates is dominated by free molecular gas conduction between 300 K and the low temperature surface. In this region the heat flow through the stack of plates is proportional to the gas pressure between the plates and inversely proportional to the number of insulation layers. More layers of multilayer insulation mean a lower heat leak just as when the heat flow is dominated by radiation. Only when one is in the continuum gas conduction region does the number of insulation layers make only a small difference in the heat transfer across the insulation.

Above 1000 Pa, ordinary continuum gas conduction dominates. Continuum gas conduction is independent of the pressure between the plates. The four moment transition region solution smoothly connects the free molecular gas conduction region and the continuum gas conduction region. The accuracy of the heat transfer calculation is probably at its worst in middle of the transition region (at pressures between at about 300 Pa). The calculations shown in Figure 1 assume that the space between the warm wall and the cold wall is filled with layers of multilayer insulation. In reality, the space between the cold wall and the warm wall is not completely filled with multilayer insulation. There is often a gap between the warm wall and the insulation or the shield and the insulation. Under normal operating conditions, when the vacuum is good, these gaps are of no real concern. When the pressure in the vacuum space is high, continuum gas conduction breaks down; the large gaps become convection cells. Continuum gas conduction only occurs in the multilayer insulation blanket itself, so the heat transfer rate is increased as the pressure goes up. The pressure where convection cells form is a function of the size of the gap, the gas in the vacuum space, and the orientation of the wall. If one is interested in reducing the heat flow into a cryostat when the pressure is at 0.103 MPa (1 atm), one has to fill all of the vacuum space with multilayer insulation to prevent convection cells from forming in the vacuum space.

A complete vacuum failure either from a break in the cryogen vessel or an air (nitrogen) leak from the outside is a catastrophic event. An air leak from the outside can produce high heat fluxes on a cryogen vessel that is below 77 K, because air will condense on the outside of the cryogen vessel. The initial heat flow to the cryogen vessel will be proportional the area of the hole letting air into the vacuum space, when there is no insulation on the cryogen vessel wall. Multilayer insulation on the cryogen vessel wall can reduce the initial surge of heat to the cryogen vessel by one or two orders of magnitude depending on the number of layers of insulation and other factors. Once the cryogen vessel has reached 77 K, the heat flow to the vessel is limited to the continuum gas conduction through nitrogen to the cryogen vessel. A catastrophic leak from a helium or hydrogen vessel has a lower heat leak initially but later the heat leak can be much worse than for an air leak.

## CONCLUSION

Heat transfer through layers of multilayer insulation in a vacuum is combination of radiation and gas conduction heat transfer. Radiation heat transfer across the stack of plates is proportional to the plate emissivity and inversely proportional to the number of plates in the stack. At pressures below 100 Pa, the gas conduction through the stack of plates is by free molecular gas conduction. Free molecular gas conduction is proportional to the gas pressure and the accommodation coefficient and is inversely proportional to the number of layers. In both the radiation and free molecular gas conduction regimes, more layers in the multilayer insulation means less heat is transferred. In the transition and continuum regimes, the number of insulation layers is of less concern. In these regimes, conduction heat transfer is controlled by the distance between the vacuum wall and the cryogen vessel. At pressures above 3000 Pa, convection cells can be formed in spaces between the insulation blankets and the walls. To minimize the heat flow to the cryogen vessel in the event an insulating vacuum failure, one should fill the insulating vacuum space with multilayer insulation.

## ACKNOWLEDGMENTS

This work was supported by the Director of the Office of Basic Energy Science, High Energy Physics Division, United States Department of Energy under contract number DE-AC03-76SF00098.

## REFERENCES

1. E. M. Sparrow and R. D. Cess, Radiation Heat Transfer, Brooks/Cole Publishing Company, Belmont CA, (1966)
2. Frank Kreith, Principles of Heat Transfer, International Textbook Company, Scranton PA, (1961)
3. James Jeans, An Introduction to the Kinetic Theory of Gases, Cambridge Press, (1962)
4. V. D. Arp and R. D. McCarty, "Thermophysical Properties of Helium-4 from 0.8 K to 1500 K with Pressures to 1000 MPa," NIST Tech Note 1334, November 1989
5. V. J. Johnson editor, A Compendium of Properties of Materials at Low Temperature (Phase 1), Part 1 Properties of Gases, Pergamon Press, New York, (1961)
6. M. A. Green, "Radiation and Gas Conduction Heat Transport Across a Helium Dewar Multilayer Insulation System," The Art and Science of Magnet Design, a Festschrift in Honor of Klaus Halbach, Vol. 1, p 191, LBL PUB-754 (February 1995)
7. C. Y. Lin and L. Lees, Rarefied Gas Dynamics, p 391, Edited by L. Talbot, Academic Press, New York (1961)
8. W. P. Teagan and G. S. Springer, "Heat Transfer and Density Distribution Measurements between Parallel Plates in the Transition Regime," MIT Fluid Mechanics Laboratory Report 67-5, September 1967

# CRYOGENIC SUPERINSULATIONS WITH INCREASED EFFICIENCY

V.F. Getmanets, G.G. Zhun , R.S. Mikhalchenko, N.P. Pershin, and A.M. Borisov

Special Research & Development Bureau in Cryogenic Technologies of the Institute for Low Temperature Physics & Engineering of the National Academy of Sciences of Ukraine
47, Lenin Ave., Kharkov, 310164, Ukraine

## ABSTRACT

Methods and results of improvement of SI thermal performances of cryobiological vessels with volume 35 l are presented. Life - time of the vessel (time of complete boil - off of liquid nitrogen) was increased from 100-120 to 350-370 days by :

using new high-effective spacer materials with adsorbents;
optimization of SI mounting parameters;
selection of design of thermal protection of cryo- vessels and regimes of thermal vacuum treatment of SI materials ensuring elimination of cryocondensation of water vapor in the layers of SI and decreasing compression of insulation;
maximum using of the cold of vapors.

## INTRODUCTION

Modern superinsulations (SI), due to their extremely low thermal conductivity and mass, have found broad application in cryogenics, space-rocketry, chemical and other branches of industry. Through the last decade, research and development in this area were practically stopped, the thermal performances of cryogenic vessels of many corporations have remained at a constant level. Therefore experts have come to a conclusion about the limitation of development of this branch, as under their judgement the thermal parameters of SI have reached limited low values.

However our research and development of the last 15 years have allowed a 3 time increase the life-time of cryogenic vessels (see fig.1) and to reach the lowest level of a thermal conductivity of SI. The purpose of this work is provide an account of the scientific basis and methods achieving this goal.

## PERFECTION OF CRYOVESSELS INSULATION

Carried out by us over the last 15 years the theoretical and experimental researches[1-6] have shown the following. A basic problem on cryogenic vessels in deriving a thermal conductivity close to calorimetric data, is maintaining inside thick packages of SI (more than 3-5 cm) a required vacuum $P < 1.10^{-2}$ Pa. The principal reason is due to a low velocity of pumping out of thin (about 0.2 mm) and long (about 1 m) clearances between screens and spacers[4].

As a result gas pressure inside an average layer of packages of a thermal insulation at the first hours of using of cryogenic vessels can reach $1.10^{-1}$ Pa (see fig.2), that multiply increases a thermal conductivity of SI (see fig. 2,3). At long-term exploitation of cryogenic vessels their thermal performances are deteriorated more significantly.

The basic solution of this problem is the arrangement outgassed before hand adsorbents immediately in layers of SI, namely, in a structure of their spacer material. However, originally in England it was not possible to introduce this process engineering [7] in an industry due to adsorption of water vapor in a phase of installing of SI.

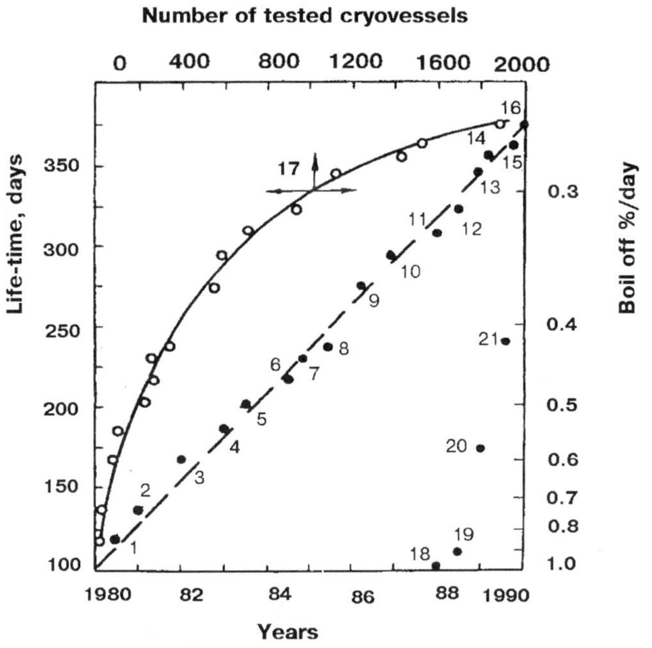

**Figure 1**. Dynamics of perfecting cryogenic vessels with liquid nitrogen (volume 35 l, diameter of a neck - 60 mm, length - 210 mm): 1 - glassveil spacer EVTI- 7; 2 - the thickness of a neck is reduced from 2.2 to 1.2 mm; 3 - optimization of the scheme of contact of SI with neck; 4 - magnification of a work part of neck from 0.13 up to 0.21 m; 5 - replacement of a thickness of screens from 12 to 5 microns; 6 - optimization of heat exchange inside neck; 7 - optimization of thermal degasation; 8 - optimization of conditions of mounting; 9 - use of the spacer USNT-10; 10 - optimization of mounting USNT-10; 11 - optimization of thermal vacuum handling of SI with USNT-10; 12 - optimization of thermal degasation USNT-10 in a package of SI and its preliminary thermal gas handling; 13 - refinement of the adsorbing pump; 14 - development of a process engineering of long maintaining of vacuum; 15 - thermoplastic handling of SI and removal of a static electricity; 16 - optimization of types of spacers; 17 - modification of number of tested cryogenic vessels; 18 - cryogenic vessels of the Messer Griesheim corporation, Germany, capacity V = 30 l, diameter of neck d= 50 mm, life-time R = 77 days; 19 - cryogenic vessels of the 5У -LAB corporation, Austria, d = 64 mm, V = 35 l, R = 109 days; 20 - cryogenic vessels of the Statebourne corporation, England, d = 51 mm, V = 35 l, R = 175 days; 21 - cryogenic vessels RSV-35A, France, d=50 mm, V=35 l, R=240 days.

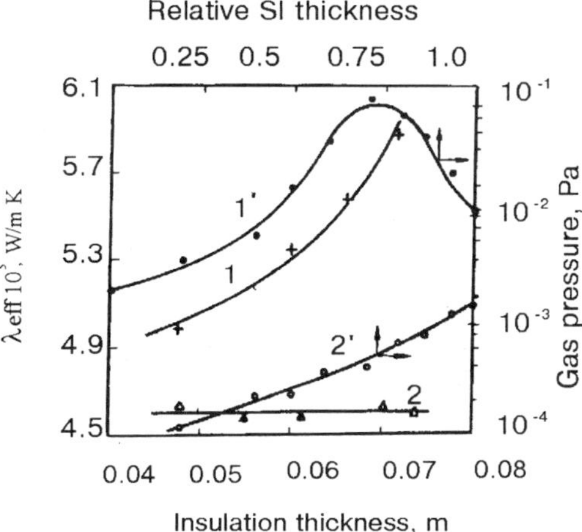


**Figure 2**. Dependence of a thermal conductivity of SI with spacers USNT-10 on a thickness of insulation and pressure profile of residual gases in layers: 1,1'- at conventional conditions of thermal and vacuum handling of SI; 2,2'- after a selection of optimum conditions of thermal and vacuum handling of SI.

We have developed and use the method of maintaining of vacuum in SI by introducing into spacers USNT - 10 a carbon adsorbent, which is outgassed after mounting of the SI in a phase of vacuum operation of the cryogenic vessel at the temperature of 100-120 °C. As an outcome of this, we managed in cryogenic biological vessels with an average thickness of a package of SI up to 80 mm to achieve and maintain vacuum better than $2.10^{-3}$ Pa. It was confirmed by direct measurements of pressure profile (see fig.2). In the total the local zones of an increased thermal conductivity in layers of SI (see fig.3, curve 3) and magnification of its thermal conductivity with an increase of thickness of package of SI (see fig.2 curves 1) was eliminated (see fig.3 curve1 and fig.2 curve2). Necessary pressure in SI ( $< 1.10^{-2}$ Pa) was maintained only during 20-25 months even with the use of USNT-10 spacers (see curves 1,2 fig. 4). Development of a new design of SI and adsorbent pumps for cryogenic vessels and introduction of special vacuum handling materials of SI before their mounting on cryogenic vessels has allowed to prolong the maintenance of required vacuum till 5 years (see curves 3,4 fig. 4).

These measures have not eliminated influence of outgassing products on a thermal conductivity of SI. We have determined, that even at vacuum at a level $1.10^{-3}$Pa in cold layers of SI (with temperature lower than $0^0$C) there is condensation of water vapor from yields of outgassing [6]. The growing layer of cryogenic condensate - ice increases the emissivity of the SI screens, which leads to gradual diminution of life-time of cryogenic vessels (for example, for SI with USNT-10 - 0,2-0,3 % in a month). The modification of the SI and method of thermal evacuation has allowed us to minimize the phenomenon.

The elimination of increased pressure of gases in layers of SI has lead to a reduction of the heat transfer between the spacer and screen, but nevertheless it remained rather significant. The experiments have shown, that the availability of "black" coal adsorbent in the spacer USNT-10 conducts to sharp increase of a thermal conductivity of SI at a raise of strength of a fillet tension during SI winding more than 0.25 kgs (see curve 2 fig. 5). As it is seen from fig. 5 (curve 1) on the spacer of a type SNT-10, which differs only by lack in it coal adsorbent, such effect is absent.

The phenomenon is easily explained with the help of open by us earlier [3,4] effects, namely - manifold amplification of radiation heat transfer between adjacent screens (see fig.6) at 3 simultaneously operating conditions: 1) optical density of the spacer make 1-

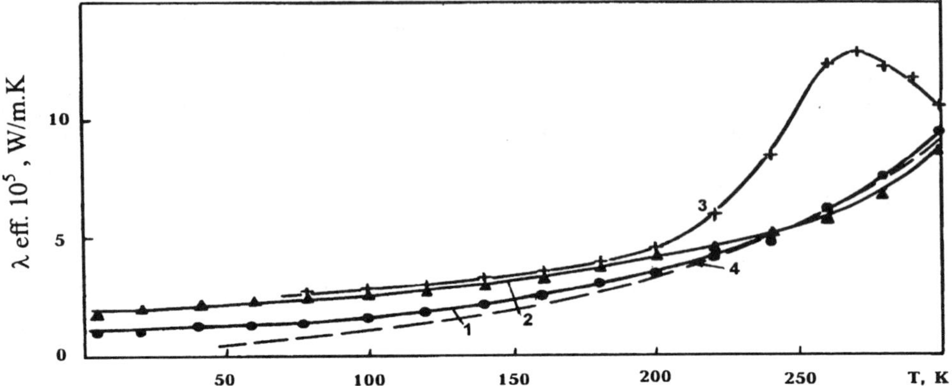


**Figure 3**. Dependence of thermal conductivity of SI on temperature: 1,3 - SI with USNT-10; 2 - SI with SNT-10; 4 - calorimetric data for SI with spacers from glass-paper SBSh - T; 1,2,4 - temperature of vacuum handling T = 390K; 3 - T = 315K.

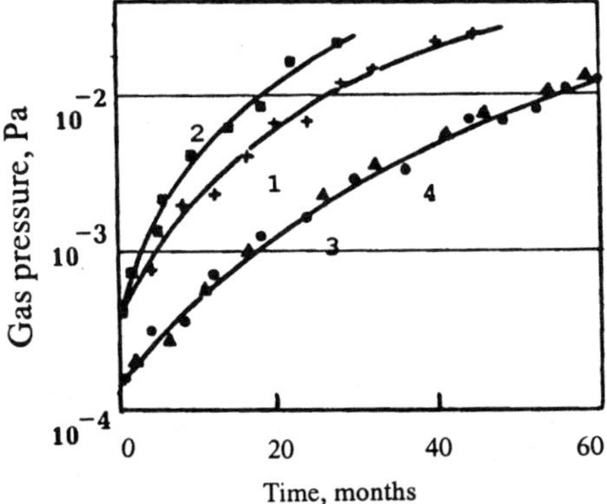


**Figure 4**. A modification of vacuum in cryogenic vessel insulation space in time: 1,2 - serial vessel; 3,4 - cryogenic vessel with improved design of adsorbing pump and special thermal and vacuum handling of SI; 1,3 - continuous operational mode; 2,4 - with heating to room temperature through 1 month.

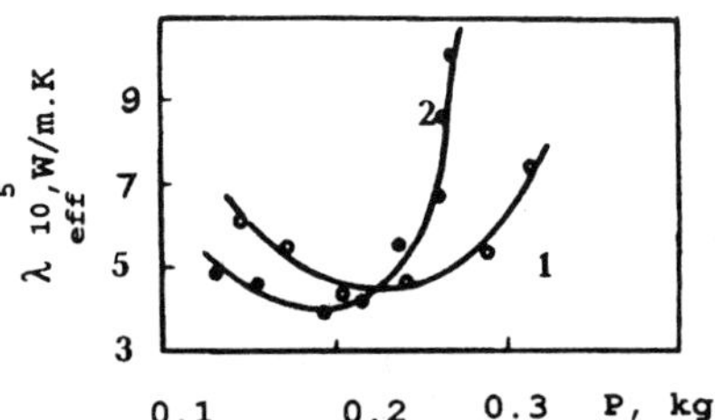


**Figure 5**. Dependence of a thermal conductivity of SI on strength of a fillet tension at SI winding: 1 - SI with SNT-10; 2 - SI with USNT-10

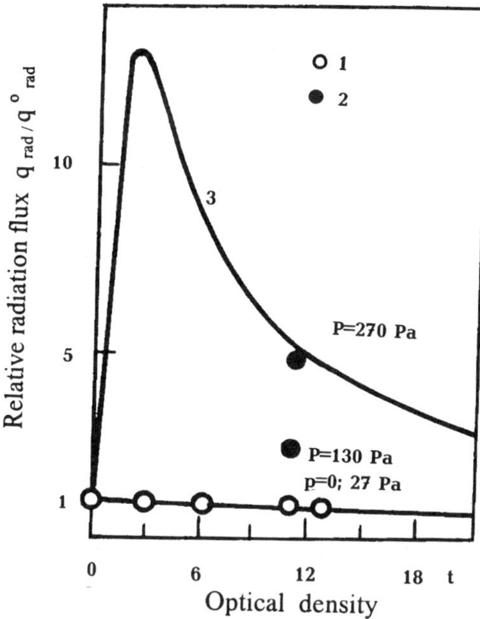


**Figure 6**. Effect of spacer optical density and compressive load on radiative heat transfer between SI shields: 1- experiment for scattering spacers or «weak» contact of adsorbing spacers; 2- experiment for scattering - adsorbing spacers ; 3 - theoretical curve for adsorbing spacers; $q^0_{rad}$ - radiation heat flux in the absence of a spacer.

10; 2) spacers have essential ability to an absorption (instead of scattering) of radiation; 3) contact heat conduction between the spacer and screen noticeably exceeds a thermal conduction of the spacer (in direction of screen to spacer). At high densities of mounting of SI with USNT-10 all conditions are fulfilled, as conducts to sharp increase of total radiation - conductive - contact heat transfer between two screens.

This problem we have successfully solved at the expense of a reduction of the optical thickness of the spacer by a drop of mass per 1 $m^2$ of spacer to 8-10 g, its thickness - up to a level of 30-50 microns, and also at the expense of monitoring gains of mounting of SI and their optimization.

It was also established earlier [6,9], that the thermal resistance of a screen - spacer zone can noticeably be reduced by removing an attraction due to an electrization of their surfaces, and also heating of SI to temperature of a beginning of plastic relaxation of a screen and spacer and reduction of their mutual compression. The resulting contribution of these components in a modification of life - time of cryogenic biological vessel is equal to 13-14% (fig.7)

The realization of all these measures (experiments on more than 2 thousand cryogenic vessels(see fig. 1) has allowed us to sharply reduce the thermal conductivity of SI of cryogenic vessels (not less than in 3-5 times) and for the first time approve to a level of calorimetric data - approximately 0.4 μW/cm.K. The values, being available in the literature, of thermal conductivity of SI of some authors on flat and cylindrical calorimeters up to 0.1-0.2 μW/cm.K are explained by large errors of a measurement, connected with ends effects - not enough by excellence guard of end faces of the packages of SI from heat exchange to outside surfaces with temperature ~ 80 K.

For deriving of extreme low performances of SI on cryogenic vessels, the problem of maximum use of a cold of cryogen vapor was solved. During a solution of this problem the mathematical model of heat exchange in a heat-shielding cryogenic vessels is created and experimentally tested and the schemes of contact of end faces of SI packages with thermal bridges are optimized.

The lowest values of a boil off of biological cryogenic vessels are obtained with the use of spacer materials of three types: USNT-10, SNT-10 and NT-10 with a mass of 1 $m^2$ not more than 10 g. Their basic specificity at a comparison with spacers, used in much countries, (for example, glass-paper, viscose papers or nylon nets) is their composite structure -2,3 or 4 various types of fibers. All of them contain various synthetic or mineral fibers with extreme small diameters (for example, cellulose, lavsan, basalt, carbon etc.), the wet paper production process, used by us, is high-duty, it is enough cheap and allows to introduce into a structure of spacer materials every possible aspects of fibers, components, to change a structure of a material, and thereby to change over a wide range thermal, optical, gasodynamical, mechanical and other performances of spacers. In an outcome on the thermal parameters on cryogenic vessels new SI exceed best known thermal insulation composition on 25-30%. Besides they have other rather essential advantages: low cost, small thickness and mass (7-10 $g/m^2$); high strength supposing the high-duty machine winding of SI; a lack in a structure of spacers (SNT-10, USNT-10) silicone particles, harmful to the person. The new materials have outgassing characteristics at a level of the best vacuum materials - glasses.

The new materials already have found a use in many items with an increased level of the requirements on thermal effectiveness, for example, space cryostats, automobile tanks with liquid methane, nitrogen transport refrigerators, helium vessels, pipelines with cryogenic fluids, magnetically transparent cryostats made of fiberglass for SQUID based equipment. The experience of work on creation and trial of new SI has shown, that there is a possibility of a further drop of a thermal conductivity SI (at least in 1,5-2 times) at the expense of improving optical properties of screens and drop of contact thermal conductivity.

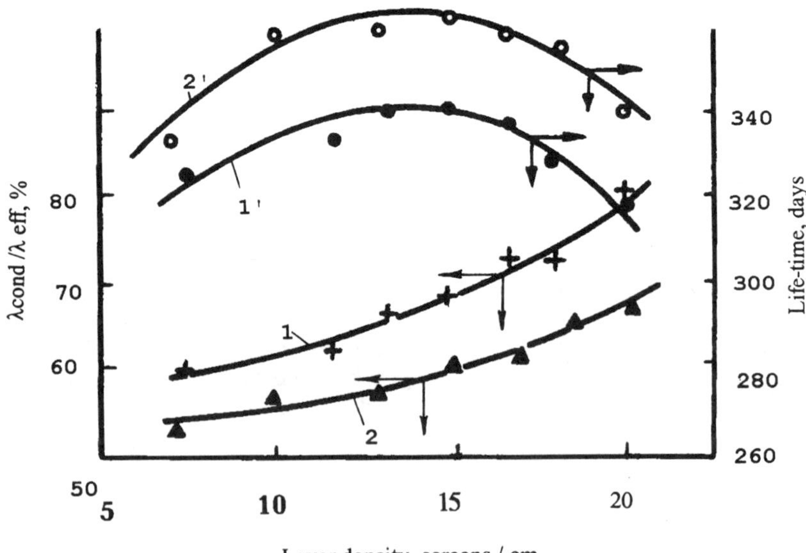


**Figure 7**. Dependence of thermal conductivity by conduction per total thermal conductivity of SI (%) and life - time of cryogenic vessels on layer density of winding of SI with USNT-10: 1,1' - without handling; 2,2' - after removal of a static electricity and thermal plastic relaxation.

## Conclusions

1. SI with composite spacer USNT-10, SNT-10 and NT-10 in cryogenic vessel on thermal parameters exceed the best world analogs by 25 - 30 %.
2. A design both process engineering of SI thermal and vacuum handling and mounting optimized on small industrial cryogenic vessels, has allowed to approach a thermal conductivity of SI close to calorimetric.
3. The techniques of calculation of a heat-shielding and design concepts ensuring an effective utilization of a cold vapor and an optimum junction of the packages of SI with thermal bridges are developed.
4. Accomplished cycles of SI research and trials have shown good reproducibility of outcomes, that allows to put a problem of creation metrologically standardized methods of SI trials on flat and spherical calorimeters.

## REFERENCES


1. Verkin B.I., Mikhalchenko R.S., Gerzhin A.G. Effect of multilayer insulation thickness on apparent thermal conductivity.Proc. of ICEC-2. Guildford, (1969). P. 123.
2. Mikhalchenko R.S., Getmanets V.F., Pershin N.P., Batozskij Yu. V. Study of heat transfer in multilayer insulations based on composite spacer materials.Cryogenics,23:309 (1983).
3. Mikhalchenko R.S., Getmanets V.F., Pershin N.P., Batozskij Yu. V. Theoretical and experimental investigation of radiative - conductive heat transfer in multilayer insulation. Cryogenics, 25: 275 (1985).
4. Verkin B.I., Mikhalchenko R.S., Getmanets V.F., Mikheev V.A. Application of multilayer insulations in cryogenic engineering and improvement of its efficiency. Proc. of ICEC-10, (1984), p. 529.
5. Mikhalchenko R.S., Getmanets V.F., Pershin N.P., Annikova T.A. New efficient composite superinsulations. Cryogenics, V.32, ICMC Supplement, (1992), p. 60.
6. Zhun G.G., Getmanets V.F., Mikhalchenko R.S., Sutula V.V., Podolski A.G. High efficiency small cryovessels. Gryogenics, V.32, ICEC Supplement, (1992), p. 199.
7. Scurlock R.G., Saull B. Development of multilayer insulations with thermal conductivities below 0,1 μW/cm.K. Cryogenics, 16 : 303 (1976).
8. Getmanets V.F., Yurchenko P.N., Dushechkin Yu. A. One - dimensional model of heat transfer in multilayer insulation under radiation heat flux to its layers. Proc. of ICEC 8, Genova, (1980), p. 695.
9. Kurskaja T.A., Mikhalchenko R.S., Getmanets V.F. An effective cryogenic thermal insulation for optical devices and pipelines. Scient.-Technol. J. Eng. Phys., (1995), V.68, N1, p.51.

# PUMPING PERFORMANCE OF CRYOPANELS COATED WITH ACTIVATED CARBON

Chr. Day, B. Kammerer, and A. Mack

Forschungszentrum Karlsruhe, Hauptabteilung Ingenieurtechnik
P.O. Box 3640, D-76021 Karlsruhe, Germany

## ABSTRACT

Within the framework of the European Fusion Technology Programme, a cryopump system for ITER is being developed. The pumping characteristics of cryosorption panels coated with activated carbon are investigated under ITER relevant operating conditions. The underlying design of the panels described in this paper is generally accepted for the cryopump system of the ITER tokamak.

Quantitative vacuum pumping performance data are obtained by the measurement of pumping speed as a function of gas load and pumping pressure for various gases. We have investigated pure gas species (helium, protium and deuterium) and their mixtures. Our measurements have been performed to determine the major factors affecting cryopumping performance, such as flow rate, charcoal saturation, panel temperature and pressure during pumping. Selectivity of mixture pumping has been monitored by the use of a gas mass spectrometer. Dynamic loading effects due to the mobility of the molecules are also investigated. The results presented demonstrate the flexibility and assess the constraints of the recommended pump concept.

## INTRODUCTION

The Karlsruhe Research Center (FZK) has been charged with the development of the cryopump system for the International Thermonuclear Experimental Reactor ITER. A reliable high-speed vacuum pump system is required for high-vacuum pumping of the torus during all phases of operation, for conditioning of the torus walls, during bake-out and leak-testing. The design is based on 16 batch-regenerating cryopumps[1]. The pump is designed for the pressure range up to 10 Pa, which can be further reduced by means of a special throttle valve at the pump inlet[2]. For sufficient helium pumping, cryosorption panels are used in the pump, which are coated with activated charcoal as sorbent material. To assess the suitability of the cryopump design, a testing program has been started. The required key scaling parameter involved in the transfer from test results to the ITER pump is the specific pumping speed (related to the cryopanel surface) of about 1 L/(s·cm$^2$) at a maximum specific gas load of about 0.23 (Pa·m$^3$)/cm$^2$ .

The pumping speed of a cryosorption pump depends to a great extent on the arrival rate of the gas, the capture probability, the cold surface, the type of sorbent used, the temperature and the gas load. The speed of a gas mixture may not only depend on the total amount of previously pumped gas, but also on the time scale and the sequence with which the gases have been admitted into the pump. The latter influence is due to the fact that the cryopump has to handle different kinds of gases, some of which are pumped by sorption, others by condensation. For this reason, the overall pumping speed is strongly determined by the interrelation between the two different pumping mechanisms, which bear different dependencies on temperature, time and gas load. In conclusion, it must be emphasized that the performance of a cryopump does not depend on a specific parameter but depends on the entire pumping history. Consequently, the experimental evaluation is necessary in order to assess the cryopump behavior for each individual application case.

One of the first decisions to be made when designing a cryopump is the appropriate selection of the sorbent material. Screening tests have been made to determine an optimum sorption method (argon frost, molecular sieve, charcoal), which allows for pumping of hydrogen isotopes and helium by the same panel[3-5]. In the literature, a significant diffusion-rate limitation has been described for molecular sieves[6] and correspondingly poor results for co-pumping[7] have been reported. Thus, an activated charcoal, CHEMVIRON SC-II type (mean particle diameter 1 mm), was selected and bonded by means of a special inorganic cement, designed to withstand thermal cycling and compatible with radiation. The latter requirement does not allow to use standard epoxy resins. An additional advantage of charcoal compared to molecular sieve is the moderate temperature of about 400 K required to achieve total regeneration. Thus, we recommend charcoal as the method of choice to operate with the large flow rates and impure gases encountered.

## DESCRIPTION OF TEST FACILITY

The tests were performed in TITAN, a versatile test facility which permits evaluation of the cryopanel pumping speed as a function of pressure, temperature and gas load. A detailed description of the panel and facility is given elsewhere[8,9].

The test panel is installed in the top part of a cylindrical vacuum vessel with 0.7 m diameter and 2.2 m height (free volume of 0.76 m$^3$). The gas to be pumped is metered into the bottom section of the vessel designed in accordance with PNEUROP standards. In the direction of the incoming flow, the cryopanel is shielded by means of a 106° chevron baffle, which is attached to the bottom of a cylindrical annular LN vessel and thus kept at 80 K. The baffle diameter is about 400 mm. Near the top, the LN volume is covered by a cooled copper plate. The specific cryopanel investigated was of a quilted design geometry (rectangular panel, 0.35 x 0.5 m, channel pitch 50 mm). This configuration is also the one seen by the ITER primary cryopump. The quilted panels are formed by line-welding two SS sheets together to form a channel pattern upon which the interior is hydrostatically inflated. The panel was coated both sides with activated charcoal. Prior to carbon coating, electric resistance heaters were attached to the panel surface to allow for regeneration of the sorbent and for operating the panels at elevated temperatures. All relevant instruments (flow meters, pressure gauges and temperature sensors) were individually calibrated. The effective pumping speed at the pump inlet was determined according to the flowmeter method using the PNEUROP standards.Throughput and pressure during the gas metering phase were measured at the dome. The reproducibility of the pumping speed results was better than 10 %. The surface gas load was calculated by integrating the flow rate vs. time.

## RESULTS

In this paper, emphasis is placed on presenting typical results for all relevant factors influencing cryopumping. Wherever possible, a comparison is made with data in the literature. However, it is often quite difficult to compare pumping results from different cryopump facilities. This is because commercial pumps do not provide PNEUROP standardized measuring devices and, the individual geometry, e.g. the type of baffles used, has a dominating influence on conductance and hence, on the pumping characteristics of the entire pump system. For this test facility, the maximum capture probability of the cryopanel (i.e., a sticking coefficient of unity) was experimentally determined to be about 0.1 and would be too low for an industrial application; it is an integrated value which combines the baffle transmission probability (~ 0.2) and the conductance of the TITAN vessel. The typical influences of cryopanel and baffle configuration are discussed elsewhere[10,11].

The statements given below, are based on a large number of tests at LHe conditions, which involve the pure gases of hydrogen, deuterium, helium, neon, nitrogen and 20 binary and ternary mixtures of these components. The pumping speed, gas load and flow rate values are referenced to 273.15 K and related to the coated cryosurface (3500 $cm^2$). Before each test run, the charcoal was activated for several hours with a vacuum bake at 400 K.

### Influence of Flow Rate and Pressure Regime on Pumping Speed

The typical dependence of measured pumping speed on flow rate is highlighted in Figure 1 for deuterium. The ITER-relevant flow rate is $2.6 \cdot 10^{-4}$ $(Pa \cdot m^3)/(s \cdot cm^2)$ (corresponding to 0.54 L/min (referenced to a standard pressure 1 atm) in the test facility). For interpretation of these results, the pressure must be considered. For the test runs at the TITAN facility, a pressure range up to approximately 0.1 Pa has been investigated, except for pure helium, where 10 Pa has been achieved. This pressure regime is related to transition flow conditions, where the mean free path becomes comparable with the aperture dimensions. In this flow regime, the pumping speed is increasing with pressure due to the increase in conductance[12-14]. Therefore, the pumping speed must always be associated with a certain value of pressure and flow rate, and shall not be regarded as an independent property.

The influence of pressure is illustrated in the lower part of Figure 1, which shows the measurement results for different runs with deuterium at various flow rate programs; the flow rate was varied as indicated by the bars. Run #1 was performed in 13 flow rate steps with consecutive equilibration breaks after each step, Run #2 was performed similarly, but starting at higher flow rates and Run #3 (see top of figure) was performed without equilibration breaks, as discussed above. The agreement is excellent, particularly since Run #3 was performed one year later after two dismantling and maintenance periods were performed. It has been shown repeatedly within the test program at LHe temperature that breaks (up to 30 min each) within the dosage steps do not significantly influence the pumping speed results.

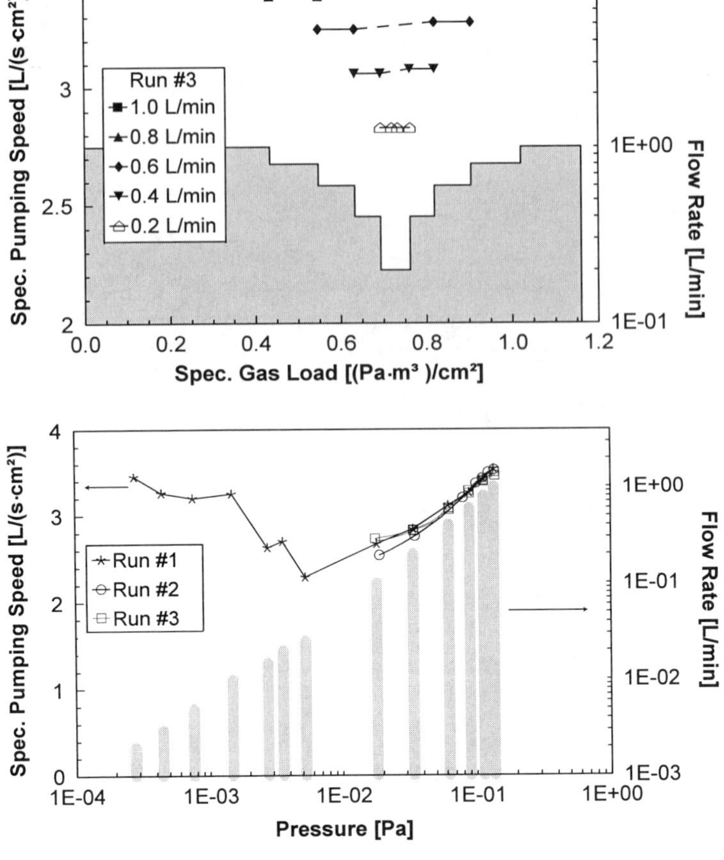


**Figure 1.** Pumping speed curves for deuterium (LHe-cooled) at different flow rates and dosage programs.

The pumping speed curve according to Run #1 illustrates, how flow rate and absolute pressure contribute to the overall pumping behavior. In principle, at LHe temperature, the pumping speed is determined by the limitation of diffusion of the sorbed gas species from the charcoal surface into the pores (see below). This leads to a decrease in pumping speed with increasing flow rate at comparable gas loads. If the resulting pressure has increased such that the transition flow regime is achieved, the effect of diffusion limitation is overcompensated and the pumping speed again rises. Another effect to be considered is the slight temperature rise associated with increasing flow rate. For the test runs from Figure 1, the panel temperature increases by almost 1 K for an increase in flow rate from 0.1 to 1 L/min. This is due to the energy transported by the molecules themselves (coming from the baffle at 80 K), the heat of adhesion of the molecules and the additional heat flux by molecular gas conduction as the pressure rises. The temperature peaks during pumping may be reduced by using supercritical helium as the coolant medium[15]. However, the impact of such a temperature rise is not that relevant, since the panel temperature remains below the corresponding gas sublimation temperature and far away from the sorption critical mobility temperature. Both mechanisms contribute to deuterium pumping[8].

## Influence of Temperature on Pumping Speed

The temperature influence is illustrated in Figure 2. The latter shows the pumping speed for different pure gases at various flow rates as a function of gas load at two temperatures, the charcoal temperature at LHe cooling and an elevated temperature of 12.2. K. The decrease in pumping speed with increasing temperature is due to a corresponding loss in sticking efficiency. For helium no sufficient steady-state pumping was achieved at temperatures above 10 K. The measured temperature influences are comparable to the results reported by Longsworth[16]. If the different capture coefficients are taken into consideration, we obtained better pumping speeds at 12 K than currently reported for hydrogen by Oka et al.[17]. The relations for the pumping speed obtained at LHe temperature for helium relative to the hydrogens are in excellent agreement with those of Sedgley et al.[18] for helium and Halama et al.[7] for cryocondensation of hydrogens.

## Pumping of Mixtures

A systematic study has been performed to gain a fundamental understanding of how mixtures are pumped and whether their pumping curves may be derived from the results for pure gases. The study investigated mixtures based on $H_2$ or $D_2$, covering a range of 5 to 20 % (in steps of 5 %) of He, Ne, Ar, or $N_2$ [8,19]. It has been shown that the most critical mixture combination is $H_2$ with He. Some of the results obtained are shown in Figure 3. Even small percentages of helium reduce the overall pumping speed almost to the level of pure helium. Nevertheless, the ITER requirements can be fulfilled. It generally has been found that helium is the predominant influence on the pumping speed of mixtures. This is due to the fact that He is competing very strongly with $H_2$ for active sites on the carbon surface, as both are pumped by sorption only. This effect is less critical for $D_2$/He mixtures.

Interrelation of He and $H_2$ has already been presumed by Halama et al.[7], but he could not verify it experimentally. Therefore, we installed a high resolution gas mass spectrometer and monitored the composition changes during mixture cryopumping[19]. A typical result is depicted in Fig. 3 (bottom) for a fusion-typical mixture composition. A strong enrichment of helium is found. However, an increase in He concentration becomes significant only after pumping periods which are considerably longer than those proposed for ITER.

## Dynamic Sorbent Capacity

**At LHe Temperature.** For our special application case, it is of considerable interest whether charcoal saturation is already attained within the designed maximum gas load of about 0.23 $(Pa{\cdot}m^3)/cm^2$. All of the tests at LHe temperature covered gas loads at least up to 0.6 $(Pa{\cdot}m^3)/cm^2$; part of the tests were even extended to more than 1 $(Pa{\cdot}m^3)/cm^2$. For pure gases, it was found that within this range, the saturation limit was never reached, except for helium. This is observed in Figure 1 for deuterium and Figure 2 for helium. Our mixture pumping speed results indicate no dependence on gas load for the mixtures which do not contain helium. The sorbent capacity for helium containing mixtures is about 1 $(Pa{\cdot}m^3)/cm^2$ over the entire concentration range investigated (up to 20 mol-% He), which is more extensive than the coverage reported by Hseuh et al.[20].

**Helium Capacity at Elevated Temperatures.** In Figure 4, the charcoal capacity measurements for pure He at different temperatures are compared with those obtained in the literature[18,20-22]. It should be noted that these data were derived from real pumping tests and thus, are dynamic in nature and quite appropriate for the design of He cryopumps.

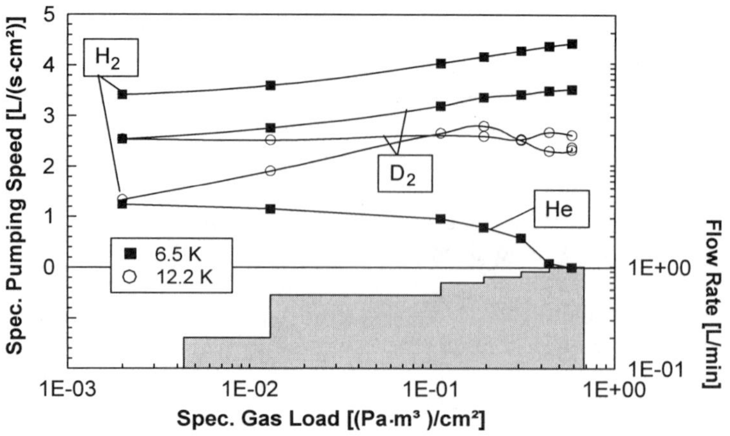


**Figure 2.** Pumping speed curves of pure gases at different temperatures.

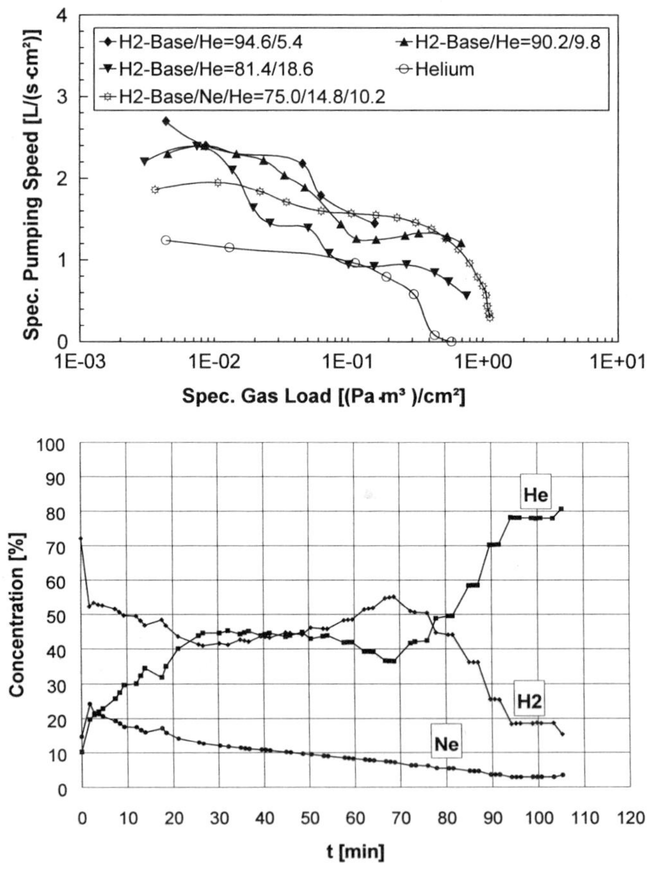


**Figure 3.** Pumping speed curves for $H_2$/He mixtures (H2-Base denotes $H_2$ containing 3.8 % fusion-typical impurities). The lower part shows the composition changes for the ternary mixture (nominally 75.0 mol-% H2-Base, 14.8 % Ne, 10.2 % He) within the selective cryopumping process.

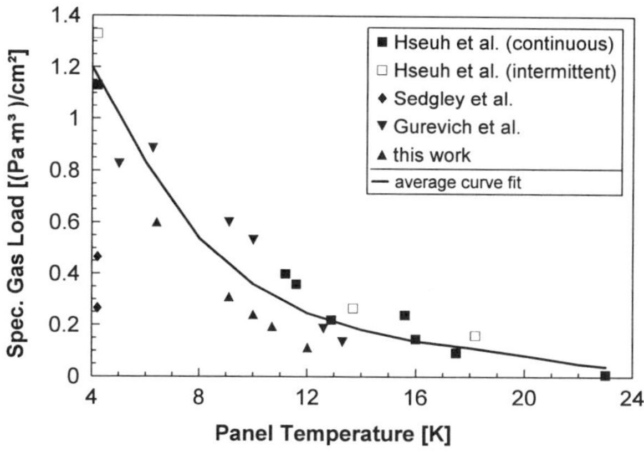


**Figure 4.** Helium sorbent capacity vs. temperature. Comparison between results and literature data [18,20-22].

The scatter of data is mainly due to the different types of charcoal used. As for all dynamic properties, the time scale, to which the maximum capacity has been related, may be a key property. The data of Hseuh et al.[20] indicate that the capacities after intermittent flow are 20 to 30 % higher than those obtained for continuous flow at the same temperature. However, the information given in the literature is not consistent[23,24].

## Mobility and Dynamic Loading

Longsworth[16,25] reports a decrease in speed with accumulated gas on the surface, which can be reversed slightly by stopping the flow for a period of time. This is due to a restricted mobility within the type of charcoal he used. The breaks occuring between the dosage steps provide additional time for a further molecule migration into the charcoal and, thus, new active sites at the outer surface are available. The necessary break time can be considerably shortened when heating occurs between the pumping cycles. Schwenterly[6] and Hseuh et al.[20] showed that the increase in pumping is more pronounced at higher temperatures. According to Lessard[26], a critical temperature of mobility exists, which has to be exceeded to benefit from this effect. We determined this critical temperature for different pure gases in heating experiments of pre-loaded panels. During these tests, the cryopanel was heated at zero flow condition to help disperse the adhered molecules more rapidly. The equilibrium pressure served as an indicator for the increased mobility of the sorbed particles. When the critical temperature was reached, the equilibrium pressure dropped significantly by more than one decade. This pressure drop indicates that the charcoal surface offers additional active sites, as the particles have migrated into the pores due to their higher mobility at the increased temperature. If the panel is cooled down again to the normal operation temperature, the equilibrium pressure was about three decades lower than before the heating treatment. By this method, the critical temperature region for $D_2$ was observed to lie between 21 and 24 K, for $H_2$ between 12 and 13 K (as compared with 13.5 K by Lessard[26] and 15-20 K by Longsworth et al.[25]). For He, we derived a temperature of about 9 K, in agreement with Gurevich et al.[22].

This effect can be used, if the pumping speed is minimized by clogging of the pores, which will probably only be the case at critical mixture compositions or at very high loadings. However, for ITER, the gas load is limited by a maximum tritium inventory, not by achieving the saturation capacity. Contrary to Lessard[26], we could not observe an increased pumping speed after exceeding the critical temperature.

## CONCLUSION

Within the performed test program, the parameters which greatly influence the cryosorption pumping and their intense interrelation could be determined and their potential for pumping performance was discussed. It was shown that the key requirements for ITER can be met. It was noted that in principle the entire cryopump history has an effect on the overall pumping performance.

## ACKNOWLEDGMENTS

The authors wish to thank D. Zimmerlin for his assistance in the operation of the test facility and A. Edinger for her careful evaluation of the measurement data. This work has been performed within the framework of the Nuclear Fusion Project of FZK, supported by the European Communities under the European Fusion Technology Programme.

## REFERENCES

1. D.K. Murdoch, A. Mack, and J.-C. Boissin, Proc. 16th IAEA Fusion Energy Conference, Montreal, Canada (1996).
2. N. Petersohn et al., Proc. 4th Int. Symp. on Fusion Nuclear Technology, Tokyo, Japan (1997).
3. A. Mack et al., *Fusion Technology* 21:902 (1992).
4. D. Perinic et al., Proc. 17th Symp. on Fusion Technology, Roma, Italy (1992), p. 334.
5. D. Perinic, H. Haas, and A. Mack, in:"Advances in Cryogenic Engineering," Vol. 39, Plenum Press, New York (1994), p. 1553.
6. S.W. Schwenterly, Proc. 7th Symp. on Engineering Problems of Fusion Research, Knoxville, TN, USA (1977), p. 1793.
7. H.J. Halama, H.C. Hseuh, and T.S. Chou, in:"Advances in Cryogenic Engineering," Vol. 27, Plenum Press, New York (1982), p. 1125.
8. Chr. Day and A. Mack, Proc. 4th Int. Symp. on Fusion Nuclear Technology, Tokyo, Japan (1997).
9. Chr. Day, B. Kammerer and A. Mack. "Tests on Fast Heating for the Regeneration Process of ITER Cryopumps", Report FZKA 5806, Forschungszentrum Karlsruhe, Germany (1996).
10. S.P. Ansley, G.H. Merz, and W.R. Simpson, Proc. 7th Symp. on Engineering Problems of Fusion Research, Knoxville, TN, USA (1977), p. 1801.
11. B.-K. Liu, J.-S. Ren, and X.-H. Cui, *J. Vac. Sci. Technol.* 20:1000 (1982).
12. P.D. Bentley, *Vacuum* 30:145 (1980).
13. M.M. Menon et al., *Fusion Technology* 27:355 (1995).
14. C. Tzemos, M. Gegenheimer and C.B. Hood Jr., in: "Advances in Cryogenic Engineering," Vol. 31, Plenum Press, New York (1986), p. 575
15. K.M. Schaubel et al., Proc. 18th Symp. on Fusion Technology, Karlsruhe, Germany (1994), p. 347.
16. R.C. Longsworth, in: "Advances in Cryogenic Engineering," Vol. 23, Plenum Press, New York (1978), p. 658.
17. Y. Oka et al., *Fusion Engineering and Design* 31:89 (1996).
18. D.W. Sedgley, T.H. Batzer, and W.R. Call, *J. Vac. Sci. Technol. A* 6:1209 (1988).
19. Chr. Day, The use of a high-resolution quadrupole gas mass spectrometer system for selective detection of helium and deuterium, accepted for publication in *Vacuum*.
20. H.C. Hseuh et al., Proc. 8th Symp. on Engineering Problems of Fusion Research, San Francisco, CA, USA (1979), p.1568.
21. D.W. Sedgley et al., *J. Vac. Sci. Technol. A* 5:2572 (1987).
22. L.S. Gurevich et al., *Plasma Devices and Operation* 2:287 (1994).
23. J.R. Porter, *J. Vac. Sci. Technol. A* 6:1214 (1988).
24. J.R. Coupland et al., *J. Vac. Sci. Technol. A* 5:2563 (1987).
25. R.C. Longsworth and R.J. Webber, *J. Vac. Sci. Technol. A* 9:2766 (1991).
26. P.A. Lessard, *J. Vac. Sci. Technol. A* 7:2373 (1989).

# A SIMPLIFIED THEORY OF CURRENT LEADS

M.L.F. Rehak

Brookhaven National Laboratory, Upton, NY 11973

## ABSTRACT

Closed form solutions for temperature, conducted heat and voltage drop are derived for a lead cooled with forced flow. The mass flow which will cause thermal runaway is expressed as a function of current. An optimal lead is defined as a lead where the conducted heat at the warm end is zero and where the total refrigerator load is minimal. Optimal mass flow and geometry at design current are derived under those two conditions. The margin between critical and optimal flow is too small: a 2% decrease in mass flow or a 10 % increase in current are sufficient to produce temperatures of 450 K. Therefore an operating mass flow is defined: it is the smallest flow for which the lead is always stable as long as the flow is proportional to the current. A curve giving the operating mass flow as a function of current is established. Mass flow, voltage drop and conducted heat under operating conditions can then be established. During operation, reading of voltage drop is the best way to monitor the leads since it is shown that voltage drop is a linear function of maximum temperature. Comparisons with leads cooled with boil-off and by conduction only are presented.

## INTRODUCTION

This work describes problems associated with current lead design and their solution from an analytical perspective rather than following a numerical treatment. The benefits of closed form solutions lie in a enhanced understanding of physical phenomena. The shortcomings are that simplifying assumptions are invoked and complete numerical solutions must be used for actual lead design. Nevertheless this analysis provides a guide for a numerical analysis. The scope of this work is as follows:

1) to state the definition of optimal design: the conducted heat at the warm end should equal to zero and the total refrigerator load should be minimized.

2) to give analytical expressions for temperature, conducted heat, voltage drop at optimal mass flow and optimal geometry for given design current.

3) to establish operating curves for mass flow and voltage drop under variable current.

## NOMENCLATURE

All units are in cgs.

$A$: lead copper cross–section, assumed to be constant, $cm^{2.}$

$A_s$: heat exchange area, $cm^{2.}$

$\beta$: one of two parameters for lead optimization, related to cooling.

*Advances in Cryogenic Engineering,* Vol. 43
Edited by P. Kittel, Plenum Press, New York, 1998

$c$: heat capacity of copper, $J/g$
$c_l$:helium latent heat of vaporization, $c_l$= $20.9\ J/g$.
$c_p$: helium heat capacity at constant pressure, $c_p = 5.2\ \ J/g$.
$f$: factor converting liquefaction load into Watts at 4 K, $f$ =80 $W/g.sec$.
$h$: heat exchange coefficient, $W/cm^3.K$.
$I$: current, $A$.
$k$: copper thermal conductivity, assumed to be constant, $k$= 6 $W/cmK$.
$Q_{tot}$ : total refrigerator load, W. $q_{tot} = Q_{tot}/I,\ W/A$.
$Q_{in}$: heat conducted at the warm end the lead,, W.$q_{in} = Q_{tot}/I,\ W/A$.
$Q_{out}$: heat conducted in at the cold end the lead,$q_{out} = Q_{out}/I,\ W/A$.
$L$: lead length, $cm$.
$L_0$: Lorenz number , $L_0 = 0.2410^7 W\Omega cm/K$.
$M$: mass flow $g/sec$. $m = M/I,\ g/\sec .A$.
$\rho$:copper electrical resistivity , $\Omega cm$.
$\sigma$:copper density , $g/cm^3$.
$\sigma_h$:helium density , $g/cm^3$.
$T$: lead temperature, $K$.
$T_h$: helium coolant temperature, $K$.
$T_1$: lead temperature at the cold end, $T_1$= 0 $K$.
$T_2$: lead temperature at the warm end, $T_2$= 300 $K$.
$V$ : voltage drop, $A$.
$X$: length along the lead $cm$, $x = X/L, .1$

## GOVERNING EQUATION

The energy conservation equation describing the lead and its coolant are

$$kA\frac{d^2T(X)}{dX^2} - \frac{hA_s\,(T-T_h)}{L} + \frac{\rho I^2}{A} = 0,\ Mc_p\frac{dT_h\,(X)}{dX} = hA_s\,(T-T_h) \tag{1}$$

In order to reduce these equations to a form that can easily be integrated analytically, the following assumptions are introduced:

- $dT/dx = dT_h/dx$. With good heat exchange this is a reasonable assumption.
- $\rho k = L_0 T$ : the material obeys the Wiedeman-Franz law. While $L_0$ does in fact vary with $T$, this dependence is small when compared to other factors.
- $k$ the heat capacity of copper is constant. The exact value used only affects the value of $LI/A$ which is shown not to be critical within a given range.
- The initial temperature $T_1$which is 4.4 K is replaced by 0 in some of the derivations in order to proceed with the analytical solution.

The following parameters are introduced:. $m = M/I$ is the mass flow per unit current, $\beta = \frac{mc_p}{2\sqrt{L_0}}$ is a measure of cooling, $\ell = \frac{LI\sqrt{L_0}}{kA}$ is a shape factor related to the aspect ratio of the lead and current. When $\beta < 1$ the parameter $\alpha = \sqrt{1-\beta^2}$ is used to simplify notation.

Using the simplifying assumptions and the definitions of parameters $\ell$ and $\beta$, the differential equation describing heat balance along the lead can be then reduced to:

$$\frac{d^2T(x)}{dx^2} - 2\beta\ell\frac{dT(x)}{dx} + \ell^2 T(x) = 0, T(0) = T_1, T(1) = T_2 \tag{2}$$

The problem of designing an optimal lead is thus reduced to finding two parameters $\ell$ and $\beta$.

## DESIGN GOALS

Leads which are insufficiently cooled experience thermal runaway where temperatures grow suddenly out of bounds. The minimal mass flow needed to prevent this condition will be called here the critical mass flow. The two parameters $\beta$ and $\ell$ must be chosen so as to satisfy the following requirements.

- The total refrigerator load defined as $q_{tot} = q_{in} + fm$ is minimal.
- $q_{in}$ must be less than a given value so as not to quench the superconductor at the cold end for instance.
- The operating mass flow must be above the critical mass flow with a sufficient margin (20%).
- The slope of the temperature profile at the warm end must be zero in an optimal lead. The lead should neither conduct or receive heat from or to the outside at the warm end. This can be stated with $q_{out} = 0$.
  The system of two equations $\partial q_{tot}/\partial\beta = 0$ (minimal total refrigerator load) , $q_{out} = 0$ and two unknowns $\beta, \ell$ is solved next.

## LEAD COOLED WITH FORCED FLOW COOLING

### Temperature profile

The nature of the solution of the governing equation depends on the roots of the characteristic equation: $r^2 - 2\beta\ell r/L + \ell^2/L^2$. There are three types of solutions depending on the discriminant $\Delta = 1 - \beta^2$.

Overcooled lead. The first case where $\beta > 1$ leads to a solution containing only exponential function. While the temperature can become high it cannot suddenly grow out of bounds, the lead is always stable and the condition for optimal design $q_{out} = 0$ cannot be satisfied.

$$T(x) = e^{\ell x\beta}(T_2 e^{\ell\alpha}(e^{2\ell x\alpha} - 1) - T_1 e^{\ell\beta}(e^{2\ell x\alpha} - e^{2\ell\alpha}))/(e^{\ell(\alpha x+\beta)}(e^{2\ell\alpha} - 1)) \tag{3}$$

Overcooled limit case. When $\beta = 1$, the mass flow is $m = 0.0596\ g/s.kA$.

$$T(x) = (e^{\ell} - xe^{\ell})e^{\ell(x-1)}T_1 + T_2 xe^{\ell(x-1)} \tag{4}$$

It is found that the slope of the temperature profile at the warm end cannot be equal to zero, consequently cases where $\beta \geq 1$ do not lend themselves to optimization.

Undercooled lead. When $\beta < 1$, the solution is of the form: .

$$T(x) = (c_1 \sin(\ell x\alpha) + c_2(\cos \ell x\alpha))e^{\ell x\beta}, c_1 = -(\cos(\ell\alpha)T_1 - T_2 e^{-\ell\beta})/\sin(\ell\alpha), c_2 = T_1$$

This solution for temperature contains trigonometric functions and can be singular when $\sin(\ell\alpha) = 0$. The condition $q_{out} = 0$ is possible, thus this is the case where optimizing is possible.

### Conducted heat $\beta < 1$

Differentiating $T(x)$ with respect to $x$ gives the conducted heat at any location along the lead $q = Q/I$ in W/A. The conducted heat per unit current at the cold end, $x = 0$ is

$$q_{in} = -\frac{\sqrt{L_0}}{\sin(\ell\alpha)}(\alpha e^{-\ell\beta}T_2 + (-\alpha\cos(\ell\alpha) + \beta\sin(\ell\alpha))T_1) \tag{5}$$

The conducted heat per unit current at the warm end is

$$q_{out} = \frac{\sqrt{L_0}}{\sin(\ell\alpha)}(\alpha e^{\ell\beta}T_1 - (\alpha\cos(\ell\alpha) - \beta\sin(\ell\alpha))T_2) \tag{6}$$

It is verified that the relation $\frac{\partial q_{in}}{\partial \ell} + \frac{q_{out} e^{-\beta\ell}\alpha}{\sin(\ell\alpha)} = 0$ holds. Therefore a lead with optimal geometry conducts a minimum amount of heat at the cold end or, equivalently, conducts no heat at the warm end.

**Voltage drop $\beta < 1$**

The lead resistance can be expressed using the Wiedeman-Franz-Lorenz relation $k\rho = L_0 T$ . With the voltage drop defined as $V = RI$, and the lead resistance $R = \int_0^1 \rho(x)L/A dx$ one has

$V = \sqrt{L_0}((-\alpha\cos(\ell\alpha)+\beta\sin(\ell\alpha)+\alpha e^{-\ell\beta})T_2+(-\alpha\cos(\ell\alpha)-\beta\sin(\ell\alpha)+\alpha e^{\ell\beta})T_1)/\sin(\ell\alpha)$

It is verified that the equation expressing energy conservation is satisfied: $V = q_{out} - q_{in} + mc_p(T_2 - T_1) = 0$ .

**Critical mass flow $\beta < 1$**

The critical mass flow is the minimum flow required to prevent thermal runaway, it corresponds to the singularity in temperature when $\sin(\alpha\ell) = 0$ at $\alpha\ell = \pi$. The case where $\alpha = 0$ corresponds to the solution $\beta = 1$ which is not under consideration here. Solving for $\beta$, one gets $\beta_{crit} = \sqrt{1 - \pi^2/\ell^2}$ and $\ell_{crit} = \pi$. $\ell$ has to be larger than $\pi$ for this solution to be valid. Then the critical mass flow is

$$m_{crit} = 2\sqrt{\frac{L_0}{c_p^2} - (\frac{\pi Ak}{c_p LI})^2} \tag{7}$$

This expression shows that when current dominates conduction the second term under the square root vanishes and $m \approx 2\sqrt{L_0}/c_p$.

**Operating flow $\beta = 1$**

The solution is valid for $\beta$ values up to 1 which corresponds to $m$ values up to $m_{oper}$ . For mass flows equal or greater than $m_{oper} = 0.0596\ g/s.kA$, the lead is stable but overcooled. The subscript *oper* is used here since it will be shown in the following that it is desirable to operate at $\beta = 1$.

**Optimal values $\beta < 1$**

One needs to determine values for the two parameters describing the problem, $\beta$ and $\ell$. The first of the two available equations, $q_{out} = 0$, yields the following relation between $\ell$ and $\beta$ when $T_1 = 0$ : $\sin(\alpha\ell)\beta + \cos(\alpha\ell)\alpha = 0$ . The solution is $\tan(\alpha\ell) = -\frac{\alpha}{\beta}$ and for the optimal value $\ell_{opt} = (\pi - \arctan(\frac{\alpha}{\beta}))/\alpha$ because $\ell$ must be positive. Moreover, since the solution lies in the second quadrant, $\sin(\alpha\ell) \geq 0$ and $\cos(\alpha\ell) \leq 0$. Thus $\sin(\alpha\ell) = \alpha$ and $\cos(\alpha\ell) = -\beta$. Replacing $\sin(\alpha\ell)$ by $\alpha$ in the expression for $T(x)$ removes the singularity in the temperature profile. This shows that at optimal mass flow and geometry the lead is stable.

The second relation is obtained by finding the value of $\beta$ which minimizes the total refrigerator load defined as $q_{tot} = q_{in} + fm$ where $f = 80$ is the factor which is used to convert liquefaction load into heat load at 4.4 K.

Upon substitution of $\ell_{opt}$ into $q_{tot}$, one finds $q_{tot}=\sqrt{L_0}e^{(\arctan(\alpha/\beta)-\pi)\beta/\alpha}T_2 + f\beta$.

Setting the derivative of $q_{tot}$ with respect to $\beta$ equal to zero gives $\beta_{opt} = 0.913$. At that point (Figure 1) the value for $q_{in}$ is negligeable and the major contribution to $q_{tot}$ comes from $fm$.The other parameters follow: $m_{opt} = 0.0544\ g/s.kA$, $\ell_{opt} = 6.68$, $LI/A_{opt} = 2.587\ 10^5\ A/cm$, $q_{in\_opt} = -0.104\ W/kA$, $q_{tot\_opt} = 4.46\ W/kA$, $V_{opt} = 85.0\ mV$. It can be noted that for an optimal lead, $q_{in} \approx 0$ , $q_{out} = 0$ thus $V_{opt} \approx mcp(T_2 - T_1)$ is constant. If $RI$ is constant at constant $m$ or $M/I$ then the resistance of the lead is $R = V_{opt}m_{opt}/M$. This states that the lead at optimal design has a variable resistance which is inversely proportional to the mass flow in $g/sec$ used to cool the lead when this mass flow is a linear function of current.

Using the flow for which the lead has a stable operation $\beta = 1$, it can be seen (Figure 2) that $q_{tot}$ has a very flat optimal value over a large range of $\ell$. Deviations from the optimal $\ell$

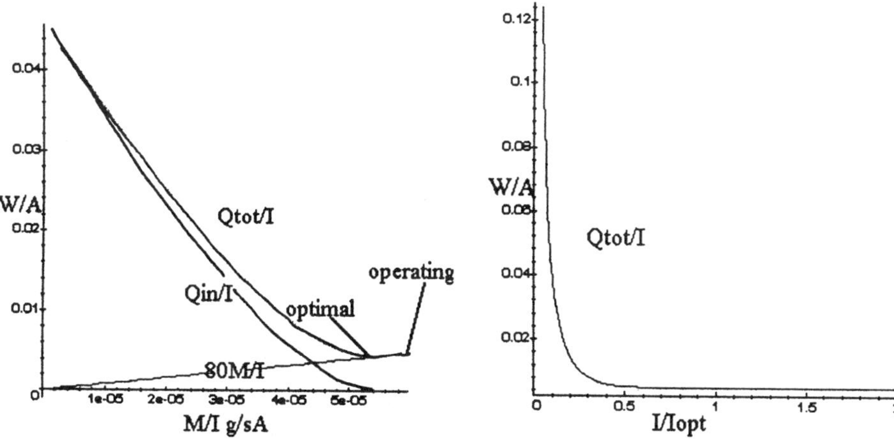


**Figure 1.** $Q_{tot}/I, Q_{in}/I, 80M/I$ *vs* $M/I$.

**Figure 2.** $Q_{tot}/I$ *vs* $\ell/\ell_{opt}$.

are permissible when the flow is equal to $m_{oper}$. The value for $LI/A$ at optimal $\ell$ will depend on the choice of a numerical value for $k$. The fact that there is a large latitude in choice of values for $\ell$ makes the initial assumption that $k$ is constant less critical.

**Operating values $\beta = 1$**

During ramp up or at operating currents less than the design current, the mass flow needs to be adjusted as a function of current. Since the operating mass flow must be above the critical mass flow with a sufficient margin, the first step is to compare the critical mass flow per unit current, $m_{crit}$, to the optimal mass flow, $m_{opt}$, and to $m_{oper}$ (which is the maximum mass flow for which this solution is valid and corresponds to the $\beta = 1$ case). The following plot shows how $m_{crit} = 2\sqrt{L_0\frac{\ell^2-\pi^2}{l^2}/c_p}$ varies with $\ell/\ell_{opt.}$ It can be seen in Figure 3 that as $\ell$ exceeds $\ell_{opt}$ by about 10%, the optimum flow is smaller than the critical flow and the lead is no longer stable. At optimal $\ell$, the optimal flow has a very small margin over the critical flow. A better choice for operating flow would be to use the limit case $\beta = 1$ case where $m_{oper} = 0.06$ g/sec/kA which is always above the critical flow. As long as the mass flow increases linearly with current, the lead will be stable.

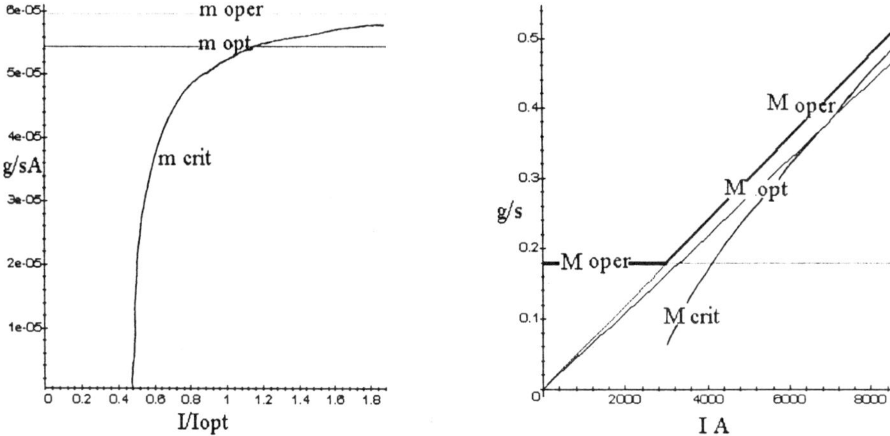


**Figure 3.** $m$ critical, optimal and operating vs $\ell/\ell_{opt}$.

**Figure 4.** Operating mass flow g/sec vs current A.

Figure 4 shows the operating flow versus current curve which was developed using half the mass flow at design current to limit conducted heat for currents up to half the current design value. At currents higher than half the design current, the flow is linear with current at the rate of 0.06 g/sec/kA.

From the operating value $\beta_{oper} = 1$ thus obtained the other parameters follow: $m_{oper} = 0.0596\ g/s.kA,\ q_{in\_oper} = 0, q_{tot\_oper} = 4.76\ W/kA,\ V_{oper} = 39.5\ mV.$

**Temperature profiles**

The temperature profile at optimal values for $\beta$ and $\ell$ is shown in Figure 5. Deviations from optimal when the mass flow is reduced by 80% or when the current is increased by 20% are also shown along with the $\beta = 1$ and $\beta > 1$ cases. The location of maximum temperature along the lead length occurs at $x_{\max} = (\arctan(\alpha/\beta) - \pi)/\ell\alpha$. At $\ell_{opt}$, it is found that $x_{\max} = I_{opt}/I$ where $I$ is the overcurrent and $I_{opt}$ is the design current.

The condition that $q_{out} = 0$ cannot be satisfied for case where $\beta \geq 1$. Furthermore $q_{in}$ is already small when $\beta = 1$ and gets exponentially smaller. In this region the lead is always stable and consumes a mass flow higher than the minimum needed to stabilize the lead and run it at optimal value.

**Voltage drop**

Monitoring the voltage drop across the lead is a better way to control a lead during operation than using temperature sensors: the voltage drop is constant while maximum temperature location varies along the lead as was just established. The operating voltage drop across the lead of 40 $mV$ is approximately half that of the optimal solution 85 mV. At operating conditions where $\beta = 1$,the lead is overcooled the area under the temperature profile is smaller that in the undercooled case, consequently the difference of a factor of two in voltage drop (which is an integral of the temperature) between optimal and operating conditions is not surprising. Figure 6 shows that the voltage drop is approximately a linear function of maximum temperature and is an appropriate way to control leads.

## LEAD COOLED BY BOIL OFF

When the lead is cooled with boil-off instead of forced flow, the mass flow is given by the latent heat of vaporization and the heat conducted at the cold end is $q_{in} = mc_l$. This equa-

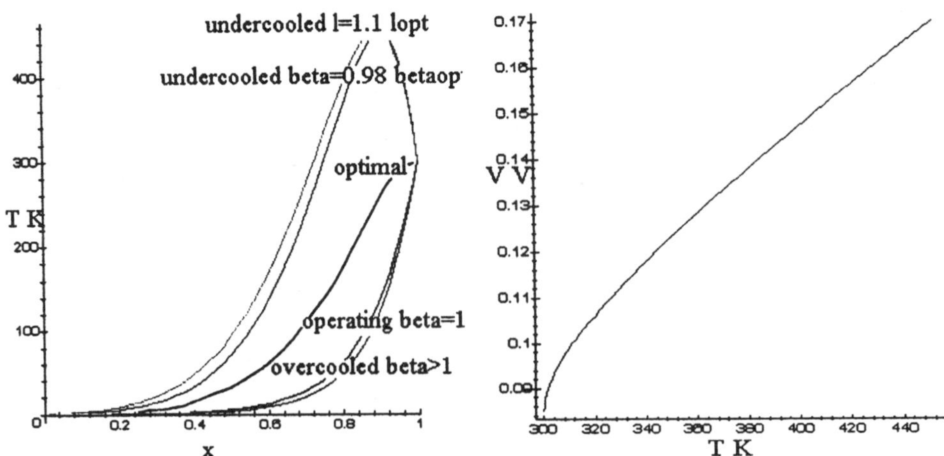


**Figure 5.** Lead temperature profiles vs lead length. **Figure 6.** Voltage drop vs maximum temperature.

**Table 1.** Summary

| Cooling method | mass flow | $LI/A$ | volt drop | $Q_{in}/I$ | $Q_{tot}/I$ |
|---|---|---|---|---|---|
| | $g/sec/kA$ | $kA/cm$ | $mV$ | $W/kA$ | $W/kA$ |
| forced @critical | 0.0526 | 260 | $\infty$ | $\infty$ | $\infty$ |
| forced @optimal | 0.0544 | 260 | 85.0 | 0.104 | 4.46 |
| forced @operating | 0.0596 | 260 | 40.0 | 0.01 | 4.76 |
| conduction | | 61 | 46.5 | 46.5 | 46.5 |
| boil-off | 0.050 | 177 | 78.2 | 1.0 | 5.0 |

tion replaces the equation minimizing $q_{tot}$ with respect to $\beta$ in the system of two equations and two unknowns for the forced flow case. Solving for $\beta$ the equation $q_{in} = mc_l$ where $q_{in} = -\sqrt{L_0}e^{-\ell\beta}T_2$, $\ell = \ell_{opt}$, one obtains for an optimized lead cooled by boil off $\beta = 0.83$, $\ell = 4.58$, $LI/A = 1.77\ 10^5 A/cm, m = 0.0495 g/s.kA, q_{in} = -1.03\ W/kA, q_{tot} = 0.50\ W/kA, V = 78.2\ mV$. With boil-off cooling, the optimal refrigerator load is slightly higher than forced-flow cooling: $5\ W/kA$ versus $4.5\ W/kA$. The contribution from conducted heat in boil-off is higher by an order of magnitude but the required mass flow is lower.

## LEAD COOLED BY CONDUCTION ONLY

For comparison purposes, the case where the lead is cooled by conduction only is examined next. The conducted heat is simply $q = -\frac{\sqrt{L_0}}{\sin\ell}(T_2 - T_1\cos\ell)$ which is minimal at $\ell = \pi/2$. The conducted heat at the cold end is $q_{in} = -0.046 W/kA$ which is an order of magnitude larger than the other cooling methods. The voltage drop is $V = 46.5\ mV$. This case would be representative of a lead cooled by heat sinking only.

## CONCLUSIONS

After defining the goals of optimal design, it was shown that the problem is reducible to finding values of only two parameters, one related to cooling and one related to geometry at a given current.

The subcooled case which is the case of interest has yielded values for optimal mass flow and geometry at design current. Since the optimum is too close to the critical flow, it is preferable to operate at that minimum flow which guarantees a stable operation at the cost of an increase of total refrigerator lead from 4.5 $W/kA$ to 4.8 $W/kA$. Measuring voltage drop is the best way to control leads during operation as it is a linear function of maximum temperature. Comparisons with leads cooled with boiloff and conduction only have been made.

Operating curve for mass flow versus current and the value of 40 $mV$ for voltage drop resulting from this analysis have been successfully applied to control and monitor 6300 A and 1600 A gas cooled current leads in RHIC[1].

Table 1 summarizes parameters for three cooling methods.


## ACKNOWLEDGMENT

Work performed under the Contract No.De-AC02-76CH00016 with the U.S. Department of Energy.


## REFERENCES


A.Nicoletti, M.Rehak. "Performance of RHIC Power Leads," to be presented at the CEC/ICMC, Portland, Oregon, 1997.

# OPTIMAL DESIGN OF SIZES IN BINARY CURRENT LEADS COOLED BY CRYOGENIC REFRIGERATOR

H.-M. Chang[1], S.J. Song[1], G.W. Hong[2], and H.J. Lee[2]

[1]Hong Ik University, Department of Mechanical Engineering
Seoul, 121-791, Korea
[2]Korea Atomic Energy Research Institute
Daejeon, 305-600, Korea

## ABSTRACT

Analysis is performed to determine the optimal lengths or cross-sectional areas of refrigerator-cooled current leads that can be applied to the conduction-cooled superconducting systems. The binary current lead is composed of the series combination of a normal metal at the upper (warm) part and a high Tc superconductor (HTS) at the lower (cold) part. The heat conduction toward the cold end of HTS part constitutes a major refrigeration load. In addition, the joint between the two parts should be cooled by a refrigerator in order to reduce the load at the low end and maintain the HTS part in a superconducting state. The sum of the work inputs required for the two refrigeration loads needs to be minimized for an optimal operation. In this design, three simple models that depict the refrigeration performance as functions of cooling temperature are developed based on some of the existing refrigerators. By solving one-dimensional conduction equation that takes into account the temperature-dependent properties of the materials, the refrigeration works are numerically calculated for various values of the joint temperatures and the sizes of two parts. The results show that for a given size of HTS, there exist the optimal values for the joint temperature and the size of the normal metal. It is also found that the refrigeration work decreases as the length of HTS increases and that the optimal size of the normal metal is quite independent of the size of HTS. For a given length of HTS, there is an optimal cross-sectional area and it increases as the length increases. The dependence of the optimal sizes on the refrigerator models employed are presented for 1 kA lead.

## INTRODUCTION

Current leads are key components in superconducting systems that supply electrical current to the cold superconducting magnet from the power source at ambient temperatures. The material for an ideal current lead should have a good electrical conductivity and a poor thermal conductivity, since both the Joule heat generation in the lead and the heat

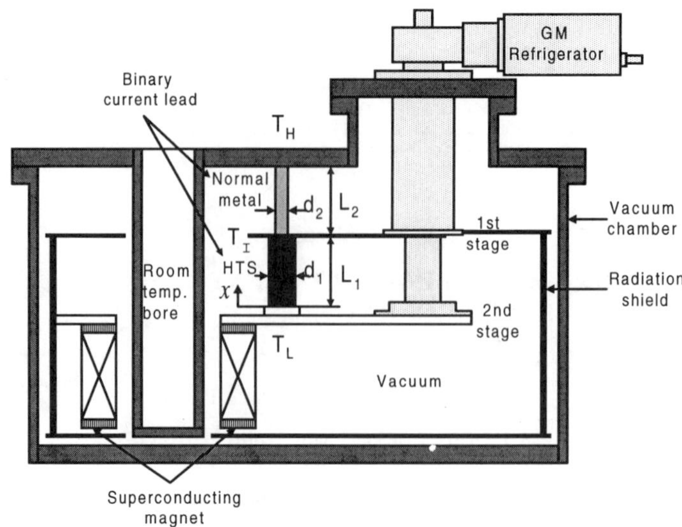


**Figure 1.** Schematic representation of refrigerator-cooled superconducting systems with binary (higher-temperature metal and lower-temperature HTS) current leads.

conduction through the lead to the cold magnet should be small. The conventional metallic leads[1-2], which basically obey the Wiedemann-Franz law, is constrained by an inherent limit in reducing the heat leak to the cold magnet. In practice, the largest cooling load of the conventional superconducting systems is due to the metallic current leads.

After the discovery of the high Tc superconductors (HTS), the earliest useful application has been as a low temperature component of the current lead. The oxidized ceramic superconductors have very low thermal conductivity and zero electrical resistance as long as they remain in a superconducting state. The binary or hybrid current lead[3-10] is the series combination of normal metal as a higher temperature part and HTS as a lower temperature part. Over the past several years, there have been many efforts directed at the practical utilization of the binary leads.

In addition, recent progress in the development of 4 K Gifford-McMahon refrigerators[11] has raised the possibility of the liquid-free or the refrigerator-cooled superconducting magnets[12-13]. Since there is no liquid helium, one of the most significant and inevitable problems in the refrigerator-cooled superconducting system is the cooling of the current lead without helium vapor in vacuum. A typical configuration of the liquid-free superconducting systems is schematically shown in Figure 1. The second stage of a 2-stage GM refrigerator cools the magnet and the cold end of the binary current leads at about 4 K, while the first stage cools the joint between the two parts of the binary lead and the radiation shield. It is a well-established fact that the intermediate cooling is necessary to reduce the amount of heat leak to the lower temperature as well as to maintain the HTS in a superconducting state. The authors think that the feasibility of the refrigerator-cooled current leads has been demonstrated by recent studies[12-13], and the next crucial step towards the practical application could be to develop energy-efficient current leads.

In a refrigerator-cooled current lead, there are two separate cooling loads at respective temperatures, and the optimal design with respect to the energy efficiency can be obtained by minimizing the sum of the work inputs for the two refrigeration loads. This study aims to determine the optimal conditions for 1 kA level of binary current leads, including the intermediate cooling temperature and the sizes (the cross-sectional area and the length) of both parts of the current lead. The analysis incorporates the temperature-dependent material properties and is based upon a few simple yet reasonable models of the refrigeration performance.

## ANALYSIS MODEL

### Cooling Load

The temperature of the current lead in vacuum as shown in Figure 1 can be determined by solving one-dimensional conduction equation

$$\frac{d}{dx}\left(k(T)\frac{dT}{dx}\right)+\frac{\rho(T)I^2}{A}=0 \tag{1}$$

where $T(x)$ is the temperature along the lead and $A$ is the cross-sectional area. The thermal conductivity, $k(T)$, and the electrical resistivity, $\rho(T)$, are functions of temperature. The coordinate $x$ is defined as the distance from the cold end of the lead. Eq.(1) can be applied both to the metal and the HTS. For the boundary conditions, the warm end of the normal metal part operates at the room temperature, $T_H$, and the cold end of the HTS part operates at the low temperature, $T_L$, which is identical to the magnet temperature. At the joint, the contact resistance is neglected, so that the two parts are assumed to have the same temperature, $T_I$. The cooling load at the joint can be expressed as

$$Q_I = k_2(T_I)A_2\frac{dT_2(L_1)}{dx}-\left(k_1(T_I)A_1\frac{dT_1(L_1)}{dx}\right) \tag{2}$$

and the cooling load at the cold end can be expressed as

$$Q_L = k_1(T_L)A_1\frac{dT_1(0)}{dx} \tag{3}$$

In Eqs.(2) and (3), the subscripts 1 and 2 denote the HTS and the normal metal, respectively.

### Properties of Materials

The thermal conductivity and the electrical resistivity of normal metals are strongly dependent on temperature. The Wiedemann-Franz law describes the general correlation of the two properties with temperature. In this paper, however, the functional relations for the two properties given by White[2] and Jones et al.[4] are selected for a more precise calculation of the temperature distribution.

Concerning the high Tc superconductors, two properties are important for this study. The first one concerns the critical property or the normal-superconducting transition property. Typically the transition property is given by the current density ($J_c$)-temperature (T)-magnetic field (B) surface[14]. The detailed phenomena are quite complicated and depend on the composition of the materials and the fabrication processes. In this study, a simple model with a linear $J_c$-T relation for zero magnetic field is selected. For HTS, the electrical resistivity in Eq.(1) is assumed to be zero, as long as the HTS material is superconducting. If this condition is satisfied, Eq.(1) is directly integrated so that the cooling load at the cold end may be written as

$$Q_L = \frac{A_1}{L_1} \int_{T_L}^{T_I} k_1(T) dT \tag{4}$$

The second property concerns the thermal conductivity of HTS as a function of temperature. A functional relation is taken from Herrmann et al.[5] for BSCCO-2212.

## Refrigerator

The most difficult and crucial part of the analysis model concerns the performance of the refrigerators. As mentioned above, in the binary current leads, there are two cooling loads at two different temperatures. Generally, the work required for refrigeration depends on the refrigeration temperature and the refrigerator type as well as on the cooling load. The sum of the works for the two cooling loads rather than the sum of the two cooling loads should be minimized for an optimal operation. The refrigeration work can be expressed as

$$\begin{aligned} W &= W_I + W_L \\ &= \frac{Q_I}{FOM_I}\left(\frac{T_H}{T_I} - 1\right) + \frac{Q_L}{FOM_L}\left(\frac{T_H}{T_L} - 1\right) \end{aligned} \tag{5}$$

where *FOM* denotes the figure of merit, defined as the ratio of the coefficient of performance (*COP* = refrigeration/work) of the actual refrigerator to that of reversible refrigerator which operates between the same refrigeration temperature and the same heat rejection temperature. In Eq.(5), the temperature at which the heat is rejected from the refrigerator is assumed to be the room temperature, $T_H$.

To construct simple yet reasonable models for the $FOM_I$ of the refrigerator, the performance of typical commercial refrigerators are surveyed and plotted in Figure 2. It is observed that for Stirling refrigerators, the COP and the FOM are strongly dependent on the size or the cooling capacity as well as on the refrigeration temperature. The required cooling capacity is closely related with the current level of the lead. For GM (Gifford-McMahon) refrigerators, however, the COP and the FOM are determined primarily by the refrigeration temperature and have values close to those for the small sized Stirling refrigerators. It is also noticed that the FOM of GM refrigerators decreases as the refrigeration temperature rises

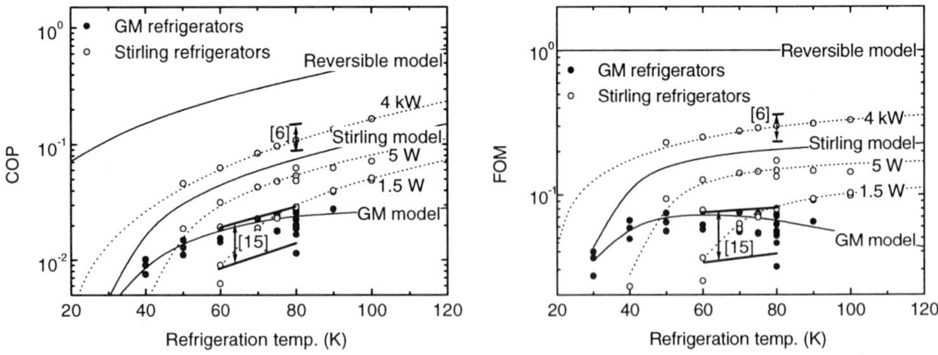


**Figure 2.** Coefficient of performance (COP) and figure of merit (FOM) for three different (Reversible, Stirling, Gifford-McMahon) refrigerator models as functions of refrigeration temperature. Hermann et al.[6] have assumed greater FOM's than practical refrigerator and Yang and Pfotenhauer[15] have assumed FOM's close to those of GM model.

above 80 K. No commercial pulse tube refrigerator could be found for the capacity and temperature range of interest.

In this paper, three different refrigerator models - reversible, Stirling, and GM - are considered as shown by solid curves in Figure 2. The reversible refrigerator model represents the thermodynamically ideal refrigeration or the case of the minimal work required for a given refrigeration load. The Stirling refrigerator model is selected for approximately 40 W range at 80 K, which is suitable for 1 kA binary leads of the optimal design case to be described below. The GM refrigerator model is closer to many commercial refrigerators. It is noted that the FOM of Stirling model is higher than that of GM model at relatively higher temperatures and nearly the same at relatively lower temperatures. Hermann et al.[6] have assumed higher FOM's than the typical commercial refrigerator and Yang and Pfotenhauer[15] have assumed FOM's close to those of GM model.

Since the cold end temperature of the binary current lead is fixed and is not a design parameter, the $FOM_L$ of the refrigerator for cooling the cold end is assumed to be a constant. Since a typical value of $FOM_L$ of the refrigerator at about 4 K is 1/15[16-17], it is used in the calculation. When a two-stage refrigerator cools the current lead as shown in Figure 1, the refrigeration performance at the second stage is related with that at the first stage in a rather complicated manner. In the present analysis model, the two performances are assumed to be independent for simplicity.

## RESULTS AND DISCUSSION

### Intermediate Cooling Temperature

The intermediate cooling temperature, $T_I$, at which the joint of the normal metal and the HTS is maintained, is important in designing the binary current lead. For all cases considered in this paper, the materials for the normal metal and the HTS in this analysis are copper and BSCCO-2212, respectively. The room temperature and the cold end temperature are 300 K and 4.2 K, respectively. The current level of the lead is 1 kA, unless specified otherwise.

Figure 3 shows the total refrigeration work as functions of the intermediate cooling temperature for three refrigeration models, all for specific sizes of the lead. The solid curves denote the sum of work for the two loads, which are shown as the dotted curves. The curves for the total work can be classified into two difference shapes; typical representations are given in Figures 3(a) and (b).

For Figure 3(a), the HTS is relatively thick and short so that the refrigeration work for the cold load is comparable with that for the intermediate load. It is clearly observed that for each model, there exists an optimal $T_I$ that minimizes the refrigeration work. At $T_I$ lower than the optimum, the total work is larger due to the work for the cold end load, and at higher $T_I$, the total work is also larger due to the work for the intermediate cooling load. In this case, it is noted that the optimal value of the intermediate cooling temperature is about 92 K for the reversible model, but is as low as about 73 K or 65 K for Stirling or GM model. The reason why the Stirling or GM model has a lower optimum than the reversible model is that the $FOM_L$ in the Stirling or GM model is much lower, and the work ($W_L$) for the cold end cooling is more significant than in the reversible model. Apparently, the highest possible value of the intermediate cooling temperature is limited by the critical current density of HTS.

For the case of Figure 3(b), the HTS is relatively thin and long so that the refrigeration work for the cold load is negligible. It is observed that the total refrigeration work decreases as the intermediate temperature increases to the critical temperature. From the viewpoint of

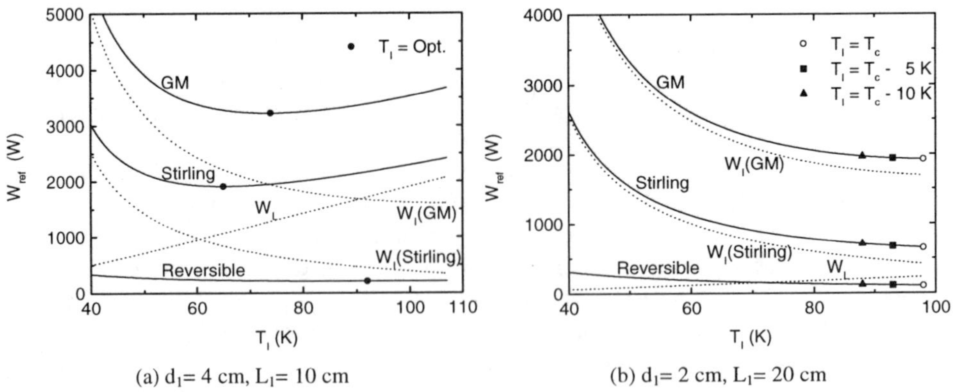


(a) $d_1$= 4 cm, $L_1$= 10 cm (b) $d_1$= 2 cm, $L_1$= 20 cm

**Figure 3.** Refrigeration work as a function of intermediate cooling temperature for three refrigerator models. (1: BSCCO-2212, 2: Cu, I= 1 kA, $d_2$= 0.5 cm, $L_2$= 12 cm)

energy efficiency, the optimal $T_I$ is the highest possible value that still keeps the HTS superconducting. In practice, when we consider the thermal stability of HTS, $T_I$ should be set at a temperature slightly lower than the critical temperature. The points where $T_I$ is lower than $T_c$ by 5 K or 10 K are indicated on the curves. The four different markers in the legends in Figure 3 will be used consistently in the subsequent graphs. It should be noted that since the analysis is one-dimensional, the diameter of the lead simply represents the size of the cross-section and the lead should not necessarily be a circular rod.

## Sizes of Normal Metal

Figure 4 shows the total refrigeration work as a function of the metal diameter for various values of the metal length. In the case of Figure 4(a), the intermediate cooling temperature has been optimized as in Figure 3(a), and in the case of Figure 4(b), the intermediate cooling temperature is the critical temperature of Figure 3(b). In both cases, it is noted that for a fixed diameter, the length has an optimum value with respect to the minimum refrigeration work required. When the length is shorter than the optimum, the cooling load due to the Joule heating by large electrical resistance entails more work. When the length is longer, the load due to large thermal conduction also causes more work.

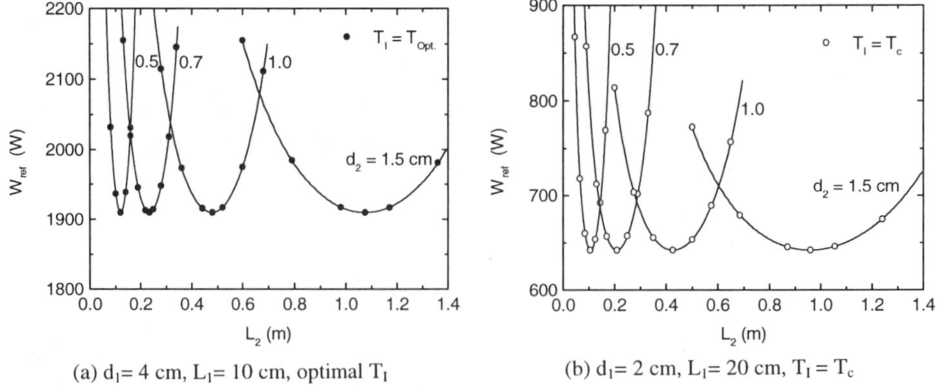


(a) $d_1$= 4 cm, $L_1$= 10 cm, optimal $T_I$ (b) $d_1$= 2 cm, $L_1$= 20 cm, $T_I = T_c$

**Figure 4.** Refrigeration work as a function of diameter of normal metal for various lengths of normal metal. (1: BSCCO-2212, 2: Cu, I= 1 kA, Stirling model)

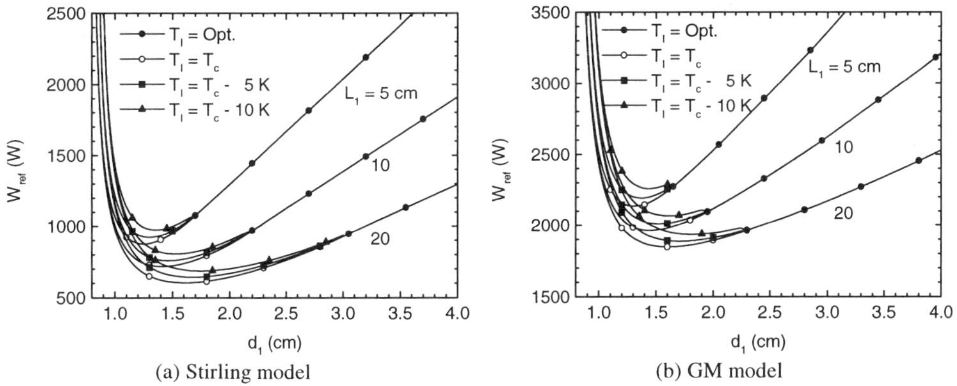

(a) Stirling model (b) GM model

**Figure 5.** Refrigeration work as a function of diameter of HTS for various values of HTS length with Stirling and GM models. The markers indicates how the joint (intermediate cooling) temperatures are selected. (1: BSÇCO-2212, 2: Cu, $d_2$ & $L_2$=optimized, I= 1 kA)

It is interesting to note that the optimal length increases but the refrigeration work remains nearly constant as the diameter increases. This behavior is quite similar to the case of the simple current lead without HTS. In practice, the length of the normal metal will be constrained by the geometric factors such as the length of the cold head of the refrigerator. With the constraints included, the optimal diameter is determined such that the total refrigeration work is minimized. While there is a close relationship between the length and the diameter of the normal metal for minimum work, the minimum work itself is dependent not on the sizes, but only on the current level. It should also be mentioned that in practical design, a thick and long design of the metal part may be desirable, because the additional work may not entail large cost even if the design deviates from the optimal.

## Sizes of HTS

The sizes of HTS part are important design factors in the practical design of the binary current lead. In particular, the sizes are quite closely related with the fabrication method of HTS. Concerning the normal metal, the sizes are important from the viewpoint of balancing the heat generation by electrical resistance and the heat conduction. However, since the HTS has no electrical resistance in superconducting state, no heat is generated and the heat conduction becomes the dominant factor in minimizing the refrigeration work. In general, a large length or small diameter of the HTS part will be preferred so far as the refrigeration is concerned. On the other hand, the cross sectional area of the HTS affects the current level which the lead can carry in a superconducting state or the temperature level below which the HTS should be cooled to.

Figure 5 shows the refrigeration work as a function of HTS diameter given various values of HTS length for Stirling and GM refrigerator models. In the calculations, the sizes of the normal metal and the intermediate cooling temperature have been optimized as described in the previous sections. It is noted that if the diameter is large enough, there exists an optimal $T_I$ as in Figure 3(a). As the diameter decreases, the optimal $T_I$ reaches the critical temperature and then the $T_I$ can be chosen as in Figure 3(b).

Regardless of the choice of $T_I$, it is found that the refrigeration work decreases as the length of HTS increases. For a given length, however, there is an optimal diameter of the HTS that requires a minimum refrigeration work. As the diameter increases above the optimum, the heat conduction through the HTS to the cold end increases, and the corresponding refrigeration work increases as well. As the diameter decreases, the current density increases and the optimal intermediate temperature reaches the critical temperature of HTS for the given current density. It is also noted that the refrigeration work is independent of the length when the diameter is much shorter than the optimum, and the optimal length-diameter combination does not significantly depend on the refrigeration models employed as well.

## CONCLUSIONS

The refrigeration work required to cool 1 kA binary current leads is calculated through analysis to determine the effect of sizes. The optimal design that minimizes the work is also discussed. The calculation of the work is based upon three different simple models that depict the refrigeration performance as functions of cooling temperature. The following main conclusions can be drawn from the analysis.

(1) There exists an optimal intermediate cooling temperature between 60 K and 100 K in which the exact value depends on a given size combination of the metal and the HTS parts if the cold end load is relatively significant. When the cold end load is small, the work decreases as the intermediate cooling temperature increases up to the critical temperature of HTS.

(2) There exists an optimal the diameter-and-length combination of the metal part, if given an optimal intermediate cooling temperature and a particular size of HTS. As the length increases, the corresponding optimal diameter increases, but the refrigeration work does not vary much.

(3) When the intermediate cooling temperature and the size of the metal are optimally designed, the refrigeration work decreases as the length of HTS increases. For a given length of HTS, there exists a corresponding optimal diameter, and it increases with the length. It is demonstrated that in the case of 1 kA binary lead, the optimal diameters of HTS fall between 1.2 and 1.8 cm when the length are from 5 to 20 cm.

## REFERENCES


1. M.N. Wilson., "Superconducting Magnets," Oxford University Press, New York (1986) pp.256-271.
2. G. K. White, "Experimental Techniques in Low-Temperature Physics," Clarendon Press, Oxford, (1979) pp.284-305, p.319.
3. R. Wesche, and A.M. Fuchs, Cryogenics, 34:145 (1994).
4. M.C. Jones et al.., Cryogenics, 18:337 (1978).
5. P.F. Herrmann et al., IEEE Trans. App. Superconductivity, 3:876, (1993).
6. P.F. Herrmann et al., Cryogenics, 33:555 (1993).
7. P.F. Herrmann et al., Cryogenics, 34:543 (1994).
8. J.R. Hull et al., Cryogenics, 32:822 (1992).
9. S.Y. Seol and J.R. Hull, Cryogenics, 33:966 (1993).
10. B. Dorri et al., IEEE Trans. Magnetics, 27:1858 (1991).
11. T. Kuriyama et al., "Advances in Cryogenic Engineering," Vol.41, Plenum Press, New York (1996) pp.1615-1622.
12. T. Hase et al., Cryogenics, 36:971 (1996).
13. K. Watanabe et al., Cryogenics, 36:1019 (1996).
14. L. Dresner, "Stability of Superconductors," Plenum Press, New York, (1995) , pp.15-34.
15. S. Yang and J.M. Pfotenhauer, "Advances in Cryogenic Engineering," Vol.41, Plenum Press, New York (1996), pp.567-572.
16. R. Li et al., "Advances in Cryogenic Engineering," Vol.41, Plenum Press, New York (1996), pp.1601-1607.
17. R. Li et al., "Advances in Cryogenic Engineering," Vol.41, Plenum Press, New York (1996), pp.1631-1637.

# EFFECT OF HEAT TRANSFER ON COOLING OF CURRENT LEADS

K. S. Iyengar and S.W. Van Sciver

FAMU/FSU College of Engineering
& National High Magnetic Field Laboratory
Tallahassee, FL 32310, USA

## ABSTRACT

Current leads supply electrical current to high-field superconducting magnets, operating at low temperatures. The heat leak flowing down the current lead must be intercepted and extracted. In the present work, we consider different methods of intercepting this heat, along with the refrigeration power required for each method. The cooling methods studied include envelope cooling, wherein the coolant flows parallel to the lead, and helical coil cooling, in which, a coolant channel is wrapped around the lead.

## INTRODUCTION

Current leads are used to transport large electrical currents from a power supply, at room temperature, to devices such as high-field magnets. Such magnets may employ superconducting windings, and are then operated at low temperatures, below the critical temperature of the material of the superconductor. The temperature gradient imposed across the current lead results in a large quantity of heat being conducted from the ambient into the magnet, via the current leads. In addition, the Ohmic heat generated within the current-carrying lead also tends to be conducted into the cryogenic system, via the current leads. A refrigeration system is then required to extract this entire quantity of heat from the cryogenic system, in order to enable steady-state operation of the magnet.

In order to minimize the heat leak due to thermal conduction via the current lead, the thermal resistance of the current lead should be high, while the electrical resistance of the lead should be low. The thermal resistance is given by

$$R_t = \frac{L}{Ak_{mean}} \tag{1}$$

where $L$ and $A$ are, respectively, the length and cross-sectional area of the current lead, and $k_{mean}$ is the mean thermal conductivity of the current lead, over the temperature range in question. The thermal resistance can be increased by increasing the lead length, decreasing the lead cross-sectional area, and selecting a lead material of low thermal conductivity. The electrical resistance of the lead is given by

$$R_e = \frac{L\rho_{mean}}{A} \tag{2}$$

where $\rho_{mean}$ is the value of the temperature-dependent electrical resistivity of the lead material, integrated over the length of the lead. The electrical resistance of the current lead can be reduced by decreasing the lead length, increasing the cross-sectional area, and by selecting a lead material with a low electrical resistivity. Recall that the Wiedemann-Franz law[1] for pure metals states that

$$k\rho = L_0 T \quad (3)$$

where $L_0$ is the Lorentz constant. We see from Equations (1) through (3) that the conditions necessary for high thermal resistance and low electrical resistance are in conflict with each other, and cannot, therefore, be simultaneously achieved. Hence, an optimum geometry must be selected for the current lead, in order to minimize the refrigeration power requirements.

## COOLING METHODS

Minimization of refrigeration power by passive means such as increasing the lead thermal resistance, and decreasing the lead electrical resistance, can lead to only a limited improvement in lead design. For efficient operation of the system, it is necessary to design the system as a whole, and minimize the total room-temperature refrigeration power requirements, rather than concentrate solely on minimizing the heat leaking into the cryogenic apparatus, at the cold end of the current lead. This involves the study of various suitable refrigeration schemes.

A factor to be considered in any refrigeration system is the *Coefficient of Performance (COP)*[2] of the refrigerator. The*COP* of an actual refrigerator is always less than that of a Carnot refrigerator operating between the same two temperatures. The*COP* of a Carnot refrigerator declines sharply with decreasing cold-space temperature. Hence, in order to remove a given quantity of heat, greater refrigerator power is required at lower cold-space temperatures. On the other hand, heat interception near the room temperature end, while allowing for greater refrigerator efficiency, causes a steeper temperature gradient at the warm end of the current lead, thereby resulting in an increased quantity of total heat to be removed. It is clear that the intermediate heat interception process must be optimized. Two different current lead cooling methods are studied here, viz. co-axial cooling with a coolant envelope, and helical coil cooling[3]. These methods are schematically depicted in Figure (1).

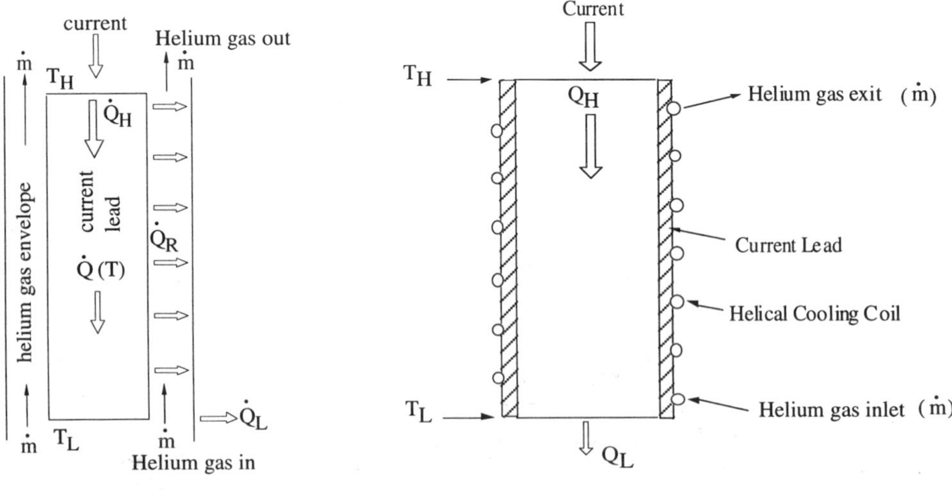


**Figure 1.** Schematic representation of two different methods of intercepting and extracting the heat being conducted along the current lead.

## CO-AXIAL COOLING BY COOLANT ENVELOPE

In this cooling method, the current lead is cooled by a stream of coolant gas, such as helium gas, flowing in a direction counter to that of the heat flow. The coolant stream may surround the current lead, and may also flow through interstitial coolant passages within the current lead. The heat leak down the current lead is intercepted by convective heat transfer to the coolant gas.

### Analysis of Coolant Envelope Method

A copper current lead, cooled by helium gas coolant is considered here. The ends of the current lead are at 300 K and 4.5 K. If the cooling surface area made available for heat exchange between coolant and current lead is extremely large, then perfect heat transfer may be assumed to exist between coolant and current lead. With the assumption of perfect heat transfer, the governing energy equation is

$$\frac{d}{dx}\left(k(T)\frac{dT}{dx}\right) - \frac{\dot{m}C_{p_f}(T)}{A}\frac{dT}{dx} + \rho(T)J^2 = 0 \tag{4}$$

where J is the current density in the current lead, and T is the temperature of both the current lead and the coolant. In non-dimensional form, we obtain

$$\frac{d}{d\overline{x}}\left(\overline{k}\frac{d\overline{T}}{d\overline{x}}\right) - m_1\overline{C}_{p_f}\frac{d\overline{T}}{d\overline{x}} + g_1\overline{\rho} = 0 \tag{5}$$

where $m_1$ and $g_1$ are two non-dimensional parameters given by

$$m_1 = \dot{m}C_{p_{f\,ref}}\frac{L}{Ak_{ref}}, \qquad g_1 = \frac{\rho_{ref}J^2}{k_{ref}}\frac{L^2}{(T_H - T_L)} \tag{6}$$

Convenient reference values are chosen as given below.

$$C_{p_{f\,ref}} = C_{p_f}(T = 0.5(300+4.5)\text{ K}),\ k_{ref} = k(T = 21.5\text{ K}),\ \rho_{ref} = \rho(T = 300\text{ K}) \tag{7}$$

The above system of equations is numerically solved, using a finite difference scheme. The results are plotted in Figure (2), for a lead made of copper with RRR = 30, in terms of the room-temperature Carnot refrigeration power requirement per Ampere of current in the current lead, as a function of $m_1$, the non-dimensional mass flow rate of coolant, with the non-dimensional Ohmic heat generation, $g_1$, as a parameter. For each value of $g_1$, the curves extend down to the lowest possible safe value of m1, below which value of $m_1$, unstable thermal runaway would occur in the current lead. The information in this figure can be used as a tool for preliminary design of current leads. Thus, for the envelope cooling of the current lead of the above material, operating between the given temperatures, with perfect heat transfer, the minimum room-temperature Carnot refrigeration power per unit current of 0.3141 W/Amp is obtained in the vicinity of $m_1 = 6.8$, $g_1 = 29.875$. The optimum values of $m_1$ and $g_1$ are a function of the material of the current lead, as well as the operating temperature range. The results of a similar analysis for a current lead made of copper, with RRR = 3.0, are plotted in Figure (3). For this case, the minimum refrigeration power of 0.3848 W/Amp is obtained in the vicinity of $m_1 = 31.2$, $g_1 = 40.0$.

The energy equations, without the assumption of perfect heat transfer are :

$$\frac{d}{dx}\left(k(T)\frac{dT}{dx}\right) + \rho(T)J^2 - \frac{hP}{A}(T - T_f) = 0 \qquad \text{: Current Lead} \tag{8}$$

$$\dot{m}C_{p_f}\frac{dT_f}{dx} - hP(T - T_f) = 0 \qquad \text{: Helium gas coolant} \tag{9}$$

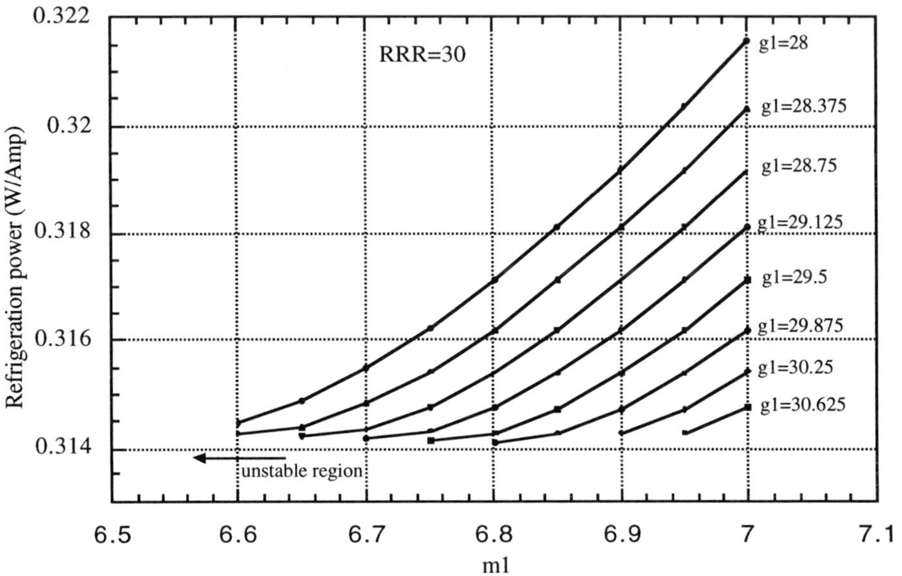


**Figure 2.** Room-Temperature Carnot refrigeration power (W/Amp of current) as a function of the non-dimensional parameter $m_1 = \dot{m}C_{p_{ref}} \frac{L}{Ak_{ref}}$, with $g_1 = \frac{\rho_{ref} J^2}{k_{ref}} \frac{L^2}{(T_H - T_L)}$ as a parameter, for a current lead with RRR = 30, cooled by a coolant envelope (assuming perfect heat transfer).

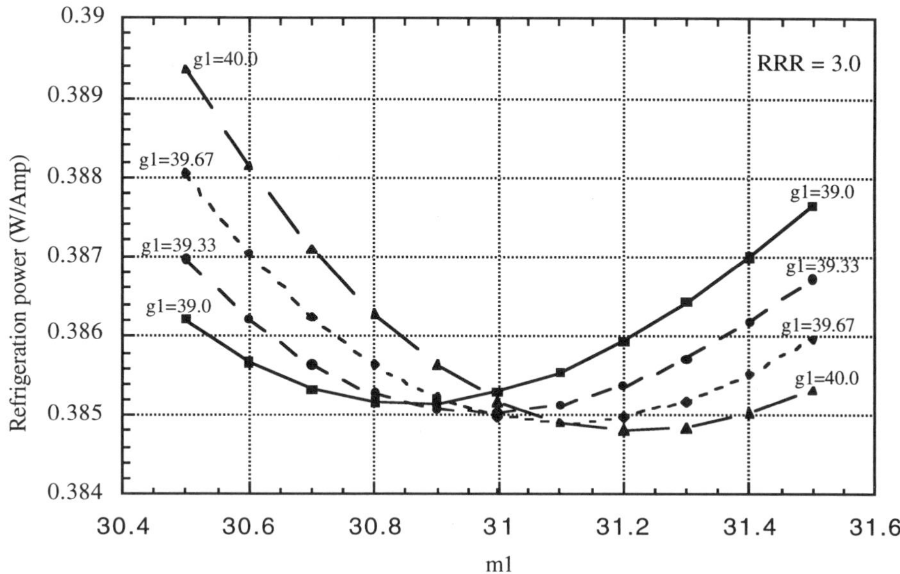


**Figure 3.** Room-Temperature Carnot refrigeration power as a function of $m_1$, with $g_1$ as a parameter, for a current lead with RRR = 3.0, cooled by a coolant envelope (assuming perfect heat transfer).

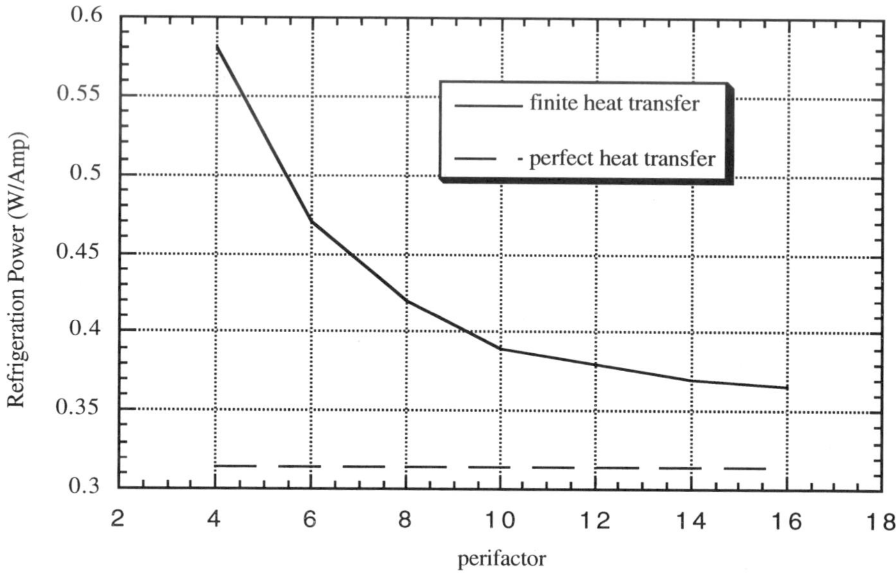


**Figure 4.** Room-Temperature Carnot refrigeration power as a function of perimeter enhancement (perifactor), for a current lead of RRR = 30, cooled by a coolant envelope with finite heat transfer is compared with the corresponding case of perfect heat transfer.

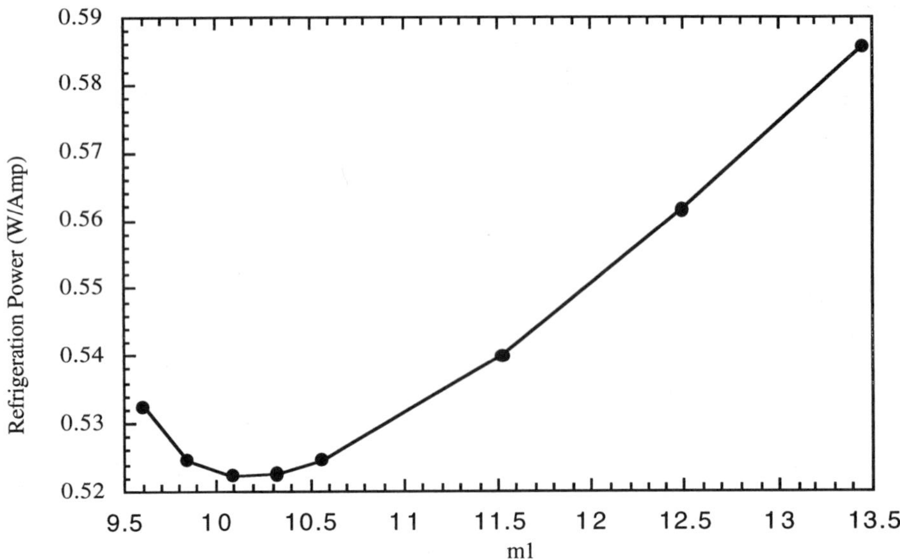


**Figure 5.** Room-Temperature Carnot refrigeration power requirement (W/Amp of current) as a function of non-dimensional mass flow rate of coolant for a current lead, with RRR = 30, cooled by a helical coil, (with finite heat transfer).

where h is the heat transfer coefficient, P is the cooled perimeter, $C_{pf}$ is the specific heat of the coolant. The above equations are also numerically solved, using a finite difference method. As a test case, a current lead with the following parameters was selected :

$L = 1.0$ m, radius $=$ 16 mm , radius of envelope $=$ 18 mm, $A = 643\ mm^2$,
solidity = 0.8, RRR $=$ 30, $T_H = 300$ K, $T_L = 4.5$ K, $T_f$ @ inlet $=$ 4.5K
$I = 13$ kA, $\dot{m} = 0.76\ g/s$, $m_l = 7.2$, $g_l = 28.9$

It was assumed that the cooled perimeter for the coolant flow could be enhanced by providing flow channels within the interior of the current lead. A measure of this perimeter enhancement has been defined :

$$\text{perifactor} = \frac{\text{actual cooled perimeter}}{2\pi(\text{outer radius} + \text{inner radius})} \qquad (10)$$

In Figure (4), the refrigerator power per unit current is plotted as a function of perifactor, for the above current lead. With increasing values of perifactor, the refrigeration power approaches that obtained for perfect heat transfer. For a perifactor of 16, the refrigeration power with finite heat transfer is 16 % more than that required with perfect heat transfer.

## HELICAL COIL COOLING

In this cooling method[3], the current lead is cooled by means of heat transfer that flows within a thin-walled tube, that is coiled around the current lead in a helical fashion. The cold coolant enters the cooling coil near the cold end of the current lead, and exits the cooling coil near the warm end of the current lead. The cooling coil is seated within, and soldered to a groove that is machined on the surface of the current lead.

This method of cooling was also analyzed numerically, using a finite difference scheme. A two-dimensional analysis is carried out, with the temperature field being allowed to vary in the axial and circumferential directions. The governing equations are

$$\frac{\partial}{\partial x}\left(k(T)\frac{\partial T}{\partial x}\right) + \frac{\partial}{\partial z}\left(k(T)\frac{\partial T}{\partial z}\right) + \rho(T)J^2 - \frac{hP}{A}\frac{(i\max)}{\sin(\alpha)}(T - T_f) = 0 \quad \text{: Current Lead} \tag{11}$$

$$\dot{m}C_{p_f}\frac{dT_f}{ds} - hP(T - T_f) = 0 \quad \text{: Helium gas coolant} \tag{12}$$

where x is the co-ordinate in the circumferential direction, z is the axial co-ordinate, s is the distance along the helical flow path of the coolant, imax is the number of divisions in the circumferential direction, and $\alpha$ is the helix angle of the cooling coil. In order to distribute the cooling effect of the helical cooling coil over the entire cross-sectional area of the current lead, the current leads are assumed to have a small wall thickness. A current lead similar to the one used for the analysis of Figure (4) was used. While the cross-sectional area of the current lead is the same in the two analyses, the current lead with the helical cooling coil has different inner and outer radii. The particulars of this current lead are as follows :

$L = 1.0$ m, outer radius $= 30$ mm, inner radius $= 26.4$ mm, $A = 643.4\ \text{mm}^2$,
$T_H = 300$ K, $T_L = 4.5$ K, RRR = 30, I = 13kA,
20 helical turns of cooling coil, $T_f$ @ inlet $= 4.5$ K

Figure (5) shows the room-temperature Carnot refrigerator power as a function of coolant mass flow rate, for the above current lead. Comparing Figures (4) and (5), it is seen that, for the current lead examined here, the helical cooling coil is approximately as effective as a envelope cooled coil, with a perimeter-enhancement factor of 5.

## CONCLUSIONS

The helium coolant envelope method of cooling current leads, with perfect heat transfer, has been analyzed, and a parametric study has been performed. The results of this exercise serve as a design tool, in the selection of the geometry of the current lead, and the mass flow rate of the coolant. In order to account for finite heat transfer between coolant and lead that exists in actual situations, a search should be done for the optimum values of the parameters $m_l$ and $g_l$, in the vicinity of the optimum values of m1 and g1, that were obtained for perfect heat transfer. Enhancement of the cooled perimeter, if possible, will reduce the optimum refrigeration power requirements. The helically-cooled current lead, with finite heat transfer, has also been analyzed. In situations which do not permit enhancement of the cooled perimeter, use of the helical cooling coil may provide an effective alternative cooling method.

## ACKNOWLEDGMENT

Research supported by the State of Florida and the National Science Foundation through NSF Cooperative Grant No. DMR9016241. The support provided by the National High Magnetic Field Laboratory, Tallahassee, Florida is gratefully acknowledged.

## REFERENCES

1. Adrian Bejan. "Entropy Generation through Heat and Fluid Flow", John Wiley & Sons, Inc., USA (1982).
2. G. J. Van Wylen and R.E. Sonntag. "Fundamentals of Classical Thermodynamics ", John Wiley & Sons, Inc., USA (1973).
3. K. Iyengar and S. W. Van Sciver, "Advances in Cryogenic Engineering", Vol. 41 B, New York, 1996.

# THERMAL RESISTANCE ACROSS A COPPER/KAPTON/COPPER INTERFACE AT CRYOGENIC TEMPERATURES

L. Zhao,[1] P.E. Phelan,[1] R.C. Niemann,[2] and B.R. Weber[3]

[1]Department of Mechanical & Aerospace Engineering
Arizona State University, Tempe, AZ 85287-6106
[2]Argonne National Laboratory, Argonne, IL 60439
[3]Superconductivity, Inc., Madison, Wisconsin 53705

## ABSTRACT

The high-$T_C$ superconductor current lead heat intercept connection, which is utilized as a thermal intercept to remove the Joule heat from the upper stage lead to a heat sink operating at 50 - 77 K, consists of a structure where a 152-$\mu$m film is sandwiched between two concentric copper cylinders. The material chosen for the insulating film is Kapton MT,* a composite film which has a relatively low thermal resistance, but yet a high voltage standoff capability. Here, the measured thermal conductance of a copper/Kapton MT/copper junction in a flat-plate geometry is compared to the results obtained from the actual heat intercept connection. Increasing the contact pressure reduces the thermal resistance to a minimum value determined by the film conduction resistance. A comparison between the resistance of the copper/Kapton MT/copper junction and a copper/G-10/copper junction demonstrates that the Kapton MT layer yields a lower thermal resistance while still providing adequate electrical isolation.

## INTRODUCTION

Thermal contact resistance arises in the region of contact where two solid specimens are pressed together. It has long been realized that surfaces are rough on a microscopic scale, which causes the real contact area to be significantly smaller compared to the nominal contact area.[1] Interstitial materials may be deliberately introduced in order to control the thermal contact conductance ($h_c$), $h_c = q/\Delta T$, where $q$ and $\Delta T$ are the heat flux and the temperature drop across the interface respectively. To enhance $h_c$, the bare metal surfaces may also be coated with soft

* Kapton MT is a registered trademark of DuPont Films.

metals of high thermal conductivity by electroplating or vacuum deposition.[2] The insert's thermal conductivity and hardness, relative to the values of the corresponding properties of the base materials forming the contact, together with the insert's thickness, dictate the resulting change in $h_c$.

The high-$T_c$ superconductor current lead heat intercept connection designed by Argonne National Laboratory and Superconductivity, Inc. consists of a structure where two concentric copper cylinders are separated by a composite (Kapton MT) film. As the purpose of the heat intercept connection is to efficiently transfer the heat generated in the copper portion of the current lead, plus that conducted in from the ambient, away from the low-temperature superconductor magnet, it is imperative that the total thermal resistance of the heat intercept connection in the radial direction be kept as small as possible. At the same time, electrical isolation must be maintained between the inner and outer copper cylinders, necessitating the use of an electrically insulating interstitial material. It was determined previously that $h_c$ at the interface between the two copper cylinders, separated by a thin-walled G-10 tube, provided the largest component of the total thermal resistance.[3] Therefore, in the most recent design, the G-10 tube is replaced with a thin Kapton film which, while still providing electrical isolation, is softer than G-10 and has a higher thermal conductivity, offering the potential for a reduced $R_c$ ($=1/h_c$) and thus a lower total thermal resistance. Detailed descriptions of previous versions of the heat intercept connection are given elsewhere.[4]

Measuring $h_c$ in a cylindrical geometry is relatively difficult. Because the contact pressure in a cylindrical joint is dependent on the initial interference and the differential expansion of the two cylinders, which is a function of radial heat flux, it is implicit and not readily controllable.[5] Furthermore, the cylindrical surfaces also involve other characteristics such as out-of-roundness, in addition to parameters like surface roughness and waviness that are important in flat junctions. Due to these complexities in the cylindrical geometry, the present work concentrates on the measurement of the thermal resistance of a copper/Kapton MT/copper interface in a flat-plate geometry at cryogenic temperatures, and correlates these results with the limited data available from the thermal performance of the actual heat intercept connection. The effects of the contact pressure and the operating temperature on the thermal resistance are also evaluated.

## HEAT INTERCEPT CONNECTION

A schematic diagram of the cross-section of the high-$T_c$ superconductor current lead heat intercept connection is presented in Fig. 1. The heat intercept connection is primarily a thermal intercept to remove the Joule heat from the upper stage lead to a heat sink operating at 50 - 77 K, but also electrically isolates the leads from the high temperature superconductor (HTS) elements mounted in the central hole. A 152-$\mu$m film is sandwiched between the two concentric ETP copper cylinders, the inner disk and the outer ring. The outer ring is in the shape of a square, with a circular hole.

The material chosen for the insulating film is Kapton MT, a polyimide film with alumina particles as a filler, giving it a relatively low thermal resistance, but yet a high voltage standoff capability. The thermal conductivity of this material at cryogenic temperatures is sufficient to result in a small temperature drop across the

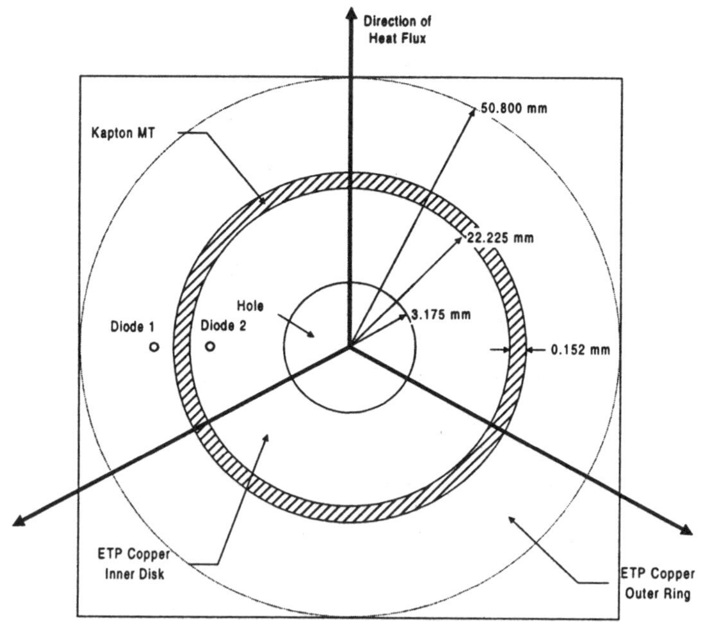


**Figure 1.** Heat intercept connection cross section.

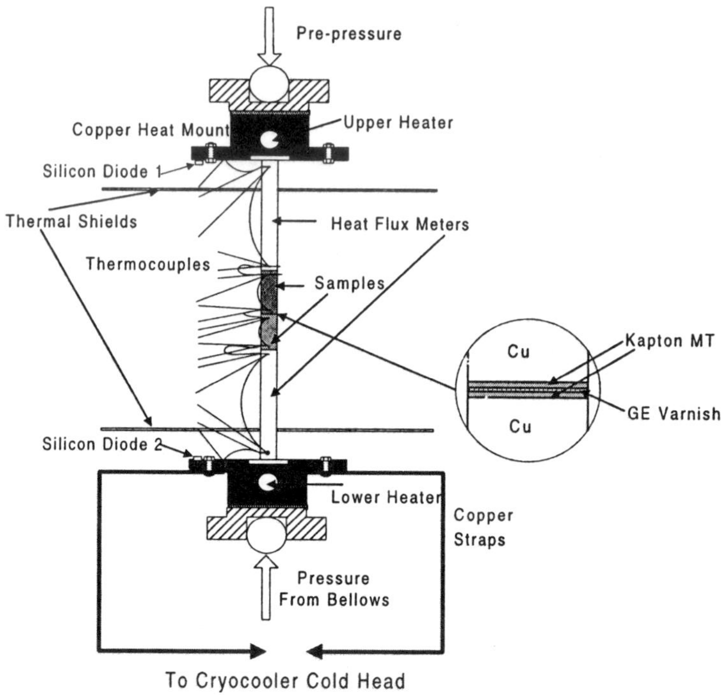


**Figure 2.** Schematic diagram of the flat-plate geometry experimental apparatus.

sheet at the desired heat flux input, while the high dielectric strength of Kapton MT (114 V/$\mu$m) allows only a thin layer to be used. The Kapton sheet is wrapped snugly around the inner disk with a double wrap. Its outside edge is sealed to the underlying layer with GE 7031 varnish, and its ends nearly meet but do not overlap.

The ETP copper disk and ring are cleansed by acid pickling, then treated by 1.27-$\mu$m nickel plating and 1.91-$\mu$m gold plating in order to reduce $h_c$ across the interface.

The inner disk, the interstitial layers, and the outer ring are held together by means of a thermal-interference fit. The radial interference is measured to be 63.5 $\mu$m, which results in a calculated contact pressure of 110.3 MPa at room temperature. The amount of interference dictates the radial contact pressure on the joint which affects the overall thermal resistance.

A heater is mounted in the center of the inner disk to simulate the Joule heat generated by the current lead, while the outer ring is thermally connected to a heat sink. The resulting radial heat flux is approximately uniformly outward.

The experimental apparatus is operated under vacuum conditions. Temperatures are measured by silicon diodes mounted as shown in Fig. 1.

## FLAT-PLATE GEOMETRY EXPERIMENTAL APPARATUS

Due to the relative difficulties in controlling and measuring the contact pressure, as well as the additional surface characteristics such as out-of-roundness, only limited data are available for the thermal performance of the actual cylindrical heat intercept connection. Therefore, a flat-plate geometry consisting of a copper/Kapton MT/copper junction is investigated instead.

The test system illustrated in Fig. 2 consists of two samples, upper and lower, and two calibrated heat flux meters which are pressed together and aligned to the center line of the apparatus. The test specimens are ETP copper, coated identically as the copper pieces in the actual heat intercept connection. The interstitial material is two layers of Kapton MT which are glued together by GE 7031 varnish and allowed to dry.

The surface treatments and roughnesses of the components in the test column, as well as the thickness of the interstitial material, are tabulated in Table 1, which provides a comparison with those in the actual heat intercept connection shown in Fig. 1 and the heat intercept connection containing a G-10 tube.[3]

The thermal conductivity of the heat flux meters was previously calibrated against a reference 304SS bar obtained from the National Institute of Science and Technology. The temperature drops along the heat flux meters and across the sample interface are measured by differential Type E thermocouples. The differential thermocouples are calibrated at a single point by immersing one junction into liquid nitrogen and the other into an ice bath, and applying the resulting percent deviation from the standard table to all measurements. The silicon diodes mounted on the copper heat mounts are utilized as the reference temperatures for the differential thermocouples. The thermocouples are inserted into 51-$\mu$m diameter holes drilled to the center lines of the samples and heat flux meters, and which are partially filled with fine copper powder in order to provide reliable thermal contact.

Indium foil is inserted at all junctions besides the sample interface in order to reduce extraneous contact resistance. Two thermal shields are placed above and

**Table 1.** Surface treatments and roughnesses of the components in the three joints

| Flat-Plate Geometry Experimental Apparatus | | |
|---|---|---|
| Component | Surface Treatment | Surface Roughness (μm) |
| ETP Copper Upper Sample | Nickel, Gold Plate | 3.575 |
| Kapton MT Sheet (185 μm) | none | 0.390 |
| ETP Copper Lower Sample | Nickel, Gold Plate | 0.994 |
| Heat Intercept Connection Using G-10 as Interstitial Material | | |
| Component | Surface Treatment | Surface Roughness (μm) |
| Cu Inner Disk | Grind, Gold Plate | 0.254 (before plating) |
| G-10CR Tube (381 μm) | none | 1.676(inner)/1.981(outer) |
| Cu Outer Ring | Grind, Gold Plate | 0.254 (before plating) |
| Heat Intercept Connection Using Kapton MT as Interstitial Material | | |
| Component | Surface Treatment | Surface Roughness (μm) |
| ETP Copper Inner Plug | Nickel, Gold Plate | 0.813 (before plating) |
| Kapton MT Sheet (152 μm) | none | 0.390 |
| ETP Copper Outer Ring | Nickel, Gold Plate | 0.813 (before plating) |

below the heat flux meters in order to reduce the radiation heat transfer. The entire test column is surrounded by a thermal shield to minimize radiation gain from the ambient.

Before each experiment, a small amount of initial pre-stress is applied to the samples via a loading screw mechanism to guarantee a good contact. The pre-pressure, which has a magnitude of 0.809 MPa around room temperature, reduces to 0.007 MPa at 123 K and vanishes below 100 K. In order to apply and control the contact pressure at cryogenic temperatures, a bellows located below the lower ball bearing is pressurized with helium gas to obtain an evenly distributed pressure, which is measured by a load cell.

The outputs of the thermocouples, silicon diodes, and the load cell amplifier are monitored and recorded by a Macintosh PowerPC computer, then simultaneously converted and analyzed by a labVIEW control program. If the temperature drops across the sample interface as well as the upper and lower heat flux meters are all within a set criterion for the last 50 data points, the system is considered to be in steady state. Starting with the reference silicon diodes located on the copper heat mounts, each temperature point is calculated sequentially from the differential thermocouple readings. The temperature drop at the sample interface is over-determined by the calculation downwards and upwards, which are averaged to determine the final $\Delta T$. The heat flux at the sample interface is considered as the average of the upper and lower heat flux meters.

All measurements are conducted under a vacuum condition of $6.5 \times 10^{-5}$ torr on average, which is measured by a cold cathode vacuum gauge.

## UNCERTAINTY ANALYSIS

The total experimental uncertainty in the thermal contact conductance, $u_h$, is calculated from:

$$u_h = \sqrt{\left(\frac{u_q}{\Delta T}\right)^2 + \left(\frac{q\, u_{\Delta T}}{(\Delta T)^2}\right)^2} \tag{1}$$

where $u_q$ and $u_{\Delta T}$ are the uncertainties in $q$ and $\Delta T$ respectively. Since $q$ is the average of the measurements of the upper and lower heat flux meters, $u_q$ is reduced to $\frac{1}{2}\left(u_{q,upper}^2 + u_{q,lower}^2\right)^{1/2}$ where $u_{q,upper}$ and $u_{q,lower}$ are the uncertainties in $q$ from the upper and lower heat flux meters, which take the form

$$u_{q,upper/lower} = \sqrt{\left(\frac{u_k \Delta T_q}{L}\right)^2 + \left(\frac{K\, u_{\Delta T_q}}{L}\right)^2} \tag{2}$$

where $L$ and $\Delta T_q$ are the length and the temperature drop across the heat flux meter respectively, and $u_k$ and $u_{\Delta T_q}$ are the uncertainties in the thermal conductivity $K$ of the heat flux meters and the differential thermocouple readings.

We consider the total experimental uncertainty, consisting of bias errors determined by instrument specifications, and precision errors (95% confidence level) caused by scatter in the data.[6] The term $u_{k,bias}$ is the material variability of the standard reference bar, given as 2% below 100 K, and $u_{\Delta T,bias}$ is the error introduced by the differential thermocouples (±1%). The corresponding uncertainties in $h_c$, reported in the figures in the form of error bars, are equal to $\left(u_{h_c,bias}^2 + u_{h_c,precision}^2\right)^{1/2}$ in which the different errors are treated separately but following the same procedure.

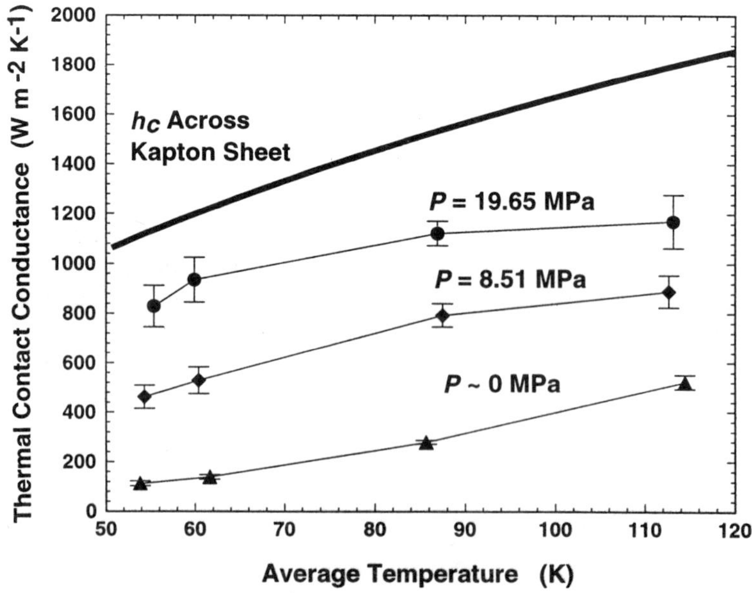


**Figure 3.** Effect of contact pressure on thermal contact conductance, as measured in a flat-plate geometry.

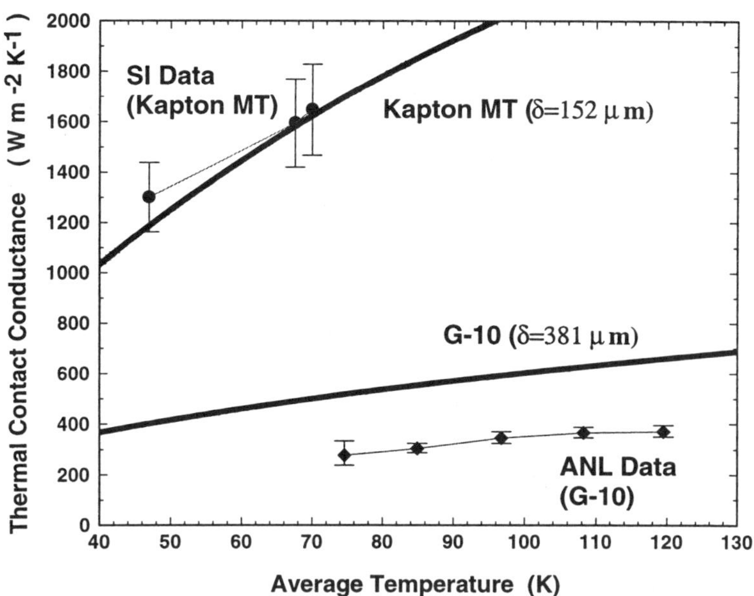


**Figure 4.** Comparison between the different interstitial materials, G-10 and Kapton MT, as measured in a cylindrical geometry.

Both junctions of the differential thermocouple positioned across the sample interface are mounted in holes located 2.413 mm from the interface. The resulting temperature drop across the copper is estimated as 1.4% of the $\Delta T$ across the interface and is considered to be negligible.

## EXPERIMENTAL RESULTS AND DISCUSSION

Measured values of $h_c$ as a function of the average temperature across the sample interface are presented in Fig. 3. For the Kapton sheet itself, $h_c$ is defined based on its thermal conductivity $K$:[7]

$$h_c = \frac{K}{\delta} \tag{3}$$

where $\delta$ includes the thickness of the GE varnish ($33\,\mu$m), which, together with the thickness of the Kapton layers ($152\,\mu$m in total) gives $\delta = 185\,\mu$m.

Figure 3 displays the experimental data over an average temperature ranging from 50 K to 120 K. Observe that $h_c$ varies with the contact pressure $P$, because when the two surfaces are pressed tighter together, the soft Kapton film is squeezed into the voids between the actual contact spots, thus increasing the effective contact area and causing the $R_c$ to diminish. From the well-known correlation for $h_c$ of two contacting randomly rough surfaces [8]

$$\frac{h_c \sigma}{k} = c\left(\frac{P}{H}\right)^n \tag{4}$$

where $\sigma$ is the surface roughness, $k$ the harmonic mean thermal conductivity and $H$ the micro hardness of the softer material in contact, $h_c$ is expected to increase with the contact pressure $P$. Moreover, the thickness of the Kapton layer probably reduces when it is pressed, therefore decreasing its contribution to the thermal resistance and resulting in an increased $h_c$. Presumably, increasing $P$ further would yield a further increase in $h_c$ up to the limit posed by the resistance of the Kapton layer itself. However, $P$ = 19.65 MPa represents the maximum allowable pressure in our flat-plate apparatus.

Figure 4 provides a comparison between the two different interstitial materials, G-10 and Kapton MT, which are utilized in the heat intercept connection. The Argonne National Laboratory (ANL) data are experimental results for $h_c$ at the interface between two concentric copper cylinders separated by a thin-walled G-10 tube. The Superconductivity, Inc. (SI) data are for a copper/Kapton MT/copper interface in the geometry shown in Fig. 1. The two sets of experimental data are obtained under comparable $q$ values, 4740 W m$^{-2}$ and 4940 W m$^{-2}$ respectively, but the SI data are at much higher contact pressure ($P$ = 91 MPa) while the ANL data are at $P$ = 43 MPa around 60 K. The two solid curves represent the contributions to $h_c$ from the Kapton and G-10 layers. In addition to being thinner, the Kapton MT layer has a higher $K$, resulting in a higher $h_c$. The Kapton MT layer, which is softer, thinner and of higher $K$ than the G-10 layer, substantially reduces the contact resistance between the Kapton MT and the copper, and subsequently the thermal resistance across the layer itself dominates. However, the thermal contact resistance at the G-10/copper interfaces is quite large, as indicated by the difference between the ANL data and the G-10 curve, which is also due to the relatively lower $P$. Thus, it is clear that Kapton MT is a better interstitial material compared to G-10, with regard to enhancing $h_c$ while still providing adequate electrical isolation.

Compared with the $h_c$ data obtained from the flat-plate geometry experimental apparatus presented in Fig. 3, the $h_c$ results are consistent in that at the much higher $P$ for the SI data, $h_c$ increases to its limiting value determined by the Kapton MT resistance.

## CONCLUSIONS

The thermal contact conductance at a copper/Kapton MT/copper interface is investigated at cryogenic temperatures. The results indicate that increasing the contact pressure increases the contact conductance, in accord with expectations. The thermal resistance of the Kapton MT layer determines the upper limit to the contact conductance. The interstitial film Kapton MT dramatically lowers $h_c$ compared with a thin G-10 layer which is relatively stiffer, thicker, and of lower thermal conductivity. Hence, the Kapton MT film utilized is a better interstitial material in the high-$T_c$ superconductor current lead heat intercept connection, because it provides relatively low thermal contact resistance, but yet sufficient electrical insulation.

## ACKNOWLEDGMENT

P.E.P. gratefully acknowledges the support of the National Science Foundation through a CAREER Award (CTS-9696003). Work at Argonne National Laboratory was supported by the U.S. Department of Energy, Energy Efficiency and Renewable Energy, as part of a program to develop electric power technology, under Contract W-31-109-Eng-38. All the authors wish to thank Lisa Debellis and Shijin Mei for their assistance in the experiment.

## REFERENCES


1. L.S. Fletcher, "Recent Developments in Contact Conductance Heat Transfer," *J. Heat Transfer* 110:1059 (1988).
2. M.A. Lambert and L.S. Fletcher, "A Review of Thermal Contact Conductance of Junctions with Metallic Coatings and Films," AIAA Paper 92-0709, AIAA 30$^{th}$ Aerospace Meeting, Reno, NV (1992).
3. P.E. Phelan, R.C. Niemann and T.H. Nicol, "Thermal Contact Resistance for a Cu/G-10 Interface in a Cylindrical Geometry," *ASME Proceedings of the 31$^{st}$ National Heat Transfer Conference*, 5:185 (1996).
4. R.C. Niemann, et al., "Design and Performance of Low-thermal-resistance, High-electrical-isolation Heat Intercept Connections," *Cryogenics* 35:829 (1995).
5. C.V. Madhusudana, "On Heat Flow across Cylindrical Joints," *Proceedings of the 8$^{th}$ International Heat Transfer Conference*, 2:651 (1986).
6. J.P. Holman, "Experimental Methods for Engineers," McGraw-Hill, Inc., New York (1994), p. 66.
7. D.L. Rule, D.R. Smith and L.L. Sparks, "Thermal Conductivity of a Polymide Film between 4.2 and 300K, with and without Alumina Particles as Filler", NISTIR 3948 (1995), p. 40.
8. J.M. Ochterbeck, G.P. Peterson, and L.S. Fletcher, "Thermal Contact Conductance of Metallic Coated BiCaSrCuO Superconductor/Copper Interfaces at Cryogenic Temperature," *Journal of Heat Transfer*, 114:21 (1992).

# THE EFFECT OF SOLID INTERFACES ON THERMAL CONTACT RESISTANCE AT LOW TEMPERATURE

L.Xu[1], J.Yang [1], J.M.Xu[2], S.M.Li[3], W.Xiong[1], T.Zhang[1]

[1]Low Temperature Center, Shanghai Jiao Tong University
Shanghai 200030, P.R.C
[2]Shanghai Sunny Research Institute of Environment and Energy
Shanghai 200042, P.R.C
[3]Cryogenic Lab. Zhejiang University
Hangzhou 310027, P.R.C

## ABSTRACT

The heat transfer of solid interfaces is analyzed in this paper. We particularly analyzed and induced the rough surfaces effect to the thermal contact resistance. During the experiment, we measured the effect of the solid interface on the thermal conductance. We got some useful conclusions.

## SURFACE TOPOGRAPHY ANALYSIS AND SURFACE TOPOGRAPHY MODEL

Although a surface is very smooth, there exists a series of concave-convex surfaces with different range and frequency, which leads to the uncompleted contact between solids, for the theropy increasing of surface layer or surface flowing during the proceeding of manufacture.

The distribution of concave-convex surfaces depends on, to a great extent, the manufacturing method and the properties of the material itself. In order to evaluate the properties of surface topography, we can use a three-dimention graph to describe it. During engineering manufacture, we usually use one of two parameters to describe surface structure: the C.L.A (Central Line Average) and the RMS(Root Mean Square) of the surface roughness.

In the view of statistics, surface structure can be described by the distribution function of its contour height:

$$F(z) = \int_{-\infty}^{+\infty} \psi(z) \cdot dz \tag{1}$$

where, z is the contour height measured from the central line and $\Psi(z)$ is the probability density function of the distribution of contour heights.

Some researchers indicate that the distribution of many surface contour heights is the Gauss Normal distribution, as shown in Fig.1. The Gauss distribution curve or its density function can be defined as follows:

$$\Psi(z) = \psi_0(z) \cdot e^{\frac{z^2}{2\sigma^2}} \tag{2}$$

From the above analyses, it can be seen that the shape of surface contour is a random phenomenon, so it is very difficult to create a universally effective model of a surface in vacuum. So we use the G-M model[5], which is used in the research of friction, to analyze the micro-contact-surface between two solids in vacuum sorting .The G-M (Greenwood and Millamson) model is based on some hypothetical conditions:

As shown in Fig. 2, the mean height of a convex summit is at the place above the mean surface $z_s$, and its standard offset is $\delta_s$. So the probability density of the convex's height is

$$f(z_s) = \frac{1}{\sigma_s (2\pi)^{\frac{1}{2}}} \cdot e^{\frac{z_s^2}{2\sigma_s^2}} \tag{3}$$

Then, the probability of height distribution between $z_s$ and $z_s$+$dz_s$ is $f(z_s)dz_s$, but the probability of height of convex greater than d is

$$(z_s > d) = \int_d^{\infty} f(z_s) dz_s = \int_d^{\infty} \frac{1}{(2\pi)^{\frac{1}{2}}} \cdot e^{-\frac{z_s^2}{2\sigma_s^2}} d\frac{z_s}{\sigma_s} = \int_{d/\sigma_s}^{\infty} \phi(x) dx = F_0\left(\frac{d}{\sigma_s}\right) \tag{4}$$

where, $\Phi(x)$ is the standard Gauss distribution function.

If the summit height $D_{sum}$ is known, then the number of actual contacts is:

$$n = D_{sum} \cdot F_0\left(\frac{d}{\sigma_s}\right) \tag{5}$$

From above, we can see that the G-M model includes three parameters: radius of curvature R, the probability density of the convex's height and the summit height $D_{sum}$.

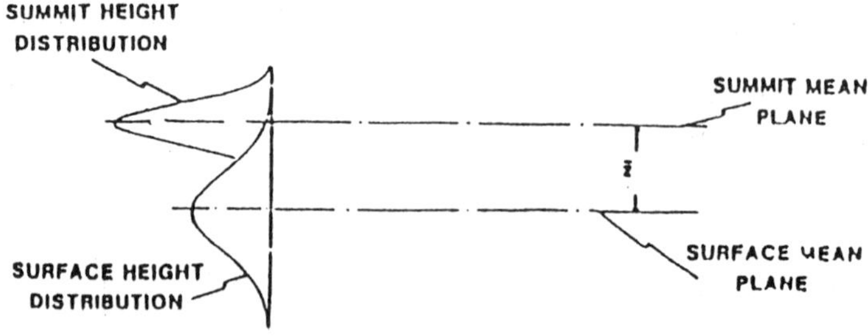


**Figure 1** Surface height distribution and distribution of the peak height[7]

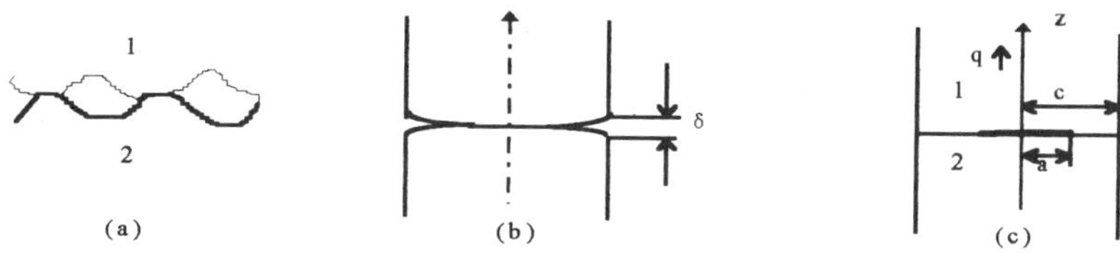


Figure 2

## THERMAL CONDUCTION RESISTANCE MODEL OF ONE CHANNEL

The heat transfer between two solid surfaces is mainly through three ways: heat conduction of the actual contact solid surface $h_s$, fluid heat transfer within the gap between the two surfaces $h_r$ and radiation heat transfer between the two surfaces $h_{rad}$. So the total heat conduction is

$$h = h_s + h_r + h_{rad} \tag{6}$$

When Grashof number Gr<2000, the heat transfer of natural convection can be restrained. So, the fluid heat transfer of the gap can be expressed by conduction. The fluid heat transfer $h_r$ between two actual contact surfaces where the mean width of the gap δ is less than 0.1mm is :

$$h_r = \frac{k_r A_r}{\delta} \tag{7}$$

When the surface temperature is not high or the difference in temperature between the two surfaces is very little, compared the radiation heat transfer between the two surfaces can be neglected. The heat transfer between the two surfaces is then determined by the $h_s$ and $h_r$.

Fluid flow must obey the mass-conservation equation. Under the condition of constant material properties and no mass source, the mass-conservation equation is

$$\frac{\partial U}{\partial x} + \frac{\partial V}{\partial y} + \frac{\partial W}{\partial z} = 0 \tag{8}$$

Where, U is axial velocity, V is radial velocity and W is circumferential velocity. For solid condution, also under the conditions of constant material properties and no heat source, there is the energy-conservation equation:

$$\frac{\partial}{\partial x}\left(\frac{\partial T}{\partial x}\right) + \frac{\partial}{\partial y}\left(\frac{\partial T}{\partial y}\right) + \frac{\partial}{\partial z}\left(\frac{\partial T}{\partial z}\right) = 0 \tag{9}$$

From equation (8) and equation (9), we can see that the temperature gradient for heat flow $\partial T/\partial x$, $\partial T/\partial y$, $\partial T/\partial z$ obey the same control equation as u,v,w.

When ΔT=0, there are many tiny channels at the joint as shown in Fig. 2(a). Fig. 2(b) is the schematic diagram of the monochannel model. Except for the conduction at the joint, there doesn't exist other means of heat transfer. When δ is very small, the heat transfer process can be described as shown in Fig.2(c). The control equation can be written as below:

$$\frac{\partial^2 T}{\partial r^2} + \frac{1}{r}\frac{\partial T}{\partial r} + \frac{\partial^2 T}{\partial z^2} = 0, \qquad \Delta T = 0 \tag{10}$$

After variable separation and a series of mathematics calculations, we find that the thermal conductance is:

$$h_1 = \frac{q/A}{\Delta T_c} = \frac{q/(\pi c^2)}{q/(4ka) \times \Psi(\varepsilon)} = \frac{4ka}{\pi c^2} \times \frac{1}{\Psi(\varepsilon)} \tag{11}$$

Because there is a thermal resistance on each of the upper and lower surfaces of the joint, the total thermal resistance is:

$$R = R_1 + R_2 \Rightarrow \frac{1}{h} = \frac{1}{h_1} + \frac{1}{h_2} \Rightarrow h = \frac{4a(1/k_1 + 1/k_2)}{\pi c^2} \frac{1}{\Psi(\varepsilon)} \tag{12}$$

We define a mean thermal conductivity k :

$$k = \frac{2}{\frac{1}{k_1} + \frac{1}{k_2}} \tag{13}$$

so the total thermal contact conductance is :

$$h = \frac{2ka}{\pi c^2} \cdot \frac{1}{\psi(\varepsilon)} \tag{14}$$

## THE THERMAL CONTACT RESISTANCE MODEL OF ROUGH SURFACE

From the above analyses, it can be seen that the thermal contact resistance is mainly caused by the constriction of the heat flow at the joint. When the rough surfaces contact with each other, there exist many small contact areas, and their sizes and distribution are very complicated, as shown in Fig. 3.

When a rough surface is extruded with a hypothetical theoretical plane, the reference planes of two surfaces are close to each other. We define the distance between mean contact plane and reference plane is d. Further, we assume that the convex whose height to reference plane is bigger than d will be contacted, the contact area is :

$$\begin{aligned} \frac{Ac}{Dsum} &= \int_d^{\infty} \pi R \delta f(Z_S) dZ_S = \int_d^{\infty} \pi R (Z_S - d) f(Z_S) dZ_S \\ &= \int_d^{\infty} \pi R \sigma_S \frac{Z_S - d}{\sigma_S} \frac{1}{(2\pi)^{1/2}} e^{-\frac{Z_S^2}{2\sigma_S^2}} d\frac{Z_S}{\sigma_S} \\ &= \int_{d/\sigma_S}^{\infty} \pi R \sigma_S \left( x - \frac{d}{\sigma_S} \right) \phi(x) dx = \pi R \sigma_S F_1\left(\frac{d}{\sigma_S}\right) \end{aligned} \tag{15}$$

where, $\phi(x) = \frac{1}{(2\pi)^{1/2}} e^{-\frac{x^2}{2}}$ , $F_1(\frac{d}{\sigma_S}) = \int_{d/\sigma_S}^{\infty} (x - \frac{d}{\sigma_S}) \phi(x) dx$

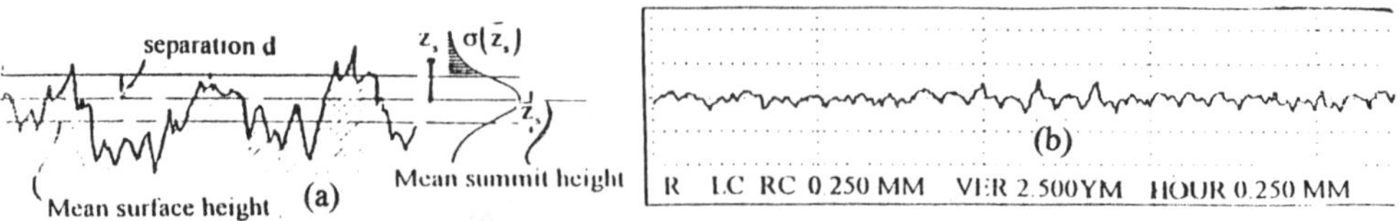


**Figure 3** Probability distribution graph of rough surface topography (a) reference[1], (b) our measured topography.

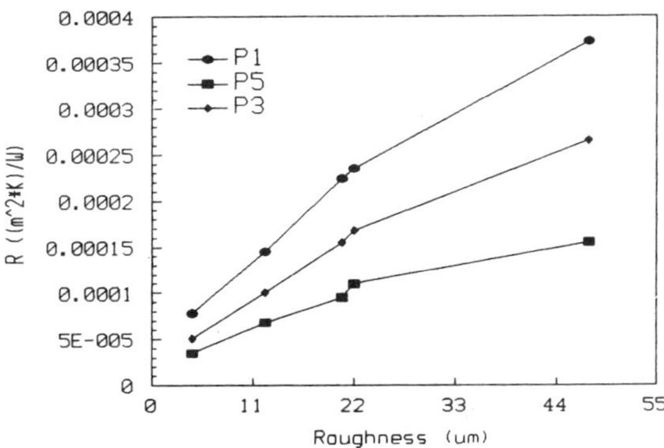


**Figure 4** Isobaric chart of TCR and roughness, Aluminum, 135K.

Using the relational expression of the pressure on the solid surface given by reference[5], and the relational expression of the pressure on the contact surface and the whole displacement given by reference[6], $F = \frac{4}{3}E^{*}R^{1/2}\delta^{3/2}$, the mean load sustained on the single rough granule is :

$$\frac{F}{D_{sum}} = \frac{4}{3}E^{*}R^{1/2}\sigma^{3/2}F_{3/2}(\frac{d}{\sigma_s}) \tag{16}$$

where $F_{3/2}(\frac{d}{\sigma_s}) = \int_{d/\sigma_s}^{\infty}(x - \frac{d}{\sigma_s})^{3/2}\phi(x)dx$ ,

so, the thermal conductance is:

$$\frac{h}{D_{sum}} = \sum_{i=1}^{D_{sum}} h_1 \Big/ D_{sum} = \frac{4kR^{1/2}\sigma_s^{1/2}}{\pi c^2}\frac{1}{\psi(\varepsilon)}F_{1/2}\left(\frac{d}{\sigma_s}\right) \tag{17}$$

where, $F_{1/2}(\frac{d}{\sigma_s}) = \int_{d/\sigma_s}^{\infty}(x - \frac{d}{\sigma_s})^{1/2}\phi(x)dx$

$$\text{so } h = 3D_{sum}\frac{k}{\sigma_s}\frac{F}{E^{X}}\frac{1}{\Psi(\varepsilon)}\frac{F_{1/2}(\frac{d}{\sigma_s})}{F_{3/2}(\frac{d}{\sigma_s})} \tag{18}$$

From above, it can be seen that the thermal contact conductance of a rough surface is a function related to the material thermal conductivity, the elastic modules, the pressure and the surface roughness etc. The roughness is an important factor. The surface topgraphy and roughness are probability distribution functions. So it is very difficult to create a universal and effective model of the actual surface.

On the basis of our studies on several surface models, we worked out the above equations and have measured the state of surface topograph and the thermal contact resistance.

## RESULTS AND ANLYSES OF THE EXPERIMENTS

Experimental results are shown in Fig. 4 to Fig. 7.

Fig. 4 and Fig. 5 show the isobaric chart of thermal contact resistance and roughness of aluminum samples at 135K and 210K. Fig. 6 is the isobaric chart of the thermal contact

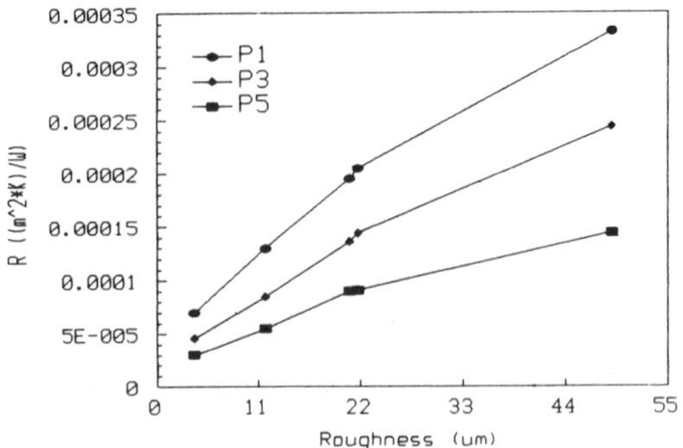


**Figure 5** Isobaric of TCR and roughness ,Aluminum, 210K .

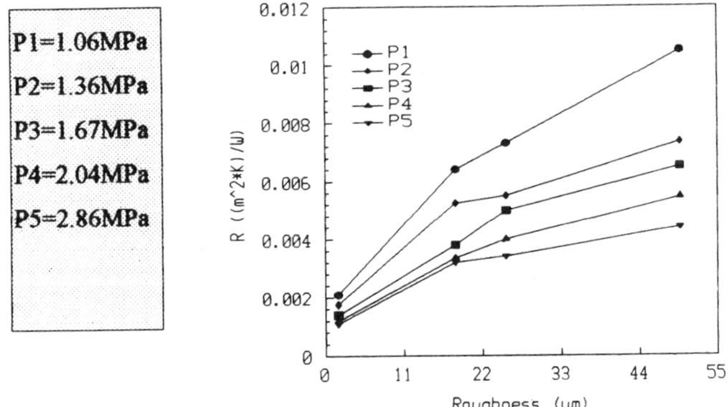


**Figure 6** Isobaric chart of TCR and roughness, stainless steel, 350K .

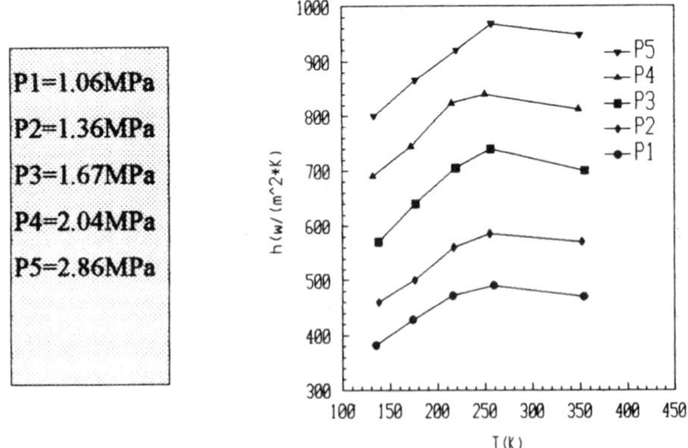


**Figure 7** The h-p graph at different temperature of stainless steel .

resistance and roughness of stainless steel samples at 350K. Fig. 7 is the h-p graph at different temperature of stainless steel. From these figures, we can get the following results:

1.When the roughness is same, with increasing of pressure, the TCR of the joint of two solids decreased.

2. When the roughness and the material are same, the TCR is smaller at high temperature.

3. With increasing roughness, the TCR between the two solid contact surfaces increases.

4. The relation of roughness and TCR is not a linear function. For example, for a stainless material sample, at the condition of 350K and 2.856MPa, when the roughness decreases from 49 μm to 18 μm, its TCR decrease 40%. On the other hand, when the roughness decreases from 18 μm to 2 μm, its TCR decreases nearly 80%.

Finally, the contact surface topography is an important factor to surface contact conductivity. Especially, when the surface roughness is small, we can greatly decreases its TCR through decreasing surface roughness. Reference[4]say the the pressure can change surface contact resistance, in other words we can say that it can be realized through changing the state of surface contactance.

## Reference


1. A.W.Bush and R.D.Gibson, "The Elastic Contact of a Rough Surface Wear," Vol 35, 1975
2. S.Song et al., "Experimental Study and Modeling of Thermal Contact Resistance across Bolted Joints," Journal of Thermophysics and Heat Transfer v8 n1 Jan-Mar , P. 159-163(1994)
3. J. Maddren et al., "Predicting Thermal Contact Resistance at Cryogenic Temperature for Spacecraft Applications," Journal of Spacrcraft and Rockets v32 n3 Mar-Jun, P. 469-474(1995)
4. Xu Lie, Yang Jun, Xu Jiamei et al., "The Effect of Solid Interfaces on Thermal Contact Resistance at Low Temperature," Cryogenics and Superconductivity(China), Vol 24, No.1, P. 53-58(1996).
5. J.A.Greenwood, "Theory of Contact Between Rough Surfaces and Flats." Trans. ASME, Journal of Lubrication Technology. 89F, P. 81-91(1967).
6. Hoolin, "The theory of friction," Machines Industry Publishing House (1981)
7. John I. McCool, "Comparison of Models for the Contract of Rough Surfaces Wear," Vol 107, P. 37-60(1986)

# MODELING SUPERFLUID PHASE SEPARATOR SYSTEMS

H. A. Snyder and A. J. Mord

Ball Aerospace & Technologies Corp. and
University of Colorado at Boulder
Boulder, Colorado, 80309, USA

## ABSTRACT

This report describes a model for the operation of superfluid helium (He II) phase separator systems. It is applied to the design of vent systems in low gravity. We used a systems approach to show the interactions of the various components of the vent system. Some of the assumptions and approximations that have been used in previous models are incorrect or inaccurate; thus, the equations used here differ somewhat from other reported models. We modeled the flow of He II with the one-dimensional steady-state Gorter-Mellink equations. The model includes several physical phenomena not found in other published analyses: the pressure jump at a gas-liquid interface due to evaporation; heat flow in the porous matrix; property variations due to temperature; and the internal structure of the porous material. When property variations and/or channel geometry are considered, the equations have to be solved as differential equations We modified the code SUPERFLOW to implement this model.

## INTRODUCTION

When He II is stored in a dewar in low gravity, it is necessary to vent the gas phase and retain the liquid regardless of bubble location. This problem is solved by using porous plugs which have been used on many space missions. The literature reports several design models, each an improvement and extension of previous ones.[1-7] We present another model in this sequence. It differs in that a system approach is used, approximations are improved, and more physical phenomena are included. The operation of a porous plug can be understood better when the interactions with the other components of the vent system are included in the analysis. We believe that some assumptions used in previously published theories are incorrect and accordingly the equations presented below differ from those used in some models.[1-7] The predictions of this model agree with experimental data from the literature.

## SYSTEM DEFINITION

Earlier published models treat the porous plug as a separate component, and the boundary conditions imposed are usually the pressure and temperature at both ends of the plug. These are not the boundary conditions determined by the operating conditions of the vent system, as a rule. The vent system consists of the following components in series:

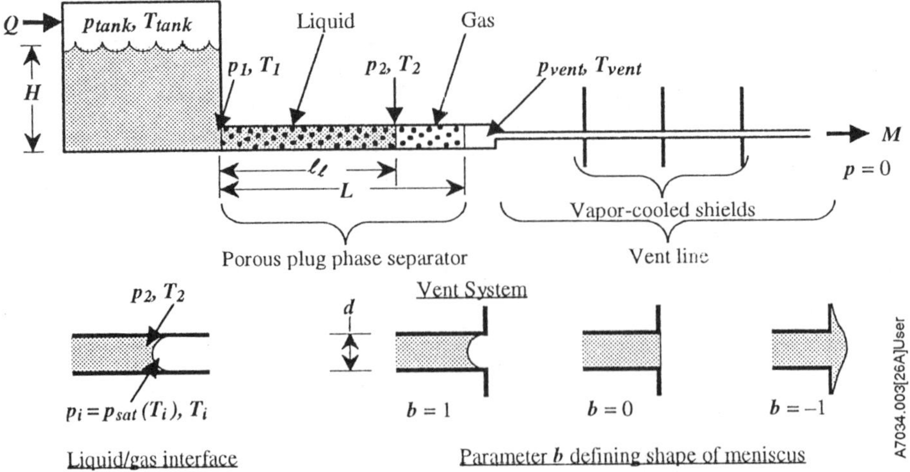


**Figure 1.** Schematic of system with definitions of symbols.

the dewar, porous plug, and the vent line. Fig. 1 shows the assumed layout and defines the symbols used below.The various pressures and temperatures are related as:

$$p_{\tan k} = p_{vent} - \nabla p_g (L - l_l) - \Delta p_{surface} - \Delta p_{evap} - \nabla p_l l_l - \rho g H \tag{1-a}$$

$$p_i = p_{sat}(T_i) = p_{vent} - \nabla p_g (L - l_l) = p_{\tan k} + \rho g H + \nabla p_l l_l + \Delta p_{surface} + \Delta p_{evap} \tag{1-b}$$

$$p_1 = p_{\tan k} + \rho g H \qquad p_2 = p_i - \Delta p_{surface} - \Delta p_{evap} \tag{1-c}$$

$$T_{\tan k} = T_i - \nabla T_l l_l \tag{1-d}$$

$$p_{sat}(T) = p_{sat}(T_r) + k_1(T - T_r) + k_2(T - T_r)^2 + \cdots \tag{1-e}$$

$\Delta p_{evap}$ is the pressure jump due to evaporation, $\Delta p_{surface}$ is the pressure jump due to surface tension, $\nabla p_g$ is the average gradient of pressure in the gas-filled portion of the porous plug, and $\nabla p_l$, is the gradient of the portion filled with liquid, while $\nabla T_l$ is the temperature gradient in the liquid. We include the gravitational pressure head, $\rho g H$, for comparison of operation on the ground and in space. Thus, $g$ is the gravitational acceleration and $\rho$ the total density of He II. Eq.(1-e) is a power series expansion of the saturation curve where $T_r$ is any reference temperature, and $k_1$ and $k_2$ are the expansion parameters. This defines our system of interest.

For the usual design problem a range of heat load to the dewar, $Q_{max}$ to $Q_{min}$, is specified and a desired $T_{tank}$ is prescribed for each $Q$. The range of the heat load that can be accommodated is determined by the porous plug and the temperature range is determined mainly by the vent line. We will use these boundary conditions; however, the same equations and methods of analysis could be used with others.

## MODEL

**Phenomena and Assumptions.** The physical phenomena included in the model are: (a) the equations of motion of He II; (b) the Darcy equations for the flow of gas in porous media; (c) the pressure jump at the gas-liquid interface due to the evaporation rate; (d) the pressure jump at a meniscus due to surface tension; (e) the fact that the gas-liquid interfaces in the dewar and in the porous plug are on the saturation line; (f) the property variations due to pressure and temperature; (g) the internal structure of the porous material such as the variability of diameter of the flow channels and transverse cross connections of the channels; and (h) the heat transfer by the porous matrix. Previous published models include some of these effects but not all.

Earlier models use one or more of the following assumptions or approximations which we believe are incorrect or inaccurate in some ranges of operation: (a) the pressure and temperature gradients in He II are connected by the London relation; (b) the heat flux is given by $\rho s v_n$ ; (c) the superfluid has a critical velocity in porous flow; and (d) the porous material can be treated as parallel soda straws with diameter based on the average diameter. In another paper we discuss these issues and our interpretation of the correct physics.[8]

There are several assumptions inherent in this model: (a) the energy flow from the dewar to the entrance of the vent line is conserved, i.e., there is no side wall heating or cooling at the porous plug; (b) the thickness of the porous plug is at least an order of magnitude larger than the average diameter of the channel or pore size; (c) the radial dependence of all variables can be averaged so that the flow dependence is one dimensional; and (d) the operation is steady state.

**Design Strategy.** The result of our model is an equation for $T_{tank}$ as a function of $Q$ and the design parameters. We evaluate each term of Eq.(1) in terms of known specified values and the design parameters to be determined. The pressure is referenced to the vacuum at the overboard end of the vent line. The temperature is referenced to the gas-liquid interface in the porous plug, $T_i$. When $Q$ is specified, the mass flow rate $M$ is also determined since energy conservation at the gas-liquid interface in the plug requires $Q = M\Lambda$, where $\Lambda$ is the latent heat. $M$ is constant throughout the system from dewar to the overboard vacuum. Since $p_{tank}$ and $T_{tank}$ are on the saturation line, it is sufficient to calculate either quantity.

The pressure drop across the vent is a function of $M$ and $T_{vent}$. Programs are available that give $p_{vent}$ in terms of the dimensions of the vent line. The gas flow through the porous plug beyond the interface is predicted accurately by Darcy's equation: $\nabla p_g = -(M/A)\nu K_p = -C_g(T_i)M$ where $A$ is the area of the plug, $\nu$ the kinematic viscosity of He gas and $K_p$ is the Darcy permeability. The gas flow in the plug is a constant enthalpy flow of an ideal gas at low pressure; there is no temperature change in the plug and $T_i = T_{vent}$. Then, $p_i = p_{vent} + MC_g(L-l_l)$. Since $p_i$ and $T_i$ are on the saturation line, we can now eliminate $T_{vent} = T_i$ in $p_{vent}$ and find $p_i$ and $T_i$ in terms of $M$, $l_l$ and the adjustable design parameters. Set $p_{vent}(M,T_l)+MC_s(T_i)(L-l_l)$ equal to the right side of Eq.(1-e) with $T = T_i$ and solve for $T_i$. Then we have $T_i$, $p_i$ and $p_{vent}$ in terms of $M$ and $l_l$ along with the adjustable design parameters. The variable properties may be included by fitting them to a curve in $T$ and setting $T = T_i$. This illustrates how the plug flow interacts with the vent design.

The next step is to consider the pressure difference across the gas-liquid interface in the porous plug. There are two effects that cause a pressure jump: evaporation and surface tension. Neither effect causes a temperature jump. The evaporative pressure jump causes the liquid side to be at a higher pressure: $\Delta p_{evap} = -(M/A)K_e = -C_e(T_i)M$ where $A_l$ is the flow area of the plug and is related to the plug area as $A_l = \varepsilon A$ with $\varepsilon$ the porosity. $K_e$ consists of products of properties of He gas evaluated at $T_i$.[9]

The pressure jump caused by surface tension can be either positive or negative. $\Delta p_{surface} = 4\sigma b/d$ where $\sigma$ is the surface tension, $d$ is the pore or channel diameter and $b$ is a number between -1 and +1. If the interface is at the surface of the plug, $l_l = L$, $b$ can take all values between -1 for a convex meniscus to +1 for a concave interface. When the interface retreats into the plug, $b$ always equals +1 and $\Delta p_{surface}$ is always positive (see Fig. 1). The range of $Q$ that can be controlled is determined by the ranges of $\Delta p_{surface}$ and $l_l$. In some porous plug materials $d$ is a rapidly varying function of $l_l$; in that case we must characterize the porous material with $d(l_l)$.

Since $\rho gH$ is given, we know all the terms in Eq.(1-a) except $\nabla p_l$ in terms of $M$, $l_l$, $b$, and the adjustable design parameters. Again, (see Fig. 1) $b$ and $l_l$ together make up one variable, since b can vary only when $l_l = L$, and $l_l$ can vary only when $b = 1$. The remaining unknowns are found by solving the equations of He II to find $\nabla p_l$ and using the fact that $p_{tank}$ and $p_i$ are both on the saturation line.

The equations of motion of He II are fourth order; there are four unknowns: $\nabla p_l$ and $\nabla T_l$ and the velocity fields, $v_n$ and $v_c$. When $Q$ is specified, then $M = Q/\Lambda$ is also known. The velocity fields $v_n$ and $v_c$ can be calculated in terms of $Q$ and $M$. Once $Q$ is chosen the gradients of $T$ and $p$ depend only on the temperature dependence of the properties of He II that appear in the equations of motion and on the channel geometry. The $p$ and $T$ at the ends of the flow channels have no effect on the gradients except for setting the temperature for property evaluation. Thus, $\nabla p_l$ and $\nabla T_l$ depend only on $Q$, the average temperature of the liquid $T_l$ and the structure of the flow channels. We can rewrite Eq.(1-a) using our above analysis as:

$$p_{\tan k} + \rho g H = p_{vent}(M, l_l) + C_g(T_i)M(L - l_l) - 4\sigma b / d(l_l) + C_e(T_i)M - \nabla p_l(\overline{T}_l, M, D)l_l \quad (2\text{-a})$$

where $T_i$ is a known function of $M$ and $l_l$, and $D$ stands for the channel structure to be defined later.

$$T_{\tan k} = T_i - \nabla T_l(\overline{T}_l, M, D)l_l \quad (2\text{-b})$$

Since $p_{tank}$ and $T_{tank}$ are on the saturation line, by Eq(1-e):

$$p_{\tan k} = p_{sat}(T_r) + k_1(T_{\tan k} - T_r) + k_2(T_{\tan k} - T_r)^2 \quad (2\text{-c})$$

We have three algebraic equations in three unknowns $p_{tank}$, $T_{tank}$ and $l_l$ and the design parameters. This is our goal, to find $T_{tank}$ vs. $Q$. We still have to find $\nabla p_l$ and $\nabla T_l$ by solving the equations for the flow of He II.

## Superfluid Helium Flow in Porous Media

**Equations of Motion.** The equations of motion of He II for one-dimensional flow are well known and have been verified experimentally for a large range of conditions. There is some controversy concerning when to use the term that accounts for turbulence in the superfluid, the Gorter-Mellink term. Since ($p_{tank}$, $T_{tank}$) and ($p_i$, $T_i$) are on the saturation line, it is not possible that $s\nabla T_l = \nabla p_l/\rho$, the necessary condition for the absence of turbulence. Accordingly, $v_s$, the superfluid component, is always turbulent in phase separator flow.

For this discussion we use the laminar friction factor in the pressure gradient term since the Reynolds number of the normal component, $v_n$, is usually less than about 1000. We do not use this assumption in computations because SUPERFLOW calculates the friction factor based on the Reynolds number. Thus, the friction factor $f$ for a circular tube is $32/d^2$. Other channel shapes have other factors. The one-dimensional steady state equations for the superfluid flow model are:

$$M = A_l \rho v_n + A_l \rho_s v_c = \text{Constant} \quad (3\text{-a})$$

$$E = Mh - A_l \rho_s s T v_c - A_p \kappa \nabla T = \text{Constant} \quad (3\text{-b})$$

$$\nabla p = -\eta f v_n \quad (3\text{-c})$$

$$s\nabla T - \frac{\nabla p}{\rho} = a\rho_n v_c^{\,3} \quad (3\text{-d})$$

$$v_c = v_s - v_n \qquad \mathrm{Q}_1 = -\rho_s s T v_c A_l \quad (3\text{-e})$$

The symbols mean: $A_l$, the flow area for the fluid; $A_p$, the porous matrix area $A_p = (1\text{-}\varepsilon)A)$; $\rho_s$, $\rho_n$,. $\rho$, the densities of the superfluid component; the normal component and the total density; $E$, the energy flow; $s$, the entropy per mass; $h$, the enthalpy per mass; $\kappa$, the thermal conductivity of the matrix; $\eta$, the viscosity of the normal component; and $a$, the Gorter-Mellink coefficient. The total heat flow in the plug is $Q = Q_l + Q_p$ where $Q_l$ refers to the liquid and $Q_p$ to the matrix of the plug. We will define $\alpha \equiv Q_l/Q$. Heat flow in the matrix is included in the model because a high $\kappa$ (low $\alpha$) sometimes improves the performance.

**General Observations.** $Q = M\Lambda$ still holds when heat flows in the porous matrix. Setting $M = Q/\Lambda$ in Eq.(3-a), and replacing $v_c$ by Eq.(3-e) we find:

$$\nabla p = -\frac{Q}{A_l}\frac{\eta}{\rho} f\left(\frac{\alpha}{sT} + \frac{1}{\Lambda}\right) = -C(T)\frac{f}{A_l}Q \quad (4\text{-a})$$

$$\nabla T = -\frac{Q}{A_l}\frac{\eta}{\rho}\left(\frac{\alpha}{sT}+\frac{1}{\Lambda}\right)\frac{f}{\rho s} - a\rho_n\left[\frac{\alpha Q}{A_l}\frac{1}{\rho_s sT}\right]^3 = -A(T)\frac{f}{A_l}Q - B(T)\frac{1}{A_l^3}Q^3 \tag{4-b}$$

$$v_n = \frac{M}{A_l\rho}(1+\frac{\alpha\Lambda}{sT}) \tag{4-c}$$

Thus, $\nabla p$, $v_n$, $v_c$, $Q$ and $M$ are all proportional. This is not usually true in superfluid flows but follows from the proportionality of $M$ and $Q$. Eq.(4-b) shows that $v_n$ is much larger than the average drift velocity $M/A_l\rho$ since $\Lambda/sT$ is large (340 at 1.2 K and 12 at 2.0 K).

As $\alpha$ decreases, more heat flows through the matrix and $\nabla p$ and $\nabla T$ decrease for fixed $Q$. A decrease in $\nabla p$ and $\nabla T$ results in a lower $T_{tank}$ for the same $Q$. The value of $\kappa$ at which the matrix transport of heat is comparable to that of the flow channels may be found as follows: set $(1-\alpha)Q = -\kappa A_p \nabla T$. Then substitute $\nabla T$ from Eq. (4-b) and solve for $\alpha$. When $\alpha > 0.2$, a noticeable increase in performance of the plug will occur. The system performance will only improve if $\nabla p_l$ and $\nabla T_l$ are an appreciable fraction of the total average gradients from $p_{vent}$ to $p_{tank}$.

The dimensionless number *Sigma,* described in our previous publications, determines the profiles along the channel of $p_l$ and $T_l$. A little algebra shows that for $M = Q/\Lambda$ flows, *Sigma* $= 1/(1+sT/\Lambda)$, it is slightly less than 1 and is independent of $M$, the length, and the parameters of the porous plug. Our previous paper contains these profiles.[10] Every flow of this sort has the same $p$ and $T$ profiles when plotted in dimensionless variables. The curvatures of the profiles are not negligible.

With Eq.(4-a) we can compare the ratio of the gradients of $p$ in the gas and liquid portion of the flow channels to see how a motion of the interface will change $p_{tank}$. At the interface the gas and liquid have the same temperature and the area of the channel is common to both. For the pressure gradient in the gas $\nabla p_g = -f(\eta_g/\rho_g)(M/A_l)$, and the gradient in the liquid is given by Eq.(4-a) with $Q = M\Lambda$. The ratio is $\nabla p_g/\nabla p_l = (\eta_g/\eta_l)(\rho_l/\rho_g)(1+\Lambda\alpha/sT)^{-1} = Pgl_{ratio}$ independent of $M$ and channel size and shape. If $\alpha=1$, the ratio is 2.5 at 1.2 K, and 5.5 at 2.0 K. Some investigators have assumed that this ratio is a large number.

**Channel Geometry and He II Property Variations**. Equation (4) applies to single channels. If the porous matrix can be modeled as "soda straws," all the same diameter with no cross links, Eq.(4) applies to each with $Q$ replaced by $Q/N$ where $N$ is the total number of channels in the plug. If we require only moderate accuracy, we can guess or iterate to a mean temperature for property evaluation and neglect the temperature dependence in Eq.(4-a and 4-b). Then, $\nabla p_l$ and $\nabla T_l$, the remaining terms needed to complete Eq.(1-a), can be found directly. For a constant property approximation the gradients are constants determined by the $p$ and $T$ conditions at the ends of the channel. They are independent of the length.

The interaction of the various terms in Eq.(1) follows algebraically; by combining all of Eq.(1) we find:

$$\nabla p_l = \frac{1}{k_1}\nabla T_l - \left(\Delta p_{evap} + \Delta p_{surface} + \rho g H\right) l_l^{-1} \tag{5-a}$$

and substituting from Eq.(4) for the gradients and solving for $l_l$:

$$l_l = \frac{\left[\dfrac{4\sigma b}{d(l_l)} + \rho g H - C_e(T_i)Q\right]}{\left[\dfrac{C(T_i)fQ}{A_l} - \dfrac{A(T_i)fQ}{k_1 A_l} - \dfrac{B(T_i)Q^3}{k_1 A_l^3}\right]} \tag{5-b}$$

where A, B and C are defined in Eqs.(4-a and 4-b). The terms containing $\sigma$ and $C$ usually dominate over the others. If Eq.(5-b) can be solved with $l_l = L$ by varying $b$ in the range $-1 < b < +1$, the meniscus is at the downstream end of the plug. If this is not possible, then set $b = 1$ and look for a solution of $0 < l_l < L$. When there is no solution, the plug leaks through to the vent or the meniscus retracts to the extent that $T_{tank}$ goes to $T_\lambda$. Eq.(5-b) determines the range $Q_{max}$ to $Q_{nmin}$.

There are two general types of porous media. The first has nearly parallel channels with slowly varying $d(l_l)$, and in some cases the length averaged $d$ varies from channel to channel. The other has a cellular pore structure where $d(l_l)$ can be modeled as $d_0 + d_1 \sin wl_l/L$ with $w>>L$. In the former case $l_l$ gradually gets smaller as $Q$ increases; it reaches a minimum and then increases. In the latter case, if $d_1$ is comparable to $d_0$, there are multiple solutions for the same $Q$ for a small range of $l_l$. This is the origin of the hysteresis that is observed with increasing and decreasing $Q$. Eq.(5-b) forms the basis for studies of the stability of $Q=\Lambda M$ flows. We are preparing another research paper dealing with this topic.

In practice, the flow channels are more complex than soda straws without cross connections. Some materials may have a range of diameters; the diameter may vary along the length; and there may be transverse cross connections. As the heat load increases, the accuracy of a constant $T$ approximation falls off rapidly. The simplified model that leads to Eqs.(4) and (5) can be extended to cover these complications.

When the channels have a random structure so that all paths have about the same impedance, we can assume that the heat flow divides equally as $Q/N$ where $N$ is the number of channels. We may still have the problem of a variable $f/A_l$ in Eq.(4). When $f$ and $A_l$ are variable, we have to solve Eqs.(3 and 4) as differential equations. When we do so, we can include the He II property variations without too much added complexity. If the channels have nearly equal impedance, we are still treating single channels. Unfortunately, we cannot use $Q/N$ as the heat load if there is an appreciable difference in the length averaged mean area $A_l$. Another complication is caused by cross connections. When there are frequent cross connections the $p$ and $T$ gradients are the same throughout the plug except for a few cells near the gas-liquid interface. Since the distance occupied by end cells is quite small, the effect on the total gradients is negligible. We will show how to deal with materials having a range of mean diameters and with the problem of cross connections.

**Variable Mean Diameter.** For this case the flow impedance of each channel is different and it is not true that $Q$ is the same in each channel. However, when there are no cross connections, steady state conditions require $Q_j = \Lambda M_j$. Here, $j$ is the channel index and $Q = \Sigma Q_j$. Also, the total pressure drop across the plug is the same for each channel. Then, Eqs.(4 and 5) apply to each channel with $Q$ replaced by $Q_j$. For every channel the ratio $\nabla p_{l,j}/\nabla p_{g,} \equiv Pgl_{ratio}$ is independent of $M$ and channel cross section. Equate the pressure drop for the full length of the plug in the first channel to that in all other channels:

$$-\nabla p_{l,j}\left[l_{l,j} + Pgl_{ratio}(L-l_{l,j})\right]+\frac{4\sigma b}{d(l_{l,j})} - C_e Q_j = \text{same expression with j} = 1 \tag{6}$$

Replace the pressure gradient with Eq.(4-a) and $l_{l,j}$ from Eq.(5-b) to get $N$-$1$ equations in $Q_j$. Solve for each $Q_j$ in terms of $Q_1$ and then use $Q = \Sigma Q_j$ to find $Q_1$. This shows how to deal with variable impedance. Once the $Q_j$ are known, each channel is treated as an independent channel.

**Cross Connections.** When there are cross connections, the assumption that $Q_j = \Lambda M_j$ in each channel fails but $\nabla p_l$ and $\nabla T_l$ are the same in all channels. Eq.(4) no longer holds and we have to return to Eq.(3). We continue as follows: eliminate $v_n$ and $v_c$ from Eqs.(3-a and 3-b) and write $Q$ and $M$ in terms of $\nabla p$, $\nabla T$, the fluid properties, areas and friction factors. Each channel has an area $A_{l,j}$ and a friction factor $f_j$ averaged along the length of the channel. $Q$ has the form $Q = -E(T)(s\nabla T-\nabla p/\rho)^{1/3}\Sigma A_{l,j}$ while $M = F(T)\,)(s\nabla T-\nabla p/\rho)^{1/3}\Sigma A_{l,j} - G(T)\nabla p\Sigma A_{l,j}/f_j$. If we set $Q = \Lambda M$, we can put these equations in the form of Eq.(4) with $\Sigma A_{l,j}$ and $\Sigma A_{l,j}/f_j$ replacing $A_l$ and $A_l/f$. Again, we can use the formalism of a single channel to deal with the more complicated situation.

**Variable Cross Section and He II Property Variations.** If the cross section along the channel varies or the temperature range is sufficient to make property variations appreciable, Eq.(3 or 4) must be solved as a set of differential equations. Our code, SUPERFLOW, solves Eq.(3) as a differential equation with the $T$ and $p$ dependence of the fluid properties included; $A_l$ and $f$ can be functions of the length, and the program computes $f$ based on $v_n$ and $A_l$.[10-11] We published comparisons of predictions with experimental data in wide tubes, narrow tubes and porous plugs.[11-13] The predictions are in very good agreement with observations.

SUPERFLOW was originally formulated with $p$ and $T$ boundary conditions at each end of the tube. For the present problem we know $Q$ and $M$, along with $p$ and $T$ at one end. We modified SUPERFLOW so that with $p$ and $T$ given at one end of the channel, it seeks a flow for given $Q$ and $Q=\Lambda M$. It solves the flow in a single channel. However, this limitation is not as serious as it appears. We showed above how to reduce most problems to flow in a single channel.

**Other Considerations.** When we calculate $Q$, we have to reduce the heat flow into the dewar by the factor $(1-\rho_g/\rho_l)$ to compensate for evaporation in the dewar. This is a very small correction for He II systems and is usually ignored.

The length variables in porous material, $L$ and $l_l$, are the actual lengths multiplied by the tortuosity.

It is easy to add a heat leak at the junction of the porous plug and the vent to the model. Set up a control zone surrounding the junction and do an energy balance. This effect is not included in the present model.

This completes a description of the model. We believe it includes all physical effects of significance in design considerations. Some of these effects are small at the low heat loads usually encountered, but may become important at higher loads required in future applications. We have included them so that the model may be applied to situations beyond those now anticipated without violating an unperceived assumption.

## EXAMPLES OF CALCULATIONS

Most of the previous publications show calculations using the constant temperature approximation and simple internal geometry. Our first calculations will compare our results with and without property variations. Then, we will show how the flow in more complicated geometries differs from that in soda straws. Finally, we present calculations to verify the model using data from the literature.

**Variable Properties.** Property variations are frequently neglected in the design of phase separators. We selected one SUPERFLOW run with $Q = M\Lambda$ to illustrate this point. The input and output are shown in Table 1. This run is reasonably representative of standard designs working at high load. When these dimensions are used in Eq.(4-a) with $Q$ replaced by $M\Lambda$ and the variables are evaluated at the mean temperature (1.70 K for a constant property mean), the mass flow rate is $M = 0.197$ µg/sec. The discrepancy with the variable property model is due to the temperature used to evaluate $sT$. It was shown above that the dimensionless variable for all $Q = M\Lambda$ flows is $Sigma = 1-sT/\Lambda \cong 1$. The $p$ and $T$ profiles for $Sigma = 1$ form Ref.(10) show that the average temperature is nearly $T_l$ for about 80% of $l_l$. It is well above 1.70 K and close to 1.73 K for the present example. When Eq.(4-a) is evaluated at 1.73 K instead of 1.70 K the mass flow is about equal to the value predicted by SUPERFLOW. The error in $M$ due to the constant property assumption is about 18% for this case. For low load situations the constant property approximation is probably satisfactory.

**Effects of Internal Structure.** Previously we reported on the effect of internal structure on flow in tubes as calculated by SUPERFLOW.[13] We choose the end conditions appropriate for a porous plug pump. We compared the mass flow, for fixed $p$ and $T$ end conditions, for tubes of increasing internal complexity. All tubes have an average diameter of 0.4 µm and variations in the range of 0.3 to 0.5 µm. As the randomness of the variation of diameter increases, the mass flow decreases and becomes asymptotic with a tube of 0.3 µm diameter. For the most random tube the flow decreases to 67% of that for a uniform tube with the same average diameter. For these runs $v_n \approx -0.04\ v_c$. We made another set of calculations using the same tube dimensions, the same internal structure and the same $p_1$ and $T_1$. But now, instead of specifying $p_2$ and $T_2$, $Q$ is set equal to $M\Lambda$. The mass flow for the most random tube is now only 30% that of a uniform tube of the same average diameter. Here, the normal velocity is much higher: $v_n \approx -0.75\ v_c$. Internal structure has a significant effect. The equivalent diameter appears to be that of the smallest passage that occurs with appreciable probability.

**Table 1.** Effect of property variations

| Diameter of channel | Length | $T_1$ | $T_2$ | $p_1$ | $p_2$ | Gravity head | Mass flow actual properties | Mass flow constant propeties |
|---|---|---|---|---|---|---|---|---|
| µm | mm | K | K | kPa | kPa | kPa | µg/sec | µg/sec |
| 10 | 2 | 1.750 | 1.650 | 1.366 | .8108 | .1200 | 0.232 | 0.197 |

**Table 2.** Comparison of calculations and observed data in uniform tubes

| $\Delta p_l$ kPa | $T_i$ K | $M$ measured mg/sec | $M$ calculated mg/sec | $\Delta T_l$ measured K | $\Delta T_l$ calculated K |
|---|---|---|---|---|---|
| 0.05 | 1.7 | 2.1 to 3.0 | 3.0 | 0.010 | 0.009 |
| 0.10 | 1.7 | 4.7 to 4.9 | 4.5 | 0.018 | 0.015 |
| 0.20 | 1.7 | 5.0 to 5.1 | 5.0 | 0.036 | 0.038 |

**Model Verification With Uniform Tubes.** We verified SUPERFLOW by comparing with experimental data for a wide range of conditions: wide and narrow tubes and porous materials.[11,13] The tests with porous materials were with boundary conditions suitable for pumps. We have now tested the program for the phase separator regime using data of DiPirro and Zahniser.[3] The porous material is a glass matrix with nearly straight uniform channels without cross connections. The channel diameter is measured as 1.8 μm, the thickness is 2 mm, $A_l = 4.9$ cm$^2$ and $T_i = 1.70$ K. We adjusted $H$ by iteration to set $\Delta p_l$ to 0.05, 0.10 and 0.20 kPa to compare to their conditions; $l_l$ is found from Eq.(5-b). The results are shown in Table 2. The agreement seems to be good.

## RESULTS

(1) Equation (4-b) shows explicitly how heat conduction through the porous plug matrix improves performance. (2) Equation (4-c) shows that $v_n$ is two orders of magnitude higher than the mean drift velocity at low temperatures. (3) It is shown that $Sigma \approx 1$ for all phase separator flows; thus, all dimensionless $p$ and $T$ profiles are the same. (4) The ratio of the pressure gradient in the gas filled portion of the plug to that of the liquid portion is not large. (5) An equation for the range of permissible heat load is presented as Eq.(5-b). (6) Equation (5-b) shows how the hysteresis with increasing and decreasing heat load is related to the internal structure. (7) Methods of dealing with the internal structure elements of variable diameter and cross connections are presented. (8) The model agrees with experimental data for uniform tubes.

## REFERENCES


1. H. Nakai and M. Murakami, Flow phenomena of superfluid helium through a porous plug phase separator, *Cryogenics*, 27:442 (1987).
2. U. Schotte, He II phase separation with slits and porous plugs for space cryogenics, *Cryogenics*, 24: 536-547 (1984).
3. M. J. DiPirro and J. Zahniser, The liquid/vapor phase boundary in a porous plug, *Adv. Cryo. Eng.* 35:173-179 (1990).
4. D. G. Elliott, Porous plug tests and flow modeling for SIRTF, LPL Publication, JPL D-11412 (June 1994).
5. C. R. Lages, R. H. Torii and D. B. DeBra, Evaporation of superfluid helium in a capillary, *Cryogenics*, 35:31-33 (1995).
6. A. Nakano, D. Petrac and C. Paine, He II liquid/vapor phase separator for large dynamic range operation, *Cryogenics*, 36:823-828 (1996).
7. S. W. K. Yuan, D. J. Frank and C. Lagas, The dependence of choked flow and break through on pore size distribution in vapor-liquid phase separators of He II using porous media, *Adv. Cryo. Eng.*41: 1189-1194 (1996).
8. H. A. Snyder and A. J. Mord, Common approximations that can invalidate calculations of the flow of superfluid helium, Presented at the Cryogenic Engineering Conference, Portland, OR (1997).
9. R. R. Rogers and M. K. Yau, "A Short Course in Cloud Physics, 3rd. ed.," Pergamon Press,Oxford (1989) Chap. 7.
10 H. A. Snyder and A. J. Mord, Calculation of He II flow in tubes, *J. Low Temp. Physics,* 86:177-209 (1992).
11. A. J. Mord, D. A. Newell and H. A. Snyder, End to end modeling of He II flow systems, *Cryogenics,* 32:291-299 (1991).
12. G. L. Mills, A. J. Mord and H. A. Snyder, Pressure maxima in the flow of superfluid He II in tubes, *Phys. Rev. B,* 49:666-669 (1994).
13. H. A. Snyder, A. J. Mord and G. L. Mills, Comparison of modeling predictions and experimental data for He II flow in porous materials, *Adv. Cryo. Eng.* 39:1821-1828 (1994).

# COMMON APPROXIMATIONS THAT CAN INVALIDATE CALCULATIONS OF THE FLOW OF SUPERFLUID HELIUM

H. A. Snyder and A. J. Mord

University of Colorado at Boulder and
Ball Aerospace & Technologies Corp.
Boulder, Colorado, 80306-1062, U.S.A.

## ABSTRACT

Calculations of the flow of He II are difficult because of its two-fluid nature and unusual boundary conditions. Investigators made many assumptions and approximations, over the years, to simplify the calculations and make them tractable. We find in scanning the literature that in a large number of situations approximations have been used beyond their range of validity. In some cases, this has resulted in substantially incorrect predictions. We have made a list of commonly used approximations and analyzed the range of validity of each. Examples of approximations are: assuming a critical velocity for the onset of turbulence in the superfluid when no critical velocity exists; attempting to specify the problem with less than four boundary conditions; neglect of the pressure gradient in evaluating the gradient of the chemical potential; neglect of property variations in one-dimensional flow; assuming the normal component flows like a Navier-Stokes fluid; attributing a dependence of the Gorter-Mellink coefficient to the structure of the material; using formulas for heat flow that apply only to counterflow situations when there is not mass flow; and assuming the normal fluid is clamped in porous flows. We made a comparison of calculations with and without the assumptions where possible.

## INTRODUCTION

Landau proposed a set of equations of motion for superfluid helium in 1941.[1] In 1949 Gorter and Mellink found discrepancies between the predictions of Landau's equations and experimental data for flow in tubes.[2] They proposed adding a term to Landau's equation, now known as the Gorter-Mellink (G-M) term that brought the observations and calculates into agreement. Hall and Vinen in 1956 showed that there is a mutual interaction between the normal and superfluid components of He II flow. They suggested that it is caused by quantized vortex lines in the superfluid.[3] They also explained how the G-M term is related to mutual friction. Others showed that the G-M term applies not only to flow in tubes but also to general flows. Khalatnikov rederived the Landau equations with great care to detail and added dissipative terms. He summarized his work in a monograph published in 1965.[4] The Landau equations, with the addition of mutual friction, appear to be adequate to explain the flow of superfluid helium. This framework is the basis of most modern interpretations and calculations of He II flow.

Experimentalists began to compare their results to the Landau equations soon after they were published. Over the years researchers made many approximations and simplifying assumptions to reduce the complexity of the analysis. In many cases, these were good choices for the problem being considered. Frequently the published reports did not state the assumptions or approximations explicitly. Others with somewhat similar applications used the same simplifications without realizing that they are approximations. Several approximations or assumptions that are used frequently in analyzing superfluid helium flow are often applied inappropriately. We list the most commonly used simplifications and indicate when they may be used without appreciable error. Most of the applications deal with one-dimensional flows and we will concentrate on these.

## DISCUSSION

### Topical Outline

After reading a large number of publications on superfluid helium flow, we identified the following eight frequently occurring simplifications that are often applied inappropriately.

(a) The pressure and temperature gradients are proportional. This implies a critical velocity below which there is no turbulence in the superfluid.

(b) The equations are less than fourth order. The problem can be posed with less than four unknowns and boundary conditions.

(c) The pressure gradient can be neglected in calculating the gradient of the chemical potential.

(d) The $p$ and $T$ dependence of the property coefficients in the equations of motion can be neglected.

(e) In most cases He II flows like water.

(f) The G-M coefficient for flows in porous materials depends on the structure.

(g) The expression for the heat flow is $\rho s T v_n$.

(h) In porous materials, it is permissible to set $v_n = 0$.

Each of these topics will be discussed and the limits of the approximation will be analyzed.

### Equations of Motion

Our analysis is based on the one-dimensional Landau equations with the G-M term added. The first four equations below are the steady state version in terms of $v_n$ and $v_c$. The last line is the time-dependent equation for the superfluid component.

Mass conservation $$M = A\rho v_n + A\rho_s v_c \tag{1-a}$$

Energy conservation $$E = Mh - A\rho_s sTv_c \tag{1-b}$$

Pressure drop $$\nabla p = -\eta f v_n \tag{1-c}$$

Gorter-Mellink equation $$s\nabla T - \frac{\nabla p}{\rho} = a\rho_n v_c^{\,3} \tag{1-d}$$

Relative velocity, Heat flow, Gradient of the chemical potential

$$v_c = v_s - v_n \qquad Q = -\rho_s sTv_c A \qquad \nabla\mu = -s\nabla T + \frac{\nabla p}{\rho} \tag{1-e}$$

Time-dependent equation of $v_s$ $$\frac{\partial v_s}{\partial t} + (v_s \cdot \nabla)v_s = -\frac{1}{\rho}\nabla p + s\nabla T - a\rho_n v_c^3 \tag{1-f}$$

The symbols mean: $M$, the mass flow rate; $E$, the energy flow rate; $A$, the flow area for the fluid; $\rho_s$, $\rho_n$, $\rho$, the densities of the superfluid component, the normal component and the total density; $v_s, v_n$, $v_c$, the velocities of the superfluid and normal components and the relative (counter flow) velocity (see Eq.(1-e)); $p$, the pressure; $T$, the temperature; $s$, the entropy per mass; $h$, the enthalpy per mass; $\eta$, the viscosity of the normal component; $a$, the Gorter-Mellink coefficient; and $Q$, the heat flow. The friction factor $f$ is a complicated function of $v_n$ and the geometry of the channel. For laminar flow it is $32/d^2$ for a circular tube with diameter $d$. Other channel shapes have different factors. When the normal fluid is turbulent, we believe that it can be calculated in nearly the same way as for an ordinary Navier-Stokes fluid.

These equations apply to a single channel. If there are several channels in parallel, such as in a porous material, they apply to each channel. For a single channel $M$ is constant and if there is no side wall heating, $E$ is also constant. For channels in parallel, the sum over all channels of $M$, and $E$ are constant.

**Dimensionless Scaling Number.** We use a dimensionless number $\sigma$ to organize and explain the limits of the approximations to be discussed. It is defined and used in previous papers to classify the $p$ and $T$ profiles as a function of parameters.[5-7] $\sigma$ is the ratio of the two terms on the right of Eq.(1-a) averaged over the length of the channel (see Eq.(2-a)). It has the remarkable property that it scales the $p$ and $T$ profiles along the channel when $p$ and $T$ are made dimensionless.[5] All one-dimensional steady flow of He II can be categorized by $\sigma$. See Ref.(5) for profiles covering a wide range of $\sigma$. For $0 < \sigma < 1$ the profiles are nearly linear. Huge maxima of $p$ and $T$ occur near the middle of the channel when $\sigma$ is large. Most of the variables in He II analysis can be scaled with $\sigma$.

We can calculate $\sigma$ from the $p$ and $T$ values at the ends of the channel, the fluid properties and the geometry of the channel:

$$\sigma \equiv - \rho_s v_c / \rho v_n \tag{2-a}$$

$$\sigma \approx \frac{-32\bar{\rho}_s \bar{\eta} \Delta\mu^{1/3} L^{2/3}}{\bar{\rho}\bar{\rho}_n^{1/3} \bar{a}^{1/3} \Delta p d^2} \quad \text{or} \quad \frac{-\bar{\rho}_s \bar{\eta}^{1/4} \Delta\mu^{1/3} L^{5/21}}{2.87\bar{\rho}^{4/7} \bar{\rho}_n^{1/3} \bar{a}^{1/3} \Delta p^{4/7} d^{5/7}} \tag{2-b}$$

The first expression applies to laminar flow and the second to Blasius type turbulent flow. The properties are evaluated at the average $p$ and $T$ of the channel. This is not the average based on the values at the ends of the channel if $\sigma > 10$. When $\sigma > 10$, we find $\sigma$ by iteration. First evaluate Eq.(2) using the end points to do the average. With that value of $\sigma$ use the profiles in Ref.(5) to find the average $p$ and $T$ and recalculate $\sigma$. Iterate if necessary. In Eq.(2) $\Delta p$ is the pressure difference between the ends of the channel, $L$ refers to the length and $d$ to the equivalent diameter.

## APPROXIMATIONS AND THEIR CONSEQUENCES

The first two simplifications are concerned with the choice of equations of motion. The first approximation is to set $a = 0$ in Eq.(1-d). In the second, fewer than four unknowns are used in the analysis. In these two cases it is difficult to express the limits of the approximation in terms of $\sigma$ only. For all the others the limits can be stated as a range of $\sigma$ independent of other variables and parameters.

### Approximation (a): The Critical Velocity and Assuming a Vanishing Gradient of the Chemical Potential

There is a critical velocity for $v_s$ that marks the onset of turbulence in the superfluid component. If $v_s < v_{s,crit}$, the G-M term, the right side of Eq.(1-d), is zero. If we set $a = 0$, the method of solution of Eq.(1) differs from the case $a \neq 0$. When $a \neq 0$, Eq.(1) can be solved as a differential equation with $p$ and $T$ as unknowns. Sufficient boundary conditions are $p$ and $T$ at both ends of the channel. Other possible boundary conditions are $E$ or $Q$ and $M$ and $p$ and $T$ at either end of the channel. If we set $a = 0$, the differential equation has only one unknown, $p$. The solution is independent of the value of $T$ at the ends of the channel; $T$ at the ends is no longer at our disposal. Although the term in $a$ may be small, the problem is not properly posed without it.

The critical velocity for $v_s$ is rather small. The experimental data roughly fit a relation:

$v_{s,crit}d^{1/4} = 1$ with the lengths measured in centimeters.[8] For a 1 μm channel $v_{s,crit} \approx 10$ cm/sec. Most flows we have measured or computed are well above the critical value. It may be that in most applications it is not significant whether we set $a = 0$ below critical or use Eq.(1) for all values of $v_s$. However, it appears that there are very few situations when $v_s$ is below the critical value. We made a comprehensive comparison of our calculations, using $a \neq 0$, with the experimental data of the literature and found good agreement.[5,9] There are physical reasons why $v_s$ is usually turbulent.

Consider Eq.(1-f) with $a = 0$ so that $v_s$ is laminar. In one-dimensional laminar flow of $v_s$ the streamlines are straight and $(v_s \cdot \nabla v_s) = 0$. If the gradient of the chemical potential $\nabla\mu \neq 0$, $\partial v_s/\partial t \neq 0$. Then, $v_s$ increases until $v_s = v_{s,crit}$ and $a \neq 0$. $\nabla p$ and $\nabla T$ are set by the boundary conditions at the ends of the channel. The only case when the flow will not accelerate until $v_s$ becomes turbulent is when $\Delta\mu = 0$ across the channel. If the $p$ and $T$ at the ends of the channel are controlled at any values for which $\Delta\mu \neq 0$, the flow is turbulent. For example, if both ends are on the saturation curve, $v_s$ is turbulent.

If $a = 0$, the superfluid component does not have a dissipative mechanism. It is not coupled to the normal component that has damping mechanisms. It is a principle of mechanics that a system needs dissipation to be stable. Thus, when $v_s$ is not turbulent, the system is marginally stable. When energy is pumped into the system by the boundary conditions, $v_s$ increases until it becomes turbulent. In most situations the assumption of laminar flow of $v_s$ is inappropriate.

There are special conditions when $\Delta\mu$ may be zero. If $p$ and $T$ at the ends of the tube are not controlled, they may adjust to this situation. Critical velocities occur in counterflow, where $M = 0$, in flows driven by a pressure head and perhaps in flows with a gas-liquid interface that can move. In the counterflow experiments $Q$ is controlled and $p$ and $T$ at one end of the channel are maintained constant. The $p$ and $T$ at the other end can adjust so that $\Delta\mu = 0$. In pressure head flow there is usually a stand pipe, with a gravitational head, attached to the channel somewhere. The height of the column of the standpipe is free to adjust and set the channel pressure so that $\Delta\mu = 0$. In other cases of pressure head flow $p$ and $T$ at one end are free and can adjust to the required condition. For flows with a meniscus at one end of the channel, the shape of the meniscus may change and the interface may recede into the channel to change the $p$ and $T$ end conditions.[10]

When $a \neq 0$ and the analysis proceeds with the G-M term omitted, $\nabla T$ is in error by $(a\rho_n/s)v_c^3$. If we know $\Delta p$ across the channel, the error can be calculated from $-(a\rho_n/s)(\rho\Delta p/32\rho_s\eta)^3(d^2/L)^3\sigma^3$ and if $M$ is given, the error is $(a\rho_n/s)(M/\rho_s A)^3(\sigma/(\sigma-1))^3$ where the properties are evaluated at a mean $T$ determined from $T$ at one end and an iterated value of $\sigma$.

If we let $a \to 0$ when solving Eq.(1) when $\Delta\mu \neq 0$, $v_c$ becomes large so as to balance Eq.(1-d). This limit, which implies $v_c \to \infty$ as $a \to 0$, is not the same as the omission of the G-M term.

## Approximation (b): Using an Incomplete Set of Equations

The four equations (1-a to 1-d) have six unknowns: $M$, $E$, $v_n$, $v_c$, $\nabla p$ and $\nabla T$. In addition, a complete solution of the problem requires a calculation not only of $\nabla p$ and $\nabla T$, but also $p$ and $T$. If $\nabla p$ and $\nabla T$ are known, we need one value of $p$ and $T$ somewhere in the integration interval to find them everywhere. It is possible to eliminate any two of $M$, $E$, $v_n$, and $v_c$ using the first two equations, and it is possible to eliminate $\nabla p$ in the last two. There is no way to eliminate $\nabla T$ in terms of the other variables. Unless $\nabla T$ is given at each point along the flow channel we always have a differential equation in $\nabla T$. There is one exception, if the property variations are neglected, $\nabla p$ and $\nabla T$ are constants and are determined by the boundary values at the ends of the channel. In this case we cannot specify $M$, $E$, $v_n$ or $v_c$ since $\nabla p$ and $\nabla T$ are set by the $p$ and $T$ at the two end points of the channel, four boundary conditions. When the differential equation is solved with any two of $M$, $E$, $v_n$ and $v_c$, we find $\nabla p$ and $\nabla T$. However, to complete the solution we still need a boundary value of $p$ and $T$ somewhere to complete the solution. Thus, in all cases four unknowns must be given. It follows that when experiments are carried out, four unknowns must be measured or controlled to specify the problem. In some situations certain conditions are implied in the problem but not always stated. For example, an experimental arrangement with a blanked off tube used to study counterflow has $M = 0$. We developed a program, called SUPERFLOW, to solve Eq.(1-a to 1-d) as a differential equation with variable properties.[5,9]

When we searched the literature to find experimental data to verify SUPERFLOW, we found that over three quarters of the reported results were useless because less than four variables were measured or controlled.

### Approximation (c): Neglecting the Pressure Gradient in the Chemical Potential

Another approximation is the neglect of $\nabla p/\rho$ when evaluating $\nabla\mu$ when $a \neq 0$. The average value of $\nabla p$ and $\nabla T$ in a channel are $\Delta p/L$ and $\Delta T/L$ where $L$ is the length. If the boundary conditions are $p$ and $T$ at the ends of the channel, then this approximation is appropriate if $\Delta p/\rho s << \Delta T$ and $0 < \sigma < 10$. For larger values of $\sigma$, $\nabla p$ can be two orders of magnitude larger than $\Delta p/L$ and the approximation fails.[5] When $\sigma > 10$, it is necessary to compare the maximum values of $\nabla p$ and $\nabla T$ from plots of the profiles to make the comparison.[5]

### Approximation (d): Ignoring He II Property Variations With $p$ and $T$

If we evaluate the fluid properties in Eq.(1) at a mean value of $p$ and $T$, obtained from the values at the ends of the channel, we can solve Eq.(1-a to 1-d) as an algebraic set. Then,$\nabla p$ and $\nabla T$ are constants and the profiles of $p$ and $T$ are straight lines connecting the end points. This procedure is valid if $\sigma \approx 0$, since only $\sigma = 0$ leads to straight profiles of $p$ and $T$. The validity of a constant property analysis can be evaluated by calculating $\sigma$ from Eq.(2). If $\sigma < 1$, the simplification is satisfactory. The error associated with the approximation may be determined by comparing the straight line profiles with those calculated for $\sigma$ as shown in Ref.(5).

### Approximation (e): Assuming Navier-Stokes Flow of $v_n$

The one-dimensional steady equations for the motion of He II can be written in dimensionless form using $\sigma$. They become the equations for the flow of a Navier-Stokes fluid as $\sigma \rightarrow 0$.[5] He II flows almost like water when $\sigma < 0.1$. When this is true, $v_n = v_s$, $\nabla p$ is constant and $\nabla T = 0$ except at the exit from the channel. Again, the error of the approximation can be checked by comparing the dimensionless profiles of $p$ and $T$ at the actual $\sigma$ with the case $\sigma = 0$.

### Approximation (f): Assuming a Dependence of G-M Coefficient on Porous Structure

Several investigators have interpreted their experimental results of flow in porous materials in terms of a structure-dependent value of the parameter $a$. They use an average value of the channel diameter to find the $A$ and $f$ of Eq.(1) and adjust $a$ to fit the results. The diameter is measured by gas flow or microscopy. SUPERFLOW can integrate Eq.(1-a to 1-d) with variations in $A$ and $f$ along the channel. We find that for the same average diameter but different internal structure the flow rates $M$ and $Q$ vary by up to 50% depending on the complexity of the internal structure.[7] As the structure becomes more random and the diameter variation increases, $M$ and $Q$ get smaller for the same boundary values of $p$ and $T$. For several runs, we selected $d$ for a uniform tube so that it has the same $M$ and $Q$ as a complex tube with the same $p$ and $T$ boundary conditions. We found that $d$ is close to the minimum diameter of the complex channel.[7] The simplest interpretation is that $a$ is not a function of the internal structure but that the flow field in complex tubes is complex. We compared the predictions of SUPERFLOW to several experimental studies of the flow in porous media, some with diameter as small as 0.25 μm, and found good agreement using the values of $a$ measured in wide tubes.[6,7]

### Approximation (g): Using a Simplified Heat Flow Expression

The approximation that appears most frequently in the literature is the expression $Q_{approx} = \rho sTv_nA$ for the heat flow. The correct expression is $Q = -\rho_s sTv_cA$, which can also be written $Q = \rho sTv_nA - MsT$. The approximation is appropriate when $\rho v_nA >> M$. The simplified version probably arose in the early days of superfluid research when most flow experiments used the counterflow configuration where $M = 0$. Khalatnikov derived the correct expression for the heat flow (his Eq.(8-20)).[4] Roberts and Donnelly reached the same expression using a different method.[11]

From the definition of $\sigma$, $Q_{approx} = (1/\sigma)Q$. The approximation is good if $\sigma \approx 1$. It overestimates at low $\sigma$ and underestimates for large $\sigma$. The error is evidently $Q_{approx}/Q = 1/\sigma$. For porous plug phase separators and for counterflow $\sigma \approx 1$ and the approximation is good, but for porous plug pumps $\sigma >> 1$ and $Q_{approx}$ is a fraction of the real heat flow. If the tube diameter is large and the tube of moderate length, $\sigma \to 0$ and again the approximation is poor.

### Approximation (h): Assuming Negligible $v_n$ in Small Tubes

Calculations on He II flow in small channels, such as occur in porous media, are sometimes simplified by assuming that $v_c = v_s$ in Eq.(1), i.e., that $v_n \approx 0$. In Eq.(1-c) we cannot set $v_n = 0$ and still fit the boundary conditions on $p$. But otherwise, it appears at first glance that $v_c \to v_s$ may be satisfactory. One way to estimate $v_n$ is to replace $\nabla p$ in Eq.(1-c) by $\Delta p/L$. For a circular channel in laminar flow $f = 32/d^2$. As $d$ gets smaller $f$ increases rapidly and $v_n \propto d^2$. This is the <u>average</u> value of $v_n$ in the channel, not the local value. We can also write this result in terms of $\sigma$. This approximation is good if $v_n/v_c << 1$. From the definition of $\sigma$, $v_n/v_c \approx -(\rho_s/\rho)(1/\sigma)$. Since $\sigma$ is based on an average along the channel, $v_n/v_c$ refers to the average values of $v_n$ and $v_c$. It seems that when $\sigma$ is large the approximation is good.

The situation is more complicated than our analysis so far suggests. In order to make $(\rho_s/\rho)(1/\sigma) << 1$, $\sigma$ has to be large. When $\sigma$ is large the $p$ profile can have a huge maximum.[5] Then, $\nabla p$ is large at the entrance and exit of the channel and the sign changes near the middle. $\nabla p$ may be two orders of magnitude larger than $\Delta p/L$ for $d = 1$ μm. The average value of $\nabla p$ is $\Delta p/L$, a small value when $\sigma$ is large, but it is large and positive at the entrance and large and negative at the exit. Since $v_n$ is proportional to $-\nabla p$, $v_n$ is large and negative at the entrance. It can be shown that $v_{n\,local}/v_{n\,average}$ is proportional to $\sigma$.[5] Thus, $v_{n\,local} \propto \sigma\, v_{n\,average} \propto \sigma\, v_c(\rho_s/\rho)(1/\sigma)$, so that $v_{n\,local}/v_c$ is nearly independent of $\sigma$. The surprising result is that local values of $v_n$ are not negligible when $d$ gets very small although the average value decreases as $d^2$. It appears that it is never a good approximation to set $v_c = v_s$ in Eq.(1).

## COMPARISON OF EXACT AND APPROXIMATE CALCULATIONS FOR THE FLOW OF HE II IN A TUBE

All of these approximations can be avoided by using the numerical solution approach used by SUPERFLOW. The calculated variables for several cases of flow in tubes are listed below. We selected the data to cover a wide range of $\sigma$. The results are displayed in Table 1. We refer to the SUPERFLOW results as "exact" although there are some approximations involved. For example, we neglect the velocity dependence of the thermodynamic properties such as $\rho_s$ on $v_s$. We believe all approximations made in SUPERFLOW are negligible for the full range of $\sigma$ reported in Ref.(5), which also discusses the experimental verification of SUPERFLOW.

The data of Table 1 are used to estimate the error of each approximation described above for each run in the table. In some situations this comparison is simple and in others the comparison is indirect. We listed some variables at the entrance, the middle and the exit of the channel indicated by $l = 0$, $l = .5L$ and $l = L$. The data of Table 1 are not precise and should be used only to compare the order of magnitude of the listed variables. Some of the variables change rapidly near $l = 0$ and $l = L$. If $\sigma$ is large, $\nabla p$ can change by two orders of magnitude along the channel. If $\sigma = 0$, the slope of $T$ is infinite at the exit. $\nabla\mu$ is the difference of two large terms of nearly equal value (Eq. 1-e). When $l = 0, L$, the velocities quoted are from the middle of the first and last cells of the computation; they are not extrapolated to the boundary. $\nabla p$, $\nabla T$ and $\nabla\mu$ are extrapolated linearly from the center of the first and last cells when $l = 0, L$. Likewise, $s$ is evaluated at the boundary when it is used for $l = 0, L$. These approximations used in reducing the output of the program result in rather large discrepancies in Table 1, but the order of magnitude comparisons are still correct.

**(a) Assuming Vanishing Gradient of the Chemical Potential.** If $v_s > v_{s,\,crit}$ and the G-M term is omitted in the calculation, the variable that is in error is $\nabla T$. Eqs. (1-a to 1-c) are

**Table 1.** Results of SUPERFOW - comparison with approximations

| | Variable | Unit | (1) | (2) | (3) | (4) | (5) | (6) | (7) |
|---|---|---|---|---|---|---|---|---|---|
| 1 | $\sigma$ | - | 1000 | 100 | 10 | 1 | 0.1 | -1 | -100 |
| 2 | $p_1$ | kPa | 0.2815 | 0.2815 | 0.2815 | 0.2815 | 0.2815 | 0.2815 | 0.2815 |
| 3 | $p_2$ | kPa | 1.635 | 1.635 | 1.635 | 1.635 | 1.635 | 1.635 | 1.635 |
| 4 | $T_1$ | K | 1.4 | 1.4 | 1.4 | 1.4 | 1.4 | 1.4 | 1.4 |
| 5 | $T_2$ | K | 1.8 | 1.8 | 1.8 | 1.8 | 1.8 | 1.8 | 1.8 |
| 6 | $d$ | μm | 0.755 | 2.49 | 8.13 | 23.8 | 879 | 28.6 | 2.55 |
| 7 | $L$ | mm | 50 | 50 | 50 | 50 | 250 | 50 | 50 |
| 8 | $Q$ | μW | -0.0140 | -0.234 | -2.67 | -14.9 | 2630 | -61.4 | -0.260 |
| 9 | $M$ | μg/sec | .0157 | 0.263 | 2.90 | 1.89 | -181000 | 90.2 | 0.300 |
| 10 | $v_n$ $l=0$ | mm/sec | -42 | -64 | -141 | -1016 | -7652 | 578 | -58 |
| 11 | $l=.5L$ | mm/sec | 0.0 | 2.0 | -30 | -289 | -2589 | 525 | 10 |
| 12 | $l=L$ | mm/sec | 19 | 31 | 17 | -207 | -2025 | 246 | 43 |
| 13 | $v_c$ $l=0$ | mm/sec | 312 | 473 | 570 | 1145 | 6396 | 422 | 502 |
| 14 | $l=.5L$ | mm/sec | 391 | 469 | 500 | 421 | 714 | 485 | 497 |
| 15 | $l=L$ | mm/sec | 333 | 505 | 543 | 348 | -46 | 1042 | 535 |
| 16 | $\nabla p$ $l=0$ | kPa/mm | 3.26 | 0.461 | 0.096 | 0.140 | 0.0002 | -0.034 | 0.401 |
| 17 | $l=.5L$ | kPa/mm | 0.0 | -0.015 | 0.019 | 0.021 | 0.0004 | -0.029 | -0.067 |
| 18 | $l=L$ | kPa/mm | -1.39 | -0.160 | -0.007 | 0.016 | 0.015 | -0.012 | -0.277 |
| 19 | $\nabla T$ $l=0$ | mK/mm | 145 | 26.4 | 10.8 | 57.2 | 122 | 0.0 | 24.0 |
| 20 | $l=.5L$ | mK/mm | 4.4 | 5.4 | 6.8 | 4.8 | 1.2 | 3.0 | 5.6 |
| 21 | $l=L$ | mK/mm | -14.8 | 6.4 | 10.8 | 3.2 | 0.16 | 74.8 | 6.8 |
| 22 | $\nabla\mu$ $l=0$ | J/g mm $10^{-4}$ | -2.6 | -5.2 | -8.2 | -62.6 | 262 | -3.2 | -6.2 |
| 23 | $l=.5L$ | J/g mm $10^{-4}$ | -3.1 | -21.4 | -19.0 | -18.8 | 5.2 | -6.4 | -24.6 |
| 24 | $l=L$ | J/g mm $10^{-4}$ | -14.0 | -3.6 | -6.0 | -1.6 | 0.0 | -392 | -5.6 |
| 25 | $MsT/Q$ | - | -0.70 | -0.50 | -0.50 | -0.06 | -31 | -0.67 | -0.52 |
| 26 | $v_n/v_c$ $l=0$ | - | -0.135 | -0.135 | -0.247 | -0.887 | -1.19 | 1.37 | -0.116 |
| 27 | $l=.5L$ | - | 0.0 | 0.004 | -0.06 | -0.686 | -3.62 | 1.08 | 0.020 |
| 28 | $l=L$ | - | 0.057 | 0.061 | 0.031 | -0.594 | 44. | 0.236 | 0.080 |
| 29 | $s\Delta T/L$ | J/g mm $10^{-4}$ | 22.8 | 22.8 | 22.8 | 22.8 | 4.56 | 22.8 | 22.8 |
| 30 | $s\nabla T$ $l=0$ | J/g mm $10^{-4}$ | 191 | 34.8 | 14.2 | 75.5 | 161 | 0 | 31.7 |
| 31 | $l=.5l$ | J/g mm $10^{-4}$ | 12.5 | 15.4 | 19.4 | 13.7 | 3.42 | 8.6 | 16.0 |
| 32 | $l=L$ | J/g mm $10^{-4}$ | -79.2 | 34.2 | 57.8 | 17.1 | 0.85 | 401 | 36.4 |
| 33 | $(1/\rho)\Delta p/L$ | J/g mm $10^{-4}$ | 1.87 | 1,87 | 1.87 | 1.87 | 0.374 | 1.87 | 1.87 |
| 34 | $(1/\rho)\nabla p$ $l=0$ | J/g mm $10^{-4}$ | 224 | 31.8 | 6.62 | 9.65 | 0.014 | -2.34 | 27.6 |
| 35 | $l=.5L$ | J/g mm $10^{-4}$ | 0 | -1.03 | 1.31 | 1.45 | 0.027 | -2.00 | -4.6 |
| 36 | $l=.5L$ | J/g mm $10^{-4}$ | -95.8 | -11.0 | -0.48 | 1.10 | 1.03 | -0.83 | -19.1 |
| 37 | $(1/\rho s)\nabla p$ $l=0$ | mK/mm | 169 | 24.1 | 5.0 | 7.3 | 0.01 | -1.78 | 20.9 |
| 38 | $l=.5L$ | mK/mm | 0 | -0.41 | 0.40 | 0.77 | 0.02 | -0.81 | -1.85 |
| 39 | $l=L$ | mK/mm | -17.9 | -2.06 | -0.09 | 0.21 | 0.19 | -0.16 | -3.6 |
| 40 | $a\rho_n v_c^3$ $l=0$ | J/g mm $10^{-4}$ | 2.6 | 5.2 | 8.2 | 62.6 | -262 | 3.2 | 6.2 |
| 41 | $l=.5L$ | J/g mm $10^{-4}$ | 3.1 | 21.4 | 19.0 | 18.8 | -5.2 | 6.4 | 24.6 |
| 42 | $l=L$ | J/g mm $*10^{-4}$ | 14.0 | 3.6 | 6.0 | 1.6 | 0.0 | 392 | 5.6 |

unchanged by the omission and $\nabla p$ is correct, except that the properties are evaluated at a slightly incorrect $T$. The error consists of omitting $a\rho_n v_c^3$ compared to $(1/\rho s)\nabla p$. Both terms are listed in the table as rows 37 to 42. There are no runs for which $(1/\rho s)\nabla p >> a\rho_n v_c^3$.

**(c) Neglecting the Pressure Gradient in the Chemical Potential.** Here, we are comparing $s\nabla T$ with $(1/\rho)\nabla p$. These terms are listed in the table in rows 30 to 36. From above we expect that $\nabla p$ is small in this sense, when $\sigma$ is small. Table 1 shows that for 0.1 $\sigma < 1$, the error is < 10%, but the error is large for other values of $\sigma$.

**(d) Ignoring He II Property Variations With *p* and *T*.** The constant property assumption leads to constant values for $\nabla p = \Delta p/L$ and $\nabla T = \Delta T/L$. For all runs except (5), $\Delta p/L$ = 0.027 kPa/mm and $\Delta T/L$ = 8 mK/mm. Reference to rows 16 to 21 shows that this is never a good approximation.

**(e) Assuming Navier-Stokes Flow of $v_n$.** This approximation implies $\nabla p = \Delta p/L$, $\nabla T = 0$ and $v_n = v_s$. or $v_c = 0$. Rows 16 to 21 of Table 1 show that $\nabla p$ and $\nabla T$ get close to this approximation as $\sigma \rightarrow 0$, but even at $\sigma = 0.1$ it is a poor approximation. Rows 13 to 15 indicate that $v_c$ does not approach zero everywhere in the channel as $\sigma$ approaches zero.

**(g) Using a Simplified Heat Flow Expression.** This approximation is satisfactory if $MsT/Q$ is negligible. This ratio is listed in the table as row 25. We expect from above that the ratio is small compared to 1 only when $\sigma \approx 1$. The data from the table confirms our conclusion.

**(h) Assuming Negligible $v_n$ in Small Tubes.** For this approximation to hold we must have $v_n/v_c << 1$. This ratio is listed in the table as rows 26 to 28. We expect that this quantity does not become small throughout the channel for any $\sigma$. We see that this prediction is correct.

## Conclusion

Every approximation considered above has a rather limited range of $\sigma$ where it can be applied without appreciable error. Our recommendation is to check the conditions and the $\sigma$ before using any of these approximations. In most cases it is just as easy to solve the equations without them.

## REFERENCES


1. L. Landau, The theory of superfluidity of helium II, *J. Physics (Moscow)*,5:71-90 (1941).
2. C. J. Gorter and J. H. Mellink, On the irreversible processes in liquid helium II, *Physica*, 15: 285-304 (1949).
3. H. E. Hall and W. F. Vinen, I. Experiments on propogation of second sound in uniformly rotating helium II, *Proc. Roy. Soc. (London)*, A238:204-214 (1956).
4. I. M. Khalatnikov, "Introduction to the Theory of Superfluidity," Benjamin, New York, (1965).
5. H. A. Snyder and A. J. Mord, Calculation of He II flow in tubes, *J. Low Temp. Physics*, 86:177-209 (1992).
6. G. L. Mills, A. J. Mord and H. A. Snyder, Pressure maxima in the flow of superfluid He II in tubes, *Phys. Rev. B*, 49:666-669 (1994).
7. H. A. Snyder, A. J. Mord and G. L. Mills, Comparison of modeling predictions and experimental data for He II flow in porous materials, *Adv. Cryo. Eng* . 39:1821-1828 (1994).
8. W. E. Keller, "Helium-3 and Helium-4," Plenum Press, New York, (1969).
9. A. J. Mord, D. A. Newell and H. A. Snyder, End to end modeling of He II flow systems, *Cryogenics*, 32:291-299 (1991).
10. H. A. Snyder and A. J. Mord, Modeling phase separator systems, Presented at the Cryogenics Engi neering Conference, Portland, OR (1997).
11. P. H. Roberts and R. J. Donnelly, in "Annual Review of Fluid Mechanics", Vol. 6, M.VanDyke, W. G. Vincenti and J. V. Wehausen, eds., Annual Reviews, Palo Alto (1974), p 179-200.

# DEVELOPMENT OF SUPERFLUID SHOCK TUBE FACILITY

H. Nagai, M. Murakami, H. Yang,
N. Takano and S. Teraoka[1]

Institute of Engineering Mechanics, University of Tsukuba
Tennoudai 1-1-1, Tsukuba, Ibaraki 305, Japan

[1]Torisha Co. Ltd. Aoki 3-5-1, Kawaguchi-city Saitama 332 Japan

## ABSTRACT

The superfluid shock tube facility has been developed as a versatile tool for general researches in low-temperature thermo-fluid dynamic phenomena. The shock tube was designed to be operated with the He II-filled test section immersed in superfluid helium. A special feature of the facility is that the gas dynamic shock tube is driven and controlled with a quick-opening valve (MO-valve). It enables a synchronized discharge of a shock wave with other target transient phenomena. Some wave phenomena such as a thermal shock and a compression shock waves in superfluid helium, and phase transition phenomena across the λ-line (He I-He II) and across the He II-solid helium phase boundary can be investigated by impinging a gas dynamic shock wave onto a vapor-He II free surface in the test section. The target physical phenomena in He II are measured by pressure transducers and superconductive temperature sensors, by applying laser beam refraction method, and with the aid of some optical visualization methods. In the present study, the general thermo-fluid dynamic performance are investigated to verify the validity in wide range of experimental researches.

## INTRODUCTION

Wide variety of thermo-fluid dynamic phenomena in superfluid helium have been investigated by many researchers. A compression shock wave equivalent to a conventional gas dynamic shock wave and a thermal shock wave as a discontinuity in the temperature are such kind of interesting phenomena [1,2]. A cryogenic shock tube, of which test section is immersed in He II, was originally developed by Liepmann et al. (1973) and Cummings (1974) [3,4,5,6] for liquid helium and high Mach number gas dynamic research. It had been utilized in the pioneering work for the study of shock waves in He II. It was found in their experiment that two modes of shock waves propagate in He II when a gas dynamic shock wave of helium vapor was made incident on a He II free surface. However, this shock tube still had some difficulties in operation. Air and water vapor migration into a shock tube during a diaphragm replacement process may cause frost contamination inside of a shock tube. Accordingly, one objective of the present development of a new type shock tube is to eliminate the difficulties. A diaphragm-free shock tube originally developed by Profs. Maeno and Oguchi [7,8], that is now called an MO-valve driven shock tube, is just in line with the demand. The diaphragm-free shock tube is capable of a frost-contamination-free operation and repeated operation with a cryogenic fluid in a short time.

In the present study, the superfluid shock tube facility has been developed and some performance test were conducted. The superfluid shock tube has some special features in addition to the contamination-free repeated operation with an MO valve. It is expected that the application can be extended to researches on phase transitions across the λ-line (HeI-HeII) and HeII-solid helium, besides researches on wave phenomena.

Two wave modes, a compression and a thermal shock waves are separately detected by pressure transducers and a superconductive temperature sensor, and by applying laser beam refraction method. The pressure history and the propagation speeds of two kinds of shock waves are measured and an x-t diagram was drawn to clarify the propagation and reflection of the two shocks.

## EXPERIMENTAL APPARATUS

### Superfluid Shock Tube Facility

An overall view picture and a schematic illustration of the facility are shown in Figures 1-(a) and (b), respectively. The shock tube facility consists of two parts, a diaphragm-free shock tube with an MO-valve and a He II cryostat with optical windows for visualization. A vertical low-pressure tube of the shock tube, called a cryogenic shock tube, is inserted in the cryostat and is maintained at cryogenic temperature about 80 K at the top portion down to 2K at the bottom portion immersed in He II. The test section of the cryogenic shock tube is filled with He II and the rest of the low-pressure tube is filled with saturated helium vapor. In most operation, high pressure driver gas and driven gas in the low pressure tube are helium gas. The pressure of helium gas in the high pressure section may be arbitrarily specified, while the pressure in the low-pressure tube is uniquely to be equal to the saturated vapor pressure of coexisting He II filled in the test section. The cryogenic shock tube portion has a rectangular cross section ( 35 mm$\times$25 mm ) and is equipped with two pair of visualization windows with a diameter of 25 mm in the test section as shown in Figure 2. Three piezoelectric pressure transducers are fixed on the wall surface of the shock tube at 260 mm (transducer A), 120 mm (transducer B) and 50 mm (transducer C) from the bottom of the test section. A superconductive temperature sensor is located at 35.5 mm from the bottom which is supported on the end plate as shown in Figure 2. A special feature of this facility in the operation is a diaphragm-free shock tube operation. An operation starts with opening the pilot valve to activate the main MO-valve between the high and low pressure sections which has a capability of extremely quick opening. A gas dynamic shock wave propagates to impinge a free surface of He II. Some portion of the shock wave penetrates into He II and the rest is reflected back into the low pressure tube. After completion of a run the inside of the shock tube need not be exposed to atmosphere because of the elimination of the diaphragm replacement process. Accordingly, significant improvement can be expected by the application of an MO-valve driven diaphragm-free shock tube to researches of cryogenic fluid. In addition to a frost-contamination-free operation, easy repeated operation in a short time is possible in the system. The shock formation time in an MO-valve driven shock tube, about 10 ms, is inevitably longer than that in conventional diaphragm type shock tubes. However, it causes no problem in operation because the shock formation is quite reproducible in an MO-valve shock tube.

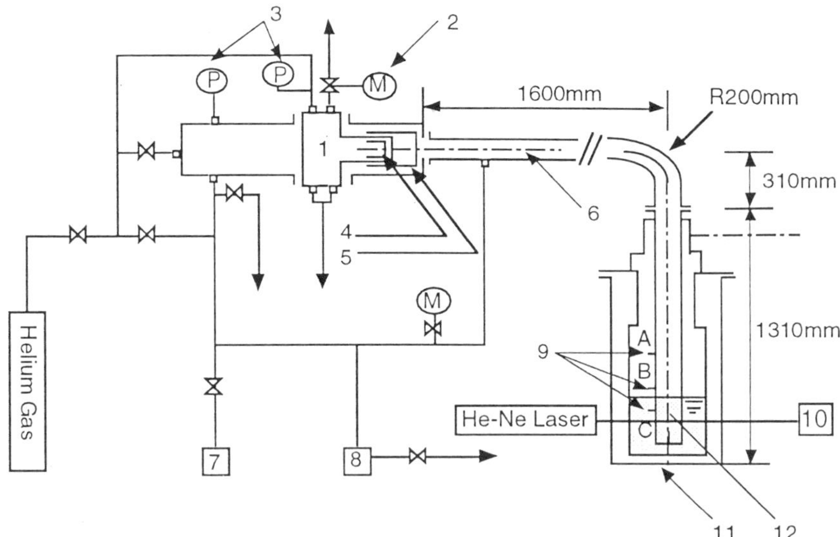


**Figure 1(a).** Schematic illustration of the superfluid shock tube facility. 1. MO-main valve 2. Electro-magnetic pilot valve 3. Pressure gauge 4. Sub-piston 5. Main-piston 6. Low pressure tube 7. Rotary pump 8. Buffer tank 9. Pressure transducers 10. Photo diode 11. Cryostat 12. Test section.

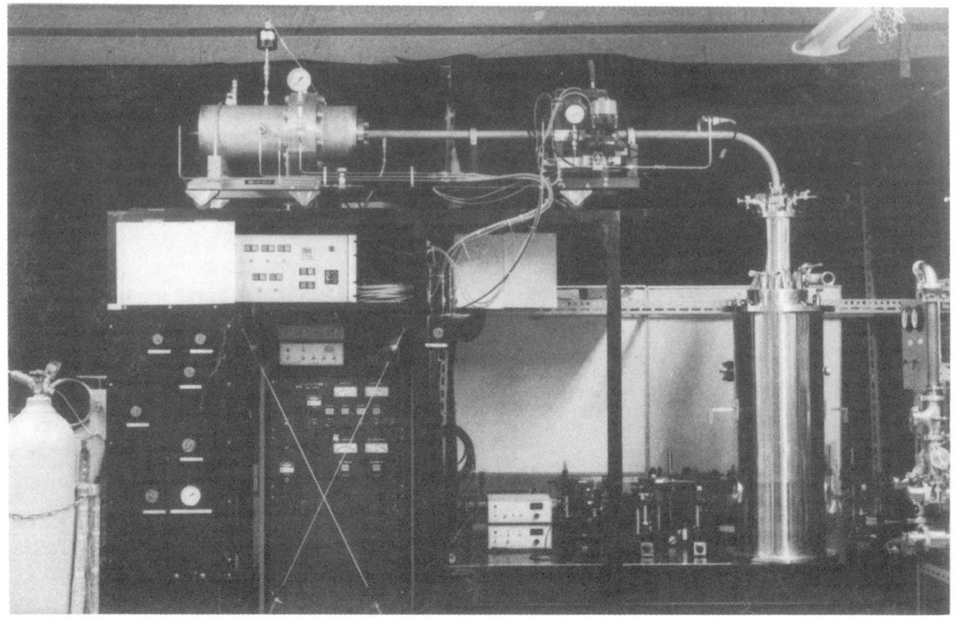

**Figure 1(b).** Overall view picture of the superfluid shock tube facility.

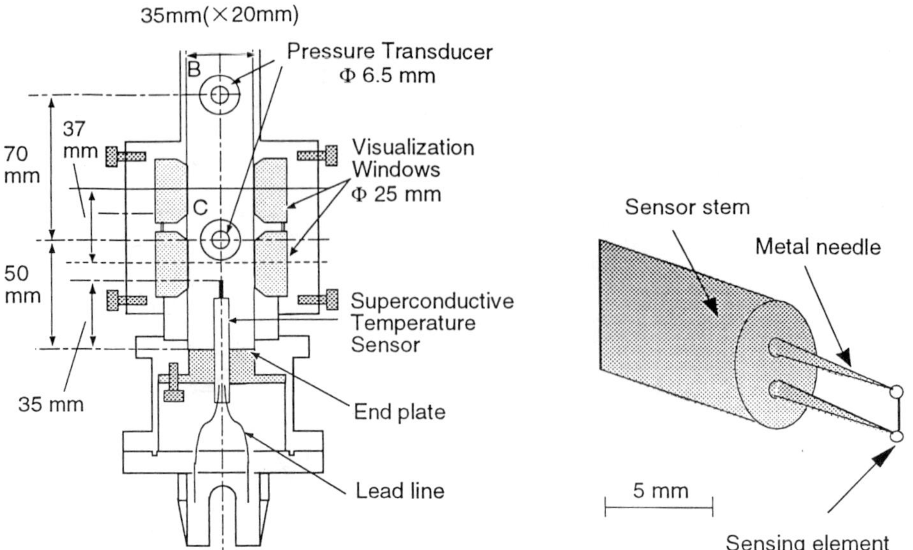


**Figure 2**. Detailed structure of the test section of the superfluid shock tube. This part, filled with He II, is completely immersed in He II.

**Figure 3**. Schematic illustration of superconductive temperature sensor.
1000Å - thick Sn on 210Å - thick Au thin film vacuum deposited on surface of Φ 40 μm quartz fiber.

### Procedure of Shock Wave Measurement

The pressure is measured by three piezoelectric pressure transducers. The incident shock Mach number or the incident shock speed in the saturated helium vapor is measured by using upper two transducers of the three. For the measurement of the shock propagation speed ( or shock Mach number ) in He II, we utilize the pressure signal detected by the pressure transducer and laser beams which are refracted by a shock front and detected by photo diodes. The temperature variation caused by both compression (first sound) and thermal (second sound) shock waves are measured by a superconductive temperature sensor shown in Figure 3, which is made of gold-tin thin film vacuum deposited on the side surface of a fine quartz rod with a diameter of 40 μm. The electric resistance changes continuously in a super-normal transition region because the sensing element behaves like a second kind superconductor. Any temperature variation as small as 1 mK results in detectable resistance change. A particular feature of this sensor is its very high sensitivity and rapid time response within a few micro seconds. Accordingly, a thermal shock wave induced by compressed high temperature zone in an incident shock wave as well as compression shock wave can be detected by this sensor.

## RESULTS AND DISCUSSION

### General Performance of a Shock Wave in He II

A typical example of detected signals of a shock wave, by the pressure transducer B and C, the superconductive temperature sensor and the photo diode is shown in Figure 4. Laser beam location is at 33 mm from the bottom of test section and a He II free surface is situated

65 mm from the bottom in the present case. Initial condition of driver and driven gas is 671 kPa (300 K) and 2.25 kPa (1.88 K), respectively. The first signal is detected by the transducer B in vapor phase. The second, third and fourth signals are detected by the transducer C, the photo diode, the superconductive temperature sensor in He II, respectively in order of location. The pressure behind the incident shock wave gradually increases owing to the strong temperature gradient developed in the cryogenic shock tube, about 80 K at the top and down to about 2K at the bottom section. The transducer C located in He II section clearly detect the arrival of a compression wave propagating in He II. The pressure rise in He II is found to be very sharp unlike shock waves propagating in other liquids. The reason for the sharp pressure rise may be that He II has considerably higher compressibility than any other liquids, and thus the impedance matching for a pressure wave penetration through a gas-liquid interface of helium is much better than conventional liquids. This means that a shock wave penetrates much efficiently into He II as compared with conventional liquids. The second pressure rise of detector signal (a) is resulted from a reflected shock wave from a He II free surface. It is of interest to note that the third pressure rise which is caused by the penetration of a shock wave in He II through He II-vapor interface into vapor phase. This indicates that a pressure wave in He II also well penetrates into vapor phase through an interface due to better impedance matching than conventional liquids. These properties are quite desirable to generate a shock wave in He II by impinging a gas dynamic shock wave on a He II free surface. The trace (b) in Figure 4 is the pressure history in He II phase. Arrivals of penetrating shock wave and a reflected shock wave from the end plate are well detected. The cause of the rapid pressure decrease between in the two arrival signals is still an open question. The trace (c) is the temperature sensor signal. First and second small decrease are the arrival of a penetrating and reflected shock wave, and a large spiky temperature rise is caused by a thermal shock wave. The detail will be described later. The trace (d) is the signal of the photo diode. The sharp decrease of signal level indicates that a shock wave arrives at the point. Figure 5 shows a x-t wave diagram for the case shown in Figure 4.

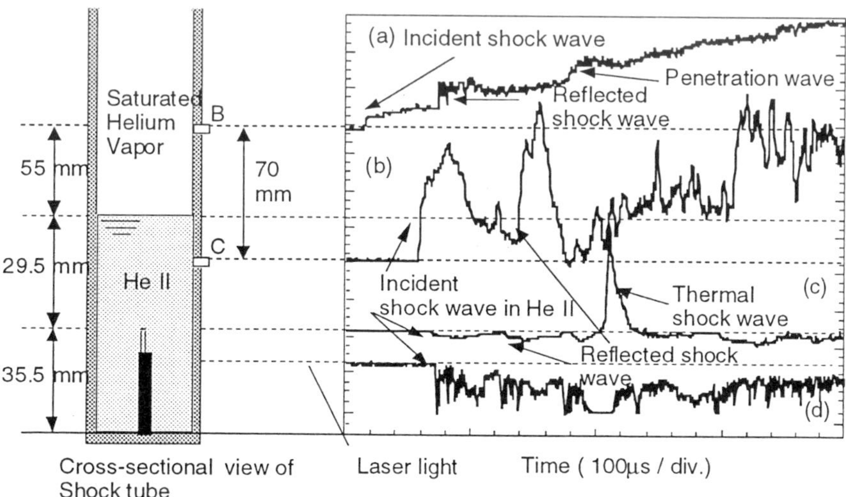


**Figure 4**. Typical example of time variation of detected signals in $P_4$=671 kPa, $P_1$=16.8 torr (2.25 kPa), $P_{41}$ (=$P_4/P_1$)=299.2, Incident shock speed = 311.1 m/s, Shock Mach number = 3.86. Driver gas is helium gas at 300 K, driven gas is saturated vapor of He II at 1.88 K.
Propagation speed in He II = 241.5 m/s, Shock Mach number in He II =1.04.
(a) Transducer B (b) Transducer C (c) Superconductive temperature sensor (d) Photo diode signal

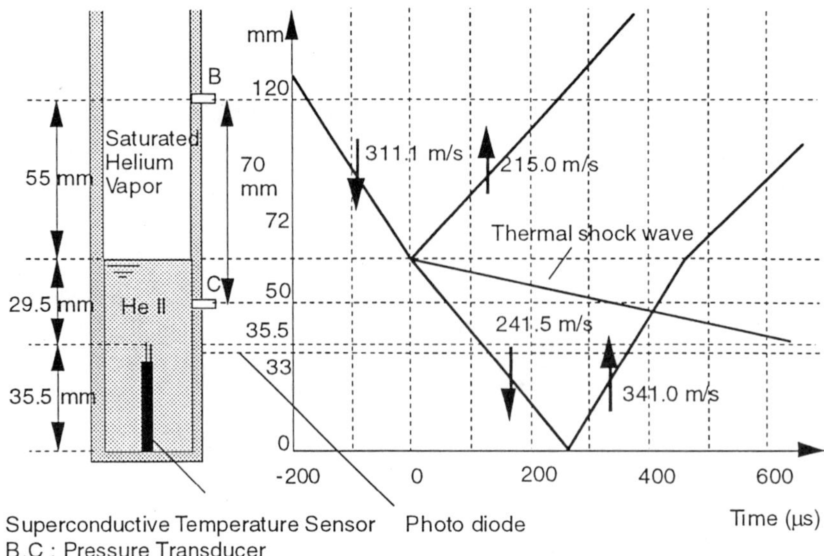


**Figure 5.** x-t diagram of an incident shock wave and a propagating shock wave in He II $P_4$=671 kPa, $P_1$=16.8 torr (2.25 kPa), $P_{41}$ (=$P_4/P_1$)=299.2.
Driver gas is helium gas at 300 K, driven gas is saturated vapor of He II at 1.88 K.

## Compression and Thermal Shock Waves in He II

Figure 6 shows the temperature variation detected by the superconductive temperature sensor in He II in the case of the initial pressure in the driver and driven section of 671 kPa (300 K) and 2.25 kPa (1.88 K), respectively. The thermal shock wave is generated by heating of He II free surface from compressed high temperature vapor upon the impingement of a gas dynamic shock wave. The first and second stepwise small temperature drops are resulted from the arrivals of a compression shock wave and a reflected one from the bottom. These variations are as small as a few mK in amplitude because of the nature of a compression ( first sound origin ) shock wave. The very sharp variations of the detected signal verifies that the superconductive temperature sensor has very high sensitivity and rapid time response. The fact that the temperature variations are negative is quite naturally understood if we consider the Maxwell relation as

$$\Delta T=\Delta P(\beta TV/C_p),$$

where $C_p$ is the heat capacity per unit volume of He II and $\beta$ is the thermal expansion coefficient that is negative in He II in the present temperature range. The relation states that a positive pressure variation causes a negative temperature variation. The quite large temperature rise, as large as 30 mK, is due to the thermal shock wave. It is, furthermore, seen that the compression shock little influences on the propagation of a thermal shock wave. This indicates that a thermal shock wave interacts only weakly with a compression shock wave.

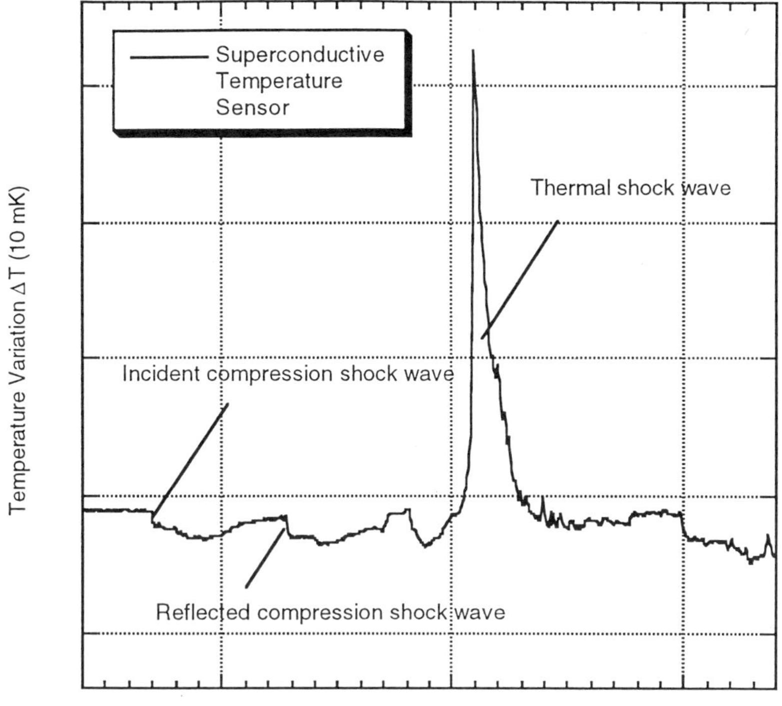


**Figure 6.** Propagating Compression and Thermal Shock Waves in He II detected by superconductive temperature sensor. $P_4$=671 kPa, $P_1$=16.83 torr ( 2.25 kPa ), $P_{41}$=299.17.
Driver gas is helium gas at 300 K, and driven gas is saturated vapor of He II at 1.88 K.
Propagation speed of compression shock wave in He II is 241.55 m/s, and shock Mach number is 1.06

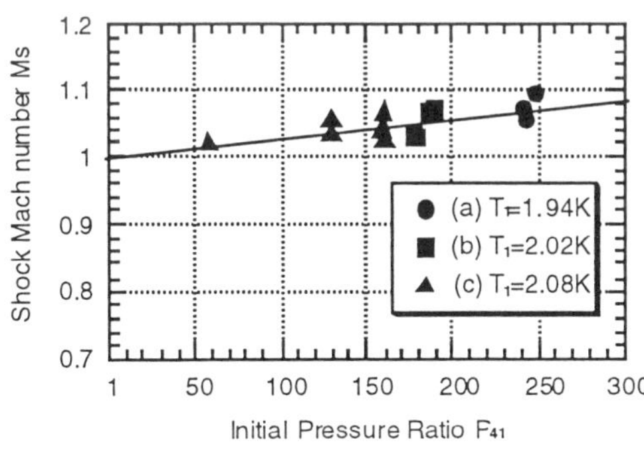


**Figure 7.** Shock Mach number in He II as a function of initial pressure ratio $P_{41}$ and bath He II temperature $T_1$. Driver gas is helium gas at 300 K, and driven gas is saturated vapor of He II.
(a) $T_1$=2.08 K (b) $T_1$=2.02 K (c) $T_1$=1.94 K

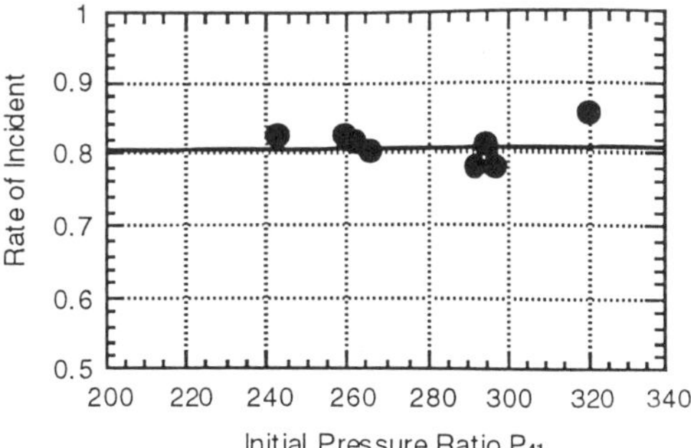


**Figure 8.** The ratio of the shock propagation speed in He II to that of incident shock wave.
( Incident shock speed in vapor / Propagating shock speed in He II )
Driver gas is helium gas at 300 K, and driven gas is saturated vapor of He II at $T_1$=1.94 K.

## Shock Mach number in He II

The propagation speed of compression shock wave in He II is investigated as an index to evaluate the performance. The measurement of the propagation speed is taken by the pressure transducers A and B in vapor phase, and by the pressure transducer C and the superconductive temperature sensor and by applying the laser beam refraction method in He II. The shock Mach number $M_S$ is plotted as a function of the initial pressure ratio $P_{41}$ ($=P_4/P_1$) in Figure 7 for three He II temperatures. The shock Mach number linearly increases a little with $P_{41}$. The dependence of $M_S$ on the bath temperature is hardly seen within the accuracy of this experiment. A more detailed experiments are needed to investigate the dependence on bath temperature. Figure 8 shows the ratio of the shock propagation speed in He II to that of incident shock wave. The result indicates that the ratio is almost constant ( about 80% of incident shock waves ) even if the initial pressure ratio increases. Accordingly, it can be concluded that the propagation speed in He II can be estimated in terms of the incident shock wave speed in vapor.

## CONCLUDING REMARKS

The performance test has been conducted for the superfluid shock tube facility. The following conclusion are drawn.

1. It is confirmed that the superfluid shock tube facility is useful for the studies of shock waves in He II and phase transitions across the $\lambda$-line and across the He II-solid helium boundary.
2. A gas dynamic shock wave efficiently penetrates into He II due to good impedance matching for gas-liquid interface and a compression shock wave in He II also does back into vapor phase through vapor-He II interface.
3. A thermal shock wave with an amplitude of about 30 mK is induced by the gas dynamic shock wave impingement.
4. Shock Mach number in He II increases a little as the initial pressure ratio $P_{41}$ increases.

## References


1. T. Shimazaki, M. Murakami and T. Iida : Temperature measurement in transient heat phenomena through a thermal boundary layer with high vortex density in He II, "Advanced in Cryogenic Engineering" (1995), vol. 41(A), pp 265-271

2. T. Iida, M. Murakami, T. Shimazaki and H. Nagai : Visualization study on the thermo-hydrodynamic phenomena induced by pulsative heating in He II by the use of a laser holographic interferometer, "Cryogenics" 36 (1996), pp 943-949

3. John C. Cummings : Development of a high-performance cryogenic shock tube, "J. Fluid, Mech. " (1974), vol. 66, part 1, pp 177-187

4. John C. Cummings : Experimental investigation of shock waves in liquid helium I and II, "J. Fluid Mech. " (1976), vol. 75, part 2, pp 373-383

5. H. W. Liepmann, G. A. Laguna : Nonlinear interactions in the fluid mechanics of helium II, "Ann. Rev. Fluid Mech. " (1984), vol. 16, pp 139-77

6. John C. Cummings : Experiments on second-sound shock waves in superfluid helium, "Phys. Fluid" (1978), vol. 21(5), pp 713-717

7. K. Maeno, Y. Shizukuda and Y. Hanaoka : Experiments on shock wave propagation in low-temperature gases and vapor bubble collapse in liquid R-12, "Memoirs of the Muroran Inst. of Tech. " vol. 36 (1986) 251

8. K. Maeno : Shock wave propagation in low temperature fluids and phase change phenomena, "IUTAM symposium", Gettingen (1989), pp 69-77

# LIQUID LEVEL CONTROL USING A POROUS PLUG IN A TWO-PHASE He II SYSTEM

J. Panek, Y. Zhao, and S. W. Van Sciver

National High Magnetic Field Laboratory
Mechanical Engineering Department
FAMU/FSU College of Engineering
Tallahassee, FL 32310

## ABSTRACT

This paper presents results of an experiment which decouples the liquid counterflow component from the evaporative component of heat transfer in the two-phase He II system. Two baths, partially filled with He II, are connected by a vapor tube at the top and a liquid tube at the bottom. Temperature and liquid level differences between the two baths are measured as a function of heat input into one bath while temperature of the other bath is held constant. A copper porous plug with 30,576 parallel channels of length 1.29 cm and diameter 30 μm is inserted in the bottom tube, which partially restricts flow of the normal fluid component. The liquid level height difference between the two baths is governed by the thermomechanical effect and saturated pressure differences. A numerical model of the combined modes of heat transport is able to predict temperature and liquid level differences accurately . A properly sized porous plug will not restrict heat flow and can reduce the undesirable liquid level differences between components at different temperatures.

## INTRODUCTION

The TESLA particle accelerator will use two-phase He II for the refrigeration system[1]. One possible concern is that saturated pressure differences between components at different temperatures could lead to undesirably low liquid levels in the warmest components, perhaps even leading to complete dryout. Placing a small second line in parallel with the components, Figure 1, and adding a fountain effect pump gives designers the ability to minimize these level differences and possibly even recirculate liquid between components.

Conventional fountain effect pumps use effective channel diameters less than 1 μm, and restrict flow of the normal fluid component. This would not be desirable in a real refrigeration system, as all flow would be blocked above $T_\lambda$. An imperfect fountain effect pump with pore size of order 10 μm can allow some leakage of normal fluid while still providing needed level difference control between the two baths.

The goal of the present experiment is to measure and predict height differences achieved by using an imperfect fountain effect pump. A simple modification of a previous experiment[2] enables us to gather the data.

## EXPERIMENT

Two baths are partially filled with He II and connected at the top and bottom with tubes, Figure 2. The top tube is open and carries vapor only, while the bottom tube is submerged completely in liquid and contains a porous plug, shown in cross section in Figure 3. The OFHC copper plug, fabricated by a process similar to that for NbTi superconductor, has 30,576 parallel channels of length 1.29 cm and diameter 30 μm[3]. Heat is added to the liquid to one bath with a carbon resistor, and vapor is removed from the other bath by a throttled vacuum pump at room temperature. More detailed information on dimensions is given in a previous paper[2].

Primary instrumentation consists of superconducting wire liquid level indicators and germanium resistance thermometers. Thermometer calibration was conducted *in situ,* against the vapor pressure read from pressure transducer located at room temperature. After analog to digital conversion, thermometer accuracy is 0.5 mK. Liquid level resolution is 1 mm, but absolute accuracy is not known due to the film creep phenomenon in He II. However, it is assumed that the creep is equal on both level indicators and relative height differences are measured accurately to 1 mm.

The procedure is to partially fill the containers, turn on the heater, and maintain constant temperature on the pumped side until steady state conditions are reached. This can sometimes take on the order of one hour, as the liquid flows through the porous plug until the hydrostatic pressure difference due to liquid level differences balances the thermomechanical and saturated pressure effects between the two sides:

$$\rho g \Delta h = \Delta P_{tm} - \Delta P_{sat} \tag{1}$$

Height difference $\Delta h$ is defined as $h_h - h_c$. The liquid level indicators are energized briefly (3-4 seconds) every 10-15 minutes. When $\Delta h$ is constant over 2 readings, steady state conditions are achieved and temperature and level data are recorded.

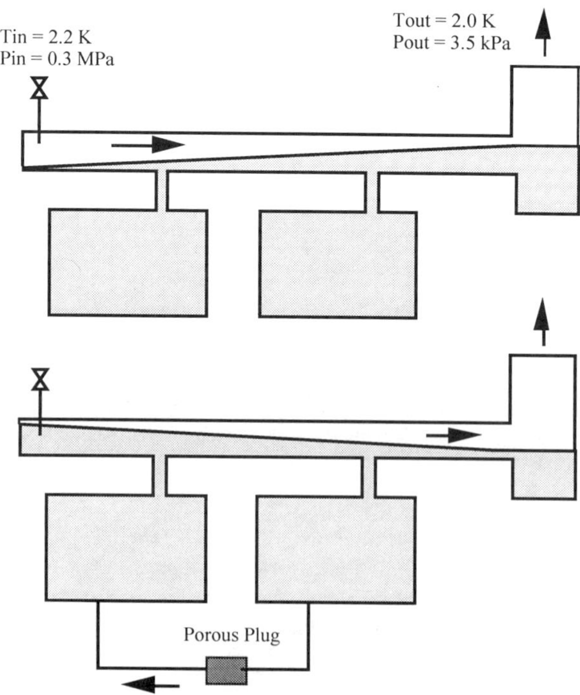


Figure 1. TESLA cooling schemes, with and without porous plug.

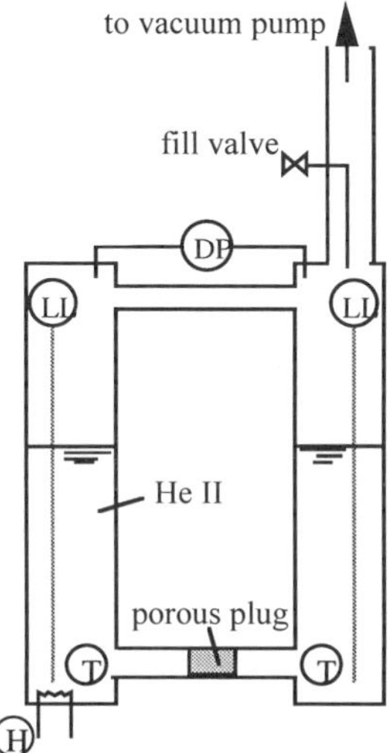


DP differential pressure transducer
LI liquid level probe
T thermometer
H heater

Figure 2. Experimental Apparatus.

Figure 3. Porous Plug.

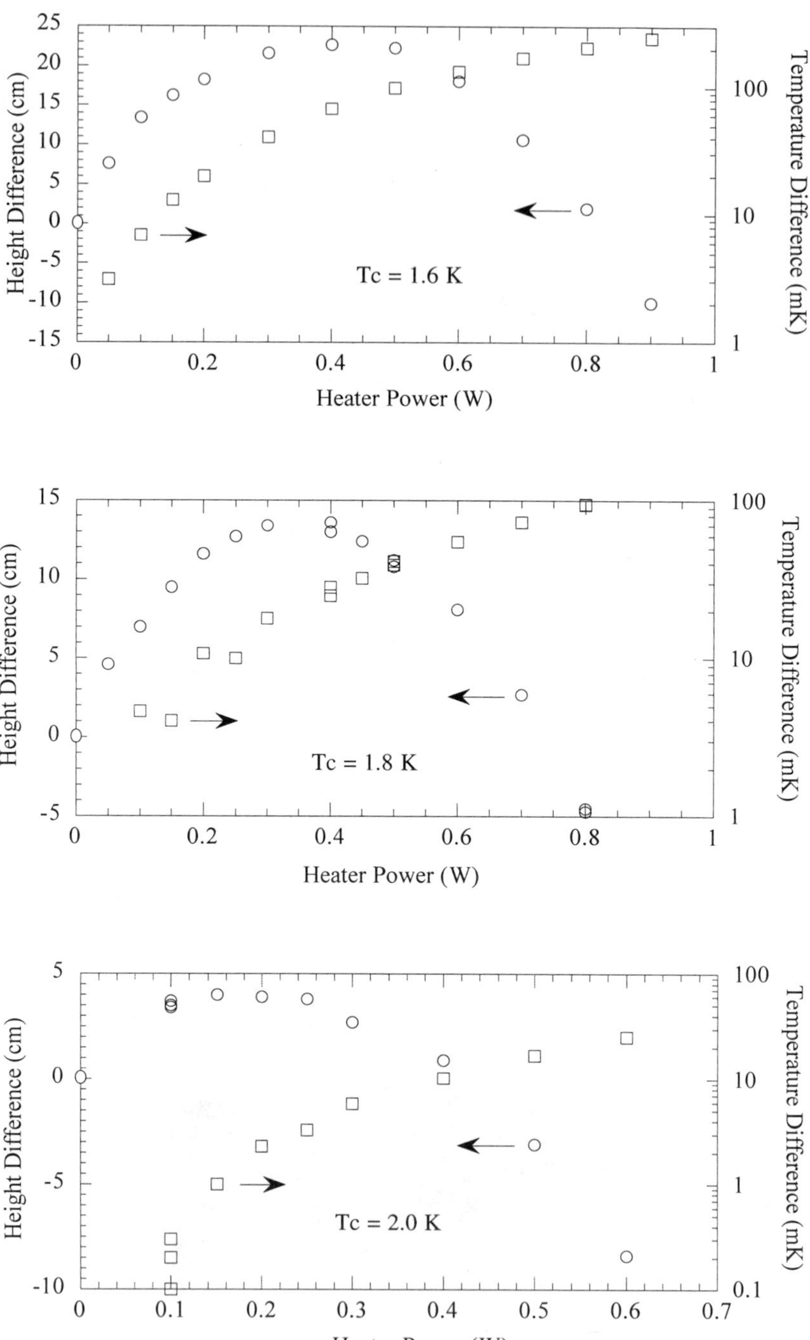


Figures 4-6. Height and Temperature Difference at Tc = 1.6, 1.8, and 2.0 K.

## RESULTS AND DISCUSSION

Data was taken for three fixed cold-side temperatures: 1.6, 1.8, and 2.0 K. Measured height and temperature differences as a function of heat input are plotted in Figures 4-6. The thermomechanical effect is at first larger than the saturated pressure effect and $\Delta h$ increases with Q, but at larger heat inputs the saturated pressure effect dominates and $\Delta h$ decreases.

A numerical model was constructed with the main assumptions of pure internal convection in the porous plug, thermodynamic equilibrium between the liquid and vapor in the warm bath, and rough pipe turbulent flow in the vapor tube. Inputs to the model are heater power, $Q_{htr}$, and cold side liquid temperature, $T_{l,c}$. At first an initial guess is made for warm side liquid temperature, $T_{l,h}$ and internal convection through the porous plug, $Q_c$, is calculated:

$$\nabla T = \frac{\beta \eta_n}{d^2 (\rho s)^2 T} q + \frac{A \rho_n}{\rho_s^3 s^4 T^3} q^3 \tag{2}$$

$$Q_c = A_{pp} q \tag{3}$$

. Evaporation and flow through the vapor tube, $Q_h$, is determined from the saturated pressure difference and friction factor as calculated by the Colebrook equation[5]:

$$2 d_v (P_{sat,h} - P_{sat,c}) = f L_v \rho_v V_v^2 \tag{4}$$

$$-0.5 f^{-0.5} = \log\left((\varepsilon / d_v)/3.7 + 2.51/(f^{0.5}\, Re_v)\right) \tag{5}$$

$$M_h = (\rho A V)_v \tag{6}$$

$$Q_h = H_{fg} M_h \tag{7}$$

$Q_c$ and $Q_h$ are the only paths for heater power removal:

$$Q_{htr} = Q_c + Q_h \tag{8}$$

At this point the calculated and experimental values of $Q_{htr}$ are compared. The guess value of $T_{l,h}$ is refined and iteration occurs until convergence is reached. After this, thermomechanical pressure and liquid level height difference are calculated:

$$\nabla P_{tm} = \frac{\beta \eta_n}{d^2 \rho s T} q + \frac{0.00494 \beta \eta_n}{\rho (sT)^{1.75} d^{1.25}} q^{1.75} \tag{9}$$

$$\rho g \Delta h = \Delta P_{tm} - \Delta P_{sat} \tag{1}$$

A comparison between the model and experimental results is shown in Figures 7-8. Agreement is fairly good at all temperatures, but only with an effective porous plug channel diameter of 26 μm, 13% smaller than the 30 μm measured by optical and scanning electron microscopy.

Several sources of error are ignored by the numerical model. The bulk mass flow rate in the porous plug is not exactly zero, so pure internal convection is not present. The true mass flow rate can be calculated after consideration of the experimental observation that in steady state, $\Delta h$ remains constant over time, i.e. liquid levels fall at the same rate, even though the evaporation rates are not equal. Ignoring the small difference in liquid densities, the He II leaving the warmer bath ($M_h + M_{pp}$) must then be the same as the amount leaving the colder bath ($M_c - M_{pp}$):

$$2 M_{pp} = M_c - M_h \tag{10}$$

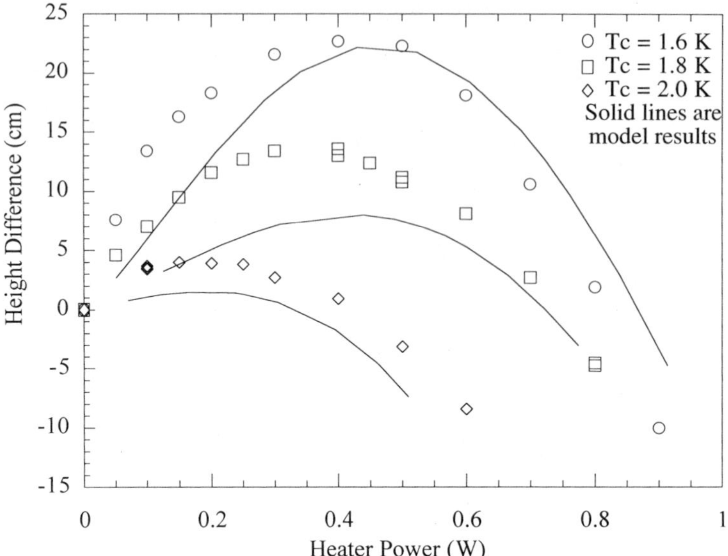


Figure 7. Height Difference vs. Heater Power.

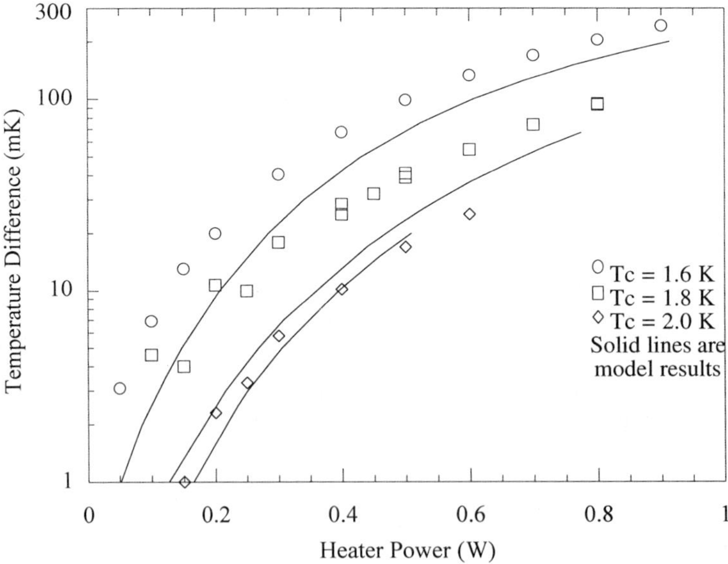


Figure 8. Temperature Difference vs. Heater Power.

The heat flow due to internal convection $Q_{ic}$ is at least 75 times larger than the enthalpy transport by bulk mass flow, $M_{pp}(H_{l,h} - H_{l,c})$, over the entire range of experimental conditions.

A further source of error lies in the simple pipe flow assumption in the vapor tube. Average values are assumed for density and velocity of the vapor, and finite difference modeling would reduce this source of error. However, even large changes to this component cannot account for the error at all experimental conditions.

One last consideration is the possibility that the porous plug itself is conducting heat in parallel with the liquid, since it is made from OFHC copper. But even at the highest heater powers, the effective thermal conductivity of the He II is never less than four times that of the copper, and usually 10-50 times larger. The Kapitza resistance also decreases the magnitude of this source of error .

## CONCLUSIONS

The thermomechanical effect in He II is a feasible way to reduce liquid level differences in two-phase baths at different temperatures. This unique transport property lessens the impact of one of the objections to using a two-phase system. Optimization of the pump channel diameter and length enables reasonable overall temperature rises and liquid level differences for a given heat load.

Simple numerical modeling predicts the overall behavior in this system. More sophisticated codes will undoubtedly lead to more accurate predictions.

A thermomechanical pump adds minimal complexity to a two-phase He II system, but would allow higher heat removal rates due to higher temperature differences without undue risk of drying out the warmest components.

## ACKNOWLEDGMENT

This research is supported in part by the U.S. Department of Energy - Division of High Energy Physics under grant DOE-FG02-96ER40952. The authors wish to thank J. B. Hendricks for supplying the porous plug, and V. Cochran for his technical support.

## NOMENCLATURE

A - area, Gorter-Mellink parameter
c - cold side
d - diameter
f - friction factor
g - gravitational acceleration constant
h - height, hot side
H - enthalpy
htr - heater
ic - internal convection in He II
L - length
M - mass flow rate
n - normal fluid component

P - pressure
pp - porous plug
q - heat flux
Q - heat power
Re - Reynolds number
s - entropy, superfluid component
tm - thermomechanical

$\beta$ - geometrical factor in Poiseville flow
$\eta$ - viscosity
$\varepsilon$ - surface roughness

## REFERENCES

1. G. Horlitz, T. Peterson, and D. Trines, A 2 Kelvin Helium II Distributed Cooling System for the 2 x 250 GeV e+e- Linear Collider TESLA, Cryogenics 1994 Vol 34 ICEC Supplement

2. J. Panek, X. Huang, and S. W. Van Sciver, Heat and Mass Transfer in Two-Phase Helium II, Proc. 1996 ASME Winter Annual Meeting

3. J. B. Hendricks, Alabama Cryogenic Engineering (private communication)

4. S. W. Van Sciver, X. Huang, and J. Panek, Heat and Mass Transfer Processes in Connected Saturated He II Baths, submitted for publication in Cryogenics

5. Streeter V. L. and Wylie E. B., 1958, Fluid Mechanics, McGraw-Hill Book Company, NY pp. 294-295.

# STEADY AND UNSTEADY HEAT TRANSFER FROM A FLAT PLATE IN SUBCOOLED HELIUM II

M. Shiotsu[1], K. Hata[2], Y. Takeuchi[2], and K. Hama[2]

[1]Department of Energy Science and Technology, Kyoto Univ.
Uji, Kyoto 611, Japan

[2]Institute of Advanced Energy, Kyoto Univ.
Uji, Kyoto 611, Japan

## ABSTRACT

Transient heat transfer caused by large stepwise heat inputs to a flat plate was measured in subcooled He II at atmospheric pressure for bulk liquid temperatures ranging from 1.9 to 2.1 K. The flat plate was made of Manganin, one side insulated, 10.3 mm in width, 40 mm in length and 0.1 mm in thickness. Steady-state heat transfer and its critical heat flux were also measured by using quasi-steadily increasing heat inputs for the same experimental conditions. The steady-state critical heat fluxes for the liquid temperatures were well expressed by the authors' correlation based on the Gorter-Mellink equations. The lifetime of quasi-steady heat flux in Kapitza conductance regime, which corresponds to that of a certain point on the extrapolation of steady-state Kapitza conductance curve, was systematically measured: the quasi-steady state rapidly changes to film boiling regime after the depletion of lifetime. Comparisons of the results on the flat plate with those on the horizontal wires with the diameters ranging from 0.08 to 0.7 mm under the same condition already reported by the authors were made to clarify the effect of heater shape.

## INTRODUCTION

Superfluid liquid helium under a subcooled condition at atmospheric pressure has excellent cooling properties and is therefore expected as a coolant for large scale superconducting magnets for superconducting magnetic energy storage systems and experimental fusion facilities. The knowledge of steady-state and transient heat transfer on a solid surface in subcooled He II is necessary for a design and safety evaluation of a He II cooled superconducting magnet.

The steady-state and transient heat transfer in He II have been mainly investigated for a one-dimensional heat flow system, namely, that for a metal heater positioned at one end of a He II filled constant cross section duct or tube whose sides are insulated and the other end is connected to a He II tank. Time delay before the onset of film boiling was observed for transient heat transfer caused by stepwise heat inputs. The results of the steady-state experiments [1–3] and transient experiments [4,5] have been analyzed in terms of two-fluid hydrodynamics and theory of mutual friction. However, two-dimensional or three dimensional heat transfer in He II, which is the case of a local thermal disturbance on a winding of a large scale superconducting magnet, have not been generally understood until now.

On the other hand, steady-state heat transfer on a wire heater was studied experimentally under saturated condition by several workers[6–10]. Effects of liquid head and wire diameter on the critical heat flux were clarified by these studies. Transient heat transfer experiments were performed for a thin wire by Gradt et al.[11] and for a cylinder by Van Sciver and Lee[9]. The existence of a time delay before the onset of film boiling was also confirmed in the two-dimensional transient heat transfer on a cylindrical test heater caused by stepwise heat inputs. The authors investigated systematically the steady-state[12–17] and transient[12–15] heat transfer from single horizontal wires with various diameter in saturated and subcooled He II for a wide range of experimental condition. They[13,14] derived a correlation for the steady-state critical heat flux on a horizontal wire by slightly modifying the solution obtained from the Gorter-Mellink equations. It was confirmed that the correlation can describe well the steady-state critical heat flux data for wire diameters ranging from 0.08 to 1.2 mm [16,17]. The transient heat transfer processes were such that the heat flux increased along the extrapolation of the steady-state Kapitza conductance curve, stopped increasing and resided on the extrapolated curve for a period of time designated as the lifetime, and then departed from the Kapitza conductance regime to film boiling. They derived the correlation of the lifetime for wide ranges of subcoolings and cylinder diameters [18].

The purpose of this study is twofold. First is to obtain the experimental data of steady-state heat transfer, peak heat flux, and the life time of quasi-steady Kapitza conductance state caused by stepwise heat input on a flat plate. Second is to clarify the effect of heater shape by comparing the results on the flat plate with those on the horizontal wires with various diameters already reported by the authors.

## EXPERIMENTAL APPARATUS AND METHOD

Details on the experimental apparatus were already reported[14]. The cryostat used is made of stainless steel and consists of the inner and outer He baths. The inner He bath is 30 cm in diameter and 140 cm in height with liquid helium content of about 60 liters. It is divided by a glass epoxy separator to constitute the double bath system to hold stable subcooled He II. The upper part of the cryostat is a saturated He I section (He I bath with a free surface). The lower part of the cryostat constitutes the He II compartment (He II bath with no free surface) with a volume of about 35 $l$.

A flat plate made of Manganin, one side insulated, 10.3 mm in width, 40 mm in length and 0.1 mm in thickness was used as the test heater. The electrical resistivity versus temperature curve for the alloy has a positive gradient over the entire experimental range from 2 K to several hundred degrees Kelvin. Two fine 30 $\mu m$ diameter platinum wires were spot welded at around 5 mm from each end of the plate. The effective lengths of the test heater between the potential taps on which heat transfer was measured was 29.3 mm. The electrical resistance versus temperature relationship was calibrated in a temperature controlled vessel immersed in liquid helium (control accuracy is 0.05 K for temperature settings below 10 K, 0.1 K from 10 to 50 K, and 0.3 K up to 300 K).

The heating current to the test heater was supplied by a power amplifier. The power amplifier can supply a direct current of up to 300 A at a power level of 6 kW; the minimum increasing time for the current from 0 to the maximum value is around 250 $\mu s$. The input signals of the power amplifier is controlled by an analog computer so that the heat generation rate in the test heater agrees with a desired value.

The average temperature of the test heater was measured by resistance thermometry using a double bridge circuit which included the test heater as a branch of the bridge. The output voltages of the bridge circuit were amplified and passed to a digital memory system. These voltages are simultaneously sampled at the shortest available time interval, 0.5 $\mu s$, and stored in a 64 k word/channel memory. The stored data were transferred to a digital computer. The average temperature of the test heater, $T_a(t)$, was calculated from its electrical resistance with the aid of the previously calibrated temperature resistance relation. The heat generation rate of the test heater, $Q(t)$, was determined from the voltage drops across the potential taps of the test heater and across the standard resistance. The heat flux from the heater surface to the liquid is the difference between the heat generation rate per unit surface area, $Q(t)\delta$, and the rate of change of energy storage in the heater, $\rho_h c_h \delta (dT_a(t)/dt)$, obtained from the faired average temperature

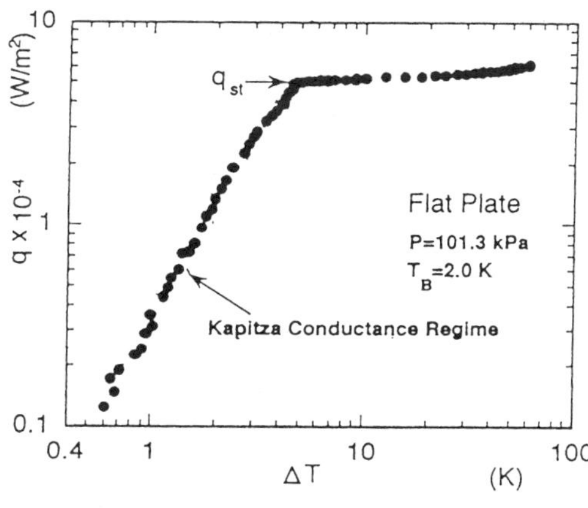


**Fig.1** Typical heat transfer process on the flat plate in a pool of subcooled He II at atmospheric pressure.

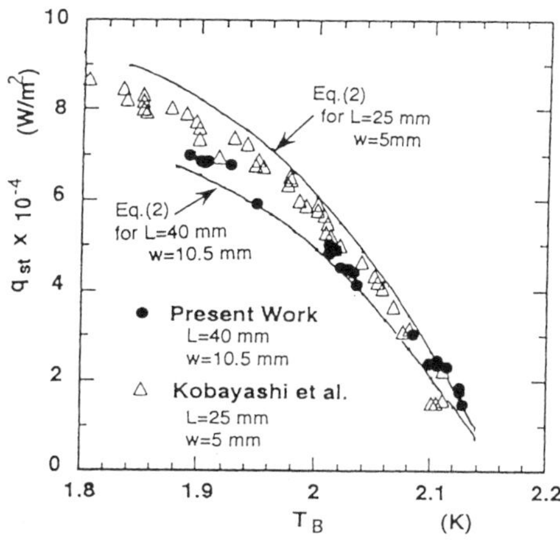


**Fig.2** Relationship between steady-state critical heat flux and bulk liquid temperature. Authors' data and the data of Kobayashi et al. [19] compared with the values predicted by Eq.(2)

versus time curve: $\rho_h$, $c_h$ and $\delta$ are the density, heat capacity and thickness of the test heater. Experimental error is estimated to be 0.2 K in the average temperature and 2 percent in the heat flux. The heater surface temperature was obtained from the measured average temperature and heat generation rate by solving the conduction equation for the test heater with the heat flux set as a boundary condition at the surface. The thermal conductivity of the test heater which becomes lower with the decrease in temperature was approximately taken to be that for the average temperature in calculating the surface temperature.

## RESULTS AND DISCUSSION

### Steady-State Critical Heat Flux

Steady-state heat transfer on the flat plate in He II was measured for the bulk liquid temperatures of 1.9, 2.0, and 2.1 K under atmospheric pressure by the quasi-steady increase of the heater's current.

Figure 1 shows the typical heat transfer process in He II at atmospheric pressure caused by the quasi-steadily increasing heat input. The heat flux gradually increases according to the heat input increase up to the steady-state critical heat flux, $q_{st}$, where the heat transfer state changes from Kapitza conductance regime to film boiling regime with little jump of the heater surface temperature.The steady-state film boiling curve obtained by quasi-steadily increasing the heat input agrees well with that by decreasing it. No appreciable temperature rise or decrease was observed with each phase change from critical heat flux to film boiling regime and from minimum film boiling heat flux to Kapitza conductance regime. These changes of phase with little temperature jump were already reported by the authors[14,15] on a test wire in He II at atmospheric pressure.

Figure 2 shows the data of $q_{st}$ versus bulk liquid temperature. As shown in the figure, the value of $q_{st}$ significantly increases with the decrease of $T_B$ from near $\lambda$ temperature. Experimental data of $q_{st}$ by Kobayashi and Yasukochi[19] on a 5 mm × 25 mm flat plate are also shown in the figure for comparison. Their data are similar in the trend of dependence on $T_B$: they are about 12 % higher than our data at each $T_B$. Kobayashi and Yasukochi[19] measured the $q_{st}$ on flat plates with the width ranging from 0.6 to 9 mm. They reported that $q_{st}$ were inversely dependent on the width. However, no theoretical correlation of $q_{st}$ on a flat plate has been reported until now as far as the authors know.

On the other hand, the authors[13,14] have already presented the following correlation for the $q_{st}$, on a horizontal test wire in subcooled He II (for $T_{sat}(P_L) > T_\lambda$) by slightly modifying the solution obtained from the Gorter-Mellink equations.

$$q_{st} = K\left[\frac{2}{r_0}\int_{T_B}^{T_\lambda} \frac{1}{f(T)} dT\right]^{\frac{1}{3}} \tag{1}$$

where

$$f(T)^{-1} = g(T_\lambda)[T_R^{6.8}(1 - T_R^{6.8})]^3 \;,$$
$$g(T_\lambda) = \rho^2 s_\lambda^4 T_\lambda^3 / A_\lambda \;, \qquad T_R = T/T_\lambda \;,$$
$$s_\lambda = 1559 \quad \mathrm{J/(kg\ K)} \;, \qquad A_\lambda \simeq 1150 \quad \mathrm{m\ s/kg} \;.$$

The modification coefficient $K$, in Eq.(1) was determined to be 0.58 by using the experimental data of $q_{st}$ for various diameter wires. The exponent of 6.8 in the equation of $f(T)^{-1}$ was changed from the original value of 5.7 for one-dimensional heat transfer in He II; the bulk liquid temperature at which the maximum of $f(T)^{-1}$ occurs was shifted from 1.923 K to 1.96 K with this modification. It was confirmed that the correlation can describe well the data of $q_{st}$ for wire diameters ranging from 0.08 to 1.2 mm [16,17].

This correlation was modified as follows to describe the $q_{st}$ for a flat plate of width $w$ and length $L$. Suppose a box of the dimension $\Delta x \times w \times L$ covers the plate, and that the heat flux is uniform on the surface of the box with the total surface area of $Lw + 2(L + w)\Delta x$. The heat flux on the box surface is $[X/(X + \Delta x)]q_0$ where $X = Lw/\{2(L+w)\}$ and $q_0$ is the heat flux on the surface of the flat plate. On the other hand, the heat flux for a cylinder at $\Delta r$ from the cylinder surface is $[r_0/(r_0 + \Delta r)]q_0$. It can be understood by comparing these expressions for a flat plate and a cylinder, the term of $X = Lw/\{2(L+w)\}$ in the former expression corresponds to the term of radius in the latter. By inserting $X$ instead of $r_0$ in Eq. (1), the equation of $q_{st}$ for a plate is given by,

$$q_{st} = K\left[\frac{2}{Lw/\{2(L+w)\}}\int_{T_B}^{T_\lambda} \frac{1}{f(T)} dT\right]^{\frac{1}{3}} \tag{2}$$

The values derived from Eq. (2) are compared with the corresponding experimental data in Fig. 2. As shown in the figure, these data are within 10 % of the predicted values: the data of this work are slightly higher than, and those by Kobayashi et al. are slightly lower than the predicted values. Figure 3 shows comparison of the $q_{st}$ predicted by Eq.(2) with the data of Kobayashi and Yasukochi[19] for 2 cm-long flat plates with widths ranging from 0.7 to 9 mm. The experimental data are within ± 10 % of the predicted values.

## Transient Heat Transfer for Stepwise Heat Inputs

To realize experimentally the ideal step heat input without any rising time is difficult. The stepwise heat input with the maximum available increasing rate (minimum rise time)

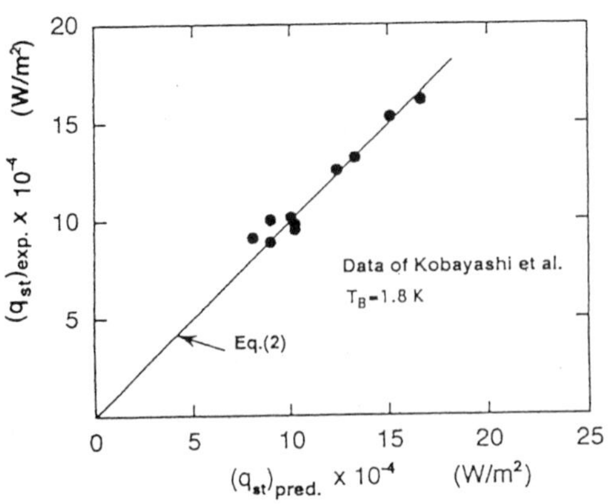


**Fig.3** Comparison of the $q_{st}$ predicted by Eq.(2) with the data of Kobayashi et al.[19] for 2 cm-long flat plates with widths ranging from 0.7 to 9 mm.

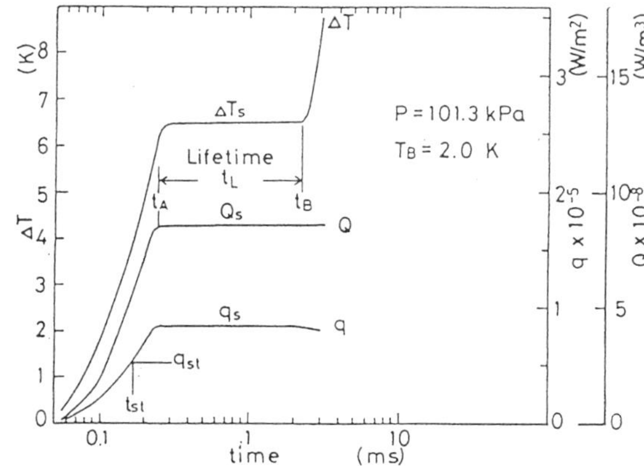


**Fig.4** Time dependencies of the heat flux and surface temperature for a pseudo- stepped heat input.

is called in this work the pseudo-step heat input. The rise time to the step height for the pseudo-step heat input is around 250 $\mu s$.

Transient heat transfer coefficients caused by pseudo-step heat inputs with the heights larger than the values corresponding to $q_{st}$ were measured for $T_B$ of 1.9, 2.0 and 2.1 K at atmospheric pressure.

Typical time traces of the pseudo-step heat input $Q$, heat flux $q$, and surface temperature difference $\Delta T(= T_w - T_B)$ are shown in Fig. 4. Initially, the heat input rapidly increases in time and then it takes a constant value, $Q_s$, after $t = t_A$. As shown in the figure, the surface temperature difference and the heat flux remain constant at $\Delta T_s$ and $q_s$, respectively, for a certain duration $(t_B - t_A)$, then they begin to increase and decrease, respectively, at $t = t_B$. The duration $t_L = t_B - t_A$ is defined as the lifetime of the quasi-steady state heat flux $q_s$. It was first found by the authors that the transient heat flux for the stepwise heat input increases along the Kapitza conductance curve and the curve's extrapolation, and then reaches a quasi-steady-state $(\Delta T_s, q_s)$ on the extrapolated curve.

The lifetimes, $t_L$, of the quasi-steady-state heat fluxes due to the pseudo-step heat inputs for $T_B$ of 1.9, 2.0 and 2.1 K are shown versus $q_s$ in Fig. 5 as a typical example of the experimental data. The lifetime is infinite for a step height corresponding to a heat flux lower than the steady-state critical heat flux which is higher for lower bulk liquid temperature. With increasing the step height and the corresponding quasi-steady heat

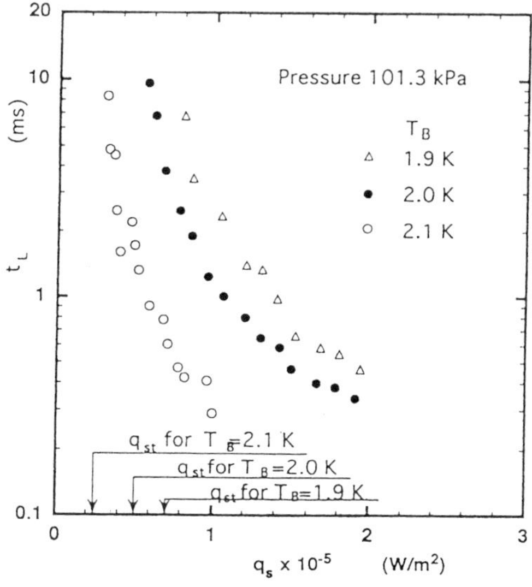


**Fig.5** Lifetime versus quasi-steady heat flux with bulk liquid temperature as a parameter.

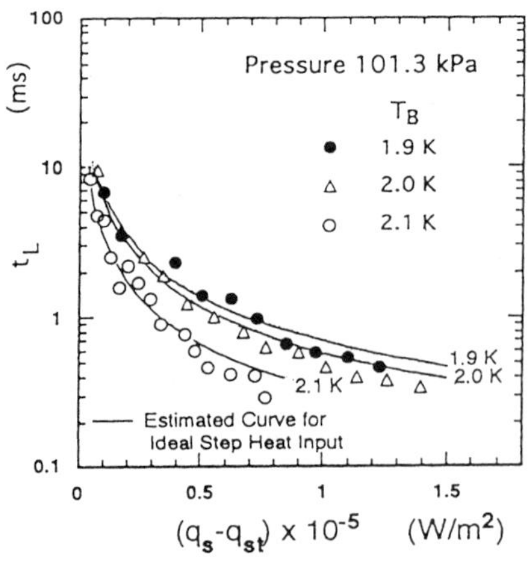


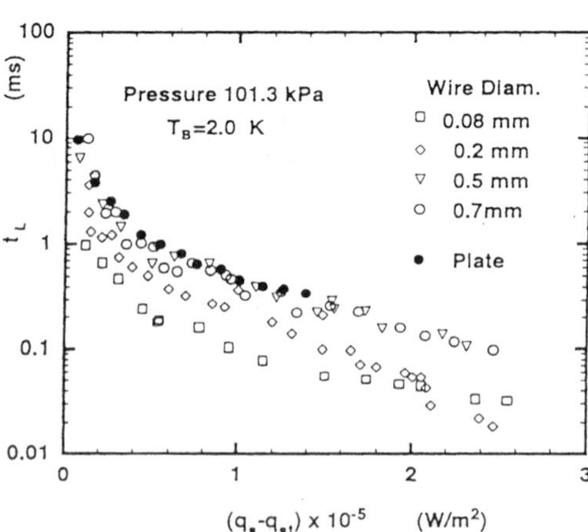


**Fig.6** Relation between the $t_L$ and $(q_s - q_{st})$ with bulk liquid temperature as a parameter.

**Fig.7** Relation between the $t_L$ and $(q_s - q_{st})$ for the flat plate compared with that for wires with various diameters.

flux $q_s$ from the steady-state critical heat flux, the lifetime of the quasi-steady state decreases.

Figure 6 shows $t_L$ versus excess heat flux beyond steady-state critical heat flux, $(q_s - q_{st})$, with $T_B$ as a parameter. As shown in the figure, the decreasing rate of $t_L$ with the increase in $(q_s - q_{st})$ is lower for lower value of $T_B$ within this experimental range.

Figure 7 shows the results of $t_L$ on the flat plate for $T_B = 2$ K compared with those on the horizontal wires with the diameters ranging from 0.08 to 0.7 mm under the same condition already reported by the authors[18]. The lifetime at an excess heat flux is longer for a larger diameter wire, though the steady-state critical heat flux which is expressed by Eq. (1) is lower for a larger diameter. The values of lifetime for the 10.5 mm-width plate used in this study are slightly higher but almost in agreement with those for 0.7 mm-diameter wire for the same value of $(q_s - q_{st})$: the value of $q_{st}$ for the plate is about 55 % lower than that for the wire. The effect of plate dimension on the lifetime will be reported elsewhere in the near future.

The integrated values, $\int_{t_{st}}^{t_B}(q - q_{st})dt$ obtained from each run for the pseudo-step heat inputs to the plate are plotted on the graph of $\int_{t_{st}}^{t_B}(q - q_{st})dt$ versus $(q_s - q_{st})$ in Fig. 8. At the time $t_{st}$, the increasing heat flux reaches the $q_{st}$, and at the time $t_B$, the lifetime for the value of $q_s$ ends as shown in Fig. 4. The integrated values for all the experimental

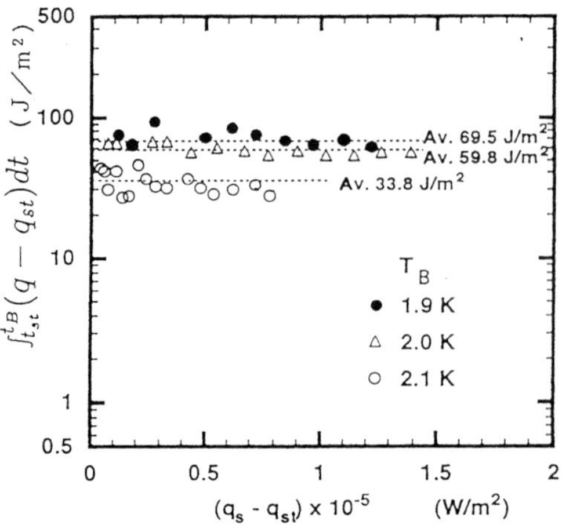


**Fig.8** Average of integrated excess heat fluxes on the flat plate for each liquid temperature.

runs obtained for each liquid temperature are almost constant and independent of the step heights. It was already clarified for the wires with various diameters that this integrated excess heat flux is independent of the initial waveform during the rise time to the step height[18]. It can be assumed that the integrated value of the excess heat flux for the plate would also be independent of the initial waveform. Then, the lifetime of stepwise heat inputs with various initial waveforms can be estimated from the integrated value. For instance, the lifetimes for the ideal step heat inputs (rise time=0) which cannot be realized experimentally can be estimated from the integrated value. They are shown in Fig. 6 in comparison with the corresponding experimental data for pseudo-step heat inputs. The experimental values of $t_L$ for pseudo-step heat inputs become lower than the evaluated values for the ideal step heat inputs for larger values of $(q_s - q_{st})$.

## CONCLUSIONS

Steady-state and transient heat transfer was measured for a flat plate heater, one side insulated, 10.3 mm in width, 40 mm in length and 0.1 mm in thickness, in subcooled He II at atmospheric pressure for bulk liquid temperatures ranging from 1.9 to 2.1 K. Experimental results lead to the following conclusions.

The correlation for steady-state critical heat flux on flat plates was obtained by modifying the authors' theoretical CHF correlation for wires based on the Gorter-Mellink equations. It was confirmed that the correlation can describe not only the authors' data but also the data for flat plates with various width by Kobayashi and Yasukochi[19].

Heat flux for a stepwise heat input increases along the Kapitza conductance curve and its extrapolation, attaining a quasi-steady state on the extrapolated curve. This quasi-steady state last a period of time which is defined as the lifetime for the quasi-steady state Kapitza conductance heat flux.

The lifetime for pseudo-step heat inputs depends on the value of $Q_s$; it becomes shorter with increased values of $Q_s$. The relationship between the quasi-steady state heat flux and its lifetime was quantitatively clarified for pseudo-step heat inputs for bulk liquid temperatures ranging from 1.9 to 2.1 K at atmospheric pressure. The decreasing rate of $t_L$ with the increase in $(q_s - q_{st})$ is lower for lower value of $T_B$ within this experimental range.

Comparisons of the results on the flat plate with those on the horizontal wires with the diameters ranging from 0.08 to 0.7 mm under the same condition already reported by the authors were made to clarify the effect of heater shape. The values of lifetime for the 10.5 mm-width plate used in this study are slightly higher but almost in agreement with those for 0.7 mm diameter wire.

The excess heat flux beyond $q_{st}$ integrated from the time $t_{st}$, when the heat flux reaches the $q_{st}$ to the time $t_B$, when the lifetime for $q_s$ ends, $\int_{t_{st}}^{t_B}(q - q_{st})dt$, is almost constant independent of the step height and of the heat input waveform up to the step height. The lifetime of quasi-steady state heat flux for a stepwise heat input with various initial waveform can be estimated from the integrated value and the correlation of $q_{st}$ presented in this work. The lifetimes for the ideal step heat inputs without any rising time up to the step heights were given.

## ACKNOWLEDGMENT

This research was supported partially by a Grant-in-Aid for Scientific Research from the Ministry of Education, Science and Culture, Japan and by the National Institute for Fusion Science, Japan.

## REFERENCES

1. Bon Mardion, G., Claudet, G., Seyfert, P., Steady State Heat Transport in Superfluid Helium at 1bar, in: "Proceedings of 7th International Cryogenic Engineering Conference", IPC Science and Technology Press, London (1978), p.441.
2. Van Sciver, S.W., Kapitza Conductance of Aluminum and Heat Transport through Subcooled He II, *Cryogenics,* 18:521 (1978).

3. Lottin, J.C., He II Experimental Facilities at Saclay, "Advances in Cryogenic Engineering", Vol.27, Plenum Press, New York, (1982), p.431.
4. Van Sciver, S.W., Transient Heat transport in He II, *Cryogenics,* 19:385 (1979).
5. Sayfert, P., Lafferanderrie, J., and Claudet, G., Time Dependent Heat Transport in Subcooled Superfluid Helium, *Cryogenics,* 22:401 (1982).
6. Frederking, T.H.K., Wärmeübergang bei der Verdampfung der verflüssigten Gase Helium und Stickstoff, *Forschung,* 27:17 (1961).
7. Goodling, J.S. and Irey, R.K., Non-Boiling and Film Boiling Heat Transfer to a Saturated Bath of Liquid Helium, "Advances in Cryogenic Engineering", Vol.14, Plenum Press, New York (1969), p.159.
8. Haben, R.L., Madsen, R.A., Leonard, A.C., and Frederking, T.H.K., Breakdown of Superfluidity for Cylinders in Saturated Liquid Helium II, "A dvances in Cryogenic Engineering", Plenum Press, New York, (1972), p.323.
9. Van Sciver, S.W., and Lee, R.L., Heat Transfer from Circular Cylinders in He II, in: "Cryogenic Process and Equipment in Energy Systems", ASME Publication No. H00164 (1981), p.147.
10. Lemieux, G.P., and Leonard, A.C., Maximum and Minimum Heat Flux in Helium II for a 76.2-$\mu$ Diameter Horizontal Wire at Depths of Immersion up to 70 centimeters, "Advances in Cryogenic Engineering", Vol.13, Plenum Press, New York (1981), p.624.
11. Gradt, T., Wang, R., Ruppert, U., and Luders, K., "Advances in Cryogenic Engineering",Vol. 35A, Plenum Press, New York (1989), p.117.
12. Shiotsu, M., Hata, K., and Sakurai, A., Transient Heat Transfer from a Horizontal wire in Superfluid Helium Caused by Exponential and Step Heat Inputs, in: "Superfluid Helium Heat Transfer", J.P. Kelly and W.J. Schneider ed., ASME Publication No. HTD-Vol.134.(1990), p.9.
13. Sakurai, A., Shiotsu, M., and Hata, K., Transient Heat Transfer for Large stepwise Heat Inputs to a Horizontal Wire in Saturated He II, "Advances in Cryogenic Engineering", Vol.37A, Plenum Press, New York (1992), p.25.
14. Shiotsu, M., Hata, K., and Sakurai, A., Transient Heat Transfer for Large Stepwise Heat Inputs to a Horizontal Wire in Subcooled He II, "Advances in Cryogenic Engineering", Vol.37A, Plenum Press, New York, (1992), p.37.
15. Shiotsu, M., Hata, K., and Sakurai, A., Steady and Unsteady Heat Transfer from a Horizontal Wire in a Pool of Subcooled He II at Pressures from Near $\lambda$-Pressure to Atmospheric, in "Heat Transfer and Superconducting Magnetic Energy Storage",J.P. Kelly and M. J. Supercznski ed., ASME Publication No. HTD-Vol.211 (1992), p.19.
16. Shiotsu, M., Hata, K., and Sakurai, A., Effect of Test Heater Diameter on Critical Heat Flux in He II, "Advances in Cryogenic Engineering", Vol.39, Plenum Press, New York (1994), p.1797.
17. Shiotsu, M., Hata, K., Takeuchi, Y., Hama, K., and Sakurai, A., Critical Heat Flux on Single Horizontal Wires in Superfluid Helium at Pressures, in: "Heat Transfer 1994 (Proc. of the 10th Internat. Heat Transf. Conf.)", G.F. Hewitt ed. Vol.5, Taylor and Francis, USA (1994), p.141.
18. Shiotsu, M., Hata, K. and Sakurai, A., Transient Heat Transfer from a Horizontal Wire in Subcooled He II at Atmospheric Pressure for a Wide Range of Wire Diameter, "Advances in Cryogenic Engineering", Vol.41A, Plenum Press, New York (1996), p.241.
19. Kobayashi, H. and Yasukochi, K., A Sumple Configuration Effect on the Heat Transfer from Metal Surfaces to Pressurized He II, "Proc. of 8th International Cryogenic Engineering Conference", IPC Science and Technology Press, Guildford (1980) p.171.

# MATERIAL AND GEOMETRY EFFECTS OF WIRE SURFACE COATING ON HEAT TRANSFER IN He II BATH

Y.Z. Li,[1] U. Ruppert,[2] I. Arend[2] and K. Lüders[2]

[1] School of Energy and Power Engineering, Xi'an Jiaotong University
Xi'an, 710049, P.R. China
[2] Department of Physics, Free University Berlin
Berlin, 14195, F.R. Germany

## ABSTRACT

Porous coating could be employed to improve heat transfer from a wire to superfluid helium. Peak and recovery heat fluxes four times the values associated with bare wire were achieved for wires cooled by saturated superfluid helium (He IIs) in our previous work. However, different materials and geometries of coatings may have different influence on heat release performance of samples. Some porous materials, different coating styles and geometries are adopted in this paper to find out their effects on heat transfer. Capillary filter, gypsum/alumina mixture and pure gypsum are taken as the alternative coating materials for copper wire (d=40μm) in direct and/or indirect contact with He II. Straight and coiled copper wires with mixture coatings are constructed and measured. The measurement shows that peak heat fluxes in excess of 2.5 times those associated with bare wire are obtained for different coatings. Copper coil with mixture coating is obviously improved by a great increase in its heat transfer performance, which is extremely significant for improving the cryostability of superconducting magnet in practical usage. It is concluded that the micro-structure geometry of porous coating rather than material and macro geometry is the main factor affecting its heat release performance.

## INTRODUCTION

Porous material is often used to enhance boiling heat transfer in normal liquid, which increases the vaporizing nucleus, decreases the potential energy of bubble breaking away from wall, and thus raises boiling heat transfer coefficient. This method is also adopted to enhance boiling heat transfer of superconductor happening in normal helium (He I).[1,2] As a better coolant, superfluid helium is widely used in many cases. However, the flow characteristics and heat transfer performance of superfluid helium are much different from those of normal liquid due to its low viscosity and high heat conductivity. Although no bubble is produced in He II, porous materials, due to their greater porosity resulting in greater heat transfer area, were still applied to a surface to enhance heat transfer.[3,4] Many researchers have paid attention to the flow of superfluid helium through porous materials in recent years. They are focused on the flow characteristics of He II in thermomechanical pump[5,6,7] and the geometrical dependence of flow characteristics.[8,9]

In our previous work,[10] it was found that heat transfer performance could be improved by using micro-porous media in He II, which is not due to its larger heat transfer area built up by higher porosity but due to the thermomechanical effect in He IIs to build up a subcooled He II zone inside porous media. Peak and recovery heat fluxes with a factor of 4 times increase were achieved for wires with porous coatings cooled by He IIs. In this paper, a systematic research on heat transfer through porous coating to He II is completed. Various wire materials and porous coating materials, different methods for coating on wire surface, and diverse sample constructions are conducted in experiments.

## EXPERIMENT AND RESULTS

### Set-up

The pressurized He II (He IIp) cryostat is cooled down by pumping, and controlled by the opening of magnet valve through initial setting of temperature. The schematic drawing of He IIp cryostat is shown in Figure 1, He IIp chamber is cooled by the surrounding He IIs, and has a pressure maintained by gaseous He (GHe) /He I chamber through thermal barrier and an auxiliary capillary tube. The adjusting rod is used to open and close the thermal barrier valve before or after the experiment. The measurement signals of testing sample are all taken from feedthrough. Filling valve is an important facility, through which helium is filled from He IIs bath to He IIp chamber before experiment and released from He IIp chamber if the pressure is over safety limit. During experiment, filling valve could be either closed for He IIp measurement or open for He IIs measurement. By adjusting the temperature of He IIs bath and the pressure of GHe/He I chamber, the cryostat has a variable range of T=1.6 K to $T_\lambda$ and $P=P_{sat}$ to 0.1 MPa.

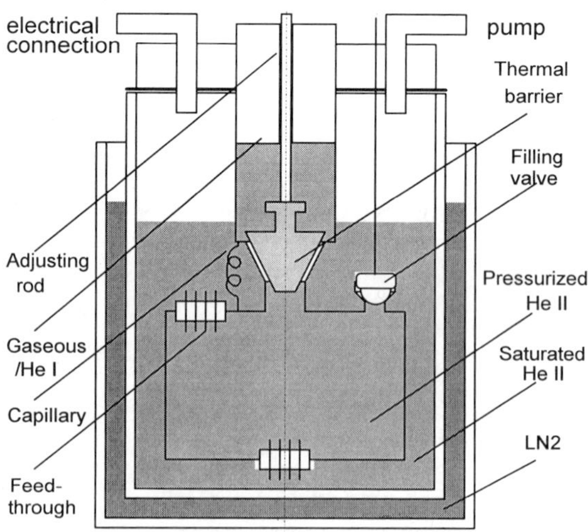


**Figure 1.** Principle of experimental set-up .

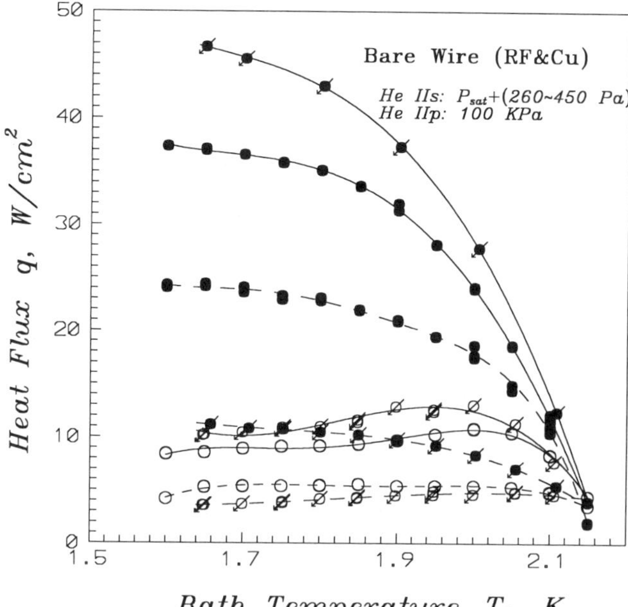


**Figure 2.** Peak (—) and recovery (---) heat flux for RhFe (○ ●) and copper (∅ ●) wires cooled by He IIs (open symbols) and He IIp (solid symbols).

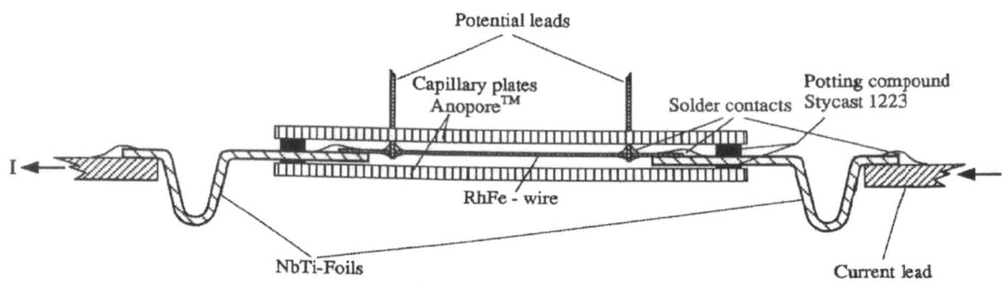


(a) Indirect-contact-coating

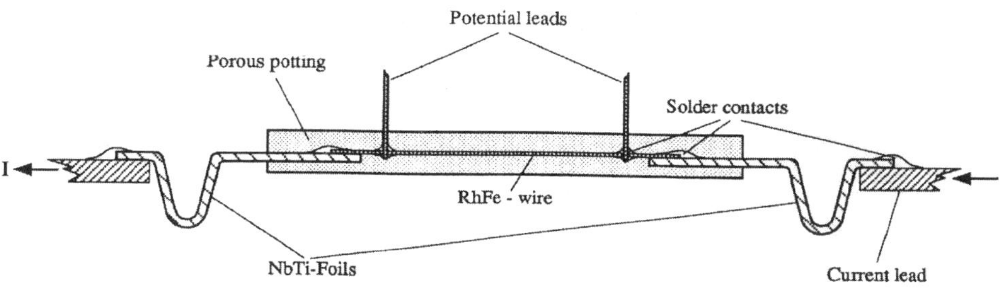


(b) Direct-contact-coating

**Figure.3** Sample construction methods .

**Table 1.** Structure parameters of coating materials

| | Capillary filter | Pure gypsum | Gypsum/Alumina mixture |
|---|---|---|---|
| Pore diameter d (μm) | 0.02 | 0.30 | 0.33 |
| porosity ε | 0.50 | 0.48 | 0.60 |

## Sample

Here two types of sample materials are used in the experiment: rhodium iron alloy wire (∅38μm) and copper wire (∅40μm). The measurements for two wires are conducted at He IIs and He IIp, respectively, and the results of the measurements are plotted in Figure 2. In the figure, peak heat flux indicates the critical surface heat flux of sample, when heat flux exceeding this value film boiling occurs. Recovery heat flux indicates the surface heat flux, at which heat transfer changes back from film boiling and the film boiling ends. It is found from the figure that copper wire has higher peak heat fluxes and lower recovery values no matter in which bath. This case is repeatedly proved, which implies that the difference of wire materials takes effect on heat transfer performance. The reason resulting in the phenomenon is not clear and the more reasonable explanation is under investigation.

The work in this paper is mostly to investigate samples with coatings. Porous coating is applied to the surface of sample and taken as a measure to improve heat release from surface to cooling liquid. There are two ways to coat wire with porous materials, which are classified as indirect- and direct-contact-coating methods. The construction schematic diagram is shown in figure 3. By using the two different coating methods, the influence of structure parameters of porous media and constructing technology on heat transfer could be reflected in experiments, which has a strong significance for considering engineering application in practical cases.

## Coating Materials

One of the purpose of the research is to find out the influence of surface coating materials. Three different materials are selected as coatings. According to the preliminary measurement, the smaller the pore size, the better the heat transfer performance. These materials used in experiments are chosen from nearly all commercially available. Three porous media and their structure characteristics are listed in Table 1.

With an indirect coating technology shown in Figure 2a, copper wires are covered by three porous materials, respectively, to construct samples. The outer cooling perimeter of sample is 20 mm and equivalent outer diameter is about 6.4 mm. The measurements for three samples are completed in He IIs and the results are plotted in Figure 4. From the figure, some important regularities have been found. With the temperature of helium bath decreasing, the peak and recovery heat fluxes of coated samples are increasing monotonically, which has same tendency as those measured in He IIp. In fact, there exists a gap between wire and porous plates inside the sample, which becomes a pressurized He II (He IIp) chamber during heat transfer. The He IIp zone can be built up in He IIs bath due to the thermo-mechanical effect through porous coating. Therefore, the quantities of heat fluxes are much greater than those of bare wires. For the three coatings listed in Table 1, Peak values with a factor of 3 through 4 times increase and recovery values with a factor 3

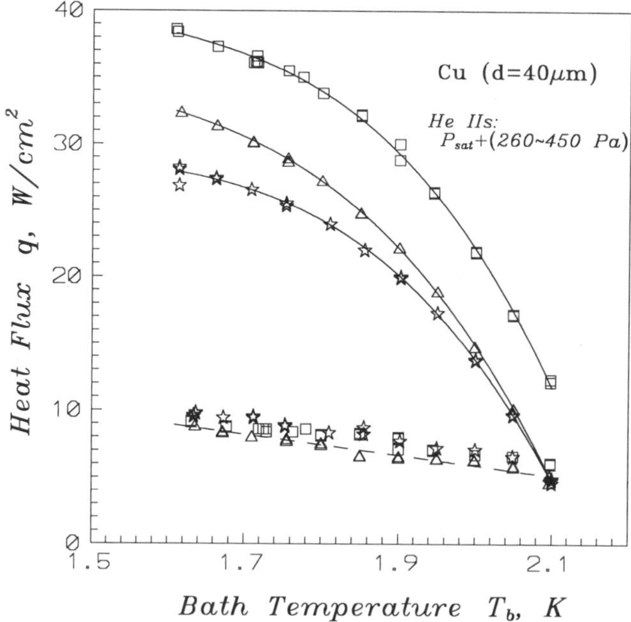


**Figure 4.** Peak (—) and recovery (---) heat flux for copper wires with different coating materials
□ Capillary filter Δ Gypsum Alumina Mixture ☆Pure Gypsum .

times increase are achieved. The results from copper wire are nearly same as those from RhFe wire.[11] The differences in peak heat fluxes for different coatings are also clearly seen in Figure 4. It is considered that the geometrical characteristics of micro-structure of porous medium are more responsible for heat transfer performance than its physical properties, because the effective thermomechanical pressure built in porous medium becomes main driving force for heat transfer.

Among three coatings, the mixture of gypsum and alumina is specially chosen to coat copper wire by two methods as shown in Figure 3. The mixture produced by mixing gypsum and alumina in a certain proportion is first filled into a cylindrical module (inner diameter: 3 mm), in which a copper wire is centrally fixed beforehand. After being dried, the sample is removed from the module. This is defined as the direct contact method. If the wet mixture is first modulated into a sheet, and a testing wire is fixed in between two parallel sheets and sealed along the rim with glue, an indirect contact sample is obtained. Two types of samples are measured respectively in He IIs, and the results are shown in Figure 5. For direct-contact coating method, better recovery values are expressed in experimental temperature range. Smaller peak heat fluxes at lower temperatures and greater heat fluxes at higher temperatures are also indicated, which is probably dependent upon constructing technology, surface situation of wire, and the relation between wire and coatings. The differences in peak and recovery heat fluxes for two samples are about 15% on average.

## Geometry Construction

Besides the consideration of porous materials and construction methods, extra attention must be paid to geometrical characteristics of samples, which here specially implies the difference in shape and dimension of sample due to the change of its

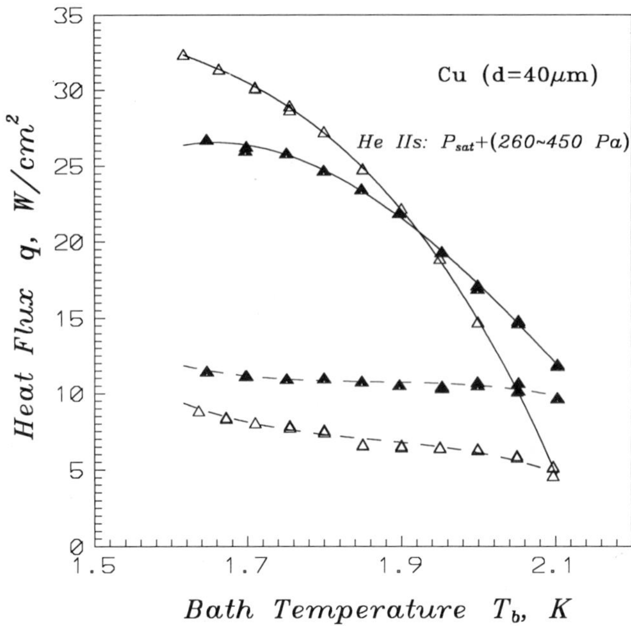


**Figure 5.** Peak (—) and recovery (---) heat flux for copper wires directly (▲) and indirectly (Δ) coated with gypsum/alumina mixture .

construction. In above mentioned experiments, samples are all constructed on the basis of a straight wire and its coatings. In order to find out the influence of sample shapes and the thickness of coating on heat transfer, other kinds of samples are also constructed. Multiple pass of straight line with mixture coating is constructed and measured, where the outer diameter of sample is the same as before (3 mm). No obvious differences in heat transfer performance between multiple-pass and one-pass samples have been found in the measurement. Copper coils with direct-contact coatings, as the case nearer to potential applications, are investigated experimentally. To construct a sample of coil, copper wire is first coiled on a former with a uniform coil pitch and then coated with mixture of gypsum and alumina (see Figure 6). The construction parameters are listed in Table 2.

Figure 7 shows the experimental results of copper coils cooled by He IIs. By comparing Figures 5 and 7, similar values for straight and coiled wires with same coatings are obtained even if the sample shapes are much different. Therefore, the influence of sample shape and its construction on heat transfer is indistinct. For the two copper coils listed in Table 2, about 10% difference in heat flux in the experimental range is found, which is regarded as the result from different cooling condition in inner side of coils. The difference in coating thickness also expresses indistinct effect on heat transfer performance. That is, the outer diameter of sample with coating is not an important factor affecting heat transfer. It is because, in this case, heat transfer between solid and liquid mainly happens inside the porous coating. The same reason results in the fact that some traditional regularities for bare wire are not yet valid for porous materials.

Copper coils coated with mixture have a 2.5 times increase in the peak heat fluxes and nearly two times increase in recovery heat fluxes compared with bare wire under same cooling condition (comparing figure 7 and 2), which are extremely significant for improving the cryostability of superconducting magnet.

**Table 2.** Geometrical properties of coiled wire with coating

| Geometrical parameters | | | Cu coil supported with "Pertinax" former | Cu coil filled with gypsum/alumina mixture |
|---|---|---|---|---|
| wire (*copper*) | diameter *d* | (μm) | 40 | 40 |
| | length *l* | (mm) | 163.5 | 152 |
| | surface area *s* | ($cm^2$) | 0.205 | 0.191 |
| coil (*copper*) | diameter *D* | (mm) | 2.7 | 3.4 |
| | length *L* | (mm) | 37 | 37 |
| | No. of circles *N* | | 18.5 | 13.8 |
| coating (*mixture*) | diameter ∅ | (mm) | 7 | 7 |
| | length *L* | (mm) | 37 | 37 |
| | surface area *A* | ($cm^2$) | 8.1 | 8.1 |

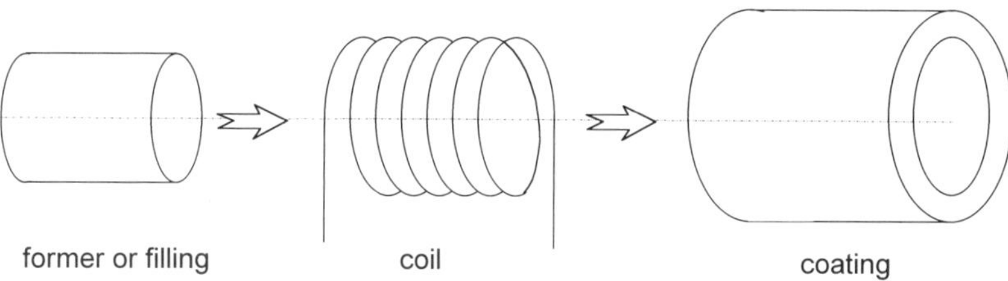


**Figure 6.** Sample formation of coiled wire with coating .

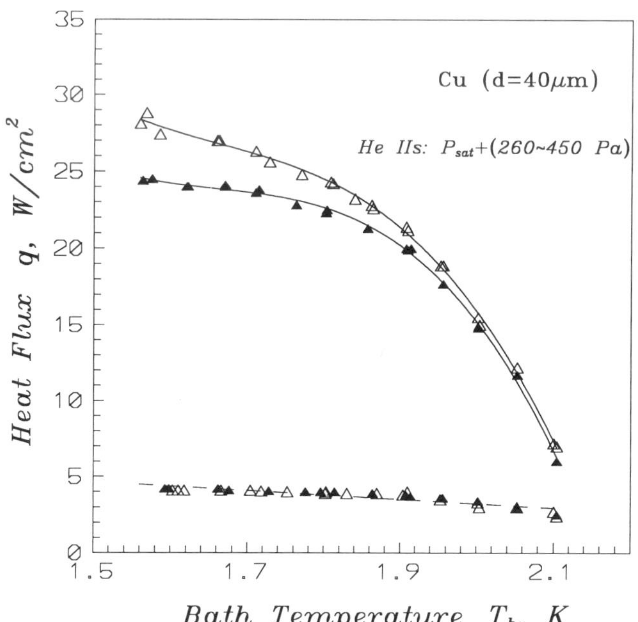


**Figure 7.** Peak (—) and recovery (---) heat fluxes for coated copper coils supported with Pertinax former (▲) and filled with mixture (Δ) .

## CONCLUSIONS

As systematic research on heat transfer from coated wire to He II, various factors affecting heat flux have been investigated. Two kinds of wire materials and three kinds of coating materials are used in the experiment. It is found through the research that the effects from material nature are slight, but the micro-structure geometries of coating materials have a great significance in improving heat transfer. With the aid of two different coating styles on wires, it is easier to identify the effects of various porous coatings and the improvement on heat transfer compared with bare wire. One-pass, multiple-pass, and coiled wire as different constructions of samples reflect indistinct influence. A maximum of ten percent scatter is found for all samples with mixture coatings. The conclusion is that appropriate porous materials greatly improve heat transfer in He II bath, the crucial factor affecting its performance being its micro-structure geometry rather than sample material, construction style, coating method, and macro geometry.

## ACKNOWLEDGMENT

The work is supported by Xi'an Jiaotong University Foundation, P. R. China. The measurements are completed in TTL, Free University Berlin, Germany.

## REFERENCES

1 Y. Kubota, T. Makiura, T. Ogasawara, and K. Yasukochi, Effects of surface coating on the cold-end recovery current of NbTi filamented conductors, *ICEC 9* (paper no. CI2-6) Bobe Japan (1982)

2 V.V. Andrianov, V.P. Baev, S.P. Malyshenko, R.G. Muchnik, and M.B. Parizh, Composed superconductors with porous coating, *Cryogenics* **29** 168 (1989)

3 N.M. Grigorenko, N.M. Volkova, A.V. Kondrashkin and E.V. Zhuravleva, Heat exchange on porous coatings in superfluid helium, *Int. Conf. on Cryo. and Ref.* Hangzhou, China (1989) p.188

4 A. Khalil, Performance of sintered copper heat exchange surfaces in helium II, *ICEC 10* (1984) p.483

5 H. Nakai, N. Kimura, M. Murakami, T. Haruyama and A. Yamamoto, Superfluid helium flow through porous media, *Cryogenics* **36:**9 667 (1996)

6 J.D. Reppy, Superfluid helium in porous media, *J Low Temp Phys* **87:**205 (1992)

7 M. Murakami and T. Hanyu, Thermomechanical flow of He II through a porous element for a fountain effect pump, *Cryogenics* **32:**371(1992)

8 N. Kimura, H. Nakai, T. Haruyama, M. Murakami and A. Yamamoto, Study of porous element for a thermomechanical pump in superfluid helium, *Adv Cryog Eng* **37A:**133 (1991)

9 H. Nakai, N. Kimura, T. Haruyama, M. Murakami and A. Yamamoto, Effect of macroscopic parameters of porous material on flow characteristics for superfluid helium thermomechanical pump, *Adv Cryog Eng* **39B:**1805 (1994)

10 Y.Z. Li, Y.Y. Wu, I. Arend, K. Lüders and U. Ruppert, Influence of porous coatings on heat transfer in superfluid helium, *Cryogenics* **34:**supp. 301 (1994)

11 I. Arend, Y.Z. Li, U. Ruppert and K. Lüders, Heat flux investigation on wires coated with porous insulation in superfluid helium, *Cryogenics* **36:**3 215 (1996)

# VARIOUS FILM BOILING STATES IN HE II AT HYDROSTATIC PRESSURES FROM SATURATED VAPOR PRESSURE TO 1 ATM

M. Murakami, M. Yamaguchi, N. Yanase , and H. Inaba

Institute of Engineering Mechanics, University of Tsukuba
Tsukuba, 305, Japan

## ABSTRACT

The characteristic feature of several film boiling states in He II is experimentally investigated by measuring pressure oscillation in a wide range of hydrostatic pressure from saturated vapor pressure to above the lambda pressure. There appear two kinds of film boiling states in He II under nearly saturated vapor pressure condition depending on the magnitude of the hydrostatic pressure. Above the lambda pressure, however, a film boiling state being different from noisy state appears in which much weaker but higher frequency audible noise is generated. The transition from one boiling state to another is clearly seen from the shift in the predominant frequency of pressure oscillation. It is found that noisy boiling does not occur in subcooled He II.

## INTRODUCTION

It has been experimentally confirmed that there appeared two kinds of film boiling states, noisy and silent boilings, in He II under nearly saturated vapor pressure condition. The occurrence of either noisy or silent film boiling state primarily depends on the magnitude of the hydrostatic pressure[1,2,3]. On the other hand, in many cryogenic cooling applications below 4 K, He II is utilized in a subcooled state in which He I may be generated in the vicinity of a heated surface and thus boiling states may be different from those in saturated He II. The real aspect of film boiling state in subcooled He II is not quite clear, though some investigations on the heat transfer performance have been carried out[4,5,6,7]. It seems that the question whether noisy boiling may appear in subcooled He II has not been positively answered.

In the present study, a number of film boiling states are experimentally investigated by measuring pressure oscillation in He II. Furthermore, the mechanism of noisy film boiling in nearly saturated He II is considered by comparing it with other film boiling states. The transition from one boiling state to another is clearly seen from the shift of the dominant frequency of pressure oscillation given by Fourier analysis applied to the pressure oscillation data. It is found that there is an obvious distinction between boiling states across the lambda pressure. Above the lambda pressure a film boiling state being different from noisy state appears, in which much weaker but higher frequency audible noise is generated. Subcooled He II film boiling may be characterized by the appearance of He I in the vicinity of a heated surface. A gradual transition between noisy and subcooled He II film boiling states takes place in a wide range of the pressure across the lambda pressure. At temperatures only slightly lower than $T_\lambda$ noisy boiling does not appear because of small pressure margin to the lambda pressure, and a direct transition from subcooled to silent boiling states takes place as the hydrostatic pressure decreases. It is, in addition, found that the pressure oscillation is not only affected by the hydrostatic pressure but also by the geometry of an experimental space because of the appearance of induced liquid column oscillation in it. It is understood that noisy boiling does not appear in subcooled He II and thus it is an unstable phenomenon which does not characteristically appear in high hydrostatic pressure situation in He II but is highly related to superfluid phenomena.

## EXPERIMENTAL SETUP AND PROCEDURE

### Experimental Setup

For the measurement of pressure oscillation in nearly saturated He II, a glass dewar was used with an inner diameter of 100 mm. A planar heater and a pressure sensor(PCB 106M122) are set in the test section near the bottom of the dewar as seen in Figure 1. The heater consists of evaporated Ni-Cr thin film (400 A° thick) on a Pyrex substrate with dimensions of 2.5 cm by 2.5 cm (in most case), which is mounted horizontally in the test section. The distance between the pressure sensor and the heater, d, is normally 5 cm. The pressure wave form and the Fourier analysis result are examined to investigate the general feature of film boiling states. Since the pressure sensor, which is sensitive only to AC component of the pressure, could not be calibrated under the liquid helium temperature condition, the output signal from the signal conditioner is analyzed. The depth of immersion h of the heater measured from a He II free surface, the saturated vapor pressure, $p_{sat}$, and the He II temperature are also

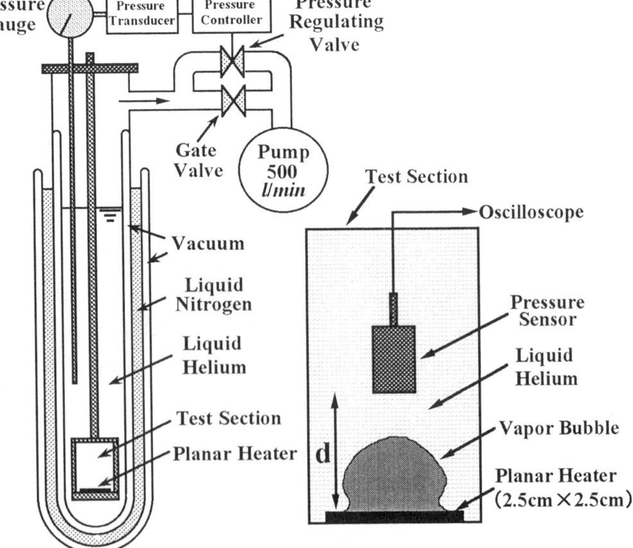


**Figure 1.** Glass dewar, evacuation system and pressure transducer mounted above planar heater.

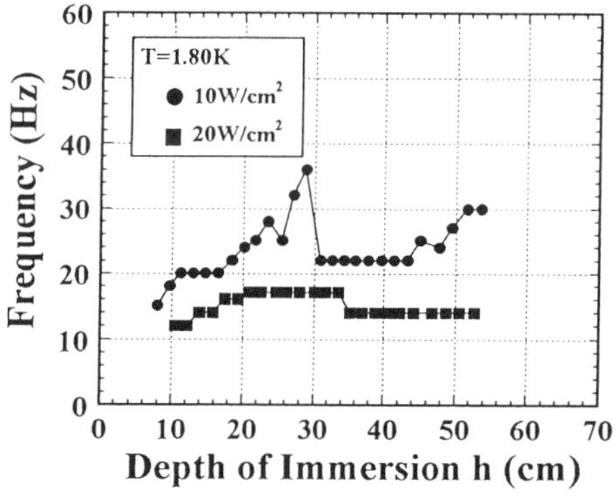


**Figure 2.** Frequency of the fundamental mode of vapor bubble oscillation during noisy film boiling, $f_{bn}$, as a function of immersion depth, h, under nearly saturated vapor pressure condition at T=1.80 K for two cases of the heat fluxes, q=10 and 20 W/cm$^2$.

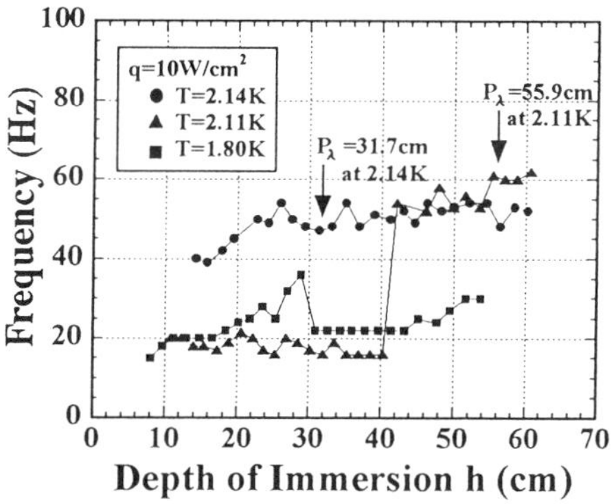

**Figure 3.** Variation of the most dominant frequency of vapor bubble oscillation with h for three temperatures,1.80 K, 2.11 K and 2.14 K. q=10 W/cm$^2$.

measured. The depth of immersion h is not only a measure of the hydrostatic pressure given by $p = p_{sat} + \rho$ gh in the experiments for nearly saturated He II, but also a measure of the characteristic length to control He II liquid column oscillation primarily excited in subcooled He II. Subcooled He II experiment is conducted in a metal cryostat, of which the subcooled He II vessel, that is the experimental space, has an inner diameter of 180 mm, a depth of 350 mm and a volume of about 5 L.

## Experimental Procedure

The pressure measurement experiments are conducted in the glass dewar under nearly saturated vapor pressure condition with a hydrostatic pressure head h up to about 60 cm, and in the metal cryostat under subcooled He II condition with a hydrostatic pressure up to about 0.03 MPa. The helium vapor pressure in the glass dewar is kept constant with the electrically controlled pressure control valve during each experimental run and thus the bath temperature is also kept so. In the case of subcooled He II experiments in the metal cryostat the vapor pressure in the He I vessel is controlled at a constant level. The heater to cause boiling is excited by stepwise electrical current for about one second. The data acquisition with a digital oscilloscope for the pressure oscillation measurement starts at 500 ms after the onset of heating when film boiling state reaches a quasi steady state.

## RESULTS AND DISCUSSION

### Effect of Heat Flux on Pressure Oscillation Frequency in Noisy Boiling

The noisy boiling is characterized by very loud acoustic noise generation induced by periodical unstable bubble formation and crush. Very large pressure spikes which are the cause of loud noise are generated by explosive burn-out-like evaporation of He II at the moment of direct contact of He II with overheated heater surface after the crush of a bubble[8]. Shown in Figure 2 is the frequency of predominant mode of pressure oscillation caused by vapor bubble oscillation during noisy film boiling, $f_{bn}$, as a function of immersion depth, h, under nearly saturated vapor pressure condition at T=1.80 K for two cases of the heat fluxes, q =10 and 20 W/cm$^2$. It is evident that the frequency is higher for smaller q. This is because the smaller vapor bubble generated by smaller q results in the higher frequency. This fact is also verified in another experiment[3] that the smaller heater causes the higher frequency at the same heat flux. It should be noted that around two values of h, about 28 and above 50 cm, the frequency becomes higher, in particular in the case of 10 W/cm$^2$. It may be reasoned that the frequency of bubble oscillation is drawn in that of the resonant liquid column oscillation, $f_c$. At higher q, 20 W/cm$^2$, the feature of noisy vapor bubble oscillation is too definite and $f_{bn}$ is too low to be drawn in $f_c$. It is also found in other experiments that the draw-in of the frequency is actually much exaggerated at smaller q, say 6 W/cm$^2$. The dependency of $f_{bn}$ on h is very weak, though it quite slightly increases with h. At smaller h than 12 cm, a transition to silent film boiling in which noisy and silent boilings appear intermittently makes the frequency lower. In silent boiling state no appreciable pressure oscillation nor acoustic noise are detected.

## Variation of Frequency of Vapor Bubble Oscillation with Temperature

The variation of the most dominant frequency of vapor bubble oscillation with h is shown in Figure 3 for three temperatures. It may be seen that the temperature dependency of the bubble frequency during noisy boiling, $f_{bn}$, is very weak at the temperatures between 1.7 and 2.05 K and h smaller than 60 cm. On the other hand, the situation is quite different for higher hydrostatic pressures than the lambda pressure ($p_\lambda$=4.97 kPa) as seen in the cases of the temperatures at 2.11 and 2.14 K, where the vapor layer oscillation frequency is higher than that of the noisy boiling family, designated as the noisy branch, by a factor of more than two. It is seen in the case of 2.11 K that the frequency is just below 20 Hz for h smaller than 40 cm and it is more than doubled for h larger than 40 cm. The frequency for the case of T=2.14 K is on the subcooled branch irrespectively of h. It can be concluded that the vapor oscillation frequency in the subcooled boiling is on a different branch from the noisy branch, which we refer to as the subcooled boiling branch of which frequency is represented by $f_{bs}$, hereafter. Subcooled film boiling can be distinguished from noisy boiling by the pressure oscillation frequency. The hydrostatic pressure reaches the lambda pressure $p_\lambda$ at h=55.9 cm in the case at T=2.11 K and 31.7 cm at 2.14 K. The transition to

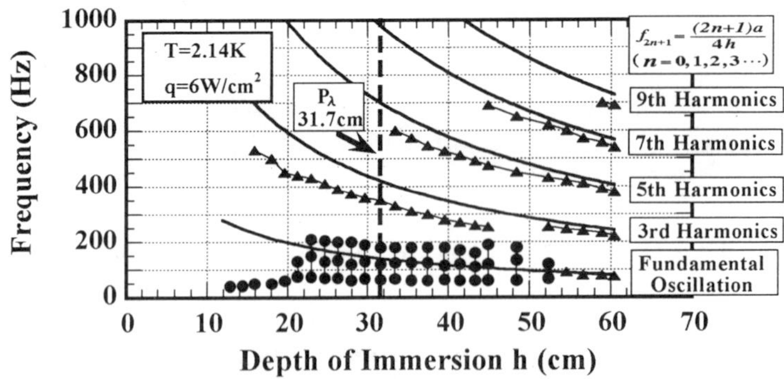


**Figure 4.** Variation of the frequencies of dominant Fourier components of the pressure oscillation with h in the case of T=2.14 K and q=6 W/cm$^2$.

the subcooled branch from the noisy branch starts at a little lower value of h than the above values because the He II temperature adjacent to a heater is higher than that of surrounding He II and thus He I phase appears at smaller h than the above values. It should be noted that at T=2.14 K no noisy boiling is observed and the subcooled boiling mode makes direct transition to the silent mode. It may be considered the difference in the vapor oscillation frequency between noisy and subcooled boilings arises from the existence of He I adjacent to a heater only in the case of subcooled boiling and the two boiling modes are essentially different. Other distinctive feature between the two boilings is also found in the loudness of acoustic noise and thus in the amplitude of the pressure oscillation discussed later.

## Liquid Column Oscillation in Cryostat

A close examination of the Fourier analysis result of the pressure oscillation data reveals their details. In Figure 4 a number of frequencies of dominant Fourier components of the pressure oscillation, including higher order harmonics, are plotted as a function of h in the case of T=2.14 K and q=6 W/cm². For h smaller than 20 cm where the static pressure is lower than $p_\lambda$ only very weak silent boiling mode is detected. For the values of h between 20 and 50 cm, the subcooled mode is dominant and thus modes with frequencies of $f_{bs}$, $2f_{bs}$ and $3f_{bs}$ are detected. The appearance of higher order modes, $2f_{bs}$ and $3f_{bs}$, results from non-linear feature of vapor layer oscillation. In addition, another family of frequencies independent of $f_{bs}$ are also detected, though the contribution is rather small. It is found that they result from liquid column oscillation of He II in the dewar. In this particular region of h, the fundamental frequency, $f_c$, can not be separated from the second harmonics of the vapor layer oscillation, $2f_{bs}$, and instead only third and fifth modes, $3f_c$ and $5f_c$, are detected. At still larger values of h than 54 cm where $p > p_\lambda$, oscillation of $f_{bs}$-family is not detected. The predominant and lowest frequency is found to belong to the liquid column oscillation family, $f_c$, and major Fourier components are found to be $3f_c$ and $5f_c$, and sometimes $7f_c$ and $9f_c$. This experimental fact that the odd number-th order harmonics dominate proves that this family results from liquid c olumn oscillation of He II with bound (bottom) and free (top) boundary conditions. The theoretical prediction for the frequencies of

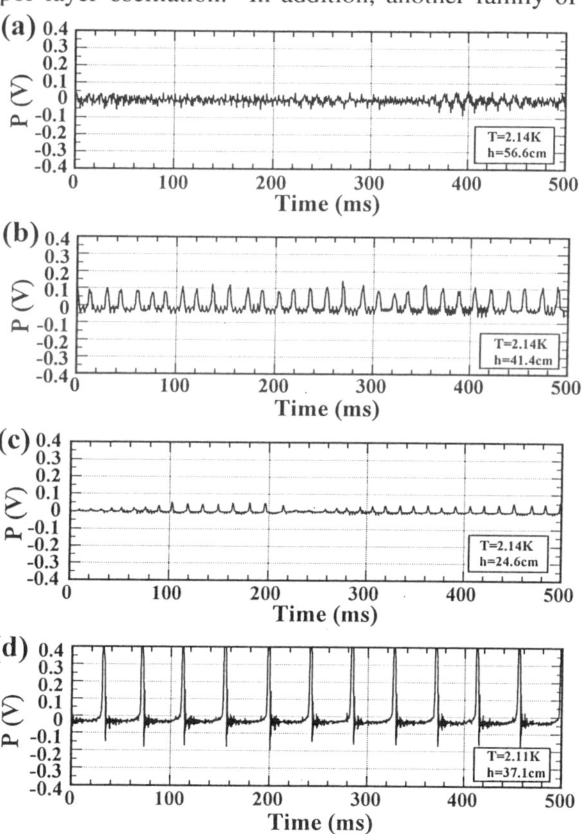


**Figure 5-a, b, c and d.** Wave forms of pressure oscillation in three film boiling states. q=6 W/cm². The temperature; 2.14 K for a, b and c and 2.11 K for d.

(2n+1)th order of harmonics;

$$f_{2n+1} = \frac{(2n+1)a}{4h} \qquad (n = 0, 1, 2, .........)$$

is also given by solid lines in the Figure 4. Here f is the frequency, $a$ is the speed of (first) sound in He II, and $h$ is the height of a liquid column in the dewar, which is approximated by h plus the distance between the heater and the bottom of the dewar, 6.5 cm. The agreement between the experimental result and the prediction is fair. The discrepancy may arise from complex geometry of He II column in the dewar and from the presence of a number of support pipes and heater holder. Actually even number-th order harmonics, $2f_c$ and $4f_c$ are also detected but the corresponding Fourier coefficients are considerably smaller. Under this condition the frequency of vapor layer oscillation, $f_{bs}$ is drawn in $f_c$ for liquid column oscillation through some feed-back mechanism. Audible noise heard in subcooled film boiling state is characterized by considerably higher frequency, several hundreds Hz, and far weaker noise as compared with noisy boiling.

## Wave Forms and Fourier Analysis Results of Pressure Oscillation

Pressure wave forms obtained in three film boiling states are compared in Figures 5-a, b, c and d, where the temperature is 2.14 K for a, b and c and 2.11 K for d, and the heat flux is all 6 W/cm$^2$. The Fourier analysis results of the above data are shown in Figures 6-a, b and c, respectively. The results shown in Figures 5-a and 6-a are for h=56.6 cm. The contribution of the fundamental and many higher order modes of liquid column oscillation makes the wave form rather complex. It is seen from Figure 5-a that the relative strength among four Fourier components, $f_c$, $3f_c$, $5f_c$ and $7f_c$, is seen to vary with time. Shown in Figures 5-b and 6-b are for h=41.4 cm. The wave form is primarily composed of $f_{bs}$ family with a larger amplitudes and of $5f_c$ with a very small amplitude. The wave form and the Fourier analysis result in the transition region between subcooled and silent boilings are shown in Figures 5-c and 6-c. The wave form in noisy boiling given in Figure 5-d is in marked contrast to the other three in far larger peak amplitude and lower frequency.

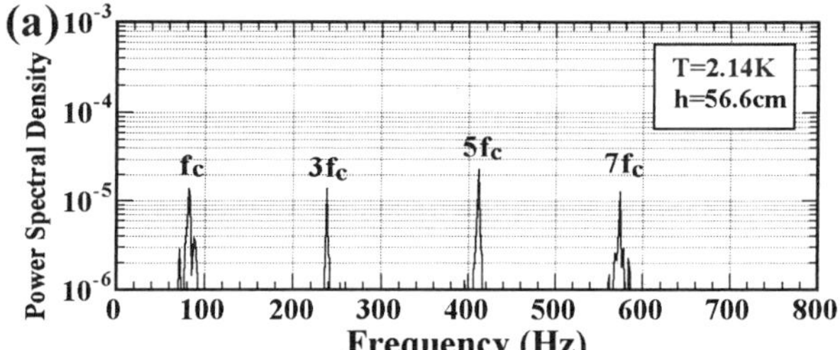


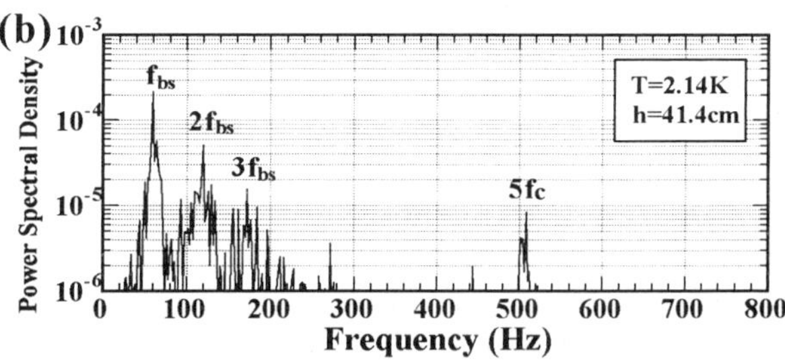


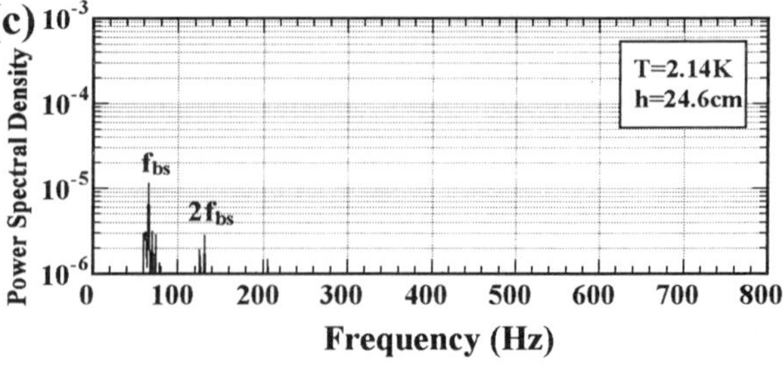


**Figure 6-a, b and c.** Fourier analysis results of the pressure oscillation data shown in Figure 5 a, b and c.

The pressure oscillation data and the Fourier analysis result in the case of subcooled He II, T=2.14 K and p = 0.016 MPa, are shown in Figures 7-a and b. This experiment was conducted in the metal cryostat dedicated to subcooled He II experiments. The vapor layer oscillation mode can be recognized, of which frequency is on the subcooled boiling branch with very small amplitude. This oscillation mode seems to diminish as the hydrostatic pressure increases. The relative contribution of the liquid column oscillation to the subcooled bubble oscillation increases as the hydrostatic pressure increases. It should be noted that the frequency of the fundamental mode of liquid column oscillation observed in the metal cryostat is higher than that observed in the glass dewar and the contribution of even number-th order modes as well as odd number-th order modes are appreciably larger. This is because the He II column in the metal cryostat is bounded by solid boundaries at both top ( $\lambda$ -plate) and bottom ends.

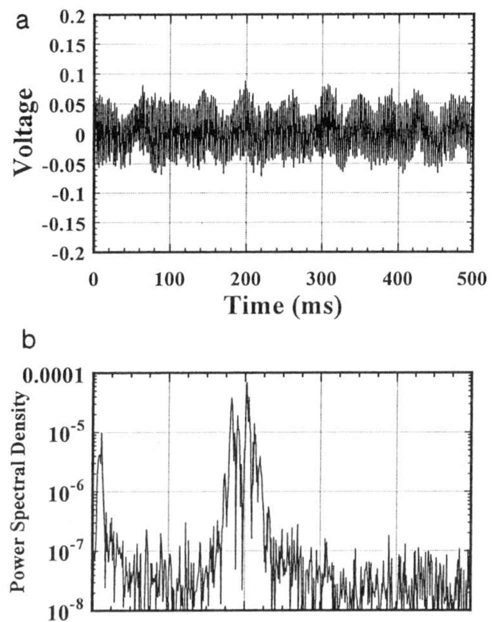


**Figure 7-a and b.** Pressure oscillation data and Fourier analysis result in the case of subcooled He II, T=2.14 K and p=0.016 MPa.

## CONCLUSION

1. In nearly saturated He II, silent or noisy film boiling mode appears. The occurrence of either of the two primarily depends on the hydrostatic pressure. The amplitude of pressure oscillation, and resultant audible noise are extremely large in noisy boiling.
2. In subcooled He II another vapor layer oscillation mode generated in subcooled boiling is excited. The amplitude of oscillation is considerably smaller and the frequency is higher, by a factor of more than two, than noisy boiling.
3. In the case of large liquid column height, in particular under large hydrostatic pressure condition liquid column oscillation is induced. It is characterized by considerably higher frequency, as high as several hundreds Hz, the appearance of many higher order harmonics and the strong column geometry dependency.
4. Noisy boiling mode does not appear in subcooled He II. The excitation of vapor bubble instability in noisy boiling originates from superfluidity.

## ACKNOWLEDGEMENT

This study was carried out as a part of "Space Utilization Frontiers Joint Research Projects" promoted by NASDA, Japan and Japan Space Utilization Promotion Center.

## REFERENCES


1. A. C. Leonard, "Helium II noisy boiling and silent boiling heat transfer coefficient values," Proc. ICEC3 109-114 (1970).
2. S. W. van Sciver, "Helium Cryogenics," Plenum Press, New York (1986).
3. M. Yamaguchi and M. Murakami, "The measurement of vapor bubble vibration during noisy film boiling in superfluid helium," Proc. ICEC16 589-592 (1997).
4. T. Okamura, T. Suzuki, N. Seki and S. Kabashima, "Heat transport in He II channel with phase transition," Cryogenics 34-3, 187-193 (1994).
5. H. Kobayashi and K. Yasukochi, "Maximum and minimum heat flux and temperature fluctuation in film boiling states in superfluid helium" Adv. Cryog. Eng., 25 372-377 (1980).
6. R. Wang, "Peak and recovery heat flux densities in bath of subcooled superfluid helium," Cryogenics 34-12, 983-990 (1994).
7. M. Shiotsu, K. Hata and A. Sakurai, "Transient heat transfer from a horizontal wire in subcooled He II at atmospheric pressure for a wide range of wire diameter," Adv. Cryog. Eng., 41A 241-248 (1996).
8. M. Yamaguchi and M. Murakami, "Study of pressure oscillation during noisy film boiling in He II," to be published in Cryogenics 34 (1997)

# HEAT TRANSFER FROM INSULATED RUTHERFORD TYPE CABLES IMMERSED IN PRESSURIZED He II

N. Kimura [1], V. Kovachev [1,2], T. Nakamoto [1], A. Yamamoto [1], T. Shintomi [1], A. Terashima [1], K. Tanaka [1], and T. Haruyama [1]

[1] Cryogenics Science Center
Applied Research Laboratory
High Energy Accelerator Research Organization, KEK
Tsukuba-shi, Ibaraki-ken, 305, Japan

[2] Superconductivity & Superconducting Materials Laboratory
Institute of Solid State Physics
Bulgarian Academy of Science
1784 Sofia, Bulgaria

## ABSTRACT

The heat load on a high field accelerator superconducting magnet induced by beam losses is a major subject to be solved for stable operation of the Large Hadron Collider (LHC) magnets. The maximum beam loss in an insertion quadrupole magnet has been estimated to be a level of $10^4$ W/m$^3$. While heating the coils to cure the epoxy resin, it will flow into gap between layers of the electrical insulation. Effects of epoxy resin permeating into the electrical insulation and the space (void) between cable strands on cooling have been studied under various conditions. A new measurement is devised to comprehend the heat transfer characteristics by using a cable that has a composite structure identical to the real one. Results of measurements on the heat transfer characteristics for several types of insulation are presented and discussed in this paper.

## INTRODUCTION

A cooperative work with CERN to develop the high gradient superconducting low $\beta$ insertion quadrupole magnet for the Large Hadron Collider (LHC) has been carried out at KEK.[1] The study of heat transfer characteristics due to the beam losses has been also carried out as a part of the R&D of the magnet. From a view point of magnet stability in the accelerator, all the heat induced by the beam losses must be removed through the electrical insulation on the cable surface of the inner coil stack to He II bath.[2] A few investigations have been carried out on the heat transfer characteristics in order to find the best solution.[3,4] Careful consideration is required when one is faced with the heat transfer problems inside coil stacks such as the composite cable structure consisting of the insulation and the cable

*Advances in Cryogenic Engineering,* Vol. 43
Edited by P. Kittel, Plenum Press, New York, 1998

strands. It is considered that the space (void) between the cable strands affects not only the electromagnetic performance but also the heat transfer characteristics. In addition, it is assumed that the epoxy resin on the surface of the insulation has a large depressing effect on the characteristics. The insulation may increase the thermal resistance between the cable strands and He II bath. For these reasons, a new measurement is devised to comprehend the heat transfer characteristics inside the cable. The temperature in the cable having a composite structure was measured directly with small thermometers in this study. He II heat transfer properties such as the second order transition were found also in the composite structure.[5] Experiments have been carried out under several conditions, in He I, in saturated He II and in pressurized He II at a pressure of 0.1 MPa.

## EXPERIMENTS

### Specimens

Since the experiments should reproduce the heat transfer characteristics of the real coil stack of the insertion quadrupole magnet, the cable of the coil stack for this experiment used the same structure of the Rutherford type superconducting cable for the magnet, but it was made of CuNi in order to generate Joule heating in the cable. The size of the cable of, 1.9 mm thick, 11 mm wide and 150 mm long, with a key-stone angle is identical to the real superconducting cable for the insertion quadrupole magnet that has been under development at KEK.[1] Figure 1 shows the preparation procedures including the location of thermometers and the wrapping of the cable with electrical insulation. The cable has two layers of a 50 % over-wrapped polyimide tape (25μm) and another polyimide tape pre-impregnated (50μm) with epoxy resin wrapped with a 3 mm gap. The specimen consists of a short cable stack and a clamp and is prepared as follows:

(1) making two holes for the thermometers on the broadside of a cable by using electrical discharge method, (thermometric cable),

(2) wrapping one drilled cable and three adjacent cables individually with electrical insulation

(3) making the cable stack with the four cables under a standard curing procedure, [1]

(4) installing the cable stack into a clamp that assures the compressive force to the stack during the experiment,

(5) inserting two uncalibrated thermometers into the holes of the thermometric cable and fixing them with Stycast.

(6) soldering voltage taps and current leads to each end of the thermometric cable,

(7) attaching two caps at both ends of the cable stack with Stycast to be free from the end effects due to longitudinal thermal conduction along the cable stack.

The cable stack consists of one thermometric cable, two small thermometers and three adjacent cables as shown in Figure 1. The three cables adjacent to the thermometric cable prevent vertical heat flow into the thermometric cable in the experiments.

It is assumed that the heat transfer from the thermometric cable is affected by parameters such as thickness of the epoxy resin on the insulation and curing conditions. Specimens used in this study have different thickness of epoxy resin and curing conditions. Some of the specimens were attached with tetlon-cloth in the curing to absorb some excessive amount of epoxy resin (No.202AA and 052AA). Two special specimens were made to compare with these specimens: one was made without the standard curing procedure (No.002BC), and the other was made of a CuNi flat plate as a thermometric cable (No.002CC-f.p.). The parameters of the prepared specimens are summarized in Table 1.

**Table 1.** Parameters of the specimens.

| Specimen No. | Epoxy resin thickness $t_{ex}$ ($\times 10^{-6}$ m) | 2nd layer thickness $t_{2nd}$ ($\times 10^{-6}$ m) | Material of insulation | Curing conditions at 50 MPa | Absorber 2) |
|---|---|---|---|---|---|
| 202AB | 20 | 50 | Polyimide-A | 130℃ 2 hours | NO |
| 202AA | 20 | 50 | Polyimide-A | 130℃ 2 hours | YES |
| 052AA | 5 | 50 | Polyimide-A | 130℃ 2 hours | YES |
| 052BC | 5 | 50 | Polyimide-A | no cure | NO |
| 002CC | 0 | 75 | Polyimide-B | 180℃ 2 hours | NO |
| 002CC-f.s. 1) | 0 | 75 | Polyimide-B | 180℃ 2 hours | NO |
| 002BC | 0 | 75 | Polyimide-B | no cure | NO |

1) A CuNi flat plate was used as a thermometric cable.
2) Amount of epoxy resin on the insulation was controlled with this absorber.

## Experimental setup

The cryostat for the experiment was 0.35 m in diameter and 1.7 m in depth. It was equipped with an insulating plate separating the He II bath and the 4.2 K helium bath. A schematic cross section of the cryostat is shown in Figure 2. Three specimens were suspended simultaneously from the insulating plate and were immersed in the He II bath. A heat exchanger with a J-T valve was equipped in the He II bath when the experiment was performed under the pressurized He II condition. A pumping system, which was used to create saturated He II in either the He II bath or the heat exchanger, consisted of a 900 $m^3/h$ Roots pump and a 390 $m^3/h$ rotary pump. The vapor pressure in either the He II bath or the heat exchanger was controlled with a pressure control valve, $V_1$. The valve was connected to an absolute pressure transducer, $P_1$, through an electric feedback loop. The transducer was mounted on the cryostat at room temperature. The temperature of the He II bath was kept constant throughout the series of measurements with the control feedback loop system and could be set between 1.65 K and 2.16 K with an accuracy of a few mK.

Calibrated carbon glass resistance thermometers (CGR-1-500, Lake Shore Cryotronic Inc.), $T_1 \sim T_6$, were attached on the surfaces of three specimens to determine the bath temperature in the cryostat and were employed for the calibration of the thermometers (CX-1030-BC, Lake Shore Cryotronic Inc.) inside the thermometric cable, $T_7 \sim T_{12}$.

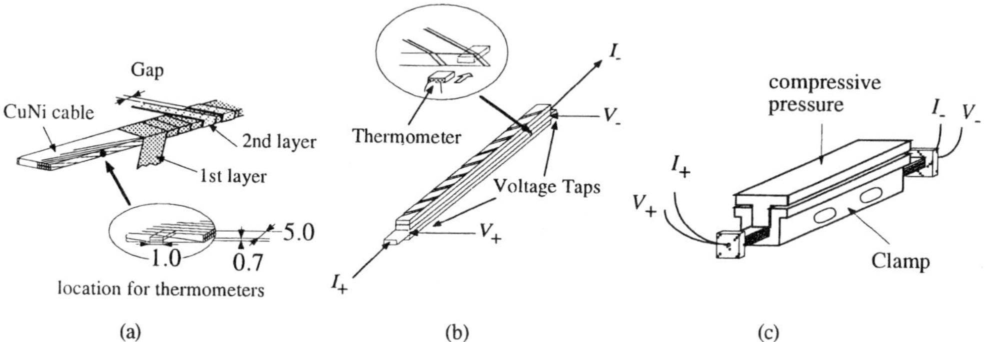


**Figure 1.** The preparation procedures, (a):wrapping a cable with electrical insulation, (b):a view of cable stack and installation of thermometers, (c):a cable stack with clamp.

Experimental data such as output voltage of the thermometers, output voltage across the thermometric cables and output voltage of the shunt resister used to measure the heating current through the cables are measured with three multichannel data recorders. The data were stored in a personal computer via a GP-IB interface from the recorders. The data acquisition system made it possible to accurately measure the temperature differences between the He II bath and the inside of cable stacks within ±5 mK.

## Experimental procedure

The experiment is separated into two procedures: the first is temperature calibration of the six uncalibrated thermometers inside the thermometric cables of the three cable stacks, and the second is measurement of the temperature difference between the bath and the inside of cable stacks as a function of heat input under several conditions.

A set of temperature calibrations of the thermometers in each experiment was made under the following two conditions. Since there should be negligible temperature differences between the bath and the thermometric cable due to high thermal conductivity and uniformity of the He II at temperatures between 1.65 K and the λ-point ($T_\lambda$=2.163K at a pressure of 0.1 MPa), the cable thermometers inside the thermometric cable were compared with secondary (calibrated) the master thermometers. The calibrations at temperatures above the λ-point were made with *in situ* thermometers under natural heating by heat leaks from the ambient atmosphere after the measurement. The specimens were calibrated and tested at different compressive pressures of 50 and 100 MPa by the clamp. The calibration was carried out for every experiment to avoid the effects of the thermal cycle and the compressive pressure.

The temperature difference between the bath and thermometric cable was developed by heating the cable with DC current. The current could be swept from 0 to 12 A with a current source. Ohmic heat which simulated the beam loss in a single layer was generated in the thermometric cable from 0 to 1.2 W. The sweep rate should be less than 0.01 A/s to maintain

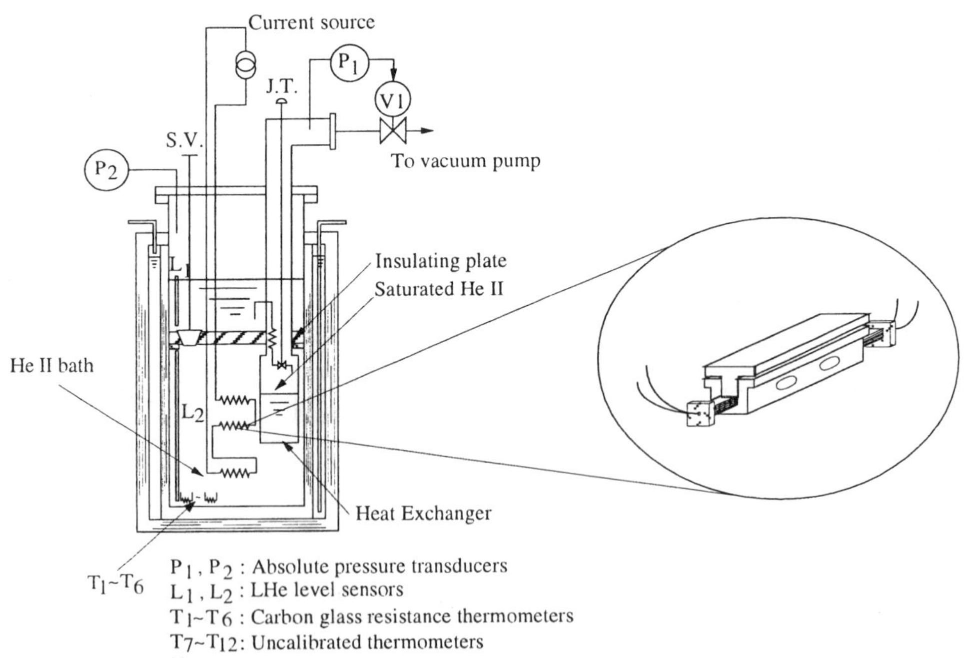


**Figure 2.** A schematic of experimental setup for heat transfer experiment.

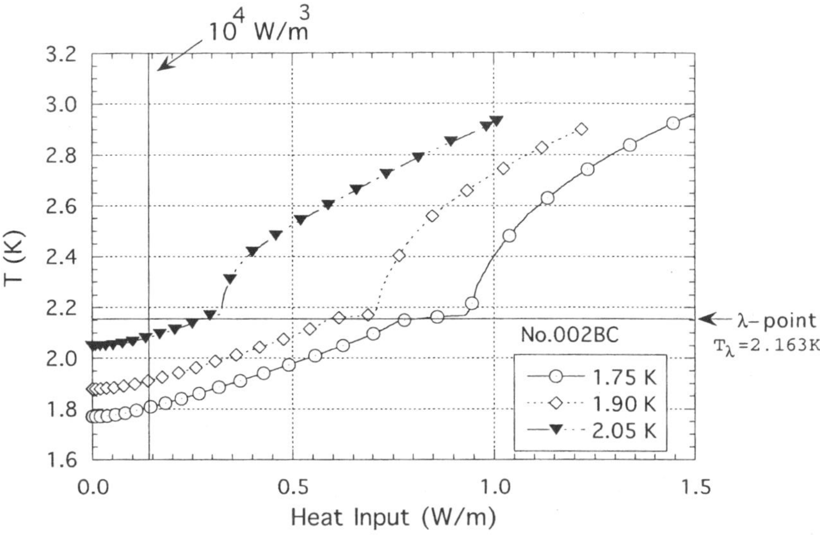


**Figure 3.** Temperature of the thermometric cable versus heat input at various temperatures (No.002BC ).

the steady state condition of the measurement. The measurements of temperature difference were explored in pressurized He II, in saturated He II, and in He I with three specimens shown in Table 1 at one experiment.

## RESULTS AND DISCUSSIONS

The measurements of temperature difference induced by the heat flux have been studied under the three bath conditions, in pressurized He II, in saturated He II and in He I. Reported in this paper are results of measurements at pressurized He II and saturated He II conditions. Specimens were tested at compressive pressures of 50 and 100 MPa. However, there were not any significant difference at the two pressures. The following discussion is made only for the results obtained at a pressure of 100 MPa.

### General heat transfer in the stack without epoxy resin effect

Typical results of the experiment in pressurized He II are shown in Figure 3, where temperature of the thermometric cable is plotted as a function of the heat input at three bath temperatures of the pressurized He II. It is consistently understood that the maximum heat input, $Q_\lambda$, at which the temperature in the thermometric cable reaches the λ-point, depends on the bath temperature. The temperature difference increases according to the heat input, and reaches a *plateau* regime near the λ-point. After reaching the *plateau* regime, the temperature in the cable shows a transition from He II to He I as the heat input increases further. These results are fairly consistent with several experiments which have been carried out within the large channels.[5,6,7] It is found that the second order transition and transient behavior can occur also in microchannels formed inside the composite structure of the cable stacks.

### Effect of permeated epoxy resin

Figure 4 shows the comparison of data sets of three specimens at the same bath temperature, 1.9 K. The results shows strong dependence of the temperature difference on the thickness of epoxy resin on the insulation tape of the second layer. The smallest temperature difference with the heat input among three specimens was obtained from the specimen (No.002CC) wrapped with the polyimide tape without resin. Thermally melted adhesive was used to shape this specimen into the cable stack in order to make a epoxy resin free condition.

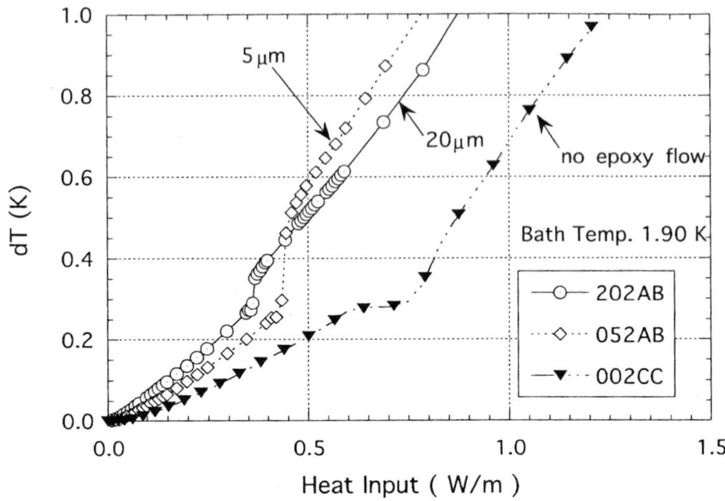


**Figure 4.** Temperature difference versus heat input of various specimens at 1.9 K.

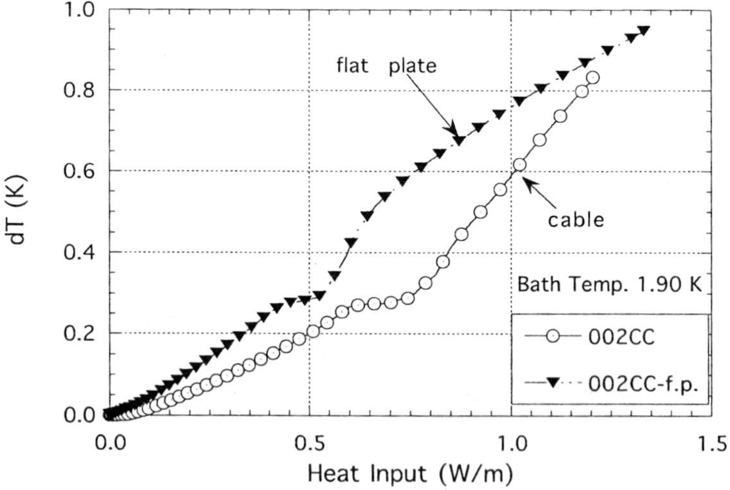


**Figure 5.** Comparison with the flat face plate specimen (No.002CC-f.s.) and the cable specimen ( No.002CC ) at 1.9 K.

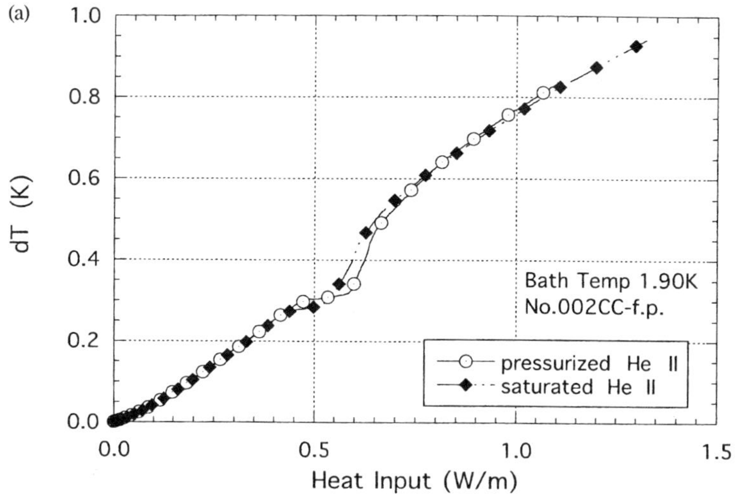


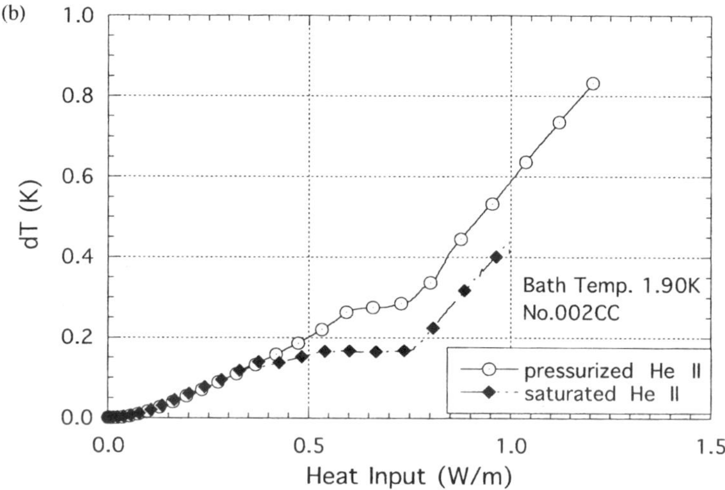


**Figure 6.** Comparison of two specimens in pressurized He II and saturated He II at 1.9 K. (a): the flat face plate specimen (No.002CC-f.s.), (b):the cable specimen (No.002CC).

From these results, it is considered that the epoxy resin flows into the insulation during the cure, seals up microchannels which are formed between the first and second layers, and reduce the effective heat transfer from the cable to the bath. In order to verify above considerations, an uncured specimen with 20μm epoxy tape was measured under the same bath conditions (see Fig.4). The cooling performance of flat specimen ( not shown on Fig.4) was fairly close to the one of specimen (No.002CC) with adhesive material and no epoxy.

**Effect of space among strands**

It is easily understood that the space among the cable strands increases the cooling performance and may enhance the heat transfer behavior in the cable. In order to confirm this, a measurement was carried out with the special specimen (No.002CC-f.p). Figure 5 shows experimental data sets of the two specimens to compare the effect of the space at the same bath temperature, 1.9 K.

The results of the experiments with each specimen in pressurized He II and in saturated He II are shown in Figure 6 (a) and (b), respectively. While the specimen with the flat plate has no significant temperature difference under both conditions, the cable specimen having composite structure shows much different behavior. This behavior is to be investigated further.

## CONCLUSIONS

The following conclusions are drawn from the study on the heat transfer measurements for the insulated Rutherford type cable under He II.

1 A new measurement is devised to comprehend the heat transfer characteristics inside the cable. Temperatures inside the composite structure which consists of the electrical insulation and the cable strands were measured directly with this measurement.

2 He II which exists among strands contributes to the heat transfer from the cable to the bath. The second order transition is also observed in the microchannels such as those formed inside the composite structure of the cable stacks.

3 The heat transfer characteristics are affected by epoxy resin permeation (during cure) which *seals* the cable insulation and blocks the cooling channels. The standard thickness of epoxy resin on the insulation should be minimized in order to keep the better heat transfer characteristics in the He II.

## ACKNOWLEDGMENT

The authors acknowledge Prof. H. Sugawara, Director General of High Energy Accelerator Research Organization, Prof. Y. Kimura, Director of Institute of Materials Structure Science, for their continuous encouragement. The KEK Mechanical Engineering Center and Cryogenics Science Center are also appreciated for their professional support and cooperation.

The authors would like to thank Prof. M. Murakami, Institute of Engineering Mechanics, University of Tsukuba and Dr. H. Nakai of KEK, for their fruitful advice and continuous encouragement.

## REFERENCES

1. A. Yamamoto, K. Tsuchya, N. Higashi, T. Nakamoto, T. Ogitsu, N. Ohuchi, T. Shintomi, A. Terashima, G. Kirby, R. Ostojic and T.M. Taylor, Design Study of a Superconducting Insertion Quadrupole Magnet for the Large Hadron Collider, *IEEE Trans. on. Appl. Superconductivity* 7: 747 (1997)
2. A. Morsch, Local Power Distribution from Particle Losses in the LHC Inner Triplet Magnet Q1, *CERN AT/94-06(DI) LHC note 265*, (1994)
3. C. Merius, Heat Transport in Insulation of Cables Cooled by Superfluid Helium, *Cryogenics* 31: 624 (1991)
4. B.J.P. Baudouy, C. Merius, L. Vieillard and M.X. François, Steady-State Heat Transferring He II Through Porous Superconducting Cable Insulation, *Advances in Cryogenics Engineering* 41: 289 (1996)
5. S. Caspi and T.H.K. Frederking, Triple-Phase Phenomena During Quenches of Superconductors Cooled by Pressurized Super-Fluid Helium II, *Cryogenics* 19: 513 (1979)
6. D. Gentile and M.X. Francois, Heat Transfer Properties in a Vertical Channel Filled with Saturated and Pressurizes Helium II, *Cryogenics* 21:234 (1981)
7. Z. Chen and S.W. Van Sciver, Channel Heat Transfer in He II - Steady State Orientation Dependence, *Cryogenics* 27: 635 (1987)

# LATEST DEVELOPMENTS ON HE II CO-CURRENT TWO-PHASE FLOW STUDIES

B. Rousset[1], A. Gauthier[1], L. Grimaud[1], R. van Weelderen[2]

[1]DRFMC/SBT/CEAG

Grenoble, 38054, France

[2]CERN, European Organization for Nuclear Research, AT/CR, CH-1211 Geneva, 23, Switzerland

## ABSTRACT

Large scale experiments were performed with the support of CERN to simulate and understand the LHC HeII cooling circuit. This paper describes the latest results obtained in HeII co-current two-phase flow configuration. First we summarize thermal and hydraulic behaviour of flows obtained in a 40 mm I.D., 86 m long tube inclined at 1.4% which resembles closely the LHC heat exchanger tube. For low vapour velocities, the flow pattern is found to be stratified. A model based on this observation has been developed and fits very well the measured pressure losses. However the wetted surface predicted by the model underestimates the measured one, notably for high vapour velocities. In this case, liquid droplets entrainment occurs. Droplets falling on the tube wall augment the wetted surface. So we infer that for higher vapour velocities, the stratified two-phase flow model cannot be applied anymore. In order to validate the range of availability of the model, and begin to draw a flow pattern map, a 20 mm I.D. horizontal test loop was built and experiments were performed. First results are presented here, including the observation of the stratified-annular flow transition.

## INTRODUCTION

For large superconductive devices (e.g. LHC[1] or TESLA[2]), HeII two-phase flow has to be employed as the cold source. Unlike the rich literature on water or petrol two-phase flow, very few studies [1-5] have already been done with HeII. As for classical fluids, the first approach consists of drawing a flow pattern map. Then, for practical reasons (stability, pressure losses,...), pipe diameter, mass flow rate and quality have to be chosen to remain if possible in the stratified flow.

In the framework of LHC cooling scheme studies, experiments were performed on a 40 mm I.D. pipe and a HeII two-phase flow model was implemented. Main results are presented here. Furthermore, in order to explore the range of validity of our stratified two-phase flow model, new experiments with a 20 mm I.D. pipe have been carried out and are commented upon here.

## DESCRIPTION OF THE EXPERIMENT

### Test Loop

The two types of tube studied and their arrangement in our test facility are shown in fig. 1a and fig. 1b.

Figure 1a for the 40 mm I.D., 86 m long pipe inclined at 1.4% has been already described in detail[3]. The test loop mainly consists of an inlet box used to create the inlet quality, the 86 m line which can be heated using the Joule effect, and an outlet saturated bath where liquid in excess is burnt.

Important measurements are total mass flow rate (performed at room temperature), saturated temperature at the extremities of the line (used to calculate pressure losses), all powers injected (in order to calculate the quality), temperature increase in the pressurized chamber when power is on (which gives access to the thermal exchange) and visualization of the flow.

### Modifications in case of 20 mm I.D. pipe

As shown in figure 1b, the 86m long line has been removed and the section between boxes B1 and B2 is now occupied by the 20 mm I.D. horizontal test section. This 1.3 m long test section contains a visual test section at each extremity and a thermal test section in the middle.

## RESULTS

### The 40 mm I.D., 86 m long tube inclined at 1.4 %

**Hydraulic results**. The flow pattern map of HeII is plotted on figure 2 using the Taitel and Dukler[6] theory. According to this map the flow should remain stratified for the

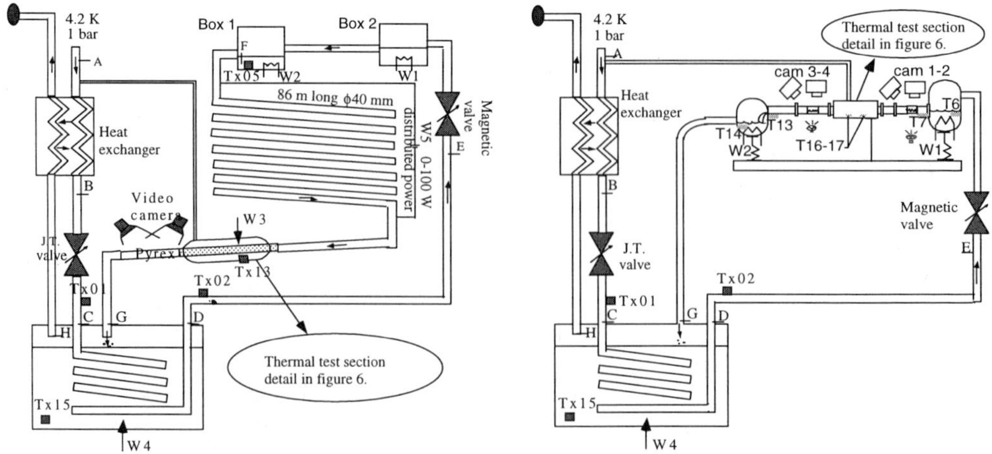


**Figure 1a**. 40mm I.D. test loop. **Figure 1b.** 20mm I.D. test sector.

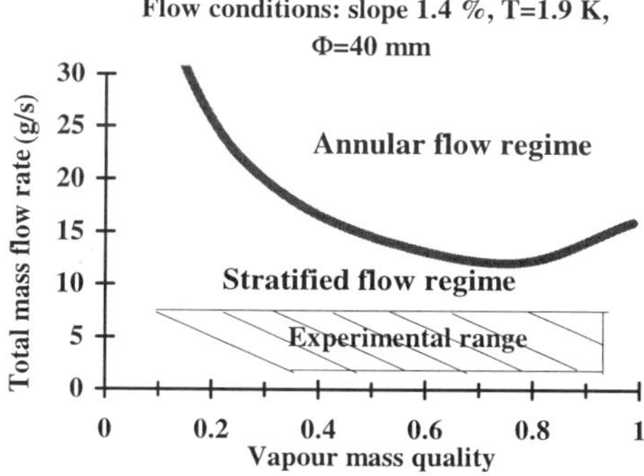


**Figure 2.** He II flow pattern map in adiabatic flow conditions using Taitel and Dukler theory (1976).

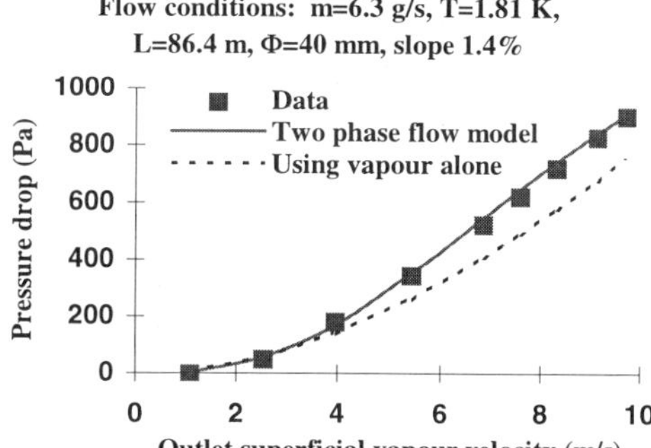


**Figure 3.** Comparison between experimental and calculated pressure losses in quasi-adiabatic flow conditions.

whole range in which we conducted our measurements. This gives us confidence in a model based on separated two-phase flow. A code solving mass, energy and momentum balance for each phase was developed. The convergence was ensured by equating pressure losses in the vapour and in the liquid phase[4]. As the prediction gives a large void fraction (mainly due to the ratio between liquid and vapour density), a simple calculation with vapour at saturation flowing alone in the whole cross section of the tube was also tested. Figure 3 shows comparison between calculated and experimental pressure losses in case of quasi adiabatic two-phase flow.

The vapour flow model has an agreement with experimental points better than 20% (compared to 2% in case of two-phase flow model), which may be sufficient for design purposes. However, if the liquid helium inventory or the wetted perimeter is to be known, the vapour model cannot be used and the two-phase flow model must be used instead. Figure 4 shows the comparison of experimental and calculated void fraction in quasi-adiabatic flow condition. With respect to the range explored, the agreement between model and experiments is excellent.

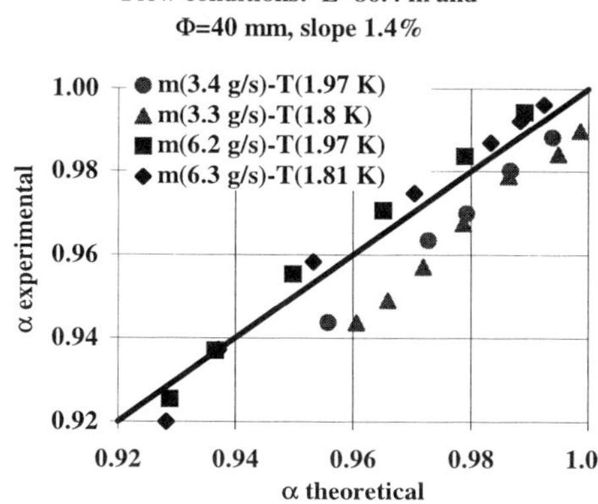


**Figure 4.** Experimental and calculated void fraction in quasi-adiabatic flow conditions.

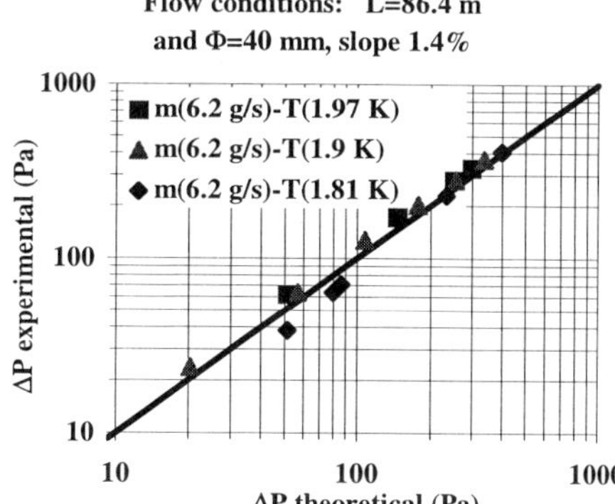


**Figure 5.** Comparison between experimental and calculated pressure losses for a linear power injected along the 86 m line.

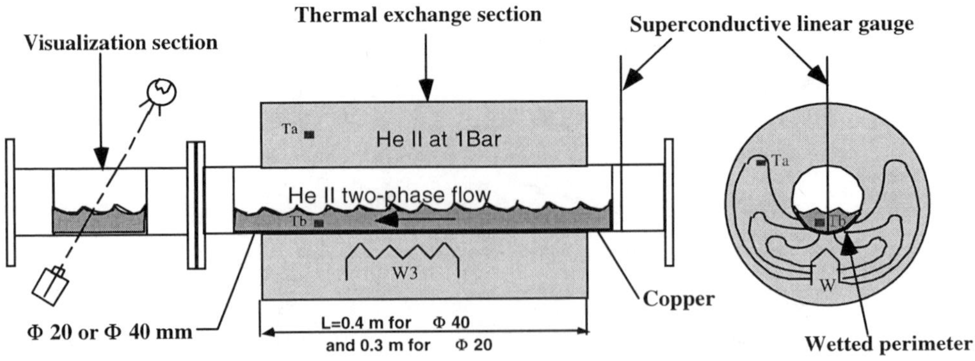


**Figure 6.** Various measurement methods of wetting.

Some experiments have been performed with linear power injected along the 86 m line. Once again, agreement between the two-phase flow model and the experimental pressure losses is very good (figure 5).

**Thermal results.** In the LHC cooling configuration, where the tube containing the He II two-phase flow is used as a heat exchanger, the wetted surface is a major parameter. In order to investigate the wetted perimeter, different measurement methods were used (figure 6).

For low vapour velocities, where interaction between liquid and vapour is weak, a simple model of open channel flow was applied and compared with the two-phase flow model and with the various experimental results (figure 7). The open channel flow model and the two-phase flow model provide about the same level of agreement with measurements. This however with the obvious exclusion of the horizontal flow configuration where the open channel flow model cannot be applied.

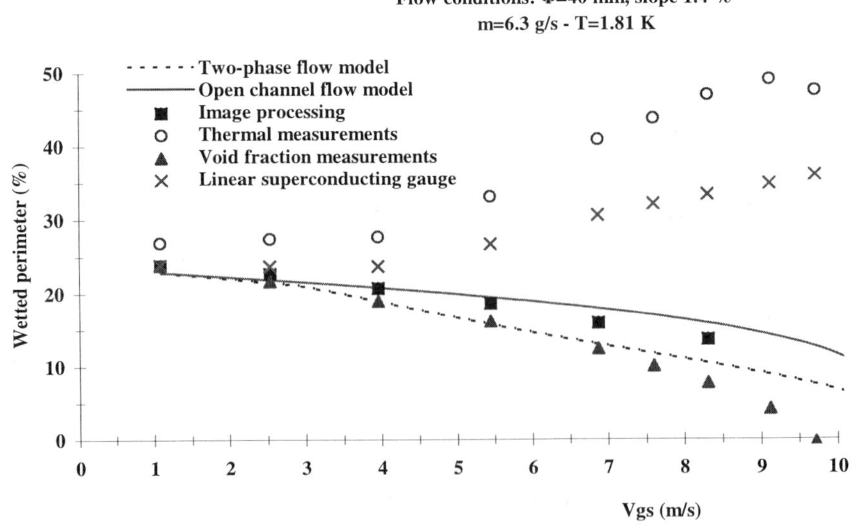


**Figure 7.** Comparison between different ways to estimate the wetted surface for high vapour velocities.

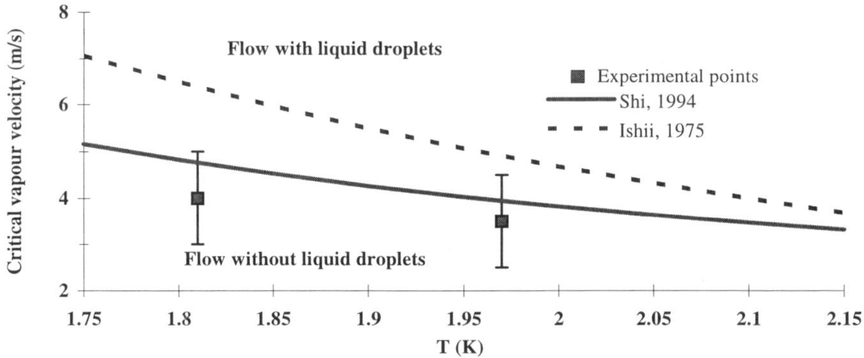


**Figure 8.** Inception criteria for liquid droplet entrainment.

For vapour velocities higher than 4 m/s, atomization occurs and the re-deposition of liquid droplet increases the wetted perimeter. The flow consists of liquid flowing at the bottom (which is always seen by the camera) surrounded by a mixture of vapour and liquid droplets above it. The density of liquid droplets increases as the vapour velocity grows. Hence, the assumption of separated two-phase flow with horizontal interface becomes invalid.

In order to check the range of validity of the separated two-phase flow model, it is important to determine the inception of liquid droplets entrainment. Some correlations which give a critical vapour velocity can be found in the literature[(7-8)] for water two-phase flow. Applying these inception criteria to HeII provides a means to predict the apparition of liquid droplets. For comparison, in the experiments, the critical vapour velocity was chosen as vapour velocity corresponding to the minimum of wetted perimeter given by the thermal measurement. Figure 8 shows the agreement between experimental and calculated (according to Shi[8]) critical vapour velocity.

This criterion is consistent with a non constant thermal exchange with power injected in the pressurized chamber (figure 9). For vapour velocities higher than the critical one, the heat exchange coefficient (slope of the curve W3( $\Delta$T)) depends on power applied at the exchange surface, as the heat load can burn liquid droplets, leaving the corresponding exchange surface dry.

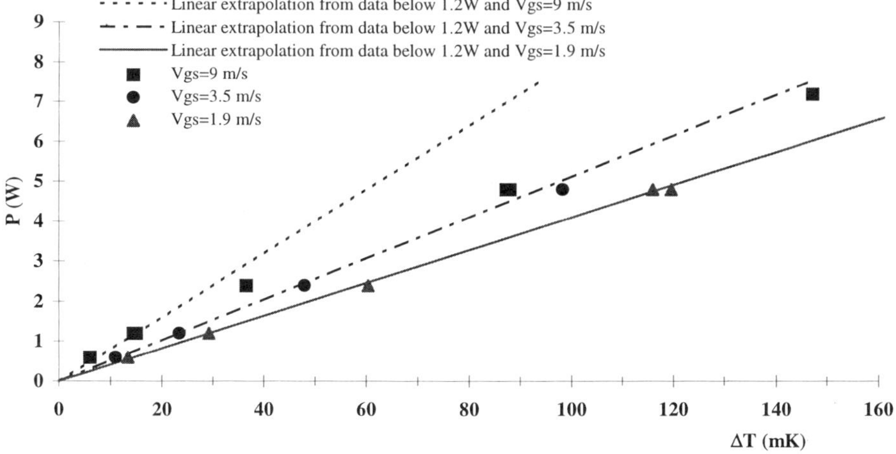


**Figure 9.** Thermal behaviour for different superficial gas velocities.

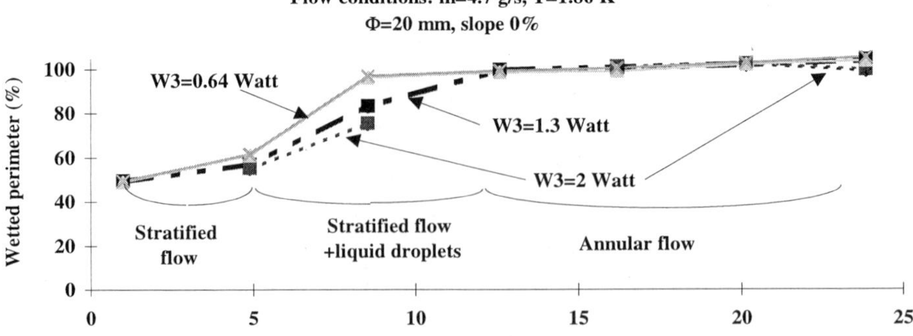


**Figure 10.** Wetting for different power injected in the thermal test section.

## The horizontal 20 mm I.D. tube

**The annular transition**. One goal of this new test circuit was the investigation of the stratified to annular flow transition boundary. In order to reach this annular flow regime, calculations using the Taitel and Dukler theory were performed and a 20 mm I.D. tube was chosen according to the mass flow rate range achievable in our test facility.

The transition from stratified to annular flow was not very clear to define as an intermediate transition to atomization was found. Consequently, the wetted surface did not exhibit a sharp step at the annular transition but grew progressively after atomization until the whole surface was wetted. Then, when the full surface seemed wetted, the discrimination between a liquid droplet re-deposition on the wall and an annular film had to be made. To solve this problem, increasing power levels were injected in the pressurized He II chamber (figure 10), and the results are analysed hereafter.

When the flow is stratified without liquid droplets, the wetted perimeter is independent of the power crossing the heat exchange surface. In the atomization regime, power crossing the wall of the pressurized chamber is able to partly burn the liquid droplets re-deposition, and the wetted perimeter decreases as the power increases. Once the annular regime is reached, the whole perimeter is wetted, whatever the injected power level may be.

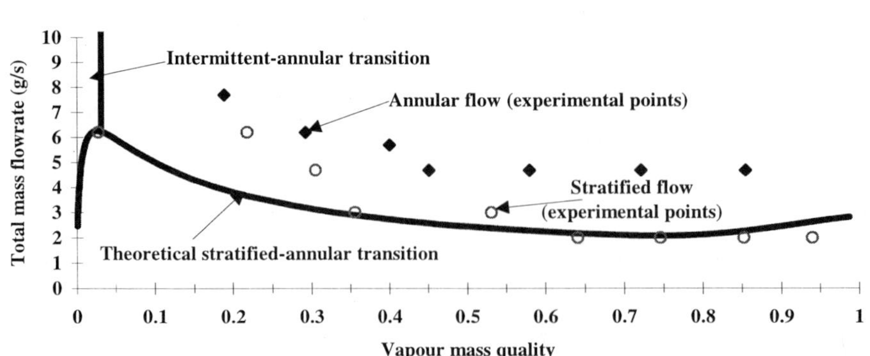


**Figure 11.** He II theoretical flow pattern map (using Taitel and Dukler 1976) and experimental data.

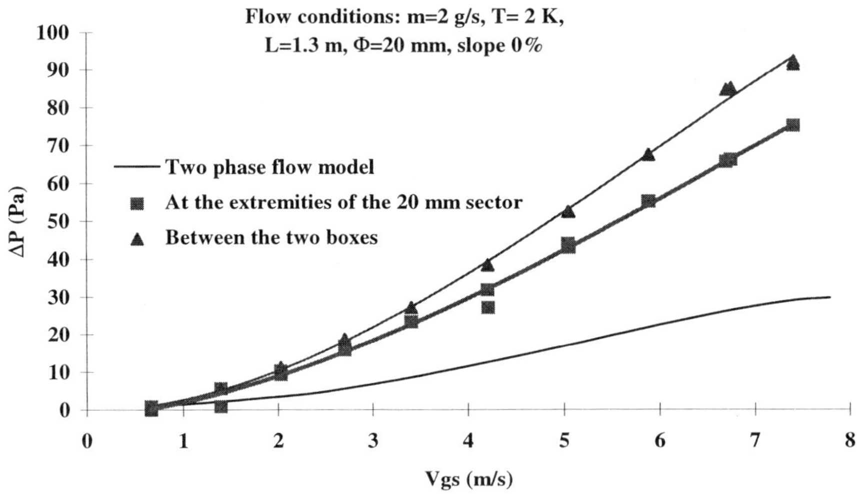


**Figure 12.** Comparison between experimental and computed pressure losses.

Using the classification previously explained, the experimental data have been classified. The last point which belongs to the atomization regime and the first point of the annular regime are placed on figure 11 to check the transition to annular flow.

**Some results in case of stratified flow**. A second goal of this series of experiments was to check the validity of scaling, provided the same flow condition (i.d. stratified flow pattern) is maintained. At 2 g/s and 2 K, the flow remains stratified and the superficial vapour velocity can be varied from 1 to 8 m/s depending on the quality. On figure 12, the experimental and computed pressure losses exhibited large discrepancies. It is important to mention that large amplitude waves were seen on the video camera. This main difference with previous observation on the 40 mm I.D. sector certainly results in an increase of pressure losses. Unfortunately, as there were many changes (slope, diameter and length) since previous experiments, it was almost impossible to find the parameter responsible for this wave flow behaviour. On figure 12, the extremity effect (difference between the experimental curves) plays an important role, which can also indicate that two-phase flow was not uniform along the 1.3 m long sector. Results presented by Spedding and Hand[9] show that for low vapour velocities, flow may remain non uniform over several tens of diameter. To solve this problem and check parameters one by one, we intend to build a new linear 40 mm I.D. test sector of 20 m in lengh and slope ranging between -/+ 1.4 %.

## CONCLUSION

We have investigated the thermohydraulic behaviour of He II two-phase flow. The general flow pattern map proposed by Taitel and Dukler[6] seems applicable in case of He II, however, more recent models[(7-8)] have to be used to predict atomization.

A separated two-phase flow model was developed in order to predict pressure losses and wetted surface. The experimental wetted surface is in agreement with the model until atomization occurs. In this last case, the actual wetted surface is always larger than the predicted one, which therefore can be considered (in term of heat transfer) as a conservative value. Concerning pressure losses, experiments done on a 40 mm I.D. tube inclined at 1.4

% show very good agreement with model, even in case of atomization. Nevertheless, big discrepancies were found using a short, horizontal, 20 mm I.D. tube. Slope, diameter and length influence will be investigated in a new 20 m long linear test facility early in 1998.

## REFERENCES


1. Lebrun P., "Superfluid Helium Cryogenics for the Large Hadron Collider Project at CERN" *Cryogenics 34* (1994)
2. Horlitz G., Peterson T. & Trines D., "A 2 Kelvin Helium II Distributed Cooling System for the 2x250 GeV $e^+ e^-$ Linear Collider TESLA" *Cryogenics 34* (1994)
3. Rousset B., Gauthier A. & Grimaud L., "Stratified Two-phase Superfluid Helium Flow-Part I" *to be published in Cryogenics*
4. Grimaud L., Gauthier A., Rousset B. & Delhaye J.M. "Stratified Two-phase Superfluid Helium Flow-Part II" *to be published in Cryogenics*
5. Filina N.N. & Weisend II, "Cryogenic Two-phase Flow - Applications to Large Scale Systems" *Cambridge university press* (1996).
6. Taitel Y. & Dukler A.E., "A Model for Predicting Flow Regime Transitions in Horizontal and Near Horizontal Gas-liquid Flow", *AIChE Journal* (1976), Vol. 22, 47-55.
7. Ishii M. & Grolmes M.A., "Inception Criteria for Droplet Entrainment in Two-phase Concurrent Film Flow" *AIChE Journal* (1975), Vol. 21 308-318
8. Shi & Kocamustafaogullari "Interfacial Measurements in horizontal Stratified Flow Patterns", *Nuclear Engineering and Design* (1994) 81-96
9 Spedding P. L. & Hand N. P., "Prediction in Stratified Gas-liquid Co-current Flow in Horizontal Pipelines" *Int. J. Heat Mass Transfer* (1997), Vol. 40, 1923-1935

# HEAT TRANSPORT IN SATURATED HE II WITH SURFACE HEAT TRANSFER IN THE TURBULENT COUNTERFLOW REGIME

S. J. Welton,[1] K. Bartholomew,[1] M. Homans[1], and S. W. Van Sciver [1,2]

[1]National High Magnetic Field Laboratory
Tallahassee, Florida, 32310, U. S. A.

[2]Mechanical Engineering Department, FAMU-FSU College of Engineering
Tallahassee, Florida, 32306-2175

## ABSTRACT

Many large cryogenic systems provide refrigeration using pressurized He II that is cooled by a saturated bath heat exchanger. These heat exchangers consist of a closed-end tube or vessel filled with saturated helium which is immersed in the pressurized bath. Significant temperature gradients are possible within the saturated helium due to high heat fluxes through the liquid. To more fully understand the temperature gradient in the saturated bath and its effects on heat exchanger performance, the relevant form of the heat equation was solved analytically and numerically. To verify the results of the mathematical model, an experiment was performed to measure the temperature gradient within the saturated helium bath of a heat exchanger at various heat loads. Good agreement between the numerical results and the experimental data was achieved.

## INTRODUCTION

There are several large superconducting magnets throughout the world which are cooled by pressurized He II. The temperature of the pressurized bath, which cools the magnet, is maintained by a heat exchanger filled with static saturated He II. Due to the heat flux across the wall of this heat exchanger a temperature difference exists between the pressurized and saturated helium baths. This temperature difference is determined by the Kapitza resistance at the liquid/solid interface. It is desirable to design a heat exchanger to minimize this temperature difference, allowing the magnet to operate at a lower temperature for the same refrigeration power.

The performance of the heat exchanger may also be affected by the temperature gradient within the saturated helium. Large heat fluxes within the fluid can cause substantial gradients. If the gradient becomes too large, the local temperature within the helium may exceed the saturation temperature and boiling will occur. The simple case of a vertically oriented tube filled with saturated helium and immersed in a pressurized bath is modeled. The purpose of the mathematical model is to determine the temperature gradient and to predict the maximum heat flux condition above which boiling will commence. An experiment was performed to verify the model by measuring the temperature profile in the saturated helium bath.

In the present paper, the refrigeration circuit is described and key design criteria for the saturated helium heat exchanger are discussed. The governing equations for the model are presented and numerical results are compared with the experimental data.

## JT REFRIGERATION CIRCUIT

The refrigeration system which maintains superconducting magnets at He II temperatures is often referred to as a JT refrigeration circuit. Helium is supplied to the system from a He I reservoir which is boiling at its saturation temperature (approximately 4.2 K). The liquid then undergoes an expansion which cools the liquid as the pressure drops to that of the saturated helium bath. The expanded fluid is two phase and its quality (mass fraction of vapor) can be reduced by use of a counterflow heat exchanger. This heat exchanger cools the liquid entering the JT valve to near $T_\lambda$, by taking advantage of the available enthalpy of the vapor boiling off from the saturated bath. The speed of the vacuum pump controls the pressure of the saturated helium bath, and the JT valve controls both the pressure and liquid level of the same. The total refrigeration power of the system depends upon the capacity of the vacuum pump. Heat is removed from the pressurized helium volume across the wall of the saturated bath. The dimensions of the saturated bath vessel must be carefully designed to optimize its performance.

The relationship between the dimensions of the saturated bath heat exchanger and its performance must be understood for optimal design. The two most important design parameters are the surface area ($A_s$) and the helium cross section ($A_c$).

To minimize the temperature difference across the wall of the heat exchanger, there must be sufficient surface area for the maximum heat load ($Q_{max}$) condition. The minimum wetted surface area requirement can be described as,

$$A_s = \frac{Q_{\max}}{U \Delta T} \tag{1}$$

where $\Delta T$ is the allowable temperature difference between the pressurized bath and the saturated bath ($T_b$ - $T_s$), and $U$ is the overall heat transfer coefficient. $U$ is defined as,

$$U = \frac{1}{h_{k_s}} + \frac{t_w}{k_w} + \frac{1}{h_{k_p}} \tag{2}$$

where the subscripts $s$ and $p$ on the Kapitza conductance terms refer to the inside (saturated helium) and outside (pressurized helium) surfaces of the heat exchanger wall respectively. The wall thickness is $t_w$ and the thermal conductivity of the wall material is $k_w$.

In addition to the surface area requirement, a sufficient cross section of helium within the saturated bath is required particularly if boiling is to be avoided. If the pressure above the liquid surface is $p_0$ then the temperature at the surface ($x = 0$) will be $T_0$. Due to the heat flux within the helium, the temperature will increase with $x$ and a maximum temperature, $T_L$, will exist at the bottom of the heat exchanger ($x = L$). Boiling will occur anywhere that the local temperature exceeds the saturation temperature ($T(x) > T_{sat}$). This means that the temperature gradient at the liquid/vapor interface exceeds the slope of the vapor pressure curve. In other words where,

$$\left.\frac{dT}{dx}\right]_{x=0} \geq \rho g \left.\frac{dT}{dp}\right]_{SVP} \tag{3}$$

where $\rho$ is the liquid density and $g$ is the acceleration due to gravity.

The vapor pressure curve is described by the Clausius-Clapeyron Equation:

$$\frac{dp}{dT} = \frac{\lambda}{Tv_{fg}} \tag{4}$$

where $\lambda$ is the latent heat of the saturated liquid, and $v_{fg}$ is the change in specific volume as the helium goes from gas to liquid ($v_{fg} = v_g - v_f$). Since $v_g >> v_f$ we can assume $v_{fg} \approx v_g$. If, in addition we take advantage of the vapor being nearly ideal, Eq. (4) becomes:

$$\frac{dp}{dT} \approx \frac{\lambda p}{RT^2} \tag{5}$$

where $R$ is the ideal gas constant and $T$ is the temperature of the saturated vapor.

For 1-D heat conduction in He II, the heat transfer rate can be described by:

$$Q = A_c \left[ f^{-1}(T) \frac{dT}{dx} \right]^{1/3} \tag{6}$$

where $f^{1}(T)$ is the He II heat conductivity function. By substituting Eq. (3) and Eq. (5) into Eq. (6), we have an approximation for the maximum heat removal rate of a heat exchanger:

$$Q_{\max} \approx A_c \left[ f^{-1}(T) \rho g \frac{RT_0^2}{\lambda p_0} \right]^{1/3} \tag{7}$$

It is evident from the above expression that the maximum cooling capacity of the heat exchanger is determined by the cross section of helium at any given operating temperature.

## MATHEMATICAL MODEL

For the heat exchanger described above, the heat transport in the saturated He II can be analyzed in terms of the non-linear turbulent counterflow equation with a surface heat transfer condition.[1,2]

$$\frac{d}{dx}\left( f^{-1}(T) \frac{dT}{dx} \right)^{1/3} + \frac{PU}{A_c}\left(T_b - T\right) = 0 \tag{8}$$

where $P$ is the wetted perimeter, $A_c$ is the helium cross section, and $T_b$ is the pressurized bath temperature.

### Analytical Solution

If the heat exchanger is carefully designed, the $\Delta T$ within the saturated liquid will be small, allowing for a constant property analytical solution of Eq. (8). The equation can be non-dimensionalized into the form

$$\frac{d}{dx}\left(\frac{d\theta}{dx}\right)^{1/3} - m^{4/3}\theta = 0 \tag{9}$$

where $\theta = \dfrac{T_b - T}{T_b - T_0}$, and $m^{4/3} = \dfrac{PU(T_b - T_0)^{2/3}}{A_c f^{-1/3}(T)}$. The $m^{4/3}$ term scales the heat transferred at the surface to that transported through the bulk fluid. This equation has an analytic solution of the form,[3]

$$Q = M\left[1 - \theta_L^2 + \alpha\theta_L^4\right]^{1/4} \tag{10}$$

where $M = 2^{1/4}\left[mf^{-1}(T_b - T_0)\right]^{1/3}$, and $\alpha = \dfrac{U^3 A_c}{2Pf^{-1}}(T_b - T_0)^2$. $\alpha$ is a parameter determined by the boundary condition at $x = L$.

**Numerical Solution**

Two of the terms in Eq. (8) are highly temperature dependent: the overall heat transfer coefficient *(U)* and the He II heat conductivity function $(f^{1})$. $U$ is dominated by the Kapitza conductance which is given by,

$$h_k = B_o T^3 \tag{11}$$

where the coefficient $B_o$ depends on the surface condition of the wall, varying from about 400 $W/m^2$-$K^4$ for dirty surfaces to about 900 $W/m^2$-$K^4$ for clean surfaces.[4] In addition to this term, the He II heat conductivity function is also strongly temperature dependent varying by a factor of 3.7 between 1.6 K and 1.8 K. At high heat fluxes, the temperature within the saturated bath might vary considerably, rendering the constant temperature assumption (and the analytical solution) less accurate. In order to solve accurately for the temperature profile in these cases a numerical solution to Eq. (8) must be used.

**Discretized Equations.** The temperature profile in the saturated helium can be calculated by discretizing Eq. (8). A series of finite-difference equations is generated in the following form for the temperature in the saturated bath,

$$T_{s_i}^* = \left(\frac{1}{a_{s0_i}}\right)\left(a_{s2_i} T_{s_{i-1}}^* + a_{s3_i} T_{s_{i+1}} + a_{s4_i} T_{w_i}\right) \tag{12}$$

where:

$$a_{s2_i} = \left(\frac{D^2}{4}\right)\left[\frac{f^{-1}(T_-)}{\Delta x\left(T_{s_{i-1}} - T_{s_i}\right)^2}\right]^{1/3} \qquad a_{s3_i} = \left(\frac{D^2}{4}\right)\left[\frac{f^{-1}(T_+)}{\Delta x\left(T_{s_{i+1}} - T_{s_i}\right)^2}\right]^{1/3}$$

$$a_{s4_i} = Dh_{k_s}\Delta x \qquad a_{s0_i} = a_{s2_i} + a_{s3_i} + a_{s4_i}$$

$$T_- = (1/2)\left(T_{s_{i-1}} + T_{s_i}\right) \qquad T_+ = (1/2)\left(T_{s_{i+1}} + T_{s_i}\right)$$

D is the tube diameter and $\Delta x$ is the spacing between adjacent nodes. The superscript (*) refers to temperatures that are generated in the current iteration step while temperature values with no (*) superscript are generated in the previous iteration step. The temperature subscripts (s) and (w) refer to saturated bath and wall temperatures respectively. The i subscript is the node counter. Close inspection of Eq. (12) reveals that one of the terms is missing $\left(a_{s_{1_i}} = 0\right)$. This is due to center line symmetry. A similar set of equations was also generated to calculate the heat exchanger wall temperatures but due to space constraints they are not shown here. To speed convergence $\Delta x$ is smaller near the liquid surface where $dT/dx$ is greatest.

**Boundary Conditions and Assumptions.** The temperature of the pressurized bath is assumed to be constant for all values of $x$ ($T_b$). This assumption is valid if the cross section of helium in the pressurized bath is large compared to the cross section of the saturated bath. The temperature at the top of the saturated bath is fixed as $T_s(0)$. The $t_w/k_w$ term in Eq. (2) is neglected as it is much smaller than the Kapitza conductance term.

The system of finite-difference equations is solved by using a stepwise iterative solution technique, beginning at the bottom of the heat exchanger and proceeding to the top. Initial values are required for all temperatures to begin the solution process. Results of the numerical model will be presented later.

The computer model calculates the temperature profiles in the saturated helium and in the heat exchanger wall, and generates the saturation temperature line as a function of $x$. Convergence is enhanced by minimizing the total number of nodes without sacrificing accuracy. Increasing the number of nodes from 100 to 1000 increases the convergence time by two orders of magnitude, yet changes the solution by only 0.5%.

## EXPERIMENT

The experimental apparatus (Figure 4) consists of a vertically oriented saturated bath heat exchanger, a pressurized He II reservoir, and a vacuum can all installed in a dewar which serves as the He I reservoir. The saturated bath heat exchanger consists of an OFHC copper tube (34.93 mm OD) with a wall thickness of 1.65 mm, and a length of 762 mm. The heat exchanger is surrounded by a bath of pressurized He II. Both reservoirs are insulated from the surrounding 4.2 K helium bath by an insulating vacuum can. The JT valve has a tapered plug for maximum flow control. A second valve is included to supply liquid to the pressurized bath. To simplify the design and reduce cost, no counterflow heat exchanger was included.

The saturated He II bath holds eight Allen-Bradley resistor temperature sensors and a superconducting liquid level indicator (LLI) mounted on a probe. The spacing between temperature sensors was measured using a caliper rule. The pressurized He II bath holds a resistive heater, a LLI, and a calibrated temperature sensor. The 10-ohm heater is connected to an adjustable power supply. The He I reservoir contains a temperature sensor and a LLI.

After filling the experiment with helium, a high capacity vacuum pump reduced the saturated bath temperature to below $T_\lambda$. During calibration, the pressure over the saturated bath was regulated using a control valve on the pumping line. The temperature sensors were calibrated at various bath pressures before and after taking experimental data.

After calibration, temperature readings in the saturated bath were taken for various heat loads. Heat was applied to the pressurized helium bath using the heater. Once temperature equilibrium was achieved ($dT/dt \approx 0$ for all sensors), the temperature profile within the saturated bath was recorded by the data acquisition computer. The saturation pressure, liquid level and pressurized bath temperature were also recorded.

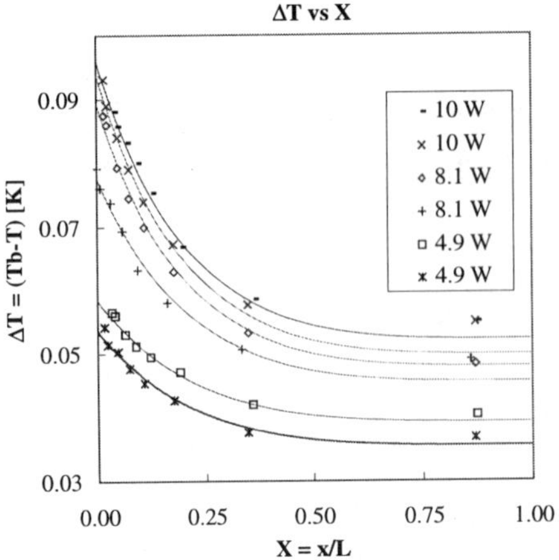


**Figure 1.** Temperature profile in saturated bath heat exchanger. Data points are experimental results. Solid lines are computer model. $T_b$ is temperature of the pressurized bath, T is temperature of the saturated bath, and x is distance below saturated liquid surface.

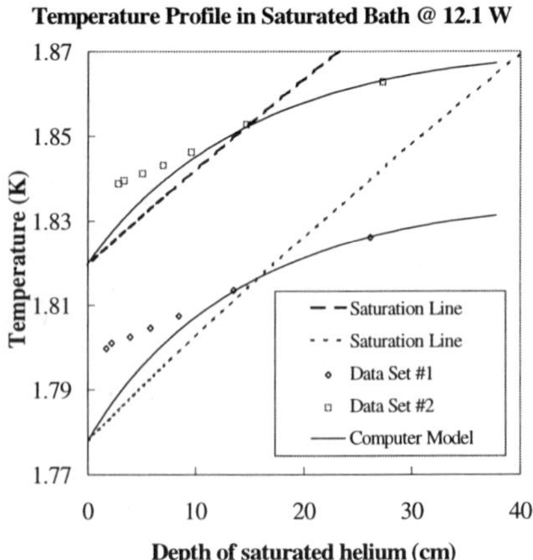


**Figure 2**. Measured temperature profiles in the saturated bath heat exchanger with a refrigeration load of 12.1 W. Two different experimental data sets are shown. Also plotted are the temperature profiles and the saturation lines from the numerical model.

## RESULTS

The experimental and computer model results are shown in Figure 1. Figure 2 plots two separate data sets with a heater power of 12.1 W. Also shown (for reference) are the model results (temperature profile of the saturated bath and the saturation line).

The maximum refrigeration capacity for the experimental heat exchanger can be calculated from Eq. (7). These theoretical data are plotted vs. the saturation temperature in Figure 3. Also shown are the data sets from Figures 1 and 2.

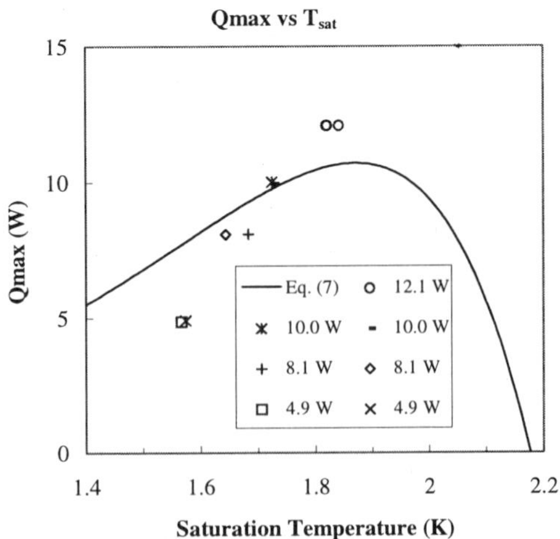


**Figure 3.** Maximum refrigeration capacity of experimental heat exchanger as a function of the saturation temperature. Also shown are the experimental data sets from Figures 1 and 2.

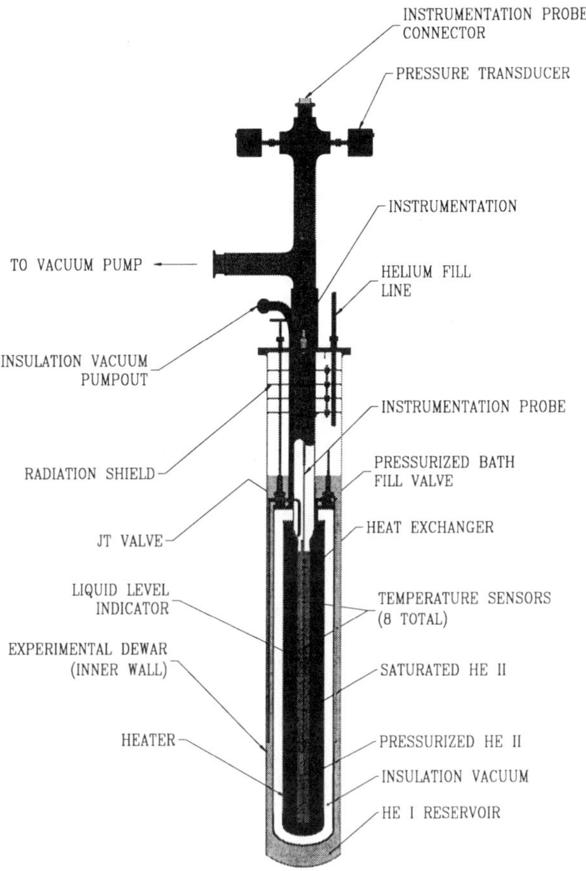


**Figure 4.** He II heat exchanger experiment.

## DISCUSSION OF RESULTS

There appears to be reasonable agreement between the computer model and the experimental results (Figure 1). The largest discrepancy appears to be at the lowest temperature sensor (maximum $x$ location in Figure 1). One possible explanation for this discrepancy is that the cross section of helium within the test section is not constant because the LLI is not long enough to reach to the bottom. Therefore, the cross section where the lowest temperature sensor is located is larger than in the rest of the saturated bath. This increased cross section reduces the heat flux in that area and reduces the temperature gradient. The computer model assumes constant cross section for all $x$.

Two outputs of the computer model are the temperature gradient in the bath and the saturation temperature of the liquid below the bath surface. When plotted together, one can view the temperature gradient relative to the slope of this saturated temperature line. In Figure 2, the temperature gradient has exceeded the slope of the saturation line. This is a prediction of boiling within the saturated helium. Boiling would be recognized experimentally by the data deviating significantly from the model. This is clearly the case for the conditions shown in Figure 2. The formation of buoyant vapor bubbles in the saturated bath renders the computer model invalid as it assumes no mass flow (Eq. (7)). The contribution to heat transport by the motion of the vapor phase is significant, and the amount of heat transported by pure conduction through the liquid is reduced. This effect can be seen by the reduction in slope of the temperature gradient. Although the model is unable to predict the temperature profile in the boiling helium, it does predict, quite well, the depth in the helium at which this boiling begins. The data begin to deviate from the numerical results very near the point where the temperature profile crosses over the saturation temperature line.

It should be noted that the saturated bath is by definition always boiling at the liquid surface. As long as the heat load within the helium is less than the value of $Q_{max}$ predicted in Eq. (6), boiling will only be at the liquid surface. However, once the heat load on the heat exchanger exceeds $Q_{max}$, boiling will begin below the surface. As the heat load increases, the boiling extends deeper into the helium. The depth of boiling in the saturated bath is predicted nicely by the model.

From Eq. (6), $Q_{max}$ may be plotted as a function of saturated bath temperature. This curve is plotted in Figure 3 and allows one to predict if boiling will occur in the heat exchanger for a specific temperature and heat load. The points at which experimental data were taken are also plotted in Figure 3. Clearly boiling is expected at the refrigeration load of 12.1 W. This was verified by the model and the experimental data in Figure 2. Additionally, one can see that boiling may occur at the 10 W refrigeration load condition as both data points are slightly above the $Q_{max}$ curve. The boiling condition in this case would only occur very near the liquid surface. The model did in fact predict boiling for these conditions, but the boiling point was still above the top temperature sensor.

## CONCLUSIONS AND FUTURE PLANS

The computer model agrees well with experimental data and is useful for calculating the temperature gradient within the saturated helium bath of a heat exchanger. In the case where $Q_{max}$ is exceeded the model is invalid for the boiling region, but it predicts well the point at which the saturation temperature is exceeded. There is good correlation between the computer model and Eq. (6) in predicting the onset of boiling.

This computer model will be used to predict temperature profiles in more complicated heat exchanger geometries such as that used in the 45-T Hybrid Magnet Cryogenic System.[5]


## ACKNOWLEDGEMENTS

This work is based upon research conducted at the National High Magnetic Field Laboratory (NHMFL), which is supported by the National Science Foundation under Award No. DMR-9527035.

The authors would also like to thank Dr. Yehia Eyssa for useful discussions on the numerical model.


## REFERENCES


1. A. Shajii, et al., Design and performance of forced flow He II heat exchangers, "Advances in Cryogenic Engineering", 35, Plenum Press, New York, (1990), pp. 1165-72.
2. Y. Huang, and S. W. Van Sciver, Forced Flow He II Heat Exchangers, *Cryogenics,* 36 (1996) 535-545.
3. S. W. Van Sciver, Heat Transfer Through Extended Surfaces Containing He II, 8th Inter. Energy Week Conf., Jan. 28-30, 1997, *ASME Journal of Heat Transfer* (submitted).
4. S. W. Van Sciver, "Helium Cryogenics", K. D. Timmerhaus, ed., Plenum Press, New York, (1986), p.178.
5. S. J. Welton, et al., Design, Development, and Testing of the JT Refrigerators for the 45-T Hybrid Magnet, "Advances in Cryogenic Engineering", 41, Plenum Press, New York, (1996), pp. 1283-9.

# HE II HEAT TRANSPORT IN CYLINDRICAL GEOMETRY

D. Pearson[†] and S.W. Van Sciver[§]

Applied Superconductivity Center
University of WI-Madison
Madison, WI, 53706

[†] Present address: Jet Propulsion Laboratory, Pasadena, CA 91109
[§] Present address: National High Magnetic Laboratory, Tallahassee FL 32310

## ABSTRACT

Temperature difference measurements are presented for two different sized radius cylindrical heaters, r = 1.6 and 4.8 mm, and at different radial positions in the fluid adjacent to the heater. Previous measurements of the peak heat flux in cylindrical geometry are found to be smaller than the value predicted from Gorter-Mellink mutual friction theory when generalized to the cylindrical coordinate system; and this deviation increases systematically as the radius of the cylindrical heater is decreased. Our measurements are compared to the theory for linear heat transport generalized to the cylindrical geometry and are found to be lower in magnitude although, functionally, the radial dependence is given correctly. The deviations appear to scale with heater radius but in a sense that is opposite to the behavior observed for the peak heat flux measurements.

## INTRODUCTION

Use of He II as a coolant for superconducting magnets is one application to which the study of He II heat transfer has been directed. Magnets used in superconductive magnetic energy storage (SMES) and particle accelerators are two particular instances where He II is proposed as a coolant. To date, almost all experiments involving He II heat transport have been concerned with heat transfer from systems where the heat flux is a constant, i.e., linear channels with constant cross section. In a cylindrical system, the heat flux is spatially varying, and this spatial variation scales with the size of the cylinder radius. Because the conductors used to construct the above superconducting magnets are commonly of a cylindrical cross-section, it is of technical interest to study the heat transfer from cylinders.

Of particular interest to a He II cooled magnet is the peak heat flux. The peak heat flux, q*, is the value of the heat flux at which boiling commences on the surface of a heater. Measurements of the peak heat flux for cylindrical heaters in He II have been found to depart from the result predicted by He II heat transport theory. Investigators have measured the peak heat flux from cylindrical heaters of various sizes ranging from 10 μm to 6.6 mm. Van Sciver and Lee[1] compared experimental results for the peak heat flux from cylinders with the result derived from the Gorter-Mellink theory. They observed that all experimental results for the peak heat flux are lower than the theory and that the deviation varied roughly as $r_h^{1/2}$, where $r_h$ is the heater radius. Because the peak flux is depressed, this implies that the temperature at the surface is higher than predicted by the above theory. Further, this means the thermal resistance of the fluid is enhanced somewhere, and that this enhancement is inversely proportional to $r_h^{1/2}$. Van Sciver and Lee did perform some temperature difference measurements to try and observe the variation from theory, but within the resolution of their experiment found no deviation. It was the intent of the present experiment to obtain improved temperature difference measurements around cylinders to understand the cylindrical peak heat flux measurements.

## TEMPERATURE DIFFERENCE IN CYLINDRICAL COORDINATES

The radial heat flux from an axially symmetric cylindrical heater is given by

$$q = q_s\left(\frac{r_h}{r}\right) \tag{1}$$

where $q_s$ is the heat flux at the surface, $r_h$ is the heater radius, and r is the radial distance from the center of the cylinder. The temperature difference of the fluid surrounding the cylinder relative to the bath is obtained from the Gorter-Mellink[2] mutual friction expression for the temperature gradient and is given by

$$\Delta T = \frac{f(T)}{2}\, q_s^3 \frac{r_h^3}{r^2}\ , \tag{2}$$

where $f(T) = A\rho_n/\rho_s^3 s^4 T^3$. A is the Gorter-Mellink parameter, $\rho_n$ and $\rho_s$ are the two fluid component densities, s the specific entropy, and T is temperature. The origin of this temperature difference is due to a dissapative interaction involving quantized vortex filaments which form in He II above a critical heat flux. The radial extent of the temperature difference scales strongly with the size of the heater radius. For a radial position one heater radius from the surface, the temperature difference is one fourth that at the surface, so most of the temperature difference occurs within one radius of the heater surface. As the cylinder size is made smaller, the extent of the temperature difference becomes smaller. This is to be contrasted with a linear channel where the temperature difference is the same, for a given distance, at any position along the channel.

The mutual friction theory applied to linear channels of constant cross section is found to explain temperature difference and peak heat flux measurements fairly well. Commonly though, the value of the heat flux exponent varies from 3 to 4. Swanson[3] and Schwarz[4] have developed similar theories where the exponent naturally varies depending on the heat flux. Our results will be compared to both Equation (2) and the Schwarz theory.

## EXPERIMENTAL APPARATUS

Since almost the entire temperature difference around a heated cylinder in He II occurs within one heater radius and it is expected that departures from theory will occur for small cylinders (<10 mm), a small temperature sensor with sub-millikelvin resolution was required for these experiments. For the above reasons, a thermocouple was selected as the temperature sensor. Thermocouple junctions can be made to a size on the order of the combined diameters of the two wires used to construct the junction. This gives a probe four times smaller then a typical resistance thermometer. With the availability of commercial SQUID systems capable of pico-volt resolution the traditional problems of using a thermocouple at liquid helium temperatures are eliminated. The pico-volt resolution of the SQUID along with the use of Au-Fe thermocouple wire, provides μK temperature sensitivity. Also, since the SQUID is essentially a low temperature amplifier, the problem of spurious voltage signals induced along the thermocouple wires passing from the low temperature environment to room temperature is resolved. For our experiments, the SQUID is coupled to the bath with exchange gas and its temperature is maintained with a PID controller.

The heaters for these experiments are OFHC copper cylinders with 100 Ω metal film resistor epoxied inside. Two heaters of radius 1.59 and 4.76 mm are used. To ensure purely radial flow, the cylindrical heaters are mounted between two G-10 plates that function as heat insulators. The plates are 76 mm by 64 mm and 6.4 mm thick. In order to lower the heat input to the bath, a wedge-shaped insert G-10 insert is placed between the two plates. This insert exposes one quarter of the cylinder surface. Constant plate spacing of 3.2 mm is maintained by invar spacers and the insert. Six 700 μm through holes are drilled into one of the plates at different radial locations relative to the heater. These holes serve to locate the thermocouple position. The hole positions are chosen based on practical limitations: the desire to measure as close to the heater as possible and to have most of the measurements within a distance of one heater radius from the surface. The holes used for these experiment were located, for the 1.5 mm heater, a distance of 380 μm and 760 μm from the heater surface, and for the 4.8 mm heater, 380 μm and 1.52 mm from the surface. A schematic of the heater assembly is shown in Figure (1).

The thermocouples are constructed using a Au 0.07 at% Fe wire as the active element. This alloy is one of the class of Kondo metals which have anomalously high thermopower at low temperatures[5]. The other element of the thermocouple is niobium wire which has zero thermopower since it is a superconductor. The thermocouple is inserted into a stainless steel (non-magnetic) tube. The tube has an outer diameter of 635 μm, an inner diameter of 345 μm, and a length of 1 cm. The inner diameter will determine the amount of fluid being probed. The junction tip is flush with one end of the tube while the other end, where the wire enters the tube, is sealed with epoxy. The sealed tube and thermocouple were modeled as a fin system. The length of the stainless steel tube is about ten fin lengths, which confirms that the junction should be at the heated fluid temperature. Because of our desire to obtain precise temperature difference measurements, we devised a simple calibration scheme for the thermocouples. The junction not used to probe the fluid near the cylinder is thermally attached inside an OFHC copper plug which also contains a heater and 100 Ω Allen-Bradley resistor to serve as a thermometer. The copper plug is covered with epoxy to isolate it from the bath. By heating the block and simultaneously measuring the SQUID output along with the thermometer resistance, the thermocouple can be calibrated. The calibration plug is shown in Figure (1).

The SQUID and its temperature control system, thermocouple, and heater assemblies form main parts of the experiment. A schematic of the experiment is shown is Figure (1). The bath temperature is maintained by a control system consisting of a capacitive pressure sensor whose output is used to control an exhaust valve. Temperature control of better than 1 mK is easily obtained. Upon cooling to He II temperatures, once the SQUID signal and bath temperature are stable, the thermocouple is calibrated and temperature difference measurements are made. The cylindrical heater is powered by a Kepco 100V, 1A DC power supply with all four wires leading to the heater RF filtered. The only other electronics used during data collection is a RF filtered bath thermometer. Before adding heat, the SQUID value (this will serve as the zero for all other data) and the bath temperature are recorded. The SQUID output voltage is averaged with a Hewlett Packard model 34401A multimeter. This averaging allows resolution of 100-300 μV even though the noise signal is on the order of 300 μV. Output

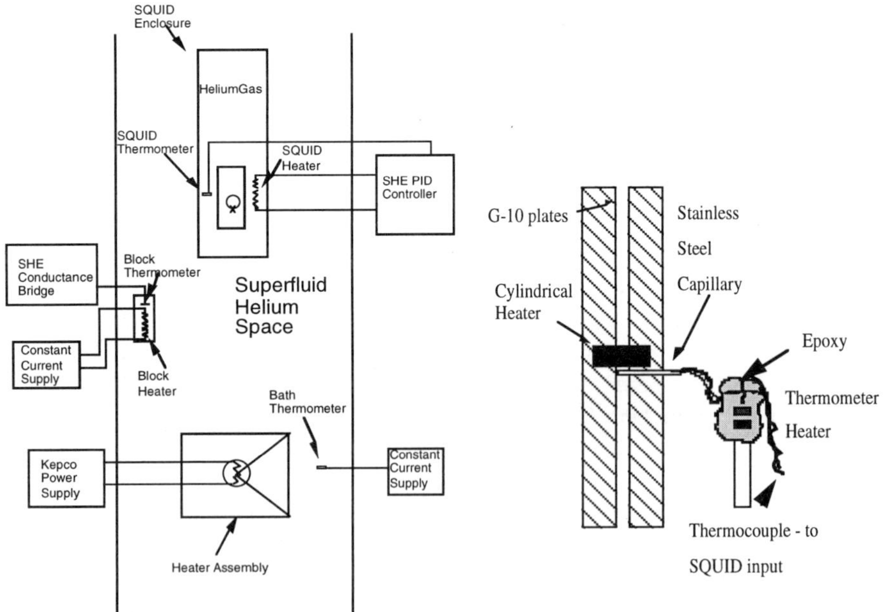


**Figure 1.** Schematic for cylindrical heat transport experiment and heater assembly with thermocouple and calibration block.

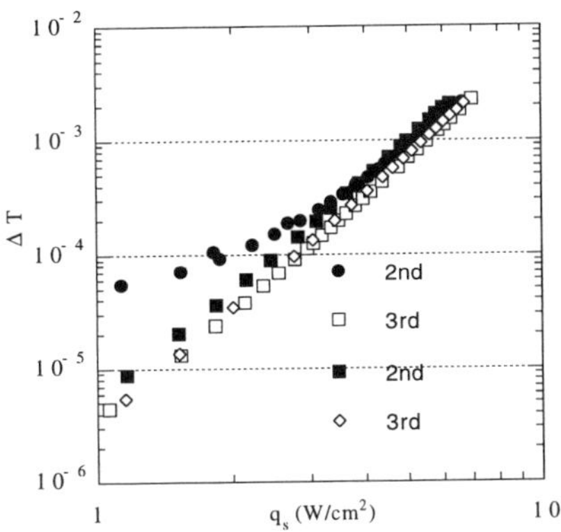


**Figure 2a** Typical data for 1.6mm heater.

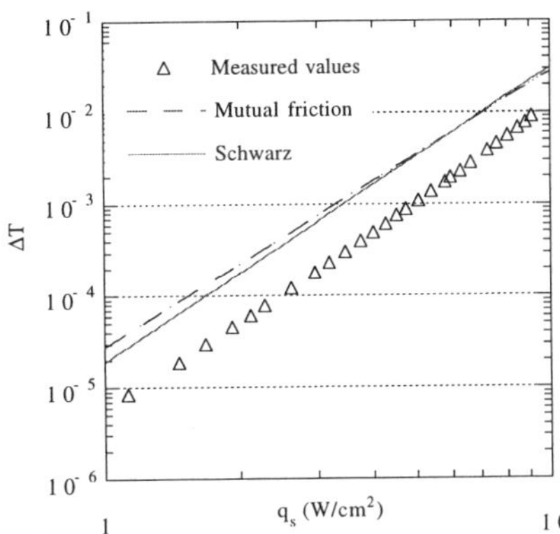


**Figure 2b** Comparison between theory and second hole data for 1.6 mm heater.

voltages typically range from 500 μV to 1 V, which corresponds to temperature differences of 5 μK to 10 mK. The process of adding heat, while recording heater current and voltage, bath temperature, and SQUID output, constitutes the primary measurement. To obtain data for another radial position or for another heater, the cryostat must be warmed to room temperature and the changes made. The collected raw data is converted to heat flux and temperature difference values with a simple spread sheet program. These experimental results are compared to the mutual friction theory as described in the next section.

## RESULTS

Figures (2a) shows temperature difference measurements versus heat flux for the 1.6 mm radius heaters at a bath temperature of 1.96 K. Similar looking data was obtained for the 4.8 mm heater. The 1.6 mm heater measurements were made with the thermocouple located in the second or third hole. Again for this heater the two holes are 380 μm and 780 μm from the heater surface. The heat flux for this heater ranged from 1 to 10 W/cm$^2$ producing temperature differences of 5 μK to 10 mK. Two separate measurements for each hole position were performed, and, as shown in Figure (1a), the data are reproducible for this heater. The 4.8 mm heater data were taken in the second or sixth hole positions. These two holes are 380 μm and 1.5 mm from the surface. For the 4.8 mm heater, the heat flux varied from 0.7 to 3.5 W/cm$^2$ with temperature differences of 5 μK to 3 mK. In one data set of Figure (2a), there is a tail that diverges from the straight line behavior of all other data at small heat flux. This is believed to be due to a jump in the SQUID output due to electromagnetic noise from the exhaust valve as it opens to keep the bath constant. This was a common problem which introduced systematic errors into the low heat flux data. The offset at the lowest heat flux is about 50 μK.

Figure (2b) shows experimental results for the second hole position of the 1.6 mm radius heater compared to temperature difference values calculated from Equation (2) and those from the Schwarz theory. As is seen clearly, the experimental data lies systematically lower than that calculated from either theory. The difference between the measured values and those predicted from theory range from about 2.5 times lower, at the highest heat flux, to almost four times smaller at the lower heat flux values. This particular set of data along with similar data for the third hole were fit to a power law of the form

$$\Delta T = \tilde{A}\, q_S{}^{m},$$

where $q_S$ is the heat flux at the surface of the heater. From the mutual friction theory, m is predicted to be three and $\tilde{A}$ is equal to temperature dependent parameters and geometrical factors. For comparison both the second and third hole measurements give a best fit heat flux exponent of 3.35±0.05. For heat fluxes in the range of 1-10 W/cm$^2$, the Schwarz theory can be fit in the same way as the experimental data. A best fit to the theory for the heat flux exponent of 3.21 and 3.19 is obtained for the second and third hole positions, respectively.

Since both data sets were obtained at the same bath temperature, in terms of the simpler mutual friction theory, the ratio of the $\tilde{A}$ values will depend only on the radial position and is given by

$$\frac{\tilde{A}_2}{A_3} = \left(\frac{r_3}{r_2}\right)^2,$$

where the subscripts refer to the hole positions. For the 1.6 mm radius heater this ratio is 1.35. Fitting the experimental values results in a ratio of 1.36. For the Schwarz theory, the predicted relationship between temperature difference measurements at different thermocouple positions is not precisely constant but over a limited range of heat flux values this is approximately true. Over the range of heat flux values used in the fitting procedure, the Schwarz theory predicts that the second hole position values will be greater than the third hole values by a factor of 1.34. The fact that the experimentally determined ratio between the values in different hole positions is consistent with theory indicates that the theoretical radial dependence is correct in form.

The same analysis was applied to the 4.8 mm heater data, and similar trends are observed. As for the 1.6 mm heater, the measured temperature difference is much lower than the predicted value, varying from four times smaller at low heat flux to about two at the highest heat flux. The heat flux exponent, m, is found again to be larger than the Schwarz value. Fitting the data for both hole positions yields for the heat flux exponent a value of 3.7 for the second hole results and 3.6 for the sixth hole. The ratio between the fits for the second and sixth hole, which serves as a check of the radial dependence of the temperature difference, is found to be three. The mutual friction theory predicts a ratio of 3 while the Schwarz theory predicts 3.2 for this ratio. Again, the radial dependence of the temperature difference is found to be functionally the same as theory.

A summary of the deviations from the Schwarz theory is given in Figure (3). This plot is obtained by dividing the Schwarz prediction by the experimental value and plotting the results versus heat flux. The data used includes all of the various hole positions. At low heat flux, there is large scatter in the deviations, but at high heat flux the deviations for

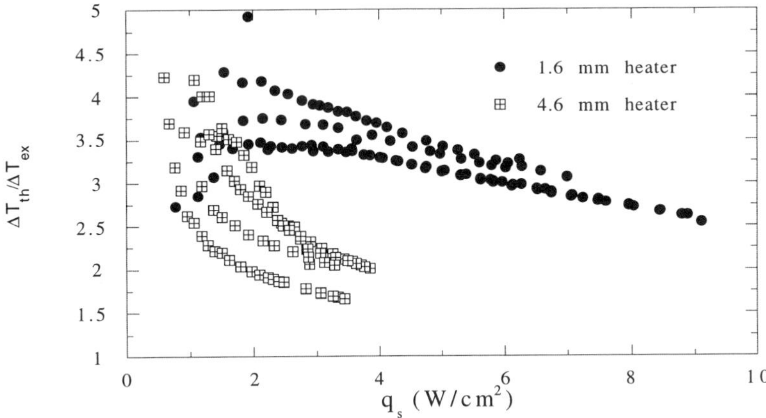


**Figure 3.** Plot of the ratio between temperature differences predicted by theory to measured values for different heater assemblies.

each heater appear to converge. The problem at low heat flux is probably due to the SQUID jumps discussed earlier. Note, that the problem is greater for the large radius heat which requires larger amounts of total heat to produce the same heat flux. The higher value of the heat flux exponent for this heater is probably due to this problem. When only the high heat flux data is used to obtain an exponent for this heater, the value is 3.4 which is consistent with the 1.6 mm heater result and the work of other investigators. In spite of the scatter in the data, Figure (3) clearly showsthat the deviations are larger for the 1.6 mm heater. Although there is a radius dependence to the deviations these results are in contradiction to the peak heat flux measurements of Van Sciver and Lee. As discussed, Van Sciver and Lee examined values of the peak heat flux and found deviations from theory that gave a smaller peak heat flux as the radius decreased. Since the peak heat flux depends on the thermal impedance of the surrounding fluid, a smaller peak heat flux indicates an increase in the thermal impedance which should result in increased values for temperature difference measurements of the surrounding fluid as the cylinder radius is decreased. Our data indicates that the thermal impedance is depressed and that the depression increases as the cylinder size is made smaller.

## CONCLUSIONS

Temperature difference measurements were made for two different cylinders in He II. Results are lower than predictions from theory. Further, these deviations were found to increase as the cylinder radius decreases. The exponent of the heat flux is found to be larger than that predicted by the Schwarz theory. For the linear channel and 1.6 mm the exponent is found to consistently lie between 3.3 and 3.4. For the 4.6 mm and 2.4 mm heater, the heat flux exponent is found to larger, but problems in the low heat flux measurements are believed to be a source of these larger values. Determining the exponent for only the higher heat flux data resulted in a value closer to that found for 1.6 mm and linear heaters. Comparison of results for different radial positions for the same heater assembly are found to agree with the theory. This is most striking in the case of the 1.6

mm heater where the two positions are separated by only 380 μm. The fact that the radial dependence is functionally correct seems to rule out effects due to a spatially varying heat flux.

Our data are in conflict with the peak flux measurements where the results imply that the thermal impedance is enhanced while our data show the opposite. A possible explanation for the apparent contradiction is that the cylinder itself is strongly influencing the heat transport. Again, the origin of the mutual friction temperature difference is due to a dissapative interaction involving quantized vortex tangle filaments. The amount of vortices present determines the magnitude of the temperature difference. As can be seen from Equation (2), for a cylindrical system, the spatial extent of the temperature difference decreases as the cylinder is made smaller. Therefore, the spatial extent of the vortex tangle will also decrease. If the tangle is assumed to be contained in an annular volume of some radius, this volume will decrease more rapidly than the surface area of the cylinder surface thus increasing the effects of the boundary on the tangle. What effects the boundary has on vortex tangle is not known. More work on this problem is clearly needed.

## ACKNOWLEDGMENT

Research supported in part by NSF Grant No. CTS 9110420.

## REFERENCES

1. Van Sciver, S.W. & Lee, R.L. *Cryogenic Processes and Equipment in Energy Systems* 147 (1981).
2. Van Sciver, S.W. & Lee, R.L. *Adv. Cryog. Eng.* **25**, 363 (1980).
3. Swanson, C.E. & Donnelly, R.J. *J. Low Temp. Phys.* **61**, 363 (1985).
4. Schwarz, K.W. *Physical Review B* **38**, 2398 (1988).
5. Armbruster, H. & Kirk, W.P. *Physica* **107B**, 335 (1981).

# EXPERIMENTAL STUDY FOR DYNAMICAL CHARACTERISTICS OF HE I–HE II INTERFACE UNDER SATURATED VAPOR PRESSURE CONDITION

K. Kamiya, M. Murakami, and N. Yanase

Institute of Engineering Mechanics, University of Tsukuba
Tennoudai 1-1-1, Ibaraki, Tsukuba 305, Japan

## ABSTRACT

We investigated appearance and propagation characteristics of He I-He II interface caused by cooling of He I. When liquid helium is cooled or heated near λ-point in a homogeneous situation, a He I-He II transition may occur without well-defined interface between the two phases. On the contrary, in such inhomogeneous situation as in the gravitational field or under a heat flow, bulk liquid helium system can not pass the λ-line as a whole at once and instead, a He I-He II interface appears. We generated He I-He II interface by imposing a cooling heat flow and measured the temperature field by a newly developed superconductive temperature sensor. We revealed some dynamical characteristics of He I-He II interface.

## INTRODUCTION

The investigation of a heat transfer problem of liquid $^4He$ near $T_\lambda$ is indispensable for designing a cooling apparatus for superconductive magnet at He II temperature. Especially in large scale He II systems in which subcooled He II is utilized or the hydrostatic pressure in He II exceeds the lambda pressure, He I-He II interface may appear around heated surface in He II. It is well known in a sample in which 2nd order phase transition takes place phase transition occurs at the critical temperature without a phase boundary in homogeneous system due to the continuity of the entropy. In the case of $^4He$, the hydrostatic pressure difference arises in an experimental cell with a finite height under the gravitational field. Accordingly, there appears He I-He II interface in the cell due to the λ-line inclined negative relative to the temperature, if the temperature of the liquid helium surface is slightly below $T_\lambda$. Ahlers[1] experimentally showed the existence of He I-He II interface in the manner mentioned above by measuring the temperature in the cell.

A heat flow can also cause inhomogeneity in liquid helium and can generate a He I-He II interface. In this research we tried to generate a He I-He II interface in this manner. A heat flow is given by cooling of a He I liquid surface at a temperature initially slightly higher than

$T_\lambda$. As shown in Figure 1, an interface appears when He I under saturated vapor pressure condition is suddenly cooled from an initial state at slightly higher temperature than $T_\lambda$ by vapor evacuation and propagates downward from a liquid helium free surface in a cryostat. In particular, a heat flow in He I region creates a temperature gradient and, on the contrary, no thermal gradient develops in He II region because of super thermal conduction. The interface is detected by the temperature measurement with a newly developed superconductive temperature sensor.

## EXPERIMENTAL

### Cryostat and Evacuation System

A glass dewar connected to the evacuation system shown in Figure 2 is used for the temperature measurement. A constant heat leak into the test section from outside is estimated to be below 0.00126W per $cm^2$ of liquid helium free surface. In addition to an ordinary evacuation system for the bath temperature control, an additional evacuation line for rapid cooling is connected in parallel to the ordinary one. The latter evacuation system consists of an electric valve for stepwise cooling and a flow rate regulating valve to control the cooling rate. The cooling rate is calculated from the evaporated vapor flow rate from a free surface measured by the gas flow meter connected to the exit of the vacuum pump.

### Interface Detection Method

A He I-He II interface can be detected by measuring the temperature on the basis of the fact that the temperature of an interface is $T_\lambda$. For the measurement of the propagation speed and the temperature profile in a transition layer, we utilize a superconductive temperature sensor[7,8] as shown in Figure 3 made of gold-tin thin film vacuum deposited on the side wall of a fine quartz rod with a diameter of 40μm.

The resistance changes continuously in a super-normal transition region, because the element behaves like a second kind superconductor. Any temperature variation can be detected as a resistance change. A particular feature of this sensor is its very high sensitivity and a rapid time response within a few micro seconds. The temperature range with a high sensitivity can be adjusted so as to include the lambda point within it by controlling the thickness ratio between gold and tin layers. The double probe type sensor of which two sensing elements are fixed apart by 5mm gaps along a propagation direction makes it possible to measure an interface speed.

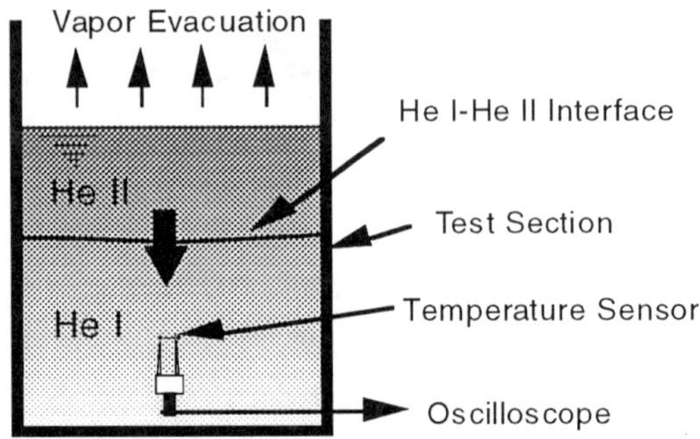


**Figure 1.** Schematic drawing of an interface propagating downward.

and a superconductive temperature sensor

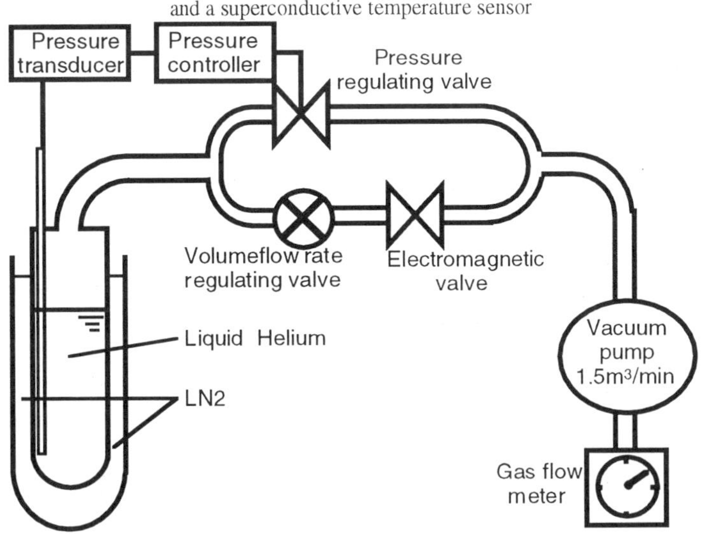


**Figure 2.** Schematic illustration of the evacuation system.

## RESULT AND DISCUSSION

### Output from Pressure Transducer and Superconductive Temperature Sensor

Typical data from the superconductive temperature sensor located 20 cm below the free surface and the pressure transducer to measure the vapor pressure during an interface propagation is shown in Figure 4. It is found that the data from the superconductive temperature sensor does not exactly indicate the temperature variation in particular in the transition region across an interface. The reason for this is that this variation is resulted from the temperature variation, however, also from the effect of the heat transfer coefficient between the sensing element and surrounding medium(He I or He II). The sensing element generates small amount of heat due to Joule heating, and thus the temperature of the element itself must be higher than that of surrounding medium when a sensor is immersed in He I because of poor heat transfer coefficient. However in He II the element temperature rapidly

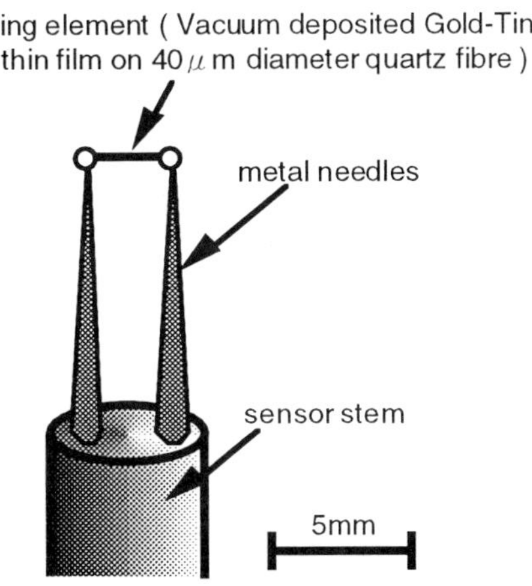


**Figure 3.** Illustration of superconductive temperature sensor.

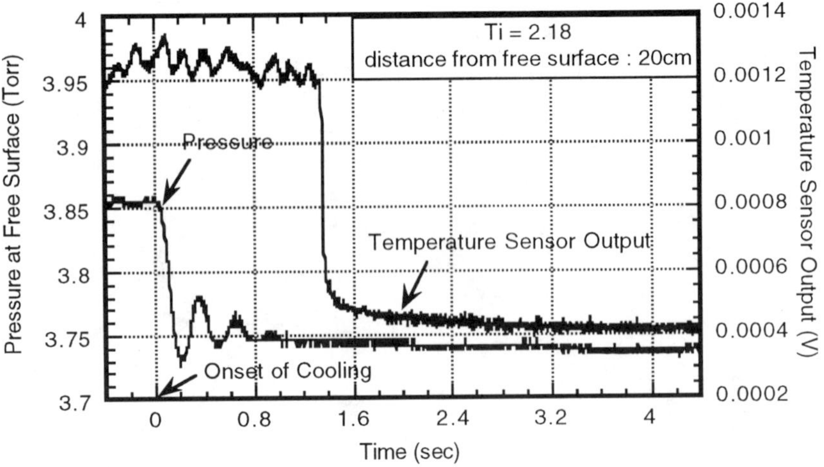


**Figure 4.** Typical data from the superconductive temperature sensor located at 20cm below the free surface and the pressure transducer to measure the vapor pressure during an interface propagation.

drops down to the surrounding He II temperature because of so called super thermal conduction. Thus, the variation of the output shown in Figure 4 is much exaggerated as compared with that of the temperature in particular in the transition region. On the other hand, it is fair to consider that the variation of the data except in transition region is that of the temperature.

After the onset of cooling, it is seen that the initial pressure at 38.6 Torr(2.18K) drops by 1 Torr across the λ-pressure at SVP(=38.0 Torr) with pressure oscillation caused by the surface oscillation. A sudden temperature drop down to $T_\lambda$ which may correspond to a pressure drop at the liquid surface due to rapid vapor evacuation is seen. Much larger temperature fluctuation detected in He I region than in He II region shows difference in the heat transfer coefficient between two phases. It is, furthermore, the appearance of a thin thermal boundary layer in He I region and a temperature decrease in He II region due to a thermal resistance caused by quantum vortex tangles.

## Initial Stage of Cooling

The initial temperature ( $T_i$ ) in the whole test section is adjusted slightly above $T_\lambda$. After the onset of cooling, the temperature in He I region in the liquid surface region first drops down to $T_\lambda$ with the thickness of thermal boundary layer increasing as illustrated in Figure 5.

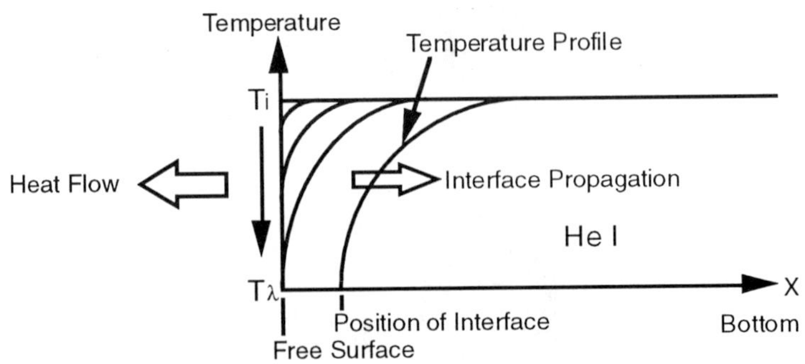


**Figure 5.** Illustration of the development of an interface layer at a free surface with time.

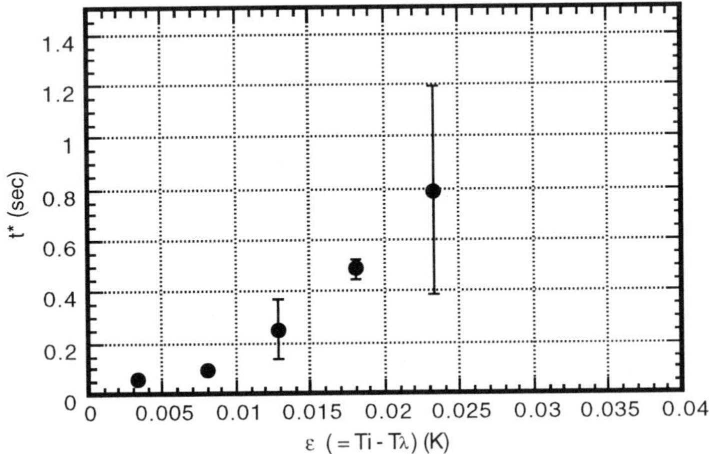


**Figure 6.** Time required for the temperature at a liquid surface reaching Tλ.

Since the temperature at a phase interface is Tλ, the onset time of an interface propagation downward from a liquid surface must be different from that of cooling. We measured the time($t^*$) when the temperature at a liquid surface reaches Tλ by a superconductive sensor located 1 mm below a liquid surface, and the result is shown in Figure 6. For large $\varepsilon = T_i - T_\lambda$ ( >0.03K ), convection and boiling arise in He I region and prevent stable heat conduction to some extent. This consequence suggests that we should select 2.18K ( $\varepsilon$ = 0.008K ) as an initial temperature in the following experiments to avoid negative effects mentioned above.

## Propagation Characteristics of an Interface

After the initial cooling stage, an interface which separates He II from He I is created first at a liquid surface and propagates downward in the test section. If super thermal conduction is realized in He II region, propagation speed should be constant and independent of the distance from the free liquid surface. The measurement result of the propagation speed is plotted against the distance from the free surface in Figure 7. The interface speed of

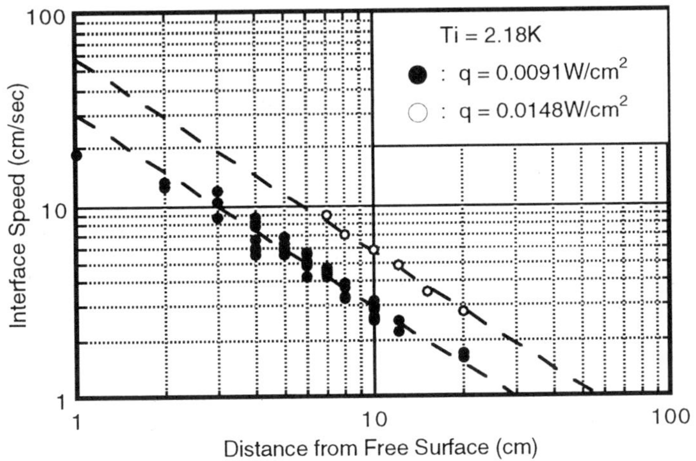


**Figure 7.** Interface speed versus distance from free liquid surface.
Dotted line indicates relation in inverse proportion to distance.

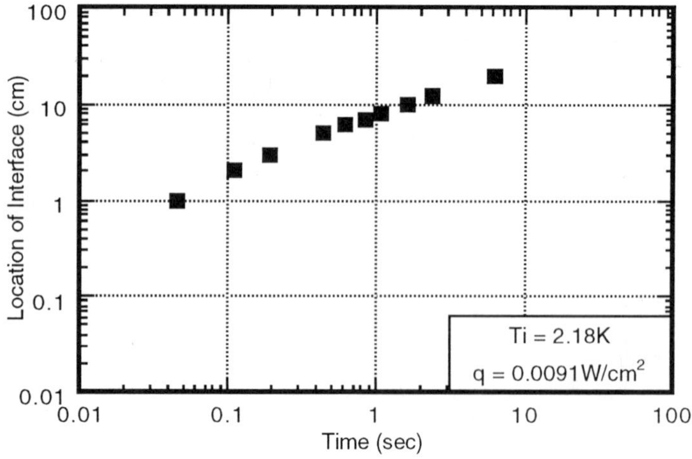


**Figure 8.** x-t diagram for an interface propagation.

0.0091W/cm$^2$ and 0.0148W/cm$^2$ decreases in inverse proportion to the distance from a free liquid surface in both cases. Interestingly this tendency does not depend on the cooling rate. These results suggest there appears the thermal resistance in He II region. Some deviation of data from the linear relation indicated by dotted line is seen at the beginning of the propagation in the case of 0.0091W/cm$^2$. The deviation means the interface speed can not be infinity but converges on some finite value. At a final stage of propagation with a speed of below 1cm/sec, an interface can not be observed due to thermal convection in He I phase. It is confirmed that the disturbance from density instability breaks a planar He I-He II interface in gravitational field.

An x-t diagram for an interface propagation in the case given in Figure 7 is drawn in Figure 8, where the initial time correction for an interface creation at liquid surface is made as discussed above. It is seen the propagated distance an interface is nearly proportional to the square root of the time.

The relation between He I-He II interface speed and the cooling rate is shown in Figure 9. Data is obtained at 20cm, 10cm and 8cm from a free liquid surface. It is seen in Figure 9 that the interface speed increases linearly with the cooling rate. The cross mark on the

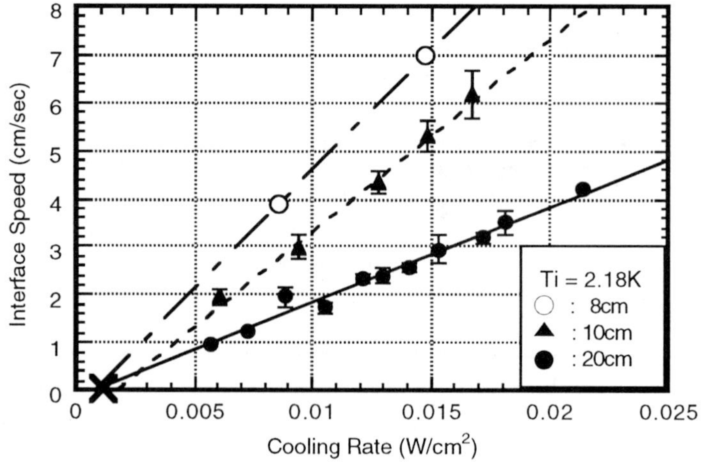


**Figure 9.** Interface speed versus cooling rate.

abscissa is the cooling rate required to cancel the background heat leak into the dewar from outside. The fact that the cross marks coincide with the intersection of the solid line with the abscissa suggests an interface is generated when the cooling rate overcomes the background heat leak.

## SUMMARY

It was revealed that He I-He II interface plays an important role in the cooling process near $T_\lambda$. The dynamic characteristics were also examined to conclude that the speed of an interface is directly proportional to the cooling rate at a speed of several cm/sec and decreases inversely proportional to the distance from the free surface. The relation between the interface speed and the cooling rate is understood by a heat balance at the interface. A decreasing interface speed, however, can not be explained unless the thermal resistance due to quantum vortex in He II region is taken into account. Therefore, a governing equation in He II region including Golter-Mellink term, a thermal resistance factor, should be solved numerically since the equation becomes nonlinear, so called Dresner equation. Results from numerical analysis will be compared with experimental data as shown in Figures 7, 8 and 9 and then the validity of the theory will be examined.

## ACKNOWLEDGMENT

This work was partly supported by a Grant-in-Aid for Scientific Research from the Ministry of Education, Science and Culture (09875247).

## REFERENCES

1. G.Ahlers, Phys. Rev. Lett. 21, 1159 (1968)
2. A.Onuki, J. Low. Temp. Phys 50, 433 (1983)
3. A.Onuki, J. Low. Temp. Phys 55, 309 (1984)
4. K.Kamiya, M.Murakami, N.Yanagise, ICEC 16 (1996), pp.531-534
5. K.Kamiya, M.Murakami, N.Yanagise, Czech. J. Phys., Vol.46-S1(1996), pp. 193-194
6. K.Kamiya, M.Murakami, S.Nemirovskii, J. of Cryogenic Society of Japan, Vol.32 No.6 (1997), pp. 5-10
7. Borner,H., Schmeling,T, and Schmidt, D.W. J Low Temp Phys (1983) 50 405-426
8. Schwerdtner, M.,Poppe, W., and Schmidt, D.W. Cryogenics (1989) 29 132-134
9. S.Caspi, ICEC 8, Genova (1980) pp. 238-242

# MEASUREMENT OF THE PRESSURE DISTRIBUTION AND DRAG ON A SPHERE IN FLOWING HELIUM I AND HELIUM II

M. R. Smith[1] and S. W. Van Sciver[1,2]

[1]National High Magnetic Field Laboratory
[2]FAMU-FSU College of Engineering
Tallahassee, Florida 32310, USA

## ABSTRACT

We have designed and constructed an apparatus for measuring the pressure distribution around a sphere as a function of Reynolds number (mass flow rate) in both He I and He II. Liquid helium possesses kinematic viscosities nearly three orders of magnitude smaller than that of air. This offers the possibility of correspondingly higher Reynolds numbers to engineers carrying out dynamic similarity studies of modern high performance aircraft and marine vessels. Important questions remain however, about the fundamental nature of turbulence in liquid helium. Additionally, many techniques in common use for room temperature fluid dynamics experiments are not well developed for low temperatures. Our goal in this work was to demonstrate one technique for measuring the pressure distribution and drag on a sphere in flowing liquid helium. A further goal was to compare our findings with classical results in the hope of assessing the degree to which helium II and helium I behave classically within the context of a classical fluid dynamics experiment.

## INTRODUCTION

With the advent of modern high performance aircraft and marine vessels, direct measurement of flow field information and dynamic loads upon models at scaled velocities has gone beyond the reach of current wind tunnel facilities. In an effort to overcome these limitations, some attention has recently focused upon liquid helium as a potential test fluid. Because its kinematic viscosity ($\nu=\eta/\rho$, where $\eta$ and $\rho$ are dynamic viscosity and density, respectively) is nearly three orders of magnitude smaller than that of air, it offers the possibility of much higher Reynolds numbers ($Re=Ud/\nu$ acts as a non-dimensional velocity, where U and d are characteristic of the flow field) for the same or smaller scale model, while staying clear of transonic effects.

So far as we know, helium above 2.176 K (He I) behaves the same as any other classical fluid. In contrast, He II (the fluid phase below 2.176 K) is a quantum fluid with a range of phenomenology which makes it particularly attractive as a coolant, and years of research continue to be directed at understanding this very special behavior. Though recent research[1] suggests that classically generated flow fields in He II may behave classically in many circumstances, the relative lack of fundamental research in this area has created doubt in many minds about the suitability of He II (and even He I) as a test fluid. More importantly, the techniques in use at room temperature (and even in cryogenic nitrogen tunnels[1]), which have been developed over many decades, are not generally adapted to liquid helium. Applications currently using hot-wire anemometers, flow visualization, or even pressure measurement via small tubes are not well explored in either helium I or II. There is currently no off the shelf equipment base available.

In this paper, we describe an experiment for measuring the pressure distribution over the surface of a sphere, and ultimately deriving the drag as a function of Reynolds number. Because the drag is so intimately tied to the boundary layer, we hope the results will provide us with important clues about the behavior of liquid helium at the small length scales characteristic of this layer. We pay particular attention to some of the instrumentation issues which came up in the course of designing the apparatus, together with the data which we have observed thus far.

## DESIGN AND INSTRUMENTATION ISSUES

In measuring the pressure distribution over the surface of a sphere, the experimentalist designing a room temperature apparatus may take several things for granted. First, a modern wind tunnel continuously supplies relatively high flow velocities through large test sections. Flow quality is carefully controlled by a series of contractions, screens and honeycombs. Furthermore, large models (in this case, a sphere) provide higher Reynolds numbers (fixed flow velocity) and ample space for instrumentation. Frequently this on board instrumentation often takes the form of small capillary tubes and scan valves connecting the array of pressure taps on the surface to pressure sensors outside the tunnel. Large models with complicated geometries might well contain a hundred or more such pressure taps[2].

In our experiment, mass flow was supplied by a single stroke stainless steel bellows[3], with an effective cross sectional area of $1.813\text{x}10^{-2}$ $m^2$. The bellows is driven by a stepper motor located at room temperature, capable of linear stroke velocities up to 0.0254 m/s. The bellows offers higher flow quality than the alternative centrifugal pump[4], in addition to being easier to use. The inherent disadvantage of a single stroke bellows is the limited time for measurement. The bellows has a finite displacement volume and stroke length; 0.16 m and $2.90\text{x}10^{-3}$ $m^3$, respectively. Thus, for a given mass flow rate, the duration of the stroke/experiment is,

$$t = \frac{\rho V}{\dot{m}} \qquad (1)$$

where $\dot{m}$ is the mass flow rate, and V is the bellows displacement volume. Once it reaches this limit, the experiment comes to an end and the bellows must be reset for another run.

The positive displacement bellows pump offers a continuous range of mass flow rates up to 65 g/s, relatively independent of test section geometry. This gives rise to the startling result that by shrinking the scale of the test section (for fixed mass flow, and assuming the model shrinks proportionally with the test section), one actually increases the Reynolds number. Specifically,

$$\text{Re} = \frac{4\dot{m}}{\pi \rho \nu d_{sphere}(X^2 - 1)} \qquad (2)$$

where $d_{sphere}$ is the diameter of the sphere, and $X$ is an aspect ratio ($X \equiv d_{tunnel}/d_{sphere}$). For fixed mass flow rate, the largest Reynolds number is achieved for small $X$ and $d_{sphere}$ (note that velocity is not constant in this situation). The practical limits to this exercise are the difficulty in instrumenting and mounting a very small model, cavitation behind such a model subjected to very high flow velocities, and excessive pressure within the bellows. In view of the materials available and keeping in mind the requirement for a pressure tap, a reasonable lower limit to the size of the sphere was $d_{sphere}$ = 10 mm.

Due to the small size of the sphere, multiple pressure taps would have required smaller tubes and greater complexity than a single pressure tap. More importantly, their response time[5] would have been unacceptably slow within the context of the short duration experiment supported by our bellows pump. The single tap alternative requires that the sphere be rotated over a range of at least ±90° so that the pressure tap may effectively traverse the meridian between the upstream and downstream points of the sphere. A strut spans the test section, perpendicular to the flow, to support the sphere in the center of the test section (see Figure 1). The pressure is then transmitted from the sphere to the sensor located outside the tunnel via the supporting strut and a coil of capillary tubing which winds or unwinds as the sphere is rotated, similar to a watch spring.

However many pressure taps and capillary tubes one uses, one must face several challenges in performing such a measurement in liquid helium. Principal among these are

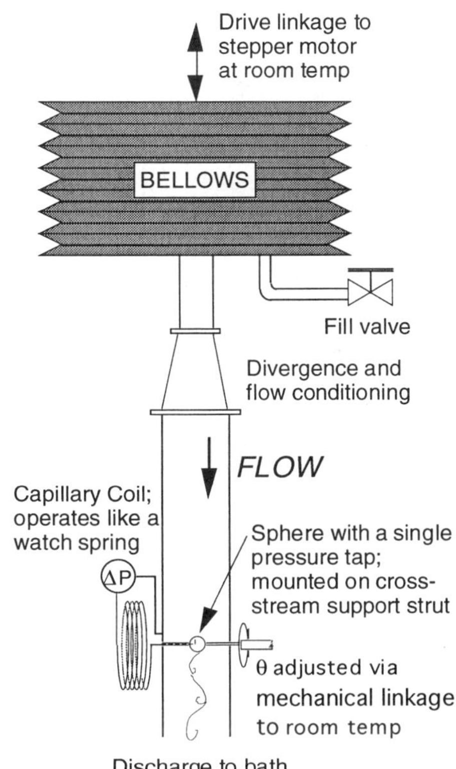


Figure 1. Schematic of the experimental apparatus. The struts provide a means of rotating the sphere so that a single pressure tap is sufficient to map out the pressure distribution as a function of $\theta$. Surface pressure is communicated outside the test section through one of the struts.

problems of vapor lock and system cleanliness. Since the experiment is carried out in saturated liquid, vapor trapped in the capillary tubes often persists until the bath is pumped down below the superfluid transition. Impurities are more difficult to deal with, and may lodge themselves in pressure taps and/or the fine mesh screens used for flow conditioning. The best cure for this is thorough purging of contamination prior to cool down, and meticulous care in preventing contaminants from entering thereafter. The pressure sensors themselves must operate in the liquid in order to avoid thermoacoustic effects. Though we have found the line of pressure sensors produced by Validyne to be most robust and reliable in our applications, one must remember that the zero drifts slightly, and is a function of both temperature and absolute pressure. The drift does not pose a problem, since the actual experiment is short in duration. Further, the response of the sensor is linear, so that one is principally interested in the voltage difference between flow and no-flow conditions. To avoid the absolute temperature dependence affecting measurements, we discharged the test section directly into the bath, as illustrated in Figure 1.

Operated in this open bath arrangement, cavitation is a continuous concern. Devices which would increase the back pressure in the tunnel cannot be tolerated, because the absolute pressure change would throw off the differential pressure measurement between the surface of the sphere and the free stream pressure. Further, cavitation occurs first in the diffuser around the various flow conditioning screens, and not behind the sphere. The entire experiment then, must be kept below enough liquid that the local total pressure at any point, including the hydrostatic head, is above the saturated vapor pressure.

## APPARATUS AND PROTOCOL

The current experiment consists of a 10 mm sphere suspended within a 40 mm circular tunnel via a thin strut ($d_{strut}$=1.07 mm), oriented perpendicular to the flow, which spans the tunnel (see Figure 2). One leg of the strut is used to communicate pressure from the single tap on the surface of the sphere to a transducer located outside the test section. This sensor measures the pressure difference between the local pressure on the surface of the sphere

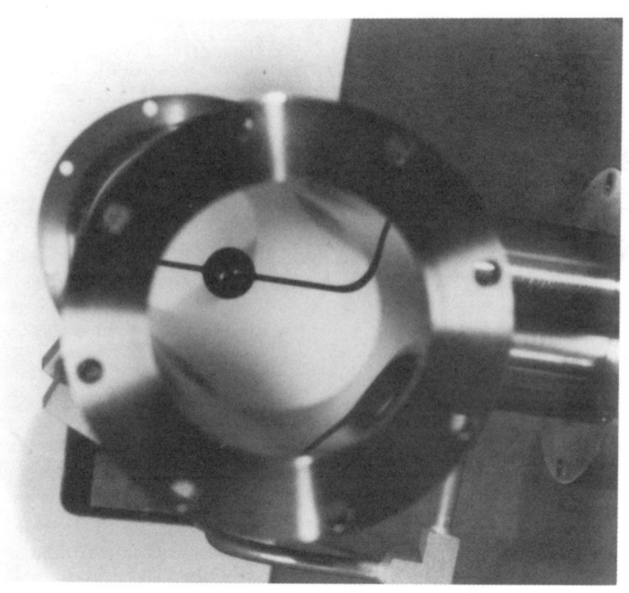

Figure 2. 10 mm sphere mounted within the test section. The 1 mm strut supports the sphere and transmits surface pressure to an external sensor (not visible).

and the free stream pressure $p_0$, accessed through a wall penetration adjacent to the sphere. The sensor in use can easily resolve pressure differences of the order 0.05 Pa. The strut is connected to a knob and drum dial/index located outside the cryostat.

The bellows pump is driven by a stepper motor, located at room temperature, and provides mass flow rates up to 65 g/s for a period of several seconds (see Equation 1), allowing us to achieve Reynolds numbers of order $10^5$ on the sphere. Flow coming out of the bellows is expanded through a divergence section with a series of intermediate screens in place to prevent separation of the flow away from the wall and the associated generation of large scale turbulence as it is slowed down and expanded. A fine stainless steel honeycomb is used to establish laminar flow and stop any large scale turbulence which may have made it through the divergence from entering the test section. A 100 mesh screen establishes a fine scale homogeneous turbulence, further breaking up any structure or pressure variations across the tunnel, which decays to quiet flow by the time it reaches the sphere.

During the initial investigation, significant experimental events (initiating the stroke, opening and closing the fill valve, setting the sphere angle, etc.) were all initiated entirely by hand. The angle, or location of the pressure tap relative to the oncoming flow, was set via the mechanical linkage. Then the fill valve (see Figure 1) was opened and the bellows was filled slowly. The fill valve was then closed, and data were recorded as the bellows was stroked downward, causing helium to flow through the test section. Final Reynolds numbers were calculated using the stroke velocity, and the recorded temperature. This procedure was repeated for each Reynolds number and angle.

The principal drawbacks to manual operation of this experiment are the relatively small data throughput and the difficulty in maintaining a uniform protocol over the course of several days. To this end, we have recently completed modifications to fully automate operations. The fill valve, bellows stroke, angle setting, and data recording/saving will all take place under LabVIEW. Besides increasing the efficiency and uniformity of the protocol, this allows us more precise control of Reynolds numbers we wish to study. Though steady state experiments may be beyond the capabilities of any bellows apparatus, automation offers the potential of an unlimited number of runs under identical circumstances. This may help us to understand the relative stability of different flow regimes.

## RESULTS AND DISCUSSION

At Reynolds numbers of $1\text{-}3 \times 10^5$ (depending upon the surface roughness), the boundary layer over the surface of a sphere transitions from laminar to turbulent. This so

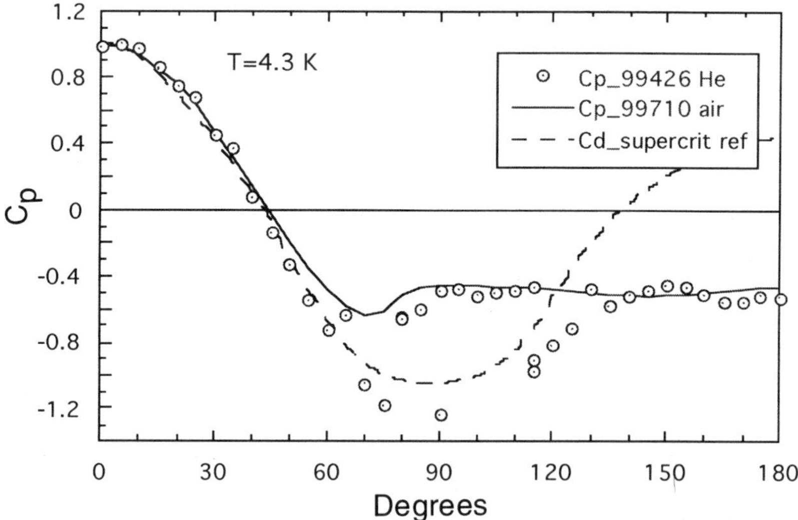


Figure 3. Observed Cp vs. angle for He I (T=4.3 K) compared with air. The solid line represents data taken for air around a sphere[7] at Re=99710. The data for He I were taken for Re=99426. The dashed line is a generalized representation of a supercritical pressure distribution[8].

called "critical transition", or "drag crisis" is characterized by a sudden shift in the point of boundary layer separation, and an associated reduction in the form drag. This shows up in a sudden drop in the coefficient of drag. The transition is particularly important in assessing boundary layer development and the behavior of turbulence at these small length scales[6].

Typical data taken manually with the present apparatus are shown in Figure 3. Here the coefficient of pressure is plotted as a function of position over the surface of the sphere for at Re=99426. The coefficient of pressure is defined as

$$Cp = (p - p_0)/\tfrac{1}{2}\rho U^2 \tag{3}$$

where p is the local total pressure, $p_0$ is the free stream pressure, and $\frac{1}{2}\rho U^2$ is the dynamic head, calculated using the oncoming velocity, U at $\infty$. The data in Figure 3 were recorded in He I. For comparison, the solid line represents data taken in the FSU Fluid Mechanics Laboratory[7] for air around a sphere in a continuous wind-tunnel for the sub-critical (below the drag crisis) Reynolds number Re=99710. The dashed line is a generalized representation of supercritical behavior ordinarily presented in textbooks[8]. Note that much of the data agrees well with the sub-critical data taken in air. The He I data presented here displays a feature common to all of our He I data taken to date; an apparent bifurcation between two possible flow regimes for the same Reynolds number (over the range of angles 60-120° in Figure 3). This bifurcation is characterized by a high degree of repeatability, and seems to alternate between sub critical and supercritical behavior. The relative stability of the two flows could not be determined with this apparatus, given the fixed displacement of the existing bellows, which limits test duration at the highest flow rates. One possibility under consideration is that the pressure tap, though small ($d_{tap}$=0.25 mm), occasionally trips the boundary layer.

Assuming azimuthal symmetry of the pressure distribution (and therefore the coefficient of pressure Cp) with respect to the oncoming flow, together with the known spherical shape of the surface, we may integrate to calculate the total drag from the measured pressure;

$$C_D = 2\int C_P \cos\theta \sin\theta d\theta \tag{4}$$

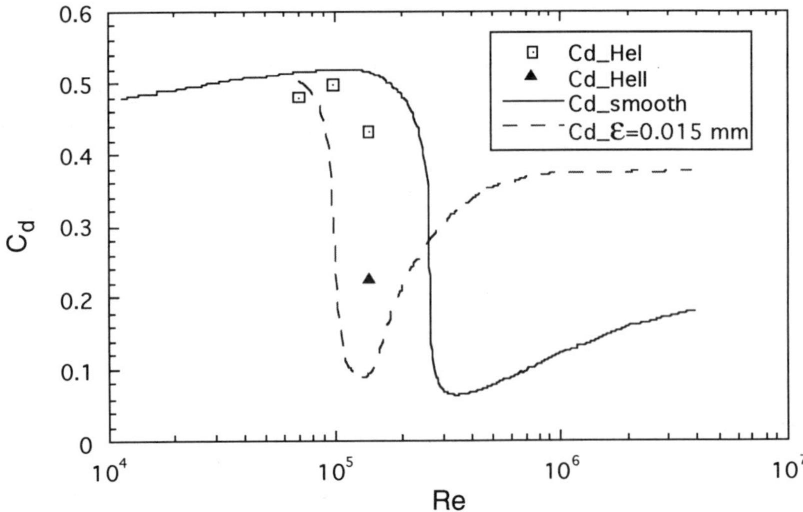


Figure 4. Calculated values for $C_d$, compared to predictions for spheres with smooth and rough surfaces[6]. Note that the He II drag coefficient, corresponding to a higher Reynolds number, drops sharply near the anticipated drag crisis. Further investigation is clearly required.

where

$$C_D = \frac{D}{\frac{1}{2}\rho U^2} \tag{5}$$

is the coefficient of drag, and D is the total drag. For the discrete data here, the contribution to the drag is calculated for each angle and summed to give the total drag, from which the drag coefficient follows quite easily. Some drag coefficients calculated by this method are plotted as a function of Reynolds number in Figure 4. The solid line shows published results for a smooth sphere, while the dashed line shows the effect a surface roughness $\varepsilon/d_{sphere}=0.0015$, or $\varepsilon=0.015$ mm would have on shifting the drag crisis to lower Reynolds numbers[8]. The sphere used in this experiment was polished stainless steel, though it is conceivable that scratches smaller than this may be present. Note that the He II drag coefficient, corresponding to a slightly higher Reynolds number, drops sharply in the neighborhood of the drag crisis. At this time, we do not have enough data to confirm whether this represents the onset of the drag crisis or an artifact of the experiment.

## FUTURE WORK

Future work will investigate smaller spheres in an effort to understand any blockage effects. The current sphere is one fourth the size of the tunnel. This may cause some local acceleration around $\theta=90°$, throwing off the measurement. We may also experiment with smaller pressure taps, to study the role of hole size in the onset of the drag crisis.

Although the present apparatus effectively probes the flow around a sphere in helium II, we are currently building a new horizontal cryostat which could house larger fluid dynamics with greater flexibility. A bellows pump would still be the most attractive choice for flow quality and ease of operation, but a larger version could provide higher flow rates as well as longer duration experiments[9]. In addition to higher Reynolds numbers, this facility will allow flow visualization studies and serve as a platform for instrumentation development, two important frontiers in this growing field.

## ACKNOWLEDGMENTS

This work was supported in part by the Naval Undersea Warfare Center, under grant N66604-96-C-A093. The National High Magnetic Field Laboratory is supported by NSF grant DMR-9527035 and the state of Florida. In addition, we wish to thank Mr. Keith Bartholomew for the many times he worked late with us, and Mr. Vince Cochran for his technical assistance.

## REFERENCES

1. R.J. Donnelly. Liquid and gaseous helium as a test fluid, in: "High Reynolds Number Flows Using Liquid and Gaseous Helium," R.J. Donnelly, ed., Springer-Verlag, New York (1991).
2. W.H. Rae and A. Pope. "Low Speed Wind Tunnel Testing, 2nd Edition," John Wiley and Sons, NY (1984).
3. X. Huang and S.W. Van Sciver. An investigation into the performance of a venturi in two-phase helium flow, in: "Advances in Cryogenic Engineering," vol 39, Plenum Press, New York (1994), p. 1065.
4. J.G. Weisend II and S.W. Van Sciver. Characterization of a centrifugal pump in he II, in: "Advances in Cryogenic Engineering," vol 33, Plenum Press, New York (1988), p. 507.
5. E. Doebelin. "Measurement Systems: Application and Design," McGraw Hill, Inc. New York (1990).
6. H. Schlichting. "Boundary Layer Theory," McGraw Hill, New York (1979).
7. C. Shih, FSU Fluid Mechanics group, private communication
8. B.R. Munson, D.F. Young and T.H. Okiishi. "Fundamentals of Fluid Mechanics, 2nd Ed.," John Wiley and Sons, Inc. (1990).
9. S.W. Van Sciver and M.R. Smith. Realization of a $10^7$ Reynolds number helium flow facility, in proceedings, International Workshop on Ultra-High Reynolds Number Flows, Brookhaven National Laboratory, 1996 (in preparation).

# STEADY STATE HEAT TRANSFER CHARACTERISTICS IN He I WITH DIFFERENT SURFACE AREA

A. Iwamoto, R. Maekawa, T. Mito, and J. Yamamoto

National Institute for Fusion Science
322-6 Oroshi, Toki, Gifu 509-52, Japan

## ABSTRACT

Accuracy of stability analyses for pool boiling superconductors depends on heat transfer from the conductor surface to liquid helium. Previously, many heat transfer measurements from copper plates have been reported.[1-4] These heat transfer surfaces were different in size and in condition. At the same heat flux, as surface area increases, the quantity of bubbles is also increasing. Free convection in the liquid helium around the surface is considered to be influenced by bubbles. It is assumed that various surface areas take different heat transfer characteristics. In this paper, the heat transfer of copper plates has been measured as a function of heat transfer surface area. We studied four types of heat transfer surface areas including : a) 10 × 18 mm, b) 18 × 18 mm, c) 40 × 18 mm, and d) 76 × 18 mm. The surface roughness is less than 10 μm. It is well-known that the heat transfer characteristics depend on surface orientation. To estimate the dependence for these samples in this study, the surface orientation was changed from 0°(horizontal, heat transfer surface upward), through 90°(vertical), to 180°(horizontal, downward) at the interval of 15°. The dependence of the heat transfer characteristics on the surface areas is discussed.

## INTRODUCTION

A lot of studies on heat transfer have been conducted; such as, dependence of the heat transfer on surface orientation,[1,2] surface treatment[3,4] and the cooling channel[3]. It is well-known that the heat transfer characteristics depend on the surface orientation because of various buoyancy effects for the free convection.[5] In the case of the channel heat transfer, the critical heat flux depends on the quality in the cooling channel.[5] The channel heat transfer also depends on the buoyancy effect for the convection. According to these studies, it is clear that the heat transfer is influenced by the condition of the free convection around the surface. In the view point of the free convection, various surface areas give some effects on heat transfer. If surface area is varied, quantity of bubbles generated from the area is changed. Convection of liquid helium around the surface is influenced by a various quantity of bubbles. Especially, at high heat flux region in nucleate boiling, a large quantity of vapor may affect free convection. It is thought that various areas of heat transfer surfaces lead the different characteristics. An investigation of the effect is important for stability analyses of pool boiling superconductors with different wetted surface areas. The study for dependence of the heat transfer characteristics on surface area has been rare.

In this study, two factors of surface orientation and surface area to change the buoyancy effect were noted. Heat transfer from copper plate with various surface areas in liquid helium was measured as changing the surface orientation. Dependence of the heat transfer characteristics, which were the critical heat flux, the minimum film boiling heat flux and heat transfer coefficient in film boiling region, on the surface area was discussed.

## EXPERIMENTAL PROCEDURE

The heat transfer of liquid helium was measured as changing the surface area and its orientation. Measurements were carried out in a glass dewar with a view window. The schematic illustration of the sample is shown in Fig.1. To thermally insulate the sample from LHe, it was covered with an FRP holder, except for the heat transfer surface. Figure 2 shows four types of the samples with the location of thermocouples, which are (a)18 mm (in width) x 10 mm (in length), (b)18 mm x 18 mm, (c)18 mm x 40 mm and (d)18 mm x 76 mm, and all copper blocks have 7.5 mm in thickness. Surfaces were polished to be less than 10 μm in roughness. The surface was heated uniformly by a thermofoil heater attached with Apiezon grease on the reverse side of the heat transfer surface. Temperature difference between the surface and liquid helium was measured with AuFe-Chromel thermocouples which were mounted 1 mm below the surface. The bath temperature was monitored, using a germanium resistance thermometer. During measurements, the bath temperature was kept at 4.22 K in the atmospheric pressure. The schematic illustration of the apparatus to vary the surface orientation is shown in Fig.3. Control of the orientation can be made from room temperature using a stainless steel rod and gears. The surface orientation was varied from 0°, that is facing upward, through 90°(vertical) to 180°, (downward) with 15° intervals.

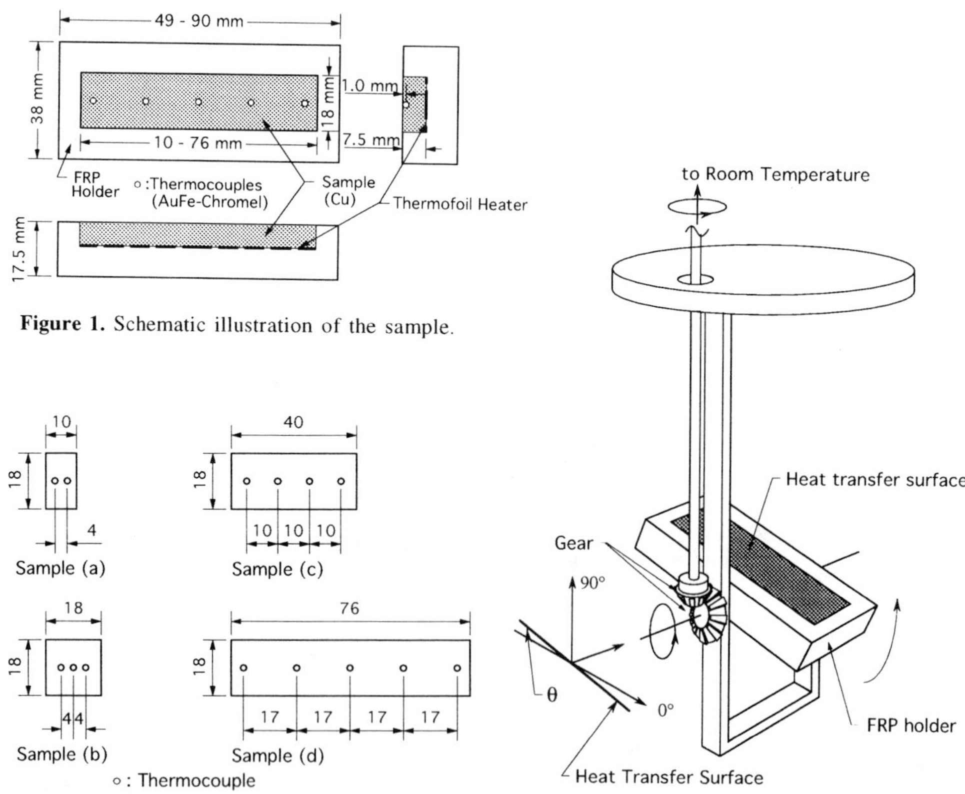


**Figure 1.** Schematic illustration of the sample.

**Figure 2.** Four types of samples with the location of thermocouples.

**Figure 3.** Schematic illustration of the apparatus to vary the surface orientation.

## RESULTS AND DISCUSSION

### Experimental Results

Figure 4 shows the orientation dependence of the critical heat flux (CHF) and the minimum film boiling heat flux (MFBHF) compared with other studies. Dependence of the critical heat flux on the surface orientation has been studied by Lyon[1] and Chandratilleke[2]. The plate sizes were 10 mm in diameter from Lyon and 20 mm in diameter from Chandratilleke. Their samples are almost the same length as the samples (a) and (b). The critical heat fluxes from Lyon and Chandratilleke almost agree with those of samples (a) and (b), respectively. According to Fig.4, it is obvious that the critical heat flux depends on the sample length. Figures 5(a) and (b) show the dependence of the critical heat flux and the minimum film boiling heat flux on the sample length at several surface orientations. The critical heat flux tends to increase at all surface orientations with decreasing sample length. At between 0° and 45°, the critical heat flux with the sample length of 18 mm is maximum. Above 45°, the critical heat flux is improved as decreasing the sample length. It is assumed that free convection around the surface is influenced by changing the quantity of bubbles generated from different surface areas. Similarly, the minimum film boiling heat flux is also influenced by the sample length. There is no difference of the sample length dependence with varying orientation. Difference of the minimum film boiling heat fluxes with various sample lengths is small.

Figures 6(a)-(d) show typical data of temperature difference versus heat flux at 0°, 45°, 90° and 135°. Temperature difference among thermocouples could not be observed. In nucleate boiling region, it is not certain that heat transfer characteristics, except for the critical heat flux, depend on sample length. Figure 7 shows the dependence of the heat transfer coefficient at 3.0 K temperature difference on the sample length at some orientations. In film boiling region, heat transfer coefficient increases with decreasing sample length. The length dependence of the heat transfer coefficient is not influenced by the surface orientation.

### Dependence of Critical Heat Flux on Sample Length

The orientation of 90° was taken as an example to illustrate the dependence. Figures 8(a) and (b) show the assumption for the sample length dependence at 90°. There is the boundary which separates liquid helium bath into two areas, as shown in Fig.8(a). The assumed heat transfer is similar to the channel heat transfer. We can define the quality (the mass ratio of vapor to liquid plus vapor mass) inside the boundary. We estimate the sample length dependence of the critical heat flux based on the concept of channel heat transfer. In the defined channel, flow velocity and quality are uniform. The channel width was decided as shown in Fig.8(b) and the channel length was equal to the sample length. The channel width depends on the sample length. It is well-known that the critical heat flux depends on the channel length and width in the case of the channel heat transfer.[5] In the channel, the quality to remain the nucleate boiling state is limited. It is called the critical quality. The critical quality ranges from 0.26 to 0.33. Nucleate boiling changes to film boiling if the quality becomes critical. Vapor filled in the channel causes the degraded critical heat flux.

In the channel, mass of vapor generated from the surface per second , $m_G$, is given by

$$m_G = qLw_{sample} \Big/ \left[\lambda + C\left(T_{vapor} - T_{liquid}\right)\right] \tag{1}$$

where q, L, $w_{sample}$, $T_{vapor}$, $T_{liquid}$, C and $\lambda$ are heat flux, sample length, sample width, vapor temperature, liquid temperature, specific heat and latent heat, respectively. Mass of vapor released from the channel per second, $m_R$, is given by

$$m_R = xvw_{sample}w_{channel}\left[x\rho_{vapor} + (1-x)\rho_{liquid}\right] \tag{2}$$

where x, $w_{channel}$, v, $\rho_{vapor}$ and $\rho_{liquid}$ are quality, channel width, velocity of flow in the channel, vapor density and liquid density. Using equations (1) and (2), the quality in the channel can be calculated, and then the critical heat flux can be estimated.

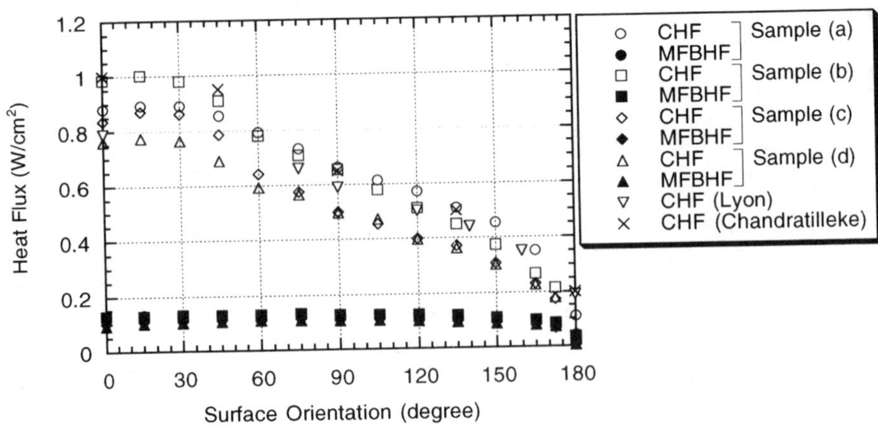


**Figure 4.** Dependence of the critical heat flux and the minimum film boiling heat flux on the surface orientation.

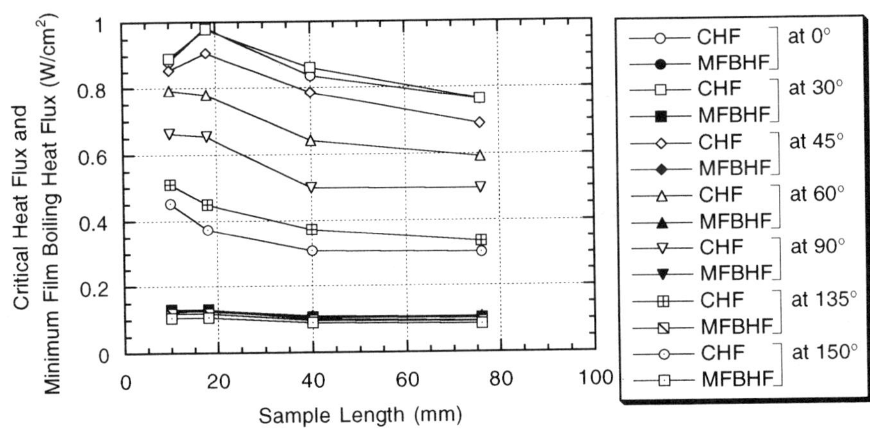


(a) Critical heat flux and minimum film boiling heat flux

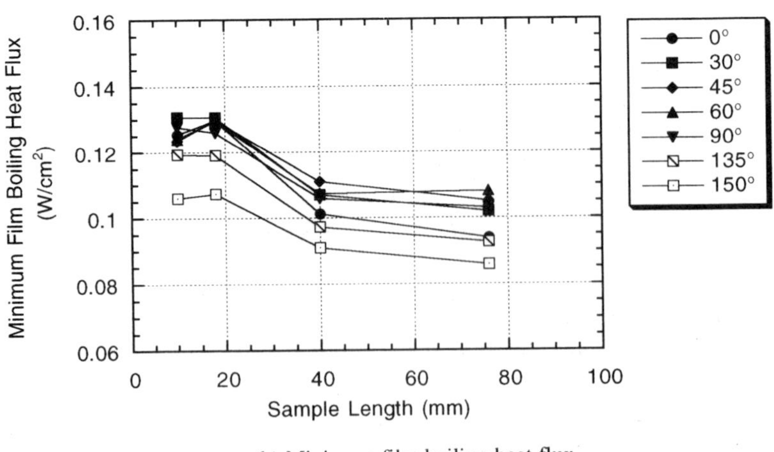


(b) Minimum film boiling heat flux

**Figure 5.** Dependence of the critical heat flux and the minimum film boiling heat flux on the sample length.

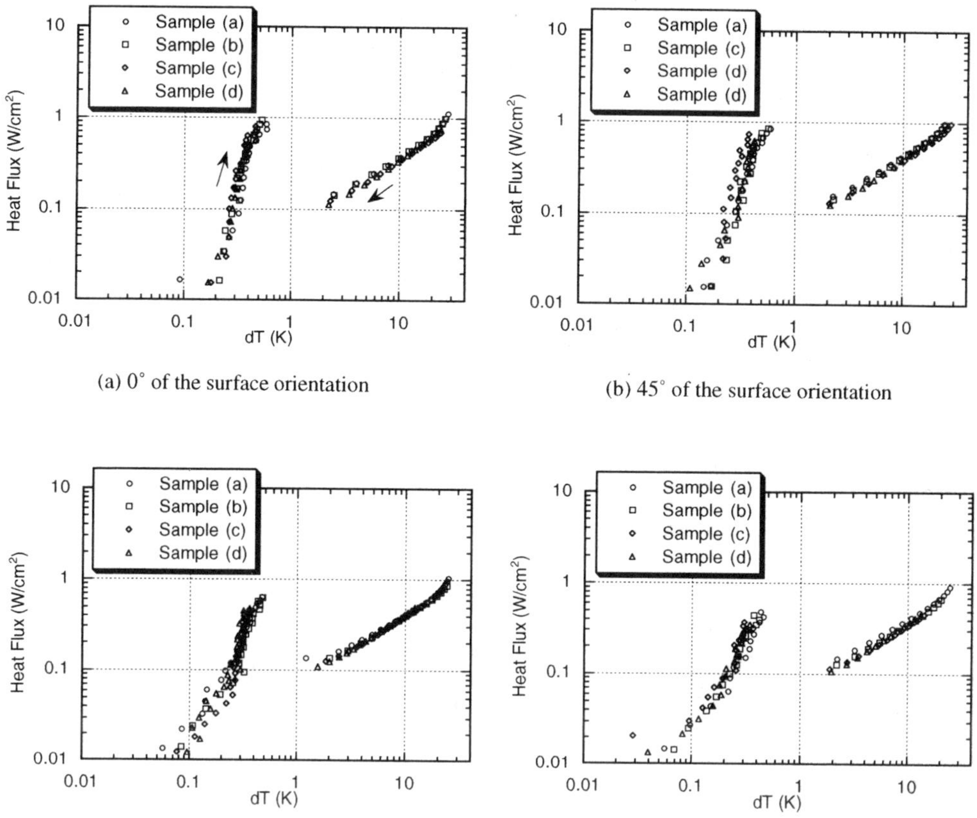

(a) 0° of the surface orientation

(b) 45° of the surface orientation

(c) 90° of the surface orientation

(d) 135° of the surface orientation

**Figure 6.** Temperature difference versus heat flux.

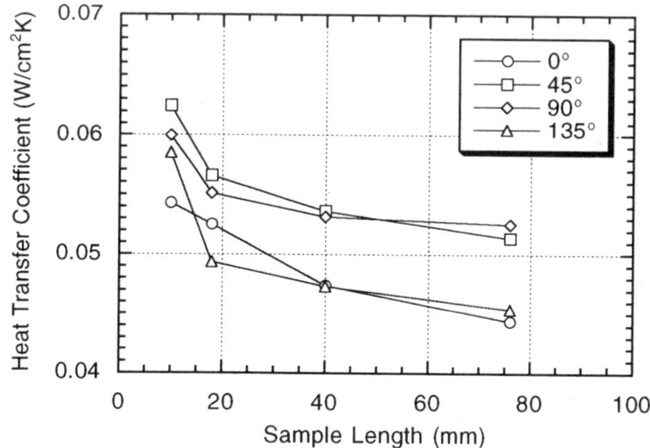


**Figure 7.** Sample length dependence of heat transfer coefficient at 3.0 K temperature difference in film boiling region.

We tried calculating the critical heat flux under the following imaginary condition. The channel widths were specified by the equation, $w_{channel}$=0.1L+1.0. The equation of the channel width was considered from our observation during experiments. When the sample lengths were 10 mm, 18 mm, 40 mm and 76 mm, the channel widths were assumed as 2 mm, 2.8 mm, 5 mm and 8.6 mm, respectively. Temperatures of vapor and liquid and the flow velocity were assumed as 4.23 K, 4.22 K and 150 mm/sec (from our observation), respectively. The critical quality was changed from 0.26 to 0.33. Figure 9 shows the calculated critical heat flux compared with those from our experimental result, Lyon and Chandratilleke. The calculated critical heat flux with 0.26 of the critical quality almost agrees with the measured critical heat flux. Around 10 mm of the sample length, the calculated critical heat flux is higher than the measured heat flux. The critical heat flux with the short sample length is not dominated by the critical quality. It is considered that quantity of vapor generated from the surface is too small to achieve the critical quality and that the critical heat flux without any effects is around 0.65 W/cm$^2$. It follows from a comparison between the measured and the calculated critical heat fluxes that the length dependence of the critical heat flux is due to the critical quality.

The length dependence of the critical heat flux is varied as changing the surface orientation because of various free convections. According to our experimental results, the length dependence with the surface that is facing downward is caused by the critical quality. The length dependence with the surface that is facing upward is also influenced by the free convection.

## Dependence of Minimum Film Boiling Heat Flux on Sample Length

According to Fig.5(b), the minimum film boiling heat flux depends on the sample length. However, the length dependence becomes negligibly small for the sample length of more than 40 mm. The length dependence of film boiling heat transfer in liquid nitrogen has been already discussed.[6] It is found that heat transfer characteristics of $LN_2$ at 90° of surface orientation with more than 15 mm of sample length do not depend on the sample length. The dependence of the minimum film boiling heat flux on the sample length does not qualitatively change by various fluids. According to Figs.5(b) and 7, there is no difference of the sample length dependence in film boiling region with varying orientation. The free convection around the surface in film boiling region has less influence for the characteristics than that in nucleate boiling region.

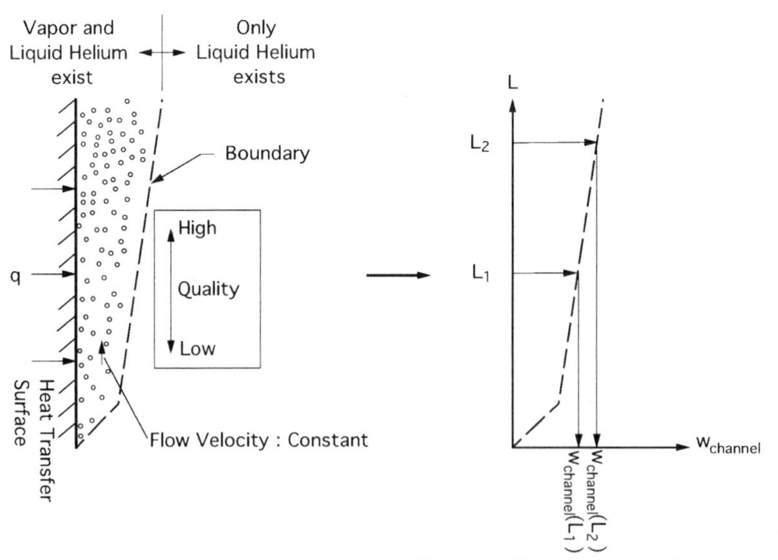


**Figure 8.** Assumption for the sample length dependence at 90°.

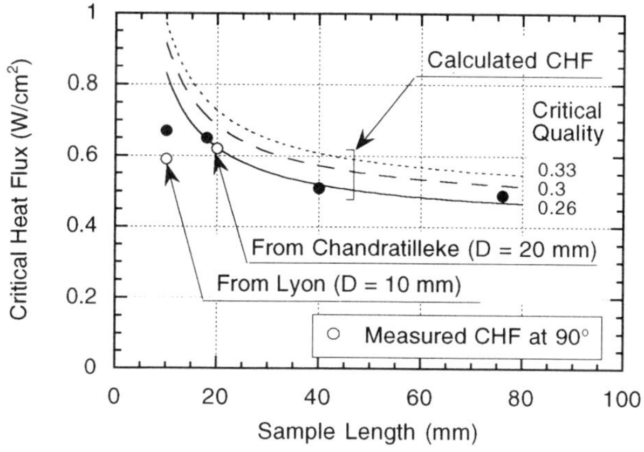


**Figure 9.** Calculated critical heat flux at 90° of the surface orientation.

## CONCLUSIONS

Heat transfer of liquid helium with various surface areas was measured as changing the surface orientation. Dependence of the heat transfer characteristics on the sample length was discussed. We obtained the following conclusions;

The critical heat flux and the minimum film boiling heat flux depend on the sample length. The critical heat flux and the minimum film boiling heat flux tend to increase at all surface orientations with decreasing the sample length. The length dependence of the critical heat flux varies with changing the surface orientation. On the other hand, the length dependence of the minimum film boiling heat flux is not affected by the varied surface orientation. There is no difference of the sample length dependence in film boiling region with varying surface orientation.

The heat transfer coefficient also depends on the sample length. The heat transfer coefficient in nucleate boiling region is not influenced by various sample lengths. On the other hand, the heat transfer coefficient in film boiling region is increased with decreasing sample length.

We could estimate the length dependence of the critical heat flux at 90° on the assumption that the critical quality around the surface dominates the dependence of the critical heat flux on the sample length. When the surface faces vertical and downward, the dependence of the critical heat flux on the sample length is caused by the critical quality, such as the channel heat transfer.

## REFERENCES

1. D. N. Lyon, Boiling heat transfer and peak nucleate boiling fluxes in saturated liquid helium between the l and critical temperature, in: "Advances in Cryogenic Engineering Vol.10," Plenum Press, New York(1965), p.371.
2. Z. Chen, et al. Channel heat transfer in He II - steady state orientation dependence, Cryogenics Vol.27:635 (1987).
3. G. R. Chandratilleke, et al., Pool boiling heat transfer to saturated liquid helium from coated surface, Cryogenics Vol.29:588 (1989).
4. A. Iwamoto, et al., Heat transfer from an oxidized large copper surface to liquid helium: dependence on surface orientation and treatment, in: "Advances in Cryogenic Engineering Vol.41," Plenum Press, New York(1996), p.217.
5. Steven W. Van Sciver, Pool boiling HeI heat transfer, in: "Helium Cryogenics", Plenum Press, New York (1986), p.199.
6. S. Nishio, et al., Natural-convection film boiling heat transfer (1st report, saturated film boiling with long vapor film), Transaction of The Japan Society of Mechanical Engineers B Vol.56:1484 (1990) (in Japanese).

# TRANSIENT HEAT TRANSFER CHARACTERISTICS OF LIQUID HELIUM IN CENTRIFUGAL ACCELERATION FIELD

O.Tsukamoto,[1] M.Furuse,[1] T.Takao,[2]
N.Tamada,[3] S.Fuchino,[3] I.Ishii,[3] N.Higuchi[3]

[1]Yokohama National University
Yokohmana, Kanagawa, 240, Japan
[2]Sophia University
Chiyodaku, Tokyo, 102, Japan
[3]Electrotechnical Laboratory
Tsukuba, Ibaraki, 305, Japan

## ABSTRACT

Transient heat transfer of liquid state helium in centrifugal acceleration fields up to 1300G have been investigated for studying the stability characteristics of the rotor field windings of a superconducting generator. In the experiment, temperature resistive platinum-0.5% cobalt (Pt-Co) thin wire was used as the temperature sensor and heater to measure the heat transfer characteristics. The thermal mass of the Pt-Co wire was small enough to observe the temperature fluctuation due to bubbling in the boiling heat transfer. In this paper, influence of the centrifugal acceleration on the transient heat transfer and its mechanism are discussed based on the experimental results. It was observed that, for step heating with heat flux higher than maximum heat flux at steady state nucleate boiling, quasi-nucleate boiling appeared also in centrifugal acceleration field at the beginning of the heating, and that duration of the quasi-nucleate boiling was not much dependent on the acceleration field for relative high heat flux. This phenomenon was explained by bubble formation process in the boiling heat transfer. In this study, a high speed video camera was used to observe the relation of the bubble formation process and the temperature fluctuation of the Pt-Co wire in the normal gravity.

## INTRODUCTION

One of the most important issues for designing the rotor field windings of a superconducting generator is estimation of the stability. A major cause of an unexpected quench of the superconducting windings is disturbance due to an abrupt wire movement.

Duration time of such a disturbance is very short, less than 1 msec. The cryogenic helium coolant of the rotor windings is subject to high centrifugal acceleration field. Therefore, for the stability analysis, transient heat transfer characteristics of helium coolant in high centrifugal acceleration field should be studied. Significant amount of data on steady state heat transfer of the cryogenic helium in centrifugal acceleration field have been accumulated [1,2], but there is a little knowledge on the transient characteristics. To measure the transient heat transfer characteristics, a quick-response temperature sensor is necessary and we used thin Pt-Co wire of 20μm diameter as the temperature sensor and heater. In the experiment, the measurement sample with the Pt-Co wire was placed in a rotating cryostat and the transient heat transfer to the helium coolant in the centrifugal acceleration up to 1300G was investigated by putting a step current to the Pt-Co wire. In the range of the experiment, the helium stayed in liquid state.

It is known for the saturated liquid helium in normal gravity (1G) that just after the step heating with heat flux higher than maximum steady state heat flux in the nucleate boiling region, vapor bubbles begin to appear around the heating surface, while the temperature rise of the heating surface is kept low as in the nucleate boiling region in the steady state boiling. This quasi-nucleate boiling is followed by transition to the film boiling where the heating surface temperature rises rapidly.[3,4] It is also known that in the nucleate boiling region thin liquid helium layer appears underneath the bubble as is shown in Figure 1(a) and that, once a bubble starts to be formed, liquid is not supplied to the thin layer till the bubble departs. Helium vapor is supplied to the bubble from this thin liquid layer and the surface temperature rise is kept low because the surface is cooled by the latent heat of the evaporation of the thin layer.[5] When the heat flux is large, the liquid in the thin layer is consumed until the bubble departs from the surface (this is the quasi-nucleate boiling region). Then, the heating surface is covered with vapor and the film boiling takes place (Figure 1(b)). In the film boiling, the heat transfer to the liquid helium is deteriorated because of the vapor layer. The vapor layer grows to bubbles and the bubbles are departed from the heating surface by their buoyancy. The buoyancy of the bubble is very large in the high centrifugal acceleration field. Therefore, bubbles depart from the heating surface more frequently and higher heat transfer flux is obtained in the higher centrifugal field. As is understood from above discussions, the heat transfer is dependent on the bubbling process affected by the centrifugal acceleration.

The Pt-Co wire in our experiment could detect temperature fluctuations due to bubbling in boiling heat transfer. In our study, high speed video pictures of the bubbling process around the wire synchronized with time trace of the temperature fluctuation of the wire were taken in normal gravity (1G). Based on these results, we studied the relation of the transient heat transfer characteristics and the bubble growing and departure processes in the high centrifugal acceleration field.

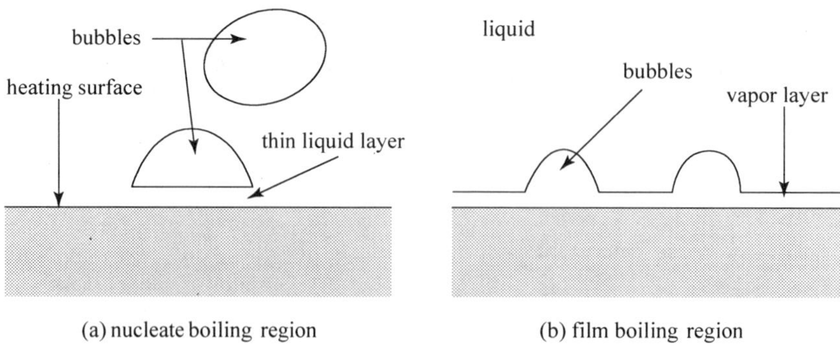


**Figure 1**. Vapor bubble formation in boiling heat transfer.

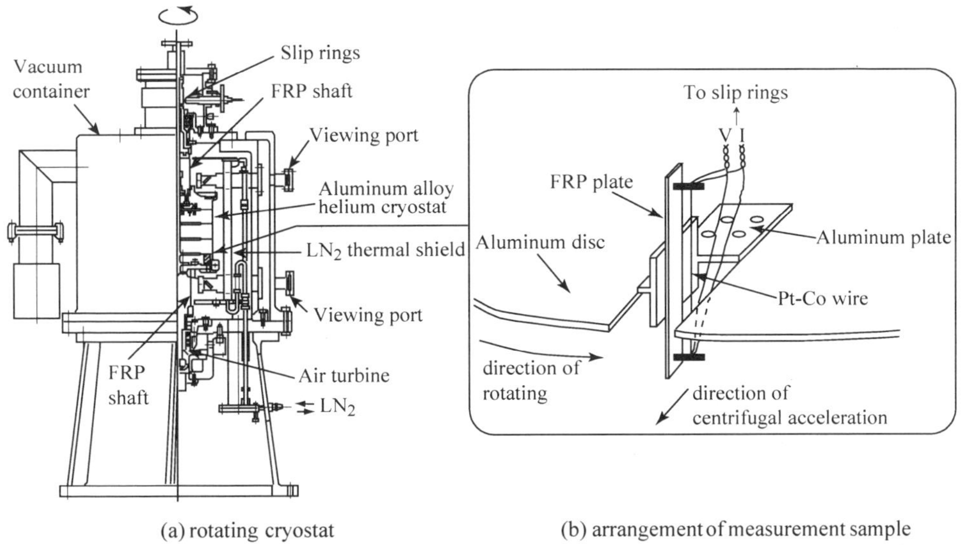


**Figure 2.** Cut-away view of rotational cryostat and schematic drawing of the measurement sample.

## EXPERIMENTAL ARRANGEMENT

### Heat Transfer Measurement in Centrifugal Acceleration Field

The experimental arrangement is shown in Figure 2. A heat transfer measurement sample with a Pt-Co wire of 20μm diameter and 48mm length (Figure 2(b)) was placed in the rotating cryostat illustrated in Figure 2(a). Details of the rotating cryostat are explained in the reference [6].

The temperature-resistance characteristic of the Pt-Co wire is shown in Figure 3. The thermal time constant of the wire was estimated as $\tau = (a^2c)/4\lambda$ , where $a$ is the radius and $c$ the heat capacity (~$9\times10^4$ J/m$^3$K) of the wire. The thermal conductivity of the wire was estimated as $\lambda$=~11W/mK from its electrical resistivity and the Wiedman-Franz relation. The estimated value of $\tau$ is $6\times10^{-8}$sec at 4.2K and $2\times10^{-7}$sec at 10K. The temperature was determined from the resistance of the wire measured by four terminal method and Figure 3. The response of Pt-Co wire was quick enough for the purpose of our experiment. However, the temperature resistivity of the Pt-Co wire is not so good as a carbon resistor and semiconductor resistor which are usually used in the heat transfer measurement in liquid helium.

A step current of $I(t)$ was applied to the Pt-Co wire and the transient wave form of the voltage $V(t)$ between the voltage tap on the wire was recorded by a transient recorder together with the current wave form. The surface heat flux $q(t)$ was calculated as $q(t)=I\cdot V/S$ where $S$ is the surface area of the wire between the voltage taps. The Pt-Co wire was so thin that we could assume that the wire surface temperature was same as the wire temperature. During the step heating, the current was kept constant. $V(t)$ changed due to the temperature change of the wire, therefore, $q(t)$ changed. Our main concern is on the transient heat transfer characteristics for a few msec after the step heating and in this time range, we assumed $q(t)$ did not change. We call $q(t)$ at the beginning of the heating the initial heat flux. Data were taken by changing the rotational speed of the cryostat from 0rpm(1G) to 3600rpm(1300G). By monitoring the liquid helium level, the pressure of the helium

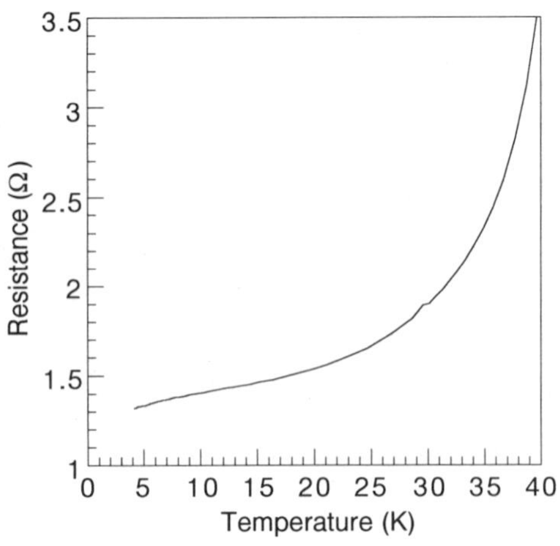


**Figure 3**. Temperature-resistance characteristic of the Pt-Co wire.

around the Pt-Co wire was calculated. In the experiment, the maximum pressure of the helium was $1.5\times10^5$Pa at 3600rpm and the helium was in the liquid state.

### Observation of Bubble Formation by High Speed Video

We observed the bubble formation around the Pt-Co wire in liquid helium of normal gravity by a high speed video camera (frame speed was 9000 frames/sec) and measured the transient heat transfer characteristics at the same time. A measurement sample arrangement similar to that shown in Figure 2(b) was used in this experiment.

## EXPERIMENTAL RESULTS

### Bubble Formation Process and Heat Transfer Characteristics in Normal Gravity

Figure 4 shows time trace of the temperature of the Pt-Co wire and high speed video pictures of the vapor bubbles around the wire, where initial heat flux $q_0$=35kW/m$^2$ and the gravity is 1G. The pictures (1)~(8) in Figure 4(a) show the bubbles at the times denoted by (1)~(8) in the temperature time trace shown in Figure 4(b).

There is a remarkable overshoot in the temperature trace at (2) in Figure 4(b), which is followed by a temperature dip at (3) and (4). At (2), the wire is covered by vapor sheath and heat transfer to the liquid helium is deteriorated because of the vapor sheath. After that, the vapor sheath swells and starts to undulate and be divided into bubbles, while convection of the vapor around the wire begins and the wire temperature goes down. When these bubbles depart from the wire, liquid helium around the wire is replenished and the wire is cooled well having the temperature dip around at (3) and (4). After these relatively large initial bubbles depart from the wire, relatively small vapor bubbles are formed and depart. The cycle of this formation and departure of the bubble is repeated as shown in Figure 4(a)(6)~(8). Obviously from Figure 4(b), there is cyclic fluctuation of the wire temperature between (6) and (8) coinciding with the cycle of the bubble formation and departure. About 100msec after the start of the step heating, the steady state film boiling is established.

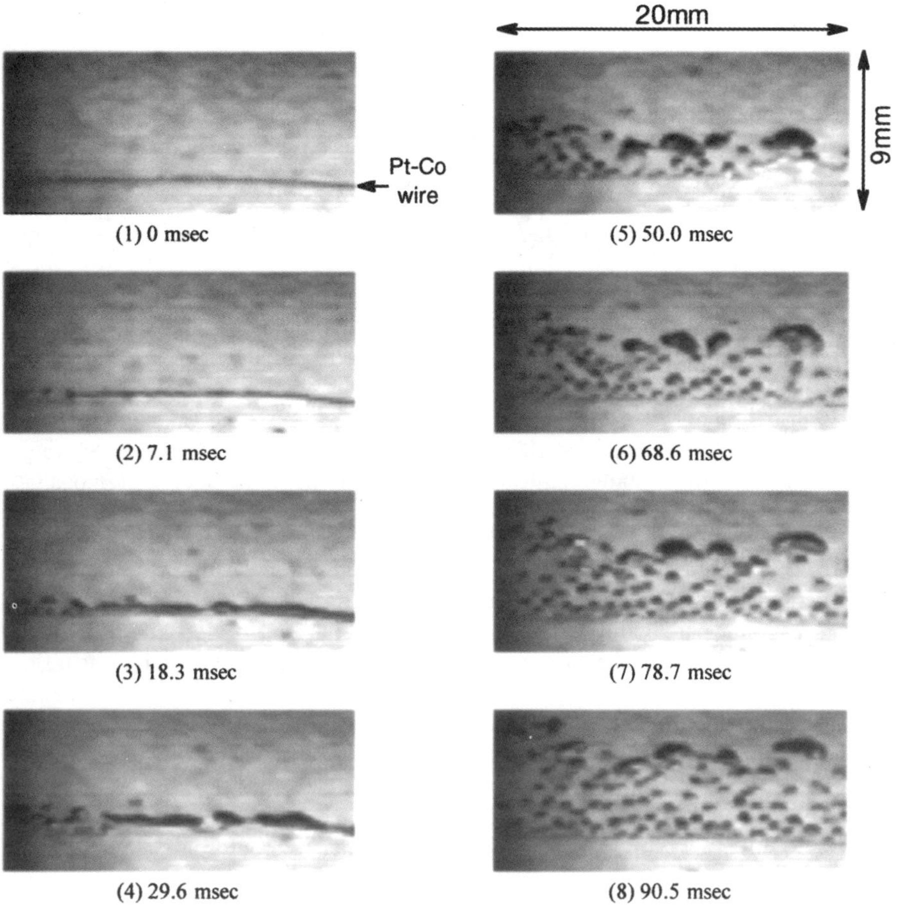

20mm

9mm

Pt-Co wire

(1) 0 msec

(5) 50.0 msec

(2) 7.1 msec

(6) 68.6 msec

(3) 18.3 msec

(7) 78.7 msec

(4) 29.6 msec

(8) 90.5 msec

(a) high speed video pictures of bubbles around the Pt-Co wire

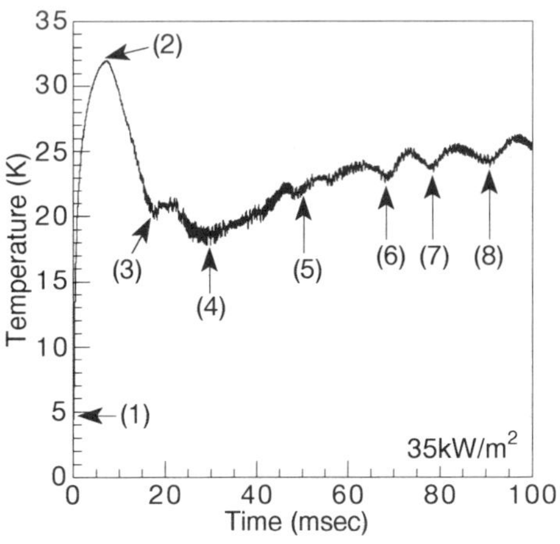


(b) temperature traces of the Pt-Co wire

**Figure 4**. High speed video pictures and temperature trace of the Pt-Co wire for step heating of $q_0$=35kW/m$^2$. The step heating started at $t$=0.

**Transient Heat Transfer in Centrifugal Acceleration Field**

Figure 5 shows temperature traces of the Pt-Co wire for step heating of $25kW/m^2 \sim 42kW/m^2$ in various centrifugal acceleration field together with the case of 1G. The quasi-nucleate boiling was obviously observed also in the centrifugal field. Based on the results shown in Figure 5, specific characteristics of the transient heat transfer in the centrifugal field are summarized as follows.

**1**. In the range of the experiment, there are three modes in the heat transfer characteristics in the centrifugal field;

(i) steady state in the nucleate boiling region (Figure 5(a)).

(ii) gradual transition from the nucleate to the film boiling regions (Figure 5(b)and(c)).

(iii) take off from the quasi-nucleate boiling to the film boiling regions (Figure 5(c)~(e)).

whereas, there is one mode for the case of 1G, where taking off from the quasi-nucleate boiling to the film boiling. Occurrence of the mode depends on the value of G and $q(t)$. It is considered that, in the mode(ii), local nucleate boiling regions and local film boiling regions coexist and that the film boiling regions gradually overwhelm the nucleate regions.

**2**. Take-off time is plotted against the heat flux for various values of G in Figure 6. The take-off time is defined as the time from the start of the step heating to the time when the wire temperature takes off due to the transition to film boiling. Obviously from the figure, when the quasi-nucleate boiling appears, the take-off time is not dependent on the value of G and data can be fitted to the line $q_0=2.6\times10^4 t^{-0.25}$, where $q_0$ is the initial heat flux. This can be explained as follows, by assuming that the thickness of thin liquid layer underneath the vapor bubble (Figure 1(a)) is not affected by the value of G. When the thin liquid layer is consumed before the initial bubbles depart, the wire temperature takes off and the film boiling begins. The time to consume the thin liquid layer, that is the take-off time, is not dependent on the value of G because the thin liquid layer thickness does not depend on the value of G. When the vapor bubbles depart from the wire before the thin liquid layer is consumed, the nucleate boiling continues. In the high centrifugal field, the bubbles depart earlier than in normal gravity because of larger buoyancy of the bubble, and nucleate boiling is sustained in higher heat flux.

**3**. In the case of the mode (iii), the steady state film boiling is established in less than 1msec for the centrifugal field higher than 400G, which is much faster than the case for 1G.

**4**. In the steady state film boiling region, there were cyclic fluctuations in the wire temperature. These fluctuations are supposed to be caused by cyclic departure of vapor bubbles as was observed by the video camera in the normal gravity experiment. Average values of the cycle times of the fluctuations were not much dependent on $q(t)$ and are plotted against G in Figure 7. The average cycle time is dependent on $G^{-1/3}$. This dependence on $G^{-1/3}$ was also observed by a direct optical observation of the cyclic bubble departure time in other experiment conducted by using a flat heating surface.[5]

**5**. A temperature overshoot as was observed in the case of 1G was not observed at the beginning of the step heat in the centrifugal field except in relative low value of G (100G). The reason for this is considered as follows. In the high centrifugal field, the buoyancy of the bubble is much higher than that in 1G, therefore, convections of the liquid and vapor around the wire are established much faster and the initial bubbles depart from the wire before growing big. Without the big initial bubbles accompanying temperature overshoot, the film boiling reaches steady state right after the take-off in high centrifugal field. This supports the result described in the specific characteristic **3**.

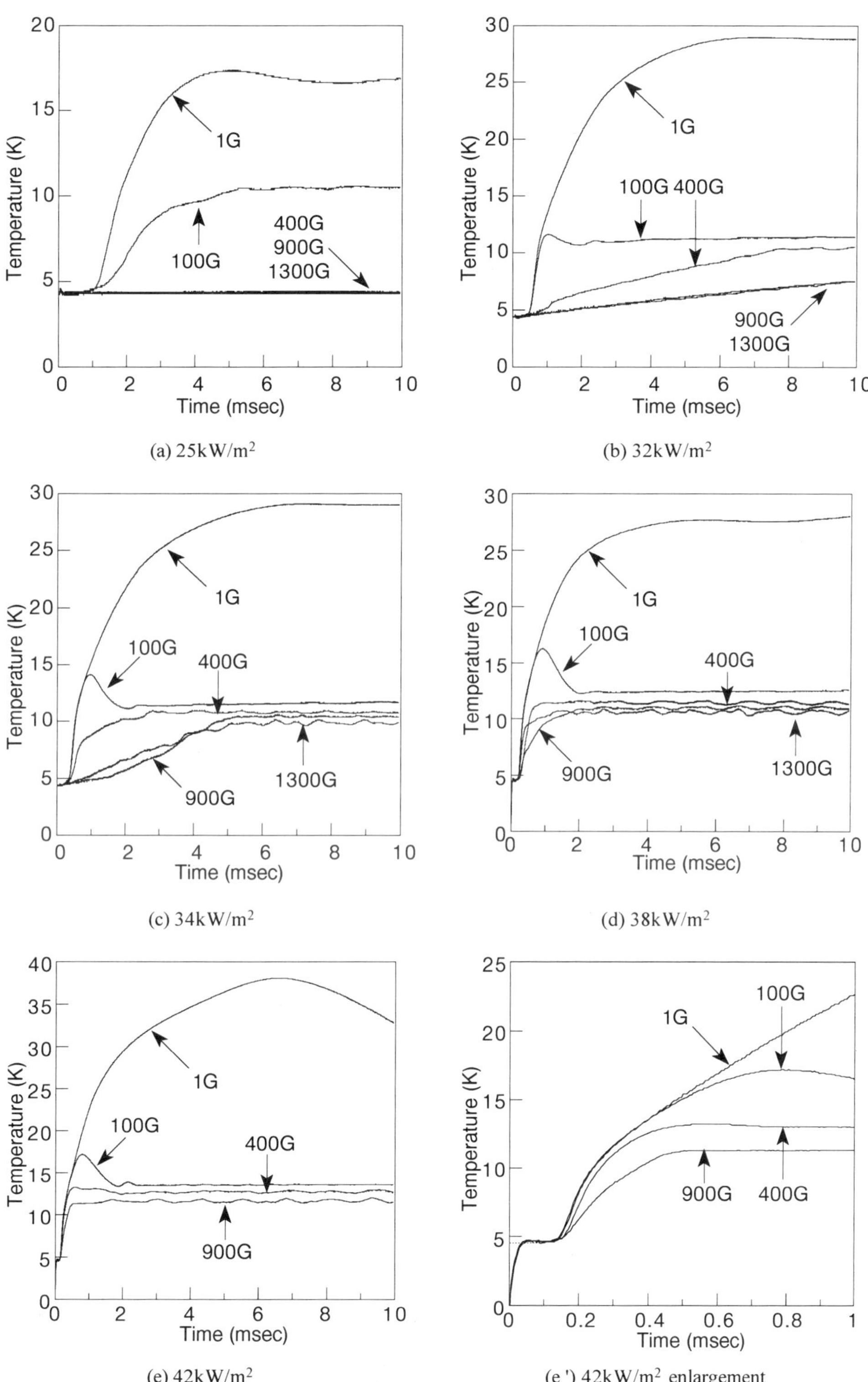


**Figure 5.** Transient traces of the Pt-Co wire for step heating in centrifugal field.

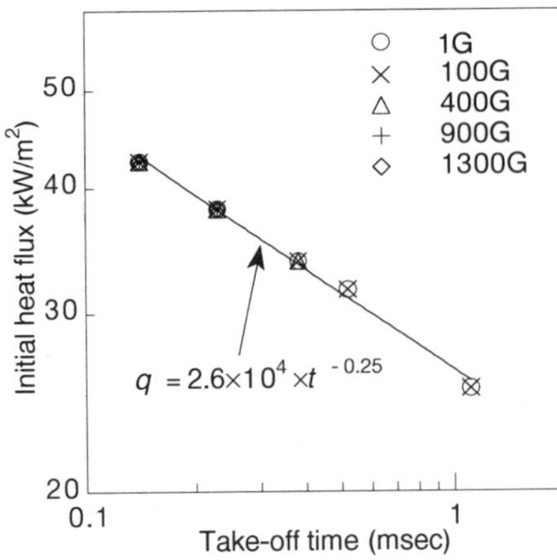


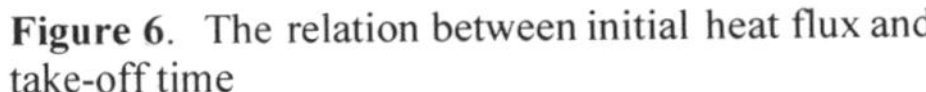

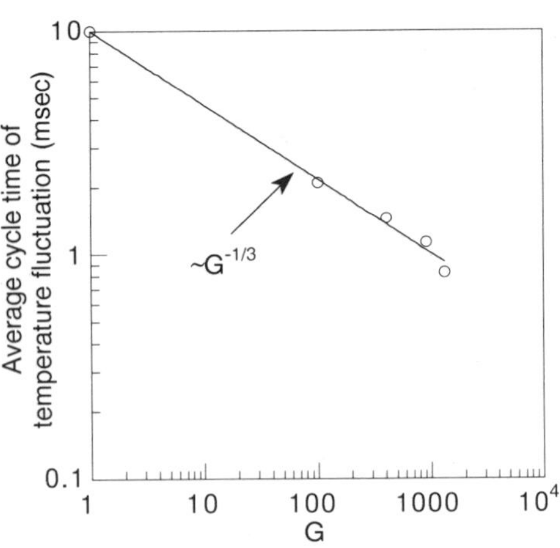


**Figure 6**. The relation between initial heat flux and take-off time

**Figure 7**. Plots of bubble departure time vs. centrifugal acceleration. Bubble departure time is proportional to the -1/3 power of centrifugal acceleration.

## CONCLUDING REMARKS

To the knowledge of the authors, transient heat transfer characteristics of liquid helium have been measured for the first time. In the experiment, it was pointed out that the quasi-nucleate boiling was observed also in the liquid helium in centrifugal acceleration field and that the relation between the take-off time and the heat flux was same as that in the normal gravity.

Due to the limitations in the rotating speed of the rotating cryostat, the centrifugal acceleration was limited up to 1300G and the pressure of the helium was up to $1.5\times10^5$Pa where the helium stayed in liquid state. Whereas, in the actual superconducting generator, state of the coolant of the field windings is supercritical helium of several atms and the centrifugal acceleration field is in the range of several thousand G. Moreover, in our experiment, the heater was a thin wire. As is known, the heat transfer characteristics of a thin wire is somewhat different from those of a flat surface and thick wire. Therefore, we can not directly use our experimental results as the heat transfer data to analyze the stability of the field windings. However, our results give the following important information on the stability analysis. The film boiling heat transfer reached the steady state in very short time less than 1msec for high heat flux in 1300G because of the high centrifugal acceleration enhanced the convection of the coolant. Therefore, we may use the steady state heat transfer data measured in high centrifugal acceleration field for the stability analysis which needs the heat transfer characteristics in the time range of less than 1msec.

## REFERENCES


1. R.G.Scurlock and G.K.Thornton, *Int J Heat Mass Transfer* Vol.20, pp.31-40 (1977).
2. S.Fuchino, H.Fukuda, T.Ogawa, K.Shimizu, Y.Nakabayashi, Y.Kobayashi, M.Ogihara, Y.Kamisada, N. Maki, A.Ueda, T.matsuda, *IEEE Trans. Mag* Vol.28, No.1, pp279-282 (1992)
3. O.Tsukamoto, S.Kobayashi, *J Applied Physics* Vol.46, No.3, pp1359-1364 (1975)
4. O.Tsukamoto, T.Uyemura, Y.Ishida, *Proc. of ICEC8*, pp.251-255 (1980)
5. O.Tsukamoto, *Cryogenic Engineering* Vol.11, No.3, pp.23-33 (1976) *IN JAPANESE*
6. N.Tamada, S.Fuchino, N.Natori, I.Ishii and M.Okano, *Cryogenics*, Vol.33, No.11, pp.1023-1027 (1993)

# A LIQUID HELIUM FILM HEAT PIPE/HEAT SWITCH

M. J. DiPirro, P. J. Shirron, and D.C. McHugh

Code 713, NASA/Goddard Space Flight Center
Greenbelt, MD 20771, USA

## ABSTRACT

We report on the development of a compact, high conductance heat pipe/heat switch using liquid helium and a wick material. As a heat pipe its operation is similar to that of capillary pumped loop heat pipes except that it is very small. The device is also used as a heat switch by attaching a getter which can be heated. The helium is desorbed from the getter heating to ~35K by using a small (< 3mW) power and the switch is then on. The on state conductance of the first device tested was over 400 mW per Kelvin at 1.8K. Unheated, the getter quickly cools and adsorbs enough helium so that the heat switch turns off. For the first device tested the switch turned off at a getter temperature of 24K. The off conductance was 3.5 microwatts per Kelvin at 1.5K. This first version had a maximum heat flow capability of 4 mW. Future switches will be designed for a larger temperature and heat flow operating range and decreased off state conductance. This device was designed to replace gas-gap heat switches currently used in adiabatic demagnetization refrigerators.

## INTRODUCTION

There are a variety of uses for heat switches and thermal links which operate at liquid helium temperatures. While this device was designed specifically for use with an adiabatic demagnetization refrigerator, its design is flexible and may be used in many other applications. In addition, in the on state the heat switch becomes such a good thermal conductor that it may be used in place of metal heat straps.

An adiabatic demagnetization refrigerator requires a heat switch to alternately make good thermal contact between the paramagnetic salt and a low temperature heat sink (such as a superfluid helium reservoir) during magnetization and to isolate the salt from the heat sink during demagnetization. Currently gas-gap heat switches[1,2], mechanical heat switches, and $^3He/^4He$ diodes[3] are used for this purpose.

Key parameters in the design of any heat switch are size, flexibility, ease of operation, heat leaks, on and off state conductances, amount of heat dissipated in activating the switch, and turn off and turn on times.

The gas-gap heat switch consists of a sealed volume containing closely spaced copper plates separated by a small gap. Gaseous helium fills the gap in the "on" state and is pumped out leaving a vacuum between the conductors in the "off" state. Because of the limited conductivity of the helium gas, the conductor area must be large (~50 cm$^2$) and the gap between conductors small (~0.25 mm) to achieve a conductance[2] of 75 mW/K. The copper must also be substantial to carry the heat with little temperature drop and to maintain the gap without deflecting under normal loads. This leads to a large, heavy heat switch with a relatively high "off" state conductance due to the surrounding structure[2]. On the other hand this type of switch has a reasonable conductance at higher temperatures which can decrease the cool down time of a large isolated mass from room temperature.

In contrast mechanical heat switches have a very low "off" state conductance. The "off" state conductance is not exactly zero due to the need to stiffen the suspended salt against any uncompensated mechanical forces during actuation). On the other hand their "on" state conductances tend to be small. Applying the necessary large force required to make a reasonable "on" state conductance requires an electromagnetic actuator, or a hydraulic or mechanical connection to outside the dewar. This will usually result in a larger parasitic heat into the cryostat either from tubes or higher current wires into the dewar. Of course, mechanical heat switches can also be used to speed cool down from room temperature if the actuator mechanism and any wiring allow room temperature operation.

The $^3He/^4He$ diode heat switch first used by Fairbank[3] is a passive device which operates through the "heat flush" of $^3He$. Because it is passive it is inherently easy to operate. It has a large on/off ratio, but cannot be used when the hot end temperature exceeds 1.4K. It also typically requires an external reservoir of the $^3He/^4He$ mixture for storage between cool downs unless the switch itself is designed to hold a very high pressure (~80 MPa).

The liquid helium film heat switch uses a similar getter on/off activation to the gas-gap heat switch, but a film is actually condensed on a substrate within a tube. This substrate allows film flow from the cold to hot end of the tube and subsequent evaporation and reflux of the helium vapor. The mass flow of liquid helium provides a very large conductance in the "on" state. The effective conductivity of this type of switch can be tens or hundreds of watts per cm K at low temperature. A drawback to this type of switch is that it is not efficient at temperatures above the point where liquid condensation takes place, and thus will not aid cool down from room temperature.

This heat switch can also be adapted for use as a heat pipe. Liquid helium mass flow is a much more efficient heat transporter than copper. It will therefore be smaller and lighter than a copper heat strap. It can be contained in very small, flexible tubes which allows ease of use in tight spaces or where mechanical isolation is important. When used as a heat pipe the tube would be sealed helium at room temperature. The pressure of helium used and the substrate area will determine the temperature range of use and the maximum heat flow capability of the heat pipe.

## THEORY

In this heat switch the heat is conducted by mass transport of liquid from the cold to hot end of the switch, followed by evaporation, then mass transport of the vapor back to the cold end, followed by condensation. Two mechanisms may be used to transport the liquid: surface tension or superfluid film flow.

### Bulk Transport

Bulk liquid is transported through surface tension. This is the type of phenomenon at work in the capillary pumped loop form of heat pipe which was first introduced in the 1960's[4]. Cryogenic capillary pumped loops are currently being developed for use with mechanical coolers at higher temperatures using liquid nitrogen and liquid neon. When liquid is evaporated at the warm end, a meniscus forms which draws more liquid into the area. Meanwhile, the evaporated vapor recondenses at the cold end of the switch/pipe due to vapor pressure differences. The limiting process to the heat flow is usually the velocity of liquid transport from the cold to warm end. The balancing forces driving the liquid flow are the surface tension and viscosity of the liquid. Either normal or superfluid may be used in this mode. The surface tension force is given by the radius of curvature, $r_c$, of the meniscus and the surface tension of the liquid, $\sigma$:

$$P_s = \frac{2\sigma}{r_c} \tag{1}$$

In a tube of circular cross section the maximum radius of curvature is the radius of the tube. The viscous drag on the liquid is given by

$$\Delta P_d = V\eta Z \quad (2)$$

where V is the volumetric flow rate, $\eta$ is the liquid viscosity, L is the tube length, and Z is the impedance, which for a circular cross section tube of radius, r, is

$$Z = \frac{8L}{\pi r^4} \quad (3)$$

One can see that in the absence of other forces, one would like as large a diameter as possible since the viscous drag increases as the inverse fourth power of the tube radius where the surface tension force only increases as the inverse of the radius. However, gravitational or other acceleration forces tend to play a big role. For instance, for superfluid helium in Earth's gravity a tube radius of 0.005 mm or smaller is required to raise the liquid 10 cm by surface tension. If we then use 5 microns as a typical pore size in a wick, the viscous losses will be given by

$$\Delta P = 8v\eta L/r^2 \quad (4)$$

where v is the flow velocity and the net driving force will be the surface tension force minus the gravitational force giving the relation for equilibrium flow velocity of:

$$v = \frac{r^2}{8\eta L}(2\sigma/r-\rho gh) \quad (5)$$

For a tube of length L = h = 5 cm, a liquid viscosity of 3.252 microPa s (using conservatively large 4.2K value), the velocity of flow would be 1.25 mm/second. For a single 5 micron radius tube this would only be a volume flow rate of about $10^{-7}cm^3$ per second. There would therefore need to be more than 3300 of these tubes in parallel to make a device capable of carrying a milliwatt of heat. Of course, such large numbers would be possible in a wick material.

**Film Transport**

For superfluid film flow, the flow is driven by a gradient in the chemical potential caused by the temperature gradient. The flow rate is usually limited by the critical velocity and the film thickness. The film thickness is determined from the van der Waals forces of the substrate, the vapor pressure, gravitational potential, and temperature. It has been experimentally shown that for superfluid film thicknesses between 2 and 35 nm, the critical velocity of the film is inversely proportional to the thickness[5]. This means that the mass transport rate per perimeter of flow tube is constant over a wide range of film thicknesses. For thicker films the mass transport rate will be higher as the critical velocity falls much slower than the inverse of the thickness. For thinner films the critical velocity stops increasing and eventually falls to zero at film thicknesses less than 1 nm. The flow rate constant for 2 to 35 nm thick films was predicted to be $7.9x10^{-5}$ $cm^3$ per second per centimeter of perimeter of the surfaces available to the film in reference 5 and has more recently been measured in a ultra low temperature apparatus and in a sensitive film flow test to be about $1.25x10^{-4}$ $cm^3$ per second per centimeter of perimeter[6,7].

Meanwhile, the thickness of film in an acceleration environment, such as that produced by a gravitational field, is determined by the balance of gravitational and van der Walls forces. For a range of materials the van der Waals force is such that the thickness of a liquid helium film is about 30 nm at a height of 1 cm above a liquid reservoir, and decreases as the height to the 1/3 power. This means that when operated vertically, if sufficient helium exists, the film thickness will be in the range of constant mass transport given above and gravity will have no effect. For just one centimeter of perimeter, then, one can expect a mass transport rate of $(0.145 g/cm^3)X(1.25x10^{-4} cm^3/s) = 1.82x10^{-5}$ g/s or a heat flow rate of 0.4mW.

Up to the limiting mass transport rate, the "on" state conductance can be so high that the overall heat switch conductance will most likely be limited by the joint conductances between the heat switch and the thermal reservoir and the device to be cooled. Above the critical film velocity, large temperature differences will be developed between the two ends of the heat switch, making it a heat current limiter.

The turn off time of the heat switch is limited by the heat of adsorption of the helium gas onto the getter, the heat capacity of the getter and its container, and the thermal conductance of the getter to the helium reservoir. The heat of adsorption of helium onto zeolite is about 250 J/mole. If one has $9cm^3$ of helium at normal temperature and pressure this is then about $3.7x10^{-4}$ mole and the total heat that must be absorbed by the getter is about 91mJ. Cooling the gas that remains in the getter housing down from 35K to low temperature is a similar sized heat. The heat capacity of the getter housing turns out to be a bigger issue and it is minimized by using thin wall materials. The thermal conductance of the getter to the helium reservoir is traded off versus the amount of power required to maintain the getter at operating "on" temperature. An advantage for this type of heat switch relative to the gas-gap heat switch is that the switch is completely off when the film is thinned to a low enough value (< 1nm). This can occur at a relatively high getter temperature, which is reached relatively quickly. The gas-gap heat switch is not completely off until the pressure is reduced to a negligible value which only occurs when the getter falls to close to 4K.

It is important to put enough helium into the switch so that a sufficiently thick liquid film exists for all temperatures of operation. A superfluid film may be used up to close to the superfluid transition temperature of 2.177K. To operate at 2.15K for instance, enough helium must be present in the switch to fill the switch volume with vapor at a density of 0.001127 $g/cm^3$ (the saturated vapor pressure at 2.15K) and still have enough liquid to coat the surfaces with a 2 nm or greater film thickness. One way to achieve this is to fill the switch at room temperature with gaseous helium at a pressure of at least 7 atmospheres. However, lower pressure helium may be used in the heat switch if compression is utilized. If the getter part of the heat switch is large compared to the cold volume, then when the getter is turned on very little of the gas is left in the warm volume but is "compressed" into the cold part of the switch.

## EXPERIMENT

A sketch of the heat switch is shown in Figure 1. The proof-of-concept version of the capillary heat switch tested was constructed from parts on hand and was therefore not optimized for performance. The hot and cold ends of the switch were made of OFHC copper. The wick, which ran from the hot to cold end, was a 200 pound test braid of Kevlar 29. This braid consists of 3600 strands of 12 micron diameter fibers. The wick fibers are splayed out at the ends and mechanically attached to the copper end pieces. The wick was surrounded by a 3.35 mm outer diameter, 0.11 mm thick wall stainless steel tube. The off conductance of the switch is almost completely due to this stainless steel. The getter used was zeolite Linde 13x. The zeolite pellets were bonded to an aluminum cap with silver filled epoxy. The cap was then sealed to a copper walled cylinder with indium. The cylinder was

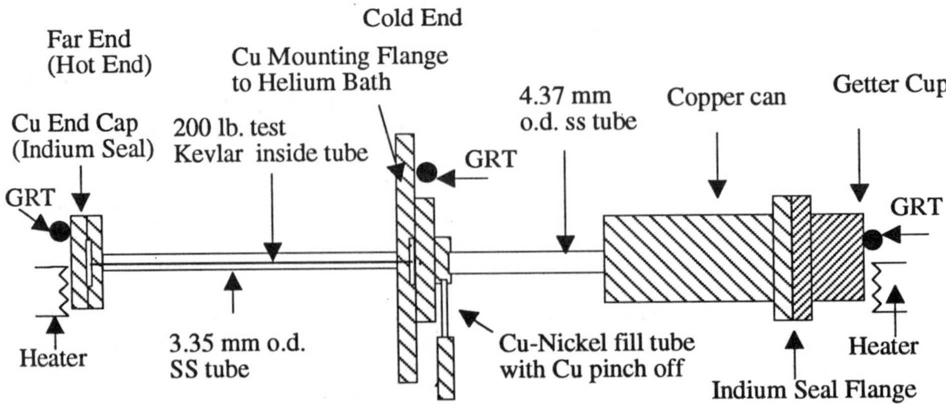


**Figure 1.** Schematic of the proof-of-concept heat switch.

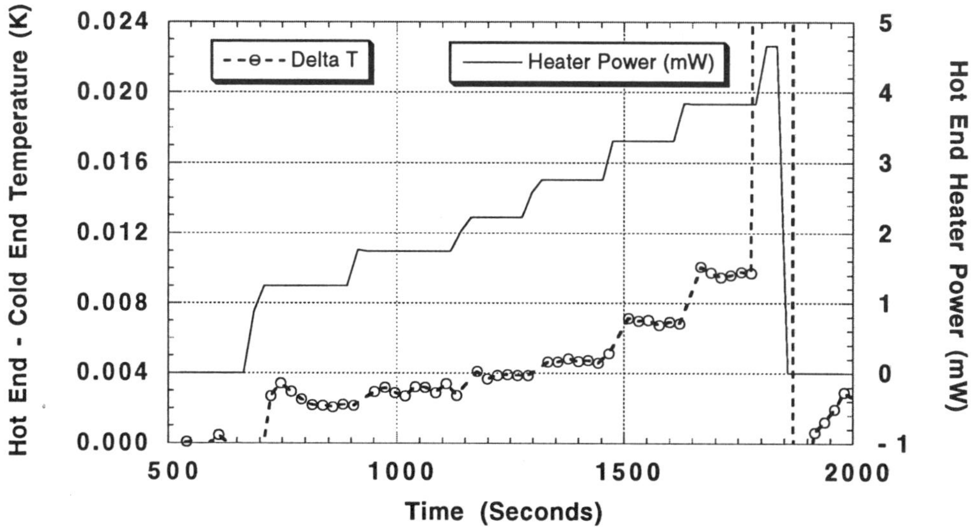


**Figure 2.** Typical data for "on" state. The heater power on the hot end is gradually increased until the heat switch stops conducting as indicated by very large temperature differences between the warm and cold ends. The heater power is then turned off and the large temperature difference quickly disappears.

sized to provide enough volume so that when heated to the getter operating temperature of 35K the helium expelled from the getter would be compressed to about 8 times its room temperature density. The entire switch was sealed with one atmosphere of GHe in it at room temperature. With the total volume of the device of 9.0 $cm^3$, this is 1.46 milligrams of helium. As calculated in the previous section, the heat of adsorption of this amount of helium is about 91mJ. Meanwhile the calculated enthalpy of the getter cup up to 35 K is about 1 J. Thinner materials will be used in future models to lower the enthalpy decreasing the cool down time and hence the time required for the switch to turn off.

A calculation of the total perimeter of the fibers gives 13.6 cm. Multiplying this by the superfluid flow rate in the previous section gives a total film mass flow rate of $1.7 \times 10^{-3}$ $cm^3$ per second. This translates into a mass flow rate of $2.47 \times 10^{-4}$ g/s and a heat transfer rate of 5.77 mW based on the latent heat of vaporization of 23.36 J/g at T=1.8K.

The volume of the heat switch from cold to hot end is minimized and the hot getter volume is maximized to ensure that more of the helium is condensed in the on state. The trade-off here is that enough space needs to be provided for the wick, and the hot getter volume must have minimum mass to allow the switch to turn off quickly when turning the switch off. Of course, the conductance of the getter to the liquid helium reservoir can be increased to shorten the cool down time at the expense of a higher heat load on the reservoir when the getter heater is on. The first heat switch had a cold volume of 0.7 $cm^3$ and a warm getter volume of 8.3 $cm^3$. Without compression, one atmosphere at room temperature would first start to condense into a liquid at 1.51K. With the compression of this first unit some liquid would be present at temperatures up to 2.19K. In addition to filling the volume with saturated vapor, the surfaces must be covered with a liquid film at least 2 nm thick. For the wick used in this test the fibers were 10 cm long (individual fibers are 1/3 longer than the braid itself). Using the perimeter above, the total surface area of fibers was 136 $cm^2$ and therefore the total liquid film volume required was > $2.7 \times 10^{-5}$ $cm^3$. At a density of 0.145 $g/cm^3$ this is 3.9 micrograms, which is a small fraction of the total helium in the system.

## RESULTS

The switch required approximately 3 mW to heat the getter to ~35K desorbing the majority of the helium. The on state conductance was measured to be over 350 mW per Kelvin at temperatures up to 1.82K. (See Figure 2, for example). Unheated, the getter

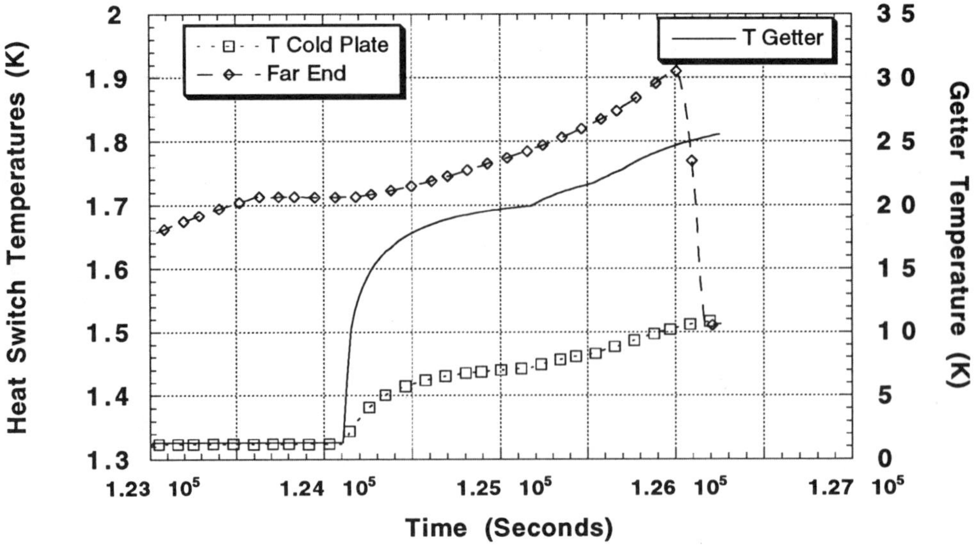


**Figure 3.** Heat switch turn on as a function of getter temperature. A power of 1.39 microwatts is applied to the far end of the heat switch. At 124,100 seconds the getter power is turned on slowly. The heat switch begins to turn on at a getter temperature of about 24K as evidenced by the far end temperature suddenly dropping.

quickly cools and adsorbs enough helium so that the heat switch turns off. The turn on/turn off temperature for the getter was about 24K. (See Figure 3). The off conductance was less than 3.5 microwatts per Kelvin at 1.5K giving an on/off ratio of over 100,000. The off conductance is equal to that calculated for the stainless steel tube and is calculated to be only 1.57 microwatts between the 0.065K adiabatic demagnetization refrigerator temperature and the helium bath operating temperature of 1.3K. This first version had a maximum heat flow capability of 4 mW at 1.82K and lower which is not too far away from the value of 5.77 mW calculated in the section above. At powers of even 4.1 mW, the hot end, also called the far end, temperature of the heat switch rapidly increased. At operating temperatures above 1.8K

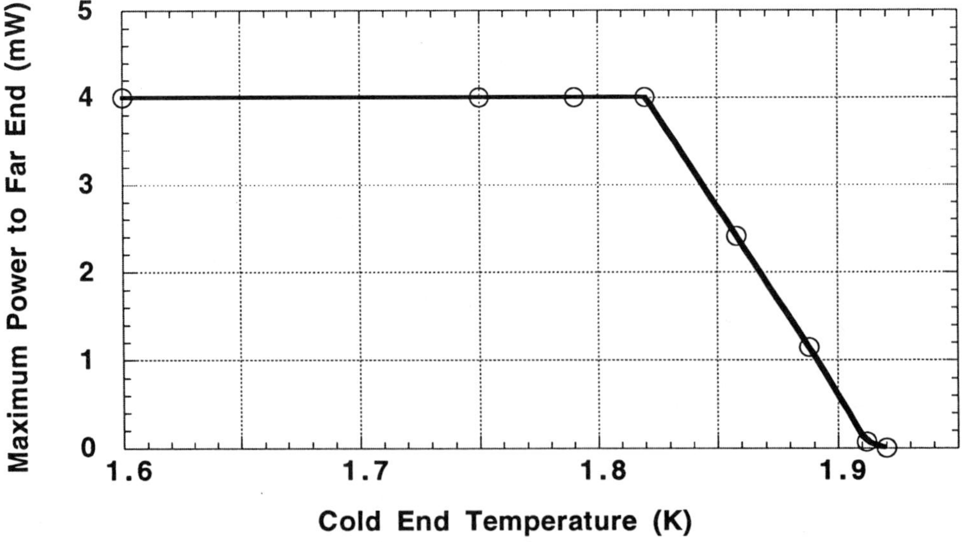


**Figure 4.** The maximum power carried by the switch at various temperatures. The warm end temperature when the heat switch was operating was never more than 11 mK hotter than the cold end.

the ability to conduct heat rapidly decreased, such that above 1.92K the switch could not conduct even 40 microwatts without a large temperature increase. (See Figure 4).

The switch stopped working at temperatures above about 1.92K. This contrasts with the density calculations in the previous section indicating that liquid would be present up to a temperature of 2.19K. Among the possible reasons for this discrepancy are that the helium did not completely desorb from the getter, only thicker films support high superfluid flow rates close to the lambda transition[8], or even that some helium had leaked out of the switch. These reasons will be explored in future testing.

The heat switch was mounted horizontally in the vacuum space of a liquid helium dewar. The dewar was tilted approximately 30° in opposite directions normal to the heat switch axis while the heat switch was in the "on" state. With either the warm or cold ends down, no change in performance was seen. For the film flow itself this was as expected since the thinning effect of gravity is so small over the few cm of heat switch length.

## SUMMARY / FUTURE DEVELOPMENT

The first heat switch tested demonstrated the predicted performance of a flowing superfluid helium film. A new heat switch test model is currently being assembled using a modular design. The new design allows remote filling of the device through a capillary tube. Various amounts of helium will be introduced into the switch to demonstrate performance in the bulk flow and normal fluid regimes as well as the superfluid film flow regime. Better heat transfer to the helium reservoir will be provided allowing the switch to be tested at lower temperatures and to be less affected by the heat dissipated at the warm end of the heat switch or by the getter heater. The new model will use a thin wall stainless steel bellows to allow more volume for more fibers to increase the maximum heat flow in the "on" state while also having a lower parasitic conductance in the "off" state. The calculated "on" state maximum heat flow for a superfluid film is 12 mW and the calculated "off" state conductance is 0.5 microwatt per Kelvin at 1.3K. This switch should have an on/off ratio of over one million.

A final device having a fixed amount of helium sealed in at room temperature will then be fabricated. Future tests may also use $^3$He, allowing operation at lower temperatures. For faster turn off, a second, smaller getter activated switch may be incorporated. Turning this switch on will short the heat switch getter to the liquid helium reservoir, cooling it quickly. This will also allow a smaller heater power to be applied to the primary switch getter in the on state.

## ACKNOWLEDGMENT

This work was supported by NASA's Office of Space Science.

## REFERENCES

1. R.P. Bywaters and R.A. Griffin, A gas-gap thermal switch for cryogenic applications, Cryogenics 13:344 (1973).
2. A.T. Serlemitsos, M. SanSebastian, and E. Kunes, Final design of the Astro-E XRS ADR, submitted for publication in Adv. Cryo. Eng. 43.
3. C.D. Fulton, C.F. Hwang, W.M. Fairbank, and J.M. Vilas, Helium heat rectifiers and a simple magnetic refrigerator, Adv. Cryo. Eng. 2 (1960) p.220.
4. F.J. Stenger, Experimental feasibility study of water-filled capillary-pumped heat-transfer loops, NASA TM-X-1310, NASA/Lewis Research Center, Cleveland, OH (1966). For a more recent review see: Jentung Ku, Overview of capillary pumped loop technology, National Heat Transfer Conference, Atlanta, Georgia, August 8-11, 1993.
5. I. Rudnick, On the thickness, Doppler shift, superfluid density and critical velocity of a flowing unsaturated helium film, in: "Liquid and Solid Helium," C. G. Kuper, S. G. Lipson, and M. Revzen, eds. Wiley and Sons, New York (1975), p. 311.

6. R. Torii, S. R. Bandler, T. More, F. S. Porter, R. E. Lanou, H. J. Maris, and G. M. Seidel, Removal of superfluid helium films from surfaces below 0.1 K, Rev. Sci. Instrum. 63:230 (1992).
7. P. J. Shirron and M. J. DiPirro, Suppression of superfluid film flow in the XRS helium dewar, submitted for publication in Adv. Cryo. Eng. 43.
8. D. Finotello, Y.Y. Yu, and F.M. Gasparini, Universal behavior of 4He films as a function of thickness near the Kosterlitz-Thouless transition, Phys. Rev. B 41:10994 (1990). The fact that the superfluid transition is suppressed for thin films is, of course, well established. In this reference a film of 5.06 nm thickness had a superfluid transition temperature of 2.011K. Even though 1.92K is lower than this temperature and a 5.06 nm film still only represents less that 1% of the total helium in the system, a much thicker film may be required to carry the relatively large heat load used in our tests.

# LIQUID HELIUM THERMOSIPHON FOR THE 4 TESLA CMS SOLENOID

J.C. Lottin and F. P. Juster

CEA-Saclay /DAPNIA-STCM
91191 Gif sur Yvette Cedex, France

## ABSTRACT

The 4 tesla superconducting solenoid for the CMS detector planned for the LHC uses a high purity aluminium stabilised conductor and the indirect refrigeration method. The coil is cooled by a network of LHe tubes which are attached to the external coil wall and are supplied in a thermosiphon circulating mode.

The design of the cooling circuit must be adapted to the restricted space and must insure proper operation in all the working conditions. The thermosiphon loop has been simulated with a dedicated code for computation of the flow distribution through parallel channels and for circuit optimisation.

## INTRODUCTION

The Compact Muon Solenoid (CMS) is one of the two large experiments planned at CERN for the LHC[1], Large Hadron Collider. The requirements of a good momentum resolution and of a high compactness have lead to the choice a magnetic field of 4 T produced by a superconducting solenoid of 12.5 m length and 5.9 m warm diameter placed inside an iron yoke.

A specific technology for the particle detector solenoids has evolved over the last 20 years through the construction and operation of several magnets[2,3]. The main features which have demonstrated the quality of these magnets are : high purity aluminium stabilised conductor, complete epoxy resin impregnation and indirect cooling. These concepts bring substantial simplification with the suppression of the coil mandrel and of the pool liquid helium vessel. The indirect cooling is provided by a set of tubes bonded to the coil external cylinder. Different methods can be used for circulating the liquid helium through the cooling tubes : forced flow from the refrigerator plant, liquid helium pump, thermally induced flow. The CMS solenoid uses this last method, called thermosiphon cooling method.

This system, where the helium flow is induced by the coil heat load, brings intrinsic reliability because it operates without any external power source nor any moving mechanical parts and so insures normal cooling conditions even in case of power shut down, provided that there is a sufficient LHe reserve[4]. This is particularly appreciable for the CMS solenoid in which fast dumping the current spontaneously induces a quench and releases about 20% of the 2.7 GJ stored energy into the winding. A period of 3 days is then necessary for re-cooling.

This method has already been used successfully for several detector solenoids.

*Advances in Cryogenic Engineering,* Vol. 43
Edited by P. Kittel, Plenum Press, New York, 1998

## THE 4 TESLA CMS SOLENOID

The superconducting coil consists of a 4 layer winding surrounded by a 12 mm thick cylinder cooled with the LHe circuit. The coil main dimensions are 6.36 m inner diameter 12.4 m length and 0.3 m thickness. The CMS solenoid magnetic induction of 4 teslas, significantly higher than the induction of all existing detector solenoids (2 teslas), has led to the concept of a fully self supporting structure for the winding. It is based on a mechanically reinforced conductor which associates a high strength aluminium alloy component to the high purity aluminium stabilised conductor [5,6].

The maximum heat load, including energising or slow dumping losses, is estimated at 400 W. The LHe cooling circuit, pictured in Figure 2, is split into 8 sub-circuits. They consist of 12 tubes welded over the external cylinder and connected to the feeding and returning manifolds installed respectively at the bottom and the top of the coil. These 8 cooling networks are connected by pair to the phase separator vessel through 4 supply lines and 4 return lines housed in the vacuum tank chimney. This arrangement has been designed in order to limit the manifold diameter because of the lack of space inside the vacuum enclosure. Also it is favourable for the flow distribution uniformity through to cooling tubes.

Another interest of a small diameter piping is the limitation of LHe volume directly associated to the coil. In case of quench only the thermosiphon helium content is evaporated and rapidly evacuated to the phase separator vessel. The smaller this volume, the lower the quench pressure rise. For the CMS cooling circuit dimension the increase in pressure is estimated at 3 to 4 bars without any helium venting out.

The liquid helium heat transfer area is 35 $m^2$ and the average heat flux is 12.4 $W/m^2$.

## THERMOSIPHON PRINCIPLE

The complete thermosiphon loop is composed of a U-shape circuit vertically oriented and a helium phase separator vessel located in elevated position. The downstream branch is

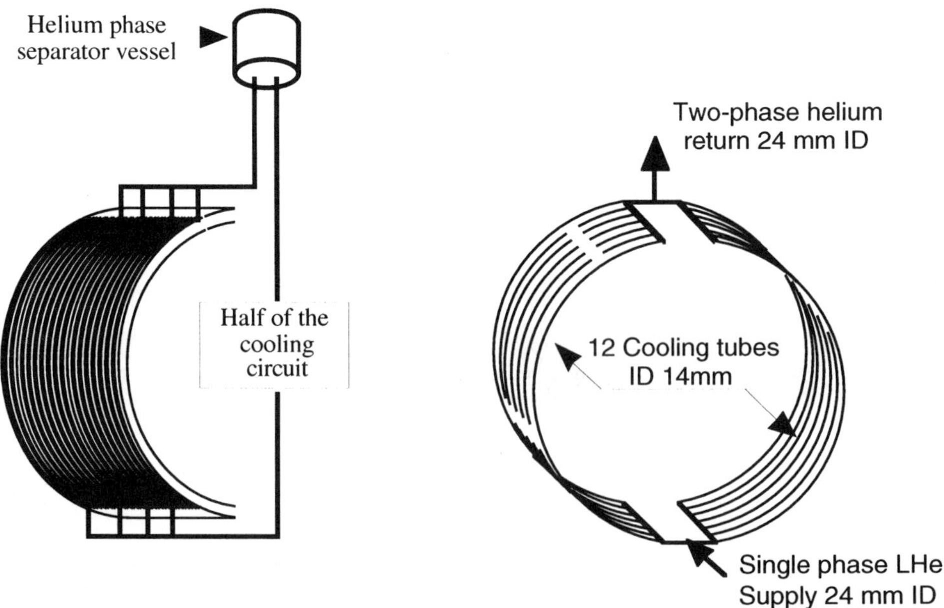


**Figure 1.** Cooling circuit of the CMS solenoid in a thermiphon mode. The 8 sub-circuits are connected by pair to the phase separator vessel. Inlet and outlet are situated at opposite side.

thermally insulated and is liquid helium supplied from the phase separator. The upstream branch consists essentially of the coil cooling loop. The cooling tubes contain two-phase helium where the vapour content results from the coil heat load. In the first part of the tubes the helium is in a sub-cooled state due to the pressure provided by the supply line.

The thermosiphon loop driving head is provided by the difference in density between the pure liquid and the two-phase mixture. It balances the resistive pressure drop due to flow friction and helium acceleration. The pressure drop across the complete loop starting and ending to the separator vessel is obviously null :

$$\Delta P = \Delta P_{gr} + \Delta P_a + \Delta P_f = 0 \tag{1}$$

where $\Delta P_{gr}$ is the gravitational pressure difference, $\Delta P_a$ the fluid acceleration pressure drop and $\Delta P_f$ the frictional pressure drop.

## THERMOSIPHON MODEL

### Two-phase Flow

The thermosiphon circuit includes necessarily two-phase flow sections with vertical, horizontal or inclined orientations. Two-phase helium flows can present various flow regimes. The parameters that determine the flow pattern include orientation, mass flow rate, vapour quality, fluid velocity, temperature. A basic principle for the present application of the thermosiphon process is to keep a low fraction of vapour in the two-phase mixture, typically lower than 10 % in mass. In this condition the flow pattern is stable enough and can be computed by using the available information.[7 8 9]

The computation of the relation (1) requires the definition of equivalent figures for the fluid parameters in the two-phase regime in order to apply the usual expressions established for single phase fluid flow. The two-phase fluid density $\rho$ is related to the void fraction $\alpha$, the ratio of the vapour area to the total cross sectional area. This parameter cannot be deduced only from the vapour quality and must also take into account the slip ratio S of the vapour to liquid velocity that determines the void fraction $\alpha$.

$$\rho = \alpha\rho_v + (1-\alpha)\rho_l \tag{2}$$

The homogeneous model assumes equal velocity for liquid and vapour and corresponds to S=1. For frictional pressure calculation it provides a fairly good agreement of about 10% with measurements both for horizontal and vertical flows up to 60 % vapour quality as verified by several experimental works[10 11]

The slip ratio value has more effect on the gravitational pressure. Experiments have shown that it ranges between 1 and 1.6 for quality lower than 10%. Several values of S have been simulated in the computation in order to evaluate its effect.

$$\Delta P_{gr}' = g\int \rho dz \tag{3}$$

The acceleration component is also dependent on void fraction and flow quality, but actually it appears negligible in the present case.

### Parallel flow paths

The 8 cooling sub-circuits are identical except for the lengths of the connecting lines to the phase separator vessel. The calculation is performed on the longest connecting lines sub-circuit.

For the calculation, the hydraulic circuit is represented as an electrical one and then is calculated by using the Kirchhoff solving method. In this analogy the hydraulic parameters are replaced by electrical parameters. The frictional pressure drop, mass flow rate, gravitational pressure and hydraulic resistance become respectively : voltage, current, electromotive force and electrical resistance. As the pressure drop is not proportional to the

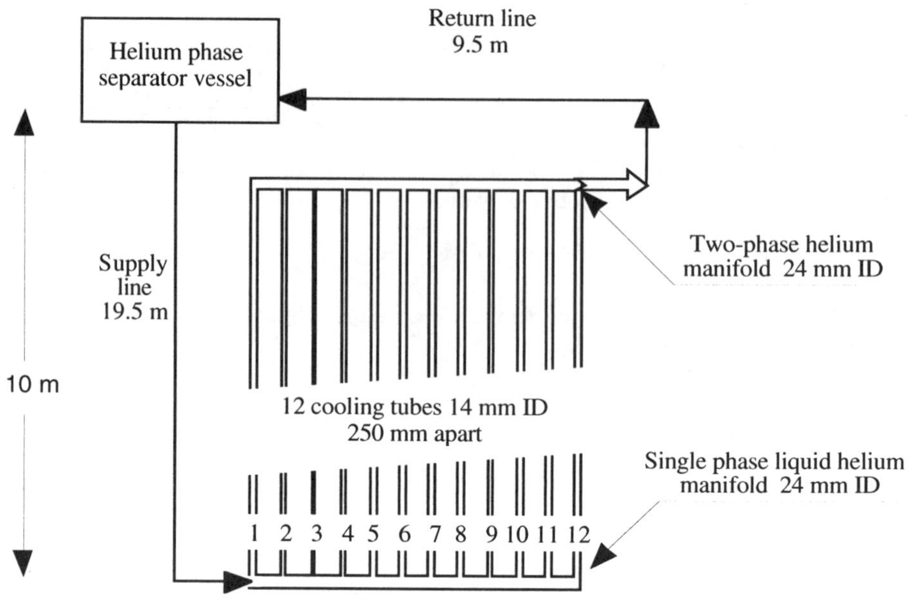


**Figure 2.** Thermosiphon sub-circuit model.

mass flow rate, the resistance is not independent of the current. An iterative calculation is used to bring successive corrections which are introduced as increments small enough to keep the convergence. The pressure distribution along the circuit and the sub-cooled state of the helium at the inlet of the cooling branches are also taken into account.

## RESULTS

### Computation criteria

The configuration of the circuit and the piping lengths being imposed together with the total heat load of 400 W, the goal of the calculation is the definition of the inner diameters of all the components. The used criteria are the following :

- maximum vapour quality in the return line : 6%
- maximum flow unbalance in the cooling tubes : 15% ( uniform heat load )

The first condition determines the mass flow rate of 0.05 kg/s per sub-circuit and governs the pipe diameters for providing low enough frictional pressure drop.

The second condition is related the ratio of the cooling tubes to the manifold diameters.

### Nominal working condition with homogeneous model

Figure 3 shows the pressure distribution along the supply and return manifolds and the pressure difference across the cooling tubes. Figure 4 gives the corresponding flow rates and qualities at the cooling tube outlets. The dimensions are given on the scheme adopted for the CMS cooling circuit and represented in Figure 2. It results from iterative computations where various values of the individual branch diameters have been simulated.

The computation has shown an optimum when using the same diameter of 24 mm for the manifolds and for the connecting lines. The cooling tube diameter has been determined at 14 mm. For larger value the total flow is not substantially increased but the flow distribution uniformity is strongly affected. Below 12 mm diameter the flow is reduced and the vapour quality increases.

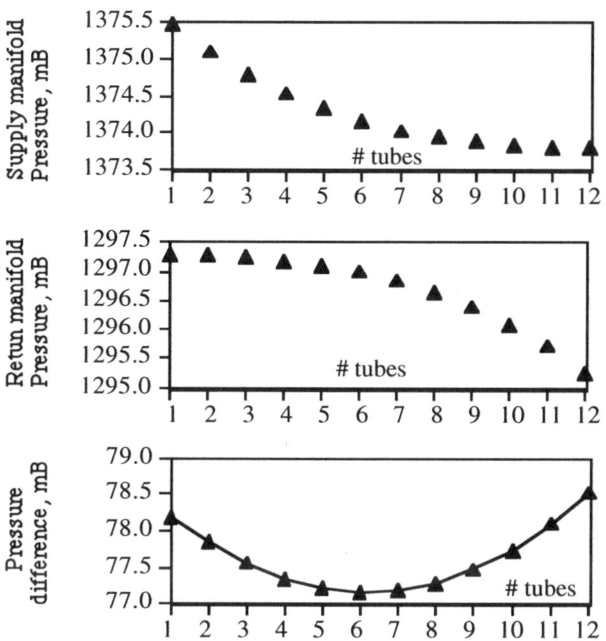


**Figure 3.** Pressure along the supply and the return manifold in mB, according to the homogeneous model, at nominal conditions.

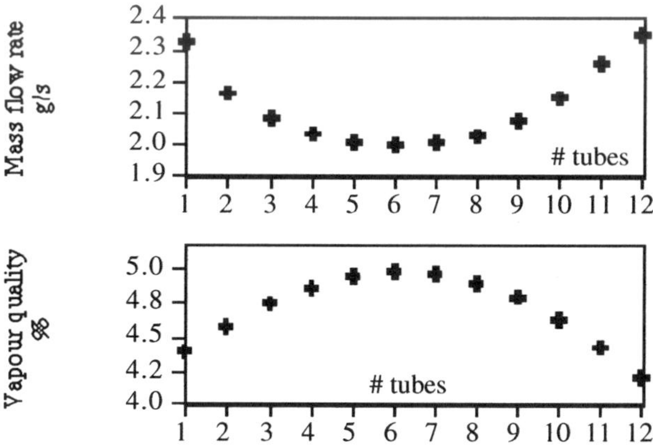


**Figure 4.** Mass flow rate and exit vapor quality in the cooling tubes, according to the homogeneous model, at nominal conditions.

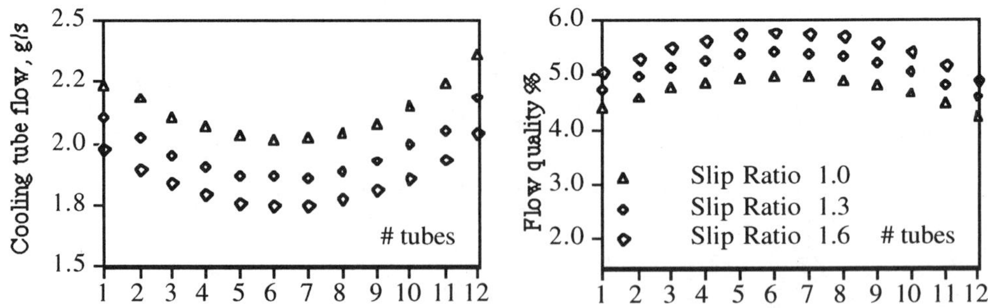


**Figure 5.** Slip ratio effect on flow rate and quality.

## Slip ratio effect

An important unknown parameter is the vapour to liquid velocity ratio. Figure 5 gives the results of computations for 3 values of S with the same circuit dimension than for the previous calculation using the homogeneous model. As expected, larger value of S reduces the flow. It must remind that the purpose of the thermosiphon circuit is cooling more than flowing process. In particular there is no velocity requirement from the LHe heat exchange at the cooling wall due the very low heat flux value, the heat transfer being mainly related to the nucleate boiling helium regime. A high vapour velocity relative to the liquid velocity clearly means that the vapour produced by the coil heat load is well evacuated to the phase separator vessel.

## Thermally induced flow

Figure 6 represents the evolution of the mass flow rate and of the vapour quality in the return line for a sub circuit as a function of the coil heat load. (The code was used for non uniform heat distribution simulation.)

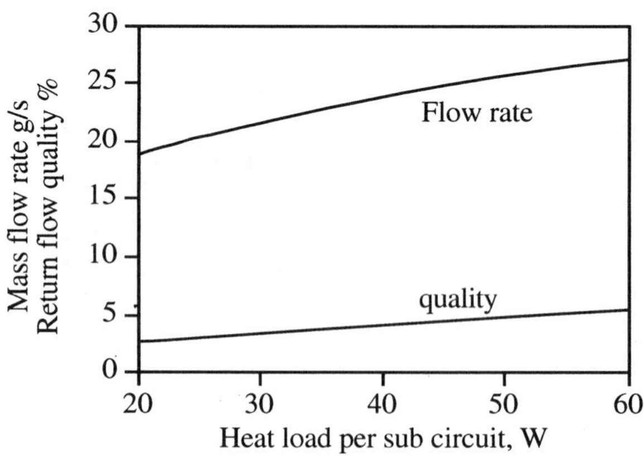


**Figure 6.** Thermosiphon flow and quality as a function of the coil heat load.

## CONCLUSION

The thermosiphon process is well adapted to the indirect cooling system used in the detector solenoids because of its intrinsic reliability and its self heat load adaptation. The low driving head that results only from the density gradient along the loop requires a proper sizing of the tubing.

The CMS solenoid cooling circuit has been calculated with simple code using the electrical analogy for kirchhoff method resolution of parallel flows. The two-phase helium, under the condition of low quality has been simulated for various slip ratio values.

More experimental work could certainly help in providing less stringent vapour quality limitation and so for the definition of more compact cooling systems.

## REFERENCES


1. CERN/LHCC 94-38 "The Compact Muon Solenoid Technical Proposal " 15 dec. 1994
2. Desportes et al. "Design, Construction and test of the large Superconducting Solenoid ALEPH ", IEEE TRANSACTIONS ON MAGNETICS, Vol. 24, N°2, March 1988
3. P.T.M. Klee and D.E. Baynham. "Toward the Realisation of Two 1.2 TESLA Superconducting Solenoids for Particle Physics Experiments " MT 11 Tsukuba 1989
4. J.C. Lottin, R. Duthil, "ALEPH Solenoid Cryogenic System " ICEC-12 Southampton 1988
5. J.C. Lottin et al., " Conceptual design of the CMS 4 tesla solenoid "Advances in Cryogenic Engineering, Vol. 41
6. D.Campi al., " Status report on the CMS magnet project at LHC," to be published at MT15
7. R. W. Boom, M.El-Walkil,G. E. McIntosh "Experimental investigation of the helium I two-phase flow pressure drop characteristics in vertical tubes," Proc ICEC 7 IPC Science and technology Press, Guildford, UK (1978)
8. A. Khalil Ph. D. Thesis and ICEC 7 (1978)
9. H. K. Zust, W. B. Bald "Experimental observations of flow boiling of liquid helium I in vertical channels," Cryogenics Nov. 1981
10. H. Katheder and M. Susser, "Results of flow experiments with two-phase helium for cooling superconductors," Proc. ICEC 12 Butterworths, Guildford, UK (1980)
11. X. Huang and S. W. Van Sciver, "Pressure drop and void fraction of two-phase helium flowing in horizontal tubes," Cryogenics 1995 Vol. 35

# DEVELOPMENT OF MULTI - LOOP HEAT PIPES FOR SUPERCONDUCTING MAGNET APPLICATIONS

G. R. Chandratilleke, Y. Ohtani, H. Hatakeyama, M. Takahashi, and H. Nakagome

R &D Center, Toshiba Corporation
4-1 Ukishima-cho, Kawasaki-ku, Kawasaki, 210, Japan

## ABSTRACT

In an early work, we reported that we have lowered the operation temperature level of a multi-loop heat pipe from the previously lowest 77 K level to 4 K level. The limiting heat transport of the heat pipe at 4 K using helium as the working fluid was however only about 0.2 W at the best. For the helium heat pipe to be able to transport 1 W, it is required to increase the heat transfer area at the heater section of the heat pipe. The area can be increased either by increasing the number of turns of the heat pipe or by increasing the heated length per turn of the heat pipe. Another factor that could have caused a low heat transport capacity in the heat pipe is the insufficiency of liquid charge. Therefore, we increased the number of turns from 10 to 52, the heated length per turn from 30 to 100 mm, and the liquid charge from 1/3 to 0.75. We have achieved nearly 1 W of heat transport for the following conditions: 52 turns, 100 mm heated length per turn, 45° heat pipe orientation and 0.45 liquid helium charge.

## INTRODUCTION

Interaction of the magnet's magnetic field with the cryocooler's magnetic materials is known to raise the temperature of the cryocooler of a conduction-cooled superconducting magnet. The cause of the temperature rise is abrasion between the crycooler's displacer and the cylinder in which it reciprocates. To eliminate the interaction, one solution is placing the cryocooler far from the magnet. Then if a heat conduction link is used between the magnet and cryocooler for simplicity of the overall magnet construction, the amount of heat that can be conducted will be limited by the cross-sectional area of this conduction member. This occurs because the temperature difference available for the heat transport has a maximum, resulting in a minimum cross-sectional area for a given heat transport. For instance, a rod of OFHC copper having a 1000 W/m/K thermal conductivity at 4 K would need at least a bulky 50 mm diameter to transport 1 W over 1 m distance for an end-to-end temperature difference of 0.5 K. The longer the heat transport distance, the larger the minimum cross-sectional area. To make the heat transport link compact, one will have to use a heat pipe because they can transport much more heat for the same temperature difference, reducing the effective cross-sectional area of the heat transport link.

We reported[1] on a loop heat pipe using helium as the working fluid, which is 30 to 60 times as effective as an electrolytic tough pitch copper rod, but its heat transport was limited to 0.2W. With applications of 1 W level cryocoolers becoming more common, it is necessary that the heat transport capacity of the helium heat pipe be increased to 1 W level.

The loop heat pipe has basically a configuration as shown in Figure 1, where a single pipe is shown coiled to form a closed loop. To add and remove heat from the heat pipe, two heat transfer plates are soldered to the turns. When the pipe is filled with a liquid at boiling point and the two plates maintained just above and below the boiling point, heat will be transported from the warm plate to the cold one by way of heat of evaporation at the warm plate and condensation at the cold plate.

In order to increase the heat transport capacity of the heat pipe, we studied the effects on the heat transport of the following factors: number of turns, length heated segment of pipe, heat pipe inclination and liquid charge.

## EXPERIMENTAL APPARATUS

Figure 2 shows a schematic diagram of the experimental apparatus. One heat transfer plate of the loop heat pipe is connected to a cryocooler and the other plate to a heater. A two - stage Gifford-McMahon cryocooler was used and the heat pipe was kept inside a thermal shield, which was cooled by the first stage of the cryocooler. Inside the thermal shield, the heat pipe was wrapped in 10 layers of multi-layer insulation. The heat pipe assembly is placed in a vacuum vessel on a rotatable support so that it can be readily inclined at cryogenic temperatures. The axis of rotation of the heat pipe is perpendicular to the plane of the paper as shown in Figure 2, and the heat pipe was rotated in the direction of the arrow to use it in "bottom - heating" position. To fill the heat pipe with an appropriate

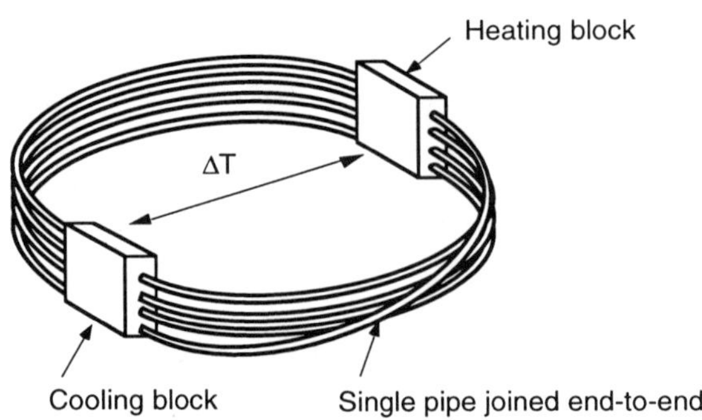


**Figure 1.** Schematic diagram of loop heat pipe.

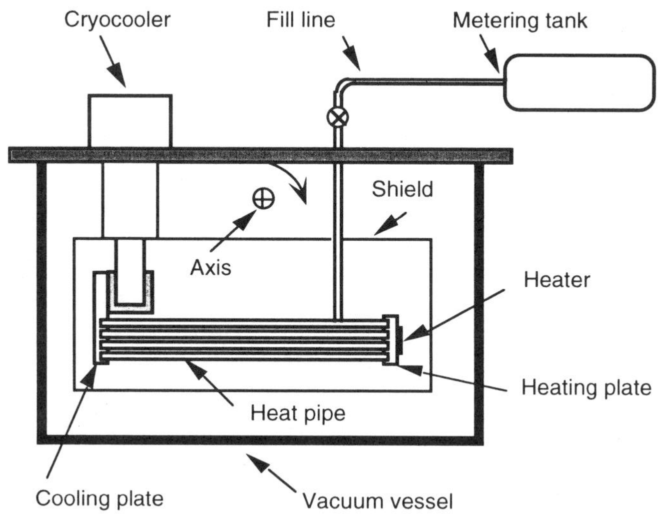


**Figure 2.** Schematic diagram of experimental apparatus for measuring heat transport .

amount of helium, we used a metering cylinder, measuring whose pressure change makes it possible to calculate the amount of helium gas liquefied. Heat pipe temperatures were measured on the heating and cooling plates with carbon - glass resistance temperature sensors; therefore, their temperature difference represents an overall temperature difference, which includes temperature differences between the pipe and the plates.

## HEAT PIPE SPECIFICATIONS

Table 1 summarizes the details of the heat pipes used in the present work. The pipe diameters were set roughly to 1.5 to 2 times Laplace constant of the working fluid as outlined elsewhere[2]. The other dimensions were roughly estimated using simple heat transfer relationships for critical heat flux[3, 4] and condensation heat transfer for pipes[5].

**Table 1.** Specifications of loop heat pipes used in the present work

| Type | Pipe diameter internal(external) (mm) | Cooling length per turn (mm) | Heating length per turn (mm) | Number of turns | Heat transport distance (mm) |
|---|---|---|---|---|---|
| 1 | 0.5 (0.8) | 100 | 30 | 10 | 516 |
| 2 | 0.5 (0.8) | 140 | 30 | 52 | 496 |
| 3 | 0.5 (0.8) | 140 | 100 | 52 | 460 |

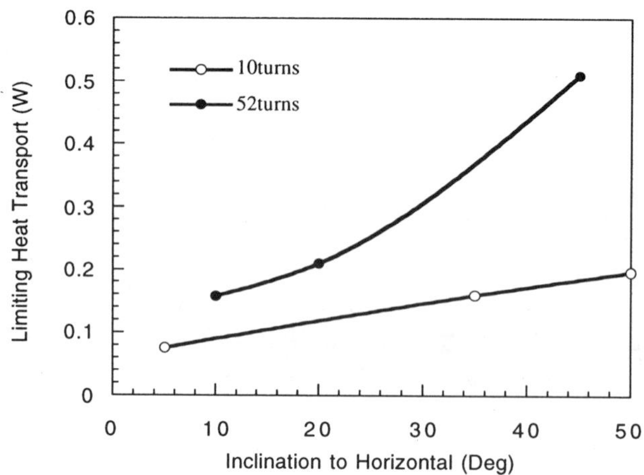


**Figure 3.** Limiting heat transport of 10 and 52 turns heat pipes as a function of heat pipe inclination.

## EXPERIMENTAL RESULTS

**Effect of Inclination and Number of Turns.** We used the following formula[3] to estimate the surface area required by the heating block for 1 W heat transfer.

$$Q_{CHF} = (\pi dnl)\ \frac{1}{1.7 + 0.125(l\ /d\ )^{0.88}} \tag{1}$$

where, $Q_{CHF}$ is the critical heat input (W) for boiling of helium in a vertical pipe, $n$ the number of turns of the heat pipe, $d$ the internal diameter of the pipe(cm), and $l$ is the heated length of the pipe(cm).

When the pipes are inclined from the vertical position, $q_{CHF}$ decreases[4] and can be roughly obtained by scaling Eq.(1) with $\sin\theta$, where $\theta$ is the inclination of the pipe to the horizontal. Therefore, it is clear that heat transfer at the heating block of the heat pipe is very small when the turns of the heat pipe are nearly horizontal as in the heat pipe position shown in Figure 2. In fact, the heat pipe using helium as the working fluid does not work in this position[1,2]. However, when the heat pipe is rotated by angle $\phi$ about a horizontal axis in the direction of the arrow in Figure 2, the heat pipe functioned and it's heat transport capacity increased as shown by open circles in Figure 3. The heat pipe in this position has its heating block at a lower height than its cooling block and is said to be in the "bottom-heated" position. For this position, the heated pipe segments are horizontal, that is $\theta = 0$, but the heat pipe works due to the difference in elevation between the cooling and heating blocks.

Figure 3 shows the limiting heat transports of type 1 and 2 heat pipes of Table 1 as a function of the heat pipe inclination $\phi$. The type 2 heat pipe, which has 5 times larger heat transfer surface than that of type 1 can provide a larger heat transport than that of type 1 heat pipe, especially at steeper inclinations. The liquid charge was roughly the same at 0.45 in both the cases.

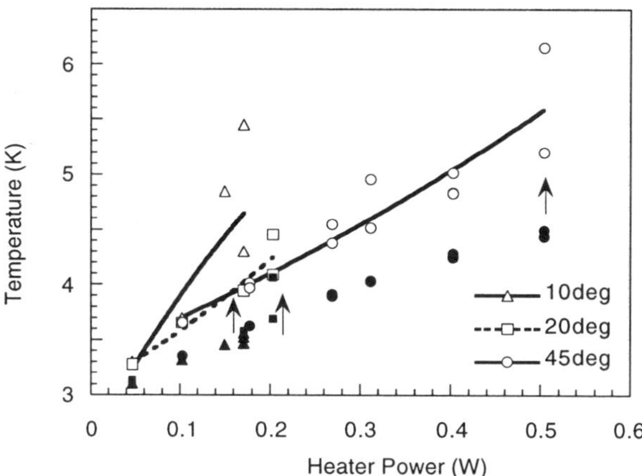


**Figure 4.** Heater and cryocooler temperatures as a function of heater power at 0.75 liquid charge for the type 2 heat pipe in Table 1 at three heat pipe orientations.

Figure 4 shows for the type 2 heat pipe the temperatures of the heating and cooling blocks as a function of the heater power for the orientations $\phi$=10°, 20° and 45° respectively by triangles, squares and circles. Corresponding solid symbols show the cooling block temperatures. The arrows indicate the heater power at which the limiting heat transport occurs. The limiting heat transport is defined here as the value of the heat transport, for a slight increase in which there is an appreciable change in the heater block temperature. Generally, the temperature difference between the heating block and cooling block is in the range 0.5 to 1 K, which correspond to 100 to 90 times heat transport compared to an electrolytic copper rod.

**Effect of Liquid Charge.** It has been reported recently that liquid charge has a significant effect on the basic heat transport mechanism of the heat pipe when water is used as the working fluid[6]. According to this work, when the liquid charge had occupied 95% of the heat pipe volume, the working fluid had been observed to be circulating in one direction due to interaction of just two elongated vapor bubbles formed in adjacent two turns. When one bubble had grown, the other bubble had diminished in size pushing the liquid in-between the bubbles in the bubble growth direction. When the former bubble had collapsed, a new bubble would form in the turn next to the expanding bubble but in the far-side turn from the collapsed bubble's. This would bring the state of the bubbles exactly to the same situation as before the collapse of the previous bubble, and the bubble formation-collapse process will repeat endlessly. For other liquid charges, it had been observed that there were multiple bubble interaction, and thus it had been difficult to establish a regular bubble formation-collapse pattern like above. The heat transport had been most effective when the regular bubble formation-collapse pattern existed.

Even though such a heat transport mechanism had not been confirmed for a cryogenic working fluid, we changed to the liquid helium charge to see whether it had an effect on the heat transport. Figure 5 shows for 45° heat pipe orientation and type 3 heat pipe, the temperatures of the heating and cooling blocks for two liquid charges: 1/3 and 0.45. We

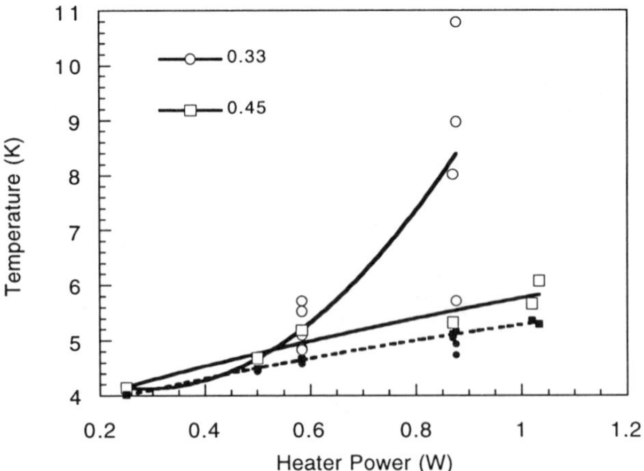


**Figure 5** Heater and cryocooler temperatures as a function of heater power for 0.33 and 0.45 gas charges at 45deg orientation to the horizontal for the type 3 heat pipe in Table 1.

also tried 0.75 liquid charge, which however increased the heat pipe pressure above helium's critical pressure on application of heat. Figure 5 shows that heat transport can be increased to 1 W level for the type 3 heat pipe with 0.45 liquid charge.

**Effect of Heated Length.** Type 2 and 3 heat pipes were tested at the same 45° orientation and 0.45 liquid charge. The results are shown in Figure 6 as a graph of temperature versus heater power. The difference in the two heat pipes are in their heated lengths. The type 2 heated pipe having 30 mm heated length shows the limiting condition at a lower

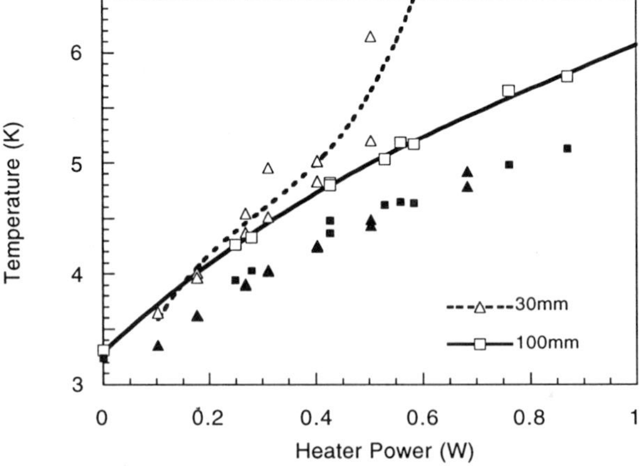


**Figure 6.** Heater and cryocooler temperatures as a function of heater power at 0.45 gas charge at 45deg orientation to the horizontal for the type 2 and type 3 heat pipe in Table 1.

heat transport than the type 3 heat pipe, which has a 100 mm heated length. For the type 3 heat pipe, the temperature difference between the heater and cooling blocks is less than 0.8 K and the heat pipe has a heat transport as 150 times as its equivalent copper rod.

## CONCLUSIONS

We changed heat pipe's number of turns, liquid charge, heated length per turn and inclination in order to obtain roughly 1 W heat transport for the helium heat pipe. We have achieved this for a heat pipe having 52 turns and a 100 mm heated length at 45° orientation and 0.45 liquid helium charge. With this increased heat transport, heater temperature exceeded 5 K, and further research is necessary to develop a heat pipe having 1 W near 4 K.

## REFERENCES


1. G. R. Chandratilleke, Y. Ohtani, H. Hatakeyama, and H. Nakagome, Development of loop heat pipes for cryogenic applications, in: "Proceedings of 16th ICEC," T. Haruyama, ed., Elsevier Science, Kitakyushu (1996) p. 501.
2. G. R. Chandratilleke, H. Hatakeyama, and H. Nakagome, Development of Cryogenic Loop Heat Pipes, Cryogenics (1997) (in print).
3. S. Lehongre, J. C. Bossin, C. Johannes, and A. De La Harpe, Critical nucleate boiling of liquid helium in narrow tubes and annuli , in: "Proceedings of 2nd ICEC," Heywood-Temple Industrial Publications, London (1968) p. 274.
4. S. Sato, and H. Ogata, Boiling heat transfer to liquid helium in inclined channels, in:" Cryophysics and Cryoengineering: Topical Problems," International. Inst. of Ref. Commission I, Tokyo (1970), p. 119.
5. M. M. Shah , A General Correlation for Heat Transfer During Film Condensation Inside Pipe, *Int. J. Heat Mass Transf.*, 22:547 (1979).
6. M. Hosoda, S. Nishio, and R. Shirakashi, Bubble-driven heat transport tube: Flow pattern and heat transport model, in:"Proceedings 34th National Heat Trans. Symposium of Japan," JSME, Sendai (1997) p. 267. (in Japanese)

## PERFORMANCE OF CRYCOOLER SHUNT HEAT PIPES

F. C. Prenger[1], D. D. Hill[1], D. E. Daney[1], M. A. Daugherty[1], G. F. Green[2], J. Chafe[2], M. Heiberger[3], and A. Langhorn[4]

[1]Los Alamos National Laboratory
Los Alamos, NM, 87545, USA

[2]Naval Surface Warfare Center
Annapolis, MD, 21402, USA

[3]General Atomics
San Diego, CA, 92186, USA

[4]Startech
Solana Beach, CA, 92075, USA

## ABSTRACT

In many important cryogenic applications the use of liquid cryogens for system cooling is either not feasible or is unsuitable. In such cases a cryogenic refrigeration system or multi stage cryocooler must be employed to provide the necessary cooling. To shorten cooldown time for such a system, especially if the thermal mass is large, a thermal shunt directly connecting the first stage of the cryocooler to the load during cooldown is desirable. This thermal shunt allows effective utilization of the greater cooling power available from the first stage of the cryocooler early in the cooldown. Upon reaching operating temperature, the thermal shunt must exhibit a high resistance to thermally isolate the first stage of the cryocooler from the load. Heat pipes are well suited to achieve these objectives. The Advanced Lightweight Influence Sweep System (ALISS), under development by the U. S. Navy for shallow water magnetic mine countermeasures, employs a large, conductively cooled, superconducting magnet that must be cooled from 300 to 4.2 K. Cryogenic heat pipes acting as cryocooler thermal shunts are used to shorten the cooldown time. Ethane, nitrogen and oxygen were evaluated as possible working fluids. A thermal model of the ALISS was developed to evaluate the cooldown performance of various heat pipe combinations. In conjunction with heat pipe performance tests, this model was used to select a suitable design for the heat pipe thermal shunts.

## INTRODUCTION

In the U.S.Navy's Advanced Lightweight Influence Sweep System[1] (ALISS), a superconducting magnet[2] will be used to sweep shallow water magnetic mines. The magnet is cooled by a pair of GM rare-earth cryocoolers. Cooldown times for these systems can be excessive because the large heat capacity magnet is connected to the low cooling power second stage of the cryocoolers. In the ALISS system heat pipes will be used as thermal shunts to interconnect the first and second stages of the cryocoolers. The heat pipe shunts are active during system cooldown providing a method for shunting the magnet cooldown heat load to the higher capacity first stages of the cryocoolers. Efficient use of this extra cooling power shortens system cooldown times significantly. During the latter stages of the cooldown as the magnet temperature decreases and falls below the triple point of the heat pipe working fluids, the heat pipes become inoperable as the working fluids freeze. The frozen heat pipes, therefore, decouple the first and second stages of the cryocoolers allowing the magnet to reach lower temperatures through connection with the second stage only.

We have previously reported[3] the results of system design studies showing the operational advantages of shunt heat pipes for this and similar applications. These studies considered various heat pipe working fluids, sizes and numbers as well as cryocooler combinations. This paper reports on the final heat pipe design and on the performance of the heat pipes during acceptance testing at NSWC.

## HEAT PIPE DESIGN REQUIREMENTS

There were two main design requirements that guided the development of cryocooler shunt heat pipe technology for the U.S. Navy. The first, a structural requirement, was that the heat pipes had to be sufficiently rugged to withstand the shock and vibration of the naval shipboard environment. Environmental simulation required that the heat pipes must survive a 100g shock loading without resulting in an excessive parasitic heat leak. The performance goal was that there be less than 100 mW of heat leak per heat pipe assembly to the cold region when the heat pipes were not operating. Use of Inconel for the heat pipe transport section provided adequate strength and sufficiently low thermal conductivity to meet this requirement.

The second requirement was that the heat pipes have sufficient thermal capacity to prevent premature freezeout of the working fluid with accompanying shutdown of the heat pipes before cooldown was completed. Undersized heat pipes are subject to subcooling by the cryocoolers and possible freezing of the working fluid in the condenser. Therefore, the heat pipes must provide adequate capacity over the full cooldown temperature range from 300 to 60 K. The heat pipes were sized to the total capacity of the 1st stage GM refrigerators. The estimated cooler capacity as a function of temperature is shown in Fig. 1. The heat pipe design curve, also shown in Fig. 1, is approximately twice the capacity of the cooler's first stage. This is the anticipated cooler capacity when operating at twice its normal speed during cooldown. It should be noted that the heat pipe design curve is similar in performance to a Cryomech, AL200 GM refrigerator.

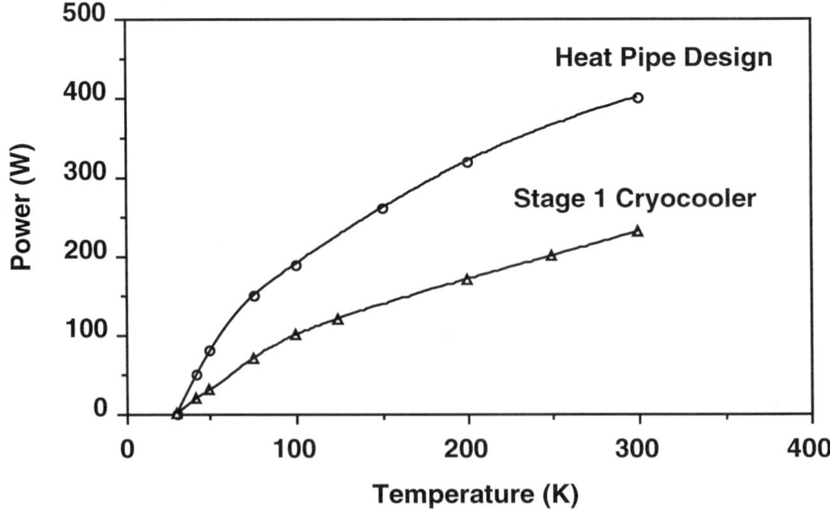


**Figure 1.** Nominal cryocooler stage 1 capacity and heat pipe design requirement.

## HEAT PIPE DESIGN

Ethane and oxygen were selected as working fluids for the heat pipes because together they provide an operational temperature range from 300 K down to approximately 55 K. The ethane and oxygen heat pipes are connected in parallel and operate in a cascading manner. At the start of cooldown, the ethane heat pipes are operational and the oxygen heat pipes are not because the oxygen working fluid is supercritical preventing formation of the necessary two-phase interface. Upon reaching 150 K, condensation begins in the oxygen heat pipes and they begin operation. The ethane heat pipes remain operational down to approximately 100 K where freezing of the ethane begins. At this point the oxygen heat pipes are fully operational and remain so until the temperatures approach 55 K where the oxygen begins to freeze. Upon further cooldown, both sets of heat pipes remain frozen preventing any internal circulation of the working fluid. From this point on, all heat pipes are inoperative and the first and second stages of the cryocoolers remain thermally decoupled. Because the optimum liquid fraction within the heat pipe is small, the freezing and thawing of the working fluid does not damage the heat pipe.

The heat pipes were fabricated from Inconel tubing that was brazed at each end into copper evaporators and condensers which are an integral part of the thermal buswork. The nominal outside tube diameter is 16 mm with a nominal wall thickness of 0.75 mm and is thickened at each end in the vicinity of the braze joint. The high thermal conductivity copper evaporators and condensers result in low radial temperature gradients in the heat transfer zones. The low thermal conductivity of the Inconel minimizes the axial parasitic heat leak when the heat pipes are inoperative. The evaporator inside diameter was increased to 23.5 mm and the condenser diameter was increased to 44.5 mm. Larger internal diameters were used to provide additional surface area and reduce thermal resistance across the heat pipe wall. Additionally, the internal surface of the condenser was axially grooved to provide enhanced drainage of condensate minimizing the liquid film thickness. A reduced liquid film thickness lowers the thermal resistance in this region.

The predicted performance limits were calculated using the LANL heat pipe model using ethane and oxygen working fluids (which can be seen in Fig. 3). Both working fluids have a pronounced peak in their limit curve which is shifted toward the corresponding critical temperature. The limit curves shown are for flooding of the heat pipe condenser. Not shown are the sonic limits for these heat pipes which are always greater than the flooding limits except at temperatures very close to the triple point. Therefore, the flooding limit defines the operating envelope for these heat pipes. There is a small region of overlapping performance between the ethane and oxygen heat pipes. Because of the pointed shape of the limit curves, it was necessary to oversize the heat pipes to get a sufficiently broad range of operating temperature for each working fluid with sufficient performance overlap.

## HEAT PIPE TESTING

Heat pipe performance testing was conducted at the Carderock Division of the Naval Surface Warfare Center, Annapolis Detachment (CDNSWC/A). The heat pipe assembly tested consisted of an ethane and an oxygen heat pipe. The heat pipe assemblies were fabricated and filled by Thermacore, Inc.

For testing the heat pipe assembly was connected to a Cryomech, model AL200, single stage GM cooler that provided cooling of 200 watts at 80K. A 12.7-mm-thick copper plate, 203 mm wide was configured in a U-shape with the AL200 cooler located at the center of the horizontal leg and the vertical sides of the plate bolted to the condenser end of the heat pipe assembly. Indium gaskets were used to provide a thermal interface between the cooler and the copper plate, and between the copper plate and the heat pipe assembly. Figure 2 shows the cooler, copper plate, and heat pipe assembly as tested.

Eight calibrated Lake Shore silicon diodes thermometers were used to instrument the heat pipe assembly. One sensor was attached to the copper plate directly under the GM cooler. A second sensor was mounted on the copper plate adjacent to the condenser of the heat pipes. Sensors were also mounted on the condenser, the evaporator, and the mid point of each heat pipe. In addition, two heaters were installed. One heater was mounted on the

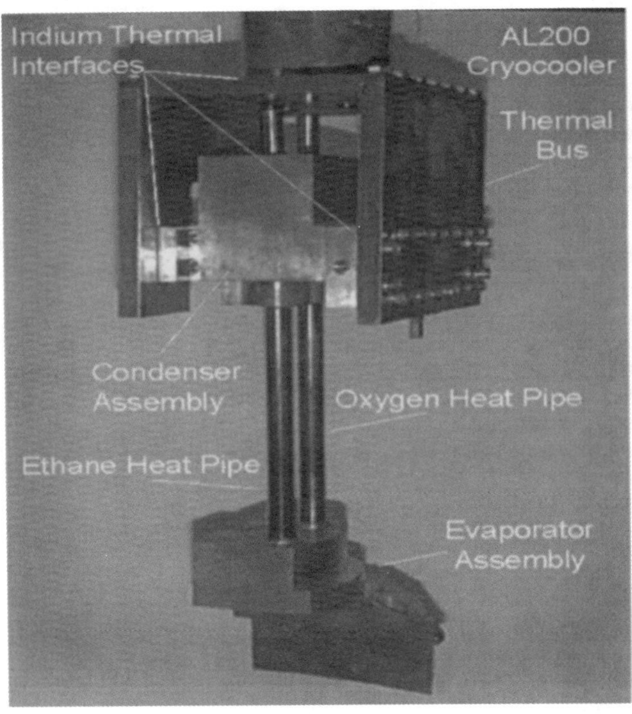


**Figure 2.** Heat pipe test assembly showing cryocooler mounting plate and thermal interfaces.

copper plate directly beneath the cold head and another was mounted on the evaporator assembly between the two heat pipes.

A data acquisition system was used to monitor the eight silicon diodes. Computer controlled power supplies were used for each of the heaters. This control system provided either a constant power to the heaters or maintained a constant temperature.

Steady state operating conditions were measured by setting the power level on the evaporator and then allowing the heat pipes to reach steady state, which was approximately 4-5 hours. The power input to the evaporator was monitored as well as all the temperature sensors.

## TEST RESULTS

The heat pipe subsystem assembly was tested across the entire operating temperature range anticipated by the ALISS system. The heat rejection limit resulting from the use of a single cryocooler in the test rig prevented operation of the heat pipe subsystem at prototypical heat loads. However, sufficient data were collected to verify the adequacy of the heat pipe subsystem performance. Figure 3 shows the steady state performance data collected for the subassembly. The total heat transferred by the subassembly is shown as a function of the temperature of the active heat pipes. In most cases either the ethane or oxygen heat pipes were active but not both. Exceptions occurred in a relatively narrow region near 150 K. Thermometers mounted on the individual heat pipes were used to determine which heat pipeswere operating. Each data point in Fig. 3 represents a steady state operating point for the subsystem. The filled data points indicate conditions where the heat pipes were operating at their respective performance limits. Performance limits were determined by observing the temperature profiles of the heat pipes and the temperature response to heat load changes on the subassembly. At the low temperature limits the heat pipe operation was stable but resulted in significant axial temperature gradients along the heat pipes.

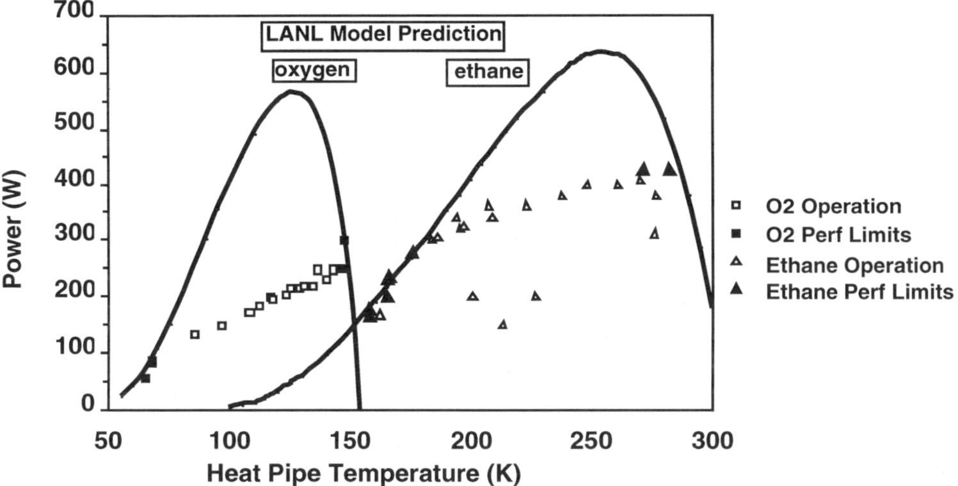


**Figure 3.** Test data for ethane and oxygen heat pipe assembly where performance limits are shown as filled points and can be compared with LANL heat pipe model predictions.

At the high temperature limits, the heat pipe operation was unstable under the test conditions of constant heat addition to the evaporators. Upon further increase in the applied heat load, the evaporators experienced dryout and, with continued heat addition, the evaporator temperature increased rapidly. Performance limits under these conditions were easily identified. In order to stabilize heat pipe operation the heat load had to be temporarily reduced to a level substantially below the observed limit. The test conditions of constant heat addition via electric heaters to the subassembly is a more severe operating condition than will exist in the actual application. The cooldown heat load results from sensible heat removal from the system and creates a constant temperature boundary condition at the heat pipe evaporators. Under these conditions the heat pipe evaporators will not overheat when a performance limit is reached and a heat pipe recovery step is unnecessary. The performance limits measured are in good agreement with predictions from the LANL heat pipe model which are also shown in Fig. 3.

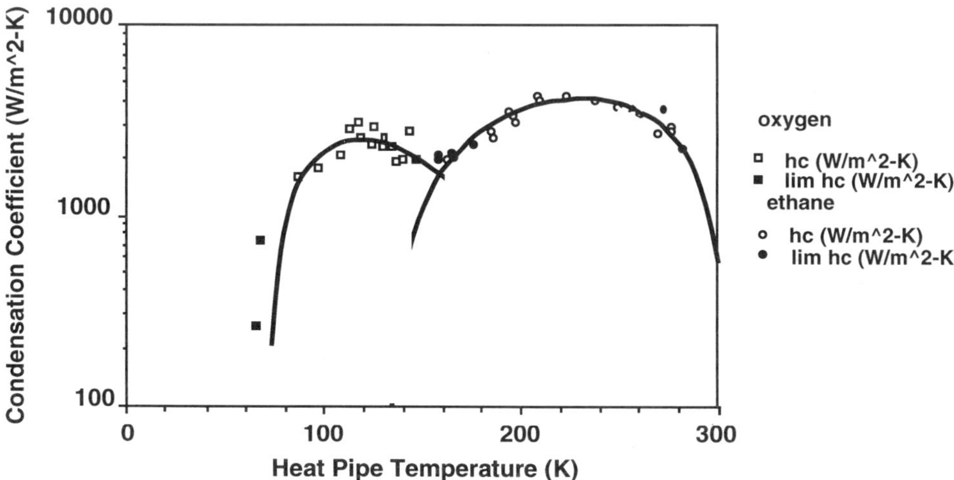


**Figure 4.** Condenser heat transfer coefficient as a function of heat pipe operating temperature for ethane and oxygen working fluids.

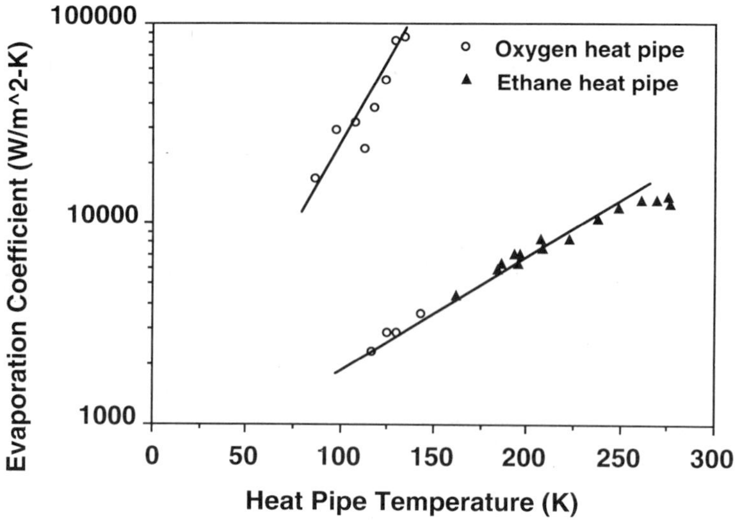


**Figure 5.** Evaporator heat transfer coefficient as a function of heat pipe operating temperature for ethane and oxygen working fluids.

Measured temperature differences in the heat pipe evaporators and condensers were used to determine heat transfer coefficients in the ethane and oxygen heat pipes. At operating conditions away from performance limits, the heat pipes were nearly isothermal and the internal evaporator and condenser temperatures were assumed to be similar to the measured adiabatic wall temperature. However, near the high temperature dryout limit and the low temperature sonic limit, axial temperature gradients in the heat pipes could be substantial, especially in the evaporator. Therefore, performance limit data were not used in determining the evaporator heat transfer coefficients. The difference between the heat pipe wall temperature and the measured external evaporator and condenser temperatures gives a reasonable approximation to the radial temperature drops in these areas. These data were used to calculate the respective heat transfer coefficients and are shown in Fig. 4 for the condenser and Fig. 5 for the evaporator. The condenser coefficients reach a maximum with

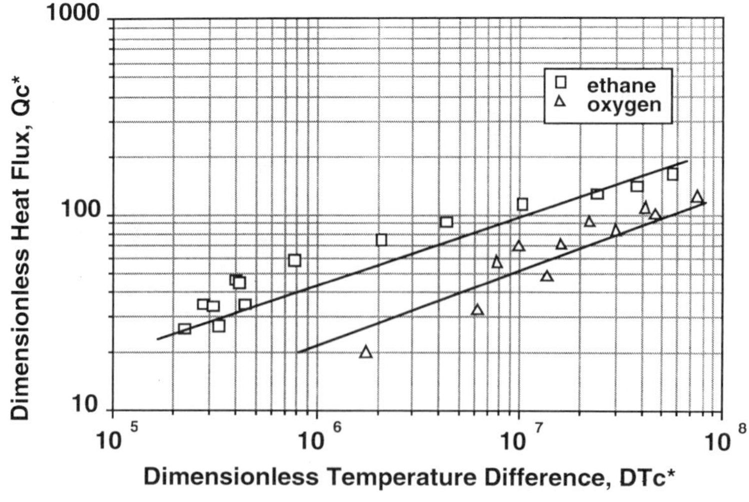


**Figure 6.** Dimensionless heat flux defined as $Q_c$* (Eq. 1) dimensionless temperature difference defined as $\Delta T_c$* (Eq. 3).

increasing temperature and then decrease as the film thickness begins to dominate. The presence of grooves in the condenser interior retards the film buildup by providing a collection area for the liquid. The net effect is to delay the onset of degradation by the liquid film.

The heat transfer coefficient results were compared with existing boiling and condensation[5] correlations. The heat flux was nondimensionalized in terms of the latent heat of vaporization, the liquid viscosity and a characteristic length as follows:

$$Q^* = \frac{q}{A} \cdot \frac{L}{h_{fg}\mu_f} \tag{1}$$

where $L = L_c$ for the condenser, the actual condenser length, and for the evaporator, the characteristic length $L = L_e$ is defined in terms of the surface tension as:

$$L_e = \left[\frac{\sigma}{g(\rho_f - \rho_g)}\right]^{0.5} \tag{2}$$

For the condenser the temperature difference was nondimensionalized using Nusselt's falling film analysis as:

$$\Delta T_c^* = g\rho_f \cdot (\rho_f - \rho_v) \cdot \left[\frac{kL_c\Delta T_c}{h_{fg}\mu_f}\right]^3 \cdot \frac{1}{\mu_f^2} \tag{3}$$

where $h_{fg}$ is modified by the factor $(1 + c_p{*}\Delta T_e)$. Q* as a function of ΔT* for the condenser is shown in Fig. 6. giving distinct correlations for oxygen and ethane.

For the evaporator the dimensionless temperature difference is defined after Rohsenow[6] as:

$$\Delta T_e^* = \Delta T_e \cdot \left[\frac{c_p}{h_{fg}}\right] \cdot \frac{1}{Pr^{1.7}} \tag{4}$$

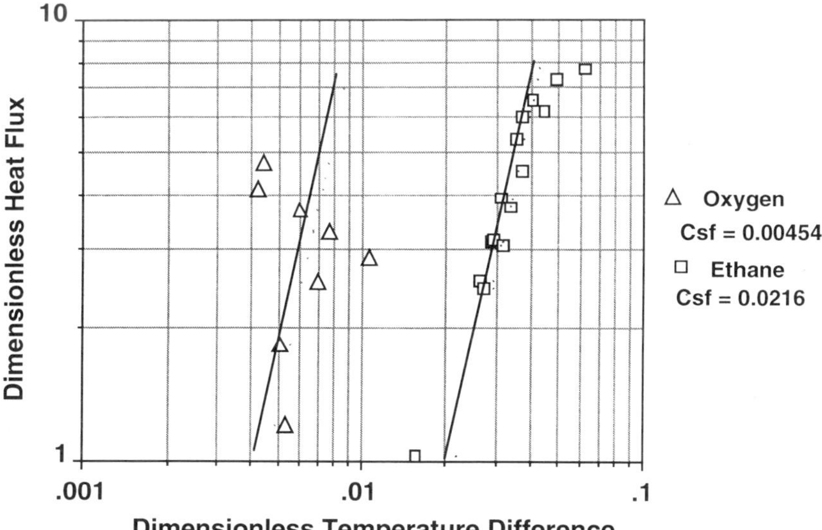


**Figure 7.** Dimensionless heat flux defined as $Q_e^*$ (Eq. 5) versus dimensionless temperature difference defined as $\Delta T_e^*$ (Eq. 4).

Figure 7 compares our test data with the Rohsenow correlation of the form:

$$\Delta T_e^* = Q_{sf}^* \left[\Delta T_e^*\right]^{0.33} Pr^{1.7} \quad (5)$$

where $C_{sf}$ was found to be equal to 0.00454 for the oxygen-copper system and equal to 0.0216 for the ethane-copper system.

## CONCLUSIONS

Heat pipe thermal shunts are shown to be an effective method for reducing cooldown times in large, conductively cooled, cryogenic systems. Analysis of the system performance predicts that cooldown times can be reduced by nearly a factor of two. We have tested prototype heat pipe shunts using oxygen, nitrogen and ethane and have shown that startup and freezeout behavior is both predictable and reliable. A shunt system using ethane and oxygen heat pipes has been fabricated and tested for the ALISS program. The test results were in good agreement with predictions from the LANL heat pipe model.

Heat transfer coefficients in the evaporator and condenser were measured for oxygen and ethane. Correlations for these data are presented using conventional correlating parameters.

## REFERENCES


1. E. M. Golda, J. D. Walters and G. F. Green; "Applications of Superconductivity to Very Shallow Water Mine Sweeping," Naval Engineers Journal, May 1992.
2. M. Heiberger, et al., "A Light-Weight Rugged Conduction-Cooled NbTi Superconducting Magnet for U.S. Navy Minesweeper Applications," Cryogenic Engineering Conference, Columbus, OH, July, 1995.
3. F. C. Prenger, et al., "Heat pipes for Advanced Cooldown of Cryogenic Systems," 9th Cryocooler Conference, Waterville Valley, NH, June, 1996.
4. K. A. Woloshun, et al., "HTPIPE: A Steady-State Heat Pipe Analysis Program", Los Alamos National Laboratory internal report no. LA-11324-M, Nov. 1988.
5. Rohsenow, W.H., and Choi, H; Heat Mass and Momentum Transfer; Prentice-Hall, Inc., Englewood Cliffs NJ, 1963.
6. Rohsenow, W.M.; "A Method of Correlating Heat Transfer Data for Surface Boiling of Liquids," Trans ASME, July 1952.

# CRYOPUMP COOLED WITH A DUAL HEAT PIPE

S. Buhler, A. Horbowa, J.P. Thermeau

Institut de Physique Nucléaire
91406 Orsay cedex, France

## ABSTRACT

A cryopumping system providing a large pumping speed in the centre of a cyclotron accelerating chamber has been built. The cooling of the cryopanels supplied from cryorefrigerators located at some distance is efficiently transfered with a dual heat pipe operating with $LN_2$ and $LH_2$ circulating by gravity or even with He at moderate pressure in a natural convection mode. The design of the cryopanel is described. Performances for cooldown and steady state operation are given.

## INTRODUCTION

The vacuum in the accelerating chamber of the SPIRAL cyclotron (1) is produced with two cryopanels in situ and requires a pumping speed of 2 x 12 000 L/s for air and 2 x 20 000 L/s for hydrogen down to a pressure of $5.10^{-6}$ Pa. Each cryopanel is connected to a twin cold head (2 x Leybold RGD 1245), located for convenience at some 2 m distance, by means of a dual heat pipe normally working with $LN_2$ and $LH_2$ circulation by gravity, an arrangement which has already been employed elsewhere (2), (3).

Condensing and vapourising of a cryofluid which circulates in a closed loop enables a very efficient and almost isothermal heat transfer under a working pressure which depends of the temperature of the saturated liquid. However, if ever the cooling power of an installed cryorefrigerator is larger than necessary the temperature at its heat stations will decrease causing the vapour pressure in the heat pipe to eventually drop below atmospheric, a situation which might induce some hazards from air and moisture leaking into the cooling loop. For security reasons it is then advisable to avoid "floating" of the these temperatures below these critical values by means of an extra heat load on the cold heads. If the possible temperature difference increases notably, however, one may switch over to the natural convection mode of helium gas at moderate pressure ($\approx 10^6$ Pa) which has been proven to perform remarkably well at the 20 K level. It also looks promising to replace the pure fluid in the heat pipe with an admixture of helium in order to keep the total pressure permanently above atmospheric in spite of a gradually lower partial pressure of the condensing $N_2$ or $H_2$.

## DESCRIPTION

The cryopumping system consists of 4 modules, the essential part is shown in Figure 1.

- The cooling module includes two cold heads in a common vaccum tank with a nominal cooling power of 2 x 45 W at 80 K and 2 x 12 W at 20 K (or some 20 % less if both heads are fed from a single compressor unit). Each heat station is equipped with a condenser block made of copper, connected in pairs to a common line which supplies the liquid or returns the vapours. The bundle of these four carefully insulated flexible lines terminates in a fourfold bayonet coupling.

A self-contained insulating vacuum is maintained with activated charcoal in good thermal contact with the 80 and 20 K heat stations. An 80 K radiation shield and extensive superinsulation further reduces the heat influxes.

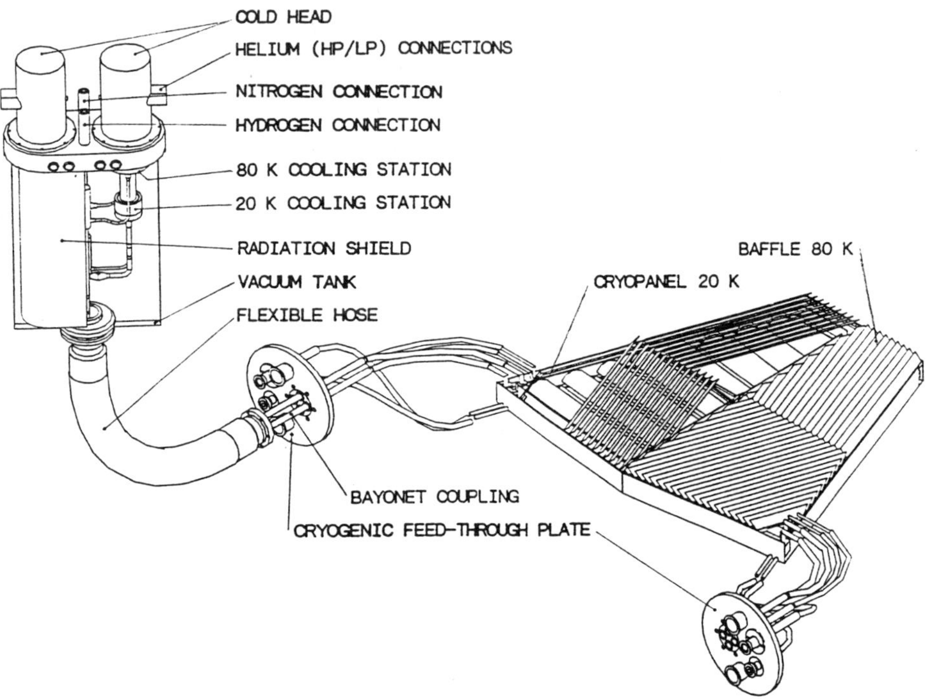


**Figure 1.** The"SPIRAL" Cryopump : cooling module with heat pipes and cryopanels.

- Each cryopump consists of two cryopanels in an arrangement shown schematically in Figure 2. A flat box (C) of a trapezoidal shape equipped with a louvre baffle (D) on its upper face is cooled with LN supplied from the cooling module to some 80 K. It encloses a second box (E) of a similar geometry which is $LH_2$ cooled to some 20 K. In order to reduce substantially the radiative heat load from the warm and relatively "grey" surroundings (A) a highly reflective radiation shield (B) is provided at ambient temperature which allows a wide open louvre baffle and thus an optimal pumping for the air which is cryocondensed on the bare surface of the convex gutters, whereas the not yet evacuated hydrogen is eventually adsorbed on the charcoal grains (F) glued onto the concave side of the same surface.

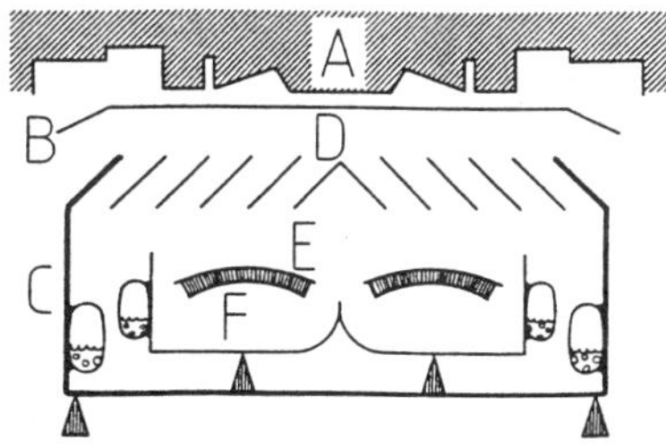


Figure 2. Principle of the pumping array.

To operate these cryopanels two ancillary modules are necessary :

- a helium compressor unit with its flexible connections to the cold heads of each cryogenerator,
- a buffer volume of 20 L for temporary storage and supply of the condensing cryofluid ($N_2$ or $H_2$ at moderate pressure (up to $2.10^6$ Pa maximum).

## PERFORMANCES

The performances of the cryogenic system depend widely on its operational mode. Refrigeration is at maximum if one compressor supplies one cold head. Feeding of two heads from the same unit reduces the available cooling power about 20 %. The cryogenic performance also depends on the filling and operation of both heat pipes. For the 80 K loop condensing and evaporation of nitrogen is indispensable for an efficient heat transfer, whereas convection cooling with helium gas only would require a substantial temperature difference. On the contrary, for the 20 K loop a liquid hydrogen filling stabilised at $10^5$ Pa or pure helium in natural convection produces identical cooling of the cryopanel. Heat pipe fillings from mixtures of $H_2$/He might perform better but are still under investigation.

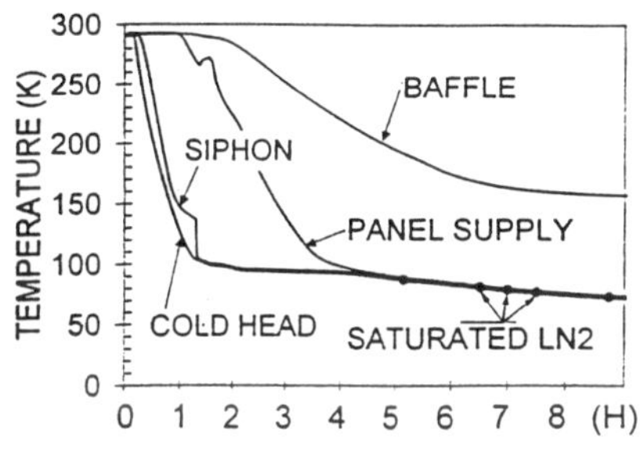


Figure 3a. Cooldown of the 80K cryopanel.

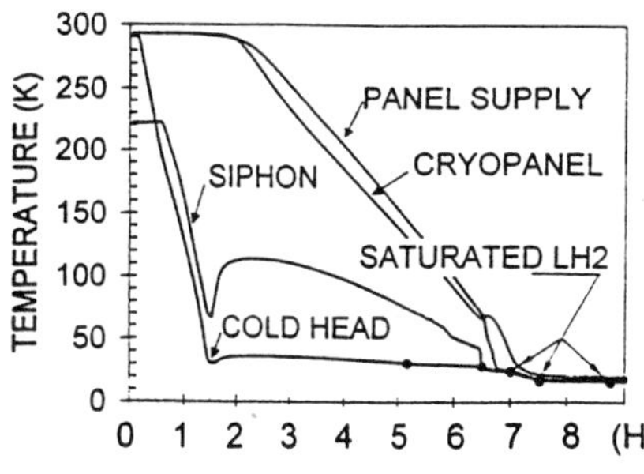


Figure 3b. Cooldown of the 20K cryopanel.

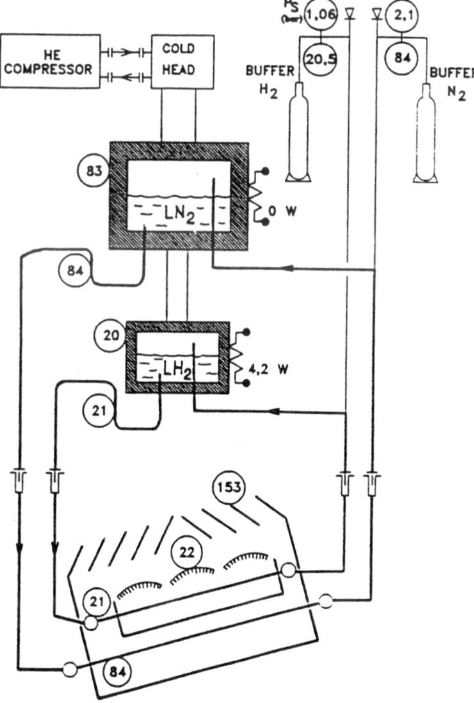


Figure 3c. Temperature distribution at steady state with $LN_2$/$LH_2$ heat pipes.

Figures 3a and 3b show the cooldown caracteristics, Figure 3c the steady state temperature distribution for a typical run with a stabilised hydrogen and a "floating" nitrogen filling of the heat pipes.

The pumping speeds of the cryopanels are not measured precisely in our first runs with a test chamber. An ultimate vaccuum of $2.10^{-6}$ Pa allows a rough estimate of 15 000 L/s for air with one cryopanel.

## CONCLUSIONS

Cryogenic heat pipes for cooling a cryopump are relatively simple and very efficient devices for almost isothermal heat transfer. Cryofluid circulation by gravity performs well even with a moderate level difference between the cooling heads and the cryopanel. For better vacuum performance and easy maintenance the cooling unit may be self-contained in a separate vacuum tank and connected to the cryopanel with an appropriate bayonet coupling

For security and reliability however the vapour pressure in the heat pipes should be permanently maintained above atmospheric pressure, if necessary, by means of some additional heating. Natural convection of He gas under moderate pressure may sometimes replace condensing hydrogen. The use of hydrogen-helium mixtures looks promising and will be investigated.

## ACKNOWLEDGMENTS

The authors appreciated the confidence of the SPIRAL group who accepted the unusual cooling concept for their machine. They are indebted for the expertise and valuable help they got from their teams, in particular P. Blache, G. Roger for design, R. Chevrollier, D. Grolet, J. Languiller, A. Pilot, A. Serafini for construction and assembly.

## REFERENCES

1 The SPIRAL. "Radioactive Ion Beam Facility", GANIL Report R9402 may 1994.

2 S. Buhler, A. Horbowa. "The AGOR Cryopump", Proc. 12th Intern. Conf. on Cyclotrons, Berlin, May 1989.

3 A. Horbowa, S. Buhler. "Le Système de cryopompage du cyclotron AGOR", Vide et Couches Minces. No 252, 1990.

# DEVELOPMENT OF INTERMETALLIC COMPOUNDS FOR USE AS MAGNETIC REFRIGERATORS OR REGENERATORS

M. Foldeaki, A. Giguere, R. Chahine, and T. K. Bose

Institute de Recherche sur l'hydrogene, UQTR
Trois-Rivieres, Quebec G9A 5H7 Canada

## ABSTRACT

Materials with complex magnetic order and complex phase transitions are our best hope in search for the giant magnetocaloric effect. Members of the $RT_2$ family of intermetallic compounds (R=Dy, Gd, Er Ho, T=Co, Ni, Al) are known to display high magnetocaloric effects. In this study, we investigate, how the classical magnetic properties correlate with the presence of high magnetocaloric effects. This knowledge provides a useful tool to identify promising materials from literature surveys and simultaneously helps the researcher to a better understanding of the material's behavior and offer useful additional data to the refrigerator designer. We report here beyond the usual magnetic entropy change measurements on the compounds' response to field and temperature cycling as well as on their time-dependent response to an instantaneous change of the external field. It will be shown that unusually high magnetic entropy changes are coupled with increased irreversibility, with respect to both field and temperature, and with a fast, asymmetrical relaxation. For the best materials, the irreversibility persists up to fields in the tesla range. Simultaneously, it is shown that slight irregularities in processing and consequently in structure, result in dramatic changes in the nature of the transition and reduces the value of the observed magnetocaloric effect as well.

## INTRODUCTION

Magnetic refrigeration produces cooling by exploitation of the magnetocaloric effect (isothermal magnetic entropy change or adiabatic temperature change of solids upon application of an external magnetic field), with the potential of replacing or completing usual gas liquefaction techniques at a later stage of technological development[1]. Beyond very low temperatures, only ordered magnetic materials at their phase transition temperatures display significant magnetocaloric effects. The size of the effect depends on the nature of the transition and on the composition. In materials with localized spins, only the magnetic (spin) state changes at the magnetic order-disorder transitions. Thus the magnetocaloric effect is determined by the number and moment of the magnetic atoms present and can be approximately calculated from the molecular field model. In case of complex magnetic order, where the transition is partially or completely order-order (i.e. between different types of magnetic order), the transition affects not only the spins, but the conduction electrons as well, or sometimes even lattice structure changes can be driven via magnetic field. The magnetism in these materials is better described by the collective or itinerant electron model[2]. Due to the involvement of the conduction electrons in magnetism, the total entropy change upon field application might be significantly higher in these materials than for the order-disorder transitions with localized spins. Unfortunately, the effect cannot be predicted on the basis of composition, like for localized spins. Revealing the correlation of observed high magnetocaloric effects with other magnetic properties provides a useful search engine for promising materials in the vast database on rare-earth magnetism and provides useful knowledge for the designer as well.

## MAGNETIC TRANSITIONS IN $RT_2$ MATERIALS

The ordering of the T atom is in $RT_2$ compounds is governed by specific condicitons[3]. Iron is always ordered and the transition temperatures are high, while no magnetism has been observed on nickel. Cobalt plays an intermediate role: cobalt atoms possess ordered magnetic moments only if the alloyed rare-earth itself is magnetically ordered. Magnetic transitions are generally of second order, except for R=Dy, Ho, Er. The cobalt moment is not intrinsic but induced in the d band by the exchange field due to the rare-earth. The moment appears if the exchange field exceeds the critical field necessary to induce the critical moment (about 1 $\mu_B$), as calculated from collective electron metamagnetism[3]. The zero-field transition temperatures for Co moment ordering can be determined this way from the Brillouin curve of the rare-earth atom, as calculated from the molecular field model, and the temperature dependence of the cobalt moment, determined from the behavior of $YCo_2$. The calculation predicts that transitions below 200 K are of first order[3]. The ordering of the rare-earth moment itself is of second order. Those materials are of specific interest in which the two transitions are fairly close. For temperatures higher than the ordering temperature of the rare-earth, magnetic field can still make cobalt magnetic (indirectly, by inducing order in the R sublattice.) The field necessary for this increases with increasing temperature, and the observed effects will depend on the applied maximal field. It should be noted that the direction of the molecular field acting on the Co sublattice is antiparallel to the R moments. This indicates that the effective field acting on the Co sublattice decreases with increasing applied field in the ground state. The Co order will collapse at a critical field, resulting in a metamagnetic transition[4].

The complex magnetic behavior well explains, that this materials display unusual transport properties as well[5]. Above the transition, the electrical resistivity is governed by the scattering of conduction electrons on spin fluctuations arising mainly from local splittings of the Co 3d band. When lowering the temperature, the 4f moments which order ferromagnetically at $T_c$ yield a strong exchange field acting on the Co 3d states. When this field is sufficient to split the 3d majority and minority sub-bands, the spin fluctuations at the Co sites are quenched, and the 3d metamagnetic state with moments on the Co sites is induced by a first order transition. Consequently, the spin fluctuation scattering vanishes, yielding a

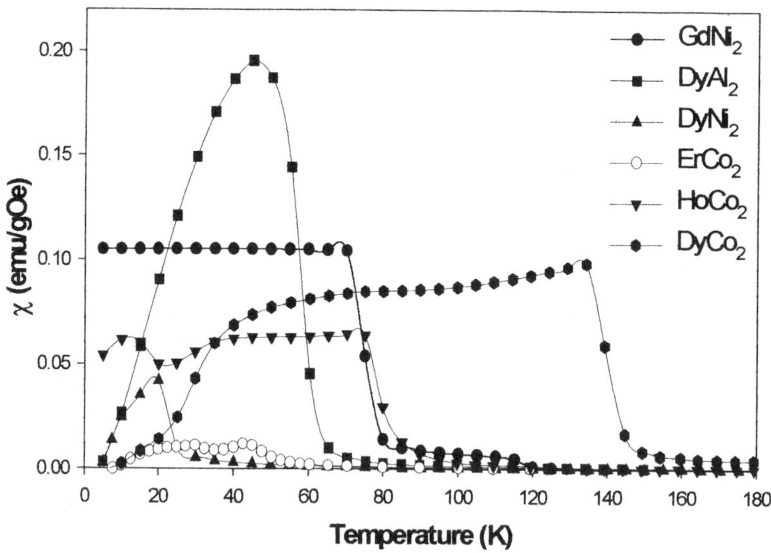


**Figure 1**. Typical behavior of the low-field (100 Oe) magnetic susceptibility after zfc.

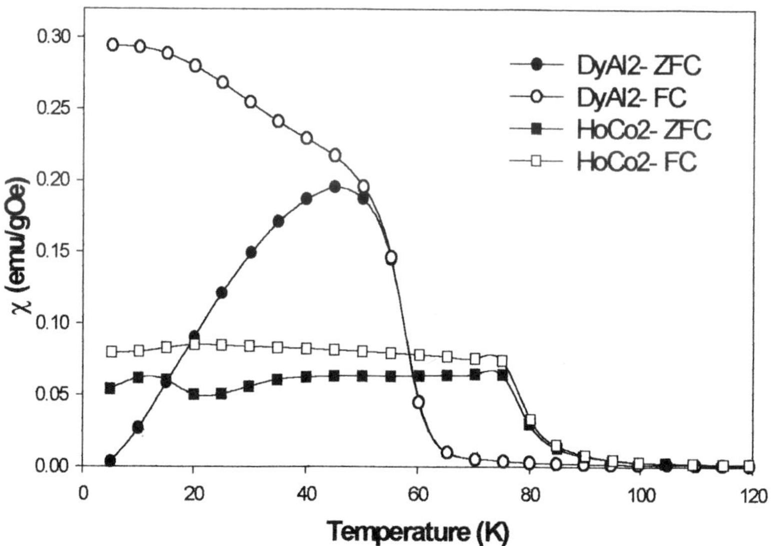


**Figure 2**. Strong temperature dependence is lost after fc in 100 Oe.

resistivity drop. This process may occur at $T_0$ in zero magnetic field or at higher temperature if the exchange field is assisted by a sufficiently large external magnetic field.

Measurements described below will display that unusual magnetic properties -- together with high magnetocaloric effects -- are observed in most investigated $RT_2$ alloys, not only in $RCo_2$.

## EXPERIMENTAL

Materials were prepared in a table-top electric arc furnace by melting nominal quantities of the constituents together. The ingot was flipped and remelted several times to ensure homogeneity. For small samples (up to 10g), this method worked well and resulted in proper, homogeneous alloys, as confirmed via X-ray diffraction. For a larger $ErCo_2$ sample (20 g), X-ray diffraction revealed minute quantities of oxides, in spite of the argon atmosphere, and lines representing the matrix slightly broadened, indicating minor distortions of the structure.

Magnetic measurements were carried out in a conventional SQUID magnetometer. Magnetization curves were always measured after zero-field cooling, subsequent temperatures in increasing order. Magnetic susceptibility was always measured upon warming, either after zero field cooling (zfc), with the field turned on at the minimum temperature, or after field cooling (fc), with the measuring field on all the time. Magnetic hysteresis was always measured with the field decreasing from its positive maximum; either after zero-field cooling or field cooling in the same maximum field. Time dependent response was measured after the field was turned on or off, with the SQUID scanning continuously, resulting in a magnetization reading at about every 20 sec.

From the high-field susceptibility above the transition, the Curie-Weiss parameters were calculated the usual way[2]. From the magnetization curves in the vicinity of the transition, the magnetic entropy change was calculated, as described in detail elsewhere[6]. Simultaneously, the Arrot plots[7] were calculated in order to determine the nature of the transition and the transition temperature.

## RESULTS AND DISCUSSION

The high field/high temperature susceptibility is consistent with the atomic moment of the rare-earth atom present, indicating a good agreement between the nominal and actual compositions.

Figure 1. reveals typical behavior of the susceptibility below the transition. After zfc, the behavior of the low field susceptibility ranges from almost temperature independent ($GdNi_2$) to a sharp cusp ($DyAl_2$) with almost zero susceptibility at the lowest measuring temperature. The high field susceptibility (not shown) is always almost temperature independent.

Field cooling does not influence samples with flat temperature dependence. The samples with the sharp cusp, on the other hand, always display a strong irreversibility, as demonstrated for $HoCo_2$ and $ErCo_2$ in Figure 2. The cusp displays a maximum with increasing fields, and disappears at about the same field (0.5 T for $ErCo_2$) as the irreversibility.

For the $RCo_2$ alloys, the observed irreversibility can be well explained on the basis of the induced order on the cobalt sublattice. In case of zero field cooling the ordering of the Co moment (antiferromagnetic with R) results in a sharp decrease of the susceptibility, as usual for antiferromagnetic transitions. Field cooling seems to align the (paramagnetic) cobalt moments parallel with the R moments at temperatures above this transition, and this order

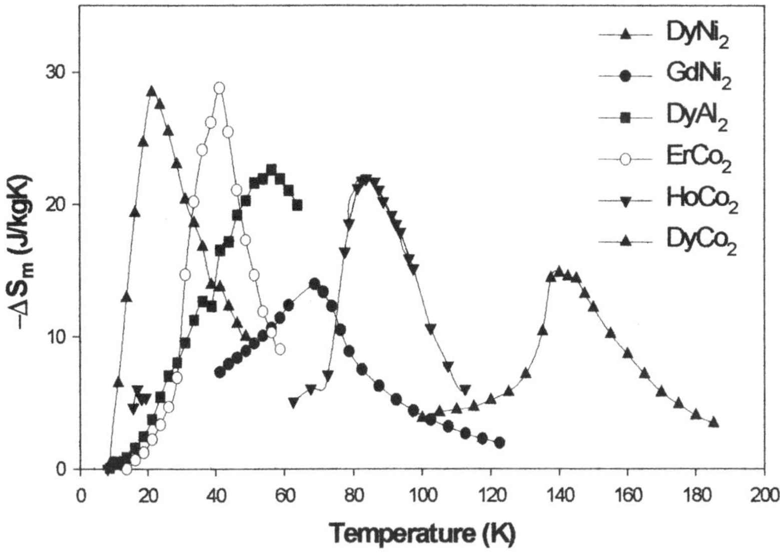


**Figure 3**. Magnetic entropy change of the investigated materials for 7 T applied field.

remains frozen (probably due to the high anisotropy of the compound) resulting in an almost field-independent susceptibility. Higher applied fields reduce the exchange field at the Co site, resulting in an increase of the ferromagnetic order and disappearing difference. An explanation along this lines cannot be given for the irreversibility in $DyAl_2$ and $DyNi_2$, where T is not magnetic. It would also be hard to argue for a spin glass state in these crystallographically ordered compounds. It is more likely, that the non-collinear magnetic order in these compounds is affected by the cooling field and due to the presence of crystal field effects the difference is more pronounced than in pure dysprosium. These anisotropy effects are likely to be responsible for the hysteresis as well. The non-irreversible samples do not display significant hysteresis, while for the other samples the hysteresis increases with increased splitting of the susceptibility curves.

The magnetic entropy change, as calculated from the magnetization curves, is displayed in Figure 3 for a group of the investigated alloys. Comparison of $GdNi_2$ and $DyNi_2$ clearly shows, that the irreversible alloy performs by a factor of two better. Investigating the nature of the phase transition in both of them might provide an explanation. Figure 4 shows the Arrott plots for both alloys. $GdNi_2$ displays nearly classical ferromagnetic behavior, with

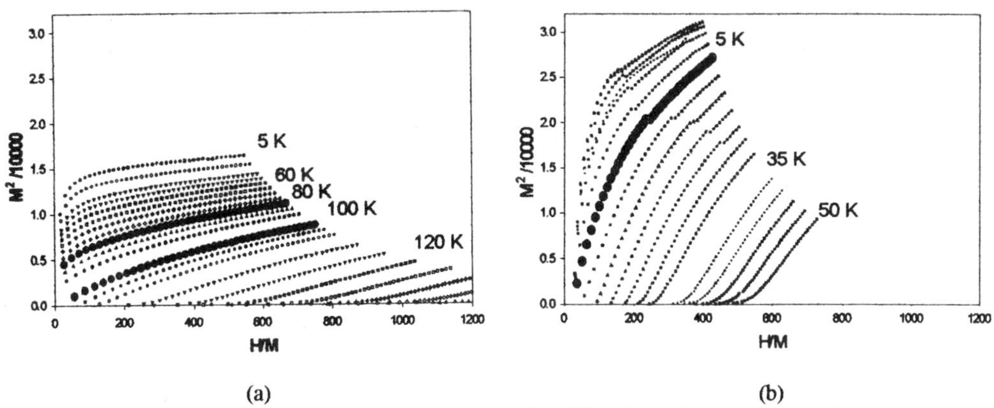


(a) (b)

**Figure 4**. Comparison of the Arrott plots for $GdNi_2$ (a) and $DyNi_2$ (b) samples.

opposite curvature below and above the transition. The curves would allow for an evaluation according to the Landau theory[6], as they do have linear sections at high field/high magnetization, which goes through the origin at the transition. In summary, the transition in $GdNi_2$ can be classified with good approximation as a second-order transition. On the contrary, $DyNi_2$ does not display linear sections at any temperature range. No curve goes through the origin, as characteristic for materials with random fields or random anisotropy[9]. Although the curvature changes at the transition, the Landau theory could not be applied for evaluation, as there are no linear sections. In summary, this is a complex magnetic transition, first order in least one of the sublattices, on the contrary of theoretical predictions[3]. We cannot decide whether this means an ordered moment on the Ni site, or simply the Dy's behavior has changed in the presence of Ni.

Figure 5 displays the magnetic entropy changes for the $ErCo_2$ composition, one for the "good" and the other for the "wrong" $ErCo_2$. The difference is about 30%, a significant change as compared the slight, easy-to-overlook changes observed with the diffractometer. Comparison of the Arrott plots (Figure 6) demonstrates, that the nature of the magnetic order and the characteristics of the transition have significantly changed in the "wrong" sample. The curvature changes at the transition, but this change is accompanied by a change in the curve's shape, similar to that observed in Dy-Zr nanocomposites[10]. This indicates that the slower cooling of a larger sample introduced a multi-phase structure in the nano-level, and proves the extreme structure (thus processing) sensitivity of the magnetic properties in rare-earth intermetallics, problems so well known to the hard-magnet manufacturer. Although not investigated in detail, but the decreased magnetic entropy change observed in $DyCo_2$ might be attributed to similar effects as well.

As shown in detail in $ErCo_2$ (Figure 7), the materials display a strong time dependent response, following a logarithmic time law. The time constant is field and temperature dependent, and goes through a maximum below the transition temperature. Near the transition, the logarithmic time law is no more valid, and the time dependence decreases and becomes unobservable with increasing temperature. Detailed measurements would allow to determine the nature of the activation (eventually calculate the activation energy) and would give a better the understanding of the process.

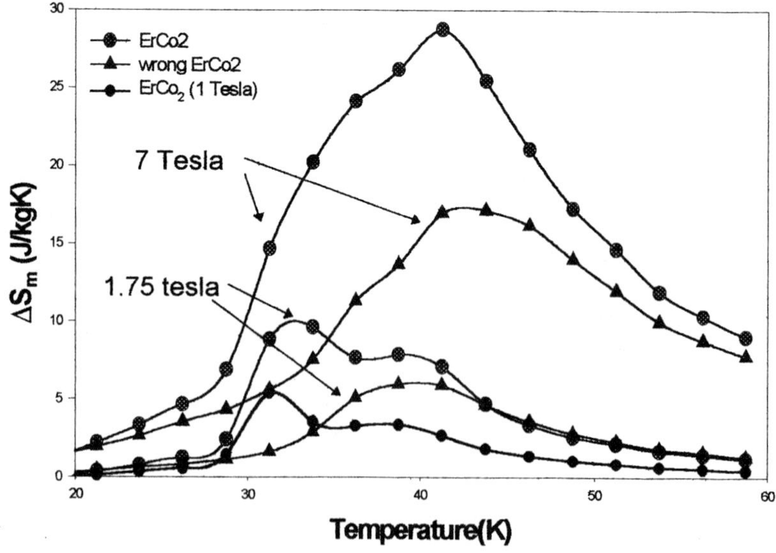


**Figure 5.** Comparison of the magnetic entropy changes in the "good" and "wrong" $ErCo_2$ samples.

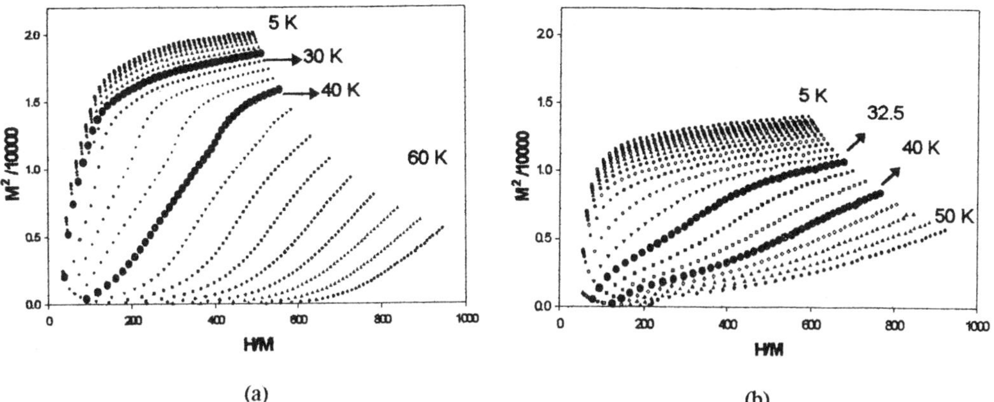


(a) (b)

**Figure 6**. Comparison of the Arrott plots in the "good" (a) and "wrong" (b) $ErCo_2$ samples.

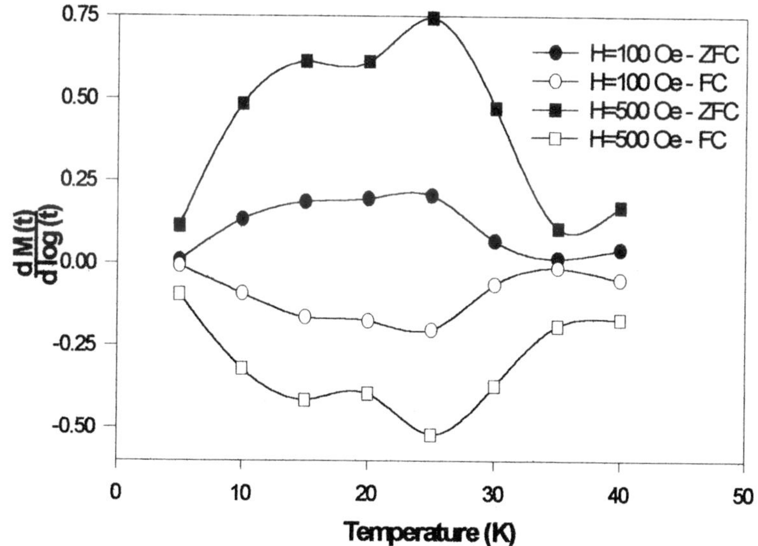


**Figure 7**. Time dependent response of $ErCo_2$.

In summary, the investigations show that

- Increased irreversibility results in increased magnetocaloric effects
- Among samples with similar irreversibilities, the material with the higher $T_f/T_c$ (with $T_f$ as the freezing, $T_c$ the transition temperature) ratio would perform better in a magnetic refrigerator.
- Slight structural imperfections dramatically affect the magnetic properties.

## CONCLUSIONS

Rare-earth transition metal alloys are promising candidates for magnetic refrigeration applications. This paper displays that the best magnetocaloric properties are systematically coupled to certain classical magnetic properties, like increased irreversibility in thermal or magnetic cycling and time dependent response. The best materials display first-order or combined transitions. This correlation facilitates the search for new, promising candidate materials. The results stress the importance of proper processing and nonforgiving quality control. It is the refrigerator designer's task to investigate the effect of irreversibity on machine operation and perform the necessary (if any) adjustments.

## ACKNOWLEDGEMENTS

This work was supported by Natural Resources Canada and Ressources naturelles du Quebec.

## REFERENCES

1. J. A. Barclay, Prospects for magnetic liquefaction of hydrogen, Proc. XVIII. International Congress on Refrigration, Montreal, 1991, Vol I., p. 297.
2. B. D. Cullity: "Introduction to Magnetic Materials," Addison-Wesley, Reading, MA (1972).
3. M. Cyrot and M.Lavagna, Density of states and magnetic properties of rare-earth compound $RFe_2$, $RCo_2$ and $RNi_2$, J. Phys. 40:763 (1979)
4. M. I. Bartashevich, H. Aruga Katori, T. Goto, H. Wada, T. Maeda, T. Mori and M. Shiga, Collapse of the itinerant Co moment in $Er_{1-x}Lu_xCo_2$ by the application of high magnetic fields, Physica B 229: 315 (1997)
5. T. D. Kuong, L. Havela, V. Sechovsky, A.V. Andreev, Z. Arnold, J. Kamarad and N. H. Duc, Band metamagnetism and related phenomena in $Er(Co_{1-x}Si_x)_2$, J. Appl. Phys. 81: 4221 (1997)
6. M. Foldeaki, R. Chahine, and T. K. Bose, Magnetic meausurements: A powerful tool in magnetic refrigerator design, J. Appl. Phys. 77: 3528 (1995)
7. A. Arrott, Criterion for Ferromagnetism from Observations of Magnetic Isotherms, Phys. Rev. 108: 1394 (1957)
8. K. P. Belov: "Magnetic Transitions" Boston Technical, Cambridge (1995)
9. A. Aharony and E. Pytte, Infinite susceptibility phase in random uniaxial anisotropy magnets, Phys. Rev. Letters, 45: 1583 (1980)
10. A. Giguere, M. Foldeaki, R. A. Dunlap, R. Chahine, Magnetic properties of Dy-Zr nanocomposites, in preparation

# RECENT PROGRESS IN THE APPLICATION OF MAGNETIC REGENERATOR MATERIALS

T. Hashimoto[1], H. Nakane[2], T. Tsukagoshi[3], and H. Nakagome[4]

[1]Tokyo Institute of Technology (Prof. Emeritus)
Meguro-ku, Tokyo, 152, Japan

[2]Kogakuin University
Shinjuku-ku, Tokyo, 163-91, Japan

[3]NTT Corporation
Yokohama-shi, Kanagawa, 220-81, Japan

[4]Toshiba Corporation
Kawasaki-shi, Kanagawa, 210, Japan

## ABSTRACT

Various magnetic materials have a large peak in heat capacity $C(T)$ in the vicinity of liquid helium temperature at $T_c$ (magnetic phase transition temperature). However, a single magnetic material is not sufficient for regenerative operations since the peak in $C(T)$ is narrow. Therefore, an important consideration in the development of a magnetic regenerator is the sequencing of the magnetic materials so as to produce the best $C(T)$ distribution. To find out the highest heat exchange effectiveness of a regenerator (regenerator effectiveness), a study was conducted into the effectiveness of multi-layer regenerators composed of several rare-earth compounds with different $T_c$ by computer simulation.

The simulation clearly indicated that regenerator effectiveness increases as the number of layers is increased. Among all the different compounds tested, maximum effectiveness was obtained in the case of a two-layer regenerator of $Er_{0.9}Yb_{0.1}Ni$ and $Er_3Co$; a three-layer regenerator of $ErNi_{0.8}Co_{0.2}$, $Er_{0.9}Yb_{0.1}Ni$ and $Er_3Co$; and a four-layer regenerator of $ErNi_{0.8}Co_{0.2}$, $Er_{0.9}Yb_{0.1}Ni$, ErNi and $Er_3Co$ inserted in sequence from cold to hot.

The refrigeration power of a four-layer regenerator is expected to be twice as much as that of a single-layer $Er_3Ni$.

## INTRODUCTION

The development of a smaller and more efficient cryogenic refrigerator is a major concern for the progress in cryogenic engineering. One such refrigerator is the Gifford-McMahon refrigerator (GM refrigerator) which is easy to use. However, the effectiveness of these types of cryogenic refrigerators largely depends on the effectiveness of the heat-exchange component. In order to obtain higher effectiveness, it is considered that the heat capacity of the regenerator material be larger than that of the working gas (helium). The specific heat of helium peaks at the supercritical state at temperatures below 10K as shown in Fig. 1. On the other hand, the specific heat of copper or lead rapidly decreases at low temperatures, which makes these materials inappropriate for use at helium temperature (4.2K).

Due to the peak caused by the magnetic phase transition, the specific heat of various rare-earth compounds is higher than that of non-magnetic materials such as copper or lead below 10K. However, the integration of the magnetic specific heat is finite because the specific heat arises from the entropy change of the spin system. Therefore, the higher the specific heat, the narrower the half-width, and vice versa.

For example, the half width of the specific heat of $Er_3Ni$ is wide, but the peak is low. On the other hand, the specific heat peak of $ErNi_{0.8}Co_{0.2}$, $Er_{0.9}Yb_{0.1}Ni$ and ErNi is close to that of helium, but the half-width is approximately half as much. Obviously neither magnetic material can cover the specific heat peak of helium.

The concept of multi-layer regenerators was developed to compensate for this deficiency in the magnetic materials and increase the regenerator effectiveness. Fig. 2 is the schematic of a multi-layer regenerator [1,2]. In a multi-layer regenerator, the magnetic materials, whose magnetic phase transition temperatures are different, are packed in order of its phase transition temperature. By doing this, it is expected that a heat capacity comparable with helium can be obtained over the whole temperature range.

## COMPUTER SIMULATION OF LAYER-STRUCTURED MAGNETIC REGENERATORS

The pressurized He gas in a GM refrigerator is under 20 atm. Magnetic materials contain a large specific heat below 10K in comparison with non-magnetic materials due

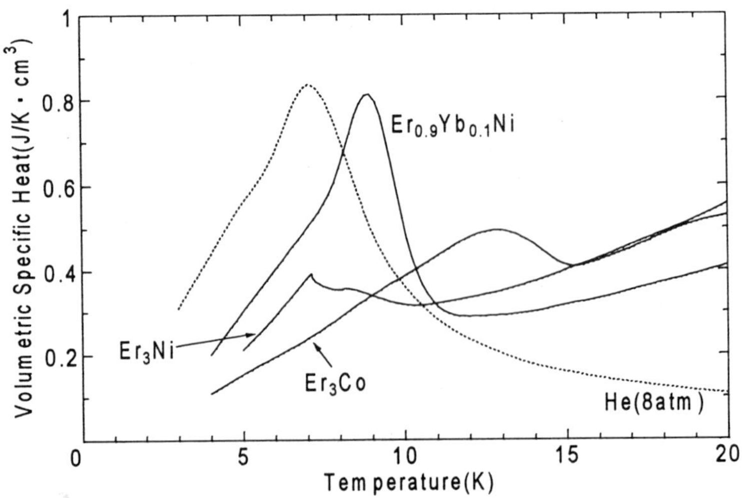


Figure 1. Specific heat of magnetic materials.

**Table 1.** The phase transition temperature $T_C$ and half-width of the peek of the magnetic specific heat

| compound | $T_C$(K) | $\Delta T$(K) | $\Delta T/T_C$ |
|---|---|---|---|
| He(8 atm) | 7.00 | 3.9 | 0.56 |
| $Er_3Ni$ | 7.16 | 2.1 | 0.29 |
| $ErNi_{0.8}Co_{0.2}$ | 6.40 | 1.9 | 0.30 |
| $Er_{0.9}Yb_{0.1}Ni$ | 9.00 | 2.7 | 0.30 |

to the phase transition. However, the peak of specific heat due to the magnetic phase transition is very sharp in comparison with that of the pressurized He gas as can be seen from the comparison of the half-width peaks of a magnetic material and of He gas at 8 atm as shown in Table 1 [2, 3]. Therefore, it is difficult to increase the effectiveness of a regenerator by using only one rare-earth element such as Gd, Er, Dy or only one compound. In order to get the best distribution of the heat capacity for the regenerative operation in a regenerator, it is necessary to analyze the multi-layer structure of magnetic materials. Fig. 3 is a schematic of a three-layer regenerator packed with materials A, B and C. To verify the superiority of a multi-layer magnetic regenerator at low temperature, the effectiveness has been compared by computer simulation. In the simulation, the maximum effectiveness of a regenerator (regenerator effectiveness) was calculated by changing the number of layers, the kind of materials and packing ratio of the materials in the regenerator. The regenerator effectiveness, which means the effectiveness of heat exchange, is defined as the effectiveness of the enthalpy ($\eta$) and is given by [4]

$$\eta = \Delta H_{real} / \Delta H_{ideal} \qquad (1)$$

where $\Delta H_{real}$ and $\Delta H_{ideal}$ are the actual and ideal enthalpy changes of the gas, respectively. The $\eta$ was calculated with a computer using the magnetic regenerator materials as shown in Fig. 4. Their peaks of the specific heats near the magnetic phase transition point are at nearly the same level as that of He gas at 8 atm. From all the curves showing the specific heats, only the curve representing the magnetic specific heat is applied in Table 1. The half-width ($\Delta T$) of the peak is also shown in Table 1. Although $\Delta T$ for each material is different, the ratio of $\Delta T$ to $T_C$ (magnetic phase transition temperature) was near 0.3 for the magnetic materials and 0.56 for He gas at 8 atm. This indicates that specific heat of the pressurized He gas cannot be covered with only a single material. This fact means that, below 15K, several kinds of magnetic materials with different $T_C$

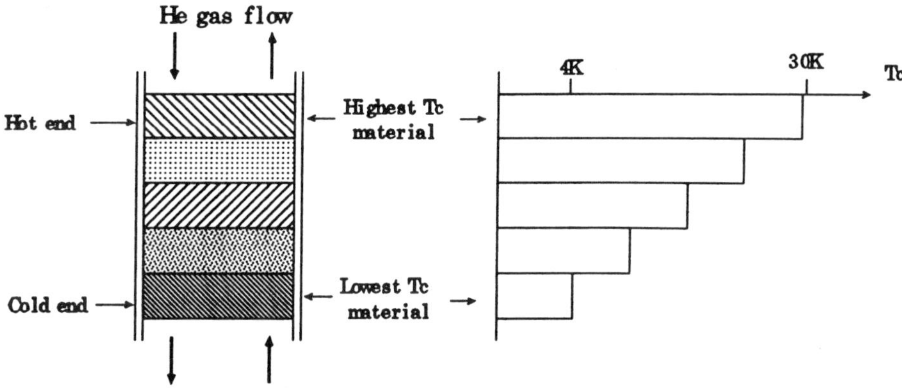


**Figure 2.** Schematic of the multi-layer regenerator.

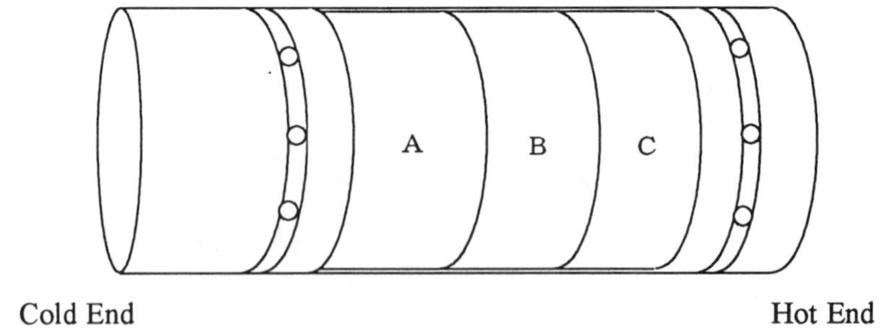


**Figure 3.** Schematic of multi-layer regenerator structure.

**Table 2.** Result of computer simulation

| | Cold end side (4K) | | | Hot end side (30K) | Regenerator Effectiveness |
|---|---|---|---|---|---|
| (A) | $Er_3Ni$(100%) | | | | 99.700 % |
| (B) | $Er_{0.9}Yb_{0.1}Ni$(50%) | | $Er_3Co$(50%) | | 99.880 % |
| (C) | $ErNi_{0.8}Co_{0.2}$(20%) | $Er_{0.9}Yb_{0.1}Ni$(30%) | | $Er_3Co$(50%) | 99.922 % |
| (D) | $ErNi_{0.8}Co_{0.2}$(15%) | $Er_{0.9}Yb_{0.1}Ni$(5%) | ErNi(20%) | $Er_3Co$(60%) | 99.979 % |

**Table 3.** Comparison of experimental and simulation results

| | Refrigeration Capacity at 4.2K(W) | Regenerator Effectiveness |
|---|---|---|
| (A) | 0.59 | 99.700 % |
| (B) | 1.00 | 99.880 % |
| (C) | 1.07 | 99.922 % |
| (D) | — | 99.979 % |

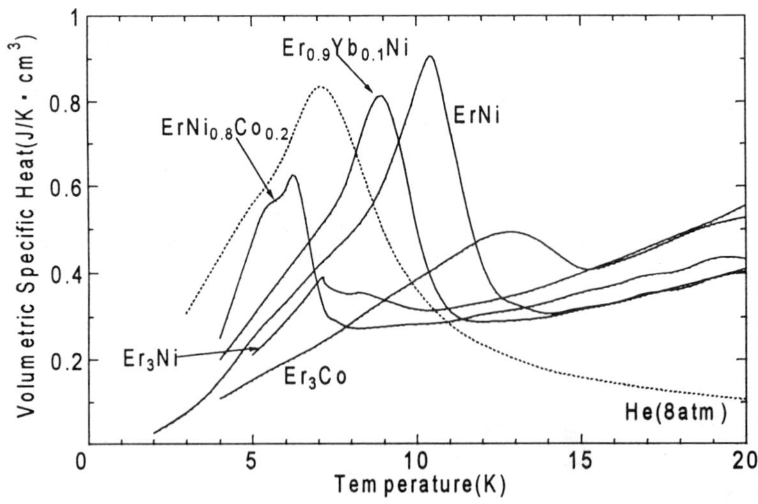


Figure 4. Specific heat of magnetic materials.

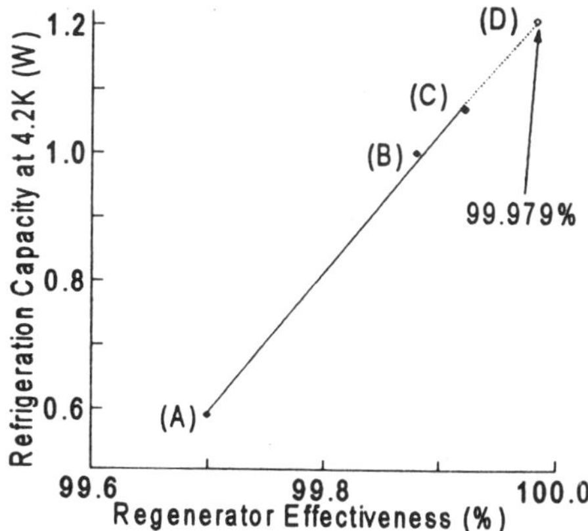


**Fig. 5.** The relation between regenerator effectiveness and refrigeration capacity.

are necessary to cover the entire property of the pressurized He gas. This is the basis for the proposal that multi-layer magnetic regenerating materials be used.

In order to quantitatively find out the effectiveness of a layer-structured regenerator, a computer simulation was done using the ratio of packing regenerating materials as shown in Table 2 to estimate $\eta$. In order to find out the relation between the $\eta$ obtained by computer simulation and the refrigeration capacity of a GM refrigerator, the experimental values, which were drawn from published literature [3], were used as the refrigeration capacities for the materials and packed according to the ratio shown in Table 2. Although the maximum value of the regenerator effectiveness ($\eta$) is not expressed, materials having $\eta$ values close to the maximum were used. The simulation indicated that the regenerator effectiveness increases with the increase in the number of layers in the regenerator.

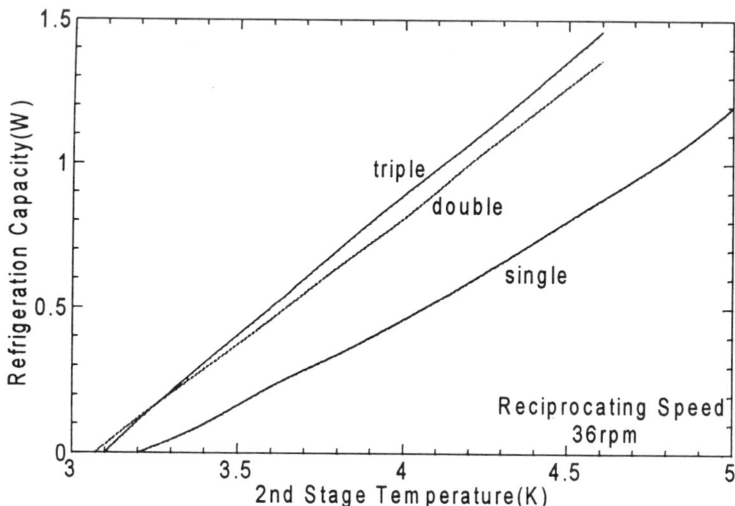


Figure 6. Refrigeration capacity of three kinds of regenerators obtained as a function of 2nd stage temperature.

## COMPARISON OF REGENERATOR EFFECTIVENESS AND REFRIGERATION CAPACITY

The most desirable composition of two-, three- and four-layer regenerators was designed by computer simulation. In this section, the results calculated with a computer are compared with the experimental values of the refrigeration capacity of a GM refrigerator. The refrigeration capacity of a two-stage single-, double- and triple-layer GM refrigerator was obtained by a conventional experiment. Table 3 [3] shows the comparison of the values obtained by computer simulation and the conventional experiment at the refrigeration capacity of 4.2K. The same packing ratio of materials was used in comparing the computer simulation (regenerator effectiveness) with the experiment (refrigeration capacity). For a four-layer structure, experimental values were not obtained and only the calculated values are shown. Obviously, a three-layer regenerator is superior to other fewer-layer regenerators. The solid lines in Fig. 5 shows the relation between regenerator effectiveness and refrigeration capacity at 4.2K for the case of one, two and three layers. From Fig. 5, the property can be assumed to be linear and the refrigeration power of a four-layer is expected to be 1.2W. The value is twice as much as that of a single-layer $Er_3Ni$. Fig. 6 indicates that the no-load temperature is not a simple relation of the refrigeration capacity at 4.2K [2, 5]. Exact comparison of the regenerator effectiveness and the refrigeration capacity at 4.2K was not possible due to the discrepancy between the conditions of the experiment and simulation. Table 3 and Fig. 6 clearly show that the refrigeration capacity increases in multi-layer structures since the heat exchange rate improves.

## CONCLUSION

To verify the superiority of multi-layer magnetic regenerators in the low temperature range, the regenerator effectiveness of several kinds of materials was estimated by computer simulation and was compared with the refrigeration capacity of a second stage GM refrigerator.

Computer simulation clearly indicated that the regenerator effectiveness increases as the number of layers in the regenerator is increased. Among all the different compound tested, maximum effectiveness can be obtained in the case of a two-layer regenerator of $Er_{0.9}Yb_{0.1}Ni$ and $Er_3Co$; a three-layer regenerator of $ErNi_{0.8}Co_{0.2}$, $Er_{0.9}Yb_{0.1}Ni$ and $Er_3Co$; and a four-layer regenerator of $ErNi_{0.8}Co_{0.2}$, $Er_{0.9}Yb_{0.1}Ni$, ErNi and $Er_3Co$ when the materials are inserted in sequence from cold to hot. Assuming the relation between the effectiveness of the regenerator and the refrigeration power in each case to be linear, the refrigeration power of a four-layer regenerator is expected to be twice that of a single-layer $Er_3Ni$.


## ACKNOWLEDGMENT

The authors would like to thank S. Yamazaki, K. Anzai and Y. Miyata for their assistance in calculations and preparation of figures and table, and T. Takahashi for his helpful discussions.

## REFERENCES


1. T. Tsukagoshi et al. "Optimum structure of multi-layer regenerator with magnetic materials", Teion-Kougaku 31 (1996) 168
2. T. Hashimoto. "Recent progress in magnetic regenerator materials and their application", Trans. of the JAR 10 (1993) 357
3. T. Hashimoto et al. "Excellent character of multi-layer type magnetic regenerator near 4.2K", Cryocoolers 8 (1995) 677
4. H. Seshake et al. "Analysis of rare earth compound regenerators operating at 4K", Adv. Cryog. Eng. 37B (1997) 995
5. T. Hashimoto et al. "Effect of high entropy magnetic regenerator materials on power of the GM refrigerator", Adv. Cryog. Eng. 40A (1994) 655

# MEAN-FIELD APPROXIMATION ANALYSIS OF MAGNETOCALORIC EFFECT IN RARE-EARTH MATERIALS

A.M.Tishin

Faculty of Physics, M.V.Lomonosov Moscow State University, Moscow,119899, Russia

## ABSTRACT

This paper describes the results of a computer simulation analysis of the magnetocaloric effect (MCE) in rare earth metals and rare earth based materials. The analysis was previously developed for calculating peak values of the MCE in rare earth magnets using a mean-field approximation. This study concentrates on the variation of the key thermodynamic parameters such as magnetic field, temperature, Curie and Debye temperatures of the magnets in wide range of their values. The results of theoretical studies of the MCE in large magnetic fields are described. The possible mechanisms of magnetic entropy change in a paraprocess region are discussed. The study shows strong influence of the Debye temperature of the magnets on the MCE value in the wide temperature region (from 4.2 K to 200-230 K).

## INTRODUCTION

One of the fast developing areas of prospective application of rare-earth metals (REM) is the design of new effective magnetic solid state refrigerators. The particular interest in magnetic refrigerators is associated with the possibility of their wide commercialization. Active prototypes of the latter are known by their high efficiency and notable life-in-service time. Hence, the practical use of the magnetocaloric effect (MCE) has potentially a great future (see, for instance, [1]).

The latest results of the experimental studies of the magnetocaloric effect are described in details in works.[2,3] Nonetheless, selection of magnetic materials, that are good candidates for cooling media (magnetic refrigerant) is not straightforward, since the theory which would permit an accurate prediction of the magnetocaloric effect does not exist yet. Hence, the purposes of this work are: 1) the numerical simulation of the magnetocaloric effect in the extended region of temperatures and magnetic fields; 2) the study of the correlation between the MCE and the magnetic entropy; 3) to find out the extent to which the MCE value is affected by the Debye temperature. The result of this research is a general-purpose computer program for numerical calculation (based on the theory of molecular field) of thermodynamic parameters for different classes of magnetic materials.

Furthermore, modern experimental investigations in high magnetic fields up to 500T and more are no longer something "exotic". Since the MCE value in such field can be extremely high, the possibility of its influence on the obtained experimental results needed to be accounted for.

## PEAK VALUES OF THE MAGNETOCALORIC EFFECT AND THE ENTROPY

Based on the molecular field theory, we performed the numerical simulation of the MCE value, $\Delta T$, and the entropy, S, of heavy lanthanides. The used numerical approach is described in details in.[4] Considered magnetic fields were assumed to significantly exceed the critical magnetic fields needed to destroy antiferromagnetic (AFM) state in lanthanides, and therefore, all calculations were performed for magnetic phase transition from ferromagnetism (FM) to paramagnetism (PM). Throughout the paper, the symbols Tb, Dy, Er, Ho and Tm are used to designate a magnetic material with such specific parameters as spectroscopic splitting factor, $g_J$, total angular moment, J, Curie temperature, $T_C$, Debye temperature, $T_D$, and electronic heat capacity coefficient, $\gamma$. All results of numerical simulation should agree with experimental data for heavy lanthanides quite well only at large magnetic field, i.e. such field where the contribution of AFM-FM phase transition to the magnetic entropy is rather small. During the simulation, the assumed values of magnetic field ranged from 0 to 4000T and the initial temperature ranged from 10K to 1000K. The energy spectrum can be extremely distorted in such high magnetic fields, therefore this naturally may affect the real field dependence of magnetic entropy, $S_m$, and $\Delta T$. The phenomena like crossover is considered by the author to yield the "staircase"-like $\Delta T(B)$ and $S_m$ (B) curves. But I acknowledge that its contribution to the overall MCE and the entropy should be quite small. It should be noted that the results of the work can be applied to the real magnetic materials only outside the critical region.

The calculations reveal the fact that if the temperature is high (near 1000K), saturation of the MCE and magnetic entropy is absent even in the highest magnetic fields (up to 4000T), while near the temperature of the phase transition to the paramagnetic state the value of the magnetic entropy saturates in a field B ~ 1000T. In the high fields (B>1000T), the magnetic entropy practically equals to zero up to the point $T \approx T_C$, i.e. there is no magnetic contribution to the entropy. This result leads to the following conclusion: in the fields higher than 1000T the magnetic moments of rare earth ions must be nearly completely aligned in a wide range of temperatures - up to the Curie temperature.

Our calculations point to the fact that the general behavior of the MCE as a function of temperature is transformed significantly under the influence of magnetic field. The anomalies on the MCE curves when $T > T_C$ are observed in magnetic fields up to 4000 T, but the MCE is increasing with the temperature, when B>500T and $T>T_C$. When B<500T, the MCE increases up to $T= T_C$ and then decreases smoothly when $T>T_C$. The temperature and field dependencies of the MCE in REM behave similarly. Figure 1 presents temperature dependence of the maximum MCE for heavy lanthanides (in the saturation field). Gadolinium has the minimum MCE at all considered temperatures.

Our analysis leads to the conclusion that at low temperatures the maximum MCE value (between heavy lanthanides) is determined by the Debye temperature, $T_D$, (i.e. by low lattice heat capacity). When temperature exceed 230K the maximum MCE in all REM follows the total angular moments, J, and/or effective magnetic moment, $\mu_{eff}$.

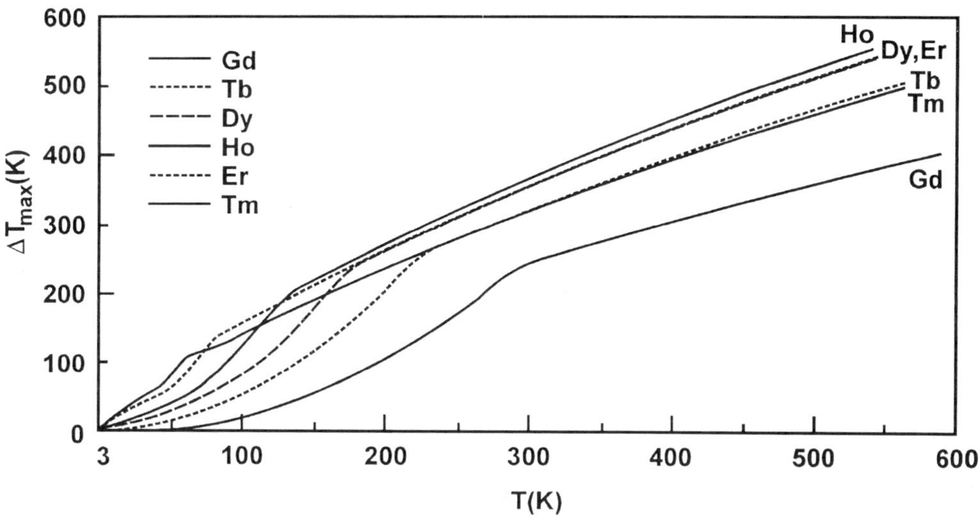


**Figure 1**. Temperature dependence of the peak MCE values for heavy lanthanides.

The calculations show, that Tb has the largest MCE in the fields up to 200T, while Er has the lowest. In higher fields Ho has the maximum value of MCE, and Gd - the minimum. Furthermore, in high fields (although not in the saturation ones) the MCE is defined not only by the magnetic entropy, but also by a number of another parameters of magnetic and crystal structures of lanthanides. Figure 2 presents field dependencies of the magnetic entropy in the vicinity of phase transition to paramagnetic state. The MCE curves are seen to be arranged in order of increasing of the Curie temperatures of lanthanides, except Gd. The calculations reveal that the MCE in REM continues to

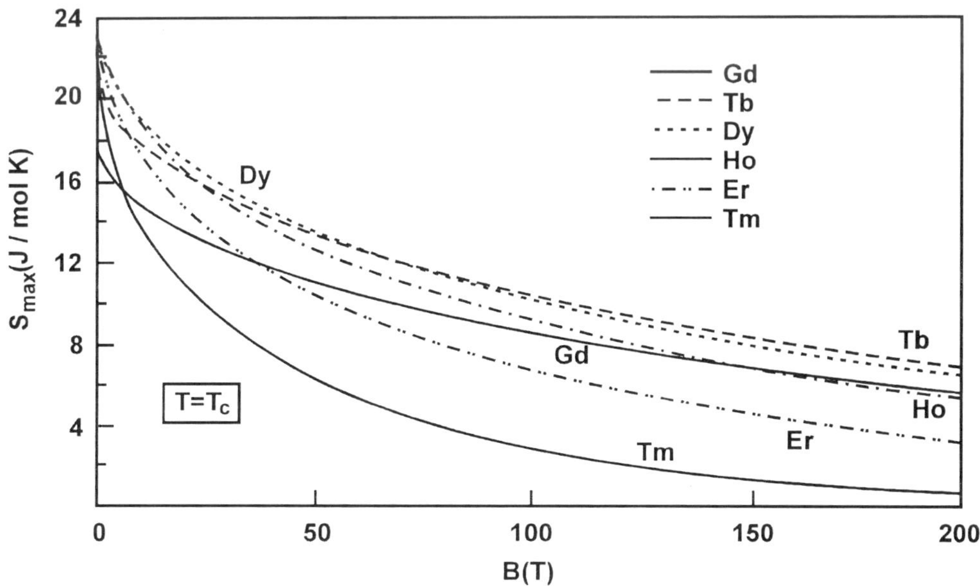


**Figure 2**. Field dependence of the magnetic entropy in the vicinity of the transition to paramagnetic state.

increase even when the magnetic entropy becomes almost equal to zero. The percentage of field dependence of the MCE and the magnetic part of entropy in Gd at $T=T_C$ is shown in the Figure 3. The correlation between the MCE and the magnetic entropy change is seen quite clearly. Thus, for instance, in a 1000T field the MCE value totals to ~88% of maximum possible, $\Delta T_{max}$=234K, and $\Delta S_m$ totals ~98% of maximum possible change, $\Delta S_{max}$=17.3 J/mol K. Hence the question is: what is the nature of continuing increase of the MCE value in high magnetic fields with practically no associated changes in the magnetic entropy? From classic point of view the change of $S_M$ value is usually related to the rotation of the vectors of atomic magnetic moments under influence of the magnetic field (i.e. the $S_M$ decreases with the increase in magnetization in classic magnetic materials). It has been experimentally established that the value of $\Delta S_m$ in rare-earth metals and their alloys in the vicinity of their magnetic phase transition totals to approximately 8 to 10% from $\Delta S_{max}$ in a 6T field.[5] In rare-earth intermetallic compounds the change of magnetic entropy can reach much higher values, close to about 90% in fields up to 10 T and at low temperatures.[6] Along with that the magnetization of Gd at the Curie point practically does not change in 10 to 30T magnetic fields (e.g. see Ref. 7). Measured in a 10T field MCE value in Gd equals $\Delta T$=19K.[8] But from our calculations it follows that $\Delta T_{max}$ in Gd is expected to be 234K at $T=T_C$. Thus, it appears that most of the sample's temperature increase accounts for the region of paraprocess (i.e. the region where the sample has already single domain structure and the atomic magnetic moments are aligned parallel with the field). So the change of MCE and the magnetic entropy is clearly determined not only by rotation of magnetization vectors, i.e. the experimental data can not be explained within the classic theory, which associates the magnetic entropy decrease at the expense of the rising order of magnetic moments in the system. Thus, the use of probability (quantum) approach to the entropy seems to be the right choice. I.e. the magnetic entropy (and consequently sample temperature) continues to change even in the region of paraprocess, just because of the fact that the probability of the atomic magnetic moments deflection from the field direction (due to heat fluctuations) remains non-zero even in high magnetic fields. It is possible, that the values of $\Delta S_m$, (which have been observed in rare-earth intermetallic compounds by Gschneidner et al [6]) are large due to the suppressing of heat fluctuations in low temperature region.

## INFLUENCE OF THE DEBYE TEMPERATURE ON THE MAGNITUDE OF MAGNETOCALORIC EFFECT

Considering that the Debye temperature, $T_D$, influences the MCE was not too much studied yet neither theoretically nor experimentally, it is of interest to define numerically the general rules of the phenomena. In general, the relation between the MCE and the entropy change is well known. To investigate the impact of the Debye temperature on the MCE we need to take into account the effect of the temperature on the lattice entropy ( it is worth noting that the concept of total entropy as a sum of different parts can be used only for magnets with localized magnetic moment, like REM). The entropy of solid is known to increase significantly when the Debye temperature decreases. Thus, for instance, it is easy to show that at 300K the decrease of the Debye temperature from 184K (Gd) down to 50K leads to the lattice entropy, $S_l$, increase on ≈33 J/mol K (70%), while when the Debye temperature increases from 184K up to 500K leads to the $S_l$ change only of ≈23 J/mol K (50%). As well known the maximum value of magnetic entropy change is defined by $\Delta S_{max}$= -Rln(2J+1). So, from one point of view the maximum of MCE is limited by total quantum moment and measurement temperature. On the other hand, in general the MCE can vary due to the change of magnetic, lattice and electronic contributions to heat capacity. Electronic part of heat capacity gives a constant contribution to heat capacity and does not change the MCE if T= const. The character of a magnetic contribution to a heat

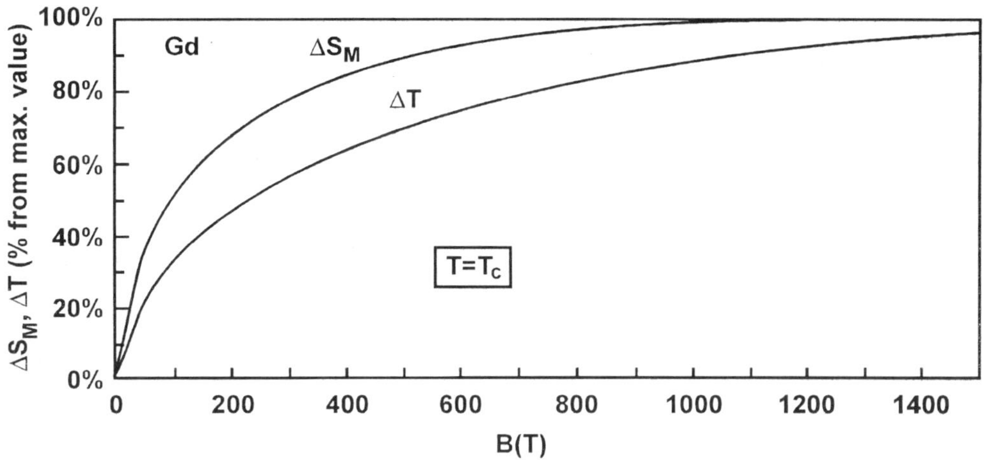


**Figure 3.** Field dependence of the MCE and magnetic part of the entropy in Gd at Curie temperature.

capacity change (generally a decrease for ferromagnets) under exposure to the field is quite complicated one and can be predicted only in general. It has been known that according to Debye theory, the heat capacity can be considered as a constant at $T>>T_D/4$ and proportional to $T^3$ at $T<<T_D/4$. On a basis of our calculations we can maintain, that the Debye temperature change significantly affects the magnetocaloric effect only above 200-230K, i.e. where $T_D$ value impacts the lattice heat capacity significantly.

As an example we can refer to temperature dependencies of the MCE in Gd (see Fig. 4). The Curie temperature (and, consequently, the MCE peak) in Gd accounts for the temperature range where the heat capacity is practically constant and does not change with the variations in $T_D$. Thus, the change in $T_D$ from 184K down to 20K at the Curie temperature leads to MCE decrease only by 0.07K at B=2T, and $T_D$ change from 184K up to 500K leads to MCE increase by 0.45K. As can be seen, the change in the Debye

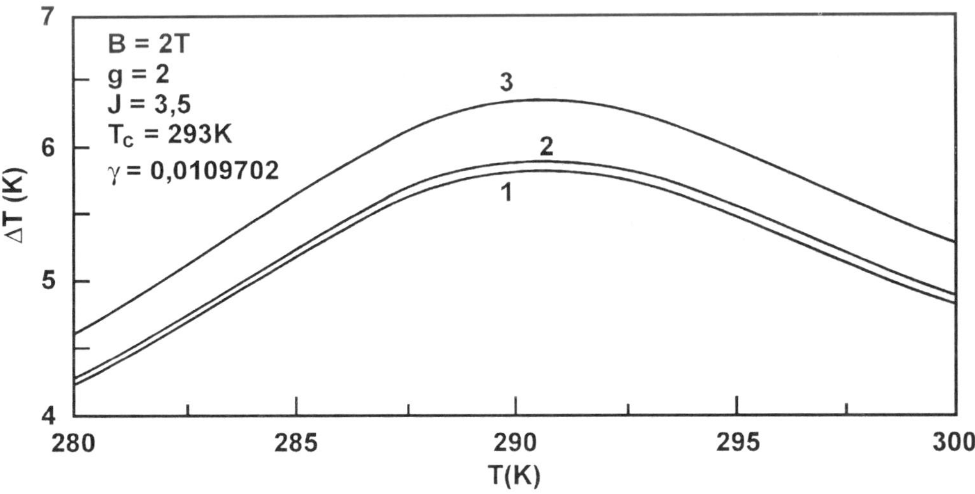


**Figure 4.** Temperature dependencies of the MCE in Gd at different Debye temperatures: 1- $T_D$=20K, 2- $T_D$=184K(Gd), 3- $T_D$=500K. Parameters used in calculations are shown in the Figure.

temperature, $T_D$, has almost no impact on the MCE value. Values of $T_D$ between 20 and 500K are intended to be taken only as limiting parameters of the simulation and are used to show overall tendency of how the $T_D$ influences MCE value in room temperature region.

The MCE in Tm is well investigated experimentally (e.g. see the data in [2,9]). Figure 5 presents the temperature dependence of the MCE for ferromagnet, that has the values of main magnetic constants close to that of Tm, at different Debye temperatures. When $T_D$ changes from 190K (Tm) up to 500K, the magnetocaloric effect value increases from 6.4K up to 12.6K in a 10T field. Thus, the change in $T_D$ can significantly influence the magnetocaloric effect value. Consequently, the change in the Debye temperature has different impact on the MCE value in the low ($T < 200$-230K) and high temperature range. However, the problem of what are the ways and how the Debye temperature can be changed in real magnetic materials still remains extremely little researched. According to general concepts, the following can be suggested: the Debye temperature can be varied by the change of the number of unit cells per unit volume or by the change of the number of particles per unit cell and, in principle, it may impact the value of the magnetocaloric effect at low temperatures.

Hence, from qualitative point of view, the Debye temperature can be treated as "rigidity" of the crystal lattice. The rigidity of a material can be changed, for instance, during the process of fabrication - by making it amorphous or by the other processing means. It is the author belief, that the Debye temperature can be affected significantly by addition of different elements to magnetic compounds. For example additions of C, N or B - are known to lead to the unit cell volume increase ("swelling") of magnetic material.

## CONCLUSIONS

Numerical simulation of thermodynamic features of rare-earth based magnetic materials in the context of molecular field theory has been performed in the wide range of temperatures and magnetic fields. The peak values of the magnetocaloric effect value were calculated. The maximum value of the MCE at low temperatures in heavy rare earth metals was found to be determined by the magnitude of the Debye temperature (i.e. by the lattice heat capacity). At higher temperatures, above 230 K, all maximum values of the MCE are

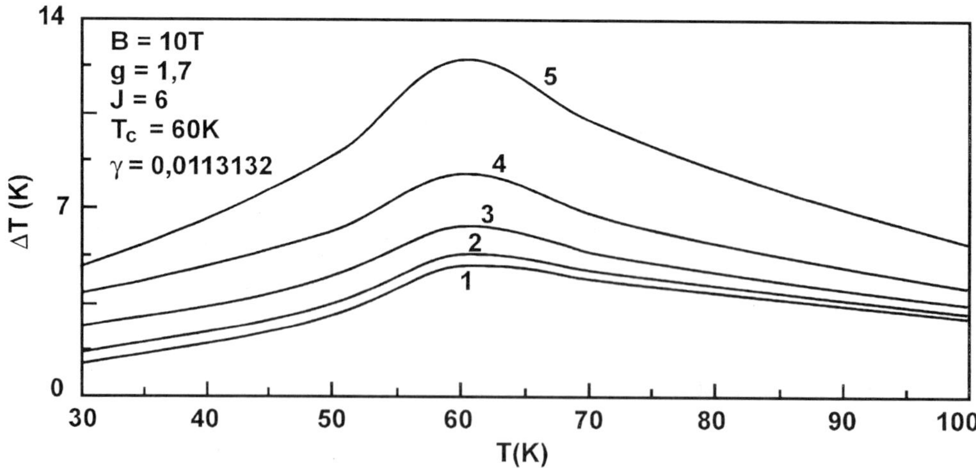


**Figure 5.** Temperature dependence of the MCE at different Debye temperatures: 1- $T_D$=20K, 2-$T_D$=100K, 3 - $T_D$=190K(Tm), 4- $T_D$=300K, 5- $T_D$=500K. Parameters used in calculations are shown in the Figure.

ordered following effective magnetic moment. Numerical simulation of correlation between the MCE and the magnetic entropy was done. The majority of the magnetic entropy change and sample's temperature increases as the result of magnetocaloric effect was found to exist in the region of paraprocess. The influence of the heat fluctuations and Debye temperature on the $\Delta T$ value was studied. The value of Debye temperature may impact the magnetocaloric effect only at lower temperatures, below 200-230K.

According to the present analysis, to increase the MCE value of the magnet (to found a good magnetic refrigerant from this point of view), the following can be suggested: try to look for the magnet with large value of magnetic moment and with such heat fluctuations which can be almost suppressed by not too large magnetic field in the vicinity of transition temperature (by the field approximately equal to value of field inside of the magnetic refrigerator).

## ACKNOWLEDGEMENT

Author thanks to Dr. V.K. Pecharsky for valuable comments during preparation of the manuscript and L.Bozkova for help in some of calculations. This work was partially supported by the NATO Linkage Grant No. 950700 and by the CEC/ICMC 1997.

## REFERENCES

1. K.A. Gschneidner, Jr., and V.K. Pecharsky, "Magnetic Refrigeration", in: "Rare Earths Science, Technology & Applications III", R.G. Bautista, C.O. Bounds, T.W. Ellis, and B.T. Kilbourn, Eds., TMS, Warrendale, PA (1997), 209.
2. M.D. Kuz'min and A.M. Tishin, Magnetocaloric effect. Part 1: An introduction to various aspects of theory and practice, Cryogenics. 32:545 (1992). Part 2 : Magnetocaloric effect in heavy rare earth metals and their alloys and application to magnetic refrigeration, Cryogenics. 33:868 (1993).
3. A.M. Tishin, Magnetocaloric effect in lanthanide materials. Invited paper submitted to the 21st Rare Earth Research Conference, July 7-12, 1996, Duluth, Minnesota (to be published in J. Alloys Comp.). v.250 (1997).
4. A.M. Tishin, Magnetocaloric effect in high magnetic field, Cryogenics 30:127 (1990).
5. A.M. Tishin, Dr. Sci. Thesis, Moscow State University, Moscow, Russia, 1994.
6. K.A. Gschneider, Jr., H.Takeya, J.O. Moorman, V.K. Pecharsky, S.K. Malik, and C.B. Zimm, New magnetic refrigeration materials for the liquefaction of hydrogen, Adv. Cryo. Eng. 39:1457 (1994).
7. B.K. Ponomarev, Magnetic properties of gadolinium in the region of paraprocess, J.Magn. Magn. Mater. 61:1291 (1986).
8. K.A. Gschneidner, Jr. and V.K Pecharsky, Unpublished result.
9. C.B. Zimm, J.A. Barclay, H.H. Harkness, G.F.Green, and W.G. Patton, Magnetocaloric effect in thulium, Cryogenics. 29:937 (1989).

# LOW TEMPERATURE ELASTICITY AND MAGNETIC SUSCEPTIBILITY OF Gd-Y ALLOYS

A.B.Beznosov, P.P.Pal-Val, and E.L.Fertman

B.Verkin Institute for Low Temperature Physics and Engineering NASU
Kharkov, 310164, Ukraine

## ABSTRACT

The longitudinal and torsional complex elastic modulus of the polycrystalline alloys $Gd_{100-x}Y_x$ (x = 0, 5.5, 7.5, 10.2, 12.0) have been measured by ultrasonic technique at the frequencies near 91 and 55 kHz correspondingly in the temperature range 5-300 K. The effective *a.c.* magnetic susceptibility $\chi_{eff}$ at the frequency 100 Hz was being measured simultaneously.

It was revealed an increase of elasticity between $x = 0$ and $x = 7.5$ in the whole temperature range, as well as four anomalies of the temperature dependences of elastic modules for yttrium containing alloys, which were interpreted as: a) conventional features in the region of Curie points $T_c$; b) first order magnetic phase transition (evidently from ferromagnetic phase to a sort of helicoid) located for all the alloys in the narrow temperature range 245.5 - 248 K); c) two adjacent second order phase transitions from helicoid tofinal angle ferromagnetic phase, with transitions temperatures monotonously decreasing in the interval $83 < T < 146$ K when increasing the yttrium contents.

## INTRODUCTION

The complex electronic structure of rare-earth metals leads to numerous features of the low temperature physical properties of related metallic systems, which are of great interest in the low temperature applications. Most of these features are caused by the interaction between the magnetic and lattice system in solids. The rare earths based alloys and compounds exhibit complex magnetic structures whose temperature transformations are correlated in some way with these of crystal lattice. These correlations are caused by the magneto-elastic interaction. The variety of magnetic and crystal phases realized in the rare-earth based compounds and alloys[1] gives a lucky opportunity to study the nature of this interaction in solids.

The alloys of gadolinium with other rare earth metals, as well as with related elements-"diluents" (Sc, Y, La) are interesting due to the features of electronic as well as magnetic systems of gadolinium. Gadolinium is being the central rare earth element, it has the half-filled electron 4f-shell and to zero angular momentum of the three-valent ion. At $226 < T < 292.5$ K it is a collinear ferromagnet with the easy magnetization axis directed along to hexagonal *c* axis of crystal structure. At $T_s = 226$ K there is the second-order spinreorientational phase transition, so that at lower temperatures the easy axis of magnetization tilts away from $c$ axis by the angle $\phi$ depending nonmonotonically on temperature in accordance with the temperature dependence of the anisotropy constants.[2-4]

Gd and Gd-based alloys attract also much attention as materials of modern technology, for example in a connection with the problem of magnetic refrigeration [5]. From this point of view special interest represents possibility of low temperature magnetic phase transitions in the alloys with high Gd contents.[6]

Yttrium is an ideal "diluent" for gadolinium, because it possesses the same crystal structure type (hcp), the similar structure of the outer electron shell (one $d$- and two $s$-electrons in a free atom, and correspondingly nearly two $d$- and one $s$-electron in metal )[7], the valence (+3), nearly equal room temperature crystal lattice parameters, and form with Gd a continuous series of solid solutions.[8] When increasing the yttrium concentration in the Gd-Y alloys, it is seen a tendency to a stabilization of an antiferromagnetic spiral structure, which appears in these alloys with the yttrium contents more than 20 at. %.[8,9] So, in the alloy $Gd_{80}Y_{20}$ it is found the ferromagnetic structure only, however unlike the pure gadolinium, the magnetic moment is tilted by the angle 70° from the hexagonal $c$ axis at all temperatures lower than Curie temperature $T_c$. This is caused by the different from pure Gd values of anisotropy constants. In the alloy $Gd_{60}Y_{40}$ there is an antiferromagnetic (the Neel temperature $T_N = 196$ K), as well as ferromagnetic ($T_c = 96$ K) transitions, and in the alloy $Gd_{50}Y_{50}$ it is observed an antiferromagnetic transition only.[8-10]

## GOAL OF PAPER

Weakly diluted gadolinium alloys with yttrium contents up to 10 at. % have not been studied thoroughly to the present time, and it was supposed, that in magnetic relation these alloys are similar to pure Gd. The goal of the present paper was to study magnetic and elastic properties of polycrystalline alloys Gd-Y with yttrium content 0, 5.5, 7.5, 10.2 and 12.0 at. % at temperatures 5-300 K. We have revealed the temperature behavior of the studied alloys, which is different from that of pure Gd as well as from that of the Gd-Y alloys with yttrium concentration $x \geq 20$ at.%.

## EXPERIMENTAL

The alloys $Gd_{100-x}Y_x$ ($x = 0$, 5.5, 7.5, 10.2 and 12.0) were prepared by melting with crucible from the raw materials of 99.9 % purity in the inert gas atmosphere at the OKMZ GIREDMET (Podol'sk, Russia). The specimens were annealed at the temperature 923 K in a vacuum of $10^{-6}$ Torr for 10 hours. By the X-ray analysis at room temperature the samples represent homogeneous single-phase solid solutions. By the metallographic data the alloys represent single-phase solid solutions with equal-axis grains of 40-130 μ size.

The samples represent rods of size 3x3x15 $mm^3$. Their demagnetization factor was approximately 0.55 along the main sample axis.

The temperature behavior of complex elastic moduli $E$ (Young's) and $G$ (shear) was studied by a long wave ultrasonic technique.[11] The sound velocities $v_{l(t)}$ and attenuation $\delta_{l(t)}$ of longitudinal and torsional acoustic waves in the samples have been measured by the two component composite oscillator method at the frequencies 54 kHz < $\nu$ < 56 kHz for transversal mode and 89 kHz < $\nu$ < 93 kHz for longitudinal one between 5 and 300 K. The ultrasonic measurements were accompanied with the measurements of the effective $a.c.$ magnetic susceptibility $\chi_{eff}$ of the samples at a frequency 100 Hz.

## RESULTS

The temperature dependences of elastic and magnetic properties of the studied systems have revealed the series of anomalies, as one can see at Figures 1-3: Fig.1 represents the

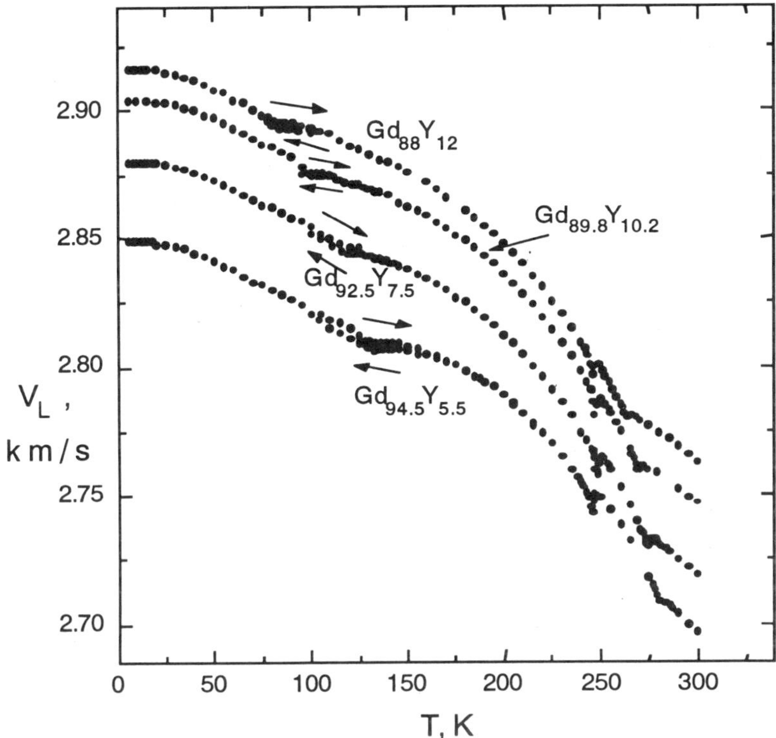


**Figure 1.** Temperature dependences of the longitudinal sound velocity in $Gd_{94.5}Y_{5.5}$, $Gd_{92.5}Y_{7.5}$, $Gd_{89.8}Y_{10.2}$ and $Gd_{88}Y_{12}$ alloys; the arrows show the temperature variation direction.

temperature dependences of longitudinal sound velocities $v_l$ for alloys. The temperature dependences of Young's modulus of Gd, $Gd_{94.5}Y_{5.5}$ and $Gd_{92.5}Y_{7.5}$, which are calculated as $E = \rho v^2$ ($\rho$ is density of alloy) are represented at Fig.2, where the dependences for $x =$ 10.2 and 12.0 are not shown because they lie very tightly to that of $Gd_{92.5}Y_{7.5}$, with some exception in the region 83 K $\leq T \leq$ 146 K. Finally, Fig.3 shows the temperature dependence of the effective magnetic susceptibility $\chi_{eff}$ of the alloy $Gd_{92.5}Y_{7.5}$ at the frequency 100 Hz.

**Phase Transitions.** The torsional wave velocity and attenuation exhibit similar temperature behavior, which differs only in some details from that of longitudinal wave characteristics. It is not the subject of the present paper, so further for definiteness we shall keep in mind the longitudinal sound behavior.

Excluding the conventional type anomalies in the region of the Curie points $T_c$ of all studied systems (it should be noted that the maxima of these ultrasound anomalies correspond to some temperatures $T_c^*$ which are somewhat lower than $T_c$)[12] , there are two temperature regions with anomalies of rather unusual type. First of them is located at temperatures $T_h$ ($T_h$ is equal to 245.5, 248, 248, 245.7 K for $x =$ 5.5, 7.5, 19.2 and 12.0, respectively) which is higher than the temperature $T_s =$ 226 K of spin reorientation in pure Gd and lower than Curie temperatures $T_c$ of alloys. The second anomalous region is represented by the interval $T_{s1} \leq T \leq T_{s2}$ , where both $T_{s1}$ and $T_{s2} \sim T_{s1}$ +10 K lie substantially lower than $T_c$ of alloys as well as $T_s$ in Gd ($T_{s1}$ is equal to 134, 116, 100 and 83 K, $T_{s2}$ is equal to 146, 126, 111 and 92 K for $x =$ 5.5, 7.5, 10.2 and 12.0, respectively).

**Elasticity.** As it is clearly seen from Fig. 2, the elasticity goes up 3% between $x = 0$ and $x =$ 7.5, in the paramagnetic region, and slightly more (3.3%) at 5 K for this concentration range. At 300 K Young's modulus increases 0.9 % when increasing Y contents from $x = 7.5$ to $x = 10.2$ and then it does not change between $x = 10.2$ and 12.0. Below $T_c$ all three curves for $x = 7.5$, 10.2 and 12.0 are close to another.

We suppose that this is the first observation of both of these two types of anomalies (at $T_h$ and $T_{s1}$, $T_{s2}$) and growth of $E$ in the studied alloys system.

## DISCUSSION

### Elasticity and Electronic Structure

The increase of $E$ between $x = 0$ and 7.5, main part of which is formed in the paramagnetic state and thus characterizes the lattice only, indicates the change in the electronic structure of alloys (since we consider the paramagnetic state, we assume, that this increase reflects changes in the interatomic cohesion forces, which are determined by the electron-ion interaction). In the simple "jelly" model of metal [13] the Young's modulus of the system is being expressed completely via following electron-ion characteristics:

$$E = \frac{(1+\sigma)(1-2\sigma)}{(1-\sigma)} \frac{Z^2}{V_a N_a(E_F)}. \tag{1}$$

Here $\sigma$ is the Poisson coefficient ($\sigma \approx 0.24$ for these alloys), $Z$ is the ion charge, $V_a$ is the atomic volume, $N_a(E_F)$ is the electron density of states at the Fermi-level per atom per pair of spin orientations. We can conclude, that 3% increasing of $E$ is being connected mainly with the decrease of $N_a(E_F)$. The decrease of $N_a(E_F)$ can be caused, according to (1), by some increase of the density of $d$-like states under the Fermi level (quasilocalization) at low Y concentrations, whereas beginning from $x = 1/(Z+1) = 7.7$ (Z =

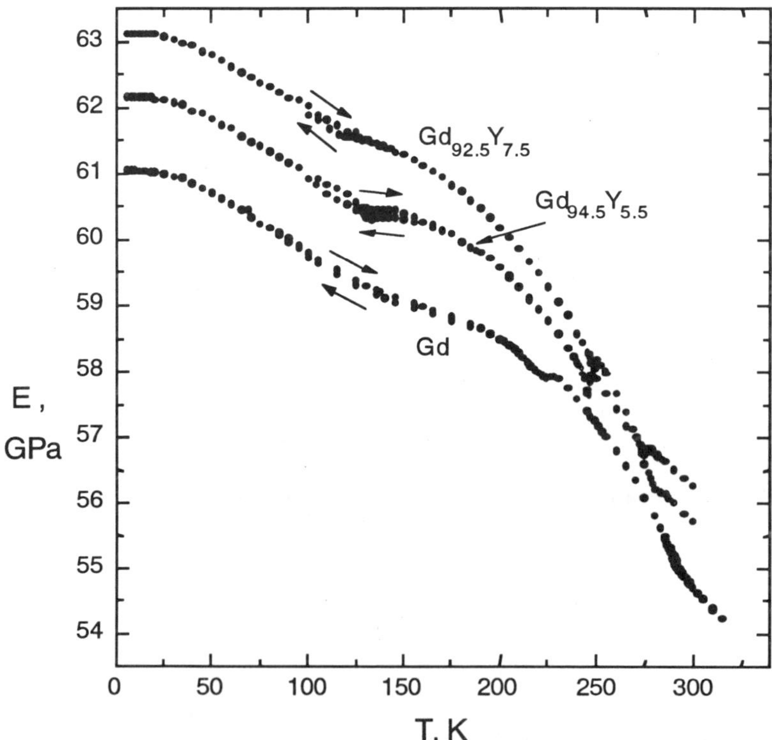


**Figure 2.** Temperature dependences of the Young's modulus in Gd, $Gd_{94.5}Y_{5.5}$ and $Gd_{92.5}Y_{7.5}$ alloys; the arrows show the temperature variation direction.

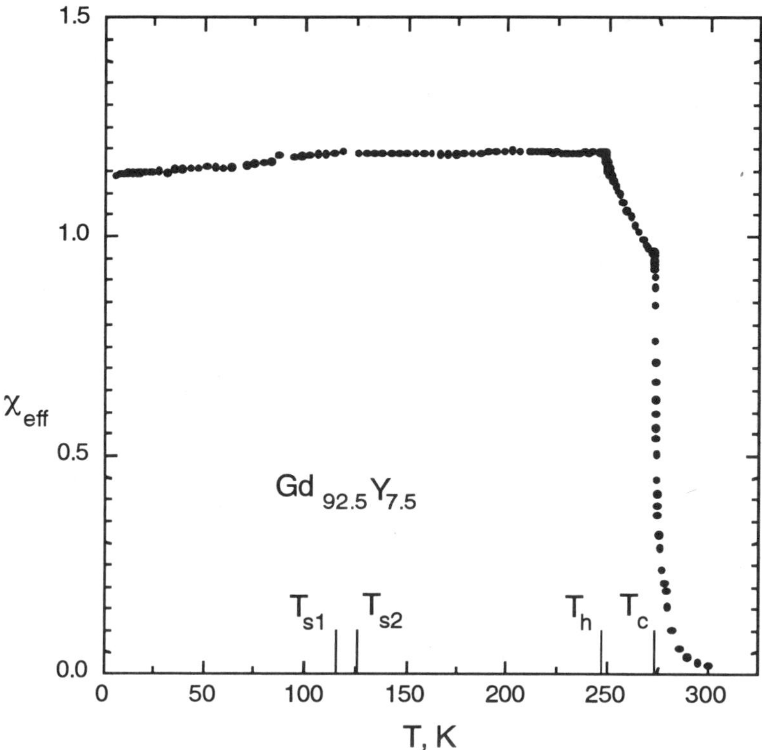


**Figure 3.** Temperature dependence of the effective magnetic susceptibility of $Gd_{92.5}Y_{7.5}$ alloy (cooling). The bars indicate the phase transitions points.

12 is the coordination number of hcp structure) this process should be stopped.[14] This is in qualitative accordance with the concentration dependence of Young's modulus obtained in the work.

## Second Order Transition at the Curie Points

For all the alloys studied the anomalies are characteristic of the temperature dependences of sound velocity $v_l\,(T)$ and attenuation $\delta_l\,(T)$ in the vicinity of Curie points. Elasticity and $v_l\,(T)$ steady increase with decreasing temperature, and in the Curie point region the attenuation exhibits $\lambda$-type anomaly, and the velocity and Young's modulus exhibit corresponding anomalies, which have, generally speaking, the shape depending on frequency.

## First Order Magnetic Phase Transition

When cooling each of the alloys (except pure Gd) below the Curie temperature, the sharp narrow pits and corresponding peaks are observed. over the background of monotonically increasing velocity and decreasing attenuation. The attenuation (as well as velocity when subtracting the background) have nearly symmetrical Gauss-function like shape, that differs radically from one at the second order phase transition (e.g. at the Curie point), where the attenuation have $\lambda$-type anomaly and velocity have a step at low enough frequencies (50 kHz is evidently much enough to observe clearly such a step). Since the Gauss-function represent the derivative of step-like function in real conditions (ideally it is delta-function), we assume that there are the first-order phase transitions in corresponding points $T_h$.

The temperatures $T_h$ weakly nonmonotonic depend on yttrium contents in alloy. The experimental points can be approximated by parabola with maximum near $x = 9$. Similar dependence from yttrium concentration exhibit also the intensities of peaks $\delta_l (T)$. So one can assume that there is a concentration threshold and limit for this transition.

The phase transitions similar to ones found at the temperatures $T_h$ are absent in the pure gadolinium,[10] they were not observed as well in the previously investigated alloys Gd-Y with various contents of "diluent".[8-10]

**Magnetic Structure Hypothesis.** The transition at $T_h$ leads to the nearly constant value of a.c. magnetic susceptibility of the samples with demagnetization factor $N = 0.55$ in a wide temperature range 5 - $T_h$ K. So there is transition tightly bound with magnetic system of alloy. Comparatively small difference between the elasticities at the branches before and after $T_h$ allow to suppose that magnetic interactions are the main ones for the transition, whereas the crystal lattice is connected with this transition indirectly via magneto-elastic coupling.

The analysis performed of the lattice and magnetic anomalies in the temperature points of observed phase transitions permits to suppose, that at temperatures $T_h$ the studied alloys go over from the collinear ferromagnetic structure to noncollinear one, being evidently by a sort of ferromagnetic helicoid, that is ferromagnetic realization of more general type of magnetic structure.

**Magnetic Interactions Analysis.** The cause of the phase transitions at $T_h$ is evidently the growth of intrinsic strains in the crystal lattice when cooling the alloys below the Curie temperature, that is caused by an increasing of the discrepancy between interatomic interactions between the magnetic Gd and nonmagnetic Y with lowering the temperature from $T_c$ down to approximately 200 K. The interactions between the magnetic gadolinium ions should lead to the increasing of the $c/a$ ratio of the parameters of hcp crystal lattice in the vicinity of Gd ions when cooling in this temperature range, whereas this ratio for yttrium in the vicinity of Y ions should decrease. When cooling the inhomogeneous field of intrinsic strains increases also in the samples due to anisotropic deformation of crystallites of the studied polycrystalline materials, which is added to the field of intrinsic strains created due to presence of yttrium ions in the alloy. Sharp transition to noncollinear magnetic phase happens when these strains achieve certain threshold value. The surprising feature of this transition is not only its presence, since it is accepted that there is no phase transformations in this temperature region in these systems for $x \leq 0.2$,[2] but its very weak nonmonotonous dependence of $x$ unlike the Curie and spin reorientation temperatures $T_{s1}$ and $T_{s2}$. The reason is in the competition of two oppositely depending on temperature factors: the increasing of yttrium contents in the alloy should lead to increasing $T_h$ , and the magnetization curve and maximum of $c/a$ ratio should shift to lower temperatures.

At the temperatures below $T_h$ down to approximately 200 K all the studied elastic and magnetic parameters of the alloys and pure Gd exhibit temperature hysteresis, which is caused, as we suppose, by the relaxation of intrinsic strains which appear when cooling and heating due to chaotic orientation of anisotropic polycrystalline grains (that is not shown in Figs. 1-3).

## Pair Phase Transition

As it is seen in the temperature dependences in Figs. 1 and 2, there are the regions where sharp decrease of the absolute value of temperature derivative of elastic modulus down to zero at temperature $T_{s2}$ changes when cooling to the similar sharp increase of this value at temperature $T_{s1}$. The difference between the temperature branches of elastic parameters higher than $T_{s2}$ and below $T_{s1}$ is close to one in pure Gd at the spin-reorientation transition at $T_s = 226$ K. In this sense, the double transition in alloys as a whole is similar to the spin-reorientation transition in Gd. Besides, as one can see from the temperature behavior of anisotropy constants of Gd-Y alloys, which may be calculated from the experimental data [10], the angle ferromagnetic phase in these systems is possible at the temperature range below $T_{s1}$.

**Hypothesis of the Sequence of Magnetic Structures.** We suppose that the order of the phase transitions at $T_{s2}$ and $T_{s1}$ is higher than one, that is proved by the absence of the symmetrical Gauss-function like peaks of sound attenuation $\delta_l$ (T) and pits of velocity $v_l$ (T). Probably the order of the transition is two or, may be, higher. The absence of the steps of velocities in the points $T_{s1}$ and $T_{s2}$ in the case if the order of transition is two may be due to the large time of establishing of thermodynamic equilibrium in the magnetic system of alloys comparatively with the sound wave period. The simplest model of the sequence of magnetic structures in this temperature range is the following one: when cooling at $T_{s2}$ the ferromagnetic helicoid structure changes to some "deformed" ferromagnetic helicoid structure, and then at $T_{s1}$ there is transition to angle ferromagnetic phase.

## CONCLUSION

So, the increasing of elasticity between $x = 0$ and $x = 7.5$ has been revealed in the $Gd_{100-x}Y_x$ alloys. The first order magnetic phase transition at yttrium contents 5.5, 7.5, 10.2 and 12.0 at temperatures $T_h$ is revealed, that manifests, as we suppose, the transition of the collinear ferromagnetic structure to the ferromagnetic helicoid. Two second or higher order phase transitions are revealed at the temperatures $T_{s1}$ and $T_{s2}$ that manifests, as we suppose, the transition at $T_{s2}$ from helicoid, when cooling, to some intermediate phase (a sort of "deformed" helicoid), and then at $T_{s1}$ from this intermediate phase to angle ferromagnetic structure.

## ACKNOWLEDGMENT

The research is supported in part by International Science Foundation, Grant No U9L000/U9L200.

## REFERENCES

1. K.N.R.Tailor and M.T.Darby. "Physics of Rare Earth Solids", Chapman and Hall Ltd, London (1972).
2. V.A.Finkel. "Structura Redkozemel'nykh Metallov", Metallurgiya, Moskva (1978) - ("Structure of Rare Earth Metals", in Russian).
3. K.P.Belov. "Redkozemel'nyie Magnetiki i Ikh Primeneniye", Nauka, Moskva, (1980) - ("Rare-Earth Magnets and Their Applications", in Russian).
4. K.P.Belov, A.K.Zvezdin, A.M.Kadomtseva and R.Z.Levitin. "Orientatsionnyie Perekhody v Redkozemel'nykh Magnetikakh", Nauka, Moskva (1979) -("Orientational Transitions in Rare-Earth Magnets", in Russian).
5. M.D.Kuz'min and A.M.Tishin, Magnetocaloric Effect in Heavy Rare Earth Metals and Their Alloys and Application to Magnetic Refrigeration, Cryogenics, 33:868 (1993).
6. A.Yu.Romanov and V.P.Silin, On Magnetocalorical Effect of Heterogeneous Ferromagnets, Fizika Metallov i Metallovedenie, 83:5 (1997) -(Metal Physics and Material Science, in Russian).
7. A.B.Beznosov and G.S.Nikol'ski, Electron interactions and crystal lattice of rare earth metals, Sov. J. Low Temp. Phys. 10:1336 (1976).
8. W.C.Thoburn, S.Legvold, and F.H.Spedding, Magnetic Properties of the Gd-La and Gd-Y Alloys, Phys.Rev., 110:1298 (1958).
9. H.R.Child and J.W.Cable, Magnetic structure properties of Gd-Y and Gd-Sc alloys, J.Appl.Phys., 40:1003 (1969).
10. K.Tohyama and S.Chikazumi, Magnetocrystalline Anisotropy of Gadolinium-Yttrium Alloys, J.Phys.Soc.Japan, 35:47 (1973).

11. P.P.Pal-Val and H.-J.Kaufmann, Influence of Microplastic Deformation on the Electronic Absorption of Ultrasound in High-Purity Molybdenum Single Crystals, Sov. J. Low Temp. Phys., 9:163 (1983).
12. M.Rosen, Elastic Moduli and Ultrasonic Attenuation of Gadolinium, Terbium, Dysprosium, Holmium, and Erbium from 4.2 to 300 K, Phys.Rev., 174:504 (1968).
13. A.O.E.Animalu. "Intermediate Quantum Theory of Crystalline Solids", Prentice-Hall, Inc., Englewood Cliffs, New Jersey, (1977).
14. A.B.Beznosov, V.N.Gorbach, and S.A.Ivakhnenko, Band Structure and Atomic Radius of Gd-La Alloys, Solid State Physics, 20:3146(1978).

# EDDY CURRENT POWER DISSIPATION IN RARE EARTH PACKED PARTICLE BED REGENERATORS FOR AMRR APPLICATIONS

**C. E. Reid, J. A. Barclay, and D. A. Leboe**

**Cryofuel Systems Group**
**University of Victoria**
**Victoria, BC, V8W 3P3, Canada**

## ABSTRACT

The magnetic regenerators in an active magnetic regenerative refrigerator (AMRR) are exposed to high fluctuating magnetic fields during operation. The rapid change of field over time can cause large eddy currents to be formed in any conducting material used in the regenerator and cause parasitic heating loads. The magnitude of eddy currents formed depends not only on the field change with time but also the shape factor of the material, its bulk resistivity, and its geometry. To determine the power dissipation due to eddy currents, specifications from the prototype AMRR being built by the Cryofuel Systems Group at University of Victoria were used. The regenerator consisted of 100 micron diameter dysprosium particles packed into a regenerator bed (120 mm long by 125 mm wide and 50 mm deep), and cycled through a magnetic field fluctuation of 8 Tesla. Eddy current generation estimates were performed based on three models. The first assumed that the particles were electrically isolated from each other. For the specified regenerator, the power dissipated was less than 1 W for frequencies up to 10 Hz. The second model assumed that the regenerator is a solid porous block. This model resulted in power dissipations 3 orders of magnitude greater than model 1. The third model assumed some degree of electrical contact among the particles. Bulk resistivity experiments were conducted to assess the degree of electrical contact between particles and to estimate the power dissipated due to eddy currents. The results show dissipated powers several orders of magnitude less than the insulated particle model over a large range of bed packing pressures. It was expected that this model would be the most realistic and show power dissipations between that shown by the first two model estimates. This discrepancy is explained by the variable packing pressure with bed length characteristic of a packed particle bed. Dissipated end plate pressure leaves particles in the bed middle relatively unloaded and leads to high bulk resistivities. Tests performed with short beds and high packing pressures lead to similar power dissipations and suggests that the insulated particle model is reasonable approximation to assess the magnitude of eddy currents formed in the prototype AMRR packed particle bed.

## INTRODUCTION

The Cryofuel Systems Group (CFS) at the University of Victoria is designing and building an active magnetic regenerative refrigerator (AMRR) prototype device to liquefy natural gas. At the core of the AMRR is the active magnetic regenerator. This device uses porous magnetic material to provide the passive regeneration necessary (as in conventional regenerative refrigerators) as well as the active magnetic work that allows the refrigerator operation. When selecting the optimum AMRR regenerator design, all of the conventional passive regenerator parameters must be considered as well as those particular to the magnetic operation of the device and the fabrication difficulties associated with magnetic materials. Characteristics of excellent AMRR regenerator geometries include: low fluid pressure drops, high heat transfer, low longitudinal heat conduction, low eddy current generation, low porosity, high thermal mass, structurally robust, and simple and inexpensive to manufacture. There are several choices when selecting optimum AMRR regenerator designs. These include packed particle beds, wire screens, parallel-flow plates, and perforated normal-flow plates. CFS has chosen packed particle beds of heavy rare earth materials as the regenerator for its prototype AMRR device. This geometry offers the best compromise among the above design characteristics. Established correlations for packed beds provided good estimates for the heat transfer, longitudinal conduction, and pressure drop characteristics. However, an estimate of the degree of parasitic power dissipated by eddy current generation could not be obtained through the literature. Experiments were performed to obtain an order of magnitude estimate for this critical eddy current design data for packed particle beds made from magnetic rare earth materials.

## THEORY

Packed particle bed AMRR regenerators have good passive regenerator characteristics. They also have the potential for magnetic eddy current generation. Eddy currents are formed when a magnetic material undergoes an applied field change over time and electric currents are induced in the material. The steeper the gradient (dB/dt), the greater the eddy currents generated. The power dissipated by the current flowing in a loop of resistance can be expressed by the equation[1]:

$$P = \frac{(dB/dt)^2}{32\rho} \Gamma A V \qquad (1)$$

where $\rho$ is the resistivity, $\Gamma$ is a shape factor, A is the area enclosed by the largest current loop, and V is the volume of the material.

Although, the power dissipated is a squared function of the fluctuating field, it is also dependent on the magnetic material's geometry, shape factor and its bulk resistivity. The shape factor depends on the material's projection area perpendicular to the applied field lines. Smaller projection areas will minimize the size of the currents generated. In the case of estimating eddy current generation in packed particle beds, three separate models were investigated. The first model assumed that each particle is electrically insulated from its neighbor. This provides an estimate of the lower limit of the power dissipated due to eddy current effects. The second model assumes that the packed bed is a solid porous block of magnetic material. This provides an upper limit to the power dissipated. The third model used a more realistic approach and assumes that the particles are in some electrical contact and that the entire bed can be treated as a porous mass with a modified bulk resistivity. The bulk resistivity of the particle bed will be less than the resistivity of a solid block of the same material. This reduction is caused by the separation of each particle by a single point contact and by the fact that this contact point is somewhat insulated by allowing the growth of thin oxide layers over the particle surface (heavy rare earth materials are reactive and will naturally grow insulating oxide layers of a few microns). This model was expected to provide a power dissipation estimate somewhere between that of the first two models. To examine the power dissipation associated with the third approach, the bulk resistivity of the packed particle bed is required.

### Bulk Resistivity Measurements

The bulk resistivity of a packed particle bed can be determined by measuring the electrical resistance of the particle bed as a whole then calculating the bulk resistivity of the regenerator:

$$\rho = \frac{RA}{L} \tag{2}$$

where R is the bed resistance, A is bed cross sectional area with the porosity accounted for, and L is the bed length.

When calculating the bed's electrical resistance, several factors must be considered:

- Particle size - As the particle size increases, the layers of particles in a given bed length decreases. This means that there are fewer point contacts between layers, the insulating factor is reduced, and the bed electrical resistance decreases.

- Regenerator bed length - As the bed length decreases, the packed bed resistance decreases for the same arguments as above (for a fixed particle diameter).

- Material reactivity - All heavy rare earth materials (Gd through Yb) will oxidize. The oxide layer that forms on the surface of the particle will act is a dielectric material that will help insulate the particles from each other.

- Packing pressure - The packed particle bed regenerator in the prototype AMRR will be under an axial load to keep the particles in place as they undergo fluctuating magnetic forces. With increasing load, the bulk resistance of the particles will decrease because the point contacts will plastically (and then elastically) flatten effectively increasing the contact area and possibly cracking the brittle surface oxide. The load distribution along the bed length will vary as energy is dissipated in particle to particle and particle to container friction.

### Active Regenerator Specifications

To estimate the power dissipation due to eddy current generation, the authors used a set of specifications for the rotary AMRR prototype being built by the Cryofuel Systems Group at the University of Victoria. These specifications are:

- regenerator media - average 100 micron diameter heavy rare earth particles (for this work, dysprosium particles are assumed)
- regenerator dimensions – circular wheel 120 mm long and 50 mm deep, with magnetic field lines aligned axially.
- bed porosity – 0.4
- magnetic field fluctuation – 0 to 8 Tesla enacted linearly over 5 degrees of the wheel shaped regenerator
- end plate pressure (packing pressure) – 150 psi
- wheel frequency – variable parameter

## METHODOLOGY AND TEST APPARATUS

Three models were used to estimate the power dissipation caused by eddy currents formed in the regenerator bed. The first model assumes that each individual particle in the regenerator bed is electrically isolated from its neighbors. In this case, the power dissipated due to eddy current generation in the volume of each particle is calculated using published values for the resistivity of dysprosium, particle geometry and equation 1. The total power is then summed over all the particles in the prototype bed. This figure is then doubled to represent magnetic field ramp up and ramp down sections of the rotating wheel.

The second model assumes that a solid porous block of dysprosium is used as the regenerator. Power dissipation is calculated as in the first model but with appropriate geometry.

The third model assumes that the particles in the bed have some electrical contact with each other. To test the particle regenerator beds for bulk resistivity, a voltage was set across the bed, the current recorded and the resistance calculated by Ohm's Law (R=V/I). The resistivity is then determined by Equation 2. The experimental apparatus used to calculate bulk resistance measurements of the particle beds is shown in Figure 1.

100 and 162.5 micron diameter dysprosium particles were loaded into the apparatus to create a predetermined bed length of both one and two centimeters. Abraded and degreased copper plates were placed on both sides of the particle bed to conduct the current to the bed. A picoammeter was used to record the low regenerator current. A simple two wire method was used instead of a four wire method because the resistance of the copper plate/lead wire assembly is insignificant compared to the large expected resistances across the particle bed. The particle bed was loaded incrementally to a maximum pressure of about 20 MPa (3000 psi) by an Ingstrom material testing machine. At each load setting, the bed deformation was measured to account for the true bed length and the regenerator resistance was recorded. Although packing pressures in the actual AMRR prototype are only 150 psi, larger loads were required to get stable current readings. Also, the bed lengths are shorter for similar reasons.

**Particle Media**

The dysprosium particles were manufactured by Nuclear Metals Inc. using a rotating electron process (REP)[2]. In general, all particles were spherical in shape (minor satelliting did occur) and the production and packaging occurred in an inert atmosphere. Before testing, the particles were unpacked and exposed to ambient air for a significant amount of time to insure that surface oxidation had occurred.

## EXPERIMENTAL RESULTS

Figure 2 shows the power dissipated in the prototype AMRR as a function of frequency for each of the proposed models. Results for Model 3 in Figure 2 were obtained using packed bed resistivity data for 1500 psi packing pressure and a bed length of 1 cm (see results presented in Figure 3).

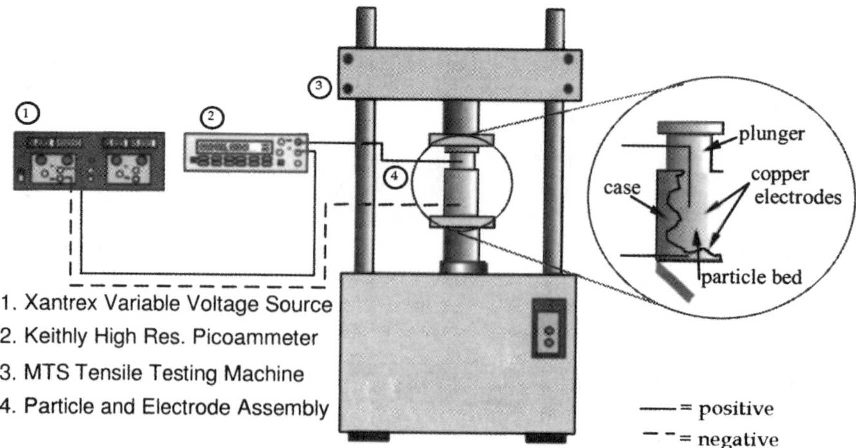


Figure 1. Schematic of regenerator resistance testing device.

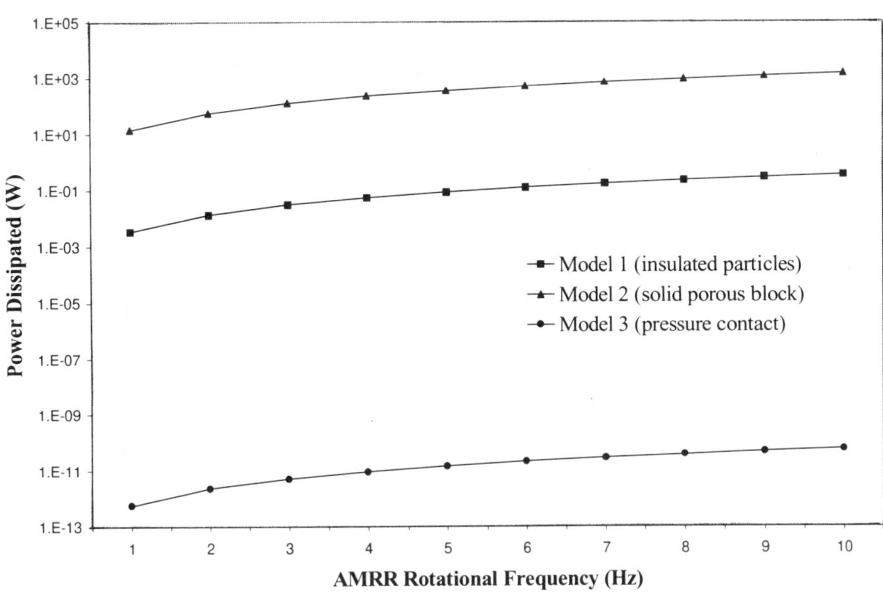


Figure 2. Eddy current power dissipated as a function of regenerator frequency

Figure 3 shows the resistivity of the dysprosium particle bed versus the endplate or packing pressure for two separate bed lengths and two different particle diameters.

Factors affecting the accuracy of these results include:

- mean particle size – The particle charge was screened to create an average of 100 μm and 162.5 μm. The allowed deviation from the mean is ±25 μm.

- contact and lead wire resistance effects – The particle bed is contacted on the top and bottom with abraded and degreased copper plates. There will be contact resistance associated with the layer of particles that press against the plate that will, at least, introduce a different resistance effect. At the same time, the total resistance of the lead wires and the copper plates will add to the net resistance

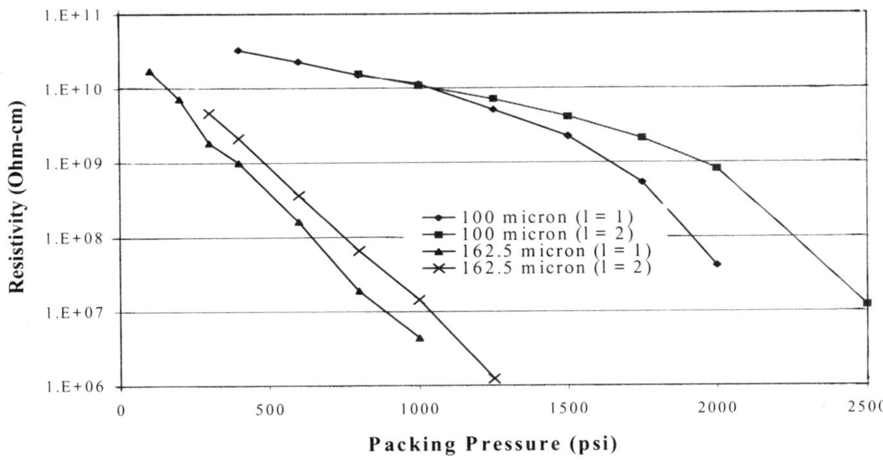


Figure 3. Resistivity versus plate pressure for various diameter dysprosium particle beds.

measured. However, this resistance is insignificant compared to the large resistances measured across the bed (less than 0.01% in most cases).

- settling effects – When the load is placed on the bed, each measurement takes several minutes for the bed to settle and the resistance reading to stabilize. In all cases, the expediency of the measurement dictated that the measurement be taken before complete steady state was achieved.

## DISCUSSION

Figure 2 shows that the insulated particle model (model 1) leads to lower power dissipation due to eddy currents than the solid porous block model (model 2) as expected. With the bed finely divided into particles, the volume available to the eddy currents is small and this leads to reduced power dissipation. Figure 2 also shows that when the particles are assumed to have some degree of electrical contact (model 3), the eddy current power dissipation is orders of magnitude less than the electrically insulated model. This is counter intuitive. We expected that when the particles are allowed some electrical contact, the eddy current loops would be larger than those assumed in model 1 but not as large as model 3. Accordingly, the dissipated power should also fall between those extremes. The reason for the low power estimates is related to the estimate of the bulk resistivity of the packed particle bed.

The lowest resistivity reading in Figure 3 is still eleven orders of magnitude greater than that of pure dysprosium ($\rho_{dy}$ = 83 $\mu\Omega$.cm)[3]. The resistivity value used to calculate the power dissipation in model 3 was for a 1 cm bed length with 1500 psi packing pressure ($2.24X10^9$ $\Omega$.cm). These high resistivity values are caused by a packing pressure gradient that is formed along the bed length and by the particle's thin insulating oxide layer.

As the end plate pressure is increased, the resistivity decreases. This is effectively caused by the increase in contact area between the particles and a break down of the thin insulating oxide layer. When a force is initially applied to the end plates, the pressure between the particles is essentially infinite because of the point contact and plastic deformation occurs. As the contact area increases, the particle pressure decreases until elastic deformation processes dominate. As the end plate force increases, the particles deform more and contact area continues to increase until the force required for further significant area increase becomes very high. During this process, the thin brittle oxide layer is also cracked enhancing electrical contact. However, the bed's packing pressure is not uniformly transmitted throughout the bed length. During testing, when the particle beds were unloaded and the charges removed, the volumes directly adjacent to the end plates were generally compacted into a solid pack while the particle volume in the middle was loose. The particle to particle and particle to wall friction dissipates the packing force and creates a packing pressure gradient that is greatest adjacent to the end plates and lowest in the bed middle. This causes large bulk resistivity numbers because the middle particles retain low contact area and brittle oxide layers might not be broken. Figure 3 shows that bed resistivity decreases with increasing particle size and decreases with bed length. This is expected because increasing particle size leads to fewer friction contacts while increased bed length leads to more. In the case of the latter, it is interesting to note that this effect overcomes the fact that bed length is inversely proportional to resistivity.

Because of the variable packing that occurs in the bed due to the pressure gradient, middle particles have very little electrical contact even under these significant end plate pressures. This creates a high overall bed resistivity and leads to insignificant power dissipation estimates. To accurately determine effect of electrical contact among particles in the bed, the variable bed resistivity as a function of length must be measured. The resistivity will vary with the packing pressure and ideally, this value should be integrated along the bed length to get an accurate model. The resistivity value used in model 3 is based on a regenerator that is 1/5 the bed length and had 10 times the packing pressure than the AMRR prototype (1 cm vs. 5 cm and 1500 psi vs. 150 psi). Even so, the power dissipation was insignificant. These results indicate that significant enhancement of the eddy current power dissipation will not occur over that predicted by the electrically insulated particle model. Although it is recognized that the actual power will be greater in the real case, model 1 provides a reasonable estimate.

## CONCLUSIONS

The assumption that each of the particles is electrically insulated from each other (model 1) provides a reasonable estimate for the eddy current power dissipated in the rare earth particle bed regenerator. This power will be less than 1 W over a frequency range of 1 – 10 Hz of AMRR operation. The second model assumes that the packed bed can be treated as a solid porous block. This model estimates power dissipations three orders of magnitude greater than model 1. This is unrealistic because each particle will have some degree of electrical isolation caused by point contact and insulating oxide layers. The third model assumes a degree of electrical contact and used the bulk resistivity of the particle bed. This model estimates power dissipations 10 orders of magnitude less than model one. It was expected that this model's power estimates would be the most realistic and fall between the first two. The power estimates were low because the packing pressure gradient along the length of the regenerator bed caused by particle to particle and particle to wall friction and by insulating oxide layers on the particle surface lead to artificially low bulk resistivities. To get a more realistic estimate using this model, the resistivity of the bed would have to be integrated along the bed length to take into account the packing gradient. However, because the tested bed length was already 1/5 that of the AMRR prototype and the packing pressure was an order of magnitude greater, there appears to be negligible enhancement of the dissipated power over that estimated by the first model.

## REFERENCES


1. P. Kittel, "Eddy Current Heating in Magnetic Refrigerators", Advances in Cryogenic Engineering, Vol. 35B, pp. 1141-1148

2. C.E. Reid, Development of Magnetic Refrigerants for Active Magnetic Regenerative Materials, M.A.Sc. Thesis, Unviersity of Victoria, 1995

3. Gschneidner and Eyring (Eds.), Handbook of the Physics and Chemistry of Rare Earths, 1, 1978

# SYSTEMATIC INVESTIGATIONS OF REGENERATORS FOR 4.2K-REFRIGERATORS

A. Lang,[1] H.-U. Häfner,[1] and C. Heiden[2]

[1]Leybold Vacuum GmbH
Bonner Str. 498, D-50968 Cologne, Germany
[2]Institute of Applied Physics
University of Giessen, 35392 Giessen, Germany

## ABSTRACT

Using Rare Earth materials and compounds as regenerator, several groups have reached temperatures below T=4.2K with Gifford-McMahon (GM), as well as with pulse tube refrigerators. Their performance is substantially determined by the choice of Rare Earth material, especially its specific heat. Based on measured pressures and displacements of an existing mechanically driven GM coldhead, cooling capacities are predicted theoretically for different regenerator matrices. This will be compared with corresponding experimental results which include measurements of the temperature profile in the regenerator. The results are discussed in terms of mass flow rate through the regenerator and the influence of different loss mechanisms. So far, an optimum capacity of >1W was achieved at T=4.2 K, using a 6 kW compressor unit for operation. Different modifications of the displacer design however were necessary in order to achieve a very good repeatability of results and high stability during operation.

## INTRODUCTION

With the employment of Rare Earth materials and different intermetallic compounds, it was possible to extend the temperature range of regenerative refrigerators to the liquid helium range. These temperatures were realized not only on the basis of two stage GM refrigerators[4,11] but also with two stage Pulse Tube coolers[9].

Compact two stage 4.2K-GM refrigerators are already employed in different cryogenic systems, as for example for cooling of MRI-systems, for direct cooling of superconducting magnets (for example NMR-spectroscopy, magnetic separation) and for sample cooling in cryostats.

The optimization of the cooling performance in this temperature range requires knowledge about the right choice of the regenerator. Different Rare Earth compounds show at temperatures below *T=20K* magnetic transitions which are accompanied by a considerable rise of specific heat. This effect can be taken advantage of - according to the temperature distribution of the regenerator - to arrange the materials with the highest specific heats in an optimal layer structure[3]. Besides the appropriate choice of the low temperature regenerator it became clear that the knowledge of different operational parameters are also of high importance for the optimization of 4.2K-GM refrigerators. This is mainly caused by the deviation of the working fluid ($^4$He) from an ideal gas below $T=15K$, leading to a partly dramatic rise of loss mechanisms[12].

In this document, investigations on the regenerator will be presented assessing the optimal regenerator arrangement while different regenerator arrangements were simulated as well as experimentally investigated. Apart from the optimization steps which are necessary for the investigations also measurements of the temperature profiles of different regenerator arrangements will be presented and discussed.

## LOW TEMPERATURE GM–REFRIGERATION

The theoretical description of the GM-cycle is quite complex[1] and shall not be treated in detail in this paper. The overall system of a GM-machine is composed of different - open - part systems (expansion- and compression space as well as the compressor). This is the reason why generally in a simplified way the part system of the expansion space is regarded for the description of the cooling cycle. On the basis of an adiabatic expansion process, it can be shown that the enthalpy change of the system just equals the pV-work[1]. However, with temperatures $T < 40K$ an additional enthalpy change has to be regarded due to real gas properties of the working fluid[12]. The ideal cooling capacity then is calculated as the sum of pV-work and the enthalpy change multiplied by the cycle frequency:

$$\dot{Q}_i = (\oint pdV + \oint hdm) \times f \qquad (1)$$

denoting V as swept volume, $m$ the mass, $h$ the enthalpy and $p$ the pressure of the working fluid ($^4$He). The term $\oint hdm$ is negative and disappears for temperatures $T>20\text{-}40K$, so that then the ideal cooling capacity corresponds again to the pV-work of the system.

Another peculiarity of the 4.2K-refrigerators is also based on the real gas properties of the working fluid. The pressure dependency of the density of $^4$He is considerably reduced below T=10K and disappears almost completely for temperatures $T \leq 4.2K$ (see figure 2). This means that the mass flow in the expansion space is almost independent from the operating pressure and is exclusively determined by the swept volume. Figure 1 shows the mass change in the expansion space for different temperatures within one cycle. Here it becomes clear that on the one hand the mass flow rate rises steeply with reduced temperature and on the other hand the so induced transition from the pressure dependent curve (asymmetric) to the pressure

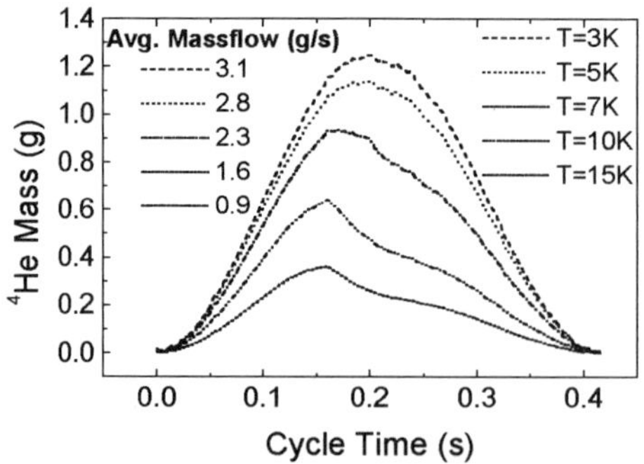


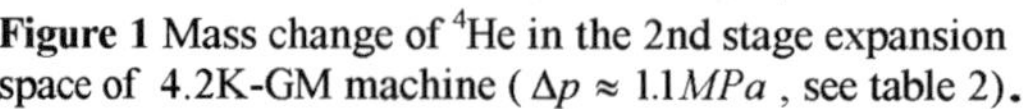

**Figure 1** Mass change of $^4$He in the 2nd stage expansion space of 4.2K-GM machine ( $\Delta p \approx 1.1MPa$ , see table 2).

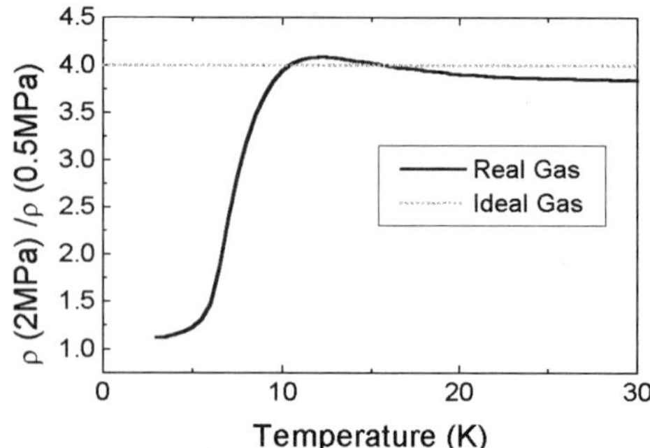


**Figure 2** Density ratio for 5bar and 20bar helium gas compared to ideal gas .

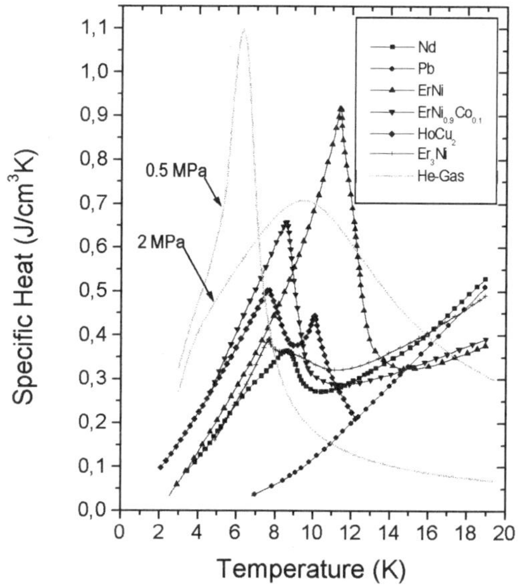


**Figure 3** Volumetric Specific Heat of Rare Earth materials compared to Pb and $^4He$ at low temperatures.

independent (symmetric) curve of the mass change in the swept volume occurred. The rising of the mass flow rate causes an increase of the loss mechanisms in the regenerator (regenerator inefficiency and temperature swing), an effect which has less impact on the cooling performance at higher temperatures.

Therefore, the crucial component of the low temperature refrigerators is - besides the operational parameters (e.g. valve timing, stroke) - the 2nd stage regenerator. The specific heat of the working fluid increases considerably at temperatures lower than $T=10\text{-}15K$ (see figure 3). This is the reason why it is still particularly important to look for regenerator materials with high specific heat for the 4.2K-refrigerators. The decisive breakthrough towards lower temperatures took place when materials with magnetic phase transitions were employed as a replacement for the lead regenerators which until then had been used for low temperature refrigerators ($T_{end}$=6-7K)[4]. The different magnetic phase transitions of these materials with the so induced different temperature dependence of the specific heat allow the choice of appropriate regenerator in form of multi-layer-regenerators[3]. In this connection, the exact knowledge on the temperature distribution within the regenerator is a necessary condition for analyzing the layer structure of the different regenerator materials.

## MODELING

Starting point of our investigation to optimize the regenerator for the 4.2K-GM refrigerator were calculations on the basis of computer models by M. Atrey[2]. On the basis of experimentally obtained pressure-displacement charts, the model calculates the ideal cooling performance, the mass flow rate in expansion spaces and different losses of the machine. Apart from the shift to a more comfortable Windows platform, the program was extended by the use of real gas properties of the working fluid[13]. In this way the variation of different system- and operational parameters (regenerator diameter and length, stroke, valve timing) as well as different regenerator arrangements (materials, layers) were simulated. Figure 4 shows the gross, loss calculations and the resulting net cooling effect of the second stage for different regenerator arrangements (see table 1). The two dominating loss mechanisms are the temperature swing loss and the losses due to the pressure drop. A reduction of the temperature swing and a rise of net cooling capacity by a factor 2.7 for the arrangement $Pb/Er_3Ni/ErNi_{0.9}Co_{0.1}$ in comparison to the initial regenerator Pb/Nd results from this model. The cooling power is already improved by a factor 2.3 only by employing 30% $HoCu_2$ at the cold end of the regenerator. The comparison of these calculated cooling performances with the experimental data will be discussed in the next chapter.

**Table 1** Arrangements of regenerators for modeling

| Model-Type | Regenerator Arrangement (from hot to cold end, Vol.%) |
|---|---|
| 1 | 25 % Pb + 75 % Nd |
| 2 | 25 % Pb + 45 % Nd + 30% $HoCu_2$ |
| 3 | 25 % Pb + 45 % ErNi + 30% $HoCu_2$ |
| 4 | 25 % Pb + 45 % Nd + 30% $ErNi_{0.9}Co_{0.1}$ |
| 5 | 25 % Pb + 45 % ErNi + 30% $ErNi_{0.9}Co_{0.1}$ |
| 6 | 25 % Pb + 45 % $Er_3Ni$ + 30% $ErNi_{0.9}Co_{0.1}$ |

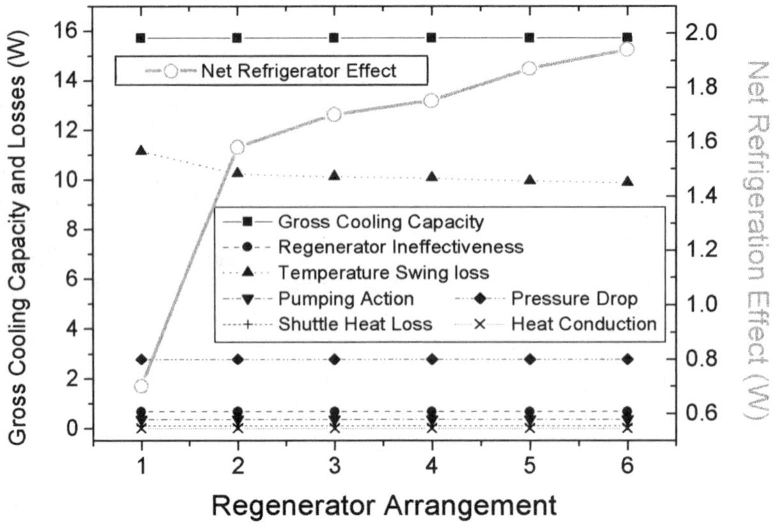


**Figure 4** Modeling results of Gross Cooling Capacity, losses and the net refrigeration effect of the second stage at T=4.2K for different regenerator arrangements: 1–6 (see table 1).

## EXPERIMENTALS AND RESULTS

### Improvement of 4.2K-Refrigerator

For the investigation of the 4.2K-regenerators, a mechanically driven 4.2K-GM refrigerator derived from standard production[11] with a (6kW, 50Hz) Scroll-compressor was used (filling pressure $p = 1.4MPa$). A few geometric data of this machine are shown in table 2. In order to adjust exact settings of the different operational parameters and to obtain the temperatures as well as the cooling performances, the machine was equipped with temperature - and pressure sensors, heaters and a pick-up-system for getting the displacement data.

Necessary requirements for a systematic analysis of the regenerator are - apart from the accuracy of the measuring sensors - the reproducibility and temperature stability of the test apparatus. The second stage sealing represents in this case a particular source for instability. Therefore, several sealing systems were investigated with regard to performance- and temperature stability. With the employment of a closed piston seal a clear reduction of temperature fluctuations within one cooling cycle and an increased stability during the complete test time (a few days respectively) were achieved. Furthermore, the feasibility of using clearance seals for 2nd stage sealing was investigated. Figure 5 and 6 show the results of the different seals. It became clear that with an appropriate adjustment of the 2nd stage displacer diameter no load temperatures of $T \cong 4.2K$ could be achieved (clearance seal). While designing the clearance seal with additional radial grooves, cooling performance increased to $Q > 0.5W$ with $T = 4.2K$ with the appropriate choice of cycle frequency (see figure 6). An important requirement for this was the employment of a stainless steel displacer coated with PTFE-foil. Especially for use as cold

**Table 2** Geometry data of 4.2K-GM-Machine

| | | 1st stage | 2nd stage |
|---|---|---|---|
| **Cylinder** | Diameter | 76.2 mm | 30.1 mm |
| | Length | 242.2 mm | 169.7 mm |
| **Regenerator** | Diameter | 44.5 mm | 25.5 mm |
| | Length | 111.2 mm | 115.8 mm |
| **Stroke** | 11.2 mm | | |

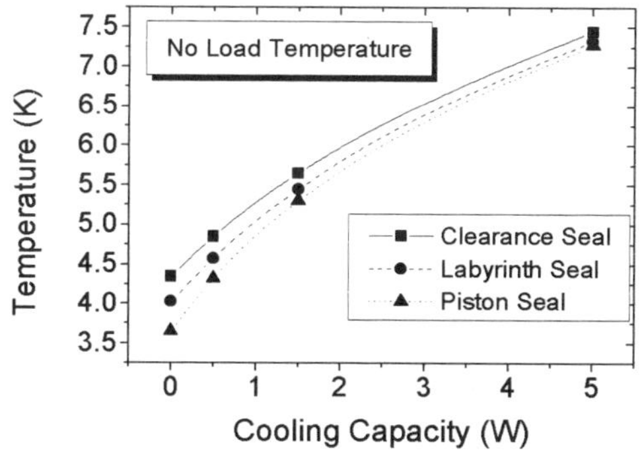


**Figure 5** Comparison of cooling performance of different 2nd stage seal systems.

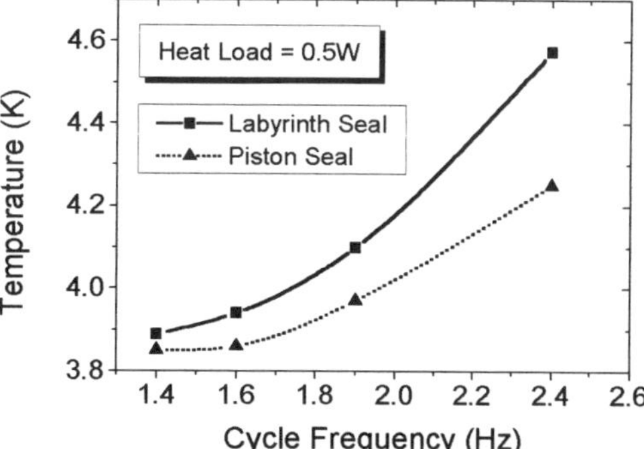


**Figure 6** Frequency dependence of end temperature for of different seal systems.

stage displacer material, stainless steel has considerable advantages compared to usually used fiber reinforced plastics. This is on the one hand the isotropic thermal expansion properties which correspond exactly to those of the cylinder (same material), and on the other hand the absence of hygroscopic properties. The PTFE coating also allows a considerable reduction of frictional losses which even were further reduced by adding solid-state lubricants. This specific coating technique has already been established in different applications (i.e. dry compressors, stirling-machines) and turned out to be very reliable in the 4.2K-GM application.

The isotropic thermal expansion of the stainless steel displacer proves to be advantageous also for a controlled tight regenerator packing. The stability of material shape guarantees constant packing densities of the regenerator in particular when installing and removing the displacer for the different cool down cycles. This was supported by a spring energized packing of the regenerator in the warm end of the second stage, especially with regard to the long-term stability of the cooling performance. Depending on the quality of the regenerator material, there might occur a loosening of the regenerator packing during operation due to small material decompositions. Owing to the spring packing a constant packing density can be guaranteed over the total running time. With all the different elements of improvement (see figure 9) it was possible to enhance considerably the temperature and performance stability. ($\Delta T_{pp} \cong 0.3K$ at $T = 4.2K$). Furthermore, a clearly reduced orientation dependency of the cold head was achieved with these improvements ($\Delta Q_{\max} \cong 30\%$ at $T = 4.2K$, cold end down to cold end up operation).

## Different Regenerator Arrangements

Implementing the different elements of improvement, it was possible to define and reproduce also small performance differences of various regenerator arrangements. Table 3 lists the different regenerator arrangements which are presented in this paper. The first stage regenerator (phosphor bronze mesh) is identical for all measurements and achieves no load temperatures of $T \cong 30K$ and cooling performances of $Q \cong 50W$ at $T = 50K$. Diameters and lengths of the regenerators are identical for all regenerator types (see table 2), the cycle frequency is $f = 2.4Hz$. Figure 7 shows both regenerator types (type 1 and type 3) with the most different cooling performance below $T = 10K$. The difference between these two regenerator arrangements at $T = 4.2K$ is about $Q \cong 0.35W$ (corresponds to an increase of 50%). The cooling performance differences become negligible for higher temperatures and with $T = 9.5K$ they are almost equal ($Q \cong 10W$). The comparison of the cooling performance at $T = 4.2K$ and the no load temperatures of the different regenerator types are shown in figure 8. Here one can also see the connection between the drop of the no load temperatures and the rise of the cooling performance at $T = 4.2K$. The lowest temperature which can be achieved with the given op-

**Table 3** Arrangements of different regenerator types

| Type | Regenerator Arrangement (from hot to cold end, Vol.%) |
|---|---|
| 1 | 25 % Pb + 75 % Nd |
| 2 | 25 % Pb + 45 % ErNi + 30% $ErNi_{0.9}Co_{0.1}$ |
| 3 | 25 % Pb + 45 % ErNi + 30% $HoCu_2$ |
| 4 | 25 % Pb + 45 % Nd + 30% $HoCu_2$ |

erational parameters (table 2, $f = 2.4Hz$) comes down to $T = 2.5K$ (type 4) with a cooling performance of $Q = 1.0W$ at $T = 4.2K$.

The dependency of the cooling performance on the cycle frequency, the first stage temperature as well as the filling pressure shall not be presented in detail (see for example Ref.[10]). The maximum cooling capacity of the regenerator type 3 is $Q_2 = 1.2W$ at $T = 4.2K$ with an increased filling pressure of 1.7 MPa (refilled during operation) and a higher first stage temperature of $T \cong 45K$ ($Q_1 = 50W$).

These measurements confirm that with the concept of the multi-layer-regenerators with different temperature dependencies of the specific heat it is not only possible to increase considerably the cooling performance (up to 50%) but also to reduce notably the no load temperatures. In particular, the pronounced influence of the enlarged specific heat regenerator material in only the cold third of the regenerator is remarkable. This is surprising, particularly in view of the flat temperature distribution in the cold end of the regenerator (see next chapter).

## Temperature Profile Measurements

The efficiency of the multi-layer regenerators depends notably on the actual temperature distribution of the gas and the matrix in the regenerator. With the help of these data it is possible to give statements not only on the best possible arrangement of the different regenerator materials in the second stage but also to assess the validity of theoretic models of low temperature regenerators[5,6,8]. For the measurement of the temperature profile in a 4.2K-GM refrigerator, the displacer of the second stage was equipped with five equidistant miniature Si-diodes (absolute accuracy $T = \pm 0.25K$). The temperature sensors are mounted in the displacer wall in the way that the temperature-sensitive side has direct contact with the regenerator (see figure 9). Due to the arrangement of the temperature sensor it can be assumed that the measured temperature corresponds to the matrix temperature of the regenerator. The dimensions of both stages as well as the set of operation parameters correspond to the settings in the previous measurements. Figure 10 shows the different regenerator arrangements which were investigated with this apparatus. In the lower part of the picture, the positions of the Si-diodes within the regenerator are indicated. The first measurements of the different regenerator types showed only a slight influence of the measuring facility on the cooling performance and on the no load temperatures of the second stage ($Q_{max}=0.8W$ at $T=4.2K$ and $T_{min}=2.9K$).

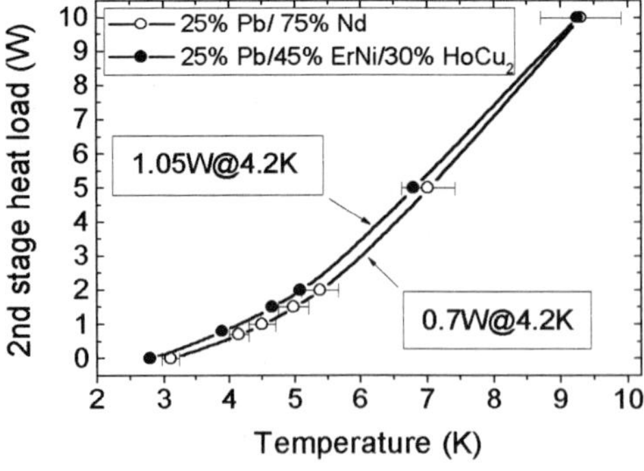


**Figure 7** 2nd stage cooling performance of two regenerator arrangements.

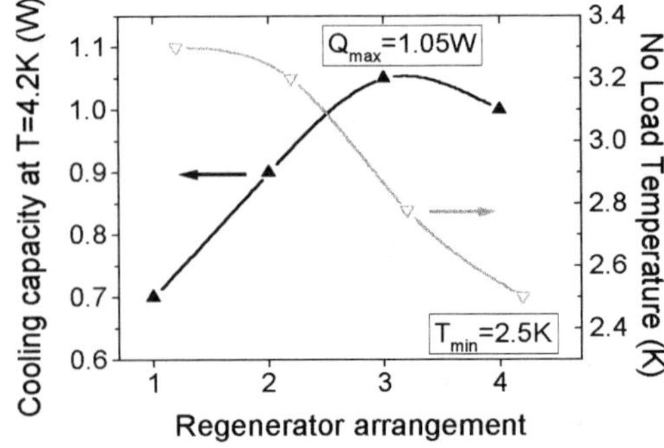


**Figure 8** Cooling capacity and no load temperatures of the different investigated regenerator arrangements.

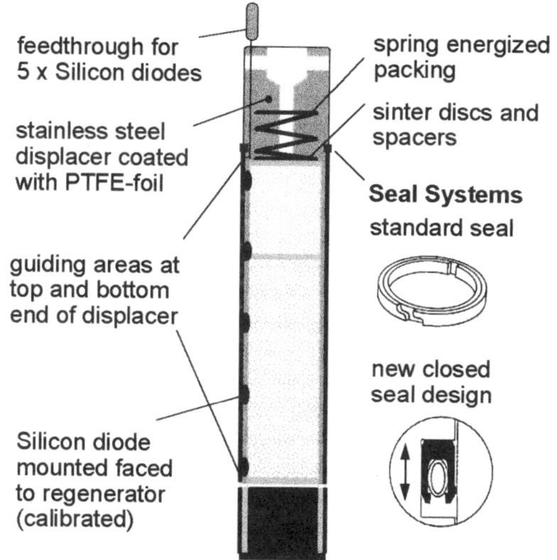


**Figure 9** Schematic experimental set-up of 2nd stage for temperature profile measurements. The different elements of improvement are shown.

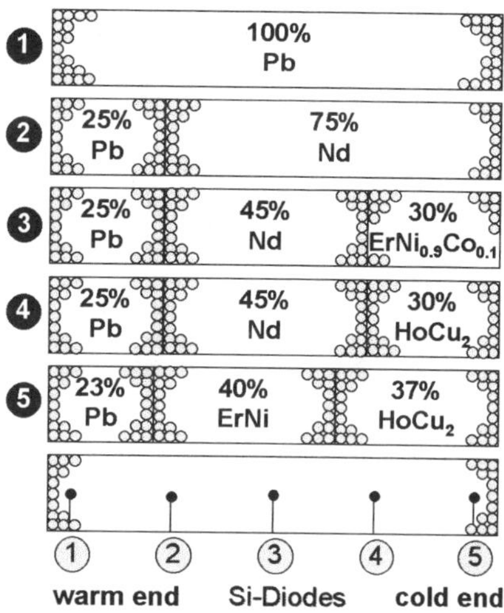


**Figure 10** Regenerator arrangements with position of Diodes.

When regarding the temperature profile of the different regenerator types (①②③④⑤, see figure 11) it can be seen that there is a steep temperature drop in the first quarter and a comparably flat profile in the cold area of the regenerator. In the first quarter of the regenerator, approx. 70% of the total temperature change occurs here. This ratio remains more or less the same for different 2nd stage inlet temperatures. The difference of the individual cooling performances between regenerator ②and ③④⑤ amounts to max. $\dot{Q} \cong 0.35W$, which is roughly in accordance with the first regenerator arrangement measurements (see section *Different Regenerator Arrangements*). The differences among the regenerators ③④⑤ are not very pronounced, however, the regenerator ⑤ with the max. proportion of enlarged specific heat material shows lower temperatures inside of the regenerator, but with no significant effect on the cooling capacity at the cold end. By comparison of model and experimental data it becomes clear that the predicted increase of cooling capacity has not been confirmed. A better agreement between modeling and experiment is expected by taking into account the real temperature distributions in the regenerator as well as the enthalpy deficit in ideal cooling capacity calculation.

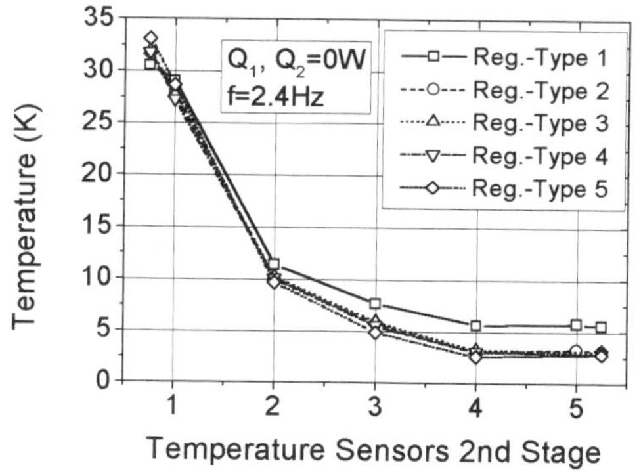


**Figure 11** Temperature profile of regenerator matrix for different regenerator arrangements. 0.75 and 5.25 indicates the 2nd stage inlet and the bottom end temperature, respectively.

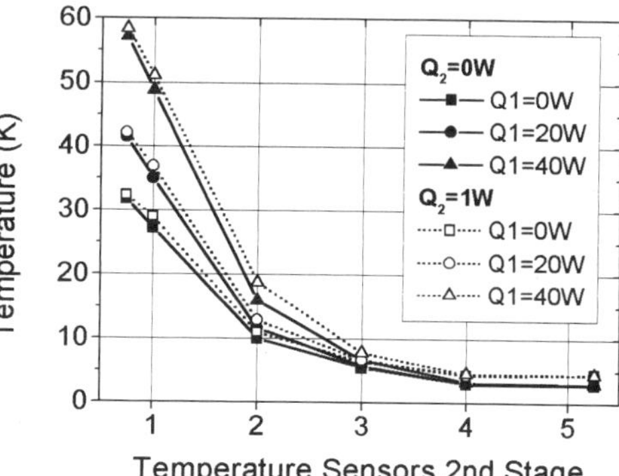


**Figure 12** Temperature profile of regenerator type ④ ($Pb/Nd/HoCu_2$) with different 2nd stage inlet temperatures ($Q_1$=0, 20, 40W).

## CONCLUSION

It has been shown that by the use of a special adapted design of critical components, i.e. 2nd stage seal, controlled regenerator packing, displacer material with defined mechanical properties respectively, a stable and reproducible cooling performance of a 4.2K-GM refrigerator was achieved.

The best regenerator arrangement we have found so far was a three layer regenerator with a composition of 25%Pb/ 45%ErNi/30%$HoCu_2$. Compared to the starting point with a two layer regenerator (25%Pb/ 75%Nd) there has been a 50% increase in cooling capacity. The maximum cooling capacity of this system with the given dimensions is 1.2W@T=4.2K and a simultaneous cooling capacity of 50W@T=45K at 1st stage.

The temperature profile of the moving regenerator was measured with special equipped displacer during operation. There is a steep temperature drop in the first quarter of the regenerator (70% of the 2nd stage temperature occurs here) with a comparably flat slope in the cold region of the regenerator.

## ACKNOWLEDGMENTS

Supports by the German BMBF (contract No. 13N6578) are gratefully acknowledged.

## REFERENCES

1. W.E. Gifford and T.E. Hoffman, "A new refrigeration system for 4.2K", *Adv. Cryo. Eng.*, 6:82 (1961)
2. M.D. Atrey, H.U. Häfner and C. Heiden, "Development of a computer model for two-stage Gifford-McMahon cryocooler", *Adv. Cryo. Eng.*, 41:1593 (1996)
3. T. Hashimoto, T. Tsukagoshi, H. Nitta, M. Yabuki, T. Kuriyama and H. Nakagome, "Excellent character of multi-layer type magnetic regenerator near 4.2K", *Cryocoolers* 8:677 (1995)
4. T. Kuriyama et al. "High efficient two stage GM refrigerator with magnetic material in the liquid helium temperature region", *Adv. Cryo. Eng.* 35:1261 (1990)
5. C. Wang, University of Giessen, *priv. comm.*
6. T. Kuriyama, M. Takahashi, H. Nakagome, T. Hashimoto, T. Eda and M. Yabuki, "Regenerator performance and refrigeration mechanism for 4K-GM refrigerator using Rare Earth compound regenerator materials", *Cryocooler* 7:429 (1992)
7. P.Y. Wu et. al., "Dynamic simulation of magnetic material regenerator and comparison with experiment", *Cryogenics* 36:259 (1996)
8. C. S. Hong and X.D. Xu, "On the thermodynamic cycle of the low temperature GM refrigerator", *Cryogenics* 34:183 (1994)
9. C. Wang, G. Thummes and C. Heiden, "A two stage pulse tube cooler operating below 4K", *Cryogenics* 37:159 (1997)
10. T. Kuriyama, et al., "Optimization of operational parameters for 4K-GM refrigerator", *Adv. Cryo. Eng.*, 41:1615 (1996)
11. J. Chafe, G. Green and C. Riedy, "Neodymium regenerator test results in a standard Gifford-McMahon refrigerator", *Cryocooler* 7:1157 (1992)
12. Z. Zhu and L. Zhang, "The theoretical cooling capacity and heat loss analysis of a 4.2K -GM refrigerator", *Adv. Cryo. Eng.*, 41:1639 (1996)
13. GASPAK 3.20 Cryodata Inc, Niwot, Colorado, USA (1992)

# NIOBIUM CLAD NEODYMIUM IN WIRE AND STRIP FORM FOR USE AS A REGENERATOR

T. Wong and M. K. Rudziak

Supercon Inc.
Shrewsbury, MA 01545

## ABSTRACT

Neodymium is a regenerator material used in Gifford-McMahon cryocoolers at temperatures <10 K due to its high heat capacity resulting from the magnetic order transformation at the Curie temperature. Typically the Nd is used as a powder or small spheres, but it is preferable to use alternative configurations that allow more efficient gas flow and heat transfer. However Nd is difficult to fabricate into wire or strips because it work hardens rapidly and also has a high affinity for oxygen. By cladding the Nd with niobium, it is possible to protect the Nd from oxidation and allow drawing and rolling to small dimensions. This work reports on the development of a process to form Nb clad Nd into fine wire, rolling the fine wire to strip, and the forming of small ridges in the strip. The strip is then wound into a pancake coil, and stacked for use in a regenerator.

## INTRODUCTION

There has been an increased level of interest in the development of conduction cooled superconducting magnets for use where liquid helium is either not available or prohibitively expensive. These magnets are typically wound from NbTi superconductor for cost and robustness reasons, and operate at a temperature between 4-6 K. Cooling is provided by cryocoolers utilizing a Gifford-McMahon cycle. One example of a conduction cooled magnet is the Advanced Lightweight Influence Sweep System (ALISS) magnet being developed by the Navy (Naval Surface Warfare Center, Carderock Division) for mine sweeping applications[1]. This magnet uses an aluminum stabilized NbTi conductor and two G-M cryocoolers with rare earth (RE) regenerators. The RE material is chosen for its high heat capacity at the temperature range of interest (< 10K) due to a magnetic phase transformation at the curie temperature.

In conjunction with the magnet development and testing program, the Navy has also been investigating methods to increase the cooling capacity of the G-M cryocoolers to provide greater performance or larger operating temperature margins[2]. One avenue of investigation is to examine alternative regenerator configurations[3] in order to increase volumetric and heat transfer efficiency. This paper details the processing necessary for the manufacture of a coiled, embossed neodymium regenerator ribbon geometry.

## BACKGROUND

Neodymium is used in the lowest temperature region of a multi-stage GM cryocooler where it has high heat capacity compared to traditional regenerator materials such as lead and copper. Other researchers have utilized intermetallic compounds such as $Er_3Ni$, $Er_3Co$, and ErYbNi compounds[4] that also demonstrate magnetic transformations that result in higher heat capacity in the temperature range of interest. Most of these alternatives are used as spheres produced by melt spinning because the compounds lack the ductility necessary for alternative forming techniques.

The use of melt spun spheres has many potential drawbacks such as contamination, non-uniform shapes, porosity and low yields[5]. It is possible to pick up enough oxygen and/or nitrogen in the final product to decrease the heat capacity of the material. The low yields result from the need to sort the melt spun material into uniform size ranges of 50-200 microns. Any spheres too large or small may not be recyclable due to contamination during processing. In addition, spheres are limited to a 35% void space when optimally packed.

The form of most regenerator material has been limited to spheres due to the processing limitations of the brittle compounds, however Nd is ductile to a limited extent. Nd will work harden rapidly to the point where further reduction is not possible without a recrystallization anneal. However, Nd has a high affinity for oxygen so during the anneal, oxygen levels are raised to a level that embrittles the material. These limits are why Nd has not been manufactured into forms other than spheres in the past. Ideally the regenerator should be of a form that results in high volumetric heat transfer efficiency and low pressure drop of the flowing gas.

More recently Nd has been manufactured into a perforated disk[3], however this approach is limited by high cost, low yields, and poor void fraction control. In this approach, Nd is bundled with a sacrificial material into a rod of the desired diameter. The rod is compacted by extrusion or other process to bond the elements together. The rod is sliced into disks and then chemically treated to remove the sacrificial material. An alternative approach, described in this paper, is to manufacture a thin Nd ribbon with ridges created by surface embossing, perpendicular to the axis of the ribbon. The ribbon is then tightly coiled to form a disk, with the ridges serving to separate the turns of ribbon as shown in Figure 1. The void fraction can easily be controlled by adjusting the relative height of the ridges to the ribbon thickness. Because standard wire drawing and ribbon rolling practices are used, it is possible to have high yields and low costs with this approach.

The use of niobium as a diffusion barrier is well known in the production of NbTi low temperature superconductors to prevent the formation of brittle CuTi intermetallics. Nb is a suitable barrier material because it is not soluble in copper and is sufficiently ductile to be compatible with both the copper and the NbTi alloy during extrusion and wire drawing. These factors make niobium a good material to clad Nd regenerator wire, as it can serve to protect the neodymium from contamination during processing while not reacting or alloying with Nd or copper during the thermal anneals received during processing. Niobium is also a very ductile metal, and thus will deform uniformly with the other materials during extrusion and drawing. Neodymium rapidly hardens due to cold work and must be annealed repeatedly during wire drawing to recrystallize the structure and restore ductility to allow further reduction without breakage. The Nb barrier around the Nd prevents rapid pickup of

**Figure 1.** A schematic of a coiled, embossed ribbon regenerator disk.

**Table 1.** Chemical Analysis of 99.9% pure Nd Billets (all values are in wt%, "-" indicates not detected)

| Ce | Y | La | Sm | Pr | Si | Fe | Mg | Ca | Al | Ni | Nd |
|---|---|---|---|---|---|---|---|---|---|---|---|
| - | <.01 | <.05 | <.05 | <.01 | <.01 | <.01 | <.01 | <.01 | <.01 | <.01 | Bal |

oxygen during the anneals, which would otherwise make the Nd become too brittle for further processing.

## PROCESSING

To produce the ridged Nd tape, the following technical approach was designed and implemented[6]. Because Nd is not available commercially in sufficiently long wire or strip form, processing was done from a starting as-cast 99.9% pure Nd ingot. The ingot was reduced to the proper size for encapsulation and hot extrusion. This occurred in a copper extrusion can with niobium foil as a diffusion barrier between the Cu and the Nd. It was then drawn to the correct starting size for rolling to final size. The process is described in more detailed below.

Concurrent Technologies Corp. (CTC) purchased from Rhône Poulenc Inc. approximately five kilograms of neodymium in the shape of two cylindrical billets 50.8 mm in diameter and 203.2 mm in length, in the as-cast condition. CTC staff machined the billets to remove surface casting roughness and then hydrostatically extruded to 47.625 mm in diameter. This also allowed higher material yields by reducing machining losses. The initial hydrostatic extrusion also provided some strain hardening and grain refinement that would promote more uniform core deformation of Nd billets during the encapsulation, extrusion, and wire drawing processes. After hydrostatic extrusion, the surface of each billet was cleaned and finish machined to 44.45 mm. Table 1 details the spectrographic chemical analysis of the billets.

Supercon received the extruded and machined billets and assembled the Nd into containers as shown in Figure 2. A Nb sheet of 0.762 mm thickness was used for the barrier, accounting for 6.6% of the non-copper area. At the time of billet assembly we did not know what the final wire size would be, so the niobium thickness was based on past experience with other composites. The Nd ingots were wire brushed and wiped with acetone

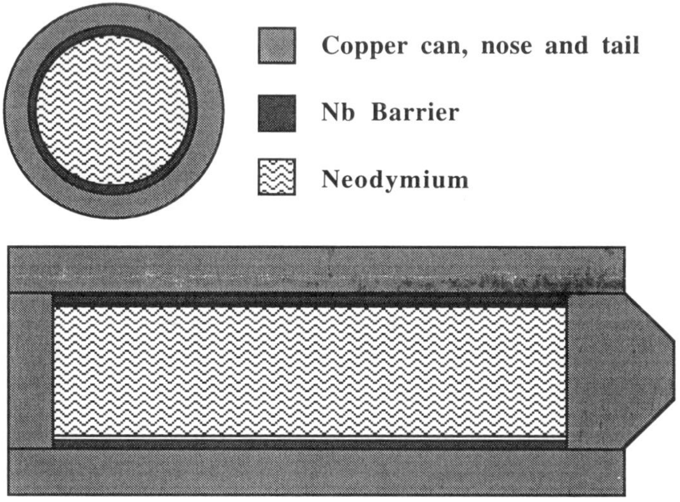


**Figure 2.** Schematic of extrusion can with niobium barrier material (drawing not to scale).

prior to assembly to remove as much surface oxide and contamination as possible to help promote good bonding of the components. The Nb sheets and Nd ingots were then inserted into 50.8 mm OD, 46.482 mm ID copper tubes and a copper nose and tail were TIG welded to enclose each billet. The enclosed billets were evacuated and then sealed for conventional hot extrusion.

The billets were hot extruded to 12.7 mm diameter with no difficulty. After extrusion the rods were cropped until the Nb barrier was complete around the Nd. Samples of each extrusion were drawn using a 20% reduction die line until 2.362 mm diameter without any intermediate anneals. After this size one billet had increased frequency of point breaks and final wire breakage at 1.892 mm, while the other sample suffered point breaks but made it to 1.684 mm without breaking. Short lengths of the 1.684 mm and 1.892 mm wire were etched in nitric acid to remove the copper jacket. A small defect in the Nb barrier in the 1.684 mm wire allowed the nitric acid to attack the Nd, resulting in wire breakage. At the time we felt this was due to the extensive cold work leading to non-uniform deformation of the Nb barrier.

The sample pieces showed much better drawability than initially hoped for, so a drawing schedule was established with anneals every half diameter in order to lesson the severity of point breaks during drawing and hopefully improve barrier uniformity. Anneals were conducted in a strand annealing furnace to limit the time exposed to high temperature to a minimum. Material was successfully drawn to sizes ranging from 1.615 mm to 0.942 mm diameter with only one wire break. The wire break occurred at a larger size and is thought to have been caused by an inclusion rather than excessive cold work.

Several lengths of the wire were etched in nitric acid to remove the copper jacket. After removal from the acid it became apparent that the Nb barrier had failed to prevent the acid from attacking the Nd in all the sizes, including the largest diameter, where the Nb was nominally the thickest, and getting increasingly worse at the smaller diameters. Some of the longer lengths of wire, with the Cu completely removed, were delivered to CTC for rolling trials. Later, additional wire was delivered with the Cu partially removed to avoid the breakage due to barrier non-uniformity. CTC was more successful at rolling the wire with some Cu left on the surface than the bare Nb coated composite. However, the Cu clad composite was not tested in a cryocooler to see the effect of the lower volume percentage of Nd in the regenerator.

The neodymium ribbon rolling and embossing was conducted at CTC facilities[6] using a Fenn rolling mill with two high and four high roll configurations where the diameter of the work rolls is 213.4 mm and 60.325 mm respectively. The roll speed was maintained at 18 rpm throughout the operations. During the entire rolling process, 4.5 kg of tension was applied to the neodymium ribbon using a digitally controlled pay-off and take-up reel system.

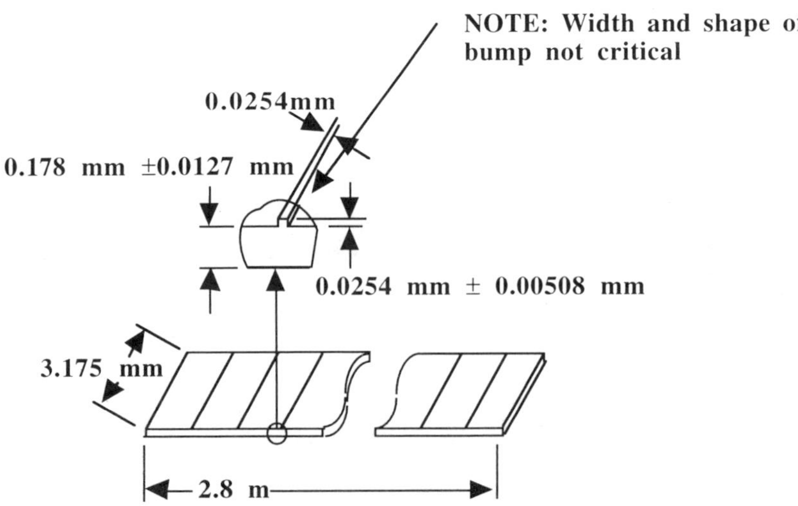


**Figure 3.** Dimensions of embossed ribbon showing typical thickness and ridge size.

The two high roll configuration was used for the first two passes to impart a large reduction in thickness, approximately 50% to 70% reduction per pass. Since neodymium hardens very rapidly and becomes brittle, an anneal was used after the second pass to soften the material. The annealing process was carried out at 450°C temperature, with one hour holding time at the temperature, and a continuous argon gas flow during heating and cooling cycles.

The four high roll configuration was used for the third through sixth passes, with an anneal after the sixth pass. Reduction was approximately 20 percent in thickness per pass. Embossing of the ribbon was done during the seventh (final) pass utilizing one 60.325 mm diameter knurled roll which imparts the 3.175 mm wide ribbon with 0.0254 mm or 0.0508 mm high ridges, 1.588 mm apart (Figure 3). A final annealing completed the fabrication of the neodymium ribbons.

For Navy evaluation, this project successfully produced 18.5 m of 0.254 mm thick ribbon with a 0.0508 mm ridge height. This would yield a theoretical void volume of 20 percent when the tape was coiled. CTC also produced 46.2 m of 0.178 mm ribbon with a 0.0254 mm ridge height. The void volume of this sample was calculated to be 15 percent. The tolerances as specified by the Navy were ±0.0127 mm for the thickness of the tape and ±0.00508 mm for the ridge height itself. Both these tolerances were met. However, the edge finish of the tapes was not optimal as unevenness and edge cracking was prevalent. Better process design, with optimization of the frequency of anneals would alleviate this problem. Figure 4 shows a cross section of the 0.254 mm high tape. This micrograph shows a section of wire rolled to tape without copper removal. The polish of the sample is not optimal due to the reactive nature of the Nd, which does not allow water based cleaners or polishes to be used. The dark spots are surmised to be pits formed by harder particles of Nd falling out of the surface. Those particles may be neodymium oxide particles formed during polishing, as any temperature increase will cause a sharp rise in oxidation rates.

## DISCUSSION

The project was successful in producing the first embossed Nd ribbon for regenerator applications. However there is room for further improvements to be made. The Nb barrier thickness is not optimized for a final ribbon size. As it was not possible to maintain a uniform barrier to allow batch etching, it will be necessary to increase the volume percentage of Nb in future work. The amount of additional Nb will depend on what final size wire proves to be the best for rolling. This in turn is dependent on additional pressure drop experiments and testing of the coiled disks in a cryocooler to determine optimal thickness and void fraction.

It was difficult to remove the copper by batch etching due to the thin spots in the Nb barrier. Some work was done that allowed partial copper removal by strand etching. Strand etching is not a preferable method as it is difficult to control uniform etching rates along the length of the wire, and is also time consuming. As an alternative, a slower more controllable etch was used in a batch process, and was successful in removing 50% of the copper without burn through to the Nd. Any further copper removal was risky as the Nd wire is not totally round, and thus some of the Nb is already exposed at 50% Cu removal, increasing the possibility of the acid attacking the Nd. The Nd is so reactive with the acid that any exposure will result in wire breakage.

CTC did not have the edge uniformity and cracking problems when it rolled samples with the copper left on the ribbon. The copper also made controlling the oxidation of the Nd during annealing much easier. Leaving the copper on the sample simplifies the ribbon rolling and avoids the difficulties of copper removal with a thin Nb barrier. However it has yet to be seen what effect the presence of copper in the regenerator has on the cooling performance of the cryocooler.

In all the work, the limit on piece length was caused by the etching process. With improvements in the barrier, or avoiding etching altogether, we expect that piece length should not be limited in future work. The ductility of the Cu/Nb/Nd composite was much better than expected. The process can have yields of 80-85% if the etching problems are bypassed or solved, a much higher yield than alternative Nd forming techniques.

There are several outstanding technical problems which need to be addressed before the optimum manufacturing technology can be developed and validated. Neodymium strain

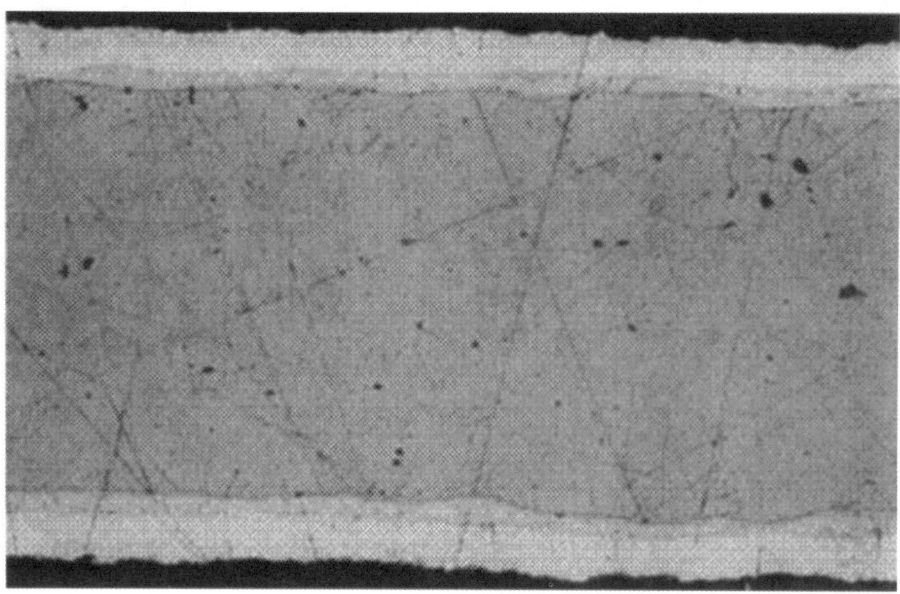

**Figure 4.** Nd ribbon longitudinal section, 0.254 mm thick, 200X.

hardens in the deformation processing quite significantly and requires frequent annealing. To solve this problem, simulation modeling needs to be employed in order to accurately predict insertion of required annealing steps during the wire drawing and tape rolling operations. In addition, removal or retention of the copper layer may have a significant impact on the quality and the cost of Nd tape. Attention also has to be paid to manufacturing process control in order to produce tapes with required geometry and dimensional tolerances.

## CONCLUSION

Continuous lengths of ridged ribbon were successfully produced in several thicknesses and volume void fractions. The use of commercial pure Nd was proven to have sufficient ductility for this processing. The Nb barrier was successful at preventing oxygen pickup during the frequent in-process anneals.

Test results obtained at the Naval Surface Warfare Center, (NSWC), Annapolis, MD, showed that Nd ribbon geometry can significantly improved regenerative efficiency and increased cooling capacity by lowering achievable temperature at one-half watt load by 0.5°C in the two-stage Gifford-McMahon cryocooler as compared to Nd spheres. Also, the ribbon configuration can offer a lower pressure drop, reducing compressor power requirements. The cooling capacity increase could be sufficient enough to meet performance requirements for the ALISS system of operating with a single cryocooler, thus providing a back-up cryocooler to improve mission reliability. In addition, the ribbon manufacturing technology which was developed here, results in a higher material yield (85% vs. 20% for Nd spheres) which, at the Nd price of $250 per pound would lead to significant cost savings.

## ACKNOWLEDGMENT

The authors express thanks to Thomas Felmley, CTC for his support of this work and paper. This work supported by the National Center for Excellence in Metalworking Technology, operated by Concurrent Technologies Corp. under contract N00140-92-C-BC49 to the US Navy as part of the US Navy Manufacturing Science and Technology Program.

## REFERENCES


1. M. Heiberger, A. Langhorn, T.W. Hurn, D.G. Morris, J.D. Walters, L.F. Aprigliano, D. Waltman, and G.F. Green, A lightweight rugged conduction-cooled NbTi superconducting magnet for US Navy minesweeper applications, in: "Adv.in Cryo Eng." Vol. 41b, Pg. 1087-1095
2. J.N. Chafe, G.F. Green, and J.W. Stevens, Geometric scaling of a 4.2K Gifford-McMahon refrigerator, in: "Cryocoolers 9,"R.G. Ross, Jr., ed., Plenum Press, New York (1997), p. 627-636.
3. J.N. Chafe, G.F. Green, and J.B. Hendricks, A neodymium plate regenerator for low temperature Gifford-McMahon cycle refrigerators, in: "Cryocoolers 9,"R.G. Ross, Jr., ed., Plenum Press, New York (1997), p. 653-662.
4. V.K. Pecharsky, K.A. Gschneidner, Jr., R.W. McCallum, and K.W. Dennis, Influence of alloying on the behavior and properties of Er3Ni, in: "Cryocoolers 9,"R.G. Ross, Jr., ed., Plenum Press, New York (1997), p. 663-668.
5. T. Wong and J. Seuntjens, "Development of rare earth regenerator materials in fine wire form, in: "Adv. in Cryo. Eng.," Vol. 42a, Pg. 439-444.
6. J. Kajuch and T. Felmley, Rapid Response for Fabrication Issues of Neodymium Ribbon Regenerators for Use in Cryogenic Refrigerators, Final Report, NCEMT TR No. 97-110, Concurrent Technologies Corporation, Johnstown, PA, July, 1997.

# NEODYMIUM-RIBBON-REGENERATOR COOLING PERFORMANCE IN A TWO-STAGE GIFFORD-MCMAHON REFRIGERATOR

J.N. Chafe and G.F. Green

Naval Surface Warfare Center, Carderock Division
Annapolis, MD 21402

## ABSTRACT

This work is part of an ongoing study into the way material configuration (i.e. flow geometry) can affect the performance of low-temperature regenerators. This work demonstrates a new regenerator configuration and fabrication technique for Gifford-McMahon (GM) refrigerators that operate at temperatures near 4.5 K. This new style regenerator promises to be less expensive, more reliable and perhaps provide better cooling power.

A regenerator was assembled from Neodymium ribbon that was approximately 0.32 cm (1/8 in.) wide and 0.025 cm (0.010 in.) thick. Neodymium was chosen because it has a relatively high volumetric heat capacity at temperatures below 10 K and it is a pure metal that can be drawn and formed with relative ease. Small ridges (approximately 0.0025 cm (0.001 in.) high) were embossed on one face of the ribbon. The ribbon was wound into coils similar to the way adhesive tape is wound into rolls. The small ridges on the ribbon provide spacing between the wound layers creating flow passages for gas. Several coils of ribbon were stacked one on top of another inside a stainless steel tube. This assembly formed the second stage regenerator. By selecting ribbon with different thickness and ridge heights the void fraction as well as the amount of heat transfer surface area of the regenerator could be controlled.

Test results are presented for two regenerators with ribbon geometry, one with a void fraction of about 20% and another with 15%. Test results include measurements of the pressure drop that occurs when gas flows through each regenerator as well as the cooling performance of a GM refrigerator while utilizing these regenerators. This data is compared to data for the more conventionally designed regenerator constructed from spherical Nd powder that has a void fraction of about 35%.

## INTRODUCTION

Operating superconducting magnets is usually expensive especially over an extended period of time. Even laboratories with expertise in this area find the current methods of cooling superconductors prohibitively expensive for long term experiments. The usual method of cooling superconductors is to fill a dewar which contains the superconductor with liquid helium. Clearly, a liquid helium plant on site would help to reduce the cost of purchasing liquid helium from local suppliers. However, this requires a large initial investment plus trained personnel to run the plant. A promising solution to this problem is using low temperature Gifford-McMahon (GM) cycle refrigerators to cool superconductors.

Gifford-McMahon refrigerators are easy to operate and maintain, and are highly reliable. This method of cooling superconductors would make them easier to use since there would no longer be the need for personnel trained in handling liquid cryogens or in running a helium plant. Superconductors could then become more marketable.

The use of GM refrigerators to cool superconductors would not have been practical five or so years ago since they could not achieve the extremely low temperatures required by the most commonly used superconductors. Only relatively recently has this become an option with several manufacturers building GM refrigerators capable of producing 0.5 W or more of cooling at 4.5 K. Although these refrigerators can now be used to cool superconductors and other types of devices, more powerful and efficient GM refrigerators would have several advantages over the currently available refrigerators. Initial start-up costs could be less since perhaps only one powerful refrigerator could do the job of two less powerful ones. In addition, a more powerful refrigerator could mean more design options. Therefore, the research for this paper deals with design changes to increase the operating efficiency and cooling power of GM refrigerators.

The most critical requirement for GM refrigerators to operate at temperatures below 7 K is a well designed low temperature regenerator. Current GM refrigerators utilize regenerative or periodic flow heat exchangers that cyclically store energy in specially selected material (referred to as regenerator matrix material) by allowing a gas to flow back and forth through a tube containing the matrix material. There are many design factors for low temperature regenerators that affect how well the regenerator performs as a heat exchanger. Varying several aspects of the currently available GM refrigerator's low temperature regenerator design should help heat exchange and limit the thermodynamic losses that occur in the regenerator. For some commercially available GM refrigerators simply replacing the lead (Pb) matrix material with a rare-earth metal such as neodymium (Nd) has lowered cooling temperatures from around 6 or 7 K to below 2.5 K. In an effort to further this improvement (allow more cooling power at the temperature or temperature range of interest e.g. 4.5 K), a regenerator was constructed with a significantly less porous matrix geometry than the commonly used spherical particles.

The regenerator design concept presented here consists of fabricating a regenerator from long strips of neodymium metal. These strips or ribbons are assembled in such a way that they form flow passages between many flat parallel plates. The design, fabrication and test results for this type of regenerator are discussed in this paper and ways to improve the regenerator's performance are presented.

## RIBBON FABRICATION AND DESIGN

One of the critical factors to be considered when designing a regenerator that operates below about 7 K is the choice of matrix material. In order for a regenerative heat exchanger to function effectively, it must have the capability to store energy with small temperature changes. There are two material properties that strongly affect this ability to store heat;

volumetric heat capacity and thermal conductivity. Typically, these properties are strongly dependent on the operating temperature and it is important that both these properties remain high over the operating range of the regenerator. If needed, moderately low values of thermal conductivity can be compensated for through the design of the regenerator, but it is crucial to use a material with high volumetric heat capacity. Unfortunately most materials do not have the high volumetric heat capacity needed for operation at temperatures below 7 K. However a few materials have been identified that do have this property. One of these materials is the element neodymium (Nd). Neodymium is used in this research because in addition to its high volumetric heat capacity, it is ductile and formable while the other materials suitable for low temperature regenerator are not. In order to form the regenerator, the matrix material was extruded, drawn and rolled extensively. These processes required that the selected matrix material be ductile and malleable and have reasonably high tensile strength. Nd is the only known material suitable for use in a low temperature regenerator that has all these properties.

The design of the ribbon regenerator is an attempt to modify a rectangular parallel plate style heat exchanger so that it can be packed it into a cylindrical tube. Using parallel plates increases greatly the flexibility in controlling the surface area of the heat exchanger, the size of the flow passages and the percentage of regenerator that is void space. A sketch showing a section of a rectangular parallel plate regenerative heat exchanger is shown in fig. 1. Shown are rectangular duct like flow passages between walls that have a specified thickness. The important design parameters of this type of geometry is the spacing between each plate and the thickness of each plate. The plate spacing determines the flow passage size or how easily gas can flow through the heat exchanger. The plate thickness controls the volume of heat storage material that is available to the gas as it passes through that particular passage. The relationship:

$$f = \frac{s}{s+t}, \tag{1}$$

between the plate spacing and the thickness determines the void fraction or the fraction of the heat exchanger volume that is occupied by gas. Here $f$ is the void fraction and $s$ and $t$ are indicated in fig. 1. The surface area available for heat transfer per unit volume of heat exchanger ($A_v$) is given by the following expression:

$$A_v = \frac{1}{s+t}. \tag{2}$$

This relation indicates that in order to make a lot of surface area available for heat transfer the spacing and the thickness should be small. The pressure drop for laminar flow can be estimated by[1]:

$$\Delta p = -\frac{3\mu\dot{V}(s+t)}{A_c s^3}L, \tag{3}$$

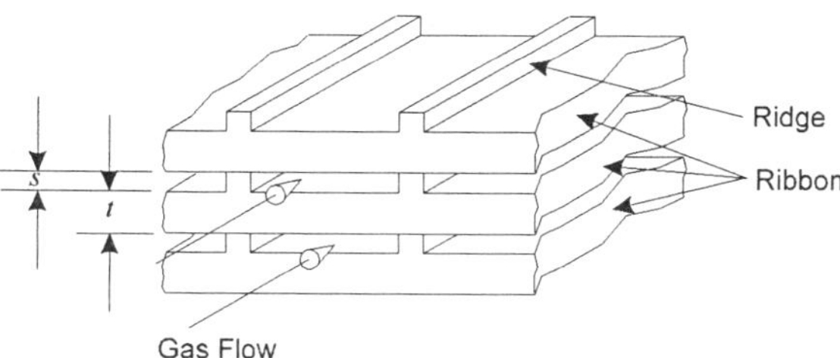


**Figure 1.** Rectangular parallel plate regenerator. The plates are separated by ridges so that uniform flow passages are formed.

where $\Delta p$ is the pressure drop, $\mu$ is the viscosity, $\dot{V}$ is the average velocity, $A_c$ is the cross sectional area of regenerator and $L$ is the length of regenerator.

This equation indicates how crucial the spacing ($s$) is in determining the pressure drop of a parallel plate heat exchanger. Therefore, if the spacing ($s$) is made too small in hopes of increasing the heat transfer by increasing the surface area, it can actually have the opposite effect by causing the pressure drop to increase to a point to where very little gas can flow through the heat exchanger.

For a GM refrigerator, the regenerator needs to be cylindrical, since that way it can be efficiently incorporated into the refrigerator. To accomplish this a long flat ribbon is coiled up similar to a watch spring or a roll of tape. Small ridges or bumps on one side of the ribbon create flow passages when the ribbon is wound as shown in fig. 2.

Another important parameter in regenerator design is length. There are actually two lengths that are important for the ribbon regenerator design. The first length is the length of the individual sections of ribbon or parallel plate. The second is the overall length of the regenerator. The way that the regenerator is constructed is to use a series of rolled ribbons placed one on top of another. This is shown in fig. 3. There is a small zone of separation between each section of rolled ribbon. This zone is comprised of a piece of fabric with randomly oriented fibers. The zone acts as a header region where the gas can collect and mix as it exits one section of plates and enters the next section. The lengths of each section is approximately 0.32 cm (1/8 in). As the gas first enters a flow passage, the velocity profiles are not fully developed which causes the gas to flow differently in this entrance region than it does further into the flow passage. This means the pressure drop and heat transfer characteristics are different in the entrance regions. Fortunately, for the parameters pertinent to a GM refrigerator, the entrance region is quite short compared with the 1/8" length of each section, so any difference in the behavior of the fluid in that region can be ignored.

A regenerator should be designed so that it can be fabricated with consistency and precision and be done at a low cost. Several researchers have teamed up to develop a process for fabricating the Nd ribbon[2]. First, wire from a Nd ingot was formed and then rolled into a flat ribbon with the small ridges on it. The rolling process used a rolling mill that flattened the wire by squeezing it between two cylindrical rolls. One roll had a smooth surface while the other had a series of grooves running axially along its surface. As it was rolled some of the Nd was extruded into the grooves and formed the spacing ridges that later controlled the size of the regenerator flow passages. Using this process, very long lengths of ribbon can be produced with very little waste material making efficient use of the raw material. By varying the wire diameter, the depths of the grooves in the roll and spacing between the rolls, the thickness , the height of the spacing ridges and the width of the ribbon can be controlled. The process produced high quality ribbons with only some minor flaws such as cracking along the edges of the ribbon. The fabrication process continues to be refined and shows promise of being an inexpensive method of producing regenerators.

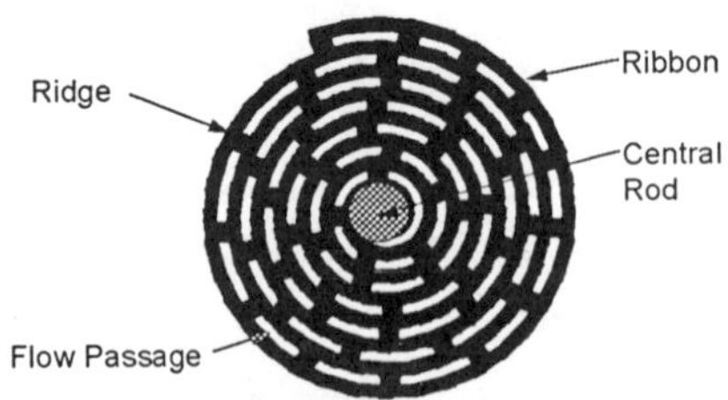


**Figure 2.** Sketch of ribbon regenerator showing how flow passages are formed by wrapping the ribbon around a cental rod. The ridges formed on the ribbon space the layers to form flow passages.

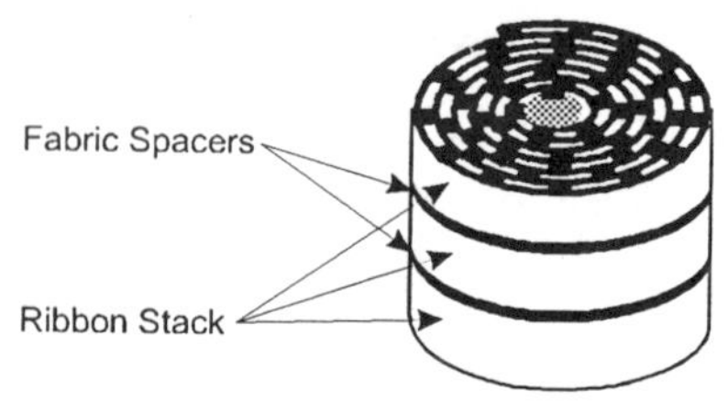


**Figure 3.** A stack of coiled ribbons that make up a regenerator. Also shown are the fabric spacer disks between each coil of ribbon.

## REGENERATOR CONSTRUCTION

Two test regenerators were constructed each using ribbon with different thickness and spacing ridges. The first regenerator, refereed to as regenerator #1, was constructed from ribbon that is 0.025 cm (0.010 in) thick with a 0.005 cm (0.002 in) spacing ridge. The second regenerator, referred to as regenerator #2, was constructed from ribbon that is 0.018 cm (0.007 in.) thick with a 0.0025 cm (0.001 in.) spacing ridge. In addition to the Nd ribbon, each of these regenerators had a quantity of Pb spherical powder (approx. 0.025 cm (0.010 in.) diameter spheres) that was also used as regenerator matrix material. Each regenerator was made with two halves. The top or warm half of the regenerator was packed with the lead spheres, while the bottom or cold end used Nd ribbon. The displacer tube which housed these materials was designed so that each end of the regenerator was packed independently of the other. The warm section of the regenerator required specially fabricated end pieces that kept the Pb spheres contained with fine mesh screens that also applied significant packing pressure to keep spheres firmly in place. The ribbon material does not require fine screens to contain it, however, it does require a solid rod that passes through each coil of ribbon. This rod is fiberglass, is 0.32 cm (1/8 in.) in diameter and fills the central hole in each coil. Each coil was separated from the coils above and below it by a porous fabric disk about 0.005 cm (0.002 in.) thick. A sketch of the cross section of the regenerator is shown in fig. 4.

To construct the ribbon half of each regenerator, pieces of ribbon were wound into spiral coils around a steel shaft that was slotted. One end of the ribbon was inserted into a slot in the shaft and the shaft was rotated while tension was applied to the ribbon. The ribbon was wound on the shaft until the coil had a diameter equal to the inside of the regenerator tube. The winding shaft was then removed and the end tab of the ribbon that fitted into the shaft was broken off. The whole coil was then pushed into the regenerator tube that had the central fiberglass rod affixed to it. A special winding fixture was used to accomplish these tasks.

The regenerator was packed in a stainless steel tube whose dimensions are shown in fig. 4. The outside of the tube was covered in plastic to prevent the regenerator from galling as the refrigerator operated.

## TESTING

For this research, the two Nd ribbon regenerators were evaluated to determine their pressure drop characteristics and their void fraction. Then the regenerators were installed in a refrigerator to determine their effectiveness in a low temperature refrigerator.

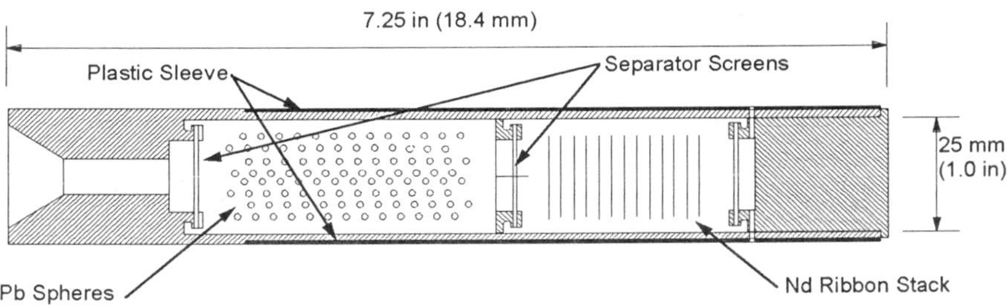


Figure 4. Cross section of complete regenerator. The left end contains Pb spheres and the right end contains the Nd ribbon.

**Table 1**. Geometric parameters for Nd ribbon regenerators

| | | Regenerator #1 | Regenerator #2 |
|---|---|---|---|
| Pb Section | Length | 7.35 cm (2.893 in) | 5.18 cm (2.038 in) |
| | Mass | 258.2 g | 195.2 g |
| | Measured Void Fraction | 0.36 | 0.31 |
| | Pressure Drop* | 1.2 kPa-min/L (5 psi/scfm) | 1.2 kPa-min/L (5 psi/scfm) |
| Nd Section | Ribbon Thickness ($t$) | 0.025 cm (0.010 in) | 0.018 cm (0.007 in) |
| | Spacing Ridge (s) | 0.005 cm (0.002 in) | 0.0025 cm (0.001 in) |
| | Theoretical Void Fraction ($f$) | 0.166 | 0.125 |
| | Mass | 131.9 g | 184.5 g |
| | Length | 4.7 cm (1.856 in) | 6.2 cm (2.442 in) |
| | Measured Void Fraction | 0.199 | 0.148 |
| | Pressure Drop* | 0.65 kPa-min/L (2.7 psi/scfm) | 5.2 kPa-min/L (21.4 psi/scfm) |

* Estimated from measurements made on fully assembled regenerators.

## *Regenerator Flow Testing*

The most critical assessment of a regenerator is its performance in a cooler but simply measuring cooler performance gives little insight into how to improve a regenerator. Understanding why a particular regenerator works well or doesn't requires a combination of analysis and careful measurements of the regenerator's components and its flow characteristics. For this reason, the pressure drop of each regenerator is measured. To do this helium gas is forced through the regenerator and the flow rate and pressure drop are recorded. These measurements are made at room temperature and are intended only to indicate the relative flow characteristics among different regenerators.

This testing allows the comparison of the flow characteristics of various types of regenerators as well as monitoring the similarity between two regenerators that should be exactly the same. In addition to measuring the pressure drop, the mass of matrix material loaded into the regenerator and the volume of space that it occupies is also measured. This allows the calculation of the void fraction for each regenerator. Then pressure drop and void fraction can be correlated to the performance of a regenerator in a refrigerator. This should help develop an empirical understanding on the characteristics of a regenerator that will give the best performance. Table 1 lists some parameters for the two ribbon regenerators used in this research. Included are the theoretical void fraction as well as the measured void fractions, estimated pressure drops, and other physical data. The measured void fraction is higher than it should theoretically be for the ribbon portion of both regenerators. This could

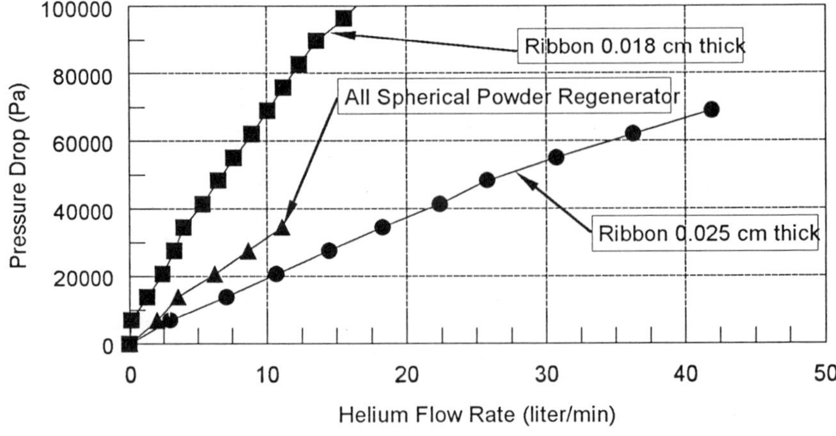


**Figure 5**. Flow measurements indicating relative pressure drops for the two ribbon regenerators as well as an all spherical powder regenerator.

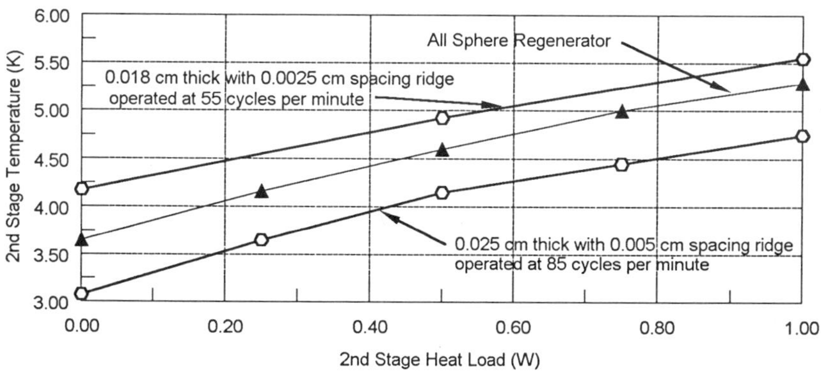


**Figure 6.** Second stage cooling performance for the ribbon regenerators. No heat was applied to the first stage for these data.

be attributed to winding the ribbon too loosely. Other contributing factors could be the fact the ribbon has a finite thickness so at the beginning and end of the coil the ribbon stops abruptly leaving somewhat tapered regions that contain no matrix material. Also, since the ribbons are not perfectly straight and they vary somewhat in the width, when they are formed into coils the tops and bottoms of the coils will not be perfectly flat but rather slightly wavy. This waviness will result in some areas devoid of Nd.

Figure 5 shows the results of flow measurements made on the two ribbon regenerators Also included for comparison is the flow measurements for an all spherical powder regenerator. This figure shows how the pressure drop in regenerator #1 is significantly lower than that for the spherical powder. On the other hand, the pressure drop for regenerator #2 is significantly higher than the spherical powder regenerator. This is evidence of how dramatically the design of the ribbon can affect the pressure drop in a regenerator. This effect is one reason the parallel plate design was adopted. Regenerators can be constructed with a specified void fraction and flow passage size that does not overly restrict flow.

## *Refrigerator Performance Testing*

The refrigerator used to test how well the regenerators performed was a Leybold Kelcool refrigerator. This refrigerator is commercially available and is typically operated with Nd and Pb spherical powder as its second stage regenerator. Unaltered, it typically produces about 0.5 W of cooling at 4.5 K.

The refrigerator was operated with the ribbon regenerators in the second stage at cycle speeds that ranged from 30 to 144 cycles per minute. The gas pressures were set nominally to 2.0 Mpa (300 psig) on the inlet and 0.69 Mpa (100 psig) at the outlet. Heat could be applied to each stage of the refrigerator to measure the cooling power. The best test results for the two regenerators are presented in fig. 6. The data in this figure are for the case where no heat is applied to the first stage of the refrigerator. Regenerator #1 performed considerably better than regenerator #2. In addition, regenerator #2 had to be operated at abnormally low refrigerator cycle speeds (the data shown in fig. 6 was taken at 55 cycles per minute) to reach temperatures below 5 K. The best performance for regenerator #1 occurred with the refrigerator operating at 85 cycles per minute. Also shown in fig. 6 is data for the refrigerator with an all spherical powder regenerator, these data were taken imediately before the ribbon regenerators were installed, with as few modifications to the refrigerator as possible. This is the most direct comparison of spherical powder regenerators to ribbon regenerators that we could obtain. In general, the cooling capacity afforded by regenerator #1 was measureably better than the spherical powder regenerator.

## CONCLUSIONS

This research is part of a continuing effort to improve low temperature closed cycle refrigerators. The data presented here is an update on the present status of ribbon regenerator fabricating and testing. There are more tests to be done and different ribbon thickness and spacing will be fabricated and tested. At this point, a method of producing the parallel plate regenerator geometry using Nd metal has been developed. This material has proven in the past to be quite satisfactory as a low temperature regenerator matrix material. The technique used to produce the ribbons is accurate and relative simple to perform once the proper equipment is in place. It should be possible to use this technique to produce Nd ribbon with a wide range of geometric parameters.

The testing results indicate that performance gains can be achieved with ribbon. This is indicated not only by performance gains of regenerator #1 but also by the pressure drop tests and the measurements of void fraction. The two ribbons that have been fabricated indicate that low void fractions are obtainable. The pressure drop test show that with the proper spacing of the ribbon, low pressure drops are also possible.

So far only two ribbon configurations have been tested and more work is required to optimize the configuration. It is possible that the amount of surface area for heat transfer was too low in regenerator #1 while the pressure drop in regenerator #2 was too high. Finding a balance between pressure drop, heat transfer surface area and void fraction needs to be the focus of continuing research.

In addition to cooling performance, two advantages that ribbon has over spheres is cost and reliability. While a commercial source for regenerator ribbon does not yet exist so that the actual manufacturing costs is not fully established, it does appear that producing ribbon should be less expensive than producing spheres. This is primarily due to the fact that so little Nd is wasted when producing ribbon. The waste that does occur happens because the wire drawing and rolling processes require a mechanical fixture to grip the ends of the Nd. Because of this some short end pieces are not properly formed. In contrast, in the process of manufacturing spheres only about 15 to 30% of the spheres produced are of useful size

The other advantage that ribbon has over spheres is reliability while operating in a regenerator. When using spheres, it is critical that the regenerator be constructed so that none of spheres can "leak" out. When this happens, not only does the refrigerator stop cooling properly, but serious damage can occur to the complete refrigeration system. This problem is not present when working with ribbons because of their relatively large size.

For these reasons, ribbon regenerators are a viable alternate to sphere regenerators. They should be further studied and improved upon. More effort in optimizing their size parameters should help them exceed the cooling capability of spherical powder.

## REFERENCES

1. W. Rohsenow and H.Choi, "Heat,Mass and Momentum Transfer", Prentice Hall, Englewood Cliffs (1961).

2. T.Wong, M.Rudsiak, and T.Felmley, Neodymium clad with Niobium in wire and strip form for use as a regenerator, in: "Advances in Crygenics Vol. 43", P.Kittel, ed., Plenum Press, New York (1998).

# MONOLITHIC REGENERATOR TECHNOLOGY FOR LOW TEMPERATURE (4 K) GIFFORD-MCMAHON CRYOCOOLERS

W. R. Mérida and J. A. Barclay

Cryofuel Systems Group, Institute for Integrated Energy Systems
University of Victoria, Victoria, British Columbia
V8W 3P6, Canada

## ABSTRACT

A two-stage Gifford-McMahon (GM) cryocooler has been selected to produce and maintain the low temperatures required by the superconducting magnet system in an active magnetic regenerative liquefier (AMRL). The operation of practical AMRLs requires relatively large magnetic fields (e.g., 8 T). Currently, these fields can only be produced via low-temperature superconducting magnets that typically operate at liquid helium temperatures (4.2 K).

Until a few years ago, obtaining non-zero cooling powers below 10 K was impossible with regenerative cryocoolers. This is mainly because the volumetric heat capacity of pressurized helium (the working fluid) is much higher than the corresponding value for lead (the material commonly used in low temperature regenerators). Several teams have successfully modified GM cryocoolers to reach temperatures < 4 K with specially manufactured second-stage regenerators. The authors have successfully modified a commercial two-stage GM cryogenic refrigerator to reduce its minimum, no-load temperature from 6.1 ± 0.1 K to 3.42 ± 0.05 K at a nominal operating frequency of 1.2 Hz. This was accomplished by using a monolithic regenerator that is relatively simple to manufacture. The cooling power at 4.2 ± 0.1 K was measured to be 0.43 W with zero thermal load at the first stage. The superconducting magnets in our AMRL have been designed to operate at 4.5 K. The refrigeration power available at this temperature was measured to be 0.50 ± 0.01 W with a simultaneous first-stage load of approximately 20 W at 42.8 ± 0.5 K.

## INTRODUCTION

The performance of GM cryocoolers depends to a large extent on the effectiveness of the thermal regenerators used. Conventional two-stage GM cryocoolers use lead as the second stage regenerator material. These devices can reach temperatures as low as ~6 K with zero external thermal load at the second stage.[1] Below this temperature the volumetric heat capacity of lead decreases rapidly while the corresponding values for pressurized helium increase (see Figure 1). This thermal-mass imbalance between the regenerator and the pressurized helium makes it impossible for the regenerator matrix to absorb enough heat to cool the gas sufficiently before the expansion step of the cycle.

One possible solution to the low thermal mass problem was proposed by Buschow et al. more than 20 years ago.[2] They suggested that lanthanide materials with low magnetic ordering temperatures (< 10 K) could be used as cryogenic regenerators. However, the high price of the proposed materials (e.g., Rh) prevented the commercialization of this design. Hashimoto et al. revisited Buschow's idea at the Tokyo Institute of Technology.[3] These authors developed a series of rare-earth/transition metal (RE/TM) alloys such as $(Er_{1-x}Dy_x)Ni_2$ and $Er(Ni_{1-x}Co_x)$ as potential low temperature regenerator materials. One of the most important results of this investigation was the design of a hybrid regenerator with layers of different materials. The layered structure attempted to match the temperature profile in the regenerator matrix with the relatively narrow heat capacity peaks of each material (see Figure 1).

The development of new regenerator materials such as $ErNi_{0.9}Co_{0.1}$, [4] $Er_3Co$, and $Er_{0.9}Yb_{0.1}Ni$ [5] has made it possible to build two-stage GM cryocoolers with larger cooling powers at 4.2 K. Examples of such devices include the cryocoolers developed by Kuriyama et al. [6] and Tsukagoshi et al. [7] which can provide slightly over 1W of cooling power at 4.2 K.

Recent developments in GM cryocooler technology have also included the work by Satoh et al. [8] Their device uses a $Er_3Ni/Er_{0.9}Yb_{0.1}Ni$ hybrid regenerator, and it can provide 1.5 W of cooling power at 4.2 K. The major modifications in this device consisted of optimizing the valve timing, increasing the size of the second-stage expansion volume, and using 500 g of $Er_3Ni$ and 500 g of $Er_{0.9}Yb_{0.1}Ni$ for the second-stage regenerator.

The largest cooling capacity for a two-stage GM cryocooler has been recently reported by Takashi and co-workers.[9] This cryocooler uses the same $Er_3Ni/Er_{0.9}Yb_{0.1}Ni$

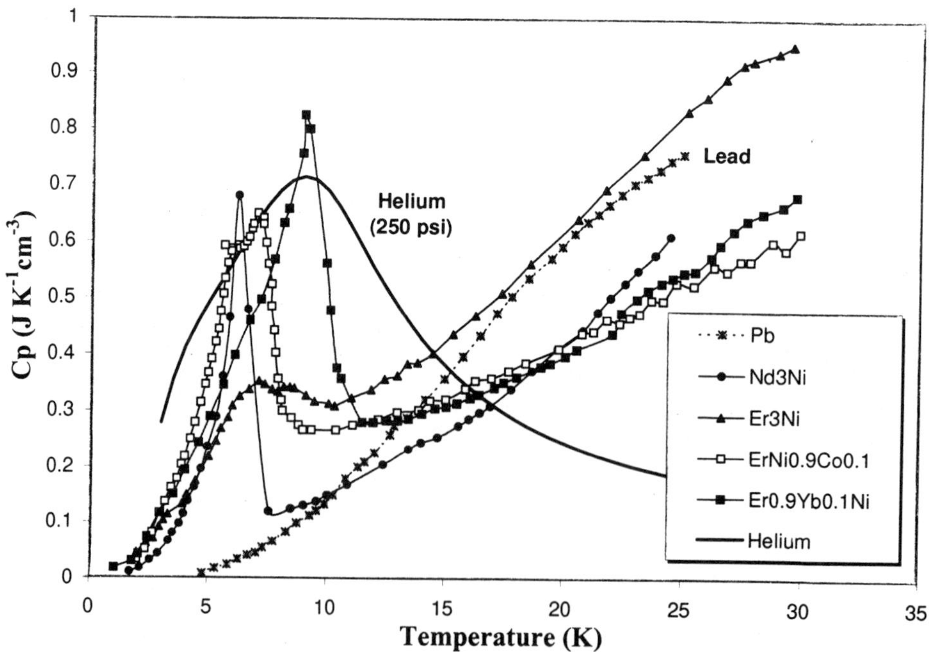


Figure 1. Volumetric heat capacities of some RE/TM alloys compared to lead and pressurized helium.

hybrid regenerator, but it can provide 2.2 W of cooling power at 4.2 K. The two major modifications required to obtain such a large cooling capacity consisted of i) increasing the expansion space volume at the second stage (the diameter was changed from 25.4 mm to 60 mm), and ii) using a multilayer regenerator separated by metallic screens placed between two felt mats.

The variety of possible regenerator materials have been developed in the last few years,[10] all use brittle, intermetallic compounds in the second stage regenerator. The brittle nature of these alloys is a serious limitation that can compromise the structural integrity of the regenerator matrix. During normal operation, the regenerator is exposed to a periodic pressure wave that can cause particle fragmentation. Unless particle disintegration is prevented, the resulting microscopic metallic powders (< 40 μm) will be carried by the moving gas thereby contaminating other parts of the system. This contamination can seriously damage the sliding seals, valving mechanism and other close tolerance components. The presence of these powders outside the regenerator produces a parasitic thermal load due to friction. In addition, small particles can fill the interstitial sites between larger ones thereby increasing the pressure drop across the regenerator and decreasing the flow of gas to the expansion spaces.

## TRADITIONAL MANUFACTURING TECHNIQUES FOR CRYOGENIC REGENERATORS

The structural limitations of RE/TM-based intermetallic compounds can be circumvented in a number of ways. One possibility consists of using pure RE metals instead of the RE/TM alloys. Pure neodymium, for example, is a ductile material that has been successfully used to replace the erbium-based alloys. Chafe et al. [11] modified a conventional GM cryocooler using this material and obtained approximately 500 mW of cooling power at 4.2 K. Although the cooling powers are modest, devices using Nd regenerators are more reliable than their Er-based counterparts.

Other approaches include the use of spherical particles to increase the tensile strength and fracture toughness of the RE/TM materials, which provides more reliable regenerators. The most common manufacturing processed used to produce spherical particles are gas and centrifugal atomization. These processes have been described elsewhere.[12, 13, 14] With this type of geometry, the regenerator beds are much more stable over time and the contamination problems are drastically reduced (spheres can withstand larger mechanical loads without breaking). Because particle disintegration is prevented, the possibility of clogging the regenerator is also minimized. However, the fabrication costs associated with spherical particles in the appropriate size range (100 to 400 μm) are high. Most of the costs are related to manufacturing techniques that require sophisticated and expensive hardware.

### Other Manufacturing Techniques and Geometries

Although beds of spherical or semi spherical particles have been the most practical choice for cryogenic regenerators, alternative geometries have received some attention recently. A Nd-based cryocooler with large cooling capacity has been recently reported. [15] The second-stage regenerator in this device uses a stack of perforated Nd plates instead of a packed bed of particles. This geometry can drastically reduce the effective porosity in a regenerator thereby increasing the thermal mass per unit volume. Chafe et al. [16] have recently reported on the use of Nd plates in the second stage regenerator of a two-stage GM cryocooler.

Wire screens have been successfully used as regenerator materials even though their porosity values (~70%) are generally higher than the corresponding values for beds of packed spheres (~40%). To improve the performance of a bed of packed screens, it is necessary to increase the mesh number. This implies that the wire itself must be drawn to very small diameters (for a standard 400 mesh screen, the wire diameter is about 25 μm). Alloys or compounds containing heavy rare earths tend to be extremely brittle and drawing them into thin wires is very difficult. Reid [17] has reported successfully drawing dysprosium metal into thin wires.

Although perforated plates and wire screens show lots of promise as effective regenerator geometries, producing thin plates or wires from RE/TM alloys will be difficult and expensive. The work described in this paper includes an alternative technique to ensure structural integrity in regenerators, while simultaneously reducing the associated manufacturing costs.

## Monolithic Regenerators

The brittleness of intermetallic compounds simplifies the production of fine, irregularly-shaped particles (by simple crushing and sieving), but also implies the need for the structural integrity considerations discussed before.

A novel manufacturing technique developed by the authors is not focused on the individual particles. Instead, the regenerator bed is considered as a whole: the main reason for particle disintegration can be related to the relative motion of individual particles and the ensuing collisions or frictional wear as they come in contact with the nearest neighbours. A monolithic bed (i.e., one in which each particle is held at a fixed position) should drastically reduce the possibility of particle disintegration. This was accomplished by bonding the entire regenerator with a diluted epoxy mixture.

After simply arc-melting of starting materials to obtain intermetallic compounds, the materials are easily ground to produce irregularly-shaped particles with an effective diameter between 200 and 300 μm. These particles are packed into a regenerator container and treated with a cryogenic cryogenic epoxy coating matrix that prevents their relative motion during operation. Three important regenerator characteristics are provided by this design:

(i) The specific surface area and total pressure drop are not compromised because only a small amount of epoxy is used,

(ii) The intrinsic brittle character of the materials is overcome by making the regenerator monolithic. Particle disintegration is eliminated producing a robust and reliable passive regenerator, and

(iii) Low values of longitudinal thermal conductivity are preserved because the point contacts between neighbouring particles do not provide a good conductive path.

The bonding of the entire regenerator into a monolithic yet porous bed makes it highly stable and resistant to structural and thermal stresses. In addition, the high cost associated with the production of spherical particles has been eliminated.

To our knowledge, this is the first time that such a technique has been used to improve the second-stage performance of a GM cryocooler. The details of this technique have been described elsewhere.[18]

## EXPERIMENTAL PROCEDURES AND RESULTS

A calibrated ruthenium oxide temperature sensor, and uncalibrated platinum resistance thermometers (PRTs) were use to measure the temperature at the second and first

Table 1. Summary of the results obtained with different regenerators at the second stage of the modified GM cryocooler

| | Regenerator Materials {size in μm}, (mass in g), [layer length in mm] | | | Stycast® 1266 Epoxy solution (A and B in g, Toluene in mL) | | | Minimum Temperature[b] (K) |
|---|---|---|---|---|---|---|---|
| | Pb {300} | $Er_3Ni$ {250-300} | $ErNi_{0.9}Co_{0.1}$ {250-300} | Part A | Part B | Toluene | $T2_{min}$ |
| REG0[a] | (191.23) [51] | – | – | – | – | – | 6.1 ± 0.1 |
| REG1 | (50.048) [12] | (65.643) [28] | (59.998) [26] | – | – | – | 3.78 ± 0.05 |
| REG2 | (60.314) [16] | (60.838) [25] | (57.952) [25] | 5.002 | 1.401 | 40.0 | 3.42 ± 0.05 |

a Original design using 300 μm lead particles only.
b At the second stage with zero thermal load.

stages of the GM cryocooler, respectively. Thermal loads were simulated by delivering a known amount of power to electrical heaters located at the first and second stages.

The minimum, no-load temperatures obtained with the GM cryocooler immediately after it was purchased were approximately 6 ± 0.1 K and 37 ± 0.5 K at the second and first stages. These results are close to the ultimate temperatures specified by the manufacturer, and to the performance of similar units using a lead particle regenerator for the second stage.

A hybrid, monolithic regenerator was built to replace the original bed of lead particles. The new bed consisted of three layers of particles made of different materials (see Table 1). The spherical lead particles used are available commercially.[19] The irregularly shaped intermetallic compound particles were produced by mechanically crushing the appropriate alloys using an automated mortar and pestle apparatus.

The new regenerator was first tested without adding any binder to the particles (REG1), and then a similar monolithic bed was built by adding a small amount of epoxy diluted in toluene (REG2). The total amount of epoxy after curing was approximately 1-2 g. The frequency of operation of the GM cryocooler was varied from 2.4 to 0.93 Hz.

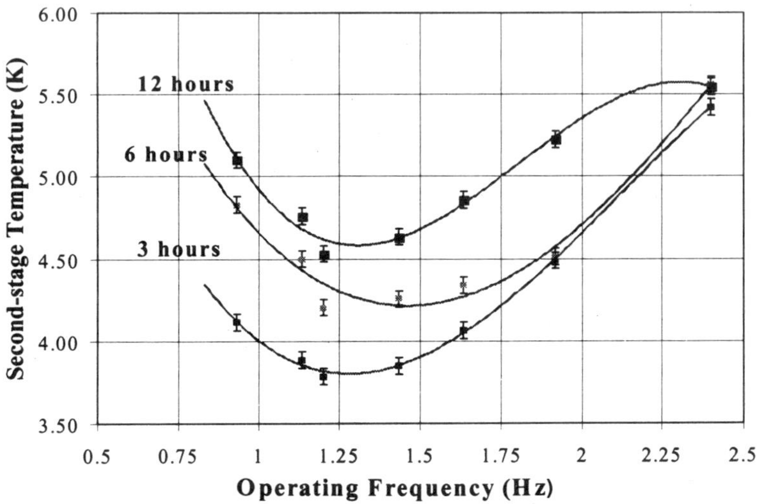


Figure 2. The performance of REG2 degraded rapidly with time. The operation of the GM Cryocooler was halted after approximately 12 hours of continuos operation.

Figure 2 shows the minimum, no-load temperatures obtained with REG1 at different operating frequencies. These temperatures became increasingly higher and, after approximately 12 hours of continuous operation, the lowest temperature had increased by almost a full degree.

To prevent possible contamination of the entire refrigerator and helium gas circuit, the operation of the GM cryocooler was halted and REG1 was detached from the displacer assembly. Inspection of the magnetic material removed from the regenerator revealed that a significant fraction of the original RE/TM materials (~ 20% by weight) had been pulverized into particles of smaller size. These powders filled the interstitial spaces between the original particles thereby obstructing the passage of gas into the second-stage expansion space. This reduction in the total gas flow resulted in a reduction in the cooling power and an increase in the minimum temperature at the second stage.

These results confirmed that, although the thermal mass of hybrid regenerator was enough to produce temperatures below 4 K, particle disintegration was still a serious limiting factor for regenerators made with brittle, irregularly shaped particles. REG2 was essentially a replica of REG1 with similar total masses for each of the materials (see Table 1). This bed of packed particles, however, was converted into a monolithic regenerator using our novel diluted epoxy technique. Figure 3 illustrates the ultimate temperature as a function of frequency.

The performance of REG2 did not degrade with time and it was still repeatable after 10 days of continuous operation. This corresponds to approximately $10^6$ cycles. The GM cryocooler was warmed and cooled again to the same final temperature. The regenerator was removed, inspected, and found to be perfectly integral. The pressure drop and other parameters were independently measured and found to be close to the values found in the original regenerator (i.e., the Pb particle bead).

## Refrigeration Power

Different cooling loads were applied to the second stage of the GM cryocooler. Figure 4 shows the temperature at the second stage as a function of cooling power.

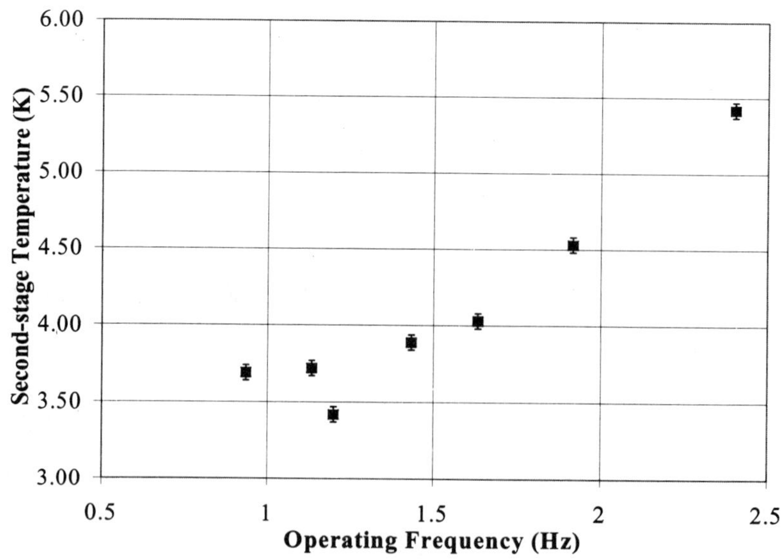


Figure 3. Measured minimum no-load temperatures obtained with REG2.

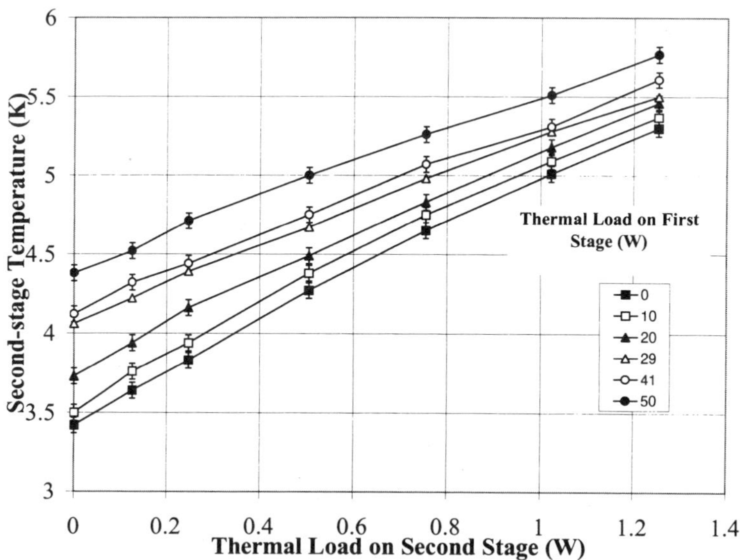

Figure 4. Measured temperature at the second stage of the GM cryocooler as a function of cooling load.

The superconducting magnets in the AMRL have been designed to operate at 4.5 K. As can be seen, the available cooling power at this temperature was measured to be 0.504 W with a simultaneous first-stage load of 20 W at 42.8 ± 0.5 K.

## CONCLUSIONS

A commercial two-stage Gifford-McMahon cryogenic refrigerator has been successfully modified to reduce its minimum, no-load temperature from 6.1 ± 0.1 K to 3.42 ± 0.05 K at a nominal operating frequency of 1.2 Hz. The cooling power at 4.2 ± 0.1 K was measured to be 0.430 W with zero thermal load at the first stage.

A novel manufacturing technique was developed to ensure the structural integrity of the second-stage regenerator without the use of spherical particles. This technique used irregularly shaped particles made of rare-earth/transition metal alloys held together by a cryogenic epoxy matrix. The bonding of the entire regenerator into a monolithic yet porous bed makes it highly stable and resistant to mechanical and thermal stresses. In addition, the high cost associated with the production of spherical particles has been eliminated.

## ACKNOWLEDGEMENTS

The authors would like to acknowledge the Natural Sciences and Engineering Council of Canada (NSERC), Centra Gas, B. C., and Natural Resources Canada for their funding of this work and the projects associated with it.

## REFERENCES

1. Riedy, R. C., "Low Temperature; High Performance G-M Refrigerator", Cryogenics, 33, pp. 653-658 (1993).
2. Buschow, K. H. J., Olijhoek, J. F and Miedema, A. R., "Extremely large Heat capacities between 4 and 10 K", Cryogenics, 5, pp. 261-264 (1975).
3. T. Hashimoto, R. Li and K. Matsumoto, "Recent Progress in the Magnetic Materials for Regenerator in the Range from 4.2 K to 20 K", Proceedings of the International Cryogenic Materials Conference, China (1988).

4. Onishi, A., Li, R., Satoh, T., Kanawaza, H., Aikawa, S., Hashimoto, T., "A 4 K GM Cryocooler with Hybrid Regenerator of Magnetic Materials", Proceedings of the 7th International Cryocooler Conference, Santa Fe, NM (November 17-19, 1992).
5. Yabuki, M., Nitta, H., Hashimot, T., Kuriyama, T., Takahashi, M., Nakagome, H., "Magnetic Regenerator material in the Liquid Helium Temperature Range — Recent Advances and its possibility", Proceedings of the 4th Joint Sino-Japanese Seminar on Cryocoolers and Concerned Topics", Beijing, P.R. China (October 19-23, 1993).
6. Kuriyama, T., Takahashi, M., Nakagome, H., Hashimoto, H., Nitta, H., Yabuki, M., "Development of 1Watt Class 4 K GM Refrigerator with Magnetic Materials", Advances in Cryogenic Engineering, 39b, pp. 1335-1342, Plenum Press, New York, USA (1994).
7. Tsukagoshi, T., Nitta, H., Yoshida, A., Matsumoto, K., Hashimoto, T., Kuriyama, T., Takahashi, M., Ohtani, Y., Nakagome, H., "Refrigeration Capacity of a GM Refrigerator with Magnetic Regenerator Materials", Cryogenic Engineering Conference, Columbus, OH, USA (July 17-21, 1995).
8. Satoh, T., Onishi, A., Li, R., Asami, H., Kanazawa, Y., "Development of 1.5W 4K GM Cryocooler with Magnetic Regenerator Material", Cryogenic Engineering Conference, Columbus, OH, USA (July 17-21, 1995).
9. I. Takashi, N. Masashi, N. Kouki and Y. Hideto, "Development of 2W Class 4 K Gifford-McMahon Cycle Cryocooler" (to be published).
10. K. A. Gschneidner, Jr., V. K. Pecharsky and M. Gailloux, "New Ternary Lanthanide Regenerator Materials for the Low Temperature Stage of a Gifford-McMahon (G-M) Cryocooler", Paper presented at the 8th International Cryocooler Conference, Vail, Colorado, USA (June 28-30, 1994).
11. Chafe, J., Green, G., Riedy, R. C., "Neodymium Regenerator Test Results in a Standard Gifford-McMahon Refrigerator", Proceedings of the 7th International Cryocooler conference, Santa Fe, NM, USA (November 17-19, 1992).
12. E. M. Ludeman and C. B. Zimm, "Production of Spherical Powders of Rare Earth Intermetallic Compounds for Use in Cryocooler Regenerators", Proceedings of the Cryogenic Engineering Conference, Huntsville, AL, USA (1991).
13. Lee, Peter W. (ed.), Powder Metallurgy, American Society for Metals, Metals Handbook Desk Edition, pp. 25.1-25.24 (1985).
14. M. G. Osborne, I. E. Anderson, K. A. Gschneidner, Jr., M. J. Gailloux, and T. W. Ellis, "Centrifugal Atomization of Neodymium and $Er_3Ni$ Regenerator Particulate", Advances in Cryogenic Engineering, 39, Plenum Press, New York, USA (1994).
15. J. N. Chafe, G. F., Green, and J. B. Hendricks, "A Neodymium Plate Regenerator for Low Temperature Gifford-McMahon Cycle refrigerators", to be published.
16. J. N. Chafe, G. Green, and P. Gifford, "The Low Temperature Performance of a Three Stage Gifford-Mcmahon Cryocooler", Advances in Cryogenic Engineering, 37B, p. 1011, Plenum Press, New York, USA (1992).
17. Reid, C. E. J., Personal communication, Department of Mechanical Engineering, University of Victoria, Victoria, BC, Canada (September 1995).
18. W.R. Mérida and J. A. Barclay, "Nouvelle Manufacturing Technique for Low Temperature Regenerators", Submitted for publication in Cryogenics.
19. Available from CLAD Metal Industries, 40 Edison Av., Oakland, NJ 07436, USA.

# ANALYSIS ON THE COLDER REGENERATOR OF 4.2K G-M REFRIGERATOR

X. D. Xu, L. Wang, L. H. Gong, and L. Zhang

Cryogenic Laboratory, Chinese Academy of Sciences
P. O. Box 2711, Beijing 100080, P. R. China

## ABSTRACT

The real gas properties of helium, especially the enthalpy change between 30K and 4.2K for the 2.0MPa helium is lower than that for the 0.6MPa helium, result in that the temperature distribution in the colder regenerator of a multistage 4.2K G-M refrigerator forms a plateau, which is measured, and also verified by a numerical calculation. Based on the calculation result, a picture of helium mass in the regenerator void is presented in this paper. According to the movement of the helium, a three-cycle assumption is suggested.

## INTRODUCTION

Since the magnetic regenerator materials were discovered at the end of eighties, great progress on G-M refrigerators has been made. The no-load temperature of a two-stage G-M refrigerator has reached 2.4 K[1], and the cooling power has reached 1.5W at 4.2 K[2]. But there are few reports on theoretical analysis and numerical calculation of 4.2 K G-M refrigerators except in[3][4]. In order to understand the thermodynamic process in more detail, in this paper, an analysis on the working process of the colder regenerator is presented on the basis of the movement of helium.

## THE WORKING PROCESS OF THE LOW TEMPERATURE REGENERATOR

### The temperature distribution characteristics of the helium in the low temperature regenerator

The G-M refrigerator cycle consists of four processes: expansion process, exhaust process, compression process, admission process. The low temperature regenerator of the 4.2K G-M refrigerator is also involved in the four processes. According to the numerical calculation of the low temperature regenerator[3][4], the results shown in Fig. 1 are obtained. Considering the practical working condition of a G-M refrigerator, calculation parameters are shown in Table 1.

Fig. 1a-d presents the temperature distribution of the helium in the low temperature regenerator during the expansion process, exhaust process, compression process, admission process respectively. Fig. 2a-d presents the temperature distribution of the helium in the conventional regenerator during the four processes respectively. Calculation parameters are also shown in Table 1. From Fig. 1 and Fig. 2 we can see that the temperature distribution of the low temperature regenerator is obviously different from that of the conventional

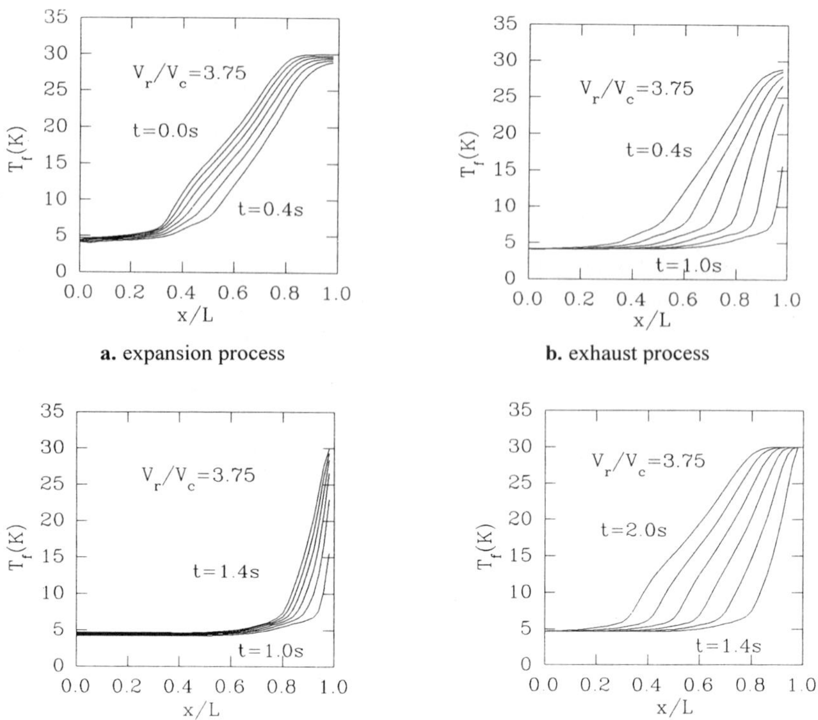


**Figure 1.** The temperature distribution of the helium in the regenerator, $Er_3Ni$ as matrix, $T_c$=4.2K, $T_h$=30K.

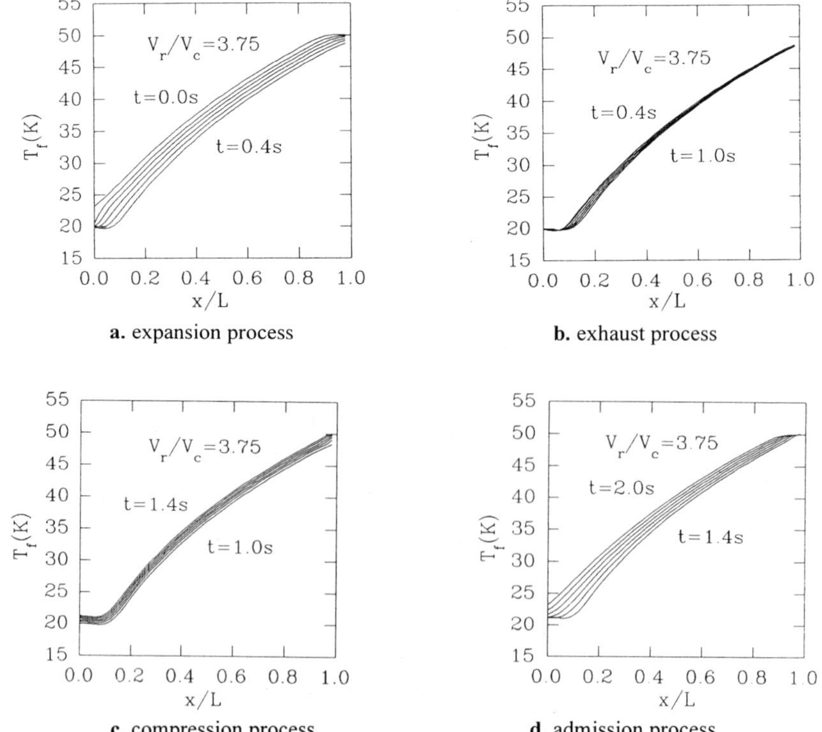


**Figure 2.** The temperature distribution of the helium in the regenerator. Pb as matrix. $T_c$=20K, $T_h$=50K.

**Table 1.** Calculation parameters of regenerators

| Parameters | Low temperature regenerator | Conventional regenerator |
|---|---|---|
| High pressure $P_h$ (MPa) | 2.0 | 2.0 |
| Low pressure $P_l$ (MPa) | 0.6 | 0.6 |
| Hot end temperature $T_h$ (K) | 30 | 50 |
| Cold end temperature $T_c$ (K) | 4.2 | 20 |
| Regenerator porosity ($\psi$) | 0.4 | 0.4 |
| Regenerator matrix | $Er_3Ni$ | Pb |
| Cycle period T (s) | 2.0 | 2.0 |
| The ratio of the total volume of the regenerator to the maximum volume of the cold chamber $V_r/V_c$ | 3.75 | 3.75 |

regenerator. A long low temperature plateau appears at the cold end of the low temperature regenerator. It is because the physical property of helium is far away from that of ideal gas at low temperature region and the disequilibrium of the isobaric enthalpy difference between 4.2K and 30K of helium appears.

## The mass distribution characteristics of the helium in the low temperature regenerator

In order to understand the working characteristics in more detail we integrated the helium in the regenerator void and obtained the mass integration distribution curves which are shown in Fig. 3. Mc is the helium mass in the cold chamber at 4.2K, 2.0MPa and Mx is the helium mass in the regenerator void from the cold end to the distance of x. L is the length of the regenerator. There are the following characteristics on the mass distribution of the helium in the low temperature regenerator: The first is the change of the total mass during the expansion and compression processes is smaller than that during the exhaust and admission processes. This is different from the conventional regenerator, in which the helium mass integration distribution is shown in Fig. 4. Table 2 presents the data of the helium mass change. The second is the helium mass distribution at the cold end of the low temperature regenerator changes less than that at the hot end. For instance, during the admission process, at the cold end the mass change $(M_{x=0.3}-M_{x=0.0})_{t=2.0s}-(M_{x=0.3}-M_{x=0.3})_{t=1.4s}=0.0$, and at the hot end the mass change$(M_{x=1.0}-M_{x=0.7})_{t=2.0s}-(M_{x=1.0}-M_{x=0.7})_{t=1.4s}=-0.15Mc$. This is mainly caused by the special temperature distribution of the low temperature regenerator.

## The flow characteristics of the helium in the low temperature regenerator

In Fig. 3 the point A presents the helium mass point between the regenerator and the cold chamber at the beginning of the expansion process and the point B presents the helium mass point between the regenerator and the hot chamber at the end of the exhaust process. Part AB is the helium between A and B. During the expansion process the part AB helium expands from $A_1B_1$ to $A_2B_2$. During the exhaust process the part AB helium moves from $A_2B_2$ to $A_3B_3$. During the compression process the part AB helium is compressed from $A_3B_3$ to $A_4B_4$. During the admission process the part AB helium moves from $A_4B_4$ to $A_1B_1$. Thus, the cycle of the part AB helium is finished and the part AB helium (about 0.2Mc) only oscillates within the regenerator and does not leave it.

However, during the expansion and exhaust processes, when the part AB helium moves from $A_1B_1$ (shown in Fig. 3a) to $A_3B_3$ (shown in Fig. 3b), the helium (about 1.0Mc seen from Fig. 3b) on the left of the part AB flows from the cold chamber to the regenerator. During the compression and admission processes, when the part AB helium moves from $A_3B_3$ (shown in Fig. 3c) to $A_1B_1$ (shown in Fig. 3d), the helium on the left of the part AB flows from the regenerator to the cold chamber. Therefore, the helium on the left of the part AB only flows between the cold chamber and the regenerator during the whole refrigeration cycle.

It can also be seen in Fig. 3 that the helium (about 0.6Mc seen from Fig. 3d) on the right of the part AB only flows between the regenerator and the hot chamber during the whole refrigeration cycle.

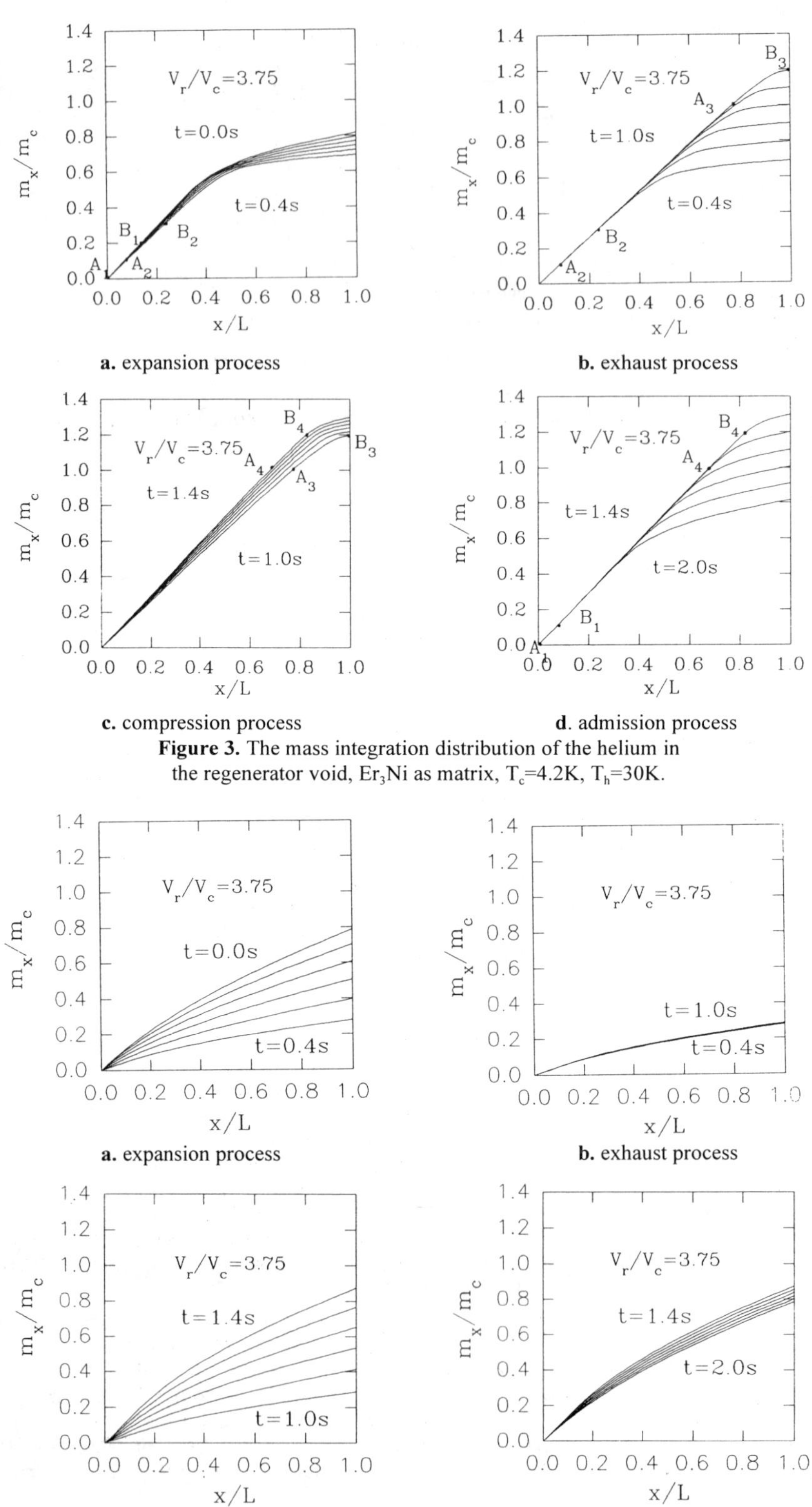


**Figure 3.** The mass integration distribution of the helium in the regenerator void, $Er_3Ni$ as matrix, $T_c$=4.2K, $T_h$=30K.

**Figure 4.** The mass integration distribution of the helium in the regenerator void, Pb as matrix. $T_c$=20K, $T_h$=50K.

**Table 2.** The changes of the total helium mass in the regenerator void

| Working processes | Low temperature regenerator | Conventional regenerator |
|---|---|---|
| Expansion process ($M_{Lt=0.0\sim0.4}$) | 0.12Mc | 0.50Mc |
| Exhaust process ($M_{Lt=0.4\sim1.0}$) | 0.52Mc | 0.00Mc |
| Compression process ($M_{Lt=1.0\sim1.4}$) | 0.10Mc | 0.60Mc |
| Admission process ($M_{Lt=1.4\sim2.0}$) | 0.50Mc | 0.10Mc |

According to the region of the movement of the helium, the helium working in a 4.2K refrigerator can be divided into three parts:

a. The part (1.0Mc) moving between the cold chamber and the regenerator.
b. The part (about 0.2Mc) moving only in the regenerator.
c. The part (about 0.6Mc) moving between the hot chamber and the regenerator.

By comparison with the 4.2K G-M refrigerator, almost all of the helium working in a conventional G-M refrigerator moves between the cold chamber and the hot chamber through the conventional regenerator, shown in Fig. 4.

## DISCUSSION AND ANALYSIS

### The cycle characteristics of the helium in the low temperature regenerator

According to the characteristics of the movement of the helium in the low temperature regenerator, the whole thermodynamic cycle can be divided into three parts.

**The "cold end" cycle** The helium in this cycle moves between the cold chamber and the regenerator. In this cycle the most of the helium remaining in the cold chamber is made a Simon expansion and all of the helium is compressed in the regenerator. This cycle produces the cooling capacity. Compared with this, Fig. 4b shows that there is much less helium still remaining in the conventional regenerator at the end of the exhaust process and most of the helium producing cooling capacity mainly flows through the regenerator and is compressed at the hot end.

**The "gas piston" cycle** The helium in this cycle oscillates only within the regenerator. The expansion, regeneration, compression processes of the cycle are all completed in the regenerator. The helium during the expansion and compression processes also exchanges heat with the matrix. The cycle pumps heat to higher temperature region and drives the "cold end" cycle. This part of the helium is like a gas piston, so the cycle is called "gas piston" cycle.

**The "hot end" cycle** The helium in this cycle moves between the hot chamber and the regenerator. In this cycle about 0.4Mc helium (on the right of $B_2$ in Fig. 3a) expands and about 0.1Mc helium (on the right of B4 in Fig. 3c) is compressed in the regenerator. About 0.5Mc helium is compressed in the hot chamber. This cycle also pumps heat to higher temperature region and drives the "cold end" cycle and "gas piston" cycle. Most of the helium in the cycle is compressed in the hot chamber and then goes into the regenerator.

The three cycles are coupled by the heat exchange between the helium and the matrix. Thus, the function of the pumping heat upward during the whole refrigeration cycle is realized.

### The function of the low temperature regenerator

From the working process of the low temperature regenerator it can be seen that the regenerator has not only the function of regeneration but also that of a compressor and a expander. It means the regenerator being as the compressor of the cold end cycle and the expander of the hot end cycle.

## CONCLUSIONS

1. The cycle of the helium in the low temperature regenerator is divided into three cycles: the "cold end" cycle, the "gas piston" cycle and the "hot end" cycle;
2. The regenerator have not only the function of regeneration but also that of a compressor and a expander. That is, as the compressor of the cold end cycle and the expander of the hot end cycle.

## ACKNOWLEDGMENT

This research was supported by the National Natural Science Foundation of China and the National Center for Research and Development on Superconductivity.

## REFERENCES

1. T.Inaguchi etc. : "Two-stage Gifford-McMahon Cycle Cryocooler Operation at about 2K", The 6th Cryocooler, Vol.2, p. 25-28, 1990.
2. Toshimi Satoh etc. : "Development of 1.5W 4K G-M Cryocooler with Magnetic Regenerator Material" Adv. in Cryo. Eng., Vol. 41, 1996.
3. Hong Chao-shen and Xu Xiangdong: "On the unique Features of the Low Temperature Regenerator ", Proceedings of 4th Joint Sino-Japanese on Cryocoolers and Concerned Topics, p. 6-13, 1993, Beijing.
4. Hong Chao-shen and Xu Xiangdong: "On the Thermodynamic cycle of the Low Temperature G-M Refrigerator", Proceedings of ICEC15, 1994, Genova, Italy.

# MEASUREMENT OF HEAT CONDUCTION THROUGH STACKED SCREENS*

M. A. Lewis, T. Kuriyama,[1] F.Kuriyama,[2] and R.Radebaugh

National Institute of Standards and Technology
Boulder, Colorado, USA 80303

[1]Guest Researcher from Toshiba Co., Japan

[2]Guest Researcher from EBARA Research Co., Japan

## ABSTRACT

This paper describes the experimental apparatus for the measurement of heat conduction through stacked screens as well as some experimental results taken with the apparatus. Screens are stacked in a fiberglass-epoxy cylinder, which is 24.4 mm in diameter and 55 mm in length. The cold end of the stacked screens is cooled by a Gifford-McMahon (GM) cryocooler at cryogenic temperature, and the hot end is maintained at room temperature. Heat conduction through the screens is determined from the temperature gradient in a calibrated heat flow sensor mounted between the cold end of the stacked screens and the GM cryocooler. The samples used for these experiments consisted of 400-mesh stainless steel screens, 400-mesh phosphor bronze screens, and two different porosities of 325-mesh stainless steel screens. The wire diameter of the 400-mesh stainless steel and phosphor bronze screens was 25.4 μm and the 325-mesh stainless steel screen wire diameters were 22.9 μm and 27.9 μm. Standard porosity values were used for the experimental data with additional porosity values used on selected experiments. The experimental results showed that the helium gas between each screen enhanced the heat conduction through the stacked screens by several orders of magnitude compared to that in vacuum. The conduction degradation factor is the ratio of actual heat conduction to the heat conduction where the regenerator material is assumed to be a solid rod of the same cross-sectional area as the metal fraction of the screen. This factor was about 0.1 for the stainless steel and 0.022 for the phosphor bronze, and almost constant for the temperature range of 40 to 80 K at the cold end.

---

## INTRODUCTION

Stacked screens are commonly used as a cryocooler regenerator for the approximate temperature range between 80 and 300 K.[1] Because of the large temperature gradient in the regenerator, heat conduction through the stacked screens is a significant loss. Some experimental studies were reported in which an apparent thermal conductivity for stacked screens was estimated from electrical conductivity.[2,3] However, in an actual regenerator, the working fluid of helium gas can transport a large fraction of the heat between screens. As far as we know, only Schumann and Voss investigated experimentally the thermal conductivity of packed beds without fluid flow.[4] Both experimental and analytical research data for heat conduction loss, however, are not available for practical use in the design of a cryocooler regenerator. The purpose of this study is to directly measure the heat conduction loss from room temperature to cryogenic temperatures through stacked screens.

## EXPERIMENTAL APPARATUS AND PROCEDURE

Figure 1 shows the experimental apparatus used for the study. The apparatus consists mainly of a test section, a two-stage Gifford-McMahon (GM) cryocooler, a heat flow sensor, and a vacuum vessel (not shown in Fig. 1).

The details of the test section are shown in Fig. 2. Two identical regenerators are used in this apparatus. Regenerator cylinders are made of fiberglass-epoxy, with an inner diameter of 24.4 mm and a length of 55 mm. The wall thickness of each cylinder is 1 mm. Heat conduction along the length of the cylinder wall of a single regenerator from room temperature to 80 K is estimated to be 0.21 W from published thermal conductivity data.[5] Stainless steel screen of 400-mesh with 0.0254 mm wire diameter, 325-mesh with 0.0279 and 0.0229 mm wire diameters, and phosphor bronze 400-mesh with 0.0254 mm diameter wire were used in the study. The screens were stacked in the regenerator cylinder to a height of approximately 45 mm. Helium gas lines are connected to the regenerator to change the filling pressure in the regenerators. The pressure can be varied from vacuum to 2.0 MPa. Multi-layer insulation was wrapped around the test section to reduce radiation heat loss.

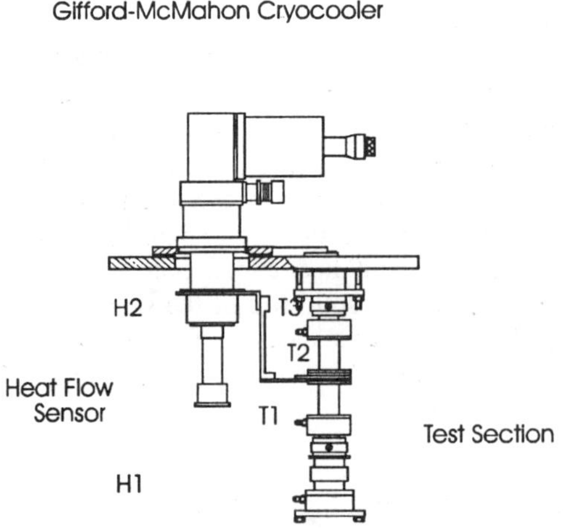


**Figure 1.** Experimental test apparatus.

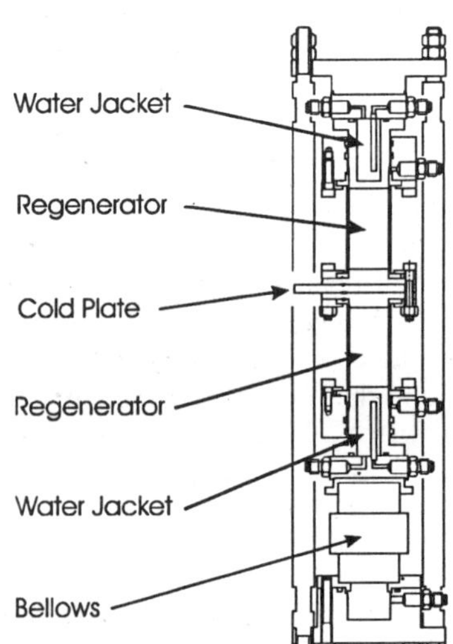


**Figure 2.** Experimental test section.

**Table 1.** Characteristics of thermal conduction screens.

| Material | Stainless Steel | Phosphor Bronze | Stainless Steel | Stainless Steel |
|---|---|---|---|---|
| Mesh | 400 | 400 | 325 | 325 |
| Wire Dia. (μm) | 25.4 | 25.4 | 27.9 | 22.9 |
| Theor. Porosity | 0.6858 | 0.6858 | 0.7703 | 0.6858 |

The cold ends of both regenerators are connected to a cold plate, which is cooled by the GM cryocooler by the heat flow sensor. The hot ends of both regenerators are capped by piston-shaped water jackets. Flowing water maintains the hot end temperature at room temperature. A bellows is attached to the lower water jacket. Changing the filling pressure of the helium in the bellows can move the lower water-cooled piston. The cold plate and the two regenerators are free to move with respect to the water jackets, so the force exerted by the bellows is applied equally to the two regenerator screen stacks.

The heat flow sensor is mounted between the cold plate and the first stage of the GM cryocooler. A flexible thermal link between the cold plate and the heat flow sensor allows for movement of the cold plate when the bellows pressure is changed. The heat flow sensor consists of a copper bar and two silicon diode thermometers. The copper bar is made of oxygen-free copper with a cross-sectional area of 72 $mm^2$ and a length of 135 mm. The distance between the two thermometers is 91.3 mm. The relationship between heat flow through the copper bar and temperature difference was calibrated before these experiments by using a heater attached to the cold end.

The experimental procedure is as follows. After pumping the vacuum vessel, the two-stage GM cryocooler is turned on. Both the cold plate and the heat flow sensor are cooled by the first stage of the GM cryocooler. The cold plate temperature is kept at a constant temperature by the temperature controller using a silicon diode thermometer at the cold plate and an electric heater mounted on the GM cryocooler first stage. The cold plate temperature can be varied from 40 to 80 K, and the temperature stability within ±0.05 K. Once the cold plate temperature is set and the temperature difference at the heat flow sensor is measured, an additional heat load is supplied to a heater mounted on the cold plate, and its effect on the temperature difference at the copper bar is measured. The heat loads, the temperature differences, and the calibration curve obtained previously are used to calculate the heat flow through the heat flow sensor. The calculated heat flow here includes the heat conduction through both stacked screens, the two fiberglass-epoxy cylinders, and other heat losses, such as radiation loss and heat conduction loss through instrument wires. In a separate experimental run, the heat flow through the regenerators without stacked screens was measured to provide information needed to determine the heat flow through the stacked screens only.

Characteristics of the screens tested in this study are given in Table 1. The theoretical porosity, for each screen shown in this table is calculated from a standard simplified equation $1\text{-}n_g = (\pi/4)(d/s)$, where $n_g$ is the theoretical porosity, $d$ is the wire diameter and $s$ is the spacing from one wire to the next. We have found from experience in packing regenerators that this porosity value is close to the experimental value whenever the screens are tamped into place a few at a time. It is a close approximation to porosity achieved when each screen layer occupies a height equal to twice the wire diameter.

In this study the thermal conductance was measured as a function of screen porosity, this ranged over lower values and higher values than the theoretical calculations. Screen porosity can be varied in two ways. Method 1 or condition 1 involves tamping the screens into the regenerator tube or using a hydraulic press with a plunger to press the screens into place with a known force. With this method, no residual force remains on the screen packing from the outside. Method 2 or condition 2 makes use of the plunger attached to the bellows in the experimental apparatus. With pressure applied to the bellows, a proportional pressure applied to the screen remains applied during the thermal conduction measurements. In the measurements reported here an applied pressure of 5.1 MPa on the screen stack was used in method 2. When the pressure in the bellows is released, the screen packing expands back to a slightly higher porosity, but yet the porosity is lower than that which existed before the application of pressure. With the pressure released, the screen packing reverts back to condition 1 where no pressure is applied during the measurements. Most measurements reported here are with the screen in condition 2, although a few measurements were made with the screen in condition 1 to investigate the effects of pressure being applied to the screen during measurements. The helium gas working fluid around the screens contributes to the heat flow from one layer to the next. For that reason measurements were made with the helium gas pressures ranging from vacuum up to 2 MPa in order to determine the contribution due to the presence of helium gas in the screen packing. Cold end temperatures were controlled at 40, 60, and 80 K to provide information on the temperature dependence of the conductivity degradation factor.

## EXPERIMENTAL RESULTS AND DISCUSSION

The first measurement was of the heat leak through the two regenerators with no screens inside and with vacuum inside the regenerator tubes. The cold plate temperature was 80 K and the hot plate temperature was 285 K. The measured total heat leak was 0.57 W, with 0.42 W the calculated heat conduction through the cylinder wall of the two regenerators. The 0.57 W of heat flow was then subtracted from subsequent measurements with screen in the regenerators to obtain the heat conduction through the screen.

The effect of helium gas pressure inside the regenerator on the total measured heat leak is shown in Figure 3. These measurements were made with the cold end held at 80 K. As this figure shows, the heat leak increases rapidly with increasing helium pressure until there is little significant change for pressures above 0.5 MPa. Figure 4 shows the pressure dependence of mean free path and thermal conductivity of helium gas at 80 K, 200 K, and 300 K. Though the thermal conductivity of bulk helium at pressures to 5 MPa is almost independent of pressure, the heat leak varies with pressure conditions at pressures below 0.5 MPa. In order to understand the test results, we discuss the heat transfer mechanism in the regions of molecular flow, viscous flow, and intermediate flow. Generally gas flow can be treated as molecular flow when the mean free path $\ell$ is larger than 10 times the distance $d$ between two solid plates. On the other hand, when $\ell$ is less than 0.01 $d$, the gas flow can be treated as viscous flow. In the viscous flow region, heat transfer between two solid plates is proportional to the thermal conductivity and is independent of pressure for an ideal gas. In the molecular flow region, the heat transfer is proportional to the gas pressure. Figure 3 shows that at approximately 0.5 MPa the thermal conduction through the helium gas has reached the viscous flow region where the mean free path of the helium atoms has become less than 0.01 $d$. Because of the temperature range variations from 80 to 285 K along the length of the regenerator, the mean free path of helium at 0.5 MPa according to Figure 4 varies from 0.010 to 0.039 μm. Therefore, the effective distance between each

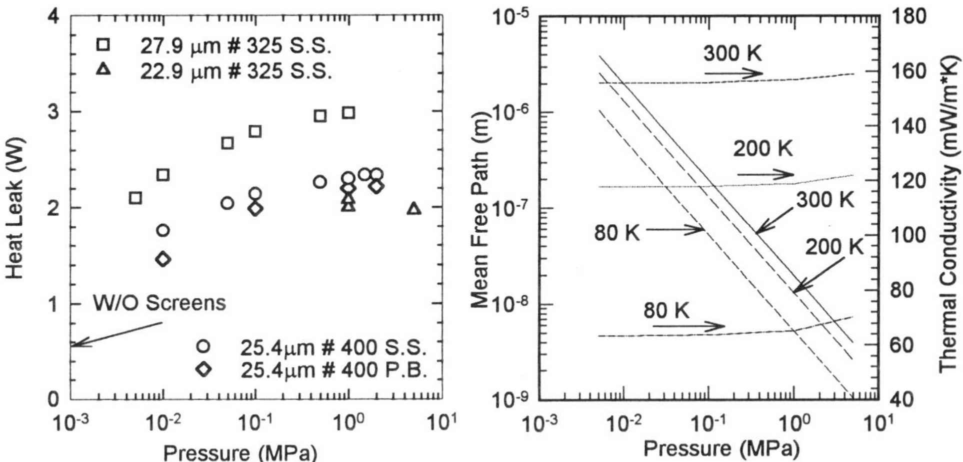


**Figure 3.** Heat leak vs. helium gas pressure.

**Figure 4.** Mean free path and thermal conductivity vs. pressure dependency over temperature ranges.

screen layer contributing to the heat flow is about 1 to 4 μm, or 100 times the mean free path. Since the wire diameter is in the range of 25 to 28 μm's, most of the heat is transported by the helium gas in a region very close to the individual contacts between wires. According to Fig. 3, the heat leak at very low helium gas pressure approaches a value very near the background value. Such behavior indicates that there is very little electronic heat leak conduction associated with direct metallic contact between the wires, in agreement with electrical resistance measurements.[2,3] The electrical resistance measurements[2,3] indicate a heat conduction of about 4 mW for our system.

Because of the many layers in a screen stack, the thermal conduction through the stack is reduced compared to a solid bar of the same material and same cross-sectional area as the metal in the screen. Therefore to estimate this heat leak through the stacked screen, a conduction degradation factor f is applied. The actual conduction through a stacked screen bed is then given as

$$\dot{Q}_c = f_c(A_t/L)(1-n_g)\int_{T_c}^{T_h} kdT, \tag{1}$$

where

$\dot{Q}_c$: Heat flow through stacked screens
$f_c$: Conductivity degradation factor
$A_t$: Total cross-sectional area of screens
$L$: Length of stacked screens
$n_g$: Porosity of stacked screens
$T_c$: Temperature at cold end of regenerator
$T_h$: Temperature at hot end of regenerator
$k$: Thermal conductivity of regenerator matrix material

Figures 5 and 6 show how the conductivity degradation factor (CDF) varies with the cold end temperature for the 400-mesh stainless steel and the 400-mesh phosphor bronze screen at the theoretical values of the porosity. The values of 0.1 and 0.022 for the degradation factors of stainless steel and phosphor bronze material, respectively, represent

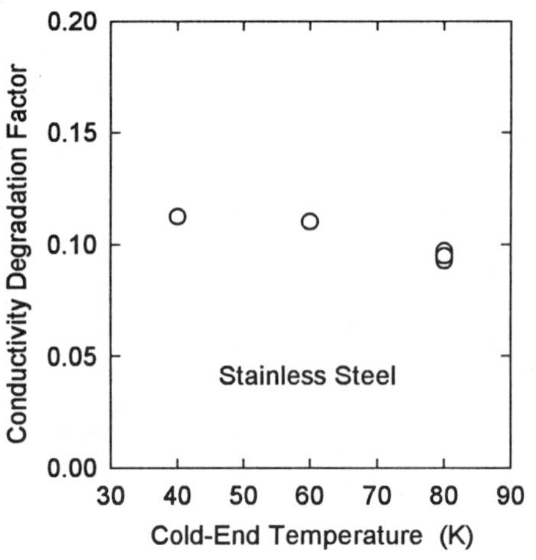


**Figure 5.** Temp vs. CDF for stainless steel.

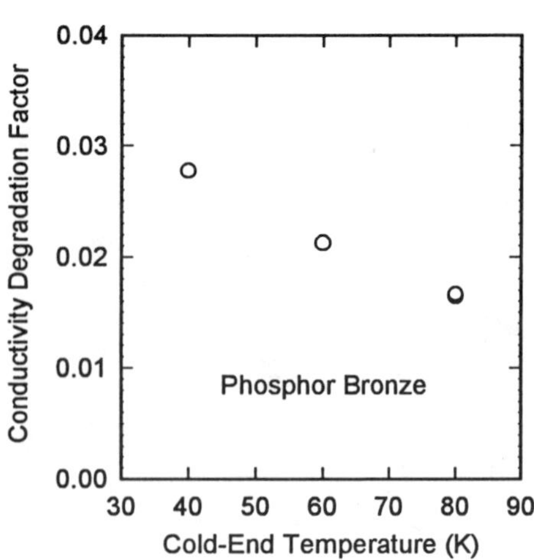


**Figure 6.** Temp vs. CDF for phosphor bronze.

our data well over the cold-end temperature range of 40 to 80 K. Figure 7 show the conductivity degradation factor for stainless steel 400-mesh, 25.4 μm wire; 325-mesh 27.9 μm wire; and 325-mesh, 22.9 μm wire over a wide porosity range. These porosities were obtained using method 2 with a static load. These data indicate that a value of 0.1 is a good representation of the conductivity degradation factor for the stainless steel screen material. This value is significantly less than the value of 0.3 that we have been using in the past for regenerator optimization, which was based on a very crude measurement made in our laboratory several years ago. The much lower value will have a significant impact on the optimization of regenerator geometry and can lead to a reduction in regenerator loss that is significantly larger than the reduced thermal conduction. Figure 8 shows the conductivity

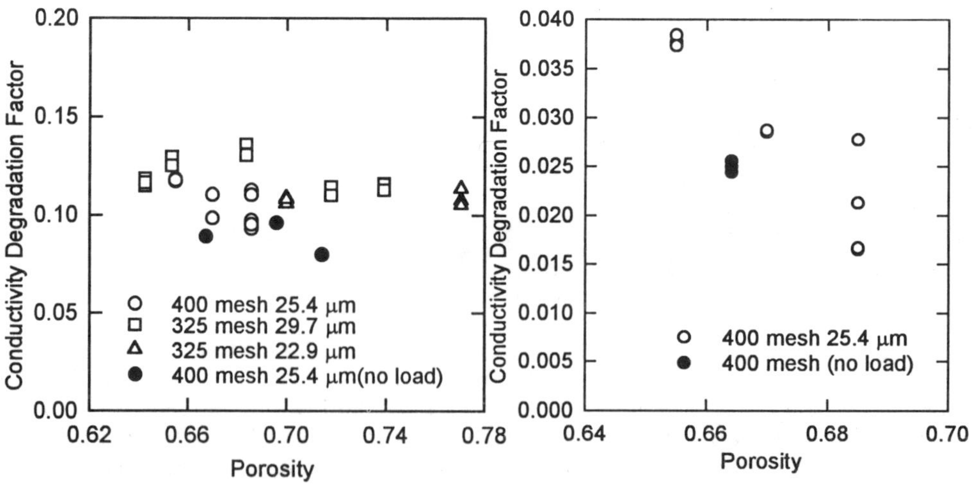


**Figure 7.** Porosity vs. CDF for stainless steel.

**Figure 8** Porosity vs. CDF for phosphor bronze.

degradation factor for 400-mesh phosphor bronze screen material as a function of porosity range. Both methods for varying the porosity are represented in Fig. 8. The application of a static load of 5.1 MPa increases the conduction by about 24 %. Although these data have more variability, a close approximation for the conductivity degradation factor for the phosphor bronze material can be achieved. For the theoretical value of porosity with phosphor bronze screen, we find a conductivity degradation factor of about 0.016 with no applied load and 0.022 with an applied load of 5.1 MPa.

## CONCLUSIONS

The heat conduction loss through stacked screens from room temperature to cryogenic temperature was measured experimentally. The experimental apparatus allows for a change in the screens, the porosity, the cold end temperature, and the helium gas pressure in the regenerator. The measurements were performed using stainless steel and phosphor bronze screen material of different mesh sizes and a wide range of porosities.

The experimental results showed that the helium gas between each screen play an important role in transporting the heat. The heat conduction through the stacked screens was enhanced by helium gas by several orders of magnitude compared to vacuum conditions inside the regenerator. The heat conduction reached a constant maximum value for helium pressures above 0.5 MPa. From a knowledge of the helium mean free path, this result indicates that most of the heat is transported a distance of the order of 3 μm from one wire to the next. There was little difference in the heat conduction between stainless steel and phosphor bronze at the theoretical porosity values. Such results indicate that the thermal resistance of the helium conduction path dominates the thermal resistance within the wire, but that it is much lower than the thermal resistance due to metallic contact between wires.

The conduction degradation factor, which is the ratio of actual heat conduction to the heat conduction where the regenerator material is assumed to be bulk, was about 0.1 for the stainless steel screen materials. This factor was relatively constant for the temperature range between 40 and 80 K at the cold end location and over the range of screen porosities tested. For the phosphor bronze material at the theoretical value of porosity, the conduction degradation factor was about 0.022 with a more pronounced negative slope with respect to cold end temperature range and screen porosity. The increased degradation factor for low porosity phosphor bronze screen may be due to its increased deformation, which leads to a greater area within about 3 μm of an adjacent wire. These new conductivity degradation factors for calculating the thermal conduction through stacked screen material gives valuable information for regenerator optimization. With the NIST regenerator optimization software REGEN3.1,[6,7] calculating improved coefficient of performance values for regenerators will be useful in designing more efficient pulse tube refrigerators for future applications.

## ACKNOWLEDGMENT

This study was funded in part by EBARA Research Co., Ltd. and by NASA Life Sciences Division.

## REFERENCES

1. Walker, G., Crycoolers, Plenum Press, New York (1983)
2. Organ, A.J., "Conductivity for quantifying regenerator thermal shorting," to be published.

3. Lee, A.C., "Contact resistance of 500 mesh regenerator screens," Cryogenics, Vol 34, No. 5 (1994), pp. 451-456.
4. Schumann, T.E.W., and Voss, V., "Heat Flow Through Granulated Material," Fuel, Vol. 13 (1934), pp. 249-256.
5. Takeno., "Thermal and Mechanical Properties of Advanced Cryogenic Materials at Low Temperatures," Cryogenic Engineering, Journal of the Cryogenic Society of Japan, Vol. 12, No. 3 (1986), pp. 182-187.
6. Gary, J., Daney,D.E., and Radebaugh, R., "A Computational Model for a Regenerator,"Proc. Third Cryocooler Conference, NBS Special Publication 698, (1985), p. 199.
7. Gary, J. and Radebaugh, R., "An Improved Numerical Model for Calculation of Regenerator Performance (REGEN3.1)," Proc. Fourth Interagency Meeting on Cryocoolers, David Taylor Research Center, Report DTRC-91/003, (1991), p. 165.

# PRESSURE DROP OVER REGENERATORS IN OSCILLATING FLOW

B. P. M. Helvensteijn,[1] A. Kashani,[1] A. L. Spivak,[1]
P. R. Roach,[2] J. M. Lee,[2] and P. Kittel[2]

[1] Atlas Scientific
Sunnyvale, CA 94086

[2] NASA, Ames Research Center
Moffett Field, CA 94035-1000, USA

## ABSTRACT

Modeling the performance of pulse tube coolers and stirling coolers involves estimating the pressure drop over the regenerator. Such pressure drop estimates are generally based on steady flow correlations which do not necessarily apply to oscillating flow. The present paper contains experimental data on the pressure drop over regenerators subjected to an oscillating flow of helium gas at 300K and a charge pressure of 11 atm. Each regenerator tested consists of a thin wall stainless steel tube packed with just one of the following materials: stainless steel screen of mesh-size 250, 300 or 400, or stainless steel felt made of wires 12 μm or 30 μm in diameter. The oscillating flow is established by means of a linear compressor (15 cc) operating between 40 Hz and 70 Hz connected to the inlet of the regenerator. The mass flow is derived from the measured pressure oscillations in a reservoir connected to the regenerator outlet. A differential pressure gauge is applied over the regenerator. Heat exchangers around and in-line with the regenerator stabilize its temperature. The pressure drop and mass flow data are converted into correlations for the friction factor as a function of Reynolds number at given void fraction and frequency. Steady flow measurements have been applied as a reference in order to note the extent to which the oscillations contribute to the pressure drop.

## INTRODUCTION

Performance predictions are as central to the design of pulse tube coolers (PTCs) as to the design of any advanced instrumentation. Computational models[1,2] for the PTC readily become very complex due to the intricate interplay between mass flow and heat transfer in the apparatus. Nearly independent of the model's complexity is the need for mass flow correlations that apply to the regenerator. The work by Kays and London[3] (K&L) is often cited as the source for such a correlation, however, K&L considered steady flow measurements only. The extent to which oscillating flow agrees with the observed steady flow behavior has not yet been studied in detail. The present work is an attempt at closing this gap in the database.

This paper discusses mass flow impedances of stainless steel screen and stainless steel felt regenerators in terms of the relation between the Fanning friction factor (fF) and the Reynolds number (Re). Our interest in measuring the pressure drop over regenerators spans both steady flow and oscillating flow. Steady flow (dc) measurements are needed in order to enable comparison with the K&L benchmark while such data will also serve as a reference frame for analyzing oscillating flow (ac).

The discussion will commence with an explanation of the various test facilities used in this work. After that, the data processing including the conversion from pressure data to friction factors follows. The results thus obtained lead to the sought-after ac to dc comparison. Finally, the deviations from steady flow observed for oscillating flow are correlated with the flow excitation (maximum Re).

## EXPERIMENT CONFIGURATIONS

Two experimental setups have been used to collect steady flow data. Figure 1a is a sketch of our initial steady flow experiment (SFE). A cylinder of compressed helium gas is used to drive the flow. The gas flows through the flowmeter and the regenerator while two differential pressure tranducers (100 Torr; 1000 Torr) determine the pressure drop. During an experiment the absolute pressure upstream of the device is also measured. The helium gas leaving the regenerator is vented to air. This setup provides a straightforward result of regenerator pressure drop versus mass flow which is then converted to a fF(Re) relation.

A variation on the SFE is the ramp flow experiment (RFE; Fig. 1b). This setup allows short bursts of high mass flow in the regenerator at pressures up to 10 atm. The automatic valve (A) is closed at first. The 1 liter reservoir is evacuated while the regenerator and the 11.4 liter buffer volume are pressurized. Then the valve A is opened and gas starts flowing through the orifice (at A) into the 1 liter reservoir. The reservoir contains copper wool which keeps it isothermal until the pressure reaches 1 atm. At this pressure a vent valve activates that opens the vessel to air. The mass flow is derived from the change in pressure in the receiving 1 liter vessel. A fast-response high-pressure differential pressure

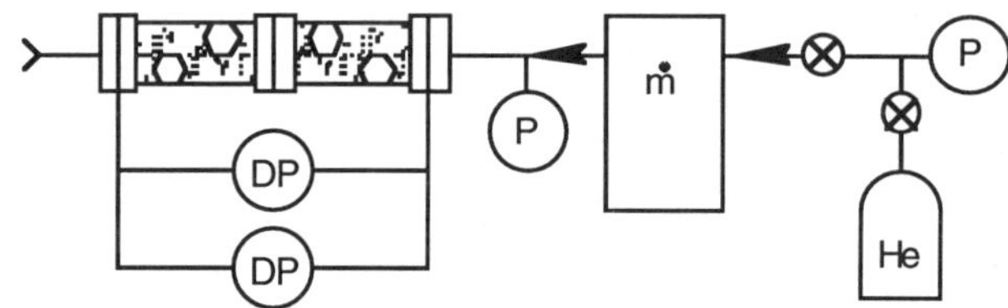


**Figure 1a.** Steady Flow Experiment (SFE).

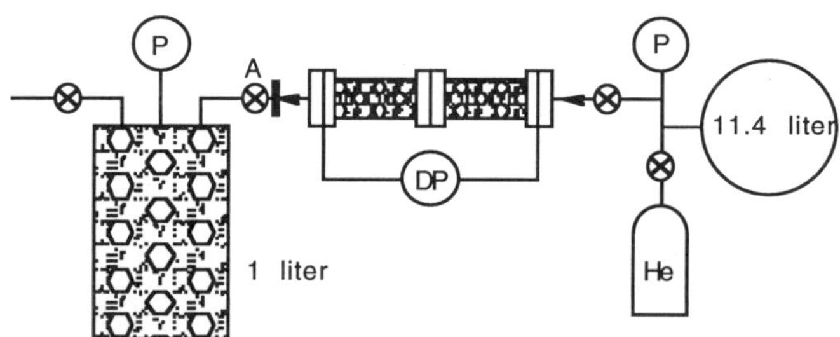


**Figure 1b.** Ramp Flow Experiment (RFE).

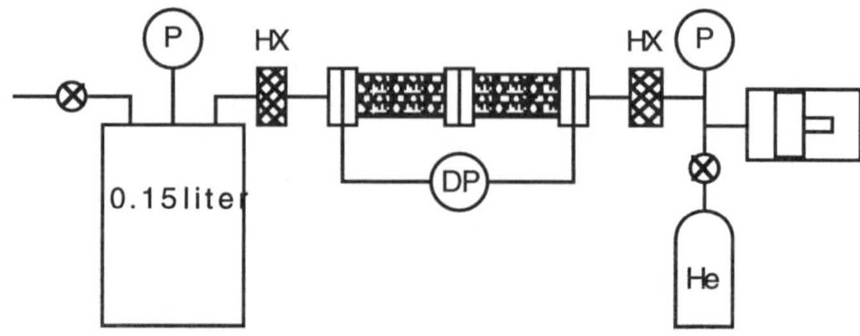


**Figure 1c.** Oscillating Flow Apparatus (OFA).

transducer determines the regenerator pressure drop. The measurements just before the vent valve opens are included in our steady flow data.

The oscillating flow apparatus (OFA) is another variation on the SFE (Fig. 1c). A charging pressure is applied to the system and is measured (≈ 11 atm absolute). Then pressure waves are created by a linear compressor (15 cc), running at 60 Hz (unless noted otherwise). The flow into and out of the regenerator occurs through heat exchangers containing 40 mesh copper screen. The side of the regenerator opposite the compressor connection is attached to a 150 cc adiabatic reservoir. The adiabatic reservoir pressure was measured with a piezo-resistive transducer. The regenerator pressure drop was measured using a strain gauge differential pressure transducer. In experiments at frequencies other than 60 Hz individual piezo-resistive devices were applied to assess the pressure drop. The relations, used to derive the mass flow from the reservoir pressure, are presented in the next section. All ac pressure sensors used in the RFE as well as the OFA are capable of reading signals at 1 kHz or faster.

Each regenerator tested consists of a thin wall stainless steel tube packed with just one of the following materials: stainless steel screen of 250, 300 or 400 mesh, or stainless steel felt with a wire-size of 12 μm or 30 μm. The tube's inner diameter is 1.49 cm; the diameter of the punched out screens is 1.54 cm (for a tight fit). Table 1 contains all further regenerator specifications. The parameters in the table are explained in the following sections.

## FRICTION FACTOR

The friction factor (fF) may be derived from the flow velocity (u) and the regenerator pressure drop ($DP_{rgn}$) using the following expression[4]:

$$fF = (D_h / 4L)\, 2\, DP_{rgn} \,/\, (\rho_{rgn}\, u^2) \tag{1}$$

where $D_h$ is the hydraulic diameter, L is the regenerator length and $\rho_{rgn}$ is the gas density inside the regenerator. The hydraulic diameter is defined as: $D_h$ = 4 (gas volume / solid surface) . Assuming the regenerator to have a void fraction $v_f$ and the matrix to consist of round wire of constant diameter d, one finds:

$$D_h = d\, v_f / (1 - v_f) \tag{2}$$

Under the stated assumptions this equation is exact for both screen and felt matrices. For the screens d was measured while for the felts nominal values were taken.

The flow velocity equals:

$$u = \dot{m} \,/\, (\rho_{rgn}\, A_{flow}) \tag{3}$$

with $\dot{m}$ being the mass flow; $A_{flow}$ is the minimum flow area, i.e. $A_{flow} = v_f\, A_{tb}$ ; $A_{tb}$ is the flow cross-section of the empty tube. For the regenerator void fraction we have used the lower of two values: the first one $v_m$ is based on the measured matrix mass and volume; the second value $v_s$ is the void fraction for the uncompressed screen. When $v_s < v_m$ there is dead space between screens, which must be eliminated from consideration. This requires a corresponding correction to the regenerator length.

The mass flow $\dot{m}$ is either directly measured (SFE) or is derived from the reservoir pressure. In deriving the mass flow two distinct cases are relevant: the isothermal reservoir and the adiabatic reservoir. In the RFE the reservoir is isothermal, its volume $vol_{rsvr}$ contains the mass: $m_{rsvr} = \rho_{ref}\, vol_{rsvr}\, P / P_{ref}$ ($\rho_{ref}$ and $P_{ref}$ are fixed reference values). Replacing the reservoir pressure P by its time derivative gives the mass flow into the isothermal reservoir. In the OFA the reservoir is adiabatic containing gas at temperature T with gas entering/exiting at temperature $T_e$ . In this case the following equation holds:

$$(1/\gamma)(dP/P) = (T_e/T)(dm/m) \tag{4}$$

(where $\gamma$ is the heat capacity ratio). Assuming $T_e \approx T$, we find: $(m / m_o) = (P / P_o)^{1/\gamma}$ (with the index "o" referring to cycle averages). As a result:

$$\dot{m}_{ad} = (dm / dt) = (m_o / P_o^{1/\gamma}) (1/\gamma) P^{(1/\gamma - 1)} (dP / dt) \quad (5)$$

The friction factor is to be considered in relation to the Reynolds number:

$$Re = \dot{m} D_h / (\mu A_{flow}) \quad (6)$$

in which $\mu$ represents the fluid's viscosity.

In order to compare the experimental results with K&L, the void fraction dependence of the K&L friction factor was characterized. The following formulation was found to adequately represent the void fraction dependence in the K&L data:

$$fF_{K\&L\ fit} = (m1 / Re) + (m2 / Re^{m3}) \quad (7)$$

for which the coefficients are:

$$m1 = 1200.8417 - (3380.7134*Re) + (2441.7295*Re*Re) \quad (7a)$$
$$m2 = 2.6490093 - (2.7316366*Re) \quad (7b)$$
$$m3 = -2.1181513 + (6.6616427*Re) - (4.945803*Re*Re) \quad (7c)$$

Taking the general trend in the K&L data, essentially eliminating the void fraction dependence, the following fitting formula has been found:

$$fF_{no\ vf} = S_{rgn} \{ (40 / Re) + (0.75 / Re^{0.1}) \} \quad (8)$$

where the "sliding factor" $S_{rgn}$ is used to attain the best fit.[#] Eqn. (8) facilitates comparisons between our ac and dc data, and K&L (Eq. (7)) as well. K&L sliding factors are established by applying Eq. (7) to the void fractions listed in Table 1 after which Eq. (8) is applied as a fitting function to the Eq. (7) results. The K&L sliding factors obtained in this manner for each of our regenerators are listed in column 6 in Table 1.

## EXPERIMENTAL RESULTS

The steady flow data were analyzed by converting the raw pressure and mass flow data into fF(Re) relations as outlined above. The results are plotted in Fig. 2. Comparison with K&L becomes feasible through curve fitting of the data using Eqn. (8) giving the steady flow $S_{rgn}$ values listed in Table 1, column 7. The agreement between the results in column 6 (K&L) and column 7 varies greatly. Considering the slight variation in void

**Table 1.** Regenerator specifications (ascending order of void fraction)

| Regenerator | $v_f$ [-] | d [μm] | $D_h$[μm] | L [cm] | $S_{rgn}$ (K&L) | $S_{rgn}$ (StFl) | 1000 * (1-C)/Re (-) | 1000 * (1-C)/Re (+) | 1000 * (1-C)/Re (<>) |
|---|---|---|---|---|---|---|---|---|---|
| 400m 5g | 0.664 | 25.4 | 50.2 | 1.15 | 0.862 | 0.840 | 15.0 | 27.2 | 21 |
| 325m 15g | 0.666 | 27.9 | 55.6 | 3.39 | 0.856 | 0.556 | 8.5 | 14.5 | 11 |
| 325m 25g | 0.669 | 27.9 | 55.6 | 5.73 | 0.846 | 0.600 | 11.1 | 18.5 | 15 |
| 325m 10g | 0.674 | 27.9 | 55.6 | 2.37 | 0.832 | 0.677 | 6.4 | 14.6 | 11 |
| 80m 10g | 0.704 | 94.0 | 224 | 2.61 | 0.801 | 0.800 | – | – | – |
| 80m 5g | 0.719 | 94.0 | 241 | 1.36 | 0.817 | 0.925 | – | – | – |
| 250m 9g | 0.725 | 35.6 | 93.5 | 2.36 | 0.830 | 0.930 | 5.3 | 14.8 | 10 |
| 12f 5g | 0.827 | 12.0 | 57.3 | 2.96 | 1.269 | 0.727 | 12.1 | 34.1 | 23 |
| 30f 5g | 0.836 | 30.0 | 153 | 2.36 | 1.664 | 1.165 | 4.9 | 14.7 | 10 |

[#] Before executing such a fit, both sides of Eq. (8) are multiplied with Re. This approach allows the fitting routine to handle the wide range in friction factors so as to attain an appropriately fitted value for $S_{rgn}$.

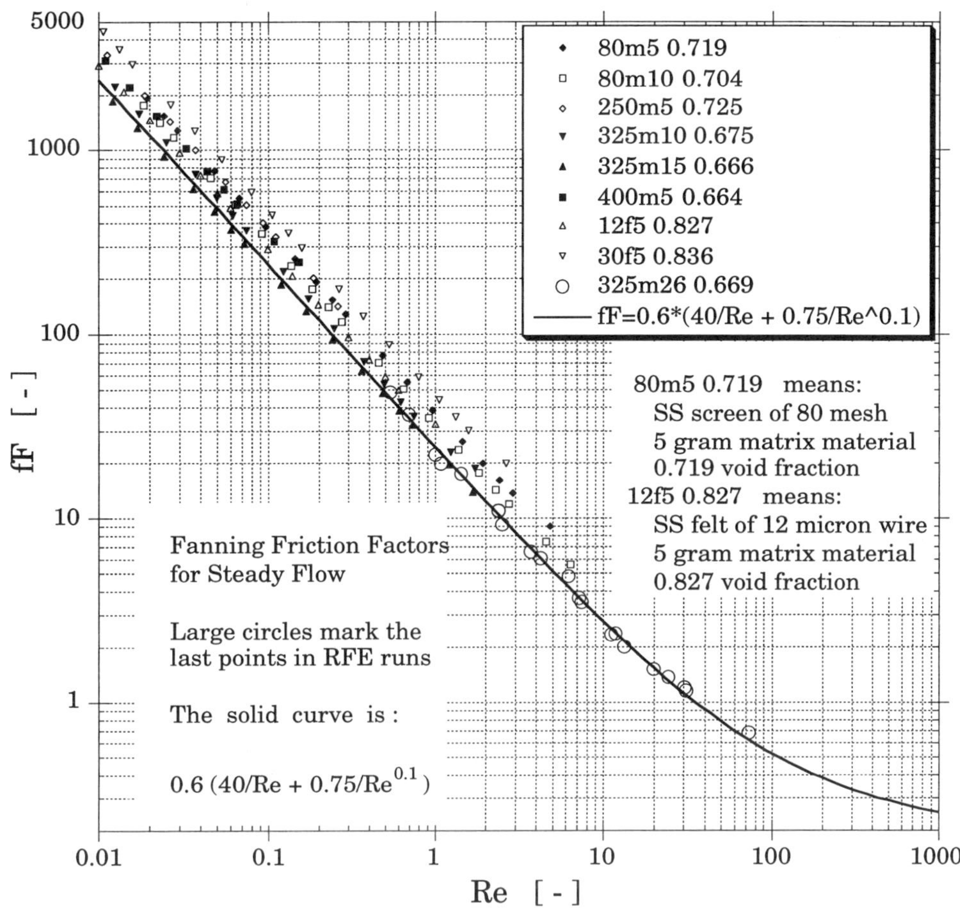


**Figure 2.** Steady flow friction factor as a function of Reynolds number.

fractions for our regenerators and the range in values observed in $S_{rgn}$, it appears that some factor other than void fraction contributes to the mass flow impedance.

We will now discuss the oscillating flow data. The oscillating flow data require various manipulations. In order to filter out noise and to establish the time derivative of the reservoir pressure we have curve-fitted all ac pressure data with a function of the form:

$$Y = A+Bt+C\sin\{([1+\alpha]\omega t)+\phi 1\}+D\sin\{(2[1+\alpha]\omega t)+\phi 2\}+E\sin\{(3[1+\alpha]\omega t)+\phi 3\} \quad (9)$$

Here t is the time and $\omega$ the nominal angular frequency. The curve fitting establishes signal offset and drift (A;B), deviation from the nominal frequency ($\alpha$), amplitudes of the harmonics in the measured pressure signals (C; D; E) and their phase shifts ($\phi 1$; $\phi 2$; $\phi 3$). The oscillating flow fF(Re) relations are then determined by applying the above equations to the pressure data minus offset and drift.

A correction is applied to the OFA mass flow obtained from Eq. (5) since it gives the value at the reservoir inlet, not inside the regenerator. To this purpose our in-house ARCOPTR[2] pulse tube model has been applied. By matching ARCOPTR results with measured amplitudes of the regenerator differential pressure and the reservoir mass flow, the regenerator mass flow was found to be roughly 1.08 times the flow into the reservoir.

An example of raw oscillating flow data and the initial data fit is presented in Fig. 3a; corresponding friction factors are displayed as open squares in Fig. 3b. As can be seen, the voltages are not true sinusoids corresponding to asymmetry in the pressure waves (the ac pressure is proportional to the measured voltage). This is typical for the data obtained in our ac experiments. The asymmetry apparent in Fig. 3a indicates that the flow out of the reservoir (decrease in PX94 voltage) takes more time than the flow into the reservoir.

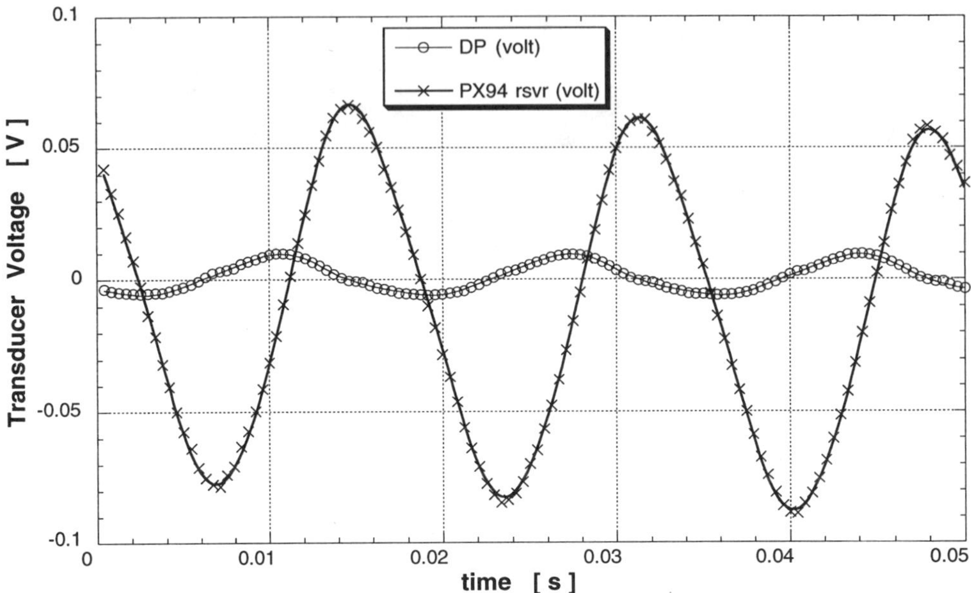


**Figure 3a.** Fits to raw data (tranducer output voltages).

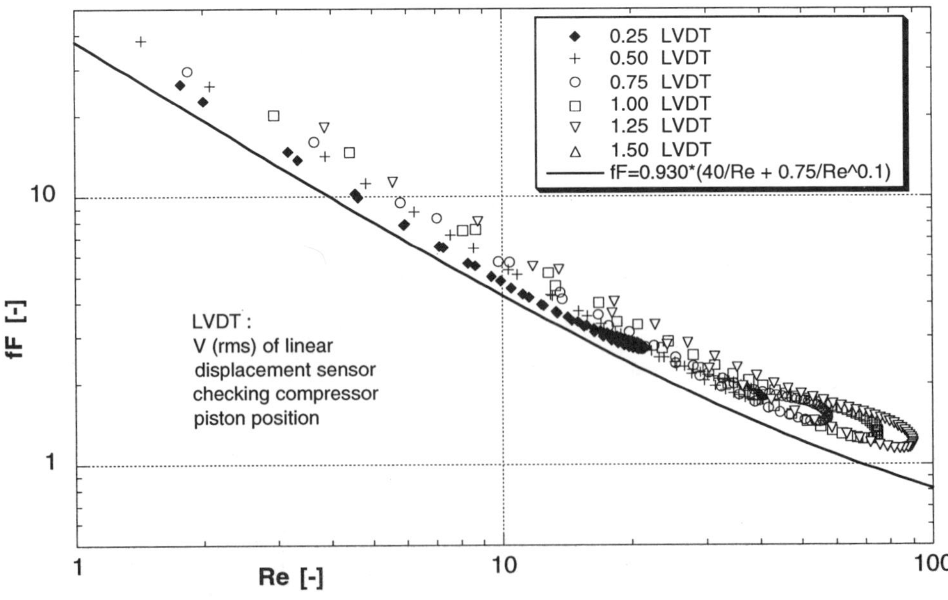


**Figure 3b.** Oscillating flow friction factor as a function of Reynolds number for various compressor strokes.

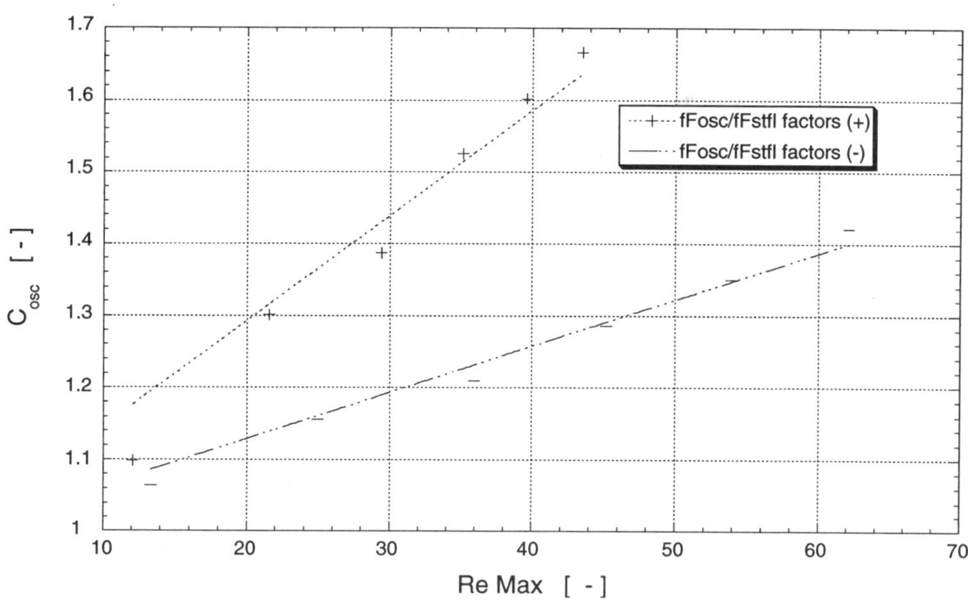


**Figure 4a.** Experimental values of $C_{osc}$ for 60 Hz operation; 325 mesh.

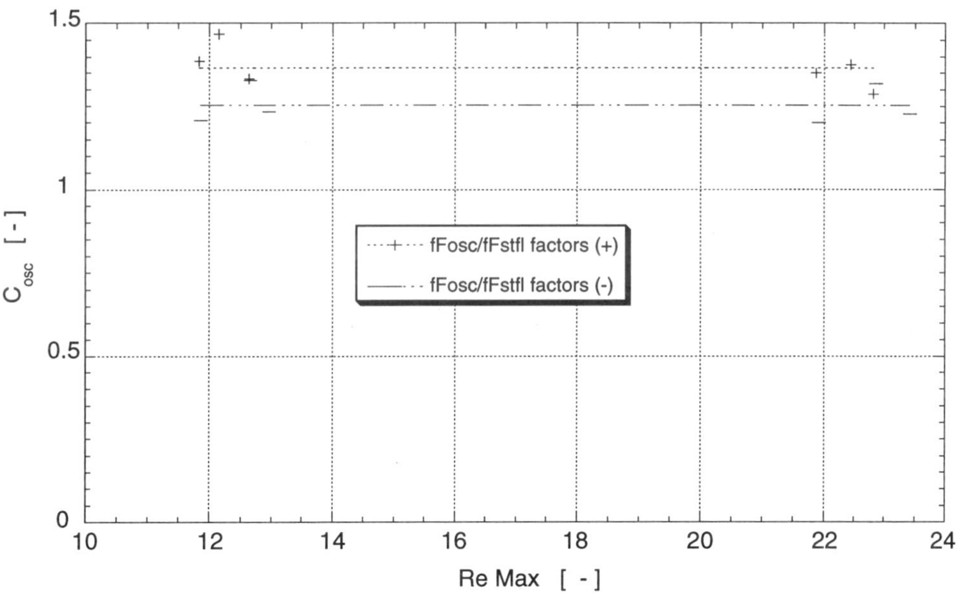


**Figure 4b.** Experimental values of $C_{osc}$ for 36 Hz operation; 325 mesh.

Another asymmetry is apparent in Fig. 3b. Friction factors for increasing absolute flow velocity exceed those for decreasing absolute flow velocity. The effect occurs in all OFA fF(Re) traces recorded in this work. Each trace starts in the upper left hand corner, travels to the right and downward dipping below the earlier part of the trace and then stays relatively low while moving back up to the upper left corner. Further study is suggested to establish the cause of this phenomenon.

For each of the OFA tests on each regenerator the ratio of the oscillating flow friction factor to the steady flow fit (Eq. (8)) was taken at 120 data points spanning one whole cycle of the oscillation. The result is the oscillating friction factor correlation:

$$C_{osc} = fF_{osc} / fF_{StFl} \tag{10a}$$

The values for this parameter that apply for flow into the reservoir (labeled "+") tend to be greater than the values for flow out of the reservoir (labeled "–"). Compiling the data on $C_{osc}$ for various strokes of the compressor shows a linear proportionality with the maximum Reynolds number (see Fig. 4a) :

$$C_{osc} \approx 1 + (a * Re_{max}) \tag{10b}$$

Values on $a = (1 - C_{osc}) / Re$ (on the order of 0.01) are contained in Table 1 including the cycle average (labeled <>).

Our data collection and analysis also has included testing 325 mesh regenerators with the compressor running at 36 Hz, 50 Hz and 70 Hz. Some of these results are shown in Fig. 4b. These data are more prone to noise since the pressure drop is established by subtracting the measurements taken with two absolute pressure transducers. The data shown are all located near $C_{osc} \approx 1.3$ and $Re \approx 17$, giving no clear indication for a frequency dependence of the friction factor over the tested frequency range.

A final note on error analysis is in order. Considering the various operations involving curve fitting, taking derivatives and then collecting the derived variables in order to arrive at the linear coefficient "a" in Eq. (10b), the errror in this final parameter is estimated to be on the order of ±20%. The sliding parameter $S_{rgn}$ has an estimated error bar of ±10%.

## CONCLUSIONS

An extensive set of data has been collected to characterize pressure drop over regenerators for both steady flow and oscillating flow. The analysis of our steady flow experimental data verifies that the Kays and London benchmark friction factor predicts friction factors of the correct order in magnitude. However, yet undetermined factors come into play in addition to void fraction and Reynolds number affecting the mass flow impedance by up to a factor of 2.

A correlation was found for predicting oscillating flow friction factors once the rather easily obtained steady flow values are known.

Oscillating flow friction factors deviate from steady flow values in the systems studied in three ways. An increase is observed relative to steady flow at any Reynolds number. There are asymmetries in the ac friction factor, being greater for increasing mass flow than for decreasing mass flow and being greater for flow into the reservoir than for flow out of the reservoir.

No clear indication was found for a frequency dependence of the friction factor over the tested frequency range (36 Hz to 70 Hz).

## REFERENCES


1. J. M. Lee, P. Kittel, K. D. Timmerhaus, R. Radebaugh, Useful scaling parameters for the pulse tube, "Advances in Cryogenic Engineering, Vol. 41B", P. Kittel, ed., Plenum Press, New York ( 1995), p. 1347.
2. P. R. Roach, A. Kashani, J. M. Lee, Theoretical analysis of a pulse tube regenerator, "Advances in Cryogenic Engineering, Vol. 41B", P. Kittel, ed., Plenum Press, New York ( 1995), p. 1357.
3. W. M. Kays, A. L. London. "Compact Heat Exchangers, 2nd Ed.", McGraw-Hill Book Company, New York (1964).
4. P. D. Roach, K. J. Bell. "Analysis of Pressure Drop and Heat Transfer Data from the Reversing Flow Test Facility", ANL/MCT-88-2, Argonne Natl. Lab., Argonne (1989).

# PARALLEL FLOW REGENERATOR FOR PULSE TUBE COOLER APPLICATION

A. Hofmann[1], S. Wild[2], and L.R. Oellrich[2]

[1] Forschungszentrum Karlsruhe (FZK)/ITP
[2] Universität Karlsruhe, Inst. für Technische Thermodynamik und Kältetechnik, Germany

## ABSTRACT

A single stage double inlet pulse tube refrigerator has been operated with two different types of regenerators, a conventional one, made from 200 mesh stainless steel screens, and a new type made from a stack of parallel channels arranged in 0.1 mm thick stainless steel foils. The overall dimensions of both are the same, but the parallel flow system has only half the void fraction as the mesh type. The initial expectation to get better performance of the cooler with the parallel flow regenerator has not been fulfilled. The no-load temperature of the mesh type system is 26 K, whereas only 33 K could be achieved with the parallel flow regenerator. A more detailed analysis based on data of Kays and London and on numeric studies based on the thermoacoustic theory is given.

## INTRODUCTION

Parallel flow regenerators are expected by several authors to provide higher efficiencies of cryocoolers than it can be obtained with packages of mesh. Respective investigations may have been motivated by the very favorable "compactness factor", $\alpha = StPr^{2/3}/f$, as given by Kays and London [1]. This factor characterizes the ratio of heat transfer and pressure drop. Hendricks [2] has redrawn a diagram of $\alpha$ vs. Reynolds number and he has added data for mesh packages. From this diagram one might draw the misleading conclusion that the compactness factor of an array of circular ducts is about 4 times that of stacked screens. All experiments done so far gave disappointing results. Rawling et al. [3] have used structured polyimid foils. The Sierra Regenerator Co. [4] has developed a technique for etching metal foils to be applied together with spacers, but no results have been published. Alabama Cryogenic Engineering [2] has developed a technique to produce perforated disks which are stacked together with spacers. To our knowledge, no results on cryocoolers which use these novel specimens have been published.

A different technique for producing parallel flow regenerators has been developed at FZK [5]. Extremely high heat flows can be obtained with cross flow arrangements [6]. Based on this technique, we have fabricated a regenerator which has been operated in a pulse tube cooler where its performance has been compared with a conventional regenerator made of stainless steel mesh. The results will be analysed with both, the Hausen theory [7] and with numerics based on the thermoacoustic model [8].

## THE EXPERIMENT

The pulse tube refrigerator is of double inlet type with magnetic valve pressure wave generator. The mesh type regenerator is made from 200 mesh stainless steel sheets stacked into a SS tube with 19 mm i. d., 0.5 mm wall thickness, and 230 mm length. The void factor is 67%. The pulse tube has the same diameter and it is 200 mm long. The structure of the parallel flow regenerator (we call it micro-regenerator, because of the fabricational technique having been developed for micro technology components) is shown in Fig. 1. It has been made from a stack of 0.1 mm thick structured SS. Rectangular flow channels with 0.07 mm width and 0.1 mm depth have been cut with high precision diamond tools. After sintering, three 70 mm long stacks are machined to fit into a SS tube with the same size as that of the mesh regenerator. The effective heat transfer area is about 1.4 $m^2$, and the porosity is 50%. Pressures are measured at both warm ends of regenerator and of pulse tube, and 10 temperature sensors have been attached to both components. The flow scheme of the cooler is shown in Fig. 2, but the experiment is U-shaped. More details are given in Ref. [9] and [10].

The refrigeration powers of both systems have been measured for 2.5 and for 3.5 Hz pulsation frequency. The results are plotted in Fig. 3. With the mesh type regenerator, the optimum performance was obtained for 2.5 Hz. Here, a no-load temperature of 26 K was obtained, and about 50 W of refrigeration power at 80 K. The parallel flow system has its optimum at slightly higer frequency, namely 3.5 Hz. The no-load temperature is 33 K. Also at higher temperatures, the parallel flow system is always worse than the conventional one.

The pressure drops of both systems are compared in Fig. 4 where the pressure waves measured at the warm ends of the regenerator and of the pulse tube are plotted. The pressure drop, the difference of both curves, proves to be 30% to 50% smaller for the parallel flow regenerator. This advantage, however is not reflected in higher refrigeration power. So far, the first expectations have not been fulfilled. A more detailed analysis must be done for explaining the result.

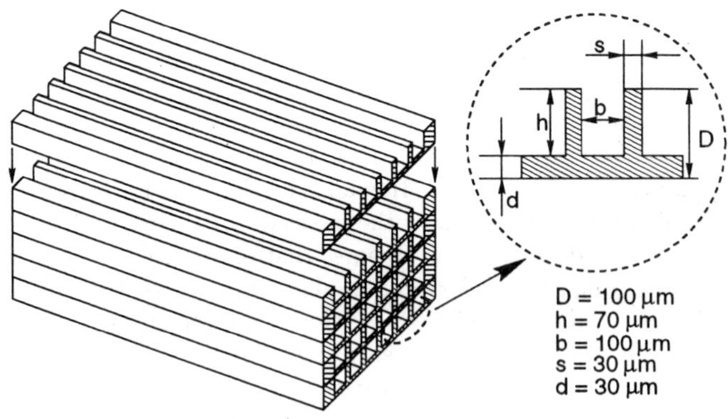


**Parallel Flow Regenerator: L = 3 * 70 mm, d = 19 mm, stainless steel, 50 % void**

**Figure** 1. Design of the "micro-regenerator".

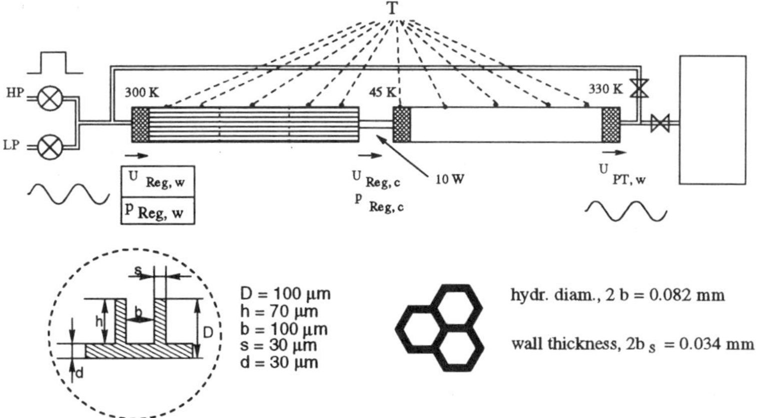


**Figure** 2. Scheme of the double inlet pulse tube refrigerator (inserts describe the adaptation to the thermoacoustic model).

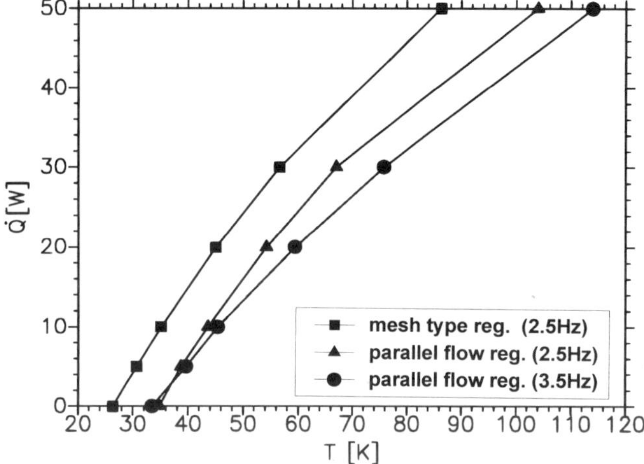


**Figure** 3. Refrigeration powers of the PTR operated with mesh type and with parallel flow type regenerators.

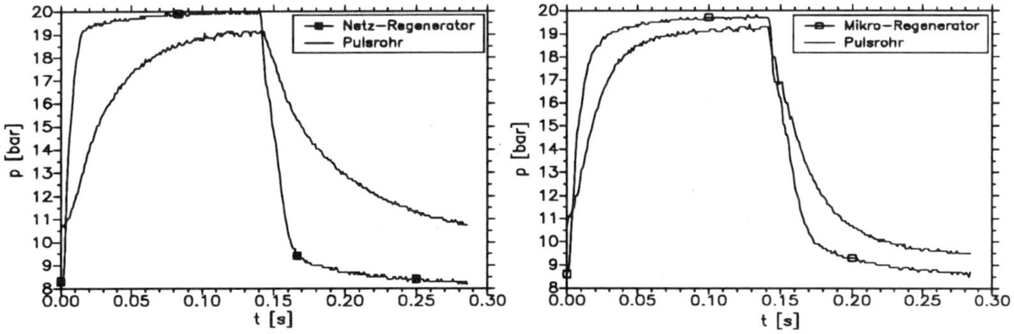


**Figure** 4. Pressures at regenerator inlet and in the pulse tube for operation of both systems at 3.5 Hz.

## CHECK OF THE COMPACTNESS FACTOR

First, a rough estimation on heat transfer and pressure drop was done, and it was found [9] that this can never predict the size of the compactness factor as it results from the diagrams of Kays/London and Hendricks. In our opinion, the discrepancy may be explained by the fact, that in the diagrams Fig.2-42 of Ref.[1] and Fig.2 of Ref.[2] compactness factors calculated with different definitions of friction factors are being compared.For parallel plates and for circular ducts f has been calculated from the hydraulic radius whereas the hydraulic diameter has been used to describe pourous beds (mesh and spheres). The Reynold's number is based on the hydraulic diameter in both cases. Recalculation with f based on the hydraulic diameter yields $f = 64/Re$, and $Nu = 3.66$ for laminar flow in circular tubes and

$$\alpha = \frac{Nu}{f Re Pr^{1/3}} = 0.0657 \tag{1}$$

Hence, the compactness factor for the circular duct is constant, and it nearby equals that of the stacked screens. In contrast to that, a four times higher value, $\alpha = 0.25$, is plotted in the mentioned diagrams. Under this aspect, the fact that both systems, parallel channels and stacked mesh, give comparable experimental results is no longer surprising. In contrary, one might expect, that the larger cross section area of the parallel flow matrix (50%) and its non-interrupted structure would cause more deterioration than detected.

The simple geometry of the parallel (laminar) flow device, however, is advantageous for theoretical studies. That makes this regenerator a valuable object for testing such models and to analyze its losses in some more details.

## EVALUATION OF LOSSES WITH THE THERMOACOUSTIC THEORY

Oscillating laminar flow with thermal and mechanical interaction to circular ducts may be described by the thermo-acoustic model developed by Xiao [8]. This results in a set of 6 linear differential equations with the complex variables of pressure, $\tilde{p}$, and volume flow, $\tilde{U}$, and the time averaged temperature, $T_0(x)$, as well as the time averaged enthalpy flow, $E(x)$. This model has been applied to the main components of the pulse tube refigerator, namely to the regenerator,to the pulse tube, and to the connection tube between both. By numeric experiments, we obtain a solution with $T_0(x)$ which fits the measured temperature distribution.

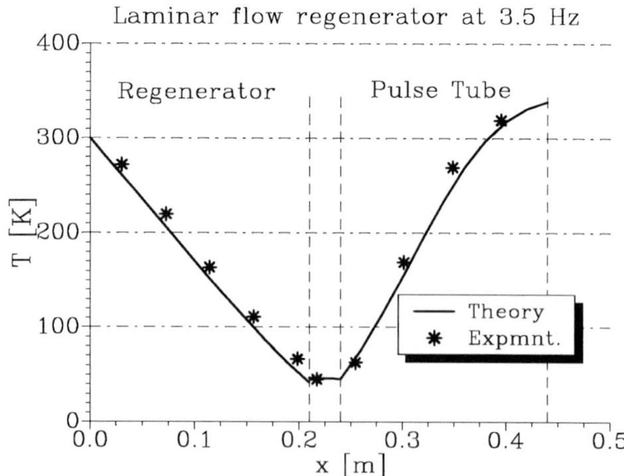


**Figure 5.** Measured and calculated temperature distribution in regenerator and and in pulse tube.

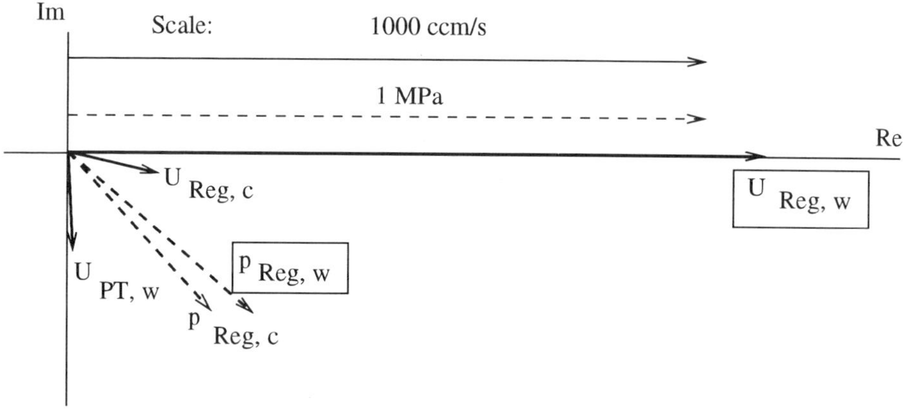


**Figure 6.** Calculated volume flows and pressure waves of the 10W/45K operarational state.

Together with $T_0$, we also get the enthalpy flow, $E_x$ as well as $\tilde{p}$ and $\tilde{U}$ at any position. Those quantities will be used to analyse the losses in both, the regenerator and in the pulse tube.

Some simplifications, however, have to be done first (Fig. 2): The rectangular flow channels are replaced by hexagons with the same hydraulic radius and the same void fraction. Those hexagons are treated as circles with inner radius $b$ and wall thickness $b_s$. The more or less square wave pressure fluctuation as caused by the valved compressor is replaced by a harmonic function. Also the volume flow is assumed harmonic. The size of the pulse tube is not modified. By numeric experiments we have found that the best performance of the refrigerator is obtained for $\tilde{p}$ delayed by 45 degrees behind $\tilde{U}$ at the warm end of the regenerator. A typical operational state obtained for 10 W heat load applied to the 3.5 Hz parallel flow regenerator system is being considered. The amplitudes of pressure and temperature waves have been varied until a reasonable agreement between the calculated temperature $T_0(x)$ and the measured temperature distribution in both components, the regenerator and the pulse tube, was obtained. The result is shown in Fig. 5. The associated pressure and volume flows, both complex quantities, are displayed in the phasor diagram (Fig. 6). The plotted phasors are: the flow rates at the warm end of the regenerator ($U_{R,w}$), and at both ends of the pulse tube, $U_{PT,c}$ and $U_{PT,w}$, respectively. The dashed lines display the oscillating fraction of pressures at the same positions. The time averaged work flow is

$$W_x = \frac{1}{2}Re[\tilde{U}\tilde{p}^*] \tag{2}$$

Additionally, the calculation yields the time averaged enthalpy flows $E_x$. In the regenerator, this is the total loss caused be all irreversible effects such as friction, thermal conduction, and non ideal heat transfer. The calculation yields $E_{Reg} = 7.3W$. In the pulse tube, the enthalpy flow is increased by the 10 W of refrigeration power and by additional irreversibilities. This is to be removed by the "pV-work" at the cold end of the regenerator, namely $W_{Reg,c} = 22W$.

## Decomposition of losses

By definition, the enthalpy flow in the adiabatic components is

$$E_x = W_x + Q_x \tag{3}$$

where $Q_x$ is the axial heat flow. In the theoretical model [8], this quantity has been composed of different fractions. Here it is broken apart in order to analyse the different sources of losses. This results in

$$E_x = E_{acoustic} + [A_f K_e^s + A_f K_0 + A_s K_{s0}]\frac{dT_0}{dx} \quad (4)$$

$$= E_{acoustic} + E(A_f, K_e^s) + E(A_f, K_0) + E(A_s, K_{s0}) \quad (5)$$

The first of the four fractions

$$E_{acoustic} = W_x + Q_x = \frac{1}{2} Re[\tilde{U}\tilde{p}^*(1 - T_0 \beta_0 f_{qx})] \quad (6)$$

does not depend on the temperature gradient. It can be positive or negative, dependent on the size of the factor $f_{qx}$ which is a function of channel dimensions and of the thermal and viscous penetration lengths (see Ref. [8] for details). The third and fourth terms are more convenient, namely the molecular thermal conduction of the fluid with conductivity $K_0$ and the cross section area $A_f$, and the Fourier conduction through the wall with conductivity $A_{s0}$ and total area $A_s$. The second term, finally, is also proportional to the fluid area, but $K_e^s$ is the effective thermal conductivity caused by fluid oscillation. It is a given function of mainly the same parameters as mentioned for the term $f_{qx}$.

The result of this decomposition for both refrigerator parts is shown in Fig. 7. Here is plotted the total enthalpy flow together with the the mentioned 4 components. Additionally, the "pV-work flow", $W_x$ in the pulse tube is displayed. The "shuttle losses", $E(A_f, K_s^e)$ prove to contribute most. It has about the same size at the cold end of both parts (the negative sign in the tube is due to inversed temperature gradient). Both term contribute to the final enthalpy flow in the pulse tube. The term $E_{acoustic}$ contributes with about 30% in the regenerator. but it is negligible in the pulse tube. This term describes losses by heat transfer between the fluid and the matrix. In the pulse tube, $E_{acoustic}$ is very high, but this is desired. It should be close to the "pV-work". The losses caused by axial heat conduction in the matrix, $E(A_s, K_s)$, are also negligible in the pulse tube, but not in the regenerator where its contribution is the range of 10% at the cold end. The losses by molecular conduction of the gas, $E(A_f, K_0)$, is extremely small in both cases.

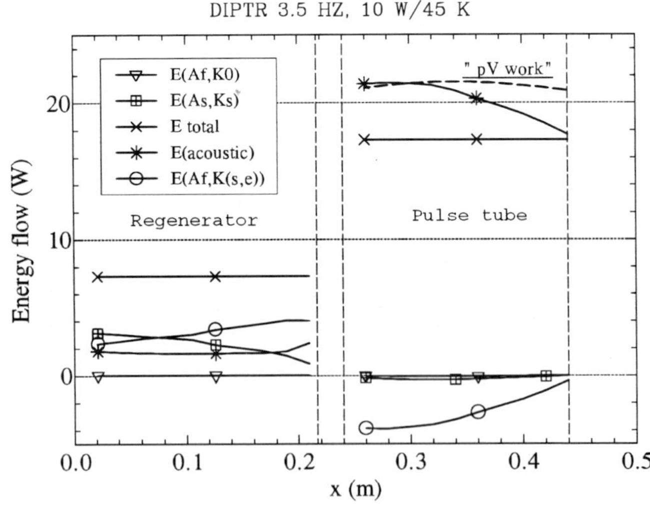


**Figure 7.** Energy flows in regenerator and in pulse tube.

## CONCLUSION

A novel type of "micro structure" parallel flow regenerator has been tested in a pulse tube refrigerator and its performance has been compared with a mesh type regenerator. Initially, the micro structure device had been developed for high power cross flow heat exchangers design. The modified parallel flow unit has been fabricated with about the same specific surface as a 200 mesh regenerator. The usability of this type of regenerator could be demonstrated. The refrigeration power obtained at 2.5 and 3.5 Hz operational frequency is rather good (about 30 W at 80 K), but it is by about 10 W smaller than it could be obtained with a conventional mesh type regenerator of the same overall size.

Nevertheless, the parallel flow device is a valuable object for more detailed analysis with laminar flow model calculations. This gives access to different factors contributing to regenerator and pulse tube losses. It is shown, that thermal conduction through the large cross section area of the matrix is still much smaller than other terms associated with the oscillation of the gas. This shows that further optimization of the laminar flow structure is possible. A final assessment of such structures cannot yet be made.

## ACKNOWLEDGMENT

The parallel flow heat exchangers have been fabricated by the FZK/HTV workshop. The willingness of Dr. Bier Dr. Linder, and Dr. Schubert to modify their facilities for meeting our special demands is greatly appreciated. Financial support by the German BMBF (grant No. 13 N 6593) is also acknowledged.

## References


[1] Kays, W.M., London, A.L.: *Compact heat exchangers.* McGraw-Hill, New York, 1984, 3rd edition.

[2] Hendricks, J.B.: *A new method for producing perforated plate recuperators.* Advances in Cryogenic Engineering, Vol. 41 B, Plenum Press, New York, 1996, p. 1329-1337.

[3] Rawlins, W., Timmerhaus, K.D., Radebaugh, R., Daney, D.E.: *Measurement of the performance of a spiral wound polyimide regenerator in a pulse tube refrigerator.* Advances in Cryogenic Engineering, Vol. 37 B, Plenum Press, New York, 1992, p. 947-952.

[4] Mitchell, M.P., Yaron, R., Shokralla, S., Yuan, J., Bradley, P.E., Radebaugh, R.: *Etched foil regenerator.* Advances in Cryogenic Engineering, Vol. 41 B, Plenum Press, New York, 1996, p. 1339-1346.

[5] Bier, W., Schubert, K.: *Herstellung und Test von kompakten Mikrowärmeübertragern.* Chem.-Ing.-Tech. 61 (1989) Nr.2, S. 172-173 Germany.

[6] Wild, S.: *Untersuchung von Mikrowärmeübertragern bei LHe-Temperaturen.* Diplomarbeit Universität Karlsruhe, 1992 Germany

[7] Hausen, H.: *Wärmeübertragung im Gegenstrom, Gleichstrom und Kreuzstrom.* Springer-Verlag, Berlin, 1976, 2. Auflage Germany

[8] Xiao, J.H.: *Thermoacoustic heat transportation and energy transformation; Part 1: Formulation of the problem.* Cryogenics 1995, Vol. 35, p. 15-19.

[9] Wild, S.: *Untersuchungen ein- und zweistufiger Pulsrohrkühler.* Dissertation Universität Karlsruhe, 1997 (german), and VDI-Berichte, Reihe 19, Nr 105, ISBN 3-18-310519-5, VDI Düsseldorf (1997).

[10] Wild, S., Oellrich, L.R., and Hofmann, A.: *Mikroregenerator für Gaskältemaschinen,* DKV-Tagungsbericht, 23. Jahrgang (1996) (Deutscher Kälte- und Klimatechnischer Verein, Stuttgart), ISBN 3 922 429 73 4 (1996), p. 137-145 Germany

# A STUDY OF OSCILLATING FLOW CHARACTRISTICS OF REGENERATORS IN A HIGH FREQUENCY PULSE TUBE REFRIGERATOR

**Y. Jiang , Y. L. Ju , and Y. Zhou**

Cryogenic Laboratory, Chinese Academy of Sciences
P.O.Box 2711, Beijing, 100080, China

## ABSTRACT

In this paper, an experiment apparatus was designed and constructed. Oscillating flow characteristics and pressure drops of regenerators in a high frequency pulse tube refrigerator are investigated. The time-averaged pressure drop of the regenerator in 50Hz oscillating flow is two to three times higher than that in steady flow. Pressure drops and phase shifts of the regenerators are correlated with the structure and the working conditions. The correlation is based on the experimental data of a 50Hz pulse tube refrigerator connected to an orifice valve and a reservoir.

## INTRODUCTION

Interests in the pulse tube refrigerator have grown rapidly in recent years because of its merits in mechanical simplicity, high reliability, low vibration and low cost. This type of cryogenic refrigerator has promising application potential in the cooling of in infrared devices and sensors, superconducting electronic devices, etc.

As a necessary component, the regenerator plays an important role as to the performance of the cryocooler. In the past, people have taken it as a kind of heat exchanger, which may similar to the low frequency regenerator. However, in a high frequency regenerator, oscillating flow characteristics influence its performance significantly and present both pressure drops and phase lags. Thus, it is important in the analysis and design to predict pressure and phase lags accurately.

There are some correlation of the pressure drop in a regenerator, but most of them are based on a steady flow through a stack of screens. Relatively little research has been performed on the pressure drop over a stack of screens subjected to an oscillating flow while considering the phase lag. In addition, the experimental conditions on which those correlation equations are based far different from those in a high frequency pulse tube refrigerator.

The objectives of the present work are: (1) to investigate experimentally pressure drops and phase lags across a regenerator packed with stainless-steel wire screens subjected to an oscillating flow; (2) to obtain the correlation equation for the friction factor of 50Hz oscillating flow in regenerators.

## EXPERIMENTAL APPARATUS

The schematic diagram of the experimental apparatus is shown in Figure1. A 50Hz valveless compressor supplies a sinusoidal pressure wave. The test section consists of a packed column (70mm in length and 10.5mm inner diameter) with each of its two ends connected to velocity straighteners (60mm in length and 9mm inner diameter). A reservoir with an orifice valve is connected to the end of the second velocity straightener.

At each end of the test section, the instant gas velocity and pressure are measured. The pressure is measured by a pressure transducer(KISTLER, TYPE 601A) connected to a charge amplifier (KISTLER, TYPE 5011) and the velocity is measured by a hot-wire probe(NANTEC, TYPE 55P11) connected to a hot-wire anemometer (NANTEC, MODEL 90N10). Both pressure and velocity voltage signals are sampled and converted to digital signals simultaneously through a multi-channel simultaneous sample and hold front end connected to an A/D conversion board (KEITHLEY, DAS-1600).

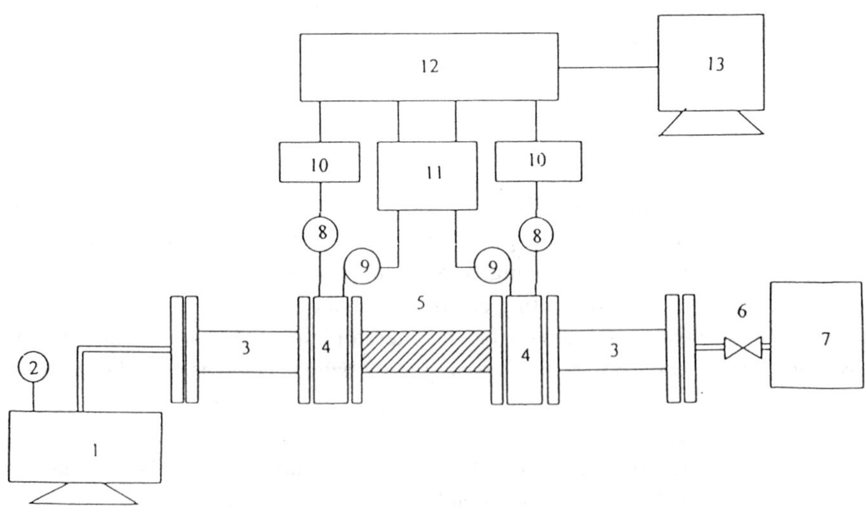


**Figure 1.** Schematic diagram of experimental apparatus.
1. compressor; 2. pressure gauge; 3. velocity straightener; 4. connector; 5. regenerator; 6. orifice valve; 7.reservoir; 8. pressure transducer; 9. hot-wire probe; 10. charge amplifier 11. hot-wire anemometer; 12.A/D conversion board; 13. computer

## EXPERIMENTAL CONDITIONS

The experimental measurements are based on :

1. Helium gas is used as working medium;
2. The operating frequency is 50Hz;
3. System pressure varies from 0.6MPa to 0.9MPa;
4. Stainless-steel screens are packed in the test section. Five different mesh sizes of screens are tested: 80, 200, 250, 300, 400 and their properties are listed in Table 1.
5. The orifice valve is adjustable and meets different end conditions.

**Table 1.** Properties of stainless-steel wire screens

| Mesh size | Number of screens, n | Wire diameter $D_w$ (mm) | Porosity $\varphi$ | Mesh distance $\beta$ (mm) | Hydraulic diameter $D_h$ (mm) |
|---|---|---|---|---|---|
| 80 | 428 | 0.102 | 0.735 | 0.140 | 0.2828 |
| 150 | 500 | 0.061 | 0.6993 | 0.120 | 0.1418 |
| 250 | 600 | 0.041 | 0.6582 | 0.100 | 0.0788 |
| 300 | 690 | 0.031 | 0.6938 | 0.087 | 0.0704 |
| 400 | 720 | 0.025 | 0.6677 | 0.083 | 0.0504 |

## RESULTS AND DISCUSSIONS

The typical measurements of gas velocity and pressure entering and leaving the test section are shown as Figure 2, from which we learn that there exists an amplitude pressure drop and phase shift of both velocity and pressure through the regenerator. We refer to the amplitude drop and phase shift as the flow characteristic.

The experimental results show that the flow characteristic depends on many factors such as matrix properties, gas properties(in this experiment, we only discuss those related to pressure), the exit conditions realized by an orifice valve, etc. The flow characteristics are also related to the frequency of the oscillating flow as some researchers have studied[1]. However, our experiments focus on 50Hz oscillating flow as an example to discuss the flow characteristics of a high frequency regenerator.

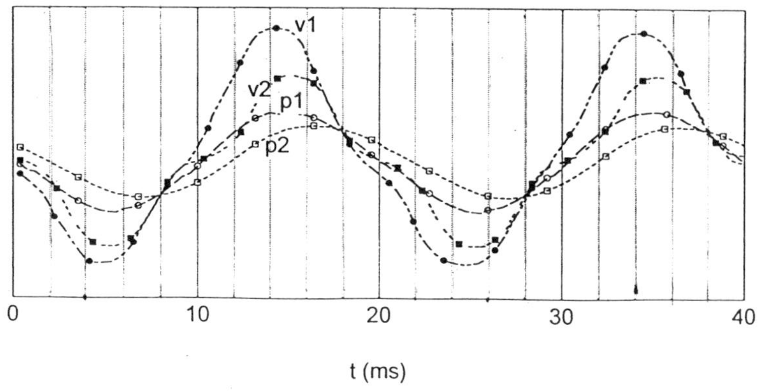


**Figure 2.** Typical measured velocity and pressure waves: input (v1,p1) and output (v2, p2) to the regenerator.

## INFLUENCE OF THE ORIFICE

In our tests, we performed experiments with different orifice valve conditions: position 1(valve closed), 2(valve opened a little), and 3(fully opened). Figure 3 indicates that the orifice influents the phase shift between velocity and pressure wave across the regenerator. We find that at position 2, the phase shift between the velocity and pressure wave across the regenerator reaches its minimum.

Our measurements indicate that the orifice setting affects the performance of the regenerator. The results below are based on tests done at position 2.

## PRESSURE DROP AND PRESSURE DROP FACTOR

Pressure drop measurements through the regenerator in 50Hz oscillating flow under different experimental conditions including five mesh sizes and several system pressures were obtained. To compare the pressure drops over the regenerator in oscillating flow with those in steady flow, we used cycle-averaged pressure drops $\Delta p_{mean}$ from our experiment and the respective pressure drops predicted by equation (1) given by Miyabe et al.[2] for pressure drop of a steady flow through a stack of woven screen:

$$f_{st} = \frac{33.6}{\mathrm{Re}_\beta} + 0.337 \tag{1}$$

with

$$f_{st} = \frac{\Delta p_{st} / n}{\frac{1}{2}\rho (u_{st})_p^2} \tag{2}$$

$$\mathrm{Re}_\beta = \frac{(u_{st})_p \beta}{\nu} \tag{3}$$

where $\Delta p_{st}$ is the steady flow pressure drop, $(u_{st})_p$ is the cross-sectional mean flow velocity in the packed column, n is the number of screens, β is the distance between meshes and $\mathrm{Re}_\beta$ is the Reynolds number based on β and $(u_{st})_p$.

Here, we use cycle-averaged velocity $u_{mean}$ instead of $(u_{st})_p$ and equation(1), (2 ) and (3) to predict the steady flow pressure drops. Pressure drops of steady flow and oscillating flow are compared in Table 2.

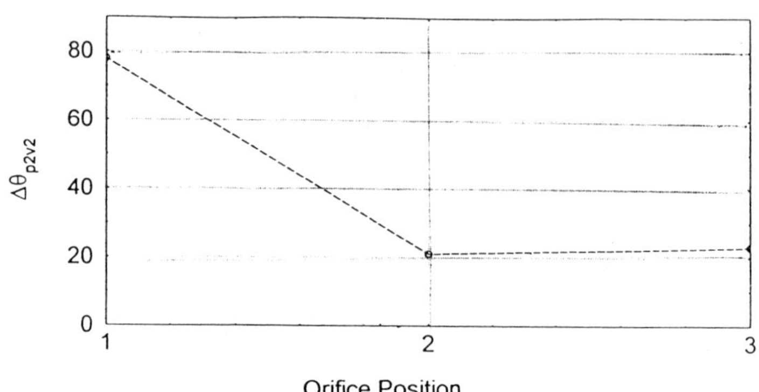


**Figure 3.** Regenerator flow characteristics influenced by orifice valve.

**Table 2.** Pressure drops of oscillatory flow compared with those of steady flow

| Mesh size | $Re_\beta$ | $\Delta P_{mean}$ (kPa) | $\Delta P_{st}$ (kPa) | $\Delta P_{mean} / \Delta P_{st}$ |
|---|---|---|---|---|
| 80 | 5.74 | 9.93 | 4.58 | 2.20 |
| 150 | 5.57 | 12.67 | 5.98 | 2.28 |
| 250 | 5.17 | 17.21 | 7.42 | 2.32 |
| 300 | 5.09 | 20.44 | 8.69 | 2.35 |
| 400 | 4.73 | 27.97 | 9.38 | 2.98 |

Table 2 shows that the pressure drops over the regenerator in 50 Hz oscillating flow are two to three times higher than that in steady flow. Thus, the equations for steady flow can not be used to predict the pressure drop over regenerators in oscillating flow.

To predict the pressure drop across the regenerator of a 50 Hz pulse tube refrigerator, we investigate the friction factor defined as follows:

$$f_{\max} = \frac{\Delta p_{\max} D_h}{\frac{1}{2}\rho (u_{\max})^2 L} \tag{4}$$

and define the Reynolds number as:

$$\mathrm{Re} = \frac{u_{\max} D_h}{\nu} \tag{5}$$

where $\Delta p_{\max}$ is the maximum pressure drop in one cycle, $u_{\max}$ is the maximum cross-sectional mean flow velocity in the regenerator, L is the length of the tested regenerator and $D_h$ is the hydraulic diameter of the screen, which is defined as follows[4]:

$$D_h = \frac{\varphi D_w}{1-\varphi} \tag{6}$$

with φ being the screen porosity defined as[3]:

$$\varphi = 1 - \frac{\pi D_w \sqrt{\lambda^2 + D_w^2}}{4\lambda^2} \tag{7}$$

where λ is the pitch of the mesh. The experimental friction factor $f_{\max}$ related to Re is shown in Figure 4. The figure shows that the results depend on screen size.

To obtain a normalized curve, we define the dimensionless distance:

$$X = \frac{1}{2}\frac{u_{\max} T}{D_h} \tag{8}$$

The normalized curve of $f_{\max}$ is shown in Figure 5. Thus, the correlation equation of the pressure drop factor of the regenerator used in 50 Hz pulse tube refrigerator in terms of Re and X is obtained:

$$f_{\max} = \frac{1}{X}(2.76 \times 10^4 + \frac{1.22 \times 10^6}{\mathrm{Re}}) \tag{9}$$

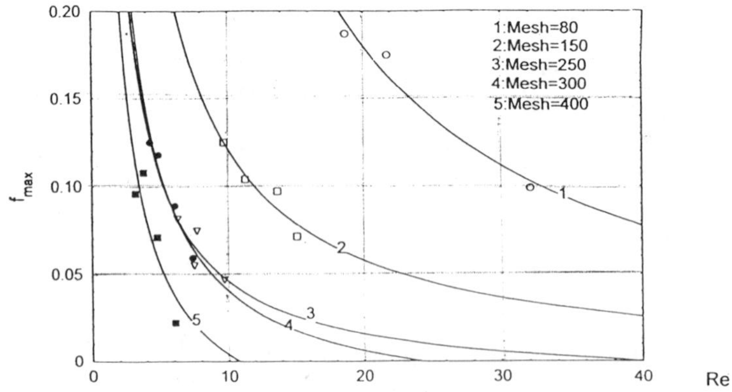


**Figure 4.** Experimental friction factor related to Re.

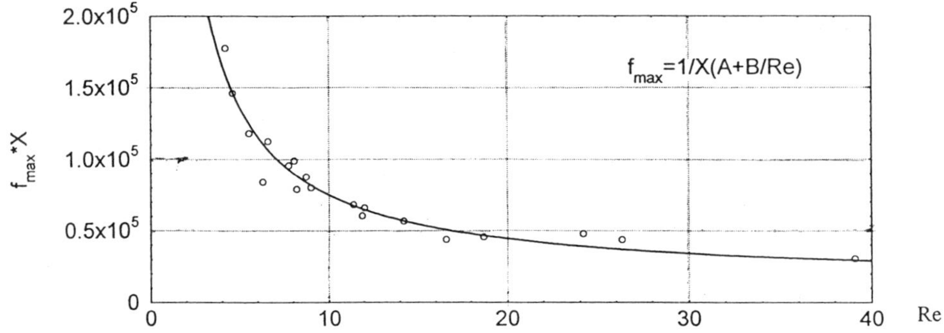


**Figure 5.** Normalized friction factor related to Re.

## PHASE SHIFT

Phase shifts are determined by the screen stack and the Re number of the working gas besides the orifice. Now we list the phase shift of velocity and pressure wave at the outlet end of the regenerator related to Reynolds numbers, as shown in Figure 6.

Firstly, Figure 6 shows that for each mesh size of screens, there exists a minimum value of phase shift, which corresponds to a special Reynolds numbers. Figure 7 indicates the relationship between the mesh size and the special Reynolds numbers.

Secondly, Figure 6 shows a tendency that the higher mesh size of screens is, the more sensitive the phase shift between the outlet pressure and gas velocity is to Reynolds number.

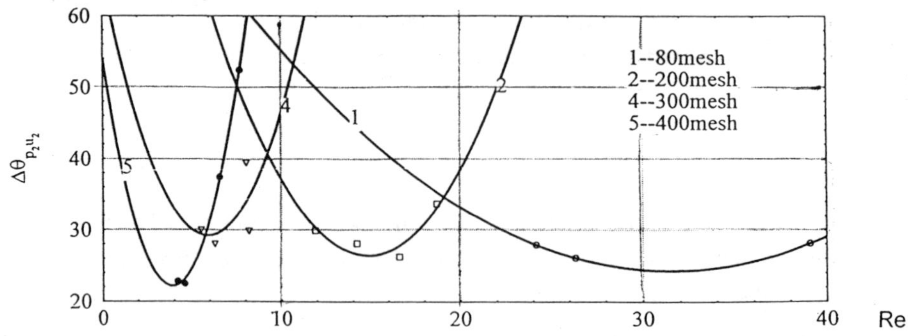


**Figure 6.** Phase shift between outlet velocity and pressure wave related to Reynolds numbers.

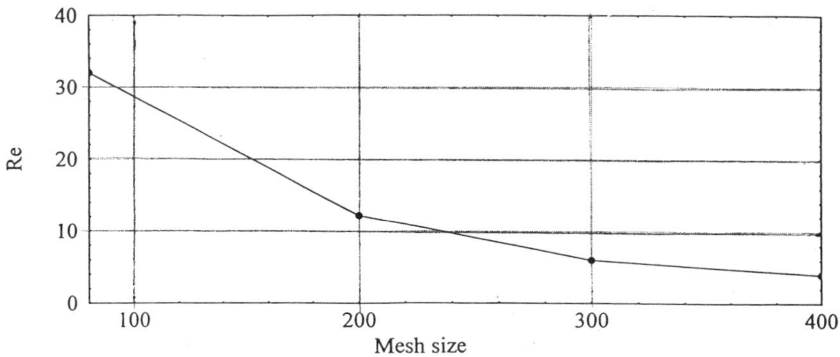


**Figure. 7** Special Reynolds number related to screen mesh size.

## CCONCLUSIONS

Some experimental results on oscillating flow characteristics of regenerators in a high frequency pulse tube refrigerator are reported in this paper. It is shown that the orifice influents the flow characteristics of the regenerator. The oscillating flow characteristics appear as pressure drops and phase lags. The value of cycle-averaged pressure drop of the regenerator in 50 Hz oscillating flow is several times higher than that in steady flow. A correlation equation for the friction factor in terms of the Reynolds number and the dimensionless distance was presented in this paper. The relationship of phase shift between the outlet gas velocity and pressure was also investigated.

## REFERENCES


1. Guo, F. Z. et al. Flow characteristics of a cyclic flow regenerator, Cryogenics, 27:152 (1987)
2. Miyabe, H., Takahashi, S. and Hamaguchi, K. An approach to the design of stirling engine regenerator matrix using packs of wire gauzes , Proc 17th IECEC, 1839 (1982)
3. Tanaka, M., Yamshita, I. and Chrisaka, F. Flow and heat transfer characteristics of the Stirling engine regenerator in an oscillating flow, JSME Int. J. 33:283 (1990)
4. Zhao, T. S. and Cheng, P. Oscillatory pressure drops through a woven-screen packed column subjected to a cycle flow, Cryogenics (1996) 36 333-341

# A STRAIGHTFORWARD METHOD FOR THE SIZING OF PERFORATED PLATE MATRIX HEAT EXCHANGERS

G. Venkatarathnam

Department of Mechanical Engineering
Indian Institute of Technology
Chennai 600 036
India

## ABSTRACT

The heat transfer between fluid streams in a matrix heat exchanger is controlled by different resistances namely conduction resistance in perforated plates, convective resistance, axial conduction through the wall and the matrix in the exchanger. In this paper closed-form expressions are presented for the effectiveness of matrix heat exchangers in terms of these different resistances. These expressions can be used to determine the effectiveness of a matrix heat exchanger of any shape (circular, rectangular etc.). Based on these expressions, a procedure is developed for the optimum sizing of balanced flow matrix heat exchangers. The procedure is illustrated with a numerical example.

## INTRODUCTION

A matrix heat exchanger (MHE), (Fig. 1) essentially consists of a stack of perforated plates of high conductivity metal (e.g. copper, aluminum) alternating with spacers made of a low conductivity material (e.g. plastics, stainless steel). The spacers are cut out of solid sheets in such a way that two or more passages are created for the fluid streams. The alternate layers of perforated plates and spacers when bonded with a header at each end result in a monoblock unit containing two or more passages through which fluid streams flow, exchanging heat with each other through the conducting plates. The spacers serve the multiple purpose of providing seals between the channels, reducing axial heat conduction, and eliminating flow maldistribution by intermediate headering[1].

An MHE is characterized by many geometrical parameters, which can be grouped into two broad headings :

- the gross or macroscopic dimensions such as width and height of the plate in each channel and the number of plates.
- the finer or microscopic features such as porosity of plates, diameter of perforations, thickness of plates and of spacers.

The latter group of parameters determine the heat transfer and flow friction characteristics and can be considered the intrinsic properties of the chosen surface. For any chosen surface, a judicious choice of the gross dimensions (W, $H_1$, $H_2$ and n) can lead to a small exchanger delivering the desired performance in terms of heat duty and pressure drop. There is no straightforward method for choosing these dimensions as the heat transfer phenomenon in a MHE is quite complex. In this paper we present closed-form expressions for the effectiveness of balanced flow MHEs and develop a simple procedure for the optimum design.

## EFFECTIVENESS-NTU RELATIONSHIP IN MATRIX HEAT EXCHANGERS

The heat transfer process in a matrix heat exchanger is quite complex. The total heat transfer resistance between the two fluid streams is composed of the following :

- thermal resistance offered by the perforated plate in either channel
- convective resistance between the perforated plate and the fluid streams in either channel
- lateral as well as axial thermal resistance of the separating wall.

The effectiveness of a MHE is a function of all the above resistances acting together. Sarangi and Barclay[2] first showed that the maximum attainable effectiveness depends on the number of plate-spacer pairs in the exchanger. A number of authors have developed numerical procedures to predict the performance of MHEs subject to the above resistances.[1,3] Recently Venkatarathnam[4] presented closed-form solutions for the effectiveness of a rectangular shaped MHE taking into account all the resistances listed above, as well as the finite number of plate-spacer pairs in the exchanger. What was not appreciated at that time was that general expressions can also be derived for MHEs of any geometry (e.g. circular cross section). The effectiveness of a MHE of any geometry can be evaluated using the following expressions[4]:

$$\varepsilon = \frac{N_{tu,eff}}{N_{tu,eff} + 1} \qquad (\nu = 1) \tag{1}$$

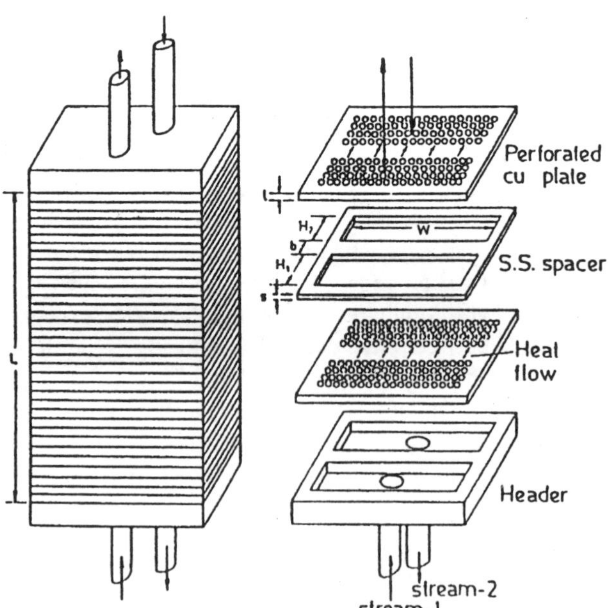


Figure 1. Schematic of perforated plate matrix heat exchanger.

$$\varepsilon = \frac{1 - \exp[-N_{tu,eff}(1-\nu)]}{1 - \nu \exp[-N_{tu,eff}(1-\nu)]} \qquad (\nu < 1) \tag{2}$$

$$N_{tu,eff} = \frac{n(1-\alpha_1)(1-\alpha_2)}{\lambda n(1-\alpha_1)(1-\alpha_2) + 1-\alpha_1\alpha_2 + (1-\alpha_1)(1-\alpha_2)/n_{tu,p,o}} \qquad (\nu = 1) \tag{3}$$

$$N_{tu,eff} = \frac{n\ln(\eta)}{n\lambda \ln(\eta) + \nu - 1} \qquad (\nu < 1) \tag{4}$$

$$\eta = \frac{\kappa + \alpha_1 + (1-\alpha_1)/n_{tu,p,o}}{1 + \kappa\alpha_2 + (1-\alpha_1)/n_{tu,p,o}},\ \kappa = \frac{1-\alpha_1}{1-\alpha_2}\nu,\ \alpha = e^{-n_{tu,f}}\ n_{tu,f} = \frac{hA}{\dot{m}c_p},\ \lambda = \frac{k_s A_c/sn}{(\dot{m}c_{p_1})} \tag{5}$$

$n_{tu,f,i}$ is the conventional "ntu" on each side (channel), $\lambda$ is the axial conductance parameter of the separating wall and $n_{tu,p,o}$ is the dimensionless conductance of the entire plate forming the two flow channels and the wall and is defined as follows:

$$\frac{1}{n_{tu,p,o}} = \frac{1}{F n_{tu,p,1}} + \frac{1}{\lambda_p} + \frac{\nu}{F n_{tu,p,2}} \tag{6}$$

$n_{tu,p,i}$ is the dimensionless plate conductance and $\lambda_p$ is the lateral dimensionless conductance of the separating wall. F is a factor dependent on the geometry of the perforated plates. The expressions for F and $n_{tu,p}$ have been derived for different plate geometries and presented in Table 1. With the effectiveness ntu relationship expressed in terms of different resistances as closed form expressions, it is possible to derive a simple procedure for determining the optimum geometry of matrix heat exchangers.

## OPTIMIZATION OF MATRIX HEAT EXCHANGER GEOMETRY

Ideally, a MHE should have a small cross-section to reduce lateral temperature differences and a large length to reduce axial heat conduction. The width and height of the exchanger should therefore be chosen to minimize the lateral plate conduction resistance, while providing enough area for convective heat transfer. The MHE length is chosen to maximize the axial resistance. The resulting heat exchanger volume depends on the combination of different parameters chosen.

In general, cryogenic exchangers need to be compact to reduce the area exposed to the ambient, and to reduce the heat leak. A small exchanger volume (mass) also helps in quick cool down of cryogenic refrigeration systems. It is therefore desirable to choose the geometry of the exchanger such that the volume (mass) of the exchanger is minimum. In this section, the procedure for the sizing of a balanced flow ($\nu$ =1) MHE of rectangular shape is developed. The formalism, however, is applicable to MHEs of any shape. The design problem can be expressed mathematically as follows:

*min (volume)* or *min* $(n\,W\,(H_1 + H_2))$

*subject to constraints:* $\varepsilon = \varepsilon$ *(specified)* and $\Delta p \leq \Delta p$ *(allowable)*

Normally the pressure drop of the low pressure stream is more critical than that of the high pressure stream in all cryogenic heat exchangers including MHEs. The flow over a stack of perforated plates separated from each other is somewhat similar to that over a bank of

Table 1. Expressions for the dimensionless plate conductance (ntu,f) and geometry factor F

| Shape | ntu,p | F |
|---|---|---|
| Rectangula plate | $\frac{Wlk}{H\dot{m}c_p}$ | 6 |
| Circular disc | $\frac{\pi lk}{\dot{m}c_p}$ | 8 |
| annular disc | $\frac{\pi lk}{\dot{m}c_p}$ | $\frac{(r_o^2-r_i^2)^2}{\frac{r_o^2 r_i^2}{2}-\frac{3}{8}r_o^4-\frac{3}{8}r_i^4+\frac{r_o^4}{2}\ln\left(\frac{r_o}{r_i}\right)}$ |

tubes. The pressure drop is a function of the number of plates (n), and can be determined using the following expression[1]:

$$\Delta p = nC_f\frac{G_d^2}{\rho} \tag{7}$$

The drag coefficient $C_f$ (not same as the fanning friction factor) is a function of the plate porosity[1].

## Design Variables and constants

The design variables are the number of plates (n), the width of the exchanger (W) and the height of the plate in either channel ($H_1$,$H_2$). The design constants are the plate porosity, plate thickness, perforation diameter, spacer thickness and the mean thermophysical properties for the two streams.

## Assumptions

(i) the number of plates in the exchanger is large and can be treated as a continuous variable (ii) the contribution of the separating wall to the total lateral heat transfer resistance is negligible i.e., ( $\lambda_p = \infty$) and (iii) the exchanger will be designed such that the allowable pressure drop on the low pressure side is utilized in full

## OPTIMIZATION PROCEDURE

The mathematical analysis consists of the following steps:

1. Expression of the design variables in terms of relevant dimensionless groups
2. Expression of the objective function (volume) and constraints (pressure drop, and effectiveness required) in terms of dimensionless groups
3. Formulation of the lagrangian equations and solution of the equations to derive algebraic relationships between the different parameters.

It is customary to use the design variables in dimensional form in optimization problems. All analytical and experimental results, on the other hand, are normally expressed in terms of dimensionless variables. In order to use the information generated in performance studies effectively, it will be beneficial to use the same set of dimensionless parameters in optimization problems as well. In this analysis, the dimensional design variables $H_1$, $H_2$ and W have been replaced by dimensionless parameters $n_{tu,f,1}$, $n_{tu,f,2}$ and $n_{tu,p,1}$, the thermophysical properties of the fluid and the convective heat transfer characteristics of the surface (Nu = C $Re^{\gamma}$) as follows:

$$W = (K_i' K_i'')^{0.5} (n_{tu,p,i})^{0.5} (n_{tu,f,i})^{0.5\beta_i} \tag{8}$$

$$H_i = \left(\frac{K_i'}{K_i''}\right)^{0.5} (n_{tu,p,i})^{-0.5} (n_{tu,f,i})^{0.5\beta_i} \tag{9}$$

where βi, $K_i'$ and $K_i''$ are dimensionless constants made up of surface characteristics and thermophysical constants and mass flow rate alone and are defined as follows:

$$\beta_i = \frac{1}{1-\gamma_i} \quad K_i' = \left[\left(\frac{p\mu}{\dot{m}d}\right)^{\gamma}\left(\frac{\dot{m}c_p d}{C_i A''' l k_f}\right)\right]_i^{\beta} \quad K_i'' = \left(\frac{\dot{m}c_p}{k_p l}\right)_i \tag{10}$$

With the dimensional variables expressed in terms of dimensionless variables, other dimensionless parameters such as the overall axial conduction parameter (λ) and the overall plate conductance ($n_{tu,p,o}$) can be expressed in terms of $n_{tu,f,1}$, $n_{tu,f,2}$ and $n_{tu,p,1}$ as follows:

$$\lambda_i = \frac{1}{n}\xi\left(n_{tu,p,1}\right)^{0.5}\left(n_{tu,f,1}\right)^{0.5\beta_1} \quad \frac{1}{n_{tu,p,o}} = \frac{1}{3\,n_{tu,p,1}}\left(1+\zeta\frac{\left(n_{tu,f,2}\right)^{\beta_2}}{\left(n_{tu,f,1}\right)^{\beta_1}}\right) \tag{11}$$

In the above expressions ξ and ς are constants and are defined as follows:

$$\zeta = \frac{K_2' \; K_2''}{K_1' \; K_1''} \quad \xi = \left(K_1' \; K_1''\right)^{0.5} \frac{k_s b/s}{\dot{m}c_{p_1}} \tag{12}$$

Assuming that the allowable pressure drop on the low pressure side will be fully utilized (assumption iii), the pressure drop constraint can be expressed in terms of dimensionless parameters as follows:

$$n - \Delta p_1 \; \phi_1\left(n_{tu,f,1}\right)^{2\beta_1} = 0 \quad \left(with \;\; \phi_1 = \frac{\rho\, p^2 (K_1')^2}{C_f \dot{m}^2}\right) \tag{13}$$

Using equations (11) and (13), the effectiveness of a MHE can be expressed in terms of three dimensionless parameters $n_{tu,f,1}$, $n_{tu,f,2}$ and $n_{tu,p,1}$ and the thermophysical properties. Similarly, the objective function (volume V) can be expressed in terms of dimensionless variables as follows:

$$Volume(V) = \left(n_{tu,f,1}\right)^{2\beta_1}\left[K_1'\left(n_{tu,f,2}\right)^{\beta_1} + K_2'\left(n_{tu,f,2}\right)^{\beta_2}\right] \tag{14}$$

## Solution Procedure

The design problem has now been fully expressed in terms of the dimensionless parameters $n_{tu,f,1}$, $n_{tu,f,2}$, $n_{tu,p,1}$. The constrained optimization problem can be converted to an unconstrained one by the Lagrangian Multiplier technique and a new Lagrangian objective function (L) can be defined as follows:

$$L = V + Z(\varepsilon - \varepsilon_{specified}) \tag{15}$$

L is a function of variables $n_{tu,f,1}$, $n_{tu,f,2}$, $n_{tu,p,1}$ and Z. The necessary conditions for an optimum to exist are as follows:

$$\frac{\partial L}{\partial n_{tu,f,1}} = 0 \qquad \frac{\partial L}{\partial n_{tu,f,2}} = 0 \qquad \frac{\partial L}{\partial n_{tu,p,1}} = 0 \qquad \frac{\partial L}{\partial n_{tu,p,2}} = 0 \tag{16}$$

Four non-linear equations result from the above conditions. A simultaneous solution of these equations gives the optimum values of design variables $n_{tu,f,1}$, $n_{tu,f,2}$, $n_{tu,p,1}$. One of the solutions of these equations results in a simple relationship between the overall axial conduction, the overall plate conductance and the number of plates for the volume of the exchanger to be minimum as follows:

$$n\lambda = \frac{2}{n_{tu,p,o}} \tag{17}$$

It can also be shown that the equation is valid even when the allowable pressure drop on the low pressure side is not fully utilized. The expression for $N_{tu,eff}$ for an optimum balanced flow MHE design can now be expressed as follows:

$$N_{tu,eff,opt} = \frac{n(1-\alpha_1)(1-\alpha_2)}{3(1-\alpha_1)(1-\alpha_2)/n_{tu,po} + 1 - \alpha_1\alpha_2} \qquad (\nu = 1) \tag{18}$$

The optimum values of variables $n_{tu,f,1}$, $n_{tu,f,2}$ and $n_{tu,p,1}$ can be obtained by solving the equations (16) simultaneously. Since the effectiveness ($\varepsilon$) has been expressed as a closed form expression, the solution procedure is quite straightforward when a symbolic programming tool (e.g. Mathematica, Maple) is used. It can be shown that one of the positive roots occurs at a very large value of $n_{tu,f,1}$, while one occurs at values less than 1.0 in most cases. Since the optimization criterion is to minimize the volume, the minimum of the positive roots should be adopted. Earlier numerical studies[3] showed that in general, it is desirable to design most MHEs with a "per plate ntu" less than 2.0. In most cases, the solutions converge rapidly when $n_{tu,f,1} = n_{tu,f,2} = n_{tu,p,1} = 1.0$ are used as starting values. No separate search for the larger root is necessary when the above procedure is followed. The Mathematica program written by us for the optimum design of MHEs consists of just 20 executable statements.

## Optimum Design Procedure

Based on the above formalism, the following procedure is suggested for an optimized design of matrix heat exchangers.

1. For a required effectiveness, allowable $\Delta p$, and other input parameters, compute the dimensionless $K_1'$, $K_1''$, $K_2'$, $K_2''$, $\xi$, $\varsigma$ and $\phi_1$ using equations (10), (12) and (13).
2. Solve equations (16) simultaneously for $n_{tu,f,1}$, $n_{tu,f,2}$, $n_{tu,p,1}$ and Z using a standard procedure such as Newton-Raphson technique. A starting value of $n_{tu,f,1} = n_{tu,f,2} = n_{tu,p,1} = 1.0$ can be used.
3. With $n_{tu,f,1}$, $n_{tu,f,2}$ and $n_{tu,p,1}$ known, determine the major dimensions of the exchanger (W, $H_1$, $H_2$) using equations (8) and (9).
4. Estimate the pressure drop on the high pressure side.

## RESULTS AND DISCUSSION

The optimum design procedure presented above has been used to evaluate the size of a matrix heat exchanger operating at an effectiveness of 95% with helium as the working fluid

Table 2. Different variables of a MHE estimated using the procedure presented in the paper. The exchanger operates with an effectiveness of 95%, flow rate of 0.2 g/s helium in either channel and an allowable pressure drop of 1500 Pa. ($T_{h,in}$ = 300 K and $T_{c,in}$ = 80 K)

| variable | cold fluid channel | hot fluid channel | variable | cold fluid channel | hot fluid channel |
|---|---|---|---|---|---|
| hole dia (input) | 0.5 mm | 0.2 mm | plate porosity (input) | 22% | 22% |
| plates | 94 | | pressure drop | 1500 Pa | 7250 Pa |
| Width | 56.35 mm | | Re | 105 | 232 |
| Plate height | 5.19 mm | 0.99 mm | Nu | 5.87 | 7.93 |
| $n_{tu,f}$ | 0.907 | 0.872 | $\lambda$ | 0.027 | |
| $n_{tu,p}$ | 1.03 | 5.64 | Fin eff[5] | 77 % | 95 % |

in either channel. The hot stream inlet temperature has been assumed to be 300 K and the cold stream inlet temperature to be 80 K. Perforated plates and spacers with a thickness of 0.5 mm have been chosen for the study. The exchanger has been designed for a helium flow rate of 0.2 g/s, with the working pressure of the hot and cold streams being 15 bar and 1.2 bar respectively. The allowable pressure drop on the low pressure stream has been assumed to be 1500 Pa. The height of the wall separating the two channels has been assumed to be 1.0 mm, while the height of the wall separating the two channels from the environment has been assumed to be 0.5 mm each.

The geometry of the exchanger estimated, along with the different dimensionless variables is presented in Table 2. The exchanger required to meet the design specifications needs a width of about 57 mm and a plate height of 6.19 mm and 0.99 mm respectively for the low and high pressure stream channels. A total of 94 plates are required for the exchanger. The equivalent fin efficiency of the plates[5] in the low pressure and high pressure fluid channels is 73% and 95 % respectively. The results clearly show that in spite of the use of stainless steel spacers, the overall performance of the exchanger is still controlled by the axial heat conduction ($\lambda$ = 0.027) through the walls.

Under certain conditions (e.g. large flow rates), the aspect ratio of the exchanger (width/plate height) designed using the above procedure may turn out to be large and unwieldy. In such cases, multiple channels such as those used by Fleming[5] may have to be used for the low pressure stream. The following procedure may be adopted for such cases:

1 The width of the multichannel exchanger may be taken as half of that of the estimated width of the optimum heat exchanger.
2 The height of the high pressure fluid channel may be taken as double the corresponding height in the optimum heat exchanger.
3 The height of each of the low pressure fluid channels may be taken the same as that in the optimum heat exchanger.

When the above procedure is adopted, the overall pressure drop and volume will not be much different from that of the optimum exchanger. The performance will be slightly lower because of larger axial conduction (wall area) in the multichannel heat exchangers. The exchanger will need to be rated and the geometry adjusted slightly (e.g. use some additional plates).

## CONCLUSIONS

- The effectiveness-ntu relations presented are useful in determining the effectiveness of matrix heat exchangers in terms of all the thermal resistances.
- A certain combinations of plate conduction resistance and the axial conduction resistance (equation 16) will result in a small matrix heat exchanger volume.
- The closed-form effectiveness expressions make it possible to derive a simple procedure for the design of MHEs.

## REFERENCES


1. G. Venkatarathnam and S. Sarangi, Matrix Heat Exchangers and Their application in Cryogenic Systems, Cryogenics, 30: 907 (1990)
2. S. Sarangi and J. Barclay, An Analysis of Compact Heat Exchanger Performance, "Cryogenic Processes and Equipment-1984", ASME, 37-44 (1984)
3. G. Venkatarathnam and Sunil Sarangi, Analysis of Matrix Heat Exchanger Performance, ASME J Heat Transfer, 113:830 (1991)
4. G. Venkatarathnam, Effectiveness-Ntu relationship in Perforated Plate Matrix Heat Exchangers, Cryogenics, 36:235 (1996)
5. R. B. Fleming, A Compact Perforated Plate Heat Exchanger, Advances in Cryogenic Engineering, 14:197 (1969)


## NOMENCLATURE

| | |
|---|---|
| A | Convective heat transfer area |
| $A_c$ | Area of Cross section |
| A''' | Heat transfer area per unit volume |
| b | separating wall width |
| C | constant in Nusselt number correlation |
| $C_f$ | drag coefficient |
| $c_p$ | specific heat at constant pressure |
| $G_d$ | mass flux through the perforation |
| h | heat transfer coefficient |
| H | height of the plate in each channel |
| k | thermal conductivity |
| K'' | intermediate parameter |
| K' | intermediate parameter |
| l | thickness of each plate |
| m | mass flow rate |
| n | number of plates |
| $n_{tu}$ | number of transfer units = dimensionless ratio of thermal conductance over ($mc_p$) |
| p | plate porosity (fraction of open area) |
| s | thickness of spacer |
| v | volume of the exchanger |
| $v_d$ | velocity through the perforation |
| W | width of the exchanger |
| Z | constant |
| $\beta$ | intermediate parameter |
| $\Delta p$ | pressure drop |
| $\Phi$ | intermediate parameter |
| $\gamma$ | exponent in nusselt number correlation |
| $\eta$ | intermediate parameter |
| $\kappa$ | intermediate parameter |
| $\lambda$ | overall axial conduction parameter |
| $\lambda_p$ | dimensionless lateral conductance of the separating wall |
| $\mu$ | viscosity |
| $\nu$ | heat capacity rate ratio |
| $\rho$ | density |
| $\tau$ | shear stress |
| $\xi$ | intermediate parameter |
| $\zeta$ | intermediate parameter |

*Subscripts*

| | | | |
|---|---|---|---|
| 1 | cold fluid i | 1,2 | |
| 2 | hot fluid | o | overall |
| eff | effective | p | plate |
| f | fluid | s | spacer |

# PHILLIPS LABORATORY SPACE CRYOCOOLER DEVELOPMENT AND TEST PROGRAM

J. Bruning [1] and T. Pilson[2]

[1]Nichols Research Corporation
Leasburg, Missouri 65535
[2]Air Force Phillips Laboratory
Kirtland AFB, New Mexico 87117

## ABSTRACT

This paper examines the results of long-life space cryocooler development and testing for various defense needs. For well over a decade, both the Air Force and the Strategic Defense Initiative Organization (now Ballistic Missile Defense Organization, or BMDO) recognized the need for long-life space cryocoolers capable of handling a variety of cooling loads spanning the temperature range between 8.5-150 Kelvin. Space borne cryocoolers had a multitude of defense system-level applications, including cryogenic cooling of chilled electronics/CMOS, infrared sensors, missile seeker precooling, launch platform systems checkout and health monitoring, maintaining stored cryogens, and high temperature superconducting device operation. Efforts to develop and fully demonstrate cryocoolers for long-term space operations have had mixed results, with several aspects of designs still preventing true long-life performance of these machines. This paper presents a summary of the results and conclusions of efforts to develop long-life cryocoolers. These results include findings from historical developments, current research areas, and Phillips Laboratory's extensive ongoing cryocooler test and evaluation efforts. Information presented emphasizes key findings and lessons learned that may help improve existing designs and concepts so that these machines can finally realize their long-life potential. These lessons learned are also examined in light of newly emerging space system requirements that will set the pace for long-life, high efficiency, larger capacity cryocoolers well into the next decade and beyond.

## INTRODUCTION

Current and projected Air Force and Ballistic Missile Defense architectures include a variety of space systems that require cryogenic refrigeration to support operations. Some spacecraft have even identified cryocoolers as the critical enabling technology essential for system development and deployment. Defense system cooler applications include infrared surveillance, acquisition, tracking, and pointing sensors, cryo-chilled electronics, stored cryogen maintenance and conditioning, low/high temperature superconducting devices, radar receivers and transmitters, and laser wavelength tuning.[1,2] Successful establishment of effective defense satellite networks, force projection systems, and space transportation elements that include any of these applications depends on the availability of a strong cryocooler technology base.[3] Phillips Laboratory (PL) has been actively pursuing long-life cryocooler technology development since 1984 to help meet these current and newly emerging requirements. PL's cryocooler development and test program has historically focused on five major thrust areas:

1) Development of new cryocoolers showing promise for achieving 95% reliability for a 5-10 year on-orbit operational life while exhibiting highly efficient and sustained thermodynamic performance as needed and defined by near-term systems within the Air Force and BMD architectures.

**Table 1.** Cryocooler Characteristics

| Project | Temp. | Cap. | Life | Rel.$_a$ | Input Power$_b$ | Induced Vibration | Mass$_c$ | Status |
|---|---|---|---|---|---|---|---|---|
| **Prototype Flight Cooler[4] (R-Cube)** | 8.5 K<br>25 K<br>80 K | 2.5 W<br>9 W<br>70 W | 5 Yrs Contin-uous | 97% | 4550 | TBD | 923 Kg | Design Complete w/ Key Parts Built |
| **Prototype Flight Cooler[5] (Tubo-Brayton)** | 8.5 K<br>25 K<br>80 K | 2.5 W<br>9 W<br>70 W | 5 Yrs Contin-uous | 96% | 5860 | - | 212 Kg | Design Complete w/ Key Parts Built |
| **Single Stage Turbo Cooler[6]** | 65 K | 5 W | 5 Yrs Contin | 97% | 226W | - | 24 Kg | EDM Built T& E Ongoing |
| **65 K Standard Spacecraft Cooler[6] (Split Stirling)** | 65 K | 1.2-2.2 W | 10 Yrs Contin-uous | 96% | 79 W | (C)>1 Nrms<br>(D)>1 Nrms | 25 Kg | 2 EDMs w/ Refurbishment T&E Ongoing |
| **65 K Standard Spacecraft Cooler[6] (Diaphragm Stirling)** | 65 K | 1 W | 10 Yrs Contin | 95% | ~85 W | (C)>1 Nrms<br>(D)>1 Nrms | 18 Kg | EDM Built T&E Ongoing |
| **65 K SPAS III Stirling Cryocooler[7]** | 65 K | 2 W | 5 Yrs | 95% | 125 W | (C)>1 Nrms<br>(D)0.5 Nrms | 25 Kg | Flight Unit Built / Qualified |
| **Multi-Stage Ten Kelvin Stirling Cooler[8]** | 10 K<br>35 K<br>80 K | 0.15 W<br>2 W<br>5 W | 10 Yrs Contin-uous | 95% | 530 W | TBD | 33 Kg | Conceptual Design Incomplete |
| **65 K Sorption J-T Cooler[9] (Chemisorption)** | 65-75 K | 2-5 W | 5-10 Yrs Contin-uous | 95% | 77-200 W | - | 17-32 Kg | Preliminary Design w/ Part Reliability T&E |
| **Intermittent 10 K Sorption$_d$ Cooler[10]** | 10 K<br>30 K<br>65 K | 0.15 W<br>1.5 W<br>8 W | 7-10 Yrs Low Duty Cycle | 95% | 125-140 W | - | <55 Kg | Proof-of-Principle Test/ STS Test Pkg. |
| **Multi-Stage 10 K Sorption Cooler[8] (Hydride/Turbo Hybrid)** | 10 K<br>35 K<br>80 K | 0.15 W<br>2 W<br>5 W | 10 Yrs Contin-uous | 95% | 752 W | - | 56 Kg | Conceptual Design Incomplete |
| **35 K/1 Watt Pulse Tube Cooler[8]** | 35 K | 1 W | 5-10 Yrs Contin | 95% | 235 W | TBD | 27.2 Kg | EDM Built T&E Ongoing |
| **35 K/0.3 Watt Pulse Tube Cooler[8]** | 35 K | 0.3 W | 5-10 Yrs Contin-uous | 95% | 172 W | 0.1 Nrms | 19.5 Kg | Protoflight Unit Built T&E Ongoing |
| **60 K/2 Watt Pulse Tube Cooler[8]** | 60 K | 2.0 W | 5-10 Yrs Contin-uous | 95% | 108 W | 0.4 Nrms | 19.5 Kg | Protoflight Unit Built T&E Ongoing |
| **150 K Protoflight Spacecraft Cryocooler[8]** | 150 K | 1 W | 7-10 Yrs Contin-uous | 95% | 19.3 W | TBD | 10.4 Kg | Protoflight Unit Built T&E Ongoing |
| **3 Stage 10 K Pulse Tube Cooler[8]** | 10 K<br>35 K<br>80 K | 0.15 W<br>2 W<br>5 W | 10 Yrs Contin-uous | 95% | 830 W | TBD | 41 Kg | Conceptual Design Incomplete |
| **Quick Cooldown J-T Coolers (3 units)[6]** | 60 K<br>30 K<br>10 K | 1 W<br>1 W<br>0.5 W | Up to 10 Yrs Storage | 99% | - | N/A | <100 grams each | Brassboard Units For Inter-ceptor Program |

a. Reliability figures are design estimates based on analysis (not demonstrated).
b. Total input power from spacecraft bus, including flight control electronics (or equivalent estimates).
c. Total cryocooler mass, including flight electronics package (actual or design estimates).
d. Includes mass and input power for an 8 W/65 K SSC-class cryocooler for 1st stage load requirements.

2) Demonstration and validation of emerging technologies developed under the AF/BMDO cryocooler technology program by conducting extensive test and evaluation (T&E) of new coolers, their key components, and related thermal management and integration technologies.

3) Exploring advanced concepts that have the potential to significantly enhance cryocooler capabilities well beyond the state-of-the-art by leveraging new advances in science, physics, and engineering.

4) Ensuring the availability of essential components to help effectively integrate the coolers with a spacecraft and its cooled instruments or other specific application. This includes not only the physical mounts, but vibration suppression, thermal isolation and insulation, EMI/RFI shielding, thermal storage units, heat transport devices, thermal straps, and heat switch technology.

5) Developing producibility technologies and strategies for cost effectively fielding cryocoolers and their integration and support components. The focus of this area is reducing the per-unit costs for coolers while reducing the unit-to-unit performance variation among like coolers in a production lot by enhancing manufacturing methods and process technology.

Recent emphasis at PL has been on the first two major thrust areas. This is due to decreasing funding availability and changing programmatic priorities within both the Air Force and BMDO. This paper summarizes the status and trends of the current cryocooler development and testing program at PL while citing more detailed references that explore various program elements in greater depth.

## CRYOCOOLER DEVELOPMENT

### Recently Completed Projects

Ongoing space systems requirements analyses and technical assessments identified numerous gaps in available cryocooler technology necessary for meeting Department of Defense (DoD) cooling needs. PL formulated a program that addressed these technological deficiencies by seeking to develop a wide range of cryocoolers for various system-level applications. Table 1 presents a summary of the cryocooler projects recently completed under PL's management purview.

### Current Development Efforts

PL has several ongoing cryocooler development projects that build on the successes and lessons learned while pursuing the technology base efforts summarized in Table 1. Further refinement of existing system requirements, as well as the emergence of backup technology strategies and pre-planned product improvement initiatives, are the drivers in the current mix of developmental programs underway. Because of the ever increasing demands of system users on cooler performance characteristics, PL has focused more recent development activities on several key aspects of cooler design and operation. Current project drivers including induced vibration reduction, multistage capabilities in the 35 K and 60 K operating temperature ranges, improved thermodynamic efficiencies of newly emerging technology, extending the life of tactical units, and establishing a more mature 10 Kelvin cooler technology base.

*60 K Protoflight Spacecraft Cryocooler (PSC):* The PSC (figure 1a) is a continuation of the Standard Spacecraft Cryocooler (SSC) family of Stirling cycle coolers developed by PL for BMDO. The PSC project builds on the lessons learned on the SSC engineering model from its design and fabrication, through its modifications during the "Improved SSC" multiple cooler production run, to the current T&E phase at PL. The thrust of the development activities on the PSC are meant to evolve the SSC-class cooler to a flight-ready protoflight qualification level[11]. Mass and system penalty are reduced while improving reliability by using higher efficiency motors, lightweighting the housings of both the compressor and expander units, hermetically sealing the unit, and improving thermodynamic efficiency by using titanium elements in the expander to improve thermal performance while reducing parasitic losses. The PSC is designed to provide over 2 Watts of cooling at 60 K with less than 100 Watts input power to the machine. This class of cooler must be able to operate with 95% reliability for over seven years of either continuous or intermittent operation. The cooler is currently in developmental performance testing at Hughes Aircraft Company and is scheduled for delivery to PL for further characterization during late-FY97.

*35 K/60 K Multi-Stage Stirling Cryocooler:* This cooler (figure 1b) addresses one potential DoD requirements set for space-based surveillance where dual load cooling is necessary for long wavelength infrared sensor operations. The objectives of this program are to develop a low vibration, highly reliable, seven year life three-stage Stirling cycle cooler that provides simultaneous 0.4 Watts at 35 K and 0.6 Watts at 60 K. The thrust of this project is to scale the compressor initially built under a NASA/Goddard

(a) Hughes PSC

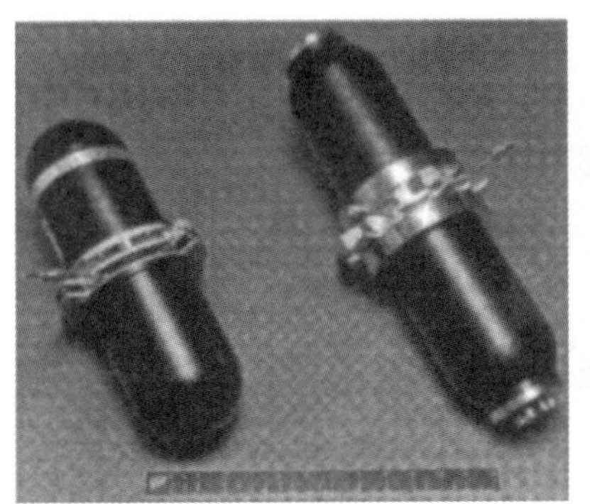
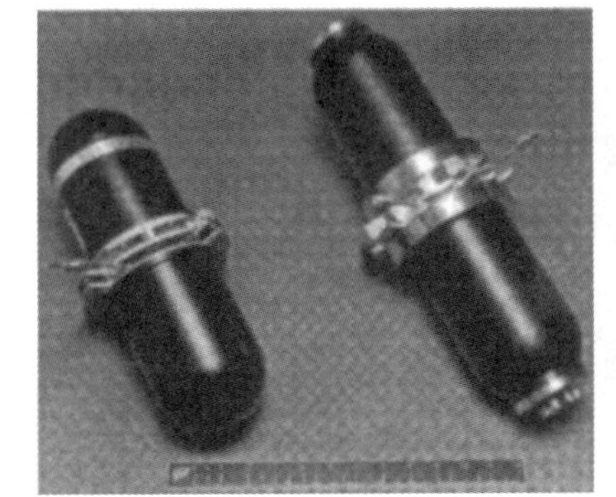

(b) Ball Multi-Stage Stirling Cooler

(c) Creare MSRB

**Figure 1.** Current cryocooler development efforts by PL.

Space Flight Center 30 K two-stage cooler program to effectively double its capacity at the dual load design points. The cooler is currently in fabrication at Ball Aerospace in Boulder, Colorado and is scheduled for delivery early in FY98.

*35 K/60 K Multi-Stage Reverse Brayton Turbocooler (MSRB):* The MSRB (figure 1c) is an alternate approach to providing the long-term cryogenic cooling necessary for the 35 K/60 K requirements previously described for the multi-stage Stirling cooler. This effort is a continuation of the evolution of the Single-Stage Turbo Cryocooler (SSTC) concept shown in Table 1. The SSTC engineering development model is currently in endurance characterization at PL where long-term performance data is being provided back to its developer, Creare, Inc. of Hanover, New Hampshire, for design improvements. The MSRB is a two stage version of the SSTC that provides 0.4 Watts of cooling at 35 K and 0.6 Watts at 60 K. Efforts are underway to improve the efficiency of the turbo cooler concept over the earlier generation EDM machine by incorporating a turboalternator at the cold end, reducing parasitic leaks into the expander section through incorporation of cryogenic cold bearings, and using more effective rare earth magnets in the compressor rotor shaft. The machine is being lightweighted further by using a compact, high efficiency radial flow heat exchanger in place of the slotted copper plate design on the EDM. Better reliability is being addressed through improvements in contamination control techniques. Component improvement efforts on the MSRB program are being baselined by a new cryocooler initiative for use on the Hubble Space Telescope (HST) NICMOS instrument.

*10 K Multi-Stage Stirling Cryocooler:* Matra Marconi Space is under contract by PL to investigate the potential for 10-12 K operation with 0.05-0.1 Watts capacity utilizing an existing 20 K Stirling cycle cryocooler compressor. A new three stage cold head is being developed under this effort and is key to realizing the 10 K performance. Consultants from Rutherford B. Appleton will assist in scaling the existing component technology through modeling and simulation techniques in order to help achieve 10 K operation of the cooler.

*65 K Mini-Stirling Cooler:* The British Defense Research Agency (DRA) is under contract with BMDO through PL to improve the performance of tactical cryocoolers by incorporating long-life cooler component technology with the promise of increasing life and reliability. The DRA cooler is a small integral Stirling machine capable of 0.25 Watts cooling at 65 K for less than 30 Watts input power to the unit. Its design life is 20,000 hours. The unit is nearing completion and is scheduled for delivery at PL's cryocooler characterization laboratory in the June-July 1997 timeframe.

## CRYOCOOLER TEST AND EVALUATION (T&E) PROGRAM

Very little detailed performance, life, and reliability data exists for the thermal technologies of interest to near-term space system developers. Cryocooler developers and users need accurate, in-depth data on the performance of coolers to better understand their inherent nature and long-life potential. Ascertaining design strengths and limitations helps the technology developers refine their designs while also helping the user community select the best designs to meet their cryocooling needs. In addition to meeting contractual requirements for thermal performance, weight, volume, material compatibility, voltages, and other physical and functional constraints, coolers must be further characterized to determine the specific power, time-dependent capacity drift, cold end temperature stability, vibration, EMI/EMC, operating life, and reliability. These critical parameters and cooler attributes form the basis for evaluating the suitability of coolers to meet DoD mission requirements. PL established extensive capabilities to measure and monitor the essential elements of analysis associated with cryocoolers. PL then structured their overall characterization T&E program into three separate focus areas-performance characterization, endurance or life testing, and system-level integrated technology demonstrations. Together, these characterization activities serve to complete the R&D phase of newly emerging cryocooler technology.

### Performance Characterization

Before space system developers baseline a cryocooler to satisfy long-term mission requirements, an accurate and complete understanding of cooler performance and operational sensitivities is essential. PL established a broad performance characterization program to gather, analyze, and disseminate data on recently developed long-life space coolers. These PL T&E activities encompass a wide array of standardized tests that serve as a basis for normalized comparisons among different cooler designs. Performance characterization at PL includes contractual acceptance testing that demonstrates conformance to specifications, off-design point performance mapping, thermal and mechanical functional tests, thermal cycling, induced vibration, and flight qualification testing according to accepted DoD and Air Force standards[11] as well as accepted practice within the cryocooler development community. Performance data gathered at PL is used to establish a detailed database of cryocooler information invaluable to space system developers during their preliminary tradeoffs and detailed design activities. Figure 2 shows the different

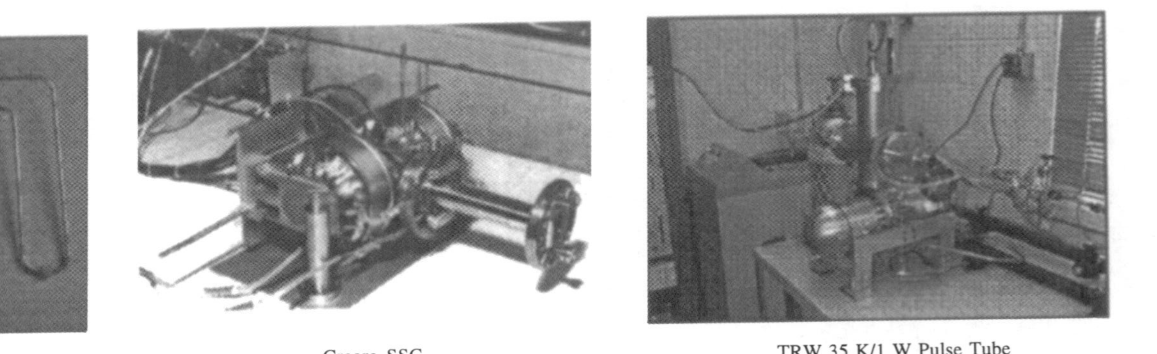

Creare SSTC

Creare SSC

TRW 35 K/1 W Pulse Tube

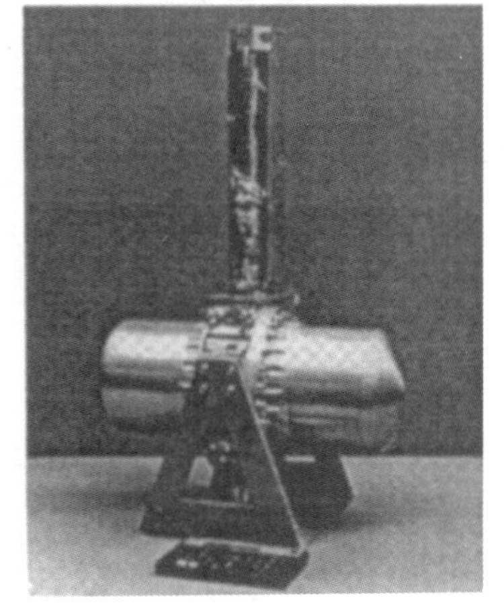

TRW 35 K/0.3 W Pulse Tube

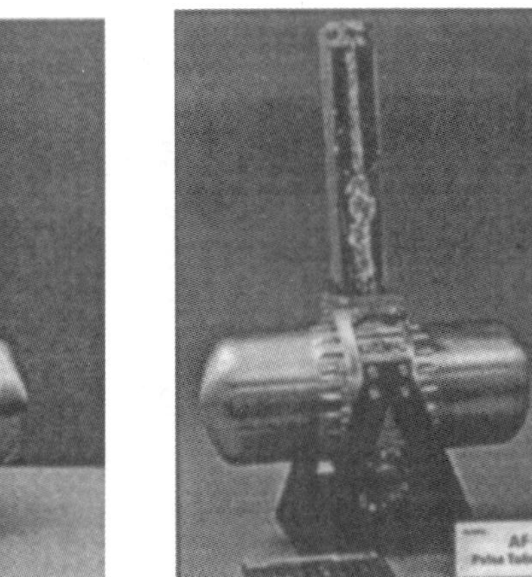

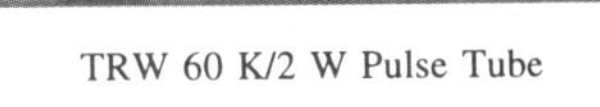

TRW 60 K/2 W Pulse Tube

Hughes SSC (2 Units)

**Figure 2.** Cryocoolers undergoing performance characterization at PL.

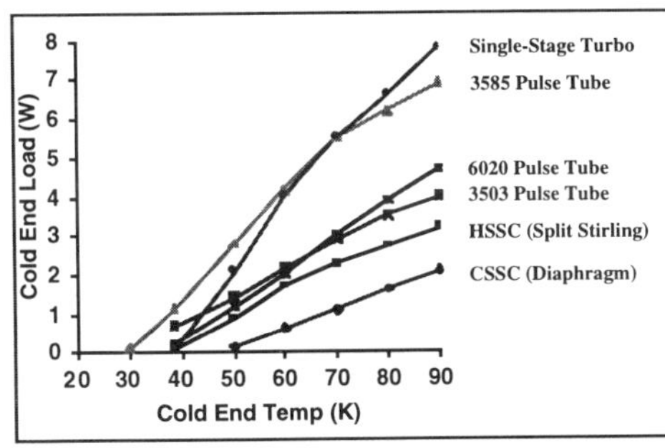


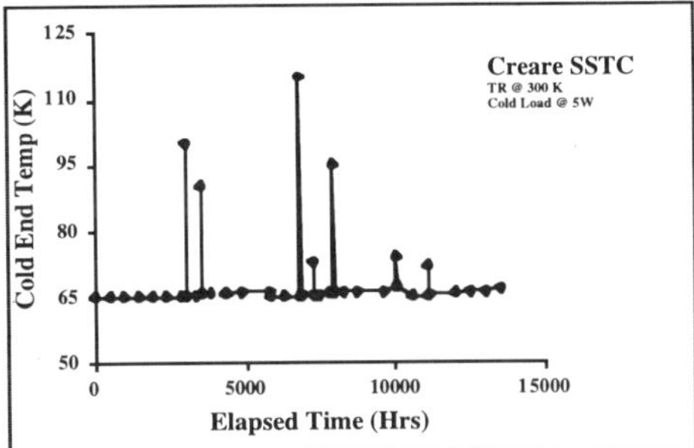


**Figure 3.** Nominal performance curves and sample time-dependent performance trends of coolers in test at PL.

coolers at PL currently in various stages of performance characterization. Of those shown, the SSTC and metal diaphragm 65 K SSC have completed performance characterization and are now undergoing endurance testing.

## Endurance Characterization

Of the parameters used to monitor and evaluate the suitability of coolers for DoD spacecraft use, *life and reliability* delineates between the capabilities of short-lived tactical units and long-life space cryocoolers. Verifying life and reliability (at user specified confidence levels) is the current primary focus of PL's cryocooler characterization T&E activities. Contributing factors to cooler unreliability are as important to understand as the inherent life and reliability of the machines themselves. Some of the potential contributors to cryocooler unreliability include wearout, drift, fatigue, material creep, gaseous contamination, particulate/compound contamination and clogging, material and workmanship defects, inadequate machining process development, assembly errors, material thermal expansion mismatch, and long-term alignment instability. However, all potential failure modes and life limiting mechanisms are not identified, and those that have been are not completely understood. Also, various cryocooler concepts can have vastly different inherent failure modes and effects, as well as "graceful degradation" characteristics. Because of the criticality of validating cryocoolers as long-life spacecraft components, a very high premium is placed on gathering cooler-specific life, reliability, and long-term performance trend data. Test facilities, procedures, and equipment/instrumentation all serve to assist in running long-duration endurance and life tests in environmental conditions simulating actual space use. Once underway, endurance and life tests are normally run until cooler failure, but not less than their design life if the test article continues to exhibit adequate performance. Figure 3 shows an example of the type of time-temperature history data emanating from endurance characterization. The other coolers shown in figure 2 will all eventually enter into similar tests.

## Space System-Level Demonstrations

This category of technology characterization at PL is similar to simulation testing, except that it is performed at the space system level. Major subsystems are integrated together and tested in the same or similar configuration planned for the eventual spacecraft using the component technology involved. Demonstration tests are high level, high fidelity "proof tests" showing that the major subsystems work well together and could enable effective spacecraft operation and enhance mission success probability determinations. Ideally, these tests can range from ambient bench testing to complex spaceflight experiments. Recently, PL assisted Sandia National Laboratories in evaluating their Multispectral Thermal Imager sensor operation using a pulse tube cryocooler of the same flight heritage as their spacecraft's selected unit. Integration constraints, thermal analysis and validation, cooler/focal plane array compatibility, and operational mode characterization were all conducted to exacting spacecraft specifications and mission profiles. Figure 4 shows these activities in progress at PL, as well as an example of the data resulting from such demonstration tests.

# RESULTS AND LESSONS LEARNED

With the large number of cooler development and characterization efforts pursued over the past decade, significant lessons have been learned that will help in the drive to achieve 5-10 years of on-orbit

MTI Sensor Demonstration Integration Activities

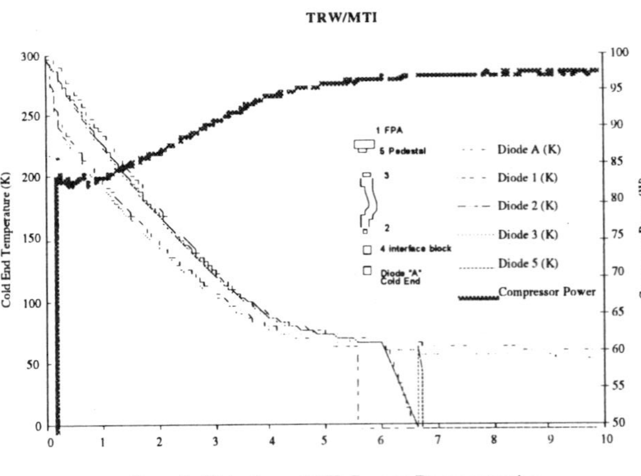


Sample Data from MTI Sensor Demonstration

**Figure 4.** Some activities and results from system-level demonstrations at PL.

operational life in these machines. While impossible to list all design issues here, some of the major results, conclusion, and recommendations from recent developments are listed below.

1) Better alignment techniques are generally needed to improve assembly time, reduce cost, and decrease the unit-to-unit performance variations experienced in like coolers. This can also help achieve non-contacting running clearances for seals and close tolerance parts. Better alignment methods and dimensional stability techniques would enable a true all-metal design cooler.

2) The trend towards increasingly allowing exposure of internal working fluid volumes to organic components is not commensurate with reliability growth. Significant data exists on previous developments that show unit life increases dramatically when exposure to organics inside the cooler are minimized or eliminated altogether.

3) Contamination continues to cause concern, whether from outgassing from internal elements of the cooler, poor hermeticity, improper or poorly defined cleanliness techniques during fabrication, inadequate fill procedures, or a combination of these. Some contamination problems are procedural in nature and can be corrected with better engineering practice and technical discipline. Others are inherent problems associated with design weaknesses in parts processing and materials selection.

4) Better manufacturing process validation is needed prior to fabrication and assembly. This includes better on-site inspection and tracking methods for vendors and part suppliers.

5) Configuration control among evolving designs in the same cooler family are essential to ascertaining reliability growth prospects.[12, 13] Loss of heritage is a real concern among some coolers where this design evolution process is allowed to proceed without proper documentation and formal engineering review.

6) Unit costs must be reduced to demonstrate the producibility of long-life coolers and allow multiple cryocooler production for statistical reliability determinations. This is essential for further life and reliability improvements before high cost space systems can baseline these machines.

As future DoD space systems identify new and more demanding requirements for cryocoolers, the technical hurdles listed here may be significantly augmented by new challenges that must be met as well. Ideas on how to solve some of these deficiencies may already exist within the cryogenic research and engineering community. Some ideas may emanate from science and technology endeavors not even related to cryocooler development. In either case, PL remains committed to continuing in its efforts to field long-life, highly reliable cryocoolers for both near and far-term space system applications. The key to making this possible is continuing to work closely with others in government, industry, and academia who share this common goal.

## ACKNOWLEDGMENTS

PL is conducting the research, development, test, and evaluation activities outlined in this presentation under the primary sponsorship of the Ballistic Missile Defense Organization, BMDO/TRS, The Pentagon, Washington, DC. Other major contributors to PL's work include the Air Force Space and Missiles System Center, Los Angeles AFB, CA and Sandia National Laboratories in Albuquerque, NM. Special mention must also be made of the technical and programmatic support from NASA/Goddard Space Flight Center and the Jet Propulsion Laboratory, without which the quality and level of cooler work at PL could not have been possible.

## REFERENCES

1. "Military and Commercial Applications for Low Cost Cryocoolers", *Proceedings of the Electronics Industries Assoc. Workshop*, San Diego, CA, 31 January-1 February 1996, EIA, Arlington, VA (1996).
2. K. P. Hyde, J. R. McCoy, C. S. Naiman, "Applications and Packaging of Semiconductor Devices for Use at Cryogenic Temperatures", *7th International Cryocooler Conference Proceedings*, Air Force Phillips Laboratory Report PL-CP--93-1001, Kirtland AFB, NM (April 1993) p. 1064-1066.
3. "Aerospace Thermal Systems", *Military Critical Technologies List*, Office of the Under Secretary of Defense for Acquisition, Washington, DC, Government Printing Office (October 1992).
4. R. W. Breckenridge, et al., "Prototype Flight Cooler Design and Component Demonstration", Arthur D. Little, AFSTC-TR-88-01, Kirtland AFB, NM (July 1988).
5. P. G. Wapato, et al., "Prototype Flight Cryocooler Design and Component Demonstration", Allied-Signal Aerospace Co., AFSTC-TR-87-14, Kirtland AFB, NM (December 1987).
6. W. Wyche, J. Bruning, "The AFSTC Cryocooler Technology Development Program Plan for Meeting SDIO Requirements", *Proceedings of the Fourth Interagency Meeting on Cryocoolers*, David Taylor Research Center Report DTRC-91/003, Bethesda, MD (January 1991), pp. 3-15.

7. I. E. Spradley, W. G. Foster, "Space Cryogenic Refrigerator System (SCRS) Thermal Performance Test Results", *Cryocoolers 8-Proceedings of the 8th International Cryocooler Conference*, Vail, CO, 1994, Plenum Press, New York (1995), pp. 13-22.
8. P. J. Thomas, "SDIO and Air Force Cryocooler Technology Development at USAF Phillips Laboratory", *7th International Cryocooler Conference Proceedings*, Air Force Phillips Laboratory Report PL-CP--93-1001, Kirtland AFB, NM (April 1993) p. 3-13.
9. S. Bard, "JPL/AFSTC Sorption Cryocooler Development Project", JPL Design Study Report for AFSTC/SDIO, JPL D-7467, Pasadena, CA (May 1990).
10. R. C. Bowman, L. A. Wade, et al., "Fabrication and Testing of the Metal Hydride Sorbent Bed Assembly for a Periodic 10 K Sorption Cryocooler", *Cryocoolers 8-Proceedings of the 8th International Cryocooler Conference*, Vail, CO 1994, Plenum Press, New York (1995), pp. 601-608.
11. MIL-STD-1540B (USAF), "Military Standard Test Requirements for Space Vehicles", 10 October 1982.
12. W. G. Ireson, C. F. Coombs, Jr., "Handbook of Reliability Engineering and Management", McGraw-Hill Book Co., St. Louis, (1988).
13. MIL-HDBK-189, "Reliability Growth Management", Def. Tech. Information Center, Alexandria, VA.

# AN UPDATE ON ESTIMATING THE COST OF CRYOGENIC REFRIGERATION

R. A. Byrns and M. A. Green

E. O. Lawrence Berkeley National Laboratory
University of California
Berkeley, CA 94720

## ABSTRACT

This paper is an update of a report written in 1991[1] concerning cost estimating of refrigeration for superconducting magnets. This report updates cost data collected by Strobridge[2,3,4] over 20 years ago and data collected by these authors. The actual costs have been inflated to 1997 dollars and are plotted. A correlation function between cost and refrigeration at 4.5 K and 1.8 K is given. An annual cost for refrigeration can also be estimated based on the refrigeration at 4.5 K or 1.8 K and the cost of electrical energy. A correlation function for estimating input power to the compressors to the refrigeration produced at 4.5 K and 1.8 K is also given.

## INTRODUCTION

This report updates costs given in a 1991 paper concerning the cost of superconducting magnets and helium refrigeration[1]. This report is restricted to the estimating of cost for helium refrigeration only. The magnet costs are not included.

In 1966, Strobridge, Mann and Chelton[2] developed a technique for estimating the cost of helium refrigerators based on a limited number of cost data points available at the time. In 1969, Strobridge[3] updated his study to include cryogenic refrigerators of all types. The cost of refrigeration was estimated based on the input power to the compressor. The 1969 Strobridge study was expanded in 1974[4] to include a number of newer refrigerators being built at that time. From 1966 to 1974, the cost of helium refrigeration did not change appreciably. From 1974 to the present, the capital cost of refrigeration appears to have escalated at the nominal rate of inflation. The 1991 report extended the number of refrigerators used to make the refrigeration cost estimate.

This report includes a number of the new small cryocoolers that have appeared on the market during the last five years. There is a trend to use these small cryocoolers to cool superconducting magnets for MRI and other magnets that operate in persistent mode.

One of the difficulties of estimating the cost of helium refrigeration is the choice of the appropriate parameter from which to calculate cost. For most of the helium refrigeration systems in this report, the refrigeration capacity at 4.5 K is used as a scaling factor. Thus the cost of small cryocoolers is based on the refrigeration generated at 4.5 K. Helium refrigeration at other temperatures is scaled to 4.5 K using the Carnot ratio. Liquefaction is converted to 4.5K refrigeration by the use of the refrigeration liquefaction coefficient (This coefficient is between 75 to 125 J per gram depending on how the cycle has been optimized).

## THE THERMODYNAMIC EFFICIENCY OF HELIUM REFRIGERATORS

Strobridge in his papers[2,3,4] discussed the efficiency of various kinds of refrigerators. Efficiency was defined as the input power of a perfect Carnot cycle refrigerator over the real compressor power that goes into the refrigerator. Figure 1 is an efficiency plot that contains both the Strobridge data and newer data. The efficiency data shown in Figure 1 shows a great deal of scatter as the original Strobridge data did. Many of the refrigerators shown in Figure 1 have liquid nitrogen precooling. The nitrogen precooler can have the effect of enhancing the apparent efficiency of the machine. Apparent efficiency increase can be a factor from 1.2 to almost 2.0, if the input power used to generate the liquid nitrogen is not used in the input power calculation. Large refrigerators use less liquid nitrogen per unit refrigeration produced than do small refrigerators. When one considers the power needed to produce liquid nitrogen, the efficiency advantage of liquid nitrogen precooled machines often disappears. The capital cost of machines using nitrogen precooling is often lower.

The effects of plant size on efficiency can be seen in Figure 1. Efficiencies range from less than 1 percent of Carnot for 1 watt plants to greater than 30 percent of Carnot for 10 kW plants. Increased plant size requires larger pistons or rotors with fixed clearances. The leakage is reduced with respect to total flow through the expander or compressor. Smaller cold boxes can only have one or two expanders, whereas the large machines can have up to seven staged expanders[5]. As the size of the plant grows, the turbines can have efficiencies that are competitive with piston expanders. Turbines are often chosen because they are thought to be more reliable, but this improved reliability may be illusionary in plants with quickly varying loads. For plant sizes under 1000 W, a piston expansion system is often the best choice. Since a major part of the plant inefficiency is in the compressors, the larger plants will often have staged compressors with intercooling between stages.

The newer data points shown in Figure 1 show that on average the overall efficiency of helium refrigerators has not increased. There are a number of reasons for this: 1) A number of the newer machines, (particularly those clustered between 80 and 400 W) have no liquid nitrogen precooling. 2) Many new machines use screw compressors. Screw compressors are reliable, but they can be less efficient (particularly if they are small single stage machines). 3) There are more turbine expanders in small machines. Small turbines are less efficient than small piston expanders. It should be pointed out that refrigeration plants that are more efficient will have a higher capital cost. If one considers cost of operation when buying a refrigeration plant, that plant will have a higher capital cost and efficientcy.

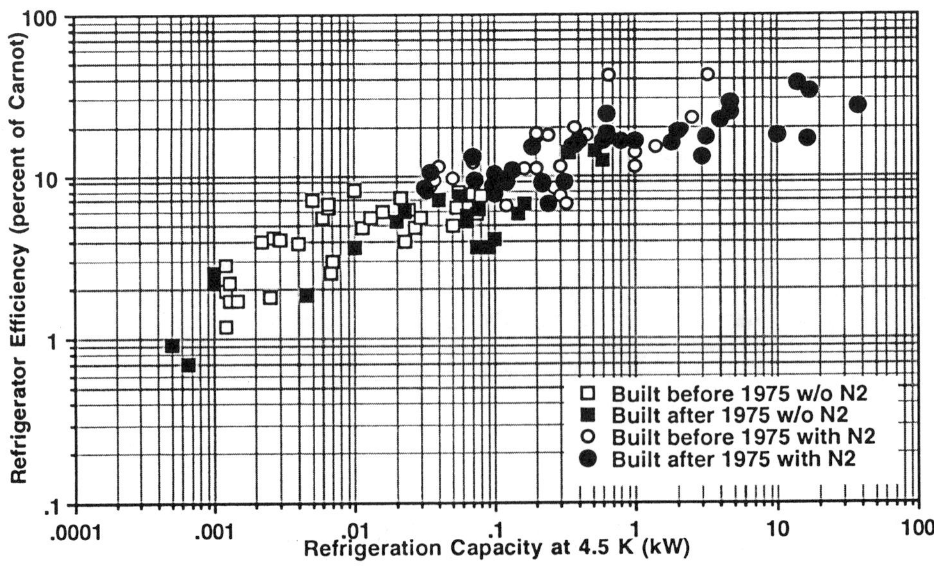


**Figure 1.** The Efficiency of Helium Refrigerators as a Function of 4.5 K Refrigeration.

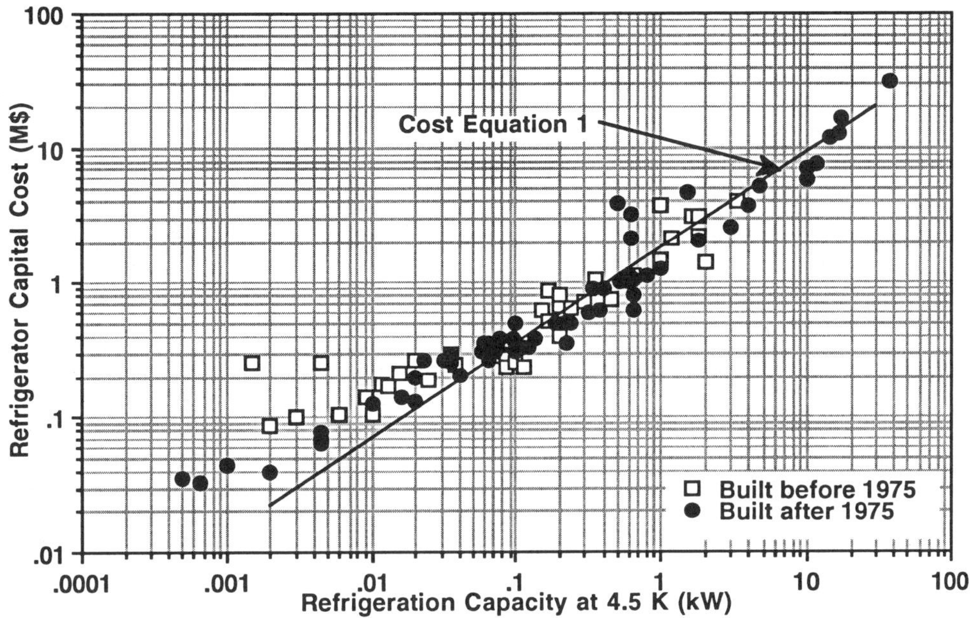


**Figure 2.** The Cost of Helium Refrigerators versus Refrigeration at 4.5 K.

## LARGE HELIUM REFRIGERATOR CAPITAL COST

Figure 2 shows the cost of various 4.5K refrigerators (compressors and cold boxes) escalated into 1997 dollars as a function of refrigeration at 4.5 K. Liquefaction was converted to refrigeration using the refrigeration-liquefaction coefficient for the machine. Refrigeration at temperatures different from 4.5 K have been converted to 4.5 K refrigeration by using the Carnot ratio. The cost of foreign made machines was converted to dollars at the exchange rate of the year of manufacture. The dollars were escalated from the year of manufacture to 1997 dollars (See Reference 6).

From Figure 2, one can see that escalated cost of refrigerators made before 1975 is about the same in 1997 dollars as the escalated cost of refrigerators made after 1975. In Figure 2 there is a line plotted with the cost points. This line represents the average capital cost in 1997 dollars for 4.5 K helium refrigerators that produce refrigeration in the range from 0.1 to 30 kW. The equation for this line is:

$$C(M\$) = 1.77\,[R(kW)]^{0.7} \tag{1}$$

where the capital cost is given in millions of 1997 dollars and R is given in kilowatts of refrigeration at 4.5 K.

The largest plants shown in Figure 2 can be quite complex. Some of these plants have several cold boxes tied together. Other plants may include helium pump systems to circulate subcooled helium. Liquid helium storage and distribution systems can cost almost as much as the basic cold box and compressor system. On the room temperature side of the plant, helium gas storage, recovery, purification and distribution can add significantly to the cost of the plant. Other factors that increase the cost of refrigeration include: computer control systems (In small machines computer controls are usually not needed.), extra purification within the cold box (sometimes a blessing, sometimes a curse), and the extra documentation that may be required by military type specifications. In some systems, the computer control system can cost up to one third the cost cold box and compressor. The costs of computer software and the debugging of that software are often grossly underestimated. If one is interested in minimizing refrigerator capital cost, one should seriously consider whether a computer control system is needed.

## SMALL HELIUM REFRIGERATOR (CRYOCOOLER) CAPITAL COST

Small refrigerators are increasingly being used for various types of superconducting magnets. These magnets are often operated in persistent mode. Magnet leads are either retracted or they are low loss leads using high Tc superconductors. In general, small cryocoolers can not produce liquid helium. As a result, this type of refrigerator can not be used to supply gas to electrical leads or shields. These machines are often staged so that heat can be pulled off (by conduction to the cold heads) at more than one temperature. A machine that produces cooling at a higher temperature will produce less cooling at 4.5 K. The amount of refrigeration available at the higher temperature is approximately proportional to the refrigeration lost (not available) at 4.5 K times the Carnot ratio. As an example, a refrigerator that produces 2 watts of refrigeration at 4.5 K will produce about 11.5 watts at 45 K plus 1 watt at 4.5 K.

Small refrigerators (less than 40 W) will cost more than is given by Equation 1 (see Figure 2). These machines do not have liquid nitrogen precooling. Even if they did have nitrogen precooling, they would be inherently less efficient. Therefore, these machines require larger compressors with respect to their refrigeration output at 4.5 K. It can be argued that small plant cost should be plotted separately and that this type of plant should have a separate cost equation. Figure 3 shows a plot of the cost of various small helium refrigerators (given in thousands of 1997 US dollars) versus the maximum refrigeration available at 4.5 K (given in watts).

From Figure 3, one can see that escalated cost of small refrigerators fits a different cost equation than is given by Equation 1. In Figure 3 there is a line plotted with the cost points. This line represents the average cost in 1997 dollars of modern helium refrigerators (built after 1975) which produce refrigeration from 0.5 to 60 W. The equation for this line is:

$$C(k\$) = 45\,[R(W)]^{0.5} \qquad (2)$$

where the capital cost C is given in thousands of 1997 US dollars and R is given in watts of refrigeration at 4.5 K.

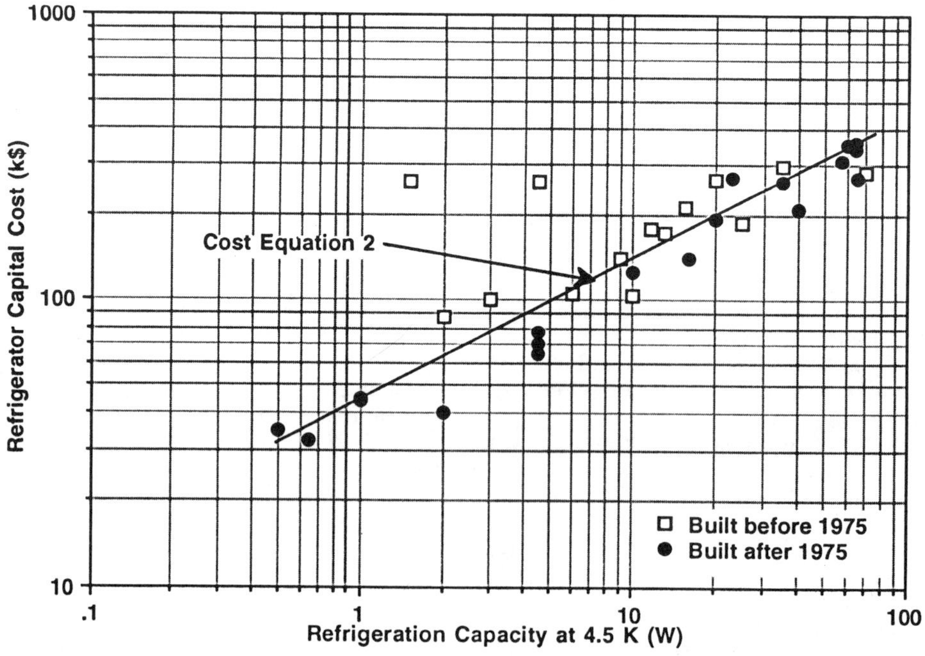


**Figure 3.** The Cost of Small Helium Refrigerators versus Refrigeration at 4.5 K.

## 1.8 TO 2.0 K REFRIGERATOR CAPITAL COST

The cost of superfluid helium refrigeration depends on the cycle used to produce refrigeration at temperatures in the range of 1.8 to 2.0 K. Over the last 30 years, two general methods have been used to produce refrigeration at 1.8 K. One method (the refrigeration mode) recovers the sensible heat from the helium gas from 1.8 to 300 K. The second method (liquefier mode) wastes the sensible heat and only uses the heat of vaporization and the sensible heat from 1.8 to 4.5 K.

The first method, uses a three pass heat exchanger in the refrigerator. One circuit carries helium at high pressure (1.5 to 1.8 MPa) helium from the compressor to the engines and J-T expansion system (a cold wet engine may also be used). The return flow is split between the flow from the engines that is returned through the heat exchanger at a pressure about 0.11 MPa and the flow from the J-T circuit that is returned through the heat exchanger at a pressure as low as 1.3 kPa (10 torr). The J-T circuit flow is pumped from 1.3 kPa to 110 kPa through a series of pumps. (These pumps may be either at low temperature or at room temperature or a combination of the two.) This method produces the most refrigeration at 1.8 K for a given input power, but sub-atmospheric pressure in the cycle can be troublesome. For this cycle, the ratio of refrigeration produced at 4.5 K to refrigeration produced at 1.8 K is about 2.5 (the Carnot ratio).

The capital cost for a refrigerator employing a three pass heat exchanger and a series of vacuum pumps can be approximately calculated using the following expression:

$$C(M\$) = 3.4\ [R(kW)]^{0.7} \tag{3}$$

where R is refrigeration at 1.8 K given in kilowatts and C is the capital cost of refrigeration given in millions of 1997 dollars. For the refrigerator described by Equation 3, the cost is 1.9 times higher at 1.8 K than the same refrigeration at 4.5 K (see Equation 1).

The second method uses a standard 4.5 K two pass heat exchanger refrigerator with a separate circuit that expands liquid helium from a pot at 4.5 K and 0.12 MPa through a heat exchanger to two phase helium at 1.8 K and 1.3 kPa. Pumping at 1.3 kPa is provided by room temperature vacuum pumps. The expanded helium at 1.3 kPa provides the cooling for material at 1.8 K and the heat exchangers between the 4.5 K region and the 1.8 K region of the cryostat. The second method produces a little less than half the refrigeration at 1.8 K for a given input power as compared to the first method, but that refrigeration can be produced in a helium bath at one atmosphere. There is very little helium that is at sub-atmospheric pressure. The second method is suited for systems that must produce refrigeration at 1.8 K, and gas at 4.5 K for cooling leads and shields (liquefaction). For many magnet applications where the heat leak into the 1.8 K region can be kept low, the second method is desirable. A refrigerator using the second method for generating 1.8 K refrigeration will produce about 5.2 W of refrigeration at 4.5 K for every watt that can be produced at 1.8 K (greater than the Carnot ratio). When using the second method of refrigeration at 1.8 K one wants to intercept as much heat as possible at 4.5 K, and gas cooled leads should use helium from the 4.5 K region.

The capital cost for a refrigerator employing a 4.5 K refrigerator (two pass heat exchangers) and a small vacuum pumped circuit can be approximately calculated using the following expression:

$$C(M\$) = 5.7\ [R(kW)]^{0.7} \tag{4}$$

where R is refrigeration at 1.8 K given in kilowatts and C is the capital cost of refrigeration given in millions of 1997 dollars. For the refrigerator described by Equation 4, the cost is 3.2 times higher at 1.8 K than the same refrigeration at 4.5 K (see Equation 1).

## ANNUALIZED COST OF REFRIGERATION AT 4.5 K

The annual cost of refrigeration can also be estimated as a function of the amount of refrigeration (R is given in kW) delivered at 4.5 K and the cost of electric power (P is given in dollars per kWh). The annual cost of refrigeration includes; amortization of the refrigerator, depreciation, operation and maintenance labor, electric power, liquid nitrogen cooling and compressor cooling. If one assumes that 22 percent of the refrigeration plant capital cost goes for the annual cost of amortization, depreciation, operation and maintenance, the following equation can be used to estimate the annual cost of providing helium refrigeration at 4.5 K:

$$\text{Annual Cost(M\$/yr)} = 3.00\,[\,R(kW)]^{0.78}\,[P(\$/kWh)]^{0.56} \tag{5}$$

Equation 5 is applicable over a range of refrigeration from 50 W to 30 kW and a range of electrical energy costs from 0.04 to 0.18 dollars per kilowatt hour. Do not use Equation 5 for estimating the annual cost for a refrigeration system operating at 1.8 K. About half of the annual cost of refrigeration shown in Equation 5 is related to the cost of energy and cooling for the compressor. The remainder of the annual cost is amortization, depreciation and maintenance. Organizations which do not amortize or depreciate their equipment can expect an annual refrigeration cost about two thirds of that given by Equation 5. The annual cost given in Equation 5 will escalate at a rate about 60 percent of the rate of inflation.

## DISCUSSION

The cost of liquid helium refrigeration expressed by Equations 1 through 5 represents the average cost for the compressors, minimal piping between the compressors and the cold box, the cold box and the expansion engines. Water cooling for the compressors, helium dewars, external helium distribution, and computer controls will cost extra. The price for any given helium refrigerator depends on the level of competition at the time the order is placed. The trend toward electronic monitoring and control is part of the reason that the cost of refrigerators manufactured after 1975 has not decreased (when escalated to 1997 dollars) when compared to refrigerators manufactured before 1975.


## ACKNOWLEDGMENT

This work was performed at the Lawrence Berkeley National Laboratory with the support of the Director, Office of Energy Research, Office of High Energy and Nuclear Physics, High Energy Physics Division, U. S. Department of Energy under Contract No. DE-AC03-76SF00098.


## REFERENCES


1. M. A. Green, R. A. Byrns and S. J. St. Lorant, "Estimating the Cost of Superconducting Magnets and the Refrigerators Needed to Keep Them Cold," Advances in Cryogenic Engineering **37**, p 637, (1991)
2. T. R. Strobridge, D. B. Mann, and D. B. Chelton, "Preliminary Analysis of Refrigeration Requirements for Superconducting Magnets in the Experimental Area of the 200 BeV Accelerator," National Bureau of Standards, Report 9259, October 1966.
3. T. R. Strobridge, IEEE Transactions on Nuclear Science **NS-16**, No. 2, p. 1104 (1969).
4. T. R. Strobridge, "Cryogenic Refrigerators, An Updated Survey," NBS Technical Note 655, June 1974.
5. R. A. Byrns, "Large Helium Refrigerators and Liquefiers," Chapter 15, Cryogenic Engineering, B. Hands, Ed., Academic Press, Inc., London (1984).
6. American Inst. Economic Resources, Economic Education Bulletin, November 1996

# FURTHER DEVELOPMENT OF A MIXED GAS JOULE THOMSON REFRIGERATOR

A. Alexeev, Ch. Haberstroh, and H. Quack

Technische Universitaet Dresden,
Lehrstuhl für Kaelte- und Kryotechnik
01062 Dresden, Germany

## ABSTRACT

A mixed gas Joule Thomson refrigerator has distinct advantages over other cryocooler for certain applications. The system is compact, has good thermodynamic efficiency and low levels of vibration and noise. Further development of this refrigeration machine is possible by optimization of the gas mixture composition. A method was developed which permits a reduction by a large factor in the number of simulation calculations and thus shortens the optimization time considerably.

## INTRODUCTION

General requirements for a cooler are reliability (long maintenance-free time) and low cost. Further criteria are connected with applications for high temperature superconductors (HTS). Good temperature stability and low levels of mechanical vibration and electromagnetic interference are necessary.

Measurements have been published[1,2] that show that temperature fluctuations as well as mechanical disturbances caused by a mixed gas Joule Thomson refrigerator are considerably lower compared with regenerative refrigeration machines (Gifford-McMahon and Stirling refrigerator) which work with pulsation of the gas. For these reasons the recuperative mixture cycles have a better chance for use in SQUID cooling than regenerative coolers.

## MIXTURE JOULE THOMSON CYCLE

In a Joule Thomson system throttling expansion of the gas is combined with recuperative precooling. The main cycle components are the compressor, heat exchanger, throttle valve and evaporator.

In the "classic" Joule Thomson cycle with a pure refrigerant the gas is discharged from a compressor and passes through the counterflow heat exchanger, where it is cooled by the returning low pressure gas. In the throttle valve the high pressure gas expands into the vapor-liquid area; the liquid evaporates in the evaporator with absorption of energy, the vapor is warmed in the heat exchanger and compressed again (Fig. 1).

In contrast to the Joule Thomson process with a pure refrigerant the mixture cycle almost completely works in the vapor-liquid region. The first liquid in the high pressure stream appears almost at ambient temperature. The high pressure stream is cooled down in the heat exchanger and further condensed at a sliding temperature, so that the refrigerant is completely in the liquid phase before entering the throttle valve. In the evaporator the refrigerant only partially evaporates. The low pressure stream continues to boil in the heat exchanger at continually rising temperature (Fig. 1).

These gas mixtures consist of a combination of gases with normal boiling temperature below 320 K. Nitrogen, hydrocarbons and their halogen derivatives, inert gases and other chemical substances can be used.

The gas is typically discharged at $p_{high}$ = 15-20 bar with a suction pressure $p_{low}$ = 1-3 bar. A low cost oil lubricated hermetic compressor can be used. The maintenance-free time of these machines exceeds by far 20,000 hours. Thereby the high reliability of the cooler is guaranteed, because no other mechanically moving components are necessary. The use of an oil lubricated compressor however has the disadvantage of oil contamination in the high pressure stream. This would result in clogging in the cold box. Therefore an oil separation in the high pressure gas after the compressor is required.

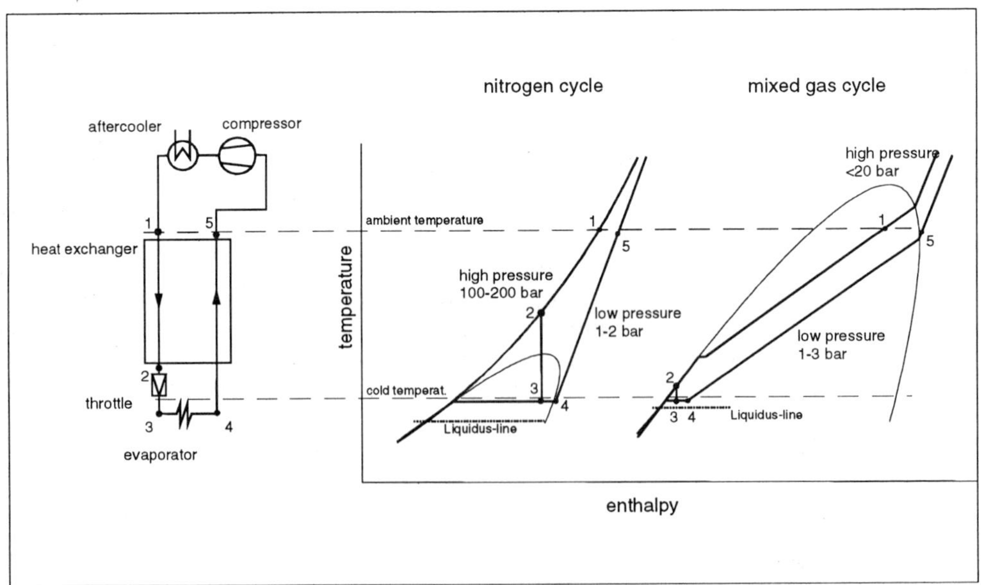


**Fig. 1** Joule Thomson cycles.

A prototype of a mixed refrigerant Joule Thomson cooler with 2 W cooling power at 77 K was developed and built at the Technische Universitaet Dresden.

The cold box consists of a cryostat 160 mm in diameter and approximately 300 mm in length (Fig. 2). In the upper flange connections for pressure lines and electric feedthroughs are placed. A multi- tube heat exchanger was used.

The oil lubricated compressor of a conventional air conditioning system was used. The electric power consumption of this machine is about 0.4 kW. The pressure lines are made of nylon. This way compressor and cold box are electrically separated. The mechanical vibrations are considerably reduced too.

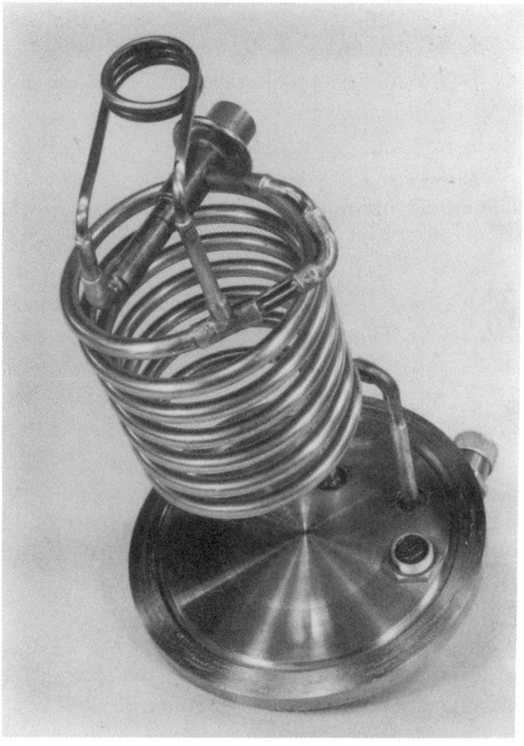

**Fig. 2** A prototype of a mixture Joule Thomson cooler developed at the TU Dresden.

## POSSIBLE FURTHER DEVELOPMENTS

The desired parameters could be reached already with the described cooler[3]. From the investigations it could be seen that this JT cooler has considerable potential for optimization. Certain technical losses can be reduced by further development of the compressor unit and heat exchanger. Considerable improvements can be reached by adjustment of working parameters such as the high and low pressure. The largest optimization potential, however, is connected with optimization of the gas mixture composition as it was shown by previous calculations[4].

The aim of the optimization is to optimum the mixture composition and determine the optimum high and low pressures for the required cooling temperature and other given parameters. It is a set of parameters, at which the efficiency of the whole system is maximal. Mathematically it can be defined as follows:

$$CF = \frac{Q_0}{P} \cdot \frac{T_{amb} - T_0}{T_0} = CF\,(T_{amb}, \Delta T_{\min}, T_0, p_h, p_n, z_i) \Rightarrow MAXIMUM \text{ , whereby}$$

$CF$ - Carnot Fraction
$Q_0$ - cooling power
$P$ - electric power consumption of the compressor
$T_{amb}$ - ambient temperature
$T_0$ - cold temperature
$\Delta T_{min}$ - minimal temperature difference in the heat exchanger
$p_h, p_l$ - high and low pressures
$z_i$ - mixture composition (molar %), this parameter contains 4 free parameters if a five component mixture is used.

For simplification we assumed as follows:

- the mixture does not form a solid phase
- the mixture composition does not change in the cycle
- no heat load from the insulation or by heat conduction
- no pressure losses in the heat exchanger

## Thermodynamic Considerations

To understand the thermodynamic behavior better we have made a series of calculations. For a set of input parameters the specific refrigeration power of the cycle was calculated. The concentrations of components were varied and changes in efficiency were determined. From these results a systematic understanding was developed.

The calculations were accomplished for a mixture of nitrogen, methane, ethane, propane and butane. The software package PROVISION[5] was used.

**Results of the Simulation.** By systematic variation of the mixture composition and numerical calculation of the cycle the following relationships were found:

a) the mixture composition has a dominant influence on the efficiency;

b) in a 3D diagram, with the concentrations of the high volatile components on the horizontal axes and the cooling power as the vertical axis, a three-sided asymmetrical pyramid is obtained (Fig. 3); the lateral surfaces differ in their gradients; the two low volatile components are kept constant;

c) by variation of the concentration of the low volatile components a similar result as in b) is obtained only with other maxima;

d) the optimal relation between nitrogen and methane concentrations for the assumed conditions ($p_h = 18$ *bar*, $p_l = 1$ *bar*, $T_0 = 80$ *K*) can be described by the equation $z_{CH4} = 1.3659\, z_{N2} - 17.962$ *mol %* for all compositions of low volatile components;

e) in a 3D diagram with the concentrations of the low volatile components on the horizontal axes and the cooling power as vertical axis a surface in the form of a three-sided asymmetrical pyramid with a maximum by $z_{C3H8} \approx 10$, $z_{C4H10} \approx 20$ *mol %* is obtained (Fig. 4); one side of this figure has a break along a straight line.

The change of the CF can be described essentially as a linear function of certain mixture

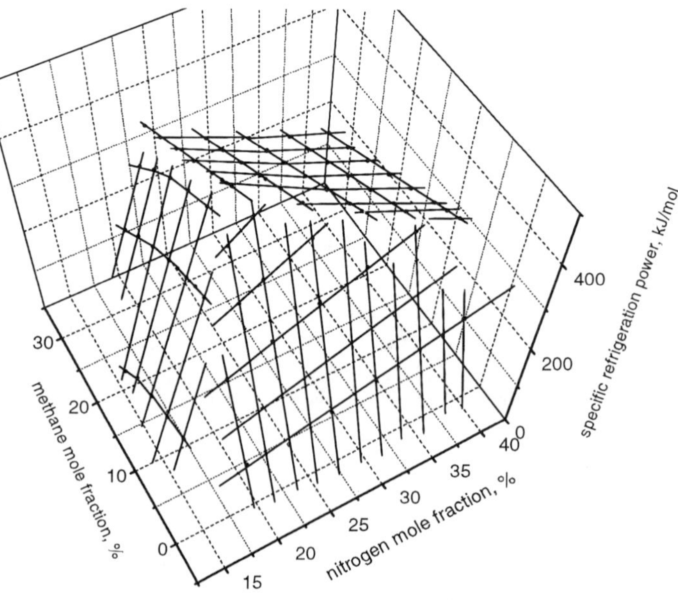


**Fig. 3** Specific refrigeration power of the mixture JT cycle, $T_0 = 80$ K, $p_h = 18$ bar, $p_l = 1$ bar, $z_{propane} = 15$ mol %, $z_{butane} = 1$ mol %.

components; there are no indications of additional local maxima, which could be caused by the complicated thermodynamic behavior.

This allows one, with some caution, to draw the conclusion, that it is possible to describe the efficiency of the cycle relatively simply, and to find an optimal mixture configuration for a given requirement very fast.

**Analysis of the Results.** To understand the observations described above the typical temperature-duty (*T-q*) diagram for the heat exchanger should be considered (Fig. 5).

Three relevant points (at $T \approx 80\ K$, $T \approx 270\ K$ and $T \approx 120\ K$) can be distinguished here. The reason for the existence of these points can be found in the thermodynamic behavior of the phases.

The first break in the *T-Q* diagram corresponds to the first liquid phase. At the example considered here the condensation of low volatile components begins at about $T \approx 270\ K$. The second critical break is connected with the liquefaction of the second liquid phase and is very important for the description of the mixture JT cycle.

The specific cooling power of the JT process is given by the enthalpy difference between the high and low pressure streams in each point of the heat exchanger with the condition that no pinch point in the heat exchanger exists:

$$q_0 = \Delta h_{n-h}$$

A pinch point can appear at one of the three critical points in the recuperator: at the warm end, at the cold end and at the point, where the second liquid phase starts to condense in the high pressure stream (split point):

$$\Delta T_{\min} = \min \begin{Bmatrix} \Delta T_{\min}^{warm\ end} \\ \Delta T_{\min}^{split\ point} \\ \Delta T_{\min}^{cold\ end} \end{Bmatrix}$$

This relationship can be applied to understand the lateral surfaces of the pyramid visible in Fig. 3.

First surface

The mixture compositions correspond to a pinch point at the warm end of the heat exchanger creating a softly bent surface (in Fig. 3 in the back). The gradient corresponds to the concentration of the low volatile components in the mixture.

surface 1:

$$\max\{q_0\} = q_0^{(1)} = \Delta h_{l-h}^{we}$$

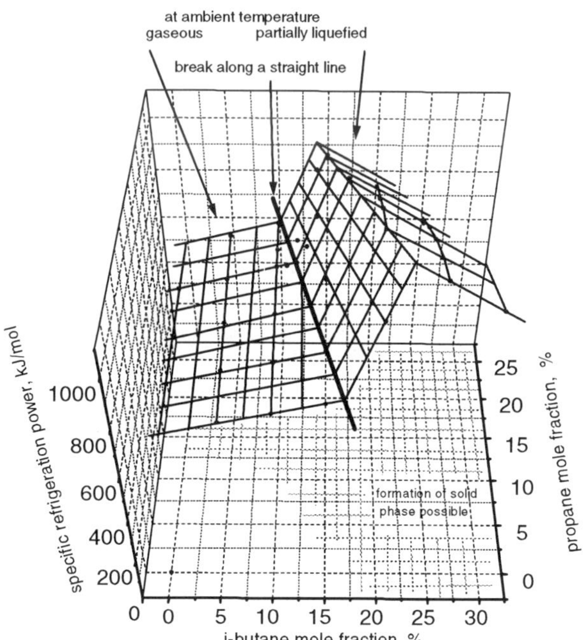


**Fig. 4** Specific refrigeration power of the mixture JT cycle, $T_0 = 80$ K, $p_h = 18$ bar, $p_l = 1$ bar.

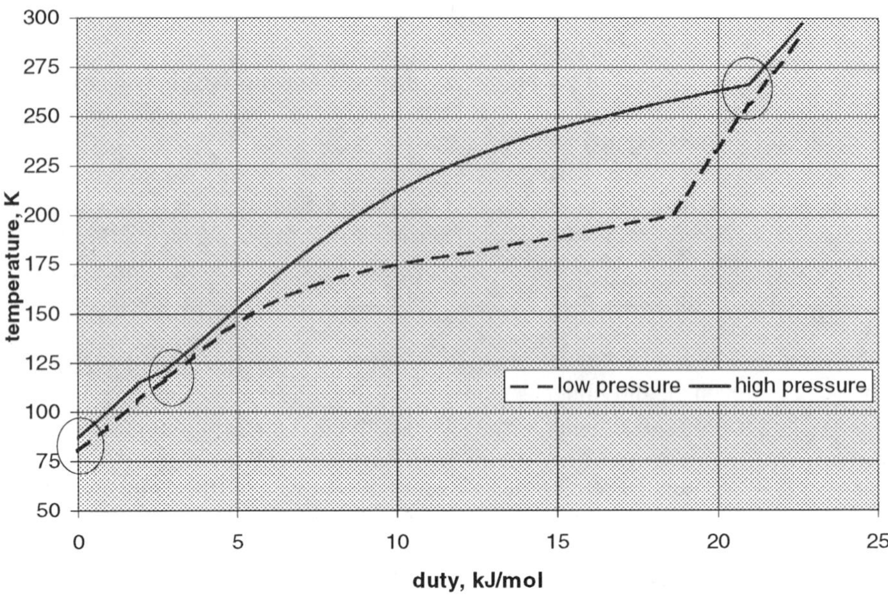


**Fig. 5** Typical temperature duty diagram for heat exchanger, nitrogen-hydrocarbons mixture, $T_{amb}$=297 K, $T_0$ = 80 K, $\Delta T_{min}$= 5 K .

1: DATA INPUT:

$$z_1^{(1)}; \quad z_1^{(2)} = z_1^{(1)} + \Delta z_1; \quad z_1^{(3)} = z_1^{(1)};$$

$$z_2^{(1)}; \quad z_2^{(2)} = z_2^{(1)}; \quad z_2^{(3)} = z_2^{(1)} + \Delta z_2;$$

$$z_3^{(i)} = 1 - z_1^{(i)} - z_2^{(i)} \quad i = 1..3$$

2: CALCULATION OF ENTHALPY DIFFERENCES AT CRITICAL POINTS OF HE:

$$\Delta h_1^{(i)} = \left| h^{P_h}_{T_{amb}} - h^{P_l}_{T_{amb}-\Delta T_{min}} \right|$$

$$\Delta h_2^{(i)} = \left| h^{P_h}_{T^{Ph}_{split}} - h^{P_l}_{T^{Ph}_{split}-\Delta T_{min}} \right| \quad \text{i=1.. 3}$$

$$\Delta h_3^{(i)} = \left| h^{P_h}_{T_0+\Delta T_{min}} - h^{P_l}_{T_0} \right|$$

3: INTERPOLATION OF CALCULATED DATA BY PLAN EQUATION:

$$A_i \quad B_i \quad C_i$$

$$A_i \cdot z_1^{(1)} + B_i \cdot z_2^{(1)} + C_i \cdot z_3^{(1)} = \Delta h_i^{(1)}$$

$$A_i \cdot z_1^{(2)} + B_i \cdot z_2^{(2)} + C_i \cdot z_3^{(2)} = \Delta h_i^{(2)} \quad \text{i=1.. 3}$$

$$A_i \cdot z_1^{(3)} + B_i \cdot z_2^{(3)} + C_i \cdot z_3^{(3)} = \Delta h_i^{(3)}$$

4: CALCULATION OF MAXIMUM AS INTERSECTION OF THE THREE SURFACES:

$$z_1^{MAX} \quad z_2^{MAX} \quad z_3^{MAX} = 1 - z_1^{MAX} - z_2^{MAX}$$

$$A_1 \cdot z_1^{MAX} + B_1 \cdot z_2^{MAX} + C_1 \cdot z_3^{MAX} = A_2 \cdot z_1^{MAX} + B_2 \cdot z_2^{MAX} + C_2 \cdot z_3^{MAX}$$

$$A_2 \cdot z_1^{MAX} + B_2 \cdot z_2^{MAX} + C_2 \cdot z_3^{MAX} = A_3 \cdot z_1^{MAX} + B_3 \cdot z_2^{MAX} + C_3 \cdot z_3^{MAX}$$

IF $\left| \left(\overline{z^{MAX}}\right)^{new} - \left(\overline{z^{MAX}}\right)^{old} \right| > \varepsilon$ DO: $\overline{z^{(1)}} = \overline{z^{MAX}}$ AND GOTO 1

ELSE: END

**Fig. 6** Algorithm to search of optimal concentration of high volatile components .

This can be described as:

$$q_0^{(1)} = h\left(p_l, T_{amb} - \Delta T_{\min}\right) - h\left(p_h, T_{amb}\right) \tag{1}$$

The specific cooling power can be described here as a linear function of the concentration of the low volatile components:

$$q_0^{(1)} \approx Const_1 + \sum_{i=3}^{n} z_i \cdot Coefficient_i \tag{2}$$

Second surface

The mixture compositions causing a pinch point at the cold end of the heat exchanger create the very steep surface (in Fig. 3 left side):

surface 2: $\max\{q_0\} = q_0^{(2)} = \Delta h_{l-h}^{cold\ end}$

The specific refrigeration power in this case depends almost completely on the concentration of the highest volatile component (nitrogen).

$$q_0^{(2)} \approx z_1 \cdot Coefficient_1 + Const_2 \tag{3}$$

Third surface

Other mixture compositions corresponding to a pinch point in the middle of the heat exchanger present the steep surface (in Fig. 3 in front):

surface 3: $\max\{q_0\} = q_o^{(3)} = \Delta h_{l-h}^{split\ point}$

The specific refrigeration power in this case depends on the concentration of the second highest volatile component (methane):

For that reason surfaces 2 and 3 correspond to the concentrations of the nitrogen ($z_1$) and methane ($z_2$) respectively. The intersection of these surfaces is a straight line $z_{CH4} = A\ z_{N2} - B$.

$$q_0^{(3)} \approx z_2 \cdot Coeffcient_2 + Const_3 \tag{4}$$

Break Line

The break line of the plane in front in Fig. 4 corresponds to the phase change in the high pressure stream at ambient temperature. Mixtures above this line have an increased concentration of low volatile components. The refrigerant is partially liquefied at the warm end of the heat exchanger in this case. These mixtures are preferable from the thermodynamic point of view. The specific refrigeration power achievable in this case is approximately two times better compared with the gaseous mixtures at ambient temperature.

## Optimization Strategy

An optimization strategy was developed based on the knowledge described above. The global maximum is determined in a two-dimensional field only of the concentrations of low volatile components $z_5=z_5^{(begin)}...z_5^{(end)}$ and $z_4=z_4^{(begin)}...z_4^{(end)}$. The optimization procedure is considerably simplified that way and can be done in a short time.

This is done by regarding only the optimal concentrations of the high volatile components (instead of all possible) for each alleged pair $(z_4, z_5)$. The search of these optimal concentrations of the high volatile components is realized by an effective and very fast algorithm. This is presented in Fig. 6 schematically.

The optimization begins with the calculation of the enthalpy differences at critical points of the heat exchanger for three alleged concentrations of the high volatile components. The calculated data interpolated are by plan equation. The optimum is defined as intersection of the three surfaces.

Because the surfaces are assumed as even surfaces, an error in the result is possible. The result should therefore be corrected with the same algorithm.

This method allows a fast maximum search in the whole area of the components concentrations.

## OUTLOOK

The mixed gas Joule Thomson refrigerator offers distinct advantages for the cooling of supersensitive components. The machine is simple in construction and is compact. The maintenance-free time amounts to far more than 20,000 operating hours. Magnetic and mechanical disturbances at the cold finger are less than for regenerative cooling system.

A further development of this cooler is possible by optimization of the gas mixtures. A strategy was developed which allows one to shorten the calculation time and to accelerate the optimization considerably.

The accomplished calculations show, that mixtures with increased concentrations of low volatile components like propane or butane are thermodynamically preferable. The refrigerant is partially liquefied at the warm end of the heat exchanger in this case. However, use of these mixtures is connected with two problems:

- formation of solid phase at low temperatures, and clogging in cold box is possible
- local mixture composition can be changed by transport of the two-phase fluid from the compressor to the cold box

But these problems can be overcome in principle. The hardware must be adjusted accordingly. Future work should focus on this point.

## REFERENCES


1. Hohmann, H. Soltner, H.-J. Krause, H. Bousack u.a., Integration of HTS SQUIDs with Portable Cooling Devices for the Detection of Materials Defects in Non-Destructive Evaluation, Proceedings of the 9th International Cryocooler Conference (ICC9), Waterville Valley, NH, USA, 25.-27.6.1996
2. Dennis Hill, Throttle Cycle Cooler Vibration Characterization, Proceedings of the 9th International Cryocooler Conference (ICC9), Waterville Valley, NH, USA, 25.-27.6.1996
3. Alexeev A., Quack H., Haberstroh Ch., Siegert J., Weiterentwicklung eines Gemisch-Joule-Thomson-Kryokuehlers, DKV-Tagungsbericht, Band 1, 1996
4. Alexeev A., Quack H., Haberstroh Ch., Low cost mixture Joule Thomson Refrigerator, Cryogenics, Proceedings of the 16th International Cryogenic Engineering Conference, Kitakyushu, Japan, 1996
5. PROVISION 1.0, Simulation Sciences Inc., 601 S Valencia Ave., Brea, California 92621, USA

# MIXED-REFRIGERANT JOULE-THOMPSON CRYOCOOLER DRIVEN BY R22/R12 COMPRESSOR

E.C. Luo, Y. Zhou, and M.Q. Gong

Cryogenic Laboratory of Chinese Academy of Sciences
Beijing 100080, China

## ABSTRACT

A reliable and low-cost miniature cryocooler for cooling infrared detectors, superconducting devices and other cryoelectronic instruments is a very challenging research topic in the cryogenic engineering field. Among different cryocoolers, a miniature, Joule-Thompson cryocooler with special mixtures is now a possible candidate for the mentioned applications due to recent developments. In this paper, the performance of a miniature mixed-refrigerant Joule-Thompson cryocooler operating around 80K and working at supply pressures of 1.0MPa to 2.5MPa has been studied experimentally. Our experimental results show that our mixed-refrigerant Joule-Thompson cooler can achieve 80K within the normal working pressure of R22 compressors and even down to the normal working pressure of R12 compressors. Thus , a 80K Joule-Thompson cryocooler driven by a R12 compressor with lower weight, smaller size and lower input electrical power can possibly be developed for cooling infrared detectors in the near future.

## INTRODUCTION

Many electronic devices operating in the range of around liquid nitrogen temperature are in the phase of practical application out of laboratory. However, an obstacle to this application is the lack of a reliable and low-cost cryogenic source. Thanks to recent developments of a close-cycle mixed-refrigerant Joule-Thompson cryocooler driven by a commercially single stage oil-lubricated R22/R12 compressor, this deadlock seems now to be overcome.

The use of multicomponent refrigerants in a Joule-Thompson cycle in the range of temperatures $T>100K$ has a rather long but obscure history. Using mixtures for the liquid nitrogen temperature range began in the seventies in the former Soviet Union [1]. However, before the nineties, a commercialized, single stage, oil-lubricated compressor had not been used in the cryocooler, thus a commercial product of mixed-refrigerant, Joule-Thompson cryocooler with high reliability and low cost had not yet been available [2,3,4]. From the beginning of the nineties, mixed-refrigerant, throttle refrigerators driven by commercial, single stage , oil-lubricated compressors have achieved liquid nitrogen temperatures, and today this technology is getting more and more attention of cryogenic scientists [5].

A compact Joule-Thompson microcooler using a special mixed -refrigerant which can operates in the temperature range from 70K to 95K has been developed successfully in our laboratory. The thermodynamic performance and operating parameters of this cryocooler are reported here.

## A MIXED-REFRIGERANT JOULE-THOMPSON MINIATURE CRYOCOOLER

Figure 1 is a schematic of a typical mixed-refrigerant Joule-Thompson Cryocooler. Some variants of this cycle type are possible for further improving the efficiency of mixed-refrigerant Joule-Thompson cryocooler. The compressor unit generally consists of a single stage, oil-lubricated compressor, an aftercooler with fan and oil separation unit. Gas lines connect the compressor to the Joule-Thompson cryostat. The cryostat is made up of a counter-flow heat exchanger, a throttle valve and an evaporator. The counter-flow heat exchanger of the cryostat is very compact. Two types of heat exchanger were tested. One is a classic Hampson -type, miniature heat exchanger. This type of heat exchanger can successfully achieve temperatures below liquid nitrogen, but it requires a special technique for fabricating it. Therefore, later, we develop a new type of counter -flow heat exchanger★, the perforated heat exchanger with a simple heat- coupling technique. The experiment shows that this type of heat exchanger can be more compact, more effective and cheaper.

The gas is typically discharged at 1.2MPa to 2.5Mpa with suction pressure from 0.1Mpa to 0.4Mpa and then flows through the aftercooler and the oil separation unit. The oil from oil separation unit is returned to the compressor. The high pressure gas then goes to the cryostat. In the cryostat the gas passes through the counter-flow heat exchanger, where the high pressure mixture gas is cooled by the returning low pressure gas mixture. The high pressure gas mixture expands when passing through a throttle valve and produces the cooling effect in the evaporator.

The mixed-refrigerant used can be chosen from some combination of inert gas, nitrogen, argon, freons and hydrocarbons. For a mixed-refrigerant Joule-Thompson cryocooler, each component plays an important role in a different temperature range. In our experiment, some mixtures on the basis of neon, nitrogen and hydrocarbons were tested. The thermodynamic analysis and optimization of the mixed-refrigerant used for the Joule-Thompson cryocooler is reported separately in another paper in this conference.

★Patent pending

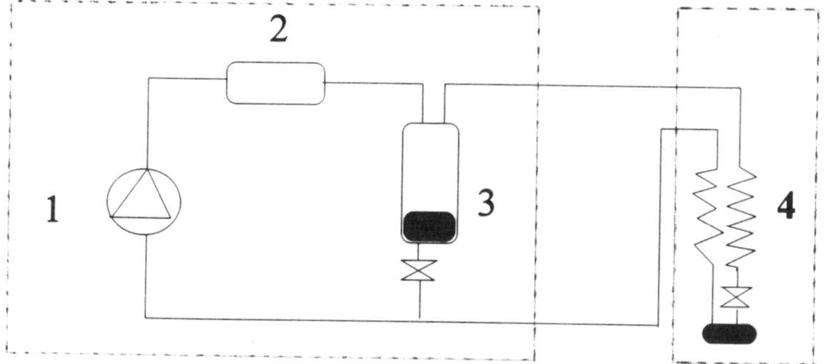


**Figure 1.** Schematic of a typical single-stage, oil-lubricated, mixed-refrigerant JT cryocooler :
1. Compressor , 2. Aftercooler, 3. Oil separator, 4. Joule-Thompson cooler unit .

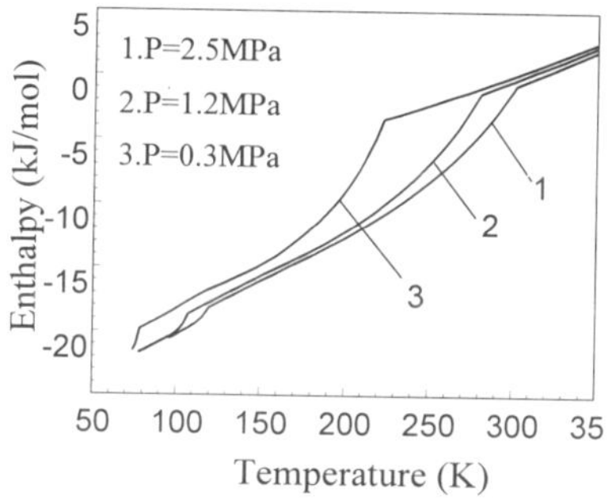


**Figure 2**. Enthalpy-Temperature Diagram of Mixture.

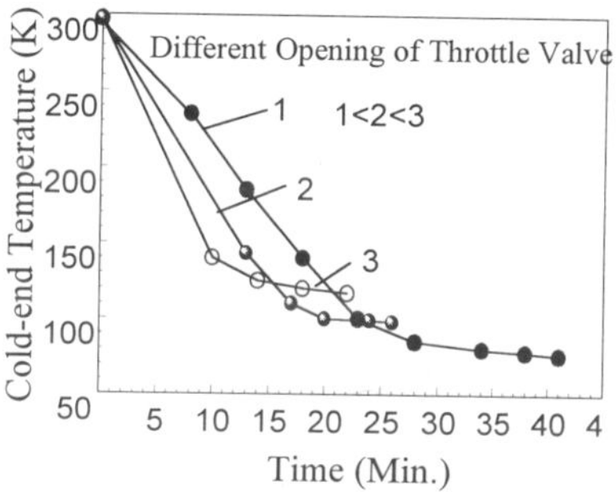


**Figure 3**. Cool-down Curve of the Cryocooler.

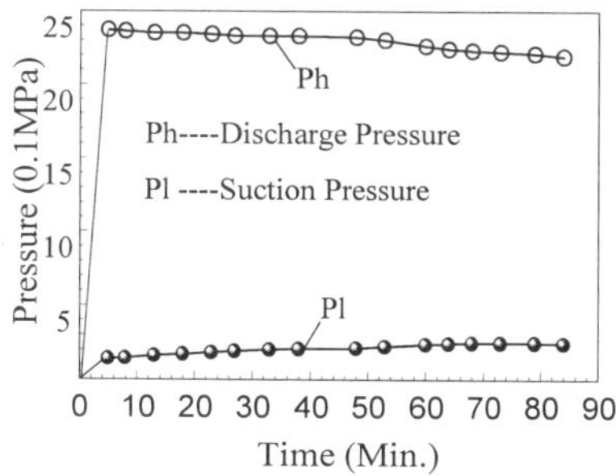


**Figure 4.** Working Pressure vs. Cool-down Time.

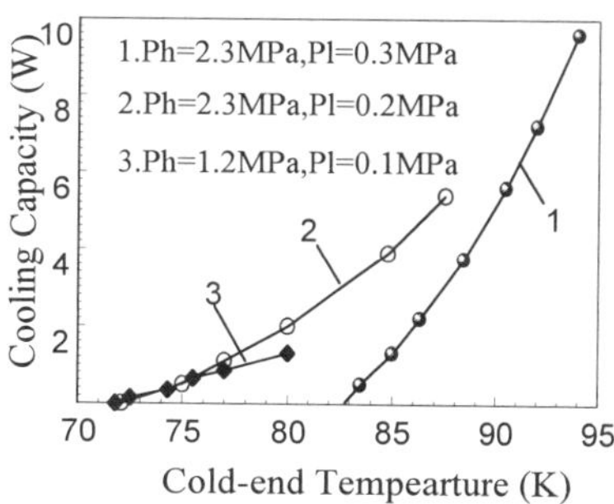


**Figure 5**. Cooling Capacity vs. Temperature.

## OPERATING PARAMETERS AND THERMODYNAMIC PERFORMANCE

Six components are included in the mixture used. Figure 2 shows the enthalpy-temperature diagram of the mixture based on the Peng-Robinson equation of state. Figure 3 gives the cool-down curve at different settings of throttling valve. Generally, the lowest temperatures can be achieved within 30 to 60 minutes. Figure 4 shows the typical pressure change during the cool-down process.The suction pressure gradually increases while the discharge pressure slowly decreases towards a steady state. Figure 5 shows the cooling capacity plot of this JT cooler for different operating pressures. Even if the discharge pressure goes down to below 1.2MPa, the cryocooler still achieves a temperature near 70K. In the experiment, a lowest temperature of 67.5K was achieved. The total electrical power input including fan ranged from 200W to 350W. In addition, a very small , R12 1/8HP compressor was also tested, and a no-load temperature of 130K was obtained. We think that the present heat exchanger does not match the compressor and a more efficient mixture is required to achieve 80K temperature range at the lower operating pressures.

## SUMMARY

A mixed-refrigerant Joule-Thompson microcooler working in the temperature range from 67.5K to 95K suitable for the cooling of infrared detectors, high temperature superconducting devices and other electronic devices has been successfully developed. A mixed-refrigerant Joule-Thompson cryocooler has good potential for application in the temperature range around 80K. A more compact Joule-Thompson cryocooler driven by a R12 compressor with smaller electrical input power(<100W) and a more efficient mixture is possible with further study.

## ACKNOWLEDGMENTS

This research work is financially supported by the National Natural Sciences Foundation of China under contract number 59476009. Also the authors are indebted to discussion with Professor Hong C.S. and help from Professor Chang Z.Z. and Professor Zhou Z.X..

## REFERENCES

1.V.M.Brodianski, V.M. Yagodin, V.A.Nikolsky, and A.V.Ivantsov. The Use of Mixtures as the Working Gas in Throttle "Joule-Thompson Cryogenic Refrigerator, The 13th international refrigeration conference, New York, 1973

2. V.N.Alfeev, V.M.Brodyanski, V.M. Yagodin, V.A.Nikolsky, and A.V.Ivantsov. "Refrigerant for a Cryogenic Throttling Unit", UK Patent 1336892(1973)

3.W.A.Little. Recent Developments in Joule-Thompson Cooling: Gases, Coolers, and Compressor, Proceedings of 5th International Cryocooler Conference, Monterey(1989), p.3

4.E.C.Luo, J.T.Liang, Y.Zhou, V.V.Yakuba, and M.P.Lobko. Experimental Investigation of an Efficient Closed-Cycle, Mixed-refrigerant J-T Cooler,in:"Cryocooler 9", ed R.G.Ross, New York(1997),p.553

5. R.C. Lonsworth, M.J.Boiarski, and L.A.Klusmier. 80K Closed Cycle Throttle Refrigerator, in: "Cryocooler 8", R.G.Ross, ed., Plenum Press, New York(1995), p.553

# THERMODYNAMIC ANALYSIS AND OPTIMIZATION OF 80K CLOSED-CYCLE JOULE-THOMPSON CRYOCOOLER WITH GAS MIXTURE

E.C. Luo, Y. Zhou, and M.Q. Gong

Cryogenic Laboratory
Chinese Academy of Science
Beijing 100080, China

## ABSTRACT

Due to using a special mixture as the working substance of a Joule-Thompson cryocooler, the thermodynamic performance of the cryocooler can be greatly improved. In recent years, this kind of cryocooler has seen some new developments, using air-conditioning compressors, and consequently it can now be developed as a highly reliable and low-cost cryocooler. However, some fundamental problems are not well understood. One problem is how to avoid the solidification of high- boiling -temperature -point components. Another problem is how to choose the mixture components and optimize their compositions. In this paper, the thermodynamic aspects of these two problems are described. In addition, a more efficient mixed refrigerant is proposed for such Joule-Thompson cryocoolers.

## INTRODUCTION

A mixed-refrigerant Joule-Thompson cryocooler for liquid nitrogen temperature range based on a one single stage oil-lubricated compressor has been getting more and more attention in recent years [1] .There are many factors which affect the performance of this type of cryocooler. Among all of these factors, the key factor obviously is the mixtures used. Theoretically, there are a large number of component which can be included in the mixed-refrigerants. However, which components are better candidates of mixed-refrigerants for obtaining liquid nitrogen temperature range? How can one optimizes the components and the compositions of the mixtures? This paper will give the answers to these questions.

## SOLUBILITY LIMITS OF MIXTURE COMPONENTS

The currently used and investigated mixtures include many components having higher freezing points than that of pure nitrogen. Typically the freezing point temperature of these components are above 90K. To avoid the solidification of these components when they pass through the throttling valve and the low temperature passage of the counter-flow heat exchanger, their concentration in the mixtures should be within their solubility limits. From thermodynamics, the soluble limit of a component in mixtures depend mainly upon its freezing point temperature , its melting enthalpy and the interactive force of the different components.

For the mixtures including only hydrocarbons, fluorinated hydrocarbons , the solids formed are nearly immisible . However, nitrogen, argon and methane are soluble in the solid state. Based on the Regular Solution Model [2], the liquid-solid equilibrium of some binary and ternary mixtures are theoretically predicted. Figure 1 to figure 6 are liquid-solid equilibrium diagrams of several binary mixtures. The predicted values reproduce the experimental data of figure 1, figure 3 and figure 4 but have bigger errors for figure 2 [3]. Figure 7 and figure 8 are the liquid-solid phase diagrams of two ternary mixtures at 80K. One is made of methane, ethane and propane, and another is made of nitrogen, methane and ethane. Within the shaded region, the mixtures do not form solid and can normally operate in a mixed-refrigerant Joule-Thompson cryocooler around 80K. In addition, table 1 gives the predicted result of the solubility limit of several hydrocarbons in their mixtures(solvent) at different temperatures. From this table, one can see that lighter hydrocarbons can vary contents in a large range without solidification , on the contrary the heavier hydrocarbons can change contents only within a smaller range without solidification. However, the heavier hydrocarbons can substantially increase isothermal effect of mixtures in high temperature region and improve thermodynamic efficiency of Joule-Thompson cryocooler.

In short, the solubility limit of each component in the mixtures at low temperatures determine the selection and the maximum contents of components in the mixtures.

## OPTIMIZATION OF MIXTURES

The thermodynamic performance of a mixed-refrigerant Joule-Thompson cryocooler depend heavily on the mixtures used. There are three kinds of throttling effects, differential effect, integral effect and isothermal effect. Among them, the isothermal effect is the most important parameter for analyzing the Joule-Thompson cycle. The isothermal effect is determined by the working substance used, the operating pressures and the temperatures. To improve the thermodynamic performance of a mixed-refrigerant Joule-Thompson

Table 1 The Solubility Limit of Hydrocarbons (Based on Ideal Solution Model)

| | 65K | 70K | 75K | 80K | 85K | 90K |
|---|---|---|---|---|---|---|
| CH4 | 0.62 | 0.70 | 0.78 | 0.85 | 0.92 | 0.99 |
| C2H6 | 0.23 | 0.33 | 0.47 | 0.62 | 0.80 | 1.00 |
| C3H8 | 0.21 | 0.33 | 0.50 | 0.71 | 0.97 | 1.00 |
| iC4H10 | 0.03 | 0.05 | 0.08 | 0.13 | 0.20 | 0.29 |
| IC5H12 | 0.01 | 0.02 | 0.04 | 0.07 | 0.10 | 0.16 |

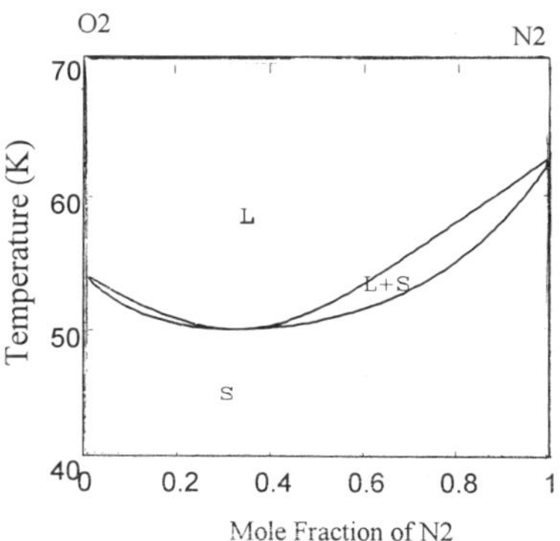


**Figure 1.** Liquid-solid phase diagram of N2 and O2.

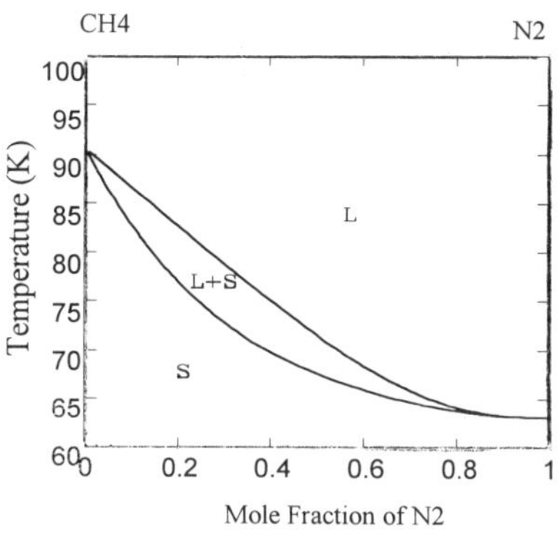


**Figure 2.** Liquid-solid phase diagram of N2 and CH4.

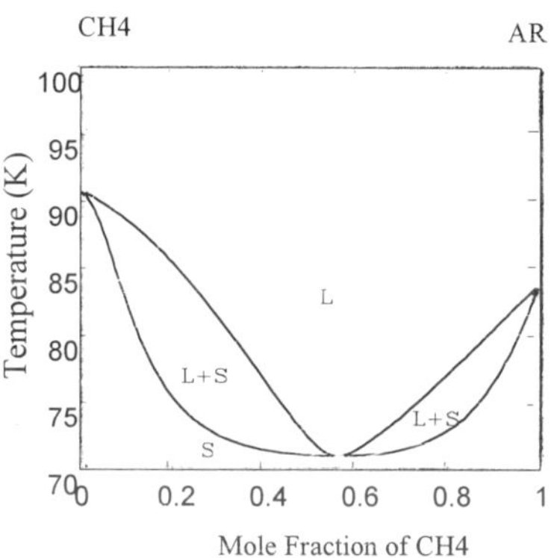


**Figure 3.** Liquid-solid phase diagram of CH4 and Ar.

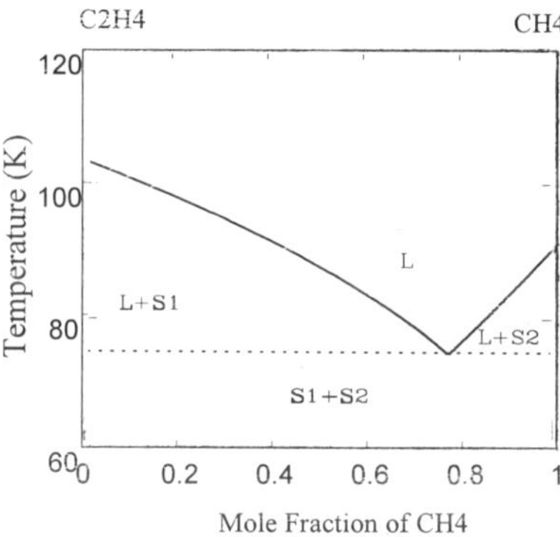


**Figure 4.** Liquid-solid phase diagram of CH4 and C2H4.

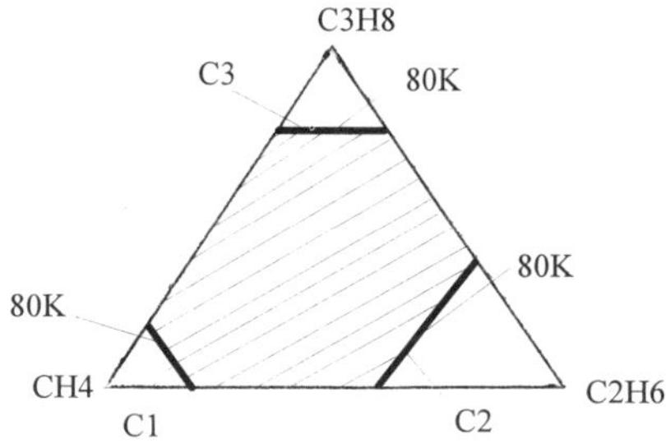


**Figure 5.** Liquid-solid phase diagram of ternary mixture (CH4+C2H6+C3H8).

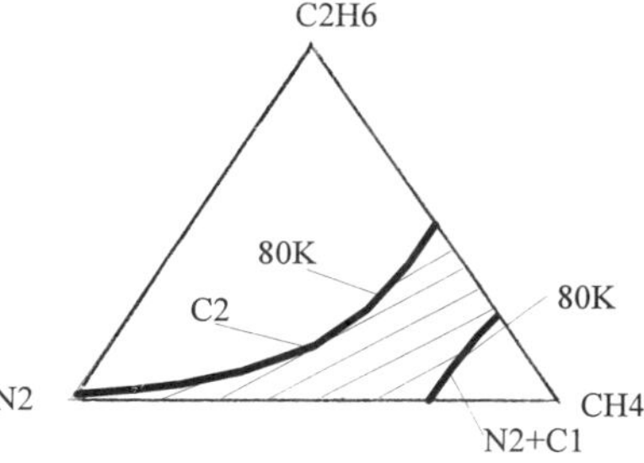


**Figure 6.** Liquid-solid phase diagram of ternary mixture(N2+CH4+C2H6).

cryocooler, it is necessary to increase the isothermal throttle effects over the whole operating temperature range. Thus, the first thing is to select the components of the mixtures and then optimize their concentration in the mixture. Based on thermodynamics, the isothermal effect is directly related to the boiling temperature and the liquid-vapor latent heat of the selected components. From liquid nitrogen temperature to room temperature (e.g.300K), there are a large number of suitable substances. However, considering the solubility limit of the mixture composition, one can further determine the possibly selected components. Furthermore, if the other factors (e.g.,safety factor and economic factor)are considered, only a small number of pure substances are the preferred candidates. These components are classified into eight groups showed in table 2 according to their boiling temperatures. Figure 7 and figure 8 show the isothermal effects of two types of mixtures at different pressures and temperatures. The first mixture (N2+CH4+C2H4+C3H8+iC4H10, here refereed as No.1) is reported in the published document [4].From a widely predicted results, the mixture(N2 + R14 + CH4 +C2H6 +C3H8 +iC5H12, here refereed as No.2 *) is found to be the best one for the Joule-Thompson cryocooler. This mixtures can give a larger isothermal throttling effect and a more reasonable distribution of enthalpy vs. temperature. In addition, it can keep a rather high efficiency even at a low working pressure(e.g., down to the normal working pressures of R12 compressor). From the predicted results, No.2 mixture will increase the thermodynamic efficiency by 50% to 100% over that of the No.1 mixture.

To form an efficient mixed refrigerant, we can choose one component from each group in table 2. Group 5 and group 6 in table 2 both include more components than the other groups. The component in the same group means that they can replace one another. For a JT cryocooler driven by oil-free compressor, the hydrocarbons and the fluoridated hydrocarbons listed in the table 2 are preferred component. However, for oil-lubricated compressor, the hydrocarbons are the more preferred components of the mixed refrigerant due to their good solubility in oil.

For a closed cycle mixed-refrigerant Joule-Thompson cryocooler, the thermodynamic performance of the whole refrigerating system is directly related to the mixture used, the

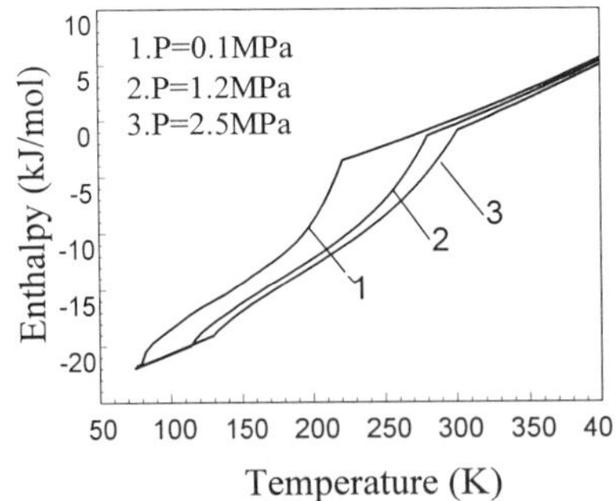


**Figure 7.** Enthalpy-temperature diagram of No.1 mixture.

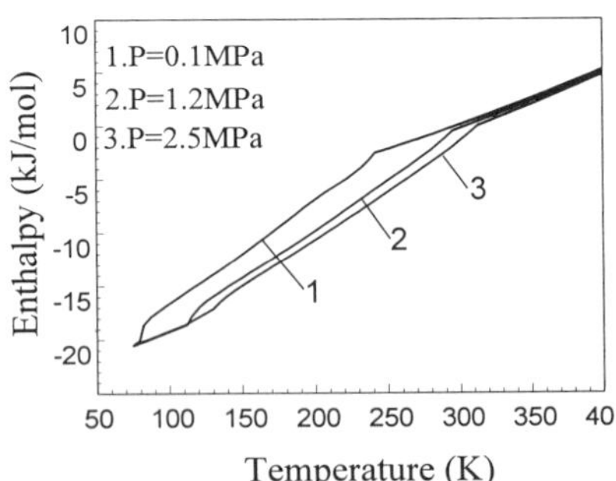


**Figure 8.** Enthalpy-temperature diagram of No.2 mixture.

* Patent Pending

**Table 2** The preferably selected components of the mixed -refrigerant for JT coolers

| | Components | Boiling Temperature (K) |
|---|---|---|
| Group 1 | He,Ne | 4.2-27.0 |
| Group 2 | N2,Ar | 77.3-87.3 |
| Group 3 | CH4 | 111.7 |
| Group 4 | R14 | 145.2 |
| Group 5 | C2H6,C2H4,R13,R23,R116 | 169.4-194.9 |
| Group 6 | C3H8,C3H6,R22,R12,R218 | 225.4-243.4 |
| Group 7 | iC4H10,iC4H8 | 261.3-266.9 |
| Group 8 | iC5H12,iC5H10,... | 293.3-303.1 |

heat exchanger, and the compressor. The optimization of a mixed-refrigerant Joule-Thompson cryocooler is a complicated thermodynamic task, but it may be developed in stages. The first step is to optimize the mixture itself from thermodynamic cycle, and the second step is to comprehensively optimize the mixture used, the heat exchanger, as well as the compressor.

## CONCLUSIONS

(1) The solubility limits of mixture components determine the possible selection of the components of the mixture used in a mixed-refrigerant Joule-Thompson cryocooler. The heavier hydrocarbons can obviously increase the isothermal effects in the high temperature region , but their maximum concentration in the low temperature region are limited.

(2) To improve the thermodynamic efficiency of a mixed-refrigerant Joule-Thompson cryocooler, the isothermal effects over the whole operating temperature range should be increased. Different components play important roles in different temperature ranges. The components and composition of the mixture should optimized.

(3) The thermodynamic performance of a mixed-refrigerant Joule-Thompson cryocooler is manly determined by the mixture used. A more efficient mixture is proposed for a closed cycle Joule-Thompson cryocooler based a single stage compressor.

## ACKNOWLEDGMENTS

We are indebted to the National Natural Sciences Foundation of China for support of this work under Contract number 59476009.

## REFERENCES

1. R.Longsworth, M. Boiarski and L. Klusmier, 80K Closed Cycle Throttle Cycle Refrigerator, in: "Cryocooler 8", R.G.Ross, ed., Plenum Press, New York(1995), p.537
2. M.W.Stanley. "Phase Equilibrium in Chemical Engineering", Butterworth, New York(1985).
3. M.P.Markov et al. "Handbook of Cryogenic Engineering",Moskva Atom Energy Press, Moskva(1985).
4. R.Longsworth, Cryogenic Refrigerator with Single Stage Compressor, US Patent 5,337,571, Aug.,1994.

# DEVELOPMENT OF A LOW VIBRATION THROTTLE CYCLE COOLER

D. H. Hill

APD Cryogenics
Allentown, Pennsylvania 18103 USA

## ABSTRACT

Cryocoolers with low vibration are desirable in a number of applications including X-ray detectors, superconducting filters, and SQUID magnetometers. Throttle cycle coolers are attractive for these applications because they operate with very low vibration. The continuing pursuit toward higher detector resolution has made it desirable to push the throttle cycle cooler to lower vibration levels.

The largest reduction in cold end vibration is achieved through development of a new throttling mechanism. Further reductions are realized through redesign of the cold tip support, heat exchanger, and heat exchanger supports. In addition to hardware changes, the composition of the mixed refrigerant is shown to affect the magnitude of vibration.

Vibration under 1 kHz is reduced below that of the standard cooler by 350 percent and by an order of magnitude in the axial and radial directions respectively. Cold tip vibration is shown to be lower than that of Gifford-McMahon coolers but somewhat higher than that of boiling liquid nitrogen. Reaction forces at the cooler interface are more than an order of magnitude lower than those of Stirling coolers and at least a factor of 4 below those of a pulse tube expander.

## INTRODUCTION

Cryogenic cooling is often employed to enhance the sensitivity of various detectors. These detectors must be held fixed in the earth's frame of reference; small perturbations in position add noise to the signal and reduce the resolution of the data being measured. The vibration imposed by the method of cooling must be considered in the system design so that resolution will not be compromised.

Liquid nitrogen has long been used to cool detectors. It is relatively inexpensive and the only source of vibration is the slow boiling of the fluid as heat is absorbed from the detector and the environment. Gifford-McMahon and Stirling coolers operate on closed cycles, but produce higher vibrations because they employ moving masses inside their respective cold ends. Pulse tube coolers have lower cold end vibration than Gifford-McMahon and Stirling coolers because there are no moving parts inside the pulse tube cold end.

The throttle cycle cooler makes use of the Joule-Thomson effect with a mixed refrigerant. This has been discussed in detail in other works.[1,2] The system consists of an air cooled compressor package, a pair of interconnecting lines, and a cold end. The compressor supplies high pressure mixed refrigerant via one of the interconnect lines to the cold end, where it flows through a counter flow heat exchanger, through the throttling device. The refrigerant expands through the throttling device, and flows through a cold tip where it absorbs heat from the device to be cooled. The low pressure refrigerant is then circulated back through the counter flow heat exchanger, the interconnect line, and back into the compressor.

The compressor package produces forcing functions in the form of mechanical vibration and pressure pulses at 58 Hz and higher harmonics, which are carried to the cold end by the interconnect lines. This was discussed in more detail in an earlier paper.[3] Compressor mechanical vibration can be prevented from having a measurable influence on the cold end by clamping the interconnect lines to a block weighing approximately 9 Kg or more. The pressure pulses can be reduced by using 15 m interconnect lines instead of the standard 3 m lines. The greater internal volume of the long lines reduces the amplitude of the pulses reaching the cold end.

The focus of the current project was vibration at the cold end, which is in direct contact with the vibration-sensitive detector. Vibration inside the cold end is produced by throttling the refrigerant from high pressure to low pressure, and by the flow of the refrigerant through the other parts of the cold end.

## COLD END DESIGN

The goal of this program was to minimize cold end vibration while cooling 2W under 82 K and maintaining the same cool-down performance, reliability, and life as the standard throttle cycle cold end. The most straightforward way to reduce vibration is to reduce the steady state flow rate. This is illustrated by Figure 1. Data was taken with an adjustable needle valve throttle mechanism, where vibration as measured by an accelerometer on the cold interface is clearly proportional to flow rate.

Other performance parameters which are affected by flow rate are refrigeration capacity, susceptibility to contaminants, and cool-down performance. Refrigeration capacity is proportional to flow rate, but sufficient margin existed such that flow rate could be reduced significantly. Generally, susceptibility to contamination increases with reduced flow rate. Although the boundary between acceptable and unacceptable is rather gray in this case, past experience and contamination tests were used to set the minimum allowable flow rate from a contamination susceptibility standpoint. In the past, a reduction in steady state flow rate necessitated a reduction in the flow rate during cool down resulting in a longer cool-down time. Ongoing research in this area allowed the development of a new type of throttling device which does not compromise cool down-time for reduced steady state flow rate. Details of this throttling device remain proprietary at this time.

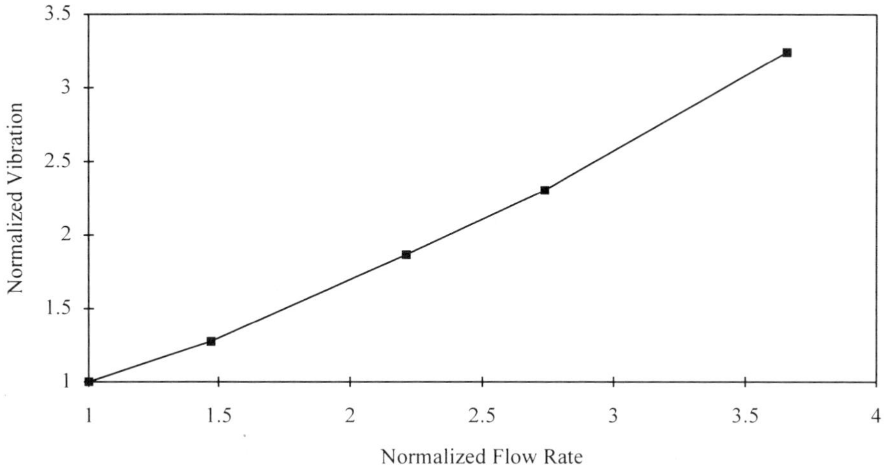


**Figure 1.** Tests with adjustable-throttle cold end show vibration is proportional to refrigerant flow rate.

State of the art detectors are often so sensitive that the minute movement of the cooler structure as the mixed refrigerant circulates through the cold end can be discerned in the detector signal. Motion of the heat exchanger, depicted in Figure 2 is measurable as acceleration at both the cold tip and warm flange. Since the low vibration cold end was not required to carry as much refrigeration capacity as the standard cold end, the heat exchanger was shortened substantially producing a stiffer structure. In addition a stiff heat exchanger support was added at the outboard end of the heat exchanger, near the cold tip.

The cold tip support tube on the standard cold end was designed to support up to 1.3 Kg with a minimum thermal loss through the support tube. In the new design, the radial stiffness of the support tube was increased by a factor of 10. Through optimization of the tube diameter and wall thickness, the thermal loss through the support tube was held to just over twice that of the standard cold end support tube.

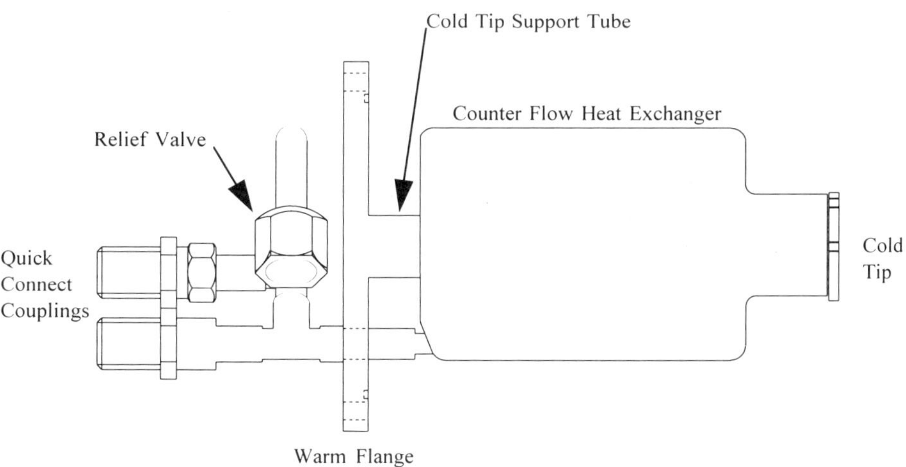


**Figure 2.** Components of the low vibration cold end.

## CHARACTERIZATION TEST SETUP

Cold end vibration will be influenced by how the cold end is mounted in the system. Rigid mounting minimizes vibration, but maximizes the interface force transmitted through the supports. The throttle cycle cold end was characterized in the "free-free" condition as depicted in Figure 3. The cold end weighed 1.6 Kg of the total 4.1 Kg supported weight which included the cold end, vacuum canister, and instrumentation portals. The canister was supported by two soft springs such that the system had a fundamental frequency of 1.0 Hz along the axis of the canister and 3.3 Hz along the axis of the springs. The "free-free" condition is a worst case because it yields the highest cold end vibrations. It is also the most repeatable condition to perform testing because there is no question of interaction of the cold end with the support structure.

The cold end was instrumeted with a PCB Model 351B41 cryogenic accelerometer on the cold tip. Acceleration data was collected one axis at a time with a Stanford Research Systems Model SR770 spectral analyzer. The axial direction was defined as the direction perpendicular to the mounting plane at the cold tip. The two orthogonal axes in the cold tip mounting plane are taken to be similar since the cold end is constructed from cylindrical geometry. The radial direction was defined as any axis in the mounting plane perpendicular to the axial direction.

Interface forces were measured with a PCB 208A03 force transducer. The cold end was suspended through its center of gravity from a 100 Kg rigid mounting structure by a stinger with the force transducer in the load path.

## CHARACTERIZATION TEST RESULTS

"Free-free" cold tip acceleration spectra of the standard and low vibration cold ends are shown in Figure 4. All cold end vibration measurements were recorded at no-load operating temperature. Radial vibration is predominant in both cold ends below 1 kHz. Comparing the peaks of the two radial vibration curves, acceleration of the low vibration

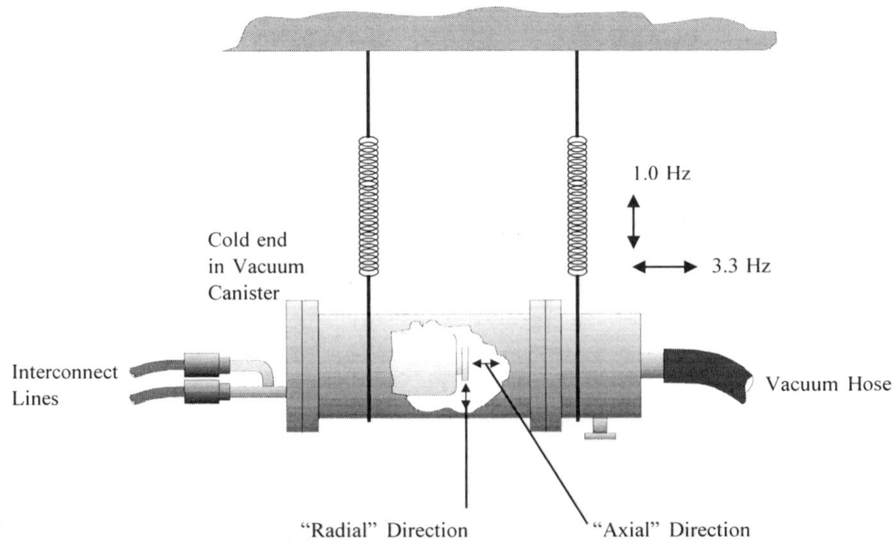


**Figure 3.** "Free-Free" Cold End Vibration Measurement Test Setup.

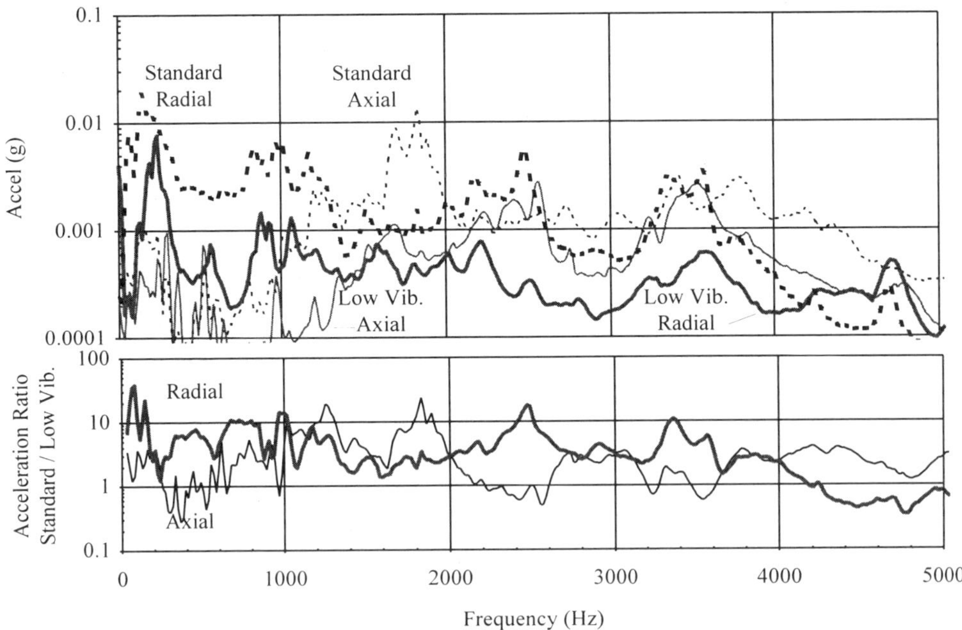


**Figure 4.** "Free-free" acceleration spectrum comparison - standard cold end and low vibration cold end.

cold end is 250 percent lower than that of the standard cold end. The dominant peaks on both radial acceleration curves (at 125 Hz for the standard cold end and 235 Hz for the low vibration cold end) correspond to the fundamental bending mode of the cold tip support tube. The upper portion of Figure 4 shows the rms average acceleration while the lower part of the figure shows the ratio of the standard cold end acceleration to the low vibration cold end acceleration. On average, cold tip vibration of the low vibration cold end under 1 kHz is reduced below that of the standard cooler by 350 percent and an order of magnitude in the axial and radial directions respectively.

Since the flow of refrigerant through the cold end causes vibration, it may be expected that the refrigerant mixture influences the cold end response. In Figure 5, the "free-free" axial acceleration response of three different mixtures is compared. Mixture PT-13 is optimized for low temperature, reaching 72 K with no load. Mixture PT-14 was designed to have higher heat lift than PT-13. PT-16 is a high capacity mixture for which the standard cold end provides 22 W of refrigeration at 120 K.

## VIBRATION COMPARISON WITH OTHER FORMS OF COOLING

The interface force produced by several types of Stirling coolers and one pulse tube expander is compared with that of the low vibration cold end in Figure 6. The data was condensed from axial and radial measurements to the resultant force, defined as

$$F_R = \sqrt{F_a^2 + F_{r1}^2 + F_{r2}^2} \tag{1}$$

where $F_R$ is the resultant force, $F_a$ is the force in the direction along the axis of the cold tip, and $F_{r1}$ and $F_{r2}$ are forces in the two orthogonal directions perpendicular to the axial direction. This equation is accurate for forces which are well correlated as those of Stirling

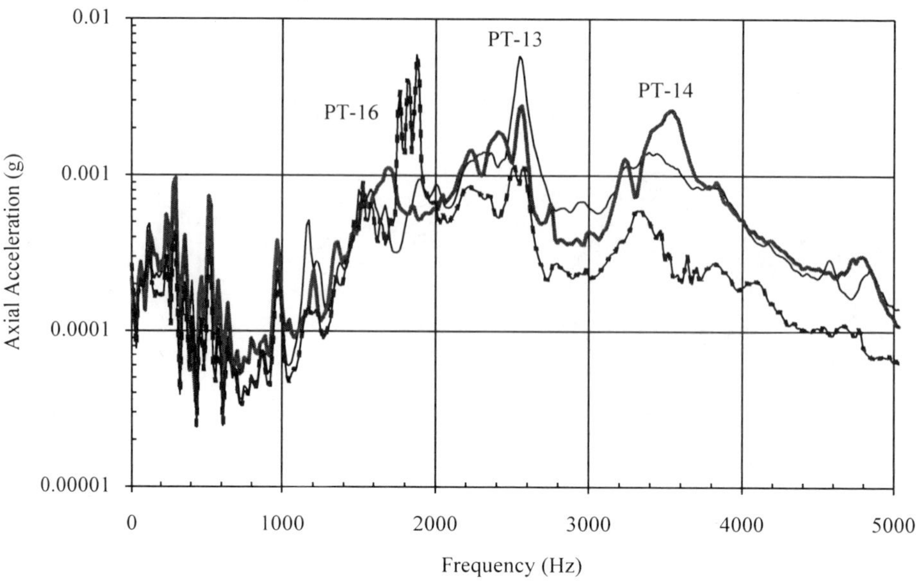


**Figure 5.** "Free-free" acceleration spectrum of low vibration cold end with different refrigerant mixtures.

coolers, but generally yields conservative (higher) resultant forces for forces which are not as well correlated such as those of the throttle cycle cooler.

The filled data points in Figure 6 are compressor vibration data while the unfilled data points are displacer data. Generally, compressor forces are higher than the displacer forces. The continuous spectra are the throttle cycle cooler interface forces. The low

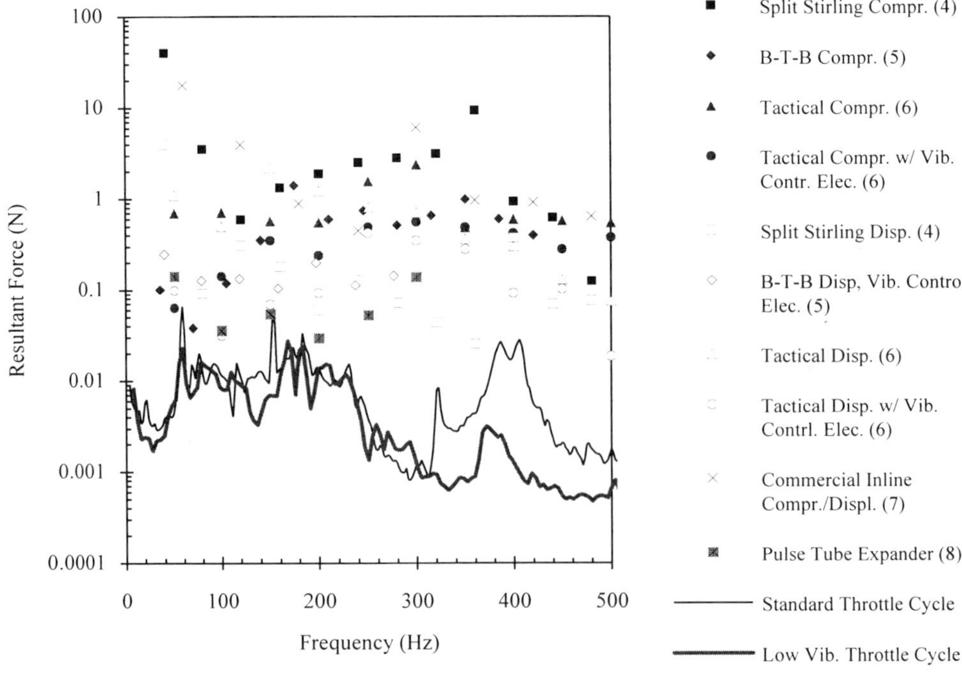


**Figure 6.** Stirling cooler, pulse tube, and throttle cycle cooler interface force comparison.

vibration throttle cycle cooler spectrum is at least an order of magnitude below the Stirling cooler data and a factor of 4 below the pulse tube data.

Gifford-McMahon deflection data is typically measured with the expander in an insulating vacuum vessel resting on the floor or clamped in place. This data is relevant for experiments which may be mounted on the expander, but lacks a consistent basis for comparison because the deflection response of the expander / vacuum vessel system is largely dependent on the mass of the vacuum vessel and the stiffness of the load path from the expander to the floor. The axial and radial deflection of the throttle cycle cooler cold end was measured with the vacuum canister rigidly supported from a 40 Kg mass to simulate the conditions under which a Gifford-McMahon cooler would be measured. The throttle cycle cooler cold end deflection data may be compared with that of Gifford-McMahon coolers in Figure 7. The throttle cycle cooler deflection data is an order of magnitude lower than the Gifford-McMahon deflection data.

One of the simplest methods of reducing the vibration of a cryocooler is to add mass. Figure 8 illustrates this technique showing "free-free" acceleration spectra of the low vibration cold end with 1.2 Kg of copper mounted on the cold tip with data from a 0.5 liter liquid nitrogen test dewar, used as a test bed for high temperature superconducting electronics. The two spectra are comparable in the axial direction, but radial acceleration of the throttle cycle cooler is somewhat higher than that of the liquid nitrogen dewar.

## CONCLUSIONS

A new throttle cycle cooler cold end has been designed for ultra-low vibration applications. A new type of throttling device and structural stiffening of the heat exchanger and cold tip support tube resulted in 350 percent and order of magnitude reductions in axial and radial vibration, respectively.

Vibration of the throttle cycle cooler has been compared with that of other forms of cooling. Stirling cycle, pulse tube, and Gifford-McMahon coolers produce higher levels of vibration. Boiling liquid nitrogen produces a lower level of vibration than the throttle cycle cooler cold end.

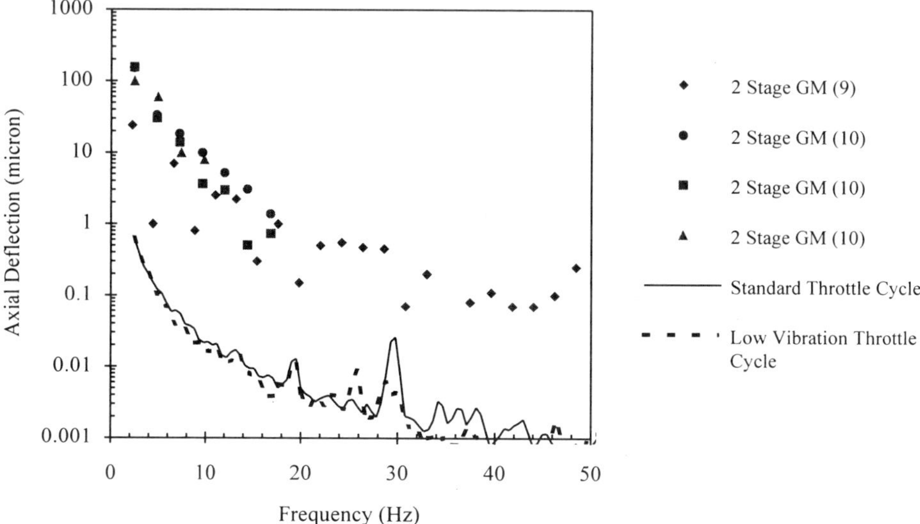


**Figure 7.** Gifford-McMahon and throttle cycle cooler axial deflection comparison.

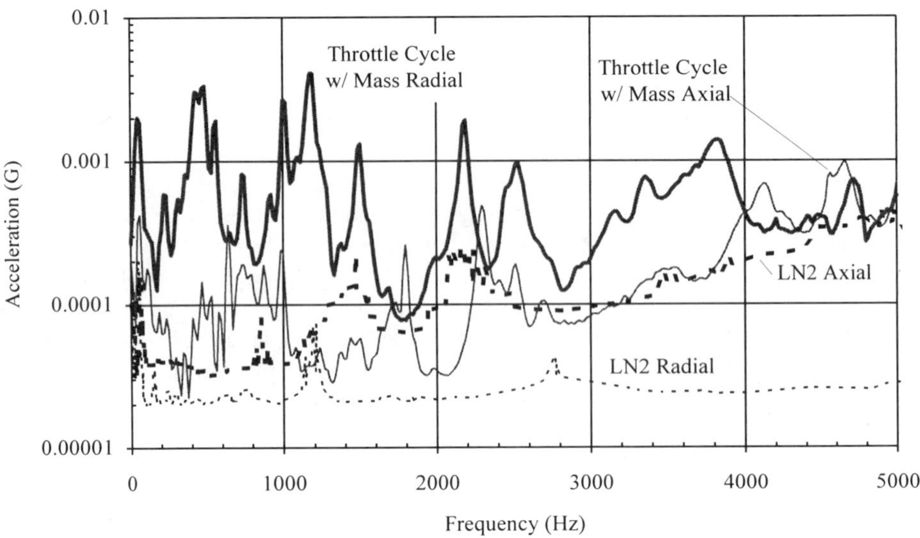


**Figure 8.** Liquid nitrogen dewar and throttle cycle cooler vibration comparison.

## ACKNOWLEDGMENTS

The author wishes to thank Hank Fasnacht for many invaluable contributions which include pivotal design ideas, manufacture of incredibly intricate parts on an R&D lathe, and assembly and test of numerous prototypes. Boris Yuden and Mikhail Boiarski are credited for their pioneering work in this area.

## REFERENCES

1. Khatri, Ajay, "A Throttle Cycle Refrigerator Operating Below 77 K," *Advances in Cryogenic Engineering*, vol. 41, Plenum Press, New York (1996), pp. 1123-1129.
2. Longsworth, R. C., Boiarski, M. J., Klusmier, L. A., "80 K Closed-Cycle Throttle Refrigerator," *Cryocoolers 8*, Plenum Press, New York (1995), pp. 537-542.
3. Hill, D. H., "Throttle Cycle Cooler Vibration Characterization," *Cryocoolers 9*, Plenum Press, New York (1997), pp. 737-745.
4. Mon, G. R., Smedley, G. T., Johnson, D. L., and Ross, R. G., "Vibration Characteristics of Stirling Cycle Cryocoolers for Space Applications," *Cryocoolers 8*, Plenum Press, New York (1995), pp. 199-200.
5. Sparr, Leroy, et al., "NASA/GFSC Cryocooler Test Program Results for FY93/94," *Cryocoolers 8*, Plenum Press, New York (1995), pp. 223-225.
6. Sparr, Leroy, et al., "Adaptation of Tactical Cryocoolers for Short Duration Space-Flight Missions," *Cryocoolers 8*, Plenum Press, New York (1995), pp. 168-170.
7. Smedley, G. T., Ross, R. G., Berchowithz, D. M., "Performance Characterization of the Sunpower Cryocooler," *Cryocoolers 8*, Plenum Press, New York (1995), pp. 155-157.
8. Collins, S. A., et al., "Performance Characterization of the TRW 35K Pulse Tube Cooler," *Advances in Cryogenic Engineering*, vol. 41, Plenum Press, New York (1996), pp. 1475-1476.
9. Kang, Yoon-Myung, et al., "4 K GM/JT Cryocooler For Cryogenic Sensors," Proceedings of the Sixth International Cryocoolers Conference, David Taylor Research Center, Maryland (1991), pp. 19-21.
10. APD Cryogenics Inc. vibration data archives, Allentown, Pennsylvania.

# WATER TRAP REFRIGERATED BY A THROTTLE-CYCLE COOLER USING MIXED GAS REFRIGERANT

**Ajay Khatri,[1] M.Boiarski ,[1] and S. Nesterov[2]**

[1]APD Cryogenics, Inc.
Allentown, PA 18103 USA
[2]Moscow Power Engineering Institute
Moscow, Russia

## ABSTRACT

The design features and the performance of the cooler are discussed to emphasize characteristics of this refrigeration system compared to other cryocoolers used for water traps. A throttle-cycle cooler based on one-stage oil-lubricated compressor provides water trap temperature in a narrow range of 115 K to 130 K depending upon the heat load on the water trap surface.

The water trap surface connected with the cooler evaporator was configured to provide an optimal correlation between the nitrogen conductance and water trapping parameters. Ability to simulate the pumping performance, the properties of mixed gas refrigerants and systems performance combined with extensive experience in designing reliable coolers will make it possible to use this development in many other applications.

## INTRODUCTION

Turbomolecular vacuum pumps are widely used in many applications in semiconductor industry to evacuate process chambers. The gas mixture to be evacuated consists mostly of water vapor which is hard to remove from the chamber by turbomolecular pumps. To capture water at the inlet of the turbomolecular pump a refrigeration system should maintain the water trapping surface in a narrow temperature range, typically from 110 K to 135 K. This temperature range relates to a partial pressure of about $10^{-9}$ Torr to $10^{-13}$ Torr for the water vapor. It also prevents condensation of other gases exhausting from the manufacturing process.

A water trap refrigerated by liquid nitrogen was proposed by Meissner[1] in 1955. Later closed-cycle coolers were developed to provide refrigeration for a water trap. These coolers may be based on the Gifford-MacMahon (GM) cycle as described elsewhere[2]. However , coolers of this type are expensive and need a special regulation system to provide a given temperature range. Polycold System, Inc. developed coolers based on the mixed refrigerant Klimenko auto-cascade cycle with one-stage compressor[3]. They are designed for trapping water vapor in a temperature range higher than 140 K.

Recent achievements in developing of mixed-refrigerant throttle-cycle based on one-stage compressors allow us to build[4] highly reliable small scale coolers providing refrigeration in a temperature range from 70 K to 130 K depending upon the mixed refrigerant (MR) composition[5]. Later it was proposed[6] to apply such coolers to water trap refrigeration.

A target of this development was to build relatively low cost highly reliable water trap system applicable to turbomolecular pumps. A closed-cycle cooler should provide a trap surface temperature in a narrow range from 115 K to 135 K independent of the water trap size. The configuration of the water trapping surface should not reduce the nitrogen pumping speed to less than 80 % providing the maximum trapping of water vapor in the given temperature range.

## CRYOCOOLER DESIGN AND PERFORMANCE

The flow schematic of the cooler which consists of a compressor, an after-cooler, a counter flow heat exchanger, a throttle device and an evaporator is presented at figure 1.

The cryocooler design for the water trap application should provide as much capacity $Q_r$=max as possible at $T_{max}$ for the use with a turbomolecular pump of a high capacity. If at a given refrigeration temperature $T_r$ an actual heat load supplied to the cooler is less than $Q_r(T_r)$ then the mixed refrigerant (MR) accumulation would take place in the heat exchanger because of an excess refrigeration capacity. In turn it will reduce pressure values in both a supply line -high pressure ($P_H$) and in a return line - low pressure - ($P_L$) . For example, processes in the evaporator are presented at fig.1 in the temperature-enthalpy diagram T-h based on our experimental data obtained with the "mix-1" MR. At $T_{max}$ a refrigeration capacity $Q_r'' = G''*q_r'' = G''*(h_3'' - h_2'')$ supports both an electrical heater load and the parasitic heat intakes ($Q_{par}$). As the heater was turned off it reduced both $P_H$, $P_L$ and $T_r$ until a lower value of $Q_r'$ supports $Q_{par}$ : $Q_r' = G'*(h_3' - h_2') = Q_{par}$. In this case the value of $T_{min}$ depends on not only the $Q_{par}$ but the efficiency of the heat exchanger and the volume of the system as well. That is why it is important to optimize components of the cooler all together including the MR composition, the compressor volumetric efficiency and the heat exchanger efficiency.

In turn, the efficiency of the heat exchanger is to be defined taking into account both the minimal temperature difference in the heat exchanger $DT_{min}$ ,which influences the specific cooling capacity $q_r$ value, and a pressure drop in the return line $DP_L$ , which influences on the mass flow rate G of the compressor and, in turn, on the overall refrigeration capacity $Q_r$ .

In order to design a cooler which meets all the requirements for a water trap application a computer software was used which allowed to predict the refrigeration map for a given MR and compressor, based on limited data on the heat transfer and hydraulic pressure drop coefficients. Comparison of the experimental and calculated data for some applications is given elsewhere [5]. The experiments show that the cryocooler of the standard configuration provides a refrigeration capacity $Q_r$ = 30 W at $T_r$ =130 K which can support an expected heat load for a water trap size up to 250 mm inner diameter.

Presented data combined with the exergy analysis[7] show some opportunities for further efficiency improvements. It is known that the exergy analysis allows to present all the losses taking place in the system in terms of work performed with the reversible process.

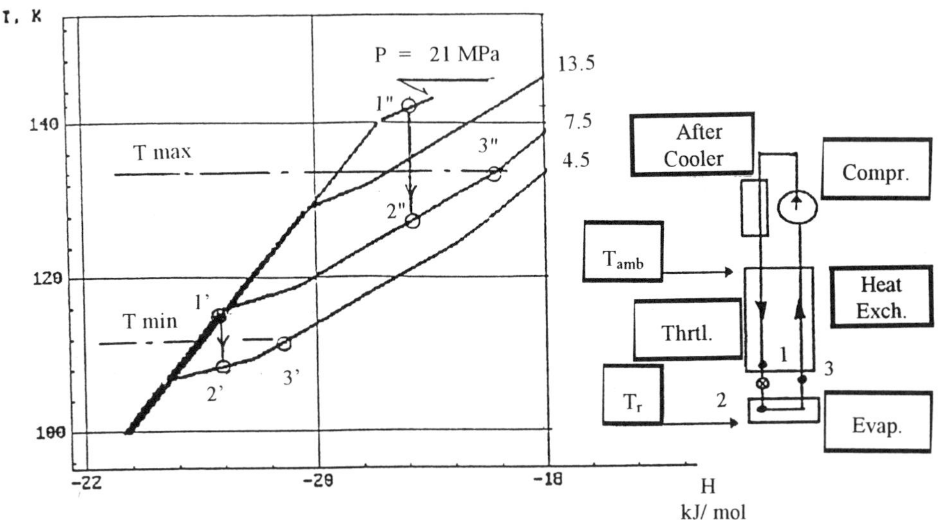


**Figure 1** Schematic of cryocooler and evaporator process.

The following simple equations have been used to define an exergy efficiency which is equal to the Carnot efficiency (CEF) of the cooler: $CEF = E'' / E'$, where $E' = P_{cm}$ compressor power consumption, $E'' = Q_r (T_a / T_r - 1)$.

The results of calculations at $T_r = 133$ K are presented at Table 1 for the blend which is shown at figure1 in T-h diagram. The table includes both an absolute value of the exergy losses Di (Watts) for each of the constituents and a dimensionless value of the losses related to the compressor power consumption $di = Di / E'_{cm}$.

**Table 1**. Exergy losses in the cooler based on 28 L/m compressor at $T_r = 133$ K

| | Compressor | | | Lines | Heat Exchanger | | | | Throttle | Evaporator | | |
|---|---|---|---|---|---|---|---|---|---|---|---|---|
| Di, | $E'_{cm}$ | $D_{cm}$ | $E''_{cm}$ | $D_{DP}$ | $E'_{he}$ | $D_{DT}$ | $D_{DP}$ | $E''_{he}$ | $D_{TR}$ | E' | $D_{DT}$ | E'' |
| Watts | 515 | 271 | 244 | 40 | 204 | 47 | 56 | 101 | 57 | 44 | 1.5 | 42.5 |
| d | 0.52 | | | 0.08 | 0.09 | | 0.11 | | 0.11 | <0.01 | | |

Based on these data it is possible to define both the cooler power efficiency as $CEF_{CL} = E''_{EV} / E'_{CM} = 0.08$ and the cryoblock (CB) efficiency which include the heat exchanger, the throttle and the evaporator losses $CEF_{CB} = E''_{EV} / E'_{HE} = 42.5/204= 0.21$. This correlation is not quite rigorous. More accurate relation is $CEF_{CB}=E''_{EV}/(E1\text{-}E5)$. This efficiency is high if taken into account a relatively small refrigeration capacity of the system. Also the hydraulic loss $D_{DP}$ in the heat exchanger is relatively high, which shows a way of the further heat exchanger optimization.

The optimization has been conducted under the constraint that the cryoblock dimensions must be identical for all the variants. That is why a number of tubes and the inner tube diameter were selected as variable parameters. In addition the refrigerant composition, supply and return pressure of the cycle provide a gaseous state of the MR if the ambient temperature is greater than Ta = 280 K. This allows us to use a single compressor attached with a few water traps if it is necessary. An optimized system produced as much as Qr = 35 watts at Tr = 135 K.

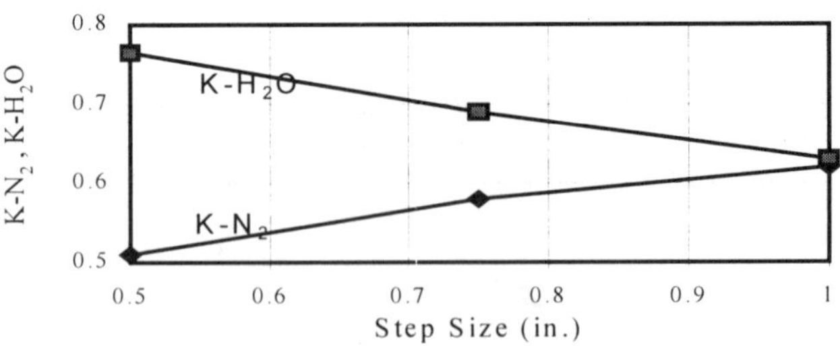


**Figure 2** Water trapping coefficient and nitrogen conductance coefficient as a function of the step size for 160 mm size AquaTrap.

## TRAP CONFIGURATION DESIGN

A water trap cold panel assembly is located in the flange installed on the top of the vacuum pump inlet. An optically penetrated cold panel assembly is formed by flat parallel fins connected with a flattened cold finger which is installed in the middle of the flange. The refrigerant flows through the cold finger which is thermally isolated from the flange at the ambient temperature $T_{amb}$ . In order to reduce the temperature gradient along the finned plates they are connected with an outer ring, which increases the cold panel rigidity as well. Such a configuration of the cold panel has been optimized to provide a reasonable correlation between the water capture coefficient and the nitrogen conductance. Step size (s) between the plates was varied as an optimization parameter.

Simulation of gas flow in vacuum systems by test particle method ( Monte-Carlo ) form the basis for the calculations[8]. Initial data to model a stationary gas flow through the baffle have been taken as follows. Entrance conditions at the inlet correspond to infinitely large volume. Molecules are equally distributed across the inlet area and have a cosine angle distribution. It is assumed that on the inner element surface walls the molecules are reflected according to the cosine law as well. Molecule interaction with the surface is modeled in the following way. A random number RN is generated by the random number generator. If RN is less than or equal to the given sticking coefficient $B_i$ ( $RN <= B_i$ ) on the I-surface, then a molecule is considered to be absorbed. If after many reflections a molecule reach the outlet opening, then it is considered to have left the baffle. A trapping coefficient is defined as a ratio of absorbed molecules $N_a$ and those which crossed the baffle inlet $N_{in}$ . In turn, $N_{in}$ equals the number of tested molecules which is always more than 15,000 .

A trapping coefficient of water $K_{H2O}$ and a conduction coefficient of nitrogen $K_{N2}$ as a function of step size is presented in figure 2 for the 160 mm water trap. The selected step size s = 0.5" provides 1.2 $m^3/s$ of nitrogen pumping speed and 2.28 $m^3/s$ of water pumping speed. For nitrogen pumping speed of a TMP rated at 500 L/s, by adding the water trap it was calculated that the nitrogen pumping speed would be 450 L/s resulting in about 10% loss. Experimental data were obtained with Osaka TG550 TMP combined with the water trap. Comparison of calculated and experimental data on nitrogen pumping speed shows that deviations are less than 5% which makes it possible to use the software for further optimization.

## SYSTEM PERFORMANCE PARAMETERS

A typical system setup is as shown in figure 3. The water trap (AquaTrap ™) is setup on top of a turbomolecular pump. A gate valve is placed below the vacuum chamber and on top of the AquaTrap. The AquaTrap is connected to the compressor with two flexible gas

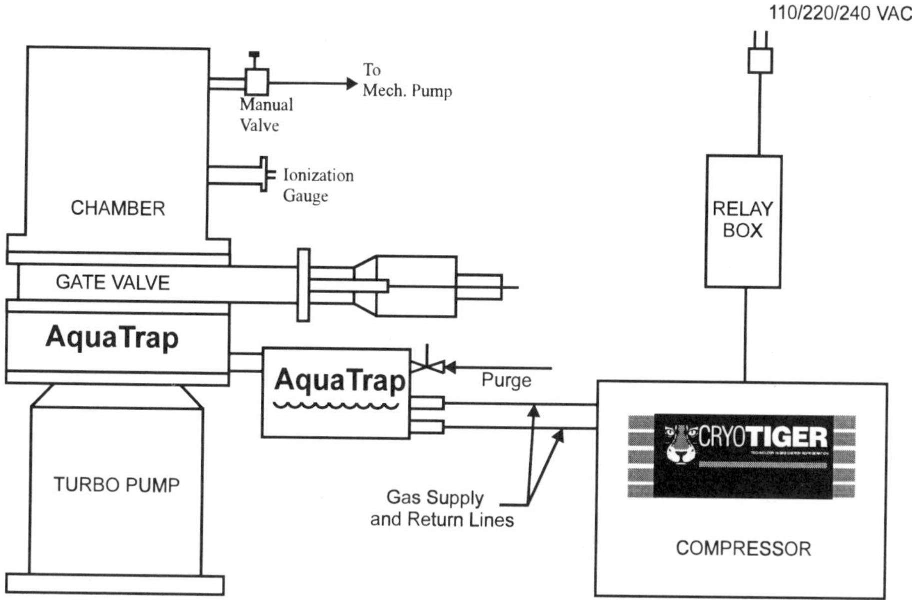


**Figure 3** AquaTrap test setup.

lines. The regeneration controller is connected to the heater/diode port on the AquaTrap, the purge solenoid valve and compressor remote on/off module. The testing of the system was carried out using a mixed gas refrigerant blend "mix-1". 160 mm and 250 mm ID water traps were tested using this refrigerant. The refrigerant blend was designed for optimum performance in the temperature range of 115 K to 135 K from no load to full load condition.

Figure 4 shows the typical performance of the 160 mm and 250 mm ID AquaTraps during cooldown. The system is considered at operational temperature and ready to run when the sensor on the cold array reads below 130 K. The mass of the cold array increases with the larger ID AquaTrap which reflects on the longer cooldown time it takes to get below 130 K for 250 mm size AquaTrap compared to 160 mm AquaTrap.

The cooldown time to 130 K for the cold array was about 30 minutes for 160 mm AquaTrap and 60 minutes for the 250 mm AquaTrap. The rate at which refrigeration is produced does not change for different sizes of AquaTrap. All the components are kept the same except for the cold array assembly size. A good starting vacuum was provided before the cooldown was started.

Figure 5 shows regeneration performance of the same AquaTraps. It shows how long it would take for the cold panel assembly to warm up from 120 K to 310 K and then cooldown again to less than 130 K temperature. During the warm up phase the heater located on the cold panel assembly is turned on and at the same time the purge solenoid is opened to start flowing nitrogen. This quickly warms up the cold panel to 310 K at which point the controller turns off the heater. It is possible to extend the purge time further if so desired at 310 K temperature. For these tests the purge time was not extended. Since the heat exchanger is still cold when the cooldown part of the cycle is started, the cooldown to 130 K takes less time compared to the time it takes for normal cooldown when the heat exchanger is warm at the start. The 250 mm AquaTrap had cool to cool cycle time of less than 60 minutes while 160 mm AquaTrap took less than 25 minutes. If the purge time was

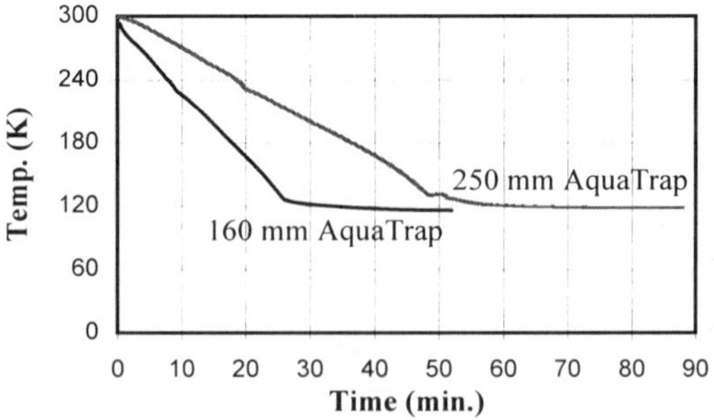


**Figure 4** Cooldown performance of AquaTrap for 160 mm and 250 mm sizes.

extended and the heat exchanger was fully warmed up, this time would be longer accordingly.

Figure 6 shows the refrigeration performance of the AquaTrap system when a heat load is applied on the cold panel assembly. The data was taken while maintaining 1 x10-3 Torr or better vacuum in the system. The AquaTrap has a minimum temperature of about 116 K to 117 K while supporting normal parasitic heat loads. As the heat load is increased the temperature on the cold panel increases as the figure shows. At 30 watts of applied heat load the cold panel temperature was about 126 K. This capacity was found to be sufficient to support heat loads on all the different sizes (100 mm to 250 mm) AquaTraps. Figure 6 also shows performance of the AquaTraps (160 mm and 250 mm) when the vacuum condition is degraded. These tests simulate how the system would behave when exposed to process environments where the vacuum level may be several microns. As the AquaTrap size gets larger and hence the corresponding array size gets larger then for the same vacuum level the heat load increases on the refrigeration system. This will cause the temperature to be little higher for 250 mm AquaTrap than 160 mm AquaTrap at the same vacuum level. It is shown that the system will run stable even when the vacuum has degraded to 500 micron.

The system has also gone through parametric testing including system static charge pressure, ambient temperature, 60 Hz and 50 Hz operation, radiation heat load etc.

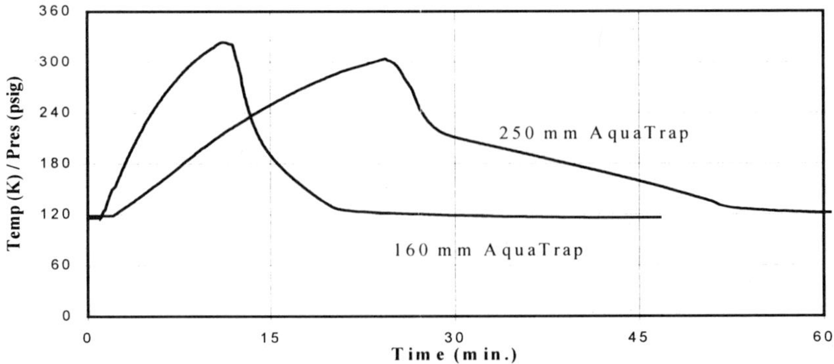


**Figure 5** Regeneration performance of AquaTrap for 160 mm and 250 mm sizes.

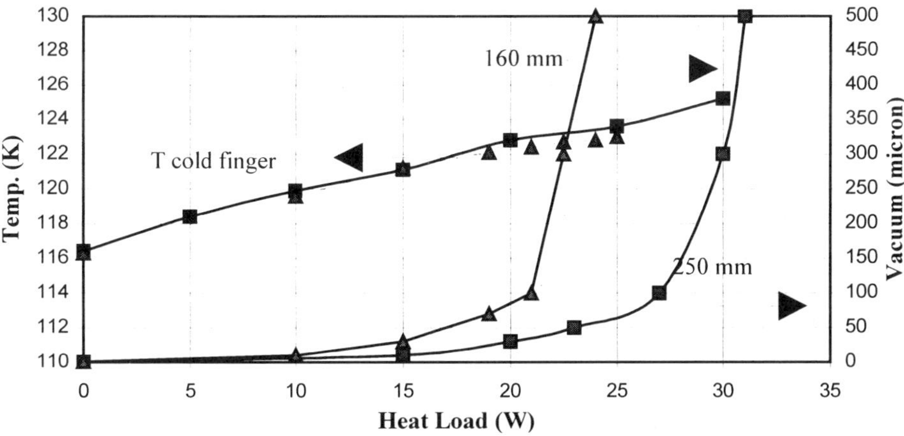


**Figure 6** Refrigeration capacity of 160 mm and 250 mm AquaTrap with applied heat load and degraded vacuum level.

The operating temperature range is a function of the system hardware design and the mixed gas refrigerant composition. Temperature range of 110 K to 135 K is considered desirable for the water trap application. The minimum temperature of about 115 K to 117 K is reached in the system and it does not require any active temperature controller to maintain it. The system will not cooldown below that point. When the heat load is applied the temperature will rise over a limited range. If the heat load is applied beyond what can be supported then the system will continue warming up as long as the heat load is maintained. If the heat load is removed or reduced then the system will cooldown again. The controller is there only to read the temperature of the system and for the regeneration cycle. Further opportunities exist where a single compressor with an adequate capacity could run multiple cold ends.

## CONCLUSION

•This type of water trap offers a simple system design. It has no moving parts in the cold end so it has extremely low vibration characteristics compared to Stirling or GM cryocooler water traps. This is quite important in the semiconductor industry where the trend is towards finer geometry. The cold finger design is flexible so that it is also possible to customize the water trap configuration.

•Gas lines are at ambient so they are easy to handle and can have longer lengths compared to the insulated refrigerated lines used in a water trap using liquid nitrogen.

•The cryocooler can be produced at lower cost in high quantity compared to any other type of cryocooler for this application.

•It has no maintenance, high reliability, long life and separation of the cold end from the compressor.

•Basic hardware can be used for different temperature ranges (70 K to 200 K) by simply recharging it with an appropriate refrigerant blend. This unique flexibility will allow this cryocooler to be used in applications such as infra-red detectors, gamma detectors, computer chip cooling, HTS filters, medical applications, electron microscope, material science studies, nitrogen recondenser, non-destructive testing and many more.

## ACKNOWLEDGMENT

The authors are grateful to Stephen Harold (APD CRYOGENICS, INC.) for help in testing the systems and Anton Hasanov (Moscow Power Engineering Institute , Russia) for help in developing the computer software to analyze the system performance.

## REFERENCES

1 Meissner, C.R., 1955, Liquid nitrogen Cold Traps, Review of Scientific Instrumentation, Vol.26, p. 305, American Institute of Physics, New York, NY.

2 K. Okumura, F. Kuriyama, Y. Murai, M. Tsujimura, H. Sobukawa; Evacuation Apparatus and Evacuation method, 1991, USA Patent 5,062,271.

3 Missimer,D.J., 1984, Fast-Cycle Pump Forecast: Cold & Dry, Photonics Spectra, The Optical Publishing Company, Pittsfield, MA.

4 Logsworth,R., Boiarski,M., Klusmier, L., 1995, 80 K Closed Throttle Cycle Refrigerator, Cryocoolers 8, Plenum Press NY, p.537.

5 Khatri , A., Boiarski, M. ,1996, A Throttle Cycle Cryocooler Operating with Mixed Gas Refrigerants in 70 K to 120 K Temperature Range, Cryocoolers 9, Plenum Press NY.

6 Longsworth, R., 1996, Throttle Cycle Cryopumping System for Group I Gases, USA Patent Pending.

7 Brodyansky V.M., Sorin M.V. and LeGoff P., 1994, Exergy Analysis and Optimization, Elsevier, Amsterdam.

8 Bird, G.A., 1994, Molecular Gas Dynamics and the Direct Simulation of Gas Flows, Oxford.

# RETROSPECTIVE OF MIXED-REFRIGERANT TECHNOLOGY AND MODERN STATUS OF CRYOCOOLERS BASED ON ONE-STAGE, OIL-LUBRICATED COMPRESSORS

M.J. Boiarski, [1] V.M. Brodianski, [2] and R.C. Longsworth [1]

[1]APD Cryogenics Inc.,
Allentown PA, 18103
[2]Moscow Power Engineering Institute
Moscow, Russia

## ABSTRACT

Advancement of applied thermodynamics made it possible starting in the late 1960's to tackle the problem of developing a highly-efficient single circuit throttling refrigeration cycle. It was shown by means of the exergy analysis that the Carnot efficiency of the well-known throttling refrigeration cycle may be essentially improved by using multi-component refrigerants (MR). Further, based on both thermodynamics of multi-component mixtures and experimental data on multi-phase vapor-liquid and solid-liquid equilibrium, highly-efficient MRs were proposed for different applications down to cryogenic temperatures. A combination of thermodynamics and technology allows us to design and build highly-reliable cryocoolers which are now commercially-available, based on one-stage oil-lubricated compressors. These low-cost coolers provide refrigeration in the temperature range from 65 K to more than 130 K depending on the MR composition. The development of this technology along with present applications, are discussed.

## INTRODUCTION

Industrial refrigeration technology was developed in the 19$^{th}$ century based on understanding the physical and chemical properties of substances, Carnot's contribution to the background of technical thermodynamics[1], and advances in mechanical engineering[2]. Two different types of refrigeration systems have been developed. One which uses a combination of absorption and mixing processes is known as thermo-chemical refrigeration. The other type which is based on a combination of mechanically driven compressors and different types of expansion processes is known as thermo-mechanical refrigeration. In the field of mechanically driven refrigerators Gorrie, G. (1845)[2] obtained the first US patent for an air-expander refrigerator. Later contributions made in particular by Siemens, V. and Kirk, A. [2] led to industrial production of air-expander refrigerators.

A different type of system based on a throttle-expansion process[2] was described by Perkins, J. (1834). He proposed a vapor-compression cycle in a closed system which provides refrigeration at steady state conditions. The throttle expansion process was investigated, described and named later as the Joule-Thomson effect. The first prototype of a system based on the vapor-compression cycle was presented by Twinning, A. (1853) and industrial scale units of this type were built by Garrison, D. (1856). Later, Linde, C. and Hampson, V. (1895) working independently developed a gas-compression cycle for air liquefaction. Throttle refrigeration systems (TRS) based on these cycles were well developed and investigated for different applications. In 1908 Onnes, H. K. used a five stage cascade TRS to liquefy helium for the first time.

It was found that the vapor-compression cycle, (VCC), is very efficient when applied at relatively high refrigeration temperatures, $T_r > 230...240$ K. That is why this is the most common cycle in use today for commercial and residential air conditioning and refrigeration. At temperatures below about 230 K the VCC running a pure refrigerant in a single circuit is not able to provide efficiencies as high as other cycles. In cryogenics for a long time it was used only in those cases when system simplicity was a major objective. Only recently, starting in 1965-1970, was it found that this cycle may be energy efficient by running a mixed gas refrigerant. Working independently Funder, A. (1964)[3], Brodianski, V. et. al. (1966)[4], Missimer, D. (1973)[5] and other researchers started investigations with MRs for different applications[6,7,8]. Brodianski referred to this as a throttle regenerative system which has the same abbreviation as TRS.

The simplicity of using mixed refrigerants in a system with a single compressor and no mechanical expander provides opportunities to build relatively low cost and highly reliable cryocoolers for lower temperatures which broadens their commercial applications. This paper discusses trends and opportunities of this technology in cryogenic applications.

## EVOLUTION OF MIXED REFRIGERANT TECHNOLOGY

Schematics and temperature-enthalpy diagrams of the VCC and TRS are presented in Fig.1. VCCs typically use refrigerants which condense at constant temperature at the high pressure and evaporate at constant temperature at the low pressure thus the process is inherently very efficient. The two phase liquid-vapor region of operation typically extends from 0.1 MPa to about 2.5 MPa. The upper part of Figure 1 illustrates a VCC with R22 to produce cooling at 233 K (-40 °C). The lower part of Figure 1 illustrates the TRS process for a refrigerator operating at 80 K with the same type of oil lubricated compressor which is designed for VCCs but used with a MR. Many mass produced oil lubricated compressors are available for applications in this range. At temperatures below about 180 K an oil separator is needed to avoid having oil freeze in the cold end. Other mixtures which might have partial condensation at ambient temperature can provide cooling at any intermediate temperature. In effect the use of a mixture and a counter-flow heat exchanger enables one to extend the range of operation of this technology to cryogenic temperatures.

The evolution of the TRS is presented in Fig.2. Early systems relied on substances for refrigerants which were available from other branches of industry[2]. Despite the synthesis of chlorofluorocarbons by F. Swarts at the end of the $19^{th}$ century, their industrial applications as refrigerants was not started until the 1930's by the Du Pont Corp. It increased the number of available refrigerants and provided in 1946 the opportunity to use azeotropic blends which behave similar to pure refrigerants.

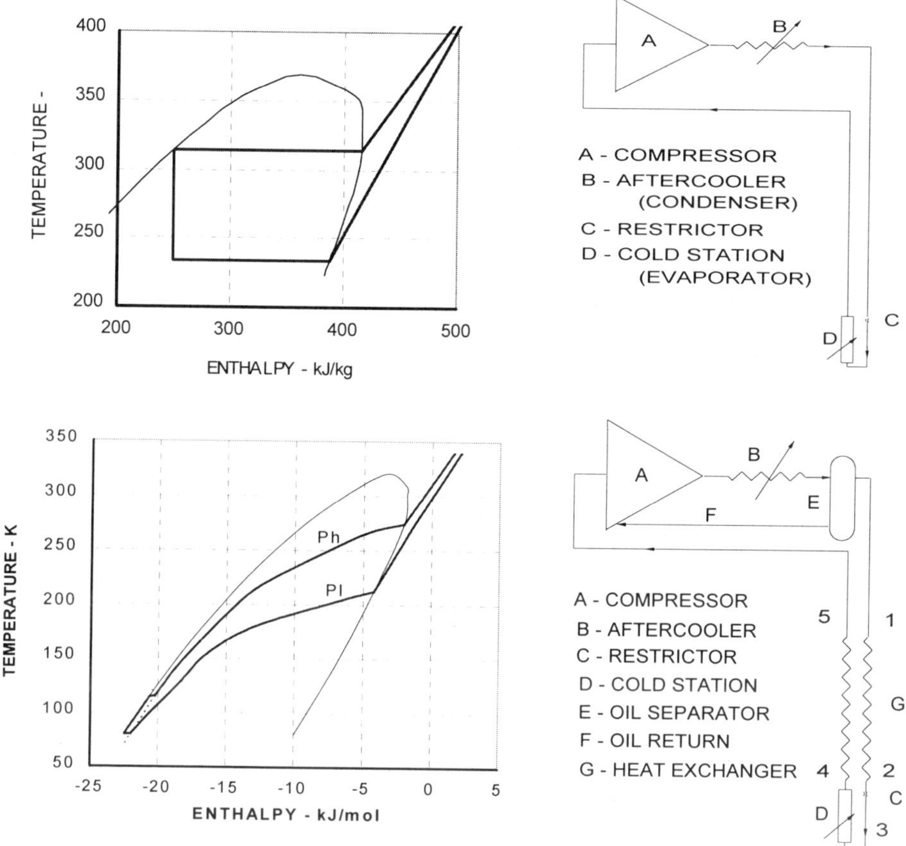


**Figure 1** Comparison of Vapor Compression Cycle cooling to -40 C with R22 and a Throttle Refrigeration System cooling to 80 K with a N2 based mixture.

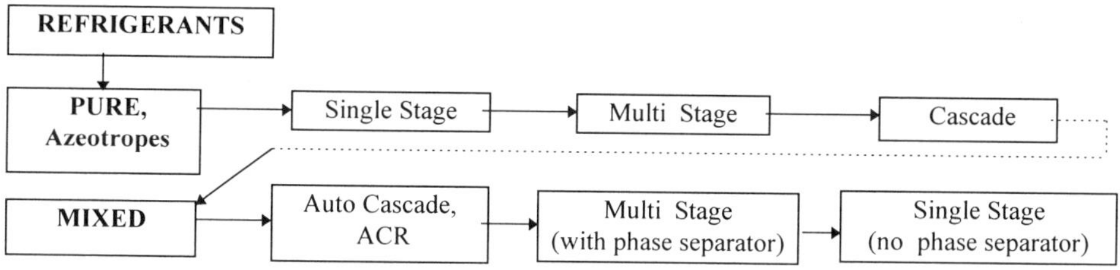


Figure 2 Evolutionary sequence of the development of Throttle Refrigeration Systems.

The first attempts to improve refrigerant properties by designing a mixed refrigerant were made as early as the $19^{th}$ century : Perkins, J. (1834) [2], who used a blend of sulfur dioxide,$SO_2$ and ethyl ether, $C_2H_5$-O-$C_2H_5$; Pikte, R. (1880) [2] unsuccessfully tried to improve a cascade refrigeration process with a MR comprised of $SO_2$ and $CO_2$. Only later, Klimenko, A.P. (1959) [9] developed a single compressor cascade cycle for natural gas liquefaction. The heat exchanger was divided into a few sections with intermediate phase separators to avoid circulation of high boiling components through the low-temperature sections of the heat exchanger. This solution, was used by Missimer, D.[5] in developing an Auto-Cascade Refrigerator (ACR) systems for $T_r > 120$ K. Later coolers of this type have been developed for cryogenic applications, for example Little, W. A. (1988)[10]. Actually the configuration of low temperature equipment in the ACR represents a modified multistage cascade system. It is more complex to build and to control the operational parameters compared to the single stage refrigeration system presented in Fig.1. It is interesting to note as illustrated in Fig. 2 that the development of refrigerators with MR went in the opposite sequence from pure refrigerants.

The idea of using MR refrigerants in single stage refrigerators for cryogenic applications was pursued by Russian researchers starting at the end of the 1960's under the leadership of V. Brodianski. The main peculiarities of the cycles which use MR were explained in the beginning of the 1970's[11]. It has been found[12] that the MRs used in single stage systems may provide idealized efficiencies as high as 0.7 in the broad range of $80 \text{ K} < T_r < 300$ K. It matches the level of efficiency of VCCs with pure refrigerants at $T_r > 230$ K. This is due to unique properties of MRs. They have a temperature drop when throttled at ambient temperature and the heat capacity ratio for the supply flow at high pressure, $P_H$, and return flow at low pressure, $P_L$, may be < 1. For pure refrigerants this ratio is always > 1. This results from the phase changes which take place in the counter-flow heat exchanger.

The idealized efficiency related to the Carnot cycle assumes the use of an isothermal compressor, and heat exchangers with an infinitely small value of minimal temperature difference and pressure drop. The value of the idealized efficiency depends on the number of MR components, their mole percentage, and the values of $P_H$ and $P_L$. Composition and the idealized efficiency of several Nitrogen based refrigerants proposed in [13] for $P_L = 0.1$ MPa and $P_H = 5$ MPa are given in Table 1. This table also includes data on a MR which was designed to be efficient at $P_H = 2$ MPa[14].

**Table 1** Composition and efficiency of mixtures for 0.1 MPa return pressure and 5 and 2 Mpa supply pressure. Mixtures #1, #2, and #3 are cited in [13] at 5 Mpa and mixture #4 is cited in [14 ] at 2 Mpa

| | Mixture Composition - % | | | | | | | |
|---|---|---|---|---|---|---|---|---|
| Mixture | #1 | | #2 | | #3 | | #4 | |
| Neon | - | | - | | 20 | | - | |
| Nitrogen | 40 | | 30 | | 25 | | 30 | |
| Methane | - | | 30 | | 25 | | 15 | |
| Ethane | 30 | | 20 | | 15 | | 30 | |
| Propane | 30 | | 20 | | 15 | | 30 | |
| Supply Pressure - MPa | **5** | 2 | **5** | 2 | **5** | 2 | 5 | **2** |
| Min. Temperature - K | **78** | 78 | **80** | 80 | **67** | 67 | 78 | **78** |
| Load Temperature - K | **78** | 78 | **99** | 80 | **71** | 70 | 83 | **80** |
| Refrigeration for .01 Mole/s - W | **9.4** | 4.8 | **17** | 6.1 | **5.6** | 3.6 | 11 | **8.0** |
| Idealized Carnot Effeciency - % | **30** | 19 | **37** | 23 | **19** | 16 | 34 | **31** |

Most mixtures such as #2 have the characteristic of boiling at increasing temperature as the mixture flows through the evaporator. Mixture #1 exhibits unusual behavior at pressures below about 6 MPa, evaporating at near constant temperature. This is because the miscibility of liquid Nitrogen in hydrocarbons at extremely low temperatures is limited. In this case one of the liquid phases, which evaporates at process 3-4 consists of almost pure Nitrogen. Boiling at near constant temperature tends to improve the cycle efficiency. Those MRs, which comprise nitrogen and methane evaporate at variable temperature even though they have a miscibility gap. Both liquid phases include methane which provides temperature distributed evaporation. By introducing Neon (or Helium) into the blend, #3, it is possible to obtain Tr = 70 K without vacuum in the return line. Mixture #4 provides an example of the ability to obtain high efficiency at relatively low supply pressure, $P_H = 2$ MPa. Freons® and halocarbon derivatives have also been studied[6] and used in mixtures at 80 K < Tr < 200 K.

Experimental and theoretical investigations have been conducted to study not only vapor-liquid but vapor-liquid-liquid phase equilibrium[6,7,8] for different types of the MR to be used at different $T_r$. It was important to accumulate experimental data and select an equation of state in order to calculate the thermodynamic properties of the mixtures, which show strong positive deviations from the behavior of the ideal solutions. Highly efficient computer software has been developed to predict properties of the MRs which include supercritical components such as Ne or He[15].

In the field of solid-liquid equilibrium it was found that properly designed MRs may be used at temperatures which are far below the triple point of the pure components. For example pure iso-butane freezes at 113.5 K, but a specific mixture with only 10 % of it does not freeze at 80 K.

The use of MRs allows a lot of freedom to optimize the composition and cycle parameters such as $P_H$ and $P_L$ in order to build high efficiency single stage refrigerators for different applications.

## COOLERS BASED ON ONE-STAGE OIL LUBRUCATED COMPRESSORS

Use of pure refrigerants such as N2 at cryogenic temperatures requires relatively high pressure, e.g. 10 MPa, and very pure gas. An example of a 4.4 K cascade refrigerator built in the early days of what is now APD Cryogenics is described by Lashmet, P. and Geist, J.[16] It used three diaphragm compressors to provide high pressure and clean gas. Recently it was recognized that mixtures can be used with good efficiency at pressures within the range of mass produced oil lubricated air conditioning compressors and that oil removal technology from GM refrigerators could be applied to purify the gas. Rolling piston type compressors are particularly well suited because they typically have volumetric efficiencies in excess of 50 % at pressure ratios above 5:1[17]. Their use makes it possible to build single stage refrigerators that have long life and require no maintenance for temperatures down to about 70 K[18].

Thermodynamic analysis shows that reducing $P_H$ causes a reduction in specific cooling capacity, $q_r$. Table 2 presents data on cycle performance where $q_r$ is given in W for 0.01 mole/s at different values of $P_H$. This data was calculated for MR #4 listed in Table 1 and another refrigerant, #5, which is obtained by adding 5% Neon to MR #4. MR #5 allows the return pressure to be increased to 0.13 MPa at $T_r = 80$ K. The data illustrates that in a real system with $DT_{min} = 5$ K the $P_H$ reduction would cause a relatively greater penalty on the performance compared to that for an idealized system with $DT_{min} = 0$.

**Table 2**. Supply pressure influence on cycle performance

| Supply Press. - MPa | $P_H$ | 1.5 | | 2.0 | | 2.5 | | 3.0 | | 4.0 | |
|---|---|---|---|---|---|---|---|---|---|---|---|
| Mixed Refrigerant | | #4 | #5 | #4 | #5 | #4 | #5 | #4 | #5 | #4 | #5 |
| Ref. for .01 mole/s - W | $Dt_{min}$ = 0 K | 6 | 5 | 8 | 7 | 10.5 | 9 | 12 | 11 | 9.5 | 12 |
| " | $Dt_{min}$ = 5 K | 3.5 | 2.7 | 5.7 | 4.7 | 8.2 | 6.7 | 9.2 | 8.6 | 9 | 9 |
| Carnot Eff. - % | $Dt_{min}$ = 0 K | 24 | 23 | 31 | 29 | 38 | 35 | 41 | 41 | 29 | 41 |
| " | $Dt_{min}$ = 5 K | 14 | 12 | 22 | 19 | 29 | 26 | 32 | 31 | 28 | 30 |

A one-stage compressor operating at $P_H$ = 2 Mpa has to have a higher flow rate to provide a given refrigeration capacity than a multistage compressor operating at $P_H$ = 4 MPa. That is why such technical factors as heat exchanger performance and pressure drop have a greater influence on the system performance. The heat exchanger for example is relatively larger in order to handle the higher flow rate and takes longer to cool down.

## DEVELOPMENTS TO COMMERCIALIZE SINGLE STAGE MR CRYOCOOLERS

Computer software has been designed to conduct optimization of the cooler based on a one stage compressor which takes into account the interactions of the compressor, a counter flow heat exchanger, a restrictor, and evaporator. Combined with the equation of state and limited experimental data it describes a refrigeration map of a given system configuration with acceptable accuracy[17].

The early markets that were envisioned for this technology required a few watts of cooling to temperatures near 77 K so this became the focus for mixture selection and cooler designs. Tube-in-tube heat exchangers were adopted and a 28 L/min (1 CFM) displacement rolling piston type oil lubricated compressor with air-cooled aftercooler and oil separator was developed. A manually adjustable needle valve was used to control the flow rate. It was found as reported[19] that a fixed setting of the needle valve which gave fast cooldown time resulted in a relatively high $P_L$ and a temperature of about 100 K. Further closing of the valve was required to reach 80 K. Much effort was expended to explore different designs of automatic variable orifice valves that would have a large orifice during cooldown and a smaller orifice for steady state[20]. Pneumatically actuated valves with active control were also developed to maintain a constant temperature and alternately minimize gas flow rate[21]. After grappling with this problem for a while it was realized that operation at lower pressures than had been used previously resulted in a larger change in

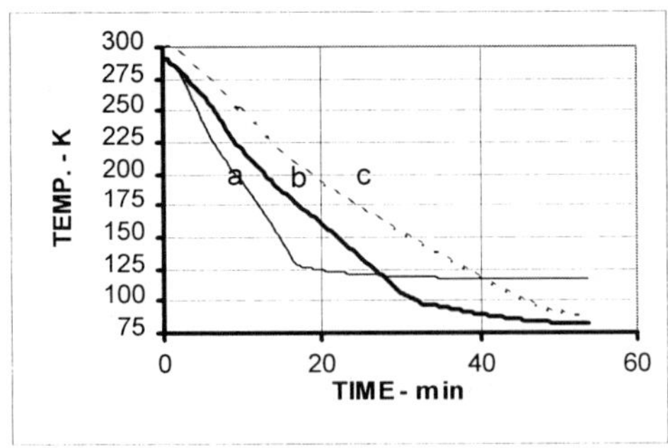


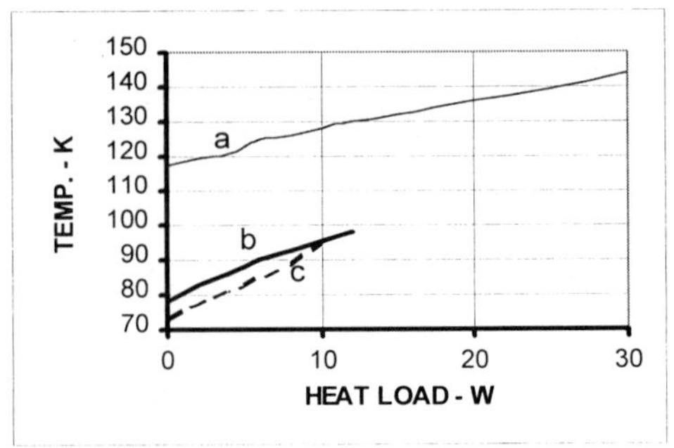


**Figure 3** Cooldown time and cooling capacity of three different mixtures in a standard system with a 1 CFM displacement oil lubricated compressor which draws less than 500 W.

the density of the gas flowing through the restrictor during cooldown. It was found that this density change could be reduced by adding a non condensable gas, e.g. He, to the mixture so that a capillary tube could be used which provided reasonably fast cooldown and low final temperature[22].

Many instruments such as X ray and gamma ray detectors which had been using LN2 as a coolant have been converted to closed cycle cooling with single stage MR cryocoolers marketed under the trade name TIGER™. Even though the inherent vibration of the throttle cycle is low, additional vibration reduction has been necessary[22].

It has been found possible to design mixtures for specific temperature ranges. In general as the base temperature is increased the maximum cooling capacity is increased and the cooldown time is decreased. Figure 3 shows plots of cooldown and capacity for three different mixtures that illustrate this capability. A mixture similar to the highest temperature one that is plotted is being used in the latest product, a cold trap[24] marketed under the trade name Aquatrap™.

## SUMMARY

Single stage Throttle Refrigeration Systems which use mixed refrigerants and mass produced oil lubricated compressors have been developed to provide refrigeration down to 70 K. This development can be seen as another step in the evolution of mixed refrigerants technology. This step was made possible with the study of throttle cycle thermodynamics and mixed refrigerants properties which had been started in Russia. Advances in thermodynamics combined with optimal system design has allowed the building of highly efficient and reliable cryocoolers based on one stage compressors. Commercialization of the technology is advancing as systems are being designed for specific applications.

## REFERENCES


1. V.M. Brodianski. "Sadi Carnot -- Biography," Nauka, Moscow, Russia [in Russian] (1996).
2. V.M. Brodianski, "From solid water to liquid helium, history of low-temperature engineering," *Energoatomizdat*, Moscow, Russia [in Russian] (1995).
3. A. Fuderer. "Compression Process for Refrigeration" U. S. Patent 3,203,194 (August 1965).
4. V.M. Brodianski, et al. "The use of mixtures as working gas in throttle (J-T) cryogenic refrigerators," *Communications of XIII Congress IIR*/ Washington (1971).
5. D.J. Missimer, "Refrigerant conversion of auto-refrigerating cascade (ARC) systems," *Polycold Systems International*/ San Rafael, CA (1973).
6. G.K. Lavrenchenko, "Formation of optimal mixed refrigerants for throttle refrigerators," *Choiodilnaia Technika and Technology* Vol. 34/ Kiev, Ukrain (1982).
7. M.J. Boiarski, et al, "Autonomic cryorefrigerators of small capacity," *Energoatomizdat*, Moscow, Russia (1984).
8. N.D. Zakharov, et al. "Investigations of enthalpy of nitrogen-freon mixtures," *Chemichesikoe Neftianoe Mashinosroenie*, No. 5 [in Russian] (1984).
9. A.P. Klimenko, "One-flow cascade cycle," *IIR Int'l Congress of Refrig.*/ Copenhagen, Denmark (1959).
10. W.A. Little, "Recent developments in Joule-Thomson cooling: gases, coolers and compressors," *Proceedings of the International Cryocoolers Conference 5* (1988).
11. V.M. Brodianski, A.K. Gresin, "Highly efficient throttle refrigerators on multi-component mixtures," *Chemichesikoe Neftianoe Mashinosroenie*, No. 12 [in Russian] (1971).
12. V.M. Brodianski, M.J. Boiarski and A.I. Lunin, "The exergy analysis of the throttle refrigerating systems on pure and mixed refrigerants," *Proceedings of the ESDA*, PD-Vol.64-3, Eng. Sys. Des.Anl., V.3, ASME (1994).
13. V.N. Alfeev, et al. Great Britain Patent 1,336,892 (1973).
14. M.J. Boiarski, et al. U.S. Patent 5,441,658 (1995).

15. M.J. Boiarski and O.N. Podcherniaev, "Methods of calculations of phase equilibria and thermodynamic properties of mixed refrigerants for analyzing throttle refrigeration cycles," *Visocotemperaturnaia Sverkhprovodimost*, Vol. 3-4/ Moscow, Russia [in Russian] (1990).
16 P. K. Lashmet and J. M. Geist, "A closed cycle cascade helium refrigerator" Adv. In Cryogenic Engr. Vol. 8, Plenum Press, NY 1963.
17. R.C. Longsworth, U.S. Patent No.5,337,572 (1994).
18. Khatri, A. and Boiarski,M.J. (1996) "A throttle cycle cryocooler operating with mixed gas refrigerants in 70 K to 120 K temperature range", Cryocoolers 9, Plenum Press,NY, (1997).
19. R.C. Longsworth, M. J. Boiarski, and L. A. Klusmier, "80 K closed-cycle throttle refrigerator", Cryocoolers 8, Plenum Press, NY, (1995), p. 537.
20. M. J. Boiarski et al. U.S. Patent No. 5,595,065 (1997)
21. R.C. Longsworth and D. Hill, " 65 K two stage MR/O2 throttle cycle refrigerator" Cryocoolers 9, Plenum Press, NY, (1997).
22. R.C. Longsworth, M. J. Boiarski and A. Khatri, U.S. Patent No. 5,579,654 (1996).
23. D. Hill, " Throttle cycle cooler vibration characterization", Cryocoolers 9, Plenum Press, NY, (1997).
24. M. J. Boiarski and A. Khatri, "Water trap refrigerated by a throttle cycle cryocooler using mixed gas refrigerant", Presented at 1997 CEC.

# CLOSED-CYCLE COOLING SYSTEM FOR A SUPERCONDUCTING MAGNET AND 2K – 400K VARIABLE TEMPERATURE INSERT

C. Winter, I. Cormie, D. Girard, and S. Wolfe

Quantum Technology Corp., 1370 Alpha Lk. Rd., Unit 15
Whistler, B.C., V0N 1B1 Canada

## ABSTRACT

A closed-cycle cooling system based on a combined Gifford-McMahon and Joule-Thomson cycle refrigerator has been successfully coupled to an 8 Tesla laboratory magnet resulting in completely closed-cycle operation. The G-M/J-T cooler is used both as a reliquefier and as a liquefier. The J-T loop cools a heat exchanger in the liquid bath surrounding the superconducting magnet eliminating the liquid helium boiloff. In addition, room temperature helium gas is injected, cooled to 80K by the first G-M stage, then to 15K by the second G-M stage and finally liquefied at 4.0K by the J-T stage. Some liquid is used by the vapour cooled magnet current leads and the remaining liquid is used by a fourth stage of cooling for the sample. J-T expansion of liquid helium into the variable temperature sample insert allows continuous operation at any sample temperature between 2K and 400K, while simultaneously allowing the superconducting magnet to operate at 4.3K. This modular cooling technology has also been applied to cooling Magnetic Resonance Imaging (MRI) machines, Dilution Refrigerators, SQUID Magnetometers and small Helium Liquefiers.

## INTRODUCTION

Many laboratory experiments involve testing a sample over a range of temperatures and magnetic fields. The apparatus which produces these environmental conditions should be as reliable, as simple to use and as inexpensive as possible. The majority of present day laboratory apparatus for this type of work requires liquid helium from an external source. The purpose of this paper is to describe a system which provides these experimental conditions without an external supply of liquid cryogen. The QUANTUMCOOLER™ system is closed-cycle both for the magnet cooling and for the sample cooling.

---

- QUANTUMCOOLER is a trademark of Quantum Technology Corp.

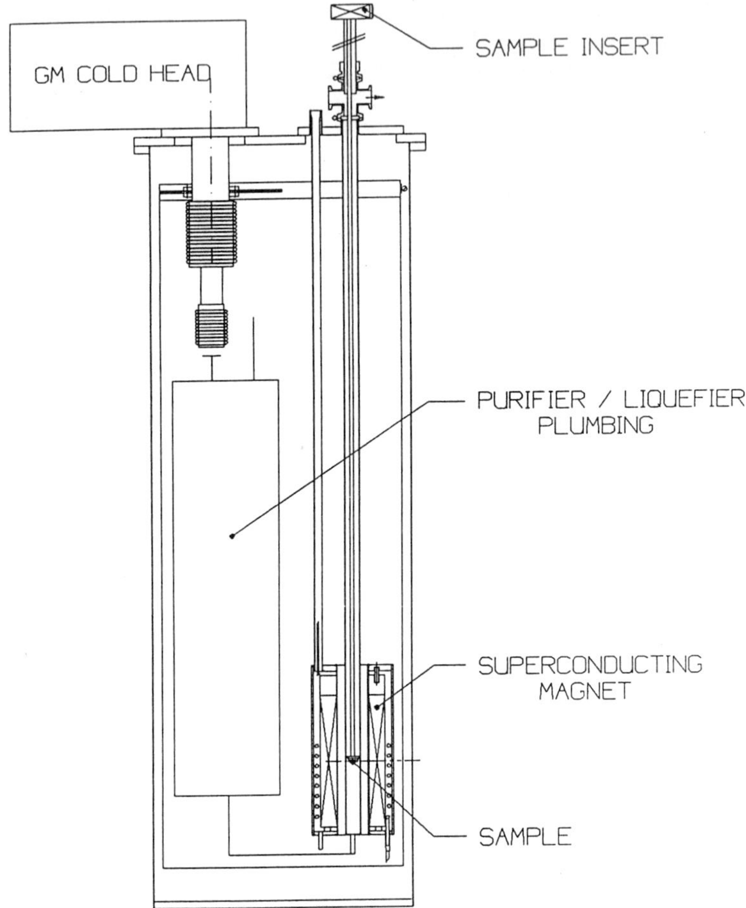


**Figure 1** Cross-sectional view of Quantumcooler, superconducting magnet and sample space. A single vacuum insulating space separates the Quantumcooler, magnet and sample space from each other and ambient temperature. The magnet is welded into an annular stainless steel vessel which is filled by liquid helium produced by the Quantumcooler. A separate flow of liquid helium is used to control the sample space temperature between 2K and 400K. The samples are top loading and the entire sample space tube can be warmed to 300K allowing samples to be changed in minutes.

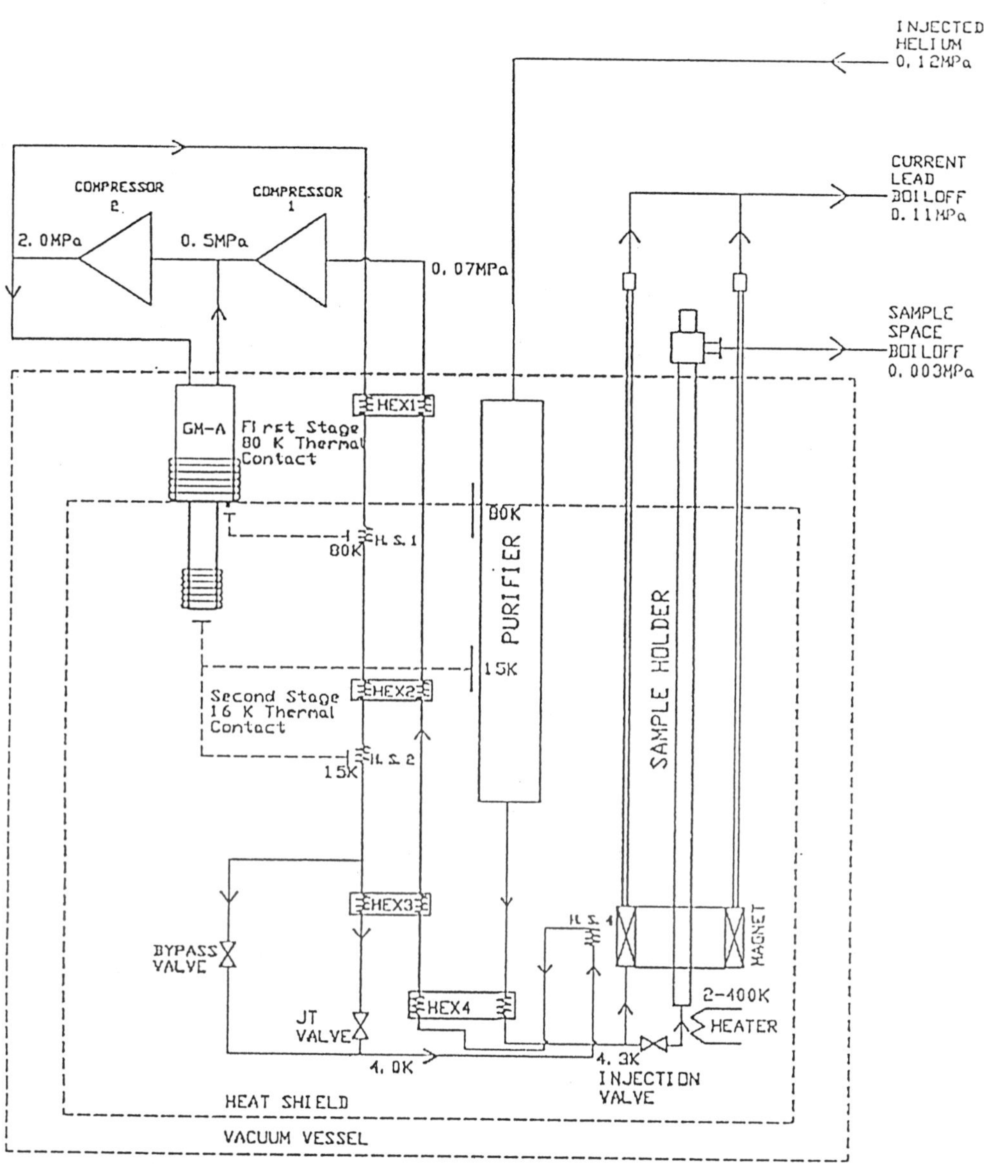


**Figure 2** Schematic of Quantumcooler cooling loop. The closed-cycle GM/JT refrigeration loop is separate from the closed-cycle injected helium/magnet bath liquid helium/sample holder/recovery helium loop. The GM moving displacer system provides cooling at 80K and 15K to precool the JT loop flow. Counterflow heat exchangers HEX1, HEX2 and HEX3 allow the JT loop to operate efficiently. The JT loop provides cooling at 4.0K to cool the HEX4 and liquefy 15K gas coming from the purifier. This liquid produced fills the superconducting magnet bath. Liquid in the magnet bath is used to cool the magnet current leads. A manually controlled INJECTION VALVE allows some liquid from the magnet bath to be withdrawn for cooling the sample space. When a vacuum pump is used on the sample space then this INJECTION VALVE acts as a second JT valve allowing sample temperatures as low as 2K to be reached.

## PRINCIPLE OF OPERATION

Four stages of refrigeration are utilized[1,2,3]. The first two stages are a conventional Gifford-McMahon (GM) cooler system nominally providing 35W of capacity at 80K and 7W of capacity at 15K. The third stage of refrigeration is a Joule-Thomson (JT) expansion of high pressure helium which has been precooled by the GM cooler. Expansion of 2MPa (20bar) helium to 0.07MPa (0.7bar) provides nominal 4W of cooling capacity at 4.0K. A separate cooling loop filled with helium at 0.12MPa (1.2bar) is cooled by the GM and JT stages and liquefies. The liquid goes through a separate JT expander to subatmospheric pressure 0.003MPa (0.03bar) cooling to 2K. Estimated heat loads are given in Table 1.

## DESCRIPTION

The Quantumcooler system is shown in cross-sectional view in Figure 1. A CTI model 1020 GM cold head was connected to the Quantumcooler JT cooling loop. The 8T superconducting magnet was welded into an annular stainless steel bath vessel supported by 3 tubes (only 1 shown in the drawing). Two of these support tubes contained the (non-detachable) current leads. The center of the magnet is hollow to allow the sample insert to be vacuum insulated from the magnet. This insulation is very important, because if the user warms the sample to 400K the superconducting magnet must remain at 4.3K.

Figure 2 shows a plumbing schematic of the cooling loop. Two Quantumcooler J20 compressors are used to provide closed-cycle refrigerant helium circulation from 0.07MPa to 2.0MPa for the JT loop and from 0.5MPa to 2.0MPa for the GM loop. Due to extensive experience, proprietary design features and exacting quality control procedures no difficulty was encountered with helium gas contamination in the closed-cycle refrigerant loop.

Note that the sample space helium is a separate system. It is in thermal contact at 80K and 15K and at 4.3K, but the refrigerant helium is not mixed with the helium to be liquefied. A built in helium purifier allows the user to contaminate this recovered helium with air, for example, by an error during changing samples, and still operate. The principle of operation of the helium purifier is shown in figure 3. As the incoming helium is cooled to 15K any air in it will be frozen solid. This can then be removed by heating the purifier and backflowing clean helium gas.

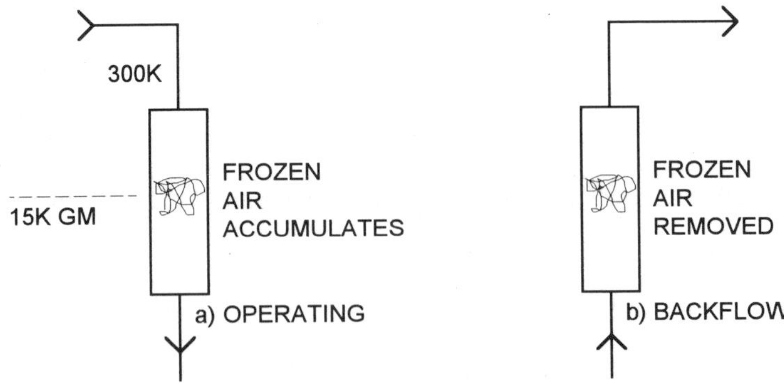


**Figure 3** Cryogenic purifier - Principle of operation. During operation (a) helium gas which may be contaminated with air (typically less than 1% air) is cooled to 80K and 15K by the GM cooler. At about 50K the air freezes and is trapped in the purifier. Periodically the flow through the purifier is reversed and heat is applied to allow the trapped contaminants to be vented as shown in (b) backflow.

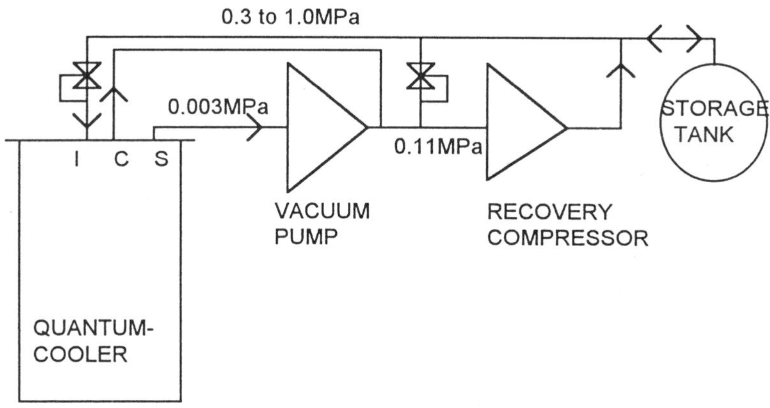


**Figure 4** Helium gas recovery circuit (simplified) plumbing schematic "I" refers to the injection of helium at 0.12MPa into the helium purifier. "C" is the boiloff helium gas from the current leads at 0.11MPa. "S" is the sample space helium gas which flows to the (optional) vacuum pump. Helium gas from the current leads and from the sample space is compressed and stored in a recovery tank at up to 1.0MPa (10bar) by a recovery compressor. Valving (not shown) is provided to allow the sample space to be backfilled with room temperature helium prior to removing a sample. The optional vacuum pump is required for sample temperatures below 4.3K. 2K is attainable using the vacuum pump.

**TABLE 1 HEAT LOADS (Estimated)**

| ITEM | TEMPERATURE | LOAD | FLOW |
|---|---|---|---|
| Heat shields & liquefier gas | 80K | 25W | |
| Precooling JT & liquefier gas | 15K | 5W | |
| Superconducting magnet | 4.3K | 0.01W | |
| Superconducting magnet leads | | | 5Liters/day |
| Sample | 2K-400K | 0.1W | |
| Heat load by radiation on the superconducting magnet when the sample is at 400K | 4.5K | 1W | |

Figure 4 shows a plumbing schematic of the helium gas recovery circuit. This allows helium gas from the current leads and from the sample space to be compressed and stored in a 450L storage tank. Flow meters and flow controls are used to allow the operator to balance the flows required for the magnet current lead cooling and sample cooling.

## OPERATION

Automatic controls which are used on the Quantumcooler system for Magnetic Resonance Imaging superconducting magnets were not yet adapted to this laboratory system so an operator adjustment during cooldown is required. This is to close the bypass valve when the magnet has been cooled to below 20K. Due to the large size of the magnet this takes about 10 hours. A few hours later the system is starting to accumulate liquid at a rate of about 3 liters/day in the magnet vessel. Care must also be taken to set the magnet current lead flow to the correct value. If the flow is set too high or too low the magnet will not be able to cool down. If the current lead flow is too high the excessive heat load from trying to cool all that gas from 300K will warm up the JT refrigerator loop above 4K. If the current lead flow is too low then the JT loop will be cold, but the heat conducted down the current leads will keep the magnet above 10K.

During operation the sample temperature can be cooling by injecting liquid helium by partially opening the injection valve and can be warmed by simultaneously heating with an electrical heater. A very nice feature of this design is the ability to quickly cool a sample by allowing a large flow of liquid through the sample space. However, care must be taken, not to leave the liquid injection valve open too wide for too long as this can result in using liquid helium faster than the liquefier can produce it. Since it is a closed system no helium is lost, and the operator can easily monitor the liquid level in the magnet by monitoring the gas pressure in the room temperature helium gas storage tank.

Samples can be changed without warming the magnet by closing the injection valve, warming the sample space to 300K, backfilling with room temperature helium and removing the sample. This allows very rapid sample change in a few minutes.

## CONCLUSIONS

A laboratory sized experimental variable temperature system has been built and operates reliably without the use of external cryogens. This Quantumcooler system allows a sample temperature variable between 2K and 400K in the field of a superconducting magnet variable from -8T to +8T. This system operates closed-cycle without any additional externally supplied cryogens.


## ACKNOWLEDGMENT

Financial assistance from the Scientific Research & Experimental Development ITC Program is gratefully acknowledged.


## REFERENCES


1. W.H. Higa and E. Wiebe, One million hours at 4.5Kelvin, NBS Special Publication Number 508 (1978)
2. C. Winter and S. Gygax, Closed-cycle Liquid Helium Dip Cryocooler, Proceedings of the 4th International Cryocooler Conference “Crycooler-4” (1986).
3. C. Winter and A. Swiecicki, Safety Considerations in Operating Dip Tube Access 10K and 3K Closed-Cycle Cryocoolers, International Cryocooler Conference Proceedings “Cryocoolers 5” (1988)

# AN HYBRID CYCLE HELIUM LIQUEFIER

J.M. Poncet and A. Ravex

C.E.A. / Grenoble - Département de Recherche Fondamentale sur la Matière Condensée - Service des Basses Températures
17, avenue des Martyrs, 38054 Grenoble Cedex 9, France

## ABSTRACT

A simple and robust helium liquefier has been developed at CEA / SBT. It is a Collins type liquefier operated at atmospheric pressure in a standard commercial helium dewar. To increase reliability, the liquefaction circuit and the refrigerator are separated. Precooling and condensation temperature are provided by an hybrid closed cycle mechanical cryocooler : a commercial Gifford Mc Mahon cryocooler (Cryomech GB 37) is used as a precooler for helium gas to be liquefied and for the Joule Thomson expansion circuit (4.5 W at 4.5 K).

The liquefaction circuit is designed to admit low purity helium. A prototype of this liquefier has been successfully operated for 6 months with a typical liquefaction rate of 1.1 liter per hour.

## INTRODUCTION

Gifford Mc Mahon (GM) / Joule Thomson (JT) hybrid cycle refrigerators are the most commonly used type of closed cycle refrigerators for applications requiring 0.5 to 4 watts of refrigeration at liquid helium temperature. A review of the technological evolutions and potential applications of this type of refrigerator has been proposed by R.C. Longsworth[1] a few years ago. During the last decade CEA/SBT has built several apparatus of this type[2] with cooling power ranging from 100 mW up to 5 W and for temperatures ranging from 2.8 K up to 4.5 K. Fully automatic cool down and operation as well as trouble free operation with several warm up and cool down cycles have been proven over about 10000 hours. Main applications are for sample or detector cooling and for helium recondensation in superconducting magnet cryostats.

Another goal for the closed-cycle refrigerators is the supply of liquid helium in a laboratory. On the basis of a 5 W / 4.5 K refrigerator, CEA / SBT has designed for this purpose a small helium liquefier with a liquefaction rate of about 1.1 liter per hour.

## DESCRIPTION OF THE HELIUM LIQUEFIER

A picture of the liquefier associated with a 100 liter storage tank is shown on Figure 1. On Figure 2 is presented a view of the inner parts of the liquefier. A scheme of the liquefier circuitery is shown on Figure 3.

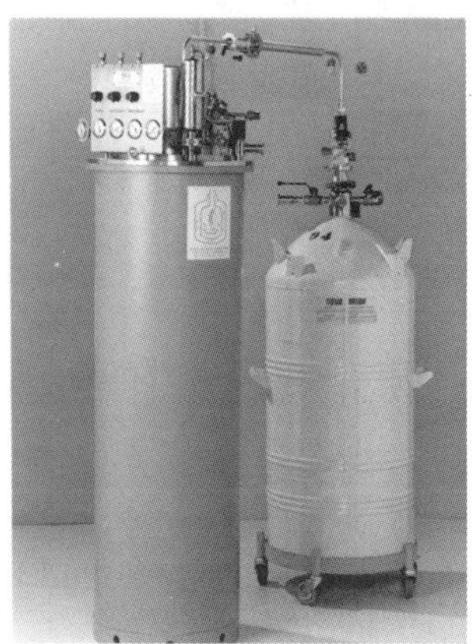

**Figure 1** . Liquefier fitting a 100 l storage tank

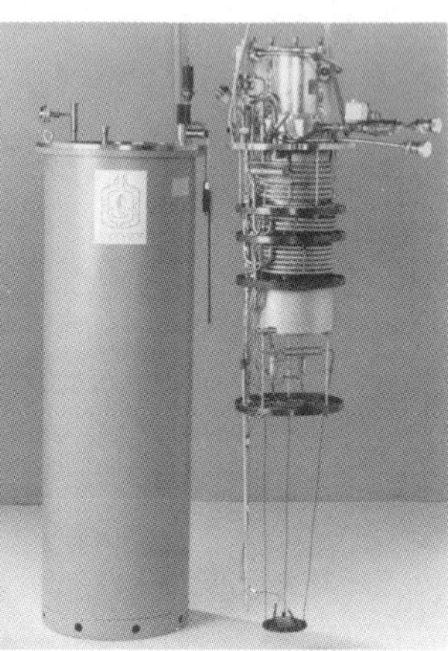

**Figure 2** . Inner parts of the liquefier

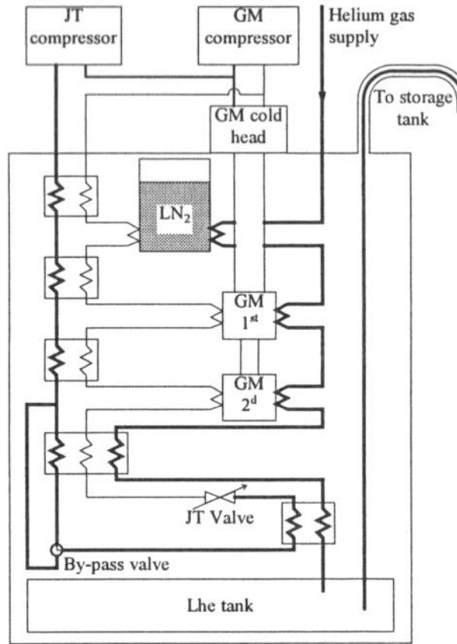


**Figure 3** . Scheme of the liquefier circuitery

To ensure high reliability, the helium liquefaction loop is totally independant from the closed-cycle refrigerator used to produce the precooling and liquefaction cooling power.

To achieve the large cooling power required at precooling and liquefaction levels, we use a liquid nitrogen precooling first stage. This first precooling stage could be replaced in the future by a single stage Gifford Mc Mahon cryocooler to suppress any cryogen consumption.

A large cooling power, commercially available, two stage Gifford Mc Mahon cryocooler (Cryomech GB37 : typical simultaneous cooling power 20 W at 20 K + 60 W at 80 K) is used for the second and third level precooling. A small fraction of the high pressure helium mass flow delivered by the GM compressor is derivated to a JT circuit. The JT compressor used to recompress helium from JT loop low pressure up to the GM compressor suction pressure is a modified GM compressor (APD HC8).

The JT circuit is automatically operated by the use of two pneumatically driven control valves ensuring fast cooling down from room temperature and stable operation without any manual interaction. During cool down, the last stage JT heat exchanger is by-passed by the pneumatic by-pass valve (Figure 4) as long as its temperature remains higher than about 15 K. This by-pass is actuated by helium gas pressure in a metallic bellows acting as a gas thermometer. The JT expansion valve (Figure 5) is also controlled pneumatically by means of helium gas in a bellows to maintain the expansion pressure at a constant value.

The helium gas to be liquefied (recovered directly from an experiment cryostat or from an external supply of gas at room temperature) is precooled in a counterflow heat exchanger

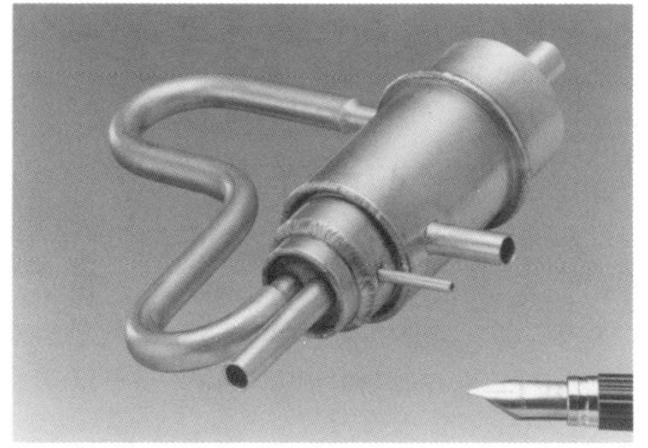

**Figure 4** . By-pass pneumatic valve

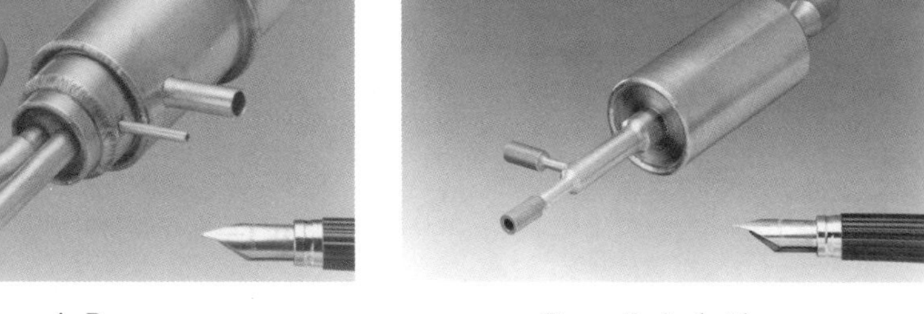

**Figure 5** . Joule Thomson expansion pneumatic valve

from 300 K down to 80 K using nitrogen vapor enthalpy and by thermal exchange at the three precooling stages ($LN_2$, and GM cooler cold stations). Between 80 K and each cold station, helium gas is only in thermal contact with the JT loop heat exchangers. This liquefaction circuit is designed to be able to admit highly polluted helium. Its large cross section allows storage of large amount of condensed impurities for a long operating period before warming up for regeneration. Between the GM second stage cold station and the JT expansion valve, a specific 3 way heat exchanger has been developed to take advantage of the enthalpy unbalance between high pressure and low pressure sides of the JT loop.

This liquefier is of Collins type : all the cryogenic assembly (Figure 3 : GM / JT hybrid cycle refrigerator and liquefaction circuit) is installed inside a commercial dewar at atmospheric pressure. Stratification occurs in the dewar and reduces parasitic heat input at 4 K. This arrangement is favourable for reliability. The liquefied helium is stored in the bottom of the dewar (storage capacity about 40 liters) and can be easily transfered to a storage dewar or an experiment through a vacuum isolated transfer line. Typical pressure in the liquefier dewar is roughly 1.3 $10^5$ Pa.

## EXPERIMENTAL RESULTS

A prototype of the liquefier has been built and extensively characterised. Table 1 gives the equilibrium heat loads and temperatures during typical operation.

The prototype has been successfully operated for more than 6 months continuously with a measured liquefaction rate of 1.1 liter per hour. The required electrical input power for operation is about 15 kW. $LN_2$ consumption is less than 2 liter per hour.

**Table 1** . Typical heat loads and temperatures

| | Temperature | Parasitic heat input | Joule Thomson loop | Liquefaction |
|---|---|---|---|---|
| $1^{st}$ stage - $LN_2$ | 80 K | 47 W | 10 W | 42 W |
| $2^d$ stage GM - $1^{st}$ stage | 42 K | 4.5 W | 5 W | 8 W |
| $3^d$ stage GM - $2^d$ stage | 16 K | 3 W | 8 W | 5 W |
| Liquefaction stage | 4.5 K | 0.5 W | | 3.5 W |

## CONCLUSION

A helium liquefier has been designed and built associated to a mechanical GM / JT hybrid cycle refrigerator. This simple Collins type helium liquefier has been successfully operated. The typical liquefaction rate of 1.1 liter per hour is well adapted for laboratory use. The reliability and low operation and investment cost of this type of liquefier are well situated for laboratory application.

## REFERENCES

1. R.C. Longsworth, 4 K Gifford Mac Mahon / Joule Thomson Cycle Refrigerators, in : « Advances in Cryogenics Engineering », Vol. 33, Plenium Press, New York (1987), pp 689-698
2. G. Claudet, R. Lagnier and A. Ravex, Closed cycle liquid helium refrigerators, Cryogenics (1992), Vol. 32, ICEC Supplement

# ANALYZING MAGNETIC REFRIGERATION EFFICIENCY: A ROTARY AMR - REVERSE BRAYTON CASE STUDY

J. L. Hall and J. A. Barclay

Cryofuel Systems
University of Victoria
Victoria, BC, Canada V8W 3P6

## ABSTRACT

Recent advances in magnetic refrigerants, regenerative magnetic cycle theory and conduction-cooled superconducting magnet systems have improved the outlook for developing practical, cost-effective magnetic refrigerators. This has prompted an effort to provide an updated estimate of achievable magnetic refrigerator thermodynamic performance as compared to conventional gas-cycle devices. Calculations were performed as part of a case study involving one candidate refrigerator from each technology: rotary active magnetic regeneration and reverse-Brayton gas compression. The study shows that the two systems are very analogous, allowing for a fair comparison. For refrigeration at 100 K and heat rejection at 300 K, both are capable of producing second law efficiencies of over 60% given a similar set of optimistic component performance assumptions. Excellent heat transfer at low-pressure drop was found critical to the success of both systems. The rotary magnetic refrigerator was found to have a clear advantage in work minimization because it relies on efficient electric drives and inherent magnetic force balancing to minimize the net cycle work, rather than coupling isothermal compressors to mechanical expanders to recover work. Several key problems are discussed that must be solved to realize the potential high thermodynamic efficiencies in a cost-effective manner.

## INTRODUCTION

Early comparisons between magnetic and gas refrigeration cycles were based on theoretical calculations for a generic magnetic cycle[1,2,3] that compared either the components required to execute the cycles, or postulated hypothetical magnetic devices based on the knowledge of the magnetic cycles at that time. Considerable progress has been made in magnetic refrigeration since then on both the theoretical and experimental fronts. In particular, recent breakthroughs in magnetic refrigerants provide almost twice the adiabatic temperature changes previously available for comparable magnetic field changes.[4,5] Advances have also been made in high temperature superconductor magnets[6] and conduction cooling of magnet systems with improved cryocoolers[7]. Several experimental devices have been constructed and tested[8,9,10]. Further, significant progress has been made toward a detailed understanding of the active magnetic regenerative cycle[11]. These advances, combined with recent work in rotary, active magnetic regenerative liquefiers (AMRLs) for natural gas[12] have provided an impetus to re-evaluate the perceived advantages and disadvantages of magnetic refrigeration versus conventional gas compression refrigeration cycles.

Early work[13,14] identified several potential advantages of magnetic refrigeration: the essentially reversible magnetocaloric effect can theoretically provide near-Carnot efficiencies; magnetic refrigerators avoid the problem of large entropy generation in aftercoolers of gas compressors; and the fluid-solid heat transfer in magnetic regenerators is inherently more efficient than fluid-solid-fluid heat transfer in recuperative gas compression refrigerators. Set against these advantages are several disadvantages: the magnetocaloric effect is small, thereby requiring very efficient regenerators and multi-Tesla magnetic fields to pump heat in practical devices; such high fields can only be produced by expensive superconducting magnet technology; and magnetic regenerators are difficult to optimize because of the complex interaction of many fluid mechanic, thermodynamic and electromagnetic parameters.

Superior thermodynamic efficiency has not yet been demonstrated by experimental magnetic refrigerators operating across large load to environment temperature spans. The Astronautics Corporation of America (ACA) reciprocating AMR[10] achieved 40% of Carnot efficiency near 77 K for a temperature span of approximately 20 K, with lower efficiencies up to a maximum temperature span of approximately 50 K. Not only did performance degrade quickly as the temperature span increased, but this efficiency value did not include the energy cost of cooling the superconducting magnet. Recently ACA has completed six months of successful operation of an AMR refrigerator operating near room temperature, although details have not yet been published[15]. The room temperature AMRs developed by the Navy[8,9] achieved maximum temperature spans in the range of 20 to 40 K, with no reported values for thermodynamic efficiencies. It should be noted that the earlier ACA and Navy devices operated at low frequencies (1 to 6 rpm) and very low cooling powers (a few Watts). This is in contrast to the prototype AMRL being developed at the University of Victoria (UVic) for the liquefaction of natural gas[12]. The UVic AMRL is designed to provide a few hundred watts of cooling at 110 K with heat rejection at 240 K, and operate at a frequency of up to 5 Hz.; however, it is undergoing testing at the time of this writing and no performance data is yet available.

It can be argued that the poor to modest performance of experimental magnetic refrigerators to date is a function of the immaturity of the technology rather than inherent deficiencies in magnetic refrigeration itself. With the recent improvements in magnetic materials, magnet systems, and understanding of the AMR cycle, there is renewed promise that a mature magnetic refrigeration technology will produce comparable or higher thermodynamic efficiencies to the best available gas compression devices. To assess this promise, we have conducted a theoretical analysis that compares similar magnetic and gas compression refrigerators on the basis of entropy production mechanisms. Due to the large number of possible cycles in both technologies, we restricted the analysis to a case study using a rotary AMR and a reverse-Brayton (RB) gas compression cycle between 300 K and 100 K. Although this was not an exhaustive evaluation, it did yield several results that provide insight into the issue of relative advantages and problems of the competing technologies.

## BACKGROUND THEORY

The rotary AMR was chosen to represent magnetic refrigeration in the case study because it is considered to be the most promising embodiment of the technology for high-power, low-cost applications. The reverse-Brayton cycle was chosen to represent gas compression and

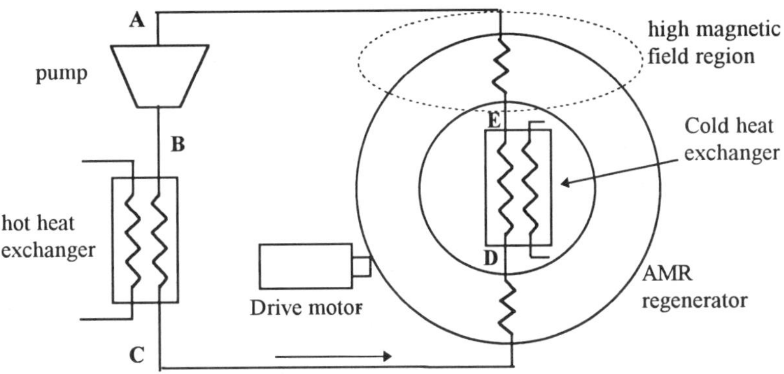


Fig. 1. Rotary AMR Flow Schematic.

expansion technology because it is the most similar cycle to the rotary AMR that could be found. This similarity allows for a fairer comparison and helps to pinpoint the respective advantages and disadvantages of the two technologies.

## Rotary AMR Description

The rotary AMR refrigerator is shown schematically in Fig. 1. The major components are a rotating ring filled with magnetic regenerator material, a superconducting magnet system that magnetizes part of the ring, a drive motor, a gaseous helium loop with load and environment heat exchangers, and a pump to push the helium through the loop. Heat pumping is provided by the magnetocaloric effect of the regenerator material that is alternately magnetized and demagnetized as it rotates into and out of the magnetic field. The helium gas thermally couples the magnetic material to the load and environment by flowing from the outer to inner rim in the demagnetized region and from the inner to outer rim in the magnetized region. Each element of solid regenerator material executes a magnetic Brayton cycle (Fig. 2a), while the helium executes a more complex cycle as shown in Fig. 2b. Note that the helium cycle in Fig. 2b is somewhat idealized because for high rotation rates of the AMR ring, the helium gas will not traverse the entire flow loop. Instead it will oscillate between different parts of the loop depending on the starting location. Further rotary AMR theoretical details can be found in Ref. 11, and prototype design details of the UVic rotary AMR liquefier can be found in Ref. 12.

## Reverse Brayton Cycle

The reverse Brayton cycle is a simple refrigerator that has seen application in the fields of miniature space-based cryocoolers[16] and larger-scale liquefaction[17]. A simple, one-stage version of this cycle is shown schematically in Fig. 3. It typically operates on helium gas, although other gases can be used in some applications. In operation, the gas is first compressed then passed through a hot heat exchanger to reject the compression and load heat to the environment. It then goes through a counter-flow heat exchanger to drop its temperature close to that of the load. It depressurizes and cools in an expander, then absorbs heat from the load in the cold heat exchanger. A swing back through the counterflow heat exchanger returns it to the compressor and completes the cycle. The T-S diagram for this process is shown in Fig. 4.

## Qualitative Comparisons of the Two Refrigeration Systems

Some of the similarities between the rotary AMR and reverse-Brayton refrigerators are immediately obvious. Both use a continuously flowing gas in a loop to absorb heat from the load in a cold heat exchanger and reject it to the environment in a hot heat exchanger. Both employ a compressor to make up the pressure drop in the flow loop. Also, both require heat exchange to span the load-to-environment temperature difference, although the rotary AMR uses a regenerator while the reverse-Brayton uses a counter-flow heat exchanger.

Other similarities are either less obvious or can be specified. The temperature drop across the expander can be made equal to the temperature change of the magnetocaloric effect at the cold temperature. This will result in the two systems having the same temperature rise across the cold heat exchanger; therefore, provided that the same gas is used in each case, the mass flow rates will be identical for equal cooling loads. The need to minimize pressure drop through the flow loops will motivate each system to use high pressure gas to increase the density and reduce the velocity. Also, the intermediate heat exchanger for the RB system can, in principle, have the same heat transfer effectiveness as the regenerator in the rotary AMR.

The differences between the two systems are four-fold: first, the rotary AMR requires a superconducting magnet system and properly selected magnetic regenerator material to work on; second, the RB system requires an expander to extract work from the flowing gas and provide cooling; third, the rotary AMR uses a simple drive motor to rotate the regenerator ring; and fourth, although both systems use a compressor, the work input is very different. The rotary AMR compressor does only enough work to overcome the pressure drop in the flow loop; the RB compressor does this plus an amount equal to the pressure drop across the expander. This difference is reflected in the enclosed area on respective T-S planes (Figs. 2b and 4): the area of the RB cycle is significantly greater, indicating that its compressor does substantially more work. Note that the primary work input for the rotary AMR is the regenerator ring drive motor, not the compressor.

It is the premise of this case study that these four differences are inherent to magnetic and gas compression refrigeration and therefore any fundamental differences in overall thermodynamic performance must result from them. Our methodology is to compute the entropy production or work input associated with these different components for the same operating conditions. This allows for quantitative comparisons of overall efficiency.

## CASE STUDY METHODOLOGY

The refrigeration load, $\dot{Q}_c$, for this case study was set to 500 W at a cold temperature ($T_c$) of 100 K. Given a specified environment temperature, $T_h$, of 300 K, this results in a ideal

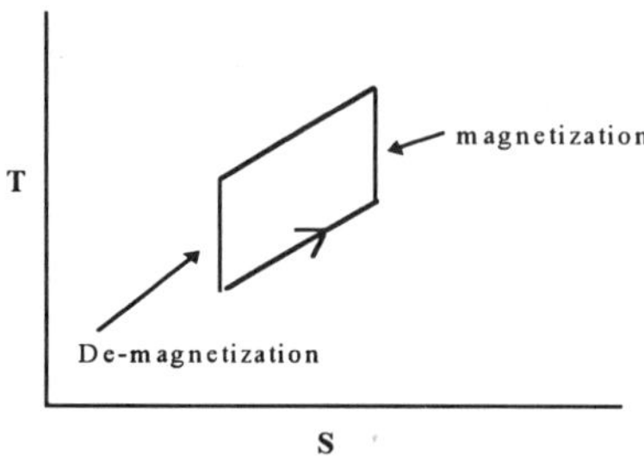


Fig. 2a. Magnetic Brayton cycle for rotary AMR.

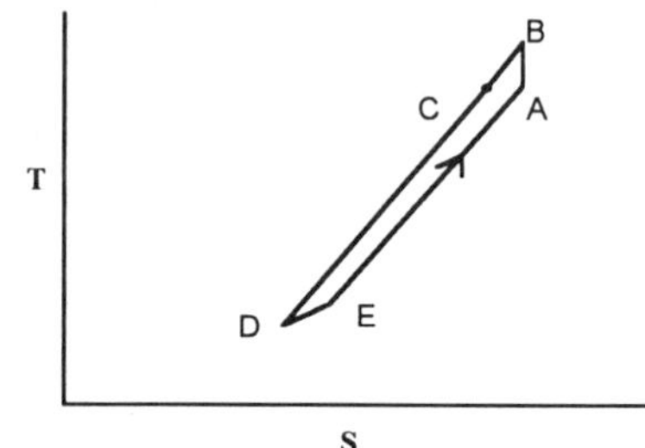


Fig. 2b. Idealized helium cycle for rotary AMR.

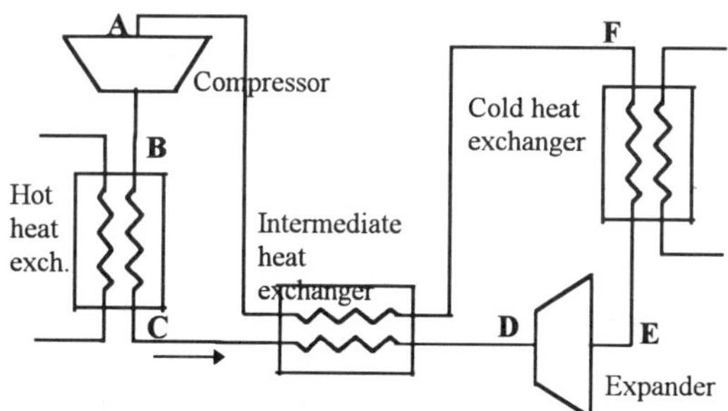


Fig. 3. Flow schematic of a reverse Brayton refrigerator.

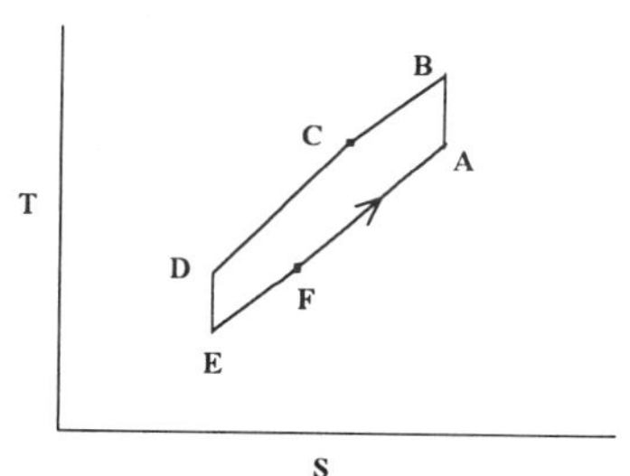


Fig. 4. T-s diagram of RB system.

(Carnot) work input, $\dot{W}_{id}$, of 1000 W. Also, if the temperature rise across the cold heat exchanger ($\Delta T_c$) is set at 5 K and helium gas is used, then the mass flow rate, $\dot{m}_f$, is 19 g/s.

The second law efficiency is defined by,

$$\eta_{II} = \frac{\dot{W}_{id}}{\dot{W}_{act}} \tag{1}$$

where $\dot{W}_{act}$ is the actual work input to the refrigerator. Although the same entropy production mechanisms are accounted for in the two systems, the presence of different components leads to slightly different actual work equations. For the rotary AMR we have,

$$\dot{W}_{act} = \dot{W}_{id} + T_h * \sum_i \dot{S}_{ht_i} + \dot{W}_P + \dot{W}_{mag} \tag{2}$$

where $\dot{S}_{ht_i}$ are the heat transfer entropy production mechanisms in the system, $\dot{W}_P$ is the pumping work required to overcome the pressure drop in the flow loop, and $\dot{W}_{mag}$ is the refrigeration work required to remove the parasitic heat that leaks into the superconducting magnet. Note that in an ideal system with no pressure drop, no heat leak into the magnet and perfect heat transfer, $\dot{W}_{id}$ equals the work done by the regenerator ring drive motor. For the reverse-Brayton system we have,

$$\dot{W}_{act} = \dot{W}_c - \dot{W}_e \tag{3}$$

where $\dot{W}_e$ is the actual expander work, and the actual compressor work is given by,

$$\dot{W}_c = \left(\frac{1}{\eta_c}\right)\left[\dot{W}_{id} + \frac{\dot{W}e}{\eta_e} + T_h * \sum_i \dot{S}_{ht_i} + \dot{W}_P\right] \tag{4}$$

Here, $\eta_c$ is the compressor efficiency and $\eta_e$ is the expander efficiency. Note that $\dot{W}_{id}$ is the ideal *cycle* work that equals the difference between the compressor input and the expander output.

Numerical calculation of refrigerator efficiencies requires explicit equations for the entropy production, pressure drop, magnet cooling and expander work terms. In their most accurate form, these will be dependent upon the particular characteristics of the device used. This results in a large parameter space because many different kinds of heat exchangers, regenerators, expanders and cryocoolers are available, each with different operating characteristics. Here we have opted for a simplified approach that ignores the mechanical details and models the devices with only one or two key performance parameters. For example, the compressors and expanders are represented solely by their overall efficiencies, $\eta_c$ and $\eta_e$. The work required to keep the magnet cold is defined to be a certain multiple ($\kappa_{mag}$) of the cooling load, $\dot{Q}_c$. The pressure drop work is defined to be a certain fraction ($\kappa_p$) of the total heat transferred in the heat exchangers and regenerators. Finally, the heat transfer entropy generation terms are defined to be functions of the number of thermal units ($N_{tu}$) of each heat exchanger, where we use the nominal definition of $N_{tu}$,

$$N_{tu} = \frac{hA_s}{\dot{m}_f C_P} \tag{5}$$

Here h is the heat transfer coefficient, $A_s$ is the total surface area, and $C_p$ is the fluid heat capacity. The underlying premise is that as long as these approximations are the same for both refrigerators, a fair comparison will result. Moreover, parametric variations of the modeling parameters $\eta_c$, $\kappa_{mag}$, $\kappa_p$, and $N_{tu}$ will indicate the sensitivity of the results to these assumptions.

The entropy generation equation can be derived as follows. For counter-flow heat exchangers (and balanced regenerators), there is a direct relationship between $N_{tu}$ and the temperature approach ($\Delta T_{app}$) between the two gas streams (or between the solid and the gas):

$$\frac{\Delta T_{app}}{T_{hot} - T_{cold}} = \frac{1}{1 + N_{tu}} \tag{6}$$

Heat flow across this finite temperature difference produces entropy:

$$\dot{S}_{ht} = \frac{\dot{Q} * \Delta T_{app}}{T_{hot} * T_{cold}} \tag{7}$$

where $\dot{Q}$ is the total heat transfer rate in the heat exchanger or regenerator. Note that $T_{hot}$ and $T_{cold}$ will be essentially equal to environment and load temperatures ($T_h$ and $T_c$) only in the magnetic regenerator and intermediate heat exchanger. For the hot and cold heat exchangers these temperatures will correspond to the much smaller inlet to outlet temperature span across these devices. Combining Eqs. 6 and 7 produces the desired relationship between entropy generation and $N_{tu}$:

$$\dot{S}_{ht} = \frac{\dot{Q}}{1 + N_{tu}} \left( \frac{T_{hot} - T_{cold}}{T_{hot} * T_{cold}} \right) \tag{8}$$

In reality, heat exchangers and regenerators possess additional entropy generation mechanisms. (Note that the pressure drop is accounted for separately.) The two most important are parasitic heat conduction from the hot to cold end of the device, and internal mixing of warmer and colder gas due to dynamic seal leakage in the rotary AMR. Nevertheless, we can adopt the concept of an "effective" $N_{tu}$ to account for these effects within the above entropy production framework. In other words, the $N_{tu}$ used in this study

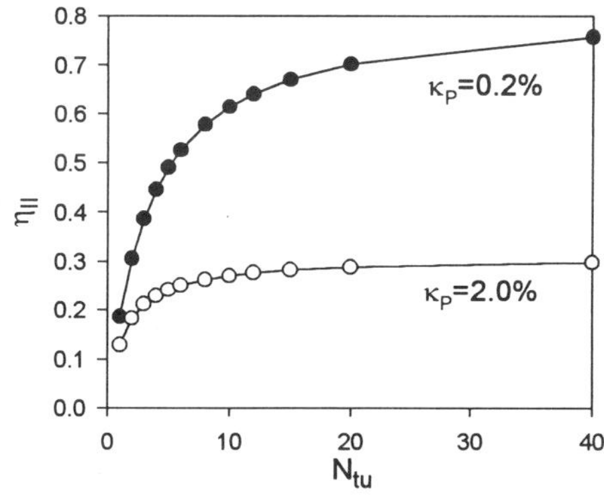


Fig. 5. AMR Performance for $\kappa_{mag}$=0.

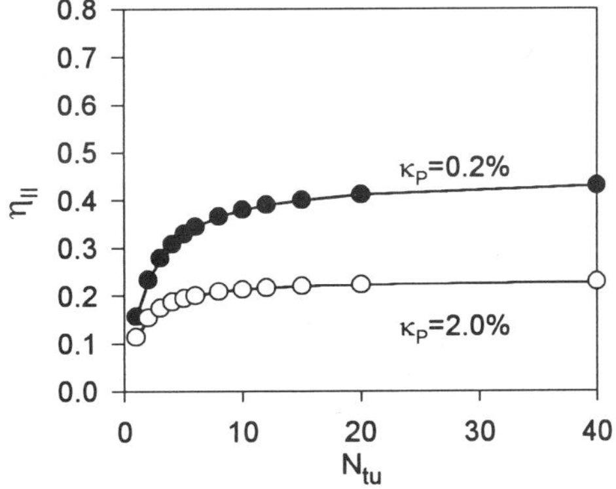


Fig. 6. AMR Performance for $\kappa_{mag}$=2 ·

will be the nominal $N_{tu}$ of Eq. 5 adjusted to a smaller value such that the entropy production computed by Eq. 8 will be equal to that of the actual device. With this approach, we can treat $N_{tu}$ as a variable parameter and use Eq. 8 to compute the entropy production for each heat exchanger. These values can then be inserted into either Eq. 2 or Eq. 4 as appropriate.

The case study employs a few additional assumptions. The compressor efficiency is arbitrarily set equal to the expander efficiency. The AMR drive motor efficiency is 100%. The hot and cold heat exchanger $N_{tu}$ are also equated. However, since the magnetic regenerator and intermediate heat exchanger transfer considerably more heat than either the cold or hot heat exchangers, their $N_{tu}$ is set twenty times higher, reflecting the need to have better heat transfer in those components. The problem of eddy current generation in the rotary AMR is ignored here, mostly on the basis of the UVic prototype design that indicated minimal eddy current generation for particle bed regenerators.[12] However, the restriction to particle beds does entail a pressure drop penalty versus the shell and tube or plate-fin geometries available for the RB intermediate heat exchanger. Ref.18 indicates that for a given heat transfer rate, particle beds can have up to 3 or 4 times the pressure drop. Here we assessed a penalty factor of only two on the assumption that more sophisticated regenerators will be developed in the future to mitigate the problem.

## RESULTS AND DISCUSSION

The calculations were performed on an Excel spreadsheet running on a MacIntosh platform. For the rotary AMR, Eqs. 2 and 8 lead to a quadratic equation in drive motor work input (or heat rejection $\dot{Q}_h$) that was solved explicitly. For the reverse-Brayton, a more complex equation for $\dot{Q}_h$ resulted that was solved iteratively. A summary of the results is presented in Figures 5 through 8.

Figure 5 shows the AMR second law efficiency as a function of $N_{tu}$ for 75% compressor efficiency, zero magnet cooling ($k_{mag}$ = 0) and two pressure drops ($k_p$ = 0.2% and 2%). The choice of values for $k_p$ spans the range of most heat transfer devices. For the low pressure drop, very high values of $h_{II}$ are attainable (up to 70%), but this value drops substantially (to 30%) for the high pressure drop case. Note that the flatness of the $k_p$ =2% curve for $N_{tu}$ > 20 indicates that entropy generation in the heat exchangers is dwarfed by the pressure drop loss. This is not true for the low pressure drop case where increasing $N_{tu}$ improves the performance even at $N_{tu}$ = 40 (which corresponds to a value of 800 in the regenerator).

Figure 6 repeats the conditions of Figure 5 except that a magnet cooling requirement of $k_{mag}$=2 has been added. This value is much smaller than that found on the UVic AMR prototype[12], but is probably reasonable for larger devices having smaller surface area to volume ratios.[19] The reduction in performance is dramatic, particularly for the low pressure drop case at high $N_{tu}$. Clearly, the magnet cooling requirement must be significantly less than $k_{mag}$ = 2 to yield overall efficiencies in the range of 60% or more.

Although not explicitly shown in either of these two figures, approximately 95% of the heat transfer entropy generation occurs in the regenerator, even though its $N_{tu}$ is twenty times that of the heat exchangers. This highlights the critical importance of using high effectiveness regenerators in AMR devices.

Figure 7 shows the reverse-Brayton performance at $k_p$ = 0.2% and 2% for a compressor and expander efficiency of 85%. This high isothermal compressor efficiency is used to represent the probable upper limit of helium turbomachine technology. The resultant low-pressure $h_{II}$ of approximately 60% is almost as high as the zero magnet cooling AMR result in Fig. 5. Figure 8 shows results for the same simulation conditions except that the compressor efficiency has been reduced to 65%. The drop in $h_{II}$ is substantial, falling from nearly 60% to 40% in the low pressure drop scenario.

Unlike the AMR where the regenerator heat transfer and pressure drop losses dominate, the reverse-Brayton cycle losses are sometimes higher in the compressor than in the intermediate heat exchanger. In particular, this is true for the low pressure drop, high $N_{tu}$ conditions that produce the highest overall efficiencies. Note that these results assume that the expander output is used to offset the compressor power input. Without it, the efficiency results drop by roughly 10% at the high end and 2% to 3% at the low end.

Several conclusions can be drawn from these results. Perhaps the most striking is the fact that both technologies produce extremely high efficiencies (60%-70%) under the most optimistic assumptions. Neither rotary AMR nor reverse-Brayton emerges as a clear winner in this sense. The key therefore becomes how successful the two technologies are at achieving the optimistic component performance targets in actual devices. For the rotary AMR these targets include low leakage, low pressure drop and high heat transfer magnetic regenerators, and minimal cooling requirements for the superconducting magnets. For reverse-Brayton these targets include high efficiency isothermal compressors, low-pressure drop, high $N_{tu}$ intermediate heat exchangers, and efficient utilization of the expander output.

It is important to realize that the work input mechanisms for the two technologies are different in a way that favors rotary AMR. A simple electric drive motor provides most of the work input to the rotary AMR. These motors generally have very high efficiencies near 100%.

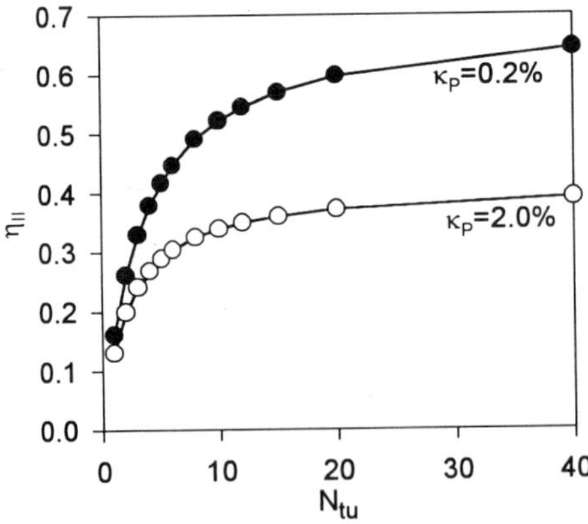


Fig. 7. Reverse-Brayton performance for $\eta_c$= 0.85.

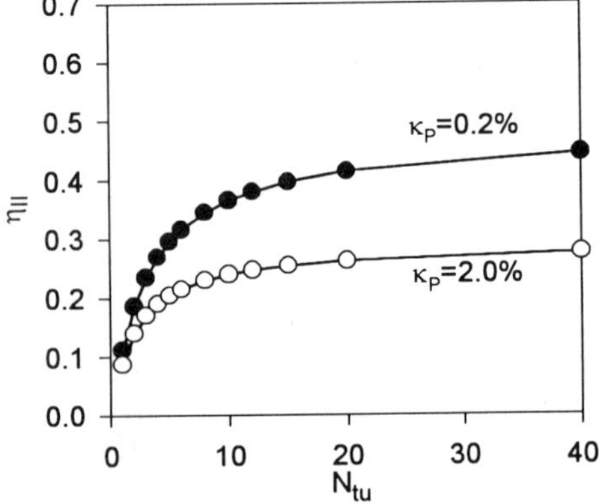


Fig. 8. Reverse-Brayton performance for $\eta_c$= 0.65.

Also, there is no expander whose output must be integrated back into the drive system. In a sense, the magnetic coupling between the superconducting magnet and the regenerator material provides "inherent" work recovery for this system: the magnet does work to draw the material in, and the motor recovers this work and does a little more to draw the material out. The reverse-Brayton system, by contrast, has separate devices for work input and work recovery, devices that are usually difficult to couple in practical refrigerators. There is also the overriding problem that highly efficient isothermal compression is difficult to achieve, especially in the high frequency devices required for practical machines. Certainly compressor efficiencies cannot approach those of simple electric motors. This work input advantage of rotary AMRs is probably not shared by reciprocating AMRs because force balancing and work recovery is much harder to accomplish in high frequency reciprocating devices.

The perceived advantage of fluid-to-solid regenerator heat transfer postulated in Ref. 2 was not supported by this study. It was realized that the helium gas has to make *two* passes through the magnetic regenerator, which results in the same fluid-to-solid-to-fluid heat transfer seen in the intermediate heat exchanger of the RB system. Both the AMR regenerator and RB intermediate heat exchanger generated the same amount of entropy at a given $N_{tu}$, although the AMR did suffer from a larger pressure drop. However, note that this study ignores the significant mechanical difficulties of producing counter-flow heat exchangers that match the high heat transfer coefficients and high surface areas readily found in particle bed or screen regenerators.

Although this study is not directly concerned with capital cost, some comments can be made based on these case-study results and the previous UVic experience with its rotary AMR prototype.[12] Magnetic refrigeration suffers greatly from the high cost of superconducting magnet technology. In fact, roughly two-thirds of the cost of the LNG prototype was in the magnet subsystem. Although maturing AMR technology will be able to reduce the cost of magnet systems in the future , it is plausible that the overall capital cost will remain too high without significant breakthroughs in either the manufacture of existing low-temperature superconducting magnets or in the commercialization of high-temperature superconducting magnets. The latter prospect is also attractive from the point of view of reducing the cryocooler power input currently required to maintain 4 K magnet temperatures.

The issue of comparing magnetic refrigeration to other gas compression cycles merits a few final comments. Currently, LNG production is based on either the Claude, cascade or mixed-refrigerant cycles. These cycles are used because they can provide moderately high second law efficiencies (close to 40% for baseload mixed-refrigerant cycle plants) at acceptable capital cost. This case study suggests the potential for both rotary AMR and reverse-Brayton to exceed this efficiency, provided that the associated technology problems are solved. Direct comparisons between these gas compression cycles and rotary AMR have not yet been attempted; however, the other cycles are not expected to have as high an efficiency as fully developed reverse-Brayton devices because they suffer from the additional loss mechanisms of J-T throttling valves and, in the cascade and mixed refrigerant cycles, phase separators. Whether or not rotary AMR or reverse-Brayton devices can be made as cheaply as the other gas compression cycles has yet to be determined.

## CONCLUSIONS

This theoretical case study compared rotary active magnetic regenerative refrigeration to reverse-Brayton gas compression refrigeration in an attempt to re-evaluate the perceived advantages and disadvantages of the competing technologies. The two refrigeration systems were found to be close analogies of each other. For refrigeration at 100 K and heat rejection at 300 K, both were capable of producing second law efficiencies of over 60% given a similar set of optimistic component performance assumptions. High effectiveness magnetic regenerators were found to be critical to the success of rotary AMR, while high effectiveness intermediate heat exchangers and compressors were necessary for the success of RB systems. Rotary AMR was found to have a clear advantage in work input since it relies on simple electric motors rather than isothermal compressors, and it avoids the mechanical complexities of coupling an expander to the compressor to utilize its output. However, rotary AMR is currently an immature technology that requires considerable technical progress to realize these potential efficiencies. Finally, the high capital cost of superconducting magnet systems is a serious concern for magnetic refrigeration that must be satisfactorily addressed to enable widespread adoption of the technology.

## ACKNOWLEDGEMENTS

The financial support from NSERC, Canada, Centra Gas B.C., Inc. and Westcoast Energy, Inc. is very much appreciated.

## REFERENCES

[1] Barclay, J. A. "An Analysis of Liquefaction of Helium Using Magnetic Refrigeration," *Los Alamos Scientific Laboratory Report*, No. LA-8991. (December 1981)
[2] Barclay, J. A. "A Comparison of the Efficiency of Gas and Magnetic Refrigerators", Proc. of 22$^{nd}$ National Heat Transfer Conference, Niagara Falls, NY, (Aug. 1984)
[3] Barclay, J.A. "Can Magnetic Refrigerators Liquefy Hydrogen at High Efficiency?" Paper presented at 20th National Heat Transfer Conference, Milwaukee, WI. Paper No. 81-HT-82. (August 2-5, 1981)
[4] Pecharsky, V. K. and Gschneidner, K. A. "Tunable Magnetic Regenerator Alloys with a Giant Magnetocaloric Effect for Magnetic Refrigeration from ~20 K to ~290 K", J. Appl. Phys. (February, 1997)
[5] Zhang, X. X., Tejada, J., Xin, Y., Sun, G. F., Wong, K. W. and Bohigas, X. "Magnetocaloric effect in $La_{0.67}Ca_{0.33}MnO$ and $La_{0.60}Y_{0.07}Ca_{0.33}MnO$ bulk materials"; Appl. Phys. Lett. 69(23), pp. 3596-3598 (1996).
[6] Russo, C., American Superconductor Company, private communication. (1997)
[7] Merida Donis, W. "Development of Monolithic Regenerator Technology for Low Temperature (4 K) Gifford McMahon Cryocoolers", M. A. Sc. thesis, Department of Mechanical Engineering, University of Victoria, *and references therein*. (December 1996)
[8] Patton, G., Green, G., Stevens, J. and Humphrey, J. "Reciprocating Magnetic Refrigerator", Proc. 4th Intl. Cryocoolers Conf., Easton, Md. (1986).
[9] Green, G.,Chafe, J., Stevens, J. and Humphrey, J. "A Gadolinium-Terbium Active Regenerator", *Adv. Cryo. Eng.*, Vol. 35, pp 1165-1174, Plenum Press, NY. (1990).
[10] Wang, A. A., Johnson, J. W., Niemi, R. W. Sternberg, A. A. and Zimm, C. B. Experimental Results of an Efficient Active Magnetic Regenerator Refrigerator, 8th Intl. Cryocooler Conf., Vail, CO. (1994)
[11] Hall, J. L., Reid, C. E., Spearing, I. G. and Barclay, J. A. "Thermodynamic Considerations For The Design Of Active Magnetic Regenerative Refrigerators", *Adv. Cryo. Eng.*, Vol. 41, Plenum Press, N. Y. (1996).
[12] Hall, J. L. and Barclay, J. A. "A Prototype Active Magnetic Regenerative Liquefier For Natural Gas, Part 1: Design and Construction", *manuscript in progress.* (1997)
[13] Barclay, J. A. "The Theory Of An Active Magnetic Regenerative Refrigerator", Los Alamos Report LA-UR-82-1792 (1982). (Also NASA-CP-2287, 1983)
[14] Barclay, J. A. "Magnetic Refrigeration: the Promise and the Problems", Proceedings of the 21st Space Congress, Kennedy Spaceflight Center, FL (1984).
[15] Zimm, C. B. Astronautics Corporation of America, private communication (1997).
[16] Swift, W.L. " Single-Stage Reverse-Brayton Cryocooler: Performance of the Engineering Model", *Cryocoolers 8*, pp 499-506, Plenum Press, NY. (1995)
[17] Ankermann, T. "Small scale liquefaction of hydrogen utilizing Reverse Brayton and modified Claude cycle helium gas refrigeration cycles", Proc. of the XVIIIth Int'l Congress of Refrigeration, Montreal, PQ, Canada, pgs 306-313. (August 10-17, 1991)
[18] Nilles, M.J., Calkins, M. E., Dingus, M. L. and Hendricks, J. B. "Heat Transfer and Flow Friction in Perforated Plate Heat Exchangers", *Aerospace Heat Exchanger Technology*, R. K. Shah and A.Hashemi, eds., Elsevier Science Publishers B. V. (1993)
[19] Barclay, J. A. and Stewart, W. F. "The Effect of Parasitic Refrigeration on the Efficiency of Magnetic Liquefiers, Proceedings IECEC'82, The 17th Intersociety Energy Conversion Engineering Conf., Aug. (1982)

# THE GIANT MAGNETOCALORIC EFFECT IN $Gd_5(Si_xGe_{1-x})_4$ MATERIALS FOR MAGNETIC REFRIGERATION

V.K.Pecharsky and K.A.Gschneidner, Jr.

Ames Laboratory and Department
of Materials Science and Engineering
Iowa State University, Ames, IA, 50011-3020, USA

## ABSTRACT

The $Gd_5(Si_xGe_{1-x})_4$ alloys, where $0 \leq x \leq 0.5$, exhibit a giant magnetocaloric effect. This extremely large magnetocaloric effect should elevate magnetic refrigeration technology to new heights allowing it to become even more competitive with other existing refrigeration technologies. The temperature of the giant magnetocaloric effect is easily tunable between ~30 and ~275 K by changing the Si:Ge ratio. Furthermore, by alloying with Ga, the giant magnetocaloric effect temperature increases to ~290 K. The reversible magnetic field induced magnetic entropy change, $\Delta S_{mag}$, and the adiabatic temperature rise, $\Delta T_{ad}$, for a low to moderate magnetic field change (from 0 to 2-10 T) are the largest ever observed at the corresponding Curie temperatures. The $\Delta S_{mag}$ is 100 to 400%, and the $\Delta T_{ad}$ is 25 to 200% larger than that for the best known prototypes.

## INTRODUCTION

Magnetic refrigeration utilizes the magnetocaloric effect (MCE), a distinct property of magnetic materials, which is perceived as the change of materials' magnetic entropy upon the isothermal application/removal of a magnetic field, or as the change of their temperature when a magnetic field is altered adiabatically. Although the MCE was discovered almost 120 years ago,[1] it was not until 45 years later when Debye[2] and Giauque[3] suggested that it could be used for cooling below ~1.5 K, and it took another 50 years for Brown[4] to show that the MCE can be used for cooling near room temperature.

The magnetic entropy usually constitutes just a fraction of the total entropy of a solid. For a magnetic lanthanide ion the molar theoretical upper limit is given as $S_{mag} = R\ln(2J+1)$, where $R$ is the universal gas constant and $J$ is the total angular momentum. Unlike the lattice entropy, which invariably rises with increasing temperature, the major portion of the magnetic entropy is localized near the magnetic ordering temperature. Hence, it was commonly thought that magnetic refrigeration is practical only at low temperatures, i.e. where the lattice entropy is small compared to the magnetic entropy, and cooling by adiabatic demagnetization was successfully employed to reach mK temperatures. Until recently it was unclear whether or not the MCE could be employed efficiently at higher temperatures, although theoretical estimates were quite favorable. A recently completed study by the Astronautics Corporation of America and the Ames Laboratory has proven that sub-room temperature active magnetic regenerator magnetic refrigeration (using Gd as a regenerator/refrigerant) is indeed energy efficient and is competitive with the gas compression technology.[5] This makes the search for materials with a large MCE more important than before, since the refrigerant performance would have a crucial impact on Carnot efficiency of the near room temperature refrigeration devices. In the past few years there has been an increased activity in the study of new magnetocaloric materials such as nanocomposites,[6] amorphous alloys,[7] intermetallic compounds,[8] and perovskite-type oxides.[9,10] This paper presents the results of our experimental study of the magnetic and thermal properties of the $Gd_5(Si_xGe_{1-x})_4$ alloys.

## EXPERIMENTAL

The $Gd_5(Si_xGe_{1-x})_4$ alloys were arc-melted from 99.99 wt.% pure Gd (prepared by the Materials Preparation Center, Ames Laboratory) and 99.99+ wt.% pure Si and Ge (purchased from CERAC, Inc.) in an argon atmosphere. The heat capacity was measured from ~3.5 to 350 K in 0 to 10 T magnetic fields using a heat-pulse calorimeter.[11] The ac and dc susceptibility, and the dc magnetization, were measured from ~4.5 to 325 K in 2.5 to 10 Oe ac fields (55 to 1000 Hz frequencies), and dc fields from 0 to 5.5 Tesla using a Lake Shore ac/dc susceptometer. The magnetocaloric effect was calculated from the total entropy evaluated from the measured heat capacity[12] (both the magnetic entropy change, $\Delta S_{mag}$, and the adiabatic temperature rise, $\Delta T_{ad}$) and from magnetization measurements using the Maxwell relation ($\Delta S_{mag}$).

## MAGNETIC PHASE DIAGRAM OF THE $Gd_5Si_4$-$Gd_5Ge_4$ SYSTEM

Holtzberg *et al.*[13] reported three distinct phase regions in the $Gd_5(Si_xGe_{1-x})_4$ system: the $Gd_5Si_4$-based solid solution ($0.5 \le x \le 1$), the $Gd_5Ge_4$-based solid solution ($0 \le x \le 0.25$), and the unidentified ternary phase ($0.3 \le x < 0.5$). When $0.5 \le x \le 1$, the alloys are ferromagnetic and the Curie temperature lowers from 335 to 295 K upon the substitution of Ge for Si. On the other hand, Ge-rich alloys order in two steps. It was reported that they order antiferromagnetically with an upper ordering temperature of ~40 K to 300 K, and upon cooling the magnetic structure changes to a ferromagnet with a lower ordering temperature which varies from ~20 K to 120 K with decreasing Ge content for $0 \le x \le 0.25$. No magnetic properties of the unidentified intermediate ternary phase was reported.

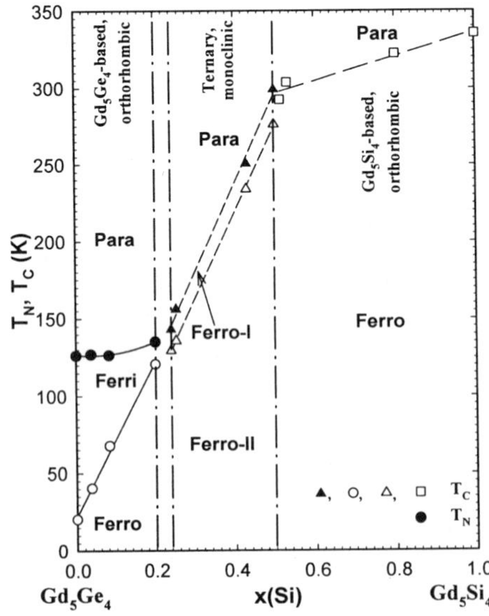


**Figure 1**. The zero magnetic field phase diagram of the $Gd_5Si_4$ - $Gd_5Ge_4$ pseudobinary system.

The magnetic phase diagram, shown in Fig.1, was determined from our study of the crystallography, phase relationships,[14] magnetic

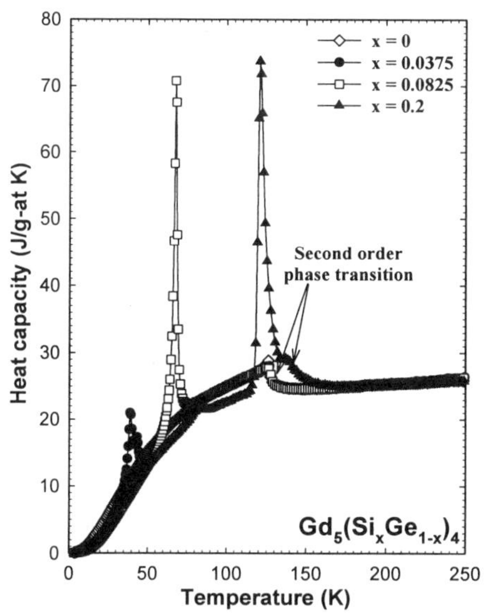


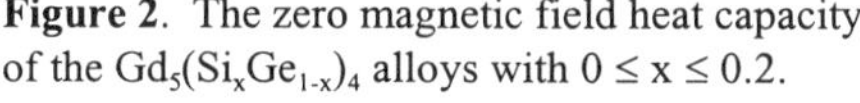

Figure 2. The zero magnetic field heat capacity of the $Gd_5(Si_xGe_{1-x})_4$ alloys with $0 \leq x \leq 0.2$.

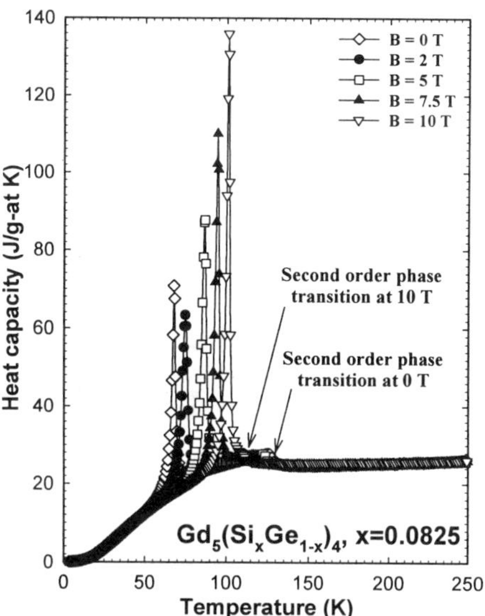


**Figure 3**. The heat capacity of the $Gd_5(Si_xGe_{1-x})_4$ with x=0.0825 as a function of magnetic field.

and magnetocaloric properties[15,16] of the $Gd_5(Si_xGe_{1-x})_4$ alloys. The alloys with $0.5 < x \leq 1$ are normal ferromagnets, and their ordering temperatures agree well with the results by Holtzberg *et al.*[13] The $Gd_5Ge_4$-based solid solution extends from $0 \leq x \leq 0.2$, and we have confirmed the two-step ordering process in this region with the upper transition being from a paramagnet to a ferrimagnet (not an antiferromagnet as previously reported), and the lower one from a ferrimagnet to a ferromagnet. The upper ordering temperature in the alloys with $0 \leq x \leq 0.2$ is almost independent of composition (Fig.1), while the lower transition temperature agrees quite well with the earlier results.[13] The intermediate ternary solid solution phase $Gd_5(Si_xGe_{1-x})_4$ extends from $0.24 \leq x \leq 0.5$ and has the monoclinically distorted lattice derivative from the parent orthorhombic $Sm_5Ge_4$-type structure. The monoclinic distortion is caused by the ordering of the bonds of the (Si+Ge) pairs, which are partially covalent.[14] The alloys in this phase region also order magnetically in two steps. Upon lowering the temperature they order ferromagnetically and later a second transition from ferromagnet-I to ferromagnet-II occurs. For all the alloys with $x \leq 0.5$ the upper transition is a second order transition and the lower one is a first order phase transition. The bulk of magnetic entropy is associated with the lower temperature first order phase transition and this brings about the giant magnetocaloric effect.

## HEAT CAPACITY OF THE $Gd_5(Si_xGe_{1-x})_4$ ALLOYS

The zero magnetic field heat capacity of the alloys from the $Gd_5Ge_4$-based solid solution region (orthorhombic structure) is shown in Fig.2. The distinct feature of the heat capacity of all the alloys, except x = 0, is the presence of a λ-type anomaly (in the ~125-140 K region) which corresponds to the second order magnetic phase transition, and a sharp maximum at a lower temperature indicating the first order magnetic phase transformation (see previous section). The magnetic field shifts the lower temperature heat capacity maximum towards higher temperatures without broadening it (Fig.3). The increase of heat capacity peak value with field indicates that magnetic field just shifts the transition and that it remains a first order phase transition having the same (or approximately the same) entropy of transformation. The upper λ-type anomaly, on the contrary, is suppressed to

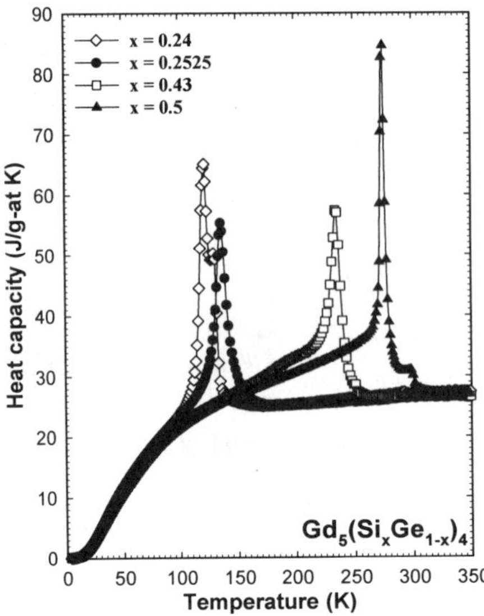


**Figure 4**. The zero magnetic field heat capacity of the $Gd_5(Si_xGe_{1-x})_4$ alloys with $0.24 \le x \le 0.5$.

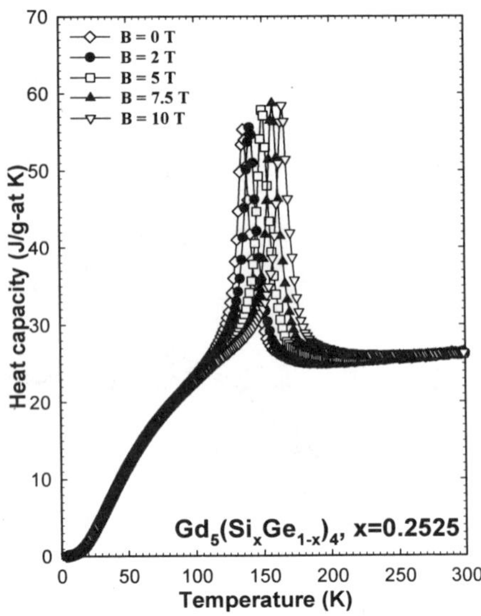


**Figure 5**. The heat capacity of the $Gd_5(Si_xGe_{1-x})_4$ with x=0.2525 as a function of magnetic field.

lower temperatures by magnetic field. It is clear that the bulk of magnetic entropy is associated with the lower magnetic ordering (Figs.2,3).

The zero magnetic field heat capacity of the $Gd_5(Si_xGe_{1-x})_4$ alloys with $0.24 \le x \le 0.5$ (i.e. those with the monoclinically distorted crystal structure) is shown in Fig.4, and the effect of magnetic field, which is similar for all of the alloys, is shown in Fig.5 for $Gd_5(Si_{1.01}Ge_{2.99})$. The behavior is quite similar to that exhibited by the alloys from the $0 \le x \le 0.2$ range of compositions.

## MAGNETIC PROPERTIES OF THE $Gd_5(Si_xGe_{1-x})_4$ ALLOYS

The magnetic susceptibility above the ordering temperatures of the $Gd_5(Si_xGe_{1-x})_4$ alloys (Fig.6) obeys the Curie-Weiss law. The effective magnetic moment ($p_{eff}$) calculated from the least squares fit of the linear regions for the alloys with $x \le 0.43$ (Table 1) is close to but slightly lower than that of the free $Gd^{3+}$ ion ($p_{eff}$ = 7.94). The ordered magnetic moment at 4.5 K

**Table 1.** Magnetic properties of the $Gd_5(Si_xGe_{1-x})_4$ alloys

| x | $p_{eff}$, $\mu_B$ paramagnet | $\Theta_p$, K | $p_{eff}$, $\mu_B$ ordered |
|---|---|---|---|
| 0 | 7.14(3) | 120.7(6) | 7.41(5) |
| 0.0825 | 7.52(2) | 111.9(3) | 7.36(5) |
| 0.2525 | 7.51(2) | 133.9(4) | 7.39(5) |
| 0.43 | 7.89(6) | 210.8(2) | 7.46(5) |
| 0.5 | - | - | 7.36(5) |

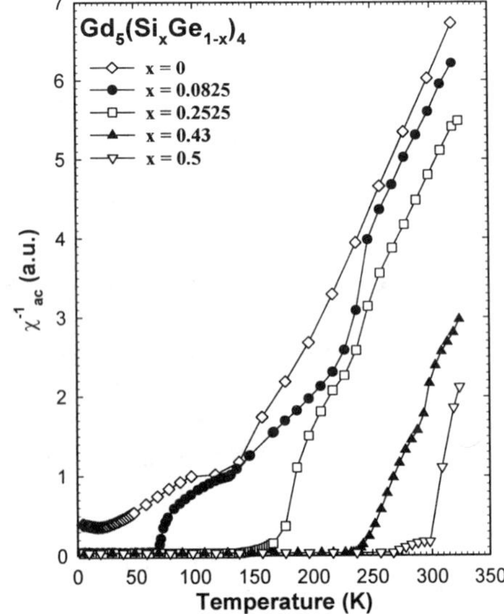


**Figure 6**. The inverse ac susceptibility of the $Gd_5(Si_xGe_{1-x})_4$ alloys with $0 \le x \le 0.5$. The solid lines are guides for the eye.

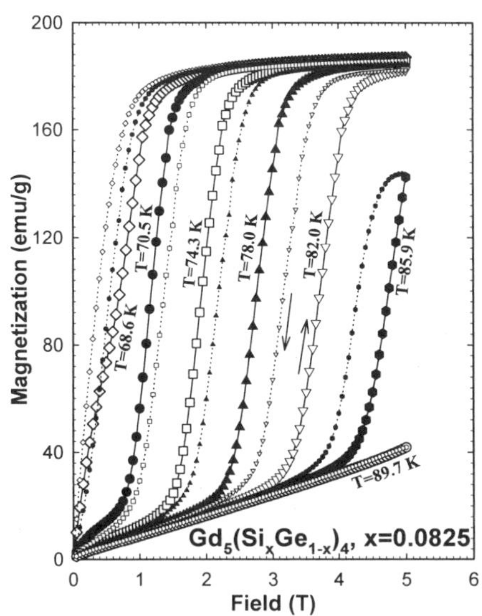


**Figure 7**. The magnetization isotherms of the $Gd_5(Si_{0.33}Ge_{3.67})$ alloy in the vicinity of the first order phase transition.

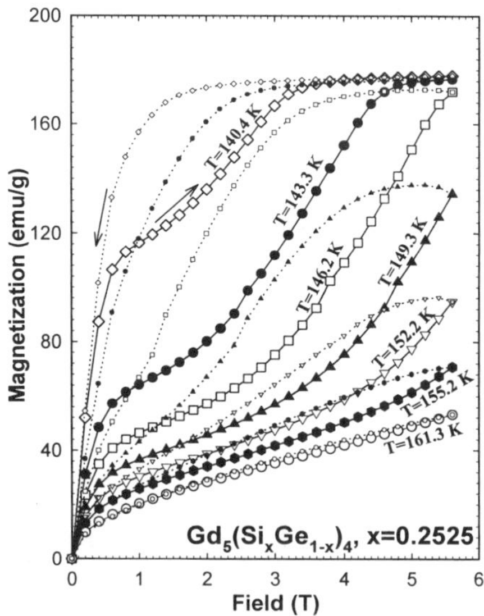


**Figure 8**. The magnetization isotherms of the $Gd_5(Si_{1.01}Ge_{2.99})$ alloy in the vicinity of the first order phase transition.

for all the alloys with $x \leq 0.5$, as determined from magnetization isotherms in a dc field of 5.6 T, is slightly larger than the theoretical value of $gJ = 7\ \mu_B$. The ac susceptibility is independent of ac field (1.25 to 10 Oe) and frequency (55 to 1000 Hz). Most of the alloys show negative deviation of the inverse susceptibility from linearity above the magnetic phase transition, which is most likely associated with high temperature spin fluctuations.

The magnetization isotherms of $Gd_5(Si_{0.33}Ge_{3.67})$ and $Gd_5(Si_{1.01}Ge_{2.99})$ in the vicinity of the lower temperature phase transitions are shown in Figs. 7 and 8, respectively. The first one is representative of the alloys from the $Gd_5Ge_4$-based (orthorhombic) solid solution, and the second one is typical for the alloys from the intermediate ternary (monoclinic) solid solution. In both cases the magnetization behavior shows that the magnetic field induces a spin flip transition from one magnetically ordered phase to another, and that this transition is both temperature and magnetic field dependent. For the alloys with $x \leq 0.20$, the spin flip transition is from a ferrimagnet to a ferromagnet, and for the alloys with $0.24 \leq x \leq 0.5$ it is from one ferromagnetic state to a different ferromagnetic phase. The partially hysteretic behavior, observed in both cases, is typical for a first order phase transition.

## MAGNETOCALORIC EFFECT AND REGENERATOR EFFICIENCY

An accurate calculation of the magnetocaloric effect from the heat capacity data for the materials which undergo a first order magnetic phase transition is complicated by the uncertainty in heat capacity measurements caused by the non-zero enthalpy of transformation and non-zero heat capacity of the calorimeter's sample holder.[17] Hence, the calculation of the isothermal magnetic entropy change, $\Delta S_{mag}$, using magnetization measurements seems to be the most accurate approach in this case. Nonetheless, it is still possible to calculate the $\Delta T_{ad}$ using the heat capacity combined with the $\Delta S_{mag}$ from the magnetization measurements. The $\Delta S_{mag}$ from magnetization is used to correct the total entropy curves from $T \geq T_C$, where $T_C$ is the Curie temperature, and then these corrected total entropy curves are used to calculate the correct $\Delta T_{ad}$. This is illustrated in Fig.9, where $\Delta S_{mag}$ for a magnetic field change from 0 to 5 T of the $Gd_5(Si_{0.33}Ge_{3.67})$ alloy is shown as calculated from magnetization data measured in an increasing field (filled diamonds) and from the as-measured heat capacity (open circles). As one can see, the difference between the two results is quite large between 70 and 90 K, well outside the estimated error limits, which in this range of temperatures equals ~2-3 J/kg K. After the raw heat capacity data were modified (i.e. corrected using the accurate $\Delta S_{mag}$ values from magnetization measurements), the agreement between the two sets of values obtained from different experimental data becomes excellent (see Fig.9, filled diamonds and open squares).

Thus, the $\Delta T_{ad}$ determined from the experimentally measured and corrected heat capacity is shown in Fig.10 for a magnetic field change from 0 to 5 T. The magnetocaloric effect reaches values as large as ~15 to ~18 K for x = 0.0825, 0.5, and 0.43, and it becomes somewhat lower for the other compositions, but still remains quite large. It is worthwhile to note that for the composition $Gd_5(Si_2Ge_2)$ the MCE is almost 30% higher than that in pure Gd, although the weight percent of gadolinium is 26% lower in $Gd_5(Si_2Ge_2)$ compared to pure Gd. Hence it is reasonable to conclude that the presence of the giant magnetocaloric effect in the ternary $Gd_5(Si_xGe_{1-x})_4$ alloys with $0 \leq x \leq 0.5$ is caused by the fact that the lower temperature magnetic phase transition in these alloys is a first order phase transformation. The first order phase transition provides extremely sharp and large changes in magnetization as a function of temperature (see Figs. 7 and 8 ), and this in turn, brings about the large $\Delta S_{mag}$ and $\Delta T_{ad}$.

The giant magnetocaloric effect in the vicinity of a first order phase transition was earlier reported in $Fe_{0.49}Rh_{0.51}$.[18,19] However, the transition in this alloy is irreversible with respect to changing magnetic field. To the contrary, the first order phase transition in $Gd_5(Si_xGe_{1-x})_4$ is reversible, and the magnetic structure is restored after the magnetic field is removed (Figs.7 and 8), which makes the $Gd_5(Si_xGe_{1-x})_4$ alloys with $0 \leq x \leq 0.5$ excellent magnetic refrigerants.

Another important characteristic of the magnetic refrigerant alloys is their, refrigeration capacity, which is given as

$$q = \int_{T_1}^{T_2} \Delta S_{mag}(T)dT$$

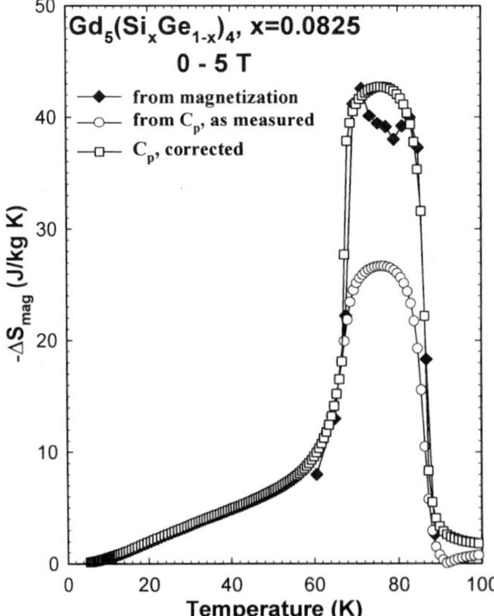


**Figure 9**. Comparison of $\Delta S_{mag}$ calculated from magnetization data, and the as-measured and corrected heat capacities.

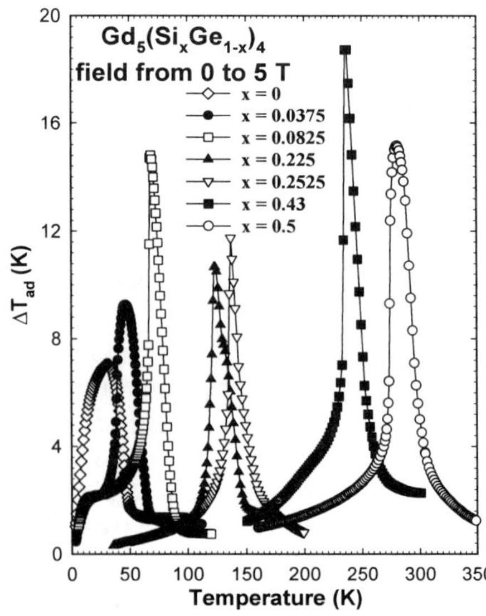


**Figure 10**. The magnetocaloric effect, $\Delta T_{ad}$, for a magnetic field change from 0 to 5 T calculated from the corrected heat capacity data.

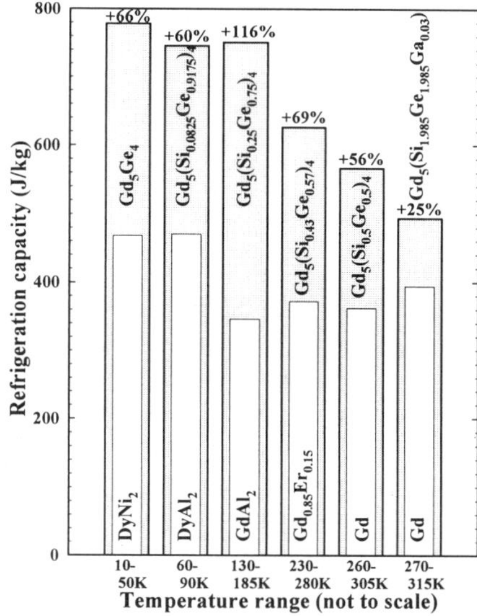


**Figure 11**. The refrigeration capacity of several $Gd_5(Si_xGe_{1-x})_4$ alloys compared with that of the best known prototype materials in different temperature ranges for a magnetic field change from 0 to 5 T.

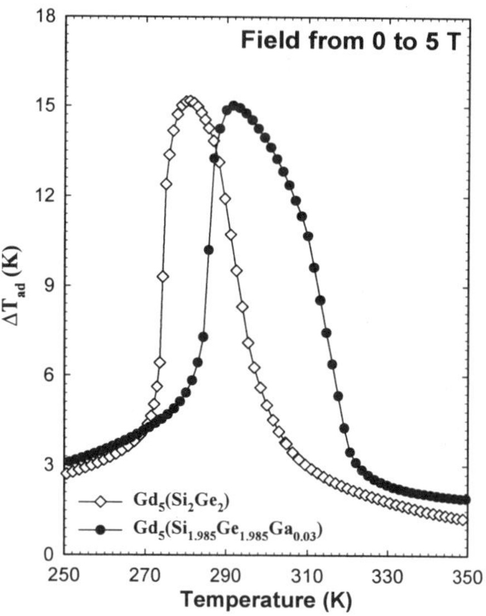


**Figure 12**. The magnetocaloric effect in $Gd_5(Si_2Ge_2)$ and $Gd_5(Si_{1.985}Ge_{1.985}Ga_{0.03})$ for magnetic field change from 0 to 5 T. The lines drawn are guides for the eye.

Here $T_1$ and $T_2$ are the temperatures of the hot and cold sinks, respectively, and $\Delta S_{mag}(T)$ is the refrigerant's magnetic entropy change as a function of temperature. The refrigerant capacity, therefore, is a measure of how much heat can be transferred between cold and hot sinks in one ideal refrigeration cycle. A comparison of some of the $Gd_5(Si_xGe_{1-x})_4$ alloys with the best known refrigerants in the corresponding ranges of temperature shows that the new materials should have ~25 to ~120% larger refrigeration capacity than any known magnetic refrigerant (Fig.11).

## EFFECT OF SMALL ALLOYING ADDITIONS

As already described above, $Gd_5(Si_2Ge_2)$ represents the terminal composition where the magnetic properties change abruptly upon a further increase of the silicon content (Fig.1). The lower ordering temperature for this alloy is 276 K and the giant magnetocaloric effect extends to just above ~290 K in a magnetic field of 5 T (see Figs. 10 and 11). This alloy would make an excellent refrigerant for a refrigerator rejecting heat to chilled water. However, if one is to build a magnetic refrigerator rejecting heat to the ambient (i.e. the temperature of the hot sink would be approximately equal to 25-35°C, or 298-308 K), then one needs a material with a higher ordering temperature than 276 K. The detailed knowledge of the mechanism of the crystal structure change in the $Gd_5Si_4$-$Gd_5Ge_4$ pseudobinary system,[14] allowed us to selectively modify the properties of $Gd_5(Si_2Ge_2)$ by alloying it with different elements substituting for Si and Ge.[20] The best alloying addition was found to be trivalent Ga, which contains just 3 valent electrons compared to 4 in tetravalent Si and Ge. The more metallic nature of Ga is thought to affect the distribution of electrons between the valence and conduction bands, and small alloying additions of Ga (0.333 at.%) retain the monoclinic crystal structure, which is required to preserve the first order magnetic phase transition and to maintain the giant magnetocaloric effect. Simultaneously, it probably enhances the exchange (RKKY) interactions which causes the increase of both the upper and the lower ordering temperatures in $Gd_5(Si_{1.985}Ge_{1.985}Ga_{0.03})$. In fact, the lower (first order phase transition) occurs at ~286 K and a 0 to 5 T magnetic field change extends the region of the giant magnetocaloric effect from ~290 to 310 K, as shown in Fig.12.

## CONCLUSION

The series of $Gd_5(Si_xGe_{1-x})_4$ alloys, where $0 \leq x \leq 0.5$, plus the $Gd_5(Si_{1.985}Ge_{1.985}Ga_{0.03})$ alloy are shown to have an extremely large magnetocaloric effect which exceeds that in known materials by as much as a factor of 2 in terms of $\Delta T_{ad}$ and by a factor of 4 in terms of $\Delta S_{mag}$. Another unique feature of this series of materials is that the magnetic ordering temperature can be easily tuned between room temperature and the boiling point of hydrogen gas. Therefore, we believe that our discovery of the large magnetic cooling power of the $Gd_5(Si_{1-x}Ge_x)_4$ materials offers the long awaited breakthrough in magnetic refrigeration technology. This breakthrough in regenerator materials can enable temperature-tuned magnetic refrigerators. If these new materials are coupled with the practical magnetic refrigeration system designs that now exist, broad commercialization of this cooling technology may result in large energy savings and in the reduction of global warming and in preservation of atmospheric quality.

## ACKNOWLEDGMENT

This work was supported by both the Office of Basic Energy Sciences, Materials Sciences Division and the Office of Computational and Technology Research, Advanced Energy Projects and Technology Research Division of the US DOE.

## REFERENCES

1. E. Warburg, *Ann. Phys. Chem.*, 13:141(1881).
2. P. Debye, *Ann. Physik*, 81:1154 (1926).
3. W. F. Giauque, *J. Am. Chem. Soc.*, 49:1864 (1927).
4. G. V. Brown, *J. Appl. Phys.*, 47:3673 (1976).
5. C.B. Zimm, A. Jastrab, A. Steinberg, V.K. Pecharsky, K.A. Gschneidner, Jr., M.G. Osborne, and I.E. Anderson, "Description and Performance of a Sub-Room Temperature Magnetic Refrigerator", paper JD3, presented at CEC/ICMC-97, Portland, OR, July 28-August 1, 1997.
6. Y.Z. Shao, J.K.L. Lai, and C.H. Shek. *J. Magn. Magn. Mater.*, 163:103 (1996).
7. X.Y. Liu, J.A. Barclay, R.B. Gopal, M. Földeàki, R. Chahine, T.K. Bose, P.J. Schurer, and J.L. LaCombe, *J. Appl. Phys.*, 79:1630 (1996).
8. J.F. Herbst, C.D. Fuerst, R.D. McMichael, *J. Appl. Phys.*, 79:5998 (1996).
9. X.X. Zhang, J. Tejada, Y. Xin, G.F. Sun, K.W. Wong, and X. Bohigas, *Appl. Phys. Lett.*, 69:3596 (1996).
10. Z.B. Guo, Y.W. Du, J.S. Zhu, H. Huang, W.P. Ding, and D. Feng, *Phys. Rev. Lett.*, 78:1142 (1997).
11. V.K. Pecharsky, J.O. Moorman, and K.A. Gschneidner, Jr., *Rev. Sci. Instr.*, to be published in November 1997.
12. V.K.Pecharsky and K.A.Gschneidner, Jr., *Adv. Cryogen. Eng.*, 42A:423 (1996).
13. F. Holtzberg, R.J. Gambino, and T.R. McGuire, *J. Phys. Chem. Solids*, 28:2283 (1967).
14. V.K. Pecharsky and K.A. Gschneidner, Jr., *J. Alloys. Comp.*, in press.
15. V.K. Pecharsky and K.A. Gschneidner, Jr., *Phys. Rev. Lett.*, 78:4494 (1997).
16. V.K. Pecharsky and K.A. Gschneidner, Jr., *Appl. Phys. Lett.*, 70:3299 (1997).
17. V.K. Pecharsky, to be published.
18. M.P.Annaorazov, K.A.Asatryan, G.Myalikgulyev, S.A.Nikitin, A.M.Tishin, and L.A.Tyurin, *Cryogenics*, 32:867 (1992).
19. M.P.Annaorazov, S.A.Nikitin, A.L.Tyurin, K.A.Asatryan, and A.Kh.Dovletov, *J. Appl. Phys.*, 79:1689 (1996).
20. V.K. Pecharsky and K.A. Gschneidner, Jr., *J. Magn. Magn. Mater.*, 167:L179 (1997).

# THE INFLUENCE OF MULTIPLE MAGNETIC ORDERING ON THE MAGNETOCALORIC EFFECT IN RNiAl ALLOYS

B. J. Korte,[1] V. K. Pecharsky,[1]
and K. A. Gschneidner, Jr.[1,2]

[1]Ames Laboratory
Iowa State University
Ames, IA 50011-3020, USA

[2]Department of Materials Science and Engineering
Iowa State University
Ames, IA 50011-3020, USA

## ABSTRACT

Both the change in magnetic entropy ($\Delta S_{mag}$) and the adiabatic temperature rise ($\Delta T_{ad}$) induced by a change of magnetic field were determined for a series of $(Gd_{1-x}Er_x)NiAl$ alloys (where x = 0, 0.30, 0.46, 0.60, 0.80 and 1.00). Through magnetic and heat capacity measurements, these materials were found to possess multiple antiferromagnetic and ferromagnetic ordering processes. The value of $\Delta S_{mag}$ was observed to peak around the antiferromagnetic transition temperature for ErNiAl, with the maximum gradually broadening and shifting upward toward the Curie temperature as the Gd content is increased. For Gd-rich alloys, a significant contribution to the magnetic entropy change is observed at both the low and high-temperature transitions, resulting in a skewed caret-like $\Delta S_{mag}$ vs. temperature profile. Broad zero-field transitions coupled with opposing shifts in entropy around the ferromagnetic and antiferromagnetic ordering temperatures induced by the application of a magnetic field are believed to contribute to this effect.

## INTRODUCTION

The specific material properties required for the operation of magnetic refrigerator devices at optimum efficiency vary with the refrigeration cycle employed. For example, refrigeration devices utilizing the Ericsson cycle[1] require that the entropy-temperature curves for the magnetic refrigerant must be parallel at the minimum and maximum applied fields, i. e. the change in magnetic entropy ($\Delta S_{mag}$) must remain constant for the full temperature span of the cycle. To achieve this, several composite or multilayer systems have been studied,[2-4] where ferromagnetic materials with varying Curie temperatures are layered according to their transition temperatures, thereby spanning the full temperature range of the thermodynamic cycle.

For ease of processibility, a single material with an appropriate $\Delta S_{mag}$ temperature profile might be more desirable. The properties of such a material were previously reported.[5] The compound $(Gd_{0.54}Er_{0.46})NiAl$ was found to possess a nearly temperature-independent magnetic entropy change (over approximately a 30 K temperature span) with a large $\Delta S_{mag}$ of -13 J/kg-K for a field change from 0 T to 5.32 T, comparable to that observed in some of the best composite materials. In order to better understand the nature of this exceptional table-like magnetocaloric effect, a study of the magnetic properties of a series of $(Gd_{1-x}Er_x)NiAl$ pseudo-ternary alloys (where x = 0, 0.30, 0.46, 0.60, 0.80, and 1.00) was undertaken. The results of this study are presented below.

## EXPERIMENTAL

The $(Gd_{1-x}Er_x)NiAl$ samples were synthesized from 99.8+ at.% pure Gd and Er metals prepared at the Materials Preparation Center of the Ames Laboratory. The major impurities were previously reported.[5] The Al and Ni were purchased from a commercial vendor with a stated purity >99.99%. The alloys were prepared by first arc-melting in an argon atmosphere the Gd and Er and (separately) the Al and Ni in the correct proportions on a water-cooled copper hearth, followed by arc-melting of the two resulting binary alloys. The $(Gd_{1-x}Er_x)NiAl$ alloys were found to be essentially single phase (< 1 % second phase) by metallography and X-ray diffraction (using a Scintag automated x-ray powder diffractometer) and therefore were not heat-treated. All alloys were found to have the hexagonal ZrNiAl-type crystal structure. The lattice parameters for the parent compounds GdNiAl and ErNiAl agree well with literature values,[6] and those for intermediate compositions decrease linearly with increasing Er content.

The heat capacity of the $(Gd_{1-x}Er_x)NiAl$ alloys was measured from approximately 3.5 K to 350 K in fields of 0, 2, 5, 7.5 and 10 T using an adiabatic heat-pulse calorimeter[7], with an error of less than 1 % over the entire measurement region. The ac and dc magnetic measurements on the as-cast polycrystalline samples were performed on a Lake Shore Susceptometer/Magnetometer Model No. 7225 at temperatures ranging from 2 K to 325 K in applied fields up to 5.6 T. The ac experiments were performed at frequencies ranging from 10 Hz to 1000 Hz and ac fields of 1.25 Oe to 5 Oe.

## RESULTS

### Magnetic and heat capacity measurements

The results of low-temperature ac susceptibility measurements in zero bias dc field are shown in Figure 1. For ErNiAl, a single sharp peak in the real component of the susceptibility is observed at 6 K; this peak is observed to shift to lower temperatures in bias dc fields up to 1 T. Above 1 T the sharp maximum splits into two peaks, with the low temperature peak shifting to lower temperatures and the high-temperature peak shifting

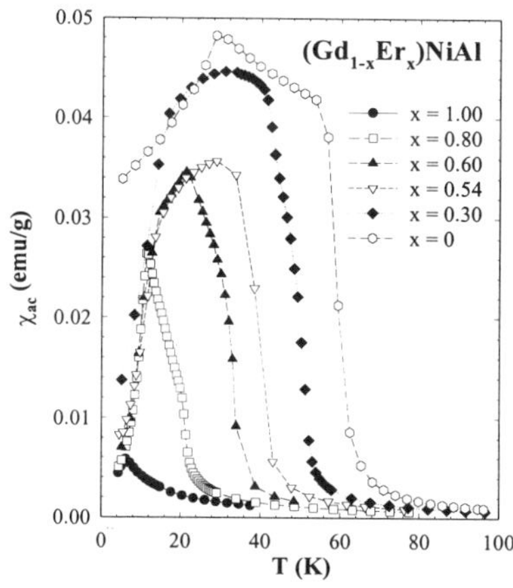


**Figure 1**. Real component of the ac susceptibility of the $(Gd_{1-x}Er_x)NiAl$ alloys as a function of temperature in zero dc field ($H_{ac}$ = 1.25 Oe, $\nu$ = 125 Hz).

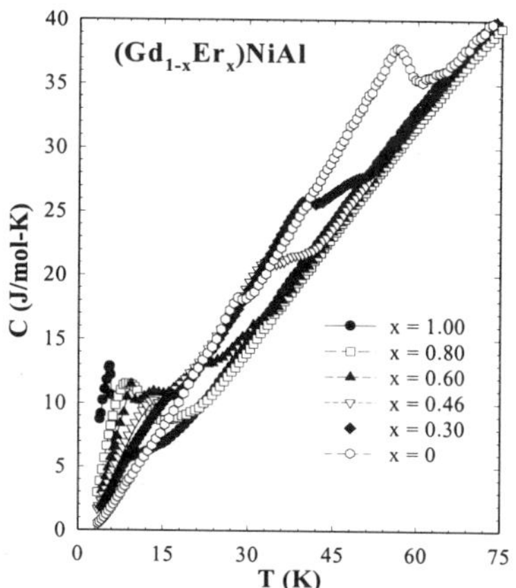


**Figure 2**. Total heat capacity of some $(Gd_{1-x}Er_x)NiAl$ alloys as a function of temperature in zero applied field.

upward. These results indicate the presence of an antiferromagnetic ground state, with a metamagnetic transition occurring at approximately 1 T as is also observed in the magnetization as a function of field below ~ 6 K. These findings are consistent with previous results.[8] For the mixed alloys, broad maxima are observed in the ac susceptibility that split into two peaks in an applied dc field above ~ 0.1 T. As with ErNiAl, the low-temperature maximum shifts to lower temperatures, with the high-temperature peak shifting upward with increasing applied field, indicating the presence of both antiferromagnetic and ferromagnetic ordering processes. These results are supported by low-temperature dc susceptibility measurements and magnetization isotherms indicating metamagnetic transitions for x = 0.80 and 0.60, with ferromagnetic-like saturation occurring for the other alloys. The estimated ordering temperatures using the ac data, as well as the observed effective magnetic moments and paramagnetic Curie temperatures derived from the temperature dependence of the inverse susceptibility are found in Table I. Good agreement between the observed moments and theoretical free-ion values is found.

The zero-field heat capacity (Figure 2) supports the presence of multiple transitions in all alloys except ErNiAl (which possesses a single antiferromagnetic transition at 5.6 K in zero field). The assignment of these ordering temperatures is given in Table I. These values were determined from peaks in the temperature-dependent magnetic heat capacity found by subtracting from the total heat capacity the lattice and electronic contributions, estimated from the prorated heat capacities of LaNiAl and LuNiAl. The Debye temperatures for LaNiAl and LuNiAl are 228(1) K and 247(1.5) K, respectively, with electronic specific heat coefficients of 1.9(2) mJ/mol-$K^2$ and 6.4(2) mJ/mol-$K^2$. The values for LaNiAl were erroneously reported as $\theta_D$ = 260(14) K and $\gamma$ = 5.5(5) mJ.mol-$K^2$ in ref. 5. The highest temperature anomalies in the mixed alloys appear as only shoulders on the $C_{mag}$ vs. T curves or as changes in the slope of the steep rise in the ac susceptibility curves. Similar behavior has been observed in related RNiAl compounds,[9] where powder neutron diffraction results have attributed such features to magnetic ordering. From the shift in the heat capacity anomalies in increasing applied field, it is observed that the low-temperature transitions are antiferromagnetic in nature whereas the upper transitions are ferromagnetic (with magnetic entropy shifting to lower and higher temperatures, respectively).

**Table I**. Summary of magnetic properties for $(Gd_{1-x}Er_x)NiAl$ alloys

| Compound | $T_{ord}$ (K) (0 T) | | Effective moment ($\mu_B$) | | $\theta_p$ |
|---|---|---|---|---|---|
| | $\chi_{ac}$ | $C_{mag}$ | (observed) | (theoretical) | (K) |
| ErNiAl | 6 | 5.6 | 9.53(1) | 9.58 | 6.0(2) |
| $(Gd_{0.20}Er_{0.80})NiAl$ | 11 | 9 | 9.21(1) | 9.28 | 13.5(2) |
| | 21 | --- | --- | --- | --- |
| $(Gd_{0.40}Er_{0.60})NiAl$ | --- | 9 | 8.86(7) | 8.96 | 25.3(7) |
| | --- | 13 | --- | --- | --- |
| | 21* | 22 | --- | --- | --- |
| | --- | 32 | --- | --- | --- |
| $(Gd_{0.54}Er_{0.46})NiAl$ | --- | 13 | 8.63(7) | 8.73 | 36.4(7) |
| | 29* | 32 | --- | --- | --- |
| | --- | 37 | | | |
| $(Gd_{0.70}Er_{0.30})NiAl$ | --- | 16 | 8.39(2) | 8.47 | 44.0(2) |
| | 31* | 37 | --- | --- | --- |
| | --- | 47 | --- | --- | --- |
| GdNiAl | --- | 23 | 8.04(3) | 7.94 | 61.7(2) |
| | 29 | 28 | --- | --- | --- |
| | 59 | 58 | --- | --- | --- |

*Broad peaks centered around 21 K, 29 K, and 31 K with full widths at half maximum of 22, 30, and 40 K are observed for alloys with x = 0.60, 0.46, and 0.30, respectively (See Figure 1).

## Magnetocaloric Effect

The complex magnetic structure described above for the $(Gd_{1-x}Er_x)NiAl$ series of alloys leads to a magnetic entropy change that exhibits a skewed caret-like dependence on temperature for Gd-rich compounds (as shown in Figure 3 for $(Gd_{0.70}Er_{0.30})NiAl$, with a comparison between values derived from both heat capacity and magnetic data). A comparison of the temperature dependence of the magnetic entropy change for all $(Gd_{1-x}Er_x)NiAl$ alloys is shown in Figure 4. For all compounds, the values of $\Delta S_{mag}$ derived from magnetic measurements are in good agreement with those calculated from the heat capacity. For ErNiAl and $(Gd_{0.20}Er_{0.80})NiAl$, $\Delta S_{mag}$ peaks around the antiferromagnetic ordering temperature; this maximum gradually shifts to the Curie temperature (the upper ordering temperature) as the Gd content is increased. For the Gd-rich compounds there remains a substantial contribution from the lower antiferromagnetic process leading to the broad maxima observed. This creates a skewed caret-like

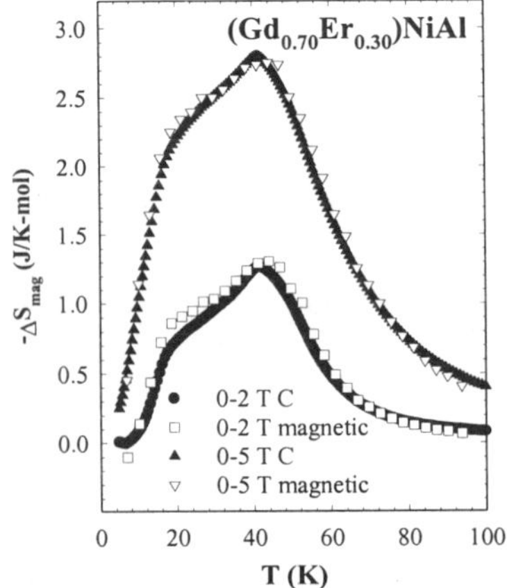


**Figure 3**. Comparison of values of $\Delta S_{mag}$ for $(Gd_{0.70}Er_{0.30})NiAl$ derived from heat capacity and magnetic measurements.

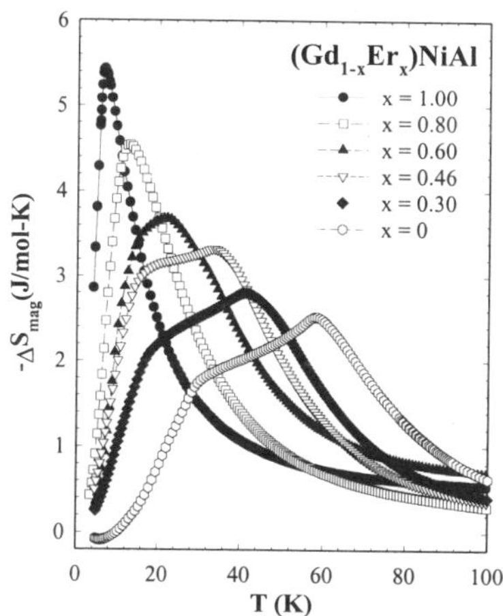


**Figure 4**. Magnetic entropy change for the $(Gd_{1-x}Er_x)NiAl$ alloys as a function of temperature for a field change from 0 T to 5 T (derived from heat capacity measurements).

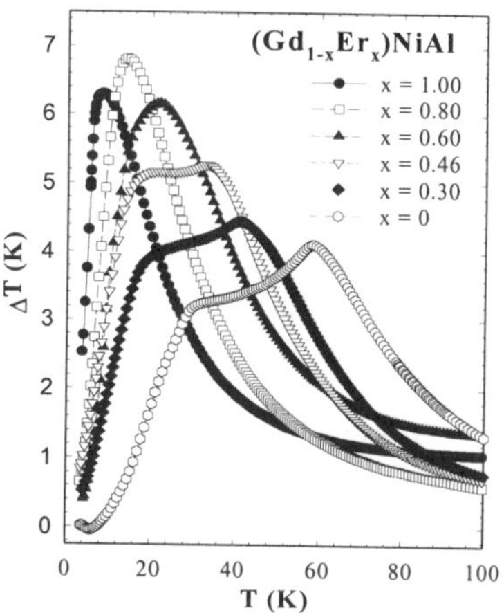


**Figure 5**. Adiabatic temperature change (MCE) for the $(Gd_{1-x}Er_x)NiAl$ alloys as a function of temperature for a field change from 0 T to 5 T.

temperature dependence of the magnetic entropy change, where the difference between the maximum values of $\Delta S_{mag}$ at the lower and higher peaks increases with increasing Gd content. The adiabatic temperature rise (MCE) has a similar temperature profile (Figure 5). For $(Gd_{0.54}Er_{0.46})NiAl$, the magnetic entropy changes at both the antiferromagnetic and ferromagnetic transition temperatures are of comparable magnitude. Between these two temperatures, the total entropy curves for different fields are close to linear and almost parallel (Figure 6, lower plot), resulting in a constant entropy change in this temperature region and therefore producing a nearly flat maximum in the $\Delta S_{mag}$ vs. T curve. For $(Gd_{0.70}Er_{0.30})NiAl$ and GdNiAl, nearly linear $S_{tot}$ curves are also observed between the transition temperatures, although the slopes of these curves differ for each applied field. This results in a steady increase in the magnetic entropy change from the lowest temperature maximum, ultimately peaking near or at the Curie temperature. Although this temperature profile does not satisfy the criterion for the Ericsson cycle (i. e. a constant $\Delta S_{mag}$ as a function of temperature), a linear dependence of the MCE (or $\Delta T_{ad}$) on temperature has been shown to be ideal for active magnetic regenerator refrigerants[9] and is therefore potentially useful for the AMR cycle as well.

## DISCUSSION

Several characteristics of the Gd-rich $(Gd_{1-x}Er_x)NiAl$ alloys lead to the skewed caret-like magnetic entropy change as a function of temperature. Broad transitions produce total entropy curves that increase slowly with increasing temperature, even in zero field (e.g. in $(Gd_{0.54}Er_{0.46})NiAl$, Figure 6, lower plot). For sharper transitions, the entropy rises quickly below the ordering temperature and more slowly above, resulting in breaks in the $S_{tot}$ vs. T curves in zero field as observed for $(Gd_{0.20}Er_{0.80})NiAl$ (Figure 6, upper plot). A second criterion for the skewed caret-like magnetic entropy change is the presence of multiple ordering processes in a relatively narrow temperature region. In the $(Gd_{1-x}Er_x)NiAl$ series, there exists up to three ordering processes spanning a temperature range (from the minimum to maximum transition temperatures) of approximately 10 K for

$(Gd_{0.20}Er_{0.80})NiAl$ and up to 35 K for GdNiAl. These processes contribute to a nonzero $\Delta S_{mag}$ over a wider temperature range than is typically seen in materials with only a single magnetic transition. Whether or not a flat maximum is observed depends in part upon the effect of the magnetic field on the shift in the entropy around these transition temperatures; this can be understood using plots of $C_{tot}/T$ vs. T at different fields, where the area under the curve is the total entropy (Figure 7). For the alloys studied, the low temperature transitions were found to be antiferromagnetic whereas the upper transitions are ferromagnetic. In an applied field, the entropy around the upper and lower transitions shifts in opposite directions, the result being that the heat capacity in a nonzero field remains below that in zero field until above the upper ordering temperature. For the Gd-rich alloys, this opposing shift in entropy creates regions between the two transition temperatures where the total entropy (the area under the curves in Figure 7) increases approximately linearly in both zero field (due to broad transitions) and nonzero applied field. The temperature range over which this occurs and the relative rate of change of entropy as a function of temperature in zero and nonzero applied fields may depend in part upon the contribution to the heat capacity of each ordering process as well as the temperature difference between transitions. These factors most likely lead to the broadening of the MCE curve with increasing substitution of Er by Gd.

For comparison, the magnetic entropy changes for $(Gd_{0.54}Er_{0.46})NiAl$, DyNiAl, $(Dy_{0.25}Er_{0.75})Al_2$ and $(Dy_{0.40}Er_{0.60})Al_2$ as a function of temperature are shown in Figure 8. These compounds all possess at least two features due to magnetic ordering or Schottky processes. Ferromagnetic ordering in DyNiAl leads to a sharper, more prominent peak in the ac susceptibility[9] and heat capacity[10] at approximately 28 K; this translates into the typical caret-like $\Delta S_{mag}$ vs. T curve that peaks around $T_c$. Although antiferromagnetic ordering does occur at 15 K (as indicated in ac susceptibility and powder neutron diffraction results[10]), the overall contribution from this process is substantially less than the high temperature feature. At most, this low temperature transition creates a shoulder on the MCE curve, and the table-like dependence found for $(Gd_{0.54}Er_{0.46})NiAl$ is not observed in DyNiAl. The $(Dy,Er)Al_2$ compounds also possess two features in the heat capacity and susceptibility,[11] where the low temperature feature around 10 K has been attributed to Schottky effects, with ferromagnetic ordering occurring at 24 K and 32 K for

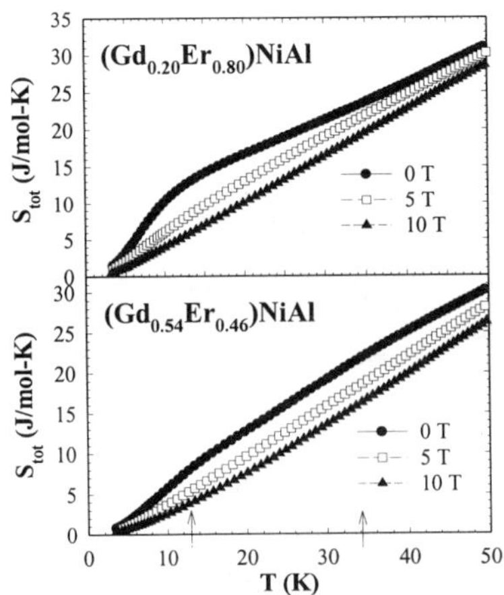


**Figure 6**. Total entropy as a function of temperature at several applied fields for $(Gd_{0.20}Er_{0.80})NiAl$ and $(Gd_{0.54}Er_{0.46})NiAl$.

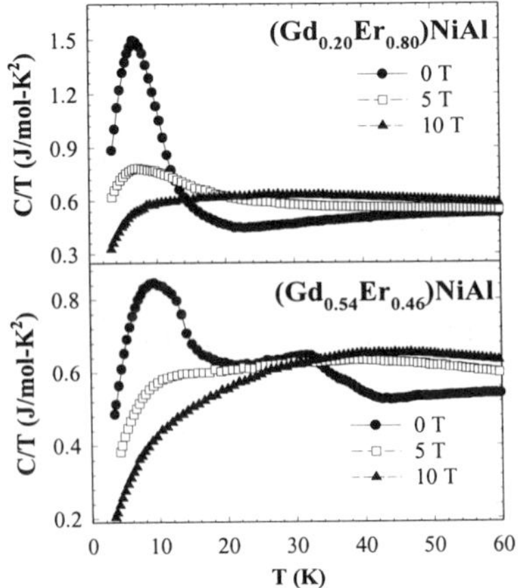


**Figure 7**. Temperature dependence of C/T at several applied fields for the alloys $(Gd_{0.20}Er_{0.80})NiAl$ and $(Gd_{0.54}Er_{0.46})NiAl$.

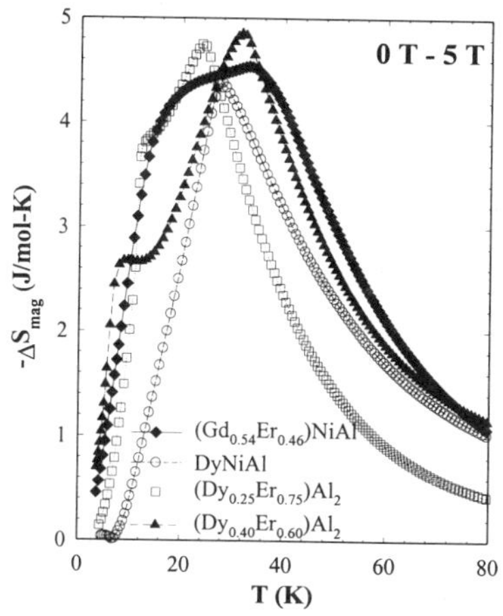


**Figure 8**. Magnetic entropy change (for a field change from 0 T to 5 T) as a function of temperature for several related Ni- and Al-based compounds.

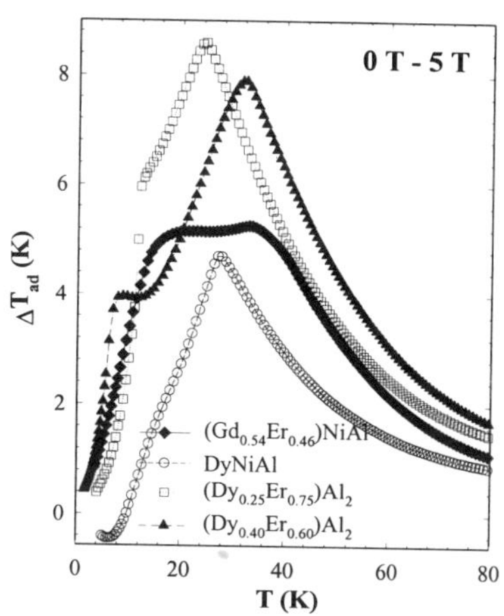


**Figure 9**. Magnetocaloric effect (for a field change from 0 T to 5 T) for several related Ni- and Al-based compounds.

$(Dy_{0.25}Er_{0.75})Al_2$ and $(Dy_{0.40}Er_{0.60})Al_2$, respectively. The magnetic entropy associated with both features shifts to higher temperature in applied fields, resulting in magnetic entropy changes that peak around the ordering temperatures. For $(Dy_{0.40}Er_{0.60})Al_2$, the width of the $\Delta S_{mag}$ peak is comparable to that observed for $(Gd_{0.54}Er_{0.46})NiAl$, although the entropy change associated with the Schottky peak is much smaller than at the Curie temperature, resulting in a caret-like appearance.

Even though the magnitude of the maximum magnetic entropy change is comparable for these four compounds, the adiabatic temperature rise (or MCE) is significantly larger in the $RAl_2$ mixed-binary compounds (Figure 9). This is the result of the larger lattice heat capacity for RNiAl compounds. The largest lattice heat capacity at the Curie temperature is found for DyNiAl (8.6 J/mol-K), while the lowest (3.5 J/mol-K) is found for $(Dy_{0.25}Er_{0.75})Al_2$; this is inversely correlated with the MCE, which increases from 4.7 K (at $T_c$) for DyNiAl up to 8.6 K for $(Dy_{0.25}Er_{0.75})Al_2$. Within the series of compounds studied, only $(Gd_{0.54}Er_{0.46})NiAl$ possesses a $\Delta S_{mag}$ temperature profile suitable for use in an Ericsson cycle refrigerator. Although the maximum adiabatic temperature rise is only about 1K/T, it is still comparable to that observed for several of the composite materials (e.g. composites consisting of layered $RAl_2$ compounds).[2-4]

## CONCLUSIONS

The table-like magnetic entropy change observed for $(Gd_{0.54}Er_{0.46})NiAl$ is unique among related RNiAl, $RAl_2$, and $RNi_2$ compounds. Several factors seem to contribute to this temperature dependence. In all compounds with multiple transitions, broader MCE peaks are observed, where the width of each peak is correlated with the relative contribution from each process and the temperature span between the features. The flattest maxima observed in the $\Delta S_{mag}$ vs T curves for the $(Gd_{1-x}Er_x)NiAl$ alloys result from transitions which span up to 35 K with comparable magnetic entropy changes at each feature. This is the result of the field dependence of low-temperature antiferromagnetic processes coupled with high-temperature ferromagnetic ordering. The opposing shifts in entropy in applied fields allow for slowly increasing total entropy between the transition temperatures; when coupled with broad zero-field transitions, this creates nearly linear total entropy curves in both zero and nonzero applied fields. For $(Gd_{0.54}Er_{0.46})NiAl$, the total entropy curves are nearly parallel, thereby leading to a table-like magnetic entropy change. The broadness of the MCE peaks observed in the Gd-rich $(Gd_{1-x}Er_x)NiAl$ alloys make these materials useful as refrigerants when specific temperature profiles of the magnetocaloric effect (such as constant $\Delta S_{mag}$ or linear $\Delta T_{ad}$) are needed over the full span of the magnetic refrigeration device. Unfortunately, this broadening of the MCE peak in general translates into smaller maximum peak values (of < 1 K/T), limiting the usefulness of these materials as magnetic refrigerants operating within small field changes. More related systems must be studied before all the factors leading to this unique behavior are understood.

## ACKNOWLEDGMENTS

The Ames Laboratory is operated for the U. S. Department of Energy by Iowa State University under contract no. W-7405-ENG-82. This work was supported by the Office of Basic Energy Sciences.

## REFERENCES

1. C. R. Cross, J. A. Barclay, A. J. DeGregoria, S. R. Jaeger and J. W. Johnson, "Optimal temperature-entropy curves for magnetic refrigeration", *Adv.Cryog. Eng.* 33:767 (1988).
2. T. Hashimoto, T. Kuzuhura, M. Sahashi, K. Inomata, A. Tomokiyo and H. Yayama, "New application of complex magnetic materials to the magnetic refrigerant in an Ericsson magnetic refrigerator", *J. Appl. Phys.* 62:3873 (1987).
3. A. Smaïli and R. Chahine, "Composite magnetic refrigerants for an Ericsson cycle: New method of selection using a numerical approach", *Adv Cryog Eng.* 42:445 (1996).
4. A. Smaïli and R. Chahine, "Composite materials for Ericsson-like magnetic refrigeration cycle", *J. Appl. Phys* 81:824 (1997).
5. H. Takeya, V. K. Pecharsky, K. A. Gschneidner, Jr. and J. O. Moorman, "New type of magnetocaloric effect: Implications on low-temperature magnetic refrigeration using an Ericsson cycle", *Appl. Phys. Lett.* 64:2739 (1994).
6. H. Oesterreicher, "Structural and magnetic studies on rare-earth compounds RNiAl and RCuAl", *J. Less-Common Metals* 30:225 (1973).
7. V. K. Pecharsky, K. A. Gschneidner, Jr. and J. O. Moorman, "A 3 to 350 K automatic small sample calorimeter", *Rev. Sci Instr.*,to be published Nov 1997 issue.
8. P. Javorský, P. Burlet, E. Ressouche, V. Sechovský, H, Michor and G. Lapertot, "Magnetic structure study of ErCuAl and ErNiAl", *Physica B* 225:230 (1996).
9. C. E. Reid, J. A. Barclay, J. L. Hall and S. Sarangi, "Selection of magnetic materials for an active magnetic regenerative refrigerator", *J. Alloys Compds* 207/208:366 (1994).
10. G. Ehlers and H. Maletta, "Spin structures with frustrated moments in RNiAl intermetallic compounds", *Z. Phys. B* 101:317 (1996).
11. K. A Gschneidner, Jr, V. K. Pecharsky, and S. K. Malik, "The $(Dy_{1-x}Er_x)Al_2$ alloys as active magnetic regenerators for magnetic refrigeration", *Adv. Cryog. Eng.* 42:475 (1996).

# EFFECT OF HEAT REJECTION CONDITIONS ON CRYOCOOLER OPERATIONAL STABILITY

**R.G. Ross, Jr. and D.L. Johnson**

Jet Propulsion Laboratory
California Institute of Technology
Pasadena, CA 91109

## ABSTRACT

It is well known that cryocooler thermal efficiency is a strong function of heat rejection temperature, roughly following the dependency described by Carnot. An equally important and generally overlooked implication of cryocooler heat-rejection thermodynamics is the effect of the heat rejection temperature control mode on cryocooler performance and operational stability. Example heat rejection temperature control modes include constant reject temperature (generally maintained via closed-loop temperature control), heat rejection temperature rising linearly with power dissipation (typical of conduction/convection to a constant temperature heat sink), and heat rejection dependent on the fourth power of reject temperature (typical of radiation to deep space). This paper presents a useful algorithm for computing the effect of changing heatsink temperature on cryocooler performance and uses the algorithm to examine the implications of various heat rejection temperature control modes on cryocooler operation. A useful system-level thermal performance map is developed to display the stability boundaries and available stable operating space for coolers of interest for various typical heat rejection control modes.

## INTRODUCTION

Increasing numbers of space instruments are using mechanical cryocoolers to enable improved scientific measurements to be made via cryogenically cooled infrared, submillimeter and gamma-ray detectors. One example is JPL's Atmospheric Infrared Sounder (AIRS) instrument, which uses a 55K cryocooler with closed-loop temperature control to maintain the HgCdTe focal plane at 58 K throughout the mission lifetime. The complete cryocooler system includes heat rejection to spacecraft-mounted radiators via a system of conductive elements and heatpipes.[1] Of particular interest is the overall thermal performance and closed-loop stability of the complete cryocooler system including the heat rejection subsystem.

To achieve the necessary understanding of overall thermal performance and closed-loop stability it is first necessary to accurately quantify the dependency of cryocooler performance on heatsink temperature. This has been done for a wide variety of coolers by characterizing their operational performance over an extensive operating range using the techniques developed at JPL for general cryocooler characterization.[2] Next, a special parametric performance plot has been developed to display the measured performance including temperature dependency, and used to note the stability boundaries and available stable operating space for the cooler with a given heat-rejection temperature control mode.

For this paper, the analysis originally done for AIRS has been generalized to provide a universal approach to examining cryocooler heatsink temperature sensitivity and operational stability boundaries. To demonstrate the techniques, data are presented for various typical heat rejection temperature control modes including constant-temperature, conduction-dominated, and radiation-dominated operation.

## PERFORMANCE SENSITIVITY TO HEATSINK TEMPERATURE

Following the dependency described by Carnot, i.e.

$$COP_{Carnot} = T_c/(T_h - T_c) \tag{1}$$

where $T_c$ is the cryogenic temperature and $T_h$ is the heatsink temperature, it is well known that cryocooler thermal efficiency is a strong function of heat rejection temperature. To analyze the thermal performance and stability implications of various cryocooler heat-rejection systems, it is necessary to first accurately quantify the actual temperature dependency of non-ideal operational cooler performance and to reduce the dependency to equation form, if possible. One means of accomplishing this is to examine overall cryocooler thermal performance dependency with respect to the principal operational variables exclusive of heatsink temperature, and then to note the change specifically associated with change in heatsink temperature. For the purposes of this paper, the TRW 3503 pulse tube cooler and the STC 80K Stirling cooler are used as examples.[3,4]

Figure 1 describes the general parameter dependency of Stirling and pulse tube cooler performance on input power, refrigeration load, refrigeration temperature, and compressor piston stroke — all for a heatsink temperature of 20°C; these particular data are for the TRW 3503 pulse tube cooler. Next, Fig. 2 displays the shift in the cooler performance caused by

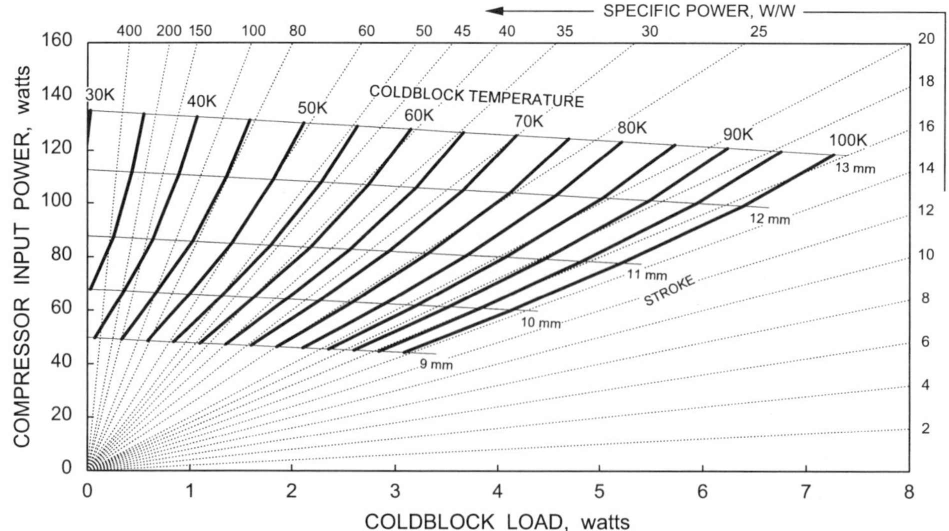


**Figure 1**. Measured thermal performance of the TRW 3503 pulse tube cooler with a 20°C heat rejection temperature .

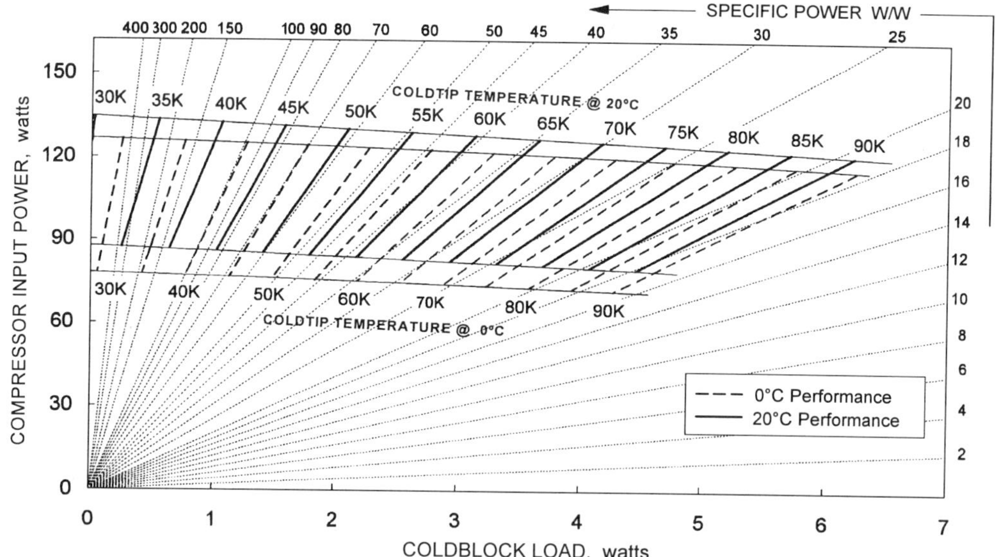


**Figure 2.** Measured sensitivity of thermal performance of the TRW 3503 pulse tube cooler to a change in heat rejection temperature from 20°C to 0°C.

lowering the heatsink temperature to 0°C. Note that the new isotherms (lines of constant refrigeration temperature) for a 0°C heatsink temperature are positioned on top of similar 20°C-isotherms corresponding to refrigeration temperatures approximately 3 K warmer; i.e. the performance at 55 K with a 0°C heatsink temperature is the same as the performance at 58 K with a 20°C heatsink temperature. Note that this shift is quite constant over the complete range of refrigeration temperatures plotted, from no-load on up. This fixed change in coldtip temperature per change in heatsink temperature has been found to be approximately true for a wide variety of cryocoolers, not just the TRW 3503 pulse tube.[2,3,5] However, the proportionality constant ($\Re$) between heatsink temperature and coldend temperature varies modestly from cooler to cooler. As noted in Table 1, values of $\Re$ measured at JPL range from 2 to 10°C/K. As an example of a higher sensitivity to heat sink temperature, Fig. 3 shows the 3.5°C/K performance of the STC Stirling cooler.

**Table 1.** Change in heatsink temperature required to cause a 1 K shift in coldend temperature for typical Stirling and pulse tube cryocoolers

| Cryocooler Model | $\frac{\text{Heatsink Temperature Change}}{\text{Coldend Temperature Shift}}$ $\Re$ (°C/K) |
|---|---|
| TRW 1W-35K Pulse Tube | 6 |
| TRW 3503 Pulse Tube | 7 |
| TRW 6020 Pulse Tube | 5 |
| STC 80K Stirling | 3.5 |
| Sunpower 140K Stirling | 2 |
| MMS 80K Stirling | 4 |
| MMS 50-80K Stirling | 4 -10 |

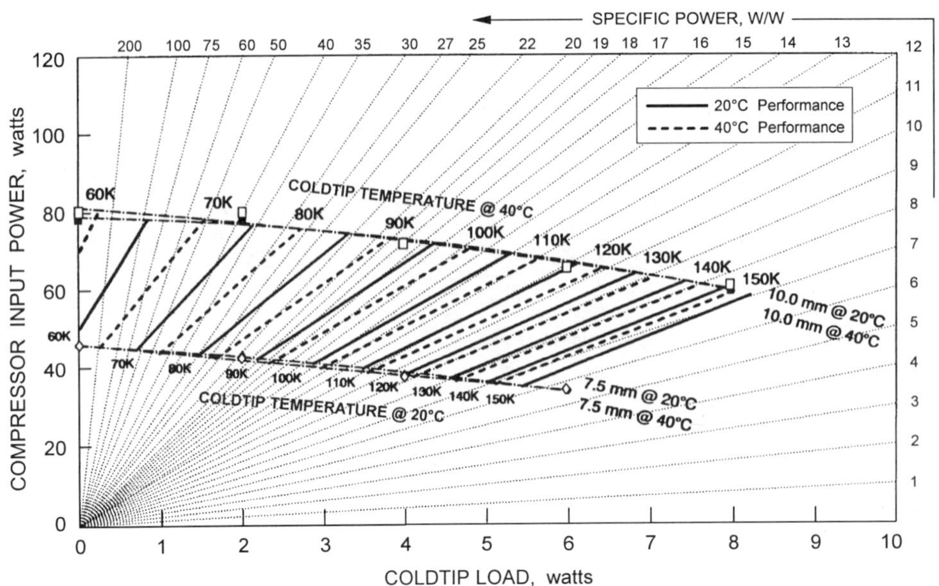


**Figure 3**. Measured sensitivity of thermal performance of the 80K STC Stirling cooler to change in heat rejection temperature from 20°C to 40°C.

Based on this empirically derived finding, one can construct the following algorithm for computing the refrigeration performance P at coldend temperature K and heatsink temperature T2 in terms of the performance of the same refrigerator at a baseline heatsink temperature T1, i.e.

$$P_{T2}(K) = P_{T1}\left(K - \frac{(T2 - T1)}{\Re}\right) \tag{2}$$

where $\Re$ is the measured change in heatsink temperature required to shift the coldend performance by 1 K. Equation (2) allows a performance plot such as Fig. 1 to be used to determine refrigeration performance for a range of heatsink temperatures (T2) different from the measuring conditions (T1) for which the plot was constructed.

As an example, consider the change in input power and compressor stroke required to maintain a 1.5 watt cryogenic load at 55 K with the TRW 3503 cooler when the heatsink temperature changes from 20°C to 40°C. Using Eq. (2) with $\Re = 7°C/K$, yields

$$P_{40}(55) = P_{20}\left(55 - \frac{(40 - 20)}{7}\right) = P_{20}(52.2) \tag{3}$$

As displayed in Fig. 4, which is a replot of Fig. 1, one finds that the power increases from 72 watts to approximately 82 watts, and the stroke increases from approximately 10.3 mm to 10.8 mm.

As a second example, Fig. 4 also notes the change in input power and compressor stroke required to maintain the 1.5 watt cryogenic load at 55 K when the heatsink temperature increases to 60°C. The corresponding calculation using Eq. (2) is

$$P_{60}(55) = P_{20}\left(55 - \frac{(60 - 20)}{7}\right) = P_{20}(49.3) \tag{4}$$

Thus, performance at 55 K with a 60°C reject corresponds to performance to 49.3 K with a 20°C reject.

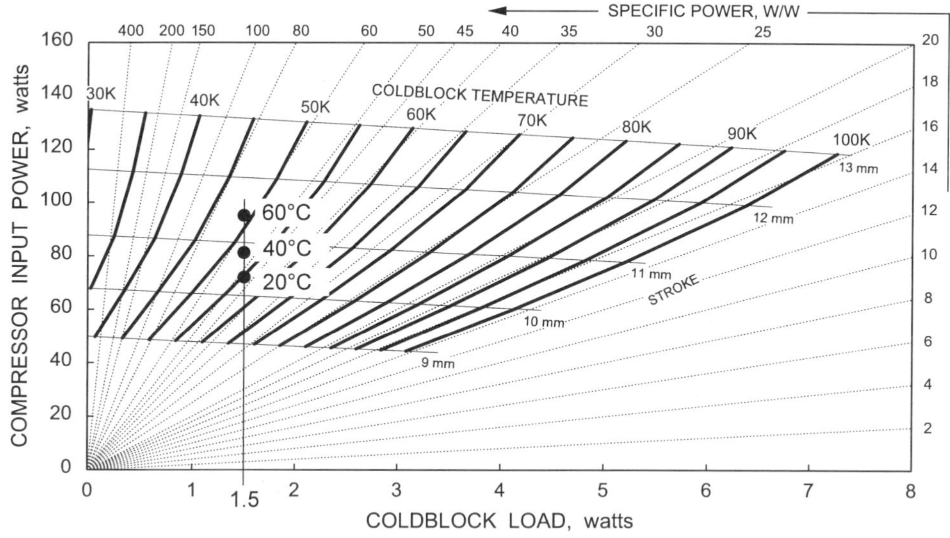


**Figure 4**. Example computation of the effect of changing the heatsink temperature from 20°C to 40°C and 60°C on the input power required to maintain a load of 1.5 watts at 55 K.

This strong coupling between heatsink temperature and cryocooler performance, as noted in Fig. 4, draws attention to the important role of the cryocooler's heat rejection system and its temperature control mode.

## IMPLICATIONS OF CRYOCOOLER TEMPERATURE-CONTROL MODE

Example heat rejection temperature control modes include constant reject temperature (generally maintained via closed-loop temperature control), heat rejection temperature rising linearly with power dissipation (typical of conduction/convection to a constant-temperature heatsink), and heat rejection temperature dependent on the fourth root of power dissipation (typical of radiation to deep space). Any mode other than constant heatsink temperature can significantly alter the performance attributes of the cooler as mapped in Fig. 1, and can raise the possibility of thermal runaway, whereby increased input power required at elevated heatsink temperatures further increases the heatsink temperature in an unstable spiral until maximum cryocooler stroke is exceeded.

### System-level Cooler Performance Plot

To understand the system-level thermal implications of the cryocooler's operation it is necessary to analyze not just the cryocooler by itself, but the complete cryocooler system including its heat rejection system. An effective means of accomplishing this is to note that the vertical axis of the parametric performance plot shown in Fig. 1 is power dissipation, and power dissipation, for a given heat-rejection temperature control system, directly determines the heatsink temperature. Thus, for a given heat-rejection temperature control system, one can modify the parametric performance plot to include the corresponding heatsink temperature for each compressor input power on the vertical axis. Going further, Eq. (2) can be used to re-map the cryocooler performance data for the specific heatsink temperature associated with each vertical position on the plot. The result is a system-level parametric plot of cooler performance for the given heat rejection system, as opposed to for a fixed heatsink temperature, as was done in Fig. 1.

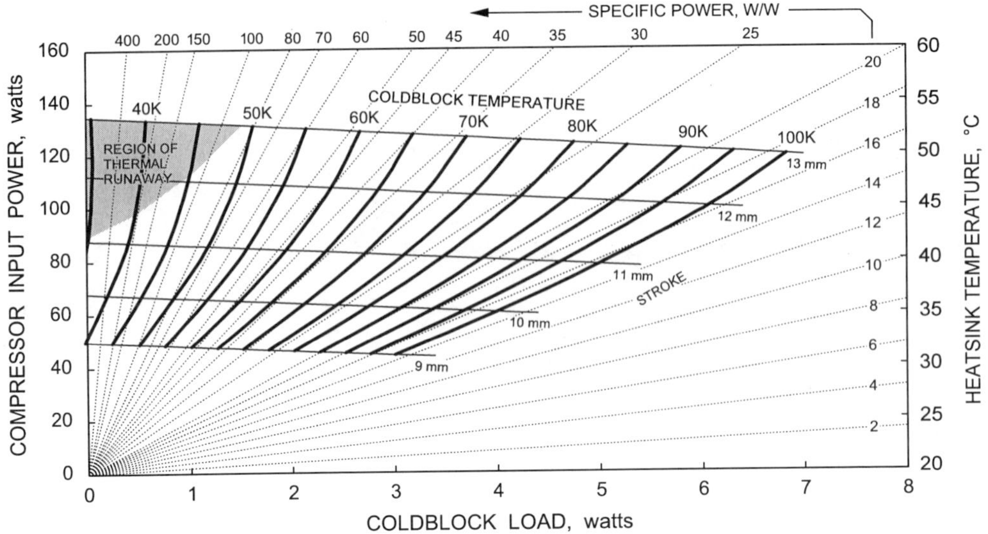


**Figure 5**. System-level performance of the TRW 3503 pulse tube cooler with a heatsink temperature defined by T = 20°C + P/4, where P is the compressor input power in watts.

**Conductive Heat Transfer**. For a condition where the cooler is coupled to a constant-temperature sink through a constant-resistance thermal link, the cryocooler temperature will increase linearly above the sink temperature with increased power dissipation (P). An example would be the operation of a cooler in a constant-temperature room-ambient environment with conductive and convective heat transfer to the room.

To understand the implications of such an environment on the TRW 3503 cooler, Fig. 5 is a system-level plot based on Fig. 1, with the heatsink temperature defined by:

$$T2 = 20°C + P/4 \tag{5}$$

where P is the cooler input power in watts. For this heat-rejection system the cooler heatsink temperature rises linearly from 20°C to 45°C when the compressor input power increases from 0 to 100 watts. This is representative of a fairly typical bench-top test condition.

Note that the increasing reject temperature with increased input power results in the isotherm lines bending toward the upper left as compared to those for the constant 20°C heatsink temperature shown in Fig. 1. Where the isotherm lines approach a vertical slope defines the onset of thermal runaway. At this point, an attempt to serve an increased cooling load results instead in increased heating and no improvement in refrigeration capacity.

To examine the effect of different $\Re$ values, Fig. 6 is a re-plot of Fig. 5 using Eq. (2) and $\Re$ = 3.5°C/K (corresponding to the temperature sensitivity of the STC 80K Stirling cooler) instead of $\Re$ = 7°C/K, as was measured for the TRW 3503 cooler. Note the increased bending over of the performance at higher power levels, and the larger thermal-runaway keepout region associated with this performance degradation at higher temperatures.

**Radiation Heat Transfer**. For a condition where the cooler is directly coupled to deep space using a high-conductivity radiator, the cryocooler absolute temperature will increase roughly proportional to the fourth root of the power dissipation, i.e. $P \propto \Theta^4$, where $\Theta$ is the absolute heatsink temperature in Kelvin.

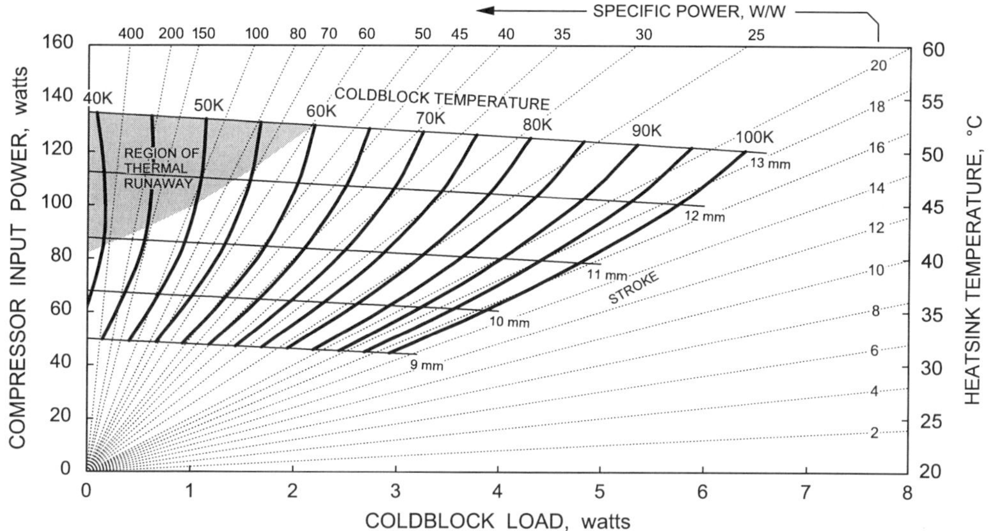


**Figure 6.** System-level performance of the TRW 3503 pulse tube cooler if it were to have a temperature sensitivity of $\Re = 3.5$ (instead of its measured $\Re = 7$) and a heatsink temperature defined by $T = 20°C + P/4$, where P is input power in watts.

As an example, Fig. 7 presents the system-level performance plot for the TRW 3503 cooler for a system where the heatsink temperature is defined by:

$$T2(°C) = 318\ (P/100)^{0.25} - 273 \qquad (6)$$

i.e., the cooler absolute heat-rejection temperature rises proportional to $P^{0.25}$, reaching 318 K (45°C) when the compressor input power reaches 100 watts. This is representative of many spacecraft applications.

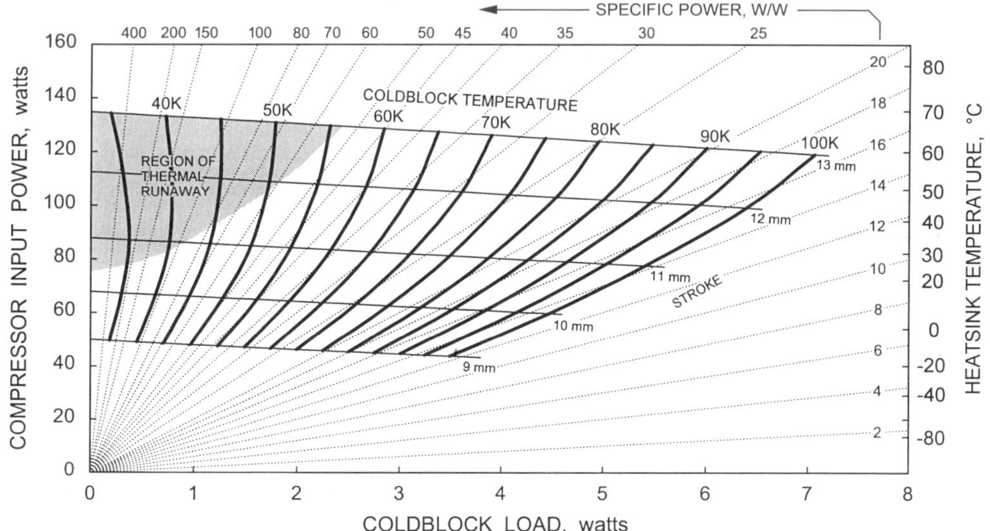


**Figure 7.** System-level performance of the TRW 3503 pulse tube cooler with a radiatively coupled heatsink temperature defined by $T(°C) = 318\ (P/100)^{0.25} - 273$, where P is the compressor input power in watts.

Note that the sharply increasing reject temperature with increasing input power results in the isotherm lines bending extensively toward the upper left as compared to those for the constant 20°C heatsink temperature shown in Fig. 1. Figure 7 delineates the resulting large thermal-runaway keepout region associated with these radiation heat-rejection conditions as defined by Eq. (6).

## SUMMARY

It is well known that cryocooler thermal efficiency is a strong function of heat rejection temperature, roughly following the dependency described by Carnot. Measurements made on a variety of coolers indicate that this performance temperature sensitivity can be well modeled by an equation of the sort: $P_{T2}(K) = P_{T1}(K - [T2-T1]/\Re)$, i.e. the performance at cryogenic temperature K with and a heatsink temperature T2 is the same as that at heatsink temperature T1 and cryogenic temperature K-(T2-T1)/$\Re$, where $\Re$ is a constant with a value in the range 2 to 7°C/K. This relationship has been used to develop a system-level cooler performance plot that displays the effect of the heat rejection temperature control mode on overall cryocooler performance. The plot also allows easy visualization of the stability boundaries and available stable operating space for the selected cooler and heat rejection control mode.

## ACKNOWLEDGMENT

The work described in this paper was carried out by the Jet Propulsion Laboratory, California Institute of Technology and was sponsored by the NASA EOS AIRS Project through an agreement with the National Aeronautics and Space Administration.

## REFERENCES

1. Ross, R.G., Jr. and Green, K.E., "AIRS Cryocooler System Design and Development," *Cryocoolers 9*, Plenum Publishing Corp., New York, 1997, pp. 885-894.

2. Smedley, G.T., Mon, G.R., Johnson, D. L. and Ross, R.G., Jr., "Thermal Performance of Stirling-Cycle Cryocoolers: A Comparison of JPL-Tested Coolers," *Cryocoolers 8*, Plenum Publishing Corp., New York, 1995, pp. 185-195.

3. Johnson, D.L., Collins, S.A., Heun, M.K. and Ross, R.G., Jr., "Performance Characterization of the TRW 3503 and 6020 Pulse Tube Coolers," *Cryocoolers 9*, Plenum Publishing Corp., New York, 1997, pp. 183-193.

4. Burt, W.W. and Chan, C.K., "New Mid-Size High Efficiency Pulse Tube Coolers," *Cryocoolers 9*, Plenum Publishing Corp., New York, 1997, pp. 173-182.

5. Collins, S.A., Johnson, D.L., Smedley, G.T. and Ross, R.G., Jr., "Performance Characterization of the TRW 35K Pulse Tube Cooler," *Advances in Cryogenic Engineering*, Vol. 41, Plenum Publishing Corp., New York, 1996, pp. 1471-1478.

# TEST AND ANALYSIS OF FIVE TYPES OF REGENERATIVE CRYOCOOLERS

Shimo Li, Z. Huang, J. Yang, S. Ge, and H. Pan

Cryo-Lab. Zhejiang University
Hangzhou 310027 China

## ABSTRACT

In this paper, the experimental results of five types of regenerative Cryocoolers: Stirling (St), Vuilleumier (V-M), Gifford-McMahon (G-M), Solvay (SV), and Pulse Tube (PTR) obtained by our research group within the last 10 years have been introduced. The thermodynamics, working conditions, and the choice of regenerative matrix have been comprehensively analyzed and discussed.

## EXPERIMENTAL RESULTS

Since 1980, our research group has studied experimentally of five types of common regenerative cryocoolers, including single-stage, two-stage, four-stage plastic St, $LN_2$ precooling V-M, two-stage SV, two-stage G-M, and single-stage PTR. The regenerative materials which have been used are copper wire velvet, stainless wire net, lead spheres, magnetic materials of $Er_3Ni$, $Er(Ni_{0.79}Co_{0.21})_2$, rare earth metal hydrides (non hydriding), active carbon and molecular sieve, etc. The main performance of these cryocoolers are described in Table 1.

Since the refrigeration capacity of our two-stage G-M is very small, so the first-stage temperature was over 40K, and the second-stage (with no load) temperature could not be reduced to 4K, it just reached 5.6K.

The S-V used is CSA-202, only for the student's experimental measurements.

The tests of PTR is composed of St compression system, G-M compression system and its gas distribution system, so we called it "St type PTR (no valve) and G-M type PTR (with valve)". The test conditions of G-M type PTR are in the following: ① There are three kinds of regenerator construction with the same regenerator matrix-rare earth metal hydrides (non hydriding), ② These three kinds of refrigerator have the same size of pulse tube. Each construction has three working states. They are basic (BPTR), orifice (OPTR) and double-inlet (DPTR). The test results show that the optimal condition is DPTR with $V_r/V_p=2.0$ and $P_h/P_l=3.0$, see Table 2.

The test conditions of St type PTR are in the following: ① Regenerator and pulse tube are coaxial arranged. ② The regenerator and pulse tube are arranged to be U-shape and the temperature defference is 70K.

**Table 1.** The main parameters of cryocoolers

| Type | Regenerator | Qr W/K | Ns | Speed | Tmin |
|---|---|---|---|---|---|
| 1−stage St 3Ly0. 8/194 | copper wire velvet | 930W/77. 4K | 12W/W 77. 4K | 24HZ | 33K |
| 2-stage St A3040 | Stainless steel wire net 250～300mesh | 2W/37K | 275W/W 37K | 24HZ | 30K |
| 4-Stage St plastic expander | gap regenerator | / | / | 1HZ | 8K |
| V-M $LN_2$ precooler | Pb+ $Er(Ni_{0.79}Co_{0.21})_2$ | 1. 25W/20K | / | 2HZ | 11. 7K |
| S-V CSA-202 | Stainless steel wire net. Pb spheres | 1. 75W/15K | 970W/W 15K | 2HZ | 9K |
| 2-Stage G-M | Pb+$Er_3Ni$ Pb+MH | 3W/15K 2W/15W | 1000W/W 15K 1500W/W 15K | 1HZ | 5. 6K 8. 4K |
| 1-Stage PTR G-M type | MH non hydriding | / | / | 1HZ | 91K* |
| 1-Stage PTR St type | Stainless steel wire net. Pb | / | / | 24HZ | △T=70K** |

*: double-inlet PTR  **: basic PTR

According to the study by our experiments[1-6] and the references[7], in this paper, some discussions are given in the following section.

## DISCUSSION

### 1. Themodynamics of the Processes in Regenerative Cryocoolers

It is known that, there are two main kinds methods achieving low temperature with a working gas: Joule-Thomson expansion and isentropy expansion. The basic thermodynamic cycles of cryogenic systems are gas compression and expansion. All the common regenerative cryocoolers are designed with suitable structures to let the working gas be compressed and expansed at a certain phase angle to achieve low temperature.

The J-T inversion temperature of helium is about 40K. However, at any temperature, the irreversible loss of the expansion of He in the expandor is less than that of J-T

**Table 2.** Test results of G-M type PTR

| Regenerator | Pulse tube | $V_r/V_p$ | Tmin K | | |
|---|---|---|---|---|---|
| | | | BPTR | OPTR | DPTR |
| Ø17mm×L 170mm | Ø10mm×L 380mm | 1. 16 | 199 | 182 | 121 |
| Ø19mm×L 170mm +Ø22mm×L 30mm | Ø10mm×L 380mm | 2. 0 | 167 | 147 | 91 |
| Ø19mm×L 170mm +Ø22mm×L 80mm | Ø10mm×L 380mm | 2. 63 | 181 | 159 | 105 |

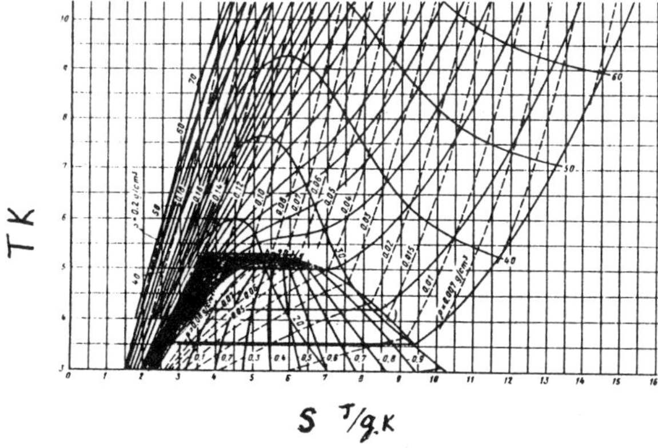

Fig. 1 LHe Temperature level cryocoolers on T-S diagram.

expansion. Therefore, on H-S diagram under normal condition with a suitable phase angle, the value of expansion efficiency of these cryocoolers is always between the isentropic curve and the isenthalpic curve, but the difference in manufacturing techniques and operating conditions will cause a different efficiency.

## 2. Experimental Study and Theoretical Design

Among these five cryocoolers the design of St and G-M are mature, there are some references, but the design of the whole machine is still being improved rapidly, because the performance are influenced by many parameters, such as structure, operation, hydrodynamics and thermodynamics. However, it is difficult to simulate. The promising design method comes from practice only. The machine will be improved along with improvement in techniques. The optimal pilot plant has to be improved by experiments, that is why there are hundreds of test PTR with large differences. It is difficult to find out the optimal design because the parameters, for example-the range of regenerator volume / pulse tube volume (=0.5~5), the variance of the regenerator length / diameter, pulse tube length / diameter, reservior volume etc., have large differences in different devices. The theoretical study is complicated, including surface heat pumping, enthalpy flow analysis, isothermal model, adiabatic model and thermoacustics mechnism etc. Besides, in several different experimental results the difference of the specific power consumption may be very large, even as large as an order of magnitude.

Therefore, the real criterion is: test practice—theoretical design—test practice, including excellent simulation and CAD design, to improve the performance of the cryocooler.

## 3. 4K Cryocooler

The critical point of helium is 5.2K and 0.227MPa. Its temperature of λ point is 2.17K, and the corresponding pressure is 0.005MPa. Therefore, the cryocooler with temperature range 5.20K ~ 2.17K, is called "LHe temperature level cryocooler". G. Thummes and his coworkers[8] have reported that refrigeration temperature is 2.1K with no load, then the corresponding pressure should be higher than 1.47MPa or its corresponding vapor—liquid equilibrium pressure is far less than 0.1MPa, that is not the normal working pressure. From the reports presented: the working pressure of the LHe temperature range is over supercritical pressure, and vapor and liquid coexist in the cool chamber, there is no phase interface, i. e. supercritical state, Fig. 1. The properties changed much near critical point, so the expansion frequency should not be high, otherwise there would occur slugging. If expansion pressure is lower than critical pressure, the expansion chamber would be as the wet expandor of the helium liquifier.

In order to obtain a practical 4K G-M with enough cooling capacity, its working pressure should be higher than the critical pressure, but in the cold chamber of G-M or

PTR, helium would be at a supercritical unstable state, therefore, the irreversible loss of the expansion chamber would increase. The specific power consumption at 4K will exceed 2500 W/W.

Besides, if the expansion process approaches the Simon isentropic expansion, the reduction of temperature below 4K is very small, Fig. 2. But the thermodynamic cycle of G-M/J-T complex cryocooler is reasonable, because near critical point, the isothalpy curve shows that the efficiency of reduction of temperature reaches around 80%. Furthermore, the cold end of J-T expansion is an independent wet, cold chamber which is working at a thermodynamics steady state, so the irreversible loss is relatively small, the thermodynamic efficiency is higher, and the specific power consumption would be reduced to 1800~1500 W/W below 4K.

## 4. Regenerator Matrix Selection

In recent years, the development and application of rare earth magnetic matrices is a great breakthrough in regenerative cryocooler techniques, which makes both G-M and PTR to work in LHe temperature level directly. Manufacturers try to produce 4K cryocoolers without a J-T cycle. In fact, the percentage of the novel magnetic matrix in second stage or third stage regenerators is not large. The range of the specific heat peak of the matrix is quite narrow. And it is known that when the temperature becomes higher than 15K, the specific heat (volume) of the novel matrix will be lower than that of lead. Therefore, using complex regenerator structure with the range of 4 ~ 20K would be promising.

The test shows that over 10K, using cheap metal hydrides (non hydriding) would be suitable.

In general, there are some alloys of copper which are located on the cylinder cold end, but at 4K, the specific heat of copper is too low, so we suggest the use of a sintered magnetic matrix to replace copper on the cylinder cold end, Fig. 3. In this case the magnetic matrix on the regenerator cold end can be neglected.

## 5. Choice of the Type of Regenerative Cryocoolers

So far, there are five regenerative cryocoolers, however, Boreas refrigerator is the newest development with gap regenerative and recuperative exchanger. Along with the technique improvements, the applications of refrigerators are changing. In the 80's, peo-

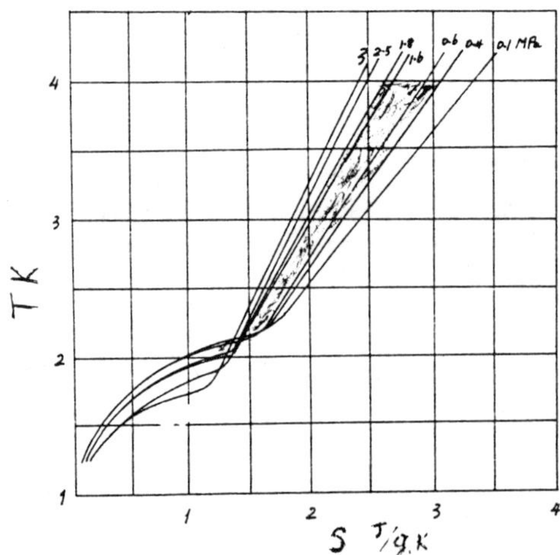

Fig. 2 The temperature drop of isentropic expansion is very small at T≤4K.

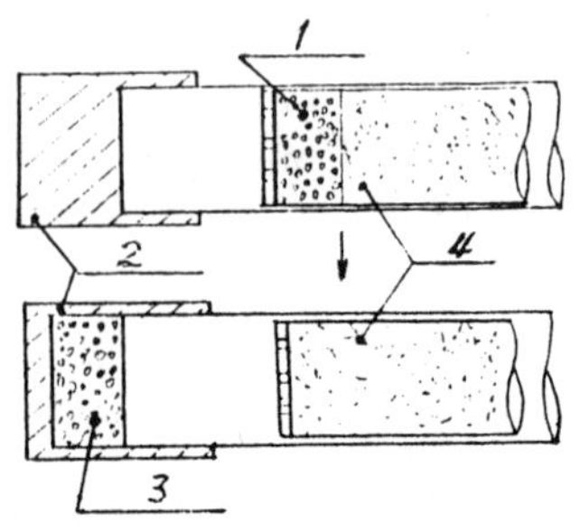

1. powder magenetic matrix
2. copper
3. sintered magenetic matrix
4. regenerator & displacer

Fig. 3 A sintered magnetic matrix to replace copper on the cold end.

ple generally recognized that PTR is difficult in practice, because of its low thermodynamic efficiency, but nowadays it becomes competitive with St. Also, plastic non-magnetic cryocooler are generally recognized to be very good, but its use has become minimal, because of its inherent disadvantage and impracticality.

Using St with magnetic bearing to replace normal bearing has been recognized to be a great improvement, but nowadays using flexure bearing or gas bearing to replace normal bearing gives St longer life. Besides, other improvements such as using linear motor to replace rotary motor drives will much improve its comprehensive ability. So, St refrigerator is still the main cryocooler in space techniques. Recently, PTR has been the most popular in study, it may be the optimal cryocooler in space technique.

G-M with magnetic matrix of large specific heat can reduce its working temperature to liquid He directly. So, it is the main commercial cryocooler, however, it meets a new challenge by the Boreas cryocooler.

## FUTURE WORK

We have studied experimentally five types of regenerative cryocoolers and have obtained some good results which can help us understand the mechnism quite well. But, perhaps some of our view is not perfect, especially the behavior of working gas in the cold chamber is not clear enough. The optimal design of PTR and the study of long life refrigerator etc, should be further developed.

## REFERENCES


1. Shimo Li, G. Chen, C. Cui, and J. Li, An experimental study for a miniature Stirling refrigerator, proceedings of the 3rd cryocooler conference, NBS special publication 698 (1984) p. 71—79
2. Shimo Li, Pan Jinwen, Ge S., The study of the improvement of a 3Ly0.8/194 Stirling cryocooler, Advances in Cryogenic Engineering. Vol. 35, (1989), p1401—1406
3. Shimo Li, H. Pan, S. Ge, Experimental tests for improving refrigerator efficiency of a V-M cryocooler, ACE. Vol. 37, (1991), p1079—1084
4. Shimo Li, Y. Mao, H. Zhao, H. Pan, Z. Huang and S. Ge, Theory and experiment for rare-earth magnetic regenerative materials, ACE. Vol. 40, (1993), p625—630
5. Shimo Li, H. Pan, Y. Chen, Q. Li and H. Zhao, Operation characteristics and analysis of a two stage G-M cryocooler, ACE, Vol. 41, (1995), p1609—1614
6. Jihong Yang, Analysis and experimental investigation on refrigeration mechanism of the valved PTR, Master thesis (1996), p. 84
7. Ray Radebaugh, Advances in cryocoolers, 16th ICEC/ICMC, proceeding (1996), p33—44
8. G. Thummes, S. Bender, and C. Heiden, Nitrogen precooled multi-stage pulse tube refrigerator reaching 2.1K, 16th ICEC/ICMC, proceedings (1996), p311-314

# DESCRIPTION AND PERFORMANCE OF A NEAR-ROOM TEMPERATURE MAGNETIC REFRIGERATOR

C. Zimm[1], A. Jastrab[1], A. Sternberg[1], V. Pecharsky[2], K. Gschneidner,Jr.[2], M. Osborne[2], and I. Anderson[2]

[1]Astronautics Technology Center, Madison WI 53716
[2]Ames Laboratory, Iowa State University, Ames IA 50011

## ABSTRACT

Magnetic refrigeration has been viewed as primarily a cryogenic technology because the necessary high magnetic fields are most easily provided by superconducting magnets. However, some of the largest magnetocaloric effects are exhibited by gadolinium-based alloys near room temperature. Ames Laboratory and Astronautics Corporation of America have been collaborating to apply such materials to large-scale commercial and industrial cooling near room temperature. Astronautics has designed and operated a reciprocating magnetic refrigerator that uses water as a heat transfer fluid. The device uses the active magnetic regeneration concept of recent cryogenic devices, but in contrast to the cryogenic case, the heat capacity of the fluid in the pores of the regenerator bed is comparable to that of the solid matrix. Using a 5 T field, the refrigerator reliably produces cooling powers exceeding 500 watts at coefficients of performance of 6 or more. This record performance puts magnetic refrigeration in a class with the best of current technology, vapor cycle refrigeration, without having to use volatile, environmentally hazardous fluids.

## INTRODUCTION

Magnetic refrigeration is based on the magnetocaloric effect, which in the case of ferromagnetic materials is a warming as the magnetic moments of the atoms are aligned by the application of a magnetic field, and the corresponding cooling upon removal of the magnetic field. This magnetic warming and cooling process can have very low losses for soft ferromagnets such as gadolinium (Gd, a rare-earth element), allowing for the possibility of an extremely efficient refrigeration process.

Two major difficulties arise in the design of a magnetic refrigerator. First, the magnetocaloric effect is fairly small; among the strongest effects is seen in gadolinium, for which the application of a 5 T magnetic field produces a maximum adiabatic temperature change of 11 K.[1] Second, the refrigerant is a solid, and is thus not easily pumped through heat exchangers, as is the case for gas and vapor cycle refrigerants.

The problems of heat transfer and temperature span can be overcome with the introduction of a heat transfer fluid and use of regeneration or recuperation. Regeneration can be accomplished by blowing fluid in a reciprocating fashion through a porous bed of magnetocaloric material that is alternately magnetized and demagnetized.

In one regenerator implementation limit, the heat capacity of the fluid blown through the bed is much larger than that of the bed, and the magnetocaloric material serves to maintain a temperature gradient in a column of fluid. In this case, the fluid behaves as a conventional regenerator, and the magnetic material can be conveniently represented on a T-S diagram as following a conventional cycle such as the Ericsson or Brayton[2]. A device was built by NASA using Gd metal plates as the magnetocaloric material, a water-ethanol mixture as the heat transfer fluid, and a 7 T water cooled charge-discharge solenoid to apply a changing magnetic field.[3,4] This device was able to achieve a remarkable temperature span of 80 K near room temperature, but the cooling power was low because churning of the fluid column shorted out the temperature gradient, except at very low frequency. Even if the churning could be somehow suppressed, a theoretical analysis[5] concluded that high power operation would suffer large pressure drop losses due to the high fluid flow rates required.

In the opposite regenerator limit, the heat capacity of the fluid blown through the magnetocaloric bed is much less than that of the bed. In this case, the temperature gradient is maintained inside the magnetocaloric bed. The magnetocaloric bed generates refrigeration that appears to regenerate the bed itself using the thermal linkage of the fluid. The fluid also carries heat to and from external heat exchangers. Much lower fluid mass flow rates are required for a given cooling power than the case described in the previous paragraph. This peculiar configuration has been called the Active Magnetic Regenerator Refrigerator (AMRR). The AMRR does not appear to have a gas cycle analogue and it is hard to describe the state of the magnetocaloric material on a T-S diagram.[6] However, a theoretical model of the AMRR has been developed[7,8] and tested against a cryogenic experimental device. At cryogenic temperatures, the AMRR, using helium as a heat transfer fluid, has operated at good efficiency.[9] An AMRR operating near room temperature using nitrogen gas achieved a 50 K temperature span, but the cooling power was low.[10] The cooling power was limited in part by the long time needed to charge and discharge a superconducting magnet, but also by the relatively low volumetric heat capacity of nitrogen at room temperature.

The ideal fluid to use in an AMRR would have a high thermal conductivity ($K_t$) to enhance heat transfer, a high volumetric heat capacity (C) to minimize volumetric flow rate, and a low viscosity ($\mu$) to minimize flow work. On this basis, water is by far the best fluid. The value of $K_tC/\mu$ in MKS units at 25 C for water is $2.85 \times 10^9$; for ethylbenzene (a liquid hydrocarbon), $0.308 \times 10^9$; for helium at 2 MPa, $0.1305 \times 10^9$; and nitrogen at 2 MPa, $0.051 \times 10^9$.

Use of water as the heat transfer fluid in an AMRR appeared problematic given the assumption that the heat capacity of the fluid blown through the bed should be less than that of the bed itself. Water trapped in the pores of a bed of gadolinium spherical particles of 36% porosity has a heat capacity equal to that of the gadolinium. However, the volume of fluid blown through the bed can be held to less than that of the pore volume. Our experimental results show that a highly successful device can be constructed on this basis. The theoretical analysis of the effects of pore fluid heat capacity and the engineering details of handling the pore fluid heat capacity will be described in a future paper.

## APPARATUS CONFIGURATION

We have tested a AMRR containing two 1.5 kg beds composed of Gd spheres. The magnetic field is provided by a conventional liquid helium immersed NbTi solenoid mounted in a warm bore dewar. The magnet was run in persistent mode. The AMR beds were mounted on a G-10 composite carrier assembly and alternately inserted in the bore of the magnet using an air cylinder drive (Figure 1). The maximum net magnetic force was about 2500 N. The insertion process takes 1 second.

The flow in the primary loop of the device starts with water being cooled as it is pumped from the hot end to the cold end of the demagnetized bed. Next, the water picks up a thermal load as it passes through the cold heat exchanger (CHEX). The water then absorbs heat as it travels from the cold to the hot end of the magnetized bed. The heat is given up as the water passes through the hot heat exchanger (HHEX) and returns to the pump inlet. After an interval of time set by a timer, the magnetized bed is removed from the magnet and the formerly demagnetized bed is inserted. Then the flow direction in each bed is reversed. No flow occurs in the beds while they are being moved. The thermal load is supplied to the CHEX by a secondary flow loop with a centrifugal pump and a temperature-controlled electric heater. The heat is removed from the HHEX by a secondary flow path using building cooling water. Flow rates were measured in the primary loop and each secondary loop using turbine flow meters with 1% of reading accuracy. The flow rate in the primary flow loop was controlled by varying the speed of a positive displacement gear pump. Throttling valves controlled the flow rates in the secondary flow loops. Temperature differences at the HEX's were measured using thermocouples and thermistors installed at each inlet and outlet. These thermometers were interchangeable to 0.1 K, i.e., when held at the same temperature, their readings would agree to 0.1 K (their absolute accuracy may not be as high). Thermocouples were also installed in five locations in each AMR bed. Knowledge of the flow rates and inlet and outlet temperatures of the HEXs allows computation of the cooling power, $Q_c$, and the heat rejected, $Q_h$.

## REFRIGERATION MATERIAL

The magnetic refrigerant was gadolinium metal. The material was purchased in chunk form from Baotou Steel and Rare Earth Company of China. The vendor rated the material as 99.9% pure. Chemical analysis indicated that it was only 93 at.% (99.7 wt.%) pure, with the major impurities being interstitial elements (in at.% - 3.2 H, 1.2 O, 1.0 C, 0.8 F, 0.4 N and 0.1 Ca). 13.6 kg of the metal chunks were vacuum-melted in a tantalum crucible at 1600° C for one hour to reduce calcium and hydrogen impurities by vaporization. The molten metal was bottom-poured to form 6.5 cm diameter rods. The rods were finish machined and shipped to a commercial vendor to prepare gadolinium spheres by the plasma-rotating electrode process (PREP). The yields were: 3.6 kg of spheres with diameter < 150 μm, 4.4 kg of spheres between 150 and 300 μm, and 0.7 kg of spheres greater than 300 μm. The 150 to 300 μm size spheres were used in the AMRR beds. The yield of suitably sized spheres was 33 % as based on the initial 13.6 kg of metal. The other major losses of material (other than the unwanted size spheres) were volatiles in the melting operation (minor), slag in the casting step (intermediate), and machining (major) and the end stubs from the PREP process (intermediate). The yield of usable spheres from a second run was increased to 51%. Chemical analysis of the spheres revealed that the hydrogen content was reduced to 0.08 at.% and the calcium to <0.005 at.%, while the other interstitial impurity contents remained essentially the same.

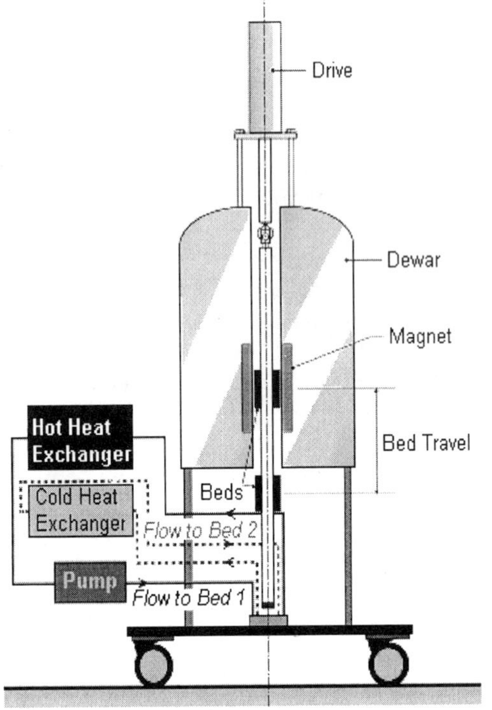


Figure 1. Configuration of AMR refrigerator .

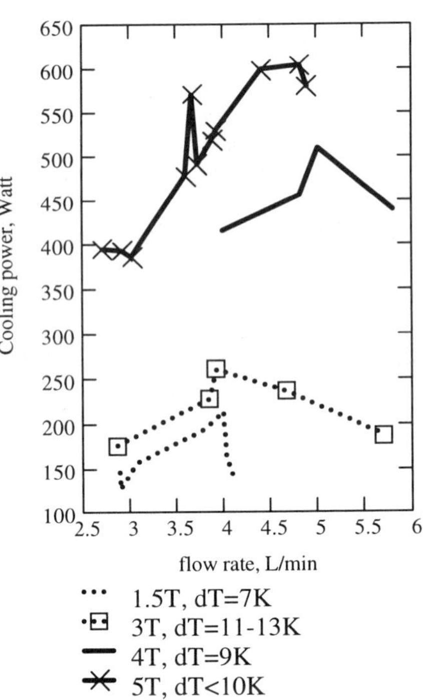


Figure 2. Cooling power vs. flow rate .

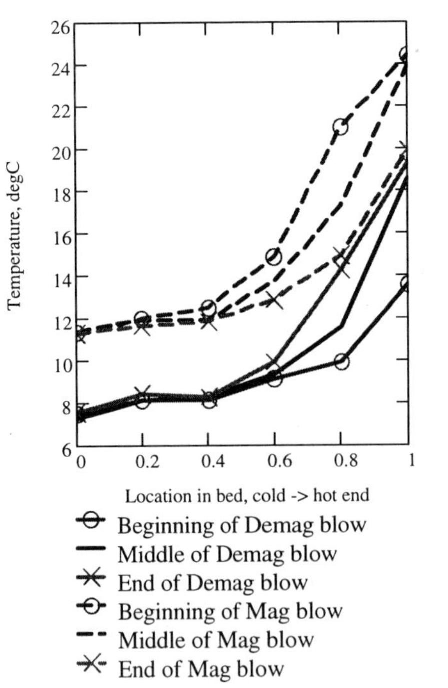


Figure 3. Bed temperature profiles, 0.05 L/s .

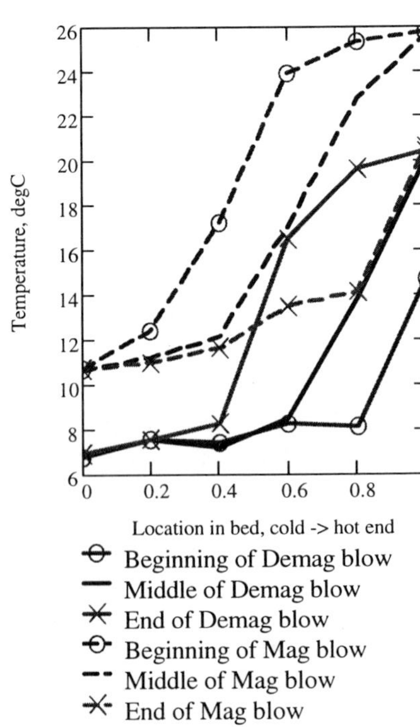


Figure 4. Bed temperature profiles, 0.1 L/s .

Heat capacity of the as-received Baotou gadolinium was about the same as that of high purity Ames Laboratory metal from 3.5 to 275 K, but near the Curie temperature, 293 K, it was ~ 10% smaller; and it also had a slightly lower Curie temperature, 287 K vs 293 K. As a result, the magnetocaloric effect of the Baotou gadolinium was ~5% larger below 285 K, ~5% smaller above 300 K, and 10% smaller at 293 K than the Ames Laboratory material. Although the magnetocaloric effect of the Chinese gadolinium is inferior to the high purity material, the large cost difference, \$110/kg vs \$23,000/kg, justifies the use of the less pure material in a magnetic refrigerator. Magnetization measurements on the gadolinium spheres indicated that the magnetocaloric effect was essentially unchanged during the processing of the ingot material into spheres.

## EXPERIMENTAL RESULTS

Figure 2 shows the measured cooling power as a function of water flow rate for a total cycle time of 6 s and a fluid flow time of 2 s in each direction. Note that the cooling power increases with magnetic field as a result of the increase of the adiabatic temperature change with field. The cooling power is initially proportional to the fluid flow rate. As the flow rate increases, the viscous flow losses increase as the square of the flow rate (the losses are proportional to the product of flow rate and pressure drop, and the pressure drop is proportional to flow rate in the laminar regime). The viscous flow losses can be calculated from the measured pressure drops and flow rates, but are not sufficient to account for the loss in cooling power at higher flow rates seen in Figure 2. More important is the effect of too large of a flow on the temperature profile in the magnetocaloric beds. When the heat capacity of the fluid blown through a regenerator bed exceeds that of the solid regenerator matrix, the temperature profile of a regenerator is severely disturbed, causing a loss in regenerator effectiveness.[11] A similar effect is seen in the AMRR. Figures 3 and 4 show the temperature profiles inside one of the magnetocaloric beds as measured by fine thermocouples inserted in the bed matrix. For low flow (Figure 3), the temperature profile in the beds changes substantially as the field changes due to the magnetocaloric effect, but the temperatures of the bed does not change much during the blow period. At a higher flow (Figure 4), the temperature profile changes substantially during the blow period. Any additional flow would not accomplish useful heat transfer.

Figure 5 shows the effect of temperature span ($T_h$-$T_c$) on the cooling power, with $T_h$ held nearly constant. A decrease in cooling power with increasing temperature span is to be expected because the adiabatic temperature change of gadolinium decreases with temperature below its Curie point. However, the drop in cooling power with decrease in cold end temperature is faster than the drop in adiabatic temperature, probably as a result of losses produced by deviations of the magnetocaloric properties of gadolinium from that of an ideal magnetic refrigerant.[2,12] Experimental verification of this point requires performance comparisons to other magnetocaloric refrigerants, which will be the subject of a future paper.

Figure 6 shows the coefficient of performance (COP) as a function of temperature span. The COP is a dimensionless quantity calculated from the measured cooling power, $Q_c$, and heat rejected, $Q_h$, by COP = $Q_c/(Q_h$-$Q_c)$ which by energy balance = $Q_c/W$, where W is the work input to the refrigeration device. $Q_c$ is calculated using the measured inlet and outlet temperatures at the CHEX and the measured water flow rate. Either W or $Q_h$ could be measured to determine COP. We chose to use $Q_h$ because measuring the net work W is difficult because the magnitude and sign of the magnetic forces changes rapidly with displacement for this reciprocating design. When the COP is larger than one, variations in experimental conditions or measurement errors tend to get magnified by the subtraction of the comparable quantities $Q_h$ and $Q_c$, hence the scatter in the plot of COP is larger than

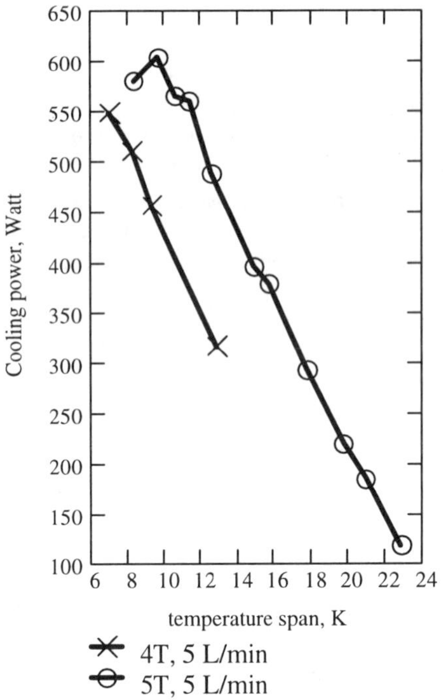


Figure 5. Cooling power vs. temperature span.

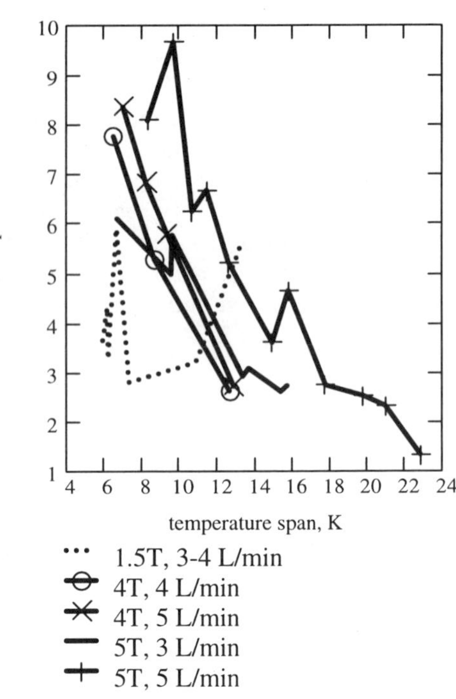


Figure 6. COP vs. temperature span.

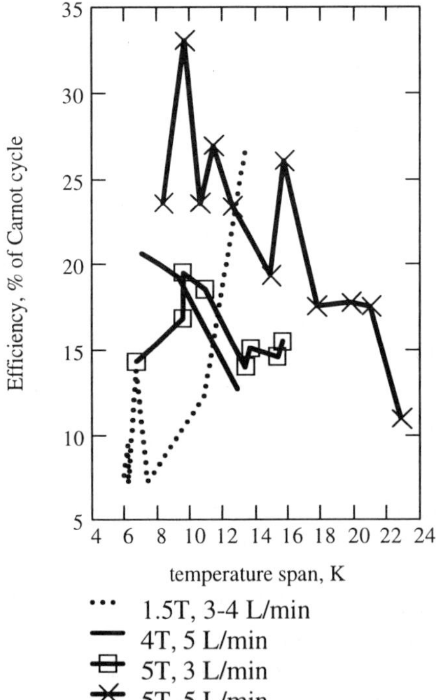


Figure 7. Percent of Carnot efficiency.

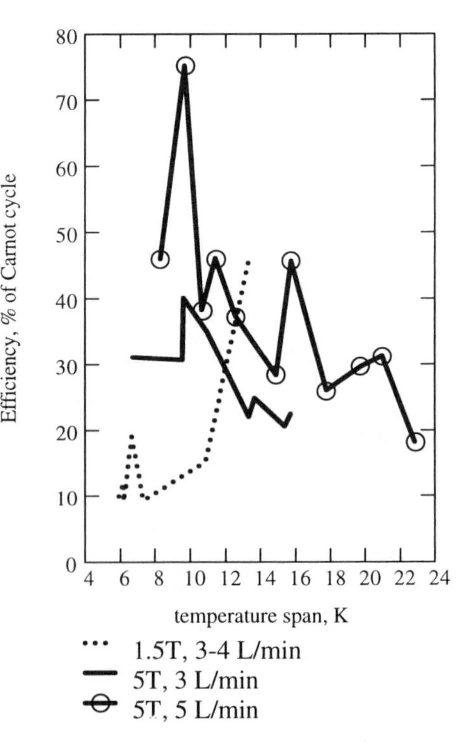


Figure 8. Efficiency with seal friction subtracted.

that observed in the plot of $Q_c$. $Q_h$ is measured using inlet and outlet temperatures and the flow rate similarly to $Q_c$, except that the frictional heat added by the pump seals is subtracted off. The reason we subtract off the pump friction is that the variable speed gear pump used here is only 2 to 10% efficient, which is much less than that of centrifugal pumps of 50 to 80% efficiency normally used for commercial applications. The gear pump was used because it is a positive displacement pump able to provide stable flow rates over a wide range of impedances. The $Q_h$ also does not include AC losses in the superconducting magnet due to changing bore permeability, which is reflected in helium boiloff in the magnet dewar. The present magnet has relatively high AC loss characteristics leading to a maximum liquid helium boiloff of 2L/hr. Such losses will not occur if a permanent magnet is used or if a rotary bed design is used. The $Q_h$ reported here does include all the losses expected to be encountered in practical devices, such as the viscous flow work in the piping, HEX's and beds, seal friction associated with the movement of the beds, hysteresis and eddy current heating in the beds, and the effects of non-ideal heat transfer in the beds or HEXs and non-ideal magnetocaloric properties of the beds.

The COP ranges from 2 to 8, which compares favorably with commercial vapor cycle refrigerators and greatly exceeds that obtainable from alternate technologies such as thermoelectric coolers. The Carnot limit to the COP of a refrigerator is COP (limit) = $T_c/(T_h-T_c)$. For example, if the refrigerator pumps heat from $T_c$ = 280 K to $T_h$ = 300 K, the limiting COP is 280/20 = 14. Figure 7 plots the ratio of the measured COP to that of the limiting Carnot COP at the same temperature span, i.e. the efficiency as a percent of that of the Carnot cycle. The percentage of Carnot efficiency is generally less than 50%, as is generally true for real refrigerators. In addition, the efficiency drops with increasing temperature span, presumably because the properties of gadolinium deviate increasingly from that of an ideal refrigerant as the temperature span is increased.[2,12]

The efficiency should start to drop as the flow rate is increased because of increased flow work and finite heat transfer, but until large flow rates are reached, the opposite is seen. In addition, the efficiency at 1.5 T appears to increase with increasing temperature span, in contrast to the higher field results. These two trends occur because the dominant loss turns out to be sliding friction in elastomer seals around the magnetocaloric beds. The frictional heat load was estimated to be about 30 watts by measuring the force needed to move the beds in the absence of magnetic field. We expect the seal friction can be greatly reduced as the device is scaled up and as the stock O-rings are replaced by more suitable seals. Figure 8 shows what the efficiency would be with the seal friction subtracted off from Qh. The efficiency is between 30 and 50% of Carnot over a range of flow rates, temperature spans and magnetic fields. Note the efficiency of 20% with a 1.5 T field that can be obtained using commercially available permanent magnets. Further performance improvements should be achievable with newly discovered refrigerants.[13]

## CONCLUSIONS

The search for a practical magnetocaloric cooling configuration for the large near-room temperature refrigeration and air conditioning market has been conducted by many groups in different nations over the past two decades. The results reported here appear to justify the effort. Significant cooling power of 500 watts has been achieved at a COP of 5 over a 12 K temperature span using a readily available material, gadolinium, readily achieved magnetic field, 5 T, and an environmentally safe heat transfer fluid, namely water. Users of refrigeration systems containing volatile fluids that endanger the ozone layer or increase the greenhouse effect are becoming concerned with the increasing expense and complexity associated with the safeguarding and recycling of refrigerants. Controversy over how to cut fossil fuel use in order to reduce greenhouse gas emissions underscores the importance of improving the efficiency of energy-intensive refrigeration and air conditioning systems. A magnetic refrigeration technology that uses no hazardous fluids and has the potential for very high efficiency should thus be of great current interest.

## ACKNOWLEDGEMENTS

Ames Laboratory is operated for the U.S. Department of Energy (DOE) by Iowa State University under Contract No. W-7405-ENG-82. The work was supported by the Advanced Energy Projects and Technology Research Division, Office of Computational and Technology Research of DOE. This work also benefitted from the expert contributions of T.A. Henning, J.F. Laatsch, J.W. Johnson and L.M. Lawton, Jr.

## REFERENCES

1. V.K. Pecharsky and K.A. Gschneidner, Jr., Comparison of the magnetocaloric effect derived from heat capacity, direct, and magnetization measurements, in: "Advances in Cryogenic Engineering", Vol 42, Plenum, New York (1996), p. 423
2. F.C. Chen, R.W. Murphy, V.C. Mei, and G.L. Chen, Thermodynamic analysis of four magnetic heat-pump cycles, *J. of Eng. for Gas Turbines and Power*, 114:715 (1992)
3. G.V. Brown, Magnetic heat pumping near room temperature, *J. Appl. Phys.* 47:3673 (1976)
4. G.V. Brown and S.S. Papell, Regeneration tests of a room temperature magnetic refrigerator and heat pump, unpublished (1978)
5. J.G. Purnell, Performance predictions for a room temperature, Ericsson cycle magnetic heat pump, DTNSRDC Report PAS-82-11, David Taylor Ship R&D Center, Annapolis (1982)
6. J.L. Hall, C.E. Reid, I.G. Spearing and J.A. Barclay, Thermodynamic considerations for the design of active magnetic regenerative refrigerators, in: "Advances in Cryogenic Engineering", Vol 41, Plenum, New York (1996), p. 1653
7. A.J. DeGregoria, Modeling the active magnetic regenerator, in: "Advances in Cryogenic Engineering", Vol 37, Plenum, New York (1992), p. 867
8. J.W. Johnson and C.B. Zimm, Performance modeling of a 4 K active magnetic regenerative refrigerator, *J. Appl. Phys.*, 79:2171 (1996)
9. A.A. Wang, J.W. Johnson, R.W. Niemi, A.A. Sternberg, and C.B. Zimm, Experimental Results of an efficient active magnetic regenerator refrigerator, "Proceedings of the 8th International Cryocooler Conference", Plenum, New York, (1995), p. 665
10. G. Green, J. Chafe, J.Stevens and J. Humphrey, A gadolinium-terbium active regenerator, in: "Advances in Cryogenic Engineering", Vol 35, Plenum, New York (1990), p. 1165
11. F.W. Schmidt and A.J. Willmot, "Thermal Energy Storage and Regeneration", McGraw-Hill, New York (1981)
12. C.R. Cross, J.A. Barclay, A.J. DeGregoria, S.R. Jaeger, and J.W. Johnson, Optimal temperature-entropy curves for magnetic refrigeration, in: "Advances in Cryogenic Engineering", Vol 33, Plenum, New York (1987), p. 767
13. V.K. Pecharsky and K.A. Gschneidner, Jr, Giant magnetocaloric effect in $Gd_5(Si_2Ge_2)$, Phys Rev Lett, 78:4494, 1997

# AN EXPERIMENTAL GM/MAGNETIC REFRIGERATOR

G. F. Nellis [1] and J. L. Smith, Jr. [2]

[1]Creare, Inc.
Hanover, NH, 03755, U.S.
[2]M.I.T. Cryogenic Engineering Lab
Cambridge, MA, 02139, U.S.

## ABSTRACT

This paper describes the results obtained with an experimental prototype of a new type of cryocooler. The cycle employed by the device (a GM/magnetic cycle) uses two thermodynamically active substances: a compressible gas (helium) and a paramagnetic solid (Gadolinium Gallium Garnet or GGG). The two substances alternately regenerate one another, thereby overcoming the low temperature thermal saturation associated with a GM cycle while producing more refrigeration than a regenerative magnetic cycle. The experimental prototype utilizes a unique regenerative upper stage device to maintain the warm reservoir temperature seen by the GM/magnetic stage at the required 10-20 K. The pressure in the apparatus is controlled using a room temperature compressor and a system of valves. The device operates at a pressure level which is on the order of 4 atmospheres and uses a pressure ratio of approximately 2. The applied field in the GM/magnetic stage is controlled with a superconducting solenoid and a power supply. The maximum applied field imposed on the paramagnetic regenerator bed is 4 tesla.

The GM/magnetic refrigerator ultimately reached a no load temperature of 4.5 K while rejecting heat at 10.9 K. The device supplied 0.36 W of refrigeration at 6 K while rejecting heat at 15.4 K. The prototype clearly shows that the GM/magnetic cycle is an effective refrigeration cycle below 15 K.

## INTRODUCTION

There are many applications which must be kept at very low temperatures. The most important of these are superconducting devices which must be kept at or near 4.2 K in order to remain in their superconducting state. These devices typically are cooled by immersion in a liquid helium bath. While this technique is reliable, it is inefficient and inconvenient for remotely operating devices [1].

In order to provide the required cooling at 4.2 K without relying on liquid helium, several types of closed cycle cryocoolers have been developed. The most important of these include Stirling or GM type cryocoolers and Joule-Thomson type cryocoolers [2]. Both of these techniques fail to reliably provide significant refrigeration at temperatures below 15 K. Cryocoolers which employ a Joule-Thomson expansion valve suffer from poor reliability [3] while Stirling and GM cycles must overcome regenerator thermal saturation [2].

The thermal and magnetic properties of some paramagnetic substances are highly coupled in this temperature range, making them attractive alternative low temperature refrigerants. Several cryocoolers based upon the concept of pumping heat via a magnetic work input (i.e. a magnetic cycle) have been developed [4]. These cryocoolers have taken many different forms including: reciprocating, rotating, and regenerative. These devices have successfully overcome many of the limitations associated with GM and Joule-Thomson cycles.

Unfortunately, magnetic cycle cryocoolers use isobaric helium as a heat transfer medium for the paramagnetic substance without taking advantage of the refrigeration potential associated with expansion of the helium. It has been shown computationally that a cryocooler in which both the pressure and applied magnetic field are appropriately controlled should achieve 50% more refrigeration than either a GM or a magnetic cryocooler [5]. Furthermore, the pressure ratio at which such a cryocooler (subsequently referred to as a GM/magnetic cryocooler) operates most effectively is on the same order as the pressure ratio employed by a GM device. Therefore, a GM/magnetic cryocooler is ideally suited for the lowest temperature stage of a conventional cryocooler, extending its temperature range and increasing its refrigeration power.

In order to investigate the properties of a GM/magnetic cycle, an experimental prototype has been designed and constructed. This paper briefly explains the thermodynamics of a GM/magnetic cycle, describes the experimental apparatus, and presents some of the results obtained with the device.

## GM/MAGNETIC CYCLE

The GM/magnetic cycle allows both the compressible fluid and the paramagnetic material to undergo thermodynamic cycles and produce refrigeration. This is a fundamentally new type of cycle which has not previously been tested. The GM/magnetic cycle was first proposed by Jeong, et. al. [6] and subsequently analyzed by the authors [7].

An idealized version of a GM/magnetic device is illustrated in Figure 1. The device is identical to a conventional GM apparatus except that the regenerator is composed of a paramagnetic material which is surrounded by a superconducting solenoid. The solenoid can impose a controlled, time varying applied field on the paramagnetic matrix. The four processes which make up an ideal GM/magnetic cycle are illustrated in Figure 2 and described below.

*hot-to-cold blow/demagnetization*

During the hot-to-cold blow/demagnetization process, the displacer moves upwards, pushing gas from the hot space, through the regenerator, into the cold space. This process occurs at constant maximum system pressure. The applied field in the paramagnetic regenerator goes from the maximum value to the minimum value seen by the cycle. The regenerator absorbs entropy from the gas in order to reduce its temperature. Because the regenerator is being demagnetized, it can absorb this entropy even as its temperature decreases. Therefore, the fluid may be delivered to the cold space at a temperature *below* the temperature of the cold reservoir. This is in contrast to a conventional regenerator where superheated fluid presents an additional

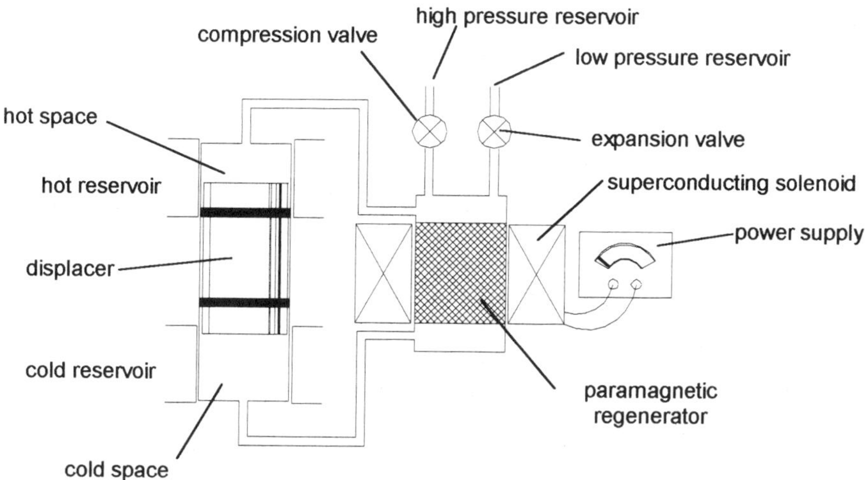


**Figure 1.** Schematic of GM/Magnetic device.

heat leak due to regenerator ineffectiveness. To first order, the refrigeration due to the flow of sub-cooled fluid into the cold space may be thought of as the 'magnetic' component of the total refrigeration ($Q_{C,mag}$).

*expansion*

During the expansion process, the displacer is not moved and all of the gas remains in the cold space of the device. The compression valve is closed and the expansion valve is opened so that the pressure in the device falls to the minimum pressure seen by the system. The pressure drop induces a heat transfer from the cold reservoir to the gas, producing refrigeration in the same manner as a GM cycle. To first order, the refrigeration associated with the expansion of the compressible fluid in the cold space may be thought of as the 'mechanical' component of the total refrigeration ($Q_{C,mech}$).

*cold-to-hot blow/magnetization*

During the cold-to-hot blow/magnetization process, the displacer moves downwards, pushing gas from the cold space, through the regenerator, into the hot space. This process occurs at constant minimum system pressure. The applied field in the paramagnetic regenerator goes from a low value to the maximum applied field seen by the cycle. The magnetization of the regenerator causes the paramagnetic material to reject entropy to the gas. Consequently the gas may enter the hot reservoir at a temperature above the hot reservoir temperature, inducing a heat rejection.

*compression*

During the compression process, the displacer is not moved so that all of the gas remains in the hot space of the device. The compression valve is opened and the expansion valve is closed so that the pressure increases in the device. The pressure rise induces a heat rejection to the hot reservoir.

## EXPERIMENTAL APPARATUS

The experimental prototype is illustrated schematically in Figure 3. The design of the experimental apparatus has been described elsewhere [8,9]. The prototype can be divided into the magnetic stage and the upper stage. The magnetic stage imposes the GM/magnetic cycle on a paramagnetic bed of GGG and is representative of a practical lowest temperature stage of a GM type device. Because the magneto-caloric effect associated with GGG disappears above approximately 20 K (and is drastically reduced above 12 K), the warm end of the paramagnetic bed must be maintained in the 10 - 20 K range. The upper stage is essentially an experimentally expedient technique to maintain this temperature without purchasing or constructing a room temperature to 15 K refrigeration device.

The upper stage system is a regenerative device which is made more effective by utilizing a stream of cold helium vapor (boil-off from the superconducting magnet). The helium vapor is forced to flow in close thermal contact with the upper stage regenerator material (lead shot). The upper stage configuration is a shell and tube heat exchanger. The working fluid for the GM/magnetic cycle flows within the tubes which are filled with lead shot. The stream of cold helium vapor flows within the shell through a series of baffles.

The paramagnetic regenerator matrix is a bed of crushed GGG particles contained inside a superconducting solenoid. The large specific area associated with the packed bed matrix creates good thermal contact between the helium working fluid and the GGG. The applied field within the GGG is controlled by varying the current within the superconducting solenoid. GGG was selected due to its large magneto-caloric effect below 10 K [11] and its well known thermodynamic properties [12].

The refrigeration load is generated by a resistive Kapton heater placed within the cold heat exchanger. This heat exchanger is affixed to the cold end of the magnetic regenerator.

The experimental apparatus is placed within a double vacuum jacketed dewar with a liquid nitrogen radiation shield. The upper stage, magnetic stage, cold heat exchanger, and most of the displacer are vacuum insulated from the liquid helium pool which cools the superconducting magnet. This ensures that the magnetic stage is thermally isolated from the 4.2 K reservoir.

The displacer is driven sinusoidally by a cam mechanism. When the displacer hits top-dead-center: the expansion valve is opened, the compression valve is closed, and the superconducting solenoid begins to magnetize. When the displacer hits bottom-dead-center, the compression valve is opened, the expansion valve is closed, and the superconducting solenoid begins to demagnetize. The resulting time variation of pressure at the warm and cold ends of the magnetic stage are shown for a typical cycle in Figure 4. The displacer volume

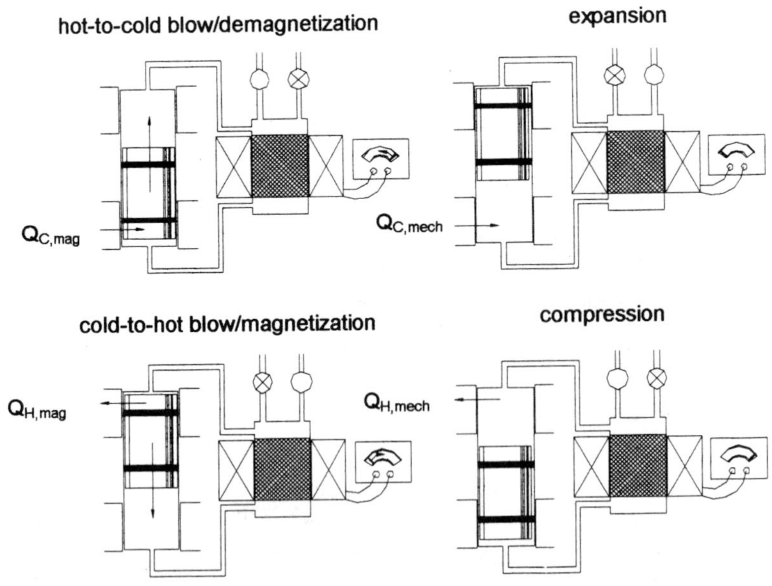


**Figure 2.** GM/Magnetic cycle.

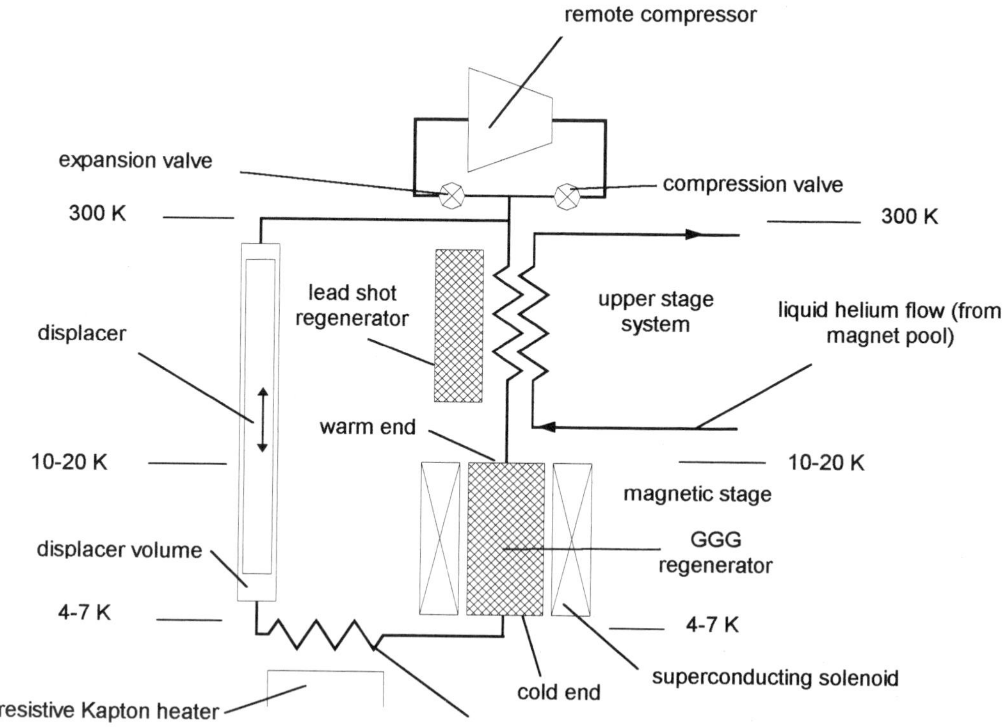


**Figure 3.** Schematic of experimental apparatus.

and applied field variation are shown in Figure 5. The solenoid voltage is limited by the quench voltage resulting in a total cycle time between 12 and 24 seconds.

## EXPERIMENTAL RESULTS

The temperature variation at the warm end, the midpoint, and the cold end of the magnetic stage are illustrated in Figure 6 for a typical steady-state cycle. Although not measured explicitly, the mass flow rate at the cold end of the magnetic stage and the temperature of the fluid within the displacer volume may be inferred by applying mass and energy balances to the displacer volume. A similar technique is used to determine the mass flow rate at other axial positions in the paramagnetic regenerator bed. Figure 7 illustrates the mass flow rate variation for a typical cycle. The dramatic decrease in the mass flow rate with axial position illustrates the effect of regenerator porosity on the cycle - the warm end of the regenerator has to deal with nearly 3 times as much helium as enters the cold end.

The temperature at the warm and cold ends of the magnetic regenerator bed vary with time, making it difficult to identify the appropriate refrigeration and warm reservoir temperature associated with a particular cycle. The warm reservoir temperature is taken to be the temperature associated with the mass average enthalpy and entropy of the flow into the warm end of the regenerator during the hot-to-cold blow portion of the cycle. The refrigeration temperature is taken to be the temperature associated with the mass average enthalpy and entropy of the flow into the cold end of the regenerator during the cold-to-hot blow portion of the cycle.

Using these definitions, the refrigeration temperature as a function of warm reservoir temperature can be plotted for various applied field changes and refrigeration loads. These

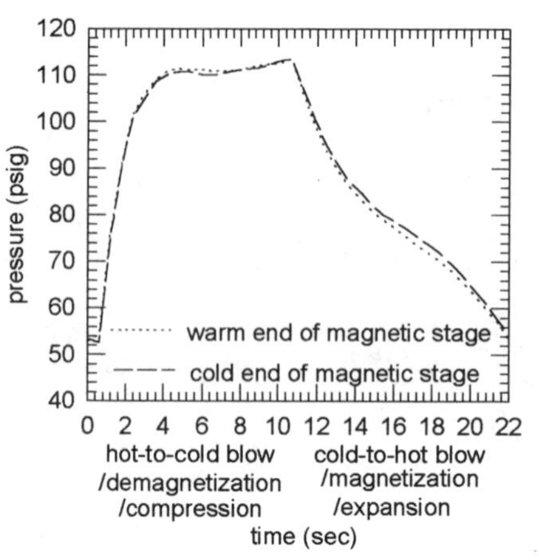


**Figure 4.** Pressure variation for a typical cycle.

**Figure 5.** Displacer volume and applied field variation.

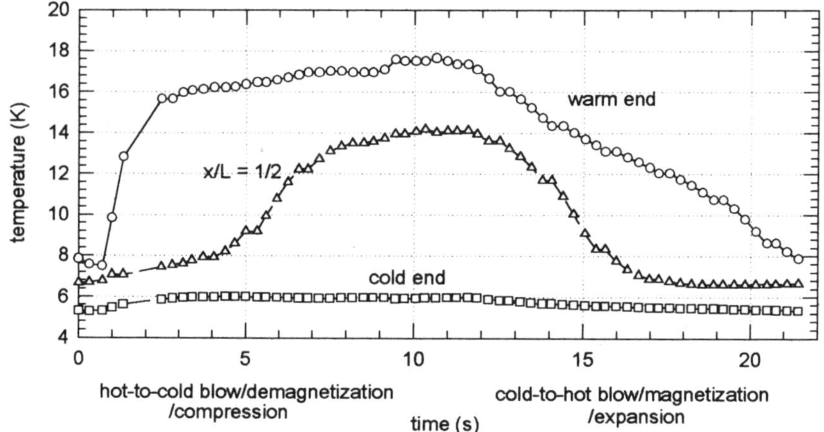


**Figure 6.** Temperature variation for a typical cycle.

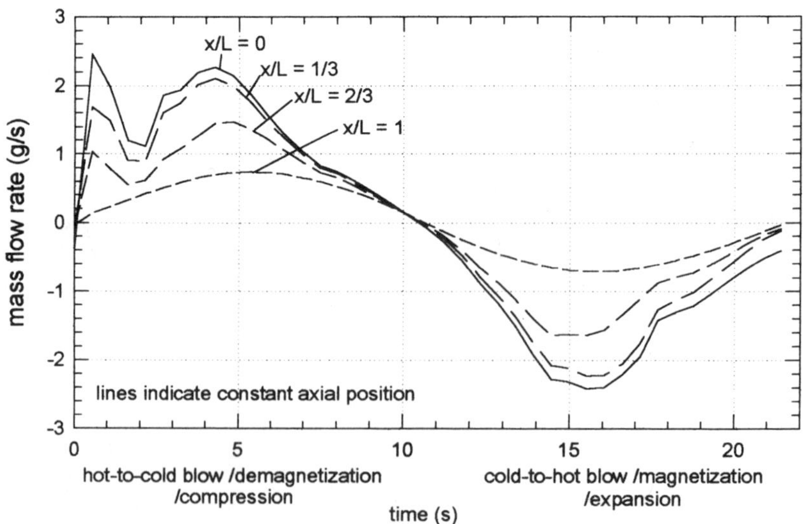


**Figure 7.** Estimated mass flow rate as a function of time for a typical cycle.

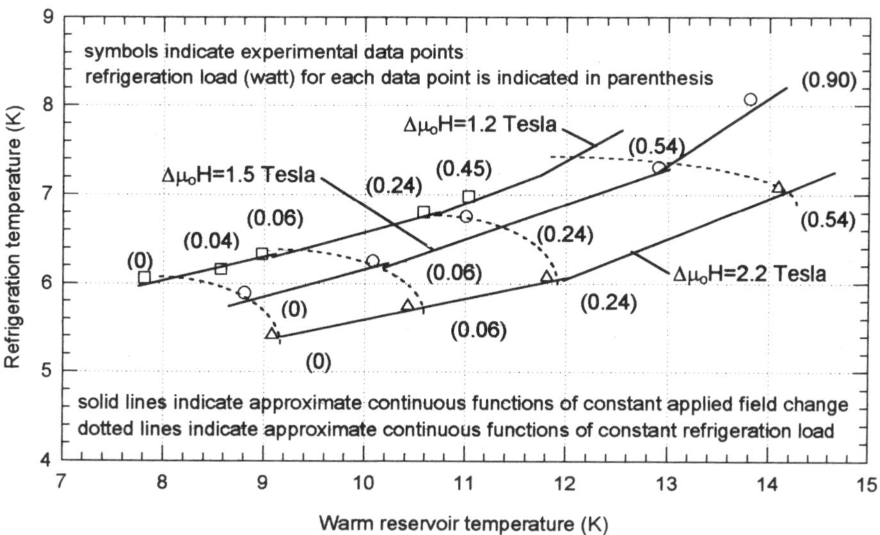


**Figure 8.** Refrigeration temperature as a function of warm reservoir temperature for various levels of applied field swing and refrigeration load.

curves are shown in Figure 8. Figure 9 illustrates the refrigeration temperature as a function of warm reservoir temperature for two different displacer volumes as the refrigeration load is varied. In each of these figures, the refrigeration load is taken to be the load on the resistive heater, parasitic heat loads are not included.

## CONCLUSIONS

Some of the more important conclusions associated with the GM/magnetic cycle in general and this project in particular are listed below.

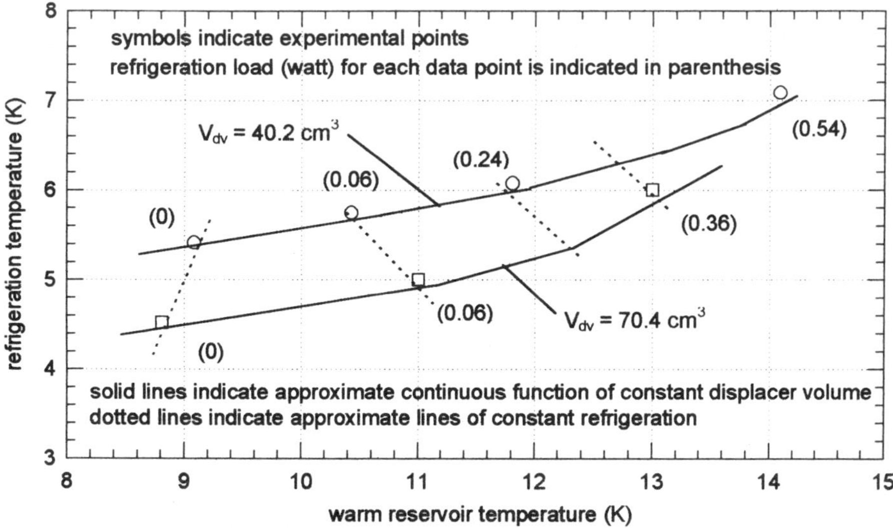


**Figure 9.** Refrigeration temperature as a function of warm reservoir temperature at different values of displacer volume and refrigeration load.

- The GM/magnetic cycle has been experimentally shown to be an effective refrigeration cycle. The GM/magnetic stage ultimately reached a no load temperature of 4.5 K while rejecting heat at 10.9 K. The device supplied 0.36 W of refrigeration at 6 K while rejecting heat at 15.4 K. This dual working fluid cycle has definite advantages over either a Gifford-McMahon or a magnetic cycle.
- Gadolinium Gallium Garnet is not a particularly effective refrigeration material at temperatures above approximately 10 K. The magneto-caloric effect exhibited by GGG decreases rapidly in this temperature range, resulting in a regenerator which is only marginally more effective than a conventional lead regenerator above 10 K.
- The void volume which is unavoidably present in any real regenerator matrix has a particularly large effect on a GM/magnetic cycle. Not only does the entrained helium prevent the magnetic material from undergoing true adiabatic processes, but the pressurization of this helium acts as an additional heat load on the regenerator.
- The time constant associated with the discharge of the superconducting magnet was the limiting time constant for the experimental apparatus. The cycle could be made more effective if the cycle time could be reduced.

## ACKNOWLEDGMENTS

The authors would like to acknowledge the Naval Surface Warfare Center for their generous financial support on this project.

## REFERENCES

1. K. D. Timmerhaus, "Recent Trends in Cryocooler Development", *Proceedings of the XVII'th International Congress of Refrigeration*, **1** : 233, (1991).

2. J. L. Smith, Jr., "Cryogenic refrigeration for space exploration - a challenge for the future", *Proceedings of the XVII'th International Congress of Refrigeration*, **1** : 2, (1991).

3. G. Walker and E. R. Bingham, "Low-Capacity Cryogenic Refrigeration", Clarendon Press, Oxford (1994).

4. J. A. Barclay, "Magnetic refrigeration for low-temperature applications", *Proceedings of the Third Cryocooler Conference*, **1**:20, (1984).

5. G. F. Nellis and J. L. Smith, Jr. , "Investigation of a magnetically augmented cryogenic refrigerator", *Proceedings of the Eighth Cryocooler Conference*, **1**:647, (1995).

6. S. Jeong and J. L. Smith, Jr. , "Magnetically augmented regeneration in Stirling Cryocooler", *Advances in Cryogenic Engineering,* **39B**:1399, (1994).

7. G. F. Nellis, "Magnetically augmented cryogenic refrigeration", M. I. T. (M.S. Thesis), Cambridge, (1995).

8. G.F. Nellis and J. L. Smith, Jr., "Design of an Experimental Apparatus for Investigation of Magnetically Active Regeneration in the Presence of Cyclic Pressure Variations", *Advances in Cryogenic Engineering*, **41**: (1995).

9. G. F. Nellis, "Stirling/magnetic cryocooler", M. I. T. (Ph.D. Thesis), Cambridge, (1997).

10. R.A. Fisher, G. E. Brodale, E. W. Hornung, and W. F. Giauque, "Magnetothermodynamics of gadolinium gallium garnet. I. Heat capacity, entropy, magnetic moment from 0.5 to 4.2 K, with fields to 90 kG along the [100] axis", *J. Chem. Phys.* , **59** : 4652, (1972).

11. G. R. Ghallagher, "Analysis of a Magnetically Active Regnerator", M.I.T. (M.S. Thesis), Cambridge, (1986).

# PERFORMANCE AND INTERNAL PROCESS OF A 4 K GM COOLER

C. Wang[1], H.U. Häfner[2], and C. Heiden[1]

[1]Institute of Applied physics, University of Giessen
D-35392 Giessen, Germany
[2]Leybold Vakuum GmbH
D-50968 Köln, Germany

## ABSTRACT

A numerical simulation of 4 K GM cooler, which takes into account properties of real gas and magnetic regenerative material as well as heat transfer in the heat exchanger and regenerator, has been developed. The source code of NIST12 helium database is incorporated within the simulation program. The predicted temperature profiles along the regenerator are in reasonable agreement with experimental data. Predictions show some features in the low temperature regenerator, such as a flat temperature region, increased mass flow rate at cold end and a part of the helium fluid never leaving the regenerator. Optimization of structure parameters deals with the effects of the displacer phase as well as the length and porosity of regenerator on the cooling performance.

## INTRODUCTION

There is a growing number of superconducting devices and detectors that require cryogenic cooling at liquid helium temperatures. An obstacle towards reaching temperatures of 4 K and below was overcome by introducing magnetic regenerator materials[1] in the coldest part of the system which exhibit a magnetic phase transition at temperatures near 10 K, accompanied by a corresponding strong enhancement of their heat capacity. Liquid helium temperatures were first achieved with two-stage GM coolers by using $Er_3Ni$[2,3]. Significant cooling power of more than 1.5 W at 4.2 K has recently been reported with multi-layer regenerative materials in the 2nd stage regenerator[4,5].

Theoretical analyses[6,7,8,9] have shown some features in 4 K regenerators that differ much from the corresponding ones at higher temperatures because of the real gas properties of $^4He$ and the strong decrease of specific heat of regenerative materials below 20 K.

The present work is based on numerical simulations developed to aid the design of 4 K pulse tube cooler and to understand its working mechanism[8,9]. Similar numerical techniques are applied here to simulate the performance of 4 K GM coolers. The NIST helium database for real gas properties of $^4$He is incorporated within the simulation program. The predictions are verified by experiments. The characteristics and optimization results of a 4 K G-M cooler are presented in this paper.

## PHYSICAL MODEL

Figure 1 shows the components of a two-stage GM cooler configuration that is used for the simulation. As can be seen the focus is on the low temperature stage. The junction between the upper stage regenerator and low temperature regenerator is taken as the left boundary in the simulation model. The right boundary is the surface of the displacer. The motion of the displacer is given for the different simulations.

The model includes regenerator 1, cold heat exchanger 2 and displacer 3. The basic assumptions, governing equations and the numerical method for simulation are the same as those in reference[8,9]. The calculation accuracy and speed are improved in the present study by further developing the numerical technique. The enthalpy flow difference at the two ends of regenerator (unbalanced energy in the regenerator ) is less than 0.02 W in all calculations.

## COMPARISON WITH EXPERIMENT

Measurements have been performed on a Leybold RGD 5/100 two-stage GM cooler, which has pneumatic dispalcers. Five Silicon-Diode sensors are fixed in a stainless steel tube in the axis of the 2nd stage regenerator to measure the temperature distribution along it. The stainless steel tube has an outer diameter of 3 mm and a thickness of 0.2 mm. The second stage regenerator is filled with Pb spheres. The measured pressure variation at the inlet of the 2nd stage regenerator and the motion of the displacer are given in Figure 2. These are the boundary condition for the simulation.

Figure 3 shows the comparison between prediction and experiment. Four experimental and four model curves are obtained respectively at the temperatures on the 1st and 2nd stage cold head of 8 K and 27 K, 10 K and 40 K, 12 K and 50 K as well as 20 K and 80 K. There is reasonable agreement. Some differences may be caused by the strong temperature variation in the cold chamber of the 1st stage. As can be seen in Figure 3 the temperature

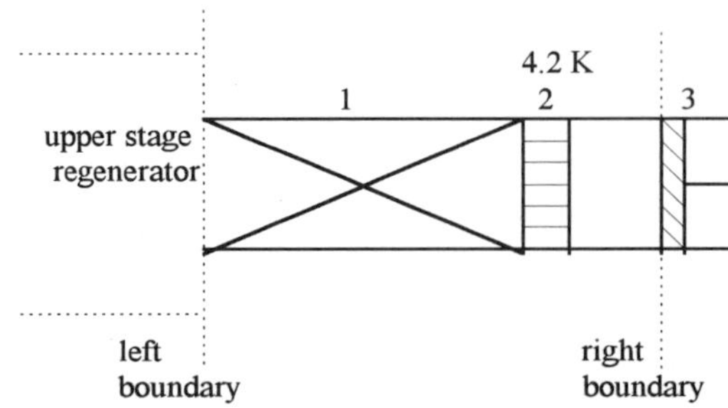


**Figure 1.** Physical model of GM cooler. 1.coldest regenerator; 2. cold heat exchanger; 3. displacer.

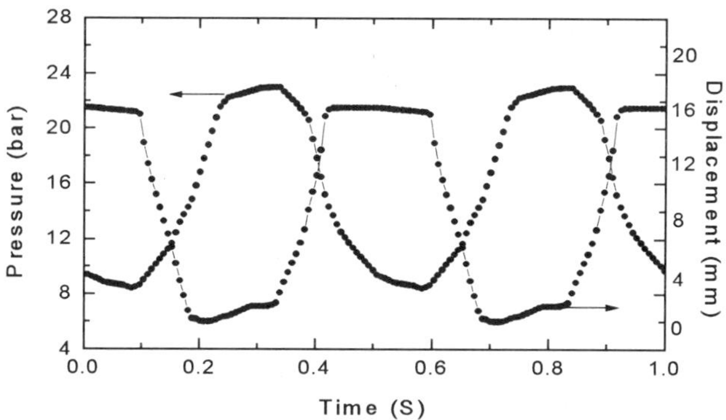


**Figure 2.** Measured pressure variation and displacer motion (pneumatically driven) in RGD/100 GM cooler.

of the first stage cold head is 40 K, however, the temperature measured at inlet position of the second stage is 46 K. The inlet temperature for modelling is assumed to be constant at 40 K.

## NUMERICAL RESULTS AND DISCUSSIONS

The GM cooler with mechanical driven dispalcers is used for the following analysis since most existed 4 K GM coolers are this type and it is also convenient to change the timing of displacer for analysis. The structure parameters of the GM cooler used for the typical calculation are: Regenerator: 29 mm inner diameter, filled with $Er_3Ni$ spheres (diameter of 0.3 mm). Without special note the length and porosity of the regenerator are 160 mm and 0.4. Cold head is assumed as a gap heat exchanger with 32 mm O.D. and 31.8 mm I.D.. The length of it is 60 mm. The displacer diameter is 32 mm and the stroke is 32 mm.

The temperatures are assumed to be 40 K at the inlet of regenerator and 4.2 K at the cold heat exchanger. The operating frequency is 1 Hz.

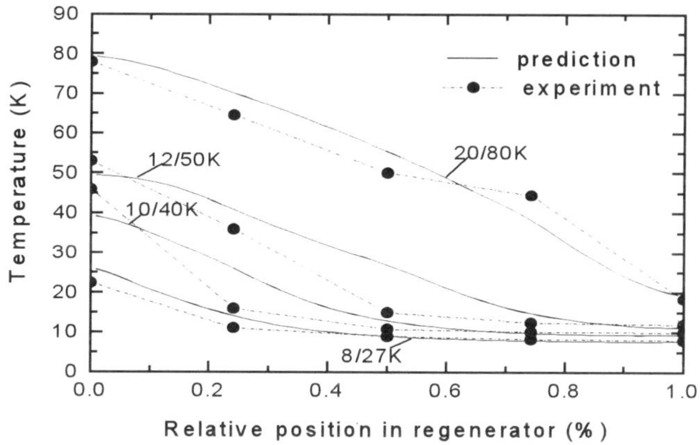


**Figure 3.** Comparison between the predictions and experiments. The 2nd stage regenerator is packed with Pb, and operating frequency is 2 Hz.

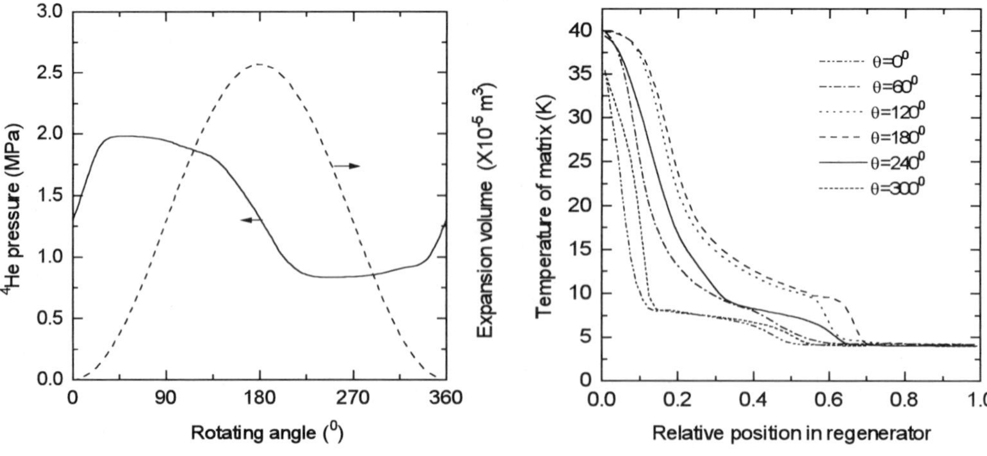


**Figure 4.** Typical input pressure wave and displacer motion for simulation.

**Figure 5.** Matrix temperature along the regenerator.

The inlet pressure wave and variation of expansion for the present simulation are shown in Figure 4. The equation of displacer motion is given by:

$$V(\theta) = V_e(1 - \cos(\theta + \phi)) / 2 \tag{1}$$

$V_e$ is the volume swept by the displacer. $\theta$ is the rotational angle of the displacer. $\phi$ is the initial angle of the displacer as a cycle-constant that controls timing, and is assumed as zero without special note in the paper.

## 1. Some features of the 4 K regenerator

The behavior of the low temperature regenerator reaching temperatures in the liquid $^4$He range, as revealed by the simulation, is in several respects completely different from

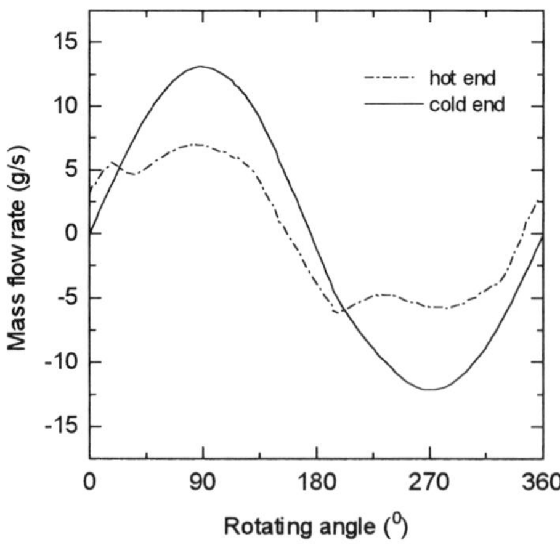


**Figure 6.** Dynamic mass flow rate at two ends of the regenerator.

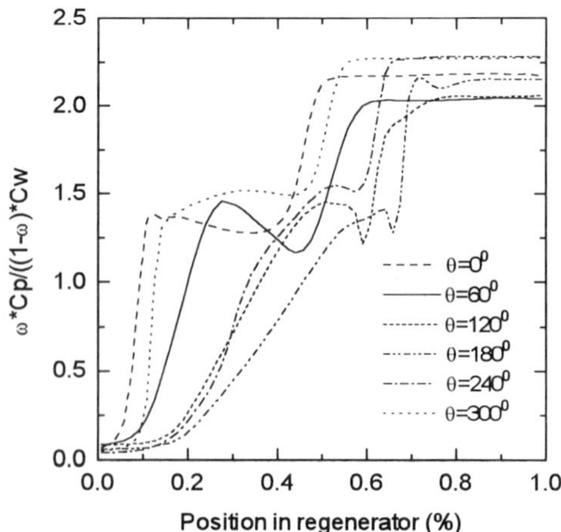


**Figure 7.** Variations of the local heat capacity ratio.

that of a conventional regenerator at higher temperatures in which the helium gas behavior to good approximation can be assumed to be ideal.

*(a) temperature plateau near cold end*

As is well known, the temperature distribution along the axis of a conventional regenerator is a monotonous, almost linear function of position. However, when using real gas properties for $^4$He in the model, a region of almost constant temperature appears near the cold end of a 4 K regenerator, as can be seen in Figure 5. About 35 percent of the regenerative material is near 4 K. This is caused by the large heat capacity of helium in this temperature range, large compared to that of the regenerative material. It is important to consider this flat temperature region for selecting the regenerative materials.

*(b) increased mass flow rate at cold end*

Figure 6 shows the mass flow rate at the hot and cold end of the regenerator. The average mass flow rates for half a cycle for both ends are 2.34 g/s and 4.05 g/s. The mass flow rate at the hot end is only 58% of that at the cold end. This behaviour is opposite to that of a conventional cooler in which the average mass flow rate at hot end is higher than that at cold end. The authors reason that the larger mass flow rate at the cold end is caused by moving of the incompressible $^4$He fluid with high density driven by gas at hot end (see Figure 8). The result also shows us a possibility to increase the refrigeration capacity by slightly increasing the mass flow rate at the inlet of regenerator.

*(c) large heat capacity ratio of helium and the matrix*

The efficiency of the regenerator can be judged by looking at the local ratio $r_c$ of heat capacities of the helium fluid and the matrix material. Taking into account the porosity $\omega$ of the regenerative material this ratio for a given control volume $V_i$ can be expressed as follows:

$$r_c = \frac{C_p \cdot \omega \cdot V_i}{(1-\omega) \cdot C_w \cdot V_i} \tag{2}$$

where $C_p$ is the specific heat of helium, and $C_w$ the specific heat of the matrix.

The local ratio of heat capacity inside the regenerator is illustrated in Figure 7. The largest value occurs in the 4K region, where the local heat capacity of helium is about 2.2 times that of the matrix.

*(d) helium partition inside the regenerator*

The predicted motion of helium particles and the input motion of displacer for simulation are illustrated in Figure 8 (a) and (b). It can be seen that for the chosen working condition a part of helium never leaves the regenerator. The helium inside the regenerator therefore can be divided into three parts (c.f. Figure 8(a)): part I moves back and forth between regenerator and upper stage regenerator, part II never leaves the regenerator, and part III moves back and forth between the regenerator and the cold chamber. This result also verifies a deduction[6] that the circulation of helium in the low temperature GM cooler can break up into two coupled circulations: the low temperature circulation between the expansion space and the low temperature regenerator, and the high temperature circulation between the low temperature regenerator and the ambient temperature compressor.

The helium partition inside the regenerator depends on pressure oscillation, length of the regenerator and swept volume of the dispalcer. For example for a short regenerator (i.g. less than 120 mm in the present simulation), all the working fluid will flow out of the cold or hot end of the regenerator. The motion of helium particles in the 2nd stage regenerator

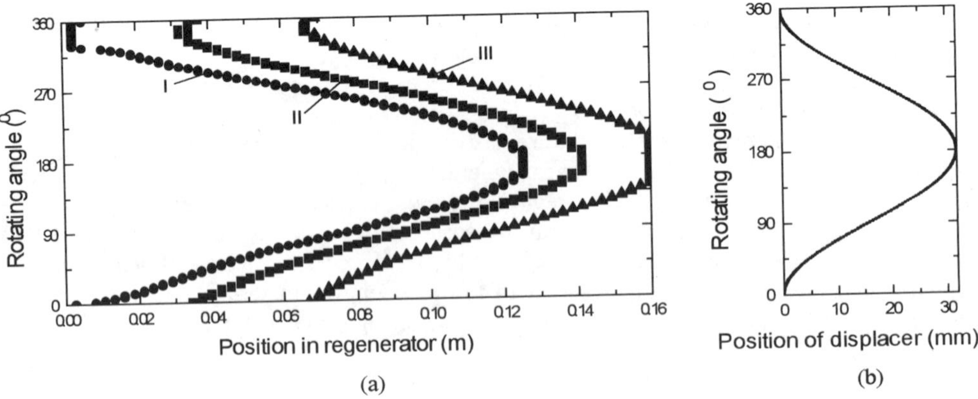


**Figure 8.** Predicted motion of helium particles and the input motion of the displacer for a 4 K GM cooler. (a) helium particles; (b) displacer

and displacer in the Leybold RGD 5/100 GM cooler is illustrated in Figure 9 (a) and (b), which operates with a 1st stage temperature of 40 K and a 2nd stage of 10 K.

## 2. Effect of structure parameters on cooler performance

The effects of initial angle $\phi$ of the displacer on the cooling performance are given in Table 1. There is an optimum value for highest P-V work that leads to maximum cooling power at 4.2 K. The regenerator efficiency increases and the mass flow rates decrease as the angle $\phi$ increases.

Variations of the cooling power at 4.2 K with the length of regenerator are illustrated in Figure 10. The cooling power increases rapidly with increasing length and then remains

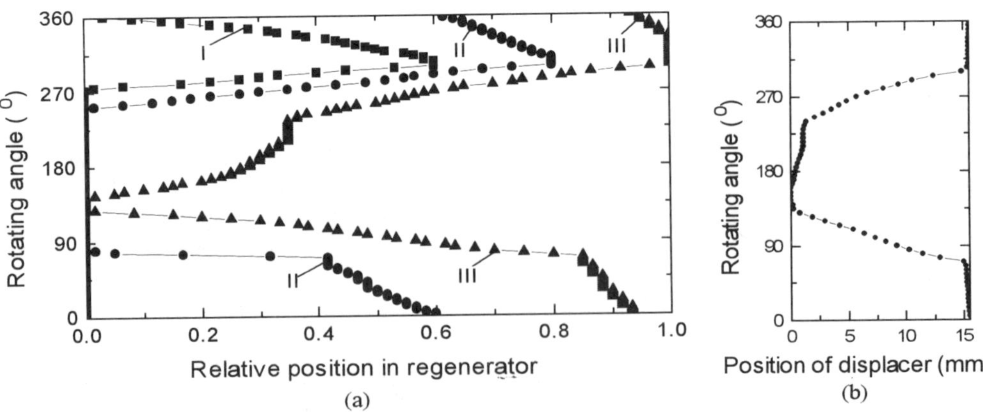


**Figure 9.** Predicted motion of helium particles in the 2nd stage regenerator and the ı easured motion of the displacer in RGD5/100 GM cooler. (a) helium articles; (b) displacer

**Table 1.** Effects of $\phi$ on cooling performance

| | $\phi = -30^0$ | $\phi = -15^0$ | $\phi = 0^0$ | $\phi = 15^0$ | $\phi = 30^0$ |
|---|---|---|---|---|---|
| $\dot{Q}_c$ (W) | 1.88 | 2.13 | 2.19 | 2.11 | 1.91 |
| $\dot{H}_r$ (W) | 21.42 | 23.26 | 23.52 | 22.15 | 19.26 |
| $\dot{H}_c$ (W) | 23.30 | 25.39 | 25.71 | 24.27 | 21.17 |
| $\eta$ | 92.19 | 93.29 | 94.66 | 96.56 | 98.63 |
| $\overline{\dot{m}_h}$ (g/s) | 2.67 | 2.53 | 2.34 | 2.10 | 1.80 |
| $\overline{\dot{m}_c}$ (g/s) | 4.14 | 4.10 | 4.05 | 3.99 | 3.92 |

Note: $\dot{Q}_c$ is net cooling power; $\dot{H}_r$ enthalpy flow in regenerator; $\dot{H}_c$ enthalpy flow into cold chamber (P-V Work); $\eta$ regenerator efficiency; $\overline{\dot{m}_h}$, $\overline{\dot{m}_c}$ mass flow rate at hot and cold end of regenerator.

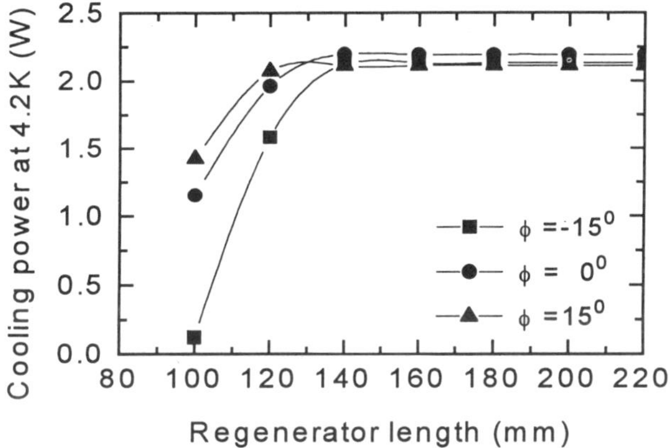


**Figure 10.** Variation of cooling power with regenerator length and initial angle of displacer.

almost constant after reaching a maximum. The most suitable length for this cooler therefore is about 140 mm in the present study. The volume ratio of regenerator and expansion space here is 3.6. Further increasing the regenerator length will not lead to improved refrigeration but only to larger mass flow rate through the hot end. Significant reduction of cooling power with decreasing regenerator length occurs when the initial angle $\phi$ is negative. This may be due to the lower regenerator efficiency with negative initial angle $\phi$. The low temperature plateau should be long enough to avoid large compression heat into the cold head in this case.

**Table 2.** Effects of regenerator porosity $\omega$ on cooling performance

| | $\omega = 0.35$ | $\omega = 0.40$ | $\omega = 0.45$ | $\omega = 0.50$ |
|---|---|---|---|---|
| $\dot{Q}_c$ (W) | 2.40 | 2.19 | 1.97 | 1.76 |
| $\dot{H}_r$ (W) | 23.31 | 23.52 | 23.74 | 23.95 |
| $\dot{H}_c$ (W) | 25.71 | 25.71 | 25.71 | 25.71 |
| $\eta$ | 95.99 | 94.66 | 93.17 | 91.37 |
| $\overline{\dot{m}_h}$ (g/s) | 2.49 | 2.34 | 2.21 | 2.09 |
| $\overline{\dot{m}_c}$ (g/s) | 4.05 | 4.05 | 4.05 | 4.05 |

The effects of regenerator porosity on the performance are shown in Table 2. Increasing the porosity of the regenerator greatly decreases the cooling power and regenerator efficiency. For a regenerator with higher porosity (i.e. 0.45), refrigeration power can not be further improved by increasing the regenerator length as shown by the simulation. Therefore the regenerative materials should be well packed for 4 K GM coolers.

## CONCLUSION

Numerical simulation has been developed for improving the understanding of the internal processes and the performance of 4 K G-M cooler. The predictions are in reasonable agreement with the experiment.

Several features of the low temperature regenerator are completely different from the behavior of regenerator working at higher temperature, such as a temperature plateau near the cold end and increased mass flow rate at cold end.

There are optimum values for phase of displacer and length of regenerator for maximum cooling power. Variation of cooling power with regenerator length depends to some degree on the displacer phase. Slight increases in porosity of regenerator will cause strong reductions of cooling power.

## ACKNOWLEDGMENT

We thank C. Beckers for carrying out the measurements in Figure 2 and 3. Financial support by the Alexander von Humboldt Foundation for one of us (C.W.) is gratefully acknowledged.

## REFERENCES

1. T.Hashimoto, M. Ogawa and R. Li, Recent advances in magnetic regenerator material, *Cryogenics (supplement)* 30: 192 (1990).
2. T, Kuriyama, R. Hakamada, H. Nakagome, et al., High efficient two-stage GM refrigerator with magnetic material in liquid helium temperature region, in: "Advances in Cryogenic Engineering" , Vol.35, Plenum Press, New York (1990) p.1261
3. T.Inaguchi, M.Nagao and H. Yoshimura, Two-stage Gifford-McMahon cycle cryocooler operating at about 2K, in: "Cryocoolers 6", Vol.II, Plenum Press, New York (1990), p.25
4. T. Satoh, A. Onishi and R. Li, Development of 1.5W 4K G-M Cryocooler with magnetic regenerative material, in: "Advances in Cryogenic Engineering", Vol.41B, Plenum Press, New York (1996), p.1631.
5. T.Inaguchi, M.Nagao, K. Naka and H. Yoshimura, Development of 2W Class Gifford-McMahon cycle cryocooler, Proc. of the sixteenth International Cryogenic Engineering Conference and International Cryogenic Materials Conference, Kitakyushu, Japan, May 1996, p.335.
6. C. Hong and X. Xu, On the thermodynamic cycle of the low temperature G-M refrigerator, *Cryogenic (Supplement)* 34: 183 (1994).
7. P. Wu, W. Huang, S.L. Hu, et al. Dynamic simulation of magnetic material regenerator and comparison with experiment, *Cryogenic* 36: 259 (1996).
8. C. Wang, Numerical analysis of 4 K pulse tube coolers Part I: Numerical simulation, *Cryogenics* 37: 207 (1997).
9. C. Wang, Numerical analysis of 4 K pulse tube coolers Part II: Performances and internal processes, *Cryogenics* 37: 215 (1997).

# PERFORMANCE OF A LOW-TEMPERATURE GIFFORD-McMAHON REFRIGERATOR UTILIZING A NEODYMIUM DISK REGENERATOR

J.N. Chafe and G.F. Green

Naval Surface Warfare Center, Carderock Division
Annapolis, MD 21402

## ABSTRACT

In this study, the cooling capability of a Gifford-McMahon (GM) refrigerator has been dramatically increased at temperatures near 4.5 K with no significant increase in input power. This increase in low temperature cooling power can be attributed to a change in design in the low temperature regenerative heat exchanger (regenerator). This regenerator was constructed by precisely forming flow holes through disks of neodymium (Nd) which allowed the reduction of the void fraction from about 35%, for conventionally designed regenerators that utilize spheres, to approximately 10%. Neodymium was chosen because it has a relatively high volumetric heat capacity at temperatures below 10 K and it is a pure metal that can be extruded and drawn with relative ease. The flow holes in the disks were formed using a technique similar to the one used to fabricate multi-filamentary NbTi superconducting wire. Approximately fifty disks, each separated from the next by a porous plastic cloth material, were stacked in a stainless steel tube to form the regenerator. The cooling performance of a Leybold Kelcool 4.2® refrigerator with this regenerator installed as its second stage heat exchanger is presented here and compared to its performance with the factory standard regenerator which uses Nd spheres. The new design improved cooling capacity by more than two fold at temperatures near 4.5 K.

## INTRODUCTION

The long term operation of superconducting magnets has historically been expensive. Short term operation of superconductors in the laboratory could be afforded by purchasing liquid helium from a local supplier and filling a dewar that housed the superconductor. This is relatively inexpensive for short duration experiments run by personnel trained in the handling of liquid cryogens. However for long term operation, even in a laboratory environment, this method of cooling a superconductor becomes prohibitively expensive. In the past, a liquid helium plant might be used to produce liquid helium on site and thus keep the cost of long term operation low. This required a large initial investment in a liquid helium plant and a trained, dedicated staff to operate the plant.

The use of low temperature Gifford-McMahon (GM) cycle refrigerators promises to allow the long term operation of superconductors at a much reduced cost. In addition, the easy operation, high reliability and low maintenance requirements of this type of refrigerator means that superconductors will be available to a more diverse group of users. It will no longer be necessary to have technicians trained in the handling of liquid cryogens or in the operation of complicated liquid helium plants.

Until about 1990, the use of GM refrigerators to cool superconductors was impractical since they could not achieve low enough operating temperatures for most superconductors. Presently, there are several manufacturers offering GM refrigerators that produce at least 0.5 W of refrigeration at 4.5 K. Refrigerators such as these are useful for cooling superconducting magnets and other types of devices, however, more powerful and more efficient refrigerators would have many advantages. One powerful cooler might take the place of two less powerful ones, thus reducing initial costs. In addition, a more powerful cooler would make very carefully designed insulation systems less critical. Increasing the operating efficiency and cooling power of GM refrigerators is the focus of this paper.

Gifford McMahon refrigerators can now reach low temperatures (below 4.5 K) primarily due to the use of specially selected materials for the lowest temperature regenerative heat exchanger. GM refrigerators utilize regenerative or periodic heat exchangers that cyclically store energy in some solid material. This energy storage is accomplished by allowing a gas to flow through a tube that is packed with heat storage material. This material (referred to as the regenerator matrix material) must be porous to allow the gas to pass through it. The gas and matrix material start off at different temperatures but as the gas flows through the matrix material the temperature of the gas begins to approach the temperature of the matrix as energy is thermally exchanged between the two. At a later time the gas flows in the reverse direction through the matrix material and again energy is transferred between the two. There are many design factors that affect the performance of regenerator, especially if it must operate at temperatures (below 7 K).

The most critical requirement for GM refrigerators to operate below 7 K is a well designed low temperature regenerator. Although some GM refrigerators only have one regenerator, all GM coolers that reach temperatures below 4.5 K have either two or three regenerators with each regenerator operating over different temperature spans. The coldest part of the first stage of these refrigerators usually reaches a temperature between 20 and 50 K. Each remaining stage cools the gas further until it eventually reaches its lowest temperature. The warmer stage or stages usually don't require any special treatment because they are not the limiting element in achieving low temperatures. It is the final stage (in our case the second stage) that requires special attention. The goal is to construct this regenerator from materials having high heat capacity and to assemble it in a way that allows good heat exchange and limits the regenerator's porosity (highly porous low temperature regenerators tend to have higher thermodynamic losses associated with them). For some GM refrigerators simply replacing the lead (Pb) matrix material with a rare-earth metal such as neodymium (Nd) has lowered cooling temperatures from around 7 K to below 2.5 K. In an effort to further improve these coolers (allow more cooling power at some temperature or temperature range of interest e.g. 4.5 K), a regenerator was constructed with a significantly less porous matrix geometry than the commonly used spherical particles.

The initial phase of this work was presented at an earlier conference[1]. That paper discussed the results of experiments performed with a regenerator constructed from Nd disks with many small holes passing through them. This paper discusses Nd disks formed by the same manufacturing process except that the configuration and quality of these disks are much improved and the test results show improved cooling power.

## THE TEST COOLER

The test refrigerator is a two stage GM cooler. Its major components are a model Kelcool 4.2™ cold head module and a model UCC 110S™ compressor module presently manufactured by Leybold, Inc. While the compressor module is unaltered from its factory configured state, the cold head module is modified. Some of the matrix material in the first stage (originally all phosphor bronze screens) has been replaced with Pb spheres. Additionally the sliding seal, located at the top of the second stage regenerator, has been reconfigured.

## SECOND STAGE REGENERATOR AND THE DISKS

One of the critical factors to be considered when designing a regenerator that operates below about 7 K is the choice of matrix material. In order for a regenerative heat exchanger to function effectively, it must have the capability to store energy with small temperature changes. There are two material properties that strongly affect this ability to store heat; volumetric heat capacity and thermal conductivity. Typically these properties are strongly dependent on the operating temperature. It is important that both these properties remain high over the operating range of the regenerator. If needed, moderately low values of thermal conductivity can be compensated for through the design of the regenerator, but it is crucial to use a material with a high volumetric heat capacity. Unfortunately most materials do not have the high volumetric heat capacity needed for operation at temperatures below 7 K. However a few materials have been identified that do have this property. One of these materials is the neodymium (Nd). Neodymium is used in this research because in addition to its high volumetric heat capacity, it is ductile and formable while various other materials suitable for low temperature regenerators are not. In order to form the regenerator, the matrix material was swaged and drawn extensively. These processes required that the selected matrix material be ductile and malleable.

### *Geometry*

The regenerators of commercially available GM refrigerators are assembled either by stacking screens in a tube or by filling the tube with small spherical particles. The first stage regenerator typically uses screens made from very fine phosphorous bronze wire (0.005 cm (0.002 in) diameter.) woven in a 180 or 200 mesh pattern. The second stage regenerator is most often made from lead (Pb) spheres of around 0.025 cm (0.010 in) diameter. These materials are loaded into their respective regenerator tubes and each end of the tubes is fitted with an end piece that keep the matrix material in place while allowing gas to flow through the regenerator.

The use of spheres or screens as a regenerator matrix material is generally advantageous, yet there are some limitations associated with this geometry. Spheres and screens are widely available and offer an easy way to construct a regenerator with continuous flow passages and uniform porosity. The uniformity that can be obtained by controlling sphere or screen size provides the ability to produce consistent performance from regenerator to regenerator. However, a limitation of both screens and spheres is that they form matrices that are too porous and this porosity cannot be easily controlled. For example, a typical regenerator made from uniform spheres is about 35% porous regardless of the size of the spheres. It would be preferable to have a regenerator that is less porous, perhaps 15% to 20%. The only way to reduce the porosity in a sphere regenerator is to use spheres of more than one size, this is difficult to do and the resulting flow characteristics vary significantly from regenerator to regenerator. It is similarly difficult to tailor the porosity of screens.

There are two major effects that lowering porosity has on the performance of a GM refrigerator. First, low porosity allows for more matrix material to be placed in the regenerator resulting in more heat storage medium in the regenerator. Second, the amount of gas contained by the regenerator is reduced and there is significant loss associated with the pressure cycling of the gas.

An alternative to spheres or screens is the use of a stack of disks that have flow holes through their faces. These are used in much the same way as screens are used. The number, size and spacing of these holes control the porosity, pressure drop and heat exchange in a regenerator. The primary obstacle to the widespread use of this geometry is the difficult fabrication process. In order to construct a regenerator that out-performs spheres or screens, each disk must have tens of thousands of small (0.002 cm (0.001 in) to 0.025 cm (0.010 in) in diameter) holes that are evenly spaced and uniform in size and shape. A technique that can accomplish this difficult task is under development.

### *Fabrication Technique*

The technique commonly used to fabricate superconducting wire was modified to produce precisely formed holes through neodymium disks. Superconducting wire has many small and precisely formed NbTi filaments in a copper matrix. For the manufacture of regenerator disks, Nd replaces the copper matrix and the NbTi filaments are replaced by a proprietary material that is eventually etched away. As in the manufacture of superconducting wire, a composite billet is constructed from the different materials. This billet is extruded and drawn and then cut into short lengths. The short lengths are then reassembled into a new composite billet that is extruded, drawn and cut. This process is repeated until a rod of Nd, with many fine filaments uniformly distributed within it, is obtained. This rod is sliced into disks approximately 0.1 cm (0.040 in) thick. The sacrificial filaments that pass through each disk are etched away to form holes. A more detailed description of the fabrication technique for the disks is presented in reference 1.

### *Finished Disks*

Several of the completed disks are shown in Fig. 1. The diameter of each disk is about 2.5 cm (1 in) with a thickness of about 0.1 cm (0.040 in) and an average mass of about 2.8 grams. Figure 2 is a photomicrograph of one of the disks taken by a scanning electron microscope. Each hole is 0.005 to 0.008 cm (0.002 to 0.003 in) in diameter and the porosity of each disk, estimated graphically from photomicrographs similar to fig. 2, is about 12.5%. There are roughly 16,500 holes in each disk. Figure 2 also shows how the size and shape of the holes varies within a disk.

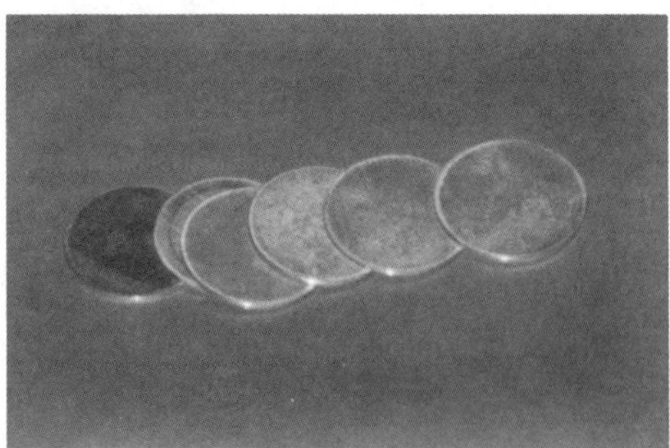

**Figure 1.** Several Nd disks, ready for assembly into the regenerator.

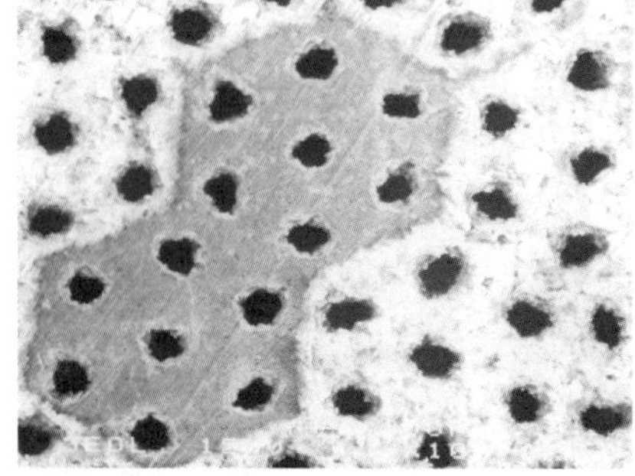

**Figure 2.** Photomicrograph of a Nd disk. There are about 16,500 holes per disk.

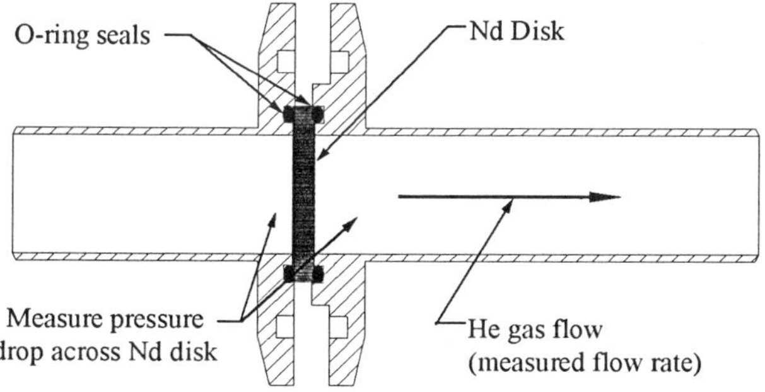


**Figure 3.** Pressure drop measurement apparatus for the individual Nd disks. Helium gas was supplied to the left side of the apparatus and flow rate was measure at the right.

Seventy-five disks manufactured as described were available for use in this study. The flow characteristics of each disk was measured because some of the disks varied in appearance from the rest and it was feared that some were damaged during fabrication. Figure 3 depicts the apparatus that measured the pressure drop across each disk. Each disk was mounted in the apparatus and pressurized helium was applied to one side of the disk. The applied helium pressure was adjusted to produce a flow rate of 57 L/min (2.0 SCFM). Results of this testing is shown in Fig. 4 as a bar graph. The length of each bar is proportional to the pressure drop across the corresponding disk for a fixed flow rate. Figure 4 highlights the fact that while most disks have a pressure drop per unit flow rate of 0.122 to 0.146 kPa-min/L (0.5 to 0.6 psi/SCFM), several of the disks have a considerably higher pressure drop per unit flow rate. The disks with a pressure drop per unit flow rate greater than 0.15 kPa-min/L (0.6 psi/SCFM) were discarded as faulty and were not used in the assembly of the test regenerator.

### *Assembled Regenerator*

The second stage regenerator tube has an internal cavity that houses the regenerator matrix material. This cavity is about 10 cm (4 in) long and has a diameter of approximately 2.5 cm (1.0 in). In the standard, factory supplied configuration, the cavity is filled with roughly equal volumes of Pb and Nd spheres. The Pb spheres are loaded into the top (warm) half of the regenerator while the Nd spheres are loaded into the bottom (cold) half. For the experiment, Pb spheres were still loaded into the warm half, but Nd disks instead of the spheres were used in the cold half. There were two reasons why the regenerator matrix was not solely Nd disks. First, Pb works better at the higher temperatures found at the warm

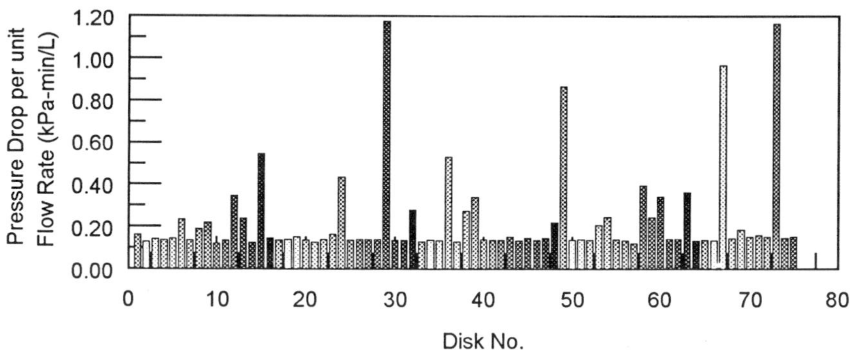


**Figure 4.** Measured pressure drop for each individual Nd disk. Twenty-three disks with the largest pressure drops were disqualified for regenerator construction.

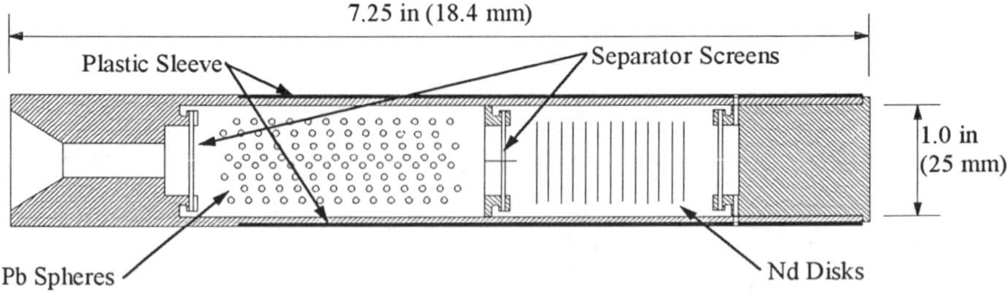


**Figure 5.** Sketch of the assembled second stage regenerator. The left side of the figure depicts the warm end of the regenerator. The right side of the figure, shown housing the Nd disks, depicts the cold end of the regenerator.

end of the regenerator because its heat capacity and thermal diffusivity are greater than Nd's at these temperatures. Second, due to manufacturing difficulties, there were not enough high quality disks available to construct an all Nd regenerator.

A sketch of the regenerator is shown in Fig. 5. The regenerator is comprised of a stainless steel tube that has a plastic sleeve covering most of its length. This sleeve prevents the stainless steel regenerator tube from rubbing against and galling its corresponding cylinder. About a dozen 200-mesh stainless steel screens were welded to the warm end of the regenerator tube and then about 200 g of 0.23 mm (0.009 in) diameter Pb spheres were loaded in. Next came a short length of tube with more screens welded to it. This separator keeps the Pb spheres captive in their section of the regenerator. Fifty-one Nd disks (approximately 140 g total) were loaded and finally a stainless steel plug was welded onto the end to keep the disks in place. No attempt was made to align the holes from one Nd disk to the next. Instead each Nd disk was separated from its neighbor by circular disks cut from porous, spun-plastic-fiber fabric. These fabric circles provided space for the gas to exit and enter neighboring disks and prevented one disk from blocking the flow holes of its neighbor. The fabric also acted as a thermal insulator between each Nd disk keeping the thermal conduction along the length of the regenerator small.

The assembled regenerator was flow tested in a similar manner to the testing of each disk. However, for the complete regenerator, pressure drop measurements were made for several flow rates. The results of the tests are shown in Fig. 6. Also shown in the figure are

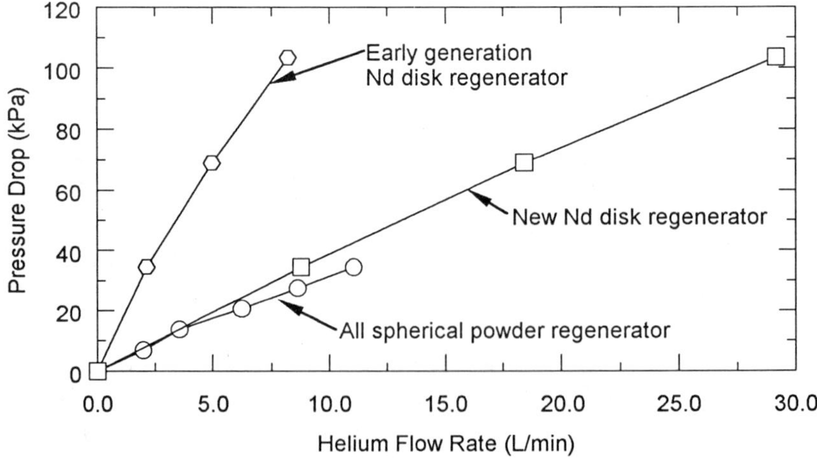


**Figure 6.** Pressure drop measurements in assembled regenerators. The curve labeled "New Disks" represent the data for the Nd disk reported on in this paper. The curve labeled "All Sphere Regenerator" represent a regenerator that was constructed using all spheres and no disks. The other curve represents data for Nd disks that were of inferior quality.

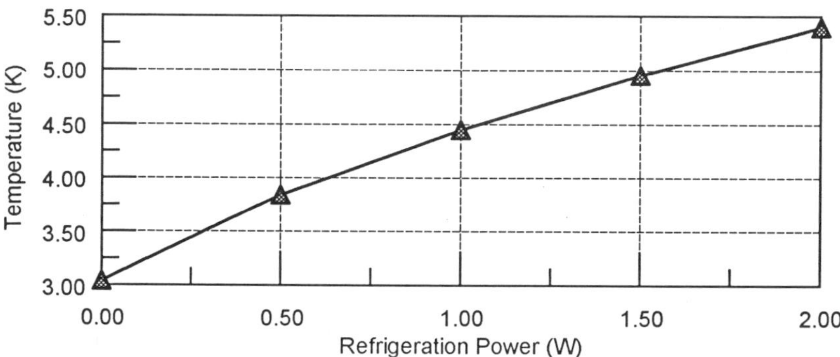


**Figure 7.** Second stage operating temperature as a function of cooling power. The first stage is unloaded for these data.

flow measurements for an all sphere regenerator and a regenerator made with an earlier generation of Nd disks that had smaller and less well formed holes. This figure demonstrates that the new disk regenerator, while having less than half the porosity of a sphere regenerator, has only a slightly greater pressure drop. This means that the thermodynamic losses associated with large porosity should be much lower for the disk regenerator and the loss associated with pressure drop should be only slightly higher.

## TEST RESULTS

The refrigerator was operated under normal pressure levels (0.7 to 2.0 MPa (100 to 300 psig)) and speeds (approximately 2 cycles per second) while various heat loads were applied to both the first and second stages using electric heating elements. The temperature of both stages of the refrigerator were then measured using calibrated silicon diodes.

The second stage temperature is plotted as a function of applied heat load to the second stage in Fig. 7. There was no heat applied to the first stage for the data shown in this figure. The refrigerator reached a low temperature of about 3 K and it supplied 1 W of cooling at 4.5 K and 1.5 W at 5 K. This is a significant improvement over the standard production Kelcool 4.5™ that uses Nd spheres. That cooler typically only produces 0.5 to 0.7 W at 4.5 K. The use of Nd disks in the second stage regenerator increased the cooling power by more than 40% at 4.5 K while not requiring any significant increase in input power to the system.

A heat load map of the test refrigerator is shown in Fig. 8. The graph presents refrigerator temperatures when both the first and second stages are simultaneously loaded with heat. Such a load map gives visual cues of how heat loads applied to one stage affects the cooling performance of the other stage. The more rectangular the map is, i.e. the individual load lines are close to horizontal and vertical, the less affected one stage is by the other. An example of this can be seen by noting that the line representing 0 W of heat input to the second stage is roughly vertical for first stage temperatures less than 40 K. For first stage temperatures above 40 K, the temperature of the second stage also rises due only to the increased heat load on the first stage. This is indicated by the slanted part of the line. It is interesting that when the second stage heat load is between 1 and 2 W, the second stage is less affected by the heat load on the first stage.

## CONCLUSIONS AND FUTURE WORK

Improving the configuration of the low temperature regenerator can result in significant increases in cooling power at low temperatures. This has been demonstrated by using a

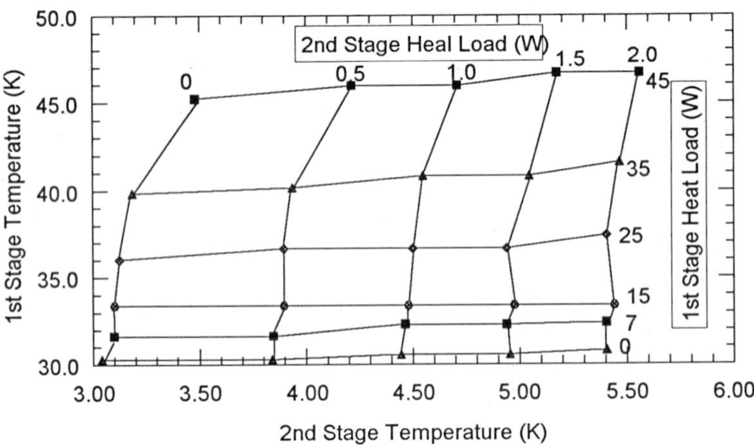


**Figure 8.** Load map for the refrigerator operating with the Nd disk regenerator. The operating temperatures for the two stages are given for conditions when heat is applied to one or both stages.

regenerator that has approximately 1/3 the void volume of the commonly used spherical powder regenerator. The pressure drop in this regenerator was only slightly higher than the conventional regenerator. When installed in the test refrigerator over 1 W of cooling was produced at 4.5 K. This is a significant improvement over the 0.5 W that is typically available with the spherical powder regenerator. This work demonstrates the importance of reducing the void volume in low temperature regenerators.

The regenerator used in this research demonstrated only one of many possible geometries that could be used to improve low-temperature refrigerator performance. However for a regenerator to be commercially viable, it must not only perform well but it must also be affordable. The experience so far with producing Nd disk has shown there to be two major impediments to making such a regenerator affordable. These are: an extremely low yield of usable disks; and the fact that only large quantities of materials can be processed at a time. In the attempts to produce Nd disks, several production runs were made that started with large quantities of Nd (45.5 kg) that resulted in only a very small number of usable disks. It may be possible to refine the fabrication process, but until this happens widespread, use of Nd disks is not practical.

This work focused on the use of Nd as a matrix material primarily because it is a ductile material that has high specific heat at low temperatures. However, other materials also make excellent low temperature regenerators. The study presented here shows that if ways were established of fabricating these materials into low void-volume geometries then improved performance could be expected. In addition, work clearly needs to continue in identifying new materials that could be useful for low temperature regenerators.

As an alternative to the process used to produce the Nd disks used in this study perhaps other processes to produce disks with precise holes could be devised. Other geometries that have low void volumes and low pressure drops should also be investigated. Another area of investigation that might be useful is the use of low void volume or low pressure drop regenerators for higher temperature use.

## REFERENCES


1 J. Chafe, G. Green, and J. Hendricks, A neodymium plate regenerator for low-temperature Gifford-McMahon refrigerators, in: "Cryocoolers 9", R.G. Ross, Plenum Press, New York, 1997.

# NUMERICAL AND EXPERIMENTAL STUDY OF A 4 K MODIFIED-SOLVAY CYCLE CRYOCOOLER

T. Kurihara,[1] M. Okamoto,[1] K. Sakitani,[1] H. Torii,[2] and H. Morishita[2]

[1]Mechanical Engineering Laboratory, Daikin Industries Ltd.
1304 Kanaoka-cho, Sakai, Osaka 591, Japan
[2]VC Division, Daikin Industries Ltd.
3-12 Chikko-shinmachi, Sakai, Osaka 592, Japan

## ABSTRACT

A two-stage 4 K Modified-Solvay cycle cryocooler[1], which is a pneumatically driven Gifford - McMahon cryocooler and will be used for the conduction-cooled low temperature superconducting magnets, has been optimized by numerical and experimental study. After the optimization, the maximum cooling capacity of 36.5 W at 35 K on the first stage and 1.08 W at 4.2 K on the second stage with 6.7 kW input power has been obtained. This is the world's first pneumatically driven G-M cryocooler which has over 1 W cooling capacity at 4.2 K. In this paper, experimental and numerical results of the influence of the valve opening timing and the displacer motion concerning the cooling capacity are described. The experimentally obtained optimum P-V diagram and valve timing are in good agreement with the numerical simulation results. And the reliability tests (over 100 times of cool-down and warm-up cycle, and over 3,000 hours continuous operation) were successfully conducted. It can be expected that this 4 K cryocooler has a high reliability equivalent to the commercial G-M cryocoolers which operate above 10 K on the 2nd stage temperature.

## INTRODUCTION

In recent years, the performance of 4 K two stage G-M cryocoolers with mechanical driving mechanism of displacer has been improved by the several experimental approaches [2-5], which are the development of the new magnetic regenerative materials and the optimization of the several design parameters such as the displacer reciprocating frequency, the displacer stroke, the valve opening timing, the 2nd stage cylinder diameter, the packing quantity of the 2nd stage regenerator materials and the multi-layer regenerator. And several numerical optimization analyses of the regenerator operating at 4 K also have been reported [6-11]. At present time, the 4 K two stage G-M cryocooler with the cooling capacity of 1 W at 4.2 K is commercialized [12]. However for the pneumatically driven 4 K G-M cryocoolers, there are only a few reports existing [13].

We have already commercialized the 4 K two-stage Modified-Solvay cycle cryocooler with the cooling capacity of 0.5 W at 4.2 K, and now we have been developing the high performance 4 K two-stage cryocoolers in order to cool the MRI magnet and the other comparatively large superconducting magnets.

For the pneumatically driven 4 K G-M cryocoolers, the most difficult point is the optimization of the displacer motion, that is very sensitive to several conditions. However we believe that if the displacer motion is optimized , its performance will be equivalent to mechanically driven G-M cryocoolers (from numerical simulation results but not mentioned in this paper) and the good points of Modified-Solvay cycle cryocooler reflect to the 4 K systems. For the optimization purpose, we developed numerical simulation of the pneumatically driven 4 K two-stage G-M cryocooler[14], and accomplished the refrigeration capacity of 0.93W at 4.2 K by numerical and experimental optimization study of 2nd regenerator materials[15]. Furthermore by optimizing the valve opening timing, the displacer motion, cylinder diameter and the quantity of regenerator materials and by using the new magnetic regenerator material $HoCu_2$, over 1 W cooling capacity at 4.2 K has been obtained, and the 1st stage cooling capacity has been considerably improved [16].

It has been already demonstrated that our numerical simulation has good performance to calculate the influence of the specific heat of regenerator materials[15]. In this paper, the main purpose is the demonstration of the validation of our numerical simulation for estimating the influence of the valve opening timing and the displacer motion or the shape of the P-V diagram.

## EXPERIMENTAL APPARATUS AND PROCEDURE

The schematic diagram of the experimental apparatus is shown in Figure 1. In order to measure the displacer motion, the valve motor head is separated from the expander cold head, and the displacer motion is measured by Laser Doppler Displacement Sensor. The

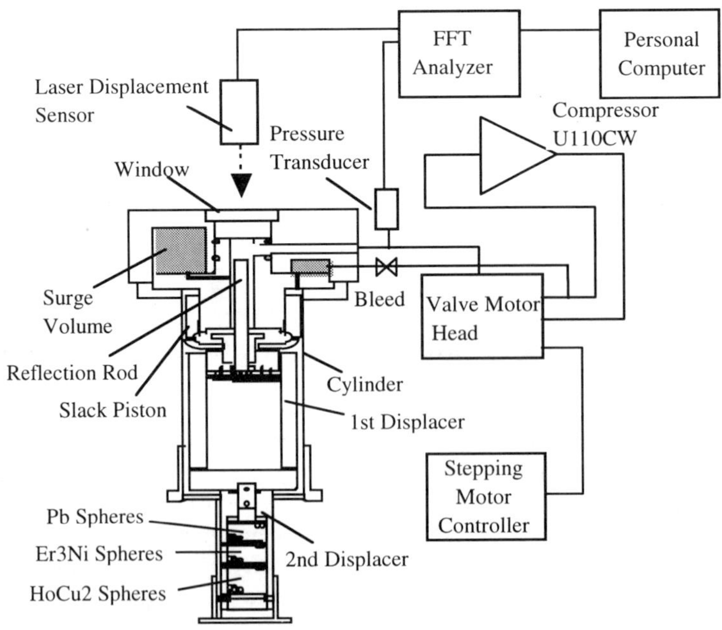


**Figure 1.** Schematic diagram of experimental apparatus .

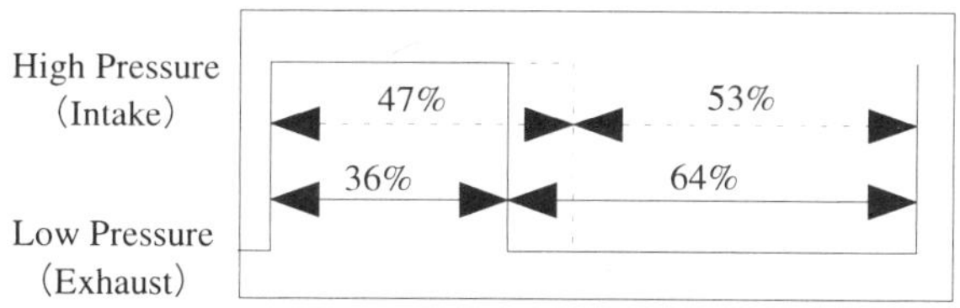


**Figure 2.** The definition of the valve opning timing.

warm end pressure is measured and its value is used to estimate the P-V work of 2nd stage expansion volume. In this case, as the pressure drop in the regenerators is not considered, the obtained P-V work is larger than the actual one.

The valve opening timing is changed by means of the stepping motor speed control in the one rotation cycle and the displacer motion is changed by the orifice and the bleed valve shown in Figure 1. The definition of the valve opening timing is shown in Figure 2 and as shown in Figure 2, the exhaust period is longer than the intake period, and there is no closing period.

The bleed has the effect to change the displacer movement by means of the pressure level control of the surge volume. When the bleed is connected to low pressure side, the surge volume pressure is lower than its ordinary value, so the displacer has a tendency to move on the warm end side, and the bleed is connected to high pressure side, the displacer has a tendency to move on the cold end side.

## NUMERICAL SIMULATION

The one-dimensional governing equations (conservation of mass, momentum and energy of helium, conservation of energy of regenerator matrix, and motion equation of the displacer and the slack piston) are solved using finite difference method considering variable physical properties of helium and regenerator materials [14]. The heat conduction loss of the cylinder and shuttle heat loss are estimated from the literature [17].

The numerical optimization process is as follows. First of all, the materials and the packing quantity of the multilayer regenerator have been optimized. Next the displacer stroke and the operating frequency have been optimized. Finally the influence of the valve opening timing and the displacer motion on the cooling capacity has been investigated to clear the optimum operating conditions. The expander's specifications obtained from the numerical optimization analysis are shown in Table 1.

For example, the calculated temperature distribution in the 2nd stage regenerator is shown in Figure 3. This temperature distribution is considerably similar to the experimental measurement[18] and typical for the 4 K G-M cryocoolers, because the enthalpy difference between hot end and cold end in the 2nd regenerator of the high pressure helium is much smaller than that of the low pressure helium.

**Table 1.** Specifications of 4 K Modified-Solvay Cycle Cryocooler

| | |
|---|---|
| Cylinder diameter | 77 mm/1st 36 mm/2nd |
| 2nd regenarater materials | Pb(103g) + Er3Ni(86g) + HoCu2(170g) |
| Stroke | 12 mm |
| Frequency | 2 Hz |
| Valve Intake/Exhaust ratio | from 47/53 to 36/64 % |

Input power 6.7 kW (DAIKIN U110CW 60Hz)

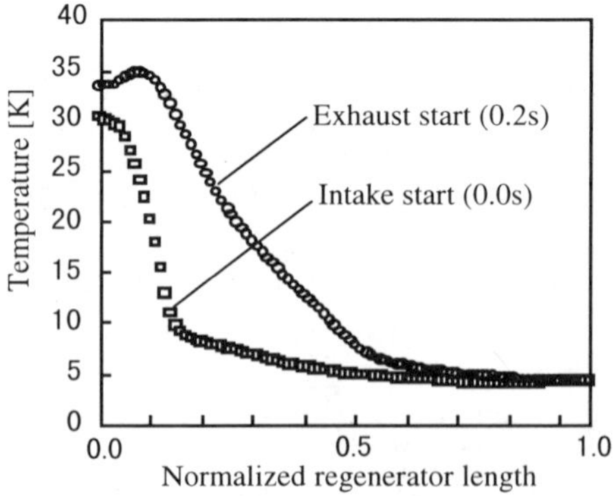

**Figure 3.** Temperature distribution of 2nd regenerator (Intake/Exhaust ratio: 40/60).

## COMPARISON BETWEEN THE EXPERIMENTAL AND NUMERICAL RESULTS

From the experimental P-V measurements, it is found that the displacer motion is sensitive to the characteristics of the seal leakage of the slack piston. However by adjusting the bleed valve, the displacer motion can be changed and optimized. The numerical simulation also can calculate the effect of slack piston seal leakage characteristics, but the comparison to the experimental results is difficult because of the lack of actual seal leakage data in the moving condition of the slack piston. So only the influence of the displacer motion or the shape of P-V diagram and the valve opening timing is compared.

### The relationship between the valve opening timing and the cooling capacity

Figure 4 shows the relationship between the valve opening timing and the cooling capacity of 1st and 2nd stage obtained from the experimental measurements and the numerical calculations. In this case, the displacer motion is optimized by adjusting the orifice valve and the bleed valve for both experimental measurement and numerical analysis. The agreement between experimental and numerical results is good and as shown in Figure 4, the optimum valve Intake/Exhaust opening ratio lies from 40/60 to 36/64.

### The relationship between the shape of P-V diagram and the cooling capacity

Figure 5 and Table 2 show the relationship between the shape of P-V diagram and the cooling capacity, in case of the same valve Intake/Exhaust opening ratio(40/60) and the

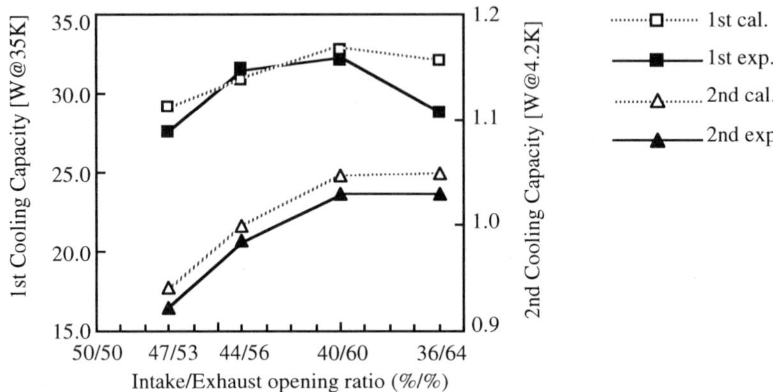

**Figure 4.** Relationship between Intake/Exhaust opening ratio and cooling capacity.

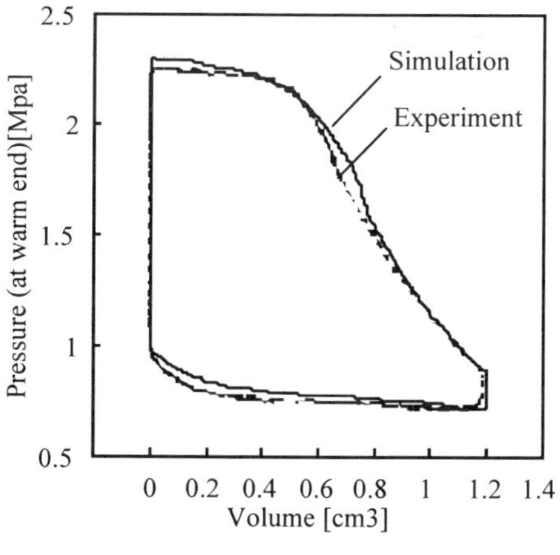


(a) In case of the displacer moving in the cold end side

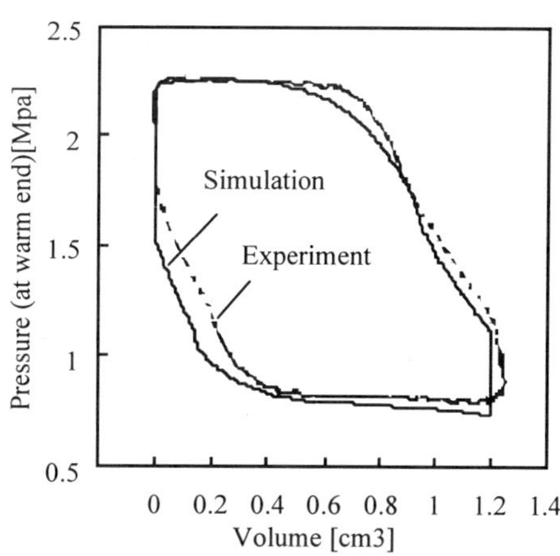


(b) In case of the displacer moving in centor of expansion space : optimum case

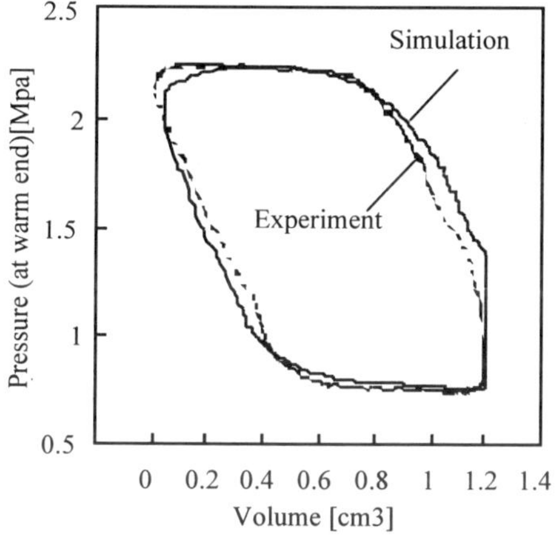


(c) In case of the displacer moving in the warm end side

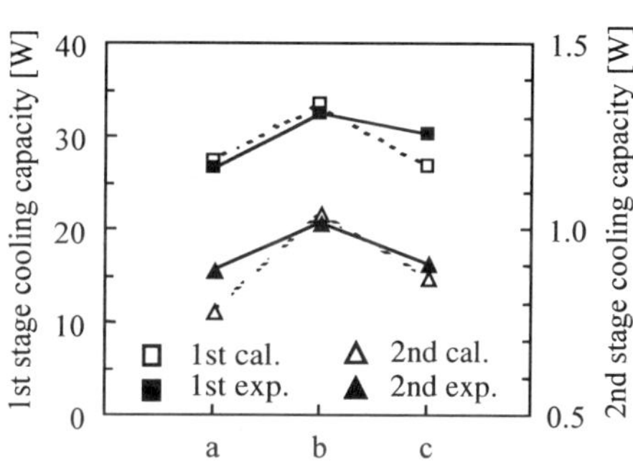


(d) Cooling capacity in case of (a) ~ (c)

**Figure 5.** Relationship between the shape of P-V diagram and the cooling capacity (35K/4.2K) at Intake/Exhaust ratio 40/60.

**Table 2.** Experimental and numerical results for P-V work and cooling capacity correspond to the P-V diagrams in Figure 5

| | Experimental results | | | Simulation results | | |
|---|---|---|---|---|---|---|
| | (a) | (b) | (c) | (a) | (b) | (c) |
| 2nd stage P-V work (using warm end Pressure) | 24.6W | 26.4W | 25.3W | 24.6W | 26.1W | 24.8W |
| 2nd stage P-V work (using cold end Pressure) | | | | 19.9W | 22.1W | 21.1W |
| 1st stage cooling capacity@35K | 26.6W | 32.5W | 30.2W | 27.2W | 33.5W | 26.8W |
| 2nd stage cooling capacity@4.2K | 0.89W | 1.02W | 0.91W | 0.780W | 1.045W | 0.871W |

different bleed opening rate. The agreement between the experimental and the numerical results is comparatively good, and the optimum P-V diagram is shown in Figure 5(b). And the 2nd stage P-V work calculated using the cold end pressure obtained from the simulation is also indicated in Table 2. The 2nd stage P-V work is reduced about 15% by the pressure drop in the regenerator. In case of the other valve opening timings , which are not shown, the shape of the optimum P-V diagram is almost same.

Figure 6 shows the pressure change and the displacer displacement change in one cycle for the optimum case (correspond to Figure 5(b)). In the optimum case, the displacer oscillates around the center of its stroke and it is moved rapidly in the regions near the upper and lower limit of the displacer stroke by the force due to the pressure difference in the both side of the displacer, when the valve connection is switched from high pressure to low pressure (or from low pressure to high pressure) of the compressor. This displacer movement produces the nearly adiabatic pressure-volume change of the upper right and the lower left in the P-V diagram. This nearly adiabatic pressure-volume change gives nearly the maximum P-V work per unit mass flow rate in the expander, so the balance among the P-V work, the regenerator efficiency and the mass flow rate is nearly best.

## The influence of the 1st stage temperature for the cooling capacity and P-V diagram

Figure 7 shows influence of the 1st stage temperature for the cooling capacity of the 1st and 2nd stage. From the experimental results, the 2nd stage cooling capacity increases with increasing the 1st stage temperature from 25 K to 55 K. However from the numerical simulation, the maximum 2nd stage cooling capacity is obtained at from 35 K to 45 K temperature range of the 1st stage. The 4 K G-M cryocooler has the characteristics that when the 1st stage temperature is over the critical value ranged from 50 K to 60 K, the 2nd stage cooling capacity is drastically reduced to zero watt[5]. Therefore it is difficult to accurately calculate the temperature dependent cooling capacity in this critical temperature range, and so we think that the difference between experimental measurements and simulation results is acceptable for practical use.

Figure 8 shows the influence of the 1st stage temperature for the P-V diagram, and the agreement between experimental and numerical results are considerably good. When the 1st

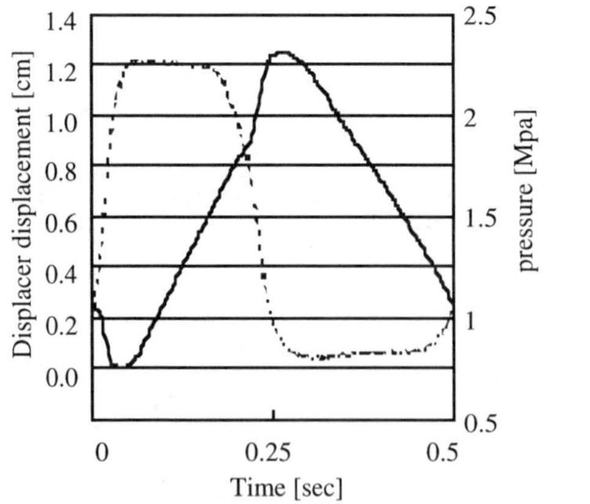


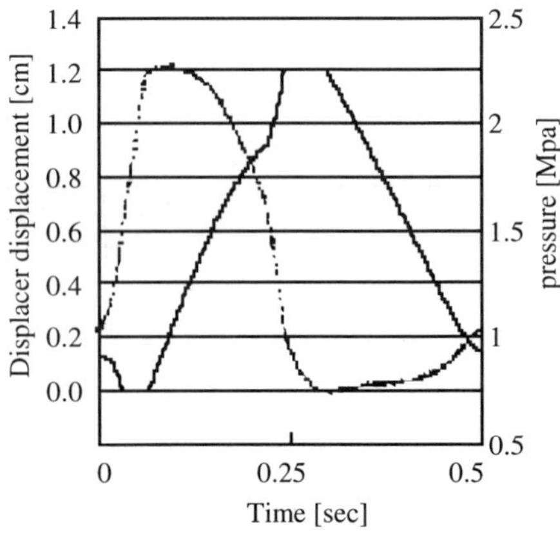


**Figure 6.** Pressure and displacer displacement in one cycle (In the case of Figure 5(b) ). Left: Experimental measurement, Right: Numerical simulation.

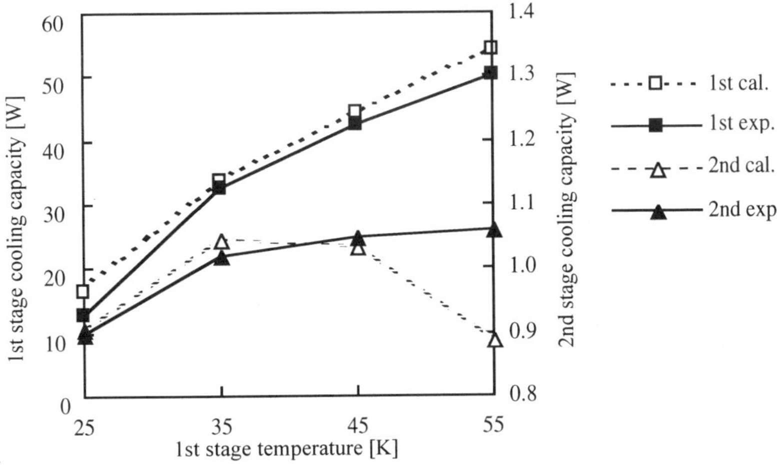


**Figure 7.** Influence of 1st stage temperature for the cooling capacity.

stage temperature rises from 25 K to 55 K, the P-V work of the 2nd stage increases considerably (8W from the experiment), but cooling capacity slightly increases (0.16W from the experiment) due to mainly the increase of the regenerator loss.

## ADDITIONAL REMARKS

In the previous section, the experimental results are obtained by using valve motor head separate type expander, which have dead volume in the connecting tube, so the cooling capacity is slightly reduced the normal type expander. For the normal type expander, the maximum cooling capacity of 36.5 W at 35 K on the first stage and 1.08 W at 4.2 K on the second stage with 6.7 kW input power has been obtained. And the 100 cycle on/off test for one expander and over 3,000 hours continuous operation test for the several expanders had been successfully conducted. No obvious cooling capacity degradation had been found.

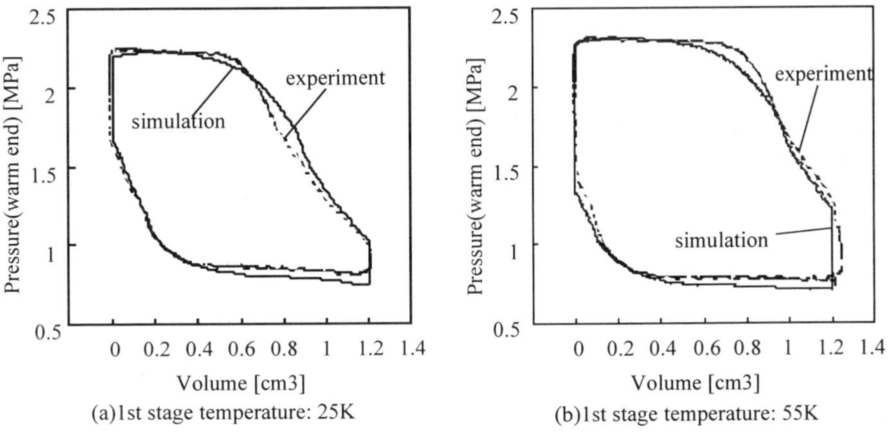


**Figure 8.** Influence of 1st stage temperature for the P-V diagram.

## CONCLUSION

For the two-stage 4 K Modified-Solvay cycle cryocooler, numerical and experimental study has been performed concerning the valve opening timing and the displacer motion. The numerical simulation results have been in good agreement with experimental results, and the optimum valve Intake/Exhaust opening ratio and the optimum displacer motion have been obtained. It is concluded that our numerical simulation is useful for the design of pneumatically driven 4 K G-M cryocoolers. Of course as the sinusoidal displacer motion can be given, it is expected that this simulation is available for the design of the mechanically driven 4 K G-M cryocoolers.

Furthermore, the developed 4 K cryocooler has good efficiency on the performance which is as good as the mechanically driven 4 K G-M cryocooler reported in the literature [3], and has a high reliability equivalent to the commercial G-M cryocoolers which operate above 10 K on the 2nd stage temperature.

## REFERENCES


1. R.C.Longsworth, A modified Solvay-cycle cryogenic refrigerator, Adv. Cryog. Eng., Vol.16:195(1971).
2. M.Sahashi et al., New magnetic material $R_3T$ system with extremely large heat capacities used as heat regenerators, Adv. Cryog. Eng., Vol.35B:1175(1990).
3. R.Li et al., Influence of valve open timing and interval on performance of 4 K Gifford-McMahon cycle cryocooler, Adv. Cryog. Eng., Vol.41B:1601(1996).
4. T.Kuriyama et al., Optimization of operational parameters for a 4 K-GM refrigerator, Adv. Cryog. Eng., Vol.41B:1615(1996).
5. T.Satoh et al., Development of 1.5W 4K G-M cryocooler with magnetic regenerator material, Adv. Cryog. Eng., Vol.41B:1631(1996).
6. H. Seshake et al., Analysis of rare earth compound regenerators operating at 4 K, Adv. Cryog. Eng., Vol.37B:995(1992).
7. T.Tsukagoshi et al., Refrigeration capacity of a GM refrigerator with magnetic regenerator materials, Adv. Cryog. Eng., Vol.41B:1623(1996).
8. Zhang Zhu and Liang Zhang, The theoretical cooling capacity and heat loss analysis of a 4.2K G-M refrigerator, Adv. Cryog. Eng., Vol.41B:1639(1996).
9. L.Zang et al., Study of effect of heat transfer in the cold head to the performance of a 4.2K G-M refrigerator, Adv. Cryog. Eng., Vol.41B:1645(1996).
10. P.Y.Wu et al., Dynamic simulation of magnetic material regenerator and comparison with experiment, Cryogenics, Vol.36:259(1996).
11. T.Inaguchi et al., Proceedings of 56th Meeting on Cryogenics and Superconductivity:3(1997).
12. Sumitomo Heavy Industries, Ltd. "Sumitomo Heavy Industries Technical Review", Vol.45 No.133:50(1997).
13. G.chen et al., Investigation on a two-stage Solvay refrigerator with magnetic material regenerator, Cryogenics ICEC Supplement, Vol.32:5(1992).
14. T.Kurihara et al., Numerical analysis of the performance of pneumatic driven 4 K-GM refrigerator, Cryogenic Engineering, Vol.31:197(1996).
15. H.Torii et al., Development of two-stage 4K cryocooler with modified Solvay cycle, Proceedings of ICEC 16 (1996).
16. M.Okamoto et al., Proceedings of 55th Meeting on Cryogenics and Superconductivity:26(1996).
17. R.Radebaugh and J.E.Zimmerman, NBS Special Publication 508:67(1978).
18. T.Kuriyama et al., Temperature profile and mass flow rate distribution in regenerator of Gifford-McMahon refrigerator using magnetic regenerator materials, Cryogenic Engineering, Vol.31:203(1996).

# A NUMERICAL SIMULATION METHOD OF A TWO-STAGE G-M REFRIGERATOR AND COMPARISON WITH EXPERIMENT

L. Wang, X.D.Xu, and L. Zhang

Cryogenic Laboratory, Chinese Academy of Sciences
P.O.Box 2711, Beijing, 100080, China

## ABSTRACT

An improved numerical model and calculation method to simulate the dynamic process of a two-stage G-M refrigerator has been developed in this paper. A variety of factors, such as variable properties of regenerator materials and helium, existence of helium in a void space of a regenerator, were taken into account in the model. It can predict the distributions of temperature and mass flow, respectively with time, in the first and second stage regenerators, the regenerator efficiency and refrigeration capacity. Some analyses are made to study the effects of some parameters on refrigerator performance. The values predicted by computing agree with experimental results. The numerical method can provide useful information for the design of a high efficiency refrigerator.

## NOMENCLATURE

| | |
|---|---|
| A | Heat transfer area |
| C | Volumetric specific heat |
| $D_h$ | Equivalent diameter |
| f | Resistance factor |
| h | Specific enthalpy of fluid |
| K | Isolation heat index |
| m | Mass of control volume |
| $\dot{m}$ | Mass flow rate |
| P | Pressure |
| $P_H$ | Inlet high pressure |
| $P_L$ | Outlet low pressure |
| V | Volume |
| x | Coordinate of flow direction |

*Greek letters*

| | |
|---|---|
| $\alpha$ | Heat transfer coefficient |
| $\theta$ | Crank angle |
| $\lambda$ | Friction factor |
| $\mu$ | Mass flow coefficient |
| $\rho$ | Density |

*Subscripts*

| | | | |
|---|---|---|---|
| Q | Heat exchange with surroundings | | |
| S | Displacer movement distance | f | Boundary of control volume |
| T | Temperature | i | The i-th control volume |
| t | Time | w | Matrix or wall of control volume |
| u | Gas flow velocity | 0 | Nozzle inlet position |

## INTRODUCTION

In recent years, with the introduction of magnetic regenerator materials, the performances of multistage G-M refrigerators at low temperatures have been greatly improved. The lowest cooling temperature reached 2.1K[1]. A cooling capacity larger than 1.5W/4.2K has been obtained by optimizing operating parameters and using hybrid multilayer regenerators with magnetic materials, although the cooling efficiency was very low[2]. Since experimental investigation is very difficult to conduct in the 4K region, theoretical analysis is very important to understand the dynamic process and the work mechanism of a 4K multistage G-M refrigerator in order to improve cooling capacity and efficiency. Some work has been done by C. S. Hong et al[3][4] and P. Y. Wu et al[5] to study the features of low temperature regenerators using numerical methods. A dynamic simulation of one-stage G-M refrigerator was presented by P. Y. Wu et al[6]. Heat balance analysis is still the most common method for analyzing the performance of a 4K multistage G-M refrigerator. This method can not provide proper understanding and predict the dynamic characteristics of the refrigerator.

In this paper, An improved numerical model and calculation method to simulate the dynamic process of a two-stage G-M refrigerator has been introduced. A variety of factors , such as the variable properties of regenerator materials and helium, existence of helium in the void space of a regenerator, were taken into account. A set of differential equations based on the conservation of mass, energy and momentum for gas flow were solved by a

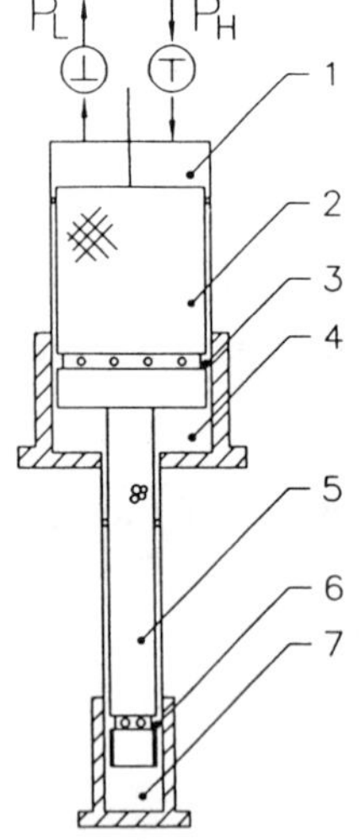


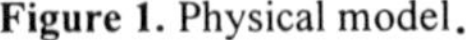

Figure 1. Physical model.

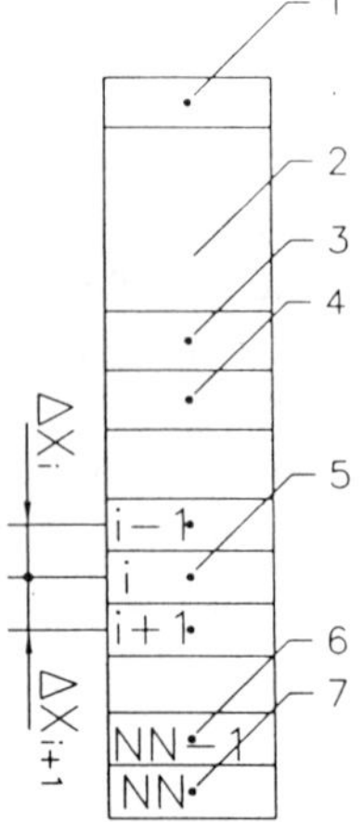


Figure 2. Domain discretization.

1. Hot space 2. 5. regenerators

3. 6. cold end heat exchangers 4. 7. cold spaces

numerical method. The results show the changes of pressure, temperature, and mass flow, respectively with time, in the first and second stage regenerators. Some work was made to study the effects of some parameters on refrigerator performance. Some values predicted by computing agree with the experimental results.

## PHYSICAL MODEL AND GOVERNING EQUATIONS

Figure 1 shows the physical model of a two-stage G-M refrigerator. The simulation system includes regenerators, hot and cold spaces, and cold end heat exchangers. The right boundary is the surface of the second-stage cold space. The left boundary is the intake/exhaust valves. The basic assumptions made in this model are as follows:

(1) one-dimensional flow of gas
(2) negligible axial heat conduction
(3) constant wall temperatures of the second-stage cold space and cold end heat exchanger

The whole model is divided into many subsystems. Each subsystem is considered as a uniform system which can exchange work, heat and mass with the surroundings through its boundaries. Its parameters vary with time. The governing equations for each subsystem are given as follows:

Continuity equation for gas flow:

$$\frac{\partial m_i}{\partial t} = \dot{m}_{f_i} - \dot{m}_{f_{i+1}} \tag{1}$$

Conservation of momentum for gas( by simplification):

$$\frac{dP_i}{dx} = -f_{f_i} \cdot \frac{1}{D_{hf_i}} \cdot \frac{(\rho u^2)_{f_i}}{2} \tag{2}$$

Conservation of energy for gas:

$$\frac{\partial (mh - PV)_i}{\partial t} = (\dot{m}h)_{f_i} - (\dot{m}h)_{f_{i+1}} + \frac{\delta Q}{dt} - P_i \frac{dV_i}{dt} \tag{3}$$

Where $\delta Q = \alpha_i \cdot A_i \cdot (T_{wi} - T_i) \cdot dt$ (4)

Conservation of energy for the matrix in regenerators, for the cold end heat exchangers and for the walls of cold spaces:

$$C_{mi} \cdot V_{mi} \cdot \frac{\partial T_{wi}}{\partial T} = \alpha_i \cdot A_i \cdot (T_i - T_{wi}) \tag{5}$$

Properties of helium and matrix:

h=f ( T, P )  and  Cm=f (Tm)

Movement of the displacer assembly:

$$S = S_{\max} \cdot \frac{1 - \cos\theta}{2} \tag{6}$$

It should be noted that the mass flow rate is defined as positive if the gas flows from the hot space to the cold one and as negative for the opposite direction. The real properties of helium were obtained from NIST TN-1334. The relation of the volumetric specific heat of matrix with temperature was obtained by using test curve fitting.

Both the intake and exhaust valves are considered as a combination of a resistance component and an ideal convergent nozzle, as shown in Figure1. The pressure $P_0$ at the inlet of the nozzles is determined from the pressure drop $\Delta P$ through the resistance component. The resistance factor $\lambda$ is taken from reference[7].

$$\Delta P = \lambda(P_H - P) \qquad (\text{if } \dot{m} > 0) \qquad (7)$$

$$\Delta P = \lambda(P - P_L) \qquad (\text{if } \dot{m} < 0) \qquad (8)$$

$$P_0 = \max(\frac{\dot{m}}{|\dot{m}|},0)(P_H - \Delta P) + \max((-\frac{\dot{m}}{|\dot{m}|},0)(P - \Delta P) \qquad (9)$$

The mass flow rate through the intake/exhaust valves is determined by using a formula for the convergent nozzle with a correction factor.

For subsonic flow, P/Po (if $\dot{m} > 0$) or PL/Po (if $\dot{m} < 0$) is larger than $\left(\frac{2}{k+1}\right)^{k/(k+1)}$

$$\dot{m} = \mu A \sqrt{\left(\frac{2k}{k-1}\right)\left(\frac{P_0}{V_0}\right)\left[\left(\frac{P}{P_0}\right)^{2/k} - \left(\frac{P}{P_0}\right)^{(k+1)/k}\right]} \qquad (\text{if } \dot{m} > 0)$$

$$\dot{m} = -\mu A \sqrt{\left(\frac{2k}{k-1}\right)\left(\frac{P_0}{V_0}\right)\left[\left(\frac{P_L}{P_0}\right)^{2/k} - \left(\frac{P_L}{P_0}\right)^{(k+1)/k}\right]} \qquad (\text{if } \dot{m} < 0) \qquad (10)$$

For choked flow, P/Po(if $\dot{m} > 0$) or PL/Po(if $\dot{m} < 0$) is equal to or less than $\left(\frac{2}{k+1}\right)^{k/(k+1)}$

$$\dot{m} = \pm \mu A \sqrt{\left(\frac{2k}{k+1}\right)\left(\frac{P_0}{V_0}\right)\left(\frac{2}{k+1}\right)^{2/(k-1)}} \qquad (11)$$

μ is correction factor, which is determined by the experimental data.

Since we could not find a friction factor and the heat transfer coefficient suitable for the oscillating gas flow, and the cycle frequency of the G-M cryocooler is low (usually 1 Hz), we adopted the steady state friction factor and the heat transfer coefficient from reference[8].

All the above equations make up the set of govering equations used for numerical simulation of a two-stage G-M refrigerator.

## DOMAIN DISCRETIZATION AND EQUATION DISCRETIZATION

The above equations are non-linear and unsteady. Some efforts were made to improve the calculation accuracy and increase the calculation speed.

Considering the non-continuity of temperature in a G-M refrigerator, the large temperature gradient, and the large change of the physical properties of the gas and matrix with the temperature in the regenerators, the domain discretization of the system was made by a method of internal points shown in Figure 2. The non-uniform grid for space discretization was used to reduce the number of nodes. A uniform grid for time discretization and an implicit scheme for time items in the governing equations are used for stability.

The staggered grid approach was adopted, shown in Figure 3, to eliminate the mistaken pressure profiles for discretization of the momentum equation. This approach uses control volumes for the velocities that are staggered with respect to those for the temperature and the pressure. The physical properties of helium at the interface are obtained by interpolation.

The convection items in the energy equation for gas were discreted by using a second-order upwind difference scheme in order to ensure the transportive property of the discretization equation, its higher order accuracy, and fewer number of nodes.

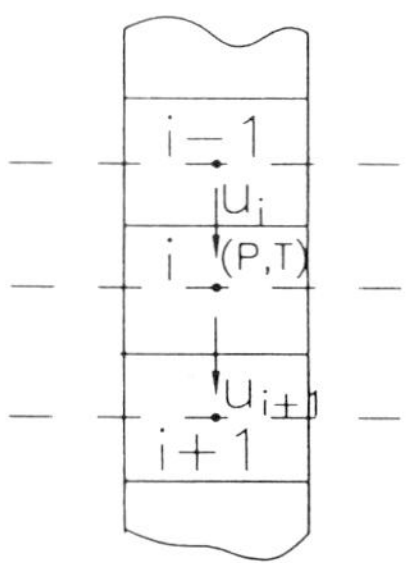


**Figure 3.** Staggered grid.

Discretization of continuity equation:

$$\frac{m_i - m_i^{k-1}}{\Delta t} = \dot{m}_{f_i} - \dot{m}_{f_{i+1}} \tag{12}$$

Discretization of energy equation for matrix:

$$C_{mi} \cdot V_{mi} \cdot (T_{wi} - T_{wi}^{k-1}) = \alpha_i A_i (T_i - T_{wi}) \Delta t \tag{13}$$

Discretization of momentum equation:

$$P_{i-1} - P_i = f_{f_i} (\frac{x_i - x_{i-1}}{D_{hf_i}})(\frac{\rho u^2}{2})_{f_i} \tag{14}$$

Discretization of energy equation for gas can be written:

$$A_i h_{i-1} + B_i h_i + C_i h_{i+1} = D_i \tag{15}$$

Where

$$A_i = -\max(\dot{m}_{f_i}, 0)(1 + e_1)\Delta t - \max(\dot{m}_{f_{i+1}}, 0) e_3 \Delta t$$

$$B_i = m_i + \max(-\dot{m}_{f_i}, 0)(1 + e_2)\Delta t + \max(\dot{m}_{f_{i+1}}, 0)(1 + e_3)\Delta t$$

$$C_i = -\max(-\dot{m}_{f_i}, 0) e_2 \Delta t - \max(-\dot{m}_{f_{i+1}}, 0)(1 + e_4)\Delta t$$

$$D_i = \alpha_i A_i (T_{wi} - T_i)\Delta t - P_i (V_i - V_i^{k-1}) + (m_i h_i - P_i V_i)^{k-1} + P_i V_i$$

$$-\max(\dot{m}_{f_i}, 0) e_1 h_{i-2} \Delta t - \max(-\dot{m}_{f_{i+1}}, 0) e_4 h_{i+2} \Delta t$$

Here $e_1 = 0.5\frac{\Delta x_i}{\Delta x_{i-1}}$ $\quad e_2 = 0.5\frac{\Delta x_i}{\Delta x_{i+1}}$ $\quad e_3 = 0.5\frac{\Delta x_{i+1}}{\Delta x_i}$ $\quad e_4 = 0.5\frac{\Delta x_{i+1}}{\Delta x_{i+2}}$

Note that the superscript k-1 means variables at k-1 time interval while variables without a superscript stand for the value at k time interval.

## NUMERICAL METHODS

The discretization energy equation for gas can be solved by the TDMA (tridiagonal matrix algorithm method). All governing equations were solved by using an under-relaxation iteration. The iteration results should be satisfied by periodic stable conditions for all the time-dependent parameters and the energy balance conditions during a cycle.

## COMPUTATIONAL RESULTS

A typical calculation has been made for a two-stage G-M refrigerator. The main parameters are as shown in the Table 1. The operating temperature are respectively 4.2K and 300K at the walls of the cold space and the hot one. The operating inlet high pressure and outlet low pressure values are 1.9MPa and 0.5MPa respectively[9].

**Table 1.** Main parameters of a two-stage G-M refrigerator

| | First stage | Second stage |
|---|---|---|
| Cylinder diameter (mm) | 53 | 26 |
| Regenerator diameter (mm) | 45 | 22 |
| length (mm) | 70 | 190 |
| material | phosphor bronze meshes | $Er_3Ni$ grains |
| porosity | 0.69 | 0.41 |
| Stroke (mm) | 25 | |
| Frequency (Hz) | 1 | |

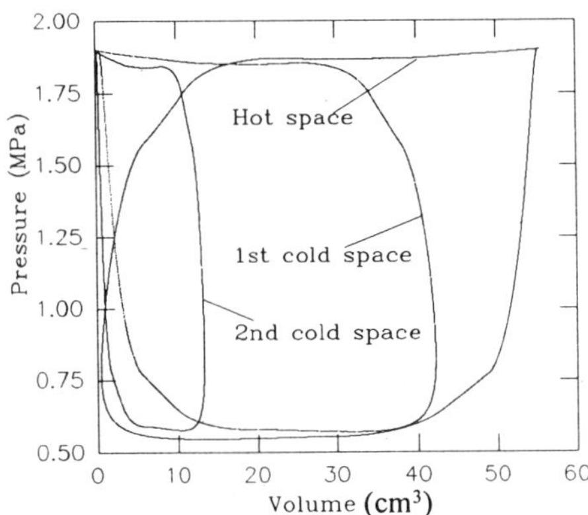


**Figure 4.** P-V diagrams.

**Figure 5.** Variation of pressure and volume with time.

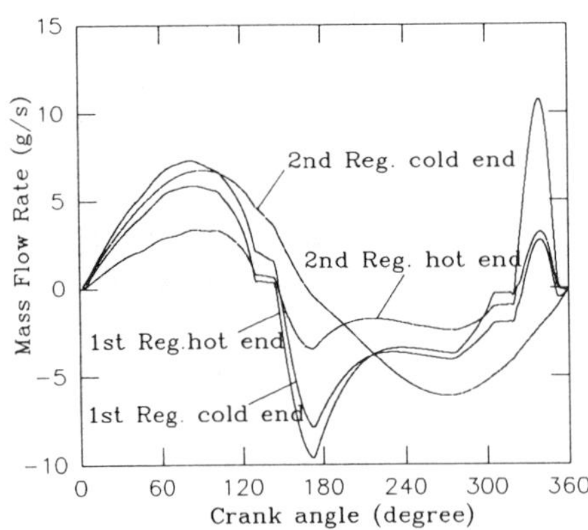


**Figure 6.** Variation of mass flow rate of gas with time.

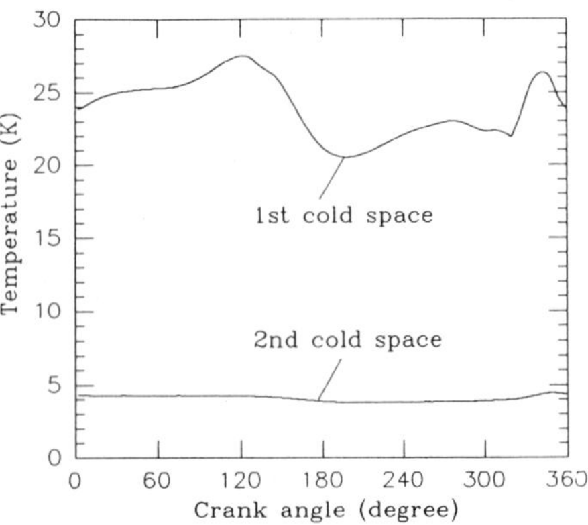


**Figure 7.** Variation of gas temperature with time.

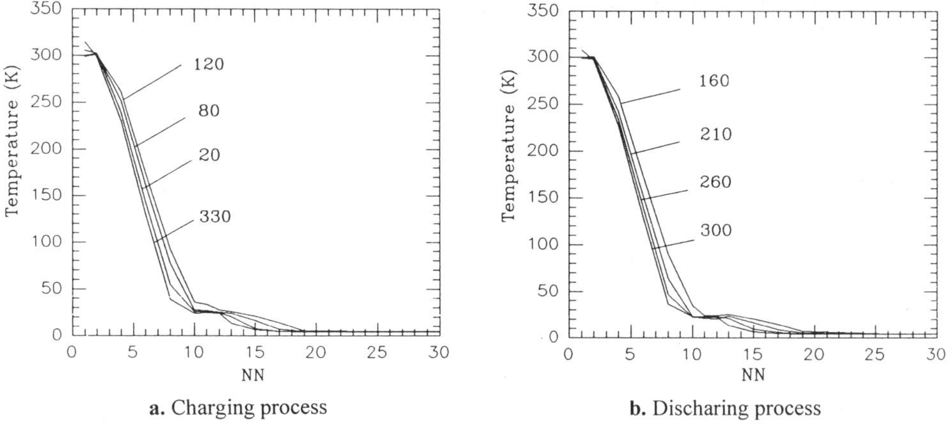


**a.** Charging process  **b.** Discharing process

**Figure 8.** Variation of gas temperature with time and the location in the refrigerator. 20-300 — crank angle(degree), NN — node number, NN=1 — hot space, NN=12, 30 — 1st and 2nd cold spaces.

The dynamic characteristics, including instantaneous pressure, temperature, mass flow rate in the refrigerator, and the P-V diagrams in hot and cold spaces are shown in Figures 4-8. This gives a good understanding of the internal processes occurring in a two-stage G-M refrigerator. Figure 8 shows the axial temperature distributions with time in the refrigerator during a cycle. The linear temperature distributions in the first-stage regenerator and non-linear temperature distributions caused by the properties of helium at low temperature in the second-stage regenerator are indicated respectively by the node number NN from 2 to 10 and from 13 to 27. Figure 9 shows the effect of the gas pressure on the cooling capacity at 4.2K. The inlet high pressure is 1.9MPa. The cooling capacity decreases with an increasing of the outlet low pressure from 0.5MPa to 0.8MPa. The maximum pressure difference between the hot and cold space during a cycle is lower than 0.15MPa which agrees with the experimental result[10].

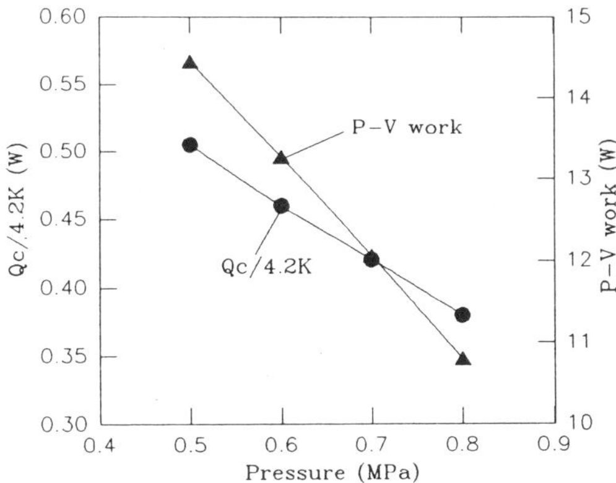


**Figure 9.** Variation of the cooling capacity at 4.2K with pressure.

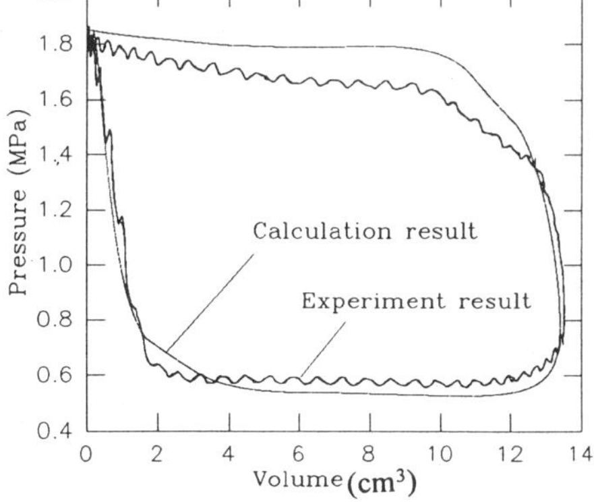


**Figure 10.** P-V diagram and comparison between the calculation and the experiment results.

## COMPARISON WITH EXPERIMENT

The tested machine is a two-stage 4K G-M refrigerator which was designed and made by ourselves. The configuration of the experimental apparatus is the same as that of the numerical model. Figure 10 shows the P-V diagram of the second-stage cold space and comparison between the experiment and calculation. The operating high and low pressures are 1.9MPa and 0.48MPa. The cooling temperature is 3K. They basically agree with each other. The deviation of the P-V work between them is 5% which is caused by some factors such as the steady friction factor and some negliged heat losses.

## CONCLUSIONS

A numerical simulation method which can predict the dynamic process and cooling performance of a two stage G-M refrigerator with given structural dimensions and operating conditions is presented. Good agreement between the experiment and calculation has been found. The calculation program developed can be run on a personal computer. The numerical model is powerful for studying the influent mechanism of various parameters on refrigerator performance, and can give some useful information for the design of a high efficiency refrigerator.

## ACKNOWLEDGMENTS

This research was supported by the National Natural Science Foundation of China.

## REFERENCES

1. M. Nagao et al. Generation of superfluid Helium by a G-M Cycle Cryocooler, The 6th Cryocooler 2, 37-47(1990)
2. A. Onishi et al. Improvement of a 1.5W-class 4K Gifford-McMahon cryocooler, ICEC16/ICMC Proceedings, Part 1, 355-358 (1996)
3. C. Hong and X. D. Xu, On the unique features of the low temperature regenerator in a 4.2K G-M refrigerator, Proc.4th Joint Sino-Japanese Seminar on Cryocoolers and Concerned Topics, 6-13 (1993)
4. C. Hong and X. D. Xu, On the thermodynamic cycle of the low temperature G-M refrigerator,Cryogenics, Vol.34, ICEC Supplement, 183-186 (1994)
5. P. Y. Wu, W.Huang, et al. Dynamic simulation of magnetic material regenerator and comparison with experiment, Cryogenics, Vol.36, No.4, 259-261(1996)
6. W. Huang, P. Y. Wu, et al, Dynamic simulation of one-stage G-M refrigerator and comparison with experiment, Cryogenics, Vol.36, No.9, 643-647 (1996)
7. Haselden, G.G., Cryogenic Fundamentals Academic Press, London and New York 269 (1971)
8. Kays, W.M. and London, A.L., Compact Heat Exchangers 2nd Edn, McGraw-Hill, New York, USA (1964)
9. L.Zhang et al. Study of effect of heat transfer in the cold head to the performance of a 4.2K G-M refrigerator, Advances in Cryogenic Engineering, Vol 41, 1645-1652 (1996)
10. X.D.Xu et al. Investigation of the performance of a 4.2K G-M refrigerator, ICEC16/ICMC Proceedings, Part 1, 343-346 (1996)

# EFFECTS OF THERMAL CONDUCTANCE IN THE COOLING STAGE OF A 4K-GM REFRIGERATOR ON REFRIGERATION CAPACITY

T. Inaguchi, M. Nagao, K. Naka, and H. Yoshimura

Mitsubishi Electric Corporation , Advanced Technology R&D Center
8-1-1, Tsukaguchi Honmachi, Amagasaki, Hyogo, 661 Japan

## ABSTRACT

This paper describes the effects of thermal conductance in the cooling stage of a 4K-GM refrigerator on its refrigeration capacity. Calculation of the temperature difference between the gas and cooling stage in the expansion space of a 4K-GM refrigerator has shown that the thermal conductance between the gas and cooling stage is inadequate, and the heat transfer loss is too large to be ignored. Hence, we have made improvements in the shape and material of the cooling stage to increase thermal conductance, and have been successful in increasing refrigeration capacity at 4.2 K from 2.2 W to 3.0 W.

## INTRODUCTION

Since the successful liquefaction of helium in a 4K-GM refrigerator in 1988[1], the refrigeration capacity at 4.2 K has improved continually, with the present refrigeration capacity being above 2 W.[2] With the refrigeration capacity getting larger, the heat transfer loss between the gas and cooling stage in the expansion space, which was estimated to be small until now, has turned into a problem that cannot be ignored.

There have been reports[3] about improvements in the cooling stage of refrigerators operating at 77 K. Also, improved heat conduction in the cylinder in the low temperature zone in a 4K-GM refrigerator has also been reported[4]. But no reports have so far been made on the heat transfer loss between the gas and the cooling stage in the expansion space at about 4 K.

We calculated the temperature difference between the gas and cooling stage in the expansion space of a 4K-GM refrigerator with a refrigeration capacity of 2.2 W at 4.2 K, and have found that the temperature difference between the gas and cooling stage is excessively large and the thermal conductance is inadequate.

Therefore, we increased thermal conductance by improving the material and shape of the cooling stage, and examined the effects on refrigeration capacity.

In this paper, we first calculate the temperature difference between the gas and cooling stage of the expansion space in a machine with a refrigeration capacity of 2.2 W at 4.2 K. Then we show the characteristics of cooling stages made of improved material and having better shapes. We also show to what extent the temperature difference between the gas and cooling stage of the expansion space has been reduced and how much the refrigeration capacity has increased, when the improved cooling stages are used. Finally, we talk a bit about how cooling stages of a 4K-GM refrigerator is designed based on the results of this experiment.

## TEMPERATURE DIFFERENCE BETWEEN THE GAS AND COOLING STAGE IN THE EXPANSION SPACE

Fig. 1 shows a schematic diagram of a two-stage 4K-GM refrigerator whose refrigeration capacity at 4.2 K is 2.2 W. The inner diameter of the cylinder at the first stage is 90 mm and that of the cylinder at the second stage is 60 mm, while the stroke is 32 mm. Phosphor bronze screens are used as regenerator materials in the high temperature zone and lead shots are used for the same purpose in the low temperature zone in the first stage. $Er_3Ni$ is used similarly in the high temperature zone and $ErNi_{0.9}Co_{0.1}$ in the low temperature zone in the second stage.

Fig. 2 shows a schematic of the experimental apparatus. The cooling stages of the 4K-GM refrigerator are set in a vacuum chamber. A radiation shield is attached to the first stage, and it encloses the second stage to prevent radiant heat from the room from entering the second stage. The temperature of the first stage was measured using a Pt-Co resistance sensor and the temperature of the second stage was measured using a carbon glass resistance sensor. Cartridge heaters were installed at each stage to measure refrigeration capacity.

In order to secure enough flux to the expander, two compressors, each with a rated input power of 6 kW totalling a rated input power of 12 kW, are arranged in a row.

A pressure transducer is set into the room temperature space of the expander and a linear displacement converter is set up on the drive mechanism of the displacers in order to measure the indicated work.

The refrigeration capacity is given in Fig. 3, where the -□- mark indicates the measured refrigeration capacity. The no-load temperature is 2.35 K and the refrigeration capacity at 4.2 K is 2.2 W.

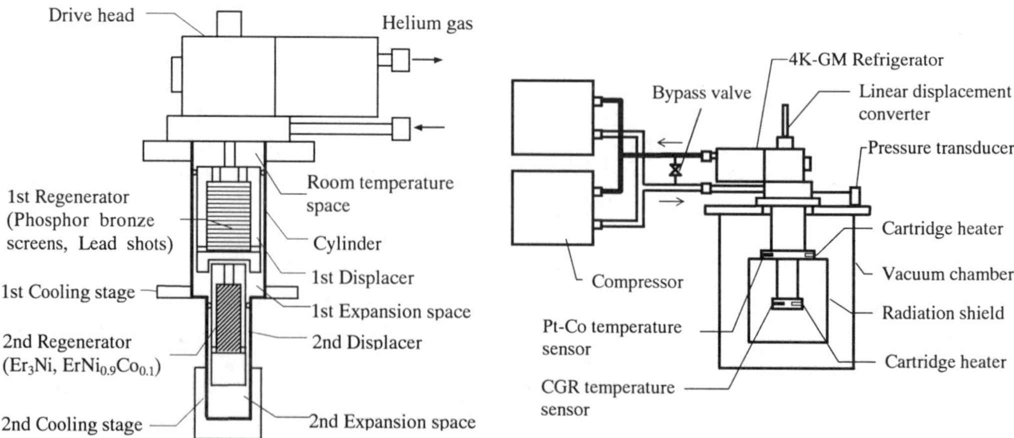


Fig.1 Schematic diagram of 4K-GM refrigerator.

Fig.2 Schematic diagram of experimental apparatus.

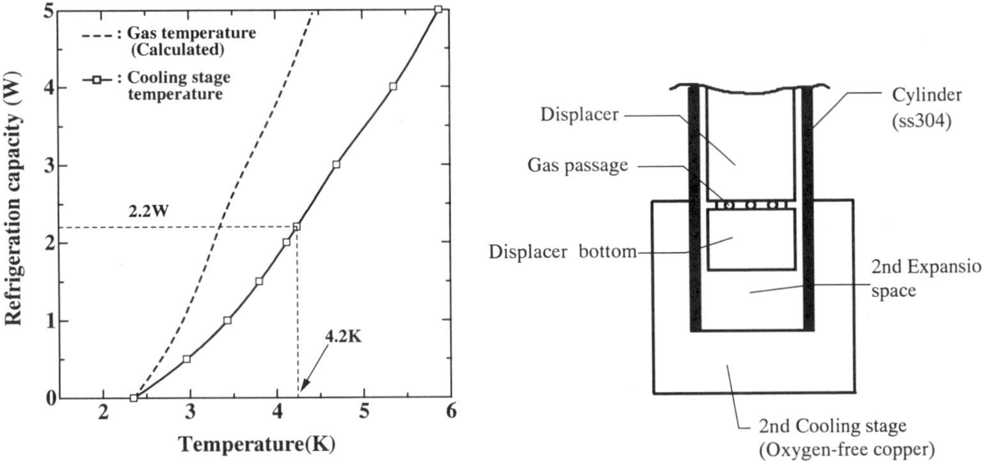


Fig.3 Refrigeration capacity and calculated gas temperature in the expansion space of the 4K-GM refrigerator.

Fig.4 Enlarged diagram of second cooling stage (Conventional cooling stage).

Fig. 4 is an enlarged view of the second cooling stage. This cooling stage is a pipe-shaped cylinder made of ss304, and was covered with Oxygen-free copper before being brazed. Thus the cooling stage is composed of oxygen-free copper and ss304.

Helium which flows out from gas passages of the displacer goes through the gap between the cylinder and the displacer bottom, and helium reaches the second expansion space. Heat transfer between the gas and cooling stage differs depending on whether the section has a displacer bottom.

The temperature difference between the gas and the cooling stage in the expansion space is calculated under the following assumptions.

[1] The temperature of the cooling stage section composed of oxygen-free copper is uniform.

[2] The temperature of the gas in the expansion space is uniform.

The heat transfer from cooling stage to gas in the expansion space can be expressed by equation (1), and the thermal conductance by equation (2).

$$Q = \alpha \left(T_w - T_f\right) \tag{1}$$

$$\alpha = \frac{1}{\dfrac{l_b}{\lambda_s A_b} + \dfrac{1}{h_b A_b}} + \frac{1}{\dfrac{l_s}{\lambda_s A_s} + \dfrac{1}{h_d A_d + h_b A_e}} \tag{2}$$

where $Q$ = quantity of heat transfer, $\alpha$ = thermal conductance, $T_w$ = temperature of cooling stage (oxygen-free copper section), $T_f$ = temperature of gas in expansion space, $\lambda_s$ = thermal conductivity at the ss304 section of the cooling stage, $l_s$ = length of heat conduction at the side (ss304 section) of the cooling stage, $l_b$ = length of heat conduction at the bottom of the cooling stage (equal to zero in Fig. 4), $A_b$ = heat transfer area at the bottom of the cooling stage, $A_s$ = heat conduction area at the side (ss304 section) of the cylinder, $A_e$ = heat transfer area at the side of the expansion space (the place where the displacer bottom does not exist), $A_d$ = heat transfer area at the side of the expansion space (the place where the displacer bottom exists), $h_b$ = heat transfer rate at the bottom, or side (where the displacer bottom does not exist) of the cooling stage, $h_d$ = heat transfer rate at the side (where the displacer bottom exists) of the expansion space.

The heat transfer rate in the expansion space was obtained by using the Petukhov-Gnielinski equation (3)[5] when the Reynolds' number was over 3000, and by using heat transfer equation(4)[6] of the laminar flow when the Reynolds number was under 3000.

$$h_b = h_d = \frac{\lambda}{4R_h}\frac{(f/2)(\mathrm{Re}-1000)\mathrm{Pr}}{1+12.7\sqrt{f/2}\left(\mathrm{Pr}^{2/3}-1\right)} \tag{3}$$

$$h_b = h_d = \frac{\lambda}{4R_h}\times 4.36 \tag{4}$$

$$\mathrm{Re} = \frac{4R_h\rho u}{\mu} \tag{5}$$

where Re =Reynolds number, Pr =Prandtle number, $\lambda$ =coefficient of thermal conductivity of helium, $R_h$ = hydraulic radius, $\rho$ = density, $u$ = velocity, $\mu$ = coefficient of viscosity, $f$ is the friction factor and we used the friction law of Blasius.

The hydraulic radius, which differs depending on whether the section has a displacer bottom, is evaluated as follows.

$$R_h = \frac{D_{in}}{4} \quad \text{(section without displacer bottom)} \tag{6}$$

$$R_h = \frac{D_{in}^2 - D_d^2}{4D_{in}} \quad \text{(section with displacer bottom)} \tag{7}$$

Where $D_{in}$ = inner diameter of cylinder, $D_d$ = diameter of displacer bottom.

The velocity u was obtained by using the law of conservation of mass at an expansion space. The velocity u also differs depending on whether the section has a displacer bottom, and these velocities are expressed in eq.(8) and eq.(9) respectively.

$$u = \frac{1}{\rho A}\frac{d(\rho V)}{dt} \quad \text{(section with no displacer bottom)} \tag{8}$$

$$u = \frac{D_{in}^2}{D_{in}^2 - D_d^2}\frac{1}{\rho A}\frac{d(\rho V)}{dt} \quad \text{(section with displacer bottom)} \tag{9}$$

Where $V$ = volume of expansion space, $A$ = sectional area of expansion space.

The density $\rho$ in the expansion space is obtained, using thermophysical properties software[7] for helium, if the temperature and pressure are given. We used the temperature of the cooling stage and the pressure which was measured in the room temperature space.

The broken line in Fig. 3 indicates the temperature of gas in the expansion space obtained with equation (1). Under a heat load of 2.2 W, the temperature of the cooling stage is 4.2 K while the temperature of gas in the expansion space is 3.3 K. The temperature difference is 0.9 K. If the heat exchange between the gas and cooling stage goes to completion, a refrigeration capacity of 2.2 W can be obtained at 3.3 K and the refrigeration capacity at 4.2 K becomes 4.3 W. This means the heat transfer loss at the cooling stage is 2.1 W in the present machine under a heat load of 2.2 W at 4.2 K. Because the loss is extremely big, we changed the material and shape of the cooling stage to improve the thermal conductance, and investigated the effects of this on the refrigeration capacity.

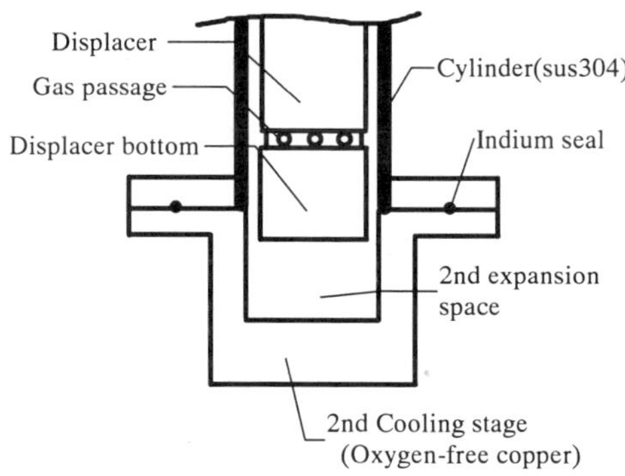

Fig.5 Copper cooling stage.

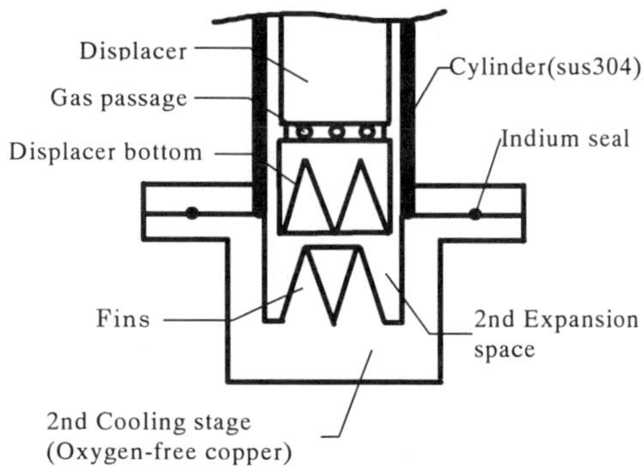


Fig.6 Copper cooling stage with fins.

## EXPERIMENTAL METHOD

We used three types of cooling stages and investigated their effects on the refrigeration capacity. First, we used the cooling stage described in Fig.4. We will hereafter call the cooling stages in Fig.4 "the conventional cooling stage".

Next, we used the cooling stage made of only oxygen-free copper (C1020) in Fig.5. The cooling stage has a flange structure and is sealed with an indium seal. In the conventional cooling stage the ss304 cylinder extends up to the expansion space. However, the thermal conductivity of the ss304 (0.3 W/m-K) at 4.2 K is about 1/3000$^{th}$ of the thermal conductivity of oxygen-free copper (950 W/m-K). Therefore, we decided to construct the cooling stage wholly from oxygen-free copper. We will hereafter call the cooling stage in Fig. 5 "the copper cooling stage".

Finally, we used the cooling stages made of oxygen-free copper and attached fins to the bottom of the expansion space of the cooling stage to increase the heat transfer area. This is shown in Fig.6. Just like in the copper cooling stage, the cooling stage has a flange structure and is sealed with an indium seal. The fins are designed to intermesh with each other in a cusped shape on the cooling stage side and at the bottom of the displacer in order to minimize dead volume. We will hereafter call the cooling stage in Fig. 6 "the copper cooling stage with fins".

Table 1 shows the material, heat transfer area and thermal conductance at 4.2 K (equation (2)) of each cooling stage. It indicates that the copper cooling stage has a thermal conductance 4.2 times higher than that of the conventional cooling stage. A great improvement in thermal conductance is gained by simply changing the material of the conventional stage entirely into oxygen-free copper, while leaving the heat transfer area the same.

Table 1 Thermal conductance of the cooling stage

| | Conventional cooling stage | Copper cooling stage | Copper cooling stage with fins |
|---|---|---|---|
| Material | ss304+ Oxygen-free copper | Oxygen-free copper | Oxygen-free copper |
| Heat transfer area | 150 cm$^2$ | 150 cm$^2$ | 412 cm$^2$ |
| Thermal conductance at 4.2 K (W/K) (eq.(2)) | 2.5 | 10.5 | 26.7 |

Table 2 Indicated work

| | Conventional cooling stage | Copper cooling stage | Copper cooling stage with fins |
|---|---|---|---|
| Indicated work (W) at 4.2 K | 71.2 | 69.6 | 68.3 |

The heat transfer area becomes 2.7 times larger, when the copper cooling stage with fins is used, providing a thermal conductance 2.5 times higher than that of the copper cooling stage and 10.7 times higher than that of the conventional cooling stage.

## RESULTS OF THE EXPERIMENT

The measured cooling capacities of the three types of cooling stages show above are given in Fig. 7, with the cycle frequency in all cases set to 42 rpm, and the indicated work of the second expansion space (the pressure was measured in the room temperature space) set at from 68.3 W to 71.2 W, which is almost constant, as shown in Table 2.

The refrigeration capacity in the case of the conventional cooling stage was 2.2 W at 4.2 K and the no-load temperature was 2.35 K as has already been shown in Fig.3, while the refrigeration capacity in the case of the copper cooling stage was 2.7 W at 4.2 K and the no-load temperature was 2.52 K. The refrigeration capacity is 0.5 W higher than that of the conventional cooling stage.

In the case of the copper cooling stage with fins, the refrigeration capacity was 3.0 W at 4.2 K and the no-load temperature was 2.72 K. The refrigeration capacity is 0.8 W higher than that of the conventional cooling stage.

The copper cooling stage with fins has a no-load temperature 0.37 K higher than that the conventional cooling stage. This is why the copper cooling stage with fins is of a complicated fin shape and the fins at the bottom of the displacer and in the cooling stage are poorly intermeshed. It is necessary to make further studies on methods for improving refrigeration capacity and quality control by making the shape simpler and increasing the heat transfer area.

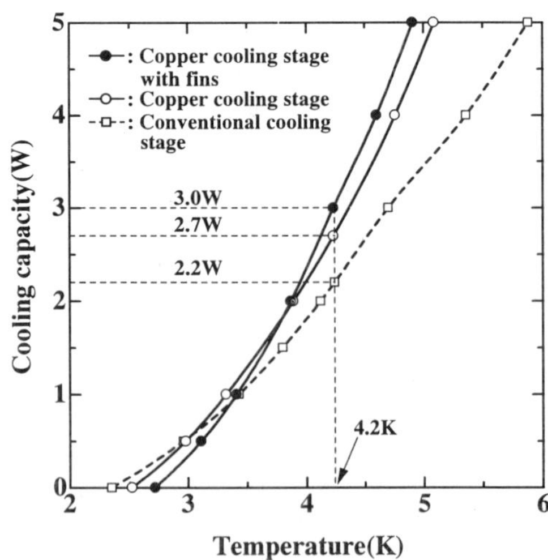


Fig.7 Refrigeration capacity for the three types of cooling stages.

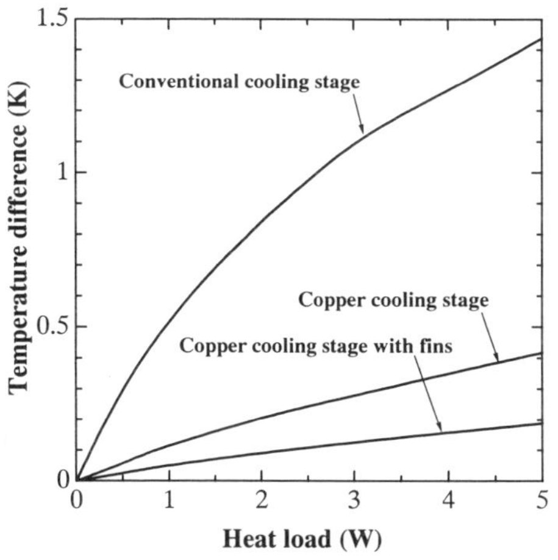


Fig.8 Calculated temperature difference between the gas and cooling stages.

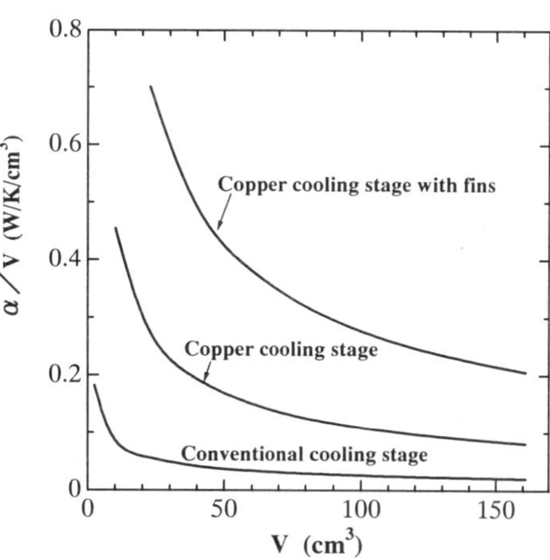


Fig.9 Thermal conductance per unit of volume of expansion space.

The temperature differences between the gas and cooling stages obtained with equation (1) are shown in Fig. 8. Under a heat load of 3 W, the temperature differences between the gas and cooling stage in the case of the copper cooling stage and in the case of the copper cooling stage with fins are respectively 0.28 K and 0.13 K, while that in the case of the conventional stage is 1.1 K. The copper cooling stage and the copper cooling stage with fins show a remarkable improvement over the conventional cooling stage.

## DISCUSSION

Items requiring caution, when the cooling stage of a refrigerator is designed based on the above experimental results, are described below.

When the volume of an expansion space increases, the refrigeration capacity increases. The thermal conductance increases, too, but is not as much as the refrigeration capacity. The thermal conductance that must be obtained increases as the refrigeration capacity increases, so we need to pay attention to this in our design.

The refrigeration capacity $Q$ of a GM refrigerator is proportional to the volume of the expansion space, and thus can be expressed as

$$Q \sim V \tag{10}$$

When the cooling capacity $Q$ is transferred outside, the temperature difference between the gas and cooling stage is expressed as $\Delta T$, and $Q$ can be also expressed as

$$Q = \alpha \ \Delta T \tag{11}$$

From equations (10) and (11),

$$\frac{\alpha}{V} \sim \frac{1}{\Delta T} \tag{12}$$

Hence, when $\alpha / V$ is large, $\Delta T$ gets small and the heat transfer loss at a cooling stage decreases. Conversely, when $\alpha / V$ is small, $\Delta T$ gets large and the heat transfer loss at a cooling stage increases. $\alpha / V$ is the thermal conductance per unit of volume of an expansion space and this parameter is a good index for use in the design of a cooling stage. We therefore examined how $\alpha / V$ changed as the volume of the expansion space changed.

Fig.9 shows the thermal conductance per unit of volume of an expansion space. During calculation, the inner diameter of an expansion space was varied while the stroke was kept constant. The diameter of the displacer bottom was varied while the ratio of it to the diameter of expansion space was kept constant. The other elements were kept constant.

When the volume ($V$) of an expansion space increases, the thermal conductance ($\alpha / V$) per unit of volume decreases in the three types of cooling stages.

Particularly in the case of the conventional cooling stage, $\alpha / V$ decreases conspicuously. For example, the $\alpha / V$ of the copper cooling stage in this experimental machine(the volume of expansion space = 90.5 $cm^3$, the inner diameter = 60 mm) is 0.12 $W/K/cm^3$. If we are going to get this $\alpha / V$ equivalent or better in a conventional cooling stage, the volume of the expansion space must be less than 7.7 $cm^3$ (inner diameter = 17.5 mm).

Therefore, unless the refrigerator has a sufficiently small expansion space volume, it is desirable that oxygen-free copper be wholly used as the material for the cooling stages.

## CONCLUSIONS

The investigation into the effect of the thermal conductance of cooling stages in 4K-GM refrigerators on refrigeration capacity made the following points clear.

(1) The refrigeration capacity can be improved from 2.2 W to 3 W at 4.2 K by improving the thermal conductance of the cooling stages.

(2) By using wholly oxygen-free copper for cooling stages, the thermal conductance can be made 4.2 times larger than that of the conventional cooling stages and the refrigeration capacity at 4.2 K can be increased by 0.5 W. By using wholly oxygen-free copper for the cooling stages and making the heat transfer area 2.7 times larger, the thermal conductance can be made 10.7 times larger than that of the conventional cooling stage, and the refrigeration capacity at 4.2 K can be increased by 0.8 W.

(3) Oxygen-free copper should be wholly used as the material of cooling stages unless volume of an expansion space is sufficiently small.

## REFERENCES


1. H.Yoshimura, M.Nagao, T.Inaguchi, T.Yamada, M.Iwamoto, "Helium Liquefaction by a Gifford-McMahon Cycle Cryogenic Refrigerator," Rev.Sci.Instrum., Vol.60(1989), p.3533
2. T.Inaguchi, M.Nagao, K.Naka, H.Yoshimura, "Development of 2W Class 4K Gifford-McMahon Cycle Cryocooler," ICEC16(1996), p.335
3. F.F.Chellis, T.P.Hosmer, and E.Keller, "Closed-cycle Refrigeration for an Airborne Illuminator," Advances in Cryogenic Engineering, Vol.16(1970), p.214
4. A.Onishi, R.Li, H.Asami, T.Satoh, and Y.Kanazawa, "Development of a 1.5W-class 4K Gifford-McMahon Cryocooler," Cryogenic Engineering, Vol.31(1996), p.162
5. V. Gnielinski, "New Equations for Heat and Mass Transfer in Turbulent Pipe and Channel Flow," Int.Chem.Eng.,Vol.16, No.2(1976),p359
6. JSME,"Heat Transfer Handbook," Maruzen(1993),p.69
7. Cryodata Inc.,"Thermophysical Properties Software Hepak v.3.0",(1989)

# LONG-TERM PERFORMANCE OF 4 K GIFFORD-McMAHON REFRIGERATORS ON THE BIMA ARRAY

R.L. Plambeck[1]

[1]Radio Astronomy Lab, University of California
Berkeley, CA 94720

## ABSTRACT

Eight radio telescopes in the Berkeley-Illinois-Maryland Array (BIMA) are outfitted with 4 K Gifford-McMahon refrigerators. The refrigerators cool superconducting detectors (SIS mixers) on the radio receivers; the heat load at 4 K is 10 to 20 mW. The refrigerators were constructed by welding third stages onto standard CTI model 1020 cryocoolers; $Er_3Ni$ or ErNi spheres are used as the low temperature regenerators. The refrigerators are operated almost continuously; one has been in service for 4 years. Maintenance consists of disassembling the cold heads every 1 to 2 years to clean out water and dust.

On the telescopes the refrigerators' third stage temperatures range from 2.7 to ~5 K. Often these temperatures are stable within $\pm$ 0.1 K for weeks, even while the refrigerators are tipped from 0 to 90 degrees as the telescopes track radio sources. Sometimes, however, the temperature creeps up by $>$ 1 K over a period of days or weeks, then jumps back to its original value following a power interruption or "defrost" cycle. Contaminants within the cold head may possibly cause this behavior. With active temperature regulation, the SIS mixer temperatures can be stabilized to $\pm$0.01 K despite the refrigerator fluctuations.

## INTRODUCTION

Three-stage Gifford-McMahon refrigerators are used to cool SIS (superconductor–insulator-superconductor) mixers on the Berkeley-Illinois-Maryland array (BIMA), an aperture synthesis instrument consisting of 10 radio telescopes[1]. Each BIMA dewar accommodates up to 4 SIS mixers, cooled by a single refrigerator. The heat loads on the refrigerators' first, second, and third stages are estimated to be 6 watts (from the 50 K outer radiation shield), 0.5 watts (from the 12 K inner radiation shield and I.F. amplifiers), and 10 to 20 milliwatts (from conduction losses and SIS mixer bias circuits), respectively.

The BIMA cryocooler design has been described previously[2,3]. We machine the end plug off a standard CTI model 1020 cryocooler, then weld on a 1.27 cm I.D. $\times$ 15 cm long third stage cylinder. The third stage displacer consists of a stainless steel tube, with an 0.5 mm thick plastic[4] sleeve epoxied to the outside to serve as a bearing surface. Spheres[5] of $Er_3Ni$ or ErNi, diameter range 180–450 $\mu$m, are used as the third stage regenerator. The third stage swept volume is 4 cm$^3$, while the regenerator volume is 10 to 12 cm$^3$. A key difference from most cryocoolers is that the third stage seal[6]

**Table 1.** Lab test results for all cryocoolers

| Serial Number | Regenerator | T (0 mW) | T (50 mW) |
|---|---|---|---|
| 5093 | 53.0 g $Er_3Ni$ | 2.50 K | 3.80 K |
| 5747 | 68.8 g $Er_3Ni$ | 2.50 K | 3.73 K |
| 6154 | 70 g $Er_3Ni$ | 2.60 K | 3.62 K |
| 5746 | 68.5 g $Er_3Ni$ | 2.40 K | 3.64 K |
| 6418 | 67.3 g $Er_3Ni$ | 2.56 K | 3.77 K |
| 6584 | 50.5 g $Er_3Ni$ | 3.28 K | 4.00 K |
| 6585 | 60.7 g ErNi | 2.62 K | 3.70 K |
| 5346 | 60.6 g $Er_3Ni$ | 2.71 K | 3.71 K |
| 7528 | 60.6 g ErNi | 2.65 K | 3.67 K |

is clamped to the warm end of the cylinder, not mounted on the displacer, so that it shrinks more tightly onto the displacer when cold. A digital power inverter controls the refrigerator's cycle frequency, which must be reduced below the normal value of 72 rpm in order to achieve third stage temperatures below ~4.5 K. On the telescopes the refrigerators normally operate at 31 rpm.

As of June 1997 we have built 9 such cryocoolers (the first of the series, #5093, differs from the rest in that it uses a conventional split seal mounted on the displacer). As indicated in Table 1, all performed comparably in a laboratory test dewar, where we measured the third stage temperatures with applied heat loads of 0 and 50 mW; cryocoolers with larger regenerators reached slightly lower temperatures.

## LONG TERM PERFORMANCE ON THE TELESCOPES

As of June 1997, 8 BIMA telescopes were outfitted with 4 K cryocoolers, for a total accumulation of 12 cryocooler-years operation; one refrigerator has been in service since March 1993. Over the past 3 years only one cryocooler suffered a catastrophic failure – a shorted motor winding, possibly caused by voltage spikes from the digital power inverter. Maintenance is done once per year: we take apart each cold head, wipe dirt and water off the displacer and cylinder, then reassemble and purge the unit. Thus far it has not been necessary to replace any of the third stage seals.

The cryocoolers are tipped over an elevation angle range of 0–150 degrees as the telescopes track radio sources. The cold tip points straight up at elevation 90 degrees. Typically, as shown in Figure 1, the third stage temperature varies by less than 0.1–0.2 K over the full range of elevation angles.

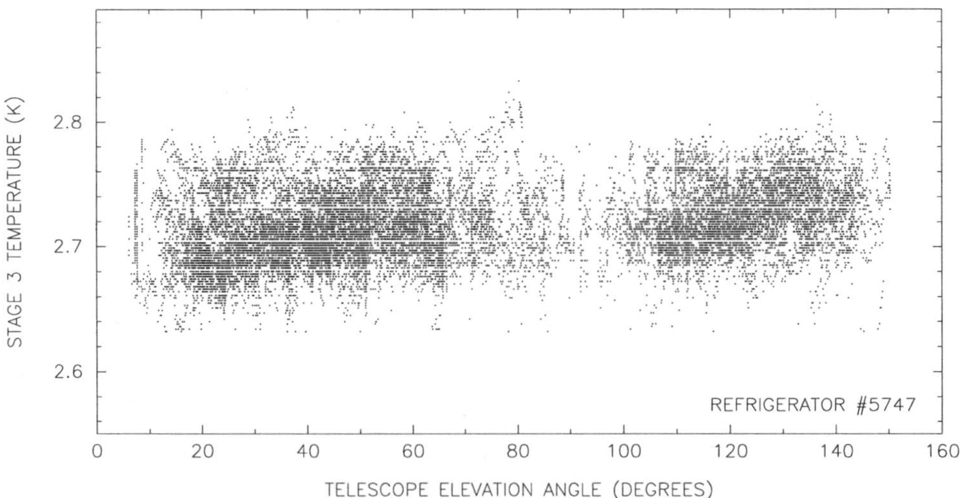


**Figure 1.** Cryocooler third stage temperature as a function of telescope elevation angle. At 90 degrees, the cryocooler's cold end is up. Data were measured every 1 minute over an 18 day period.

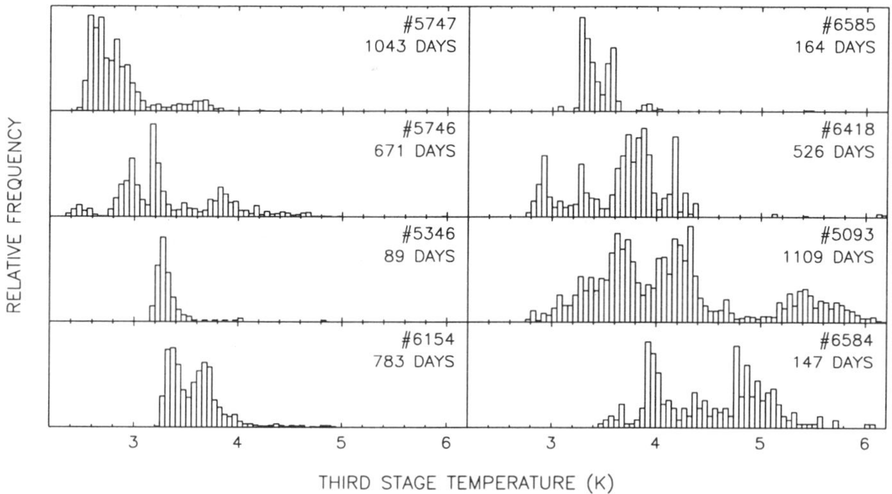


**Figure 2.** Temperature behavior of all cryocoolers from January 1994 to June 1997, based on measurements made every 12 hours. Histograms show the relative number of data points in each 0.05 K temperature interval; the time accumulated on a telescope is indicated for each cryocooler.

The refrigerator temperatures are recorded twice per day, averaged over 20 second intervals to smooth over temperature variations caused by the 1.92 second refrigerator cycle. Histograms of these data are shown in Figure 2. The third stage temperature is most erratic on refrigerator #5093, which uses a conventional split seal on the displacer.

Figure 3 shows the time series of temperatures for two of the oldest cryocoolers. Typically, the third stage temperature is stable to ± 0.1 K for weeks or months, then drifts upward over a period of days or weeks, sometimes becoming erratic. Then, following a power interruption, it jumps back to its original value. Brief power interruptions are rarely apparent in the long term record because the refrigerators recool in minutes. However, they are guaranteed to occur about every 2 months, when the telescopes are unplugged for a few minutes in order to move them to new locations.

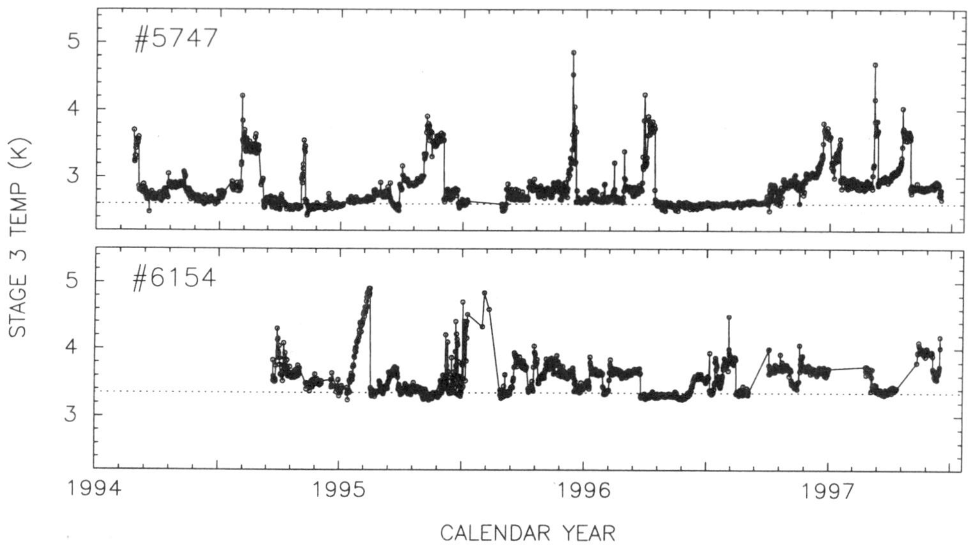


**Figure 3.** Third stage temperatures sampled twice per day over 3.3 years for cryocoolers #5747 and #6154.

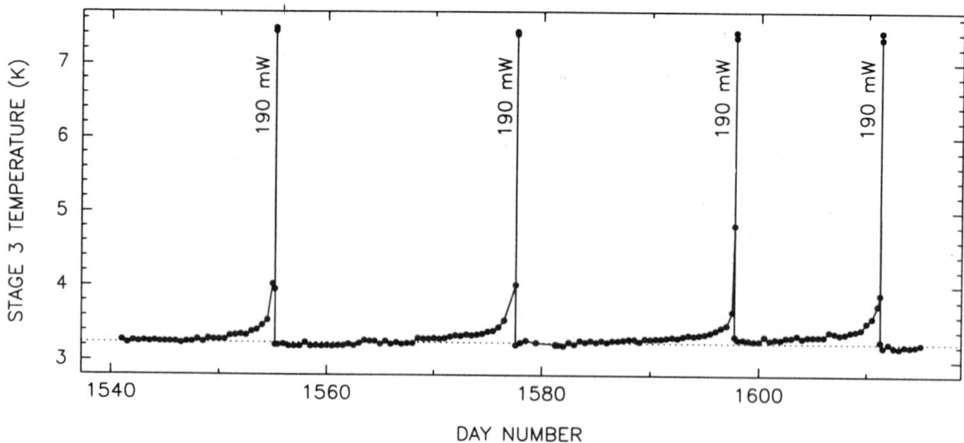


**Figure 4.** Temperature history of refrigerator #5346 over a 3 month period. The stage 3 temperature drifts upward gradually, but is driven back to its original value with the "defrost" cycle.

Instead of turning off the refrigerator, one often can succeed in driving the third stage temperature back to a lower value using a "defrost" cycle. One allows the refrigerator to run normally, but uses a heater resistor to apply about 200 mW to the third stage, raising its temperature to 7 or 8 K. After 15 minutes the heater is turned off and the third stage temperature drops back to its original value in a few minutes. An example of this behavior is shown in Figure 4.

Although the cause of the upward temperature drift is uncertain, 3 possibilities suggest themselves:

1. Gas leakage past the third stage seal. We are doubtful about this possibility – it isn't clear how the "defrost" cycle could restore the seal's original performance.
2. Development of flow channels within the regenerator. As a cryocooler's exhaust valve opens and the helium pressure drops from 18 to 6 atmospheres, gas rushes out of the third stage expansion space through the regenerator. Below 6 K, the helium density changes little from 18 to 6 atmospheres, so the mass flow and pressure drop through the regenerator is small. Perhaps this allows the gradual development of preferred flow channels through the regenerator, which degrade its performance. Figure 5 shows that the pressure drop rises rapidly above 6 K, perhaps explaining how the "defrost" cycle could disrupt the flow channels and drive the cryocooler back to its original temperature.

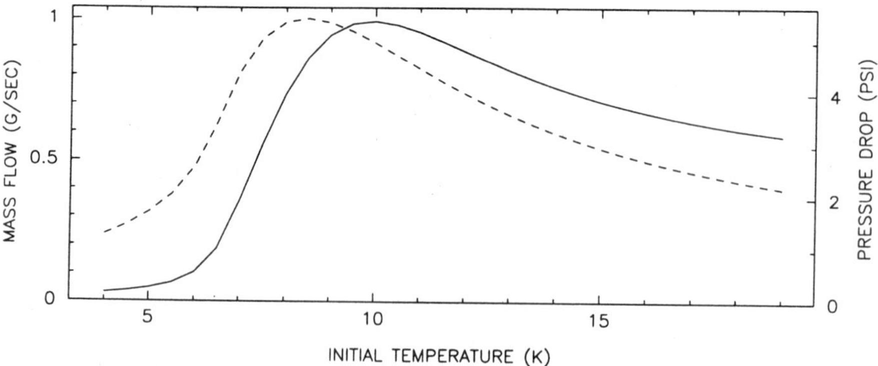


**Figure 5.** Calculated mass flow (dashed line) and pressure drop through a 1 cm long section of regenerator as a cryocooler's exhaust valve opens. The computation assumes that helium from a 4 $cm^3$ working volume flows through the regenerator in 0.3 seconds; the regenerator, filled with 0.018 cm diameter spheres with a packing factor of 0.6, has a cross section of 0.71 $cm^2$.

3. Buildup of contaminants within the third stage might increase friction between the displacer and cylinder wall. Since the "defrost" cycle raises the temperature only to 7 K, hydrogen is the most plausible contaminant.

## ACTIVE TEMPERATURE REGULATION

The conversion gain of an SIS mixer is a sensitive function of the DC bias voltage applied to the superconducting junction. Normally one sets the bias voltage to maximize the output power from the receiver. As the physical temperature of the junction increases, the superconductor energy gap decreases and so does the optimum bias point, as shown in Figure 6. Thus, to obtain stable, well calibrated performance from the mixer, one must stabilize its temperature within a few milliKelvin over periods of hours. This requires active temperature regulation.

A calibrated silicon diode sensor and a heater resistor are attached to each mixer block in the BIMA dewars. The mixers are attached to the third stage heat station with flexible copper straps. Typically the thermal resistance from the mixer to the cold station is 0.15 K/mW, so one may regulate the mixer temperature over a range of ~0.5 K with just a few milliwatts of heater power; such a small heat load has little effect on the cryocooler's third stage temperature. Each mixer temperature is measured every 0.32 seconds by a control computer, which averages groups of 6 samples to compute the mean temperature over the past refrigerator cycle, then adjusts the heater power as follows:

$$P = G_p \times (T - T_{set}) + G_i \times \int_0^t (T - T_{set})\, dt'.$$

Here P is the heater power, $T_{set}$ is the desired setpoint, and $G_p$ and $G_i$ are the proportional and integral gains, which are optimized empirically. Figure 7 shows that this technique succeeds in stabilizing the cycle-averaged mixer temperature to 1.8 mK r.m.s. – a peak-to-peak variation of 15 to 20 mK.

While the temperature regulation scheme described above works adequately, it does not remove the 10 to 20 mK swing within each refrigerator cycle. As shown in Figure 8, this causes the receiver's output power to vary by a few percent, which is

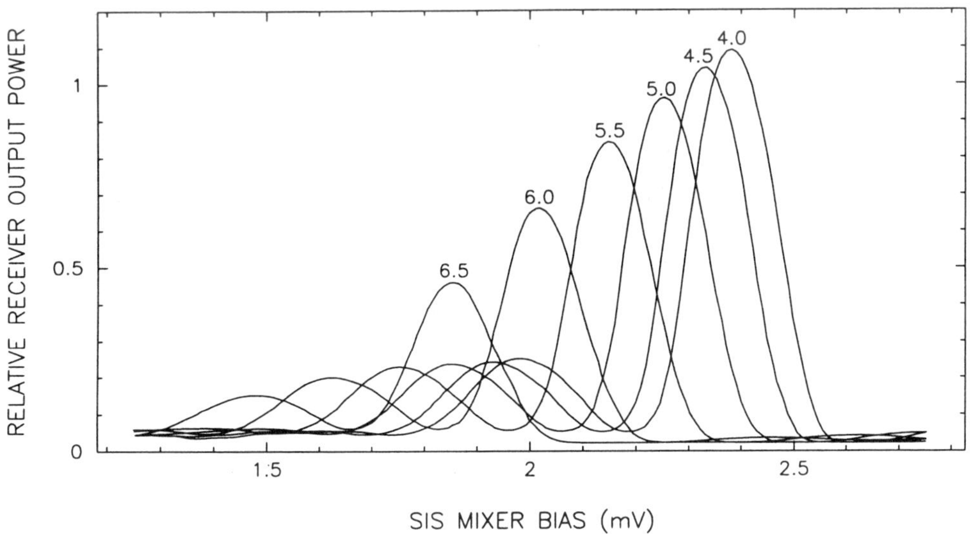


**Figure 6.** Receiver output power vs. DC bias voltage applied to a Nb/AlOx/Nb SIS junction, with the junction held at 5 temperatures. The local oscillator frequency is 107 GHz in this example. Generally the bias voltage is adjusted to maximize the output power. The optimum bias is a sensitive function of temperature.

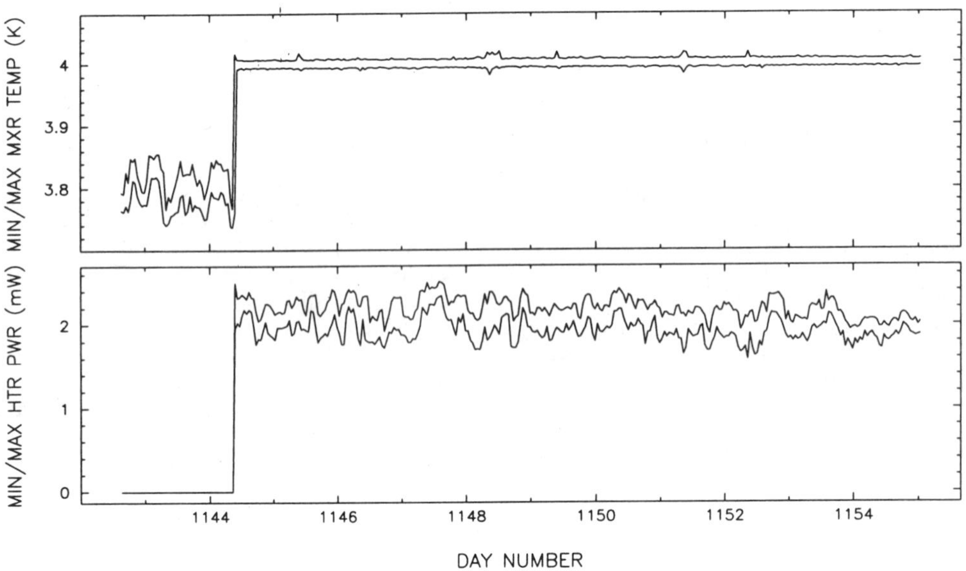


**Figure 7.** SIS mixer block temperature monitored over a 12 day period, showing the effect of temperature regulation. The top panel shows the minimum and maximum mixer temperatures (averaged over the 1.92 sec refrigerator cycle) measured in 53-minute intervals. The bottom panel shows the minimum and maximum heater powers in the corresponding intervals. Temperature regulation lowers the peak-to-peak mixer temperature fluctuations from 100 mK to <20 mK.

not generally serious. If necessary, the mixer block temperature can be controlled more tightly by pulsing the heater power within each refrigerator cycle. Lab tests, shown in Figure 9, demonstrate that this can reduce the mixer's peak-to-peak temperature swing to $< 2$ mK.

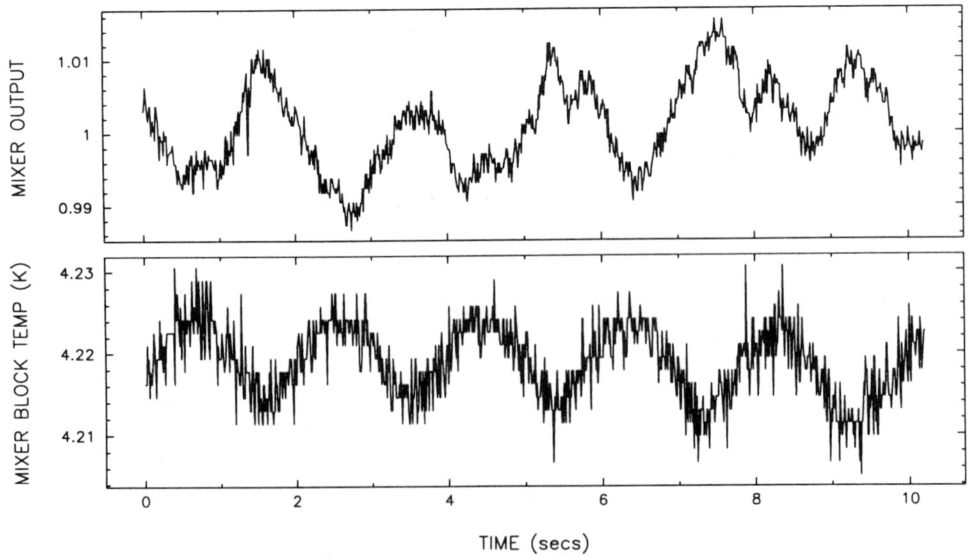


**Figure 8.** Mixer block temperature and receiver output power monitored every 0.01 sec for 10 seconds. The mixer block temperature varies by 10 mK with each refrigerator cycle. This causes the receiver output power to vary by 1.5%.

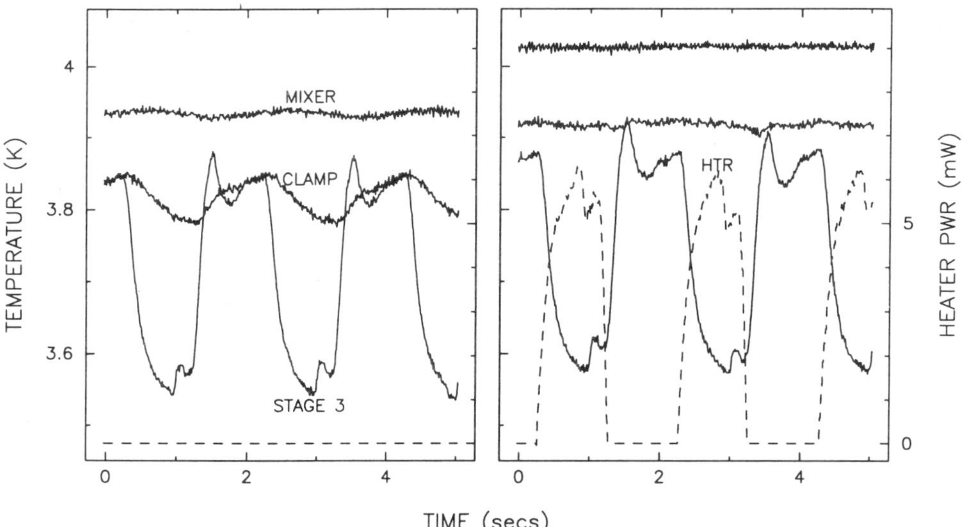


**Figure 9.** The left panel shows the stage 3, mixer clamp ring, and mixer temperatures with no heater current applied. The right panel shows the same temperatures with the heater pulsed to smooth out the cyclical temperature variations. The peak-to-peak temperature swing on the mixer is reduced from 15 mK to less than 2 mK. In this test the heater was mounted on the clamp ring, not the mixer block.

## CONCLUSIONS

Gifford-McMahon cryocoolers operating at ~4 K have performed reliably on radio telescopes in the BIMA array for several years with no long-term degradation. The third stage temperatures are insensitive to the refrigerator orientation, and typically are stable to ± 0.1 K for periods of weeks. Over time the third stage temperatures tend to drift upward, but often can be caused to jump back to their original values using a 15-minute "defrost" cycle. SIS mixers attached to the cryocoolers can be temperature stabilized to a few milliKelvin r.m.s. with a straightforward regulation scheme.

## ACKNOWLEDGEMENTS

We thank Dr. Toru Kuriyama for the loan of ErNi spheres for testing. This work was supported in part by grant AST-9613998 from the National Science Foundation.

## REFERENCES

1. Welch, W.J., et al., The Berkeley-Illinois-Maryland-Association millimeter array, Publications of the Astronomical Society of the Pacific, 108:93 (1996).
2. Plambeck, R., Thatte, N., and Sykes, P., A 4 K Gifford-McMahon refrigerator for radio astronomy, in "Proc. 7th International Cryocooler Conf.," Phillips Laboratory report PL-CP-93-1001, Kirtland AFB, NM (1993), p. 401.
3. Plambeck, R.L., Improved seal for a 4 K Gifford-McMahon cryocooler, in: "Cryocoolers 8," R.G. Ross, Jr., ed., Plenum Press, New York (1995), p. 795.
4. Vespel SP-21, DuPont Engineering Polymers, Newark, DE.
5. Purchased from Toshiba America Electronics, Westboro, MA.
6. Part number R310MB-014-SP31, Bal Seal Engineering Co., Santa Ana, CA.

# EFFICIENT SINGLE STAGE GIFFORD-McMAHON REFRIGERATOR OPERATING AT 20 K

A. Fiedler, J. Gerban, and H.U.Haefner

Leybold Vakuum GmbH, Cologne, Germany

## ABSTRACT

In order to operate $HT_c$-wires under high magnetic field conditions (for example: SMES) cooling with liquid nitrogen is not sufficient since a temperature between 20-30 K has to be maintained continuously in order to produce high fields. Using commercial Gifford-McMahon refrigerators this range till now was only been achieved by dual stage machines. It will be shown that such temperatures can be achieved even more efficiently by improved single stage coldheads, providing capacities of more than 25 W at 20K.

## INTRODUCTION

For many years now, the Gifford-McMahon ("GM") refrigerators have been established in commercial applications requiring cooling capacities of a few Watts up to above 100W. In this context, their employment in refrigerator cryopumps is particularly emphasised (especially in semiconducting and coating processes, space simulation chambers and more) as well as in the so-called shield cooling for the MRI business. The overall market volume amounts to approx. 8,000 - 10,000 coolers per year. Service intervals occur typically between 9,000 and 18,000 operating hours, and often even intervals of 27,000 hours are realized.

GM refrigerators are normally built with two stages, and the above mentioned applications have determined the performance, design, and typical operating temperatures of both stages in the following way:

| | Cryopumps | Shield Cooling |
|---|---|---|
| 1st stage | 40 - 100 K<br>$H_2O$ condensation<br>thermal shielding | 40 - 60 K<br>1st thermal radiation shield |
| 2nd stage | 10 - max. 20 K<br>condensation/adsorption<br>$N_2$,$O_2$,Ar,$H_2$ and others | 12 - 16 K<br>2nd thermal radiation shield |

*Advances in Cryogenic Engineering*, Vol. 43
Edited by P. Kittel, Plenum Press, New York, 1998

Another typical characteristic of two stage machines is that the ratio of the cooling performance of the 1st/2nd stage at reference temperatures of 80 K / 20 K is generally greater than 3 do to the construction design of the refrigerator and also the thermal needs of these applications.

Beyond this, the increasing implementation of the superconductor technology required new divergent demands concerning the refrigeration technology. For direct cooling of low temperature-superconductors, concepts were develcped for refrigerators with operating temperatures between 4 K and 8 K. In this context, the choice of the regenerative heat exchanger appropriate for these temperatures is a decisive element. Similar to this, with high temperature-superconductor applications becoming increasingly of interest, it appears useful to adjust the design of cryo refrigerators for this application as well.

For instance, the employment of $HT_c$-wires in magnets demands for cooling of the superconducting material to at least approximately 1/3 of the superconducting transition temperature, $T_c$. In the case of Bi-2223 filament wires, and similar materials, this means operation at approximately 20 - 30 K. It is necessary to provide a cooling performance of 10 W or more for larger applications charged with variable operating conditions.

Do to the fact that this area is not covered by present single stage GM refrigerators (lowest zero Watt temperature typically at 25 - 27 K) this demand can only be fulfilled by operating a couple of two stage machines in parallel. This automatically results in a considerably higher capacity at the first stage than is required in this application.

In the following it will be shown that, similar to the low temperature optimization of two stage refrigerators, it is also possible to modify single stage machines in such a way that a new temperature range of 20 to 30 K can be *optimally* operated.

## FUNCTIONAL PRINCIPLE OF THE SINGLE STAGE GM-COOLER

Single stage cold heads are products which were developed by almost every GM-refrigerator supplier out of the dominant two stage variants by eliminating the second stage. Figure 1 shows the components of such a cooler.

Operating the compressor unit, a high pressure (HP) and a low pressure (LP) level is established permanently and is transferred to the coldhead via flexible pressure lines. The GM-cooling process itself is controlled by a valve mechanism at the warm end of the coldhead. In many cases this is realized by a rotating valve driven by a synchroneous motor, which controls the sequence of pressurization and expansion in the main working volume as well as the pneumatic operation of the displacer. The working fluid Helium thus can enter voulme 1 (warm end) and after flowing through the regenerator it enters volume 2, the expansion volume. When the displacer is moved the ratio of the two volumes is changed under constant pressure, resulting in the pV-cycle shown in its ideal form in Figure 1.

The ideal theoretical cooling performance of each cycle is calculated as follows:

$$Q_{ideal} = (V_{max} - V_{min})\,(P_H - P_L)$$

just according to the rectangular surface in the PV-diagram. More detailed desriptions of the cycle can be found in the cryogenic literature.

## ARRANGEMENT OF SINGLE STAGE GM-REFRIGERATORS

Figure 2 shows the typical arrangement of a single stage refrigerator for the Leybold RGS120 coldhead.

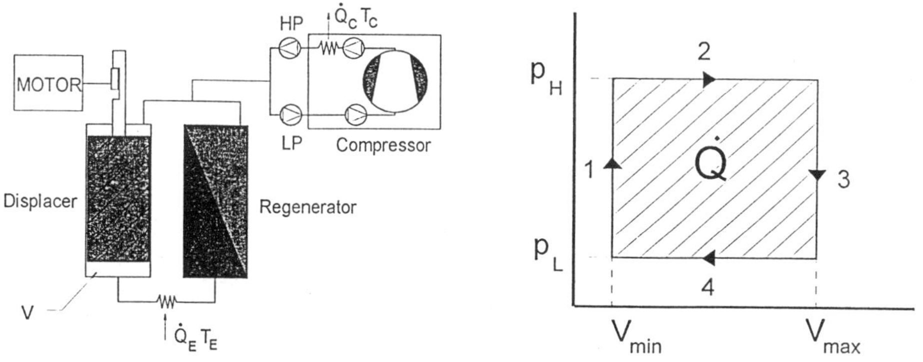


**Figure 1.** Sketch of functional principle of GM cryocooler and pV-diagram.

The compressor unit, an RW 6000, provides high pressure Helium of 21-22 bar. After it passes through a water cooled heat exchanger, the oil separators and an adsorber in the compressor ensure that only high purity helium reaches the cold head through the high pressure line. The control unit with the motor and rotating valve is directly attached to the expander part with the coldhead cylinder and displacer. The cycle frequency typically is 2 Hz.

The object to be cooled is connected to the copper flange of the cold stage. This coldhead reaches a minimum temperature of approximately 25 K, and at 80K a capacity of approximately 120 W at 80 K is achieved. After expansion the gas is returned to the compressor through the low pressure line.

## OPTIMIZATION FOR LOWER TEMPERATURES

The characteristics of regenerator materials has a considerable impact on the temperature range that can be achieved with GM coolers, with regards to heat capacity, thermal conductivity, throttling of the gas flow etc.

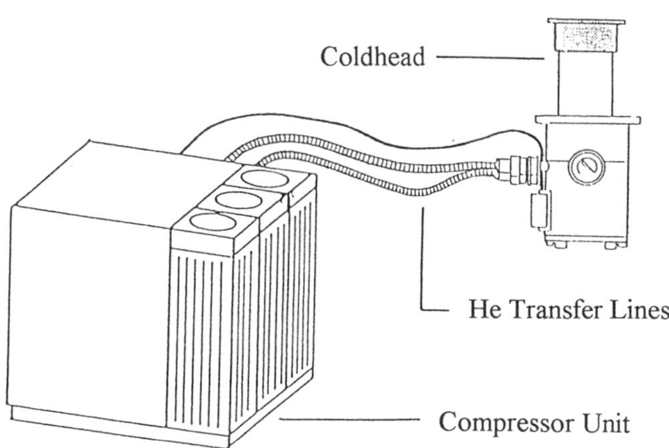


**Figure 2.** Components of the single stage refrigerator system RGS120.

Figure 3 shows the curve of the heat capacities of the regenerator materials Cu-bronze ($CuSn_6$), stainless steel (CrNi) and lead (Pb). For Cu-bronze and steel which are generally employed as regenerators in single stage machines the heat capacity drops below 30-40 K to a few % of the ambient temperature value, thus limiting the bottom temperature as well as the cooling performance that can be achieved in this area. Lead, however, serves much better as a heat storing material below 50 K.

The cooling requirements of a $HT_c$-magnet are determined by the fact, that for high magnetic field applications the coil should be cooled down to arround 1/3 of the material transition temperature, i.e. to 20 - 30 K. Based on experiences with low temperature optimization of two stage coldheads the same approach was applied to a single stage head RGS 120. A prototype has been built for test purposes which was lengthened by a few inches to provide additional lead in the regenerator. A sectional drawing of this variant called „RGS 120-T" is shown in Figure 4.

The regenerator is made up out of three layers: stainless steel screens and/or Cu-bronze screens of different meshes at the warm end as well as a filling of fine lead grains at the cold end. The cooling capacity curve resulting from this modification is shown in Figure 5 which, for comparative reasons, also points out the performance curves of a "normal" single stage cold head RGS 120 as well as for the second stage of a two stage cold head (Leybold RGD 1245).

The range which is required for the $HT_c$-magnet is covered best by the modified single stage cold head. Its cooling performance of around 25 W at 20 K is twice as high as the performance of a two stage cold head RGD 1245. Thus, it is possible to minimize the total number of cold heads for cooling the $HT_c$-conductor. Another advantage is that the higher cooling capacity above 30K (120 W at 80 K, approx. 260 W at 300 K) is of considerable help in achieving a faster initial cooldown of large components such as the $HT_c$-magnet. This is much faster than it would be possible even with a multiplicity of two stage machines.

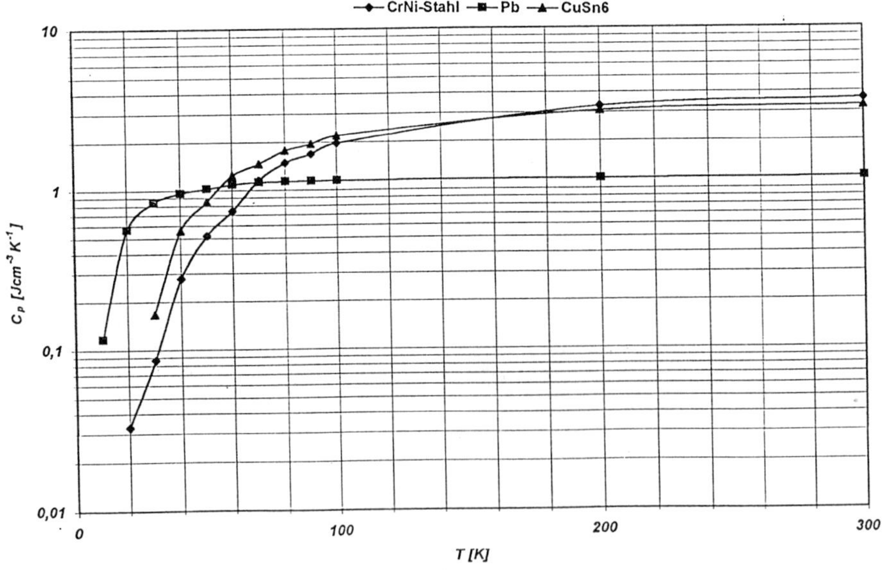


**Figure 3.** Specific heat capacity of stainless steel, lead and Cu-bronze.

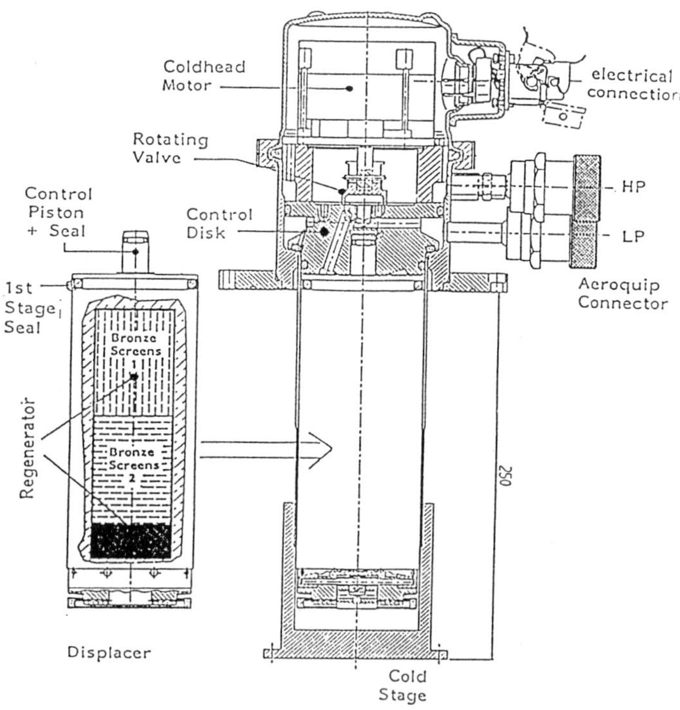


**Figure 4.** Sectional drawing of „low T single stage cold head" RGS 120-T.

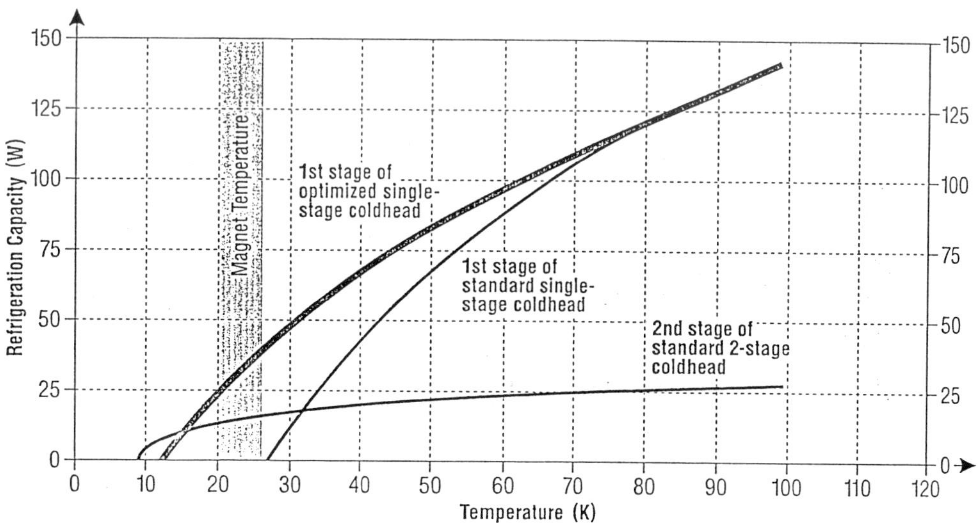


**Figure 5.** Refrigeration capacities of the standard and low temperature version of the RGS 120 cold head and the two stage head RGD 1245.

## $HT_c$-SMES COOLING WITH SINGLE STAGE REFRIGERATOR

The first application of the single stage 20 K-refrigerator was used in a demonstrator-prototype of a $HT_c$-superconductive magnetic energy storage ("SMES") which was recently developed by American Superconductor Corp., Westborough/MA[1]. Figure 6 shows the arrangement of this demonstrator designed for an energy storage of 5 kJ.

The RGS 120-T is directly linked to the magnet coil, made from Bi-2223 superconductor material, and another smaller cold head (single stage standard type, temperature range approx. 60 - 70 K) is operated for thermal intercept of the supply wires (also partly out of $HT_c$-conductor) and for cooling of the thermal radiation shield. The magnet coil is the largest in the world which has been designed with the $HT_c$-technique and has the following data:

| | |
|---|---|
| diameter: | approx. 44 cm / 17.3 inches |
| length: | ca. 10 cm / 3.9 inches |
| max. field intensity: | 0.97 Tesla |
| inductivity: | 1.00 H |
| max. current: | 100 A |
| max. voltage (DC) | 200 V |

The demonstrator serves as a field trial model to investigate the use of $HT_c$-SMES in a local network to protect against short-term power failures and irregularities of the main supply. In this case, the magnet can be charged within 2 seconds, from 0 to a total 100A, and discharged in the same way.

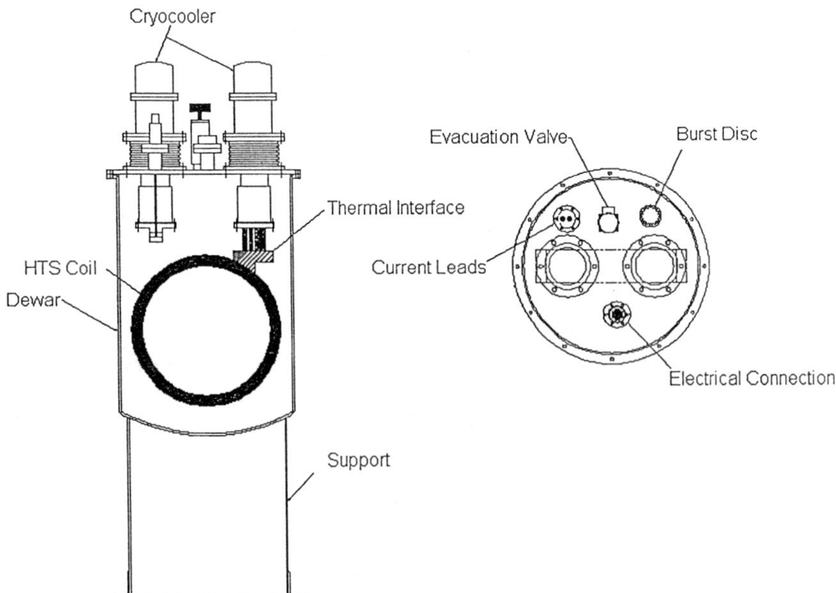


**Figure 6** $HT_c$-SMES with 5 kJ energy storage demonstrator [2].

**Figure 7.** $HT_c$-SMES with two single stage cryocoolers (RGS120-T and RGS70).

The short-term repeating of such cycles represents a considerable heat load generated by eddy currents which also have to be compensated for by the cryorefrigerator. From that perspective the new single stage cold head turns out to be particularly advantageous as it shows a clearly steeper increase of cooling performance over 20 K than is the case with two stage cold heads. This allows faster return of the $HT_c$-coil to its lower operating temperatures after being ramped up or down. The initial cooldown to T<25K in 12 hours, whereas cooling with two dual stage coldheads (such as the RGD 1245) takes considerably longer times by a factor of 2 to 3.

This first $HT_c$-SMES with its 5 kJ energy storage does not yet meet the targets for commercially employable energy storages, i.e. planned high-scaled versions certainly require the employment of a couple of these coolers. It is clear that in the $HT_c$-SMES application the *low temperature parts,* at 20-30 K, require a considerably *higher cooling capacity* than is the case for the 60-80 K stage (intercept of the supply lead, thermal shielding).This is an *inversion* relative to the case for normal two stage coldheads. Thus the application of high capacity, low temperature single stage heads (plus a much smaller 70K-intercept cooler) considerably improves the cooling efficiency as well as the cost perspectives of this cooling application.

## SUMMARY AND OUTLOOK

The cooling requirements of $HT_c$-magnets can be currently satisfied with single stage Gifford-McMahon cold heads optimized to 20-30 K operating temperature, since:

* the reliability of Gifford-McMahon coolers has been well established from other commercial applications (vacuum technique, MRI shield cooling) on the basis of high quantity statistics
* single stage coolers are less complex than two stage coolers (only one displacer seal at the "warm side") and therefore allowing one to expect additional long-term reliability

* the special cooling requirements of $HT_c$-magnets (less “first stage“ than “second stage“ performance) can be realized more effectively by operation, in parallel, of single stage cold heads optimized to the required operating temperature.

The 25 W at 20 K (6 kW compressors) which have been realized so far are more than sufficient for the above described SMES-demonstrator. Additional tests meanwhile have proved that even >30 W at 20 K can be achieved with the same system when additional modifications in the regenerator area and operating pressure ratio are carried out.

## REFERENCES

1. S.S.Kalsi et al., Proc.Applied.Supercond. Conf., Pittsburgh.(Aug.1996)
2. by kind permission of the American Superconductor Corporation, Westborough/MA

# THERMOFLUID ANALYSIS OF SHUTTLE LOSS ON REGENERATIVE CRYOCOOLERS

K. Naka, T. Inaguchi, M. Nagao, and H. Yoshimura

Advanced Technology R&D Center, Mitsubishi Electric Corp.
Amagasaki City Hyogo 661, Japan

## ABSTRACT

Shuttle loss is one of major losses of regenerative cryocoolers. In order to improve the efficiency of the cryocoolers, we have to estimate the shuttle loss precisely and reduce it. However, it has been difficult to calculate shuttle loss precisely because the hydrodynamic behavior of the working fluid has not been considered in the analyses and the temperature gradient of the working fluid was given as linear. Therefore, using a general purpose thermofluid analysis software STAR-CD[1], we made a model in which the hydrodynamic behavior of the working fluid was considered. By using this model we did a thermofluid analysis which calculated the movement of the working fluid and the temperature change. From this, we calculated the shuttle loss. As a result of these calculations, the preciseness of the calculation model was confirmed by comparing with previous analyses. This means that acceptable analysis is possible with general purpose software, and various models. This paper reports on the analysis model and the results.

## INTRODUCTION

There are many reports on the analysis of the shuttle loss. Zimmerman and Longsworth[2] calculated the shuttle loss assuming that the thermal capacities of the cylinder and displacer are infinite. In that case, the temperature of the cylinder and displacer is held constant and the shuttle loss depends only on the heat resistance of the working fluid. Radebaugh and Zimmerman[3] reported the shuttle losses as a problem of heat conduction of half infinity solids which cause a sinusoidal temperature fluctuation of the surface. These two analyses consider the solid and working fluid separately. But it is not realistic to consider them separetly. Nishio[4] considered the solid and the working fluid at the same time and considered the temperature gradient in the radial direction of the cylinder and displacer.

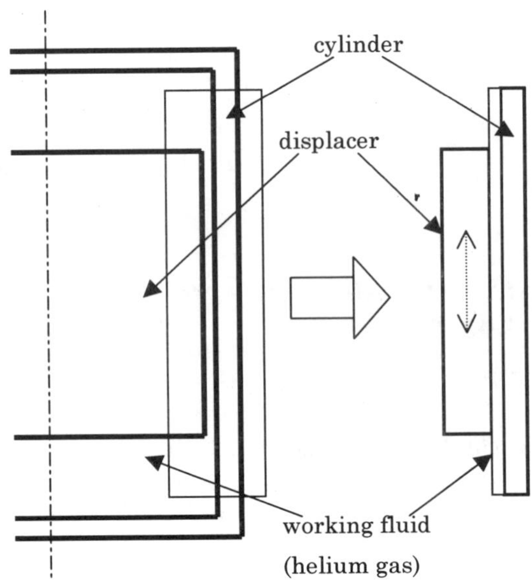


**Figure 1.** Analysis model

However all these analyses gave the temperature gradient between the cylinder and displacer as linear. It is well known that shuttle loss originates in the working fluid between the cylinder and displacer, however no analysis has considered the hydrodynamic behavior of the working fluid. Therefore, we made a model which calculated the hydrodynamic behavior of the working fluid between the cylinder and displacer. Using this model we did a thermofluid analysis and calculated the shuttle loss for sinusoidal motion and compared it with these former analyses.

## MODEL OF ANALYSIS

This time we made a model of shuttle loss. The analysis model is shown in Figure 1. This analysis model considers only the cylinder, displacer and the working fluid, expansion space and regenerator are not considered. We made a model using general purpose thermofluid analysis software, STAR-CD. In particular this model utilized the sliding mesh function of the software and expressed the movement of the displacer. It is a characteristic of this software that a moving mesh, which is able to contain solids, can be analyzed. The analysis mesh is shown in Figure 2. Two meshes are the displacer in the

Table 1. Main parameters of the analysis model

| | | |
|---|---|---|
| Cylinder (SUS304) | Inner diameter | 61.9 mm |
| | Thickness | 0.9 mm |
| Displacer (Bakelite) | Outer diameter | 60.5 mm |
| | Thickness | 6.5 mm |
| | Length from hot end to cold end | 140.8 mm |
| | Stroke | 32 mm |
| | Frequency | 1 Hz |
| Working fluid (Gas Helium) | Clearance | 0.7 mm |
| | Pressure | 0.1 MPa |

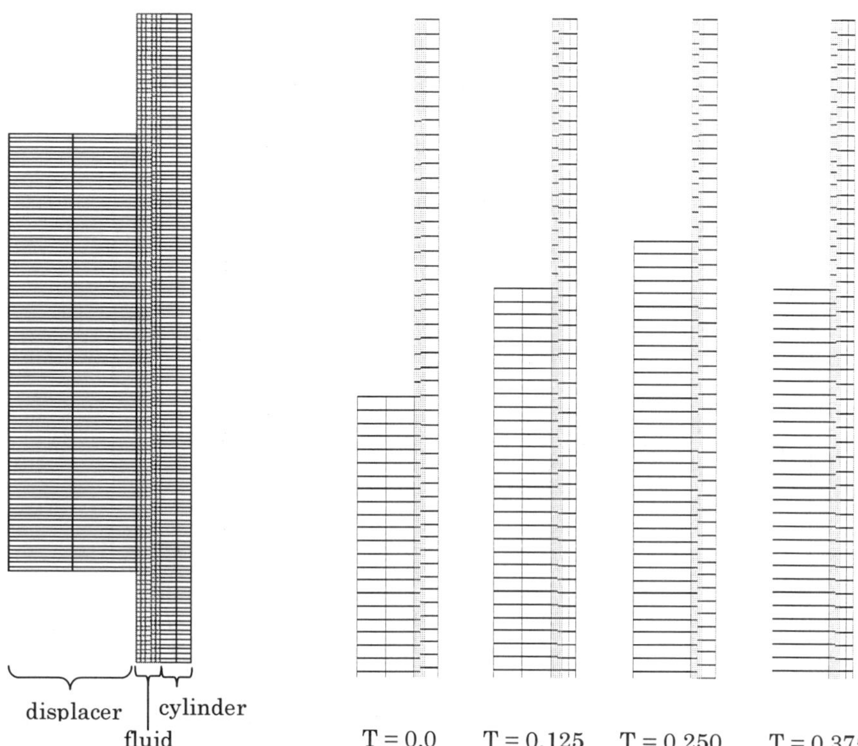


**Figure 2.** Mesh division.

**Figure 3.** Mesh division of upper half part of this analysis model.

radius direction on the left of the figure, and the two meshes on the right are the cylinder. The remaining meshes are the fluid. The upper part of this model is shown in Figure 3. In this model, we express the movement of the displacer using a sliding mesh. A mesh usually becomes coarse and minute when it expresses the movement of displacer. STAR-CD can use a completely discontinuous mesh. Whenever the displacer moves, we remake the mesh of the fluid for the displacer and reduce the coarseness and minuteness of the mesh and raise the convergence of calculation.

## METHOD OF ANALYSIS

A difference of temperature occurs between the cylinder and displacer because the displacer, which has a temperature gradient, moves up and down, which we analyzed. We assumed that properties (density, specific heat, heat conductivity, viscosity coefficient) did not depend on temperature and an average temperature value was used. Only the density of helium gas was calculated using an equation for the state of an ideal gas. The initial condition is shown in Figure 4 and the boundary condition is shown in Figure 5. We calculated the temperature change of the cylinder, displacer and the working fluid, and the flow of the working fluid, to calculate the shuttle loss. The Reynolds number of this model is about 10. Therefore the analysis of the working fluid was done with a laminar flow. Only 10 cycles were analyzed, because calculation time was very long.

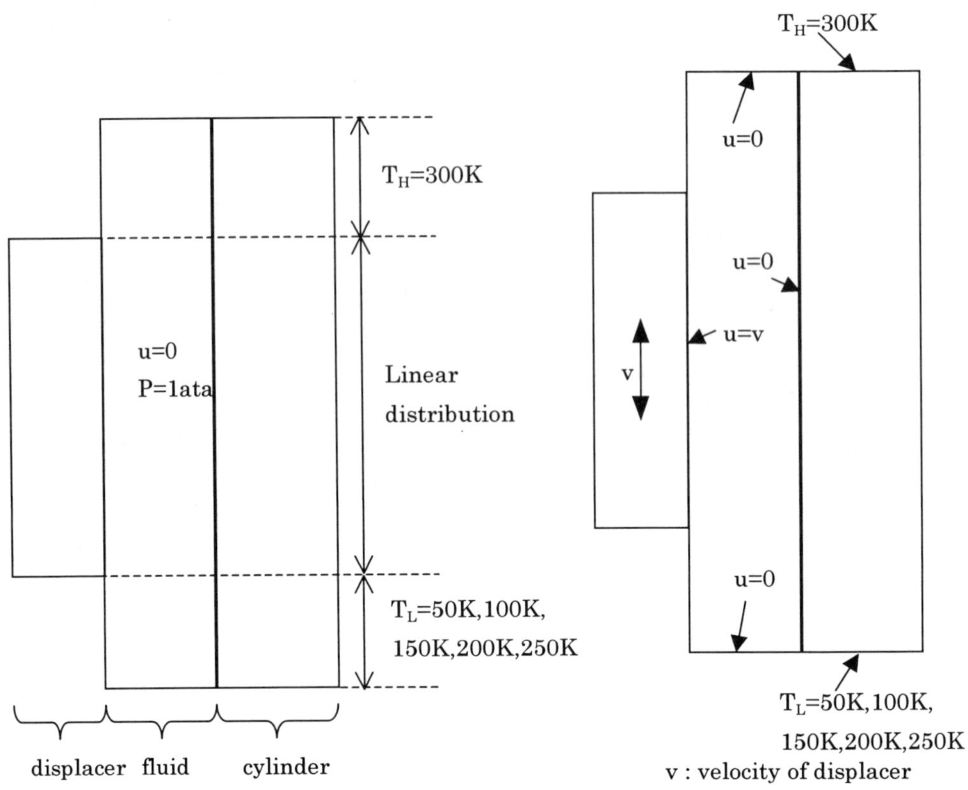


**Figure 4.** Initial condition.

**Figure 5.** Boundary condition.

## ANALYSIS RESULT AND CONSIDERATION

The analysis of the temperature change in one cycle is shown in Figure 6. Low end temperature $T_L$ is 200K. We represent the position of the displacer using phase angle. It is defined that the phase angle is zero when displacer is positioned at the bottom end, and phase angle is 180 degrees when the displacer is positioned at the top end. Figure 6 shows the temperature change every 45 degrees from 90 degrees. Because the displacer has a temperature gradient and moves, a difference in temperature occurs between the cylinder and displacer, and a temperature gradient is produced in the working fluid in a radial direction. We show the temperature change of the central part of the working fluid in Figure 7. The temperature gradient in the working fluid is not linear and becomes larger at the displacer side in this figure. This is regarded as the influence of the flow of the working fluid. The velocity in the working fluid is shown in Figure 8. In this figure, the displacer is at a position of 90 degrees, displacer then moves above. Therefore the fluid at the displacer side moves upward. The space of the working fluid in this analysis was closed. Therefore the fluid at the cylinder side flows adversely underneath. We show the velocity of the central part of the working fluid in Figure 9. In this figure, we can understand that the flow of the working fluid at the displacer side and the flow at the cylinder side flows underneath adversely.

The temperature change of the cylinder is shown in Figure 10. From this figure, we can see that the temperature of the central part of cylinder hardly changes, but the temperatures of the upper and lower parts do change. The shuttle loss was calculated from the change of internal energy of the upper cylinder and the change of internal energy of lower cylinder. The change of internal energy for every cycle of the upper and lower

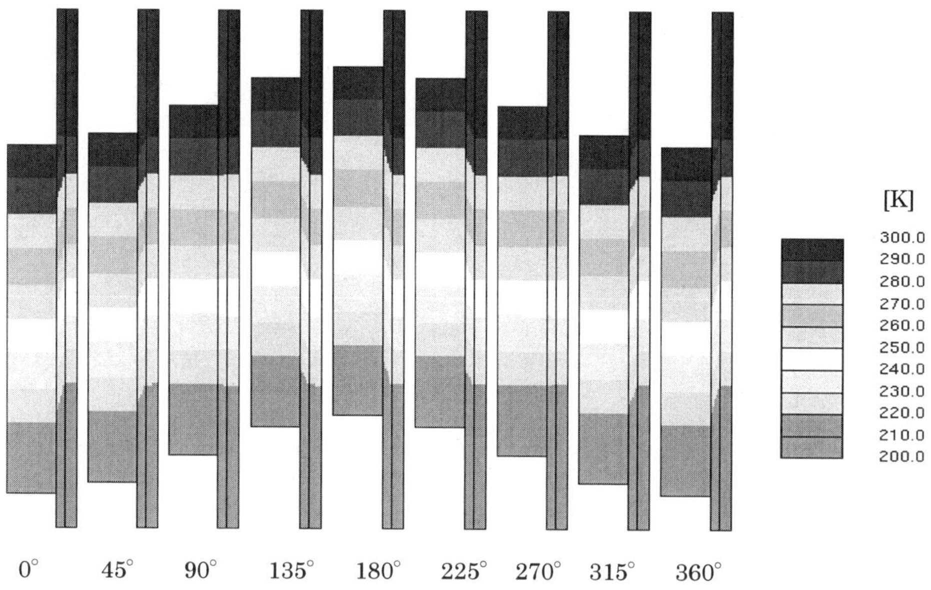


**Figure 6.** Temperature change in 1 cycle ($T_L$=200K).

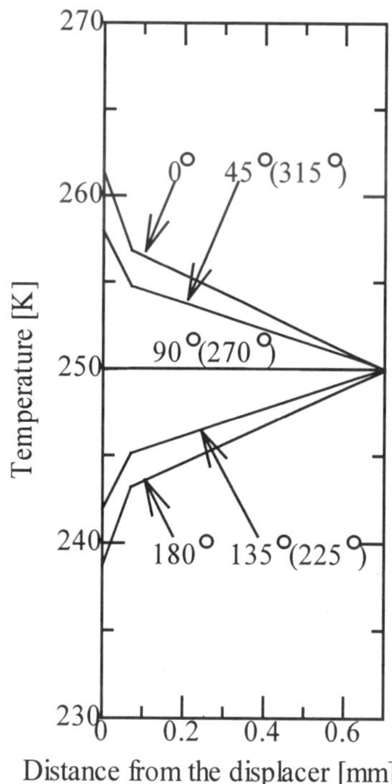


**Figure 7.** Temperature distribution of the working fluid ($T_L$=200K).

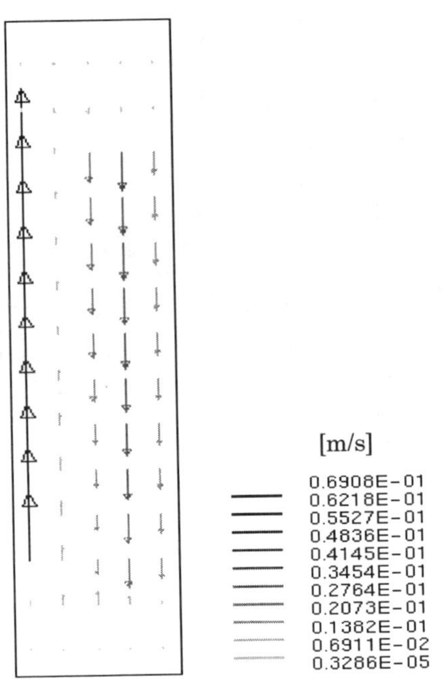


**Figure 8.** Velocity distribution of the working fluid when the displacer is at 90 degrees.

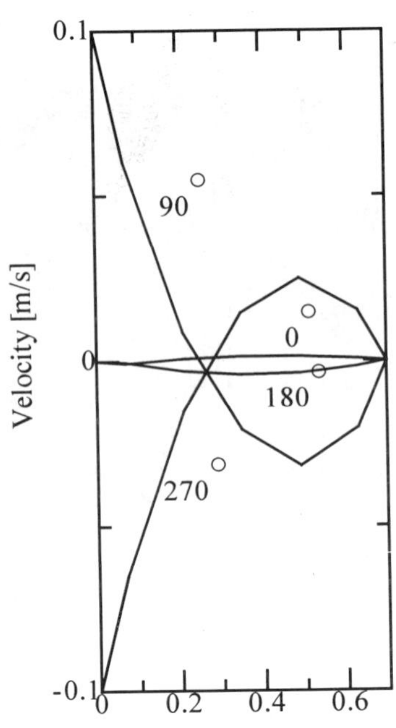


**Figure 9.** Velocity distribution of the working fluid ($T_L$=200K).

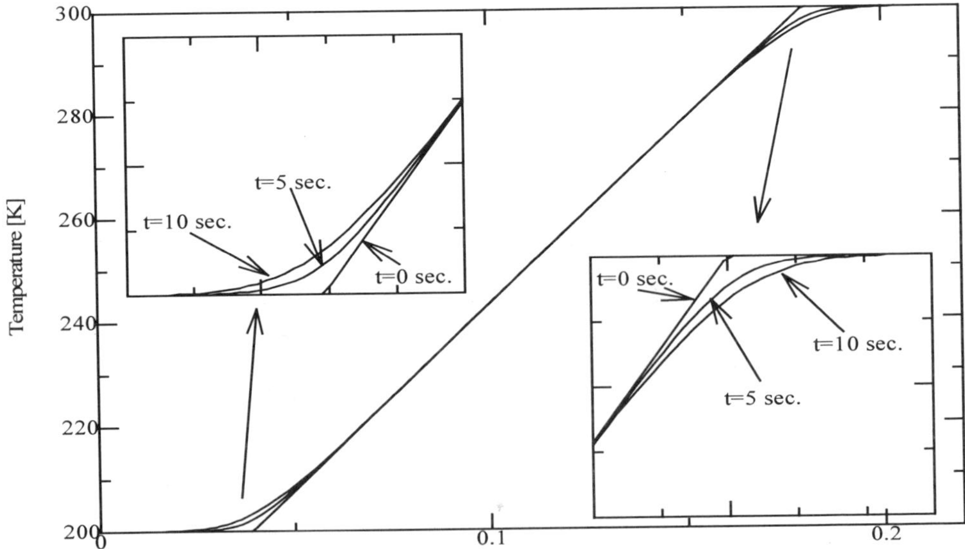


**Figure 10.** The change of longitudinal temperature distribution of cylinder ($T_L$=200K).

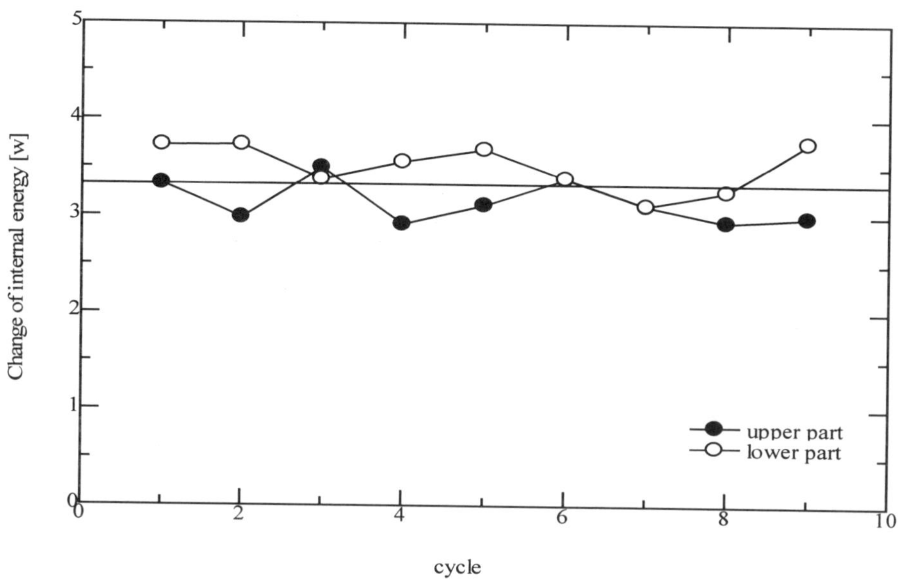


**Figure 11.** Change of internal energy of cylinder upper part and lower part at every cycle ($T_L$=200K).

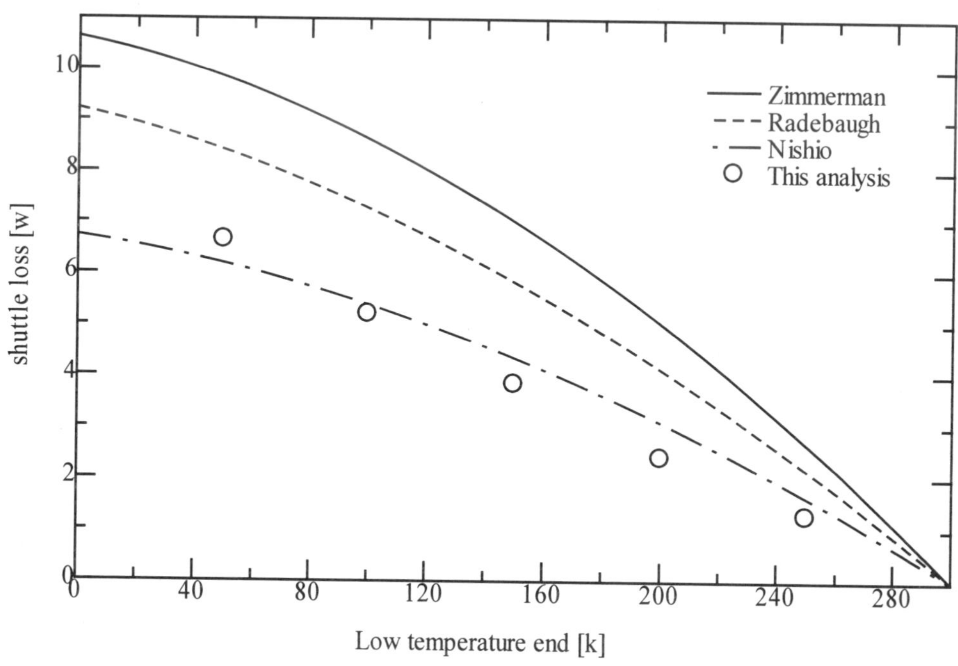


**Figure 12.** Result of this analysis and comparison with previous analyses.

cylinder is shown in Figure 11. Shuttle loss defined as a difference of the mean of these internal energies and the energy of conduction. As a result shuttle loss at 200K became 2.41W.

We changed low end temperature, analyzed shuttle loss, and compared it with previous analyses. The result is shown in Figure 12. The circles in the figure are the analysis results, the solid line shows Zimmerman's results, the dashed line Radebaugh's, and the dot dash line Nishio's.

The difference in temperature of the cylinder and the working fluid became smaller than the linear temperature gradient of previous analyses. As a result, shuttle loss becomes small. The precision of analysis was verified by comparing it to previous analyses.

## CONCLUSION

We made a model that considered the hydrodynamic behavior of the working fluid. Using this model we did a thermofluid analysis which calculated the movement of the working fluid and the temperature change. From this analysis, we calculated the shuttle loss. The precision of this analysis was confirmed by comparing it with previous analyses. This means that acceptable analysis is possible with general purpose software. Also it is possible to analyze various models easily.

## REFERENCES


1. Computational Dynamics Ltd.: "STAR-CD Ver 3.0"
2. F. J. Zimmerman and R. C. Longsworth: "Shuttle Heat Transfer" Adv. Cryog. Eng. 16 (1970) 342
3. R. Radebaugh and J. E. Zimmerman: "Shuttle Heat Transfer in Plastic Displacer at Low Speeds" NBS Special Publication 508 (1978) 67
4. S. Nishio, T. Inada and H. Nakagome: "Shuttle Heat Transfer in Refrigerators" Journal of the Cryogenic Society of Japan 27 (1992) 507

# PERFORMANCE OF HUGHES' "SBIRS LOW" FDS CRYOCOOLER

M. C. Barr and D. C. Gilman

Hughes Aircraft Company
El Segundo, California 90245

## ABSTRACT

A cryocooler for use in long-term space environment adapted from earlier designs has been chosen as the baseline cryocooler for SBIRS Low FDS. First two of four units have been designed, assembled and acceptance tested. The first unit has been placed into a simulated space life test environment. This cryocooler design operates on the Stirling cycle in the split configuration. It utilizes "heat intercept" concept to enhance its performance. Heat Intercept has been demonstrated at JPL previously by Dr. Dean Johnson and Dr. Ron Ross and its use here and on an earlier prototype bears out the concept. The cooler is intended to be used at two discrete cold end temperatures for two different focal plane arrays in a space-based infrared telescope. Intended loads are: a) 0.9 Watt heat lift at 35K rejecting to 293K for 80 Watts of input power with the aid of a 170K heat intercept of less than 3 Watts and b) 0.9 Watts heat lift at 60K rejecting to 293K for 60W of input power with the aid of a 170K heat intercept of less than 3 Watts. Specified performance has been demonstrated and both units have survived a simulated space launch environment. Description of cryocooler geometry, test methods and performance data are presented.

## INTRODUCTION

The Hughes SBIRS Low Flight Demonstration System (FDS) cryocooler is designed for the very specialized space-based sensor environment. Some of the design objectives and the Hughes cooler performance are summarized in Table 1. These requirements lead to the use of precise tolerances and specialized materials and processes. Extensive work in electronics and control algorithms has also been done at Hughes but is not covered in this paper.

| Table 1. Design Objectives for SBIRS Low FDS Cryocooler | |
|---|---|
| Objective | SBIRS Low FDS Cryocooler Performance |
| Spacecraft Environment | All conductive heat transfer from cooler surfaces |
| Efficient | See performance |
| Compact | Two modules ~150 mm diameter by ~420 mm long |
| Lightweight | Total mass of 17.2 kg |
| Low Residual Vibration | Control first 5 harmonics below .1 N, none > .2 N |
| Survive Launch | Random Vibe tested > 5.4 $G_{rms}$ |
| Long Life | Designed for > 10 years operation |
| Simple Spacecraft Interface | All flat reject surfaces, single cold finger, 4 connectors |
| Performance: | |
| 0.9 W @ 35K for <80 W in* <3 W to 170K intercept | 1.19 W @ 35K for 80 W in*, 2.3 W to 170K intercept |
| 0.9 W @ 60K for <60 W in* <3 W to 170K intercept | 2.8 W @ 60K for 60 W in*, 2.9 W to 170K intercept |

* 293K reject temperature

## DESIGN HISTORY

Hughes has been designing cryocoolers for space and tactical applications for over 30 years. This latest line of low-vibration split Stirling cryocoolers started with R&D efforts in 1988. Through the generous funding of the BMDO and the Air Force SMC we have developed two new generations of more efficient and powerful coolers. The history of this line of coolers is summarized in Table 2.

Table 2. Design History of Hughes Stirling Space Cryocoolers

| Program | Year | Units | Performance* | Mass | Funding Source |
|---|---|---|---|---|---|
| HAC IR&D | '89-'90 | 1 | 0.35 W @ 65K for 65 W in | 13 Kg | Hughes |
| SSC | '90-'94 | 2 | 2.0 W @ 65K for 58 W in | 18.5 Kg | BMDO/PL |
| ISSC | '93-'97 | 4 | 2.1 W @ 60K for 80 W in | 18.2 Kg | Hughes/SMC |
| PSC | '94-'97 | 1 | 3.1 W @ 60K for 93 W in | 12.6 Kg | BMDO/PL |
| SBIRS Low | '95-'97 | 4 | 3.0 W @ 60K for 65 W in | 17.2 Kg | SMC |
| Mini Cooler | '96-'97 | 1 | 0.5 W @ 60 K for 20 W in** | 6 Kg** | Hughes |

* 293K reject temperature

** Projected

The SBIRS Low FDS coolers are based on the SSC and ISSC cryocooler designs and materials but utilize the user-friendly outline and mounting features developed on the PSC program. The housings are stainless steel with the expander heat-reject flange of aluminum and the cryogenic interface of copper. Materials and construction are consistent with minimizing contamination or leaking of the ultra-high purity helium working fluid.

## USE OF HEAT INTERCEPT

The use of heat intercept to enhance cryocooler performance was pioneered by Dr. Dean Johnson and Dr. Ron Ross at JPL [1]. Their work was done primarily on a BAe 80K cryocooler and demonstrated increases in cold end capacity of as much as 50% (using an intercept of 1.5 watts at 150K) over performance demonstrated without heat intercept. The concept pre-supposes that there is available a source of cooling that can draw off a combination of waste heat and parasitic heat loads to free up cold end PV work to do more cooling. In the case of the SBIRS Low FDS cooler, we have such a source of cooling in our cryoradiator, a device used to cool the focal plane of one sensor and the radiation shields for both sensors. Our thermal analysts calculated that we could build into the spacecraft extra cryoradiator capacity which could be used for cryocooler heat intercept purposes. The budget they set for this was 3.0 Watts at 170K. Given this number, the cryocooler design team used the ISSC 4 cryocooler in an intercept experiment to determine a suitable intercept location to optimize thermodynamic performance while keeping the intercept load below 3.0 Watts at 170K [2]. This was done and the data was used to design the cold finger housings for the SBIRS Low FDS coolers.

The cryocooler cold finger design is depicted in figure 1. At the base of the coldfinger is the expander heat reject flange which is maintained at a reject temperature of 293K to 302K by heat pipes in the SBIRS Low FDS satellite. The coldfinger steps down to a thin-walled section to minimize parasitic heat loads and then steps to the integral machined heat intercept flange. The coldfinger steps back down to the thin-walled section out to the cold tip. The solid cylindrical copper cold tip is welded in place and is the thermal interface that pulls heat out of the SBIRS Low FDS Track Sensor Infrared Focal Plane Arrays. The sensor's two modes of operation require that the refrigerator be able to operate the FPAs at either 65K or 40K (5K is allotted for thermal strap temperature deltas).

To test the SBIRS Low FDS cooler in a representative environment, we designed a custom test dewar that accommodated the SBIRS Low FDS coldfinger as well as an additional tactical cooler coldfinger to simulate the sensor's cryoradiator. The layout is depicted in figure 2. The tactical cooler's cold end is strapped to the heat intercept flange on the SBIRS Low cryocooler. The tactical cooler has a temperature control hybrid built in and its feedback is from a temp sensor on the intercept flange. An instrument ring (with a load resistor and two precision temp sensors) attached to the cold tip of the SBIRS Low FDS cryocooler simulates the sensor heat load to the cooler. Multi-Layer Insulation (MLI) envelopes the cold finger and lines the test dewar to minimize extraneous parasitic loads and permit accurate calorimetry. The strap attached from the intercept flange to the tactical cooler is instrumented for delta T and calibrated to be used as a measure of intercept heat load. Other temp sensors measure warm end temperature and reject flange temperature.

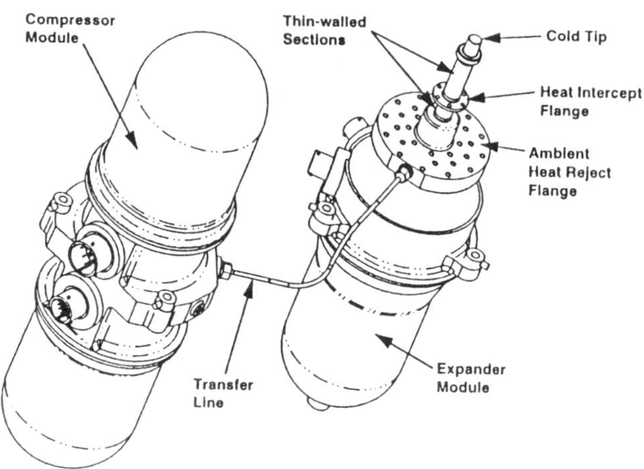


Figure 1 The configuration of the SBIRS Low FDS cryocooler coldfinger facilitates easy attachment of the heat intercept cold strap and cold end load.

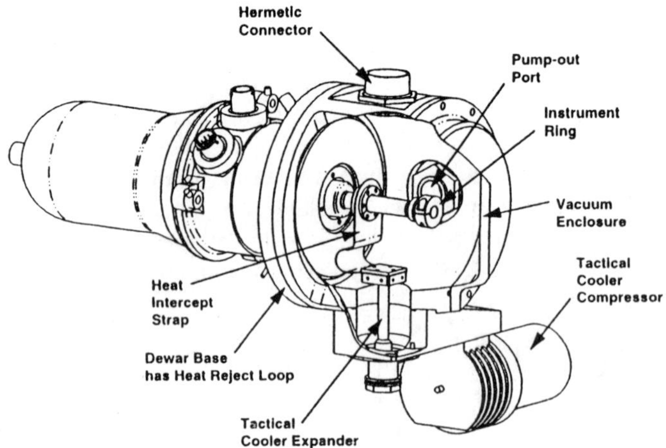


Figure 2 Our heat intercept dewar is designed to implement the heat intercept scheme and facilitate measurement of cold end and heat intercept heat loads.

On the actual SBIRS Low FDS Sensor, there is a flexible thermal strap from the FPA mount clamped to the copper cold tip of the cryocooler. Another flexible strap is clamped to the intercept flange connecting it to the sensor radiation shield and a second strap connects this to the cryo heat pipe attached to the satellite's cryoradiator. Ambient heat pipes are attached directly to the expander and compressor heat reject flanges with two pipes per device carrying the rejected heat to the satellite's ambient radiator. The modules are anchored to the Track Sensor Baseplate by brackets and are connected by a stainless steel transfer line. Load washers at the module mounts provide feedback of residual vibration forces to the cryocooler electronics to drive the harmonic cancellation algorithms. Temperature feedback for the refrigerator comes from temp sensors on the FPAs.

The outline and mounting dimensions for the SBIRS Low cryocooler are shown in figure 3. Transfer line length used in this testing is 270 mm. On the flight coolers the transfer line length will be slightly shorter (~250 mm).

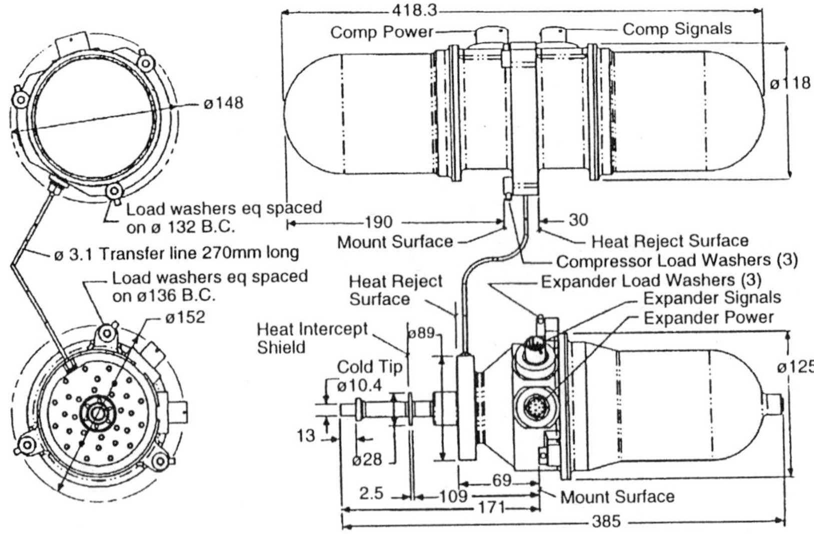


Figure 3 Hughes' SBIRS Low cryocooler outline and mounting.

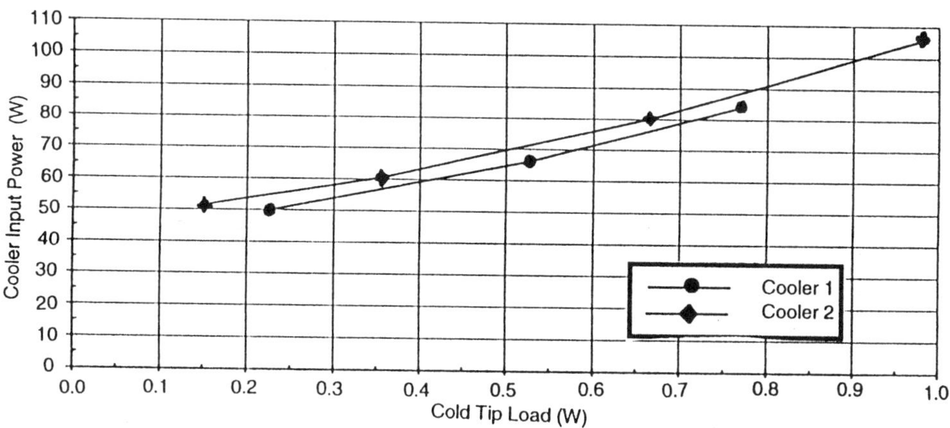


Figure 4 35K Baseline (no heat intercept) performance of Coolers 1 & 2 at 295K reject temperature.

## HEAT INTERCEPT PERFORMANCE DATA

The baseline performance of the SBIRS Low FDS Cryocooler is taken with nothing attached to the intercept flange except a temp sensor to document flange temperature. This testing is done throughout the integration and test cycle as a monitor of cooler health and status. Testing is done at a 295K reject temperature for both the 35K load point and the 60K load point. Figure 4 shows typical performance of coolers 1 & 2 at 35K and figure 5 shows typical performance at 60K. Cooler 2 was built as a pathfinder using a slightly less efficient compressor motor resulting in a slightly lower overall cooler efficiency.

Both coolers were tested with the heat intercept test dewar described above. An intercept temperature of 170K was used at three different heat reject temperatures: 290K, 300K and 310K at the 35K set point and at a single reject temperature, 290K, for the 60K set point. Figures 6 through 9 show the performance of the two coolers under the various reject temperatures at 35K and 60K load points. Figure 10 shows the load curve for cooler 2 with a 170K intercept temperature and a 290K reject temperature.

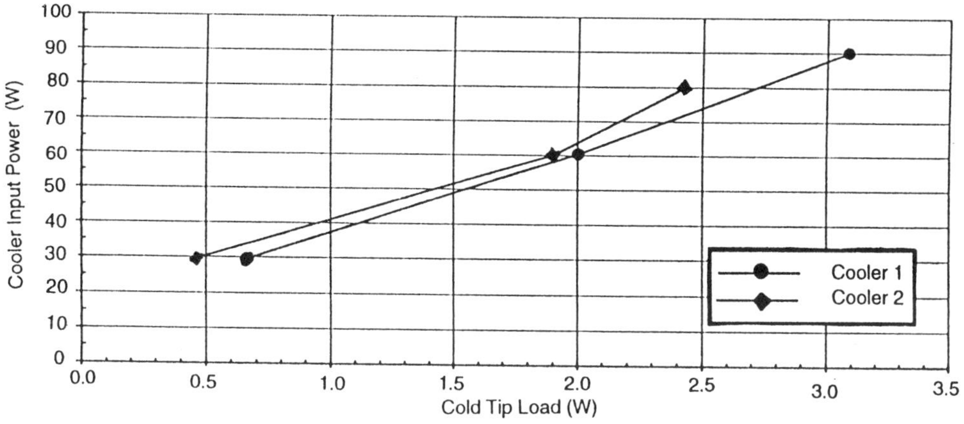


Figure 5 60K Baseline (no heat intercept) performance of Coolers 1 & 2 at 295K reject temperature.

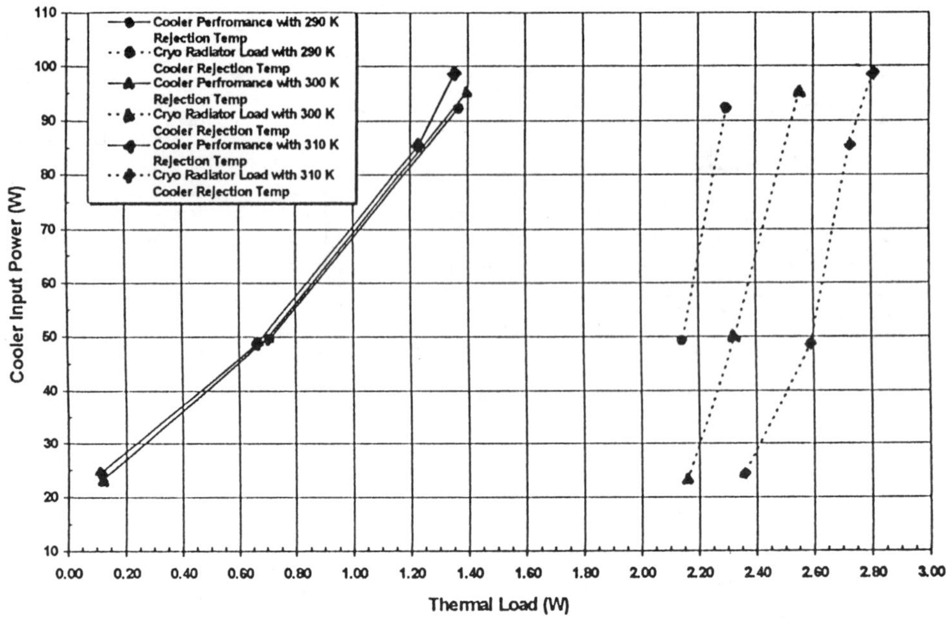


Figure 6 35K performance of Cooler 1 at 290K, 300K and 310K reject temperature with 170K intercept temperature.

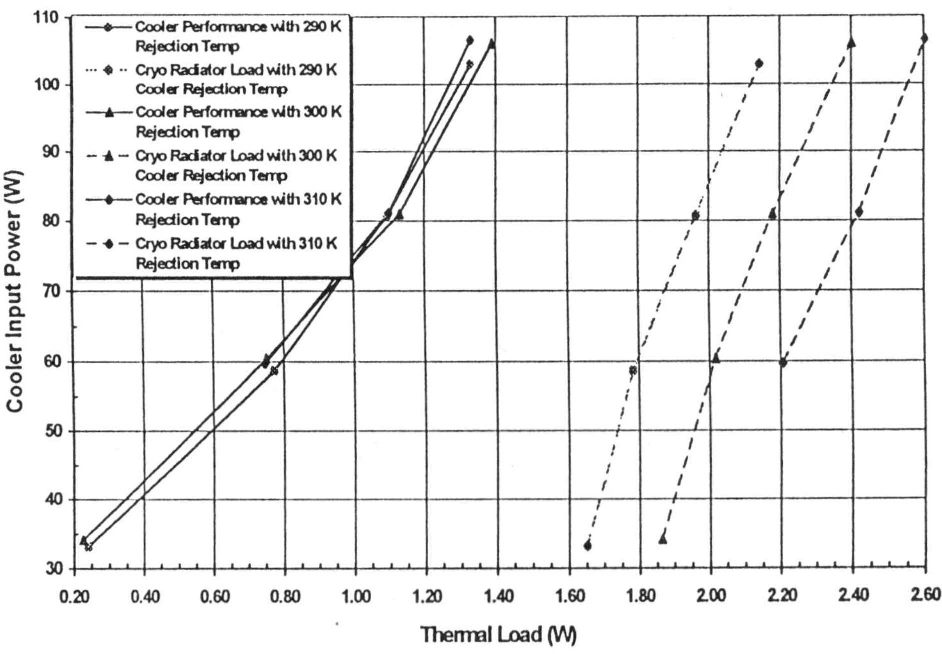


Figure 7 35K performance of Cooler 2 at 290K, 300K and 310K reject temperature with 170K intercept temperature.

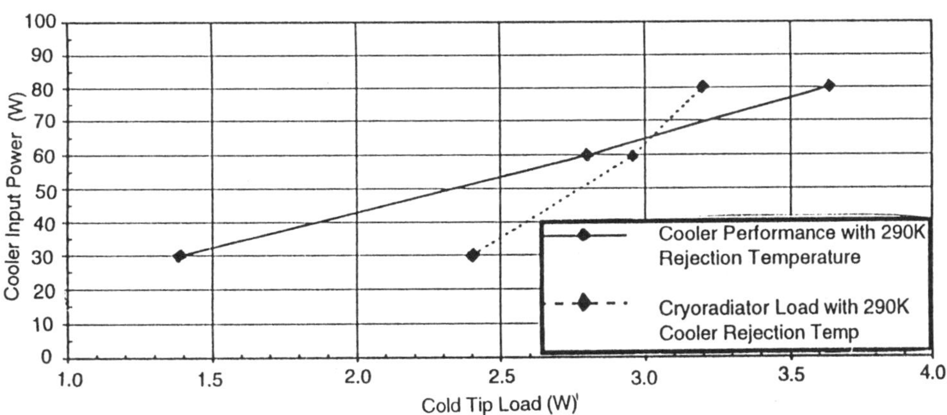


Figure 8 60K performance of Cooler 1 at 290K reject temperature with 170K intercept temperature.

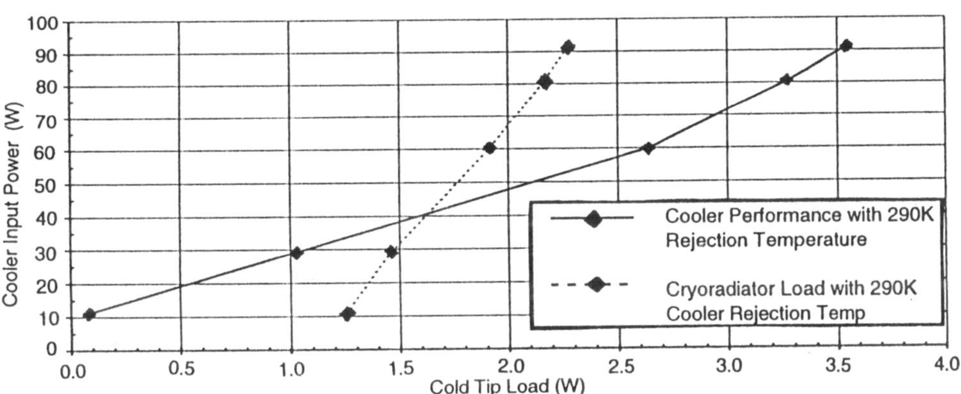


Figure 9 60K performance of Cooler 2 at 290K reject temperature with 170K intercept temperature.

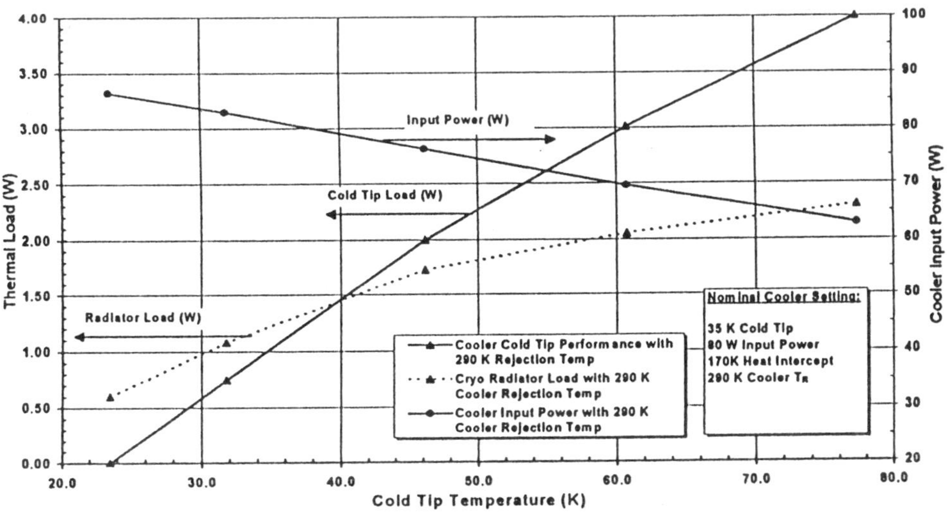


Figure 10 Standard load line of Cooler 2 with a 170K intercept temperature and a 290K reject temperature.

## OTHER TESTING

Further tests are pending on the pathfinder cooler and the two flight units. ;Tests of residual vibration levels and testing in a thermal-vacuum environment will be conducted on each cooler. The life test unit is currently running in a thermal-vacuum environment; in alternate months we run the 60K mode rejecting to 280K and the 35K mode rejecting to 305K, both with a 170K intercept temperature. It has accumulated over 4000 hours of operation to date with no degradation of performance: 35K cold end capacity is 1.05W for 74W in (170K load of 2.5W); 60K cold end capacity is 1.05W for 26W in (170K load of 1.8W). Life test will continue for 1 year under current program funding and probably longer.

## CONCLUSIONS

The cryocoolers designed for the SBIRS Low FDS mission meet design thermodynamic performance goals. By utilizing the heat intercept concept at 170K the 35K cold end capacity of the machines is enhanced by over 40% and the sensitivity of the cold end capacity to reject temperature is greatly reduced. Intercept heat load limits were not exceeded and target cold end capacities were attained within specified power limits over the specified rejection temperature range. Performance at 35K is better than 80W/W from 0.6W to 1.3W loads and as good as 72W/W at a 1.1W load.

By utilizing the heat intercept concept at 170K the 60K cold end capacity of the machines is enhanced by over 39%. Intercept heat load limits were not exceeded and target cold end capacities were attained within specified power limits over the specified rejection temperature range. Performance at 60K is better than 22W/W for loads from 1.5W to 3.5W.

## ACKNOWLEDGMENT

Financial support provided by the USAF Space and Missile Systems Center to Hughes Aircraft Company (subcontract HA2758GR3S) through TRW (contract F04701-92-C-0062). The authors would like to thank Lt. Charles Light, USAF SMC, Drs. Ron Ross and Dean Johnson, NASA JPL, and Dr. David Curran, Aerospace Corporation for their support. Thanks also to Ken Price of Hughes for design support and overview.

## REFERENCES

1. D.L. Johnson and R.G. Ross, Jr., Cryocooler coldfinger heat interceptor, in: "Cryocoolers 8", Plenum Publishing Corp., New York (1995), pp 709-718.
2. D.C. Gilman, "Cryocooler heat interceptor test for the SMTS Program", presented at the 9th International Cryocooler Conference, Waterville, NH, June '96.

# ENHANCED PERFORMANCE OF THE BEI 0.5 WATT MINI-LINEAR STIRLING COOLER

S.W.K.Yuan, D.T.Kuo, and A.S.Loc

BEI Technologies
Sylmar, CA, 91342, USA

## ABSTRACT

The BEI B500C Mini-Linear Cooler delivers 0.5 Watt of refrigeration at 80 K with an input power of less than 21W (at 23C case temperature). To meet more stringent requirements, the design of this cooler has been modified to enhance its performance. The new refrigerator, known as the B512C Mini-Linear Cooler has a more efficient motor with less dead volume and parasitic heat loads down the coldfinger. It can provide ~0.6 Watt of cooling at 80K. The cooldown time from room temperature to 73K is less than 8 minutes with 250 joules of thermal mass and a case temperature of 60C. At the same demanding case temperature of 60C, the refrigeration capacity is >0.25 Watt (with a cap of 11 Volt or 22 Watt), and the cooler can deliver 0.15 Watt of cooling with less than 13 Watt of input power. In this paper, the enhanced performance of this cooler is discussed in detail together with some of the analyses which lead to the final product.

## INTRODUCTION

The progressively more demanding requirements on the 0.5 Watt Linear Cooler have prompted BEI to modify the design of the mini-linear cooler. The enhanced performance of this new design is presented in this paper.

## THE OLD DESIGN OF THE BEI B500C COOLER

The typical performance of the old baseline BEI B500C cooler is listed in the second column of Table 1. In order to enhance the performance of the B500C coolers to meet the stringent requirements of some of our customers, BEI has investigated ways to improve the following areas of the cooler, namely, the effectiveness of the regenerator, efficiency of the motor, and parasitic heat loss.

**Table 1** Summary on the Enhancement of Cooler Performance

| | Old Baseline | Improved Regenerator Efficiency | Improved Motor Efficiency | Reduced Parasitic Heat Loss |
|---|---|---|---|---|
| Cooldown Time to 73K | 3:56 | S1- 4:14<br>S2-4:47<br>S3-4:35 (Avg. 4:45)<br>S4-4:39<br>S5-5:06 | C1751- 4:26<br>C1752- 4:33<br>C1753- 4:47<br>( Avg. 4:35) | C1751- 4:10<br>C1752- 4:34<br>C1753- 4:17<br>(Avg. 4:20) |
| Input Power @ 73K 150 mW Ambient: 60 C | 14.6 W | S1- 14.7 W<br>S2- 14.1 W<br>S3- 13.7 W(Avg. 14W)<br>S4- 13.3 W<br>S5- 14.2 W | C1751- 10.2 W<br>C1752- 10.3 W<br>C1753- 12.5 W<br>(Avg. 11W) | C1751- 9.5 W<br>C1752- 10.6 W<br>C1753- 11.37 W<br>(Avg.10.5W) |
| Input Power @ 73K 150 mW Ambient: 23 C | 10.7 W | S1- 9.8 W<br>S2- 10 W<br>S3- 10.4 W(Avg. 10.4W)<br>S4- 10.4 w<br>S5- 11.4 W | C1751- 7.97 W<br>C1752- 7.55 W<br>C1753- 8.9 W<br>(Avg.8.1W) | C1751- 6.85 W<br>C1752- 6.7 W<br>C1753- 7.83 W<br>(Avg. 7.1W) |
| Input Power @ 78K 150 mW Ambient: 23 C | 9.1 W | S1- 8.7 W<br>S2- 8.4 W<br>S3- 8.8 W (Avg. 8.92 W)<br>S4- 8.7 W<br>S5- 10 W | C1751- 6.99 W<br>C1752- 6.67 W<br>C1753- 8.25 W<br>(Avg.7.3W) | C1751- 6.25 W<br>C1752- 5.9 W<br>C1753- 6.76 W<br>(Avg. 6.3W) |
| Lowest Temp Reached | ~ 44 K | ~ 40 K | ~ 37 K | ~ 34 K |

## REGENERATOR EFFECTIVENESS

The regenerator is one of the most critical components of the cryocooler. It absorbs the heat energy from the working gas as it passes from the hot end of the displacer to the cold end and releases the energy to the gas stream as the flow reverses in direction. By improving the effectiveness of the regenerator, one can enhance the performance of the cooler by cutting down the enthalpy flow, which decreases the PV work of the thermodynamic cycle. There are three major considerations in the design of a regenerator.

### Regenerator configuration

The conventional method of using screens as the regenerator material is only one of the many possible configurations, including parallel plates and bundled tubes. The effectiveness of a regenerator configuration is generally characterized by the amount of heat transfer divided by the pressure drop across the regenerator. Recently, etched-foil regenerators, based on the parallel plate concept, stirred up quite a bit of interest in the cryocooler field. Some believe that by keeping the transport within the regenerator in the developing flow regime, one can maximize the heat transfer and minimize the pressure drop. Etched-foil regenerators were built and tested at BEI. It was concluded that one should be cautious in applying the etched-foil technique to pneumatically driven split Stirling coolers, for the extremely low pressure drop in the regenerator may change the optimum phase angle between the displacer and piston motion, and lead to adverse effects.

### Regenerator dimensions

It is quite obvious that each cooler has an optimum dimension of the regenerator for the best performance. Bigger and longer regenerators result in good heat transfer with the working gas and reduced axial heat conduction, but the presence of extra dead volume

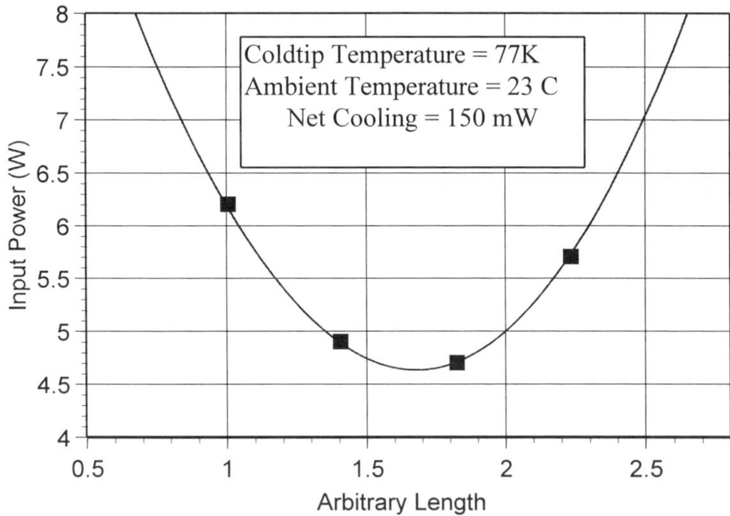


**Figure 1**. Effect of Regenerator Length.

degrades the performance. On the other hand, too small a regenerator might have limited dead volume, but the adverse heat transfer effects are not favorable. Figure 1 shows the input power as a function of regenerator length as predicted by a Hybrid Computer Model (reference 1 to 8), showing the optimum length for the lowest input power.

### Regenerator material

The choice of material is crucial for the success of a regenerator. The material used should have a high heat capacity for heat transfer and low thermal conductivity to minimize parasitic heat loss. Common material used nowadays are stainless steel and phosphor bronze. Etched foil regenerators made of nickel were also fabricated at BEI.

Five prototype expanders (S1, S2, S3, S4 and S5) with improved screen-regenerators were built and tested, and the results are summarized in the third column of Table 1. The new regenerator design increased the cooldown time (from room temperature to 73K) by approximately 50 seconds (at 4 min. 45 sec), but a moderate gain in cooling capacity was observed. The benefit of cooling capacity is more pronounced at low coldtip temperature and high ambient. Since in most applications, the required cooldown time is around 6 minutes, BEI decided to adopt the new regenerator design to take advantage of the extra cooling capacity.

## MOTOR PERFORMANCE

The performance of a motor depends on the motor efficiency, the presence of dead volume, and the size of the clearance gap of the piston. Excessive dead volume decreases the pressure ratio thus degrades the performance. This effect is shown in Figure 2, and is more pronounced at small input powers.

It is a well-known fact that blow-by losses across the piston clearance gap tend to reduce the useful work of the compressor.

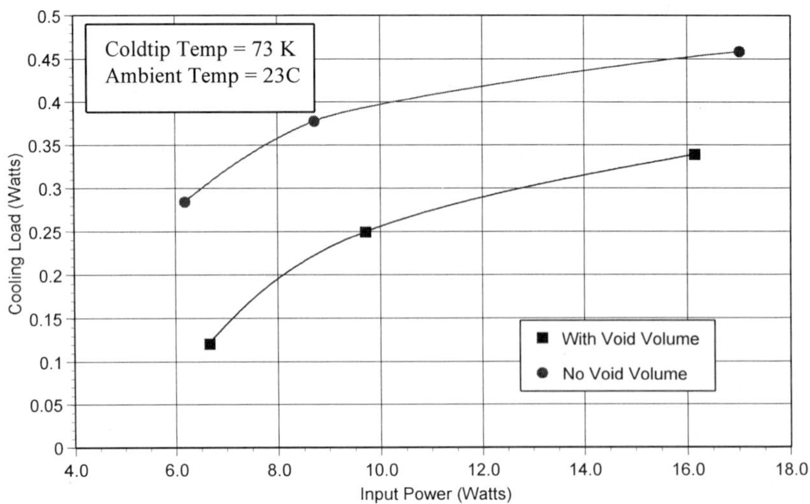


**Figure 2.** Effect of Dead Volume.

The mass flow of the helium gas in the piston clearance gap is governed by the following equation,

$$\dot{m} = \frac{\pi D S^3}{32} \frac{\rho \Delta P}{\eta L} \tag{1}$$

where S is the radial gap of the piston clearance, D is the piston diameter, L is the gap length, $\Delta P$ is the pressure drop, $\rho$ is the density and $\eta$ is the viscosity of the helium gas. From the above equation, one can see that the blow-by flow across the clearance gap is a strong function of the gap size.

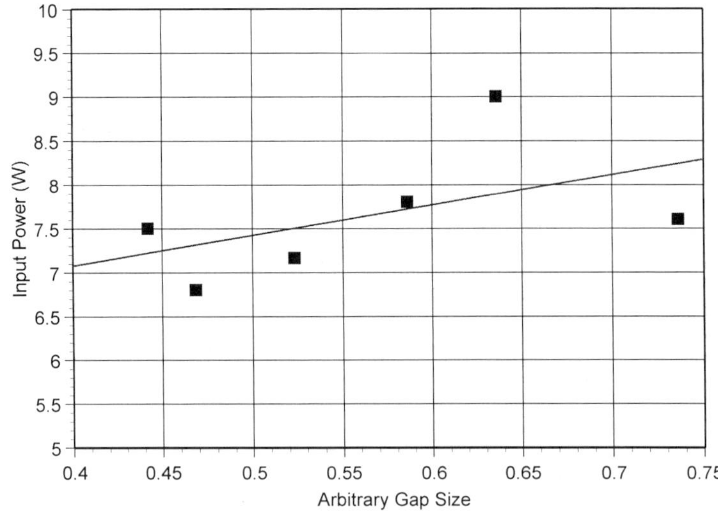


**Figure 3**. Input power versus piston gap size.

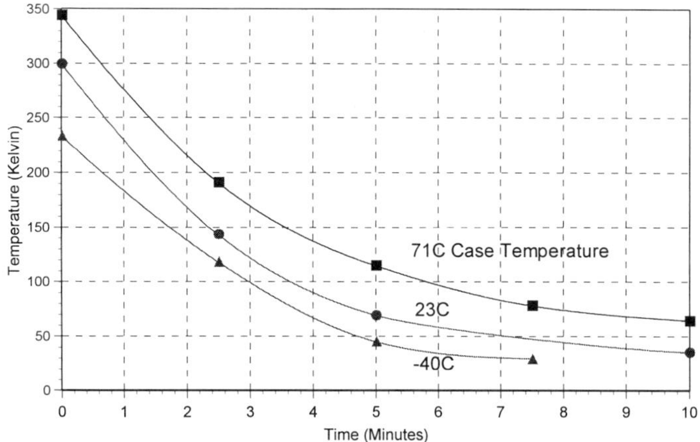


**Figure 4.** Cooldown time as a function of time.

Figure 3 shows the input power of the cooler as a function of the piston clearance gap. Within data scatter and limited data studied, a correlation can be seen between the gap size and input power, favoring smaller gap size for better performance.

Three prototype high performance compressors (C1751, C1752, and C1753) were built and tested. The result of these tests can be found in Column 4 of Table 1. The new compressor design improves the cooldown time (to 73K) by approximately 10 seconds and consumes about 3 watts less power at 73K, with an ambient temperature of 60C. The decrease in input power is less at room temperature (~2.3 Watts).

## REDUCE PARASITIC HEAT LOSSES

Another important area in enhancing cooler performance is the reduction of parasitic heat losses. Parasitic heat losses include heat conduction along coldfinger and regenerator shell, pressure drop loss, shuttle and/or blow-by losses in the displacer gap, and pumping loss. The BEI high-performance cooler contains enhancement in one or more of the above areas.

Comparing the fifth column to the fourth column, one can see that approximately 1 Watt decrease in input power is resulted by reducing the parasitic heat loss.

## RESULTS

The experimental results of the high-performance cooler with enhanced regenerator, motor efficiency, and reduced parasitic heat loss are listed in the fifth column of Table 1. From left to right, Table 1 shows the progressional enhancement of experimental performance due to the above study. The new cooler is approximately 3 Watts lower in input power with 150 mW heat load at room temperature, when compared to the old baseline. The enhancement is approximately 4 Watts at high ambient.

The typical cooldown characteristic of the enhanced cooler is shown in Figure 4 for ambient temperatures of -41C, 23C and 71C. The cooldown time to 73K is less than 5 minutes at room temperature and approximately 10 minutes at 71C.

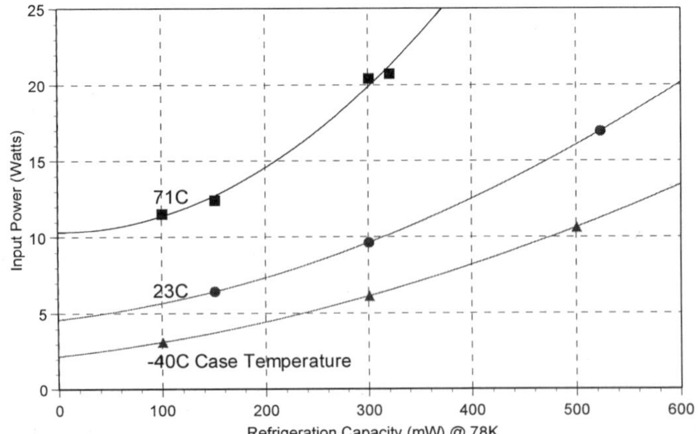


**Figure 5.** Input power as a function of refrigeration capacity.

Figure 5 shows the input power as a function of refrigeration capacity. The cooler provides close to 500mW of cooling at 78K with 15 Watts of input power. The coldtip temperature as a function of refrigeration capacity is plotted in Figure 6.

## CONCLUSIONS

A 73K High-Performance Cooler has been analyzed, designed, fabricated and tested at BEI, which delivers 150 mW of cooling with a mere input power of 10.5 W at 60C ambient, and 7.1 W at 23C. The cooldown time to 73K is approximately 4 minutes and 20 seconds. At 78K, the input power is 6.3W for a heat load of 150mW at 23C ambient.

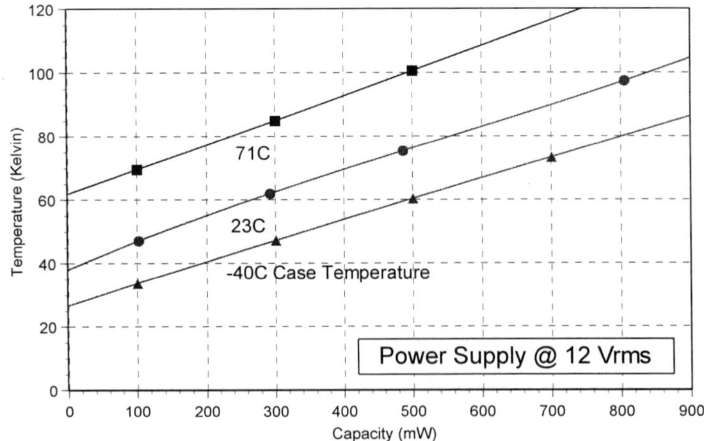


**Figure 6**. Coldtip temperature as a function of refrigeration capacity.

## REFERENCES


1. D.T. Kuo, A.S. Loc, and S.W.K. Yuan, Experimental and Predicted Performance of the BEI Mini-linear Cooler, Cryocoolers 9, Plenum Press, New York (1997) p.119.
2. S.W.K. Yuan and R. Radebaugh, A Blind Test on the Pulse Tube Refrigerator Model, in "Advances in Cryogenic Engineering", Vol. 41, (1996) p.1383.
3. S.W.K. Yuan, Validation of the Pulse Tube Refrigerator Model Against a Lockheed Built Pulse Tube Cooler, in "Cryogenics", Vol. 36, No.10, (1996) p.871.
4. S.W.K. Yuan, L.G. Naes, and T.C. Nast, Prediction of Natural Frequency of the NASA 80 K Cooler by the Stirling Refrigerator Performance Model, in: "Cryogenics", Vol. 34, No.5, (1994) p.383.
5. S.W.K. Yuan, I.E. Spradley, and W.G. Foster, Validation of the Stirling Refrigerator Performance Model Against the Oxford Refrigerator, in: "Advances in Cryogenic Engineering", Vol. 39, (1994) p.1359.
6. S.W.K. Yuan and I.E. Spradley, Validation of the Stirling Refrigerator Performance Model Against the Philips/NASA Magnetic Bearing Refrigerator, in: "Proc. 7th Int. Cryocooler Conf.", Vol. 1, Phillips Lab, USA (1993) p.280.
7. S.W.K. Yuan, I.E. Spradley, P.M. Yang, and T.C. Nast, Computer Simulation Model for Lucas Stirling Refrigerators, in: "Cryogenics", Vol. 32, (1992) p.143.
8. S.W.K. Yuan and I.E. Spradley, A Third Order Computer Model for Stirling Refrigerators, in: "Advances in Cryogenic Engineering", Vol. 37 (1992) p.1055.

# EXPERIMENTAL PERFORMANCE OF THE BEI ONE WATT LINEAR (OWL) STIRLING COOLER

S.W.K.Yuan, D.T.Kuo, A.S.Loc, and T.D. Lody

BEI Technologies
Sylmar, CA, 91342, USA

## ABSTRACT

The experimental performance of the BEI One Watt Linear (OWL) Cooler is presented in this paper. In particular, the compliance with the Army specification will be discussed in detail. The cooldown time of the BEI OWL Cooler (to 77K) is less than 7 minutes with a thermal mass of 1465 Joules (at 23C ambient temperature). Its refrigeration capacity at 77K is 1.8W with 40W of input power, and 1W with 25.8W input power (at 23C ambient temperature).

## INTRODUCTION

The BEI OWL cooler is one of many in a family of linear Stirling-cycle cooler designs developed by the company.

BEI's involvement in cryocoolers began with a corporate sponsored project in 1991 to develop a closed-cycle Joule-Thomson cooler for high-temperature-superconductivity application. After achieving limited success with a mixed gas refrigerant, the company leveraged its expertise in linear compressor technology to develop a miniature Stirling-cycle refrigerator for IR detector cooling.

The first miniature cooler designed for 150mW capacity at 78K has been well-received in the infrared user community [1,2]. BEI has recently enhanced the performance of this cooler by as much as 30% (Reference 1). A third design in BEI's family of coolers, in addition to the OWL Cooler, is the 350mW Linear Cooler. This design has recently undergone a flight test and received high-marks on its performance. BEI has undertaken

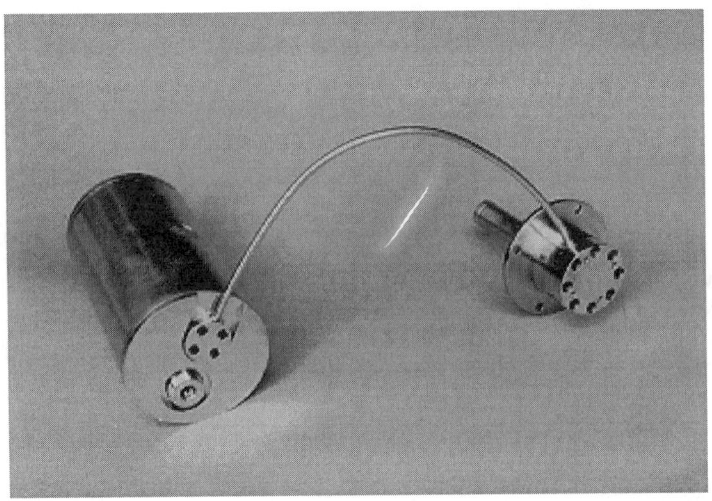

**Figure 1.** The BEI OWL Cooler.

another development effort in the 1.75 Watt linear cooler. All these coolers are designed to satisfy their respective U.S. Army specifications. BEI is also in the process of developing a Dual Use Long-Life Low-Cost Pulse Tube Cooler[3].

## THE ONE WATT LINEAR (OWL) COOLER

Figure 1 shows the external view of the OWL Cooler. The dimensions of this cooler are summarized in Table 1. The entire unit weighs only 3.8 lbs (meeting the Army requirement of 4.2 lbs).

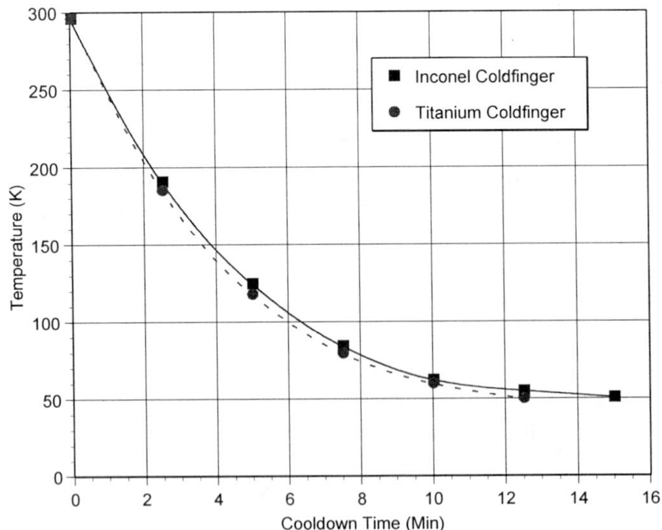


**Figure 2.** Cooldown characteristics of the BEI OWL Cooler.

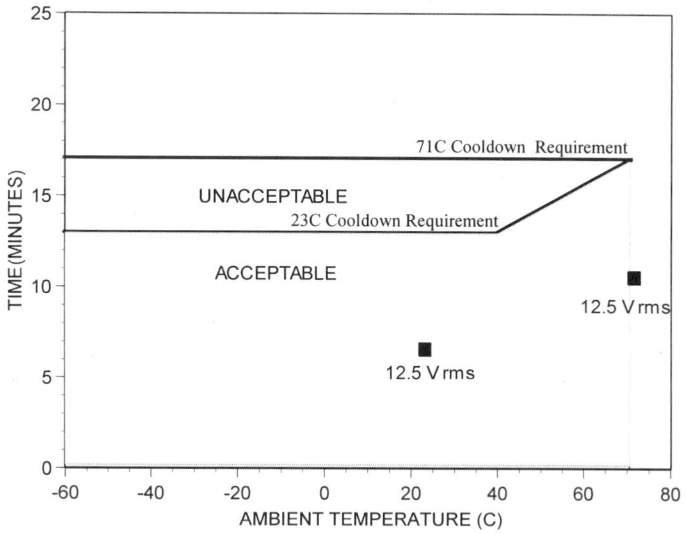


**Figure 3**. Cooldown time versus ambient temperature.

## COOLDOWN TIME

The cooldown time of a cooler is a function of the thermal mass and input power. Figure 2 shows a typical plot of coldtip temperature as a function of cooldown time (12.5 Volt rms, 23C ambient temperature) for the BEI OWL Cooler with an 8 inch transfer line and a thermal mass of 1465 Joules. The effect of using titanium as the coldfinger material as versus inconel is also shown, with a shorter cooldown time for the former, due to less static heat conduction.

With a titanium coldfinger, the typical cooldown time to 77K is 6.8 minutes at 23C ambient temperature and 10.5 minutes at 71C ambient temperature, meeting the requirements of 13 minutes and 17 minutes respectively as depicted in Figure 3.

## COOLING CAPACITY

Figure 4 is a plot of cooling capacity as a function of ambient temperature. The data points in the figure represent that of a BEI OWL Cooler with a titanium coldfinger (with 40 Watts of input power). As one can see, the refrigeration capacity of the cooler (at 1.8 Watts) far exceeds that of the requirement (1 Watt) at 23C ambient. At a high ambient of 71C, the cooler provides 1.77 Watts of cooling at the maximum allowable input power of 60W.

Figure 5 is a plot of refrigeration capacity vs. input power. The BEI OWL Cooler can deliver as much as 2.25 Watt at 77K with an input power of 50 W, making this cooler a good candidate for the High Tc Superconductor cooling. This cooler is also suitable for other commercial applications which require large cooling capacities. To increase the reliability of this cooler, flexure bearings or gas bearings can be incorporated into the compressor. The load curves of the BEI OWL cooler are plotted in Figure 6.

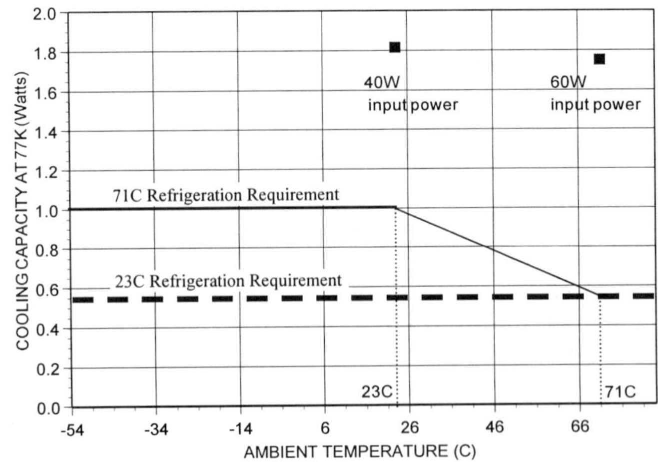


**Figure 4**. Cooling capacity as a function of ambient temperature.

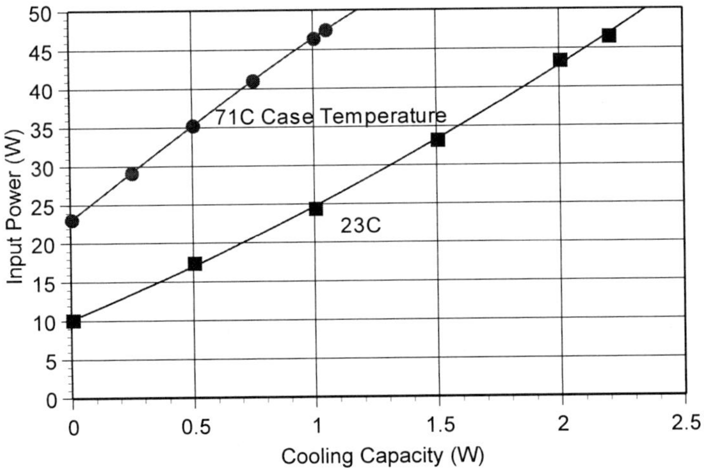


**Figure 5.** Cooling capacity vs. input power.

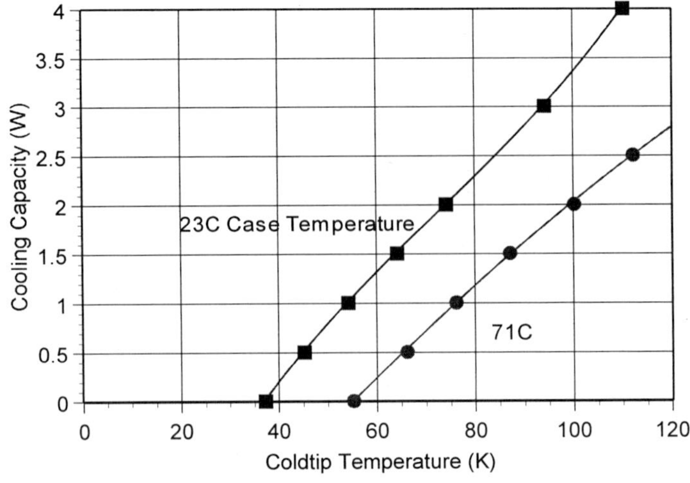


**Figure 6.** Cooling capacity vs. coldtip temperature.

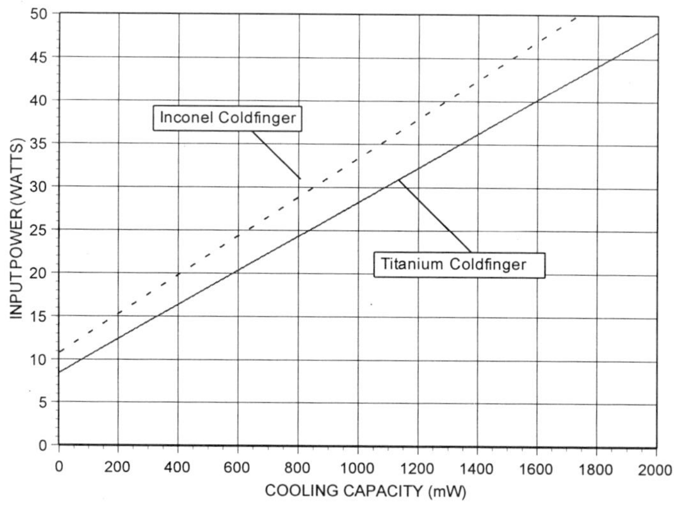


**Figure 7**. Input power as a function of cooling capacity.

## INPUT POWER

Input power versus heat load of the BEI OWL Cooler (at 23C ambient temperature) is plotted in Figure 7 for both the inconel and titanium coldfingers. As expected, the latter out-performed the former due to less parasitic heat conduction along the coldfinger.

Maximum input power as a function of ambient temperature is shown in Figure 8. At 23C ambient temperature, the cooler provides 1 Watt of cooling with an input power of 25.8 W (more than 14W below the requirement), and at 71C ambient temperature, the cooler delivers 0.55 Watt of cooling with an input power of 32.4W (more than 27W below the requirement).

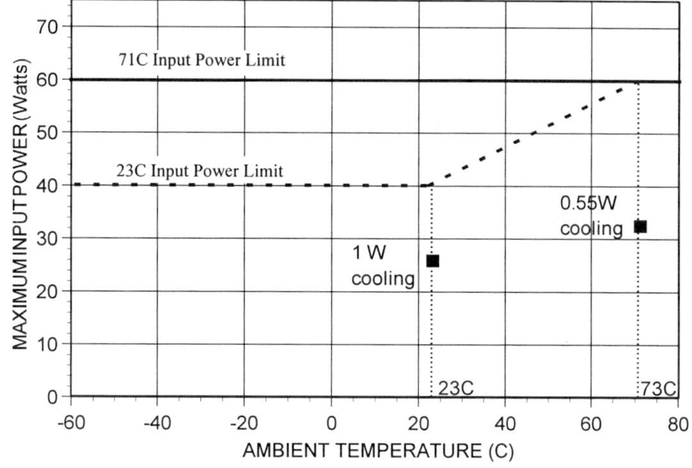


**Figure 8**. Maximum power versus ambient temperature.

**Table 1.** Specification Compliance

| Ambient Temperature | -54C | | 23C | | 71C | |
|---|---|---|---|---|---|---|
| | Specification | Performance | Specification | Performance | Specification | Performance |
| **Cooldown Time to 77K** | 13 min. | 4.5 min (12.5 V) | 13 min. | 6.8 min. (12.5 V) | 17 min. | 10.5 min. (12.5 V) |
| **Cooling Capacity @77K** | 1.0 W | 1.61W @ 40W Input Power | 1.0 W | 1.8W @ 40W Input Power | 0.55 W | 1.75 W @ 60W Input Power |
| **Maximum Input Power @77K** | 40 W | 24 W (1W cooling) | 40 W | 25.8 W (1W cooling) | 60 W | 32.4 W (0.55W cooling) |

## RELIABILITY

According to the Army specification, a failure is characterized by failure in performance in one of the five areas, namely, minimum refrigeration, cooldown time, input power, leak rate, and vibration. BEI has ample experience in the life-test of a similar cooler from which the current proposed cooler was designed, based on scaling. Experimental results show that within 4000 hours of operation, the input power of the cooler did not exceed the specification. Figure 9 shows the life projection of the BEI OWL cooler based on analysis assuming that the life of the cooler is dictated by the wear within the compressor. After 4000 hours, the cooling capacity of the proposed cooler is predicted to be 1.4W at 23C and 0.8W at 71C, exceeding the requirement of the U.S. Army.

## QUALIFICATION AND LIFE TESTS

Three BEI OWL Coolers have been fabricated and sent to the U.S. Army for environmental test to be followed by another three units for life tests. These units will be tested for vibration output, audible noise, and electromagnetic radiation. The results will be published in another paper.

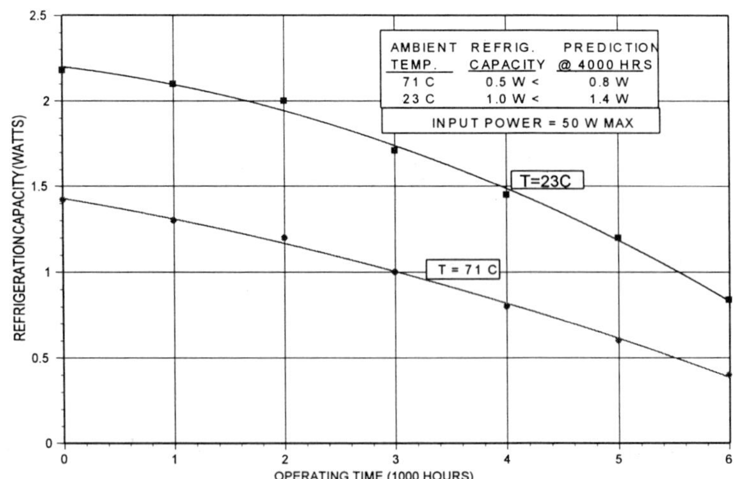


**Figure 9.** Predicted life time of the BEI OWL Cooler.

## CONCLUSIONS

Experimental performance of the BEI One Watt Linear (OWL) Cooler is presented in this paper (as summarized in Table 1). The performance of this OWL Cooler far exceeds the U.S. Army's specification. Six units are scheduled to undergo environmental and life tests, and their results will be reported elsewhere. The BEI OWL Cooler is also suitable for other commercial applications which require large cooling capacities. To increase the reliability of this cooler, flexure bearings or gas bearings can be incorporated into the compressor.

## REFERENCES


1. D.T. Kuo, A.S. Loc, and S.W.K. Yuan, Enhanced Performance of the BEI 0.5 Watt Mini-Linear Stirling Cooler, parallel paper in Cryogenic Engineering Conference, Portland, Oregon, 1997.
2. D.T. Kuo, A.S. Loc, and S.W.K. Yuan, Experimental and Predicted Performance of the BEI Mini-linear Cooler, in the Proc. of the 9th International Cryocooler Conference (1997) p.119.
3. D.T. Kuo, A.S. Loc, and S.W.K. Yuan, Design of a 0.5 Watt Dual Use Long-Life Low-Cost Pulse Tube Cooler, parallel paper in Cryogenic Engineering Conference, Portland, Oregon, 1997.

# EXERGY ANALYSIS OF STIRLING CYCLE CRYOGENERATOR

K.G. Narayankhedkar

Indian Institute of Technology, Bombay
Mumbai-400 076, India

## ABSTRACT

Exergy or the available work energy function is a useful concept for analysing various systems. This paper attempts to present exergy analysis for Stirling cycle cryogenerator. The cycle deviates from the ideal cycle due to (1) finite temperature difference ($\Delta t$) during heat transfer processes and (2) loss of refrigerating effect due to regenerator ineffectiveness. The analysis shows that at lower temperature ratios the influence of $\Delta t$ on exergetic efficiency is marginal. However, the influence of regenerator effectiveness on the exergetic efficiency even at lower temperature ratios is considerable. The combined effect of $\Delta t$ and regenerator effectiveness leads to the practical case and the same also has been analysed.

## INTRODUCTION

The cryogenerators based on Stirling cycle have been successfully utilised for liquefaction of gases like nitrogen for laboratory applications. A gas, such as hydrogen or helium, is used in this type of Stirling cryogenerator. The gas performs a closed cycle, during which it is alternately compressed at ambient temperature in a compression space and expanded at the desired low temperature in the expansion space. An ideal Stirling cycle consisting of two isotherms and two isochores produces refrigeration.

Exergy method of analysis has been reported by Parulekar and Narayankhedkar[1] for air liquefaction processes, Rangrej and Narayankhedkar[2] for low temperature heat pump, Thirumaleshwar[3] for helium cryorefrigerator, Thirumaleshwar and Subramanyam[4] for Gifford-McMahon cycle cryorefrigerator, Banerjee et al.[5] for pressure swing adsorption process for air separation. This paper attempts to present exergy analysis for Stirling cycle cryogenerator. The actual Stirling cycle deviates from the ideal cycle due to (1) finite temperature difference ($\Delta t$) during isothermal heat transfer processes and (2) loss of refrigerating effect due to regenerator ineffectiveness. The combined effect of $\Delta t$ and regenerator ineffectiveness leads to the practical case and the same also has been analysed.

*Advances in Cryogenic Engineering,* Vol. 43
Edited by P. Kittel, Plenum Press, New York, 1998

## EXERGY

'Exergy' simply means - the available work energy function. By definition, the exergy of a system at a given state is the maximum work that would be obtained in taking the system to the 'dead state' (normally, the ambient) for heat engine. For a refrigeration system, exergy is the minimum work required to take the system to a given state from its dead state. The methodology of exergy analysis can be useful for selecting optimal operating parameters and for comparing different process configurations. Exergy balance leads to the concept of exergetic efficiency which takes into consideration the quality of energy.

It is known that for a heat engine, when heat supplied is q, the availability or exergy of this amount of heat varies with temperature at which it is supplied. Thus,

$$e_q = q\,(T - T_o)/T \tag{1}$$

Similarly, for a refrigeration system, the minimum work required (or exergy) to remove a given amount of heat increases as the temperature is decreased.

The exergy for flow process is the availability of work due to the energy content of a flow of mass flux. It can be shown that for a flow process, the maximum available work per unit mass between a state point 1 and the ambient state is given by

$$e_{f1} = (h_1 - h_o) - T_o\,(s_1 - s_o) \tag{2}$$

Thus, it is obvious that exergy depends on the parameters of the state point 1. Equation (2) can be rearranged as

$$e_{f1} = (h_1 - T_o s_1) - (h_o - T_o s_o)$$
$$= h_1 - T_o s_1 + C_1 \tag{3}$$

where $C_1$ is a constant for given ambient conditions.

In fact, one can choose any value for $C_1$ since in most of the cases one is interested only in the difference between the exergy values and not in the absolute value. So, equation for exergy function can be expressed as

$$e = h - T_o s + C_a \tag{4}$$

where $C_a$ is an arbitrary constant. For the convenience the subscript for e is omitted. Thus here after e represent $e_{f1}$.

### Coefficient of Performance (COP) and Exergetic Efficiency ($\eta_e$)

Fig. 1 shows Carnot cycle and Stirling cycle operating with the same temperature limits of $T_o$ and $T_r$. It is seen that for the ideal cycle, COP expression is the same for both the cycles and is given by

$$COP_i = T_r/(T_o - T_r)$$

Although for ideal case the COP of Stirling cycle is the same as that of Carnot cycle, in actual practice Stirling cycle deviates considerably from the ideal one. When the cycle is used for providing refrigeration at constant temperature, 'relative efficiency' is defined as

$$\eta_{re} = COP/COP_i \tag{5}$$

For constant temperature refrigeration applications, the relative efficiency is also termed as figure of merit (FOM) or exergetic efficiency. Exergetic efficiency, $\eta_e$ is the ratio of ideal exergy required to actual exergy supplied for the same cold produced.

$$\eta_e = e_i/e_s = COP/COP_i \tag{6}$$

Fig. 2 shows Carnot cycle and Stirling cycle with the same temperature limits of $T_o$ and $T_r$ on exergy-enthalpy diagram.

**For liquefaction cycle exergetic efficiency, $\eta_e$, is defined as the ratio of minimum exergy as required by ideal liquefaction cycle to actual exergy supplied .**

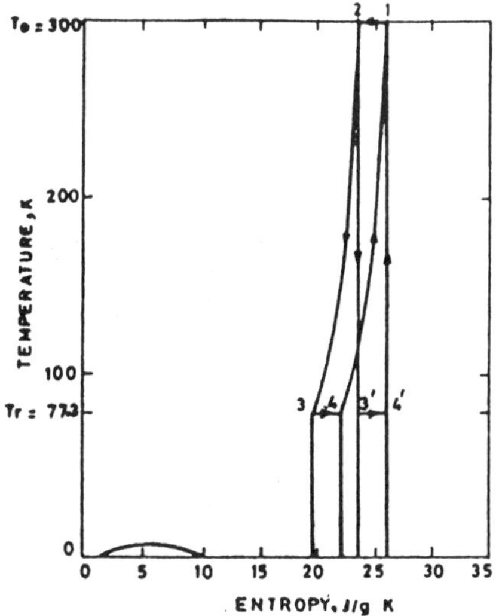


Fig. 1 Ideal Carnot cycle (1-2-3'-4'-1) and ideal Stirling cycle (1-2-3-4-1) on T-s diagram.

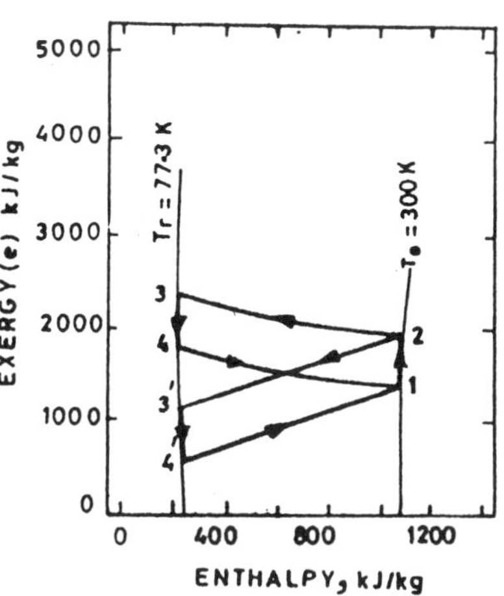


Fig.2 Ideal Carnot cycle (1-2-3'-4'-1) and ideal Stirling Cycle (1-2-3-4-1) on exergy enthalpy diagram.

Fig. 3 illustrates the above definition by comparing ideal liquefaction cycle and ideal Carnot cycle on T-s diagram. Since COP of ideal Stirling cycle is the same as that of ideal Carnot cycle, $W_{Carnot}$ represented on Fig. 3 also represents work required for Stirling cycle. Fig. 4 shows exergy diagram for nitrogen. It is seen that minimum exergy required for ideal liquefaction cycle for changing the state of nitrogen gas from 1 atm, 300 K to 1 atm, 77.3 K in saturated liquid state is 747.9 kJ . The refrigerating effect (RE) to be provided is $(h_1 - h_f)$ and is equal to 433.14 kJ/kg. Thus if ideal Stirling cycle is employed for liquefaction of nitrogen, exergy to be supplied is given by

$$e_{S,i} = RE / COP_i = (h_1 - h_f) / (T_r / (T_o - T_r))$$
$$= 433.14 / (77.3/300-77.3) = 1248.24 \text{ kJ}$$

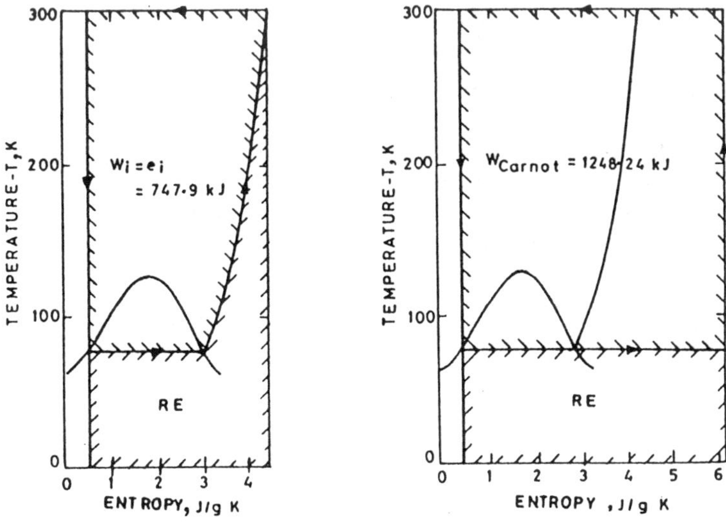


Fig. 3 Liquefaction cycles : a) Ideal cycle and b) Carnot cycle.

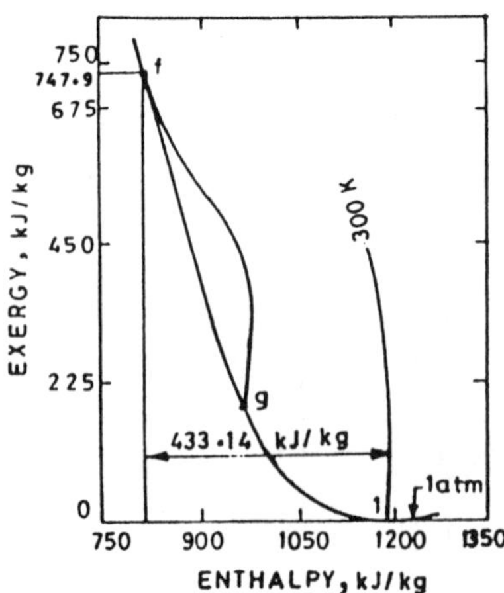


Fig. 4 Exergy diagram for nitrogen .

Thus exergetic efficiency of ideal Stirling cycle for liquefaction of nitrogen gas ($\eta_{e,s}$) is given by

$$\eta_{e,s} = \text{ideal exergy required/ exergy supplied}$$
$$= (747.9 \ / \ 1248.24) = 0.6$$

Thus for ideal Stirling cycle, energy consumption works out as 0.347 kWh per kg of nitrogen liquefied.

With isothermal efficiency of compressor $\eta_{i,c}$ and isothermal efficiency of expander $\eta_{i,e}$, the expression for relative efficiency $\eta_{re}$ for the Stirling cycle is given by

$$\eta_{re} = \eta_{i,c}\eta_{i,e}\,(r_t - 1)/(r_t - \eta_{i,c}\,\eta_{i,e}) \tag{7}$$

For $T_o = 300$ K, $T_r = 77.3$ K, $\eta_{i,c} = 0.6$, $\eta_{i,e} = 0.7$ , relative efficiency works out as 0.349. The exergetic efficiency is given by

$$\eta_e = \eta_{e,s} \times \eta_{re}$$
$$= 0.6 \times 0.349 = 0.209 \tag{8}$$

Thus energy consumption works out as 0.994 kWh per kg of nitrogen liquefied.

**Deviation from Ideal Cycle.** The Stirling cycle used for gas liquefaction deviates from the ideal cycle in the following respects:

1. Finite temperature difference during heat transfer process.
2. Loss of refrigeration due to regenerator ineffectiveness.

**Finite Temperature Difference.** Fig. 5 illustrates Stirling cycle with finite temperature difference during isothermal heat transfer processes. Assuming $(\Delta t)_c = (\Delta t)_r = \Delta t$, COP can be expressed as

$$\text{COP} = \frac{\eta_{i,c}\,\eta_{i,e}}{(T_o + \Delta t)\,/\,(T_o/r_t - \Delta t) - \eta_{i,c}\eta_{i,e}} \tag{9}$$

and relative efficiency, $\eta_{re}$ can be expressed as

$$\eta_{re} = \frac{\eta_{i,c}\,\eta_{i,e}\,(r_t - 1)}{(T_o + \Delta t)/(T_o/r_t - \Delta t) - \eta_{i,c}\,\eta_{i,e}} \tag{10}$$

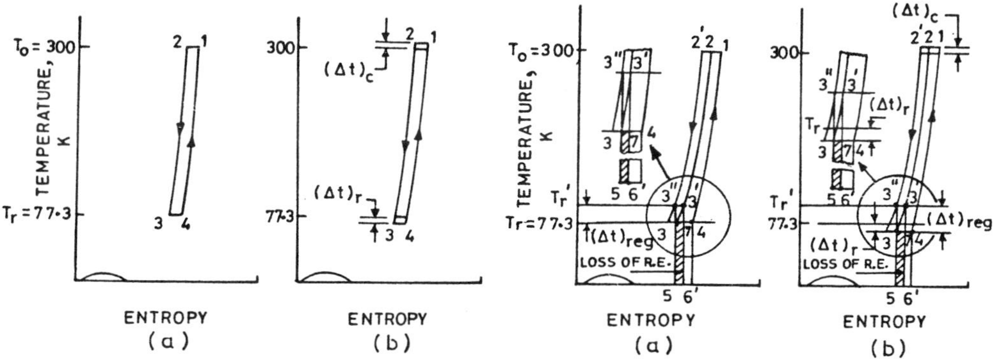


Fig.5 Stirling cycle: a) Ideal
b) Finite temperature difference.

Fig. 6 Effect of regenerator effectiveness
a) $\Delta t = 0$, b) $\Delta t \neq 0$.

Since the exergetic efficiency of Carnot/Stirling cycle with $T_o = 300$ K and $T_r = 77.3$ K is equal to 0.6, exergetic efficiency of the Stirling cycle with finite temperature difference, ($\Delta t = 3^0$C) is given by

$$\eta_e|_{S,\Delta t} = \eta_e|_{S,\Delta t=0} \cdot \eta_{re}$$
$$= 0.6 \times 0.33 = 0.198$$

Thus the energy consumption per kg of nitrogen liquefied is 1.05 kWh.

**Regenerator Ineffectiveness.** The success of the Stirling cycle liquefier depends to a large extent on the effectiveness of the regenerator used in the system. As the heat transferred in the regenerator is 30 to 50 times more than that in the expansion space, a small change in the effectiveness largely affects the performance of the system. It has been shown by a simple analysis[6] that the loss of refrigeration is 21 per cent even for a regenerator having effectiveness of 99 per cent .

**Case 1, $(\Delta t) = 0$.** Since the regenerator effectiveness is not equal to 100 per cent, temperature attained during regeneration is $T_{r'}$ instead of $T_r$ as shown in Fig. 6(a). Thus to attain temperature equal to $T_r$, it is assumed that the gas is isentropically expanded from 3' to 3 followed by isothermal expansion. However, this results in loss of refrigerating effect which is equal to area 3 - 5 - 6' - 7 - 3 as shown in Fig. 6 (a). Thus to obtain the same refrigerating effect, isothermal compression is continued till state 2' so that isentropic expansion after regeneration results in state 3. This increases work input and is given by

$$RT_o \ln (P_{2'}/P_2) = RT_o (\gamma/\gamma\text{-}1) \ln (T_{r'}/T_r) \tag{11}$$

and the ratio of extra work to initial work, $r_w$ is given by

$$r_w = \frac{\text{Extra work}}{\text{Initial work}} = \frac{(\gamma/\gamma\text{-}1) \ln (T_{r'}/T_r)}{\ln (P_2/P_1)} \tag{12}$$

The expression for relative efficiency is given by

$$\eta_{re} = \frac{\eta_{i,c}\,\eta_{i,e}\,(r_t - 1)}{r_t\left[1 + \dfrac{\gamma}{\gamma - 1}\,\dfrac{\ln (T_r'/T_r)}{\ln (r_p)}\right] - \eta_{i,c}\,\eta_{i,e}} \tag{13}$$

**Case 2, $(\Delta t) \neq 0$.** Fig. 6 (b) illustrates the cycle with finite temperature difference $(\Delta t)$ during isothermal processes and also considering regenerator ineffectiveness by defining $(\Delta t)_{reg}$ as $T_{r'} - (T_r - \Delta t)$.

The expression for relative efficiency is given by

$$\eta_{re} = \frac{\eta_{i,c}\,\eta_{i,e}\,(r_t - 1)}{\dfrac{T_o + \Delta t}{T_o/r_t - \Delta t}\left[1 + \dfrac{\gamma}{\gamma - 1}\,\dfrac{\ln\left(T_{r'} / (T_r - \Delta t)\right)}{\ln(r_p)}\right] - \eta_{i,c}\,\eta_{i,e}} \tag{14}$$

For $(\Delta t)_{reg} = 3^0$ C and $(\Delta t) = 3^0$ C, $\eta_{i,c} = 0.6$ and $\eta_{i,e} = 0.7$, $\eta_{re}$ works out as 0.285. Thus the exergetic efficiency works out as 0.171. The energy consumption per kg of nitrogen liquefied works out 1.22 kWh. The corresponding values of $\eta_e$ and energy consumption for $(\Delta t)_{reg} = 5^0$C and $(\Delta t) = 5^0$C are 0.151 and 1.382 kWh respectively.

## DISCUSSION

Fig. 7 illustrates four cases. It is seen that at lower temperature ratios, the influence on $\eta_e$ is marginal. However, at higher temperature ratios, the effect is considerable. For the cycle considering both $(\Delta t)_{reg} = 5^0$C as well as $\Delta t = 5^0$C, Case d, $\eta_{re}$ is the lowest. The cyclic analysis reported by Atrey et al.[7] shows that the Stirling cycle based liquid nitrogen plant (PLN 106) provides refrigerating effect of 787.87 W and energy requirement is 8986.34 W. Thus COP is 0.0876. By using concepts of relative efficiency and exergetic efficiency as discussed in this paper, $\eta_{re}$ and $\eta_e$ work out as 0.252 and 0.151 respectively.

Table 1 summarises the results in terms of exergetic efficiency.

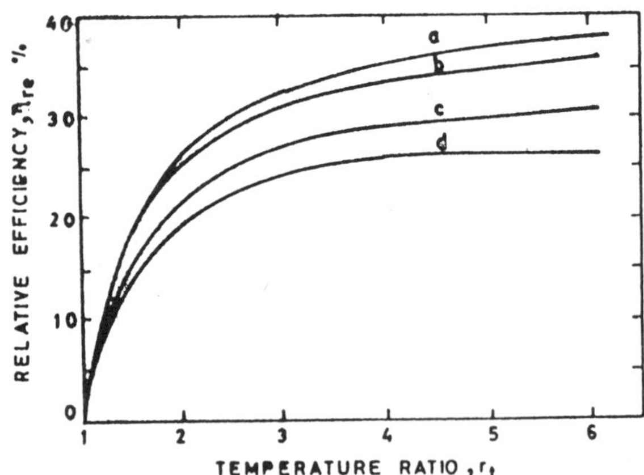


Fig.7 Effect of $\Delta t$ and regenerator ineffectiveness on relative efficiency. $\eta_{i,c} = 0.6$, $\eta_{i,e} = 0.7$, $(\Delta t)_o = (\Delta t)_r = \Delta t$ ; Case a- $\Delta t = 0, \eta_{reg} = 1$ ; Case b- $\Delta t = 3^0$C, $\eta_{reg} = 1$; Case c- $\Delta t = 3^0$C, $r_{t'} = 77.3/74.3$; Case d- $\Delta t = 5^0$ C, $r_{t'} = 77.3/72.3$.

**Table 1.** Exergetic efficiency for Stirling Cycle liquefaction system. ($T_o$ = 300 K, $T_r$ =77.3 K, $r_p$ = 2, $\gamma$ = 1.67)

| Cycle | Ideal Cycle | $\eta_{i,c} = 0.6$, $\eta_{i,e} = 0.7$ | | | | PLN 106 Cyclic Analysis[7] |
|---|---|---|---|---|---|---|
| | | $\Delta t = 0$, $(\Delta t)_{reg} = 0$ | $\Delta t = 3°C$ $(\Delta t)_{reg} = 0$ | $\Delta t = 3°C$ $(\Delta t)_{reg} = 3°C$ | $\Delta t = 5°C$ $(\Delta t)_{reg} = 5°C$ | |
| COP | 0.347 | 0.121 | 0.115 | 0.098 | 0.087 | 0.0876 |
| $\eta_{re}$ | 1.000 | 0.349 | 0.330 | 0.285 | 0.251 | 0.252 |
| $\eta_e$ | 0.600 | 0.209 | 0.198 | 0.171 | 0151 | 0.151 |
| kWh/kgl * | 0.347 | 0.994 | 1.05 | 1.22 | 1.382 | 1.377 |

* Energy consumption per kg of nitrogen liquefied.

## CONCLUSIONS

The approach of exergy analysis is illustrated for Stirling cycle cryogenerator used for nitrogen liquefaction. The effect of finite temperature difference ($\Delta t$) and regenerator ineffectiveness represented by $(\Delta t)_{reg}$ are also considered. The exergetic efficiency for $\Delta t$ = $5^0$ C and $(\Delta t)_{reg}$ = $5^0$ C is 0.151 which is same as the exergetic efficiency of PLN 106 as per the cyclic analysis[7]. Thus exergy analysis is a useful tool for predicting the performance of the systems, the influence of various parameters and the losses.

## NOMENCLATURE

- C - Constant
- $e_{fl}$ - Exergy of flow process
- $e_q$ - Exergy of heat energy
- $es_{,i}$ - Exergy supplied for ideal Stirling cycle
- P - Pressure
- R - Gas constant
- $r_p$ - Pressure ratio
- $r_t$ - Temperature ratio ($T_o/T_r$)
- $r_{t'}$ - Temperature ratio ($T_{r'}/T_o$)
- $r_w$ - Ratio of extra work to initial work
- T - Temperature, Source temperature
- $T_o$ - Ambient temperature
- Tr - Refrigeration temperature
- $T_{r'}$ - Temperature of gas after regeneration process

Greek Symbols

- $\Delta t$ - Finite temperature difference during isothermal heat transfer process
- $(\Delta t)_c$ - Finite temperature difference during isothermal heat rejection process
- $(\Delta t)_r$ - Finite temperature difference during isothermal heat reception (refrigeration) process
- $\eta$ - Efficiency

$\eta_e$ - Exergetic efficiency
$\eta_{re}$ - Relative efficiency
$\gamma$ - Specific heat ratio

Subscripts

a - Arbitrary
c - Compressor
e - Expander
fl - Flow process
i - Ideal/Isothermal
o - Ambient condition
reg - Regenerator
s - Supplied

## REFERENCES


1. B.B. Parulekar and K.G. Narayankhedkar, Use of Exergy Chart for the Study of Air Liquefaction Processes, Indian Journal of Cryogenics, 4:78 (1979).
2. H.A. Rangrej and K. G. Narayankhedkar, Exergy Analysis of a Low Temperature Heat Pump, Cryogenics, 23:148 (1983).
3. M. Thirumaleshwar, Exergy Method of Analysis and its Application to a Helium Cryorefrigerator, Cryogenics, 19:355 (1979).
4. M. Thirumaleshwar and S.V. Subramanyam, Exergy Analysis of a Gifford-McMahon Cycle Cryorefrigerator, Cryogenics, 19:355(1979).
5. R. Banerjee, K.G. Narayankhedkar and S.P. Sukhatme, Exergy Analysis of Pressure Swing Adsorption Processes for Air Separation, Chemical Engg. Science, 45: 467 (1990).
6. R.Barron , Cryogenic Systems, McGraw Hill, New York (1966).
7. M.D. Atrey , S.L.Bapat and K.G. Narayankhedkar, Cyclic Simulation of Stirling Cryocoolers, Cryogneics 30:341 (1990).

# THEORETICAL PERFORMANCE OF SINGLE STAGE AND TWO STAGE SUPERFLUID STIRLING REFRIGERATORS USING KAPTON RECUPERATORS

Ashok B. Patel and J. G. Brisson

Massachusetts Institute of Technology
Cambridge, MA 02139

## ABSTRACT

Previous superfluid Stirling refrigerators (SSR) using CuNi recuperators have experimentally achieved temperatures of 0.168 K and 0.296 K operating from high temperatures of 0.383 K and 1.05 K respectively. This article describes the theoretical performance of a single stage SSR and a two stage SSR which use Kapton recuperators. Specifically, a single stage SSR operating from a high temperature of 1.0 K should achieve a low temperature of 0.22 K and deliver 420 μW of cooling power at 0.3 K. A two stage SSR operating from a high temperature of 1.0 K should reach a low temperature of 0.12 K and deliver 750 μW of cooling power at 0.30 K.

## INTRODUCTION

The superfluid Stirling refrigerator (SSR) is a Stirling cycle refrigerator which uses a $^3He$-$^4He$ mixture as the working fluid. The basic components of a SSR are hot and cold pistons connected by a regenerator. The refrigerator operates below 2 K where the $^3He$-$^4He$ mixture behaves to first approximation as an ideal gas of $^3He$ in an inert background of superfluid $^4He$. Each piston is bypassed with a superleak which allows the superfluid $^4He$ to flow freely through the pistons. These superleaks enable the thermodynamically active $^3He$ to be expanded and compressed through the Stirling cycle which results in cooling. The SSR was first demonstrated by Kotsubo and Swift[1] in 1990. In 1992, Brisson and Swift[2-5] improved the SSR design by operating two refrigerators 180 degrees out of phase and using a counterflow recuperator as the regenerator. They achieved a low temperature of 0.296 K operating from a high temperature of 1.05 K. Using the same machine, Watanabe, Swift, and Brisson[6] later reached 0.168 K operating from a high temperature of 0.383 K.

This article estimates the theoretical performance of two SSR's of the Brisson and Swift type that are currently being built at the Massachusetts Institute of Technology. These SSR's differ from the Brisson and Swift SSR in that they are significantly larger

(91.6 $cm^3$ and 140 $cm^3$ versus 14.3 $cm^3$) and use recuperators made of Kapton instead of CuNi. The overall performance of Kapton heat exchangers is expected to be better than that of metal heat exchangers at low temperature due to the high Kapitza conductance of Kapton-helium interfaces relative to that of metal-helium interfaces. Using a simple model, an estimate of the cooling power of these SSR's for a range of cold piston temperatures and $^3$He-$^4$He mixture concentrations is calculated. A comparison between the estimated performance for SSR's using Kapton and CuNi recuperators of the same size is also provided.

The first SSR, called the single stage SSR, is shown in Figure 1. This SSR is basically a Brisson and Swift type SSR using a Kapton recuperator. Both the cold and hot platforms are made from solid blocks of copper and are connected by the recuperator. The pistons are made from stainless steel bellows which have nested convolutions to minimize void volume. The hot platform pistons are rigidly connected together and driven by a rigid push rod from a room temperature drive. The cold piston platforms are connected and driven in a similar manner. The superleaks between the pistons are made from porous Vycor glass.

The two stage SSR, shown in Figure 2, is basically the single stage SSR with an additional expansion platform and another Kapton recuperator. The cold platform of the single stage SSR now serves as the intermediate platform of the two stage SSR. The intermediate platform pistons and the cold platform pistons are rigidly connected together and are driven by the same drive rod. As the analysis will show, the second expansion stage should enable the SSR to reach lower ultimate temperatures and deliver higher cooling powers at these temperatures than a single stage SSR.

Specifications for both the single stage SSR and two stage SSR are provided in Table 1 in the results section. The two Kapton recuperators, described in Table 1, respectively contain 170 and 190 counterflow heat exchanger passages (0.238 cm x 0.0127 cm x 20 cm) separated by 12.7 μm thick Kapton film. The hypothetical CuNi recuperators, described in Table 1, consist of 607 and 668 CuNi tubes ( 0.308 mm O.D. x 0.038 mm wall x 20 cm) arranged in the same counterflow heater exchanger geometry as the Brisson and Swift recuperator.[4]

## ANALYSIS

In this section, we show how the cooling power of a two stage SSR is calculated. The cooling power of the single stage machine follows directly from this analysis.

The cooling power of a SSR can be predicted by using the Schmidt model[7] of the Stirling cycle coupled with a model which estimates the heating of the cold platform of the SSR due to imperfect regeneration. Fermi gas effects and fountain pressure effects are neglected. The Schmidt model assumes that the working fluid is an ideal gas, the regenerator is perfect, the temperature distribution along the regenerator is linear, the compression and expansion of the pistons are isothermal, and there are no other losses. For one side of a two stage SSR, the volumes of the pistons are assumed to vary sinusoidally as

$$V_c = V_{clc} + \frac{V_{swc}}{2}\left[1+\cos\left(\theta-\frac{\pi}{2}\right)\right],\quad V_i = V_{cli} + \frac{V_{swi}}{2}\left[1+\cos\left(\theta-\frac{\pi}{2}\right)\right],\ \text{and}\ V_h = V_{clh} + \frac{V_{swh}}{2}\left[1+\cos\theta\right] \qquad (1)$$

where $V_c$, $V_i$, and $V_h$ are the cold, intermediate, and hot platform volumes, $V_{clc}$, $V_{cli}$, and $V_{clh}$ are the cold, intermediate, and hot platform clearance volumes, and $V_{swc}$, $V_{swi}$, and $V_{swh}$ are the cold, intermediate, and hot platform swept volumes. Using the ideal gas law, the pressure in the SSR is given by

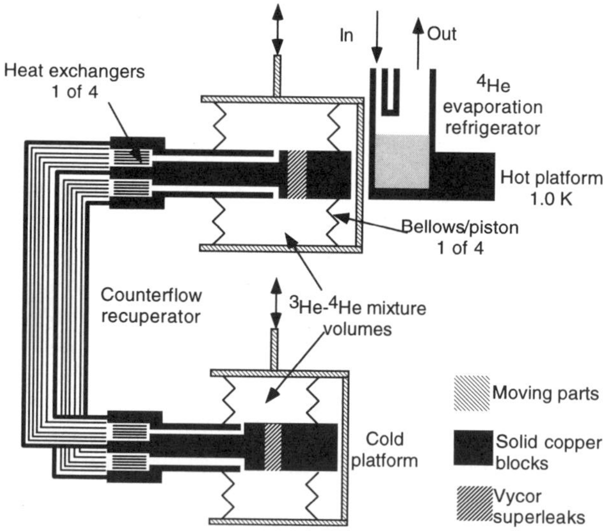


**Figure 1.** Single Stage SSR.

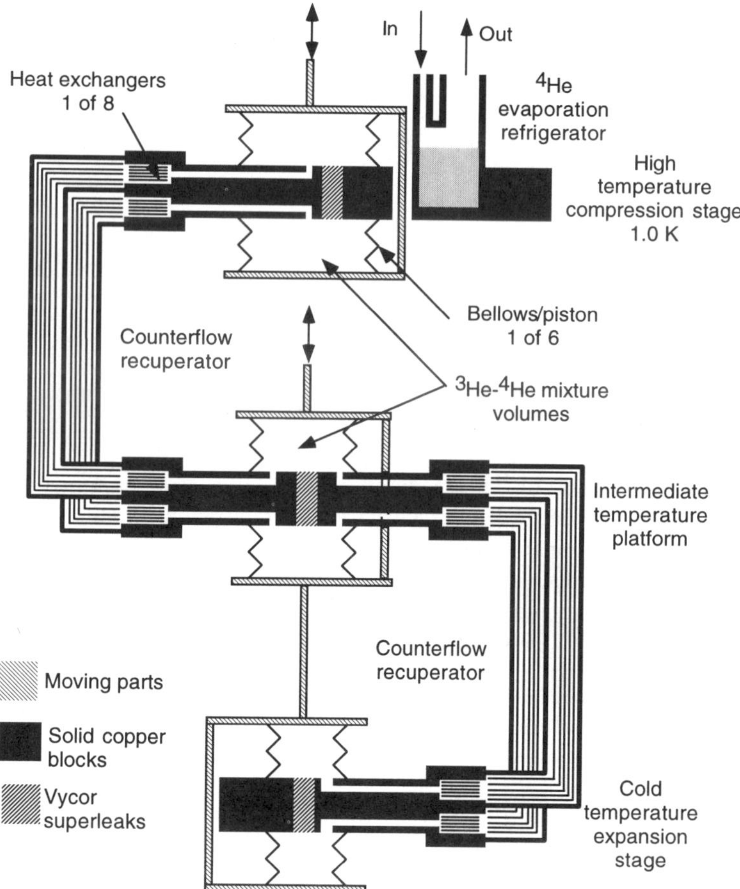


**Figure 2.** Two Stage SSR.

$$P = nR\left[\frac{V_c}{T_c} + \frac{V_i}{T_i} + \frac{V_h}{T_h} + \ln\left(\frac{T_h}{T_i}\right)\frac{V_{top}}{T_h - T_i} + \ln\left(\frac{T_i}{T_c}\right)\frac{V_{bottom}}{T_i - T_c}\right]^{-1} \tag{2}$$

where n is the number of moles of $^3He$ on one side of the SSR, R is the universal gas constant, $T_c$, $T_i$, and $T_h$ are the cold, intermediate, and hot platform temperatures, and $V_{top}$ and $V_{bottom}$ are the volumes of regenerators between the hot and intermediate platforms and the intermediate and cold platforms. The Schmidt cooling powers at the cold platform, $Q_{sc}$, and intermediate platform, $Q_{si}$, are then given by

$$Q_{sc} = 2f \oint_{cycle} p\frac{dV_c}{d\theta}d\theta, \quad Q_{si} = 2f \oint_{cycle} p\frac{dV_i}{d\theta}d\theta \tag{3}$$

where f is the frequency of operation and θ is the crank angle. The factor of 2 is present because the SSR contains two refrigerators.

An estimate of the heating of the cold and intermediate platforms due to imperfect regeneration in a two stage SSR can be made using several simplifying assumptions. First, we assume that the $^3He$ flow on both sides of the SSR is symmetric. Second, we assume that a worst case estimate of the heating can be made by assuming a steady $^3He$ flow at the peak flow rate through the recuperator. Third, we neglect effects such as viscous heating and axial conduction of the fluid which are small compared to the heat exchanged.

The geometry of the flow passages of the Kapton recuperator used in the SSR is shown in Figure 3. An energy balance of the elemental control volume results in the following set of equations:

hot stream
$$\dot{n}c_p\frac{dT_{hot}}{dx} = -h_h P\left(T_{hot} - T_{wall}\right) \tag{4}$$

wall
$$k_w A_w \frac{d^2T_{wall}}{dx^2} = -h_h P\left(T_{hot} - T_{wall}\right) + h_c P\left(T_{wall} - T_{cold}\right) \tag{5}$$

cold stream
$$\dot{n}c_p\frac{dT_{cold}}{dx} = -h_c P\left(T_{wall} - T_{cold}\right) \tag{6}$$

b.c.
$$T_{hot}\Big|_{x=0} = T_{high},\ T_{cold}\Big|_{x=20cm} = T_{low},\ \frac{dT_{wall}}{dx}\Big|_{x=0} = 0,\ \frac{dT_{wall}}{dx}\Big|_{x=20cm} = 0 \tag{7}$$

where $\dot{n}$ is the molar flow rate of $^3He$ through the recuperator, $c_p$ is the molar heat capacity of $^3He$ (assumed to be 3R/2), $h_h$ is the heat transfer coefficient on the hot side of the recuperator, $h_c$ is the heat transfer coefficient on the cold side of the recuperator, P is the perimeter of the heat exchange surface, and $A_w$ is the transverse cross sectional area of the wall. $T_{high}$ and $T_{low}$ refer to platform temperatures at the hot and cold ends of the heat exchanger. The heat transfer coefficients, $h_h$ and $h_c$, are given by

$$h_h^{-1} = 2.5U_{Kap}^{-1}(T_h) + U_{wall}^{-1}(T_h) + U_{fluid}^{-1} \tag{8}$$

$$h_c^{-1} = 2.5U_{Kap}^{-1}(T_c) + U_{wall}^{-1}(T_c) + U_{fluid}^{-1} \tag{9}$$

where $U_{Kap}(T)$ is the Kapitza conductance of the wall material, $U_{wall}(T) = 2k_{material}$/thickness $W/(K\ m^2)$ is the heat transfer coefficient through half the wall, and $U_{fluid}$ is the convective heat transfer coefficient for the $^3He$-$^4He$ mixture assuming fully developed flow. The factor of 2.5 is an empirical scaling factor whose use results in good agreement over a wide range of operating conditions between predictions made using this model and the experimental data of Brisson and Swift[2-5] and Watanabe, Swift, and Brisson.[6] The Kapitza conductance of Kapton[8] is given by $U_{Kapton}(T) = 1428.6\ T^3$ W/(K m$^2$). Although this expression is valid for temperatures ranging from 0.030 K to 0.150 K, we use this expression for temperatures up to 1.0 K because we are interested in a worst case estimate of heating and because we have not been able to find Kapitza conductances for any plastics above 0.2 K. The Kapitza conductance of CuNi is given by $U_{CuNi}(T) = 1000\ T^{4.431}$ W/(K m$^2$) and is valid for temperatures ranging from 0.15 K to 1.0 K.[5] The thermal conductivity of Kapton, $k_{Kapton}$, is assumed to obey the relation $k_{Kapton}(T) = 0.002\ T^{1.431}$ W/(K m) which is obtained by using the value of $k_{Kapton}$ at 5 K and assuming that Kapton's thermal conductivity behaves similarly to that of nylon.[9] The thermal conductivity of CuNi obeys the relation $k_{CuNi}(T) = 0.093\ T^{1.23}$ W/(K m).[5] $U_{fluid}$ obeys the relation $U_{fluid} = 5.385 k_{fluid}/(gap/2)$ W/(K m$^2$) for the Kapton recuperator geometry and $U_{fluid} = 4.364\ k_{fluid}/(D/2)$ W/(K m$^2$) for CuNi recuperator geometry where D is the inner diameter of the CuNi tubes. The thermal conductivity of $^3$He-$^4$He mixtures, $k_{fluid}$, is assumed constant at 0.04 W/(K m).[5] Eq.'s (4-9) are solved using a finite element scheme for the temperature defect, $\theta = T_{hot}|_{x=20cm} - T_{low} = T_{high} - T_{cold}|_{x=0cm}$, which is used to calculate the heating due to imperfect regeneration, $Q_{loss} = \dot{n}\, c_p \theta$.

Assuming $Q_{loss\ top}$ and $Q_{loss\ bottom}$ are the heating losses of the top and bottom recuperators, the cooling power at the cold platform of a two stage SSR is determined by iterating over different values of $T_i$ until the energy balance at the intermediate platform, $Q_{si} - Q_{loss\ top} + Q_{loss\ bottom} = 0$, is satisfied. Then, the cooling power at the cold platform, $Q_{cold}$, is given by $Q_{cold} = Q_{sc} - Q_{loss\ bottom}$.

## RESULTS AND DISCUSSION

The cold platform cooling powers for both the single stage SSR and two stage SSR are calculated using the SSR specifications in Table 1. The calculations assume that both types of SSR's are operated from a high temperature of 1.0 K, have a mean operating pressure of 1 atmosphere, and have an operating frequency of 0.1 Hz.

Figures 4a and 4b show the cold platform cooling power of a single stage SSR for the specifications given in Table 1. The plots show that operating from a high temperature of 1.0 K, a single stage SSR should be able to deliver significant cooling power (4-8 mW) between 0.5-0.8 K. At lower temperatures, a single stage SSR becomes less effective mainly due to regenerator losses. A single stage SSR should achieve 0.22 K using a 1.5 % $^3$He-$^4$He mixture as the working fluid. Using a 6% mixture, it should deliver 420 μW of cooling power at 0.3 K.

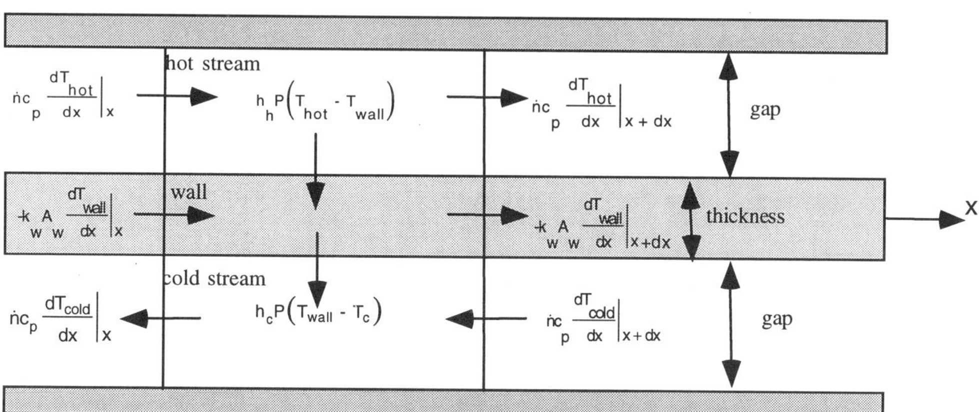


**Figure 3.** Elemental control volume and geometry of Kapton counterflow heat exchanger; gap = 0.127 mm and thickness = 12.7 μm; length of heat exchanger passage = 20 cm.

Figures 4c and 4d provide the cold platform cooling power of a two stage SSR. For cold platform temperatures below 0.3 K, the two stage SSR should produce almost double the cooling power of a single stage SSR. Our two stage SSR should achieve an ultimate low temperature of 0.12 K using a 1.5% $^3$He-$^4$He mixture as working fluid and should deliver 750 μW of cooling power at 0.3 K using a 6% mixture. At higher temperatures, the two stage SSR does not deliver as much cold platform cooling power as a single stage SSR for two reasons: 1) as shown in Figure 4e, a significant portion of the intermediate platform cooling power is used to cool the intermediate platform below the cold platform temperature; and 2) the expansion and compression ratios for our two stage SSR are lower than that of the single stage SSR. Figure 4f shows that the low temperature performance of the two stage SSR using $^3$He-$^4$He mixture concentrations above 3% is not limited by regenerator losses but by the phase separation curve. However, because the solubility of $^3$He in $^4$He can be increased significantly by increasing the pressure,[10] the low temperature performance of a two stage SSR will be improved by operating the SSR at higher mean pressures.

Figure 5a compares the estimated cooling power of a single stage SSR using a 5 cm$^3$ Kapton recuperator with one using a 5 cm$^3$ CuNi recuperator. The plots show that the performance of these recuperators is almost identical. However, we expect that the actual performance of a Kapton recuperator will be significantly better than a CuNi recuperator due to the very conservative expression we used for the Kapitza conductance of Kapton. Figure 5b compares the performance of a two stage SSR using Kapton recuperators with one using CuNi recuperators. At high cold platform temperatures, the CuNi recuperators outperform the Kapton recuperators. However, at temperatures below 0.4 K, the two stage SSR using Kapton heat exchangers should deliver significantly more cooling power and achieve lower ultimate temperatures.

It should be noted that these calculations assume that the $^3$He component of the mixture behaves as a Boltzmann gas; however, this component will begin to behave as a Fermi gas for temperatures well below 0.3 K. The calculated cooling powers of this model are optimistic for temperatures below 0.2 K. Furthermore, this analysis does not include mechanical heating (due to bellows flexure), conduction heating (through structural

**Table 1.** Single stage SSR and two stage SSR specification for one side of SSR

| Single Stage SSR | | Two Stage SSR | |
|---|---|---|---|
| Total Volume (cm$^3$) | 45.8 | Total Volume (cm$^3$) | 70.0 |
| Hot Platform | | Hot Platform | |
| Swept Volume (cm$^3$) | 17.74 | Swept Volume (cm$^3$) | 17.74 |
| Clearance Volume (cm$^3$) | 12.50 | Clearance Volume (cm$^3$) | 12.50 |
| Heat Exchanger Area (cm$^2$) | 61 | Heat Exchanger Area (cm$^2$) | 61 |
| Cold Platform | | Intermediate Platform | |
| Swept Volume (cm$^3$) | 13.16 | Swept Volume (cm$^3$) | 13.16 |
| Clearance Volume (cm$^3$) | 11.91 | Clearance Volume (cm$^3$) | 12.81 |
| Heat Exchanger Area (cm$^2$) | 114 | Heat Exchanger Area (cm$^2$) | 228 |
| Kapton Recuperator | | Cold Platform | |
| Total Volume (cm$^3$) | 5 | Swept Volume (cm$^3$) | 7.68 |
| Perimeter (cm) | 32.1 | Clearance Volume (cm$^3$) | 14.67 |
| Wall Cross Sectional Area (cm$^2$) | 0.025 | Heat Exchanger Area (cm$^2$) | 600 |
| Top CuNi Recuperator | | Top Kapton Recuperator | |
| Total Volume (cm$^3$) | 5 | Total Volume (cm$^3$) | 5 |
| Perimeter (cm$^3$) | 43.7 | Perimeter (cm) | 32.1 |
| Wall Cross Sectional Area (cm$^2$) | 0.655 | Wall Cross Sectional Area (cm$^2$) | 0.025 |
| Bottom CuNi Recuperator | | Bottom Kapton Recuperator | |
| Total Volume (cm$^3$) | 5.5 | Total Volume (cm$^3$) | 5.5 |
| Perimeter (cm) | 48.0 | Perimeter (cm) | 36.1 |
| Wall Cross Sectional Area (cm$^2$) | 0.773 | Wall Cross Sectional Area (cm$^2$) | 0.0275 |

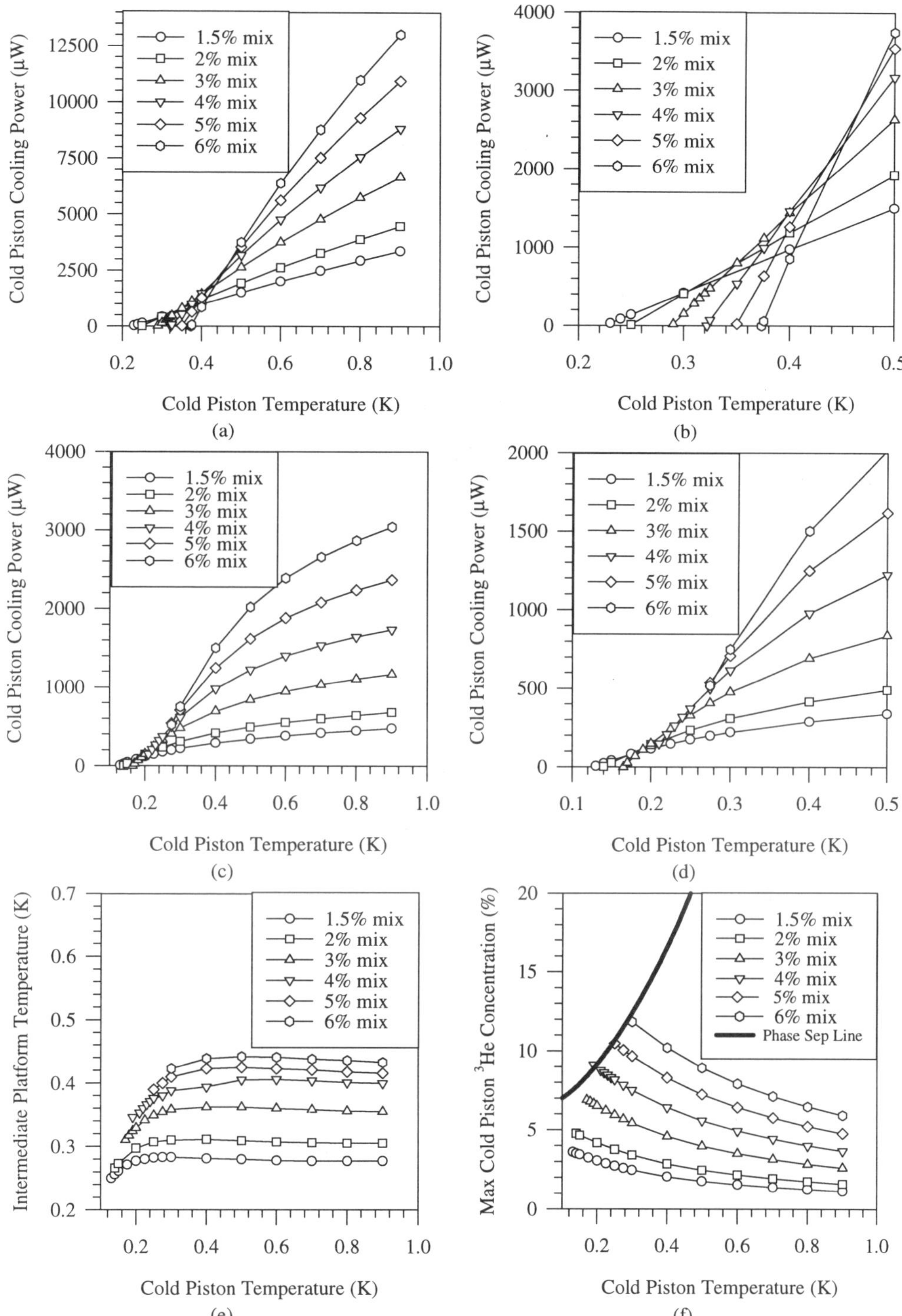


**Figure 4**. (a) and (b) Single stage SSR cold platform cooling powers; (c) and (d) two stage SSR cold platform cooling powers; (e) two stage SSR intermediate platform temperatures and (f) two stage SSR maximum cold platform $^3$He concentration profiles for various $^3$He-$^4$He mixtures.

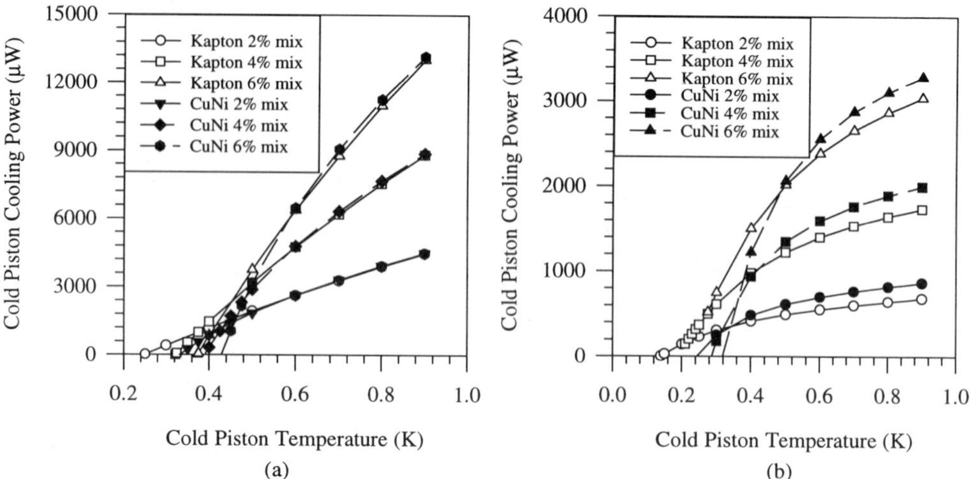


**Figure 5**. Comparison of SSR performance using Kapton and CuNi recuperators; (a) single stage SSR and (b) two stage SSR.

elements) and radiation heating to the cold platform. We estimate these losses to be of order 5 μW, 40 μW and 1μW, respectively.

## CONCLUSIONS

The SSR's currently being constructed at the Massachusetts Institute of Technology should deliver significant cooling power between 0.3 K and 1.0 K. At high cold platform temperatures, a single stage SSR should outperform a two stage SSR. However, at cold platform temperatures below 0.4 K, a two stage SSR should provide more cooling power and lower ultimate temperatures. The results also show that Kapton recuperators should perform on par with similarly sized CuNi recuperators at high temperatures and outperform them at lower temperatures.

## ACKNOWLEDGEMENTS

This work has been supported by National Science Foundation grant CTS-9416689.

## REFERENCES

1. V. Kotsubo and G. W. Swift. Superfluid Stirling-cycle refrigeration below 1 Kelvin, *J. Low Temp. Phys.* 83:217 (1991).
2. J. G. Brisson and G. W. Swift. Superfluid Stirling refrigerator with a counterflow regenerator, in: "Proceeding of the Seventh International Cryocoolers Conference," Santa Fe, NM (1992).
3. J.G. Brisson and G. W. Swift. A recuperative superfluid Stirling refrigerator, in: "Advances in Cryogenic Engineering," 39B (1994) p.1393.
4. J. G. Brisson and G. W. Swift. High Temperature Cooling Power of the Superfluid Stirling Refrigerator, *J. Low Temp. Phys.* 98:141 (1995).
5. J. G. Brisson and G. W. Swift. Measurements and modeling of a recuperator for the superfluid Stirling refrigerator, *Cryogenics* 34:971 (1994).
6. A. Watanabe, G. W. Swift, and J. G. Brisson. Measurements with a recuperative superfluid Stirling refrigerator, in: "Advances in Cryogenic Engineering," 41 (1996) p. 1527.
7. I. Urieli and D. M. Berchowitz. "Stirling Engine Analysis," Adam Hilger Ltd., Bristol (1984).
8. P. J. Nacher, P. Schleger, I. Shinkoda, and W. N. Hardy. Heat exchange in liquid helium through plastic foils, *Cryogenics* 32:353 (1992).
9. H. P. Gush. A compact low thermal conductivity support for cryogenic use, *Rev. Sci. Instrum.* 62:1106 (1991).
10. F. Pobell. "Matter and Methods at Low Temperatures," Springer-Verlag, Berlin (1996).

# DEVELOPMENT OF A LINEAR DRIVE CRYOCOOLER SYSTEM INCORPORATING BOTH STIRLING AND ORIFICE PULSE TUBE REFRIGERATION CYCLES

L. B. Penswick, D. C. Lewis, and R. W. Olan

Stirling Technology Company
Kennewick WA 99336-2626 USA

## ABSTRACT

This paper describes the development, testing, and field trial results of a free-piston, linear drive cryocooler system designed to meet various commercial market requirements. The cryocooler units are used to cool high performance computer system microprocessors, high temperature superconductor (HTS) devices, and macromolecular X-ray crystallography equipment. Three distinct cold head configurations are utilized: 1) a Stirling cycle unit for HTS applications in the 60 to 65 K temperature range with a capacity up to 10 watts, 2) a Stirling cycle computer cooling unit providing approximately 25 watts capacity at 90 K, and 3) an orifice Pulse Tube refrigeration cycle operating over a temperature range of 75 to 90 K with a nominal capacity of 7.5 watts. All of these coolers incorporate a common free-piston, linear drive gas compressor utilizing flexural bearings to eliminate potential degradation due to contamination or seal wear and insure a long, maintenance-free operating life. A description of the cooler operating requirements along with mechanical details of the gas compressor, cold heads, heat rejection system, and supporting power electronics is provided. Issues concerning the use of a common gas compressor for both Stirling and Pulse Tube coolers are discussed. An emphasis is placed on cooler test results gathered under the unique operating conditions experienced by the coolers when integrated into their respective system configurations. A detailed comparison between actual cooler operating characteristics and predicted values based on refrigeration cycle simulation models developed at Stirling Technology Company (STC) is presented.

## INTRODUCTION

In recent years there has been a significant increase in potential applications of cryogenic temperatures in: 1) low temperature cooling of electronics—for example, high performance computers and LNA's (Low Noise Amplifiers), 2) biomedical laboratory test equipment, and 3) HTS (High Temperature Superconducting) materials for a wide range of applications such as passive and active communication system components, high performance sensors, and various types of power conditioning equipment. To satisfy these requirements, the mechanical cryogenic refrigeration systems must be reliable, have long, maintenance-free operating lives, and most importantly be practical from the cost point of view.

This paper describes the technical aspects of the development of a free-piston Stirling cycle cryocooler to meet many of the above end user requirements. Specific issues addressed include the use of a modular design concept which reduces cooler system costs, the application of flexural bearings in a cost effective manner to insure high operating reliability, and use of a number of innovative concepts to provide good overall performance at acceptable costs.

Performance of the cooler and a comparison with various simulation model results are presented. The impact of using a Pulse Tube to replace the Stirling cycle cold head is also discussed.

## GENERAL DESIGN PHILOSOPHY

The current family of STC coolers evolved based on the concept that key cooler system components should be as modular as possible since a number of potential markets were in their early stages and in some cases had not developed firm technical specifications for their final product. This approach would allow various relatively high cost components such as the linear drive gas compressor to be used in multiple applications, thus increasing the production volumes and driving down costs.

The specific cooler addressed in this paper began with the hardware developed in a NASA Goddard Phase II SBIR project. This hardware, shown in Figure 1, incorporated a free-piston Stirling cycle cryocooler with a two-stage expander (60 K/20 K) which was used to cool the working fluid of a JT expander. The JT compressors were attached directly to the linear drive compressor and are contained within a common pressure vessel [1]. The potential for a modular system is evident in the areas of the linear drive gas compressors and their associated linear motors, single- or dual-opposed piston configurations, and a straightforward approach to incorporate various cold head configurations. To provide the necessary long operating life, STC incorporated its flexural bearing technology into all of the system's moving components to insure that no contact occurred, essentially eliminating wear as a failure mode.

By removing the JT gas compressors, using a single drive compressor, and switching to a single stage cold head, the cooler system shown in Figure 2 was developed. Nominal capacity was in the range of 15 watts at 80 K with an electrical input power to the drive motor on the order of 400 watts. The ability to incorporate dual drives and a wide range of cold head configurations (Stirling and Pulse Tube) allows the basic hardware components to cover the temperature range of 20 K to 200 K with cooling capacities from 1 watt to over 100 watts. The prototype unit shown in Figure 2 has been further refined for manufacturing and is currently in low volume production. The following sections provide further technical details concerning the various key components making up the cooler system depicted in Figure 2.

### Flexural Bearings

The use of flexural bearings in free-piston Stirling and Pulse Tube hardware represents a key factor in the improved operating life and reliability of newer cooler systems. As indicated in Figure 3, the flexural bearing allows the moving components to be accurately aligned within their respective cylinders in a manner which insures that the moving components never contact

**Figure 1.** Integrated Stirling Cooler/JT System.

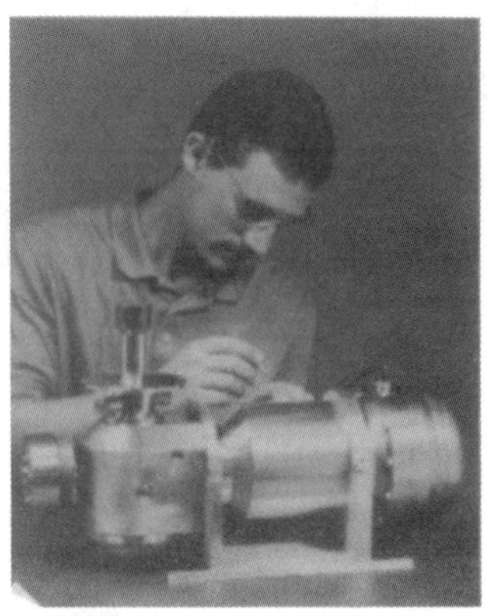

**Figure 2.** Prototype Cooler System.

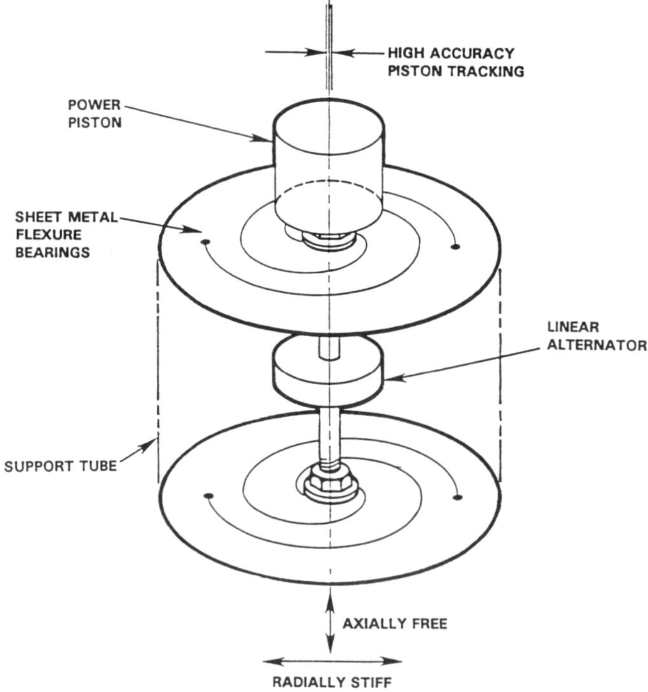


**Figure 3.** Flexural Bearing Concept.

the cylinders even when small radial gas clearance seal gaps are utilized (<25 microns). The flexure itself has a high radial stiffness to provide the necessary radial alignment while at the same time providing a significant portion of the axial spring effect necessary to operate the moving components in a near resonant condition. Flexures developed and used by STC range in diameter from greater than 150 mm to under 20 mm and are shown in Figure 4. STC has carried out extensive flexure life testing in both special test rigs and actual hardware. Testing has included operation in highly over stressed conditions to define the actual flexures' ultimate operating limits and accuracy of modeling employed in the design of the flexures. Many hundreds of flexures have been on continuous life test for over 4 years without any failures.

**Figure 4.** Flexure Hardware.

The flexure design utilized for the displacer is primarily governed by the spring effect required to resonate the displacer since there is little gas spring effect provided by the cycle gas forces. The number of flexures, therefore, is generally greater than that required to align the displacer in its cylinder. A critical factor in the compressor flexure design is the magnetic side load induced between the moving iron and the magnets located on the stator in higher power cooling systems. Overcoming this side load under worst case conditions represents a major design issue for the flexures since even small radial deflections due to this load can lead to piston contact with its cylinders. In lower power coolers it is possible to effectively employ moving coil motors which do not have magnetic side loads or provide an axial magnetic spring effect.

**Linear Drive Motor**

The linear drive motor for the compressor is of the moving iron type, incorporating stationary high performance rare earth magnets. This particular design, shown in Figure 5, is essentially made up of an individual fundamental motor element which is stacked to provide the necessary total input power for the compressor. For the current cooler, an individual motor element provides approximately 75 watts of PV power with five elements stacked into the complete motor package providing a total of 375 watts of PV power to the Stirling cycle. The stator structure (Figure 5C) contains the stator laminations, coils, and the stationary magnets. The laminations are of a simple geometry and orientated perpendicular to the axis of motion. This configuration allows the stator laminations to be assembled in a manner very similar to that of conventional rotating electric motors. Coils are fabricated separately and installed into the stator lamination as noted in Figure 5A. The permanent magnets are also attached directly to the bore of the stator lamination assembly, Figure 5C. The moving portion of the motor is attached directly to the piston rod which is supported by the flexures and has the actual compressor piston mounted on one end (see Figure 5B). The mover is made up of stacks of laminations (one stack for each motor element) separated by a non-magnetic spacer. The assembled motor is shown in Figure 5D.

The stator, mover, piston rod, and flexures are assembled and aligned in the forward portion of the compressor pressure vessel which contains the piston cylinder. A series of mechanical and electrical motoring tests are performed on this integrated assembly to insure that all critical functions are within specification. Once these tests are completed, the rear portion of the compressor pressure vessel, which contains the power feedthrough, is attached and

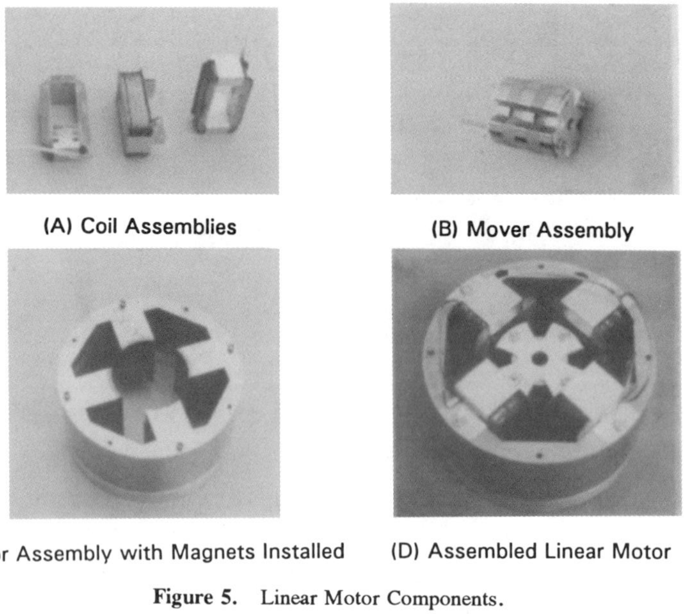

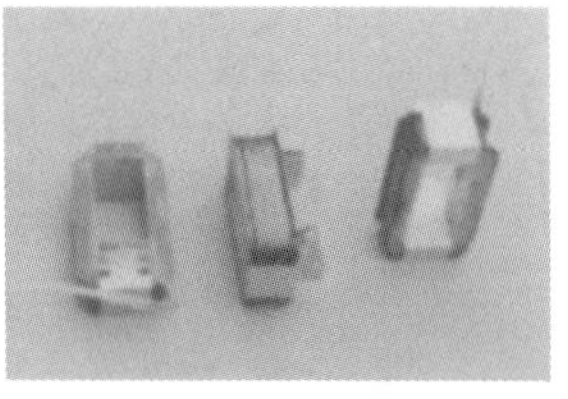

(A) Coil Assemblies

(B) Mover Assembly

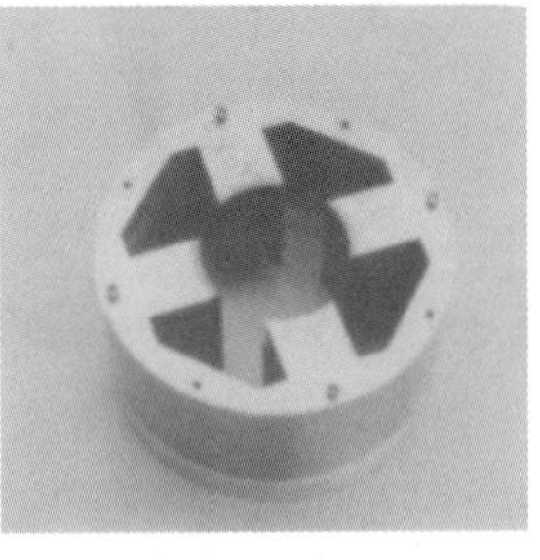

(C) Stator Assembly with Magnets Installed

(D) Assembled Linear Motor

**Figure 5.** Linear Motor Components.

**Figure 6.** In-Line Cold Head Configuration.

closeout welding performed. At this time the compressor enters the bakeout/contamination control processing.

**Stirling Cycle Cold Head Configurations**

As noted previously, the original cold head configuration for the cryocooler was a two-stage displacer assembly optimized for operation at 65 K/6 watts and 20 K/0.5 watt for use as a pre-cooler for a JT expander. Due to market driven requirements, the primary focus for the Stirling cycle cold head has been a single stage unit which can operate over the temperature range of 80 to 150 K with a nominal cooling capacity of 25 watts at 90 K. In all cases the cold head/displacer assembly axis is mounted at 90 degrees to the axis of the piston motion. Both the single- and two-stage displacers for this cold head configuration have the regenerator mounted within the moving displacer. Flexural bearings align the displacer within its cylinder which contains small radial gap clearance seals which insure that the working gas flows through the various heat exchangers rather than around the exterior of the displacer.

A second cold head configuration (see Figure 6) also incorporates the same linear drive gas compressor. The design of this cold head was driven by the requirements for better performance at lower operating temperatures in support of possible HTS markets. The nominal operating point is 60 K/7 watts capacity; however, the unit can be operated over the range of 45 to 110 K. Due to end user requirements, the cold head is mounted integrally to the compressor, with the displacer motion in line with that of the power piston. This configuration dictated that the flexural bearings for the displacer be mounted within the displacer body and the annular regenerator stationary.

**Pulse Tube Configuration**

The orifice Pulse Tube with various by-pass configurations represents a cold head assembly that may have distinct advantages and disadvantages compared to conventional Stirling cycle systems. For the laboratory and specialty sensor market, the key attribute of the Pulse Tube is the potential for very low vibration at the cold head. Other advantages include potentially higher reliability due to the removal of the moving displacer and its associated components, and possibly a lower cost cold head assembly. It is important to note that the basic reliability of a Pulse Tube system will still be dictated by the reliability of the gas compressor. Based on these advantages, STC is actively pursuing Pulse Tubes which incorporate the same linear drive gas compressor as the previously discussed Stirling cycle coolers. The specific design point of interest has a capacity of 7 to 10 watts at 80 K.

From the negative viewpoint, Pulse Tubes of the size under consideration have somewhat poorer performance than equivalent Stirling systems. This is particularly true if non-conventional Pulse Tube configurations are utilized. For example, the current Pulse Tube cold head (see Figure 7), is of the standard 'in-line' configuration in which the cold head is located essentially in the middle of the device. This location can make packaging quite cumbersome. By folding the Pulse Tube back on itself either U-shaped tube or concentric Pulse Tube configurations are possible. In both of these configurations, the cold head is located on the end of the device similar to that of a Stirling cycle systems, allowing conventional packaging techniques to be employed; however, the performance of the Pulse Tube is generally reduced.

Due to the fact that the hardware must interface with the existing linear drive gas compressor, the Pulse Tube makes use of a number of the previously discussed Stirling cycle cooler components in the areas of heat rejection and regenerator. The orifice and reservoir are located on the end of the Pulse Tube assembly downstream of the warm end rejector. Due to the robust nature of the hardware, it will be possible to evaluate a number of by-pass configurations during the prototype test program.

## COOLER SYSTEM PERFORMANCE

### Power Supply and Control System

The cooler system employs a variable voltage, fixed frequency power supply to drive the linear gas compressor. To insure reliable operation of the overall cooler system while offering as many possible cooling options, a microprocessor-based control system is employed. Signals from the moving components (piston and displacer) and temperatures of the cold head, heat rejection system, and linear motor housing are used as input to the control system.

During cooldown, the power supply voltage is increased until either the displacer or power piston reaches its maximum value at which time the power supply is commanded to hold the voltage constant. If the component amplitude continues to increase, the voltage is reduced until the component is within its design range. Generally during the initial phase of the cooldown process the displacer amplitude limits the level of the applied voltage. At low temperatures the situation is reversed with the power piston amplitude defining the quantity of power available to the drive motor. The measurement of the component motions is made by Hall effect sensors mounted on the exterior of the pressure vessel to allow easy sensor replacement if necessary.

The cooler control system generally transitions from the cooldown phase to the constant operating temperature mode once the cold head temperature reaches the desired set point. During this phase, the applied voltage is varied in response to the cold head temperature signal. This loop is tailored to provide approximately $\pm 1\,^{\circ}C$ variation from the set point temperature when in the temperature control mode.

The controller safety system compares component amplitudes (Hall effect sensors) and the rejector and motor temperatures (Thermistors) to set limits. If these values are exceeded, the cooler system can be shut down by reducing the applied voltage to a specified lower value or disconnecting the power supply completely from the drive motor.

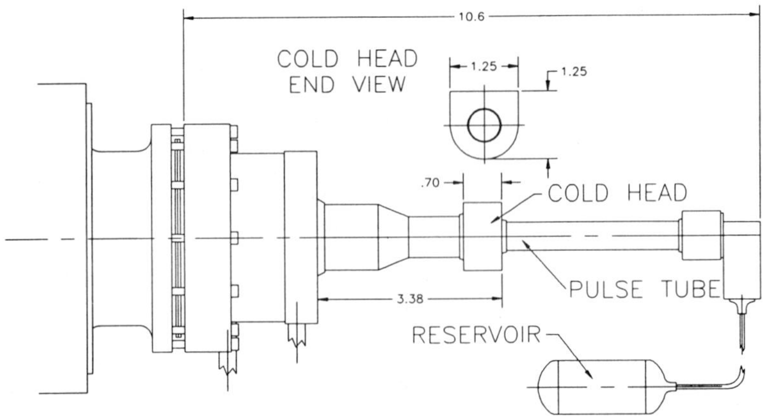


**Figure 7.** Pulse Tube Cryocooler/Cold Head Assembly.

**Figure 8.** Cooler System Package and Key Components.

Figure 8 depicts the basic layout of a cooler package which is utilized in a laboratory environment. The power supply and controller are located within the enclosure and mounted to the package frame. The specific heat rejection system employed in the unit shown incorporates a closed, pumped coolant loop which utilizes a small magnetically coupled circulating pump and an extended surface radiator/fan combination to reject the waste heat to the surrounding air. An alternative configuration of the rejection system is also available in which a pool boiler heat transfer loop is employed rather than a pumped loop. The working fluid R134a is contained within the area surrounding the Stirling cycle heat rejector which serves as the boiler portion of the system. The vapors rise through manifolds to an extended surface radiator/fan in which they condense and flow under the influence of gravity back to the boiler portion of the system. This configuration provides good overall cooling efficiency and eliminates the need for a circulation pump; however, due to the buoyancy/gravity induced flows, the lowest portion of the air-cooled condenser must be above the boiler for proper operation.

Vibration isolation for the majority of the cooler systems is provided by a passive mass balance system which can be seen mounted to the left end of the cooler in Figure 2. This type of balancer is made up of a mass which is supported by flexural springs and tuned to operate at a frequency slightly below the operating frequency of the power supply. Further improvements in vibration isolation can be attained by employing a small linear motor on a passive balance system. This motor is driven via a voltage defined by the measured acceleration of the cooler housing. When coupled with the proper control algorithm, it is possible to further reduce vibration loads. To further reduce vibration loads, it is also possible to separate the cold head/expander portion of the cooler from the compressor. This particular configuration was used for a cooler utilized in X-ray crystallography equipment. The small diameter tube is used to separate a standard cold head and compressor module. A passive balance unit can then be employed along the displacer axis to further minimize vibration levels. In this particular hardware, the addition of the small diameter, approximately 0.5 meter long tube reduced cooling capacity of the system by slightly more than one-third.

## Cooling Performance

A number of Stirling cycle and Pulse Tube simulation models are employed at STC during hardware design, development, and testing. The primary thermodynamic model is based on the GLIMPS software developed by Gedeon Associates and used by STC for over 10 years. Recently this software has been supplemented by another Gedeon Associates package, SAGE, which reflects the various improvements in thermodynamic cycle modeling that have occurred based on a better understanding of Stirling cycle and Pulse Tube operating characteristics.

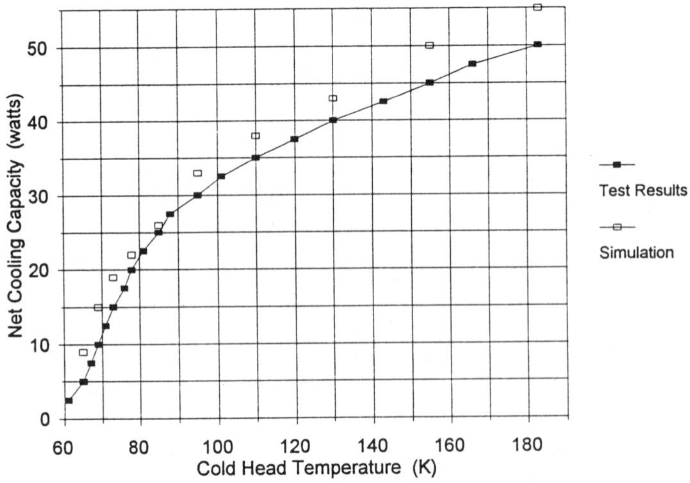


**Figure 9.** Comparison of Cooler Test Results and Simulation Predictions.

Cooler testing performed at STC employs an instrumentation suite that can fully characterize the cooler's operating conditions. Information from these sensors is collected by a high speed data acquisition system which carries out the various post processing necessary in the data evaluation process. The system (National Instruments/Lab View™-based) is easily adapted to carry out tests on specific components or on the overall system.

Figure 9 depicts the measured cooling capacity of one cooler system which, as previously mentioned, is optimized for cooling at 90 K (nominal capacity of approximately 25 watts). Peak electrical input power occurs in the 80 to 90 K range with lower input power levels occurring due to displacer amplitude limits at higher temperatures and piston amplitude limits at lower temperatures. Also shown in Figure 9 are the results of simulations at a wide range of operating temperatures for this cooler. As can be seen, the models track the actual test data over the entire temperature range of interest. The relatively rapid falloff in cooling capacity at temperatures below about 75 K is attributed by the modeling primarily to regenerator and acceptor heat exchanger operating characteristics (effectiveness). Tests with modified regenerator and acceptor heat exchanger hardware have confirmed these simulation model predictions.

## CONCLUSIONS

The cryocooler system described above is one member of a family of cryocoolers which represent an innovative approach to long life, cryogenic refrigeration systems tailored to meet market requirements. The hardware has demonstrated that the modular component concept is compatible with good cooling performance at cooler costs which are competitive with emerging market needs. The flexural bearing system and innovative linear drive motor configuration will allow this portion of the system to be utilized on single- and two-stage Stirling cycle cold heads as well as in coolers based on Pulse Tubes.

## ACKNOWLEDGEMENTS

We wish to acknowledge and express our gratitude for the support provided by NASA Goddard Space Flight Center, the technical expertise from Dr. Syed Nasar, and the technical support from STC personnel, Ian Williford, David Underwood, and Shawn Simons.

## REFERENCES

1. L. B. Penswick and D. C. Lewis, "Design and Testing of a Combined Stirling Cycle, Joule Thomson Cryocooler System," Cryocoolers 9, Plenum Press, NY, (1997).

# A NEW APPROACH FOR THERMOACOUSTIC REFRIGERATION*

J. H. Xiao[1] * and Ray Radebaugh[2]

[1]Cryogenic Laboratory, Chinese Academy of Sciences, PO Box 2711 Beijing 100080, People's Republic of China

[2]National Institute of Standards and Technology, 325 Broadway, MS 838.09, Boulder, Colorado 80303, USA

## ABSTRACT

We suggest a new design configuration for thermoacoustic refrigeration that optimizes the refrigeration effect caused by both the standing wave and the traveling wave. In current thermoacoustic refrigerators the refrigeration effect comes mainly from the standing wave component, in which the phase angle between the pressure and the velocity is 90°. In the orifice pulse tube refrigerator the refrigeration effect comes from the traveling wave component, in which the pressure and the velocity are in phase. By combining these two devices and optimizing the refrigeration effect caused by both the traveling wave and the standing wave components, we may be able to extend the temperature range of the thermoacoustic refrigerator down to the cryogenic range. The resulting device is a thermoacoustic refrigerator with an orifice pulse tube as a cold end expander, or an orifice pulse tube cooler operating near the resonance condition. The new design configuration may have a higher efficiency than that of the orifice pulse tube refrigerator or the thermoacoustic refrigerator in the intermediate temperature range. Another advantage over the pulse tube is the ability to use a driver with a small displacement, since the device is operating near resonance. This paper will discuss the working process, the new configuration, and some preliminary modeling results of the new type of thermoacoustic refrigerator.

## INTRODUCTION

Thermoacoustic refrigeration was developed during the past decade, and it is attractive as a potential alternative cooling method for room temperature refrigeration[1]. The

*Also guest researcher at Cryogenic Technologies Group, NIST, Boulder, Colorado, USA

operation of a thermoacoustic refrigerator is based upon the time-averaged thermoacoustic effect between the stacks and the standing wave created in a half wavelength resonator. Thermoacoustic refrigeration is reasonably efficient for temperatures near ambient, but cannot be used to achieve cryogenic temperatures. This is mainly because the refrigeration is based on the irreversible (standing wave) part of the thermoacoustic effect, which requires a moderate heat transfer between the gaseous fluid and the solid stack materials[2]. In thermoacoustic refrigeration the device operates near the resonance condition. This enables the use of a small displacement driver to deliver a relatively large amount of energy to drive the cooler. A thermoacoustic refrigerator is mechanically simple. It needs no moving parts at the cold end and does not use environmentally harmful working fluids or materials. Existing thermoacoustic refrigerators lacks a mechanism to impose an acoustic traveling wave, which could possibly increase the efficiency of the device. Because the entire resonator of the existing thermoacoustic refrigerator is at cold temperature during operation, frictional dissipation and imperfect heat transfer between the fluid and wall generates entropy and becomes a heat load to the cold end heat exchanger.

Orifice pulse tube refrigeration was also developed during the past decade[3]. Because of its mechanical simplicity (no moving parts at the cold end) and relatively high efficiency, it is one of the most promising technologies for the next generation of cryocoolers. The operation of the orifice pulse tube refrigerator is based upon the time averaged thermoacoustic effects taking place between the regenerator matrix and the traveling wave (which is created by the orifice and reservoir structure in the pulse tube cooler) [4-8]. The orifice pulse tube cooler operates based on the reversible (traveling wave) thermoacoustic effect, which requires excellent heat transfer between the gaseous fluid and the solid regenerator materials. These factors make the device efficient at cryogenic temperatures. However, the orifice pulse tube does not operate efficiently over small temperature differences for two reasons: first, the thermoacoustic effect caused by the standing wave component introduces a loss to the system, and second, the expansion work in the pulse tube is not recovered. Existing orifice pulse tube refrigerators operate far below the resonance condition of the gas column in the system. They also lack the mechanism to take advantage of the energy inherent in the always-present standing wave, which by default introduces losses to the system.

It is the purpose of this paper to introduce a new design configuration that combines some features of both the thermoacoustic and the pulse tube refrigerators to improve the performance. Depending upon the temperature range and the way the device operates, we may have two varieties here. The first is a pulse tube refrigerator operating near resonance condition, whose operation is mainly based on the thermoacoustic effect caused by the traveling wave component (small phase angle and good heat transfer between the gas and the regenerator matrix). The second is a thermoacoustic refrigerator with a pulse tube expander, whose operation is mainly based on the thermoacoustic effect caused by the standing wave component (large phase angle and moderate heat transfer between gas and stack matrix). The resonance orifice pulse tube refrigerator exploits the unused energy in the standing wave to improve efficiency. Similarly, the thermoacoustic refrigerator with a pulse tube expander introduces a traveling wave component to realize higher efficiency.

## THEORETICAL ASPECTS

Thermoacoustic theory [4,5] gives the longitudinal heat flux (heat-pumping rate when the direction of the heat flux is the same as that of the temperature gradient) of any cross section:

$$Q_x = -\frac{1}{2} T_o \beta_o \,\mathrm{Re}\,[U p^* f_{qx}] - A_f K_e \frac{\mathrm{d}\,T_o}{\mathrm{d}\,x} \qquad (1)$$

which can be rewritten as:

$$Q_x = -\frac{1}{2} T_o \beta_o |p||U||f_{qx}| \cos(\theta + \theta_q) - A_f K_e \frac{\mathrm{d}\, T_o}{\mathrm{d}\, x} \tag{2}$$

where: $\theta$ is the phase angle by which the velocity leads the pressure, $\theta_q$ is the phase angle of $f_{qx}$ , $\theta_q = \tan^{-1}\frac{\mathrm{Im}[f_{qx}]}{\mathrm{Re}[f_{qx}]}$ and $|f_{qx}|$ is the magnitude of $f_{qx}$, a complex constant determined by the material and structure of the regenerator or stacks. For parallel-plate stacks and helium gas, $f_{qx}$ is illustrated in Figure 1 (see references [4,5] for details of the equations).

If we assume that the positive direction is the direction in which temperature decreases in the regenerator or stacks, the heat-pumping effect exists when $Q_x < 0$ . This is satisfied if both the following conditions are true:

1) $\cos(\theta + \theta_q) > 0$ or $-\pi/2 < \theta + \theta_q < \pi/2$

2) $\left|\frac{\mathrm{d}\, T_o}{\mathrm{d}\, x}\right| < \frac{1}{2}\frac{T_o \beta_o}{A_f K_e} |p|\ |U| \cos(\theta + \theta_q)$

The first condition guarantees that the heat-pumping effect exists, while the second condition guarantees that the heat-pumping effect is larger than the conductive losses.

The first condition can be further expanded:

$$\cos(\theta + \theta_q) = \cos(\theta)\cos(\theta_q) - \sin(\theta)\sin(\theta_q) \tag{3}$$

On the right hand side of Equation (3), $\mathrm{Cos}(\theta)$ in the first term represents the traveling wave component of the acoustic oscillations, $\mathrm{Sin}(\theta)$ in the second term represents the standing wave component of the acoustic oscillations. To have both the traveling wave and the standing wave contribute to the heat-pumping effect requires:

$$\cos(\theta) > 0 \ \text{ and } \ \sin(\theta) < 0. \tag{4}$$

This yields:

$$-\pi/2 < \theta < 0, \tag{5}$$

which means the phase angle of the pressure oscillation leads that of the velocity oscillation by 0 to 90 degrees.

To have the heat-pumping rate caused by the traveling wave component be larger than that caused by the standing wave component requires:

$$\cos(\theta)\cos(\theta_q) > -\sin(\theta)\sin(\theta_q) \tag{6}$$

This means the phase angle θ should satisfy:

$$-\pi/2 + \theta_q < \theta < 0 \tag{7}$$

This discussion applies to the pulse tube as well. If the pulse tube is to pump heat, then the phase angle θ must satisfy:

$$\pi/2 - \theta_q < \theta < 3\pi/2 - \theta_q \tag{8}$$

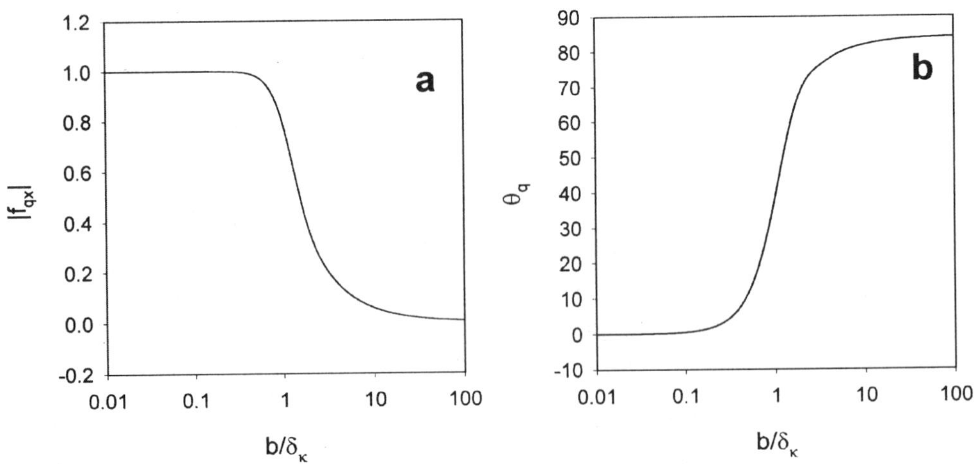


**Figure 1**. Illustration of $f_{qx}$. a, the magnitude; b, the phase angle [°].

## THE NEW APPROACH FOR THERMOACOUSTIC REFRIGERATION

We propose two new configurations to realize the design requirements. They are the resonance orifice pulse tube refrigerator and the thermoacoustic refrigerator with pulse tube expander. Both the traveling wave and standing wave contribute to the heat-pumping effect in the new designs. In the resonance pulse tube refrigerator, the heat-pumping effect caused by the traveling wave component is larger than that caused by the standing wave component. In the thermoacoustic refrigerator with pulse tube expander, the heat-pumping effect caused by the standing wave component is larger than that caused by the traveling wave component.

### The Resonance Orifice Pulse Tube Refrigerator

The configuration of the resonance orifice pulse tube refrigerator looks very similar to that of the normal orifice pulse tube refrigerator, except that with the new design, the refrigerator operates near the resonance condition of the whole system (not only that of the compressor or other driver). Thus, for the same operating frequency the device must have a longer pulse tube and possibly a longer transfer tube. Alternatively, a system of normal length requires a higher operating frequency. The length of the whole system must be somewhere between a quarter and a half of the wavelength to insure the proper phase angle of the pressure and velocity oscillations. A Helmholtz resonator (orifice, reservoir and a neck tube) attached at the warm end of the pulse tube provides a mechanism to create the traveling wave component in the oscillations. A fine matrix type regenerator is necessary here because we mainly want to utilize the heat-pumping effect caused by the traveling wave component. This also insures that the refrigeration effect caused by the traveling wave component of the system is larger than that caused by the standing wave, i.e. the system operates mainly on the reversible thermoacoustic effect.

The new design configuration eliminates the loss caused by the standing wave component, but still retains the main advantages of the orifice pulse tube refrigerator. The new device may operate efficiently at relatively high frequencies (though the heat transfer in the regenerator may not be as good as that in the lower operating frequency case) because the device uses both the traveling wave and the standing wave for heat pumping. The new device is operating near the resonance condition of the system, so it may use a small displacement driver to deliver a relatively large amount of energy for heat pumping. The device should be efficient over a wider temperature range.

The disadvantages related to the new configuration may be as follows: The longer pulse tube used here will need to be thermally insulated. For low operating frequencies, it may be difficult to build a small system due to the length requirement. Otherwise, a higher operating frequency must be used.

### Thermoacoustic Refrigerator with a Pulse Tube Expander

The configuration of the thermoacoustic refrigerator with pulse tube expander replaces the cold end resonator of the current thermoacoustic refrigerator with a pulse tube expander. It looks just like an orifice pulse tube refrigerator. However in contrast to the resonance orifice pulse tube refrigerator, where a fine matrix regenerator is used, a coarse matrix type regenerator or a stack is required here because we only want to have moderate heat transfer coupling between the gaseous fluid and the solid matrix. In this way, we mainly utilize the heat-pumping effect caused by the standing wave component in the system. The use of an orifice (possibly with a neck tube to provide inertance) and reservoir structure (more precisely a Helmholtz resonator) here provides a mechanism to impose the traveling wave in the system. The introduction of a traveling wave gives us another dimension to further optimize the refrigeration effect.

The new design eliminates most of the entropy generation at the cold end resonator of the current system, provides a way to make use of the heat-pumping effect caused by the traveling wave component and still retains the main feature of the current thermoacoustic refrigerator. The new device may operate efficiently at lower temperatures, because its operation is based upon both irreversible and reversible thermoacoustic effects; i.e. the device uses the thermoacoustic effect caused by both the standing wave and traveling wave for heat pumping. The new device is still operating near the resonance condition of the system, so it still may use a small displacement driver to deliver a relatively large amount of energy for heat pumping. We also expect the device to be more efficient over a wider temperature range.

The possible disadvantages related to the new configuration are as follows: A longer pulse tube will be used here which needs to be thermally insulated. A high operating frequency must be used for small systems.

## PRELIMINARY MODELING RESULTS

To demonstrate the feasibility of the new configuration, a resonance pulse tube refrigerator was studied. The pulse tube refrigerator used for the study is a primitive design that is described in the pulse tube calculation section of reference [9]. This pulse tube refrigerator design was not the same as that developed by Godshalk *et al.*[10] with an inertance tube, although they are related. The goal for this modeling study is to show that by changing the operating conditions of the pulse tube refrigerator to near resonance condition, at least theoretically, the performance of the refrigerator could be improved.

The original geometric dimensions and the operating conditions of the sample pulse tube refrigerator are listed in reference [9]. The cooler is operating at 300 Hz with a pulse tube length of 0.2 m. The pressure oscillation of the drive end is fixed at 8 % of the mean pressure. By the DeltaE Model [9], the cooler provides 5.34 W of cooling power at 150 K with a power input of 546 W. The relative Carnot efficiency of the cooler (COP/COPc) is about 1 %.

The operating frequency and the length of the pulse tube are changed to show the improvement in the performance of the cooler. All the other parameters of the cooler remain the unchanged. According to our discussion above, as the length of the system approaches half of the wavelength, an improvement in performance is expected.

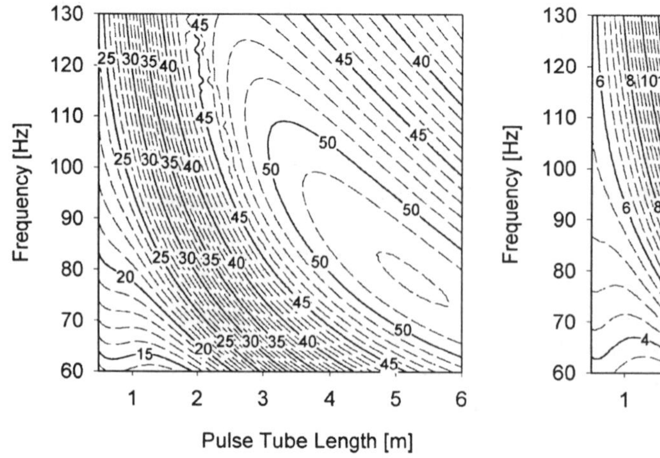


Figure 2. Contour of the refrigeration power [W].

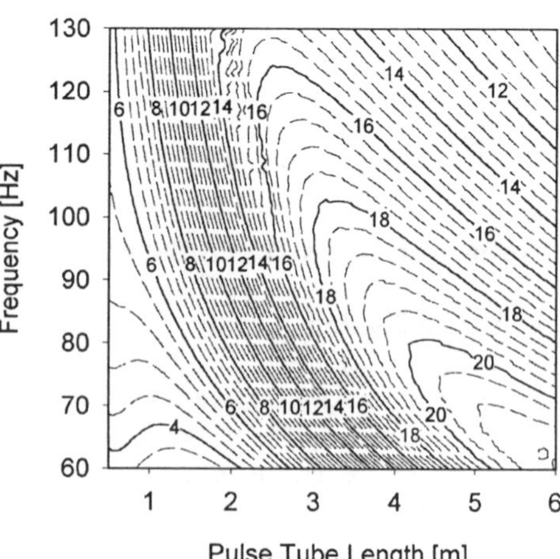


Figure 3. Contour of the COP/COPc [%].

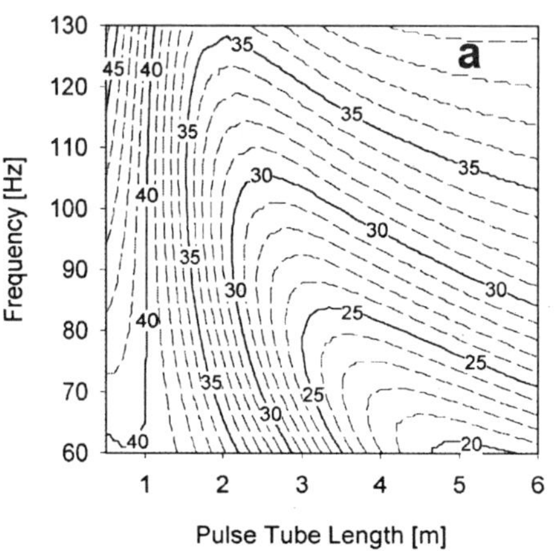


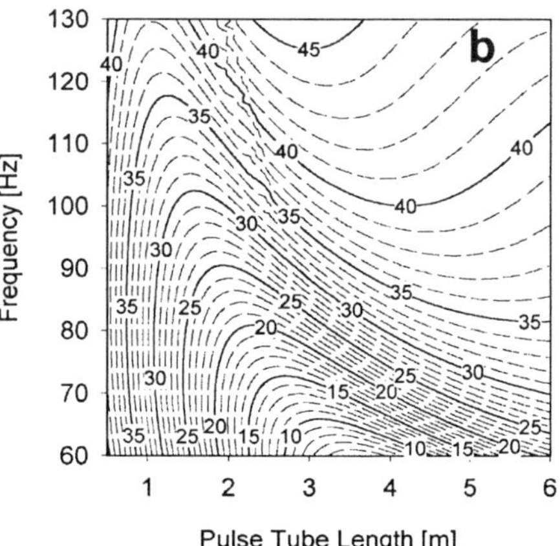


**Figure 4.** Contour of the volumetric velocity at the driver end of the refrigerator. a, the magnitude [$10^{-4}$ $m^3/s$]; b, the phase angle [°].

**Table 1** Comparison of parameters of the two configurations

| Locations | Phase difference[°] Orig./New | Total energy flux[W] Orig./New | Work flux[W] Orig./New |
|---|---|---|---|
| Driver | 50.4 / 33.5 | 546.2 / 263.9 | 546.2 / 263.9 |
| After cooler | 47.8 / 27.4 | 54.7 / 22.0 | 465.2 / 256.4 |
| Regenerator | -3.3 /-20.2 | 54.7 / 22.0 | 109.3 / 110.7 |
| Cold heat exchanger | -5.3 /-26.3 | 60.0 / 74.1 | 62.8 / 82.6 |
| Pulse tube | 12.5 / 41.1 | 60.0 / 74.1 | 62.5 / 67.2 |
| Hot heat exchanger | 60.0 /-81.2 | 8.4 / 9.0 | 8.4 / 9.0 |

The speed of sound in helium $a_o$ in the pulse tube operating between 150 K and 300 K is approximately 880 m/s. With a frequency of 300 Hz, the half wavelength of the acoustic wave is about 1.5 m. With a frequency of 70 Hz, the half wavelength is about 6.3 m. Figure 2 shows the contour of refrigeration power as a function of the frequency and the pulse tube length. The refrigeration power of the system is optimized to 52.1 W at a frequency of 79 Hz and a pulse tube length of 5.1 m (very close to the half wavelength of 5.6 m at 79 Hz). It can also be seen that, near the optimal point, the refrigeration power is less sensitive to changes in pulse tube length and frequency near the curve $L = 0.46\frac{a_o}{f}$.

Figure 3 shows the contour of the relative Carnot efficiency of the refrigerator. The efficiency of the refrigerator is optimized to about 21 % at a frequency of 62 Hz and pulse tube length of 5.85 m. Near the optimal point, the efficiency is also less sensitive to changes in pulse tube length and frequency near the curve $L = 0.46\frac{a_o}{f}$.

Figure 4 shows the contour of the volumetric velocity at the driver of the refrigerator. Note that the amplitude and the phase angle of the pressure oscillation remain constant at the driver, the amplitude is 0.24 MPa and the phase angle is set to be the reference ($0^o$). Figures 4a and 4b show the amplitude and the phase angle of the velocity, respectively. As the operating condition approaches resonance, both the amplitude and the phase angle of the volumetric velocity are much smaller than those in the original design.

Table 1 shows a comparison of values for some of the most important parameters of the original design with those of the new design. One thing we need to mention here is that the heat exchanger at the warm end of the pulse tube for this design also acts as the flow impedance for the orifice reservoir structure. This means that most of the pressure drop is generated across the heat exchanger. This also accounts for the big change of the phase angle between the velocity and pressure.

The data in Table 1 show that by using the new design, the regenerator loss is much less than that of the original design (from 54.7 W to 22.0 W, an improvement of 32.7 W). The total energy flux (enthalpy flux) in the pulse tube becomes much bigger than that in the original design (from 60 W to 74.1 W, a gain of 14.1 W). Also, in the original configuration, a pulse tube loss of 2.8 W is introduced (compare the value of the work flow with the total energy flow). In the new configuration the pulse tube is pumping the heat at a rate of 6.9 W. Those changes occur because the phase relationship between the pressure and velocity oscillations has been changed dramatically in the new configuration due to resonance operation.

## CONCLUSIONS

The new design configuration for thermoacoustic refrigeration proposed in this paper is a resonance orifice pulse tube refrigerator, or a thermoacoustic refrigerator with an orifice pulse tube as a cold end expander. The resonance orifice pulse tube refrigerator uses a fine matrix type regenerator. It mainly uses the heat-pumping effect caused by the traveling wave component. The phase angle of the velocity is leading that of the pressure oscillation, and is close to 0 degrees. The refrigeration effect caused by the traveling wave component of the system is larger than that caused by the standing wave component. The thermoacoustic refrigerator with pulse tube expander uses a much coarser matrix type regenerator or stacks. It mainly uses the heat-pumping effect caused by the standing wave component. The phase angle of the velocity is leading that of the pressure oscillation, and is much closer to 90 degrees. The refrigeration effect caused by the standing wave component of the system is larger than that caused by the traveling wave component.

The new design configuration may have an improved efficiency over the existing orifice pulse tube refrigerator and the thermoacoustic refrigerator because it uses the thermoacoustic effects of both the traveling wave and standing wave components.

## ACKNOWLEDGMENTS

The authors would like to thank Matthew Hill for his assistance in preparing this paper.

## REFERENCES


1 G.W. Swift, Thermoacoustic engines and refrigerators, *Physics Today* (July 1995), p.22
2 G. W. Swift, Thermoacoustic engine, *J. Acoust. Soc. Am.* 84:1145 (1988)
3 Ray Radebaugh, Advances in Cryocoolers, “Proc. 16th CEC/ICMC”, Elsevier Science Inc., Oxford, England(1996), p.33
4 J. H. Xiao, Thermoacoustic theory for cyclic flow regenerators, *Cryogenics*, 32:895 (1992)
5 J. H. Xiao, Thermoacoustic heat transportation and energy transformation, Part I: Formulation of the problem. *Cryogenics*, 35:15(1995)
6 J. H. Xiao, Thermoacoustic heat transportation and energy transformation, Part II: Isothermal wall thermoacoustic effects. *Cryogenics*, 35:21 (1995)
7 J. H. Xiao, Thermoacoustic heat transportation and energy transformation, Part III: Adiabatic wall thermoacoustic effects. *Cryogenics*, 35:27 (1995)
8 J. H. Xiao, Thermoacoustic theory for regenerative cryocoolers: A case study for a pulse tube refrigerator, “Proc. 7th Int’l Cryocooler Conf.”, Airforce Report PL-CP-93-1001(1993), p.305
9 Ward and G. Swift, Design environment for low-amplitude thermoacoustic engines, *Tutorial and User’s Guide, V2.0*, Los Alamos National Laboratory (1995), (email to ww@lanl.gov for a copy)
10 K. M. Godshalk, C. Jin, Y. K. Kwong, E. L. Hershberg, G.W. Swift and R. Radebaugh, Characterization of 350 Hz thermoacoustic driven orifice pulse tube refrigerator with measurements of the phase of the mass flow and pressure, “Advances in Cryogenic Engineering”, Plenum Press, New York (1997), Vol. 41 p.1411

# PULSE TUBE COOLERS WITH AN INERTANCE TUBE: THEORY, MODELING,AND PRACTICE

Pat R. Roach[1] and Ali Kashani[2]

[1]NASA Ames Research Center
Moffett Field, CA 94035
[2]Atlas Scientific
Sunnyvale, CA 94086

## ABSTRACT

We have studied the advantages to be gained by replacing the orifice of a pulse tube cooler with an inertance tube—a long thin tube that introduces the possibility for additional phase shift between pressure and mass flow in the pulse tube section. A case for using an inertance tube is made by employing an electrical analogy, where the 'inductance' added by the inertance tube allows for an improved power transfer efficiency at the cold end of the pulse tube. Detailed computer modeling of pulse tube systems with inertance tubes confirms these advantages. Comparison between a laboratory cooler with an orifice and with two inertance tubes is presented; the inertance tubes yield dramatic improvements over the use of the orifice.

## INTRODUCTION

Pulse tube coolers have a considerable advantage over Stirling coolers because they have no moving parts in the low-temperature region. This leads to much more reliable operation and longer life times for the low-temperature components and also much lower vibration in the cold region. However, Stirling coolers have demonstrated better thermal efficiency than pulse tube coolers because the low-temperature displacer of the Stirling cooler can be driven at a phase that is adjusted to give the best performance. Pulse tube coolers have used either an orifice to create the phase shift between pressure and mass flow in the pulse tube or both an orifice and a double-inlet connection that allows some of the flow to bypass the regenerator and pulse tube. In neither case does the efficiency quite reach that of Stirling coolers.

Recent studies[1,2] have suggested that there is a simple way to generate the phase shift needed to make pulse tube coolers operate as efficiently as Stirling coolers. This is to replace the orifice in the system with an 'inertance' tube. The inertance tube is a long, thin tube that provides a complex impedance at the warm end of the pulse tube rather than a simple resistive impedance that the orifice provides. The inertance tube adds a reactive impedance, analogous to inductance in electrical circuits, that allows the phase between the pressure and mass flow in the pulse tube to be adjusted to an extent that was not previously possible. In principle, it might now be possible to adjust the complex impedance of the inertance tube to achieve the maximum cooling efficiency that the system is capable of.

## ELECTRICAL ANALOGY

Considerable insight to the inertance tube can be gained from an analogy to electrical circuits. Consider the relations between electrical current, I, and voltage, V, for the cases of a resistor of resistance, R, an inductor of inductance, L, and a capacitor of capacitance, C:

$$\text{Resistor:} \quad V = I R \tag{1}$$

$$\text{Inductor:} \quad V = L \frac{dI}{dt} \tag{2}$$

$$\text{Capacitor:} \quad \frac{dV}{dt} = \frac{I}{C} \tag{3}$$

### Fluid flow

Similar relationships can be found for gas flow and pressure in the elements of a pulse tube cooler. Consider the 1-dimensional momentum conservation equation[3] for the flow in a tube of radius, r:

$$\rho \frac{\partial q}{\partial t} = -\frac{\partial P}{\partial z} - \frac{\mu q}{K_p}, \tag{4}$$

where P is the pressure, $\rho$ is the gas density, q is the average velocity in the tube, $\mu$ is the viscosity and $K_p$ is the Darcy permeability.

Substituting $q = U/\pi r^2$ and rearranging the terms yields:

$$-\frac{\partial P}{\partial z} = \frac{\rho}{\pi r^2}\frac{\partial U}{\partial t} + \frac{\mu U}{\pi r^2 K_p}, \tag{5}$$

where U is the volume flow rate. If these parameters are independent of the distance, z, along the tube, then the equation can be integrated along the tube for its entire length, $\lambda$, and one gets:

$$-\Delta P = \frac{\rho \lambda}{\pi r^2}\frac{\partial U}{\partial t} + \frac{\mu \lambda U}{\pi r^2 K_p}. \tag{6}$$

If the pressure, P, is taken to be the analog of voltage, V, and the volume flow, U, is taken to be the analog of electrical current, I, then comparison with eqs. (1) and (2) shows that eq. (6) is analogous to a resistance of value

$$R = \mu\lambda/(\pi r^2 K_p), \quad (K_p = r^2/8 \text{ for laminar flow}) \tag{7}$$

in series with an inductance of value

$$L = \rho\lambda/\pi r^2 . \tag{8}$$

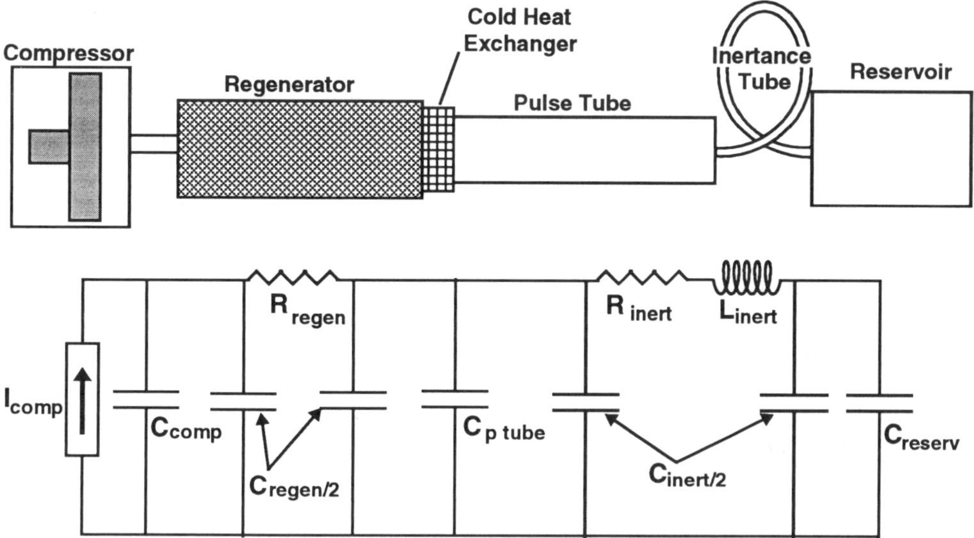


**Fig. 1.** A simple pulse tube cooler with an inertance tube, and the analogous electrical circuit, using lumped circuit elements. $I_{comp}$ is a current source representing the piston stroke; $C_{comp}$ represents the compressor volume; $R_{regen}$ represents the flow impedance of the regenerator and the two $C_{regen/2}$ are approximations to the distributed volume of the regenerator; $C_{p\ tube}$ represents the volume of the pulse tube; $R_{inert}$ represents the dissipative resistance of the inertance tube; $L_{inert}$ represents the inductive 'inertance' of the inertance tube; $C_{inert/2}$ are approximations to the distributed volume of the inertance tube; and $C_{reserv}$ represents the volume of the reservoir (assumed to be large so that $C_{reserv}$ and the capacitance in parallel with it can be ignored).

This 'inductance' analog gets the name of inertance for the gas flow case.

For an isothermal volume, $V_t$, with volume flow, U, into it, the pressure rise will be:

$$\frac{\partial P}{\partial t} = \frac{P_{av}\, U}{V_t}\ . \tag{9}$$

This is exactly analogous to eq. (3) with a capacitance of value

$$C = V_t / P_{av}. \tag{10}$$

Equations (6) and (9) show that gas flow in tubes and volumes is analogous to electrical current flow in resistors, inductors and capacitances.

## Electrical analogy for a pulse tube cooler

In a pulse tube cooler it is important to maximize the cooling efficiency: the ratio of the cooling at the cold end of the pulse tube to the work of the compressor. Figure 1 shows a simple pulse tube cooler with an inertance tube along with an equivalent circuit that approximates the main features of the cooler. In this circuit, all the power dissipation in the system takes place in the resistive elements $R_{regen}$ and $R_{inert}$. The electrical power that flows through $R_{inert}$ is the analog of the cooling power that flows through the pulse tube and into the inertance tube. The ratio of this power to the sum of this power plus the power dissipated in $R_{regen}$ is the efficiency. If one assumes that all the other parameters are fixed, then only $R_{inert}$ and $L_{inert}$ are to be adjusted. Equations (7) and (8) were used to express $R_{inert}$ and $L_{inert}$ in terms of the length, $\lambda$, and the radius, r, of the inertance tube. Figure 2 shows analytical results, using values that are typical for a pulse tube cooler. The contours are constant values of efficiency. Clearly, the efficiency is greatest for $\lambda$ and $r = \infty$. However, the efficiency can be quite close to the optimum value (1.0 for this case) even with finite values of $\lambda$ and r. For any given value of r, the value of $\lambda$ that gives the best efficiency is:

$$\lambda_{opt} = \frac{\pi r^2}{\rho \, \omega^2 C_2} \frac{k}{\sqrt{k^2+1}}, \tag{11}$$

where $\omega$ is the angular frequency, $k = \omega \rho r^2/8\mu$ and $C_2 = C_{regen/2} + C_{p\ tube} + C_{inert/2}$. For this value of $\lambda$:

$$\text{efficiency}(\lambda_{opt}) = \frac{1}{1 + 2\,\omega\, C_2 R_1\left(\sqrt{k^2+1} - k\right)} \quad . \tag{12}$$

For $k >> 1$, $\lambda_{opt} = \pi r^2/(\rho\, \omega^2 C_2)$ or

$$L_{opt} = 1/(\omega^2 C_2) \text{ by eq. (8)}, \tag{13}$$

and the efficiency = 1. For the case analyzed in Fig. 2 and for $r = 0.13$ cm, then $k = 11.5$, $\lambda_{opt} = 265$ cm and the efficiency = 0.94.

By way of comparison, for the case where the inertance tube is replaced by an orifice (a simple resistance in the electrical analog) of optimum value, the efficiency becomes

$$\text{efficiency}(\text{orifice}_{opt}) = \frac{1}{1 + 2\,\omega\, C_2 R_1} \tag{14}$$

which reduces to an efficiency = 0.42 for the same parameters used above.

This is only suggestive of the way that the inertance tube benefits a pulse tube cooler. Many of the features of a cooler cannot be modeled by such an analog: the presence of temperature gradients in the actual cooler are one of the obvious effects that an electrical analog can't deal with. Nevertheless, it is clear that this treatment captures the primary way the inertance tube yields the benefits it does.

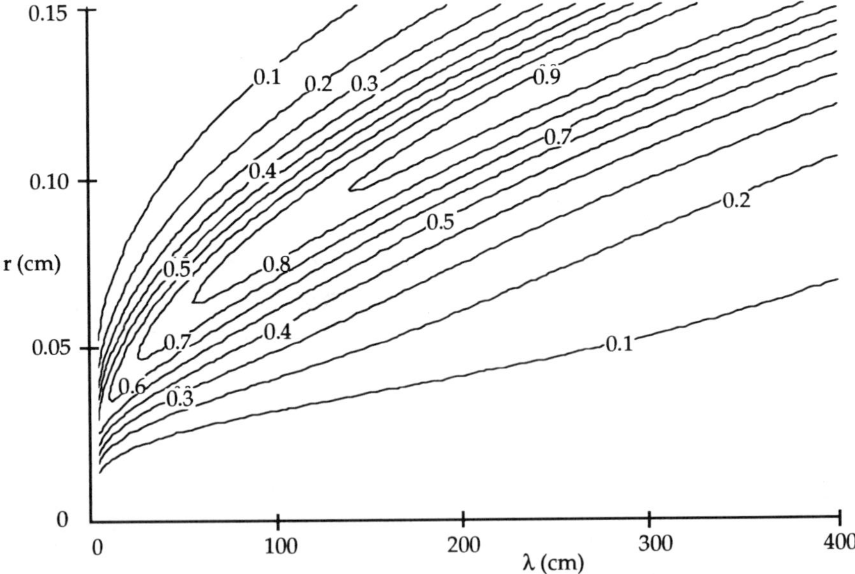


Fig. 2. The efficiency of the electrical analog of an inertance pulse tube as a function of the length, λ, and the radius, r, of the inertance tube. The lines are contours of constant efficiency.

**Table 1.** Orifice pulse tube coolers, modeled for $T_{hot}$=300 K and $T_{cold}$=80 K freq. =55 Hz, Pressure =2.0 x$10^6$ Pa, pressure ratio in pulse tube =1.20

| | Regenerator | | | | Pulse Tube | | Orifice | Cooling Power | Compress. PV work | Efficiency |
|---|---|---|---|---|---|---|---|---|---|---|
| | ID (cm) | L (cm) | Mesh (/in) | Wire (cm) | ID (cm) | L (cm) | (kg/s Pa) | (W) | (W) | |
| Pulse Tube 1 | 0.615 | 5.00 | 400 | .0025 | 0.87 | 2.01 | 8.4 | 0.82 | 64.7 | 1.27% |
| Pulse Tube 2 | 1.545 | 4.68 | 400 | .0025 | 1.10 | 3.00 | 21.8 | 3.37 | 93.2 | 3.63% |
| Pulse Tube 3 | 3.77 | 4.30 | 400 | .0025 | 1.55 | 5.02 | 90.0 | 15.95 | 333.8 | 4.78% |

## COMPUTER MODELING

We have adapted our ARCOPTR computer model[4] to include an inertance tube. This allows us to easily and quickly study the effect on a given pulse tube design of changing the orifice to an inertance tube. This modification to the model was readily implemented since the model already contained a detailed analytic solution for flow in the hot heat exchanger; it was only necessary to take out the screen mesh, make the heat exchanger long and thin, and it thus becomes an isothermal inertance tube. This solution includes all the effects that arise for lengths comparable to the acoustic wavelength. In electrical terms, it treats the inertance tube as a lossy transmission line.

It was also necessary to treat the effects of turbulent flow since the velocities tend to be rather high in the inertance tube. To do this, we used a Reynolds number, $N_\delta$, that is appropriate for oscillating flow[5]:

$$N_\delta = \frac{q\,\delta_v\,\rho}{\mu} \tag{15}$$

where $\delta_v$ is the boundary layer thickness (or Stokes length): $\delta_v = \sqrt{\frac{\mu\,\tau}{\pi\,\rho}}$ where $\tau$ is the period of oscillation (q, $\rho$ and $\mu$ are defined for Eq. 4). This is appropriate because the transition to turbulent flow occurs first in the boundary layer; it is the boundary layer dimension rather than the tube diameter that determines the critical Reynolds number for oscillatory flow. The friction factor for turbulent flow is determined from the Blasius eq.:

$$f = C\,N_D^{-1/4} \tag{16}$$

where $N_D$ is the Reynolds number using the tube diameter, D: $N_D = qD\,\rho/\mu$, and C is a constant near 0.1.

### Three model pulse tube coolers

**Orifice versions.** We have modeled three orifice pulse tube coolers of rather different sizes; their dimensions are shown in Table 1. The net cooling power is the sum of the enthalpy flow at the cold end of the pulse tube and the various losses in the system (the residual enthalpy flow from the regenerator and the conduction losses from the regenerator wall and matrix and from the pulse tube wall). The efficiency is the ratio of this net cooling power to the compressor PV work. In each case the orifice setting used was that which maximized the efficiency.

**Table 2.** The same pulse tube coolers as in Table 1 with the orifice replaced by an inertance tube in the model calculation; pressure ratio in pulse tube =1.20

| | Inertance Tube | Cooling Power | Compress. PV work | Efficiency | Eff. (Inert. tube) / Eff. (Orifice) |
|---|---|---|---|---|---|
| | length x dia.(cm) | (W) | (W) | | |
| Pulse Tube 1 | 100 x 0.104 | 0.98 | 48.3 | 2.02% | 1.59 |
| Pulse Tube 2 | 250 x 0.22 | 4.72 | 81.4 | 5.79% | 1.60 |
| Pulse Tube 3 | 400 x 0.58 | 17.9 | 215 | 8.34% | 1.74 |

**Inertance tube versions.** We then replaced the orifice of each cooler with an inertance tube that was chosen to give the best cooling performance. The results of this model calculation are shown in Table 2. For all three pulse tubes the inertance tube increases the efficiency by about a factor of 1.6 to 1.7 under similar operating conditions. Part of the increase in efficiency is brought about by an increase in cooling power but most of the increase comes from a reduction in the required compressor PV work. This is one of the main benefits of an inertance tube, that it significantly lowers the compressor PV work required for the same system with an orifice.

**Comments on modeling.** Figure 3 shows the results of more detailed calculations for pulse tube 1. For a wide range of inertance tube lengths, from 60 cm to 150 cm, there is a value of inertance tube radius, r, that brings the efficiency to about 2%. The values of $\lambda/r^2$ (which is proportional to the inertance) at which these maximum efficiencies occur are clustered around 9000 $cm^{-1}$ for this example. For a given value of $\lambda$, however, the efficiency is sharply peaked at a particular value of r.

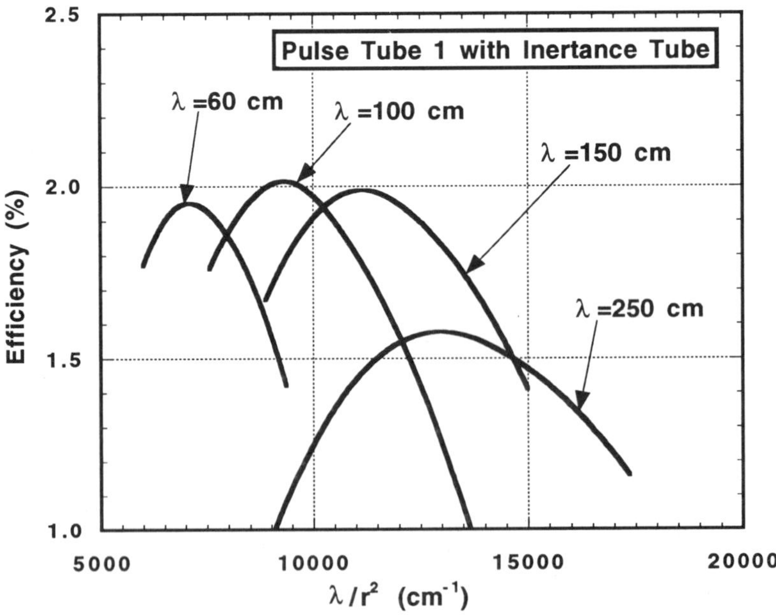


Fig. 3. Efficiency of pulse tube 1 from model calculation for different values of inertance tube length, $\lambda$, as a function of $\lambda/r^2$, where r is the inertance tube radius.

**Table 3.** Experimental orifice pulse tube cooler, with $T_{hot} = 303$ K, freq. = 65 Hz and pressure = 2.27 x10$^6$ Pa. No heat is applied to the cold heat exchanger

| | Regenerator | | | | Pulse Tube | | Press. ratio in PT | Calculated compressor PV work | T(cold) |
|---|---|---|---|---|---|---|---|---|---|
| | ID (cm) | L (cm) | Mesh (/in) | Wire (cm) | ID (cm) | L (cm) | | (W) | (K) |
| Experimental pulse tube | 1.54 | 6.73 | 400 | .0025 | 1.27 | 5.69 | 1.13 | 113 | 111 |

## EXPERIMENTAL TESTING

We have performed a similar experimental comparison by measuring the lowest temperature reached by a laboratory pulse tube with an orifice and then replacing the orifice with an inertance tube and measuring the change in lowest temperature reached.

### Orifice version

Table 3 gives the dimensions of the experimental pulse tube cooler we used and the results of the measurements with the orifice set to give optimum cooling. The pressure ratio in the pulse tube is the maximum that we can obtain from our compressor for this cooler. The calculated compressor work comes from a fit to the experimental data by our model. The model determines the mass flows through the various parts of the cooler in order to match the experimental pressures at the compressor, the pulse tube and the reservoir. From this information the model calculates the PV work of the compressor, assuming adiabatic compression.

### Inertance tube version

Table 4 shows the results of measurements on the same experimental cooler with the orifice replaced by two different inertance tubes. Again, the pressure ratio in the pulse tube is the maximum that we can obtain from our compressor. The fact that the pressure ratio in the pulse tube is higher for the inertance tube versions than for the orifice version is one of the advantages of an inertance tube. For inertance #1 the minimum temperature, 118 K, isn't quite as low as that with the orifice, 111 K. Note,

**Table 4.** The same experimental pulse tube cooler as Table 3 but with two different inertance tubes replacing the orifice; $T_{hot} = 303$ K, freq. = 65 Hz and pressure = 2.3 x10$^6$ Pa. No heat is applied to the cold heat exchanger

| | Inertance Tube | Press. ratio in PT | Calculated compressor PV work | T(cold) |
|---|---|---|---|---|
| | length x dia.(cm) | | (W) | (K) |
| Inertance 1 | 105 x 0.17 | 1.17 | 82 | 118 |
| Inertance 2 | 226 x 0.28 | 1.17 | 76 | 92 |

however, that the calculated compressor PV work is only 73% of that for the orifice version. For inertance #2 the minimum temperature, 92 K, is considerably lower than it is with the orifice, and the compressor PV work is even lower, at 67% of that for the orifice version. Clearly, the inertance tubes provide considerable benefit in both cases.

## CONCLUSIONS

The advantages of employing an inertance tube in the design of a pulse tube cooler have been shown by a simplified electrical analogy, by detailed computer modeling and by experimental measurements.

The electrical analogy reveals that the optimum inertance (the electrical inductance) obeys a resonance condition (eq. (13)) in terms of the volume in the middle of the cooler (half the regenerator volume plus the pulse tube volume plus half the inertance tube volume). This shows that an optimum inertance tube depends primarily on the volumes in the system (except the compressor volume) and not on the impedance of the regenerator.

Computer modeling for three different size pulse tubes shows that a large benefit of about a factor of 1.6 can be gained in the cooling efficiency for a pulse tube cooler with an optimized inertance tube over one with an optimized orifice. It also shows thaat there is a sharp maximum in cooling efficiency as a function of r for a fixed value of $\lambda$. However, for a wide range of $\lambda$ there are values of r that bring the efficiency to about the same maximum value.

Experimental comparisons of a pulse tube with an orifice and with two different inertance tubes likewise show substantial benefits either in the lowest temperature reached, the compressor PV power needed, or both, when employing an inertance tube.

## REFERENCES

1. S. W. Zhu, S. L. Zhou, N. Yoshimura and Y. Matsubara, Phase shift effect of the long neck tube for the pulse tube refrigerator, Proceedings of the 9th International Cryocoolers Conference, Waterville Valley, New Hampshire, June 1996, p. 269.
2. D. L. Gardner and G. W. Swift, Use of inertance in orifice pulse tube refrigerators, *Cryogenics* (to be published).
3. Pat R. Roach, Ali Kashani and J.M.Lee, Theoretical analysis of a pulse tube regenerator, in: "Advances in Cryogenic Engineering", 41B (1995) p.1357, eq. (5).
4. Pat R. Roach and Ali Kashani, A Simple modeling program for orifice pulse tube coolers, Proceedings of the 9th International Cryocoolers Conference, Waterville Valley, New Hampshire, June 1996, p. 327.
5. David M. Eckmann and James B. Grotberg, Experiments on transition to turbulence in oscillatory pipe flow, J. Fluid Mech. 222:329 (1991).

# AN OPTIMIZATION PROGRAM FOR MODELING PULSE TUBE COOLERS

Ali Kashani[1] and Pat R. Roach[2]

[1]Atlas Scientific
Sunnyvale, CA 94086

[2]NASA Ames Research Center
Moffett Field, CA 94037-1000

## ABSTRACT

ARCOPTR is a program developed at NASA-Ames Research Center for modeling pulse tube coolers. The ARCOPTR is a 1-D model that treats all the components of an orifice pulse tube cooler or a pulse tube cooler employing an inertance tube. It analyzes the thermodynamic equations for the regenerator and assumes that all mass flows, pressure oscillations and temperature oscillations are small and sinusoidal. The resulting mass flows and pressures are matched at the boundaries with the other components of the cooler: compressor, aftercooler, cold heat exchanger, pulse tube, hot heat exchanger, orifice/inertance tube and reservoir. The results of the calculation are oscillating pressures, mass flows and enthalpy flows in the main components of the cooler. An optimizer has been built into the ARCOPTR that allows optimization of the cooler parameters to attain maximum efficiency. The parameters that can be optimized are the length and the diameter of the regenerator and those of the pulse tube, the wire diameter and the mesh size of the regenerator screen, and the orifice opening or in the case of an inertance tube, its length and diameter. In this study, pulse tube cooler configurations, employing an orifice or an inertance tube, have been optimized using the ARCOPTR. The optimum parameters are plotted as a function of the regenerator diameter. In the optimization the pressure ratio in the pulse tube is maintained at 1.2 by adjusting the compressor stroke.

## INTRODUCTION

Pulse Tube coolers are complex systems that require careful optimization of their many components to achieve maximum efficiency. In particular, the regenerator, where a large surface area and high heat capacity are needed to damp out temperature oscillations, is a component that must be designed to maximize heat exchange with the gas passing through it while minimizing the pressure drop over its length. The fact that the gas flow is oscillating back and forth while the gas pressure is also oscillating with a different phase makes the system too complex for a simple analysis.

To understand this interaction in the regenerator and to design pulse tube coolers a computer model[1] (ARCOPTR) has been developed which captures the fundamental behavior of the gas-matrix interaction. In optimizing the performance of the regenerator the performance of the other parts of the cooler is also affected. Since it is only the efficiency of the system that is ultimately important, the model is extended to include the other major components of a typical cooler. The model treats the entire system starting with the compressor and includes the heat exchangers, the regenerator, the pulse tube section itself and also the orifice/inertance tube and reservoir that provide the phase shift of the mass flow that is necessary for cooling to occur. To simplify the calculations, the model is limited to the consideration of only the fundamental frequency component of the oscillating parameters and to the limiting case of infinitesimal amplitude oscillations. This allows the study of all the fundamental processes that affect the performance of the pulse tube cooler, however, highly accurate estimates of performance of actual systems having large amplitudes of pressure oscillation or mass flow are not expected. Nevertheless, the model provides a useful guide to the optimum trade-off between the many conflicting requirements, especially since higher accuracy is hard to justify without a better understanding of some of the loss mechanisms that occur.

To make the optimization of pulse tube coolers easier using the ARCOPTR an optimizer has been developed. The optimizer provides the user with the optimum values for the critical parameters of a pulse tube cooler to achieve maximum efficiency. The present is a brief overview of the ARCOPTR and its optimization routine.

## FEATURES OF ARCOPTR

The ARCOPTR is a 1-D model that treats all components of an orifice pulse tube cooler or one employing an inertance tube. Schematics of both these configurations are shown in Figure 1. The compressor is taken to be adiabatic. All temperature oscillations are assumed to be completely damped in the aftercooler. The regenerator matrix is made up of stacked SS screens having a void fraction of 2/3. The equations derived for the regenerator are solved to arrive at a fundamental-frequency amplitude for the pressure; temperature and velocity are then derived from this pressure solution. An initial linear temperature gradient from the hot end to the cold end is used to evaluate parameters that are temperature dependent. The heat flow to the solid matrix is taken to be in phase with the temperature difference between the gas and matrix. This assumes that the dimensions of the pores containing the gas are small enough that the gas is in fairly good contact with the matrix; in other words, the hydraulic radius of the pores is less than the diffusion length in the gas.

The heat transfer coefficient used in the regenerator is from Kays and London[2] and the friction factor is based on the oscillating flow measurements performed at NASA-ARC and reported elsewhere.[3] Since these data show considerable non-linearities at higher velocities, it is important to include some high-velocity effects. An initial guess to the velocity in the regenerator is used to estimate the heat transfer coefficient and the friction factor; the regenerator equations are then solved and much more accurate values of the velocity are obtained. From these velocities the final values for the heat transfer coefficient and the friction factor are found and the calculation is repeated. This is the only instance where some allowance for finite-velocity effects is included. In calculating the effect of axial thermal conduction through the regenerator matrix, the bulk conductivity of SS is reduced by a factor of 10 to reflect the poor contact between adjacent screens.[4] The boundary conditions for the regenerator solution are that the pressures and mass flows at the ends of the regenerator match those of the compressor and pulse tube (as modified by the aftercooler and cold heat exchanger, respectively).

The aftercooler and the cold and hot heat exchangers are made up of stacked screens and assumed to be isothermal. The primary effect they have on the modeling is to introduce a pressure drop due to their impedance and a phase shift in the mass flow due to their void volume. No attempt is made to assess their adequacy for heat transfer or to calculate the amount of heat that flows through them. It is assumed that all temperature oscillations are completely damped in passing through them. The equations for the heat exchangers are solved analytically.

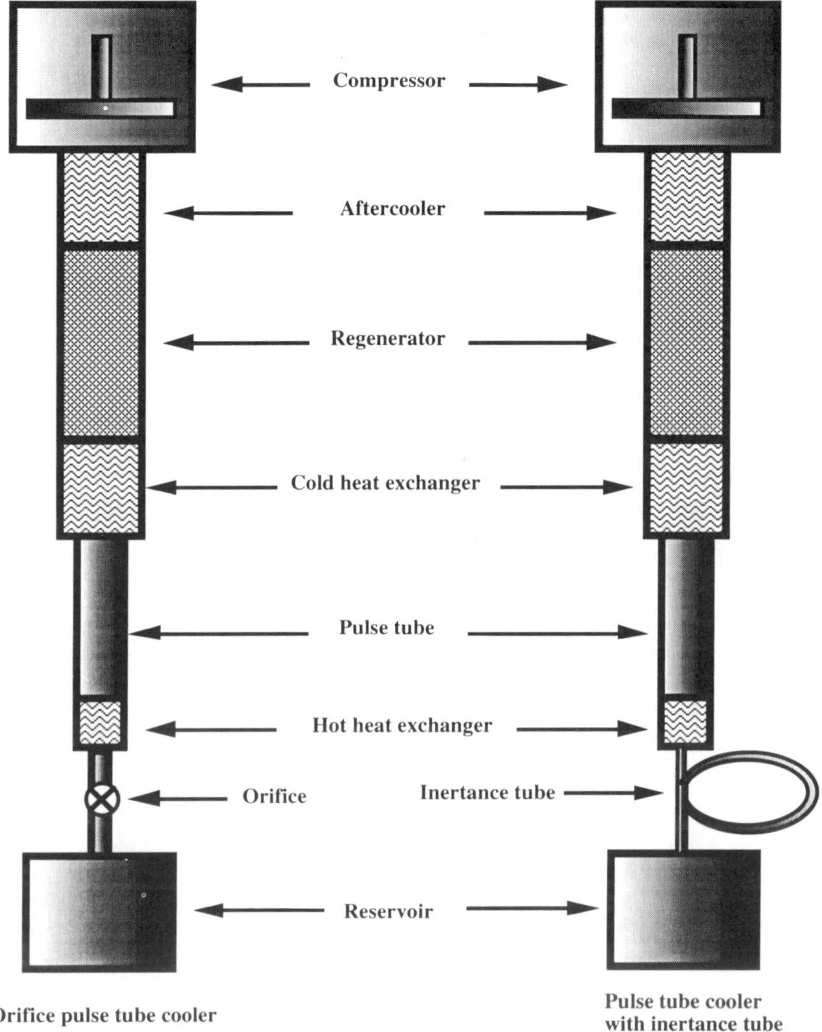


**Figure 1.** Schematics of an orifice pulse tube cooler and one with an inertance tube.

In the pulse tube there is no pressure gradient. The mass conservation and the energy conservation equations are used to arrive at an equation for the velocity. This equation is solved analytically assuming a linear steady temperature profile. However, the assumption, made for the regenerator, that the heat transfer to the wall is in phase with the temperature difference between the gas and the wall is no longer adequate since the gas in the center of the pulse tube can be many diffusion lengths from the wall. When that is the case it is possible to have the heat transfer occurring at a phase different from that of the phase of the temperature difference between the wall and the average temperature of the gas. The heat transfer coefficient in the pulse tube is based on a complex Nusselt number as discussed in an earlier report.[5]

In a basic pulse tube with no wall interaction, all the mass flow into the cold end goes into compressing the gas in the pulse tube. This mass flow will be 90° out of phase with both the pressure and the temperature and there will be no work flow or enthalpy flow at the cold end. Presence of an orifice adds a component of flow throughout the pulse tube that is in phase with the pressure and leads to an enthalpy flow that is the basis for the cooling in an orifice pulse tube cooler. If the orifice flow is too large, however, the pressure drop in the regenerator will yield a small pressure oscillation in the pulse tube, and result in little cooling. Clearly, there is an optimum orifice setting that produces the best cooling. ARCOPTR treats the orifice as an impedance with flow proportional to pressure; the flow is symmetrical with flow direction and has no dependence on velocity. Replacing the orifice with an inertance tube (a long and narrow tube) replaces a resistive impedance with a reactive one allowing more adjustment in the phase difference between the pressure and the mass flow in the pulse tube. This configuration is found to increase the pulse tube cooler efficiency.[6] In the present version of the ARCOPTR a pulse tube cooler with an isothermal inertance tube can be modeled for both the laminar and the turbulent flow regimes.[7] The mass flow rate and pressure oscillations in the reservoir are solved for analytically assuming a finite-volume adiabatic reservoir.

The equation for the regenerator is solved numerically using a 'shooting' method where the slope of the pressure at the warm end of the regenerator is adjusted until the pressure and mass flow at the cold end, as determined by integrating the differential equation, agree with the boundary conditions. The analytic solutions for the compressor and for the pulse tube provide relationships between pressure and mass flow at each end of the regenerator; these relationships (rather than fixed values of pressure or flow) make up the boundary conditions at each end of the regenerator. The analytic solutions for the heat exchangers modify these boundary conditions in a way that accounts for the pressure drop and the phase shift in the mass flow that occurs in the heat exchangers. The general form of the relationship between pressure and mass flow is not changed by the heat exchangers.

ARCOPTR is an easy program to use. It has a graphical interface and requires no guesses for the operating pressures or mass flows.

## OPTIMIZATION ROUTINE

To model a pulse tube cooler using the ARCOPTR it is required that all the cooler parameters be specified. Often, a user desires to know the optimum values of the cooler parameters that achieve maximum efficiency. To this end, an optimization routine has been built into the ARCOPTR that finds the optimum values for the critical parameters of a pulse tube cooler which maximize its efficiency. Based on the manual optimizations of pulse tube coolers performed using the ARCOPTR it is concluded that the efficiency of a pulse tube cooler is a continuous function of its parameters and has no local maxima. However, the efficiency is calculated by solving a non-analytic multi-variable function which in general is non-trivial to optimize. This is why the number of parameters varied in the optimization are limited to those considered critical in the design of a pulse tube cooler. These are the length and diameter of the regenerator, the length and diameter of the pulse tube, the mesh size and wire diameter of the regenerator screen, the orifice opening or the length and diameter of the inertance tube.

Steepest-ascent, a first-order optimization routine, was selected for the optimization problem at hand. This routine requires evaluation of the partial derivatives of the function with respect to its variables to obtain the direction of steepest ascent. The partial derivatives are numerically evaluated using a forward differencing technique. The finite-difference intervals used in these calculations are made proportional to the magnitudes of the variables. Since these variables are several orders different in magnitude the partial derivatives are scaled by appropriate scaling factors. Steps taken in the direction of ascent are adjusted automatically to increase the optimization speed.

The optimizer requires that for each variable an initial value, an upper limit, and a lower limit be specified. If both limits of a variable are set equal to its initial value that variable will become invariant. If during optimization a step takes a variable outside its specified limits the optimizer reduces the step length appropriately to remedy this problem. As the maximum is approached the difference intervals are reduced to increase the accuracy of calculations. The optimization terminates when the difference in the efficiencies calculated from two consecutive iterations is less than a preset limit.

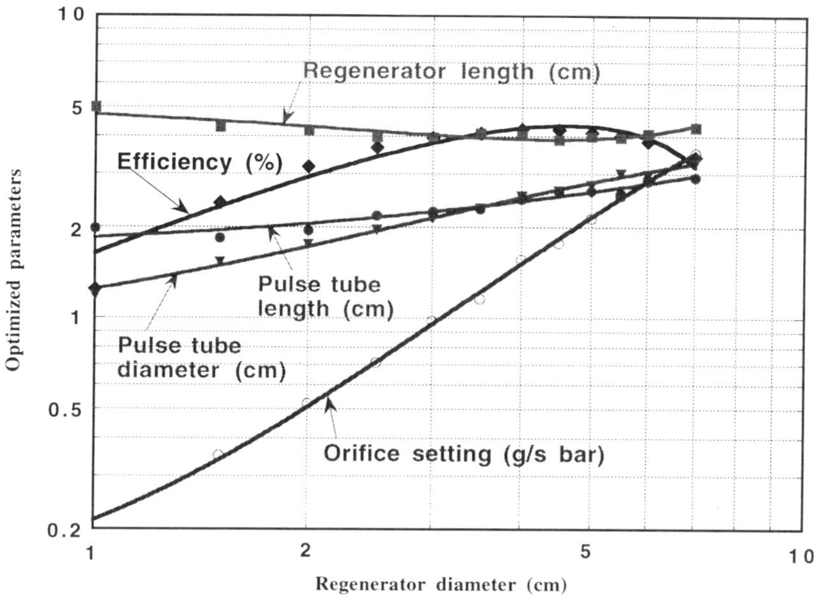


**Figure 2.** Optimized parameters for orifice pulse tube coolers.

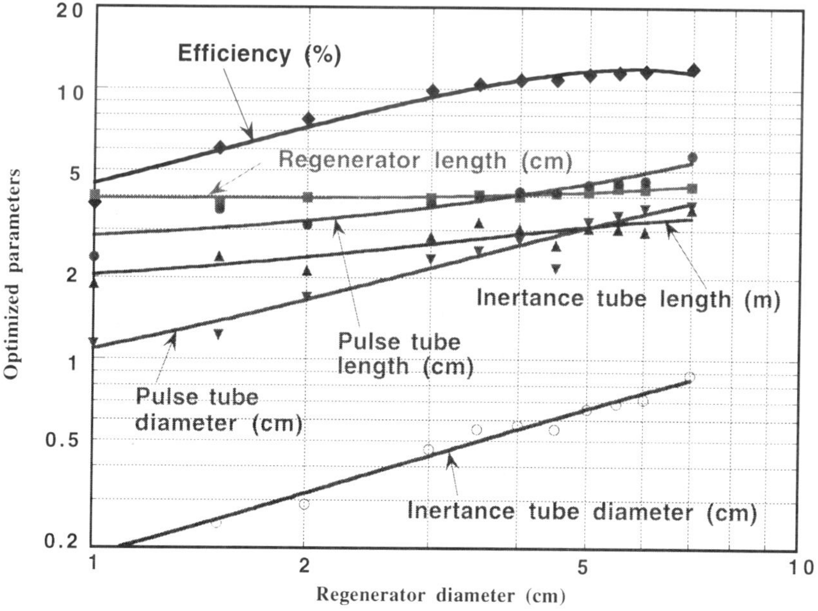


**Figure 3.** Optimized parameters for pulse tube coolers with an inertance tube.

**Table 1.** Parameters used in the calculations of optimized pulse tube coolers

| Parameter | Value |
|---|---|
| Frequency | 60 (Hz) |
| Ambient pressure | 20 (bar) |
| Hot end temperature | 300 (K) |
| Cold end temperature | 80 (K) |
| Aftercooler length | 7.0 (cm) |
| Aftercooler diameter | 0.4 * Regenerator diameter + 0.7 (cm) |
| Aftercooler screen wire diameter | 0.025 (cm) |
| Aftercooler screen mesh size | 40 (#/in) |
| Regenerator wall thickness | 0.0052 * Regenerator diameter + 0.0026 (cm) |
| Cold heat exchanger length | 1.5 cm |
| Cold heat exchanger diameter | 0.4 * Regenerator diameter + 0.7 (cm) |
| Cold heat exchanger screen wire diameter | 0.025 (cm) |
| Cold heat exchanger screen mesh size | 40 (#/in) |
| Pulse tube wall thickness | 0.0052 * Regenerator diameter + 0.0026 (cm) |
| Hot heat exchanger length* | 1.5 cm |
| Hot heat exchanger diameter* | 0.4 * Regenerator diameter + 0.7 (cm) |
| Hot heat exchanger screen wire diameter* | 0.025 (cm) |
| Hot heat exchanger screen mesh size* | 40 (#/in) |
| Orifice opening for coolers with an inertance tube | 50 (g/s bar), i.e., large opening |
| Reservoir volume | 300 (cc) |

* For orifice pulse tube coolers only. No hot heat exchanger is used for coolers with an inertance tube.

## RESULTS FROM THE OPTIMIZER

The ARCOPTR program with the built in optimizer is used to calculate optimum efficiencies for different size orifice pulse tube coolers and those employing an inertance tube. Results of the optimization calculations are presented graphically for fixed values of the regenerator diameter in Figures 2 and 3 for the orifice and the inertance tube configurations, respectively.

The regenerator screen mesh size is set at discrete values of 250, 325, and 400. Of these the 400 mesh size screen yields the highest efficiency for all of the regenerator diameters considered. The screen wire diameter is related to the screen mesh size as: screen wire diameter (cm ) = 1 / screen mesh size ( #/in.).

The pressure ratio in the pulse tube is maintained at 1.2 during the optimization by adjusting the compressor stroke. This is to limit the compressor stroke and thus the amplitude of the oscillations.

The fixed parameters used in all the cases considered are listed in Table 1. The diameters of the heat exchangers and the wall thickness of the regenerator and that of the pulse tube are scaled proportional to the regenerator diameter for each case.

## CONCLUSIONS

The pulse tube modeling program ARCOPTR with the built in optimizer can be used in the optimization of pulse tube coolers which employ an orifice or an inertance tube. The optimizer allows optimization of the regenerator and the pulse tube parameters along with those of the orifice or the inertance tube to achieve maximum efficiency. The optimizer has been employed to determine the optimum parameters of several pulse tube coolers with different size regenerator diameters. It is noted that the efficiency function is fairly flat and only weakly dependent on its variables as its peak is approached. In the cases considered the use of an inertance tube more than doubles the calculated efficiency of an orifice pulse tube cooler.

## REFERENCES


1. P.R. Roach, A. Kashani, and J.M. Lee, Theoretical analysis of a pulse tube regenerator, in: "Advances in Cryogenic Engineering," P. Kittel, ed., Plenum Press, New York (1995), p. 1357.
2. W.M. Kays, and A.L. London, "Compact Heat Exchangers," McGraw-Hill, New York (1955).
3. B.P.M. Helvensteijn, A. Kashani, A.L. Spivak, P.R. Roach, J.M. Lee, and P. Kittel, Pressure drop over regenerators in oscillating flow, Presented at CEC 1997.
4. T. Kuriyama, F. Kuriyama, M. Lewis, and R. Radebaugh, Measurement of heat conduction through stacked screens, in: "Cryocoolers 9," R.G. Ross, ed., Plenum Press, New York (1997), p. 459.
5. P.R. Roach, A. Kashani, A simple modeling program for orifice pulse tube cooler, in: "Cryocoolers 9," R.G. Ross, ed., Plenum Press, New York (1997), p. 327.
6. S.W. Zhu, S.L. Zhou, N. Yoshimura, and Y. Matsubara, Phase shift effect of the long neck tube for the pulse tube refrigerator, in: "Cryocoolers 9," R.G. Ross, ed., Plenum Press, New York (1997) ), p. 269.
7. P.R. Roach, A. Kashani, Pulse tube coolers with an inertance tube, Presented at CEC 1997.

# PULSE TUBE REFRIGERATOR BASED ON FLUID INERTIA

**K. V. Ravikumar and Y. Matsubara**

Atomic Energy Research Institute, Nihon University
7-24-1 Narashino-dai, Funabashi, Chiba, 274
Japan

## ABSTRACT

Refrigeration effect in pulse tube coolers can be explained/understood in terms of phase shift between oscillations of pressure and displacement of gas particle at the cold end of the pulse tube. There are different mechanisms (means) to create the phase shift. Surface heat pumping and resistance to fluid flow in orifice are two examples of commonly used mechanisms to create needed phase shifts in basic pulse tube cooler and orifice pulse tube cooler respectively. At high frequency operation there is yet another mechanism possible due to fluid inertia that could effectively be used to create favorable phase shift. Results of experimental investigations of the new mechanism are presented. It is found in the case double inlet pulse tube cooler that effectiveness of second (double) inlet progressively diminished as operating frequency increased. It is also observed as the operating frequency increased that fluid inertia based phase shifter achieved lower temperatures than simple orifice and double inlet configurations in a certain frequency range.

## INTRODUCTION

At present there exists three main variations of pulse tube coolers (PTC), viz., basic pulse tube cooler (BPTC), orifice pulse tube cooler (OPTC) and double inlet orifice pulse tube cooler (DOPTC). There are several approaches by which one can explain or understand the operation of PTC. One such method, known as Phasor Analysis [1] tries to explain the functioning of PTC in terms of phase difference between oscillating quantities at the cold end of the pulse tube. In accordance with this view point heat transfer between the fluid and pulse tube wall, known as surface heat pumping effect [2], is the significant factor in creating the required phase difference in BPTC. In cases of OPTC and DOPTC, the needed phase shift, however, is produced by combined effects of resistance to fluid flow by orifice(s) and of capacitance of reservoir (buffer) volume [1,3].

Another type of phase shift mechanism based on fluid inertia is possible for PTC. Inertial effects of fluid become important either for high frequency operation or when the size of the cooler is increased to achieve higher cooling capacity. The above mentioned Phasor Analysis in its present formulation does not explicitly account for fluid inertia. Zhu et al. [4] studied the effects of fluid inertia on PTC performance by carrying out numerical simulations. Using fluid inertia effects to control the phase, it was shown in their studies, it is possible to achieve at the cold end of the pulse tube equivalent PV diagrams similar to that of Stirling cooler. Tominaga [3] and Gardner and Swift [5] studied using electrical analogy the fluid inertia effects on improving the phase between pressure and displacement/velocity. Godshalk et al. [6] operated OPTC with a thermo-acoustic pressure wave generator operating at 350Hz. In their design fluid inertia in the pulse tube and in the connecting tube between pulse tube and buffer played a important role in the performance of the cooler. Kanao et al. [7] reported high frequency operation of OPTC with capillary tube as phase shifter. During these experiments they noted that double inlet configuration did not yield better performance. Haruyama and Inoue [8] reported dependence of no load refrigeration temperature on frequency with length of capillary phase shifter as a parameter.

We present here experimental results demonstrating the effects of fluid inertia at high frequency on the operation of pulse tube refrigerator. In these experiments orifice of the PTC is replaced by a long tube. This long tube, hereafter, is referred as "neck-tube or NT" in the following discussion.

## PRELIMINARY STUDIES

Here a comparison is made of two groups of experiments with the purpose of demonstrating fluid inertia effects. In the first group of experiments a stainless steel thin walled pulse tube (ϕ19mmx120mm) is included between the regenerator (ϕ38mmx92mm, phosphor bronze, 200 mess size) and the copper neck-tube (ϕ4mmx2m). Also whole of the neck-tube is placed outside the vacuum space in thermal communication with the surroundings. A reservoir of 2250cc is connected at the other end of the neck tube. A linear compressor is used as a pressure wave generator. In the second group of experiments a

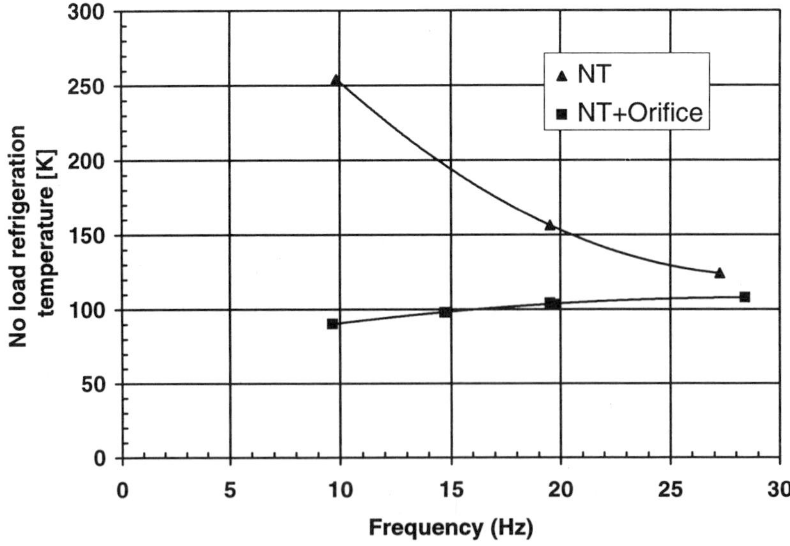


Figure 1. No load refrigeration temperature versus frequency - comparison of neck-tube phase shifter versus orifice + neck-tube combination as phase shifter showing that fluid inertia may be effectively used as phase shifter at higher frequencies. Input P-V power at the inlet of the regenerator is 150 Watts.

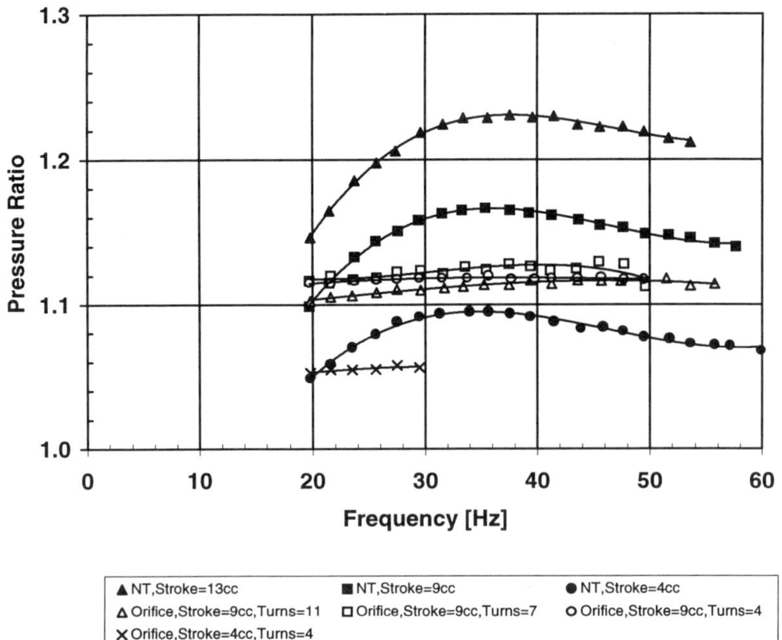


Figure 2. Pressure ratio versus frequency characteristics of neck-tube phase shifter and orifice phase shifter. Neck-tube size: $\phi$4mm x 4m. Orifice: needle valve (Nupro Company), $C_V$=0.03. Charge pressure: 1.58MPa.

valve is added between the pulse tube and neck-tube, every thing else remaining same as in the Group-I. The Group-I of experiments represent, in terms of electrical analogy, distributed resistance in regenerator and neck-tube, distributed inductance in pulse tube and neck-tube, and lumped capacitance in pulse tube and reservoir volumes. In addition to these the Group-II of experiments contain lumped resistance arising from a valve.

Figure (1) presents no load refrigeration temperature versus frequency for a fixed P-V power input at the inlet of the regenerator. The upper line is obtained by the use of neck-tube alone working as the phase shifter in Group-I experiments. The lower line is obtained by optimizing the valve opening for each frequency in Group-II experiments. It may be observed from the Figure (1) that at lower frequencies addition of an orifice, i.e., a pure resistive component, has greater influence in lowering the temperature than at higher frequencies. This data suggests by analogy to electrical circuits that at higher frequencies the inertia of the fluid is effective and is playing a greater role in achieving lower temperatures.

## CHARACTERISTICS OF TWO TYPES OF PHASE SHIFTERS

Refrigeration temperature and refrigeration capacity are influenced by pressure ratio, phase between oscillations of pressure and gas displacement and equivalent cold end PV work per unit of fluid mass displaced through the regenerator. The two type of phase shifters studied are neck-tube and orifice. In each case one end of the pulse tube is connected directly to a linear compressor with an after-cooler between them. The other end of the pulse tube is connected either to a neck tube phase shifter or to a adjustable orifice phase shifter. And a reservoir of 2250cc is connected to the other end of the phase shifter. Pressure at the compressor end of the pulse tube is recorded. Displacement of the compressor piston is measured using LVDT position sensor. Since the compressor end of the pulse tube represents the cold end of the pulse tube in an actual pulse tube cooler the

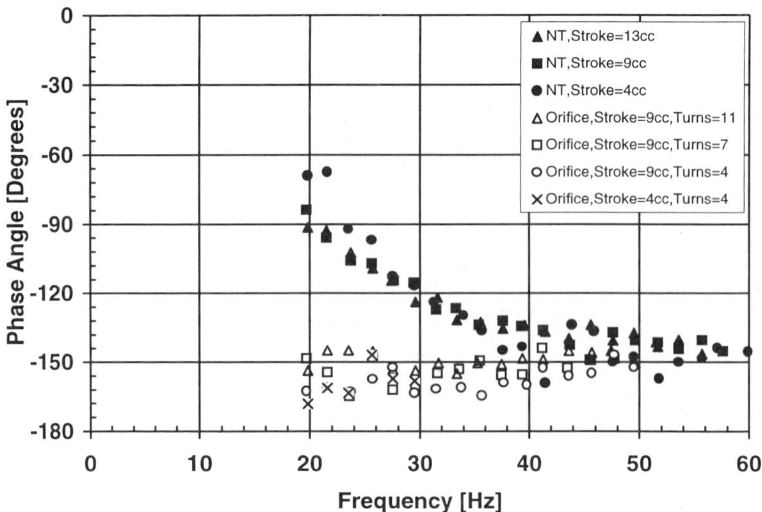


Figure 3. Phase angle versus frequency characteristics of neck-tube phase shifter and orifice phase shifter. Neck-tube size: ϕ4mm x 4m. Orifice: needle valve (Nupro Company), $C_V$=0.03. Charge pressure: 1.58MPa.

measured pressure and compressor piston displacement are indicative of the pressure and gas displacement at the cold end of the pulse tube. The fluid mass displaced is computed from the measured pressure and volume changes and assuming near isothermal compression process corresponding to after-cooler temperature. The phase is also derived from the measured piston displacement and pressure. In this work phase, $\phi$, is defined as the difference between the angles corresponding to swept volume peak, $\theta_v$, and pressure peak, $\theta_p$, i.e. $\phi = \theta_v - \theta_p$.

Figures (2), (3) and (4) show the quantitative comparison of the two type of phase shifters in terms of pressure ratio, phase between oscillations of pressure and gas displacement and PV work per unit of fluid mass displaced. Inner diameter and length of the

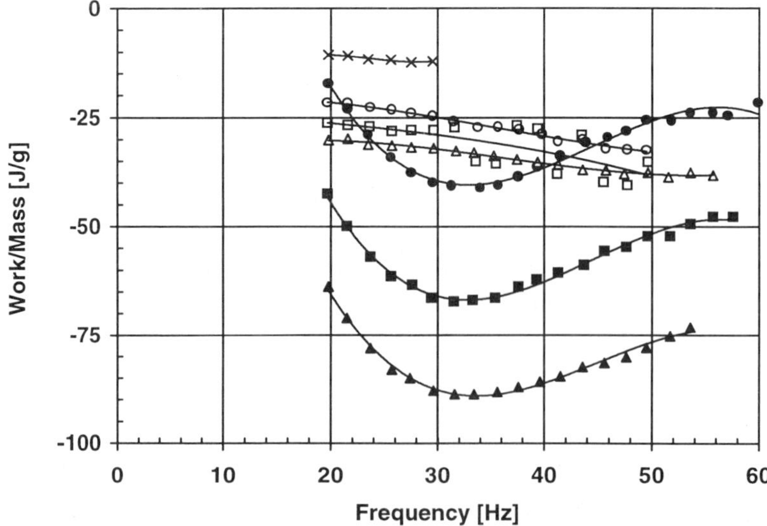


Figure 4. Work/unit mass displaced versus frequency characteristics of neck-tube phase shifter and orifice phase shifter. Neck-tube size: ϕ4mm x 4m. Orifice: needle valve (Nupro Company), $C_V$=0.03. Charge pressure: 1.58mpa. Please refer to the above figure for symbol identification.

neck-tube are 4mm and 4m respectively. The tube is made of copper. The pulse tube size is same as that given in the previous section. Figure (2) shows that in case of orifice phase shifter both the frequency and the orifice opening did not show strong influence on pressure ratio in the frequency range studied. This must be so can be understood from the considerations of volumes involved in the system. They are: 20cc of compressor dead volume, 34cc of pulse tube volume and 9cc of swept volume. These experiments were done at room temperature and all the components of the system were in good thermal communication with the surroundings. Assumptions of isothermal pressure change, negligible fluid inertia and completely closed orifice will cause a maximum pressure ratio of 1.166. Any amount of orifice opening will result in lower pressure ratio due to the gas flow into the 2250cc reservoir through orifice.

In case of the neck-tube phase shifter the pressure ratio has a maximum around 35Hz. It can also be noted by comparing the data corresponding to compressor stroke volume of 9cc that above the frequency of 20Hz there is a marked gain in pressure ratio for neck-tube phase shifter over that of orifice phase shifter. Swept volume has strong effect on the magnitude of pressure ratio. The following viewpoint may qualitatively explain the observed behavior. This viewpoint originates from the observed similarities of the processes in the neck-tube system and forced oscillations of a damped spring mass system. The oscillating gas displacer is acted upon by forces due to gas spring in the reservoir volume, viscous forces in the neck-tube and compressor piston. In the case of spring-mass system the displacement amplitude of the mass as a function of frequency peaks at a frequency different from the natural frequency of the system when the damping factor, i.e., ratio of damping coefficient to critical damping coefficient is less than the critical value of $0.5^{0.5}$ and different from zero. Under these conditions the phase is different from -90 degrees. Phase angle of -90 degrees represent the operation of the system when forcing frequency equals natural frequency of the system, and this condition also represents where external force is exactly balanced by viscous damping forces in the system. Therefore, the observed variations of pressure ratio in Figure (2) and phase angle in Figure (3) of the neck-tube system suggests that the damping factor in the system is less than the critical value and the forcing frequencies corresponding to the pressure peaks are different from the natural frequency of the system.

Equivalent PV work at the cold end of the pulse tube per unit of mass displaced through the regenerator is an indicator of the performance of the PTC. Bigger is this number the better. As can be noted from Figure (4) the neck-tube system promises better performance when operated in a certain restricted frequency range.

## PERFORMANCE COMPARISON

In this section performance of a pulse tube cooler with neck-tube as phase shifter is compared with that of OPTC and DOPTC. For a realistic comparison all the experiments are performed with the same regenerator and pulse tube changing only the phase shifter. A rotary valve was used as pressure wave generator. The data points were taken at optimum orifice opening in OPTC and DOPTC experiments. The neck-tube has length and inner diameter of 4m and 4mm respectively. Regenerator and pulse tube sizes are same as those given in the earlier sections. Figure (5) compares no load refrigeration temperature as a function of operating frequency of the cooler. It is generally believed that double inlet configuration improves the performance of the cooler over single orifice configuration at high frequencies. Double inlet configuration has no doubt greatly improved the performance of the cooler over the single orifice configuration in a range of low frequencies as shown in the Figure (5). The effectiveness of the double inlet configuration, however, quickly

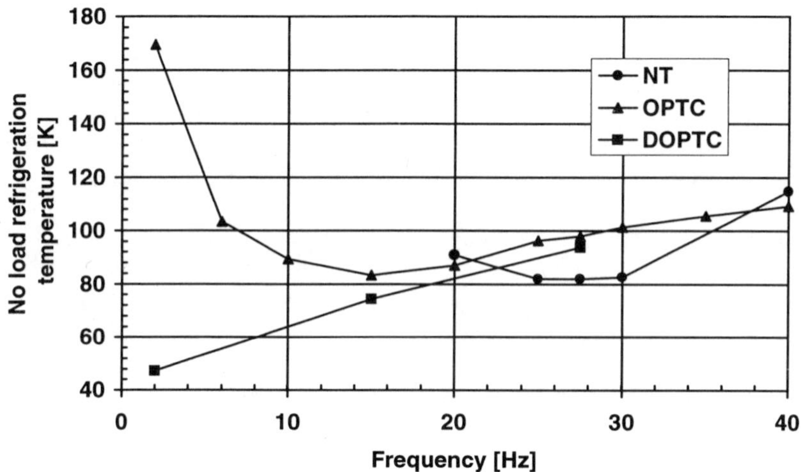


Figure 5. No load refrigeration temperature versus frequency - comparison of OPTC, DOPTC and PTC with neck-tube as phase shifter.

dropped as the operating frequency increased. This kind of behavior was also noted by Ravex et al [9]. These results, therefore, call for careful studies to determine the extent of benefits of double inlet configuration over single inlet configuration at high frequencies. As noted above in the introduction Kanao et al [7] observed similar behavior at high frequency operation of their cooler.

As the operating frequency increased neck-tube phase shifter started to show better performance. For the particular system studied as shown in the Figure (5) there is a range of frequencies where neck-tube phase shifter is better than the other two phase shifters. The optimum frequency in Figure (5) is for practical purposes same as the frequency from Figure (4) corresponding to the maximum equivalent PV work per unit of mass displaced. Slight difference is believed to come from the addition of regenerator volume, adiabatic versus isothermal wall conditions on pulse tube and pressure drop and heat transfer in the regenerator.

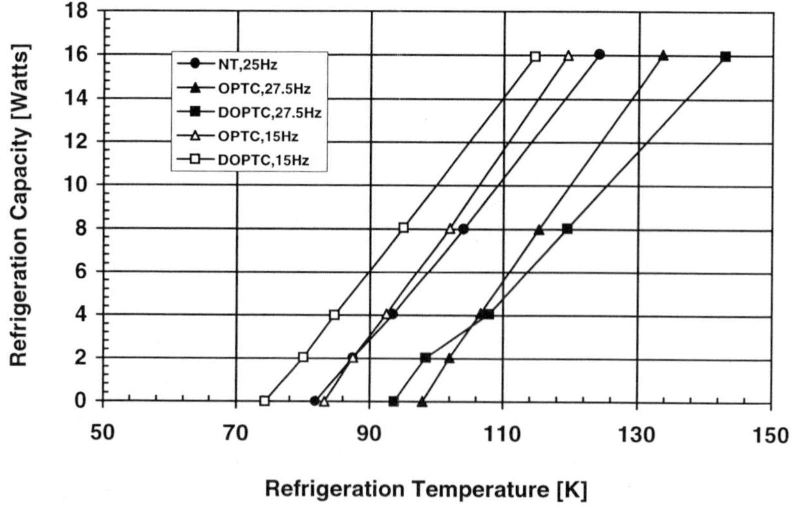


Figure 6. Refrigeration capacity versus refrigeration temperature - comparison of OPTC, DOPTC and PTC with neck-tube as phase shifter.

For a given pulse tube cooler system as the operating frequency is increased the thermal penetration depth in the regenerator material decreases. As a result the effective thermal capacity in the regenerator is reduced. Further, mass flow rate and pressure drop in the regenerator increases with frequency. Another effect is that as the frequency increased the pressure ratio at the inlet of the regenerator dropped irrespective of whether the system is OPTC, DOPTC or PTC with neck-tube as phase shifter. This drop in pressure ratio as frequency increased is a characteristic of rotary valve arising from its poor performance at high frequency. The experimental results suggest that in a limited range of high frequency the fluid inertial effects arising from the neck-tube phase shifter could more than offset the losses due to the regenerator poor performance and due to the reduced pressure ratio at the inlet of the regenerator. The observed improved performance of the neck-tube phase shifter over the other two phase shifters in the limited frequency range is believed to come from the increased pressure ratio at the cold end of pulse tube and increased work per unit mass displaced through the regenerator as evidenced in Figures (2) and (4).

Figure (6) presents refrigeration capacity versus refrigeration temperature for different phase shifters. In case of OPTC and DOPTC data are obtained at two different frequencies, viz., 15Hz and 27.5Hz. First frequency corresponds to the optimum frequency of OPTC system and the latter corresponds to the optimum frequency of PTC with neck-tube phase shifter.

## CONCLUSIONS

It is observed that at higher frequencies the benefits of DOPTC over OPTC are not as great as those at low frequencies. These results, therefore, call for careful studies to determine the actual extent of benefits of double inlet configuration over single inlet configuration at high frequencies.

It is shown that fluid inertia could be effectively used as a phase shifter for PTC operation. From simple considerations it can be shown that fluid inertia is proportional to square of the neck-tube inner diameter and inversely proportional to its length. The range of frequencies where fluid inertia becomes effectively useful over other phase shifters can, in principle, be selected by choosing proper length and inner diameter of the neck-tube. Further studies are needed to understand the fluid inertia effects, and its extent of usefulness to high frequency operation of pulse tube cooler.

**ACKNOWLEDGMENT.** First author thankfully acknowledges the National Science Foundation for supporting the research with a fellowship under Center for Global Partnership program.

## REFERENCES

1. P.J. Storch and R. Radebaugh, Development and Experimental Test of an Analytical Model of the Orifice Pulse Tube Refrigerator, Advances in Cryogenic Engineering, Vol. 33, p: 851 (1987)
2. W.E. Gifford and R.C. Longsworth, Surface Heat Pumping, Advances in Cryogenic Engineering, Vol. 11, p: 171 (1965)
3. A. Tominaga, Phase Controls for Pulse-Tube Refrigerator of the Third Generation, Cryogenic Engineering, Vol. 27, No. 2, p: 146 (1992) (in Japanese)
4. S.W. Zhu, S.L. Zhou, N. Yoshimura and Y. Matsubara, Phase Shift Effect of the Long Neck Tube for the Pulse Tube Refrigerator, in Proceedings of the 9th International Cryocooler Conference, held June 1996, in New Hampshire (to be published)

5. D.L. Gardner and G.W. Swift, Use of Inertance in Orifice Pulse Tube Refrigerators, (submitted to Cryogenics)
6. K.M. Godshalk, C. Jin, Y.K. Kwong, E.L. Hershberg, G.W. Shift and R. Radebaugh, Characterization of 350Hz Thermoacoustic Driven Orifice Pulse Tube Refrigerator with Measurements of the Phase of Mass Flow and Pressure, Advances in Cryogenic Engineering, Vol. 41B, p: 1411 (1996)
7. K. Kanao, N. Watanabe and Y. Kanazawa, A Miniature Pulse Tube Refrigerator for temperature Below 100K, Cryogenics, Vol. 34, ICEC Supplement, p:167 (1994)
8. T. Haruyama and H. Inoue, Cooling Characyeristics of a Modified Miniature Pulse Tube Refrigerator, Advances in Cryogenic Engineering, Vol. 41B, p: 1427 (1996)
9. A. Ravex, P. Rolland and J. Liang, Experimental Study and Modelisation of a Pulse Tube Refrigerator, Cryogenics, Vol. 32, ICEC Supplement, p:9 (1992)

# THERMODYNAMICAL ASPECTS OF PULSE TUBES

A.T.A.M. de Waele and P.P. Steijaert

Department of Physics, Eindhoven University of Technology
P.O. Box 513, NL-5600 MB Eindhoven, The Netherlands

## ABSTRACT

In this paper we treat the fundamental aspects of the single-orifice pulse tube. This is also the basis for the operation of pulse tubes with more valves and/or inlets. We discuss the dynamic behaviour of the temperature profiles in the regenerator and in the gas in the tube near the hot and cold ends. Furthermore the operation of single-orifice pulse tubes is decribed in general thermodynamical terms.

## INTRODUCTION

In this paper we will consider the single-orifice pulse tube, which is schematically represented in Fig.1, with a harmonic time dependence of the pressure

$$p = p_0 + p_1 \sin \omega t \quad \text{with} \quad p_0 \gg p_1 . \tag{1}$$

We will use various expressions derived in our paper in Cryogenics.[1] We assume that the fluid is an ideal gas, so the molar entropy is given by

$$S_{\mathrm{m}}(p,T) = S_{\mathrm{m}}(p_0, T_0) + \frac{C_{\mathrm{p}}}{T}(T - T_0) - R\frac{p_1}{p_0} \sin \omega t, \tag{2}$$

the molar enthalpy by

$$H_{\mathrm{m}}(p,T) = C_{\mathrm{p}} T, \tag{3}$$

and the molar flow rate density by

$$j = \frac{\dot{n}}{A} = \frac{p_0 v}{RT} \tag{4}$$

where $A$ is the area of the cross section of the flow channel and $v$ the velocity.

## REGENERATOR DYNAMICS

From conservation of energy the following relation can be obtained (see Ref.1. Eqs.(31) and (35))

$$\frac{(1-f)C_{\mathrm{p}}}{V_{\mathrm{m}}}\frac{\partial T_{\mathrm{g}}}{\partial t}+C_{\mathrm{r}}\frac{\partial T_{\mathrm{r}}}{\partial t}+jC_{\mathrm{p}}\frac{\partial T_{\mathrm{g}}}{\partial l}+j\left(V_{\mathrm{m}}-T_{\mathrm{g}}\left(\frac{\partial V_{\mathrm{m}}}{\partial T_{\mathrm{g}}}\right)_{p}\right)\frac{\partial p}{\partial l}$$

$$-\frac{(1-f)T_{\mathrm{g}}}{V_{\mathrm{m}}}\left(\frac{\partial V_{\mathrm{m}}}{\partial T_{\mathrm{g}}}\right)_{p}\frac{\partial p}{\partial t}-\frac{\partial}{\partial l}\left(\kappa_{\mathrm{g}}\frac{\partial T_{\mathrm{g}}}{\partial l}\right)-\frac{\partial}{\partial l}\left(\kappa_{\mathrm{r}}\frac{\partial T_{\mathrm{r}}}{\partial l}\right)=0, \quad (5)$$

where $f$ is the matrix filling factor, $C_{\mathrm{r}}$ the matrix specific heat per unit volume, the gas and matrix temperatures are $T_{\mathrm{g}}$ and $T_{\mathrm{r}}$ respectively and the other parameters have their usual meaning. We consider an ideal regenerator in which the thermal conductivities are zero (neglect 6th and 7th term), the flow resistance is zero (neglect 4th term), the volumetric heat capacity of the gas much smaller than $C_{\mathrm{r}}$ (neglect first and 5th term), and, finally, there is ideal heat exchange so $T_{\mathrm{g}} = T_{\mathrm{r}} = T$. In this way in Eq.(5) only the second and third term remain. It can be further simplified with the minus-one rule to

$$\left(\frac{\partial l}{\partial t}\right)_{T}=j\frac{C_{\mathrm{p}}}{C_{\mathrm{r}}}=v_{\mathrm{T}}(T). \quad (6)$$

The velocity $v_{\mathrm{T}}$ is the velocity with which a temperature $T$ moves through the regenerator. Its value is proportional to the molar flow rate density and the ratio of the specific heats of the gas and the matrix, which is a function of $T$ only. So if $j$ is constant in the regenerator, all points with the same temperature move through the matrix with the same speed. If $\oint j dt = 0$ the $T$-profile only moves back and forth.

At the hot end the temperature of the gas entering the regenerator is $T_{\mathrm{H}}$ during half the cycle. The temperature $T_{\mathrm{H}}$ penetrates the regenerator over a certain distance. The next half of the cycle it is pushed back out again. At the cold end the same happens with $T_{\mathrm{L}}$. Usually $C_{\mathrm{r}}(T_{\mathrm{L}}) \ll C_{\mathrm{r}}(T_{\mathrm{H}})$ so the distance travelled during a half cycle will be larger at low temperature. As a result the temperature profile in the regenerator moves back and forth as schematically represented in Fig 2.

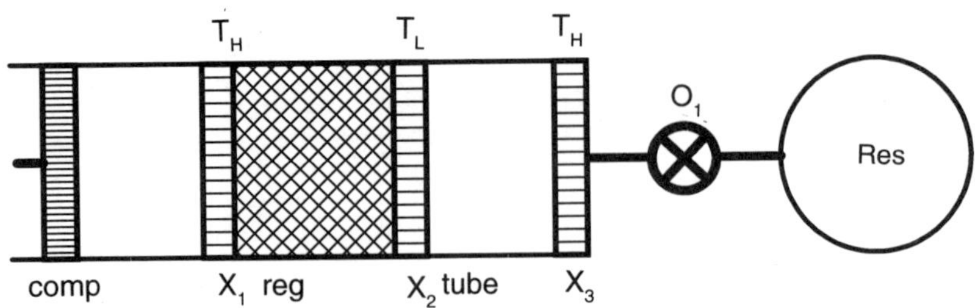


**Figure 1.** Schematic diagram of a Stirling type single-orifice pulse tube. From left to right the system consists of a compressor (comp), the after cooler ($X_1$), a regenerator (reg), a low-temperature heat exchanger ($X_2$), a pulse tube (tube), a second room-temperature heat exchanger ($X_3$), an orifice ($O_1$), and a reservoir (Res). The cooling power is generated at the low temperature $T_{\mathrm{L}}$. Room temperature is $T_{\mathrm{H}}$.

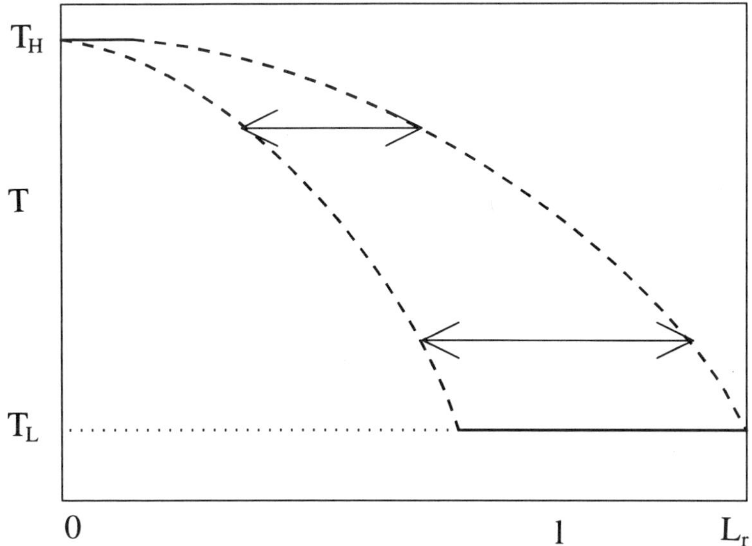


**Figure 2.** Schematic representation of the time dependence of the temperature profile in the regenerator. At the warm end a region of constant temperature $T_H$ moves into and out of the regenerator. At the cold end the same holds for $T_L$. The dotted curve is a qualitative continuous connection between these two regions.

## TEMPERATURE PROFILES INSIDE THE TUBE

The amplitude of the volume flow through the first orifice is

$$V_1 = \frac{p_1}{R_1}, \tag{7}$$

where $R_1$ is the flow resistance of the orifice. The velocity $v_H$ of the gas at the warm end of the tube is

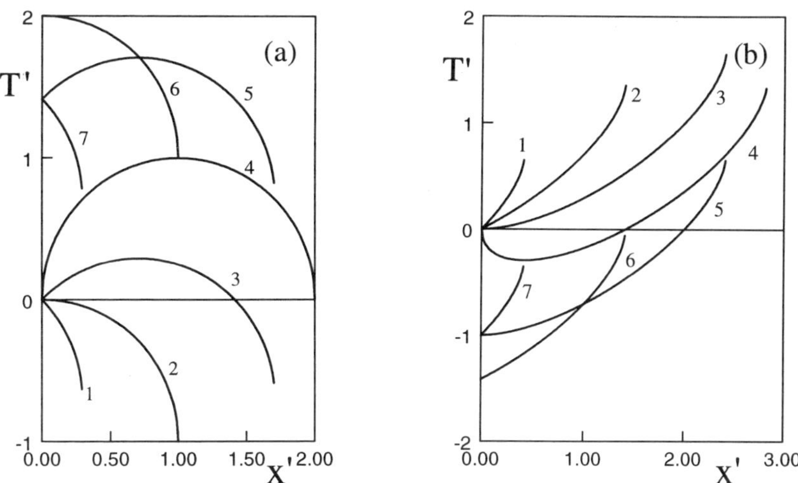


**Figure 3.** $T' - x'$ profiles in the tube at eight time intervals (a) at the warm end; (b) at the cold end.

$$v_{\mathrm{H}} = \frac{V_1}{A} \sin \omega t. \tag{8}$$

The position of a gas element, leaving the heat exchanger at $x = L$ on a time $t_0$, is given by

$$x = L - \frac{V_1}{A\omega}(\cos \omega t - \cos \omega t_0). \tag{9}$$

If $v_{\mathrm{H}} < 0$ then the gas element leaves $X_3$ with a temperature $T_{\mathrm{H}}$ and moves with constant entropy afterwards (Eq.(2)). Its temperature at time $t$ is

$$T = T_{\mathrm{H}} + T_{\mathrm{H}} \frac{R}{C_{\mathrm{p}}} \frac{p_1}{p_0}(\sin \omega t - \sin \omega t_0). \tag{10}$$

The essence of the time dependencies is contained in the dimensionless factors for the position

$$x' = \cos \omega t - \cos \omega t_0 \tag{11}$$

and for the temperature

$$T' = \sin \omega t - \sin \omega t_0. \tag{12}$$

Elimination of $t_0$ gives the temperature profile in the tube in dimensionless form

$$1 = (T' - \sin \omega t)^2 + (x' - \cos \omega t)^2. \tag{13}$$

The $T'$-profile is the positive branch of the solution of Eq.(13) for $x' > 0$. It is a segment of a circle with its centre moving on a circle as represented in Fig.3a. This property has been discussed before by Radebaugh[2] and Kittel[3]. The dimensionless temperature $T'(x')$ at the warm end ($x' = 0$) is

$$T'(0) = 0 \text{ for } v_{\mathrm{H}} < 0 \quad \text{and} \quad T'(0) = 2 \sin \omega t \text{ for } v_{\mathrm{H}} > 0. \tag{14}$$

The average heat removed per second in $X_3$ is

$$\dot{Q}_{\mathrm{H}} = \frac{1}{2} p_1 V_1. \tag{15}$$

The velocity of the gas near $X_2$ is the sum of $v_{\mathrm{H}}$ and the length variation of the gas piston

$$v_{\mathrm{L}} = v_{\mathrm{H}} + \frac{C_{\mathrm{V}}}{C_{\mathrm{p}}} \frac{L}{p_0} \frac{\mathrm{d}p}{\mathrm{d}t} = \frac{V_1}{A}(\sin \omega t + \alpha \cos \omega t) \quad \text{with} \quad \alpha = R_1 A \frac{C_{\mathrm{V}}}{C_{\mathrm{p}}} \frac{L}{p_0} \omega. \tag{16}$$

The position of a gas element, leaving $X_2$ on $t = t_0$, is

$$x = \frac{V_1}{A\omega}(-\cos \omega t + \cos \omega t_0 + \alpha(\sin \omega t - \sin \omega t_0)). \tag{17}$$

If $v_{\mathrm{L}} > 0$ the temperature of the gas element at time $t$ (Eq.(2)) is

$$T = T_{\mathrm{L}} + T_{\mathrm{L}} \frac{R}{C_{\mathrm{p}}} \frac{p_1}{p_0}(\sin \omega t - \sin \omega t_0). \tag{18}$$

With Eq.(12) and

$$x' = -\cos \omega t + \cos \omega t_0 + \alpha T', \tag{19}$$

eliminating $t_0$ as before, we obtain

$$1 = (T' - \sin \omega t)^2 + (x' + \cos \omega t - \alpha T')^2. \tag{20}$$

This $T'$-profile is a segment of a tilted ellipse through the origin with its centre moving over an ellipse (Fig.3b). The dimensionless temperature of the gas near $X_2$ $(x' = 0)$ is

$$T' = 0 \quad \text{for} \quad v_{\rm L} > 0 \quad \text{and} \quad T' = \frac{2}{1+\alpha^2}\frac{A}{V_1}v_{\rm L} \quad \text{for} \quad v_{\rm L} < 0. \tag{21}$$

From Eq.(16) and Eq.(21) the cooling power can be calculated to be

$$\dot{Q}_{\rm L} = \frac{1}{2}p_1 V_1 . \tag{22}$$

## GENERAL THERMODYNAMIC CONSIDERATIONS

The entropy production rate in the orifice (Ref.1, Eq.(123)) is

$$\dot{S}_{\rm O1} = \frac{p_1 V_1}{2T_{\rm H}}. \tag{23}$$

Mass conservation in the regenerator gives that the velocity of the compressor piston is

$$v_{\rm p} = \frac{T_{\rm H}}{T_{\rm L}}v_{\rm L} . \tag{24}$$

So the work performed by the piston per second is

$$\dot{W}_{\rm p} = \frac{1}{t_{\rm c}}\int_0^{t_{\rm c}} p_1 \sin\omega t \; v_{\rm p} A dt, \tag{25}$$

where $t_{\rm c}$ is the cycle time of the cooler. With Eq.(16) and Eq.(24)

$$\dot{W}_{\rm p} = \frac{T_{\rm H}}{T_{\rm L}}\frac{1}{2}p_1 V_1 . \tag{26}$$

In the following discussion we will frequently use the first and second law for open systems in the *steady state* ($\dot{U} = 0$ and $\dot{S} = 0$)

$$0 = \Sigma\dot{Q} + \Sigma H^*, \tag{27}$$

where $\Sigma\dot{Q}$ is the sum of the heat flows into the system. The average net enthalpy flow into the system, $\Sigma H^*$, is due to gas flow through the system boundaries. For one boundary it is given by

$$H^* = \frac{1}{t_{\rm c}}\int_0^{t_{\rm c}} \dot{n} H_{\rm m} \, {\rm d}t. \tag{28}$$

The second law reads

$$0 = \Sigma\frac{\dot{Q}}{T} + \Sigma S^* + \dot{S}_{\rm i} . \tag{29}$$

Here $\Sigma S^*$ is the sum of the average entropy flow into the system, due to gas flow, defined analogous to Eq.(28), and $\dot{S}_{\rm i}$ the entropy production rate. We will split up the pulse tube in subsystems in such a way that only two terms play a role in Eq.(27) and Eq.(29). In our approximation entropy is produced *only in the orifice*. In all the other subsystems $\dot{S}_{\rm i} = 0$. The heat flows $\dot{Q}$ are only nonzero in the heat exchangers.

We will discuss the various subsystems of the pulse tube going from right to left (Fig.4). At the reservoir side of the orifice $H^* = 0$ and $S^* = 0$ since $p$ and $T$ are constant here. In between the orifice and $X_3$ the temperature is constant so $H^* = 0$ but $S^* \neq 0$. In the orifice entropy is produced with rate $\dot{S}_{O1}$ . In the gas piston of the pulse tube the gas moves back and forth isentropically, so $S^* = 0$ here. The second law, for the system consisting of the orifice and $X_3$, gives

$$\dot{Q}_H = T_H \dot{S}_{O1} . \tag{30}$$

The gas temperature in the tube is high when it flows to the right and low when it moves to the left so enthalpy flows from $X_2$ towards $X_3$. The first law, for the system containing the tube and its two heat exchangers, gives

$$\dot{Q}_L = \dot{Q}_H . \tag{31}$$

If the heat capacity of the regenerator material is so large that $v_T \approx 0$ then the temperature in the regenerator is constant in time. As a result of Eq.(3) $H^* = 0$. However, the entropy flow in the regenerator $S^*_r$ is nonzero. It is directed from the *cold* to the *warm* end since the gas moves to the right when the pressure is high (low entropy, see Eq.(2)) and to the left when the pressure is low (entropy high). In the ideal case $S^*_r$ is constant. The second law for $X_2$ gives

$$\dot{Q}_L = T_L S^*_r . \tag{32}$$

The gas in the adiabatic compressor moves back and forth isentropically, so $S^* = 0$ here. As the gas temperature is high when it flows to the right and low when is moves to the left enthalpy flows from the piston to $X_1$. In $X_1$ heat is removed ($\dot{Q}_1 < 0$), so enthalpy (Eq.(27)) and entropy (Eq.(29)) have to flow towards $X_1$. The second law for $X_1$ prescribes

$$\dot{Q}_1 = T_H S^*_r . \tag{33}$$

The first law for the system containing the compressor and $X_1$

$$\dot{Q}_1 = \dot{W}_p . \tag{34}$$

Combining these equations gives the well known result for the COP of pulse tubes[4]

$$\frac{\dot{Q}_L}{\dot{W}_p} = \frac{T_L}{T_H} . \tag{35}$$

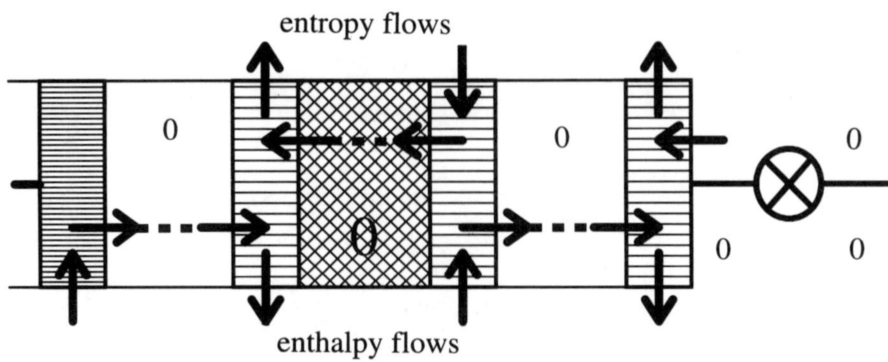


**Figure 4.** Diagram with the entropy flow (upper half) and the enthalpy flows (lower half). The zeros indicate that the flow in question is zero in this part of the system.

The general picture is summarised in Fig.4. The operation of the pulse tube can be understood by splitting it up in a number of subsystems of which the heat flows and the entropy and enthalpy flows are prescribed by the first and second law. One can start the argument at any point of the system. The generality of the first and second law prescribe that many of the expression given here will also be valid in the case of a nonideal gas.

## ACKNOWLEDGMENT

This work is supported by the Dutch Technology Foundation (STW).

## REFERENCES

1. A.T.A.M. de Waele, P.P. Steijaert, and J. Gijzen, *Cryogenics* 37: 313 (1997).
2. P.J. Storch, R. Radebaugh, and J.E. Zimmerman; NIST Technical Note 1343 (1990).
3. P. Kittel, Advances in Cryogenic Engineering, Vol. 42A.
4. P. Kittel, *Cryogenics* 32: 843 (1992).

# THE TEMPERATURE PROFILE WITHIN PULSE TUBES

Peter Kittel

NASA, Ames Research Center
Moffett Field, CA 94035-1000, USA

## ABSTRACT

A simple model is developed of the temperature profile within the pulse tube section of Pulse Tube Refrigerators. Using the Lagrangian reference frame, this simplified model directly gives the temperature overshoot at the hot and cold ends. The overshoots are the result of the interaction of the gas flow and the heat exchangers. They result from the work flow in the gas being converted to heat flow prior to reaching the heat exchangers. The mean temperature overshoot at the hot end is $\Delta T \approx {}^{\pi}/_{2}\, T_1$, where $T_1$ is the amplitude of the temperature oscillation within the pulse tube. At the cold end, the mean temperature overshoot is $\Delta T \approx -({}^{\pi}/_{2})\,(\dot{Q}_{net}/\dot{Q}_{gross})\, T_1$, where $\dot{Q}_{net}$ and $\dot{Q}_{gross}$ refer to the net and gross cooling powers respectively. Away from the ends, the temperature gradient is primarily due to thermal conductivity. The temperature overshoot effects are excluded in many models developed in the Eulerian reference frame because these models often explicitly exclude the changes in mean gas temperature that occur at the entrance to the heat exchangers.

## INTRODUCTION

It has been observed that the gas temperature near the ends of pulse tubes overshoot the temperature of the heat exchanger.[1] Near the hot heat exchanger, the gas temperature exceeds the heat exchanger temperature; while, near the cold end, the gas temperature is lower than the heat exchanger temperature. These features are usually missing from models developed in the Eulerian reference frame.[2] In this reference frame, the observer is fixed in space while the fluid flows past. Models developed in the Lagrangian reference frame usually show these features.[3] In the Lagrangian reference frame, the observer moves with an element of gas. There is a one-to-one correspondence between the two approaches. Complete models developed in either reference frame will produce the same results. However, the models emphasize different thermodynamic quantities. Eulerian models deal with enthalpy and entropy flows, while Lagrangian models emphasizes work and heat flows. The biggest differences appear in incomplete models. Practical models are based on making approximations such as truncating expansion series, ignoring harmonics, or assuming small amplitudes. Such approximations have different effects in different reference frames. Thus, practical Eulerian and Lagrangian models do not show all the same features.

A common assumption in Eulerian models of pulse tubes is to ignore the details of the heat exchanger to pulse tube interface. Such models inherently ignore the temperature overshoots. The models typically have a constant or near constant enthalpy flow in the pulse tube and an equivalent heat flow between the heat exchanger and the external world. The mean temperature of the gas on the pulse tube side of the interface is assumed to equal the heat exchanger temperature. No consideration is given to how the enthalpy flow becomes heat flow. A previous paper discussed this transition in terms of entropy production.[4] The model presented here uses the Lagrangian frame which lends itself to emphasizing this transition. The model considers the evolution of gas elements that spend part of the cycle in the heat exchanger and part of the cycle in the pulse tube.

## SIMPLE MODEL

The model will be developed in the Lagrangian reference frame. Only the pulse tube and the two heat exchangers at the pulse tube ends will be included. The other components of an Orifice Pulse Tube Refrigerator are only considered to the extent that they define the boundary conditions at the heat exchangers. Initially, thermal conductivity will be ignored in the pulse tube and heat transfer within the heat exchangers is assumed perfect. The heat exchange between the gas and the pulse tube wall is ignored; i.e., the gas is adiabatic. Later, axial thermal conduction will be added to the pulse tube. To further simplify the model, all quantities will be expanded into Fourier series and approximated by keeping only the time invariant and fundamental frequency terms, a small amplitude assumption. The working fluid is assumed to be an ideal monatomic gas.

In this frame the motion of individual gas elements are tracked as they move. Because of the repetitive behavior of the system, once steady state operation is reached, each element retraces the same path on successive cycles. For the hot end, the resulting temperature-displacement path is shown in Figure 1. When an element leaves the heat exchanger, it is at the temperature of the heat exchanger. Within the pulse tube pressure changes (in the absence of heat transfer) results in an adiabatic temperature change; thus, the temperature oscillation is in phase with the pressure oscillation. The orifice at the hot heat exchanger forces the pressure and velocity to be in phase. For sinusoidal oscillations, the displacement of an element is 90° out of phase to the velocity. Thus, inside the pulse tube, the displacement and temperature are 90° out of phase. When the element returns to the heat exchanger, its temperature returns to that of the heat exchanger. (This is the result of assuming perfect heat transfer within the heat exchanger.) Finally, within the heat exchanger there is no temperature oscillation; although, the pressure and displacement continue to evolve. Even though different elements of gas leave the heat exchanger at different times, their temperature-displacement paths are all anchored to the temperature of the heat exchanger at the interface with the pulse tube.

At the cold heat exchanger, a similar process takes place. Except there the pressure and velocity are no longer in phase. There is an additional component of velocity proportional to dP/dt. This component results from the pressurization of the pulse tube.

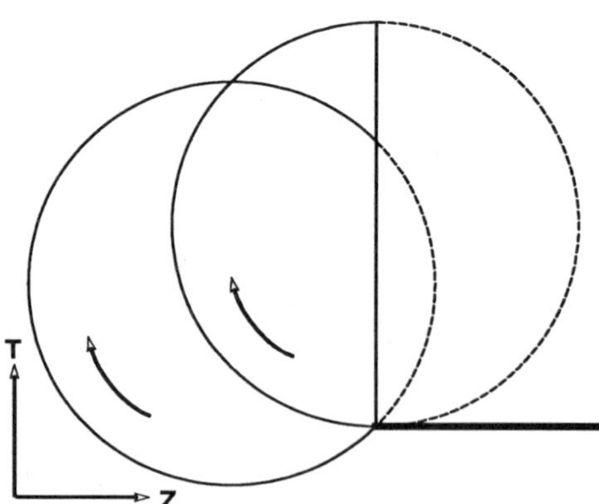


**Figure 1.** A diagram of the temperature-displacement histories of two elements of gas in contact with the hot heat exchanger. An element traces a clockwise path around the diagram. The solid line (——) represents the case in which thermal conduction with in the gas is ignored and heat transfer between the heat exchanger and the gas is perfect. The dashed line (- - - -) is the path the element would follow in the absence of a heat exchanger. To the left of the vertical line, the element is in the pulse tube; to the right, it is in the heat exchanger. The thick solid line (▬▬ ) represents the heat exchanger.

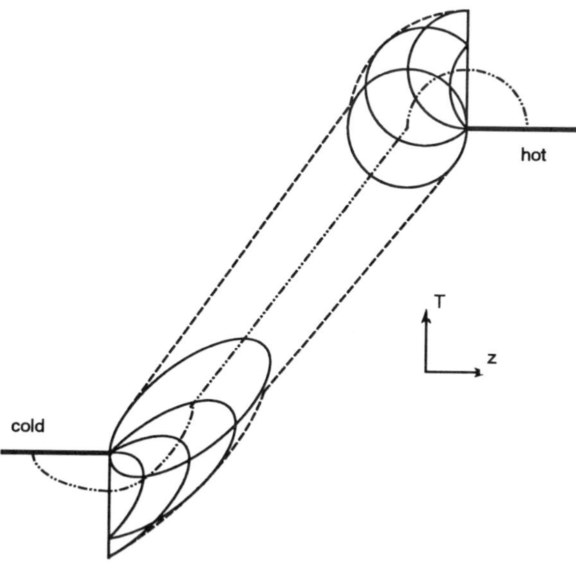


**Figure 2.** A diagram of the temperature (——) as a function of position of several elements as they move through a Pulse Tube. The motion is clockwise. Also shown are the temperature extrema ( - - - ) and the mean temperature ( - · · · - ) of all elements. The mean is only taken for the time an element spends in the Pulse Tube and is shown as a function of the mean displacement of each element. The mean temperature seen by a stationary observer (Eulerian frame) would be different. The constant temperature ends represent isothermal heat exchangers.

Thus, an element at the cold end follows an elliptical temperature-displacement path. This is shown in Figure 2. An element that does not touch either heat exchanger follows an elliptical path intermediate between those at the ends. Without heat transfer to a heat exchanger, there is nothing to set the temperature of an intermediate element. When axial thermal conduction is considered, the temperature of the intermediate elements is set by the condition: $\dot{Q}_z = -k\,\nabla T$ averaged over a cycle is constant in space; i.e. $\langle d\dot{Q}_z/dz\rangle = 0$. This condition does not mean that $\dot{Q}_z$ constant in time. Elements closer to the cold end move further during a cycle than elements near the hot end. Thus, the separation between elements varies over a cycle. For an adiabatic process, the temperature is given by

$$T(t)/T_0 = (P(t)/P_0)^{(\gamma-1)/\gamma} \tag{1}$$

Since $T_0$ varies along the length of the pulse tube; the temperature swing ($T_{max}$ - $T_{min}$) also varies along the pulse tube. Therefore, $\nabla T$ varies over a cycle. This is also shown in Figure 2. The discontinuities in $\nabla T$ shown in Figure 2 are the result of the simplicity of the model. In actuality, $\nabla T$ is a continuous function of z.

## DISCUSSION

The simple model developed above appears to fulfill the goal of predicting the temperature overshoots. It predicts that the mean temperature overshoot is $\approx T_1$,. At the hot end,

$$T_1 \approx 2\langle \dot{H}\rangle/\dot{m}_1 C_p\,, \tag{2}$$

where the right hand side is measured in the Eulerian frame. (This relation is only approximate because $T_1$ is not identical in the two reference frames.)

Unfortunately, this model contains a serious thermodynamic flaw. The process depicted in Figure 1 results in insufficient heat transfer between the gas and the heat exchanger. Consider the path of an element which starts with the element leaving the hot heat exchanger. Figure 1 shows paths for both an adiabatic case and for a perfect heat transfer case. If there were no thermal contact to the heat exchanger, the element would follow the adiabatic path. The other path can be modeled by an adiabatic section within the pulse tube followed by two heat transfer sections within the heat exchanger. The second section (vertical line in Figure 1) is an isobaric cooling. The heat transferred from the gas in this section is

$$\Delta Q_1 = \delta m\, C_p\, \Delta T \tag{3}$$

where $\delta m$ is the mass of the element. The third section is an isothermal expansion within the heat exchanger. The heat transferred here is

$$\Delta Q_2 = \delta m\, T\, \Delta S\,; \tag{4}$$

where

$$\Delta S = -R \ln\left(1 + \Delta P/P\right) . \tag{5}$$

For steady state operation, the end point for both paths is the same as the starting point. Thus the pressure and temperature change is independent of path. So, the pressure and temperature change while an element is within the heat exchanger is the same whether or not there is any heat transfer. For an adiabatic process the pressure and temperature change are related:

$$\Delta P/P \approx (C_p/R)\, \Delta T/T . \tag{6}$$

This same relation must also apply to the heat transfer process, Eqs. (4) and (5). Thus, the total heat transfer, the sum of Eqs. (3) and (4) in the small amplitude approximation, is

$$\Delta Q_t = \Delta Q_1 + \Delta Q_2 \approx 0. \tag{7}$$

(There is a small amount of net heat transfer because the amount of work contained in the two paths is not identical; they differ by a small amount. The adiabatic path does no net work. However, the other path does a small amount because the gas density and therefore displacement during the isothermal expansion is smaller than during the adiabatic expansion portion of the cycle. This is a second order effect and is insignificant in this context.)

## IMPROVED MODEL

In the above model the only heat transfer was the result of a negligible loss in work flow. To improve the model, all of the work flow must be converted to heat and transferred to the heat exchanger. This can be simplified as a two step process. First, the work flow is converted to heat and then this heat was transferred to the heat exchanger. This process is illustrated in Figure 3. The work transmitted per cycle by a gas element (work flow) is

$$W = \oint P\, A\, dx = \pi\, P_1\, A\, x_1 \tag{8}$$

The work flow vanishes within the hot heat exchanger/orifice because the volumetric displacement, $A\, x_1$, decreases as the orifice is approached and the pressure amplitude, $P_1$, vanishes at the orifice. The change in work flow, heats the working gas to a $\Delta T$ above the heat exchanger. Second, the heated gas is cooled by the heat exchanger with a heat transfer of

$$\Delta Q = m\, C_p\, \Delta T, \tag{9}$$

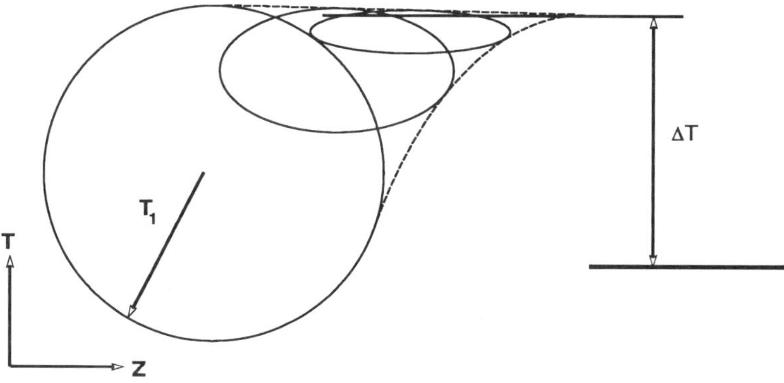


**Figure 3.** A representation of the work flow decrease for elements close to the hot heat exchanger/orifice. The paths are ellipsoidal, as the temperature and displacement changes are not linked.

where m is the mass of gas being cooled. The only mass that participates in the heat transfer comes from those elements that spent part of their cycle in the heat exchanger and part in the pulse tube. Other elements are considered to be at the heat exchanger temperature or isolated from the heat exchanger. The mass that crosses the pulse tube to heat exchanger interface during a cycle is $m = 2\ \rho\ A\ x_1$. Setting $\Delta Q = W$, using the ideal gas equation ($P = \rho RT$), and the small amplitude approximation of Eq.(1): $P_1 \approx (C_p/R)\ T_1(P/T)$; results in

$$\Delta T \approx \eta\, T_1 \tag{10}$$

where $\eta = \pi/2$.

Equation (10) is in fair agreement with reported data for the overshoot at the hot end.[1] The measured offset is slightly lower than given by Eq. (10). This is due to the simplicity of the model. The work conversion and heat transfer are not separated; rather, the work conversion and heat transfer will overlap, reducing the overshoot. Also, the preceding discussion assumed that all of the heat flows between the gas and the heat exchanger. In actuality, some of the heat is conducted along the pulse tube toward the cold heat exchanger. At the hot end, the gas conduction is usually only a small amount of the total heat flow. The $\eta$ factor is reduced by this gas conduction.

$$\eta = (\pi/2)\,(\dot{Q}_{net}/\dot{Q}_{gross}) \tag{11}$$

where $\dot{Q}_{net}$ is $\dot{Q}_{gross}$ less the heat conduction away from the heat exchanger.

At the cold end, the situation is different. There, the gas conduction and other losses can be substantial. In Eq. (11) the net heat represents the heat that flows between the heat exchanger and the gas. At the cold end this is the heat from an external load plus what ever other parasitic heat loads that are external to the gas flow. At the minimum operating temperature, gas conduction, regenerator inefficiency, and other parasitic losses account for all of the gross cooling power. Under these conditions $\eta \rightarrow 0$; as has been observed.[1]

## CONCLUSION

A simple model has been developed that describes the major features of the temperature profile in pulse tubes. The model is developed in the Lagrangian frame. It ignores secondary flows and heat pumping effects. The temperature profile in a pulse tube is dominated by heat transfer effects. At the hot end the temperature overshoot is caused by the conversion of work flow into heat which is removed by the heat exchanger. At the cold end the reverse process occurs. In the central region, the temperature gradient is determined by the bulk thermal conductivity. Once the overshoot approaches the condition given in Eqs. (10) and (11), further improvements in heat exchanger effectiveness will not significantly effect performance. In effect, a Pulse Tube Cooler with heat exchangers at $T_c$ and $T_h$ operates over an effective temperature range of $T_c$ - $\Delta T_c$ to $T_h$ - $\Delta T_h$. These overshoots represent an inherent loss. The overshoots produce an additional entropy at the rate:

$$\dot{S} \approx \dot{Q}_{gross}\left(\Delta T_h/T_h^2 + \Delta T_c/T_c^2\right) \tag{12}$$

Eq. (12) may be rewritten in terms of temperature oscillation:

$$\dot{S} \approx C_p\left(\dot{m}_h \eta_h \cos(\phi_h) T_{1h}^2/T_h^2 + \dot{m}_c \eta_c \cos(\phi_c) T_{1c}^2/T_c^2\right) \tag{13}$$

where $\dot{Q}_{gross}$ has been approximated as $\langle \dot{H} \rangle$. The most effective method of reducing this loss is to reduce $T_1$. To keep the same cooling power, this can only be done by increasing the mass flow or by changing the phase. The use of inertance tubes may make such changes practical.

## REFERENCES


1. P.J. Storch, R. Radebaugh, and J.E. Zimmerman; "Analytical Model for the Refrigeration Power of the Orifice Pulse Tube Refrigerator," NIST Technical Note 1343 (1990).
2. P. Kittel, A. Kashani, J.M. Lee, and P.R. Roach, General pulse tube theory, *Cryogenics* 36:849 (1996).
3. P.C.T. de Boer, Thermodynamic analysis of the basic pulse-tube cryocooler, *Cryogenics* 34:699 (1994).
4. P. Kittel, Enthalpy Flow transition losses in regenerative cryocoolers, Proc. 7th Int. Cryocooler Conf., USAF Phillips Lab. Report, PL-CP--93-1001 (1993) p. 1145.


## NOMENCLATURE

| | |
|---|---|
| A | cross-sectional area |
| C | heat capacity |
| H | enthalpy |
| P | pressure |
| Q | heat |
| R | gas constant |
| S | entropy |
| T | temperature |
| W | work |
| m | mass |
| z | distance / displacement |
| Δ | change / difference |
| γ | heat capacity ratio |
| η | temperature coefficient |
| ρ | density |
| φ | phase between mass flow and temperature oscillation |

<u>subscripts</u>

| | |
|---|---|
| 0 | mean |
| 1 | amplitude of oscillation |
| 1, 2, ... | series |
| c | cold end |
| gross | gross (total) |
| h | hot end |
| max | maximum |
| net | net (gross - losses) |
| min | minimum |
| p | constant pressure |
| t | total |

<u>other</u>

| | |
|---|---|
| $\cdot$ | flow, time derivative |
| $\nabla$ | gradient |
| $\langle\ \rangle$ | time averaged |

# INTERFACE LOSS IN THE SMALL AMPLITUDE ORIFICE PULSE TUBE MODEL

Luc Bauwens

The University of Calgary, Department of Mechanical Engineering
Calgary, AB T2N 1N4, Canada

## ABSTRACT

Linearized analysis of the orifice pulse-tube refrigerator yields simple expressions and key concepts such as the enthalpy flux. However, except for the irreversibility that occurs at the orifice, the linear model is ideal, and it yields the theoretical Kittel coefficient of performance, equal to the temperature ratio. One of the major losses, and one of the least-well understood, that this device shares with Stirling refrigerators, is the interface loss. This loss is due to the abrupt transition between the cold end or freezer, which is an approximately isothermal space, and the tube, in which the flow is approximately isentropic. In the linearized analysis, the loss appears as a second order correction. It is found to be strongly dependent upon the pressure amplitude and of the cosine of the phase angle.

## NOMENCLATURE

| | |
|---|---|
| $A$ | cross-section |
| $\dot{H}$ | enthalpy flux |
| $L$ | length of the tube |
| $R$ | tube radius |
| $\dot{S}$ | entropy flux |
| $T$ | temperature, dimensionless |
| $X$ | rescaled coordinate |
| $a$ | speed of sound at $T_{ref}$ |
| $i$ | $\sqrt{-1}$ |
| $p$ | pressure |
| $r$ | radial coordinate |
| $s$ | specific entropy |
| $t$ | time |
| $u$ | velocity |
| $x$ | longitudinal coordinate |
| $\Omega$ | orifice resistance |
| $\delta$ | $= L/a\tau$ |
| $\varepsilon$ | small dimensionless No., $= \Delta L_{CP}/L$ |
| $\gamma$ | ratio of specific heats, $= c_p/c_v$ |
| $\mu$ | dynamic viscosity |
| $\varphi$ | phase angle between u and p |
| $\pi$ | $= 3.141596535....$ |
| $\rho$ | density of the cycle fluid |
| $\tau$ | period |

**Subscripts**

| | |
|---|---|
| CP | compressor (piston) |
| F | interface freezer/tube |
| O | orifice |
| T | reservoir |
| 0, 1, 2 | term of order 1, $\varepsilon$, $\varepsilon^2$.etc. |

## INTRODUCTION

The orifice pulse-tube refrigerator[1] is closer to the Stirling than to the original basic pulse-tube of Gifford and Longsworth.[2] It can be described as a Stirling device in which the piston on the cold side is replaced by the tube-orifice-reservoir assembly. The functions of these components are to dissipate and reject the work recoverable at the cold end and to ensure an appropriate phase shift. The coefficient of performance is limited to a value equal to the temperature ratio,[3] but since the recoverable work vanishes as the temperature ratio approaches zero, it approaches the Carnot value. Ideally, the tube operates as an isentropic enthalpy pump between freezer and orifice; surface heat pumping[2,4] plays no role or may even be detrimental.[1-6]

The orifice pulse-tube is amenable to linearized, small amplitude analysis,[7] although the orifice needs a different limit.[8] In the linear limit, all losses except the orifice loss are either neglected or asymptotically small, yielding the ideal Kittel efficiency.[3] Besides surface heat pumping in the tube and regenerator losses, one additional loss is particularly significant. Freezer and regenerator are designed to operate at a near-constant temperature close to the wall temperature. In the tube, however, the flow as close as possible to isentropic. This results in an interface loss, which Kittel[9] calls "transition losses." A loss called adiabatic loss in the Stirling literature[10] is closely related,[11-13] although one could argue that the adiabatic loss traditionally includes not only the interface loss but also mixing losses.[14] In the linearized analysis that follows, the interface loss appears as a small, lower order correction.

In contrast with Kittel's,[9] this study requires no additional ad-hoc model or assumption besides linearization; a simple expression for the interface loss is obtained. Numerical studies[13] do not require the small amplitude assumption, but merely provide numbers and no physical insight. They have shown that, except by lowering the pressure amplitude, hence also the specific refrigeration, little can be done using other design parameters to minimize the loss. Indeed, the following linear analysis shows that, over a wide range of phase angles between pressure and velocity, the ratio loss/refrigeration is insensitive to phase angle.

A complete list of the assumptions leading to the small amplitude model follows, including some not addressed in previous work.[6-7] The derivation proceeds directly as a small amplitude approximation to the boundary value problem of gas dynamics. On the whole, the results are very similar to Storch and Radebaugh's,[7] except with respect to mass conservation: on phasor diagrams, velocity, and not the mass flow rate, is found to vary linearly with position.[7] Finally, the interface loss is found as a second order correction.

## THE LINEAR THEORY

### Formulation and Assumptions

Except for the compressor, all volumes are fixed. The device is driven by one single function: the periodic motion of the compressor. The assumptions are:

- All the fixed volumes, including, sequentially, aftercooler, regenerator, freezer, orifice and reservoir, are of similar magnitude, except for the orifice, discussed below.
- Except in the orifice, transverse gradients are neglected; cross-sections remain uniform within each volume; effects of changes in cross-section at interfaces between volumes are neglected. This one-dimensional model requires two independent variables, space x and time t. Lengths are rescaled so that x actually represents volume/tube cross-section, and fluxes are per unit tube cross-section. L is the tube length, $L_T$ is the normalized reservoir length and $L_{CP}$ is the normalized compressor dead volume at mid-stroke.
- The piston displacement is small compared with the total length. The normalized piston displacement $\Delta L_{CP}$ is thus small compared with L, which defines the small parameter ε:

$$\varepsilon = \Delta L_{CP}/L << 1 \tag{1}$$

- The flow is periodic in time.
- The period $\tau$ is much larger than the period of resonant acoustic modes of the device, so that a second small parameter, $\delta$ can be defined:

$$\delta = L/a\tau << 1 \quad (2)$$

- In the compressor and heat exchangers, the fluid temperatures are time-independent.
- In the tube and reservoir, the flow is inviscid, non-conducting, hence isentropic, but not homentropic. *Each fluid parcel traveling inside the tube retains a fixed entropy, but neighboring parcels are characterized by different entropy;* entropy gradients occur.[13]
- The orifice volume is assumed to be negligible compared with the tube volume, the aspect ratio diameter/length is negligible, but the Mach number remains negligible.
- Finally, the cycle fluid is assumed to be an ideal gas with constant specific heats.

Except in the orifice, the variables are made dimensionless uniformly, scaling velocities (normalized by the ratio $A/A_{tube}$ - these velocities are based upon rescaled length) by the ratio $\Delta L_{CP}/\tau$, length by the tube length L, time by the period $\tau$, pressure by its mean value, temperature by some reference value $T_{ref}$ and density by its value at the mean pressure and $T_{ref}$. The variables, including u, p, T, $\rho$ and s, are written in perturbation series such as:

$$u = u_0 + \varepsilon\, u_1 + \ldots \quad (3)$$

**Isentropic Flow**

In the tube and in the reservoir, the flow is one-dimensional, inviscid and non-conducting, hence, in smooth flow, isentropic. The conservation laws are:

$$\frac{\partial \rho}{\partial t} + \varepsilon \frac{\partial}{\partial x} \rho u = 0 \quad (4a)$$

$$\varepsilon\delta^2 \frac{\partial}{\partial t} \rho u + \frac{\partial}{\partial x}\left(\varepsilon^2\delta^2 \rho u^2 + \frac{1}{\gamma} p\right) = 0 \quad (4b)$$

$$\frac{\partial}{\partial t}\left(\rho T + \varepsilon^2\delta^2 \frac{\gamma(\gamma-1)}{2} \rho u^2\right) + \frac{\partial}{\partial x}\left(\varepsilon\, \gamma u p + \varepsilon^3\delta^2 \frac{\gamma(\gamma-1)}{2} \rho u^2\right) = 0 \quad (4c)$$

and the thermodynamic properties are related by the state equation $p = \rho T$. Assuming that the solution remains continuous, Eqs. (4) can be combined, yielding an entropy equation, which shows that entropy remains constant along particle paths.

From Eq. (4b), pressure is spatially uniform to order $\varepsilon\delta^2$. From Eq. (4c), leading order pressure is time-independent: $p_0$ equals 1; and $p_1$ depends upon time only, $p_1 = p_1(t)$. From Eq. (4a), leading order density is time-independent. Thus, leading order density and temperature depend upon x only. The first non-trivial set of equations is the perturbation of order $\varepsilon$:

$$\frac{\partial \rho_1}{\partial t} + \frac{\partial}{\partial x} \rho_0 u_0 = 0, \quad \frac{dp_1}{dt} + \gamma \frac{\partial u_0}{\partial x} = 0 \text{ and } p_1 = \rho_0 T_1 + \rho_1 T_0 \quad (5a\text{-}c)$$

Equation (5b) is readily integrated with respect to x.

$$\frac{x}{\gamma}\frac{dp_1}{dt} + u(0) - u(x) = 0 \quad (6)$$

On phasor diagrams, it is thus velocity that varies linearly with x, and not the mass flow rate, as assumed by Storch and Radebaugh. (Temperature, hence density, varies with x; if u is a linear function of x, the product $\rho u$ can only be linear if temperature is uniform .)

Applying Eq. (6), respectively, for the tube with unit dimensionless length, and for the reservoir with dimensionless length $L_T$,

$$\frac{dp_1}{dt} + \gamma\,(u_O^- - u_F) = 0 \quad \text{and} \quad L_T \frac{dp_1}{dt} - \gamma u_O^+ = 0 \quad (7a,b)$$

where subscript + and - indicate values respectively on the right and left side of the respective interface. Likewise, integrating Eq. (5a) will yield the density perturbation.

**Isothermal Flow**

In the compressor, the aftercooler, the regenerator and the freezer, the temperature is time-independent. Except in the regenerator, these temperatures are also spatially uniform in each space. (Admittedly, this model may not be very realistic in the compressor.) The scaling assumptions are the same as in the isentropic spaces and Eq. (5a) remains valid. Pressure is spatially uniform to order $\varepsilon\delta^2$. The key difference occurs in the energy equation, which still yields $T_0 = T_0(x)$, but $T_1 = 0$ in these spaces, because of the isothermal assumption. Thus, the Schmidt model is obtained.[15] Indeed, $p_1 = \rho_1 T_0$ and Eq. (5a) becomes:

$$\frac{1}{T_0}\frac{dp_1}{dt} + \frac{\partial}{\partial x}\frac{u_0}{T_0} = 0 \tag{8}$$

Integrating between the piston face and the cold space,

$$\frac{dp_1}{dt}\left(\frac{L_{CP}}{T_{CP}} + \int_{x_C}^{x_F}\frac{1}{T}\,dx\right) + \frac{u_F}{T_F} - \frac{u_{CP}}{T_{CP}} = 0 \tag{9}$$

**The Orifice**

In a slender orifice, longitudinal diffusion is negligible, but transverse diffusion is not. If, furthermore, the length of the orifice is short compared to the sweep by the flow, then the flow becomes quasi-steady, and all time-derivatives are negligible. In particular, mass conservation then results in a spatially uniform mass flow rate $\rho u$ and the inertia terms are negligible in the momentum equation. Finally, for velocities smaller than the speed of sound, acceleration terms are negligible. The yields the familiar Poiseuille velocity profile: [6]

$$u(r,t) = -\frac{\partial p}{\partial x}\frac{R^2}{4\mu}\left(1 - \frac{r^2}{R^2}\right) \tag{11}$$

Next, this is integrated over the cross-section and over the orifice length. Clearly there would be no point for a velocity of leading order to result in a pressure drop of magnitude larger than the magnitude of the pressure fluctuations, $\varepsilon$. Thus, it is assumed that $4\mu L_O/R_O^2$ is of order $\varepsilon$. The mass flow rate $\rho_0 u$ being uniform along the orifice, so is u, [u] =0, and

$$[p_1] = -\Omega\gamma\, u_O \tag{12}$$

The orifice resistance $\Omega$ is defined by Eq. (14). $A_O$ is the ratio orifice cross-section/tube cross-section. A small $\Omega$ corresponds to a large and short orifice offering little resistance.

$$\Omega = \frac{8\mu L L_O}{\tau\gamma P A_O R_O^2} \tag{13}$$

**Other Interfaces**

Pressure and the mass flow rate are continuous at the interfaces since there are no shocks or mass leaks or sources. Using square brackets for jumps across interfaces,

$$[p_0] = 0,\ [p_1] = 0,\ [\rho_0 u_0] = 0 \text{ and } [\rho_0 u_1 + \rho_1 u_0] = 0 \tag{14a-d}$$

The mass flow rate $\rho u$ being continuous, continuity of the enthalpy flux $\rho c_p u T$ would imply continuity of the temperature. But the enthalpy flux is not necessarily continuous at interfaces.[12-13] Indeed, no heat transfer occurs *inside* either an isothermal or an isentropic space, so that the only way that heat transfer can occur is through singularities in the heat flux, at the interfaces between these regions. Two situations are then possible. Whenever the flow is moving from an isothermal channel to an isentropic one, the flux remains continuous because there is no temperature difference that can induce heat transfer, and the fluid enters the region of isentropic flow at the temperature of the isothermal region. But fluid originating in the isentropic region reaches the interface at a temperature determined by its history, including the effect of pressure fluctuations, while traveling isentropically in the insulated channel. Typically, this temperature will not match the temperature of the isothermal region. However, the hypothesis of isothermal flow requires the temperature of the entering fluid immediately to jump to the wall temperature. This results is a finite and irreversible heat transfer at the interface, no matter how effective heat transfer is. Thus, to arbitrary order n, *if and only if the flow goes from an isothermal to an isentropic region,*

$$[T_n] = 0 \tag{15}$$

Equations (14a) and (14c) imply that $[u_0/T_0] = 0$. But both in isothermal and isentropic flow regions, $T_0$ has been found above to be time-independent. Thus, if it is continuous across the interfaces when the flow is in one direction, it is continuous, and, so is $u_0$.

$$[u_0] = 0 \text{ and } [T_0] = 0 \tag{16a,b}$$

**Solution**

All equations have already been integrated with respect to x, resulting in a set of ordinary differential equations with respect to t, with constant coefficients. In Fourier space, this system becomes a set of linear algebraic equations which is readily solved. For a sinusoidal piston displacement, $u_{CP} = 2\pi \cos 2\pi t$, and the Fourier transform $\hat{u}_{CP} = 2\pi$. The relationship between pressure and velocity at the freezer-tube interface is found:

$$\hat{u}_F = \frac{2\pi i}{\gamma}\frac{1 + L_T + 2\pi i\Omega L_T}{1 + 2\pi i\Omega L_T}\hat{p}_1 \tag{17}$$

The full solution is given by Eqs. (18) and (19). The symbol a defined by Eq. (20) is independent of the orifice and reservoir; it characterizes the compressor and heat exchangers.

$$\hat{p}_1 = \frac{\gamma}{i}\frac{T_F}{T_C}\frac{1 + 2\pi i\Omega L_T}{a + L_T + 2\pi i a\Omega L_T} \quad \text{and} \quad \hat{u}_F = \frac{T_F}{T_C}\frac{2\pi(1 + L_T + 2\pi i\Omega L_T)}{a + L_T + 2\pi i a\Omega L_T} \tag{18,19}$$

$$a = 1 + \gamma\left(L_{CP}\frac{T_F}{T_C} + \int_{x_C}^{x_F}\frac{1}{T}\,dx\right) \tag{20}$$

The compressor work and the refrigeration, which, for an ideal regenerator, equals the enthalpy flux in the duct, can now be calculated. Using asterisks to indicate complex conjugates, refrigeration (scaled by the mean pressure × piston displacement/τ) is given by

$$\dot{H} = \int_0^1 p_1 u_F dt = \frac{1}{2}\mathrm{Re}[\hat{p}_1\hat{u}_F^*] = \left(\frac{T_F}{T_C}\right)^2 \frac{2\pi^2\gamma\Omega L_T^2}{(a + L_T)^2 + 4\pi^2 a^2\Omega^2 L_T^2} \tag{21}$$

And, as expected, the compression work is found to be equal to $\dot{H}$ divided by the temperature ratio so that the COP equals the Kittel value. The phase angle between $p_1$ and $u_F$ depends only upon the tube, the orifice and the reservoir:

$$\tan\varphi_F = \frac{1 + L_T + 4\pi^2\Omega^2 L_T^2}{2\pi\Omega L_T^2} = \frac{1 + L_T}{2\pi\Omega L_T^2} + 2\pi\Omega \tag{22}$$

## INTERFACE LOSS

The solution above corresponds to the ideal orifice pulse-tube with no interface loss. Furthermore, the solution does not satisfy Eq. (15). Indeed, in the freezer, $T_1 = 0$, while inside the tube, close to inlet, $T_1$ is a sinusoidal function of time, obtained from Eqs. (5):

$$\hat{T}_1 = \frac{1}{2\pi i}\frac{dT_0}{dx}\hat{u}_F - \frac{1}{\gamma}\left((\gamma - 1)T_0 + x\frac{dT_0}{dx}\right)\hat{p}_1 \qquad (23)$$

The flow near the interface is as follows.[11-13] Flow leaving the isothermal region enters the tube at the temperature of the isothermal region. However, as the fluid travels inside the tube, its entropy remains constant, and pressure fluctuations result in temperature fluctuations. When eventually, the fluid returns to the isothermal region, its temperature normally will not match the temperature of the isothermal region. Irreversible heat transfer occurs over a small entry length determined by heat transfer effectiveness. The small amplitude model, however, assumes that the distance traveled by the fluid is negligible compared to the tube length, blurring the physics of the loss.

Since the discontinuity affects the temperature perturbation $T_1$, the loss is a fraction of order $\varepsilon$ of the enthalpy flux, hence refrigeration, and the leading order results above remain correct. But the evaluation of the interface loss will require a detailed study of the flow near the interface, taking into account the length scale of the fluid motion, $\varepsilon L$. To that effect, an inner problem is introduced, characterized by the rescaled coordinate $X = (x - x_F)/\varepsilon$. Using X, rescaled Eqs. (4) are readily obtained and pressure is found to be uniform to order $\varepsilon^2\delta^2$. Thus, the inner $p_0$ and $p_1$ are equal to their already known outer values. Likewise, since $p_0 = 1$, the rescaled energy equation results in $u_0$ being spatially uniform, hence equal to $u_F(t)$, and $\rho_0(X,t) = 1/T_F$. The first perturbation to the rescaled equations can then be written as:

$$\frac{\partial \rho_1}{\partial t} + u_F\frac{\partial \rho_1}{\partial X} + \rho_0\frac{\partial u_1}{\partial X} = 0 \qquad \text{and} \qquad \frac{dp_1}{dt} + \gamma\frac{\partial u_1}{\partial X} = 0 \qquad (24a,b)$$

Replacing $u_1$ from Eq. (24b) and $\rho_1$ from the equation of state, Eq. (5c), in Eq. (24a):

$$\rho_0\frac{\partial T_1}{\partial t} + \rho_0 u_F\frac{\partial T_1}{\partial X} = \frac{\gamma - 1}{\gamma}\frac{dp_1}{dt} \qquad (25)$$

Eq. (25) shows that, as a fluid particle moves along the tube, its temperature varies isentropically with pressure, hence the factor $(\gamma - 1)/\gamma$ on the right hand side. Since $\rho_0 = 1/T_F$, and $p_1$ and $u_F$ are given respectively by Eqs. (18) and (19), the only unknown is $T_1$. Being first order, hyperbolic, Eq. (25) only supports one boundary condition, at the extremity where the fluid originates. When $u_F$ is positive, the fluid enters the tube at $X = 0$, coming from the isothermal space, so that $T_1 = 0$. When $u_F$ is negative, the fluid reaching the interface comes from the tube. The characteristics, which originated at the interface when $u_F$ was positive, have now changed direction and they are now moving back toward the interface; no boundary condition is needed and none exists. The temperature of the fluid *returning* from the tube to the isothermal region is determined by integrating on particle paths starting at $X = 0$ at time $t = t_0$, (with $u_F > 0$) and returning to $X = 0$ at a later time $t_1$ (with $u_F < 0$). On particle paths described by Eq. (26), Eq. (25) simplifies to Eq. (27).

$$dx = u_F\,dt \qquad\qquad \rho_0\frac{dT_1}{dt} = \frac{\gamma - 1}{\gamma}\frac{dp_1}{dt} \qquad (26,27)$$

Integrating between $t_0$ and $t_1$ and noting that $T_1(0,t_0) = 0$, $t_1$ and $T_1(0,t_1)$ are found, from:

$$\int_{t_0}^{t_1} u_F\,dt = 0 \qquad\qquad \rho_0\,T_1(0,t_1) = \frac{\gamma - 1}{\gamma}[p_1(t_1) - p_1(t_0)] \qquad (28,29)$$

Calling $t^*$ the time at which $u_F$, a sinusoidal function of time, is zero and increasing, Eq. (28) implies that $t_1 = 2t^* + 1/2 - t_0$. When $u_F > 0$, $T_1 = 0$, and, from Eq. (29), when $u_F < 0$:

$$\hat{T}_1(0,t) = -\frac{\gamma - 1}{\gamma} T_F (\hat{p}_1 + \hat{p}_1^*) \exp i\varphi_F \qquad (30)$$

which is in phase with $u_F$. The refrigeration, scaled as in Eq. (21), is then given by:

$$\dot{H} = \frac{c_p}{R} \int_0^1 (\rho_0 T_1 u_0 + \rho_1 T_0 u_0 + \rho_0 T_0 u_1)\, dt = \frac{\gamma}{\gamma - 1} \int_0^1 \rho_0 T_1 u_F\, dt$$

taking into account that $\rho_1 u_0 + \rho_0 u_1$ is periodic and that $T_0 = T_F$ is constant. In the last integral, $T_1 = 0$ during one-half of the period when $u_F > 0$ and $T_1$ is given by Eq. (30) during the other half when $u_F < 0$. The contribution of the first half is zero, and the contribution from the second is one-half of the integral over an entire period of the product $u_F$ by a temperature which would be given by Eq. (30) over the entire period, times $\gamma\rho_0/(\gamma - 1)$, and the same value given by Eq. (21) is recovered:

$$\dot{H} = \frac{1}{4} (\hat{p}_1 + \hat{p}_1^*) \operatorname{Re}[\exp i\varphi_F\, \hat{u}_F^*] = \frac{1}{2} \operatorname{Re}[\hat{p}_1 \hat{u}_F^*]$$

Per unit of enthalpy flux, the instantaneous irreversibility due to the temperature mismatch $T_1$ at the interface, given by Eq. (30), equals the difference $1/(T_F + \varepsilon\, T_1 + O(\varepsilon^2)) - 1/T_F + O(\varepsilon^2)$. Multiplying by the instantaneous enthalpy flux and integrating over the period, an expression for the total irreversibility is found. $T_1$ is nonzero only when $u_F$ is negative, and $\Delta\dot{S}$ is thus intrinsically positive. $\Delta\dot{S}$ is readily evaluated, yielding the value given by Eq. (31).

$$\Delta\dot{S} = -\varepsilon \frac{\gamma - 1}{\gamma} \int_0^1 \rho_0 u_F \frac{T_1^2}{T_F^2}\, dt = -\varepsilon \frac{\gamma - 1}{\gamma} \frac{1}{T_F^3} \int_0^1 u_F T_1^2\, dt$$

$$\Delta\dot{S} = \varepsilon \frac{4(\gamma - 1)}{3\gamma\pi T_F} \dot{H} (\hat{p}_1 + \hat{p}_1^*) \cos\varphi_F \qquad (31)$$

This results in a coefficient of performance (COP) given by:

$$\frac{\text{COP}}{\text{Ideal COP}} \cong 1 - \frac{8(\gamma - 1)}{3\gamma\pi} \Delta p \cos\varphi_F \qquad (32)$$

with the pressure ratio $\Delta p = \varepsilon(\hat{p}_1 + \hat{p}_1^*)/2$ defined by $\Delta p = (p_{Max} - p_{Min})/(p_{Max} + p_{Min})$. (If the compressor were isentropic there would be an interface loss at the warm end also, and the COP would be given by the same expression replacing $\cos\varphi_F$ by the sum $\cos\varphi_F + \cos\varphi_C$.)

For instance, for a large reservoir, $L_T >> 1$ and small heat exchangers, $a \cong 1$. Then:

$$\dot{H} = \left(\frac{T_F}{T_C}\right)^2 \frac{2\pi^2\gamma\Omega}{1 + 4\pi^2\Omega^2} \qquad \frac{\text{COP}}{\text{Ideal COP}} \cong 1 - \varepsilon \frac{8(\gamma - 1)}{3\gamma\pi} \frac{T_F}{T_C} \frac{2\pi\gamma\Omega}{(1 + 4\pi^2\Omega^2)^{3/2}} \qquad (33,34)$$

$$\tan\varphi_F = 2\pi\Omega \qquad \hat{p}_1 = \frac{T_F}{T_C} \frac{2\pi\gamma\Omega}{1 + 2\pi i\Omega} \qquad \text{and} \qquad \hat{u}_F = 2\pi \frac{T_F}{T_C} \qquad (34\text{–}36)$$

Fig. 1 shows, as a function of the phase angle, the refrigeration, the relative efficiency loss and the value of $\Omega$ necessary to achieve the phase angle. The refrigeration plot is a parabola; equal values of the refrigeration are obtained for phase angles equidistant from the 45 degree centerline. However, the efficiency deficit is smaller for larger phase angles which are thus always preferable.

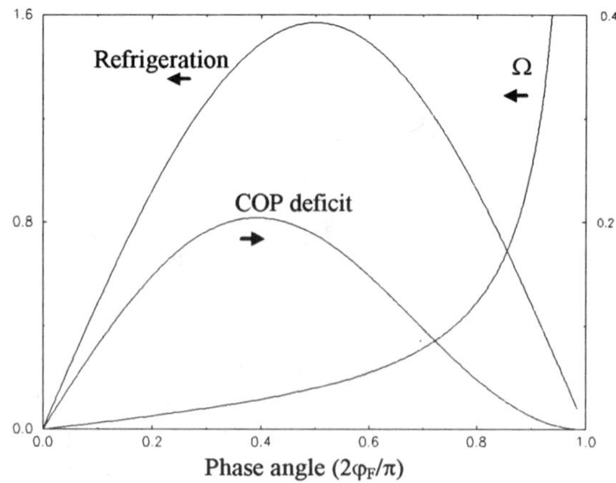


**Figure 1.** Results for large reservoir and small heat exchangers, including refrigeration $\dot{H}/\left(\frac{T_F}{T_C}\right)^2$, COP deficit (1 - COP/Ideal COP) $\times$ $T_F/\varepsilon T_C$ and orifice resistance $\Omega$

## CONCLUSION

The small amplitude, linear model was developed, starting from a complete statement of assumptions and using a rigorous procedure. The results include the effect on refrigeration of the main geometric parameters. The interface loss was determined, as a smaller order correction to the main results. The results show that, among all designs yielding the same gross refrigeration, the configuration maximizing the velocity amplitude will suffer from the smallest loss. This is because the loss is proportional to refrigeration, times the product pressure amplitude, times the cosine of the phase angle between velocity and pressure at the interface, while refrigeration is proportional to the same product, times the velocity amplitude.

## REFERENCES

1. Radebaugh, Ray, "A Review of Pulse Tube Refrigeration," *Adv. Cryo Eng,* 35 (1990), pp. 1191-1206.
2. Gifford, W.E. and Longsworth, R.C., "Surface Heat Pumping," *Adv. Cryo Eng,* 11 (1966), pp. 171-179.
3. Kittel, P., "Ideal Orifice Pulse Tube Refrigerator Performance," *Cryogenics*, 32,:9, pp. 843-844 (1992).
4. Bauwens, L., "Oscillating Flow of a Heat-Conducting Fluid in a Narrow Tube," *J. Fluid Mech,* 324, pp. 135-161 (1996).
5. Bauwens, L., "Entropy Balance and Performance Characterization of the Narrow Basic Pulse-Tube Refrigerator," *AIAA Journal of Thermophysics and Heat Transfer*, 10:4, pp. 663-671 (1996).
6. Mirels, H., "Linearized Theory for Pulse Tube Cryocooler Performance," *AIAA J.*, 32:8, pp. 1662-1669 (1994).
7. Storch, P.J. and Radebaugh, R., "Development and experimental test of an analytical model of the orifice pulse tube refrigerator," *Adv. Cryo Eng*, Plenum Press, NY, 33 (1988), pp. 851-859.
8. Mirels, H., "Effect of Orifice Flow and Heat Transfer on Gas Spring Hysteresis," *AIAA J.*,. 32:8, pp. 1656-1661 (1994).
9. Kittel, P., "Enthalpy Flow Transition Losses in Regenerative Cryocoolers," Phillips Lab Report #PL-CP-93-1001 (1993).
10. Walker, G., *Stirling Engines*, Clarendon Press, Oxford, U.K. (1980).
11. Bauwens, L., "Stirling Cryocooler Model with Stratified Cylinders and Quasisteady Heat Exchangers," *AIAA Journal of Thermophysics and Heat Transfer*, 9:1, pp. 129-135 (1995).
12. Bauwens, L., "Adiabatic Losses in Stirling Cryocoolers - A Stratified Flow Model," *Cryogenics*, 34:8, pp. 627-633 (1994).
13. Bauwens, L., "Adiabatic Losses in Stirling Refrigerators," *Transactions of the ASME, Journal of Energy Resources Technology,* 1996, 118:2, pp. 120-127 (1996).
14. Walker, G., *Stirling Cycle Machines*, Clarendon Press, Oxford, U.K. (1973).
15. Schmidt, G., "Theorie der Lehmann'schen kalorische Maschine," *Zeitschrift des Vereins Deutscher Ingenieure*, 15:1, pp. 1-12 (1871).

# AN ADIABATIC MODEL FOR A TWO-STAGE DOUBLE-ORIFICE PULSE TUBE REFRIGERATOR TO REACH LIQUID HELIUM TEMPERATURES*

Limin Qiu, Guobang Chen, and Jianyao Zheng

Cryogenics Laboratory, Zhejiang University
Hangzhou 310027, P.R.China

## ABSTRACT

An adiabatic model for 4K two-stage double-orifice pulse tube refrigerator is proposed to study the detailed fluid dynamics within the pulse tube. Numerical predictions also reveal the energy transportation in the pulse tube. The calculated results show that the double-orifice configuration developed by the authors has an obvious effect by increasing the time-averaged enthalpy flow at the cold end of the low temperature stage pulse tube while also decreasing irreversible losses compared with the double-inlet version.

## INTRODUCTION

The pulse tube refrigerator contains no moving parts at the cold end and thus has considerable advantages over most other types of cryocoolers in terms of reliability, lifetime, vibration and cost. More and more laboratories around the world are interested in pulse tube refrigerators because of its promising application potential. In order to obtain 4K by a pulse tube refrigerator, a multi-stage configuration must be employed. In 1994, a temperature of 3.6K was obtained by Matsubara and Gao with a three-stage pulse tube refrigerator which has a regenerative tube at the 3rd stage pulse tube [1]. Recently, a lowest temperature of 2.23K and cooling power of 370mW at 4.2K were obtained by C. Wang with a two-stage pulse tube refrigerator [2].

---

* The project is supported by the National Natural Science Foundation of China.

The authors developed a double-orifice configuration, in which a bypass valve connects the low temperature stage reservoir and the inlet tube of the compressor. The performance improvement test of the double-orifice configuration has been reported[3,4]. However, the theoretical analysis has not been published yet about this version until now.

Because of the absence of a solid displacer, the expansion space within the pulse tube sometimes could be expanded to the whole volume of the tube. In this case, there is a great temperature gradient along the pulse tube, so the isothermal model developed for the pulse tube refrigerator can not be used since this model assumes an isothermal equivalent cold and hot volumes. At the same time, the system is too complex to be calculated by the conventional nodal model. On the other hand, the compression and expansion processes within a pulse tube are closer to an adiabatic process rather than an isothermal one. In the present study we shall describe the detailed fluid dynamics within the 4K multi-stage pulse tube refrigerator by an adiabatic model which considers real properties of the helium gas . As pressure in the system can easily be measured, the dynamic pressure is given directly as an input data, in order to reduce the complex equation-solving process. Compared with the conventional numerical model, the adiabatic model is simpler and easier to use.

## PHYSICAL MODEL AND GOVERNING EQUATIONS

In order to give a set of Equations for the adiabatic model, the following basic assumptions are adopted.

1. one-dimensional flow;
2. real fluid;
3. negligible axial heat conduction and no heat exchange between the gas and the tube wall;
4. constant wall temperatures of the hot and cold heat exchangers respectively;

Attention is drawn only to the calculated system including the pulse tube, heat exchanger and phase shifter of the low temperature stage as shown in Fig. 1. On the basis of the above assumptions, the whole system is divided into many subsystems. Each subsystem can exchange work, heat and mass with the surroundings through its boundaries but its parameters vary with time. For each subsystem 'i', the governing equation set can be written as follows:

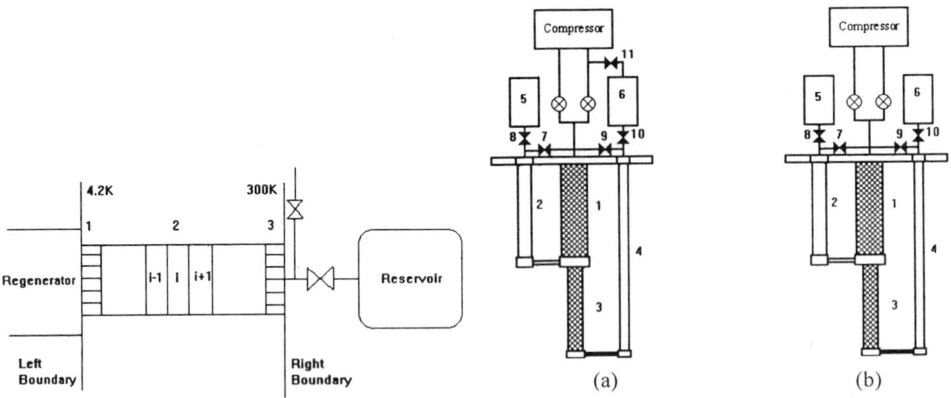


**Figure 1.** The calculated system for adiabatic model.

**Figure 2.** Schematics of the two-stage double orifice (a) and double inlet (b) pulse tube refrigerator.

Conservation of energy for a working fluid:

$$C_p\ (T_i, P)\frac{m_i T_i - m_i^{k-1}T_i^{k-1}}{\Delta t} + C_p(T_{f_{i+1}}, P)(\dot{m}T)_{f_{i+1}} - C_p(T_{f_i}, P)(\dot{m}T)_{f_i}$$
$$+\alpha_i A_i(T_i - T_{w_i}) - V_i\frac{P - P^{k-1}}{\Delta t} = 0 \qquad (1)$$

where superscript k-1 means variables at the (k-1)th time interval. Variables without a superscript stand for the value at the kth time interval. Subscript i denotes the nodal number, while subscript $f_i$ means the left boundary of the ith subsystem. $\Delta t$ is the calculated time interval. According to the assumptions, the heat transfer coefficient $\alpha_i$ equals zero.
Conservation of energy of heat exchangers:

$$\frac{T_{w_i} - T_{w_i}^{k-1}}{\Delta t} = \alpha_i\ A_i\ \frac{(T_i - T_{w_i})}{C_{w_i}(T_{w_i})m_{w_i}} \qquad (2)$$

Continuity of fluid flow:

$$\frac{m_i - m_i^{k-1}}{\Delta t} = \dot{m}_{f_i} - \dot{m}_{f_{i+1}} \qquad (3)$$

Equation of state:

$$m_i = \rho(P,T)V_i \qquad (4)$$

Pressure in the pulse tube is harmonic as follows:

$$P = P_0 + P_a\ sin\,2\pi ft \qquad (5)$$

The "M" & "L" Series Fine Metering Valves (1/8″) made in NUPRO Company U.S.A are used as valves in our experimental apparatus. Mass flow rate through valves such as orifice, second-inlet, second-orifice is calculated by the fomulars from NUPRO Corp.

$$\dot{m} = 16.05Cv\sqrt{\frac{(P_1^2 - P_2^2)}{(SG)T}} \qquad (6a)$$

$$\dot{m} = 13.61P_1Cv\sqrt{\frac{1}{(SG)T}} \qquad (6b)$$

Where $P_1$、 $P_2$ are the inlet and outlet pressures of the valve in PISA and $Cv$ is flow coefficient of the valve. If $P_2$ is greater than $0.53P_1$, Equation (6a) should be used; otherwise Equation (6b) should be used. SG is equal to 0.137 for helium.

The above set differential equations are non-linear and unsteady. Some efforts must be made to increase both the calculation accuracy and speed. The real fluid properties for helium such as specific heat , density etc. can be calculated directly by the Hepak[5]. The discretization can be solved by the TDMA (Tridiagonal Matrix Algorithm) method. The iteration results should be satisfied by periodic stable conditions for all the parameters and the energy balance condition.

## COMPUTATION CONDITIONS

A typical calculation has been made for a two-stage pulse tube refrigerator as dipicted in Fig. 2. The structure and operating parameters of the refrigerator are listed as follows:

**Table 1.** Description of the two-stage pulse tube refrigerator

| | Pulse tube (mm) | Regenerator (mm) | Regenerator Matrix |
|---|---|---|---|
| First stage | Φ 27.3 × 280 | Φ 59 × 130+ | No.250 bronze mesh + |
| | | Φ 45 × 60 | Φ 0.3 lead shot |
| Second stage | Φ 12 × 368 | Φ 19 × 190 | 0.1~0.3mm $Er_3Ni$ grains |

1. Operating parameters:
   Working fluid: helium ; Operating frequency: f=2Hz;
   Cold end temperature: $T_C$=4.2K, Hot end temperature: $T_h$=300K;
   Pressure in pulse tube: $P = P_0 + P_a \sin 2\pi ft$, where $P_0$=1.35MPa, $P_a$=0.65MPa;
2. Structure parameters are given in Table 1.

## CALCULATION RESULTS AND DISCUSSIONS

Calculations for the double-orifice pulse tube refrigerator have been carried out using the numerical model described above. Both the double-orifice configuration shown in Fig. 2(a) and a conventional double-inlet version shown in Fig. 2(b) are calculated. The results are discussed as follows.

### Double-inlet Type

Fig.3 shows the transient mass flow rate at different positions in the second stage pulse tube of the two-stage double-inlet version pulse tube refrigerator, where curves 1, 2, and 3 demonstrate positions at the cold, middle and hot ends of the pulse tube respectively. The mass flow rate at the cold end is much larger than that at the hot end, which is caused mainly by the increase in the density of helium below 20K. Large mass flow rate will lead to the reduction of regenerator efficiency.

Fig. 4 shows calculated temperature oscillation of the gas for the same three positions for

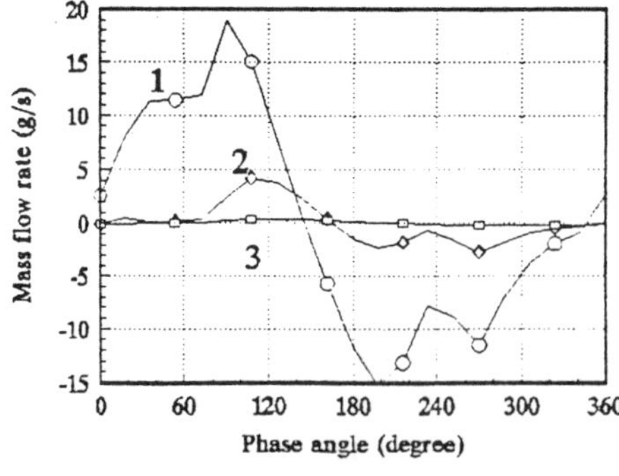


**Figure 3.** Mass flow rate of double-inlet type.

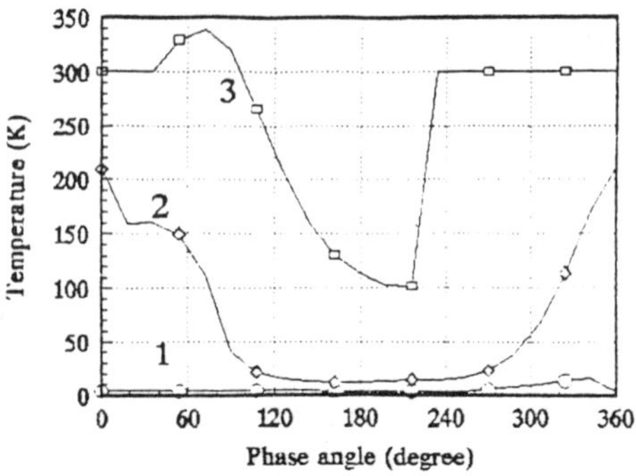


**Figure 4.** Temperature wave of double-inlet type.

the double inlet pulse tube refrigerator. We can see that the amplitude of the temperature wave at the hot end becomes several times larger than that at the cold end. The transient gas temperature for different openings of the second-inlet valve at the cold and hot ends are shown in Figs. 5(a) and (b). The amplitude of the gas temperature at the cold and hot ends increases for smaller openings of the second-inlet. However, the transient minimum temperatures of the cold end are almost the same for different openings. Therefore, the effect of the second-inlet vlave on the refrigeration temperature is not very obvious.

The calculated energy transportation in the pulse tube is shown in Fig.6, where curves 1, 2, and 3 are the time averaged enthalpy flow at the cold, middle and hot ends of the second stage pulse tube. There is an optimum open ratio for the second-inlet valve to get the maximum enthalpy flow at the cold end. However, the range of open ratio to maintain positive enthalpy flow is relatively narrow and the value of the positive enthalpy flow is also small , i.e. the cooling power of the second stage pulse tube is limited. Enthalpy flow at the hot end is negative within the whole range of openings of the second-inlet valve, while the enthalpy flow is positive at the middle of the pulse tube. This means there is a net enthalpy flow from the hot end to the cold end of the pulse tube during a cycle. That is why there is a peak temperature on tube wall near the hot end [6,7]. We can also find this point from the average gas temperature along the second stage pulse tube shown in Fig. 7, curve 1. There is a section near the hot end of curve 1 where the gas temperature is much higher than room temperature. The peak temperature increases irreversible losses in the pulse tube process. By analyzing and phenomenalizing, we have realized that it is most important to remove the anomalous temperature peak, i.e. to dissipate the heat accumulated near the hot end of the pulse tube for obtaining liquid helium temperatures. For the purpose, a novel phase shifter named second-orifice was developed for operating the pulse tube from room temperature to liquid helium temperature. The second-orifice combined with the original orifice is called double-orifice type[3].

## Double-orifice Type

Fig. 8 shows the transient mass flow rate at different positions in the second stage pulse version (Fig. 3), the average mass flow at the cold end is slightly larger than that of

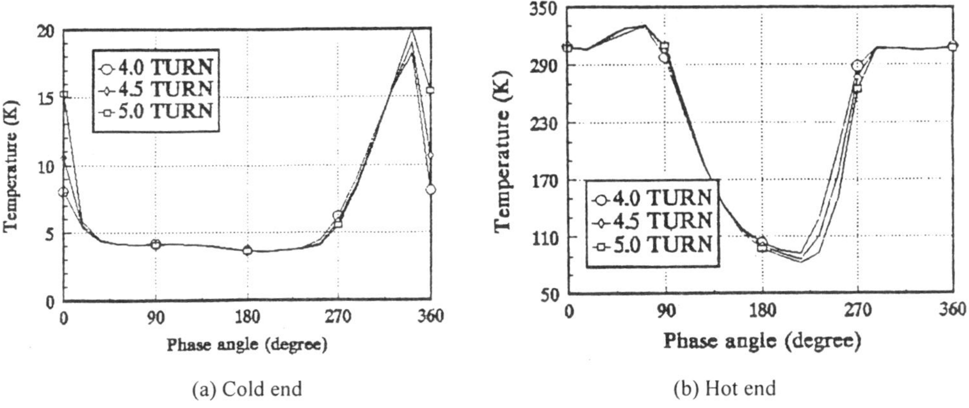


(a) Cold end

(b) Hot end

**Figure 5.** Temperature wave at different positions of double-inlet type.

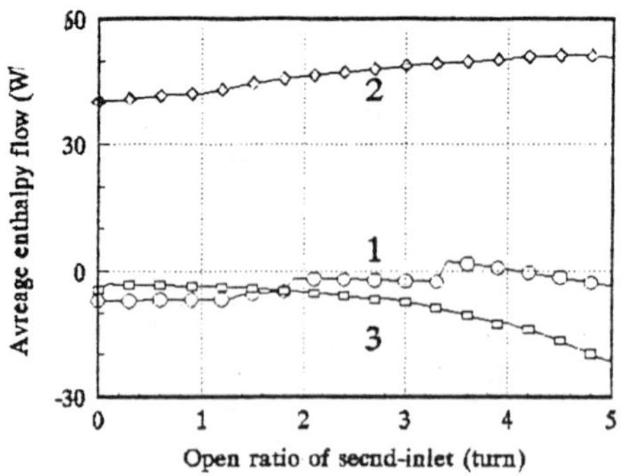


**Figure 6.** Time-averaged enthalpy flow of double-inlet type.

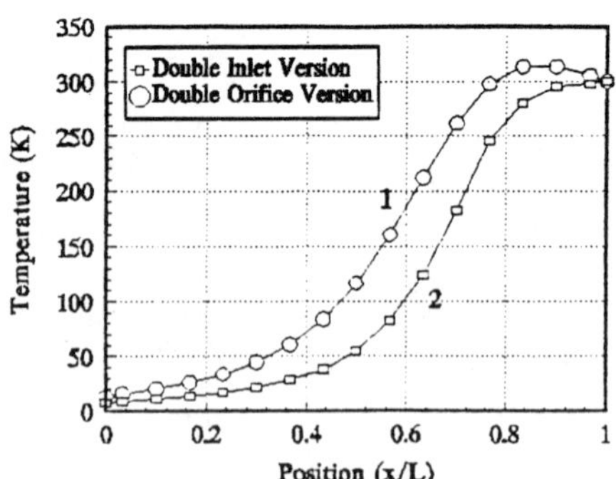


**Figure 7.** Average gas temperature in pulse tube.

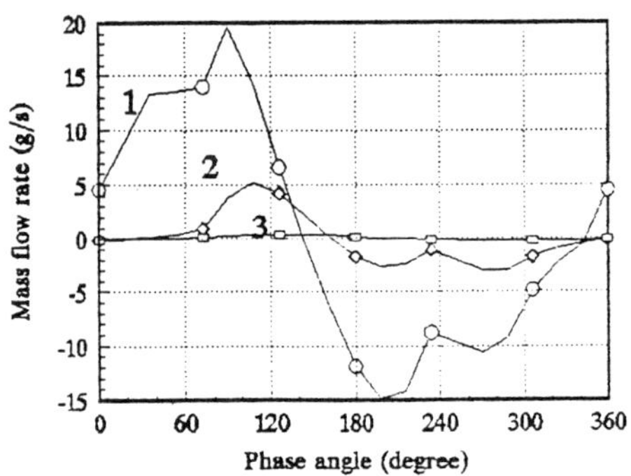


**Figure 8.** Mass flow rate of double-orifice type.

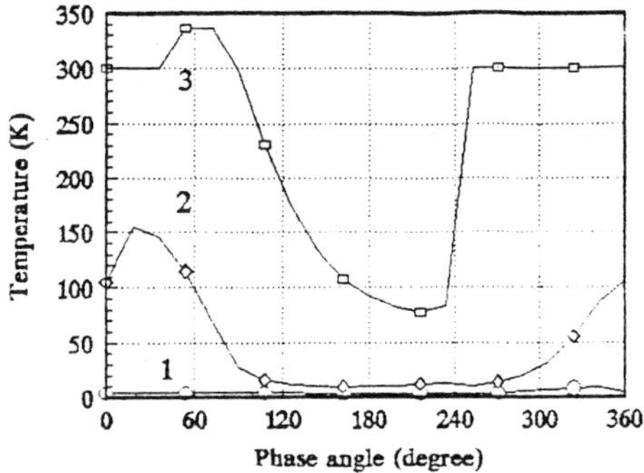


**Figure 9.** Temperature wave of double-orifice.

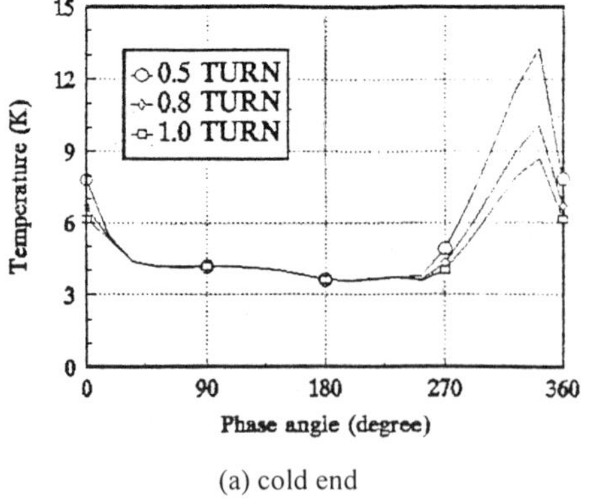


(a) cold end

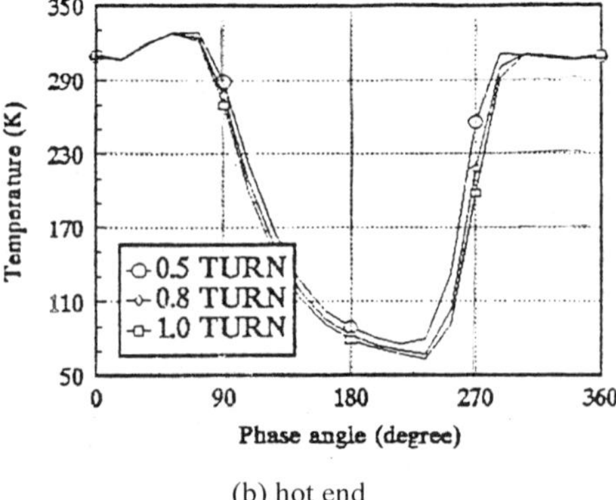


(b) hot end

**Figure 10.** Temperature wave at different positions of double-orifice type.

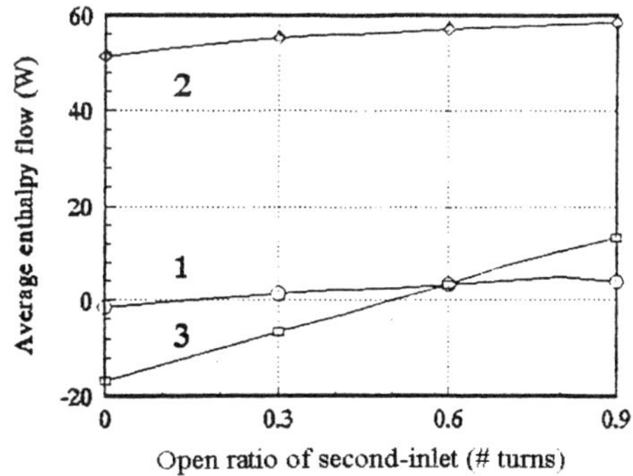


**Figure 11.** Time-averaged enthalpy flow of double-orifice type.

double-inlet version. It is caused mainly by the second orifice. Fortunately, the open ratio of the second orifice is so small that the small rise in average mass flow is acceptable.

Fig. 9 shows temperature oscillation of the gas calculated at various positions for the double-orifice pulse tube refrigerator. Compared with double-inlet version (Fig. 4), the amplitude of the temperature oscillation at the hot end increased slightly and decreased at the middle and cold ends of the second stage pulse tube. This means the effect of the hot gas flowing into pulse tube, caused by using a second-inlet valve on the temperature oscillations at the cold end and the middle of pulse tube decreased. The transient gas temperature for different open ratios of the second-inlet at the cold and hot ends are compared in Figs. 10(a) and (b). With the same open ratio of the second-inlet, the gas temperatures at the cold end are lower than that of the double-inlet version (Fig. 5). Curve 2 in Fig. 7 shows the average gas temperature along the second stage pulse tube of this version. In contrast to double-inlet version (curve 1), the double-orifice version obtains a lower average gas temperature. This is the main reason why the double-orifice version performs better.

Fig. 11 shows the time-averaged enthalpy flow in the second stage pulse tube of the double-orifice version. When open ratios of the orifice and second-inlet are fixed, there is an optimum setting of the double orifice valve to obtain the maximum enthalpy flow. With the accretion of second-orifice, the value of time-averaged enthalpy flow increases and changes from negative to positive. That means the second-orifice transfers enthalpy flow from pulse tube to the reservoir, thereby increasing the heat pumping effect of the pulse tube. Compared with the double-inlet version (Fig. 6), the double-orifice version has a larger enthalpy flow at the cold end, i.e. the double-orifice may obtain much greater cooling capacity at 4K.

## CONCLUSION

1. An adiabatic model for 4K pulse tube refrigerator has been proposed to study the heat transfer and fluid flow in the pulse tube. Compared with the conventional numerical model, the adiabatic model is simpler and easier to use.
2. Numerical predictions reveal anomalies of energy transfer and temperature distribution within the pulse tube. A double-orifice two-stage pulse tube refrigerator has been proposed to obtain liquid helium temperatures.
3. Calculation results show that the double-orifice configuration increases the enthalpy flow and decreases the temperature that can be achieved at the cold end of the pulse tube refrigerator.

## REFERENCES


1. Matsubara, Y. and Gao, J.L., Novel configuration of three-stage pulse tube refrigerator for temperatures below 4K, Cryogenics, 1994, 34, 259
2. C.Wang et al. A two-stage pulse tube cooler operating below 4K, Cryogenics, 1997, 37, 159
3. Guobang Chen, Limin Qiu, Jianyao Zheng, et al. Experimental study on a double-orifice two-stage pulse tube refrigerator, Cryogenics, 1997, 37, 271
4. Guobang Chen,Jianyao Zheng, Limin Qiu, et al. Modification test of staged pulse tube refrigerator for temperatures below 4K, Cryogenics, 1997, in press
5. V. Arp et al., Hepak User's Guide, Cryodata, 1995
6. Wang Chao, Ju Yonglin, and Zhou Yuan, Experimental Investigation of a two-stage pulse tube refrigerator, Cryogenics, 1996, 36,605
7. T. Shigi et al. Anomaly of one stage double-inlet pulse tube refrigerator, ICEC/ICMC 16, Japan, May, 1996

# NUMERICAL SIMULATION AND EXPERIMENTAL VERIFICATION OF THE OSCILLATING FLOW IN A MULTI-BYPASS PULSE TUBE REFRIGERATOR WITH AREA CHANGE

**Y. L. Ju, C. Wang, and Y. Zhou**

Cryogenic Laboratory, Chinese Academy of Sciences
P. O. Box 2711, Beijing, 100080, China

## ABSTRACT

This paper describes an improved numerical model for simulating the oscillating gas flow and detail performance of a multi-bypass pulse tube refrigerator (MPTR) with area change. The governing equations that include the pressure gradient, inertia, viscous and convection terms are based on the conservation of mass, energy and momentum for oscillating flow. The control-volume method is used to discrete the governing partial differential equations on a staggered grid. Simulation results and predicated performance are compared with experiment. Good agreement has been found between the two. The effects of the orifice, double-inlet and multi-bypass valves on the dynamic parameters and refrigeration performance of the MPTR are presented.

## INTRODUCTION

The need for cryogenic cooling of space-base electronics, infrared detectors and the associated electronics has brought about a strong requirement for high reliability and efficiency in refrigerators. The refrigerator meeting most capable of these requirements is the pulse tube refrigerator that was invented by Gifford and Longsworth[1]. Developments and innovations[2-6] in recent years have enabled the pulse tube refrigerator to achieve comparable refrigeration with Stirling or G-M machines. A detail review of the works done on the pulse tube refrigerator can be found in reference[6].

The working process of the pulse tube refrigerator is very complex due to the unsteady, oscillating compressible gas flow, the porous media in the regenerator and the addition of the orifice valve and double-inlet valve, etc. Many previous analyses of the pulse tube refrigerator are based on the method of enthalpy flow and phase analysis reported by Storch and Radebaugh[7]. Unfortunately, the quantitative accuracy of the analytical results was not satisfactory. An understanding of the mechanisms associated with pulse tube refrigerator requires a full solution of the Navier-Stokes equations. However, the complexity of these equations makes an analytical solution for pulse tube refrigerator essentially impossible. Therefore, numerical methods have usually been employed to simulate the oscillating process. The numerical modeling presented by Wu and Zhu[8] neglected the pressure drop in the regenerator, so it was impossible to obtain satisfactory quantitative results. Wang et al. published a set of numerical papers[9-10] which included the gas flow friction and heat transfer in the heat exchanger and the regenerator. However, the momentum equation for the regenerator in their model is not in the naturally characteristic conservation form and neglects the momentum equation of gas flow in the pulse tube. Therefore, an improved numerical modeling method for simulating the detail dynamic performance and characteristics of the oscillating gas flow in the MPTR has been developed in the paper.

## PHYSICAL MODEL AND GOVERNING EQUATIONS

Pulse tube and regenerator length, are much longer than the typical hydraulic diameter of the cross-section of pulse tube. The variation in the thermal properties and momentum in the radial direction seems to have little effect on the behavior of the pulse tube refrigerator. Therefore, the one-dimensional model has been adopted for numerical simulation to reduce the CPU time and avoid any difficulty involved with the two- or three-dimensional models. For the sake of simplicity, it is assumed that the orifice, double-inlet and multi-bypass valve openings remain circular as the flow area of the valves are varied. Neglecting any geometric complexity in regenerator, a simulation of the pulse tube refrigerator is carried out for the simplified one-dimensional physical model as shown in Figure 1.

The basic assumptions are as follows:

(1) one-dimensional laminar compressible flow
(2) the gas and porous medium in regenerator are in local thermal equilibrium
(3) the porous medium is incompressible, uniform porosity
(4) boundary and variable permeability effects are negligible

The one-dimensional system of conversation laws for the oscillating gas flow, according to conservative form, which is usually used in the numerical simulation of compressible fluid flow, is the following:

$$\frac{\partial E}{\partial t} + \frac{\partial F}{\partial x} + G = 0 \tag{1}$$

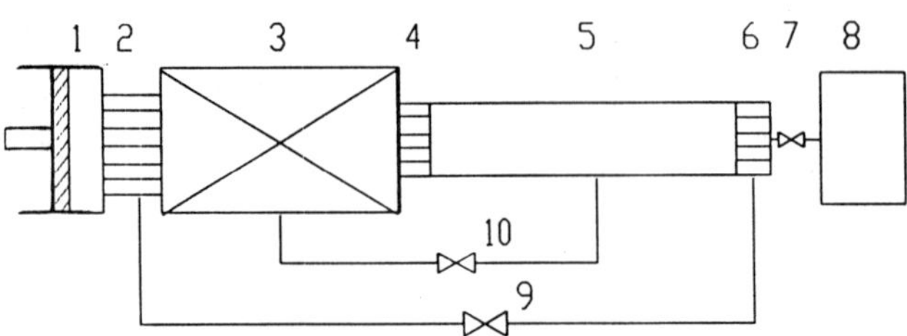


1 compressor; 2 aftercooler; 3 regenerator; 4 cold heat exchanger; 5 pulse tube; 6 hot heat exchanger; 7 orifice; 8 reservoir; 9 double-inlet valve; 10 multi-bypass valve

**Figure 1.** Schematic of the physical model for the pulse tube refrigerator.

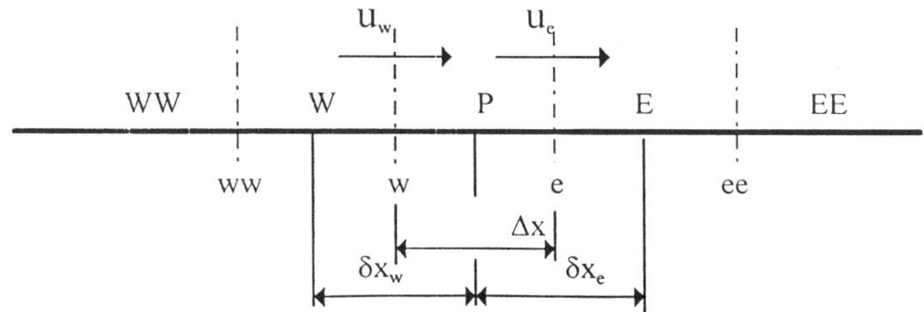


**Figure 2.** The schematic of the staggered grid.

where:

$$E = \begin{bmatrix} \rho \\ \dot{m} \\ E \end{bmatrix} \qquad F = \begin{bmatrix} \rho u \\ \dot{m}u + P \\ u(E+P) \end{bmatrix} = \begin{bmatrix} \rho u \\ \dot{m}u + P \\ \rho u h \end{bmatrix} \qquad G = \begin{bmatrix} 0 \\ \sigma \\ S \end{bmatrix}$$

$$\sigma = f\frac{\rho}{2D_h}|u|u \qquad S = \frac{\partial q}{\partial x} \qquad q = \alpha(T - T_w)$$

here u is the velocity, m is the mass flow rate, $u=\dot{m}/\rho$, E is the total internal energy, $E = \rho(e + uu/2)$, and $H = \rho h = E + P$ is the total enthalpy, h is the specific enthalpy.

Equation of state:

$$P = \rho RT \tag{2}$$

Assuming simple sinusoidal oscillations for the compressor, thus the compressor volume change with respect to crank angle is

$$V = V_d + \frac{1}{2}V_S(1 + \cos 2\pi ft) \tag{3}$$

where $V_d$ is the dead volume and $V_S$ is the swept volume.

It should be noted that the mass flow rate is defined as positive if the gas flows from the compressor to the reservoir and as negative for the back towards the compressor. The mass flow rate through the orifice and double-inlet valves are determined using the formula for a convergent nozzle with a correction factor which is determined by the experimental data.

It is resonable to assume constant wall temperature of the hot and cold heat exchangers, and the working process in the reservior to be at a constant temperature. Since we can not find the the friction factor and heat transfer coefficient for the oscillating flow, and the transition from laminar to turbulent flow is not obvious in the pulse tube refrigerators, we adopted the steady state friction factor and heat transfer coefficients given by Kays and London[11].

## NUMERICAL METHODS AND PROCEDURE

The above equations are non-linear and unsteady. Considering the non-continuity of interface temperature at various components in the pulse tube refrigerator, the complex heat and fluid flow boundary, the conjugate thermal conduction and convection between the gas flow and the solid matrix in the regenerator, etc., some effects were made to increase the calculation accuracy. The governing equations are discretized using a control-volume approach by the inter node method[12]. We adopted the staggered grid approach (Figure 2) to eliminate the mistaken pressure profiles. This approach uses control volumes for the velocities that are staggered with respect to those for temperature and pressure.

A solution is obtained from an initial guess through an iterative scheme using a line-by-line TDMA(tri-diagonal matrix algorithm) method. The governing equations are solved with the fully implicit and upwind second-order scheme for the dependent variables(as discussed later), harmonic mean formulation for the interface diffusivities. The conjugate thermal conduction and convection between the gas and the solid matrix in the regenerator are handled numerically by solving the same full set of momentum and energy equations throughout the entire pulse tube refrigerator, but with a large value of viscosity specified for the solid matrix. A general numerical scheme for Equation (1) in Figure 2 can be written as

$$E_P^{n+1} = E_P^n + \lambda\theta(F_e^{n+1} - F_w^{n+1}) - \lambda(1-\theta)(F_e^n - F_w^n) + G_P^n \tag{4}$$

where, $\lambda = \Delta t/\Delta x$, the superscripts n and n+1 represent time levels, and $\theta$ is a measure of implicitness of the scheme. We can obtain explicite, fully implicit, and Crank-Nicolson schemes for $\theta = 0$, 1, and 1/2 respectivity. $F_P$ are the numerical fluxes at the control-volume interfaces. A general expression for the upwind second-order scheme at the interface can be written as:

$$F_e^{(1)} = \frac{1}{2}(F_P + F_E) - \frac{1}{2}\Gamma[\frac{1}{2}(F_E - F_P), - \frac{1}{2}(F_E - F_P)] \tag{5}$$

$$F_e^{(2)} = F_e^{(1)} + \frac{1}{2}[(F_P - F_w) + (F_E - F_{ee})] \tag{6}$$

where the function $\Gamma$ is defined as the largest among the variables, that is $\Gamma(x,y)=\max0(x,y)$.

Eq. (5) can be differentiated and rewritten by using the TDMA method. Computational accuracy and convergence rate are affected by the orifice and double-inlet valve. In order to improve the simulation accuracy, the mass flow rate through the orifice and double-inlet valves, calculated by the mass conservation equation, is corrected by Eq.(4). The iteration results should be satisfied by periodic stable conditions for all the time-dependent parameters and the energy balance condition for the matrix of the regenerator during one cycle.

After a grid refinement study, all calculations presented here were carried out using a 200 non-uniform grid size. The calculated results with a finer grid do not reveal any noticeable variations. The solution was obtained through the entire pulse tube refrigerator, from compressor to reservoir. With the oscillating compressible gas flow and the conjugate thermal conduction and convection between the gas flow and the solid matrix in the regenerator, the solution took a long time to obtain a convergent solution, requiring 200~300 CPU min on a HP 5/90C mainframe.

## COMPARISON WITH EXPERIMENT

Verification of our numerical modeling was carried out by comparing the simulation results with experimental data. The test machine was a MPTR with two-bypass tubes and area change. The configuration of the experimental apparatus was the same as given in reference[13].

A typical calculation has been made at conditions as follows: The operating wall temperature of the cold and hot end heat exchangers are 77K and 300K, respectively. The system mean pressure is 1.2MPa, and the operating frequency is 16Hz. Predicted time-average wall temperature distributions of the pulse tube and regenerator, and dynamic pressure compared with the experimental data are shown in Figures 3a and 3b, respectively. The results from the numerical simulation show reasonable agreement with the experimental data.

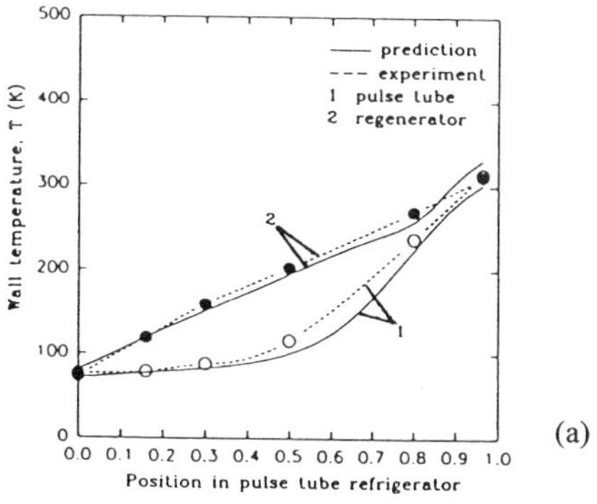


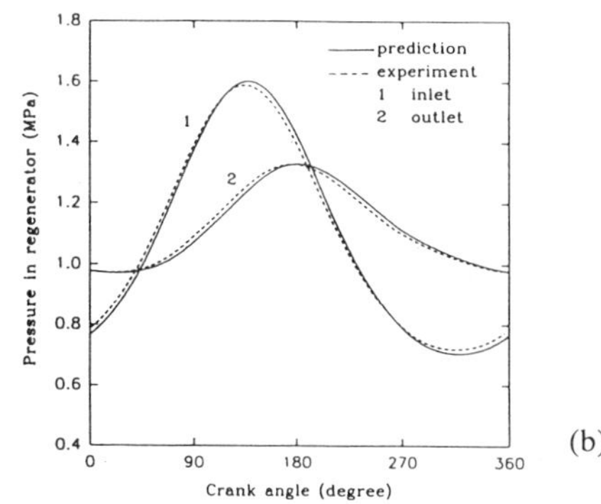


**Figure 3.** Comparison of the predicted simulation results with the experimental data (a) time-average wall temperature distributions; (b) dynamic pressure in the pulse tube.

Figure 4a, 5a and 6a are plots of the predicted refrigeration power of orifice, double-inlet and multi-bypass refrigerator, respectively. Curve 1 in the Figures is the refrigeration power obtained by neglecting pressure drop in pulse tube refrigerator; Curve 3 is the refrigeration power calculated by our numerical modeling; Curve 2 is the losses and curve 4 is the compressor work. Results at constant cold end temperature are compared to the experimental data in figures 4b, 5b and 6b, in which the cold end temperature are plotted as a function of the valve opening. It is easily observed that at the optimum valve opening, the low temperature in the experiment corresponds to the maximum refrigeration power in the numerical simulation.

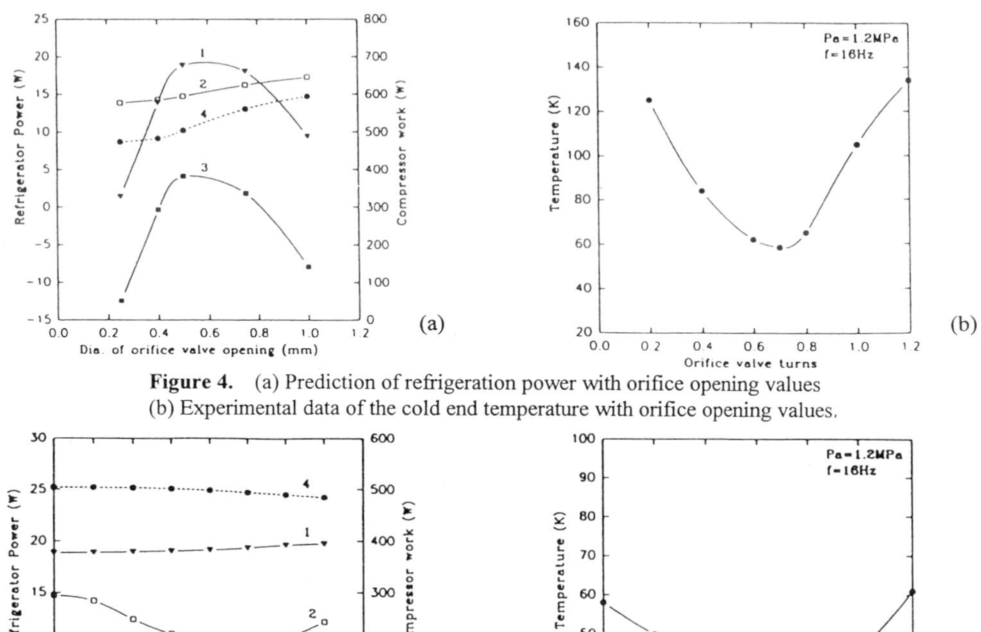


**Figure 4.** (a) Prediction of refrigeration power with orifice opening values
(b) Experimental data of the cold end temperature with orifice opening values.

**Figure 5.** (a) Prediction of refrigeration power with double-inlet opening values
(b) Experimental data of the no load cold end temperature with double-inlet values ($d_o$=0.5mm).

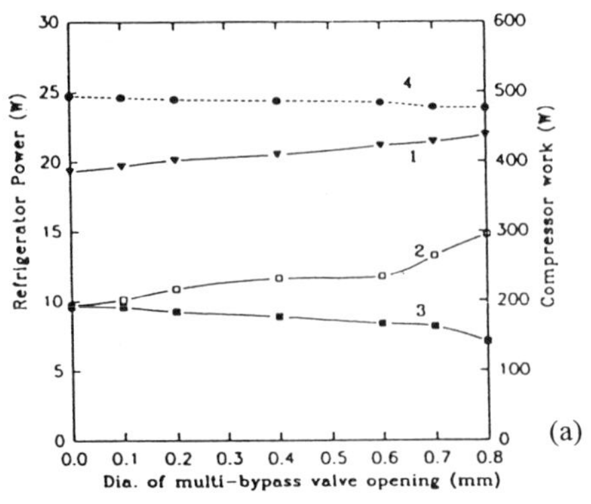

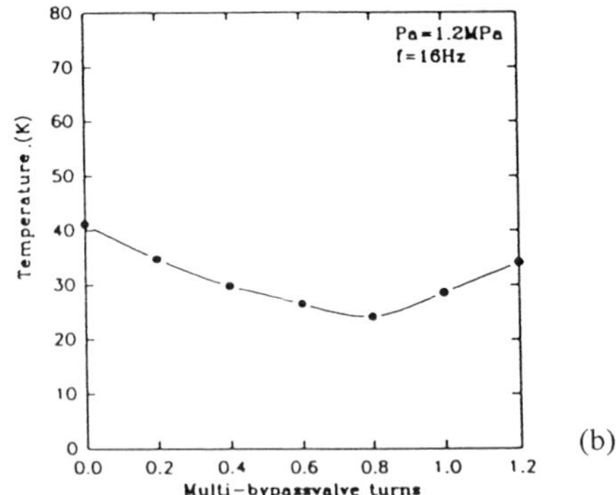


**Figure 6.** (a) Prediction of refrigeration power with multi-bypass opening values
(b) Experimental data of the no load cold end temperature with multi-bypass values ($d_o$=0.5mm, $d_1$=0.5mm).

## NUMERICAL RESULTS AND DISCUSSIONS

Table 1 shows the predicted time-average half-cycle parameters and refrigeration performance of the MPTR. It can be seen the effect of the optimum opening of orifice, double-inlet and multi-bypass valves on the performance of the pulse tube refrigerator.

**Table 1.** Predicted refrigeration performance of the MPTR at optimum orifice, double-inlet and multi-bypass opening values (Tc = 77K, $P_{mean}$ =1.2MPa, f =16Hz)

| Type | | orifice | double-inlet | multi-bypass |
|---|---|---|---|---|
| optimum opening valve | | $d_o$=0.5mm | $d_1$=0.5mm | $d_{k1}$=0.6mm, $d_{k2}$=0.6mm |
| pressure ratio in compressor | | 1.87 | 1.83 | 1.79 |
| pressure ratio in pulse tube | | 1.33 | 1.36 | 1.42 |
| mass flow rate at hot end of regenerator, | (g/s) | 1.32 | 1.32 | 1.33 |
| mass flow rate at cold end of regenerator, | (g/s) | 0.542 | 0.551 | 0.517 |
| mass flow rate at cold end of pulse tube, | (g/s) | 0.521 | 0.530 | 0.509 |
| mass flow rate at hot end of pulse tube, | (g/s) | 0.121 | 0.123 | 0.131 |
| enthalpy at hot end of regenerator, | (W) | 18.0 | 3.62 | 4.49 |
| enthalpy at cold end of regenerator, | (W) | 2.25 | 2.22 | 2.83 |
| enthalpy at cold end of pulse tube, | (W) | 21.4 | 21.8 | 24.6 |
| enthalpy at hot end of pulse tube , | (W) | 27.0 | 27.3 | 34.6 |
| mass flow rate at double-inlet, | (g/s) | 0 | 0.201 | 0.198 |
| enthalpy flow rate at double-inlet, | (W) | 0 | -12.6 | -12.5 |
| mass flow rate at multi-bypass $K_1$, | (g/s) | 0 | 0 | 0.0548 |
| enthalpy flow rate at multi-bypass $K_1$, | (W) | 0 | 0 | 1.80 |
| mass flow rate at multi-bypass $K_2$, | (g/s) | 0 | 0 | 0.0457 |
| enthalpy flow rate at multi-bypass $K_2$, | (W) | 0 | 0 | 1.94 |
| compressor work, | (W) | 504.28 | 494.55 | 479.91 |
| refrigeration power, | (W) | 4.11 | 9.60 | 8.15 |

Figure 7 shows the dynamic pressure wave and mass flow rate in regenerator and pulse tube at the optimum opening values of the orifice, double-inlet and multi-bypass valves. We can find clearly the amplitude and phase difference between the dynamic pressure and the mass flow rate at the hot and cold ends of the multi-bypass pulse tube refrigerator.

It should be noted that the phase shifts between the dynamic pressure and the mass flow rate at the cold end of the pulse tube are independent of the orifice, double-inlet and multi-bypass opening values. In this pulse tube refrigerator system, the phase difference are still about 45~50 degree (curve 3 and curve b in Figure 7), this agrees with the experiment data[13].

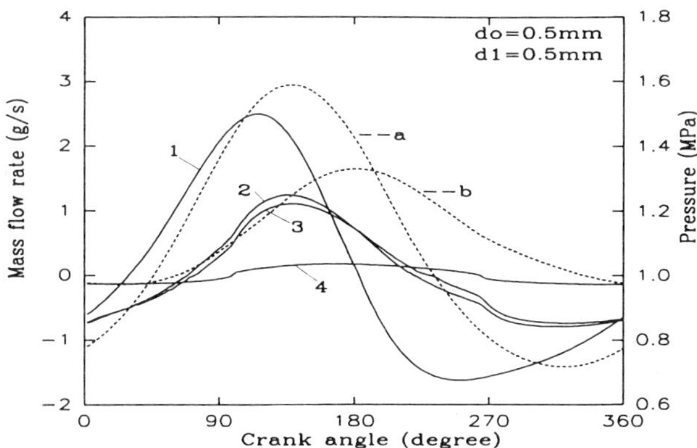


**Figure 7.** Variation of the dynamic pressure and mass flow rate in the pulse tube refrigerator
1 hot end of regenerator; 2 cold end of regenerator; 3 cold end of pulse tube; 4 hot end of pulse tube
a pressure in compressor; b pressure in pulse tube.

## CONCLUSIONS

An improved numerical modeling for simulating the oscillating flow and detail dynamic performance of a multi-bypass pulse tube refrigerator has been developed. The governing equations that include the pressure gradient, inertia, viscous and convection terms are based on the conservation of mass, energy and momentum for oscillating flow. Good agreement has been found between the numerical simulation and experimental data. The simulation model is useful for predicting the effect of the orifice, double-inlet and multi-bypass valves on the refrigeration power and efficiency of the pulse tube refrigerator.

## REFERENCES

1. Gifford, W.E. and Longsworth, R.C. Pulse tube refrigeration, Trans. ASME, No. 63- WA-290, NOV. 17-22:264 (1963).
2. Mikulin, E. I., Tarasov, A. A. and Shrebyonock, M. P. Low-temperature expansion pulse tube, Adv. Cry. Eng. 29: 629 (1984).
3. Zhu, S., Wu, P. and Chen, Z. Double inlet pulse tube refrigerator--An important improvement, Cryogenics. 30:514 (1990).
4. Zhou, Y. and Han, Y. J. Pulse tube refrigerator research, Int. Cryocooler Conf. 7:147 (1992).
5. Ishizaki, Y. and Ishizaki, E. Experimental study and modelization of pulse tube refrigerator below 80K down to 23K, Int. Refrigeration Conference, 7:140 (993).
6. Radebaugh, R. Recent developments in refrigerators, Int. Con. Ref. 19: 973 (1995).
7. Storch, P. J. and Radebaugh, R. Development and experimental test of an analysis model of the orifice pulse tube refrigerator, Adv. Cry. Eng. 33:851 (1988).
8. Wu, P. Y and Zhu, S. W. Mechanism and numerical analysis of orifice pulse tube refrigerator with a valveless compressor, Proc. Int. Conf. Cry. and Refrig. 85 (1989).
9. Wang, C., Wu, P. Y. and Chen, Z. Q. Numerical modeling of an orifice pulse tube refrigerator, Cryogenics, 32:785 (1992).
10. Wang, C., Cai, J. H., Zhou, Y. and Wang, S. Q. Numerical analysis and experimental verification of multi-bypass pulse tube refrigerator, Adv. Cry. Eng. 41B:1389 (1996).
11. Kays, W. M. and London, A. L. Compact Heat Exchangers, 2nd McGraw-Hill, New York, USA (1964).
12. Patankar, S.V. Numerical Heat Transfer and Fluid Flow, Hemisphere/2McGraw-Hill, New York, USA 1980).
13. Ju, Y. L., Wang, C. and Zhou, Y. Dynamic experimental study of the multi-bypass pulse tube refrigerator with two-bypass tubes, presented at CEC/ICMC 97, Portland, USA.

# DEVELOPMENT OF LOW FREQUENCY PULSE TUBE REFRIGERATORS

A. Ravex, J.M. Poncet, I. Charles, and P. Bleuzé

C.E.A. Grenoble/Département de Recherche Fondamentale sur la Matière Condensée-Service des Basses Températures
17, rue des Martyrs, 38054 Grenoble Cédex 9, France

## ABSTRACT

Gifford-McMahon (G.M.) like pulse tube (P.T.) refrigerators are under development at CEA/SBT. The pressure oscillation in the P.T. is generated by a commercial helium compressor associated with a rotating distribution valve. The P.T. are operated at low frequency (1-2 Hz) and their performance and potential applications are close to those of conventional G.M. cryorefrigerators.

A single stage prototype achieving an ultimate temperature of 26 K and producing a net cooling power of 100 W at 80 K has been developed. The effective performances have been compared to the prediction of a theoretical model. The influence of geometrical parameters and P.T. inclination have been extensively studied. The contribution to parasitic heat input of gas permanent flow has also been evaluated.

A double stage prototype is presently under tests with the aim to achieve a simultaneous cooling power of 7 W at 20 K and 40 W at 80 K.

To prepare the future development of a 4 K multistage P.T., the characterisation of a 20 K-4 K stage using a rare earth regenerator has been performed. Some experimental results are compared to model predictions. An ultimate temperature of 3.5 K has been achieved with a net cooling power of 30 mW at 4.5 K.

## INTRODUCTION

Various applications for mechanical cryorefrigerators are today clearly identified such as detectors, electronics and superconductor cooling, cryopumping, thermal shield cooling in large helium cryostats or even cryogenic fluids liquefaction. Low cost and high reliability are important factors for the industrial success of these applications. As a consequence of these criteria, the most commonly used mechanical closed cycle cryorefrigerators are today of the G.M. type. The P.T. can even improve the reliability and

*Advances in Cryogenic Engineering*, Vol. 43
Edited by P. Kittel, Plenum Press, New York, 1998

reduce both manufacturing and operation costs of G.M. refrigerators by the suppression of all moving parts and associated clearance and friction constraints at low temperature. For this reason extensive work is performed worldwide to develop and industrialise G.M. type P.T. Some of the developments undertaken at CEA/SBT in this area are presented hereafter. They are part of a larger program[1] including also high frequency P.T. of Stirling type.

## SINGLE STAGE PULSE TUBE : DESCRIPTION AND PERFORMANCES

We have developed a single stage P.T. for operation at 2 Hz with a goal of 100 W for net cooling power at liquid nitrogen temperature. The double inlet[2] configuration has been selected since it has been previously proven that it was improving P.T. performances[3].

The pressure oscillation in the P.T. is generated by a commercial helium compressor commonly used for G.M. refrigerators, associated with a rotating distribution valve. The measured volumetric flow rate of this CTI 8500 compressor is 10 $m^3/h$ when operated at 50 Hz with suction and exhaust pressures of typically $7.10^5$ Pa and $20.10^5$ Pa. The rotating distribution valve has been developed at CEA/SBT to allow for a variable rotation speed (0-10 Hz) during development tests. Its technology is similar to existing G.M. refrigerators rotating distribution valves.

The geometry of the P.T. has been calculated and optimised using a numerical model[1,3] previously developed at CEA/SBT. The main construction parameters are the following :

| | | |
|---|---|---|
| Pulsation tube : (stainless steel) | inner/outer diameter | 29/30 mm |
| | length | 200 mm |
| Regenerator tube : (stainless steel) | inner/outer diameter | 40/41 mm |
| | length | 104 mm |
| Regenerator mesh: (stainless steel) | wire diameter | 50 µm |
| | aperture | 71 µm |

Room temperature and cold tip heat exchangers, located at each extremity of the pulsation tube, are made of copper gauze discs stacked and soft soldered in a copper block. They also act as flow straighteners.

The orifice impedance is made of a calibrated thin walled orifice and the associated buffer volume is about 600 $cm^3$. The double inlet impedance is made of a capillary. Both orifice diameter and capillary length to diameter ratio have been experimentally adjusted to achieve optimal performances.

The P.T. prototype has been initially assembled in an in-line configuration which is known to allow for optimal performance due to the reduction of dead volumes and flow disturbances at regenerator and pulsation tube junction.

In Figure 1 is shown a photograph of the pulse tube with its distribution valve. In this picture, the P.T. is assembled in a U-shape configuration.

**Figure 1** . U-shape pulse tube with distribution valve.

The performance obtained with this single stage P.T. prototype are presented in Figures 2 and 3.

Figure 2 illustrates the behaviour of the P.T. when changing the operation frequency : the cooling temperature for a given applied heat load (80 W) is plotted versus the operation frequency. The optimal operation frequency is very close to the specified value (2 Hz) for the design, illustrating the reliability of our model. At high frequency, the compressor is not able to generate the appropriate pressure oscillation and the regenerator efficiency decreases due to the mass flow rate augmentation : it results in a degradation of performance. At low frequency the amplitude of the pressure oscillations is kept constant by an internal by pass in the compressor for safety reasons and furthermore the gas displacement in the pulsation tube increases too much, reducing the overall efficiency : cold gas is reaching the room temperature heat exchanger and warm gas the cold tip heat exchanger.

In Figure 3 is reported the net cooling power versus the cooling temperature : the ultimate temperature (no net cooling) is close to 26 K and the net cooling power of 100 W is achieved at (72 K). The dashed line represents the calculated performance : there is a good agreement for the ultimate temperature which is the most severe test for the model. The discrepency, for a given heat load, between the measured and the calculated temperatures results from the overall heat transfer coefficient at the cold tip (0.2 K/W) : the experimental temperature is measured on the outer side of the cold tip copper block, the calculated temperature is for the cycle helium gas.

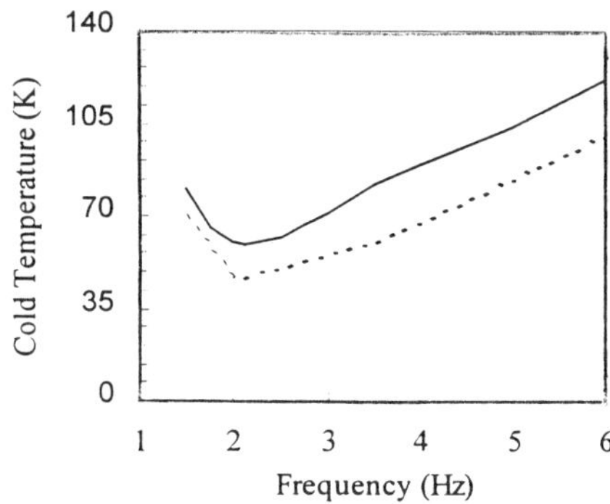


**Figure 2** . Inflence of the frequency on the cooling temperature for a constant applied load (80 W).

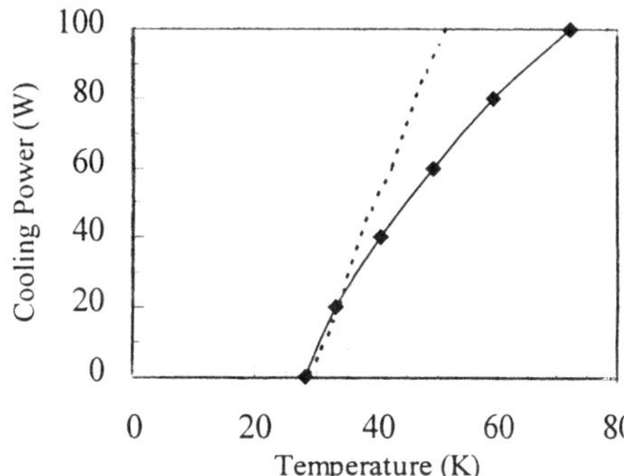


**Figure 3** . Cooling power versus cooling temperature.

(Dashed line represents a model prediction)

## SINGLE STAGE PULSE TUBE : PARAMETRIC STUDY

The influence of several parameters on the P.T. performances has been experimentally studied to test their sensitivity for potential modifications in some applications.

### Length to diameter ratio of the pulsation tube

For a given regenerator size (diameter = 45 mm, length = 82 mm) the length to diameter ratio l/∅ of the pulsation tube is modified keeping the tube volume constant. The Figure 4 illustrates the effect of this parameter on the cooling temperature for a given heat load (80 W). For l/∅ ranging between 3.5 and 7 the performance of the P.T. remains almost insensitive. The reduction of l/∅ below 3.5 rapidly decreases the performance due to the increasing importance of the perturbed flow zone at each end of the pulsation tube following the flow straighteners. The performance degradation when increasing l/∅ over 7 is less important. The increasing proportion of gas involved in wall heat transfer compared to the proportion involved in the adiabatic enthalpy flow process when the pulsation tube diameter decreases is the reason for this evolution.

### Regenerator length to diameter ratio

For a given pulsation tube (diameter = 29 mm, length = 200 mm) the length to diameter ratio l/∅ of the regenerator is modified keeping the regenerator volume and type of matrix constant. Figure 5 illustrates the effect of this parameter on the cooling temperature for a given heat load (80 W). For small values of l/∅ the heat transfer coefficient is reduced and for large values of l/∅ the pressure drop is increased. Both effects reduce the P.T. performances.

### Pulsation tube to regenerator volumes ratio

For a given regenerator, the volume of the pulsation tube has been increased. The corresponding geometries and performances are summarised in Table 1.

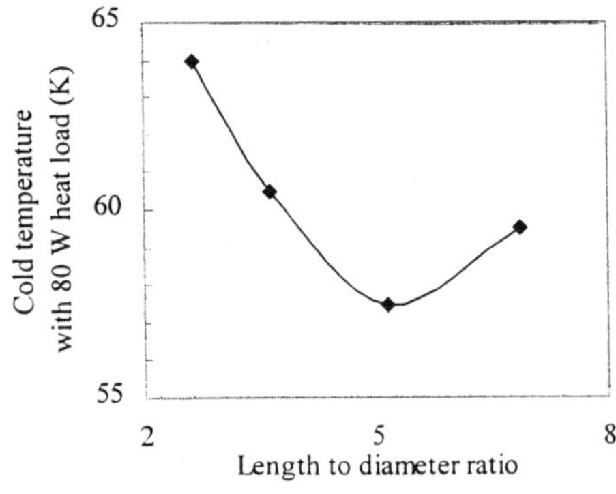


**Figure 4** . Influence of the pulsation tube length to diameter ratio.

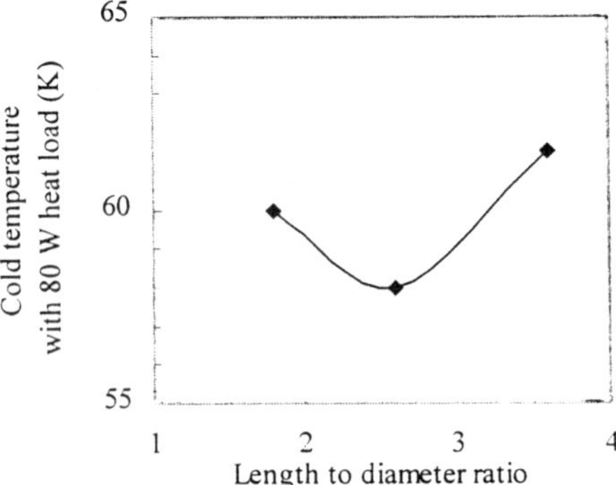


**Figure 5** . Influence of the regenerator length to diameter ratio .

**Table 1.** Influence of pulsation tube volume

| Tube volume | Tube diameter (mm) | Tube length (mm) | Cooling temperature (K) with 80 W load | Optimal operation frequency (Hz) |
|---|---|---|---|---|
| Nominal | 29 | 200 | 60 | 2.25 |
| Increased | 36 | 200 | 67 | 1.75 |

The pulsation tube volume augmentation has slightly reduced the performance : the regenerator has a larger mass flow rate to thermalise and becomes therefore less efficient. The optimal operation frequency is also shifted to a lower value as expected since the volumic flow rate of the compressor remains constant. This also confirms that the calculated geometry was effectively optimised for the 2 Hz specified operation frequency.

**Pulse tube geometrical assembly effect**

The P.T. prototype was initially assembled in a in-line configuration which is optimal from a gas flow point of view but which is not obviously suitable for all applications. We have consequently tested both a U-shape configuration which is more comparable to the traditionnal G.M. refrigerator geometry and an in-line configuration with a bent (90°) pulsation tube. In Table 2 we report the different performances obtained for these various configurations for the same operation parameters (frequency, pressure oscillation) and the same tube and regenerator volumes. We can thus experimentally observe how flow perturbations can influence the P.T. performances.

**Convective heat losses**

In a P.T. operated at low frequency, it can be easily calculated that inertia forces associated with the oscillatory mass flow and gravity forces are of the same order of magnitude. Consequently, as a function of inclination, convection may occur and drastically increase the internal heat losses. This fact has been recently observed and published[4] by other authors. We have observed similar results at the nominal operation frequency (2 Hz) as illustrated in Figure 6 where the cooling temperature under a 80 W heat load is plotted versus inclination. Attempts to weaken convective effects by the introduction of porous material in the tube were not successful due to the induced regenerative effect which is incompatible with the adiabaticity required for the enthalpy flow. It has been experimentally observed that the cooler is no longer sensitive to convective perturbation if the operation frequency exceeds about 8 Hz.

**Table 2.** Influence of geometrical configuration

| Type of assembly | Ultimate temperature (K) | Cooling power (W) at 80 K |
|---|---|---|
| in line | 26 | 105 |
| U-shape | 30 | 85 |
| in line (bent tube) | 32 | 80 |

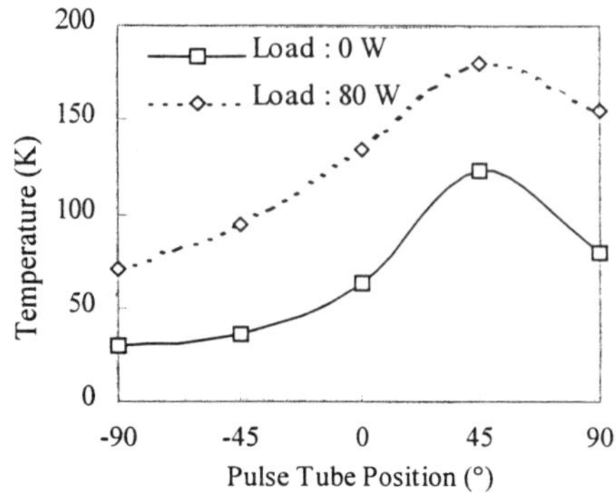


**Figure 6** . Influence of the P.T. inclination.

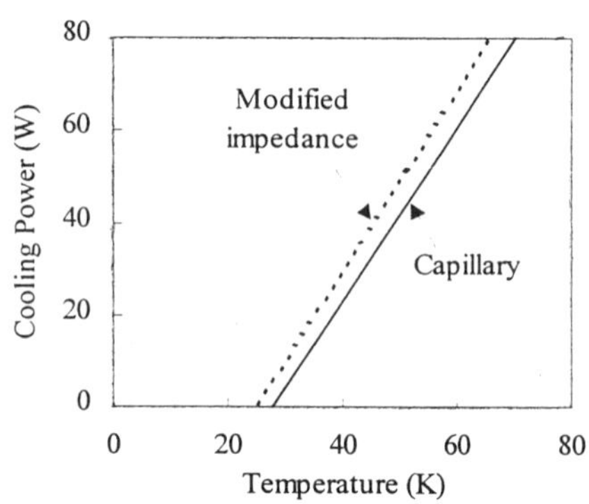


**Figure 7** . Permanent flow influence on P.T. performances .

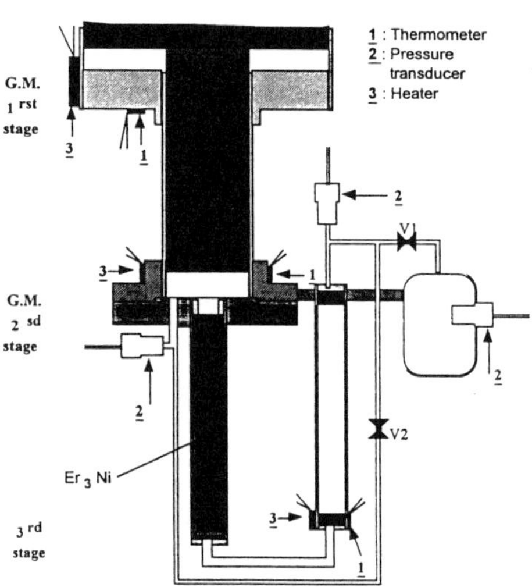


**Figure 8** . Scheme of the PT 20 K - 4 K stage associated to a GM cooler .

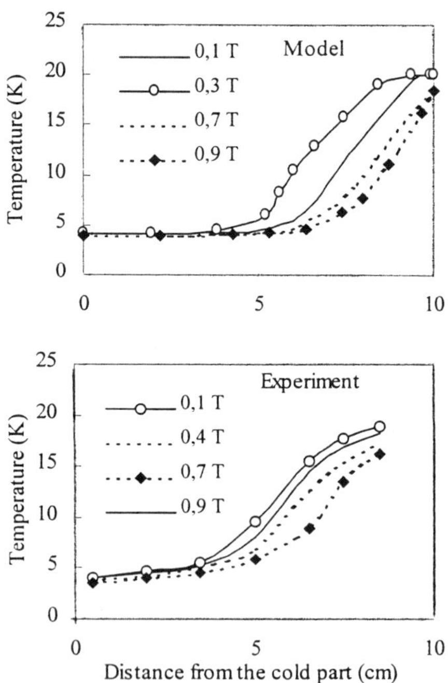


**Figure 9** . Evolution with time of the temperature profiles in a 20 K - 4 K regenerator .

## SINGLE STAGE PULSE TUBE : PHASING AND PERMANENT FLOW

It can be shown, for example using the enthalpy flow analysis[5], that for an optimal operation of the P.T., the mass flow and the pressure oscillations have to be in phase in the pulsation tube. In a single orifice configuration this is automatically achieved at the room temperature end of the pulsation tube using a resistive type impedance. But the condition is not satisfied at the cold end, due to the capacitive effect of the pulsation tube, and thus a large mass flow rate is passed through the regenerator. The use of the double inlet configuration solves this problem, allowing for the appropriate phasing at the cold end, thus reducing the mass flow rate and associated thermal load in the regenerator. An alternate solution to control this phasing has been recently suggested[6] : the active buffer system.

It is also generally accepted that a drawback of the double inlet configuration is the possibility for a permanent flow[7] through the regenerator and the pulsation tube to be superimposed on the forced alternate flow. This may induce an important parasitic heat load. We have performed extensive work recently at CEA/SBT, either on low frequency or high frequency P.T., to find a signature and measure the effect of such permanent flow. This work will be published elsewhere. Nevertheless to illustrate the permanent flow effect in the case of the present single stage prototype we have reported in Figure 7 two different cooling curves obtained using the original capillary and a modified double inlet impedance. In the temperature range explored there is no modification of the cooling power versus cooling temperature slope (2 W/K), indicating that the basic cooling process has not been modified. The difference in absolute cooling power (about 7 W) corresponds to a permanent flow effect, estimated to be of the order of a few mg/sec and which has been observed by other indicators such as temperature profile in the tube.

## MULTISTAGE PULSE TUBE DEVELOPMENTS

Using the G.M. compressor (C.T.I. 8500) previously described and used for the single stage prototype, we are developping a two stage P.T. The aim is to produce simultaneously 7 W at 20 K and 40 W at 80 K using stainless-steel gauze discs at the first stage regenerator and lead shots at the second stage. A first prototype has been built and is under testing. An ultimate temperature of 16 K at the second stage has been achieved, still limited by some remaining permanent flow. Modifications are in progress to improve the phase shift control and reduce the d.c. flow. Nevertheless, the second stage cooling power curves exhibits a slope of 1 W/K in perfect agreement with the model predictions.

In the course of the development of a multistage 4 K P.T., we have performed numerical simulation and experimental tests of a P.T. stage in the temperature range 20 K-4 K associated to a G.M. cooler in the framework of a thesis[8]. The experimental set-up is shown shematically in Figure 8. In this range of temperature the main features are the unusual properties of the regenerator matrix (magnetic material - $Er_3Ni$ in our case - with large specific heat anomaly due to magnetic ordering) and of the cycle helium gas which no longer behaves as a perfect gas. Resulting from these facts is a specific evolution of the regenerator matrix temperature with time during a period (T on Figure 9), half of the regenerator remaining at 4 K and the other half undergoing large temperature oscillations. In Figure 9 both calculated and experimental temperature profile variations are reported : the good agreement between these curves gives us great confidence in our model. Similar behaviour has been predicted[9,10] and observed[11] by other authors. With this experimental set-up an ultimate temperature of 3.5 K has been achieved with a cooling power of 30 mW at 4.5 K. It is intended to improve this performance in the future either by using new magnetic materials for the regenerator matrix or by replacing the regenerator by a counterflow heat exchanger[8,12,13].

## CONCLUSION

A single stage P.T. with a cooling power of 100 W at liquid nitrogen temperature, as initially specified, has been designed and successfully operated at the nominal 2 Hz frequency. The influence of various geometrical parameters has been tested allowing for future modifications for specific applications and showing the relevance of the initial design and optimisation performed with our thermal model. D.c. flow in double inlet configuration has been observed and cancelled.

Multistage prototypes are undergoing performance adjustment tests.

From a technological and a performance point of view, low frequency P.T. have now achieved the required maturity to complete with G.M. refrigerators.

## REFERENCES


1. A. Ravex, P. Bleuzé, L. Duband, J.M. Poncet and P. Rolland, Pulse tube cooler development at CEA/SBT, in : "Proceedings of 19th International Congress of Refrigeration", IIR, Paris (1995) p.1209.
2. S. Zhu, P. Wu and Z. Chen, Double inlet pulse tube refrigerators : an important improvement, *Cryogenics*, Vol. 30 : 514 (1990).
3. J. Liang, "Development and Experimental Verification of a Theoretical Model for Pulse Tube Refrigeration", Thesis, Grenoble (1993).
4. G. Thummes, M. Schreiber, R. Landgraf and C. Heiden, Convective heat losses in pulse tube coolers : effect of pulse tube inclination, in : *"Cryocoolers 9"*, R.G. Ross Jr., ed., Plenum Press, New York (1997), p. 393.
5. P.J. Storch and R. Radebaugh, Development and experimental test of an analytical model of the orifice pulse tube refrigerator, in : *"Advances in Cryogenic Engineering"*, Plenum Press, New York, Vol. 33 (1988), p. 851.
6. Y. Kakimi, S.W. Zhu, T. Ishiga, K. Fujioka and Y. Matsubara, Pulse tube refrigerator and nitrogen liquefier with active buffer system, in : *"Cryocoolers 9"*, R.G. Ross Jr., ed., Plenum Press, New York (1997), p. 247.
7. D. Gedeon, DC gas flows in Stirling and pulse tube cryocoolers, in : *"Cryocoolers 9"*, R.G. Ross Jr., ed., Plenum Press, New York (1997), p. 385.
8. P. Bleuzé, "Etude et modélisation d'un étage 20 K-4 K de tube à gaz pulsé", Thesis, Grenoble (1996).
9. C. Hong and X. Xu, On the unique features of the low temperature regenerator in a 4.2 K G.M. refrigerator, in : "Proceedings of the 4th Joint Sino-Japanese Seminar on Cryocoolers and concerned topics", Beijing (1993), p. 6.
10. C. Wang, Numerical analysis of 4 K pulse tube coolers, *Cryogenics*, Vol. 37 : 207 (1997).
11. P.Y. Wu, W. Huang, S.L. Hu, Y.L. He, L. Zhang and G.B. Cheng, *Cryogenics*, Vol. 36 : 259 (1996).
12. A. Watanabe, G.W. Swift and J.G. Brisson, Superfluid orifice pulse tube refrigerator below 2 Kelvin,in: "Advances in Cryogenics Engineering", Plenum Press, New York, Vol. 41 (1996), p. 1519.
13. J.L. Gao and Y. Matsubara, An inter-phasing pulse tube refrigerator for high refrigeration efficiency, in : "Proceedings of the 16th Cryogenic Engineering Conference", T. Haruyama, T. Mitsui and K. Yamafriji, ed., Elsevier Science, Oxford (1997), p. 295.

# EXPERIMENTAL RESULTS OF A SINGLE STAGE LINEAR ORIFICE PULSE TUBE REFRIGERATOR

M. Ishimura and Erik Bodegom

Department of Physics and
Environmental Sciences and Resources Program
Portland State University
Portland, OR 97207

## ABSTRACT

A single stage linear Orifice Pulse Tube Refrigerator (OPTR) was designed, and tested. Gas pressure, gas/wall temperature, heat load, phase shift, and frequency were measured and a simple discrete model of the pulse tube was developed to describe the cooling power. Our model treats the gas in the tube as a set of elements. The summation over these small gas elements permits the calculation of the cooling power. Based on the experimental data, the cooling power was evaluated. An analysis of the differences between our experimental and theoretical results yields a better understanding of the features of the OPTR system.

## INTRODUCTION

There is a tremendous need for a low cost and reliable small size cryocooler. The applications range from cooling computers to operating detectors in space. The orifice pulse tube refrigerator (OPTR) is a candidate to fulfill this need. After the idea of an orifice pulse tube refrigerator was introduced, many experimental papers were written. Some analytical models have been developed such as the enthalpy flow model by Storch et al.[1] and the thermoacoustic model based on Rott's work.[2-3] However, there is as yet no adequate theory for the OPTR system because the basic cooling mechanisms are not well understood. Therefore, a new approach was developed based on the conservation of energy of the system. The cooling power (the rate of heat flow out of the system) was calculated by this model and compared with experimental results. In this model the cooling power is separated into two different contributors which are: 1) the amount of heat carried by the surface heat pumping effect, and 2) the amount of heat that is exchanged at the hot-end heat exchanger due to the direct contact of the working fluid and the heat exchange elements.

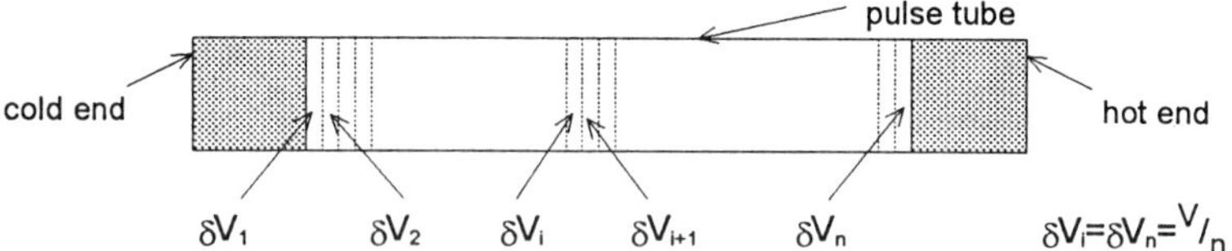


**Figure 1.** Individual gas elements considered in the model.

## THEORY

The amount of gas in the pulse tube is a function of time due to the oscillatory pressure change of the gas in the tube. It is assumed that the gas pressure in the tube is uniform, i.e., there is no pressure difference in the tube at any moment. The gas temperature within the tube is a function of time and position because of the temperature gradient caused by shuttle heat transfer.

The volume of the pulse tube can be divided into $n$ tiny equal volume elements as shown in Figure 1. It is assumed that the gas in each tiny element is in thermal equilibrium at any moment and the ideal gas law can then be applied.

### The Cooling Power of the System

The hot-end heat exchanger is the only place where the system can reject heat to the outside of the system. When the system is in steady state, the amount of heat rejected through the hot-end heat exchanger (cooling power) must be equal to the heat that comes into the system. The heat that comes into (and is rejected through) the hot-end heat exchanger, is separated into two parts. One is the heat which moves into the hot-end heat exchanger by the surface heat pumping mechanism. The other is the heat which is rejected because of the direct contact with the heat exchanging elements (copper screens) in the hot-end heat exchanger. This heat exchanger is water cooled. By calculating these two heats which come into the hot-end heat exchanger separately, the cooling power of the system can be determined.

### Contributions due to the Heat Carried by Surface Heat Pumping

The amount of heat which is carried out of the pulse tube by the surface heat pumping effect is one part of the cooling power of the system. After the compression period and at the highest gas pressure, $P_H$, the $i^{th}$ gas element in the pulse tube has the enthalpy of:

$$h_{i,H} = C_P m_i T_{i,H} = C_P \frac{P_H}{R}\left(\frac{V}{n}\right)\frac{1}{T_{i,H}}(M.W.)\,T_{i,H} = C_p \frac{P_H}{R}\left(\frac{V}{n}\right)(M.W.) \tag{1}$$

where $h_{i,H}$, $m_i$, and $T_{i,H}$ are the enthalpy, the mass, and the temperature of the $i^{th}$ gas element respectively. $C_P$, $R$, $V$, $n$ and $(M.W.)$ are the specific heat at constant pressure, the gas constant, the volume of the pulse tube, the number of elements in the tube, and the molecular weight of working gas respectively. Eq. (1) is the enthalpy of the $i^{th}$ gas element at $P_H$ with no heat load. The total enthalpy is:

$$H = \sum_i h_{i,H}' = C_P \frac{P_H}{R} V(M.W.)\,. \tag{2}$$

Then the gas element gives heat to the wall and the temperature of the gas element drops from $T_{i,H}$ to $T'_i$ which is equal to the wall temperature corresponding to the $i^{th}$ element of gas. Therefore, assuming $m_i$ remains constant, the enthalpy of the $i^{th}$ gas element drops to:

$$h'_{i,H} = C_P \, m_i \, T'_i = C_p \frac{P_H}{R} \left(\frac{V}{n}\right) \frac{1}{T_{i,H}} (M.W.) \, T'_i \,. \tag{3}$$

The total enthalpy of gas in the pulse tube (after heat is transferred to the wall) is then,

$$H' = \sum_i h'_i = C_P \frac{P_H}{R}\left(\frac{V}{n}\right)(M.W.)\left(\frac{T'_1}{T_{1,H}} + \frac{T'_2}{T_{2,H}} + \cdots + \frac{T'_n}{T_{n,H}}\right). \tag{4}$$

If the heat transfer from the gas to the wall occurs for all of the gas, then the total amount of heat transferred to the wall is the difference of eq. (2) and eq. (4) because for this isobaric process $\Delta H$ is the energy exchanged between gas and wall.

$$\Delta H = H - H' = C_P \frac{P_H}{R} V (M.W.) - C_P \frac{P_H}{R}\left(\frac{V}{n}\right)(M.W.)\left(\frac{T'_1}{T_{1,H}} + \frac{T'_2}{T_{2,H}} + \cdots + \frac{T'_n}{T_{n,H}}\right). \tag{5}$$

However, the heat transfer from gas elements to tube wall occurs only for the gas that is inside of the thermal boundary layer of the gas. If the ratio of the cross sectional area for the region inside the thermal boundary layer to the whole cross sectional area is $\alpha$, and if the efficiency of heat transfer from gas to wall is $\beta$, then the net heat transfer to the wall is:

$$\Delta H_{Surf} = \alpha \, \beta \, C_P \frac{P_H}{R} V (M.W.) \left[1 - \frac{1}{n}\left(\frac{T'_1}{T_{1,H}} + \frac{T'_2}{T_{2,H}} + \cdots + \frac{T'_n}{T_{n,H}}\right)\right]. \tag{6}$$

$\Delta H_{Surf}$ is the amount of heat which is transferred per cycle from the gas inside of the thermal boundary layer to the tube wall when the temperatures of the gas and wall at the hot-end are $T_{n,H}$, $T'_n$, and the temperature of gas and wall at the cold-end are $T_{1,H}$ and $T'_1$ respectively. $\Delta H_{Surf}$ can be calculated from the experimentally obtained values for the temperature as a function of position in the pulse tube. For a fixed wall position the amount of heat transferred from a gas element to the wall during the $P_H$ must be equal to the amount of heat transferred from the wall to the gas element during $P_L$ because the wall temperature is stable when the system is in steady state. Eq. (6) is the amount of heat that is pumped by the surface heat pumping effect at the condition of zero heat load. The factor $\alpha$ for a system depends on the diameter of the pulse tube. The factor $\beta$ is unknown at this moment, but in principle can be obtained from experimental observations.

Similarly, $\Delta H_{Surf}$ can be expressed in terms of the lowest gas pressure, $P_L$, and the gas temperatures at $P_L$.

$$\Delta H_{Surf} = \alpha \, \beta \, C_P \frac{P_L}{R} V (M.W.) \left[\frac{1}{n}\left(\frac{T'_1}{T_{1,L}} + \frac{T'_2}{T_{2,L}} + \cdots + \frac{T'_n}{T_{n,L}}\right) - 1\right] \tag{7}$$

where $T_{i,L}$ is the temperature of the $i^{th}$ gas element when the pressure is $P_L$, which gives the same result as eq. (6).

When a heat load ($Q_c$) is applied to the system, the temperatures of gas and wall increase. We define $T_{i,H,Q}$, $T_{i,L,Q}$, and $T'_{i,Q}$ to be the highest gas temperature, the lowest gas temperature, and the wall temperature of the $i^{th}$ gas element respectively. Then, at $P_H$ the $i^{th}$ gas element has the enthalpy of:

$$h_{i,H,Q} = C_p m_i T_{i,H,Q} = C_p \frac{P_H}{R}\left(\frac{V}{n}\right)\frac{1}{T_{i,H,Q}}\ (M.W.)\ T_{i,H,Q} = C_p \frac{P_H}{R}\left(\frac{V}{n}\right)(M.W.) \tag{8}$$

The total enthalpy of gas in the tube (before the heat transfer occurs) is:

$$H_{H,Q} = \sum_i h_{i,H,Q} = C_P \frac{P_H}{R} V(M.W.). \tag{9}$$

After the heat transfer occurs, the enthalpy of the $i^{th}$ gas element is:

$$h'_{i,H,Q} = C_p \frac{P_H}{R}\left(\frac{V}{n}\right)\frac{1}{T_{i,H,Q}}\ (M.W.)\ T'_{i,Q}. \tag{10}$$

and the total enthalpy of gas in the tube is:

$$H'_{H,Q} = \sum_i h'_{i,H,Q} = C_P \frac{P_H}{R}\left(\frac{V}{n}\right)(M.W.)\left(\frac{T'_{1,Q}}{T_{1,H,Q}} + \frac{T'_{2,Q}}{T_{2,H,Q}} + \cdots + \frac{T'_{n,Q}}{T_{n,H,Q}}\right). \tag{11}$$

But since only a small layer near the wall plays a roll, the effective change in enthalpy, $\Delta H_{Surf,Q}$ is

$$\Delta H_{Surf,Q} = \alpha\beta\ (H_{H,Q} - H'_{H,Q})$$

$$= \alpha\beta C_P \frac{P_H}{R} V(M.W.)\left[1 - \frac{1}{n}\left(\frac{T'_{1,Q}}{T_{1,H,Q}} + \frac{T'_{2,Q}}{T_{2,H,Q}} + \cdots\cdots + \frac{T'_{n,Q}}{T_{n,H,Q}}\right)\right] \tag{12}$$

Eq. (12) gives the amount of heat pumped in the tube for each cycle when a heat load is applied to the system and after $T_{i,H,Q}$, $T_{i,L,Q}$, and $T'_{i,Q}$ are established.

## Contribution due to Direct Contact with the Hot-End Heat Exchanger

The gas which shuttles between the pulse tube and the reservoir passes the hot-end heat exchanger and the heat transfer from gas to the heat exchanging elements (copper screens) occurs in the hot-end heat exchanger. The gas pressure and the gas temperature in the reservoir are assumed to be uniform in the entire reservoir at any moment. In other words, there is no pressure or temperature gradient in the reservoir at any time. The total mass of the gas in the reservoir is a function of time. The mass of the gas can be calculated from the gas pressure and temperature. The mass change with respect to time is the mass flow rate into (or out of) the reservoir. The gas temperature before it enters the hot-end heat exchanger is $T_H$ and the temperature in the reservoir is $T_R$. Experimentally $T_H$ is a function of the time but $T_R$ is not. $T_R$ is almost a constant because of the much larger volume of the reservoir. The enthalpy change at the hot-end heat exchanger, brought about by a small amount of gas which passes through the heat exchanger in a time interval $\Delta t$, is calculated as follows:

$$\Delta h = C_P\ \dot{m}(t)\ (T_H(t) - T_R)\ \Delta t \tag{13}$$

where $\dot{m}(t)$ is the instantaneous mass flow rate. Since the pressure change is small over a short period of time, as a good approximation we can consider that $\Delta h$ is equal to the heat exchanged. Thus for a complete cycle, we have:

$$Q \approx \Delta H_{Exch} = \gamma C_P \int_{t}^{t+\tau} \dot{m}(t)\left[T_H(t) - T_R\right]dt \tag{14}$$

where $Q$ is the heat exchanged at the heat exchanger, $\gamma$ is the coefficient of heat transfer between gas and heat exchanging elements, and $\tau$ is the time for one cycle. The coefficient $\gamma$ is introduced to deal with the fact that we know the temperature of the gas moving into the reservoir ($\dot{m}(t)>0$) from the experimental measurements, but during the expansion of the gas from reservoir into the pulse tube we do not know the temperature right after the passage through the heat exchanger. We assume that $T_H$ is equal to the temperature of the cooling water when $\dot{m}(t)<0$. This overestimates $Q$ and hence we introduce the coefficient $\gamma$. We estimate that $\gamma$ is about 0.7 in our case.[6] So, eq. (14) permits us to calculate $Q$ from our experimental observations.

The total amount of heat which is carried out of the system is the sum of eq. (7) and eq. (14) for the case when no heat load is applied, and the sum of eq. (12) and eq. (14) for the case when a heat load is applied. The latter case is for a system under normal operating conditions.

## EXPERIMENTAL SET-UP

The experiments are conducted to measure gas and wall temperatures, and gas pressures as functions of time and location in the system. Figure 2 shows our experimental set-up. The gas used for the experiments is 99.995 % purified helium gas. At nominal operation the gas pressure (supply) is between 180 and 200 psi. The gas is supplied to the rotary valve from the compressor and then is supplied to the OPTR system (and returned to the compressor) through the rotary valve. The rotary valve is connected to a DC motor. The DC motor speed is controlled by a speed controller. The maximum frequency of gas supply and return is 12 Hz.

The regenerator is located just in front of the pulse tube in the OPTR system. It is made of stainless steel. 1103 screens of 150 mesh stainless steel are packed inside of the regenerator.

The pulse tube itself is simply a hollow stainless steel tube. The cold-end heat exchanger is attached between the regenerator and the pulse tube. It is made of copper. Thirty copper screens (25 mesh) are inserted into the cold heat exchanger in order to increase heat exchange capability. An electric heater (nichrome wire) is wrapped around the cold-end heat exchanger for supplying the heat load to the OPTR system.

The hot-end heat exchanger is placed on the other end of the tube. It is made of copper. Forty-five copper screens (25 mesh) are packed inside the hot-end heat exchanger to enhance heat exchange. The heat exchanger is cooled by water that flows through a copper tube wrapped around the hot-end heat exchanger. The heat exchanger is just beyond the vacuum jacket.

A needle valve is set between the hot-end heat exchanger and the reservoir. This valve acts as an orifice. The gas flow through the valve is controlled by changing the valve opening.

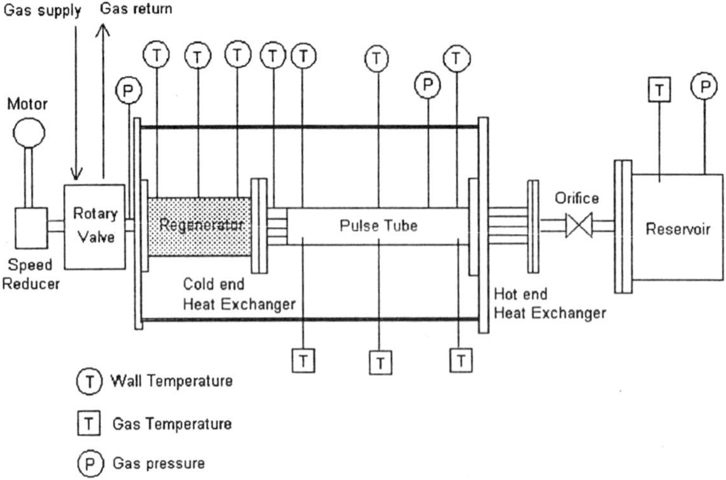


**Figure 2.** Experimental set-up.

The reservoir is made of stainless steel. The reservoir is designed to have a much larger volume than that of the pulse tube. This makes the gas pressure change in the reservoir small. Our system has a reservoir to a pulse tube volume ratio of 13:1.

A stainless steel vacuum jacket covers the regenerator, cold end heat exchanger, and the pulse tube. The vacuum jacket minimizes the heat leak from the outside to the OPTR system. For the experimental set-up, the maximum heat leak to the system is calculated to be 8.5 W. This heat leak is due to conduction through the residual gas, and by conduction along the pulse tube wall, and by radiation.

## RESULTS AND DISCUSSION

The lowest achieved temperatures were measured as a function of the orifice opening. The only parameter actually changed was the orifice opening. The orifice opening controls gas flow that directly contacts the hot-end heat exchanger located between the pulse tube and the reservoir. Therefore, this opening is an important factor for generating low temperatures. The temperature, 161 K (wall), was obtained with a vacuum jacket pressure of 20 ~ 30 microns at f = 8.5 Hz as shown in Figure 3. This pressure was not low enough to allow the generation of lower temperatures because of the relatively large heat leak due to conduction through the residual gas in the vacuum jacket. This temperature is close to the reported value by Shiraishi et al.[4] even though our gas pressure change is much smaller (from a high pressure ($P_H$) of 11.7 atm to a low pressure ($P_L$) of 10.4 atm).

The minimum temperatures were measured as a function of frequency with fixed orifice opening (1/6 turn) and with no heat load. The minimum temperature was obtained when the frequency was about 6 ~ 7 Hz. The minimum temperature was not very sensitive to the operational frequency of the system.

The relationship between the lowest wall temperature and heat load was measured. This experiment was conducted with fixed frequency (f =6.5 Hz) and fixed orifice opening (1/6 turn). The minimum wall temperature was 161 K with no heat load and was 234 K with 30 W of heat load as shown in Figure 4. 30 W is a relatively large heat load for a small size cryocooler. Usually, the heat load would be smaller. The minimum temperature should be lower with a higher vacuum and no heat load.

The optimum orifice opening and operational frequency for zero heat load were 1/6 turn and 6.5 Hz respectively. The optimum condition for each heat load was not

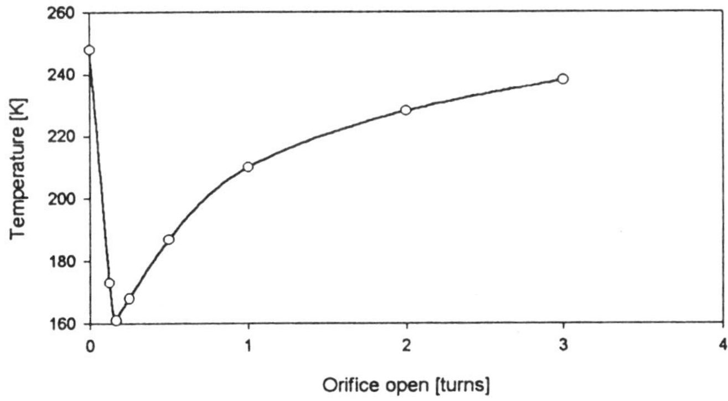


**Figure 3.** Minimum temperatures as a function of orifice opening (f=8.5 Hz).

determined. It might be interesting to determine the optimum orifice opening and frequency for each heat load.

The factor $\alpha$ in eq. (6) was found to be 0.157 with a thermal penetration depth in the pulse tube of about 1 mm.[5] The results are shown in Table 1. The variation of $\Delta H_{Surf}$ is due to uncertainties in the temperature measurements. Two different values of $\Delta H_{Surf}$ were put in the table because the value of $\beta$ is not known at the moment. Consequently, two different values of $\Delta H_{Total}$ were presented in the table. It seems reasonable to assume that $\beta$ is less than 0.5 from the data. Based on a rough calculation of the time available for the heat transfer and the temperature gradient in the penetration depth, a value of 0.5 seems to be on the high side for $\beta$. The factor $\gamma$ was assumed to be about 0.7 based on our experimental setup.[6]

## CONCLUSIONS

An orifice pulse tube refrigerator was built and data were taken. A theoretical model was presented that, in principle, clearly delineates the contribution due to surface heat pumping and the contribution due to the direct contact between the gas and the hot-end heat exchanger. Three factors come into play to completely understand surface heat pumping. One is the thickness of the thermal penetration depth; this thickness is not exactly known in our case. A more elaborate theory needs to be developed that incorporates the temperature and position dependence of this thickness. The other factor is the efficiency of the heat transfer from the gas in the boundary layer to the wall. This factor can be determined. Currently, this was not done.

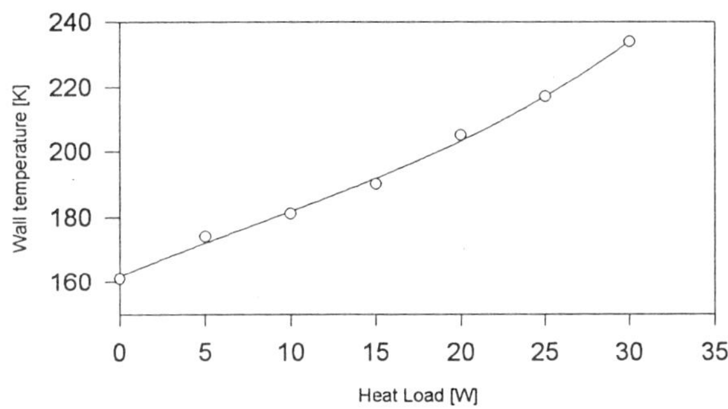


**Figure 4.** The cold-end wall temperature as a function of heat load.

**Table 1.** Calculated and measured heat as a function of $Q_C$
(f = 6.5 Hz, orifice opening = 1/6 turn )

| $Q_C$ (W) | $\Delta H_{surf}$ (W) ($\beta$ =1 ) | $\Delta H_{surf}$ (W) ($\beta$ =0.5 ) | $\Delta H_{exch}$ (W) ($\gamma$ = 0.7) | $\Delta H$ (W) total ($\beta$ =1 ) | $\Delta H$ (W) total ($\beta$ = 0.5 ) | $Q_{OUT}$ (W) by measurements |
|---|---|---|---|---|---|---|
| 0 | 16.5 | 8.3 | 16.6 | 33.1 | 24.9 | 22.3 |
| 5 | 15.5 | 7.8 | 23.4 | 38.9 | 31.2 | 27.8 |
| 10 | 17.0 | 8.5 | 28.0 | 45.0 | 36.5 | 33.4 |
| 15 | 15.9 | 8.05 | 34.6 | 50.5 | 42.6 | 40.5 |
| 20 | 15.4 | 7.75 | 39.5 | 54.9 | 47.2 | 44.3 |
| 25 | 14.4 | 7.2 | 44.5 | 58.9 | 51.7 | 48.2 |
| 30 | 11.9 | 6.0 | 50.1 | 62.0 | 56.1 | 51.9 |

The third uncertainty arises because of the approximation in assuming a constant mass in each of the volume elements in the tube. This assumption needs to be explored in more detail. The contribution to the cooling power due to the direct interaction between gas and hot-end heat exchanger was determined. A more complete model would need to incorporate the efficiency of the heat exchanger. The calculated cooling power of the system was compared with the direct measurement of the cooling power. The calculated cooling power is larger than the experimental result by a factor of 1.19 to 1.48 (if $\beta = 1$) or 1.05 to 1.12 (if $\beta = 0.5$). These results look very promising. It is hoped that, with further fine-tuning, this approach will lead to a better description of the OPTR.

## ACKNOWLEDGMENT

This research has been made possible by an equipment grant from Tektronix.

## REFERENCES


1. P.J. Storch, and R. Radebaugh, Development and experimental test of an analytical model of the orifice pulse tube refrigerator, in: *Advances Cryogenic Engineering* **33:851** (1988)
2. N. Rott, Damped and Thermally Driven Acoustic Oscillations. Part VI: Excitation and Power, *ZAMP (J Appl Math Physics)* **34:609** (1983)
3. N. Rott, Thermoacoustic heating at the closed end of an oscillating gas column, *J. Fluid Mech* **145:1** (1984)
4. M. Shiraishi, K, Seo, and M. Murakami, Pressure and Temperature Oscillations of Working Gas in a Pulse Tube Refrigerator, Cryocoolers **8**:403 (1995)
5. M. David, J.-C. Marechal, Y. Simon, and C. Guilpin, Theory of Ideal Orifice Pulse Tube Refrigerator *Cryogenics*, **33:154** (1993)
6. H. Martin, "Heat Exchangers," Hemisphere Publishing Company, New York (1992)

# EXPERIMENTS ON THE THERMODYNAMIC PERFORMANCE OF A "U-TUBE" PULSE TUBE EXPANDER

C. S. Kirkconnell

Hughes Aircraft Company
El Segundo, California, 92648, USA

## ABSTRACT

As has been extensively discussed in the literature, pulse tube cryocoolers have numerous advantages over other closed-cycle coolers which arise from the absence of moving parts in the pulse tube expander. Advantages in the areas of cost and reliability are of interest to the user community in general, and the additional advantage of cold-tip vibration minimization is of particular importance for cooling applications in electro-optical (E-O) systems for it leads to unique system integration advantages. For example, the low vibration of the expander may make it possible to mount the focal plane directly on the cold tip, eliminating the thermal strap required when using other types of expanders. The 'U-Tube' configuration is particularly convenient from the standpoint of system integration because the cold region extends as a tip from a single warm end heat exchanger, unlike the linear pulse tube (LPT) in which the cold region is bounded on either side by warm regions of the expander. A U-Tube expander was recently constructed and tested, and the thermodynamic efficiency was seen to be comparable to that of the LPT design. At an operating temperature of 77K, a cooling capacity of 2.9 W was measured for an input P-V power of 84 W. The expander design and experimental data are discussed.

## INTRODUCTION

Pulse tube refrigerators (PTRs) are typically categorized according to the method used to attain the desired phase shift between the mass flow rate and pressure waves within the cold region of the refrigerator. The term "Basic Pulse Tube" (BPT) is used to describe the original Gifford and Longsworth design in which oscillatory heat transfer between the gas and pulse tube walls is the dominant phase shift mechanism.[1] Other types of pulse tube refrigerators include the "Orifice Pulse Tube" (OPT),[2] the "Double-Inlet Pulse Tube" (DIPT),[3] and, most recently, the "Multi-Bypass Pulse Tube" (MBPT).[4] The phase shift in the OPT is controlled by an orifice-surge volume arrangement at the warm end of the pulse tube in which an oscillating pressure differential is developed between the gas within the pulse tube and the surge volume gas. Kittel *et al.* describe this as an RC phase shifter.[5] The DIPT and MBPT are variations on the OPT design; the performance of the refrigerator is improved, largely by optimizing the phase angle between the mass flow and pressure wave in the cryogenic region of the cooler, with the use of one or more flow bypasses. The common and distinctive features of these types of PTRs are discussed in detail in the references provided.

A second categorization describing the physical configuration of the refrigerator is also informative. The most common physical configuration, the Linear Pulse Tube (LPT) configuration, features the pulse tube expander components (heat exchangers, pulse tube, and regenerator) connected end-to-end along a single axis. The U-Tube, however, includes a $180^{\circ}$ turn in the flow path at the cold end of the refrigerator. This configuration offers significant advantages for system integration because the expander has a true cold tip, as in a Stirling refrigerator, as opposed to a cold region bounded on either side by warmer components, as in the LPT. With respect to the design of electro-optical (E-O) systems, the combined advantages of low vibration and an easily-accessible cold tip may eliminate the need for a flexible thermal strap to connect the expander to the focal plane assembly or optical bench. This simplifies the mechanical design and reduces the required refrigeration power because a thermal strap temperature delta need not be overcome. Cryocooler efficiency also being a critical system integration criterion, the purpose of this paper is to evaluate the performance of a U-Tube expander to determine if the unique physical features of the U-Tube, such as the porting of the gas at the cold end, may have a deleterious effect on performance.

## EXPERIMENTAL APPARATUS AND INSTRUMENTATION

The U-Tube configuration has been used previously.[6,7] Figure 1 shows this particular U-Tube expander, which is of DIPT design, in cross-section. The U-Tube is particularly convenient for implementing the DIPT design because the compressor and surge tank ports are in close physical proximity. The bypass and surge tank flow restrictors are both housed within the warm-end heat exchanger manifold, but for testing purposes, the surge orifice was removed and the flow regulated with an external metering valve. This allowed changing the setting during test without shut down and disassembly. The warm-end regenerator and pulse tube isothermalizers are also housed within the

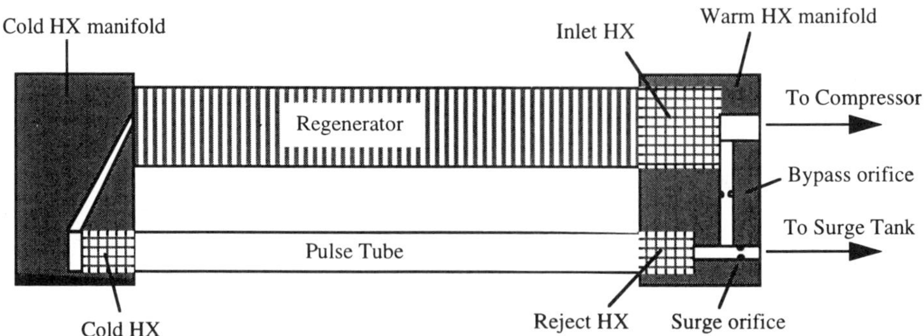


**Figure 1**. Schematic of U-Tube expander. Many design features, such as coupling the two warm heat exchangers into a single manifold, were included for cost reduction purposes.

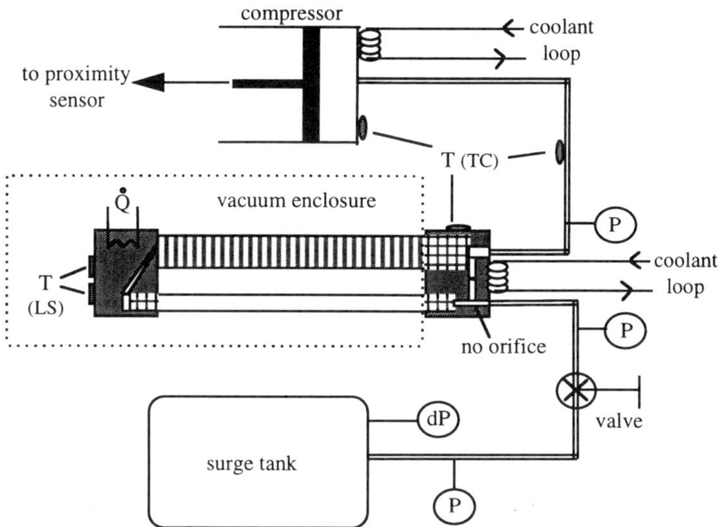


**Figure 2**. Experimental test setup and instrumentation. P = absolute pressure, dP = differential pressure, T = temperature, LS = LakeShore diode, TC = type-T thermocouple, Q = electrical heat load.

warm-end heat exchanger manifold, which is cooled by a circulating glycol-water flow circuit. The oscillating flow is turned 180° within the load heat exchanger at the cold end of the expander. Both heat exchanger housings are of copper construction, the pulse tube and regenerator tube are inconel, and the regenerator and heat exchangers are filled with porous matrices of stainless steel and copper, respectively. The design of this expander was discussed briefly in a previous paper.[8]

The pressure wave was supplied by a wall-powered, rotary-drive laboratory compressor with fixed swept volume (20 cc) and operating frequency (21 Hz). The expander was sized for use with a 10 cc, 30 Hz compressor, so it is somewhat undersized for the laboratory compressor used for these experiments. The expander was made to work with this compressor by increasing the flow restriction at the surge volume entrance to reduce the mean amplitude of the mass flow rate wave within the cooler. To get a more meaningful assessment of the expander efficiency, future experiments may be conducted in which an inlet surge volume will be used as a shunt for the excess input P-V power. At present the compressor is coupled directly to the expander with a 7 cm transfer line.

The cryocooler instrumentation and test configuration are shown in Figure 2. Paine strain gauges and amplifier circuits are used to measure the DC pressure waves at the transfer line, on the expander side of the metering valve, and within the surge tank. A differential AC pressure transducer is used to obtain a precise measurement of the amplitude of the small pressure fluctuation within the surge volume. The compressor is equipped with a proximity sensor which indicates the position of a cam mounted directly off the drive shaft. The mechanical linkage is such that the cam is in phase with the compressor piston, so the proximity sensor provides an accurate measurement of the piston motion itself. The cold tip is instrumented with an 80 Ω resistive heater for load capacity measurements, and two LakeShore DT470 temperature diodes, one for redundancy, are used to accurately measure the refrigeration temperature. A vacuum environment ($< 10^{-5}$ torr) is maintained around the expander using a Leybold turbo pumping station. Glycol-water coolant circuits are used to cool the system at two

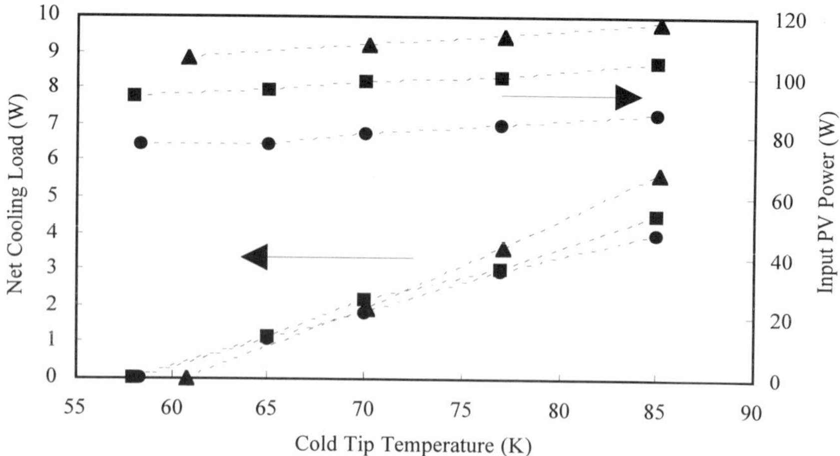


**Figure 3.** Experimental load and power curves. ▲ --> P=28.2 atm, ■ --> P=24.8 atm, ● --> P=20.6 atm.

locations, the compressor head and the expander warm-end manifold. Type-T thermocouples on the heat exchangers are used to measure the rejection temperatures.

## RESULTS

Experiments were performed at three different mean pressures: 20.6 atm, 24.8 atm, and 28.2 atm. As described in the previous section, the compressor frequency and swept volume cannot be varied, so these values were constant at 21 Hz and 20 cc for all three mean pressures. The bypass flow restriction is built into the warm heat exchanger manifold and requires some disassembly and reassembly to change. The test schedule did not allow for the amount of time required to properly optimize this setting, so it was held constant as well. The surge volume orifice setting was adjusted for each mean pressure such that the minimum no load temperature was obtained. Once this orifice setting was

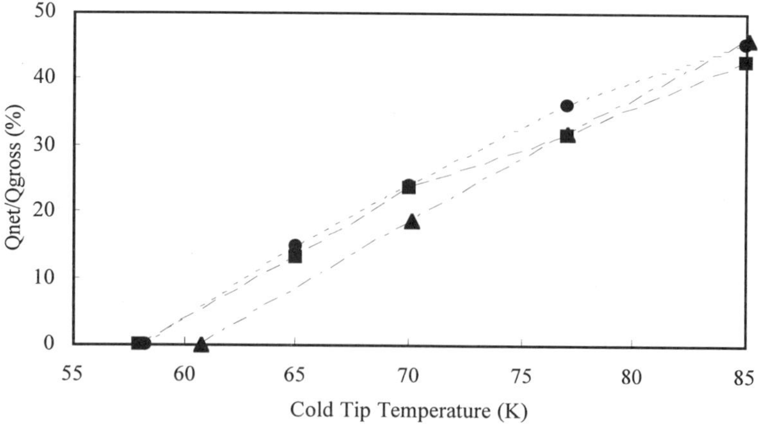


**Figure 4.** Expander efficiency versus mean operating pressure. ▲ --> P=28.2 atm, ■ --> P=24.8 atm, ● --> P=20.6 atm.

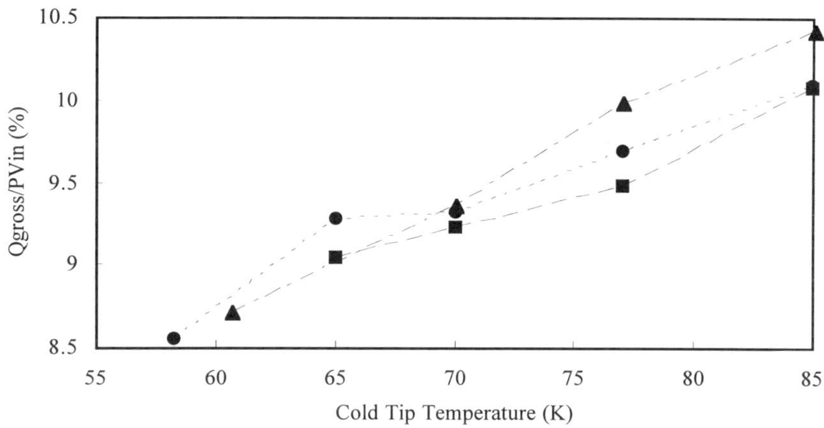


**Figure 5.** P-V power efficiency versus mean operating pressure. ▲ --> P=28.2 atm, ■ --> P=24.8 atm, ● --> P=20.6 atm.

determined for a given pressure, it was held constant for all measurements taken at that pressure.

The net refrigeration load and input P-V power curves for all three cases are shown in Figure 3. The coldest no load temperature achieved was 58.0 K, obtained with the mean pressure set at 24.8 atm. The 20.6 atm case yielded an almost identical 58.2 K no load temperature, but the 28.2 atm case had a minimum temperature almost 3 K higher. This is evidence that the parasitic loads around 60 K were significantly larger for the high pressure case.

Figure 4, a plot of the expander efficiency, defined as the ratio of the net refrigeration to the gross refrigeration, i.e., $\eta_{exp} = \dot{Q}net/\dot{Q}gross$, further illustrates this fact. Note that the expander efficiency for the high pressure case lags the other two cases at low temperatures, trailing by about 5% until crossing over with the intermediate pressure curve at 77 K. The curvature of the traces towards the high temperature end of the data is also of interest. While the slopes of the 20.6 and 24.8 atm curves decrease, the slope of the high pressure curve exhibits no such obvious drop off. Since the radiation and

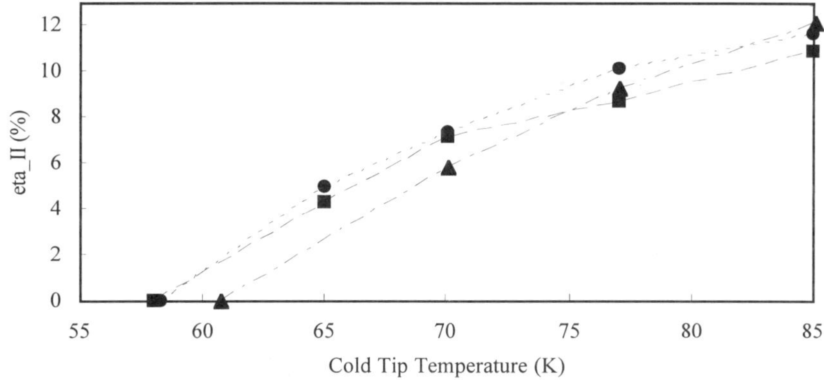


**Figure 6.** Second Law thermodynamic efficiency versus mean operating pressure. These curves reveal that the optimum operating pressure is a function of the desired operating temperature. ▲ --> P=28.2 atm, ■ --> P=24.8 atm, ● --> P=20.6 atm.

conduction losses are (essentially) independent of the charge pressure, a flow loss or combination of flow losses, such as regenerator enthalpy flux or pulse tube secondary flow losses, must be greater for the two lower pressure cases at higher load temperatures. Regenerator inefficiency losses have been discussed extensively, and examples and discussions of pulse tube flow losses can be found in the recent literature.[9,10,11] Though beyond the scope of this paper to evaluate the extent and nature of these contributing losses, this issue of the effect of mean pressure on expander efficiency does warrant further investigation.

Another measure of expander performance is the gross P-V power efficiency, $\eta_g$, defined as the percentage of input P-V power which gets converted to expansion P-V power at the cold tip. Figure 5 is a plot of $\eta_g$ versus cold tip temperature. The curves are relatively indistinguishable below 77 K, above which point the high pressure case separates from the other two.

The ultimate figure of merit for cryocooler expander thermodynamic performance is the second law efficiency, given by the following equations:

Carnot efficiency: $$\eta_C = T_L / (T_H - T_L) \quad (1)$$

First Law efficiency: $$\eta_I = \dot{Q}_{net} / \dot{PV}_{comp} \quad (2)$$

Second Law efficiency: $$\eta_{II} = \eta_I / \eta_C \quad (3)$$

$T_L$ is the cold tip temperature, and the temperature of the warm end manifold is used for $T_H$ since, as verified experimentally, roughly 80% of the heat rejection occurs there. Figure 6 contains the $\eta_{II}$ versus cold tip temperature plots for all three cases. This figure reveals that the optimum charge pressure is not an absolute value, rather it is a function of the desired operating point. For example, the highest second law efficiency at 77 K was achieved with a mean pressure of 20.6 atm (2.9 W, $\eta_{II}$ = 10.1%), while the high pressure setting of 28.2 atm proved to be the optimum at 85 K (5.7 W, $\eta_{II}$ = 12.1%). More extensive experimentation over a wider range of pressures and temperatures is required to state with more precision the effect of the mean pressure on the values of $\eta_g$, $\eta_{exp}$, and $\eta_{II}$.

## DISCUSSION

The arrangement of the regenerator and pulse tube in the U-Tube configuration allows for more intimate thermal coupling between these two components as compared to the linear design. The two parallel tubes are radiatively coupled along the entire length of the expander, though the effect was reduced with a set of simple multi-layer insulation (MLI) blankets, and they are conductively linked at the warm end through the heat exchanger manifold. In addition to these thermal characteristics, another unique feature of the U-Tube is that the oscillating flow must turn 180° in the cold end heat exchanger. If the flow profile entering the pulse tube exhibits a great deal of non-uniformity as a result of this turn, vorticity could be introduced, creating unpredictable, but most likely deleterious, consequences. If necessary, all of these phenomena can be controlled through careful design. The radiative coupling can be further reduced with the use of more extensive MLI shielding, and the conductive coupling can be eliminated by using two

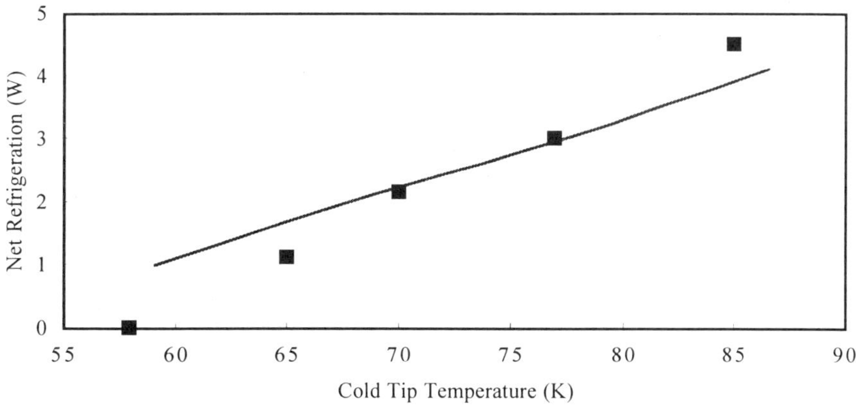


**Figure 7.** Comparison of experimental and numerically-predicted data. Numerical calculations assume linear configuration. P=24.8 atm data shown as typical.

distinct warm-end heat exchangers. Furthermore, the effect of the 180° turn on the inlet pulse tube flow dynamics can be controlled by lengthening the cold-end screen pack.

To determine if these types of design modifications are necessary to achieve comparable performance in the U-Tube as in the linear expander, the experimental data were compared to numerical predictions which were obtained assuming no performance impact as a result of the unique configurational characteristics of the U-Tube. (However, the fact that the rejection temperature is the same for both warm heat exchangers is part of the input data file, so that effect is built into the numerical data.) The numerical model used has been previously correlated against the performance of a range of linear pulse tube expanders.[12] Figure 7 is a comparison of the numerical and experimental load curves for the 24.8 atm case, selected as a representative example of the data set. The experimental data points are seen to agree reasonably well with the numerical prediction. One minor discrepancy is that the slope of the experimental load curve is slightly steeper than that predicted analytically, but this is probably because the input P-V power to the expander drifted up with load temperature (see Fig. 3) due to the imperfect mean pressure modulation provided by the test setup (93 W @ 58 K vs. 105 W @ 85 K). Regardless, it appears evident from this data that the thermodynamic efficiency of the U-Tube expander is comparable, if not indistinguishable, from that of the linear expander.

## CONCLUSION

A U-Tube pulse tube expander has been built and tested. The optimum charge pressure, as determined by maximizing the second law efficiency, was seen to be a function of the desired operating point, the trade-off arising largely because of the effect of the mean pressure on the expander efficiency. Further analysis of the data is planned to investigate the comparative roles of the pulse tube and regenerator flow losses in the observed pressure vs. efficiency effects. In general, the performance has been shown to be essentially identical to that of the more common linear expander. The U-Tube has potentially significant system integration advantages over the linear, so this finding is of significance in the cryocooler applications arena.

## APPENDIX: DATA ANALYSIS

Several direct measurements were taken during these experiments, such as temperatures, pressures, and external heat load (net refrigeration capacity). The accuracy of each direct measurement is dominated by the physical limitations of the measurement method, and the accuracy of the method is primarily a function of the component accuracies of the devices used to implement the method. These measurement device accuracies are typically provided by the manufacturer. The accuracy of the measurement devices used for these experiments ranges from 0.5% to 5%.

In addition to the direct measurements, two important derived measurements were also obtained, input P-V power and expansion P-V power (= gross refrigeration capacity). These quantities were estimated from the direct measurements using the following relationships derived by K.D. Price:[13]

$$PV_{comp} = f\left(\frac{\pi}{2}\right)\overline{P}\,\frac{PR_{comp}-1}{PR_{comp}+1}V_{swept}\sin(\Phi) \qquad \text{(A.1)}$$

$$PV_{ex} = f\pi\frac{(PR_{ex}-1)}{k(PR_{ex}+1)}dP_{surge}V_{surge}\sin(\theta) \qquad \text{(A.2)}$$

(PV = P-V power, comp = compressor, ex = expander, PR = max-to-min pressure ratio, dP = mean-to-peak pressure difference, $\overline{P}$ = mean pressure, Vswept = compressor swept volume, Vsurge = surge tank volume, k = ratio of specific heats, $\phi$ = phase angle between piston motion and pressure wave in the compressor, and $\theta$ = phase angle between mass flow rate and pressure wave at the orifice.) In the case of the input P-V power, Eq. A.1 was used instead of the preferred method of numerically integrating the exact equation, i.e.,

$$\dot{P}V_{comp} = f\oint P_{comp}dV_{comp} \qquad \text{(A.3)}$$

because the data acquisition system used did not have the capability to digitize the piston motion and pressure waves, so numerical integration was not possible. The waveforms were simply displayed on an oscilloscope, and the phase angles and signal amplitudes were read manually from the display. With respect to the expansion power calculation, determining this quantity in a pulse tube refrigerator is always a complicated proposition because the expansion "piston" is only a conceptualization, not a physical entity, as in a Stirling or Vuilleumier cooler. The assumption implicit in Eq. A.2, which is based upon the 1-D pulse tube model, is that the compression power absorbed at the warm end of the pulse tube is equal to the expansion power produced at the cold end. Equation A.2 is therefore actually an estimate of the warm-end compression power.

The uncertainty in these derived measurements is the sum of the uncertainty of the direct measurements from which they are calculated and the inaccuracy of the form of the equations used to describe the actual P-V integral spaces. As stated above, the former are on the order of 1% to 5%. The latter were estimated by utilizing a pulse tube refrigerator model which determines the input P-V power and expansion power by direct integration.[11] (The expansion power calculation is facilitated by modeling the pulse tube gas piston as an adiabatic, constant mass control volume.) Using the physical geometry of the test article and a typical operating point for these experiments, the pressure waves throughout the device were determined using the model. The direct integration calculations for the input and expansion power were then compared to the values obtained manually from Eqs. A.1 and A.2 using the numerical pressure solution to provide the inputs to the equations. For both calculations, the values obtained from the equations were higher than the model's direct integration calculations, 127 W versus 107 W for the input power and 16.5 W versus 14.6 W for the expansion power. These 10-20% errors are significantly larger than the measurement device errors, thus they dominate the total error in these calculations.

Since the same equations were used for all cases and the same measurement devices were used throughout the course of the experiments, the relative values of these derived measurements between data points are believed to be reliable. However, one should bear in mind that the absolute values reported for both input power and gross refrigeration are likely 10-20% high. The present plan calls for the implementation of a data acquisition system which will digitize and store the pressure and piston stroke data to facilitate more accurate P-V power measurements.

## ACKNOWLEDGMENT

The contributions of the following individuals were essential elements of this overall effort: Kenneth Price (LCC Technical Director), Samuel Russo (LCC Program Manager), Alan Rattray (Mechanical Designer), and William Croft (Test Engineer). The author would also like to acknowledge Dr. Martin Nisenoff and the Naval Research Laboratory for their generous loan of the LCC expander used in these experiments.

## REFERENCES

1. W.E. Gifford and R.C. Longsworth, Pulse tube refrigeration progress, in: "Advances in Cryogenic Engineering," vol. 10, (1965), p. 69.
2. E.I. Mikulin, A.A. Tarasov, and M.P. Shkrebyonock, Low-temperature expansion pulse tubes, in: "Advances in Cryogenic Engineering," vol. 29, (1965), p. 629.
3. S. Zhu, P. Wu, Z. Chen, W. Zhu, and Y. Zhou, A single stage double inlet pulse tube refrigerator capable of reaching 42 K, *Cryogenics* 30 suppl., p. 257, (1990).
4. C. Wang, S.Q. Wang, J.H. Cai, and Z. Yuan, Experimental study of multi-bypass pulse-tube refrigerator, *Cryogenics* 35:555, (1995).
5. P. Kittel, A. Kashani, J.M. Lee, and P.R. Roach, General pulse tube theory, *Cryogenics* 36:849, (1996).
6. P. Curlier, Cryocooler technologies: 2. Pulse tube miniature cryocoolers, *SPIE* 2552:791.
7. J. Liang, A. Ravex, and P. Rolland, Study on pulse tube refrigeration, Part 3: Experimental verification, *Cryogenics* 36:101, (1996).
8. S.C. Russo and G.R. Pruitt, Development of a Low Cost Cryocooler for HTS applications, accepted for publication in "Cryocoolers 9."
9. J.M. Lee, P. Kittel, K.D. Timmerhaus, and R. Radebaugh, Steady secondary momentum and enthalpy streaming in the pulse tube refrigerator, in: "Cryocoolers 8", R.G. Ross, Jr., ed., Plenum Press, NY (1995), p. 359.
10. J.R. Olson and G.W. Swift, Acoustic streaming in pulse tube refrigerators: Tapered pulse tubes, to be published.
11. C.S. Kirkconnell, S.C. Soloski, and K.D. Price, Experiments on the effects of pulse tube geometry on PTR performance, accepted for publication in "Cryocoolers 9."
12. C.S. Kirkconnell. "Numerical Analysis of the Mass Flow and Thermal Behavior in High-Frequency Pulse Tubes," Ph. D. thesis, Georgia Institute of Technology, Atlanta, GA (1995).
13. K.D. Price, private communication (1994).

# A SINGLE STAGE FIVE VALVE PULSE TUBE REFRIGERATOR REACHING 32 K

**J. Yuan and J.M. Pfotenhauer**

Applied Superconductivity Center
University of Wisconsin - Madison
Madison, WI 53706 USA

## ABSTRACT

In order to enhance the performance of a single stage pulse tube refrigerator, a new G-M type pulse tube refrigerator with five valves is introduced in this paper. In this novel configuration, five valves with different flow coefficients are used to connect the compressor, regenerator, pulse tube and reservoir. The compression and expansion processes inside the pulse tube are realized by carefully controlling the opening and closing time of the five valves. The system is optimized by changing two intermediate pressures during the compression and expansion processes. A detailed description of the operating procedure and test results of a prototype five-valve pulse tube is given in the paper. Additionally, a simplified thermodynamic simulation is presented and compared with test results. Measurements of the refrigeration power and minimum temperature indicate that the performance of the new configuration is better than that of a standard two-valve double inlet configuration. A cooling capacity of 8 watts at 60 K is achieved with an input power of 1 kW, equivalent to an efficiency of 3.2% of Carnot. A minimum temperature of 32 K was reached with an operating frequency of 1.75 Hz.

## INTRODUCTION

In the last decade, a number of methods have been introduced to improve the performance of orifice pulse tube refrigerators due to their significant advantages over other cryocoolers. In the case of the GM type (or controlled-valve type) pulse tube refrigerator, the most efficient methods include double inlet [1], four valve [2] and active buffer [3]. In general, these modifications attempt to optimize the area of a P-V diagram traced out by the cold end control volume during one cycle, and at same time reduce system losses. Optimizing the area of the P-V diagram at the cold end can be accomplished by manipulating two intermediate pressures and the duration of the isobaric processes. One of the intermediate pressures is produced by gas flowing into the warm end of the pulse tube during the compression process. The other intermediate pressure occurs during the expansion process when gas flows out of the warm end of the pulse tube. The isobaric processes occur when gas moves from the compressor, through the regenerator and pulse tube, to the reservoir, and again when gas moves from the reservoir, through the pulse tube and regenerator, to the compressor.

The shortcoming of the standard double inlet configuration is that there is no control over the two intermediate pressures and additional compressor work is wasted due to gas flow directly between compressor and reservoir through the double inlet valve. The active buffer configuration successfully solves the problems associated with the double inlet configuration, but system complexity is increased due to the introduction of additional reservoirs. In the present study, we introduce a new method called a five-valve pulse tube refrigerator to improve the efficiency of pulse tube refrigerator. In this configuration, active valves with different flow coefficients are used to connect the compressor, regenerator, pulse tube and reservoir. The compression and expansion processes inside the pulse tube are realized by carefully controlling the opening and closing times of the five valves.

As will be detailed below, the performance of the pulse tube is theoretically understood on the basis of a control volume at the cold end of the pulse tube. The boundaries of this control volume are defined by the walls of the pulse tube on the side, by a movable boundary between the top of the cold end volume and the gas piston, and by a fixed boundary at the upper end of the cold end heat exchanger. The control volume can do work on the gas piston, but exchanges no mass or heat with this piston. The side walls are also adiabatic. Mass flow is allowed between the cold end control volume and the cold end heat exchanger. To maximize the area of the P-V diagram associated with the cold end control volume, two intermediate pressures and the duration of the isobaric process are adjusted by manipulating the timing and flow coefficient of the five valves.

To verify the analytical results, a prototype single-stage five-valve pulse tube refrigerator has been built and tested. A cooling capacity of 8 watts at 60 K was achieved with input power of 1.0 kW; equivalent to an efficiency of 3.2% of Carnot. The minimum temperature of 32 K was reached with operating frequency of 1.75 Hz.

## SYSTEM DESCRIPTION

### Configuration

The configuration of the five-valve pulse tube is illustrated schematically in Fig. 1. The system consists of five active valves, a rotary vane compressor, a regenerator, a cold end heat exchanger, a pulse tube, a warm end heat exchanger and a reservoir. Here three active valves are used to control the gas flow at the warm end of the pulse tube instead of the two passive orifice valves in a standard two-valve GM-type double-inlet pulse tube refrigerator. Valves 1 and 3 are connected to the high pressure side of the compressor, while valves 2 and 4 are connected to the low pressure side. A reservoir is connected to the warm end heat exchanger of the pulse tube by valve 5. Valves 1 and 2 have the largest flow coefficients while valves 3 and 4 have smaller flow coefficients and valve 5 has the smallest flow coefficient. The flow coefficients of valves 3, 4 and 5 are adjusted by opening or closing the needle valves connected in series with these valves. Gas compression, expansion and displacement within the pulse tube are controlled by opening and closing the five valves.

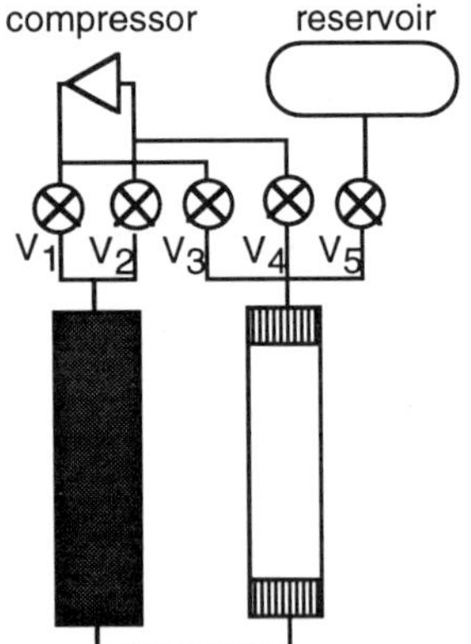


**Figure 1.** Configuration of the five valve pulse tube refrigerator.

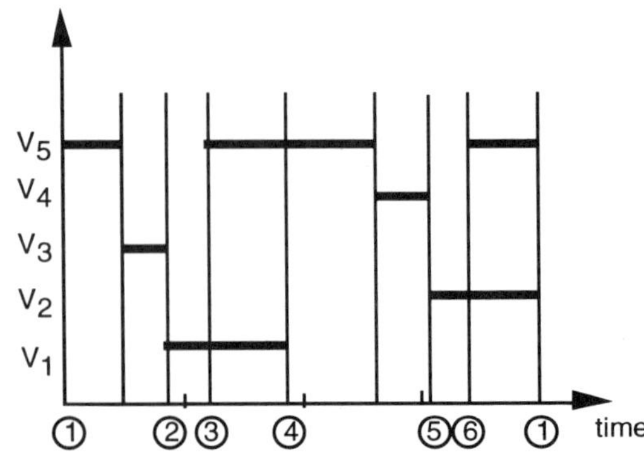


**Figure 2**. A typical valve timing of the five valve pulse tube refrigerator over one cycle.

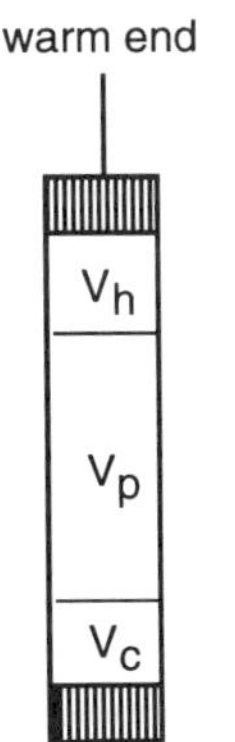


**Figure 3**. Gas distribution in the pulse tube

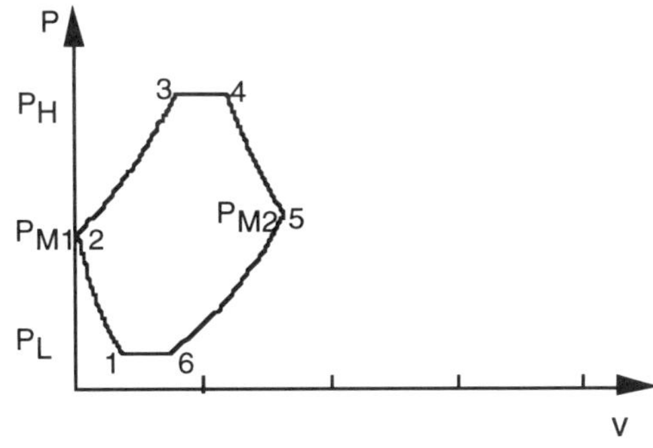


**Figure 4**. Equivalent P-V diagram of cold end control volume .

## Operating Procedure

A typical valve operating sequence is shown in Fig. 2. The working process of the five valve pulse tube refrigerator consists of 6 steps. The thermodynamic cycle of the five valve pulse tube can be described by considering three different gas segments within pulse tube. Figure 3 illustrates a typical gas distribution at one state. Here $V_h$ represents the gas column at the warm end of pulse tube, $V_c$ represents the gas column at cold end of pulse tube, and $V_p$ is the gas column called the gas piston. The distinctive feature of the gas piston is that none of the mass within this volume ever leaves the pulse tube. Figure 4 displays the equivalent P-V diagram of the gas column $V_c$. At the beginning of the cycle, all valves are closed. The system is at low pressure state $P_L$ and the gas column $V_c$ has volume $V_1$.

**Step 1.** (state 1 to state 2): open valve 5 and then valve 3. The gas will flow first from the reservoir and then from the compressor into the pulse tube warm end through valve 5 and valve 3 respectively. The gas pressure will increase from $P_L$ to $P_{M1}$. Note that $P_{M1}$ can be varied from $P_L$ to $P_H$ depending on the timing and flow coefficients of valves 5 and 3. The volume of cold gas changes from $V_1$ to $V_2$ as gas flows from the pulse tube into the dead volume of regenerator.

**Step 2.** (state 2 to state 3): close valve 3 and open valve 1. The system pressure increases to $P_H$ and the gas in the pulse tube is further compressed. The pulse tube cold end will be filled with cold gas through the cold end of the regenerator. The volume of $V_c$ changes from $V_2$ to $V_3$.

**Step 3.** (state 3 to state 4): open valve 5 and keep valve 1 open. The gas in the pulse tube moves towards the reservoir. The cold space volume changes from $V_3$ to $V_4$ which is determined by the timing of valves 1 and 5. This is an isobaric process.

**Step 4.** (state 4 to state 5): close valve 1 and keep valve 5 open. After some time close valve 5 and open valve 4. During this step an adiabatic expansion takes place in the pulse tube. The system pressure drops from $P_H$ to $P_{M2}$. Note that $P_{M2}$ can be varied from $P_H$ to $P_L$ depending on the timing and flow coefficients of valves 5 and 4. The volume of cold gas changes from $V_4$ to $V_5$.

**Step 5.** (state 5 to state 6): close valve 4 and open valve 2. The gas in the pulse tube further expands. The system pressure decreases from $P_{M2}$ to $P_L$. The volume of $V_c$ changes from $V_5$ to $V_6$.

**Step 6.** (state 6 to state 1): open valve 5 and keep valve 2 open. The expanded cold gas will move back towards the cold end heat exchanger. This also is an isobaric process. The system state will return to the initial state, completing one cycle.

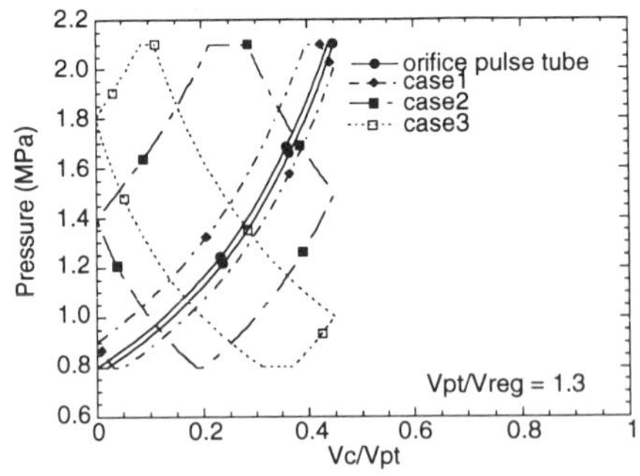


**Figure 5.** An equivalent P-V diagram of the cold space of a small pulse tube.

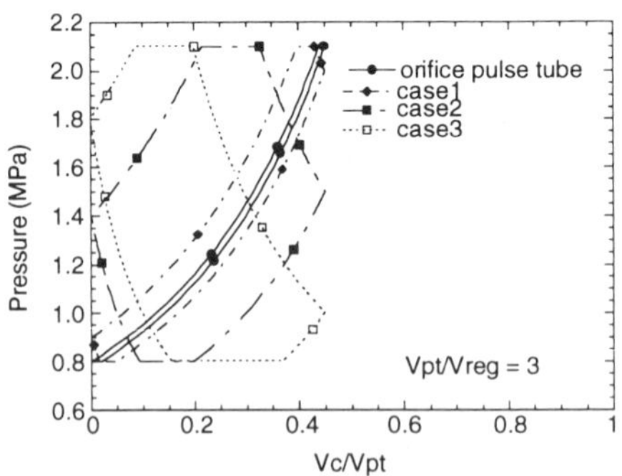


**Figure 6**. An equivalent P-V diagram of the cold space of a large pulse tube.

## Thermodynamic Simulation

A complete and detailed description of the thermodynamic analysis will be presented in a forthcoming publication, however the salient features of the approach are summarized here. In order to predict the optimized intermediate pressures, a thermodynamic simulation was carried out based on the gas distribution shown in figure 3. To simplify the analysis, The following idealizations are made:

1. The ineffectiveness of regenerator is zero.
2. The compressor low and high pressures are constant.
3. No heat transfer or gas mixing occurs between the three gas columns.
4. At any given time, a uniform temperature exists in $V_c$ and $V_h$.
5. The compression processes within pulse tube are adiabatic.

Using the first law of thermodynamics for an open system, it follows that the energy balance for $V_c$ and cold end heat exchanger over one cycle is given by

$$q_c = w_c \tag{1}$$

where $q_c$ is the cooling capacity (heat transferred to cold end heat exchanger) per cycle and $w_c$ is the net work $V_c$ does on the gas piston in one period. From equation (1), it can be seen that the cooling capacity equals the area of cold gas P-V diagram. It should be noted that optimizing the area of the cold gas P-V diagram can be realized by manipulating the following three parameters for a given regenerator, pulse tube, and compressor system: $P_{M1}$, $P_{M2}$ and the timing of process 3-4. Considering the loss caused by the gas shuttle within the pulse tube, there is a limitation on the timing of process 3-4. Although the longer timing of process 3-4 is desirable for maximizing the area of cold gas P-V diagram, the shuttle loss will increase with an increasing duration of process 3-4.

In order to determine the effect of $P_{M1}$ and $P_{M2}$ on the area of the P-V diagram, two series of thermodynamic simulations are carried out. Both sets use the same regenerator, and low and high pressures, but different pulse tube sizes. The ratio of the pulse tube volume to the dead volume of regenerator in the first series is 1.3, while the other is 3. For each series, four cases with different $P_{M1}$ and $P_{M2}$ combinations are investigated. Table 1 shows the

**Table 1**. Calculation Parameters

| case | $P_{M1}$ (MPa) | $P_{M1}$ (MPa) |
|---|---|---|
| orifice pulse tube | 0.8 | 2.1 |
| 1 | 0.9 | 2.0 |
| 2 | 1.4 | 1.5 |
| 3 | 1.8 | 1.0 |

parameters for the four different cases. In the first case, the arrangement is a simple orifice pulse tube, while the remaining three cases utilize the five controlled valves. The maximum volume of $V_c$ is approximately 45% of the pulse tube volume for all calculations. Figures 5 and 6 show the simulation results, and display that the optimized intermediate pressure in fact varies with the pulse tube size. With the small pulse tube, the optimized intermediate pressures are approximately equal to the average pressure of system, while for the larger pulse tube, the optimized intermediate pressures $P_{M1}$ and $P_{M2}$ are respectively close to the high and low pressures of the compressor. The compressor input power also increases with increasing pulse tube size.

## EXPERIMENTAL VERIFICATION

### Apparatus

To verify the qualitative analysis presented above, a prototype single stage five valve refrigerator has been constructed and tested. A GM compressor (Cryomech model CP510) with a nominal input power of 1 kW is used to generate the low and high pressures. The valve operation timing is controlled by a self designed electrical circuit which allow ease in testing different valve timings. The operating frequency can be adjusted over a wide range. The pulse tube is fabricated from a thin wall stainless steel tube with 17.5 mm id, 140 mm length, and 0.5 mm wall thickness: the same volume as the first series simulation. The regenerator is fabricated from a thin wall G-10 tube 100 mm in length, 19 mm in diameter and with a 3.3 mm wall thickness. About 1000 pieces of 200 mesh bronze screens are packed inside the regenerator tube yielding a regenerator porosity of about 65%. The regenerator and pulse tube are arranged side by side but connected at cold end heat exchanger. The pulse tube warm end is equipped with a water-cooled heat exchanger. Both ends of the pulse tube are filled with copper screens with 250 mesh to provide for both a large heat exchanger surface area and flow straightening. The volume of the reservoir is 1000 $cm^3$.

A 30 watt heater is placed at cold end to measure the cooling capacity. Three separate piezoresistive pressure transducers which allow measurement of the pressure oscillation are mounted on the warm ends of the pulse tube, regenerator, and reservoir, respectively. The temperature at the cold end of the pulse tube is measured with a silicon diode manufactured by Lake Shore Cryotronics. Five solenoid valves, manufactured by CO-AX, are used as control valves.

### Results and Discussion

All tests are conducted under low and high pressures of 0.8 MPa and 2.1 MPa respectively. The optimization procedure to minimize the cold end temperature is carried out by changing the two intermediate pressures and the timing of the isobaric process. This is

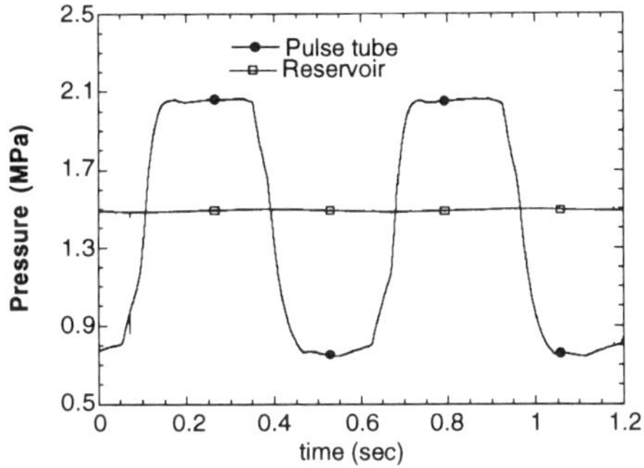


**Figure 7**. Pressure waves at the pulse tube and reservoir.

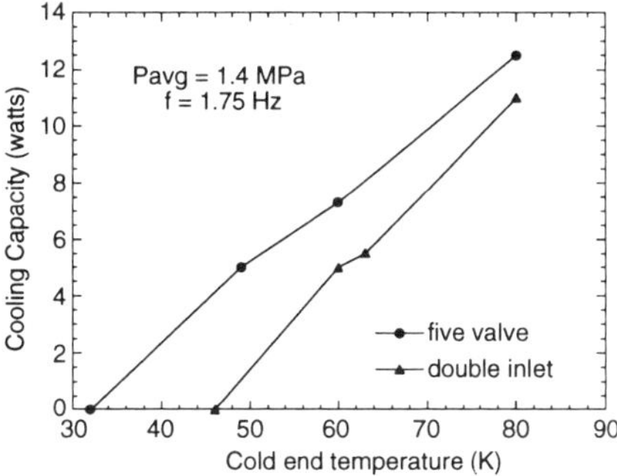


**Figure 8**. Cooling power for the five valve and double inlet configurations.

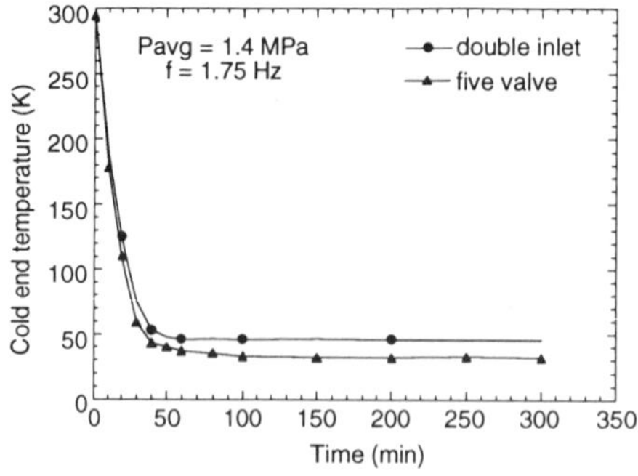


**Figure** 9. Cooldown curve for five valve and double inlet configurations.

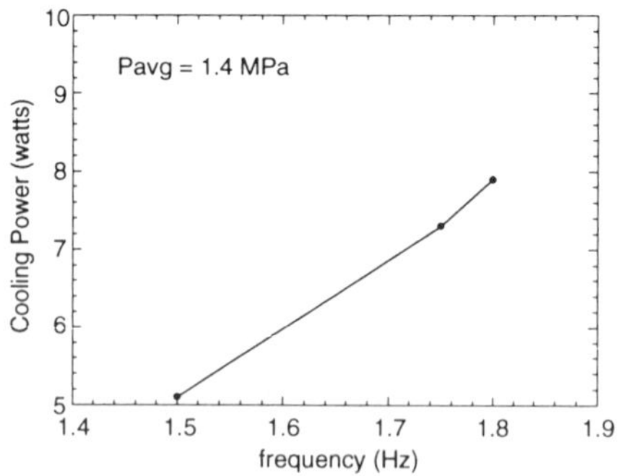


**Figure** 10. Effect of operating frequency on cooling power at 60 K.

realized by varying the timing and flow coefficients of the five valves. A lowest temperature of 32 K is achieved with $P_{M1}$ = 1.3 MPa and $P_{M1}$ = 1.7 MPa and at a frequency of 1.75 Hz. The intermediate pressures obtained by experiments are in reasonable agreement with our simulation results. Figure 7 displays the pressure wave in the pulse tube and reservoir. As indicated before, there are two approximately isobaric processes during one cycle.

As a reference for comparison, performance results are also displayed from a standard two valve double inlet configuration tested under the same operating conditions. Fig. 9 illustrates the typical cooldown curve for the double inlet and the five valve configuration. The working frequencies for both cases are 1.75 Hz. The figure clearly demonstrates that the cooling rate of the five valve configuration is better than that of the double inlet configuration. The minimum temperature reached with the five valve configuration is 32 K while the minimum temperature reached using the double inlet configuration is only 46 K. A comparison of the cooling capacity for these two configurations is shown in Fig. 8. For the five valve configuration, the cooling power at 60 K is 7.3 watts while for the double inlet it is 5.0 watts. Although the warm end heat exchanger of pulse tube was cooled by water in the cases presented above, only a slight performance degradation was observed when using air cooling.

The effect of the operating frequency on the cooling capacity at 60 K is illustrated in Fig. 10. In this case all tests are done using the same valve timing. The tests suggest that the cooling power strongly depends on the operating frequency. The highest frequency we have tested so far is 1.8 Hz due to the limitation of valve response time and the capacity of compressor.

## CONCLUSIONS

A single stage five valve pulse tube refrigerator has been constructed and tested. Measurements of the refrigeration power and minimum temperature indicate that two intermediate pressures play a significant role in system performance. The optimized intermediate pressures depend on cooling system parameters including the size of pulse tube, regenerator dead volume, compressor capacity, and the low and high source pressures. Tests show that the performance of the new configuration is better than that of a standard two-valve double inlet configuration. A cooling capacity of 8 watts at 60 K was achieved with input power of 1 kW. A minimum temperature of 32 K was reached with operating frequency of 1.75 Hz. The overall performance of the pulse tube refrigerator is shown to correlate well with the PV area defined by the cold end control volume, and this in turn can be easily optimized through the active control of the valve timings.

## ACKNOWLEDGMENTS

This work was funded in part by the Naval Research Laboratory, contract #N00014-95-C-2017, through CryoMech, Inc.

## REFERENCES

1.Zhu, S., Wu, P. and Chen, Z. "Double Inlet Pulse Tube Refrigerator: An Important Improvement", *Cryogenics* Vol. 30, 1990, p.514

2.Matsubara, Y., Gao, J.L., Tanida, K., Hiresaki, Y. and Kaneko, M., "An Experimental and Analytical Investigation of 4 K Pulse Tube Refrigerator", *Proc. Seventh Intl. Cryocooler Conf.*, 1992, p.166

3.Zhu, S., Kakimi, Y., Fujioka, K. and Matsubara, Y. " Active-Buffer Pulse Tube Refrigerator", *Proc. Sixteenth Intl Cryogenic Engr. Conf. / Intl Cryogenics Matl. Conf.* 1996, p.291

# A FOUR-VALVE PULSE TUBE CRYOCOOLER WITH A COOLING POWER OVER 30 W AT 80 K

R. Li, K. Kanao, N. Watanabe, and Y. Kanazawa

Research and Development Center
Sumitomo Heavy Industries Limited
63-30, Yuhigaoka, Hiratsuka, Kanagawa, 254, JAPAN

## ABSTRACT

A single stage four-valve pulse tube cryocooler with a large cooling power at 80 K has been built and tested. The cryocooler has a regenerator stacked with stainless steel screen disks of 250 mesh, and a pulse tube of 200 mm long. A rotary valve unit is employed to control the mass flow at the hot ends of regenerator and pulse tube. Because the size of pulse tube is an important parameter of the cryocooler, three kinds of pulse tube, 18, 28 and 38 mm in diameter, were prepared for investigating the influence of pulse tube size on cooling performance. Initial test demonstrated that the cryocooler with the pulse tube of 28 mm in diameter delivers a cooling power of 33.5 W at 80 K. Cycle frequency also shows a great effect on cooling performance, and the cryocooler reached its terminal temperature of 20.5 K with a cycle frequency of 1.8 Hz. This paper describes details of the cryocooler design and the experimental results.

## INTRODUCTION

Since W. E. Gifford and R. C. Longsworth invented pulse tube cryocooler at 1963[1], a number of kinds of the cryocooler have been developed. The major types of them are basic type[1], orifice type[2], double inlet type[3], four-valve type[4], active buffer type[5], and so on.

The four-valve type pulse tube cryocooler was proposed by Y. Matsubara et al. in 1993, and shows impressive results of cooling performance. The most attractive feature of the four-valve pulse tube cryocooler is the absence of a reservoir. Almost all the pulse tube cryocooler systems have reservoirs connected with the hot end of pulse tube, but in the four-valve pulse tube cryocooler the reservoir is replaced by a controlled valve unit. The four-valve pulse tube cryocooler accordingly has a potential for being very compact in size. Furthermore, the four valves can easily be compactly designed with a high reliability similar to the valve unit of Gifford-McMahon (G-M) cycle cryocoolers.

In the present investigation, we designed and constructed a single stage four-valve cryocooler with a rotary valve unit directly mounted on the hot ends of the regenerator and pulse tube. The cryocooler is able to deliver a large cooling power at 80 K, and the initial experimental results are given below.

## CRYOCOOLER DESIGN

As reported by other researches, the hot end of the cold head is usually connected with controlled valves by connection tubes in G-M type pulse tube cryocoolers, and there always is a needle valve inserted in the connection tube for adjusting the mass flow at the hot end of the pulse tube. The main feature of our cryocooler, however, is a rotary valve unit directly united to the hot ends of regenerator and pulse tube. The purpose of this configuration is to build the cryocooler more compactly and to take away the influence of the connection tube on cooling performance. J. Blaurock et al.[6] have reported a four-valve pulse tube cryocooler with such compact design, but their cryocooler had a rather poor cooling power of 5 W at 80 K with an input power of 2.2 kW for compressor.

Figure 1 shows the sketch of the four-valve pulse tube cryocooler built by this work. The main parts of the rotary valve unit are a rotor and a stator. They are similar to those used in G-M cryocoolers, but are specially designed to be double speed for intake and exhaust. In other words, the intake and exhaust of the cryocooler are done twice per a revolution of the rotor. The rotor is driven by a synchronous motor, and its revolution speed, in effect, the cycle frequency is easily changed by using an electric inverter.

Between the rotary valve and the hot end of pulse tube, a removable part with a pin hole orifice, is set in the valve unit for controlling the mass flow at the hot end of the pulse tube.

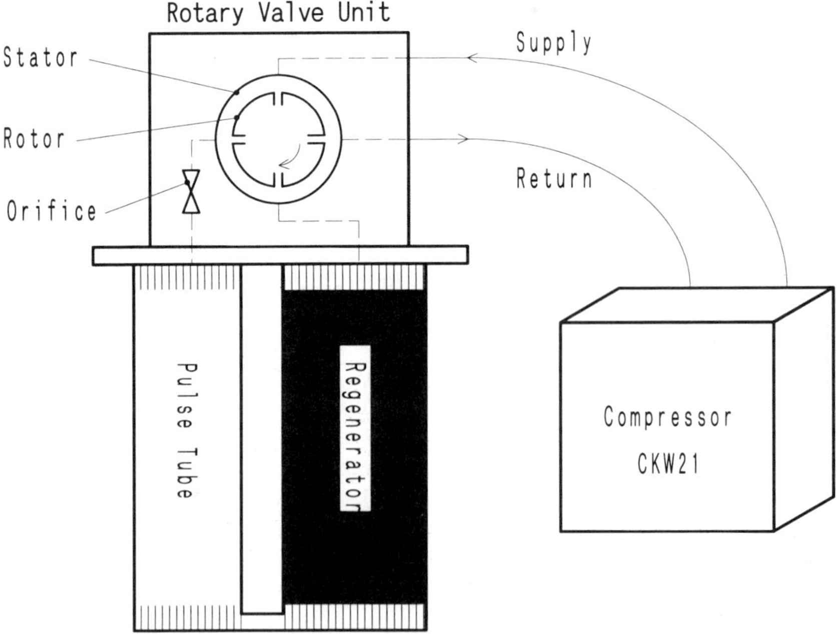


**Figure 1.** Schematic of the four-valve pulse cryocooler.

**Table 1.** Dimensions of the pulse tubes

| | Pulse tube A | Pulse tube B | Pulse tube C |
|---|---|---|---|
| Inner diameter (mm) | 18 | 28 | 38 |
| Length (mm) | 200 | 200 | 200 |
| Thickness (mm) | 1 | 1 | 1 |
| Volume (cm3) | 50.9 | 123.2 | 226.8 |

The regenerator consists of a stainless steel tube and a stack of stainless steel screen disks. The stainless steel tube is 170 mm long with an inner diameter of 38 mm and a thickness of 1 mm. The disks of stainless steel screen are 250 mesh, and their total number is 2000. At both ends of the regenerator, 30 copper screen disks of 80 mesh are installed as heat exchangers and flow straighteners.

The pulse tube is made of a stainless steel tube of 200 mm long with a thickness of 1 mm. There are three pulse tubes prepared for investigating the influence of pulse tube size on cooling performance. The dimensions of the pulse tubes are given in table 1. It is the same as the regenerator that 30 copper screen disks are also stacked at both ends of the pulse tubes as heat exchangers and straighteners. The cold ends of the regenerator and the pulse tube are connected by a copper tube with an outer diameter of 1/4 inch.

The compressor unit (model CKW21) applied in the cryocooler system is manufactured by Sumitomo Heavy Industries Limited and is mainly used for G-M cryocooler system. The rated input power of the compressor is 2.4 kW with 50 Hz.

## EXPERIMENTAL DETAILS

The temperature of the cold stage was measured by a platinum-cobalt resistance thermometer. The wall temperatures of the regenerator were monitored by four thermocouples of T-type at regular intervals, and those of the pulse tube were also obtained by the same way.

The pressure measurement was done at five points: (1) the supply line of the compressor, (2) the return line of the compressor, (3) the hot end of the regenerator, (4) the hot end of the pulse tube, and (5) the point between the rotary valve and orifice.

In the present initial investigation, we kept the initial pressure of the compressor to be 1.67 MPa, varied cycle frequency by using an electric inverter, looked for the optimum orifice by exchanging the removable orifice, and estimated net cooling power by an electric heater of manganin wire.

Only one set of stator and rotor was employed in the present experiment. The valve timing and the phase shift at the hot end of the cold head are consequently fixed for the following results. The pressure waves generated by the rotary valve at the hot ends of pulse tube and regenerator are illustrated in figure 2. The phase of the pressure wave at the hot end of regenerator is almost the same as that at the hot end of pulse tube, but is shifted about 60 degrees (360 degrees per one cycle) from that between the orifice and the rotary valve.

## RESULTS AND DISCUSSIONS

### Pulse Tube Size

The cooling performance of the pulse tube cryocooler is strongly dependent on the size of pulse tube. The performance of the pulse tubes with different sizes listed in table 1 were measured.

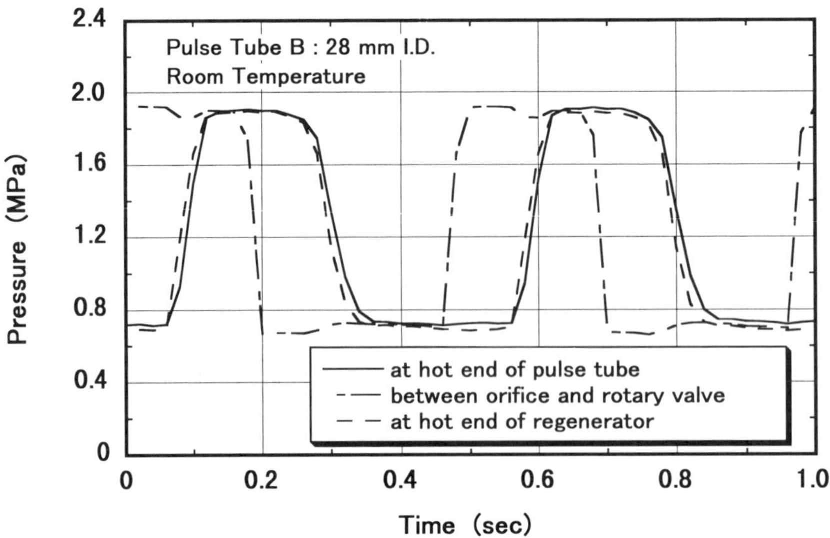


**Figure 2.** Pressure waves at three points : (1) the hot end of pulse tube, (2) the point between rotary valve and orifice, and (3) the hot end of regenerator. The cycle frequency is 2 Hz.

Figure 3 shows the terminal temperatures achieved by the three pulse tubes with a cycle frequency of 2 Hz. The pulse tube of 28 mm reaches very low terminal temperature of 21.8 K, while they are above 35 K for the pulse tubes of 18 mm and 38 mm.

The difference of cooling performance is also apparent in cooling power. The relationship of cooling power versus cold stage temperature for the three pulse tubes are given in figure 4. The 28 mm pulse tube has good cooling performance, and delivers cooling powers of 20 W at 52.7 K and 30 W at 71.8 K. For a cold stage temperature of 80 K the cooling power is 33.5 W, whereas the pressure ratio of supply pressure to return pressure is 2.51 and the input power of compressor is 2.93 kW. Compared with the results of J. Blaurock et al., the cooling performance is much improved in our cryocooler.

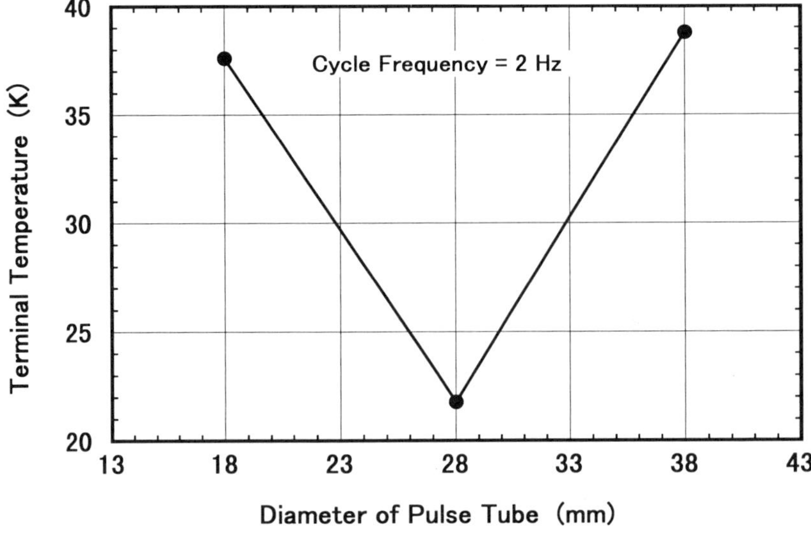


**Figure 3.** Terminal temperatures for the three pulse tubes with a cycle frequency of 2 Hz.

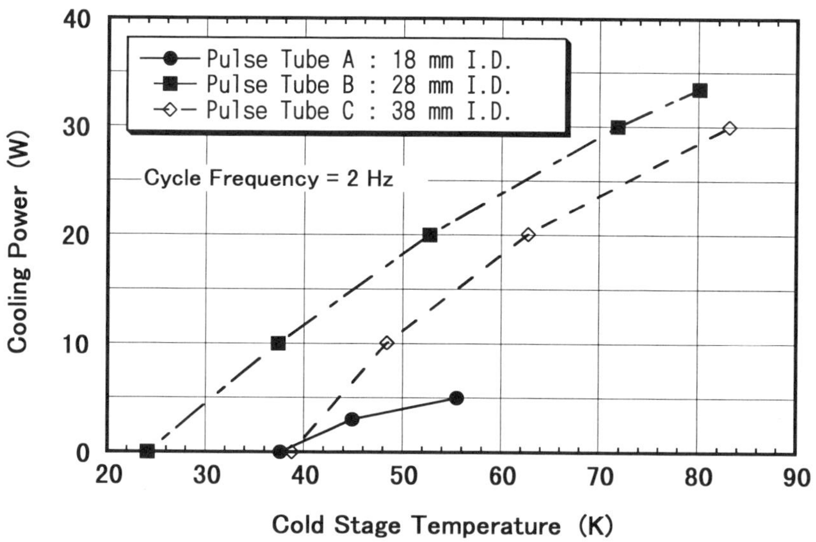


**Figure 4.** Cooling Powers for the three pulse tubes with a cycle frequency of 2 Hz.

The trend of cooling power curve for the 38 mm pulse tube is similar to that of the 28 mm pulse tube, but the cooling power is inferior to that of the latter for temperatures up to 80 K. For the 18 mm pulse tube, the cooling power is more poor though its terminal temperature is almost the same to that of 38 mm pulse tube.

It is a matter of common sense that the cooling performance would be conspicuously poor if the size of pulse tube is too large or too small compared with that of regenerator. As an empirical knowledge, the volume of the optimum pulse tube is almost equal to the void volume of regenerator. In our cryocooler, the void volume of regenerator is 135 $cm^3$ if a packing factor of 0.3 is assumed for the stainless steel screen disks. By using the pulse tube volumes listed in table 1, the ratios of pulse tube volume to regenerator void volume are calculated to be 0.38, 0.91 and 1.68 for the 18mm, 28 mm and 38 mm pulse tube respectively. The ratio for the 28 mm pulse tube is the closest to 1, and this is the reason that the cooling performance of the pulse tube is superior to those of the others.

## Orifice

The orifice between the pulse tube and rotary valve plays an important role in control of the mass flow at the hot end of pulse tube. In order to find out the optimum orifice, we exchanged the removable part of orifice having different diameters of the pin hole, and evaluated them with terminal temperatures. Figure 5 gives the experimental results. The terminal temperature is considerably influenced by the orifice, and the results indicate that the orifice with a diameter of $\phi$ 0.7 mm is the best for the pulse tube of 28 mm. All of the following results are therefore obtained with the orifice of $\phi$ 0.7 mm diameter.

## Cycle Frequency

It is suggested by figure 5 that cycle frequency gives great effect on cooling performance. Figure 6 shows another result about the influence of cycle frequency. The terminal temperature drastically depends on cycle frequency, and the lowest terminal temperature achieved in the present investigation is 20.5 K for a cycle frequency of 1.8 Hz.

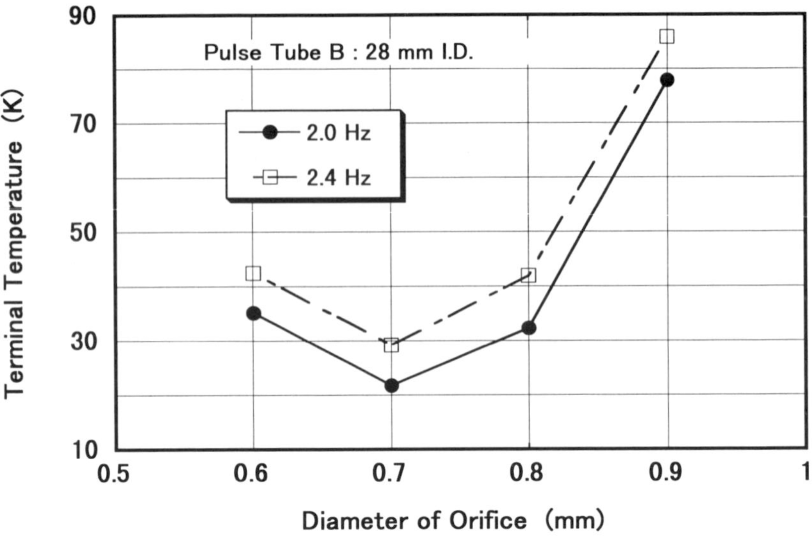


**Figure 5.** Terminal temperatures versus the diameter of orifice.

### Cool Down Curve

The cool down curve of the cryocooler is illustrated in figure 7. The cryocooler is rapidly cooled from room temperature down to 80 K with 15 minutes and down to 50 K with 20 minutes. A quite long time more than 3 hours, however, is necessary to reach the terminal temperature.

### Profile of Wall Temperature

Finally, the temperature profiles measured by thermocouples on the walls of regenerator and pulse tube are given in figure 8. The horizontal axis presents the distance from the hot

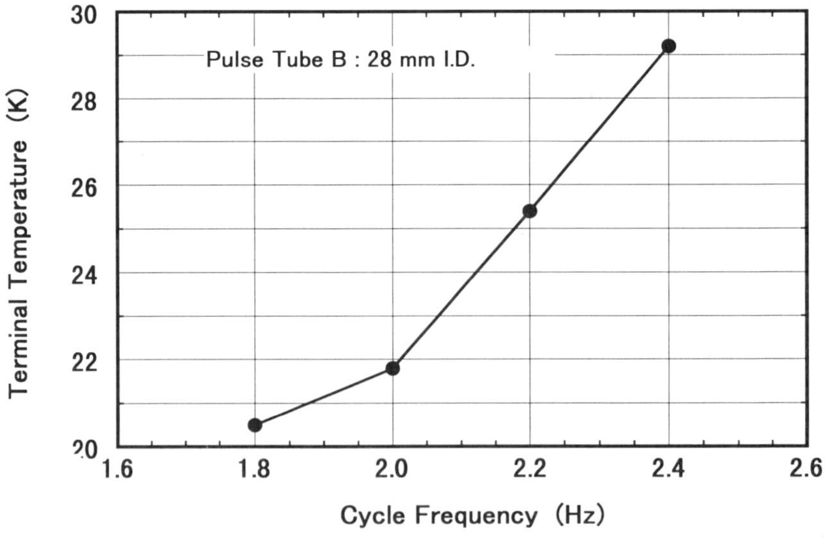


**Figure 6.** Cycle frequency dependence of terminal temperature for the 28 mm pulse tube.

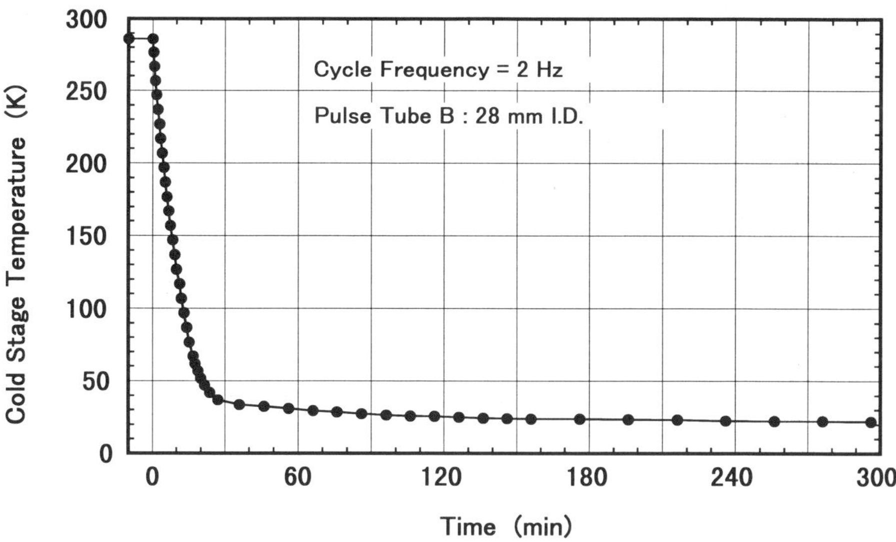


**Figure 7.** Cool down curve for the 28 mm pulse tube.

end of regenerator to the hot end of pulse tube via the cold stage. When there is no heat load supplied to the cold stage, both the regenerator and the pulse tube has a wall temperature profile close to linear functions. Under the conditions of heat load input to the cold stage, the wall temperatures of regenerator and pulse tube considerably rise, especially, the increase of wall temperature is more than 100 K in the middle position of regenerator and pulse tube for larger heat loads. The temperature profile of the pulse tube has a peak at the position near to the hot end for a heat load of 30 W. The reason is that the hot end of the pulse tube is always cooled by natural convection at room temperature. The real temperature of working gas in the hot end is considered to be higher.

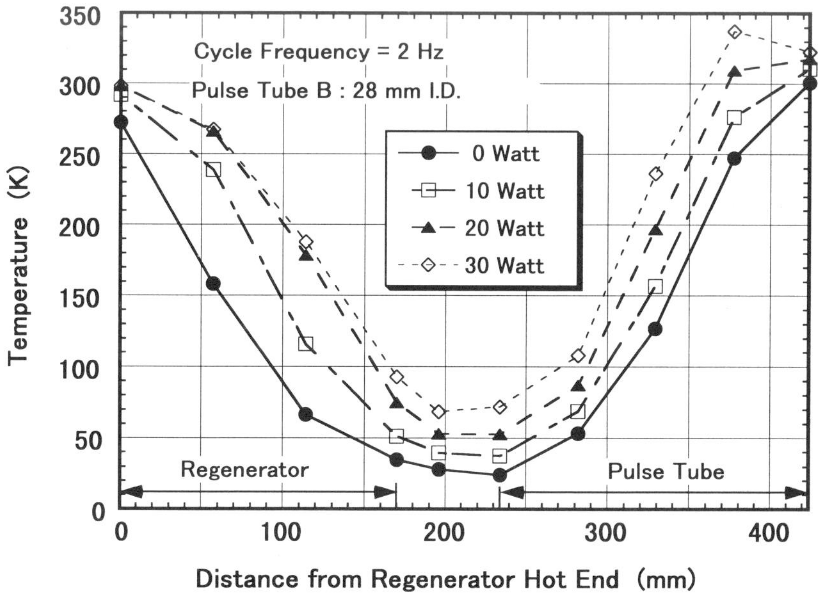


**Figure 8.** Profiles of wall temperature for regenerator and pulse tube.

## SUMMARY

A single stage four-valve pulse tube cryocooler with a rotary valve unit directly united to the cold head has been designed and constructed. Because there is no connection tube between the valve unit and the hot end of cold head, the cryocooler has the advantages of no influence of connection tube and compact size. As the initial experiment, the cooling performance of the cryocooler was investigated. The major results obtained are the following:

1. A terminal temperature of 20.5 K is achieved with a cycle frequency of 1.8 Hz.
2. The cryocooler with 28 mm pulse tube delivers a cooling power of 33.5 W at 80 K.
3. For a given regenerator, there is the optimum pulse tube which volume almost equals the void volume of regenerator.

The cryocooler will be investigated further. The main works we scheduled are to find out the optimum valve timing, and to examine the influence of pulse tubes which have a constant volume but have different combinations of diameter and length.

## REFERENCES

1. W.E. Gifford and R.C. Longsworth, Pulse-tube refrigeration, ASME paper No. 63-WA-290 presented at Annual Meeting of the American Society of Mechanical Engineering, Philadelphia, Pennsylvania Nov. 17-22, (1963).
2. E.I. Mikulin, A.A. Tasarov and M.P. Shkrebyonock, Low temperature expansion pulse tubes, in: "Advances in Cryogenic Engineering", Vol. 29, Plenum Press, New York (1984), p. 629.
3. S. Zhu, P. Wu, Z. Chen, W. Zhu and Y. Zhou, A single stage double inlet pulse tube refrigerator capable of reaching 42 K, Cryogenics **30** ICEC 13 Supplement: 257 (1990).
4. Y. Matsubara, K. Tanida, J.L. Gao, Y. Hiresaki and M. Kaneko, Four-valve pulse tube refrigerator, in: "Proceedings of Fourth Joint Sino-Japanese Seminar on Cryocoolers and Concerned Topics", Beijing (1993), p. 54.
5. S.W. Zhu, Y. Kakimi, K. Fujioka and Y. Matusbara, Active-buffer pulse-tube refrigerator, in: "Proceedings of ICEC 16 / ICMC", Kitakyushu (1996), p. 291.
6. J. Blaurock, R. Hackenberger, P. Seidel and M. Thurk, Compact four-valve pulse tube refrigerator in coaxial configuration, in: "Cryocoolers 8", R.G. Ross, Jr., ed., Plenum Press, New York (1995), p. 359.

# EFFECTS OF REGENERATOR GEOMETRY ON PULSE TUBE REFRIGERATOR PERFORMANCE*

M.Lewis, T.Kuriyama,[1] J. H. Xiao,[2] and R.Radebaugh

National Institute of Standards and Technology
Boulder, Colorado, USA 80303

[1]Guest Researcher from Toshiba Co., Kawasaki, Japan

[2]Guest Researcher from Chinese Academy of Sciences
Beijing, China

## ABSTRACT

This paper gives results of the cooling performance of a double-inlet pulse tube refrigerator using various regenerators. The same pulse tube was used for all the experiments and measured 4.76 mm in diameter and 46.2 mm in length. A commercial linear compressor with a swept volume of 4 $cm^3$ was used in these experiments. The operating conditions were held constant at a mean pressure of 2.0 MPa and a frequency of 54 Hz. Using finite difference software called REGEN3.1, developed at NIST, and recent experiment results, we optimized a series of regenerators based on dimensions, materials and screen packing. The values used for calculating the thermal conduction through stacked screens by REGEN3.1 were based on recent experimental results from NIST. The regenerator tubes were designed using 316 stainless steel and titanium materials. The regenerator matrices investigated were 400-mesh and 500-mesh stainless steel screen. The valve settings for both orifices were adjusted to minimize the no-load temperature for all regenerators. A cooling capacity curve from 0 to 3 W was also determined. The performance of the pulse tube refrigerator using the different regenerators is discussed. The experimental results from the various regenerators are evaluated and compared with their corresponding numerically calculated coefficient of performance (COP) and regenerator design as determined by REGEN3.1.

---

## INTRODUCTION

The pulse tube refrigerator has gained significant interest over the past few years because of its simplified mechanical properties, overall reliability, and lower costs. Since the performance of the pulse tube refrigerator has nearly equaled that of the Stirling refrigerator, the interest in the pulse tube has increased because there are no moving parts at the cold end as there are in the Stirling machine. High efficiency is demanded in many applications such as cooling of infrared sensors, superconducting electronic components, and semiconducting devices. In our efforts to improve the pulse tube performance, we focused our efforts on the cold head, where the most significant difference is between the two refrigerators. We have essentially ignored the compressor losses since they would be comparable for both refrigerators.

This study addresses one of the largest losses in the pulse tube refrigerator, which are those attributed to the regenerator. The axial thermal conduction in the regenerator is part of the overall regenerator loss which must be considered. At NIST, we recently completed research on the axial thermal conduction in regenerators with stacked screens.[1] There were no previous data for axial thermal conduction other than an estimated measurement obtained years ago at NIST. The recent measurements allowed us to improve the optimization of the regenerator geometry by using the new thermal conduction degradation factors of 0.1 for stainless steel and 0.022 for phosphor bronze instead of the old values of 0.3 and 0.05.

This paper shows the results of numerical COP values calculated by using the NIST software REGEN3.11,[2,3] on six different regenerator configurations with stacked screens. REGEN3.1 calculated a range of COP values for configurations based on the new research data on thermal conduction through stacked screens. From these calculations an optimized regenerator length, diameter, and mesh size for a specific tube material and matrix material were identified. With this information six regenerators were designed and fabricated and a series of experiments was performed in which the regenerators were inserted in a pulse tube refrigerator and the load curves were measured. This paper also shows the experimental results of four of these optimized regenerator designs, specifically the low-end temperature values, and their respective cooling capacities.

## NUMERICAL CALCULATIONS

REGEN3.1 has been used at NIST as a tool to optimize pulse tube regenerators on many pulse tube refrigerator applications. One input variable required is the conduction degradation factor used to calculate the thermal conduction through the stacked screen beds. With recent research performed at NIST on a more accurate coefficient for this input variable, we ran calculations for the coefficient of performance for a series of regenerators using various lengths, regenerator tube materials, and matrix materials. The lengths of these regenerators were optimized using stainless steel and titanium for the regenerator wall material. The matrix materials used were 500-mesh stainless steel screen, 400-mesh stainless steel screen, and 400-mesh phosphor bronze screen. Figure 1 shows the results of these calculations for COP values versus regenerator length for all the conditions observed. As a result of these data we designed and constructed regenerators based on the optimized COP values for specific regenerator materials and dimensions. Table 1 defines the calculation parameters used by REGEN3.1 for six specific conditions and the calculated values for the coefficient of performance (COP) for the pulse tube refrigerator using each of these regenerators. Heat exchanger and compressor losses were ignored in these COP calculations. The pressure ratio of 1.4 is

**Table 1**. Calculation parameters for COP

| Values for numerical calculations | | | | | | |
|---|---|---|---|---|---|---|
| REGEN3.1 | #1 | #2 | #3 | #4 | #5 | #6 |
| COP | 0.0737 | 0.0698 | 0.0680 | 0.0639 | 0.0646 | 0.0612 |
| Thermodynamic | | | | | | |
| Temp (Cold) K | 80 | 80 | 80 | 80 | 80 | 80 |
| Temp (Warm) K | 300 | 300 | 300 | 300 | 300 | 300 |
| Pressure (avg) MPa | 2.0 | 2.0 | 2.0 | 2.0 | 2.0 | 2.0 |
| Pressure ratio MPa | 1.4 | 1.4 | 1.4 | 1.4 | 1.4 | 1.4 |
| Test Fluid | Helium | Helium | Helium | Helium | Helium | Helium |
| Geometry | | | | | | |
| Hydraulic diameter m | $0.45\times10^{-4}$ | $0.56\times10^{-4}$ | $0.45\times10^{-4}$ | $0.56\times10^{-4}$ | $0.45\times10^{-4}$ | $0.56\times10^{-4}$ |
| Area (cross sect.) $m^2$ | $0.67\times10^{-4}$ | $0.67\times10^{-4}$ | $0.67\times10^{-4}$ | $0.67\times10^{-4}$ | $0.67\times10^{-4}$ | $0.67\times10^{-4}$ |
| Length mm | 32 | 32 | 36 | 40 | 40 | 40 |
| Material | | | | | | |
| Regenerator tube | Titanium | Titanium | Stainless | Stainless | Stainless | Stainless |
| Screen | S.S. | S.S. | S.S. | S.S. | S.S. | PhBr |
| Mesh | 500 | 400 | 500 | 400 | 500 | 400 |
| Porosity | 0.694 | 0.686 | 0.694 | 0.694 | 0.686 | 0.686 |
| Cond. Degrad. Factor | 0.10 | 0.10 | 0.10 | 0.10 | 0.10 | 0.022 |
| Flow | | | | | | |
| Frequency Hz | 54 | 54 | 54 | 54 | 54 | 54 |
| Mass flow (cold) kg/s | $0.50\times10^{-3}$ | $0.50\times10^{-3}$ | $0.50\times10^{-3}$ | $0.50\times10^{-3}$ | $0.50\times10^{-3}$ | $0.50\times10^{-3}$ |
| Mass flow (warm)kg/s | $0.64\times10^{-3}$ | $0.64\times10^{-3}$ | $0.65\times10^{-3}$ | $0.67\times10^{-3}$ | $0.69\times10^{-3}$ | $0.67\times10^{-3}$ |
| Phase angle ° | 38 | 42 | 40 | 43 | 46 | 43 |

for the cold end. The pressure ratio at the warm end was about 1.47. Figure 1 shows that the best regenerator should be #1 when a thermal conduction degradation factor of 0.1 is used for S.S. screen. In contrast, a degradation factor of 0.3 suggests regenerator #4 worked the best, although those calculations are not shown here.

## EXPERIMENTAL TECHNIQUES

The tests were performed using a small pulse tube[4] that measured 4.76 mm in diameter and 46.2 mm in length and a wall thickness of 0.254 mm. The average operating pressure was maintained at 2.0 MPa using helium gas for all test data. A commercial linear compressor was used in which the volumetric displacement could be varied from 0 to 4 $cm^3$. The compressor was driven at a fixed frequency of 54 Hz with a power amplifier that was continually adjusted to operate the compressor at the maximum stroke. There were no position sensors on the compressor; therefore, the compressor was operated at maximum stroke, which resulted in a known 4 $cm^3$ swept volume. With the operating conditions used in these experiments, the compressor was operating far from resonance conditions, which resulted in low efficiency. The pulse tube was designed with a primary orifice valve associated with the reservoir volume and a secondary valve between the compressor and the pulse tube warm end. Pressure measurements at the reservoir volume, pulse tube, and compressor were recorded using piezoresistive transducers. Temperature measurements at the cold end of the pulse tube and the aftercooler using silicon diodes were also recorded. An electrical heater was located a the cold end of the pulse tube to perform the cooling capacity tests. The pulse tube and regenerator assembly were located in a stainless steel vacuum vessel and were wrapped in a multi-layer insulation material during testing to minimize the effects of radiation heat leak. Table 2 summarizes these experimental conditions. The pulse tube pressure ratio given is an average for the cold-end temperatures obtained.

**Table 2.** Experimental conditions

| | |
|---|---|
| Temperature at cold-end K | 36-62 |
| Temperature at warm-end K | 300 |
| Pressure MPa | 2.0 |
| Pressure ratio (Pulse Tube) | 1.42 |
| Test fluid | Helium |
| Pulse tube length mm | 46.2 |
| Pulse tube diameter mm | 4.76 |
| Pulse tube wall thickness mm | 0.254 |
| Compressor swept volume $cm^3$ | 4.0 |
| Frequency Hz | 54 |

On the basis of the numerical calculations, six optimized regenerator configurations were chosen for the experimental apparatus. The experimental work on four of the regenerators has been completed. The summaries of the characteristics of these four regenerators are given in Table 3. The identification number in the first column corresponds to the numerical data from Figure 1. The experimental regenerators were built with the input parameters used in REGEN3.1 where possible. Parameters such as porosity, mass flow, and pressure ratios cannot be achieved precisely in the fabrication.

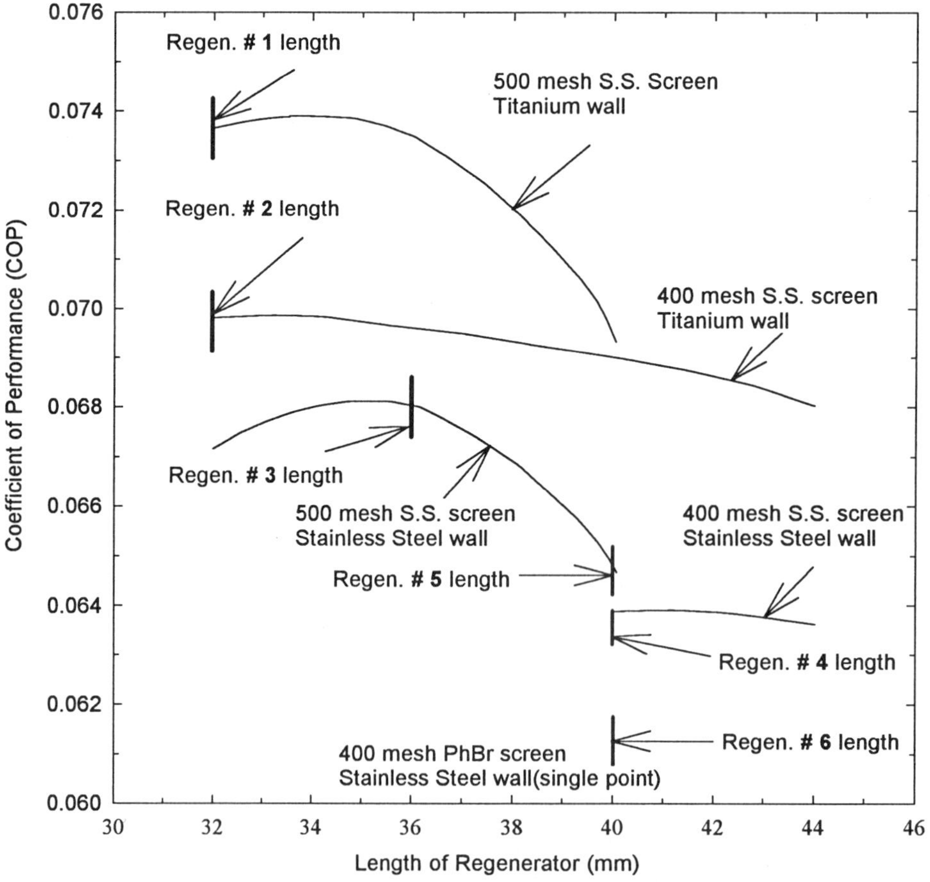


Figure 1. Calculated COP vs. regenerator length.

**Table 3.** Regenerator characteristics

| | Regen. Length (mm) | Tube Material | O.D / Wall Dimensions (mm) | Screen Material | Mesh Size | Wire Dia. (μm) | Actual Porosity | Theoretical Porosity |
|---|---|---|---|---|---|---|---|---|
| #1 | 32 | Titanium | 9.525 / 0.152 | S.S. | 500 | 25.4 | 0.6148 | 0.6073 |
| #3 | 36 | S.S. | 9.525 / 0.152 | S.S. | 500 | 25.4 | 0.6385 | 0.6073 |
| #4 | 40 | S.S. | 9.525 / 0.152 | S.S. | 400 | 25.4 | 0.6477 | 0.6858 |
| #5 | 40 | S.S. | 9.525 / 0.152 | S.S. | 500 | 25.4 | 0.6300 | 0.6073 |

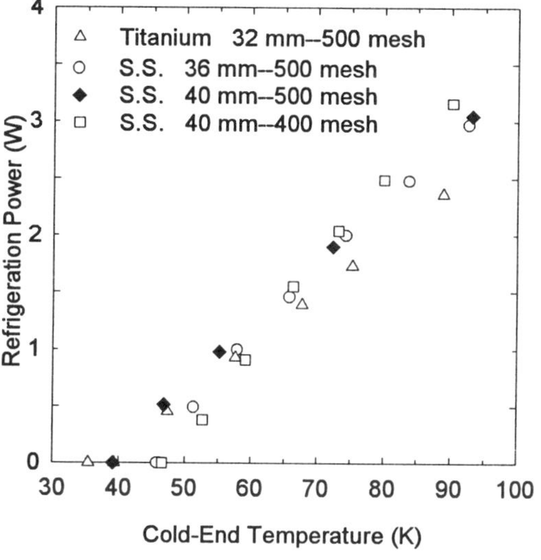


**Figure 2.** Refrigeration power for orifices optimized for minimum no-load temperatures.

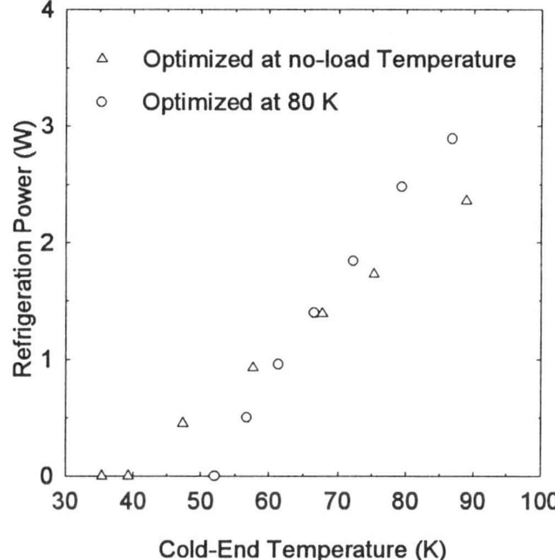


**Figure 3.** Refrigeration power for orifices optimized at a specific temperature.

The theoretical porosity, used by REGEN3.1, is calculated from the simplified equation $1-n_g = (\pi/4)(d/s)$, where $d$ is the wire diameter and $s$ is the spacing from one wire to the next. It is a close approximation to porosity achieved when each screen layer occupies a height equal to twice the wire diameter.

## EXPERIMENTAL RESULTS

The regenerators were installed into the pulse tube apparatus and operated at the described test conditions. The primary and secondary valves were adjusted to give the minimum no-load temperature at the cold end. Once this lowest pulse tube cold end temperature was obtained and allowed to stabilize, a series of cooling capacity measurements were recorded with the same valve settings from 0 to about 3 W to the cold-end electrical heater. The results of these measurements appear in Figure 2. The

**Table 4.** Regenerator test results

| Fig 1 | Regenerator Characteristics | No-load temperature(K) | Calculated COP from REGEN 3.1 | Calculated no-load temperature(K) |
|---|---|---|---|---|
| #1 | Titanium(32 mm--500 mesh) | 35.41 | 0.0737 | 36.0 |
| #3 | S.S.(36 mm—500 mesh) | 45.76 | 0.0680 | 38.8 |
| #4 | S.S.(40 mm--400 mesh) | 46.56 | 0.0639 | 42.3 |
| #5 | S.S.(40 mm--500 mesh) | 39.14 | 0.0646 | 39.0 |

lowest no-load temperature was achieved using the 32 mm long titanium regenerator with 500 mesh stainless steel screens where a temperature of 35.4 K was obtained. Pressure ratios of 1.41 at the warm end of the pulse tube and 1.59 at the compressor were measured under these conditions.

We performed a second set of tests using this same regenerator and matrix material that optimized the cooling capacity of the pulse tube at a cold end temperature of 80 K. These results obtained a much higher cooling capacity at 80 K than the previous method of optimizing at the no-load conditions but also resulted in a higher cold end temperature. We performed a series of tests using this method of optimization; one set of actual test results is given in Figure 3. The best cooling capacity achieved for these experiments was 3 W at 80 K. This data point is not shown in Figure 3.

## DISCUSSION

A nearly exact agreement for # 1 and # 5 and a fairly close agreement for # 3 and # 4 are observed between the calculated no-load low-end temperature values from our mathematical modeling with the actual no-load low-end temperatures obtained through the experiments. Table 4 documents the experimental no-load optimized low-end temperatures of the pulse tube as it relates to the corresponding no-load low-end temperature calculation derived using REGEN3.1. It also shows the numerically calculated COP values from REGEN3.1. The numerical values in column 1 are related to the corresponding regenerators in the previously shown figure 1. Figure 4 shows a plot

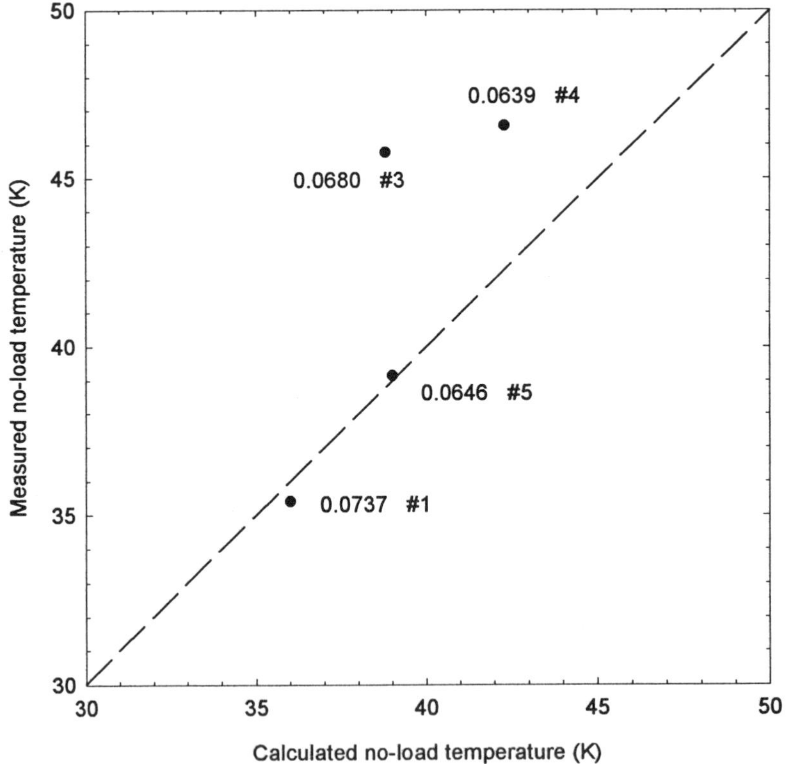


**Figure 4** Measured versus calculated no-load temperature values. The numbers are the associated calculated COP.

of these data. The general tendency represents improved no-load low-end pulse tube temperatures with larger regenerator COP values. The experimental no-load temperatures are in good agreement with the calculated values. The optimization calculations showed regenerator #1 to be the best. The experiments confirmed the predicted behavior and support the use of the value 0.1 for the thermal conduction factor for stacked stainless steel screen.

## CONCLUSIONS

The experimental results for the no-load temperatures with four different regenerators in a small double-inlet pulse tube refrigerator agree well with the values calculated from REGEN3.1. The lowest measured temperature of 35.4 K occurred with the regenerator that was calculated to give the highest COP at 80 K from optimization calculations using 0.1 for the thermal conduction factor for stainless steel screen. These results show how significant improvements to the performance of pulse tube refrigerators can be obtained when the correct thermal conduction degradation factor is used in optimization calculations to find the best regenerator geometry.

## ACKNOWLEDGMENTS

This study was funded in part by NASA Life Sciences Division, NASA headquarters. We thank R. Rawlings of Texas Instruments, Inc. for the loan of the compressor used in these experiments.

## REFERENCES

1. Lewis, M., Kuriyama, T., Kuriyama, F., and Radebaugh, R., "Measurement of Heat Conduction Through Stacked Screens," presented at CEC Conference in Portland Oregon, 1997,to be published.
2. Gary,J., Daney, D.E., and Radebaugh, R., "A Computational Model for a Regenerator," Proc. Third Cryocooler Conference, National Bureau of Standards(U.S.) Special Publication 698, (1985), p. 199.
3. Gary, J., and Radebaugh, R., "An Improved Numerical Model for Calculation of Regenerator Performance (REGEN3.1)," Proc. Fourth Interagency Meeting on Cryocoolers, David Taylor Research Center, Report DTRC-91/003, (1991), p.165.
4. Yaron, R., Shokralla, S., Yuan, J., Bradley, P., and Radebaugh, R., "Etched Foil Regenerators," Advances in Cryogenic Engineering, Vol, 41, Part B, p. 1342-1346, Plenum Press, New York (1996).

# INVESTIGATION ON THE HEAT TRANSFER CHARACTERISTIC IN PULSE TUBE REFRIGERATOR

S. Jeong and K. Nam

Cryogenic Engineering Laboratory
Korea Advanced Institute of Science and Technology
Taejon, Korea

## ABSTRACT

It is one of the critical issues to understand heat transfer characteristic of the pulse tube in developing a more efficient pulse tube refrigerator at cryogenic temperature. This paper describes the experimental investigation on the heat transfer mechanism during both oscillating pressure and oscillating flow. The gas temperature and the tube wall temperature along the pulse tube were measured with fast response thermocouples for various cycle conditions. The real time measurement of the heat flux using the heat flux transducers was also made at the warm and the cold end heat exchangers. The pressurization period of a pulse cycle was varied by controlling the stepper motor in the rotary valve system to study the effect of pressure wave shape. The experimental results were compared with the developed theoretical analysis. To improve the heat-pumping characteristic of the pulse tube refrigerator, a unique design of the warm end heat exchanger was also proposed.

## INTRODUCTION

There have been great number of research studies on the pulse tube refrigerator(PTR) since it was first conceived by Gifford and Longsworth[1]. The PTR has a unique advantage that it doesn't have any moving components at the cold end, which results in mechanical simplicity, small vibration and long term reliability. In the early stages of its development, the PTR was not as successful as other well commercialized cryocoolers such as Stirling and G-M. Continuous research on the PTR, however, has improved its cooling capacity and lowered the cold end temperature[2]. Even though theoretical analyses of PTR have been developed[(3),(4)], the experimental studies of the heat transfer were mostly for the sinusoidal oscillating pressure and oscillating flows. As the various types of PTR have been introduced, not only sinusoidal but also step-wise pressure variation cases need great attention, because this phenomenon occurs inherently in most PTRs. A large scale experimental apparatus of a pulse tube was constructed in our laboratory with extensive instrumentation. This paper describes an experimental investigation on the pulse tube to see how the heat is transferred during the oscillating pressures. A phase difference between the gas temperature and the wall heat flux was clearly observed and further discussed. The heat flux was measured at the cold end and the warm end heat exchangers. The cooling capacity, however, was insignificant since the regenerator was not installed in the apparatus. The experimental results are also presented for the different pulsing frequencies and the pressurized period.

## EXPERIMENTAL APPARATUS

A schematic diagram of the experimental setup is shown in Figure 1. High pressure helium gas is generated by a commercial helium compressor(CTI Model 8200), which is then converted into pulse waves by the rotary valve system. The rotary valve is actuated by a 5-phase stepper motor. The rotation of the stepper motor for various pressure waves is controlled by D/A operation of the PC data acquisition and control board. The pulse tube system is located in the vacuum chamber($\sim 10^{-3}$ torr) to prevent any convection heat transfer. The pressure wave during the cycle is measured by a strain gauge type pressure transducer. Real time data acquisition of the heat flux is performed during the cycle by the thermofoil type gauge(RdF model 20455-1) which is installed at the outside walls of the cold and warm end heat exchangers. The response time of the heat flux sensor is about 60 ms for step input function. The gas and the wall temperatures are also measured by type E(chromel-constantan) thermocouples at the same location where the heat flux is located. The gas temperature instrumentation is shown in Figure 2. Very fine-wire thermocouples (0.05 mm

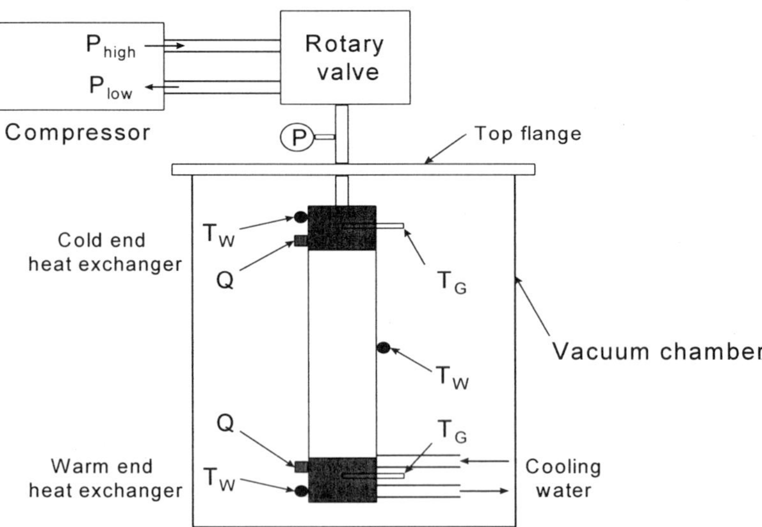


**Figure 1.** Schematic diagram of the pulse tube experiment apparatus. $T_W$ = Wall temperature thermocouple, $T_G$ = Gas temperature thermocouple, Q = Heat flux sensor, P = Pressure transducer.

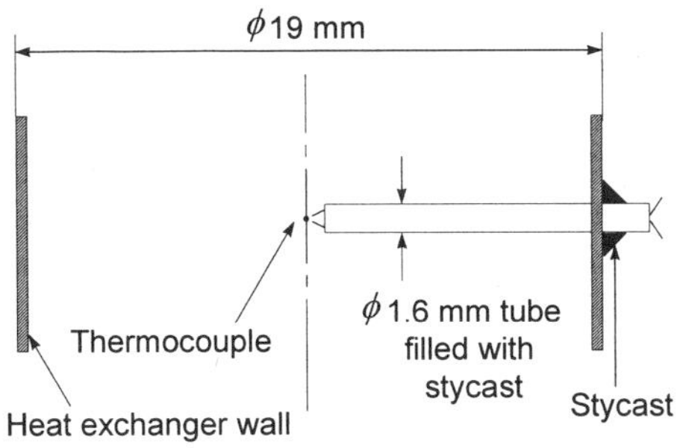

**Figure 2.** Gas temperature instrumentation.

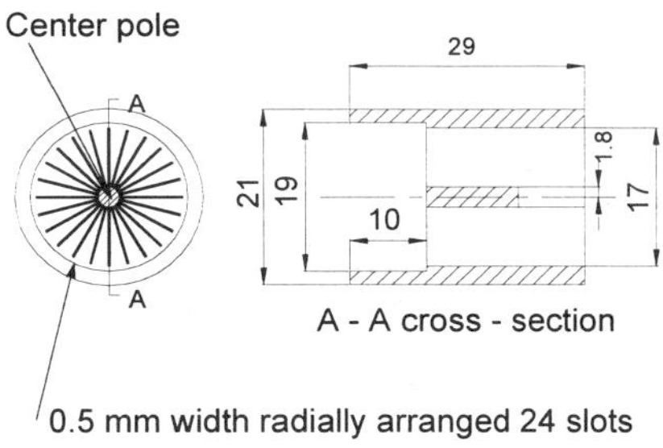


**Figure 3.** Internally finned heat exchanger (unit : mm).

diameter) were inserted into a 1.6 mm(1/16 inch) stainless steel tube and the tube was filled with epoxy (Stycast 2850FT) to prevent any leakage due to high oscillating pressure. The gap between the thermocouple probe and the heat exchanger was filled with Stycast, too. The pulse tube is made of thin-walled stainless steel tube (0.25 mm thick) to minimize axial thermal conduction. Its length is 200 mm and the outside diameter is 19 mm (3/4 inch). The heat exchangers of this experiment are fabricated as shown in Figure 3. The warm and the cold end heat exchangers are identically 21 mm diameter and 29 mm long except that the warm end one has a water-cooling jacket around it. The heat exchanger is made of OFHC(Oxygen Free High purity Copper) that has high thermal conductivity. The 24 triangular-shape internal fins were arranged radially by EDM(Electric Discharge Machining). As the gas is compressed and flows through the thin slots, it rejects its thermal energy to the copper fins. This design concept of one-piece heat exchanger without any joining (such as brazing) was for the purpose of enhancing radial heat transfer between the warm gas flow and the heat exchanger outside wall without any thermal contact resistance.

## EXPERIMENTAL RESULTS AND ANALYSIS

The cold end wall temperature was monitored during the transient process to determine when the pulse tube reached the cyclic steady state and all data were recorded by the A/D data board during the cycle. The flow rate of the cooling water for the warm end heat exchanger was maintained at 100 cc/min. The frequency of the tested cycle was 0.2, 0.5, 0.7, 1.0 Hz, respectively. The variation of the pressure wave form was only tested for 0.2 Hz case.

### Variation of frequency

As shown in Figure 4, there exists a clear phase shift between the gas temperature and the heat flux at all experiments. The heat flux at the inside wall of the heat exchanger can be calculated from the outside one which is measured by the heat flux sensor.[5] The instrumental phase shift is about 10 degrees for 1 Hz case from this calculation, which is due to the capacitive effect of the heat exchangers. Since the wall temperature is almost constant, it can be said that the instantaneous heat flux is not proportional to the wall and the gas temperature difference. The pressure is nearly a square wave due to a sudden opening of high and low pressure ports at the rotary valve, which results in sharp peak in the gas temperature and the heat flux. Since the thermal transport from the gas to the heat exchanger results in time delay, the gas temperature looks broader as the cycle frequency becomes higher. The wall temperature does not change very much since the heat capacity of the tube wall is much larger than that of the gas in the experimental temperature range. The positive heat flux data mean that the heat flows into the pulse tube and the negative one means the opposite direction. The heat flux always lags the temperature difference between the wall and the gas as the internal energy of the gas is stored instantaneously. As shown in Figure 5, the phase difference increases with the cycle frequency. This fact agrees with the prediction of the theoretical investigations of oscillating pressure heat transfer [(6),(7)]. Since the conventional steady-state heat transfer model can not explain this phenomenon, a complex Nusselt number was defined to predict the heat flux quantity. The complex Nusselt number was defined as[(6)] ;

$$q'' = \frac{k}{D}\left[ Nu_r \cdot \Delta T + \frac{Nu_i}{\omega} \cdot \frac{dT}{dt} \right] \tag{1}$$

where $k$ is the thermal conductivity of the gas, $D$ is the tube inside diameter, $\omega$ is the angular frequency, $Nu_r$ is the real part of the complex Nusselt number, and $Nu_i$ is its imaginary part.

The first term on the right hand side represents the heat transfer due to the temperature difference only like Newton's law of cooling. The second term indicates the additional

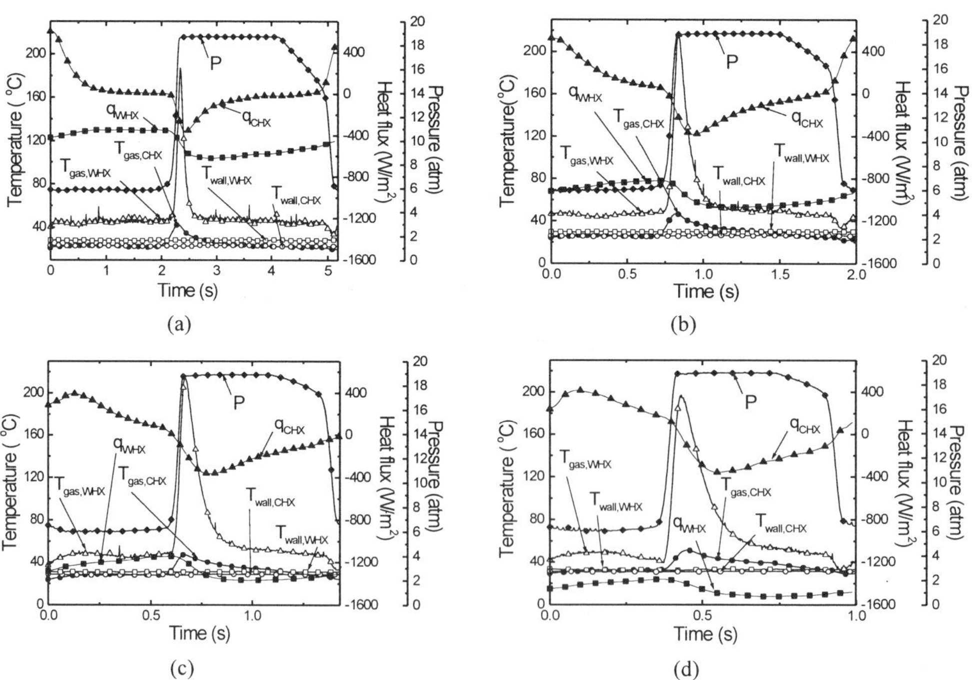


**Figure 4.** The experimental data of the gas temperature, the wall temperature, the pressure and the heat flux. (a) 0.2 Hz (b) 0.5 Hz (c) 0.7 Hz (d) 1.0 Hz (WHX means the warm end heat exchanger. CHX means the cold end heat exchanger.)

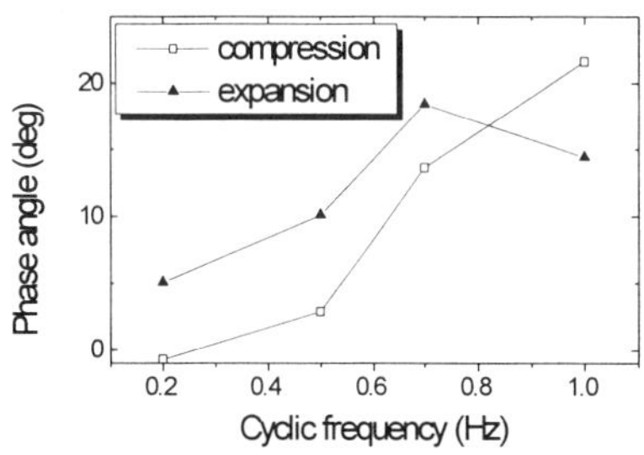


**Figure 5.** Phase lag angles of the peak heat flux with respect to the peak temperature difference at the cold end.

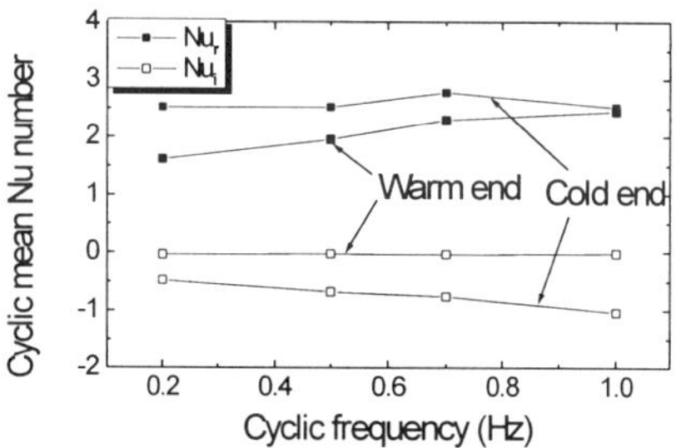


**Figure 6.** Cyclic mean Nu number at the cold and the warm ends.

contribution of the heat transfer by temperature difference rate, which explains the phase shift between the heat flux and the temperature difference. The cyclic mean Nusselt number was computed by a least square method for all experimental data points ($q''$, $\Delta T$, $d\Delta T/dt$) with the relation as shown in Eq. (1). The calculation result is as follows[6].

$$Nu_r = \left(\frac{D}{k}\right) \cdot \frac{\Sigma q''\Delta T \Sigma \Delta T'^2 - \Sigma q''\Delta T' \Sigma \Delta T \Delta T'}{\Sigma \Delta T^2 \Sigma \Delta T'^2 - \left(\Sigma \Delta T \Delta T'\right)^2} \tag{2}$$

$$Nu_i = \left(\omega \frac{D}{k}\right) \cdot \frac{\Sigma q''\Delta T' \Sigma \Delta T^2 - \Sigma q''\Delta T \Sigma \Delta T \Delta T'}{\Sigma \Delta T^2 \Sigma \Delta T'^2 - \left(\Sigma \Delta T \Delta T'\right)^2} \tag{3}$$

where $\Sigma = \sum_{n=1}^{m}$, $\Delta T' = \frac{d\Delta T}{dt}$, and $q'', \Delta T, \Delta T' = q''_n, \Delta T_n, \Delta T'_n$

m: number of data points, $\Delta T = T_{wall} - T_{gas}$

As shown in Figure 6, the absolute value of $Nu_i$ at the cold end increases slightly as the cyclic frequency goes up, which is reasonable in consideration of the phase shift results in Figure 5. We also calculated the cyclic mean heat flux at the warm end heat exchanger by integrating the experimental heat flux data. The cyclic mean heat flux is defined as ;

$$q''_{mean} = \frac{\oint q'' dt}{T} \tag{4}$$

where $T$ is the cycle period.

The cyclic mean heat flux represents the energy intensity extracted by the warm end heat exchanger. Table 1 shows that the high frequency pressure wave has a large amount of heat

**Table 1.** Cyclic average heat flux amount at the warm end.
(Negative sign indicates the heat flux out of the warm end )

| Cycle frequency (Hz) | 0.2 | 0.5 | 0.7 | 1.0 |
|---|---|---|---|---|
| Heat flux (W/m$^2$) | -475.5 | -956.5 | -1296.4 | -1448.6 |

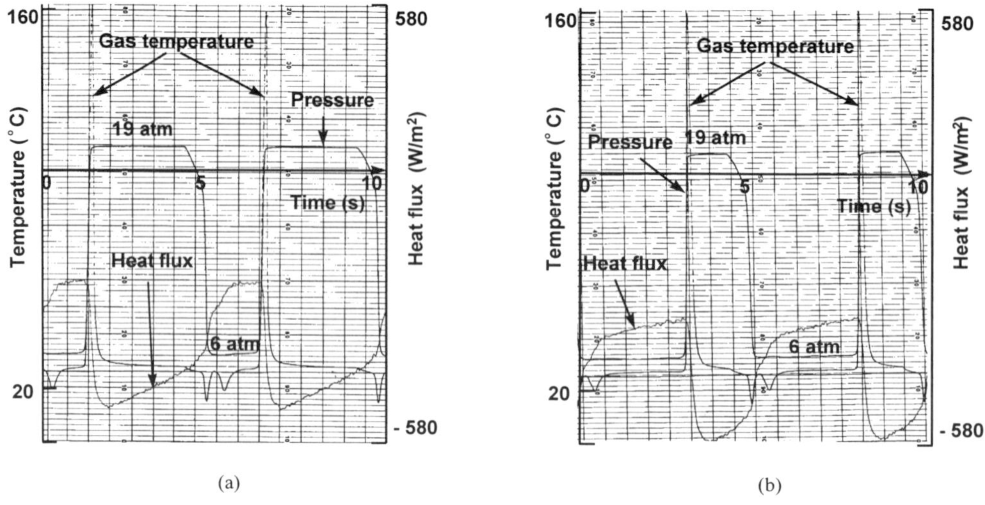


**Figure 7.** Experimental data at the warm end.(0.2 Hz) with (a) Long high pressure period, (b) Short high pressure period.

flux at the warm end. The total pumping heat per cycle, however, decreases at high frequency.

### Variation of pressure wave form

The stepper motor for the rotary valve was controlled by the PC. Therefore, any arbitrary pressure wave could be generated theoretically by adjusting the pulse intervals of the D/A board for the stepper motor. The duration of the high pressure period was changed to examine its typical effect on the heat transfer characteristic. As shown in Figure 7, the gas temperature always increased very sharply at the beginning of the pressurization process of the square wave. This result was different from that of Seo[(8)], in that the gas temperature dropped very quickly. The gas temperature was measured at the center of the heat exchanger and the effective heat transfer always occurred between the center and the warm end heat exchanger that is well cooled by cooling water. It is interesting to observe that the amount of the total heat flux at the warm end heat exchanger becomes larger as the high pressure period is shorter. Relatively long expansion period in a cycle helps the warm end heat exchanger pump more heat[(9)]. It is believed that this type of the pulse tube refrigeration cycle also enhances the cooling performance.

## SUMMARY

(1) The experimental pulse tube apparatus was fabricated to investigate the heat transfer with the typical pressure waves of pulse tube refrigerators.
(2) The heat flux always lags the gas-wall temperature difference. As the cyclic frequency increases, the absolute value of the imaginary part of the cyclic averaged complex Nusselt number becomes large.
(3) The extended low pressure period in a cycle can enhance the heat pumping capability at the warm end heat exchanger.

## REFERENCES


1. W.E. Gifford and R.C. Longsworth, Pulse tube refrigeration, ASME paper No.63-WA-290, (1963).
2. E.I. Mikulin, A.A. Tarasov and M.P. Shkrebyonock, Low temperature expansion pulse tubes, "*Advances in Cryogenic Engineering,*" Vol. 29, Plenum Press, New York, (1984), p. 629.
3. P.C.T., de Boer, Thermodynamic analysis of the basic pulse tube refrigerator, *Cryogenics*, 34:699(1994).
4. E.S. Jeong, Secondary flow in basic pulse tube refrigerator, *Cryogenics*, 36:317(1995).
5. O.R. Burggraf, An exact solution of the inverse problem in heat conduction theory and applications, Journal of heat transfer, Trans. ASME, 86:373(1964).
6. A.A. Kornhauser and J.L. Smith, Jr., Application of a complex Nusselt number to heat transfer during compression and expansion, On flows in internal combustion engine-IV, ASME, (1988b), p. 1.
7. E.S. Jeong and J.L. Smith, Jr., An analytic model of heat transfer with oscillating pressure, General papers in heat transfer, ASME, (1992), p. 97.
8. K. Seo, M. Shiraishi, N. Nakamura and M. Murakami, Investigation of radial temperature and velocity profiles in oscillating flows inside a pulse tube refrigerator, in: "Cryocoolers 9," R.G. Ross, Jr. Ed., Plenum Press, New York (1997), p. 365.
9. G. Thummes, F. Giebeler and C. Heiden, Effect of pressure wave form on pulse tube refrigerator performance, in: "Cryocoolers 8," R.G. Ross, Jr. Ed., Plenum Press, New York (1995), p. 383.

# OSCILLATING FLOW BEHAVIOR IN A PULSE TUBE REFRIGERATOR UNDER OPTIMIZED CONDITIONS

M. Shiraishi[1], K. Seo[2], N. Nakamura[2], and M. Murakami[2]

[1]Mechanical Engineering Laboratory, MITI
Namiki 1-2, Tsukuba, Ibaraki, 305 Japan

[2]Institute of Engineering Mechanics, University of Tsukuba
Tennodai 1-1-1, Tsukuba, Ibaraki, 305 Japan

## ABSTRACT

Optimum conditions, which are for the lowest value of minimum cold end temperatures, are investigated by changing the frequency and the opening of the orifice and double inlet valves. Moreover, oscillations of velocity are obtained directly by using a hot-wire anemometer and the P-V work in the pulse tube is estimated. The velocity oscillation under the optimum conditions and the effects of valve opening on the P-V work are discussed. It is found that for the orifice configuration the P-V work increases with increasing valve opening and reaches a maximum value at a valve opening larger than that for the optimum condition. For the double inlet configuration the dependency of the P-V work on the valve opening is negligibly small except in the range of very small opening. It is suggested from the results of P-V work that the losses in the pulse tube have significant effects in the range of large valve opening.

## INTRODUCTION

The performance of pulse tube refrigerators has been improved in the recent years. Several approaches have been done to research and develop a practical pulse tube refrigerator[1-6]. The modelling of pulse tubes has become of major interest lately[7,8,9]. Although these models are convenient for making approximate predictions of basic trends or dependencies of performance, there are still significant shortcomings. For developing a more realistic analytical model, it is required to clarify the thermo-fluiddynamic phenomena experimentally and to evaluate whether the assumptions used in the model are adequate or not.

The fundamental flow phenomena in the pulse tube should be described by four variables, such as pressure, temperature, density and velocity. Several experimental investigations on pressure and temperature oscillations have been reported[7,10,11], but there are very few reports about a velocity oscillation due to difficulty of a velocity measurement[12,13]. In this study, we investigate the velocity oscillation by using a hot-wire anemometer and discuss the effects of valve opening on the velocity oscillation in the orifice and the double inlet pulse tube refrigerators. Moreover, we estimate the displacement of fluid elements from the velocity oscillation and discuss P-V work of these element for better understanding the optimum conditions for both pulse tube refrigerators.

## EXPERIMENT

### Experimental Apparatus

Figure 1 shows a schematic of the experimental apparatus. A helium compressor unit for a Gifford-Macmahon refrigerator and a rotary valve are used to generate pressure oscillations with frequencies varying from below 1Hz up to 20 Hz. The regenerator is composed of a stainless steel housing and a bakelite tube, 35 mm in diameter by 160 mm long, packed with 865 discs of 100 mesh stainless steel screen. The pulse tube, which has a total length of 288 mm with an inner diameter of 15.6 mm, is made of stainless steel pipe brazed to copper pipe at both ends as the heat exchangers. The reservoir has a volume of about 11 times that of the pulse tube volume. The orifice and the double inlet valves, which have a flow coefficient($C_V$) of 0.30 at full opening of 6 turns, are installed to find the optimum conditions. Tubes for introducing the hot-wire probe are attached near both ends and the middle point of the pulse tube. The hot-wire probe with a thermocouple was designed to measure the velocity of oscillating helium gas flow with the variations of its temperature. A platinum wire of 25 μm in diameter and a mineral insulated thermocouple(Chromel/Alumel) of 150 μm in diameter were used (see top right figure). Although the nominal 63 % response time of the thermocouple is 5 msec(200 Hz), the practical response time is found to be 15 msec(70 Hz) at most. Strain gauge type pressure transducers with a response time of about 0.25 msec(4 kHz) are installed at the entrance of the regenerator and the reservoir. Thermocouples(Copper/Constantan) are also soldered on the outer surface of the pulse tube at seven points to obtain the wall temperature distribution. The pulse tube and the regenerator are insulated by using a polyurethane foam insulation to provide convenient measurements for the velocity so that less attention is paid to minimize the heat leak from the environment at room temperature.

### Experimental Procedures

After evacuation of the apparatus, helium gas is charged at the pressure level of 1.4 MPa. The configuration of a pulse tube is provided by setting the orifice and the double inlet valves. Upon supplying the pressure oscillation, a frequency, a compression ratio(ratio of high to low pressures) and openings of both valves are adjusted at prescribed levels. In this experiment the compression ratio is fixed at the level of about 1.4. The hot end is kept at room temperature by circulating cooling water. The released heat from the hot end is derived from

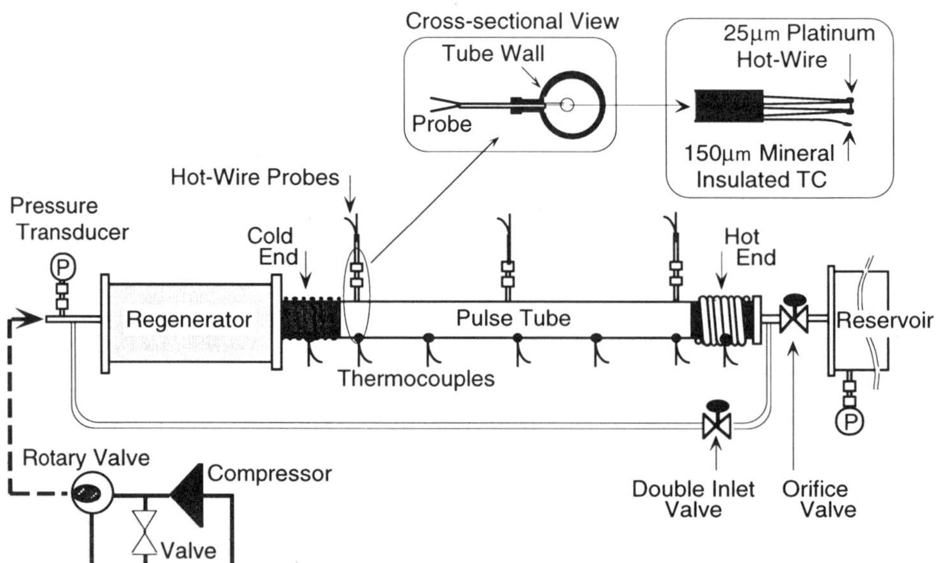


**Figure 1**. Experimental apparatus.

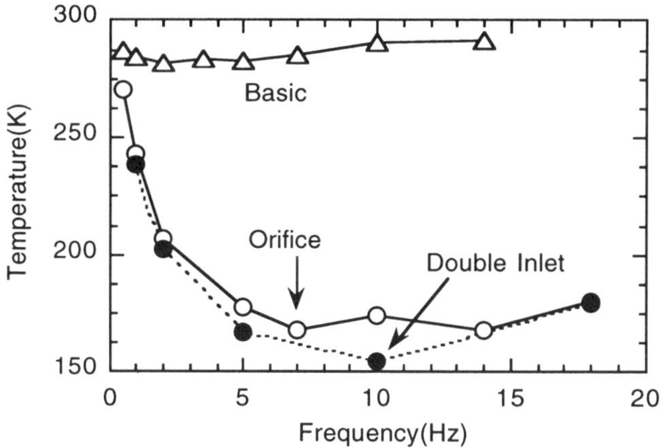


**Figure 2.** The minimum cold end temperature as a function of frequency.

the mass flow rate and the temperature increase of circulating water. After both the wall temperatures at the cold end and the released heat reach steady values, data are stored. By changing the frequency and the valve opening, the optimum conditions are found from the lowest value of minimum cold end temperatures.

## EXPERIMENTAL RESULTS AND DISCUSSION

Figure 2 shows the minimum cold end temperature without net refrigeration power at various operating frequencies. For the orifice and double inlet configurations the valve openings are optimized at each frequency. For the basic configuration the optimum frequency is found to be around 2 Hz. For other configurations it seems to be in the range of 5 Hz or more, but at 18 Hz the temperature increases slightly.

From the former results, we selected 10 Hz as the typical frequency in the optimum range to investigate the effects of valve opening. Figure 3 shows the effects of the orifice and the double inlet valve openings($V_o$, $V_d$) on the minimum cold end temperature under the fixed frequency of 10 Hz. For the double inlet configuration the orifice valve is kept at the optimized opening for the orifice configuration($V_o$=0.36) and only the double inlet valve is changed. The lowest temperature is only obtained under the optimized orifice valve opening and not

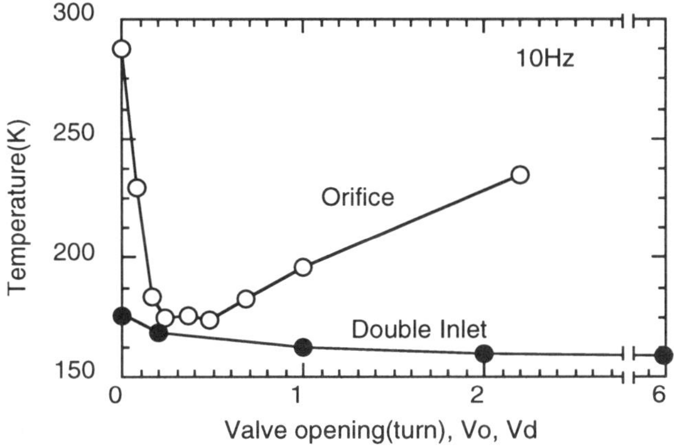


**Figure 3.** The minimum cold end temperature as a function of opening of the orifice and the double inlet valves.

any other opening. It should be noted that the orifice valve has a very narrow range for the optimization while the double inlet valve has a wider range except in the range of very small opening. The frequency is an essential parameter to the optimum operation, especially for the basic configuration, but for the orifice and the double inlet one the valve opening is a more effective parameter. The effects of frequency for the basic configuration were discussed in the previous study[14]. Here, we shall concentrate on the effects of valve opening on the inside flow phenomena for the other configurations.

Figure 4 shows the typical velocity oscillation at the center line of the pulse tube at each position. The valve openings are optimized as shown in the figure for the orifice and the double inlet configurations(the basic one is for a comparison). In the basic configuration, the

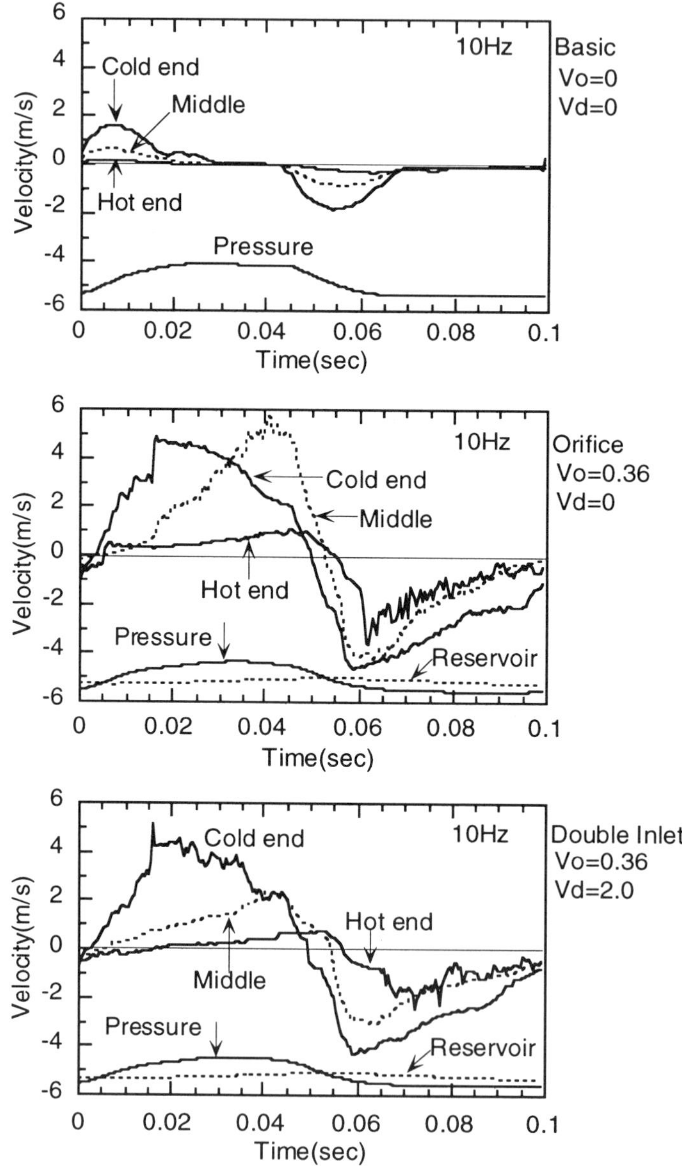


**Figure 4.** Variation of velocity during one cycle at both ends and the middle point of the basic, the orifice and the double inlet pulse tube refrigerators. For the orifice and double inlet configurations each valve is optimized.

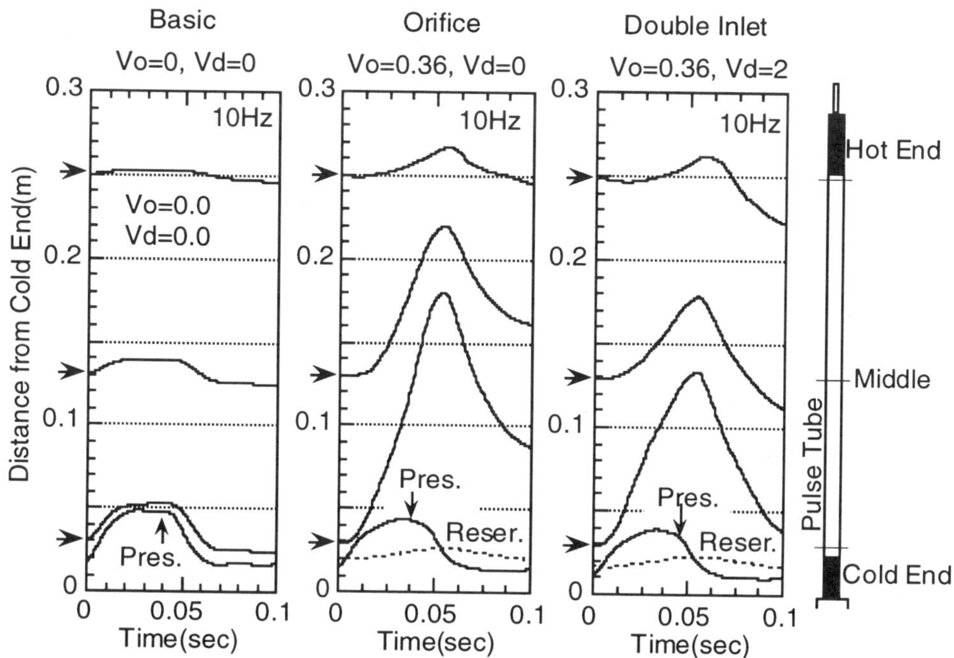


**Figure 5.** Variation of displacement of a fluid element during one cycle at both ends and the middle point of the basic, the orifice and the double inlet pulse tube refrigerators.

velocity varies as the variation of the time-derivative of the pressure(dP/dt), while in others the velocity does as the variation of (dP/dt) and the pressure difference between the pulse tube and the reservoir. This pressure difference can be controlled by the valve opening. It can clearly be seen that the orifice valve has large effects on the velocity oscillations at every point compared to the basic one, and that the double inlet valve has considerable influences at the middle point, but not so large at the cold end compared to the orifice one. It can be also seen that the velocity at the hot end is disturbed during the expansion process. This is explained as the influences of the jet injecting into the hot end from the reservoir.

Figure 5 shows the change of the displacement of fluid element during one cycle. They are derived from the velocity oscillations in figure 4 by using a linear approximation. For the basic configuration the oscillation is very similar to and almost in phase with the pressure oscillation. For other configurations they are very different from the basic one and its phase shifts about 90° from the pressure oscillation. In the case of the orifice configuration there is a large difference in the displacement between before and after one cycle. It indicates that a fluid element does not return to the initial position. Even if the error in the linear approximation is considered, the difference is considerably large compared to other cases. Now, the reason is not understood whether it originates in the flow phenomena such as a secondary flow[15,16] or the estimation from experimental data. This is the subject for a future study.

The P-V work is estimated by using the previous results of the displacement. Figure 6 shows the variation of the P-V work at the cold end as a function of the valve opening. For the orifice configuration the P-V work increases with the increase of the valve opening up to 1 turn, and then decreases over 1 turn. The optimum opening, which is around 0.36, does not correspond to that of the maximum value of P-V work. In the range less than the optimum opening, the minimum cold end temperature decreases with the increases of the valve opening as shown in figure 3. It can be explained that the refrigeration power increases with the valve opening due to the increase of P-V work. On the other hand, in the range over the optimum opening, the temperature increases with valve opening, but P-V work is still kept at larger value compared to that at the optimum opening. It may be explained as the increase of losses in the pulse tube due to the increase of velocity[17].

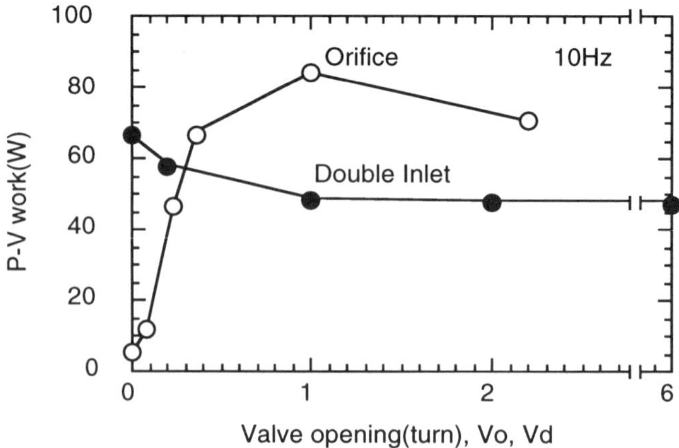


**Figure 6.** Estimated P-V work as a function of opening of the orifice and the double inlet valves.

For the double inlet configuration the dependency of the P-V work on the valve opening is negligible small except in the range of very small opening. Although the P-V work is smaller than that of the orifice configuration, the minimum cold end temperature is lower than that of orifice one as shown in figure 3. The velocity decreases from the level in the orifice one by opening the double inlet valve as shown in figure 4 so that the main reason of the improvement of refrigeration power is also explained as the decrease of the losses.

## SUMMARY

We investigate the velocity oscillation by using a hot-wire anemometer and discuss the effects of valve opening on the velocity oscillation and P-V work in pulse tube refrigerators.

It is found that the velocity increases and its phase shifts about 90° by the opening of the orifice valve compared to the basic configuration, and that the velocity at the middle point of the pulse tube decreases by the opening of the double inlet valve compared to the orifice configuration.

It is also found that for the orifice configuration the P-V work increases with the increase of the valve opening and then decreases, and that the optimum opening does not correspond to that of the maximum value of P-V work; for the double inlet configuration the dependency of the P-V work on the valve opening is negligibly small except in the range of very small opening.

It is suggested from the results of P-V work that the losses in the pulse tube have significant effects in the range of large valve opening. The clarification of details of losses through experiments is left for future studies.

## REFERENCES

1. W.E.Gifford and R.C.Longsworth, Pulse Tube Refrigeration progress, Advances in Cryogenic Engineering, Plenum Press, New York, 10B (1965) p.69.
2. E.L.Mikulin, A.A.Trasov and M.P.Shkerbyonock, Low-Temperature Expansion Pulse Tubes, Advances in Cryogenic Engineering, Plenum Press, New York, 29 (1984) p.629.
3 R.Radebaugh, J.Zimmerman, D.R.Smith and B.Louie, A Comparison of Three Types of Pulse Tube Refrigerators : New Methods for Reaching 60 K, Advances in Cryogenic Engineering, Plenum Press, New York, 31 (1986) p.779.
4. S.Zhu, P.Wu, Z.Chen, W.Zhu, and Y.Zhou, A Single Stage Double Inlet Pulse Tube Refrigerator Capable of Reaching 42 K, Cryogenics, 30 (1990) p.257.

5. R.Radebaugh, A Review of Pulse Tube Refrigerator, Advances in Cryogenic Engineering,Plenum Press, New York, 35 (1990) p.1191.
6. J.L.Gao and Y.Matsubara, Experimental Investigation of 4 K Pulse Tube Refrigerator, Cryogenics, 34 (1994) p.25.
7. P.J.Storch, R.Radebaugh and J. Zimmerman, Analytical model for the refrigeration power of the orifice pulse tube refrigerator, NIST Technical Notes 1343(1990).
8. A.Tominaga, Thermodynamic aspects of thermoacoustic theory, Cryogenics, 35, (1995), p.427.
9. P. Kittel, A.Kashani, J.M. Lee and P.R. Roach, General pulse tube theory, Cryogenics, 36, (1996), p.849.
10. M. Shiraishi, K.Seo and M. Murakami, Pressure and temperature oscillations of working gas in a pulse tube refrigerator, Cryocooler 8, Plenum Press, New York(1995), p.403.
11. J. Liang, A. Ravex and P. Rolland, Study on pulse tube refrigeration, Part3: Experimental verification, Cryogenics, 36, (1996), p.101.
12. W.Rawlins, R.Ragebaugh and K.D.Timmerhaus, Thermal anemometry for mass flow measurement in oscillating cryogenic gas flows, Rev. Sci. Instrum. 64(1993), P.3229.
13. K.Seo, M.Shiraishi, N.Nakamura and M.Masahide, Radial temperature and velocity profiles of oscillating flows in a pulse tube refrigerator, Proc.of the 16th ICEC/ICMC, Elsevier Science, Oxford(1997), p. 251.
14. K.Seo, M.Shiraishi and M.Murakami, Temperature oscillations of working gas inside a basic, an orifice and double inlet pulse tube refrigerators, Advances in Cryogenic Engineering, Plenum Press, New York, 41 (1996) p.1395.
15. J.M.Lee, P.Kittel, K.D.Timmerhaus and R.Radebaugh, Steady Secondary Momentum and Enthalpy Streaming in the Pulse Tube Refrigerator, Cryocooler 8, Plenum Press, New York, (1995), p.359.
16. E.S.Jeong, Secondary Flow in Basic Pulse Tube Refrigerators, Cryogenics, 36, (1996), p.317.
17. S.A.Colgate, Regenerator optimization for Stirling cycle refrigeration II, Cryocooler 8, Plenum Press, New York, (1995), p.247.

# VISUALIZATION STUDY OF OSCILLATING FLOW INSIDE A PULSE TUBE REFRIGERATOR

N. Nakamura,[1] M. Shiraishi,[2] K. Seo,[3] and M. Murakami[1]

[1] University of Tsukuba, Institute of Engineering Mechanics
Tennoudai 1-1, Tsukuba, Ibaraki, 305, Japan

[2] Mechanical Engineering Laboratory
Ministry of International Trade and Industry
Namiki 1-2, Tsukuba, Ibaraki, 305, Japan

[3] Yamagata University, Faculty of Education
Kojirakawa-machi 1-4-12, Yamagata, 990, Japan

## ABSTRACT

The flow inside typical pulse tube refrigerators which are a basic, an orifice and a double inlet type, has been observed by the visualization technique to investigate the effects of opening of the orifice and the bypass valves. A smoke-wire and a tuft flow visualization techniques are adopted. The displacement and the velocity of the working gas at cold end, middle of the pulse tube and in the bypass tube are measured by changing the valve openings and the flow is classified into 8 typical flow patterns. It is found that the flow pattern in the pulse tube is changed by the opening of the orifice and bypass valves. In the orifice type, phase difference between the pressure and the displacement increases as the increase of the orifice valve opening. It has maximum value at the optimum opening of the orifice valve. In the double inlet type, phase difference does not so change, but the flow behavior in the pulse tube is different from that of the orifice type.

## INTRODUCTION

A pulse tube refrigerator is a new type refrigerator discovered by Gifford and Longsworth[1]. It has been developed and investigated by many researchers to improve the performance. Because of many improvements, in recent years the performance of the pulse tube refrigerator is comparable with other cryo-coolers such as Striling, G-M refrigerator and so on. Therefore, it is possible to expect many applications which are not only in special fields such as space, but also in commercial fields[2].

Many researchers have been investigating the refrigeration mechanism. There are many researches related to analytical approaches, but not as many experimental studies about flow behaviors have been done. Lee et al.[3] made visualization studies of the internal flow pattern in the basic and orifice types, but the flow pattern of the whole pulse tube during one cycle was not clear. In a recent study, we visualized the flow in the pulse tube and investigated the characteristics of the oscillating flow[4,5].

The objective of this study is to understand the whole flow behaviors throughout the pulse tube refrigerator during one cycle by the visualization and to discuss the effect of the flow behavior on the refrigeration performance.

## EXPERIMENT

A schematic of the experimental apparatus is shown in figure 1. The pulse tube which is made of a transparent acrylate tube oriented in vertical position, having a length of 315 mm, an inner diameter of 16 mm and a wall thickness of 4 mm. It has three feedthroughs at the middle and near both ends of the tube to thread a smoke-wire across the tube and to insert a thermocouple. The regenerator is made of a bakelite tube with an inner diameter 18 mm and length of 168 mm, into which 770 discs of 100-mesh stainless steel screen are compressed. The reservoir has a volume of about 0.3 liters which is about 5 times that of the pulse tube volume. The hot end is connected to the reservoir through the orifice valve and to the warm side of the regenerator by the bypass tube through the bypass valve. The tee connecting the hot end, the reservoir and the bypass tube are made of a transparent acrylate tube to visualize the direction of the inside flow by using the tuft flow visualization as shown in the figure. The bypass tube is made of a transparent acrylate tube, having a length of 600 mm, an inner diameter of 4 mm. The smoke-wire with a diameter of 0.1 mm is made of a tungsten wire. It is connected with copper tubes at both ends and fixed by the tube fittings. Air is used as the working gas and the pressure oscillation is generated by introducing the compressed air of about 0.2 MPa into the rotary valve and releasing to the atmosphere.

The compression ratio is adjusted to about 1.2 and this value is smaller than the previous experiment[4,5]. The reason is to prevent the smoke emitted from the middle of the pulse tube from flowing over the hot end. Paraffin is used to make a smoke line. It is coated on the smoke wire surface and heated by instantaneous electrical heating. The smoke is emitted at

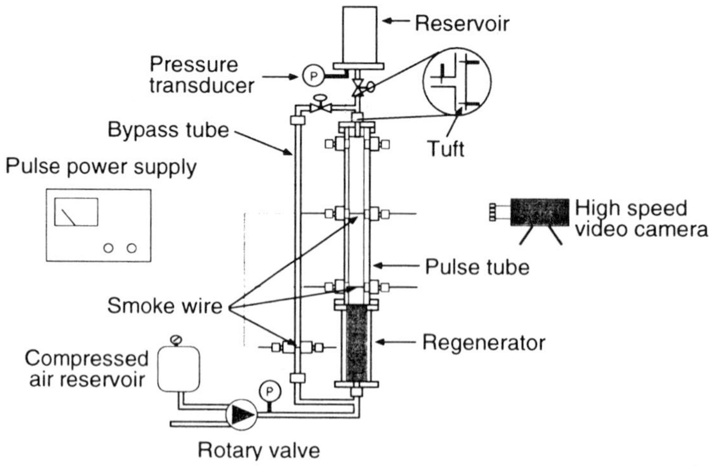


**Figure 1.** Experimental apparatus.

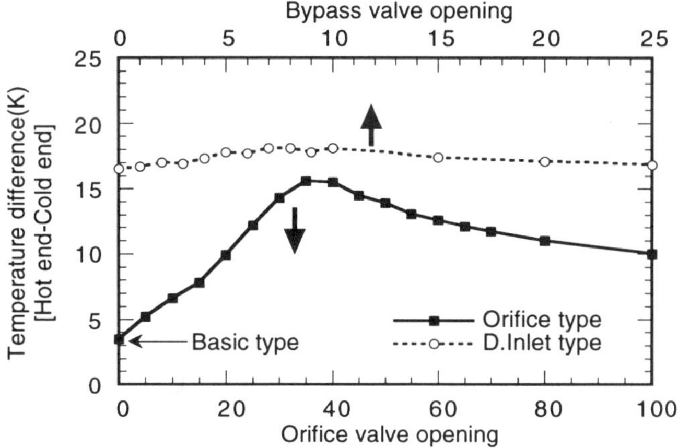


**Figure 2.** The change of the temperature difference between the hot and cold ends at 5 Hz by the orifice and bypass valve openings.

the moment of the beginning of the compression process and recorded by a high speed video camera. The video frames are processed by a computer. The displacement($x$) of the gas are measured from the distance between the smoke wire and the smoke line. The velocity at the displacement($x$) is measured from the difference of the displacement between a frame and next frame. In this study, we present the typical results based on the flow at the center axis of the pulse tube.

## RESULTS AND DISCUSSION

### Observation of Oscillating Flow

Firstly we investigate the performance of our pulse tube apparatus to find the optimum experimental conditions by changing the frequency and the valve openings. The difference of the gas temperature between the cold and hot ends is obtained as a function of frequency and the opening of the valves. The results show that the maximum difference is at 5 Hz for the orifice and double inlet types so that 5 Hz is determined as the frequency for the observation. Figure 2 shows the change of the temperature difference between the hot and cold ends at 5 Hz by the valve openings. In the orifice type, the temperature difference increases as the increase of the orifice valve opening, but it decreases in the range over the opening of about 40. In the double inlet type, the temperature difference increases slightly upon increasing the bypass valve opening to about 10, but it decreases slightly with larger valve openings.

Figures 3 and 4 show the photographs of the flow patterns at 5 Hz during one cycle in the basic type and the orifice type at the optimum opening of the valve, respectively. The interval between photographs is 25 msec, and the pressure oscillation in the pulse tube and the reservoir are shown below. In the basic type, variation of both smoke lines are almost in phase with the pressure oscillation. The maximum displacement at the cold end is larger than that at the middle of the pulse tube. In the orifice type, variation of smoke lines are not in phase with the pressure oscillation and there is a large phase difference. The difference in displacement between the cold end and at the middle is not so large compared to the former case.

It is also found that flow patterns in the double inlet type are very similar to that of the orifice type except the displacement of the smoke line is smaller.

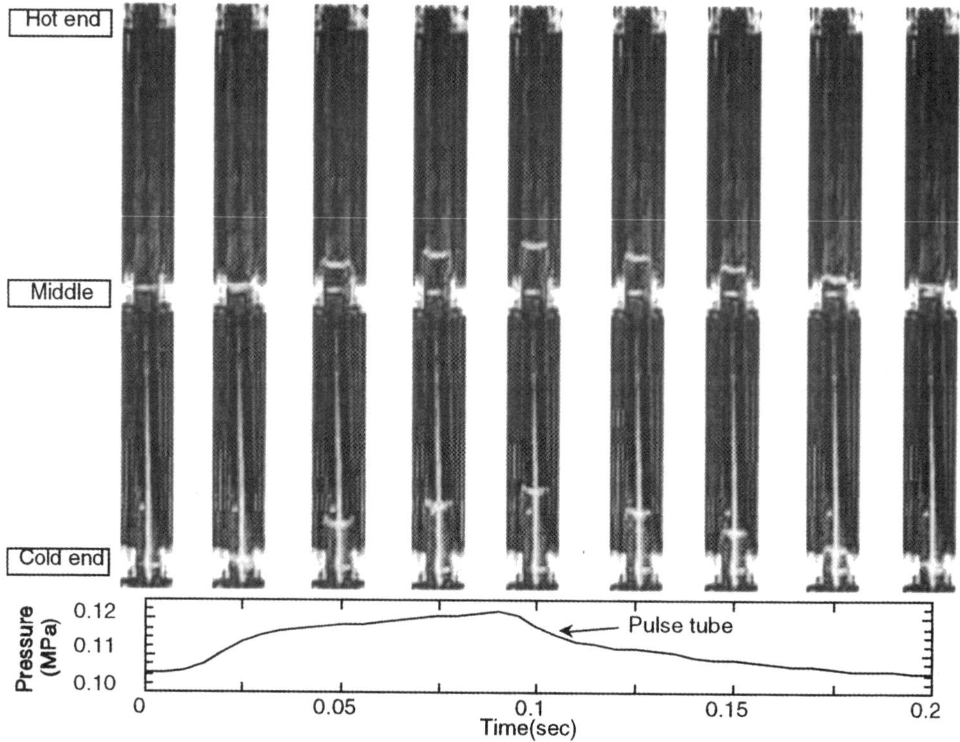


**Figure 3.** Photographs and the pressure oscillation in the basic type.

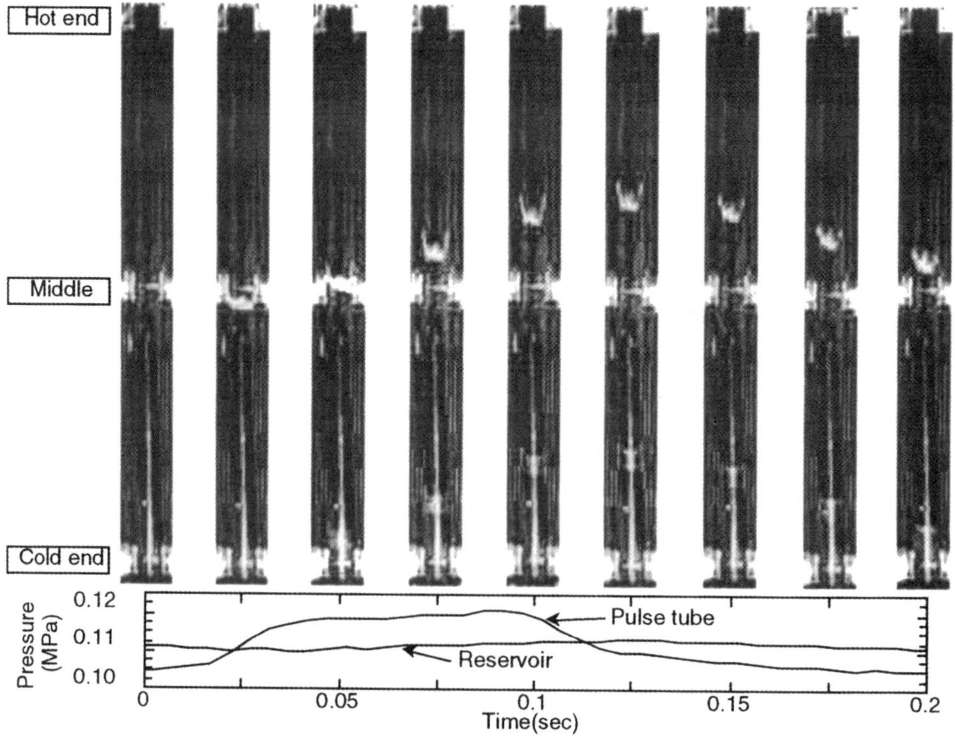


**Figure 4.** Photographs and the pressure oscillation in the orifice type.

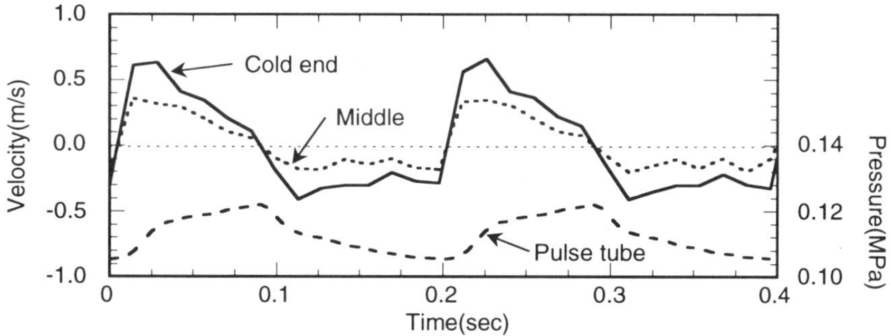


**Figure 5.** Velocity of the smoke and the pressure oscillation during two cycles in the basic type.

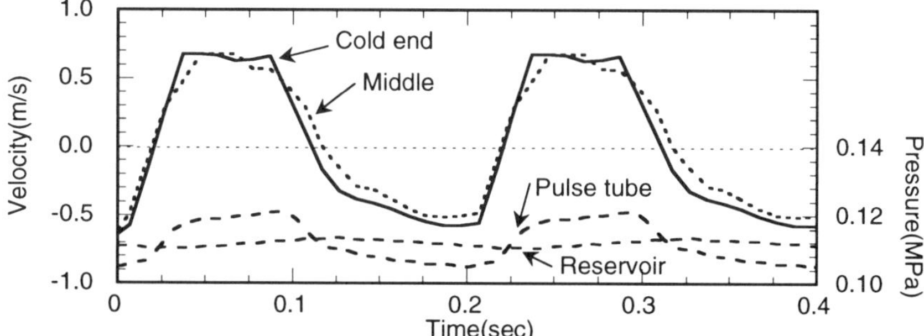


**Figure 6.** Velocity of the smoke and the pressure oscillation during two cycles in the orifice type.

## Velocity Profiles

The velocity on the center axis of the pulse tube are obtained from the results of visualization. Figures 5, 6 and 7 show the results for the basic, the orifice and the double inlet types at the optimum valve openings, respectively. The positive velocity means the flow direction to the hot end and the negative means that to the cold end. The solid and the dotted lines show the velocity of the smoke line emitted from the cold end and the middle of the pulse tube, respectively. The gray line in the double inlet type shows the velocity in the bypass tube. The broken lines show the pressure in the pulse tube and the reservoir.

The velocity in the basic type increases immediately as the increase of the pressure and decreases gradually in the compression process. The variation of velocity corresponds to the time-derivative of the pressure oscillation[4,5]. The time of inversion, when the flow

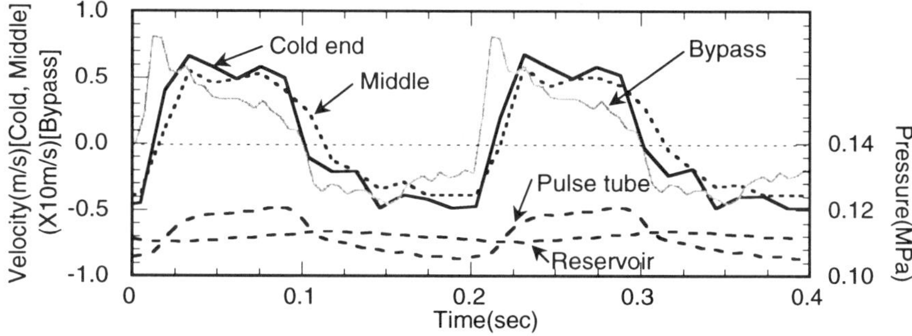


**Figure 7.** Velocity of the smoke and the pressure oscillation during two cycles in the double inlet type. The unit of velocity for the bypass is ten times of that for others.

direction changes, corresponds to the peak and the bottom of the pressure oscillation, which means the change of the process. There is no difference in the time between both positions because in the basic type, the change of the flow direction occurs at the same time in the pulse tube.

The variation of the velocity in the orifice type are different from in the basic type. When the pressure decreases, the velocity in the pulse tube decreases, but the flow direction does not change. Because the flow caused by the difference of the pressure between in the pulse tube and the reservoir still remains. The flow direction change at the time when the pressure in the pulse tube is equal to that in the reservoir. The time of inversion is almost the same at both points in the compression process, but it is different in the expansion process. The time of inversion at the cold end is faster than at the middle of the pulse tube. It means that the change of flow direction depends on the position in the pulse tube.

In the double inlet type, the velocity in the pulse tube is a little smaller and there is a larger difference in the time of inversion between at the cold end and the middle point compared to that in the orifice type. It is remarkable that the velocity in the bypass tube increases immediately with beginning of the compression process. Its maximum velocity is about ten times faster than that in the pulse tube. Then, the velocity begins to decrease when the pressure in the pulse tube is equal to that in the reservoir.

## Flow Pattern in the Pulse Tube

The change of flow direction during one cycle is obtained from results of the smoke-wire visualization at the cold end and the middle, and of the tuft visualization at the hot end.

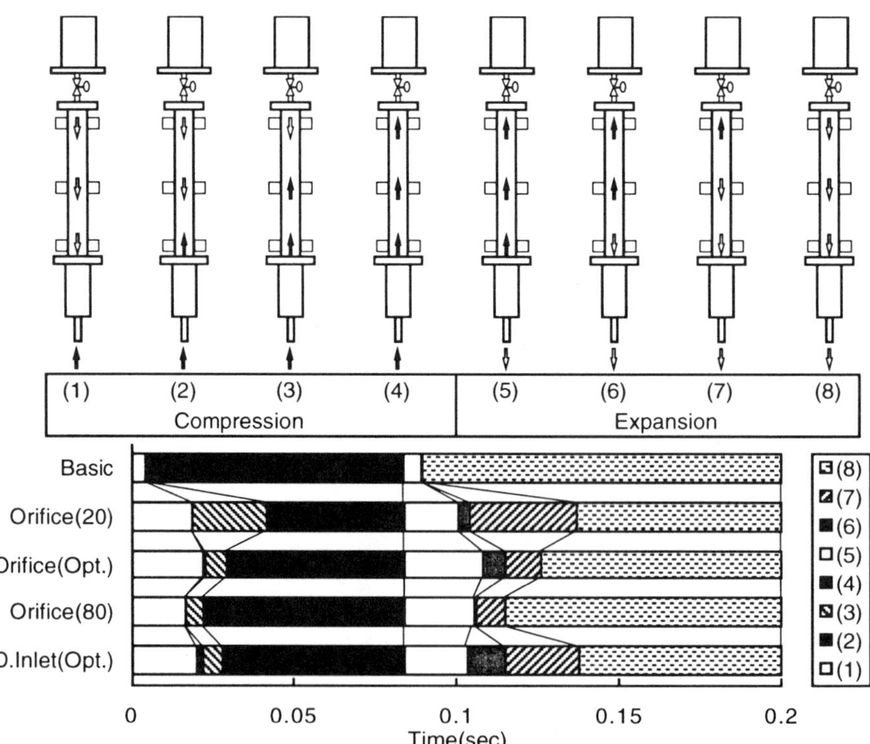


**Figure 8.** Time charts of the flow patterns in the pulse tube during one cycle for the valve openings. The value after the name of the type show the orifice valve opening and opt. means the optimum valve openings.

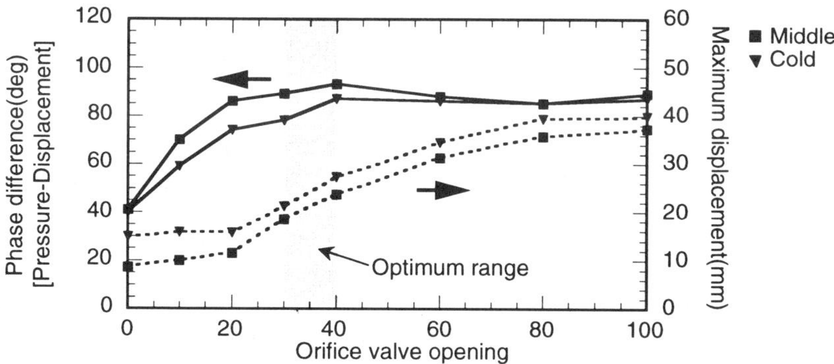


**Figure 9.** Phase difference between the pressure and the displacement, and the maximum displacement at each point for the orifice valve opening.

The flow pattern can be classified into typical 8 patterns. Figure 8 shows the proportion of flow patterns during one cycle for various openings of the orifice and bypass valves. The top figure shows the flow patterns and its number corresponds to the number of legends in the graphs.

In the basic type, the flow direction changes almost at the same time at any point in the pulse tube. It must be noted that as the opening of the orifice valve, the noticeable flow patterns, in which the direction of the flow depends on the position in the pulse tube such as (2), (3), (6) and (7), appear in a cycle. In the expansion process, as the increase of the interval of (6) increases, the interval of (7) decreases. The two intervals are comparable at the optimum opening. However, interval of (6) disappears as the opening exceeds the optimum one. As the result, it is supposed that flow pattern of (6) has the significant effects on the improvement of the performance. Although the velocity profile in the double inlet type is similar to that in the orifice type under the optimum conditions, the time intervals of (6) and (7) are longer. The increase of these patterns, especially (6), is thought to be one of the reasons for the improvement of performance in the double inlet type.

## Effect of the Orifice and the Bypass Valve Openings

In order to evaluate quantitatively for the difference of the flow behavior in the axial position, the phase difference between the pressure and the displacement are calculated.

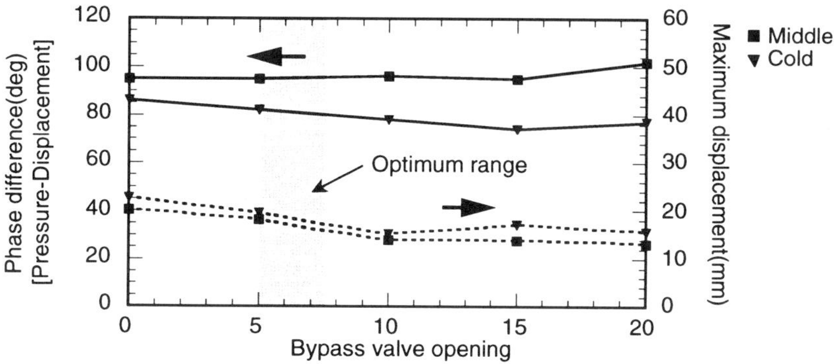


**Figure 10.** Phase difference between the pressure and the displacement, and the maximum displacement at each point for the bypass valve opening.

Figure 9 shows the dependency of the phase difference and the maximum displacement on the orifice valve opening at the cold end and the middle point. As the increase of the opening of the orifice valve, the phase difference increases and it has the maximum value within a range of the optimum condition shown by gray. Moreover, the values of the phase difference are different in the position along the pulse tube up to the range of the optimum opening. The difference between the values are negligibly small in the range of the large valve opening.

Figure 10 shows the changes of the phase difference and the maximum displacement with the bypass valve opening. The phase difference does not depend on the opening of the bypass valve compared to the former case, but it strongly depends on the position. These fact suggests that the dependency of the performance on the opening of the bypass valve is not so large and the gas in the pulse tube is more compressed or expanded compared to other case. The maximum displacement slightly decrease from that of the orifice type so that indicates a decrease of the losses in the pulse tube. But sufficient information to evaluate losses is not obtained in this experiment.

## SUMMARY

The summary is as follows:

The typical flow phenomena in the pulse tube refrigerator are visualized by using the smoke-wire and the tuft techniques.

The details of the flow patterns during one cycle are made clear and the effects of opening of the orifice and the bypass on the flow patterns are discussed.

It is confirmed directly from the visualization that the phase difference between the pressure and the displacement in the orifice and the double inlet types increases compared to the basic one.

## REFERENCES


1. Gifford. W. E. and Longsworth. R. C., Pulse-tube refrigeration, *ASME paper*, No.63-WA-290 (1963).
2. M. Nisenoff, F. Patten and S. A. Wolf, And What about Cryogenic Refrigeration?, "Cryocoolers 9," Plenum Press, New York (1997), p. 25.
3. Lee. J. M., Kittel. P. Timmerhaus. K. D. and Radebough. R., Flow patterns intrinsic to the pulse tube refrigerator, "Proceedings of the 7th International Cryocooler Conference," PL-CP-93-1001, Santa Fe (1993), p. 125.
4. M. Shiraishi., N. Nakamura., K. Seo. and M. Murakami., Investigation of Velocity Profiles in Oscillating Flows Inside a Pulse Tube Refrigerator, "Proceedings of the 16th International Cryogenic Engineering Conference," Elsevier, Kitakyushu Japan (1996), pp. 255-259.
5. M. Shiraishi., N. Nakamura., K. Seo. and M. Murakami., Visualization Study of Velocity Profiles and Displacements of Working Gas Inside a Pulse Tube Refrigerator, "Cryocoolers 9," Plenum Press, New York (1997), p. 355.

# DYNAMIC EXPERIMENTAL STUDY OF THE MULTI-BYPASS PULSE TUBE REFRIGERATOR WITH TWO-BYPASS TUBES

**Y. L. Ju, C. Wang , and Y. Zhou**

Cryogenic Laboratory, Chinese Academy of Sciences
P.O.Box 2711, Beijing, 100080, China

## ABSTRACT

A dynamic experimental apparatus has been constructed to investigate the instantaneous velocity and pressure in the multi-bypass pulse tube refrigerator(MPTR). Some important experimental results of the instantaneous measurements of the velocity and the pressure in the MPTR with two-bypass tubes during actual operation are presented. The effects of the middle-bypass version on the dynamic pressure and mass flow rate at the cold end of the pulse tube are evaluated from experimental measurements. DC-flow phenomena are observed in this MPTR. The reasons of the multi-bypass version improved the performance of pulse tube refrigerator are given.

## INTRODUCTION

Developments in the pulse tube refrigerators have progressed through three important stages from basic [1], orifice [2] to double-inlet [3]. Other modifications were presented by Ishizaki et al.[4] and Matsubara et al.[5] Interest in pulse tube refrigerator has grown rapidly due to the fact that it has no moving parts at the cold end. Therefore, it has considerable system advantages over most other types of refrigerators[6-7].

The pulse tube refrigerator is a variation of Stirling refrigerator, where the moving displacer is substituted by a gas piston. The refrigeration substance in the pulse tube refrigerator includes three part: hot gas, cold gas and medium gas. The medium gas acts as a gas piston. The refrigerating cycle in the pulse tube refrigerator is the same as in the Stirling machine. To raise the efficiency of the pulse tube refrigerator, we attempt to increase the pressure ratio in the pulse tube refrigerator and decrease the proportion of the gas acting as the gas piston; which is the principle of the double-inlet method. Actually, the double-inlet version allows some room-temperature gas to enter the pulse tube directly without passing through the regenerator. The main contribution of the double-inlet to the pulse tube is to adjust the phase shift between the pressure wave and the mass flow at the hot end of the pulse tube refrigerator, and also to increase their amplitudes[8].

To make further improvement, the multi-bypass pulse tube refrigerator[9] was suggested in 1992. A bypass tube and a valve connect the middle parts of the pulse tube and the regenerator. A part of the gas flow in or out of the pulse tube is due to the pressure drop. Therefore, there is an enthalpy flow from the regenerator to the pulse tube brought about by the middle-bypass tube. This means that refrigeration capacity is generated at the segment connected between the multi-bypass tube and the pulse tube.The improvement resulting from the addition of the middle bypass to the pulse tube refrigerator was initially verified experimentally by Cai et al[10], Wang et al[11]. However, due to the lack of the experimental measurements of the instantaneous mass flow rate and temperature in the MPTR during actual operation, there has been little experimental and theoretical analysis of the effects brought about by the introduction of the multi-bypass on the refrigeration performance of the pulse tube refrigerator. Som believe that the MPTR is actually a kind of multi-stage pulse tube refrigerator, and that it produces multi-stage expansion processes. Others think that the effect of the multi-bypass version is to adjusted the phase shift between the pressure wave and mass flow rate.

A better understanding of pulse tube refrigerators requires measurements of the velocity, temperature and pressure waves at different locations in the pulse tube. The difficulty here lies in obtaining the velocity measurement independent of temperature and with minimal disturbance. David et al.[12] reported a method of measuring the instantaneous velocity in an orifice pulse tube (OPTR), a hybrid pulse tube (HPT) and a double inlet pulse tube (DPT) using hot wire anemometry. Rawlins et al.[13] adapted constant temperature anemometers (CTA) and resistance temperature detectors (RTD) to measure the instantaneous mass flow rates and temperature during actual operation of an OPTR. They presented the values of enthalpy, entropy, and work flow at the cold end of the pulse tube evaluated from their measurements. Cai et al.[8] in our laboratory measured the amplitudes and phase difference between the pressure wave and the mass flow by using CTA.

To explain the reason why the multi-bypass version can improve the performance of the pulse tube refrigerator, the dynamic experimental apparatus of the MPTR with two-bypass tubes was designed and constructed. The experimental data of the amplitudes and phase difference between the pressure wave and mass flow rate at the cold end in the MPTR during actual operation are presented. DC-flow phenomena has been observed and is discussed in this paper.

## EXPERIMENTAL APPARATUS

A schematic of the experimental apparatus is shown in Figure 1. It is designed to measure the dynamic pressure, temperature and mass flow rate of the working gas at the cold end of the pulse tube. The swept volume of the valveless compressor is 55.6cm$^3$; the operating frequency is 16Hz; The pulse tube and the refrigerator walls are made of stainless steel with a thickness of 0.25mm. A reservoir of 300cm$^3$ volume is connected to the pulse tube hot end after an orifice valve. A double-inlet valve is placed between the hot end of the pulse tube and the hot end of the regenerator. The orifice valve and the double-inlet valves are all adjustable needle valves. This pulse tube refrigerator is a new type multi-bypass version with two-bypass tubes. Considering that the gas density and viscosity vary greatly with the temperature in the refrigerator, the pulse tube and the regenerator are separated into three segments, respectively. The dimensions of each of the segments of the pulse tube and the regenerator are given in table 1. The regenerators are packed with phosphor bronze screens and stainless steel screens. In the connecting parts of the pulse tube, we added 20 sheet of 100 mesh copper screens to act as a gas flow straightener to reduce the gas flow disturbance caused by gas flow in or out of the pulse tube[10]. The multi-bypass valves 9 and 10 are capillary tubes whose flow area can be adjusted, the length of the capillary tubes is 25mm.

**Table 1.** Dimensions of each segments of pulse tube and refrigerator

| components | pulse tube | regenerator |
|---|---|---|
| first segment | 12mm × 85mm | 17mm × 80mm |
| second segment | 10mm × 60mm | 14mm × 58mm |
| third segment | 7.5mm × 50mm | 10mm × 55mm |

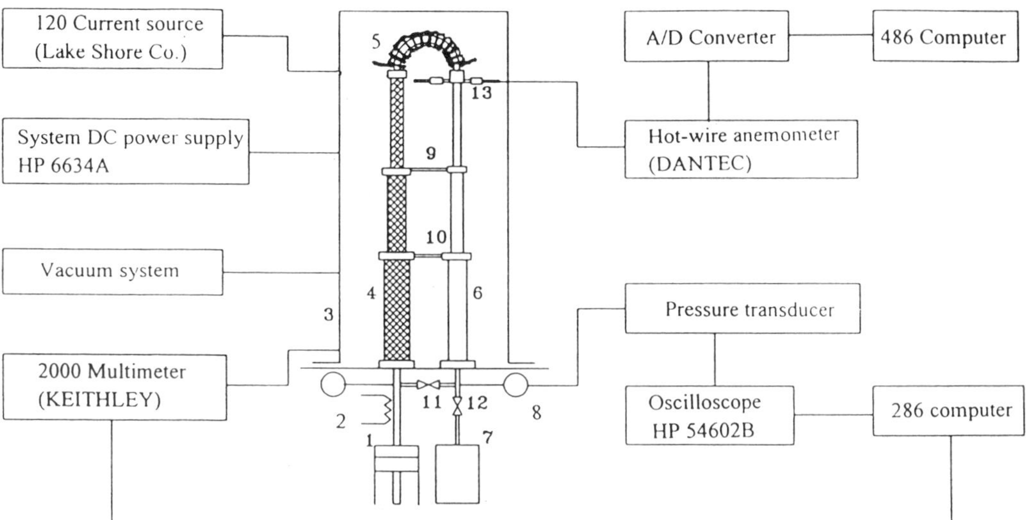


**Figure 1.** Schematic diagram of the dynamic experimental apparatus:
1. compressor; 2. aftercooler; 3. vacuum chamber; 4. regenerator; 5. cold head; 6. pulse tube; 7. reservoir; 8. pressure transducer; 9. middle-bypass valve (K2); 10. middle-bypass valve(K1) 11. double-inlet valve; 12. orifice

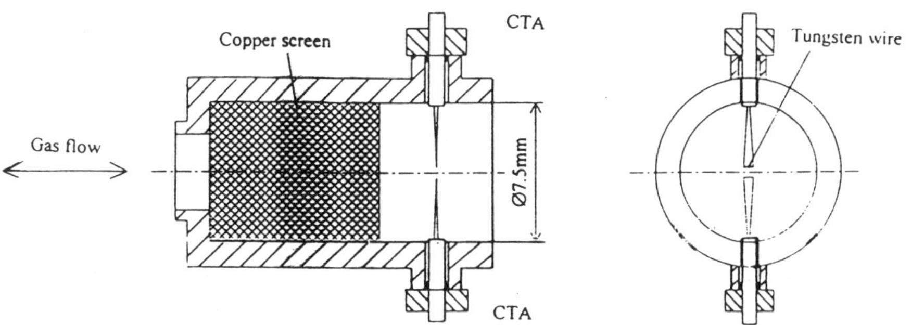


**Figure 2.** Experimental arrangement for the hot wire measurement.

Ten copper-constantan thermocouples are used to measure the wall temperature of the pulse tube and the regenerator. A resistance heater is provided at the cold end to measure the refrigeration capacity at a particular temperature. The refrigeration power is supplied by a system DC power supply (HP 6634A).Two small quartz pressure transducers (KISTLER Co.) with a high natural frequency (70KHz) are used to measure the transient gas pressure wave at the hot ends of the regenerator (after the compressor) and the pulse tube, respectively.

We used the hot wire anemometry (DANTEC, Model 90N10) to measure the instantaneous gas velocity and temperature. The anemometer works according to the constant temperature anemometer (CTA) principle where the sensor probe forms an arm of a Wheatstone bridge. It is kept in balance by the deviation signal across the bridge diagonal so that the probe resistance, and hence its temperature is kept constant independently of the cooling from the flowing medium. The solution for fast varying temperature is to operate two identical probes placed in very close to each other with different overheat ratios. This is called the power drop for double sensors method. That is, we can obtain the velocity independently of the temperature according to power drops of two identical probes operated with different overheat ratios.

Two hot wire probes (DANTEC, Model 9055) which are verified precisely identical to each other are located at center in the cold head of the pulse tube. We assume that the velocity is homogeneous in a cross-sectional area. The probes are made of 5μm diameter tungsten wire between the probe supports of 2mm length The wires are suspended in the gas stream by probe supports which are electrical conductors and serve as the electrical connections for the sense. The CTA are operated at approximately 200K warmer than the environment fluid. Both hot wires are placed inside of the cold head of the pulse tube. The response times of the CTA in this system are 20μs. Figure 2 shows the experimental arrangement for the hot wire measurement.

Analogy-to-digital conversions are carried out by a conversion board (KEITHLEY, DAS 1600) which is plugged into a 486 personal computer. A 4-channel simultaneous sample and hold front end are employed so that both the dynamic pressure and velocity voltage signals can be sampled simultaneously. An oscilloscope (HP 5402B) is also employed to simultaneously observe the wave signals.

## EXPERIMENTAL RESULTS AND DISCUSSIONS

To compare the effects of the multi-bypass version on the performance of the pulse tube refrigerator, we studied experimentally the refrigeration performance of the DPTR and MPTR. In our experiments, both multi-bypass tubes 9 and 10 were changed simultaneously. From our experimental measurement, the multi-bypass version can improve the refrigeration performance of the pulse tube refrigerator. The optimum flow area of the two multi-bypass tubes was $0.75mm^2$. The temperature that the MPTR achieved was 10~15K lower than that of the DPTR. However, the refrigeration temperature increased with further increase of the flow area of the multi-bypass valves. We will explain the reasons through dynamic experimental results.

The raw experimental data measured from the hot wire anemometry and pressure transducers needs data processing which include velocity transformation and correction, high frequency harmonic filtration, Fourier analysis, and so on. Figures 3 (a) and (b) show the mass flow rate and pressure wave as a function of time at the cold end for the DPTR and MPTR at the same operating conditions, respectively. They demonstrated that the amplitudes and phase relationship between the mass flow rate and dynamic pressure in the pulse tube refrigerator were not truly sinusoidal. It should be noticed that the peak flows and half-cycle frequencies are not equivalent in the Figures. Therefore, the experimentally measured phase angles were averaged between the two half-cycles. From the experimental results, we found that the phase relationship between the mass flow rate and

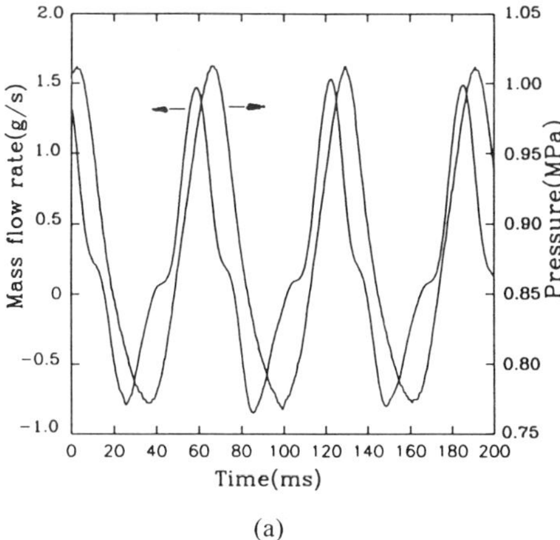


(a)

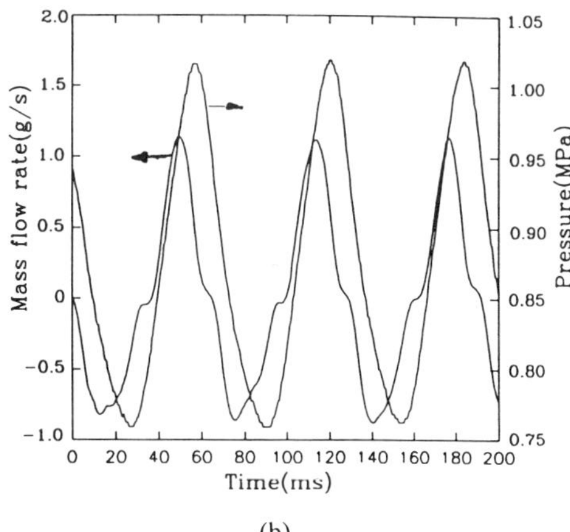


(b)

**Figure 3.** The mass flow rate and pressure wave as a function of time at the cold end of pulse tube refrigerator (Tc=77K, Pa = 0.9MPa, f =16Hz) (a) Double-inlet version; (b) multi-bypass version

pressure wave at the cold end of the pulse tube are independent of the multi-bypass flow area and the double-inlet opening values. It is different from the phase shift at hot end of the pulse tube refrigerator[8]. In this pulse tube refrigerator system, the average phase differences determined from the experimental data was 45 degrees with a standard deviation of 4.2 degrees.

We evaluated the two half-cycles time-averaged mass flow rate from the experimental data and found that the average mass flow rate which flow past the sensor in the positive direction (away from the compressor) was not equal to that in the negative direction (towards the compressor). The results are different from the experimental results for the OPTR in reference[13]. Some reasons may be the effects of the double-inlet and multi-bypass version. The uncertainty of the mass flow rate from experimental measurements is 2%.

Table 2 shows the two half-cycles time-averaged mass flow rate evaluated by integrating the instantaneous velocity values from the experimental data for different opening values of the double-inlet valve. The average mass flow rate in the positive direction is not equal to that in the negative direction. Therefore, there is a net mass flow rate which flows to one direction for fixed opening values of the double-inlet valve. This is a kind of DC-flow phenomenon which is not analogous to the AC electrical system described in reference[6]. It is natural to guess that the DC-flow is some kind of circulating flow (The maximum ratio of the DC-flow rate to the mass flow rate at cold end of the pulse tube refrigerator is about 3.5%). The path of the circulating flow is between the cold end and the double-inlet valve. The larger the opening values of the double-inlet valve, the larger the DC-flow rate at the cold end of the pulse tube refrigerator. As we know, the DC-flow introduces loss to the pulse tube refrigerator. Although the double-inlet version can improve the refrigeration performance of the pulse tube refrigerator by adjusting the phase shift between the pressure wave and the mass flow rate, the opening values of the double-inlet valve has an optimum value.

Table 3 shows the two half-cycles time averaged mass flow rate for different flow area of the multi-bypass tubes. The DC-flow rate decreased with the increase of the flow area of the multi-bypass tubes. When the flow area of the multi-bypass values tubes are adjusted to an optimum value, the DC-flow rate is zero. The lowest temperature at the cold end of the multi-bypass pulse tube corresponds to a DC-flow rate of zero. The optimum values of the two middle-bypass tubes in the pulse tube refrigerator were both 0.75mm$^2$. By further increasing the flow area of multi-bypass, a DC-flow in the negative direction observed and resulted in decreasing the refrigeration performance of the MPTR. The temperature at the cold end increased and might have been higher than that at the double-inlet version. Because there are two multi-bypass tubes in our MPTR system, we think there are two paths of circulating flow between the two multi-bypass tubes and the cold end of the pulse tube refrigerator.

**Table 2.** Time average mass flow rate and DC-flow rate at the cold end of DPTR (Tc=77K, Pa= 0.9MPa, f=16Hz)

| Double-inlet valve (turns) | 0 | 0.2 | 0.4 | 0.6 | 0.8 | 1 |
|---|---|---|---|---|---|---|
| mass flow rate in position direction (g/s) | 0.498 | 0.504 | 0.513 | 0.521 | 0.530 | 0.539 |
| mass flow rate in negative direction(g/s) | 0.496 | 0.498 | 0.504 | 0.508 | 0.512 | 0.516 |
| DC-flow rate (g/s) | 0.002 | 0.006 | 0.009 | 0.013 | 0.018 | 0.023 |

**Table 3.** Time average mass flow rate and DC-flow rate at the cold end of MPTR (Tc=77K, Pa= 0.9MPa, f=16Hz)

| Multi-bypass values ($mm^2$) | 0 | 0.25 | 0.51 | 0.75 | 1.05 | 1.25 |
|---|---|---|---|---|---|---|
| mass flow rate in position direction (g/s) | 0.530 | 0.515 | 0.493 | 0.466 | 0.502 | 0.497 |
| mass flow rate in negative direction(g/s) | 0.512 | 0.504 | 0.487 | 0.443 | 0.509 | 0.516 |
| DC-flow rate (g/s) | 0.018 | 0.011 | 0.006 | 0.003 | -0.007 | -0.019 |

The DC-flow introduces losses to the pulse tube refrigerator. From experimental data, we found that it may occur in the DPTR and MPTR if the flow area of the double-inlet and multi-bypass tubes are not adjusted to optimum values. Therefore, the ratio of the DC-flow rate to the average mass flow rate is an important parameter for the design of the pulse tube refrigerator. In addition, we found that the multi-bypass version can reduce the DC-flow rate occurring in the DPTR. Therefore, the multi-bypass version can improve the refrigeration performance of the pulse tube refrigerator if the DC-flow rate in MPTR is zero.

## CONCLUSIONS

Measurements of instantaneous mass flow rate and pressure were presented at the cold end of the MPTR with two-bypass tubes during actual refrigeration operation. Some important experimental results of the instantaneous measurements were given. Results revealed that DC-flow phenomena occurred in the DPTR. We found that the multi-bypass version can reduce the DC-flow rate that occurs in the DPTR. The DC-flow introduces losses which can be reduced to zero by adjusting the flow area of the multi-bypass tubes to an optimum value. Comparison of experiments verifies that the multi-bypass version can improve the performance of the pulse tube refrigerator.


## ACKNOWLEDGEMENT

The research is supported by the National Natural Sciences Foundation of China. The authors wish to thank Professor C. S. Hong for useful discussion on this paper


## REFERENCES


1. Gifford, W.E. and Longsworth, R.C. Pulse tube refrigeration, Trans of the ASME. 63-WA-290:264 (1964).
2. Mikulin, E. I., Tarasov, A. A. and Shrebyonock, M. P. Low-temperature expansion pulse tube , Adv. Cryo. Eng. 29:629 (1984).
3. Zhu, S., Wu, P. and Chen, Z. Double inlet pulse tube refrigerator--An important improvement, Cryogenics. 30:514 (1990).

4. Ishizaki,Y. and Ishizaki, E. Experimental study and modelization of pulse tube refrigerator below 80K down to 23K, Proc. 7th. Int. Cryo. Conf., USA. 7:140 (1993).
5. Matsubara, et al. Four-valve pulse tube refrigerator, Proc. of JSJS-IV, Beijing, 4:54 (1993).
6. Radebaugh, R. A Review of Pulse Tube Refrigeration, Adv. Cryo. Eng. 35B:1191 (1990).
7. Radebaugh, R. Recent development of Pulse Tube Refrigeration, 19th. Int. Conf. (1995)
8. Cai, J.H. et al. Experimental analysis of the double-inlet principle in pulse tube refrigerator, Cryogenics. 33:522 (1993).
9. Zhou, Y. and Han, Y.J. Pulse tube refrigerator research, Proc. 7th. Int. Cryo. Conf., Santa Fe, NM, USA(1992)
10. Cai, J.H., Wang, J.J., Zhu,W.X., and Zhou, Y. Experimental analysis of the multi-bypass principle in pulse tube refrigerator, Cryogenics. 34:713 (1994).
11. Wang,C. et al. Experimental study of multi-bypass pulse-tube refrigerator, Cryogenics. 35:555 (1995).
12. David, M., Marechal, J.C. and Encrenaz, P. Measurements of instantaneous gas velocity and temperature in a pulse tube refrigerator, Adv. Cryo. Eng. 37:939 (1992).
13. Rawlins, W. et al. Energy flows in an orifice pulse tube refrigerator, Adv. Cryo. Eng. 39:1449 (1994).

# DESIGN OF A 0.5 WATT DUAL USE LONG-LIFE LOW-COST PULSE TUBE COOLER

D.T.Kuo[1], A.S.Loc[1], S.W.K.Yuan[1], and A.L. Johnson[2]

[1]BEI Technologies, Inc.,
Sylmar, CA, 91342, USA
[2]Electro Thermo Associates
Manhattan Beach, CA, 90266, USA

## ABSTRACT

The design of a 0.5 Watt Dual Use Pulse Tube Cooler discussed in this paper is based on the analyses of a Hybrid Refrigerator Model (HRM) which is similar in nature to a third order model that has been validated extensively against various Stirling and Pulse Tube Coolers in the literature[1-8]. Together with BEI's experience with the Stirling refrigerators and the introduction of flexure bearings, the design of a long-life and low-cost cooler emerges. The cooler should have a lifetime of at least five years with high reliability and is capable of producing 0.5 Watt of cooling power at 80 Kelvin with a total input power of less than 15 Watt and with a cooler heat sink temperature of 310 Kelvin. The system should also deliver 0.3 Watt of cooling at 65 Kelvin with the same input power and heat sink temperature. The cooler is also light-weight, weighing less than 1 Kg.

## INTRODUCTION

BEI has developed a family of "tactical" cryocoolers based on the U.S. Army Night Vision Electronics Directorate (NVESD) specifications. These units are of split Stirling cycle variety, employing opposed piston compressors and reciprocating regenerator displacers. The first step in designing the Long-Life Low-Cost Cooler ($L^3C^2$) required the identification of the compressor needed to drive the Pulse Tube. The current design of the BEI B512 cooler meets all the thermodynamic requirements of the $L^3C^2$ cooler described in this paper (see Reference 9). However there are two major concerns in using the B512 Compressor as the driver. First, in order to meet the 5+ year of life time requirement, the springs of the compressor have to be replaced by flexure bearings. Due to the size limitation, the incorporation of flexure bearings in the B512 compressor might present a problem. Second, thermodynamic analyses showed that the B512 compressor does not have enough power to drive a 0.5W pulse tube. Thus it was concluded to use the B600 compressor instead.

A schematic diagram of the Dual Use $L^3C^2$ Cooler is presented in Figure 1. The pulse tube and regenerator are designed to fit within the current design of an existing BEI coldfinger. For the sake of easy packaging, a concentric design of the pulse tube is adopted. Four hundred mesh screens made of stainless steel are used for the regenerator material. For reasons to be discussed later, etched foils might be used as regenerator material instead. In this paper, the results of an orifice pulse tube are discussed. Due to the fact that concentric pulse tubes are less efficient than their linear cousins, the double-inlet and/or multi-bypass configuration might be used to make up the difference between the theoretical and actual performance of the cooler.

## PREDICTED PERFORMANCE

Due to the stringent requirements of this cooler, (0.5 W cooling per 15W input power at 80K), it is important to perform a thorough thermodynamic analysis to determine the configuration of the cooler and the operating conditions. The model used for the design of this $L^3C^2$ cooler is similar to a third order model that has been validated extensively against various Stirling and Pulse Tube coolers in the literature[2-8]. This new model is know as the Hybrid Refrigerator Model[1] (HRM) which uses second order analysis for heat transport and third order analysis for mass flow. For a detailed description of the HRM model, please refer to Reference 1. Some of the key parameters of the $L^3C^2$ cooler are tabulated in Table 1.

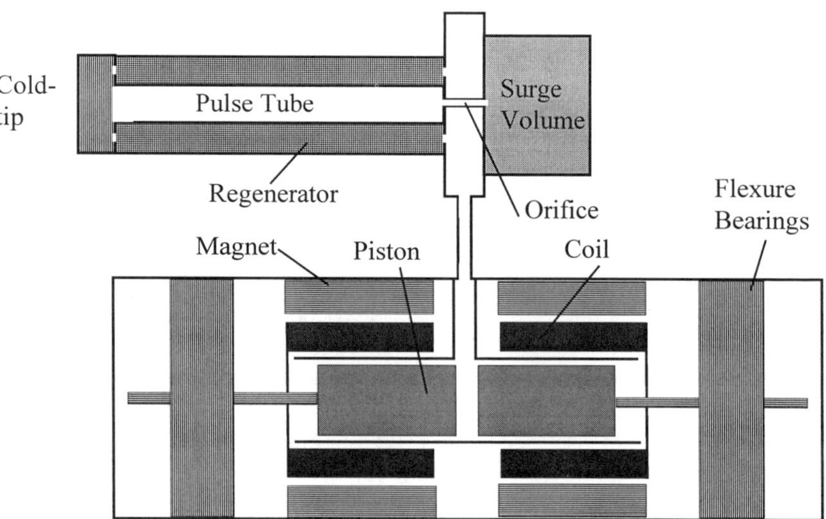


**Figure 1**. A schematic diagram of the Dual Use $L^3C^2$ Cooler.

**Table 1**. Key parameters of the $L^3C^2$ Cooler

| | |
|---|---|
| Compressor Stroke | 0.9 cm |
| Pulse Tube Length | 3.58 cm |
| Pulse Tube Diameter | 0.3 cm |
| Regenerator Length | 3.58 cm |
| Regenerator Diameter | 0.5 cm |
| Orifice Diameter | 0.018 cm |
| Surge Volume | 25 $cm^3$ |

Figure 2 shows the Cooling Capacity of the cooler as a function of the Specific Power (W/W) for different frequencies. The design point of this cooler is marked by the intersection of 0.5W cooling and 30 W/W specific power. As one can see, the optimum frequency to drive the cooler is between 40 and 50 Hz, agreeing with the operating frequency (at 44.625 Hz) of the TRW's AIRS Pulse Tube Cooler[10-12]. The natural frequency of the BEI compressor is around 60 Hz. To match the desired operating frequency of the compressor to its natural frequency, one can either increase the moving mass of the piston to decrease the natural frequency, or one can use etched-foil regenerators. Experimental data show that due to the less pressure drop of the etched-foil regenerator, one can operate the cooler at a higher frequency.

Figure 3 is a plot of Cooling Capacity versus input power. Unlike Figure 2, the orifice diameter here is 0.018cm. The thermodynamic analysis shows that at 40Hz, the cooler is more efficient with an orifice diameter of 0.018cm, and at 60Hz, the cooler is more efficient with an orifice diameter of 0.016cm.

The effect of the pulse tube / regenerator length on performance is plotted in Figures 4a and 4b. Within the range studied, the model shows a slight increase in the cooling capacity with a shorter regenerator length. The specific power also favors a shorter length. However, the gain in performance is not significant and since the pulse tube is designed to fit inside an existing coldfinger, the length of the existing BEI coldfinger has been chosen.

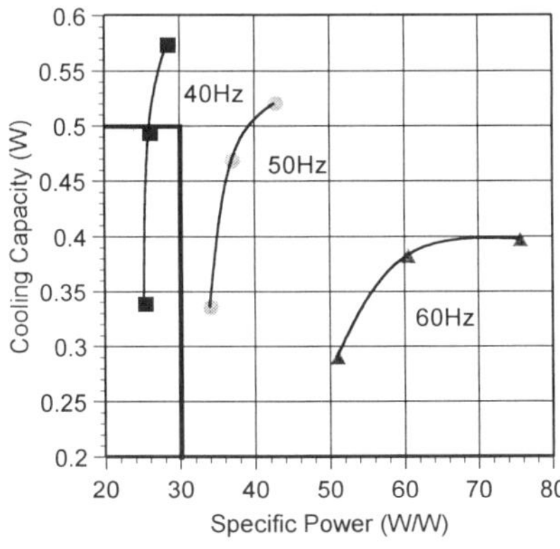


**Figure 2**. Refrigeration vs. specific power.

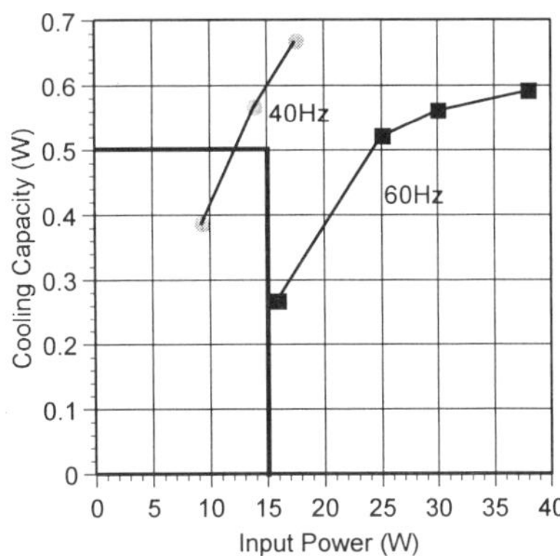


**Figure 3.** Refrigeration vs. input power.

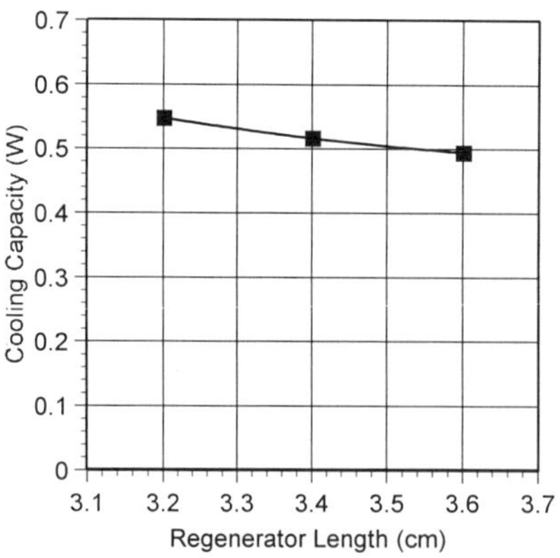


**Figure 4a.** Cooling capacity versus regenerator length.

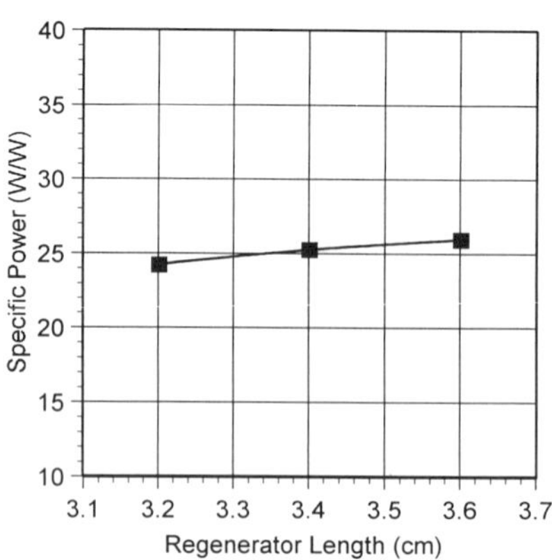


**Figure 4b**. Specific power versus regenerator length.

The effect of the regenerator diameter on performance is studied in Figures 5a and 5b. While the cooling capacity favors a smaller regenerator, an optimum specific power is found at a regenerator diameter of around 0.5cm. Since the input power poses a bigger concern than the absolute cooling capacity, a regenerator diameter of 0.55cm has been selected.

The cooling capacity and specific power of the $L^3C^2$ cooler is plotted as a function of the pulse tube diameter in Figures 6a and 6b respectively. In both cases, the optimum performance of the cooler is found at a pulse tube diameter of 0.3cm.

The effect of orifice diameter on performance is depicted in Figures 7a and 7b, with the optimum value at around 0.019cm.

In order to satisfy the long-life requirement of the cooler, the compressor springs have to be replaced by flexure bearings. A schematic diagram of this compressor can be found in Figure 1.

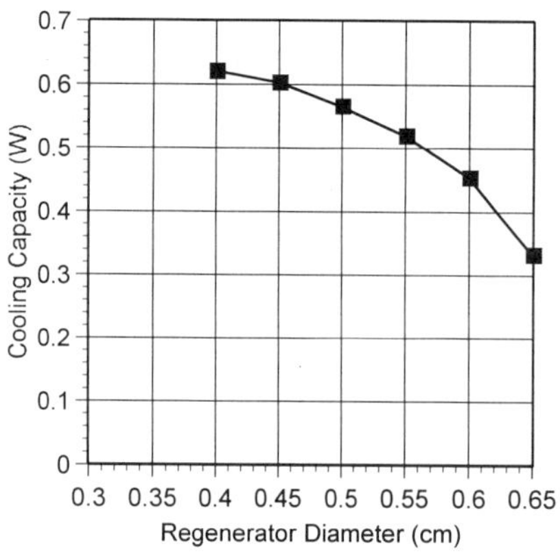


**Figure 5a.** Cooling capacity versus regenerator diameter.

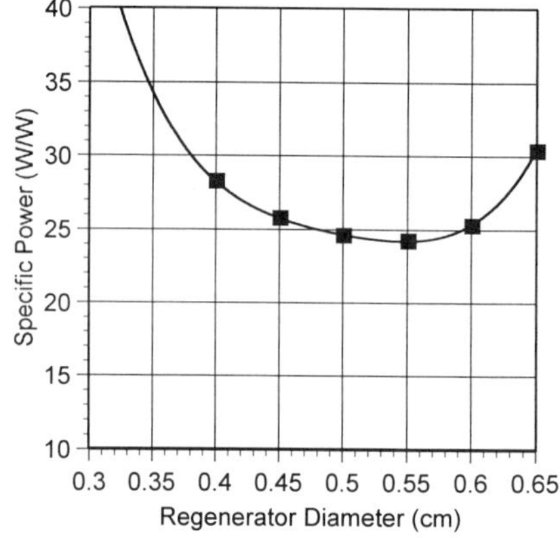


**Figure 5b**. Specific Power versus regenerator diameter.

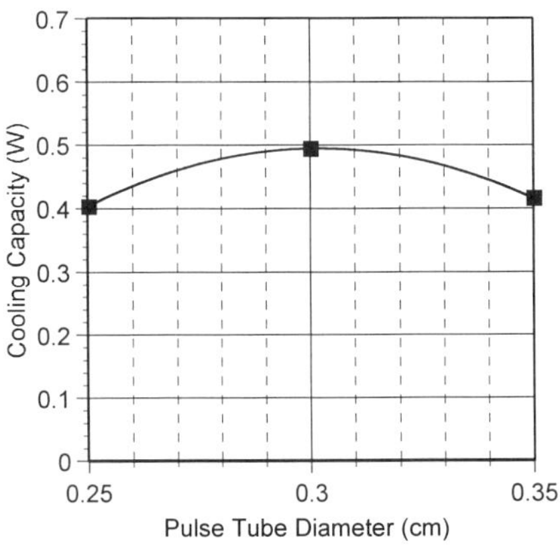


**Figure 6a.** Cooling capacity versus pulse tube diameter.

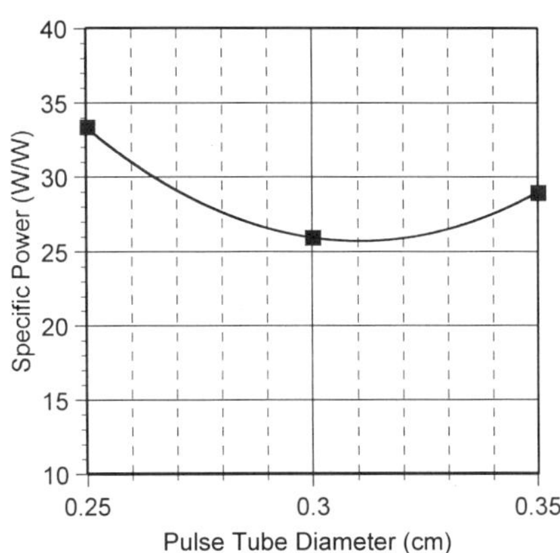


**Figure 6b.** Specific power versus pulse tube diameter.

## FLEXURE BEARINGS

To allow operation in any gravitational orientation, the non-wearing bearings must provide a high radial stiffness while maintaining a close piston to cylinder concentricity throughout the stroke length. Electro-magnetic bearings, linear flexure bearings, hydrostatic gas bearings and hydrodynamic gas bearings have all been successfully used in this type of application, however, for the sake of simplicity, cost effectiveness, and maximum reliability, only the linear flexure bearing and the hydrostatic gas bearing are practical for the cooler discussed in this paper.

The best designs are the Aerospace and "delta" flexures[13]. These configurations result in the lowest radial noise, the highest radial stiffness to axial stiffness ratio when compared to all the other configurations when the characteristics are normalized to the infinite cycle fatigue life limited maximum local stress.

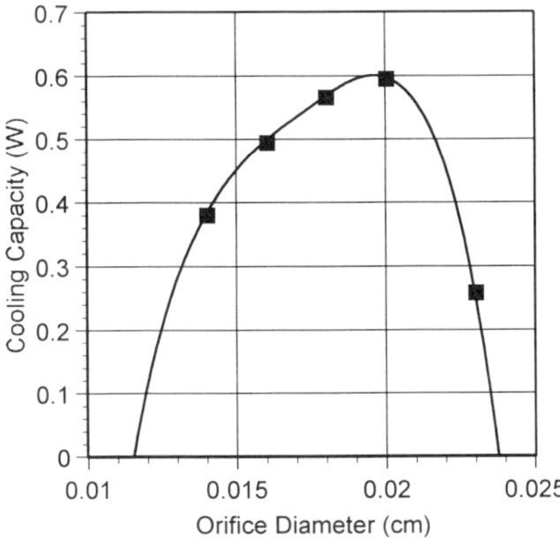


**Figure 7a.** Cooling capacity versus orifice diameter.

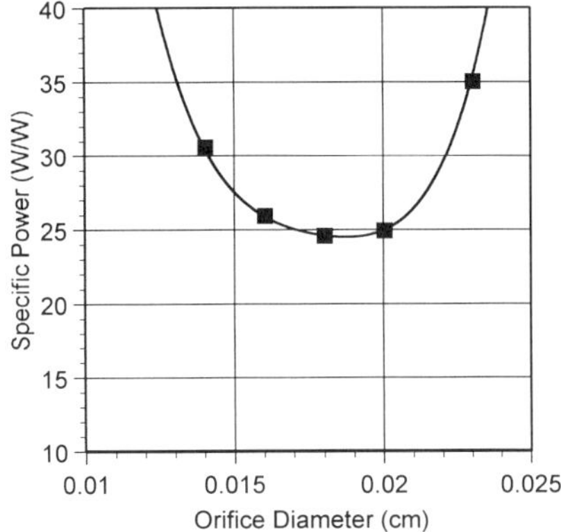


**Figure 7b.** Specific power versus orifice diameter.

The assembly, alignment, and attachment of the stack of flexure diaphragms and spacers required to create a linear flexure bearing has proven to be feasible but costly. Skilled workers, complex fixturing, and multiple trial and error fit-ups have been required to obtain an acceptable installation. A potential solution to this problem is the use of an integrated flexure bearing cartridge which can be installed in a manner analogous to that used in roller and ball bearing cartridges.

The Peckham version of the linear flexure bearing cartridge is composed of a bonded assembly of diaphragm flexure springs, inner diameter spacers, outer diameter spacers, and a retainer system which maintains the assembly as a single entity with a precise outside diameter, inside diameter and length. The inner and outer diameters are precision ground to the tolerance and concentricity required by the specific design. The ends of the assembly are parallel ground perpendicular to the axis of the ID/OD in order to produce the part in a configuration that can be assembled with the final product without using shims.

A study of the total fabrication / assembly process utilizing the Peckham cartridge verified the potential cost savings of this approach, however, this study also showed that an alternative approach which located the outer fixed rim by the use of three locating pins rather than a precision outside diameter surface reduced the cost of both the cartridge and the assembly housing. Additional finite element analysis showed that the flexure could be improved by shaping the outer edge of the flexure leaf. In addition, the use of the locating pins to position the cartridge eliminated the need for a circular outer rim, thereby reducing the material content of a flexure. Johnson (one of the authors) incorporated all these changes into an alternate flexure bearing cartridge, identified as the ETA flexure bearing.

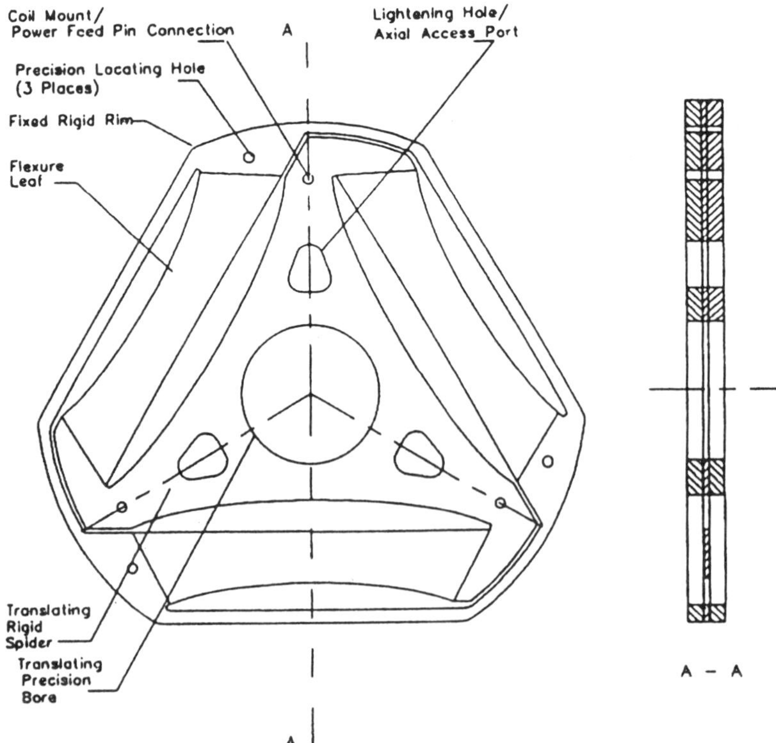


**Figure 8**. The ETA flexure bearing cartridge.

Figure 8 shows the ETA flexure bearing cartridge. Peckham now manufactures both the Aerospace and the ETA type flexure bearing cartridges. These "Delta" flexure cartridges have other advantageous characteristics: The tangential leaf configuration virtually eliminates radial noise since it is not subjected to the many natural vibration modes of the spiral bladed flexure; the flexure cartridge is easily integrated into a moving sleeve compressor configuration; and the individual cartridges can be electrically insulated so that the flexure can also serve as the electrical power lead for the moving coil.

Hydrostatic gas bearings have the advantage of high radial stiffness, but they have the disadvantage of parasitic high pressure working gas flow rate required to provide this high stiffness. Due to the power limitation of this high efficient $L^3C^2$ cooler, the use of flexure bearings is favored.

## CONCLUSIONS

The design of a Long-Life Low-Cost Pulse Tube Cooler is discussed in this paper. Using a well-validated computer model, the $L^3C^2$ cooler was predicted to deliver 0.5W cooling at 80K, with an input power of less than 15W. The predicted frequency agrees with the operating frequency of the TRW AIRS Pulse Tube Cooler. The pulse tube and the regenerator are designed to fit within the design of an existing BEI coldfinger. To increase reliability of the cooler, flexure bearings are to be incorporated into the cooler to extend the life time to beyond 5 years. This cooler will be fabricated and tested in fall of 1997.

## REFERENCES

1. D. Kuo, T. Loc, and S.W.K. Yuan, Experimental and Predicted Performance of the BEI Mini-linear Cooler, in Proc. the 9th International Cryocooler Conference, (1997) p. 119.

2. S.W.K. Yuan and R. Radebaugh, A Blind Test on the Pulse Tube Refrigerator Model, in "Advances in Cryogenic Engineering", Vol. 41, (1996) p.1383.

3. S.W.K. Yuan, Validation of the Pulse Tube Refrigerator Model Against a Lockheed Built Pulse Tube Cooler, in "Cryogenics", Vol. 36, No.10, (1996) p.871.

4. S.W.K. Yuan, L.G. Naes, and T.C. Nast, Prediction of Natural Frequency of the NASA 80 K Cooler by the Stirling Refrigerator Performance Model, in: "Cryogenics", Vol. 34, No.5, (1994) p.383.

5. S.W.K. Yuan, I.E. Spradley, and W.G. Foster, Validation of the Stirling Refrigerator Performance Model Against the Oxford Refrigerator, in: "Advances in Cryogenic Engineering", Vol. 39, (1994) p.1359.

6. S.W.K. Yuan and I.E. Spradley, Validation of the Stirling Refrigerator Performance Model Against the Philips/NASA Magnetic Bearing Refrigerator, in: "Proc. 7th Int. Cryocooler Conf.", Vol. 1, Phillips Lab, USA (1993) p.280.

7. S.W.K. Yuan, I.E. Spradley, P.M. Yang, and T.C. Nast, Computer Simulation Model for Lucas Stirling Refrigerators, in: "Cryogenics", Vol. 32, (1992) p.143.

8. S.W.K. Yuan and I.E. Spradley, A Third Order Computer Model for Stirling Refrigerators, in: "Advances in Cryogenic Engineering", Vol. 37 (1992) p.1055.

9. D.T. Kuo, A.S. Loc, and S.W.K. Yuan, Enhanced Performance of the BEI 0.5 Watt Mini-Linear Stirling Cooler, a parallel paper at Cryogenic Engineering Conference, 1997.

10. C.K. Chan, et al., Performance of the AIRS Pulse Tube Engineering Model Cooler, in Proc. of the 9th International Cryocooler Conference, (1997) p. 195.

11. R.G. Ross, Jr., and K.E. Green, AIRS Cryocooler System Design and Development, in Proc. of the 9th International Cryocooler Conference, (1997) p. 885.

12. C.K. Chan, et al., AIRS Pulse Tube Cryocooler System, in Proc. of the 9th International Cryocooler Conference, (1997) p. 895.

13. Electro Thermo Associates, Manhattan Beach, CA, 90266, USA

# DEVELOPMENT OF A LOW-COST UNLIMITED-LIFE PULSE-TUBE CRYOCOOLER FOR COMMERCIAL APPLICATIONS

T. Nast, P. Champagne, and V. Kotsubo

Lockheed Martin Missiles & Space, Palo Alto, CA 94304

## ABSTRACT

Our low-cost unlimited-life pulse-tube cryocooler was developed under a NASA cooperative program for high-temperature superconducting (HTS) cellular phone applications. Team members include the Lockheed Martin Advanced Technology Center, Superconductor Technology Inc., the National Institute of Standards and Technology, and the NASA Goddard Space Flight Center. This paper reports on the development of the pulse tube and one of the two compressor types developed. The Lockheed Martin (LM) compressor described utilizes a moving magnet design combined with self-aligning flexure supports in combination with a co-axial pulse tube. This approach evolved specifically to reduce production costs while maintaining the unlimited lifetime and high reliability of the spaceborne flexure systems. Early testing of an in-line pulse tube shows 3.7 W of cooling at 77 K and no life limitation. The coaxial version was also developed and provided cooling at the same level. Production costs, based largely on actual parts quotes, are estimated at under $2,000, including the electronic controller, for quantities of 10,000 units per year.

## INTRODUCTION

High-temperature superconductor (HTS) materials have been proven useful for ultraselective radio-frequency filters for communication systems such as cellular phones.

The keys to the commercial success of this technology are the cost and reliability of the cooler. The mean time to failure (MTTF) must exceed 40,000 hours. Because traditional Stirling coolers have MTTF levels of less than 10,000 hours, cooler development is a major focus for technology development.

Under a NASA Aerospace Industry Technology Program (AITP), a consortium of federal and industry partners has developed prototype cryocooler systems for low-cost commercial applications. The consortium is led by the Advanced Technology Division of Lockheed Martin Missiles & Space in partnership with Superconductor Technology, Inc. (STI), a small business in Santa Barbara, California, the National Institute of Standards and Technology (NIST) in Colorado, and NASA's Goddard Space Flight Center in Maryland.

The pulse tube approach increases the reliability and reduces the cost of the cryocooler by eliminating one of the few moving parts (the displacer) from the system. Two low-cost commercial approaches for the compressor that drives the pulse tube are being developed under this NASA cooperative program: the current STI approach used in their Stirling-cycle machine, which uses gas bearings, and a flexure bearing approach being developed at the Lockheed Martin Advanced Technology Center.

The development of the flexure-bearing/pulse-tube system is described in this paper.

## DESIGN APPROACH

### Goals

The principal goals of this program were to develop a cryocooler system with a lifetime greater than 40,000 hours and with approximately 4 W of cooling at 80 K. The cost goals were $1,000 in quantities of 1,000 units. There were no specific requirements for power, size and weight.

## FLEXURE BEARING COMPRESSOR

The low-cost compressor is based on the same three technologies that give traditional space-based systems their long life and high reliability: a clearance-sealed piston, guided by a spiral flexure linear bearing, and actuated by a direct-drive linear motor. This combination eliminates all sliding and rolling contact between moving parts, thereby removing theoretical limits on operational life.

However, clearance seals require such tight tolerances that these high-precision systems have thus far been too expensive for commercial applications. Therefore the primary focus of this project is to drastically reduce the cost of manufacture while still retaining the base technologies. This is accomplished by:

1. Devising a simplified mechanical architecture with fewer parts
2. Configuring these parts to be adaptable to mass production processes
3. Reducing assembly time by simplifying and automating the alignment procedure

Figure 1 shows the layout of the cryocooler in a typical air-cooled application with the integral coaxial pulse tube. This system weighs 8.1 kg (excluding the electronic controller) and is 10 cm in diameter and 31 cm long. Figure 2 is a photograph of the compressor.

### Architecture

A new mechanical architecture was devised for the compressor which simplifies the design, reduces part count, and allows for easier assembly. Major areas of change included a moving magnet motor, modular assembly, and external sensors.

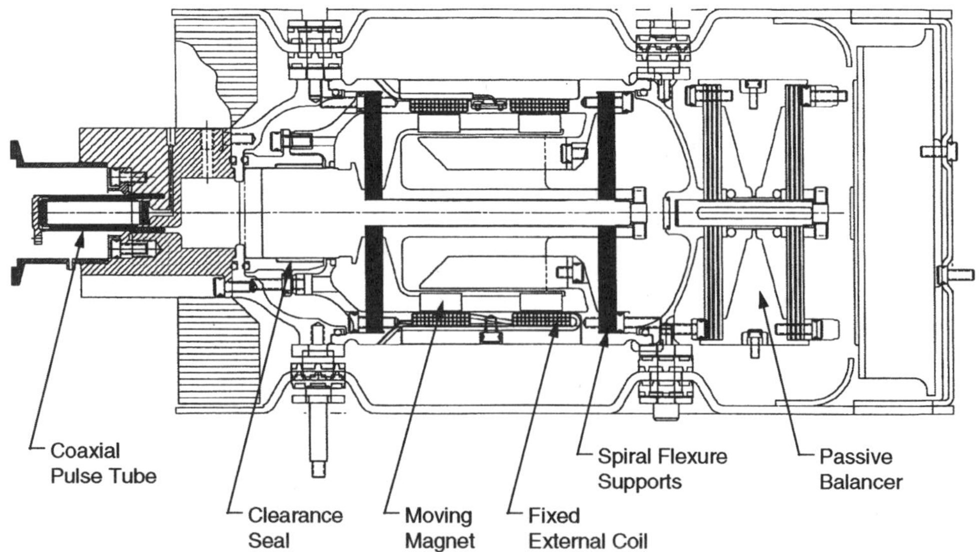


Figure 1. Layout of low-cost pulse-tube cryocooler.

Figure 2. Our low-cost compressor.

**Moving Magnet Motor.** The moving magnet motor reverses the traditional motor design to use internal moving magnets with a stationary coil outside the working fluid containment. While the added moving mass of this type of motor can be a problem in fast-response systems, it is not necessarily a disadvantage in tuned resonant reciprocating systems such as this. Moving magnet motors eliminate the need for problem-prone flexing motor leads and their electrical feedthroughs in the pressure wall, thereby greatly simplify sealing. In addition, magnets are less likely than epoxy-bonded wire coils to contaminate the working fluid. Concerns about unbalanced radial forces resulting from nonuniformities in the magnets proved to be unfounded because of the relatively large effective air gaps in the magnetic circuit and the radial stiffness of the bearing system.

**Modular Assembly.** An all-steel piston/shaft, flexures, and housing components are combined into a modular assembly, which constitutes the complete linear bearing system and can be preassembled and checked prior to integration into the compressor. Because similar materials are used, thermal expansions do not distort the structure and affect the linearity of the bearing system. This is essential to maintain the fit of the clearance seals during temperature excursions.

**External Sensors.** External sensors are used to sense piston position from outside the pressure wall. Ultraprecision control of position, which is required for vibration cancellation in satellite applications, is not required for ground-based HTS applications. This makes possible other, lower cost approaches to position sensing. External detection of the magnetic fields from the moving magnets is one possibility with this type of motor. Magnetic field sensors are small, inexpensive, and can work from outside a nonmagnetic pressure wall.

**Piece Parts Manufacturing**

Manufacturing methods for higher production quantities were evaluated for each piece part. With some investment in tooling for mass-production processes, such as investment casting and forging, parts costs can be dramatically reduced. With these higher-production processes, parts are produced to near net shape inexpensively, significantly reducing raw material and machining operations—the primary components of manufacturing costs. Other possible manufacturing methods include chemical milling and laser cutting, both of which are repeatable and accurate, and also eliminate machining time.

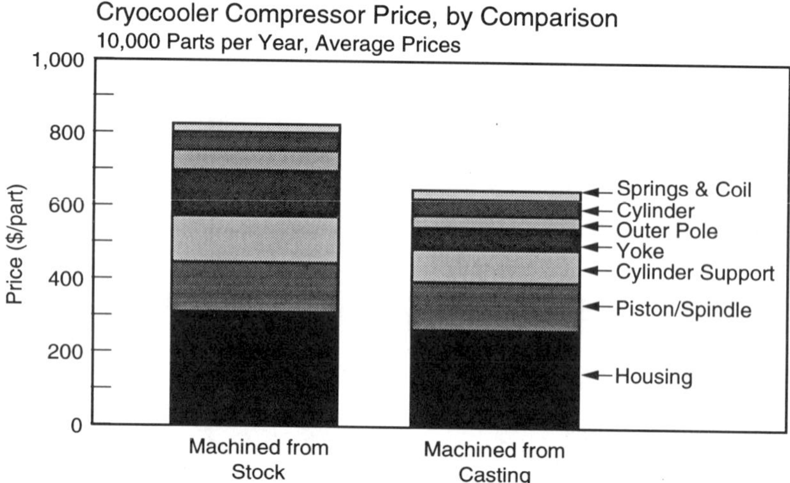


Figure 3. Composite prices by component.

To verify that design goals had been achieved and to estimate actual manufacturing costs in volume production, a study of piece part fabrication methods was commissioned by an outside consulting firm. The report from this study indicates that the major parts cost for the compressor would range from $700 to $900 (depending on tooling investment) for build rates of 10,000 units/year (see Figure 3).

This study focused on the seven major piece parts that account for about 90% of the total parts costs. While cost per unit generally drops with increased volume, there was found to be a full 30% savings realized when casting processes are used to reduce machining labor (see Figure 4).

## Assembly and Alignment

Assembly and alignment procedures now constitute a large part of the compressor manufacturing costs. Trueing the linear bearing system and centering the clearance seal has traditionally been a long and tedious procedure, requiring highly skilled technicians who

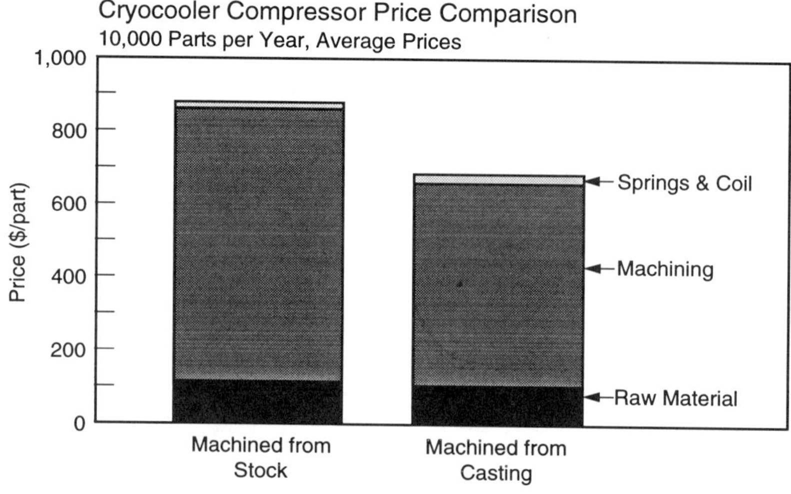


Figure 4. Composite prices, comparing manufacturing approach.

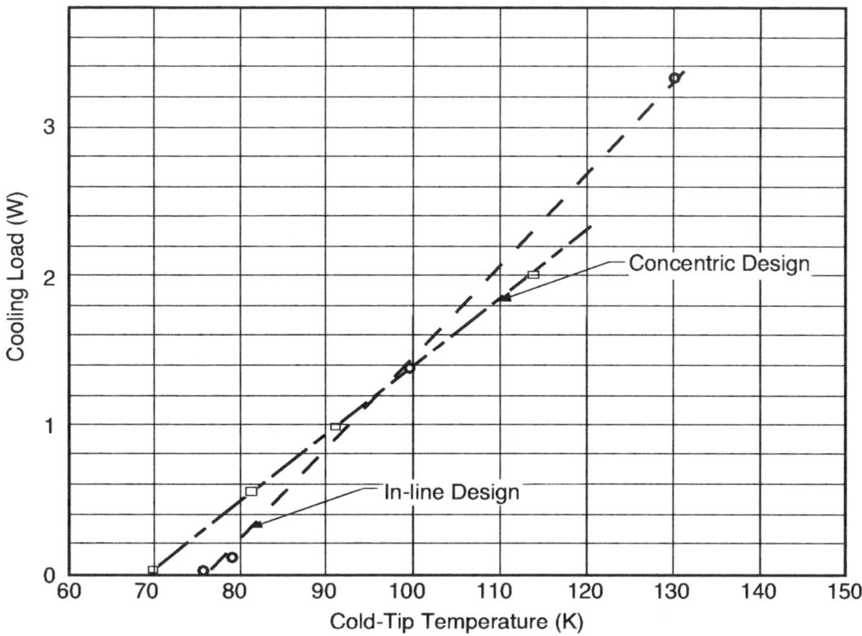


Figure 5. Comparative testing on concentric and in-line configurations (first design).

often resort to trial and error methods. Reducing this assembly to a simple, step-by-step procedure was necessary to meet cost goals. This is being accomplished in two stages, by simplifying assembly and by automating alignment.

**Simplified Assembly.** Assembly is simplified by eliminating adjustments in the linear bearing. The flexure bearing system is designed to be self-aligning upon assembly. Features integrated into the piece parts cause the piston/shaft to be centered in the single-piece housing, producing the required precision linear motion of the piston without the necessity for adjustments.

**Automated Alignment.** Automated alignment eliminates the need for operator skill. The final critical alignment of the piston and cylinder is accomplished by following a set procedure that determines the relative position of the parts and displays the results graphically in real time. The operator uses this display in conjunction with adjusting screws to move the cylinder into final alignment. High-resolution fiber-optic position sensors, which track part position, are already in use at Lockheed Martin. Microactuators with force feedback that can be adapted to computer control are also used. Computer control of the entire procedure will ultimately eliminate the need for skilled operators, resulting in improved repeatability and yields.

## PULSE TUBE

Two pulse tube designs were tested under this program. Both were of the double inlet configuration, using adjustable metering valves to tune the flow impedances for laboratory testing. Each design was tested in both in-line and concentric configurations in an attempt to quantify any excess losses in the concentric configuration due to effects such as lateral heat transfer between the pulse tube and the regenerator.

The first pulse tube design was tested using our L1710C flexure bearing compressor with a compressor swept volume of 3.6 cm$^3$. The load lines for both the in-line and concentric configurations are shown in Figure 5. The data show similar cooler load lines for the in-line and concentric configurations. The second pulse tube design was tested with the low-cost compressor described in this paper. The load lines for both the in-line and concentric pulse tubes are shown in Figure 6. The in-line data was with about 190 W of compressor power, while the concentric data was taken with about 170 W of compressor power. Both versions produced about 3 W of cooling at 80 K, close to the original goal and well within the current requirements of the HTS filter system.

To compare the in-line and concentric configurations directly, we attempted to match the regenerator volumes, pulse tube volumes, and dimensions of the transfer line and other flow passages. In practice, differences in the actual hardware occurred, some which were unavoidable due to the basic arrangement difference between the in-line and concentric configurations and others which were intentional. For example, for the second design, the particular arrangement of the bypass valve resulted in an overly long transfer line, which we estimated to have dissipated 30 W of compressor power. In addition, the concentric configuration for the second design used titanium tubing, while the in-line version used 304 stainless steel tubing. We estimated that the stainless steel tubing contributed about 0.5 W of additional parasitics compared to the titanium tubing. Adjusting the data in Figure 6 for these two differences suggests slightly better performance for the in-line configuration.

In general, these and other hardware differences between the in-line and concentric configurations for both designs prevent precise quantitative determination of any losses due to lateral heat transfer. The data does suggest, however, that the loss is not overwhelming and that concentric pulse tubes are feasible, particularly for commercial applications where the concentric design can significantly reduce packaging and integration costs.

System tests with the concentric pulse tube integrated with the STI filter package are in process at the time of this paper (July 1997).

## SYSTEM COST STUDIES

The majority of the cost studies conducted on this program were focused on the compressor, since the parts cost was 54% of the projected total system cost. Figure 7 summarizes the various costs for the system. The complete system cost is $1648 each in quantities of 10,000/year, including tooling and setup costs. The electronic controller system utilized feedback loops to control temperature and stroke, along with a high-efficiency pulse width

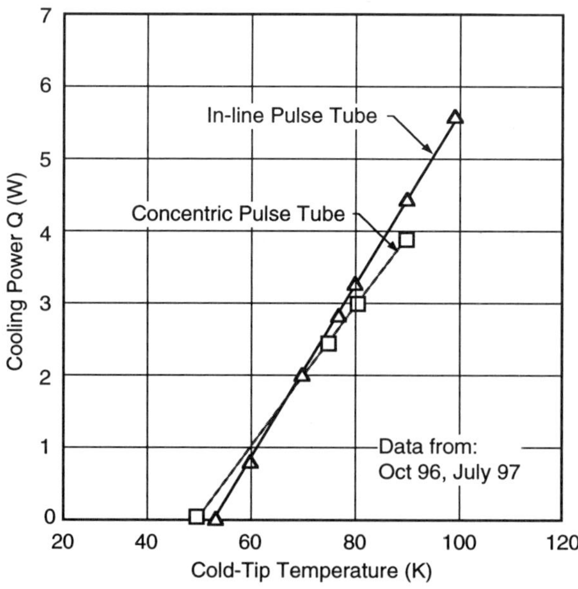


Figure 6. Comparative testing on concentric and in-line configuration (second design).

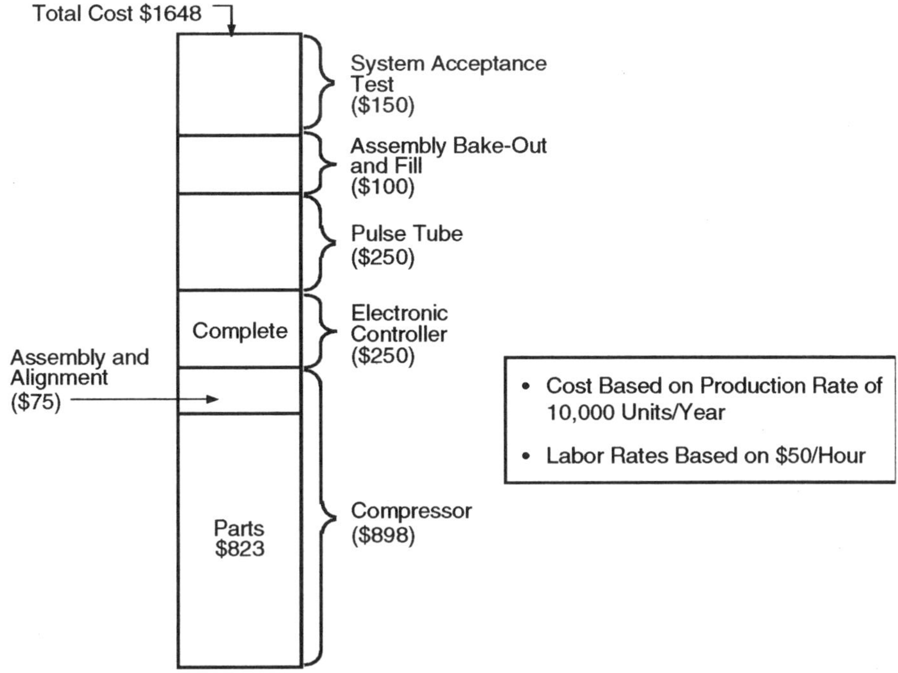


Figure 7. Current commercial cryocooler cost estimate.

modulator. Vendor quotes of $250 each were received in 10,000/year quantities. Other labor costs were calculated at $50/hour.

## SUMMARY AND CONCLUSIONS

The consortium has teamed up to produce a low-cost pulse-tube cryocooler system which has an unlimited lifetime. There are no identified failure modes of the only moving part, the compressor, and this is well established by the heritage from the space-based flexure bearing compressors. The pulse tube produced sufficient cooling to meet the filter requirements, and the concentric version allows for an easy change out from the Stirling cycle to the pulse tube in the HTS filter package.

One of these systems has been put under life testing and will be subjected to temperature cycling over the range of interest to verify adequate operation in the operating environment.

The basic approach, which resulted in the low-cost cryocooler system is usable in space-based systems and is currently undergoing modification and evaluation for that environment and will be tested for vibration, vacuum operation, and compatibility with launch environments.

Current pulse tube technology introduced into this system would produce nearly 10 W of cooling at 80 K.

## ACKNOWLEDGMENTS

This research was carried out under partial funding from NASA under a NASA Aerospace Industry Technology Program (AITP). The authors would like to thank Dr. Steve Castles and Vincent Bly of NASA/GSFC for their support and advice, Dr. Ray Radebaugh of NIST for his efforts and support in the pulse tube area, and Vince Loung, Tim James, and their technical staff in numerous areas. At Lockheed Martin, excellent support was provided by Bob Fujimoto in the electronic controller, Paul Huang in the system testing, and Bobby Evtimov and Andre Callaco in the design, fabrication, and assembly of the hardware.

# PERFORMANCE STUDY ON A TWO-STAGE 4 K PULSE TUBE COOLER

C. Wang, G. Thummes, and C. Heiden

Institute of Applied Physics, University of Giessen
D-35392 Giessen, Germany

## ABSTRACT

Systematic investigations towards performance improvement of a two-stage 4 K pulse tube cooler, concerning structural parameters, staging configuration and compressor input power, are presented in this paper. Theoretical analysis and experiment reveal that a proper negative DC flow through the double-inlet bypass has positive effects on cooling performance. With a modified cooler design a highest cooling power of 0.5 W at 4.2 K and 1 W at 5 K have been achieved. With heating load of 20 W at 67 K on the first stage, the 2nd stage can provide 0.42 W at 4.2 K. All these results have been obtained by using a 6 kW G-M compressor combined with a redesigned GM rotary valve. By using a compressor with measured input power of 1.7 kW the same cooler has a net cooling power of 170 mW at 4.2 K. The corresponding COP at 4.2 K is $1.0\times10^{-4}$, which reaches the same order of magnitude as that of GM or GM+JT coolers at the same temperature range.

## INTRODUCTION

After a quiet period of two decades following the introduction of the basic pulse tube cooler by Gifford and Longsworth in the mid sixties[1], the introduction of a reservoir connected to the hot end of the pulse tube via an orifice produced a phase shift between mass flow and pressure of the working fluid thereby leading to an essential step in cooler performance[2,3]. A further significant improvement resulted from adding a second orifice connecting the hot end of the pulse tube to the main gas inlet[4] ( so called double-inlet pulse tube cooler).

Owing to these advances in cooler design and to the development of magnetic regenerative materials, a cooling temperature below 4 K (3.6 K) was first obtained by Matsubara and Gao[5] in 1994 with a three stage pulse tube cooler. Using precooling with liquid nitrogen it was possible to reach a low temperature of 2.07 K with two pulse tube stages[6]. The net cooling power of the above coolers at 4.2 K of around 30 mW was comparatively low, corresponding to a COP of the order of $5\times10^{-6}$. This low coefficient at present appears as the main obstacle against a more widespread acceptance of this cooler type at liquid $^4$He temperatures, despite its inherent advantages with respect to reliability, lifetime, vibration and cost. Recently we designed and constructed a two-stage pulse tube cooler[7] that reached a minimum temperature of 2.23 K with a net cooling power of 370 mW at 4.2 K. The comparatively high efficiency of that cooler was achieved by design optimization based on numerical analysis[8,9] and by employing a three-layer 2nd-stage regenerator filled with $ErNi_{0.9}Co_{0.1}$, ErNi and Pb.

In order to further improve the cooler performance at 4.2 K, structure parameters and compressor input power have been further optimized in this paper. Negative DC flow, which was found to have positive effect on cooling performance[10], is introduced into the cooler design.

## EXPERIMENTAL SET-UP

A schematic of our first two-stage pulse tube cooler operating below 4 K is shown in Figure 1. The pressure oscillation is generated by a valved compressor (Leybold RW6000) in combination with a GM rotary valve redesigned with regard to proper timing. If not otherwise noted the following experiments are performed with this 6 kW compressor. The hot end of the 1st and 2nd stage pulse tube are connected to the phase shifting assembly at room temperature. A common reservoir of 1200 $cm^3$ is used for both two stages.

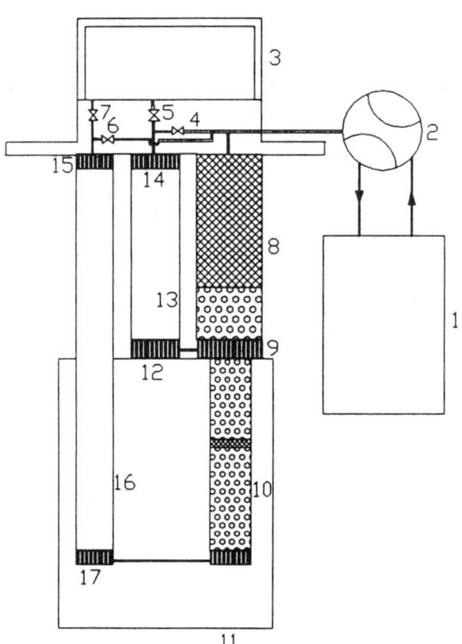


**Figure 1.** Schematic of a two-stage pulse tube cooler. 1: compressor; 2: rotary valve; 3: reservoir; 4,6: double-inlet valves; 5,7: orifice valves; 8: 1st stage regenerator; 9: flow straightener and heat exchanger; 10: 2nd stage regenerator; 11: radiation shield; 12: 1st stage cold head; 13: 1st stage pulse tube; 14, 15: heat exchangers; 16: 2nd stage pulse tube; 17: 2nd stage cold head.

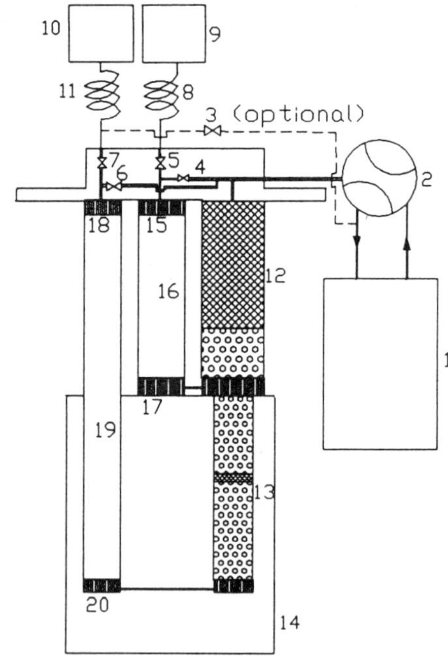


**Figure 2.** Schematic of a modified two-stage pulse tube cooler. 1: compressor; 2: rotary valve; 3:optional DC flow valve; 4,6: double-inlet valves; 5,7: orifice valves; 8: 1st stage coiled tube; 9,10: 1st and 2nd stage reservoirs; 11: 2nd stage coiled tube; 12: 1st stage regenerator; 13: 2nd stage regenerator; 14: radiation shield; 15: 1st stage heat exchanger; 16: 1st stage pulse tube; 17: 1st stage cold head; 18: 2nd stage heat exchanger; 19: 2nd stage pulse tube; 20: 2nd stage cold head.

The first stage regenerator has an inner diameter of 58 mm and a length of 140 mm. It is filled in its upper part with a sequence of 1280 disks of 250 mesh stainless steel and phosphor bronze screens, and with lead spheres (diameter: 0.18-0.23 mm) in its lower part over a length of 30 mm. Outer diameter, wall thickness and length of the 2nd stage regenerator are: 30 mm×0.5 mm×180 mm. The matrix of it contains three layers consisting of $ErNi_{0.9}Co_{0.1}$ grains (size: 0.18-0.25 mm) in the lower 40%, ErNi spheres (diameter: 0.25-0.5 mm) in the middle 27% and Pb spheres (diameter: 0.18-0.23 mm) in the upper 33% of the tube length. The dimensions of the 1st and 2nd stage pulse tube are optimized.

A copper radiation shield is fixed at the 1st stage cold head to reduce the radiation loss of the 2nd stage. The temperatures of the 1st and 2nd stage cold heads are measured by means of a rhodium-iron and a carbon-glass (Lake Shore, model CGR-1-1000) resistance thermometer, respectively. The net cooling powers of the 1st and 2nd stage are measured using resistive heaters.

## RESULTS OF IMPROVEMENT AND DISCUSSION

To increase the cooling performance of the 1st and 2nd stage, we tried to enlarge the pulse tube volumes. Table 1 gives three sets of diameters of pulse tubes as well as the corresponding cooling performance at 4.2 K. As seen from Table 1, increase in the diameter of pulse tubes leads to a lower cooling power. The main obstacle against using larger diameter of the 2nd stage pulse tube is a temperature maximum in the middle of the 2nd pulse tube, which means that the heat dissipation through the hot end is not efficient in the present system. The valve timing here was only optimized for pulse tube (A)[7].

DC flow through the double-inlet bypass was found from simulation in Ref. 9 and was studied theoretically and experimentally in Ref. 10. Interestingly a negative DC flow from the hot end of pulse tube to the hot end of regenerator has two positive effects on cooling performance[10]: 1. increasing cooling capacity of pulse tube; 2. dissipating the enthalpy flow at the hot end of pulse tube. There are two possibilities to create DC flow in a double-inlet pulse tube cooler. DC flow can first be caused by the pressure wave form. For a valved pulse tube system, DC flow can be obtained by changing the timing of the rotary valve as was done in our first two-stage 4 K cooler[7] by increasing the exhaust time of the valve. Second, DC flow can be caused by the asymmetric geometry of the double-inlet orifice.

**Table 1.** Effects of pulse tube diameters on the cooling power at 4.2 K for the cooler in Figure1

| Pulse tube type | (A) (Ref. 7) | (B) | (C) |
|---|---|---|---|
| 1st stage pulse tube O.D.×thickness× length | 35mm ×0.3mm × 140mm | 35mm ×0.3mm × 140mm | 41.7mm× 0.35mm ×140mm |
| 2nd stage pulse tube O.D.×thickness× length | 15.5mm × 0.25mm ×330mm | 20mm × 0.4mm ×330mm | 20mm × 0.4mm ×330mm |
| cooling power at 4.2K | 370 mW | 250 mW | 230 mW |

Note: one common reservoir of 1.2 L for the two stages; frequency is 1.0 Hz.

**Table 2.** Effects of DC flow on the cooling power at 4.2 K

| | 1. without DC flow valve | 2. DC flow valve open, optimizing with fixed valve settings of case 1 | 3. DC flow valve open, optimizing all valve settings |
|---|---|---|---|
| cooling power at 4.2K | 260mW | 310 mW | 420 mW |

Note: Cooler with pulse tube (B) and coiled tube (a). Frequency is 1.0 Hz

A modified two-stage pulse tube cooler combined with negative DC flow and two copper coiled tubes has been constructed and tested. Its configuration is shown in Figure 2. Two split reservoirs of 2 liters and 1 liters are connected to the hot ends of the 1st and 2nd stage pulse tube, respectively. Two coiled tubes here are expected to perform as flow resistances and to a minor degree as inertances[11]. Because optimization of the rotary valve timing and of the asymmetric double-inlet valve for proper DC flow for different systems is a very time consuming task, we adopted a simple method to introduce a DC flow by adding an optional valve between the hot end of the 2nd stage pulse tube and the low pressure line (see dashed line in Figure 2). A similar method connecting an orifice between the second stage reservoir and low pressure line was used by Chen et al[12]. The difference between the work in Ref.12 and the present work is that the former introduces a DC flow from the reservoir to the low pressure line, while here a negative DC flow through the double-inlet bypass from the hot end of pulse tube to the hot end of regenerator is to be expected. With a proper system design the DC flow valve is not needed for a two-stage cooler. The following tests of the cooler driven by a 2 kW compressor were performed without the DC flow valve.

Effects of DC flow on the cooling performance at 4.2 K are given in Table 2. Here, pulse tube (B) and coiled tube (a) in Table 3 are used for the comparison. The experiments verify that a proper negative DC flow increases the cooling performance greatly as predicted in Ref.10. By optimizing all the valve settings the cooling power of the cooler at 4.2 K increases from 260 mW to 420 mW by aid of the negative DC flow.

The operating frequency of the cooler obtained with the increased diameter of pulse tube (B) is optimized for cooling power at 4.2 K, as shown in Figure 3. The optimum frequency is now 0.8 Hz, not 1.0 Hz as for pulse tube (A)[7]. The following results are obtained with a frequency of 0.8 Hz.

The cooling powers of this modified cooler obtained with different pulse tube diameters are shown in Figure 4. Pulse tube (C) with the 1st stage pulse tube diameter of 41 mm and the 2nd stage pulse tube diameter of 19.2 mm achieves the highest cooling power of 480 mW at 4.2 K. Pulse tube (B) has a cooling power of 450 mW at 4.2 K. The original cooler with pulse tube (A) had a cooling power of 370 mW at 4.2 K. Therefore the following results are all obtained with pulse tube (C).

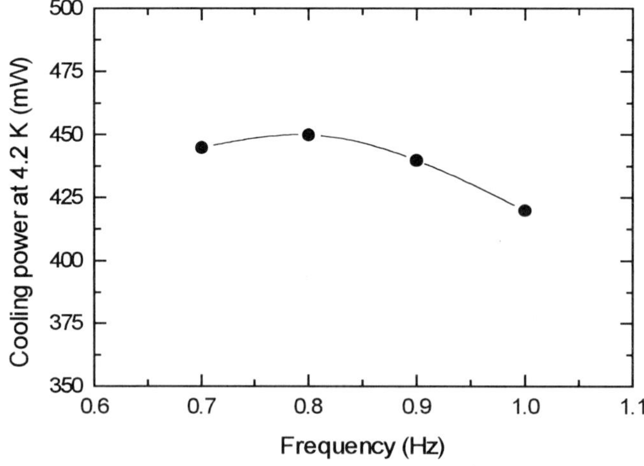


**Figure 3.** Effect of frequency on the cooling power at 4.2 K. Cooler with pulse tube (B) and coiled tube (a).

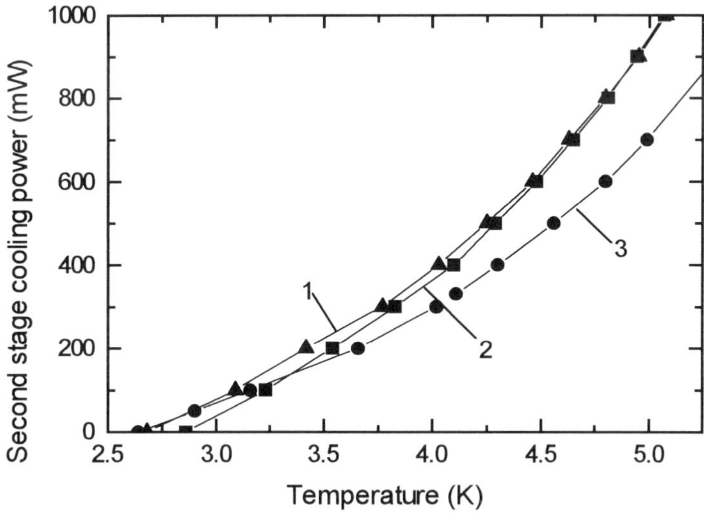


**Figure 4.** Comparison of the cooling power obtained with different pulse tubes. Curve 1: modified cooler with pulse tube (C) and coiled tube (a), f=0.8 Hz; Curve 2: modified cooler with pulse tube (B) and coiled tube (a), f=0.8 Hz; Curve 3: original cooler (see Fig.1) with pulse tube (A), f=1.0 Hz.

Table 3 shows the effect of different coiled tubes on the cooler performance at 4.2 K. The coiled tubes used here lead to some improvement indeed. The best cooling performance of 0.5 W at 4.2 K has been obtained with coiled tube (b). However, it is not clear up to now, which effect is mainly responsible for the obtained improvement. It may (most likely) be the combined effects of the flow resistance (capillary tube ) and negative DC flow, or be, to a lesser degree, also due to phase shift caused by the inertance of the capillary.

Also with this modified configuration the two-stage pulse tube cooler can be optimized to provide suitable cooling power at given temperature on both stages for practical use. The valve settings here are optimized for the highest cooling power at 4.2 K with 20 W heating on the 1st stage. With these fixed valve settings a corresponding cooling load map for the 1st and 2nd stage is given in Figure 5. The cooling power of the 2nd stage here is 380 mW at 4.2 K with the 1st stage heated to 53 K by a 10 W heat load. With the heat load of 20 W at 67 K on the first stage, the 2nd stage can provide 420 mW at 4.2 K. The observed increase in cooling power at 4.2 K upon increasing the 1st stage heat load in Figure 5 is similar to that in Ref. 7.

**Table 3.** Effect of coiled tubes on the cooling power at 4.2K

| Type of coiled tube | (a) | (b) | (c) | (d) |
|---|---|---|---|---|
| 1st stage tube coil | 3 mm i.d. and 0.45 m long | 3 mm i.d. and 1.5 m long | 3 m i.d. and 1.5 m long | 3 mm i.d. and 3 m long |
| 2nd stage tube coil | 2 mm i.d. and 0.45 m long | 2 mm i.d. and 1.5 m long | 3 mm i.d. and 1.5 m long | 2mm i.d. and 3m long |
| cooling power @ 4.2 K | 480 mW | 500 mW | 450 mW | 400 mW |

Note: Cooler with pulse tube (C), f=0.8 Hz.

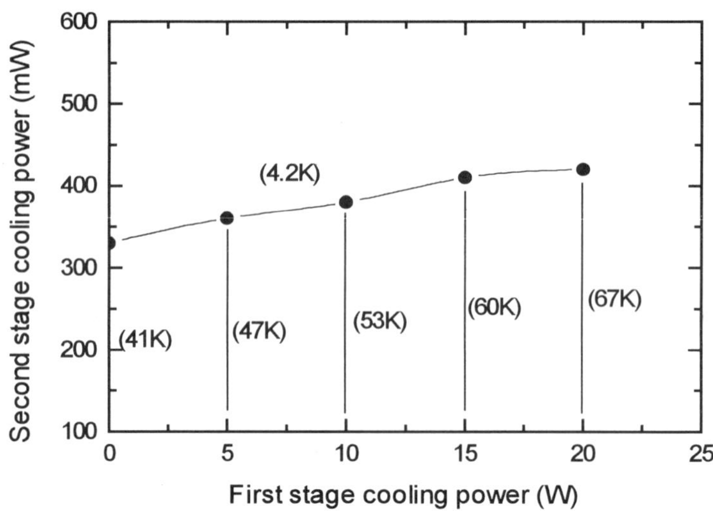


**Figure 5.** Cooling load map for the 1st and 2nd stages. Cooler with pulse tube (C) and coiled tube (b), f=0.8 Hz.

The two-stage pulse tube cooler was initially designed for operating with a 6 kW compressor. All above results have been obtained by using the 6 kW G-M compressor. With regard to some applications requiring less cooling power at 4.2 K, we also tested the cooler with 4 kW and 2 kW compressors. The 2nd stage cooling power of the cooler with three different compressors is shown in Figure 6. The 2nd stage net cooling power with 6 kW compressor is 0.5 W at 4.2 K and 1.0 W at 5 K, that with 4 kW compressor is 290 mW at 4.2 K and 580 mW at 5 K. Cooling powers of 170 mW at 4.2 K and 370 mW at 5 K are obtained with the 2 kW compressor. The actual input power of this compressor was measured to be 1.7 kW. This results is encouraging to develop a small and low cost two-stage pulse tube cooler for cooling below 4 K.

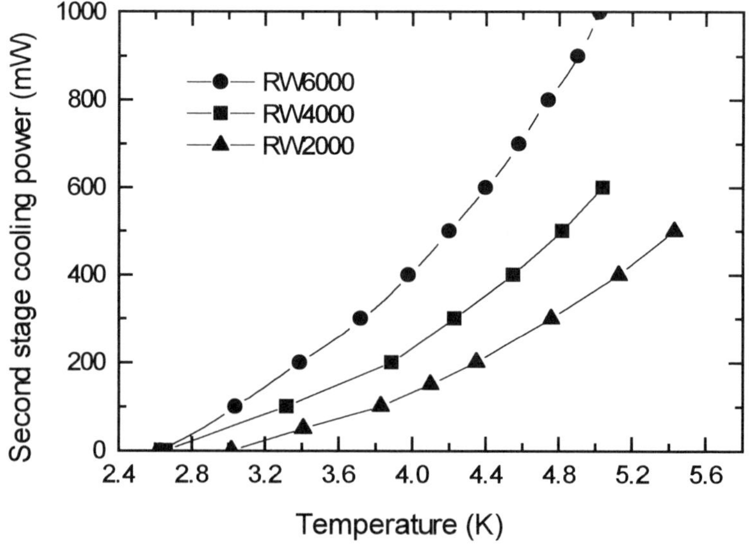


**Figure 6.** The second stage cooling power obtained from three different compressors with nominal input powers of 2, 4 and 6 kW. Cooler with pulse tube (C) and coiled tube (b); f=0.8 Hz.

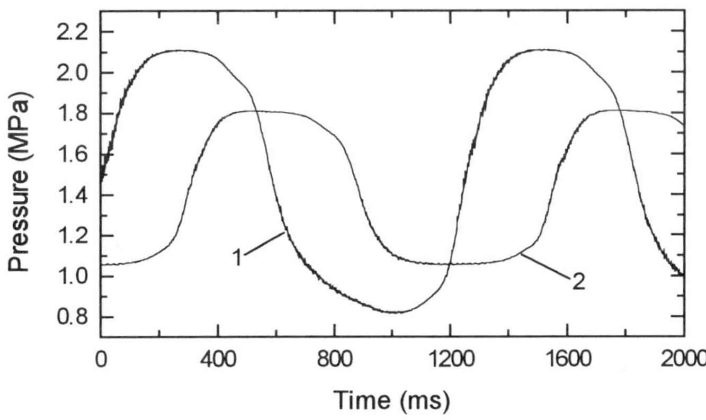


**Figure 7.** Pressure waves at the main gas inlet. f=0.8 Hz. Curve 1: 6 kW compressor. Curve 2: 2 kW compressor.

Typical pressure oscillations at the 1st stage regenerator inlet with 6 kW and 2 kW compressor are illustrated in Figure 7. In both cases the cooler works with the same rotary valve. The pressure loss through the rotary valve for the 6 kW compressor is larger than that for the 2 kW compressor. This may be an important reason for the fact that the cooler with 2 kW compressor achieved a higher efficiency.

The opening value of the DC flow valve decreases with decreasing input power of compressor. It is worth to note that the DC flow valve is completely closed when the cooler is operated with the 2 kW compressor. As can be seen from Figure 7 the time ratio of low pressure to high pressure interval for the 2 kW compressor is larger than that for the 6 kW compressor. Therefore there is a larger negative DC flow through the double-inlet valve for the system driven by the 2 kW compressor. In the present study the opening of the DC flow valve decreases the cooler performance for the 2 kW compressor. It means that negative DC flow caused by the pressure wave and double-inlet valve is sufficient for operation, and the additional DC flow valve is not needed.

The performances of the cooler with three different compressors are compared in Table 4. So far the cooler driven by the 2 kW compressor achieved the highest efficiency. Since the peak-to-peak pressure variation of $\Delta P$=0.76 MPa is rather low the actual measured input power of the compressor is only 1.7 kW. Its COP of $1.0 \times 10^{-4}$ at 4.2 K reaches the same order of magnitude of the present GM or GM+JT coolers at the same temperature range. The lowest measured temperatures are 3.02 K with the 2 kW compressor and 2.43 K with the 4 kW compressor. The COP of the cooler with the 4 kW compressor is lower in the present study.

**Table 4.** Performance comparison of the cooler with three different compressors. Cooler with pulse tube (C) and coiled tube (b)

| compressor type | Leybold RW2000 | Leybold RW4000 | Leybold RW6000 |
|---|---|---|---|
| measured input power ( kW ) | 1.7 | 4.2 | 6.3 |
| $\Delta P$ ( MPa ) | 0.76 | —— | 1.29 |
| $T_{min}$ ( K ) | 3.02 | 2.43 | $\leq 2.6$ |
| Cooling power @ 4.2 K ( mW ) | 170 | 290 | 500 |
| COP @ 4.2 K | $1.0 \times 10^{-4}$ | $0.69 \times 10^{-4}$ | $0.79 \times 10^{-4}$ |

Note: $\Delta P$ is peak-to-peak pressure variation at the main gas inlet; $T_{min}$ lowest achievable temperature.

## CONCLUSION

Structure parameters and compressor input power of a two-stage pulse tube cooler have been optimized to increase the refrigeration efficiency. Negative DC flow through the double-inlet bypass has been applied to improve the cooling performance of the 2nd stage at 4.2 K. Moreover, it is found that a long tube between orifice and reservoir leads to further system improvement.

A net cooling power of 0.5 W at 4.2 K and 1 W at 5 K are obtained with the cooler driven by a 6 kW compressor. With a heat load of 20 W on the 1st stage at 67 K the 2nd stage can provide 0.42 W at 4.2 K.

The highest efficiency of the cooler has been achieved by using a compressor with measured input power of 1.7 kW. The cooler then achieves 170 mW at 4.2 K corresponding to a COP of $1.0 \times 10^{-4}$.

The results obtained so far demonstrate that the two-stage pulse tube cooler developed in the present work can be used for operating small scale conventional superconducting systems like magnets or for small scale $^4$He liquefaction.

## ACKNOWLEDGMENTS

Supports by Alexander von Humboldt Foundation and the German BMBF (contract no. 13N6513) are gratefully acknowledged. We also would like to thank H.U.Häfner (Leybold, GmbH, Cologne) for support.

## REFERENCES

1. W.E. Gifford and R.C. Longsworth, Pulse tube refrigeration, *Trans. ASME, J. Eng. Ind.* 86 Series B: 264 (1964).
2. E.I. Mikulin, A.A. Tarasov, M.P. Shkrebyonoch, Low temperature expansion pulse tubes, in: "Adv. in Cryog. Eng.", Vol. 29, Plenum Press, New York(1984), p.629.
3. R. Radebaugh, J.E. Zimmerman, R.S. David, and L. Beverly, A comparison of three types of pulse tube refrigerator: New methods of reaching 60 K, in: "Adv. in Cryog. Eng.", Vol.31, Plenum Press, New York (1986), p.779.
4. S.W. Zhu, P.Y. Wu, and Z.Q. Chen, Double inlet pulse tube refrigerator: an important improvement, *Cryogenics* 30: 514 (1990).
5. Y. Matsubara and J.L. Gao, Novel configuration of three-stage pulse tube refrigerator for temperatures below 4 K, *Cryogenics* 34: 259 (1994).
6. G. Thummes, S. Bender and C. Heiden, Approaching the $^4$He lambda-line with a liquid nitrogen precooled two-stage pulse tube refrigerator, *Cryogenics* 36: 709 (1996).
7. C. Wang, G. Thummes and C. Heiden, A two stage pulse tube cooler operating below 4 K, *Cryogenics* 37: 159 (1997),.
8. C. Wang, Numerical analysis of 4 K pulse tube coolers, Part I: Numerical simulation, *Cryogenic* 37: 207 (1997).
9. C. Wang, Numerical analysis of 4 K pulse tube coolers, Part II: Performances and internal processes, *Cryogenics* 37: 215 (1997).
10. C. Wang, G. Thummes and C. Heiden, Effects of DC flow on performance of a two-stage 4K pulse tube cooler, to be published.
11. S.W. Zhu, S.L. Zhou, N. Yoshimura, and Y. Matsubara, Phase shift effect of the long neck tube for the pulse tube refrigerator, in: "Cryocooler 9" ,R.G. Ross, Jr., ed., Plenum Press, New York (1997), p.269
12. G.B. Chen, L.M. Qiu and J.Y. Zhen, et al., Experimental study on a double-orifice two-stage pulse tube refrigerator, *Cryogenics* 37: 271 (1997).

# 3K DOUBLE-ORIFICE TWO-STAGE PULSE TUBE REFRIGERATOR*

G.B. Chen, L.M. Qiu, J.Y. Zheng, P.D. Yan,
Z.H Gan, X. Bai, and T. Jin

Cryogenics Laboratory, Zhejiang University
Hangzhou 310027, China

## ABSTRACT

Based on the analysis of heat transfer and fluid flow in the staged pulse tube, a novel phase-shifter named secondary(minor) orifice was firstly introduced between the reservoir of the final stage pulse tube and the inlet line of the helium compressor, so as to eliminate the anomalous temperature distribution along the pulse tube. The influence of construction and operating parameters on cooling performance of the machine was experimentally studied. Finally a refrigerating temperature as low as 3.0K has been obtained by a two-staged pulse tube refrigerator with the double-orifice configuration.

## INTRODUCTION

The cooling performance of pulse tube refrigerators has been greatly improved by introducing the second inlet phase shifter. A refrigerating temperature of 22K was reported by a single stage pulse tube refrigerator[1]. In order to obtain liquid helium temperature, it is necessary to use a multistaged configuration for the pulse tube refrigerator. Matsubara and Gao successfully reached 2.8K by means of a three-staged pulse tube refrigerator with a regenerative tube, which was used for removing heat from the warm end of the final stage pulse tube to the reservoir[2].

Recently, Wang[3] reported that a temperature as low as 2.23K had been achieved by

* The project was supported by The National Natural Science Foundation of China

means of a two-stage pulse tube refrigerator without the regenerative tube. However, a rather powerful helium compressor of 6kW and high-grade complex magnetic regenerative materials of $ErNi_{0.9}Co_{0.1}$ and ErNi as well as $Er_3Ni$ were used in his experiments. Almost at the same time, we constructed a two-stage pulse tube refrigerator, in which a double-orifice phase shifter was adopted to replace the regenerative tube. A low temperature of 3.0K was obtained employing only a 3.7kW helium compressor and a single regenerative material of $Er_3Ni$ in the second stage regenerator[4].

## THEORETICAL ANALYSIS

In recent years, a number of authors have reported that an anomalous temperature profile near the warm end of the pulse tube in double inlet configuration has been observed[5,6]. There is a temperature peak on the tube wall near the hot end. This peak temperature is much higher than room temperature and increases with the opening of the second inlet valve,which is at room temperature. By analyzing and phenomenalizing, we have recognized that this is a rather serious problem and a key issue to be solved to improve the performance of multistaged pulse tube refrigerators in order to reach liquid helium temperatures.

Numerical simulation of pulse tube refrigeration allows clear visualization of heat transfer and fluid flow. With the help of a simplified adiabatic model presented in a separate paper[7], the relationship of the temperature of the working fluid with respect to position along the tube has been obtained(Fig.1). We can see from the curve 1 that there exists an inefficient tube length L, where the gas temperature is higher than 300K. This is nearly a quarter of the entire tube length. This will cause an accumulation of heat near the hot end of the tube. Then compressed gas from the cold end hardly flows through this inefficient volume to transfer heat to the heat exchanger at room temperature. Obviously, the second inlet plays the roles of gas mixing and introducing turbulence in this volume.

In order to remove the anomalous temperature peak, that is to dissipate the heat accumulated near the hot end of the pulse tube, we recommended the introdution of a

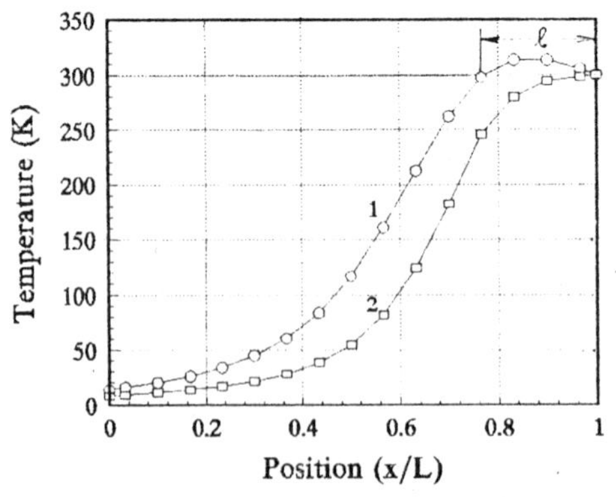

**Fig.1.** Gas temperature profile along pulse tube.

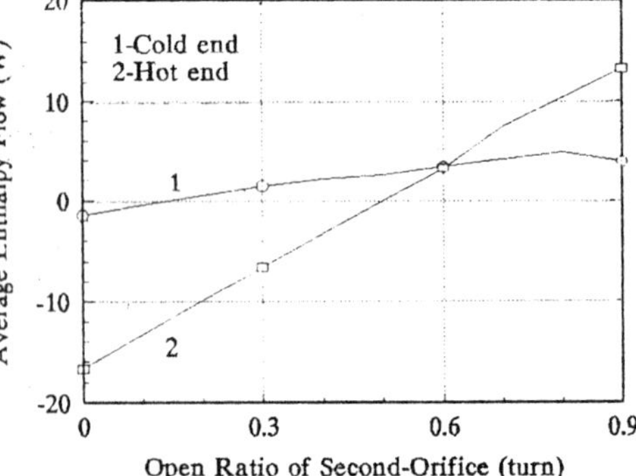

**Fig.2.** Enthalpy flow vs. open ratio of second orifice.

second orifice between the second stage reservoir and the return line of the compressor. This is termed a double-orifice with the primary one, to operate the two-stage pulse tube from room temperature to the liquid helium temperature region. The numerical simulation calculation indicates that the temperature distribution along the tube is much improved and the inefficient length of the pulse tube is reduced by use of this double-orifice configuration (curve 2 in Fig.1).

Fig.2 shows the variation of enthalpy flow within the pulse tube versus the open ratio of the second orifice. The enthalpy flows at both the cold and hot ends are negatives when the open ratio of the second orifice is equal to zero. It means that any gas mixing or turbulence caused by the second inlet within the tube volume not only effect the heat pumping to the hot end of the tube, but also introduce an additional heat load to the cold end. Fortunately, enthalpy flows gradually become positive either at the cold or hot end of the tube with an increase in the second orifice opening. Meanwhile, the temperature profile of the gas along the tube becomes reasonable (see curve 2, Fig.1).

## EXPERIMENTAL APPARATUS

A modified experimental apparatus of the two-stage pulse tube refrigerator with double orifices is shown in Fig 3, which consists of a helium compressor, regenerators(1,3), pulse tubes(2,4), heat exchangers, reservoirs(5,6), primary orifices(8,10), second inlets(7,9) and a second(minor) orifice(11). The first stage regenerator consists of two parts, one part of $\Phi59\times130$mm filled with No.250 bronze mesh and the second part of $\Phi45\times60$mm filled with $\Phi0.3$mm lead shot. Dimensions of the second stage regenerator are $\Phi19\times190$mm filled with magnetic regenerative material of $Er_3Ni$ grains. All final dimensions of the pulse tubes were derived from experiments. A valved helium compressor with only about 3.7kW of input power is used in the system. The second(minor) orifice is located between the second stage reservoir and the inlet line of the compressor.

A self-made on-off timing computer control system is used to accurately adjust the operating frequency and time ratio between the compression and the expansion processes. This device is very useful for obtaining an optimum operating condition, not only for pulse tube refrigerators, but also for conventional cryocoolers such as the G-M refrigerator etc..

In the measuring system, rhodium-iron resistance thermometers with 0.1K accuracy are used to measure temperatures at the cold ends of each stage. A number of thermocouples are set onto the pulse tubes to monitor the temperature distribution along the tube. Any anomalous temperature distribution phenomenon can be eliminated by adjusting the opening of the minor orifice, so as to approach an optimum operating condition. A pressure gauge is installed at each reservoir, each having a volume of 1L.

## STRUCTURE AND OPERATING PARAMETERS

The effect of the volume ratio between the regenerator and the corresponding pulse tube on the refrigerator performance has been tested. The dimensions of the regenerators are fixed, while the pulse tubes are gradually changed in search of optimum construction parameters.

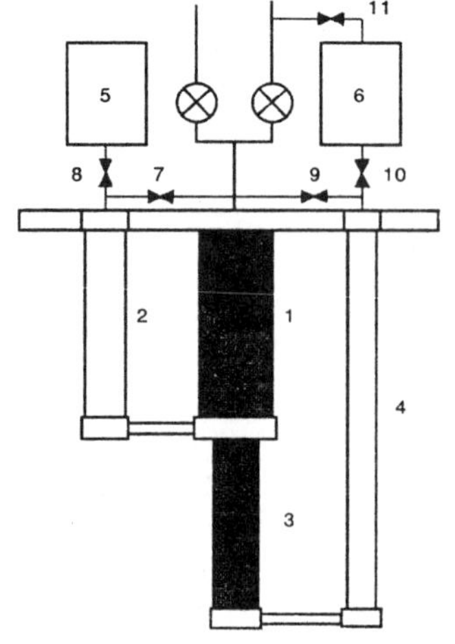


**Fig.3.** Schematic diagram of two-stage pulse tube refrigerator.

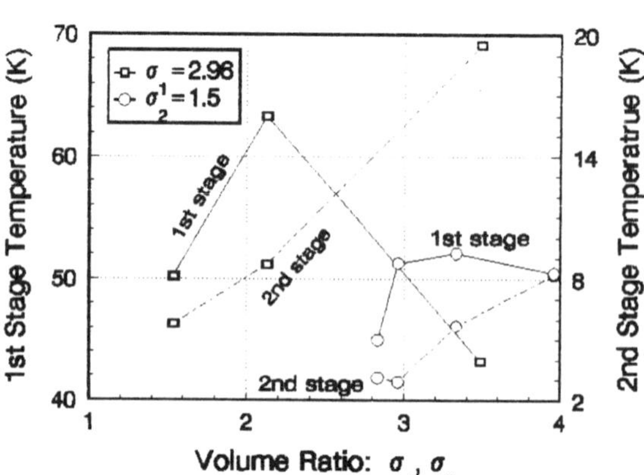


**Fig.4.** Influence of volume ratio on cooling temperature.

First, the influence of the second stage pulse tube geometry on the refrigerator performance is investigated, while the geometry of the first stage pulse tube is fixed (i.e. $\sigma_1$=const). Then, the influence of the first stage pulse tube geometry on the refrigerating performance is also tested, while the second stage pulse tube geometry is fixed $\sigma_2$=const. The volume ratio of the first stage is expressed by $\sigma_1=Vr_1/Vp_1$ and the second stage by $\sigma_2=Vr_2/Vp_2$, where Vr and Vp represent volumes of the regenerator and the pulse tube respectively. Fig.4 shows the test results. We can see that reasonable values of $\sigma_1=2.96$ and $\sigma_2=1.5$ are preliminary confirmed for our apparatus[8]. The corresponding lowest temperatures of 52K and 3.0K for the first and second stages respectively are obtained. Table 1 gives the dimensions of the tested refrigerator. Fig.5 shows cool-down curves of the double-orifice two-stage pulse tube refrigerator. The charge pressure of the system is about 1.3MPa. And maximum and minimum pressures are 2.0 MPa and 0.8MPa, respectively. The operating frequency is about 1.5Hz.

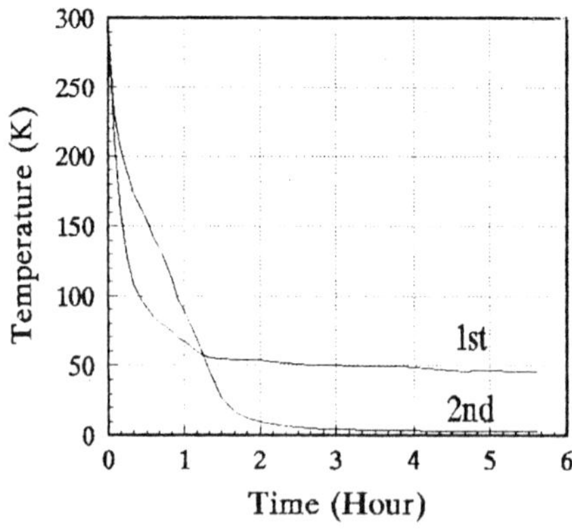


**Fig.5.** Cool-down curves of two-stage pulse tube cooler,

**Table 1.** Dimensions of two-stage pulse tube refrigerator(mm)

| Stage | Pulse Tube | Regenerator |
|---|---|---|
| 1st | Φ27.3×280 | Φ59×130+Φ45×60 |
| 2nd | Φ12×368 | Φ19×190 |

## DOUBLE-ORIFICE EFFECT

The primary orifice proposed by Mikulin in 1985, together with the secondery orifice developed by the authors are called double orifices. The function of the minor orifice is to enhance the heat transmitting mechanism of the working gas from the cold end to the hot end of the second stage pulse tube. Figure 6 gives the influence of the secondary orifice opening on temperatures at various measuring points. The curve $T_1$ shows cold temperatures of the first stage. Curve $T_2$ and $T_4$ show the cold and warm end temperatures of the second stage respectively. The curve $T_3$ expresses the anomalous temperature peak near the warm end of the second stage pulse tube. We can see that the curves $T_2$ and $T_3$ strongly depend on the opening of the second orifice until an optimum point is approached. At this condition, the temperature distribution of the second stage pulse tube is obviously improved. As a result, the temperature of the second stage is gradually reduced to a minimum value. However, when the second orifice opens above the optimum value, the temperature of the second stage increases since too much of the cold helium gas flows to the warm end. Meanwhile, the first stage temperature only rises slightly with an increase in the second orifice opening due to its drawing effect.

Fig.7 shows the temperature distribution along the second stage pulse tube at a steady state condition. Curve 1 represents the case of a single double inlet configuration. Curve 2 shows the temperature distribution for a double-orifice configuration with a small opening of the second orifice, for the same operating condition as that for curve 1. Clearly, the temperature distribution of the pulse tube is obviously improved for the double-orifice system, i.e. the irreversible loss in the expansion process is greatly reduced.

## CONCLUSIONS

1. The double-orifice configuration has been successfully used in a two-stage pulse tube refrigerator to obtain liquid helium temperatures. A low refrigerator temperature of 3.0K was achieved by the double-orifice two-stage pulse tube refrigerator with a 3.7kW helium compressor.

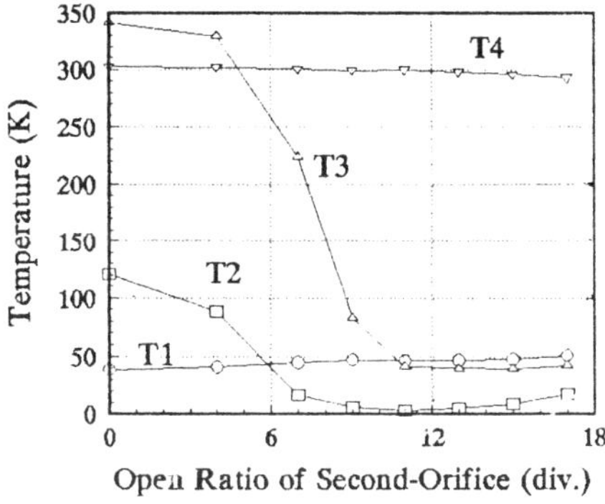


**Fig.6.** Influence of open ratio of 2nd orifice on temperature.

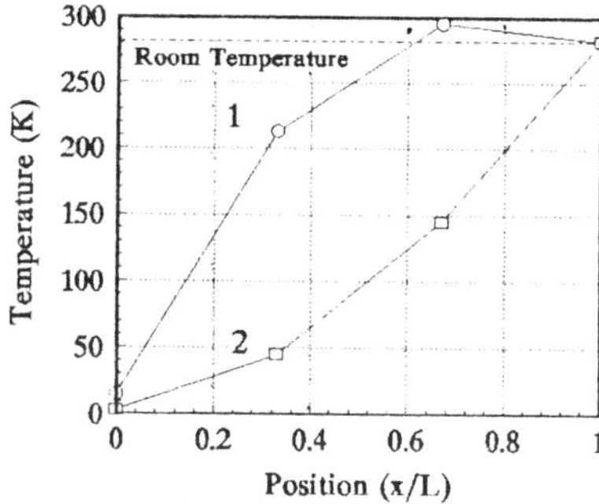


**Fig.7.** Temperature profile along pulse tube wall of 2nd stage.

2. The second orifice can effectively eliminate the anomalous temperature peak near the warm end of the pulse tube, which is caused by the second inlet.
3. The volume ratio of the regenerator to the corresponding pulse tube significantly influences the temperature of pulse tube refrigerator. In this study, the typical volume ratio values of 2.96 and 1.56 are confirmed for the first and second stages of the refrigerator respectively.

## REFERENCES


1. Y. Ishizaki, et al. Experimental performance of modified pulse tube refrigerator below 80K down 23K, 7th Intl. Cryocooler Conf. Proceeding, Part 1, (1993), p.140
2. Y. Matsubara and J.L. Gao, Novel configuration of three-stage pulse tube refrigerator for temperatures below 4K, Cryogenics, 34:259 (1994)
3. C. Wang, et al. A two-stage pulse tube cooler operating below 4K, Cryogenics, 37:159 (1997)
4. G.B. Chen, L.M. Qiu, et al. Experimental study on a double-orifice two-stage pulse tube refrigerator, Cryogenics, 37:271 (1997)
5. T. Shigi, et al, Anomaly of one stage double-inlet pulse tube refrigerator, Proceedings of ICEC/ICMC 16, Elsevier Science, Part 1, (1997), p.263
6. C. Wang, Y. Ju and Y. Zhou, Experimental investigation of a two-stage pulse tube refrigerator, Cryogenics, 36:606 (1996)
7. L.M. Qiu and G.B. Chen, Adiabatic model and experimental verification of a two-stage double-orifice pulse tube refrigerator reached liquid helium temperatures, CEC/ICMC, Portland, USA(1997)
8. G.B. Chen, J.Y. Zheng, L.M. Qiu, et al. Modification test of staged pulse tube refrigerator for temperatures below 4K, Cryogenics, (1997), in Press

# DESIGN AND TESTING OF A PULSE TUBE BASED COOLING SYSTEM FOR HIGH PURITY GERMANIUM DETECTORS

P. E. Tissot[1], L. Crowe[1], J. Colvin[1], T. L. Mann[1], and A. Guerra[2]

[1]Houston Advanced Research Center, Technology Development Laboratory
The Woodlands, TX 77381
[2]Kelvin International Corporation
Hampton, VA 23664

## ABSTRACT

A cooling package based on a new type of production cryocooler was designed and tested at the Technology Development Laboratory (TDL) of the Houston Advanced Research Center (HARC) in collaboration with the company Kelvin International. The prototype is based on an Iwatani CryoMini® Pulse Tube Cryocooler provided by Kelvin International. The CryoMini cryocooler was integrated with an EG&G Pop-Top High Purity Germanium (HPGe) detector within a TDL designed cooling package. One of the key features of this new type of cryocooler is the small size of its cold head allowing for the combination of the cooling package prototype and an EG&G Pop-Top detector to fit within a cylindrical volume with a diameter slightly larger than 3 inches and a length of less than 2 1/2 feet long. The compact design of the cryocooler is attractive for environmental surveys, shallow hole investigations, and other applications that require long uninterrupted operating time, minimal space and ease of operation. The Cooling package temperature histories were recorded during operation, and gamma ray spectra were acquired with the prototype. A steady state temperature of 78 K was achieved in the Pop-Top capsule. This temperature is well within the operating range of a standard HPGe Pop-Top detector and provides a comfortable margin to convert the prototype into a more rugged field device. The energy resolution of an HPGe detector is compared for spectra acquired with the prototype and with a standard liquid nitrogen dewar for calibration sources and Naturally Occurring Radioactive Material (NORM) samples. The energy resolution of the prototype was measured at 2.08 keV for the 1.332 MeV line of a Cobalt-60 gamma ray source. Although somewhat higher than the resolution measured for the same detector cooled with a liquid nitrogen cryostat (1.73 keV) the measured energy resolution is within the warranted range for the HPGe detector used in this study.

*Advances in Cryogenic Engineering,* Vol. 43
Edited by P. Kittel, Plenum Press, New York, 1998

## INTRODUCTION

The number of applications for cryocoolers has been steadily increasing as these refrigerators are becoming more cost effective with the decrease of both capital expenditures and maintenance costs. For certain types of applications such as semiconductor detector cooling[1] and low temperature electronics[2], considerations other than simple economics determine the success of these refrigeration systems. For these applications, characteristics such as the size of the cooling package and the vibrational field generated by the cryocooler are key parameters. The purpose of this study is to determine if a new type of cryocooler, an Iwatani CryoMini Pulse Tube Cryocooler provided by Kelvin International, has the appropriate thermal and vibrational characteristics to be used as the cooling mechanism for a High Purity Germanium (HPGe) gamma ray detector.

High Purity Germanium detectors are widely used as gamma ray detectors because of their excellent energy resolution. Some of the main applications of these detectors are Neutron Activation Analysis (NAA)[3], monitoring of the natural gamma ray environmental and geological formations[4] and Prompt Gamma Neutron Activation Analysis (PGNAA)[5]. These detectors are used both in the laboratory environment and in the field and must be cooled to the liquid nitrogen temperature range during operations. Liquid nitrogen dewars are currently the method of choice for cooling HPGe detection systems. Their main advantages are the low capital investment of the cooling package, the reliability and simplicity of operation, and the inherent temperature stability provided by the liquid-vapor phase change. With the growing use of HPGe detection systems outside the laboratory environment, there is an increasing number of applications where either liquid nitrogen is difficult to obtain or simply impossible to use. For field surveys in remote areas, liquid nitrogen is sometimes difficult to obtain and for some applications, it is desirable to have the detection system operational for extended periods of time without an operator being present to periodically refill the dewars. In the down hole environment, for oil and gas exploration or for deep geological surveys, liquid nitrogen cannot be used because of its thermodynamic properties: liquid nitrogen contained in a closed system will generate an unmanageable pressure rise[6]. Field surveys and shallow hole measurements are two of the applications for which the prototype described in this paper is designed.

An alternative solution to liquid nitrogen is the use of cryocoolers. EG&G Ortec provides a line of cryocooler based solutions, Electricool®[7,8]. The first cryocoolers provided by EG&G (Electricool I and II[7]) were adequate for a number of applications, long unattended monitoring tasks in particular, and had performances nearly equivalent to that obtained with liquid nitrogen cooling. The set ups were however of relatively large size as compared to a small liquid nitrogen dewar and were therefore not well suited for field and down hole applications. Recently[8], EG&G has made available a new cryocooler in the electricool family. The *New Electricool* has features similar to the prototype described in this article. In particular, it has small overall dimensions and a light weight. At the time at which this article is written, the authors have not been able to gather significant information about the *New Electricool* and therefore are not able to give a thorough comparison between the two cooling packages.

The advantages of the new Iwatani Pulse Tube Cryocooler are its light weight and the small size of its cold head. The combination of the cooling package prototype presented here and an EG&G Pop-Top detector could fit within a cylindrical volume slightly larger than 3 inches in diameter and less than 2 1/2 feet long. The size of the device makes it highly attractive for environmental surveys, shallow hole investigations, and for other applications that require the type of energy resolution of a HPGe detection system as well as long uninterrupted operating time, ease of operation and small size. The absence of moving parts in the cold head, characteristic of pulse tube cryocoolers, potentially reduces the

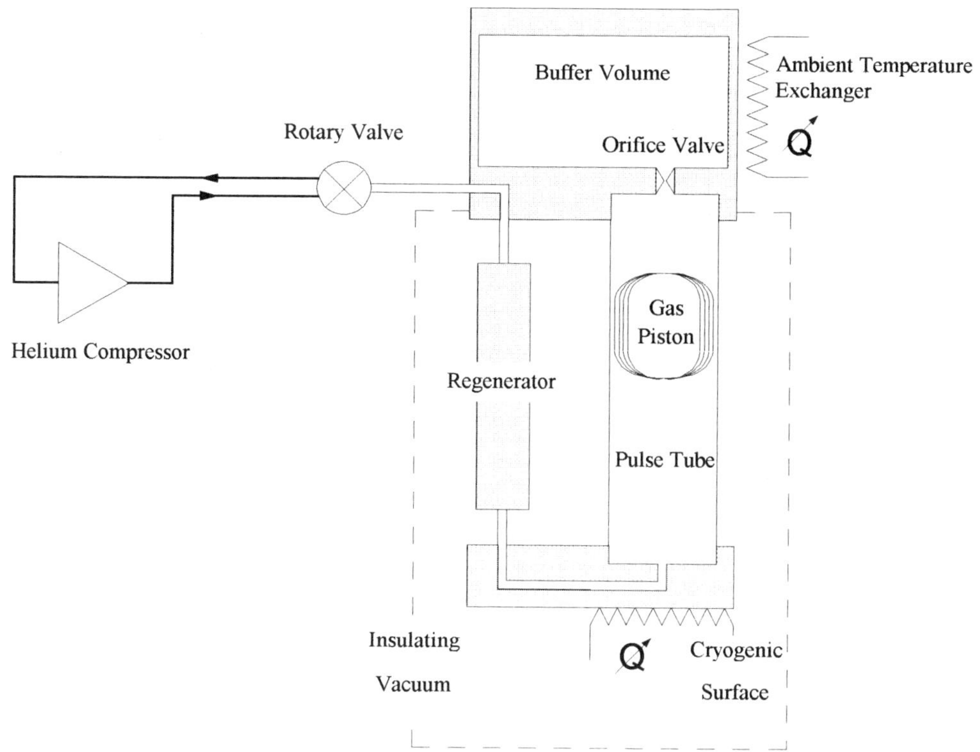


**Figure 1.** Schematic of a pulse tube Cryocooler such as the Iwatani Cryomini Pulse Tube Cryocooler used during these tests.

degradation of the energy resolution of the system due to microphonics. The mechanical vibrations generated during the operation of most cryocoolers make them unsuitable for operation with a semiconductor detector unless a special interface such as the ones included in the first generation of EG&G Electricool cryocoolers is added to dampen the vibrations. While other types of cryocoolers have been evaluated for their vibrational characteristics[9], to our knowledge, pulse tube cryocoolers have yet to be studied extensively for their inherent vibrational properties. The effect on the energy resolution of a high purity germanium detector cooled by a pulse tube cryocooler is also a direct test of its vibrational properties in the microphonics range.

The design of this prototype is part of a collaborative effort between HARC TDL and Kelvin International to take advantage and test the unique features of this new type of production cryocooler for various cryogenics applications. This article describes the design characteristics of the prototype and discusses its performances as far as the thermal properties and the energy resolution of the overall tool.

## IWATANI PULSE TUBE REFRIGERATOR

The Iwatani CryoMini Pulse Tube Refrigerator is a regenerative cryocooler operating at a constant speed. Thermodynamically similar to a Stirling refrigerator, shaped gas pulses (piston) remove heat by expansion in a closed volume. A schematic of the operating principle is presented in Figure 1. Compressed helium gas enters the rotary valve where a pulse of gas is allowed to pass to the regenerator. After giving up its sensible heat, the gas expands in the pulse tube chamber providing cooling to the cold head. A moving gas wave (piston) travels to the other end of this chamber and is compressed against the warm end as heat is transferred through the tube wall and rejected. Lastly, the gas piston reverses its course, cools the regenerator, and returns to the compressor through the rotary valve.

The above mentioned process is repeated several times per second providing a stable temperature at a constant heat lift. Higher efficiencies are achieved by tuning the piston phase angle through the adjustment of the size of the buffer volume and its orifice. Ultimate temperatures of 55K and heat lifts of 2 watts at 77K are achieved with standard single stage units. Several cold heads can be manifolded with a larger single compressor to provide greater heat lifts and temperature distribution. No maintenance is required within the cold head as there are no moving parts. The rotary valve and compressors will require a few minutes of preventative maintenance (MTBM) every 15,000 hours.

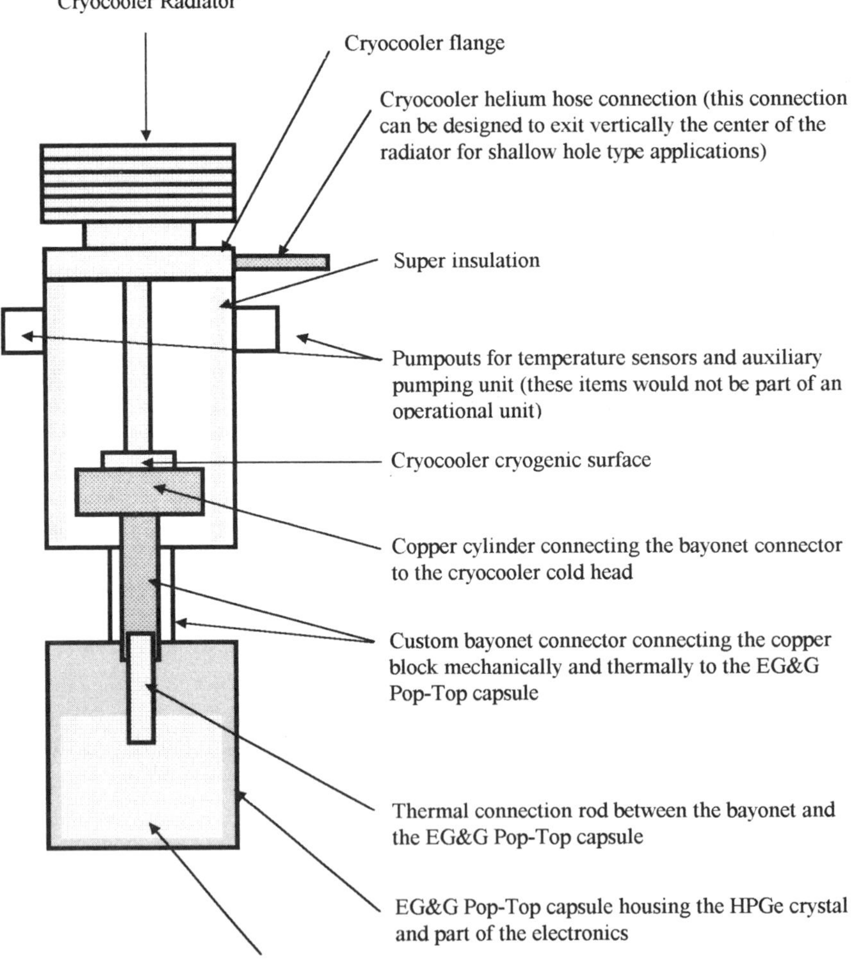


**Figure 2.** Schematic of the cooling package developed to test the use of the pulse tube cryocooler to cool the high purity germanium detector.

## EXPERIMENTAL DESIGN

The prototype is based on an Iwatani CryoMini Pulse Tube Cryocooler provided by Kelvin International and is designed to be integrated with off-the-shelve EG&G Pop-Top Detectors. The detector used during this study was a 50 mm diameter, 50 mm long High Purity Germanium crystal enclosed in a EG&G Ortec GEM series Pop-Top capsule. A schematic of the design is presented in Figure 2. The prototype consists of a vacuum chamber enclosing the Iwatani CryoMini cryocooler, and a custom copper bayonet connecting the cryocooler cold head and the EG&G Pop-Top detector. Superinsulation and other cryogenics design features were used to minimize the heat load on the cryocooler. The vacuum chamber was fitted with two pumpouts to allow the use of sensors within the chamber and as a connection to an auxiliary vacuum system. These pumpouts, as well as some of the other features allowing for a good monitoring of the experiment, would be eliminated or replaced by more compact solutions in the design of a field prototype. The cold head assembly of the cryocooler is connected to a rotary valve which itself is connected to a compressor. For the sake of clarity, the rotary valve and the air cooled compressor are not shown on Figure 2.

A series of temperature sensors was used to record the temperatures during operation at several key locations of the overall set-up. These locations are the surface of the radiator, the top of the rotary valve, the cryogenic surface of the cryocooler, and the two ends of the custom bayonet connector. The Pop-Top capsule temperature was recorded using the platinum resistor which is an integral part of the detector package. The vacuum in the cooling package was achieved with a 60 l/sec turbomolecular pump. The vacuum was maintained below the $10^{-4}$ Torr range during operations. For a field prototype these vacuum systems could probably be replaced by a getterer system attached to the cryogenic surface of the cold head. Foam pads and flexible vacuum tubing lines were used to avoid the propagation of vibrations from the mechanical vacuum pumps to the detection system set-up. The possibility of interference from the mechanical pumps was tested by intermittently replacing these pumps with a sorption pump. No changes in energy resolution were detected indicating the adequacy of the vibrational isolation of the system from the mechanical vacuum pumps.

## THERMAL HISTORIES

The cooling cycle is initiated by starting the compressor. The thermal histories for each of the sensors are displayed in Figure 3. The cooldown time of approximately 6 hours to reach steady state temperatures is very similar to the cooldown time associated with the use of a liquid nitrogen dewar or an EG&G Electricool system for this type of detector. The steady state temperatures reached for the Pop-Top capsule as well as for the different components of the prototype are also displayed in Figure 3.

The 78 K steady state temperature reached by the Pop-Top is well within the operating range of a Pop-Top capsule (~ 55K to 115K) and will allow for the necessary modifications to design a more rugged prototype for field operations. The Pop-Top temperature was very stable over extended periods of time. On several occasions, the Pop-Top capsule was maintained at operating temperature for more than a week without any noticeable temperature change. Temperature control is however an important issue for the design of a truly functional cooling package. Changes in environmental temperatures during measurements or increased heat loads resulting from very high count rates such as the one encountered in PGNAA[5] applications will no doubt affect the Pop-Top capsule temperature and possibly the energy resolution of the system.

## HPGE SPECTRA

To test the vibrational characteristics of the prototype, Cobalt-60 gamma ray spectra were acquired while cooling the same HPGe Pop-Top detector with both the standard liquid nitrogen cryostat and the pulse tube based cooling system. The energy resolution was assessed by fitting the 1.332 MeV line of the Co-60 spectrum with a Gaussian curve[1]. The Full Width at Half Maximum was then computed based on the parameters of the curve fit. The experimental conditions such as the count rate and the electronics settings were set following EG&G recommended procedures[10]. The energy resolutions for both the benchmark liquid nitrogen cryostat and the pulse tube refrigerator are displayed in Figure 4. The 2.08 keV energy resolution measured for the prototype is higher than the 1.73 keV benchmark, but is still within the range of the warranted performance for this detector (2.09 keV[10]). The 2.08 keV resolution was obtained using a pulse shaping time constant of 2 μs rather than the recommended 6 μs, 6 μs was used for the benchmark measurement. The fact that the optimal pulse shaping time is low is often correlated to system vibrations (microphonics). Although there are no moving parts in the pulse tube there could be small vibrations generated as the gas flows through the cold head. Another possibility is that vibrations from the compressor are propagated through the flexible hoses connecting the compressor to the rotary valve and then from the rotary valve to the cold head through the connecting tube. An improvement of the interface between the cryocooler cold head and the Pop-Top capsule would probably reduce the difference in energy resolution between the

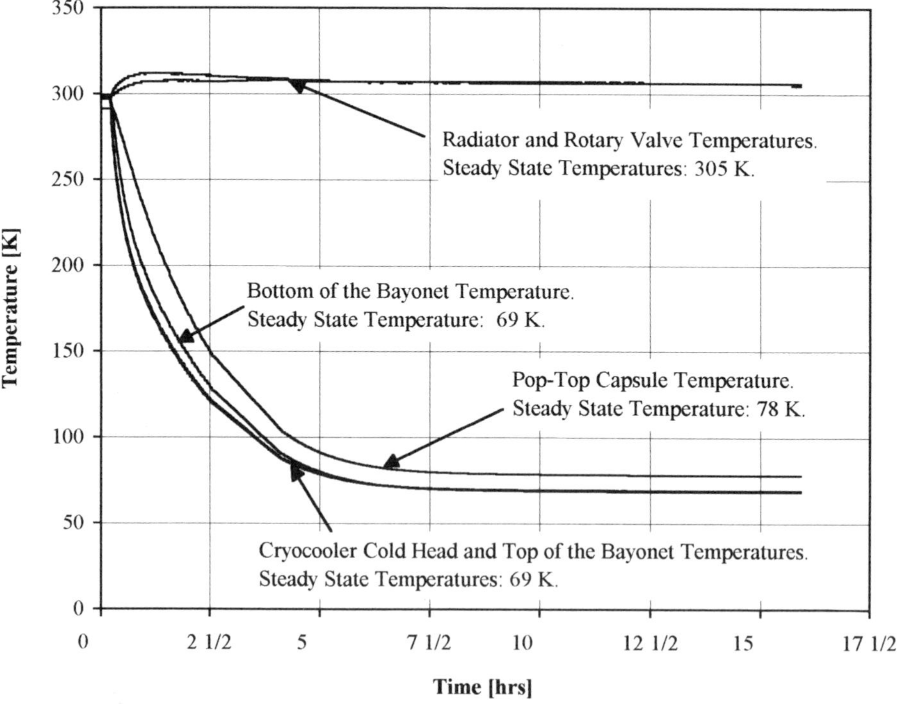


**Figure 3.** Temperature histories at key locations in the cooling package during cool down and operational steady state temperatures.

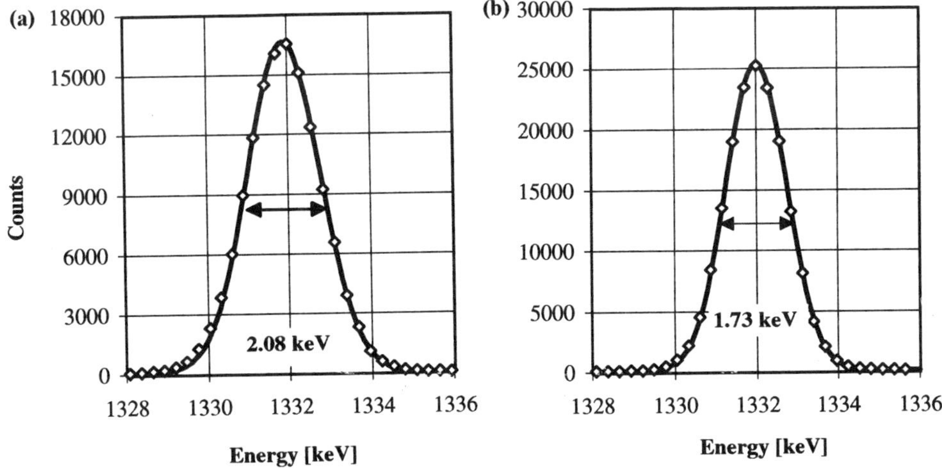


**Figure 4.** Quantitative measurements of the energy resolution of a HPGe Pop-Top detector when cooled with a liquid nitrogen cryostat (a) and with the prototype cooling package based on a pulse tube cryocooler (b). The energy resolution is assessed by measuring the Full Width at Half Maximum of the 1.332 MeV line of a Cobalt 60 gamma ray source.

two configurations. Nevertheless, the measured energy resolution of 2.08 keV makes the system fully operational for almost all current applications.

To test the performance of the cooling package for a practical application, the spectrum of a Naturally Occurring Radioactive Material (NORM) sample was measured with the prototype. The spectrum displayed in Figure 5 shows that the prototype is well suited for environmental measurements such as NORM monitoring.

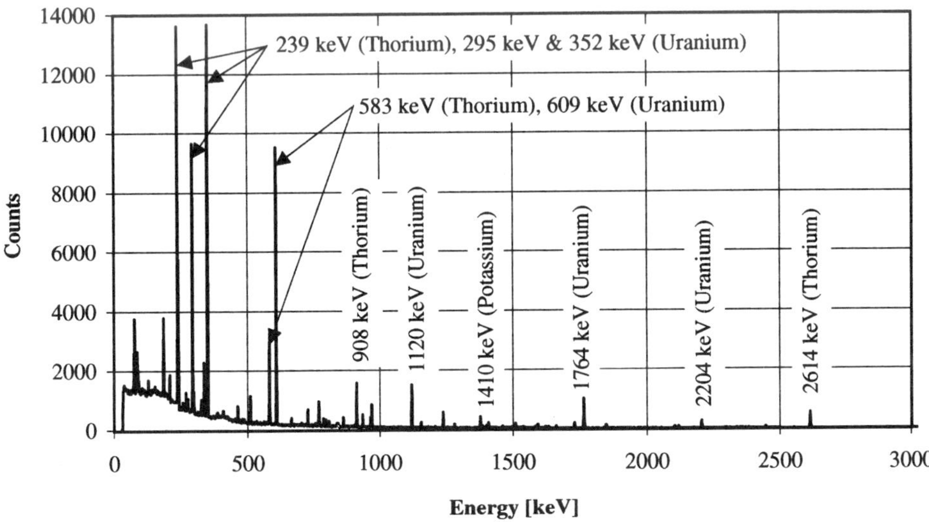


**Figure 5.** Naturally Occurring Radioactive Material (NORM) gamma ray spectrum obtained with the prototype cooling package. NORM measurement is one of the potential application for this type of tool. The natural radioactive nuclei series of the gamma ray lines are indicated in parenthesis.

## CONCLUSION

A cooling system based on a Pulse Tube Cryocooler was designed to operate a high purity germanium detector. The design was successfully implemented and tested on an EG&G Pop-Top HPGe detector. The thermal performances were such that the detector can operate within the temperature range usually provided by a cooling system based on liquid nitrogen, and even at lower temperatures. The cooldown time is similar to that of a liquid nitrogen based system and the system can be operated for long periods of time without interruption. Gamma ray spectra were acquired for several gamma ray sources. The energy resolution of the prototype at 1.332 MeV was measured at 2.08 MeV, within the warranted energy resolution for this particular detector. The energy resolution of the prototype also compared well to the energy resolution of 1.73 keV obtained for the same detector cooled by a standard liquid nitrogen system. The difference is attributed to microphonics and could be reduced, if necessary, by improving the interface between the cryocooler cold head and the detector. Nevertheless, the prototype is fully functional for most applications including environmental monitoring. The spectrum of a NORM sample was recorded with the prototype as an example of the performance for an environmental monitoring application. For the applications mentioned in this study and for downhole cryogenics in particular, a pressurized helium canister could replace the compressor. The result would be a small size self contained cooling system with great promise for down hole cryogenics.

## ACKNOWLEDGMENTS

This project was supported in part by the Houston Environmental Initiative Program, and by T. Nishitami and M. Yanai of Iwatani Corporations. We would also like to thank the EG&G Ortec technical staff for their assistance during this project as well as B. Freeman of Centrus Plasma Technologies.

## REFERENCES

1. G.F. Knoll. "Radiation Detection and Measurement," John Wiley & Sons, New York (1989).
2. T. Mizutani, K. Hirata, M. Hiramaya, and A. Ishida, Ultra-high speed semiconductor devices and low temperature electronics, *Cryogenics* 30:1024 (1990).
3. Z.B. Alfassi. "Activation Analysis," CRC Press, Boca Raton (1990).
4. R.M. Kogan, I.M. Nazarov, and Sh.D. Fridman. "Gamma Spectroscopy of Natural Environment and Formations," The National Science Foundation (1971).
5. Z.B. Alfassi, and C. Chung. "Prompt Gamma Neutron Activation Analysis," CRC Press, Boca Raton (1995).
6. G.R. Boynton, Canister cryogenic system for cooling germanium semiconductor detectors in borehole and marine probes, *Nucl. Inst. Meth.* 123:599 (1975).
7. "Modular Pulse-Processing Electronics and Semiconductor Radiation Detectors", EG&G Ortec, Oak Ridge (1995), p. 526 .
8. "Modular Pulse-Processing Electronics and Semiconductor Radiation Detectors", EG&G Ortec, Oak Ridge (1997-1998), p. 1.103.
9. D. Hill, Throttle cycle cooler vibration characterization, in: "Cryocoolers 9," Plenum Press, New York (1997), p. 737.
10. "Solid-State Photon Detector, Operators Manual, GEM Series HPGE Coaxial Detector System," EG&G Ortec, Oak Ridge, (1991).

# LOW NOISE COLD HEAD OF A FOUR-VALVE PULSE TUBE REFRIGERATOR

J. Gerster, G. Kaiser, L. Reißig, M. Thürk, and P. Seidel

Institut für Festkörperphysik
Friedrich-Schiller-Universität Jena, Lessingstraße 8, 07743 Jena
Germany

## ABSTRACT

We have designed and constructed a split type of a four-valve pulse tube refrigerator (FVPTR) in order to cool high-$T_c$ Superconducting Quantum Interference Devices[1] (SQUIDs). A SQUID, in particular, places rigorous upper limits on the tolerable levels of magnetic interference, vibrations and temperature fluctuations. The pressure wave in the system is generated by means of a commercial He-compressor in combination with a rotary valve. In order to protect the cold head from disturbances generated by the valve gear and the compressor, it is useful to separate those parts spatially. Reduction of interference signals from the compressor and the rotary valve is accomplished using flexible tubes between the rotary valve and the cold head. Most parts of the cold head, including the regenerator, were made of non-metallic, electrically insulating, and non-magnetic materials to minimize eddy currents and local magnetic fields, which interfere the sensor directly.

In this contribution we show our setup of a low noise split FVPTR. Some fundamental changes have been done in this setup compared with the integral FVPTR[2]. These changes will be discussed in detail. Furthermore, we will present the non-metallic components of the cold finger. The thermodynamic consequences and the magnetic noise reductions which result from the material substitution will be considered.

## INTRODUCTION

High-$T_c$ Superconducting Quantum Interference Devices[1] (SQUIDs) are usually operated in a temperature range between 50 and 80 K and require a cooling power of about 100 mW. Mostly they are cooled by liquid nitrogen, at its boiling point (77.3 K at 1.013 bar) or in the temperature range above the triple point (63.2 K). For a user-friendly high-$T_c$ magnetometer system, it is useful to cool without liquefied gases. One-stage cryocoolers can reach every temperature in this desired temperature range.

In order to detect anomalies in the magnetocardiogram (MCG) the magnetic noise level has to be less[3] than 100 fT/√Hz in a frequency range between 0.1 and 100 Hz. The cryocooler has to meet this requirement also. This demand is an enormous challenge for a cryocooler.

Compared to other regenerative cryocoolers like Gifford-McMahon and Stirling refrigerators, pulse tube refrigerators have no moving parts in the cold finger. This feature has the advantage that the cold finger is easy to construct and has a low level of mechanical vibration. For that reason the pulse tube refrigerator is very attractive for SQUID cooling. In order to protect the cold head from disturbances generated by the compressor and the rotary valve it is necessary to separate these two parts from the cold head spatially[4]. If the construction of the cold head of entirely non-metallic, non-magnetic, and electrically insulating materials is successful, the magnetic noise of the pulse tube refrigerator will be reduced strongly. In this contribution we present our efforts minimizing the magnetic noise which is generated by the pulse tube refrigerator.

## SETUP OF A LOW NOISE FVPTR

The distance of the compressor and the rotary valve to the cold head should be as long as necessary. The compressor is connected to the cold head via the rotary valve. The distance between the compressor and the rotary valve can be made very long, because it only depends on the pressure drop in the high and low pressure pipes. If the diameters of these pipes are large enough the pressure drop will be low. Small distances must be chosen between the rotary valve and the cold head , because the cooling power and the lowest temperature decrease with an increase of distance.

Figure 1 shows the schematic setup of the split four-valve pulse tube refrigerator (FVPTR). A precooler has been inserted in the main inlet at the hot end of the regenerator in contrast to the integral type of the FVPTR[2]. Furthermore, it can be recognized that the second inlet pipe is replaced by two pipes, called second inlet and second outlet. The working gas in the second inlet pipe flows from the valve unit into the pulse tube. The gas in the second outlet pipe flows in the opposite direction.

**Precooler.** As the working gas is compressed nearly adiabatically the temperature of the gas inside the pipe will increase. In order to avoid the loss due to the flow of hot gas into the regenerator a precooler is necessary which keeps the gas at ambient temperature.

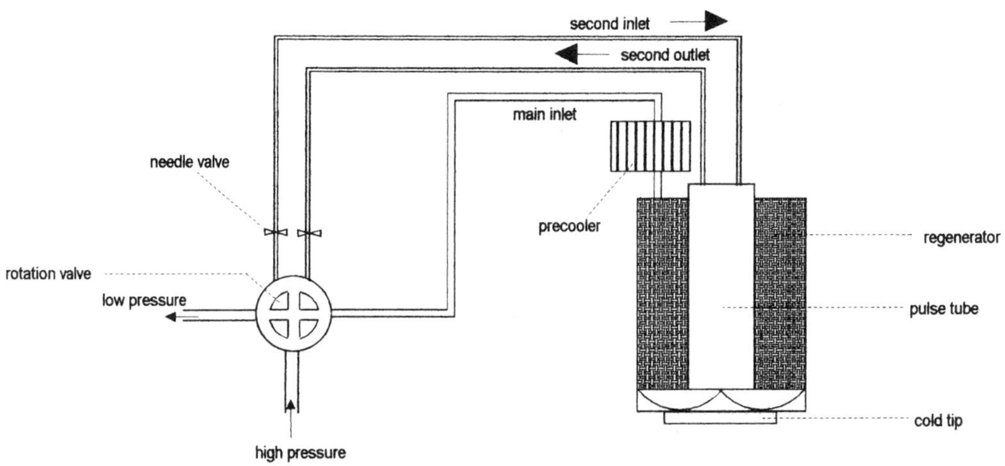


**Figure 1.** Setup of a split four-valve pulse tube refrigerator (FVPTR).

**Second inlet / second outlet.** For our first experiments with the FVPTR we used a single pressure pipe connecting the hot end of the pulse tube to the control valves. Operating in this setup we observed a reduced cooling performance because of regeneration losses acting between the gas and the pipe wall.

After compression, helium gas coming from the pulse tube hot end with a high temperature $T_h$ flows through the pipe into the pressure sink heating up the inner wall surface of the pipe. A particular amount of heat is supplied to the pipe depending on the heat transfer resistance and the temperature difference between gas and pipe wall.

Afterward the expansion gas at ambient temperature $T_{amb}$ coming from the pressure source returns through the same pipe into the pulse tube. It enters the pulse tube hot end with a temperature higher than ambient temperature $T = T_{amb} + \Delta T$, because a part of the heat supplied to the wall will heat up the returning gas again.

The energy exchange H with the environment of one cycle provided by mass flow enthalpies

$$H \propto (T_h - T_{amb})$$

will be reduced by the hot end loss $\Delta H$ due to the temperature shift $\Delta T$

$$H \propto \Delta T.$$

Taking this into account we have changed our experimental setup, see Fig. 1. Now the hot end of the pulse tube is connected to the control valves using two pressure pipes for gas exit and gas enterance, respectively. With this setup the gas enters the pulse tube hot end at ambient temperature.

## NON-MAGNETIC AND NON-METALLIC COLD FINGER

### Tubes, flanges and adhesive connections

The tubes of the cold finger and the flanges were made of glassfilled epoxy resins. At 200 K, resins have a thermal conductivity k = 0.002 W/(cm K) [5] compared to stainless steel k = 0.13 W/(cm K). Due to this fact the axial thermal conduction loss from the hot end to the cold end is low not only for thin walls of the tubes.

We built up a split FVPTR in coaxial configuration. The regenerator tube has an inner diameter of 20 mm, a wall thickness of 2.5 mm and a length of 250 mm. So the axial thermal conduction loss by the regenerator tube can be reduced down to 35 mW.

The flanges at the cold and at the hot ends are connected to the regenerator tube by special adhesive technology with an epoxy resin. The connections meet the requirements of the tensile strength due to the high pressure of the working gas.

No diffusion of helium through the adhesive connections and the epoxy resin tubes into the vacuum chamber could be observed.

### Cold Tip

The cold tip has to seal the regenerator tube against high pressure helium gas leaking to the vacuum dewar. Additionally, it has to transfer the cooling power to the device.

For the first task we made a solid cold tip of polycarbonate. Due to the fact[6] that polycarbonate has a low thermal conductivity k = 0.001 W/(cm K) at 50 K it is necessary to insert a material with a high thermal conductivity into the cold tip. For the heat transfer we glued 17 pieces M1.6 threaded pin made of aluminum into the polycarbonate cold tip. The pins act as a cold end heat exchanger. The large surface of the threaded pins guarantee a sufficient heat transportation. A comparison with a conventional solid copper cold tip shows no differences of the lowest temperature. Even the slope in the cool-down curve of the polycarbonate / Al-pins has a stronger decrease compared to that of the copper cold tip, because of the lower volumetric heat capacity of the polycarbonate / Al-pins cold tip.

In the first cool-down experiments the split FVPTR with the polycarbonate / Al-pins cold tip, the glassfilled epoxy resin tubes and flanges reached the lowest temperature of below 90 K. The length of the pulse tube is 280 mm and the diameter 9.5 mm. The main inlet pipe is 2.5 m and the second inlet and outlet pipes are 0.5 m long.

Figure 2 shows the magnetic noise of the copper cold tip at 50 K. The cold tip has a diameter of 48 mm and a mass of 64 g. This noise spectrum was measured with a low-$T_c$ SQUID gradiometer in a shielded room. The magnetic noise of the polycarbonate / Al-pins cold tip is lower than the intrinsic noise of the low-Tc SQUID measurement system (<10 fT/√Hz). Therefore, the cryocooler with a copper cold tip cannot be used for a direct mounted SQUID magnetometer in a MCG measurement system.

## Regenerator material

Figure 3 shows the magnetic noise of the split FVPTR operating at 8 Hz. The noise was measured with a magnetoresistive sensor in parallel orientation to the cryocooler axis. The noise with a bronze / stainless steel regenerator is illustrated in figure 3 a). One can see clearly the peaks at the working frequency and its harmonics. Without regenerator material these peaks are vanished, see figure 3 b). The noise of the cryocooler is below the background noise.

These considerable magnetic disturbances are caused by magnetic impurities, induced eddy currents, and thermally activated currents in the metallic regenerator. These interferences can be removed by the use of a non-magnetic, non-metallic and electrically insulating regenerator material. The main problem is to find an adequate material which does not reduce the efficiency of the cryocooler with respect to the dimensions of the regenerator housing.

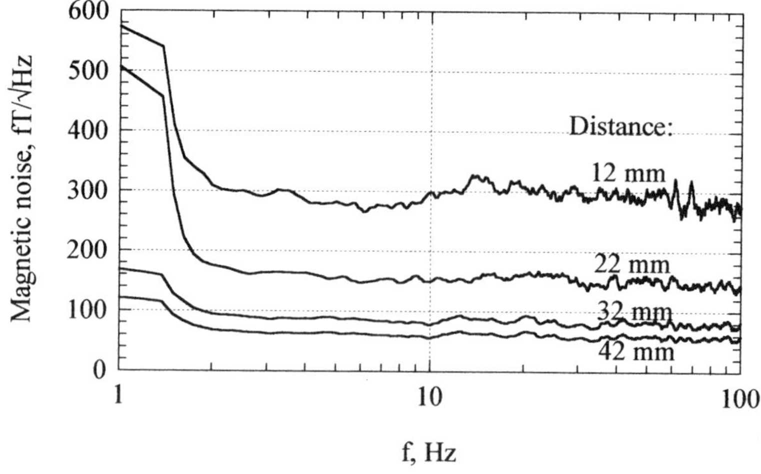


**Figure 2.** Noise of a copper cold tip at 50 K at several distances of the sensor.

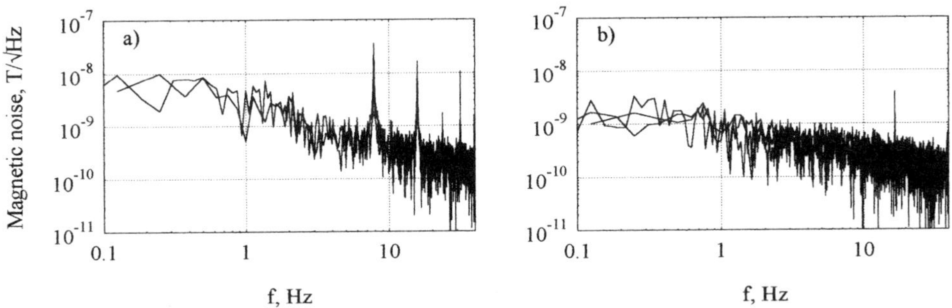


**Figure 3.** Magnetic noise of a non-metallic split FVPTR with a bronze mesh regenerator a) and without regenerator material b) at a working frequency of 8 Hz measured with a magnetoresistive sensor.

The volumetric specific heat c of a regenerator material should be as high as possible. Figure 4 shows c versus temperature for copper and for two polymer materials. Copper has a 1.8 to 2.3 times higher c compared to Nylon or Teflon in the concerning temperature range.

Further problems occur in relation with the low temperature conductivity k/c of the polymer materials. Nylon and Teflon have about two order of magnitude lower temperature conductivities compared to bronze. Due to this fact, the temperature wave penetrates worse into the material. The ratio of the temperature wave amplitude inside $\vartheta_x$ to the surface $\vartheta_d$ of a slab with the thickness 2d can be expressed[7] as

$$\frac{\vartheta_x}{\vartheta_d} = \sqrt{\frac{(\cosh(\frac{x}{d}M)\cos(\frac{x}{d}M))^2+(\sinh(\frac{x}{d}M)\sin(\frac{x}{d}M))^2}{(\cosh(M)\cos(M))^2+(\sinh(M)\sin(M))^2}} \quad (1)$$

M is the ratio of d to the thermal penetration depth λ:

$$M = \frac{d}{\lambda} = d\sqrt{\frac{\pi f c}{k}} \quad (2)$$

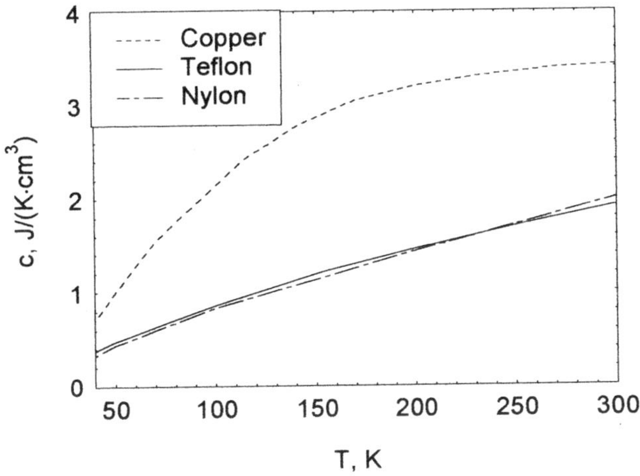


**Figure 4.** Volumetric specific heat c of two polymer materials in comparison to copper.

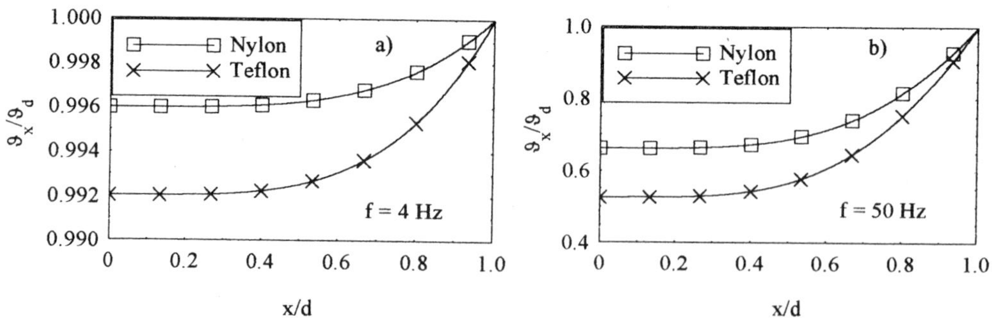


**Figure 5.** Ratio of the temperature wave amplitudes inside to the surface of a slab (thickness = 75 μm) made of Nylon and Teflon at 4 Hz a) and 50 Hz b) at 300 K.

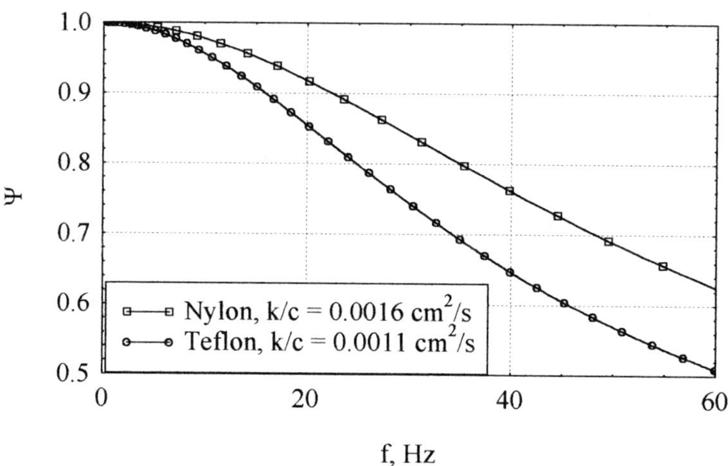


**Figure 6.** Utilization rate Ψ of the regenerator at 300 K and a slab thickness 2d = 75 μm as a function of the frequency for two polymers.

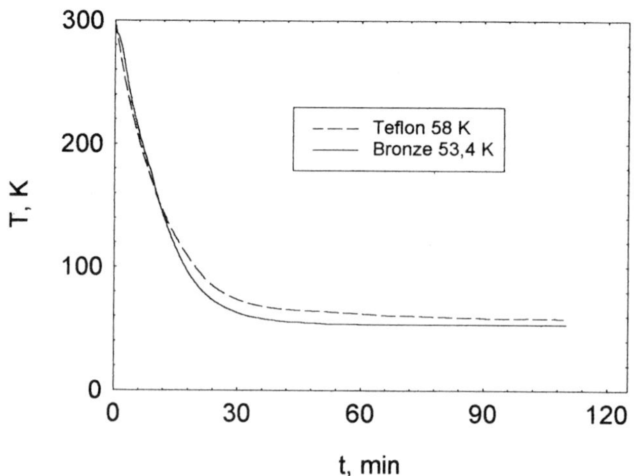


**Figure 7.** Cool-down curve of a FVPTR at a frequency of 4 Hz with a regenerator made of Teflon or bronze, respectively.

M depends on the frequency f and therefore $\vartheta_x/\vartheta_0$ does also. In figure 5 the ratio $\vartheta_x/\vartheta_0$ for 4 Hz and for 50 Hz with a slab thickness 2d = 75 μm at 300 K is shown. The amplitudes of the temperature wave decrease strongly inside the material.

The utilization rate Ψ of a regenerator is defined as

$$\Psi = \frac{Q_{\tau/2}}{Q_{max}}. \tag{3}$$

$Q_{max}$ is the maximum heat which can be stored and $Q_{\tau/2}$ is the real amount of heat which is stored in a half of a period. For a slab, Ψ can be written[7] as

$$\Psi(M) = \frac{1}{2M}\sqrt{2\frac{\cosh^2 M - \cos^2 M}{\cosh^2 M - \sin^2 M}}. \tag{4}$$

In figure 6 the utilization rate Ψ versus frequency is shown for 2d = 75 μm at 300 K. One can see that at high frequencies Teflon and Nylon exhibit a low Ψ. In order to avoid a low Ψ one should minimize the slab thickness 2d or the diameter of the wire 2d'. Due to the fact that the pressure drop Δp is proportional to $f_r/d'$ and the friction factor $f_r$ strongly increases at small d' [8], a reduction of d' would cause a dramatical increase of the pressure loss. For comparison, at 50 Hz bronze has a value of Ψ = 0.9999 at the same slab thickness. Using a non-metallic regenerator material the cryocooler has to operate at a moderately low frequency.

First investigations have been done using our integral FVPTR[2] with Teflon as the regenerator material. We have selected a matter of Teflon fabric made of a multifilament yarn consisting of 29 monofilaments (diameter ≈ 40μm) per wire. With this fabric we have achieved a minimum temperature of 58 K. The working frequency of the FVPTR was 4 Hz.

Figure 7 shows the cool-down curves for Teflon and bronze regenerator material, respectively. The lowest temperature using Teflon is only 4.6 K higher than in the case of bronze. Therefore we could show that it is possible to operate a one-stage cryocooler with a non-metallic regenerator material.

## CONCLUSIONS

Highly sensitive SQUIDs place a rigorous upper limit on the allowable level of the magnetic interferences. In order to limit the magnetic disturbances of a FVPTR, a low noise cryocooler was designed.

Both the compressor and the rotary valve generate disturbances and were separated spatially from the cold finger. For that purpose, we had inserted two new parts in the split FVPTR, a precooler and an additionally pipe at the pulse tube hot end.

For the construction of the cold finger most components were made of non-metallic materials. Tubes and flanges were constructed of glassfilled resins and were glued together. So they meet the requirements of the cryocooler. The cold tip was made of polycarbonate with aluminum threaded pins. In contrast to a comparable copper cold tip no thermal magnetic could be observed. The cold tip shows no disadvantages in the cool-down behavior compared with the copper cold tip. Noise measurements show that the main disturbances are generated by the metallic regenerator material. In our integral type FVPTR we have achieved 58 K with a Teflon regenerator. Due to the small penetration depth of polymers, the frequency of a cryocooler with a polymer regenerator should not be too high.

Our further work will concentrate on the reduction of the vibrations, and the temperature fluctuation in order to enable the cryocooler operation close to the SQUID sensor during the measurement.

## ACKNOWLEDGMENT

We thank F. Giesler (Biomagnetic center of FSU Jena) for the low-$T_c$ SQUID measurements. Financial support by a BMBF grant (No. 13 N 6618) is gratefully acknowledged.

## REFERENCES

1. E. Il'chev, L. Dörrer, G. Hildebrandt, F. Schmidl, V. Zakosarenko and P. Seidel, Current resolution, noise, and inductance measurements on high-Tc dc SQUID galvanometers, Appl. Phys. Lett. 68, 5 (1996), p. 708
2. J. Blaurock, R. Hackenberger, P. Seidel and M. Thürk, Compact four-valve pulse tube refrigerator in coaxial configuration, "Cryocoolers 8," R. G. Ross, ed., Plenum Press, New York (1995), p. 395
3. G. Kaiser, S. Wunderlich, M. Thürk and P. Seidel, Cryocoolers for high-Tc superconducting sensors: case study on a planar SQUID gradiometer, "Applied Superconductivity 1995" in IOP Conf. Series 148, D. Dew-Hughes, ed., IOP Publ., Bristol (1995), p. 1637
4. G. Thummes, R. Landgraf, F. Giebeler, M. Mück, and C. Heiden, Pulse tube refrigerator for high-Tc SQUID operation, "Adv. Cryo. Eng., Vol. 41," P. Kittel, ed., Plenum Press, New York (1996), p. 1463
5. G. Hartwig, Tieftemperatureigenschaften von Epoxidharzen, Progr. Colloid & Polymer Sci. 64, 56-67 (1978), p. 660
6. G. Hartwig, "Polymer Properties at Room and Cryogenics Temperatur", Plenum Press, New York (1994)
7. H. Gröber, S. Erk and U. Grigull, "Die Grundgesetze der Wärmeübertragung," 3. Aufl., Springer, Berlin (1954)
8. W. M. Kays and A. L. London, "Compact Heat Exchangers," 3rd Ed., McGraw-Hill, New York (1985)

# USE OF A PULSE TUBE REFRIGERATOR FOR COOLING A HTS-ANTENNA FOR MAGNETIC RESONANCE IMAGING

K. Klundt[1], C. Lienerth[1], G. Thummes[1], F. Steinmeyer[2], M. Vester[2], W. Renz[2], and C. Heiden[1]

[1]Institute of Applied Physics, Justus-Liebig-University
D-35392 Giessen, Germany
[2]Siemens Corporate Technology
D-91050 Erlangen, Germany

## ABSTRACT

We report on the first operation of a YBCO receiving antenna for magnetic resonance imaging by use of a specially designed double-inlet pulse tube refrigerator. The pressure oscillation in the cooler is generated by means of a rotary valve in combination with a 2 kW helium-compressor. To reduce noise at the cold head, the oscillating gas flow from the rotary valve is supplied to the cold head via a thin flexible tube with a length of 4 m. Optimum cooler performance is achieved with the pulse tube in vertical position (cold end facing downwards). Then, at an operating frequency of 4.6 Hz, a minimum no-load temperature of 31 K is achieved, and a net cooling power of 2 W is available at 50 K. Tilting of the pulse tube leads to enhanced heat losses through free convection of helium in the tube. By increasing the operating frequency to 8 Hz the orientation dependence of the cooler performance could be reduced. This made possible the testing of the PTR-cooled antenna in a Siemens Magnetom Open 0.2 T MRI-system with a horizontal pulse tube and with the cold block at 55 K. Compared to a standard room temperature antenna the improved signal to noise ratio that is achieved with the superconducting antenna is clearly reflected in the higher quality of MR-images taken from parts of the human body.

## INTRODUCTION

Superconducting receiving antennas offer improved image quality in a number of magnetic resonance imaging (MRI) applications, e.g. at low magnetic fields[1]. The use of such high-quality antennas in clinical environment, where handling with cryogenic liquids should be avoided, requires a cooler that is electromagnetically compatible with the MRI detection system. The pulse tube refrigerator (PTR) is a very attractive candidate for cryogen-free operation of highly sensitive superconducting devices, since the absence of moving parts in the cold stage results in low vibrations and low electromagnetic interference signals[2].

In a previous study, the inherent low noise level of the PTR has been proven by operation of a SQUID gradiometer consisting of two YBCO rf-SQUIDs, which were mounted on the cold head of a specially designed double-inlet PTR[3,4]. To reduce noise from the rotary valve and helium compressor, the cold head was connected to the pressure-wave generating assembly by means of a long flexible tube.

Also, a first test of that PTR to cool a YBCO MRI-antenna showed rather promising results[5]. One disadvantage of that cooler was the pronounced dependence of cooling temperature on orientation of the pulse tube with respect to gravity, which made impossible operation in a horizontal position, as would be desirable for highest resolution of the antenna[5]. The orientation dependence of cooler performance has been related to convective heat losses by free convection of He-gas which is superimposed on the oscillatory gas displacement in the pulse tube, occurring with a frequency of 2 Hz in that PTR[6].

On the basis of these results, we have constructed a new PTR specially designed for cooling a HTS-antenna in a C-shape open magnet MRI-system with static magnetic field $B_0$ in vertical direction.

## EXPERIMENTAL SET-UP

### Refrigerator Description

For practical operation in a MRI-system the PTR has to meet the following requirements: low mechanical vibration and no magnetic-field distortion from the cold head, no metallic parts near the antenna, spatial separation of cold head and pressure-wave generator, feasibility of operation in horizontal position of the pulse tube.

A cross section of the PTR is shown in Figure 1. A U-shaped configuration of pulse tube and regenerator is used because of ease of construction and to avoid radial heat transfer losses. These tubes, made of stainless steel, have outer diameter, wall thickness and length as follows: pulse tube 12 mm × 0.3 mm × 145 mm, regenerator tube 18 mm × 0.3 mm × 140 mm. The volume ratio of regenerator to pulse tube is 2.2, which is close to the optimum ratio of about 2 that has been found empirically[7]. The pulse tube volume is chosen to give an optimum working frequency around 5 Hz, following previous experience[3,4].

The regenerator matrix consists of a stack of 1500 phosphorus bronze screens of mesh no. 247 (porosity = 0.673). The screens are fabricated with a slight oversize of 0.26 % in order to avoid flow leakage along the regenerator tube wall[8].

The part of the refrigerator that is located outside the vacuum insulation, is made of two cylindrical aluminum blocks. The first block, to which the regenerator and pulse tube are attached, serves as vacuum flange and carries the heat exchanger at the warm end of the pulse tube and the aftercooler at the regenerator inlet. Both heat exchangers consist of axial flow channels ($\phi$1.4 mm) which were drilled into the aluminum. The second block contains the gas inlets and the orifice- and double-inlet valves that serve for adjusting the phase shift between gas flow rate and pressure oscillation in the cooler.

The cold ends of regenerator and pulse tube are soldered into a cylindrical copper block ($\phi$47 mm) containing the cold heat exchanger, which consists of a copper cylinder with 64 axial flow channels ($\phi$1 mm, length 13.5 mm). The antenna is cooled via a sapphire rod (length 3 cm, cross section 1 $cm^2$) that is glued to the cold block.

The vacuum vessel (outer diameter 69 mm) is fabricated from acrylic glass. The end face of the vessel is formed by an alumina window (1 mm thick) with an O-ring seal. Ten layers of superinsulating foil are wrapped around the cooler. In the 1 mm-gap between antenna and ceramic window no superinsulation is used in order to avoid degradation of the quality-factor of the superconducting antenna. The calculated heat load by radiation from ambient temperature to the cold antenna surface is then about 0.6 W.

A valved compressor with nominal input power of 2 kW (Leybold, model RW 2) provides high pressure helium gas as working fluid. The pressure oscillation in the cooler is

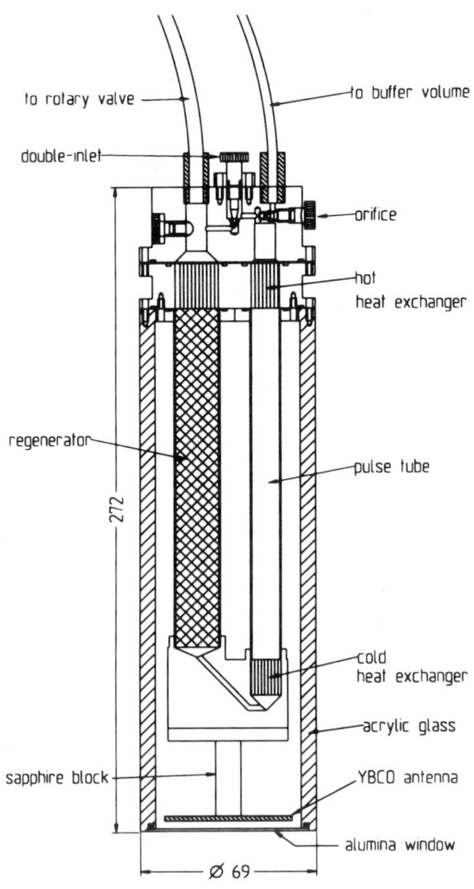


**Figure 1.** Cross section of the double-inlet pulse tube refrigerator for cooling a HTS-antenna for magnetic resonance imaging.

generated by means of a motor-driven rotary valve, as used for Gifford-McMahon coolers, which connects periodically the high and low pressure side of the compressor to the cooler inlet. The valve has been modified to make possible the operation of the cooler at different frequencies.

## Arrangement of Cooler in MRI-System

A schematic of the PTR arrangement in the MRI-system (Siemens Magnetom Open) is displayed in Figure 2. The C-shaped conventional magnet with a static induction of $B_0$ = 0.2 T has a patient gap of 41 cm.

To reduce magnetic noise and mechanical vibrations, the cold head with the antenna is connected to the rotary valve by means of a flexible polyamide tube with an inner diameter of 4 mm and with a length of 4 m. This split-type arrangement allows placing the compressor and rotary valve outside the screened room, which is a key condition for low-noise operation of the HTS-antenna by use of the PTR cold head. To facilitate the handling of the cold head, the buffer volume (reservoir) of 200 $cm^3$ also is connected to the orifice valve (see Figure 1) by a 4 m polyamide tube with an inner diameter of 2 mm.

## YBCO Antenna

The HTS receiving antenna is designed as a pair of coupled planar spiral coils operating in self resonance[5,9]. The planar coils (12 windings each, effective diameter 3.4 cm) are made

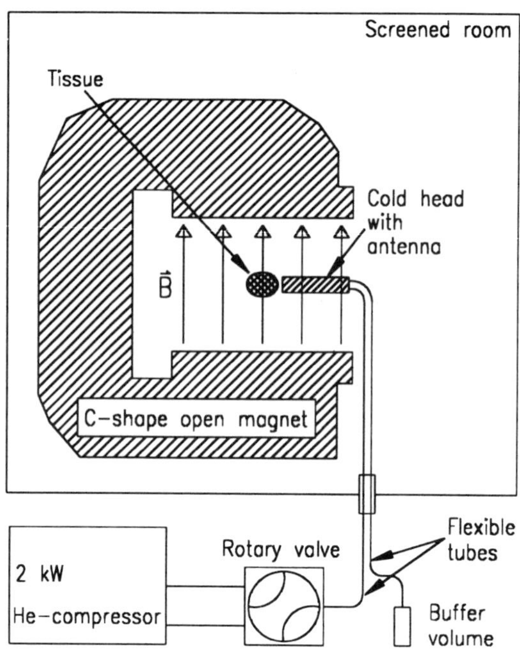


**Figure 2.** Schematic of the MRI-system with PTR-cooled receiving antenna.

by photolithographic patterning of two epitaxial YBCO films. The films were deposited on the opposite surfaces of a 2-inch $LaAlO_3$ wafer (1mm thick) by laser ablation[10]. At the resonance frequency of 8.25 MHz, the unloaded Q-factor of the antenna, as measured at 77 K in a static field of 0.2 T, is about $7 \times 10^4$. Inductive coupling of the antenna to the room temperature preamplifier is made by means of a single loop of copper wire and a miniature coaxial cable. For highest resolution, parallel alignment of the coil plane to the static magnetic field is desirable; this also minimizes field distortion by the diamagnetic superconducting windings. Therefore, a horizontal orientation of the cold head is demanded, as indicated in Figure 2.

## EXPERIMENTAL RESULTS

### Cooler Performance

The performance of the PTR in Figure 1 was investigated before installation of the HTS-antenna. The cold block temperature was measured by means of a Pt100 resistance thermometer. A resistive heater attached to the cold block served for measuring the net cooling power. Dynamic pressures were monitored by use of piezoresistive sensors. In all tests the helium charging pressure was 17 bar.

**Minimum temperature and pressure wave.** Figure 3 displays the cool-down curves at three different operating frequencies with the cold end of the pulse tube facing downwards. The settings of the orifice and double-inlet needle valves were optimized here for minimum temperature. The lowest stationary temperature of $T_{min}$ = 31 K is achieved with a frequency of f = 4.6 Hz after operation of 100 minutes. Fastest cooling is observed for f = 8 Hz where the cold block attains $T_{min}$ = 36 K already after one hour. At 12 Hz only $T_{min}$ = 43 K is achieved. This behavior can be qualitatively attributed to the variation of dynamic pressure amplitude in the cooler with frequency.

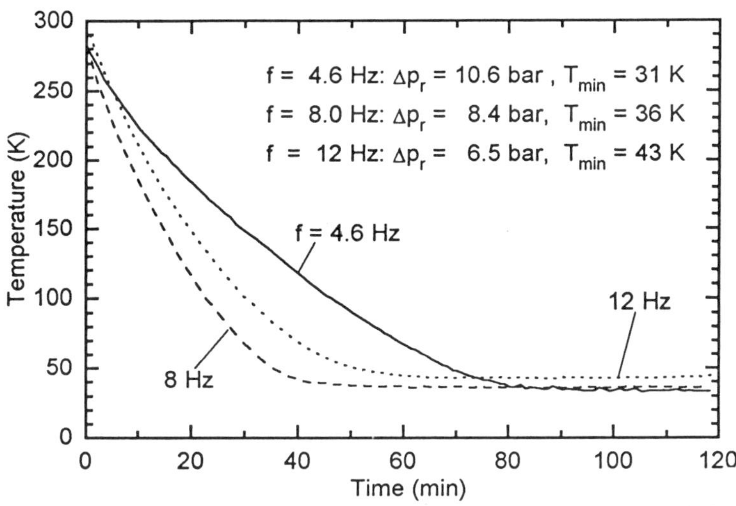


**Figure 3.** Cool-down curves at different frequencies. Vertical pulse tube with cold end facing downwards.

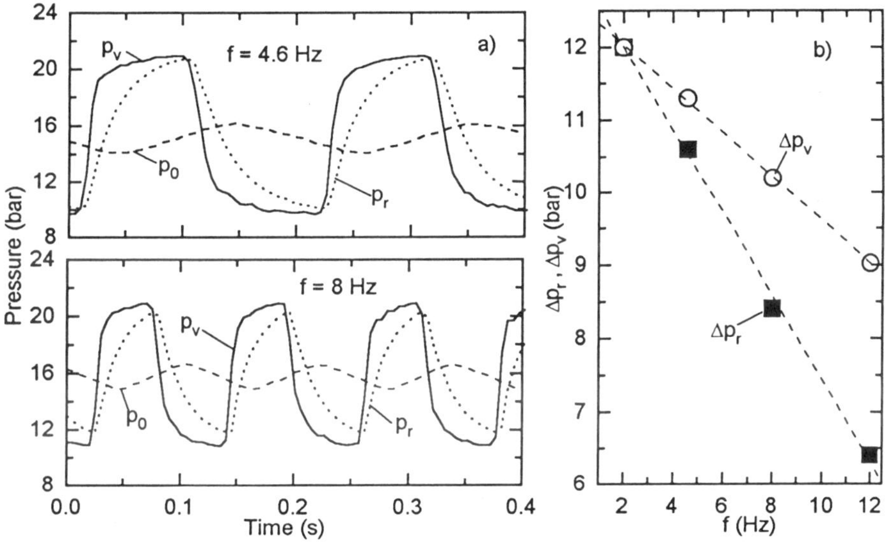


**Figure 4. a)** Pressure waves at rotary valve ($p_v$), regenerator inlet ($p_r$), and reservoir ($p_o$) at 4.6 Hz and 8 Hz. The amplitude of $p_0$ is magnified by a factor of 5. **b)** Peak to peak pressure difference $\Delta p_{v,r}$ vs frequency.

Examples of pressure oscillations monitored at the rotary valve ($p_v$), the regenerator inlet ($p_r$) and in the reservoir ($p_o$) are shown in Figure 4a. While $p_v(t)$ is approximately rectangular in shape (crest factor $\approx$ 1), $p_r(t)$ is given by alternate exponentials (crest factor $\approx$ 1.4). This change in wave form can be related to the low-pass filter characteristic of the 4 m connecting line between rotary valve and cold head. Consequently, the peak to peak pressure difference $\Delta p_r$, which approximately equals that in the pulse tube, decreases rapidly with increasing frequency, as seen from Figure 4b.

**Inclination effect**. Figure 5 displays the variation of the minimum no-load temperature upon tilting the pulse tube with respect to the vertical position ($\theta$ = 0°) at distinct working frequencies. The dependence of temperature on tilt angle $\theta$ resembles the formerly observed variation for a 2 Hz-PTR[4], which has been discussed in terms of gravity-induced convective heat losses[6]. Due to the higher working frequencies of the present PTR the tilt effect is less

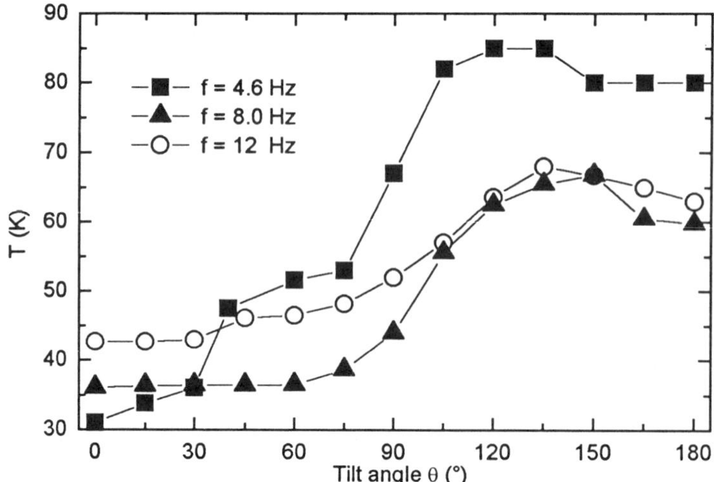


**Figure 5.** Variation of minimum no-load temperature with angle θ of inclination of the pulse tube at distinct frequencies of the pressure oscillation.

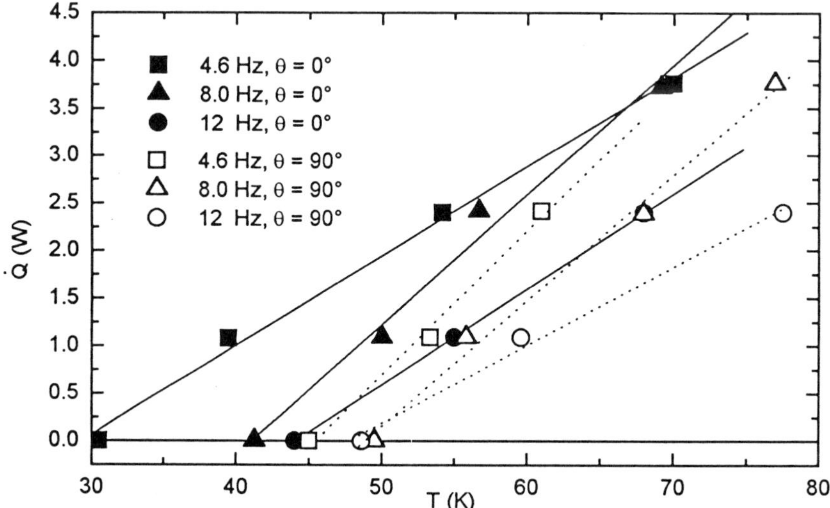


**Figure 6.** Net cooling powers vs cold end temperature at different frequencies for vertical (θ= 0°) and horizontal (θ = 90°) position of pulse tube. Straight lines represent linear fits.

dramatic, since the oscillatory gas displacement attentuates the superimposed natural convection[6]. This is reflected in Figure 5, where the maximum change of temperature with θ amounts to 55, 30, and 25 K for 4.6, 8, and 12 Hz, respectively.

**Cooling power.** The net cooling powers as a function of cold block temperature are shown in Figure 6 for vertical and horizontal orientation at different frequencies. Here, in each run the settings of the needle valves were optimized for highest cooling power at about 70 K. Optimum performance below 70 K is obtained with f = 4.6 Hz. Then at 50 K cooling powers of 2 W and 0.75 W are available in vertical and horizontal position, respectively. The disadvantage of operating at 4.6 Hz is the rather strong orientation dependence of cold block temperature at small heat loads (Figures 5 and 6). Therefore, for the required operation of the HTS-coil with horizontal pulse tube at a cold block temperature of 50-60 K

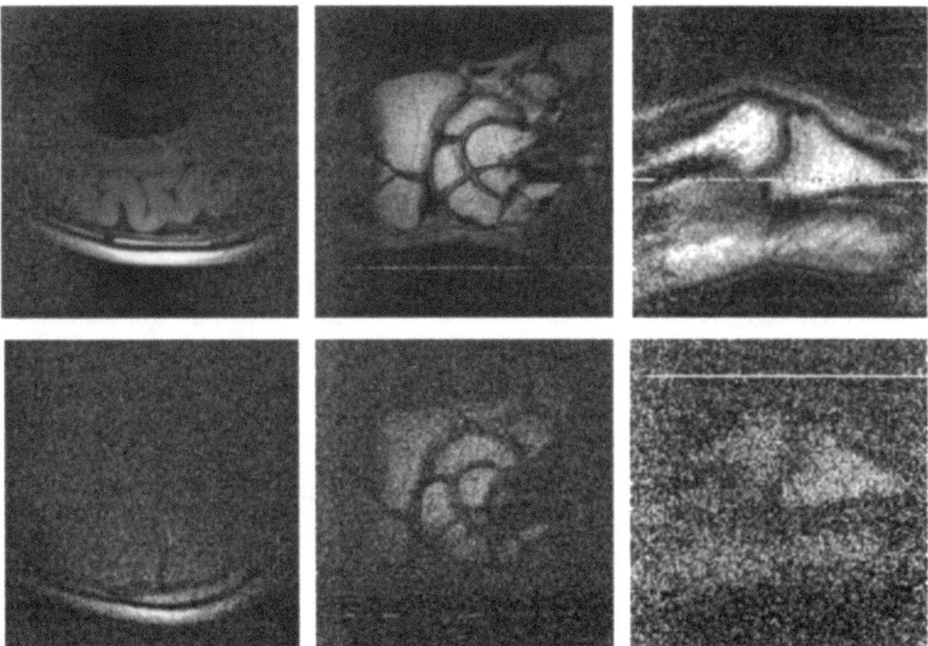

**Figure 7.** Examples of MR-images taken with PTR-cooled HTS-antenna (top) and with room temperature silver coil (bottom) in a Siemens Magnetom Open 0.2 T system. From left to right: back of human head (sample spacing: 5 cm, field of view: 15 × 15 $cm^2$, S/N gain: 6.4), wrist (spacing 1.2 cm, field of view: 10 × 10 $cm^2$, S/N gain: 3.6), and little finger (spacing 1.2 cm, field of view: 3 × 3 $cm^2$, S/N gain: 4.1).

the frequency was set to 8 Hz. This working frequency also provides a short cool-down time (Figure 3).

## YBCO Antenna Test

The tests of the HTS-antenna were conducted in the arrangement sketched in Figure 2 with the coil surface aligned parallel to the static field. With the antenna coil installed, the cold block attains a temperature of 55 K at $\theta = 90°$. At the frequency of 8 Hz this corresponds to a heat load of 0.8 W (see Figure 6) which is mainly caused by heat radiation to the coil surface. The YBCO coil temperature was estimated to be about 60 K.

For comparison, a room temperature silver spiral coil of the same size (3.4 cm diameter) was used. The unloaded Q-factors of the two coils, including tuning capacitors and matching circuit to the preamplifier, were 10573 and 214 for the HTS-coil and silver coil, respectively. The higher Q-factor of the HTS-coil results in an improved signal-to-noise (S/N) ratio, the magnitude of which depends on the coil loading. For sample to coil spacings of 5 cm and above, the S/N gain (relative to the silver coil) is greater than 6. With decreasing spacing the S/N gain decreases to about 2 at the minimum possible distance of 2 mm that is given by the present set up (see Figure 1).

Figure 7 shows MR-images taken from human head, wrist, and little finger using the superconducting coil in comparison to those taken with the room temperature coil under identical conditions. Acquisition times ranged from 5.6 minutes (head) to 42 s (little finger). The S/N gain of 3.6 to 6.4 that is achieved with the YBCO coil is clearly visible in the improved image quality.

It should be noted that imaging artifacts, which could arise from residual vibrations of the cold head and the diamagnetic antenna, were not observed in theses imaging sequences.

A source of residual vibration of the cold block with antenna is given by pressure-induced elastic length oscillations of pulse tube and regenerator tube. For a pressure variation of $\Delta p_r = 8.4$ bar at 8 Hz (Figure 4) the resulting length variation is calculated to be about 7 μm.

## CONCLUSIONS

The integration of a PTR-system into a low-field MR-imager has been successfully demonstrated.

One important feature of the present PTR-system is the spatial separation of cold head and the pressure-wave generating assembly that was placed outside the shielded room. Due to the long interconnecting tube, which acts as a low-pass filter, the operating frequency of the PTR is limited to below about 12 Hz. Therefore miniature PTRs, operating at considerable higher frequencies, would be unsuitable for the present application.

With respect to practical use, further reduction of the orientation dependence of cooler performance would be desirable. Since operating frequency is limited, this reduction could possibly be achieved by introducing an internal structure in the pulse tube that hinders the development of the convective flow pattern, as already pointed out in Ref. 6.

Concerning the HTS antenna, further improvement of signal-to-noise ratio is likely to be achieved by using a very low-noise cooled preamplifier.

## ACKNOWLEDGMENT

We would like to thank U. Häfner (Leybold, Cologne) for support.

## REFERENCES

1. D.G. Smith, P. Gerner, and C.P. Zuhoski, A study of the applicabilty of high temperature superconducting materials to the MRI industry, *IEEE Trans. Appl. Supercond.* 7: 3710 (1997)
2. R. Radebaugh, Recent developments in cryocoolers, in: "Proceedings of the 19th International Congress of Refrigeration", The Hague, Vol. IIIb (1995), p. 973
3. G. Thummes, R. Landgraf, F. Giebeler, M. Mück, and C. Heiden, Pulse tube refrigerator for high-$T_c$ SQUID operation, in: "Advances in Cryogenic Engineering", Vol. 41B, Plenum Press, New York (1996), p. 1463
4. G. Thummes, R. Landgraf, M. Mück, K. Klundt, and C. Heiden, Operation of a high-$T_c$ SQUID gradiometer by use of a pulse tube refrigerator, in: "Proceedings of ICEC16/ICMC", T. Haruyama, T. Mitsui, and K. Yamafuji, eds., Elsevier Science, Oxford, New York, Tokyo (1997), p. 283
5. M. Vester, F. Steinmeyer, B. Roas, G. Thummes, and K. Klundt, High temperture superconducting surface coils with liquid nitrogen or pulse tube refrigeration, in: "Proceedings of the 5th Scientific Meeting of the International Society for Magnetic Resonace in Medicine", Vancouver (B.C.) (1997), p. 1528
6. G.Thummes, M. Schreiber, R. Landgraf, and C. Heiden, Convective heat losses in pulse tube coolers: effect of pulse tube inclination, in: "Cryocoolers 9", R.G. Ross, Jr., ed., Plenum Press, New York (1997), p. 393
7. J. Liang, Y. Zhou, and W. Zhu, Development of a single-stage pulse tube refrigerator capable of reaching 49 K, *Cryogenics* 30: 514 (1990)
8. P.-H. Chen, Z.-C. Chang, and B.-J. Huang, Effect of oversize in wire-screen matrix to the matrix-holding tube on regenerator thermal performance, *Cryogenics* 36: 365 (1996)
9. R.S. Withers, G.-C. Liang, B.F. Cole, and M. Johansson, Thin-film HTS probe coils for magnetic-resonance imaging, *IEEE Trans. Appl. Supercond.* 3: 2450 (1993)
10. B. Roas, L. Schultz, and G. Enders, Epitaxial growth of $YBa_2Cu_3O_{7-x}$ thin films by a laser evaporation process, *Appl. Phys. Lett.* 53: 1557 (1988)

# AUTHOR INDEX, Vol. 43

## SUBJECT INDEX, Vol. 43